D1753645

Dictionary of Electronics, Computing and Telecommunications
Wörterbuch der Elektronik, Datentechnik und Telekommunikation

English-German
Englisch-Deutsch

Vittorio Ferretti

Dictionary of Electronics, Computing and Telecommunications

English-German

Wörterbuch der Elektronik, Datentechnik und Telekommunikation

Englisch-Deutsch

Springer-Verlag
Berlin Heidelberg NewYork
London Paris Tokyo
HongKong Barcelona Budapest

Dipl.-Ing. Vittorio Ferretti
Frillenseestraße 6
W-8000 München 70
Germany

ISBN 3-540-54202-7 Springer-Verlag Berlin Heidelberg New York
ISBN 0-387-54202-7 Springer-Verlag New York Berlin Heidelberg

Library of Congress Cataloging-in-Publication Data
Ferretti, Vittorio:
Dictionary of electronics, computing and telecommunications, English-German =
Wörterbuch der Elektronik, Datentechnik und Telekommunikation, Englisch-Deutsch / Vittorio Ferretti.
Includes bibliographical references.
ISBN 0-387-54202-7 (U.S.)
1. Electronics-Dictionaries-German. 2. Electronics-Dictionaries. 3. Telecommunication-Dictionaries-German.
4. Telecommunication-Dictionaries. 5. Electronic data processing-Dictionaries-German. 6. Electronic data processing-Dictionaries.
7. German language-Dictionaries-English. 8. English language-Dictionaries-German.
I. Title. II. Title: Wörterbuch der Elektronik, Datentechnik und Telekommunikation.
TK7804.F47 1992   621.38·03-dc20   92-17916 CIP

This dictionary has been compiled with the utmost care and to the best of the author's knowledge. Neither the author nor the publisher can assume any liability or legal responsibility for omissions or errors.
It would have exceeded the scope of this work to go into the restrictions on the use of the terms in all English- and German-speeking countries.
Suggestions as to how the dictionary might be improved will be much appreciated. They can be sent to the publisher, Springer Verlag, Planung Technik, Tiergartenstr.17, D-6900 Heidelberg 1/Germany; Telefax + 496221/487150 [e.g. on a copy of the appropriate page(s) with the suggested correction(s) and amendment(s) marked]

This work is subject to copyright. All rights are reserved, whether the whole or part of the material is concerned, specifically the rights of translation, reprinting, reuse of illustrations, recitation, broadcasting, reproduction on microfilm or in other ways, and storage in data banks. Duplication of this publication or parts thereof is permitted only under the provisions of the German Copyright Law of September 9, 1965, in its current version, and permission for use must always be obtained from Springer-Verlag. Violations are liable for prosecution act under the German Copyright Law.

© Springer-Verlag Berlin Heidelberg 1992
Printed in Germany

The use of general desriptive names, registered names, trademarks, etc. in this publication does not imply, even in the absence of a specific statement, that such names are exempt from the relevant protective laws and regulations and therefore free for general use.

Typesetting: Data conversion by Druckerei Appl, W-8853 Wemding (Germany)
Printing: Mercedes-Druck, W-1000 Berlin (Germany); Book binding: Lüderitz & Bauer,
W-1000 Berlin (Germany)
60/3020 5 4 3 2 1 0 Printed on acid-free paper

Preface

This dictionary covers the entire field of electronics in the widest sense. This used to be called "Low Current Engineering," as opposed to "Power Current Engineering." The modern term is "Information Technology." Nevertheless, I have chosen the title "Electronics, Computing and Telecommunications" in order to circumscribe more explicitly the over fifty specialties of this hemisphere of electric engineering. Why a single dictionary about such a vast field?

First of all, we are nowadays confronted with an increasing number of specialties. Electronics, computers and telecommunications are now concurrently involved anywhere.

On the other hand, the technical terminologies of the different subject areas are coming to overlap more and more. This is the result of the aforemencioned electronization and computerization of almost all technologies. As an example we can observe that modern communications terminology is bristling with computing terms; computing terminology itself is largely derived by from microelectronics and can hardly be separated from data communications, i.e. telecommunications.

International communication has progressed to the point where an expanding circle of persons is faced with terminologies in foreign languages. This makes new demands on multilingual dictionaries.

Conventional dictionaries give only term-to-term equivalents; in the present work **user friendliness** has been enhanced in the following ways:

– Every entry is assigned to a **subject field**. This reassures users that they are searching under the right term; in spite of ongoing integration of terminologies, the same term may still have very different meanings and translations in different specialities. The subject field codes can be understood without reference to a coding list.

– The entries contain exhaustive listings of **synonyms** for both languages. Also this makes the dictionary easier to use; it adds to users' certainty that they are searching within the correct field and allows them to deploy their passive knowledge of the target language to select the most appropriate synonym, as from a menu.

Vorwort

Dieses Wörterbuch deckt den gesamten Bereich der Elektronik im weitesten Sinne ab. Ursprünglich war hierfür der Begriff Schwachstromtechnik gebräuchlich, im Gegensatz zur Starkstromtechnik. Er wurde dann vom Terminus Nachrichtentechnik abgelöst, und letztens ist der aus dem Englischen entlehnte Ausdruck Informationstechnologie aufgekommen. Ich habe der Eindeutigkeit halber den weitschweifigeren Arbeitstitel „Elektronik, Datentechnik und Telekommunikation" gewählt. Diese eine Hälfte der Elektrotechnik fächert sich heute in über fünfzig Spezialgebiete auf. Warum ein Fachwörterbuch für ein derart weitgespanntes Feld?

Zum ersten kommen wir mit immer mehr Fachgebieten gleichzeitig in Berührung. Allerorten kommen heutzutage elektronische Technologien, rechnergestützte Verfahren und Kommunikationsmittel simultan zum Einsatz.

Andererseits hat der Überlappungsgrad der Fachsprachen stark zugenommen. Dies hängt mit der bereits erwähnten "Elektronisierung" und "Computerisierung" zusammen. Die Terminologie der modernen Telekommunikation ist beispielsweise mit Begriffen der Datentechnik gespickt; ebenso baut die Fachsprache der Datenverarbeitung auf der Mikroelektronik auf und ist kaum noch von der Datenübermittlung – und damit der Telekommunikation – zu trennen.

Die wachsende internationale Verflechtung stellt einen immer größeren Personenkreis vor die Aufgabe, selbständig mit fremdsprachigen Fachwörtern – vor allem des Englischen – umzugehen. Daraus ergeben sich neue Anforderungen an Fachwörterbücher.

Bei der Gestaltung dieses Wörterbuchs wurde ein höchstmöglicher **Benutzerkomfort** angestrebt. Gegenüber traditionellen Äquivalenzwörterbüchern, die ohne weitere Hinweise lediglich eine punktuelle Entsprechung wiedergeben, bietet es folgende Zusatzinformationen:

– Für jeden Eintrag wird grundsätzlich das **Fachgebiet** genannt. Der Benutzer erhält dadurch die Sicherheit, sich im korrekten Begriffsfeld zu bewegen; denn aller Uniformierung der Fachsprachen zum Trotz kann ein und derselbe Terminus in verschiedenen Fachgebieten völlig unterschiedliche

– Additional information is given in the case of basic or difficult terms, in order to explain the meaning and make the translation easier: **short definitions**, **quasi-synonyms**, antonyms (opposites), **general terms** and **derivative terms**.

Thanks to its lexicographic information and cross-referencing, this dictionary is **also very helpful in monolingual tasks,** e.g. when searching for appropriate terms or synonyms.

The above mentioned features could hardly have been implemented without computer support. It was a piece of great good fortune for me that Siemens-Nixdorf released its software "Term-PC," which is dedicated to terminology management, just in time for my working schedule. I am very grateful to Mr. Vollnhals, head of the Department for Computer-Aided Lexicography with Siemens-Nixdorf, for his valuable help.

I am also much obliged to the publisher Springer-Verlag, for the confidence placed in my work and, especially, for supporting my innovative dictionary concept. Finally many thanks to all lexicographers, terminologists, authors of monographs and standards committees: without their preliminary work of digesting the terminologies, a comprehensive work of this type would hardly have been possible.

I hope this book will prove a faithful and trustworthy helpmate to all who opt for it!

Munich, August 1992

Bedeutungen und Übersetzungen haben. Die angewandten Fachgebietskürzel sind in sich – ohne Nachschlagen in einer Schlüsselliste – verständlich.
– Systematische Auflistung aller **Synonyme** auf beiden Sprachseiten. Die Übersicht der bedeutungsgleichen Fachwörter ist für den Benutzer ungemein wertvoll: Auch sie trägt zur Gewißheit bei, daß er sich im gewünschten Eintrag befindet; außerdem kann er sein passives Wissen einsetzen, um in der Zielsprache sozusagen " à la carte " das geeignete Synonym auszuwählen.
– Bei grundlegenden oder diffizilen Begriffen werden weitere Zusatzinformationen gegeben, um die Bedeutung und Übersetzung zu präzisieren: **Kurzdefinitionen**, Hinweise auf **Quasisynonyme** (Nebenbegriffe), auf **Antonyme** (Gegensätze), auf **Oberbegriffe** sowie auf **Unterbegriffe**.

Dank der lexikalischen Zusatzinformationen und der Vernetzung der Einträge ist dieses Wörterbuch **auch für einsprachiges Arbeiten von großem Nutzen,** zum Beispiel bei der Suche nach einem Fachausdruck oder nach einem Synonym bei einer Textredaktion.

Die erwähnten Gestaltungsmerkmale hätten ohne Rechnerunterstützung schwerlich implementiert werden können. Durch eine glückliche Fügung konnte ich rechtzeitig das Terminologie-Verwaltungsprogramm "Term-PC" von Siemens-Nixdorf einsetzen. Herrn Vollnhals, dem Leiter der Fachabteilung "Computerunterstützte Lexikographie" von Siemens-Nixdorf, danke ich für diesbezügliche Ratschläge und Unterstützung.

Dem Springer-Verlag bin ich für das entgegengebrachte Vertrauen und Eingehen auf meine innovative Wörterbuchkonzeption in außerordentlichem Maße verbunden. Dank gebührt schließlich auch der Vielzahl von Lexikographen, Terminologen, Fachbuchautoren und Normungsgremien beider Sprachen, ohne deren Vorarbeit bei der Sichtung der Fachsprachen ein zusammenfassendes Werk dieser Art ebenfalls kaum hätte gelingen können.

Sollten Sie sich, lieber Leser, für die Benutzung dieses Wörterbuchs entschieden haben, so wünsche ich, daß es sich als Ihr getreuer und vertrauenswürdiger Gehilfe bewähren möge!

München, im August 1992

Contents

Inhalt

Preface	V	Vorwort		V
Explanatory Notes	IX	Hinweise zur Benutzung		IX
Alphabetical List of Abbreviations	XVI	Alphabetische Liste der Abkürzungen		XVI
Alphabetical List of Subject Fields	XVII	Alphabetische Liste der Fachgebiete		XVII
Applied Morphology of Subject Fields	XX	Angewandte Morphologie der Fachgebiete		XX
Bibliography	XXIV	Literaturhinweise		XXIV
Dictionary English – German	1–669	Wörterverzeichnis Englisch – Deutsch		1–669
Phonetic Lists	Back cover inside	Buchstabierlisten		Rückendeckel-Innenseite

MEIS QUIBUS

HUIC LABORE INTENTUS

NIMIUM TEMPUS ABSTULI

Explanatory Notes

1 Structure of Entries

This dictionary contains two types of entries: main entries and synonym entries. A **main entry** is made of: the main term and its translation, a subject field code and a comprehensive listing of any synonyms existing. If appropriate, additional information is supplied for both languages: grammatical notes, reference to usage, a short definition, quasi-synonyms, antonyms, general terms and derivative terms. A **synonym entry** contains a reference to the main entry in both languages, besides the subject field code.

The main entries are structured as follows:

Main term (grammatical notes) (references to usage)
[short definition]
= synonym 1; synonym 2; ...
≈ quasi-synonym 1; quasi-synonym 2; ...
≠ antonym 1; antonym 2; ...
↑ generic term 1; generic term 2; ...
↓ derivative term 1; derivative term 2; ...

1.1 Grammatical Notes

Grammatical information is given only when it is indispensable to specify the type of word, or when knowledge of it cannot be assumed from colloquial fluency. The former case is frequent in English, where nouns, verbs, adjectives and adverbs can have the identical spelling (homographs). In German sometimes the gender of a noun gives rise to doubts. The appendix **Alphabetical List of Abbreviations** defines all codes used for grammatical notes.

1.2 References to Usage

Within this rubric, indications of regional or institutional preferences are given. In English there are regional differences both in spelling and in predilection of some synonym instead of others. The main difference is that between British and American usage, as well known. International organizations

Hinweise zur Benutzung

1 Aufbau der Einträge

In diesem Wörterbuch gibt es zwei Arten von Einträgen: Haupteinträge und Synonymeinträge. Ein **Haupteintrag** enthält – neben der Fachgebietsangabe – auf beiden Sprachseiten eine Auflistung aller Synonyme. Bedarfsweise werden im Haupteintrag zweisprachig noch folgende Zusatzinformationen geboten: grammatische Angaben, Hinweis zum Sprachgebrauch, Kurzdefinition, Quasisynonyme, Antonyme, Oberbegriffe und Unterbegriffe. Ein **Synonymeintrag** enthält, neben der Fachgebietsangabe, in beiden Sprachen den Verweis auf den entsprechenden Haupteintrag.

Die Haupteinträge sind folgendermaßen strukturiert:

Hauptstichwort (grammatische Angaben)(Hinweis zum Sprachgebrauch)
[Kurzdefinition]
= Synonym 1; Synonym 2; ...
≈ Quasisynonym 1; Quasisynonym 2; ...
≠ Antonym 1; Antonym 2; ...
↑ Oberbegriff 1; Oberbegriff 2; ...
↓ Unterbegriff 1; Unterbegriff 2; ...

1.1 Grammatische Angaben

Grammatische Angaben werden nur in den Fällen gemacht, in denen sie zur eindeutigen Kennzeichnung der Wortart unerläßlich sind oder in denen sie aus der Beherrschung der Umgangssprache nicht vorausgesetzt werden können. Bei englischen Fachwörtern ist ersteres häufig der Fall, da Substantive, Verben, Adjektive und Adverbien vielfach gleichgeschrieben werden. Im Deutschen treten hingegen gelegentlich Zweifel über das anzuwendende Geschlecht von Substantiven auf. Die für grammatische Angaben angewandten Abkürzungen werden im Anhang **Alphabetische Liste der Abkürzungen** erklärt.

tend to a hybrid English. All these variations are treated as synonyms (see item 1.5). The author hopes that British readers will excuse his tendency to give precedence to American usage.

The abbreviations used to mark regional or institutional usage are listed in the annex **Alphabetical List of Abbreviations.**

1.3 Subject Field

Every entry is assigned to a subject field by a self-explanatory code, thus avoiding the reference to a coding list. The subdivision of specialities is only as fine as strictly necessary. With increasing distance from the central topics of the dictionary, the differentiation becomes coarser. Mathematics, for instance, is treated as single subject field.

Coverage and delimitation of the subject fields can be taken from the appendix **Applied Morphology of Subject Fields.** Besides the different specialties within information technology, this dictionary also covers the basic terms of some fundamental sciences and some remote disciplines which impinge on electronic terminology. In texts about word processing, for instance, terms relating to typography and linguistics come up very commonly. In all, about 70 subject areas are considered in this dictionary, which are listed in the appendix **Alphabetical List of Subject Fields.**

1.4 Short Definitions

For fundamental or difficult terms short definitions are given. The aim is to explain a complex word by means of more familiar ones. No scientific precision or exhaustiveness is claimed in these definitions.

1.5 Synonyms

Synonyms are defined in this book as expressions having the same essential meaning; variants of spelling are also treated as synonyms.

For many technical concepts different terms are used in different countries, specialities or even corporations. In each case one of these equivalent terms has been elected by the author, to the best of his knowledge and belief, as the basic term for the concept concerned. The remaining terms are listed as its synonyms, in decreasing order of frequency and precision of use. Together with other additional information, the basic term constitutes the main entry mentioned under item 1.

1.2 Hinweise zum Sprachgebrauch

Unter dieser Rubrik werden hauptsächlich Hinweise auf regionalsprachliche Präferenzen gegeben oder auf von Institutionen bevorzugte Sprachregelungen. Im Englischen bestehen regionalsprachliche Unterschiede der Schreibweise aber auch der Wortwahl, vor allem zwischen dem britischen und dem nordamerikanischen Sprachraum. Internationale Fachgremien neigen zu einem hybriden Fachenglisch. Regionalsprachliche Varianten werden hier lexikographisch wie Synonyme behandelt (siehe Abschnitt 1.5). Der britische Leser möge es dem Autor nachsehen, wenn er die nordamerikanischen Ausdrücke tendenziell an erster Stelle anführt.

Die für Regionalsprachen und Quellenangaben angewandten Abkürzungen werden im Anhang **Alphabetische Liste der Abkürzungen** erklärt.

1.3 Fachgebiet

Für jeden Eintrag wird das Fachgebiet angegeben. Die dafür angewandten Kürzel sind so bemessen, daß sie in sich verständlich sind, d. h., daß ein Nachschlagen in einer Schlüsselliste nicht erforderlich ist. Da sich die englische Abkürzung in vielen Fällen von einer deutschen kaum unterscheiden würde, werden nur englische Fachgebietskürzel verwendet. Der deutschsprachige Leser möge dies im Sinne einer besseren Übersichtlichkeit in Kauf nehmen. Bei der Auswahl der englischen Bezeichnungen wurde auf die Verständlichkeit für deutschsprachige Benutzer Rücksicht genommen.

Die Fachgebiete wurden nur so weit aufgegliedert, wie es für eine Unterscheidung der Fachsprachen unbedingt notwendig ist; mit zunehmender Entfernung vom Zentralthema des Wörterbuchs wird die Differenzierung immer gröber. So wird zum Beispiel die gesamte Mathematik hier als ein einziges Fachgebiet behandelt. Der Umfang und die Abgrenzung der Fachgebietsklassifizierung ist in der Anlage **Angewandte Morphologie der Fachgebiete** wiedergegeben. Außer den Grundlagenwissenschaften wurden auch Fachgebiete weit außerhalb der Schwachstromtechnik berücksichtigt, deren Grundbegriffe für die Kernthemen des Wörterbuchs relevant sind. So treten z. B. in der Datenverarbeitung (v.a. in der Textverarbeitung) viele Fachwörter des Druckwesens oder der Sprachwissenschaft auf. Insgesamt werden etwa 70 Fachgebiete unterschieden. In der Anlage **Alphabetische Liste der Fachgebiete** sind sie in der genannten Sortierung aufgelistet.

Every synonym is also listed separately at the appropriate point in the alphabet, where a cross-reference indicated by the symbol → is given to the main entry in both languages. Together with the subject field code, this constitutes the synonym entry mentioned under item 1. In this way, the hasty user finds a translation at the first attempt; the user looking for synonyms or supplementary information can find them in a second step by cross-consulting the referenced main entry.

1.6 Quasi-synonyms

Terms that are very close in meaning, especially those likely to be erroneously taken as synonyms, are listed as quasi-synonyms. Under this heading reference is also made to diverging meanings given to very specific terms within the same area of technology. This is generally due to a multiple use given to the term in a metaphorical sense, which may not however coincide in different languages.

Example: Under "hard disk memory [TERM&PER]" (a specific type of magnetic layer memory) cross-reference is made to the quasi-synonym "hard disk memory [DATA PROC]," as this same term is used figuratively in PC jargon to designate an "external, high-capacity random-access memory." For this latter concept German PC jargon has a different metaphoric term, namely "Festplattenspeicher [DATA PROC]" which really means "fixed disk memory."

1.7 Antonyms

In the rubric of antonyms references are given to technical terms of opposite or complementary meaning. Sometimes the most concise way to explain the meaning of a word is to give its antonym. It can also happen that the user is searching for a term, while actually having only its opposite in mind.

1.8 General Terms and Derivative Terms

General and derivative terms are often helpful in the elucidation of a technical term. Furthermore, specialists often use a general term or a derivative term in lieu of a technical term. This liberty of expression may not, however, concide in different languages and must be handled with care. References to general and derivative terms are intended to help in selecting the correct conceptual level when translating.

1.4 Kurzdefinition

Die bei grundsätzlichen oder diffizilen Begriffen gegebenen Kurzdefinitionen sind als Hinweise zur Bedeutung des Begriffs gedacht nach dem Prinzip, Schwieriges mit leichter Verständlichem zu erläutern. Dabei kann leider keine wissenschaftliche Präzision geboten werden.

1.5 Synonyme

Synonyme sind in diesem Wörterbuch als gleichbedeutende Fachbegriffe definiert; darunter fallen auch Variationen der Schreibweise.

Für viele Fachbegriffe werden je nach Land, Spezialistenkreis oder Firma unterschiedliche Synonyme bevorzugt. Der Autor hat nach bestem Wissen und Gewissen für jeden Begriff einen Terminus als Hauptstichwort erwählt. Beim Hauptstichwort werden auf beiden Sprachseiten alle Synonyme aufgeführt, und zwar in der Rangfolge abnehmender Gebräuchlichkeit und Korrektheit. Zusammen mit allen weitern anfallenden Zusatzinformationen bildet das Hauptstichwort den unter Abschnitt 1 erwähnten Haupteintrag.

Jedes Synonym tritt darüber hinaus an seinem alphabetischen Platz als Synonymeintrag auf. Von dort wird auf beiden Sprachseiten durch das Symbol → auf den Haupteintrag verwiesen. Der eilige Benutzer findet so auf Anhieb eine Übersetzung; der weiter ausholende Benutzer kann in einem zweiten Schritt beim Haupteintrag alle weiteren Synonyme und Zusatzinformationen finden.

1.6 Quasisynonyme

Als Quasisynonyme werden hier bedeutungsverwandte Termini, insbesonders die leicht verwechselbaren, verstanden. Unter dieser Rubrik werden auch Fachausdrücke angeführt, die, obwohl sehr spezifisch, im gleichen Fachbereich mit unterschiedlichen Bedeutungen benutzt werden (Nebenbegriffe). Dies ist meist auf eine metaphorische Mehrfachbelegung eines Terminus zurückzuführen, die jedoch in verschiedenen Sprachen unterschiedlich sein kann.

Beispiel: Unter "Festplattenspeicher [TERM&PER]" versteht man primär einen bestimmten Typ von Magnetschichtspeicher (mit nicht auswechselbaren Magnetplatten). Bei diesem Eintrag wird auf das Quasisynonym "Festplattenspeicher [DATA PROC]" hingewiesen, welches im PC-Jargon metaphorisch für "externer Massenspeicher mit direktem Zugriff" an-

2 Indexed Terms

Sometimes an identical term is used within the same speciality with several different meanings. These are distinguished by indexes and treated as independent entries.

Example:

    CAE 1 [DATA PROC] = computer-aided engineering;
    CAE 2 [DATA PROC] = computer-aided education.

3 Composite Terms

English and German technical terms are mostly composite, with the attributive word first and the main word in second place. Example: circulating storage = Umlaufspeicher. The alphabetic sorting of composite words in their natural sequence is therefore determined in both languages by secondary aspects. This lexographic disadvantage can hardly be counteracted.

3.1 In German

German is a highly agglutinative language with composite terms normally fused into single words. It has not been possible to break such words up and to quote them in their inverted sequence (in the example: speicher, Umlauf-), as this would have almost doubled the size of the dictionary.

3.2 In English

English is much less prone to fuse composite words into single words than German. This applies especially to British English. A particular difficulty with English composite technical terms is the hyphenation. The questions of whether or not a term should be hyphenated or even written in one word are complicated, and any rule is subject to countless exceptions. In many cases all three variants are found in the technical literature, e.g.: standby, stand-by, stand by. The type of lexicographic sorting used in this book, however, ensures that the three versions are shown side by side. As two out of three English technical terms are composite words, it has not been feasable to include additional entries with the inverted sequences.

gewandt wird. Für diesen PC-Fachausdruck verwendet man aber im Englischen, ebenfalls im übertragenen Sinne, den Begriff "hard disk memory [DATA PROC]", d. h. "Hartplattenspeicher" (mit harten Magnetplatten).

1.7 Antonyme

Unter der Rubrik Antonyme (Gegenwörter) wird auf Termini gegensätzlicher oder komplementärer Bedeutung hingewiesen. Gelegentlich wird ja ein Begriff durch ein Antonym am besten charakterisiert. Außerdem kommt es vor, daß man nach einem Fachausdruck sucht, von dem einem nur das Gegenteil oder Komplement gegenwärtig ist.

1.8 Oberbegriffe und Unterbegriffe

Oberbegriffe (Hypernyme) und Unterbegriffe (Hyponyme) können dazu beitragen, die Bedeutung eines Fachausdrucks besser zu beleuchten. Hinzu kommt, daß Fachleute sich vielfach die Freiheit erlauben, stellvertretend für einen Terminus den Oberbegriff oder einen Unterbegriff zu nennen. Dieser Wechsel der Begriffsebene ist aber in unterschiedlichen Sprachen nicht gleichläufig, so daß er bei einer Übersetzung bereinigt werden muß; Hinweise auf Oberbegriffe und Unterbegriffe sollen dabei hilfreich sein.

2 Indizierte Stichwörter

In manchen Fällen hat ein Fachausdruck selbst innerhalb eines engen Fachgebiets mehrere Bedeutungen und demzufolge verschiedene Übersetzungen. Derartige Mehrfachbelegungen werden durch numerische Indizes auseinandergehalten und bilden jeweils eigene Einträge.

Beispiel aus der Physik:

    Remanenz 1 = Restmagnetisierung (ein Effekt);
                    Engl. = residual magnetism
    Remanenz 2 = remanente Induktion (ein Wert);
                      Engl. = residual induction

3 Mehrgliedrige Stichwörter

Fachausdrücke sind sowohl im Deutschen als auch im Englischen meist mehrgliedrige Wörter, wobei das Attribut (Bestimmungswort) an erster Stelle und der Hauptbedeutungsträger (Grundwort) an letzter Stelle steht (Beispiel: Umlaufspeicher; circulating storage).

4 Sequence of Terms

4.1 In German

The entries are sorted according to the system used for well-known German dictionaries such as "Duden," which obeys to the following priorities:

1. Blank space
2. Period
3. Small letter
4. Capital letter
5. Parenthesis

Beyond that, the following criteria apply:

6. Numeric characters are disregarded
7. Hyphens and other punctuation marks are disregarded
8. Special characters (e.g. &, ^{2}, $, %, *, #) are treated as blank spaces
9. Umlauts (mutated vowels, e.g. ä,ö,ü) are treated as the corresponding basic vowels (a,o,u)
10. Accented vowels are treated as the corresponding basic vowels

Example:

```
°C
a
A
A h
A s
A-Ader
abätzen
dB
D-Betrieb
3-dB-Koppler
Durchlaßbereich
durchlässig
Durchlaßrichtung
E&M-Signalisierung
E²CL
E²PROM
EA
E/Λ
EAPROM
Ruhe vor dem Telefon
Ruheenergie
Ruhe-und-Arbeitskontakt
```

4.2 In English

The sorting criteria applied here are the same as those used for well-known English dictionaries, which are subject to the following priorities:

1. Capital and small characters with equal priority
2. Numeric characters disregarded

Die alphabetische Sortierung der Wörter in ihrer natürlichen Zusammensetzung orientiert sich in beiden Sprachen folglich leider nach sekundären Attributen. Diesem lexikographischen Nachteil kann mit vertretbarem Aufwand nicht entgegengewirkt werden.

3.1 Im Deutschen

Im Deutschen werden mehrgliedrige Wörter meist zusammengeschrieben. Es mußte davon abgesehen werden, solche Wörter zusätzlich auch in aufgebrochener und umgestellter Form (z.B. speicher, Umlauf-) anzuführen, da sich der Buchumfang dabei fast verdoppelt hätte.

3.2 Im Englischen

Die Wortverschmelzungen finden im Englischen, vor allem im britischen Englisch, in einem weit beschränkteren und langsameren Maße statt als im Deutschen. Eine spezielle Schwierigkeit des Englischen ist dabei die Unsicherheit, ob man mit oder ohne Bindestrich trennt oder doch zusammenschreibt. Diesbezügliche Regeln sind von geringem Nutzen, weil voller Ausnahmen. In vielen Fällen findet man alle Varianten vor (Beispiel: standby, stand-by, stand by). Die lexikographische Sortierung (siehe Abschnitt 4) stellt sicher, daß die verschiedenen Schreibarten nebeneinander aufgelistet erscheinen. Da dem Autor in Einzelfällen die eine oder andere Abart der Orthographie entgangen sein könnte, wird bei Mißerfolg dem Leser empfohlen, den Eintrag auch in den anderen Schreibarten zu suchen.

Auf eine zusätzliche Anführung der mehrgliedrigen Fachwörter in ihrer umgestellten Form (z. B. memory, circulating) mußte auch im Englischen aus Platz- und Preisgründen verzichtet werden.

4 Reihenfolge der Stichwörter

4.1 Im Deutschen

Die Stichwörter wurden nach dem sogenannten "lexikographischen Kriterium, numerische Zeichen ignoriert" sortiert, entspechend der im Duden angewandten Systematik. Es gilt dabei folgende Reihenfolge:

1. Leerstelle
2. Komma
3. Kleinbuchstabe
4. Großbuchstabe
5. Klammer

In addition:

3. Blank spaces are disregarded
4. Period, hyphen and other punctuation marks are disregarded
5. Special characters are disregarded
6. Umlauts are treated like the corresponding basic vowels
7. Accented vowels are treated like the corresponding basic vowels

Example:

```
A
a
abampere
A h
°C
dB
3-dB coupler
E²CL
E²PROM
EA
electronics
empty
E&M signaling
inflexion
3¹/₄ in. floppy disk
stand-by
stand by
standby
stand-by button
```

Außerdem gilt:

6. Numerische Zeichen ignoriert
7. Bindestrich und sonstige Satzzeichen ignoriert
8. Sonderzeichen (z. B. &, ², $, %, §, *, #) wie Leerstelle
8. Umlaut (z. B. ä) wie Grundvokal (z. B. a)
9. Vokal mit Akzent wie Vokal ohne Akzent
10. ß wie ss

Beispiel:

```
°C
a
A
A h
A s
A-Ader
abätzen
dB
D-Betrieb
3-dB-Koppler
Durchlaßbereich
durchlässig
Durchlaßrichtung
E&M-Signalisierung
E²CL
E²PROM
EA
E/A
EAPROM
Ruhe vor dem Telefon
Ruheenergie
Ruhe-und-Arbeitskontakt
```

4.2 Im Englischen

Es wurde das Sortierkriterium " durchgehend, numerische Zeichen berücksichtigt" angewandt, entsprechend der Systematik namhafter englischsprachiger Wörterbücher. Dieses Kriterium befolgt folgende Rangfolge:

1. Großbuchstaben wie Kleinbuchstaben
2. Numerische Zeichen ignoriert

Ansonsten gilt:

3. Leerstellen ignoriert
4. Komma, Bindestrich und sonstige Satzzeichen ignoriert
5. Sonderzeichen ignoriert
6. Umlaut wie Grundvokal
7. Vokal mit Akzent wie Grundvokal

Beispiel:

```
A
a
abampere
A h
°C
```

dB
3-dB coupler
E^2CL
E^2PROM
EA
electronics
empty
E&M signaling
inflexion
$3^1/_4$ in. floppy disk
stand-by
stand by
standby
stand-by button

Alphabetical List of Abbreviations
Alphabetische Liste der Abkürzungen

AM	see NAM	siehe NAM
adj.	Adjective	Adjektiv (Eigenschaftswort)
adv.	Adverb	Adverb (Umstandswort)
ANSI	American National Standards Institute	American National Standards Institute
BRD	Usage of the Federal Republic of Germany	Sprachgebrauch in der Bundesrepublik Deutschland
BRI	British-English usage	Britischer Sprachgebrauch
BSI	British Standards Institute	British Standards Institute
CCIR	International Radio Consultative Committee	Comité Consultatif International des Radiocommunications
CCITT	International Telegraph and Telephone Consultative Committee	Comité Consultatif International Téléfonique et Télégraphique
DDR	Usage in the former German Democratic Republic	Sprachgebrauch in der ehemaligen DDR
dim.	Diminutive	Diminutiv (Verkleinerungsform)
DIN	German Standards Institute	Deutsches Institut für Normung
DUDEN	The most authoritative dictionary for the German language	Deutsches Universalwörterbuch
fig.	Figurative meaning	Figurative (bildliche) Bedeutung
IEC	International Electric Committe	
IEEE	Institute of Electrical and Electronics Engineers (USA)	Institute of Electrical and Electronics Engineers (USA)
ISO	International Organization for Standardization	International Organization for Standardization
NAM	American-English usage	Nordamerikanischer Sprachgebrauch
n.	Noun	Nomen (Substantiv, Hauptwort)
n.f.	Feminine noun	Femininum (weibliches Hauptwort)
n.m.	Masculine noun	Maskulinum (männliches Hauptwort)
n.n.	Neuter noun	Neutrum (sächliches Hauptwort)
n.plt.	Noun used only in the plural form	Pluraletantum (nur in der Mehrzahl gebräuchliches Hauptwort)
n.slt.	Noun used only in the singular form	Singularetantum (nur in der Einzahl gebräuchliches Hauptwort)
NTG	German Information Technology Association	Nachrichtentechnische Gesellschaft
obs.	Obsolete usage	Veraltete (obsolete) Form
OES	Austrian usage	Österreichischer Sprachgebrauch
pl.	Plural	Plural (Mehrzahl)
sl.	Slang, nonstandard language	Saloppe Ausdrucksweise
sup.	Superlatíve	Superlativ (Höchststufe)
SWZ	Swiss usage	Schweizer Sprachgebrauch
VDE	Association of German Electrotechnicians	Verband Deutscher Elektrotechniker
VDI	Association of German Engineers	Verein Deutscher Ingenieure
v.i.	Intransitive verb	Intransitives Verb (nichtzielendes Tätigkeitswort)
v.r.	Reflexive verb	Reflexives Verb (rückbezügliches Tätigkeitswort)
v.t.	Transitive verb	Transitives Verb (zielendes Tätigkeitswort)

Alphabetical List of Subject Areas
Alphabetische Liste der Fachgebiete

ACOUS	Acoustics	Akustik
AERON	Aeronautics	Luftfahrt
	Astronautics	Raumfahrt
ANT	Antennas	Antennen
ASTROPHYS	Astrophysics	Astrophysik
	Astronomy	Astronomie
BROADC	Broadcasting	Rundfunktechnik
	Cable TV	Kabelfernsehen
CHEM	Chemistry	Chemie
CIRC.ENG	Circuit Engineering	Schaltkreistechnik
	Pulse Technique	Impulstechnik
CIV.ENG	Civil Engineering	Bautechnik
CODING	Coding	Codierung
COLLOQ	Colloquial	Umgangssprache
COMM.CABLE	Communications Cables	Nachrichtenkabel
COMPON	Components and Devices	Bauelemente u. Bauteile
	Electronic Tubes	Elektronenröhren
CONS.EL	Consumer Electronics	Konsumelektronik
	Entertainment Electronics	Unterhaltungselektronik
	Car Electronics	Fahrzeugelektronik
CONTROL	Control Systems	Steuer-u.Regelungstechnik
	Cybernetics	Kybernetik
	Robotics	Robotik
	Automation	Automatisierung
DATA COMM	Data Communication	Datenkommunikation
	Data Transmission	Datenübertragung
	Data Switching	Datenvermittlung
	Data Networks (LAN,WAN,..)	Datennetze (LAN,WAN,..)
DATA PROC	Data Processing	Datenverarbeitung
	Computer Theory	Rechnertheorie
	Hardware	Hardware
	Software	Software
	Computer Applications (CAD,CAM,..)	Computeranwendungen (CAD,CAM,..)
DOC	Documentation	Dokumentationswesen
ECON	Economics	Wirtschaft
EL.ACOUS	Electroacoustics	Elektroakustik
ELECTRON	Electronics	Elektronik
EL.INST	Electrical Installation	Elektroinstallation
EL.TECH	Electrical Engineering *(general terminology)*	Elektrotechnik *(allgemein)*
	Electrical Science	Elektrizitätslehre
ENG.DRAW	Engineering Drawing	Technisches Zeichnen
ENG.LOG	Engineering Logic	Schaltalgebra
EQUIP.ENG	Equipment Engineering and Design	Gerätetechnik und Konstruktion
GEOPHYS	Geophysics	Geophysik ·
HF	High Frequency Engineering *(general terminology)*	Hochfrequenztechnik *(allgemein)*

INF.TECH	Information Technology	Nachrichtentechnik
INF	Information Theory	Informationstheorie
INSTR	Test, Measuring and Recording Instruments	Meß-, Prüf- u. Registriertechnik
LINE TH	Line Theory	Leitungstheorie
LING	Linguistics	Sprachlehre
MANUF	Manufacturing	Fertigungstechnik
MATH	Mathematics	Mathematik
MECH	Mechanics	Mechanik
	Mechanical Engineering	Maschinenbau
METAL	Metallurgy	Metallurgie
METEOR	Meteorology	Meteorologie
MICROEL	Microelectronics	Mikroelektronik
	Semiconductor Devices	Halbleiterbauelemente
MICROW	Microwave Engineering	Mikrowellentechnik
MIL.COMM	Military Communications	Militärische Nachrichtentechnik
MOB.COMM	Mobile Communications	Mobilfunk
MODUL	Modulation	Modulation
NETW.TH	Network Theory	Netzwerktheorie
		Theorie der Wechselstromschaltungen
OFFICE	Office Systems	Bürowirtschaft
OPT.COMM	Optical Communications	Optische Nachrichtentechnik
OPTOEL	Optoelectronics	Optoelektronik
OUTS.PLANT	Outside Plant	Linientechnik
PHYS	Physics	Physik
POST	Postal Services	Postwesen
POWER ENG	Electrical Power Engineering	Starkstromtechnik Elektrische Energietechnik
POWER SYS	Power Supply Systems	Stromversorgungsanlagen
QUAL	Quality Assurance Reliability	Qualitätssicherung Zuverlässigkeit
RADIO	Radio Technique *(general terminology)*	Funktechnik *(allgemein)*
	Radio Communications	Funkwesen
RADIO LOC	Radio Location	Funrortung
	Radar	Radar
	Radio Monitoring	Funküberwachung
RADIO NAV	Radio Navigation	Funknavigation
RADIO PROP	Radio Propagation	Funkwellenausbreitung
RADIO REL	Radio Relay Systems	Richtfunktechnik
RAILW.SIGN	Railway Signaling	Eisenbahnsignaltechnik
SAT.COMM	Satellite Communications	Nachrichtensatelliten
SCIE	Science *(general terminology)*	Wissenschaft *(allgemein)*
SIGN.ENG	Signal Engineering	Signal- u. Sicherungstechnik
SIGN.TH	Signal Theory	Signaltheorie
SWITCH	Communications Switching	Nachrichtenvermittlungstechnik
SYS.INST	System Installation, Station Installation	Anlagen- u. Amtsbautechnik
TECH	Technology *(general terminology)*	Technik *(allgemein)*
TELEC	Telecommunications	Telekommunikation
	Communications	Fernmeldetechnik
TELECONTR	Telecontrol	Fernwirktechnik
TELEGR	Telegraphy	Fernschreibtechnik
	Facsimile	Faksimile

TELEPH	Telephony	Fernsprechwesen
	PABX	Nebenstellentechnik
TERM&PER	Terminals and Peripherals	Endgeräte und Peripheriegeräte
TRANS	Communications Transmission	Nachrichtenübertragungstechnik
TV	Television	Fernsehtechnik
TYPOGR	Typography and Printing	Druckwesen

Applied Morphology of Subject Fields
Angewandte Morphologie der Fachgebiete

COLLOQ	COLLOQUIAL	UMGANGSSPRACHE
SCIE	SCIENCE	WISSENSCHAFT
	• HUMAN SCIENCES	• GEISTESWISSENSCHAFTEN
	• •	• •
	• •	• •
	• •	• •
LING	• • Linguistics	• • Sprachwissenschaften
ECON	• • Economics	• • Wirtschaftswissenschaften
	• • • Business Administration	• • • Betriebswirtschaft
	• • • Political Economy	• • • Volkswirtschaft
MATH	• MATHEMATICS	• MATHEMATIK
	• • Elementary Mathematics	• • Elementarmathematik
	• • • Algebra	• • • Algebra
	• • • Geometry	• • • Geometrie
	• • • Trigonometry	• • • Trigonometrie
	• • Higher Mathematics	• • Höhere Mathematik
	• • • Set Theory	• • • Mengenlehre
	• • • Mathematical Logic	• • • Mathematische Logik
	• • • Linear Algebra	• • • Lineare Algebra
	• • • • Matrix Calculus	• • • • Matrizenrechnung
	• • • • Vector Calculus	• • • • Vektorrechnung
	• • • Analysis	• • • Analysis
	• • • • Series	• • • • Folgen und Reihen
	• • • • Differential Calculus	• • • • Differentialrechnung
	• • • • Integral Calculus	• • • • Integralrechnung
	• • • • Differential Equations	• • • • Differentialgleich.
	• • • • Harmonic Analysis	• • • • Harmonische Analyse
	• • • • Calculus of Variations	• • • • Variationsrechnung
	• • • Analytical Geometry	• • • Analytische Geometrie
	• • • Differential Geometry	• • • Differentialgeometrie
	• • • Integral Geometry	• • • Integralgeometrie
	• • • Statistics	• • • Statistik
	• • • Numerical Mathematics	• • • Numerische Mathematik
	• NATURAL SCIENCES	• NATURWISSENSCHAFTEN
PHYS	• • PHYSICS	• • PHYSIK
	• • • Mechanics	• • • Mechanik
	• • • • Statics	• • • • Statik
	• • • • Dynamics	• • • • Dynamik
	• • • • Hydrodynamics	• • • • Hydrodynamik
ACOUS	• • • Acoustics	• • • Akustik

	• • • Thermodynamics	• • • Thermodynamik
	• • • Electrical Science	• • • Elektrizitätslehre
	(→Electrical Engineering)	(→Elektrotechnik)
	• • • Optics	• • • Optik
	• • • Atomical Physics	• • • Atomphysik
	• • • Nuclear Physics	• • • Kernphysik
ASTROPHYS	• • Astrophysics	• • Astrophysik
CHEM	• • CHEMISTRY	• • CHEMIE
	• • • Anorganic Chemistry	• • • Anorganische Chemie
	• • • Organic Chemistry	• • • Organische Chemie
GEOPHYS	• • Geophysics	• • Geophysik
METEOR	• • Meteorology	• • Meteorologie
TECH	**TECHNOLOGY**	**TECHNIK**
EL.TECH	• **ELECTRICAL ENGINEERING**	• **ELEKTROTECHNIK**
EL.SCIE	• • ELECTRICAL SCIENCE	• • ELEKTRIZITÄTSLEHRE
	• • • Electrostatics	• • • Elektrostatik
	• • • Electrodynamics	• • • Elektrodynamik
	• • • Magnetism	• • • Magnetismus
NETW.TH	• • • Network Theory	• • • Netzwerktheorie
LINE.TH	• • • Line Theory	• • • Leitungstheorie
HF	• • • High Frequency	• • • Hochfrequenztechnik
MICROW	• • • Microwaves	• • • Mikrowellentechnik
ELECTRON	• • ELECTRONICS	• • ELEKTRONIK
COMPON	• • • Components and Devices	• • • Bauelemente u. Bauteile
	• • • • Passive Components	• • • • Passive Bauelemente
	• • • • Electron Tubes	• • • • Elektronenröhren
MICROEL	• • • Semiconductors and Microelectronics	• • • Halbleitertechnik und Mikroelektronik
OPTOEL	• • • Optoelectronics	• • • Optoelektronik
CIRC.ENG	• • • Circuit Engineering and	• • • Schaltkreistechnik und
	• • • Pulse Technique	• • • Impulstechnik
POWER EL	• • • Power Electronics	• • • Leistungselektronik
CONS.EL	• • • Consumer Electronics	• • • Konsumelektronik
	• • • • Entertainment Electron.	• • • • Unterhaltungselektr.
	• • • • Household Electronics	• • • • Haushaltselektronik
	• • • • Car Electronics	• • • • Fahrzeugelektronik
EL.ACOUS	• • ELECTROACOUSTICS	• • ELEKTROAKUSTIK
SIGN.ENG	• • SIGNALING ENGINEERING	• • SIGNALISIERUNGSTECHNIK
	• • • Security Electronics	• • • Sicherheitselektronik
RAILW.SIGN	• • • Railway Signaling	• • • Eisenbahnsignaltechnik
	• • • Traffic Signaling	• • • Straßenverkehrstechnik
CONTROL	• • CONTROL ENGINEERING	• • STEUER-u.REGELUNGSTECH.
	• • • Cybernetics	• • • Kybernetik
	• • • Automation	• • • Automation
	• • • Robotics	• • • Robotik

INSTR	•• ELECTRIC INSTRUMENTATION		•• ELEKTRI.MESSTECHNIK
EQUIP.ENG	•• EQUIPMENT ENGINEERING AND DESIGN		•• GERÄTETECHNIK UND KONSTRUKTION
INF.TECH	•• INFORMATION TECHNOLOGY		•• NACHRICHTENTECHNIK
	••• FUNDAMENTALS OF IT		••• THEOR. NACHRICHTEN-TECHNIK
	•••• Signal Theory		•••• Signaltheorie
INF	•••• Information Theory		•••• Informationstheorie
CODING	•••• Coding		•••• Codierung
MODUL	•••• Modulation		•••• Modulation
ENG.LOG	•••• Engineering Logic		•••• Schaltalgebra
	••• COMPUTING		••• DATENTECHNIK
DATA PROC	•••• Data Processing		•••• Datenverarbeitung
DATA COMM	•••• Data Communications		•••• Datenübermittlung
	••••• Data Switching		••••• Datenvermittlung
	••••• Data Transmission		••••• Datenübertragung
	••••• Data Networks (LAN, WAN,..)		••••• Datennetze (LAN, WAN,..)
TERM&PER	••• TERMINALS & PERIPHERALS		••• ENDGERÄTE UND PERIPHERIEGERÄTE
	•••• Telephone Terminals		•••• Fernsprechendgeräte
	•••• Data Terminals		•••• Datenendgeräte
	•••• Text Terminals		•••• Textendgeräte
TELEC	••• TELECOMMUNICATIONS		••• TELEKOMMUNIKATION
TELEPH	•••• Telephony		•••• Fernsprechtechnik
TELEGR	•••• Telegraphy		•••• Fernschreibtechnik
TV	•••• Television		•••• Fernsehtechnik
RADIO	•••• Radio Engineering		•••• Funktechnik
ANT	••••• Antennas		••••• Antennentechnik
RADIO PROP	••••• Radio Propagation		••••• Funkausbreitung
HF.COMM	••••• HF Communication		••••• Kurzwellenfunk
BROADC	••••• Broadcasting Engineering		••••• Rundfunktechnik
RADIO REL	••••• Radio Relay		••••• Richtfunktechnik
RADIO LOC	••••• Radio Location and Radio Monitoring		••••• Funkortung und Funküberwachung
RADIO NAV	••••• Radio Navigation		••••• Funknavigation
SAT.COMM	••••• Satellite Communication		••••• Satellitenfunk
MOB.COMM	••••• Mobile Communication		••••• Mobilfunk
	•••••• Cellular Radio		•••••• Zellularfunk
	•••••• Land Mobile Telephony		•••••• Mobiler Landfunk
COMM.CABLE	•••• Communications Cables		•••• Nachrichtenkabel
OPT.COMM	•••• Optical Communications		•••• Opt.Nachrichtentechnik
OUTS.PLANT	•••• Outside Plant		•••• Linientechnik
SWITCH	•••• Communications Switching		•••• Nachrichten-Vermittlungstechnik
TRANS	•••• Communications Transmission		•••• Nachrichten-Übertragungstechnik
TELECONTR	•••• Telecontrol Engineering		•••• Fernwirktechnik
	••••• Telemonitoring		••••• Fernüberwachen
	••••• Telecommand		••••• Fernsteuern

MIL.COMM	• • • • Military Communications	• • • • Mil.Nachrichtentechnik

SYS.INST	• • • • System and
	  Station Installation
		• • • • Anlagen- und
		  Amtsbautechnik

MANUF	• • MANUFACTURING ENGINEERING	• • FERTIGUNGSTECHNIK
POWER ENG	• • ELECTRIC POWER ENGINEERING • • • Electric Machines Engin. • • • • Generators • • • • Transformers • • • • Electric Motors • • • Industrial Engineering • • • • Power Stations • • • • Power Distribution • • • • Electric Traction	• • ELEKTRI. ENERGIETECHNIK • • • Elektromaschinenbau • • • • Generatoren • • • • Transformatoren • • • • Elektromotoren • • • Elektri. Anlagentechnik • • • • Kraftwerkstechnik • • • • Elektrische Netze • • • • Elektrische Antriebe
POWER.SYS	• • • • Power Systems • • • • Pumps Engineering	• • • • Stromversorgungsanlagen • • • • Pumptechnik
EL.INST	• • • El.Installation Engineering • • • • Power Cables • • • • Lighting Engineering • • • • Air Conditioning • • • • Ventilating Engineering • • • Domestic El. Appliances • • • Electrochemistry and • • • Electrometallurgy • • • Indus.Electrical Heating • • • • Induct.Electroheating • • • • Electric Welding	• • • Elektri.Istallationstechnik • • • • Energiekabeltechnik • • • • Beleuchtungstechnik • • • • Klimatechnik • • • • Lüftungstechnik • • • Haushaltgerätetechnik • • • Elektrochemie und • • • Elektrometallurgie • • • Industr. Heiztechnik • • • • Induktionsheizung • • • • Elektri.Schweißtechnik
MECH	• MECHANICAL ENGINEERING • • Machine Elements • • Moving Machines • • Machine Tools • • Apparatus Engineering • • Transportation Engineering	• MASCHINENBAU • • Maschinenelemente • • Kraftmaschinen • • Werkzeugmaschinen • • Apparatebau • • Transporttechnik
ENG.DRAW	• Engineering Drawing	• Technisches Zeichnen
TYPOGR	• Typography and Printing	• Druckwesen
AERON	• Aeronautics	• Luft-u.Raumfahrt
CIV.ENG	• CIVIL ENGINEERING • • Building Technology • • Underground Construction	• BAUWESEN • • Hochbau • • Tiefbau
POST	Postal Services	Postwesen
DOC	Documentation	Dokumentationswesen
OFFICE	Office Systems	Bürowitschaft

Bibliography
Literaturhinweise

A. Dictionaries and Handbooks
Lexika und Handbücher

Bachmann, B.: Großes Lexikon der Computerfachbegriffe, 1. Auflage 1990, IWT Verlag, Vaterstetten bei München

Bosch, R.; Daemisch, K.: Stichwörter der Telekommunikation, 1988, Fachverlag Schiele & Schön, Berlin

Collin, S. M. H.: Dictionary of Computing, 1988, Peter Collin Publishing, Teddington (Middlesex)

Drosdowski, G. (Herausgeber): DUDEN, Deutsches Universalwörterbuch, 1983, Dudenverlag, Mannheim-Wien-Zürich

Falkner, R. (Bearb.): Lexikon der modernen Elektronik, 1980, Verlag Markt & Technik, München

Feichtinger, H.: Fachlexikon der Computer, 1988, Franzis Verlag, München

Fellbaum, K.; Hartlep, R.: Lexikon der Telekommunikation, 2. Auflage, 1985, VDE-Verlag, Berlin-Offenbach

Frank, J. (Editor in Chief); Goetz, J. A. (Chairman): IEEE Standard Dictionary of Electrical and Electronic Terms, 4th Edition, 1988, The Institute of Electrical and Electronics Engineers, Inc., New York, NJ

Gusbeth, H.: Mobilfunk-Lexikon, 1. Auflage, 1990, Franzis-Verlag, München

Hille, K.; Krischke, A.: Das Antennen-Lexikon, 1988, Verlag für Technik und Handwerk, München

IEEE Standard Computer Dictionary: A Compilation of IEEE Standard Computer Glossaries, 1990, The Institute of Electrical and Electronics Engineers, New York

Kaltenbach, T.; Reetz, U.; Woerrlein, H.: Das große Computer Lexikon, 2. Auflage, 1990, Markt&Technik Verlag AG, Haar bei München

Lange, R.; Watzlawik, G.: Glossar CAD/CAM, 2. Auflage 1987, Siemens AG, Berlin-München

Langley, G.: telephony's Dictionary, 2nd Edition, 1986, telephony Publishing Company, Chichago

Mache, W.: Lexikon der Text- und Datenkommunikation, 1980, R.Oldenburg Verlag, München-Wien

Markus, J.: Electronics Dictionary, 4th Edition, 1978, McGraw-Hill Book Company, New York-St.Louis-San Francisco

Müller, P.; Löbel, G.; Schmidt, H.: Lexikon der Datenverarbeitung, 10. Auflage, 1989, Verlag Moderne Industrie, Landsberg am Lech

Musiol, A.: Glossar Bürokommunikation, 1984, Siemens AG, Berlin-München

Neufang, O. (Herausgeber): Lexikon der Elektronik, 1983, Friedrich Vierweg & Sohn, Braunschweig-Wiesbaden

Parker, S. (Editor in chief): McGraw-Hill Dictionary of Electrical and Electronic Engineering, 1984, McGraw-Hill Book Company, New York-St.Louis-San Francisco

Pietsch, H.-J.: Amateurfunk-Lexikon, 3. Auflage, 1990, Franzis-Verlag, München

Pooch, H.; Heydel, J.; Schlolaut, R.: Fachwörterbuch des Nachrichtenwesens, 1976, Fachverlag Schiele & Schön, Berlin

Rentsch, S. B.: Begriffe der Elektronik, 1989, Franzis-Verlag, München

Schneider, H. J.: Lexikon der Informatik und Datenverarbeitung, 3. Auflage, 1990, R.Oldenbourg Verlag, München-Wien

Schulze, H. H.: Das roro Computer-Lexikon, 3. Auflage, 1988, Rowolt Taschenbuch Verlag, Reinbeck bei Hamburg

Schulz, K.-P.; Steinke, E.: Rundfunk-und Fernsehtechnik, Verlag Technik, Berlin

SYBEX: Mikrocomputer Lexikon, 7. Auflage, 1987, SYBEX-Verlag, Düsseldorf

tau-tron: Tau-tron's Pocket Dictionary, of Terms, Measurements, Acronyms, and Abbreviations, 2nd edition 1991, Tau-tron, Westford, MA

Telefunken-Fachbuch: Halbleiter-Lexikon, 1. Auflage 1965, Franzis-Verlag, München

The British Computer Society: A Glossary of Computer Terms, 6th Edition, 1989, Cambridge University Press, Cambridge-New York-Melbourne

Tiemeyer, E.(Schriftleiter): Lexikon Büro und Sekretariat, 1990, Orbis Verlag, München

Tornow, W.: Lexikon der Nachrichtentechnik, 1990, Fachverlag Schiele & Schön, Berlin

Ulrich, W.: Linguistische Grundbegriffe, 4. Auflage, 1987, Verlag Ferdinand Hirt AG, Unterägeri

Voltz, H.: Das große Wörterbuch der Computer-Fachbegriffe, 1989, Signum Medien Verlag, München

Wahrig, G.: Deutsches Wörterbuch, Ausgabe 1986, Bertelsmann Lexikon Verlag GmbH, Gütersloh/München

Walker, P. (General editor): Chambers Science and Technology Dictionary, 1988, Chambers/Cambridge, Cambridge-Edinburgh-New York-New Rochelle-Melbourne-Sydney

Webster's New World: Dictionary of Computer Terms, 3rd Edition, 1988, Webster's New World, New York

Webster's Seventh New Collegiate Dictionary, 1967, G.&C.Merriam Company Publishers, Springfield, Mass.

B. Monographs
Fachbücher

Bergmann, K. (Begründer): Lehrbuch der Fernmeldetechnik, Band I, Band II, 5. Auflage, 1986, Schiele & Schön

Bergmann-Schäfer: Lehrbuch der Experimentalphysik, Band I – Mechanik, Akustik, Wärmelehre, 3. Auflage, 1945, Walter de Gruyter, Berlin-New York

Bergmann-Schäfer: Lehrbuch der Experimentalphysik, Band II – Elektrizität, Magnetismus, 7. Auflage, 1987, Walter de Gruyter, Berlin-New York

Bergmann-Schäfer: Lehrbuch der Experimentalphysik, Band III – Optik, 8. Auflage, 1987, Walter de Gruyter, Berlin-New York

Bidlingmaier, M.; Haag, A.; Kühnemann, K.: Einheiten – Grundbegriffe – Meßverfahren der Nachrichten-Übertragungstechnik, 4. Auflage, 1973, Siemens AG, Berlin-München

Bremberger, M.: Konstruktion und Fertigung, Design and Production, 1960, R.Oldenbourg, München

Brodhage, H.; Hormuth, W.: Planung und Berechnung von Richtfunkverbindungen, 10. Auflage, 1977, Siemens AG, Berlin-München

Bronstein, I.N.; Semendjajew, K.A.: Taschenbuch der Mathematik, 2. Auflage, 1962, Harri Deutsch, Frankfurt a.M.

Freeman, R.L.: Radio System Design for Telecommunications (1–100 GHz), 1987, John Wiley and Sons, New York-Chichester-Brisbane-Toronto-Singapore

Freeman, R.L.: Reference Manual for Telecommunications Engineering, 1985, John Wiley and Sons, New York-Chichester-Brisbane-Toronto-Singapore

Gottwald, S.; Küstner, H.; Hellwich, M.; Kästner, H.: Handbuch der Mathematik, 1986, Buch und Zeit Verlagsgesellschaft, Köln

Heinrich, W.: Richtfunktechnik, 1988, R.v.Decker's Verlag G.Schenk, Heidelberg

Holleman, A.F.; Wiberg, E.: Lehrbuch der anorganischen Chemie, 37.–39. Auflage, 1956, Walter de Gruyter & Co, Berlin

Hölzler, E.; Thierbach, D.: Nachrichtenübertragung, 1966, Springer-Verlag, Berlin-Heidelberg-New York

ICC: Incoterms, 1980, International Chamber of Commerce, Paris

Jackson, J.D.: Classical Electrodynamics, 1962, John Wiley & Sons, New York-London

Kreyszig, E.: Statistische Methoden und ihre Anwendungen, 3. Auflage, 1968, Vanderhoeck & Ruprecht, Göttingen

Küpfmüller, K.: Einführung in die theoretische Elektrotechnik, 6. Auflage, 1959, Springer-Verlag, Berlin-Göttingen-Heidelberg

Lindner, H.; Bauer, H.; Lehmann, C.: Elektrotechnik – Elektronik, Formeln und Gesetze, 1988, Buch- und Zeit-Verlagsgesellschaft mbH, Köln

Mahlke, G.; Gössing, P.: Lichtwellenleiterkabel, 1986, Siemens AG, Berlin-München

Meadows, R.G.: Electric Network Analysis, 1972, Penguin Books, Hamondsworth-Baltimore-Victoria

Neuhaus, K.; Haltern, M.: Euro Business – Sprachführer Englisch, 1. Auflage, 1990, ILT-Verlag, Bochum

Ohmann, F.: Kommunikations-Endgeräte, 1983, Springer-Verlag, Berlin-Heidelberg-New York-Tokyo

Philippow E.(Herausgeber): Taschenbuch Elektrotechnik, Band 3, Nachrichtentechnik, 1967, VEB Verlag Technik, Berlin

Pooch, H.; Köhler, K.; Gräber, H.-J.: Richtfunktechnik, 1974, Fachverlag Schiele & Schön, Berlin

Pribich, K.; Haslinger, H.: Bauelemente Nachrichtentechnik, 1969, Bohmann-Verlag, Heidelberg

Rahmig, G.: Niederfrequenz-Übertragungstechnik, 1972, Verlag Berliner Union, Stuttgart

Robbin, J.: Das DOS 4.0 Buch, 1989, Sybex, Düsseldorf-San Francisco-Paris-London-Arnheim

Schubert, W.: Nachrichtenkabel und Übertragungssysteme, 2. Auflage, 1980, Siemens AG, Berlin-München

Siemens: Technical Tables, Quantities, Formulas, Definitions, 1986, Siemens AG, Berlin-München

Siemens: Technische Tabellen, Größen, Formeln, Begriffe, 1986, Siemens AG, Berlin-München

Steinbuch, K.; Rupprecht, W.: Nachrichtentechnik, 3. Auflage, 1982, Springer-Verlag, Berlin-Heidelberg-New York

Turner, L.W.: Electronics Engineer's Reference Book, 4th Edition, 1976, Newnes-Butterworth, London

Woller, H.: Neuzeitliche Fernsprechvermittlungstechnik, 1968, Telekosmos-Verlag, Stuttgart

C. Industrial Catalogues
Industriekataloge

Andrew: Catalog 33, 1986, Andrew Corporation, Orland Park, IL

Farnell: Electronic Components, 1990, Deisenhofen b.München

Grundig: revue'88, 1988, Grundig AG, Fürth

Hewlett-Packard: Test & Measuring Catalog, 1989, Hewlett-Packard Company, Palo Alto, CA

Hewlett-Packard: Test- und Meßtechnik, Hauptkatalog, 1989, Hewlett-Packard GmbH, Böblingen

RADIO-RIM: RIM electronic 88, 1988, RADIO-RIM GmbH, München

Rhode&Schwarz: Measuring Equipment Catalog 90/91, 1990, Rhode&Schwarz, München

Rhode&Schwarz: Meßgerätekatalog 90/91, 1990, Rhode&Schwarz, München

Salhöfer Elektronik: Spezial-Katalog, 1987, Kulmbach

Sennheiser: bestseller, 1987, Sennheiser electronic KG, Wedemark

D. Standards
Normen

DIN 1301: Einheiten, Einheitennamen, Einheitenzeichen

DIN 1320: Allgemeine Benennungen in der Akustik

DIN 2140: Textautomaten, Begriffe

DIN 2148: Tastaturen, Begriffe

DIN 9784: Druckeinrichtungen, Begriffe

DIN 40146: Begriffe der Nachrichtenübertragung, Grundbegriffe

DIN 40148: Übertragungssysteme und Vierpole

DIN 44300: Informationsverarbeitung, Begriffe

DIN 44301: Informationstheorie, Begriffe

DIN 44302: Datenübertragung, Datenübermittlung, Begriffe

DIN 44330: Telegrafentechnik und Telegrafie-Einrichtungen für Datenübertragung, Begriffe

DIN 66010: Magnetbänder, Begriffe

DIN 66200: Betrieb von Rechnern, Begriffe

DIN 66201: Prozeßrechner, Begriffe

DIN 66214: Lochstreifentechnik, Begriffe, NORM

DIN 66233: Bildschirmarbeitsplätze, Begriffe,

DIN 66257: NC-Maschinen, Begriffe

DIN 69901: Projektmanagement, Begriffe

NTG 0101: Modulationstechnik, Begriffe, 1971

NTG 0103: Impuls- und Modulationstechnik, Begriffe

NTG 0104: Codierung, Grundbegriffe, 1982

NTG 0902: Nachrichtenvermittlungstechnik, Begriffe, 1982

NTG 0903: Nachrichtenverkehrstheorie, Begriffe, 1983

NTG 1203: Daten- und Textkommunikation, Begriffe, 1981

NTG 2001: Fernwirktechnik, Begriffe, 1978

E. Multilingual Dictionaries
Mehrsprachige Fachwörterbücher

AT&T: Telecommunications Glossary, English, Dutch, German, French, Spanish, Italian, Japonese, 1987, AT&T Costumers Information Center, Indianapolis

Attiyate, Y.H.; Shah, R.R.: Wörterbuch der Mikroelektronik und Mikrorechnertechnik, Deutsch-Englisch/Englisch-Deutsch, Dictionary of Microelectronics and Microcomputer Technology, German-English/English-German, 1984, VDI-Verlag, Düsseldorf

Bezner, H. (Bearbeiter): Fachwörterbuch Energie- und Automatisierungstechnik; Band 1: Deutsch-Englisch, Dictionary of Power Engineering and Automation; Vol.1: German-English, 2. Auflage, 1988, Siemens AG, Berlin-München

Bezner, H. (Bearbeiter): Fachwörterbuch der Energie- und Automatisierungstechnik; Band 2: Englisch-Deutsch, Dictionary of Power Engineering and Automation; Vol.2: English-German, 2. Auflage, 1988, Siemens AG, Berlin-München

Bezner, H.: Elektromaschinen-Wörterbuch, Deutsch-Englisch/Englisch-Deutsch, Electrical Machines Dictionary, German-English/English-German, 1978, Oscar Brandstetter Verlag, Wiesbaden

Bindmann, W.: Fachwörterbuch der Mikroelektronik, Englisch-Deutsch/Deutsch-Englisch, 2. Auflage, 1987, Dr. Alfred Hüthig Verlag, Heidelberg

Böhss, G.: Fachwörterbuch Leiterplatten, Deutsch-Englisch/Englisch-Deutsch, 2. Auflage, 1990, Dr. Alfred Hüthig Verlag, Heidelberg

Brendel, H.; Grießmann, A.: Großes iwt-Wörterbuch der Computertechnik und der Wirtschaftsinformatik, Deutsch-Englisch/Englisch-Deutsch, 1. Auflage, 1987, IWT-Verlag, Vaterstetten bei München

Brinkmann, K. H.: Wörterbuch der Daten- und Kommunikationstechnik; Deutsch-Englisch/Englisch-Deutsch, Dictionary of Data Systems and Communications; German-English/English-German, 4. Auflage, 1989, Oscar Brandstetter Verlag, Wiesbaden

Budig, P.-K.: Fachwörterbuch Elektrotechnik, Elektronik, Band 1: Englisch-Deutsch, 5. Auflage, 1989, Dr. Alfred Hüthig Verlag, Heidelberg

Carl-Amkreutz: Wörterbuch der Datenverarbeitung; 2 Bände; Deutsch-Englisch-Französisch, Dictionary od Data Processing; 2 Volumes; German-English-French, 4. Auflage, 1987, Datakontekt Verlag, Köln

Daubach, G.: Wörterbuch der Computerei, Englisch-Deutsch/Deutsch-Englisch, 8. Auflage, 1989, IWT-Verlag, Vaterstetten bei München

De Vries, L.; Clason, W. E.: Wörterbuch der Reinen und Angewandten Physik; Band 1: Deutsch-Englisch, Dictionary of Pure and Applied Physics; Vol.1: German-English, 1964, R. Oldenbourg Verlag, München-Wien

De Vries, L.; Clason, W. E.: Wörterbuch der Reinen und Angewandten Physik; Band 2: Englisch-Deutsch, Dictionary of Pure and Applied Physics; Vol.2: English-German, 1964, R. Oldenbourg Verlag, München-Wien

Dorian, A. F.: Handwörterbuch der Naturwissenschaft und Technik; Deutsch-Englisch, Dictionary of Science and Technology; German-English, 2. Auflage, 1981, Elsevier Scientific Publishing Company, Amsterdam-Oxford-New York

Falkner, R.: Lexikon der modernen Elektronik, Englisch-Deutsch, 1980, Verlag Markt&Technik, München

Goedecke, W.: Wörterbuch der Elektrotechnik, Fernmeldetechnik und Elektronik; Deutsch-Englisch-Französich, 2. Aufl. 1968, Oscar Brandstetter Verlag, Wiesbaden

IBM: Fachausdrücke der Informationsverarbeitung, Wörterbuch und Glossar, Englisch-Deutsch/Deutsch-Englisch, 1990, IBM

Jauß, D.; Villani, C.: Großes IWT-Wörterbuch Datenverarbeitung und Programmiertechnik; Englisch-Deutsch/Deutsch-Englisch, 1. Auflage 1989, IWT-Verlag, Vaterstetten bei München

Junge, H.-D.: parat Dictionary of Information Technology; English-German, parat Wörterbuch Informationstechnologie; Englisch-Deutsch, 1. Auflage, 1989, VCH, Weinheim

Junge, H.-D.: parat Wörterbuch Informationstechnologie; Deutsch-Englisch, parat Dictionary of Information Technology; German-English, 1. Auflage, 1990, VCH, Weinheim

Klause, G.: CAD, CAE, CAM, CIM Lexikon, Deutsch-Englisch/Englisch-Deutsch, 1987, expert-Verlag, Vogel-Verlag, Ehningen

Kucera, A.: The Compact Dictionary of Exact Science and Technology; English-German, Compact Wörterbuch der exakten Naturwissenschaften und Technik; Englisch-Deutsch, 2. Auflage, 1989, Oscar Brandstetter Verlag, Wiesbaden

Kucera, A.: The Compact Dictionary of Exact Science and Technology; German-English, Compact Wörterbuch der exakten Naturwissenschaften und Technik; Deutsch-Englisch, 1. Auflage, 1982, Oscar Brandstetter Verlag, Wiesbaden

Metzger, R. M.: Fachwörtersammlung moderner Telekommunikation; Deutsch-Englisch-Französisch, Glossary of Telecommunication Terms; German-English-French, Standard Telephon und Radio AG, Zürich

Neufang, O.; Rühl, H.: Elektronik-Wörterbuch, Englisch-Deutsch, 3. Auflage, 1988, Verlag Zimmermann-Neufang, Ulmen

Oppermann, A.: Wörterbuch der Elektronik; Deutsch-Englisch, 1. Auflage, 1980, Alfred Oppermann, Aeronautischer Verlag, Baldham bei München

Profos, P. (Herausgeber): Lexikon und Wörterbuch der industriellen Meßtechnik, Deutsch-Englisch-Französisch, 2. Auflage, 1986, Vulkan-Verlag, Essen

Renouard, H. v.: Fachwörterbuch Neue Informations- und Kommunikationsdienste, Deutsch-Englisch/Englisch-Deutsch, 1989, Dr. Alfred Hüthig Verlag, Heidelberg

Schwenkhagen, H. F.; Meinhold, H.: Dictionary of Electrical and Electronic Engineering; German-English/English-German, Wörterbuch Elektrotechnik und Elektronik; Deutsch-Englisch/Englisch-Deutsch, 5. Auflage, 1989, Verlag W. Girardet, Essen

Tietz, W.: Wörterbuch der Datenkommunikation; Englisch-Deutsch/Deutsch-Englisch, Dictionary of Data Communication Terms; English-German/German-English, 1. Auflage, 1984, R. v. Decker's Verlag, G. Schenck, Heidelberg

UIT: CCIR XVth Plenary Assembly, Yellow Book, Vol.XIII, Vocabulary (CMV) – English, 1982, UIT, Geneve

UIT: CCITT VIIIth Plenary Assembly, Málaga 1984, Red Book, Vol.X, Fasc. X.1, Terms and Definitions – English, 1985, UIT, Geneve

Vollnhals, O.: Elsevier's Dictionary of Personal and Office Computing; English-German-French-Italian-Portuguese, 1st Edition, 1984, Elsevier Science Publishers, Amsterdam-Oxford-New York-Tokyo

Vollnhals, O.: Wörterbuch Desktop Publishing; Englisch-Deutsch/Deutsch-Englisch, 1. Auflage, 1989, Siemens AG, Berlin-München

Wennrich, P.: Wörterbuch der Elektronik und Informationsverarbeitung; Band 1: Englisch-Deutsch, Dictionary of Electronics and Information Processing; Vol.1: English-German, 1. Auflage, 1990, K.G.Saur, München-London-New York-Paris

Wennrich, P.: Wörterbuch der Elektronik und Informationsverarbeitung; Band 2: Deutsch-Englisch, Dictionary of Electronics and Information Processing; Vol.2: German-English, 1. Auflage, 1990, K.G.Saur, München-London-New York-Paris

Wernicke, H.: Lexikon der Elektronik, Nachrichten- und Elektrotechnik; Band 1: Englisch-Deutsch, Dictionary of Electronics, Communications and Electrical Engineering; Vol.1: English-German, 4. Auflage, 1985, H.Wernicke Verlag, Deisenhofen

Wernicke, H.: Lexikon der Elektronik, Nachrichten- und Elektrotechnik; Band 2: Deutsch-Englisch, Dictionary of Electronics, Communications and Electrical Engineering; Vol.2: German-English, 4. Auflage, 1985, Verlag H.Wernicke, Deisenhofen

Wittmann, A.; Klos, J.: Wörterbuch der Datenverarbeitung; Deutsch-Englisch-Französisch, Dictionary of Data Processing; German-English-French, 5. Auf., 1987, R.Oldenbourg Verlag, München-Wien

A

' (→minute)	PHYS	(→Minute)	
A (→ampere)	PHYS	(→Ampere)	
A (→area)	MATH	(→Fläche)	
-a) (→cube 2)	MATH	(→Würfel)	
Å (→Ångström)	PHYS	(→Ångström)	
a (→are)	PHYS	(→Ar)	
a (→year)	PHYS	(→Jahr)	
A and B signaling (→AB signaling)	TRANS	(→AB-Signalisierung)	

abampere PHYS **Abampere**
[CGS unit for current; = 10 A] [CGS-Einheit für Strom; = 10 A]

abandon call DATA COMM **erfolglose Wahl**
AB bits TRANS **AB-Bits**
abbreviated address calling SWITCH (→Kurzwahl)
 (→abbreviated dialing)
abbreviated addressing DATA PROC **verkürzte Adressierung**
abbreviated calling SWITCH (→Kurzwahl)
 (→abbreviated dialing)
abbreviated code 2 SWITCH **verkürzte Kennzahl**
[abbreviation of region/country code] ≈ Kurzrufnummer
≈ abbreviated number
abbreviated code 1 SWITCH (→Kurzrufnummer)
 (→abbreviated number)
abbreviated code dialing SWITCH (→Kurzwahl)
 (→abbreviated dialing)
abbreviated dialing SWITCH **Kurzwahl**
= abbreviated calling; instant dialing; = Kurzrufnummernwahl;
 abbreviated code dialing; short code Schnellwahl
 dialing; abbreviated address calling;
 speed calling
abbreviated number SWITCH **Kurzrufnummer**
= abbreviated code 1; short code ≈ verkürzte Kennzahl
≈ abbreviated number
abbreviated-number generator SWITCH **Kurzrufnummern-Geber**
abbreviated specification TECH **Kurzspezifikation**
= short spec
abbreviation LING **Abkürzung**
 = Abbreviation; Kurzbezeichnung
abcd parameter (→chain NETW.TH (→Kettenparameter)
 parameter)
abcoulomb PHYS **Abcoulomb**
[CGS unit for el. charge; = 10 C] [CGS-Einheit für el. Ladung; = 10 C]
abend (→abnormal end) DATA PROC (→Absturz)
aberration 1 PHYS **Abbildungsfehler**
 = Aberration; Bildfehler
aberration 2 (→deviation) PHYS (→Abweichung)
aberration function (→error MATH (→Fehlerfunktion)
 function)
abhenry PHYS **Abhenry**
[CGS unit for inductance; = 10 H] [CGS-Einheit für Induktivität; = 10 H]
ABM (→asynchronous DATA COMM (→gleichberechtigter Spontbalanced mode) anbetrieb)
abnormal (→erratic 2) TECH (→anormal)
abnormal end DATA PROC **Absturz**
[unwanted stoppage] [ungewollter]
= abend; abnormal termination; forced = Zusammenbruch
 termination; cancel (n.); truncation; = Abbruch
 program crash; crash; drop dead halt; ↓ Systemabsturz; Programmdead halt; close down (n.); bomb (n.) absturz
↓ system crash; program crash
abnormal termination DATA PROC (→Absturz)
 (→abnormal end)
abohm PHYS **Abohm**
[CGS unit for el. resistance; = 10 [CGS-Einheit für el. Widerohm] stand; = 10 Ohm]
abolition (→cessation) COLLOQ (→Wegfall)
abort (v.t.) DATA PROC **abbrechen**
= kill = anhalten
≈ reset (v.t.) ≈ rücksetzen
abort DATA PROC **Abbruch**
[wanted termination] [gewollter]
= abortion; termination = Programmabbruch
≈ abnormal end ≈ Absturz
abort (→block abort) DATA COMM (→Blockabbruch)

abortion (→abort) DATA PROC (→Abbruch)
above average COLLOQ **überdurchschnittlich**
≈ extraordinary ≈ außerordentlich
above ground TECH **oberirdisch**
= overhead
abrade TECH **abschleifen**
= wear down = abreiben
↑ grind ↑ schleifen
abrasion TECH **Abschleifen**
 = Abrieb
abrasion-proof MECH **abriebfest**
abrasion resistance MECH **Abriebfestigkeit**
abrasive compound TECH **Schleifmittel**
= abrasive n. = Schmirgel
abrasive n. (→abrasive TECH (→Schleifmittel)
 compound)
abrasive paper (→emery paper) TECH (→Schmirgelpapier)
abrasive paste TECH **Schmirgelpaste**
 = Polierpaste
abrasive powder TECH **Schleifpulver**
 = Polierpulver
abrasive tool TECH **Schleifwerkzeug**
 = Polierwerkzeug
abrasive trimming MICROEL **Schleiftrimmen**
abridge LING **kürzen**
= condense ≈ zusammenfassen
≈ summarize
abridge (→cut) ECON (→kürzen)
abridgment (BRI) (→abridgment) LING (→Kürzung)
abridgment (NAM) LING **Kürzung**
= abridgement (BRI); condensation
abridgment (→cut) ECON (→Kürzen)
abrupt TECH **schlagartig**
≈ radical = abrupt; plötzlich
 ≈ radikal
abrupt junction MICROEL **abrupter Übergang**
[abrupt transition from one impurity [von einer Störstellenkonzentration to other] zentration zur anderen]
= step junction = Stufensperrschicht
abrupt varactor diode MICROW **Abrupt-Varaktordiode**
abscissa MATH **Abszisse**
= X-coordinate; X-axis = X-Achse
absence of discontinuities ELECTRON **Sprungfreiheit**
absence of flux leakage EL.TECH **Streuungsfreiheit**
= zero-flux leakage; absence of stray
absence of stray (→absence of EL.TECH (→Streuungsfreiheit)
 flux leakage)
absent-subscriber service TELEPH (→Abwesenheitsdienst)
 (→telephone answering service)
AB signaling TRANS **AB-Signalisierung**
["bit stealing" signaling in 1,5 Mbit/s ["bitstehlende" Signalisierung in 1,5-Mbit/s-Systemen]
 (T1) systems]
= A and B signaling
absolute address DATA PROC **absolute Adresse**
[allows direct store access, without [erlaubt, ohne Umrechmodifications address] nung, direkten Zugriff auf
= real address; machine address; spe- Speicherplatz indirekte
 cific address; physical address; actual Adresse]
 address 2 = Absolutadresse; Maschi≈ direct address; current address nenadresse; echte Adresse;
≠ floating address; virtual address; sym- reale Adresse; Realadresbolic address; indirect se; tatsächliche Adresse;
 wirkliche Adresse; physikalische Adresse; effektive
 Adresse; Effektivadresse
 ≈ direkte Adresse; Momentanadresse
 ≠ relative Adresse; virtuelle
 Adresse; symbolische
 Adresse;
absolute addressing DATA PROC **absolute Adressierung**
[indication of real store place] [Angabe des echten Speicherplatzes]
= actual addressing; real addressing; = Absolutadressierung;
 machine addressing; physical address- Maschinenadressierung;
 ing; specific addressing echte Adressierung; reale
 Adressierung; Realadressierung; effektive Adressierung; Effektivadressierung;
 aktuelle Adressierung

absolute assembler	DATA PROC	**Absolutassemlierer** = absoluter Assemblierer; Absolutassembler; absoluter Assembler	abstract (→compendium)	LING	(→Abriß)
			abstract automat	ENG.LOG	abstrakter Automat
			abstract mathematics	MATH	reine Mathematik
absolute code (→machine code 1)	DATA PROC	(→Maschinencode 1)	abstract word ≠ concrete word	LING	**Begriffswort** = Abstraktum ≠ Konkretum
absolute coding = absolute programming ≠ symbolic coding	DATA PROC	**absolute Codierung** = Absolutcodierung; absolute Programmierung ≠ symbolische Codierung	abutment	MECH	**Widerlager** = Stützlager
			abvolt [CGS unit for el. voltage; = 10 V]	PHYS	**Abvolt** [CGS-Einheit für el. Spannung; = 10 V]
absolute error	INSTR	**absoluter Fehler** = Absolutfehler	AC (→alternating current)	EL.TECH	(→Wechselstrom)
absolute frequency	MATH	absolute Häufigkeit	ac (IEEE) (→alternating current)	EL.TECH	(→Wechselstrom)
absolute humidity	PHYS	**absolute Feuchtigkeit** = absolute Feuchte	a-c (→alternating current)	EL.TECH	(→Wechselstrom)
absolute instruction [requiring no other data]	DATA PROC	**vollständiger Befehl** = vollständige Anweisung	a.c. (IEC) (→alternating current)	EL.TECH	(→Wechselstrom)
absolute jump	DATA PROC	absoluter Sprung	Ac (→actinium)	CHEM	(→Actinium)
absolute level = actual level (BRI)	TELEC	absoluter Pegel	academic engineer = graduated engineer	SCIE	**Hochschulingenieur** = Diplom-Ingenieur
absolute loader [programm to load programs into the main memory, with fixed loading address] ≠ relocatable loader	DATA PROC	**Absolutlader** [Programm zum Laden von Programmen in den Arbeitsspeicher, mit festgelegter Ladeadresse] ≠ Relativlader	academic maturity	SCIE	Hochschulreife
			ac bridge = alternating current bridge	EL.TECH	Wechselstrombrücke
			accelerating algorithm (→acceleration algorithm)	MATH	(→Beschleunigungsalgorithmus)
absolute maximum ratings	COMPON	**Grenzdaten** = Grenzwerte	accelerating factor	QUAL	Zeitraffungsfaktor
absolute power level	TELEC	absoluter Leistungspegel	accelerating field (→acceleration field)	ELECTRON	(→Beschleunigungsfeld)
absolute program [written in absolute code] ≈ object program	DATA PROC	**Absolutprogramm** [in Absolutcode geschrieben] ≈ Objektprogramm	accelerating force	PHYS	Beschleunigungskraft
			accelerating grid	ELECTRON	Beschleunigungsgitter
			accelerating lens	ANT	Beschleunigungslinse
absolute programming (→absolute coding)	DATA PROC	(→absolute Codierung)	accelerating lens (→acceleration lens)	ELECTRON	(→Beschleunigungslinse)
absolute standard	INSTR	**Primärnormal** = absolutes Normal	accelerating voltage	ELECTRON	Beschleunigungsspannung
absolute temperature (→thermodynamic temperature)	PHYS	(→thermodynamische Temperatur)	acceleration [increment of velocity in the unit of time]	PHYS	**Beschleunigung** [Geschwindigkeitszuwachs in der Zeiteinheit]
absolute value = magnitude	MATH	**Absolutwert** = Betrag	acceleration ↑ time scaling	TECH	**Zeitraffung** ↑ Zeitmaßstabsänderung
absolute value of admittance (→magnitude of admittance)	NETW.TH	(→Betrag des Scheinleitwerts)	acceleration algorithm = accelerating algorithm	MATH	Beschleunigungsalgorithmus
absolute value of impedance (→magnitude of impedance)	NETW.TH	(→Betrag des Scheinwiderstandes)	acceleration distance	PHYS	Beschleunigungsweg
			acceleration electrode	ELECTRON	Beschleunigungselektrode
absolute value resonance	NETW.TH	Betragsresonanz	acceleration field = accelerating field	ELECTRON	Beschleunigungsfeld
absolute voltage level	TELEC	absoluter Spannungspegel			
absolute zero	PHYS	absoluter Nullpunkt	acceleration lens = accelerating lens; immersion lens	ELECTRON	**Beschleunigungslinse** = Immersionslinse
absorb	TECH	aufsaugen			
absorbed power (→power input)	EL.TECH	(→Leistungsaufnahme)	acceleration of gravity [SI unit: m/s²] = g ≈ gravitational constant	PHYS	**Fallbeschleunigung** [SI-Einheit: m/s²] = Erdanziehungskonstante; Erdbeschleunigungskonstante; g ≈ Gravitationskonstante
absorber	ANT	Absorber
absorptance (→absorption factor)	PHYS	(→Absorptionsfaktor)
absorption	PHYS	Absorption
absorption area (→effective aperture)	ANT	(→Antennenwirkfläche)	acceleration resistant	TECH	beschleunigungsfest
			acceleration room	ELECTRON	Beschleunigungsraum
absorption band	PHYS	Absorptionsband	acceleration sensor	COMPON	**Beschleunigungssensor** = Beschleunigungsfühler
absorption capacity (→absorptivity)	PHYS	(→Absorptionskapazität)			
			acceleration time	TERM&PER	Anlaufzeit
absorption clamp	INSTR	Absorptions-Meßwandlerzange	accelerator	PHYS	Beschleuniger
			accelerometer	INSTR	Beschleunigungsmesser
absorption coefficient	PHYS	Absorptionskoeffizient	accent 1 (n.) ≈ accentuation ↑ diacritic mark ↓ acute accent; grave accent; circumflex; accent 2	LING	**Akzent** = Akzentzeichen; Tonzeichen; Betonungszeichen ≈ Betonung ↑ diakritisches Zeichen ↓ Akut; Gravis; Zirkumflex; Apex 1
absorption edge	PHYS	Absorptionskante
absorption factor = absorptance	PHYS	**Absorptionsfaktor** = Absorptionsgrad
absorption fading	RADIO PROP	**Absorptionsschwund** = Dämpfungsschwund
absorption loss	PHYS	Absorptionsverlust	accent 2 (n.) [a mark ^ or ' indicating long vocals] ↑ accent 1	LING	**Apex 1** [lange Selbstlaute markierendes Zeichen ^ oder ,] ↑ Akzent
absorption maximum	PHYS	**Absorptionsspitze** = Absorptionsmaximum
absorption power meter	INSTR	**Absorptionsleistungsmesser** = Endleistungsmesser
absorption spectrum	PHYS	Absorptionsspektrum	accent 3 (n.) (→acute accent)	LING	(→Akut)
absorption trap	PHYS	Absorptionsfalle	accent aigu (→acute accent)	LING	(→Akut)
absorptivity = absorption capacity	PHYS	**Absorptionskapazität** = Absorptionsvermögen; Aufnahmefähigkeit	accented	LING	akzentuiert
			accent grave (→grave accent)	LING	(→Gravis)
			accent key	TERM&PER	Akzenttaste
AB stereophony	EL.ACOUS	AB-Stereophonie	accept (v.t.)	TECH	**annehmen** = akzeptieren

acceptable quality level = AQL	QUAL	**Annahmegrenze** = annehmbare Qualitätsgrenzlage; AQL
acceptance	TECH	**Abnahme** ≈ Übergabe
acceptance	COLLOQ	**Akzeptanz** = Aufnahme
acceptance angle	ELECTRON	**Einfangwinkel** = Eintrittswinkel
acceptance certificate = protocol of acceptance	ECON	**Abnahmeprotokoll** = Übergabeprotokoll; Abnahmeakte; Übergabeakte
acceptance characteristic	QUAL	**Annahmekennlinie**
acceptance condition ≈ acceptance criterion	TECH	**Abnahmebedingung** = Abnahmekriterium
acceptance criterion (→acceptance condition)	TECH	(→Abnahmebedingung)
acceptance inspection ≈ acceptance test ≈ quality inspection	QUAL	**Abnahmeprüfung** ≈ Qualitätsprüfung
acceptance number	QUAL	**Annahmezahl** = Gutzahl
acceptance probability	QUAL	**Annahmewahrscheinlichkeit**
acceptance region	QUAL	**Annahmebereich**
acceptance test (→acceptance inspection)	QUAL	(→Abnahmeprüfung)
accepted in the trade	ECON	**handelsüblich 2** [im Handel üblich]
acceptor [atom or impurity accepting electrons] ↓ acceptor atom; acceptor ion	MICROEL	**Akzeptor** [Elektronen aufnehmendes Atom oder Störstelle] ↓ Akzeptor-Atom; Akzeptor-Ion
acceptor (→series-resonant circuit)	NETW.TH	(→Reihenschwingkreis)
acceptor atom [atom accepting electrons]	PHYS	**Akzeptor-Atom** [Elektronen aufnehmendes Atom]
acceptor circuit (→series-resonant circuit)	NETW.TH	(→Reihenschwingkreis)
acceptor density	PHYS	**Akzeptorendichte**
acceptor distribution	PHYS	**Akzeptorverteilung**
acceptor exhaustion	PHYS	**Akzeptorenerschöpfung**
acceptor impurity	PHYS	**Akzeptorverunreinigung**
acceptor ion = ionized acceptor atom; dissociated acceptor atom ↑ acceptor	MICROEL	**Akzeptor-Ion** = ionisiertes Akzeptor-Atom; dissoziiertes Akzeptor-Ion ↑ Akzeptor
acceptor level	PHYS	**Akzeptorniveau** = Akzeptorterm
accesories supplied (→accessories furnished)	TECH	(→mitgeliefertes Zubehör)
access (v.t.) [data, files,..]	DATA PROC	**zugreifen auf** [Daten, Dateien,..]
access (v.t.)	TECH	**zugreifen auf**
access (n.) = admission ≈ grip	TECH	**Zugang** = Zulassung; Zutritt ≈ Zugriff
access (n.) [localization of a defined storage position, to store or read data] = admission ↓ direct access; sequential access; random access	DATA PROC	**Zugriff** [Auffinden einer gesuchten Speicherstelle, zum Speichern oder Lesen von Daten] = Zugang ↓ direkter Zugriff; sequentieller Zugriff; wahlloser Zugriff
access ↓ multiple access	TRANS	**Zugriff** ↓ Vielfachzugriff
access (n.) (→network access)	TELEC	(→Teilnehmerzugang)
access arm [magnetic disk memory] = actuator	TERM&PER	**Zugriffsarm** [Magnetplattenspeicher] = Aktuator
access authorization = access permission	DATA PROC	**Zugriffsberechtigung** = Zugangsberechtigung
access barred (→call barred)	DATA COMM	(→Verbindung gesperrt)
access bus	DATA PROC	**Zugriffsbus**
access category (→access mode)	DATA COMM	(→Zugangsart)
access charge (→access fee)	DATA COMM	(→Zugangsgebühr)
access circuit	SWITCH	**Zugangssatz**
access code (→prefix)	SWITCH	(→Verkehrsausscheidungszahl)
access contention	DATA COMM	**Zugangskonflikt** = Zugriffskonflikt
access control = entry control	SIGN.ENG	**Zutrittskontrolle**
access control	DATA PROC	**Zugriffskontrolle**
access control register	DATA PROC	**Zugriffskontrollregister**
access control server	DATA COMM	**Zugangskontrollserver**
access control system = entry control system	SIGN.ENG	**Zutrittskontrollsystem**
access control system	DATA PROC	**Zugriffskontrollsystem**
access control terminal = entry control terminal	TERM&PER	**Zutrittskontrollterminal**
access fee = access charge	DATA COMM	**Zugangsgebühr** = Zugriffsgebühr
accessibility [number of outlets which can be reached from an inlet] = availability	SWITCH	**Erreichbarkeit** [Zahl der Abnehmerleitungen die von einer Zubringerleitung erreicht werden können] = Verfügbarkeit
accessibility	TECH	**Zugänglichkeit**
accessibility time = availability time	SWITCH	**Verfügbarkeitsdauer**
accessible	TECH	**zugänglich**
accessible price	ECON	**erschwinglicher Preis**
access information path	DATA COMM	**Zugangspfad**
access level [file]	DATA PROC	**Zugriffsebene** [Datenbank]
access line (→subscriber line)	TELEC	(→Teilnehmerleitung)
access mechanism	TERM&PER	**Zugriffsmechanismus**
access method (→access procedure)	DATA PROC	(→Zugriffsverfahren)
access mode = access category	DATA COMM	**Zugangsart** = Zugriffsart
access mode ≈ access technique	TRANS	**Zugriffsverfahren**
access mode (→access procedure)	DATA PROC	(→Zugriffsverfahren)
access network = access switching network	SWITCH	**Anschaltenetz**
access network (→subscriber network)	TELEC	(→Teilnehmernetz 1)
accessories furnished = accesories supplied	TECH	**mitgeliefertes Zubehör**
accessory [necessary for normal service] = accessory part; fitting; add-on 2 ≈ accessory	TECH	**Zubehör** [für die Funktionserfüllung notwendig, „gehört dazu"] = Zubehör
accessory part (→accessory)	TECH	(→Zubehör)
access path ↑ path	DATA PROC	**Zugriffspfad** ↑ Pfad
access permission (→access authorization)	DATA PROC	(→Zugriffsberechtigung)
access point ≈ interface [TELEC]	DATA COMM	**Zugriffspunkt** ≈ Schnittstelle [TELEC]
access procedure = access method; access mode	DATA PROC	**Zugriffsverfahren** = Zugriffsmethode; Zugriffsmodus; Zugriffsroutine
access protocol	DATA COMM	**Zugangsprotokoll** = Zugriffsprotokoll
access road	CIV.ENG	**Zufahrtstraße** = Zufahrtsweg
access security	DATA PROC	**Zugriffssicherheit** [gegen unbefugtem Zugriff]
access speed (→access time)	DATA PROC	**Zugriffszeit**
access switching network (→access network)	SWITCH	(→Anschaltenetz)
access system [transfers informations between central and peripheral control units]	SWITCH	**Zugriffssystem** [tauscht Informationen zwischen zentralen und dezentralen Steuerungseinrichtungen aus]

access technique

access technique (→access mode)	TRANS	(→Zugriffsverfahren)
access time [time between activation and completion of an access] = access speed; search time; searching time ↓ positioning time; latency time	DATA PROC	**Zugriffszeit** [Zeit zwischen Veranlassung und Durchführung eines Zugriffs] = Zugriffsgeschwindigkeit; Suchzeit ↓ Positionierzeit; Latenzzeit
accident	COLLOQ	**Unfall**
accidental (→random)	TECH	(→zufällig)
accidental error (→statistical error)	MATH	(→statistischer Fehler)
accidental ground (→ground fault)	EL.TECH	(→Erdschluß)
accident-sensitive (→susceptible)	TECH	(→störanfällig)
accomodate (→house)	EQUIP.ENG	(→unterbringen)
accomodation [eye]	PHYS	**Akkomodation** [Auge]
accomodation (→lodging)	ECON	(→Unterkunft)
accomplish	COLLOQ	**vollziehen**
accomplishment (→achievement)	COLLOQ	(→Errungenschaft)
ac component = periodic and random deviation; PARD	EL.TECH	**Wechselstromkomponente** = Wechselstromanteil; Wechselstromgehalt
ac converter [converts from one ac system into another] = a.c. convertor (IEC); electronic a.c. convertor (IEC); ↑ static power converter; voltage system converter	POWER ENG	**Wechselstromumrichter** [wandelt von einem Wechselstromsystem in ein anderes] ↑ Stromrichter; Umrichter
a.c. convertor (IEC) (→ac converter)	POWER ENG	(→Wechselstromumrichter)
accordance = consonance; consonancy; concordance; agreement 2	COLLOQ	**Übereinstimmung**
according to plan (→scheduled)	TECH	(→planmäßig)
accordion fold (n.) (→fan folding)	TERM&PER	(→Leporellofalzung)
accordion-fold (adj.) (→fanfold)	TERM&PER	(→zickzackgefaltet)
account	ECON	**Konto**
account (→report)	LING	(→Bericht)
account (n.) (→invoice)	ECON	(→Rechnung)
account (→accounting 2)	ECON	(→Rechnungslegung)
account (→account code)	DATA COMM	(→Kontierungscode)
accountant ≈ bookkeeper	ECON	**Fachmann des Rechnungswesens** [in USA und Lateinamerika ein akademischer Titel] ≈ Buchhalter
accountant (BRI) (→auditor)	ECON	(→Wirtschaftsprüfer)
account code	ECON	**Kontonummer**
account code = account	DATA COMM	**Kontierungscode**
accounting 1 = account	ECON	**Rechnungslegung** = Abrechnung
accounting 2 (→bookkeeping)	ECON	(→Buchführung)
accounting machine	TERM&PER	**Buchungsmaschine**
accounting management [of inventories and billing] = inventory record keeping	TELEC	**Bestandsverwaltung** = Bestandsmanagement
accounting period (→billing period)	ECON	(→Abrechnungszeitraum)
accounting program	DATA PROC	**Finanzbuchhaltungsprogramm**
accounts payable (→liabilities)	ECON	(→Verbindlichkeiten)
ac-coupled	EL.TECH	**wechselstromgekoppelt**
accourtments (→equipment 1)	TECH	(→Ausstattung)
accretion (→increase 1)	TECH	(→Zunahme)
accrual (→provision 2)	ECON	(→Rückstellung)
accumulate (→enrich)	MICROEL	(→anreichern)
accumulate (→sum-up)	MATH	(→aufsummieren)
accumulated	COLLOQ	**aufgelaufen** = akkumuliert
accumulated experience	COLLOQ	**Erfahrungsschatz**
accumulation (→enrichment)	MICROEL	(→Anreicherung)
accumulation (→concentration)	PHYS	(→Stau)
accumulation layer (→enhancement layer)	MICROEL	(→Anreicherungszone)
accumulator [dc-chargeable energy store] = storage cell; secondary cell; cell ↑ galvanic cell	POWER SYS	**Akkumulator** [durch Gleichstrom ladbarer Energiespeicher] = Akku; Sammler; Sekundärelement; sekundäres Element ↑ galvanisches Element
accumulator [a register of ALU where arithmetic operations are carried out] ↑ arithmetic and logic unit	DATA PROC	**Akkumulator** [ein Register des Rechenwerks für arithmetische Operationen] = A-Register ↑ Rechenwerk
accumulator-driven	EQUIP.ENG	**akkubetrieben**
accumulator pack	POWER SYS	**Akkumulatorsatz** = Akkusatz
accumulator register	DATA PROC	**Akkumulatorregister**
accumulator shift instruction	DATA PROC	**Akkumulatorverschiebebefehl**
accuracy = precision; exactness; fidelity	TECH	**Genauigkeit** = Präzision; Exaktheit; Treue
accuracy	INF	**Sicherheit**
accuracy to size (→size permanency)	TECH	(→Maßbeständigkeit)
accuracy window	INSTR	**Genauigkeitsfenster**
accurate (→correct)	TECH	(→genau)
accurate alignment (→congruent matching)	TECH	(→Deckungsgleichheit)
accurately aligned (→congruent)	TECH	(→deckungsgleich)
ac current (→alternating current)	EL.TECH	(→Wechselstrom)
accusative	LING	**Akkusativ** = Wenfall; 4. Fall; vierter Fall
AC/DC (→all mains)	POWER SYS	(→Allstrom)
acess unit	DATA COMM	**Zugangseinheit**
acetate	CHEM	**Azetat**
acetobutyrate	CHEM	**Acetobutyrat**
ac generator = alternator; alternating current generator (IEC); ac machine ↑ generator	POWER SYS	**Wechselstromgenerator** = Wechselstrommaschine ↑ Stromgenerator
achievable	TECH	**erzielbar** = erreichbar
achieve = reach; obtain	TECH	**erzielen** = erreichen
achievement = accomplishment	COLLOQ	**Errungenschaft**
achromatic (→colorless)	PHYS	(→farblos)
achromatic locus = achromatic region	TV	**achromatischer Bereich** = Unbunt-Bereich
achromatic region (→achromatic locus)	TV	(→achromatischer Bereich)
aci (→alternating current)	EL.TECH	(→Wechselstrom)
ACIA = asynchronous communications interface adapter	DATA PROC	**asynchroner Schnittstellenadapter** = ACIA
acid (n.)	CHEM	**Säure**
acid-resistant ≈ non-corroding	CHEM	**säurebeständig** ≈ korrosionsfest
acid solution	CHEM	**saure Lösung**
ACK (→acknowledge)	DATA COMM	(→positive Rückmeldung)
Ackerman's function	DATA PROC	**Ackerman-Funktion**
acknowledge	TELEC	**quittieren**
acknowledge = ACK	DATA COMM	**positive Rückmeldung** [Code] = ACK
acknowledge bit	DATA COMM	**Rückmeldebit** = Quittungsbit
acknowledgement (BRI) (→acknowledgment)	ECON	(→Bestätigung)
acknowledgement (BRI) (→acknowledgment)	TELEC	(→Quittierung)

acknowledgement (BRI) TELECONTR (→Quittung)
 (→acknowledgment)
acknowledgement request DATA COMM (→Quittungsaufforderung)
 (BRI) (→acknowledgment request)
acknowledgement signal (BRI) TELEC (→Quittungszeichen)
 (→acknowlegment signal)
acknowledgement telegram TELECONTR (→Quittungstelegramm)
 (BRI) (→acknowledgment telegram)
acknowledge signal TELECONTR **Quittungssignal**
acknowledgment (NAM) ECON **Bestätigung**
 = acknowledgement (BRI); confirmation
acknowledgment (NAM) TELEC **Quittierung**
 = acknowledgement (BRI); receipt confirmation; receipt = Quittung; Rückmeldung; Bestätigung
acknowledgment (NAM) TELECONTR **Quittung**
 = acknowledgement (BRI)
acknowledgment request DATA COMM **Quittungsaufforderung**
 (NAM) = Quittungsanforderung
 = acknowledgment request (BRI)
acknowledgment telegram TELECONTR **Quittungstelegramm**
 (NAM)
 = acknowledgment telegram (BRI)
acknowlegment signal (NAM) TELEC **Quittungszeichen**
 = acknowledgement signal (BRI); receipt signal; wink pulse; wink = Quittierungszeichen; Bestätigungszeichen
 ↑ switching signal ↑ Kennzeichen
ac line frequency (→mains POWER SYS (→Netzfrequenz)
 frequency)
AC-line-supply cable EQUIP.ENG (→Netzanschlußkabel)
 (→power cable)
AC line voltage (→line POWER ENG (→Netzspannung)
 voltage)
ac machine (→ac POWER SYS (→Wechselstromgenerator)
 generator)
ac measuring bridge INSTR **Wechselstrom-Meßbrücke**
 = alternating current measuring bridge
ac motor (→alternating POWER ENG (→Wechselstrommotor)
 current motor)
ACO (→alarm cut-off) EQUIP.ENG (→Alarmabschaltung)
acorn nut MECH **Hutmutter**
acorn tube ELECTRON **Kleinströhre**
 = subminiature tube
acoustic PHYS **akustisch**
 = audible
acoustic absorption ACOUS **Schallschluckgrad**
 coefficient
acoustical alarm EQUIP.ENG **akustischer Alarm**
 = audible alarm
acoustical sound enclosure TERM&PER (→Schallschluckhaube)
 (→noise-absorbing cover)
acoustical transmission factor EL.ACOUS (→Schalltransmissionsgrad)
 (→sound transmission factor)
acoustic aperture ACOUS **Schallöffnung**
acoustic beam (→acoustic ray) ACOUS (→Schallstrahl)
acoustic conductor ACOUS **Schalleiter**
 = akustischer Leiter
acoustic coupler TERM&PER **Akustikkoppler**
 ↑ modem = akustischer Koppler
 ↑ Modem
acoustic damping material EL.ACOUS **Dämpfungsmaterial**
acoustic feedback EL.ACOUS **akustische Rückkopplung**
acoustic hood TERM&PER (→Schallschluckhaube)
 (→noise-absorbing cover)
acoustic impedance ACOUS **Schallimpedanz**
acoustic input DATA PROC **akustische Eingabe**
 [speech recognition] [Spracherkennung]
acoustic lens EL.ACOUS **akustische Linse**
acoustic memory (→acoustic EL.ACOUS (→Schallspeicher)
 storage)
acoustic power ACOUS **Schalleistung**
acoustic power level ACOUS **Schalleistungspegel**
acoustic radiator (→sound EL.ACOUS (→Schallgeber)
 generator)
acoustic ray ACOUS **Schallstrahl**
 = acoustic beam ↑ Strahl [PHYS]
 ↑ ray [PHYS]
acoustic receiver EL.ACOUS **Schallempfänger**
 = sound receiver = Schallaufnehmer
 ↓ microphone ↓ Mikrophon

acoustic resistance ACOUS **akustische Resistanz**
 [real component of acoustic impedance] [Realteil der Schallimpedanz]
acoustics PHYS **Akustik**
 = Lautlehre; Schallehre
acoustic sensor COMPON **Schallsensor**
 = sound sensor = Schallfühler
acoustic shadow ACOUS **Schallschatten**
acoustic short-circuit ACOUS **akustischer Kurzschluß**
acoustics of room ACOUS **Raumakustik**
 = room acoustics
acoustic source ACOUS **Schallquelle**
 = sound source = Schallgeber; akustische Quelle
acoustic source (→sound EL.ACOUS (→Schallgeber)
 generator)
acoustic storage EL.ACOUS **Schallspeicher**
 = acoustic memory = akustischer Speicher
 ↓ Schallplatte; Tonband
acoustic velocity (→sound particle ACOUS (→Schallschnelle)
 velocity)
acoustoelectric PHYS **akustoelektrisch**
 ≈ electroacoustic ≈ elektroakustisch
acoustoelectric amplifier EL.ACOUS **akustoelektrischer Verstärker**
acoustoelectric effect PHYS **akustoelektrischer Effekt**
acoustoelectric index TELEPH **Sendeübertragungsfaktor**
acousto-electric index EL.ACOUS (→Übertragungsfaktor)
 (→electroacoustic index)
acousto-optic (adj.) PHYS **akusto-optisch**
ac plate current ELECTRON **Anodenwechselstrom**
ac plate voltage ELECTRON **Anodenwechselspannung**
ac power distribution POWER ENG **Netzverteilertafel**
 panel
ac powered EL.TECH **wechselstromgespeist**
ac power source EL.TECH **Wechselstromquelle**
ac power source (→mains EQUIP.ENG (→Netzstromversorgungsgerät)
 power supply)
ac power supply POWER SYS **Netzanschluß**
 = mains connection
ac power supply (→mains EQUIP.ENG (→Netzstromversorgungsgerät)
 power supply)
acquisition CIRC.ENG **Akquisition**
acquisition (→procurement) ECON (→Beschaffung)
acquisition and tracking RADIO LOC **Erfassungs-und Folgeradar**
 radar
acquisition costs ECON **Anschaffungskosten**
 ≈ first costs ≈ Investitionskosten
acquisition radar RADIO LOC **Erfassungsradar**
acquisition time INSTR **Erfassungszeit**
ac relay COMPON **Wechselstromrelais**
ac remote power feeding TRANS **Wechselstromfernspeisung**
ac resistance NETW.TH (→komplexer Scheinwiderstand)
 (→impedance)
acronym LING **Akronym**
 [artificial word formed by single letters of several words, mostly initial letters, e.g. RADAR] [aus einzelnen Buchstaben mehrerer Wörter, meist Anfangsbuchstaben, gebildetes Kunstwort, z.B. RADAR]
 ≈ abbreviation
 ↑ artificial term = Initialwort; Buchstabenwort
 ≈ Abkürzung
 ↑ Kunstwort
ac signaling (NAM) SWITCH **Wechselstromsignalisierung**
 = ac signalling (BRI)
ac signalling (BRI) (→ac SWITCH (→Wechselstromsignalisierung)
 signaling)
ac telegraphy (→voice frequency TELEGR (→Wechselstromtelegrafie)
 telegraphy)
AC tension (→alternating EL.TECH (→Wechselspannung)
 voltage)
a-c tension (→alternating EL.TECH (→Wechselspannung)
 voltage)
ac tension (→alternating EL.TECH (→Wechselspannung)
 voltage)
AC thyristor (→triac) MICROEL (→Triac)
actinide CHEM **Aktinid**
actinium CHEM **Actinium**
 = Ac = Ac
action TECH **Einwirkung**
 ≈ effect ≈ Effekt

action

action (→operating mode)	TECH	(→Betriebsart)
action cycle	DATA PROC	Aktionszyklus
action key (→release key)	TERM&PER	(→Auslösetaste)
action message	DATA PROC	Aktionsaufforderung
action paper (→pressure-sensitive paper)	TERM&PER	(→druckempfindliches Papier)
action report (→activity report)	ECON	(→Tätigkeitsbericht)
action statement (→statement 1)	DATA PROC	(→Anweisung 1)
action window	DATA PROC	Arbeitsfenster
activate (→trigger)	DATA PROC	(→auslösen)
activate (→trigger)	ELECTRON	(→auslösen)
activate (→arrange)	COLLOQ	(→veranlassen)
activated charcoal	CHEM	Aktivkohle
activation	PHYS	Aktivieren
activation (→triggering)	ELECTRON	(→Auslösung)
activation current (→trigger current)	ELECTRON	(→Zündstrom)

activation energy PHYS **Aktivierungsenergie**
[necessary to reach another state] [Anregungsenergie zum Erreichen eines neuen Zustandes]
↑ excitation energy ↑ Anregungsenergie
↓ ionization energy ↓ Ionisierungsenergie

activation period (→on period)	ELECTRON	(→Einschaltzeit)

active (adj.) TECH **aktiv**
= working; running = in Betrieb befindlich; laufend
≠ stand-by

active DATA PROC **aktiv**
= running; working

active aerial (→active antenna) ANT (→aktive Antenne)

active antenna ANT **aktive Antenne**
= direct-fed antenna; integrated antenna; active aerial; direct-fed aerial; integrated aerial
= direktgespeiste Antenne; elektronische Antenne

active balance return loss (→structural balance return loss) NETW.TH (→komplexes Echodämpfungsmaß)

active cell	DATA PROC	aktiviertes Tabellenfeld
active component	COMPON	aktives Bauelement
= active device		

active current EL.TECH **Wirkstrom**
[active power to active voltage ratio] [Wirkleistung durch Wirkspannung]
= effective current = Effektivstrom

active decoding	RADIO LOC	aktive Decodierung
active device (→active component)	COMPON	(→aktives Bauelement)

active field ANT **Wirkfeld**
≈ far field ≈ Fernfeld

active file	MOB.COMM	Aktivdatei

active file DATA PROC **Aktivdatei**
[the just beeing used one] [gerade benutzte Datei]

active file change MOB.COMM **Umbuchen**
[Wechsel des Aktivregisters]

active filter	NETW.TH	aktives Filter

active function NETW.TH **Wirkungsfunktion**
[frequency response of transmission coefficient] [Übertragungsfaktor als Funktion der komplexen Frequenz]
↓ transfer function ↓ Transferfunktion

active high data (→positive logic) CIRC.ENG (→positive Logik)

active information system	DATA PROC	aktives Informationssystem
active line	DATA COMM	Beschriftungszeile
active material	ELECTRON	Emissionssubstanz
active network	NETW.TH	aktives Netzwerk
active partition (→boot partition)	DATA PROC	(→Boot-Partition)

active power EL.TECH **Wirkleistung**
[time average of power] [zeitlicher Mittelwert der Leistung]
= effective power; real power; true power = mittlere Leistung; Effektivleistung
↓ dissipation power ↓ Verlustleistung

active power match	EL.TECH	(→Wirkleistungsanpassung)
(→active power matching)		

active power matching EL.TECH **Wirkleistungsanpassung**
= effective power matching; true power matching; active power match; effective power match; true power match ↑ Anpassung 1
↑ matching

active power meter INSTR **Wirkleistungsmesser**
= real power meter; effective power meter = Wirkverbrauchszähler
↑ watthour meter ↑ Leistungsmesser; Wechselstromzähler

active probe	INSTR	Aktivtastkopf

active program DATA PROC **Aktivprogramm**
[the just running one] [das gerade ablaufende]

active quadripole (→active two-port) NETW.TH (→aktiver Vierpol)

active radiator (→primary radiator) ANT (→Primärstrahler)

active region MICROEL **aktiver Bereich**
= normaler Bereich

active repair time QUAL **Instandsetzungsdauer**
= Reparaturdauer

active resistance NETW.TH **Wirkwiderstand**
[real part of impedance] [Realteil des komplexen Scheinwiderstandes]
= effective resistance; resistance = Resistanz
≈ true resistance; dc resistance [EL.TECH]; ohmic resistance [EL.TECH] ≈ reeller Widerstand; Gleichstromwiderstand [EL.TECH]; ohmscher Widerstand [EL.TECH]
≠ reactance ≠ Blindwiderstand

active return loss NETW.TH **Reflexionsdämpfung**
= return loss 2; matching loss; balance return loss; matching attenuation; reflection loss 2; echo loss; echo attenuation [Realteil des Echodämpfungsmaßes]
= Echodämpfung; Fehlerdämpfung; Fehlanpassungsdämpfung; Anpassungsdämpfung; Stoßdämpfung
↓ composite return loss ↓ Betriebsreflexionsdämpfung

active state	ELECTRON	Aktivzustand

active two-port NETW.TH **aktiver Vierpol**
= active quadripole

active two-terminal (→active two-terminal network) NETW.TH (→aktiver Zweipol)

active two-terminal network NETW.TH **aktiver Zweipol**
[has a voltage between its terminals, with no current flowing] [weist eine Spannung auf, ohne daß ein Strom fließt]
= active two-terminal = Zweipolquelle

active voice LING **Aktiv**
[e.g. I listen, I listened, I have listened] [z.B. ich höre, ich hörte, ich habe gehört]
≠ passive voice = Aktivform; Tatform; Tätigkeitsform
≠ Passiv

active voltage EL.TECH **Wirkspannung**
[active power to active current ratio] [Wirkleistung durch Wirkstrom]
= effective voltage = Effektivspannung

active window DATA PROC **aktives Fenster**
[currently in use] [das gerade bearbeitete]

activity PHYS **Aktivität**
[SI unit: Bequerel] [SI-Einheit: Bequerel]

activity ECON **Tätigkeit**
≈ occupation ≈ Aktivität; Beschäftigung
≈ profession ≈ Beruf

activity DATA PROC **Aktivität**
[subunit of a task] [Untereinheit eines Task]

activity TECH **Aktivität**
= Tätigkeit

activity field	ECON	Tätigkeitsbereich
activity file (→transaction file)	DATA PROC	(→Bewegungsdatei)
activity plan (→timing diagram)	TECH	(→Ablaufdiagramm)
activity rate	DATA PROC	Bewegungshäufigkeit

activity ratio DATA PROC **Bewegungsindex**
= file activity ratio = Dateibewegungsindex; Bewegungsquotient; Dateibewegungsquotient

activity report ECON **Tätigkeitsbericht**
= action report; progress report
actor MICROEL **Aktor**
 [an add-on component] [nachgeschaltetes Bauelement]
 = actuator = Wirkungselement
actor COMPON **Aktor**
 [transforms electrical signals into mechanical movement] [wandelt elektrische Signale in mechanische Bewegungen]
 = Wirkungselement
ACT plumbicon ELECTRON (→ACT-Plumbicon)
 (→ anti-comet-tail plumbicon)
actual COLLOQ **aktuell**
 = topical
actual (→ real) COLLOQ (→tatsächlich)
actual address 2 (→absolute address) DATA PROC (→absolute Adresse)
actual address 1 (→current address) DATA PROC (→Momentanadresse)
actual addressing (→absolute addressing) DATA PROC (→absolute Adressierung)
actual balance DATA PROC **Tagfertigkeit**
 [ready on day] [am selben Tag fertig]
actual condition (→actual state) TECH (→Ist-Zustand)
actual costs ECON **Ist-Kosten**
 = effektive Kosten
actual deviation CONTROL **Ist-Abweichung**
actual frequency (→working frequency) ELECTRON (→Arbeitsfrequenz)
actual indication INSTR **Istwertanzeige**
actualization (→update) TECH (→Aktualisierung)
actual level (BRI) (→absolute level) TELEC (→absoluter Pegel)
actual signal CONTROL **Ist-Signal**
actual size MECH **Ist-Maß**
actual state TECH **Ist-Zustand**
 = actual condition
actual value CONTROL **Ist-Wert**
actual value (→instant value) PHYS (→Augenblickswert)
actual-value signal ELECTRON **Istwertsignal**
actuate TECH **betätigen**
 ≈ operate 1 ≈ bedienen
actuate message TELECONTR **Ausgabebefehl**
 = execute command
actuating (→control-) CONTROL (→Steuer-)
actuating element (→final control element) CONTROL (→Stellglied)
actuating force (→actuation force) TECH (→Betätigungskraft)
actuating signal (→deviation) CONTROL (→Regelabweichung)
actuating signal (→control signal) CONTROL (→Steuersignal)
actuating unit (→final control element) CONTROL (→Stellglied)
actuating variable (→control variable) CONTROL (→Stellgröße)
actuation TECH **Betätigung**
 ≈ operation ≈ Bedienung
actuation force TECH **Betätigungkraft**
 = actuating force
actuator POWER ENG **Aktuator**
 [an electromagnetic servomechanism with two positions] [elektromechanisches Stellglied mit zwei Stellungen]
actuator CONTROL **Aktuator**
 [signal-controllable mechanical device] [durch Signale steuerbare mechanische Vorrichtung]
 ↑ final control element = Aktor
 ↑ Stellglied
actuator (→actor) MICROEL (→Aktor)
actuator 1 (→final control element) CONTROL (→Stellglied)
actuator (→access arm) TERM&PER (→Zugriffsarm)
actuator (→effector) DATA PROC (→Effektor)
actuator (→control) ELECTRON (→Bedienelement)
acute accent LING **Akut**
 [symbol ′] [Symbol ′]
 = accent aigu; accent 3 (n.) = Accent aigu; Apex 2
 ↑ diacritic mark; accent ≈ scharfer Akzent
 ↑ diakritisches Zeichen; Akzent

acute angle MATH **spitzer Winkel**
acute-angled MATH **spitzwinklig**
 ≠ obtuse-angled ≠ stumpfwinkling
acute-angled triangle MATH **spitzwinkliges Dreieck**
acv (→alternating voltage) EL.TECH (→Wechselspannung)
ac voltage (IEEE) (→alternating voltage) EL.TECH (→Wechselspannung)
a-c voltage (→alternating voltage) EL.TECH (→Wechselspannung)
AC voltage (→alternating voltage) EL.TECH (→Wechselspannung)
a.c. voltage (IEC) (→alternating voltage) EL.TECH (→Wechselspannung)
ac voltmeter INSTR **Wechselspannungsmesser**
 = alternating current voltmeter = Wechselspannungsvoltmeter
ACW (→aircraft control and warning service) RADIO LOC (→Luftraum-Kontroll-und-Warndienst)
acyclic TECH **azyklisch**
A/D (→analog/digital) INF.TECH (→analog/digital)
ad (→advertising) ECON (→Werbung)
ad (→advertisement) ECON (→Zeitungsinserat)
ADA (→angular diversity antenna) ANT (→Winkeldiversityantenne)
adamantine PHYS **diamantartig**
adapt 1 (v.t.) TECH **anpassen 1** (v.t.)
 ≈ convert; conform ≈ umrüsten; angleichen
adapt 2 (v.r.) TECH **anpassen 2** (v.r.)
adaptable (→convertible) TECH (→umrüstbar)
adaptation TECH **Anpassung**
 ≈ conversion ≈ Umrüstung
adaptation 2 (→power matching) NETW.TH (→Leistungsanpassung)
adaptation 1 (→matching 1) NETW.TH (→Anpassung 1)
adapter MECH **Anpassungsglied**
adapter 1 COMPON **Übergangsstecker**
 [matches non-compatible connectors] [stellt den Übergang zwischen nichtkorrespondierenden Steckern her]
 = adapter plug; adaptor = Steckerübergang; Adapterstecker; Adapter 1; Zwischenstecker; Übergangsstück; Übergangsverbinder 1; Übergang
 ↓ in-series adapter; inter-series adapter; reductor
 ↓ Kupplung 1; Adapter 2; Reduzierstück
adapter 2 COMPON **Zwischensockel**
adapter (→board extender) INSTR (→Adapter)
adapter (→matching circuit) CIRC.ENG (→Anpassungsschaltung)
adapter board DATA PROC **Anpassungskarte**
 = adapter card = Adapterkarte
adapter bushing MECH **Reduzierbuchse**
adapter card (→adapter board) DATA PROC (→Anpassungskarte)
adapter plug (→adapter 1) COMPON (→Übergangsstecker)
adapter sleeve MECH **Reduzierhülse**
 = Spannhülse
adapting equipment EQUIP.ENG **Vorschaltgerät**
adaption CONTROL **Adaption**
adaptive (→self-adapting) TECH (→selbstanpassend)
adaptive allocation (→demand assignment) TELEC (→bedarfsweise Zuteilung)
adaptive channel allocation (→demand assignment of channel) TELEC (→bedarfsweise Kanalzuteilung)
adaptive control CONTROL **adaptive Regelung**
 = adaptive Steuerung
adaptive design (→adaptive development) TECH (→Anpassungsentwicklung)
adaptive development TECH **Anpassungsentwicklung**
 = adaptive design; adaptive engineering ≈ kundenspezifische Anpassung
 ≈ customization
adaptive differential pulse code modulation (→ADPCM) CODING (→ADPCM)
adaptive engineering (→adaptive development) TECH (→Anpassungsentwicklung)
adaptive equalizer NETW.TH **adaptiver Entzerrer**
adaptive feedback CIRC.ENG **nachgebende Rückkopplung**

adaptive PCM

adaptive PCM (→ APCM)	CODING	(→ APCM)
adaptive routing	SWITCH	adaptive Verkehrslenkung
adaptive scanning	ANT	Nachlaufsteuerung
adaptive system	CONTROL	adaptives System
adaptive time domain equalizer = ATDE	NETW.TH	adaptiver Zeitbereichsentzerrer
adaptive transversal equalizer	NETW.TH	adaptiver Transversalentzerrer
adaptor (→ adapter 1)	COMPON	(→ Übergangsstecker)
ADB technique (→ alloy diffusion technique)	MICROEL	(→ AD-Technik)
ADC (→ analog-to-digital converter)	CODING	(→ A/D-Wandler)
Adcock antenna ↑ directional antenna ↓ H-Adcock antenna; U-Adcock antenna; rotary Adcock antenna; fixed Adcock antenna	ANT	Adcock-Antenne ↑ Richtantenne ↓ H-Adcock-Antenne; U-Adcock-Antenne; Drehadcock-Antenne; Fest-Adcock-Antenne
Adcock direction finder	RADIO LOC	Adcock-Peiler
A/D conversion (→ analog-to-digital conversion)	TELEC	(→ Analog-Digital-Umsetzung)
A/D converter (→ analog-digital converter)	CIRC.ENG	(→ Analog-Digital-Umsetzer)
A/D converter (→ analog-to-digital converter)	CODING	(→ A/D-Wandler)
add (v.t.) = sum (v.t.); totalize; total (v.t.)	MATH	addieren = summieren; zusammenzählen; summen
add (→ addition 1)	MATH	(→ Addition)
add-drop (→ drop/insert)	TRANS	(→ Abzweig und Wiederbelegung)
add/drop multiplexer (→ drop/insert multiplexer)	TRANS	(→ Abzweigmultiplexer)
add-drop traffic (→ way-side traffic)	TRANS	(→ Unterwegsverkehr)
added feature = special feature	TECH	Komfortleistungsmerkmal
added-feature keyboard ↑ special keyboard	TERM&PER	Komforttastatur ↑ Spezialtastatur
added-feature telephone = special-feature telephone	TELEPH	Komfortfernsprecher = Comforttelefon
added-main-line system (→ subscriber loop carrier)	TRANS	(→ Teilnehmermultiplexsystem)
added value = value added	ECON	Wertschöpfung = Mehrwert
addend 2 [a number to be added to another] = addendum (pl.-a) ↓ augend; addend 1	MATH	Summand 2 [zu addierende Zahl] = Addend 2; Addende ↓ erster Summand; zweiter Summand
addend 1 [number to be added to the augend] = addendum; augmenter ↑ addend 2	MATH	zweiter Summand [dem ersten Summanden hinzuzuzählende Zahl] = Summand 1; Addend 1 ↑ Summand 2
addendum [gear]	MECH	Zahnkopfhöhe [Zahnrad]
addendum (→ supplement)	LING	(→ Nachtrag)
addendum (pl.-a) (→ addend 2)	MATH	(→ Summand 2)
addendum (→ addend 1)	MATH	(→ zweiter Summand)
adder = adder circuit; summing amplifier ≈ integration circuit ↓ half adder; full adder	CIRC.ENG	Addierer = Addierschaltung; Addierglied; Addierwerk; Addierverstärker; Adder; Summierglied; Summierer; Umkehraddierer; Summationsverstärker; Summierverstärker ≈ Integrierglied ↓ Halbaddierer; Volladdierer
adder circuit (→ adder)	CIRC.ENG	(→ Addierer)
adder-subtractor	CIRC.ENG	Addierer-Subtrahierer
add-in (n.)	DATA PROC	Einbauzubehör = Aufsteckzubehör
add-in (n.) (→ attachment)	EQUIP.ENG	(→ Zusatz)
add-in board (→ expansion board)	EQUIP.ENG	(→ Erweiterungskarte)
add-in card (→ expansion board)	EQUIP.ENG	(→ Erweiterungskarte)
adding counter	CIRC.ENG	Addierzähler
adding machine	OFFICE	Addiermaschine
adding machine keypad (→ numeric keypad)	TERM&PER	(→ Zifferntastenblock)
add-in module (→ expansion board)	EQUIP.ENG	(→ Erweiterungskarte)
add-in PCB (→ expansion board)	EQUIP.ENG	(→ Erweiterungskarte)
addition 1 [the act of adding] = summation; add ≈ sum	MATH	Addition = Addieren; Summierung; Aufsummierung; Summation ≈ Summe
addition (→ supplement)	TECH	(→ Zusatz)
addition 2 (→ sum)	MATH	(→ Summe)
addition agent (→ dopant)	MICROEL	(→ Dotierungsmaterial)
additional = extra (adj)	TECH	zusätzlich = hinzukommend
additional board (→ expansion board)	EQUIP.ENG	(→ Erweiterungskarte)
additional card (→ expansion board)	EQUIP.ENG	(→ Erweiterungskarte)
additional charge (→ addition price)	ECON	(→ Aufpreis)
additional condition (→ supplementary condition)	TECH	(→ Zusatzbedingung)
additional consumption (→ production scrap)	MANUF	(→ Mehrverbrauch)
additional device (→ add-on device)	TECH	(→ Zusatzeinrichtung)
additional equipment (→ add-on equipment)	EQUIP.ENG	(→ Zusatzgerät)
additional expenses	ECON	Mehraufwand
additional feature = add-on feature	TECH	Zusatzmerkmal = Zusatzfunktion
additional function	TERM&PER	Zusatzfunktion
additional load	ECON	Zuladung
additional module (→ expansion board)	EQUIP.ENG	(→ Erweiterungskarte)
additional need	ECON	Mehrbedarf
additional noise	MICROEL	Zusatzrauschen
additional PCB (→ expansion board)	EQUIP.ENG	(→ Erweiterungskarte)
addition of energy	PHYS	Energiezufuhr
addition price = surplus price; additional charge	ECON	Aufpreis
addition record [record with changes] = modification record; change record	DATA PROC	Änderungssatz = Modifikationssatz
addition roll	TERM&PER	Additionsrolle
addition time (→ add time)	ELECTRON	(→ Additionszeit)
additive	MATH	additiv
additive color composition = additive colour composition	TV	additive Farbmischung
additive colour composition (→ additive color composition)	TV	(→ additive Farbmischung)
additive conversion (→ additive mixing)	HF	(→ additive Mischung)
additive mixing = single-input mixing; additive conversion	HF	additive Mischung
add-on (v.t.)	TELEC	zuschalten = aufschalten
add-on (n.)	DATA PROC	Nachrüstzubehör
add-on (n.)	TELEC	Zuschalten = Aufschalten; Zuschaltung; Aufschaltung
add-on (adj.) ≠ built-in	TECH	angebaut = Anbau- ≠ eingebaut
add-on (→ retrofit)	EQUIP.ENG	(→ nachrüsten)
add-on 2 (→ accessory)	TECH	(→ Zubehör)
add-on (n.) (→ attachment)	EQUIP.ENG	(→ Zusatz)
add-on 1 (→ add-on device)	TECH	(→ Zusatzeinrichtung)
add-on board (→ expansion board)	EQUIP.ENG	(→ Erweiterungskarte)

German-English dictionary entries			
add-on card (→expansion board)	EQUIP.ENG	(→Erweiterungskarte)	
add-on conference (→tripartite conference)	TELEPH	(→Dreierkonferenz)	
add-on connection	EL.TECH	Zuschaltung	
add-on device	TECH	Zusatzeinrichtung	
= additional device; add-on 1; ancillary device; applique devoice ≈ accessory		≈ Zubehör	
add-on equipment	EQUIP.ENG	Zusatzgerät	
= additional equipment; applique equipment		= Erweiterungsgerät	
add-on feature (→additional feature)	TECH	(→Zusatzmerkmal)	
add-on kit (→retrofit kit)	EQUIP.ENG	(→Nachrüstsatz)	
add-on module (→expansion board)	EQUIP.ENG	(→Erweiterungskarte)	
add-on PCB (→expansion board)	EQUIP.ENG	(→Erweiterungskarte)	
address (v.t.)	DATA PROC	adressieren	
		= ansteuern	
address (n.)	TELECONTR	Adresse	
address (n.)	DATA PROC	Adresse	
= data address; label 3		= Datenadresse	
address	TELEC	Anschrift	
		= Adresse	
address (→speech)	COLLOQ	(→Ansprache)	
address (→postal address)	POST	(→Postanschrift)	
addressability	DATA PROC	Adressierbarkeit	
addressable	DATA PROC	adressierbar	
		= ansteuerbar	
addressable storage		Adreßraum	
= address space			
address administration program	DATA PROC	Adressenverwaltungsprogramm	
address batch	SWITCH	Adressenschub	
		= Adreßschub	
address bit	CODING	Adressenbit	
address bus	DATA PROC	Adreßbus	
= address highway; highway 2 ↑ bus		= Adressenbus; Adreßpfad ↑ Bus	
address capacity	DATA PROC	Adreßbreite	
		= Adreßkapazität	
address check	DATA PROC	Adreßprüfung	
address coder	DATA PROC	Adressencodierer	
address computation	DATA PROC	Adreßrechnung	
address computation instruction	DATA PROC	Adreßrechnungsbefehl	
address conversion (→address mapping)	DATA COMM	(→Adreßabbildung)	
address conversion (→address translation)	DATA PROC	(→Adreßumsetzung)	
address converter	DATA PROC	Adreßkonverter	
= address convertor			
address convertor (→address converter)	DATA PROC	(→Adreßkonverter)	
address counter (→instruction counter)	DATA PROC	(→Befehlszähler)	
address decoder	DATA PROC	Adressendecodierer	
addressee	POST	Empfänger	
		= Adressat	
addresser (→sender)	INF.TECH	(→Absender)	
address field	DATA PROC	Adreßfeld	
= operand field			
address file	SWITCH	Adressendatei	
		= Adreßdatei; Anschriftendatei	
address file up-dating	DATA COMM	Anschriftenverwaltung	
address format	DATA PROC	Adressenformat	
address highway (→address bus)	DATA PROC	(→Adreßbus)	
address information	SWITCH	Adreßinformation	
addressing	DATA PROC	Adressierung	
addressing capability	DATA PROC	Adressierungsmöglichkeit	
addressing capacity	DATA PROC	Adressierfähigkeit	
addressing level	DATA PROC	Adressierebene	
addressing machine	OFFICE	Adressiermaschine	
addressing method	DATA PROC	Adressierungsmethode	
= addressing mode		= Adressierungsart	
addressing mode (→addressing method)	DATA PROC	(→Adressierungsmethode)	
4-address instruction (→four-address instruction)	DATA PROC	(→Vieradreßbefehl)	
address latch	SWITCH	Adressenspeicher	
address length	DATA COMM	Adreßlänge	
addressless	DATA PROC	adressenlos	
= no-address			
addressless instruction	DATA PROC	adressenloser Befehl	
address location (→instruction address)	DATA PROC	(→Befehlsadresse)	
address management	DATA PROC	Adressenverwaltung	
address mapping	DATA COMM	Adreßabbildung	
= address conversion		= Adreßumsetzung	
address mark	DATA PROC	Adreßmarkierung	
address message	DATA COMM	Adreßblock	
address modification	DATA PROC	Adreßmodifikation	
address part	DATA PROC	Adressenteil	
address range	DATA PROC	Adreßbereich	
address register [part of main memory]	DATA PROC	Adreßregister [Teil des Zentralspeichers] = Adressenregister; Adressierungsregister	
address signal	DATA COMM	Adreßkennzeichen	
address space (→addressable storage)	DATA PROC	(→Adreßraum)	
address strobe [signal indicating a valid address being transmitted on a bus]	DATA PROC	Adreßhinweissignal [weist auf Gültigkeit der Adresse hin, die auf einem Bus gerade übertragen wird]	
address substitution ↑ address modification	DATA PROC	Adressensubstitution ↑ Adressenmodifikation	
address translation	DATA PROC	Adreßumsetzung	
= address conversion		= Adreßkonvertierung; Adreßübersetzung	
address word	DATA PROC	Adreßwort	
add time	ELECTRON	Additionszeit	
= addition time			
adequacy	COLLOQ	Angemessenheit	
		= Adäquatheit	
adequate (adj.)	SCIE	angemessen	
= fair		= adäquat	
≈ appropriate; fitting		≈ geeignet; passend	
adhere	TECH	haften	
		= anhaften	
		≈ kleben	
adherent point (→contact point)	TECH	(→Berührungspunkt)	
adhesion [between molecules of different bodies or materials] ≈ cohesion ↓ adsorption	PHYS	Adhäsion [zwischen Molekülen verschiedener Körper oder Stoffe] = Haftung; Adhäsionskraft ≈ Kohäsion ↓ Adsorption	
adhesive	TECH	Kleber	
= glue; bonding agent		= Klebstoff; Klebemittel; Leim	
≈ cement		≈ Kitt	
↓ rubber cement		↓ Alleskleber	
adhesive film	TECH	Klebefolie	
adhesive label	TERM&PER	Haftetikette	
≈ stick-on label		≈ Klebeetikette	
adhesive stamp	ECON	Wertmarke	
≈ postal stamp [POST]		≈ Briefmarke [POST]	
↑ stamp		↑ Wertzeichen	
adhesive tape	TECH	Klebeband	
≈ scotchtape		↓ Selbstklebeband	
↓ self-adhesive tape			
adhesive tape dispenser	OFFICE	Klebestreifenspender	
adiabatic [without change of heat content] ≈ isentropic	PHYS	adiabatisch [ohne Änderung des Wärmeinhalts] ≈ isentropisch	
adjacency	MATH	Adjazenz	
adjacent	TECH	benachbart	
= adjoining; contiguous; neighboured ≈ connected		= angrenzend; Nachbar-; anliegend ≈ zusammenhängend	

adjacent angle

adjacent angle	MECH	Nebenwinkel	
adjacent band	RADIO	Nachbarband	
= near band			
adjacent carrier trap (→adjacent video carrier trap)	TV	(→Bildträgersperre)	
adjacent channel	TELEC	Nachbarkanal	
adjacent-channel interference	MODUL	Nachbarkanalbeeinflussung	
= near-channel interference		= Nachbarkanalstörung	
adjacent-channel operation	RADIO	Nachbarkanalbetrieb	
= near-channel operation			
adjacent-channel power meter	INSTR	Nachbarkanal-Leistungsmesser	
adjacent-channel rejection	MODUL	Nachbarkanalunterdrückung	
adjacent-channel selection	MODUL	Nachbarkanalentkopplung	
= near-channel selection; adjacent-channel selectivity; near-channel selectivity		= Nachbarkanalselektion; Nahselektion; Trennschärfe [RADIO]	
adjacent-channel selectivity (→adjacent-channel selection)	MODUL	(→Nachbarkanalentkopplung)	
adjacent channel translation	BROADC	Nachbarkanalumsetzung	
adjacent picture carrier	TV	Nachbarbildträger	
adjacent video carrier trap	TV	Bildträgersperre	
= adjacent carrier trap		= Bildfalle; Nachbarbildfalle	
adjective (n.)	LING	Adjektiv	
		= Adjektivum; Eigenschaftswort; Beiwort; Artwort	
		↑ Nomen 2	
adjoining (→adjacent)	TECH	(→benachbart)	
adjoint	MATH	adjungiert	
adjoint expression	ENG.LOG	adjungierter Ausdruck	
adjourn (→postpone)	COLLOQ	(→vertagen)	
adjournment (→postponement)	COLLOQ	(→Vertagung)	
adjunction (→OR operation)	ENG.LOG	(→ODER-Verknüpfung)	
adjunction circuit (→OR gate)	CIRC.ENG	(→ODER-Glied)	
adjunction element (→OR gate)	CIRC.ENG	(→ODER-Glied)	
adjunction gate (→OR gate)	CIRC.ENG	(→ODER-Glied)	
adjunct professor	SCIE	Hilfsprofessor	
adjust	TECH	einstellen	
= set-up		= regeln; justieren; abgleichen	
adjust	ECON	bereinigen	
[figures]		[Zahlenreihen]	
adjustability	TECH	Einstellbarkeit	
= settability		= Regelbarkeit	
adjustable	TECH	einstellbar	
= settable		= regelbar 1	
≈ variable		≈ veränderbar	
adjustable base	EQUIP.ENG	Einstellfuß	
adjustable in height	TECH	höhenverstellbar	
= height-adjustable			
adjustable point	DATA PROC	einstellbares Komma	
adjustable resistance (→adjustable resistor)	COMPON	(→veränderbarer Widerstand)	
adjustable resistor	COMPON	veränderbarer Widerstand	
= variable resistor; adjustable resistance; variable resistance		= veränderlicher Widerstand; variabler Widerstand; einstellbarer Widerstand	
≈ balancing resistor [ELECTRON]		≈ Abgleichwiderstand [ELECTRON]	
≠ fixed resistor		≠ Festwiderstand	
↑ resistor		↑ Widerstand	
↓ potentiometer; thermistor; varistor		↓ Regelwiderstand; Thermistor; Varistor	
adjustable threshold metal oxide semiconductor (→ATMOS)	MICROEL	(→ATMOS)	
adjustable transformer	PHYS	Drehtransformator	
adjusted angle	TECH	Einstellwinkel	
= setting angle; indicated angle			
adjusted value	ELECTRON	Einstellwert	
= setting value		= Abgleichwert	
adjusting appliance (→setting gauge)	MECH	(→Paßlehre)	
adjusting command	TELECONTR	Stellbefehl	
adjusting device	TECH	Einstellvorrichtung	
adjusting drive (→servo drive)	CONTROL	(→Stellantrieb)	
adjusting gage (→setting gauge)	MECH	(→Paßlehre)	
adjusting instructions (→adjustment instructions)	DOC	(→Einstellanleitung)	
adjusting knob (→setting knob)	EQUIP.ENG	(→Einstellknopf)	
adjusting pliers	ELECTRON	Justierzange	
= snipe nose pliers			
adjusting potentiometer	ELECTRON	Abgleichpotentiometer	
= balancing potentiometer			
adjusting screw (→setscrew)	MECH	(→Stellschraube)	
adjusting time (→setting time)	TECH	(→Einstellzeit)	
adjusting torque	TECH	Einstellmoment	
= controlling torque			
adjustment	TECH	Einstellung	
= variation		= Justage; Justierung; Abgleich	
≈ positioning; alignment		≈ Positionierung; Ausrichtung	
adjustment	ELECTRON	Einstellung	
= setting; setup; control; variation			
adjustment (→alignment)	ANT	(→Ausrichtung)	
adjustment (→intervention)	TECH	(→Eingriff)	
adjustment instructions	DOC	Einstellanleitung	
= adjusting instructions		= Einstellangabe; Einstellanweisung	
adjustment mark (→setting mark)	INSTR	(→Einstellmarke)	
adjustment range	TECH	Einstellbereich	
= setting range			
adjustment screwdriver	ELECTRON	Justierschraubendreher	
adjustment specification	DOC	Einstellvorschrift	
ADM (→asynchronous disconnected mode)	DATA COMM	(→unabhängiger Wartezustand)	
ADM (→drop/insert multiplexer)	TRANS	(→Abzweigmultiplexer)	
administer (v.t.)	ECON	verwalten	
= manage		= administrieren	
administer (→dispense)	COLLOQ	(→austeilen)	
administration 1	ECON	Verwaltung 1	
[activity]		[Tätigkeit]	
= management 3		= Administration	
administration 2	ECON	Verwaltung 2	
[organization]		[Organisation]	
≈ authority		≈ Behörde	
administration (→data management)	DATA PROC	(→Datenverwaltung)	
administration and data server	TELEC	Betriebs-und Datenserver	
[ISDN]		[ISDN]	
administration expenses	ECON	(→Verwaltungskosten)	
(→administrative expenses)			
administrative building	ECON	Verwaltungsgebäude	
administrative costs	ECON	(→Verwaltungskosten)	
(→administrative expenses)			
administrative council	ECON	Verwaltungsrat	
[supervisory board for public institutions]		[Gremium zur Überwachung von Institutionen des öffentlichen Rechts]	
≈ advisory board		≈ Verwaltungsbeirat	
≈ supervisory board; board of directors		≈ Aufsichtsrat; Vorstand	
administrative expenses	ECON	Verwaltungskosten	
= administrative costs; administration expenses		≈ Verwaltungsgemeinkosten	
administrative headquartes (→headquarters)	ECON	(→Hauptverwaltung)	
administrative unit	TELEC	Verwaltungseinheit	
[SDH]		[SDH]	
= AU			
administrator	ECON	Verwalter	
= manager 2			
Admiralty mile	PHYS	englische Seemeile	
[1,853.1824 m or 6,080 ft]		[1.853,1824 m oder 6.080 ft]	
admissible solution (→feasible solution)	MATH	(→zulässige Lösung)	
admission	ECON	Zulassung	
= licence		= Lizenz	
admission (→access)	TECH	(→Zugang)	
admission (→access)	DATA PROC	(→Zugriff)	

admittance	NETW.TH	komplexer Scheinleitwert

admittance NETW.TH
[reciprocal of impedance; vectorial sum of conductance and susceptance; SI unit: Siemens]
= Y

komplexer Scheinleitwert
[Kehrwert des komplexen Scheinwiderstandes; Realteil = Wirkleitwert, Imaginärteil = Blindleitwert; SI-Einheit: Siemens]
= Admittanz; Scheinleiwert 1; Leitwert; komplexer Leitwert; Y

admittance equation NETW.TH **Leitwertgleichung**
↑ Vierpolgleichung

admittance matrix NETW.TH **Leitwertmatrix**
= Y matrix = y-Matrix; Admittanz-Matrix

admittance parameter NETW.TH (→Leitwertparameter)
(→conductance parameter)

admitted (→licensed) ECON (→zugelassen)
admix (v.t.) TECH **beimengen**
admixed air TECH **Nebenluft**
ADP (→electronic data processing) INF.TECH (→elektronische Datenverarbeitung)

ADPCM CODING **ADPCM**
[PCM coding reduced to 4 bits per sample, to save storage or transmission capacity]
= adaptive differential pulse code modulation
≈ transcoder [TRANS]; APCM

[auf 4 Bit pro Abtastwert reduzierte PCM-Codierung, zur Einsparung von Speicher- oder Übertragungskapazität]
= adaptive differentielle Pulscodemodulation
≈ Transcoder [TRANS]; APCM

adsorption PHYS **Adsorption**
[adhesion of gas on solid]
↑ adhesion

[Adhäsion von gasförmigem auf festen Stoff]
↑ Adhäsion

adsorptive (n.) (→adsorptive material) TECH (→Absorptionsmaterial)
adsorptive material TECH **Absorptionsmaterial**
= adsorptive (n.) = absorbierendes Material
AD technique (→alloy diffusion technique) MICROEL (→AD-Technik)

advance ELECTRON **Fortschaltung**
= increment
advance (n.) (→progress) TECH (→Fortschritt)
advance (→constantan) METAL (→Konstantan)
advance- (→preliminary) COLLOQ (→Vorab-)
advance (→prepayment) ECON (→Vorauszahlung)
advance (n.) (→feed) TERM&PER (→Vorschub)
advance (n.) (→lead) COLLOQ (→Vorsprung)
advance (→antedate v.t.) ECON (→vorziehen)
advance control (→feed control) TERM&PER (→Vorschubsteuerung)
advanced (→progressive 2) TECH (→fortschrittlich)
advanced development (→front-end development) TECH (→Vorfeldentwicklung)
advance delivery ECON **Vorauslieferung**
≈ partial delivery ≈ Teillieferung
advanced level exam (BRI) SCIE **Abitur**
= A-level exam; A-Level = Oberschulreife; Bakkalaureat; Hochschulreifeprüfung
advanced standard-buried collector technology MICROEL **ASBC-Technik**
= ASBC technology
advanced technology TECH **hochentwickelte Technik**
= hochentwickelte Technologie; fortgeschrittene Technik
advance increment TERM&PER **Vorschubschritt**
= Vorschubweg
advance information TECH (→Vorabinformation)
(→preliminary information)
advance instruction (→feed instruction) DATA PROC (→Vorschubbefehl)
advance in the art (→technological progress) TECH (→technischer Fortschritt)
advance mechanism (→moving mechanism) MECH (→Transporteinrichtung)
advance of works TECH **Baufortschritt**

advance order ECON **Vorausbestellung**
= Vorbestellung; Vorabbestellung; Reservierung
advance payment ECON (→Vorauszahlung)
(→prepayment)
advance-payment bond ECON **Anzahlungsgarantie**
advance performance ECON **Vorleistung 1**
advance planning TECH **Vorausplanung**
advantage COLLOQ **Vorteil**
= payoff; benefit

advection METEOR **Advektion**
[horizontal movement of air causing change in physical parameters]
[horizontale Luftbewegung mit Veränderung physikalischer Parameter]

adventure (→adventure game) DATA PROC (→Abenteuerspiel)
adventure game DATA PROC **Abenteuerspiel**
= adventure

adverb LING **Adverb** (pl.-ien)
[word to specify circumstances of an action or process; e.g. today, rapidly]
[Wortart zur Umstandsbestimmung; nicht flektierbar; z.B. heute, schnell]
= Umstandswort

adverbial (n.) (→adverbial specification) LING (→Adverbial) (n.n.; pl.-e)

adverbial sentence LING **Adverbialsatz**
= Umstandssatz

adverbial specification LING **Adverbial** (n.n.; pl.-e)
= adverbial (n.)
[Wortkombination zur Bestimmung des Umstandes einer Handlung]
= Adverbiale (n.n.; pl.-ien); adverbiale Bestimmung; Umstandsbestimmung

adverb of place LING **Lokaladverb**
adversative LING **adversativ**
= entgegensetzend
adverse (→unfavorable) COLLOQ (→ungünstig)
adverse consequence COLLOQ (→Nachteil)
(→disadvantage)
advert (→advertisement) ECON (→Zeitungsinserat)
advertisement ECON **Zeitungsinserat**
= advert; ad
= Zeitungsannonce; Annonce; Inserat; Anzeige
advertising ECON **Werbung**
= ad; propaganda; publicity; canvassing; solicitation
= Propaganda; Reklame; Publizität
↓ institutional advertising; product advertising
↓ Repräsentativwerbung; Produktwerbung
advertising agency ECON **Werbefirma**
= publicity agency = Werbeagentur
advertising brochure DOC **Werbebroschüre**
≈ data sheet; leaflet ≈ Datenblatt; Prospekt
advertising budget ECON **Werbeetat**
advertising costs ECON **Werbekosten**
= publicity costs; advertising expenditures; publicity expenditures
= Werbungskosten 1
advertising department ECON **Werbeabteilung**
= publicity department
advertising expenditures ECON (→Werbekosten)
(→advertising costs)
advertising gift ECON **Werbegeschenk**
adviser (→consultant) ECON (→Berater)
advisory board (→advisory council) ECON (→Beirat)
advisory board (→administrative council) ECON (→Verwaltungsrat)
advisory council ECON **Beirat**
= advisory board; board (n.)
A-encoding law CODING **A-Gesetz**
= A-law
A.E.N. value TELEPH **Ersatzdämpfung**
= A.E.N.-Wert
aerial OUTS.PLANT **oberirdisch**
[line] [Leitung]
aerial (BRI) (→antenna) RADIO (→Antenne)
aerial amplifier (→antenna booster) HF (→Antennenverstärker)
aerial base impedance (→antenna input impedance) ANT (→Fußpunktwiderstand)

aerial booster

aerial booster (→antenna booster)	HF	(→Antennenverstärker)	
aerial cable	COMM.CABLE	**Luftkabel**	
aerial cable (→antenna cable)	RADIO	(→Antennenanschlußkabel)	
aerial cable (→antenna cable)	CONS.EL	(→Antennenkabel)	
aerial cable line = cable line	OUTS.PLANT	**Luftkabellinie**	
aerial cable system	OUTS.PLANT	**oberirdische Kabelanlage**	
aerial choke (→antenna choke)	ANT	(→Antennendrossel)	
aerial coil (→antenna coil)	ANT	(→Antennenspule)	
aerial conductor	PHYS	**Luftleiter**	
aerial coupling capacitor (→antenna coupling condenser)	ANT	(→Antennenkopplungskondensator)	
aerial coupling condenser (→antenna coupling condenser)	ANT	(→Antennenkopplungskondensator)	
aerial drive (→antenna drive)	ANT	(→Antennendrehvorrichtung)	
aerial exciting (→antenna feeding)	ANT	(→Antennenspeisung)	
aerial feed (→antenna feeding)	ANT	(→Antennenspeisung)	
aerial feeder (→antenna feeding line)	ANT	(→Antennenspeiseleitung)	
aerial feeder line (→antenna feeding line)	ANT	(→Antennenspeiseleitung)	
aerial feeding (→antenna feeding)	ANT	(→Antennenspeisung)	
aerial feeding line (→antenna feeding line)	ANT	(→Antennenspeiseleitung)	
aerial feed line (→antenna feeding line)	ANT	(→Antennenspeiseleitung)	
aerial height (→antenna height)	ANT	(→Antennenlänge)	
aerial impedance (→antenna impedance)	ANT	(→Antennenwiderstand)	
aerial input impedance (→antenna input impedance)	ANT	(→Fußpunktwiderstand)	
aerial jack (→antenna socket)	CONS.EL	(→Antennenbuchse)	
aerial mast (→antenna mast)	ANT	(→Antennenmast)	
aerial mount (→antenna mount)	ANT	(→Antennenbefestigung)	
aerial outlet (→antenna outlet)	CONS.EL	(→Antennensteckdose)	
aerial resistance (→antenna impedance)	ANT	(→Antennenwiderstand)	
aerial rotator (→antenna rotator)	ANT	(→Antennenrotor)	
aerial rotor (→antenna rotor)	ANT	(→Antennenrotor)	
aerial selector (→antenna selector)	ANT	(→Antennenwahlschalter)	
aerial series capacitor (→antenna series capacitor)	ANT	(→Antennenverkürzungskondensator)	
aerial socket (→antenna socket)	CONS.EL	(→Antennenbuchse)	
aerial spacing (→antenna spacing)	RADIO	(→Antennenabstand)	
aerial support (→antenna support)	ANT	(→Antennenträger)	
aerial system (→antenna system)	ANT	(→Antennenanlage)	
aerial tuning coil (→antenna tuning coil)	ANT	(→Antennenabstimmspule)	
aerial tuning unit (→antenna tuning unit)	ANT	(→Antennenanpaßgerät)	
aerodrome (BRI) (→airport)	AERON	(→Flughafen)	
aerodrome beacon (→airport beacon)	RADIONAV	(→Flughafenbake)	
aerodrome surveillance radar (→ASR)	RADIONAV	(→Flughafen-Überwachungsradar)	
aeronautical engineering ↑ aerospace engineering	AERON	**Luftfahrttechnik** ↑ Luft- und Raumfahrttechnik	
aeronautical mobile satellite service	SAT.COMM	**beweglicher Flugfunkdienst über Satelliten**	
aeronautical radio	RADIO	**Flugfunk**	
aeronautical radionavigation system	RADIONAV	**Flugnavigationsfunkdienst**	
aeronautical radio service	TELEC	**Flugfunkdienst**	
aeronautical station	RADIONAV	**Bodenfunkstelle**	
aeronautics	TECH	**Luftfahrt**	
aerophore (→airport beacon)	RADIONAV	(→Flughafenbake)	
aerospace engineering ↓ aeronautical engineering; space-flight engineering	TECH	**Luft- und Raumfahrttechnik** ↓ Luftfahrttechnik; Raumfahrttechnik	
aesthetic (adj.)	COLLOQ	**ästhetisch** = ansprechend ≈ formschön	
AF (→voice frequency)	TELEC	(→Niederfrequenz)	
AFC (→automatic frequency control)	CIRC.ENG	(→Frequenznachregelung)	
affect (v.t.) = influence ≈ impair	TECH	**beinflussen** ≈ beinträchtigen	
affiliate (→subsidiary)	ECON	(→Tochtergesellschaft)	
affiliate company (→subsidiary)	ECON	(→Tochtergesellschaft)	
affirmative ≠ negative	LING	**bejahend** ≠ verneinend	
affix [dependent word element aggregated to a word kernel] ↓ prefix; infix; suffix	LING	**Affix** [einem Wortkern angefügtes unselbständiges Wortelement] ↑ gebundenes Morphem ↓ Präfix; Infix; Suffix	
A-fixture	OUTS.PLANT	**A-förmiger Mast** ↑ Telefonmast	
aftereffect	TECH	**Nachwirkung**	
afterglow	PHYS	**Nachglimmen**	
afterglow = afterglow persistence; after image; screen persistence	TV	**Nachleuchten** = Nachleuchteffekt	
afterglow persistence (→afterglow)	TV	(→Nachleuchten)	
afterglow screen (→persistent screen)	TERM&PER	(→nachleuchtender Bildschirm)	
afterimage	TERM&PER	**nachleuchtendes Bild** = Nachbild	
after image (→afterglow)	TV	(→Nachleuchten)	
afterpulse	ELECTRON	**Nachimpuls**	
after-sales service = costumer service; service; costumer support; support; assistance ≈ maintenance service	ECON	**Kundendienst** = Service; Kundenbetreuung; Betreuung ≈ Wartungsdienst	
Ag (→silver)	CHEM	(→Silber)	
AGA [Advanced Graphics Adapter; emulator of Commodore for other graphic standards]	TERM&PER	**AGA** [Emulator der Fa. Commodore für andere Graphikstandards]	
AGA board ↑ graphics board	TERM&PER	**AGA-Karte** = AGA-Adapter ↑ Graphikkarte	
Agate (BRI) (→type size 5 1/2 point)	TYPOGR	(→Schriftgröße 5 1/2 Punkt)	
AGC (→automatic gain control)	CONTROL	(→automatische Verstärkungsregelung)	
AGC amplifier (→regulating amplifier)	CIRC.ENG	(→Regelverstärker)	
aged	QUAL	**gealtert**	
aged data	DATA PROC	**veraltete Daten**	
ageing = aging	QUAL	**Alterung**	
ageing (→obsolete)	TECH	(→veraltet)	
ageing failure = aging failure; wearout failure ≈ late failure	QUAL	**Verschleißausfall** = Alterungsausfall; Ermüdungsausfall ≈ Spätausfall	
ageing properties	QUAL	**Alterungseigenschaften**	
ageing rate = aging rate; drift characteristic	QUAL	**Alterungsrate** = Alterungszahl	
agency (→office 1)	ECON	(→Amt)	
agency mask (→standard mask)	INSTR	(→Standardmaske)	
agenda	ECON	**Tagesordnung**	
agent 1	CHEM	**Mittel** = Agens	
agent	ECON	**Beauftragter** = Agent	
agent 2 (→binding agent)	CHEM	(→Bindemittel)	
aggregate bit rate (→transmission rate)	TELEC	(→Übertragungsgeschwindigkeit)	
aggregate peak level	TELEGR	**Summenspitzenpegel**	
agile (adj.)	MIL.COMM	**agil**	
agile bandwidth	MIL.COMM	**agile Bandbreite**	
agile communication	MIL.COMM	**agile Kommunikation**	
agile signal generator	INSTR	**Agil-Signal-Generator** = Hochgeschwindigkeitssignal-Generator	
agility	MIL.COMM	**Agilität**	
agil signal [changing its parameters with high speed]	MIL.COMM	**Hochgeschwindigkeitssignal** = agiles Signal	
agil transceiver	MIL.COMM	**agiles Funkgerät**	

aging (→ageing)	QUAL	(→Alterung)	**air-core electromagnet**	EL.TECH	**eisenfreier Elektromagnet**
aging failure (→ageing failure)	QUAL	(→Verschleißausfall)	= air-core magnet		= eisenfreier Magnet; eisenloser Elektromagnet; eisenloser Magnet
aging rate (→ageing rate)	QUAL	(→Alterungsrate)	**air-core magnet** (→air-core electromagnet)	EL.TECH	(→eisenfreier Elektromagnet)
agree	COLLOQ	**vereinbaren**	**aircraft**	AERON	**Flugzeug**
= stipulate		= abmachen; übereinkommem; stipulieren	= airplain		
agreed penalty (→contractual penalty)	ECON	(→Konventionalstrafe)	**aircraft antenna**	ANT	**Flugzeugantenne**
			= airplain antenna; airborne antenna		
agreement 1	COLLOQ	**Vereinbarung**	**aircraft control and warning service**	RADIO LOC	**Luftraum-Kontroll-und-Warndienst**
= stipulation; deal; arrangement; convention		= Abmachung; Übereinkommen; Übereinkunft; Abkommen; Abrede; Stipulation; Verständigung	= ACW		
			aircraft pilot	AERON	**Flugzeugpilot**
agreement 2 (→accordance)	COLLOQ	(→Übereinstimmung)	= airplane pilot; pilot		= Pilot
agricultural machine engineering	TECH	**Landmaschinenbau**	**aircraft radar**	RADIO LOC	**Flugzeugradar**
A h (→ampere-hour)	PHYS	(→Amperestunde)	= airplain radar; airborne radar		
AI (→artificial intelligence)	DATA PROC	(→künstliche Intelligenz)	**aircraft stand taxilane**	AERON	**Standplatzrollgasse**
			aircraft transceiver	RADIO NAV	**Flugfunkgerät**
aid	TECH	**Hilfsmittel**	**aircraft transmitter**	RADIO NAV	**Flugzeugbordsender**
= tool			= airplain transmitter; airborne transmitter		
Aiken code	CODING	**Aiken-Code**	**aircraft warning lightning** (→lightning)	AERON	(→Befeuerung)
AI language [program language]	DATA PROC	**KI-Sprache** [Programmiersprache]	**air damping** (→pneumatic damping)	TECH	(→Luftdämpfung)
AIM (→avalanche-induced migration)	MICROEL	(→AIM)	**air defense**	MIL.COMM	**Luftabwehr**
					= Flugabwehr
aim at each other	TECH	**gegenseitig ausrichten**	**air defense system**	RADIO LOC	**Flugabwehrsystem**
air-activated (→pneumatically operated)	TECH	(→druckluftbetätigt)	**air density**	PHYS	**Luftdichte**
air-actuated (→pneumatically operated)	TECH	(→druckluftbetätigt)	**air dielectric trimmer**	COMPON	**Lufttrimmer**
			= air trimmer		
air aspiration	TECH	**Luftansaugung**	**air discharge**	PHYS	**Luftentladung**
= air input		≈ Lufteinlaß	**air duct**	TECH	**Luftkanal**
airbase	AERON	**Militärflughafen**	**air extraction**	TECH	**Lufterneuerung**
airborne	AERON	**Flugzeug-**	**airfield**	AERON	**Rollfeld**
airborne (→flying)	TECH	(→fliegend)	[total surface accessible to airplaines]		[Gesamtheit der von Flugzeugen befahrbaren Flächen]
airborne antenna (→aircraft antenna)	ANT	(→Flugzeugantenne)			
airborne beacon	RADIO NAV	**fliegendes Funkfeuer**	**airfield surveillance radar**	RADIO LOC	**Rollfeld-Überwachungsradar**
airborne computer (aircraft) (→onboard computer)	DATA PROC	(→Bordcomputer)	**air filter 1**	TECH	**Luftfilter**
			= air cleaner		
airborne radar (→aircraft radar)	RADIO LOC	(→Flugzeugradar)	**air filter 2**	TECH	**Druckluftfilter**
			air flow 1	TECH	**Luftfluß**
airborne transmitter (→aircraft transmitter)	RADIO NAV	(→Flugzeugbordsender)	= airflow		
			air flow 2 (→air throughput)	TECH	(→Luftdurchsatz)
air brake	INSTR	**Luftdämpfung**	**airflow** (→air flow 1)	TECH	(→Luftfluß)
air capacitor	COMPON	**Luftkondensator**	**airflow sheet** (→air conducting sheet)	EQUIP.ENG	(→Luftablenkblech)
= air-spaced capacitor					
air cargo (→air freight)	ECON	(→Luftfracht)	**air freight**	ECON	**Luftfracht**
air circulation	TECH	**Luftzirkulation**	= air cargo		
air circulator (→fan)	TECH	(→Ventilator)	**air freighter**	AERON	**Frachtflugzeug**
air cleaner (→air filter 1)	TECH	(→Luftfilter)	**air-friction damping** (→pneumatic damping)	TECH	(→Luftdämpfung)
airconditioned	TECH	**klimatisiert**			
= with controlled ambient			**air-gap**	MECH	**Luftspalt**
air-condition equipment (→air conditioner)	TECH	(→Klimaanlage)	**air gap** (→magnetic gap)	PHYS	(→Magnetluftspalt)
			air gap (→head distance)	TERM & PER	(→Flughöhe)
air conditioner	TECH	**Klimaanlage**	**air-gap induction**	EL.TECH	**Luftspaltinduktion**
= air conditioning 2; air-condition equipment			**air-gap lightning arrestor**	COMPON	**Luftblitzableiter**
			air humidity	PHYS	**Luftfeuchtigkeit**
air conditioning 1	TECH	**Klimatisierung**	= atmosferic humidity		= Luftfeuchte
air conditioning 2 (→air conditioner)	TECH	(→Klimaanlage)	**air inlet**	MICROEL	**Lufteinfüllstück**
			air inlet	TECH	**Lufteinlass**
air conditioning engineering	TECH	**Klimatechnik**	= air intake		= Luftzuführung
		= Versorgungstechnik			≈ Luftansaugung
air conducting sheet	EQUIP.ENG	**Luftablenkblech**	**air input** (→air aspiration)	TECH	(→Luftansaugung)
= airflow sheet		= Wärmeleitblech	**air interface**	RADIO	**Luftschnittstelle**
air control (→pneumatic control)	TECH	(→pneumatische Steuerung)	**airline**	AERON	**Fluggesellschaft**
					= Fluglinie
air-controlled (→pneumatically operated)	TECH	(→druckluftbetätigt)	**airline pilot**	AERON	**Linienflugzeugführer**
			airmail	POST	**Luftpost**
air cooled	TECH	**luftgekühlt**	**air motion transformer**	EL.ACOUS	**Folientöner**
air cooling	TECH	**Luftkühlung**	**air-operated** (→pneumatically operated)	TECH	(→druckluftbetätigt)
air-core [EL.TECH] (→iron-free)	TECH	(→eisenfrei)	**air outlet**	TECH	**Luftauslass**
					= Luftaustritt
air-core cable	COMM.CABLE	**ungefülltes Kabel**	**air permeability**	TECH	**Luftdurchlässigkeit**
≠ jelly-filled cable			**airplain** (→aircraft)	AERON	(→Flugzeug)
air core choke (→air coil)	COMPON	(→Luftspule)	**airplain antenna** (→aircraft antenna)	ANT	(→Flugzeugantenne)
air core coil	COMPON	**Luftspule**			
= air core choke					

airplain radar (→aircraft radar)		RADIO LOC	(→Flugzeugradar)	**alarm indication** (→visual alarm)	EQUIP.ENG	(→optischer Alarm)
airplain transmitter (→aircraft transmitter)		RADIO NAV	(→Flugzeugbordsender)	**alarm indication signal** = AIS; blue signal	TRANS	**Alarm-Meldesignal** = AIS-Signal; Alarmmeldewort
airplane flutter		AERON	**Flugzeugstörung**	**alarm message** (→alarm signal)	TELEC	(→Alarmsignal)
airplane pilot (→aircraft pilot)		AERON	(→Flugzeugpilot)	**alarm processing** = alarm treatment	EQUIP.ENG	**Alarmbehandlung**
airport = aerodrome (BRI) ≈ airfield		AERON	**Flughafen** ≈ Rollfeld	**alarm signal** = alarm message	TELEC	**Alarmsignal** = Alarmmeldung
airport beacon = aerodrome beacon; aerophore		RADIO NAV	**Flughafenbake** = Flugplatzbake	**alarm signal distributor**	SYS.INST	**Alarmverteiler**
airport beaconing		RADIO NAV	**Flughafen-Befeuerung**	**alarm signal panel**	EQUIP.ENG	**Alarmsignalfeld**
airport of destination		AERON	**Bestimmungsflughafen**	**alarm stop**	TECH	**Alarmstopp** = Notstopp
airport surface detection equipment		RADIO LOC	**Flughafen-Oberflächenradar**	**alarm timing delay**	TELEC	**Alarmverzögerung**
airport surveillance radar (→ASR)		RADIO NAV	(→Flughafen-Überwachungsradar)	**alarm treatment** (→alarm processing)	EQUIP.ENG	(→Alarmbehandlung)
air-position indicator		RADIO NAV	**Lagebestimmungsgerät** = Positionsbestimmungsgerät	**A-law** (→A-encoding law)	CODING	(→A-Gesetz)
				album [container for phogram records]	CONS.EL	**Plattenalbum** = Schallplattenalbum
air pressure (→atmospheric pressure)		PHYS	(→Luftdruck)	**ALC** (→automatic level control)	ELECTRON	(→automatische Pegelregelung)
air pump		TECH	**Luftpumpe** ↓ Vakuumpumpe	**alcali metall**	CHEM	**Alkalimetall**
air purging		TECH	**Luftspülung**	**alcaline photoelectric cell**	POWER SYS	**Alkalizelle**
air recycling		TECH	**Umluft**	**alcatron** ↑ FET	MICROEL	**Alcatron** ↑ FET
air resistance		TECH	**Luftwiderstand**	**alcohol**	CHEM	**Alkohol**
air space		AERON	**Luftraum**	**A-lead** (→tip wire)	TELEPH	(→A-Ader)
air-spaced capacitor (→air capacitor)		COMPON	(→Luftkondensator)	**aleatory** (→random)	TECH	(→zufällig)
				alert (n.) (→warning)	COLLOQ	(→Warnung)
air space surveillance		RADIO LOC	**Luftraumüberwachung**	**alert box**	DATA PROC	**Warntafel**
airstrip [runway without further airport facilities] ≈ runway		AERON	**Landestreifen** ≈ Landebahn	**alerting paint** (→warning paint)	TECH	(→Warnbemalung)
				alerting system (→danger detection system)	SIGN.ENG	(→Gefahrenmeldeanlage)
air tank		TECH	**Lufttank**	**A-Level** (→advanced level exam)	SCIE	(→Abitur)
air temperature		PHYS	**Lufttemperatur**	**A-level exam** (→advanced level exam)	SCIE	(→Abitur)
air throughput = air flow 2		TECH	**Luftdurchsatz**	**Alexanderson antenna** = multiple-tuned antenna 2 ≈ multiple-tuned antenna 1	ANT	**Alexanderson-Antenne** ≈ Mehrfachresonanz-Antenne
airtight = hermetic		TECH	**luftdicht** = hermetisch ↑ gasdicht	**Alford loop antenna**	ANT	**Alford-Ringantenne**
				algebra ↑ elementary mathematics	MATH	**Algebra** ↑ Elementarmathematik
airtight packing		TECH	**luftdichte Verpackung**	**algebraic**	MATH	**algebraisch**
air traffic = flight traffic ↓ line air traffic; charter air traffic; military air traffic		AERON	**Luftverkehr** = Flugverkehr ↓ Linienluftverkehr; Bedarfsluftverkehr; Militärluftverkehr	**algebraic complement** [determinant] = cofactor	MATH	**algebraisches Komplement** [Determinate] = Adjunkte
				algebraic expression	MATH	**algebraischer Ausdruck**
air traffic control = air traffic management		RADIO NAV	**Flugsicherung**	**algebraic language** (→algebraic program language)	DATA PROC	(→algebraische Programmiersprache)
air traffic controller		RADIO NAV	**Fluglotse**	**algebraic number** ≠ transcendental number	MATH	**algebraische Zahl** ≠ transzendente Zahl
air traffic management (→air traffic control)		RADIO NAV	(→Flugsicherung)	**algebraic program language** = algebraic language	DATA PROC	**algebraische Programmiersprache** = algebraische Sprache
air trimmer (→air dielectric trimmer)		COMPON	(→Lufttrimmer)	**algebraic sign** (→sign)	MATH	(→Vorzeichen)
air valve		TECH	**Luftventil**	**ALGOL** [ALGOrithmic Language] ↑ high-level programming language	DATA PROC	**ALGOL** ↑ problemorientierte Programmiersprache
AIS (→alarm indication signal)		TRANS	(→Alarm-Meldesignal)			
ajustment accuracy = setting accuracy; settability		ELECTRON	**Einstellgenauigkeit**	**algorithm** [rule to solve a problem] ↓ solution algorithm; decision algorithm; iteration algorithm; optimizing algorithm; prediction algorithm; acceleration algorithm; correlation algorithm	MATH	**Algorithmus** [Verfahren zur Lösung eines Problems] ↓ Lösungsalgorithmus; Entscheidungsalgoritmus; Iterationsalgorithmus; Optimierungsalgorithmus; Vorhersagealgorithmus; Beschleunigungsalgorithmus; Korrelationsalgorithmus
A key (→alternate coding key)		TERM&PER	(→ALT-Taste)			
akin ≈ similar		COLLOQ	**artverwandt** ≈ ähnlich			
Al (→aluminum)		CHEM	(→Aluminium)			
alarm		TELECONTR	**Alarm** = Warnmeldung			
alarm activation		EQUIP.ENG	**Alarmauslösung**			
alarm and alerting system		SIGN.ENG	**Alarm- und Meldeanlage**			
alarm call (→alarm-clock calling)		TELEPH	(→Weckdienst)			
alarm-clock calling = alarm call ↑ telephone notification system		TELEPH	**Weckdienst** ↑ Fernsprechauftragdienst	**algorithmic** (adj.)	MATH	**algorithmisch**
				algorithmic language (→algorithmic programming language)	DATA PROC	(→algorithmische Programmiersprache)
alarm collector		TELEC	**Alarmsammler**	**algorithmic programming language** = algorithmic language	DATA PROC	**algorithmische Programmiersprache** = algorithmische Sprache
alarm cut-off = ACO		EQUIP.ENG	**Alarmabschaltung**			
alarm detector		SIGN.ENG	**Alarmgerät**			
alarm horn		TECH	**Alarmhupe**			

aliasing		MODUL	**Rückfaltung**	**allocation**		TECH	**Zuordnung**

aliasing MODUL **Rückfaltung**
[causes generally undesired optical effects] [führt i.a. zu unerwünschten optischen Effekten]
= foldover distortion; overlap distortion = Überfaltung; Alias-Effekt; Faltungsverzerrung; Überlappungsverzerrung
≈ staircaising

aliasing DATA PROC **Bildunschönheit**
[computer graphics; unwanted visual defects due to limited display resolution] [Computergraphik; auflösungsbedingte Darstellungsfehler]
↓ staircase effect ↓ Aliasing
↓ Treppeneffekt

alien (adj.) TECH **fremd**
alien (v.t.) (→alienate) ECON (→veräußern)
alienable ECON **veräußerbar**
≈ salable ≈ verkäuflich
alienate ECON **veräußern**
= alien (v.t.) = verkaufen
≈ sale

alienation ECON **Veräußerung**
≈ sale ≈ Verkauf 1
alien disk DATA PROC **Fremdplatte**
[in a non-readable format] [in nicht lesbarem Format]
alien disk reader TERM&PER **Fremdplattenleser**
align TECH **ausrichten**
= collimate = fluchten
aligned (→collimated) TECH (→ausgerichtet)
aligning edge TECH **Ausrichtkante**
aligning edge (→reference edge) TERM&PER (→Bezugskante)
aligning pin MECH **Paßstift**
aligning plug (→guide pin) MECH (→Führungsstift)
alignment ANT **Ausrichtung**
= adjustment; collimation; orientation
alignment TECH **Ausrichtung**
= straightening; collimation ≈ Einstellung
≈ adjustment
alignment (→synchronism) TECH (→Synchronismus)
alignment bit (→synchronization bit) TELEC (→Synchronisierbit)
alignment diskette TERM&PER **Abgleichdiskette**
alignment error TELEC (→Synchronisationsfehler)
(→synchronization failure)
alignment line (→convergence line) MATH (→Fluchtlinie)
alignment mark MICROEL **Justiermarke**
alignment point (→convergence point) MATH (→Fluchtpunkt)
alignment tool ELECTRON **Abgleichwerkzeug**
= trimmtool
alive (→live) EL.TECH (→heiß)
alive (→live) BROADC (→direktübertragen)
alkali lye CHEM **Alkalilauge**
alkaline battery POWER SYS **alkalische Batterie**
alkaline manganese battery POWER SYS **Alkali-Mangan-Zelle**
alkaline solution CHEM **alkalische Lösung**
Allan variance MATH **Allan-Varianz**
all-digital ELECTRON **volldigital**
all-electronic (→fully electronic) ELECTRON (→vollelektronisch)
all-embracing COLLOQ **allumfassend**
Allen screw MECH **Innensechskantschraube**
↓ Inbusschraube
Allen wrench MECH **Innensechskantschlüssel**
↓ L-shaped hexagon key ↓ Inbusschlüssel
alligator clip (→crocodil clip) ELECTRON (→Abgreifklemme)
all mains POWER SYS **Allstrom**
= AC/DC
all-metal METAL **Ganzmetall-**
allocate (→assign) TECH (→zuordnen)
allocated (→connected) SWITCH (→beschaltet)
allocation SWITCH **Beschaltung**
= connection (NAM); connexion (BRI); configuration = Anschaltung
allocation ECON **Umlage**
= appointment; levy; charge = Zuweisung
↓ costs appointment; loss appointment ↓ Kostenumlage; Verlustumlage

allocation TECH **Zuordnung**
= assignment; allotment; appropriationing; grant (n.) = Zuteilung; Zuweisung
≈ occupation; positioning ≈ Belegung; Aufteilung
≠ deallocation ≠ Freigabe
allocation list SWITCH **Beschaltungsliste**
= assignment list
allocation of overheads ECON **Gemeinkostenumlage**
allocation plan (→allocation scheme) EQUIP.ENG (→Belegungsplan)
allocation routine DATA PROC **Zuteilungsroutine**
allocation scheme EQUIP.ENG **Belegungsplan**
= allocation plan; equipment configuration plan; equipment plan = Beschaltungsplan
≈ assembly drawing; circuit diagram ≈ Zusammenbauzeichnung; Stromlaufplan
allocator DATA PROC **Zuordner**
[allocates defined input signals to defined output signals] [ordnet definierte Eingangssignale definierten Ausgangssignalen zu]
= coder; translator
all-ones (→continuous one) DATA COMM (→Dauereins)
allot (→assign) TECH (→zuordnen)
allotment (→allocation) TECH (→Zuordnung)
all over TECH **allseitig**
allowance ECON **Gutschrift**
allowance (→tightest fit) ENG.DRAW (→Kleinstsitz)
allowance (→discount) ECON (→Preisnachlaß)
allowance (→margin) COLLOQ (→Spielraum)
allowance (→daily allowance) ECON (→Tagegeld)
allowance of charge ECON **Gebührenermäßigung**
= reduction of charge; tariff discount; reduced tariff = Gebührenvergünstigung
allow deferment ECON **stunden**
= grant time
allowed band PHYS **erlaubtes Band**
alloy (v.t.) METAL **legieren**
alloy (n.) METAL **Legierung**
alloy bulk diffusion technique MICROEL (→AD-Technik)
(→alloy diffusion technique)
alloy-diffused transistor MICROEL (→Legierungstransistor)
(→alloy-junction transistor)
alloy diffusion technique MICROEL **AD-Technik**
= AD technique; alloy bulk diffusion technique; ADB technique; post-alloy diffusion technique; PAD technique; push-out base technique; POB technique = ABD-Technik; PAD-Technik; POB-Technik
alloyed diode MICROEL **legierte Diode**
alloyed transistor MICROEL (→Legierungstransistor)
(→alloy-junction transistor)
alloy front PHYS **Legierungsfront**
alloying process (→alloying technique) MICROEL (→Legierungstechnik)
alloying technique MICROEL **Legierungstechnik**
= alloying process = Legierungsverfahren
alloy junction MICROEL **Legierungsschicht**
alloy-junction transistor MICROEL **Legierungstransistor**
= alloy-diffused transistor; alloyed transistor = legierter Transistor
↑ bipolar transistor; junction transistor ↑ Bipolartransistor; Flächentransistor
alloy-treated METAL **niedriglegiert**
all-pass (→all-pass filter) NETW.TH (→Allpaß)
all-pass element NETW.TH **Allpaßglied**
all-pass filter NETW.TH **Allpaß**
= all-pass; lattice filter; phase filter = Allpaßfilter
all-pass network NETW.TH **Allpaßnetzwerk**
all-purpose TECH **Allzweck-**
all-purpose computer DATA PROC (→Universalrechner)
(→general-purpose computer)
all-purpose matchbox ANT **Universalabstimmgerät**
all-round knowledge SCIE **Allgemeinwissen**
≈ basic education ≈ Allgemeinbildung
all-terrain TECH **geländegängig**
all-transistorized ELECTRON (→transistorisiert)
(→transistorized)
all trunks busy (→external blocking) SWITCH (→äußere Blockierung)
all-trunks-busy condition SWITCH (→Gassenbesetztzustand)
(→congestion)

all-trunks-busy time SWITCH **Blockierungsdauer**
= busy period
all-trunks-busy tone TELEPH **Gassenbesetztton**
all-wave antenna ANT **Allwellenantenne**
 = Allbereichantenne
all-wave receiver RADIO **Allwellenempfänger**
 = Universalempfänger; Mehrwellenempfänger
all-weather landing RADIO NAV **Schlechtwetterlandung**
all-zeroes (→continuous DATA COMM (→Dauernull)
 zero)
almost periodic PHYS (→quasiperiodisch)
 (→quasiperiodic)
alphabet LING **Alphabet**
 [ordered sequence of a character set] [geordnete Folge eines Buchstabenvorrats]
alphabet (→character INF.TECH (→Zeichenvorrat)
 set)
alphabetic CODING **alphabetisch**
 = alphabetical ≠ numerisch; alphanumerisch
 ≠ numeric; alphanumeric
alphabetical (→alphabetic) CODING (→alphabetisch)
alphabetical code (→alphabetic CODING (→alphabetischer Code)
 code)
alphabetical data DATA PROC **alphabetische Daten**
 = alphabetic data
alphabetical key TERM&PER **Buchstabentaste**
 = alphabetic key
alphabetical keyboard TERM&PER **Buchstabentastatur**
 = alphabetic keyboard
alphabetical order LING **alphabetische Reihenfolge**
 = alphabetic order
alphabetical position DATA PROC (→Buchstabenstelle)
 (→alphabetic position)
alphabetic character LING (→Buchstabe)
 (→letter)
alphabetic code CODING **alphabetischer Code**
 = alphabetical code
alphabetic data DATA PROC (→alphabetische Daten)
 (→alphabetical data)
alphabetic key (→alphabetical TERM&PER (→Buchstabentaste)
 key)
alphabetic keyboard TERM&PER (→Buchstabentastatur)
 (→alphabetical keyboard)
alphabetic order (→alphabetical LING (→alphabetische Reihenfolge)
 order)
alphabetic position DATA PROC **Buchstabenstelle**
 = alphabetical position = Alphastelle
 ↑ position ↑ Stelle
alphabetic sort DATA PROC **alphabetische Sortierung**
alphabetic string DATA PROC **Buchstabenkette**
alphabet of lines ENG.DRAW **Linienarten**
alpha cutoff frequency MICROEL **Alpha-Grenzfrequenz**
 [where transistor current gain falls to [Transistor-Stromverstärkung auf 0,707 des NF-Wertes]
 0.707 of LF figure]
alpha decay PHYS **Alphazerfall**
alpha emission PHYS **Alpha-Emission**
alpha emitter PHYS **Alphastrahler**
 = alpha radiator
alphameric (→alphanumeric) CODING (→alphanumerisch)
alphanumeric CODING **alphanumerisch**
 = alphameric ≠ numerisch; alphabetisch
 ≠ numeric; alphabetic
alphanumeric code CODING **alphanumerischer Code**
 = Schreibcode
alphanumeric display TERM&PER **alphanumerisches Datensichtgerät**
 terminal
alphanumeric instruction DATA PROC **alphanumerischer Befehl**
alphanumeric key TERM&PER **Datentaste**
 = character key; data key = alphanumerische Taste
alphanumeric keyboard TERM&PER **alphanumerische Tastatur**
alphanumeric position DATA PROC **alphanumerische Stelle**
alphanumeric sort DATA PROC **alphanumerische Sortierung**
alphanumeric string DATA PROC (→Zeichenfolge)
 (→character string)
alpha particle PHYS **Alphateilchen**
alpha radiation PHYS **Alphastrahlung**
alpha radiator (→alpha emitter) PHYS (→Alphastrahler)
alpha ray PHYS **Alphastrahl**

alpha test (→alpha DATA PROC (→firmeninterner Feldversuch)
 testing)
alpha testing DATA PROC **firmeninterner Feldversuch**
 = alpha test; in-house testing = hausinterner Probebetrieb
alter (→change) TECH (→ändern)
alterable TECH **veränderbar**
 [permits alteration] [sich verändern lassend]
 ≈ adjustable; variable = verstellbar
 ≠ fixed ≈ einstellbar; veränderlich
 ≠ festeingestellt
alterable memory DATA PROC **veränderbarer Speicher**
 = alterable storage; alterable store; programmable memory; programmable storage; programmable store = programmierbarer Speicher
alterable storage (→alterable DATA PROC (→veränderbarer Speicher)
 memory)
alterable store (→alterable DATA PROC (→veränderbarer Speicher)
 memory)
alteration (→change) TECH (→Wechsel)
alternate (v.t.) TELEPH **makeln**
 = split (v.t.)
alternate (adj.) TECH **abwechselnd**
 = wechselnd; alternierend
alternate channel RADIO **übernächster Kanal**
 = dritter Kanal
alternate coding key TERM&PER **ALT-Taste**
 [to generate characters not considered on the keyboard] [zur Generierung von auf der Tastatur nicht vorhandenen Zeichen]
 = alternate key; ALT key; A key = Codetaste; A-Taste
alternate key (→alternate TERM&PER (→ALT-Taste)
 coding key)
alternate keyboard TERM&PER (→intelligente Tastatur)
 (→intelligent keyboard)
alternate mark inversion CODING (→AMI)
 (→AMI)
alternate position ENG.DRAW **Wechselstellung**
alternate routing SWITCH **Leitweglenkung**
 [assignment of most favourable route, taking into account the occupation of the alternatives] [Ansteuern des günstigsten Vermittlungsweges, unter Berücksichtigung des Belegungszustandes der Alternativen]
 = traffic routing; alternative routing; route selection; route administration; routing = Wegewahl; Verkehrslenkung
alternate sweep INSTR **alternierende Wobbelung**
alternate track TERM&PER **Ersatzspur**
 = backup track
alternate type COMPON **Ausweichtyp**
alternating (→alternation between TELEPH (→Makeln)
 lines)
alternating burst TV **Wackelburst**
 [PAL] [PAL]
 = alternierender Burst
alternating component EL.TECH **Wechselanteil**
 ≠ continuous value ≠ Gleichwert
alternating control CONTROL **Wechselregelung**
 [main control temporarily substituted by an auxiliary control] [zeitweise Ablösung der Hauptregelung durch eine Hilfsregelung]
 = Ablöseregelung
alternating current EL.TECH **Wechselstrom**
 = ac current; ac (IEEE); a.c. (IEC); a-c; AC; aci ≠ Gleichstrom
 ↓ Drehstrom
 ≠ direct current
 ↓ three-phase current
alternating current bridge EL.TECH (→Wechselstrombrücke)
 (→ac bridge)
alternating current generator POWER SYS (→Wechselstromgenerator)
 (IEC) (→ac generator)
alternating current measuring INSTR (→Wechselstrom-Meßbrücke)
 bridge (→ac measuring bridge)
alternating current motor POWER ENG **Wechselstrommotor**
 = ac motor
alternating current tension EL.TECH (→Wechselspannung)
 (→alternating voltage)
alternating current voltage EL.TECH (→Wechselspannung)
 (→alternating voltage)
alternating current voltmeter INSTR (→Wechselspannungsmesser)
 (→ac voltmeter)

alternating field PHYS **Wechselfeld**
= pulsating field; variable field = veränderliches Feld
alternating magnetization PHYS **Wechselmagnetisierung**
= periodische Magnetisierung
alternating magnitude EL.TECH (→Wechselgröße)
(→alternating value)
alternating parameter EL.TECH (→Wechselgröße)
(→alternating value)
alternating permeability PHYS **Wechselfeldpermeabilität**
alternating tension EL.TECH (→Wechselspannung)
(→alternating voltage)
alternating value EL.TECH **Wechselgröße**
= alternating magnitude; alternating = Wellengröße; Wellenparameter
parameter; wave parameter
alternating voltage EL.TECH **Wechselspannung**
= ac voltage (IEEE); a.c. voltage ≠ Gleichspannung
(IEC); a-c voltage; AC voltage; alternating current voltage; alternating tension; ac tension; a-c tension; AC tension; alternating current tension; acv
≠ direct voltage
alternation (→alternation between TELEPH (→Makeln)
lines)
alternation (→OR ENG.LOG (→ODER-Verknüpfung)
operation)
alternation (→reversal) PHYS (→Wechsel)
alternation (→reversal) TELEC (→Wechselzeichen)
alternation between lines TELEPH **Makeln**
[discretionary swith-over between [beliebiges Umschalten zwischen rückgefragtem Teilnehmer und wartendem, ursprünglich angerufenem Teilnehmer]
consuted and waiting, first called station]
= alternation; alternating; two-way splitting; splitting; broker's call
alternative (adj.) ECON (→wahlweise)
(→optional)
alternative denial ENG.LOG **Alternativverneinung**
[output false if all inputs true, and [Ausgang = 0 wenn A und B = 1, = 1 wenn A und/oder B = 0]
true if any input false]
= dispersion
alternative hypothesis MATH **Gegenhypothese**
alternative power supply POWER SYS **Ersatzstromquelle**
alternative route TELEC **Ersatzweg**
alternative routing (→alternate SWITCH (→Leitweglenkung)
routing)
alternative supplier (→second ECON (→Zweitlieferant)
source)
alternator (→ac POWER SYS (→Wechselstromgenerator)
generator)
ALT GR key TERM&PER **Taste ALT GR**
[shift key for lower right case characters] [Umschalttaste für rechts unten dargestellte Zeichen]
altimeter INSTR **Höhenmesser**
altitude GEOPHYS **Höhe**
[vertical elevation above sea level] [Entfernung über Meeresspiegel]
= elevation = Höhenlage; Kote
altitude ASTROPHYS **Höhe**
[angular elevation above horizon] [Winkelabstand vom Horizont]
= elevation
altitude (→height) MATH (→Höhe)
ALT key (→alternate coding TERM&PER (→ALT-Taste)
key)
ALU (→arithmetic-logic DATA PROC (→Rechenwerk)
unit)
alumina CHEM **Alaunerde**
aluminium (BRI) (→aluminum) CHEM (→Aluminium)
aluminum (AM) CHEM **Aluminium**
= aluminium (BRI); Al = Al
aluminum bronze METAL **Aluminiumbronze**
aluminum carrying case TECH **Aluminiumkoffer**
aluminum electrolytic capacitor COMPON **Aluminium-Elektrolytkondensator**
aluminum solder METAL **Aluminiumlot**
AL value EL.TECH **Induktivitätsfaktor**
= AL-Wert
AM (→amplitude modulation) MODUL (→Amplitudenmodulation)
AM) (→despatch) ECON (→Versand)
AM) (→printout) DATA PROC (→Ausdruck)
AM (→americium) CHEM (→Americium)

amateur (→radio amateur) RADIO (→Funkamateur)
amateur (→tinker) TECH (→Bastler)
amateur radio RADIO **Amateurfunk**
amateur radio communication RADIO **Amateurfunkverbindung**
amateur radio operator (→radio RADIO (→Funkamateur)
amateur)
amateur satellite service SAT.COMM **Amateurfunkdienst über Satelliten**
amateur station RADIO **Amateurfunkstelle**
amateur tape recordist EL.ACOUS **Tonband-Amateur**
ambassadorial radio network TELEC **Botschafter-Funknetz**
amber (adj.) PHYS **bernsteinfarben**
[color]
amber CHEM **Bernstein**
amber monitor TERM&PER **Bernstein-Monitor**
ambient climate (→environmental TECH (→Umweltklima)
climate)
ambient condition TECH (→Umweltbedingung)
(→environmental condition)
ambient environment TECH **Umwelt**
= environment = Umgebung; Umfeld
ambient-induced TECH (→umgebungsbedingt)
(→environment-induced)
ambient light TECH **Raumbeleuchtung**
= Vorlicht
ambient noise (→room noise) ACOUS (→Raumgeräusch)
ambient temperature TECH **Umgebungstemperatur**
(→environmental temperature)
ambiguity (→equivocality) MATH (→Mehrdeutigkeit)
ambiguous MATH **mehrdeutig**
= many-valued = vieldeutig
↓ zweideutig
ambiguous function MATH **mehrdeutige Funktion**
ambipolar ELECTRON **ambipolar**
amend TECH **berichtigen**
= put right; correct (v.t.) = verbessern; bessern; richtigstellen; korrigieren
≈ complement ≈ ergänzen
amendment LING **Berichtigung**
= correction; rectification = Korrektur; Besserung; Verbesserung
≈ compementation ≈ Ergänzung
amendment ECON **Änderung**
amendment record DATA PROC **Ergänzungseintrag**
= update record = Korrektureintrag; Aktualisierungseintrag
American cloth (BRI) (→oil-cloth TECH (→Wachstuch)
lining)
American Military Standard MIL.COMM **US-amerikanische Militärnorm**
= AMS
american sectional view ENG.DRAW **amerikanische Schnittdarstellung**
American Society for Testing Materials QUAL **Amerikanische Gesellschaft für Materialprüfung**
= ASTM
American Standarts Association TECH **Amerikanische Normengesellschaft**
= ASA
American Wire Gauge METAL **amerikanische Drahtlehre**
= AWG = AWG
americium CHEM **Americium**
= AM = Am
AMH CONTROL **AMH**
= automated materials handling = rechnergestützte Materialhandhabung
AMI CODING **AMI**
= alternate mark inversion
ammeter (→current meter) INSTR (→Strommesser)
amorphous PHYS **amorph**
amorphous semiconductor PHYS **amorpher Halbleiter**
amount ECON **Betrag**
≈ sum; value = Geldsumme; Höhe
≈ Summe; Wert
amount (n.) (→quantity) SCIE (→Menge)
amount of substance PHYS **Stoffmenge**
[SI unit: Mole] [SI-Einheit: Mol]
amp (→ampere) PHYS (→Ampere)
ampacity (→current carrying POWER ENG (→Strombelastbarkeit)
capacity)
ampacity table POWER ENG **Strombelastbarkeitstabelle**
= load capacity table

amperage

amperage [POWER ENG] EL.TECH (→Stromstärke)
(→current intensity)
ampere PHYS **Ampere**
[SI unit for elctric current, electric [SI-Basiseinheit für elektri-
flux and magnetic potencial] sche Stromstärke, elektri-
= amp; A sche Durchflutung und
magnetische Spannung]
= A
ampere-hour PHYS **Amperestunde**
= A h [abgeleitete SI-Einheit der
Elektrizitätsmenge]
= A h
ampere-second PHYS **Amperesekunde**
= A s [abgeleitete SI-Einheit der
Elektrizitätsmenge]
= A s
ampere-turn 1 PHYS **Amperewindung**
ampere turns 2 (→ number of PHYS (→ Amperewindungszahl)
ampere turns)
ampersand ECON **Und-Zeichen**
[&] [&]
= and sign; and symbol; and character = Et-Zeichen; kommerzielles
↑ sign Und; Ampersand; kauf-
männisches Und
↑ Zeichen
amplification (→enlargement) TECH (→Vergrößerung)
amplification (→gain) NETW.TH (→Verstärkung 1)
amplification drift ELECTRON **Verstärkungsdrift**
[Verstärkungsänderung
durch Temperatureffekt]
amplification factor NETW.TH (→Verstärkung 1)
(→gain)
amplification stage (→amplifier CIRC.ENG (→Verstärkerstufe)
stage)
amplification tube ELECTRON (→Verstärkerröhre)
(→amplifying tube)
amplification valve ELECTRON (→Verstärkerröhre)
(→amplifying tube)
amplifier CIRC.ENG **Verstärker**
amplifier (→ line repeater) TRANS (→Leitungsverstärker)
amplifier bridge CIRC.ENG **Verstärkerbrücke**
amplifier circuit CIRC.ENG **Verstärkerschaltung**
↓ basic amplifier connection ↓ Verstärkergrundschaltung
amplifier engineering CIRC.ENG **Verstärkertechnik**
amplifier noise CIRC.ENG **Verstärkerrauschen**
≈ basic noise [TELEC] ≈ Grundgeräusch [TELEC]
amplifier point BROADC **Verstärkerpunkt**
amplifier stage CIRC.ENG **Verstärkerstufe**
= amplifying stage; amplification stage
amplify ELECTRON **verstärken**
= enhance
amplify (→enlarge) TECH (→vergrößern)
amplifying stage (→amplifier CIRC.ENG (→Verstärkerstufe)
stage)
amplifying tube ELECTRON **Verstärkerröhre**
= amplification tube; amplifying valve;
amplification valve
amplifying valve (→amplifying ELECTRON (→Verstärkerröhre)
tube)
amplitude PHYS **Amplitude**
= magnitude = Schwingweite
amplitude accuracy PHYS **Amplitudengenauigkeit**
amplitude compensation method INSTR **Amplitudenverfahren**
amplitude discriminator CIRC.ENG **Amplitudendiskriminator**
= Amplitudenentscheider
amplitude distortion EL.TECH **Amplitudenverzerrung**
= Amplitudenschräglage
amplitude excursion PHYS **Schwingwegamplitude**
amplitude fading (→flat RADIO PROP (→Flachschwund)
fading)
amplitude gain ELECTRON **Amplitudenverstärkung**
amplitude keying MODUL **Amplitudentastung**
= amplitude shift keying; ASK
amplitude limiter (→amplitude CIRC.ENG (→Amplitudenbegrenzer-
limiter circuit) schaltung)
amplitude limiter circuit CIRC.ENG **Amplitudenbegrenzerschal-**
= amplitude limiter; peak limiter; peak **tung**
clipper; clipper circuit; clipping cir- = Spitzenbegrenzer
cuit; clipper; amplitude lopper ↑ Begrenzerschaltung
↑ limiter circuit

amplitude linearity ELECTRON **Amplitudenlinearität**
amplitude lopper (→amplitude CIRC.ENG (→Amplitudenbegrenzer-
limiter circuit) schaltung)
amplitude measurement INSTR **Amplitudenmessung**
amplitude modulation MODUL **Amplitudenmodulation**
= AM
amplitude modulation with MODUL **Rauschamplituden-Modula-**
a noise carrier **tion**
amplitude-pure ELECTRON **amplitudenrein**
amplitude range PHYS **Amplitudenbereich**
amplitude regulation ELECTRON **Amplitudenregelung**
amplitude response (→frequency TELEC (→Amplitudenfrequenzgang)
response of amplitude)
amplitude separator TV **Amplitudensieb**
= synchron separator
amplitude shift keying MODUL (→Amplitudentastung)
(→amplitude keying)
amplitude slice MODUL **Amplitudenscheibe**
amplitude spectrum EL.TECH **Amplitudenspektrum**
amplitude step MODUL **Amplitudenstufe**
amplitude swing MODUL **Amplitudenhub**
amplitude time value EL.TECH **Amplitudenzeitwert**
amplitude variation PHYS **Amplitudenschwankung**
AMS (→ American Military MIL.COMM (→US-amerikanische Militär-
Standard) norm)
analog (adj.) INF.TECH **analog** (adj.)
[continuous in amplitude and time] = zeit- und wertkontinuierlich
= analogue ≈ wertkontinuierlich
≈ value-continuous ≠ digital
≠ digital
analog channel TELEC **Analogkanal**
= analogue channel = analoger Kanal
analog circuit CIRC.ENG **Analogschaltung**
analog color monitor TERM&PER **analoger Farbbildschirm**
= analog colour monitor; analog RGB = analoger RGB-Monitor
monitor
analog colour monitor TERM&PER (→analoger Farbbildschirm)
(→analog color monitor)
analog comparator CIRC.ENG **analoger Vergleicher**
analog computer DATA PROC **Analogrechner**
≠ digital computer ≠ Digitalrechner
analog conclusion (→analogical SCIE (→Analogieschluß)
reasoning)
analog data INF.TECH **Analogdaten**
= analoge Daten
analog/digital INF.TECH **analog/digital**
= A/D = A/D
analog-digital converter CIRC.ENG **Analog-Digital-Umsetzer**
= A/D converter = Analog-Digital-Wandler;
A/D-Wandler
analog display EQUIP.ENG **Analoganzeige**
= analog indication
analog divider CIRC.ENG **Analogdividierer**
analog driver CIRC.ENG **Analogtreiber**
analog filter NETW.TH **Analogfilter**
≠ digital filter ≠ Digitalfilter
analogical reasoning SCIE **Analogieschluß**
= analog conclusion; analogue conclusion
analog indication (→analog EQUIP.ENG (→Analoganzeige)
display)
analog interface TELEC **Analogschnittstelle**
= analogue interface = analoge Schnittstelle
analog line SWITCH **Analoganschluß**
analog line TELEC **Analogleitung**
analog matched filter SIGN.TH (→Optimalfilter)
(→optimum matched filter)
analog measuring instrument INSTR **analoges Meßgerät**
= analogue measuring instrument; = analoges Meßinstrument
analog meter; analogue meter
analog memory (→analog DATA PROC (→Analogspeicher)
storage)
analog meter (→analog measuring INSTR (→analoges Meßgerät)
instrument)
analog multiplexer DATA PROC **Analogmultiplexer**
analog multiplier DATA PROC **Analogmultiplizierer**
analog multiplier CIRC.ENG **analoger Multiplizierer**
= Analogmultiplizierer
analog RGB monitor TERM&PER (→analoger Farbbildschirm)
(→analog color monitor)

analog signal	INF	**Analogsignal**	
= continuous signal		= analoges Signal; zeit- und wertkontinuierliches Signal; stetiges Signal	
analog storage	DATA PROC	**Analogspeicher**	
= analog memory			
analog switch	COMPON	**analoger Schalter**	
analog switching system	SWITCH	**Analogvermittlungssystem**	
		= analoges Vermittlungssystem	
analog technique	TELEC	**Analogtechnik**	
analog-to-digital conversion	TELEC	**Analog-Digital-Umsetzung**	
= A/D conversion		= A/D-Umsetzung	
analog-to-digital converter	CODING	**A/D-Wandler**	
= A-to-D converter; A/D converter; ADC; digitizer		= A/D-Umsetzer; A/D-Umwandler; Analog-Digital-Wandler; Analog-Digital-Umsetzer; Analog-Digital-Umwandler; Digitalisierer	
≠ digital-to-analog converter		≈ D/A-Wandler	
		≠ D/A-Wandler	
analog transistor	MICROEL	**Analog-Transistor**	
[concentric transistor with features similar to an electronic tube]		[konzentrisch aufgebauter Transitor mit Eigenschaften ähnlich einer Elektronenröhre]	
↓ spacistor		↓ Spacistor	
analogue (→ analog)	INF.TECH	(→ analog) (adj.)	
analogue channel (→ analog channel)	TELEC	(→ Analogkanal)	
analogue conclusion (→ analogical reasoning)	SCIE	(→ Analogieschluß)	
analogue interface (→ analog interface)	TELEC	(→ Analogschnittstelle)	
analogue measuring instrument (→ analog measuring instrument)	INSTR	(→ analoges Meßgerät)	
analogue meter (→ analog measuring instrument)	INSTR	(→ analoges Meßgerät)	
analogy	SCIE	**Analogie**	
analyse (v.t.)(BRI)	SCIE	(→ analysieren)	
(→ analyze)			
analysis (pl.-es)	SCIE	**Analyse**	
analysis	MATH	**Analysis**	
↑ higher mathematics		↑ Höhere Mathematik	
↓ calculus; differential calculus; integral calculus; differential equations; harmonic analysis; calculus of variations		↓ Infinitesimalrechnung; Differentialrechnung; Integralrechnung; Differentialgleichungen; Harmonische Analyse; Variationsrechnung	
analysis graphic	DOC	**Schaubild**	
analysis problem	NETW.TH	**Analyseproblem**	
analyst	SCIE	**Analytiker**	
analytic (adj.)	SCIE	**analytisch**	
= analytical			
analytical (→ analytic)	SCIE	(→ analytisch)	
analytical function generator	DATA PROC	**analytischer Funktionsgenerator**	
analytical process model	DATA PROC	**analytisches Prozeßmodell**	
analyze (v.t)	SCIE	**analysieren**	
= analyse (v.t.)(BRI)			
analyzer	ELECTRON	**Analysator**	
analyzer	DATA PROC	**Analysator**	
[analog computer]		[Analogrechner]	
↓ differential analyzer		↓ Differentialanalysator	
analyzer	INSTR	**Analysator**	
ancestral file (→ generation principle)	DATA PROC	(→ Generationen-Prinzip)	
anchor (v.t.)	TECH	**verankern**	
[to fasten securely]			
anchor (→ stay)	OUTS.PLANT	(→ Anker)	
anchorage	CIV.ENG	**Verankerung**	
anchor bolt (→ foundation bolt)	MECH	(→ Ankerschraube)	
anchor guy (→ guy rope)	OUTS.PLANT	(→ Abspannseil)	
anchoring point	OUTS.PLANT	**Abspannpunkt**	
		[Freileitung]	
anchoring rail	EQUIP.ENG	**Ankerschiene**	
ancillary (→ auxiliary)	TECH	(→ Hilfs-)	
ancillary device (→ add-on device)	TECH	(→ Zusatzeinrichtung)	
ancillary industry	ECON	**Zulieferindustrie**	
ancillary memory (→ secondary memory)	DATA PROC	(→ Sekundärspeicher)	
ancillary storage (→ secondary memory)	DATA PROC	(→ Sekundärspeicher)	
ancillary store (→ secondary memory)	DATA PROC	(→ Sekundärspeicher)	
ancillary system	TECH	**Hilfssystem**	
= auxiliary system		≈ Untersystem	
≈ subsystem			
and (→ plus)	MATH	(→ plus)	
and character (→ ampersand)	ECON	(→ Und-Zeichen)	
AND circuit (→ AND gate)	CIRC.ENG	(→ UND-Glied)	
AND element (→ AND gate)	CIRC.ENG	(→ UND-Glied)	
Anderson bridge	EL.TECH	**Anderson-Brücke**	
AND function (→ AND operation)	ENG.LOG	(→ UND-Verknüpfung)	
AND gate	CIRC.ENG	**UND-Glied**	
= AND element; AND circuit; conjunction gate; conjunction element; conjunction circuit; coincidence gate; coincidence element; coincidence circuit		= UND-Gatter; UND-Schaltung; UND-Tor; Konjunktionsglied; Konjunktionsgatter; Konjunktionsschaltung; Konjunktionstor; Koinzidenzglied; Koinzidenzgatter; Koinzidenzschaltung; Koinzididenztor	
↑ logic gate		↑ Verknüpfungsglied	
AND operation	ENG.LOG	**UND-Verknüpfung**	
[output = 1 if simultaneously A = 1 and B = 1]		[Ausgang = 1 wenn gleichzeitig A = 1 und B = 1]	
= AND function; conjunction operation; conjunction function; conjunction; coincidence operation; coincidence function; coincidence; logical multiply; logical product; intersection		= UND-Funktion; Konjunktionsverknüpfung; Konjunktionsfunktion; Konjunktion; Koinzidenzverknüpfung; Koinzidenzfunktion; Koinzidenz	
↑ logic operation		↑ logische Verknüpfung	
android	DATA PROC	**Android**	
[human-like robot]		[menschliches Verhalten imitierender Robot]	
and sign (→ ampersand)	ECON	(→ Und-Zeichen)	
and symbol (→ ampersand)	ECON	(→ Und-Zeichen)	
anechoic (→ dead)	ACOUS	(→ schalltot)	
anechoic chamber	ACOUS	**schalltoter Raum**	
≠ reverberation room		≠ Nachhallraum	
anemometer	METEOR	**Windmesser**	
		= Anemometer	
angle	MATH	**Winkel**	
↓ plain angle; solid angle		↓ ebener Winkel; Raumwinkel	
angle at a point (→ complete angle)	MATH	(→ Vollwinkel)	
angle bar	MECH	**Winkelschiene**	
angle bracket	MATH	**spitze Klammer**	
[symbol ⟨ ⟩]		[Symbol ⟨ ⟩]	
↑ bracket 1		= Winkelklammer	
		↑ Klammer	
angle characteristic (→ angular characteristic)	PHYS	(→ Winkelcharakteristik)	
angled	MECH	**abgewinkelt**	
angled circumflex	LING	**Hochpfeil**	
≈ caret [TYPOGR]			
↑ circumflex			
angle diversity (→ angular diversity)	RADIO REL	(→ Winkeldiversity)	
angle gage	MECH	**Winkellehre**	
angle iron	MECH	**Winkeleisen**	
angle modulation	MODUL	**Winkelmodulation**	
↓ phase modulation; frequency modulation		↓ Phasenmodulation; Frequenzmodulation	
angle of deflection	PHYS	**Ablenkwinkel**	
angle of incidence	PHYS	**Einfallswinkel**	
= wave angle			
angle of inclination	MATH	**Neigungswinkel**	
angle of polarization	PHYS	**Polarisationswinkel**	
= Brewster angle		= Brewsterscher Winkel; Brewster Winkel	

angle of radiation (→radiation angle) PHYS (→Strahlungswinkel)
angle of reflection PHYS **Reflexionswinkel**
angle of refraction PHYS **Brechungswinkel**
= refractive angle
angle of rotation PHYS **Drehwinkel**
= Drehungswinkel
angle of sight PHYS **Sehwinkel**
= visual angle
angle-position encoder COMPON **Winkelcodierer**
[encodes angular values] [setzt Winkelwerte in Codewörter um]
≈ angular position pickup = Drehwinkelcodierer
≈ Winkelwertgeber
angle-preserving MATH **winkelgetreu**
= isogonal 2 = winkeltreu; isogonal 2
angle quotation mark LING **spitzes Anführungszeichen**
angle sign MATH **Winkelzeichen**
= angle symbol
angle symbol (→angle sign) MATH (→Winkelzeichen)
angle tolerance TECH **Winkeltoleranz**
angle value (→angular position) TECH (→Winkelstellung)
Ångström PHYS **Ångström**
[unit for length, 0.1 nm] [Längeneinheit; 0,1 nm]
= Å = Å
angular TECH **winkelförmig**
= eckig
angular acceleration PHYS **Winkelbeschleunigung**
angular adapter COMPON **Winkeladapter**
angular aperture PHYS **Winkelöffnung**
angular attenuation ANT **Winkeldämpfung**
[relative to main direction] [bezogen auf Hauptrichtung]
= angular decoupling; angular discrimination
= Winkelentkopplung
angular characteristic PHYS **Winkelcharakteristik**
= angle characteristic
angular decoupling (→angular attenuation) ANT (→Winkeldämpfung)
angular dependence MATH **Winkelabhängigkeit**
= angular function
angular-dependent MATH **winkelabhängig**
angular dimension MATH **Winkelmaß**
↓ angular measure; radian measure ↓ Gradmaß; Bogenmaß
angular dipole ANT **Winkeldipol**
= V dipole; quadrant antenna = Quadrantantenne; V-Dipol
angular discrimination (→angular attenuation) ANT (→Winkeldämpfung)
angular displacement MATH **Winkelabweichung**
angular distance (→scatter angle) RADIO PROP (→Streustrahlwinkel)
angular distribution PHYS **Winkelverteilung**
angular diversity RADIO REL **Winkeldiversity**
= angle diversity
angular diversity antenna ANT **Winkeldiversityantenne**
= ADA
angular drive MECH **Winkeltrieb**
[gear] [Zahnrad]
angular error TECH **Winkelfehler**
= angularity error
angular frequency PHYS **Kreisfrequenz**
[number of oscillations per 2π seconds] [Zahl der in 2π Sekunden ausgeführten Schwingungen]
= angular velocity 2; radian frequency; radial frequency; circular frequency; pulsation; gyrofrequency
= Schwingungsfrequenz; Pulsation; Winkelgeschwindigkeit 2
angular frequency deviation MODUL (→Kreisfrequenzhub)
(→angular frequency shift)
angular frequency shift MODUL **Kreisfrequenzhub**
= angular frequency swing; angular frequency deviation; angular frequency sweep; radian frequency swing; radian frequency deviation; radian frequency sweep; radial frequency swing; radial frequency deviation; radial frequency sweep
= Schwingungsfrequenzhub; Pulsationshub
angular frequency sweep MODUL (→Kreisfrequenzhub)
(→angular frequency shift)
angular frequency swing MODUL (→Kreisfrequenzhub)
(→angular frequency shift)

angular function MATH **Winkelfunktion**
= trigonometric function = trigonometrische Funktion
angular function (→angular dependence) MATH (→Winkelabhängigkeit)
angularity error (→angular error) TECH (→Winkelfehler)
angular measure MATH **Gradmaß**
≈ radian measure ≈ Bogenmaß
↑ Winkelmaß
↓ Grad
angular misalignmemt (→misalignment) TECH (→Fluchtungsfehler)
angular momentum PHYS **Drehimpuls**
= moment of momentum = Impulsmoment; Drall 2
angular momentum quantum number (→secondary quantum number) PHYS (→Nebenquantenzahl)
angular offset screw driver (→angular screw driver) TECH (→Winkelschraubendreher)
angular point OUTS.PLANT **Winkelpunkt**
angular position TECH **Winkelstellung**
= angle value = Winkelwert
angular position pickup COMPON **Winkelwertgeber**
≈ angular position encoder ≈ Winkelcodierer
angular range MATH **Winkelbereich**
angular resolution PHYS **Winkelauflösung**
angular screw driver TECH **Winkelschraubendreher**
= angular offset screw driver = Winkelschraubenzieher
angular section analysis TERM&PER **Winkelschnittanalyse**
angular velocity 1 PHYS **Rotationsgeschwindigkeit**
[SI unit: radia per second] [SI-Einheit: Radiant durch Sekunde]
≈ rotational frequency = Winkelgeschwindigkeit 1
≈ Drehzahl
angular velocity 2 (→angular frequency) PHYS (→Kreisfrequenz)
angular velocity pickup COMPON **Winkelgeschwindigkeits-Aufnehmer**
= Drehgeschwindigkeits-Aufnehmer
anharmonic PHYS **nichtharmonisch**
= nonharmonic = unharmonisch
animation (→computer animation) DATA PROC (→Computeranimation)
anion PHYS **Anion**
[negatively charged ion] [negativ geladenes Ion]
≠ cation ≠ Kation
↑ ion ↑ Ion
anisochronous TELEC **anisochron**
anisotropic PHYS **anisotrop**
[different properties in different directions] [unterschiedliche Eigenschaften in verschiedenen Richtungen]
≠ isotropic ≠ isotrop
anisotropic conductivity PHYS **anisotrope Leitfähigkeit**
anisotropy PHYS **Anisotropie**
anneal (v.t.) METAL **glühen**
[heat treatment] [Warmbehandlung]
= ausglühen
anneal (n.) METAL **Glühung**
[heat treatment] [Warmbehandlung]
= annealing = Ausglühung
annealed METAL **geglüht**
annealing (→anneal) METAL (→Glühung)
annex (n.) LING **Anlage**
= enclosure; enc; encs; appendix; inclosure; inclose
= Anhang
≈ Beilage
annex (n.) CIV.ENG (→Anbau)
(→sidebuilding)
annotation (→comment) DATA PROC (→Kommentar)
annotation field (→comment file) DATA PROC (→Kommentarfeld)
annotation line (→comment line) DATA PROC (→Kommentarzeile)
annotation symbol (→comment symbol) DATA PROC (→Kommentarsymbol)
announcement service TELEPH **Ansagedienst**
= message service
announcer DATA COMM **Ankündigungszeichen**
annoyance call trap (→tracing switch) SWITCH (→Fangschaltung)

annual accounts (BRI) (→financial ECON (→Jahresabschluß)
statement)
annual fee ECON **Jahresgebühr**
annual output MANUF **Jahresausstoß**
= annual production = Jahresproduktion
annual production (→annual MANUF (→Jahresausstoß)
output)
annul (v.t.) (→override) TECH (→lahmlegen)
annular TECH **ringförmig**
= ring-shaped ≈ kreisförmig; rund
≈ circular; round
annular code COMM.CABLE **Ringkennzeichnung**
annular core (→toroidal COMPON (→Ringkern)
core)
annular lens PHYS **Ringlinse**
= Fresnelsche Ringlinse
annular slot ANT **Ringspaltantenne**
= Kreisschlitzantenne
annulment (→cancellation) ECON (→Annullierung)
annulus (pl.-i) MATH **Kreisring**
[surface bound by two concentric [durch zwei konzentrische
circles] Kreise begrenzte Fläche]
= ring 2 ≈ Torus
≈ torus
annunciator SIGN.ENG **Alarmierungsfeld**
= security monitor
annunciator (→display ELECTRON (→Anzeigeelement)
element)
annunciator (→indicator) INSTR (→Anzeiger)
anode PHYS **Anode**
= positive electrode = positive Elektrode
↑ electrode ↑ Elektrode
anode ELECTRON **Anode**
[tube] [Röhre]
= plate (AM)
anode POWER SYS **Anode**
[battery] [Batterie]
= plate (AM) ≠ Katode
≠ cathode
anode COMPON **Anode**
[capacitor] [Kondensator]
= plate (AM) ≠ Katode
≠ cathode
anode MICROEL **Anode**
≠ base; collector ≠ Basis; Kollektor
anode battery (→plate POWER SYS (→Anodenbatterie)
battery)
anode-B modulation ELECTRON **Anoden-B-Modulation**
↑ anode modulation ↑ Anodenmodulation
anode choke ELECTRON **Anodendrossel**
anode current PHYS **Anodenstrom**
anode current ELECTRON **Anodenstrom**
= plate current
anode-current limiter ELECTRON **Anodenstrombegrenzer**
anode fall PHYS **Anodenfall**
anode hum ELECTRON **Anodenrauschen**
anode modulation ELECTRON **Anodenmodulation**
anode ray PHYS **Anodenstrahl**
anode region PHYS **Anodengebiet**
anode sputtering PHYS **Anodenzerstäubung**
anode voltage PHYS **Anodenspannung**
anode voltage ELECTRON **Anodenspannung**
= plate voltage
anodic oxidation CHEM **Eloxalverfahren**
anodize (→eloxadize) METAL (→eloxieren)
anormal dispersion PHYS **anormale Dispersion**
ANSI keyboard TERM&PER **ANSI-Tastatur**
answer (v.t.) TELEPH **antworten**
[a call] [auf einen Anruf]
= receive
answer (n.) TECH **Antwort**
= reply = Antwortgabe
≈ solution ≈ Lösung
answer (n.) SWITCH **Antwort**
answerback code (→station DATA COMM (→Stationskennung)
identification)
answerback code request DATA COMM (→Kennungsabfrage)
(→station identification request)
answerback code storage DATA COMM **Kennungsspeicher**
= identifier storage

answerback device DATA COMM (→Kennungsgeber)
(→answerback generator)
answerback exchange DATA COMM **Kennungstausch**
= Kennungsaustausch
answerback generator DATA COMM **Kennungsgeber**
= identifier generator; answerback de- = Namengeber
vice
answered call SWITCH **Belegung mit Melden**
answering SWITCH **Melden**
[connection of called terminal to the [Anschließen der gerufe-
upset circuit] nen Gegenstelle an die auf-
gebaute Leitung]
answering device TRANS **Abfrageeinrichtung**
answering jack TELEPH **Abfrageklinke**
answering key TERM&PER (→Abfragetaste)
(→interrogation key)
answering machine TERM&PER (→Anrufbeantworter)
(→automatic answering equipment)
answering service (→telephone TELEPH (→Abwesenheitsdienst)
answering service)
answering station (→polling TELEC (→Abfragestation)
station)
answer mode DATA COMM **Antwortbetrieb**
[automatic management of incoming [automatische Abwicklung
calls] ankommender Anrufe]
= auto-answer
answer signal (BRI) (→off-hook SWITCH (→Beginnzeichen)
signal)
answer tone DATA COMM **Antwortton**
answer-to-seize ratio SWITCH **Erfolgsquote**
= ASR
antedate v.t. ECON **vorziehen**
= advance [einen Termin]
antenna (pl.-as, -ae) RADIO **Antenne**
= aerial (BRI) = Luftleiter (obs.); Luftdraht
≈ radiator (obs.)
≈ Strahler
antenna allocation ANT **Antennenbeschaltung**
antenna amplifier (→antenna HF (→Antennenverstärker)
booster)
antenna array (→array ANT (→Strahlerfeld)
antenna)
antenna base impedance (→antenna ANT (→Fußpunktwiderstand)
input impedance)
antenna beamwidth (→half-power ANT (→Halbwertsbreite)
beamwidth)
antenna booster HF **Antennenverstärker**
= booster; antenna amplifier; aerial = Booster
booster; aerial amplifier
antenna cable RADIO **Antennenanschlußkabel**
= aerial cable = Antennenkabel
antenna cable CONS.EL **Antennenkabel**
= aerial cable = Antennenanschlußkabel
antenna change-over switch RADIO (→Antennenumschalter)
(→duplexer 2)
antenna choke ANT **Antennendrossel**
= aerial choke
antenna coil ANT **Antennenspule**
= aerial coil
antenna coupling capacitor ANT (→Antennenkopplungskon-
(→antenna coupling condenser) densator)
antenna coupling condenser ANT **Antennenkopplungskonden-**
= aerial coupling condenser; antenna **sator**
coupling capacitor; aerial coupling capacitor
antenna curtain (→curtain ANT (→Vorhangantenne)
antenna)
antenna diversity (→space RADIO (→Raumdiversity)
diversity)
antenna drive ANT **Antennendrehvorrichtung**
= aerial drive = Antennenantrieb
antenna duplexer RADIO **Sendeantennenweiche**
[connects several transmitters to one [mehrere Sender an einer
antenna] Antenne]
= duplexer 3; transmitter combining fil- = Simultanweiche; Sender-
ter weiche
↑ Antennenweiche
antenna effect RADIO LOC **Antenneneffekt**
antenna element (→radiating ANT (→Einzelstrahler)
element)

antenna engineer	ANT	**Antennentechniker**	
antenna engineering	HF	**Antennentechnik**	
antenna exciting (→antenna feeding)	ANT	(→Antennenspeisung)	
antenna factor [receiver votage to field strength ratio] = K factor; K antenna factor	ANT	**Antennenfaktor** [Empfangsspannung zu Feldstärke] = K-Faktor	
antenna feed (→antenna feeding)	ANT	(→Antennenspeisung)	
antenna feeder line (→antenna feeding line)	ANT	(→Antennenspeiseleitung)	
antenna feeding = antenna feed; aerial feeding; aerial feed; antenna exciting; aerial exciting; feed ≈ antenna feeder ↓ top feed; center feed; base feed	ANT	**Antennenspeisung** ≈ Antennen-Erreger; Erreger ↓ Obenspeisung; Mittelpunktspeisung; Fußpunktspeisung	
antenna feeding line = antenna feed line; antenna feeder line; aerial feeding line; aerial feed line; aerial feeder line; transmission line; aerial feeder; lead-in	ANT	**Antennenspeiseleitung** = Antennenleitung; Antennenzuleitung	
antenna feed line (→antenna feeding line)	ANT	(→Antennenspeiseleitung)	
antenna filter [generel concept for filters which combine different transmitters or receiver or transmitter with receiver to one antenna] ↓ diplexer; duplexer	RADIO	**Antennenweiche** [Oberbegriff für Filter zum Betrieb mehrerer Sender, oder mehrerer Empfänger, oder von Sender und Empfänger, an einer Antenne] = Verzweigungsfilter ↓ Sendeantennenweiche; Sende-Empfangsweiche	
antenna filter (→antenna low pass filter)	ANT	(→Antennenfilter)	
antenna gain (→gain)	ANT	(→Gewinn)	
antenna height = aerial height	ANT	**Antennenlänge** = Antennenhöhe	
antenna impedance = aerial impedance; antenna resistance; aerial resistance ↓ radiation resistance; resonant impedance; antenna input impedance; antenna loss resistance	ANT	**Antennenwiderstand** ↓ Strahlungswiderstand; Resonanzwiderstand; Fußpunktwiderstand; Antennenverlustwiderstand	
antenna input impedance = aerial input impedance; antenna base impedance; aerial base impedance ↑ antenna impedance	ANT	**Fußpunktwiderstand** = Antenneneingangswiderstand; Speisepunktwiderstand ↑ Antennenwiderstand	
antenna jack (→antenna socket)	CONS.EL	(→Antennenbuchse)	
antenna litz wire	ANT	**Antennenlitze** = Antennenseil	
antenna load	ANT	**Antennenbelastung**	
antenna loss resistance = loss resistance ↑ antenna impedance	ANT	**Antennenverlustwiderstand** = Verlustwiderstand ↑ Antennenwiderstand	
antenna low pass filter = antenna filter	ANT	**Antennentiefpass** = Antennentiefpass	
antenna mast = aerial mast ≈ antenna tower	ANT	**Antennenmast** = Antennenturm ↑ Antennenträger	
antenna mount = aerial mount	ANT	**Antennenbefestigung**	
antenna noise	RADIO	**Antennenrauschen**	
antenna noise temperature = antenna temperature	HF	**Antennenrauschtemperatur** = Antennentemperatur	
antenna outlet = aerial outlet	CONS.EL	**Antennensteckdose** [an der Wand]	
antenna pattern (→radiation pattern)	ANT	(→Strahlungscharakteristik)	
antenna resistance (→antenna impedance)	ANT	(→Antennenwiderstand)	
antenna resonance = resonance ↓ series resonance; parallel resonance	ANT	**Antennenresonanz** = Resonanz ↓ Stromresonanz; Spannungsresonanz	
antenna rotator (→antenna rotor)	ANT	(→Antennenrotor)	
antenna rotor = aerial rotor; antenna rotator; aerial rotator	ANT	**Antennenrotor**	
antennascope [bridge to measure active resistance of antennae]	INSTR	**Antennascope** [Meßbrücke für Wirkwiderstandsmessungen an Antennen]	
antenna selection switch (→antenna selector)	ANT	(→Antennenwahlschalter)	
antenna selector = aerial selector; antenna selection switch	ANT	**Antennenwahlschalter**	
antenna series capacitor = aerial series capacitor	ANT	**Antennenverkürzungskondensator**	
antenna socket = antenna jack; aerial socket; aerial jack	CONS.EL	**Antennenbuchse**	
antenna spacing = aerial spacing	RADIO	**Antennenabstand**	
antenna support = aerial support	ANT	**Antennenträger** ≈ Fernmeldeturm ↓ Antennenmast; Antennenturm	
antenna system = aerial system	ANT	**Antennenanlage**	
antenna temperature (→antenna noise temperature)	HF	(→Antennenrauschtemperatur)	
antenna test equipment	INSTR	**Antennenmeßgerät**	
antenna theory	HF	**Antennentheorie**	
antenna tower	ANT	**Antennenturm** ≈ Antennenmast ↑ Antennenträger	
antenna transformer	RADIO	**Antennenübertrager**	
antenna tuning coil = aerial tuning coil	ANT	**Antennenabstimmspule**	
antenna tuning unit = aerial tuning unit	ANT	**Antennenanpaßgerät**	
antenna wire	ANT	**Antennendraht**	
anteroom	OFFICE	**Vorzimmer**	
anti-aliasing [a filtering to smooth lines in raster images]	TELEC	**Antialiasing** [eine Filterung zur Glättung von Kurven in Rasterbildern]	
anti-aliasing [computer graphics; countermeasure to reduce visual imperfections]	DATA PROC	**Antialiasing** [Computergraphik; Maßnahme zur Minderung von Bildunschönheiten]	
anticapacitance (adj.) (→low-capacitance)	EL.TECH	(→kapazitätsarm)	
anticathode	PHYS	**Antikathode**	
anticipate (→forestall)	TECH	(→zuvorkommen)	
anticipated value (→expectation value)	INF	(→Erwartungswert)	
anticipating	COLLOQ	**zuvorkommend** = antizipativ; vorwegnehmend	
anticipation	COLLOQ	**Vorwegnahme**	
anticipation = time lead ≈ lead time	TECH	**Vorlauf** ≈ Vorlaufzeit	
anticipation (→time lead)	TECH	(→Voreilung)	
anticipation value (→expectation value)	INF	(→Erwartungswert)	
anticlockwise (→counterclockwise)	TECH	(→linksdrehend)	
anticoincidence	MATH	**Antikoinzidenz**	
anticoincidence (→EXCLUSIVE-OR operation)	ENG.LOG	(→EXKLUSIV-ODER-Verknüpfung)	
anticoincidence circuit (→EXCLUSIV OR gate)	CIRC.ENG	(→EXKLUSIV-ODER-Glied)	
anticoincidence counter	CIRC.ENG	**Antikoinzidenzzähler**	
anticoincidence element (→EXCLUSIV OR gate)	CIRC.ENG	(→EXKLUSIV-ODER-Glied)	
anticoincidence gate (→EXCLUSIV OR gate)	CIRC.ENG	(→EXKLUSIV-ODER-Glied)	
anticoincidence stage ≈ exclusive OR gate	CIRC.ENG	**Antikoinzidenzstufe** ≈ Exklusiv-ODER-Glied	
anticollision radar	RADIO LOC	**Kollisionsschutzradar**	

anti-comet-tail plumbicon	ELECTRON	ACT-Plumbicon
= ACT plumbicon		
anticompetitive	ECON	wettbewerbsschädlich
		= konkurrenzschädlich
anti-fading	RADIO PROP	schwundmindernd
anti-fading aerial (→ anti-fading antenna)	ANT	(→ schwundmindernde Antenne)
anti-fading antenna	ANT	schwundmindernde Antenne
= fading-reducing antenna; anti-fading aerial; fading-reducing aerial		= schwundfreie Antenne; Antifading-Antenne
anti-fading device	RADIO	Schwundausgleichschaltung
antifading tube	ELECTRON	Exponentialröhre
anti-feedback device	ELECTRON	Rückkopplungssperre
= reaction suppressor		
antiferroelectric (adj.) [without electric hysteresis]	PHYS	antiferroelektrisch [ohne elektrische Hysterese]
antiferroelectricity	PHYS	Antiferroelektrizität
antiferroelectric material	PHYS	Antiferroelektrikum
antiferromagnetic [without overall bulk spontaneous magnetisation]	PHYS	antiferromagnetisch [weist keine Eigenmagnetisierung auf]
antiferromagnetism	PHYS	Antiferromagnetismus
antigradient (→ negative gradient)	MATH	(→ negativer Gradient)
antihole	PHYS	Antiloch
antihunting circuit	CIRC.ENG	Beruhigungskreis
		= Beruhigungsschaltung
antihunt transformer	CIRC.ENG	Beruhigungsdrossel
anti-interference (→ interference suppression)	ELECTRON	(→ Entstörung)
anti-interference capacitor	CIRC.ENG	Entstörungskondensator
= interference suppression capacitor		
anti-intrusion protection	SIGN.ENG	Intrusionsschutz
		= Einbruchschutz
anti-jamming	MIL.COMM	Störungsbeseitigung
		≈ Enstörung
antiknock	QUAL	klopffest
= knockresistant		
≈ impact-resistant [TECH]		≈ schlagfest [TECH]
antiknocking characteristic	QUAL	Klopffestigkeit
antilogarithm [the number corresponding to a given logarithm]	MATH	Antilogarithmus [die Zahl die einem Logarithmus entspricht]
antilogarithmic amplifier	CIRC.ENG	antilogarithmischer Verstärker
		= Antilogverstärker; Antilogarithmierschaltung; antilogarithmischer Funktionsumformer
antimetric (→ antimetrical)	NETW.TH	(→ antimetrisch)
antimetrical	NETW.TH	antimetrisch
= antimetric		
antimetrical filter	NETW.TH	antimetrisches Filter
antimetrical quadripole (→ antimetrical two-port)	NETW.TH	(→ antimetrischer Vierpol)
antimetrical two-port	NETW.TH	antimetrischer Vierpol
= antimetrical quadripole		
antimonide	CHEM	Antimonid
antimony	CHEM	Antimon
= Sb		= Sb
antinode	PHYS	Schwingungsbauch
= crest		= Antinode; Bauch
≠ node		≠ Schwingungsknoten
antinomy [inherent contradiction]	SCIE	Antinomie [inhärenter Widerspruch]
anti-parallax (→ parallax-free)	PHYS	(→ parallaxfrei)
anti-parallel (adj.)	MATH	antiparallel
= parallel		≈ gegensinnig [TECH]
≈ reverse [TECH]		≠ parallel
anti-parallel connection	CIRC.ENG	Gegenparallelschaltung
		= Antiparallelschaltung
antiparticle	PHYS	Antiteilchen
≠ elementary particle		≠ Elementarteilchen
antiquate (v.t.) (→ become outdated)	TECH	(→ veralten)
antireflective coat	TECH	Antireflexbelag
antireflective layer	PHYS	Antireflexionsschicht
antiresonance	NETW.TH	Antiresonanz
anti-resonant circuit (→ parallel-resonant circuit)	NETW.TH	(→ Parallelschwingkreis)
anti-resonant inverter	POWER SYS	Parallelschwingkreis-Wechselrichter
antisaturation diode	CIRC.ENG	Antisättigungsdiode
antisismic protection	TECH	Erdbebenschutz
antistatic aerial (→ antistatic antenna)	ANT	(→ geräuscharme Antenne)
antistatic antenna	ANT	geräuscharme Antenne
= antistatic aerial		= Anti-QRN-Antenne
antistatic cloth	TERM&PER	Antistatik-Tuch
antistatic mat	ELECTRON	Antistatik-Matte
antistatic system	SYST.INST	Antistatiksystem
anti-tapping (n.)	TELEC	Abhörsicherung
anti-transmission-reception switch (→ ATR switch)	RADIO LOC	(→ Sperrschalter)
anti-transmitting receiving tube [RADAR]	RADIO LOC	Sendersperröhre [RADAR]
= ATR tube		= Sperröhre
antivalence (→ EXCLUSIVE-OR operation)	ENG.LOG	(→ EXKLUSIV-ODER-Verknüpfung)
antivalence circuit (→ EXCLUSIVE OR gate)	CIRC.ENG	(→ EXKLUSIV-ODER-Glied)
antivalence element (→ EXCLUSIVE OR gate)	CIRC.ENG	(→ EXKLUSIV-ODER-Glied)
antivalence gate (→ EXCLUSIVE OR gate)	CIRC.ENG	(→ EXKLUSIV-ODER-Glied)
anti-virus program	DATA PROC	Antivirenprogramm
		= Virenschutzprogramm
antonym [word with complementary, opposite or converse meaning]	LING	Antonym [Wort komplementärer, kontrárer oder konverser Bedeutung]
≠ synonym		= Gegensatzwort; Gegenwort; Oppositionswort
		≠ Synonym
anywhere call pickup	MOB.COMM	netzweite Erreichbarkeit
AOQ (→ average outgoing quality)	QUAL	(→ mittlere Auslieferqualität)
APC (→ application program command)	DATA PROC	(→ Anwenderprogrammanweisung)
APCM [temporary reduction of bit rates during traffic peaks]	CODING	APCM [zeitweise Reduzierung der Abtastgeschwindigkeit während Verkehrsspitze]
= adaptive PCM		= adaptives PCM
≈ ADPCM		≈ ADPCM
APD (→ avalanche photodiode)	MICROEL	(→ Lawinenphotodiode)
aperiodic	PHYS	aperiodisch
		= unperiodisch
aperiodic aerial (→ aperiodic antenna)	ANT	(→ aperiodische Antenne)
aperiodic antenna	ANT	aperiodische Antenne
= nonresonant antenna; aperiodic aerial; nonresonant aerial		
aperiodic discharge	PHYS	aperiodische Entladung
aperiodicity	PHYS	Aperiodizität
aperture	PHYS	Apertur
		= Öffnungswinkel; Öffnung
aperture	ANT	Strahlaustrittsfläche
		= Öffnungsfläche
aperture angle	ANT	Öffnungsbreite
aperture antenna	ANT	Aperturantenne
= plane antenna; sheet antenna; flat-top antenna		= Aperturstrahler; Flächenstrahler; Flächenantenne 1; Flächenreflektor; Flat-top-Antenne; Kreissektorzahn-Antenne
aperture blockage	ANT	Aperturbehinderung
aperture distortion	TELEGR	Aperturverzerrung [Faxsimile]
aperture distribution (→ aperture illumination)	ANT	(→ Ausleuchtung)
aperture field distribution	ANT	Aperturfeldverteilung
aperture illumination	ANT	Ausleuchtung
= illumination; aperture distribution		= Aperturverteilung

aperture lens (→electron lens) ELECTRON (→Elektronenlinse)
aperture loading ANT **Aperturbelegung**
aperture-to-medium coupling ANT **Antennenkopplungsverlust**
loss [Überhorizontverbindungen]
= multipath coupling loss; loss in path-
antenna gain
aperura (→focal aperture) PHYS (→Blendenöffnung)
apex TYPOGR **Spitze**
[symbol ^] [Symbol ^]
= curet [DATA PROC]; carat [DATA = Zirkumflex [DATA
PROC] PROC]
apex MATH **Scheitelpunkt 1**
[of an angle] [eines Winkels]
= vertex = Scheitel 1
apex angle (→crest angle) MATH (→Scheitelwinkel)
apex drive (→center feed) ANT (→Mittelpunktspeisung)
APL DATA PROC **APL**
[A Programming Language] ↑ problemorientierte Programmiersprache
↑ high-level programming language
apogee ASTROPHYS **Apogäum**
[orbital point farest from earth] [Bahnpunkt größter Entfernung zur Erde]
↑ apsis ↑ Apside
A pole OUTS.PLANT **A-Mast**
apostilb PHYS **Apostilb**
[unit for brightness; = $1/\pi$ cd/m^2] [Einheit für Leuchtdichte; = $1/\pi$ cd/m^2]
= asb; Blondel = asb; Blondel
apostrophe LING **Apostroph**
[mark ' indicating omission] [eine Auslassung kennzeichnender Haken ']
= single quotation mark
↑ ellipsis 2 ↑ Auslassungszeichen
apparatus (pl.-us or -uses) TECH **Apparat**
[an equipment for specific operation] [ein Gerät 1 für spezifische Funktion]
≈ set of apparatus
↑ equipment 1 ≈ Apparatur
↑ Gerät 1
apparatus (→set) TERM&PER (→Gerät)
apparent capacitance COMPON **Scheinkapazität**
[electrolytic capacitor] [Elektrolytkondensator]
apparent capacitance meter INSTR **Scheinkapazitätsmesser**
apparent permeability EL.TECH **Scheinpermeabilität**
≈ wirksame Permeabilität
apparent power EL.TECH **Scheinleistung**
= complex power 2; vector power [Produkt Effektivspannung mal Effektivstrom]
apparent resistance NETW.TH (→komplexer Scheinwiderstand)
(→impedance)
appeal (→objection) ECON (→Einspruch)
append (v.t.) DATA PROC **anfügen**
append (→complement) MATH (→komplementieren)
appendix (→annex) LING (→Anlage)
appliance 1 (→gauge) MECH (→Lehre)
appliance 2 (→device) TECH (→Vorrichtung)
appliance 3 (→tool) TECH (→Werkzeug)
applicability TECH **Anwendbarkeit**
= utilization mode = Anwendungsmöglichkeit; Einsatzmöglichkeit
≈ field of application; user friendliness
≈ Anwendungsgebiet; Benutzerfreundlichkeit
applicable TECH **anwendbar**
application TELEC **Anmeldung**
[to a connection or service] [eines Anschlußes]
= subscription = Beantragung; Antragsstellung
application ECON **Antrag**
= request
application 1 TECH **Anwendung**
≈ task; use = Einsatz
= Aufgabe; Gebrauch
application DOC **Applikation**
application 2 TECH **Aufbringung**
[e.g. of a paint] [z.B. Farbe]
application case TECH **Anwendungsfall**
= Einsatzfall
application class COMPON **Anwendungsklasse**
application-dependent TECH (→anwendungsspezifisch)
(→application-specific)
application domain TECH (→Anwendungsbereich)
(→application field)

application field TECH **Anwendungsbereich**
= range of application = Einsatzbereich
application for trademark ECON **Warenzeichenanmeldung**
application guide (→application DOC (→Anwendungsrichtlinie)
note)
application-independent TECH **anwendungsneutral**
= use-independent = einsatzneutral
≈ user-independent ≈ anwenderneutral
≠ application-specific ≠ anwendungsspezifisch
application layer DATA COMM **Verarbeitungsschicht**
[7th layer of OSI] [7.Schicht im ISO-Schichtenmodell]
= Anwendungsschicht; Anwendungsebene
application note DOC **Anwendungsrichtlinie**
= application guide = Anwendungsbeschreibung; Einsatzrichtlinie; Nutzungshinweise
≈ Gebrauchsunterlage
application-orientated DATA PROC (→anwendungsorientiert)
(→application-oriented)
application-oriented DATA PROC **anwendungsorientiert**
= use-oriented; job-oriented; application-orientated; use-orientated; job-orintated = anwendungsbezogen; einsatzorientiert; einsatzbezogen; aufgabenorientiert;
≈ application-specific; user-specific; dedicated aufgabenbezogen; auftragsorientiert; auftragsbezogen
≈ anwendungsspezifisch; anwenderspezifisch
application-oriented language DATA PROC (→anwendungsorientierte Programmiersprache)
(→application-oriented programming language)
application-oriented program- DATA PROC **anwendungsorientierte Programmiersprache**
ming language
= application-oriented language = anwendungsorientierte Sprache
application place (→application TECH (→Anwendungsort)
site)
application program DATA PROC **Anwenderprogramm 1**
≠ system software [für den Anwender geschrieben]
= Anwendungsprogramm
≠ Systemsoftware
application program DATA PROC **Anwenderprogrammanweisung**
command
= APC = APC
application program DATA PROC **Anwenderprogrammschnittstelle**
interface
application programmer DATA PROC **Applikationsprogrammierer**
application programming DATA PROC **Applikationsprogrammierung**
≠ system programming ≠ Systemprogrammierung
applications engineer MICROEL **Applikationsingenieur**
application site TECH **Anwendungsort**
= application place ≈ Einsatzort
≈ operation site
application software DATA PROC **Anwendersoftware 1**
≠ system software [für den Anwender geschrieben, systembezogen]
≠ Systemsoftware
application software DATA PROC (→Anwendungssoftware)
(→special applications software)
application-specific TECH **anwendungsspezifisch**
= use-specific; job-specific; task-specific; application-dependent; use-dependent; job-dependent; task-dependent; dedicated 1; special-use (adj.) = einsatzspezifisch; aufgabenspezifisch; anwendungsbezogen; einsatzbezogen; aufgabenbezogen; zweckbestimmt; dediziert; spezialisiert; zweckbestimmt; zugeordnet
≈ user-specific; application-oriented [DATA PROC]; custom-designed
≠ application-independent ≈ anwenderspezifisch; anwendungsorientiert [DATA PROC]; kundenspezifisch
↑ dedicated 2
≠ anwendungsneutral
↑ spezifisch
application-specific IC MICROEL **anwendungsspezifische integrierte Schaltung**
= ASIC; application-specific integrated circuit
= ASIC; anwendungsspezifischer intergrierter Schaltkreis; anwendungsspezifischer IC; anwender-
≈ custom IC; full custom IC
↓ gate array

application-specific integrated circuit (→application-specific IC)	MICROEL	spezifische integrierte Schaltung; anwenderspezifischer IC; anwenderspezifischer integrierter Schaltkreis ≈ kundenspezifische integrierte Schaltung; Vollkundenschaltung ↓ Gate Array (→anwendungsspezifische integrierte Schaltung)
application-specific terminal (→applications terminal)	TERM&PER	(→Spezialendgerät)
applications programmer	DATA PROC	Anwendungsprogrammierer
applications terminal = single-purpose terminal; dedicated terminal; application-specific terminal ≠ multi-purpose terminal	TERM&PER	Spezialendgerät ≠ Mehtzweckendgerät
applied-for patent (→pending patent)	ECON	(→angemeldetes Patent)
applied informatics ≈ telematics ↑ informatics; information technology	INF.TECH	angewandte Informatik ≈ Telematik ↑ Informatik; Nachrichtentechnik
applied mathematics	MATH	angewandte Mathematik
applied physics	PHYS	angewandte Physik
applied science	SCIE	angewandte Forschung
applique (→applique circuit)	CIRC.ENG	(→Additivkreis)
applique circuit = applique	CIRC.ENG	Additivkreis
applique devoice (→add-on device)	TECH	(→Zusatzeinrichtung)
applique equipment (→add-on equipment)	EQUIP.ENG	(→Zusatzgerät)
apply [tension, signal ...] = inject	EL.TECH	anlegen [Spannung, Signal ...] = einspeisen; zuführen
appointed execution time	ECON	Fertigstellungstermin
appointed time [established moment] = due time; fixed time; time limit = term; date; deadline	COLLOQ	Termin [festgelegter Zeitpunkt] ≈ Frist 2; Datum; äußerster Termin
appointment 2	ECON	Verabredung
appointment ≈ allocation; occupation	TECH	Festlegung [fig.] = Festsetzung 2 ≈ Zuteilung; Belegung
appointment 1 (→allocation)	ECON	(→Umlage)
appointment book = date book; datebook; follow-up file; appointment scheduler; appointment calendar	OFFICE	Terminkalender = Terminplaner
appointment calendar (→appointment book)	OFFICE	(→Terminkalender)
appointment scheduler (→appointment book)	OFFICE	(→Terminkalender)
apportionment (→partitioning)	TECH	(→Aufteilung)
apposition	LING	Apposition = Beisatz
appraisal (→estimate)	COLLOQ	(→Schätzung)
appraise = estimate (n.)	COLLOQ	schätzen
appraised value = estimated value	ECON	Schätzwert
appraiser = valuer ≈ designated expert	ECON	Schätzer ≈ Gutachter
apprentice = trainee 1	ECON	Lehrling = Anlernling; Auszubildender; Azubi
apprenticeship	ECON	Lehre [praktische Berufsausbildung]
approach (n.)	RADIO NAV	Anflug = Einflug
approach (→method)	SCIE	(→Methode)
approach aid	RADIO NAV	Anfluganweisung
approach light	AERON	Landelicht
approach path (→landing path)	AERON	(→Landeweg)
approach plane	AERON	Anflugebene
approach radar	RADIO NAV	Anflugradar
approach radiobacon = landing beacon	RADIO NAV	Landebake
approach surface	AERON	Anflugfläche
approbation (→approval)	COLLOQ	(→Zustimmung)
approportioning (→allocation)	TECH	(→Zuordnung)
appropriate (→suited)	TECH	(→geeignet)
approval = release	QUAL	Freigabe = Zulassung ≈ Abnahme
approval = approbation	COLLOQ	Zustimmung = Billigung; Zulassung; Anerkennung
approved	COLLOQ	altbewährt
approximate (adj.) = approximating ≈ inaccurate	SCIE	angenähert = approximativ ≈ ungenau
approximate (adj.) = rough	MATH	näherungsweise = überschläglig; grob
approximately-equal sign [≈] = approximately-equal symbol	MATH	Üngefähr-gleich-Zeichen [≈]
approximately-equal symbol (→approximately-equal sign)	MATH	(→Üngefähr-gleich-Zeichen)
approximate method (→approximation method)	MATH	(→Näherungsverfahren)
approximate solution	MATH	Näherungslösung
approximate theory	MATH	Näherungstheorie
approximate value (→approximation value)	MATH	(→Näherungswert)
approximating (→approximate)	SCIE	(→angenähert)
approximation	MATH	Näherung = Annäherung; Approximation
approximation accuracy = approximation precision	MATH	Näherungsgenauigkeit = Approximationsgenauigkeit
approximation calculus = simplified calculus	MATH	Näherungsrechnung = Approximationsrechnung
approximation equation = simplified equation	MATH	Näherungsgleichung = Approximationsgleichung
approximation error	MATH	Näherungsfehler = Approximationsfehler
approximation formula = simplified formula	MATH	Näherungsformel = Approximationsformel
approximation function	MATH	Approximationsfunktion = Näherungsfunktion
approximation method = approximate method	MATH	Näherungsverfahren = Näherungsmethode; Approximationsverfahren; Approximationsmethode
approximation precision (→approximation accuracy)	MATH	(→Näherungsgenauigkeit)
approximation value = approximate value	MATH	Näherungswert = Approximationswert
APR (→precision approach radar)	RADIO NAV	(→Präzisions-Anflug-Radar)
apron	AERON	Vorfeld = Flugzeugabstellplatz
apron taxiway	AERON	Vorfeldrollbahn
APS (→protection switching equipment)	TRANS	(→Schutzschalteinrichtung)
apsides connecting line ≠ inter-apsides line	ASTROPHYS	Apsidenlinie
apsis (pl. apsides) [orbital point with nearest or greatest distance from center of attraction] ↓ perigee; apogee; perihelion	ASTROPHYS	Apside (pl.-en) [Bahnpunkt kleinster oder größter Entfernung zum Anziehungspunkt] ↓ Perigäum; Apogäum; Perihel
aptitude (→suitability)	TECH	(→Eignung)
AQL (→acceptable quality level)	QUAL	(→Annahmegrenze)
aqueous ≈ hydrated; hydrous	CHEM	wäßrig = wässerig ≈ hydratiert; wasserhaltig
Ar (→argon)	CHEM	(→Argon)

ARB (→arbitrary waveform)	INSTR	(→anwenderdefinierte Signalform)
arbiter	DATA PROC	**Zuteiler**
= tie breaker		
arbitrary	COLLOQ	**beliebig**
arbitrary	TECH	**willkürlich**
= at random		= beliebig
arbitrary function	INSTR	**anwenderdefinierbare Funktionsfolge**
arbitrary waveform	INSTR	**anwenderdefinierte Signalform**
[shapable by the user]		
= ARB		= benutzerdefinierbare Signalform; beliebige Signalform
arbitrary waveform synthesizer	INSTR	**Synthesizer für beliebige Signalformen**
arbitration (→decision)	INF	(→Entscheidung)
arbitration algorithm (→decision algorithm)	MATH	(→Entscheidungsalgorithmus)
arbitration award	ECON	**Schiedsspruch**
= award 2 n.		
arbitration logic	MATH	**Entscheidungslogik**
= decision logic		
arbitrator	ECON	**Schiedsrichter**
arc (→circular arc)	MATH	(→Kreisbogen)
arcade game	DATA PROC	**Arcade**
[adventure game played on coin-operated public machine]		[Geschicklichkeits-Computerspiel auf öffentlichen Münzautomaten]
arc-back	ELECTRON	**Rückzündung**
arc discharge	PHYS	**Bogenentladung**
↑ gaseous discharge		↑ Gasentladung
arc discharge tube	ELECTRON	**Bogenentladungsröhre**
↓ thyratron; gas discharge rectifier		↓ Thyratron; Gasentladungsgleichrichter
arch	TECH	**Wölbung**
= camber		= Bogen
arched	TECH	**bogenförmig**
archimedian spiral (→spiral of Archimedes)	MATH	(→archimedische Spirale)
archimedian spiral antenna	ANT	**archimedische Spiralantenne**
architecture	DATA PROC	**Architektur**
archival (adj.)	DATA PROC	**die Langzeitspeicherbarkeit betreffend**
archival (n.)	DATA PROC	**Archivierung**
[storage over a long period]		[langfristige Speicherung]
= archive (n.); archive storage		↑ Speicherung
↑ storage		
archival quality	DATA PROC	**Langzeitspeicherbarkeit**
		= Speicherlebensdauer
archive (v.t.)	DATA PROC	**archivieren**
↑ store (v.t.); file (v.t.)		↑ ablegen; speichern
archive (n.) (→archival)	DATA PROC	(→Archivierung)
archive bit	DATA PROC	**Archivbit**
archive storage	DATA PROC	(→Archivierung)
(→archival)		
archiving device	TERM&PER	**Archivierungseinrichtung**
arcing	PHYS	**Lichtbogenbildung**
arc lamp	PHYS	**Bogenlampe**
arc length	MATH	**Bogenlänge**
= radial length		≈ Bogenmaß
≈ radian measure		
arc-over (→spark-over)	PHYS	(→Funkenüberschlag)
arc spectrum	PHYS	**Bogenspektrum**
Arcus clamp	POWER SYS	**Arcus-Klemme**
arc-weld (v.t.)	METAL	**lichtbogenschweißen**
arc welding	METAL	**Lichtbogenschweißen**
= spark welding		= Bogenschweißen; Elektroschweißen
are	PHYS	**Ar**
[unit for land areas; = $10 \text{ m} \cdot 10 \text{ m} = 100 \text{ m}^2$]		[Maß für Grundstücksflächen; = $10 \text{ m} \cdot 10 \text{ m} = 100 \text{ m}^2$]
= a		= a
area	MATH	**Fläche**
[SI unit: square meter]		[SI-Einheit: Quadratmeter]
= A; q		= A; q
↓ surface		↓ Oberfläche
area	DATA PROC	**Bereich**
[of a memory]		[in einem Speicher]
area code (AM)	SWITCH	**Ortskennzahl**
= trunk code; dialling code (BRI); local network code;		[z.B. 89 für München]
		= Ortznetzkennzahl; nationale Kennzahl; Inlandsvorwählnummer
		≈ Ortsvorwählnummer
area composition [DATA PROC] (→page makeup)	TYPOGR	(→Seitenumbruch)
area control radar	RADIO LOC	**Gebietskontrollradar**
area dose product	PHYS	**Flächendosisprodukt**
area filling	DATA PROC	**Bildfülloperation**
[graphics processor]		[Graphikprozessor]
areal density (→surface density)	PHYS	(→Flächendichte)
area load	CIV. ENG	**Flächenbelastung**
= load per unit area		
area mass	PHYS	**Flächenmasse**
area of reflection	RADIO PROP	**Reflexionsfläche**
= reflecting surface		
area protection	SIGN. ENG	**Innenraumschutz**
area search	DATA PROC	**Grobrecherche**
= area searching		= Grobklassifizierung; Vorselektierung
area searching (→area search)	DATA PROC	(→Grobrecherche)
area switch-over (→subband switch-over)	RADIO	(→Teilbereichsumschalter)
area utilization factor	MICROEL	**Flächennutzungsfaktor**
argon	CHEM	**Argon**
= Ar		= Ar
argot (→jargon)	LING	(→Jargon)
argument	DATA PROC	**Argument**
[variable for either logical or numeric values]		[Variable für sowohl numerische als auch logische Werte]
≈ variable		≈ Variable
arithmetic (n.)	MATH	**Arithmetik**
arithmetical (adj.)	MATH	**arithmetisch**
		= rechnerisch
arithmetical mean (→arithmetic mean)	MATH	(→arithmetisches Mittel)
arithmetical progression	MATH	**arithmetische Reihe**
= arithmetic progression; arithmetical series; arithmetic series		= arithmetische Progression; endliche arithmetische Reihe
arithmetical series (→arithmetical progression)	MATH	(→arithmetische Reihe)
arithmetic and logic unit (→arithmetic-logic unit)	DATA PROC	(→Rechenwerk)
arithmetic data	DATA PROC	**Rechendaten**
		= arithmetische Daten
arithmetic element (→arithmetic-logic unit)	DATA PROC	(→Rechenwerk)
arithmetic function	MATH	**arithmetische Funktion**
arithmetic instruction	DATA PROC	**arithmetischer Befehl**
≠ logic instruction		= Rechenbefehl
		≠ logischer Befehl
arithmetic logical unit (→arithmetic-logic unit)	DATA PROC	(→Rechenwerk)
arithmetic-logic unit	DATA PROC	**Rechenwerk**
[part of CPU, where arithmetic and logic operations are carried out]		[Teil des Prozessors der Zentraleinheit, der die arithmetischen und logischen Operationen durchführt]
= arithmetic and logic unit; arithmetic logical unit; ALU; arithmetic unit; arithmetic element		= arithmatisch-logische Einheit; Arithmetik- und Logikeinheit; ALU; Recheneinheit; Rechenelement; Arithmetikeinheit
↑ processor; central processing unit		↑ Prozessor; Zentraleinheit
↓ accumulator; multiplier-quotient register		↓ Akkumulator; Multiplikator-Quotienten-Register
arithmetic mean	MATH	**arithmetisches Mittel**
[sum of A1 till An, divided by n]		[Summe A1 bis An, geteilt durch n]
= arithmetical mean		= lineares Mittel
arithmetic operation	MATH	**Rechenoperation**
		= arithmetische Operation

arithmetic operator	DATA PROC	arithmetischer Operator
arithmetic progression	MATH	(→arithmetische Reihe)
(→arithmetical progression)		
arithmetic series (→arithmetical progression)	MATH	(→arithmetische Reihe)
arithmetic statement	DATA PROC	arithmetische Anweisung
arithmetic unit	DATA PROC	(→Rechenwerk)
(→arithmetic-logic unit)		
arm	MECH	Arm
arm	TYPOGR	Ast
ARM (→asynchronous response mode)	DATA COMM	(→Spontanbetrieb)
arm (→leg)	MATH	(→Schenkel)
armament industry	TECH	Rüstungsindustrie
armature [carries the windings]	POWER ENG	Anker [trägt die Wicklungen der Maschine]
armature [the movable part of a magnetic circuit] ≈ yoke [EL.TECH]	COMPON	Anker [der bewegliche Teil des magnetischen Kreises] = Magnetanker ≈ Magnetjoch [EL.TECH]
armature bounce = armature rebound	COMPON	Ankerprellen
armature current	POWER ENG	Ankerstrom
armature excursion [relay] = armature travel	COMPON	Ankerhub [Relais]
armature gap [relay]	COMPON	Ankerluftspalt [Relais]
armatureless	COMPON	ankerlos
armature reaction	POWER ENG	Ankerrückwirkung
armature rebound (→armature bounce)	COMPON	(→Ankerprellen)
armature relay	COMPON	Ankerrelais
armature resetting spring (→armature restoring spring)	COMPON	(→Ankerrückzugfeder)
armature restoring spring [relay] = armature resetting spring	COMPON	Ankerrückzugfeder [Relais]
armature reversal	POWER ENG	Ankerumschaltung
armature shaft	POWER ENG	Ankerwelle
armature slot [relay]	COMPON	Ankernut [Relais]
armature stop [relay]	COMPON	Ankeranschlag [Relais]
armature travel (→armature excursion)	COMPON	(→Ankerhub)
armature winding	POWER ENG	Ankerwicklung
armor (→armouring)	COMM.CABLE	(→Bewehrung)
armored	COMM.CABLE	bewehrt
armored cable	COMM.CABLE	bewehrtes Kabel
armoring machine	COMM.CABLE	Kabelbewehrungsmaschine
armorless	COMM.CABLE	unbewehrt
armotization [of a debt or an inversion]	ECON	Amortisation [ratenweise Tilgung einer Schuld; Deckung von Investitionskosten durch Ertrag] ≈ Abschreibung
armouring = armor	COMM.CABLE	Bewehrung
armouring tape	COMM.CABLE	Bandeisenbelegung
Aron measuring circuit	INSTR	Aronschaltung
ARQ (→automatic repeat request)	DATA COMM	(→automatische Wiederholanforderung)
arrange	TECH	anordnen = aufstellen 2
arrange = activate	COLLOQ	veranlassen = aktivieren
arrangement [spacial] = configuration	TECH	Anordnung [räumliche] = Aufstellung 2
arrangement = order	MATH	Anordnung
arrangement (→circuit arrangement)	CIRC.ENG	(→Schaltungsaufbau)
arrangement 2 (→agreement 1)	COLLOQ	(→Vereinbarung)
arrangement 1 (→precaution)	COLLOQ	(→Vorkehrung)
array 1 (n.) [ordered arrangement of indexed elements] = matrix; vector	DATA PROC	Datenfeld 2 [geordnete Anordnung von indizierten Elementen] = Feld 2 ↓ Matrix; Vektor
array 2 (→data field)	DATA PROC	(→Datenfeld 1)
array (→array antenna)	ANT	(→Strahlerfeld)
array antenna = antenna array; array	ANT	Strahlerfeld = Antennengruppe; Antennenfeld; Antennennetz; Strahlfeld; Strahlergruppe; Gruppenantenne; Gruppe
array bounds	DATA PROC	Datenfeldgrenzen
array cell (→core cell)	MICROEL	(→Kernzelle)
array computer (→array processor)	DATA PROC	(→Vektorrechner)
array element	DATA PROC	Matrixelement
array element	ANT	Einheitsfeld
array factor (→space factor)	ANT	(→Gruppencharakteristik)
array logic [microprocessor]	MICROEL	Array-Logik [Mikroprozessor]
array of parallel dipoles = broadside array with parallel elements	ANT	Dipolreihe = Dipolzeile
array processor [mainframe computer dedicated to fast matrix arithmetic] = vector processor; array computer; vector computer; distributed array processor; DAP ≠ scalar processor ↑ multi-processor system ↓ SIMD processor; MIMD procesor	DATA PROC	Vektorrechner [auf schnelle Matrixrechnungen ausgelegter Großrechner] = Feldrechner; Matrizenrechner; Vektor-Prozessor; Array-Prozessor; Zellenrechner; DAP ≠ Skalarrechner ↑ Mehrprozessorsystem ↓ SIMD-Prozessor; MIMD-Prozessor
array variable (n.)	DATA PROC	Feldvariable
arrears (→backlog 1)	COLLOQ	(→Rückstand)
arrest (v.t.) = lock; stop; detain; secure	MECH	arretieren = feststellen
arrest (n.) = arresting device; binding; hesitation; locking device ≈ clamping device; detent (n.)	TECH	Hemmung = Arretierung; Arretiervorrichtung; Feststellvorrichtung; Sperrvorrichtung ≈ Klemmvorrichtung; Spannvorrichtung; Raste
arrester (→overvoltage protector)	COMPON	(→Spannungssicherung)
arresting device (→arrest)	TECH	(→Hemmung)
arrestor (→overvoltage protector)	COMPON	(→Spannungssicherung)
arrival process	SWITCH	Ankunftsprozeß = Anrufprozeß
arrival rate	DATA COMM	Empfangsrate
arrow	TYPOGR	Pfeil
arrow diagram	DOC	Pfeildiagramm
arrowed line	DATA PROC	Pfeillinie = gepfeilte Linie
arrow-head antenna	ANT	Pfeilspitzenantenne
arrow key (→directional key)	TERM&PER	(→Richtungstaste)
arsenic = As	CHEM	Arsen = As
art (→engineering)	TECH	(→Technik 1)
art director	TYPOGR	Chefgrafiker
article	LING	Artikel = Geschlechtswort
article number code	DATA PROC	Artikelnummercode
articulated lorry (BRI) (→semi-trailer truck)	TECH	(→Sattelschlepper)
articulation	LING	Artikulation = Lautbildung
articulation (→intelligibility)	TELEC	(→Verständlichkeit)
articulation for logatomes	TELEPH	Logatomverständlichkeit
artifact [error in an image]	INF.TECH	Bildfehler
artifice	TECH	Kunstgriff
artificial ≈ synthetic	TECH	künstlich ≈ synthetisch

artificial aerial (→dummy antenna) ANT (→künstliche Antenne)
artificial antenna (→dummy antenna) ANT (→künstliche Antenne)
artificial black signal TELEGR **künstliches Schwarz**
= nominal black signal
artificial dielectric ANT **künstliches Dielektrikum**
artificial ear TELEPH **künstliches Ohr**
artificial horizon AERON **künstlicher Horizont**
artificial intelligence DATA PROC **künstliche Intelligenz**
= AI = KI
artificial language INF.TECH **künstliche Sprache**
artificial line TELEC **Kunstleitung**
artificial line (→line-balancing network) TELEC (→Leitungsnachbildung)
artificial mouth TELEPH **künstlicher Mund**
= artificial voice
artificial network (→line-balancing network) TELEC (→Leitungsnachbildung)
artificial term LING **Kunstwort**
↓ acronym; portmanteau word ↓ Akronym; Schachtelwort
artificial traffic SWITCH **künstlicher Verkehr**
artificial voice (→artificial mouth) TELEPH (→künstlicher Mund)
artificial white signal TELEGR **künstliches Weiß**
= nominal white signal
artillery computer RADIO LOC **Feuerleitrechner**
artillery control system RADIO LOC **Feuerleitsystem**
artillery point PHYS **artilleristischer Strich**
[0° 3'22.5"] [0° 3'22,5"]
artillery radar RADIO LOC **Feuerleitradar**
arts COLLOQ **Kunst**
↓ fine arts; performing arts; music ↓ bildende Kunst; darstellende Kunst; Musik
artwork master MANUF **Druckvorlage**
[PCB] [Leiterplattenherstellung]
= PCB artwork; production master = Leiterplattendruckvorlage
A s (→ampere-second) PHYS (→Amperesekunde)
As (→arsenic) CHEM (→Arsen)
ASA (→American Standarts Association) TECH (→Amerikanische Normengesellschaft)
asb (→apostilb) PHYS (→Apostilb)
ASBC technology (→advanced standard-buried collector technology) MICROEL (→ASBC-Technik)
asbestos CHEM **Asbest**
= asbestus
asbestus (→asbestos) CHEM (→Asbest)
ascend (→rise) ECON (→steigen)
ascender 1 TYPOGR **Oberlänge**
[part of character above x-height] [Teil eines Buchstabens der über die Oberkante des Buchstabens x hinausragt]
≠ descender ≠ Unterlänge
ascender 2 TYPOGR **Oberlängenbuchstabe**
ascending link TELEC **aufsteigende Verbindung**
ascending order MATH **aufsteigende Reihenfolge**
ascending traffic SWITCH **aufsteigender Verkehr**
ASCII (→ASCII code) CODING **ASCII-Code**
ASCII code CODING **ASCII-Code**
[United States of America Standard Code for Information Interchange; pronounced "as-key"; originally with 7 bit, now mostly with 8 bit] [ursprünglich ein 7-Bit-Code, heute meist ein 8-Bit-Code; im Engl. als „äskäy" ausgesprochen]
= USASCII code; ASCII = ASCII-Zeichensatz; USASCII-Code
ASCII keyboard TERM&PER **ASCII-Tastatur**
[with a key to every ASCII code] [mit einer Taste für jeden ASCII-Code]
ASIC (→application-specific IC) MICROEL (→anwendungsspezifische integrierte Schaltung)
ASK (→amplitude keying) MODUL (→Amplitudentastung)
asked price ECON **Preisforderung**
aslant (→slant) TECH (→schräg)
as needed TECH **bedarfsweise**
aspect ratio TV **Seitenverhältnis**
ASR RADIO NAV **Flughafen-Überwachungsradar**
= airport surveillance radar; aerodrome surveillance radar = ASR
ASR (→answer-to-seize ratio) SWITCH (→Erfolgsquote)

ASR keyboard TERM&PER **ASR-Tastatur**
assault alarm device (→holdup alarm device) SIGN.ENG (→Überfallmelder)
assault alarm system (→holdup alarm system) SIGN.ENG (→Überfallmeldesystem)
assault button (→holdup button) SIGN.ENG (→Überfallalarmknopf)
assemble (v.t.) DATA PROC **assemblieren**
[to translate a machine-oriented code into a machine code] [von maschinenorientierter Sprache in Maschinensprache übersetzen]
↑ translate ↑ übersetzen
assemble (v.t.) TECH **zusammenbauen**
≈ install; mount = zusammenstellen
≠ disassemble ≈ installieren; montieren
 ≠ zerlegen
assembled board (→assembled PCB) ELECTRON (→bestückte Leiterplatte)
assembled PCB ELECTRON **bestückte Leiterplatte**
= assembled board; equipped PCB; populated board
assembler DATA PROC **Assembler 1**
[translates from a machine-oriented language into machine code] [übersetzt von maschinenorientierter Sprache in Maschinensprache]
= assembly program; assembly routine; assembler program = Assemblierer 1; Assemblerprogramm; Assemblierprogramm
≈ assembler language; compiler ≈ Assemblierersprache; Kompilierer
≠ disassembler ≠ Disassembler
↑ translator ↑ Übersetzer
assembler (→mechanic) TECH (→Monteur)
assembler instruction DATA PROC **Assemblerbefehl**
= assembly instruction = Assemblierbefehl
assembler language DATA PROC **Assemblersprache**
[processor-specific, mnemotechnic programming language; is translated into object code by an assembler] [Prozessor-spezifische, dem Maschinencode angepaßte, mnemotechnische Programmiersprache; wird per Assembler in Objektcode übersetzt]
= assembly language; computer-dependent language; assembler source program; base language = Assemblersprache; Assembler 2; Assemblierer 2; Eins-zu-Eins-Sprache; 1:1 Sprache
≈ assembler ≈ Assembler 1
↑ low-level programming language; computer-oriented language ↑ niedere Programmiersprache; maschinenorientierte Programmiersprache
assembler listing DATA PROC **Assemblerprotokoll**
= assembly listing = Assemblierprotokoll
↑ translator listing ↑ Übersetzungsprotokoll
assembler program DATA PROC (→Assembler 1)
(→assembler)
assembler source program DATA PROC (→Assemblersprache)
(→assembler language)
assembling DATA PROC **Assemblierung**
assembling gauge MANUF **Aufbaulehre**
assembly ECON **Versammlung**
≈ conference ≈ Besprechung
assembly TECH **Zusammenbau**
≈ installation; mounting ≈ Installierung; Montage
≠ disassembly ≠ Zerlegung
assembly (→module) EQUIP.ENG (→Baugruppe)
assembly (→packaging) MICROEL (→Montage)
assembly appliance MANUF **Montagelehre**
= assembly jig = Montagevorrichtung; Einbauvorrichtung; Zusammenbaulehre
≈ mounting frame ≈ Montagerahmen
assembly diagram (→wiring diagram) EL.TECH (→Bauschaltplan)
assembly drawing ENG.DRAW **Zusammenbauzeichnung**
≈ installation drawing = Zusammenstellungszeichnung
 ≈ Montagezeichnung
assembly instruction DATA PROC (→Assemblerbefehl)
(→assembler instruction)

assembly instructions	ENG.DRAW	Zusammenbauvorschrift	assumed mean	MATH	provisorisches Mittel
= assembly specifications		= Zusammenstellungsvorschrift	assumption (→ supposition)	SCIE	(→ Annahme)
≈ mounting instructions		≈ Montagevorschrift; Bauvorschrift	assurance (→ insurance)	ECON	(→ Versicherung)
			assure (→ insure)	ECON	(→ versichern)
assembly jig (→ assembly appliance)	MANUF	(→ Montagelehre)	assured (→ insured)	ECON	(→ Versicherter)
			assured value (→ insured value)	ECON	(→ Versicherungswert)
assembly language (→ assembler language)	DATA PROC	(→ Assemblersprache)	assurer (→ insurer)	ECON	(→ Versicherer)
			assuror (→ insurer)	ECON	(→ Versicherer)
assembly line	MANUF	Fließband	assy (→ module)	EQUIP.ENG	(→ Baugruppe)
= production line; manufacturing line		= Fertigungslinie; Fertigungsstraße; Montagelinie; Montagestraße	assymetrical feedline	ANT	unsymmetrische Speiseleitung
			assymmetry	MATH	Asymmetrie
			= dissymmetry		= Ungleichmäßigkeit
assembly listing (→ assembler listing)	DATA PROC	(→ Assemblerprotokoll)	≠ symmetry		≠ Symmetrie
			assymmetry factor	NETW.TH	Asymmetriefaktor
assembly plan	DOC	Montageplan	astable multivibrator	CIRC.ENG	astabile Kippschaltung
assembly position (→ board position)	EQUIP.ENG	(→ Baugruppenposition)	[self-oscillating, without external triggering]		[selbstschwingend, ohne externe Auslösung]
assembly program (→ assembler)	DATA PROC	(→ Assembler 1)	= free-running multivibrator; astable trigger circuit; multivibrator 2; bivibrator		= astabiler Multivibrator; Kippschwinger; selbstschwingende Kippschaltung; Bivibrator
assembly routine (→ assembler)	DATA PROC	(→ Assembler 1)	↑ toggle generator; relaxation oscillator		≠ stabile Kippschaltung
assembly specification (→ construction specification)	TECH	(→ Bauvorschrift)			↑ Kippgenerator; Relaxationsoszillator
assembly specifications (→ assembly instructions)	ENG.DRAW	(→ Zusammenbauvorschrift)	astable trigger circuit (→ astable multivibrator)	CIRC.ENG	(→ astabile Kippschaltung)
			astatic	CONTROL	astatisch
assembly time	DATA PROC	Assemblierzeit			= integralwirkend
↑ translation time		↑ Übersetzungszeit	astatic couple of needles	INSTR	astatisches Nadelpaar
assembly variant (→ board variant)	EQUIP.ENG	(→ Baugruppenvariante)	astatine	CHEM	Astat
assess (v.t.)	COLLOQ	bemessen	= At		= At
assessment	COLLOQ	Bemessung	asterisc	TYPOGR	Sternchen
assessment [taxes]	ECON	Veranlagung [Steuer]	[*]		[*]
assessment basis	ECON	Bemessungsgrundlage	= star		= Asteriskus (pl.-en); Sternsymbol; Stern; Sternzeichen
assets (→ funds)	ECON	(→ Mittel)			
assign	TECH	zuordnen	asterisk address	DATA PROC	Sternadresse
= allocate; allot		= zuteilen; zuweisen	astigmatism	PHYS	Astigmatismus
assignment (→ transfer 1)	ECON	(→ Abtretung)	ASTM (→ American Society for Testing Materials)	QUAL	(→ Amerikanische Gesellschaft für Materialprüfung)
assignment (→ partitioning)	TECH	(→ Aufteilung)			
assignment (→ division)	TECH	(→ Einteilung)	astrophysics	PHYS	Astrophysik
assignment (→ transfer)	COLLOQ	(→ Übertragung)	A subscriber (→ calling subscriber)	SWITCH	(→ rufender Teilnehmer)
assignment (→ allocation)	TECH	(→ Zuordnung)			
assignment list (→ allocation list)	SWITCH	(→ Beschaltungsliste)	asymmetrical	MATH	asymmetrisch
assignment mode	TELEC	Zuteilungsverfahren	asymmetrical feeding	ANT	Gleichtaktspeisung
assignment operator	DATA PROC	Zuweisungsoperator	asymptote	MATH	Asymptote
assignment statement	DATA PROC	Zuweisungsanweisung	[line approached by a curve without beeing touched]		[Gerade der sich eine Kurve nähert ohne erreicht zu werden]
assistance (→ after-sales service)	ECON	(→ Kundendienst)			
			asymptotic	MATH	asymptotisch
associate	ECON	Teilhaber	asymptotic distribution	MATH	Grenzverteilung
= partner		= Partner	asymptotic focal length	MATH	Asymptotenbrennweite
associated	COLLOQ	zugehörig	asymptotic point	MATH	asymptotischer Punkt
≈ respective		≈ betreffend	↑ singularity		↑ Singularität
associated mode (→ associated signaling)	SWITCH	(→ assoziierte Zeichengabe)	asymptotic stability	CONTROL	asymptotische Stabilität
			async (→ asynchronous)	PHYS	(→ asynchron)
associated signaling (AM)	SWITCH	assoziierte Zeichengabe	async (→ asynchronous)	CIRC.ENG	(→ ungetaktet)
= associated signalling (BRI); associated mode		= assoziierte Signalisierung	asynchronous	PHYS	asynchron
			= async		
associated signalling (BRI) (→ associated signaling)	SWITCH	(→ assoziierte Zeichengabe)	asynchronous	CIRC.ENG	ungetaktet
			= async		≠ taktgesteuert
association	ECON	Verband	≠ clock-pulse controlled		
= federation			asynchronous [with different clock]	TELEC	asynchron [nicht taktgleich]
associative memory (→ associative storage)	DATA PROC	(→ Assoziativspeicher)	≈ plesiochronous		= nichtsynchron
			≠ synchronous		≈ plesiochron
associative storage	DATA PROC	Assoziativspeicher			≠ synchron
[the information is retrieved with a piece of it, and not through an address]		[die Information wird nicht über Adressen, sondern durch Angabe eines Teils von ihr gefunden]	asynchronous access (→ asynchronous operation)	TELEGR	(→ Asynchronbetrieb)
			asynchronous balanced mode	DATA COMM	gleichberechtigter Spontanbetrieb
= associative memory; content-addressable storage; content-addressable memory; content-addressed storage; content-addressed memory; CAM; data-addressed storage; data-addressed memory; parallel-search memory; parallel-search storage; search memory		= inhaltsadressierter Speicher; CAM; Durchrufspeicher	= balanced mode; ABM		= ABM
			asynchronous communications (→ asynchronous operation)	TELEGR	(→ Asynchronbetrieb)
		↑ Direktzugriffsspeicher	asynchronous communications interface adapter (→ ACIA)	DATA PROC	(→ asynchroner Schnittstellenadapter)
↑ random access memory			asynchronous counter	CIRC.ENG	Asynchronzähler
assortment	ECON	Sortiment			= asynchroner Zähler
= range; line; gang		≈ Produktspektrum			≠ Synchronzähler
≈ product line					

asynchronous disconnected DATA COMM **unabhängiger Wartezustand**
mode = ADM
= ADM

asynchronous machine POWER ENG **Asynchronmotor**
= induction motor = Asynchronmaschine

asynchronous machine POWER SYS (→Induktionsmaschine)
(→induction machine)

asynchronous mode TELEGR (→Asynchronbetrieb)
(→asynchronous operation)

asynchronous motor POWER SYS (→Induktionsmaschine)
(→induction machine)

asynchronous network TELEC **asynchrones Netz**
= nichtsynchrones Netz

asynchronous operation TELEGR **Asynchronbetrieb**
[each character contains individual in- [jedes einzelne Zeichen ent-
formation for synchronism of re- hält eigene Informationen
ceiver with transmitter] für den Gleichlauf von
= asynchronous principle; asyn- Empfänger mit Sender]
chronous transmission; asynchronous = Asynchronverfahren;
mode; asynchronous communica- Asynchronübertragung;
tions; asynchronous access Nichtsynchronbetrieb;
↓ start-stop operation Nichtsynchronübertra-
gung
↓ Start-Stop-Betrieb

asynchronous operation TELEC (→asynchrones Übertra-
(→asynchronous transfer mode) gungsverfahren)

asynchronous principle TELEGR (→Asynchronbetrieb)
(→asynchronous operation)

asynchronous principle TELEC (→asynchrones Übertra-
(→asynchronous transfer mode) gungsverfahren)

asynchronous response DATA COMM **Spontanbetrieb**
mode
= ARM

asynchronous TDM TELEC (→asynchrone Multiplextech-
(→asynchronous time division multi- nik)
plexing)

asynchronous time division TELEC **asynchrone Multiplextechnik**
multiplexing = ATD
= ATD; asynchronous TDM

asynchronous transfer mode TELEC **asynchrones Übertragungs-**
= ATM; asynchronous operation; asyn- **verfahren**
chronous principle = asynchroner Übertragungs-
≈ asynchronous transmission; asyn- modus; ATM;
chronous operation [DATA COMM] Asynchronbetrieb; Asynchronverfahren; Asyn-
chronübermittlung
≈ Asynchronübertragung;
Asynchronbetrieb [DATA
COMM]

asynchronous transmission TELEC **Asynchronübertragung**
≈ asynchronous transfer mode ≈ asynchrones Übertragungsverfahren

asynchronous transmission TELEGR (→Asynchronbetrieb)
(→asynchronous operation)

at (→technical atmosphere) PHYS (→technische Atmosphäre)
At (→astatine) CHEM (→Astat)
at a favourable price ECON (→preiswert)
(→cheap)

ATB (→external blocking) SWITCH (→äußere Blockierung)
AT compatible DATA PROC **AT-kompatibel**
AT computer DATA PROC **AT-Computer**
[Advanced Technology; a computer [mit 16-Bit- oder 32-Bit-
working with 16 bit or 32 bit proces- Prozessoren arbeitender
sors] Computer]
= PCAT; PC/AT = AT-Rechner; PCAT;
PC/AT

AT cut COMPON **AT-Schnitt**
[quartz] [Quarz]

ATD (→asynchronous time TELEC (→asynchrone Multiplextech-
division multiplexing) nik)

ATDE (→adaptive time NETW.TH (→adaptiver Zeitbereichsent-
domain equalizer) zerrer)

ATE (→automatic tester) INSTR (→Prüfautomat)
at intervals (→periodic) SCIE (→periodisch)
atlas GEOPHYS **Atlas** (pl.Atlanten; Atlasse)
[bound collection of correlated maps] [gebundene Sammlung kor-
≈ map; globe relierter Landkarten]
≈ Landkarte; Globus

ATM (→asynchronous transfer TELEC (→asynchrones Übertra-
mode) gungsverfahren)

atm (→physical atmosphere) PHYS (→physische Atmosphäre)
ATM cell TELEC **ATM-Zelle**
ATME TELEPH **ATME**
[automatic transmission and signaling [automatische Meß- und
testing equipment for international Prüfeinrichtung für interna-
telephone trunks] tionale Fernsprechleitun-
gen]

ATMOS MICROEL **ATMOS**
= adjustable threshold metal oxide
semiconductor

atmosferic humidity (→air PHYS (→Luftfeuchtigkeit)
humidity)

atmosphere GEOPHYS **Atmosphäre**
atmospheric GEOPHYS **atmosphärisch**
atmospheric discharge METEOR **atmosphärische Entladung**
= thunderstroke; thunderbolt; thunder- ≈ Gewitter
stone ↓ Blitz; Donner
≈ thunderstorm
↓ lightning; thunder

atmospheric layer GEOPHYS **atmosphärische Schicht**
atmospheric noise RADIO (→atmosphärisches Ge-
(→statics) räusch)

atmospheric noise METEOR (→atmosphärische Störung)
(→atmospheric perturbation)

atmospheric perturbation METEOR **atmosphärische Störung**
= atmospheric noise

atmospheric precipitation METEOR **atmosphärischer Niederschlag**
= precipitation = Niederschlag
↓ rainfall; snowfall ↓ Regenfall; Schneefall

atmospheric pressure PHYS **Luftdruck**
= barometric pressure; air pressure

atmospheric radiowave RADIO PROP (→Raumwelle)
(→sky wave)

atmospheric whistler RADIO **Whistler**
[whistler in the VLF band] [Pfeifton im Myriameterbe-
reich]

ATM switching SWITCH **ATM-Vermittlungstechnik**
ATM switching matrix SWITCH **ATM-Koppelfeld**
ATM switching system SWITCH **ATM-Vermittlungseinrich-**
tung

at negative pole (→negative) PHYS (→negativ)
A-to-D converter CODING (→A/D-Wandler)
(→analog-to-digital converter)

atom PHYS **Atom**
atom DATA PROC **Atom**
[unreduceable element of a list or a [nicht unterteilbares Ele-
file] ment einer Liste oder Da-
tei]

atom diameter PHYS **Atomdurchmesser**
atomic bond PHYS **atomare Bindung**
atomic core PHYS **Atomkern**
= atomic nucleus; nucleus; core = Kern

atomic cross-sectional area PHYS **atomarer Wirkungsquer-**
schnitt

atomic distance PHYS **Atomabstand**
= atomic spacing

atomic frequency standard PHYS **Atomfrequenznormal**
= Atomuhr

atomic lattice PHYS **Atomgitter**
atomic mass PHYS **Atommasse**
atomic mass number PHYS **Massenzahl**
[number of protons plus neutrons] [Anzahl Protonen plus
= mass number; nuclear number; nu- Neutronen eines Kerns]
cleon number

atomic mass unit PHYS **atomare Einheitsmasse**
[Konstante]

atomic model PHYS **Atommodell**
atomic nucleus (→atomic PHYS (→Atomkern)
core)

atomic number PHYS **Atomnummer**
= Atomzahl; Ordnungszahl;
Kernladungszahl

atomic radius PHYS **Atomradius**
atomic spacing (→atomic PHYS (→Atomabstand)
distance)

atomic spectrum PHYS **Atomspektrum**
atomic standard INSTR **Atomnormal**
= atomares Normal

atomic structure PHYS **Atomaufbau**
= Atomstruktur

atomic torso	PHYS	**Atomrumpf**
atomic weight	PHYS	**Atomgewicht**
atomization (→pulverization)	TECH	(→Zerstäubung)
atomize (→pulverize)	TECH	(→zerstäuben)
ATPC (→automatic transmit power regulation)	RADIO	(→automatische Sendeleistungsregelung)
at positive pole (→positive)	PHYS	(→positiv)
at random (→arbitrary)	TECH	(→willkürlich)
at receiving side	TELEC	**empfangsseitig**
ATR switch	RADIO LOC	**Sperrschalter**
= anti-transmission-reception switch		
ATR tube (→anti-transmitting receiving tube)	RADIO LOC	(→Sendersperröhre)
AT sign	ECON	**Klammeraffe**
[@]		[@]
= AT symbol; EACH sign; EACH symbol		= Pro-Stück-Zeichen
AT symbol (→AT sign)	ECON	(→Klammeraffe)
att (→care of)	OFFICE	(→zu Händen)
attach (v.t.)	TECH	**anbringen**
= link (v.t.)		≈ befestigen
≈ fasten		
attachment	TECH	**Anbringung**
≈ fastening		≈ Befestigung
attachment	DATA COMM	**zusätzlicher Anschluß**
attachment	EQUIP.ENG	**Zusatz**
= add-on (n.); add-in (n.); extra (n.)		
attainment	COLLOQ	**Erreichung**
		[fig.]
		= Erzielung
attempt (→trial)	TECH	(→Versuch)
attendance (of persons) (→participation)	COLLOQ	(→Teilnahme)
attendant (→operator)	TELEPH	(→Beamtin)
attendant (→operator)	TECH	(→Bedienungsperson)
attendant console	SWITCH	**Abfragestelle**
= console		
attendant console (→operator's console)	TECH	(→Bedienplatz)
attendant recall (→operator recall)	SWITCH	(→Platzherbeiruf)
attendant's board (→control desk)	TECH	(→Kontrollpult)
attendant's console (→control desk)	TECH	(→Kontrollpult)
attendant's desk (→control desk)	TECH	(→Kontrollpult)
attendant's position (→operator's position)	TECH	(→Bedienplatz)
attendant's terminal (→operator's position)	TECH	(→Bedienplatz)
attendant telephone station	TELEPH	**Vermittlungsfernsprecher**
attended (→staffed)	TELEC	(→bemannt)
attendee	ECON	**Teilnehmer**
		[einer Besprechung o.dgl.]
attention (AM) (→care of)	OFFICE	(→zu Händen)
attenuation	LINE TH	**Dämpfung**
= transmission loss; loss		= Verlust; Abschwächung
attenuation (→decrement)	PHYS	(→Dämpfung)
attenuation box (→step attenuator)	INSTR	(→Stufendämpfungsglied)
attenuation characteristic	NETW.TH	**Dämpfungscharakteristik**
attenuation coefficient	PHYS	**Schwächungskoeffizient**
attenuation coefficient (BRI)	LINE TH	(→Dämpfungskonstante)
(→attenuation constant)		
attenuation constant	LINE TH	**Dämpfungskonstante**
= attenuation coefficient (BRI)		[Realteil der Fortpflanzungskonstante]
		= Dämpfungsmaß; Dämpfungsbelag; Dämpfungskoeffizient
attenuation constant	NETW.TH	(→Kettendämpfung)
(→iterative attenuation constant)		
attenuation curve	NETW.TH	**Dämpfungskurve**
attenuation diode	MICROW	**Abschwächungsdiode**
attenuation distortion	TELEC	**Dämpfungsverzerrung**
attenuation edge	NETW.TH	**Dämpfungsflanke**
attenuation equalizer	NETW.TH	**Dämpfungsentzerrer**
attenuation factor	NETW.TH	**Dämpfungsfaktor**
[input to output value; reciprocal of transmission coefficient]		[Eingangsgröße zu Ausgangsgröße; Kehrwert des Übertragungsfaktors]
= attenuation ratio; damping coefficient; damping factor; attenuation function; damping function		= Dämpfungskoeffizient; Dämpfungsfunktion
≠ transmission coefficient		≠ Übertragungsfaktor
↓ image attenuation constant; effective attenuation constant; voltage attenuation constant; current attenuation constant; power attenuation constant		↓ Wellendämpfungsfaktor; Betriebsdämpfungsfaktor; Spannungsdämpfungsfaktor; Stromdämpfungsfaktor; Leistungsdämpfungsfaktor
attenuation fluctuation	TELEC	(→Dämpfungsschwankungen)
(→attenuation variation)		
attenuation function	NETW.TH	(→Dämpfungsfaktor)
(→attenuation factor)		
attenuation measurement	INSTR	**Dämpfungsmessung**
attenuation notch	NETW.TH	**Dämpfungskerbe**
attenuation plan (→transmission plan)	TELEC	(→Dämpfungsplan)
attenuation pole	NETW.TH	**Dämpfungspol**
attenuation ratio (→attenuation factor)	NETW.TH	(→Dämpfungsfaktor)
attenuation variation	TELEC	**Dämpfungsschwankungen**
= attenuation fluctuation		
attenuator	NETW.TH	**Dämpfungsglied**
= pad; resistance pad		= Abschwächer
↓ L section; star section; delta section; balanced T section; lattice section; extension line [TELEC]		↓ L-Schaltung; T-Schaltung; Pi-Schaltung; H-Schaltung; X-Schaltung; Verlängerungsleitung [TELEC]
attenuator calibration set	INSTR	**Abschwächer-Kalibrier-System**
attenuator probe	INSTR	**Tastteiler**
↑ test probe		↑ Prüfspitze
attenuator set	INSTR	**Abschwächersatz**
attestation (→certification)	ECON	(→Beglaubigung)
attested by notary	ECON	**notariell beglaubigt**
attitude (→position 1)	TECH	(→Lage)
atto..	MATH	**atto..**
[10E-18]		[10E-18]
attorney 1 (→lawyer)	ECON	(→Rechtsanwalt)
attorney-at-law (AM) (→lawyer)	ECON	(→Rechtsanwalt)
attorneyship	ECON	**Prokura**
[legal appointment for commercial transactions]		[handelsrechtliche Vollmacht eines Angestellten]
↑ procuration		↑ Vollmacht
attraction	PHYS	**Anziehung**
attractive force	PHYS	**Anziehungskraft**
attribute	LING	**Beifügung**
≈ attibutive		= Attribut
		≈ Attributivum
attribute	DATA PROC	**Attribut**
attribute check	QUAL	**Attributprüfung**
attribute domain (→domain)	DATA PROC	(→Domäne)
attributive	LING	**Attributivum**
≈ attribute		≈ Beifügung
attributive word	LING	**Bestimmungswort**
Au (→gold)	CHEM	(→Gold)
AU (→administrative unit)	TELEC	(→Verwaltungseinheit)
auction (n.)	ECON	**Versteigerung**
auction (v.t.) (→sell by auction)	ECON	(→versteigern)
auctioneer	ECON	**Versteigerer**
audibility	ACOUS	**Hörbarkeit**
audibility acuity	ACOUS	**Hörschärfe**
audibility range	ACOUS	**Hörbarkeitsbereich**
= audible range		= Hörbereich
audibility threshold	ACOUS	**Hörbarkeitsschwelle**
= threshold of hearing		= Hörschwelle; Schallschwelle
audible	ACOUS	**hörbar**
		= vernehmbar
audible (→acoustic)	PHYS	(→akustisch)
audible alarm (→acoustical alarm)	EQUIP.ENG	(→akustischer Alarm)
audible noise (→noise)	ACOUS	(→Geräusch)

audible range

audible range (→audibility range) ACOUS (→Hörbarkeitsbereich)
audible signal TELEPH Hörton
= audible tone; tonality ↓ Wählton; Freiton; Besetztton; Hinweiston
audible signal (→sound signal) EL.ACOUS (→Schallsignal)
audible sound (→sound 2) ACOUS (→Hörschall)
audible spectrum EL.ACOUS Tonfrequenzspektrum
audible tone (→audible signal) TELEPH (→Hörton)
audience COLLOQ Zuhörerschaft
audience rating BROADC Einschaltquote
audio (adj.) TELEC tonfrequent
audio amplifier EL.ACOUS Tonfrequenzverstärker
audio amplifier (→voice frequency amplifier) TRANS (→NF-Verstärker)
audio analyzer INSTR Audioanalysator
[signal source and distortion analyzer for audio measurements] [Generator und Verzerrungsmesser für Tonsignalmessungen]
↑ signal analyzer ↑ Signalanalysator
audio band EL.ACOUS Tonfrequenzband
audio cassette EL.ACOUS Tonbandkassette
↑ magnetic tape cassette [TERM&PER] = Tonbandcassette; Tonkassette; Toncassette; Musikkassette; Musiccassette
↑ Magnetbandkassette [TERM&PER]
audio channel (→sound channel) BROADC (→Tonkanal)
audioconference (→audio conferencing) TELEC (→Fernsprechkonferenz)
audio conferencing TELEC Fernsprechkonferenz
= audioconference ↑ Telekonferenz
↑ teleconferencing
audio connecting cable CONS.EL NF-Verbindungskabel
= Niederfrequenz-Verbindungskabel
audio connector COMPON NF-Steckverbinder
= Niederfrequenz-Steckverbinder; Diodenstecker
audio console EL.ACOUS Tonregiesystem
audio control engineer EL.ACOUS (→Toningenieur)
(→sound engineer)
audio correlator EL.ACOUS NF-Korrelator
audio delay BROADC NF-Verzögerung
audio detection system SIGN.ENG (→Geräuschmeldesystem)
(→sound sensing detector system)
audio dub CONS.EL Nachvertonung
[video recorder] [Videorecorder]
audio electronics (→audio frequency electronics) ELECTRON (→NF-Elektronik)
audio engineering (→sound engineering) EL.ACOUS (→Tontechnik)
audio frequency EL.ACOUS Tonfrequenz
[a frequency between 20 Hz and 20 kHz] [eine Frequenz zwischen 20 Hz und 20 kHz]
audio frequency (→voice frequency) TELEC (→Niederfrequenz)
audio frequency amplifier TRANS NF-Verstärker
(→voice frequency amplifier)
audiofrequency; audio frequency EL.TECH Niederfrequenz
[range of frequencies which are audible when converted to acoustic waves; from 30 Hz to 20 kHz] [Bereich von Frequenzen die hörbar sind, wenn in akustische Wellen gewandelt; von 30 Hz bis 20 kHz]
≈ voice frequency [TELEC] = NF
audio frequency cable COMM.CABLE NF-Kabel
= VF cable = Niederfrequenzkabel
audio frequency electrolytic capacitor COMPON Tonfrequenz-Elko
audio frequency electronics ELECTRON NF-Elektronik
= audio electronics = Niederfrequenzelektronik
audio frequency engineering TELEC (→NF-Technik)
(→voice-frequency engineering)
audio frequency generator INSTR NF-Generator
= audio generator = Niederfrequenz-Generator
audio frequency impedance converter CIRC.ENG NF-Impedanzwandler
= Niederfrequenz-Impedanzwandler

audio frequency level TELEC (→NF-Pegel)
(→voice-frequency level)
audio frequency millivoltmeter INSTR NF-Millivoltmeter
= Niederfrequenz-Millivoltmeter
audiofrequency power amplifier CIRC.ENG NF-Endverstärker
= Niederfrequenz-Endverstärker
audio frequency preamplifier CIRC.ENG NF-Vorverstärker
= Niederfrequenz-Vorverstärker
audio frequency signal (→voice frequency signal) TELEC (→NF-Signal)
audiofrequency transistor (→audio transistor) MICROEL (→NF-Transistor)
audio generator (→audio frequency generator) INSTR (→NF-Generator)
audio level indication EL.ACOUS NF-Pegelanzeige
audio mixer (→mixing console) BROADC (→Mischpult)
audio oscillator EL.ACOUS Tonfrequenzgenerator
audio program (→sound programm) TELEC (→Tonprogramm)
audio signal TV Audiosignal
audio signal (→sound signal) EL.ACOUS (→Schallsignal)
audio squelch BROADC NF-Squelch
audio tape recorder CONS.EL Tonbandmaschine
↑ tape recorder = Spulentonbandgerät; Bandmaschine
↑ Tonbandgerät
audio transformer COMPON NF-Übertrager
= Niederfrequenzübertrager
audio transistor MICROEL NF-Transistor
[cutoff frequency < 100 kHz] [Grenzfrequenz < 100 kHz]
= audiofrequency transistor = Niederfrequenztransistor
audiovision INF.TECH Audiovision
audio wattmeter INSTR NF-Leistungsmesser
= Niederfrequenz-Leistungsmesser; NF-Wattmeter
audit (n.) ECON Revision
= auditing [Betriebsprüfung]
= Audit
audit (→re-check) TECH (→nachprüfen)
audit (n.) (→re-check) TECH (→Nachprüfung)
auditing (→audit) ECON (→Revision)
auditing firm ECON Prüfungsgesellschaft
auditor ECON Wirtschaftsprüfer
= accountant (BRI); chartered accountant = Revisor
≈ controller ≈ kaufmännischer Direktor
audit trail DATA PROC Prüfpfad
= Rückverfolgungspfad
augend MATH erster Summand
[a number to which an addend is added] [Zahl zu der eine andere dazugezählt wird]
↑ addend 2 = Augend
↑ Summand 2
Auger effect PHYS Augereffekt
augment (→enlarge) TECH (→vergrößern)
augmenter (→addend 1) MATH (→zweiter Summand)
aural carrier (→sound carrier) EL.ACOUS (→Tonträger)
aurora borealis (→northern light) GEOPHYS (→Nordlicht)
auroral scatter RADIO PROP Polarlichtstreuung
auspices COLLOQ Schirmherrschaft
austenitic steel METAL austenitischer Stahl
autarchic (→independent) TECH (→unabhängig)
autarkic (→independent) TECH (→unabhängig)
authentic (→genuine) COLLOQ (→authentisch)
authentication center MOB.COMM Berechtigungszentrum
authority ECON Behörde
≈ administration 2 ≈ Verwaltung 2
authority to sign ECON Unterschriftsberechtigung
= Unterschriftsvollmacht
authorization TYPOGR Freigabevermerk
authorization (→entitlement) ECON (→Berechtigung)
authorization control DATA PROC (→Befugniskontrolle)
(→security control)

authorized TELEPH **berechtigt**
= classmarked ↓ vollamtsberechtigt; halb-
↓ non-restricted; outward restricted; amtsberechtigt; nichtamts-
fully restricted berechtigt
authorized access DATA PROC **befugter Zugriff**
= befugter Zugang
authorized class of service TELEC **Anschlußberechtigung**
= user service category
auto-answer (→automatic TERM&PER (→Anrufbeantworter)
answering equipment)
auto-answer (→answer DATA COMM (→Antwortbetrieb)
mode)
autobahn (→freeway) CIV.ENG (→Autobahn)
auto-calibration INSTR **automatische Kalibrierung**
auto-call DATA COMM **automatischer Verbindungs-**
= automatic call set-up; automatic con- **aufbau**
nection set-up; automatic link set-up = automatische Verbindungsherstellung
autocorralation function SIGN.TH **Autokorrelationsfunktion**
autocorrelation MATH **Eigenkorrelation**
= self-correlation = Autokorrelation; Selbstkor-
relation
autocorrelation SIGN.TH **Autokorrelation**
autocorrelation analysis SIGN.TH **Autokorrelationsanalyse**
[Signalverarbeitung]
autocorrosion (→self-corrosion) CHEM (→Eigenkorrosion)
autodecremental ELECTRON **selbstzurückschaltend**
= selbstdekrementierend;
selbstzurückrückend
autodelimitation MICROEL **Selbstabgrenzung**
autodiagnostic TECH **Selbstdiagnose**
auto-dialing (adj.) (AM) TELEC **selbstwählend**
= auto-dialling (BRI); self-dialing
auto-dialling (BRI) TELEC (→selbstwählend)
(→auto-dialing)
auto-dial modem DATA COMM **Selbstwahlmodem**
autogenous welding METAL **Autogenschweißung**
autoincremental ELECTRON **selbstvorschaltend**
= selbstinkrementierend;
selbstvorrückend
autointerrupting ELECTRON (→selbstunterbrechend)
(→self-interrupting)
autointerruption ELECTRON (→Selbstunterbrechung)
(→self-interruption)
auto-load DATA PROC **Selbstladefunktion**
auto mains failure system POWER SYS (→Notstromversorgung)
(→emergency power supply)
automata (→automata theory) INF (→Automatentheorie)
automata theory INF **Automatentheorie**
= automate theory; automata
automated materials handling CONTROL (→AMH)
(→AMH)
automated office (→electronic INF.TECH (→automatisiertes Büro)
office)
automate theory (→automata INF (→Automatentheorie)
theory)
automatic TECH **automatisch**
= self-acting; self-actuated; self-opera- = selbsttätig
ting; self-operated ≈ adaptiv; alleinbetriebsfähig
≈ adaptive; independently operating
automatic answering TERM&PER **Anrufbeantworter**
equipment
= telephone answering machine;
answering machine; auto-answer;
telephone recorder
automatic balancing ELECTRON (→selbstabgleichend)
(→self-adjusting)
automatic balancing bridge INSTR **selbstabgleichende Brücke**
automatic barrier close RAILW.SIGN **Bahnselbstschlußanlage**
= Basa
automatic blanking TERM&PER **Bildschirmabschaltung**
= Schirmabschaltung
automatic breaker (→circuit COMPON (→Schutzschalter)
breaker)
automatic call TELEC **Selbstwählverbindung**
automatic callback TELEPH **automatischer Rückruf**
= camp-on
automatic call set-up DATA COMM (→automatischer Verbin-
(→auto-call) dungsaufbau)

automatic call transfer TELEPH **automatische Rufweiterlei-**
[if call is not answered within a fixed **tung**
time] [wenn nicht innerhalb ei-
≈ call forwarding ner bestimmten Zeit abge-
hoben wird]
= selbsttätige Rufweiterleitung
≈ Rufumleitung
automatic cash dispenser TERM&PER **Geldausgabeautomat**
= Geldautomat; Bankautomat
automatic check EQUIP.ENG **Selbsttest**
= built-in check
automatic circuit breaker COMPON (→Sicherungsautomat)
(→automatic cutout)
automatic coding DATA PROC (→automatische Program-
(→machine-aided programming) mierung)
automatic connection TELEPH **automatischer Verbindungs-**
= hot line **aufbau**
= Hot-line
automatic connection set-up DATA COMM (→automatischer Verbin-
(→auto-call) dungsaufbau)
automatic control CONTROL **automatische Steuerung**
= Selbststeuerung
automatic control (→control TECH (→Steuer-und Regelungs-
engineering) technik)
automatic cutout COMPON **Sicherungsautomat**
= automatic circuit breaker; cutout [thermomagnetische Strom-
↓ automatic cutout with signal contact sicherung, deren Reaktions-
zeit vom Überstrom
abhängt]
↓ Fernmelde-Schutzschalter
automatic cutout with signal COMPON **Fernmelde-Schutzschalter**
contact [Sicherungsautomat mit Si-
↑ automatic cutout cherungskontakt]
↑ Sicherungsautomat
automatic data processing INF.TECH (→elektronische Datenverar-
(→electronic data processing) beitung)
automatic decimal point DATA PROC **Komma-Automatik**
capability
automatic error correction DATA COMM **automatische Fehlerkorrektur**
automatic exchange SWITCH **Wählvermittlungsstelle**
↑ switching center = Wählvermittlung; Wählamt
↑ Vermittlungsamt
automatic fading compensation RADIO **Schwundregelung**
≈ automatic gain control
automatic film processor OFFICE **Entwicklungsautomat**
automatic formatting [DATA TYPOGR (→Blocksatz)
PROC] (→justified typesetting)
automatic frequency control CIRC.ENG **Frequenznachregelung**
= AFC = automatische Frequenz-
nachsteuerung; automat-
ische
Frequenznachstellung;
automatische Frequenzre-
gelung; AFC; Abstimmau-
tomatik; Abstimmungsrege-
lung
automatic gain control CONTROL **automatische Verstärkungsre-**
= AGC **gelung**
= automatische Verstärkungs-
nachregelung
automatic gain control amplifier CIRC.ENG (→Regelverstärker)
(→regulating amplifier)
automatic homing RADIO NAV **Eigenpeilung**
= automatischer Zielanflug
automatic hyphenation DATA PROC **automatische Silbentrennung**
= intelligent spacer
automatic insertion machine MANUF (→Bestückungsautomat)
(→automatic placement system)
automatic landing system RADIO NAV **automatisches Landesystem**
automatic lathe TECH **Werkzeugautomat**
automatic level control ELECTRON **automatische Pegelregelung**
= ALC; automatic leveling = ALC
automatic leveling ELECTRON (→automatische Pegelrege-
(→automatic level control) lung)
automatic line break DATA PROC **automatischer Zeilenumbruch**
[word processing] [Textverarbeitung]
= automatic word wrapping; automatic
word wrap; word wrap; wraparound

automatic link set-up (→auto-call) DATA COMM (→automatischer Verbindungsaufbau)
automatic loader DATA PROC **automatischer Lader**
automatic machine (→automaton) INF.TECH (→Automat)
automatic measuring equipment = measuring automat INSTR **Meßautomat**
automatic mode TECH **automatischer Betrieb**
automatic network (→switched network) TELEC (→Wählnetz)
automatic operation (→switched operation) TELEC (→Wählbetrieb)
automatic PCB artwork creation (→auto-routing) DATA PROC (→automatische Leiterplattenentflechtung)
automatic placement machine (→automatic placement system) MANUF (→Bestückungsautomat)
automatic placement system = automatic placement machine; automatic insertion machine MANUF **Bestückungsautomat** = Bestückautomat
automatic programming (→machine-aided programming) DATA PROC (→automatische Programmierung)
automatic program relocation (→dynamic program relocation) DATA PROC (→dynamische Programmverschiebung)
automatic protection switch (→protection switching equipment) TRANS (→Schutzschalteinrichtung)
automatic quality control (→CAQ) DATA PROC (→CAQ)
automatic receiver RADIO **Automatik-Empfänger**
automatic redialing key TERM&PER **Wiederwahltaste** = Wahlwiederholungstaste
automatic relocation (→dynamic program relocation) DATA PROC (→dynamische Programmverschiebung)
automatic repeat request = automatic request; ARQ; automatic request for repetition DATA COMM **automatische Wiederholanforderung** = automatische Wiederholung; ARQ-Verfahren
automatic request (→automatic repeat request) DATA COMM (→automatische Wiederholanforderung)
automatic request for repetition (→automatic repeat request) DATA COMM (→automatische Wiederholanforderung)
automatic restart (→power fail) DATA PROC (→Wiedereinschaltautomatik)
automatic restoral TRANS **automatische Zurückschaltung**
automatic sequencing (→control program) DATA PROC (→Organisationsprogramm)
automatic service (→switched operation) TELEC (→Wählbetrieb)
automatic shutdown DATA PROC **automatischer Ausschalter**
automatic station finder CONS.EL **Suchlaufautomatik** = Kanalsuchlauf
automatic switchover TELEPH **Selbstumschaltung**
automatic test equipment (→automatic tester) INSTR (→Prüfautomat)
automatic tester = test automat; automatic test equipment; ATE INSTR **Prüfautomat** = Testautomat; automatische Prüfeinrichtung
automatic track control RAILW.SIGN **induktive Zugsicherung** = Indusi; automatische Zugsicherung
automatic train stop RAILW.SIGN **Zwangsbremsung**
automatic transmit power regulation = ATPC RADIO **automatische Sendeleistungsregelung** = ATPC
automatic tuning CONS.EL **Abstimmautomatik**
automatic tuning mechanism INSTR **Abstimmautomatik**
automatic typewriter OFFICE **Schreibautomat**
automatic winding machine [coil manufacturing] ≈ winding machine MANUF **Wickelautomat** [Spulenfertigung] ≈ Wickelmaschine
automatic word wrap (→automatic line break) DATA PROC (→automatischer Zeilenumbruch)
automatic word wrapping (→automatic line break) DATA PROC (→automatischer Zeilenumbruch)
automatic zero-current controller INSTR **Nullstromregler**
automation (n.plt.) = automatization (n.plt.) ≈ robotization TECH **Automatisierung** = Automation ≈ Robotisierung

automation engineering = automatization POWER ENG **Automatisierungstechnik**
automation level TECH **Automatisierungsgrad**
automatism TECH **Automatismus**
automatization (n.plt.) (→automation) TECH (→Automatisierung)
automatization (→automation engineering) POWER ENG (→Automatisierungstechnik)
automaton (pl. automata) = automatic machine ≈ robot [CONTROL] INF.TECH **Automat** ≈ Roboter [CONTROL]
automobile (→motor vehicle) TECH (→Kraftfahrzeug)
automobile protection SIGN.ENG **Kfz-Sicherung**
automonitor (n.) DATA PROC **Selbstkontrollprogramm**
automotive TECH **selbstbeweglich**
automotive electronics (→car electronics) ELECTRON (→Fahrzeugelektronik)
autonomous (→stand-alone) DATA PROC (→allein operierend)
autonulling INSTR **automatische Nullpunkteinstellung** = automatische Nullung
auto-parallel connection POWER SYS **Auto-Parallelschaltung**
autopilot RADIO NAV **Kursgeber**
auto-poll DATA COMM **automatischer Sendeaufruf**
autoprotective (→fail-safe) TECH (→selbstschützend)
autoradio (→car radio) RADIO (→Autoradio)
autoranging INSTR **automatische Meßbereichswahl** = automatische Bereichswahl
autoregression MATH **Eigenregression** = Autoregression
auto-repeat (→repeat function) TERM&PER (→Wiederholfunktion)
auto-restart (n.) DATA PROC **Neustartfunktion**
auto-reverse CONS.EL **automatische Spurumschaltung** = Autoreverse
auto-routing [of conductor paths on PCB] = automatic PCB artwork creation DATA PROC **automatische Leiterplattenentflechtung** [von Leiterbahnen auf Leiterplatten] = automatische Entflechtung; Autorouting
auto-scaling INSTR **Autoskalierung**
autoscanning (→station search) CONS.EL (→Sendersuchlauf)
autoscore (n.) DATA PROC **Unterstreichfunktion**
autosequence programm DATA PROC **automatisches Ablaufprogramm**
auto-series connection POWER SYS **Auto-Serienschaltung**
autostore CONS.EL **Senderkralle**
autosync display (→multi-frequency monitor) TERM&PER (→Autosync-Monitor)
autosync monitor (→multi-frequency monitor) TERM&PER (→Autosync-Monitor)
autotracking ANT **automatische Antennennachführung**
autotracking CONTROL **automatische Nachführung** = Nachführung
autotransformer EL.TECH **Spartransformator** = Autotransformator; Spartrafo
auto vernier INSTR **automatische Feineinstellung**
auto-zeroing (→self-calibrating) ELECTRON (→selbsteichend)
auto-zeroing (→self-calibration) ELECTRON (→Selbsteichung)
auxiliary = ancillary TECH **Hilfs-**
auxiliary (n.) (→auxiliary verb) LING (→Hilfsverb)
auxiliary address register SWITCH **Hilfsadreßregister**
auxiliary air = supplementary air TECH **Hilfsluft**
auxiliary amplifier CIRC.ENG **Hilfsverstärker**
auxiliary antenna ≈ provisional antenna ANT **Hilfsantenne** ≈ Behelfsantenne
auxiliary bridge INSTR **Hilfsbrücke**
auxiliary capacity (→overhead capacity) TRANS (→Zusatzkapazität)

auxiliary channel	TELEC	**Hilfskanal**	**average** (adj.) (→mean)	MATH	(→durchschnittlich)
= sub-channel		≈ Rückkanal	**averaged**	MATH	**gemittelt**
≈ backward channel			**averaged measurement**	INSTR	**Messung mit Mittelwertbildung**
auxiliary circuit	CIRC.ENG	**Hilfskreis**	= averaged mode		
= fall-back circuit					= gemittelte Messung
auxiliary circuit	EL.TECH	**Hilfsstromkreis**	**averaged mode** (→averaged measurement)	INSTR	(→Messung mit Mittelwertbildung)
		= Hilfskreis			
auxiliary controlled variable	CONTROL	(→Hilfsregelgröße)	**average meter**	INSTR	**Mittelwertmesser**
(→objective variable)			**average noise factor**	HF	**Band-Rauschzahl**
auxiliary electrode	ELECTRON	**Hilfselektrode**	**average outgoing quality**	QUAL	**mittlere Auslieferqualität**
auxiliary information	TELECONTR	**Hilfsinformation**	= AOQ; outgoing fraction defective		= AOQ; Durchschlupf
auxiliary material	TECH	**Hilfsstoff**	**average picture signal level**	TV	**Bildsignalmittelwert**
auxiliary memory (→external memory)	DATA PROC	(→Externspeicher)	**average rectifier**	CIRC.ENG	**Mittelwertgleichrichter**
auxiliary quantity	PHYS	**Hilfsgröße**	**average total value**	PHYS	**Halbschwingungsmittelwert**
auxiliary register	DATA PROC	**Hilfsregister**	**average transmission rate** (→mean date rate)	DATA COMM	(→mittlere Übertragungsgeschwindigkeit)
= scratch pad		≈ Notizblockspeicher			
= scratch pad memory			**average value** (→average)	MATH	(→Mittel)
auxiliary resource manager	SWITCH	**Betriebsmittelverwaltung**	**averaging** (n.)	MATH	**Mittelwertbildung**
= resource management					= Mittelwertbestimmung; Durchschnittsbildung
auxiliary rope	OUTS.PLANT	**Hilfsseil**	**averaging** (adj.)	MATH	**mittelwertbildend**
auxiliary shaft	MECH	**Hilfswelle**			= durchschnittbildend
auxiliary store (→external memory)	DATA PROC	(→Externspeicher)	**avionics**	ELECTRON	**Luftfahrtelektronik**
auxiliary supply	CIRC.ENG	**Hilfsstromquelle**	**Avogadro constant 1**	CHEM	**Avogadro-Konstante 1**
auxiliary system (→ancillary system)	TECH	(→Hilfssystem)	[number of molecules per mole]		[Anzahl Moleküle je Mol]
			= Avogadro number 1; Loschmidt number 2		= Avogadro-Zahl 1; Loschmidtsche Zahl 2
auxiliary verb	LING	**Hilfsverb**	≈ Loschmidt number 1		≈ Loschmidtsche Zahl 1
= auxiliary (n.)		≠ Hilfszeitwort	**Avogadro number 2** (→Loschmidt number 1)	CHEM	(→Loschmidtsche Zahl 1)
		≠ Vollverb			
auxiliary view	ENG.DRAW	**Hilfsansicht**	**Avogadro number 1** (→Avogadro constant 1)	CHEM	(→Avogadro-Konstante 1)
auxiliary voltage source	CIRC.ENG	**Hilfsspannungsquelle**			
= auxiliary voltage supply			**avoirdupois**	PHYS	**Sechzehn-Unzen-System**
auxiliary voltage supply (→auxiliary voltage source)	CIRC.ENG	(→Hilfsspannungsquelle)	[series of units of mass based on the pound of 16 ounces]		[auf das Pound von 16 Unzen basierendes angelsächsisches Meßsystem für Masse]
availability	QUAL	**Verfügbarkeit**	= avoirdupois weight		
= disponibility; operation ratio; up-time ratio; UTR			**avoirdupois weight** (→avoirdupois)	PHYS	(→Sechzehn-Unzen-System)
availability (→accessibility)	SWITCH	(→Erreichbarkeit)	**award 1** (n.)	ECON	**Zuschlag 2**
availability time (→accessibility time)	SWITCH	(→Verfügbarkeitsdauer)			[Ausschreibung]
					= Zuspruch; Vergabe 1
available	COLLOQ	**verfügbar**	**award 2 n.** (→arbitration award)	ECON	(→Schiedsspruch)
≈ free; existent; in stock		≈ frei; vorhanden; vorrätig	**AWG** (→American Wire Gauge)	METAL	(→amerikanische Drahtlehre)
avalanche	PHYS	**Lawine**			
avalanche amplification	PHYS	**Lawinenverstärkung**	**A wire** (→tip wire)	TELEPH	(→A-Ader)
avalanche breakdown	MICROEL	**Lawinendurchbruch**	**axial**	MATH	**axial**
		= Avalanche-Durchbruch	≠ radial		≠ radial
avalanche breakdown voltage (→avalanche voltage)	MICROEL	(→Lawinendurchbruch-Spannung)	**axial fan**	TECH	**Axiallüfter**
			= axial flow fan		
avalanche diode (→avalanche photodiode)	MICROEL	(→Lawinenphotodiode)	**axial field**	PHYS	**axiales Feld**
					= Axialfeld
avalanche effect	PHYS	**Lawineneffekt**	**axial flow fan** (→axial fan)	TECH	(→Axiallüfter)
		= Avalanche-Effekt	**axial force**	PHYS	**Axialkraft**
avalanche-induced migration	MICROEL	**AIM**			= axiale Kraft
= AIM			**axial lead**	COMPON	**Axialanschluß**
avalanche injection	MICROEL	**Lawineneinspritzung**			= axialer Anschlußdraht
		= Stoßinjektion	**axial load**	MECH	**Axiallast**
avalanche multiplication	PHYS	**Lawinenmultiplikation**			= axiale Last
avalanche photodiode	MICROEL	**Lawinenphotodiode**	**axially parallel**	MECH	**achsparallel**
= APD; avalanche diode		= Avalanche-Photodiode; Lawinendiode; Avalanche-Diode; APD-Diode	**axially symmetric**	MATH	**axialsymmetrisch**
↓ impatt diode; trapatt diode			**axially vertical**	MECH	**achssenkrecht**
			axial ratio (→ellipticity)	MATH	(→Ellipizität)
		↓ Impatt-Diode; Trapatt-Diode	**axial section**	ENG.DRAW	**Axialschnitt**
avalanche transistor	MICROEL	**Lawinentransistor**	**axial travel**	MECH	**Axialbewegung**
avalanche transit time diode	MICROEL	(→Impatt-Diode)	**axiom**	SCIE	**Axiom**
(→impact avalanche transit diode)			**axiomatic**	SCIE	**axiomatisch**
avalanche voltage	MICROEL	**Lawinendurchbruch-Spannung**	**axis** (pl.-es)	MATH	**Achse**
= avalanche breakdown voltage			**axis of reference**	ENG.DRAW	**Bezugsachse**
↑ breakdown voltage		↑ Durchbruchspannung	= datum axis		
average (v.t.)	MATH	**mitteln**	**axle**	MECH	**Achse**
		= Mittelwert bilden; Mittelwert bestimmen; Durchschnitt bilden; Durchschnitt bestimmen	**axle coupling**	MECH	**Achskupplung**
			Ayrton shunt	INSTR	**Ayrton-Nebenwiderstand**
			= universal shunt		= Mehrfachnebenwiderstand
			AZERTY keyboard (→french keyboard)	TERM&PER	(→französische Tastatur)
average	MATH	**Mittel**	**azimuth**	MATH	**Azimut**
= average value; mean value; mean		= Mittelwert; Durchschnitt; Durchschnittswert; Schnitt 1	**azimuthal**	MATH	**azimutal**
≈ midpoint		≈ Medianwert; Mittelwert	**azimuthal analyzer**	RADIO	**Azimut-Analysator**

azimuthal angle	RADIO	**Azimutwinkel**
azimuthal diagram (→azimuthal radiation pattern)	ANT	(→Azimutaldiagramm)
azimuthal mode	ANT	**Azimutalschwingung**
azimuthal radiation pattern	ANT	**Azimutaldiagramm**
= azimuthal diagram		
azimuth indicating goniometer	RADIO LOC	**Azimutpeilvorrichtung**
azimuth marker	RADIO NAV	**Azimutgeber**
azur blue	PHYS	**azurblau**

B

B (→boron)	CHEM	(→Bor)
b (→barn)	PHYS	(→Barn)
BA (→ISDN basic access)	TELEC	(→ISDN-Basisanschluß)
Ba (→barium)	CHEM	(→Barium)
babble	TELEPH	**Babbeln**
[crosstalk from many channels]		[Nebensprechen aus mehreren Kanälen]
Babinet's principle	EL.TECH	**Babinetsches Prinzip**
baby	POWER SYS	**Baby**
		[Zellengröße ca 26 mm Durchmesser x 50 mm]
Baby Bell (→Regional Bell Operating Company)	TELEC	(→regionale Bell-Betreibergesellschaft)
baby board (→expansion board)	EQUIP.ENG	(→Erweiterungskarte)
baby module (→expansion board)	EQUIP.ENG	(→Erweiterungskarte)
back (n.)	TYPOGR	**Rücken**
		= Buchrücken
back 2 (→rear side)	TECH	(→Rückseite)
back 1 (n.) (→rear part)	TECH	(→Hinterteil)
back bearing	RADIO NAV	**Abflugpeilung**
= outbound bearing		
backbone (→backbone route)	TRANS	(→Haupttrasse)
backbone artery (→backbone route)	TRANS	(→Haupttrasse)
backbone line-of-sight (→backbone microwave)	RADIO REL	(→Weitverkehrsrichtfunk)
backbone LOS (→backbone microwave)	RADIO REL	(→Weitverkehrsrichtfunk)
backbone microwave	RADIO REL	**Weitverkehrsrichtfunk**
= backbone radio relay; backbone line-of-sight; backbone LOS; long-haul microwave; long-haul radio relay; long-haul line-of-sight; long-haul LOS; long-distance radio relay; long-distance microwave; long-distance line-of-sight; long-distance LOS		≈ Breitbandrichtfunk
≈ high-capacity radio		
backbone network (→toll network)	TELEC	(→Fernnetz)
backbone radio relay (→backbone microwave)	RADIO REL	(→Weitverkehrsrichtfunk)
backbone route	TRANS	**Haupttrasse**
= backbone artery; backbone		= Hauptarterie
≈ long-haul route		≈ Ferntrasse
backbone system (→long-haul system)	TRANS	(→Weitverkehrssystem)
back conductance	MICROEL	**Sperrleitwert**
back coupling (→feedback)	CIRC.ENG	(→Rückkopplung)
back cover	TYPOGR	**Rückendeckel**
back-diffusion	PHYS	**Rückdiffusion**
back direction (→reverse direction)	ELECTRON	(→Sperrichtung)
back electron	ELECTRON	**Rückelektron**
back-end processor	DATA PROC	**Spezialhilfsprozessor**
↑ auxiliary processor		
backfire antenna	ANT	**Backfire-Antenne**
backfire radiation	ANT	**Rückwärtswellen-Anregung**
back gate bias (→substrate bias)	MICROEL	(→Substratvorspannung)
background (n.)	TECH	**Hintergrund**
≠ foreground (n.)		≠ Vordergrund
background BER (→background bit error rate)	CODING	(→Grundbitfehlerrate)
background bit error rate	CODING	**Grundbitfehlerrate**
= background BER; residual bit error rate; residual BER		
background brightness	TV	**Grundhelligkeit**
background colour	TECH	**Hintergrundfarbe**
background noise	TELEPH	**Hintergrundrauschen**
background noise (→basic noise)	TELEC	(→Grundgeräusch)
background processing	DATA PROC	**Hintergrundverarbeitung**
[takes place when priority programs are inactive]		[findet statt wenn Programme höchster Priorität inaktiv sind]
≠ foreground processing		≠ Vordergrundverarbeitung
background program	DATA PROC	**Hintergrundprogramm 1**
[a program with low priority, or running unperceivable to the user]		[Programm niedrigster Priorität, oder vom Anwender unbemerkt ablaufend]
= low priority program		= Hintergrundprozess
≠ priority program		≈ Hauptprogramm
		≠ Prioritätsprogramm
background program memory	DATA PROC	**Hintergrundprogramm 2**
[section of main memory]		[Bereich des Hauptspeichers]
background system	DATA PROC	**Hintergrundsystem**
back heating	ELECTRON	**Rückheizung**
= backheating		
backheating (→back heating)	ELECTRON	(→Rückheizung)
backhoe fading (sl.) (→cable disruption)	OUTS.PLANT	(→Kabelbruch)
backing memory 2 (→scratch pad memory)	DATA PROC	(→Notizblockspeicher)
backing memory 1 (→external memory)	DATA PROC	(→Externspeicher)
backing storage 2 (→scratch pad memory)	DATA PROC	(→Notizblockspeicher)
backing storage 1 (→external memory)	DATA PROC	(→Externspeicher)
backing store 2 (→scratch pad memory)	DATA PROC	(→Notizblockspeicher)
backing store 1 (→external memory)	DATA PROC	(→Externspeicher)
backing-up (→saving)	DATA PROC	(→Sicherung)
backing-up area (→saving area)	DATA PROC	(→Sicherungsbereich)
backlash	MECH	**Flankenspiel**
		= mechanisches Spiel; Hystereseeffekt
backlash	ELECTRON	**Gitterrückstrom**
back-lighting	TV	**Hintergrundbeleuchtung**
= bias lightning; backlit		
backlit (adj.)	TECH	**hinterleuchtet**
backlit (→back-lighting)	TV	(→Hintergrundbeleuchtung)
backlit LCD	TERM&PER	**Backlit-LCD**
		[Flüssigkristallschirm mit Hintergrundbeleuchtung]
back lobe	ANT	**Rückwärtskeule**
↑ side lobe		= Hinterkeule
		↑ Nebenzipfel
backlog 1 (n.)	COLLOQ	**Rückstand**
[fig.]		[fig.]
= arrears		
backlog 2 (n.)	COLLOQ	**Überhang**
[fig.]		[fig.]
backlog (n.)	DATA PROC	**Verarbeitungsrückstand**
[not yet processed]		
backlog demand	ECON	**Nachholbedarf**
back-off (n.)	ELECTRON	**Unteraussteuerung**
= power derating		= Leistungsminderung; Leistungsreduzierung
back-off (→feedback)	CIRC.ENG	(→Rückkopplung)
backplane	EQUIP.ENG	**Rückwandleiterplatte**
= mother board [DATA PROC]; rear-panel board		= Rückwandplatine
backplane cable connector	COMPON	**Rückwandkabelstecker**
= rearpanel cable connector		
backplane wiring	EQUIP.ENG	**Rückwandverdrahtung**
= rearpanel wiring		
back plate	EQUIP.ENG	**Rückwand**
= rear cover; rear panel; rearpanel		
back plate (→rear cover)	TECH	(→rückseitige Abdeckung)
back porch	TV	**hintere Schwarzschulter**

back pressure PHYS **Gegendruck**
= counterpressure
backscattering PHYS **Rückstreuung**
= Rückwärtsstreuung
backscattering cross section RADIO LOC (→Radarquerschnitt)
(→radar cross section)
backslant TYPOGR **linksgeneigte Schrägschrift**
≠ slant ≠ Schrägschrift
backslash TYPOGR **verkehrter Schrägstrich**
[\] [\]
= reverse solidus; reverse slant = Backslash; inverser Schrägstrich; Schrägstrich nach links; Schrägstrich rückwärts
≠ slash ≠ Scrägstrich
backspace (v.t.) TERM&PER **rücksetzen**
= zurücksetzen
backspace DATA COMM **Rückwärtsschritt**
[code] [Code]
= backwards step; back step; BS = BS
backspace key TERM&PER **Rücksetztaste 2**
[moves the cursor one step to the left and removes the character situated there, all character at the right shifting by; generally labeled with BS or ←] [bewegt die Schreibmarke einen Schritt nach rechts und entfernt das dort vorhandene Zeichen, wobei alle Zeichen rechts davon nachrücken; meist mit BS oder ← beschriftet]
= reset key 2; INST/DEL key; insert/delete key; cancel key; erase key; clear key = Rückstelltaste 2; Rückschrittstaste; Rückwärtsschritt-Taste; Rücktaste; INST/DEL-Taste; Taste INST/DEL; BACKSPACE-Taste; Taste BACKSPACE; BS-Taste; Taste BS; Annulliertaste; Löschtaste; Radiertaste
≈ DEL key; correction key; reset key ≈ ENTF-Taste; Korrekturtaste; Rücksetztaste
backspacing TERM&PER **Rücksetzung**
= Zurücksetung; Rücksetzen; Zurücksetzen
back step (→backspace) DATA COMM (→Rückwärtsschritt)
backstop MECH **Rückanschlag**
back stroke MECH **Rückwärtshub**
back-to-back SYS.INST **Rücken-an-Rücken**
back-to-back credit ECON **Gegenakkreditiv**
back-to-back installation SYS.INST **Rücken-an-Rücken-Aufbau**
backtracing TECH **Zurückverfolgen**
= backtracking = Zurückverfolgung
backtrack (v.t.) DATA PROC **rückwärtsverarbeiten**
backtracking DATA PROC **Rückwärtsverarbeitung**
backtracking (→backtracing) TECH (→Zurückverfolgen)
backup (→data protection) DATA PROC (→Datensicherung)
back-up (→stand-by) TELEC (→Ersatz)
backup (v.t.) (→save) DATA PROC (→sichern)
back-up (→saving) DATA PROC (→Sicherung)
backup area (→saving area) DATA PROC (→Sicherungsbereich)
backup capacitor CIRC.ENG **Speicherkondensator**
backup copy DATA PROC **Sicherungskopie**
≈ backup file = Reservekopie; Backup-Kopie; Sicherheitskopie; Sicherstellungskopie
≈ Sicherungsdatei
backup disk DATA PROC **Sicherungsdiskette**
≠ work disk ≠ Arbeitsdiskette
backup file DATA PROC **Sicherungsdatei**
= security file = Sicherstellungsdatei; Bakkup-Datei
≈ backup copy ≈ Sicherungskopie
back-up line TELEC **Ersatzleitung**
backup memory DATA PROC **Sicherungsspeicher**
[for backup copies of data and programs] [zur Aufnahme einer Sicherungskopie von Daten und Programmen]
= backup storage = Backup-Speicher; Reservespeicher; Sicherstellungsspeicher
↑ external memory ↑ externer Speicher

backup power plant POWER SYS (→Notstromversorgung)
(→emergency power supply)
backup storage (→backup memory) DATA PROC (→Sicherungsspeicher)
backup track (→alternate track) TERM&PER (→Ersatzspur)
Backus-Naur form DATA PROC **Backus-Naur-Form**
= BNF = BNF
backward (adv.) TECH **rückwärts**
≠ forward adv. ≠ vorwärts
backward (→rear) TECH (→rückseitig)
backward barring (→revertive blocking) ELECTRON (→rückwärtige Sperre)
backward blocking ELECTRON (→rückwärtige Sperre)
(→revertive blocking)
backward blocking voltage ELECTRON **Rückwärtsblockierspannung**
backward busy signal SWITCH **Rückwärts-Belegtkennzeichen**
backward chaining DATA PROC **Rückwärtskettung**
backward channel TELEC **Rückkanal**
≈ reverse channel ≈ Hilfskanal
≈ auxiliary channel ≠ Hauptkanal
≠ main channel
backward clearance (→backward release) SWITCH (→Rückwärtsauslösung)
backward diode MICROEL **Rückwärtsdiode**
= uni tunnel diode = Backward-Diode; Uni-Tunneldiode
backward direction (→reverse direction) ELECTRON (→Sperrichtung)
backward indicator SWITCH **Rückwärts-Indikatorbit**
[CCITT Nr.7] [CCITT Nr.7]
backward read TERM&PER **Rückwärtslesen**
backward recovery DATA PROC **Datenrückgewinnung durch Rückwärtsverarbeitung**
backward recovery time MICROEL (→Sperrverzögerungszeit)
(→reverse recovery time)
backward release SWITCH **Rückwärtsauslösung**
= backward clearance; clear backward = Rückwärtsauslösen
↑ release ↑ Auslösung
backward scan (→rear scanning) ELECTRON (→Rückwärtsabtastung)
backward scanning (→rear scanning) ELECTRON (→Rückwärtsabtastung)
backward sequence number SWITCH **Rückwärts-Folgenummer**
[CCITT Nr.7] [CCITT Nr.7]
backward signal SWITCH **Rückwärtszeichen**
= response signal = Rückwärtskennzeichen; Antwortkennzeichen
↑ switching signal ↑ Kennzeichen
backwards step DATA COMM (→Rückwärtsschritt)
(→backspace)
backwards supervision DATA COMM (→Rückwärtssteuerung)
(→backward supervision)
backward supervision DATA COMM **Rückwärtssteuerung**
[receiver controlled] [empfängergesteuerte Übertragung]
= backwards supervision = Rückwärtsüberwachung
backward tabulator TERM&PER **Rücktabulator**
backward tone DATA COMM **Bestätigungston**
backward wave LINE TH **Rückwärtswelle**
backward wave oscillator MICROW **Rückwärtswellenröhre**
= BWO; carcinotron = Rückwärtswellenoszillator
↑ traveling-field tube ↑ Lauffeldröhre
baconing lights AERON **Flugwarnlichter**
bad acoustic ACOUS **schlechte Akustik**
= schlechte Hörsamkeit
badge TECH **Abzeichen**
= emblem = Emblem; Kennmarke
badge reader TERM&PER **Ausweisleser**
bad sector TERM&PER **unbrauchbarer Sektor**
= fehlerbehafteter Sektor
bad soldering point (→cold soldering point) ELECTRON (→kalte Lötstelle)
bad weather METEOR **Schlechtwetter**
≠ fair weather ≠ Schönwetter
baffle EQUIP.ENG **Luftleitblech**
baffle EL.ACOUS **Schallwand**
= Schallschirm

baffle

baffle		TECH	Schutzschirm		
			= Schutzwand		
bag		TECH	Sack		
= sack					
bag phone		MOB.COMM	Rucksack-Mobiltelefon		
[packed in a shoulder case]			↑ Zellulartelefon		
↑ cellular phone					
bakelite		ELECTRON	Bakelite		
bakelized paper (→laminated paper)		ELECTRON	(→Hartpapier)		
balance (v.t.)		MECH	auswuchten		
balance (v.t.)		TECH	wägen		
			= wiegen; ausgleichen (v.t.)		
balance (v.i.)		TECH	ausgleichen (v.r.)		
balance (n.)		TECH	Waage		
balance		ECON	Saldo		
balance		NETW.TH	Symmetrie		
= impedance balance; symmetry			≠ Unsymmetrie		
≠ unbalance					
balance control (→fade control)		CONS.EL	(→Überblendregler)		
balanced		EL.TECH	symmetrisch		
= symmetrical; symmetric			= erdsymmetrisch		
balanced aerial (→balanced antenna)		ANT	(→symmetrische Antenne)		
balanced amplifier (→push-pull amplifier)		CIRC.ENG	(→Gegentaktverstärker)		
balanced antenna		ANT	symmetrische Antenne		
= balanced aerial					
balanced code		CODING	symmetrischer Code		
balanced duplexer		RADIO LOC	symmetrischer Duplexer		
balanced error		DATA PROC	gleichwahrscheinlicher Fehler		
balanced feeder cable (→ribbon cable)		COMM.CABLE	(→Bandkabel)		
balanced line		LINE TH	erdsymmetrische Leitung		
= symmetrical line			= symmetrische Leitung		
balanced mixer		MICROW	Gegentaktmischer		
= push-pull mixer			= Brückenmischer		
balanced mode		DATA COMM	(→gleichberechtigter Spontanbetrieb)		
(→asynchronous balanced mode)					
balanced modulator		CIRC.ENG	Gegentaktmodulator		
= push-pull modulator			= Gegentaktumsetzer		
balanced pair		COMM.CABLE	symmetrisches Aderpaar		
= symmetrical pair			= symmetrisches Paar		
balanced pairs cable		COMM.CABLE	erdsymmetrisches Kabel		
balanced state		SCIE	Gleichgewichtszustand		
= equilibrium state					
balanced station		DATA COMM	Hybridstation		
= combined station					
balanced to ground		LINE TH	erdsymmetrisch		
balanced transistor		MICROEL	Gegentakttransistor		
balanced T section		NETW.TH	Viereckschaltung		
= twin-T network			= H-Schaltung; H-Glied; Doppel-T-Schaltung; Doppel-T-Glied		
↑ transverse-symmetric two-port; attenuator					
			↑ quersymmetrischer Vierpol; Dämpfungsglied		
balanced-unbalanced (→balun)		ANT	(→Balun)		
balance error		INSTR	Nullpunktfehler		
= zero error; zero deviation; zero variation			= Nullpunktabweichung		
balance of payments		ECON	Zahlungsbilanz		
balance out (→equalize)		TECH	(→ausgleichen)		
balance return loss (→active return loss)		NETW.TH	(→Reflexionsdämpfung)		
balance sheets		ECON	Bilanz		
= financial statement			= Bilanzierung		
balance-to-unbalance transformer (→balun)		EL.TECH	(→Symmetrietransformator)		
balancing		NETW.TH	Abgleich		
[conjugate different circuit elements]			[Angleichen der Übertragungseigenschaften verschiedener Schaltkreise]		
= equalize; compensate					
			= Ausgleich		
balancing capacitor (→trimming capacitor)		COMPON	(→Trimmerkondensator)		
balancing circuit (→line-balancing network)		TELEC	(→Leitungsnachbildung)		
balancing circuit (→simulation network)		CIRC.ENG	(→Nachbildung)		
balancing element		NETW.TH	Abgleichelement		
balancing network (→line-balancing network)		TELEC	(→Leitungsnachbildung)		
balancing potentiometer (→adjusting potentiometer)		ELECTRON	(→Abgleichpotentiometer)		
balancing resistor		ELECTRON	Abgleichwiderstand		
= trimming resistor			≈ veränderbarer Widerstand [COMPON]		
≈ adjustable resistor [COMPON]					
balck-and-white TV tube		ELECTRON	Schwarz-Weiß-Fernsehbildröhre		
↑ balck-and-white tube; picture tube					
			↑ Schwarz-Weiß-Bildröhre; Bildwiedergaberöhre		
ball		TECH	Kugel		
= sphere; globe					
ball (→sphere)		MATH	(→Kugel)		
ball-and-socket joint		MECH	Kugelgelenk		
ball bearing		MECH	Kugellager		
ball bonding (→nailhead bonding)		MICROEL	(→Nagelkopfbondierung)		
ball burnish		MECH	kugelpolieren		
ballistic		PHYS	ballistisch		
ballistic galvanometer		INSTR	ballistisches Galvanometer		
ballistic measuring instrument		INSTR	ballistisches Meßgerät		
ballistic moving-coil galvanometer		INSTR	ballistisches Drehspulgalvanometer		
			= Stromstoßgalvanometer		
ball key		MECH	Kugelpaßfeder		
ball-like (→spheroidal)		MATH	(→sphäroid)		
balloon antenna		ANT	Ballonantenne		
ball-point (→biro)		OFFICE	(→Kugelschreiber)		
ball-point pen (→biro)		OFFICE	(→Kugelschreiber)		
ball printer (→type-ball printer)		TERM&PER	(→Kugelkopfdrucker)		
Balmer series		PHYS	Balmer-Serie		
balun		ANT	Balun		
= balanced-unbalanced			= Symmetrieglied		
↓ rod-core balun; toroidal-balun transformer; coiled balun			↓ Stabkern-Balun; Ringkern-Balun; Spulen-Balun		
balun		EL.TECH	Symmetrietransformator		
[device to couple a balanced to an unbalanced system]			[zur Anpassung eines symmetrischen an ein unsymmetrisches System]		
= balance-to-unbalance transformer			= Symmetrieglied		
≈ cable choke			≈ Mantelwellensperre		
bamboo cable		COMM.CABLE	Ballontube		
↑ coaxial tube			↑ Koaxialtube		
banana pin (→banana plug)		COMPON	(→Bananenstecker)		
banana plug		COMPON	Bananenstecker		
= banana pin					
band		PHYS	Band		
↓ energy band; valence band; conduction band			↓ Energieband; Valenzband; Leitungsband		
band (→frequency band)		RADIO	(→Frequenzband)		
band antenna		ANT	Bereichsantenne		
band bending		MICROEL	Bandverbiegung		
band broadening		PHYS	Bandverbreiterung		
band center		RADIO	Bandmitte		
= band centre (BRI); midband 2			≈ Mittelband		
≈ midband 1					
band center frequency		RADIO	Bandmittenfrequenz		
= midband frequency					
band center frequency (→midband frequency)		NETW.TH	(→Bandmittenfrequenz)		
band centre (BRI) (→band center)		RADIO	(→Bandmitte)		
band change (→wave selection)		RADIO	(→Wellenbereichswahl)		
band-change switch (→wave-change switch)		RADIO	(→Wellenbereichsschalter)		
band edge		PHYS	Bandkante		
= energy band edge; band rim			= Energiebandkante		
band edge (→band limit)		RADIO	(→Bandgrenze)		
band edit		DATA PROC	Bandauszug		
band efficiency		RADIO	Bandausnutzung		
≈ band utilization			≈ Bandnutzung		
band-elimination filter (→bandstop filter)		NETW.TH	(→Bandsperrfilter)		
band filter (→bandpass filter)		NETW.TH	(→Bandpaßfilter)		

band gap = gap	PHYS	**Bandabstand**
banding (→elastic banding)	DATA PROC	(→Bildeinschnürung)
band limit = band edge	RADIO	**Bandgrenze** ↓ Bandanfang; Bandende
band limitation	PHYS	**Bandbegrenzung**
band-limited	PHYS	**bandbegrenzt**
bandpass (→bandpass filter)	NETW.TH	(→Bandpaßfilter)
bandpass filter = bandpass; band filter ↓ zigzag filter	NETW.TH	**Bandpaßfilter** = Bandfilter; Bandpaß ↓ Zickzackfilter
band printer (→belt printer)	TERM&PER	(→Banddrucker)
band rim (→band edge)	PHYS	(→Bandkante)
band selection (→wave selection)	RADIO	(→Wellenbereichswahl)
band selector = tape selector	CONS.EL	**Bandsortenschalter** = Bandsorten-Wahlschalter
band selector = band switch	RADIO	**Bereichsumschalter**
band setting oscillator	RADIO	**Bandsetzoszillator**
band sharing	RADIO	**gemeinschaftliche Bandnutzung**
band sorting	DATA PROC	**Bandsortierung**
band spreading	RADIO	**Bandspreizung**
bandstop (→bandstop filter)	NETW.TH	(→Bandsperrfilter)
bandstop filter = bandstop; band-elimination filter; tuned trap [HF]; trap [HF]; notched filter [INSTR]; slot filter [INSTR]	NETW.TH	**Bandsperrfilter** = Bandsperre; Trap [HF]
band switch (→band selector)	RADIO	(→Bereichsumschalter)
band-translated	NETW.TH	**bandübersetzt**
band utilization	RADIO	**Bandnutzung** = Bandbenutzung ≈ Bandausnutzung
bandwidth	PHYS	**Bandbreite**
bandwidth = band width	NETW.TH	**Bandbreite** = Durchlaßbreite
band width (→bandwidth)	NETW.TH	(→Bandbreite)
bandwidth-efficient	INF.TECH	**bandbreitenwirksam**
bandwidth-limited	NETW.TH	**bandbreitenbegrenzt**
bandwidth on demand	TELEC	**Bandbreite nach Wunsch**
bandwidth switching	INSTR	**Bandbreitenumschaltung**
bandwith-saving	INF.TECH	**bandbreitensparend**
bank ↑ financial institution	ECON	**Bank** ↑ Geldinstitut
bank ↑ memory unit	DATA PROC	**Bank** ↑ Speichereinheit
bank balance equiry terminal	TERM&PER	**Kontostand-Abfrageterminal**
bank code	ECON	**Bankleitzahl**
banked winding (→bank winding)	EL.TECH	(→verschachtelte Windung)
bank guarantee (→bank guaranty)	ECON	(→Bankgarantie)
bank guaranty = bank guarantee	ECON	**Bankgarantie**
bank money	ECON	**Girageld** = Buchgeld
banknote counting machine	OFFICE	**Banknotenzählmaschine**
banknote sorting machine	OFFICE	**Banknotensortiermaschine**
bankruptcy = liquidation	ECON	**Konkurs** = Bankrott; Pleite
bank select	DATA PROC	**Bankauswahl** [Speichererweiterung]
bank switching	DATA PROC	**Speicherblockumschaltung**
bank winding = banked winding ≈ series winding	EL.TECH	**verschachtelte Windung** ≈ Serienwicklung
banner headline (AM) (→headline 2)	LING	(→Schlagzeile)
bantam tube	ELECTRON	**Miniaturröhre**
bar	TV	**Balken**
bar 2 ≈ panel	MECH	**Schiene** = Blende
bar 1 = staff; stick ↓ rod	MECH	**Stab** = Stange ↓ Rundstab
bar [= 10,000 Pascal = 0.1 N/mm^2]	PHYS	**Bar** [= 10.000 Pascal = 0,1 N/mm^2] = bar
bar (→slash)	LING	(→Schrägstrich)
bar chart [representation by rectangles, with length proportional to the variable represented, mostly in vertical and side-by-side] = bar graph; column chart; columnar graph ≈ rod chart ↓ histogram	MATH	**Balkendiagramm** [Darstellung mit Rechtecken, deren Länge proportional zur zu veranschaulichenden Variable ist, meist vertikal und aneinander anstoßend] = Säulendiagramm; Säulenschaubild; Balkengraph; Säulengraphik ≈ Stabdiagramm ↓ Histogramm
bar clamp	MECH	**Schraubzwinge**
bar code = barcode; bar graphics (NAM)	DATA PROC	**Balkencode** = Strichcode; Balkenkode; Strichkode ↓ Artikelnummercode
barcode (→bar code)	DATA PROC	(→Balkencode)
bar code manual reader	TERM&PER	**Strichcode-Handleser** = Barcode-Handleser
bar code printer	TERM&PER	**Strichcode-Drucker** = Barcode-Drucker
bar-code reader = bar-code scanner	TERM&PER	**Strichcodeleser** = Strichcode-Lesegerät; Balkencodeleser; Barcode-Leser; Barcode-Scanner; Strichkodeleser; Strichkode-Lesegerät; Balkenkodeleser; Strichcode-Abtaster
bar-code scanner (→bar-code reader)	TERM&PER	(→Strichcodeleser)
bar controller	CONTROL	**Stabregler**
bare [wire] = uncoated; non-insulated	COMM.CABLE	**blank** [Draht] = unisoliert; unbeschichtet
bare board (→unassembled PCB)	ELECTRON	(→unbestückte Leiterplatte)
bare PCB (→unassembled PCB)	ELECTRON	(→unbestückte Leiterplatte)
bare wire (→bare wiring)	EL.TECH	(→Blankverdrahtung)
bare wiring = bare wire	EL.TECH	**Blankverdrahtung**
bargain (→negotiate)	ECON	(→verhandeln)
bargaining (→negotiation)	ECON	(→Verhandlung)
bar generator	TV	**Balkengenerator**
bar graph (→bar chart)	MATH	(→Balkendiagramm)
bar graphics (NAM) (→bar code)	DATA PROC	(→Balkencode)
bar iron	METAL	**Stangeneisen**
BARITT diode = barrier-injected transit time diode	MICROEL	**BARRIT-Diode**
barium = Ba	CHEM	**Barium** = Ba
Barker code	CODING	**Barker-Code**
Barker coded radar pulse	RADIO LOC	**Barker-codierter Radarimpuls**
Barkhausen effect = magnetic fluctuation noise	PHYS	**Barkhausen-Effekt**
Barkhausen formula	ELECTRON	**Barkhausensche Röhrenformel** [SRD = 1] = Barkhausen-Formel; Barkhausensche Röhrengleichung
Barkhausen jump	PHYS	**Barkhausen-Sprung**
Barlow's wheel	PHYS	**Barlowsches Rad**
barn [unit for atomic cross sectional areas; = 10 to the power of minus 24 cm^2] = b	PHYS	**Barn** [Maßeinheit für atomare Wirkungsquerschnitte; = 10 hoch minus 24 cm^2] = b
barometric pressure (→atmospheric pressure)	PHYS	(→Luftdruck)
bar pattern (→bar test pattern)	TV	(→Balkentestbild)
bar printer (→typebar printer)	TERM&PER	(→Typenstabdrucker)

barred zero

barred zero [ɸ] = slashed zero	DATA PROC	durchgestrichene Null [ɸ]	
barrel = vat ↑ receptacle	TECH	Tonne = Faß ↑ Behälter	
barrel distortion [optics] = positive distortion	PHYS	tonnenförmige Verzeichnung [Optik]	
barrel printer (→drum printer)	TERM&PER	(→Walzendrucker)	
barretter (→bolometer)	MICROW	(→Bolometer)	
barretter (→hydrogen-iron resistance)	PHYS	(→Eisen-Wasserstoff-Widerstand)	
barrier	RAILW.SIGN	Bahnschranke = Schranke	
barrier (obsol.) (→depletion layer)	MICROEL	(→Sperrschicht)	
barrier (→threshold value)	TECH	(→Schwelle)	
barrier-injected transit time diode (→BARITT diode)	MICROEL	(→BARRIT-Diode)	
barrier layer (obsol.) (→depletion layer)	MICROEL	(→Sperrschicht)	
barrier-layer capacitor (→junction capacitor)	COMPON	(→Sperrschichtkondensator)	
barrier-layer rectifier = dry rectifier	COMPON	Sperrschicht-Gleichrichter	
barrier system	RAILW.SIGN	Schrankensystem	
barring (→restriction)	SWITCH	(→Sperre)	
barring (→interlock)	DATA PROC	(→Verriegelung)	
barrister (→lawyer)	ECON	(→Rechtsanwalt)	
barrister-at-law (BRI) (→lawyer)	ECON	(→Rechtsanwalt)	
bar steel	METAL	Stabstahl	
barter (n.) (→barter business)	ECON	(→Tauschgeschäft)	
barter business = barter trade; barter transaction; barter (n.); counter-trading; switch transaction; exchange (n.)	ECON	Tauschgeschäft = Bartergeschäft; Tausch-Kompensationsgeschäft	
barter trade (→barter business)	ECON	(→Tauschgeschäft)	
barter transaction (→barter business)	ECON	(→Tauschgeschäft)	
bar test pattern = bar pattern	TV	Balkentestbild	
Barth key	MECH	Trapezpaßfeder	
Bartlett's bisection theorem (→Bartlett's theorem)	NETW.TH	(→Theorem von Bartlett)	
Bartlett's theorem = Bartlett's bisection theorem	NETW.TH	Theorem von Bartlett	
base 1 (n.) = radix; number base ≠ exponent ↑ power	MATH	Basis = Grundzahl 2; Radix ≠ Exponent ↑ Potenz	
base (n.) [main place of a company]	ECON	Hauptsitz = Hauptstandort	
base 1 (n.)	MATH	Grundfläche	
base [controls the carrier flux from emitter to collector] ↓ base zone; base region; base terminal	MICROEL	Basis [steuert den Ladungsträgerfluß zwischen Emitter und Kollektor] ↓ Basiszone; Basisbereich; Basisanschluß	
base = terminal base; lower end	ANT	Fußpunkt = Aufpunkt	
base	TECH	Grundfläche	
base 2 [number of digits used in a number system]	MATH	Basiszahl [Anzahl der im Zahlensystem verwendeten Zahlen]	
base (→socket)	COMPON	(→Fassung)	
base (→stand)	EQUIP.ENG	(→Fuß)	
base address = basic address; main address; reference address; referencial address	DATA PROC	Grundadresse = Basisadresse; Bezugsadresse; Referenzadresse	
base addressing (→relative addressing)	DATA PROC	(→relative Adressierung)	
base address register	DATA PROC	Basisadreßregister	
baseband = base band; BB	TRANS	Basisband = BB	
base band (→baseband)	TRANS	(→Basisband)	
baseband module	TRANS	Basisbandbaugruppe	
baseband transmission [transmission of pulse-shaped unmodulated signals] = limited distance modem transmission	DATA COMM	Basisbandtechnik [Übertragung pulsgeformter unmodulierter Datenströme] = Basisbandübertragung	
base bias	CIRC.ENG	Basisvorspannung	
base bulk resistance (→extrinsic base resistance)	MICROEL	(→Basisbahnwiderstand)	
base coil (→base loading coil)	ANT	(→Fußpunktspule)	
base connection (→base terminal)	MICROEL	(→Basisanschluß)	
base current	CIRC.ENG	Basisstrom	
base diffusion	MICROEL	Basisdiffusion	
base diffusion isolation process (→BDI process)	MICROEL	(→BDI-Verfahren)	
based in	ECON	mit Sitz in	
base doping	MICROEL	Basis-Dotierung	
base electrode	MICROEL	Basiselektrode	
base-emitter diode	MICROEL	Basis-Emitter-Diode	
base emitter voltage	MICROEL	Basis-Emitter-Spannung	
base excess current	CIRC.ENG	Basisüberschußstrom	
base feed = series feed	ANT	Fußpunktspeisung	
base isolating ring	MICROEL	Basis-Isolationskranz	
base language (→assembler language)	DATA PROC	(→Assemblersprache)	
base layer (→base zone)	MICROEL	(→Basiszone)	
base line	RADIO LOC	Standlinie	
base line [the imaginary line on which the caracters rest, only exceeded by descenders] ≠ cap line	TYPOGR	Schriftlinie [nur durch Unterlängen unterschrittene, imaginäre untere Begrenzungslinie der Buchstaben] = Grundlinie ≠ Oberlinie	
baseline offset	ELECTRON	Grundlinien-Offset	
baseline overshoot = trailing-edge overshoot ↑ transient	ELECTRON	Nachschwingung = Nachschwinger; Unterschwingung; Unterschwinger; nachlaufender Unterschwinger ↑ Übergangsvorgang	
base load	TECH	Grundlast	
base-load antenna	ANT	Grundlastantenne	
base loading (→base loading coil)	ANT	(→Fußpunktspule)	
base loading coil = base coil; base loading	ANT	Fußpunktspule	
base load power set	POWER SYS	Dauerstromzentrale	
base material = Grundstoff	TECH	Ausgangsstoff	
base material	MICROEL	Basismaterial	
basement (→cellar)	CIV.ENG	(→Keller)	
base metal	METAL	unedles Metall	
base notation (→radix notation)	MATH	(→Radixschreibweise)	
base on (v.i.)	SCIE	basieren = bestehen auf	
base open (→collector-emitter cut-off current)	MICROEL	(→Kollektor-Emitter-Reststrom)	
base pan	SYS.INST	Bodenwanne	
base part (→lower part)	TECH	(→Unterteil)	
base plate = bottom plate; mounting plate	MECH	Grundplatte = Trägerplatte; Bodenplatte	
base plate (→floor plate)	SYS.INST	(→Bodenplatte)	
base potential divider	CIRC.ENG	Basisspannungsteiler	
base processor	DATA COMM	Basisprozessor	
base region [base zone without both junctions] ≈ base zone	MICROEL	Basisbereich [Basiszone ohne die beidseitigen Sperrschichten] = Basisraum = Basiszone	
base register [contains address of program start]	DATA PROC	Basisregister [enthält Programmstartadresse]	
base SI unit (→SI base unit)	PHYS	(→SI-Basiseinheit)	

base station (→radio base station)	MOB.COMM	(→Funk-Basisstation)
base technology	TECH	Basistechnologie
base terminal	MICROEL	Basisanschluß
= base connection		
base thickness 1 (→base zone thickness)	MICROEL	(→Basisbreite)
base thickness 2 (→base width)	MICROEL	(→Basisweite)
base time constant	MICROEL	Basiszeitkonstante
base voltage	CIRC.ENG	Basisspannung
base width	MICROEL	Basisweite
[distance of emitter- to collector junction]		[Abstand Emitter-zu Kollektor-Sperrschicht]
= base thickness 2		= Basisdicke 2
≈ base zone thickness		
base zone	MICROEL	Basiszone
[region between emitter and collector zone, including both junctions]		[Bereich zwischen Emitterzone und Kollektorzone, einschl. beider Sperrschichten]
= base layer		= Basisschicht
≈ base region		≈ Basisbereich
↑ base		↑ Basis
base zone thickness	MICROEL	Basisbreite
= base thickness 1		[Dicke des Basisbereichs]
≈ base width		= Basisdicke 1
		≈ Basisweite
BASIC	DATA PROC	BASIC
[Beginners All-purpose Symbolic Instruction Code]		↑ problemorientierte Programmiersprache
↑ high-level programming language		
basic access (→ISDN basic access)	TELEC	(→ISDN-Basisanschluß)
basic access method	DATA PROC	Basiszugriffsmethode
= elementary access method		
basic address (→base address)	DATA PROC	(→Grundadresse)
basically	COLLOQ	grundsätzlich
= principally		= im Grunde; prinzipiell
≈ fundamentally		≈ fundamental
basic amplifier circuit (→basic amplifier connection)	CIRC.ENG	(→Verstärkergrundschaltung)
basic amplifier connection	CIRC.ENG	Verstärkergrundschaltung
= basic amplifier circuit; basic connection; basic circuit		= Grundschaltung
↓ basic transistor connection; basic tube connection		↓ Transistorgrundschaltung; Röhrengrundschaltung
basic arithmetic operation	MATH	Grundrechnungsart
↓ addition; subtraction; multiplication; division		= Grundrechenart; Grundrechenoperation
		↓ Addition; Subtraktion; Multiplikation; Division
basic cell (→core cell)	MICROEL	(→Kernzelle)
basic channel	TELEC	Basiskanal
[ISDN]		[ISDN]
= B channel; payload channel; bearer channel		= B-Kanal; Nutzkanal
basic characteristic (→fundamental property)	TECH	(→Grundeigenschaft)
basic character set	TELEGR	Primärzeichensatz
basic circuit (→basic amplifier connection)	CIRC.ENG	(→Verstärkergrundschaltung)
basic clock	TELEC	Grundtakt
basic code	DATA PROC	Grundcode
[the code operating directly a CPU]		[der die Zentraleinheit direkt steuernde Code]
basic concept	SCIE	Grundprinzip
basic configuration	EQUIP.ENG	Grundaufbau
		= Grundausführung; Grundausbau
basic connection (→basic amplifier connection)	CIRC.ENG	(→Verstärkergrundschaltung)
basic control loop	CONTROL	Grundregelkreis
= basic loop		
basic-coverage transmitter	BROADC	Grundnetzsender
basic dimension (→basic size)	ENG.DRAW	(→Grundmaß)
basic electronics	PHYS	Grundlagen-Elektronik
basic element	SCIE	Grundelement
		= Grundbaustein
basic entry	LING	Grundeintrag
[dictionary]		[Wörterbuch]
= main entry		= Haupteintrag
basic equipment	EQUIP.ENG	Grundgerät
= mainframe		
basic feature	TECH	Kernleistungsmerkmale
		= Hauptleistungsmerkmal; Grundleistungsmerkmal
basic features (→standard equipment)	TECH	(→Grundausstattung)
basic frequency (→fundamental frequency)	PHYS	(→Grundfrequenz)
basic group	TRANS	Grund-Primärgruppe
[FDM]		[TF-Technik]
= basic primary group; L group (NAM)		= Grundgruppe
≈ primary group		≈ Primärgruppe
basic instruction	DATA PROC	Grundbefehl
basic instruction key	TERM&PER	Grundbefehlstaste
basic logic operations	ENG.LOG	Grundverknüpfungsarten
basic loop (→basic control loop)	CONTROL	(→Grundregelkreis)
basic mastergroup	TRANS	Grund-Tertiärgruppe
[FDM]		[TF-Technik]
basic mode	DATA COMM	Grundmodus
		= Basismodus
basic model	TECH	Grundbauform
= basic version; standard model; standard version		= Standardbauform; Grundversion; Standardversion
≈ serial model; standard finish		≈ Seriengerät; Standardausführung
basic model	SCIE	Grundmodell
= fundamental model		
basic mode link control	DATA COMM	Grundverbindungssteuerung
basic noise	TELEC	Grundgeräusch
= idle noise; intrinsic noise; load-invariant noise; background noise; self-noise		= Wärmegeräusch
		≈ thermisches Rauschen; Verstärkerrauschen
≈ thermal noise; amplifier noise [CIRC.ENG]; intrinsic noise; noise floor		[CIRC.ENG]; Eigenrauschen; Störpegel
basic operating system	DATA PROC	Grundbetriebssystem
[basic version]		[Grundausstattung]
= BOS		= Basisbetriebsystem
basic oscillation (→fundamental wave)	PHYS	(→Grundschwingung)
basic pattern	RADIO REL	Grundraster
basic primary group (→basic group)	TRANS	(→Grund-Primärgruppe)
basic processing unit 2 (→processor 1)	DATA PROC	(→Prozessor 1)
basic property (→fundamental property)	TECH	(→Grundeigenschaft)
basic rate	TRANS	Basisrate
[64 kb/s]		[64 kbit/s]
= DS1; T1		
basic rate (→basic rental)	TELEC	(→Grundgebühr)
basic R&D	TECH	Grundsatzentwicklung
≈ basic research [SCIE]		≈ vorwettbewerbliche FuE
		≈ Grundlagenforschung [SCIE]
basic rental (AM)	TELEC	Grundgebühr
= rental charge; basic rate; basic subscription price; unit charge		≠ Gesprächsgebühr
≠ call fee		
basic research (→fundamental research)	SCIE	(→Grundlagenforschung)
basics (→principles)	SCIE	(→Grundlagen)
basic section	NETW.TH	Grundglied
basic service	TELEC	Grunddienst
= fundamental service; bearer service 2		= Basisdienst
↓ POT service		↓ konventioneller Fernsprechdienst
basic set	MATH	Grundmenge
basic setting	ELECTRON	Grundeinstellung
= basic setup		
basic setup (→basic setting)	ELECTRON	(→Grundeinstellung)
basic size	ENG.DRAW	Grundmaß
= basic dimension; basic value		
basic size	ENG.DRAW	Passungsgrundmaß

basic software

basic software (→system software) DATA PROC (→Systemsoftware)
basic structure (→framework) SCIE (→Gerüst)
basic subscription price (→basic rental) TELEC (→Grundgebühr)
basic supergroup [FDM] TRANS **Grund-Sekundärgruppe** [TF-Technik]
basic supermastergroup [FDM] TRANS **Grund-Quartärgruppe** [TF-Technik]
basic transmission loss RADIO PROP (→Freiraumdämpfung)
(→free-space attenuation)
basic tube circuit (→basic tube connection) CIRC.ENG (→Röhrengrundschaltung)
basic tube connection CIRC.ENG **Röhrengrundschaltung**
= basic tube circuit ↑ Verstärkergrundschaltung
↑ basic amplifier connection
basic unit PHYS **Grundeinheit**
basic unit DATA PROC (→Grundelement)
(→primitive 1)
basic value (→basic size) ENG.DRAW (→Grundmaß)
basic value (→initial value) DATA PROC (→Anfangswert)
basic version (→basic model) TECH (→Grundbauform)
basis charge MICROEL **Basisladung**
bass EL.ACOUS **Baß**
= Tiefen; Tiefton
bass control EL.ACOUS **Baßregler**
bass reflex box EL.ACOUS **Baßreflexbox**
bass reflex tube EL.ACOUS **Baßreflexrohr**
bass switch EL.ACOUS **Baßschalter**
batch (n.) DATA PROC **Stapel 1**
[a set of data or programs considered as a single processing unit] [als Verarbeitungseinheit betrachteter Satz von Daten und Programmen]
= Schub
batch (→lot 1) ECON (→Los)
batch (→series) SCIE (→Serie)
batch-bulk processing DATA PROC (→Stapelverarbeitung)
(→batch processing)
batch-conversational mode DATA COMM (→gemischter Verkehr)
(→mixed traffic)
batched communication DATA COMM (→Stapelübertragung)
(→batch transmission)
batch file DATA PROC **Stapeldatei**
[file with stored system commands] [enthält Systembefehle]
= Batch-Datei
batching (→stacking) COLLOQ (→Stapelung)
batch job DATA PROC **Stapelverarbeitungsautrag**
batch-mode processing DATA PROC (→Stapelverarbeitung)
(→batch processing)
batch number DATA PROC **Stapelnummer**
batch processing DATA PROC **Stapelverarbeitung**
[processing occurs groupwise after accumulation of tasks] [schubweise Verarbeitung nach vorangegangener Ansammlung]
= batch-mode processing; batch-bulk processing = Stapelbetrieb; Schubbearbeitung; Schubbetrieb; Schubverarbeitung; Monitorbetrieb; Batch-Verarbeitung; Batch-Prozessing; Sukzessivverarbeitung
≠ interactive processing ≠ Dialogverarbeitung
↓ Ortsstapelverarbeitung; Fernstapelverarbeitung
batch region DATA PROC **Stapelbereich**
batch size (→lot size) QUAL (→Losgröße)
batch system DATA PROC **Stapelsystem**
batch terminal DATA COMM **Stapelstation**
↑ terminalstation = Stapelendgerät
↑ Datenstation
batch total DATA PROC **Stapelsumme**
batch traffic DATA COMM **Stapelverkehr**
batch transmission DATA COMM **Stapelübertragung**
[by large data blocks] [mit großen Datenblöcken]
= batched communication
bathing-tub diagram (→bathtub curve) QUAL (→Badewannenkurve)
bathtub curve QUAL **Badewannenkurve**
= bathing-tub diagram

battery PHYS **Batterie**
[a series-connection of capacitors, cells] [Hintereinanderschaltung von Kondensatoren oder galvanischen Elementen]
↑ bank ↑ Satz
battery POWER SYS **Batterie 1**
[series connection of galvanic elements] [Hintereinanderschaltung von galvanischen Elementen]
= voltaic battery; galvanic battery = galvanische Batterie
↓ Sammlerbatterie
battery-backed POWER SYS **batteriegeschützt**
= batteriegesichert
battery backup POWER SYS **Batteriereserve**
= Batteriesicherung
battery bank POWER SYS **Batteriesatz**
battery cell POWER SYS **Batteriezelle**
= cell = Zelle; Batterie 2
battery charger POWER SYS (→Ladegerät)
(→charger)
battery charging control POWER SYS **Batterieladekontrolle**
battery clamp COMPON **Batterieklemme**
battery holder COMPON **Batteriehalter**
battery low annunciator EQUIP.ENG **Batterie-leer-Anzeige**
= battery low indication
battery low indication EQUIP.ENG (→Batterie-leer-Anzeige)
(→battery low annunciator)
battery magazine POWER SYS **Batterieschublade**
battery-operated EQUIP.ENG **batteriebetrieben**
= battery-powered; self-powered = batteriegespeist
battery operation EQUIP.ENG **Batteriebetrieb**
battery pack POWER SYS **Batteriesäule**
battery panel POWER SYS **Batterieschaltfeld**
= battery rack = Batteriefeld
battery-powered EQUIP.ENG (→batteriebetrieben)
(→battery-operated)
battery rack (→battery panel) POWER SYS (→Batterieschaltfeld)
battery receptacle EQUIP.ENG **Batteriefach**
battery room POWER SYS **Batterieraum**
battery save switch EQUIP.ENG **Batteriesparschalter**
battery separator POWER SYS **Scheider**
[Akkumulator]
= Separator
battery tester (→battery testing instrument) INSTR (→Batterieprüfgerät)
battery testing instrument INSTR **Batterieprüfgerät**
= battery tester = Batterietester; Elementeprüfer
battery voltage (→supply voltage) TELEPH (→Speisespannung)
batwing antenna ANT **Schmetterlingsantenne**
= butterfly antenna; bow-tie antenna = Fledermausantenne; Flächendipol; Batwing-Antenne
↓ paralleled antenna
↑ parallelgeschaltete Antenne
baud TELEC **Baud**
[mesuring unit for telegraph speed; 1 baud = 1 signal element per second] [Maßeinheit der Schrittgeschwindigkeit; 1 Baud = 1 Schritt / Sekunde]
Baudot code (1) TELEGR **Baudot-Code** (1)
= CCITT code no.1 = Baudot-Kode; CCITT-Code Nr.1; CCITT-Alphabet Nr.1
≈ interrnational alphabet no.2 ≈ internationales Telegraphenalphabet Nr.2
↑ telegraph code ↑ Fernschreibcode
Baudot code no.2 (→international telegraph alphabet no.2) TELEGR (→internationales Telegraphenalphabet Nr.2)
baud rate (→telegraph speed) TELEC (→Schrittgeschwindigkeit)
baud rate generator DATA COMM **Baudratengenerator**
= signaling rate generator; telegraph speed generator = Schrittgeschwindigkeitsgenerator
baud-spaced filter CIRC.ENG **Baud-Spaced-Filter**
bay (→rack) EQUIP.ENG (→Gestell)
bay face TELEC **Gestellansichtsplan**
= frame layout
bay line (→rack row) SYS.INST (→Gestellreihe)
bayonet base ELECTRON **Bajonettsockel**
= bayonet socket = Bajonettfassung

bayonet lock	MECH	Bayonettverschluß	
bayonet socket (→bayonet base)	ELECTRON	(→Bajonettsockel)	
bazooka	ANT	Viertelwellen-Sperrtopf	
= folder top; detuning sleeve		= Sperrtopf; Bazooka ≈ Symmetriertopf	
bazooka dipole antenna	ANT	Halbwellen-Sperrtopfantenne	
BB (→baseband)	TRANS	(→Basisband)	
B-battery (NAM) (→plate battery)	POWER SYS	(→Anodenbatterie)	
BBD (→bucked brigade device)	MICROEL	(→Eimerkettenschaltung)	
B box	DATA PROC	B-Register	
= B register			
BBS (→bulletin board system)	DATA COMM	(→Schwarzes-Brett-System)	
BC (→blind copy)	DATA COMM	(→Blindkopie)	
BCC (→block check)	CODING	(→Blockprüfung)	
BCC (→block check character)	DATA COMM	(→Blockprüfzeichen)	
BCD code (→binary-coded decimal code)	CODING	(→binärer Dezimalcode)	
BCD switch	COMPON	BCD-Codierschalter	
B channel (→basic channel)	TELEC	(→Basiskanal)	
BCH code (→Bose-Chandhuri-Hocquenghem code)	CODING	(→Bose-Chandhuri-Hocquenghem-Code)	
BCS (→block check sequence)	DATA COMM	(→Blockprüfzeichenfolge)	
BDI process	MICROEL	BDI-Verfahren	
= base diffusion isolation process			
Be (→beryllium)	CHEM	(→Beryllium)	
beach manhole	OUTS.PLANT	Strandschacht	
beacon	RADIO NAV	Bake	
↓ radiobacon; identification beacon; approach radiobeacon; marker beacon; radar beacon		↓ Funkfeuer; Kennbake; Landebake; Einflugzeichen; Markierungsfunkfeuer	
beaconing	RADIO NAV	Befeuerung	
bead (v.t.)	MECH	sicken	
bead (n.) [tiny ball] = pearl	TECH	Kügelchen = Perlenkugel	
bead (n.)	MECH	Sicke	
bead (n.) [programm section related to a task]	DATA PROC	Programmelement	
bead [coaxial tube]	COMM.CABLE	Abstandshalter [Koaxialtube]	
beaded	METAL	gesickt	
bead seal	COMPON	Glastropfendurchführung = Tropfendurchführung	
beam	PHYS	Bündel	
beam (→directional antenna)	ANT	(→Richtantenne)	
beam (→ray)	PHYS	(→Strahl)	
beam (→support)	CIV.ENG	(→Träger)	
beam aerial (→directional antenna)	ANT	(→Richtantenne)	
beam angle (→radiation angle)	PHYS	(→Strahlungswinkel)	
beam antenna (→directional antenna)	ANT	(→Richtantenne)	
beam aperture	PHYS	Bündelquerschnitt	
beam axis	PHYS	Strahlachse	
beam axis (→boresight)	ANT	(→Hauptstrahlrichtung)	
beam bending	PHYS	Strahlenkrümmung	
beam catcher = deflecting plate; deflector plate; deflector	ELECTRON	Ablenkplatte	
beam compression	ELECTRON	Strahlverdichtung	
beam control	ELECTRON	Strahlführung	
beam deflection	PHYS	Strahlungsablenkung	
beam deflection system (→electron beam deflection system)	ELECTRON	(→Elektronenstrahl-Ablenksystem)	
beam diameter	PHYS	Strahldurchmesser	
beam divergence	PHYS	Strahldivergenz	
beam generating system ≈ electron gun	ELECTRON	Strahlerzeugungssystem ≈ Elektronenkanone	
beamlead	MICROEL	Fahnenanschluß = Streifenanschluß	
beam-lead bonding	MICROEL	Beam-Lead-Lead-Kontaktierung	
beam-lead diode	MICROEL	Beam-Lead-Diode	
beam-lead technique	MICROEL	Balken-Leiter-Technik	
beam power = radiated power; radiated intensity	ANT	Strahlstärke = abgestrahlte Leistung; Strahlungsleistung	
beam return (→reverse action)	ELECTRON	(→Rücklauf)	
beam splitter	OPT.COMM	Strahlteiler	
beam spot	ELECTRON	Strahlfleck	
beam squint	ANT	Strahlauslenkung	
beam steering [main lobe modelling by electric means]	ANT	Keulenschwenkung [Modifikation der Hauptkeule mit elektrischen Mitteln]	
beam waveguide system	SAT.COMM	Strahlwellenleitersystem	
beam width	ANT	Keulenbreite	
bearer (→carrier)	TELEC	(→Träger)	
bearer channel (→carrier channel)	DATA COMM	(→Trägerkanal)	
bearer channel (→basic channel)	TELEC	(→Basiskanal)	
bearer channel rate (→carrier channel rate)	DATA COMM	(→Trägergeschwindigkeit)	
bearer control	TELEC	Nutzkanalsteuerung	
bearer identification code = BIC	DATA COMM	Trägerkennungscode	
bearer service 1 = carrier service	TELEC	Übermittlungsdienst = Trägerdienst	
bearer service 2 (→basic service)	TELEC	(→Grunddienst)	
bearing	MECH	Lager	
bearing	RADIO LOC	Peilung	
bearing correction	RADIO LOC	Peilungskorrektur	
bearing error	RADIO LOC	Peilfehler	
bearing shell	MECH	Lagerbuchse [Gleitlager]	
bearing surface	MECH	Auflagefläche	
beat [amplitude pulsation of the sum of two oscillations with different frequencies] ≈ pulsation	PHYS	Schwebung [Amplitudenschwankung bei Überlagerung zweier Schwingungen unterschiedlicher Frequenz] ≈ Kreisfrequenz	
beat (n.) (→heterodyning)	HF	(→Überlagerung)	
beat (→hit)	TECH	(→schlagen)	
beat detection (→beat reception)	HF	(→Überlagerungsempfang)	
beat frequency	PHYS	Schwebungsfrequenz	
beat frequency (→heterodyne frequency)	INSTR	(→Schwebungsfrequenz)	
beat-frequency generator = beat generator	INSTR	Schwebungsgenerator = Schwebungsoszillator	
beat frequency oscillator (→BFO)	TELEGR	(→BFO)	
beat generator (→beat-frequency generator)	INSTR	(→Schwebungsgenerator)	
beat reception = beat detection; heterodyne reception; heterodyne detection	HF	Überlagerungsempfang = Heterodynempfang	
Beck's arc lamp	PHYS	Beck-Bogenlampe	
become antiquated (→become outdated)	TECH	(→veralten)	
become obsolete (→become outdated)	TECH	(→veralten)	
become outdated = become obsolete; become antiquated; antiquate (v.t.); become outmoded; outmode (v.t.)	TECH	veralten = altmodisch werden	
become outmoded (→become outdated)	TECH	(→veralten)	
becomes symbol (→colons equal)	MATH	(→Ergibtsymbol)	
bedspring antenna	ANT	Matrazenfeder-Antenne = Bettgestellantenne	
bee line ≈ line of sight	TECH	Luftlinie ≈ Sichtlinie	
beep (v.i.) ≈ bleep (v.i.)	ACOUS	piepsen	
beep (n.) = bleep (n.); feep (n.)	ACOUS	Pieps	
beeper (→radio paging)	MOB.COMM	(→Funkruf)	
begin (v.t.) = start ≈ start-up (v.t.); initiate	TECH	beginnen = starten ≈ anlaufen; initiieren	

begin

begin (n.)	TECH	**Beginn**	**belt drive**	MECH	**Riemenantrieb**
= inception; start (n.); beginning; commencement		= Anfang; Start; Anlauf; Anbeginn			= Riementrieb
≈ onset		≈ Einsetzen	**belting**	MANUF	**Gurtung**
beginner	ECON	**Berufsanfänger**	[of components]		[von Bauteilen]
= inceptor			**belt printer**	TERM&PER	**Banddrucker**
beginner (→novice)	TECH	(→Anfänger)	[types on a rotating metallic belt]		[Drucktypen auf rotierendem Metallband]
beginning (adj.)	TECH	**beginnend** (adj.)	= band printer		
= starting; commencing (BRI); incipient; initialing; initialling		= anfangend	↑ impact printer		↑ Anschlagdrucker
		≈ anfänglich	**belt pulley**	MECH	**Riemenrolle**
≈ initial			**belt tightener**	MECH	**Riemenspanner**
beginning (→begin)	TECH	(→Beginn)	**bench**	MECH	**Werkbank**
beginning of conversation	SWITCH	**Gesprächsbeginn**	**bench** (→laboratory bench)	TECH	(→Labortisch)
= start of conversation; time-on			**bench** (→test department)	MANUF	(→Prüffeld)
beginning of file (→header label)	DATA PROC	(→Dateianfangs-Etikett)	**bench application**	ELECTRON	**Laboranwendung**
			= laboratory application		
beginning-of-information mark	DATA PROC	**Datenstrom-Anfangszeichen**	**bench instrument** (→table instrument)	INSTR	(→Tischinstrument)
= beginning-of-information marker; BIM; BIM mark; BIM marker		= Datenfluß-Beginnzeichen	**bench-light**	TECH	**Arbeitsplatzleuchte**
			bench mark	TECH	**Fixpunkt**
beginning-of-information marker (→beginning-of-information mark)	DATA PROC	(→Datenstrom-Anfangszeichen)			= fester Bezugspunkt
			benchmark (→benchmark program)	DATA PROC	(→Bewertungsprogramm)
beginning of line	TYPOGR	**Zeilenanfang**	**benchmarking** (→benchmark test)	DATA PROC	(→Bewertungstest)
= line start		= Zeilenbeginn			
beginning-of-tape mark	TERM&PER	**Bandanfangsmarke**	**benchmark problem**	DATA PROC	**Bewertungsaufgabe**
= beginning-of-tape marker; BOF; BOF mark; BOF marker		[Magnetband]			= Bewertungsproblem
		↑ Bandmarke	**benchmark program**	DATA PROC	**Bewertungsprogramm**
↑ tape mark			= benchmark		= Benchmark
beginning-of-tape marker	TERM&PER	(→Bandanfangsmarke)	**benchmark test**	DATA PROC	**Bewertungstest**
(→beginning-of-tape mark)			= benchmarking		
beginning stroke	TYPOGR	**Aufstrich**	**bench supply**	POWER SYS	**Tisch-Stromversorgung**
≠ end stroke		≠ Abstrich	**bench-top model** (→desktop model)	EQUIP.ENG	(→Tischgerät)
begin-of-paper detector	TERM&PER	**Papieranfangabtaster**			
begin of signal	TELEC	**Signalanfang**	**bend** (v.t.)	MECH	**knicken**
behaviour (n.)	ELECTRON	(→Antwort)	= buckle (v.i.)		
(→response)			**bend** (n.)	MICROW	**Knick**
behaviour (→response)	TECH	(→Verhalten)	**bend**	MATH	**Biegung**
beige	PHYS	**beige**	= curvature		[einer Kurve]
belated	TECH	**verspätet**			= Krümmung
= late		≈ verzögert	**bend**	MECH	**Knick**
≈ delayed		↑ unzeitig	= knee; buckling; jog; knuckle		= Knickung
↑ untimely			**bending**	METAL	**Biegen**
bell (n.)	TERM&PER	**Klingel**			[Stanzen]
= ringer; clock; call bell		= Glocke; Wecker; Anrufwecker	**bending** (→diffraction)	PHYS	(→Beugung)
			bending (→flection)	PHYS	(→Biegung)
bell (→bell character)	DATA COMM	(→Klingel-Zeichen)	**bending bulge**	METAL	**Biegewulst**
bell character	DATA COMM	**Klingel-Zeichen**	**bending couple**	PHYS	**Biegemoment**
= bell		= Klingel	= bending moment; bending torque		
bell gong	COMPON	**Glockenschale**	**bending die**	METAL	**Biegestanze**
bell hammer	COMPON	**Glockenköppel**	**bending diffraction**	PHYS	(→Beugung)
= hammer		= Köppel	(→diffraction)		
Bell hierarchy (→North American hierarchy)	TELEC	(→nordamerikanische Hierarchie)	**bending edge**	MECH	**Biegekante**
			bending limit	MECH	**Biegezahl**
bell load	TELEPH	**Weckerlast**	**bending load**	MECH	**Biegebelastung**
bellow	TECH	**Balg**	**bending loss**	OPT.COMM	**Biegungsverlust**
bell shaped	MATH	**glockenförmig**	= bend loss		
bell-shaped curve	MATH	**Glockenkurve**	**bending moment** (→bending couple)	PHYS	(→Biegemoment)
bell transformer (→clock transformer)	EL.INST	(→Klingeltransformator)			
			bending radius (→bend radius)	MATH	(→Biegeradius)
bellville spring	MECH	**Tellerfeder**	**bending stiffness** (→bending strength)	PHYS	(→Biegefestigkeit)
bell volume	TERM&PER	**Weckerlautstärke**			
below ground (→buried)	TECH	(→unterirdisch)	**bending strength**	PHYS	**Biegefestigkeit**
below-ground cable (AM)	COMM.CABLE	**unterirdisches Kabel**	= flexure; bending stiffness		= Biegungsfestigkeit
= underground cable (BRI)		↑ Außenkabel	**bending stress**	MECH	**Biegespannung**
↑ outside cable		↓ Erdkabel; Röhrenkabel	**bending tool**	TECH	**Biegewerkzeug**
↓ earth cable; duct cable			**bending torque** (→bending couple)	PHYS	(→Biegemoment)
below value	TECH	**unterwertig**			
[below standard value]		[unter Normalwert]	**bend loss** (→bending loss)	OPT.COMM	(→Biegungsverlust)
= inferior		≈ minderwertig [ECON]			
= low-quality [ECON]			**bend permanently**	TECH	**verbiegen**
belt	TECH	**Gurt**	[deform permanently from straight shape]		[durch Biegen dauerhaft aus der Normalform bringen]
belt	MECH	**Riemen**			
= strap			≈ bend; deform; distort		≈ biegen; verformen; verdrehen
beltbed plotter	TERM&PER	**Vertikalplotter**			
[paper fixed od a belt]		[Papier auf einem Band befestigt]	**bend radius**	MATH	**Biegeradius**
= vertical potter			= bending radius		= Krümmungsradius
		= Beltbed-Plotter; Bandunterlage-Plotter	**beneficial**	COLLOQ	**fördernd**
belt conveyer (→conveyor belt)	TECH	(→Förderband)	**benefit** (→privilege)	ECON	(→Vergünstigung)

benefit (→advantage)	COLLOQ	(→Vorteil)		bias (→biasing)	PHYS	(→Vorpolung)
benificiary	ECON	Begünstigter		bias (→bias voltage)	ELECTRON	(→Vorspannung)
bent	TECH	gebogen		bias address (→displacement address)	DATA PROC	(→Distanzadresse)
≈ curved		≈ gekrümmt		bias current [magnetic tape]	TERM&PER	Bias-Strom [Magnetband]
benzene = benzol	CHEM	Benzol		bias current	ELECTRON	Vorstrom
benzol (→benzene)	CHEM	(→Benzol)		bias distortion	DATA COMM	einseitige Verzerrung
BER (→bit error rate)	TELEC	(→Bitfehlerrate)		biased [statistics]	MATH	nicht erwartungstreu [Statistik]
BER (→error ratio)	INF	(→Fehlerquote)		= Bk		= vorgepolt
berkelium = Bk	CHEM	Berkelium = Bk		biased exponent = floating-point-number exponent	DATA PROC	Gleitkommaexponent = Gleitpunktexponent
BER meter (→bit error counter)	INSTR	(→Bitfehlermeßplatz)		biasing = bias; polarization	PHYS	Vorpolung ↓ Vormagnetisierung; Vor-
Bernoulli box [mass storage with exchangable and transportable data carrier for PC]	TERM&PER	Bernoulli-Box [Massenspeicher für PC's mit wechselbarem Datenträger]		↓ premagnetization; pre-electrification bias lightning (→back-lighting)	TV	elektrisierung (→Hintergrundbeleuchtung)
Bernoulli distribution (→binomial distribution)	MATH	(→Binomialverteilung)		bias point (→operating point)	ELECTRON	(→Arbeitspunkt)
BERT (→bit error counter)	INSTR	(→Bitfehlermeßplatz)		bias point adjustment (→operating point adjustment)	ELECTRON	(→Arbeitspunkteinstellung)
BERTS (→bit error counter)	INSTR	(→Bitfehlermeßplatz)		bias voltage = bias	ELECTRON	Vorspannung = Bias-Spannung
beryllium = Be	CHEM	Beryllium = Be		bias winding	ELECTRON	Polarisationswicklung
Bessel filter	NETW.TH	Besselfilter		biaxial	MATH	zweiachsig
Bessel function	MATH	Besselfunktion		BIC (→bearer identification code)	DATA COMM	(→Trägerkennungscode)
Bessel polynomial	MATH	Bessel-Polynom		BICMOS [combination of bipolar and CMOS technology]	MICROEL	BICMOS [Kombination von Bipolar- und CMOS-Technologie]
best fit (n.) = nearest match	MATH	beste Anpassung		biconcave [optics]	PHYS	bikonkav [Optik]
best-fit analysis (→curve fitting)	MATH	(→Kurvenermittlung)		biconical	MATH	doppelkonisch
beta cutoff (→ß cutoff)	MICROEL	(→Beta-Grenzfrequenz)		biconical antenna ↑ geometrically thick antenna ↓ discone antenna	ANT	Doppelkonusantenne = Doppelkegelantenne ↑ geometrisch dicke Antenne ↓ Discone-Antenne
beta decay	PHYS	Betazerfall		biconical horn	ANT	Doppelkonusstrahler = Doppelkonustrichter
beta distribution	MATH	Betaverteilung		bid (v.i.)	DATA COMM	ansuchen
beta emitter (→beta ray emitter)	PHYS	(→Betastrahler)		bid 2 (AM) (→offer)	ECON	(→Angebot)
beta gain	MICROEL	Betaverstärkung		bid (→request for service)	SWITCH	(→Vermittlungswunsch)
beta match	ANT	Beta-Anpassung ↓ L-Anpassung; C-Anpassung		bid closing date (→due date)	ECON	(→Abgabetermin)
beta particle ≈ electron	PHYS	Betateilchen ≈ Elektron		bidder (AM) = tenderer; offeror	ECON	Anbieter = Bieter; Submittent; Offerent
beta ray	PHYS	Betastrahl		bidirectional (→both-way)	TELEC	(→doppeltgerichtet)
beta ray emission	PHYS	Betastrahlung		bidirectional antenna	ANT	zweiseitige Richtantenne = zweiseitig gerichtete Antenne
beta ray emitter = beta emitter	PHYS	Betastrahler		bidirectional counter (→up-down counter)	CIRC.ENG	(→Vor-Rückwärts-Zähler)
beta testing [with selected users]	DATA PROC	externer Feldversuch [mit ausgesuchtem Teilnehmerkreis] = externer Probebetrieb		bidirectional good transmission (→five-by-five)	RADIO	(→beidseitig guter Empfang)
betatron	PHYS	Betatron		bidirectional mode (→duplex operation)	TELEC	(→Duplexbetrieb)
between-series adapter (→inter-series adapter)	COMPON	(→Adapter 2)		bidirectional operaration (→duplex operation)	TELEC	(→Duplexbetrieb)
bevel 1 (n.) = inclination angle	MECH	Schrägungswinkel = Neigungswinkel		bidirectional printer [from left to right and viceversa]	TERM&PER	bidirektionaler Drucker [druckt im Hin- und Rücklauf]
bevel (v.t.) ≈ chamfer; fold	MECH	abschrägen ≈ abkanten 2; abfasen		bidirectional printing	TERM&PER	bidirektionales Drucken
bevel 2 (n.)	MECH	Abschrägung		bidirectional transistor	MICROEL	Zweirichtungstransistor = bidirektionaler Transistor; Bidirektional-Transistor
beveled = tapered ≈ folded; chamfered	MECH	abgeschrägt ≈ abgekantet; abgefast		bidirectional triode thyristor (→triac)	MICROEL	(→Triac)
bevel gear [gear]	MECH	Kegelrad [Zahnrad] ↑ Kegelrad		BIFET = bipolar field-effect transistor technology	MICROEL	BIFET-Technologie
bevel gear drive	MECH	Kegelradtrieb		bifilar	ELECTRON	doppelfädig = bifilar
Beverage antenna = wave antenna 2 ↑ ground antenna; wave antenna 1	ANT	Beverage-Antenne ↑ Erdantenne 1; Wellenantenne		bifilar coil balun	ANT	Guanella-Übertrager
bezel [LCD]	COMPON	Frontrahmen [LCD]		bifilar insulated gate field-effect transistor (→BIGFET)	MICROEL	(→BIGFET)
Bezier curve	MATH	Bezier-Kurve		bifilar winding = double winding; double-spiral winding; noninductive winding	COMPON	Bifilarwicklung = bifilare Wicklung
Bezier surface	MATH	Bezier-Fläche				
BFO = beat frequency oscillator	TELEGR	BFO				
Bi (→bismuth)	CHEM	(→Wismut)				
bias (n.) = reference value ↓ bias voltage; bias current	EL.TECH	Bezugswert ↓ Vorspannung; Vorstrom				
bias (→displacement)	DATA PROC	(→Distanz)				
bias (n.) (→systematic error)	MATH	(→systematischer Fehler)		bifurcated contact	COMPON	Gabelkontakt

bifurcation 46

bifurcation	DATA PROC	**Binärverzweigung**
[with only two possible results]		[mit nur zwei möglichen Ergebnissen]
BIGFET	MICROEL	**BIGFET**
= bifilar insulated gate field-effect transistor		
big wheel antenna	ANT	**Big-wheel-Antenne**
bijection	MATH	**Bijektion**
[set theory]		[Mengenlehre]
bilateral	TECH	**zweiseitig**
= two-faced; two-sided; double-sided; on both sides		= doppelseitig; beidseitig
≠ unilateral		≠ einseitig
bilateral aerial (→bilateral antenna)	ANT	(→zweiseitig gerichtete Antenne)
bilateral antenna	ANT	**zweiseitig gerichtete Antenne**
= bilateral aerial		
bilateral characteristic	EL.ACOUS	**Achtercharakteristik**
= octagonal characteristic; figure-eight pattern; figure-eight characteristic		
bilateral transistor	MICROEL	**Bilateral-Transistor**
bill (n.) (→invoice)	ECON	(→Rechnung)
bill (→bill of exchange)	ECON	(→Wechsel)
bill (→paper currency)	ECON	(→Banknote)
billable (→chargeable)	TELEC	(→gebührenpflichtig)
bill endorser	ECON	**Wechselgirant**
[transfers a bill of exchange to another person]		[überträgt sein Bezugsrecht eines Wechsels auf einen anderen Empfänger]
= endorser		= Girant; Begeber; Wechselindossant; Indossant; Indossent
billet	METAL	**Knüppel**
bill feed (→single-sheet feed)	TERM&PER	(→Einzelblattzuführung)
billing	TELEC	**Rechnungserstellung**
		= Verrechnung
billing administration	SWITCH	**Gebührendatenverwaltung**
billing counter	SWITCH	**Verrechnungszähler**
billing data (→call-charge data)	SWITCH	(→Gebührendaten)
billing period	ECON	**Abrechnungszeitraum**
= accounting period		
billing record	TELEC	**Gebührenprotokoll**
billion (BRI) (→trillion)	MATH	(→Billion)
billion (AM)	MATH	**Milliarde**
[thousand millions]		= Md.; Mrd.
= million (BRI); bn		
≈ billion (BRI)		
bill of exchange	ECON	**Wechsel**
[an unconditional written promise to pay the indicated sum to the owner of the bill]		[schuldrechtliches Forderungspapier, in dem die Zahlung einer Summe an den Wechselinhaber versprochen wird]
= bill		
↓ promissory note; draft		↓ Eigenwechsel; Tratte
bill of materials	ECON	**Stückliste**
bill of materials (→parts list)	MANUF	(→Stückliste)
BIM	DATA PROC	(→Datenstrom-Anfangszeichen)
(→beginning-of-information mark)		
bimetal	PHYS	**Bimetall**
bimetallic	PHYS	**bimetallisch**
bimetallic instrument	INSTR	**Bimetallinstrument**
bimetallic release	POWER ENG	**Bimetallauslöser**
bimetallic thermal switch	COMPON	**Bimetall-Thermostat**
bimetallic thermometer	INSTR	**Bimetallthermometer**
bimetal relay	COMPON	**Bimetallrelais**
BIM mark	DATA PROC	(→Datenstrom-Anfangszeichen)
(→beginning-of-information mark)		
BIM marker	DATA PROC	(→Datenstrom-Anfangszeichen)
(→beginning-of-information mark)		
bin	ELECTRON	**Kriterium**
bin (→vessel)	TECH	(→Behälter)
bin (→supply bin)	TERM&PER	(→Vorratsbehälter)
binarization (→digitizing)	DATA PROC	(→Digitalisierung)
binary (adj.)	MATH	**binär**
[relative to a number system of radix 2]		[zu einem Zahlensystem mit Basis 2 gehörend]
= dual; dyadic		= zweiwertig 1; dual; dyadisch
binary addition	MATH	**binäre Addition**
binary arithmetics	MATH	**Binärarithmetik**
= dual arithmetics		= Dualarithmetic
binary bit (→bit)	INF	(→Bit)
binary cell	DATA PROC	**Binärzelle**
= binary storage element		= binäre Speicherzelle; Binärelement; binäres Speicherelement
binary channel	INF	**Binärkanal**
binary character (→binary signal)	INF	(→Binärsignal)
binary chop (→binary search)	DATA PROC	(→Binärsuche)
binary code	CODING	**Binärcode**
[with binary code elements]		[Code mit zweistufigen Codeelementen]
↓ dual code		↓ Dualcode
binary coded	CODING	**binärcodiert**
		= binär codiert
binary coded character (→binary signal)	INF	(→Binärsignal)
binary-coded decimal code	CODING	**binärer Dezimalcode**
= BCD code		= BCD-Code
binary coding	CODING	**Binärcodierung**
= binary encoding		= Dualcodierung
binary counter	CIRC.ENG	**Dualzähler**
		= Binärzähler
binary-decimal conversion	DATA PROC	(→Binär-Dezimal-Wandlung)
(→binary-to-decimal conversion)		
binary-decimal converter	DATA PROC	(→Binär-Dezimal-Umsetzer)
(→binary-to-decimal converter)		
binary digit (→bit)	INF	(→Bit)
binary division	MATH	**binäre Division**
binary dump	DATA PROC	**Binärausdruck**
		[in Binärcode]
binary encoding (→binary coding)	CODING	(→Binärcodierung)
binary error correcting code	CODING	**binärer Fehlerkorrekturcode**
binary exponent	DATA PROC	**Binärexponent**
binary file	DATA PROC	**Binärdatei**
[contains machine programs]		[enthält Maschinenprogramme]
binary fraction	DATA PROC	**Binärbruch**
[decimal fraction in binary form]		[Binärdarstellung eines Dezimalbruchs]
binary information	INF	**Binärinformation**
binary information meaning	TELECONTR	**Meldungsinhalt**
binary loader	DATA PROC	**Binärlader**
binary mantissa	DATA PROC	**Binärmantisse**
binary multiplication	MATH	**binäre Multiplikation**
binary notation (→binary number system)	MATH	(→Dualsystem)
binary number	MATH	**Binärzahl**
= dual number		= Dualzahl
binary number system	MATH	**Dualsystem**
= binary notation; binary representation; binary system		[Zahlendarstellung mit 0 und 1]
		= duales Zahlensystem; binäres Zahlensystem; Binärsystem
		↑ Stellenwertsystem
binary operation	ENG.LOG	**binäre Funktion**
= dyadic operation		= Binäroperation; binäre Operation; dyadische Operation
binary point	INF	**Binärkomma**
		= binäres Komma; Binärpunkt
binary representation (→binary number system)	MATH	(→Dualsystem)
binary scale (→bit position)	CODING	(→Binärstelle)
binary search	DATA PROC	**Binärsuche**
= dichotomizing search; binary chop		= Halbierungssuchverfahren; dichotomische Suche
binary sequence	CODING	**Binärfolge**
binary signal	INF	**Binärsignal**
= binary character; binary coded character		= Binärzeichen; binäres Signal
binary signalling (→binary transmission)	TELEC	(→Binärzeichenübertragung)

binary source		INF	Binärquelle	bipartite (→two-part)	TECH	(→zweiteilig)
binary state		DATA PROC	binärer Zustand	bipartition (→bisection)	TECH	(→Halbierung)
			= Binärzustand; zweiwertiger Zustand	bipartition angle	MATH	Halbierungswinkel
binary storage element		DATA PROC	(→Binärzelle)	biplug (→male-male adapter)	COMPON	(→Doppelstecker)
(→binary cell)				bipolar	PHYS	zweipolig
binary subtraction		MATH	binäre Subtraktion	= two-pole		= bipolar
binary synchronous communications		DATA COMM	binäre Synchronübetragung	bipolar circuit	MICROEL	Bipolarschaltung
= BSC						= bipolare Schaltung
binary system (→binary number system)		MATH	(→Dualsystem)	bipolar code [binary cero = 0 V, binary one = alternating positive and negative voltage]	DATA COMM	Bipolarcode [binäre Null = 0 V, binäre Eins = alternierend positive oder negative Spannung]
binary-to-decimal conversion		DATA PROC	Binär-Dezimal-Wandlung = Binär-Dezimal-Umsetzung			= Doppelstromcode
= binary-decimal conversion				bipolar field-effect transistor technology (→BIFET)	MICROEL	(→BIFET-Technologie)
binary-to-decimal converter		DATA PROC	Binär-Dezimal-Umsetzer = Binär-Dezimal-Wandler	bipolar impedance chart (→rectangular transmission line chart)	LINE TH	(→Schmidt-Buschbeck-Diagramm)
= binary-decimal converter						
binary transfer (→binary transmission)		TELEC	(→Binärzeichenübetragung)	bipolar modulation	MODUL	Bipolartastung
binary transmission		TELEC	Binärzeichenübetragung	bipolar power supply	POWER SYS	bipolare Netzstromversorgung
= binary signalling; binary transfer			= Binärübetragung			= bipolares Netzgerät
binary tree		DATA PROC	Binärbaum	bipolar semiconductor	MICROEL	bipolarer Halbleiter
[two-branched data structure]			= binärer Baum	bipolar semiconductor memory	MICROEL	Bipolarspeicher
binary variable		DATA PROC	Binärvariable = binäre Variable	bipolar signal [with positive and negative significant conditions]	DATA COMM	Bipolarsignal [mit positiven und negativen Kennzuständen]
binaural		ACOUS	zweiohrig			
≈ stereophonic			≈ stereophonisch	bipolar technology	MICROEL	Bipolar-Technologie
≠ monaural			≠ monoaural	bipolar transistor	MICROEL	Bipolartransistor
bind (v.t.)		EQUIP.ENG	abbinden	[transistor build-up of p-type as well as of n-type material]		[Transistor der sowohl mit P- als auch mit N-Halbleitern aufgebaut ist]
[a cable harness]			[eines Kabelbaumes]	≠ unipolar transistor		
= tie-up				↓ point-contact transistor; junction transistor; alloy-junction transistor; mesa transistor; diffusion transistor; planar transistor		= bipolarer Transistor
bind (v.t.) (→link-edit)		DATA PROC	(→binden)			≠ Unipolartransistor
binder equipment		TERM&PER	Bindegerät			↓ Spitzentransistor; Flächentransistor; Legierungstransistor; Mesatransistor; Diffusionstransistor; Planartransistor
binder file		OFFICE	Heftmappe			
binding		EQUIP.ENG	Abbindung			
[a cable harness]			[eines Kabelbaums]			
= tying-up						
binding		CHEM	Bindung	bipolar transmission line chart (→rectangular transmission line chart)	LINE TH	(→Schmidt-Buschbeck-Diagramm)
= bond						
binding (adj.)		ECON	verbindlich			
= obliging				biprism	PHYS	Biprisma
binding		TYPOGR	Einband	biquadrate	MATH	Biquadrat
binding (→arrest)		TECH	(→Hemmung)	biquadratic	MATH	biquadratisch
binding (→binding margin)		TYPOGR	(→Heftrand)	biquadratic equation	MATH	biquadratische Gleichung
binding agent		CHEM	Bindemittel	biquinary	MATH	biquinär
= agent 2						= biquintal
binding electron (→bonding electron)		PHYS	(→Valenzelektron)	biquinary code [decimal 1–2–3–4–5–6 etc. as 00–01–02–03–04–50–51]	CODING	Biquinärcode [dezimal 1–2–3–4–5–6 u.s.f. dargestellt als 00–01–02–03–04–50–51]
binding energy		CHEM	Bindungsenergie			
binding-head screw		MECH	Hemmkopfschraube			= biquinärer Code; Biquintalcode; biquintaler Code
binding margin		TYPOGR	Heftrand			
= binding				bird cage antenna	ANT	Vogelkäfigantenne
binding post (→terminal)		COMPON	(→Anschlußklemme)			= Bird-cage-Antenne
binding time		DATA PROC	Bindevorgang	birefringence	PHYS	Doppelbrechung
binding wire (→tie wire)		TECH	(→Bindedraht)	= double refraction; twofold refraction		= doppelte Brechung
binistor		MICROEL	Binistor	Birmingham wire gage	METAL	Birmingham Drahtlehre
[injector-controlled tetrode]			[über Injektor gesteuerte Tetrode]	= BWG		= BWG
≈ PNPN diode			≈ Vierschichtdiode	biro	OFFICE	Kugelschreiber
binoculars		PHYS	Feldstecher	= ball-point pen; ball-point		= Kugelstift
binomial		MATH	binomisch	bisect (→halve)	TECH	(→halbieren)
binomial antenna array		ANT	Binomialantennenfeld	bisection	TECH	Halbierung
binomial coefficient		MATH	Binomialkoeffizient	= bipartition		= Zweiteilung
binomial distribution		MATH	Binomialverteilung	↑ division		↑ Teilung
= Bernoulli distribution			= Bernoulli-Verteilung	bisectional line (→center line)	ENG.DRAW	(→Mittellinie)
biometric identification system		SIGN.ENG	Biometrisches Identifikationssystem			
				bisectional plane (→central plane)	ENG.DRAW	(→Mittelebene)
bionics		INF.TECH	Bionik			
[analysis of biology by engineering science]			[ingenieurwissenschaftliches Studium der Biologie]	bisector	MATH	Winkelhalbierende
				bisector (→center line)	ENG.DRAW	(→Mittellinie)
BIOS		DATA PROC	BIOS	bisectrix	PHYS	Bisektrix
[Basic Input/Output System; in some operating systems as MS-DOS a set of booting instructions to customize it to a specific system]			[systemspezifischer Befehlssatz zum Laden bestimmter Betriebssysteme wie MS-DOS]	bisectrix (→center line)	ENG.DRAW	(→Mittellinie)
				bismuth	CHEM	Wismut
				= Bi		= Bismut; Bi; Bismutum
≈ input/output processor; ROS			≈ Ein-Ausgabe-Werk; ROS	bisquare antenna	ANT	Bisquare-Antenne
Biot-Savart law		PHYS	Biot-Savartsches Gesetz	bistability	TECH	Bistabilität
bipack (→two-layered)		TECH	(→zweischichtig)			

bistability effect

bistability effect	PHYS	Bistabilitätseffektffekt	
		= Hysterese-Effekt	
bistable	TECH	bistabil	
bistable flip-flop	CIRC.ENG	(→bistabile Kippschaltung)	
(→flip-flop)			
bistable multivibrator	CIRC.ENG	(→bistabile Kippschaltung)	
(→flip-flop)			
bistable relay	COMPON	bistabiles Relais	
bistable trigger circuit	CIRC.ENG	(→bistabile Kippschaltung)	
(→flip-flop)			
bistatic cross section	RADIO LOC	bistatische Reflexionsfläche	
[scatters back to other directions]		[strahlt in andere Richtungen zurück]	
bistatic-scattering cross section (→radar cross section)	RADIO LOC	(→Radarquerschnitt)	
bistre	PHYS	bister	
biswitch (→trigger diode)	MICROEL	(→Triggerdiode)	
bit	INF	Bit	
[0 or 1]		[0 oder 1]	
= binary digit; binary bit		= Binärziffer; Dualziffer; Binärelement	
bit	TECH	Werkzeugspitze	
bit addressing	DATA PROC	Bitadressierung	
bit-at-a-time (→bit-serial)	TELEC	(→bit-seriell)	
bit-by-bit	INF.TECH	bitweise	
≈ bitwise		≈ bitseriell	
bit-by-bit (→bit-serial)	TELEC	(→bit-seriell)	
bit-by-bit operation	DATA PROC	bitweise Operation	
		= bitweiser Betrieb	
bit-by-bit stuffing	CODING	bitweises Stopfen	
bit combination	INF	Bitkombination	
bit configuration (→bit pattern)	INF	(→Bitmuster)	
bit counter	DATA PROC	Bitzähler	
bit density (→recording density)	DATA PROC	(→Speicherdichte)	
bit error	TELEC	Bitfehler	
≈ erroned bit		≈ fehlerhaftes Bit	
bit error counter	INSTR	Bitfehlermeßplatz	
= bit error measuring set; bit error rate tester; BERT; BER meter; bit error rate test set; BERTS		= Bitfehlermeßgerät; Bitfehlerraten-Meßplatz	
bit error measuring set (→bit error counter)	INSTR	(→Bitfehlermeßplatz)	
bit error probability	INF	Bitfehlerwahrscheinlichkeit	
bit error rate	TELEC	Bitfehlerrate	
= BER		= Bitfehlerhäufigkeit; Bitfehlerquote; Pulsfehlerate	
bit error rate tester (→bit error counter)	INSTR	(→Bitfehlermeßplatz)	
bit error rate test set (→bit error counter)	INSTR	(→Bitfehlermeßplatz)	
bit error ratio (→error ratio)	INF	(→Fehlerquote)	
bit flipping (→bit manipulation)	DATA PROC	(→Bithantierung)	
bit group	TELEC	Bitgruppe	
bit-group processor	DATA PROC	Bitgruppen-Prozessor	
bit handling	DATA PROC	Bithandhabung	
bit image	DATA PROC	Bitbild	
bit-interleaved	TELEC	bitverschachtelt	
bit interleaving	TELEC	Bitverschachtelung	
bit inversion	CODING	Bitinversion	
		= Bitinvertierung	
bit location (→bit position)	CODING	(→Binärstelle)	
bit manipulation	DATA PROC	Bithantierung	
= bit flipping			
bit-map (v.t.)	DATA PROC	mit Einzelbits darstellen	
[to represent by single bits]			
bit map 2	TERM&PER	Pixelmuster	
[a graphics dissolved into an array of dots]		[in ein Punktraster aufgelöstes Bild]	
= pixel map; pixel array		= Pixelraster; Punktraster 2; Punktmuster; Bit-Map	
bit map 1 (→graphics memory)	DATA PROC	(→Graphikspeicher)	
bit-mapped	DATA PROC	einzelbitweise	
bit-mapped console	TERM&PER	Bit-map-Konsole	
bit-mapped font	TERM&PER	Rasterschrift	
= bit-mapped type		= Bit-map-Schrift	
bit-mapped graphic (→raster graphics)	DATA PROC	(→Rastergraphik)	
bit-mapped graphics (→memory-mapped video)	DATA PROC	(→speicherkonforme Bildschirmanzeige)	
bit-mapped screen	TERM&PER	Bit-map-Bildschirm	
[every pixel controlled by one bit]		[Bildpunkte mit je einem Bit gesteuert]	
bit-mapped screen (→memory-mapped video)	DATA PROC	(→speicherkonforme Bildschirmanzeige)	
bit-mapped terminal	TERM&PER	Bit-map-Terminal	
bit-mapped type (→bit-mapped font)	TERM&PER	(→Rasterschrift)	
bit-mapped video (→memory-mapped video)	DATA PROC	(→speicherkonforme Bildschirmanzeige)	
bit-mapping (→memory-mapped video)	DATA PROC	(→speicherkonforme Bildschirmanzeige)	
bit map screen (→raster screen)	TERM&PER	(→Rasterbildschirm)	
bit-oriented	INF.TECH	bitorientiert	
bit-parallel	INF.TECH	bitparallel	
[simultaneously over several lines]		[über mehrere Leitungen gleichzeitig]	
≠ bit-serial		= schrittparallel	
		≠ bitseriell	
bit parity violation (→parity violation)	CODING	(→Paritätsverletzung)	
bit pattern	INF	Bitmuster	
= bit configuration			
bit position	CODING	Binärstelle	
= bit location; binary scale		= Bitposition; Bitstelle	
bit rate (→transmission rate)	TELEC	(→Übertragungsgeschwindigkeit)	
bit-rate matching	TELEC	Bitratenanpassung	
↓ stuffing mode		↓ Stopfverfahren	
bit-rate reduction	TELEC	Bitratenreduktion	
bit-rate-variant	TELEC	bitratenvariabel	
bit-rate-variant communications	TELEC	bitratenvariable Kommunikation	
bit repetition	TELEC	Bitwiederholung	
bit robbery (→bit stealing)	TELEC	(→Bitstehlen)	
bit/s (→bits per second)	TELEC	(→Bit/s)	
bit sequence	TELEC	Bitfolge	
= bit string; bit train		= Bitsequenz	
bit sequence independence	TELEC	Bitfolgeunabhängigkeit	
bit sequence plan	TELEC	Bitfahrplan	
		= Bitfolgeplan	
bit-serial	TELEC	bit-seriell	
[sequencically on a single line]		[auf einer Leitung nacheinander]	
= bit-by-bit; bit-at-a-time		= schrittseriell	
≠ bit-parallel		≈ bitweise	
		≠ bit-parallel	
bit slice	MICROEL	Bit-Slice	
		= Scheibe	
bit-sliced	DATA PROC	bitgeteilt	
bit-slice processor	MICROEL	Bit-slice-Prozessor	
bit slip	TELEC	Bitschlupf	
bits per inch	TERM&PER	Bits pro Zoll	
= bpi		= bpi	
bits per second	TELEC	Bit/s	
= bit/s; bp/s; BPS; bps		= Bits pro Sekunde	
bit stealing	TELEC	Bitstehlen	
= bit robbery			
bit stream	INF	Bitstrom	
= digital stream			
bit-stream font	TERM&PER	bitorientierter Zeichensatz	
[a font defined by unscalable dot arrays, for laser printer]		[als Punktmuster definierter, nicht skalierbarer Zeichensatz für Laserdrucker]	
≠ outline font		≠ Outline-Font	
bit string (→bit sequence)	TELEC	(→Bitfolge)	
Bitter pattern	PHYS	Bitter-Streifen	
bit timing	TRANS	Bittakt	
bit track (→information track)	TERM&PER	(→Informationsspur)	
bit train (→bit sequence)	TELEC	(→Bitfolge)	
bit transfer rate	DATA COMM	Bitübertragungsgeschwindigkeit	
bit violation	TELEC	Bitverletzung	
bitwise (→bit-by-bit)	INF.TECH	(→bitweise)	

16-bit word	DATA PROC	**Langwort**	
		= 16-Bit-Wort	
bivalent	MATH	**zweiwertig 2**	
= divalent; two-valued		= bivalent; zweideutig	
bivariate (→two-dimensional)	MATH	(→zweidimensional)	
bivariate normal distribution	MATH	**zweidimensionale Normalverteilung**	
bivector	MATH	**Bivektor**	
bivibrator (→astable multivibrator)	CIRC.ENG	(→astabile Kippschaltung)	
Bk (→berkelium)	CHEM	(→Berkelium)	
black	PHYS	**schwarz**	
black (→bold)	TYPOGR	(→fett)	
black and white	TV	**Schwarz-Weiß**	
= monochromatic			
black-and-white reception	TV	**Schwarz-Weiß-Empfang**	
black-and-white television	TV	**Schwarz-Weiß-Fernsehen**	
≠ colour television		≠ Farbfernsehen	
black-and-white tube	ELECTRON	**Schwarz-Weiß-Bildröhre**	
= monochrome tube; monochrome display		= monochrome Bildröhre	
blackboard (→bulletin board)	COLLOQ	(→Anzeigetafel)	
black body (→black-body radiator)	PHYS	(→schwarzer Strahler)	
blackbody radiation	PHYS	**schwarze Strahlung**	
black-body radiator	PHYS	**schwarzer Strahler**	
= black body		= schwarzer Körper	
black box	TECH	**schwarzer Kasten**	
[unit of unknown functioning]		[Einheit unbekannter Funktionsweise]	
≠ white box		≠ weißer Kasten	
black compression	TV	**Schwarzkompression**	
= black crushing			
black crushing (→black compression)	TV	(→Schwarzkompression)	
blacken	TECH	**schwärzen**	
blacken	TYPOGR	**schwärzen**	
blacker-than-black	TV	**Ultraschwarz**	
black letter	TYPOGR	**gotische Schrift**	
[a font with heavy face and angular outline]		= Gotisch	
= Gothic			
black level	TV	**Schwarzpegel**	
[75% modulation]		[75% Modulation]	
≈ porch		= Schwarzwert	
		≈ Schwarzschulter	
black level control (→black level restoration)	TV	(→Schwarzwerthaltung)	
black-level lift (→black-level set-up)	TV	(→Schwarzabhebung)	
black level restoration	TV	**Schwarzwerthaltung**	
= black level control; dc reinsertion; dc restoring		= Schwarzwertsteuerung; Schwarzpegel-Wiederherstellung; Schwarzpegelregelung	
black-level set-up	TV	**Schwarzabhebung**	
= black-level lift; lift			
blacklistener (→unlicensed listener)	BROADC	(→Schwarzhörer)	
black-out (n.)	POWER SYS	**Spannungsausfall**	
[accidental]		[unbeabsichtigt]	
= blackout; power failure; voltage brakdown		= Stromausfall	
≈ power cut; brown-out		≈ Stromabschaltung; Spannungsmangel	
↑ power outage		↑ Stromunterbrechung	
↓ mains failure		↓ Netzausfall	
black-out (→blanking)	ELECTRON	(→Austastung)	
blackout (→black-out)	POWER SYS	(→Spannungsausfall)	
blackout (→catastrophic failure)	QUAL	(→Totalausfall)	
black-out (→interruption)	EL.TECH	(→Unterbrechung)	
black-out level (→blanking level)	TV	(→Austastwert)	
black printer	TERM&PER	**Schwarzdrucker**	
= black writer		↑ Laserdrucker	
↑ laser printer			
black screen	TV	**Grauscheibe**	
black writer (→black printer)	TERM&PER	(→Schwarzdrucker)	
black-write technique	TERM&PER	**Black-write-Verfahren**	
blade	TECH	**Klinge**	
blade antenna	ANT	**Klingenantenne**	
↑ monopole antenna		= Blattantenne	
		↑ Monopolantenne	
blade-connector strip	COMPON	**Messerleiste**	
= blade-contact strip			
blade-contact strip	COMPON	(→Messerleiste)	
(→blade-connector strip)			
blade type pointer	INSTR	**Messerzeiger**	
↑ pointer		↑ Zeiger	
blank (→blanking)	ELECTRON	(→Austastung)	
blank (→blanking)	TERM&PER	(→Dunkelsteuerung)	
blank (→standard form)	OFFICE	(→Formular)	
blank (n.)	CODING	**Leerstelle**	
= void (n.); space (n.)		= Leerzeichen; Füllzeichen; Zwischenraum; Blank	
blank (→blank space)	TYPOGR	(→Leerraum)	
blank (→empty)	TERM&PER	(→leer)	
blank address	DATA PROC	**Leeradresse**	
blank bit (→dummy bit)	CODING	(→Leerbit)	
blank character	DATA COMM	**Leerzeichen**	
= idle character		≈ Füllzeichen	
≈ filler			
blank character (→space character)	TELEGR	(→Zwischenraumzeichen)	
blank command (→blank instruction)	DATA PROC	(→Leeranweisung)	
blank disk	DATA PROC	**Leerplatte**	
[with no data]		[ohne Daten]	
= empty disk			
blank diskette	DATA PROC	**Leerdiskette**	
= empty diskette; clean diskette			
blanked picture signal	TV	**BA-Signal**	
		= ausgetastetes Videosignal; unterdrücktes Bildsignal	
blank flange	MECH	**Blindflansch**	
blank form	TERM&PER	**Leerformular**	
		= Blankoformular	
blanking	METAL	**Ausschneiden**	
		[Stanzen]	
blanking	ELECTRON	**Austastung**	
= blank; black-out; masking-off; masking-out		= Ausblendung	
		↓ Impulsausblendung	
↓ strobing			
blanking	TERM&PER	**Dunkelsteuerung**	
= blank		= Dunkeltastung	
blanking interval	TV	**Austastintervall**	
↓ vertical blanking interval; horizontal blanking interval		= Austastlücke	
		↓ Bildaustastlücke; Zeilenaustastlücke	
blanking level	TV	**Austastwert**	
= black-out level		= Austastpegel	
blanking pulse		**Austastimpuls**	
[suppresses the image signal; consists of front porch, sync signal and back porch]		[unterdrückt das Bildsignal; enthält die vordere Schwarzschulter, den Synchronimpuls und die hintere Schwarzschulter]	
blanking signal	TV	**Austastsignal**	
		= A-Signal	
blank instruction	DATA PROC	**Leeranweisung**	
[does nothing]		[löst nichts aus]	
= do-nothing instruction; dummy instruction; skip instruction; null instruction; no-operation instruction; no-op instruction; NOP instruction; blank instruction; blank command; do-nothing command; dummy command; skip command; null command; no-op command; blank statement; do-nothing statement; dummy statement; skip statement; null statement; no-op statement		= Blindanweisung; Scheinanweisung; Leerbefehl; Blindbefehl; Scheinbefehl; Nulloperationsanweisung; Nulloperationsbefehl; Nulloperation; NOP	
blank instruction (→blank instruction)	DATA PROC	(→Leeranweisung)	
blank line	TYPOGR	**Leerzeile**	
		= Blindzeile; Leergang; Leerdruck	
blank line skipping	TELEGR	**Leerzeilensprung**	
[fax]		[Fax]	

blank module

blank module (→dummy module)	EQUIP.ENG	(→Blindbaugruppe)
blank-out	RADIO LOC	Bildausfall
blank page	TYPOGR	Leerseite
		= Vakatseite
blank paper	TERM&PER	Blankopapier
blank set (→empty set)	MATH	(→Leermenge)
blank space	EQUIP.ENG	Leerplatz
blank space	TYPOGR	Leerraum
= space; blank		= Zwischenraum; Leerstelle
↓ left blank; right blank		↓ Vorbreite; Nachbreite
blank statement (→blank instruction)	DATA PROC	(→Leeranweisung)
blank string (→null string)	DATA PROC	(→Leerkette)
blank tape (→unpunched tape)	TERM&PER	(→ungelochter Lochstreifen)
blank track	TERM&PER	Leerspur
blast (v.t.)	DATA PROC	freimachen
[to free memory sections]		[von Speicherplatz]
		= säubern
blast (n.)	TECH	Gebläse
= high-pressure fan ; blower		= Hochdrucklüfter
↑ fan		↑ Ventilator
blast	PHYS	Druckwelle 2
		[Gas]
blast (v.t.) (→program)	MICROEL	(→schießen)
blast (→windflaw)	METEOR	(→Windstoß)
blast (→blow-up)	TECH	(→sprengen 1)
blaze (v.i.)	TECH	lodern
		= auflodern
B-lead (→ring wire)	TELEPH	(→B-Ader)
bleed (v.i.)	ENG.DRAW	verlaufen
[ink]		[Tinte]
		= zerfließen
bleed (v.t.)	TECH	ablassen
[pressure]		[Druck]
bleed (n.)	TERM&PER	Zeilenüberlauf
[line beyond paper edge]		
bleeder (→bleeder resistance)	POWER SYS	(→Entladungswiderstand)
bleeder resistance	POWER SYS	Entladungswiderstand
= bleeder		
bleeder resistance (→leck resistance)	EL.TECH	(→Ableitungswiderstand)
bleeder valve (→drain valve)	TECH	(→Ablaßventil)
bleeding	TV	Farbsaum
= color fringing (AM)		≈ Farbrand
bleep (n.) (→beep)	ACOUS	(→Pieps)
bleep (v.i.) (→beep)	ACOUS	(→piepsen)
blend (→portmanteau word)	LING	(→Schachtelwort)
blendable	TECH	wischbar
[paint]		[Farbe]
blending valve (→mixing valve)	TECH	(→Mischventil)
BLER (→block error rate)	DATA COMM	(→Blockfehlerhäufigkeit)
blind approach (→blind landing)	RADIO NAV	(→Blindlandung)
blind copy	DATA COMM	Blindkopie
= BC		
blind flying (→instrument navigation)	RADIO NAV	(→Instrumentennavigation)
blind hole	MECH	Sackloch
blinding-out (→extraction)	DATA PROC	(→Ausblendung)
blind keyboard	TERM&PER	Blindtastatur
[without visual check of keyed input]		[ohne Sichtkontrolle des Eingetippten]
		= blinde Tastatur
blind landing	RADIO NAV	Blindlandung
= blind approach		
blind landing system (→instrument landing system)	RADIO NAV	(→Instrumentenlandesystem)
blind navigation (→instrument navigation)	RADIO NAV	(→Instrumentennavigation)
blind-out (→mask-off)	ELECTRON	(→ausblenden)
blind search	DATA PROC	Blindsuche
blink (v.i.) (→flicker)	TECH	(→blinken)
blinking (→flashing)	TERM&PER	(→blinkend)
blinking cursor	TERM&PER	blinkende Schreibmarke
		= Blinkcursor
blip (→pip)	RADIO LOC	(→Bilschirmmarkierung)
blister steel	METAL	Blasenstahl
blob (n.)	COLLOQ	Farbfleck
Bloch wall	PHYS	Blochwand
block (v.t.)	DATA PROC	blocken
[to group several records to a block]		[mehrere Sätze zu einem Block zusammenfassen]
block (v.t.)	SWITCH	sperren
block (v.t.)	MECH	sperren
block (n.)	TECH	Klotz
= log (n.)		
block (n.)	MECH	Sperre
block (→note pad)	OFFICE	(→Notizblock)
block (→data block)	DATA PROC	(→Datenblock)
block (v.t.) (→disable)	ELECTRON	(→sperren)
block abort	DATA COMM	Blockabbruch
= abort		
block address register	DATA PROC	Blockadressregister
block antenna	BROADC	Gemeinschaftsantenne
= common antenna; community antenna		
block capacitor (→blocking capacitor)	CIRC.ENG	(→Sperrkondensator)
block capital (→capital character)	TYPOGR	(→Großbuchstabe)
block character check (→block check)	CODING	(→Blockprüfung)
block check	CODING	Blockprüfung
= block character check; BCC; longitudinal parity check; longitudinal redundancy check; LRC		= Blockparitätsprüfung; Blocksicherung; Longitudinalprüfung; Längsparitätsprüfung; Längsparitätskontrolle; Längsprüfung
block check character	DATA COMM	Blockprüfzeichen
= BCC; block parity control character		= Blockparitätszeichen; Blockparitäts-Kontrollzeichen
↑ longitudinal redundancy check character		↑ Längsparitätszeichen
block check sequence	DATA COMM	Blockprüfzeichenfolge
= BCS; frame check sequence; FCS		= Fehlersicherungsteil
block code	CODING	Blockcode
block command	DATA COMM	Blockbefehl
block compaction (→block compression)	DATA PROC	(→Blockkompression)
block compression	DATA PROC	Blockkompression
= block compaction		
block control	DATA COMM	Blocksicherung
[the control information is within the block]		[die Prüfinformation ist im Block enthalten]
= block securing; block parity check; redundancy check		
block copy	DATA PROC	Blockkopie
block diagram	TELEC	Blockdiagramm
= functional diagram		= Blockschaltbild; Übersichtsschaltplan; Blockschema; Blockschaltung; Funktionsbild
blocked record	DATA PROC	geblockter Satz
block error probability	TELEGR	Blockfehlerwahrscheinlichkeit
block error rate	DATA COMM	Blockfehlerhäufigkeit
= BLER		= Blockfehlerquote
block factor	DATA PROC	Blockfaktor
[number of records grouped to a block]		[Anzahl de Sätze die zu je einem Block zusammengefaßt werden]
= blocking factor		= Blockungsfaktor
block flag (n.)	DATA COMM	Blockbegrenzer
		= Blockbegrenzung; Flag
block format	DATA PROC	Blockformat
block gap (→interblock gap)	TERM&PER	(→Blockzwischenraum)
block gluing equipment	OFFICE	Blockverleimgerät
block graphics	DATA PROC	Blockgraphik
[with a fixed set of graphic characters]		[mit definiertem Satz von Graphikzeichen]
block header	DATA COMM	Blockkopf
		= Blockvorspann

block identification field	DATA PROC	Blockkennungsfeld
= block identifier		
block identifier (→block identification field)	DATA PROC	(→Blockkennungsfeld)
block-ignore character	DATA COMM	Blockungültigkeitszeichen
blocking	SWITCH	Blockierung
= congestion		↓ äußere Blockierung
↓ all trunks busy		
blocking [combination of records]	DATA PROC	Blockung [Zusammenfassung von Datensätzen]
blocking	ELECTRON	Sperre
= lock		
blocking	NETW.TH	Sperrung
= rejection 1		
blocking acknowledgement signal	DATA COMM	Sperrbestätigungs-Kennzeichen
blocking bias (→reverse voltage)	ELECTRON	(→Sperrspannung)
blocking capacitance	MICROEL	Sperrkapazität
blocking capacitor	CIRC.ENG	Sperrkondensator
= block capacitor		= Blockkondensator
blocking characteristic (→reverse characteristic)	ELECTRON	(→Rückwärtskennlinie)
blocking circuit	CIRC.ENG	Sperrschaltung
= paralysis circuit		
blocking circuit (→parallel-resonant circuit)	NETW.TH	(→Parallelschwingkreis)
blocking code	SWITCH	Kennzahlsperre
blocking factor (→block factor)	DATA PROC	(→Blockungsfaktor)
blocking filter	NETW.TH	Sperrfilter
= stop filter; notch filter		= Sperre
blocking ground potential	ELECTRON	Sperrerde
blocking layer (→depletion layer)	MICROEL	(→Sperrschicht)
blocking magnet	POWER ENG	Sperrmagnet
blocking mechanism (→bolt)	TECH	(→Riegel)
blocking oscillator	CIRC.ENG	Sperrschwinger
= self-blocking oscillator; squegging oscillator		
blocking period (→blocking time)	ELECTRON	(→Sperrzeit)
blocking probability	SWITCH	Blockierungswahrscheinlichkeit
blocking signal	DATA COMM	Sperrkennzeichen
blocking signal	ELECTRON	Sperrsignal
blocking state	ELECTRON	Sperrzustand
= inhibiting condition		
blocking-state region (→cutoff region)	ELECTRON	(→Sperrbereich)
blocking time	ELECTRON	Sperrzeit
= off time; guard time; paralysis time; blocking period; off period ≈ deadtime		= Schutzzeit ≈ Totzeit
blocking valve	TECH	Sperrventil
= gating valve; stop valve		= Absperrventil
blocking voltage	COMPON	Blockierspannung
blocking voltage (→reverse voltage)	ELECTRON	(→Sperrspannung)
block layout	MICROEL	Block Layout
= floor plan		
block length	DATA PROC	Blocklänge
block letter (→capital character)	TYPOGR	(→Großbuchstabe)
block letters [a sans serif roman type]	TYPOGR	Blockschrift [eine serifenlose lateinische Druckschrift]
block list	DATA PROC	Blockliste
block mark	DATA PROC	Blockmarke
		= Blockendemarke
block marker	DATA PROC	Blockmarkierer
block multiplex channel [interconnects a computer with several fast peripherals] ≠ byte multiplex channel ↑ multiplex channel	DATA PROC	Blockmultiplexkanal [verbindet einen Computer mit mehreren schnellen Peripheriegeräten] ≈ Selektorkanal ≠ Bytemultiplexkanal ↑ Multiplexkanal
block multiplex mode	DATA COMM	Blockmultiplexbetrieb
block name [FORTRAN]	DATA PROC	Speicherbereichsname [FORTRAN]
block operation	DATA PROC	Blockmanipulation
block-oriented RAM (→BORAM)	DATA PROC	(→blockorientierter Schreib-Lese-Speicher)
block parity	DATA COMM	Blockparität
↑ longitudinal parity		↑ Längsparität
block parity check (→block control)	DATA COMM	(→Blocksicherung)
block parity control character (→block check character)	DATA COMM	(→Blockprüfzeichen)
block retrieval	DATA PROC	Blockrückgewinnung
block securing (→block control)	DATA COMM	(→Blocksicherung)
block sequence	DATA COMM	Blockfolge
block-stapling machine	OFFICE	Blockheftmaschine
block structure	DATA COMM	Blockstruktur
block synchronization	TELEGR	Blocksynchronisierung
		= Blocksynchronisation
block transfer	DATA PROC	Blocktransfer
		= Blockverlagerung
Blondel (→apostilb)	PHYS	(→Apostilb)
bloom (n.)	TV	Bildweichheit
		= Überstrahlung
bloom	METAL	Vorblock
		= Luppe
blooming (n.) [increase of disoplay size due to overload]	ELECTRON	Überstrahlung [Elektronenröhre]
blooming mill	METAL	Vorwalzwerk
= slabbing mill		= Luppenwalzwerk
bloop (n.) [unpleasing sound caused by a tape splice]	EL.ACOUS	Klebstellengeräusch
blow (v.t.) [to activate a protecting device]	COMPON	auslösen [einer Sicherung] = ansprechen 2
blow (v.t.) (→program)	MICROEL	(→schießen)
blower (→fan)	TECH	(→Ventilator)
blower (→blast)	TECH	(→Gebläse)
blowing current [of a overcurrent fuse] = tripping current; release current	COMPON	Auslösestrom [einer Stromsicherung]
blow torch	METAL	Lötlampe
blow-up (v.t.)	TECH	sprengen 1
= blast		[mit Explosionsmittel]
blow-up (→enlarge)	TECH	(→vergrößern)
blue	PHYS	blau
blue glow (→glow light)	PHYS	(→Glimmlicht)
blueish (→bluish)	PHYS	(→bläulich)
blueprint [copy on photosensitive paper, from transparent original] = diazo print ↑ copy (N.)	DOC	Lichtpause [Pause einer Transparentvorlage, auf fotoempfindlichem Papier] = Blaupause ↑ Pause
blue ribbon program [works properly on first try]	DATA PROC	Blaues-Band-Programm [auf Anhieb fehlerfrei]
blue screen	TV	Blaustanze
blue signal (→alarm indication signal)	TRANS	(→Alarm-Meldesignal)
bluish	PHYS	bläulich
= blueish		
blunt 1 [not pricking] ≠ pointed	TECH	stumpf 1 [nicht stechend] ≠ spitz
blunt 2 [not cutting] ≠ sharp	TECH	stumpf 2 [nicht schneidend] ≠ scharf
blunt (v.t.) (→round-off)	MECH	(→abkanten 1)
blur (v.t.)	TERM&PER	unscharf machen
blur (n.)	TERM&PER	Unschärfe
= smear (n.)		= Verwaschung
blur (n.) (→blurred picture)	TERM&PER	(→unscharfes Bild)
blurr (v.t.)	PHYS	verschmieren
blurred	TV	unscharf
= confused; diffuse; out of focus; fuzzy		= verwaschen
blurred picture	TERM&PER	unscharfes Bild
= blur (n.)		

blurring

blurring	PHYS	**Verschmierung**	
[of a pulse, a spectral line]		[eines Impulses, einer Spektrallinie u.dgl.]	
blurring	DATA PROC	**Unschärfeerzeugung**	
[computer graphics; to make fuzzy as an anti-aliasing measure]		[Computergraphik; zur Minderung von Bildunschönheiten]	
		= Blurring	
bn (→billion)	MATH	(→Milliarde)	
BNC connector	COMPON	**BNC-Stecker**	
BNF (→Backus-Naur form)	DATA PROC	(→Backus-Naur-Form)	
board	SWITCH	**Platz**	
board 2 (→committee)	ECON	(→Ausschuß)	
board (→module)	EQUIP.ENG	(→Baugruppe)	
board 3 (n.) (→advisory council)	ECON	(→Beirat)	
board 1 (→body)	ECON	(→Gremium)	
board (→printed circuit board)	ELECTRON	(→Leiterplatte)	
board (→console)	EQUIP.ENG	(→Pult)	
board computer	DATA PROC	(→Einkartenrechner)	
(→single-board computer)			
board-exchange warranty	DATA PROC	**Kartentausch-Garantie**	
board extender	INSTR	**Adapter**	
= extender board; card extender; adapter		[zur Prüfung herausgezogener Steckmodule]	
		= Baugruppenadapter; Adapterbaugruppe	
board guide bar (→module guide bar)	EQUIP.ENG	(→Baugruppenführung)	
board position	EQUIP.ENG	**Baugruppenposition**	
= module position; assembly position; subassembly position; plug-in position; insert position		= Steckposition; Modulposition	
board tester	MANUF	**Leiterplatten-Testgerät**	
board test system	INSTR	**Leiterplatten-Prüfsystem**	
= PCB test system			
board variant	EQUIP.ENG	**Baugruppenvariante**	
= module variant; assembly variant; subassembly variant		= Modulvariante	
bobbin	COMPON	**Wickelkörper**	
= coil former		= Spulenkörper	
bobbin (→reel)	TECH	(→Spule 1)	
bobtail curtain	ANT	**Bobtail-Vorhang-Antenne**	
		[bobtail = engl. Stutzschwanz]	
Bode diagram (→frequency characteristic)	TELEC	(→Frequenzkennlinie)	
body	ECON	**Gremium**	
= board			
body	COMPON	**Griffkörper**	
[connector]		[Stecker]	
body	DATA COMM	**Hauptteil**	
[of a message]		[einer Mitteilung]	
body (→body text)	TYPOGR	(→Grundschrifttext)	
body (→current text)	TYPOGR	(→Lauftext)	
body antenna	ANT	**Flächenantenne 2**	
[a radiating body]		[ein strahlender Körper]	
body-centered	PHYS	**raumzentriert**	
[crystal]		[Kristall]	
body dose	PHYS	**Körperdosis**	
body heat detector	SIGN.ENG	**Körperwärmemelder**	
body height	TYPOGR	**Kegel**	
[maximum vertical span of a typefont, from maximum descender to maximum ascender]		[maximale Spannweite einer Schriftart, von größter Unter- zu größter Oberlänge]	
= body size		= Schriftkegel; Kegelmaß; Kegelgröße; Kegelhöhe	
≈ point size		≈ Punktgröße	
body of book	TYPOGR	**Buchblock**	
[part of book without cover]		[Teil des Buches ohne Deckel]	
body size (→body height)	TYPOGR	(→Kegel)	
body text	TYPOGR	**Grundschrifttext**	
= body			
body text type (→body type)	TYPOGR	(→Grundschrift)	
body type	TYPOGR	**Grundschrift**	
[font used for the main text]		[für den Haupttext verwendeter Zeichensatz]	
= body text type			
BOF (→beginning-of-tape mark)	TERM&PER	(→Bandanfangsmarke)	
BOF (→header label)	DATA PROC	(→Dateianfangs-Etikett)	
BOF mark	TERM&PER	(→Bandanfangsmarke)	
(→beginning-of-tape mark)			
BOF marker	TERM&PER	(→Bandanfangsmarke)	
(→beginning-of-tape mark)			
Bohr magneton	PHYS	**Bohr'sches Magneton**	
[unit]		[Konstante]	
Bohr radius	PHYS	**Bohr'scher Radius**	
[a constant]		[Konstante]	
boil	PHYS	**sieden**	
≈ evaporate		= kochen	
		≈ verdampfen	
boiler plate	METAL	**Kesselblech**	
boilerplate	DATA PROC	**Normtext**	
= prerecorded text element		= Textbaustein; vorformulierter Text	
boilerplate- (→preformulated)	DOC	(→vorformuliert)	
boilerplate letter	OFFICE	**Standardbrief**	
= form letter; standard letter		= vorformulierter Brief; Ganzbrief; Schemabrief	
boiler work	MECH	**Kesselbau**	
boiling point (→evaporation point)	PHYS	(→Verdampfungspunkt)	
boiling temperature (→evaporation point)	PHYS	(→Verdampfungspunkt)	
bold	TYPOGR	**fett**	
= bold-faced; bold face; black		= fettgedruckt; kräftig	
↑ font attribute		↑ Schriftattribut	
boldface (→boldface printing)	TYPOGR	(→Fettschrift)	
bold face (→bold)	TYPOGR	(→fett)	
bold-faced (→bold)	TYPOGR	(→fett)	
boldface printing	TYPOGR	**Fettschrift**	
= boldface type; boldface; bold tape; bold print		= Fettdruck	
boldface type (→boldface printing)	TYPOGR	(→Fettschrift)	
bold print (→boldface printing)	TYPOGR	(→Fettschrift)	
boldprinting	DATA PROC	**Fettdruckfunktion**	
[wordprocessing]		[Textverarbeitung]	
bold tape (→boldface printing)	TYPOGR	(→Fettschrift)	
bolometer	MICROW	**Bolometer**	
= barretter			
bolt (v.t.)	MECH	**schrauben 2**	
		[mit Mutter]	
		= verschrauben 2	
bolt (n.)	TECH	**Riegel**	
= blocking mechanism			
bolt 2 (n.)	MECH	**Schraube 2**	
≠ nut		[mit Mutter zu verwenden]	
		≠ Mutter	
bolt 1 (n.)	MECH	**Bolzen**	
= pin			
bolted	MECH	**geschraubt 2**	
		[mit Mutter]	
bolted joint	MECH	**Schraubverbindung 2**	
		[mit Mutter]	
bolt head (→screw head)	MECH	(→Schraubenkopf)	
Boltzmann constant	PHYS	**Boltzmann-Konstante**	
Boltzmann equation	PHYS	**Boltzmann-Gleichung**	
Boltzmann factor	PHYS	**Boltzmann-Faktor**	
bomb (n.) (→abnormal end)	DATA PROC	(→Absturz)	
bomb (v.i.) (→crash)	DATA PROC	(→abstürzen)	
BOMOS	MICROEL	**BOMOS-Technologie**	
= buried oxide metal-oxide semiconductor			
bond (v.t.)	TECH	**kleben**	
≈ cement		≈ kitten	
bond	MICROEL	**kontaktieren**	
		= bonden	
bond (→binding)	CHEM	(→Bindung)	
bond (→elbow)	MICROW	(→Bogen)	
bonded (→cemented)	TECH	(→gekittet)	
bonding	MICROEL	**Kontaktierung**	
		= Bonden	
bonding agent (→adhesive)	TECH	(→Kleber)	

bonding device		MANUF	**Klebevorrichtung**	**boom** (v.i.) (→roar)	ACOUS	(→dröhnen)
bonding diagram		MICROEL	**Bond-Plan**	**boost** (v.t.)	COLLOQ	**hochtreiben** [fig.]
bonding electron		PHYS	**Valenzelektron**	**boost** (n.)	COLLOQ	**Auftrieb** [fig.]
= valence electron; outer-shell electron; pripheral electron; binding electron			= Bindungselektron	**booster**	PHYS	**Booster** [Vakuumpumpe]
bonding method		MICROEL	**Kontaktierungsverfahren**	**booster** (→antenna booster)	HF	(→Antennenverstärker)
book (v.t.)		ECON	**buchen**	**booster amplifier**	TRANS	**NF-Zwischenverstärker**
= post (v.t.)			= verbuchen	= voice frequency amplifier; VF amplifier		
book (n.) [typefont]		TYPOGR	**Buchschrift** [Schriftart]	**booster diode** [TV]	TV	**Schalterdiode** = Boosterdiode
book		LING	**Buch**	= damper		
booking [call queuing]		SWITCH	**Anmelden** [Warten]	**booster oscillator** (→control oscillator)	CIRC.ENG	(→Steueroszillator)
booking = entry; posting; reservation		ECON	**Buchung**	**booster station** (→radio relay station)	RADIO	(→Relaisfunkstelle)
booking (→call booking)		SWITCH	(→Gesprächsanmeldung)	**booster transformer**	EL.TECH	**Saugtransformator**
booking system		DATA PROC	**Buchungssystem**	= negative boosting transformer; draining transformer		
booking terminal = reservation terminal		TERM&PER	**Buchungsplatz** = Buchungsstation	**boot** (v.t.) (→bootstrap)	DATA PROC	(→urladen)
book jacket = dust jacket; wrapper		TYPOGR	**Buchumschlag** = Umschlaghülle; Schutzumschlag	**boot address** (→start address)	DATA PROC	(→Startadresse)
bookkeeper ≈ accountant		ECON	**Buchhalter** ≈ Fachmann für Rechnungswesen	**boot drive** [from which the operating system is loaded automatically after start]	DATA PROC	**Boot-Laufwerk** [von dem nach Einschalten das Betriebssystem automatisch geladen wird]
bookkeeping = accounting 2		ECON	**Buchführung** = Buchhaltung	**booth** (n.) = stand (n.); stall (n.); exhibition stand	ECON	**Messestand** = Stand; Ausstellungsstand
bookkeeping (→housekeeping)		DATA PROC	(→Systemverwaltung)	**booth** = cabinet	CIV.ENG	**Zelle** = Kabine
bookkeeping operation (→housekeeping sequence)		DATA PROC	(→Organisationsablauf)	**booting** (→bootstrapping)	DATA PROC	(→Urladen)
booklet (→brochure)		LING	(→Broschüre)	**bootlace lens**	ANT	**Schuhbandlinse**
book value		ECON	**Buchwert**	**bootleg** (→pirated copy)	DATA PROC	(→Raubkopie)
bookware [book with data medium]		DATA PROC	**Bookware** [Buch mit Datenträger]	**boot partition** [partition of fixed disk memory, reserved for booting routines]	DATA PROC	**Boot-Partition** [Speicherbereich der Festplatte für Startroutinen]
boolean = Boolean		MATH	**boolesch** ↓ aussagenlogisch	= active partition		= aktive Partition
Boolean (→boolean)		MATH	(→boolesch)	**boot record**	DATA PROC	**Startroutine**
Boolean algebra [algebra of binary variables] = logic algebra; symbolic algebra; Boolean logic ↑ propositional calculus ↓ set theory; propositional calculus; engineering logic [INF]		MATH	**Boolesche Algebra** [Algebra binärer Variablen] = Bool'sche Algebra; logische Algebra; Algebra der Logik; symbolische Algebra; zweiwertiger Aussagenkalkül; Boolesche Logik; Bool'sche Logik; Boolsche Algebra; Boolsche Logik ↑ Aussagenlogik ↓ Mengenlehre; Aussagenlogik; Schaltalgebra [INF]	**boot ROM** [contains primary boot routines]	DATA PROC	**Boot-ROM** [enthält die primären Startroutinen]
				bootstrap (v.t.) [to activate a computer by loading a program from a storage] = to load initial program; boot (v.t.); boot-up (v.t.)	DATA PROC	**urladen** [einen Computer durch Einlesen eines Programms von einem Speicher in Betrieb setzen] = booten
				bootstrap (n.) [the very first input into a main memory]	DATA PROC	**Ureingabe** [allererste Eingabe in einen Arbeitsspeicher; bootstrap = engl. „Schnürriemen"] = Kernanlauf
Boolean function (→Boolean operation)		ENG.LOG	(→Boolesche Verknüpfung)	**bootstrap** (adj.) (→self-loading)	DATA PROC	(→selbstladend)
Boolean instruction ↑ logic instruction		DATA PROC	**Boolescher Befehl** = Bool'scher Befehl; Boolscher Befehl ↑ logischer Befehl	**bootstrap generator** (→Miller integrator)	CIRC.ENG	(→Miller-Integrator)
Boolean logic (→Boolean algebra)	MATH		(→Boolesche Algebra)	**bootstrap initialization key**	DATA PROC	**Urladetaster**
Boolean operation = Boolean function		ENG.LOG	**Boolesche Verknüpfung** = Boolesche Funktion; Boolesche Operation; Boole'sche Verknüpfung; Boole'sche Funktion; Boole'sche Operation; Boolsche Verknüpfung; Boolsche Funktion; Boolsche Operation	**bootstrap initialization program**	DATA PROC	**Urladeschalter**
				bootstrap loader (→initial program loader)	DATA PROC	(→Urlader)
				bootstrapping = initial program loading; IPL; boot-up (n.); booting	DATA PROC	**Urladen** = Booten
Boolean operator ↑ logic operator ↓ AND; OR; XOR; NOR; NOT		DATA PROC	**Verknüpfungsoperator** ↑ logischer Operator; Boole'scher Operator ↓ UND; ODER; EXKLUSIV ODER; WEDER-NOCH	**bootstrap program** ≈ initial program loader	DATA PROC	**Ureingabeprogramm** [Programm für erste Ladung in einen Arbeitsspeicher] ≈ Urlader
				boot-up (v.t.) (→bootstrap)	DATA PROC	(→urladen)
Boolean variable		ENG.LOG	**Boolesche Variable** = Boole'sche Variale; Boolsche Variable	**boot-up** (n.) (→bootstrapping)	DATA PROC	(→Urladen)
boom [yagi antenna] = longitudinal support		ANT	**Längsträger** [Yagi-Uda-Antenne] = Boom	**BORAM** = block-oriented RAM	DATA PROC	**blockorientierter Schreib-Lese-Speicher** = blockorientierter RAM; BORAM

border

border (n.)		TYPOGR	seitlicher Papierrand
border (n.)		TYPOGR	Umrahmung
= box			= Umrandung; Einrahmung
border (→margin)		TECH	(→Rand)
borderline (→boundary)		TECH	(→Grenze)
borderline (→borderline case)		COLLOQ	(→Grenzfall)
borderline case		COLLOQ	Grenzfall
= borderline			↓ bester Fall; schlimmster Fall
↓ best case; worst case			
border-punched card		TERM&PER	(→Randlochkarte)
(→marginal punched card)			
bore 2 (n.)		MECH	Lochdurchmesser
			= Bohrungsdurchmesser
bore (v.t.) (→drill)		MECH	(→bohren)
bore 1 (n.) (→drill hole)		MECH	(→Bohrloch)
bored (→drilled)		MECH	(→gebohrt)
borehole (→drill hole)		MECH	(→Bohrloch)
Borel set		MATH	Borel-Menge
bore-out (v.t.) (→drill-out)		MECH	(→ausbohren)
boresight (n.)		TECH	Zielrichtung
= bore sight			
boresight		ANT	Hauptstrahlrichtung
= bore sight; maximum-radiation direction; main-radiation direction; beam axis			= Keulenachse
			≈ Zielrichtung
≈ reference boresight			
bore sight (→boresight)		ANT	(→Hauptstrahlrichtung)
bore sight (→boresight)		TECH	(→Zielrichtung)
boresight antenna		ANT	Einmeßantenne
boring (→drill hole)		MECH	(→Bohrloch)
boring tool		MECH	Bohrspitze
boron		CHEM	Bor
= B			= B
borrow (v.t.)		MATH	borgen
borrower		ECON	Entleiher
≈ debtor			= Darlehensnehmer
≠ lender			≈ Schuldner
			≠ Verleiher
borrower (→creditor)		ECON	(→Gläubiger)
borrowing		ECON	Schuldenaufnahme
≈ indebtedness			= Verschuldung
BOS (→basic operating system)		DATA PROC	(→Grundbetriebssystem)
Bose-Chandhuri-Hocquenghem code		CODING	Bose-Chandhuri-Hocquenghem-Code
= BCH code			= BCH-Code
Bose-Einstein statistics		PHYS	Bose-Einstein-Statistik
boss (v.t.)		MECH	punzen
boss (n.)		MECH	Punze
both-way		TELEC	doppeltgerichtet
≈ two-way; bidirectional			= Zweiweg-; Zweirichtungs-; bidirektional
≈ duplex			
≠ one-way			≠ einfachgerichtet
Bott-Duffin procedure		NETW.TH	Bott-Duffin-Verfahren
bottle		TECH	Flasche
= flask			
bottleneck		COLLOQ	Engpaß
= defile (n.)			= Flaschenhals (fig.)
bottom		TECH	Boden 2
[of a vessel]			[eines Behälters]
bottom (→floor)		CIV.ENG	(→Sohle)
bottom die (→female die)		MECH	(→Matrize)
bottom edge		TECH	Unterkante
bottom plate (→base plate)		MECH	(→Grundplatte)
bottom-up method		DATA PROC	(→Bottom-up-Programmierung)
(→bottom-up programming)			
bottom-up programming		DATA PROC	Bottom-up-Programmierung
[starting from partial problems]			[bei Teilproblemen beginnend]
= bottom-up technique; bottom-up method			
bottom-up technique		DATA PROC	(→Bottom-up-Programmierung)
(→bottom-up programming)			
bottom view		ENG.DRAW	Untersicht
Botzmann distribution		PHYS	Boltzmann-Verteilung
boule (→preform)		OPT.COMM	(→Vorform)
bounce (v.i.)		MECH	prellen
= rebound (v.i.)			
bounce (n.)		MECH	Prellen
= rebound (n.)			
bounce (→contact bounce)		COMPON	(→Kontaktprellen)
bounce-free switch		COMPON	prellfreier Schalter
bounce time		COMPON	Prellzeit
bound (n.)		DATA PROC	Beschränkung
= limitation			
bound 1 (n.)		MATH	Schranke
[set theory]			[Mengenlehre]
bound 2		MATH	Werteschranke
[error analysis]			[Fehlerrechnung]
bound		TECH	gebunden
[fig.]			[fig.]
			= belegt
bound (→restriction)		TECH	(→Beschränkung)
bound (→limit)		TECH	(→begrenzen)
boundary		TECH	Grenze
= limit; borderline			≈ Rand
boundary address		SWITCH	Grenzadresse
boundary condition		MATH	Randbedingung
= constraint; limiting condition			= Grenzbedingung; Nebenbedingung; Zwangsbedingung
boundary current		MICROEL	Randströmung
= random current			
boundary curve		MATH	Randkurve
boundary layer		PHYS	Grenzschicht
			= Unstetigkeitsschicht
boundary layer (→surface layer)		MICROEL	(→Randschicht)
boundary light		RADIO NAV	Grenzlichtbake
boundary marker		RADIO NAV	(→Platz-Einflugzeichen)
(→boundary marker beacon)			
boundary marker beacon		RADIO NAV	Platz-Einflugzeichen
= boundary marker			= Grenzmarkierungs-Funkfeuer; Grenzfunkbake
boundary protection		DATA PROC	Bereichschutz
boundary punctuation		DATA PROC	Bereichsinterpunktion
boundary register		DATA PROC	Bereichsregister
boundary representation		DATA PROC	Umrißdarstellung
[computer graphics; volume modelling with data on distance, orientation and hiding]			[Computergraphik; Darstellung von Festkörperbildern aus Angaben über Entfernung, Orientierung und Verdeckungen]
boundary surface		MATH	Grenzfläche
boundary value		MATH	Randwert
boundary value problem		MATH	Randwertproblem
boundary wavelength (→cutoff wavelength)		MICROEL	(→Grenzwellenlänge)
bound mode (→core mode)		OPT.COMM	(→Kernmodus)
bow compass (→drop pen)		ENG.DRAW	(→Nullenzirkel)
bowl (→shell 1)		TECH	(→Schale)
bow-tie antenna (→batwing antenna)		ANT	(→Schmetterlingsantenne)
box		TECH	Kasten
≈ light box; vessel			= Kiste
↑ vessel			≈ Schachtel
			↑ Behälter
box (→border)		TYPOGR	(→Umrahmung)
box (→frame)		DOC	(→Kästchen)
boxed		TECH	in Schachtel verpackt
box process		MICROEL	Box-Verfahren
b.p. (→evaporation point)		PHYS	(→Verdampfungspunkt)
bpi (→bits per inch)		TERM&PER	(→Bits pro Zoll)
bps (→bits per second)		TELEC	(→Bit/s)
BPS (→bits per second)		TELEC	(→Bit/s)
bp/s (→bits per second)		TELEC	(→Bit/s)
BPU (→processor 1)		DATA PROC	(→Prozessor 1)
Br (→bromine)		CHEM	(→Brom)
brace 1 (v.t.)		TECH	verspannen
= guy (v.t.)			= verstreben
brace (→curly bracket)		MATH	(→geschweifte Klammer)
bracing		TECH	Verstrebung
bracket 1 (n.)		MATH	Klammer
↓ parenthesis; square bracket; angle bracket; curly bracking			↓ runde Klammer; eckige Klammer; spitze Klammer; geschweifte Klammer
bracket		MECH	Ausleger
bracket (v.t.) (→enclose in brackets)		MATH	(→einklammern)
bracket 2 (n.) (→square bracket)		MATH	(→eckige Klammer)

bracket (→console table)	TECH	(→Konsole)	
bracket (→support)	TECH	(→Stütze)	
bracket (→console)	EQUIP.ENG	(→Pult)	
bracket-free	MATH	klammerlos	
= bracketless			
bracketless (→bracket-free)	MATH	(→klammerlos)	
bracket relation	MATH	**Klammerausdruck**	
brackets 3 (n.)	MATH	(→runde Klammer)	
(→parenthesis)			
Bragg effect	PHYS	**Bragg-Effekt**	
		= Farbfilter-Effekt	
braid breaker (→RF blocking transformer)	ANT	(→HF-Trenntransformator)	
braided conductor	COMM.CABLE	(→Litzendraht)	
(→stranded wire)			
braided shield	COMM.CABLE	**Schirmgeflecht**	
brake (v.t.; v.i.)	MECH	bremsen	
brake (n.)	MECH	**Bremse**	
branch (n.)	MECH	**Abzweig**	
		= Verzweigung	
branch (n.)	MICROW	**Arm**	
branch 1 (n.)	ECON	**Branche**	
= business line; kind of business		= Sparte	
branch (n.)	ECON	**Zweigwerk**	
= division (n.)			
branch	SCIE	**Fachgebiet**	
= special branch; speciality		= Fach; Spezialdisziplin; Spezialgebiet; Spezialfach	
branch	NETW.TH	**Zweig**	
branch 3 (→branch office)	ECON	(→Filiale)	
branch (→program branch)	DATA PROC	(→Programmzweig)	
branch (→subdivision)	ECON	(→Unterabteilung)	
branch cable (→distribution cable)	OUTS.PLANT	(→Sekundärkabel)	
branch condition (→jump condition)	DATA PROC	(→Sprungbedingung)	
branch current	POWER ENG	**Zweigstrom**	
branch distribution cable	POWER SYS	**Zweigverbraucherleitung**	
branched	TECH	**verzweigt**	
≈ tree-like		≈ baumähnlich	
branched duplexer	NETW.TH	**Abzweig-Duplexer**	
branched program	DATA PROC	**verzweigtes Programm**	
branched star connection	DATA COMM	(→passive Sternanschaltung)	
(→passive bus connection)			
branch exchange (→private branch exchange)	SWITCH	(→Nebenstellenanlage)	
branch feeder cable	OUTS.PLANT	**Verteilkabel** [zweigt vom Hauptkabel ab]	
branching	NETW.TH	**Verzweigung**	
= fork		≈ Zweig	
branching (→coupler)	ELECTRON	(→Koppler)	
branching circulator	MICROW	**Verzweigungszirkulator**	
branching closure (→drop sleeve)	OUTS.PLANT	(→Abzweigmuffe)	
branching filter (→directional filter)	RADIO REL	(→Richtungsweiche)	
branching filter (→duplexer filter)	RADIO	(→Sende-Empfangs-Weiche)	
branching jack (→bridging jack)	TELEPH	(→Parallelklinke)	
branching network	OUTS.PLANT	**Verzweigungsnetz**	
branch instruction (→jump instruction)	DATA PROC	(→Sprungbefehl)	
branch jack (→bridging jack)	TELEPH	(→Parallelklinke)	
branch line [between a front-end equipment and the subscriber]	TELEC	**Zweigleitung** [zwischen Vorfeldeinrichtung und Teilnehmer]	
= main line		= Hauptleitung	
branch-off	NETW.TH	**abzweigen**	
branch-off	TRANS	**abzweigen**	
= drop (v.t.)			
branch-off amplifier	TELEGR	**Staffelverstärker**	
branch office	ECON	**Filiale**	
= branch 3; regional office; sales and support office		= Niederlassung; Zweigstelle; Zweigniederlassung; Geschäftsstelle	
≈ off-premises		≈ Außenstelle	
branch-off sleeve (→drop sleeve)	OUTS.PLANT	(→Abzweigmuffe)	
branch-off traffic	TELEGR	**Staffelverkehr**	
branch-on condition	DATA PROC	**bedingte Verarbeitung**	
branch point	DATA PROC	**Verzweigungspunkt**	
branch statement (→jump instruction)	DATA PROC	(→Sprungbefehl)	
brand	ECON	**Marke**	
brand (→brand name)	ECON	(→Markenname)	
brand (→trademark)	ECON	(→Markenzeichen)	
brand name	ECON	**Markenname**	
= brand		= Ma1rke	
≈ trademark		≈ Markenzeichen	
brandname merchandise	ECON	**Markenware**	
= brandname product			
brandname product (→brandname merchandise)	ECON	(→Markenware)	
brand-new (adj.)	COLLOQ	**funkelnagelneu**	
		= nagelneu; brandneu	
brand-new	ECON	**fabrikneu**	
= factory-new			
brand policy	ECON	**Markenpolitik**	
brass	METAL	**Messing**	
brass casting	METAL	**Messingguß**	
brass wire	METAL	**Messingdraht**	
Braun tube (→cathode ray tube)	ELECTRON	(→Kathodenstrahlröhre)	
braze 1	METAL	**messinghartlöten**	
↑ hard-solder		= messinglöten	
		↑ hartlöten	
braze 2 (→hard-solder)	METAL	(→hartlöten)	
brazing	METAL	**Messinghartlötung**	
↑ hard soldering		↑ Hartlötung	
brazing alloy (→brazing solder)	METAL	(→Messinghartlot)	
brazing paste	METAL	**Hartlötpaste**	
brazing solder	METAL	**Messinghartlot**	
= brazing alloy; brazing spelter		= Schlaglot; Messinglot	
brazing spelter (→brazing solder)	METAL	(→Messinghartlot)	
breadboard (→breadboard circuit)	ELECTRON	(→Brettschaltung)	
breadboard circuit	ELECTRON	**Brettschaltung**	
= breadboard		= Laborplatine; Experimentierplatine	
breadboard engineer (→circuit engineer)	ELECTRON	(→Schaltungsentwickler)	
break (v.t.) [a difficult code]	TELEC	**knacken** [einen schwierigen Code]	
↑ decipher		↑ entziffern	
break (n.) [point of]	TECH	**Bruchstelle**	
break	ENG.DRAW	**Bruchkante**	
		↑ Kante	
break (→pause)	COLLOQ	(→Pause)	
break (→dot)	ENG.DRAW	(→stricheln)	
break (v.t.) (→interrupt)	EL.TECH	(→unterbrechen)	
break (n.) (→interruption)	EL.TECH	(→Unterbrechung)	
breakable (→fragile)	TECH	(→zerbrechlich)	
breakage (→breaking 2)	MECH	(→Bruch)	
break-before-make contact ≠ make-before-break contact	COMPON	**Folgekontakt Öffnen-vor-Schließen** [Relais] ≠ Folgeumschaltekontakt	
break-break contact ↑ relay contact	COMPON	**Zwillingsöffner** = Zwillingsruhekontakt ↑ Relaiskontakt	
break character (→underline character)	TYPOGR	(→Unterstreichungszeichen)	
break contact [open when operated, closed when disactivated]	COMPON	**Ruhekontakt** [offen bei Betätigung, geschlossen in Ruhestellung]	
= resting contact; interrupter; normally closed contact; N/C contact; breaker contact; normal contact		= Trennkontakt; Öffnerkontakt; Öffner; Unterbrechungskontrakt; Unterbrecherkontakt; Unterbrecher	
≠ make contact		≠ Arbeitskontakt	
↑ relay contact		↑ Relaiskontakt	

breakdown 2 | | | 56

breakdown 2 (n.)	TECH	**Betriebsstörung**	
≈ service interruption		≈ Ausfall	
≈ failure			
breakdown (n.) (NAM)	PHYS	**Durchschlag**	
[discharge through an isolator]		[Entladung durch einen Isolierstoff]	
= break-down (BRI); disruption; disruptive discharge; puncture; punch-through breakdown		= Durchbruch	
		↑ Entladung	
↑ discharge		↓ Überschlag	
↓ spark-over			
breakdown (→itemized breakdown)	COLLOQ	(→Aufschlüsselung)	
breakdown (n.) (→failure)	QUAL	(→Ausfall)	
break down (→loss)	TELEC	(→Ausfall)	
break-down (BRI)	PHYS	(→Durchschlag)	
(→breakdown)			
break down (→partition)	TECH	(→aufteilen)	
breakdown (v.t.) (→collapse)	TECH	(→kollabieren)	
breakdown 1 (→collapse)	TECH	(→Zusammenbruch)	
breakdown voltage	PHYS	**Durchbruchspannung**	
= disruptive voltage; puncture voltage		= Durchschlagspannung	
breakdown voltage	MICROEL	**Durchbruchspannung**	
= disruptive voltage		≈ Umkehrspannung	
≈ reverse isolation		↓ Zenerspannung; Lawinendurchbruchspannung	
↓ Zener voltage; avalanche voltage			
breaker (→circuit breaker)	COMPON	(→Schutzschalter)	
breaker contact (→break contact)	COMPON	(→Ruhekontakt)	
breaker plug (→splitting plug)	COMPON	(→Trennstecker)	
break-field (→retarding field)	PHYS	(→Bremsfeld)	
break-field tube (→retarding-field tube)	ELECTRON	(→Bremsfeldröhre)	
break-in (n.)	DATA COMM	**Gegenschreiben**	
break-in (→override)	TELEPH	(→aufschalten)	
break-in (→override)	TELEPH	(→Aufschalten)	
break-in detection	DATA COMM	**Gegenschreibauswertung**	
breaking 1	MECH	**Bremsung**	
= stopping; retardation; deceleration			
breaking 2	MECH	**Bruch**	
= breakage			
breaking load	MECH	**Bruchlast**	
= ultimate load		= Bruchbelastung	
breaking loss	MICROEL	**Abschaltverlust**	
breaking loss (→turn-off loss)	MICROEL	(→Ausschaltverlust)	
breaking power	MECH	**Bremsvermögen**	
= stopping power			
break key	TERM&PER	**Unterbrechungstaste**	
break-make contact (→changeover contact)	COMPON	(→Umschaltekontakt)	
break-off (v.t.) (→disconnect 2)	TELEC	(→abbrechen)	
breakover point [thyristor]	MICROEL	**Kippunkt** [Thyristor]	
		= Schaltpunkt	
breakover voltage [thyristor]	MICROEL	**Kippspannung** [Thyristor]	
		= Schaltspannung	
break point	MATH	**Knickpunkt**	
[where a curve changes abruptly its sense, the tangents beeing different]		[Kurve ändert sprunghaft den Sinn, Kurventangenten verschieden]	
≈ invertive point		≈ Rückkehrpunkt	
↑ singularity		↑ Singularität	
breakpoint	DATA PROC	**Unterbrechungspunkt**	
[address stopping the processing and permitting an external intervention]		[Adresse die ein Anhalten des Programms bedingt, und somit externe Eingriffe im Ablauf ermöglicht]	
= haltpoint			
≈ checkpoint		= Haltepunkt; Interrupt-Punkt; Breakpoint	
		≈ Fixpunkt	
breakpoint instruction (→halt)	DATA PROC	(→Halt)	
breakpoint interruption	DATA PROC	**Haltepunktanweisung**	
breakpoint symbol	DATA PROC	**Haltepunktsymbol**	
break-proof (→unbreakable 2)	TECH	(→unzerbrechlich)	
break signal	DATA COMM	**Anhaltesignal**	
break strip	COMPON	**Trennleiste**	
		= Trennstreifen	
break-through (→cutout)	TECH	(→Durchbruch)	
break-through (→wall opening)	CIV.ENG	(→Wanddurchbruch)	
breakup (→signal breakup)	ELECTRON	(→Signalverfälschung)	
B register (→B box)	DATA PROC	(→B-Register)	
bremsspectrum (→retardation spectrum)	PHYS	(→Bremsspektrum)	
bremsstrahlung	PHYS	**Bremsstrahlung**	
= retardation radiation; continuous radiation			
breve [phonetic symbol]	LING	**Kürzezeichen** [phonetisches Zeichen]	
Brevier (BRI) (→type size 8 point)	TYPOGR	(→Schriftgröße 8 Punkt)	
Brewster angle	PHYS	**Brewster-Winkel**	
Brewster angle (→angle of polarization)	PHYS	(→Polarisationswinkel)	
brick	CIV.ENG	**Ziegel 2**	
		= Ziegelstein	
bridge (v.t.)	EL.TECH	**überbrücken**	
bridge (→strap)	ELECTRON	(→Brücke)	
bridge (→bridge circuit)	EL.TECH	(→Brückenschaltung)	
bridge (→bridge processor)	DATA COMM	(→Bridge)	
bridge amplifier	CIRC.ENG	**Brückenverstärker**	
bridge arm	NETW.TH	**Brückenzweig**	
= ratio arm		= Quotientenzweig	
bridge balance	INSTR	**Brückenabgleich**	
		= Brückengleichgewicht	
bridge cable	COMM.CABLE	**Brückenkabel**	
bridge circuit	EL.TECH	**Brückenschaltung**	
= bridge connection; bridge		= Brücke	
bridge connection (→bridge circuit)	EL.TECH	(→Brückenschaltung)	
bridged taps (→multiple teeing)	OUTS.PLANT	(→Parallelschalten)	
bridged T section	NETW.TH	**überbrückte T-Schaltung**	
bridge feedback	CIRC.ENG	**Brückenrückkopplung**	
bridge filter	NETW.TH	**Brückenweiche**	
bridge lattice [impedance transformer]	ANT	**Boucherot-Brücke** [Impedanzwanler]	
bridge measurement	INSTR	**Brückenmeßverfahren**	
bridge method	INSTR	**Brückenverfahren**	
bridge modulator	TV	**Brückenmodulator**	
bridge network	NETW.TH	**Vierpolkreuzglied**	
bridge processor	DATA COMM	**Bridge**	
[connects data networks of same type]		[verbindet gleichartige Datennetze]	
= bridge		↑ Verbindungsrechner	
bridge rectifier	CIRC.ENG	**Brückengleichrichter**	
bridge transformer	COMPON	**Brückenübertager**	
		= Ausgleichübertrager	
bridgeware	DATA PROC	**Bridgeware**	
[hardware and software to transfer data between different computer types]		[Hardware und Software für den Datentransfer zwischen verschiedenen Computertypen]	
bridging	EL.TECH	**Überbrückung**	
		= Überkopplung	
bridging jack	TELEPH	**Parallelklinke**	
= branching jack; branch jack			
briefcase computer (→laptop computer)	DATA PROC	(→Aktentaschencomputer)	
brief description (→data sheet)	TECH	(→Datenblatt)	
brief interruption	TELEC	**Kurzzeitunterbrechung**	
= drop-out		= kurzzeitige Unterbrechung	
bright (adj.)	PHYS	**hell**	
= light 2 (adj.)			
bright (adj.)	TECH	**glänzend**	
= lustrous; brillant; shining			
bright annealed	METAL	**blankgeglüht**	
bright drawn	METAL	**blankgezogen**	
brightness	TECH	**Glanz**	
brightness	TV	**Helligkeit**	
= luminosity; luminance			
brightness (→luminance)	PHYS	(→Leuchtdichte)	

brightness control	TV	**Helligkeitsregler**
		= Bildhelligkeitsregler
bright rolled	METAL	**blankgewalzt**
bright steel	METAL	**Blankstahl**
brillance	PHYS	**Brillanz**
[sparkling brightness]		[funkelnde Helligkeit]
= brillancy; brillantness		
brillance	TV	**Brillanz**
brillancy (→brillance)	PHYS	(→Brillanz)
brillant	COLLOQ	**glanzvoll**
[fig.]		[fig.]
		= brillant
brillant (→bright)	TECH	(→glänzend)
brillant color (NAM)	PHYS	**glänzende Farbe**
= brillant colour		= Glanzfarbe
brillant colour (→brillant color)	PHYS	(→glänzende Farbe)
brillantness (→brillance)	PHYS	(→Brillanz)
Brillouin scattering	OPTOEL	**Brillouin-Effekt**
		= Brillouin-Streuung
Brillouin zone	PHYS	**Brillouin-Zone**
Brinell hardness	METAL	**Brinell-Härte**
brittle	TECH	**spröde**
[hard and frail, as e.g. glass]		[hart und leicht brechend, z.B. Glas]
≈ frail		= spröd
		≈ brüchig
broadband (n.)	TELEC	**Breitband**
= wideband (n.)		
broadband (adj.)	TELEC	**breitbandig**
= wideband adj.; wide-bandwidth		
broadband aerial (→broadband antenna)	ANT	(→Breitbandantenne)
broadband amplifier	CIRC.ENG	**Breitbandverstärker**
= wideband amplifier		
broadband antenna	ANT	**Breitbandantenne**
= broadband aerial; wideband antenna; wideband aerial		↓ frequenzunabhängige Antenne; Wanderwellenantenne; geometrisch dicke Antenne; parallelgeschaltete Antenne; widerstandsbelastete Antenne
↓ frequency-indipendent antenna; traveling-wave antenna; geometrically thick antenna; paralleled antenna; resistance-loaded antenna		
broadband aperture antenna	ANT	**Breitband-Flächenantenne**
= wideband aperture antenna		
broadband cable network	TELEC	**Breitband-Kabelnetz**
= wideband cable network		
broadband cable system	BROADC	**Breitbandkabelanlage**
↑ TV distribution system		= BK-Anlage
		↑ Fernsehverteilanlage
broadband cage antenna	ANT	**Breitbandreuse**
= wideband cage antenna		↑ geometrisch dicke Antenne; Vertikalantenne; Monopol
↑ geometrically thick antenna; vertical antenna; monopole antenna		
broadband channel	TELEC	**Breitbandkanal**
= wideband channel		
broadband communication	TELEC	**Breitbandkommunikation**
= wideband communication		
broadband compensation	ANT	**Breitbandkompensation**
= wideband compensation		
broadband conical monopole	ANT	**Breitband-Kegelantenne**
= wideband conical monopole		= Doppelkegel-Breitbandantenne
broadband dipole	ANT	**Breitband-Dipol**
= wideband dipole		
broadband feeding	ANT	**Breitbandspeisung**
= wideband feeding		≠ Schmalbandspeisung
≠ narrowband feeding		
broadband filter	NETW.TH	**Breitbandfilter**
= wideband filter		
broadband folded unipole	ANT	**Breitband-Faltunipol**
↑ unipole		↑ Monopolantenne
broadband ISDN	TELEC	**Breitband-ISDN**
= wideband ISDN		
broadband ISDN switch	SWITCH	**Breitband-ISDN-Vermittlung**
broadband line-of-sight radio (→high-capacity radio)	TRANS	(→Breitbandrichtfunk)
broadband link (→high capacity link)	TRANS	(→Breitbandstrecke)
broadband LOS (→high-capacity radio)	TRANS	(→Breitbandrichtfunk)
broadband loudspeaker	EL.ACOUS	**Breitbandlautsprecher**
= wideband loudspeaker		= Vollbereichslautsprecher
broadband measuring	INSTR	**Breitbandmessung**
= wideband measuring		
broadband microwave	TRANS	(→Breitbandrichtfunk)
(→high-capacity radio)		
broadband modem	DATA COMM	**Breitbandmodem**
broadband modulation	MODUL	**Breitbandmodulation**
= wideband modulation		
broadband network	TELEC	**Breitbandnetz**
= wideband network		
broadband noise (→wideband noise)	TELEC	(→Breitbandrauschen)
broadband path	TELEGR	**Breitbandstromweg**
broadband polarization diplexer	RADIO REL	**Breitband-Polarisationsweiche**
= wideband polarization filter		
broadband radio (→high-capacity radio)	TRANS	(→Breitbandrichtfunk)
broadband radio-relay (→high-capacity radio)	TRANS	(→Breitbandrichtfunk)
broadband rhombic antenna	ANT	**Breitband-Rhombusantenne**
= wideband rhombic antenna		= Wanderwellen-Rhombus
↑ travelling wave antenna		↑ Wanderwellenantenne
broadband route (→high capacity link)	TRANS	(→Breitbandstrecke)
broadband service	TELEC	**Breitbanddienst**
= wideband service		
broadband sweep	INSTR	**Breitbandwobbeln**
= wideband sweep		= Breitbandwobbelung
broadband synthesizer	INSTR	**Breitbandsynthesizer**
broadband system	TRANS	(→Breitbandsystem)
(→high-capacity system)		
broadband transmission	DATA COMM	**Breitbandtechnik**
[over a defined sector of baseband]		[ein definierter Bereich des Basisbandes steht zur Verfügung]
broadcast (v.t.)	BROADC	**senden**
		= ausstrahlen
broadcast (→point-to-multipoint operation)	TELEC	(→Punkt-zu-Mehrpunkt-Betrieb)
broadcast (→radio broadcasting)	RADIO	(→Rundfunk)
broadcast (→multi-address message)	DATA COMM	(→Rundschreiben)
broadcast address	BROADC	**Rundfunkansprache**
= broadcast speech		
broadcast antenna	ANT	**Rundfunkantenne**
= broadcasting antenna		
broadcast band	RADIO	**Rundfunkband**
broadcast command	TELECONTR	(→Sammelbefehl)
(→collective command)		
broadcast communications	TELEC	(→Verteilkommunikation)
(→distributive communication)		
broadcast conference	TELEC	**Rundspruchkonferenz**
broadcaster's coverage area	BROADC	(→Rundfunk-Versorgungsbereich)
(→broadcaster's service area)		
broadcaster's service area	BROADC	**Rundfunk-Versorgungsbereich**
= broadcaster's coverage area		
broadcasting (→radio broadcasting)	RADIO	(→Rundfunk)
broadcasting (→multi-address message)	DATA COMM	(→Rundschreiben)
broadcasting antenna (→broadcast antenna)	ANT	(→Rundfunkantenne)
broadcasting company	BROADC	**Rundfunkanstalt**
= broadcasting corporation		
broadcasting corporation	BROADC	(→Rundfunkanstalt)
(→broadcasting company)		
broadcasting network	BROADC	**Rundfunknetz**
↓ TV network		↓ Fernsehnetz
broadcasting satellite	SAT.COMM	(→Rundfunksatellit)
(→broadcast satellite)		
broadcasting-satellite service	SAT.COMM	**Rundfunkdienst über Satellit**
broadcasting service	TELEC	**Rundfunkddienst**
broadcasting system	TELEPH	**Rundsprechanlage**
broadcasting transmitter	BROADC	(→Rundfunksender)
(→broadcast transmitter)		
broadcast interview	BROADC	**Rundfunkinterview**

broadcast n.

broadcast n. (→ broadcast transmission) BROADC (→ Rundfunksendung 1)
broadcast operation TELEC (→ Punkt-zu-Mehrpunkt-Betrieb)
(→ point-to-multipoint operation)
broadcast program (NAM) BROADC Rundfunkprogramm
= broadcast programme (BRI); program (NAM); programme (BRI) ↓ Programm
↓ sound broadcast program; television broadcast programm ↓ Hörfunkprogramm; Fernsehprogramm
broadcast programme (BRI) BROADC (→ Rundfunkprogramm)
(→ broadcast program)
broadcast receiver RADIO Rundfunkempfänger
= Rundfunkempfangsgerät
broadcast satellite SAT.COMM Rundfunksatellit
= broadcasting satellite ↓ Hörfunksatellit; Fernsehsatellit
↓ sound broadcast satellite; television broadcast satellite
broadcast service user BROADC Rundfunkteilnehmer
↓ sound broadcast service user; television broadcast service user ↓ Hörfunkteilnehmer; Fernsehteilnehmer
broadcast speech (→ broadcast address) BROADC (→ Rundfunkansprache)
broadcast transmission BROADC Rundfunksendung 1
= radio broadcast; broadcast n.; transmission = Radiosendung; Sendung
↓ television broadcast; sound broadcast transmission ↓ Fernsehsendung; Hörfunksendung
broadcast transmitter BROADC Rundfunksender
= broadcasting transmitter ↓ Fernsehsender; Hörfunksender
↓ television transmitter; sound transmitter
broadcast videotex TELEC (→ Teletext)
(→ teletext)
broad dimension MECH Breitseite
≠ narrow dimension ≠ Schmalseite
broaden (→ widen) TECH (→ verbreitern)
broadening (→ widening) TECH (→ Verbreiterung)
broadly diversified COLLOQ breitgestreut
broadside (adj.) TECH breitseitig
≈ transversal ≈ transversal
broadside antenna (→ broadside array) ANT (→ Dipolebene)
broadside array ANT Dipolebene
= broadside antenna; broadside radiator; stacked dipole = gestockter Dipol; Querstrahler; Breitseitenantenne
≈ lazy H antenna; curtain array ≈ Lazy-H-Antenne; Vorhangantenne
broadside array with parallel elements (→ array of parallel dipoles) ANT (→ Dipolreihe)
broadside radiator (→ broadside array) ANT (→ Dipolebene)
brochure LING Broschüre
= booklet
broken (→ dotted 2) ENG.DRAW (→ gestrichelt)
broken interest ECON Bruchzins
broken line (→ dotted line 2) ENG.DRAW (→ gestrichelte Linie)
broken-out section ENG.DRAW Teilschnitt
= cut-away (n.)
broken wire (→ wire breakage) COMM.CABLE (→ Drahtbruch)
broker DATA COMM Informationsvermittler
= Broker
broker's call (→ alternation between lines) TELEPH (→ Makeln)
bromine CHEM Brom
= Br = Br
bronze METAL Bronze
bronze bushing METAL Bronzebuchse
bronze wire METAL Bronzedraht
broom TV Auflösungskeil
[Testbild]
= Besen
brown PHYS braun
brownish PHYS bräunlich
brownish yellow (→ ocher yellow) PHYS (→ ockergelb)
brown-out (n.) POWER SYS Spannungsmangel
= voltage insufficiency ≈ Spannungsausfall
≈ black-out (n.)

browsing (n.) DATA PROC Stöbern
[superficial, mostly unauthorized search in files] [oberflächliche, meist unerlaubte Suche in Dateien]
= Schnüffeln; Schmökern
browsw (v.i.) COLLOQ schmökern
= stöbern
Bruce antenna ANT Bruce-Antenne
= Bruce array; grecian antenna ↑ Mäanderantenne
↑ meander-line antenna
Bruce array (→ Bruce antenna) ANT (→ Bruce-Antenne)
Brune network NETW.TH Brune-Netzwerk
Brune procedure NETW.TH Brune-Verfahren
brush (v.t.) METAL bürsten
brush (n.) POWER ENG Bürste
brush (n.) DATA PROC Graphikpinsel
[computer graphics] = Pinsel; Brush
brush collar POWER ENG Bürstenträger
brush discharge PHYS Büschelentladung
= corona discharge [POWER ENG] = Koronaentladung [POWER ENG]
↑ gaseous discharge ↑ Gasentladung
brush discharge [PHYS] POWER SYS (→ Koronaentladung)
(→ corona discharge)
brute-force method DATA PROC (→ Brechstangenprinzip)
(→ brute-force technique)
brute-force technique DATA PROC Brechstangenprinzip
[solving problems in an akward way, with exaggereted deployment of means] [unelegante Lösung mit unverhältnismäßig hohem Aufwand]
= brute-force method
BS (→ backspace) DATA COMM (→ Rückwärtsschritt)
BSC (→ binary synchronous communications) DATA COMM (→ binäre Synchronübetragung)
B subscriber (→ called suscriber) SWITCH (→ gerufener Teilnehmer)
BTAM DATA COMM BTAM-Methode
[Basic Telecommunications Access Method]
bubble (→ magnetic bubble) PHYS (→ Magnetblase)
bubble chamber PHYS Blasenkammer
bubble memory (→ magnetic bubble memory) TERM&PER (→ Magnetblasenspeicher)
bubble memory cassette TERM&PER Blasenspeicherkassette
bubble sort (→ bubble sorting) DATA PROC (→ Bubble-Sortieren)
bubble sorting DATA PROC Bubble-Sortieren
= bubble sort; ripple sort; exchange sort ↑ Sortieren
↑ sorting
bubble technique TERM&PER Bubble-Verfahren
[ink-jet printer] [Tintenstrahldrucker]
bucked brigade device MICROEL Eimerkettenschaltung
= BBD; CCD shift register = Eimerkettenspeicher; CCD-Schieberegister; BBD; Pumpenbrigade-Schaltung; Schiebekettenspeicher
↑ Ladungstransferelement
bucket (→ memory area) DATA PROC (→ Speicherbereich)
bucking circuit CIRC.ENG (→ Ausgleichsschaltung)
(→ compensating circuit)
buckle (v.i.) (→ bend) MECH (→ knicken)
buckled rhombic antenna ANT geknickte Rhombusantenne
buckling (→ bend) MECH (→ Knick)
buckling load MECH Knicklast
buckling strength MECH Knickfestigkeit
buckling stress MECH Knickspannung
budget (n.) ECON Etat
= economic plan = Wirtschaftsplan; Budget
budget costs ECON Plankosten
= Sollkosten
budgeting ECON Wirtschaftsplanung
= economic planning
buff (v.t.) TECH schwabbeln
[polish] [Oberflächenbehandlung]
buffer (v.t.) DATA PROC zwischenspeichern
= store temporarily; store intermediately
buffer (n.) CIRC.ENG Puffer
[device absorbing discontinuities] [eine Vorrichtung die Unregelmäßigkeiten ausgleicht]
↓ Pufferspeicher

buffer (→buffer store) DATA PROC (→Pufferspeicher)
buffer (→driver) CIRC.ENG (→Treiber)
buffer amplifier CIRC.ENG **Pufferverstärker**
= Trennverstärker
buffer capacitor EL.TECH **Pufferkondensator**
buffer circuit EL.TECH **Pufferschaltung**
buffer coating OPT.COMM **Polsterbeschichtung**
buffered POWER SYS **gepuffert**
buffered battery POWER SYS **gepufferte Batterie**
buffered from mains POWER SYS **netzgepuffert**
buffered from mains battery POWER SYS **netzgepufferte Batterie**
buffered from mains operation POWER SYS **netzgepufferter Betrieb**
buffered memory DATA PROC **gepufferter Speicher**
= buffered storage; buffered store
buffered power supply POWER SYS **gepufferte Stromversorgung**
buffered storage (→buffered memory) DATA PROC (→gepufferter Speicher)
buffered store (→buffered memory) DATA PROC (→gepufferter Speicher)
buffering POWER SYS **Pufferbetrieb**
= float
buffering DATA PROC **Pufferbetrieb**
buffer jacket (→buffer loose tube) OPT.COMM (→Hohlader)
buffer loose tube OPT.COMM **Hohlader**
= buffer tube; buffer jacket = Schutzhülle
≠ tight buffer tube ≠ Vollader
buffer memory (→buffer store) DATA PROC (→Pufferspeicher)
buffer rectifier POWER SYS **Puffergleichrichter**
buffer register DATA PROC **Zwischenregister**
= intermediate register = Pufferregister
buffer size DATA PROC **Pufferlänge**
buffer stage CIRC.ENG **Pufferstufe**
≈ buffer amplifier = Trennstufe
≈ Pufferverstärker
buffer store DATA PROC **Pufferspeicher**
[transitory store to compensate different data velocities] [kurzfristiger Speicher zum Ausgleich unterschiedlicher Datengeschwindigkeiten]
= buffer memory; buffer; temporary memory = Zwischenspeicher 3; elastischer Speicher; Puffer
↓ cache storage ↓ Cache-Speicher
buffer tube (→buffer loose tube) OPT.COMM (→Hohlader)
bug (n.) DATA PROC **Fehler**
[a malfunction] ↓ Programmfehler; Hardwarefehler
= error; fault
↓ program error; hardware fault
bug EL.ACOUS **vestecktes Mikrophon**
bug (→eavesdrop) TELEPH (→abhören)
bugging device SIGN.ENG (→Abhörvorrichtung)
(→eavesdropping device)
build (v.t.) TECH **bauen**
= construct
build-in command (→resident command) DATA PROC (→residentes Kommando)
building CIV.ENG **Gebäude**
= edifice = Bauwerk
≈ premises
building (AM) (→shelter) SYS.INST (→Shelter)
building authority ECON **Baubehörde**
building block TECH **Systembaustein**
= Baustein
building construction (→building technology) CIV.ENG (→Hochbau)
building-out capacitor COMM.CABLE **Ergänzungskondensator**
building-out network CIRC.ENG **Ergänzungsnetzwerk**
building-out section OUTS.PLANT **Anlaufflänge**
building security system SIGN.ENG **Gebäudesicherheitsanlage**
building site [CIV.ENG] (→works) TECH (→Baustelle)
building technology CIV.ENG **Hochbau**
= building construction ↑ Bauwesen
↑ civil engineering
building-up transient PHYS **Einschwingvorgang**
= transient ≈ Relaxation
≈ relaxation

buildup (→rise) ELECTRON (→Anstieg)
build-up (→set-up) TELEC (→Aufbau)
build-up a message (→generate a message) SWITCH (→eine Meldung erstellen)
build-up behaviour ELECTRON (→Übergangsverhalten)
(→transient response)
build-up characteristics ELECTRON (→Übergangsverhalten)
(→transient response)
build-up period (→build-up time) ELECTRON (→Einschwingzeit)
build-up process ELECTRON (→Übergangsvorgang)
(→transient 1)
build-up response (→transient response) ELECTRON (→Übergangsverhalten)
build-up time 1 ELECTRON **Einschwingzeit**
= settling time; transient time; buil-up period; settling period; transient period
build-up time 2 (→rise time) ELECTRON (→Anstiegzeit)
built-in (adj.) TECH **eingebaut**
= inbuild; built-into; integrated; incorporated; self-contained = integriert
≠ angebaut
≠ add-on
built-in DATA PROC (→speicherresident)
(→memory-resident)
built-in antenna ANT **Gehäuseantenne**
= Einbauantenne
built-in check (→automatic check) EQUIP.ENG (→Selbsttest)
built-in font TERM&PER **integrierter Zeichensatz**
built-in microphone EL.ACOUS **Einbaumikrofon**
built-into (→built-in) TECH (→eingebaut)
bulb ELECTRON **Kolben**
[electronic tube] [Elektronenröhre]
bulb (→incandescent lamp) EL.INST (→Glühlampe)
bulb blackening ELECTRON **Kolbenschwärzung**
bulb neck ELECTRON **Kolbenhals**
[electron tube] [Bildröhre]
bulb resistance (→bulb resistor) INSTR (→Widerstandskopf)
bulb resistor INSTR **Widerstandskopf**
= bulb resistance [Meßfühler]
= Thermowiderstand
bulge MECH **Ausbauchung**
bulge (n.) (→protuberance) TECH (→Protuberanz)
bulging METAL **Wulstanstauchen**
bulk (n.) COLLOQ **Unmenge**
= large quantity
bulk billing (→bulk registration) SWITCH (→Summengebührenerfassung)
bulk changeover (→large-scale changeover) SWITCH (→Massenumschaltung)
bulk core memory DATA PROC **Großraumkernspeicher**
= bulk core storage
bulk core storage (→bulk core memory) DATA PROC (→Großraumkernspeicher)
bulk erase DATA PROC **Totallöschung**
= Globallöschung
bulk erase head TERM&PER **Blocklöschkopf**
↑ Löschkopf
bulk haulage ECON **Massengüterverkehr**
bulkhead (→circular mounting) COMPON (→Zentralbefestigung)
bulkhead jack (→bulkhead socket) COMPON (→Einbaubuchse mit Zentralbefestigung)
bulkhead plug COMPON **Einbaustecker mit Zentralbefestigung**
≠ Flanschstecker
↑ Einbaustecker
bulkhead receptacle COMPON (→Einbaubuchse mit Zentralbefestigung)
(→bulkhead socket)
bulkhead socket COMPON **Einbaubuchse mit Zentralbefestigung**
= bulkhead jack; bulkhead receptacle
≠ Flanschbuchse
bulk memory (→bulk storage) DATA PROC (→Großraumspeicher)
bulk memory (→mass memory) DATA PROC (→Massenspeicher)
bulk modulus PHYS **Kompressionsmodul**

bulk order		ECON	**Großauftrag**	burglar alarm		SIGN.ENG	**Einbruchalarm**
			= Riesenauftrag	= intrusion alarm			= Intrusionsalarm
bulk RAM		DATA PROC	**Großspeicher 1**	burglar alarm system		SIGN.ENG	(→Einbruchmeldeanlage)
			[Direktzugriffsspeicher gro-	(→intrusion detection system)			
			ßer Kapazität]	burial		OUTS.PLANT	**Vergrabung**
bulk registration		SWITCH	**Summengebührenerfassung**	buried		OUTS.PLANT	**erdverlegt**
= bulk billing			= Summenerfassung				= erdgelegt
≠ toll ticketing			≠ Einzelgebührenerfassung	buried		TECH	**unterirdisch**
bulk resistance		MICROEL	**Bahnwiderstand**	= below ground; underground; subter-			
bulk silicon (→substrate)		MICROEL	(→Substrat)	ranean			
bulk storage (→mass		DATA PROC	(→Massenspeicher)	buried antenna		ANT	**Erdantenne 2**
memory)							= vergrabene Antenne
bulk vibration		MECH	**Körperschwingung**	buried cable (→earth cable)		COMM.CABLE	(→Erdkabel)
↑ mechanical vibration			↑ mechanische Schwingung	buried cable marker		OUTS.PLANT	**Kabelmerkstein**
bulk wave		PHYS	**Körperwelle**	= marking post; marker			= Merkstein
bulky		TECH	**unhandlich**	buried contact		MICROEL	**vergrabener Kontakt**
			= sperrig	buried diffused layer (→buried		MICROEL	(→vergrabene Schicht)
bullet (→eye-catcher)		DOC	(→Blickfang)	layer)			
bulletin board		COLLOQ	**Anzeigetafel**	buried layer		MICROEL	**vergrabene Schicht**
= notice board; blackboard			= Anzeigebrett; schwarzes	= buried diffused layer			
			Brett; Anschlagbrett; Be-	buried oxide metal-oxide		MICROEL	(→BOMOS-Technologie)
			kanntmachungstafel	semiconductor (→BOMOS)			
bulletin board (→bulletin		DATA COMM	(→Schwarzes-Brett-System)	buried repeater (→underground		TRANS	(→Unterflurverstärker)
board system)				repeater)			
bulletin board system		DATA COMM	**Schwarzes-Brett-System**	burner (→programming		MICROEL	(→Programmiergerät)
= BBS; electronic bulletin board; bul-			[zum Ablesen oder Hinter-	device)			
letin board			lassen von Nachrichten]	burn-in (v.t.)		QUAL	**einbrennen**
↑ message handling system			= schwarzes Brett;	burn-in (n.)		QUAL	**Voralterung**
			Anschlagbrettsystem; elek-				= Voraltern; Einbrennen;
			tronische Anschlagtafel				Burn-in
			↑ Mitteilungssystem	burn-in (v.t.) (→program)		MICROEL	(→schießen)
bullet-proof		TECH	**kugelsicher**	burning question		COLLOQ	**brennende Frage**
			= schußsicher	burning time		QUAL	**Brenndauer**
bump		TRANS	**Buckel**	burn-in period		QUAL	**Einlaufphase**
[in a frequency respose curve]			[in einem Frequenzgang]	= debugging time			= Einbrennphase
bump		MICROEL	**Bump**	burn-in process		QUAL	**Einbrennvorgang**
[a raised contact-pad]			[erhöhter Kontaktierungs-	burnish		METAL	**preßpolieren**
			flecken]	[surface termination]			[Oberflächenbehandlung]
bump		DATA PROC	**nicht adressierbarer Hilfsspei-**				= brünieren
			cher	burnisher		ELECTRON	**Kontaktreinigungsmittel**
			= Bump	burn-out level (→overload		ELECTRON	(→Überlastungsgrenze)
bump equalizer		TRANS	**Buckelentzerrer**	level)			
bumper strip (→skirting		EQUIP.ENG	(→Sockelleiste)	burnt carmin		PHYS	**karmin gebrannt**
board)				burnt sienna		PHYS	**siena gebrannt**
bumpless		TECH	**stoßfrei**	burr		MECH	**Grat**
= hitless				burrer (→buzzer)		COMPON	(→Summer)
bunch (AM) (→trunk group)		SWITCH	(→Leitungsbündel)	burring		METAL	**Düsenziehen**
bunch (→ray bunch)		PHYS	(→Strahlenbüschel)	= nozzle-drawing			[Stanzen]
bunch pin plug		COMPON	**Büschelstecker**	Burrus diode		OPTOEL	**Burrus-Diode**
bundle (v.t.)		TELEC	**bündeln**	[a high-intensity LED]			[Leuchtdiode hoher Strah-
≈ multiplex			≈ multiplexen				lungsdichte]
bundle		COMM.CABLE	**Bündel**	bursary (→scholarship)		SCIE	(→Stipendium)
bundle (→ray bundle)		PHYS	(→Strahlenbündel)	burst (v.t.)		TERM&PER	**auftrennen**
bundled		DATA PROC	**im Preis inbegriffen**	[to rip-up continuous-form paper]			[Endlospapier]
= included in the price				burst (n.)		ELECTRON	**Burst** (n.)
bundle structure		COMM.CABLE	**Bündelaufbau**				= Bündel; Anhäufung
bungle (v.t.)		TECH	**murksen**	burst		TV	**Farbsynchronsignal**
≈ tinker			= basteln	= color burst (AM); colour burst (BRI)			= Farb-Synchronisier-Puls;
bungler		TECH	**Murkser**				Burst
bungling		TECH	**Murks**	burst [DATA COMM]		DATA PROC	(→Datenblock)
Bunsen burner		PHYS	**Bunsenbrenner**	(→data block)			
buoyancy		PHYS	**Auftriebskraft**	burst characteristics		ELECTRON	**Burst-Kenndaten**
			= Auftrieb	burster		TER&PER	**Einzelblattaufbereiter**
burden		INSTR	**Bürde**	[separates continuous-form paper			= Reißer
burden (AM) (→indirect		ECON	(→Gemeinkosten)	into single sheets]			
costs)				burster		TERM&PER	**Schlagschere**
burden effective resistance		INSTR	**Bürdenwiderstand**	≈ edge cutter			≈ Führungsstreifen-Abtren-
bureau (pl.-s, or -x)		OFFICE	**Büro**				ner
[facility]			[Einrichtung]	burst error		TELEC	**Büschelfehler**
= office 2							= Burstfehler
bureau equipment (→office		TERM&PER	(→Bürogerät)	burst gate		TV	**Farbsynchronsignal-Abtren-**
equipment)							**nung**
bureau machine (→office		TERM&PER	(→Bürogerät)	burst interference		TELEC	**Büschelstörung**
equipment)				≈ burst			= Impulsgruppe
bureau software (→office		DATA PROC	(→Bürosoftware)	burst mode		DATA COMM	**Stoßbetrieb**
software)				burst mode (→time-division		TELEC	(→Zeitgetrenntlage-Verfah-
bureau typewriter		OFFICE	**Büroschreibmaschine**	mode)			ren)
= office typewriter			= bureau typewriter	burst modem		SAT.COMM	**Burst-Modem**
Burgeois (BRI) (→type size 9		TYPOGR	(→Schriftgröße 9 Punkt)	burst open		TECH	**aufbrechen**
point)				= force open			= sprengen 2

burst operation	TRANS	Burstverfahren	business for own account	ECON	Eigengeschäft
= burst transmission; grouped time operation		= Bitbündelübertragung			≠ Provisionsgeschäft
burst operation (→time-division mode)	TELEC	(→Zeitgetrenntlage-Verfahren)	business goal	ECON	Unternehmensziel
			business graphics	DATA PROC	Geschäftsgraphik
burst transmission (→burst operation)	TRANS	(→Burstverfahren)	= management graphics		= Geschäftsgrafik; Bürografik; Bürografik
			≈ presentation graphics		≈ Präsentationsgraphik
bursty (adj.)	TELEC	diskontinuierlich	business hours	ECON	Geschäftszeit
= bursty-traffic		≠ kontinuierlich			= Geschäftsstunden
≠ streamy			business letter	OFFICE	Geschäftsbrief
bursty traffic	TELEC	diskontinuierlicher Verkehr	business line (→branch 1)	ECON	(→Branche)
bursty-traffic (→bursty)	TELEC	(→diskontinuierlich)	business machine (→office equipment)	TERM&PER	(→Bürogerät)
bursty-traffic service	TELEC	Dienst diskontinuierlichen Bitstroms	businessman (→entrepreneur)	ECON	(→Unternehmer)
			business paper	ECON	Wirtschaftszeitung
bury	OUTS.PLANT	eingraben	business prospects (n.plt.)	ECON	Geschäftsaussichten (n.plt.)
≈ lay		= vergraben	business software	DATA PROC	Geschäftssoftware
		≈ verlegen	business subscriber	TELEC	Geschäftsteilnehmer
burying depth (→laying depth)	OUTS.PLANT	(→Verlegetiefe)	= business costumer; business user		≈ Großteilnehmer
			≈ large user		≠ Privatteilnehmer
bus (pl.-es,-ses)	DATA PROC	Bus (pl. Busse)	≠ private subscriber		↓ Firmenkunde
[signal exchange line between functional units]		[Signalaustauschleitung zwischen Funktionseinheiten]	↓ corporate costumer		
			business telephone	TELEPH	Geschäftsanschluß
= highway 1 (BRI); computer bus		= Busleitung; Rechnerbus; Computerbus; Sammelleitung; Sammelweg; Pfad	= commercial telephone		≠ Privatanschluß
↓ control bus; data bus; address bus; serial bus; parallel bus			≠ private telephone		
			business transaction (→trade)	ECON	(→Handel)
		↓ Steuerbus; Datenbus; Adreßbus; serieller Bus; Parallelbus	business unit (→corporate division 1)	ECON	(→Unternehmensbereich)
bus (pl.-es, -ses)	ELECTRON	Sammelschiene	business user (→business subscriber)	TELEC	(→Geschäftsteilnehmer)
↓ bus [DATA PROC]		= Sammelleitung; Bus			
		↓ Bus [DATA PROC]	bus interface	DATA PROC	Busanschluß
bus A	DATA PROC	A-Bus	[directly to computer bus]		[direkt an Rechnerbus]
bus arbiter	DATA PROC	Buszuteiler			= Busschnittstelle
		= Bus-Arbiter	bus master (→bus controller)	DATA PROC	(→Buskontroller)
bus arbitration	DATA PROC	Buszuteilung			
bus architecture	DATA PROC	Busarchitektur	bus mouse	TERM&PER	Bus-Maus
busbar	POWER ENG	Sammelschiene	[directly connectable to the computer bus]		[direkt am Rechnerbus anschließbar]
↓ grounding bus; current bus		↓ Erdsammelschiene; Stromschiene	bus network	DATA COMM	Busnetz
bus board	DATA PROC	Buskarte	bus register	DATA PROC	Busregister
bus controller	DATA PROC	Buskontroller			= Sammelleitungsregister
= bus master		= bussteuerndes Gerät	bus slave	DATA PROC	Busklave
bus distributor	CIRC.ENG	Busverteiler	≠ bus master		≠ Buscontroller
bus driver	CIRC.ENG	Bustreiber	bus structure (→bus topology)	DATA COMM	(→Bus-Topologie)
bus enable (→enable signal)	ELECTRON	(→Freigabeimpuls)	bus system	DATA PROC	Bussystem
			bus termination	DATA PROC	Busabschluß
bus extender	DATA PROC	Buserweiterung	bus topology	DATA COMM	Bus-Topologie
bushing (→sleeve)	MECH	(→Buchse)	= bus structure		= Busstruktur
business (pl.-es)	ECON	Geschäft	bus user	DATA COMM	Busteilnehmer
= deal; transaction; trading			bus width	DATA PROC	Busbreite
business	ECON	Gewerbe	[number of parallel lines]		[Anzahl der parallelen Leitungen]
= industry					
business (→company)	ECON	(→Gesellschaft)	busy (adj.)	SWITCH	belegt
business administration	ECON	(→Betriebswirtschaftslehre)	= engaged		= besetzt
(→business administration economics)			≠ idle		≠ unbelegt
			busy (v.t.) (→seize)	SWITCH	(→belegen)
business administration economics	ECON	Betriebswirtschaftslehre	busy condition (→busy state)	SWITCH	(→Besetztzustand)
		= Betriebswirtschaft	busy flash (→busy flash signal)	SWITCH	(→Besetztzeichen)
= business administration		↑ Wirtschaftswissenschaften			
↑ economics		↓ Buchhaltungs- und Bilanzwesen	busy flash signal	SWITCH	Besetztzeichen
↓ accounting			= busy flash		
business area (→corporate division 1)	ECON	(→Unternehmensbereich)	busy hour call attempt	SWITCH	Anrufversuch in der Hauptverkehrsstunde
business communication	TELEC	Geschäftskommunikation	busy-hour to day ratio	SWITCH	Konzentration 2
business computer	DATA PROC	Geschäftscomputer	busy indication	TERM&PER	Belegtanzeige
= commercial-use computer		= kommerzieller Rechner	busy lamp	TERM&PER	Besetztlampe
business correspondence	DOC	Geschäftskorrespondenz	= engaged lamp		
business costumer (→business subscriber)	TELEC	(→Geschäftsteilnehmer)	busy line	TELEC	besetzte Leitung
			busy period (→all-trunks-busy time)	SWITCH	(→Blockierungsdauer)
business cycle	ECON	Konjunktur			
business efficiency exhibition	ECON	Büroartikelmesse	busy signal (→busy tone)	TELEPH	(→Besetztton)
		= Büroartikelausstellung	busy state	SWITCH	Besetztzustand
business enterprise (→company)	ECON	(→Gesellschaft)	= busy condition; occupation state		= Belegtzustand
business equipment (→office equipment)	TERM&PER	(→Bürogerät)	≠ idle state		≠ Ruhezustand
			busy time (→holding time)	SWITCH	(→Belegungsdauer)
business equipment (→office equipment)	OFFICE	(→Bürotechnik)	busy time (→heavy-traffic period)	TELEC	(→verkehrsstarke Zeit)
business espionage	ECON	Wirtschaftsspionage			
↓ industrila espionage		↓ Industriespionage			

busy tone

busy tone	TELEPH	**Besetztton**	
[indicating called line is busy]		[gewählter Anschluß oder Leitung besetzt; schneller Takt mit 425 Hz]	
= engaged tone (BRI); line busy tone (AM); busy signal			
≠ idle tone		= Besetztzeichen; Belegtton; Belegtzeichen	
↑ audible signal		≠ Freiton	
		↑ Hörton	
busy verification	SWITCH	**Besetztkontrolle**	
butt	MECH	**stumpfes Ende**	
butt end (→pole socket)	OUTS.PLANT	(→Stangenfuß)	
butterfly antenna (→batwing antenna)	ANT	(→Schmetterlingsantenne)	
butterfly capacitor	COMPON	**Schmetterlings-Drehkondensator**	
butterfly circuit	CIRC.ENG	**Schmetterlingskreis**	
Butterworth filter (→maximally flat filter)	NETW.TH	(→Potenzfilter)	
butt joint	MECH	**Stumpfverbindung**	
butt-joint riveting	MECH	**Laschennietung**	
butt-joint welding	METAL	**Stumpfstoßschweißung**	
bottom-up-design	MICROEL	**Bottom-up-Entwurf**	
button (→key)	TERM&PER	(→Taste)	
button cell	POWER SYS	**Knopfzelle**	
button head	MECH	**Halbrundkopf 1** [Niet]	
button-headed screw (→round-head screw)	MECH	(→Halbrundkopfschraube)	
button microphone (→lapel microphone)	EL.ACOUS	(→Knopflochmikrophon)	
butt welding	METAL	**Stumpfschweißen**	
buy back (→repurchase)	ECON	(→zurückkaufen)	
buyer (→purchaser)	ECON	(→Käufer)	
buying (→procurement)	ECON	(→Beschaffung)	
buying price (→cost price)	ECON	(→Einstandspreis)	
buzz	ACOUS	**summen**	
buzzer	COMPON	**Summer**	
[electric device for loud hum]		= Schnarre	
= burrer			
buzzer (→horn)	EL.ACOUS	(→Hupe)	
buzz saw (→circular saw)	TECH	(→Kreissäge)	
buzz tone	TELEPH	**Summton**	
= purring			
buzzword (→keyword)	LING	(→Schlagwort)	
bW (→reactive watt)	PHYS	(→Blindwatt)	
BWG (→Birmingham wire gage)	PHYS	(→Birmingham Drahtlehre)	
B wire (→ring wire)	TELEPH	(→B-Ader)	
BWO: carcinotron (→backward wave oscillator)	MICROW	(→Rückwärtswellenröhre)	
by hand (→manual)	TECH	(→manuell)	
bypass (n.)	ELECTRON	**umgehen**	
bypass capacitor	EL.TECH	**Ableitkondensator**	
bypass capacitor	NETW.TH	**Überbrückungskondensator**	
≈ speed-up capacitor [CIRC.ENG]		= Nebenschlußkondensator; Bypasskondensator	
		≈ Überhöhungskondensator [CIRC.ENG]	
bypass mode	TERM&PER	**Bypassbetrieb**	
bypass valve	TECH	**Umgehungsventil**	
by phone	TELEC	**fernmündlich** (adv.)	
= by telephone; telephonically		= telefonisch (adv.)	
byproduct	TECH	**Abfallprodukt**	
= by-product		= Nebenprodukt	
by-product (→byproduct)	TECH	(→Abfallprodukt)	
by return of mail (→by return of post)	ECON	(→postwendend)	
by return of post	ECON	**postwendend**	
= by return of mail			
byte	DATA PROC	**Byte**	
[a standard word lenth, mostly 8 bits, sometimes 16 bits]		[genormte Wortlänge, meist 8 bit, seltener 16 bit]	
≈ word		≈ Wort	
↓ octet		↓ Oktett; Langwort	
byte address	DATA PROC	**Byteadresse**	
byte-at-a-time (→byte-serial)	DATA PROC	(→byteseriell)	
byte-by-byte	INF	**byteweise**	
byte-by-byte (→byte-serial)	DATA PROC	(→byteseriell)	
byte-by-byte stuffing	CODING	**byteweises Stopfen**	
byte counter	DATA PROC	**Bytezähler**	
by telephone (→by phone)	TELEC	(→fernmündlich) (adv.)	
byte machine	DATA PROC	**Bytemaschine**	
byte manipulation	DATA PROC	**Byteeingiff**	
byte multiplex channel	DATA PROC	**Bytemultiplexkanal**	
[to interconnect simultaneously several slow peripherals]		[zur gleichzeitigen Anbindung mehrerer langsamer Peripheriegeräte]	
↑ multiplex channel		↑ Multiplexkanal	
byte-parallel	DATA PROC	**byteparallel**	
		= Byte-parallel	
byte-serial	DATA PROC	**byteseriell**	
= byte-by-byte; byte-at-a-time		= Byte-seriell	
byte-serial mode (→byte-serial transmission)	DATA COMM	(→byteserielle Übertragung)	
byte-serial transmission	DATA COMM	**byteserielle Übertragung**	
= byte-serial mode			
byte timing	DATA COMM	**Bytetakt**	
by the day	COLLOQ	**tageweise**	
by the hour	COLLOQ	**stundenweise**	

C

°C (→degree Celsius)	PHYS	(→Grad Celsius 1)	
C (→carbon)	CHEM	(→Kohlenstoff)	
C (→electric capacitance)	PHYS	(→elektrische Kapazität)	
C (→Coulomb)	PHYS	(→Coulomb)	
Ca (→calcium)	CHEM	(→Kalzium)	
cabinet	EQUIP.ENG	**Schrank**	
= cubicle			
cabinet (→control cubicle)	POWER SYS	(→Schaltschrank)	
cabinet (→booth)	CIV.ENG	(→Zelle)	
cabinet (→computer cabinet)	DATA PROC	(→Computergehäuse)	
cabinet connecting cable	EQUIP.ENG	**Schrankverbindungskabel**	
cabinet construction	EQUIP.ENG	**Schrankbauweise**	
cabinet foot	EQUIP.ENG	**Aufstellfuß**	
cabinet row	SYS.INST	**Schrankreihe**	
≈ rack row		≈ Gestellreihe	
cable (v.t.)	TELEC	**verkabeln**	
cable 1 (v.t.)	EQUIP.ENG	**verkabeln**	
≈ wire (v.t.)		≈ verdrahten; beschalten	
cable (n.)	TECH	**Kabel** (pl.-n)	
[a strong rope]		[dickes Tau]	
≈ cord; rope		≈ Leine; Seil; Strick; Tau	
cable (n.)	COMM.CABLE	**Kabel** (pl.-n)	
↓ earth cable; duct cable; submarine cable; aerial cable; paired cable; optical fiber cable		↓ Erdkabel; Röhrenkabel; Seekabel; Luftkabel; Paarkabel; Lichtwellenleiterkabel	
cable 2 (→wire 2)	EQUIP.ENG	(→beschalten)	
cable (v.t.) (→telegraph)	TELEC	(→telegrafieren)	
cable (n.) (→telegram)	POST	(→Telegramm)	
cable access	BROADC	**Kabelanschluß**	
cable address (→telegraphic address)	POST	(→Telegrammadresse)	
cable armoring	COMM.CABLE	**Kabelbewehrung**	
cable assembly list	SWITC	**Leitungsnamenliste**	
cable assignment	EQUIP.ENG	**Kabelbelegung**	
= cabling code; cable code			
cable bridge	SYS.INST	**Kabelbrücke**	
cable chamber	OUTS.PLANT	**Kabelschacht**	
= jointing chamber (BRI); splicing chamber (BRI)			
cable choke	ANT	**Mantelwellendrossel**	
↓ coiled-up cable choke		↓ Kabeldrossel	
cable chute (→cable shaft)	OUTS.PLANT	(→Aufstiegkanal)	
cable cladding (→cable sheath)	COMM.CABLE	(→Kabelmantel)	
cable clamp	TECH	**Kabelschelle**	
= cable clamping; cable clip		= Kabelabfangung; Leitungsschelle; Bügelschelle	
cable clamping (→cable clamp)	TECH	(→Kabelschelle)	
cable clip (→cable clamp)	TECH	(→Kabelschelle)	
cable code (→cable assignment)	EQUIP.ENG	(→Kabelbelegung)	

cable color code (AM) COMM.CABLE
= cable colour code (BRI)
cable colour code (BRI) COMM.CABLE
(→cable color code)
cable comb EQUIP.ENG
cable conduit OUTS.PLANT
[one or more ducts in the same trench]
= conduits; duct nest; duct run; duct bank; conduit run
≈ cable duct
↓ single-duct conduit; multiple-duct conduit
cable connection DATA COMM
cable connector EQUIP.ENG
cable construction COMM.CABLE
= cable design; cable make-up
cable core COMM.CABLE
= core
cable cut (→cable OUTS.PLANT disruption)
cabled EQUIP.ENG
≈ wired
≠ uncabled
cable design (→cable COMM.CABLE construction)
cable detector OUTS.PLANT
cable diagram (→cable SYS.INST layout)
cable dispenser OUTS.PLANT
[cable laying]
= wire dispenser
cable disruption OUTS.PLANT
= cable cut; backhoe fading (sl.)
cable distribution OUTS.PLANT
cable distribution box OUTS.PLANT
(→distribution cabinet)
cable distribution OUTS.PLANT
cellar
cable distribution rack OUTS.PLANT
cable ditch (→cable OUTS.PLANT trench)
cable drum COMM.CABLE
= cable reel; drum; reel
cable drum OUTS.PLANT
cable duct OUTS.PLANT
[a single pipe]
≈ cable conduit; single-duct conduit
cable end OUTS.PLANT
[component to seal outside cables]
= cable termination; cable terminal head
cable-end forming COMM.CABLE
cable entrance (→cable OUTS.PLANT inlet)
cable entry COMPON
[connector]
cable entry EQUIP.ENG
cable fastener COMPON
= cable strap; cable ties
cable feed-trough SYS.INST
cable fittings COMM.CABLE
cable form (→cable EQUIP.ENG harness)
cable forming plan DOC
= harness wiring plan
cable funnel OUTS.PLANT
cable gland SYS.INST
cable grid (→planar cable SYS.INST grid)

Farbkennzeichnung
= Farbcode
(→Farbkennzeichnung)

Kabelkamm
Kabelkanal
[ein oder mehrere Kabelkanalzüge auf derselben Trasse]
= Kanal; KK; Kabelkanalverband; Kabelkanalanlage; Röhrenzug; Rohrzug
≈ Kabelkanalzug
↓ Einrohrkanal; Mehrröhrenkanal

Kabelanschluß
Kabelverbinder
Kabelaufbau
= Kabelausführung
Kabelseele
[Verseilelemente mit deren Bewicklung]
= Kabelkern
(→Kabelbruch)

verkabelt
≈ verdrahtet
≠ unverkabelt
(→Kabelaufbau)

Kabelsuchgerät
= Trassensuchgerät
(→Kabelführungsplan)

Abrollvorrichtung
[Kabelverlegung]

Kabelbruch

Kabelaufteilung
(→Kabelverzweigergehäuse)

Kabelaufteilungsraum

Kabelaufteilungsgestell
≈ Kabelendgestell
(→Kabelgraben)

Kabeltrommel
= Trommel
Kabelwinde
= Kabelziehwinde; KZW
Kabelkanalzug
[einzelne durchlaufende Röhre]
= Kabelkanal; Einrohrkanal
Kabelendverschluß
[Bauteil für den luftdichten Abschluß von Außenkabeln]
= Endverschluß; EVs; Kabelabschlußeinrichtung
Kabelausformung
(→Kabeleinführung)

Kabeldurchlaß
[Stecker]
Kabeleinfall
Kabelbinder
= Kabelband
Kabeldurchführung
Kabelgarnitur
(→Kabelform)

Kabelformplan

Kabeltrichter
Kabelverschraubung
(→Flächenkabelrost)

cable grip OUTS.PLANT
= pulling eyer
cable grip (→thimble) COMPON
cable harness EQUIP.ENG
= cable form
cable harness SYS.INST
cable holder SYS.INST
= cable support
cable inlet OUTS.PLANT
= cable entrance
cable jack COMPON
= female cable connector; free socket (BRI)
cable jacket (→cable COMM.CABLE sheath)
cable ladder (→cable SYS.INST rack)
cable lasher OUTS.PLANT
cable layer (→cable OUTS.PLANT plow)
cable layer (→cable ship) TELEC
cable laying OUTS.PLANT
= laying; cable running
cable laying machine OUTS.PLANT
cable laying plough (→cable OUTS.PLANT plow)
cable laying plow (→cable OUTS.PLANT plow)
cable layout SYS.INST
= cable routing plan; cable diagram
cable layout EQUIP.ENG
cable length COMM.CABLE
cable line (→aerial cable OUTS.PLANT line)
cable locating system OUTS.PLANT
(→cable locator system)
cable locator system OUTS.PLANT
= cable locating system
cable lug (→thimble) COMPON
cable make-up (→cable COMM.CABLE construction)
cableman OUTS.PLANT
= cable splicer
cable marker ELECTRON
cable measuring equipment INSTR
= cable test equipment
cable plant (→cable OUTS.PLANT system)
cable plough (→cable OUTS.PLANT plow)
cable plow OUTS.PLANT
= cable laying plow; cable layer; cable plough; cable laying plough
↓ submarine cable plow
cable plug COMPON
= free plug (BRI)
cable protection tube SYS.INST
cable protector (→cable SYS.INST trough)
cable proyect OUTS.PLANT
cable rack SYS.INST
= rack (IEC); cable shelf; cable runway; cable ladder
≈ cable trough
cable rack (→cable support OUTS.PLANT rack)
cable reel OUTS.PLANT
cable reel (→cable COMM.CABLE drum)
cable rerouting tube EQUIP.ENG
cable routing (→cable SYS.INST run)
cable routing plan (→cable SYS.INST layout)

Ziehstrumpf
= Kabelziehstrumpf
(→Kabelschuh)
Kabelform
= Kabelbaum; Kabelstamm; Leitersystem
Kabelpaket
Kabelhalter
= Kabelhalterung
Kabeleinführung

Kabelbuchse
= Kabelkupplung; Kabeldose
≠ Einbaubuchse; Kabelstecker
(→Kabelmantel)

(→Kabelrost)

Laschvorrichtung
(→Kabelpflug)

(→Kabelschiff)
Kabelverlegung
= Verlegung; Kabellegung; Legung
Kabelverlegemaschine
(→Kabelpflug)

(→Kabelpflug)

Kabelführungsplan
= Kabellaufplan; Kabelplan; Verkabelungsplan
Kabelplan
Kabellänge
(→Luftkabellinie)

Kabelortungssystem

Kabelortungssystem

(→Kabelschuh)
(→Kabelaufbau)

Kabelmonteur

Kabelmarke
Kabelmeßgerät

(→Kabelanlage)

(→Kabelpflug)

Kabelpflug
↓ Unterwasserkabelpflug

Kabelstecker
≠ Kabelbuchse; Einbaustecker
Kabelschutzrohr
= Schutzrohr
(→Kabelwanne)

Kabelplanung
Kabelrost
≈ Kabelwanne

(→Kabelendgestell)

Kabelrolle
(→Kabeltrommel)

Kabelumkehrrohr
(→Kabelführung)

(→Kabelführungsplan)

cable run	SYS.INST	**Kabelführung**	
= cable routing; running of cables			
cable run [lattice mast]	ANT	**Kabelleiter** [Gittermast]	
cable run	OUTS.PLANT	**Kabelverlauf**	
cable rung	SYS.INST	**Kabelsprosse**	
cable running (→cable laying)	OUTS.PLANT	(→Kabelverlegung)	
cable runway (→cable rack)	SYS.INST	(→Kabelrost)	
cable screen	COMM.CABLE	**Kabelschirm**	
= screen		= Schirm	
cable section [between sleeves]	OUTS.PLANT	**Kabelstück** [Länge zwischen Muffen] = Kabelstrang; Kabelabschnitt	
cable shaft	OUTS.PLANT	**Aufstiegkanal**	
= cable chute			
cable sheath	COMM.CABLE	**Kabelmantel**	
= cable jacket; cable cladding; sheath; cladding		= Mantel	
cable sheathing	COMM.CABLE	**Kabelhülle**	
cable shelf (→cable rack)	SYS.INST	(→Kabelrost)	
cable ship	TELEC	**Kabelschiff**	
= cable layer; laying ship		= Kabelleger; Verlegeschiff; Legeschiff	
cable sleeve	COMPON	**Kabeltülle**	
cable socket (→thimble)	COMPON	(→Kabelschuh)	
cable splice	COMM.CABLE	**Kabelspleiß**	
= splice; splicing; junction		= Spleiß; Spleißung	
cable splicer (→cableman)	OUTS.PLANT	(→Kabelmonteur)	
cable strap (→cable fastener)	COMPON	(→Kabelbinder)	
cable subway	OUTS.PLANT	**begehbarer Kabelkanal**	
= tunnel; gallery		= Kabeltunnel	
cable supplier	TELEC	**Kabellieferant**	
= cable vendor			
cable support (→cable holder)	SYS.INST	(→Kabelhalter)	
cable support rack	OUTS.PLANT	**Kabelendgestell**	
= cable rack		= Kabelgestell ≈ Kabelaufteilungsgestell	
cable switch	COMPON	**Zwischenschalter**	
= intermediate switch			
cable system	OUTS.PLANT	**Kabelanlage**	
= cable plant			
cable tail	OUTS.PLANT	**Kabelschwanz**	
cable television (→cable TV)	BROADC	(→Kabelfernsehen)	
cable television system	BROADC	**Kabelfernsehanlage**	
= cable TV system; CATV system			
cable terminal head (→cable end)	OUTS.PLANT	(→Kabelendverschluß)	
cable termination (→cable end)	OUTS.PLANT	(→Kabelendverschluß)	
cable test equipment (→cable measuring equipment)	INSTR	(→Kabelmeßgerät)	
cable text	TELEC	**Kabeltext**	
cable ties (→cable fastener)	COMPON	(→Kabelbinder)	
cable trench	OUTS.PLANT	**Kabelgraben**	
= cable ditch			
cable trough	SYS.INST	**Kabelwanne**	
= cable protector ≈ cable rack		= Kabelrinne ≈ Kabelrost	
cable TV	BROADC	**Kabelfernsehen**	
= cable television ≈ CATV		= KTV; Drahtfernsehen ≈ Großgemeinschaftsanlage	
cable TV system (→cable television system)	BROADC	(→Kabelfernsehanlage)	
cable vault [room under main distribution frame, where outside plant cables enter] = splicing chamber (AM)	OUTS.PLANT	**Kabelkeller**	
cable vendor (→cable supplier)	TELEC	(→Kabellieferant)	
cabling	TELEC	**Verkabelung**	
cabling ≈ wiring	EQUIP.ENG	**Verkabelung** ≈ Verdrahtung	
cabling code (→cable assignment)	EQUIP.ENG	(→Kabelbelegung)	
cabling list	OUTS.PLANT	**Kabellegeliste** = Kabelliste	
cabling space	EQUIP.ENG	**Verkabelungsraum**	
cache (→cache memory)	DATA PROC	(→Cache-Speicher)	
cache memory [speeds-up by storing also, at chance, neighbouring memory contents] = cache storage; cache store; cache ≈ scratch-pad memory ↑ buffer store	DATA PROC	**Cache-Speicher** [beschleunigt durch auf-Verdacht-Mitspeichern benachbarter Bereiche; vom engl. „cash" = Geheimlager; Aussprache „käsch"] = Cash; schneller Pufferspeicher; Hintergrundspeicher 2 ≈ Notizblockspeicher ↑ Pufferspeicher	
cache storage (→cache memory)	DATA PROC	(→Cache-Speicher)	
cache store (→cache memory)	DATA PROC	(→Cache-Speicher)	
CAD = computer aided design; computer aided development	DATA PROC	**CAD** = rechnergestützte Entwicklung; rechnergestütztes Konstruieren; rechnergestützte Konstruktion; rechnergestütztes Zeichnen; computergestützte Entwicklung; computergestütztes Konstruieren; computergestützte Konstruktion; computergestütztes Zeichnen; computergestützter Entwurf; computergestützter Entwurf	
CAD complete solution	DATA PROC	**CAD-Komplettlösung**	
cadence (n.) [regular sequence]	TECH	**Kadenz** [regelmäßige Abfolge]	
cadence (→clock)	ELECTRON	(→Takt)	
cadmium = Cd	CHEM	**Kadmium** (n.n.) = Cadmium; Cd	
cadmium-plate	METAL	**kadmieren**	
cadmium plating	METAL	**Kadmierung**	
cadmium sulfide cell	POWER SYS	**Cadmiumsulfid-Zelle**	
CAD tool	DATA PROC	**CAD-Werkzeug**	
CAD workstation	DATA PROC	**CAD-Arbeitsplatz**	
CAE 1 = computer aided engineering	DATA PROC	**CAE 1** = rechnergestütztes Ingenieurtechnik; rechnergestütztes Engineering; computergestützte Ingenieurtechnik	
CAE 2 (→computer-aided education)	DATA PROC	(→rechnergestützter Unterricht)	
caesium beam frequency standard (→cesium frequency standard)	INSTR	(→Cäsium-Frequenzstandard)	
cage	MECH	**Käfig** [Kugellager]	
cage aerial (→cage antenna)	ANT	(→Käfigantenne)	
cage antenna = cage aerial; squirrel cage antenna	ANT	**Käfigantenne**	
cage dipole = sausage dipole ↑ geometrically thick antenna	ANT	**Reusendipol** = Käfigdipol ↑ geometrisch dicke Antenne	
cage monopole antenna ↑ vertical antenna	ANT	**Reusen-Monopol** ↑ Vertikalantenne	
cage winding (→squirrel cage winding)	EL.TECH	(→Käfigwicklung)	
CAI (→computer-aided education)	DATA PROC	(→rechnergestützter Unterricht)	
CAIM = computer aided inventory and maintenance	DATA PROC	**CAIM** = rechnergestützte Bestandskontrolle und Wartung; computergestützte Bestandskontrolle und Wartung	
cal (→calory)	PHYS	(→Kalorie)	
CAL (→computer-aided education)	DATA PROC	(→rechnergestützter Unterricht)	
cal adj (→calibration adjustment)	INSTR	(→Kalibriereinstellung)	
calandria	TECH	**Schlangenkühler**	

calcite	CHEM	**Kalkspat**
calcium	CHEM	**Kalzium**
= Ca		= Calcium; Ca
calculable	SCIE	**berechenbar**
≈ computable		≈ kalkulierbar
calculate	MATH	**rechnen**
= compute		= berechnen
≈ compute [DATA PROC]		≈ berechnen [DATA PROC]
calculate costs	ECON	**kalkulieren**
calculated curve	TECH	**gerechnete Kurve**
		= theoretische Kurve
calculating machine	DATA PROC	**Rechner 1**
[machine to solve mathematical problems]		[Maschine zur Lösung mathematischer Aufgaben]
= calculator		= Rechenmaschine 1; Rechenautomat
≈ computer; data processing equipment		≈ Computer; Datenverarbeitungsanlage
↓ mechanical calculator; electronic calculator		↓ mechanischer Rechner; Elektronenrechner
calculation	MATH	**Rechnung**
= computation		= Berechnung
≈ counting; computation [DATA PROC]		≈ Kalkulation [ECON]; Zählung; Berechnung [DATA PROPC]
calculation (→cost calculation)	ECON	(→Kalkulation)
calculation error	MATH	**Rechenfehler**
= computational error		= Berechnungsfehler
calculation method (→mode of calculation)	MATH	(→Berechnungsart)
calculator (→calculating machine)	DATA PROC	(→Rechner 1)
calculator chip (→computer chip)	MICROEL	(→Rechner-Chip)
calculator mode	DATA PROC	**Rechenbetrieb**
calculus	MATH	**Infinitesimalrechnung**
↑ analysis		↑ Analysis
↓ differential calculus; integral calculus		↓ Differentialrechnung; Integralrechnung
calculus (→mode of calculation)	MATH	(→Berechnungsart)
calculus of variations (→variational calculus)	MATH	(→Variationsrechnung)
calendar of events	ECON	**Veranstaltungskalender**
cal factor (→calibration factor)	INSTR	(→Kalibrierfaktor)
calibrate	INSTR	**eichen**
= gauge (v.t.)		= kalibrieren; einmessen
calibrating accuracy	INSTR	**Eichgenauigkeit**
calibrating curve	INSTR	**Eichkennlinie**
= calibration curve		= Eichkurve
calibration (→gauging)	INSTR	(→Eichung)
calibration adjustment	INSTR	**Kalibriereinstellung**
= cal adj		
calibration curve (→calibrating curve)	INSTR	(→Eichkennlinie)
calibration factor	INSTR	**Kalibrierfaktor**
= cal factor; instrumentation correction		= Meßgerätekorrektur
calibration kit	INSTR	**Kalibriersatz**
calibration range	INSTR	**Kalibrierbereich**
calibration series	INSTR	**Eichreihe**
calibration signal	DATA COMM	**Eichsignal**
calibration specification	INSTR	**Eichvorschrift**
calibration uncertainty	INSTR	**Kalibrierunsicherheit**
= calibratot uncertainty		
calibrator	INSTR	**Kalibrator**
		= Kalibriereinrichtung; Kalibrierschaltung
calibratot uncertainty (→calibration uncertainty)	INSTR	(→Kalibrierunsicherheit)
californium	CHEM	**Californium**
= Cf		= Cf
call (v.t.)	DATA PROC	**aufrufen**
= excite; invoke (v.t.); invocate		
call (n.)	ECON	**Abruf**
= requisition		[Lieferung, Geld]
call (n.)	TELEPH	**Anruf**
= calling; ring n.		= Ruf
≈ conversation; connection [TELEC]		≈ Gespräch; Verbindung [TELEC]
call	DATA PROC	**Aufruf**
[sequence of instructions to activate defined hardware or software functions]		[Befehlsfolge zum Auslösen bestimmter Hardware- oder Softwarefunktionen]
= call instruction; excitation; invocation		= Aufrufbefehl; Rufanweisung
↓ macro instruction		↓ Makroaufruf
call (→ring)	TELEC	(→rufen)
call (v.t.) (→telephone)	TELEPH	(→telefonieren)
call-... (→switching-...)	SWITCH	(→vermittlungstechnisch)
call accepted	DATA COMM	**Rufannahme**
call accepted condition	DATA COMM	**Rufannahme-Zustand**
call accepted message	DATA COMM	**Rufannahme-Block**
= CAM		
call accepted signal	DATA COMM	**Antwortkennzeichen für Rufannahme**
call accept signal	DATA COMM	**Rufannahmesignal**
call attempt	SWITCH	**Wählversuch**
= offered call		= Belegungsversuch; Verbindungsversuch; Anrufversuch; angebotene Belegung
≈ request		≈ Aufforderung
call back 2 (v.t.)	TELEPH	**rückfragen**
[the PABX facility to contact a third party during a call]		[bei Nebenstellenanlagen gebotene Möglichkeit während eines Gesprächs bei einem Dritten rückzufragen]
= ring back; recall		
call back 1 (v.i.)	TELEPH	**rückrufen**
[to call a subscriber who could not reach you]		[einen Teilnehmer anrufen, der einen nicht erreichen konnte]
call-back 2 (n.)	TELEPH	**Rückfrage**
[PABX]		[Nebenstellenanlagen]
= callback 2; consultation call; ring-back; ring back; recall; inquiry		= Rückruf; Internrückfrage
callback 1 (n.)	TELEPH	**Rückruf**
[to a party which couldn't reach you]		[an einen Teilnehmer der einen nicht erreichen konnte]
callback (→call-back)	TELEPH	(→Rückfrage)
call barred	DATA COMM	**Verbindung gesperrt**
= access barred		= Zugang verhindert
call bell (→bell)	TERM&PER	(→Klingel)
call booking	SWITCH	**Gesprächsanmeldung**
= booking		
call box (→telephone booth)	TELEPH	(→Fernsprechzelle)
call by name	DATA PROC	**Namensaufruf**
call by reference	DATA PROC	**Referenzaufruf**
[ALGOL]		[ALGOL]
call by value	DATA PROC	**Wertaufruf**
call charge [TELEPH]	TELEC	(→Verbindungsgebühr)
(→connection fee)		
call-charge computer (→tax computer)	SWITCH	(→Gebührenrechner)
call-charge data	SWITCH	**Gebührendaten**
= billing data; charging information		= Gebühreninformation
call-charge meter (NAM) (→charge indicator)	TERM&PER	(→Gebührenanzeige)
call-charge pulse (→tax-metering pulse)	SWITCH	(→Gebührenimpuls)
call-charge registration	SWITCH	**Gebührenspeicherung**
call clear-down time (→call release time)	SWITCH	(→Verbindungsauslösedauer)
call collision	DATA COMM	**Verbindungszusammenstoß**
= head-on collision		= Rufzusammenstoß; Belegungszusammenstoß; Zusammenstoß
call completion (→connection set-up)	SWITCH	(→Verbindungsaufbau)
call condition (→call state)	SWITCH	(→Verbindungszustand)
call confirmation protocol	DATA COMM	**Bestätigungsprotokoll**
call confirmation signal	DATA COMM	**Anrufbestätigung**
= reception confirmation signal		
call-confirmation signal (→dial tone)	TELEPH	(→Wählton)
call connected	TELEC	**Verbindung hergestellt**
call connected signal	DATA COMM	**Freizeichen**
		= Verbunden-Kennzeichen; Verbundensignal

call control TELEC **Rufbehandlung**
call control SWITCH **Verbindungssteuerung**
= connection control; link control
call control character SWITCH **Verbindungssteuerungszeichen**
= call control signal; connection control character; connection control signal; link control character; link control signal
= Steuerzeichen; Steuerkennzeichen
call control procedure SWITCH **Verbindungssteuerungsverfahren**
= connection control procedure; link control procedure; LCP
call control signal (→call control SWITCH (→Verbindungssteuerungszeichen)
character)
call counting SWITCH **Gesprächszählung**
call-count meter SWITCH **Belegungszähler**
= Verkehrsmeßgerät
call data DATA COMM **Benutzerdaten**
= communication data
= Benutzerangaben; Bedienerdaten; Bedienerangaben
call data SWITCH **Verbindungsdaten**
= link data
= Verbindungsangaben; Gesprächsdaten; Anrufdaten
call delay SWITCH **Wartedauer**
= delay time; waiting delay
call distribution SWITCH **Anrufverteilung**
call distributor SWITCH **Anrufverteiler**
call diversion (→call forwarding) TELEPH (→Rufumleitung)
call duration TELEC **Verbindungsdauer**
= connection time; communication time
= Anrufdauer
↓ conversation time [TELEPH]
↓ Gesprächsdauer [TELEPH]
called line identity (→call DATA COMM (→Anschlußkennung)
line identification)
called party (→called suscriber) SWITCH (→gerufener Teilnehmer)
called station DATA COMM **gerufene Station**
= called subscriber; called user
= gerufener Teilnehmer
called subscriber (→called DATA COMM (→gerufene Station)
station)
called suscriber SWITCH **gerufener Teilnehmer**
= called party; B subscriber; terminating station
= B-Teilnehmer; ferner Teilnehmer; Angerufener
called user (→called station) DATA COMM (→gerufene Station)
caller identification (→call SWITCH (→Fangen)
tracing)
call establishment (→connection SWITCH (→Verbindungsaufbau)
set-up)
call failure SWITCH **Verbindungsstörung**
= connection failure; link failure
call fee [TELEPH] (→connection TELEC (→Verbindungsgebühr)
fee)
call fee indicator (BRI) TERM&PER (→Gebührenanzeige)
(→charge indicator)
call fill (→call fill percentage) SWITCH (→Wirkungsgrad)
call fill pecentage SWITCH **Wirkungsgrad**
= call fill
call finder (→line finder) SWITCH (→Anrufsucher)
call for bids (AM) (→invitation ECON (→Ausschreibung)
to tender)
call for tenders (→invitation to ECON (→Ausschreibung)
tender)
call forwarding TELEPH **Rufumleitung**
[automatic rerouting of calls to other number or to attendant]
[automatische Weiterleitung eines Anrufs an andere Nummer oder an Beamtin]
= call diversion; call redirection; call transfer; call routing
≈ automatic call transfer
= Rufweiterleitung; Rufumlenkung; Anrufweiterleitung; Anrufumlenkung; Anrufweiterschaltung; Rufweiterschaltung; Gesprächsumleitung;Gesprächsweiterleitung; Gesprächsumlenkung; Gesprächsweiterschaltung; Umlegen; Rufweitergabe; Anrufweitergabe; Rufzuweisung; Anrufzuweisung; Gesprächszuweisung
≈ selbsttätige Rufweiterleitung

call handling TELEPH **Vermittlung**
[reference in telephone number to a manually operated private exchange]
[bei Rufnummern der Hinweis auf handbediente Nebenstellenanlage]
= operator
call identifier DATA COMM **Verbindungskennung**
= connection identifier; link identifier
= Rufkennung
calligraphy (→letter quality) TERM&PER (→Schönschrift)
call indicator DATA COMM **Unterscheidungskennzeichen**
= Rufkennzeichnung
call information service DATA COMM **Auskunftdienst**
= information service
call information service signal DATA COMM **Auskunftdienst-Kennzeichen**
calling (→call) TELEPH (→Anruf)
calling card (AM) (→visiting ECON (→Visitenkarte)
card)
calling condition (→ringing SWITCH (→Rufzustand)
condition)
calling cord TELEPH **Rufschnur**
calling frequency (→ringing TELEC (→Rufstromfrequenz)
frequency)
calling key TELEGR **Anruftaste**
= starting key
≠ Schlußtaste
≠ clearing key
calling line identity (→call DATA COMM (→Anschlußkennung)
line identification)
calling method TELEC **Wählverfahren**
calling number (→dial number) TELEC (→Rufnummer)
calling-number memory TERM&PER **Rufnummernspeicher**
= call-number memory
= Nummernspeicher
calling party (→calling DATA COMM (→rufende Station)
station)
calling party (→calling SWITCH (→rufender Teilnehmer)
subscriber)
calling plug TELEPH **Rufstöpsel**
calling sequence DATA PROC **Abrufsequenz**
≈ concatenation
≈ Verkettung
calling signal (→ringing signal) TELEC (→Rufsignal)
calling state (→ringing SWITCH (→Rufzustand)
condition)
calling station DATA COMM **rufende Station**
= calling party; originating call; calling subscriber; calling user
= rufender Teilnehmer
calling subscriber SWITCH **rufender Teilnehmer**
= calling party; A subscriber; outgoing subscriber
= A-Teilnehmer; Anrufer; Anrufender
calling subscriber (→calling DATA COMM (→rufende Station)
station)
calling tone (→idle tone) TELEPH (→Freiton)
calling user (→calling DATA COMM (→rufende Station)
station)
call in progress TELEPH **laufendes Gespräch**
call instruction (→call) DATA PROC (→Aufruf)
call line identification DATA COMM **Anschlußkennung**
= calling line identity; called line identity; line identification; CDI
[Sicherung gegen Manipulation von Kennungen]
call load DATA COMM **Ruflast**
call logging DATA COMM **Verbindungsprotokollierung**
= connection logging; link logging
call meter TELEPH **Gesprächszähler**
call metering (→tax SWITCH (→Gebührenzählung)
metering)
call not accepted (→call DATA COMM (→Rufabweisung)
rejection)
call number (→dial number) TELEC (→Rufnummer)
call-number display TERM&PER **Rufnummernanzeige**
call-number generator TERM&PER **Rufnummerngeber**
call number identification TELEC **Rufnummeridentifizierung**
= number identification
call-number memory TERM&PER (→Rufnummernspeicher)
(→calling-number memory)
call-off program MANUF **Abrufprogramm**
call parking TELEPH **Parken**
[holding of a temporarily suspended call]
[Festhalten einer vorübergehend unterbrochenen Verbindung]
= parking
call pickup TELEPH **Anrufübernahme**
call processing SWITCH **Verbindungsabwicklung**
= Gesprächsabwicklung; Verbindungsbearbeitung

call-processing ... (→switching-...)	SWITCH	(→vermittlungstechnisch)	**calory** [old unit for heat quantity; ≈ 4.1868 Joule] = cal	PHYS	**Kalorie** [alte Einheit für Wärmemenge; ≈ 4,1868 Joule] = cal
call-processing function (→switching task)	SWITCH	(→vermittlungstechnische Aufgabe)	**CAM** = computer aided manufacturing ≈ CIM	DATA PROC	**CAM** = rechnergestützte Fertigung; rechnergestützte Herstellung; Ferigungsleittechnik; rechnergestützte Produktion; computergestützte Fertigung ≈ CIM
call processing program (→switching program)	SWITCH	(→Vermittlungsprogramm)			
call processor (→switching processor)	SWITCH	(→Vermittlungsprozessor)			
call progress signal (→service signal)	DATA COMM	(→Dienstsignal)			
call queuing system (→queuing system)	SWITCH	(→Wartesystem)			
call receiver	TELEC	**Rufempfänger**	**cam**	MECH	**Nocken**
call recorder = speech recorder; voice recorder	TERM&PER	**Sprachaufzeichnungsgerät** = Anrufaufzeichnungsgerät	**CAM** (→associative storage)	DATA PROC	(→Assoziativspeicher)
call recording	TELEPH	**Gesprächsaufzeichnung**	**CAM** (→call accepted message)	DATA COMM	(→Rufannahme-Block)
call recording equipment	TELEPH	**Gesprächsaufzeichnungsgerät**	**camber** (→arch)	TECH	(→Wölbung)
call redirection (→call forwarding)	TELEPH	(→Rufumleitung)	**cambered** = convex	MECH	**gewölbt** = konvex ≈ erhaben
call register = connection register; link register; carry register	SWITCH	**Verbindungsspeicher**			
call rejection = call not accepted	DATA COMM	**Rufabweisung** = Rüfzurückweisung	**cam contact**	POWER SYS	**Nockenkontakt**
			cam control	MECH	**Nockensteuerung**
call release time = call clear-down time	SWITCH	**Verbindungsauslösedauer** = Verbindungsauslösungsdauer; Auslösedauer; Auslösungsdauer; Auslöseverzug; Auslösungsverzug	**cam-controlled** (→cam-operated)	TECH	(→nockengesteuert)
			cam disk 1	MECH	**Kurvenscheibe**
			cam disk 2 (→jumping cam)	MECH	(→Nockenscheibe)
			camera	TV	**Kamera**
			camera dolly = dolly	TV	**Fahrstativ**
call request = connection request; link request; request	SWITCH	**Verbindungsanforderung** = Verbindungsaufforderung; Verbindungswunsch	**camera monitoring system**	SIGN.ENG	**Kameraüberwachungssystem**
			camera recorder	CONS.EL	**Camera Recorder**
			camera tube ↑ cathode ray tube	ELECTRON	**Bildaufnahmeröhre** = Aufnahmeröhre; Kameraröhre; Bildaufnahme-Elektronenröhre ↑ Kathodenstrahlröhre
call request time	SWITCH	**Vorwahlzeit**			
call routing (→call forwarding)	TELEPH	(→Rufumleitung)			
call sequence (→connection sequence)	SWITCH	(→Verbindungsablauf)			
call setting-up time (→call set-up time)	DATA COMM	(→Verbindungsaufbaudauer)	**cam-follower lever**	MECH	**Nockenhebel**
			cam-operated = cam-controlled	TECH	**nockengesteuert** = nockenbetätigt
call set-up	TELEPH	**Gesprächsaufbau**	**camped-on call**	TELEPH	**Gespräch in Wartestellung**
call set-up (→connection set-up)	SWITCH	(→Verbindungsaufbau)	**camped-on condition** (→disconnected mode)	DATA PROC	(→Wartestellung)
call set-up time = connecting delay; link set-up time; set-up time; call setting-up time; setting-up time; setting time	DATA COMM	**Verbindungsaufbaudauer** = Verbindungsherstellungsdauer; Vorbereitungszeit	**camp-on** (→automatic callback)	TELEPH	(→automatischer Rückruf)
			camshaft	MECH	**Nockenwelle** = Steuerwelle
call sharing	SWITCH	**Anrufteilung**	**cam switch**	COMPON	**Nockenschalter**
call signal (→ringing signal)	TELEC	(→Rufsignal)	**can** (n.) = metal container	TECH	**Kanister**
call simulator	SWITCH	**Rufsimulator**	**can** (→case)	COMPON	(→Gehäuse)
call splitting = splitting	SWITCH	**Verbindungsaufspaltung**	**can** (→sleeve)	TECH	(→Hülse)
			CAN (→cancel)	DATA COMM	(→ungültig)
call state = connection state; link state; call condition	SWITCH	**Verbindungszustand** = Gesprächszustand	**canal** (n.) [artificial waterway]	CIV.ENG	**Kanal** [künstlicher Wasserweg] = Wasserkanal
call status indicator	TELEPH	**Besetzanzeigenterminal**	**canal ray** = positive ray	ELECTRON	**Kanalstrahl**
call tear-down (→connection tear-down)	SWITCH	(→Verbindungsabbau)	**can anode** = cylinder anode ↑ cylinder electrode	ELECTRON	**Topfanode** = Zylinderanode ↑ Zylinderelektrode
call throughput (→through switching)	SWITCH	(→Durchschaltung)			
call tracing = malicious call tracing; malicious call identification; caller identification	SWITCH	**Fangen**	**canary yellow** = golden yellow	PHYS	**goldgelb**
			cancel (v.t.) = rescind; countermand (v.t.); withdraw ≈ revoke	ECON	**rückgängigmachen** = annullieren; stornieren ≈ widerrufen
call tracing criterion	SWITCH	**Fangkriterium**			
call tracing order	SWITCH	**Fangauftrag**			
call transfer	TELEPH	**Gesprächsweitergabe** = Gesprächszuweisung	**cancel** [code] = CAN; invalid	DATA COMM	**ungültig** [Code] = CAN
call transfer (→call forwarding)	TELEPH	(→Rufumleitung)	**cancel** (n.) (→abnormal end)	DATA PROC	(→Absturz)
call up (→request)	DATA COMM	(→abrufen)	**cancel** (→erase)	DATA PROC	(→löschen)
call waiting [busy station is made aware of a waiting call by a call-waiting tone] ≈ override	TELEPH	**Anklopfen** [besetzter Teilnehmer wird durch Anklopfton auf eine wartende Verbindung aufmerksam gemacht] ≈ Aufschalten	**cancel** (→erase)	ELECTRON	(→löschen)
			cancel character = CCH; invalid character	TERM&PER	**ungültiges Zeichen** = CCH
			cancel character (→ignore character)	DATA COMM	(→Ungültigkeitszeichen)
call waiting security	TELEPH	**Anklopfschutz**	**cancel command**	TELECONTR	**Rücknahmebefehl**
call-waiting tone	TELEPH	**Anklopfton**	**cancel command** (→delete statement)	DATA PROC	(→Löschanweisung)
calorimetric temperature (→color temperature)	PHYS	(→Farbtemperatur)			

cancel input (→erase input) CIRC.ENG (→Löscheingang)
cancel instruction (→delete statement) DATA PROC (→Löschanweisung)
cancel job (→delete statement) DATA PROC (→Löschanweisung)
cancel key (→backspace key) TERM&PER (→Rücksetztaste 2)
cancellation (→withdrawal) ECON (→Widerruf 2)
cancellation MOB.COMM **Ausbuchen**
cancellation DATA PROC **Annullierung**
cancellation completed DATA COMM **Löschungsvollzug**
= Annullierungsvollzug
cancellation completed signal DATA COMM **Löschungsvollzugzeichen**
= Anullierungsvollzugszeichen
cancellation request DATA COMM **Löschungsanforderung**
= Annulllierungsanforderung
cancellation-request signal DATA COMM **Löschungsanforderungszeichen**
= Löschforderungszeichen; Annullierungsanforderungszeichen
cancel statement (→delete statement) DATA PROC (→Löschanweisung)
candela PHYS **Candela**
[SI unit for luminous intensity] [SI-Basiseinheit für Lichtstärke]
= cd = cd
canned (adj.) TECH **konfektioniert**
[fig.] [fig.]
= precanned (adj.) = seriengefertigt; aus Massenproduktion
canned software (→standard software) DATA PROC (→Standard-Software)
cannibalization TECH **Ausschlachtung**
= Querentnahme
cannibalize TECH **ausschlachten**
[to maintain operation taking parts of intact spares] [Betriebserhaltung durch Entnahme von Teilen intakter Beständen]
= querentnehmen
canonic NETW.TH **kanonisch**
canonical MATH **kanonisch**
canonical schema DATA PROC **kanonisches Schema**
[independent of hardware or software] = von Hardware oder Software unabhängig
canonical transformation MATH **kanonische Transformation**
canonic quadripole (→canonic two-port) NETW.TH (→kanonischer Vierpol)
canonic two-port NETW.TH **kanonischer Vierpol**
[with the minimum of circuit elements] [mit einem Minimum von Schaltelementen]
= canonic quadripole
cantenna ANT **Cantenna**
[an oil filled can as antenna termination] [Antennennachbildung]
cantilever CIV.ENG **freitragender Träger**
cantilever spring MECH **Biegefeder**
[unilaterally clamped] [einseitig festgeklemmt]
canvassing (→advertising) ECON (→Werbung)
CAP 2 DATA PROC **CAP 2**
= computer aided publishing = rechnerunterstütztes Publizieren; computergestützte Dokumentenerstellung
CAP 1 (→computer aided planning) DATA PROC (→CAP 1)
cap (→capital character) TYPOGR (→Großbuchstabe)
cap (→hood) TECH (→Haube)
cap (→cover 1) TYPOGR (→Kappe)
capability TECH **Leistungsfähigkeit**
= performance; functionality = Fähigkeit; Befähigung
≈ efficiency ≈ Eignung
capable (→suited) TECH (→geeignet)
capacitance PHYS **Kapazität**
= capacity ↓ elektrische Kapazität; Wärmekapazität
↓ electric capacitance; heat capacity
capacitance NETW.TH **Kapazitanz**
= condensance = Kondensanz

capacitance (→electric capacitance) PHYS (→elektrische Kapazität)
capacitance alarm system (→proximity alarm system) SIGN.ENG (→Annäherungsmeldesystem)
capacitance bridge INSTR **Kapazitätsmessbrücke**
capacitance decade INSTR **Stufenkondensator**
capacitance detector (→proximity detector) SIGN.ENG (→Annäherungsmelder)
capacitance diode (→varactor diode) MICROEL (→Varaktordiode)
capacitance measurement INSTR **Kapazitätsmessung**
capacitance meter INSTR **Kapazitätsmesser**
capacitance per unit length (→distributed capacitance) LINE TH (→Kapazitätsbelag)
capacitance to ground (AM) EL.TECH **Erdkapazität**
= earth capacitance (BRI)
capacitative (→capacitive) EL.TECH (→kapazitiv)
capacitive EL.TECH **kapazitiv**
= capacitative
capacitive coupling EL.TECH **kapazitive Kopplung**
= Kondensatorkopplung
capacitive load EL.TECH **kapazitive Last**
= Kapazitivlast
capacitive pick-up INSTR **kapazitiver Meßfühler**
capacitive reactance NETW.TH **kapazitiver Widerstand**
[1/ΩC] [1/ΩC]
= capacitive resistance
capacitive resistance NETW.TH (→kapazitiver Widerstand)
(→capacitive reactance)
capacitor COMPON **Kondensator**
= condenser (BRI) ≈ Kapazität
≈ capacitance ↓ Festkondensator; einstellbarer Kondensator
↓ fixed capacitor; variable capacitor
capacitor aerial (→capacitor antenna) ANT (→kapazitive Antenne)
capacitor antenna ANT **kapazitive Antenne**
= capacitor aerial
capacitor bank EL.TECH **Kondensatorbatterie**
capacitor loss angle EL.TECH **Kondensatorverlustwinkel**
capacitor loudspeaker EL.ACOUS **Kondensator-Lautsprecher**
= condenser loudspeaker
capacitor memory TERM&PER (→kapazitiver Speicher)
(→capacitor storage)
capacitor microphone EL.ACOUS **Kondensatormikrophon**
= condenser microphone; electrostatic microphone
capacitor pickup EL.ACOUS **Kondensator-Tonabnehmer**
capacitor quenching PHYS **Kondensatorlöschung**
capacitor storage TERM&PER **kapazitiver Speicher**
= capacitor memory
capacity TECH **Kapazität**
= Fassungsvermögen
capacity (→capacitance) PHYS (→Kapazität)
capacity balancing EL.TECH **Kapazitätsausgleich**
capacity bottleneck MANUF **Kapazitätsengpass**
capacity cutback (→capacity reduction) MANUF (→Kapazitätsabbau)
capacity-flat EL.TECH **kapazitätsgerade**
capacity reduction MANUF **Kapazitätsabbau**
= capacity cutback = Kapazitätsverringerung
capacity stage (→construction stage) TECH (→Ausbaustufe)
capacity unbalance EL.TECH **Kapazitätsunsymmetrie**
capillary TECH **kapillar**
capillary electrometer INSTR **Kapillarelektrometer**
capillary jointing METAL **Kapillarschweißung**
capillary tube TECH **Kapillarrohr**
capital ECON **Kapital**
capital case printing TYPOGR (→Großschreibung)
(→capitalization)
capital character TYPOGR **Großbuchstabe**
= capital letter; uppercase character; uppercase letter; cap; majuscule; block letter; block capital = Versalbuchstabe; Versal (n.m.,pl.-ien); Versalie (n.f.,pl.-ien); Majuskel; Blockbuchstabe
≈ initial (n.) ≈ Anfangsbuchstabe
capital character font TYPOGR **Versalschrift**
[typeface in capital letters] [in Großbuchstaben]
= titling ≈ Kapitälchen
≈ small capitals
capital cost ECON **Kapitalkosten**

capital expenditure	ECON	**Kapitalaufwand**
capital height	TYPOGR	**Versalhöhe**
[between base line and cap line]		[zwischen Schriftlinie und Oberlinie]
= caps size		
capitalization	TYPOGR	**Großschreibung**
= capital case printing		
capitalize (v.t.)	TYPOGR	**großschreiben**
capital letter (→capital character)	TYPOGR	(→Großbuchstabe)
capitals lock key (→SHIFT LOCK key)	TERM&PER	(→Umschaltfeststelltaste)
cap line	TYPOGR	**Oberlinie**
[the immaginary upper limiting line of the characters]		[immaginäre obere Begrenzungslinie der Buchstaben]
≠ base line		≠ Schriftlinie
capping system	TECH	**Abdichtungsvorrichtung**
= sealing device		
cap screw (→ head screw)	MECH	(→Kopfschraube)
CAPS-LOCK (→SHIFT LOCK key)	TERM&PER	(→Umschaltfeststelltaste)
caps size (→capital height)	TYPOGR	(→Versalhöhe)
capstan	TERM&PER	**Capstan**
[shaft for magnetic tape reel]		[Welle in Tonbandgerät]
↑ shaft		= Bandantriebsachse
		↑ Welle
capstan bearing	TERM&PER	**Capstan-Lagerung**
capstan drive	TERM&PER	**Capstan-Antrieb**
= capstan servo		
capstan motor	TERM&PER	**Capstan-Motor**
capstan servo (→capstan drive)	TERM&PER	(→Capstan-Antrieb)
capsulated (→encapsulated)	TECH	(→gekapselt)
capsule	TECH	**Kapsel**
caption	TV	**Kennung**
		= Insert; Caption
caption 2	TYPOGR	**Bildlegende**
[explanatory note to an illustration]		= Bildunterschrift
≈ legend		≈ Zeichenerklärung
caption 1 (→heading)	LING	(→Überschrift)
capture	ELECTRON	**Erfassung**
capture area (→effective aperture)	ANT	(→Antennenwirkfläche)
capture range	ELECTRON	**Fangbereich**
= lock-in range		
CAQ	DATA PROC	**CAQ**
= computer aided quality control; automatic quality control		= rechnergestützte Qualitätssicherung; rechnergestützte Qualitätskontrolle; computergestützte Qualitätssicherung
car (→vehicle)	TECH	(→Fahrzeug)
CAR (→current address register)	DATA PROC	(→Momentanadressenregister)
car aerial (→car antenna)	ANT	(→Fahrzeugantenne)
car antenna	ANT	**Fahrzeugantenne**
= car aerial; car radio antenna; car radio aerial; vehicle-mounted antenna; vehicle-mounted aerial		= Autoantenne
carat [DATA PROC] (→apex)	TYPOGR	(→Spitze)
carbide	CHEM	**Karbid**
carbide drill	METAL	**Hartmetallbohrer**
carbide metal	METAL	**Hartmetall**
carbon	CHEM	**Kohlenstoff**
= C		= C
carbon	COMPON	**Kohlespannungsableiter**
↑ fine overvoltage protection		↑ Spannungsfeinsicherung
carbon arc	PHYS	**Kohlebogen**
carbonate	CHEM	**Karbonat**
carbon copy	TERM&PER	**Durchschlag**
		= Durchschrift
carbon filament	PHYS	**Kohlefaden**
carbon film potentiometer	COMPON	**Kohleschichtpotentiometer**
carbon film resistor	COMPON	**Kohleschichtwiderstand**
= carbon resistor		= Kohlewiderstand
carbon microphone	EL.ACOUS	**Kohlemikrophon**
		= Kohlekörnermikrophon
carbon microphone	TELEPH	**Kohlesprechkapsel**
		= Kohlemikrophon
carbon paper	OFFICE	**Kohlepapier**
= flimsy paper		= Durchschlagpapier
carbon resistor (→carbon film resistor)	COMPON	(→Kohleschichtwiderstand)
carbon ribbon	TERM&PER	**Kohlefarbband**
= carbon tape		= Kohleband; Karbonfarbband; Karbonband; Carbonfarbband; Carbonband
carbon ribbon shredder	OFFICE	**Karbonbandvernichter**
= carbon tape shredder		= Carbonbandvernichter
carbon silk	TERM&PER	**Farbtuch**
carbon tape (→carbon ribbon)	TERM&PER	(→Kohlefarbband)
carbon tape shredder (→carbon ribbon shredder)	OFFICE	(→Karbonbandvernichter)
carbon tetrachloride	CHEM	**Tetrachlorkohlenstoff**
card	TERM&PER	**Karte**
↓ punched card; magnetic card		↓ Lochkarte; Magnetkarte
card	ECON	**Ausweis**
↑ identity card; corporate identification card		↑ Personalausweis; Firmenausweis
card aligner	TERM&PER	**Kartenanschlag**
		= Kartenausrichter
card-based data entry terminal	TERM&PER	**Lochkartenerfassungsgerät**
cardboard	TECH	**Pappe**
= paperboard		= Karton 1
cardboard box	TECH	**Pappschachtel**
= carton		= Kartonschachtel; Karton 2
card box	TERM&PER	**Kartenkasten**
= card rack		
card cage (→module frame)	EQUIP.ENG	(→Baugruppenrahmen)
card chassis (→module frame)	EQUIP.ENG	(→Baugruppenrahmen)
card code	TERM&PER	**Lochkartencode**
card collator (→punched card collator)	TERM&PER	(→Lochkartenmischer)
card column	TERM&PER	**Lochkartenspalte**
[vertical, parallel to the narrow card side]		[senkrecht, parallel zur Kartenschmalseite]
		= Kartenspalte
card container	EQUIP.ENG	**Kartenfach**
		= Kartentasche
card controlled	DATA PROC	**kartengesteuert**
card counter	TERM&PER	**Kartenzähler**
card cycle	TERM&PER	**Kartenzyklus**
[for transport and processing]		[Transport und Bearbeitung]
		= Kartengang
card deck (→punched card deck)	TERM&PER	(→Lochkartenstapel)
card edge connector (→direct plug connector)	COMPON	(→Direktsteckverbinder)
card ejection (→card stacker)	TERM&PER	(→Lochkartenstapler)
card extender (→board extender)	INSTR	(→Adapter)
card feed	TERM&PER	**Kartenvorschub**
= card transport		= Kartentransport
card field	TERM&PER	**Lochkartenfeld**
card file	OFFICE	**Kartei**
= card-index		↓ Steilkartei; Staffelkartei
card format	TERM&PER	**Kartenformat**
card frame (→module frame)	EQUIP.ENG	(→Baugruppenrahmen)
card gripper (→card picker)	TERM&PER	(→Kartenaufnehmer)
card handle	EQUIP.ENG	**Kartengriff**
card holder	TERM&PER	**Postkartenhalter**
[typewriter]		[Shreibmaschine]
card hopper	TERM&PER	**Kartenstapler**
[holds and feeds punched cards]		= Eingabefach
= hopper		
card image	DATA PROC	**Kartenabbild**
[exact card-content image in a memory]		[in einem Speicher ein exaktes Abbild eines Karteninhalts]
cardinal number	MATH	**Kardinalzahl**
[any number obtainable by addition of 1 to 1]		[jede aus 1 durch Addition von 1 erhältliche Zahl]
= natural number; nonnegative integer		= natürliche Zahl; Grundzahl 1
≈ ordinal number		≈ Ordinalzahl

cardinal point		MECH	**Angelpunkt**	care of (bri) [addressing] = c/o; attention (AM); att; for the attention of	OFFICE	**zu Händen** [Adresse] = z.Hdn.
card-index (→card file)		OFFICE	(→Kartei)			
card-index cabinet = filing cabinet		OFFICE	**Karteikasten**			
card-index file		OFFICE	**Karteiregister**	caret (→upward arrow)	TYPOGR	(→Pfeil nach oben)
card-index set		OFFICE	**Leitregister**	cargo = load ≈ freight	ECON	**Ladung** ≈ Fracht
card-index tab		OFFICE	**Karteireiter**			
card-index table		OFFICE	**Karteitisch**			
card-index tray		OFFICE	**Karteitrog**	cargo (→freight 2)	ECON	(→Fracht 1)
card interpreter (→punched card scanner)		TERM&PER	(→Lochkartenleser)	cargo ship = Transportschiff	TECH	**Frachtschiff**
cardioid (n.) = Herzkurve		MATH	**Kardioide** (n.f.)	car measuring instrument	ELECTRON	**Cockpit-Instrument**
				carmine red = carminrot	PHYS	**karminrot**
cardioid (adj.) = kardioid		MATH	**herzförmig**			
				caron	DOC	**Caron**
cardioid (→cardioid pattern)		PHYS	(→Kardioidendiagramm)	carpenter's pincers (→nipper pliers)	TECH	(→Beißzange)
cardioid antenna		ANT	**Kardioid-Antenne**			
cardioid characteristic = Nierencharakteristik		EL.ACOUS	**Nieren-Richtcharakteristik**	car phone = car telephone; mobile cellular phone ↑ cellular phone; cellular telephone	MOB.COMM	**Autotelefon** ↑ Mobiltelefon; Zellulartelefon
cardioid diagram (→cardioid pattern)		PHYS	(→Kardioidendiagramm)			
cardioide microphone		EL.ACOUS	**Kardioidenmikrofon**	car radio = autoradio	RADIO	**Autoradio**
cardioid pattern = cardioid diagram; cardioid		PHYS	**Kardioidendiagramm** = Herzkurvendiagramm; Kardioide; Herzkurve; Herzcharakteristik			
				car radio aerial (→car antenna)	ANT	(→Fahrzeugantenne)
				car radio antenna (→car antenna)	ANT	(→Fahrzeugantenne)
cardioid shape		COMPON	**Nierenplattenschnitt**			
carditioner (→card reconditioner)		TERM&PER	(→Kartenbügler)	carriage [typewriter] = Schreibwagen; Schlitten	TERM&PER	**Wagen** [Schreibmaschine]
card jam = Kartensalat		TERM&PER	**Kartenstau**			
				carriage (→freight 2)	ECON	(→Fracht 1)
card loader [a program]		DATA PROC	**Kartenlader** [ein Programm]	carriage (→freight 1)	ECON	(→Frachtkosten)
				carriage (→transport)	ECON	(→Transport)
card output		TERM&PER	**Kartenausgabe**	carriage bolt = Schloßschraube	MECH	**Flachrundschraube**
cardphone		TELEPH	**Kartentelefon**			
card picker = card gripper		TERM&PER	**Kartenaufnehmer** = Kartengreifer	carriage control (→feed control)	TERM&PER	(→Vorschubsteuerung)
card punch (→card puncher)		TERM&PER	(→Kartenlocher)	carriage lock	TERM&PER	**Wagenverriegelung**
card puncher = card punch; punch card machine; keyboard punch; keyboard perforator; keypunch; handpunch		TERM&PER	**Kartenlocher** = Lochkartenlocher; Kartenstanzer; Lochkartenstanzer; Tastenlocher; Handlocher; Tastaturlocher	carriage release = Wagenauslöser	TERM&PER	**Wagenlöser**
				carriage return = CR	TELEGR	**Wagenrücklauf** = CR; WR; Zeilenrücklauf
				carriage return key = CR key; RETURN key ≈ ENTER key [TERM&PER]	TELEGR	**Wagenrücklauftaste** = WR-Taste; Zeilenschalttaste ≈ Eingabetaste [TERM&PER]
card rack (→card box)		TERM&PER	(→Kartenkasten)			
card reader (→punched card scanner)		TERM&PER	(→Lochkartenleser)			
card reconditioner = carditioner		TERM&PER	**Kartenbügler**	carriage tape	TERM&PER	**Vorschublochband**
				carried call = verarbeitete Belegung	SWITCH	**angenommene Belegung**
card reproducer (→punched card reproducer)		TERM&PER	(→Lochkartendoppler)			
				carried traffic (→traffic intensity)	SWITCH	(→Verkehrswert)
card reverser = reverser		TERM&PER	**Kartenwender**			
				carrier [system or organization conveying information] = bearer ↓ telecommunication carrier	TELEC	**Träger** [Nachrichten übermittelndes System oder Organisation] ↓ Fernmeldegesellschaft
card row (→punched card row)		TERM&PER	(→Lochkartenzeile)			
card run		TERM&PER	**Kartendurchlauf**			
card scanner (→punched card scanner)		TERM&PER	(→Lochkartenleser)			
				carrier (→substrate)	MICROEL	(→Substrat)
card sensing = Kartenablesung		TERM&PER	**Lochkartenablesung**	carrier (→carrier wave)	MODUL	(→Trägerschwingung)
				carrier (→transmission medium)	TELEC	(→Übertragungsmedium)
card stacker = card ejection ↑ stacker		TERM&PER	**Lochkartenstapler** ↑ Stapler			
				carrier amplifier	TELEC	**Trägerverstärker**
				carrier avalanche = charge carrier avalanche	MICROEL	**Trägerlawine** = Ladungsträgerlawine
card transport (→card feed)		TERM&PER	(→Kartenvorschub)			
card verification = card verifying		TERM&PER	**Lochkartenprüfung** = Kartenprüfung	carrier beat	MODUL	**Trägerschwebung**
				carrier channel = FDM channel	TRANS	**TF-Kanal** = Trägerfrequenzkanal
card verifier (→punched card verifier)		TERM&PER	(→Kartenprüfer)			
card verifying (→card verification)		TERM&PER	(→Lochkartenprüfung)	carrier channel = bearer channel	DATA COMM	**Trägerkanal**
				carrier channel rate = carrier transfer rate; bearer channel rate	DATA COMM	**Trägergeschwindigkeit**
care (n.) = diligence		COLLOQ	**Sorgfalt**			
careful		TECH	**schonend**	carrier concentration layer (→enhancement layer)	MICROEL	(→Anreicherungszone)
careful		TECH	**schonend** (adj.) = sorgfältig	carrier density	MICROEL	**Trägerdichte**
				carrier diffusion (→charge carrier diffusion)	PHYS	(→Ladungsträgerdiffusion)
carefully = diligently		COLLOQ	**sorgfältig**			
car electronics = automotive electronics ↑ consumer electronics		ELECTRON	**Fahrzeugelektronik** = Kfz-Elektronik; Autoelektronik ↑ Konsumelektronik	carrier dispersal	RADIO	**Trägerverwischung**
				carrier dispersal gain = dispersal gain	RADIO	**Trägerverwischungsgewinn** = Verwischungsgewinn
				carrier extraction (→carrier recovery)	MODUL	(→Trägerrückgewinnung)

carrier fading	HF	Trägerschwund	
		= Trägerfading	
carrier frequency	MODUL	Trägerfrequenz	
≈ carrier wave		= TF	
		≈ Trägerschwingung	
carrier-frequency bridge	INSTR	Trägerfrequenzmeßbrücke	
		= TF-Meßbrücke; Trägerfrequenzbrücke; TF-Brücke	
carrier frequency generator [FDM]	TRANS	Trägerfrequenzgenerator [TF-Technik]	
= carrier generator			
carrier frequency measuring technique	INSTR	TF-Meßtechnik	
carrier frequency voltmeter (→selective level meter)	INSTR	(→selektiver Pegelmesser)	
carrier generator (→carrier frequency generator)	TRANS	(→Trägerfrequenzgenerator)	
carrier injection (→charge carrier injection)	MICROEL	(→Ladungsträgerinjektion)	
carrier interference	RADIO	Trägerinterferenz	
carrier jam effect (→hole-storage effect)	MICROEL	(→Trägerstaueffekt)	
carrier leak	MODUL	Trägerrest	
carrier level	TRANS	Trägerfrequenzpegel	
		= Trägerpegel	
carrier lifetime	MICROEL	Trägerlebensdauer	
= charge carrier lifetime		= Ladungsträgerlebensdauer	
carrier link line	TRANS	TF-Grundleitung	
= multichannel circuit			
carrier mobility (→charge carrier mobility)	PHYS	(→Ladungsträgerbeweglichkeit)	
carrier multiplication	MICROEL	Ladungsträger-Vervielfachung	
= charge carrier multiplication		= Trägervervielfachung; Ladungsträgermultiplikation; Trägermultiplikation	
carrier noise test set	INSTR	Trägerrausch-Prüfplatz	
		= Trägerrausch-Testset	
carrier offset	RADIO	Offsetbetrieb	
carrier pair	MICROEL	Trägerpaar	
carrier phase duplex	MODUL	Amplituden- und Phasentastung	
carrier phase noise	MODUL	Träger-Phasenrauschen	
carrier power	MODUL	Trägerleistung	
carrier pulse [electron]	PHYS	Trägerimpuls [Elektron]	
carrier recovery	MODUL	Trägerrückgewinnung	
= carrier extraction			
carrier replenishment	MICROEL	Trägerauffüllung	
= carrier support			
carrier sense multiple access	TRANS	Vielfachzugriff mit Leitungsüberwachung	
= CSMA		= CSMA	
carrier service (→bearer service 1)	TELEC	(→Übermittlungsdienst)	
carrier shift (→frequency shift keying)	MODUL	(→Frequenzumtastung)	
carrier signal	TRANS	Trägersignal	
carrier signaling	TELEC	Trägersignalisierung	
= carrier signalling		= Trägertastung	
carrier signalling (→carrier signaling)	TELEC	(→Trägersignalisierung)	
carrier storage time	MICROEL	Speicherzeit	
= storage time; pulse storage time			
carrier supply [FDM]	TRANS	Trägerversorgung [TF-Technik]	
carrier support (→carrier replenishment)	MICROEL	(→Trägerauffüllung)	
carrier suppression	MODUL	Trägerunterdrückung	
carrier swing	MODUL	Trägerhub	
carrier telegraphy (→voice frequency telegraphy)	TELEGR	(→Wechselstromtelegrafie)	
carrier telephony	TELEC	Trägerfrequenz-Telephonie	
carrier-to-noise ratio	MODUL	Träger-Rausch-Verhältnis	
≠ C/N			
carrier transfer rate (→carrier channel rate)	DATA COMM	(→Trägergeschwindigkeit)	
carrier transmission over power lines	TRANS	TFH-Technik	
= powerline carrier transmission		= Trägerfrequenztechnik auf Hochspannungsleitungen	
carrier-transport mechanism (→conduction mechanism)	PHYS	(→Leitungsmechanismus)	
carrier wave	MODUL	Trägerschwingung	
= modulation carrier; carrier		= Trägerwelle; Modulationsträger; Träger	
≈ carrier frequency		≈ Trägerfrequenz	
≠ modulating wave		≠ Modulationsschwingung	
carry	MATH	Übertrag	
[increment if result of addition at preciding digit position exceeds maximum of the numbering system]		[Zuschlag wenn Additionsergebnis in der vorangegangenen Stelle den Höchstwert des Zahlensystems überschritten hat]	
= transfer; overflow; carry-over		= Überlauf	
≈ overflow [DATA PROC]		≈ Überlauf [DATA PROC]	
↓ complete carry [DATA PROC]		↓ Vollübertrag [DATA PROC]	
carry (→carry bit)	DATA PROC	(→Übertragbit)	
carry bit	DATA PROC	Übertragbit	
= transfer bit; carry flag; carry; transfer flag; overflow flag; overflow bit		= Übertragsbit; Übertragmerker	
carry complete signal	DATA PROC	Übertrag-Ausführungssignal	
carry digit	MATH	Übertragerziffer	
= overflow digit; overflow 2		= Übertragsziffer	
carry flag (→carry bit)	DATA PROC	(→Übertragbit)	
carrying (→conductive)	PHYS	(→leitend)	
carrying handle	EQUIP.ENG	Tragegriff	
↑ handle		↑ Tragebügel	
		↑ Handgriff 1	
carry look-ahead (n.)	DATA PROC	Parallelübertrag	
[speeds-up predicting the carry]		[beschleunigt durch Vorhersage des Übertrags]	
		= Carry-look-ahead	
carry out (→execute)	DATA PROC	(→ausführen)	
carry-over (→carry)	MATH	(→Übertrag)	
carry register (→call register)	SWITCH	(→Verbindungsspeicher)	
carry time	DATA PROC	Übertragszeit	
car security system	ELECTRON	Auto-Alarmanlage	
car speaker	EL.ACOUS	Autolautsprecher	
carted goods	ECON	Rollgut	
car telephone (→car phone)	MOB.COMM	(→Autotelefon)	
Carter chart	LINE TH	Carter-Diagramm	
↑ transmission line chart		↑ Leitungsdiagramm	
Carter stub	ANT	Carter-Schleife	
↑ stub		↑ Stichleitung	
cartesian coordenate	MATH	kartesische Koordinate	
= rectangular coordenate		= rechtwinklige Parallelkoordinate	
cartesian coordinate system	MATH	kartesisches Koordinatensystem	
= rectangular coordinate system			
cartesian product [set theory]	MATH	kartesisches Produkt [Mengenlehre]	
		= Kreuzprodukt	
cartesian space	MATH	kartesischer Raum	
carton (→cardboard box)	TECH	(→Pappschachtel)	
cartridge	TECH	Patrone	
cartridge	TERM&PER	Einschubkassette	
↓ magnetic tape cassette		↓ Magnetbandkassette	
cartridge (→phonograph pickup)	EL.ACOUS	(→Tonabnehmer)	
cartridge drive	TERM&PER	Kassettenlaufwerk	
cartridge font	DATA PROC	Zeichensatzkassette	
cartridge fuse	COMPON	Glasrohr-Feinsicherung	
= fuse cartridge		= G-Sicherung; Glasröhrensicherung	
		↑ Stromfeinsicherung	
cartridge loader	OFFICE	Kassetten-Füllgerät	
cartridge ribbon	TERM&PER	Kassettenfarbband	
cartridge streamer (→streamer)	TERM&PER	(→Streamer-Magnetbandgerät)	
cartridge tape drive (→magnetic tape cassette drive)	DATA PROC	(→Magnetbandkassetten-Laufwerk)	
cartwheel antenna	ANT	Radantenne	
CAS (→column address)	MICROEL	(→Spaltenadresse)	
CAS (→channel-oriented signaling)	SWITCH	(→kanalgebundene Zeichengabe)	
cascadable microprocessor	MICROEL	kaskadierbarer Mikroprozessor	

cascade

cascade (v.t.)	EL.TECH	**hintereinanderschalten**
		= in Reihe schalten; kaskadieren
cascade (→series connection)	EL.TECH	(→Reihenschaltung)
cascade amplifier	CIRC.ENG	**Kaskadenverstärker**
= multistage amplifier; distributed amplifier		= Kettenverstärker; mehrstufiger Verstärker; Loftin-White-Verstärker
cascade carry	DATA PROC	**Kaskadenübertrag**
cascade circuit (→series connection)	EL.TECH	(→Reihenschaltung)
cascade connection (→series connection)	EL.TECH	(→Reihenschaltung)
cascade control (→sequence control)	CONTROL	(→Folgesteuerung)
cascade equation	MATH	**Kaskadengleichung**
cascade generator	CIRC.ENG	**Kaskadengenerator**
cascade network	TELEC	**Kaskadennetz**
[star-connected star networks]		[sternförmig zusammengefaßte Sternnetze]
cascade of cascades	CIRC.ENG	**Kaskadenanreihung**
cascade rectifier	CIRC.ENG	**Kaskadengleichrichter**
cascade sorting	DATA PROC	**Kaskadensortieren**
↑ sorting		↑ Sortieren
cascade tube	ELECTRON	**Kaskadenröhre**
cascading (→series connection)	EL.TECH	(→Reihenschaltung)
cascode	ELECTRON	**Kaskode**
		= Cascode
cascode circuit	CIRC.ENG	**Kaskode-Schaltung**
↑ amplifier circuit		= Cascode-Schaltung; Kaskodenschaltung; Cascodenschaltung
		↑ Verstärkerschaltung
cascode difference amplifier	CIRC.ENG	**Kaskode-Differenzverstärker**
↑ direct current amplifier		↑ Gleichspannungsverstärker
cascode FET current source	CIRC.ENG	**Kaskode-FET-Stromquelle**
↑ constant current power supply		↑ Konstantstromquelle
case (n.)	EQUIP.ENG	**Gehäuse**
= enclosure; housing		≈ Schrank
≈ cabinet		
case	ECON	**Fall**
case	TECH	**Gehäuse**
= housing; chassis; frame		≈ Umhüllung
≈ cladding		
case	COMPON	**Gehäuse**
= holder; package; enclosure; can		= Verpackung
CASE	DATA PROC	**CASE**
= computer-aided software engineering		= rechnergestützte Programmierung
case	LING	**Kasus**
[inflectional form of a noun 2]		[Beugungsform eines Nomens 2]
↓ nominative; genitive; dative; accusative; vocative; locative; instrumental		= Fall
		↓ Nominativ; Genitiv; Dativ; Akkusativ; Vokativ; Lokativ; Instrumental
case (→casing)	TECH	(→Hülle 1)
case (→portable case)	EQUIP.ENG	(→Koffer)
CASE (→machine-aided programming)	DATA PROC	(→automatische Programmierung)
case (→proceedings 1)	ECON	(→Verfahren)
case (→computer cabinet)	DATA PROC	(→Computergehäuse)
case (→disk jacket)	TERM&PER	(→Diskettenhülle)
case antenna	ANT	**Kofferantenne**
case by case	COLLOQ	**fallweise**
= case to case		
case-harden	METAL	**einsatzhärten**
case history (→case study)	SCIE	(→Fallstudie)
case shift	TERM&PER	**Groß-Klein-Umschaltung**
= upper/lower case shift; uppercase shift; shift; case change;upper/lower case change; upper/lower change; change		= Klein-Groß-Umschaltung; Umschaltung; Zeichenwechsel; Großbuchstabenumschaltung
≈ shift key		≈ Umschaltetaste
case shift key (→SHIFT key)	TERM&PER	(→Umschalttaste 1)
case study	SCIE	**Fallstudie**
= case history		
case temperature	COMPON	**Gehäusetemperatur**
case to case (→case by case)	COLLOQ	(→fallweise)
cash	ECON	**bar**
≠ cashless		≠ unbar
cash-card (→prepaid phonecard)	TERM&PER	(→Guthabenkarte)
cashless	ECON	**unbar**
≠ cash		≠ bar
cash on shipment	ECON	**zahlbar bei Verschiffung**
= COS		
cash price	ECON	**Barpreis**
cash register	TERM&PER	**Registrierkasse**
cash testing equipment (→money testing equipment)	OFFICE	(→Geldprüfgerät)
casing	TECH	**Hülle 1**
[solid protective box]		[festes Gebilde]
= case		≈ Gehäuse
		↓ Schutzhülle
casket	TECH	**Transportbehälter**
= coffin		
Cassegrain antenna (→Cassegrain reflector antenna)	ANT	(→Cassegrain-Antenne)
Cassegrain arrangement	ANT	**Cassegrain-Anordnung**
Cassegrain feed	ANT	**Cassegrain-Erreger**
Cassegrain reflector antenna	ANT	**Cassegrain-Antenne**
= Cassegrain antenna		= Cassegrain-Reflektor-Antenne
cassette (→magnetic tape cassette)	TERM&PER	(→Magnetbandkassette)
cassette autoradio (→cassette car radio)	CONS.EL	(→Kassetten-Autoradio)
cassette car radio	CONS.EL	**Kassetten-Autoradio**
= cassette autoradio		= Cassetten-Autoradio
cassette deck	CONS.EL	**Kassettendeck**
[cassette recorder without amplifier]		[Kassettenrecorder ohne Verstärker]
≈ cassette recorder		= Cassettendeck
		≈ Kassettenrecorder
cassette ejector (→eject)	CONS.EL	(→Kassettenausschub)
cassette magnetic tape	TERM&PER	**Kassetten-Magnetband**
cassette player	CONS.EL	**Kassettenspieler**
[permits playing only]		[nur zum Abspielen]
		= Cassettenspieler
		↑ Kassettenrecorder
cassette receptacle	CONS.EL	**Kassettenfach**
		= Cassettenfach
cassette recorder	CONS.EL	**Kassettenrecorder**
[for record and play]		[zum Abspielen und Bespielen]
≈ cassette deck		= Cassettenrecorder
		≈ Kassettendeck
		≈ Kassettenspieler
cassette tape	TERM&PER	**Kassettenband**
cassette tape memory	TERM&PER	**Magnetbandkassetten-Speicher**
= cassette tape storage		
↑ magnetic tape memory		↑ Magnetbandspeicher
cassette tape recorder	TERM&PER	**Kassettenrecorder**
= magnetic tape cassette recorder		= Kassettengerät; Cassettenrecorder; Cassettengerät; Datasette 2; Magnetkassettenstation
cassette tape storage (→cassette tape memory)	TERM&PER	(→Magnetbandkassetten-Speicher)
cass horn antenna	ANT	**Casshorn-Antenne**
↑ reflector antenna		↑ Reflektorantenne
cast	METAL	**gegossen**
= cast-metal		
cast (→found)	METAL	(→gießen)
cast alloy	METAL	**Gußlegierung**
cast aluminum	METAL	**Gußaluminium**
casting 1	METAL	**Guß**
casting 2	METAL	**Gußstück**
casting-out-nines	CODING	**Neunerprüfung**
[longitudinal parity check with decimal numbers]		[Längsparitätskontrolle bei Dezimalzahlen]
casting resin	CHEM	**Gießharz**
cast iron (n.)	METAL	**Gußeisen**
cast-iron ...	METAL	**gußeisern**
= cast-metal ...		

English	Domain	German
castle nut	MECH	Kronenmutter
cast-metal (→cast)	METAL	(→gegossen)
cast-metal ... (→cast-iron ...)	METAL	(→gußeisern)
cast-on (adj.)	TECH	angegossen
cast-resin transformer	POWER ENG	Gießharztransformator
cast steel	METAL	Gußstahl
casual (→random)	TECH	(→zufällig)
casual user (→occasional user)	TECH	(→Wenigbenutzer)
CAT = computer aided testing	DATA PROC	CAT = rechnergestütztes Prüfen; computergestütztes Prüfen
catalog (v.t.) (NAM) = catalogue (BRI)	DATA PROC	katalogisieren
catalog [list in order] = catalogue	DOC	Katalog [geordnete Liste]
catalog (→directory)	DATA PROC	(→Verzeichnis)
catalogue (→catalog)	DOC	(→Katalog)
catalogue (n.) (→directory)	DATA PROC	(→Verzeichnis)
catalogue (BRI) (→catalog)	DATA PROC	(→katalogisieren)
catastrophic failure = complete failure; total failure; blackout	QUAL	Totalausfall = Vollausfall
catch (n.) ≈ latch	MECH	Schnapper ≈ Klinke
catcher space = output gap	MICROW	Auskoppelraum
catching nozzle	TECH	Fangdüse
catchline (→headline 2)	LING	(→Schlagzeile)
catchment area (→servicing area)	TELEC	(→Versorgungsbereich)
catch pan	OUTS.PLANT	Schmutzfänger
catchphrase (→keyword)	LING	(→Schlagwort)
category	SCIE	Kategorie
catena = chain	DATA PROC	Kette
catenary	MATH	Kettenlinie
catenary wire (→messenger wire)	OUTS.PLANT	(→Tragseil)
catenary wire aerial cable (→integral messenger cable)	COMM.CABLE	(→Tragseil-Luftkabel)
catenate = concatenate	DATA PROC	verketten
caternary (→overhead line)	POWER ENG	(→Fahrleitung)
caterpillar tractor	TECH	Raupenschlepper
cathete [side of a right angle]	MATH	Kathete [einen rechten Winkel bildende Seite] ↓ Ankathete; Gegenkathete
cathode = negative electrode ↑ electrode	PHYS	Kathode = Katode; negative Elektrode ↑ Elektrode
cathode base circuit = grounded cathode circuit; grounded cathode amplifier	CIRC.ENG	Kathodenbasisschaltung = Katodenbasisschaltung; Kathodenschaltung; Katodenschaltung; KB-Schaltung; Kathodengrundschaltung; Katodengrundschaltung; Kathodenbasisverstärker; Katodenbasisverstärker
cathode bias	ELECTRON	Kathodenvorspannung = Katodenvorspannung
cathode characteristic	ELECTRON	Emissionskennlinie
cathode coating impedance	ELECTRON	Schichtwiderstand [einer Kathode]
cathode-coupled circuit (→grounded-anode circuit)	CIRC.ENG	(→Anoden-Basis-Schaltung)
cathode cup	ELECTRON	Kathodenbecher = Katodenbecher
cathode current	ELECTRON	Kathodenstrom = Katodenstrom
cathode desintegration (→cathode sputtering)	ELECTRON	(→Kathodenzerstäubung)
cathode evaporation (→cathode sputtering)	ELECTRON	(→Kathodenzerstäubung)
cathode fall	PHYS	Kathodenfall = Katodenfall
cathode follower (→grounded-anode circuit)	CIRC.ENG	(→Anoden-Basis-Schaltung)
cathode hum = filament hum	ELECTRON	Kathodenrauschen = Katodenrauschen
cathode poisoning	PHYS	Kathodenvergiftung = Katodenvergiftung
cathode ray = electron ray	PHYS	Kathodenstrahl = Katodenstrahl; Elektronenstrahl
cathode-ray direction finder	RADIO LOC	Sichtfunkpeiler
cathode ray display ↓ image store display; image repeat display	TERM&PER	Kathodenstrahlbildschirm ↓ Speicherbildschirm; Bildwiederholschirm
cathode ray oscillograph [records rapidly varying magnitudes] ↑ oscillograph	INSTR	Elektronenstrahl-Oszillograph [registriert schnelle Größen] = Elektronenstrahl-Oszillograf ↑ Oszillograph
cathode ray oscilloscope [displays rapidly varying magnitudes] ↑ oscilloscope	INSTR	Elektronenstrahl-Oszilloskop [stellt schnelle Größen am Bildschirm dar] = Kathodenstrahloszilloskop; Universaloszilloskop ↑ Oszilloskop
cathode ray tube = CRT; Braun tube ↓ picture tube; camera tube; velocity-modulated tube	ELECTRON	Kathodenstrahlröhre = Katodenstrahlröhre; Elektronenstrahlröhre; Braunsche Röhre ↓ Bildwiedergaberöhre; Bildaufnahmeröhre; Laufzeitröhre
cathode spot	ELECTRON	Kathodenfleck = Katodenfleck
cathode sputtering = sputtering; cathode desintegration; cathode evaporation	ELECTRON	Kathodenzerstäubung = Katodenzerstäubung
cathodoluminescence [induced by impact of electrons]	PHYS	Kathodolumineszenz [durch Elektronenaufprall induziert] = Katodolumineszenz; Kathodenlumineszenz; Katodenlumineszenz
cathodophosphorescence	PHYS	Kathodophosphoreszenz = Katodophosphoreszenz
cation [positively charged ion] ≠ anion ↑ ion	PHYS	Kation [positiv geladenes Ion] ≠ Anion ↑ Ion
cat's eye	TECH	Katzenauge
CATV = community antenna television ≈ cable TV ↑ television distribution system	BROADC	Großgemeinschaftsanlage = GGA; Groß-Gemeinschaftsantennen-Anlage ≈ Kabelfernsehen ↑ Fernsehverteilanlage
CATV analyzer	INSTR	CATV-Analysator
CATV head station	BROADC	Kabelkopfstation
CATV system (→cable television system)	BROADC	(→Kabelfernsehanlage)
Cauchy distribution	MATH	Cauchy-Verteilung
Cauer filter	NETW.TH	Cauer-Filter
caulk (v.t.)	MECH	verstemmen
causal = causative	LING	kausal
causal sentence ↑ conjunctional sentence ↓ conditional sentence; consecutive sentence; purposive sentence; concessive sentence	LING	Kausalsatz = Begründungssatz ↓ Konditionalsatz; Konsekutivsatz; Finalsatz; Konzessivsatz; Instrumentalsatz
causative (→causal)	LING	(→kausal)
cause (v.t.) = originate ≈ produce	TECH	verursachen ≈ erzeugen
cause (n.)	TECH	Ursache = Grund
causer = originator ≈ producer	TECH	Verursacher ≈ Erzeuger
caustic surface	PHYS	Brennfläche = kaustische Fläche

causting solution

causting solution (→lye)	CHEM	(→Lauge)	
cauterize	CHEM	ausätzen	
caution marks	ECON	Vorsichtsmarkierung	
cavitation (→cavity)	TECH	(→Hohlraum)	
cavity	TECH	Hohlraum	
= cavitation; hollow		= Höhlung; Hohlstelle	
cavity	MICROW	Hohlraum	
		= Kammer	
cavity	METAL	Lunker	
cavity-coupled	MICROW	hohlraumgekoppelt	
cavity-coupled laser	OPTOEL	hohlraumgekoppelter Laser	
cavity field	MICROW	Hohlraumfeld	
cavity ion dose	PHYS	Hohlraumdosis	
cavity resonance	MICROW	Hohlraumresonanz	
cavity resonator (→coaxial cavity resonator)	MICROW	(→Topfkreis)	
cavity-tuned	MICROW	hohlraumabgestimmt	
cavity-tuned oscillator	CIRC.ENG	hohlraumabgestimmter Oszillator	
CAV method (→CAV mode)	TERM&PER	(→CAV-Verfahren)	
CAV mode	TERM&PER	CAV-Verfahren	
[Constant Angular Velocity; magnetic head run with same angular velocity on all tracks]		[Magnetkopf auf allen Spuren mit gleicher Winkelgeschwindigkeit]	
= CAV method			
CB (→citizens band)	RADIO	(→CB)	
CB (→central battery)	TELEPH	(→Zentralbatterie)	
CB antenna	ANT	CB-Antenne	
CBL (→computer-aided education)	DATA PROC	(→rechnergestützter Unterricht)	
cbm (obs.) (→cubic meter)	PHYS	(→Kubikmeter)	
CBMS (→message handling system)	DATA COMM	(→Mitteilungsübermittlungssystem)	
CB radio	RADIO	CB-Funk	
CB radio (→CB radio equipment)	RADIO	(→CB-Funksprechgerät)	
CB radio engineering	RADIO	CB-Funktechnik	
CB radio equipment	RADIO	CB-Funksprechgerät	
= CB radio; citizen band radio		= CB-Funkgerät	
CBT (→cursor backward tabulation)	TERM&PER	(→Schreibmarke Rücktabulation)	
c.c. (→cubic centimeter)	PHYS	(→Kubikzentimeter)	
CCCL	MICROEL	CCCL	
= complementary constant current logic			
CCD (→charge coupled device)	MICROEL	(→ladungsgekoppelte Schaltung)	
CCD diode (→charge storage diode)	MICROEL	(→Speicherschaltdiode)	
CCD filter	CIRC.ENG	CCD-Filter	
CCD memory (→CCD store)	MICROEL	(→CCD-Speicher)	
CCD register	MICROEL	CCD-Register	
CCD shift register (→bucked brigade device)	MICROEL	(→Eimerkettenschaltung)	
CCD store	MICROEL	CCD-Speicher	
= CCD memory			
CCH (→cancel character)	TERM&PER	(→ungültiges Zeichen)	
CCIR	TELEC	CCIR	
[international radio consultative committee of ITU]		[Normungsinstitution der UIT] = Comité Consultatif International des Radio-Communications	
CCIS (AM) (→common-channel signaling)	SWITCH	(→zentrale Zeichengabe)	
CCITT	TELEC	CCITT	
[international telephone and telegraph consultive committee of ITU]		[Normungsorganisation der UIT] = Comité Consultatif International de Téléfonique et Télégraphique	
CCITT code no.1 (→Baudot code)	TELEGR	(→Baudot-Code (1))	
CCITT high level programming language (→CHILL)	SWITCH	(→CHILL)	
CCM (→cross-connect multiplexer)	TRANS	(→Verteilmultiplexer)	
C compiler	DATA PROC	C-Kompilierer = C-Compiler	
CCS (→current-controlled current source)	NETW.TH	(→stromgesteuerte Stromquelle)	
CCS (→common-channel signaling)	SWITCH	(→zentrale Zeichengabe)	
CCSL	MICROEL	CCSL	
= compatible current-sinking logic			
CCTL	MICROEL	CCTL	
= collector coupled transistor logic			
CCTV (→closed-circuit television)	TV	(→nichtöffentliches Fernsehen)	
CCU (→communications control unit)	DATA PROC	(→Datenübertragungseinheit)	
CCV (→control-configured vehicle)	RADIO NAV	(→CCV)	
cd (→candela)	PHYS	(→Candela)	
CD (→compact disk)	CONS.EL	(→Compact Disc)	
Cd (→cadmium)	CHEM	(→Kadmium) (n.n.)	
CDI (→call line identification)	DATA COMM	(→Anschlußkennung)	
CDI (→collector diffusion insulation)	MICROEL	(→Kollektordiffusionsisolation)	
CDI process	MICROEL	CDI-Verfahren	
= collector diffusion isolation process			
CDMA (→code-division multiple access)	TELEC	(→Codemultiplex-Vielfachzugriff)	
CD player	CONS.EL	CD-Spieler	
= compact disk player		= CD-Player	
CDR (→constant-density recording)	TERM&PER	(→lineare Schreibdichte)	
CD record changer	CONS.EL	CD-Plattenwechsler	
CD-ROM	TERM&PER	CD-ROM	
[on standard 4 2/3 " disks] = compact-disk ROM; compact disk read-only memory; read-only optical store ↑ optical disk		[nur lesbarer optischer Speicher auf 4,75"-Platten] ↑ Bildplatte	
Ce (→cerium)	CHEM	(→Cer)	
ceasium beam standard (→cesium frequency standard)	INSTR	(→Cäsium-Frequenzstandard)	
CECUA	DATA PROC	CECUA	
= Confederation of European Computer Users's Associations		[Vereinigung von Verbänden zum Schutz von Computeranwendern der EWG]	
cedilla	LING	Cedille	
[hooked mark placed under a letter, e.g. with ç] ↑ diacritic mark		[Komma-ähnliches diakritisches Zeichen unterhalb eines Buchstabens, z.B. ç] = Häkchen ↑ diakritisches Zeichen	
Ceefax (GBR) (→teletext)	TELEC	(→Teletext)	
CEI (→IEC)	EL.TECH	(→IEC)	
ceiling	CIV.ENG	Decke	
ceiling height	CIV.ENG	Raumhöhe	
ceiling height indicator (→cloud ceilometer)	RADIO NAV	(→Wolkenhöhenmesser)	
ceiling mounting (→ceiling suspension)	SYS.INST	(→Deckenaufhängung)	
ceiling opening	SYS.INST	Deckendurchbruch	
ceiling speaker	EL.ACOUS	Deckenlautsprecher	
ceiling suspension	SYS.INST	Deckenaufhängung	
= ceiling mounting		= Deckenbefestigung	
ceilometer (→cloud ceilometer)	RADIO NAV	(→Wolkenhöhenmesser)	
celestial sphere	ASTROPHYS	Himmelkuppel	
cell	ANT	Elementarbereich	
cell	MOB.COMM	Zelle	
cell	MICROEL	Zelle	
cell [ATM]	TELEC	Zelle [ATM] = Paket	
cell 1 (→data element)	DATA PROC	(→Datenelement)	
cell 2	DATA PROC	Tabellenfeld	
cell (→accumulator)	POWER SYS	(→Akkumulator)	
cell (→battery cell)	POWER SYS	(→Batteriezelle)	
cell (→galvanic cell)	POWER SYS	(→galvanisches Element)	
cell (→radio cell)	MOB.COMM	(→Funkzelle)	
cell (→class interval)	MATH	(→Klassenintervall)	
cell allocation	MOB.COMM	Zellenzuteilung	
cellar	CIV.ENG	Keller	
= basement		= Kellergeschoß; Untergeschoß	
cellar (→push-down storage)	DATA PROC	(→Kellerspeicher)	

cell boundary (→radio cell boundary)	MOB.COMM	(→Funkzellgrenze)
cell catalog (NAM)	MICROEL	Zellenkatalog
= cell catalogue (BRI)		
cell catalogue (BRI) (→cell catalog)	MICROEL	(→Zellenkatalog)
cell-end voltage	POWER SYS	Entladeschlußspannung [Akkumulator]
cell group (→cluster)	MOB.COMM	(→Cluster)
cell header [ATM]	TELEC	Zellenkopf [ATM]
= header		= Paketkopf; Zellkopf
cell jar	POWER SYS	Batteriegefäß
cell library	MICROEL	Zellenbibliothek
cell limit (→radio cell boundary)	MOB.COMM	(→Funkzellgrenze)
cellophane	CHEM	Zellophan
cell-oriented switch	TELEC	zellorientierte Vermittlung
cell reference variable	DATA PROC	Bezugszellenvariable
cell split	MOB.COMM	Zellenteilung
cell splitting	MOB.COMM	Kleinzellenbildung
cell stream [ATM]	TELEC	Zellenstrom [ATM]
cell stream policing [ATM]	TELEC	Zellstromkontrolle [ATM]
= policing		
cellular carrier company	MOB.COMM	Mobilfunk-Betreibergesellschaft
cellular insulation	COMM.CABLE	Zellisolierung
cellular licence	MOB.COMM	Mobilfunkkonzession
cellular network (→mobile telephony network)	MOB.COMM	(→Mobilfunknetz)
cellular phone	MOB.COMM	Zellulartelefon
= cellular telephone		↑ Funktelefon
↑ radiotelephone		↓ Autotelefon; transportierbares Mobiltelefon;Rucksack-Mobiltelefon;
↓ car phone; transportable cellular phone; bag phone; hand-held cellular phone		
cellular radio	MOB.COMM	Zellenfunk
		= Kleinzonentechnik
cellular system	MOB.COMM	Zellularsystem
cellular telephone (→cellular phone)	MOB.COMM	(→Zellulartelefon)
celluloid	CHEM	Zelluloid
cellulose	TECH	Zellstoff
= pulp		
Celsius temperature [SI unit: degree Celsius]	PHYS	Celsius-Temperatur [SI-Einheit: Grad Celsius]
= centigrade temperature		
cement ≈ concrete	CIV.ENG	Zement ≈ Beton
cement (n.) ≈ adhesive	TECH	Kitt ≈ Kleber
cement (v.t.) ≈ bond (v.t.)	TECH	kitten ≈ kleben
cemented = bonded	TECH	gekittet = geklebt
CENELEC [European electrotechnical standards coordinating committee]	EL.TECH	CENELEC [europäisches Komitee für elektrotechnische Normung] = Comité Européen de Normalisations Electrotechniques
census = inquiry; survey	MATH	Erhebung = Ermittlung
center (v.t.)	TECH	zentrieren
center (n.) (AM) = centre (BRI)	MATH	Mitte = Zentrum; Mittelpunkt
center (NAM) = centre (BRI)	TELEC	Zentrale = Zentrum
center [keyboard function]	TERM&PER	Zentrierfunktion [einer Tastatur]
center angle [formed by two radii]	MATH	Zentriwinkel [von zwei Radien gebildeter Winkel]
center axis = center line	TECH	Mittelachse
center conductor (→inner conductor)	COMM.CABLE	(→Innenleiter)

center deviation (AM) ≈ centre deviation (BRI)	MECH	Mittenabweichung
center distance (AM) = centre distance (BRI)	MECH	Mittenabstand
center drive (AM) (→center feed)	ANT	(→Mittelpunktspeisung)
centered (→centric)	TECH	(→zentrisch)
center feed (AM) = centre feed (BRI); apex drive; center drive (AM); centre drive (BRI); ; symmetrical feed	ANT	Mittelpunktspeisung [Mittelanschluß] = Mittenspeisung
center frequency (AM) = centre frequency (BRI) ≠ limit frequency	NETW.TH	Mittenfrequenz ≠ Eckfrequenz
center frequency (AM) = centre frequency (BRI)	RADIO	Mittenfrequenz
center frequency (AM) (→mid frequency)	TELEC	(→Mittenfrequenz)
center frequency (AM) (→central frequency)	PHYS	(→Mittenfrequenz)
center frequency tracking (AM) = centre frequency tracking (BRI); signal track ≈ signal track	INSTR	Mittenfrequenznachführung ≈ Signalgleichlauf
center gap (AM) (→central gap)	RADIO REL	(→Mittellücke)
center-gap channel (AM) = centre-gap channel (BRI)	RADIO REL	Mittellückenkanal
center hole (→feed hole)	TERM&PER	(→Vorschubloch)
centering	TECH	Zentrierung
centering pin	TECH	Zentrierstift
centering plate	TECH	Zentrierplatte
centering square (→center square)	TECH	(→Zentrierwinkel)
centering strip	TECH	Zentrierleiste
center line (AM) = centre line (BRI); bisector; bisectrix; bisectional line	ENG.DRAW	Mittellinie
center line (→center axis)	TECH	(→Mittelachse)
center loading coil	ANT	Zentralspule
center of current distribution	ANT	Stromverteilungs-Schwerpunkt
center of curvature	MATH	Krümmungsmittelpunkt
center of gravity = center of mass	PHYS	Schwerpunkt = Massenmittelpunkt
center of mass (→center of gravity)	PHYS	(→Schwerpunkt)
center of rotation	PHYS	Drehpol
center square = centering square	TECH	Zentrierwinkel
center tap = center-tap connection; center tapping; electrical midpoint	EL.TECH	Mittelabgriff = Mittelanzapfung; Mittenanzapfung
center-tap connection (→center tap)	EL.TECH	(→Mittelabgriff)
center tapping (→center tap)	EL.TECH	(→Mittelabgriff)
center tolerance	MECH	Mittentoleranz
center wave	ANT	Zentralwelle
centigrade (→degree)	PHYS	(→Grad Celsius 2)
centigrade temperature (→Celsius temperature)	PHYS	(→Celsius-Temperatur)
centimeter (NAM) [0.01 m] = centimetre (BRI); cm	PHYS	Zentimeter [0,01 m] = cm
centimetre (BRI) (→centimeter)	PHYS	(→Zentimeter)
centimetric waves [0.1 m-0.01 m; 3 GHz- 30 GHz] = microwaves; super-high frequency; SHF	RADIO	Zentimeterwellen [0,1 m-0,01 m; 3 GHz-30 GHz] = Mikrowellen; Höchstfrequenzbereich; SHF
centisecond	PHYS	Hunderstelsekunde
cent mark [¢] = cent sign	ECON	Cent-Zeichen [¢]
central (adj.) = centralized; centralised (BRI) ≈ common	TECH	zentral = zentralisiert ≈ gemeinsam
central (n.) (→operator)	TELEPH	(→Beamtin)

central bank

central bank	ECON	**Zentralbank**	
≈ bank of issue		≈ Notenbank	
central bank money	ECON	**Zentralbankgeld**	
central battery	TELEPH	**Zentralbatterie**	
= common battery; CB; exchange battery; central office battery; station battery		= ZB; Wählerbatterie; Amtsbatterie	
≠ local battery		≠ Ortsbatterie	
central bus	DATA PROC	**Zentralbus**	
= unibus		= zentraler Bus; Unibus	
central channel	SWITCH	**Zentralkanal**	
central clock	CIRC.ENG	**Taktzentrale**	
= clock pulse generator central		≈ Taktversorgung	
central clock	TELEC	**Zentraltakt**	
central computer	DATA PROC	**Zentralrechner**	
= host computer			
central computer (→host computer)	DATA COMM	(→Hauptrechner)	
central computer network	DATA COMM	**Zentralcomputernetz**	
		= zentrales Computernetz	
central control unit (→switching processor)	SWITCH	(→Vermittlungsprozessor)	
central deposit	ECON	**Zentrallager**	
= central store			
central dictating facility	OFFICE	**Zentraldiktieranlage**	
central dictating system	OFFICE	**Zentrale Diktieranlage**	
central division	ECON	**Zentralabteilung**	
= corporate division 2			
central force	PHYS	**Zentralkraft**	
[acting in the direction of the connecting line of two mass points]		[in Richtung der Verbindungslinie zweier Massen]	
central frequency	PHYS	**Mittenfrequenz**	
= center frequency (AM); centre frequency (BRI)			
central gap	RADIO REL	**Mittellücke**	
= center gap (AM); centre gap (BRI)		= Zwischenbandlücke	
central gap decoupling	RADIO REL	**Mittellückenentkopplung**	
central hole	TERM&PER	**Spindelloch**	
[floppy disk]		[Diskette]	
centralised (BRI) (→central)	TECH	(→zentral)	
centralize	TECH	**zentralisieren**	
≈ concentrate		≈ konzentrieren	
centralized (→central)	TECH	(→zentral)	
centralized control	TELEC	**Zentralsteuerung**	
= common control			
centralized data acquisition	DATA PROC	**zentrale Datenerfassung**	
centrally controlled call set-up	SWITCH	**vollversetzter Verbindungsaufbau**	
= centrally controlled connection set-up; centrally controlled link set-up		= vollversetzte Verbindungsherstellung	
centrally controlled connection set-up (→centrally controlled call set-up)	SWITCH	(→vollversetzter Verbindungsaufbau)	
centrally controlled link set-up (→centrally controlled call set-up)	SWITCH	(→vollversetzter Verbindungsaufbau)	
central member	COMM.CABLE	**Zentralelement**	
central memory (→main memory 1)	DATA PROC	(→Hauptspeicher 1)	
central memory address (→main memory address)	DATA PROC	(→Hauptspeicheradresse)	
central office (NAM)	SWITCH	**Fernsprechvermittlungsstelle**	
= telephone central office; office; telephone switching exchange; telephone exchange (BRI) ↑ exchange		= Fernsprechamt; Fernsprechzentrale; Telefonzentrale; Telephonzentrale; Telefonamt; Telephonamt; Amt; Zentrale ↑ Vermittlungsstelle	
central office (→head office)	ECON	(→Zentralverwaltung)	
central office battery (→central battery)	TELEPH	(→Zentralbatterie)	
central office equipment (→exchange equipment)	TELEC	(→Amtsausrüstung)	
central-office switching (→public switching)	SWITCH	(→öffentliche Vermittlungstechnik)	
central plane	ENG.DRAW	**Mittelebene**	
= midplane; bisectional plane			
central processing element = CPE	DATA PROC	**Zentralverarbeitungselement** = zentrales Verarbeitungselement	
central processing unit 1	DATA PROC	**Zentraleinheit 1**	
[central control of a computer, consisting of arithmetic and logic unit, processing unit, input-output unit, main memory]		[zentrale Steuereinheit eines Rechners; besteht aus Rechenwerk, Steuerwerk = Prozessor, Ein-/Ausgabewerk, Speicherwerk = Hauptspeicher 1]	
= CPU; central processor; processing unit; central unit		= Zentralprozessor; CPU; Hauptprozessor; ZE	
↓ processor; arithmetic and logic unit; control unit; input-output unit; main memory		↓ Rechenwerk; Steuerwerk; Ein-/Ausgabewerk; Hauptspeicher 1	
central processing unit 2 (→processor 1)	DATA PROC	(→Prozessor 1)	
central processor (→host computer)	DATA COMM	(→Hauptrechner)	
central processor (→central processing unit 1)	DATA PROC	(→Zentraleinheit 1)	
central pulse	CODING	**Mittenimpuls**	
central sound	EL.ACOUS	**Mittelschall**	
central storage (→main memory 1)	DATA PROC	(→Hauptspeicher 1)	
central store (→central deposit)	ECON	(→Zentrallager)	
central studio (→main studio)	BROADC	(→Hauptstudio)	
central tower	ANT	**Mittelmast**	
central typing pool	OFFICE	**zentraler Schreibdienst**	
central unit	EQUIP.ENG	**Zentraleinheit**	
= common unit			
central unit (→central processing unit)	DATA PROC	(→Zentraleinheit)	
central zoning	SWITCH	**zentrale Verzonung**	
centre (BRI) (→center)	MATH	(→Mitte)	
centre (BRI) (→center)	TELEC	(→Zentrale)	
centre conductor (→inner conductor)	COMM.CABLE	(→Innenleiter)	
centre distance (BRI) (→center distance)	MECH	(→Mittenabstand)	
centre drive (BRI) (→center feed)	ANT	(→Mittelpunktspeisung)	
centre feed (BRI) (→center feed)	ANT	(→Mittelpunktspeisung)	
centre frequency (BRI) (→center frequency)	NETW.TH	(→Mittenfrequenz)	
centre frequency (BRI) (→mid frequency)	TELEC	(→Mittenfrequenz)	
centre frequency (BRI) (→center frequency)	RADIO	(→Mittenfrequenz)	
centre frequency (BRI) (→central frequency)	PHYS	(→Mittenfrequenz)	
centre frequency tracking (BRI) (→center frequency tracking)	INSTR	(→Mittenfrequenznachführung)	
centre gap (BRI) (→central gap)	RADIO REL	(→Mittellücke)	
centre-gap channel (BRI) (→center-gap channel)	RADIO REL	(→Mittellückenkanal)	
centre hole (→feed hole)	TERM&PER	(→Vorschubloch)	
centre line (BRI) (→center line)	ENG.DRAW	(→Mittellinie)	
Centrex	TELEC	**Centrex**	
[USA, PABX services by central office]		[USA, Nebenstellendienste durch Fernsprechamt]	
centric	TECH	**zentrisch**	
= centered		= mittig	
≈ central		≈ zentral	
centrifugal	PHYS	**zentrifugal**	
≠ centripetal		≠ zentripetal	
centrifugal acceleration	PHYS	**Zentrifugalbeschleunigung**	
centrifugal clutch	MECH	**Fliehkraftkupplung**	
centrifugal force	PHYS	**Fliehkraft**	
≠ centripetal force		= Zentrifugalkraft; Schwungkraft	
		≈ Anstrebkraft	
centrifugally cast concrete	CIV.ENG	**Schleuderbeton**	
centrifugal switch	POWER ENG	**Fliehkraftschalter**	
centrifuge	TECH	**Schleuder**	
		= Zentrifuge	
centripetal	PHYS	**zentripetal**	
≠ centrifugal		≠ zentrifugal	

centripetal force		PHYS	**Anstrebkraft**	ch (→hyperbolic cosine)	MATH	(→Cosinus hyperbolicus)
≠ centrifugal force			= Zentripetalkraft	CHA (→cursor horizontal)	TERM&PER	(→Schreibmarke horizontal)
			≠ Fliehkraft	chad (n.)	TERM&PER	**Schnitzel**
centroid		PHYS	**Flächenschwerpunkt**			= Schnipsel; Schnippel; Stanzrest
Centronix connector		COMPON	**Centronix-Schnittstellenverbinder**	chadbox (→chad container)	TERM&PER	(→Schnitzelkasten)
Centronix interface		TERM&PER	**Centronix-Schnittstelle**	chad container	TERM&PER	**Schnitzelkasten**
[a parallel type interface for printer]			[Parallelschnittstelle für Druckeranschluß]	= chadbox		= Schnipselkasten
				chadded	TERM&PER	**durchgelocht**
cent sign (→cent mark)		ECON	(→Cent-Zeichen)	chadded tape	TERM&PER	**perforierter Lochstreifen**
CEPT		TELEC	**CEPT**			= durchlochter Lochstreifen
[union of European PTT's]			[Vereinigung der europäischen PTT-Anstalten]	chadless [formular]	TERM&PER	**angelocht** [Formular]
			= Conférence Européenne des Administrations des Postes et des Télécommunications	chadless perforation	TERM&PER	**Schuppenlochung**
				chadless tape	TERM&PER	**Schuppenlochstreifen**
				= partially perforated tape		= geprägter Lochstreifen; angelochter Lochstreifen
CEPT hierarchy		TELEC	**CEPT-Hierarchie**	chaff	RADIO LOC	**Düppel**
ceramic (n.)		CHEM	**Keramik**	[silver tinsel for radar jamming]		[Lametta zur Radarstörung]
ceramic (adj.)		CHEM	**keramisch**			
ceramic capacitor		COMPON	**Keramikkondensator**	chain	TECH	**Kette**
ceramic chip capacitor		COMPON	**Keramik-Chip-Kondensator**	chain (→catena)	DATA PROC	(→Kette)
ceramic dual-in-line package (→CERDIP)		MICROEL	(→CERDIP)	chain code	CODING	**Kettencode**
ceramic filter		COMPON	**keramisches Filter**	= recurrent code		= rekurrenter Code
↑ piezoelectric filter			= Keramikfilter	chain-dotted	ENG.DRAW	**strichpunktiert**
			↑ piezoelektrisches Filter	[.–.–.–]		[.–.–.–]
ceramic microphone		EL.ACOUS	**Keramikmikrophon**	= dot-dashed		
ceramic package		MICROEL	**Keramikgehäuse**	chain drive	MECH	**Kettenantrieb**
ceramic substrate		MICROEL	**Keramiksubstrat**			= Kettentrieb
ceramic trimmer		COMPON	**Keramik-Trimmwiderstand**	chained address	DATA PROC	**gekettete Adresse**
CERDIP		MICROEL	**CERDIP**	chained data record (→chained record)	DATA PROC	(→verketteter Datensatz)
= ceramic dual-in-line package			[zweireihiges keramisches IC-Gehäuse]	chained file	DATA PROC	**verkettete Datei**
ceriph (→serif)		TYPOGR	(→Serife) (n.f. pl.-n)	= threaded file; linked file		= Verkettungsdatei
ceriph font (→serif font)		TYPOGR	(→Serifenschrift)	chained list	DATA PROC	**verkettete Liste**
ceriph font style (→serif font)		TYPOGR	(→Serifenschrift)	= threaded list; linked list		= Verkettungsliste
cerium		CHEM	**Cer**	chained record	DATA PROC	**verketteter Datensatz**
= Ce			= Ce	= chained data record		
cermet (→powder metal)		METAL	(→Pulvermetall)	chain field	DATA PROC	**Kettfeld**
cermet resistor		COMPON	**Cermet-Widerstand**	[data field containing the address of logically related data records]		[Datenfeld welches die Adresse logisch zusammenhängender Datensätze enthält]
			= Metallglasur-Festwiderstand			
cermet trimmer		COMPON	**Cermet-Trimmer**	= pointer		= Zeiger
			= Cermet-Trimmpotentiometer	chaining	DATA PROC	**Kettung**
certificate 1 (n.)		ECON	**Bescheinigung**	[linking of records or programs]		[von Daten oder Programmen]
			= Zertifikat	= concatenation; linkage; linking		= Verkettung; Verknüpfung
certificate		QUAL	**Protokoll**	= chain; calling sequnzy		≈ Kette; Abruffrequenz
			= Zertifika	↓ simple chaining; multiple chaining; forward chaining; backward chaining; data chaining		↓ Einfachkettung; Mehrfachkettung; Vorwärtskettung; Rückwärtskettung; Datenverkettung
certificate 2 (→testimonial)		ECON	(→Dienstzeugnis)			
certificate of indebtness (→promissory note)		ECON	(→Eigenwechsel)			
certification		ECON	**Beglaubigung**			
= attestation				chaining address (→reference address)	DATA PROC	(→Verweisadresse)
certification		QUAL	**Zertifizierung**	chaining search	DATA PROC	**Suche in geketteter Datei**
cesium		CHEM	**Cäsium**	chain name	DATA PROC	**Kettenname**
= Cs			= Cs	chain number	DATA PROC	**Kettennummer**
cesium frequency standard		INSTR	**Cäsium-Frequenzstandard**	chain of television line	BROADC	**Fernsehleitungskette**
= cesium standard; caesium beam frequency standard; ceasium beam standard			= Cäsium-Frequenznormal; Cäsiumnormal; Cäsiumstrahl-Frequenznormal; Cäsiumstrahlnormal	= chain of TV line		= TV-Leitungskette
				chain of TV line (→chain of television line)	BROADC	(→Fernsehleitungskette)
				chain parameter	NETW.TH	**Kettenparameter**
cesium standard (→cesium frequency standard)		INSTR	(→Cäsium-Frequenzstandard)	= abcd parameter		= Kettenkoeffizient
				chain parameter filter	NETW.TH	**Kettenfilter**
cessation		COLLOQ	**Wegfall**	chain parameter matrix	NETW.TH	**Kettenmatrix**
= abolition			[figurativ]	chain parameter matrix form	NETW.TH	**Kettenform**
≈ omission			≈ Weglassung	chain parameter relations	NETW.TH	**Kettengleichungen**
cession (→transfer 1)		ECON	(→Abtretung)			= Primärgleichungen
Cf (→californium)		CHEM	(→Californium)	chain printer	TERM&PER	**Kettendrucker**
CGA		TERM&PER	**CGA**	↑ type printer		↑ Typendrucker
[Color Graphics Adapter; graphics standard with 320x200 pixels and 4 out of 16 colours or 640x200 pixels and 2 colors]			[Graphikstandard mit 320x200 Punkten und 4 aus 16 Farben oder 640x200 Punkten und 2 Farben]	chain reaction	PHYS	**Kettenreaktion**
				chairman	ECON	**Vorsitzender**
				= president; chairperson		= Präsident
				chairmanship	ECON	**Vorsitz**
CGA board		TERM&PER	**CGA-Karte**	= presidency		
[establishes CGA resolution on RGB monitors]			[ermöglicht CGA-Auflösung auf RGB-Monitoren]	chairperson (→chairman)	ECON	(→Vorsitzender)
↑ colour graphics board			= CGA-Adapter	challenge	TECH	**Herausforderung**
			↑ Farbgraphikkarte			

chalnicon	TERM&PER	Chalnicon	change parameter	DATA PROC	(→Änderungsparameter)
chamber	CIV.ENG	Kammer	(→modification parameter)		
↑ room 1		↑ Raum	change record (→addition	DATA PROC	(→Änderungssatz)
chamber	OUTS.PLANT	Schacht	record)		
≈ manhole		= Unterflurkammer	change run (→modification	DATA PROC	(→Änderungslauf)
		≈ Mannloch	run)		
chamfer (v.t.)	MECH	abfasen	change tape	DATA PROC	Änderungsband
[to cut a slant surface to the inside]		[nach innen abschrägen]	= modification tape; modifications tape		= Modifikationsband
≈ bevel		= fasen	changing device	TERM&PER	Wechseleinrichtung
		≈ abschrägen	changeover to standby	TRANS	(→Ersatzschalttechnik)
chamfer (n.)	MECH	Fase	(→protection switching)		
[beveled edge]		[nach innen gehende Abschrägung einer Kante]	channel (n.)	CIV.ENG	Flußbett
			[bed of a natural waterway]		
		= Abfasung	channel	TELEC	Kanal
chamfer angle	MECH	Abfasungswinkel	[smallest subdivision of a transmission path in one sense of transmission]		[kleinste Unterteilung eines Übertragungsweges in einer Übertragungsrichtung]
chamfered	MECH	abgefast			
[beveled to the inside]		[nach innen abgekantet]			
change (v.t.)	TECH	ändern	≈ transmission path		≈ Übertragungsweg
= vary; modify; alter		= verändern; variieren; modifizieren	↓ data channel		↓ Datenkanal
= fluctuate; remodel			channel	DATA PROC	Kanal
		≈ schwanken; umbauen	[transmission path between CPU and peripherals]		[Übertragungsweg zwischen Zentraleinheit und Peripheriegeräten]
change (n.)	TECH	Wechsel			
= change-out (n.); alteration		= Veränderung	↓ data channel; input channel; output channel; selector channel; multiplex channel; control channel; main channel; auxiliary channel		↓ Datenkanal; Eingabekanal; Ausgabekanal; Selektorkanal; Multiplexkanal; Steuerkanal; Hauptkanal; Hilfskanal
changeable memory	DATA PROC	wechselbarer Speicher			
= changeable storage					
changeable storage	DATA PROC	(→wechselbarer Speicher)			
(→changeable memory)					
change block (→modification block)	DATA PROC	(→Änderungsblock)	channel	MICROEL	Kanal
change-climates	QUAL	Wechselklima	channel	TERM&PER	Lochspur
change data (→transaction data)	DATA PROC	(→Bewegungsdaten)	[longitudinal row of perforations]		[in Streifenrichtung]
			↑ hole row; track		↑ Lochreihe; Spur
changed number	TELEC	Rufnummer geändert	channel (→channel profile)	METAL	(→U-Profil)
= transferred subscriber			channel adapter	DATA COMM	Kanaladapter
change dump	DATA PROC	Änderungsauszug	channel address	SWITCH	Kanaladresse
		= Speicheränderungsabzug	channel addressing	SWITCH	Kanaladressierung
change entry (→modification entry)	DATA PROC	(→Änderungseintrag)	channel address register	DATA PROC	Kanaladressregister
			channel allocation (→channel assignment)	TELEC	(→Kanalzuteilung)
change file (→transaction file)	DATA PROC	(→Bewegungsdatei)			
			channel allocation plan	DOC	Kanalaufteilungsplan
change information (→update information)	TECH	(→Änderungsmitteilung)	channel arrangement (→channel configuration)	RADIO	(→Kanalraster)
change log (→modification log)	DATA PROC	(→Änderungsprotokoll)	channel arrangement plan	TELEC	Kanalbelegungsplan
			= channelization plan; channeling plan		= Kanalplan 1; Kanalisierungsplan
change loop (→modification loop)	DATA PROC	(→Änderungsschleife)	≈ channel routing plan		
					≈ Kanalführungsplan
change mode (→modification mode)	DATA PROC	(→Änderungsmodus)	channel-assigned signaling	SWITCH	(→kanalgebundene Zeichengabe)
			(→channel-oriented signaling)		
change notice (→update information)	TECH	(→Änderungsmitteilung)	channel assignment	TELEC	Kanalzuteilung
			= channel allocation		= Kanalzuordnung
change notification (→update information)	TECH	(→Änderungsmitteilung)	≈ channel occupancy		≈ Kanalbelegung
			channel-associated	TELEC	kanalgebunden
change of picture	TV	Bildwechsel			≈ kanalindividuell
change of technology	TECH	Technologiewechsel	channel bank (NAM) (→channel modulator)	TRANS	(→Kanalumsetzer)
change-out (n.) (→change)	TECH	(→Wechsel)			
changeover (→switchover logic)	ELECTRON	(→Umschaltelogik)	channel bank (AM) (→primary multiplexer)	TELEC	(→Primärmultiplexer)
changeover (v.t.) (→switchover)	ELECTRON	(→umschalten)	channel block (→channel command block)	SWITCH	(→Kanalblock)
changeover (→switchover)	ELECTRON	(→Umschaltung)	channel block register	DATA COMM	Kanalblockregister
changeover contact	COMPON	Umschaltekontakt	channel board (→channel unit card)	TRANS	(→Kanalkarte)
[when activated, one contact opens and another closes]		[bei Betätigung wird ein Kontakt unterbrochen und ein anderer hergestellt]			
			channel bundling (→channel packing)	TELEC	(→Kanalbündelung)
= break-make contact; switchover contact		= Wechsler; Wechselkontakt; Umschalter 2			
			channel capacity	INF	Kanalkapazität
↑ relay contact		↑ Relaiskontakt	120 channel carrier frequency system	TRANS	V120
changeover criterion	ELECTRON	(→Umschaltekriterium)			[TF-Technik]
(→switchover criterion)			= 120 channel FDM multiplex system		
change-over switch	COMPON	Wechselschalter	channel characteristic	INF	Kanaleigenschaft
		= Umschalter 1			= Kanalcharakteristik
		↑ Niederspannungsschalter; Schalter	channel coding	TV	Kanalcodierung
			channel command	DATA PROC	Kanalbefehl
changeover threshold (→switching threshold)	ELECTRON	(→Schaltschwelle)	channel command address	DATA PROC	Kanalbefehlsadresse
			channel command block	SWITCH	Kanalblock
changeover threshold (→protection switching threshold)	TRANS	(→Umschaltschwelle)	= channel block		
			channel command word	DATA PROC	Kanalbefehlswort
changeover time (→switchover time)	ELECTRON	(→Umschaltzeit)	[to control data input and output]		[zur Steuerung von Daten-ein- und ausgabe]
			channel command word sequence	DATA COMM	Kanalsbefehlskette
changeover time (→transit time)	COMPON	(→Umschlagzeit)			

channel configuration		RADIO	Kanalraster	channel redirection (→channel rerouting)		TELEC	(→Kanalumleitung)

channel configuration RADIO **Kanalraster**
= channel arrangement; channel pattern = Kanalanordnung
≈ radio frequency pattern ≈ Radiofrequenzraster
channel converter (→channel TRANS (→Kanalumsetzer)
modulator)
channel current MICROEL **Kanalstrom**
[FET] [FET]
channel D (→signal channel) TELEC (→Signalkanal)
channel decoupling (→channel TELEC (→Kanalentkopplung)
isolation)
channel derivation equipment TRANS **Abzweigeinrichtung**
channel discharge PHYS **Kanalentladung**
↑ gaseous discharge ↑ Gasentladung
channeled (→U-shaped) TECH (→U-förmig)
channel electron multiplier INSTR **Kanalelektronenvervielfacher**
channel encoding INF **Kanalcodierung**
[coding occurs in the transmission [Codierung erfolgt im Si-
channel] gnalkanal]
≠ primary encoding ≠ Quellencodierung
channel error DATA PROC **Kanalfehler**
120 channel FDM multiplex TRANS (→V120)
system (→120 channel carrier frequency system)
channel filter TRANS **Kanalfilter**
[FDM] [TF-Technik]
channel filter RADIO REL **Kanalweiche**
= Kanalfilter; Kanalweichen-
filter
channel filter TV **Kanalweiche**
channel filter chain RADIO REL **Kanalweichenkette**
channel group (AM) (→primary TRANS (→Primärgruppe)
group)
channel grouping (→channel TELEC (→Kanalbündelung)
packing)
channel group number DATA COMM **Kanalgruppennummer**
channeling plan (→channel TELEC (→Kanalbelegungsplan)
arrangement plan)
channel isolation TELEC **Kanalentkopplung**
= channel decoupling
channelization 2 (→multichannel TELEC (→Merkanalübertragung)
transmission)
channelization 1 (→channel TELEC (→Kanalführung)
routing)
channelization plan (→channel TELEC (→Kanalbelegungsplan)
arrangement plan)
channel length MICROEL **Kanallänge**
channel level TRANS **Kanalpegel**
channel level adjustment TRANS **Kanalpegeleinstellung**
channel loading (→channel TELEC (→Kanalbelegung)
occupancy)
channel modulator TRANS **Kanalumsetzer**
[FDM] [TF-Technik]
= channel bank (NAM); channel trans- ↑ Primärmultiplexer
lator; channel converter
↑ primary multiplexer
channel monitoring TELEC **Kanalüberwachung**
= channel supervision
channel noise INF **Kanalrauschen**
channel occupancy TELEC **Kanalbelegung**
= channel loading; usage of channels ≈ Kanalzuteilung
≈ channel assignment
channel-oriented mode SWITCH (→kanalgebundene Zeichen-
(→channel-oriented signaling) gabe)
channel-oriented signaling (AM) SWITCH **kanalgebundene Zeichengabe**
= channel-oriented signalling (BRI); = kanalgebundene Signalisie-
channel-oriented mode; channel-as- rung; kanalzugehörige Si-
signed signaling; CAS gnalisierung
channel-oriented signalling (BRI) SWITCH (→kanalgebundene Zeichen-
(→channel-oriented signaling) gabe)
channel oveload DATA COMM **Kanalüberlastung**
channel packing TELEC **Kanalbündelung**
= channel bundling; channel grouping = Kanalgruppierung
channel pattern (→channel RADIO (→Kanalraster)
configuration)
channel profile METAL **U-Profil**
= channel
channel program start DATA COMM **Kanalprogramm-Startadresse**
address
channel queue DATA COMM **Kanalschlange**

channel redirection (→channel TELEC (→Kanalumleitung)
rerouting)
channel rerouting TELEC **Kanalumleitung**
= channel redirection = Kanalumlenkung
channel routing TELEC **Kanalführung**
= channelization 1
channel routing plan TELEC **Kanalführungsplan**
≈ channel arrangement plan = Kanalplan 2
≈ Kanalbelegungsplan
channel sample CODING **Kanalinformation**
channel selector CONS.EL **Kanalwähler**
= tuner 2 = Tuner 2
channel separation TRANS **Kanalabstand**
= channel spacing
channel separation RADIO **Rasterabstand**
= Kanalabstand
channel separation (→stereo EL.ACOUS (→Stereo-Übersprechdämp-
separation) fung)
channel spacing (→channel TRANS (→Kanalabstand)
separation)
channel-specific TELEC **kanalindividuell**
≈ channel-associated = kanalspezifisch; kanalzuge-
hörig
≈ kanalgebunden
channel speed (→line DATA COMM (→Leitungsgeschwindigkeit)
speed)
channel stop MICROEL **Kanalstopper**
channel supervision (→channel TELEC (→Kanalüberwachung)
monitoring)
channel synchronizer DATA PROC **Kanalsynchronisiereinrich-**
tung
channel timing CODING **Pausentakt**
channel-to-channel DATA PROC **Kanal-zu-Kanal-Verbindung**
connection
channel translator (→channel TRANS (→Kanalumsetzer)
modulator)
channel type INF **Kanalart**
channel unit card TRANS **Kanalkarte**
= channel board
channel width TELEC **Kanalbreite**
channel width MICROEL **Kanalbreite**
chapter LING **Kapitel**
chapter DATA PROC **Programmkapitel**
[executable without the rest of the [ohne die übrigen Pro-
program] grammteile ausführbar]
= program chapter
chapter heading (→heading) LING (→Überschrift)
character INF **Zeichen**
[element of a character set, to convey [Element eines Zeichenvor-
information; signals are carriers for rats, zur Informationsüber-
characters] mittlung; Signale sind
= symbol Träger von Zeichen]
≈ signal ≈ Signal
character alignment DATA COMM **Zeichensynchronisierung**
= character synchronization; symbol = Zeichenbildung
alignment
character arithmetics DATA PROC **Zeichenarithmetik**
≠ fixed-word-length arithmetics ≠ Festwortarithmetik
character assembly DATA PROC **Zeichenformung**
character blink DATA PROC **Zeichenflackern**
character block DATA PROC **Zeichenblock**
character-bound INF.TECH (→zeichenorientiert)
(→character-oriented)
character-by-character INF.TECH **zeichenweise**
= symbol-by-symbol = symbolweise
≈ character-oriented ≈ zeichenorientiert
character byte DATA PROC **Zeichenbyte**
character check DATA COMM **Zeichenprüfung**
= character checking; symbol check; = Symbolprüfung
symbol checking
character checking DATA COMM (→Zeichenprüfung)
(→character check)
character code CODING **Zeichencode**
character counter TELEGR **Zeichenzähler**
= symbol counter
character delay TELEC **Zeichendauer**
= character period; character duration
character delete TERM&PER **Zeichenlöschung**
= character erase = Zeichenentfernung

character density TYPOGR **Zeichendichte**
[number of characters per unit length; unit: pitch/inch] [Zeichen pro Längeneinheit; Maßeinheit: Zeichen/Zoll]
= horizontal print density = Schreibdichte; Typendichte
≈ character spacing ≈ Zeichenabstand

character density DATA PROC
(→recording density) (→Speicherdichte)

character display TERM&PER **Zeichenbildschirm**
[characters are assembled of stored picture elements] [Zeichen werden aus gespeicherten Elementen zusammengefügt]
= character map screen
↑ display terminal ↑ Sichtgerät

character duration (→character delay) TELEC (→Zeichendauer)

character erase (→character delete) TERM&PER (→Zeichenlöschung)

character error INF.TECH **Zeichenfehler**
= symbol error

character error frequency DATA COMM (→Zeichenfehlerhäufigkeit)
(→character error rate)

character error probability DATA COMM **Zeichenfehlerwahrscheinlichkeit**
= symbol error probability ≈ Bitfehlerwahrscheinlichkeit [TELEC]
≈ bit error probability [TELEC]

character error rate DATA COMM **Zeichenfehlerhäufigkeit**
= character error frequency; symbol error rate; symbol error frequency

character escapement TYPOGR (→Zeichenabstand)
(→character spacing)

character fill DATA PROC **Zeichenauffüllung**
[to every memory location] [ein Zeichen in jeden Speicherplatz]

character font (→font) TYPOGR (→Schriftzeichensatz)

character frequency DATA COMM (→Zeichengeschwindigkeit)
(→character rate)

character generator DATA PROC **Zeichengeber**
= symbol generator = Zeichengenerator; Bildgenerator; Buchstabengenerator

character imaging equipment TERM&PER **zeichendarstellendes Gerät**

characteristic 3 (n.) TECH **Kennlinie**
= characteristic curve; diagram; graph

characteristic 1 (n.) TECH **Merkmal**
= feature 1; property; identification mark; mark; sign = Eigenschaft; Charakteristik; Kennzeichen
≈ pecularity ≈ Besonderheit
↓ feature 2 ↓ Leistungsmerkmal

characteristic (adj.) TECH **charakteristisch**
= peculiar; particular; distinctive = kennzeichnend; eigentümlich; eigenartig
≈ special ≈ besonders

characteristic (n.) (→floating point exponent) MATH (→Gleitpunktexponent)

characteristic 2 (n.) TECH (→Kenngröße)
(→characteristic quantity)

characteristic curve TECH (→Kennlinie)
(→characteristic 3)

characteristic data TECH (→Kenndaten) (n.n.pl.t)
(→characteristics)

characteristic distortion DATA COMM **charakteristische Verzerrung**
= Einschwingverzerrung

characteristic dose rate PHYS **Kenndosisleistung**

characteristic equation MATH **charakteristische Gleichung**
= Eigenwertgleichung

characteristic frequency ACOUS **Formant**
= formant; fundamental frequency

characteristic frequency TRANS (→Kennfrequenz)
(→identification frequency)

characteristic frequency EL.TECH (→Nennfrequenz)
(→nominal frequency)

characteristic function MATH **Eigenfunktion**
= eigenfunction = charakteristische Funktion

characteristic impedance NETW.TH **Wellenwiderstand**
= wave resistance; surge impedance = charakteristischer Widerstand

characteristic noise parameter MICROEL **Rauschkenngröße**

characteristic quantity (n.) TECH **Kenngröße**
= characteristic 2 (n.); variable = Kennwert
≈ figure of merit; parameter; characteristics ≈ Gütezahl; Parameter; Kenndaten

characteristics (n.pl.t.) TECH **Kenndaten** (n.n.pl.t)
= characteristic data = Kennwerte (n.m.pl.t.)

characteristic sound impedance ACOUS (→Schallwellenwiderstand)
(→sound radiation impedance)

characteristic value MATH (→Eigenwert)
(→eigenvalue)

characteristic vector MATH (→Eigenvektor)
(→eigenvector)

character key DATA PROC **Einzelzeichentaste**
[for characterwise word processing mode] [für zeichenweise Textverarbeitung]

character key TERM&PER (→Datentaste)
(→alphanumeric key)

character map screen TERM&PER (→Zeichenbildschirm)
(→character display)

character multiplexer INF.TECH **Zeichenmultiplexer**
= symbol multiplexer; signal multiplexer

character-orientated computer DATA PROC (→Stellenmaschine)
(→character-oriented computer)

character-orientated machine DATA PROC (→Stellenmaschine)
(→character-oriented computer)

character-oriented INF.TECH **zeichenorientiert**
= symbol-oriented; character-bound; symbol-bound = zeichengebunden; symbolorientiert
≈ character-by-character ≈ zeichenweise

character-oriented computer DATA PROC **Stellenmaschine**
[computer with characters as smallest addressable units] [Computer mit Stelle als kleinste adressierbare Einheit]
= digit computer; character-oriented machine; character-orientated computer; character-orientated machine = Zeichenmaschine; zeichenorientierter Computer; zeichengebundener Computer; Stellencomputer; Zeichencomputer
≠ word-oriented computer ≠ Wortmaschine

character-oriented machine DATA PROC (→Stellenmaschine)
(→character-oriented computer)

character-parallel INF.TECH **zeichenparallel**
= parallel by character

character period (→character delay) TELEC (→Zeichendauer)

character pitch (→character spacing) TYPOGR (→Zeichenabstand)

character position TERM&PER **Schreibstelle**

character position DATA COMM **Zeichenstelle**
= symbol position

character range (→character set) INF.TECH (→Zeichenvorrat)

character rate DATA COMM **Zeichengeschwindigkeit**
= character frequency; symbol rate; symbol frequency

character reader TERM&PER **Klarschriftleser**
[reads characters perceivables to the human eye] [liest für das menschliche Auge lesbare Schrift]
= character scanner; symbol reader; symbol scanner = Schriftleser; Schriftenleser; Zeichenleser
↓ optical character reader; magnetic character reader ↓ optischer Leser; Magnetschriftleser

character recognition TERM&PER **Klarschriftlesen**
= symbol recognition; text recognition; font recognition = Zeichenerkennung; Zeichenlesung; Schrifterkennung; Texterkennung
↓ optical character recognition; omnifont character recognition ↓ optische Zeichenerkennung; universelle Schrifterkennung

character recognition INF.TECH **Zeichenerkennung**
= Zeichenlesung

character repertoire INF.TECH (→Zeichenvorrat)
(→character set)

character repetition DATA COMM **Zeichenwiederholung**

character representation DATA PROC **Zeichendarstellung**

character rounding DATA PROC **Zeichenabrundung**
= Zeichenretusche

character scanner (→character reader)	TERM&PER	(→Klarschriftleser)	
character sensing	TELEGR	**Zeichenüberprüfung**	
character-serial	DATA COMM	**zeichenseriell**	
character set	INF.TECH	**Zeichenvorrat**	
[an agreed set of signal configurations to convey information]		[vereinbarter Satz von unterschiedlichen Signalkonfigurationen zur Informationsübermittlung]	
= character repertoire; character supply; character range; alphabet; symbol set; symbol repertoire; symbol supply; symbol range		= Symbolvorrat; Vorrat; Zeichenrepertoire; Symbolrepertoire; Repertoire; Zeichensatz; Symbolsatz; Alphabet	
character signal (→codeword)	CODING	(→Codewort)	
character skew	TYPOGR	**Schrägstellung**	
character spacing	TYPOGR	**Zeichenabstand**	
= character pitch; character escapement; pitch; intercharacter spacing; spacing 2; lettersapce; letterspacing		= Abstand; Spatium (pl.-ien); Buchstabenzwischenraum; Typenabstand; Spatiierung; Spationierung; Sperrung; Sperren	
≈ character density		≈ Zeichendichte	
characters per inch	TERM&PER	**Zeichen pro Zoll**	
= cpi		= cpi; Zeichen pro Inch	
≈ pitch		≈ Teilung	
characters per second	DATA PROC	**Zeichen pro Sekunde**	
= cps		= cps	
character string	DATA PROC	**Zeichenfolge**	
= symbol string; alphanumeric string; string		= Folge; Zeichenkette; String 2	
character stuffing	DATA PROC	**Zeichenstopfen**	
character supply (→character set)	INF.TECH	(→Zeichenvorrat)	
character synchronization (→character alignment)	DATA COMM	(→Zeichensynchronisierung)	
character table	DATA PROC	**Zeichentabelle**	
character template	ENG.DRAW	**Zeichenschablone**	
= symbol stencil			
character timing	SWITCH	**Zeichentakt**	
character timing control	SWITCH	**Zeichentaktsteuerung**	
character wheel (→typewheel)	TERM&PER	(→Typenrad)	
character-wheel printer (→typew-heel printer)	TERM&PER	(→Typenraddrucker)	
character width	TYPOGR	**Dickte**	
[width of type including the blank on both sides]		[Buchstabenbreite einschließlich des beidseitigen Leerraums]	
= set width; width		= Zeichendickte; Zeichenbreite; Schriftdickte; Schriftbreite; Buchstabenbreite	
charge 1 (n.)	ECON	**Belastung 1**	
[of an account]		[eines Kontos]	
= debit		≈ Rechnung	
≈ invoice			
charge 2 (n.)	ECON	**Gebühr**	
[imposed pecunary burden]		= Abgabe; Tarif	
= duty; rate (proportional to measured service); levy; fee (a fixed charge); tariff (a schedule of charges or rates); rental (a fixed periodical pecunary burden)		= Honorar; Steuer	
		↓ Zollgebühr; Maut	
≈ tax			
↓ customs fee; toll			
charge (n.)	PHYS	**Ladung**	
charge	POWER SYS	**Ladung**	
[process of charge transfer]		[Vorgang]	
charge (→stress)	QUAL	(→Beanspruchung)	
charge 3 (→allocation)	ECON	(→Umlage)	
charge (→invoice)	ECON	(→verrechnen)	
chargeable	TELEC	**gebührenpflichtig**	
= billable; subject to fee; toll-			
chargeable	POWER SYS	**ladbar**	
		= aufladbar	
chargeable-time device (→timing register)	SWITCH	(→Gesprächszeitmesser)	
charge accumulation	TELEC	**Gebührenstand**	
charge carrier	PHYS	**Ladungsträger**	

charge carrier avalanche (→carrier avalanche)	MICROEL	(→Trägerlawine)	
charge carrier current	MICROEL	**Ladungsträgerstrom**	
↓ drift current; diffusion current		↓ Feldstrom; Diffusionsstrom	
charge carrier diffusion	PHYS	**Ladungsträgerdiffusion**	
= carrier diffusion		= Trägerdiffusion	
charge carrier injection	MICROEL	**Ladungsträgerinjektion**	
= carrier injection		= Trägerinjektion	
charge carrier lifetime	MICROEL	(→Trägerlebensdauer)	
(→carrier lifetime)			
charge carrier mobility	PHYS	**Ladungsträgerbeweglichkeit**	
= carrier mobility		= Trägerbeweglichkeit	
charge carrier multiplication (→carrier multiplication)	MICROEL	(→Ladungsträger-Vervielfachung)	
charge carrier transport (→charge transport)	PHYS	(→Ladungstransport)	
charge computer (→tax computer)	SWITCH	(→Gebührenrechner)	
charge coupled device	MICROEL	**ladungsgekoppelte Schaltung**	
= CCD		= ladungsgekoppeltes Bauelement; CCD; ladungsgekoppelter Baustein; Ladungsspeicherbaustein	
		↑ Ladungstransferelement	
charge coupling	MICROEL	**Ladungskopplung**	
charge density	PHYS	**Ladungsdichte**	
charge distribution	PHYS	**Ladungsverteilung**	
charge exchange	PHYS	**Umladung**	
		= Ladungsaustausch	
charge indicator	TERM&PER	**Gebührenanzeige**	
= call-charge meter (NAM); call fee indicator (BRI)		= Gebührenanzeiger	
charge injection device (→CID)	MICROEL	(→CID-Element)	
charge jack	EQUIP.ENG	**Ladebuchse**	
charge maintaining current	POWER SYS	**Ladehaltestrom**	
[for an accumulator]		[für einen Akkumulator]	
charge/mass ratio (→specific load)	PHYS	(→spezifische Ladung)	
charge meter (→tax meter)	SWITCH	(→Gebührenzähler)	
charge observation	SWITCH	**Zählvergleich**	
charge printing (→tax printing)	SWITCH	(→Gebührenausdruck)	
charger	POWER SYS	**Ladegerät**	
[accumulator]		[Akkumulatoren]	
= charging set; battery charger; charging rectifier		= Batterieladegerät; Ladegleichrichter	
charge regulator	POWER SYS	**Laderegler**	
[accumulator]		[Akkumulatoren]	
charge retention	EL.TECH	**Ladungserhaltung**	
charge shifting	PHYS	**Ladungsverschiebung**	
≈ charge transfer; charge transport		≈ Ladungstransfer; Ladungstransport	
charge storage	MICROEL	**Ladungsspeicherung**	
charge storage diode	MICROEL	**Speicherschaltdiode**	
= CCD diode		= Speichervaraktor; Speichervaractor; Speicherdiode; Ladungsspeicherdiode	
↑ varactor diode			
↓ step-recovery diode; snap-off diode		↑ Varaktordiode	
		↓ Step-Recovery-Diode; Snap-off-Diode	
charge-storage tube (→storage tube)	ELECTRON	(→Speicherröhre)	
charge-storing tube (→storage tube)	ELECTRON	(→Speicherröhre)	
charge transfer	PHYS	**Ladungstransfer**	
≈ charge transport; charge shifting		≈ Ladungstransport; Ladungsverschiebung	
charge transfer device	MICROEL	**Ladungstransferelement**	
= CTD		= Ladungsverschiebeschaltung	
↓ charge coupled device; bucked brigade device		↓ ladungsgekoppelte Schaltung; Eimerkettenschaltung	
charge transport	PHYS	**Ladungstransport**	
= charge carrier transport		= Ladungsträgertransport	
≈ charge transfer; charge shifting		≈ Ladungstransfer; Ladungsverschiebung	
charge unit (→unit-fee)	SWITCH	(→Gebühreneinheit)	
charging	PHYS	**Aufladung**	
charging	ECON	**Verrechnung**	
= clearing; settlement		= Gebührenverrechnung	
≈ invoicing		≈ Fakturierung; Belastung 2	

charging

charging (→tax metering)	SWITCH	(→Gebührenzählung)	
charging area (→metering zone)	SWITCH	(→Gebührenzone)	
charging circuit	POWER SYS	**Ladestromkreis**	
charging current [accumulator]	POWER SYS	**Ladestrom** [Akkumulator]	
charging information	DATA COMM	**Gebührenzuschreiben** [Mitteilung der Gegühr nach Verbindungsende]	
charging information (→call-charge data)	SWITCH	(→Gebührendaten)	
charging of capacitor	PHYS	**Kondensatoraufladung**	
charging policy	TELEC	**Gebührenpolitik**	
charging rectifier (→charger)	POWER SYS	(→Ladegerät)	
charging set (→charger)	POWER SYS	(→Ladegerät)	
charging time	POWER SYS	**Ladezeit**	
charging unit (→unit-fee)	SWITCH	(→Gebühreneinheit)	
charging voltage [accumulator]	POWER SYS	**Ladespannung** [Akkumulator]	
charging zone (→metering zone)	SWITCH	(→Gebührenzone)	
charging zone list = zoning list	SWITCH	**Verzonungsliste**	
chart (→diagram 1)	TECH	(→Diagramm)	
chart (→graphic)	DOC	(→Graphik)	
chart (→graph)	DOC	(→Kurvenblatt)	
chartered accountant (→auditor)	ECON	(→Wirtschaftsprüfer)	
charter flight	AERON	**Charterflug**	
charter flight traffic = charter traffic ≠ line air taffic; military air traffic ↑ air traffic	AERON	**Bedarfsflugverkehr** = Bedarfsluftverkehr; Charterflugverkehr; Charterflugverkehr ≠ Linienflugverkehr; Militärluftverkehr ↑ Luftverkehr	
charter traffic (→charter flight traffic)	AERON	(→Bedarfsflugverkehr)	
chart graticule	NETW.TH	**Diagrammgitter**	
chase	METAL	**gewindestrehlen** = strehlen	
chased	METAL	**gestrehlt**	
chassis (→case)	TECH	(→Gehäuse)	
chassis ground = frame ground; ground; earth	ELECTRON	**Masse** = Erde	
chassis ground bus	EQUIP.ENG	**Masseschiene**	
chassis grounding (→frame grounding)	EQUIP.ENG	(→Gehäuseerde)	
chassis illumination	EQUIP.ENG	**Gehäusebeleuchtung**	
chassis receptacle (→mounting jack)	COMPON	(→Einbaubuchse)	
chatter (→contact bounce)	COMPON	(→Kontaktprellen)	
cheap = at a favourable price ≈ economic; profitable	ECON	**preiswert** = billig; preisgünstig ≈ wirtschaftlich 2; rentabel	
cheapen ≠ increase in price	ECON	**verbilligen** ≠ verteuern	
cheapened = reduced-rate	ECON	**verbilligt**	
Chebyshev filter	NETW.TH	**Tschebyscheff-Filter**	
Chebyshev polynomial	NETW.TH	**Tschebyscheff-Polynom**	
check (v.t.) = inspect; test; control; probe; examine ≈ re-examine	TECH	**prüfen** = testen; inspizieren; kontrollieren; examinieren ≈ nachprüfen	
check (n.) = examination; inspection; test; observation ≈ re-examination; verification; investigation	TECH	**Prüfung** = Test; Inspektion; Kontrolle ≈ Nachprüfung; Bestätigung; Untersuchung	
check (n.) (NAM) = cheque (BRI) ↑ payment means	ECON	**Scheck** ↑ Zahlungsmittel	
check (→inspection)	QUAL	(→Prüfung)	
check (n.) (→revision)	TECH	(→Überprüfung)	
check (→data validation)	DATA PROC	(→Datenüberprüfung)	
check a calculation (→recalculate)	MATH	(→nachrechnen)	
check bit = option bit; test bit ↓ parity bit	CODING	**Prüfbit** = Kontrollbit ↓ Paritätsbit	
check character = control character 1 ≈ control character 2	INF.TECH	**Kontrollzeichen** = Sicherungszeichen; Prüfzeichen ≈ Steuerzeichen	
check command	TELECONTR	**Prüfbefehl**	
check digit = control digit; check number	CODING	**Kontrollziffer** = Kontrollzahl; Kontrollnummer; Sicherheitsziffer; Prüfziffer	
checked = proved; proven ≈ tested ≠ unchecked	TECH	**geprüft** ≈ erprobt ≠ ungeprüft	
checker (v.t.) ≈ hatch (v.t.); shade (v.t.)	ENG.DRAW	**riffeln** ≈ schraffieren; schattieren	
checker (v.t.) ≈ corrugate	MECH	**riffeln** [mit Riffeln versehen] ≈ wellen	
checker (n.) ≈ hatching	ENG.DRAW	**Riffelung** ≈ Schraffur	
checker (n.) [alternation of grooves and fins] ≈ ondulation	MECH	**Riffel** [abwechselnde rillenförmige Vertiefungen bzw. rippenförmige Erhöhungen] = Riffelung ≈ Wellung	
checkerboarding (n.) [waisting memory by unusable gaps]	DATA PROC	**Mosaikspeicherung** [Speicherverschwendung mit unbrauchbaren Lücken]	
checkered	ENG.DRAW	**geriffelt**	
checkered = corrugated	MECH	**geriffelt**	
check-in	AERON	**Passagierabfertigung** = Abfertigung	
check indicator	DATA PROC	**Prüfanzeige**	
check information	TELECONTR	**Prüfinformation**	
checking (→inspection)	QUAL	(→Prüfung)	
checking appliance ≈ gage	MECH	**Prüflehre** ≈ Meßlehre	
checking device	TECH	**Kontrolleinrichtung**	
check key [word processing]	DATA PROC	**Kontrollschlüssel** [Textverarbeitung]	
check lamp (→pilot lamp)	EQUIP.ENG	(→Kontrollampe)	
checklist	TECH	**Prüfliste**	
check mark	DOC	**Haken**	
check number (→check digit)	CODING	(→Kontrollziffer)	
check-out (v.t.) (→debug)	DATA PROC	(→austesten)	
check out (n.) (→debugging)	DATA PROC	(→Austesten)	
check plot	DATA PROC	**Probezeichnung**	
checkpoint [prepared program point, from which the program can be restarted after an interrupt] = check point; restart point; rerun point ≈ breakpoint	DATA PROC	**Fixpunkt** [definierter Programmpunkt, ab dem nach einer Unterbrechung der Programmablauf wieder aufgenommen werden kann] = Wiederanlaufpunkt; Stützpunkt; Wiederanlaufkennzeichen; Wiederholpunkt; Checkpoint; Anhaltepunkt ≈ Unterbrechungspunkt	
checkpoint (→test point)	ELECTRON	(→Meßpunkt)	
check point (→checkpoint)	DATA PROC	(→Fixpunkt)	
checkpoint character [data block marking a checkpoint] ≈ checkpoint character ↑ label	DATA PROC	**Fixpunktsatz** [einen Fixpunkt markierender Datenblock]	
checkpoint dump	DATA PROC	**Fixpunktausdruck**	
checkpoint label [identifies checkpoints] ≈ checkpoint character ↑ label	DATA PROC	**Fixpunkt-Etikett** [kennzeichnet Fixpunkte] = Fixpunkt-Kennsatz; Checkpoint-Etikett ≈ Fixpunktsatz ↑ Etikett	

checkpoint routine	DATA PROC	**Fixpunktroutine**
≈ restart routine		≈ Wiederanlaufroutine
checkpoint selector (→ test point selector)	EQUIP.ENG	(→ Meßstellenwahlschalter)
checkpoint technique	DATA PROC	**Fixpunkttechnik**
[automatic back-up of main memory at fixed instants]		[automatische Erstellung von Sicherungskopien des Hauptspeichers zu festen Zeitpunkten]
check problem	DATA PROC	**Prüfproblem**
check register	DATA PROC	**Prüfregister**
checksum (→ running digital sum)	DATA PROC	(→ **Prüfsumme**)
check sum (→ running digital sum)	DATA PROC	(→ **Prüfsumme**)
check total (→ running digital sum)	DATA PROC	(→ **Prüfsumme**)
cheese (→ cheese antenna)	ANT	(→ **Käseanetenne**)
cheese antenna	ANT	**Käseanetenne**
= cheese		= Cheese-Antenne
↑ segment antenna		↑ Segment-Antenne
cheese-head screw (BRI) (→ pan-head screw)	MECH	(→ **Zylinderkopfschraube**)
chemical	TECH	**chemisch**
chemical bond	CHEM	**chemische Bindung**
chemical composition (→ composition)	CHEM	(→ **Zusammensetzung**)
chemical energy	PHYS	**chemische Energie**
chemical formula	CHEM	**chemische Formel**
		= Formel
chemical industry	ECON	**chemische Industrie**
		= Chemieindustrie
chemical machining	METAL	**Formteilätzen**
chemist	ECON	**Chemiker**
chemistry	SCIE	**Chemie**
chemoluminiscence	PHYS	**Chemolumineszenz**
[induced by chemical reaction]		[durch chemische Reaktion induziert]
↑ luminescence		↑ Lumineszenz
cheque (BRI) (→ check)	ECON	(→ **Scheck**)
cheque printer	TERM&PER	**Scheckdrucker**
cheque reader	TERM&PER	**Scheckleser**
chess board frequency	TV	**Schachbrettfrequenz**
chess board pattern	TECH	**Schachbrettmuster**
chess-playing computer	DATA PROC	**Schachcomputer**
chest microphone	EL.ACOUS	**Umhängemikropohon**
chief engineer	TECH	**Chefingenieur**
chief function (→ main function)	TECH	(→ **Hauptfunktion**)
chief operator	SWITCH	**Vermittlungsstellenleiter**
chief programmer	DATA PROC	**Chefprogrammierer**
child record	DATA PROC	**abgeleiteter Datensatz**
≠ parent record		≠ Stammdatensatz
CHILL	SWITCH	**CHILL**
= CCITT high level programming language		
chinch	COMPON	**Cinch**
[coaxial hifi connector]		[koaxiales Stecksystem für HiFi]
chinch angular adapter	COMPON	**Chinch-Winkeladapter**
chinch angular jack	COMPON	**Chinch-Winkelbuchse**
chinch angular plug	COMPON	**Chinch-Winkelstecker**
chinch female	COMPON	**Chinch-Kupplung**
		= Chinch-Kabelkupplung
chinch jack	COMPON	**Chinch-Buchse**
chinch mounting jack	COMPON	**Chinch-Einbaubuchse**
chinch plug	COMPON	**Chinch-Stecker**
Chinese-binary	TERM&PER	**spaltenbinär**
[punching and reading cards in columns]		[Lochkartenlochung]
Chinese ink (→ drawing ink)	ENG.DRAW	(→ **Tusche**)
chink (→ crack)	TECH	(→ **Sprung**)
chip (n.)	METAL	**Span**
chip (n.)	TECH	**Schnitzel**
[small flat piece worked-off]		[kleines abgetrenntes Stück]
= shred; snip		= Schnipsel; Schnippel
chip	MICROEL	**Chip**
[slice of semiconductor material, some millimeters of size, containing an integrated circuit or a semiconductor device, cutted from a "wafer"]		[millimetergroßes Plättchen mit einem aufdiffundierten intergrierten Schaltkreis oder Halbleiterbaustein, durch Zerkleinerung einer „Kristallscheibe" hergestellt]
= microchip; semiconductor chip; die (pl. dice or dies); semiconductor die; silicon chip; silicon die		= Mikrobaustein; Halbleiterchip; Mikrochip; Halbleiterplättchen; Siliziumplättchen
≈ wafer; integrated circuit; flip chip; semiconductor device		≈ Kristallscheibe; integrierte Schaltung; Flip-chip; Halbleiterbauelement
↓ transistor chip; IC chip		↓ Transistorchip; IC-Baustein
chip (v.t.) (→ shred)	TECH	(→ **zerschnitzeln**)
chip area	MICROEL	**Chip-Fläche**
chip capacitor	MICROEL	**Chip-Kondensator**
chip card	TERM&PER	**Chip-Karte**
[magnetic card with embedded microprocessor]		[Magnetkarte mit integriertem Mikroprozessor]
= telephone chip card; smart card; processor chip card		= Chipkarte; Telefon-Chipkarte
≈ magnetic card		≈ Magnetkarte
↓ prepaid phonecard; credit phonecard		↓ Guthabenkarte; Buchungskarte
chip-card telephone	TERM&PER	**Chipkartentelefon**
		= Chipkartenfernsprecher
chip carrier	MICROEL	**Chip-Träger**
		= Chip Carrier
chip-disk	TERM&PER	**Chip-Diskette**
[made with memory chips]		[mit Speicherbausteinen aufgebaut]
chip enable (→ chip enable signal)	ELECTRON	(→ **Chip-Freigabesignal**)
chip enable signal	ELECTRON	**Chip-Freigabesignal**
= chip enable		= Bausteinfreigabesignal
chip family	MICROEL	**Chip-Familie**
chipping knife	OUTS.PLANT	**Kabelmesser**
chip production	MICROEL	**Chip-Herstellung**
chip resistor	MICROEL	**Chip-Widerstand**
chip select	MICROEL	**Chip-Auswahl**
		= Bausteinauswahl; Chip Select
chip select line	CIRC.ENG	**Chip-Freigabeleitung**
chip set	CIRC.ENG	**Chip-Satz**
chip slice	MICROEL	**Chip-Scheibe**
chip technology	MICROEL	**Chip-Technik**
chip topopology	MICROEL	**Chip-Topologie**
Chireix-Mesny antenna (→ zigzag antenna)	ANT	(→ **Zickzackantenne**)
chirp	RADIO LOC	**Chirp**
[pulse compression by FM]		[Impulskompression durch FM]
chirped radar	RADIO LOC	**Chirp-Radar**
chirped signal (→ chirp signal)	RADIO LOC	(→ **Chirp-Signal**)
chirp linearity	RADIO LOC	**Chirp-Linearität**
chirp signal	RADIO LOC	**Chirp-Signal**
= chirped signal		
chisel	TECH	**Meißel**
chisel	TECH	**Stemmeisen**
		= Stechbeitel; Beitel; Stecheisen
chi-square function	MATH	**Chi-Quadrat-Verteilung**
↑ test distribution		↑ Testverteilung
chi-square test	MATH	**Chi-Quadrat-Test**
chlorine	CHEM	**Chlor**
= CL		= Cl
↑ halogen		↑ Halogen
choise	SWITCH	**Suchstellung**
choke	COMPON	**Drossel**
= choke coil; reactor; retard coil; inductor 2		= Drosselspule
choke coil (→ choke)	COMPON	(→ **Drossel**)
chop (v.t.)	DATA PROC	**verwerfen**
[to destroy useless data]		[unbrauchbare Daten]
chopper	CIRC.ENG	**Zerhacker**
= vibrator		= Chopper
chopper amplifier	CIRC.ENG	**Zerhackerverstärker**
		[wandelt Gleichspannung in Rechteckwechsel]
		= Chopperverstärker

chopper bar	TERM&PER	**Schreibstange**	↑ clock		↑ Uhr
		= Fallbügel	↓ stop watch		↓ Stoppuhr
chopper interference (→electronic hash)	ELECTRON	(→Zerhackergeräusch)	**chronometry** (→time measurement)	INSTR	(→Zeitmessung)
chopper noise (→electronic hash)	ELECTRON	(→Zerhackergeräusch)	**chuck** (n.)	TECH	**Spannvorrichtung**
			chuck jaw	MECH	**Spannbake**
chopper stabilized	CIRC.ENG	**chopperstabilisiert**	= jaw		
chopper stabilized operational amplifier	CIRC.ENG	**chopperstabilisierter Operationsverstärker**	**churning** (→thrashing)	DATA PROC	(→Zeitverschwendung)
			chute (n.)	TECH	**Sammelbehälter**
chopping	CIRC.ENG	**Zerhackung**	**chute blade**	TERM&PER	**Sortierschiene**
chord	MATH	**Sehne**	**Ci** (→Curie)	PHYS	(→Curie)
[line connecting two points of a curve]		[Gerade die zwei Punkte einer Kurve verbindet]	**CIC** (→circuit identification code)	DATA COMM	(→Leitungskennzeichnungs-Code)
chord keying	DATA PROC	**Mehr-Tasten-Betätigung**	**cicero**	TYPOGR	**Cicero**
[pressing several keys simultaneously]			[typographic measuring unit; = 12 points, corresponding to 4,51 mm]		[typographische Maßeinheit; = 12 Punkte, entsprechend 4,51 mm]
christmastree antenna (→fishbone antenna)	ANT	(→Fischgrätenantenne)			
chroma 3	PHYS	**Farbeigenschaft**	**CID**	MICROEL	**CID-Element**
= color attribute			= charge injection device		
chroma 2 (→color fastness)	PHYS	(→Farbechtheit)	**CIF**	ECON	**CIF**
chroma 1 (→color intensity)	PHYS	(→Farbstärke)	= cost, insurance, freight		= Kosten, Versicherung und Fracht
chroma control (→color intensity control)	TV	(→Farbstärkeregler)	**cigar antenna**	ANT	**Zigarrenantenne**
			CIM	DATA PROC	**CIM**
chromatic (→colored)	PHYS	(→farbig)	= computer integrated manufacturing		= rechnerunterstützte integrierte Produktion
chromatic aberration	PHYS	**chromatische Aberration**	≈ CAM		≈ CAM
		= Farbfehler	**CIM**	TERM&PER	**Mikrofilmeingabe**
chromatic dispersion (→dispersion)	PHYS	(→Dispersion)	= computer input from microfilm		≈ CIM
			cinematographic camera	TECH	**Filmkamera**
chromaticity	PHYS	**Farbart**	**cinetics**	PHYS	**Kinetik**
[dominant color and its purity]		[dominierende Farbe und deren Reinheit]	[theory of force-dependence of movements]		[Theorie der Bewegungen in Abhängigkeit der Kräfte]
≈ chromatic purity		= Farbechtheit	≈ cinematics		≈ Kinematik
chromaticity coordinates	PHYS	**Farbwertanteil**	**cipher** (v.t.)	TELEC	**verschlüsseln**
chromaticity diagram	TV	**Farbtafel**	= enciper; encrypt; encode; sramble; code (v.t.)		= chiffrieren; codieren
chromaticity diagram (→color triangle)	PHYS	(→Farbdreieck)	≠ decipher (v.t.)		≠ entschlüsseln
chromatic purity	PHYS	**Farbenreinheit**	**cipher** (v.t.)	ECON	**beziffern**
≈ color fastness		≈ Farbechtheit	= cypher		
chromatic rendering (→color rendition)	PHYS	(→Farbwiedergabe)	**cipher** (→numeral)	MATH	(→numerisches Zeichen)
			cipher (n.) (→cipher key)	INF.TECH	(→Schlüssel)
chromatic rendition (→color rendition)	PHYS	(→Farbwiedergabe)	**ciphered text** (→ciphertext)	INF.TECH	(→verschlüsselter Text)
chromatic temperature (→color temperature)	PHYS	(→Farbtemperatur)	**cipher key**	INF.TECH	**Schlüssel**
chromatron (→chromatron tube)	ELECTRON	(→Gitterablenkröhre)	= key; cipher (n.); code		= Code
			↓ encryption key; decryption key		↓ Verschlüsselungscode; Entschlüsselungscode
chromatron tube	ELECTRON	**Gitterablenkröhre**			
= chromatron; Lawrence tube		= Chromatron-Röhre; Chromatron	**cipher method** (→cipher system)	INF.TECH	(→Verschlüsselungssystem)
chrominance	TV	**Farbwert**	**cipher system**	INF.TECH	**Verschlüsselungssystem**
[colometric difference to e reference color]		[chromatische Differenz zu einer Bezugsfarbe]	= cipher method; encryption system; encryption method		= Verschlüsselungsverfahren; Verschlüsselungsmethode
≈ tristimulus value		= Chrominanz	**ciphertext** (n.)	INF.TECH	**verschlüsselter Text**
chrominance carrier	TV	**Farbträger**	= ciphered text; coded text		= Schlüsseltext
= chrominance subcarrier			≠ cleartext		≠ Klartext
chrominance modulator	TV	**Farbmodulator**	**ciphony** (→speech encryption)	INF.TECH	(→Sprachverschlüsselung)
		= Chrominanzmodulator	**CIR** (→current instruction register)	DATA PROC	(→Momentanbefehlsregister)
chrominance signal	TV	**Chrominanzsignal**	**circle**	MATH	**Kreis**
[carrier modulated with the primary signal]		[mit Primär-Farbartsignal modulierter Träger]	↑ conical section		= Zirkel
≈ primary signal; color difference signal		= Farbartsignal; Farbwertsignal			↑ Kegelschnitt
		≈ Primär-Farbartsignal; Farbdifferenzsignal	**circle chart** (→transmission line chart)	LINE TH	(→Leitungsdiagramm)
chrominance subcarrier (→chrominance carrier)	TV	(→Farbträger)	**circle cutter**	TECH	**Kreisschneider**
			circle of curvature	MATH	**Krümmungskreis**
chromium	CHEM	**Chrom**	**circuit**	SWITCH	**Satz**
= Cr		= Cr	[functional unit of a switching system, assigned to a call with switching or interfacing functions, generally only during the task period]		[Funktionseinheit eines Wählsystems, die einer Verbindung zur Erfüllung wähltechnischer Funktionen oder für Schnittstellenaufgaben zugeordnet wird, meist nur vorübergehend]
chromium-plate	METAL	**verchromen**			
chromium plating	METAL	**Verchromung**			
chromoscope	ELECTRON	**Chromoskop**			
chronogram	TECH	**Chronogramm**			
		≈ Ablaufplan			
chronologic (adj.)	SCIE	**chronologisch**	↓ subscriber line circuit; signaling circuit; junctor; trunk circuit; test circuit		↓ Teilnehmersatz; Signalisierungssatz; Verbindungssatz; Leitungssatz; Prüfsatz
= chronological		= zeitlich geordnet			
chronological (→chronologic)	SCIE	(→chronologisch)			
chronologically interleaved	TECH	**zeitlich verschachtelt**			
chronological order	SCIE	**chronologische Ordnung**			
chronometer	INSTR	**Chronometer**			
[high precision clock]		[Uhr hoher Genauigkeit]			

English	Domain	German
circuit	NETW.TH	**Schaltkreis**
[network containing one or more closed electrical paths]		[Netzwerk mit einer oder mehreren Stromschleifen]
= electric circuit; current circuit; connection; section		= Schaltung; elektrischer Stromkreis; Stromkreis; Kreis
↑ electric network		↑ Netzwerk
circuit	AERON	**Platzrunde**
circuit (→line)	TELEC	(→Leitung)
circuit (→cycle)	TECH	(→Umlauf)
circuit (→connection)	TELEC	(→Verbindung)
circuital erogation	CIRC.ENG	**Schaltungsaufwand**
circuit algebra (→engineering logic)	INF	(→Schaltalgebra)
circuit analyzer	INSTR	**Schaltkreisanalysator**
circuit arrangement	CIRC.ENG	**Schaltungsaufbau**
= arrangement		
circuit board (→printed circuit board)	ELECTRON	(→Leiterplatte)
circuit breaker	CIRC.ENG	**Aus-Schalter**
≠ circuit closer		= Ausschalter; Stromunterbrecher
		≠ Einschalter
circuit breaker	COMPON	**Schutzschalter**
= safety switch; fuse-disconnector (IEC); fuse-isolator (IEC); automatic breaker; breaker		= Sicherungstrennschalter; Sicherungsschalter
		↑ Stromgrobsicherung
↑ high-current fuse		
circuit breaker	POWER ENG	**Trennschalter**
= power switch		= Leistungsschalter; Netzschalter
circuit closer	CIRC.ENG	**EIN-Schalter**
= switch-on		≠ AUS-Schalter
≠ circuit breaker		
circuit complexity	CIRC.ENG	**Schaltungskomplexität**
circuit connector	SWITCH	**Anschaltesatz**
= connecting relay set		
circuit description	DOC	**Stromlaufbeschreibung**
circuit design	CIRC.ENG	**Schaltungsentwicklung**
≈ circuit engineering		= Schaltungsentwurf
		≈ Schaltkreistechnik
circuit design engineer (→circuit engineer)	ELECTRON	(→Schaltungsentwickler)
circuit diagram	ELECTRON	**Stromlaufplan**
[a schematic representation, whether in true position nor in true shape, of the interconnection of components]		[weder lagerichtige noch maßstabgerechte Darstellung der Zusammenschaltung der Bauelemente]
= schematic circuit diagram; schematic diagram; wiring scheme; circuit schematic		= Stromlauf; Schaltplan; Prinzipschaltbild; Prinzipschaltung; Prinzipstromlauf; Stromlaufzeichnung; Schaltbild; Verdrahtungsschema; Schaltschema; Wirkschaltplan
≈ wiring diagram		≈ Bauschaltplan
circuit element	NETW.TH	**Schaltungselement**
[element of an electrical circuit]		[Element eines elektrischen Schaltkreises]
= component [COMPON]		= Schaltelement; Schaltorgan; Bauelement [COMPON]
circuit element [NETW.TH] (→component)	COMPON	(→Bauelement)
circuit end	TELEC	**Leitungsende**
= line end		
circuit engineer	ELECTRON	**Schaltungsentwickler**
= circuit design engineer; breadboard engineer; squeezer		
circuit engineering	EL.TECH	**Schaltkreistechnik**
≈ circuitry		= Schaltungstechnik
circuit group	SWITCH	**Satzgruppe**
circuit identification code	DATA COMM	**Leitungskennzeichnungs-Code**
= CIC		
circuit implementation	CIRC.ENG	**Schaltungsrealisierung**
circuit module	SWITCH	**Satzbaugruppe**
circuit multiplication system	TELEC	**Sprechkreis-Vermehrungssystem**
circuit noise (→line noise)	TELEC	(→Leitungsgeräusch)
circuit-oriented	MICROEL	**schaltungsorientiert**
circuit periphery	SWITCH	**Satzperipherie**
circuit polling (→line polling)	DATA COMM	(→Leitungsabruf)
circuit-polling operation (→line-polling operation)	DATA COMM	(→Leitungsabrufbetrieb)
circuit position	SWITCH	**Satzposition**
circuit-related	TELEC	**leitungsbezogen**
circuit release (→connection tear-down)	SWITCH	(→Verbindungsabbau)
circuitry (n.; s.t.)	CIRC.ENG	**Schaltungskomplex**
		= Schaltungsanordnung
circuitry (→circuit engineering)	EL.TECH	(→Schaltkreistechnik)
circuit schematic (→circuit diagram)	ELECTRON	(→Stromlaufplan)
circuit side	SWITCH	**Satzseite**
circuit simulation	MICROEL	**Schaltkreis-Simulation**
circuit stage	CIRC.ENG	**Stufe**
= stage		= Schaltungsstufe; Schaltkreisstufe
circuit state	DATA COMM	**Leitungsstatus**
circuit switched	SWITCH	**leitungsvermittelt**
circuit switching	SWITCH	**Durchschaltevermittlung**
[connection via through switched lines, without intermediate stores]		[Verbindung über durchgeschaltete Leitungen, ohne Zwischenspeicher]
= line switching		= Leitungsvermittlung
≠ store-and-forward switching		≠ Speichervermittlung
circuit switching network	DATA COMM	**Durchschaltenetz**
circuit symbol	EL.TECH	**Schaltkennzeichen**
		= Schaltungssymbol
circuit technology	MICROEL	**Schaltkreistechnologie**
		= Schaltungstechnologie
circuit terminating equipment (→line terminating unit)	DATA COMM	(→Leitungsabschlußeinrichtung)
circuit tester (→in-circuit tester)	INSTR	(→Schaltkreisprüfgerät)
circuit testing	SWITCH	**Satzprüfung**
circuit type	SWITCH	**Satztyp**
circular (n.)	ECON	**Rundschreiben**
= circular letter; newsletter		= Zirkular
circular	MATH	**kreisförmig**
↑ round		= kreisartig; kreisrund; zirkular; zirkulär
		↑ rund
circular aperture	TV	**Kreislochblende**
circular aperture	PHYS	**Kreisblende**
circular arc	MATH	**Kreisbogen**
[embraces 360°]		[umfasst 360°]
= arc		= Bogen; Arcus
circular area	MATH	**Kreisfläche**
circular array (→circular array antenna)	ANT	(→Kreisgruppenantenne)
circular array antenna	ANT	**Kreisgruppenantenne**
= circular array; ring array		= Kreisgruppe
↑ planar array		↓ Wullenweber-Antenne
circular bent dipole (→circular dipole)	ANT	(→Ringdipol)
circular characteristic	EL.ACOUS	**Kreischarakteristik**
circular chart	NETW.TH	**Kreisblatt**
= round chart		
circular chart diagram	NETW.TH	**Kreisblattdiagramm**
= round chart diagram		
circular chart recorder (→pie recorder)	INSTR	(→Kreisblatt-Schreiber)
circular cross section	MATH	**Kreisquerschnitt**
circular cylinder	MATH	**Walze**
		[Zylinder mit kreisförmigem Querschnitt]
		= Kreiszylinder
circular cylindrical	MATH	**kreiszylinrisch**
circular dipole	ANT	**Ringdipol**
= circular bent dipole		≈ Halo-Antenne
≈ half-wave loop antenna		
circular file	DATA PROC	**Ringdatei**
circular flange	MECH	**Rundflansch**
≠ rectangular flange		≠ Rechteckflansch
circular frequency (→angular frequency)	PHYS	(→Kreisfrequenz)

circular fringe PHYS
[optics]
circularity MATH
circular letter (→circular) ECON
circular list DATA PROC
[cyclic arrangement of data]
= ring (n.)
circularly polarized PHYS
circular motion (→circular PHYS
movement)
circular mounting COMPON
= bulkhead
≠ flange mounting

circular movement PHYS
= circular motion; circulation
↓ rotary movement
↓ twist 1

circular nut MECH
circular orbit PHYS
= circular path
↑ trajectory
circular path (→circular orbit) PHYS
circular pitch MECH

circular plate shape COMPON
[plate capacitor]
circular polarization PHYS

circular polarized wave PHYS
circular saw TECH
= buzz saw
circular scanning ELECTRON
circular shift DATA PROC
circular shift (→end-around DATA PROC
shift)
circular slide rule MATH
circular test pattern TV
≈ circular test chart
circular wave PHYS

circular waveguide MICROW

circulate PHYS
[to move in a circle]
≈ turn 2; rotate
circulate SCIE
≈ rotate

circulating PHYS
≈ rotating

circulating assets ECON
[balance]
= current assets
circulating current (→ring POWER ENG
current)
circulating memory DATA PROC
[consists of looped data channels
where informations circulate continuously, which can be accessed cyclically without erasing]
= cyclic memory; circulating register;
cyclic storage; circulating storage; recirculating memory; recirculating
storage; recirculating store

circulating pump TECH
circulating register (→cyclic CIRC.ENG
shift register)
circulating register DATA PROC
(→circulating memory)

Ring
[Optik]
Kreisform
= Zirkularität
(→Rundschreiben)
Ringliste
[zyklische Anordnung von
Daten]

zirkular polarisiert
(→Kreisbewegung)

Zentralbefestigung
[von Einbausteckverbindern]
≠ Flanschbefestigung
Kreisbewegung
= Rotation; Rotationsbewegung
≈ Drehbewegung
↓ Drall
Ringmutter
Kreisbahn
= Kreisumlaufbahn
↑ Bahn
(→Kreisbahn)
Kreisteilung
[Zahnrad]
Kreisplattenschnitt
[Plattenkondensator]
zirkulare Polarisation
= Zirkularpolarisation; Kreispolarisation
zirkular polarisierte Welle
Kreissäge

Kreisabtastung
Kreisverschiebung
(→Ringschieben)

Rechenscheibe
Kreismuster

Kreiswelle
= Zirkularwelle
Rundhohlleiter
= Kreishohlleiter; kreisförmiger Hohlleiter
kreisen
[sich im Kreis bewegen]
≈ zirkulieren
≈ drehen; rotieren
umlaufen
≈ kreisen; zirkulieren
≈ drehen
umlaufend
≈ kreisend; zirkulierend
≈ drehend
Umlaufvermögen
[Bilanz]

(→Kreisstrom)

Umlaufspeicher
[besteht aus geschlossenen
Datenkanälen auf denen
Informationen ständig umlaufen, und auf die zyklisch
zugegriffen werden kann,
ohne den Inhalt zu löschen]
= zyklischer Speicher; periodischer Speicher
↓ Magnettrommelspeicher;
Magnetplattenspeicher;
CCD-Speicher; Magnetblasenspeicher
Umwälzpumpe
(→Ringschieberegister)

(→Umlaufspeicher)

circulating storage DATA PROC
(→circulating memory)
circulation PHYS

circulation MATH
[line integral over a closed integration path]
= contour integral
↑ line integral
circulation (→circular PHYS
movement)
circulation list (→distributor) DOC
circulation quantum PHYS
circulator NETW.TH
[multiport network, transmits only
from a port to the next one]

circulator MICROW

circumference 1 MATH
↑ circumference 2

circumference 2 (→perimeter) MATH
circumflex LING
[e.g. in ñ]
= circumflex accent
↑ diacritic mark; accent

circumflex accent LING
(→circumflex)
circumvention SIGN.ENG
[of an alarm system]
≈ spoofing
CISC (→complex instruction DATA PROC
set computer)
CISC processor DATA PROC
[Complex Instruction Set Code]
CISPR EMI receiver INSTR
citizen band (→citizens RADIO
band)
citizen band radio (→CB radio RADIO
equipment)
citizens band RADIO
[at 27 MHz]
= citizen band; CB

city (BRI) (→downtown) ECON
city rail (→tramway) TECH
civil aviation AERON
civil engineering TECH
= constructional engineering
↑ engineering
civil project TECH
= civil works
↑ project
civil servant (→official) ECON
civil works (→civil project) TECH
CKSM (→running digital DATA PROC
sum)
CL (→chlorine) CHEM
clad (v.t.) TECH
[to cover with a film of another material, most metallic]
↑ coat

clad (n.) (→cladding) TECH
clad board (→clad PCB) ELECTRON
cladded (adj.) TECH
↑ coated
cladding TECH
= clad (n.)
≈ wrapping
↑ coating
cladding METAL
↑ plating
cladding (→cable sheath) COMM.CABLE
clad PCB ELECTRON
= clad board
claim 1 (n.) ECON
= right; title

(→Umlaufspeicher)

Kreislauf
= Zirkulation
Umlaufintegral
[Kurvenintegral über geschlossenen Integrationsweg]
↑ Kurvenintegral
(→Kreisbewegung)

(→Verteiler)
Zirkulationsquant
Zirkulator
[mehrtoriges Netzwerk,
überträgt nur von Tor zu
nachfolgendem Tor]
Zirkulator
= Richtungsgabel
Kreisumfang
= Zirkumferenz
↑ Umfang
(→Umfang)
Zirkumflex
[z.B. in ñ]
= Dehnungszeichen
↑ diakritisches Zeichen; Akzent
(→Zirkumflex)

Umgehung
[einer Warnanlage]
≈ Ausschaltung
(→CISC-Computer)

CISC-Prozessor

CISPR-Empfänger
(→CB)

(→CB-Funksprechgerät)

CB
[„Bürgerfrequenzband"
um 27 MHz]
= Citizen-Band
(→Geschäftszentrum)
(→Straßenbahn)
Zivilluftfahrt
Bautechnik
= Bauwesen
↑ Technik
Bauvorhaben
= Bau
↑ Projekt
(→Beamter)
(→Bauvorhaben)
(→Prüfsumme)

(→Chlor)
kaschieren
[mit einem Überzug eines
anderen, meist metallischen, Materials versehen]
↑ beschichten
(→Kaschierung)
(→kaschierte Leiterplatte)
kaschiert
↑ beschichtet
Kaschierung
≈ Umhüllung
↑ Beschichtung
mechanische Plattierung
↑ Plattierung
(→Kabelmantel)
kaschierte Leiterplatte

Anspruch
= Anspruch

claim 2 (n.)		ECON	**Beanstandung**	clapper armature magnet		COMPON	(→Klappankermagnet)

claim 2 (n.) ECON **Beanstandung**
= complaint; objection   = Reklamation; Beschwerde
claim 3 (n.) ECON **Forderung**
= demand; requirement   ≈ Schuld
≈ debt
claim 4 (n.) ECON **Schadensfall**
clamp (v.t.) TECH **abfangen**
[of a cable]   [ein Kabel]
clamp 1 (v.t.) MECH **festklemmen**
[to fix with a vise-type device]   = festspannen; einspannen; aufspannen
clamp (v.t.) EL.TECH **klemmen**
clamp 2 (v.t.) MECH **schellen**
[to fix by contraction of a metal tape]   = abschellen; anschellen
clamp (n.) MECH **Schelle**
[a fixing device by contraction of a metal tape]   ↓ Rohrschelle
↓ pipe clamp
clamp INSTR **Meßwandlerzange**
clamp (→clamping) TECH (→Abfangung)
clamp (v.t.) (→jam) TECH (→klemmen 1) (v.t.)
clamp (→lamp socket) COMPON (→Lampenfassung)
clamp ammeter (→pliers ammeter) INSTR (→Zangenstrommeter)
clamp coupling MECH **Klemmschalenkupplung**
clamped amplifier CIRC.ENG **Klemmverstärker**
clamper (→clamping circuit) CIRC.ENG (→Klemmschaltung)
clamper noise TV **Zeilenrauschen**
    = Zeilenstreifigkeit
clamp handle MECH **Knebel**
clamping MECH **Aufspannung**
[of a machine tool]   [Werkzeugmaschine]
clamping TECH **Abfangung**
[of a cable]   [eines Kabels]
= clamp
clamping TV **Klemmung**
clamping appliance (→clamping device) MECH (→Spannelement)
clamping circuit CIRC.ENG **Klemmschaltung**
= clamper   = Klammerschaltung; Clamping-Schaltung
clamping collar EL.INST **Klemmtülle**
[for a cable]   [Kabel]
clamping device MECH **Spannelement**
= fixture; clamping appliance   = Spannvorrichtung; Klemmvorrichtung; Anspannvorrichtung; Abspannvorrichtung
≈ arrest   ≈ Hemmung
clamping device TECH **Anschellmaschine**
clamping diode 1 ELECTRON **Kappdiode**
[levels pulse peaks]   [kappt Impulsspitzen]
≈ limiter diode   = Abfangdiode
↑ protective diode   = Begrenzerdiode
    ↑ Schutzdiode
clamping diode 2 CIRC.ENG **Klemmdiode**
[restores dc component]   [stellt Gleichstromanteil wieder her]
    = Klammerdiode; Clamping-Diode
clamping distortion TV **Klemmverzerrung**
clamping pulse TV **Klemmimpuls**
clamping ring MECH **Klemmring**
clamping screw MECH **Klemmschraube**
    = Spannschraube
clamping sleeve COMM.CABLE **Klemm-Muffe**
clamping spring MECH **Klemmfeder**
clamping washer MECH **Spannscheibe**
clamp meter (→pliers meter) INSTR (→Zangenmeßgerät)
clamp-on ac probe INSTR **Wechselstromzange**
clamp-on ammeter INSTR **Stromzange**
= clip-on ammeter
≈ clamp-on current amplifier   ≈ Stromzangenverstärker
clamp-on current amplifier INSTR **Stromzangenverstärker**
= clip-on current amplifier
clamp-on dc current probe INSTR **Gleichstromzange**
clamp-on meter (→pliers meter) INSTR (→Zangenmeßgerät)
clamp power meter (→pliers power meter) INSTR (→Zangenleistungsmesser)

clapper armature magnet COMPON (→Klappankermagnet)
 (→hinged-armature magnet)
clapper armature relay COMPON (→Klappankerrelais)
 (→hinged-armature relay)
Clapp oscillator CIRC.ENG **Clapp-Ostillator**
↑ crystal oscillator   ↑ Quarzoszillator
clarification COLLOQ **Verdeutlichung**
≈ ilustration   ≈ Veranschaulichung
clarify COLLOQ **verdeutlichen**
≈ illustrate   ≈ veranschaulichen
clasp handle TECH **Klappgriff**
= hinged handle; fold-in handle   ↑ Handgriff 1
↑ handle 1
class (n.) MATH **Klasse**
class A (→class A operation) CIRC.ENG (→A-Betrieb)
class A amplifier CIRC.ENG **Klasse-A-Verstärker**
    = A-Verstärker; Eintakt-A-Verstärker
class AB (→class AB operation) CIRC.ENG (→AB-Betrieb)
class AB amplifier CIRC.ENG **Klasse-AB-Verstärker**
    = AB-Verstärker
class AB mode (→class AB operation) CIRC.ENG (→AB-Betrieb)
class AB operation CIRC.ENG **AB-Betrieb**
= class AB mode; class AB
class AB push-pull operation CIRC.ENG **Gegentakt-AB-Betrieb**
class AB stage CIRC.ENG **AB-Stufe**
    [Verstärker]
class A mode (→class A operation) CIRC.ENG (→A-Betrieb)
class A operation CIRC.ENG **A-Betrieb**
[amplifier]   [Verstärker]
= class A mode; class A
class B (→class B operation) CIRC.ENG (→B-Betrieb)
class B amplifier CIRC.ENG **Klasse-B-Verstärker**
    = B-Verstärker; Gegentakt-B-Verstärker
class B mode (→class B operation) CIRC.ENG (→B-Betrieb)
class B operation CIRC.ENG **B-Betrieb**
= class B mode; class B
class B push-pull operation CIRC.ENG **Gegentakt-B-Betrieb**
class C (→class C operation) CIRC.ENG (→C-Betrieb)
class C amplifier CIRC.ENG **Klasse-C-Verstärker**
    = C-Verstärker
class C mode (→class C operation) CIRC.ENG (→C-Betrieb)
class C operation CIRC.ENG **C-Betrieb**
= class C mode; class C
class D (→class D operation) CIRC.ENG (→D-Betrieb)
class D amplifier CIRC.ENG **Klasse-D-Verstärker**
    = D-Verstärker
class D mode (→class D operation) CIRC.ENG (→D-Betrieb)
class D operation CIRC.ENG **D-Betrieb**
= class D mode; class D
class frequency MATH **Klassenhäufigkeit**
[statistics]   [Statistik]
classical physics PHYS **klassische Physik**
classification SCIE **Klassifizierung**
    = Einstufung
classification (→division) TECH (→Einteilung)
classification letter DATA PROC **Kennbuchstabe**
classified room (→clean room) MANUF (→Reinraum)
classify (v.t.) SCIE **klassifizieren**
    = einstufen
class interval MATH **Klassenintervall**
= cell
class mark (→class midpoint) MATH (→Klassenmitte)
classmarked (→authorized) TELEPH (→berechtigt)
class midpoint MATH **Klassenmitte**
[statistics]   [Statistik]
= class mark

class of data signalling rate   88

class of data signalling rate	DATA COMM	**Geschwindigkeitsklasse**	
class of emission	RADIO	**Sendeart**	
class of fit	ENG.DRAW	**Passungsklasse**	
= class of fits		= Sitzklasse	
class of fits (→class of fit)	ENG.DRAW	(→Passungsklasse)	
class of line (→user group)	DATA COMM	(→Teilnehmerbetriebsklasse)	
class of operation	ELECTRON	**Betriebsart**	
= operation mode; operation; mode		= Modus; Betrieb	
↓ class A; class B; class C		↓ A-Betrieb; B-Betrieb; C-Betrieb	
class of service (→user group)	DATA COMM	(→Teilnehmerbetriebsklasse)	
classroom	SCIE	**Schulungsraum**	
= training room		= Unterrichtsraum; Klassenraum	
class S amplifier	CIRC.ENG	**Klasse-S-Verstärker**	
		= S-Verstärker	
clause	ECON	**Klausel**	
clause (→subordinate clause)	LING	(→Nebensatz)	
claw (n.) (→grip 4)	TECH	(→Greifer)	
C lead (→tip wire)	TELEPH	(→C-Ader)	
clean (v.t.)	TECH	**säubern**	
= scavenge		= reinigen	
clean (adj.)	TECH	**sauber**	
= scavenged		= gereinigt	
clean (→error-free)	INF	(→fehlerfrei)	
clean diskette (→blank diskette)	DATA PROC	(→Leerdiskette)	
cleaning	TECH	**Reinigung**	
		= Säuberung	
cleaning agent	TECH	**Reinigungsmittel**	
= detergent (n.)		= Waschmittel	
≈ solnent		≈ Spülmittel; Lösungsmittel	
cleaning cassette	CONS.EL	**Reinigungskassette**	
↓ video cleaning cassette		= Reinigungscassette	
		= Video-Reinigungskassette	
cleaning diskette (→head cleaning disk)	TERM&PER	(→Reinigungsdiskette)	
cleaning equipment	TERM&PER	**Reinigungsgerät**	
cleaning kit	TECH	**Reinigungssatz**	
clean room	MANUF	**Reinraum**	
= classified room		= Reinstraum; staubfreier Raum	
clean-room engineering	TECH	**Reinstraumtechnik**	
clear 2 (v.t.)	TERM&PER	**aufheben**	
[e.g. a flag]		[z.B. eine Markierung]	
clear (v.t.)	TECH	**beheben**	
[a fault]		[Fehler, Störung]	
clear 1 (v.t.)	TERM&PER	**löschen**	
[tape]		[Band]	
clear (v.t.)	ECON	**verzollen**	
= pay duty			
clear (v.t.)	DATA PROC	**löschen**	
[a display]		[eines Bildschirminhalts]	
clear (→release)	SWITCH	(→auslösen)	
clear (→erase)	DATA PROC	(→löschen)	
clear (→erase)	ELECTRON	(→löschen)	
clearable (→erasable)	DATA PROC	(→löschbar)	
clearance	TECH	**Räumung**	
clearance	RADIO PROP	**Sichtfreiheit**	
≠ obstruction		≠ Sichtbehinderung	
clearance (→release)	SWITCH	(→Auslösung)	
clearance (→entitlement)	ECON	(→Berechtigung)	
clearance (→path clearance)	RADIO PROP	(→Hindernisfreiheit)	
clearance (→play)	ENG.DRAW	(→Spielraum)	
clearance (→custom payment)	ECON	(→Verzollung)	
clearance fit	ENG.DRAW	**Spielpassung**	
↑ class of fit		= Spielsitz	
		↑ Passungsklasse	
clearance papers	ECON	**Verzollungspapiere**	
clear-back signal	SWITCH	**Schlußzeichen**	
= clearing signal; disconnect signal; clearing			
clear backward (→backward release)	SWITCH	(→Rückwärtsauslösung)	
clear command (→delete statement)	DATA PROC	(→Löschanweisung)	
clear confirmation	DATA COMM	**Auslösebestätigung**	
		= Auslösungsbestätigung	
clear down (→release)	SWITCH	(→auslösen)	
clear down (→trigger)	DATA PROC	(→auslösen)	
clear forward (→forward release)	SWITCH	(→Vorwärtsauslösung)	
clear forward signal (→release signal)	SWITCH	(→Auslösezeichen)	
clear indication	DATA COMM	**Auslöseanzeige**	
		= Auslösungsanzeige	
clearing	SWITCH	**freischalten**	
= trunk release			
clearing (→clear-back signal)	SWITCH	(→Schlußzeichen)	
clearing (→triggering)	ELECTRON	(→Auslösung)	
clearing (→charging)	ECON	(→Verrechnung)	
clearing-cause field	DATA COMM	**Auslösungsgrund-Feld**	
clearing forward (→release)	SWITCH	(→Auslösung)	
clearing key	TELEGR	**Schlußtaste**	
≠ calling key		≠ Anruftaste	
clearing pulse	ELECTRON	**Abschaltstromstoß**	
clearing signal (→clear-back signal)	SWITCH	(→Schlußzeichen)	
clear input (→erase input)	CIRC.ENG	(→Löscheingang)	
clear instruction (→delete statement)	DATA PROC	(→Löschanweisung)	
clear job (→delete statement)	DATA PROC	(→Löschanweisung)	
clear key (→backspace key)	TERM&PER	(→Rücksetztaste 2)	
clear message	DATA COMM	**Auslösungsblock**	
clearness of tuning (→selectivity)	NETW.TH	(→Selektion)	
clear pulse (→reset pulse)	ELECTRON	(→Rücksetzimpuls)	
clear request	DATA COMM	**Auslöseanforderung**	
= invitation to clear		= Auslösungsanforderung	
clear statement (→delete statement)	DATA PROC	(→Löschanweisung)	
clear text (→plaintext)	INF.TECH	(→Klartext)	
cleavage (→fragmentation)	TECH	(→Zerstückelung)	
cleave (→fragment)	TECH	(→zerstückeln)	
cleft (→split)	TECH	(→Spalte)	
clerck 1	ECON	**Büroangestellter**	
clerck 2	ECON	**Verkäufer**	
clerical error	DOC	**Schreibfehler**	
= typing error		= Tippfehler	
clerical staff	OFFICE	**Büropersonal**	
clerk (→employee)	ECON	(→Angestellter)	
clerk-typist (→typist)	OFFICE	(→Schreibkraft)	
click (v.t.)	TERM&PER	**klicken**	
[mouse]		[Maus]	
click (n.)	TELEPH	**Knackgeräusch**	
= click interference; crackle; crackle interference		= Knacken; Knackstörung	
click (→hard keying)	MODUL	(→Harttastung)	
click (→impulsive noise)	TELEC	(→Impulsgeräusch)	
click (→clicking)	TERM&PER	(→Klicken)	
click absorber	TELEPH	**Gehörschutzgleichrichter**	
click absorption	TELEPH	**Gehörschutz**	
click filter	TELEPH	**Knackfilter**	
clicking (n.)	TERM&PER	**Klicken**	
[short depression of mouse key]		[kurzes Drücken der Mausetaste]	
= click		= Mausklick	
clicking on (n.)	DATA PROC	**Anklicken**	
click interference (→click)	TELEPH	(→Knackgeräusch)	
click on (v.t.)	DATA PROC	**anklicken**	
client	ECON	**Kunde**	
= costumer			
client (→requester)	DATA COMM	(→Requester)	
clientele	ECON	**Kundschaft**	
		= Klientel; Kundenstamm	
clima	QUAL	**Klima**	
= climatic conditions		= Klimabedingungen	
climate	METEOR	**Klima**	
[typical yearly course of weather conditions in a region]		[typischer Jahresverlauf der Witterung in einem Gebiet]	
≈ weather; weather conditions		≈ Wetter; Witterung	

climate-proof (→climate resistant)		QUAL	(→klimabeständig)	**clocked pulse** (→clock pulse)	ELECTRON	(→Schrittpuls)
climate resistant		QUAL	**klimabeständig**	**clock error**	ELECTRON	**Taktfehler**
= climate-proof			= klimafest	**clock error alarm**	EQUIP.ENG	**Taktfehlermeldung**
↓ tropicalized			↓ tropenfest	**clock extraction**	CODING	**Taktausblendung**
climatic box		QUAL	**Klimaschrank**	**clock frequency**	ELECTRON	**Taktfrequenz**
climatic conditions (→clima)		QUAL	(→Klima)	= clock rate; timing frequency		= Taktgeschwindigkeit
climatic factor		RADIO PROP	**Klimafaktor**	≈ pulse repetition rate		≈ Pulsrate
climatic test		QUAL	**Klimaprüfung**	**clock frequency**	DATA PROC	**Taktfrequenz**
climatic test chamber		QUAL	**Klimakammer**	= clock rate; strobe frequency; strobe 2 (n.)		= Taktrate
climatogram		QUAL	**Klimatogramm**	**clock generation**	ELECTRON	**Takterzeugung**
climbing aid [mast]		OUTS.PLANT	**Steighilfe** [Mast]	**clock generator** (→timing generator)	CIRC.ENG	(→Taktgeber)
climbing facility		OUTS.PLANT	**Klettervorrichtung**	**clock inhibit**	ELECTRON	**Taktsperre**
climbing grid		SYS.INST	**Steigrost**	**clock input**	CIRC.ENG	**Takteingang**
climbing irons (→hooks)		OUTS.PLANT	(→Steigeisen)	**clock interface**	CIRC.ENG	**Taktschnittstelle**
clinch (v.t.)		MECH	**einbördeln**	**clock mode**	CIRC.ENG	**Taktbetriebsart**
clip (v.t.)		TECH	**klammern**	**clock multiplexer**	CIRC.ENG	**Taktmultiplexer**
			= anklammern; festklammern	**clock offset** (→clock pulse offset)	ELECTRON	(→Taktversatz)
clip (n.)		TECH	**Klammer**	**clock oscillator** (→timing generator)	CIRC.ENG	(→Taktgeber)
clip (n.) (→pressure clamp)		EL.INST	(→Klemme)	**clock output**	CIRC.ENG	**Taktabgabe**
clip (→paper clip)		OFFICE	(→Büroklammer)	**clock pulse**	ELECTRON	**Schrittpuls**
clip art		DATA PROC	**Clip-Art**	= clocked pulse; synchronizing pulse		[Takt gebende Impulsfolge]
clipboard [transitional storage]		DATA PROC	**Ablagefläche** [Zwischenspeicherung]	≈ clock; timing signal [CODING]; timing pulse		= Taktpuls; Taktimpuls; Synchronimpuls
clip bolt (→hook bolt)		MECH	(→Hakenkopfschraube)			≈ Takt; Taktsignal [CODING]; Zeitsteuertakt
clip connector		COMPON	**Andruckleiste**			
			= Andruckverbinder	**clock pulse** (→timing signal)	CODING	(→Taktsignal)
clip-on ammeter (→clamp-on ammeter)		INSTR	(→Stromzange)	**clock-pulse controlled**	CIRC.ENG	**taktgesteuert**
clip-on current amplifier (→clamp-on current amplifier)		INSTR	(→Stromzangenverstärker)	= clocked		= getaktet; taktgebunden
				≠ asynchronous		≠ ungetaktet
clip-on microphone		EL.ACOUS	**Ansteckmikrofon**	↓ clock-pulse-state controlled		↓ taktzustandsgesteuert
= lavalier clip-on microphone				**clock-pulse-controlled flip-flop**	CIRC.ENG	**taktzustandsgesteuerte Kippschaltung**
clipper (→limiter circuit)		CIRC.ENG	(→Begrenzerschaltung)	= clocked flip-flop		= taktzustandsgesteuertes Flipflop
clipper circuit (→limiter circuit)		CIRC.ENG	(→Begrenzerschaltung)			↑ dynamische Kippschaltung
clipper tube		ELECTRON	**Begrenzerröhre**	**clock pulse edge**	ELECTRON	**Taktflanke**
clipping		CIRC.ENG	**Begrenzung**	**clock-pulse-edge controlled**	CIRC.ENG	**taktflankengesteuert**
= limitation				**clock pulse generator** (→timing generator)	CIRC.ENG	(→Taktgeber)
clipping [computer graphics; – of a picture sector]		DATA PROC	**Abschneiden** [Computergraphik; – eines Bildausschnittes]	**clock pulse generator central** (→central clock)	CIRC.ENG	(→Taktzentrale)
			= Abtrennen; Clipping; Clippen	**clock pulse offset**	ELECTRON	**Taktversatz**
clipping circuit (→limiter circuit)		CIRC.ENG	(→Begrenzerschaltung)	= clock offset; clock difference; clock time difference		
clobber (v.t.)		DATA PROC	**unabsichtlich überschreiben**	**clock pulse period**	ELECTRON	**Taktperiode**
[to overwrite inadvertently good data]			≈ überschreiben	**clock pulse rate** (→timing pulse rate)	CODING	(→Taktfolge)
≈ overwrite				**clock pulse signal** (→timing signal)	CODING	(→Taktsignal)
clock (n.)		ELECTRON	**Takt**	**clock-pulse-state controlled**	CIRC.ENG	**taktzustandsgesteuert**
[sequence]			[Sequenz]	↑ clocked		↑ taktgesteuert
= cadence			= Schrittakt; Synchrontakt; Kadenz	**clock pulse supply**	CIRC.ENG	**Taktversorgung**
			≈ Schrittpuls	= clock supply		≈ Taktzentrale; Taktgeber
clock		INSTR	**Uhr**	≈ timing generator		
↓ chronometer			↓ Chronometer	**clock pulse system**	DATA COMM	**Taktsystem**
clock (→bell)		TERM&PER	(→Klingel)	**clock Q factor**	CIRC.ENG	**Taktkreisgüte**
clock (→timing generator)		CIRC.ENG	(→Taktgeber)	**clock radio** (→digital clock radio)	CONS.EL	(→Radio-Digitaluhr)
clock alignment (→timing alignment)		CODING	(→Taktanpassung)	**clock rate** (→clock frequency)	ELECTRON	(→Taktfrequenz)
clock amplifier		CIRC.ENG	**Taktverstärker**	**clock rate** (→clock frequency)	DATA PROC	(→Taktfrequenz)
clock card		TERM&PER	**Stechkarte**	**clock recovery**	CODING	**Taktrückgewinnung**
clock controlled (→clock synchronized)		ELECTRON	(→taktsynchron)	= timing recovery; timing extraction		
clock cycle		ELECTRON	**Taktzyklus**	**clock signal** (→timing signal)	CODING	(→Taktsignal)
clock difference (→clock pulse offset)		ELECTRON	(→Taktversatz)	**clock source**	CIRC.ENG	**Taktquelle**
clock distributor		CIRC.ENG	**Taktverteiler**	≈ timing generator		≈ Taktgeber
clocked		ELECTRON	**getaktet**	**clock stage**	ELECTRON	**Taktstufe**
[synchronized with a clock pulse]				**clock supplied**	CIRC.ENG	**taktversorgt**
clocked (→clock-pulse controlled)		CIRC.ENG	(→taktgesteuert)	**clock supply** (→clock pulse supply)	CIRC.ENG	(→Taktversorgung)
clocked flip-flop (→clock-pulse-controlled flip-flop)		CIRC.ENG	(→taktzustandsgesteuerte Kippschaltung)			
clocked production		MANUF	**Taktfertigung**			

clock synchronized	ELECTRON	**taktsynchron**	**closed shop operation**	DATA PROC	**Closed-shop-Betrieb**
= clock controlled			[user doesn't have acess to the computer]		[Anwender hat kein Zugang zur DVA]
clock time difference (→clock pulse offset)	ELECTRON	(→Taktversatz)	= CS operation; CS mode		= CS-Betrieb
clock track	TERM&PER	**Taktspur**	≠ open shop operation		≠ Open-shop-Betrieb
↑ track		↑ Spur	**closed stop pulse** (→stop pulse)	TELEGR	(→Sperrschritt)
clock transformer	EL.INST	**Klingeltransformator**	**closed subroutine**	DATA PROC	**geschlossenes Unterprogramm**
= bell transformer			= linked subroutine		
clock trap	TERM&PER	**Taktsperre**	**closed system**	INF.TH	**abgeschlossenes System**
clockwise	TECH	**rechtsdrehend**	**closed tube process**	MICROEL	**Ampullendiffusion**
= dextrorotatory; right-hand ...		= im Uhrzeigersinn	**closed user group**	DATA COMM	**geschlossene Teilnehmerbetriebsklasse**
≠ counterclockwise		≠ linksdrehend	= CUG		
clockwork	MECH	**Federwerk**			= geschlossene Benutzergruppe
clog (v.t.)	TECH	**verstopfen**			
= obstruct		= zustopfen	**closed user network**	TELEC	**geschlossenes Teilnehmernetz**
clone (n.)	DATA PROC	**Klon** (n.m.)	[services of a public switched network, accessible to a limited group of subscribers]		[auf bestimmte Teilnehmer beschränkte, vermittelte Dienste eines öffentlichen Netzes, z.B. Btx]
[fully compatible hardware imitation; from "clone" = asexually procreated living being]		[völlig kompatible Hardware-Imitation; von „Klon" = ungeschlechtlich vermehrtes Lebewesen]			
≈ no-name		= Clonus; Clone; Hardware-Imitation; Imitation; Hardware-Plagiat; Plagiat; Nachbau	= CUN		
			≈ private network		≈ Privatnetz
			close-field EMC analyzer	INSTR	**Nahfeld-EMV-Analysator**
			close-field probe	INSTR	**Nahfeldsonde**
		≈ Noname	**close-in** (→near)	COLLOQ	(→nah)
close (v.t.)	EL.TECH	**schließen**	**close loop signal**	DATA COMM	**Schleifenschlußsignal**
[a circuit]		[Stromkreis]	**close-meshed** (→fine-meshed)	TECH	(→engmaschig)
close (v.t.)	DATA PROC	**schließen**	**closeness** (→proximity)	TECH	(→Nähe)
[a file]		[eine Datei]	**close-range action**	PHYS	**Nahewirkung**
close (→terminate)	TECH	(→beenden)	**close scanning**	ELECTRON	**Feinabtastung**
close (→make)	COMPON	(→schließen)	**close-talking microphone**	EL.ACOUS	(→Handmikrophon)
close (→near)	COLLOQ	(→nah)	(→hand microphone)		
close bracket (→right bracket)	MATH	(→Klammer zu)	**close to ground** (→close to surface)	RADIO	(→bodennah)
close coupling	PHYS	**feste Kopplung**	**close-to-ground aerial**	ANT	(→bodennahe Antenne)
closed-circuit current	ELECTRON	(→Ruhestrom)	(→close-to-ground antenna)		
(→quiescent current)			**close-to-ground antenna**	ANT	**bodennahe Antenne**
closed-circuit operation	TELEGR	(→Ruhestrombetrieb)	= close-to-ground aerial		
(→closed-circuit working)			**close-tolerance**	TECH	**engtoleriert**
closed-circuit signaling (→tone-on idle)	TRANS	(→Ruhestromverfahren)	**close to surface**	RADIO	**bodennah**
			= close to ground		= in Bodennähe
closed-circuit state	ELECTRON	(→Ruhezustand)	**closure** (→shut-down)	ECON	(→Stillegung)
(→quiescent state)			**closure** (→eye closure)	INSTR	(→Augenschließen)
closed-circuit television	TV	**nichtöffentliches Fernsehen**	**closure sealing cord**	TECH	**Dichtungsschnur**
= CCTV		↓ Betriebsfernsehen; Industriefernsehen	= sealing cord		
↓ plant TV; industrial TV			**clothoid**	MATH	**Klothoide**
closed-circuit working	TELEGR	**Ruhestrombetrieb**			= Spiralkurve
[current flows in idle condition]		≠ Arbeitsstrombetrieb	**cloud ceilometer**	RADIO NAV	**Wolkenhöhenmesser**
= closed-circuit operation			= ceilometer; ceiling height indicator		
≠ open-circuit working			**cloud chamber**	PHYS	**Nebelkammer**
closed code	SWITCH	**verdeckte Kennzahl**	= expansion chamber		
closed contact resistance	ELECTRON	**Schließungswiderstand**	**cloudless**	METEOR	**wolkenfrei**
closed file	DATA PROC	**geschlossene Datei**	**clover leaf antenna**	ANT	**Kleeblattantenne**
		= unzugängliche Datei	**clover leaf pattern**	ANT	**Kleeblattdiagramm**
closed loop	TELEC	**geschlossene Schleife**	**cluster** (v.t.)	DATA PROC	**gruppieren**
closed loop	CONTROL	**geschlossener Regelkreis**	[to group similar things]		
		= geschlossene Regelschleife	**cluster** (n.)	MATH	**Klumpen**
closed-loop control	CONTROL	**Regelung**	≈ cluster		= Häufungsstelle
[influence on a control quantity in conformance with a control deviation]		[Beinflußung einer Steuergröße durch Regelabweichung]			≈ Gruppe
			cluster	MOB.COMM	**Cluster**
= regulation			= cell group		= Zellengruppe
		≈ Streuung	**cluster**	DATA COMM	**Cluster**
closed-loop control circuit	CONTROL	**Regelschleife**	= station group		= Stationsgruppe
= control loop		= Regelkreis	**cluster**	MICROEL	**Cluster**
closed-loop mode	DATA PROC	**geschlossen prozeßgekoppelter Betrieb**	**cluster**	TERM&PER	**Cluster**
[process control]		[Prozeßsteuerung]	[group of sectors of a memory]		[Gruppierung von Speichersektoren]
= closed-loop operation		= Closed-loop-Betrieb			
closed-loop operation	DATA PROC	(→geschlossen prozeßgekoppelter Betrieb)	**cluster** (→group)	TECH	(→Gruppe)
(→closed-loop mode)			**cluster** (n.) (→clustered devices)	DATA PROC	(→Anschlußgerätegruppe)
closed-loop tracking	CIRC.ENG	**geregelter Nachlauf**	**cluster** (→group)	TECH	(→gruppieren)
closed numbering	SWITCH	**verdeckte Numerierung**	**cluster** (→grouping)	TECH	(→Gruppierung)
= linked numbering			**cluster controller**	DATA PROC	**Anschlußgruppensteuerung**
close-down (v.t.)	ECON	**stillegen**			= Cluster-controller
[a manufacturing plant]		[einen Betrieb]	**clustered devices**	DATA PROC	**Anschlußgerätegruppe**
= shut-down (v.t.)		= schließen	[jointly connected group of devices]		= Cluster
close down (n.) (→abnormal end)	DATA PROC	(→Absturz)	= cluster (n.)		
			clustering	DATA PROC	**Gruppierung**
closed routine	DATA PROC	**geschlossene Routine**	= grouping		
			clustering (→grouping)	TECH	(→Gruppierung)
			cluster station	DATA COMM	**Mehrfachstation**

clutch (n.)		MECH	Kupplung	coaxial	LINE TH	koaxial
= coupling				coaxial	MATH	koaxial
clutch (→grip 2)		TECH	(→Zugriff)	≈ concentrical		≈ konzentrisch
clutch head screw		MECH	Spezialschraube	coaxial antenna (→coaxial-fed antenna)	ANT	(→Koaxialantenne)
[with unique head requiring uncommon removal tool]			[mit Sonderform des Kopfes, nur mit Spezialwerkzeug abschraubbar]	coaxial cable	COMM.CABLE	Koaxialkabel
				= koaxiales Kabel		
clutter (n.)		RADIO LOC	Störfleck	coaxial cavity resonator	MICROW	Topfkreis
CLV method (→CLV mode)		TERM&PER	(→CLV-Verfahren)	= cavity resonator; coaxial resonator; concentric line resonator; line resonator; resonant cavity ; tuned cavity		= Koaxialresonator; Leitungsresonator; Hohlraumresonator
CLV mode		TERM&PER	CLV-Verfahren			
[Constant Linear Velocity; magnetic head running with the same linear velocity on all tracks]			[Magnetkopf läuft auf allen Spuren mit der gleichen Lineargeschwindigkeit]	coaxial circuit	MICROW	Koaxialkreis
				coaxial connector	COMPON	Koaxialstecker
= CLV method						= Koaxstecker
cm (→centimeter)		PHYS	(→Zentimeter)	coaxial-fed antenna	ANT	Koaxialantenne
Cm (→curium)		CHEM	(→Curium)	= coaxial antenna		
CMC letter		TERM&PER	CMC-Schrift	coaxial-fed log-periodic antenna (→logarithmically periodic antenna)	ANT	(→logarithmisch-periodische Antenne)
= coded magnetic character						
CMI (→computer-aided education)		DATA PROC	(→rechnergestützter Unterricht)	coaxial filter	MICROW	Koaxialfilter
				coaxial line	LINE TH	Koaxialleitung
CMI code		CODING	CMI-Code	= concentric line		= koaxiale Leitung
[coded mark inversion]				coaxial pair	COMM.CABLE	Koaxialpaar
CML (→emitter-coupled logic)		MICROEL	(→emittergekoppelte Logik)	[inner plus outer conductor]		[Innen- und Außenleiter]
CMOS memory		MICROEL	CMOS-Speicher	= coaxial tube; tube		= Koxialtube; Tube
CMOS RAM		DATA PROC	CMOS-RAM	↑ stranding element		↑ Verseilelement
CMOS transistor		MICROEL	CMOS-Transistor	coaxial relay	COMPON	Koaxialrelais
[complementary symmetry metal-oxide semiconductor transistor]			[Komplementär-Feldeffekttransistor mit Metall-Oxid-Halbleiter-Aufbau]	coaxial resonator (→coaxial cavity resonator)	MICROW	(→Topfkreis)
				coaxial switch	COMPON	Koaxialschalter
CMS (→radiocommunication tester)		INSTR	(→Funkmeßplatz)	coaxial trap	ANT	Koaxial-Trap
				coaxial tube (→coaxial pair)	COMM.CABLE	(→Koaxialpaar)
CNC (→computerized numerical control)		CONTROL	(→CNC-Steuerung)			
				coax-waveguide transition	MICROW	Koaxial-Hohlleiter-Übergang
c/o (→care of)		OFFICE	(→zu Händen)	cobalt	CHEM	Kobalt
Co (→cobalt)		CHEM	(→Kobalt)	= Co		= Cobaltum; Co
coal (n.)			Kohle	cobalt blue	PHYS	kobaltblau
coalesce (→merge)		DATA PROC	(→mischen)	co-band mode (→equal band mode)	TRANS	(→Gleichlageverfahren)
coal gas		CHEM	Leuchtgas			
coalition		MATH	Koalition	COBOL	DATA PROC	COBOL
[theory of games]			[Spieltheorie]	[Common Business-Oriented Language]		↑ problemorientierte Programmiersprache
coal power station		POWER SYS	Kohlekraftwerk			
coarse		TECH	grob	↑ high-level programming language		
≠ fine			≈ rauh	cochannel interference	RADIO	Gleichkanalstörung
			≠ fein	= common channel interference		= Gleichkanalbeeinflussung; Gleichkanalinterferenz
coarse adjustment		ELECTRON	Grobeinstellung			
= rough adjustment				cochannel interference immunity	RADIO	Gleichkanal-Störfestigkeit
coarse-grained		TECH	grobkörnig			
coarse overvoltage protection		ELECTRON	Überspannungsgrobschutz	cochannel mode (→cochannel operation)	RADIO REL	(→Gleichkanalbetrieb)
coarse protection		CIRC.ENG	Grobschutz			
coarse radiolocation		RADIO LOC	Grobortung	cochannel operation	RADIO REL	Gleichkanalbetrieb
coarse scanning		ELECTRON	Grobabtastung	= frequency reuse; cochannel transmission; dually polarized operation; dually polarized transmission; dual-pol operation; dual-pol transmission; cochannel mode		= Gleichkanalübertragung; Gleichkanalverfahren; gleichpolarer Nachbarkanalbetrieb
coarse spark gap		COMPON	Grobfunkenstrecke			
coarse thread		MECH	Grobgewinde			
coarse tuning		ELECTRON	Grobabgleich			
= main tuning			= Hauptabstimmung			≠ kreuzpolarer Nachbarkanalbetrieb
coastal festoon		OPT. COMM	Küstengirlande	≠ interleaved operation		
[submarine optical link without submarine regenerators]			[optische Seekabelverbindung ohne Unterseeregeneratoren]	↑ frequency reuse [RADIO]		↑ Frequenzwiederbenutzung [RADIO]
				cochannel operation	RADIO	Gleichkanalbetrieb
coastal line (→shoreline)		GEOPHYS	(→Küstenlinie)	[reuse of same RF in relatively short distance]		[Wiederbelegung eines RF-Kanals im relativ kurzer Entfernung]
coastal radio station		RADIO NAV	Küstenfunkstelle			
= shore-based station			= Küstenstation			
coastal region		GEOPHYS	Küstengebiet	cochannel pattern	RADIO REL	Gleichkanalraster
coat (v.t.)		TECH	beschichten	cochannel transmission (→cochannel operation)	RADIO REL	(→Gleichkanalbetrieb)
↓ clad			= überziehen			
			↓ kaschieren	cochannel transmitter	BROADC	Gleichkanalsender
coat (n.)		TECH	Mantel	cock	TECH	Hahn
= jacket; protective wrapping			= Ummantelung; Schutzhülle 2	= faucet (AM)		
≈ cladding				cock (→valve)	TECH	(→Ventil)
			≈ Umhüllung	COD (→collect on delivery)	POST	(→Nachnahme)
coat (v.t.) (→cover 1)		TECH	(→zudecken)	code	INF	Code
coated		TECH	beschichtet	[set of rules to represent information]		[Regelwerk zur Darstellung von Informationen]
↓ cladded			= überzogen			
			↓ kaschiert			= Kode (Duden)
coated (→tarnished)		METAL	(→angelaufen)	code (v.t.) (→encode)	INF	(→codieren)
coated cathode		ELECTRON	Schichtkathode	code 1 (→code number)	SWITCH	(→Kennzahl)
coating		TECH	Beschichtung	code 2 (→number)	SWITCH	(→Nummer)
≈ layer; coat (n.)			= Überzug; Belag	code (v.t.) (→cipher)	TELEC	(→verschlüsseln)
↓ cladding			≈ Schicht; Mantel	code (→cipher key)	INF.TECH	(→Schlüssel)
			↓ Kaschierung			

code area

code area [of main memory]	DATA PROC	**Codebereich** [im Hauptspeicher]
code base	CODING	**Codebasis**
codec [coder + decoder]	CIRC.ENG	**Codec** [Codierer + Decodierer]
code change (→ESCAPE)	DATA PROC	(→Codeumschaltung)
code check	CODING	**Codeprüfung**
code combination	TELEGR	**Codekombination** = Kombination
code construction = code structure	CODING	**Codeaufbau** = Codebildung; Codestruktur
code conversion = code translation; transcoding	CODING	**Codeumsetzung** = Codeübersetzung; Codewandlung; Umcodierung
code converter	CODING	**Code-Umsetzer** = Codewandler; Codeübersetzer; Umcodierer
code-dependent	TELEC	**codeabhängig**
code-division multiple access = CDMA	TELEC	**Codemultiplex-Vielfachzugriff** = CDMA
coded magnetic character (→CMC letter)	TERM&PER	(→CMC-Schrift)
coded rotary switch = thumbwheel coding switch	COMPON	**Dreh-Codierschalter** ↑ Codierschalter
coded speech (→encrypted voice)	TELEC	(→verschlüsselte Sprache)
coded speech 1 (→coded voice signal 1)	CODING	(→codiertes Sprachsignal)
coded switch (→coding switch)	COMPON	(→Codierschalter)
coded text (→ciphertext)	INF.TECH	(→verschlüsselter Text)
coded voice (→encrypted voice)	TELEC	(→verschlüsselte Sprache)
coded voice 2 (→coded voice signal 1)	CODING	(→codiertes Sprachsignal)
coded voice signal = coded voice 1; coded speech 1 ≈ encrypted voice	CODING	**codiertes Sprachsignal** = digitalisiertes Sprachsignal ≈ verschlüsselte Sprache
code element [smallest unit to represent a code] = digit	CODING	**Codeelement** [kleinste Einheit zur Darstellung eines Codes] = Code-Element
code extension	DATA COMM	**Codeerweiterung** = Code-Erweiterung
code family	CODING	**Codefamilie**
code frame	CODING	**Coderahmen**
code gain (→coding gain)	CODING	(→Codiergewinn)
code generator	CIRC.ENG	**Codegenerator**
code holes ≠ feed holes	TERM&PER	**Informationslochung** ≠ Transportlochung
code-independent ≈ code-transparent	TELEC	**codeunabhängig** ≈ codetransparent
code injector	TERM&PER	**Schlüsseleingabegerät**
code level [number of bits per character]	CODING	**Codestufe** [Anzahl der Bits pro Zeichen]
code line [line containing an instruction]	DATA PROC	**Codezeile** [enthält einen Befehl]
code list = coding 2	DATA PROC	**Codeliste**
code mutilation	CODING	**Codeverfälschung**
code network = coding matrix	CIRC.ENG	**Codiermatrix** = Verschlüsselungsmatrix
code number = code 1; identification number ≈ prefix plus code number; number ↓ trunk code; country code; regional identity code	SWITCH	**Kennzahl** [Ziffernfolge zur Kennzeichnung des Vermittlungsamtes oder des Netzes] ≈ Vorwahlnummer; Nummer ↓ Ortskennzahl; Landeskennzahl; internationale Kennzahl
code number = part number; model number; device number; specification	TECH	**Sachnummer**
code number (→keyword 1)	DATA PROC	(→Paßwort)
code pattern	CODING	**Codemuster**
code pen = data pen; reading wand; wand ≈ scanning pistol ↑ hand-held reader; document reader	TERM&PER	**Lesestift** ≈ Lesepistole ↑ Handleser; Beleglser
code position ≠ in the code table	CODING	**Codeposition** [in der Codetabelle]
code procedure [procedure translated to machine language]	DATA PROC	**Codeprozedur** [in Maschinensprache übersetzte Prozedur]
coder = encoder	CIRC.ENG	**Codierer** = Coder; Encoder; Kodierer (Duden)
coder [forms the FBAS]	TV	**Coder** [bildet das FBAS-Signal]
coder [a person]	DATA PROC	**Codierer** [Person]
coder (→encryption equipment)	TELEC	(→Verschlüsselungsgerät)
coder (→allocator)	DATA PROC	(→Zuordner)
coder (→external label)	TECH	(→Etikett)
code receiver	SWITCH	**Codeempfänger**
code redundancy	CODING	**Coderedundanz**
code representation	CODING	**Code-Darstellung**
code review (→desk checking)	DATA PROC	(→Schreibtischtest)
code skeleton = skeletal code	DATA PROC	**Programmgerippe**
code structure (→code construction)	CODING	(→Codeaufbau)
code table	CODING	**Codetabelle**
code translation (→code conversion)	CODING	(→Codeumsetzung)
code translation (→decryption)	CODING	(→Entschlüsselung)
code transparency	TELEC	**Codetransparenz**
code-transparent	TELEC	**codetransparent**
code violation (→coding law violation)	CODING	(→Coderegelverletzung)
code violation monitoring = violation monitoring	CODING	**Coderegelüberwachung** = Codefehlermessung
codeword = character signal	CODING	**Codewort**
coding 1 = encoding	INF.TECH	**Codierung** = Codieren
coding 1 [translation into a program]	DATA PROC	**Codierung** [Umsetzung in ein Computerprogramm]
coding 3 (→programming)	DATA PROC	(→Programmierung)
coding 2 (→encryption)	INF.TECH	(→Verschlüsselung)
coding 2 (→code list)	DATA PROC	(→Codeliste)
coding form = coding sheet ≈ data sheet	DATA PROC	**Programmvordruck** = Programmformular; Codierblatt; Codierformular; Codierungsformular ≈ Dateneingabeformular
coding gain = code gain	CODING	**Codiergewinn** = Codierungsgewinn; Codegewinn
coding law = encoding law	CODING	**Coderegel** = Codierungsgesetz; Codierungskennlinie
coding law violation = code violation; violation	CODING	**Coderegelverletzung** = Codeverletzung
coding matrix (→code network)	CIRC.ENG	(→Codiermatrix)
coding sheet (→coding form)	DATA PROC	(→Programmvordruck)
coding step	CODING	**Codierschritt**
coding switch = coded switch ↓ key coding switch; coded rotary switch	COMPON	**Codierschalter** = Eingabeschalter ↓ Tast-Codierschalter; Dreh-Codierschalter
coding syntax = program syntax	DATA PROC	**Programmierungssyntax**
coding theory	INF	**Codierungstheorie**
codirectional ≈ favourably oriented; co-operative ≈ parallel [MATH]; collimated ≠ reverse	TECH	**gleichsinnig** ≈ parallel [MATH]; ausgerichtet ≠ gegensinnig

English	Domain	German
coefficient	PHYS	**Beiwert**
		= Koeffizient; Beizahl
coefficient	MATH	**Koeffizient**
[a constant factor to a variable]		[konstanter Faktor einer Variablen]
↑ factor		↑ Faktor
coefficient control	CIRC.ENG	**Koeffizientensteuerung**
coefficient of absorption	ANT	**Absorptionsgrad**
coefficient of self-inductance	PHYS	**magnetische Feldkonstante**
= coefficient of self-induction; permeability of vacuum; magnetic space constant		
coefficient of self-induction	PHYS	(→magnetische Feldkonstante)
(→coefficient of self-inductance)		
coefficient of variance	MATH	**Variationskoeffizient**
		= relative Standardabweichung
coefficient potentiometer	COMPON	(→Koeffizientenpotentiometer)
(→coefficient setting potentiometer)		
coefficient setting potentiometer	COMPON	**Koeffizientenpotentiometer**
		= Funktionspotentiometer
= coefficient potentiometer		
coefficient setting voltage	CIRC.ENG	**Koeffizientenstellspannung**
CO equipment (→exchange equipment)	TELEC	(→Amtsausrüstung)
coercimeter	INSTR	**Koerzimeter**
coercitivity (IEC) (→coercive force)	PHYS	(→Koerzitivkraft)
coercive force	PHYS	**Koerzitivkraft**
[ferromagnetism]		[Ferromagnetismus, Magnetisierungskurve]
= coercitivity (IEC)		= Koerzitivfeldstärke
cofactor (→algebraic complement)	MATH	(→algebraisches Komplement)
coffin (→casket)	TECH	(→Transportbehälter)
cog (→indent)	COLLOQ	(→verzahnen)
cogwheel (→gear)	MECH	(→Zahnrad)
coherence	SCIE	**Kohärenz**
[systematic logic connection]		[systematischer logischer Zusammenhang]
≈ consistent		≈ Konsistenz
coherence	INF.TECH	**Kohärenz**
[assumption of equal values for adjacent points]		[Annahme gleichen Wertes für Nachbarpunkte]
coherence	PHYS	**Kohärenz**
[fixed phase relation]		[konstante Phasenbeziehung]
coherent	PHYS	**kohärent**
coherent detection	HF	**kohärenter Empfang**
coherent light	PHYS	**kohärentes Licht**
coherent MTI	RADIO LOC	**kohärente Festzeichenunterdrückung**
[radar]		[Radar]
coherent optical communications	OPT.COMM	**kohärente optische Übertragung**
coherent oscillator	RADIO LOC	**Kohärenzoszillator**
[radar]		[Radar]
= coho		= Coho
coherent radar	RADIO LOC	**Kohärenzradar**
coherent radiation	PHYS	**Kohärenzstrahlung**
		= kohärente Strahlung
coherent sampling (→sequential sampling)	INSTR	(→kohärente Abtastung)
coherent scanning (→sequential sampling)	INSTR	(→kohärente Abtastung)
coherent scattering	PHYS	**kohärente Streuung**
coherer	HF	**Fritter**
		= Kohärer
cohesion	PHYS	**Kohäsion**
[between molecules of the same body]		[zwischen Molekülen desselben Körpers]
		= Kohäsionskraft
coho (→coherent oscillator)	RADIO LOC	(→Kohärenzoszillator)
coil	COMPON	**Spule**
↑ inductor		↑ induktives Bauelemt
coil (→reel)	TERM&PER	(→Spule)
coil (→winding)	EL.TECH	(→Wicklung)
coil (v.t.) (→wrap-up)	TECH	(→aufwickeln)
coil aereal (→frame antenna)	ANT	(→Rahmenantenne)
coil antenna (→frame antenna)	ANT	(→Rahmenantenne)
coil cord (→retractile cord)	TERM&PER	**Spiralschnur**
coil core	COMPON	**Spulenkern**
= core		= Kern
coildriven louspeaker (→electrodynamic loudspeaker)	EL.ACOUS	(→elektrodynamischer Lautsprecher)
coiled	TECH	**gewendelt**
coiled balun	ANT	**Spulen-Balun**
= coiled-wire balun		
coiled coaxial balun	ANT	**Balun-Doppeldrossel**
coiled-up-cable choke	ANT	**Einspeisedrossel**
coiled-wire balun (→coiled balun)	ANT	(→Spulen-Balun)
coil filter	NETW.TH	**Spulenfilter**
≠ inductorless filter		≠ spulenloses Filter
coil former (→bobbin)	COMPON	(→Wickelkörper)
coil-load (→pupinize)	COMM.CABLE	(→bespulen)
coil-loaded	COMM.CABLE	**bespult**
= loaded		= pupinisiert
≠ unloaded		≠ unbespult
coil-loading	COMM.CABLE	**Bespulung**
= Pupin-coil loading; inductive loading; lumped loading		= Pupinisierung; punktförmige Bespulung
coil-minimizing	NETW.TH	**spulensparend**
coil Q	COMPON	**Spulengüte**
coil resistance	COMPON	**Wicklungswiderstand**
		[ohmscher Widerstand einer Wicklung]
coil section	OUTS.PLANT	**Spulenfeldlänge**
= loading section; coil spacing		= Spulenabstand; Spulenfeld
coil set (→filter kit)	ELECTRON	(→Filterbausatz)
coil spacing (→coil section)	OUTS.PLANT	(→Spulenfeldlänge)
coil-type capacitor	COMPON	**Wickelkondensator**
coil winder (→winding machine)	MANUF	(→Wickelmaschine)
coil winding	COMPON	**Spulenwicklung**
↑ winding		↑ Wicklung
coin (v.t.)	METAL	**münzprägen**
↑ emboss		↑ prägen
coin (n.)	ECON	**Münzgeld**
= metal money		
coin	TERM&PER	**Münze**
[token money]		[gesetzliches Zahlungsmittel]
≈ token		≈ Einwurfmünze
coin box 1	TELEPH	**Münzbehälter**
= coin collector		
coin box 2 (BRI) (→coin telephone)	TELEPH	(→Münzfernsprecher)
coinbox telephone (BRI) (→coin telephone)	TELEPH	(→Münzfernsprecher)
coin checking device	TERM&PER	**Münzprüfer**
coincide (→concur)	TECH	(→zusammentreffen)
coincidence	PHYS	**Koinzidenz**
coincidence (→AND operation)	ENG.LOG	(→UND-Verknüpfung)
coincidence circuit (→AND gate)	CIRC.ENG	(→UND-Glied)
coincidence element (→AND gate)	CIRC.ENG	(→UND-Glied)
coincidence function (→AND operation)	ENG.LOG	(→UND-Verknüpfung)
coincidence gate (→AND gate)	CIRC.ENG	(→UND-Glied)
coincidence memory	DATA PROC	**Koinzidenzspeicher**
coincidence operation (→AND operation)	ENG.LOG	(→UND-Verknüpfung)
coincidence suppression	ELECTRON	**Koinzidenzunterdrückung**
coincident (→congruent)	TECH	(→deckungsgleich)
coincident (→concurrent 2)	TECH	(→zusammenfallend)
coin collecting box (BRI) (→coin telephone)	TELEPH	(→Münzfernsprecher)
coin collector (→coin box 1)	TELEPH	(→Münzbehälter)
coin counting machine	OFFICE	**Münzzählmaschine**
coin-operated telephone (→coin telephone)	TELEPH	(→Münzfernsprecher)

coin phone (→coin telephone)	TELEPH	(→Münzfernsprecher)	**cold-work** (v.t.)	METAL	**kaltverformen**
coin slot	TERM&PER	**Münzeinwurfschlitz**	**cold work** (n.)	METAL	**Kaltverformung**
= slot		= Einwurfschlitz; Geldeinwurf; Schlitz	**colinear** (→rectilinear)	MATH	(→geradlinig)
coin sorting machine	OFFICE	**Münzsortiermaschine**	**colinear balun**	ANT	**Halbwellen-Symmetriertopf**
coin telephone (AM)	TELEPH	**Münzfernsprecher**	**colision broadening**	PHYS	**Stoßverbreiterung**
= pay phone (AM); coin phone; coinbox telephone (BRI); coin collecting box (BRI); coin box (BRI); paystation (AM); coin-operated telephone		= Münzer; Münztelefon	**collapse** (v.t.)	TECH	**kollabieren**
			= breakdown (v.t.)		= zusammenfallen; zusammenbrechen
			collapse (n.)	TECH	**Zusammenbruch**
			= breakdown 1		= Kollabierung
coin wrapping machine	OFFICE	**Münzverpackungsmaschine**	**collar**	TECH	**Kragen**
cold (→dead)	EL.TECH	(→kalt)			= Bund 1
cold boot (→cold start)	DATA PROC	(→Kaltstart)	**collate** (→merge)	DATA PROC	(→mischen)
cold cathode	ELECTRON	**Kaltkathode**	**collate** (→compare)	COLLOQ	(→vergleichen)
= dull emitter			**collating** (→merging)	DATA PROC	(→Mischen)
cold-cathode tube	ELECTRON	**Kaltkathodenröhre**	**collating** (→file collating)	DATA PROC	(→Dateienabgleich)
↓ gas-discharge tube; flash tube		↓ Gasentladungsröhre; Blitzröhre			
			collating sort (→merge sorting)	DATA PROC	(→Mischsortieren)
cold-draw	METAL	**kaltziehen**			
cold drawing	METAL	**Kaltziehen**	**collation** (→merging)	DATA PROC	(→Mischen)
cold-drawn	MECH	**kaltgereckt**	**collation** (→comparison)	COLLOQ	(→Vergleich)
= cold-strained			**collation sequence** (→sorting sequence)	DATA PROC	(→Sortierfolge)
cold fault	DATA PROC	**Einschaltfehler**			
[appears as soon as switching on]		[erscheint gleich beim Einschalten]	**collator** (→punched card collator)	TERM&PER	(→Lochkartenmischer)
cold-forge	METAL	**kaltschlagen**	**collect**	TECH	**einsammeln**
cold forging	METAL	**Fließpressen**	= gather		= vereinnahmen
= swaging; impact molding		= Kaltschlagen	**collect call** (→transferred-charge call)	TELEPH	(→R-Gespräch)
cold grey	PHYS	**blaugrau**			
cold-hardened	METAL	**kaltgehärtet**	**collecting electrode**	ELECTRON	(→Kollektor)
cold-hardening	METAL	**kalt aushärtend**	(→collector)		
cold-head (→cold-upset)	METAL	(→kaltanstauchen)	**collection** (→data acquisition)	DATA PROC	(→Datenerfassung)
cold joint (→cold soldering point)	ELECTRON	(→kalte Lötstelle)	**collective account**	ECON	**Sammelkonto**
			collective acknowledge	TELECONTR	**Sammelquittung**
cold junction	INSTR	**Kaltlötstelle**	**collective command**	TELECONTR	**Sammelbefehl**
		= Vergleichstelle	= broadcast command		
cold lamination (→cold rolling)	METAL	(→Kaltwalzen)	**collective indication**	TELECONTR	**Sammelmeldung**
			= group information		
cold moulding (→cold pressing)	METAL	(→Kaltpressen)	**collect on delivery**	POST	**Nachnahme**
			= COD		
cold-press	METAL	**kaltpressen**	**collector**	POWER ENG	**Kollektor**
cold pressing	METAL	**Kaltpressen**	= commutator		= Kommutator
= cold moulding			**collector**	MICROEL	**Kollektor**
cold-press welding	METAL	**Kaltpreßschweißung**	[collects the carrier stream injected by the emitter and controled by the basis; with PNP on negative absorbing holes, with NPN on positive absorbing electrons]		[„Sammler"; nimmt den vom Emitter eingespeisten u. von der Basis gesteuerten Ladungsträgerfluß auf; bei PNP auf Minuspol Löcher u. bei NPN auf Pluspol Elektronen aufnehmend]
cold-reduce	METAL	**kaltreduzieren**			
cold reducing	METAL	**Kaltreduzieren**			
cold-resisting	TECH	**kältebeständig**			
cold-roll	METAL	**kaltwalzen**			
cold-rolled	METAL	**kaltgewalzt**			
cold rolling	METAL	**Kaltwalzen**	↓ collector zone; collector region; collector terminal		
= cold lamination					
cold rolling mill	METAL	**Kaltwalzwerk**			= Collector
cold setting	TECH	**kaltverfestigen**			↓ Kollektorzone; Kollektorbereich; Kollektoranschluß
cold shunt (→cold welding)	METAL	(→Kaltschweißung)			
cold soldering point	ELECTRON	**kalte Lötstelle**	**collector**	ELECTRON	**Kollektor**
= cold solder joint; cold joint; dry joint; dry contact; bad soldering point; faulty soldering point			[electron tubes]		[Elektronenröhren]
			= collector electrode; collecting electrode; pick-up electrode		= Kollektorelektrode; Auffangelektrode; Auffangelektronik; Auffänger; Sammelelektrode
cold solder joint (→cold soldering point)	ELECTRON	(→kalte Lötstelle)	↑ electrode		
			↓ electron collector		↑ Elektrode
cold stand-by	QUAL	**Kaltersatz**			↓ Elektronenkollektor
≠ hot-stand-by		≠ Heißersatz	**collector barrier** (→collector depletion layer)	MICROEL	(→Kollektor-Sperrschicht)
cold start (n.)	TECH	**Kaltstart**			
cold start	DATA PROC	**Kaltstart**	**collector barrier capacitance** (→collector transition capacitance)	MICROEL	(→Kollektor-Sperrschichtkapazität)
[activate a computer from its starting point, by power-on or pressing keys]		[Starten eines Computers von Null an, durch Einschalten oder Betätigen von Tasten]			
			collector barrier temperature (→collector junction temperature)	MICROEL	(→Kollektor-Sperrschichttemperatur)
= cold boot					
≠ warm start		≠ Warmstart	**collector-base cut-off current**	MICROEL	**Kollektor-Basis-Reststrom**
cold-strained (→cold-drawn)	MECH	(→kaltgereckt)	= cutoff collector current, emitter open		= Kollektor-Reststrom
cold type	TYPOGR	**Kaltsatz**	**collector-base voltage**	MICROEL	**Kollektor-Basis-Spannung**
↓ phototypesetting; electronic typesetting		≠ Bleisatz	**collector-basis diode** (→collector diode)	MICROEL	(→Kollektordiode)
		↓ Fotosatz; elektronischer Satz			
cold-upset (v.t.)	METAL	**kaltanstauchen**	**collector breakdown voltage**	MICROEL	**Kollektor-Durchbruchspannung**
= cold-head		= kaltstauchen			
cold upsetting (n.)	METAL	**Kaltstauchen**	**collector capacitance**	MICROEL	**Kollektorkapazität**
cold-weld (v.t.)	METAL	**kaltschweißen**	= collector feedback capacitance		= Kollektor-Basis-Kapazität; Kollektor-Rückwirkungskapazität
cold welding	METAL	**Kaltschweißung**			
= cold shunt					

collector cavity	ELECTRON	**Ausgangsresonator** [Klystron]	collision	MECH	**Zusammenstoß** = Kollision	
collector coupled transistor logic (→CCTL)	MICROEL	(→CCTL)	collision (→shock)	PHYS	(→Stoß)	
collector current	MICROEL	**Kollektorstrom**	collision detection [multiple access]	TELEC	**Kollisionserkennung** [Vielfachzugriff]	
collector depletion layer = collector junction; collector barrier	MICROEL	**Kollektor-Sperrschicht** [Grenzschicht zwischen Kollektor- und Basiszone] = Kollektorübergang; Kollektor-Grenzschicht	collision detection	DATA PROC	**Kollisionserkennung**	
			collision ionization (→impact ionization)	PHYS	(→Stoßionisierung)	
collector depletion layer capacitance (→collector transition capacitance)	MICROEL	**Kollektordiffusionsisolation** (→Kollektor-Sperrschichtkapazität)	collision strategy	DATA PROC	**Kollisionsstrategie**	
			colloidal	PHYS	**kolloidal**	
collector diffusion insulation = CDI	MICROEL	**Kollektordiffusionsisolation** = CDI	colloquial = conversional language; vernacular ≠ technical language	LING	**Umgangssprache** = Allgemeinsprache; Normalsprache; Alltagssprache ≠ Fachsprache	
collector diffusion isolation process (→CDI process)	MICROEL	(→CDI-Verfahren)	co-located	TECH	**am gleichen Aufstellungsort**	
collector diode = collector-basis diode	MICROEL	**Kollektordiode** = Kollektor-Basis-Diode	cologarithm	MATH	**negativer Logarithmus**	
			colons [symbol :] ↑ punctuation mark	LING	**Doppelpunkt** [Symbol :] = Kolon ↑ Satzzeichen	
collector dissipation power	MICROEL	**Kollektorverlustleistung**				
collector dot	MICROEL	**Kollektorpille**				
collector electrode (→collector)	ELECTRON	(→Kollektor)	colons equal [symbol : =] = becomes symbol	MATH	**Ergibtsymbol** [Symbol : =] = Ergibtzeichen	
collector-emitter cut-off current = cut-off collector current; base open	MICROEL	**Kollektor-Emitter-Reststrom**	color (n.) (AM) = colour (BRI) ≈ hue; tint 2	PHYS	**Farbe** ≈ Farbton; Farbnuance	
collector-emitter voltage	MICROEL	**Kollektor-Emitter-Spannung**	colorant	TECH	**Färbemittel** = Farbstoff	
collector feedback capacitance (→collector capacitance)	MICROEL	(→Kollektorkapazität)	color artifact ↑ artifact	INF.TECH	**Farbfehler** ↑ Bildfehler	
collector inverse current	MICROEL	**Kollektorsperrstrom**	color attribute (→chroma 3)	PHYS	(→Farbeigenschaft)	
collector junction (→collector depletion layer)	MICROEL	(→Kollektor-Sperrschicht)	color bar (AM) = colour bar (BRI)	TV	**Farbbalken**	
collector junction temperature = collector barrier temperature	MICROEL	**Kollektor-Sperrschichttemperatur**	color bar signal (AM) = colour bar signal (BRI)	TV	**Farbbalkensignal**	
			color break-up (AM) = colour break-up (BRI)	PHYS	**Farbenzerlegung**	
collector modulation	ELECTRON	**Kollektormodulation**	color burst (AM) (→burst)	TV	(→Farbsynchronsignal)	
collector region [collector zone without junction]	MICROEL	**Kollektorbereich** [Kollektorzone minus Sperrschicht]	color capability (AM) = colour capability (BRI)	TERM&PER	**Farbtüchtigkeit**	
collector resistance	ELECTRON	**Kollektorwiderstand**	color change (AM) = colour change (BRI)	TERM&PER	**Farbumschlag**	
collector saturation voltage (→knee voltage)	MICROEL	(→Kniespannung)	color code (AM) = colour code (BRI); color coding (AM); colour coding (BRI)	COMPON	**Farbcode** = Kennfarbe	
collector series resistance	MICROEL	**Kollektorbahnwiderstand**				
collector terminal	MICROEL	**Kollektoranschluß**	color coder (AM) = colour coder (BRI)	TV	**Farbcoder**	
collector time constant	MICROEL	**Kollektor-Zeitkonstante**				
collector transition capacitance = collector depletion layer capacitance; collector barrier capacitance	MICROEL	**Kollektor-Sperrschichtkapazität**	color coding (AM) (→color code)	COMPON	(→Farbcode)	
			color composition (AM) = colour composition (BRI); color synthesis; colour sintesis	TV	**Farbmischung**	
collector voltage [collector to other electrode]	CIRC.ENG	**Kollektorspannung** [Kollektor gegen andere Elektrode] ↓ Kollektor-Emitter-Spannung; Kollektor-Basis-Spannung	color cone (AM) = colour cone (BRI)	PHYS	**Farbkegel**	
			color contamination (AM) = colour contamination (BRI)	TV	**Farbverschmelzung**	
collector zone [collector region plus junction] ≈ collector region	MICROEL	**Kollektorzone** [Kollektorbereich plus Sperrschicht] ≈ Kollektorbereich	color coordinate (AM) = colour coordinate (BRI)	PHYS	**Farbkoordinate**	
			color copier (AM) = colour copier (BRI)	OFFICE	**Farbkopiergerät**	
collet	TECH	**Spannzange**	color decoder (AM) = colour decoder (BRI)	TV	**Farbdecoder**	
collimate (→align)	TECH	(→ausrichten)				
collimated ≈ aligned ≈ codirectional; parallel [MATH]	TECH	**ausgerichtet** ≈ gleichsinnig; parallel [MATH]	color difference signal (AM) = colour difference signal (BRI) ≈ chrominance signal	TV	**Farbdifferenzsignal** ≈ Chrominanzsignal ↑ Primär-Farbartsignal	
collimation [optics; to adjust by fitting two lines]	PHYS	**Kollimation** [Optik; Ausrichten durch Überdecken zweier Linien]	color disabler (AM) = colour disabler (BRI); color killer	TV	**Farbabschalter**	
			color display (AM) = colour display (BRI)	TERM&PER	**Farbanzeige**	
collimation (→alignment)	ANT	(→Ausrichtung)	color distance (AM) = colour distance (BRI)	PHYS	**Farbabstand**	
collimation (→alignment)	TECH	(→Ausrichtung)				
collimator	TECH	**Kollimator**	color distortion (AM) = colour distortion (BRI)	PHYS	**Farbverzerrung**	
collinear	MATH	**kollinear**				
collinear antenna	ANT	**Kollinearantenne**	colored (AM) = coloured (BRI); chromatic	PHYS	**farbig** ≈ chromatisch	
collinear array (→collinear dipole array)	ANT	(→Dipollinie)				
collinear dipole array = collinear array	ANT	**Dipollinie** = Dipolspalte	color edge (AM) = colour edge (BRI) ≈ bleeding	TV	**Farbrand** = Farbcontour ≈ Farbsaum	
collinear mapping	PHYS	**kollineare Abbildung**				
Collins filter ↑ pi filter	NETW.TH	**Collins-Filter** ↑ Pi-Filter	color edge sharpness (AM) = colour edge sharpness (BRI)	TV	**Farbrandschärfe**	
collision	DATA PROC	**Kollision**				

color edging (AM)   TV   **Farbrandfehler**
= colour edging (BRI)
color equation (AM)   PHYS   **Farbgleichung**
= colour equation (BRI)
color fastness (AM)   PHYS   **Farbechtheit**
= colour fastness (BRI); color fidelity (AM); colour fidelety (BRI); chroma 2
≈ chromatic purity
= Farbtreue
≈ Farbenreinheit
color fidelity (AM) (→color fastness)   PHYS   (→Farbechtheit)
color fringing (AM) (→bleeding)   TV   (→Farbsaum)
colorful (adj.) (AM)   PHYS   **farbreich**
= colourful (BRI)
color graph (AM)   TERM&PER   **Farbdiagramm**
= colour graph (BRI)
color graphics (AM)   TERM&PER   **Farbgraphik**
= colour graphics (BRI)
= Farbgrafik; farbige Graphik; farbige Grafik
color graphics board (AM)   TERM&PER   **Farbgraphikkarte**
[controls coloured display]
= colour graphics board (BRI); color graphics card; colour graphics card
↑ graphics board
↓ CGA board; EGA board; VGA board
[steuert farbige Darstellung am Bildschirm]
= Farbgrafikkarte; Color-Graphik-Karte
↑ Graphikkarte
↓ CGA-Karte; EGA-Karte; VGA-Karte
color graphics card (→color graphics board)   TERM&PER   (→Farbgraphikkarte)
color hue (AM) (→hue)   PHYS   (→Farbton)
colorimeter   INSTR   **Farbmeßgerät**
colorimetrie (→color metric)   PHYS   (→Farbenmetrik)
color ink (AM)   TERM&PER   **Farbtinte**
= colour ink (BRI)
color intensity (AM)   PHYS   **Farbstärke**
= colour intensity (BRI); color saturation; colur saturation; saturation; chroma 1
≈ hue
= Farbintensität; Farbsättigung; Sättigung; Farbkontrast
≈ Farbton
color intensity control (AM)   TV   **Farbstärkeregler**
= colour intensity control (BRI); color saturation control; colour saturation control; chroma control
≈ hue control
= Farbsättigungsregler; Sättigungsregler
≈ Farbtonregler
color killer (→color disabler)   TV   (→Farbabschalter)
colorless (AM)   PHYS   **farblos**
= colourless (BRI); achromatic
= unbunt; achromatisch
color mask (AM)   TV   **Farbmaske**
= colour mask (BRI)
color matrix unit (AM)   TV   **Farbmatrixschaltung**
= colour matrix unit (BRI)
color memory (AM)   TERM&PER   **Farbspeicher**
= colour memory (BRI)
color metric (AM)   PHYS   **Farbenmetrik**
= colour metric (BRI); colorimetrie
= Kolorimetrie; Farbmessung
color mixture (AM)   PHYS   **Farbmischung**
= colour mixture (BRI)
color monitor (AM)   TERM&PER   **Farbbildschirm**
= colour monitor (BRI); color picture screen; colour picture screen; color screen; colour screen
↓ RGB monitor; composite monitor
= Farbschirm; Farbmonitor; Multichrombildschirm
↓ RGB-Bildschirm; FBAS-Bildschirm
color picture screen (→color monitor)   TERM&PER   (→Farbbildschirm)
color picture signal (AM)   TV   **Farbbildsignal**
= colour picture signal (BRI)
= FBA-Signal
color picture transmission (AM)   TV   **Farbbildübertragung**
= colour picture transmission (BRI)
color picture tube (AM)   ELECTRON   **Farbbildröhre**
= colour picture tube (BRI)
= Farbbild-Wiedergaberöhre
color pigment (→dye)   PHYS   (→Farbstoff)
color plotter (AM)   TERM&PER   **Farbplotter**
= colour plotter (BRI)
color printer (AM)   TERM&PER   **Farbdrucker**
= colour printer (BRI)
color printing (AM)   TERM&PER   **Farbendruck**
= colour printing (BRI)
= Farbdruck
color purity error (AM)   TV   **Farbverfälschung**
= colour purity error (BRI); color registration error; colour registration error

color pyrometer (AM)   PHYS   **Farbpyrometer**
= colour pyrometer (BRI)
color range (AM)   TERM&PER   (→Farbpalette)
(→gamut)
color register (AM)   DATA PROC   **Farbregister**
= colour register (BRI)
color registration error (→color purity error)   TV   (→Farbverfälschung)
color rendering (→color rendition)   PHYS   (→Farbwiedergabe)
color rendition (AM)   PHYS   **Farbwiedergabe**
= colour rendition (BRI); chromatic rendition; color rendering; colour rendering; chromatic rendering
color saturation (→color intensity)   PHYS   (→Farbstärke)
color saturation control (→color intensity control)   TV   (→Farbstärkeregler)
color scanner (AM)   TERM&PER   **Farb-Scanner**
= colour scanner (BRI)
color screen (→color monitor)   TERM&PER   (→Farbbildschirm)
color sensation (AM)   PHYS   **Farbempfindung**
= colour sensation (BRI)
color shade (NAM) (→shade)   PHYS   (→dunkelgetönte Farbe)
color shift (AM)   PHYS   **Farbverschiebung**
= colour shift (BRI)
color space (AM)   PHYS   **Farbraum**
= colour space (BRI)
color stimulus specification (AM)   TV   **Farbvalenz**
= colour stimulus specification (BRI)
color synthesis (→color composition)   TV   (→Farbmischung)
color television (AM)   TV   **Farbfernsehen**
= colour television (BRI); color TV; colour TV
color television camera (AM)   TV   **Farbfernsehkamera**
= colour television camera (BRI); color TV camera; colour TV camera
≈ video camera
≈ Videokamera
color television standard (AM)   TV   **Farbfernsehnorm**
= colour television standard (BRI); color TV standard (AM); colour TV standard (BRI)
= Farbfernsehstandard
color television technique (AM)   TV   **Farbfernsehtechnik**
= colour television technique; color TV technique; colour TV technique; color TV engineering; colour TV engineering
color temperature (AM)   PHYS   **Farbtemperatur**
= colour temperature (BRI); chromatic temperature; calorimetric temperature
= chromatische Temperatur; kalorimetrische Temperatur
color tint (NAM) (→tint 1)   PHYS   (→Helltönung)
color tint (AM) (→tint 2)   PHYS   (→Farbnuance)
color tone (→hue)   PHYS   (→Farbton)
color triangle (AM)   PHYS   **Farbdreieck**
= colour triangle (BRI); chromaticity diagram
color TV (→color television)   TV   (→Farbfernsehen)
color TV camera (→color television camera)   TV   (→Farbfernsehkamera)
color TV engineering (→color television technique)   TV   (→Farbfernsehtechnik)
color TV standard (AM) (→color television standard)   TV   (→Farbfernsehnorm)
color TV technique (→color television technique)   TV   (→Farbfernsehtechnik)
color TV tube (AM)   ELECTRON   **Farb-Fernsehbildröhre**
= colour TV tube (BRI)
↑ color picture tube; television picture tube
↑ Farbbildröhre; Fernsehbildröhre
colour (BRI) (→color)   PHYS   (→Farbe)
colour bar (BRI) (→color bar)   TV   (→Farbbalken)
colour bar signal (BRI) (→color bar signal)   TV   (→Farbbalkensignal)
colour break-up (BRI) (→color break-up)   PHYS   (→Farbenzerlegung)
colour burst (BRI) (→burst)   TV   (→Farbsynchronsignal)
colour capability (BRI) (→color capability)   TERM&PER   (→Farbtüchtigkeit)

colour change (BRI) (→color TERM&PER (→Farbumschlag) change)
colour code (BRI) (→color COMPON (→Farbcode) code)
colour coder (BRI) (→color TV (→Farbcoder) coder)
colour coding (BRI) (→color COMPON (→Farbcode) code)
colour composition (BRI) (→color TV (→Farbmischung) composition)
colour cone (BRI) (→color PHYS (→Farbkegel) cone)
colour contamination (BRI) (→color TV (→Farbverschmelzung) contamination)
colour coordinate (BRI) (→color PHYS (→Farbkoordinate) coordinate)
colour copier (BRI) (→color OFFICE (→Farbkopiergerät) copier)
colour decoder (BRI) (→color TV (→Farbdecoder) decoder)
colour difference signal (BRI) TV (→Farbdifferenzsignal) (→color difference signal)
colour disabler (BRI) (→color TV (→Farbabschalter) disabler)
colour display (BRI) (→color TERM&PER (→Farbanzeige) display)
colour distance (BRI) (→color PHYS (→Farbabstand) distance)
colour distortion (BRI) (→color PHYS (→Farbverzerrung) distortion)
coloured (→multi-coloured) TECH (→mehrfarbig)
coloured (BRI) (→colored) PHYS (→farbig)
colour edge (BRI) (→color TV (→Farbrand) edge)
colour edge sharpness (BRI) (→color TV (→Farbrandschärfe) edge sharpness)
colour edging (BRI) (→color TV (→Farbrandfehler) edging)
colour equation (BRI) (→color PHYS (→Farbgleichung) equation)
colour fastness (BRI) (→color PHYS (→Farbechtheit) fastness)
colour fidelety (BRI) (→color PHYS (→Farbechtheit) fastness)
colourful (BRI) (→colorful) PHYS (→farbreich)
colour graph (BRI) (→color TERM&PER (→Farbdiagramm) graph)
colour graphics (BRI) TERM&PER (→Farbgraphik) (→color graphics)
colour graphics board (BRI) TERM&PER (→Farbgraphikkarte) (→color graphics board)
colour graphics card (→color TERM&PER (→Farbgraphikkarte) graphics board)
colour hue (BRI) (→hue) PHYS (→Farbton)
colour ink (BRI) (→color TERM&PER (→Farbtinte) ink)
colour intensity (BRI) (→color PHYS (→Farbstärke) intensity)
colour intensity control (BRI) TV (→Farbstärkeregler) (→color intensity control)
colourless (BRI) PHYS (→farblos) (→colorless)
colour mask (BRI) (→color TV (→Farbmaske) mask)
colour matrix unit (BRI) (→color TV (→Farbmatrixschaltung) matrix unit)
colour memory (BRI) TERM&PER (→Farbspeicher) (→color memory)
colour metric (BRI) (→color PHYS (→Farbenmetrik) metric)
colour mixture (BRI) (→color PHYS (→Farbmischung) mixture)
colour monitor (BRI) (→color TERM&PER (→Farbbildschirm) monitor)
colour picture screen (→color TERM&PER (→Farbbildschirm) monitor)
colour picture signal (BRI) (→color TV (→Farbbildsignal) picture signal)
colour picture transmission (BRI) TV (→Farbbildübertragung) (→color picture transmission)

colour picture tube (BRI) ELECTRON (→Farbbildröhre) (→color picture tube)
colour plotter (BRI) (→color TERM&PER (→Farbplotter) plotter)
colour printer (BRI) (→color TERM&PER (→Farbdrucker) printer)
colour printing (BRI) (→color TERM&PER (→Farbendruck) printing)
colour purity error (BRI) (→color TV (→Farbverfälschung) purity error)
colour pyrometer (BRI) (→color PHYS (→Farbpyrometer) pyrometer)
colour range (BRI) TERM&PER (→Farbpalette) (→gamut)
colour register (BRI) (→color DATA PROC (→Farbregister) register)
colour registration error (→color TV (→Farbverfälschung) purity error)
colour rendering (→color PHYS (→Farbwiedergabe) rendition)
colour rendition (BRI) (→color PHYS (→Farbwiedergabe) rendition)
colour saturation control (→color TV (→Farbstärkeregler) intensity control)
colour scanner (BRI) (→color TERM&PER (→Farb-Scanner) scanner)
colour screen (→color TERM&PER (→Farbbildschirm) monitor)
colour sensation (BRI) (→color PHYS (→Farbempfindung) sensation)
colour shade (BRI) (→shade) PHYS (→dunkelgetönte Farbe)
colour shift (BRI) (→color PHYS (→Farbverschiebung) shift)
colour sintesis (→color TV (→Farbmischung) composition)
colour space (BRI) (→color PHYS (→Farbraum) space)
colour stimulus specification (BRI) TV (→Farbvalenz) (→color stimulus specification)
colour television (BRI) (→color TV (→Farbfernsehen) television)
colour television camera (BRI) TV (→Farbfernsehkamera) (→color television camera)
colour television standard (BRI) TV (→Farbfernsehnorm) (→color television standard)
colour television technique (→color TV (→Farbfernsehtechnik) television technique)
colour temperature (BRI) (→color PHYS (→Farbtemperatur) temperature)
colour tint (BRI) (→tint 1) PHYS (→Helltönung)
colour tint (BRI) (→tint 2) PHYS (→Farbnuance)
colour tone (→hue) PHYS (→Farbton)
colour triangle (BRI) (→color PHYS (→Farbdreieck) triangle)
colour TV (→color television) TV (→Farbfernsehen)
colour TV camera (→color television TV (→Farbfernsehkamera) camera)
colour TV camera (→color television TV (→Farbfernsehkamera) camera)
colour TV engineering (→color TV (→Farbfernsehtechnik) television technique)
colour TV standard (BRI) (→color TV (→Farbfernsehnorm) television standard)
colour TV technique (→color TV (→Farbfernsehtechnik) television technique)
colour TV tube (BRI) ELECTRON (→Farb-Fernsehbildröhre) (→color TV tube)
colour video signal (BRI) TV (→FBAS-Signal) (→composite color picture signal)
Colpitts oscillator CIRC.ENG **Colpitts-Oszillator**
↑ LC oscillator ↑ LC-Oszillator
Colpitts oscillator CIRC.ENG **kapazitive Dreipunktschal-**
circuit **tung**
Columbian (BRI) TYPOGR (→Schriftgröße 16 Punkt) (→type size 16 point)
column TYPOGR **Spalte**
≠ line = Kolonne
≠ Zeile
column MATH **Spalte**
≠ row (n.) ≠ Zeile

English	Domain	German
column address = CAS	MICROEL	Spaltenadresse
columnar	TECH	säulenförmig
columnar graph (→bar chart)	MATH	(→Balkendiagramm)
column balance (→column compensation)	TYPOGR	(→Spaltenausgleich)
column balancing (→column compensation)	TYPOGR	(→Spaltenausgleich)
column chart	MATH	Säulendiagramm
column chart (→bar chart)	MATH	(→Balkendiagramm)
column compensation = column balancing; column balance	TYPOGR	Spaltenausgleich = Spaltabgleich
column driver	ELECTRON	Spaltentreiber
column guide	TYPOGR	Spaltenlinie
column gutter [free space between columns]	TYPOGR	Spalten-Zwischenschlag [Leerraum zwischen Spalten]
column heading	TYPOGR	Spaltenüberschrift
column path = column track ≠ line path	MICROEL	Spaltenleitung ≠ Zeilenleitung
column port ≠ line port	MICROEL	Spalteneingang ≠ Zeileneingang
column speaker	EL.ACOUS	Tonsäule
column track (→column path)	MICROEL	(→Spaltenleitung)
column width	TYPOGR	Spaltenbreite
colur saturation (→color intensity)	PHYS	(→Farbstärke)
COM [computer-output microfilm]	TERM&PER	COM [Computerausgabe auf Mikrofilm]
coma (pl.-ae) [optics, comet-shaped blur]	PHYS	Koma [Optik, kometenschweifartiger Abbildungsfehler]
COMAL [Common Algorithmic Language; a development of BASIC]	DATA PROC	COMAL [Weiterentwicklung von BASIC]
comb	TECH	Kamm 1 [mit Zinken versehenes Gerät]
comb antenna	ANT	Kammantenne
comb filter	MICROW	Kammfilter = Comb-Filter
comb generator	INSTR	Kammgenerator
combination 1 [grouping of n elements out of m elements] ↓ variation; combination 2	MATH	Kombination 1 [Zumammenstellung von n Elementen aus m Elementen] ↓ Variation; Kombination 2
combination 2 [with consideration of sequence] ↑ combination 1	MATH	Kombination 2 = Kombination ohne Berücksichtigung der Anordnung ↑ Kombination 1
combination (AM) = combine (BRI)	ECON	Konzern
combinational (adj.) = combinatorial	TECH	Kombinations- = Verknüpfungs-
combinational circuit (→combinatorial circuit)	CIRC.ENG	(→Schaltnetz)
combinational logic (→combinatorial logic)	ENG.LOG	(→Kombinationslogik)
combination cable (→composite cable)	COMM.CABLE	(→gemischtpaariges Kabel)
combination frequency (→combination tone)	ACOUS	(→Kombinationston)
combination microphone	EL.ACOUS	Kombinations-Mikrophon
combination pliers	TECH	Kombinationszange = Kombizange
combination resistor	COMPON	Kombinations-Widerstand
combination tone = complex tone; combination frequency	ACOUS	Kombinationston = Kombinationsfrequenz
combinatorial (→combinational)	TECH	(→Kombinations-)
combinatorial circuit [logic circuit without storing features] = combinational circuit; switching network; logical network; switching circuit	CIRC.ENG	Schaltnetz [logische Schaltung ohne Speichervermögen] = kombinatorisches Schaltwerk; Kombinationsschaltung; Schaltnetzwerk; Schaltsystem
combinatorial explosion [too many possibilities]	DATA PROC	Kombinationsexplosion [zu viele Kombinationen]
combinatorial logic = combinational logic	ENG.LOG	Kombinationslogik
combinatorics	MATH	Kombinatorik
combine (BRI) (→combination)	ECON	(→Konzern)
combine (→link)	ENG.LOG	(→verknüpfen)
combined branch (→combined program branch)	DATA PROC	(→kombinierte Programmverzweigung)
combined communications = mixed communications	TELEC	Mischkommunikation
combined fixed/removable disk memory	TERM&PER	Fest/Wechsel-Plattenspeicher
combined head (→read-write head)	TERM&PER	(→Schreib-Lese-Kopf)
combined jump (→combined program jump)	DATA PROC	(→kombinierter Programmsprung)
combined program branch = combined branch	DATA PROC	kombinierte Programmverzweigung = kombinierte Verzweigung
combined program jump = combined jump	DATA PROC	kombinierter Programmsprung = kombinierter Sprung
combined station (→balanced station)	DATA COMM	(→Hybridstation)
combined technology	MICROEL	Kombinationstechnik
combined winding	EL.TECH	kombinierte Wicklung
combiner	RADIO REL	Kombinator
combiner diversity	RADIO REL	Kombinationsdiversity
combustible (n.) = fuel	TECH	Brennstoff = Kraftstoff; Betriebsstoff; Treibstoff
combustible (adj.)	TECH	brennbar ≠ entflammbar
COM device = computer output microfilmer	TERM&PER	COM-Anlage
come into force = take effect	ECON	inkrafttreten
comfortable	TECH	komfortabel = bequem
coming (→forthcoming)	TECH	(→kommend)
comma [symbol ,] ↑ punctuation mark	LING	Komma (n.n.; pl.-s oder -tas) [Symbol ,] = Beistrich ↑ Satzzeichen
comma fault	LING	Kommafehler
command (n.)	TELECONTR	Befehl
command (n.) [instruction of a user to start or terminate a computer action, e.g. RUN; formerly synonym with instruction] ≈ statement 1; instruction	DATA PROC	Kommando [Anweisung eines Benutzers zum Starten oder Beenden einer Computeroperation, z.B. RUN; früher synonym zu Befehl] ≈ Anweisung 1; Befehl
command 1 ("loosely") (→instruction)	DATA PROC	(→Befehl)
command block (→instruction block)	DATA PROC	(→Befehlsblock)
command cache (→instruction cache)	DATA PROC	(→Befehls-Cache-Speicher)
command catena (→instruction chain)	DATA PROC	(→Befehlskette)
command chain (→instruction chain)	DATA PROC	(→Befehlskette)
command chaining (→instruction chaining)	DATA PROC	(→Befehlsverkettung)
command character (→instruction character)	DATA PROC	(→Befehlszeichen)
command code (→operation code)	DATA PROC	(→Operationsschlüssel)
command control language = control language; command language ↓ job control language	DATA PROC	Kommandosprache = Befehlssprache; Steuersprache; Betriebssprache ↓ Auftragssprache
command counter (→instruction counter)	DATA PROC	(→Befehlszähler)
command cycle (→instruction cycle)	DATA PROC	(→Befehlszyklus)
command cycle time (→instruction cycle time)	DATA PROC	(→Befehlszykluszeit)
command decoder (→instruction decoder)	DATA PROC	(→Befehlsdecodierer)

command decoding	DATA PROC	(→Befehlsentschlüsselung)	**commencement** (→begin)	TECH	(→Beginn)
(→instruction decoding)			**commencing** (BRI) (→beginning)	TECH	(→beginnend) (adj.)
command-driven	DATA PROC	**kommandogesteuert**	**commensurable**	MATH	**kommensurabel**
≠ menu-driven		≠ menügesteuert	[divisible by common divisor]		[mit gemeinsamen Teiler]
command-driven program	DATA PROC	**kommandogesteuertes Programm**	**comment**	DATA PROC	**Kommentar**
			= commentary; annotation		
command-driven software	DATA PROC	**kommandogesteuerte Software**	**commentary** (→comment)	DATA PROC	(→Kommentar)
			commentator line	BROADC	**Kommentarleitung**
command duration	TELECONTR	**Befehlsdauer**			[Ton- oder Fernsprechleitung]
command entry	ELECTRON	**Befehlseingabe**			
= enter command; command input			**comment file**	DATA PROC	**Kommentarfeld**
command execution time	DATA PROC	(→Operationszeit)	= annotation field		
(→operation time)			**comment line**	DATA PROC	**Kommentarzeile**
command file (→instruction file)	DATA PROC	(→Befehlsdatei)	= annotation line		
			comments (n. pl.t.)	DATA PROC	**Klartextkommentar**
command file processor	DATA PROC	**Kommandodateiprozessor**			= Textkommentar
command format	DATA PROC	(→Befehlsformat)	**comment symbol**	DATA PROC	**Kommentarsymbol**
(→instruction format)			= annotation symbol		
command function	DATA PROC	**Kommandofunktion**	**commerce** (→trade)	ECON	(→Handel)
command input	DATA PROC	**Kommandoeingabe**	**commercial** (adj.)	ECON	**handelsüblich 1**
command input (→command entry)	ELECTRON	(→Befehlseingabe)	= commercially available		[im Handel vorhanden]
					= kommerziell
command interface	DATA PROC	**Kommandoschnittstelle**	**commercial** (→commecial spot)	BROADC	(→Werbeeinblendung)
command key	TERM&PER	**Kommandotaste**	**commercial agent**	ECON	(→Vertreter)
≈ control key		= Befehlstaste	(→representative)		
↑ function key		≈ Steuertaste	**commercial communications**	TELEC	**kommerzielle Fernmeldedienste**
		↑ Funktionstaste	[common carrier + private]		[öffentliche + private]
command language	DATA PROC	(→Kommandosprache)	**commercial executive**	ECON	**Kaufmann** (f. Kauffrau; pl. Kaufleute)
(→command control language)					
command length	DATA PROC	(→Befehlslänge)	**commercialize** (→sale)	ECON	(→verkaufen)
(→instruction length)			**commercialize** (→market)	ECON	(→vertreiben)
command line (→instruction line)	DATA PROC	(→Befehlszeile)	**commercial line** (→mains)	POWER ENG	(→Starkstromnetz)
command list (→instruction list)	DATA PROC	(→Befehlsliste)	**commercial-line frequency** (→mains frequency)	POWER SYS	(→Netzfrequenz)
command meaning	TELECONTR	**Befehlsinhalt**	**commercially available** (→commercial)	ECON	(→handelsüblich 1)
command mode	DATA PROC	**Kommandoebene**			
command normalization	DATA PROC	(→Befehlsnormalisierung)	**commercial operation**	TELEC	**kommerzieller Betrieb**
(→instruction normalization)			**commercial power** (→mains)	POWER ENG	(→Starkstromnetz)
command prefetch	DATA PROC	(→Befehlsvorauslesen)			
(→instruction prefetch)			**commercial product** (→good)	ECON	(→Ware)
command procedure	DATA PROC	**Kommandoprozedur**	**commercial satellite**	SAT.COMM	**Nutzsatellit**
command processing	DATA PROC	(→Befehlsverarbeitung)	**commercial supply** (→mains)	POWER ENG	(→Starkstromnetz)
(→instruction processing)					
command processing time	DATA PROC	(→Befehlsverarbeitungszeit)	**commercial telecommunications** (→telecommunication service)	TELEC	(→Fernmeldedienst)
(→instruction processing time)					
command processor	DATA PROC	(→Befehlsprozessor)	**commercial telecommunications carrier** (→telecommunications carrier)	TELEC	(→Fernmeldegesellschaft)
(→instruction processor)					
command prompt	DATA PROC	**Kommando-Aufforderungszeichen**	**commercial telephone** (→business telephone)	TELEPH	(→Geschäftsanschluß)
command register	DATA PROC	(→Befehlsregister)	**commercial traffic**	TELEC	**Geschäftsverkehr**
(→instruction register)			**commercial-use computer** (→business computer)	DATA PROC	(→Geschäftscomputer)
command reject	DATA PROC	(→Befehlsrückweisung)			
(→instruction reject)			**commission** (v.t.)	TELEC	**einschalten**
command repertoire	DATA PROC	(→Befehlsvorrat)	= line-up; turn-up; cut-over [SWITCH]		= in Betrieb nehmen; aktivieren
(→instruction set)					
command retry (→instruction retry)	DATA PROC	(→Befehlswiederholung)	**commission**	ECON	**Provision**
			commission (→committee)	ECON	(→Ausschuß)
command sequence	DATA PROC	(→Anweisungsfolge)	**commission agent**	ECON	(→Zwischenhändler)
(→statement sequence)			(→intermediary)		
command set (→instruction set)	DATA PROC	(→Befehlsvorrat)	**commissioning**	TELEC	**Einschaltung**
command signal	DATA PROC	(→Befehlszeichen)	= turn-up; start-up; cutover [SWITCH]		= Inbetriebnahme
(→instruction character)			≈ field tests; line-up (n.)		≈ Streckenabnahme; Einmessung
command signal delimiter	DATA COMM	**Befehlstrennzeichen**			
command staticizing	DATA PROC	(→Befehlsübernahme)	**commissioning engineer**	TELEC	**Einschaltingenieur**
(→instruction staticizing)			= cut-over engineer [SWITCH]; line-up engineer		= Einschalter
command string	DATA PROC	(→Befehlsfolge)			
(→instruction sequence)			**commission of inquiry**	ECON	**Untersuchungsausschuß**
command type	TELECONTR	**Befehlsart**	**commited** (→tied)	ECON	(→zweckgebunden)
command validation	DATA PROC	**Befehlsberechtigung**	**commitment** (→obligation)	ECON	(→Verpflichtung)
command wait list	DATA PROC	(→Befehlswarteliste)	**commitment** (→promise)	ECON	(→Zusage)
(→instruction wait list)			**commitment of supply**	ECON	**Lieferantgarantie**
command window	DATA PROC	**Kommandofenster**	**committee**	ECON	**Ausschuß**
command word	DATA PROC	(→Befehlswort)	= commission; board 2		[Gremium]
(→instruction word)					= Kommission; Komitee
commecial spot	BROADC	**Werbeeinblendung**	**commodity** (→raw material)	TECH	(→Rohstoff)
= commercial		= Werbespot	**commodity** (→good)	ECON	(→Ware)
commecial TV (→sponsored television)	BROADC	(→Werbefernsehen)	**common** (→general)	COLLOQ	(→gewöhnlich)

common alarm EQUIP.ENG **Sammelalarm**
= summary alarm (BRI)
commonality TECH **Gemeinsamkeit**
common antenna (→block BROADC (→Gemeinschaftsantenne)
antenna)
common band mode (→equal TRANS (→Gleichlageverfahren)
band mode)
common base (→common base CIRC.ENG (→Basisschaltung)
connection)
common base circuit CIRC.ENG (→Basisschaltung)
(→common base connection)
common base connection CIRC.ENG **Basisschaltung**
= common base circuit; common base; = Basistransistorschaltung;
grounded base circuit; inter-base cir- Basisschaltung vorwärts;
cuit; grounded-base connection; inter- Zwischen-Basisschaltung
base connection ↑ Transistorgrundschaltung
↑ transistor basic connection
common battery (→central TELEPH (→Zentralbatterie)
battery)
common-battery station TELEPH **amtsgespeister Fernsprechap-**
parat
common carrier TELEC (→Fernmeldegesellschaft)
(→telecommunications carrier)
common carrier communications TELEC (→öffentliche Telekommuni-
(→public communications) kation)
common carrier communications TELEC (→öffentliche Nachrichten-
engineering (→public communica- technik)
tions engineering)
common carrier network TELEC (→öffentliches Netz)
(→public network)
common channel interference RADIO (→Gleichkanalstörung)
(→cochannel interference)
common-channel interoffice SWITCH (→zentrale Zeichengabe)
signaling (AM) (→common-channel
signaling)
common channel mode TRANS (→Gleichkanalverfahren)
(→cochannel mode)
common-channel signaling (AM) SWITCH **zentrale Zeichengabe**
= common-channel signaling (BRI); = zentrale Signalisierung;
CCS; common-channel interoffice sig- Zentralkanal-Signalisie-
naling (AM); CCIS (AM) rung; ZZK-Signalisierung
common-channel signalling SWITCH (→zentrale Zeichengabe)
(BRI) (→common-channel signaling)
common channel transmission TRANS (→Gleichkanalverfahren)
(→cochannel mode)
common collector (→common CIRC.ENG (→Kollektorschaltung)
collector connection 1)
common collector circuit 1 CIRC.ENG (→Kollektorschaltung)
(→common collector connection 1)
common collector circuit 2 CIRC.ENG (→Kollektorschaltung vor-
(→common collector connection 2) wärts)
common collector connection 1 CIRC.ENG **Kollektorschaltung**
= common collector circuit 1; common = Emitterfolger
collector ↑ Transistorgrundschaltung
↑ transisitor basic connection
common collector connection 2 CIRC.ENG **Kollektorschaltung vorwärts**
= common collector circuit 2
common control (→centralized TELEC (→Zentralsteuerung)
control)
common denominator MATH **gemeinsamer Nenner**
= Generalnenner
common drain connection CIRC.ENG **Drainanschluß**
= Sourcefolger
common emitter CIRC.ENG (→Emitterschaltung)
(→common-emitter connection)
common emitter circuit CIRC.ENG (→Emitterschaltung)
(→common-emitter connection)
common-emitter connection CIRC.ENG **Emitterschaltung**
= common emitter circuit; common = Emitterschaltung vorwärts
emitter; grounded emitter circuit ≈ Source-Schaltung
≈ common-source connection ↑ Transistorgrundschaltung
↑ transistor basic connection
common fraction MATH **gemeiner Bruch**
common gate CIRC.ENG **Gate-Schaltung**
[FET] [FET]
common highway (→common SWITCH (→gemeinsame Abnehmer-
trunk) leitung)
common logarithm MATH **dekadischer Logarithmus**
= logarithm to the base of 10; decadic = Logarithmus zur Basis 10;
logarithm Briggscher Logarithmus;
gewöhnlicher Logarithmus;
Zehnerlogarithmus
common mode ELECTRON **Gleichtakt**
common-mode (→in-phase) PHYS (→gleichphasig)
common mode distortion CIRC.ENG **Gleichtaktverzerrung**
common mode driving CIRC.ENG **Gleichtaktsteuerung**
common mode input CIRC.ENG **Gleichtakt-Eingangswider-**
resistance **stand**
common mode interference EL.TECH **Gleichtaktstörung**
common mode rejection CIRC.ENG **Gleichtaktunterdrückung**
common-mode suppression NETW.TH (→Unsymmetriedämpfung)
(→unbalance attenuation)
common mode voltage CIRC.ENG **Gleichtaktspannung**
common mode voltage gain CIRC.ENG **Gleichtaktverstärkung**
common return EL.TECH **Betriebserde**
= signal ground ≈ Schutzerde
common return conductor EL.TECH **Betriebserder**
common signaling channel (AM) SWITCH **zentraler Zeichenkanal**
= common signalling channel (BRI) = ZZK; Zentralzeichengabe-
kanal; zentraler Signalisie-
rungskanal;
Zentralsignalisierungskanal
common signalling channel (BRI) SWITCH (→zentraler Zeichenkanal)
(→common signaling channel)
common-source connection CIRC.ENG **Source-Schaltung**
[with FET, corresponds to the com- [mit FET, entspricht de
mon-emitter connection with transi- Emitterschaltung bei Tran-
tors] sistoren]
≈ common-emitter connection ≈ Emitterschaltung
common statement DATA PROC **Speicherbereichsanweisung**
[FORTRAN] [FORTRAN]
common stock (AM) (→ordinary ECON (→Stammaktie)
share)
common storage 1 DATA PROC **gemeinsamer Speicherbereich**
common storage 2 DATA PROC **gemeinsamer Speicherinhalt**
[accessible to all programs] [für alle Programme zu-
gänglich]
common trademark ECON **Freizeichen**
[Patentwesen]
common trunk SWITCH **gemeinsame Abnehmerlei-**
= common highway **tung**
common unit (→central unit) EQUIP.ENG (→Zentraleinheit)
common volumen RADIO PROP **Streustrahlvolumen**
common wave BROADC **Gleichwelle**
[global coverage with the same fre- [Flächendeckung mit Sen-
quency] dern gleicher Frequenz]
common-wave transmitter BROADC **Gleichwellensender**
= CW transmitter
communication TELEC **Nachrichtenübermittlung**
[general term of transmission and [Überbegriff für Übertra-
switching] gung und Vermittlung]
↓ transmission; switching = Übermittlung
↓ Übertragung; Vermittlung
communication INF.TECH (→Kommunikation)
(→communications)
communication (→message) INF (→Nachricht)
communication cable COMM.CABLE **Nachrichtenkabel**
= communications cable; telecommuni- = Kommunikationskabel;
cations cable Fernmeldekabel
↓ signal and metering cable; trans- ↓ Signal- und Meßkabel;
mission cable; connecting cable Übertragungskabel; Schalt-
kabel
communication capacity (→trans- INF (→Übertragungskapazität)
mission capacity)
communication channel INF **Übertragungskanal**
= communications channel; trans- = Nachrichtenkanal; Informa-
mission channel; information transfer tionskanal
channel; information channel
communication computer SWITCH **Kommunikationsrechner**
= communications computer; communi- = Kommunikationsprozessor
cation processor
communication computer DATA COMM (→Kommunikationsrechner)
(→communications computer)
communication control 1 DATA COMM (→Datenübertragungssteue-
(→data transmission control) rung 1)
communication control 2 DATA COMM (→Datenübertragungssteue-
(→transmission control) rung 2)
communication control DATA COMM (→Übertragungssteuerzei-
character (→transmission control chen)
character)

communications software

communication controller DATA COMM (→communications controller)
communication control unit DATA PROC (→communications control unit)
communication data (→call DATA COMM data)
communication front-end DATA COMM **Kommunikationsvorrechner** computer
= communications front-end computer
communication interlocking DATA COMM **Kommunikationsverbund**
= communications interlocking
communication lines network TELEC (→Liniennetz 1)
(→outside plant network)
communication line technique TELEC (→Linientechnik)
(→outside plant technique)
communication link TELEC **Nachrichtenverbindung**
= communications link ≈ Nachrichtenweg; Anruf
≈ communication path; call [TELEPH] [TELEPH]
↑ connection ↑ Verbindung
communication mast OUTS.PLANT **Fernmeldemast**
communication media (n.pl.t.) TELEC **Kommunikationsmedien**
= media (n.pl.t.) (n.pl.t.)
 = Medien
communication network TELEC (→Kommunikationsnetz)
(→communications network)
communication node DATA COMM (→Kommunikationsknoten)
(→communications node)
communication node com- DATA COMM (→Kommunikationsknoten-
puter (→communications node computer) rechner)
communication node pro- DATA COMM (→Kommunikationsknoten-
cessor (→communications node computer) rechner)
communication path TELEC **Nachrichtenweg**
= connection path; path = Verbindungsweg
≈ communication link ≈ Nachrichtenverbindung
communication processor DATA COMM (→Kommunikationsrechner)
(→communications computer)
communication processor SWITCH (→Kommunikationsrechner)
(→communication computer)
communication program DATA PROC (→Kommunikationsprogramm)
(→communications program)
communication protocol DATA COMM (→Kommunikationsprotokoll)
(→communications protocol)
communications INF.TECH **Kommunikation**
= communication = Nachrichtenwesen; Kommunikationswesen
↓ information transfer; telecommunications; informatics; telematics; dialog ↓ Informationsaustausch;
communication; distribution com- Informationsübermittlung;
munication; request communication Telekommunikation; Informatik; Telematik;
 Verteilkommunikation;
 Dialogkommunikation;
 Aufrufkommunikation
communications buffer DATA COMM (→Datenübertragungspuffer)
(→data communications buffer)
communications cable COMM.CABLE (→Nachrichtenkabel)
(→communication cable)
communications capacity INF (→Übertragungskapazität)
(→transmission capacity)
communications channel INF (→Übertragungskanal)
(→communication channel)
communications common carrier TELEC (→Fernmeldegesellschaft)
(→telecommunications carrier)
communications computer DATA COMM **Kommunikationsrechner**
= communication computer; communications processor; communication = Kommunikationsprozessor
processor ↑ Vorrechner
↑ front-end processor
communications computer SWITCH (→Kommunikationsrechner)
(→communication computer)
communications confidentiality INF.TECH **Nachrichtengeheimnis**
≈ privacy = Fernmeldegeheimnis
 ≈ Geheimhaltung
communications connection COMPON **Fernmelde-Anschlußdose**
socket
communications controller DATA COMM **Kommunikationssteuerung**
= communication controller = Leitzentrale

communications control unit DATA PROC **Datenübertragungseinheit**
= communication control unit; = DUET
CCU
communications co-operative TELEC (→Fernmeldegenossenschaft)
(→telecommunications co-operative)
communications coverage TELEC **Fernmeldeversorgung**
= telecommunications coverage
communications coverage grade TELEC **Fernmeldeversorgungsgrad**
communications directory TELEC **Kommunikationsverzeichnis**
≈ subscriber directory ≈ Teilnehmerverzeichnis
communications electronics INF.TECH (→Telekommunikationstechnik)
(→telecommunication engineering)
communications engineering INF.TECH (→Telekommunikationstechnik)
(→telecommunication engineering)
communications equipment TELEC (→Telekommunikationsgerät)
(→telecommunications equipment)
communication server DATA COMM **Kommunikationsserver**
= communications server
communication service TELEC (→Kommunikationsdienst)
(→communications service)
communication session DATA COMM (→Sitzung)
(→session)
communications executive DATA COMM **Kommunikations-Ausführungsprogramm**
communications front-end DATA COMM (→Kommunikationsvorrechner)
computer (→communication front-end computer)
communications front-end DATA COMM **Datenübertragungsvorrechner**
processor
[relieves the host computer from [entlastet den Verarbeitungsrechner von DÜ-Aufgaben]
communication tasks]
↑ front-end processor; communications = Frontrechner;
computer Front-end-Rechner
 ≈ Vorverarbeitungsrechner
 ↑ Vorrechner; Kommunikationsrechner
communications interlocking DATA COMM (→Kommunikationsverbund)
(→communication interlocking)
communications line TELEC (→Fernmeldeleitung)
(→telecommunication circuit)
communications link TELEC (→Nachrichtenverbindung)
(→communication link)
communications network TELEC **Kommunikationsnetz**
= communication network ≈ Nachrichtennetz
≈ telecommunication network ≈ Fernmeldenetz
communications node DATA COMM **Kommunikationsknoten**
= communication node
communications node DATA COMM **Kommunikationsknotenrechner**
computer
= communication node computer; communications node processor; communication node processor
communications node pro- DATA COMM (→Kommunikationsknoten-
cessor (→communications node computer) rechner)
communication socket TELEC **Kommunikationssteckdose**
communication software DATA COMM (→Kommunikationssoftware)
(→communications software)
communications processor DATA COMM (→Kommunikationsrechner)
(→communications computer)
communications program DATA PROC **Kommunikationsprogramm**
= communication program; trans- = Übertragungsprogramm
mission program
communications protocol DATA COMM **Kommunikationsprotokoll**
= communication protocol; trans- = Übertragungsprotokoll
mission protocol
communications receiver HF **Betriebsempfänger**
communications satellite SAT.COMM **Nachrichtensatellit**
≈ space radio station ≈ Weltraumfunkstelle
↓ telecommunication satellite ↓ Fernmeldesatellit
communications server DATA COMM (→Kommunikationsserver)
(→communication server)
communications service TELEC **Kommunikationsdienst**
= communication service
communications software DATA COMM **Kommunikationssoftware**
= communication software; trans- = Übertragungssoftware
mission software

communications switching

communications switching TELEC (→Vermittlung)
(→switching)
communications terminal TERM&PER **Kommunikationsendgerät**
equipment
= communication terminal equipment
communications transmission TELEC **Nachrichtenübertragungs-**
engineering **technik**
= communications transmission tech- = Übertragungstechnik; Fern-
nique; transmission engineering; übertragungstechnik
transmission technique; trunk trans- ↑ Telekommunikationstech-
mission engineering nik
↑ telecommunication engineering ↓ Weitverkehrstechnik
↓ long-haul communications
engineering
communications transmission TELEC (→Nachrichtenübertragungs-
technique (→communications trans- technik)
mission engineering)
communication system TELEC **Kommunikationssystem**
communication terminal DATA COMM (→Datenstation)
(→terminal station)
communication terminal TERM&PER (→Kommunikationsendge-
equipment (→communications termi- rät)
nal equipment)
communication time (→call TELEC (→Verbindungsdauer)
duration)
communication tower OUTS.PLANT **Fernmeldeturm**
= telecommunication tower ↓ Richtfunkturm; Rundfunk-
↓ radio relay tower; broadcasting turm; Fernsehturm
tower; television tower
communication transmission TELEC (→Übertragung)
(→transmission)
community antenna (→block BROADC (→Gemeinschaftsantenne)
antenna)
community antenna BROADC **Gemeinschaftsantennenanla-**
installation **ge**
[covers a multi-family house] [versogt Mehrfamilienhäu-
↑ TV distribution system ser]
= GA
↑ Fernsehverteilanlage
(→Großgemeinschaftsanlage)
community antenna television BROADC
(→CATV)
commutated antenna RADIO LOC **CA-Peiler**
commutating capacitance POWER SYS (→Löschkondensator)
(→surge-absorbing capacitor)
commutating inductance POWER SYS **Kommutierungsinduktivität**
commutating number POWER SYS **Kommutierungszahl**
commutating period POWER SYS **Kommutierungszeit**
commutation POWER SYS **Kommutierung**
commutation MATH **Vertauschung**
commutation curve (→normal PHYS (→Kommutierungskurve)
magnetization curve)
commutation rule MATH **Vertauschungsregel**
commutative group MATH **kommutative Gruppe**
= Abelsche Gruppe
commutator (→collector) POWER ENG (→Kollektor)
commutator motor POWER SYS **Kollektormotor**
commutator rectifier POWER SYS **Kontaktgleichrichter**
compact (adj.) TECH **gedrängt**
= dicht; kompakt
compact TYPOGR **englaufend**
↑ font attribute = kompakt
↑ Schriftattribut
compact cassette TERM&PER **Kompaktkassette**
compact computer DATA PROC **Kompaktrechner**
≈ small computer = Kompaktcomputer
≈ Kleinrechner
compact design EQUIP.ENG **Kompaktbauweise**
= compact structure = kompakte Konstruktion;
kompakter Aufbau
compact direction finder RADIO LOC **Kompaktpeiler**
compact disk CONS.EL **Compact Disc**
= CD = CD; Kompaktplatte
compact disk player (→CD CONS.EL (→CD-Spieler)
player)
compact disk read-only TERM&PER (→CD-ROM)
memory (→CD-ROM)
compact-disk ROM TERM&PER (→CD-ROM)
(→CD-ROM)
compact handset (→dial-in TELEPH (→Kompakttelefon)
handset)

compacting TELEC **Bundelverdichtung**
= Bündelzusammenfassung
compacting algorithm DATA PROC **Kompaktierungsalgorithmus**
compaction (→condensation) SCIE (→Verdichtung)
compactness TECH **Kompaktheit**
compact structure (→compact EQUIP.ENG (→Kompaktbauweise)
design)
compact version TECH **Kompaktversion**
companded code CODING **kompandierter Code**
compander TELEC **Kompander**
[compressor + expander] [Kompressor + Expander]
compander gain TELEC **Kompandergewinn**
= Kompandierungsgewinn
companding TELEC **Kompandierung**
companding characteristic TELEC **Kompandierungskennlinie**
companion instrument INSTR **Begleitgerät**
company ECON **Gesellschaft**
[association for economic purposes] = Firma; Betrieb; Unterneh-
= firm; business; enterprise; business men 1; Wirtschaftsunter-
enterprise; society; undertaking; nehmen; Haus
house ↓ Kommanditgesellschaft;
↓ limited partnership; patnership; cor- GmbH; Aktiengesellschaft
poration
company initials (→corporate ECON (→Firmenzeichen)
logo)
company limited by shares (BRI) ECON (→Aktiengesellschaft)
(→corporation 2)
company logo (→corporate logo) ECON (→Firmenzeichen)
company logotype (→corporate ECON (→Firmenzeichen)
logo)
company report ECON **firmenkundlicher Bericht**
company's premises ECON **Firmengelände**
= corporate premises
company standard TECH **Werknorm**
= Werksnorm; Firmenstan-
dard; Fabriknorm
company with limited liability ECON **GmbH**
= Ltd.; partnership = Gesellschaft mit beschränk-
↑ company ter Haftung
↑ Gesellschaft
comparability COLLOQ **Vergleichbarkeit**
= comparableness ≈ Ähnlichkeit
≈ similarity
comparableness COLLOQ (→Vergleichbarkeit)
(→comparability)
comparation circuit CIRC.ENG (→Vergleicher)
(→comparator)
comparative (n.) LING **Komparativ**
= Höherstufe; Steigerungsstu-
fe
comparator CIRC.ENG **Vergleicher**
= comparation circuit = Vergleichsschaltung; Kom-
≈ difference amplifier parator
≈ Differnzverstärker
comparator (→relational DATA PROC (→Vergleichsoperator)
operator)
comparator check (→comparison TECH (→Vergleichstest)
check)
comparator test (→comparison TECH (→Vergleichstest)
check)
compare COLLOQ **vergleichen**
= collate
≈ juxtapose
comparing instruction DATA PROC **Vergleichsbefehl**
↑ logic instruction ↑ logischer Befehl
comparing operation DATA PROC **Vergleichsoperation**
= comparison = Vergleich
comparison COLLOQ **Vergleich**
= collation
≈ juxtaposition [SCIE] ≈ Juxtaposition [SCIE]
comparison INSTR **Vergleichsmessung**
comparison (→comparing DATA PROC (→Vergleichsoperation)
operation)
comparison bridge INSTR **Vergleichsbrücke**
comparison check TECH **Vergleichstest**
= comparator check; comparison test; = Vergleichsprüfung
comparator test; cross validation
comparison logic CIRC.ENG **Vergleichslogik**
comparison method INSTR **Vergleichsmethode**
= reference method = Vergleichsverfahren

English	Subject	German
comparison register	DATA PROC	**Vergleichsregister**
		= Vergleicherregister
comparison result	DATA PROC	**Vergleichsergebnis**
comparison test (→comparison check)	TECH	(→Vergleichstest)
comparison value (→reference value 2)	TECH	(→Vergleichswert)
compart (→computer art)	DATA PROC	(→Computerkunst)
compasses (pl.t.)	ENG.DRAW	**Zirkel**
= pair of compasses		↓ Stechzirkel; Nullzirkel
↓ dividers		
compatibility	TECH	**Kompatibilität**
= interoperability		= Verträglichkeit
≈ strength		≈ Festigkeit
compatibility service	DATA COMM	**Kompatibilitätsdienst**
compatible	TECH	**kompatibel**
		= verträglich
compatible current-sinking logic (→CCSL)	MICROEL	(→CCSL)
compatible software	DATA PROC	**kompatible Software**
compel (→force)	TECH	(→erzwingen)
compelled guidance	TECH	**Zwangsführung**
compelled signaling	SWITCH	**Zwangslaufverfahren**
= compelled system		
compelled synchronization (→despotic synchronization)	TELEC	(→Zwangssynchronisierung)
compelled system (→compelled signaling)	SWITCH	(→Zwangslaufverfahren)
compendium	LING	**Abriß**
[short description]		[knappe Darstellung]
= abstract		= Kompendium; Kurzbeschreibung
≈ summary		≈ Zusammenfassung
compendium of formulas	MATH	**Formelsammlung**
compensate (→balancing)	NETW.TH	(→Abgleich)
compensate (→equalize)	TECH	(→ausgleichen)
compensated antenna	ANT	**kompensierte Antenne**
compensated semiconductor	MICROEL	**Kompensationshalbleiter**
compensating capacitor (→equalizing capacitor)	COMM.CABLE	(→Ausgleichkondensator)
compensating charge (→float charging)	POWER SYS	(→Pufferung)
compensating circuit	CIRC.ENG	**Ausgleichsschaltung**
= corrective circuit; compensation circuit; corrector circuit; bucking circuit		= Kompensationsschaltung
compensating coil	POWER ENG	**Kompensationsspule**
compensating lead (→temperature compensating lead)	INSTR	(→Thermoausgleichsleitung)
compensating magnet	ELECTRON	**Kompensationsmagnet**
= correcting magnet		= Korrekturmagnet
compensating network (→correcting network)	CIRC.ENG	(→Ausgleichnetzwerk)
compensating recorder	INSTR	**Kompensationsschreiber**
		= Kompensograph; Y-t-Schreiber; Potentiometerschreiber
compensating turn	ELECTRON	**Ausgleichswindung**
		= Kompensationswindung
compensating voltage	EL.TECH	**Ausgleichsspannung**
compensating winding (→compensation winding)	ELECTRON	(→Ausgleichswicklung)
compensation	TECH	**Ausgleich**
= equalization		
compensation	INSTR	**Kompensation**
compensation	ECON	**Schadenersatz**
= indemnification; damages; reimbursement		= Schadenloshaltung; Kompensation; Entschädigung; Ersatz; Abfindung
compensation (→salary)	ECON	(→Gehalt)
compensation circuit (→compensating circuit)	CIRC.ENG	(→Ausgleichsschaltung)
compensation conductivity	PHYS	**Kompensationsleitfähigkeit**
compensation method	INSTR	**Kompensationsmethode**
		= Kompensationsverfahren; Ausgleichmethode; Ausgleichverfahren
compensation resistor	MICROEL	**Kompensationswiderstand**
compensation theorem	NETW.TH	**Kompensationstheorem**
		= Ausgleichtheorem
compensation thermistor	MICROEL	**Kompensationsheißleiter**
		= Ausgleichheißleiter
compensation voltage	POWER SYS	**Erhaltungsladespannung**
= float voltage		[Akkumulator]
compensation winding	ELECTRON	**Ausgleichswicklung**
= compensating winding		= Kompensationswicklung
compensation zone	MICROEL	**Kompensationszone**
compensator	INSTR	**Kompensator**
compensator (→phase shifter)	POWER ENG	(→Phasenschieber)
compensatory current	EL.TECH	**Ausgleichsstrom**
competence	ECON	**Zuständigkeit**
≈ responsibility		≈ Verantwortlichkeit
competitor	ECON	**Mitbewerber**
= rival		= Konkurrent (pl.-enz); Wettbewerber; Wettbewerbsteilnehmer
competition	ECON	**Wettbewerb**
		= Konkurrenz
competitive	ECON	**wettbewerbsfähig**
		= konkurrenzfähig
competitive component	ECON	**Wettbewerbskomponente**
competitive edge	ECON	**Wettbewerbsvorteil**
		= Wettbewerbsvorsprung
competitiveness	ECON	**Wettbewerbsfähigkeit**
		= Konkurrenzfähigkeit
compilation	DATA PROC	**Kompilierung**
[translation into machine code of the whole program before execution]		[Übersetzung in die Maschinensprache des gesamten Programms vor dessen Ausführung]
= compiling		= Kompilieren; Compilierung; Compilieren
↑ program translation		↑ Programmübersetzung
compilation error	DATA PROC	**Kompilierungsfehler**
= compiling error		= Compilierungsfehler; Compilerfehler; Compilerfehler
compilation time	DATA PROC	(→Kompilierungszeit)
(→compiling time)		
compile	DATA PROC	**kompilieren**
[to translate from high-level into machine language]		[von problemorientierter Programmiersprache in Maschinensprache übersetzen]
↑ translate		= compilieren
		↑ übersetzen
compile	LING	**zusammenstellen**
[to compose out of different sources a text or a book]		[einen Text oder ein Buch, aus verschiedenen Quellen]
		= zusammentragen
compile-and-go	DATA PROC	**kompilieren und starten**
[without operator interaction]		[ohne Bedienereingriff]
compiler	DATA PROC	**Kompilierer**
[translates high-level programming language into machine language; the whole program is translated before its execution]		[übersetzt Hochsprache in Maschinensprache; das Programm wird dabei gänzlich übersetzt bevor es ausgeführt wird]
= compiler program; compiling program; compiler routine; compiling routine		= Kompiler; Compiler; Kompilierprogramm; Compilerprogramm; kompilierendes Programm; compilierendes Programm
≠ interpreter 1		≠ Interpreter 1
↑ translator		↑ Übersetzer
compiler diagnostics	DATA PROC	**Kompilerdiagnose**
		= Compilerdiagnose
compiler language	DATA PROC	(→Kompilersprache)
(→compiler-level language)		
compiler-level language	DATA PROC	**Kompilersprache**
[any high level language which can be converted into machine code by a compiler, e.g. ALGOL, C, COBOL, FORTRAN, PASCAL]		[jede problemorientierte Programmiersprache, die per Kompilierer direkt in Maschinensprache übersetzt werden kann; z.B. ALGOL, C, COBOL, FORTRAN, PASCAL]
= compiler language		
≠ interpreter language		= Kompilierersprache; Compilersprache; Compilersprache
↑ high-level programming language		

compiler listing   DATA PROC   **Kompiliererprotokoll**
↑ translator listing   ↑ Übersetzungsprotokoll
compiler program   DATA PROC   (→Kompilierer)
(→compiler)
compiler routine   DATA PROC   (→Kompilierer)
(→compiler)
compile time (→compiling   DATA PROC   (→Kompilierungszeit)
time)
compiling   DATA PROC   (→Kompilierung)
(→compilation)
compiling error   DATA PROC   (→Kompilierungsfehler)
(→compilation error)
compiling program   DATA PROC   (→Kompilierer)
(→compiler)
compiling routine   DATA PROC   (→Kompilierer)
(→compiler)
compiling time   DATA PROC   **Kompilierungszeit**
= compilation time; compile time   = Compilierungszeit; Kompilierzeit; Compilierzeit
↑ translation time   ↑ Übersetzungszeit
complaint (→claim 2)   ECON   (→Beanstandung)
complaints board (→complaints   TELEC   (→Störungsannahmeplatz)
desk)
complaints desk   TELEC   **Störungsannahmeplatz**
= complaints board; repair desk   = Störungsannahmetisch
complement (v.t.)   MATH   **komplementieren**
= append
complement (n.)   DATA PROC   **Komplement**
[auxiliary number to represent negative values and to subtract]   [zur Darstellung von Negativwerten oder für Subtraktionen gebildete Hilfszahl]
↓ radix complement; radix-minus-one complement   = Kehrwert
   ↓ Basiskomplement; Basis-minus-Eins-Komplement
complementary   TECH   **komplementär**
complementary BCD code   CODING   **komplementärer BCD-Code**
complementary channel pair   SWITCH   **komplementäres Kanalpaar**
complementary color   PHYS   **Komplementärfarbe**
   = Ergänzungsfarbe
complementary constant current logic (→CCCL)   MICROEL   (→CCCL)
complementary Darlington pair circuit   CIRC.ENG   **Komplementär-Darlington-Schaltung**
complementary event   MATH   **entgegengesetztes Ereignis**
   = komplementäres Ereignis
complementary MOS   MICROEL   **Komplementär-MOS**
complementary operation   ENG.LOG   **Komplementäroperation**
complementary push-pull amplifier   CIRC.ENG   **Komplementär-Gegentaktverstärker**
complementary technology   MICROEL   **Komplementärtechnik**
complementary transistor   MICROEL   **Komplementärtransistor**
complementary transistor amplifier   CIRC.ENG   **Komplementärverstärker**
complementation   MATH   **Komplementierung**
= complementing   = Komplementation
complementation   DATA PROC   (→Komplementbildung)
(→complementing)
complementation representation   MATH   **Komplementärdarstellung**
complement circuit   CIRC.ENG   (→Komplementgatter)
(→complement gate)
complement gate   CIRC.ENG   **Komplementgatter**
= complement circuit   = Komplementärglied; Komplementärschaltung
complementing   DATA PROC   **Komplementbildung**
= complementation   = Komplementierung
complementing   MATH   (→Komplementierung)
(→complementation)
complement to nine   DATA PROC   **Neunerkomplement**
[of decimal numbers]   [von Dezimalzahlen]
= nine's complement   ↑ Basis-minus-Eins-Komplement
↑ radix-minus-one complement
complement to one   DATA PROC   **Einerkomplement**
[of binary numbers]   [von Dualzahlen]
= one's complement   = Einserkomplement
↑ radix-minus-one complement   ↑ B-minus-Eins-Komplement
≠ Interpretersprache
↑ problemorientierte Programmiersprache
complement to ten   DATA PROC   **Zehnerkomplement**
[of decimal numbers]   [von Dezimalzahlen]
= ten's complement   ↑ Basiskomplement
↑ radix complement
complement to two   DATA PROC   **Zweierkomplement**
= two's complement   [von Binärzahlen]
↑ radix complement   ↑ Basiskomplement
complete (adj)   COLLOQ   **vollständig**
   = komplett
complete angle   MATH   **Vollwinkel**
[unit; 360°]   [Maßeinheit; 360°]
= angle at a point
complete carry   DATA PROC   **Vollübertrag**
completed call (→effective call)   TELEPH   (→ausgeführtes Gespräch)
complete failure (→catastrophic failure)   QUAL   (→Totalausfall)
completeness   COLLOQ   **Vollständigkeit**
= completion 2   = Komplettheit
≈ completion 1   ≈ Vervollständigung
completeness error   DATA PROC   **Vollständigkeitsfehler**
complete solution   DATA PROC   **Komplettlösung**
= integral solution
complete version   DATA PROC   **Vollversion**
≠ demo version   ≠ Demo-Version
completion 1   COLLOQ   **Vervollständigung**
[act or process of completing]   = Komplettierung
≈ completion 2   ≈ Vollständigkeit
completion 2 (→completeness)   COLLOQ   (→Vollständigkeit)
complex (adj.)   MATH   **komplex**
complex (→complicated)   COLLOQ   (→kompliziert)
complex alternating current   EL.TECH   **komplexer Wechselstrom**
= vector alternating current
complex alternating voltage   EL.TECH   **komplexe Wechselspannung**
= vector alternating voltage
complex attenuation constant   NETW.TH   **komplexes Dämpfungsmaß**
[logarithm of attenuation factor; negative value of the complex transfer constant; real part = attenuation constant, imaginary part = phase angle]   [Logarithmus des Dämpfungsfaktors; Negativwert des komplexen Übertragungsmaßes; Realteil = Dämpfungsmaß, Imaginärteil = Dämpfungswinkel]
↓ complex image attenuation constant; complex effective attenuation constant   ↓ komplexes Wellendämpfungsmaß; komplexes Betriebsdämpfungsma
complex balance return loss   NETW.TH   (→komplexes Betriebsreflexionsdämpfungsmaß)
(→complex composite return loss)
complex calculus   MATH   **komplexe Rechnung**
   = symbolische Rechnung
complex cell   MICROEL   **komplexe Zelle**
complex composite return loss   NETW.TH   **komplexes Betriebsreflexionsdämpfungsmaß**
= complex balance return loss   = komplexes Fehlerdämpfungsmaß; komplexes Anpassungsdämpfungsmaß
complex-conjugated   MATH   (→konjugiert-komplex)
(→conjugate)
complex effective attenuation constant   NETW.TH   **komplexes Betriebsdämpfungsmaß**
[effective attenuation constant under operational matching conditions; negative value of complex effective transfer constant; real part = effective attenuation constant, imaginary part = effective phase angle]   [komplexes Dämpfungsmaß unter Betriebsbedingungen; Negativwert des komplexen Betriebsübertragungsmaßes; Realteil = Betriebsdämpfungsmaß, Imaginärteil = Betriebsdämpfungswinkel]
≈ complex effective transfer constant; effective attenuation constant   ≈ komplexes Betriebsübertragungsmaß; Betriebsdämpfungsmaß
↑ complex attenuation constant   ↑ komplexes Dämpfungsmaß
complex effective transfer constant   NETW.TH   **komplexes Betriebsübertragungsmaß**
[complex transfer constant under operational matching conditions; negative value of complex effective attenuation constant]   [komplexes Übertragungsmaß unter Betriebsbedingungen; Negativwert des komplexen Betriebsdämpfungsmaßes]
= composite gain constant

≈ complex effective attenuation constant
↑ complex transfer constant
↓ effective transfer constant

complex frequency PHYS
complex function MATH
complex image transfer coefficient NETW.TH (→complex image transfer constant)
complex image transfer constant NETW.TH [complex transfer constant with impedance matching on both sides; negativ value of the complex image attenuation constant]
= complex image transfer coefficient
≈ complex image attenuation constant
↑ complex transfer constant

complex instruction set computer DATA PROC [conventional computer with unrestricted set of instructions]
= CISC
≠ reduced instruction set computer
complexity TECH
= intricacy

complex mismatch factor NETW.TH (→complex return current coefficient)
complex number MATH
complex permeability PHYS
complex permittivity PHYS
complex polarization ratio RADIO PROP
complex pole NETW.TH
complex power 1 EL.TECH [complex voltage multiplied by complex current]
= phasor power
complex power 2 (→ apparent power) EL.TECH
complex power meter INSTR
complex return current coefficient NETW.TH
= complex mismatch factor
complex sentence LING [with a main and a complementary clause]
complex tone (→combination tone) ACOUS
complex transfer coefficient NETW.TH (→complex transfer constant)
complex transfer constant NETW.TH [logarithm of the transmission coefficient; the real component is the transfer constant, the imaginary one is the phase angle factor; negative value of the complex attenuation constant]
= complex transfer coefficient
≈ complex attenuation constant
↓ complex image transfer constant; complex effective transfer constant
complex variable domain MATH
complex zero point NETW.TH
complicate (v.t.) COLLOQ
≠ simplify
complicated COLLOQ
= cumbersome; complex; intricate

≈ komplexes Betriebsdämpfungsmaß
↑ komplexes Übertragungsmaß
↓ Betriebsübertragungsmaß

komplexe Frequenz
Bildfunktion
(→komplexes Wellenübertragungsmaß)

komplexes Wellenübertragungsmaß [komplexes Übertragungsmaß bei beiseitiger Wellenanpassung, Negativwert des komplexen Wellenanpassungsmaßes]
≈ komplexes Wellendämpfungsmaßes
↑ Übertragungsmaß
CISC-Computer [konventioneller Computer mit uneingeschränktem Befehlsvorrat]
≠ RISC-Computer
Komplexität
= Kompliziertheit; Verzwicktheit
(→komplexer Reflexionsfaktor)

komplexe Zahl
komplexe Permeabilität
komplexe Dielektrizitätskonstante
komplexes Polarisationsverhältnis

komplexer Pol
komplexe Wechselleistung [komplexe Wechselspannung mal komplexer Wechselstrom]
= Wechselleistung; komplexe Leistung
(→Scheinleistung)

Scheinleistungsmesser
↑ Leistungsmesser
komplexer Reflexionsfaktor
= komplexer Echofaktor

Satzgefüge [Hauptsatz mit Nebensatz]
= Satzreihe; komplexer Satz
(→Kombinationston)

(→komplexes Übertragungsmaß)
komplexes Übertragungsmaß [Logarithmus des Übertragungsfaktors; Realteil = Übertragungsmaß, Imaginärteil = Phasenmaß; Negativwert des komplexen Dämpfungsmaßes]
≈ komplexes Dämpfungsmaß
↓ komplexes Wellenübertragungsmaß; komplexes Betriebsübertragungsmaß
Bildbereich
komplexe Nullstelle
komplizieren
≠ erschweren
≠ vereinfachen
kompliziert
= umständlich; komplex; verzwickt

component COMPON
= device; circuit element [NETW.TH]
= module; chip [MICROEL]

component SCIE
≈ contribution
component (→component part) TECH
component [COMPON] NETW.TH (→circuit element)
component (→stranding element) COMM.CABLE
component analysis MATH
component assortment COMPON
component density EQUIP.ENG (→packaging density)
component encoding TV
≠ composite encoding

component lead (→ lead) COMPON
component list MANUF
≈ parts list

component measurement COMPON
component mounting hole COMPON
component part TECH
= single part; component
≈ constituent
component placement MANUF
components (→ parts and pieces) MANUF

component screening QUAL

component side ELECTRON [PCB]
= side one
≠ solder side

components manufacture MANUF
components tester (→parts-tester) INSTR
component testing QUAL
compose (v.t.) TYPOGR
≈ break
composed of ECON [delivery note]
composite TECH
= multicomponent; compound
composite attenuation NETW.TH (→effective attenuation constant)
composite attenuation constant NETW.TH (→effective transmission coefficient)
composite attenuation factor NETW.TH (→effective attenuation factor)
composite bit rate (→ transmission rate) TELEC
composite buffered fiber OPT.COMM
composite cable COMM.CABLE
= combination cable
composite circuit CIRC.ENG

composite color picture signal TV (AM,NTSC)
[color-picture signal plus all synchronizing signals]
= colour video signal (BRI); composite color signal; composite video
↑ composite video signal

composite color signal (→composite color picture signal) TV
composite current transmission coefficient NETW.TH (→effective current transmission coefficient)
composite encoding TV
≠ component encoding

Bauelement
= Bauteil; Schaltungselement [NETW.TH]
≈ Baustein; Chip [MICROEL]

Komponente
≈ Beitrag
(→Einzelteil)
(→Schaltungselement)

(→Verseilelement)

Komponentenanalyse
Bauteilesortiment
(→Packungsdichte)

Komponentencodierung
= komponentenweise Codierung
≠ geschlossene Codierung
(→Anschlußdraht)
Bauteileübersicht
= BÜ; Bauteileliste; Bauteileblatt;
≈ Stückliste
Bauelementemessung
Anschlußloch
Einzelteil
= Satzteil
= Bestandteil
Bauteilebestückung
(→Satzteile)

Bauteileselektierung
= Bauteilesortierung
Bauelementeseite [einer Leiterplatte]
= Bauteileseite; Bestückungsseite
≠ Lötseite
Teilefertigung
(→Bauelemente-Testgerät)
Bauteileprüfung
setzen
≈ umbrechen
bestehen aus [Lieferpapiere]
zusammengesetzt

(→Betriebsdämpfungsmaß)

(→Betriebsübertragungsfaktor)

(→Betriebsdämpfungsfaktor)

(→Übertragungsgeschwindigkeit)
Kompaktader
gemischtpaariges Kabel

komplexer Schaltkreis
= komplexe Schaltung
FBAS-Signal [das ausgestrahlte Gesamtsignal bei Farbfernsehen]
= Farbbildsignalgemisch; Farb-Video-Signalgemisch; Farbbild-Austastsynchronsignal
≈ BAS-Signal
↑ Videosignal
(→FBAS-Signal)

(→Betriebs-Stromübertragungsfaktor)

geschlossene Codierung
= direkte Codierung
≠ Komponentencodierung

composite gain (→effective NETW.TH
transfer constant)
composite gain constant NETW.TH
(→complex effective transfer constant)
composite gain factor NETW.TH
(→effective transmission coefficient)
composite loss (→effective NETW.TH
attenuation constant)
composite loudspeaker EL.ACOUS
= multichannel loudspeaker
composite matching loss NETW.TH
(→composite return loss)
composite monitor TERM&PER
↑ color monitor
composite parameter NETW.TH
(→effective transmission parameter)
composite phase angle NETW.TH
(→effective phase angle)
composite phase constant NETW.TH
(→effective phase angle factor)
composite picture signal (→video TV
signal)
composite reflection coefficient NETW.TH
(→composite return current coefficient)
composite return current NETW.TH
coefficient
= composite reflection coefficient; effective return current coefficient; effective reflexion coefficient
composite return loss NETW.TH
[under effective operating conditions]
= effective return loss; composite matching loss; effective matching loss
↑ active return loss
composite symbol DATA PROC
composite tape TERM&PER
composite transmission coefficient (→effective transmission coefficient) NETW.TH
composite transmission parameter (→effective transmission parameter) NETW.TH
composite video (→composite color TV
picture signal)
composite voltage transmission NETW.TH
coefficient (→effective voltage transmission coefficient)
composition CHEM
= chemical composition
≈ mixture
composition (→typesetting) TYPOGR
composition computer DATA PROC
= typesetting computer
composition formatting DATA PROC
[text processing]
composition resistor COMPON
compound CHEM
compound (→composite) TECH
compound (→compound word) LING
compound die TECH
compound filter NETW.TH
compound glass OPT.COMM
compound horn antenna ANT
compound interest ECON
compound motor EL.TECH
compound semiconductor MICROEL
[a semiconducting chemical compound, e.g. GaAs]
compound statement DATA PROC

(→Betriebsübertragungsmaß)
(→komplexes Betriebsübertragungsmaß)
(→Betriebsübertragungsfaktor)
(→Betriebsdämpfungsmaß)
Lautsprecherkombination
(→Betriebsreflexionsdämpfungsmaß)
FBAS-Bildschirm
= FBAS-Monitor
↑ Farbbildschirm
(→Betriebsübertragungsparameter)
(→Betriebsdämpfungswinkel)
(→Betriebsphasenmaß)
(→Videosignal)
(→Betriebsreflexionsfaktor)
Betriebsreflexionsfaktor
= Betriebsechofaktor
↑ Reflexionsfaktor
Betriebsreflexionsdämpfungsmaß
= Anpassungsdämpfungsmaß; Fehlerdämpfungsmaß
↑ Reflexionsdämpfung
zusammengesetztes Symbol
zusammengesetzter Lochstreifen
(→Betriebsübertragungsfaktor)
(→Betriebsübertragungsparameter)
(→FBAS-Signal)
(→Betriebs-Spannungsübertragungsfaktor)
Zusammensetzung
= chemische Zusammensetzung
≈ Gemisch
(→Satz)
Satzrechner
= Satzcomputer
Satzaufbereitung
[Textverarbeitung]
Massewiderstand
= Massefestwiderstand
Verbindung
(→zusammengesetzt)
(→zusammengesetztes Wort)
Verbundwerkzeug
zusammengesetztes Filter
Mehrkomponentenglas
Stufenhornstrahler
Zinseszins
Doppelschlußmotor
= Verbundmaschine
Verbindungshalbleiter
[eine halbleitende chemische Verbindung, z.B. GaAs]
Mehrfachanweisung
= zusammengesetzte Anweisung

compound word LING
= compound
comprehensive school (BRI) SCIE
compress (v.t.) PHYS
≈ condense
compress DATA PROC
= condense; pack
compress TYPOGR
= condense
compress (→concatenate) TECH
compressed (→condensed) TYPOGR
compressed air PHYS
= pressure air
compressibility PHYS
compression MECH
compression TELEC
compression PHYS
≠ expansion
compression (→data DATA PROC
compression)
compression (→condensation) SCIE
compressional wave PHYS
= compressive wave; pressure wave
compression load MECH
[spring]
compression radar RADIO LOC
receiver
compression spring MECH
= pressure spring
compression strength MECH
(→compressive strength)
compression stress (→compressive MECH
stress)
compressive strength MECH
= compression strength
compressive stress MECH
= compression stress
compressive wave (→compressional PHYS
wave)
compressor TELEC
≠ expander
compressor (→data DATA PROC
compressor)
compromise equalizer TELEC
compromise network TELEC
= compromize balance
compromize balance TELEC
(→compromise network)
Compton wavelength PHYS
[a constant]
compulsory (→mandatory) COLLOQ
compulsory education SCIE
compulsory hand-off MOB.COMM
compulsory justification TERM&PER
(→forced justification)
compunication INF.TECH
(→telematics)
computability DATA PROC
computable (→calculable) SCIE
computation DATA PROC
[mathematical with logical operations]
= computing
≈ calculation [MATH]; Verarbeitung
computation (→calculation) MATH
computational (adj.) DATA PROC
= computing; computer-
computational error (→calculation MATH
error)
computation effort DATA PROC

zusammengesetztes Wort
= Kompositum; Zusammensetzung
Gesamtschule
[Integration von Hauptschule, Realschule und Gymnasium]
verdichten
≈ zusammendrängen; komprimieren
≈ kondensieren
verdichten
= komprimieren; kondensieren; Redundanz reduzieren
komprimieren
= schmallegen
(→verketten)
(→schmal)
Druckluft
Kompressibilität
Flächendruck
Kompression
Kompression
= Verdichtung
≠ Expansion
(→Datenverdichtung)
(→Verdichtung)
Druckwelle 1
= Kompressionswelle
Druckbelastung
[Feder]
Kompressionsradarempfänger
Druckfeder
(→Druckfestigkeit)
(→Druckspannung)
Druckfestigkeit
Druckspannung
= Druckbeanspruchung
(→Druckwelle 1)
Kompressor
= Kompresser; Presser
≠ Dehner
(→Datenverdichter)
Kompromißentzerrer
Kompromißnachbildung
(→Kompromißnachbildung)
Compton-Wellenlänge
[Konstante]
(→vorgeschrieben)
Schulpflicht
Zwangsumschaltung
(→Zwangsausschließen)
(→Telematik)
Rechnerfreundlichkeit
= Computerfreundlichkeit
(→berechenbar)
Berechnung
[mathematisch-logische Operation]
≈ Rechnen [MATH]; processing
(→Rechnung)
Computer-
= Rechner-; Rechen-
(→Rechenfehler)
Rechenaufwand

compute	DATA PROC	**berechnen**	**computer application** DATA PROC	**Computeranwendung**
≈ calculate [MATH]; process		≈ rechnen [MATH]; verarbeiten		= Rechneranwendung
			computer architecture DATA PROC	**Rechnerarchitektur**
compute (→calculate)	MATH	(→rechnen)	**computer arithmetic** (→engineering logic) INF	(→Schaltalgebra)
compute-bound	DATA PROC	**rechnerbegrenzt**		
[limited by CPU performance]		[durch die Zentraleinheit limitiert]	**computer art** DATA PROC	**Computerkunst**
			= compart	= Rechnerkunst
compute-intensive	DATA PROC	**rechenintensiv**	**computer assistance** DATA PROC	(→Rechnerunterstützung)
≠ data-intensive		≈ arbeitsintensiv	(→computer aid)	
		≠ datenintensiv	**computer-assisted** DATA PROC	(→rechnergestützt)
computer	DATA PROC	**Computer**	(→computer-aided)	
[electronic, digital, program-controlled machine, to solve logical and mathematical problems]		[elektronische, digitale, programmgesteuerte Maschine, zur Lösung logischer und mathematischer Aufgaben]	**computer-assisted diagnosis**	**rechnergestützte Diagnose**
			= computer-aided diagnosis	= computergestützte Diagnose
≈ machine			**computer-assisted instruction** DATA PROC	(→rechnergestützter Unterricht)
≈ calculating machine		= Rechner 2; Rechenanlage 2; Maschine	(→computer-aided education)	
↑ information processing machine			**computer assisted software engineering** (→machine-aided programming) DATA PROC	(→automatische Programmierung)
↓ home computer; hobby computer; personal computer; office computer; data processing equipment; mainframe computer; universal computer; analog computer; digital computer		≈ Rechner 1		
		↑ informationsverarbeitende Maschine	**computer-augmented learning** DATA PROC (→computer-aided education)	(→rechnergestützter Unterricht)
		↓ Heimcomputer; Hobbycomputer; Personal Computer; Bürocomputer; Datenverarbeitungsanlage; Großrechner; Universalrechner; Analogrechner; Digitalrechner	**computer awareness** DATA PROC	**Computer-Verständnis**
				= Verständnis für Datenverarbeitung
			computer-based DATA PROC	(→rechnergestützt)
			(→computer-aided)	
			computer-based learning DATA PROC	(→rechnergestützter Unterricht)
computer- (→computational)	DATA PROC	(→Computer-)	(→computer-aided education)	
computer aid	DATA PROC	**Rechnerunterstützung**	**computer-based message system** (→message handling system) DATA COMM	(→Mitteilungsübermittlungssystem)
= computer assistance		= Computerunterstützung		
computer-aided	DATA PROC	**rechnergestützt**	**computer-bound** DATA PROC	(→prozessorgebunden)
= computer-assisted; computer-based; computerized		= computergestützt	(→processor-bound)	
		≈ rechnergesteuert	**computer bus** (→bus) DATA PROC	(→Bus) (pl. Busse)
≈ computer-controlled			**computer cabinet** DATA PROC	**Computergehäuse**
computer-aided coding (→machine-aided programming)	DATA PROC	(→automatische Programmierung)	= cabinet; computer case; case; computer enclosure; computer housing	
computer aided design (→CAD)	DATA PROC	(→CAD)	**computer case** (→computer cabinet) DATA PROC	(→Computergehäuse)
computer aided development (→CAD)	DATA PROC	(→CAD)	**computer cassette** TERM&PER	**Computerkassette**
			computer center DATA PROC	(→Rechenzentrum)
computer-aided diagnosis (→computer-assisted diagnosis)	DATA PROC	(→rechnergestützte Diagnose)	(→computing center)	
			computer chip MICROEL	**Rechner-Chip**
computer-aided education	DATA PROC	**rechnergestützter Unterricht**	= calculator chip	= Rechnerbaustein
= CAE 2; computer-aided instruction; computer-assisted instruction; CAI; computer-augmented learning; CAL; computer-based learning; CBL; computer-managed instruction; CMI		= computergestützter Unterricht; CAI; CAE 2; CAL; CBL; CMI	**computer code** (→machine code 1) DATA PROC	(→Maschinencode 1)
			computer communication DATA COMM	**Rechnerkommunikation**
			≈ data communication	≈ Datenübermittlung
			computer communication service (→data communication service) TELEC	(→Datenübermittlungsdienst)
computer aided engineering (→CAE 1)	DATA PROC	(→CAE 1)	**computer company** DATA PROC	**Computerfirma**
			= computer firm; computer house	
computer-aided image processing (→computer image processing)	DATA PROC	(→rechnergestützte Bildverarbeitung)	**computer conference** DATA COMM	**Computerkonferenz**
			= computer conferencing	= Rechnerkonferenz
computer-aided instruction (→computer-aided education)	DATA PROC	(→rechnergestützter Unterricht)	**computer conferencing** (→computer conference) DATA COMM	(→Computerkonferenz)
computer aided inventory and maintenance (→CAIM)	DATA PROC	(→CAIM)	**computer-controlled** DATA PROC	**rechnergesteuert**
			= computer-managed	= rechnergeführt; computergesteuert; computergeführt
computer aided manufacturing (→CAM)	DATA PROC	(→CAM)	≈ computer-aided	≈ rechnergestützt
computer aided planning	DATA PROC	**CAP 1**	**computer controlled switching system** SWITCH	**rechnergesteuertes Vermittlungssystem**
= CAP 1		= rechnergestütztes Planen; computergestütztes Planen		= rechnergesteuertes Wählsystem
computer-aided programming (→machine-aided programming)	DATA PROC	(→automatische Programmierung)	**computer crime** DATA PROC	**Computerverbrechen**
computer aided publishing (→CAP 2)	DATA PROC	(→CAP 2)	**computer criminality** DATA PROC	**Computerkriminalität**
			computer-dependent DATA PROC	(→maschinenorientiert)
computer aided quality control (→CAQ)	DATA PROC	(→CAQ)	(→computer-oriented)	
			computer-dependent language (→assembler language) DATA PROC	(→Assemblersprache)
computer-aided simulation (→simulation technique)	DATA PROC	(→Simulationstechnik)		
computer-aided software engineering (→machine-aided programming)	DATA PROC	(→automatische Programmierung)	**computer-dependent language** (→machine-oriented programming language) DATA PROC	(→maschinenorientierte Programmiersprache)
computer aided testing (→CAT)	DATA PROC	(→CAT)	**computer-dependent programming language** (→machine-oriented programming language) DATA PROC	(→maschinenorientierte Programmiersprache)
computer amateur	DATA PROC	**Computer-Amateur**	**computer design** DATA PROC	**Computerentwicklung**
= nerd				= Rechnerentwicklung
computer animation	DATA PROC	**Computeranimation**	**computer diode** (→switching diode) MICROEL	(→Schaltdiode)
= animation		= Animation		

computer drawing	DATA PROC	**Computerzeichnung**	computerized numerical control	CONTROL	**CNC-Steuerung**
computer electronics (→computer engineering)	DATA PROC	(→Computerelektronik)	= CNC		
computer enclosure (→computer cabinet)	DATA PROC	(→Computergehäuse)	computerized system	DATA PROC	**DV-Verfahren**
			computer jargon	DATA PROC	**Computer-Jargon**
computer engineer	DATA PROC	**Computeringenieur**	= computerese		
computer engineering [subject field] = computer electronics ≈ informatics	DATA PROC	**Computerelektronik** [Fachgebiet] = Rechnerelektronik; Computertechnik; Rechnertechnik ≈ Informatik	computer keyboard	TERM&PER	**Datentastatur**
			computer kit	DATA PROC	**Rechnerbausatz**
			computer language (→machine language)	DATA PROC	(→Maschinensprache)
			computer letter	DATA PROC	**Computerbrief**
			computer linguistics	LING	**Computerlinguistik**
computer error	DATA PROC	**Computerfehler**	computer listing	DATA PROC	**Rechneausdruck** = Computerausdruck
computerese (→computer jargon)	DATA PROC	(→Computer-Jargon)	computer literacy	DATA PROC	**Computerwissen** = Computerfachwissen; Rechnerfachwissen
computer expert (→computer professional)	DATA PROC	(→Computerfachmann)			
			computer literate (→computer professional)	DATA PROC	(→Computerfachmann)
computer family	DATA PROC	**Rechnerfamilie** = Computerfamilie	computer logic (→engineering logic)	INF	(→Schaltalgebra)
computer file (→data file)	DATA PROC	(→Datei)			
computer firm (→computer company)	DATA PROC	(→Computerfirma)	computer magazine = computer-oriented magazine	DATA PROC	**Computerzeitschrift** = Computermagazin
computer floor (→raised floor)	SYS.INST	(→Doppelboden)	computer mail (→electronic mail)	TELEC	(→elektronischer Briefdienst)
computer flooring (→raised floor)	SYS.INST	(→Doppelboden)	computer-managed (→computer-controlled)	DATA PROC	(→rechnergesteuert)
computer game (→video game)	DATA PROC	(→Videospiel)	computer-managed instruction (→computer-aided education)	DATA PROC	(→rechnergestützter Unterricht)
computer game program = recreational software	DATA PROC	**Spielprogramm** = Spielsoftware	computer management	DATA PROC	**Rechnerverwaltung** = Computerverwaltung
computer-generated	DATA PROC	**computererzeugt** = rechnererzeugt	computer memory = computer store	DATA PROC	**Rechnerspeicher** = Computerspeicher
computer generation	DATA PROC	**Rechnergeneration** = Computergeneration	computer music	DATA PROC	**Computermusik** = Rechnermusik
computer graphics ≈ graphical processing	DATA PROC	**Computergraphik** = Computergrafik ≈ graphische Datenverarbeitung	computer network (→multi-computer system)	DATA PROC	(→Mehrrechnersystem)
			computernik (→freak)	DATA PROC	(→Freak)
computer house (→computer company)	DATA PROC	(→Computerfirma)	computer-on-a-chip (→single-chip microcomputer)	DATA PROC	(→Ein-Chip-Mikrocomputer)
computer housing (→computer cabinet)	DATA PROC	(→Computergehäuse)	computer operation 1 = machine operation	DATA PROC	**Rechneroperation** = Computeroperation
computer illiterate ≠ computer professional	DATA PROC	**Computerlaie** = Rechnerlaie ≠ Computerfachmann	computer operation 2	DATA PROC	**Rechnerbetrieb** = Computerbetrieb
			computer operations manager = data processing manager	DATA PROC	**DV-Beauftragter** = Computerbeauftragter; DV-Verantwortlicher; Computerverantwortlicher
computer image processing = computer-aided image processing	DATA PROC	**rechnergestützte Bildverarbeitung**			
computer-independent	DATA PROC	**computerunabhängig** = rechnerunabhängig	computer operator = console operator; operator 2	DATA PROC	**Rechnerbediener** = Computerbediener; Anlagenbediener; Bediener; Operator 2; Operateur
computer-independent language	DATA PROC	**computerunabhängige Programmiersprache** = rechnerunabhängige Programmiersprache			
			computer-oriented = machine-oriented; computer-dependent; machine-dependent ≈ processor-bound	DATA PROC	**maschinenorientiert** = maschinennah; rechnerorientiert; computerorientiert ≈ prozessorgebunden
computer industry	DATA PROC	**Computerindustrie**			
computer input from microfilm (→CIM)	TERM&PER	(→Mikrofilmeingabe)			
computer instruction (→machine instruction)	DATA PROC	(→Maschinenanweisung)	computer-oriented language (→machine-oriented programming language)	DATA PROC	(→maschinenorientierte Programmiersprache)
computer-integrated	DATA PROC	**rechnerintegriert** = computerintegriert	computer-oriented magazine (→computer magazine)	DATA PROC	(→Computerzeitschrift)
computer integrated manufacturing (→CIM)	DATA PROC	(→CIM)	computer-oriented programming language (→machine-oriented programming language)	DATA PROC	(→maschinenorientierte Programmiersprache)
computer-integrated telephony	TELEC	**rechnerintegriertes Fernsprechen** = computerintegrierte Telefonie			
			computer output	DATA PROC	**Computerausgabe** = Rechnerausgabe
computer interconnection	DATA COMM	**Rechnerkopplung** = Computer-Kopplung	computer output microfilmer (→COM device)	TERM&PER	(→COM-Anlage)
computerization	DATA PROC	**Rechnerdurchdringung** = Computer-Durchdringung; Computerisierung; Umstellung auf Rechnerbetrieb	computerphobia	DATA PROC	**Computerfeindlichkeit** = Rechnerfeindlichkeit; Computerphobie
computerize		**auf Rechnerbetrieb umstellen** = computerisieren	computer power	DATA PROC	**Rechnerleistung** = Computerleistung
computerized (→computer-aided)	DATA PROC	(→rechnergestützt)	computer professional = computer expert; computer specialist; computer literate; computing professional; computing expert; computing specialist; computing literate; data processing professional; data process-	DATA PROC	**Computerfachmann** = Rechnerfachmann; DV-Fachmann; Computerexperte; Rechnerexperte; DV-Experte; Computerspezialist; Rechnerspezialist;
computerized database	DATA PROC	**Computer-Datenbank**			
computerized game (→video game)	DATA PROC	(→Videospiel)			

ing expert; data processing specialist; data processing literate			DV-Spezialist; Computerkenner; Rechnerkenner; DV-Kenner; Datenverarbeitungsfachmann; Datenverarbeitungsexperte; Datenverarbeitungsspezialist; Datenverarbeitungskenner
≈ freak			
≠ computer illiterate			
			≈ Freak
			≠ Computerlaie
computer program	DATA PROC	(→Programm)	
(→program)			
computer-readable	TERM&PER	(→maschinenlesbar)	
(→machine-readable)			
computer-readable character	TERM&PER	(→maschinenlesbares Zeichen)	
(→machine-readable character)			
computer revolution	DATA PROC	**Computerrevolution**	
computer room	SYS.INST	**Rechnerraum**	
= machine room		= Computerraum; Maschinenraum	
computer run (→program run)	DATA PROC	(→Programmlauf)	
computer science	INF.TECH	(→Informatik)	
(→informatics)			
computer scientist	INF.TECH	**Informatiker**	
computer-sensible	TERM&PER	(→maschinenlesbar)	
(→machine-readable)			
computer services company	DATA PROC	(→DV-Dienstleistungsbetrieb)	
(→computer utility)			
computer simulation	TECH	**Rechnersimulation**	
		= Computersimulation	
computer specialist	DATA PROC	(→Computerfachmann)	
(→computer professional)			
computer store	DATA PROC	**Computerladen**	
computer store (→computer memory)	DATA PROC	(→Rechnerspeicher)	
computer system (→data processing equipment)	DATA PROC	(→Datenverarbeitungsanlage)	
computer time	DATA PROC	**Maschinenzeit**	
= machine time		= Rechnerzeit	
computer time sales	DATA PROC	**Rechenzeitverkauf**	
		= Rechenzeitvertrieb	
computer training station	DATA PROC	**Computer-Lehrplatz**	
computer user	DATA PROC	**Computeranwender**	
≈ computer operator		≈ Rechneranwender	
		≈ Rechnerbetreiber	
computer utility	DATA PROC	**DV-Dienstleistungsbetrieb**	
= information utility; computer services company; service bureau; servicer		= Computer-Dienstleistungsbetrieb; Rechner-Dienstleistungsbetrieb; DV-Dienstleistungsfirma; Computer-Dienstleistungsfirma; Rechner-Dienstleistungsfirma; Datenverarbeitungsbetrieb; DV-Betrieb;	
computer virus (→virus)	DATA PROC	(→Virus)	
computer word (→word)	DATA PROC	(→Wort)	
computer word length	DATA PROC	(→Maschinenortlänge)	
(→machine word length)			
computing (→data technology)	INF.TECH	(→Datentechnik)	
computing (→computational)	DATA PROC	(→Computer-)	
computing (→computation)	DATA PROC	(→Berechnung)	
computing center	DATA PROC	**Rechenzentrum**	
= computer center; data processing center; information processing center; data center		= Datenverarbeitungs-Zentrum; DV-Zentrum; EDV-Zentrum	
computing expert (→computer professional)	DATA PROC	(→Computerfachmann)	
computing literate (→computer professional)	DATA PROC	(→Computerfachmann)	
computing mode	DATA PROC	**Rechenmodus**	
computing performance (→computing speed)	DATA PROC	(→Rechengeschwindigkeit)	
computing professional (→computer professional)	DATA PROC	(→Computerfachmann)	
computing science (BRI) (→informatics)	INF.TECH	(→Informatik)	
computing specialist (→computer professional)	DATA PROC	(→Computerfachmann)	
computing speed [unit of measure: MIPS]	DATA PROC	**Rechengeschwindigkeit** [Maßeinheit: MIPS]	
= computing performance		= Rechenleistung	
computing staff (→liveware)	DATA PROC	(→DVA-Personal)	
computing system (→data processing equipment)	DATA PROC	(→Datenverarbeitungsanlage)	
COM recorder [data on microfilm]	TERM&PER	**COM-Recorder** [Daten auf Mikrofilm]	
concatenate	TECH	**verketten**	
= link; interlink; compress		= entketten	
≠ decatenate			
concatenate (→catenate)	DATA PROC	(→verketten)	
concatenated key	DATA PROC	**verketteter Schlüssel**	
concatenated signal	CODING	**verkettetes Signal**	
concatenation	TECH	**Verkettung**	
= linkage; linking; interlinking		≠ Entkettung	
≠ decatenation			
concatenation (→chaining)	DATA PROC	(→Kettung)	
concave [hollowed inward]	MATH	**konkav** [nach innen gewölbt]
≈ hollow [TECH]		≈ hohl [TECH]	
≠ convex		≠ konvex	
concave cathode (→hollow cathode)	ELECTRON	(→Hohlkathode)	
concave mirror	PHYS	**Hohlspiegel**	
concavity	MATH	**Konkavität**	
concavo-convex (→convexo-concave)	MATH	(→konvex-konkav)	
conceal (→hide)	COLLOQ	(→verbergen)	
concealed	EL.INST	**Unterputz-**	
concentrate [to serve a larger number of users by a smaller number of facilities]	SWITCH	**konzentrieren** [eine größere Anzahl von Teinehmern mit einer kleineren Anzahl von Einrichtungen bedienen]	
≠ expand		≠ expandieren	
↓ concentrar			
concentrated load	MECH	**Punktlast**	
concentrated mass	PHYS	**Punktmasse**	
concentration	SCIE	**Konzentration**	
		= Ansammlung	
concentration [serving a larger number of users by a smaller number of facilities]	SWITCH	**Konzentration 1** [Bedienung einer größeren Anzahl von Teilnehmern durch eine geringere Anzahl von Einrichtungen]	
≠ expansion		≠ Expansion	
concentration	PHYS	**Stau**	
= accumulation; stagnation			
concentration (→data compression)	DATA PROC	(→Datenverdichtung)	
concentration cell	PHYS	**Konzentrationselement**	
concentration gradient	PHYS	**Konzentrationsgradient**	
		= Konzentrationsgefälle	
concentration stage	SWITCH	**Konzentrationsstufe**	
concentrator	SWITCH	**Konzentrator 1**	
[device to serve a number of inlets with a smaller number of outlets]		[Vorrichtung zur Bedienung einer Anzahl von Eingängen mit einer kleineren Anzahl von Ausgängen]	
↓ line concentrator; traffic concentrator		↓ Leitungskonzentrator; Verkehrskonzentrator	
concentrator panel (→jack panel)	TELEPH	(→Klinkenfeld)	
concentric	MATH	**konzentrisch**	
= concentrical		≈ koaxial	
≈ coaxial			
concentrical (→concentric)	MATH	(→konzentrisch)	
concentricity	MATH	**Konzentrizität**	
concentricity error	MECH	**Konzentrizitätsfehler**	
concentric line (→coaxial line)	LINE TH	(→Koaxialleitung)	
concentric line resonator (→coaxial cavity resonator)	MICROW	(→Topfkreis)	
conceptual	SCIE	**konzeptionell**	
concertina fold (n.) (→fan folding)	TERM&PER	(→Leporellofalzung)	

concession		ECON	**Konzession**	**condensation** (→data		DATA PROC	(→Datenverdichtung)
concession		COLLOQ	**Zugeständnis**	compression)			
concessionaire		ECON	**Konzessionär**	**condensation temperature**		PHYS	**Kondensationstemperatur**
= concessioner				**condense** (→abridge)		LING	(→kürzen)
concessionaire company		ECON	**Konzessionsgesellschaft**	**condense** (→compress)		DATA PROC	(→verdichten)
concessioner		ECON	(→Konzessionär)	**condense** (→compress)		TYPOGR	(→komprimieren)
(→concessionaire)				**condensed**		TYPOGR	**schmal**
concessive		LING	**konzessiv**	= compressed			= schlank; schmalgelegt; eng
			= einräumend	**condensed print** (→condensed		TYPOGR	(→Schmalschrift)
concessive sentence		LING	**Konzessivsatz**	type)			
↑ causal sentence			= Einräumungssatz	**condensed style** (→condensed		TYPOGR	(→Schmalschrift)
			↑ Kausalsatz	type)			
conclusion		LING	**Abschluß**	**condensed type**		TYPOGR	**Schmalschrift**
[final remarks of a text]			[Schlußbemerkungen eines	= condensed print; condensed style			= Kompreßschrift; Kompreß-
			Texts]				druck; Engdruck
conclusion		SCIE	**Schlußfolgerung**	**condensed water**		TECH	**Kondenswasser**
= inference; reasoning			= Rückschluß; Schluß; Kon-				= Schwitzwasser
↓ wrong inference; fallacy			klusion; Folgerung; Fazit	**condenser**		PHYS	**Kondensor**
			↓ Fehlschluß; Trugschluß	[optics]			[Optik]
conclusive (→final)		TECH	(→abschließend)	**condenser** (BRI)		COMPON	(→Kondensator)
concordance		SCIE	**Konkordanz**	(→capacitor)			
			= Übereinstimmung	**condenser** (→phase shifter)		POWER ENG	(→Phasenschieber)
concordance		DATA PROC	**Konkordanz**	**condenser antenna**		ANT	**Kondensatorantenne**
concordance (→accordance)		COLLOQ	(→Übereinstimmung)	**condenser loudspeaker**		EL.ACOUS	(→Kondensator-Lautspre-
concrete		CIV.ENG	**Beton**	(→capacitor loudspeaker)			cher)
≈ cement			≈ Zement	**condenser microphone**		EL.ACOUS	(→Kondensatormikrophon)
concrete pedestal		TECH	**Betonsockel**	(→capacitor microphone)			
concrete pole base		OUTS.PLANT	**Mastfuß**	**condition** (v.t.)		TECH	**aufbereiten**
[to support wooden pole]			[Betonzylinder zum Befe-	[to put into condition]			[in den Zustand versetzen]
			stigen eines Holzmastes]	**condition 1** (n.)		DATA PROC	**Bedingung**
concrete word		LING	**Konkretum**	**condition 2** (→status)		DATA PROC	(→Zustand)
≠ abstract word			= Gegenstandswort	**condition** (→prerequisite)		COLLOQ	(→Voraussetzung)
			≠ Abstraktum	**condition** (→signal		DATA COMM	(→Zeichenlage)
concur		TECH	**zusammentreffen**	condition)			
= coincide; meet				**condition** (→state)		PHYS	(→Zustand)
concurrent 2 (adj.)		TECH	**zusammenfallend**	**condition** (→state)		TECH	(→Zustand)
[in time]			[zeitlich]	**condition A** (→signal		DATA COMM	(→Startpolarität)
= coincident			= zusammentreffend; neben-	condition A)			
≈ simultaneous			läufig	**conditional**		COLLOQ	**bedingt**
			≈ gleichzeitig	**conditional**		LING	**Konditional**
concurrent 1 (→converging)		COLLOQ	(→konvergierend)	≈ subjunctive			[z.B. ich würde hören,
concurrent data processing		DATA PROC	(→verzahnte Verarbeitung)				wenn]
(→concurrent processing)							= Konditionalis; Bedingungs-
concurrent mode		DATA PROC	(→Mehrprogrammbetrieb)				form
(→multiprogramming)							≈ Konkunktiv
concurrent mode (→concurrent		DATA PROC	(→verzahnte Verarbeitung)	**conditional branch**		DATA PROC	(→bedingte Programmver-
processing)				(→conditional program branch)			zweigung)
concurrent operation		DATA PROC	(→verzahnte Verarbeitung)	**conditional branching**		DATA PROC	(→bedingte Programmver-
(→concurrent processing)				(→conditional program branch)			zweigung)
concurrent processing		DATA PROC	**verzahnte Verarbeitung**	**conditional branch instruction**		DATA PROC	(→bedingter Sprungbefehl)
[several tasks are performed in inter-			[mehrere Aufgaben wer-	(→conditional jump instruction)			
leaved intervals]			den in abwechselnden Zeit-	**conditional breakpoint**		DATA PROC	**bedingter Unterbrechungs-**
= concurrent mode; concurrent data			abschnitten verzahnt				**punkt**
processing; concurrent operation;			bearbeitet]	**conditional execution**		DATA PROC	**bedingte Befehlsausführung**
multiplex mode			= zeitlich verschachtelte Ver-	**conditional expression**		DATA PROC	**bedingter Ausdruck**
≠ simultaneous processing			arbeitung; Multiplexbetrieb	**conditional flip-flop**		CIRC.ENG	**Bedingungskippstufe**
↑ multiprogramming			≠ Simultanverarbeitung	**conditional implication**		INF	**Subjunktion**
			↑ Mehrprogrammbetrieb	**conditional information content**		INF	**Verbundinformationsentropie**
concurrent processing		DATA PROC	(→Mehrprogrammbetrieb)	**conditional information entropy**		INF	**bedingte Informationsentro-**
(→multiprogramming)							**pie**
concurrent program		DATA PROC	**simultaner Programmablauf**	**conditional instruction**		DATA PROC	**bedingter Befehl**
execution				**conditional jump**		DATA PROC	(→bedingter Programm-
= concurrent programming				(→conditional program jump)			sprung)
concurrent programming		DATA PROC	(→simultaner Programmab-	**conditional jump instruction**		DATA PROC	**bedingter Sprungbefehl**
(→concurrent program execution)			lauf)	= conditional branch instruction			= bedingter Verzweigungsbe-
concuss (v.t.)		TECH	**erschüttern**				fehl; Abfragebefehl
= jar			≈ schütteln	**conditional loop**		DATA PROC	**bedingte Schleife**
≈ shake (v.t.)				**conditional paging**		DATA PROC	**bedingter Seitenwechsel**
concussion		TECH	**Erschütterung**	**conditional probability**		MATH	**bedingte Wahrscheinlichkeit**
≈ shock [PHYS]; shake			≈ Stoß [PHYS]; Schütteln	**conditional program branch**		DATA PROC	**bedingte Programmverzwei-**
condensance		NETW.TH	(→Kapazitanz)	= conditional branch; conditional			**gung**
(→capacitance)				branching			= bedingte Verzweigung
condensate (n.)		PHYS	**Kondensat**	**conditional program jump**		DATA PROC	**bedingter Programmsprung**
condensation		PHYS	**Kondensation**	= conditional jump; conditional transfer			= bedingter Sprung
			= Kondensierung	**conditional replenishment**		TV	**bedingte Auffrischung**
condensation		SCIE	**Verdichtung**				[Codierung]
[fig.]			[fig.]	**conditional selection**		SWITCH	**bedingte Wegesuche**
= compaction; compression			= Kondensierung; Kompres-	= conjugate selection			
			sion	**conditional selection system**		SWITCH	(→Linksystem)
condensation (→abridgment)		LING	(→Kürzung)	(→link system)			

conditional sentence	LING	Konditionalsatz
↑ causal sentence; conjunctional sentence		↑ Kausalsatz; Konjunktionalsatz
conditional statement	DATA PROC	bedingte Anweisung
conditional transfer (→conditional program jump)	DATA PROC	(→bedingter Programmsprung)
condition bit 1	DATA PROC	Bedingungsbit
condition bit 2 (→flag)	DATA PROC	(→Merker)
condition byte (→status byte)	DATA PROC	(→Zustandsbyte)
condition code	DATA PROC	Bedingungsschlüssel = Zustandscode
condition code register	DATA PROC	Anzeigenregister = Bedingungsmarkenregister
condition for passivity	NETW.TH	Passivitätsbedingungen
conditioning	TECH	Aufbereitung
conditioning line	MECH	Aufbereitungsanlage
conditions (→terms)	ECON	(→Konditionen)
condition Z (→signal condition Z)	DATA COMM	(→Stoppolarität)
conductance [real part of admittance]	NETW.TH	Wirkleitwert [Realteil des komplexen Scheinleitwerts] = Konduktanz
conductance (→conductivity)	PHYS	(→Leitfähigkeit)
conductance (→electric conductivity)	PHYS	(→elektrischer Leitwert)
conductance measurement	INSTR	Leitwertmessung
conductance parameter = admittance parameter; y parameter	NETW.TH	Leitwertparameter = y-Parameter; y-Vierpolparameter; Admittanzparameter
conducted = conductor-bound; line-bound ↓ wire-bound	TELEC	leitergebunden = leitungsgebunden ↓ drahtgebunden
conducted emission (→interfering voltage)	TELEC	(→Störspannung)
conducted interference (→interfering voltage)	TELEC	(→Störspannung)
conducted susceptibility	TELEC	leitungsgebundene Störfestigkeit
conducting = in on-stage	ELECTRON	durchlässig [in leitendem Zustand]
conducting (→conductive)	PHYS	(→leitend)
conducting electron (→conduction electron)	PHYS	(→Leitungselektron)
conducting layer	PHYS	leitende Schicht
conducting state current (→forward current)	ELECTRON	(→Durchlaßstrom)
conducting track (→track conductor)	ELECTRON	(→Leiterbahn)
conducting track routing (→conductor track routing)	ELECTRON	(→Leiterbahnführung)
conducting voltage (→conduction voltage)	ELECTRON	(→Durchlaßspannung)
conduction ↓ electric conduction; thermal conduction	PHYS	Leitung ↓ Elektrizitätsleitung; Wärmeleitung; Lichtwellenleitung
conduction angle (→current flow angle)	POWER SYS	(→Stromflußwinkel)
conduction band [lowest energy band] ↑ energy band	PHYS	Leitungsband [unterstes Energieband] ↑ Energieband
conduction current ≈ dielectric current	EL.TECH	Leitungsstrom ≈ Verschiebungsstrom
conduction electron = conducting electron; free electron	PHYS	Leitungselektron
conduction mechanism = carrier-transport mechanism	PHYS	Leitungsmechanismus
conduction path (→track conductor)	ELECTRON	(→Leiterbahn)
conduction voltage = conducting voltage ≠ reverse voltage	ELECTRON	Durchlaßspannung = Vorwärtsspannung ≠ Sperrspannung
conductive = conducting; carrying ≠ insulating ↓ heat conducting; conductive [EL.TECH]	PHYS	leitend = leitfähig ≠ isolierend ↓ wärmeleitend; stromleitend [EL.TECH]
conductive ≈ current-carrying	EL.TECH	stromleitend = leitend ≈ stromführend
conductive coupling (→dc coupling)	EL.TECH	(→galvanische Kopplung)
conductive foam material	TECH	elektrisch leitfähiger Schaumstoff
conductive pattern (→wiring pattern)	ELECTRON	(→Verdrahtungsmuster)
conductivity = conductance ↓ heat conductivity; electric conductivity	PHYS	Leitfähigkeit = Leitungsfähigkeit; Leitungseigenschaft; Leitvermögen ↓ Wärmeleitfähigkeit; elektrische Leitfähigkeit
conductivity modulation	MICROEL	Leitfähigkeitsmodulation
conductivity type	PHYS	Leitfähigkeitstyp
conductometer	INSTR	Leitwertmesser
conductor	OPT.COMM	Ader
conductor ≠ insulator ↓ electrical conductor; thermal conductor	PHYS	Leiter ≠ Isolator ↓ Elektrizitätsleiter; Wärmeleiter
conductor (→wire)	COMM.CABLE	(→Ader)
conductor (→wire)	EL.TECH	(→Draht)
conductor-bound (→conducted)	TELEC	(→leitergebunden)
conductor bundle (→wire bundle)	COMM.CABLE	(→Aderbündel)
conductor cord (→connecting cord)	EQUIP.ENG	(→Anschlußschnur)
conductor earthing electrode	SYS.INST	Seilerder
conductor pair (→wire pair)	COMM.CABLE	(→Aderpaar)
conductor rail (→current bus)	POWER ENG	(→Stromschiene)
conductor resistance	ELECTRON	Leiterwiderstand
conductor short	ELECTRON	Leiterbahnschluß
conductor track routing = conducting track routing	ELECTRON	Leiterbahnführung
conduit run (→cable conduit)	OUTS.PLANT	(→Kabelkanal)
conduits (→cable conduit)	OUTS.PLANT	(→Kabelkanal)
cone ≈ taper	MATH	Kegel = Konus ≈ Verjüngung
cone antenna (→conical antenna)	ANT	(→Konusantenne)
cone cut	MECH	Blechschälbohrer
cone head	MECH	Hochkegelkopf
cone-head rivet	MECH	Hochkegelkopfniet
cone loudspeaker = cone system	EL.ACOUS	Konuslautsprecher
cone system (→cone loudspeaker)	EL.ACOUS	(→Konuslautsprecher)
Confederation of European Computer Users's Associations (→CECUA)	DATA PROC	(→CECUA)
conference = meeting; discussion ≈ assembly	ECON	Besprechung = Konferenz; Diskussion; Rundgespräch; Sammelgespräch ≈ Versammlung
conference (→session)	ECON	(→Sitzung)
conference call = conference connection; conference circuit; multiaddress call; multiaddress circuit ≠ selective call	TELEC	Konferenzverbindung = Konferenzschaltung; Rundgesprächsverbindung; Sammelruf; Sammelverbindung; Sammelgesprächsverbindung [TELEPH]; Rundschreibverbindung [DATA COMM] ≠ Selektivruf
conference circuit (→conference call)	TELEC	(→Konferenzverbindung)
conference connection (→conference call)	TELEC	(→Konferenzverbindung)
conference mode (→conference operation)	TELEC	(→Konferenzbetrieb)

conference operation

conference operation	TELEC	**Konferenzbetrieb**
= conference mode		
conference participant	ECON	**Tagungsteilnehmer**
= congressman		= Kongreßteilnehmer; Konferenzteilnehmer
conference room	OFFICE	**Konferenzraum**
conference studio	TELEC	**Konferenzstudio**
conference table	ECON	**Verhandlungstisch**
		= Konferenztisch
conference traffic	TELECONTR	**Konferenzverkehr**
confidence	COLLOQ	**Vertraulichkeit**
≈ secrecy		≈ Geheimhaltung
confidence belt (→ confidence interval)	MATH	(→ Vertrauensbereich)
confidence coefficient	MATH	**Aussagewahrscheinlichkeit**
confidence intervall	MATH	**Vertrauensbereich**
= confidence region; confidence belt		= Vertrauensintervall; Konfidenzbereich; Konfidenzintervall; Konfidenzgürtel
confidence level	MATH	**Vertrauensgrad**
		= Konfidenzzahl
confidence limit	MATH	**Vertrauensgrenze**
[limits of the confidence interval]		[Endpunkte des Vertrauensbereichs]
		= Konfidenzgrenze; Mutungsgrenze
confidence region (→ confidence interval)	MATH	(→ Vertrauensbereich)
confidential	OFFICE	**vertraulich**
≈ secret		≈ geheim
≠ unrestricted		≠ allgemein zugänglich
configurability	TECH	**Konfigurierbarkeit**
configurable	DATA PROC	**einstellbar**
= settable		= konfigurierbar
configuration	TECH	**Aufbau 2**
= layout; structure		= Struktur; Gefüge;
configuration	SCIE	**Konfiguration**
configuration	DATA PROC	**Konfiguration**
[assembly of equipment and programs adjusted to operate as a system]		[Gruppe von Geräten und Programmen die aufeinander abgestimmt sind um als System zu arbeiten]
		= Anlagenkonfiguration; Konfigurierung; Ausführung; Ausbau
configuration (→ arrangement)	TECH	(→ Anordnung)
configuration (→ type)	TECH	(→ Ausführung 1)
configuration (→ allocation)	SWITCH	(→ Beschaltung)
configuration device	DATA PROC	**Konfigurationsgerät**
		= Einstellgerät
configuration file	DATA PROC	**Konfigurationsdatei**
[stores settings for the configuration of the PC]		[speichert Einstellungen zur Konfigurierung des PC's]
		= Konfigurierungsdatei
configuration management	TELEC	**Beschaltungsverwaltung**
		= Netzausführungsverwaltung; Zuteilungsverwaltung
configuration management	DATA PROC	**Ausführungsverwaltung**
		= Konfigurationsverwaltung; Konfigurierungsverwaltung
configuration menu	DATA PROC	**Konfigurationsmenu**
= system configuration menu; set-up menu		= Konfigurationsmenu; System-Konfigurationsmenü; System-Konfigurierungsmenü; Set-up-Menü
configuration program	DATA PROC	**Konfigurationsprogramm**
= system configuration program		= Konfigurierungsprogramm; System-Konfigurationsprogramm; System-Konfigurierungsprogramm
configurator	DATA PROC	**Konfigurator**
configure (v.t.)		**konfigurieren**
[to select and make up a hardware and software configuration]		[eine spezielle Hardware- und Softwarekombination auswählen und in Betrieb nehmen]
= set 2 (v.t.)		= ausführen 2; einstellen 2; ausbauen
confine (→ limit)	TECH	(→ begrenzen)
confirmation (→ acknowledgment)	ECON	(→ Bestätigung)
conform (v.t.)	TECH	**angleichen**
≈ adapt		≈ anpassen
conformal antenna	ANT	**Konformantenne**
= conformal array		
conformal array (→ conformal antenna)	ANT	(→ Konformantenne)
conformal mapping	MATH	**konforme Abbildung**
conformance	TECH	**Normgerechtigkeit**
conformance test (→ homologation test)	QUAL	(→ Typprüfung)
conform to	COLLOQ	**entsprechen**
		= gerecht werden
confused (→ blurred)	TV	(→ unscharf)
congested	SWITCH	**gassenbesetzt**
= with congestion; with all trunks busy		
congestion	SWITCH	**Gassenbesetztzustand**
= all-trunks-busy condition		
congestion (→ blocking)	SWITCH	(→ Blockierung)
congestion control	DATA COMM	**Überlastabwehr**
↓ flow control; load control		↓ Flußsteuerung; Laststeuerung
congestion signal	DATA COMM	**Überlastungskennzeichen**
congress (→ meeting)	ECON	(→ Tagung)
congressman (→ conference participant)	ECON	(→ Tagungsteilnehmer)
congruence	MATH	**Kongruenz**
= superposability		= Deckungsgleichheit
congruent	TECH	**deckungsgleich**
= coincident; accurately aligned		
congruent	MATH	**kongruent**
= superposable		= deckungsgleich
congruent matching	TECH	**Deckungsgleichheit**
= accurate alignment		
congruent sign	MATH	**Geometrisch-kongruent-Zeichen**
↑ mathematical symbol		↑ mathematisches Zeichen
conical	MATH	**kegelförmig**
≈ tapered		≈ konisch
		≈ verjüngt
conical antenna	ANT	**Konusantenne**
= cone antenna		= Kegelantenne
conical helical spring	MECH	**kegelförmige Schraubenfeder**
conical horn	ANT	**Konushorn**
= taper		= konischer Trichter; Trichterhorn; Trichter; Kegelhorn
↑ horn radiator		↑ Hornstrahler
conical horn antenna	ANT	**Konushornantenne**
		= Kegelhornantenne; Trichterhornantenne
conical refraction	PHYS	**konische Refraktion**
conical scan (→ conical scanning)	ELECTRON	(→ konische Abtastung)
conical scanning	ELECTRON	**konische Abtastung**
= conical scan		
conical scanning	RADIO LOC	(→ Quirlen)
(→ conical-scan tracking)		
conical-scan tracking	RADIO LOC	**Quirlen**
= quirl (n.); conical scanning		= konisches Sucherverfahren
conical section	MATH	**Kegelschnitt**
↓ circle; ellipse; hyperbola; parabola		↓ Kreis; Ellipse; Hyperbel; Parabel
conical spiral antenna	ANT	**konische Spiralantenne**
conical spring	MECH	**Kegelstumpffeder**
CONIFAN antenna	ANT	**CONIFAN-Antenne**
conjugate	MATH	**konjugiert-komplex**
= complex-conjugated		= komplex-konjugiert; konjugiert
conjugate attenuation constant	LINE TH	**konjugiert-komplexe Dämpfung**
conjugate impedance	NETW.TH	**konjugiert komplexer Scheinwiderstand**
		= konjugiert komplexer Widerstand
conjugate phase constant	LINE TH	**konjugiert komplexe Phasenkonstante**
conjugate selection (→ conditional selection)	SWITCH	(→ bedingte Wegesuche)

English	Domain	German
conjugate selection system (→link system)	SWITCH	(→Linksystem)
conjugation [inflection of verbs] ↑ flection; morphology	LING	**Konjugation** [Beugung des Verbs] ↑ Flexion; Morphologie
conjunction [connects words or sentences; e.g. and; because]	LING	**Konjunktion** [verbindet Wörter oder Sätze; z.B. und, weil] = Bindewort
conjunction (→AND operation)	ENG.LOG	(→UND-Verknüpfung)
conjunctional sentence	LING	**Konjunktionalsatz** [stellvertredend für Adverb] ↓ Temporalsatz; Modalsatz; Kausalsatz
conjunction circuit (→AND gate)	CIRC.ENG	(→UND-Glied)
conjunction element (→AND gate)	CIRC.ENG	(→UND-Glied)
conjunction function (→AND operation)	ENG.LOG	(→UND-Verknüpfung)
conjunction gate (→AND gate)	CIRC.ENG	(→UND-Glied)
conjunction operation (→AND operation)	ENG.LOG	(→UND-Verknüpfung)
connect (v.t.)	TELEC	**anschließen**
connect 1 = link (v.t.)	EL.TECH	**anschließen**
connect 2 = power-up (v.t.); switch-up (v.t.); start-up (v.t.); turn-on (v.t.)	EL.TECH	**einschalten** ≠ ausschalten
connect	SWITCH	**verbinden** = beschalten
connect (→through connect)	TELEC	(→durchschalten)
connect additionally	EL.TECH	**zuschalten**
connect/disconnect relay (→isolation relay)	POWER SYS	(→Trennrelais)
connected = allocated	SWITCH	**beschaltet**
connected = related ≈ adjacent	TECH	**zusammenhängend** ≈ benachbart
connected types [like handwriting]	TYPOGR	**verbundene Zeichen** [handschriftartig]
connecting cable ≈ connecting line ≈ interconnecting cable ↓ power cable	EQUIP.ENG	**Anschlußkabel** ≈ Anschlußleitung ≈ Verbindungskabel ↓ Netzanschlußkabel
connecting cable = switchboard cable ≈ office cable [SYS.INST] ↑ communication cable	COMM.CABLE	**Schaltkabel** = Amtskabel [SYS.INST] ↑ Nachrichtenkabel
connecting charge (→subscription fee)	TELEC	(→Anschlußgebühr)
connecting contact	COMPON	**Anschaltekontakt**
connecting cord = conductor cord; cord	EQUIP.ENG	**Anschlußschnur** = Verbindungsschnur; Leitung ≈ Anschlußkabel
connecting delay (→call set-up time)	DATA COMM	(→Verbindungsaufbaudauer)
connecting jack	COMPON	**Anschlußklinke** = Anschalteklinke
connecting lead (→lead)	COMPON	(→Anschlußdraht)
connecting line = connection line; junction line	MATH	**Verbindungslinie**
connecting line (→connecting cable)	EQUIP.ENG	(→Anschlußkabel)
connecting lug	COMPON	**Kontaktfahne**
connecting matrix (→switching matrix)	SWITCH	(→Koppelvielfach)
connecting network 1 (→switching network 1)	SWITCH	(→Koppelanordnung)
connecting network 2 (→switching network 2)	SWITCH	(→Koppelnetz)
connecting piece = fitting	MECH	**Verbindungsstück**
connecting plug	COMPON	**Anschlußstecker** = Verbindungsstecker
connecting point (→terminal point)	COMPON	(→Stützpunkt)
connecting range (→exchange area)	SWITCH	(→Anschlußbereich)
connecting relay	SWITCH	**Anschalterelais**
connecting relay set (→circuit connector)	SWITCH	(→Anschaltesatz)
connecting request	TELEC	**Schaltauftrag**
connecting row (→switching row)	SWITCH	(→Koppelreihe)
connecting stage (→switching stage)	SWITCH	(→Koppelstufe)
connecting strap	MECH	**Verbindungslasche**
connecting strip	COMPON	**Anschlußstreifen** = Anschlußschiene
connecting time	SWITCH	**Beschaltungsdauer**
connecting unit	DATA COMM	**Anschalteinrichtung**
connecting wire = lead-in; lead	TELEC	**Anschlußdraht**
connect in parallel = parallel (v.t.); shunt (v.t.)	EL.TECH	**parallelschalten** = nebenschließen; überbrücken
connection (NAM) = connexion (BRI)	EL.TECH	**Anschluß** ≈ Zusammenschaltung
connection (AM) = connexion (BRI); circuit; link ≈ call [TELEPH]; conversation [TELEPH] ↓ communication link	TELEC	**Verbindung** ≈ Anschluß; Anbindung ≈ Anruf [TELEPH]; Gespräch [TELEPH] ↓ Nachrichtenverbindung
connection (NAM) (→terminal connection)	SWITCH	(→Anschluß)
connection (NAM) (→allocation)	SWITCH	(→Beschaltung)
connection (→circuit)	NETW.TH	(→Schaltkreis)
connection (NAM) (→relationship)	ECON	(→Verbindung)
connection (→interconnection)	EL.TECH	(→Zusammenschaltung)
connection (NAM) (→enabling)	ELECTRON	(→Einschaltung)
connection block (→terminal strip)	EL.INST	(→Klemmleiste)
connection board	EQUIP.ENG	**Anschlußplatte**
connection cable	OUTS.PLANT	**Anschlußkabel**
connection capacity	SWITCH	**Anschlußkapazität**
connection charge (→subscription fee)	TELEC	(→Anschlußgebühr)
connection charge (→connection fee)	TELEC	(→Verbindungsgebühr)
connection control (→call control)	SWITCH	(→Verbindungssteuerung)
connection control character (→call control character)	SWITCH	(→Verbindungssteuerungszeichen)
connection control procedure (→call control procedure)	SWITCH	(→Verbindungssteuerungsverfahren)
connection control signal (→call control character)	SWITCH	(→Verbindungssteuerungszeichen)
connection costs	TELEC	**Anschaltkosten**
connection diagram (→wiring diagram)	EL.TECH	(→Bauschaltplan)
connection endpoint	DATA COMM	**Verbindungsendpunkt**
connection establishment (→connection set-up)	SWITCH	(→Verbindungsaufbau)
connection failure (→call failure)	SWITCH	(→Verbindungsstörung)
connection fee = connection charge; link fee; link charge; call charge [TELEPH]; call fee [TELEPH] ≠ basic rental	TELEC	**Verbindungsgebühr** = Gesprächsgebühr [TELEPH] ≠ Grundgebühr
connection function	DATA COMM	**Verbindungsfunktion**
connection identifier (→call identifier)	DATA COMM	(→Verbindungskennung)
connectionless service	TELEC	**verbindungsloser Dienst**
connection line (→connecting line)	MATH	(→Verbindungslinie)
connection logging (→call logging)	DATA COMM	(→Verbindungsprotokollierung)
connection matrix = incidence matrix	DATA PROC	**Verbindungsmatrix**

connection network	OUTS.PLANT	Anschlußnetz	
connection-oriented mode [ATM]	TELEC	verbindungsorientierter Betrieb [ATM]	
connection path (→communication path)	TELEC	(→Nachrichtenweg)	
connection register (→call register)	SWITCH	(→Verbindungsspeicher)	
connection request (→call request)	SWITCH	(→Verbindungsanforderung)	
connection sequence = call sequence; link sequence	SWITCH	Verbindungsablauf = Ruffolge	
connection set-up = call set-up; connection establishment; call establishment; link set-up; link establishment; call completion	SWITCH	Verbindungsaufbau = Verbindungsherstellung	
connections observed	SWITCH	beobachtete Belegungen	
connection state (→call state)	SWITCH	(→Verbindungszustand)	
connection tear-down = call tear-down; link tear-down; disconnection 1; circuit release; dissociation	SWITCH	Verbindungsabbau = ABBAU	
connection technique	EQUIP.ENG	Anschlußtechnik	
connection time (→call duration)	TELEC	(→Verbindungsdauer)	
connection unit	SWITCH	Beschaltungseinheit	
connective (n.) (→connector symbol)	DATA PROC	(→Verknüpfungssymbol)	
connective (n.) (→operation)	ENG.LOG	(→Verknüpfung)	
connectivity	TELEC	Anschlußmöglichkeit = Anschließbarkeit	
connect node [CAD]	DATA PROC	Verknüpfungspunkt [CAD]	
connector	SWITCH	Anschalter	
connector	ANT	Armatur	
connector (→plug connector)	COMPON	(→Steckverbinder)	
connector (→connector symbol)	DATA PROC	(→Verknüpfungssymbol)	
connector circuit (→junctor)	SWITCH	(→Verbindungssatz)	
connector-ended cable (→plug-in cable)	EQUIP.ENG	(→Steckkabel)	
connector housing	COMPON	Steckergehäuse	
connectorized (→pluggable)	ELECTRON	(→steckbar)	
connectorized cable (→plug-in cable)	EQUIP.ENG	(→Steckkabel)	
connector lug	COMPON	Steckerfahne	
connector panel (→terminal field)	EQUIP.ENG	(→Anschlußfeld)	
connector pin = plug pin; contact pin	COMPON	Steckerstift	
connector speaker (→loudspeaker connector)	EL.ACOUS	(→Lautsprecher-Stecker)	
connector strip = frame connector ≠ spring contact strip	COMPON	Steckerleiste ≠ Federleiste	
connector-strip body	COMPON	Leistenkörper	
connector symbol [circle with number, to identify interrupted lines] = connector; connective (n.)	DATA PROC	Verknüpfungssymbol [Kreis mit Zahl, zur Kennzeichnung unterbrochener Linien] = Übergangsstelle; Verbinder; Koonektor	
connect single path	SWITCH	Einzelweganschaltung	
connect time	DATA PROC	Anschlußintervall	
connexion (BRI) (→connection)	EL.TECH	(→Anschluß)	
connexion (BRI) (→terminal connection)	SWITCH	(→Anschluß)	
connexion (BRI) (→allocation)	SWITCH	(→Beschaltung)	
connexion (BRI) (→relationship)	ECON	(→Verbindung)	
connexion (BRI) (→connection)	TELEC	(→Verbindung)	
connexion (BRI) (→enabling)	ELECTRON	(→Einschaltung)	
connotation [suggested emotional meaning]	LING	Konnotation [Andeutung, emotionale Nebenbedeutung]	
conscious error	DATA PROC	bewußter Fehler	
consecutive = succesive; sequential ≈ succeeding	TECH	aufeinanderfolgend = aneinandergereiht; sukzessiv; sequentiell; laufend 2 ≈ nachfolgend	
consecutive	LING	konsekutiv	
consecutive number (→sequential number)	MATH	(→laufende Zahl)	
consecutive numbering (→serial numbering)	SWITCH	(→fortlaufende Numerierung)	
consecutive sentence ↑ causal sentence	LING	Konsekutivsatz = Folgesatz ↑ Kausalsatz	
consecutive zeros	CODING	aufeinanderfolgende Nullen	
consent (n.)	ECON	Zustimmung	
consequence finding program	DATA PROC	Konsequenzen findendes Programm	
conservation	PHYS	Erhaltung [Energie]	
conservation law	PHYS	Erhaltungssatz	
conservation of energy	PHYS	Energieerhaltung	
conservation of momentum	PHYS	Impulserhaltung	
consideration	COLLOQ	Gegenleistung	
consign (→send)	ECON	(→übersenden 1)	
consignee [of a delivery] ≈ recipient	ECON	Lieferungsempfänger = Adressat; Konsignatär; Kommissionär	
consigner [of a good] = consignor; shipper	ECON	Absender [einer Ware] = Adressant; Konsignant; Versender; Kommittent	
consignment (→despatch)	ECON	(→Versand)	
consignment stock	ECON	Konsignationslager	
consignor (→consigner)	ECON	(→Absender)	
consistency	MATH	Folgerichtigkeit = Konsistenz	
consistency check	DATA PROC	Konsistenzprüfung	
Consolan antenna ≈ Consol antenna	ANT	Consolan-Antenne ≈ Consol-Antenne	
Consol antenna ≈ Consolan antenna	ANT	Consol-Antenne ≈ Consolan-Antenne	
console = board; panel; bracket ≈ attendant console	EQUIP.ENG	Pult = Konsole; Konsol; Board ≈ Bedienungsfeld	
console [peripheral mostly composed by a keyboard, monitor and printer]	DATA PROC	Konsole [meist aus Tastatur, Bildschirm und Drucker bestehende Bedieneinheit] = Konsol; Ein-/Ausgabe-Einheit	
console (→attendant console)	SWITCH	(→Abfragestelle)	
console (→control desk)	TECH	(→Kontrollpult)	
console (→control board)	BROADC	(→Kontrollpult)	
console (→operator's console)	SWITCH	(→Vermittlungsplatz)	
console operator (→computer operator)	DATA PROC	(→Rechnerbediener)	
console screen	TERM&PER	Konsolenbildschirm	
console table [a table type construction designed to fit against a wall] = bracket ≈ pulpit; desk	TECH	Konsole [an der Wand aufgestellte tischartige Konstruktion] ≈ Pult; Tisch	
console typewriter (→page printer 1)	TERM&PER	(→Blattschreiber)	
consolidation (→hubbing)	TELEC	(→Verkehrskonzentrierung)	
Consol radio beacon	RADIO NAV	Consol-Funkfeuer	
consonance (→accordance)	COLLOQ	(→Übereinstimmung)	
consonancy (→accordance)	COLLOQ	(→Übereinstimmung)	
consonant (n.) ↑ sound	LING	Konsonante = Konsonant; Mitlaut ↑ Laut	
consonant articulation	TELEPH	Konsonantenverständlichkeit	
consortium = syndicate	ECON	Konsortium = Bietergemeinschaft; Arbeitsgemeinschaft	

constancy	COLLOQ	**Konstanz**	
= fortitude		= Beständigkeit; Beharrungsvermögen	
constant	MATH	**Konstante**	
≈ invariable		≈ Invariante	
≠ variable (n.)		≠ Variable	
constant (→continuous 1)	TECH	(→kontinuierlich)	
constant (n.) (→literal constant)	DATA PROC	(→Literalkonstante)	
constant (→invariable)	TECH	(→unveränderlich)	
constant accessibility	SWITCH	**konstante Erreichbarkeit**	
= constant availability		= konstante Verfügbarkeit	
constantan	METAL	**Konstantan**	
= eureka; advance		= Eureka; Advance	
constant availability (→constant accessibility)	SWITCH	(→konstante Erreichbarkeit)	
constant bit rate mode (→fixed bit rate mode)	TELEC	(→Festbitratenbetrieb)	
constant current	EL.TECH	**Konstantstrom**	
constant-current crossover	POWER SYS	**Konstantstrom-Überkreuzung**	
constant-current measuring bridge	INSTR	**Konstantstrom-Meßbrücke**	
constant current operation	NETW.TH	**Konstantstrombetrieb**	
constant-current power supply (→stabilized current regulator)	CIRC.ENG	(→Stromstabilisator)	
constant-density recording [magnetic store] = CDR	TERM&PER	**lineare Schreibdichte** [Magnetspeicher]	
constant-failure period (→constant-failure-rate period)	QUAL	(→Konstantausfallratenzeit)	
constant-failure-rate period = constant-failure period	QUAL	**Konstantausfallratenzeit** = Konstantfehlerzeit	
constant field	PHYS	**Gleichfeld**	
constant-frequency	ELECTRON	**frequenzstarr**	
constant-length field	DATA PROC	**Datenfeld konstanter Länge**	
constant light barrier	ELECTRON	**Gleichlichtschranke**	
constant-pitch font (→constant-width font)	TYPOGR	(→Konstantschrift)	
constant-pitch print (→constant-width font)	TYPOGR	(→Konstantschrift)	
constant-pitch printing (→constant-width font)	TYPOGR	(→Konstantschrift)	
constant ratio code = constant weight code	CODING	**gleichgewichtiger Code**	
constant-value control (→fixed command control)	CONTROL	(→Festwertregelung)	
constant voltage	EL.TECH	**Konstantspannung**	
constant-voltage crossover	POWER SYS	**Konstantspannungs-Überkreuzung**	
constant-voltage/current-limiting power supply = CV/CL power supply	POWER SYS	**Konstantspannungs-/Strombegrenzungs-Stromversorgung**	
constant-voltage generator	INSTR	**Gleichspannungsgenerator**	
constant voltage operation	NETW.TH	**Konstantspannungsbetrieb**	
constant-voltage power supply (→voltage stabilizer)	POWER ENG	(→Spannungsstabilisator)	
constant-voltage source (→voltage stabilizer)	POWER ENG	(→Spannungsstabilisator)	
constant voltage transformer	ELECTRON	**magnetischer Spannungskonstanthalter**	
constant weight code (→constant ratio code)	CODING	(→gleichgewichtiger Code)	
constant-width font = constant-pitch font; constant-width print; constant-pitch print; constant-width printing; constant-pitch printing; monospaced font ≠ proportional-width font	TYPOGR	**Konstantschrift** [alle Buchstaben gleich breit] ≠ Proportionalschrift	
constant-width print (→constant-width font)	TYPOGR	(→Konstantschrift)	
constant-width printing (→constant-width font)	TYPOGR	(→Konstantschrift)	
constellation (→signal constellation)	MODUL	(→Signalkonstellation)	
constellation analysis	INSTR	**Konstellationsanalyse**	
constellation analyzer = digital radio constellation analyzer	INSTR	**Konstellationsanalysator**	
constellation diagram	MODUL	**Konstellationsdiagramm**	
constellation display	INSTR	**Konstellationsanzeige**	
constituent ≈ part ≈ component part	TECH	**Bestandteil** ≈ Einzelteil	
constituent signal (→tributary signal)	TRANS	(→Zubringersignal)	
constrain (v.t.) = restrict	TECH	**einschränken**	
constrain (→force)	TECH	(→erzwingen)	
constrained magnetization	PHYS	**erzwungene Magnetisierung**	
constraint (→boundary condition)	MATH	(→Randbedingung)	
constraint (→difficulty)	COLLOQ	(→Schwierigkeit)	
constrict ≈ contract	COLLOQ	**verengen** ≈ zusammenziehen	
constriction ≈ contraction	TECH	**Verengung** ≈ Schrumpfung	
construct (→build)	TECH	(→bauen)	
construction (→mechanical design)	EQUIP.ENG	(→Konstruktion)	
constructional characteristic (→mechanical characteristic)	TECH	(→konstruktives Merkmal)	
constructional engineering (→civil engineering)	TECH	(→Bautechnik)	
constructional feature (→mechanical characteristic)	TECH	(→konstruktives Merkmal)	
constructional unit	EQUIP.ENG	**konstruktive Einheit**	
construction document	DOC	**Konstruktionsunterlage**	
construction engineer (→draftsman)	TECH	(→Konstrukteur)	
construction practice = packaging structure; packaging system; packaging technique	EQUIP.ENG	**Bauweise** = Aufbausystem [SWITCH]; Aufbautechnik = Gerätetechnik; Konstruktionstechnik; Konstruktion	
construction specification = manufacturing specification; fabrication specification; fabricating specification; assembly specification	TECH	**Bauvorschrift** = Fabrikationsvorschrift = Montagevorschrift; Zusammenbauvorschrift	
construction stage = expansion stage; capacity stage; increment stage	TECH	**Ausbaustufe** = Erweiterungsstufe	
construction standard = packaging standard	EQUIP.ENG	**Baunorm** = Gerätenorm ≈ Bauweise	
construction suggestions	ELECTRON	**Bauvorschlag**	
construction unit (→module)	EQUIP.ENG	(→Baugruppe)	
consult (v.t.)	COLLOQ	**konsultieren**	
consultant = adviser; counselor	ECON	**Berater**	
consultant terminal	TERM&PER	**Beraterterminal**	
consultation call (→call-back)	TELEPH	(→Rückfrage)	
consultation service	ECON	**Beratungsdienst**	
consulting company = consulting firm	ECON	**Beraterfirma** = Beratungsunternehmen	
consulting firm (→consulting company)	ECON	(→Beraterfirma)	
consumables (n. pl.t.) = Verbrauchsteil	TECH	**Verschleißteil**	
consumer ≈ costumer; purcheaser	ECON	**Verbraucher** ≈ Abnehmer ≈ Kunde; Käufer	
consumer = load	EL.TECH	**Verbraucher** = Last	
consumer electronics ↓ entertainment electronics; household electronics; car electronics	ELECTRON	**Konsumelektronik** ↓ Unterhaltungselektronik; Haushaltselektronik; Fahrzeugelektronik	
consumer inquiry	ECON	**Verbraucherbefragung**	
consuming ≈ expensive	ECON	**aufwendig** ≈ teuer	
consumption = expenditure	ECON	**Kostenaufwand** = Aufwand	
consumption	TECH	**Verbrauch** = Konsum	
consumption good ≠ industrial product; investment good	ECON	**Konsumgut** ≠ Industriegut; Investitionsgut	
contact (n.) ↓ lead; terminal pin	COMPON	**Anschluß** ↓ Anschlußdraht; Anschlußstift	

contact (v.t.)		TECH	anliegen [berühren]	contact rivet		COMPON	Kontaktniet
contact		PHYS	Kontakt	contact sensor		COMPON	Berührungssensor
= touch			= Berührung	= touch sensor; tactile sensor			= Berührungsfühler; Kontaktsensor; Kontaktfühler
-contact		ELECTRON	-polig				
contact (→lead)		COMPON	(→Anschlußdraht)	contact separation (→contact clearance)		COMPON	(→Kontaktabstand)
contact area (→contact surface)		PHYS	(→Kontaktfläche)				
contact arrangement		COMPON	Kontaktanordnung = Kontaktaufbau	contact series (→electrochemical series)		PHYS	(→voltaische Spannungsreihe)
contact bank		EQUIP.ENG	Kontaktbank	contact set [relay]		COMPON	Kontaktsatz [Relais]
contact bar		COMPON	Kontaktschiene				
contact base		COMPON	Kontaktträger	contact spring		COMPON	Kontaktfeder
contact between lines		EL.TECH	Leitungsberührung = Leitungsschluß	= contact clip ↑ contact element			↑ Kontaktelement
contact blade		COMPON	Kontaktmesser	contact strip		COMPON	Kontaktleiste
contact bounce		COMPON	Kontaktprellen	↓ spring contact strip; pin contact strip			↓ Federleiste; Stiftleiste
= bounce; contact chatter; chatter; contact rebound			= Prellen	contact stroke (→contact travel)		COMPON	(→Kontakthub)
contact chatter (→contact bounce)		COMPON	(→Kontaktprellen)	contact surface = contact area		PHYS	Kontaktfläche ≈ Kontaktpunkt
contact clause		LING	Relativsatz ohne Relativpronomen	≈ contact point			
				contact thermometer		INSTR	Kontaktthermometer
contact clearance		COMPON	Kontaktabstand	contact to earth (→ground fault)		EL.TECH	(→Erdschluß)
= contact separation							
contact clip (→contact spring)		COMPON	(→Kontaktfeder)	contact travel		COMPON	Kontakthub
contact closure		COMPON	Kontaktschließung	= contact stroke; contact excursion			
contact electricity		EL.TECH	Berührungselektrizität	contact voltage		EL.TECH	Berührungsspannung
= voltaic electricity			= Kontaktelektrizität				= Kontaktspannung
contact element		COMPON	Kontaktelement				≈ Berührungspotential; Fehlerspannung
↓ contact pin; contact spring			↓ Kontaktstift; Kontaktfeder				
contact excursion (→contact travel)		COMPON	(→Kontakthub)	contact wiper		SWITCH	Kontaktarm = Schaltarm
contact float		COMPON	Kontaktspiel	contact wire (→overhead line)		POWER ENG	(→Fahrleitung)
contact follow		COMPON	Folgeweg [Kontakt]	contain = include		COLLOQ	enthalten ≈ einschließen (fig.)
contact force		COMPON	Kontaktkraft	container		OUTS.PLANT	Behälter
= contact pressure			= Kontaktdruck	container [SDH/SONET]		TELEC	Container [SDH/SONET]
contact grid		MICROEL	Kontaktgitter				
contactless		PHYS	berührungslos	container		ECON	Frachtbehälter
= noncontact (adj.)			= kontaktlos				= Container; Warenbehälter
contact material		COMPON	Kontaktträgerwerkstoff = Leiterwerkstoff	container (→vessel)		TECH	(→Behälter)
contact mattress [key pad]		TERM&PER	Kontaktmatte [Tastatur]	container (→shelter)		SYS.INST	(→Shelter)
				container installation		SYS.INST	Containeranlage
contact noise		TELEPH	Kontaktgeräusch	contaminate = pollute		TECH	verunreinigen ≈ verschmutzen
contact noise		EL.ACOUS	Kratzgeräusch	≈ foul (v.t.)			
= frying noise; frying; cracking noise; line scratches (BRI)				contamination = pollution; poisoning ≈ fouling		TECH	Verunreinigung = Vergiftung ≈ Verschmutzung
contactor		OUTS.PLANT	Drucküberwachungskontakt				
contactor		POWER ENG	Schütz	contamination (→impurity)		PHYS	(→Verunreinigung)
= magnetic switch			= Magnetschalter	content-addressable file		DATA PROC	Assoziativdatei
≈ switch [COMPON]			≈ Schalter [COMPON]	content-addressable memory (→associative storage)		DATA PROC	(→Assoziativspeicher)
contact paper		TERM&PER	Kontaktpapier				
contact pin ↑ contact element		COMPON	Kontaktstift ↑ Kontaktelement	content-addressable storage (→associative storage)		DATA PROC	(→Assoziativspeicher)
contact pin (→connector pin)		COMPON	(→Steckerstift)	content-addressed memory (→associative storage)		DATA PROC	(→Assoziativspeicher)
contact point ≈ contact surface		PHYS	Kontaktstelle = Kontaktpunkt = Kontaktfläche	content-addressed storage (→associative storage)		DATA PROC	(→Assoziativspeicher)
				content-changeable = content-variable		DATA PROC	inhaltveränderbar = inhaltvariabel
contact point = adherent point		TECH	Berührungspunkt	contention ≈ request mode		DATA COMM	Konkurrenz = Konkurrenzsituation ≈ Konkurrenzbetrieb
contact-potencial difference		PHYS	Kontaktpotentialdifferenz = Kontaktspannung; Galvanispannung				
				contention bus		DATA PROC	Konkurrenzbus
contact potential		EL.TECH	Berührungspotential = Kontaktpotential ≈ Berührungsspannung	contention delay		DATA COMM	Konkurrenzverzug
				contention mode (→request mode)		DATA COMM	(→Anforderungsbetrieb)
				contention point		DATA COMM	Konkurrenzpunkt
contact pressure (→contact force)		COMPON	(→Kontaktkraft)	contents (n. pl.t.)		COLLOQ	Inhalt
				contents		DOC	Inhaltsverzeichnis
contact printing		MICROEL	Kontaktbelichtung	= table of contents			= Inhaltsübersicht
contact rebound (→contact bounce)		COMPON	(→Kontaktprellen)	contents directory		DATA PROC	Programmverzeichnis
				content-variable (→content-changeable)		DATA PROC	(→inhaltveränderbar)
contact reed [relay] = reed		COMPON	Kontaktzunge [Relais]				
				contermand (v.t.) (→cancel)		ECON	(→rückgängigmachen)
contact resistance		COMPON	Kontaktübergangswiderstand = Kontaktwiderstand; Übergangswiderstand; Durchgangswiderstand	context (n.)		LING	Kontext = Zusammenhang
				contextual		LING	kontextual = kontextuell

English	Domain	German
contiguous (→adjacent)	TECH	(→benachbart)
contiguous data structure	DATA PROC	(→sequentielle Datenstruktur)
(→sequential data structure)		
contiguous file	DATA PROC	Nachbardatei
contingency	COLLOQ	Eventualfall
contingency	MATH	Kontingenz
contingency plan	TECH	Notplan
= emergency plan		
continious duty (→continuous operation)	TECH	(→Dauerbetrieb)
continous phase frequency shift keying (→CPFSK)	MODUL	(→CPFSK)
continual 1 (→continuous 1)	TECH	(→kontinuierlich)
continual 2 (→repetitive)	TECH	(→wiederholt)
continuation address (→reference address)	DATA PROC	(→Verweisadresse)
continuation instruction (→sequential instruction)	DATA PROC	(→Folgebefehl)
continuation page	TYPOGR	Folgeseite
= next page		
continued fraction	MATH	Kettenbruch
continued fractions arrangement	NETW.TH	Kettenbruchschaltung
continuing education (→further education)	SCIE	(→Weiterbildung)
continuity	SCIE	Kontinuität
≈ steadiness		≈ Stetigkeit
continuity check	ELECTRON	Durchgangsprüfung
continuity check	SWITCH	Durchschalteprüfung
= continuity test; cross-office check		= Durchgangsprüfung; Kontinuitätsprüfung; Verbindungsweg-Durchschalteprüfung; Stetigkeitsprüfung
continuity condition	SCIE	Kontinuitätsbedingung
= continuity state		
continuity equation	MATH	Kontinuitätsgleichung
continuity state (→continuity condition)	SCIE	(→Kontinuitätsbedingung)
continuity test (→continuity check)	SWITCH	(→Durchschalteprüfung)
continuity tester	INSTR	Durchgangsprüfgerät
continuous 1	TECH	kontinuierlich
[in time domain]		[zeitlich]
= uninterrupted; continual 1; incessant; constant		= ununterbrochen; unterbrechungsfrei; pausenlos; lückenlos (fig.)
≈ stepless; endless; repetitive		≈ stufenlos; endlos; wiederholt
continuous	MATH	stetig
= uniform; steady		= gleichförmig;: stufenlos
≈ monotonous		≈ monoton
≠ discrete		≠ diskret
continuous 2 (→endless)	TECH	(→endlos)
continuous 3 (→stepless)	TECH	(→stufenlos)
continuous-action controller	CONTROL	stetiger Regler
↓ P controller; I controller; PI controller		↓ P-Regler; I-Regler; PI-Regler
continuous cassette	TERM&PER	Endloscassette
= endless-tape cassette		
continuous casting	METAL	Strangguß
continuous charge (→continuous load)	TECH	(→Dauerbelastung)
continuous control	CONTROL	Bahnsteuerung
↑ numeric control		↑ numerische Steuerung
continuous controller	CONTROL	kontinuierlicher Regler
≠ sampled-data controller		≠ Abtastregler
continuous copying paper	TERM&PER	Endlosdurchschreibsatz
continuous current telegraphy (→direct current telegraphy)	TELEGR	(→Gleichstromtelegrafie)
continuous data print	TERM&PER	Endlosdatendruck
continuous distribution	MATH	stetige Verteilung
continuous feed (→continuous form feed)	TERM&PER	(→Endlosformularzuführung)
continuous file card	TERM&PER	Endloskarteikarte
continuous form	TERM&PER	Endlosformular
≠ single form		= Endlosvordruck
↑ continuous paper		≠ Einzelformular
↓ rollpaper form; fanfold form		↑ Endlospapier
		↓ Rollenformular; Zickzackformular
continuous form feed	TERM&PER	Endlosformularzuführung
= continuous feed		= Endlosvordruckzuführung; Endlosformulartransport; Endlosvordrucktransport; Endlosformulareingabe; Endlosvordruckeingabe; Endlosformulareinzug; Endlosvordruckeinzug; Endlosformularvorschub; Endlosvordruckvorschub
≠ single-form feed		
↑ form-feed		
		≠ Einzelformularzuführung
		↑ Formularzuführung
continuous information source	INF	kontinuierliche Informationsquelle
continuous load	TECH	Dauerbelastung
= permanent load; steady load; continuous charge; permanent charge		= Dauerlast; Dauerbeanspruchung
continuous loading	COMM.CABLE	Krarupisierung
= uniform loading; Krarup loading		[gleichmäßige Umwicklung mit Eisendraht]
continuous loop	TERM&PER	Endlosband
continuously adjustable	ELECTRON	stufenlos einstellbar
= continuously variable; infinitely variable		= stetig einstellbar; stufenlos veränderlich; stetig veränderlich
continuously loaded cable	COMM.CABLE	Krarup-Kabel
= Krarup cable		
continuously variable (→continuously adjustable)	ELECTRON	(→stufenlos einstellbar)
continuously variable slope delta modulation (→CVSDM)	MODUL	(→CVSDM)
continuous mode	DATA PROC	Zügig-Betrieb
[data are stored or transferred sequentially without interruptions]		[Daten werden ohne Unterbrechung nacheinander eingelesen, gespeichert oder übertragen]
= streaming mode		= Streaming-Betrieb
≠ start-stop mode		≠ Start-Stop-Betrieb
continuous one	DATA COMM	Dauereins
= all-ones		
continuous operation	TECH	Dauerbetrieb
= continious duty		
continuous paper	TERM&PER	Endlospapier
= continuous stationery; continuous stock; listing paper		≠ Einzelblatt
≠ single sheet		↓ Leporellopapier; Rollenpapier; Endlosformular
↓ fanfold paper; rollpaper; continuous form		
continuous play	CONS.EL	Endlosbetrieb
		= automatische Deck-zu-Deck-Umschaltung
continuous plotter paper	TERM&PER	Endlos-Plotterpapier
continuous-progression code (→cyclic code)	CODING	(→zyklischer Code)
continuous radiation (→bremsstrahlung)	PHYS	(→Bremsstrahlung)
continuous rollpaper (→rollpaper)	TERM&PER	(→Rollenpapier)
continuous signal	SWITCH	Dauerkennzeichen
↑ status identifier		↑ Zustandskennzeichen
continuous signal (→analog signal)	INF	(→Analogsignal)
continuous space bar	TERM&PER	automatische Leertaste
= continuous spacing key		
continuous spacing key (→continuous space bar)	TERM&PER	(→automatische Leertaste)
continuous spectrum	PHYS	kontinuierliches Spektrum
continuous stationery (→continuous paper)	TERM&PER	(→Endlospapier)
continuous stick-on label	TERM&PER	Endlos-Haftetikette
continuous stock (→continuous paper)	TERM&PER	(→Endlospapier)
continuous system	SYS.TH	kontinuierliches System
≠ discrete system		≠ diskretes System
continuous test (→permanent test)	QUAL	(→Dauerprüfung)
continuous tone	TYPOGR	Halbton
= contone; halftone		
continuous-tone image	DATA PROC	Mosaikbild
continuous value	EL.TECH	Gleichwert
≠ alternating component		≠ Wechselanteil
↓ dc component		↓ Gleichstromkomponente

continuous-wave operation ELECTRON (→Dauerstrichbetrieb)
(→continuous-wave mode)
continuous wave ELECTRON **Dauerstrich**
= CW
continuous-wave laser OPTOEL **Dauerstrichlaser**
= CW laser = CW-Laser
≠ pulsed laser ≠ Impulslaser
continuous-wave mode ELECTRON **Dauerstrichbetrieb**
= CW mode; continuous-wave operation; CW operation = CW-Betrieb
≠ Impulsbetrieb
≠ pulsed mode
continuous wave modulation MODUL **Gleichwellenmodulation**
= CW modulation
continuous-wave power ELECTRON **Dauerstrichleistung**
= CW power ≠ Impulsleistung
≠ pulsed power
continuous-wave radar RADIO LOC **Dauerstrichradar**
= CW radar = CW-Radar
continuous welding (→seam welding) METAL (→Nahtschweißung)
continuous zero DATA COMM **Dauernull**
= all-zeroes
continuum PHYS **Kontinuum**
contone (→continuous tone) TYPOGR (→Halbton)
contour (→outline) TECH (→Umriß)
contour analysis DATA PROC **Konturenanalyse**
contour control MECH **Formtreueprüfung**
contour error METAL **Nachformfehler**
contour fidelity (→form fidelity) TECH (→Formtreue)
contouring ANT **Konturkorrektur**
[to modify side lobes] [zur Veränderung von Nebenzipfeln]
contouring DATA PROC **Konturzeichnen**
contour integral MATH (→Umlaufintegral)
(→circulation)
contour line (→equipotential line) PHYS (→Äquipotentiallinie)
contour vibration MECH **Formschwingung**
contract (n.) ECON **Vertrag**
= Kontrakt
contract TECH **schrumpfen**
= shrink = schwinden
≈ constrict ≈ verengen
≠ expand ≠ ausdehnen
contract awarder ECON **Auftraggeber**
= contractor 2 ≠ Auftragnehmer
≠ supplier ↑ Vertragsteilnehmer
↑ contractor 1
contracted mode LING **Allegroform**
[e.g. "math" for "mathematics"] [umgangssprachliche Verkürzung, z.B. „Demo" für „Demonstration"]
≠ regular mode
= Schnellsprechform
≠ Lentoform
contractible TECH **zusammenziehbar**
contracting force MECH **Schrumpfkraft**
= shrinking force
contracting partner (→contract party) ECON (→Vertragspartner)
contracting party (→contract party) ECON (→Vertragspartner)
contraction TECH **Schrumpfung**
= shrinkage; diminution; dwindling; loss = Schwund; Schwindung; Zusammenziehung
≈ constriction ≈ Verengung
contractor 2 (→contract awarder) ECON (→Auftraggeber)
contractor 3 (→supplier 2) ECON (→Auftragnehmer)
contractor 1 (→contract party) ECON (→Vertragspartner)
contract party ECON **Vertragspartner**
= contracting party; contracting partner; contractor 1 = Vertragsteilnehmer
↓ Auftraggeber; Auftragnehmer
↓ contract awarder; supplier
contract termination ECON **Vertragsauflösung**
contractual ECON **vertraglich**
contractual penalty ECON **Konventionalstrafe**
= agreed penalty = Vertragsstrafe
↓ penalty for delay ↓ Verzugsstrafe
contradiction SCIE **Widerspruch**
= Kontradiktion
contradictory SCIE **widersprüchlich**
= kontradiktorisch
contradirectional (→reverse) TECH (→gegensinnig)
contra-dope (v.t.) MICROEL **gegendotieren**
contragredient matrix MATH **kontragrediente Matrix**
contraposition SCIE **Kontraposition**
[negative proposition derived from a positive one] [negative Aussage aus positiver]
= opposition ≈ Antithese
≈ antithesis
contra-rotating TECH (→gegenläufig)
(→countermoving)
contrast (n.) TV **Kontrast**
contrast control TV **Kontrastregler**
contrast enhancement TERM&PER **Kontrastverstärkung**
contrast fidelity (→contrast rendition) TV (→Kontrastwiedergabe)
contrast range TV **Kontrastumfang**
contrast rendition TV **Kontrastwiedergabe**
= contrast fidelity
contravariance MATH **Kontravarianz**
contravariant MATH **kontravariant**
contravention COLLOQ **Zuwiderhandlung**
contribution ECON **Beitrag**
contribution LING **Beitrag**
= Artikel
contribution level (→studio quality level) BROADC (→Studioqualität)
control (v.t.) CONTROL **steuern**
≈ regulate ≈ regeln
control (n.) ELECTRON **Bedienelement**
= actuator = Bedienungselement; Betätigungselement
control- CONTROL **Steuer-**
= actuating
control (n.) TECH **Steuerung**
control (→drive 1) ELECTRON (→Ansteuerung)
control (→adjustment) ELECTRON (→Einstellung)
control (→check) TECH (→prüfen)
control (→open-loop control) CONTROL (→Steuerung)
control accuracy CONTROL **Regelgüte**
= control precision = Regelgenauigkeit
control air TECH **Steuerluft**
= controlling air
control amplifier CONTROL **Regelverstärker**
control area CONTROL **Regelfläche**
control behaviour (→control response) CONTROL (→Regelverhalten)
control bit DATA PROC **Steuerbit**
control block 1 DATA PROC **Kennblock**
= Steuerblock
control block 2 DATA PROC **Steuerbereich**
[storage area] [im Speicher]
control board BROADC **Kontrollpult**
= control console; console = Regiepult
control board (→control desk) TECH (→Kontrollpult)
control box TECH **Schaltkasten**
control break DATA PROC **Kontrollunterbrechung**
control bus DATA PROC **Steuerbus**
↑ bus = Steuerpfad; Kontrollbus; Kontrollpfad
↑ Bus
control byte DATA PROC **Steuerbyte**
control cabinet (→control cubicle) POWER SYS (→Schaltschrank)
control card DATA PROC **Steuerlochkarte**
= parameter card = Steuerkarte; Parameterkarte
control center (AM) (→main station) TELECONTR (→Zentrale)
control centre (BRI) (→main station) TELECONTR (→Zentrale)
control channel DATA PROC **Steuerkanal**
control character DATA COMM **Steuerzeichen**
↓ format control character; device control character; transmission control character ↓ Formatsteuerzeichen; Gerätesteuerzeichen; Übertragungssteuerzeichen
control character 2 INF.TECH **Steuerzeichen**
≈ check character ≈ Kontrollzeichen

control character 1 (→check character) INF.TECH (→Kontrollzeichen)
control characteristic ELECTRON **Steuerkennlinie**
= Übertragungskennlinie
control chip MICROEL **Steuerchip**
= Steuerbaustein; Regelchip; Regelbaustein
control circuit 1 CIRC.ENG **Steuerkreis**
= driving circuit; drive circuit = Ansteuerschaltung
↓ trigger circuit 1 ↓ Triggerschaltung
control circuit 2 CIRC.ENG **Überwachungsschaltung**
= supervisory circuit; monitoring circuit
control code TELEGR **Kontrollcode**
control code DATA PROC **Steuerschlüssel**
= control key = Steuercode
control command DATA PROC **Steuerbefehl**
[from computer to peripheral device] [von Computer zu Peripheriegerät]
↓ channel command word ↓ Kanalbefehlwort
control computer CONTROL **Steuerrechner**
= embedded computer
control-configured vehicle RADIO NAV **CCV**
= CCV
control console (→control desk) TECH (→Kontrollpult)
control console (→control board) BROADC (→Kontrollpult)
control criterion CONTROL **Gütekriterium**
control cubicle POWER SYS **Schaltschrank**
= control cabinet; panel cabinet; cabinet; cubicle = Schaltkabine
control current ELECTRON **Steuerstrom**
control cylinder (→Wehnelt cylinder) ELECTRON (→Wehneltzylinder)
control data DATA PROC **Steuerdaten**
= control parameter; control information = Steuerparameter; Steuerinformation; Steuerinstruktion; Steuerungsdaten
control desk TECH **Kontrollpult**
= operating desk; switch-desk; control board; control console; console; operator's console; operator's desk; operator's board; attendant's console; attendant's desk; attendant's board = Schaltpult; Steuerpult; Bedienpult; Bedienungspult; Bedienungskonsole; Systemkonsole
≈ service panel ≈ Bedienungsfeld
↑ operator's position ↑ Bedienplatz
control deviation (→deviation) CONTROL (→Regelabweichung)
control device CONTROL **Regelglied**
control digit (→check digit) CODING (→Kontrollziffer)
control diode CIRC.ENG **Regeldiode**
control direction TELECONTR **Steuerungsrichtung**
[from control center to controlled object] [von steuernder Stelle zum gesteuerten Objekt]
control drift MICROEL **Aussteuerungsdrift**
[drive-dependent variation of parameters] [aussteuerungsbedingte Parameteränderung]
control electrode ELECTRON **Steuerelektrode**
control element CONTROL **Steuerelement**
= Schaltelement; Stellelement
control engineering TECH **Steuer-und Regelungstechnik**
= automatic control; control technology = Regelungstechnik
↓ cybernetics; robotics; automation ↓ Kybernetik; Robotik; Automatisierung
control equipment EQUIP.ENG **Steuergerät**
= controller
control experiment PHYS **Gegenversuch**
control field 1 DATA PROC **Kontrollfeld**
[memory field] [Speicherbereich]
≈ instruction register; control register ≈ Befehlsregister; Steuerregister
control field 2 (→key) DATA PROC (→Kennbegriff)
control frequency TRANS **Steuerfrequenz**
[FDM] [TF-Technik]
control function ELECTRON **Steuerungsaufgabe**
control function layer DATA COMM **Steuerfunktionsschicht**
control grid ELECTRON **Steuergitter**

control head TERM&PER **Steuerkopf**
control hysteresis CONTROL **Steuerhysterese**
control information CONTROL **Steuerinformation**
control information (→switching information) SWITCH (→vermittlungstechnische Information)
control instruction CONTROL **Steueranweisung**
[of process control computer to controlling device] [von Prozeßrechner zu Steuergerät]
control instruction (→control statement) DATA PROC (→Steueranweisung)
control key 1 TERM&PER **Steuerungstaste**
[activates functions if pressed simultaneously with other key] [löst bei gleichzeitigem Drücken anderer Taste Funktionen aus]
= CONTROL key; CTRL key; CTR key = Steuertaste 1; STRG-Taste; CONTROL-Taste; CTRL-Taste; Kontrolltaste
≈ function key; command key ≈ Funktionstaste; Kommandotaste
control key 2 (→function key) TERM&PER (→Funktionstaste)
control key (→control code) DATA PROC (→Steuerschlüssel)
CONTROL key (→control key 1) TERM&PER (→Steuerungstaste)
control keyboard TERM&PER **Bedientastatur**
= Steuertastatur
control knob ELECTRON **Bedienungsknopf**
control label DATA PROC **Steuerkennung**
controllability TECH **Steuerbarkeit**
controllable TECH **steuerbar**
≈ regulable ≈ regulierbar
control language DATA PROC (→Kommandosprache)
(→command control language)
controll-current terminal MICROEL (→Steueranschluß)
(→gate terminal)
controlled accessibility SWITCH **gesteuerte Erreichbarkeit**
= controlled availability
controlled availability SWITCH (→gesteuerte Erreichbarkeit)
(→controlled accessibility)
controlled condition CONTROL (→Regelgröße)
(→controlled magnitude)
controlled convertor POWER SYS **steuerbarer Gleichrichter**
= controlled rectifier
controlled current source NETW.TH **gesteuerte Stromquelle**
controlled magnitude CONTROL **Regelgröße**
[magnitude which has to be controlled by the control circuit] [durch einen Regelkreis zu beeinflussende Größe]
= controlled variable; controlled condition; directly controlled variable ≠ Führungsgröße
≠ reference magnitude
controlled measuring device INSTR (→Meßresponder)
(→measuring responder)
controlled rectifier POWER SYS (→steuerbarer Gleichrichter)
(→controlled convertor)
controlled slip (→frame slip) CODING (→Rahmenschlupf)
controlled source NETW.TH **gesteuerte Quelle**
controlled system CONTROL **Regelstrecke**
= directly controlled system; plant = Steuerstrecke
controlled test DATA PROC **Kontrollertest**
controlled variable CONTROL (→Regelgröße)
(→controlled magnitude)
controlled voltage source NETW.TH **gesteuerte Spannungsquelle**
controller DATA PROC **Kontroller**
[HW or SW controlling computer parts or peripherals] [Hardware oder Software die Geräteteile eines Coppmputers steuert, z.B. Peripheriegeräte]
= Controller
controller CONTROL **Regler**
= regulator; governor
controller INSTR **Kontroller**
= Controller
controller (→control unit) EQUIP.ENG (→Steuereinheit)
controller (→control equipment) EQUIP.ENG (→Steuergerät)
controller coupling CONTROL **Reglerverknüpfung**
= coupling = Verknüpfung 1

control lever

control lever	TECH	**Schalthebel**
= switch lever		
control line	SWITCH	**Steuerleitung**
= control wire		
controlling air (→control air)	TECH	(→Steuerluft)
controlling torque (→adjusting torque)	TECH	(→Einstellmoment)
controlling valve (→regulating valve)	TECH	(→Regelventil)
control logic	ELECTRON	**Ansteuerlogik**
= drive logic; trigger logic		= Steuerlogik
control logic	DATA PROC	**Steuerlogik**
		= Ansteuerlogik
control loop (→closed-loop control circuit)	CONTROL	(→Regelschleife)
control mark (→tape mark)	TERM&PER	(→Bandmarke)
control memory	SWITCH	**Haltespeicher**
control mode	DATA PROC	**Steuermode**
control module	EQUIP.ENG	**Steuerbaugruppe**
control mouse (pl. c.mice or c.mouses) [hand-held device to control cursor position on a screen]	TERM&PER	**Maus** (pl.Mäuse) [handgeführtes Gerät zur Steuerung der Schreibmarkenposition auf einem Bildschirm]
= mouse		
≈ track ball		≈ Rollkugel
↑ rollover device		↑ Abrollgerät
control oscillator	CIRC.ENG	**Steueroszillator**
= booster oscillator		= Hilfsoszillator
control packet	DATA COMM	**Steuerpaket**
control panel	TECH	**Schalttafel**
= switch panel		= Schaltbrett
≈ distribution panel		
control panel (→operating panel)	EQUIP.ENG	(→Bedienfeld)
control point	CONTROL	**Regelpunkt**
control post	TECH	**Stellwerk**
control post	CONTROL	**Leitstand**
≈ control center		≈ Leitzentrale
control power	ELECTRON	**Steuerleistung**
= driving power		
control precision (→control accuracy)	CONTROL	(→Regelgüte)
control procedure	DATA PROC	**Steuerungsverfahren**
control program [controls overall operations of CPU and peripherals]	DATA PROC	**Organisationsprogramm** [steuert allgemeine Abläufe des Rechners und seiner Peripheriegeräte]
= supervisor routine; control routine; executive control program; housekeeping register; master program; automatic sequencing		= Organisationsregister; Steuerprogramm; Programmablaufsteuerung; Monitor
↑ operating system		↑ Betriebssystem
↓ executive program; job management; task management; data management		↓ Hauptsteuerprogramm; Auftragsverwaltung; Task-Management; Datenverwaltung
control pulse	ELECTRON	**Steuerimpuls**
= drive pulse		↓ Triggerimpuls; Ansteuerimpuls; Torimpuls
↓ trigger pulse; drive pulse; gate control pulse		
control pulse (→drive pulse)	ELECTRON	(→Ansteuerimpuls)
control quantity	CONTROL	**Steuergröße**
≈ reference magnitude		≈ Führungsgröße
control range	CONTROL	**Stellbereich**
= correcting range		≈ Regelbereich
control range (→dynamic range)	ELECTRON	(→Dynamikbereich)
control range (→regulation range)	CONTROL	(→Regelbereich)
control rate	CONTROL	**Regelgeschwindigkeit**
control ratio	CONTROL	**Steuerverhältnis**
control register (→instruction register)	DATA PROC	(→Befehlsregister)
control register (→control field)	DATA PROC	(→Kontrollfeld)
control response	CONTROL	**Regelverhalten**
= control behaviour		
control room	BROADC	**Regieraum**
control room equipment	TELECONTR	**Wartenperipherie**
control routine (→control program)	DATA PROC	(→Organisationsprogramm)
control section 1	DATA PROC	**Programmabschnitt**
≈ dummy section		≈ Pseudoabschnitt
control section 2 (→control unit)	DATA PROC	(→Steuerwerk)
control sequence	DATA PROC	**Steuerungsfolge**
		= Steuerfolge; Steuerungsablauf
control signal	CONTROL	**Steuersignal**
= actuating signal		
control statement	DATA PROC	**Steueranweisung**
= control instruction		
control station (→transmitting terminal)	DATA COMM	(→Sendestation)
control station (→main station)	TELECONTR	(→Zentrale)
control string	DATA COMM	**Steuerzeichenfolge**
control sum (→control total)	ECON	(→Kontrollsumme)
control switch	POWER ENG	**Steuerschalter**
control system	CONTROL	**Regelungssystem**
control system	DATA PROC	**Steuersystem**
= control unit		
control system call (→supervisor call)	DATA PROC	(→Organisationsaufruf)
control technology (→control engineering)	TECH	(→Steuer-und Regelungstechnik)
control terminal (→control unit)	TERM&PER	(→Bediengerät)
control thermistor	COMPON	**Regelheißleiter**
control time	CONTROL	**Stellzeit**
= correcting time		
control token (→token)	DATA COMM	(→Sendeberechtigung)
control total	ECON	**Kontrollsumme**
= proof total; control sum; proof sum		
control tower	AERON	**Kontrollturm**
control track	TV	**Steuerspur**
		= Führungsspur
control transfer	DATA PROC	**Steuerungsübergabe**
control transformer (→trigger transformer)	CIRC.ENG	(→Ansteuerübertrager)
control unit	TERM&PER	**Bediengerät**
= control terminal		= Bedieneinheit; Bedienterminal
control unit	EQUIP.ENG	**Steuereinheit**
= controller		= Steuereinrichtung; Steuerteil
control unit [N + 1]	RADIO REL	**Steuereinsatz** [N + 1]
control unit	CONTROL	**Steuerwerk**
control unit [functional part of CPU, decodes the instructions and controls its execution]	DATA PROC	**Steuerwerk** [Teil des Prozessors in der Zentraleinheit, entschlüsselt die Befehle und steuert deren Ausführung]
= control section 2		= Leitwerk; Kommandowerk
↑ processor; central processing unit		↑ Prozessor; Zentraleinheit
control unit (→control system)	DATA PROC	(→Steuersystem)
control variable [generated from the forward controlling element, drives the directly controlled system]	CONTROL	**Stellgröße** [vom Stellglied erzeugt, steuert die Regelstrecke]
= actuating variable; manipulated variable; manipulation variable; correcting variable; correcting quantity		
control voltage	ELECTRON	**Steuerspannung**
= drive voltage		= Regelspannung
control winding (→drive winding)	COMPON	(→Steuerwicklung)
control wire (→control line)	SWITCH	(→Steuerleitung)
control word 1	DATA PROC	**Kontrollwort**
control word 2	DATA PROC	**Steuerwort**
controversy (→dispute)	ECON	(→Streit)
conurbation (→metropolitan area)	ECON	(→Ballungsgebiet)
convection current	PHYS	**Konvektionsstrom**
		= Konvektion

convection current		RADIO PROP	**Konvektionsstrom**	**conversation time**	TELEPH	**Gesprächsdauer**
[RF current flowing through the earth]			[durch die Erde fließender HF-Strom]	= speech time ↑ call duration [TELEC]		= Gesprächszeit ↑ Verbindungsdauer [TELEC]
≠ dielectric current [EL.SCIE]			≠ elektrischer Verschiebungsstrom [EL.SCIE]	**conversion**	PHYS	**Konversion**
convenience (→operator convenience)		TECH	(→Bedienungskomfort)	= transformation		= Umwandlung; Transformation; Transformierung
convenience feature (→operator convenience)		TECH	(→Bedienungskomfort)	**conversion**	MATH	**Umrechnung**
						= Konversion
conveniencies (→premises)		CIV.ENG	(→Räumlichkeiten)	**conversion**	TECH	**Umrüstung**
convention (→programming convention)		DATA PROC	(→Programmkonvention)	= retrofitting ≈ adaption		≈ Anpaßung
convention (→agreement 1)		COLLOQ	(→Vereinbarung)	**conversion**	INF.TECH	**Umwandlung**
conventional (→usual)		COLLOQ	(→üblich)	[of signal form]		[einer Signalform, z.B. von analog auf digital]
conventional load		TELEC	**konventionelle Belastung**			= Wandlung; Umsetzung; Konvertierung
conventional telephonic service (→POT service)		TELEC	(→konventioneller Fernsprechdienst)	**conversion** (→mixing)	HF	(→Mischung)
converge		MATH	**konvergieren**	**conversion** (→translation)	MODUL	(→Umsetzung)
convergence		MATH	**Konvergenz**	**conversion** (→transformation)	TECH	(→Umwandlung)
convergence		TV	**Konvergenz**	**conversion** (→data conversion)	DATA PROC	(→Datenkonvertierung)
convergence criterion		MATH	**Konvergenzkriterium**	**conversational language** (→colloquial)	LING	(→Umgangssprache)
convergence electrode		ELECTRON	**Konvergenzelektrode**			
[image tube]			[Bildröhren]	**conversion coefficient**	PHYS	**Konversionskoeffizient**
convergence in probability		MATH	**stochastische Konvergenz**	**conversion equipment** (→data conversion unit)	DATA PROC	(→Umsetzanlage)
convergence line		MATH	**Fluchtlinie**			
= alignment line				**conversion factor**	MATH	**Umrechnungsfaktor**
convergence magnet		ELECTRON	**Konvergenzmagnet**	**conversion kit**	TECH	**Umrüstsatz**
			[Bildröhren]	**conversion loss**	HF	**Mischdämpfung**
convergence plane		TV	**Konvergenzebene**	**conversion program**	DATA PROC	**Umsetzprogramm**
= convergence surface			= Konvergenzfläche	= conversion routine		= Konvertierungsprogramm; Umsetzroutine; Konvertierungsroutine
convergence point		MATH	**Fluchtpunkt**			
= alignment point			= Konvergenzpunkt			
convergence point		TV	**Konvergenzpunkt**	**conversion-program control**	DATA PROC	**Umsetzer-Steuerprogramm**
convergence surface (→convergence plane)		TV	(→Konvergenzebene)			
				conversion rate (→transformation rate)	TECH	(→Umwandlungsgeschwindigkeit)
convergence time		CIRC.ENG	**Konvergenzzeit**			
[e.g. of digital filters]			[z.B. bei Digitalfiltern]	**conversion routine** (→conversion program)	DATA PROC	(→Umsetzprogramm)
convergent		MATH	**konvergent**			
= converging			= konvergierend	**conversion table**	TECH	**Umrechnungstabelle**
convergent integral		MATH	**konvergentes Integral**	= translation table		
converging (fig.)		COLLOQ	**konvergierend**	**conversion unit** (→data conversion unit)	DATA PROC	(→Umsetzanlage)
= concurrent 1			[fig.]			
			= zusammenlaufend	**convert** (v.t.)	DATA PROC	**konvertieren**
converging (→convergent)		MATH	(→konvergent)			= wandeln; umwandeln
converging lens		PHYS	**Sammellinse**	**convert**	TECH	**umrüsten**
[thickest in the center]			[in der Mitte am dicksten]	= retrofit		≈ anpassen
= focusing lens			↓ bikonvexe Linse; plankonvexe Linse; konkavkonvexe Linse	≈ adapt		
				converted quadripole (→converted two-port)	NETW.TH	(→konvertierter Vierpol)
converging wave		PHYS	**konvergierende Welle**			
conversation		TELEPH	**Gespräch**	**converted two-port**	NETW.TH	**konvertierter Vierpol**
≈ call; connection [TELEC]; speech			≈ Anruf; Verbindung [TELEC]; Sprechen	[with impedance transformers on both sides]		[beidseitig mit Impedanzwandler]
conversational (→interactive)		DATA PROC	(→Dialog-)	= converted quadripole		
conversational data processing (→interactive processing)		DATA PROC	(→Dialogbetrieb)	**converter**	DATA PROC	**Konverter**
				[hardware or software to convert data]		[Hardware oder Software zur Datenwandlung]
conversational device (→interactive terminal)		DATA COMM	(→Dialogstation)	= convertor		
				converter	BROADC	**Konverter**
conversational information system		DATA PROC	**passives Informationssystem**	**converter**	CIRC.ENG	**Umsetzer**
= passive information system; interactive information system			[konversationelles Informationssystem]	[converts a frequency or a code]		[wandelt eine Frequenz oder einen Code]
				= convertor (BRI)		
conversational interaction		DATA PROC	(→Dialogbetrieb)	↓ code converter; series-parallel converter; parallel-series converter		↓ Codeumsetzer; Serien-Parallel-Umsetzer; Parallel-Serien-Umsetzer
[DATA COMM] (→interactive processing)						
				converter	RADIO	**Umsetzer**
conversational language		DATA PROC	(→natürliche Programmiersprache)	**converter** (→impedance converter)	NETW.TH	(→Impedanzkonverter)
(→conversational programming language)						
				converter (→static power converter)	POWER SYS	(→Stromrichter)
conversational mode		DATA PROC	(→Dialogbetrieb)			
(→interactive processing)				**converter** (→convertor 1)	POWER SYS	(→Umformer 1)
conversational processing		DATA PROC	(→Dialogbetrieb)	**converter** (→transducer)	COMPON	(→Wandler)
(→interactive processing)				**converter** (→data converter)	DATA PROC	(→Datenkonverter)
conversational programming language		DATA PROC	**natürliche Programmiersprache**			
= conversational language				**converter connection**	POWER SYS	**Stromrichterschaltung**
conversational terminal		DATA COMM	(→Dialogstation)	**converter feed motor**	POWER SYS	**Stromrichtermotor**
(→interactive terminal)				**converter valve**	COMPON	**Stromrichterventil**
conversation test		TELEPH	**Sprechprobe**			

convertibility

convertibility	TECH	**Konvertierbarkeit**
		≈ Umformbarkeit
convertible	TECH	**umrüstbar**
= adaptable		= konvertierbar
≈ expandable		≈ erweiterbar
convertor 1 (IEC)	POWER SYS	**Umformer 1**
[rotating machine to transform the type of current]		[umlaufende Maschine zur Umwandlung der Stromart]
= converter		
convertor 2 (IEC)	POWER SYS	(→Umrichter)
(→Voltage system converter)		
convertor (→converter)	DATA PROC	(→Konverter)
convertor (BRI)	CIRC.ENG	(→Umsetzer)
(→converter)		
convertor (→data converter)	DATA PROC	(→Datenkonverter)
convex	MATH	**konvex**
[hollowed outward]		[nach außen gewölbt]
≈ raised [TECH]		≈ erhaben [TECH]
≠ concave		≠ konkav
convex (→cambered)	MECH	(→gewölbt)
convexity	MATH	**Konvexität**
convex mirror	PHYS	**Wölbspiegel**
convexo-concave	MATH	**konvex-konkav**
= concavo-convex		= konkav-konvex; hohlerhaben
convey	TECH	**befördern**
= transport		= fördern; transportieren
conveyer belt	TECH	**Förderband**
= belt conveyer; conveyor		= F²rderanlage
conveyor (→conveyer belt)	TECH	(→Förderband)
conveyor trip (→rope traction)	MECH	(→Seilzug)
convolution	MATH	**Faltung**
= folding; fold-over		
convolution duration	MIL.COMM	**Faltungsdauer**
convolution efficiency	MIL.COMM	**Faltungseffizienz**
convolution integral	MATH	**Faltungsintegral**
convolver	MIL.COMM	**Convolver**
cookbook (fig.)	TECH	**Kochrezept** [fig.]
		= Kochbuch (fig.)
cool (v.i.)	PHYS	**erkalten**
		≈ abkühlen (v.i.)
cool (v.t.)	TECH	**kühlen**
= refrigerate		≈ abkühlen (v.t.)
coolant (→cooling liquid)	TECH	(→Kühlflüssigkeit)
coolant (→refrigerant)	TECH	(→Kühlmittel)
coolant air (→cooling air)	TECH	(→Kühlluft)
coolant liquid (→cooling liquid)	TECH	(→Kühlflüssigkeit)
coolant pump	TECH	**Kühlpumpe**
cooler (→refrigerator)	TECH	(→Kühler)
cooling (n.)	TECH	**Kühlung**
= refrigeration		= Abkühlung
cooling agent (→refrigerant)	TECH	(→Kühlmittel)
cooling air	TECH	**Kühlluft**
= coolant air		↑ Kühlmittel
↑ refrigerant		
cooling circuit	TECH	**Kühlkreislauf**
cooling clamp	MICROEL	**Kühlschelle**
cooling coil	MECH	**Kühlschlange**
cooling column (→cooling tower)	TECH	(→Kühlturm)
cooling device (→refrigerator)	TECH	(→Kühler)
cooling fan (→fan)	TECH	(→Ventilator)
cooling fin	EQUIP.ENG	**Kühlrippe**
= cooling rib		
cooling fin (→cooling vane)	COMPON	(→Kühlfahne)
cooling flange	TECH	**Kühlflansche**
cooling hole	TECH	**Lüftungsloch**
= ventilation hole		= Belüftungsloch; Ventilationsloch
cooling jacket	TECH	**Kühlmantel**
cooling liquid	TECH	**Kühlflüssigkeit**
= coolant liquid; coolant		↑ Kühlmittel
↑ refrigerant		
cooling medium (→refrigerant)	TECH	(→Kühlmittel)
cooling opening	TECH	**Lüftungsöffnung**
= ventilation opening; cooling vent		= Belüftungsöffnung
cooling plate	TECH	**Kühlplatte**
≈ cooling sheet		≈ Kühlblech
cooling rib (→cooling fin)	EQUIP.ENG	(→Kühlrippe)
cooling sheet	TECH	**Kühlblech**
≈ cooling plate		≈ Kühlplatte
cooling slit (→ventilation slit)	TECH	(→Lüftungsschlitz)
cooling surface	PHYS	**Kühlfläche**
cooling system	TECH	**Kühlsystem**
cooling tower	TECH	**Kühlturm**
= cooling column		
cooling vane	COMPON	**Kühlfahne**
= ventilating vane; cooling fin		
cooling vent (→cooling opening)	TECH	(→Lüftungsöffnung)
cooling water	TECH	**Kühlwasser**
↑ refrigerant		↑ Kühlmittel
cool light reflector lamp	TECH	**Kaltlichtspiegellampe**
cool thoroughly	TECH	**auskühlen**
cooperate (v.i.) (NAM)	ECON	**zusammenarbeiten**
= co-operate (BRI)		= kooperieren
co-operate (BRI) (→cooperate)	ECON	(→zusammenarbeiten)
cooperate (→interwork)	TECH	(→zusammenwirken)
cooperation	ECON	**Zusammenarbeit**
		= Kooperation
co-operative (→codirectional)	TECH	(→gleichsinnig)
cooperative information processing	INF.TECH	**kooperative Informationsverarbeitung**
Cooper pair	PHYS	**Cooper-Paar**
coopers rivet	MECH	**Küferniet**
		= Böttcherniet
coordinate (n.)	MATH	**Koordinate**
↓ abcissa; ordinate; polar coordinate		↓ Abszisse; Ordinate; Polarkoordinate
coordinate axis	MATH	**Koordinatenachse**
coordinate field	MATH	**Koordinatenfeld**
= coordinate lattice; coordinate frame		= Koordinatenraster; Koordinatennetz
≈ grid [TECH]		≈ Raster [TECH]
coordinate frame (→coordinate field)	MATH	(→Koordinatenfeld)
coordinate lattice	ENG.DRAW	**Koordinatenraster**
coordinate lattice (→coordinate field)	MATH	(→Koordinatenfeld)
coordinate paper	ENG.DRAW	**Koordinatenpapier**
= gridsheet		= Rasterpapier
coordinate plane	MATH	**Koordinatenebene**
coordinate space	MATH	**Koordinatenraum**
coordinate system	MATH	**Koordinatensystem**
coordinate value	MATH	**Koordinatenwert**
coordination	SCIE	**Koordinierung**
		= Koordination; Abstimmung
coordination area	RADIO	**Koordinierungsgebiet**
coordination distance	RADIO	**Koordinierungsentfernung**
coordination function	DATA COMM	**Koordinierungsfunktion**
coordination processor	SWITCH	**Koordinationsprozessor**
coordinatograph	MICROEL	**Koordinatenschreiber**
		= Koordinatengraph; Koordinatograph
cophasal (→in-phase)	PHYS	(→gleichphasig)
copier (→copying machine)	OFFICE	(→Kopiergerät)
co-polar	RADIO	**kopolar**
≠ cross-polar		≠ kreuzpolar
co-polarization	RADIO	**Kopolarisation**
≈ working polarization		= Gleichpolarisation
≠ cross-polarization		≈ Nutzpolarisation
		≠ Kreuzpolarisation
co-polarization pattern	ANT	**Kopolarisationsdiagramm**
copper	CHEM	**Kupfer**
= Cu		= Cu
copper-base alloy	METAL	**Kupferlegierung**
copper-clad (v.t.)	METAL	**kupferkaschieren**
		= kupferplattieren
copper cladding	METAL	**Kupferplattierung**
copper-clad wire	COMM.CABLE	**Stahlkupferdraht**
		= Staku-Draht; Kupfermanteldraht

copper conductor	PHYS	**Kupferleiter**
copper loss	EL.TECH	**Kupferverlust**
copper oxide rectifier	COMPON	**Kupferoxydul-Gleichrichter**
copper pair	COMM.CABLE	**Kupferaderpaar**
= copper wire pair		= Kupferadernpaar; Kupferdoppelader
copper-plate (v.t.)	METAL	**verkupfern**
↑ electroplate		↑ galvanisieren
copperplate (n.)	TYPOGR	**Kupferstich**
copper plating	METAL	**Verkupferung**
↑ galvanizing		↑ Galvanisierung
copper wire	METAL	**Kupferdraht**
copper wire pair (→copper pair)	COMM.CABLE	(→Kupferaderpaar)
coprocessor	MICROEL	**Koprozessor**
		= Coprozessor
coprocessor 1	DATA PROC	**Koprozessor 1**
[specialized processor to free main processor]		[spezialisierter Prozessor zur Entlastung Hauptprozessors]
↓ graphics coprocessor; maths coprocessor		= Coprozessor; Hilfsprozessor
		↓ Graphik-Koprozessor; mathematischer Koprozessor
coprocessor 2	DATA PROC	**Koprozessor 2**
[a CPU working in tandem]		[parallel arbeitende Zentraleinheit]
copy (v.t.)	DATA PROC	**kopieren**
[to reproduce data without change of content, however possibly of their physical form of storage]		[Daten ohne Änderung des Inhalts, jedoch eventuell der physikalischen Speicherungsform, vervielfältigen]
≈ duplicate (v.t.)		≈ duplizieren
copy 2 (n.)	OFFICE	**Pause**
= reproduction; print		= Abzug; Kopie; Vervielfältigung
≈ duplicate; facsimile		≈ Duplikat; Faksimile
≠ original		≠ Original
↑ copy 1		↑ Exemplar
↓ photocopy; blueprint		↓ Fotokopie; Lichtpause
copy 1 (n.)	OFFICE	**Exemplar**
[printed reproduction including the first one]		[Original oder Kopie]
↓ first copy		↓ Original; Kopie
copy 3 (n.)	OFFICE	**Vorlage 1**
[document subject to reproduction]		[für eine Reinschrift]
= model (n.)		= Konzept
copy (n.)	TECH	**Nachbildung**
= simulation; imitation; look-alike (n.)		= Kopie; Nachahmung; Imitation
≈ model		≈ Modell
copy (→dubb)	CONS.EL	(→Kopie)
copy (v.t.) (→print)	OFFICE	(→pausen)
copy (→dub)	CONS.EL	(→überspielen)
copy (v.t.) (→imitate)	TECH	(→nachbilden)
copy block (→note pad)	OFFICE	(→Notizblock)
copy-drill	MECH	**abbohren**
copy-grind	MECH	**kopierschleifen**
copy impact printer	TERM&PER	**Kopiendrucker 2**
[with sufficient impact for copies]		[mit ausreichendem Anschlag für Durchschläge]
copying (→dubbing)	CONS.EL	(→Überspielen)
copying lead (→dubbing lead)	CONS.EL	(→Überspielkabel)
copying machine	OFFICE	**Kopiergerät**
= copier; duplicator; duplicating equipment		= Kopierer; Vervielfältigungsmaschine; Dupliziergerät
↓ desk copier; pedestal copier		↓ Tischkopierer; Standkopierer
copying paper set	TERM&PER	**Durchschreibsatz**
copy-mill	MECH	**kopierfräsen**
copy paper	OFFICE	**Kopierpapier**
copy protect (→copy protection)	DATA PROC	(→Kopierschutz)
copy-protected	DATA PROC	**kopiergeschützt**
copy protection	DATA PROC	**Kopierschutz**
= copy protect		
↓ dongle		↓ Kopierschutzschaltung
copyright	ECON	**Urheberrecht**
		= Copyright
copyrighted	ECON	**urheberrechtlich geschützt**
copyright sign	ECON	**Copyrightzeichen**
[a C within a circle]		[eingekreistes C]
= copyright symbol		
copyright symbol (→copyright sign)	ECON	(→Copyrightzeichen)
cord	TECH	**Schnur**
[thin, consisting of several strands of filaments]		[dünn, aus mehreren Fäden oder Drähten gebildet]
= lace; string; cordon		= Schnüre; Kordel; Bindfaden
≈ thread		≈ Faden
cord (→connecting cord)	EQUIP.ENG	(→Anschlußschnur)
cord (→patch cord)	ELECTRON	(→Steckschnur)
Cordially yours (AM) (→Very sincerely yours)	ECON	(→Mit freundlichen Grüßen Ihr)
cord jack	TELEPH	**Klinkenkupplung**
		≈ Klinken-Einbaubuchse
		↑ Klinke
cordless	EQUIP.ENG	**schnurlos**
		= kabellos; drahtlos
cordless (→wireless)	TELEC	(→drahtlos)
cordless microphone system	EL.ACOUS	**drahtlose Mikrophonanlage**
cordless soldering iron	ELECTRON	**kabelloser Lötkolben**
cordless telephone	TELEPH	**schnurloser Fernsprechapparat**
= wireless telephone		
		= schnurloses Telefon; schnurloses Telephon
cordon (→cord)	TECH	(→Schnur)
cord terminal	COMPON	**Schnurstecker**
cord wood technique	EQUIP.ENG	**Bündelholzbauweise**
		= Kompaktbausteintechnik
core	TECH	**Kern**
= nucleus; kernel		= Seele
core	EL.TECH	**Kern**
[transformer]		[Transformator]
core (→atomic core)	PHYS	(→Atomkern)
core (→cable core)	COMM.CABLE	(→Kabelseele)
core (→coil core)	COMPON	(→Spulenkern)
core assembly (→laminated core)	EL.TECH	(→Schnittbandkern)
core cell	MICROEL	**Kernzelle**
= array cell; basic cell		= Grundzelle
coreless	EL.TECH	**ohne Eisenkern**
core loss	EL.TECH	**Eisenverlust**
= iron loss		
core matrix	ELECTRON	**Kernmatrix**
core memory (→magnetic core memory)	TERM&PER	(→Magnetkernspeicher)
core memory (→main memory 1)	DATA PROC	(→Hauptspeicher 1)
core mode	OPT.COMM	**Kernmodus**
= bound mode; guided mode		
core-resident	DATA PROC	**kernspeicherresident**
coresident		**koresident**
[simultaneously in main memory]		[gleichzeitig im Hauptspeicher]
core stack (→laminated core)	EL.TECH	(→Schnittbandkern)
core storage (→magnetic core memory)	TERM&PER	(→Magnetkernspeicher)
core storage (→main memory 1)	DATA PROC	(→Hauptspeicher 1)
core store (→magnetic core memory)	TERM&PER	(→Magnetkernspeicher)
core-type transformer	EL.TECH	**Kerntransformator**
Coriolis force	PHYS	**Coriolis-Kraft**
cork	TECH	**Kork**
corkscrew (→helical)	TECH	(→schraubenförmig)
corkscrew field (→helicoidal field)	PHYS	(→schraubenförmiges Feld)
corkskrew antenna (→helix antenna)	ANT	(→Wendelantenne)
cornel-reflector aerial (→corner-reflector antenna)	ANT	(→Winkelreflektor-Antenne)
corner	MECH	**Ecke**
corner	TECH	**Winkel**
		[Ecke]
corner aerial (→corner-reflector antenna)	ANT	(→Winkelreflektor-Antenne)
corner antenna (→corner-reflector antenna)	ANT	(→Winkelreflektor-Antenne)

corner fittings   TECH   **Eckbeschlag**
corner frequency (→limit frequency)   NETW.TH   (→Eckfrequenz)
corner reflector   ANT   **Winkelreflektor**
= V reflector   = Eckreflektor; Corner-Reflektor
corner-reflector antenna   ANT   **Winkelreflektor-Antenne**
= corner antenna; cornel-reflector aerial; corner aerial   = Corner-Reflektor-Antenne
corner reflector array   ANT   **Winkelreflektorfeld**
cornucopia antenna (→scimitar antenna)   ANT   (→Sichelantenne)
cornucopia shape (→scimitar cut)   COMPON   (→Sichelplattenschnitt)
corona 1 [electricity]   PHYS   **Korona** [Elektrizität]
corona 2 [optics]   PHYS   **Hof** [Optik]
≈ halo   = Kranz
   ≈ Lichthof
corona discharge   POWER SYS   **Koronaentladung**
= brush discharge [PHYS]   = Büschelentladung [PHYS]
corona discharge [POWER ENG]   PHYS   (→Büschelentladung)
(→brush discharge)
co-rotational   TECH   **mitrotierend**
co-routine (→utility program)   DATA PROC   (→Dienstprogramm)
corporate (adj.) (→proprietary 1)   ECON   (→firmeneigen)
corporate (→in-house)   ECON   (→firmenintern)
corporate business   ECON   **Verbundgeschäft**
= intercompany business; inter-group sale
corporate costumer   TELEC   **Firmenkunde**
= corporate subscriber   ≈ Großteilnehmer
≈ large user   ↑ Geschäftsteilnehmer
↑ business subscriber
corporate division 1   ECON   **Unternehmensbereich**
= division; corporate group 2; business area; business unit   = Geschäftsbereich; Geschäftsgebiet; Unternehmenseinheit
corporate division 2 (→central division)   ECON   (→Zentralabteilung)
corporate group 1   ECON   **Firmengruppe**
corporate group 2 (→corporate division 1)   ECON   (→Unternehmensbereich)
corporate headquarter   ECON   **Firmenzentrale**
≈ registered office   ≈ Firmensitz
corporate identification card   ECON   **Firmenausweis**
↑ card   = Werksausweis
   ↑ Ausweis
corporate initials (→corporate logo)   ECON   (→Firmenzeichen)
corporate logo   ECON   **Firmenzeichen**
= corporate initials; trademark; sign; company logo; company initials; company logotype   = Firmenlogo; Firmenmonogramm; Firmensymbol
   ≈ Markenzeichen
↑ logotype   ↑ Logotype
corporate mail   OFFICE   **Hauspost**
corporate management   ECON   **Unternehmensführung**
= entrepreneurship
corporate network   TELEC   **Firmennetz**
≈ enterprise network   = Firmenkundennetz
≈ local area network [DATA COMM]   ≈ lokales Netz [DATA COMM]
↑ private network   ↑ Privatnetz
corporate photography   ECON   **Werkbild**
corporate premises (→company's premises)   ECON   (→Firmengelände)
corporate security service   ECON   **Werkschutz**
corporate subscriber (→corporate costumer)   TELEC   (→Firmenkunde)
corporate telecommunications   TELEC   **Firmentelekommunikation**
corporation 1   ECON   **Körperschaft**
corporation 2 (AM)   ECON   **Aktiengesellschaft**
= company limited by shares (BRI); incorporation; stock corporation (AM); public limited company (BRI)   = AG
   ↑ Gesellschaft
↑ company
corpuscle (→particle)   PHYS   (→Teilchen)
corpuscular radiation   PHYS   **Korpuskularstrahlung**
= particle radiation

correct (adj.)   TECH   **genau**
= accurate; right   = korrekt; exakt; präzise; präzis [OES]
correct (adj.) (→valid)   DATA PROC   (→zulässig)
correct (v.t.) (→amend)   TECH   (→berichtigen)
correctable   SCIE   **korrigierbar**
correcting capacitor   CIRC.ENG   **Ausgleichkondensator**
correcting code   CODING   (→fehlerkorrigierender Code)
(→error-correcting code)
correcting displacement   CONTROL   **Stellweg**
correcting element (→final control element)   CONTROL   (→Stellglied)
correcting magnet   ELECTRON   (→Kompensationsmagnet)
(→compensating magnet)
correcting network   CIRC.ENG   **Ausgleichnetzwerk**
= corrective network; compensating network; equalizing network 2   = Ausgleichsnetzwerk; Kompensationsnetzwerk
   ↓ Entzerrernetzwerk
correcting quantity (→control variable)   CONTROL   (→Stellgröße)
correcting range (→control range)   CONTROL   (→Stellbereich)
correcting space bar   TERM&PER   **korrigierende Leertaste**
= correcting spacing key
correcting spacing key   TERM&PER   (→korrigierende Leertaste)
(→correcting space bar)
correcting term   MATH   **Korrekturglied**
correcting time (→control time)   CONTROL   (→Stellzeit)
correcting unit (→final control element)   CONTROL   (→Stellglied)
correcting variable (→control variable)   CONTROL   (→Stellgröße)
correction   SCIE   **Korrektur**
   = Verbesserung; Korrektion
correction (→amendment)   LING   (→Berichtigung)
correction factor   MATH   **Korrekturfaktor**
correction instruction   DATA PROC   **Korrekturanweisung**
= patch 4   = Korrekturbefehl
correction key   TERM&PER   **Korrekturtaste**
≈ cancel key; reset key   ≈ Löschtaste; Rücksetztaste
correction run   DATA PROC   **Korrekturlauf**
correction sign   TYPOGR   (→Korrekturzeichen)
(→proofreaders' mark)
corrective circuit   CIRC.ENG   (→Ausgleichsschaltung)
(→compensating circuit)
corrective maintenance   TECH   **korrigierende Wartung**
≈ repair   = Bedarfswartung
   ≈ Instandsetzung
corrective measure   TECH   (→Gegenmaßnahme)
(→countermeasure)
corrective network   CIRC.ENG   (→Ausgleichnetzwerk)
(→correcting network)
correctly (adv.)   COLLOQ   **korrekt** (adv.)
= properly   = richtig
correctly phased   EL.TECH   **phasenrein**
correctly phased   TELEC   (→synchron)
(→synchronous)
corrector circuit   CIRC.ENG   (→Ausgleichsschaltung)
(→compensating circuit)
correlated   MATH   **korreliert**
correlated variable   MATH   **korrelierte Variable**
correlation   MATH   **Korrelation**
correlation admittance   NETW.TH   **Korrelationsadmittanz**
   = komplexer Korrelationsleitwert
correlation algorithm   MATH   **Korrelationsalgorithmus**
correlation analysis   TELEC   **Korrelationsanalyse**
correlation coefficient   MATH   **Korrelationskoeffizient**
correlation electronics   SIGN.TH   **Korrelationselektronik**
= statistical communication theory
correlation factor   SIGN.TH   **Korrelationsfaktor**
correlation function   SIGN.TH   **Korrelationsfunktion**
correlation matrix   MATH   **Korrelationsmatrix**
correlation modulation   MODUL   **Korrelationsmodulator**
correlation-modulation with a noise carrier   MODUL   **Rauschkorrelationsmodulation**
correlation receiver   MIL.COMM   **Korrelationsempfänger**
correlator   CIRC.ENG   **Korrelationsanalysator**
   = Korrelator

correspondence 1 [Laplace transformation]	MATH	**Korrespondenz 1** [Laplace-Transformation]
correspondence	OFFICE	**Schriftverkehr** = Schriftwechsel; Korrespondenz ↓ Briefwechsel
correspondence 2 [set theory] = non-unique mapping	MATH	**Korrespondenz 2** [Mengenlehre] = mehrdeutige Abbildung
correspondence quality (→letter quality)	TERM&PER	(→Schönschrift)
correspondence quality printer (→letter-quality printer)	TERM&PER	(→Schönschreibdrucker)
corridor	CIV.ENG	**Gang**
corridor [between rack rows]	SYS.INST	**Gang** [zwischen Gestellreihen]
corrode ≠ etch	CHEM	**korrodieren** ≈ ätzen
corrosion ≈ etching	CHEM	**Korrosion** ≈ Ätzung
corrosion-proof (→non-corroding)	TECH	(→korrosionsfest)
corrosion-proof cable	COMM.CABLE	**korrosionsgeschütztes Kabel**
corrosion-protected (→non-corroding)	TECH	(→korrosionsfest)
corrosion protection ≈ corrosion resistance	TECH	**Korrosionsschutz** ≈ Korrosionsfestigkeit
corrosion resistance ≈ corrosion protection	TECH	**Korrosionsfestigkeit** = Korrosionsbeständigkeit = Korrosionsschutz
corrosion-resistant (→non-corroding)	TECH	(→korrosionsfest)
corrosive (n.)	CHEM	**Abbeizmittel**
corrugate ≈ checker	MECH	**wellen** ≈ riffeln
corrugated	MECH	**gewellt**
corrugated (→checkered)	MECH	(→gerifelt)
corrugated cardboard (→corrugated paper)	TECH	(→Wellpappe)
corrugated diaphragm	ANT	**Riffelkonus**
corrugated horn ↑ horn radiator	ANT	**Rillenhorn** ↑ Hornstrahler
corrugated paper = corrugated cardboard	TECH	**Wellpappe**
corrugated sheet metal	METAL	**Wellblech**
corrugated steel sheath	COMM.CABLE	**Stahlwellmantel**
corrugated tube	TECH	**Wellrohr**
corrugation = ondulation ≈ checker	MECH	**Wellung** ≈ Riffelung
cos (→cosine)	MATH	(→Kosinus)
COS (→user group)	DATA COMM	(→Teilnehmerbetriebsklasse)
COS (→cash on shipment)	ECON	(→zahlbar bei Verschiffung)
cosec (→cosecant)	MATH	(→Kosekans)
cosecant = cosec ↑ trigonometric function	MATH	**Kosekans** [Hypothenuse zu Gegenkathete] = cosec ↑ trigonometrische Funktion
cosecant-squared antenna	ANT	**Cosec-Quadrat-Antenne**
cosecant squared beam antenna = cosß dipole	ANT	**cosß-Dipol** = Cosecans-Beam-Antenne
coset	MATH	**Nebenklasse**
cosine = cos ↑ trigonometric function	MATH	**Kosinus** (Duden) = cos; Cosinus ↑ trigonometrische Funktion
cosine equalizer	CIRC.ENG	**Kosinusentzerrer** = cos-Entzerrer
cosine transform (→cosine transformation)	MATH	(→Kosinustransformation)
cosine transformation = cosine transform	MATH	**Kosinustransformation** = Cosinustransformation
cosinusoidal roll-of	MODUL	**Kosinus-roll-off**
cosinus roll-off shape	MODUL	**Kosinus-roll-off-Charakteristik**
cosmic noise = galactic noise	RADIO	**kosmisches Rauschen** = galaktisches Rauschen; Radiorauschen
cosmic radiation = cosmic rays	GEOPHYS	**Höhenstrahlung** = kosmische Strahlung
cosmic rays (→cosmic radiation)	GEOPHYS	(→Höhenstrahlung)
cosß dipole (→cosecant squared beam antenna)	ANT	(→cosß-Dipol)
cost (v.i.)	ECON	**kosten**
cost (n.) ≈ expense	ECON	**Kosten** (pl.t.; Singular: Kostenpunkt) ≈ Ausgaben
cost accounting ↑ accounting	ECON	**Kostenrechnung** ↑ Buchhaltungs- und Bilanzwesen
cost-adequate (→cost-conforming)	ECON	(→kostengerecht)
cost allocation	ECON	**Kostenumlage**
cost ascertainment = cost evaluation	ECON	**Kostenermittlung** = Kostenauswertung
cost-benefit analysis	ECON	**Nutzen-Kosten-Analyse**
cost calculation = costing; calculation	ECON	**Kalkulation** = Kostenberechnung
cost-conforming = cost-adequate	ECON	**kostengerecht**
cost-covering	ECON	**kostendeckend**
cost curtailment (→cost reduction)	ECON	(→Kostensenkung)
cost-effective 1 [saving costs] = economic ≈ profitable; cheap	ECON	**wirtschaftlich 2** [Kosten sparend] = ökonomisch ≈ rentabel; preiswert
cost-effective 2 [having effect on costs]	ECON	**kostenwirksam**
cost effectiveness (→profitability)	ECON	(→Wirtschaftlichkeit)
cost evaluation (→cost ascertainment)	ECON	(→Kostenermittlung)
cost expenditure (→consumption)	ECON	(→Kostenaufwand)
cost increase	ECON	**Kostensteigerung** = Kostenerhöhung
costing (→cost calculation)	ECON	(→Kalkulation)
cost,insurance,freight (→CIF)	ECON	(→CIF)
cost-intensive	ECON	**kostenintensiv**
costly (→expensive)	ECON	(→teuer)
cost-neutral [not causing additional costs]	ECON	**kostenneutral** [ohne zusätzliche Kosten]
cost price = buying price	ECON	**Einstandspreis** = Einkaufspreis; Selbstkostenpreis
cost recording equipment [PABX]	SWITCH	**Kostenerfassungsgerät** [Nebenstellenanlage]
cost recovery	ECON	**Kostendeckung**
cost reduction = cost curtailment	ECON	**Kostensenkung** = Kostenreduktion; Kostenreduzierung
cost sharing	ECON	**Kostenbeteiligung** = Kostenteilung
costume-exchangable	TECH	**vom Benutzer austauschbar**
costumer (→client)	ECON	(→Kunde)
costumer (→suscriber)	TELEC	(→Teilnehmer)
costumer acceptance (→subscriber acceptance)	TELEC	(→Teilnehmerakzeptanz)
costumer audit	QUAL	**Kundenaudit**
costumer behaviour	TELEC	**Teilnehmerverhalten**
costumer demand	ECON	**Kundenanfrage**
costumer equipment (→user terminal)	TELEC	(→Teilnehmergerät)
costumer-installable (→user-installable)	TECH	(→vom Benutzer installierbar)
costumer line (→subscriber line)	TELEC	(→Teilnehmerleitung)
costumer-located	TELEC	**teilnehmerseitig**
costumer loop (→subscriber line)	TELEC	(→Teilnehmerleitung)
costumer premises (→suscriber premises)	TELEC	(→Teilnehmerbereich)
costumer-premises equipment (→user terminal)	TELEC	(→Teilnehmergerät)
costumer-premises interface	TELEC	**Teilnehmerschnittstelle**
costumer-premises near (→user-near)	TELEC	(→teilnehmernahe)
costumer requirement	ECON	**Kundenanforderung**

costumer service (→after-sales ECON (→Kundendienst)
service)
costumer service engineer TECH **Kundendienstingenieur**
= field engineer ≈ Kundendiensttechniker;
≈ costumer service technitian Wartungsingenieur
costumer's premises (→suscriber TELEC (→Teilnehmerbereich)
premises)
costumer support (→after-sales ECON (→Kundendienst)
service)
costumer training TECH **Kundenschulung**
cot (→cotangent) MATH (→Kotangens)
cotangens hyperbolicus MATH **Cotagens hyperbolicus**
= coth; cth = Kotangens hyperbolikus;
 Hyperbelkotangens; coth;
 cth
cotangent MATH **Kotangens**
= cot; cotg; ctg = Cotangens; cot; cotg; ctg
↑ trigonometric function ↑ trigonometrische Funktion
cotg (→cotangent) MATH (→Kotangens)
coth (→cotangens MATH (→Cotagens hyperbolicus)
hyperbolicus)
cottage key people DATA PROC **Heimarbeiter**
cotter 2 (v.t.) TECH **versplinten**
cotter 1 (n.) MECH **Sicherungselement**
↓ cotter 2 (n.); cotter pin ↓ Querkeil; Splint
cotter 2 MECH **Querkeil**
↑ cotter 1 ↑ Sicherungselement
cotter pin MECH **Splint**
= slit pin ↑ Sicherungselement
↑ cotter
Cotton-Mouton effect PHYS **Cotton-Mouton-Effekt**
Coulomb PHYS **Coulomb**
[SI unit for electric charge; electric [SI-Einheit für elektrische
flux and quantity of electricity; = 1 Ladung, elektrischen Fluß
A s] und Elektrizitätsmenge; =
= C 1 A s]
 = C
Coulomb force PHYS **Coulombkraft**
Coulomb law PHYS **Coulombsches Gesetz**
coulometer INSTR **Ladungsmesser**
 = Coulombmeter; Voltameter
counselor (→consultant) ECON (→Berater)
count (v.t.) MATH **zählen**
 = abzählen
countable (n.) LING **zählbares Nomen**
count arrow NETW.TH **Zählpfeil**
≠ directional arrow = Kettenzählpfeil
 ≠ Richtungspfeil
counted measurand TELECONTR **Zählwert**
= metered measurand
counter ECON **Schalter**
 ↓ Bankschalter
counter INSTR **Zähler**
= meter ↓ Frequenzzähler; Zeitinter-
↓ frequency counter; time interval vallzähler
counter
counter CIRC.ENG **Zähler**
= counting circuit; counter circuit; = Zählschaltung
meter ↓ Vorwärtszähler; Rück-
↓ up counter; down counter; up-down wärtszähler;
counter; modulo-n counter; ring Vor-Rückwärts-Zähler;
counter; binary counter; decimal Modulo-N-Zähler; Ring-
counter zähler; Dualzähler; Dezi-
 malzähler
counter accuracy INSTR **Zählergenauigkeit**
counteract TECH **gegenwirken**
counter advance sense ELECTRON **Zählrichtung**
[sense in which counter pulses ad- [Richtung in der Zählim-
vance the counter] pulse den Zähler fortschal-
 tet]
counterbalance MECH **Gegengewicht**
= counterweight; counterpoise
counter cell POWER SYS **Gegenzelle**
= counter EMF cell; counter electromo- [Fernmeldestromversor-
tive cell gung]
counter circuit CIRC.ENG (→Zähler)
(→counter)
counterclockweise TECH **linksdrehend**
= anticlockwise; levorotatory; left-hand = gegen den Uhrzeigersinn
≠ clockwise ≠ rechtsdrehend

counter constant INSTR **Zählerkonstante**
= meter constant
countercurrent PHYS **Gegenstrom**
counter decade COMPON **Zähldekade**
[component to count in decimal code] [Baustein zur Zählung im
 Dezimalcode]
 = Zähltetrade
counter delay (→counter lag ELECTRON (→Zählerverzögerung)
time)
counter electromotive cell POWER SYS (→Gegenzelle)
(→counter cell)
counter EMF cell (→counter POWER SYS (→Gegenzelle)
cell)
counterexample COLLOQ **Gegenbeispiel**
counter field (→opposite PHYS (→Gegenfeld)
field)
counter key (→metering key) INSTR (→Zähltaste)
counter lag time ELECTRON **Zählerverzögerung**
= counter time lag; counter delay
counter-light TECH **Gegenlicht**
counter loop DATA PROC **Zählerschleife**
countermand (→cancel) ECON (→stornieren)
countermand (n.) ECON (→Widerruf 2)
(→withdrawal)
countermeasure TECH **Gegenmaßnahme**
= corrective measure; remedy; mitiga-
tion technique
countermovement MECH **Gegenbewegung**
countermoving TECH **gegenläufig**
= contra-rotating ≈ rückläufig
≈ retrogade
counter-offer ECON **Gegenangebot**
counter overflow ELECTRON **Zählerüberlauf**
counterpart TECH **Gegenstück**
= mating part; mating component
counterpoise ANT **Gegengewicht**
= ground plane; radial 3
counterpoise MECH (→Gegengewicht)
(→counterbalance)
counter preset ELECTRON **Zählervoreinstellung**
= counter presetting
counter presetting (→counter ELECTRON (→Zählervoreinstellung)
preset)
counterpressure (→back PHYS (→Gegendruck)
pressure)
counter reading 1 INSTR **Zählerablesung**
[the act of reading]
= meter reading 1
counter reading 2 INSTR **Zählerstand**
[the value indicated]
= meter reading 2
counter reading date ECON **Zählerablestag**
= meter reading day
counter register DATA PROC **Zählregister**
≈ counter ≈ Zähler
countersink (v.t.) MECH **spitzsenken**
 = ausfräsen
counter terminal TERM&PER **Schalterterminal**
counter time lag (→counter ELECTRON (→Zählerverzögerung)
lag time)
counter-torque MECH **Gegenmoment**
counter-trading (→barter ECON (→Tauschgeschäft)
business)
counter variable DATA PROC **Zählvariable**
counterweight MECH (→Gegengewicht)
(→counterbalance)
counting MATH **Zählung**
= count n. = Zählen; Abzählung
≈ calculation ≈ Rechnung
counting bundle (→meter COMM.CABLE (→Zählbündel)
bundle)
counting circuit CIRC.ENG (→Zähler)
(→counter)
counting code CODING **Zählcode**
[representation of numbers by serial [Darstellung von Zahlen
pulse sequencies] durch bitserielle Impulsfol-
= step-at-a-time code gen]
counting device COMPON **Zählbaustein**
counting element (→meter COMM.CABLE (→Zählader)
wire)

English	Domain	German
counting loop	DATA PROC	Zählschleife
counting pulse	ELECTRON	Zählimpuls
counting relay	COMPON	Zählrelais
counting system (→number system)	MATH	(→Zahlensystem)
counting track [punched card]	TERM&PER	Zählspur [Lochkarte]
counting tube	ELECTRON	Zählrohr
countless (→innumerable)	COLLOQ	(→zahllos)
count n. (→counting)	MATH	(→Zählung)
country code [up to three digits to mark the country, follows the prefix and the regional identity code, e.g. 4 for Great Britain and Northern Ireland] ≈ international code; country prefix	SWITCH	Landeskennzahl [nach der Verkehrsausscheidungszahl und nach der Weltnummerierungszone, ein bis drei Ziffern zur Kennzeichnung des Landes, z.B. 9 für BRD] ≈ Landesvorwahl; internationale Kennzahl
country prefix [international prefix + international code; e.g. 0044 for Great Britain and Northern Ireland when dialing from Germany; not appliable when dialing North American subscribers] ≈ country code; international code; international number	SWITCH	Landesvorwahl [internationale Verkehrsausscheidungszahl + internationale Kennzahl; z.B. 01049 für BRD von Großbritannien aus; bei nordamerikanischen Teilnehmernummern nicht anwendbar] ≈ Landeskennzahl; internationale Kennzahl; internationale Rufnummer
country-wide (→nationwide)	TELEC	(→landesweit)
couple	PHYS	koppeln
couple (→couple of forces)	PHYS	(→Kräftepaar)
coupled	PHYS	gekoppelt
coupled-circuit oscillation	CIRC.ENG	Kopplungsschwingung
coupled parallel resonant circuit	NETW.TH	gekoppelter Parallelschwingkreis
coupled resonant circuit	NETW.TH	Schwingkreiskopplung
couple of forces = couple	PHYS	Kräftepaar = Drehzwilling
coupler = branching	ELECTRON	Koppler
coupling	PHYS	Kopplung
coupling (→clutch)	MECH	(→Kupplung)
coupling (→sleeve)	TECH	(→Muffe)
coupling (→controller coupling)	CONTROL	(→Reglerverknüpfung)
coupling admittance (→mutual admittance)	NETW.TH	(→Kernleitwert)
coupling attenuation	COMPON	Koppeldämpfung
coupling capacitor = coupling condenser	EL.TECH	Kopplungskondensator = Koppelkapazität; Koppelkondensator
coupling coefficient = coupling factor; coupling constant	PHYS	Kopplungskoeffizient = Kopplungsfaktor; Kopplungskonstante
coupling condenser (→coupling capacitor)	EL.TECH	(→Kopplungskondensator)
coupling constant (→coupling coefficient)	PHYS	(→Kopplungskoeffizient)
coupling diode	CIRC.ENG	Koppeldiode
coupling efficiency	OPT.COMM	Kopplungsgüte
coupling element	CIRC.ENG	Koppelelement
coupling factor (→coupling coefficient)	PHYS	(→Kopplungskoeffizient)
coupling filter	NETW.TH	Kopplungsfilter
coupling hole	MICROW	Koppelöffnung = Koppelloch
coupling impedance (→mutual impedance)	NETW.TH	(→Kernwiderstand)
coupling loop	MICROW	Koppelschleife = Kopplungsschleife
coupling loss = cross-coupling ratio	EL.TECH	Kopplungsdämpfung
coupling mechanism	MICROW	Einkoppelmechanismus
coupling network	NETW.TH	Koppelnetzwerk
coupling probe	MICROW	Kopplungssonde
coupling radiator	ANT	Koppelstrahler
coupling RC	CIRC.ENG	Koppel-RC-Glied
coupling resistance	EL.TECH	Kopplungswiderstand
courier service	ECON	Kurierdienst
course 2 = route ≈ trayectory [PHYS]	TECH	Kurs ≈ Bahn [PHYS]
course [training]	SCIE	Kurs [Schulung]
course ↓ course of curve	MATH	Verlauf ↓ Kurvenverlauf
course 1 [spacial] = route ≈ trayectory [PHYS]	TECH	Kurs [räumlich] = Route; räumlicher Verlauf ≈ Bahn [PHYS]
course 2 (→temporal course)	TECH	(→zeitlicher Verlauf)
course aim	TECH	Kursziel
course calculator (→course line computer)	RADIONAV	(→Kursrechner)
course computer (→course line computer)	RADIONAV	(→Kursrechner)
course line computer = course computer; course calculator	RADIONAV	Kursrechner
course material = training material; training documentation	ECON	Schulungsunterlage = Kursunterlage
course objective	SCIE	Schulungsziel
course of lectures	SCIE	Vortragsreihe
course of ray	PHYS	Strahlengang = Strahlenverlauf
course of study	SCIE	Studienfach = akademische Fachrichtung
courseware	DATA PROC	Unterrichtssoftware
courtesy copy (→unofficial print)	DOC	(→Informationspause)
couteracting winding (→differential winding)	EL.TECH	(→Differentialwindung)
coutoff frequency (→limit frequency)	NETW.TH	(→Eckfrequenz)
covalence	CHEM	Atombindung = Kovalenz
covalent binding (→valence bond)	CHEM	(→Valenzbindung)
covalent bond	CHEM	kovalente Bindung = homöopolare Bindung
covariance [statistics]	MATH	Kovarianz [Statistik]
covariance analysis [statistics]	MATH	Kovarianzanalyse [Statistik]
covariant	MATH	kovariant
cover (v.t.) = coat (v.t.) ≈ mask (v.t.)	TECH	zudecken ≈ abdecken 2; verdecken; bedecken; verkleiden
cover 2 (v.t.) [hiding effect of a paint] = hide	TECH	decken [Farbe]
cover (n.) ≈ panel; hood; flap; masking; protective case ↓ front panel; dummy cover; insulating cover; dust cover; cover plate; protective cover; lid	TECH	Abdeckung ≈ Blende; Haube; Klappe; Verkleidung; Schutzgehäuse ↓ Frontabdeckung; Blindabdeckung; Isolierabdeckung; Staubabdeckung; Abdeckplatte; Schutzabdeckung; Deckel
cover (n.)	COMPON	Isolierkappe
cover 1 (n.) = cap; hood	TYPOGR	Kappe = Vorderdeckel
cover 2 (n.) = frontispiece	TYPOGR	Titelbild
cover (n.)	MICROEL	Deckschicht
cover 3 (→cover sheet)	TYPOGR	(→Deckblatt)
cover (→record cover)	CONS.EL	(→Schallplattenhülle)
coverage	SAT.COMM	Bedeckung
coverage = service	TELEC	Versorgung
coverage	ECON	Deckung
coverage area (→service area)	TELEC	(→Versorgungsbereich)
coverage corridor	MOB.COMM	Versorgungsstreifen = Versorgungskorridor

coverage gap	TELEC	Versorgungslücke	
coverage index	TELEC	Versorgungsgrad	
coverage radius	BROADC	Versorgungsradius	
cover flap (→cover plate)	TECH	(→Abdeckplatte)	
covering	TECH	Bedeckung	
covering [graph theory]	MATH	Überdeckung [Graphentheorie]	
covering depth ≈ laying depth	OUTS.PLANT	Überdeckung [Höhe des Erdreichs zwischen unterirdischer Kabelanlage und Erdoberfläche] ≈ Verlegetiefe	
covering letter	OFFICE	Anschreiben	
covering material = hiding material	TECH	Deckmittel [Farbe]	
covering panel ↑ panel	TECH	Abdeckblende ↑ Blende	
covering power = hiding power	TECH	Deckfähigkeit [Farbe]	
cover plate = cover flap ↑ cover (n.)	TECH	Abdeckplatte ↑ Abdeckung	
cover ring	TECH	Abdeckring	
cover sheet ↑ cover	MECH	Abdeckblech ↑ Abdeckung	
cover sheet = cover 3; front cover; lead sheat	TYPOGR	Deckblatt	
cover shortage = partial cover; deficient cover	ECON	Unterdeckung	
cover slab (of concrete) (→manhole cover)	OUTS.PLANT	(→Schachtdeckel)	
cover strip = dummy strip; ≈ dummy bar	TECH	Abdeckleisten = Abdeckstreifen ≈ Abdeckschiene	
cover strip (→dummy strip)	MECH	(→Abdeckleiste)	
CPE (→central processing element)	DATA PROC	(→Zentralverarbeitungselement)	
CPFSK = continous phase frequency shift keying	MODUL	CPFSK = phasenkontinuierliche Frequenzumtastung	
cpi (→characters per inch)	TERM&PER	(→Zeichen pro Zoll)	
CPI (→cycles per instruction)	DATA PROC	(→Takte je Befehl)	
c.p.s. (→cycles per second)	PHYS	(→Hertz)	
cps (→characters per second)	DATA PROC	(→Zeichen pro Sekunde)	
CPU 2 (→processor 1)	DATA PROC	(→Prozessor 1)	
CPU (→central processing unit 1)	DATA PROC	(→Zentraleinheit 1)	
cq call	RADIO	cq-Ruf	
Cr (→chromium)	CHEM	(→Chrom)	
CR (→carriage return)	TELEGR	(→Wagenrücklauf)	
crack (n.) [a narrow break without separation of the pieces] = crevice; fissure; rift; chink; flaw ≈ split; crevasse	TECH	Sprung [feiner Riß in sprödem Material, ohne Trennung der Teile] = Riß 1; Anriß ≈ Spalte; Kluft	
cracker [a criminal hacker 2] ≈ Hacker 2	DATA PROC	Cracker [ein krimineller Hacker 2] ≈ Hacker 2	
crack formation	TECH	Rißbildung	
cracking noise (→contact noise)	EL.ACOUS	(→Kratzgeräusch)	
crackle (→click)	TELEPH	(→Knackgeräusch)	
crackle interference (→click)	TELEPH	(→Knackgeräusch)	
crack-off (v.t.)	MECH	abknicken	
cradle = receiver rest; cradle hook; hook	TELEPH	Gabel [Hörerauflage] = Gabelachse; Keile; Haken	
cradle hook (→cradle)	TELEPH	(→Gabel)	
cradle relay	COMPON	Kammrelais	
cradle switch (→hookswitch)	TELEPH	(→Gabelumschalter)	
craftmanship (→workmanship 1)	TECH	(→Handwerkskunst)	
craftsman ≈ technician	ECON	Handwerker ≈ Techniker	
cranck (v.t.)	MECH	kröpfen	
cranck shaft	MECH	Kurbelwelle	
crane	TECH	Kran	
crank (n.)	MECH	Kurbel	
cranked	METAL	gekröpft	
crank resistance = lever type decade resistance; rotary rheostat	INSTR	Kurbelwiderstand	
crash (v.i.) = bomb (v.i.); fail (v.i.)	DATA PROC	abstürzen = zusammenbrechen	
crash (→abnormal end)	DATA PROC	(→Absturz)	
crash analysis	DATA PROC	Kollisionsanalyse = Kollisionsuntersuchung	
crash conversion [ceasing abruptly to operate the old system] = direct convertion	DATA PROC	abrupter Übergang [schlagartiges Aufgeben des alten Systems] = radikale Umstellung; schlagartige Umstellung	
crash-protected	TERM&PER	absturzgesichert	
crash simulation	DATA PROC	Kollisionssimulation	
crate (n.)	TECH	Lattenkiste	
CRC (→cyclic redundancy check)	CODING	(→zyklische Blocksicherung)	
create 1	DATA PROC	neuanlegen	
create 2	DATA PROC	festlegen	
credit 2 (n.) [amount of money placed at disposal by a bank] = loan	ECON	Kredit	
credit 1 (n.) [balance in favour]	ECON	Guthaben	
credit card	ECON	Kreditkarte	
credit card (→credit phonecard)	TERM&PER	(→Buchungskarte)	
credit card reader	TERM&PER	Kreidkartenleser	
credit check terminal	TERM&PER	Kreditprüfterminal	
credit indication [coinbox telephone]	TERM&PER	Guthabenanzeige [Münzfernsprecher]	
credit information automat	TERM&PER	Kontoauszugsautomat	
credit institution	ECON	Kreditinstitut	
creditor = borrower ≈ lender ≠ debtor	ECON	Gläubiger = Kreditor ≈ Verleiher ≠ Schuldner	
credit phonecard [with personal identification number] = credit card ↑ phonecard	TERM&PER	Buchungskarte [mit persönlicher Identifikationsnummer] = Kredit-Telefonkarte ↑ Telefonkarte	
creep [slow deformation under load]	MECH	Kriechen [langsame Verformung unter Dauerbelastung]	
creep (v.t.) (→side-scroll)	DATA PROC	(→seitlich rollen)	
creeping (→side-scrolling)	DATA PROC	(→seitliches Rollen)	
creeping current (→tracking current)	EL.TECH	(→Kriechstrom)	
creeping distance (→tracking distance)	EL.TECH	(→Kriechstrecke)	
creeping galvanometer ≈ flux meter ↑ moving-coil galvanometer	INSTR	Kriechgalvanometer ≈ Fluxmeter ↑ Drehspulvanometer	
creeping strength	MECH	Kriechfestigkeit	
crest	TECH	Kamm 2 [oberster Teil einer Erhebung]	
crest (→peak 3)	MATH	(→Scheitel 2)	
crest (→antinode)	PHYS	(→Schwingungsbauch)	
crest angle = apex angle	MATH	Scheitelwinkel	
crest clearance [thread]	MECH	Kopfspiel [Gewinde]	
crest current (→peak current)	EL.TECH	(→Scheitelstrom)	
crest deviation (→maximal deviation)	CONTROL	(→Maximalabweichung)	
crest factor [peak value to effective value] = peak factor; peak-to-average ratio	EL.TECH	Spitzenfaktor [Spitzenwert zu Effektivwert] = Scheitelfaktor; Spitzenwertfaktor; Spitzen-zu-Effektivwert; Crestfaktor	

crest inverse voltage (→peak inverse voltage) ELECTRON (→Spitzensperrspannung)
crest reverse voltage (→peak inverse voltage) ELECTRON (→Spitzensperrspannung)
crest voltmeter (→peak voltmeter) INSTR (→Scheitelspannungsmesser)
crevassse TECH **Kluft**
[a deep crevice or fissure] [tiefer Riß oder große Spalte]
≈ crack ≈ Sprung
crevice (→superficial corrosion) CHEM (→Oberflächenkorrosion)
crevice (→crack) TECH (→Sprung)
crimp cable lug COMPON **Quetschkabelschuh**
= crimp connector; crimp terminal; push-on connector; press cable lug; press connector; press terminal = Preßkabelschuh; Quetschverbinder; Preßverbinder
crimp connection COMPON **Quetschverbindung**
= wire-crimp connection; pressure connection = Quetschanschluß; Crimpverbindung
crimp connector (→crimp cable lug) COMPON (→Quetschkabelschuh)
crimping tool TECH **Kabelschuhzange**
= Quetschzange; Preßzange; Crimpwerkzeug
crimp terminal (→crimp cable lug) COMPON (→Quetschkabelschuh)
crimp tool for push-on connectors EL.INST **Kombi-Kabelschuhzange**
crimson PHYS **carmoisin**
= deep purpish red
crippled leapfrog test DATA PROC **Einzellen-Bocksprungprüfung**
[using a single memory cell]
crippled version TECH **abgemagerte Version**
= stripped-down version = abgemagerte Ausführung; Sparversion
≠ full-bore version ≠ voll ausgebaute Version
crisp (adj.) TECH **scharf**
[fig.; eg. an illustration] [fig.; z.B. eine Abbildung]
= gestochen scharf
CRISP DATA PROC **CRISP-Computer**
[Complexity Reduced Instruction Set Processor]
cristalline PHYS **kristallin**
cristallization system PHYS **Kristallsystem**
= kristallographische Punktgruppe
criterion (pl.-ia) SCIE **Kriterium** (pl.-ien)
critical TECH **kritisch**
critical angle (→limiting angle) TECH (→Grenzwinkel)
critical coupling EL.TECH **Grenzkopplung**
critical current EL.TECH **Grenzstrom**
critical current density PHYS **kritische Stromdichte**
[superconductivity] [Supraleitung]
critical dimensions TECH **Grenzabmessungen**
critical flicker frequency TV **Flimmergrenze**
critical frequency RADIO **Grenzfrequenz**
= cutoff frequency
critical frequency (→cutoff frequency) MICROW (→Grenzfrequenz)
critical limit ELECTRON **Abfallgrenze**
critical path SCIE **kritischer Pfad**
critical-path method SCIE **Kritischer-Pfad-Methode**
critical resistance EL.TECH **Grenzwiderstand**
critical temperature (→transition temperature) PHYS (→Sprungtemperatur)
critical velocity PHYS **Grenzgeschwindigkeit**
= kritische Geschwindigkeit
critical wavelength (→cutoff wavelength) MICROEL (→Grenzwellenlänge)
CR key (→carriage return key) TELEGR (→Wagenrücklauftaste)
crocodil clip ELECTRON **Abgreifklemme**
= alligator clip = Krokodilklemme
↑ test clip ↑ Prüfklemme
crooked TECH **krumm**
≈ curved ≈ gekrümmt
crop DATA PROC **beschneiden**
[of graphics] [von Graphiken]

cross 1 (v.t.) TECH **kreuzen**
[fig.] [fig.]
= intersect = überschneiden; schneiden
≈ traverse; overlap ≈ durchqueren; überlappen
cross 2 (v.t.) TECH **verschränken**
≈ interleave; interlace ≈ verschachteln
cross (→cross-section) MATH (→Querschnitt)
cross (→transversal) TECH (→transversal)
cross adding (→cross check) MATH (→Querkontrolle)
cross-application (adj.) TECH **anwendungsübergreifend**
crossarm MECH **Querträger**
= crossbar; crossbeam; traverse; pole arm = Traverse; Querschiene = Querstrebe
≈ transversal strut
cross assembler DATA PROC **Kreuzassembler**
[generates machine code for a different computer] [übersetzt in die Maschinensprache eines anderen Computertyps]
= Kreuzassemblierer; Cross-Assembler
cross-assembling DATA PROC **Kreuzassemblierung**
= Cross-Assemblierung
crossbar (→crossarm) MECH (→Querträger)
crossbar distributor ELECTRON **Kreuzschienenverteiler**
crossbar micrometer INSTR **Kreuzfadenmikrometer**
crossbar selector (→crossbar switch) SWITCH (→Kreuzschienenschalter)
crossbar switch SWITCH **Kreuzschienenschalter**
= crossbar selector = Koordinatenschalter; Crossbarschalter; Kreuzschienenwähler; Crossbarwähler
↑ Wähler
crossbar switching system SWITCH **Kreuzschienenwählsystem**
↑ indirect-control switching system ↑ indirekt gesteuertes Vermittlungssystem
crossbar transition MICROW **Kreuzbalkenübergang**
crossbeam (→crossarm) MECH (→Querträger)
cross-border ECON **grenzüberschreitend**
= transborder ≈ international
≈ international
cross bracing (→diagonal cross brace) MECH (→Kreuzverstrebung)
cross check CODING **Kreuzsicherung**
cross check MATH **Querkontrolle**
= cross adding; cross total = Querrechnung
cross coil EL.TECH **Kreuzspule**
= crossed coil
cross-coil antenna ANT **Kreuzrahmenantenne**
= crossed-loop antenna
cross-coil instrument INSTR **Kreuzspulinstrument**
cross-coil measuring system INSTR **Kreuzspulmeßwerk**
= cross-coil mechanism
cross-coil mechanism (→cross-coil measuring system) INSTR (→Kreuzspulmeßwerk)
cross compiler DATA PROC **Kreuzkompilierer**
[translates into machine code of a different computer] [übersetzt in die Maschinensprache eines anderen Computertyps]
= Kreuzkompiler; Kreuzcompiler; Cross-Compiler
cross-compiling DATA PROC **Kreuzkompilierung**
= Cross-Compilierung
cross-connect (v.t.) EQUIP.ENG **rangieren**
= strap (v.t.); jumper (v.t.)
cross connect (n.) SYS.INST **Rangierleitung**
[connects different terminal blocks] [verbindet Verteilerblöcke]
= cross-connect line; cross-connect cable; jumper; jumper line; jumper cable = Rangierverbindung; Rangierkabel
≈ strap (n.) ≈ Rangierdraht
cross-connect cable (→cross connect) SYS.INST (→Rangierleitung)
cross-connect device OUTS.PLANT **Verzweigungseinrichtung**
cross-connecting (→jumpering) SYS.INST (→Rangierung)
cross-connecting box OUTS.PLANT **Abzweigkasten**
cross connecting distributor OUTS.PLANT (→Kabelverzweigergehäuse)
(→distribution cabinet)
cross-connecting panel EQUIP.ENG (→Rangierfeld)
(→jumpering panel)

cross connecting terminal (→distribution cabinet)	OUTS.PLANT	(→Kabelverzweigergehäuse)	**crosspoint** ↓ switching element	SWITCH	**Koppelpunkt** ↓ Koppelelement	
cross connection 1	POWER ENG	**Kreuzschaltung**	**cross-pointer instrument**	INSTR	**Kreuzzeigerinstrument**	
cross connection 2	POWER ENG	**Steckverteiler**	**crosspoint switch**	SWITCH	**Koppelpunktschalter**	
cross-connection field (→patching distribution frame)	SYS.INST	(→Rangierverteiler)	**cross-polar** ≠ co-polar	RADIO	= Crosspoint-Schalter **kreuzpolar** ≠ kopolar	
cross-connect level = distribution level	TELEC	**Rangierebene** = Durchschaltebene; Verteilebene	**cross-polar discrimination** = XPD	RADIO	**Kreuzpolarisationsentkopplung** = XPD	
cross-connect line (→cross connect)	SYS.INST	(→Rangierleitung)	**crosspolar discrimination** (→depolarization loss)	ANT	(→Polarisationsentkopplung)	
cross-connect multiplexer = CCM; intelligent multiplexer; programmable multiplexer; managed multiplexer	TRANS	**Verteilmultiplexer** = Schaltmultiplexer; Steuermultiplexer; intelligenter Multiplexer; programmierbarer Multiplexer; Cross-connect-Multiplexer	**cross-polar-interference canceler** = XPIC; cross-pol interference canceller; cross-polarization canceller; cross-pol canceller	RADIO REL	**Depolarisationskompensator** = Depok; Depolarisationsentkoppler	
cross-connect point	OUTS.PLANT	**Verzweigungspunkt**	**cross polarization**	RADIO	**Kreuzpolarisation**	
cross correlation	TELEC	**Kreuzkorrelation**	≠ co-polarization		≠ Kopolarisation	
cross-correlation analysis	TELEC	**Kreuzkorrelationsanalyse**	**cross-polarization canceller**	RADIO REL	(→Depolarisationskompensator)	
cross-correlation function	SIGN.TH	**Kreuzkorrelationsfunktion**	(→cross-polar-interference canceler)			
cross-country line	TELEC	**Überlandleitung**	**cross polarized Yagi antenna**	ANT	(→Kreuz-Yagi-Antenne)	
cross-coupling	TELEC	**Kreuzkopplung**	(→crossed Yagi array)			
cross coupling (→side-to-side coupling)	COMM.CABLE	(→Übersprechkopplung)	**cross-pol canceller** (→cross-polar-interference canceler)	RADIO REL	(→Depolarisationskompensator)	
cross-coupling ratio (→coupling loss)	EL.TECH	(→Kopplungsdämpfung)	**cross-pol interference canceller** (→cross-polar-interference canceler)	RADIO REL	(→Depolarisationskompensator)	
cross current = shunt current	PHYS	**Querstrom**	**cross recess** (→indented cross)	MECH	(→Kreuzschlitz)	
cross-drilled headscrew	MECH	**Kreuzlochschraube**	**cross-recessed screw**	MECH	**Kreuzschlitzschraube**	
cross-drilled nut	MECH	**Kreuzlochmutter**	= Phillips screw			
crossed antenna (→turnstile antenna)	ANT	(→Drehkreuzantenne)	**cross-reference**	DOC	**Querverweis**	
crossed coil (→cross coil)	EL.TECH	(→Kreuzspule)	**cross-reference generator**	DATA PROC	**Querverweisgenerator**	
crossed-loop antenna (→cross-coil antenna)	ANT	(→Kreuzrahmenantenne)	**cross resistance** = shunt resistance ≠ parallel resistance	PHYS	**Querwiderstand** ≠ Parallelwiderstand	
crossed Yagi array = cross polarized Yagi antenna	ANT	**Kreuz-Yagi-Antenne**	**cross-section** [descriptive geometry] = cross	MATH	**Querschnitt** [darstellende Geometrie] ↑ Schnitt 2	
cross-fade (v.t.) [gradual transition from one audio signal to another]	EL.ACOUS	**überblenden** [gradueller Übergang von einem Tonsignal auf ein anderes]	**cross-sectional area**	MATH	**Querschnittsfläche**	
			cross-series adapter (→inter-series adapter)	COMPON	(→Adapter 2)	
cross fade (n.) = fade (n.)	EL.ACOUS	**Überblendung**	**cross stroke** (→slash)	LING	(→Schrägstrich)	
cross-fade (→lap-dissolve)	TV	(→überblenden)	**crosstalk** [interference between communication lines]	TELEC	**Nebensprechen** [gegenseitige Beinflussung von Nachrichtenleitungen]	
cross fade (n.) (→lap-dissolve)	TV	(→Überblendung)	↓ near-end crosstalk; far-end crosstalk; side-to-side crosstalk [COMM.CABLE]; side-to-phantom crosstalk [COMM.CABLE]		↓ Nahnebensprechen; Fernnebensprechen; Übersprechen [COMM.CABLE]; Mitsprechen [COMM.CABLE]	
cross force = transverse force	PHYS	**Querkraft** = Transversalkraft				
cross grating	PHYS	**Kreuzgitter**				
cross-hairs (→cross-line)	INSTR	(→Fadenkreuz)				
cross-hatch (v.t.)	ENG.DRAW	**kreuzschraffieren**	**crosstalk attenuation**	TELEC	**Nebensprechdämpfung**	
cross-hatch (n.) = cross-hatching	ENG.DRAW	**Kreuzschraffierung** = Kreuzschraffur	**crosstalk coupling**	COMM.CABLE	**Nebensprechkopplung**	
cross-hatching (→cross-hatch)	ENG.DRAW	(→Kreuzschraffierung)	**crosstalk suppression filter** [Codec]	TELEC	**Nebensprechsperre**	
cross hole	MECH	**Kreuzloch**	**cross total** (→cross check)	MATH	(→Querkontrolle)	
crossing (→transposition)	OUTS.PLANT	(→Kreuzung)	**cross under**	MICROEL	**Untertunnelung**	
cross-line = cross-hairs; reticule	INSTR	**Fadenkreuz**	**cross validation** (→comparison check)	TECH	(→Vergleichstest)	
cross-line (→slash)	LING	(→Schrägstrich)	**cross voltage** = transverse voltage	PHYS	**Querspannung** [elektrisch]	
cross modulation [unwanted modulation product with AM, causes intelligible crosstalk] = x modulation	MODUL	**Kreuzmodulation** [unerwünschte Modulationsprodukte bei AM, erzeugen verständlichen Nebensprechen]	**crosswise**	TECH	**kreuzweise** ≈ gekreuzt	
			crowbar [short-circuit or low resistance at the input] = crowbar protection circuit; overvoltage crowbar ↑ overvoltage protection [EL.TECH]	POWER SYS	**Eingangskurzschluß** = Crowbar-Schutzschaltung ↑ Überspannungsschutz [EL.TECH]	
cross-office check (→continuity check)	SWITCH	(→Durchschalteprüfung)				
crossover (n.)	ELECTRON	**Bündelknoten**				
crossover (n.)	MICROEL	**Überkreuzen** = Crossover	**crowbar protection circuit** (→crowbar)	POWER SYS	(→Eingangskurzschluß)	
crossover (n.)	POWER SYS	**Überkreuzung**	**crowfoot antenna**	ANT	**Krähenfuß-Antenne**	
crossover filter = crossover network	EL.ACOUS	**Frequenzweiche**	**crown glass**	PHYS	**Kronglas**	
cross-over frequency	EL.ACOUS	**Übergangsfrequenz** = Übernahmefrequenz	**CRT** (→cathode ray tube)	ELECTRON	(→Kathodenstrahlröhre)	
			crucible ↑ Tiegel	TECH	**Schmelztiegel**	
crossover network (→crossover filter)	EL.ACOUS	(→Frequenzweiche)	**crude oil** (→petroleum)	TECH	(→Erdöl)	
crosspiece	TECH	**Querstück**	**crude steel** ≈ pig iron	METAL	**Rohstahl** ≈ Roheisen	

crush (v.t.)		TECH	**zerquetschen**	**crystal oscillator** (→quartz oscillator)		CIRC.ENG	(→Quarzoszillator)
≈ jam			= quetschen				
			≈ zermalmen; klemmen 1	**crystal plane**		PHYS	**Kristallebene**
crushed		TECH	**gequetscht**	**crystal potential**		PHYS	**Kristallpotential**
= squeezed				**crystal-precise**		ELECTRON	**quarzgenau**
cryogenic memory		COMPON	**Tieftemperaturspeicher**	≈ crystal stabylized			≈ quarzstabilisiert
= cryogenic store			= Kryospeicher; kryogener Speicher; supraleitender Speicher	**crystal pulling**		MICROEL	**Kristallzüchtung**
				≈ crystal growth			= Kristallziehen
							≈ Kristallwachstum
cryogenics (→cryophysics)		PHYS	(→Tieftemperaturphysik)	**crystal resonator** (→quartz resonator)		COMPON	(→Schwinquarz)
cryogenic store (→cryogenic memory)		COMPON	(→Tieftemperaturspeicher)	**crystal stabilization**		ELECTRON	**Quarzstabilisierung**
				crystal-stabilized (→crystal-controlled)		ELECTRON	(→quarzstabilisiert)
cryophysics		PHYS	**Tieftemperaturphysik**				
= very low temperature physics; cryogenics				**crystal standard**		CIRC.ENG	**Quarznormal**
				= quartz standard			
cryostat		TECH	**Kälteregler**	**crystal structure**		PHYS	**Kristallstruktur**
			= Kryostat	**crystal temperature**		PHYS	**Kristalltemperatur**
cryotron		COMPON	**Kryotron**	**crystal thermostat**		ELECTRON	**Quarzthermostat**
[superconductive device]			[bei Tiefsttemperatur arbeitendes Schaltelement]	**crystal time base**		ELECTRON	**Quarzzeitbasis**
				crystal unit (→quartz resonator)		COMPON	(→Schwinquarz)
			= Cryotron				
cryptoanalysis (→decryption)		CODING	(→Entschlüsselung)	**crystal vibrator**		COMPON	**Quarzvibrator**
				Cs (→cesium)		CHEM	(→Cäsium)
cryptographic attachment		TERM&PER	**Schlüsselzusatz**	**csch** (→hyperbolic cosecant)		MATH	(→Cosecans hyperbolicus)
cryptographic equipment		TERM&PER	**Schlüsselgerät**	**CSL** (→emitter-coupled logic)		MICROEL	(→emittergekoppelte Logik)
cryptography		INF.TECH	**Kryptographie**				
= cryptology			= Kryptologie; Geheimschrift	**CSMA** (→carrier sense multiple access)		TRANS	(→Vielfachzugriff mit Leitungsüberwachung)
↓ speech encryption			↓ Sprachverschlüsselung				
cryptography (→speech encryption)		INF.TECH	(→Sprachverschlüsselung)	**CSMA/CD mode**		DATA COMM	**CSMA/CD-Verfahren**
				[Carrier Sense with Multiple Access with Collision Detection; collision avoidance procedure for LAN]			[Verfahren der Kollisionsverhinderung in LAN's]
cryptology (→cryptography)		INF.TECH	(→Kryptographie)				
crypton		CHEM	**Krypton**				
= Kr			= Kr	**CS mode** (→closed shop operation)		DATA PROC	(→Closed-shop-Betrieb)
crystal 1		PHYS	**Kristall**				
crystal 2 (→quartz)		CHEM	(→Quarz)	**CS operation** (→closed shop operation)		DATA PROC	(→Closed-shop-Betrieb)
crystal (→quartz resonator)		COMPON	(→Schwinquarz)				
				ct (→metric carat)		PHYS	(→metrisches Karat)
crystal accuracy		ELECTRON	**Quarzgenauigkeit**	**CTD** (→charge transfer device)		MICROEL	(→Ladungstransferelement)
crystal anisotropy		PHYS	**Kristallanisotropie**				
= crystalline anisotropy				**ctg** (→cotangent)		MATH	(→Kotangens)
crystal class		PHYS	**Kristallklasse**	**cth** (→hyperbolic cotangent)		MATH	(→Cotangens hyperbolicus)
			= Symmetrieklasse	**CTL**		MICROEL	**CTL**
crystal-controlled		ELECTRON	**quarzstabilisiert**	[complementary transistor logic]			
= crystal-stabilized			= quarzgesteuert	**CTR** (→current transmission coefficient)		NETW.TH	(→Stromübertragungsfaktor)
			≈ quarzgenau				
crystal-controlled oscillator (→quartz oscillator)		CIRC.ENG	(→Quarzoszillator)	**CTR key** (→control key 1)		TERM&PER	(→Steuerungstaste)
				CTRL key (→control key 1)		TERM&PER	(→Steuerungstaste)
crystal cut 1		PHYS	**Kristallschnitt**				
crystal cut 2		PHYS	**Quarzschnitt**	**Cu** (→copper)		CHEM	(→Kupfer)
crystal detector		COMPON	**Kristalldetektor**	**cube 1** (n.)		MATH	**Kubikzahl**
crystal diode		PHYS	**Kristalldiode**	= power of three			= Hoch-Drei-Zahl
crystal dislocation		PHYS	**Kristallversetzung**	**cube 2** (n.)		MATH	**Würfel**
↑ crystal imperfection			↑ Kristallbaufehler	[equilateral rectangular prism; a solid of six equal squares]			[Quader mit gleichen Kanten; von sechs gleichen Quadraten begrenzter Körper]
crystal filter		CIRC.ENG	**Quarzfilter**				
crystal generator (→quartz generator)		CIRC.ENG	(→Quarzgenerator)	= hexahedron (pl.-ons; -a)			
				↑ rectangular prism; parallelepiped; prism; polyhedron			= Kubus; Hexaeder; Sechsflächner
crystal growth		MICROEL	**Kristallwachstum**				
≈ crystal pulling			≈ Kristallzüchtung				↑ Quader; Parallelepiped; Prisma; Polyeder
crystal imperfection		PHYS	**Kristallbaufehler**				
= lattice imperfection; lattice defect			= Gitterfehler; Gitterstörstelle; Fehlordnung	**cube root**		MATH	**Kubikwurzel**
↓ crystal dislocation; lattice vacancy			↓ Kristallversetzung; Gitterfehlstelle	= third root			= dritte Wurzel
				cubic (adj.)		MATH	**kubisch**
crystal lattice		PHYS	**Kristallgitter**	**cubic** (→regular)		PHYS	(→regulär)
= space lattice			= Raumgitter	**cubical expansion coefficient**		PHYS	**Volumen-Ausdehnungskoeffizient**
crystalline anisotropy (→crystal anisotropy)		PHYS	(→Kristallanisotropie)				
				cubical quad antenna		ANT	**Cubical-quad-Antenne**
crystalline semiconductor		PHYS	**kristalliner Halbleiter**	= quad antenna; cubical quad loop; quad loop			= Quad-Antenne; Quad-Schleife
crystallinity		PHYS	**Kristallinität**				
crystallization		PHYS	**Kristallisieren**				↓ Boom-Quad; Spinnen-Quad-Antenne; Diamant-Quad
crystallize		PHYS	**kristallisieren**				
crystallographic lattice constant		PHYS	**kristallographische Gitterkonstante**				
				cubical quad loop (→cubical quad antenna)		ANT	(→Cubical-quad-Antenne)
crystal loudspeaker (→piezoelectric loudspeaker)		EL.ACOUS	(→piezoelektrischer Lautsprecher)				
				cubic centimeter (AM)		PHYS	**Kubikzentimeter**
crystal microphone		EL.ACOUS	**Kristallmikrophon**	= cubic centimetre (BRI); c.c.			
= piezoelectric microphone			= piezoelektrisches Mikrophon; Piezomikrophon	**cubic centimetre** (BRI) (→cubic centimeter)		PHYS	(→Kubikzentimeter)
crystal orientation		PHYS	**Kristallorientierung**	**cubic characteristic**		ELECTRON	**kubische Kennlinie**

cubic lattice		PHYS	**kubisches Gitter**
cubicle (→mains distribution rack)		POWER ENG	(→Netzschaltfeld)
cubicle (→control cubicle)		POWER SYS	(→Schaltschrank)
cubicle (→cabinet)		EQUIP.ENG	(→Schrank)
cubic meter (NAM) [SI unit for volume] = cubic metre (BRI); cbm (obs.)		PHYS	**Kubikmeter** [SI-Einheit für Volumen] = cbm (obs.)
cubic metre (BRI) (→cubic meter)		PHYS	(→Kubikmeter)
cue = hint		COLLOQ	**Tip** = Anstoß; Wink
cue (→queue)		COLLOQ	(→Schlange)
cue (→prompt)		DATA PROC	(→Bereitmeldung)
cue channel		BROADC	**Dienstkanal**
CUG (→closed user group)		DATA COMM	(→geschlossene Teilnehmerbetriebsklasse)
culpable ≈ responsable		ECON	**schuldhaft** ≈ verantwortlich
cumbersome (→complicated)		COLLOQ	(→kompliziert)
cumulant (n.)		MATH	**Kumulante**
cumulate (→sum-up)		MATH	(→aufsummieren)
cumulative distribution function (→distribution function)		MATH	(→Verteilungsfunktion)
cumulative envelope		ANT	**Summenhüllkurve**
cumulative failure frecuency (→failure fraction)		QUAL	(→Ausfallsatz)
cumulative frequency		MATH	**Summenhäufigkeit**
cumulative tolerance		MECH	**Summentoleranz**
CUN (→closed user network)		TELEC	(→geschlossenes Teilnehmernetz)
cup antenna ↑ geometrically antenna		ANT	**Kelchstrahler** = Kelchantenne ↑ geometrisch dicke Antenne
cup core (→pot core)		COMPON	(→Schalenkern)
cup-core coil (→pot-core coil)		COMPON	(→Schalenkernspule)
cup plug		TECH	**Verschlußkappe**
curb (n.) [edging along a street] = kerb (BRI)		CIV.ENG	**Bordstein**
curet [DATA PROC] (→apex)		TYPOGR	(→Spitze)
Curie [unit for radioactivity] = Ci		PHYS	**Curie** [Maßeinheit für Radioaktivität] = Ci
Curie law		PHYS	**Curie-Gesetz** = Curiesches Gesetz
Curie point		PHYS	**Curie-Punkt** = Curiescher Punkt; kritischer Punkt
Curie temperature		PHYS	**Curie-Temperatur** = Curiesche Temperatur
curium = Cm		CHEM	**Curium** = Cm
curl (n.) [vector analysis] = rot		MATH	**Rotation** [Vektorrechnung] = Wirbel; rot
curl (n.) ≈ eddy; vortex ≈ turbulente flow		PHYS	**Wirbel** ≈ Wirbelströmung
curled		TECH	**gerollt**
curling		METAL	**Rollstanzen**
curly bracket [symbol { }] = brace ↑ Klammer		MATH	**geschweifte Klammer** [Symbol { }] ↑ bracket 1
currency		ECON	**Währung**
currency clause = currency translation clause; foreign exchange clause		ECON	**Währungsklausel**
currency sign = currency symbol		ECON	**Währungszeichen** = Währungssymbol
currency symbol (→currency sign)		ECON	(→Währungszeichen)
currency translation clause (→currency clause)		ECON	(→Währungsklausel)
current (n.)		PHYS	**Strom**
current (→ongoing)		TECH	(→laufend 1)
current address = actual address 1 ≈ absolute address		DATA PROC	**Momentanadresse** = momentane Adresse; aktuelle Adresse ≈ absolute Adresse
current address register = CAR		DATA PROC	**Momentanadressenregister**
current amplification (→current gain)		NETW.TH	(→Stromverstärkung)
current antinode = current maximum ≠ current node		EL.TECH	**Strombauch** ≠ Stromknoten
current assets (→circulating assets)		ECON	(→Umlaufvermögen 1)
current attenuation		NETW.TH	**Stromdämpfung**
current attenuation factor [current at input to current at output]		NETW.TH	**Stromdämpfungsfaktor** [Strom am Eingang zu Strom am Ausgang]
current bus = power supply bus; power supply bar; conductor rail ↑ busbar		POWER ENG	**Stromschiene** = Stromversorgungsschiene ↑ Sammelschiene
current-carrying ≠ current-free ↑ live		EL.TECH	**stromführend** ≠ stromleitend ≠ stromlos ↑ heiß
current carrying capacity = ampacity; current rating		POWER ENG	**Strombelastbarkeit**
current changer (→rectifier)		EL.TECH	(→Gleichrichter)
current changer (→static power converter)		POWER SYS	(→Stromrichter)
current circuit (→circuit)		NETW.TH	(→Schaltkreis)
current compensation		INSTR	**Stromkompensation**
current condition (→marking pulse)		TELEGR	(→Stromschritt)
current conduction		PHYS	**Stromleitung**
current consumption ≈ power consumption; current demand		EL.TECH	**Stromaufnahme** = Stromverbrauch ≈ Leistungsaufnahme; Strombedarf
current-controlled		ELECTRON	**stromgesteuert**
current-controlled current source = CCS		NETW.TH	**stromgesteuerte Stromquelle** = CCS
current-controlled voltage source = CVS		NETW.TH	**stromgesteuerte Spannungsquelle** = CVS
current controller		CONTROL	**Stromregler**
current data (→variable data)		DATA PROC	(→variable Daten)
current demand ≈ power demand; current consumption; power consumption		EL.TECH	**Strombedarf** ≈ Leistungsbedarf; Stromaufnahme; Leistungsaufnahme
current density		PHYS	**Stromdichte**
current density modulation		ELECTRON	**Stromdichtemodulation**
current direction (→current sense)		EL.TECH	(→Stromrichtung)
current directory [where one is just working]		DATA PROC	**aktuelles Dateiverzeichnis** [in dem man gerade arbeitet]
current distribution		POWER SYS	**Stromverteilung**
current distribution		ANT	**Stromverteilung** = Strombelag
current divider		CIRC.ENG	**Stromteiler**
current drain = current sink		NETW.TH	**Stromsenke**
current drive = current source driving; current driving		CIRC.ENG	**Stromsteuerung**
current driving (adj.)		EL.TECH	**stromsteuernd**
current driving (→current drive)		CIRC.ENG	(→Stromsteuerung)
current driving magnetic amplifier (→current driving transductor amplifier)		CIRC.ENG	(→stromsteuernder Transduktorverstärker)
current driving transductor amplifier = current driving magnetic amplifier ↑ transductor amplifier		CIRC.ENG	**stromsteuernder Transduktorverstärker** = stromsteuernder Magnetverstärker ↑ Transduktorverstärker

current element (→elementary dipole)	ANT	(→Elementardipol)	
current equation	PHYS	**Stromgleichung**	
current excitation	ANT	**Stromkopplung**	
current feeding	EL.TECH	**stromspeisend**	
current flow angle	POWER SYS	**Stromflußwinkel**	
= conduction angle		= Phasenanschnittwinkel	
current follower	CIRC.ENG	(→Gitterbasisschaltung)	
(→grounded-grid circuit)			
current-free	EL.TECH	**stromlos**	
≈ voltage-free		≈ spannungsfrei	
≠ current-carrying		≠ stromführend	
↑ dead		↑ kalt	
current-frequency converter	CIRC.ENG	**Strom-Frequenz-Wandler**	
current gain	NETW.TH	**Stromverstärkung**	
= current amplification		↑ Verstärkung	
↑ gain			
current gain, output shorted	MICROEL	(→Kurzschluß-Stromverstärkung)	
(→small-signal short-circuit forward transfer ratio)			
current generator (→current source)	NETW.TH	(→Stromquelle)	
current hogging [gate]	MICROEL	**Überstromaufnahme** [Gatter]	
current impression	EL.TECH	**Stromeinprägung**	
current indicator	ANT	**Stromindikator**	
current instruction register	DATA PROC	**Momentanbefehlsregister** = momentanes Befehlsregister; CIR	
= CIR			
current intensity	EL.TECH	**Stromstärke**	
[quantity of electricity flowing through a cross section in a unit of time; SI unit: Ampere]		[in der Zeiteinheit durch einen Querschnitt hindurchgehende Elektrizitätsmenge; SI-Einheit: Ampere]	
= current strength; electric current strength; amperage [POWER ENG]; I		= Strommenge; Stromintensität; elektrische Stromstärke; Amperezahl [POWER ENG]; I	
current inverter (→inverter)	POWER SYS	(→Wechselrichter)	
current limit	EL.TECH	**Stromgrenzwert**	
current limitation	EL.TECH	**Strombegrenzung**	
= current limiting			
current limiter	EL.TECH	**Strombegrenzer**	
current limiting (→current limitation)	EL.TECH	(→Strombegrenzung)	
current limiting coil	CIRC.ENG	**Begrenzungsdrossel**	
= limiting coil			
current-limiting diode	MICROEL	**Strombegrenzerdiode**	
current limiting transistor	CIRC.ENG	**Strombegrenzungstransistor**	
current location counter	DATA PROC	**Momentanadreßzähler**	
current matrix	NETW.TH	**Strommatrix**	
current maximum (→current antinode)	EL.TECH	(→Strombauch)	
current measurement	INSTR	**Strommessung**	
current meter	INSTR	**Strommesser**	
= ammeter		= Amperemeter	
current minimum (→current node)	EL.TECH	(→Stromknoten)	
current mirror (→current mirror circuit)	CIRC.ENG	(→Stromspiegelschaltung)	
current mirror circuit	CIRC.ENG	**Stromspiegelschaltung**	
= current mirror		= Stromspiegel	
current-mode logic	MICROEL	(→emittergekoppelte Logik)	
(→emitter-coupled logic)			
current node	EL.TECH	**Stromknoten**	
= current minimum		≠ Strombauch	
≠ current antinode			
current noise	ELECTRON	**Stromrauschen**	
current of air (→draft 2)	TECH	(→Luftzug)	
current output	POWER SYS	**Stromabgabe**	
current path	EL.TECH	**Strompfad**	
		= Stromweg	
current penetration	POWER ENG	**Wirktiefe**	
current polarity (→marking pulse)	TELEGR	(→Stromschritt)	
current probe	INSTR	**Stromsonde**	
current pulse	ELECTRON	**Stromimpuls**	
current pulse (→marking pulse)	TELEGR	(→Stromschritt)	
current range	EL.TECH	**Strombereich**	
current rating (→current carrying capacity)	POWER ENG	(→Strombelastbarkeit)	
current reflection	EL.TECH	**Stromspiegelung**	
current regeneration	ELECTRON	**Stromrückgewinnung**	
current resonance (→parallel resonance)	NETW.TH	(→Parallelresonanz)	
current reversal	EL.TECH	**Stromumkehr**	
= reversal		= Stromwechsel	
current-reversing	EL.TECH	**stromumkehrend**	
current reversing key	CIRC.ENG	**Stromwender**	
current rise rate (→current slope)	ELECTRON	(→Stromsteilheit)	
current saturation	MICROEL	**Stromsättigung**	
current sense	EL.TECH	**Stromrichtung**	
= current direction			
current-sensing resistor	INSTR	**Stromfühler**	
current sharing reactor	POWER SYS	**Stromteilerdrossel**	
current shunt	INSTR	**Strommeßwiderstand**	
current shunt (→shunt)	EL.TECH	(→Nebenschluß)	
current sink (→current drain)	NETW.TH	(→Stromsenke)	
current slope	ELECTRON	**Stromsteilheit**	
= current rise rate			
current source	NETW.TH	**Stromquelle**	
= current generator		= Urstromquelle	
current source driving (→current drive)	CIRC.ENG	(→Stromsteuerung)	
current-source potential	MICROEL	**Stromquellenpotential**	
current stabilization	EL.TECH	**Stromstabilisierung**	
= current stabilizing			
current stabilizing (→current stabilization)	EL.TECH	(→Stromstabilisierung)	
current status memory	DATA PROC	**Aktualitätenspeicher**	
current strength (→current intensity)	EL.TECH	(→Stromstärke)	
current sum antenna	ANT	**Stromsummen-Antenne**	
current supply (→power supply)	CIRC.ENG	(→Stromversorgung)	
current surge	EL.TECH	**Stromstoß**	
≈ surge current		≈ Stoßstrom	
↓ pulse		↓ Impuls [ELECTRON]	
current-switch logic (→emitter-coupled logic)	MICROEL	(→emittergekoppelte Logik)	
current text	TYPOGR	**Lauftext**	
= main text; body			
current through unknown (→measurement current)	INSTR	(→Meßstrom)	
current-to-voltage converter	CIRC.ENG	**Strom-Spannungs-Wandler**	
current tracer	INSTR	**Stromtaster**	
current transfer [tube]	ELECTRON	**Stromübernahme** [Röhre]	
current transfer constant	NETW.TH	**Strom-Wellenübertragungsmaß**	
current transfer ratio (→small-signal short-circuit forward transfer ratio)	MICROEL	(→Kurzschluß-Stromverstärkung)	
current transfer ratio (→current transmission coefficient)	NETW.TH	(→Stromübertragungsfaktor)	
current transformation ratio	MICROEL	**Stromübersetzungsverhältnis**	
		= Stromübersetzung	
current transformer	CIRC.ENG	**Stromtransformator**	
		= Stromwandler	
current transformer	POWER SYS	**Stromwandler**	
current transmission coefficient [current at output to current at input]	NETW.TH	**Stromübertragungsfaktor** [Strom am Ausgang zu Strom am Eingang]	
= current transfer ratio; CTR			
≠ current attenuation factor		≠ Stromdämpfungsfaktor	
↑ transmission coefficient		↑ Übertragungsfaktor	
↓ image current transmission coefficient; effective current transmission coefficient		↓ Wellenstromübertragungsfaktor; Betriebsstromübertragungsfaktor	
current type	POWER SYS	**Stromart**	
current value (→present value)	ECON	(→Barwert)	
current vector	PHYS	**Stromvektor**	

current-voltage characteristic

current-voltage characteristic	ELECTRON	Strom-Spannungs-Charakteristik
		= Strom-Spannungs-Kennlinie
current working area	DATA PROC	aktueller Arbeitsbereich
current yield	EL.TECH	Stromausbeute
curriculum (→professional education)	ECON	(→Fachausbildung)
curriculum vitae (→résumé)	ECON	(→Lebenslauf)
curse of curve	MATH	Kurvenverlauf
cursive (→italic)	TYPOGR	(→kursiv)
cursive face (→italic face)	TYPOGR	(→Kursivschrift)
cursor	TERM&PER	Schreibmarke
[marks a relevant point on a screen]		[zeigt auf einem Bildschirm eine relevante Stelle an]
= pointer; marker; position indicator; optical pointer		= Cursor; Läufer; Laufzeichen; Positionsanzeiger; Positionsmarke; Bildschirmmarke; Positionsanzeigesymbol; Einfügemarke; Eingabezeiger; Zeiger; Lichtmarke; Leuchtmarke; Lichtzeiger; Leuchtzeiger; Blinker
≈ icon		≈ Piktogramm
↓ cursor arrow; mouse pointer; sandglass; underscore character		↓ Cursorpfeil; Mauszeiger; Sanduhr; Unterstreichungsstrich
cursor arrow	TERM&PER	Cursorpfeil
↑ cursor		↑ Schreibmarke
↓ text cursor; graphic cursor		↓ Textcursor; Graphikcursor
cursor backward tabulation	TERM&PER	Schreibmarke Rücktabulation
= CBT		= CBT
cursor control	TERM&PER	Schreibmarkensteuerung
		= Cursorsteuerung
cursor control key (→cursor key)	TERM&PER	(→Schreibmarkentaste)
cursor control pad (→cursor keypad)	TERM&PER	(→Schreibmarkentastenblock)
cursor function	DATA PROC	Cursorfunktion
		= Schreibmarkenfunktion
cursor home	DATA PROC	Schreibmarkenheimlauf
= cursor homing		= Cursorheimlauf
cursor home position	TERM&PER	Schreibmarken-Normalstellung
= home		= Cursor-Normalstellung
cursor homing (→cursor home)	DATA PROC	(→Schreibmarkenheimlauf)
cursor horizontal	TERM&PER	Schreibmarke horizontal
= CHA		= CHA
cursor key	TERM&PER	Schreibmarkentaste
= cursor control key; cursor movement key		= Cursorsteuertaste; Cursortaste
≈ HOME key		↓ HOME-Taste
cursor keypad	TERM&PER	Schreibmarkentastenblock
= cursor control pad		= Cursortastenblock; Schreibmarkensteuerungsblock; Cursorsteuerungsblock
cursor movement key (→cursor key)	TERM&PER	(→Schreibmarkentaste)
cursor pad	TERM&PER	Schreibmarkenblock
[array of cursor control keys]		[Gruppe von Cursor-Steuertasten]
= keypad 2		= Cursorblock; Schreibmarkentastenfeld; Cursortastenfeld; Schreibmarkentastatur; Cursortastatur
		↑ Tastaturfeld
cursor position	TERM&PER	Schreibmarkenposition
		= Cursorposition
curtailment (→lowering)	COLLOQ	(→Senkung)
curtain (→curtain antenna)	ANT	(→Vorhangantenne)
curtain antenna	ANT	Vorhangantenne
= curtain; antenna curtain; curtain array 1; fan antenna; harp antenna		= Vorhang; Fächerantenne
↑ planar dipole array		↑ Dipolfeld
curtain array 1 (→curtain antenna)	ANT	(→Vorhangantenne)

curvature (→bend)	MATH	(→Biegung)
curvature equation	MATH	Krümmungsgleichung
curve (v.t.)	MECH	krümmen
		≈ biegen
curve (n.)	MATH	Kurve
= curved line		= gekrümmte Linie
curve (→trace)	INSTR	(→Kurve)
curved	TECH	gekrümmt
[by a regular shape]		[nach einer Regelkurve]
≈ bent; crooked		≈ gebogen; krumm
curved line (→curve)	MATH	(→Kurve)
curve fit (→curve fitting)	MATH	(→Kurvenermittlung)
curve fitting	MATH	Kurvenermittlung
[numeric mathematics]		[numerische Mathematik]
≈ curve fit; best-fit analysis; fitting		≈ Kurvendeckungsanalyse
curve follower	TERM&PER	Kurvenleser
curve plotter (→curve recorder)	INSTR	(→Kurvenschreiber)
curve recorder	INSTR	Kurvenschreiber
= curve plotter; track recorder; track plotter		
curves (→parenthesis)	MATH	(→runde Klammer)
curve templet (→French curve)	ENG.DRAW	(→Kurvenlineal)
curve tracer	INSTR	Kennlinienschreiber
curvilinear	MATH	krummlinig
≈ nonlinear		≈ nichtlinear
curvilinear correlation	MATH	nichtlineare Korrelation
cushion (v.t.)	MECH	dämpfen
= damp (v.t.)		= polstern
cusp (→inversive point)	MATH	(→Rückkehrpunkt)
custodian	DATA PROC	Speicherbeauftragter
custody (→deposit 1)	ECON	(→Verwahrung)
custom (→usage)	COLLOQ	(→Gepflogenheit)
custom (→custom-designed)	TECH	(→kundenspezifisch)
customary (→usual)	COLLOQ	(→üblich)
custom-build (→custom-designed)	TECH	(→kundenspezifisch)
custom building (AM) (→shelter)	SYS.INST	(→Shelter)
custom circuit (→custom IC)	MICROEL	(→kundenspezifische integrierte Schaltung)
custom design	MICROEL	kundenspezifischer Entwurf
		= Custom-Design
custom-designed	TECH	kundenspezifisch
= custom-tailored; custom-made; custom-build; customized; custom; dedicated 2		= kundenindividuell
≈ application-specific; tailor-made		≈ anwenderspezifisch; anwendungsspezifisch; maßgeschneidert
custom duties exemption	ECON	Zollbefreiung
= custom duties remission		= Zollerlaß
custom duties remission (→custom duties exemption)	ECON	(→Zollbefreiung)
customer's apparatus (→subscriber set)	TELEPH	(→Teilnehmerapparat)
custom IC	MICROEL	kundenspezifische integrierte Schaltung
= custom circuit		= kundenspezifische Schaltung; kundenspezifischer IC; Kundenwunschschaltung
≈ application-specific IC		
↓ full custom IC		≈ anwendungsspezifische integrierte Schaltung
		↓ Vollkundenschaltung
customization	TECH	kundenspezifische Anpassung
= custom-specific adaptation		= Kundenanpassungsentwicklung
customize (v.t.)	TECH	kundenspezifisch anpassen
≈ personalize		≈ personalisieren
customized	TECH	(→kundenspezifisch)
(→custom-designed)		
custom-made	MANUF	fremdgefertigt
custom-made	TECH	(→kundenspezifisch)
(→custom-designed)		
custom payment	ECON	Verzollung
= duty payment; clearance		≈ Entzollung
custom production	MANUF	Fremdfertigung
customs	ECON	Zoll 1
[administration]		[Behörde]
= customs authority		= Zollbehörde

customs authority (→customs)	ECON	(→Zoll 1)
customs drawback (→drawback)	ECON	(→Rückzoll)
customs nomenclature	ECON	Zollnomenklatur
		= Nomenklatur
custom software	DATA PROC	kundenspezifische Software
≠ canned software		= maßgeschneiderte Software
		≠ Massensoftware
custom-specific adaptation	TECH	(→kundenspezifische Anpassung)
(→customization)		
customs penalty	ECON	Zollstrafe
customs service technician	TECH	Kundendiensttechniker
= field technician		= Servicetechniker
≈ field service engineer		≈ Kundendienstingenieur; Wartungstechniker
custom-tailored	TECH	(→kundenspezifisch)
(→custom-designed)		
cut (v.t.)	MECH	schneiden
cut (n.)	TECH	Einschnitt
≈ notch		= Schnitt
		≈ Kerbe
cut (n.)	ECON	Kürzung
= abridgment; reduction		= Reduzierung
cut	ECON	kürzen
= reduce; abridge		= reduzieren; beschneiden
cut (→lowering)	COLLOQ	(→Senkung)
cut-and-paste	DATA PROC	Cut-and-paste
[to move text and illustrations]		[Verschieben von Textblöcken und Graphiken]
= cut'n'paste		
cut-away (n.) (→broken-out section)	ENG.DRAW	(→Teilschnitt)
cutback 2 (→decrease)	TECH	(→Verminderung)
cutback 1 (→reduction)	TECH	(→Verkleinerung)
cutback current limiting	POWER SYS	(→Kurzschlußstrom-Rückregelung)
(→foldback current limiting)		
cutback mode (→foldback current limiting)	POWER SYS	(→Kurzschlußstrom-Rückregelung)
cut back the insulation	EL.TECH	(→abmanteln)
(→remove the sheet)		
cut free	METAL	freischneiden
[stamping]		[Stanzen]
= punch		
cut-in	SWITCH	Eintreten
cut'n'paste	DATA PROC	(→Cut-and-paste)
(→cut-and-paste)		
cut off (v.t.)	TECH	abschneiden
= cutoff (v.t.)		
cutoff (n.)	ELECTRON	Einsatzpunkt
cutoff (v.t.) (→cut off)	TECH	(→abschneiden)
cut-off (v.t.)	EL.TECH	(→ausschalten)
(→disconnect)		
cutoff (→cutoff frequency)	EL.TECH	(→Grenzfrequenz)
cut-off (→cutoff frequency)	EL.TECH	(→Grenzfrequenz)
cut-off (→truncate)	MATH	(→abstreichen)
cutoff bias (→dynamic range)	ELECTRON	(→Dynamikbereich)
cut-off collector current	MICROEL	(→Kollektor-Emitter-Reststrom)
(→collector-emitter cut-off current)		
cutoff collector current, base and emitter shorted (→residual short-circuit current)	MICROEL	(→Kurzschluß-Reststrom)
cutoff collector current, emitter open (→collector-base cut-off current)	MICROEL	(→Kollektor-Basis-Reststrom)
cutoff current 2	MICROEL	Reststrom
[transistor]		[Transistor]
= leakage current; saturation current		= Leckstrom
↓ collector-base cutoff current; collector-emitter cutoff current; emitter base cutoff current; residual short-circuit current		↓ Kollektor-Basis-Reststrom; Kollektor-Emitter-Reststrom; Emitter-Basis-Reststrom; Kurzschluß-Reststrom
cutoff current 1	MICROEL	(→Rückwärtsstrom)
(→reverse current)		
cut-off day (→key day)	ECON	(→Stichtag)
cut-off device	TERM&PER	Abschneidevorrichtung
cutoff frequency	EL.TECH	Grenzfrequenz
= cut-off frequency; cutoff; cut-off; limiting frequency		
cutoff frequency	MICROW	Grenzfrequenz
= cut-off frequency; waveguide cutoff; critical frequency		≈ Grenzwellenlänge
≈ cutoff wavelength		
cut-off frequency (→cutoff frequency)	EL.TECH	(→Grenzfrequenz)
cut-off frequency (→cutoff frequency)	MICROW	(→Grenzfrequenz)
cutoff frequency (→critical frequency)	RADIO	(→Grenzfrequenz)
cut-off jack	TELEPH	Überführungsklinke
cutoff region	ELECTRON	Sperrbereich
= cut-off region; blocking-state region		
cut-off region (→cutoff region)	ELECTRON	(→Sperrbereich)
cutoff relay	TELEPH	Trennrelais
cutoff voltage (→reverse voltage)	ELECTRON	(→Sperrspannung)
cutoff wavelength	MICROEL	Grenzwellenlänge
= cut-off wavelength; critical wavelength; boundary wavelength		≈ Grenzfrequenz
≈ cutoff frequency		
cut-off wavelength (→cutoff wavelength)	MICROEL	(→Grenzwellenlänge)
cut-out (n.)	POWER SYS	Auslöser
= release (n.)		
cutout (n.)	TECH	Durchbruch
= perforation; break-through		= Ausschnitt
= opening 1		≈ Öffnung
cut-out (→protector)	COMPON	(→Sicherung 1)
cutout (→automatic cutout)	COMPON	(→Sicherungsautomat)
cut-out (→interruption)	EL.TECH	(→Unterbrechung)
cut-over [SWITCH] (→commission)	TELEC	(→einschalten)
cutover [SWITCH] (→commissioning)	TELEC	(→Einschaltung)
cut-over engineer [SWITCH] (→commissioning engineer)	TELEC	(→Einschaltingenieur)
cut set	MATH	Schnittmenge
cut sheet feeding (→single-sheet feed)	TERM&PER	(→Einzelblattzuführung)
cut surface	MECH	Schnittfläche
cutter (→side cutting pliers)	TECH	(→Seitenschneider)
cutting (n.)	METAL	Spanabhebung
cutting (n.)	MECH	Schneiden
		= Schnitt
cutting (adj.)	METAL	spanabhebend
cutting	MECH	schneidend
= keen		↓ scharf
↓ sharp		
cutting angle	MECH	Schneidewinkel
		= Schnittwinkel
cutting die	MECH	Schneidstanze
cutting edge	METAL	Schneide
= edge		
↓ knife edge		
cutting edge	ENG.DRAW	Schnittkante
cutting-edge technology (BRI)	TECH	(→Spitzentechnologie)
(→top technology)		
cutting head	MECH	Schneidekopf
cutting line	ENG.DRAW	Schnittlinie
		= Schnittverlauf
cutting load	MECH	Schnittdruck
cutting machine	TERM&PER	Schneidemaschine
= paper cutting machine		= Papierschneidemaschine
cutting machine tool	MECH	spanabhebende Werkzeugmaschine
= cutting tool		
cutting plane	ENG.LOG	Schnittebene
cutting shaping	MECH	spanabhebende Bearbeitung
= machining 2		= spanabhebende Formung
cutting tool	TECH	Zerspanungswerkzeug
cutting tool (→cutting machine tool)	MECH	(→spanabhebende Werkzeugmaschine)
cut to fit (→cut to size)	TECH	(→zuschneiden)
cut to length	MECH	ablängen
cut to size	TECH	zuschneiden
= cut to fit		
CV/CL power supply (→constant-voltage/current-limiting power supply)	POWER SYS	(→Konstantspannungs-/Strombegrenzungs-Stromversorgung)

CVD process [chemical vapour deposition]		MICROEL	**CVD-Verfahren**		**cycle time** [time for a periodic process, e.g. a writing or reading operation in main memory]		DATA PROC	**Zykluszeit** [Durchlaufzeit eines periodischen Vorgangs, z.B. Einschreiben oder Lesen im Hauptspeicher]

CVD process MICROEL **CVD-Verfahren**
[chemical vapour deposition]
C-V meter INSTR **C-V-Meßgerät**
CVS (→ current-controlled NETW.TH (→ stromgesteuerte Spannungsquelle)
voltage source)
CVSDM MODUL **CVSDM**
= continuously variable slope delta
modulation
CW (→ continuous wave) ELECTRON (→ Dauerstrich)
C wire (→ tip wire) TELEPH (→ C-Ader)
CW laser (→ continuous-wave OPTOEL (→ Dauerstrichlaser)
laser)
CW measurement INSTR **Dauertonmessung**
CW mode (→ continuous-wave ELECTRON (→ Dauerstrichbetrieb)
mode)
CW modulation (→ continuous MODUL (→ Gleichwellenmodulation)
wave modulation)
CW operation ELECTRON (→ Dauerstrichbetrieb)
(→ continuous-wave mode)
CW power ELECTRON (→ Dauerstrichleistung)
(→ continuous-wave power)
CW radar RADIO LOC (→ Dauerstrichradar)
(→ continuous-wave radar)
CW transmitter BROADC (→ Gleichwellensender)
(→ common-wave transmitter)
cyan TV **Cyan**
↑ color ↑ Farbe
cyanide (v.t.) METAL **salzbadhärten**
cybernetic (→ cybernetical) CONTROL (→ kybernetisch)
cybernetical CONTROL **kybernetisch**
= cybernetic
cybernetics CONTROL **Kybernetik**
[theory of automatic control mechanisms] [Theorie der Regelungsmechanismen]
cycle TECH **Umlauf**
= circuit; recirculation
cycle SCIE **Zyklus**
[a periodic sequence of events] [periodische Folge von Ereignissen]
≈ period [PHYS] ≈ Periode [PHYS]
cycle DATA PROC **Zyklus**
[a cyclic sequence of processes in a computer] [zyklischer Ablauf in einem Computer]
↓ memory cycle; instruction cycle ↓ Speicherzyklus; Befehlszyklus
cycle (→ period 1) PHYS (→ Periode)
cycle (→ period length) MATH (→ Periodenlänge)
cycle accuracy ELECTRON **Ganggenauigkeit**
= cycle precision
cycle availability DATA PROC **Zyklusverfügbarkeit**
cycle-controlling TELECONTR **zyklussteuernd**
cycle count ELECTRON **Zykluszahl**
cycle counter (→ loop DATA PROC (→ Schleifenzähler)
counter)
cycle duration (→ cycle time) TELECONTR (→ Zykluszeit)
cycle frequency AERON **Taktfolge**
[light] [Feuer]
cycle index DATA PROC **Zyklenindex**
= Iterationsindex
cycle leading TELECONTR **zyklusführend**
cycle precision (→ cycle ELECTRON (→ Ganggenauigkeit)
accuracy)
cycle shift DATA PROC **Ringverschiebung**
cycles per instruction DATA PROC **Takte je Befehl**
= CPI
cycles per second PHYS **Hertz**
[SI unit for frequency; = one oscillation per second] [SI-Einheit für Frequenz; = 1 Schwingung pro Sekunde]
= Hz; hertz; c.p.s.; 1/s = Hz; 1/s
cycle stealing DATA PROC **Zyklusklau**
= Cycle-stealing
cycle-stealing mode DATA PROC **Zyklusklauverfahren**
[the CPU suspends its access mode for one cycle, in order to permit some direct memory access] [die Zentraleinheit setzt für einen Speicherzyklus aus, um einen Direktzugriffzu ermöglichen]
= Zyklusklaubetrieb; Cycle-stealing-Verfahren; Cycle-stealing-Betrieb

cycle time DATA PROC **Zykluszeit**
[time for a periodic process, e.g. a writing or reading operation in main memory] [Durchlaufzeit eines periodischen Vorgangs, z.B. Einschreiben oder Lesen im Hauptspeicher]
↓ instruction cycle time; memory cycle time = Wortzeit
↓ Befehlszykluszeit; Speicherzykluszeit
cycle time TELECONTR **Zykluszeit**
= cycle duration = Zyklusdauer
cyclic SCIE **zyklisch**
[with regular sequence of things or events] [in regelmäßiger Abfolge von Dingen oder Ereignissen]
≈ periodic ≈ periodisch
cyclic access DATA PROC **zyklischer Zugriff**
= Zykluszugriff
cyclic check DATA PROC **zyklische Kontrolle**
cyclic code CODING **zyklischer Code**
[with Hamming distance 1 between sequential numbers] [mit Hamming-Distanz 1 zwischen aufeinanderfolgenden Werten]
= cyclic permuted code; unit-distance code; continuous-progression code; reflected code = reflektierter Code; Polynomsicherung; zyklisch permutierter Code; einschrittiger Code
↓ Gray code ↓ Gray-Code
cyclic code (→ program DATA PROC (→ Programmschleife)
loop)
cyclic coding (→ cyclic DATA PROC (→ zyklische Programmierung)
programming)
cyclic decimal code CODING **zyklischer Dezimalcode**
cyclic distortion DATA COMM **zyklische Verzerrung**
cyclic error control DATA COMM **zyklische Fehlerkontrolle**
cyclic memory (→ circulating DATA PROC (→ Umlaufspeicher)
memory)
cyclic permuted code CODING (→ zyklischer Code)
(→ cyclic code)
cyclic processing (→ cyclic DATA PROC (→ zyklische Programmierung)
programming)
cyclic program (→ program DATA PROC (→ Programmschleife)
loop)
cyclic programming DATA PROC **zyklische Programmierung**
[using program loops] [mit Programmschleifen]
= cyclic processing; cyclic coding; loop coding; loop code ≠ gestreckte Programmierung
≠ straight-line programming
cyclic redundancy check CODING **zyklische Blocksicherung**
= CRC = zyklische Blockprüfung; CRC
cyclic redundancy check CODING **CRC-Prüfzeichen**
character = CRC-Zeichen; Zyklische-Blocksicherungs-Zeichen
cyclic shift (→ end-around DATA PROC (→ Ringschieben)
shift)
cyclic shift register CIRC.ENG **Ringschieberegister**
= ring shift register; endaround shift register; circulating register
cyclic storage (→ circulating DATA PROC (→ Umlaufspeicher)
memory)
cyclic transmission TELECONTR **zyklischer Betrieb**
[permanently and in a preestablished sequence] [dauernd und in vorgegebener Reihenfolge]
cycloconvertor POWER SYS **Direktumrichter**
= Steuerumrichter
cycloid MATH **Zykloide**
[curve described by the point of a circle rolling straight-on] [Kurve die der Punkt eines geradeaus rollendes Kreises beschreibt]
= Radkurve
cyclotron PHYS **Zyklotron**
cylinder MATH **Zylinder**
[a body delimited by two parallel planes and a straight lateral surface] [durch zwei parallele Ebenen und einem Mantel begrenzter Körper]
cylinder TERM&PER **Zylinder**
[vertical column of tracks of a disk pack] [übereinanderliegende Spuren eines Plattenspeichers]
cylinder anode (→ can ELECTRON (→ Topfanode)
anode)

cylinder antenna (→cylindrical antenna)	ANT	(→Zylinderantenne)	daisy chain bus	DATA COMM	Befehlskettenbus
cylinder electrode	ELECTRON	Zylinderelektrode	daisy-chain cable	COMM.CABLE	Girlandenkabel
↓ can anode		↓ Topfanode	[submarine optical cable with only land-based repeaters]		[LWL-Seekabel mit ausschließlich auf Land installierten Regeneratoren]
cylinder of revolution	MATH	Rotationszylinder	= festoon (sl.); scallop coastal cable; single-span system		= Festonkabel
		= gerader Kreiszylinder	↑ repeaterless cable		↑ repeaterloses Kabel
cylindric (→cylindrical)	MATH	(→zylindrisch)	↓ minisub cable (SIEMENS)		↓ Minisub-Kabel (SIEMENS)
cylindrical	MATH	zylindrisch	daisy chaining (→daisy chain)	DATA COMM	(→Prioritätsverkettung)
= cylindric		↓ kreiszylindrisch; parabolzylindrisch	daisy chain interrupt (→interrupt bus)	DATA PROC	(→Unterbrechungsbus)
↓ circular cylindrical; parabolic cylindrical			daisy wheel (→typewheel)	TERM&PER	(→Typenrad)
cylindrical antenna	ANT	Zylinderantenne	daisy-wheel printer (→typew-heel printer)	TERM&PER	(→Typenraddrucker)
= cylinder antenna		= zylindrische Antenne	daisy-wheel typewriter	TERM&PER	Typenradschreibmaschine
≈ tower antenna		≈ Mastantenne	dale (→valley)	GEOPHYS	(→Tal)
cylindrical armature (→drum armature)	POWER ENG	(→Trommelanker)	DAMA (→demand-assignment multiple access)	TELEC	(→bedarfsgesteuerter Vielfachzugriff)
cylindrical cathode	ELECTRON	Rundkathode	damage (n.)	TECH	Schaden
cylindrical coil	COMPON	Zylinderspule	= injury		= Beschädigung
= cylindric coil; solenoid			≈ impairment		≈ Beeinträchtigung
cylindrical coordinate	MATH	Zylinderkoordinate	damaged	TECH	beschädigt
= cylindric coordinate		= zylindrische Koordinate	= impaired		= schadhaft
cylindrical dipole	ANT	Zylinderdipol	≈ defective		≈ fehlerfaft
cylindrical earth electrode	ANT	Zylindererder	damage level	TECH	Beschädigungsschwelle
cylindrical helical spring	MECH	zylindrische Schraubenfeder	damages (→compensation)	ECON	(→Schadenersatz)
cylindrically symmetric	MATH	zylindersymmetrisch	damage to property	ECON	Sachschaden
cylindrical magnet	PHYS	Topfmagnet			≈ Materialschaden
cylindrical reflector	ANT	Zylinderreflektor			≠ Personenschaden
cylindrical-roller bearing	MECH	Zylinderrollenlager	damp (v.t.)	PHYS	dämpfen
cylindrical wave	PHYS	zylindrische Welle	= deaden		
		= Zylinderwelle	damp (v.t.) (→cushion)	MECH	(→dämpfen)
cylindric coil (→cylindrical coil)	COMPON	(→Zylinderspule)	damp (→humid)	PHYS	(→feucht)
cylindric coordinate (→cylindrical coordinate)	MATH	(→Zylinderkoordinate)	damped	PHYS	gedämpft
			damped oscillation	PHYS	gedämpfte Schwingung
cypher (→cipher)	ECON	(→beziffern)	dampen (v.t.)	TECH	anfeuchten
cyrillic	LING	kyrillisch			= befeuchten
Czochralski crystal-pulling process (→Czochralski process)	MICROEL	(→Tiegelziehverfahren)	damper (→booster diode)	TV	(→Schalterdiode)
Czochralski process	MICROEL	Tiegelziehverfahren	damping (→decrement)	PHYS	(→Dämpfung)
= Czochralski crystal-pulling process		= Czochralski-Verfahren	damping coefficient (→attenuation factor)	NETW.TH	(→Dämpfungsfaktor)
			damping constant	PHYS	Abklingkonstante
			= damping decrement		= Dämpfungsfaktor

D

DA (→device attributes)	DATA COMM	(→Gerätekennung)	damping decrement (→damping constant)	PHYS	(→Abklingkonstante)
DAC (→digital-to-analog converter)	CODING	(→D/A-Wandler)	damping factor (→attenuation factor)	NETW.TH	(→Dämpfungsfaktor)
D/A converter (→digital-to-analog converter)	CODING	(→D/A-Wandler)	damping function (→attenuation factor)	NETW.TH	(→Dämpfungsfaktor)
DACS (→digital cross-connect)	TRANS	(→Digital-Verteiler)	damping magnet	ELECTRON	Dämpfungsmagnet
dagger (n.)	TYPOGR	Kreuzzeichen	damping pad	EL.ACOUS	Dämpfungsmaterial mit Profil
[a cross-like sign]		= Kreuz	dampness (→humidity)	PHYS	(→Feuchtigkeit)
daily 1 (adj.)	COLLOQ	täglich (adj.)	dampness sensor (→humidity sensor)	COMPON	(→Feuchtesensor)
[happening every day]		≈ tagtäglich; tageweise; alltäglich	danger alarm	TELECONTR	Gefahrenmeldung
= quotidian			danger alarm engineering	SIGN.ENG	Gefahrenmeldetechnik
≈ everday; by the day; everyday			danger detection system	SIGN.ENG	Gefahrenmeldeanlage
daily 2 (adj.)	COLLOQ	werktäglich	= warning system; alerting system		= Warnanlage
[on weekdays]			↓ intrusion detection system; emergency call system; fire detection system; premises supervision system		↓ Einbruchmeldeanlage; Notrufanlage; Brandmeldeanlage; Geländeüberwachungsanlage
daily (n.) (→daily output)	MANUF	(→Tagesleistung)			
daily (→daily paper)	ECON	(→Tageszeitung)			
daily account statement	ECON	Tagesauszug			
daily allowance	ECON	Tagegeld			
= allowance; per diem (AM)		= Auslösung	danger money (→hazard bonus)	ECON	(→Gefahrenzulage)
daily output	MANUF	Tagesleistung	danger notice	TECH	Warntafel
= daily (n.)		= Tagesausbeute	= warning notice; hazard notice		= Gefahrenschild; Warnschild
daily paper	ECON	Tageszeitung	dangerous good	TECH	Gefahrengut
= daily		↑ Zeitung	dank (→humid)	PHYS	(→feucht)
↑ newspaper			DAP (→array processor)	DATA PROC	(→Vektorrechner)
daisy (→typewheel)	TERM&PER	(→Typenrad)	dark current	ELECTRON	Dunkelstrom
daisy-chain (v.t.)	DATA COMM	prioritätisch verketten	dark discharge	PHYS	Dunkelentladung
		= im Warteschlangenmodus verketten	↑ gaseous discharge		↑ Gasentladung
daisy chain (n.)	DATA COMM	Prioritätsverkettung	dark fiber [FITL]	TELEC	Reservefaser [FITL]
[devices are inserted into a bus in the sequence of their priority; daisy = "first class person or thing" in US slang]		[Geräte werden in ihrer Rangfolge in einen Bus eingefügt; daisy = „erstklassige Person oder Sache" in der nordamerikan. Umgangssprache]			= ungenutzte Faser
			dark green	PHYS	dunkelgrün
			= deep green		= sattgrün; tiefgrün
			darkness	PHYS	Dunkelheit
= daisy chaining			dark resistance	COMPON	Dunkelwiderstand
		= Daisy Chaining; Warteschlangenkette			[eines unbelichteten Fotowiderstandes]

darkroom 138

darkroom	TECH	**Dunkelkammer**
= dark space		= Dunkelraum
dark space (→darkroom)	TECH	(→Dunkelkammer)
dark-trace tube	ELECTRON	**Dunkelschriftröhre**
= skiatron		= Farbschriftröhre; Blauschriftröhre; Skiatron
dark tube	ELECTRON	**Dunkelröhre**
dark violet	PHYS	**violett dunkel**
Darlington amplifier	CIRC.ENG	**Darlington-Transistor**
= Darlington power transistor		= Transistorkaskade
Darlington circuit	CIRC.ENG	**Darlington-Schaltung**
= Darlington stage; Darlington combination		= Darlingtonstufe
Darlington combination (→Darlington circuit)	CIRC.ENG	(→Darlington-Schaltung)
Darlington differential amplifier	CIRC.ENG	**Darlington-Differenzverstärker**
Darlington pair	MICROEL	**Darlington-Paar**
Darlington power transistor (→Darlington amplifier)	CIRC.ENG	(→Darlington-Transistor)
Darlington stage (→Darlington circuit)	CIRC.ENG	(→Darlington-Schaltung)
dash	LING	**Gedankenstrich**
≈ hyphen		≈ Bindestrich
dash	MATH	**Strich**
[']		[']
dash-dot	ENG.DRAW	**Strichpunkt**
[.–]		[.–]
D/A signal source (→digital/analog signal generator)	INSTR	(→Digital/Analog-Signalgenerator)
DAT (→digital audio tape)	CONS.EL	(→digitale Musikkassette)
DAT (→dynamic address translation)	DATA PROC	(→dynamische Adreßumsetzung)
data (s.-um)	DATA PROC	**Daten** (s.-um)
[formalized representation of facts, suitable for storage, processing and transfer]		[formgerechte Wiedergabe von Fakten, geeignet für Speicherung, Verarbeitung und Übermittlung]
≈ file; data base		≈ Datei; Datenbank
data	TECH	**Angaben**
		= Daten
data (→useful information)	CODING	(→Nutzinformation)
data-above-voice modem (→DOV modem)	DATA COMM	(→DOV-Modem)
data abuse (→data misuse)	DATA PROC	(→Datenmißbrauch)
data access	DATA PROC	**Datenzugriff**
data access management	DATA PROC	**Datenzugriffsverwaltung**
		= Datenzugriffsmanagement
data acquisition	DATA PROC	**Datenerfassung**
= data collection; data gathering; data recording; data capture; data capturing; collection		= Datensammlung; Dateneinsammlung
≈ data input; data import		≈ Dateneingabe; Datenimport
data acquisition service	DATA PROC	**Datenerfassungsdienst**
data acquisition system	DATA PROC	**Datenerfassungssystem**
= data collection system		
data acquisition terminal	TERM&PER	**Datenerfassungsterminal**
data adapter unit	DATA PROC	**Datenanpassungsgerät**
data address (→address)	DATA PROC	(→Adresse)
data addressed memory (→associative storage)	DATA PROC	(→Assoziativspeicher)
data-addressed storage (→associative storage)	DATA PROC	(→Assoziativspeicher)
data administration (→data management)	DATA PROC	(→Datenverwaltung)
data administrator	DATA PROC	**Datenbankbeauftragter**
= database administrator; DBA; data librarian; file librarian		= Datenbankverwalter; Datenverwalter
≈ librarian		
data aggregate	DATA PROC	**Datengruppe**
[aggregate of data fields]		[Zusammenfassung von Datenfeldern]
= data group		↓ Datensatz
↑ data record		↓ Datenfeld; Datenelement
↓ data field; data element		
data analysis	DATA PROC	**Datenanalyse**
data analyzer	INSTR	**Datenanalysator**
data area	DATA PROC	**Datenbereich**
[in a memory]		[in einem Speicher]
data backup (→data protection)	DATA PROC	(→Datensicherung)

databank (→database)	DATA PROC	(→Datenbank) (pl.-en)
data bank (→database)	DATA PROC	(→Datenbank) (pl.-en)
data bank machine (→database machine)	DATA PROC	(→Datenbankmaschine)
database	DATA PROC	**Datenbank** (pl.-en)
[structured system of related files]		[strukturiertes System sachbezogener Dateien]
= data base; databank; data bank; file system		= Datenbasis; Dateiensystem; Dateisystem; Dateigruppe
≈ data file; data stock		≈ Datei; Datenbestand
data base (→database)	DATA PROC	(→Datenbank) (pl.-en)
database access	DATA PROC	**Datenbankzugriff**
database administration (→database management)	DATA PROC	(→Datenbankverwaltung)
database administration system (→database management system)	DATA PROC	(→Datenbank-Verwaltungssystem)
database administrator (→database manager)	DATA PROC	(→Datenbank-Verwaltungsprogramm)
database administrator (→data administrator)	DATA PROC	(→Datenbankbeauftragter)
database analysator	DATA PROC	**Datenbankanalysator**
database description language	DATA PROC	**Datenbank-Beschreibungssystem**
database language	DATA PROC	**Datenbanksprache**
↓ query language; data description language; data manipulation language		↓ Abfragesprache; Datenbeschreibungssprache; Datenmanipulationssprache
database machine	DATA PROC	**Datenbankmaschine**
= data bank machine		
database management	DATA PROC	**Datenbankverwaltung**
= database administration		
database management system	DATA PROC	**Datenbank-Verwaltungssystem**
= DBMS; database administration system; database manager; database manager system; database administrator		= Datenbank-Managementsystem; DBMS; Datenbasis-Verwaltungssystem; Datenbank-Verwaltungsprogramm
database manager (→database management system)	DATA PROC	(→Datenbank-Verwaltungsprogramm)
database manager system (→database managment system)	DATA PROC	(→Datenbank-Verwaltungsytem)
database mapping	DATA PROC	**Datenbankabbildung**
database mask	DATA PROC	**Datenbankmaske**
database organization	DATA PROC	**Datenbankorganisation**
database scheme	DATA PROC	**Datenbankkonzept**
database system	DATA PROC	**Datenbanksystem**
data bearer (→data carrier)	DATA PROC	(→Datenträger)
data bit (→useful bit)	CODING	(→Nutzbit)
data bit rate adaptation	TELEC	**Datenbitraten-Anpassung**
data block	DATA PROC	**Datenblock**
[a defined set of data handled as physical unit, in storage or transfer processes, independently from logic content]		[für Speicherung oder Übermittlung als physikalische Einheit behandelte Datenmenge, unabhängig vom logischen Gehalt]
= record block; block; physical record; burst [DATA COMM]		= Satzblock; Block; physikalischer Satz; physikalische Informationseinheit
≈ data record		≈ Datensatz
↑ information unit		↑ Informationseinheit
data book	TECH	**Datenbuch**
data break	DATA PROC	**Datenabzugspause**
data broadcasting	RADIO	**Datenrundfunk**
data buffer	DATA PROC	**Datenpuffer**
data bus	DATA COMM	**Datenbus**
= data highway; highway		= Datenpfad; Speicherbus
data byte (→useful byte)	CODING	(→Nutzbyte)
data cable	TERM&PER	**Datenkabel**
data call	DATA COMM	**Datenverbindung**
= data connection; data circuit		
data capacity	DATA PROC	**Datenbreite**
[number of bits which can be handled by each access to the main memory]		[pro Zugriff im Arbeitsspeicher erreichbare Anzahl von Bits]
≈ word length		≈ Wortlänge

data capture (→data acquisition)	DATA PROC	(→Datenerfassung)		**data communications equipment**	DATA COMM	**Datenübertragungseinrichtung**
data capture device (→data entry terminal)	TERM&PER	(→Datenerfassungsgerät)		= DCE; data circuit-terminating equipment ↑ terminal station ↓ data modem		= DÜE; Datenfernschaltgerät ↑ Datenstation ↓ Datenmodem
data capture equipment (→data entry terminal)	TERM&PER	(→Datenerfassungsgerät)		**data communication service** = computer communication service	TELEC	**Datenübermittlungsdienst** = Datentransferdienst
data capture station (→data entry terminal)	TERM&PER	(→Datenerfassungsgerät)		**data communications testing**	INSTR	**Datenkommunikations-Meßtechnik**
data capture terminal (→data entry terminal)	TERM&PER	(→Datenerfassungsgerät)		**data compacting**	DATA PROC	**Datenkompaktierung**
data capturing (→data acquisition)	DATA PROC	(→Datenerfassung)		**data compaction** (→data compression)	DATA PROC	(→Datenverdichtung)
data carrier	DATA PROC	**Datenträger**		**data compression**	DATA PROC	**Datenverdichtung**
= data medium; storage medium; memory medium; data bearer; volume ≈ information carrier; long-term storage		= Speichermedium; Datenmedium; Medium ≈ Informationsträger; Langzeitspeicher		= compression; data condensation; condensation; data reduction; reduction; data concentration; concentration; data compaction ≠ data expansion ↓ string-oriented data compression; structure-oriented data compression		= Verdichtung; Datenkompression; Kompression; Datenkomprimierung; Komprimierung; Datenkonzentration; Konzentration ≠ Datenexpansion ↓ stringorientierte Datenverdichtung; strukturorientierte Datenverdichtung
data carrier detect signal = DCD	DATA COMM	**Datenträger-Empfangssignal**				
data carrier shredder	OFFICE	**Datenträgervernichter**				
data cartridge (→data cassette)	TERM&PER	(→Datenkassette)				
data cassette	TERM&PER	**Datenkassette**		**data compressor** = compressor	DATA PROC	**Datenverdichter** = Datenkompressor
= data cartridge; datasette 1 ↑ magnetic tape cassette		= Datasette 1 ↑ Magnetbandkassette		**data concentration** (→data compression)	DATA PROC	(→Datenverdichtung)
data catalog = data directory	DATA PROC	**Datenkatalog** = Datenverzeichnis		**data concentration computer**	DATA PROC	**Datenkonzentrationsrechner**
data category (→data type)	DATA PROC	(→Datenbauart)		**data concentrator**	DATA COMM	**Datenkonzentrator**
data cell	TERM&PER	**Datenzelle**		**data condensation** (→data compression)	DATA PROC	(→Datenverdichtung)
data center (→computing center)	DATA PROC	(→Rechenzentrum)		**data confidenciality** ≈ data protection; data security and privacy	DATA PROC	**Datengeheimnis** ≈ Datensicherung; Datenschutz
data chaining	DATA PROC	**Datenverkettung**				
		= Datenkettung		**data connection** (→data call)	DATA COMM	(→Datenverbindung)
data channel = information channel	DATA PROC	**Datenkanal**		**data control**	DATA PROC	**Datensteuerung**
data check (→data validation)	DATA PROC	(→Datenüberprüfung)		**data conversion** [of format] = conversion; data translation ≈ data origination	DATA PROC	**Datenkonvertierung** [des Datenformats] = Datenkonversion; Datenübersetzung; Datenumsetzung; Konvertierung ≈ Datenaufbereitung
data circuit (→data call)	DATA COMM	(→Datenverbindung)				
data circuit-terminating equipment (→data communications equipment)	DATA COMM	(→Datenübertragungseinrichtung)				
data clearing = data correction; data purification	DATA PROC	**Datenkorrektur**				
data clerck	DATA PROC	**Datenassistent**		**data conversion program**	DATA PROC	**Datenkonvertierungsprogramm**
data code	DATA PROC	**Datencode**		**data conversion unit** = conversion unit; conversion equipment	DATA PROC	**Umsetzanlage**
data coding (→data encryption)	DATA PROC	(→Datenverschlüsselung)				
data coding program (→data encryption program)	DATA PROC	(→Datenverschlüsselungsprogramm)		**data converter** [hardware or software to convert data] = converter; convertor	DATA PROC	**Datenkonverter** [Hardware oder Software zur Datenwandlung] = Konverter
data collection (→data acquisition)	DATA PROC	(→Datenerfassung)				
data collection platform	DATA COMM	**Datensammelstelle**		**data coordination computer**	DATA PROC	**Datenkoordinierungsrechner**
data collection protocol	DATA PROC	**Datenerfassungsprotokoll**		**data correction** (→data clearing)	DATA PROC	(→Datenkorrektur)
data collection system (→data acquisition system)	DATA PROC	(→Datenerfassungssystem)		**data corruption**	DATA PROC	**Datenverfälschung**
data collection terminal (→data entry terminal)	TERM&PER	(→Datenerfassungsgerät)		**data declaration** ≈ data structure command ↑ declaration	DATA PROC	**Datendeklaration** ≈ Datenaufbauanweisung ↑ Vereinbarung
datacom (→data communications)	TELEC	(→Datenübermittlung)		**data definition** = data description language	DATA PROC	**Datendefinition** = Datenbeschreibung
data communication computer	DATA PROC	**Datenkommunikationsrechner**		**data definition language** (→data description language)	DATA PROC	(→Datenbeschreibungssprache)
data communication control (→data transmission control)	DATA COMM	(→Datenübertragungssteuerung 1)		**data delimiter**	DATA PROC	**Datenbegrenzungssymbol**
						= Datenabgrenzungssymbol
data communication processor	DATA PROC	**Datenkommunikationsprozessor**		**data description language** = data definition language; DDL ↑ data model	DATA PROC	**Datenbeschreibungssprache** = Datendefinitionssprache ↑ Datenbehandlungssprache
data communications	TELEC	**Datenübermittlung**				
= datacom ≈ information transfer; remote data processing [DATA PROC]; computer communication ↑ data technology ↓ data switching; data transmission		= Datenkommunikation ≈ Informationstransfer; Datenfernverarbeitung [DATA PROC]; Rechnerkommunikation ↑ Datentechnik ↓ Datenvermittlung; Datenübertragung		**data dictionary**	DATA PROC	**Datenlexikon**
				data direct entry equipment = data direct input equipment	TERM&PER	**Daten-Direkteingabegerät**
				data direct input equipment (→data direct entry equipment)	TERM&PER	(→Daten-Direkteingabegerät)
data communications buffer = communications buffer	DATA COMM	**Datenübertragungspuffer** = Kommunikationspuffer		**data directory** (→data catalog)	DATA PROC	(→Datenkatalog)

data disk

data disk DATA PROC **Datendiskette**
≠ program disk ≠ Programmdiskette
data display terminal TERM&PER **Datensichtgerät**
[keyboard and screen to enter and [Tastatur mit Bildschirm
display data and graphics] für Anzeige von Daten und
= video display terminal; VDT; video Graphiken]
display unit; VDU; display console; = DSG; Datensichtstation;
video terminal; video unit Sichtstation; Bildschirmsta-
↑ data terminal; display terminal tion; Bildschirmterminal;
Bildschirmkonsole
↑ Datenendgerät; Sichtgerät

data division DATA PROC **Datenteil**
↑ COBOL ↑ COBOL
data-driven DATA PROC **datengesteuert**
data editing DATA PROC **Dateneditierung**
data element DATA PROC **Datenelement**
= data item; data unit; cell 1 = Dateneinheit; Dateneintrag
≈ data level ≈ Datenposition
data encryption DATA PROC **Datenverschlüsselung**
= data coding
data encryption program DATA PROC **Datenverschlüsselungspro-**
= data coding program **gramm**
data entry (→data DATA PROC (→Dateneingabe)
input)
data entry device (→data TERM&PER (→Datenerfassungsgerät)
entry terminal)
data entry equipment (→data TERM&PER (→Datenerfassungsgerät)
entry terminal)
data entry mask (→input DATA PROC (→Eingabemaske)
mask)
data entry operator 2 DATA PROC **Datentypist**
[male]
= data entry specialist
data entry operator 1 OFFICE **Datentypistin**
data entry sheet DATA PROC **Dateneingabeformular**
= data sheet ≈ Programmvordruck
= coding form
data entry specialist (→data (→Datentypist)
entry operator 2)
data entry station (→data TERM&PER (→Datenerfassungsgerät)
entry terminal)
data entry terminal TERM&PER **Datenerfassungsgerät**
= data entry equipment; data entry sta- = Datenerfassungsstation; Er-
tion; data entry device; data capture fassungsgerät
terminal; data collection terminal;
data capture equipment; data capture
station; data capture device
data error DATA PROC **Datenfehler**
data error analyzer INSTR **Datenfehleranalysator**
data exchange (→data DATA COMM (→Datenaustausch)
interchange)
data exchange function DATA COMM **Datenaustauschfunktion**
data expansion DATA PROC **Datenexpansion**
= decompression = Dekomprimierung
≠ data compression ≠ Datenkompression
data field DATA PROC **Datenfeld 1**
[smallest significant element of a data [kleinstes selbständiges Ele-
record, e.g. "birthday"] ment eines Datensatzes,
= field; array 2 z.B. „Geburtsdatum"]
↑ data record; data aggregate = Feld 1
↓ subfield ↑ Datensatz; Datengruppe
↓ Teilfeld
data field masking DATA PROC **Datenfeldmaskierung**
data file DATA PROC **Datei**
[collection of related data with com- [Ansammlung zusammen-
patible format] hängender Daten in kom-
= computer file; record file; file; dataset patiblem Format]
1 = Datendatei
≈ data base ≈ Datenbank
data file processing DATA PROC **Dateienverarbeitung**
data file program DATA PROC **Datenbankprogramm**
data file transfer DATA PROC **Dateitransfer**
= file transfer
data flow (→data DATA PROC (→Datenfluss)
stream)
dataflow analysis DATA PROC **Datenflußanalyse**
data flowchart DATA PROC **Datenflußplan**
data flow control DATA COMM **Datenflußsteuerung**
data flow diagram DATA PROC **Datenflußdiagramm**
data flow machine DATA PROC **Datenflußmaschine**

data format DATA PROC **Datenformat**
data gathering (→data DATA PROC (→Datenerfassung)
acquisition)
data generation DATA PROC **Datengenerierung**
= Generierung
data generator DATA PROC **Datengenerator**
datagram DATA COMM **Datagramm**
[data package containing destination [Datenpaket mit Adressen-
and route] und Weginformation]
↑ packet switching ↑ Paketvermittlung
datagram mode (→virtual DATA COMM (→virtuelle Verbindung)
call)
datagram service TRANS **verbindungsloser Betrieb**
[ATM] [ATM]
data group (→data DATA PROC (→Datengruppe)
aggregate)
data handling DATA PROC **Datenbehandlung**
≈ data processing ≈ Datenverarbeitung
data hierarchy DATA PROC **Datenhierarchie**
[word-data field-record-file-data [Wort-Datenfeld-Daten-
base] satz-Datei-Datenbank]
data highway (→data DATA COMM (→Datenbus)
bus)
data import DATA PROC **Datenimport**
[entry of data developed by other sys- [Übernahme von Daten,
tems] die von anderen Systemen
≈ data input; data acquisition erfasst wurden]
≈ Dateneingabe; Datenerfas-
sung
data input DATA PROC **Dateneingabe**
= data entry; input; entry; inputting = Eingabe; Dateneintrag;
≈ read-in (n.); data acquisition; data im- Eintrag
port ≈ Einlesen; Datenerfassung;
≠ data output Datenimport
≠ Datenausgabe
data integrity DATA PROC **Datensicherheit**
= data security; information security = Datenintegrität
≈ data protection ≈ Datenschutz
data integrity (→data DATA PROC (→Datensicherung)
protection)
data-intensive DATA PROC **datenintensiv**
≠ compute-intensive ≠ rechenintensiv
data interchange DATA COMM **Datenaustausch**
= data exchange ≈ Datenübertragung
≈ data transmission
data item (→data DATA PROC (→Datenelement)
element)
data key (→alphanumeric TERM&PER (→Datentaste)
key)
data leakage DATA PROC **Datendiebstahl**
data level DATA PROC **Datenposition**
[position within database]
data librarian (→data DATA PROC (→Datenbankbeauftragter)
administrator)
data line TELEC **Datenleitung**
data line TV **Datenzeile**
dataline GEOPHYS **Datumsgrenze**
data line decoder TV **Datenzeilendecoder**
data link DATA COMM **Datenübermittlungsabschnitt**
= data transfer link = Übermittlungsabschnitt
data link control (→data DATA COMM (→Datenübertragungssteue-
transmission control) rung 1)
data link escape DATA COMM **Datenübertragungsumschal-**
[a code] **tung**
= DLE [ein Code]
= DLE
data link layer (→link layer) DATA COMM (→Sicherungsschicht)
data load DATA COMM **Datenlast**
data logger DATA PROC **Datenaufzeichnungsgerät**
data logging DATA PROC **automatische Datenerfassung**
data loss DATA PROC **Datenverlust**
data management DATA PROC **Datenverwaltung**
= data administration; administration = Verwaltung
data management system DATA PROC **Datenverwaltungssystem**
= DVS; Archivsystem; Abla-
gesystem
data manipulation language DATA PROC **Datenmanipulationssprache**
= DML = Datenmanipulierungsspra-
↑ data model che
↑ Datenbehandlungssprache

data medium (→data carrier) DATA PROC (→Datenträger)

data memory DATA PROC **Datenspeicher**
= data storage; data store; data storage device

data migration DATA PROC **Datenwanderung**
[from online to offline device]

data misuse DATA PROC **Datenmißbrauch**
= data abuse

data model DATA PROC **Datenbehandlungssprache**
↓ data description language; data manipulation language
↓ Datenbeschreibungssprache; Datenmanipulationssprache

data modem DATA COMM **Datenmodem**
[adapts data streams to voice channels]
[paßt Datenströme an Fernsprechkanäle an]
= modem; modulator-demodulator; dataset (AM)
= Modem; Datenübertragungsmodem

data movement time (→data transfer time) DATA PROC (→Transferzeit)

data multiplex DATA COMM **Datenmultiplex**
= dataplex

data multiplexer DATA COMM **Datenmultiplexer**
data mutilation DATA PROC **Datenverstümmelung**
data name DATA PROC **Datenname**
data network TELEC **Datennetz**
= Datenübermittlungsnetz

data network diagnostic equipment DATA COMM **Datennetz-Diagnoseeinrichtung**

data network identification code DATA COMM **Datennetzkennzahl**

data networking DATA COMM **Datenvernetzung**
data network terminating unit DATA COMM **Datennetz-Abschlußeinrichtung**
= DNAE

data network with fixed connection (→leased-circuit data network) DATA COMM (→Direktrufnetz)

data number DATA COMM **Datenrufnummer**
data organization DATA PROC **Datenorganisation**
= Datengliederung

data origination (→data preparation) DATA PROC (→Datenaufbereitung)

data output DATA PROC **Datenausgabe**
≈ read-out (n.) ≈ Auslesen
≠ data input ≠ Dateneingabe
↑ output (n.) ↑ Ausgabe

data package (→data packet) DATA COMM (→Datenpaket)

data packet DATA COMM **Datenpaket**
= data package; packet = Paket

data path DATA COMM **Datenweg**
data pen (→code pen) TERM&PER (→Lesestift)
dataplex (→data multiplex) DATA COMM (→Datenmultiplex)

datapoint INSTR **Datenpunkt**
datapoint duration INSTR **Datenpunktdauer**
data pointer DATA PROC **Datenzeiger**
data preparation DATA PROC **Datenaufbereitung**
[into machine-readable form] [in maschinenlesbare Form]
= data origination

data print TERM&PER **Datendruck**
= Datenausdruck

data printer TERM&PER **Datendrucker**
= dataprinter

dataprinter (→data printer) TERM&PER (→Datendrucker)

data privacy protection DATA PROC (→Datenschutz)
(→data security and privacy)

data processing INF.TECH **Datenverarbeitung**
[of analog or digital data] [Verarbeitung analoger oder digitaler Daten]
= DP; information processing 1; data processing technology
= DV; Informationsverarbeitung 1
≈ informatics ≈ Informatik
↑ message processing; data technology ↑ Nachrichtenverarbeitung; Datentechnik
↓ electronic data processing ↓ elektronische Datenverarbeitung

data processing center DATA PROC (→Rechenzentrum)
(→computing center)

data processing curriculum DATA PROC **DV-Fachausbildung**
= information processing curriculum

data processing cycle DATA PROC **Datenverarbeitungszyklus**
= DV-Zyklus

data processing equipment DATA PROC **Datenverarbeitungsanlage**
[a large computer system, mostly for business or public administration] [größere Computeranlage, meist in Witschaftsunternehmen oder Behörden eingesetzt]
= data processing system; electronic data processing machine; EDPM; electronic data processing system; electronic computer system; computing system; computer system
= DVA; EDV-Anlage; EDVA; EDV-System; Datenverarbeitungssystem; DV-Anlage; DV-System; elektronisches Datenverarbeitungssystem; elektronisches Computersystem; elektronische Rechenanlage; Rechenanlage 1; Rechneranlage
↑ computer ↑ Computer

data processing expert DATA PROC (→Computerfachmann)
(→computer professional)

data processing literate DATA PROC (→Computerfachmann)
(→computer professional)

data processing manager DATA PROC (→DV-Beauftragter)
(→computer operations manager)

data processing professional DATA PROC (→Computerfachmann)
(→computer professional)

data processing specialist DATA PROC (→Computerfachmann)
(→computer professional)

data processing system DATA PROC (→Datenverarbeitungsanlage)
(→data processing equipment)

data processing technology INF.TECH (→Datenverarbeitung)
(→data processing)

data processing terminal equipment DATA COMM (→Datenendeinrichtung)
(→data terminal equipment)

data processor DATA PROC **Datenprozessor**
data projector TERM&PER **Datenprojektor**
data proofing (→data protection) DATA PROC (→Datensicherung)

data protection DATA PROC **Datensicherung**
= data security; data integrity; data backup; backup; data proofing
= Sicherstellen
≈ data security and privacy [ECON]; data confidenciality; logging
≈ Datenschutz [ECON]; Datengeheimnis; Protokollierung

Data Protection Act DATA PROC **Datenschutzgesetz**
= Privacy Act (U.S.A.)

data protection officer DATA PROC (→Datenschutzbeauftragter)
(→data security officer)

data purification (→data clearing) DATA PROC (→Datenkorrektur)

data rate [DATA COMM] TELEC (→Übertragungsgeschwindigkeit)
(→transmission rate)

data record DATA PROC **Datensatz**
[set of data fields handled as a unit by logical criteria] [unter logischen Gesichtspunkten als Einheit behandelter Satz von Daten]
= logical record; record; data set 2
= Satz; Record
≈ data block ≈ Datenblock
↑ information unit ↑ Informationseinheit
↓ data field ↓ Datenfeld

data recording (→data acquisition) DATA PROC (→Datenerfassung)

data reduction (→data compression) DATA PROC (→Datenverdichtung)

data register DATA PROC **Datenregister**
[am Rechenwerkeingang der Zentraleingang]

data release key (→ENTER key) TERM&PER (→Eingabetaste)

data reliability DATA PROC **Datenzuverlässigkeit**
data retrieval DATA PROC **Datenwiedergewinnung**
= Datenwiederauffindung

data scope DATA COMM **Datenübertragungsmonitor**
[a display]

data security (→data integrity) DATA PROC (→Datensicherheit)

data security (→data protection) DATA PROC (→Datensicherung)

data security and privacy

data security and privacy ECON
= data proofing; data privacy protection; [DATA PROC]
≈ data confidenciality; data protection
Datenschutz
≈ Datengeheimnis; Datensicherung [DATA PROC]

data security officer DATA PROC
= data protection officer
Datenschutzbeauftragter

data separator DATA PROC **Datenfilter**
dataset 1 (→data file) DATA PROC (→Datei)
dataset (AM) (→data modem) DATA COMM (→Datenmodem)
data set 2 (→data record) DATA PROC (→Datensatz)
data set ready DATA PROC **Empfangsbereitschaft**
= DSR
datasette (→data cassette) TERM&PER (→Datenkassette)
data sharing DATA PROC **Datenverbund**
↑ multi-computer system ↑ Mehrrechnersystem
data sheet TECH **Datenblatt**
= brief description = Kennblatt; Kurzbeschreibung
≈ advertising brochure; leaflet ≈ Werbebroschüre; Prospekt
data sheet (→data entry sheet) DATA PROC (→Dateneingabeformular)
data signal DATA COMM **Datensignal**
data signal (→digital signal) TELEC (→Digitalsignal)
data signaling rate [DATA COMM] (→transmission rate) TELEC (→Übertragungsgeschwindigkeit)
data sink DATA COMM **Datensenke**
≠ data source ≠ Datenquelle
data source DATA COMM **Datenquelle**
= talker
≠ data sink ≠ Datensenke
data stack DATA PROC **Datenkeller**
[temporary storage process, where data are stored and retrieved always on the same end of file] [Zwischenspeicher bei dem Daten immer am selben Ende ein- und ausgelagert werden]
= stack
data stamp OFFICE **Tagestempel**
data station (→terminal station) DATA COMM (→Datenstation)
data stock DATA PROC **Datenbestand**
≈ data file; data base ≈ Datei; Datenbank
data storage (→data memory) DATA PROC (→Datenspeicher)
data storage and retrieval system (→documentation system) DATA PROC (→Dokumentationssystem)
data storage device (→data memory) DATA PROC (→Datenspeicher)
data storage technique DATA PROC **Datenspeicherungstechnik**
data store (→data memory) DATA PROC (→Datenspeicher)
data stream DATA PROC **Datenfluss**
= data flow = Datenstrom
data strobe DATA PROC **Datenhinweissignal**
[signal indicating validity of data being transmitted on a bus] [bestätigt die Gültigkeit von Daten die gerade über einen Bus laufen]
data strobe signal DATA PROC **Datenbestägigungssignal**
data structure DATA PROC **Datenstruktur**
data structure command DATA PROC **Datenaufbau-Anweisung**
≈ data declaration ≈ Datendeklaration
data structure tree DATA PROC **Datenstrukturbaum**
data switching DATA COMM **Datenvermittlung**
↑ data transfer ↑ Datenübermittlung
data switching engineering SWITCH **Datenvermittlungstechnik**
↑ switching engineering ↑ Vermittlungstechnik
data switching exchange DATA COMM **Datenvermittlungsstelle**
= DSE
data syntax DATA COMM **Datensyntax**
data system technology (→data technology) INF.TECH (→Datentechnik)
data tablet (→digitizing tablet) TERM&PER (→Digitalisiertablett)
data tape TERM&PER **Datenband**
↑ Magnetband
data technology INF.TECH **Datentechnik**
= data system technology; computing ↓ Datenverarbeitung; Datenübermittlung
↓ data processing; data communications

data telephone TERM&PER **Datentelefon**
= Datenfernsprecher
data terminal TERM&PER **Datenendgerät**
[for inputs and outputs on computer systems] [Gerät zur Ein-u.Ausgabe an DVA]
= terminal = Datenterminal; Terminal; Datenstation; Endgerät
↓ video display terminal ≈ Datenendeinrichtung
↓ Datensichtgerät
data terminal (→terminal station) DATA COMM (→Datenstation)
data terminal equipment DATA COMM **Datenendeinrichtung**
= DTE; data processing terminal equipment [Übermittlung steuernde Datenquelle oder -senke]
≈ data terminal [TERM&PER] = DEE; Endeinrichtung
↑ terminal station ≈ Datenendgerät [TERM&PER]
↑ Datenstation
data terminal ready DATA PROC **Datenendgerät-Bereitschaftssignal**
= DTR
data throughput DATA PROC (→Durchsatz)
(→throughput)
data thruput DATA PROC (→Durchsatz)
(→throughput)
data traffic DATA COMM **Datenverkehr**
= Verkehr
data traffic measuring equipment DATA COMM **Datenverkehrsmeßgerät**
data transaction DATA PROC **Datentranasaktion**
= Datenabarbeitung
data transfer (→data interchange) DATA COMM (→Datenaustausch)
data transfer link (→data link) DATA COMM (→Datenübermittlungsabschnitt)
data transfer operation DATA PROC **Datentransferoperation**
data transfer rate DATA COMM **Transfergeschwindigkeit**
[the useful information transmitted per time unit] [pro Zeiteinheit übertragene Nutzinformation]
↑ data signaling rate = effektive Übertragungsgeschwindigkeit; Datentransfergeschwindigkeit; Datentransferrate
↑ Datenübertragungsgeschwindigkeit
data transfer time DATA COMM (→Transferzeit)
(→transfer time)
data translation (→data conversion) DATA PROC (→Datenkonvertierung)
data transmission DATA COMM **Datenübertragung**
= data transfer = Datentransfer; DÜ
↑ data communication ≈ Datenaustausch
↓ data interchange ↑ Datenübermittlung
↓ Datenfernübertragung
data transmission block DATA COMM **Datenübertragungsblock**
= frame = DÜ-Block
data transmission cable COMM.CABLE **Datenleitung**
data transmission control DATA COMM **Datenübertragungssteuerung 1**
[a device] [ein Gerät]
= transmission control; data link control; link control; data communication control; communication control 1 = DUST; Übertragungssteuerung 1
data transmission link TELEC **Datenübertragungsleitung**
data transmission path DATA COMM **Datenübertragungsweg**
= path
data transmit key (→ENTER key) TERM&PER (→Eingabetaste)
data trunk DATA COMM **Datenstrecke**
data type DATA PROC **Datenbauart**
= data category = Datentyp; Datenkategorie
data unit (→data element) DATA COMM (→Datenelement)
data validation DATA PROC **Datenüberprüfung**
= data validity check; validation 1; data check; check; data vetting = Datenrichtigkeitsprüfung; Datenprüfung; Datenkontrolle; Datentest
data validity check (→data validation) DATA PROC (→Datenüberprüfung)
data value DATA PROC **Datenwert**

data vetting (→data validation)		DATA PROC	(→Datenüberprüfung)	d.c. component (→dc component)		EL.TECH	(→Gleichstromkomponente)
data viewing		INSTR	Datenbetrachtung	d-c component (→dc component)		EL.TECH	(→Gleichstromkomponente)
data word		DATA PROC	Datenwort	dc conversion		POWER SYS	Gleichstromumrichtung [Gleichstrom in Gleichstrom]
data word size (→word length)		DATA PROC	(→Wortlänge)				
date		COLLOQ	Datum	dc converter		INSTR	Gleichstromwandler
≈ day [PHYS]			≈ Tag [PHYS]	dc converter (→dc convertor)		POWER SYS	(→Gleichstromumrichter)
datebook (→appointment book)		OFFICE	(→Terminkalender)				
date book (→appointment book)		OFFICE	(→Terminkalender)	dc convertor (IEC)		POWER SYS	Gleichstromumrichter [Gleichstrom in Gleichstrom]
Datel service		TELEC	Datel-Dienst	= dc converter; direct current converter; dc-dc convertor; dc voltage transducer; dc transducer			
date of delivery (→delivery time)		ECON	(→Lieferfrist)				= DC-Wandler; DC-DC-Umrichter; Gleichspannungswandler
date of despatch (→shipping date)		ECON	(→Versanddatum)	↑ static power convertor			
Datex service		TELEC	Datexdienst	↓ direct dc converter			
dating		COLLOQ	Datierung				↑ Stromrichter
			= Datumsangabe				↓ Gleichstromsteller
dative		LING	Dativ	dc-coupled		EL.TECH	galvanisch gekoppelt
			= Wemfall; 3. Fall; dritter Fall	= electrically coupled			= gleichspannungsgekoppelt
DAT recorder		CONS.EL	DAT-Recorder	dc coupling		EL.TECH	galvanische Kopplung
datum (pl. -a)		DATA PROC	Datum (pl. -en) [wenig gebräuchliches Singular von Daten]	= conductive coupling; direct coupling; resistance coupling			= Gleichspannungskopplung; Gleichstromkopplung; direkte Kopplung
[rarely used singular of data]				≠ dc isolation			≠ galvanische Trennung
= piece of data				dc current (→direct current)		EL.TECH	(→Gleichstrom)
datum (→fact)		COLLOQ	(→Tatsache)	dc current amplification (→dc current gain)		MICROEL	(→Gleichstromverstärkung)
datum axis (→axis of reference)		ENG.DRAW	(→Bezugsachse)				
datum dimensioning		ENG.DRAW	Bezugskantenbemaßung	dc current gain		MICROEL	Gleichstromverstärkung
datum line		ENG.DRAW	Bezugslinie 1	= dc current amplification			
= reference line			= Ausgangslinie; Führungslinie	DCD (→data carrier detect signal)		DATA COMM	(→Datenträger-Empfangssignal)
datum plane		ENG.DRAW	Bezugsebene	dc-dc converter (→dc converter)		POWER SYS	(→Gleichspannungswandler)
			= Ausgangsebene				
datum point		ENG.DRAW	Bezugspegel	dc decoupling		EL.TECH	galvanische Trennung
			= Ausgangspunkt	= metallic isolation; electrical isolation			= galvanische Entkopplung
datum surface		ENG.DRAW	Bezugsfläche	≠ dc coupling			≠ galvanische Kopplung
			= Ausgangsfläche	dc dissipation		MICROEL	Gleichstrom-Verlustleistung
daughter product		ECON	Folgeprodukt	DCE (→data communications equipment)		DATA COMM	(→Datenübertragungseinrichtung)
day		PHYS	Tag				
[24 hours]			[24 Stunden]				
≈ date			≈ d	DCE data		DATA COMM	DÜE-Daten
			≈ Datum	DCE datagram		DATA COMM	DÜE-Datagramm
day break tariff		TELEC	Spätnachtgebühr	DCE interrupt		DATA COMM	DÜE-Unterbrechung
daylight saving time		COLLOQ	Sommerzeit	dc fault locating		TRANS	Gleichstromfehlerortung
daytime tariff		TELEC	Taggebühr	dc-free		EL.TECH	gleichstromfrei
≠ nocturnal tariff			≠ Nachttarif	dc generator (→direct-current generator)		POWER SYS	(→Gleichstromgenerator)
dB (→decibel)		PHYS	(→Dezibel)				
DBA (→data administrator)		DATA PROC	(→Datenbankbeauftragter)	dci (→direct current)		EL.TECH	(→Gleichstrom)
				dc impulse dialing		SWITCH	Gleichstromimpulswahl
DB-25 connector		COMPON	DB-25-Stecker	dc-insulated		EL.TECH	galvanisch getrennt
3-dB coupler (→three-dB coupler)		MICROW	(→Drei-dB-Koppler)	= electrically insulated			= gleichspannungsgetrennt
				DC inverter (→inverter)		POWER SYS	(→Wechselrichter)
DBMS (→database management system)		DATA PROC	(→Datenbank-Verwaltungssystem)	DCL		MICROEL	DCL
				= direct coupled logic			
DC (→device control)		DATA COMM	(→Gerätesteuerung)	dc level		ELECTRON	Gleichspannungspegel
dc (→direct voltage)		EL.TECH	(→Gleichspannung)	= direct current level			
dc (IEEE) (→direct current)		EL.TECH	(→Gleichstrom)	dc link		POWER SYS	Gleichstromzwischenkreis
d.c. (IEC) (→direct current)		EL.TECH	(→Gleichstrom)	dc load		EL.TECH	Gleichstromverbraucher
d-c (→direct current)		EL.TECH	(→Gleichstrom)				= Gleichstromlast; DC-Last
DC (→direct current)		EL.TECH	(→Gleichstrom)	DCM (→digital circuit multiplexer)		TELEC	(→digitaler Leitungsmultiplexer)
DC-AC converter (→inverter)		POWER SYS	(→Wechselrichter)				
				dc machine		POWER SYS	Gleichstrommaschine
dc ammeter		INSTR	Gleichstrommesser	dc measurement		INSTR	Gleichstrommessung
dc amplification factor		MICROEL	B-Wert	dc measuring amplifier		INSTR	Gleichspannungs-Meßverstärker
dc amplifier (→direct current amplifier)		CIRC.ENG	(→Gleichspannungsverstärker)				
				dc measuring bridge		INSTR	Gleichstrom-Meßbrücke
dc bell		COMPON	Gleichstromwecker	= direct current measuring bridge			
= trembler bell				dc microvoltmeter		INSTR	DC-Mikrovoltmeter
dc breaking		POWER ENG	Gleichstrombremsung	dc mode		EL.TECH	Gleichstrombetrieb
dc cable		POWER SYS	Gleichstromkabel	dc motor		POWER ENG	Gleichstrommotor
dc chopper (→direct d.c. convertor)		POWER SYS	(→Gleichstromsteller)	= direct current motor			
				3D computer graphics		DATA PROC	3D-Computergraphik
d.c. chopper convertor (→direct d.c. convertor)		POWER SYS	(→Gleichstromsteller)				= 3D-Computergrafik
				D-controller (→differential regulator)		CONTROL	(→Differentialregler)
dc component		EL.TECH	Gleichstromkomponente				
= d.c. component; d-c component			= Gleichspannungskomponente; Gleichstromanteil; Nullkomponente	dc output		CIRC.ENG	Gleichspannungsausgang
≠ ac component				dc power supply		EQUIP.ENG	Gleichstromversorgung
↑ continuous value				dc push-button dialing		SWITCH	Gleichstromtastwahl
			≠ Wechselstromkomponente	dc recovery		MODUL	Gleichstrom-Wiederherstellung
			↑ Gleichwert				

dc reinsertion 144

dc reinsertion (→black level restoration) TV (→Schwarzwerterhaltung)
dc relay COMPON **Gleichstromrelais**
dc remote power feeding TRANS **Gleichstromfernspeisung**
dc repeater panel TELEGR **Fernteilnehmer-Anschluß-schiene**
dc resistance EL.TECH **Gleichstromwiderstand**
  [elektrischer Widerstand für Gleichstrom]
  ≈ Wirkwiderstand; ohmscher Widerstand
dc resolver COMPON **Gleichstromdrehmelder**
dc restorer TV **Gleichstromzuschaltung**
dc restoring (→black level restoration) TV (→Schwarzwerterhaltung)
DCS (→digital cross-connect) TRANS (→Digital-Verteiler)
dc signaling SWITCH **Gleichstromsignalisierung**
dc spark-over voltage COMPON **Ansprechgleichspannung**
  [overvoltage protector] [Überspannungsableiter]
  = response dc; responding dc
dc switch COMPON **Gleichstromschalter**
dc telegraphy (→direct current telegraphy) TELEGR (→Gleichstromtelegrafie)
DC tension (→direct voltage) EL.TECH (→Gleichspannung)
dc tension (→direct voltage) EL.TECH (→Gleichspannung)
d-c- tension (→direct voltage) EL.TECH (→Gleichspannung)
dc tester INSTR **Gleichspannungsmesser**
  = dc voltmeter; direct current voltmeter
DCTL MICROEL **DCTL**
  = direct-coupled-transistor logic
dc transducer (→dc converter) POWER SYS (→Gleichspannungswandler)
dc underlay ELECTRON **Gleichstromunterlegung**
  ≈ wetting 1 [COMPON] ≈ Frittung [COMPON]
dcv (→direct voltage) EL.TECH (→Gleichspannung)
d.c. voltage (IEC) (→direct voltage) EL.TECH (→Gleichspannung)
dc voltage (→direct voltage) EL.TECH (→Gleichspannung)
DC voltage (→direct voltage) EL.TECH (→Gleichspannung)
d-c voltage (→direct voltage) EL.TECH (→Gleichspannung)
dc voltage offset CIRC.ENG **Gleichspannungs-Offset**
dc voltage transducer (→dc converter) POWER SYS (→Gleichspannungswandler)
dc voltmeter (→dc tester) INSTR (→Gleichspannungsmesser)
DD (→double density) TERM&PER (→doppelte Schreibdichte)
DDC (→direct digital control) CONTROL (→digitale Direktregelung)
DDD (AM) (→direct distance dialing) SWITCH (→Selbstwählferndienst)
DD disk (→DD diskette) TERM&PER (→DD-Diskette)
DD diskette TERM&PER **DD-Diskette**
  [5,536 bpi, 48 or 96 tpi] [2.179 Bit/cm, 19 oder 38 Spuren/cm]
  = double-density diskette; DD disk = Double-density-Diskette
DDE (→direct data acquisition) DATA PROC (→direkte Datenerfassung)
ddeg (→decadegree) PHYS (→Dez)
3D display INSTR **dreidimensionale Darstellung**
  = 3D-Darstellung
DDL (→data description language) DATA PROC (→Datenbeschreibungssprache)
DDP (→distributed data processing) DATA PROC (→verteilte Datenverarbeitung)
DDRR antenna ANT **DDRR-Antenne**
  = hula-hoop antenna; transmission line antenna = Hula-hoop-Antenne; Transmission-line-Antenne
deactivate (→disconnect) EL.TECH (→ausschalten)
dead (adj.) ACOUS **schalltot**
  = anechoic = nachhallfrei
dead (adj.) EL.TECH **kalt**
  = power-off; de-energized; cold ≠ heiß
  ≠ live ↓ stromlos; spannungslos
  ↓ current-free; voltage-free

dead angle TECH **toter Winkel**
  = dead sector
dead band CONTROL **Unempfindlichkeitsbereich**
dead beat PHYS **eigenschwingungsfrei**
deaden (→damp) PHYS (→dämpfen)
dead-end pole (→end pole) OUTS.PLANT (→Endgestänge)
deadener ACOUS **Schalldämpfer**
  = muffler; sound absorber = Schalldämmer; Schallschlucker; Schalldämmstoff; Schalldämpfstoff
dead halt DATA PROC **Blockierungsunterbrechung**
  [error or instruction causing a halt without recovery] [durch Programmfehler oder Anweisung erzeugt]
  = drop dead halt = festgefahrene Unterbrechung
  ↑ program crash ↑ Programmabsturz
dead halt (→abnormal end) DATA PROC (→Absturz)
dead key TERM&PER **Tottaste**
  [operative only if pressed simultaneously with other key, e.g. accent key with vowel key] [gibt sein Zeichen nur bei gleichzeitigem Drücken einer weiteren Taste ab, z.B. Akzenttaste mit Buchstabentaste]
dead letter box DATA COMM **Adreßlos-Datei**
  [message switching; for unaddressed messges] [Sendungsvermittlunmg; für nicht übermittelbare Nachrichten]
deadline ECON **äußerster Termin**
  ↑ due time = Grenztermin
  ↑ Frist
deadlock DATA PROC **Verklemmung**
  [mutual blocking of processes] [gegenseitige Blockierung von Prozessen]
  = deadly embrace = Blockierung; Deadlock
deadly embrace DATA PROC (→Verklemmung)
  (→deadlock)
dead sector (→dead angle) TECH (→toter Winkel)
dead spot RADIO PROP **Funkschatten**
  = radio shadow; shadow
dead time CONTROL **Totzeit**
  = retardation ≈ Verzugszeit; Schutzzeit; Verzug; Verzögerung
  ≈ guard time
dead-time correction ELECTRON **Totzeitkorrektur**
dead time element CONTROL **Totzeitelement**
dead zone CONTROL **Unempfindlichkeitszone**
  = tote Zone
dead zone RADIO PROP **tote Zone**
  = silent zone
deaf-aid amplifier TELEPH **Hörverstärker**
deaf-aid telephone TELEPH **Fernsprechapparat mit Hörverstärker**
  = hard-of-hearing telephone
deal (→business) ECON (→Geschäft)
deal (→agreement 1) COLLOQ (→Vereinbarung)
dealer ECON **Händler**
  = seller; vendor; distributor; trader; merchant = Vertreiber
  ≈ trading company ≈ Handelsunternehmen; Lieferant
dealership (→distribution network) ECON (→Vertriebsnetz)
deallocate DATA PROC **Zuweisung aufheben**
  [to free formerly allocated resources] = Zuordnung zurücknehmen
  ≈ strobe (v.t.) ≈ freigeben
deallocation TECH **Freigabe**
  ≠ allocation ≠ Zuweisung
deal out (→dispense) COLLOQ (→austeilen)
DEAP MICROEL **DEAP**
  = diffused eutectic aluminum process
dear (→expensive) ECON (→teuer)
deathnium (→recombination center) PHYS (→Rekombinationszentrum)
deathnium center (→recombination center) PHYS (→Rekombinationszentrum)
deattenuation TELEC **Entdämpfung**
debit ECON **Soll**
  [accounting] [Buchhaltung]
  ≠ Haben
debit (→charge 1) ECON (→Belastung 1)
debit card (→prepaid phonecard) TERM&PER (→Guthabenkarte)

debit sign	ECON	**Debet-Zeichen**	
= debit symbol			
debit symbol (→debit sign)	ECON	(→Debet-Zeichen)	
deblock (→unblock)	DATA PROC	(→entblocken)	
deblocking (→unblocking)	SWITCH	(→Entblockierung)	
deblocking	DATA PROC	(→Entblocken)	
(→unblocking)			
debounce (v.t.)	ELECTRON	**entprellen**	
debounced keyboard	TERM&PER	**entprellte Tastatur**	
debouncing	ELECTRON	**Entprellen**	
[of a contact]		[eines Kontaktes]	
debouncing circuit	ELECTRON	**Entprellschaltung**	
de Broglie wave	PHYS	**Materiewelle**	
= matter wave; particle wave		= Materiewelle; de-Broglie-Welle	
debt	ECON	**Schuld**	
≈ claim (v.t.)		≈ Forderung	
debt discharge	ECON	**Schuldbefreiung**	
debtor	ECON	**Schuldner**	
= obligor		≈ Entleiher	
≈ borrower		≠ Gläubiger	
≠ creditor			
debtor interest	ECON	**Sollzins**	
= receivable interest; rending rate		= Aktivzins	
debt service	ECON	**Schuldendienst**	
debug (v.t.)		**austesten**	
= check-out (v.t.); troubleshoot (v.t.)		= ausprüfen	
debugger (→diagnostic program)	DATA PROC	(→Diagnoseprogramm)	
debugging	DATA PROC	**Austesten**	
[of a program]		[eines Programms]	
= check out (n.); troubleshooting		= Ausprüfen	
debugging (→repair)	TECH	(→Instandsetzung)	
debugging aid (→test aid)	DATA PROC	(→Testhilfe)	
debugging method	DATA PROC	**Fehlersuchverfahren**	
= debug method; diagnostic method			
debugging program	DATA PROC	(→Diagnoseprogramm)	
(→diagnostic program)			
debugging routine	DATA PROC	(→Diagnoseprogramm)	
(→diagnostic program)			
debugging time (→burn-in period)	QUAL	(→Einlaufphase)	
debug method (→debugging method)	DATA PROC	(→Fehlersuchverfahren)	
deburr	MECH	**entgraten**	
deburring machine	MECH	**Entgratmaschine**	
Debye length	PHYS	**Debye-Länge**	
decade	MATH	**Zehnerpotenz**	
		= Dekade	
decade capacitance	INSTR	**Dekadenkondensator**	
decade capacitor	INSTR	**Kapazitätsdekade**	
decade counter	CIRC.ENG	**Dekadenzähler**	
decadegree	PHYS	**Dez**	
[10°]		[10°]	
= ddeg			
decade resistor	INSTR	**Dekadenwiderstand**	
decade switch	COMPON	**Dekadenschalter**	
≈ thumbwheel switch		≈ Fingerradschalter	
decadic (→decimal)	MATH	(→dezimal)	
decadic logarithm (→common logarithm)	MATH	(→dekadischer Logarithmus)	
decadic pulse dialing	SWITCH	**dekadische Impulswahl**	
decahedron	MATH	**Dekaeder**	
decaling	METAL	**Entzunderung**	
		[Oberflächenbehandlung]	
decametric waves (→short waves)	RADIO	(→Kurzwelle)	
decant	TECH	**dekantieren**	
decarbonize	TECH	**entkohlen**	
decatenate	DATA PROC	**entketten**	
≠ concatenate		≠ verketten	
decay (v.i.)	PHYS	**abklingen**	
= fade out (v.i.); evanesce			
decay (n.)	ELECTRON	**Abfall**	
[of a signal, voltage]		[eines Signals, einer Spannung,...]	
= drop (n.); decrease (n.); fall (n.); fall-off (n.)			
≠ rise (n.)		≠ Anstieg	
decay 1 (n.)	PHYS	**Abklingvorgang**	
= decay process; fall 2 (n.)		= Abfall	
decay 2 (n.)	PHYS	**Zerfall**	
= desintegration			
decay characteristic (→persistence characteristic)	PHYS	(→Nachleuchtcharakteristik)	
decay constant	PHYS	**Zerfallskonstante**	
= desintegration constant			
decay curve	PHYS	**Zerfallskurve**	
= desintegration curve			
decaying electrode	ELECTRON	(→Bremselektrode)	
(→retarding electrode)			
decaying pulse edge	ELECTRON	(→Impulsabfallflanke)	
(→trailing pulse edge)			
decay of pressure (→decrease of pressure)	PHYS	(→Druckabnahme)	
decay process (→decay 1)	PHYS	(→Abklingvorgang)	
decay rate	PHYS	**Zerfallsgeschwindigkeit**	
= desintegration rate			
decay series	PHYS	**Zerfallsreihe**	
= desintegration series			
decay time	ELECTRON	**Abfallzeit**	
[between 90% and 10% amplitude]		[zwischen 90% und 10% Amplitude]	
= fall time; release time; trailing transition time; pulse decay time		= Fallzeit; Impulsabfallzeit	
≠ rise time		≠ Anstiegzeit	
↑ ramp time		↑ Rampenzeit	
decay time	PHYS	**Abklingzeit**	
decay time	PHYS	**Zerfallszeit**	
[atomic and nuclear physics]		[Atomphysik, Kernphysik]	
= desintegration time			
DECCA navigation	RADIO NAV	**DECCA-Verfahren**	
decelerate	TECH	**verlangsamen**	
= slow-down		≈ verzögern; bremsen	
≈ delay (v.t.); brake (v.t.)			
decelerating grid (→suppressor grid)	ELECTRON	(→Bremsgitter)	
deceleration	TECH	**Verlangsamung**	
≈ delay		≈ Verzug	
deceleration (→breaking 1)	MECH	(→Bremsung)	
deceleration time	TERM&PER	**Bremszeit**	
decentralization	SCIE	**Dezentralisierung**	
decibel	PHYS	**Dezibel**	
= dB		= dB	
decidability	MATH	**Entscheidbarkeit**	
decidable	MATH	**entscheidbar**	
deciding factor	SCIE	**Entscheidungsfaktor**	
decile	MATH	**Dezil**	
decimal (adj.)	MATH	**dezimal**	
= decadic; denary		= dekadisch; zehnteilig; denär	
decimal (n.) (→decimal fraction)	MATH	(→Dezimalbruch)	
decimal carry	MATH	**Zehnerübertrag**	
decimal code	CODING	**Dezimalcode**	
		= Denärcode	
decimal counter	CIRC.ENG	**Dezimalzähler**	
↑ counter		= BCD-Zähler	
		↑ Zähler	
decimal dialing	SWITCH	**dekadische Wahl**	
= decimal switching			
decimal digit	MATH	**Dezimalziffer**	
[0, 1, 2, 3, ..]		[0, 1, 2, 3,...]	
		= dezimale Ziffer	
decimal format computer	DATA PROC	**Dezimalrechner**	
decimal fraction	MATH	**Dezimalbruch**	
= decimal (n.)			
decimalization	MATH	**Dezimalisierung**	
decimalize	MATH	**dezimalisieren**	
decimal notation (→decimal number system)	MATH	(→Dezimalsystem)	
decimal number	MATH	**Dezimalzahl**	
decimal number system	MATH	**Dezimalsystem**	
= decimal system; decimal notation; denary notation		[Zahlendarstellung mit 10 Ziffern]	
↑ radix notation		= dezimales Zahlensystem; Zehnersystem	
		↑ Stellenwertsystem	
decimal place (→decimal position)	MATH	(→Dezimalstelle)	
decimal point (→point)	MATH	(→Komma) (n.n.; pl.-s oder -tas)	
decimal position	MATH	**Dezimalstelle**	
= decimal place		↑ Ziffernstelle	
↑ digit position			

decimal position (→digit position)	DATA PROC	(→Ziffernstelle)
decimal representation	MATH	Dezimaldarstellung
decimal switching (→decimal dialing)	SWITCH	(→dekadische Wahl)
decimal switching system [electromechanical switching system]	SWITCH	dekadisches Vermittlungssystem [elektromechanisches Vermittlungssystem] = dekadisches Wählsystem
decimal system (→decimal number system)	MATH	(→Dezimalsystem)
decimal tabbing (→decimal tabulation)	DATA PROC	(→Dezimalkommatabulierung)
decimal tabulation = decimal tabbing	DATA PROC	Dezimalkommatabulierung
decimal-to-binary conversion	DATA PROC	Dezimal-Binär-Umsetzung = Dezimal-Binär-Umwandlung
decimal-to-hexadecimal conversion	DATA PROC	Dezimal-Hexadezimal-Umsetzung = Dezimal-Hexadezimal-Umwandlung
decimal-to-octal conversion	DATA PROC	Dezimal-Oktal-Umsetzung = Dezimal-Oktal-Umwandlung
decimeter (NAM) = decimetre (BRI); dm	PHYS	Dezimeter [0,1 m] = dm
decimetre (BRI) (→decimeter)	PHYS	(→Dezimeter)
decimetric waves [1 m-0,1 m; 300 MHz-3 GHz] = ultra-high frequencies; UHF	RADIO	Dezimeterwellen [1 m-0,1 m; 300 MHz-3 GHz] = UHF; Dezimalwellen
decimillimetric waves	RADIO	Dezimillimeter-Wellen
decipher (v.t.) ↓ break (v.t.)	TELEC	entziffern = dechiffrieren; entschlüsseln ↓ knacken
deciphrement (→decryption)	CODING	(→Entschlüsselung)
decision = arbitration	INF	Entscheidung
decisional algorithm (→decision algorithm)	MATH	(→Entscheidungsalgorithmus)
decision algorithm = decisional algorithm; arbitration algorithm	MATH	Entscheidungsalgorithmus
decision box (→decision symbol)	DATA PROC	(→Entscheidungssymbol)
decision circuit (→discriminator)	CIRC.ENG	(→Entscheider)
decision content	INF	Entscheidungsgehalt
decision element (→discriminator)	CIRC.ENG	(→Entscheider)
decision feedback	DATA COMM	Rückmeldung = Fehlerkorrektur
decision feedback filter	NETW.TH	Entscheidungs-Rückkopplungsfilter
decision function	MATH	Entscheidungsfunktion
decision instant	CODING	Entscheidungszeitpunkt
decision instruction [indicates conditional continuation instruction] = descrimination instruction	DATA PROC	Entscheidungsbefehl [weist auf konditionierten Folgebefehl]
decision logic (→arbitration logic)	MATH	(→Entscheidungslogik)
decision-making network (→logic circuit)	CIRC.ENG	(→Logikschaltung)
decision matrix	INF.TECH	Entscheidungsmatrix
decision structure (→selection structure)	DATA PROC	(→Selektionsstruktur)
decision-supporting	DATA PROC	entscheidungsunterstützend
decision support system = DSS	DATA PROC	Entscheidungsvorbereitungssystem
decision symbol [a diamond-shaped symbol] = decision box	DATA PROC	Entscheidungssymbol [rautenförmig] = Entscheidungskästchen
decision system	CONTROL	Entscheidungssystem
decision table	INF	Entscheidungstabelle
decision theory	SCIE	Entscheidungstheorie
decision threshold	CIRC.ENG	Entscheidungsschwelle
decision tree	INF	Entscheidungsbaum
decision unit	CIRC.ENG	Entscheidereinheit
decision value	CIRC.ENG	Entscheidungswert
decisive ≈ relevant	COLLOQ	entscheidend ≈ maßgebend
deck (→punched card deck)	TERM&PER	(→Lochkartenstapel)
deck (of cards) (→stack)	TERM&PER	(→Stapel)
deck (→switch deck)	COMPON	(→Schalterebene)
declaration = declarative statement; declarative; declare statement ↓ data declaration; variable declaration	DATA PROC	Vereinbarung = Deklaration; Konvention; Vereinbarungsanweisung; Deklarationsanweisung ↓ Datendeklaration; Variablendeklaration
declarative (COBOL) (→procedure declaration)	DATA PROC	(→Prozedurvereinbarung)
declarative (→declaration)	DATA PROC	(→Vereinbarung)
declarative language (→declarative programming language)	DATA PROC	(→deklarative Programmiersprache)
declarative programming language = descriptive programming language; declarative language; descriptive language	DATA PROC	deklarative Programmiersprache = deskriptive Programmiersprache; deklarative Sprache; deskriptive Sprache
declarative section	DATA PROC	Vereinbarungsteil
declarative statement (→declaration)	DATA PROC	(→Vereinbarung)
declarator = Deklarator	DATA PROC	Vereinbarungssymbol
declare	DATA PROC	vereinbaren
declare statement (→declaration)	DATA PROC	(→vereinbaren)
declension (→declination)	LING	(→Deklination)
declination = magnetic declination	PHYS	Mißweisung = Deklination; magnetische Deklination
declination [inflection of nouns] = declension ↑ inflection; morphology	LING	Deklination = Beugung des Substantivs ↑ Flexion; Morphologie
decline (→decrease)	TECH	(→Verminderung)
decline (→decrease)	TECH	(→vermindern)
decline (→deny)	COLLOQ	(→verweigern)
decline (→reduce 1)	TECH	(→vermindern)
declining (→retrograde)	TECH	(→rückläufig)
decode (v.t.) ≠ code	CODING	decodieren = dekodieren ≠ codieren
decode network	CIRC.ENG	Entschlüsselungsmatrix = Decodiermatrix
decoder ≠ coder	CIRC.ENG	Decodierer = Decoder; Dekodierer (Duden) ≠ Codierer
decoder network	CODING	Dekodiermatrix
decode switch	COMPON	Decodierschalter
decoding	CODING	Decodierung = Dekodierung (Duden)
decohere	ELECTRON	entfritten
decoherer	ELECTRON	Entfritterer
decollate [separate continuous form or carbon paper into single sheets]		trennen [von Endlospapier oder Durchschlagpapier in Einzelblätter]
decollator [separates continuous paper] ↑ form handling equipment	TERM&PER	Tennmaschine = Trennvorrichtung; Formulartrennmaschine; Formulartrennvorrichtung; Formulartrenner ↑ Formularhantierungsgerät
decompilation	DATA PROC	Entcompilierung
decomposability	MATH	Zerlegbarkeit
decomposable	MATH	zerlegbar
decompose	MATH	zerlegen
decomposition	MATH	Zerlegung
decompression (→data expansion)	DATA PROC	(→Datenexpansion)

decorative border (→decorative margin)		TYPOGR	(→Zierrand)		
decorative character font		TYPOGR	**Zierschrift**		
= decorative typefont; decorative typestyle			= Schmuckschrift		
decorative line		TYPOGR	**Zierlinie**		
= patterned line			= Schmucklinie		
decorative margin		TYPOGR	**Zierrand**		
= decorative border			= Schmuckrand		
decorative typefont		TYPOGR	(→Zierschrift)		
(→decorative character font)					
decorative typestyle		TYPOGR	(→Zierschrift)		
(→decorative character font)					
decouple		TELEC	**auskoppeln**		
[from a signal path]			[aus einem Signalweg]		
= extract			≈ abzweigen		
≈ derivate					
decouple		PHYS	**entkoppeln**		
decouple		EL.TECH	**entkoppeln**		
= uncouple			≈ trennen		
≈ separate					
decoupling		PHYS	**Entkopplung**		
decoupling		EL.TECH	**Entkopplung**		
≈ isolation			≈ Trennung		
decoupling capacitor		CIRC.ENG	**Entkopplungskondensator**		
= isolating capacitor; isolation capacitor; neutralizing capacitor			= Trennkondensator		
decoupling filter (→separating filter)		NETW.TH	(→Weiche)		
decoupling stub		ANT	**Entkopplungs-Stumpf**		
= isolation stub					
decoupling transformer		COMPON	(→Trennübertrager)		
(→isolating transformer)					
decrease (v.i.)		TECH	**abnehmen** (v.i.)		
= decline; lessen			= zurückgehen (fig.)		
≈ reduce; dwindle			= verkleinern; schwinden		
≠ increase			≠ zunehmen		
decrease (n.)		TECH	**Abnahme**		
= decrement; drop; decline; diminution			= Rückgang; Dekrement; Abfall 2		
≠ increase; downward trend			≠ Zunahme		
≈ reduction 1			≈ Verminderung		
decrease (n.) (→decay)		ELECTRON	(→Abfall)		
decrease of pressure		PHYS	**Druckabnahme**		
= decay of pressure					
decrement (v.t.)		MATH	**abwärtszählen**		
			= rückwärtszählen		
decrement (v.t.)		DATA PROC	**vermindern**		
= diminish			= erniedrigen; abnehmen; dekrementieren		
≠ increment (v.t)			≠ erhöhen		
decrement (v.i.)		TECH	**zurückrücken**		
= regress (v.i.)					
decrement		PHYS	**Dämpfung**		
= damping; attenuation			= Dekrement		
decrement (→decrease)		TECH	(→Abnahme)		
decremental		ELECTRON	**zurückschaltend**		
			= dekrementierend; zurückrückend		
decrementer (→down counter)		CIRC.ENG	(→Rückwärtszähler)		
decryption		CODING	**Entschlüsselung**		
= cryptanalysis; code translation; deciphrement			= Decodierung 2		
≠ encryption			≠ Verschlüsselung		
decryption code		INF.TECH	**Entschlüsselungscode**		
≠ encryption code			≠ Verschlüsselungscode		
↑ key			↑ Schlüssel		
decuple (v.t.)		MATH	**verzehnfachen**		
[to multiply by ten]			↑ vervielfachen		
↑ multiplicate					
dedendum		MECH	**Zahnfußhöhe**		
[gear]			[Zahnrad]		
dedicated 1		TECH	(→anwendungsspezifisch)		
(→application-specific)					
dedicated (→fixed)		TELEC	(→festgeschaltet)		
dedicated 2		TECH	(→kundenspezifisch)		
(→custom-designed)					
dedicated (→tied)		ECON	(→zweckgebunden)		
dedicated (→reserved)		DATA PROC	(→reserviert)		
dedicated channel		DATA PROC	**dedizierter Kanal**		
			= Standkanal		
dedicated circuit (→fixed line)		TELEC	(→Standleitung)		
dedicated computer		DATA PROC	(→Spezialrechner)		
(→special-purpose computer)					
dedicated connection (→fixed line)		TELEC	(→Standleitung)		
dedicated key (→special key)		TERM&PER	(→Sondertaste)		
dedicated line (→fixed line)		TELEC	(→Standleitung)		
dedicated network (→private network)		TELEC	(→Privatnetz)		
dedicated plant system		OUTS.PLANT	**Kabelverzweigertechnik**		
dedicated radio network		TELEC	(→Funksondernetz)		
(→special services radio network)					
dedicated register		DATA PROC	**dediziertes Register**		
			= zweckbestimmtes Register		
dedicated server		DATA COMM	**dedizierter Server**		
[exclusively for network management tasks]			[ausschließlich für Netzverwaltungsaufgaben]		
dedicated terminal		TERM&PER	(→Spezialendgerät)		
(→applications terminal)					
dedication		TECH	**Zweckbestimmung**		
			= Dedizierung		
deduce (→derive)		SCIE	(→ableiten)		
deduct (→subtract)		MATH	(→subtrahieren)		
deducting		COLLOQ	**abzüglich**		
deductive conclusion (→deductive reasoning)		SCIE	(→Deduktivschluß)		
deductive inference (→deductive reasoning)		SCIE	(→Deduktivschluß)		
deductive reasoning		SCIE	**Deduktivschluß**		
= deductive conclusion; deductive inference					
dedust		TECH	**entstauben**		
dedusting		TECH	**Entstaubung**		
deed (→document)		ECON	(→Urkunde)		
de-electrification		PHYS	**Entelektrisierung**		
de-embedding		ELECTRON	**Disintegration**		
de-emphasis		TELEC	**Rückentzerrung**		
≠ pre-emphasis			= Deemphasis		
			≠ Vorverzerrung		
de-energize (→disconnect)		EL.TECH	(→ausschalten)		
de-energize (→demagnetize)		PHYS	(→entmagnetisieren)		
de-energized (→dead)		EL.TECH	(→kalt)		
deep		TECH	**tief**		
= profound					
deep blue		PHYS	**dunkelblau**		
deep brown		PHYS	**dunkelbraun**		
deep discharge (→deep discharging)		POWER SYS	(→Tiefentladung)		
deep discharging		POWER SYS	**Tiefentladung**		
[accumulator]			[Akkumulator]		
= deep discharge					
deep-discharging protection		POWER SYS	**Tiefentladeschutz**		
[accumulator]			[Akkumulator]		
deep drawing		METAL	**Tiefziehen**		
= depth drawing			= Tiefreduzierziehen; Drahtziehen		
			↑ Ziehen		
deep-drawing sheet metal		METAL	**Tiefziehblech**		
deep green (→dark green)		PHYS	(→dunkelgrün)		
deep-groove ball bearing		MECH	**Rillenlager**		
deep purplish red (→crimson)		PHYS	(→carmoisin)		
deep red		PHYS	**dunkelrot**		
deep-sea cable		COMM.CABLE	**Tiefseekabel**		
↑ submarine cable; underwater cable			↑ Seekabel; Unterwasserkabel		
deep yellow		PHYS	**dunkelgelb**		
default (n.)		DATA PROC	**Vorgabe**		
[parameter choice of the system if no choice is made by the user]			[Parameterwahl des Systems wenn vom Anwender keine vorliegt; default = engl. „Unterlassung"]		
= default state; default value; default option; predefinition			= Vorbesetzung; Standardannahme; Zusatzparameter; Default- Wert;		

default 148

		Grundwert; Rückfallzustand; Ausgangszustand; voreingestellter Parameter; automatisch angesetzter Wert	**deferment** = postponement	ECON	**Stundung**
			deferment of payment	ECON	**Zahlungsaufschub**
			deferred (→delayed)	TECH	(→verzögert)
default (→delay)	TECH	(→Verzug)	**deferred address** (→indirect address)	DATA PROC	(→indirekte Adresse)
default (→preset)	COLLOQ	(→vorgegeben)	**deferred addressing**	DATA PROC	(→indirekte Adressierung)
default clause	ECON	**Verzugsklausel**	(→indirect addressing)		
default drive [the referred to one, if no other is specified]	DATA PROC	**Standardlaufwerk** [das, sofern nicht anders spezifiziert, angesprochene]	**deferred addressing mode** (→indirect addressing)	DATA PROC	(→indirekte Adressierung)
			deferred alarm (→non-urgent alarm)	EQUIP.ENG	(→nicht dringender Alarm)
defaulting = tardy	ECON	**säumig**	**deferred and accrued item**	ECON	**Abgrenzungsposten**
default option (→default)	DATA PROC	(→Vorgabe)	**deferred entry**	DATA PROC	**asynchroner Eintritt**
			deferred exit	DATA PROC	**asynchroner Austritt**
default profile = initial standard profile	TELEC	**Ausgangsprofil** [Signalverarbeitung] = Rückfallprofil	**deficiency** (→shortfall)	ECON	(→Fehlmenge)
			deficiency (→defect)	QUAL	(→Mangel)
			deficient (→faulty)	QUAL	(→fehlerhaft)
default state (→default)	DATA PROC	(→Vorgabe)	**deficient cover** (→cover shortage)	ECON	(→Unterdeckung)
default value	ELECTRON	**voreingestellter Wert**	**deficit**	ECON	**Defizit**
default value (→default)	DATA PROC	(→Vorgabe)			≈ Fehlbetrag
			deficit	MATH	**Fehlbetrag**
defeat (→interrupt)	EL.TECH	(→unterbrechen)	**defile** (n.) (→bottleneck)	COLLOQ	(→Engpaß)
defeatable (→disconnectable)	ELECTRON	(→abschaltbar)	**define** (v.t.)	SCIE	**definieren**
			define (v.t.)	DATA PROC	**definieren**
defeatable thyristor [for DC]	MICROEL	**abschaltbarer Thyristor** [für Gleichspannungen]	**defining argument** (→ordering argument)	DATA PROC	(→Ordnungsbegriff)
defect (n.) = deficiency; weakness	QUAL	**Mangel** = Defekt; Fehler	**defining equation**	MATH	**Definitionsgleichung** = Bestimmungsgleichung
defect (→fault)	TECH	(→Fehler)	**definite integral** [integration over a segment of a straight line] ≈ line integral	MATH	**bestimmtes Integral** [Integration über einen Abschnitt einer Geraden] ≈ Kurvenintegral
defect (→lattice vacancy)	PHYS	(→Gitterfehlstelle)			
defect (→imperfection)	MICROEL	(→Störstelle)			
defect concentration (→impurity density)	MICROEL	(→Störstellendichte)			
defect conduction (→hole conduction)	PHYS	(→Löcherleitung)	**definition**	TV	**Auflösung**
			definition	SCIE	**Definition**
defect conductor (→extrinsic conductor)	MICROEL	(→Störstellenleiter)	**definition**	DATA PROC	**Definition**
			definition (→resolution)	PHYS	(→Auflösung)
defect density (→impurity density)	MICROEL	(→Störstellendichte)	**definition file**	DATA PROC	**Definitionsdatei**
			definition range	MATH	**Definitionsbereich**
defect distribution	MICROEL	**Fehlstellenverteilung**	**deflect**	PHYS	**ablenken**
defect electron (→hole)	PHYS	(→Loch)	**deflecting electrode**	ELECTRON	**Ablenkelektrode**
defect exhaustion (→impurity exhaustion)	MICROEL	(→Störstellenerschöpfung)	**deflecting element**	COMM.CABLE	**Umlenkelement**
			deflecting field	ELECTRON	**Ablenkfeld**
defect-free zone	MICROEL	**defektfreie Zone**	**deflecting magnet** (→deflection magnet)	ELECTRON	(→Ablenkmagnet)
defect generation (→impurity generation)	MICROEL	(→Störstellenerzeugung)			
			deflecting plate (→beam catcher)	ELECTRON	(→Ablenkplatte)
defect information (→failure report)	TELECONTR	(→Fehlermeldung)			
			deflecting reflector	ELECTRON	**Ablenkreflektor**
defective (→faulty)	QUAL	(→fehlerhaft)	**deflecting system** (→deflection system)	ELECTRON	(→Ablenksystem)
defective fraction (→defective ratio)	QUAL	(→Fehleranteil)			
			deflecting voltage	ELECTRON	**Ablenkspannung**
defective ratio = defective fraction	QUAL	**Fehleranteil** = Ausschußanteil	**deflection** = deviation	PHYS	**Ablenkung**
defect level ≠ excess level	TELEC	**Unterpegel** ≠ Überpegel	**deflection** = sweep	ELECTRON	**Ablenkung**
			deflection [spring]	MECH	**Auslenkung** [Feder]
defect mobility (→impurity mobility)	MICROEL	(→Störstellenbeweglichkeit)			
			deflection [deviation of pointer from zero]	INSTR	**Ausschlag**
defect of electrons	PHYS	**Elektronenmangel**			
defect profile (→impurity profile)	MICROEL	(→Störstellenprofil)	**deflection aberration**	ELECTRON	**Ablenkfehler**
			deflection angle	INSTR	**Ausschlagwinkel** = Deflexionswinkel
defect report (→failure report)	TELECONTR	(→Fehlermeldung)			
			deflection coefficient [voltage necessary for 1 cm of deflection]	INSTR	**Ablenkkoeffizient** [für ein cm Ablenkung erforderliche Spannung] = Ablenkfaktor
defect semiconductor (→extrinsic semiconductor)	MICROEL	(→Störstellenhalbleiter)			
defence economy (BRI) (→defense economy)	ECON	(→Wehrwirtschaft)			
			deflection coil = sweeping coil ≈ deflection yoke	ELECTRON	**Ablenkspule** = Deflektionsspule ≈ Ablenkjoch
defence electronics (BRI) (→defense electronics)	ELECTRON	(→Verteidigungselektronik)			
defence technology (BRI) (→military technology)	TECH	(→Militärtechnik)	**deflection coil yoke** = deflection yoke	TV	**Ablenkspulenjoch**
defense economy (NAM) = defence economy (BRI)	ECON	**Wehrwirtschaft**	**deflection curve**	MECH	**Biegelinie**
			deflection defocussing	ELECTRON	**Ablenkdefokussierung**
defense electronics (NAM) = defence electronics (BRI); military electronics	ELECTRON	**Verteidigungselektronik** = Militärelektronik	**deflection factor**	INSTR	**Ausschlagfaktor**
			deflection magnet = deflecting magnet	ELECTRON	**Ablenkmagnet**
defense technology (NAM) (→military technology)	TECH	(→Militärtechnik)	**deflection method**	INSTR	**Ausschlagmethode**
			deflection non-linearity	ELECTRON	**Ablenk-Nichtlinearität**
defere (→postpone)	COLLOQ	(→vertagen)	**deflection sensitivity**	PHYS	**Ablenkempfindlichkeit**

deflection speed (→slewing speed)	INSTR	(→Ablenkgeschwindigkeit)
deflection system = sweeping system; deflecting system	ELECTRON	Ablenksystem = Ablenkeinheit; Deflektorsystem
deflection yoke ≈ deflection coil; sweeping yoke	ELECTRON	Ablenkjoch = Deflexionsjoch ≈ Ablenkspule
deflection yoke (→deflection coil yoke)	TV	(→Ablenkspulenjoch)
deflector (→beam catcher)	ELECTRON	(→Ablenkplatte)
deflector plate (→beam catcher)	ELECTRON	(→Ablenkplatte)
defocus (v.t.)	PHYS	defokussieren
defocussed parabolic antenna	ANT	defokussierte Parabolantenne
defocussing	PHYS	Defokussierung
deform = strain	TECH	verformen
deformation = strain	TECH	Verformung = Deformation; Gestaltsänderung
defrayal = defraying of costs	ECON	Kostenübernahme
defraying of costs (→defrayal)	ECON	(→Kostenübernahme)
degasification (→degassing)	PHYS	(→Entgasung)
degassing = degasification	PHYS	Entgasung
degauss (→demagnetize)	PHYS	(→entmagnetisieren)
degausser (→demagnetizer)	ELECTRON	(→Entmagnetisierer)
degaussing (→demagnetization)	PHYS	(→Entmagnetisierung)
degaussing coil	ELECTRON	Entmagnetisierungsspule
degaussing current (→demagnetization current)	EL.TECH	(→Entmagnetisierungsstrom)
degaussing factor (→demagnetizing factor)	PHYS	(→Entmagnetisierungsfaktor)
degeneracy = degeneration	PHYS	Entartung
degeneracy (→degeneration)	MATH	(→Degenerierung)
degeneracy temperature	PHYS	Entartungstemperatur
degenerated = entartet	MATH	degeneriert
degenerated	PHYS	entartet
degenerate mode = degenerierte Mode	MICROW	degenerierter Wellentyp
degeneration = degeneracy	MATH	Degenerierung = Entartung
degeneration (→degeneracy)	PHYS	(→Entartung)
degeneration concentration	PHYS	Entartungskonzentration
degenerative feedback (→negative feedback)	CIRC.ENG	(→Gegenkopplung)
deglitcher [suppresses interfering pulses]	ELECTRON	Deglitcher [unterdrückt Störspannungsspitzen]
degradation (→deterioration)	TECH	(→Verschlechterung)
degradation failure ≠ sudden failure	QUAL	Driftausfall ≠ Sprungausfall
degrease	TECH	entfetten
degreasing	TECH	Entfettung
degree [complete angle/360] = old degree	MATH	Grad [Vollwinkel/360] = Altgrad ↑ Gradmaß; Winkelmaß
degree [unit for temperature difference; old name for Kelvin] = centigrade	PHYS	Grad Celsius 2 [Maßeinheit für Temperaturdifferenz; alte Bezeichnung für Kelvin] ≈ Kelvin
degree	PHYS	Grad [Temperaturmessung]
degree C (→degree Celsius)	PHYS	(→Grad Celsius 1)
degree Celsius = degree C; °C	PHYS	Grad Celsius 1 [absolute Temperaturangabe] = °C
degree Fahrenheit = °F	PHYS	Grad Fahrenheit = °F
degree of contraction	MECH	Schrumpfmaß
degree of coupling	NETW.TH	Kopplungsgrad
degree of freedom	PHYS	Freiheitsgrad
degree of latitude	GEOPHYS	Breitengrad
degree of longitud	GEOPHYS	Längengrad
degree of modulation (→modulation depth)	MODUL	(→Modulationsgrad)
degree of the filter	NETW.TH	Filtergrad
degree sign [symbol °]	PHYS	Grad-Zeichen [Symbol °]
deicing system	TECH	Tauvorrichtung
deinstall (→take out of service)	TECH	(→außerbetriebsetzen)
deionization	PHYS	Entionisierung
dejagging	DATA PROC	Dünnzeichnen
dejitterize	TELEC	entjittern
dejitterizer	CIRC.ENG	Dejitterizer
delay (v.t.) = stall ≈ postpone; slow-down	TECH	verzögern = hinausziehen ≈ verschieben; verlangsamen
delay (n.) = delay time; time delay; time lag; lag; holdoff ≈ deceleration; guard time	ELECTRON	Verzögerungszeit = Verzögerung; Zeitverzögerung; Verzug ≈ Ansprechzeit; Laufzeit; Sperrzeit
delay (n.) = default ≈ time lag	TECH	Verzug = Verzögerung; Verspätung ≈ Nacheilung
delay (→propagation time)	PHYS	(→Laufzeit)
delay circuit	CIRC.ENG	Verzögerungsschaltung = Laufzeitnetzwerk; Verzögerer
delay compensator 2	TELEC	Laufzeitnachbildung
delay compensator 1 (→delay equalizer)	TELEC	(→Laufzeitentzerrer)
delay constant	PHYS	Verzögerungsmaß
delay difference	TELEC	Laufzeitunterschied
delay distortion	TELEC	Laufzeitverzerrung
delayed = deferred ≈ time-shifted; belated; lagging ↑ untimely	TECH	verzögert = verlangsamt ≈ zeitlich versetzt; verspätet; nacheilend ↑ unzeitig
delayed-action (→slow-reacting)	TECH	(→träge)
delayed-action fuse (→slow blowing fuse)	COMPON	(→träge Schmelzsicherung)
delayed burst mode	TELECONTR	Zeitstaffelbetrieb
delayed delivery	DATA COMM	verzögerte Zustellung
delay element [transversal filter]	NETW.TH	Laufzeitglied [Transversalfilter]
delay element [output delayed with respect to input e.g. by one year] = delay flipflop; d-type flipflop ≈ timer ↑ flipflop	CIRC.ENG	Verzögerungsglied [Ausgang hinkt gegenüber Eingang um z.B. einen Impuls nach] = Verzögerungskippschaltung; Verzögerungs-Flipflop; Verzögerungselement; Auffang-Flipflop; D-Flipflop ≈ Zeitglied ↑ Kippschaltung
delay equalization ≈ phase equalization; delay compensation	TELEC	Laufzeitentzerrung ≈ Laufzeitausgleich
delay equalizer = delay compensator 1 ≈ phase equalizer; phase compensator 2	TELEC	Laufzeitentzerrer ≈ Laufzeitnachbildung
delay field (→delay section)	MICROEL	(→Verzögerungsleitung)
delay filter	NETW.TH	Laufzeitfilter
delay flipflop (→delay element)	CIRC.ENG	(→Verzögerungsglied)
delay lens	ELECTRON	Verzögerungslinse
delay lens (→lens antenna)	ANT	(→Linsenantenne)
delay line	LINE TH	Laufzeitleitung = Verzögerungsleitung
delay line	ELECTRON	Verzögerungsleitung
delay-line memory = delay-line storage; delay-line store	DATA PROC	Laufzeitspeicher = Verzögerungsleitungsspeicher

delay-line storage (→delay- DATA PROC (→Laufzeitspeicher)
line memory)
delay-line store (→delay-line DATA PROC (→Laufzeitspeicher)
memory)
delay probability SWITCH **Wartewahrscheinlichkeit**
[figure of service quality for queuing [Maß der Verkehrsgüte
systems] von Wartesystemen]
= queuing probability ↑ Verkehrsgüte
↑ grade of system
delay register DATA PROC **Verzögerungsregister**
delay relay (→time-delay COMPON (→Verzögerungsrelais)
relay)
delay section MICROEL **Verzögerungsleitung**
[magnetron] [Magnetron]
= delay field
delay-slope equalizer TELEC **Laufzeitneigungsregler**
delay switch (→ time switch) COMPON (→Zeitschalter)
delay time TELEC **Verzögerungszeit**
= Verzugszeit
delay time CONTROL **Verzugszeit**
≈ dead time ≈ Totzeit
delay time (→delay) ELECTRON (→Verzögerungszeit)
delay time (→call delay) SWITCH (→Wartedauer)
delay-time characteristic TELEC **Laufzeitcharakteristik**
= delay vs. frequency characteristic = Laufzeitgang
delay-time colour decoder TV **Laufzeitdemodulator**
[PAL] [PAL]
delay vs. frequency characteristic TELEC (→Laufzeitcharakteristik)
(→delay-time characteristic)
D-element CONTROL **D-Glied**
= differential element = Differenzierglied; differen-
 zierendes Übertragungs-
 glied
delete (n.) DATA COMM **Löschen**
[code] [Code]
= DL = DEL; Ausfügen
delete DATA PROC **ausfügen**
[erase characters on the display, with- [Zeichen am Bildschirm lö-
out dislocation of subsequent charac- schen, ohne Nachrücken]
ters] ↑ löschen
↑ erase
delete (→erase) ELECTRON (→löschen)
delete accidentally (→zap) DATA PROC (→versehentlich löschen)
delete character (→erase DATA PROC (→Löschzeichen)
character)
delete command (→delete DATA PROC (→Löschanweisung)
statement)
delete input (→erase input) CIRC.ENG (→Löscheingang)
delete instruction (→delete DATA PROC (→Löschanweisung)
statement)
delete job (→delete DATA PROC (→Löschanweisung)
statement)
delete key TERM&PER **ENTF-Taste**
[deletes character under or right [entfernt das rechts oder
from cursor bar, characters at the unter der Schreibmarke be-
right shift by] findliche Zeichen,alle Zei-
= DEL key che n rechts davon rücken
≈ return key; backspace key nach]
 = DEL-Taste; DELETE-Ta-
 ste
 ≈ Rückschrittaste; Rücksetztaste 2
delete statement DATA PROC **Löschanweisung**
= delete instruction; delete command; = Löschauftrag; Löschkom-
delete job; cancel statement; cancel mando; Löschbefehl; An-
instruction; cancel command; cancel nullierungsanweisung;
job; erase statement; erase instruc- Annullierungsauftrag; An-
tion; erase command; erase job; clear nullierungskommando; An-
statement; clear instruction; clear nullierungsbefehl
command; clear job; rub-out state-
ment; rub-out instruction; rub-out
command; rub-out job
deletion (→erasing) ELECTRON (→Löschung)
deletion record DATA PROC **Überschreibeintrag**
delimit (v.t.) TECH **eingrenzen**
= pinpoint (v.t.) = eindämmen
delimit COLLOQ **abgrenzen**
delimitation TECH **Eingrenzung**
delimitation of ray PHYS **Strahlbegrenzung**
delimiter (→separator) DATA PROC (→Trennzeichen)

delimiter symbol DATA PROC (→Trennzeichen)
(→separator)
deliver (→furnish) ECON (→ausliefern)
deliverable ECON **lieferbar**
delivery ECON **Lieferung**
= handing over = Übergabe; Abgabe; Auslie-
 ferung
delivery TELEC **Weitersendung**
[of a received message to addressee] [zum Empfänger]
= retransmission 2; transfer; handing- = Zustellung; Übergabe
over ≈ Übertragungswiderholung
≈ retransmission 1
delivery confirmation DATA COMM **Übergabebestätigung**
delivery lot ECON **Lieferlos**
delivery note ECON **Lieferschein**
= dispatch note; despatch note
delivery point TELEC **Übergabepunkt**
= interchange point; point of presence; = Übergabestelle
point of termination; POT ≈ Schnittstelle
≈ interface
delivery range ECON **Lieferspektrum**
delivery state QUAL **Auslieferungszustand**
 = Lieferzustand
delivery time ECON **Lieferfrist**
= date of delivery; specified time = Liefertermin
delivery tolerance QUAL **Auslieferungstoleranz**
DEL key (→delete key) TERM&PER (→ENTF-Taste)
Delon rectifier circuit INSTR **Delon-Schaltung**
 = Greinacher-Schaltung
delta (→ delta section) NETW.TH (→Pi-Schaltung)
delta antenna ANT **Delta-Antenne**
≈ double delta antenna ≈ Doppel-Delta-Antenne
↑ long-wire antenna ↑ Langdrahtantenne
delta circuit (→delta NETW.TH (→Pi-Schaltung)
section)
delta clock DATA PROC **Deltatakt**
delta connection (→ delta NETW.TH (→Pi-Schaltung)
section)
delta function MATH **Deltafunktion**
delta function (→pulse PHYS (→Dirac-Puls)
function)
delta loop antenna ANT **Delta-loop-Antenne**
delta loop beam ANT **Delta-loop-Yagi-beam**
delta marker INSTR **Delta-Marke**
delta match LINE TH **Delta-Anpassung**
≈ T match ≈ T-Anpassung
delta-matched aerial ANT (→Y-Dipol)
(→delta-matched antenna)
delta-matched antenna ANT **Y-Dipol**
= delta-matched aerial
delta matching ANT **Anzapfanpassung**
delta modem MODUL **Delta-Modem**
delta modulation MODUL **Deltamodulation**
delta network (→delta NETW.TH (→Pi-Schaltung)
section)
delta PCM (→ differential pulse MODUL (→Differenz-Pulscodemo-
code modulation) dulation)
delta section NETW.TH **Pi-Schaltung**
= delta network; delta connection; = Pi-Glied; Dreieckschal-
delta; pi network; pi section; pi con- tung; Deltaschaltung
nection; delta circuit; triangle connec- ↑ Dämpfungsglied
tion; triangle circuit
↑ attenuator
delta-shaped element ANT **Delta-Element**
↑ radiator = Delta-Schleife
 ↑ Strahlelement
delta-star conversion POWER ENG **Dreieck-Stern-Umwandlung**
demagnetization PHYS **Entmagnetisierung**
= degaussing = Demagnetisierung; Abma-
 gnetisierung
demagnetization current EL.TECH **Entmagnetisierungsstrom**
= degaussing current = Demagnetisierungsstrom;
 Abmagnetisierungsstrom
demagnetize PHYS **entmagnetisieren**
= degauss; de-energize = demagnetisieren; abmagne-
 tisieren
demagnetizer ELECTRON **Entmagnetisierer**
= degausser = Demagnetisierer; Abma-
 gnetisierer; Entmagnetisier-
 gerät

demagnetizer cassette	TERM&PER	**Entmagnetisierungskassette**
		= Entmagnetisierungscassette
demagnetizing factor	PHYS	**Entmagnetisierungsfaktor**
= degaussing factor		
demand (n.)	ECON	**Nachfrage**
demand (→claim 3)	ECON	(→Forderung)
demand assigned	TELEC	**bedarfsgesteuert**
demand assignment	TELEC	**bedarfsweise Zuteilung**
= adaptive allocation		↓ bedarfsweise Kanalzuteilung
↓ demand assignment of channel		
demand-assignment multiple access	TELEC	**bedarfsgesteuerter Vielfachzugriff**
= DAMA		= bedarfsgesteuerter Mehrfachzugriff; DAMA
demand assignment of channel	TELEC	**bedarfsweise Kanalzuteilung**
= adaptive channel allocation		↑ bedarfsweise Zuteilung
↑ demand assignment		
demand-driven	TECH	**bedarfsgesteuert**
		= nachfragegesteuert
demand driver	ECON	**Bedarfsträger**
demand file	DATA PROC	**Abrufdatei**
demanding	TECH	**anspruchsvoll**
= exigent		
demand paging	DATA PROC	**bedarfsweiser Seitenabruf**
demand processing (→in-line processing)	DATA PROC	(→Geradewohl-Verarbeitung)
demarcation	TECH	**Abgrenzung**
		= Demarkation
demi (→semibold)	TYPOGR	(→halbfett)
democratic network	DATA COMM	**nicht-hierarchisches Netz**
(→non-hierarchical network)		
demo disk	DATA PROC	**Demodiskette**
= demonstration disk		= Demonstrationsdiskette
demodulation	MODUL	**Demodulation**
demodulation transfer function	MODUL	**Demodulations-Übertragungsfunktion**
demodulator	MODUL	**Demodulator**
= detector; rectifier		
demolish (→destroy)	TECH	(→zerstören)
demonstrate	TECH	**vorführen**
		= demonstrieren
demonstration	TECH	**Vorführung**
≈ presentation		≈ Präsentation
demonstration	SCIE	**Beweisführung**
= demonstrativeness		
demonstration disk (→demo disk)	DATA PROC	(→Demodiskette)
demonstration unit (→evaluation unit)	TECH	(→Vorführgerät)
demonstrativeness (→demonstration)	SCIE	(→Beweisführung)
demonstrative pronoun [e.g. this]	LING	**Demonstrativpronomen** [z.B. jener]
		= hinweisendes Fürwort
De Morgan rule	ENG.LOG	**De-Morgan-Regel**
demount 1 (→dismantle)	TECH	(→abbauen)
demount 2 (→disassemble)	TECH	(→zerlegen)
demountable	TECH	**zerlegbar**
≈ detachable		≈ abnehmbar
demo version	DATA PROC	**Demo-Version**
≠ complete version		≠ Vollversion
demultiplex (v.t.)	TELEC	**demultiplexieren**
		= demultiplexen
demultiplexer	TELEC	**Demultiplexer**
demultiplexing	TELEC	**Demultiplexen**
denary (→decimal)	MATH	(→dezimal)
denary notation (→decimal number system)	MATH	(→Dezimalsystem)
denial	COLLOQ	**Verweigerung**
denomination (→description)	TECH	(→Bezeichnung)
denominational notation (→positional notation)	MATH	(→Stellenwertsystem)
denominational number system (→positional notation)	MATH	(→Stellenwertsystem)
denominator (→divisor 2)	MATH	(→Divisor 2)
denormalizing	NETW.TH	**Entnormierung**
≠ normalizing		≠ Normierung
dense binary code [makes use of all patterns]	CODING	**vollständiger Binärcode** [macht von allen Mustern Gebrauch]
dense list (→sequential list)	DATA PROC	(→sequentielle Liste)
densely populated	ECON	**dichtbesiedelt**
density 1	PHYS	**Dichte 1**
↓ charge density; mass density		↓ Ladungsdichte; Massendichte
density	MICROEL	**Besetzungsdichte**
density 2 (→mass density)	PHYS	(→Massendichte)
density (→packaging density)	EQUIP.ENG	(→Packungsdichte)
density (→recording density)	DATA PROC	(→Speicherdichte)
density anisotropy	PHYS	**Dichteanisotropie**
density gradient	PHYS	**Dichtegradient**
		= Dichtegefälle
density step [facsimile]	TELEGR	**Tonwert** [Faksimile]
dent (n.) [slight depression caused by pressure]	TECH	**Delle** [durch Druckeinwirkung erzeugte leichte Vertiefung]
≈ notch		= Dalle
		≈ Kerbe
denumerable	MATH	**abzählbar**
deny (v.t.)	COLLOQ	**verweigern**
= refuse; decline		
department [of an organization]	ECON	**Abteilung** [Organisation]
= division; section		= Dienststelle; Sektion
dependability 2 (→service reliability)	QUAL	(→Betriebszuverlässigkeit)
dependability 1 (→reliability)	QUAL	(→Zuverlässigkeit)
dependable (→reliable)	QUAL	(→zuverlässig)
dependence	MATH	**Abhängigkeit**
= dependency		
dependency (→dependence)	MATH	(→Abhängigkeit)
dependent failure (→secondary failure)	QUAL	(→Folgeausfall)
depending on application	DOC	**je nach Anwendungsfall**
depict (→represent)	LING	(→darstellen)
depletion	MICROEL	**Verarmung**
≠ enrichment		= Abreicherung; Erschöpfung
		≠ Anreicherung
depletion depth (→depletion-layer thickness)	MICROEL	(→Sperrschichtdicke)
depletion layer [region at np junction with no current carriers, i.e. nonconducting, unless biased]	MICROEL	**Sperrschicht** [von Ladungsträgern freie, d.h. nichtleitende, Zone an einem NP-Übergang, sofern keine Spannung angelegt wird]
= depletion region; depletion zone; blocking layer; barrier layer (obsol.); barrier (obsol.)		= Halbleitersperrschicht; Verarmungsschicht; Verarmungsrandschicht; Ausschöpfungszone
≈ surface layer		≈ Randschicht
↑ junction		↑ Übergang
depletion layer capacity (→junction capacity)	MICROEL	(→Sperrschichtkapazität)
depletion-layer depth (→depletion-layer thickness)	MICROEL	(→Sperrschichtdicke)
depletion-layer thickness	MICROEL	**Sperrschichtdicke**
= depletion layer depth; depletion-layer width; depletion thickness; depletion depth; depletion width; junction thickness; junction depth; junction width		= Sperrschichtbreite; Übergangsdicke; Übergangsbreite
depletion-layer width (→depletion-layer thickness)	MICROEL	(→Sperrschichtdicke)
depletion mode	MICROEL	**Verarmungstyp**
		= Entblößungstyp
depletion mode FET	MICROEL	**Verarmungsisolierschicht-Feldeffekttransistor**
depletion mode operation	MICROEL	**Verarmungsbetrieb**
depletion region (→depletion layer)	MICROEL	(→Sperrschicht)
depletion thickness (→depletion-layer thickness)	MICROEL	(→Sperrschichtdicke)
depletion transistor	MICROEL	**Verarmungstransistor**
= depletion type transistor		= Depletionstransistor

depletion type transistor MICROEL (→Verarmungstransistor)
(→depletion transistor)
depletion width MICROEL (→Sperrschichtdicke)
(→depletion-layer thickness)
depletion zone (→depletion MICROEL (→Sperrschicht)
layer)
deployment (→input) ECON (→Einsatz)
deployment of personnel ECON **Personaleinsatz**
 = Personalaufwand
depolarization discrimination ANT (→Polarisationsentkopplung)
(→depolarization loss)
depolarization loss ANT **Polarisationsentkopplung**
 = depolarization discrimination; cross-
 polar discrimination
depolarizer PHYS **Depolarisator**
deposit 2 (n.) ECON **Lager**
 [place] = Depot
 = store
deposit 1 (n.) ECON **Verwahrung**
 = custody; safekeeping = Aufbewahrung; Gewahr-
 ≈ storage sam
 ≈ Lagerung
deposit (→precipitation) TECH (→Niederschlag)
deposit (n.) (→deposition) TECH (→Ablagerung)
deposited TECH **aufgebracht**
deposition TECH **Ablagerung**
 = deposit (n.); precipitation = Niederschlag
deposition method MICROEL (→Aufdampfverfahren)
(→vapor-deposition method)
depreciate ECON **entwerten**
depreciation 2 ECON **Abschreibung**
 ≈ amortization [den bilanzmäßigen Wert
 herabsetzen]
 ≈ Amortisation; Amortisie-
 rung
depreciation 1 (→wear and tear) ECON (→Wertminderung)
depress TECH **drücken**
 [key] [Knopf, Taste]
 = niederdrücken
depression MATH **Einsattelung**
 = dip (n.)
depression (→hole 2) TECH (→Loch 2)
depth TECH **Tiefe**
depth drawing (→deep METAL (→Tiefziehen)
drawing)
depth earth electrode SYS.INST **Tiefenerder**
depth gage MECH **Tiefenlehre**
depth hardening METAL **Tiefenhärtung**
depth queing DATA PROC **Bildtiefensimulation**
 ≈ shading ≈ Schattierung
deputy ECON **Stellvertreter**
 = representative
deputy director ECON **Abteilungsbevollmächtigter**
 = stellvertredender Direktor
deque (v.t.) DATA PROC **aus einer Warteschlange aus-**
 [to take from a queue] **gliedern**
deque 1 (n.) DATA PROC **Doppelendschlange**
 [items can be entered/retrieved at [beidseitig ergänzbar/les-
 both ends] bar]
 = double-ended queue
deque 2 (n.) DATA PROC **Warteschlangenausgliederung**
 [withdrawal from a queue]
derate (v.t.) QUAL **unterbelasten**
 ≈ underuse (v.t.) [TECH] = unterlasten
 ≈ unterauslasten [TECH]
derating QUAL **Unterbelastung**
 ≈ underuse [TECH] = redizierte Beanspruchung
 ≠ overload (n.) ≈ Unterauslastung [TECH]
 ≠ Überlastung
derating (→decrease) TECH (→Verminderung)
derating factor (→reduction LINE TH (→Reduktionsfaktor)
factor)
dereflection TECH **Entspiegelung**
deregulate TELEC **deregulieren**
deregulation TELEC **Deregulierung**
derivation (→derivative) MATH (→Ableitung)
derivation (→drop) TRANS (→Abzweig)
derivative MATH **Ableitung**
 = derivation ↓ Zeitableitung
 ↓ time derivative

derivative action CONTROL **Vorhalt**
 = rate action = Vorhaltewirkung
derivative control CONTROL **Differentialregelung**
derivative term LING **Unterbegriff**
 = hyponym = Hyponym
 ≠ generic term ≠ Oberbegriff
derivative time CONTROL **Vorhaltezeit**
 = rate time
derive SCIE **ableiten**
 = deduce; infer [fig.]
 = deduzieren
derived circuit NETW.TH **Abzweigstromkreis**
derived SI unit PHYS **abgeleitete SI-Einheit**
 [Kombination von SI-Ein-
 heiten]
derrick-style mast (→lattice CIV.ENG (→Gittermast)
mast)
descale METAL **entzundern**
 [surface treatment] [Oberflächenbehandlung]
descender 1 TYPOGR **Unterlänge**
 [part of character below the line] [Teil des Zeichens unter
 ≠ ascender der Zeilenlinie]
 ≠ Oberlänge
descender 2 TYPOGR **Unterlängenbuchstabe**
 [a type with descender]
descending link TELEC **absteigende Verbindung**
descending priority DATA PROC **absteigende Priorität**
descending traffic SWITCH **absteigender Verkehr**
descent TELEGR **Abfall**
 = drop (n.)
descramble CODING **entwürfeln**
 = entscrambeln
descrambler CIRC.ENG **Descrambler**
 = Entwürfler
describability SCIE **Beschreibbarkeit**
describe (→name) TECH (→bezeichnen)
describing function CONTROL **Beschreibungsfunktion**
descrimination instruction DATA PROC (→Entscheidungsbefehl)
(→decision instruction)
description LING **Beschreibung**
description TECH **Bezeichnung**
 = designation; denomination; name (n.) = Benennung; Name
 ≈ identification ≈ Kennzeichnung
description language DATA PROC **Beschreibungssprache**
 ↓ data description language ↓ Datenbeschreibungssprache
description list DATA PROC **Beschreibungsliste**
descriptive geometry MATH **Darstellende Geometrie**
descriptive language DATA PROC (→deklarative Programmier-
 (→declarative programming lan- sprache)
 guage)
descriptive programming DATA PROC (→deklarative Programmier-
language (→declarative programm- sprache)
ing language)
descriptive statistics MATH **statistische Maßzahl**
descriptor DATA PROC **Beschreiber**
 [pass or identification code for a file] [Code für Zugriffssiche-
 = keyword 2 rung oder Identifizierung
 einer Datei]
 = Deskriptor
design 2 (v.t.) TECH **konstruieren**
 = draft 2 (v.t.) ≈ entwerfen; entwickeln
 ≈ develop
design 1 (v.t.) TECH **entwerfen**
 = draft 1 (v.t.) ≈ konstruieren
 ≈ design 2 (v.t.)
design 1 (n.) TECH **Entwurf**
 = draft 1 (n.); draught 1 (n.) (BRI); out- ≈ Abfassung
 line (n.) ≈ Entwicklung
 ≈ development
design 2 (→lay-out) TECH (→Auslegung)
design 3 (v.t.) (→develop) TECH (→entwickeln)
design (→development) TECH (→Entwicklung)
design (→mechanical EQUIP.ENG (→Konstruktion)
design)
design aid (→design tool) TECH (→Entwurfswerkzeug)
designate ECON **bestimmen**
 = ernennen
designated circuit (→fixed TELEC (→Standleitung)
line)

designated expert ECON **Gutachter**
≈ appraiser ≈ Schätzer
designated line (→fixed TELEC (→Standleitung)
line)
designation ECON **Bezeichnung**
= Ernennung
designation EQUIP.ENG (→Beschriftung)
(→lettering)
designation (→description) TECH (→Bezeichnung)
designation label TECH **Beschriftungsschild**
= marking label; designation plate; let- = Bezeichnungsschild; Aus-
tering label; label (n.) zeichnungsschild; Schild
designation label (→type EQUIP.ENG (→Typenschild)
label)
designation plan (→lettering TECH (→Beschriftungsplan)
plan)
designation plate (→designation TECH (→Beschriftungsschild)
label)
designation space (→lettering EQUIP.ENG (→Beschriftungsfeld)
space)
designation strip EQUIP.ENG **Beschriftungsstreifen**
= marking strip; lettering strip; labeling = Bezeichnungsstreifen; Aus-
strip zeichnungsstreifen
design automation TECH **Entwurfsautomatisierung**
= Entwicklungsautomatisie-
rung
design characteristic (→mechanical TECH (→konstruktives Merkmal)
characteristic)
design concept TECH **Entwicklungskonzept**
design costs (→development ECON (→Entwicklungskosten)
costs)
design cycle TECH **Entwicklungszyklus**
= Entwicklungskette
design data TECH **Entwurfsdaten**
design department 1 TECH **Konstruktionsbüro**
design department 2 ECON (→Entwicklungsabteilung)
(→development department)
design drawing ENG.DRAW **Konstruktionszeichnung**
= engineering drawing 2 = technische Zeichnung
design engineer 1 TECH **Entwicklungsingenieur**
↑ designer 1 ↑ Entwickler
design engineer 2 TECH (→Konstrukteur)
(→draftsman)
designer 1 TECH **Entwickler**
= developer ≈ Forscher
≈ researcher ↓ Entwicklungsingenieur
↓ design engineer
designer 2 (→draftsman) TECH (→Konstrukteur)
designer kit (→tool) DATA PROC (→Programmierwerkzeug)
designer tool (→tool) DATA PROC (→Programmierwerkzeug)
design expenses (→development ECON (→Entwicklungskosten)
costs)
design goal (→design TECH (→Entwicklungsziel)
objective)
design lab (→design TECH (→Entwicklungslabor)
laboratory)
design laboratory TECH **Entwicklungslabor**
= design lab
design language DATA PROC **Entwicklungssprache**
design maturity TECH **Ausgereiftheit**
design method TECH **Entwurfsverfahren**
design mistake TECH **Konstruktionsfehler**
= faulty design
design modification TECH **Entwurfsänderung**
design objective TECH **Entwicklungsziel**
= design goal = Entwicklungsvorgabe; Be-
≈ design specification messungsgrundlage
≈ Entwicklungsspezifikation
design parameters (→design TECH (→Entwicklungsspezifika-
specification) tion)
design phase TECH **Entwicklungsphase**
design rating (→nominal load TECH (→Nennbelastbarkeit)
capacity)
design rule TECH **Entwurfsregel**
= Designregel
design rule check MICROEL **Entwurfregelnprüfung**
[ASIC] [ASIC]
design specification TECH **Entwicklungsspezifikation**
= design parameters = Entwicklungsziel
≈ design goal

design tool TECH **Entwurfswerkzeug**
= design aid = Entwurfshilfe
design verification TECH **Entwurfsüberprüfung**
desintegration (→decay 2) PHYS (→Zerfall)
desintegration constant (→decay PHYS (→Zerfallskonstante)
constant)
desintegration curve (→decay PHYS (→Zerfallskurve)
curve)
desintegration rate (→decay PHYS (→Zerfallgeschwindigkeit)
rate)
desintegration series (→decay PHYS (→Zerfallsreihe)
series)
desintegration time (→decay PHYS (→Zerfallszeit)
time)
desired (→scheduled) TECH (→nominell)
desired condition (→set TECH (→Sollzustand)
condition)
desired signal (→wanted TELEC (→Nutzsignal)
signal)
desired value (→nominal TECH (→Nennwert)
value)
desired value (→set CONTROL (→Sollwert)
value)
desk TERM&PER **Tisch**
≈ pulpit; console ≈ Pult; Konsole
desk OFFICE **Schreibtisch**
= writing table = Bürotisch
desk buid-in mounting TECH **Tischeinbau**
desk calculator (→desktop DATA PROC (→Tischrechner)
calculator)
desk check (→desk DATA PROC (→Schreibtischtest)
checking)
desk checking DATA PROC **Schreibtischtest**
[checking a program logic at desk- [Prüfung der Programmlo-
top] gik am Schreibtisch]
= desk check; dry run; dry running; = Blindversuch
code review
desk computer (→desktop DATA PROC (→Tischcomputer)
computer)
desk copier OFFICE **Tischkopierer**
desk instrument (→table INSTR (→Tischinstrument)
instrument)
desk model (→desktop EQUIP.ENG (→Tischgerät)
model)
desk plotter (→flatbed TERM&PER (→Flachbettplotter)
plotter)
desk telephone (→desktop TELEPH (→Fernsprech-Tischapparat)
telephone)
desktop accessory TERM&PER **Arbeitsplatzzubehör**
desktop calculator DATA PROC **Tischrechner**
[bureau machine for the four basic [Büromaschine für die vier
arithmetic opreations] Grundrechnungsarten]
= desk calculator = Tischrechenmaschine; Re-
≈ desktop computer; desk computer chenmaschine 2
≈ Tischcomputer
desktop computer DATA PROC **Tischcomputer**
= desk computer ≈ Tischrechner
≈ desktop calculator
desktop instrument (→table INSTR (→Tischinstrument)
instrument)
desktop model EQUIP.ENG **Tischgerät**
= table model; desk model; bench-top = Tischmodell; Auftischgerät
model
desktop mounting (→table TECH (→Tischaufstellung)
mounting)
desktop publishing DATA PROC **Desktop Publishing**
[production of publications with PC [Herstellen von Publikatio-
tools] nen mit PC-Mitteln]
= DTP = DTP; rechnerunterstützte
≈ electronic publishing Druckvorlagengestaltung
↑ CAP 2 ≈ Electronic Publishing
↑ CAP 2
desktop set (→table set) TERM&PER (→Tischapparat)
desktop telephone TELEPH **Fernsprech-Tischapparat**
= desk telephone; table telephone = Tischfernsprecher; Tischte-
↑ table set lefon; Tischtelephon
↑ Tischapparat
desolder ELECTRON **auslöten**
= unsolder = ablöten; entlöten
desoldering accessories ELECTRON **Auslötzubehör**

desoldering braid	ELECTRON	**Lötsauglitze**
desoldering iron	ELECTRON	**Entlötkolben**
desoldering station	ELECTRON	**Entlötstation**
desoldering tool	ELECTRON	**Entlötgerät**
		= Entlöthilfe
despatch (n.) (BRI)	ECON	**Versand**
= dispatch (n.; AM); loading; forwarding; sending-off; shipment 2; shipping 2; consignment		= Verladung; Versendung; Absendung; Übersendung; Zusendung; Sendung; Verfrachtung
≈ delivery		≈ Lieferung
↓ shipment 1; shipping 1		↓ Verschiffung
despatch (→dispatch)	TELEC	(→absetzen)
despool (v.t.)	DATA PROC	**Ausspuldateien ausdrucken**
[print-out spooled files]		
despotic	TELEC	**zwangssynchronisiert**
		= hierarchisch; despotisch
despotic network (→hierarchical network)	TELEC	(→hierarchisches Netz)
despotic synchronization	TELEC	**Zwangssynchronisierung**
= compelled synchronization		
dessicate	TECH	**entfeuchten**
≈ dry (v.t.)		≈ trocknen
dessicator	TECH	**Entfeuchter**
= dryer; drier		= Trockner
dessicator	MICROW	**Luftentfeuchter**
		= Lufttrockner
destage (→read-out)	DATA PROC	(→auslesen)
destilled water	CHEM	**destilliertes Wasser**
destination	SWITCH	**Ziel**
destination (→place of destination)	ECON	(→Bestimmungsort)
destination (→drain)	INF	(→Senke)
destination (→destination address)	DATA COMM	(→Zieladresse)
destination address	DATA COMM	**Zieladresse**
= destination		= Bestimmungsadresse
destination code	DATA COMM	**Bestimmungskennzahl**
		≈ Zieladresse
destination disk (→target disk)	DATA PROC	(→Zieldiskette)
destination exchange (→terminating exchange)	SWITCH	(→Zielvermittlung)
destination key	TERM&PER	**Zieltaste**
destination key terminal	TELEPH	**Zieltastenterminal**
destination load	SWITCH	**Zielbelastung**
destination network	DATA COMM	**Zielnetz**
		= Bestimmungsnetz
destination point	SWITCH	**Zielpunkt**
= terminating point		= Zielzeichengabepunkt
≈ terminating exchange		≈ Zielvermittlung
destination point code	SWITCH	**Zielpunktcode**
		= Zielkennung
destination station	TELEPH	**Zielsprechstelle**
destination subscriber	DATA COMM	**Zielteilnehmer**
destination track	RAILW.SIGN	**Zielgleis**
destinator	DATA COMM	**Letztempfänger**
Destriau effect	OPTOEL	**Destriau-Effekt**
		= Wechselfeldlumineszenz
destroy	TECH	**zerstören**
= demolish		= demolieren
destroy (→erase)	DATA PROC	(→löschen)
destroy (→erase)	ELECTRON	(→löschen)
destructive addition	DATA PROC	**löschende Addition**
[result overwrites an operand]		[das Ergebnis überschreibt einen der Operanden]
destructive addition (→EXCLUSIVE-OR operation)	ENG.LOG	(→EXKLUSIV-ODER-Verknüpfung)
destructive cursor	DATA PROC	**löschende Schreibmarke**
		= löschender Cursor
destructive operation	DATA PROC	**destruktiver Betrieb**
destructive read (→destructive reading)	DATA PROC	(→löschendes Lesen)
destructive reading	DATA PROC	**löschendes Lesen**
= destructive read; destructive readout; DRO		= zerstörendes Lesen
destructive readout (→destructive reading)	DATA PROC	(→löschendes Lesen)
detach (v.t.)	TECH	**lösen**
		= abnehmen
detachable	MECH	**abnehmbar**
= removable; releasable		= abtrennbar; lösbar
≈ demountable 2		≈ zerlegbar
detachable connection	EL.TECH	**lösbare Verbindung**
detachable keyboard	TERM&PER	**abgesetzte Tastatur**
= detached keyboard		
detachable optical splice	OPT.COMM	**Fingerspleiß** [lösbar]
detached	TECH	**abgesetzt**
= set-off (adj)		= abgelegen
≈ remote; separated		≈ entfernt; gesondert
detached (→remote)	TELEC	(→abgesetzt)
detached keyboard (→detachable keyboard)	TERM&PER	(→abgesetzte Tastatur)
detail (v.t.)	SCIE	**detaillieren**
detail (n.)	ENG.DRAW	**Einzelheit**
detail contrast	TV	**Feinkontrast**
detail drawing	ENG.DRAW	**Teilzeichnung**
detailed	DOC	**detailliert**
= itemized		≈ aufgeschlüßelt
detailed accounting (→toll ticketing)	SWITCH	(→Einzelgebührenerfassung)
detailed description	TECH	**Detailbeschreibung**
detailed documentation	TECH	**Detailunterlagen**
detailed message accounting (→toll ticketing)	SWITCH	(→Einzelgebührenerfassung)
detailed registration (→toll ticketing)	SWITCH	(→Einzelgebührenerfassung)
detailer	TECH	**Teilkonstrukteur**
↑ draftsman		↑ Konstrukteur
detail file (→transaction file)	DATA PROC	(→Bewegungsdatei)
detail paper (→transparent paper)	ENG.DRAW	(→Transparentpapier)
detail printing	DATA PROC	**Postendruck**
[a printing line per read record]		[eine Druckzeile pro eingelesenen Datensatz]
		= Einzeldruck
details enhancer	TV	**Einzelheitenauflöser**
detain (→arrest)	MECH	(→arretieren)
detect	INF.TECH	**erkennen**
= recognize; identify		= identifizieren
detectability	TECH	**Erkennbarkeit**
= detectivity; identifiability		= Nachweisbarkeit; Identifizierbarkeit
detectability limit	TECH	**Nachweisbarkeitsgrenze**
= detectable limit		= Nachweisgrenze; Erkennbarkeitsgrenze
detectable (→recognizable)	TECH	(→erkennbar)
detectable error	DATA PROC	**erkennbarer Fehler**
detectable limit (→detectability limit)	TECH	(→Nachweisbarkeitsgrenze)
detection	INF.TECH	**Erkennung**
= recognition; identification; reconnaissance		= Identifizierung
detection	HF	**Gleichrichtung**
detection circuit (→evaluation circuit)	CIRC.ENG	(→Auswerteschaltung)
detection logic	DATA PROC	**Erkennungslogik**
= recognition logic		
detectivity (→detectability)	TECH	(→Erkennbarkeit)
detector	PHYS	**Detektor**
detector (→evaluation circuit)	CIRC.ENG	(→Auswerteschaltung)
detector (→demodulator)	MODUL	(→Demodulator)
detector (→sensor)	SIGN.ENG	(→Melder)
detector diode	COMPON	**Detektordiode**
detector noise	ELECTRON	**Detektorrauschen**
detector tube (→rectifier tube)	ELECTRON	(→Gleichrichterröhre)
detector valve (→rectifier tube)	ELECTRON	(→Gleichrichterröhre)
detent (n.)	TECH	**Raste**
≈ locking device		[Vorrichtung zum Einrasten]
		≈ Sperrvorrichtung
detent (→stop)	MECH	(→Anschlag)
detent pin	MECH	**Arretierstift**
detergent (n.) (→cleaning agent)	TECH	(→Reinigungsmittel)

deterioration		TECH	**Verschlechterung**	**device** (→module)	COMPON	(→Baustein)
= degradation			= Qualitätsverschlechterung;	**device address**	DATA COMM	**Geräteadresse**
≈ impairment			Güteverlust	**device allocation**	DATA PROC	(→Gerätezuordnung)
			≈ Beeinträchtigung	(→equipment allocation)		
determinant (n.)		MATH	**Determinante**	**device assignment**	DATA PROC	(→Gerätezuordnung)
determinatice		COLLOQ	**maßgebend**	(→equipment allocation)		
≈ decisive			= maßgeblich	**device attributes**	DATA COMM	**Gerätekennung**
			≈ entscheidend	= DA		[Code]
deterministic		MATH	**determiniert**			= DA
≠ stochastic			≠ stochastisch	**device cluster**	DATA PROC	**Gerätegruppe**
deterministic access		DATA COMM	**deterministischer Zugang**	[group of peripherals with a common		[am selben Controller]
deterministic signal		INF	**determiniertes Signal**	controller]		= Peripheriegerätegruppe
dethnium center		MICROEL	**Reaktionshaftstelle**	**device code**	DATA PROC	**Gerätecode**
detour		COLLOQ	**Umweg**	**device control**	DATA COMM	**Gerätesteuerung**
= roundabout way				[code]		[Code]
detour connection		SWITCH	**Umwegschaltung**	= DC		= DC
detour factor		SWITCH	**Umwegfaktor**	**device control character**	DATA COMM	**Gerätesteuerzeichen**
detour line		ANT	**Umwegleitung**	**device-controlled**	DATA PROC	**gerätegesteuert**
detriment (→disadvantage)		COLLOQ	(→Nachteil)	**device coordinate**	DATA PROC	**Gerätekoordinate**
detrimental (→harmful)		TECH	(→schädlich)	**device-dependent**	DATA PROC	**geräteabhängig**
detune		ELECTRON	**verstimmen**	**device driver**	DATA PROC	**Peripheriegerätetreiber**
= untune; stagger				= peripheral driver		= Gerätetreiber
detuned		ELECTRON	**verstimmt**	**device family**	COMPON	**Bausteinfamilie**
= untuned; staggered; dumb			= unabgestimmt	**device flag**	DATA PROC	**Gerätezustandsregister**
detuning		ELECTRON	**Verstimmung**			= Gerätehinweiszeichen; Gerätezustandsmerker
= untuning; staggering						
detuning sleeve (→bazooka)		ANT	(→Viertelwellen-Sperrtopf)	**device-independent**	DATA PROC	**geräteunabhängig**
detuning-sleeve antenna		ANT	(→Sperrtopfantenne)	**device intelligence**	DATA PROC	**Geräteintelligenz**
(→folded-top antenna)				= peripheral intelligence		= dezentrale Intelligenz
Deutsche Bundespost		POST	**Deutsche Bundespost**	**device media control**	DATA PROC	**Gerätebeschreibungssprache**
[postal administration of F.R.G]			= DBP	**language**		
devaluation clause		ECON	**Abwertungsklausel**	**device name**	DATA PROC	**Gerätename**
develop (v.t.)		TECH	**entwickeln**			= Gerätebezeichnung
= design 3 (v.t.); engineer (v.t.)				**device number**	DATA PROC	**Gerätenummer**
develop		MATH	**abwickeln**	**device number** (→code number)	TECH	(→Sachnummer)
developable		MATH	**abwickelbar**	**device placement**	MANUF	**Bauteilplazierung**
developer		CHEM	**Entwickler**	**device priority**	DATA PROC	**Gerätepriorität**
[a substance]			[Substanz]	= peripherals priority		
developer (→designer 1)		TECH	(→Entwickler)	**device queue**	DATA PROC	**Geräteschlange**
development		ENG.DRAW	**Abwicklung**			= Peripheriegeräteschlange
development		TECH	**Entwicklung**	**device status report**	DATA COMM	(→Gerätestatusmeldung)
= engineering; design				(→DSR)		
development costs		ECON	**Entwicklungskosten**	**device status word**	DATA PROC	**Gerätestatuswort**
= development expenses; design costs;			= Entwicklungsausgaben	= DSW		
design expenses				**device tester**	DATA PROC	(→Prüfsystem)
development department		ECON	**Entwicklungsabteilung**	(→exerciser)		
= design department 2				**device under test** (→unit under	INSTR	(→Meßobjekt)
development expenses		ECON	(→Entwicklungskosten)	measurement)		
(→development costs)				**device under test** (→test	QUAL	(→Prüfling)
development software		DATA PROC	(→Programmierwerkzeug)	specimen)		
(→tool)				**devise** (v.t.)	COLLOQ	**ausdenken**
development strategy		TECH	**Entwicklungsstrategie**			= austüfteln; ersinnen; ausklügeln
development time		TECH	**Entwicklungszeit**			
			= Entwicklungsdauer	**dew**	METEOR	**Tau**
development tool		DATA PROC	(→Programmierwerkzeug)	**dewfall**	PHYS	**Betauung**
(→tool)				**dew point**	PHYS	**Taupunkt**
deviate		TECH	**abweichen**	≈ saturation humidity		≈ Sättigungsfeuchte
deviating (→divergent)		TECH	(→abweichend)	**dew-point hygrometer**	INSTR	**Taupunkt-Hygrometer**
deviation		MECH	**Maßabweichung**	**dextrorotatory**	PHYS	**rechtsdrehend**
≈ tolerance			= Abmaß	[optics]		[Optik]
			≈ Toleranz	≠ levorotatory		≠ linksdrehend
deviation		CONTROL	**Regelabweichung**	**dextrorotatory** (→clockwise)	TECH	(→rechtsdrehend)
[difference of controlled variable to			[Regelgröße minus Führungsgröße]	**DF** (→radio direction	RADIO LOC	(→Funkpeiler)
reference magnitude]				finder)		
= control deviation; error; offset; actuating signal			= Regeldifferenz; Abweichung	**DFB laser**	OPTOEL	**DFB-Laser**
				= distributed feedback laser		
deviation		PHYS	**Abweichung**	**DF-loop** (→loop direction	RADIO LOC	(→Rahmenpeiler)
= aberration 2; variation				finder)		
deviation		TECH	**Abweichung**	**DF-loop pattern**	ANT	**Rahmendiagramm**
= divergence			= Divergenz	**DFSP** (→digital signal	MICROEL	(→digitaler Signalprozessor)
deviation (→deflection)		PHYS	(→Ablenkung)	procesor)		
deviation (→excursion)		MODUL	(→Hub)	**DFT** (→discrete Fourier	INF.TECH	(→diskrete Fourier-Transformation)
deviation ratio		MODUL	**Hubverhältnis**	transform)		
device			**Vorrichtung**	**diac** (→trigger diode)	MICROEL	(→Triggerdiode)
[small or part of equipment for a specific function]			[kleines Gerät oder Teil davon, für spezifische Funktion]	**diacaustic**	PHYS	**Diakaustik**
				[optics]		[Optik]
= appliance 2; facility; jig; gadget			= Einrichtung	**diacritic** (adj.)	LING	**diakritisch**
≈ accessory			≈ Gerät 1; Zubehör	**diacritic** (→diacritic mark)	LING	(→diakritisches Zeichen)
				diacritical mark (→diacritic	LING	(→diakritisches Zeichen)
device (→component)		COMPON	(→Bauelement)	mark)		

diacritical symbol

diacritical symbol (→diacritic mark) LING (→diakritisches Zeichen)
diacritic mark LING diakritisches Zeichen
  [modifying phonetic or semantic value of characters] [zur Unterscheidung der Aussprache oder Bedeutung eines Buchstabens]
  = diacritical mark; diacritic symbol; diacritical symbol; diacritic (n.) ≈ Satzzeichen
  ≈ punctuation mark ↑ Zeichen
  ↑ sign ↓ Cedille; Trema; Betonungszeichen
  ↓ cedilla; diaresis; accent mark
diacritic symbol (→diacritic mark) LING (→diakritisches Zeichen)
diad (n.) DATA PROC Dyade
  [two bits or characters] [zwei Bits oder Zeichen]
  ≈ doublet
diaeresis (pl.-es) LING Trema (n.n.; pl.-ta oder -s)
  [mark over a vowel to indicated its separate pronunciation, as in "naïve"] [kennzeichnet Diäresis, z.B. im Englischen „naïve"]
  ↑ diacritic mark = Trennpunkte
  ≈ Diäresis
  ↑ diakritisches Zeichen
diagnose (v.t.) SCIE diagnostizieren
diagnose test SWITCH Diagnoseprüfung
diagnosis (pl.-es) SCIE Diagnose
diagnostic (adj.) SCIE diagnostisch
diagnostic aid (→test aid) DATA PROC (→Testhilfe)
diagnostic bus DATA PROC Diagnosebus
diagnostic capability ELECTRON Diagnosefähigkeit
diagnostic card (→test card) TERM&PER (→Prüfkarte)
diagnostic chip MICROEL Diagnosechip
  = Diagnosebaustein; Prüfchip; Prüfbaustein
diagnostic code DATA COMM Diagnoseangaben
diagnostic error message (→error message) DATA PROC (→Fehlermeldung)
diagnostic fact (→symptom) DATA PROC (→Indiz)
diagnostic feature ELECTRON Diagnoseeigenschaft
diagnostic loop DATA COMM Diagnoseschleife
diagnostic loop (→test loop) TELEC (→Prüfschleife)
diagnostic message DATA PROC Diagnosemeldung
  = error diagnostics = Fehlerursachehinweis
diagnostic method (→debugging method) DATA PROC (→Fehlersuchverfahren)
diagnostic program DATA PROC Diagnoseprogramm
  = diagnostic routine; diagnostics program; diagnostics routine; diagnostics; debugging program; debugging routine; debugger; fault location program = Fehlersuchprogramm; Debugprogramm
  ≈ Prüfhilfe; Ablaufverfolger
  ≈ test aid; fault trace
diagnostic routine (→diagnostic program) DATA PROC (→Diagnoseprogramm)
diagnostics (→diagnostic program) DATA PROC (→Diagnoseprogramm)
diagnostics (→error message) DATA PROC (→Fehlermeldung)
diagnostics flag DATA PROC Fehlerhinweiszeichen
diagnostics program (→diagnostic program) DATA PROC (→Diagnoseprogramm)
diagnostics routine (→diagnostic program) DATA PROC (→Diagnoseprogramm)
diagnostic test tape TERM&PER Prüflochstreifen
  = test perforated tape; test tape
diagnostic unit EQUIP.ENG Diagnoseeinheit
diagonal (n.) MATH Diagonale
  = diagonale Gerade
diagonal (adj.) MATH diagonal
diagonal cross brace MECH Kreuzverstrebung
  = cross bracing
diagonal horn ANT Diagonalhorn
  ↑ horn radiator ↑ Hornstrahler
diagonal pliers ELECTRON Vornschneider
diagram (n.) ECON Schaubild
  = graph = Graphik

diagram 1 TECH Diagramm
  = chart = Schemabild
  ≈ scheme
diagram (→characteristic 3) TECH (→Kennlinie)
diagrammatic form DOC Diagrammform
diagram synthesis ANT Diagrammsynthese
dial (n.) INSTR Skalenscheibe
dial (n.) TELEC wählen
  = select; mark (v.t.)
dial 1 (→rotary dial switch) TERM&PER (→Nummernschalter)
dial 2 (→number plate) TERM&PER (→Ziffernring)
dial converter SWITCH Wahlumsetzer
dialect DATA PROC Dialekt
  [variation of a programming language] [Variation einer Programmiersprache]
dial frequency TELEPH Ruffrequenz
dial gage INSTR Skalenlehre
dial-in (v.i.) SWITCH einwählen
dial information SWITCH Wählinformation
  = dialing information = Wahlinformation
dialing (AM) SWITCH Wählen
  = dialling (BRI) = Wahl
dialing (→dial pulsing) SWITCH (→Nummernschalterwahl)
dialing bit DATA COMM Wählbit
  = dialing digit; signal digit
dialing current TELEC Rufstrom
  = ringing current
dialing digit (→dialing bit) DATA COMM (→Wählbit)
dialing disk TERM&PER Wählscheibe
  = dial plate; finger wheel = Fingerlochscheibe; Nummernscheibe
  ≈ rotary dial switch ≈ Nummernschalter
dialing information (→dial information) SWITCH (→Wählinformation)
dialing network (→switched network) TELEC (→Wählnetz)
dialing pulse (→dial pulse) SWITCH (→Wählimpuls)
dialing register SWITCH Wahlspeicher
  = Wahlregister
dialing switch EQUIP.ENG Wählschalter
  = option selector ≈ Drehschalter [COMPON]
  ≈ rotary switch [COMPON]
dial-in handset TELEPH Kompakttelefon
  = compact handset = Kompakttelephon; Kompaktfernsprecher
dialling (BRI) (→dialing) SWITCH (→Wählen)
dialling code (BRI) (→area code) SWITCH (→Ortskennzahl)
dialling device TERM&PER Wählvorrichtung
  = Wählorgan
dialling exchange equipment (→switching equipment) SWITCH (→Vermittlungsanlage)
dialling tone (→dial tone) TELEPH (→Wählton)
dial mechanism INSTR Antrieb
dial number TELEC Rufnummer
  [number to control a selection process] [Nummer zur Steuerung eines Wählvorgangs]
  = call number; calling number = Wählnummer
  ↓ subscriber number; national number; international number ↓ Teilnehmerrufnummer; nationale Rufnummer; internationale Rufnummer
dialog (n.) (NAM) INF.TECH Dialog
  = dialogue (BRI)
dialog (→interactive processing) DATA PROC (→Dialogbetrieb)
dialog box DATA PROC Dialogfenster
dialog communication (→interactive communication) TELEC (→Dialogkommunikation)
dialog data processing (→interactive processing) DATA PROC (→Dialogbetrieb)
dialog field DATA PROC Dialogfeld
dialog mode (→interactive processing) DATA PROC (→Dialogbetrieb)
dialog operating system DATA PROC Dialog-Betriebssystem
dialog processing (→interactive processing) DATA PROC (→Dialogbetrieb)

dialog program (NAM) DATA PROC **Dialogprogramm**
= dialogue programme (BRI); interactive program (NAM); interactive programme (BRI); interactive routine
= interaktives Programm

dialog station (→interactive DATA COMM (→Dialogstation)
terminal)

dialog system (NAM) DATA PROC **Dialogsystem**
= dialogue system (BRI); interactive system; end-user system
= interaktives System; Endbenutzersystem

dialogue (BRI) (→dialog) INF.TECH (→Dialog)
dialogue programme (BRI) DATA PROC (→Dialogprogramm)
(→dialog program)
dialogue system (BRI) DATA PROC (→Dialogsystem)
(→dialog system)
dial-out SWITCH **hinauswählen**
dial plate (→dialing disk) TERM&PER **Wählscheibe**
dial pulse SWITCH **Wählimpuls**
= selector pulse; dialing pulse = Wahlimpuls
dial pulse storage TELEGR **Wähltonumsetzer**
= translator unit
dial pulse timing SWITCH **Wählimpulszeitsteuerung**
dial pulsing SWITCH **Nummernschalterwahl**
= dialing; rotary dial selection; rotary dial
= Nummernscheibenwahl

dial sequence TELEPH **Rufsequenz**
dial switch (→rotary dial TERM&PER (→Nummernschalter)
switch)
dial test INSTR **Skalenprüfung**
dial tone TELEPH **Wählton**
["proceed to dial"] ["bitte wählen"]
= dialling tone; proceed-to-dial signal; call-confirmation signal
= Amtszeichen
≈ Freiton; Wählzeichen
↑ ringing tone; selection signal ↑ Hörton
↑ tonalities
dial tone receiver SWITCH **Wähltonempfänger**
dial train (→selection TELEPH (→Wählzeichen)
signal)
dial-up line (→switched TELEC (→Wählleitung)
connection)
diamagnetic PHYS **diamagnetisch**
[counteracting external magnetization] [externer Magnetisierung gegenwirkend]
≠ paramagnetic ≠ paramagnetisch
diamagnetism PHYS **Diamagnetismus**
≠ paramagnetism ≠ Paramagnetismus
diameter MATH **Durchmesser**
= Diameter
diametral pitch MECH **Durchmesserteilung**
[Zahnrad]
diametric (→diametrical) MATH (→diametral)
diametrical MATH **diametral**
= diametric
diamond PHYS **Diamant**
Diamond (BRI) (→type size TYPOGR (→Schriftgröße 4 1/2 Punkt)
4 1/2 point)
diamond (→lozenge) DATA PROC (→Rhombuszeichen)
diamond antenna (→rhombic ANT (→Rhombusantenne)
antenna)
Diamond code CODING **Diamond-Code**
diamond lattice PHYS **Diamantgitter**
diamond-planish MECH **rauhplanieren**
diamond-shaped antenna ANT (→Rhombusantenne)
(→rhombic antenna)
diapason (→tuning fork) ACOUS (→Stimmgabel)
diaphanous (→translucide) PHYS (→durchscheinend)
diaphragm PHYS **Blende**
[Optik]
diaphragm EL.ACOUS **Membrane**
= membrane
diaphragm (→membrane) TECH (→Membran)
diaphragm aperture (→focal PHYS (→Blendenöffnung)
aperture)
diaphragm source EL.ACOUS **Membranstrahler**
diaresis 1 LING **Diäresis** (pl.-en)
≈ diphtong; diaresis 2 [getrennte Aussprache zweier Vokale]
= Diärese
≈ Diphtong; Trema
dia scanner TELEGR **Dia-Abtaster**
[facsimile] [Fax]

diastatic PHYS **diastatisch**
diazo print (→blueprint) DOC (→Lichtpause)
diazo print developing equipment OFFICE **Lichtpaus-Entwicklungsgerät**
diazo print exposure system OFFICE **Lichtpaus-Belichtungsanlage**
dibit INF **Dibit**
[00, 01, 10, 11] [00, 01, 10, 11]
dichotomizing search DATA PROC (→Binärsuche)
(→binary search)
dichroic (adj.) PHYS **dichroitisch**
[with different colors in different directions]
[mit verschiedenen Farben in verschiedenen Richtungen]
dichroic OPTOEL **frequenzselektiv**
dichroism PHYS **Dichroismus**
[optics] [Optik]
dichromatic PHYS **doppelfarbig**
= zweifarbig; dichromatisch
dichromatism PHYS **Doppelfarbigkeit**
= Zweifarbigkeit
dicing MICROEL **Dicing**
[wafer into chips]
disconnecting SWITCH **Aufheben**
[a subscriber line] [einen Teilnehmeranschluß]
= taking out of service
dictaphone (→dictation set) OFFICE (→Diktiergerät)
dictating equipment (→dictation OFFICE (→Diktiergerät)
set)
dictating facility OFFICE **Diktiereinrichtung**
dictating machine (→dictation OFFICE (→Diktiergerät)
set)
dictation OFFICE **Diktat**
dictation set OFFICE **Diktiergerät**
= dictating machine; dictating equipment; dictaphone
dictionary LING **Wörterbuch**
[a linguistic reference book, monolingual or plurilingual, alphabetically arranged]
[sprachliches Nachschlagewerk, ein- oder mehrsprachig, alphabetisch geordnet]
≈ lexicon = Lexikon 2 (obsol.)
= Lexikon 1
dictionary DATA PROC **Dateiliste**
[list of files] ≈ Dateiverzeichnis
≈ directory
dictionary program DATA PROC (→Rechtschreibprogramm)
(→spelling check program)
DID (AM) (→in-dialing) SWITCH (→Durchwahl)
diddle (v.t.) DATA PROC **mit Daten manipulieren**
= mittels Daten betrügen
Didot point TYPOGR **kontinentaleuropäischer**
[0,376 mm] **Punkt**
≈ pica 1 [0,376 mm]
die 1 (n.) METAL **Druckgießform**
die 2 (n.) METAL **Gesenk**
die (pl. dice or dies) (→chip) MICROEL (→Chip)
die away EL.ACOUS **ausklingen**
die-cast (→pressure die-casting) METAL (→Druckguß)
die-cast alloy METAL **Spritzgußlegierung**
= Druckgußlegierung
diecast box EQUIP.ENG **Druckgußgehäuse**
die-casting (→pressure METAL (→Druckguß)
die-casting)
dielectric (n.) PHYS **Dielektrikum**
= dielectric medium; dielectric material
dielectric (adj.) PHYS **dielektrisch**
dielectrical aerial (→dielectrical ANT (→dielektrische Antenne)
antenna)
dielectrical antenna ANT **dielektrische Antenne**
= dielectrical aerial
dielectric cable COMM.CABLE (→metallfreies Kabel)
(→metal-free cable)
dielectric coated PHYS **dielektrisch beschichtet**
dielectric constant 1 PHYS **Dielektrizitätskonstante**
= permittivity = Permittivität
≈ dielectric constant 2 ≈ elektrische Feldkonstante
↓ Dielektrizitätszahl
dielectric constant 2 PHYS **elektrische Feldkonstante**
= permittivity of vacuum; permittivity; electric space constant; dielectric constant of free space (obs.)
= Influenzkonstante; Verschiebungskonstante; Dielektrizitätskonstante des leeren Raumes (obs.)

dielectric constant of free space (obs.) (→dielectric constant 2)	PHYS	(→elektrische Feldkonstante)
dielectric current [Maxwell's equations] = displacement current ≈ conduction current	EL.TECH	elektrischer Verschiebungsstrom [Maxwellsche Gleichungen] = Verschiebungsstrom; Verschiebestrom ≈ Leitungsstrom
dielectric displacement (→electric flux density)	PHYS	(→elektrische Flußdichte)
dielectric displacement density (→electric flux density)	PHYS	(→elektrische Flußdichte)
dielectric dissipation factor (→loss factor)	EL.TECH	(→Verlustfaktor)
dielectric excitation (→electric flux density)	PHYS	(→elektrische Flußdichte)
dielectric hysteresis	PHYS	dielektrische Hysterese
dielectric insulation	MICROEL	dielektrische Isolation
dielectric layer	PHYS	dielektrische Schicht
dielectric lens	ANT	dielektrische Linse
dielectric loss factor (→loss factor)	EL.TECH	(→Verlustfaktor)
dielectric material (→dielectric)	PHYS	(→Dielektrikum)
dielectric medium (→dielectric)	PHYS	(→Dielektrikum)
dielectric relaxation	PHYS	dielektrische Nachwirkung
dielectric resonator	MICROW	dielektrischer Resonator
dielectric rod antenna = dielectric tube antenna; rod antenna; flagpole antenna; dielectric rod radiator ↑ vertical antenna	ANT	Stabantenne = dielektrischer Rohrstrahler; Rohrstrahler; dielektrischer Stielstrahler; Stielstrahler ↑ Vertikalantenne
dielectric rod radiator (→dielectric rod antenna)	ANT	(→Stabantenne)
dielectric strength = disruptive strength; withstand voltage; voltage endurance; excess voltage immunity	EL.TECH	Spannungsfestigkeit [von Material und Geometrie abhängig] = Überspannungsfestigkeit; Durchschlagfestigkeit; Durchschlagfeldstärke; Überschlagfestigkeit
dielectric tube antenna (→dielectric rod antenna)	ANT	(→Stabantenne)
dielectron ↑ FET	MICROEL	Dielektron ↑ FET
die pressing	METAL	Gesenkpressen
diesel generating set = diesel genset ↑ motor-generating set	POWER SYS	Dieselaggregat ↑ Motorgenerator
diesel genset (→diesel generating set)	POWER SYS	(→Dieselaggregat)
diesel oil	CHEM	Dieselöl
DIFAN antenna	ANT	DIFAN-Antenne
differ (v.i.) = distinguish	COLLOQ	unterscheiden (v.r.)
difference	MATH	Differenz [Ergebnis der Subtraktion]
difference = diversity; disparateness; dissimilarity; unlikeness ≈ variousness; manifoldness	COLLOQ	Verschiedenheit = Verschiedenartigkeit; Unterschiedlichkeit; Ungleichheit; Unterschied ≈ Vielfältigkeit; Mannigfaltigkeit
difference amplifier = differential amplifier ≈ comparator	CIRC.ENG	Differenzverstärker = Differentialverstärker ≈ Vergleicher
difference level	TELEC	Differenzpegel
difference measuring amplifier	INSTR	Differenzmeßverstärker
difference pattern ≠ sum pattern ↑ directional pattern	ANT	Differenzdiagramm ≠ Summendiagramm ↑ Richtdiagramm
difference voltmeter	INSTR	Differenzvoltmeter
different [of distinctive quality] = diverse 1; disparate; dissimilar; unlike; unequal ≈ various; manifold 1	COLLOQ	verschieden = verschiedenartig; unterschiedlich; ungleich; divers ≈ vielfältig; mannigfaltig
differentiability	SCIE	Unterscheidbarkeit = Differenzierbarkeit
differentiable = discernible; distinguishable; discriminable	SCIE	unterscheidbar = differenzierbar
differential (adj.)	SCIE	differenzierend
differential (adj.) = incremental	MATH	differentiell
differential (n.) ≠ integral	MATH	Differential ≠ Integral
differential amplifier (→difference amplifier)	CIRC.ENG	(→Differenzverstärker)
differential analyzer [analog computer to solve differential equations]	DATA PROC	Differentialanalysator [Analogrechner zur Lösung von Differentialgleichungen]
differential analyzer	INSTR	Differentialanalysator
differential bridge	NETW.TH	Differentialbrücke
differential bridge	INSTR	Differentialmeßbrücke
differential calculus ↑ analysis; calculus	MATH	Differentialrechnung ↑ Analysis; Infinitesimalrechnung
differential coder	CODING	Differenzcodierer
differential coding (→differential encoding)	CODING	(→Differenzcodierung)
differential coil	COMPON	Differentialspule
differential conductance = incremental conductance	MICROEL	differentieller Leitwert
differential-current switch	EL.INST	Fehlerstrom-Schutzschalter
differential decoder	CODING	Differenzdecodierer
differential decoding	CODING	Differenzdecodierung
differential discriminator	CIRC.ENG	Differenzdiskriminator [FM-Demodulation] = Gegentaktdiskriminator; Gegentaktflankendiskriminator
differential echo suppressor	TELEPH	Differentialechosperre
differential element (→D-element)	CONTROL	(→D-Glied)
differential encoding = differential coding	CODING	Differenzcodierung
differential equation	MATH	Differentialgleichung
differential fase	TV	differentielle Phase = differenteller Phasenfehler
differential filter	NETW.TH	Differentialfilter [sehr schmales Filter] ≈ Differential-Brückenschalter
differential gain	TV	differentielle Verstärkung = differentieller Verstärkungsfaktor
differential gain	NETW.TH	differentielle Verstärkung
differential linearity	INSTR	differentielle Linearität
differentially decode	CODING	differenzdecodieren
differentially encode	CODING	differenzcodieren
differential measuring amplifier	INSTR	Differentialmeßverstärker
differential microphone = push-pull microphone	EL.ACOUS	Doppelkohlemikrofon
differential mode voltage	ELECTRON	Differenzspannung
differential-mode voltage gain	ELECTRON	Differenzverstärkung
differential modulation	MODUL	Differenzmodulation = differentielle Modulation
differential operator	MATH	Differentialoperator
differential permeability	PHYS	differentielle Permeabilität
differential phase	TV	differentielle Phase = differentieller Phasenfehler
differential phase shift keying = DPSK	MODUL	Phasendifferenzmodulation = DPSK
differential pulse code modulation = delta PCM	MODUL	Differenz-Pulsecodemodulation = Delta-PCM
differential regulator = D-controller	CONTROL	Differentialregler = D-Regler
differential relay	COMPON	Differentialrelais
differential resistance = incremental resistance; dynamic resistance	MICROEL	differentieller Widerstand = dynamischer Widerstand
differential screw	MECH	Differentialschraube

differential series		POWER ENG	Gegenreihenschluß	diffusion current	MICROEL	Diffusionsstrom
differential shunt		POWER ENG	Gegennebenschluß	[caused by concentration difference of charge carriers]		[durch unterschiedliche Ladungsträgerkonzentration hervorgerufen]
differential term		CONTROL	Differentialglied	↑ charge carrier current		↑ Ladungsträgerstrom
differential thermoelectric force		PHYS	differentiale Thermokraft	diffusion current density	MICROEL	Diffusionsströmungsdichte
↑ thermoelectric force			↑ Thermospannung	diffusion depth	MICROEL	Diffusionstiefe
differential transformer		COMPON	Differentialübertrager	diffusion equation	PHYS	Diffusionsgleichung
differential winding		EL.TECH	Differentialwindung	diffusion front	PHYS	Diffusionsfront
= couteracting winding			= Gegenwicklung	diffusion furnace	MICROEL	Diffusionsofen
differentiating circuit		NETW.TH	(→Differenzierglied)	diffusion layer	MICROEL	Diffusionsschicht
(→differentiating network)				diffusion length	PHYS	Diffusionslänge
differentiating network		NETW.TH	Differenzierglied	diffusion mask	MICROEL	Diffusionsmaske
[pulse shaping network]			[impulsformendes Netzwerk]	= mask		= Halbleitermaske; Maske
= differentiating circuit; differentiator			= Differenzierschaltung; differenzierende Schaltung; differenzierendes Netzwerk; Differentiator	diffusion path	PHYS	Diffusionsweg
≠ integrating network				diffusion potential	PHYS	Diffusionspotential
				diffusion process	MICROEL	Diffusionsverfahren
				diffusion profile	MICROEL	Diffusionsprofil
			≠ Integrierglied	diffusion resistance	MICROEL	Diffusionswiderstand
differentiator (→differentiating network)		NETW.TH	(→Differenzierglied)	diffusion technique	MICROEL	Diffusionstechnik
				diffusion time	PHYS	Diffusionszeit
differntial equation of Cauchy-Riemann		MATH	Cauchy-Riemann-Differentialgleichung	diffusion transistor	MICROEL	Diffusionstransistor
				↑ bipolar transistor; junction transistor		↑ Bipolartransistor; Flächentransistor
difficulty		COLLOQ	Schwierigkeit			
= distress; constraint; trouble			≈ Engpaß	diffusion triangle	MICROEL	Diffusionsdreieck
≈ strait						= Ladungsdreieck
diffracted		PHYS	gebeugt	diffusion velocity	PHYS	Diffusionsgeschwindigkeit
diffraction		PHYS	Beugung	diffusion voltage	MICROEL	Diffusionsspannung
[deviation of the propagation of a wave from the straight path, due to obstacles]			[durch Hindernisse hervorgerufene Abweichung einer Wellenbewegung von der geradlinigen Ausbreitung]	diffusion zone	MICROEL	Diffusionszone
				digit (n.)	MICROEL	Finger
				digit (→code element)	CODING	(→Codeelement)
= bending diffraction; bending				digit (→numeral)	MATH	(→numerisches Zeichen)
≈ reflection; refraction; scattering			= Diffraktion	digital (adj.)	INF.TECH	digital (adj.)
			≈ Reflexion; Brechung; Streuung	[discrete in amplitude and time]		[zeit-u.wertdiskret]
				≈ value-discrete		= ziffernmäßig
diffraction fading		RADIO PROP	(→Beugungsschwund)	≠ analog		= wertdiskret
(→obstruction fading)						≠ analog
diffraction filter beam focussing effect		OPT OEL	Beugungsfilter-Strahlbündelungseffekt	digital (n.) (→key)	TERM&PER	(→Taste)
				digital/analog signal generator	INSTR	Digital/Analog-Signalgenerator
diffraction fringe		PHYS	Beugungssaum	= D/A signal source		
diffraction grating		PHYS	Beugungsgitter	digital audio tape	CONS.EL	digitale Musikkassette
diffraction loss		RADIO PROP	Beugungsdämpfung	= DAT		= digitale Musikkassette
= obstruction loss; irregular terrain attenuation			= Beugungsverlust; Abschattungsdämpfung; Abschattungsverlust; Hindernisdämpfung; Geländedämpfung; Zusatzdämpfung	digital automat	ENG.LOG	digitaler Automat
				digital bearer	TELEC	digitales Übertragungsmedium
				digital broadband digital radiolink	TRANS	(→Breitband-Digitalrichtfunk-Verbindung)
				(→high-capacity digital radio link)		
				digital broadband LOS	TRANS	(→Breitband-Digitalrichtfunk-Verbindung)
diffraction pattern		PHYS	Beugungsbild	(→high-capacity digital radio link)		
diffraction spectrum		PHYS	Beugungsspektrum	digital broadband LOS link	TRANS	(→Breitband-Digitalrichtfunk-Verbindung)
diffuse (v.i.)		PHYS	diffundieren	(→high-capacity digital radio link)		
diffuse (→blurred)		TV	(→unscharf)	digital broadband radiolink	TRANS	(→Breitband-Digitalrichtfunk-Verbindung)
diffused		MICROEL	diffundiert	(→high-capacity digital radio link)		
diffused alloy transistor		MICROEL	diffusionslegierter Transistor	digital button	CONS.EL	Nummerntaste
[combination of alloy process with diffusion process]			[Legierung kombiniert mit Diffusion]	digital cassette	TERM&PER	Digitalkassette
				digital channel	TELEC	Digitalkanal
			= AD-Transistor			= digitaler Kanal
diffused-base alloy technique		MICROEL	DA-Technik	digital chip	MICROEL	Digitalchip
diffused diode		MICROEL	diffundierte Diode	↓ memory chip; logic chip		↓ Speicherbaustein; Logikbaustein
diffused eutectic aluminum process (→DEAP)		MICROEL	(→DEAP)			
				digital circuit	CIRC.ENG	Digitalschaltung
diffused junction transistor		MICROEL	diffundierter Transistor			= digitale Schaltung
= diffused transistor				digital circuit multiplexer	TELEC	digitaler Leitungsmultiplexer
↓ diffused base transistor; diffused emitter-collector transistor; diffused mesa transistor				= DCM		
				digital clock (→digital watch)	ELECTRON	(→Digitaluhr)
				digital clock radio	CONS.EL	Radio-Digitaluhr
diffused reflection		PHYS	diffuse Reflexion	= clock radio		= Radiouhr; Radiowecker
			= Remission	digital closed-loop control	CONTROL	digitale Regelung
diffused transistor (→diffused junction transistor)		MICROEL	(→diffundierter Transistor)	digital color monitor	TERM&PER	digitaler Farbbildschirm
				= digital colour monitor; digital RGB monitor		= digitaler RGB-Monitor
diffuse reflection		PHYS	Streureflexion			
diffusion		PHYS	Diffusion	digital colour monitor	TERM&PER	(→digitaler Farbbildschirm)
diffusion capacitance		MICROEL	Diffusionskapazität	(→digital color monitor)		
diffusion coefficient (→diffusion constant)		PHYS	(→Diffusionskonstante)	digital computer	DATA PROC	Digitalrechner
				≠ analog computer		= digitale Rechenanlage; digitaler Rechner; digitale Datenverarbeitungsanlage; digitale DVA; Digitalcomputer
diffusion conductance		MICROEL	Diffusionsleitwert			
diffusion constant		PHYS	Diffusionskonstante			
= diffusion coefficient			= Diffusionskoeffizient; Diffusionsbeiwert			
						≠ Analogrechner
diffusion current		PHYS	Diffusionsstrom			

digital concentrator		SWITCH	**Digitalkonzentrator** = digitaler Konzentrator	digital multimeter = DMM	INSTR	**Digitalmultimeter** = digitaler Multimeter
digital control		CONTROL	**digitale Steuerung** = Digitalsteuerung	digital multiplex equipment = digital signal multiplexer; digital multiplexor	TRANS	**Digital-Multiplexeinrichtung** = Digitalsignal-Multiplexgerät; Digitalsignalmultiplexer
digital cross-connect = digital cross-connect system; DCS; DACS; digital signal distributor		TRANS	**Digital-Verteiler** = Digital-Verteileinrichtung	digital multiplexor (→digital multiplex equipment)	TRANS	(→Digital-Multiplexeinrichtung)
digital cross-connect system (→digital cross-connect)		TRANS	(→Digital-Verteiler)	digital multiplier	CIRC.ENG	**digitaler Multiplizierer**
digital data		DATA PROC	**Digitaldaten** = digitale Daten	digital network	TELEC	**Digitalnetz**
				digital optical reading (→optical disk)	TERM&PER	(→Bildplatte)
digital data test set		INSTR	**Digitaldaten-Meßplatz**	digital oscilloscope = digitizing oscilloscope	INSTR	**Digitaloszilloskop**
digital display (→numeric display)		INSTR	(→Ziffernanzeige)	digital pH meter	INSTR	**Digital-pH-Meter**
digital display tube (→numeric display tube)		ELECTRON	(→Ziffernanzeigeröhre)	digital PLL (→DPLL)	CIRC.ENG	(→DPLL)
				digital plotter (→plotter)	TERM&PER	(→Plotter)
digital divider		CIRC.ENG	**Digitaldividierer** = Dividierwerk	digital pulse duration modulation	MODUL	**Schrittdauermodulation**
digital electronics		ELECTRON	**Digitalelektronik**	digital radio constellation analyzer (→constellation analyzer)	INSTR	(→Konstellationsanalysator)
digitale phase-lock loop (→DPLL)		CIRC.ENG	(→DPLL)	digital readout (→numeric display)	INSTR	(→Ziffernanzeige)
digital exchange		SWITCH	**Digitalvermittlungsanlage** = Digitalvermittlung 2	digital recording	DATA PROC	**Digitalaufzeichnung**
digital filter		NETW.TH	**Digitalfilter**	digital repeater	DATA COMM	**Digitalpulsregenerator**
digital ground		ELECTRON	**Digitalerde**	digital representation	INF.TECH	**digitale Darstellung**
digital hierarchy		TELEC	**Digitalhierarchie**	digital resolution	CODING	**digitale Auflösung**
digital hierarchy conversion (→mapping)		TELEC	(→Digitalhierarchieumsetzung)	digital RGB monitor (→digital color monitor)	TERM&PER	(→digitaler Farbbildschirm)
digital hierarchy converter (→mapper)		TELEC	(→Digitalhierarchie-Umsetzer)	digital scale	INSTR	**Ziffernskala** = Ziffernskale
digital high-capacity line-of-sight radiolink (→high-capacity digital radio link)		TRANS	(→Breitband-Digitalrichtfunk-Verbindung)	digital signal = discrete signal; DS; data signal	TELEC	**Digitalsignal** = digitales Signal; diskretes Signal; Datensignal
digital high capacity LOS radiolink (→high-capacity digital radio link)		TRANS	(→Breitband-Digitalrichtfunk-Verbindung)	digital signal distortion	DATA COMM	**Schrittverzerrung**
				digital signal distributor (→digital cross-connect)	TRANS	(→Digital-Verteiler)
digital IC		MICROEL	**digitale Digitalschaltung**	digital signaling	TELEC	**Digitalzeichengabe** = digitale Zeichengabe
digital indicator (→numeric display)		INSTR	(→Ziffernanzeige)	digital signal multiplexer (→digital multiplex equipment)	TRANS	(→Digital-Multiplexeinrichtung)
digital input		CIRC.ENG	**Digitaleingang**	digital signal procesor = DFSP	MICROEL	**digitaler Signalprozessor**
digital interface		TELEC	**Digitalschnittstelle** = digitale Schnittstelle	digital simulation	DATA PROC	**digitale Simulation**
digitalization [of a system] ≈ digitization [INF.TECH]; quantization [CODING]		TELEC	**Digitalisierung** [eines Systems] ≈ Quantisierung [CODING]	digital sorting = radix sorting	DATA PROC	**Digitalsortierung** = Radixkommasortierung
digitalize [to change a system for digital operation] ≈ digitize [INF.TECH]; quantize [CODING]		TELEC	(→digitalisieren) [ein System auf digitalen Betrieb umstellen] ≈ quantisieren [CODING]	digital step	INSTR	**Ziffernschritt**
				digital storage (→digital memory)	ELECTRON	(→Digitalspeicher)
				digital storage oscilloscope	INSTR	**Digitalspeicher-Oszilloskop**
digitalizer (→digitizing tablet)		TERM&PER	(→Digitalisiertablett)	digital stream (→bit stream)	INF	(→Bitstrom)
				digital structure	MICROEL	**Digitalstruktur**
digital level meter		INSTR	**Digitalpegelmesser**	digital subscriber loop carrier (→digital loop carrier system)	TRANS	(→digitales Teilnehmermultiplexsystem)
digital line link (CCITT)		TRANS	**Datensignal-Grundleitung**	digital subscriber pair gain system (→digital loop carrier system)	TRANS	(→digitales Teilnehmermultiplexsystem)
digital line-of-sight radio = digital LOS		TRANS	**Digitalrichtfunk** = digitaler Richtfunk	digital switching	SWITCH	**Digitalvermittlung 1** [Technik]
digital logic (→engineering logic)		INF	(→Schaltalgebra)	digital switching network	SWITCH	**Digitalkoppelnetz**
digital loop		MICROEL	**Digitalschleife**	digital switching system	SWITCH	**Digitalvermittlungssystem** = digitales Vermittlungssystem
digital loop carrier (→digital loop carrier system)		TRANS	(→digitales Teilnehmermultiplexsystem)			
digital loop carrier system = digital loop carrier; DLC; digital subscriber loop carrier; digital subscriber pair gain system; subscriber PCM system		TRANS	**digitales Teilnehmermultiplexsystem** = digitales Teilnehmersystem; Teilnehmer-PCM	digital synthesizer (→digital measuring oscillator)	INSTR	(→Digital-Meßsender)
				digital system	ELECTRON	**Digitalsystem**
				digital system	TELEC	**Digitalsystem** = digitales System
digital LOS (→digital line-of-sight radio)		TRANS	(→Digitalrichtfunk)	digital technique	ELECTRON	**Digitaltechnik**
				digital telephone	TELEPH	**Digitalfernsprecher** = Digitaltelefon
digitally (adv.)		INF.TECH	**digital** (adv.)	digital thermometer	INSTR	**Digitalthermometer**
digital measurement		INSTR	**digitales Messen**	digital-to-analog converter = D-to-A converter; D/A converter; DAC ≠ analog-to-digital converter	CODING	**D/A-Wandler** = D/A-Umsetzer; D/A-Umwandler; Digital-Analog-Wandler; Digital-Analog-Umsetzer; Digital-Analog-Umwandler ≈ A/D-Wandler ≠ A/D-Wandler
digital measuring instrument = digital meter		INSTR	**digitales Meßgerät** = digitales Meßinstrument			
digital measuring oscillator = digital synthesizer		INSTR	**Digital-Meßsender**			
digital measuring technique		INSTR	**Digitalmeßtechnik**			
digital memory = digital storage		ELECTRON	**Digitalspeicher**			
digital meter (→digital measuring instrument)		INSTR	(→digitales Meßgerät)	digital transmission	TRANS	**Digitalübertragung**

digital transmission analyzer	INSTR	Digitalübertragungsanalysator	dilatation (→stretch)	PHYS	(→Dehnung)
digital voltmeter	INSTR	Digitalvoltmeter	dilation	PHYS	**Ausdehnung 2**
		= Digitalspannungsmesser	[increase in circumference]		[Vergrößerung der Fläche, des Umfanges]
digital watch	ELECTRON	**Digitaluhr**	≈ expansion		≈ Ausdehnung
= digital clock			dilation (→stretch)	PHYS	(→Dehnung)
digit computer	DATA PROC	(→Stellenmaschine)	**DIL connector**	COMPON	**DIL-Stecker**
(→character-oriented computer)			= dual in-line connector		= Doppelreihenanschluß-Stecker
digit driver	CIRC.ENG	**Zifferntreiber**			
digit input	SWITCH	**Wahlaufnahme**	diligence (→care)	COLLOQ	(→Sorgfalt)
digit input circuit	SWITCH	**Wahlaufnahmesatz**	diligently (→carefully)	COLLOQ	(→sorgfältig)
↑ signaling circuit		↑ Signalisierungssatz	**DIL switch** (→dual-in-line switch)	COMPON	(→DIL-Schalter)
digitiser (→digitizing tablet)	TERM&PER	(→Digitalisiertablett)			
digitization	INF.TECH	**Digitalisierung**	**dilute**	TECH	**verdünnen 1**
[of signals]		[eines Signals]	↓ water (v.t.)		[Flüssigkeit]
= digitizing		≈ Quantisierung [CODING]			↓ verwässern
≈ digitalization [TELEC]; quantization [CODING]			**dilution**	TECH	**Verdünnung 1**
					[Flüssigkeit]
digitize	INF.TECH	**digitalisieren**	**dim** (v.t.)	PHYS	**abblenden**
[to convert analog signals into digital form]		[analoge Signale in digitale Form bringen]	= iris-in		
≈ digitalize [TELEC]; quantize [CODING]		≈ quantisieren [CODING]	**dim** (v.t.)	TECH	**verdunkeln**
					= abdunkeln
digitizer (→analog-to-digital converter)	CODING	(→A/D-Wandler)	**dim-bright**	TECH	**hell-dunkel**
			dimension (v.t.)	TECH	**bemessen**
digitizer (→digitizing tablet)	TERM&PER	(→Digitalisiertablett)	= seize (v.t.); rate (v.t.)		= dimensionieren; auslegen
			≈ engineer (v.t.)		
digitizing	DATA PROC	**Digitalisierung**	**dimension** (n.)	TECH	**Abmessung**
[conversion of graphics into data]		[Umsetzung von Graphiken in Daten]	≈ size		≈ Größe
= binarization		= Binarisierung	**dimension 1** (n.)	PHYS	**Dimension 1**
			↓ height, width, depth		= Ausdehnung 4
digitizing (→digitization)	INF.TECH	(→Digitalisierung)			↓ Höhe, Breite, Tiefe
digitizing oscilloscope (→digital oscilloscope)	INSTR	(→Digitaloszilloskop)	**dimension** (n.)	PHYS	**Maßzahl**
			dimension 2 (n.)	PHYS	**Dimension 2**
digitizing pad (→digitizing tablet)	TERM&PER	(→Digitalisiertablett)	[power to which a fundamental unit is involved in a quantity]		[Potenz mit der eine Grundgröße in eine Größe eingeht]
digitizing panel (→digitizing tablet)	TERM&PER	(→Digitalisiertablett)	**dimension**	ENG.DRAW	**Maß**
			dimension (→quantity)	MATH	(→Größe)
digitizing pen	TERM&PER	**Digitalisierstift**	**dimensional equation**	PHYS	**Größengleichung**
= sensing pin		= Abtaststift; Fühlstift	= quantity equation		≠ Zahlenwertgleichung
digitizing puck	TERM&PER	**Puck**	≠ numerical-value equation		
[accessory for digitizing tablet]		[Zubehör für Digitalisiertablett]	**dimensioning**	ENG.DRAW	**Bemaßung**
			dimensioning	TECH	**Bemessung**
digitizing rate (→sampling frequency)	MODUL	(→Abtastfrequenz)	= sizing; rating		= Dimensionierung; Auslegung
digitizing tablet	TERM&PER	**Digitalisiertablett**	**dimensionless**	PHYS	**dimensionslos**
[manual input device for graphic information]		[manuelles Eingabegerät graphischer Daten]	**dimension line**	ENG.DRAW	**Maßlinie**
= data tablet; digitizing pad; digitizer; digitiser; digitalizer; graphics tablet; graphics pad; graph tablet; graphic board; graphics board 2; graphic digitizer; graphics digitizer; digitizing panel		= Digitalisiergerät; Digitalisiertisch; Digitalisiertableau; Digitalisierer; Digitizer; Graphiktablett; Grafiktablett; Graphiktableau; Grafiktableau; elektronischer Zeichentisch; Zeichentablett; Zeichentableau; Datentablett; Lokalisierer	**diminish** (→decrease)	TECH	(→vermindern)
			diminish (→decrement)	DATA PROC	(→vermindern)
			diminish (→reduce 1)	TECH	(→vermindern)
			diminished radix complement (→radix-minus-one complement)	DATA PROC	(→Basis-minus-Eins-Komplement)
			diminution (→decrease)	TECH	(→Verminderung)
			diminution (→contraction)	TECH	(→Schrumpfung)
↑ graphics input hardware			**diminution** (→reduction)	TECH	(→Verkleinerung)
		↑ Graphikeingabegerät	**diminutive** (n.)	LING	**Diminutiv**
					= Deminutiv; Verkleinerungsform
digitizing uncertainity	ELECTRON	**Digitalisierungsunsicherheit**	**dimmer**	EL.INST	**Helligkeitsregler**
digit output circuit	SWITCH	**Wahlsendesatz**	= light regulator		= Dimmer
↑ signaling circuit		↑ Signalisierungssatz	**dimming switch**	EL.INST	**Abblendschalter**
digit place (→digit position)	DATA PROC	(→Ziffernstelle)	[to adjust low luminosity]		[erlaubt Einstellung geringerer Helligkeit]
digit place (→digit position)	MATH	(→Ziffernstelle)			= Dämmerungsschalter 1
digit position	DATA PROC	**Ziffernstelle**	**din** (n.)	ACOUS	**Lärm**
= digit place; decimal position; numeric position; numeric place		= Dezimalstelle; numerische Stelle; Ziffernteil	[a loud noise]		[störende laute Geräusche]
			↑ noise		= Krach
↑ position		↑ Stelle			↑ Geräusch
digit position	MATH	**Ziffernstelle**	**DIN**	TECH	**DIN**
= digit place		↓ Dezimalstelle	[German technical standards organisation]		[früher „Deutscher Industrie-Normungsausschuß"]
↓ decimal position					= Deutsches Institut für Normung
digit rate (→transmission rate)	TELEC	(→Übertragungsgeschwindigkeit)	**diode**	MICROEL	**Diode**
digit receiver (→signaling circuit)	SWITCH	(→Signalisierungssatz)	[a two-electrode semiconductor device with non-linear voltage-current characteristic]		[zweipoliges Halbleiterbauteil mit nichtlinearer Strom-Spannungs-Kennlinie]
digroup (AM)	TELEC	**Primärmultiplexbündel**	= semiconductor diode		= Halbleiterdiode
[basic group of PCM]		[PCM24, PCM30]	≈ rectifier diode		≈ Gleichrichterdiode
= primary multiplex group; primary block			↑ rectifier [EL.TECH]		↑ Gleichrichter [EL.TECH]
DIL (→dual in line)	COMPON	(→Dual-in-line)			

diode AC switch

↓ junction diode; point-contact diode; voltage reference diode; signal diode; rectifier diode
diode AC switch (→trigger diode) MICROEL (→Triggerdiode)
diode-anode MICROEL **Dioden-Anode**
diode cathode MICROEL **Dioden-Kathode**
diode characteristic COMPON **Diodenkennlinie**
diode current MICROEL **Diodenstrom**
diode equation MICROEL **Diodengleichung**
diode function generator CIRC.ENG **Diodenfunktionsgenerator**
diode gate CIRC.ENG **Dioden-Gatter**
diode limiter CIRC.ENG **Diodenbegrenzer**
diode logic ELECTRON **Diodenlogik**
diode matrix CIRC.ENG **Diodenmatrix**
= Diodenfeld
diode mixer HF **Diodenmischer**
diode rectifier POWER SYS **Diodengleichrichter**
diode tester INSTR **Diodenprüfgerät**
diode thyristor MICROEL **Thyristordiode**
diode-transistor logic MICROEL (→DTL)
(→DTL)
diode-transistor logic with zener MICROEL (→DTLZ)
diodes (→DTLZ)
diode tuning CONS.EL **Diodenabstimmung**
[tuner] [Tuner]
diode vacuum tube ELECTRON **Hochvakuumdiode**
= Vakuumdiode
diode voltage MICROEL **Diodenspannung**
diopter PHYS **Dioptrie**
[reciprocal of focal length in meters] [Kehrwert der in Meter gemessenen Brennweite]
= dioptre
dioptre (→diopter) PHYS (→Dioptrie)
dioptric PHYS **dioptrisch**
dip (v.t.) TECH **eintauchen**
dip (n.) PHYS **Senke**
= trough ≠ Quelle
≠ source
DIP (→dual-in-line package) COMPON (→DIL-Gehäuse)
dip (n.) (→depression) MATH (→Einsattelung)
DIP FIX switch COMPON **DIP-FIX-Schalterelement**
dip-greased MANUF **tauchbefettet**
diphase TECH **zweiphasig**
= two-phase; two-phased
diphtong LING **Doppellaut**
= Diphtong; Doppelvokal; Zweilaut
diplexer BROADC **Diplexer**
diplexer (→directional filter) RADIO REL (→Richtungsweiche)
diplexer (→diplexing filter) RADIO (→Trennweiche)
diplexing filter RADIO **Trennweiche**
[filter to connect several transmitter and/or receiver to a single antenna] [Filter zum Anschluß mehrerer Sender und/oder Empfänger an eine Antenne]
= diplexer
↓ duplexer filter ↓ Sende-Empfangs-Weiche
diplex operation TELEC **Doppelsprechbetrieb**
[simultaneous transmission or reception via a common medium] [gleichzeitig in beiden Richtungen über ein gemeinsames Medium]
= diplex reception
↑ Diplexbetrieb
diplex reception (→diplex operation) TELEC (→Doppelsprechbetrieb)
dip meter (→resonance frequency meter) INSTR (→Resonanzfrequenzmesser)
dipole (n.) ANT **Dipol** (n.m.)
= doublet = Doublet
≠ monopole ≠ Monopol
↓ half-wave dipole; full-wave dipole ↓ Halbwellendipol; Ganzwellendipol
dipole PHYS **Dipol**
dipole antenna ANT **Dipolantenne**
= doublet antenna
dipole array (→planar dipole array) ANT (→Dipolfeld)
dipole curtain array (→planar dipole array) ANT (→Dipolfeld)
dipole layer PHYS **Dipolschicht**
dipole moment PHYS **Dipolmoment**
dipole radiation PHYS **Dipolstrahlung**
dipole rod ANT **Dipolstab**
dipole source PHYS **Dipolquelle**
dipole surface density PHYS **Dipolflächendichte**
dipping varnish TECH **Tauchlack**
= dip varnish
dip procedure MANUF **Tauchverfahren**
dip-solder METAL **tauchlöten**
dip soldering METAL **Tauchlöten**
= Tauchlötung
DIP switch COMPON **DIP-Schalter**
[Dual In Line Package]
dip-tin (v.t.) METAL **tauchverzinnen**
dip-tinning METAL **Tauchverzinnung**
= Tauchverzinnen
dip varnish (→dipping varnish) TECH (→Tauchlack)
dip varnishing TECH **Tauchlackierung**
= Tauchlackieren
Dirac delta function (→pulse function) PHYS (→Dirac-Puls)
Dirac delta pulse (→pulse function) PHYS (→Dirac-Puls)
Dirac pulse function (→pulse function) PHYS (→Dirac-Puls)
direccional circulator RADIO REL **Sende-/Empfangs-Zirkulator**
direccional element ANT **Richtstrahlelement**
direct (adj.) COLLOQ **direkt**
direct-acces processing DATA PROC **wahlfreie Verarbeitung**
= random processing = Direktverarbeitung
≠ sequential processing ≠ sequentielle Verarbeitung
direct access DATA PROC **Direktzugriff**
[the procured data are localized by address informations, like disk side and track number] [die gesuchten Daten werden mittels Adreßinformationen, wie Plattenseite und Spur, lokalisiert]
= random access = direkter Zugriff; wahlfreier Zugriff; wahlloser Zugriff; beliebiger Zugriff
≠ sequential access ≠ sequentieller Zugriff
direct address DATA PROC **direkte Adresse**
[specifies in the address part of the instruction directly the storage location] [gibt im Adreßteil des Befehls die Speicheradresse direkt an]
= first-level address; explicit address = Direktadresse; explizite Adresse
≈ absolute address; immediate address ≈ absolute Adresse; unmittelbare Adresse
≠ indirect address ≠ indirekte Adresse
direct addressing DATA PROC **direkte Adressierung**
[with address directly indicated in the address part of the instruction] [mit direkter Adreßangabe im Adreßteil des Befehls]
= direct addressing mode; explicit addressing = Direktadressierung; explizite Adressierung
≈ absolute addressing; immediate addressing ≈ absolute Adressierung; unmittelbare Adressierung
≠ indirect addressing ≠ indirekte Adressierung
direct addressing mode DATA PROC (→direkte Adressierung)
(→direct addressing)
direct award ECON **Direktvergabe**
direct broadcast satellite SAT.COMM **Direkt-Rundstrahlsatellit**
direct-buried cable (→earth cable) COMM.CABLE (→Erdkabel)
direct call TELEC **Direktruf**
= direct station selection [Ruf auf Knopfdruck]
direct-call line SWITCH **Hauptanschluß für Direktruf**
= HfD
direct-call network DATA COMM (→Direktrufnetz)
(→leased-circuit data network)
direct call set-up SWITCH **nicht versetzter Verbindungsaufbau**
= direct connection set-up; direct link set-up
= nicht versetzte Verbindungsherstellung
direct circuit TELEC **Vorzugsleitung**
direct code (→machine code 1) DATA PROC (→Maschinencode 1)
direct connection TELEC **Querverbindung**
= tie line [TELEPH]; tie trunk [DATA COMM] = Querverbindungsleitung

direct connection set-up (→direct call set-up)		SWITCH	(→nicht versetzter Verbindungsaufbau)	**direct-fed aerial** (→active antenna)		ANT	(→aktive Antenne)

direct connection set-up (→direct SWITCH (→nicht versetzter Verbincall set-up) dungsaufbau)
direct-connect modem DATA COMM **Direktanschlußmodem**
[pluggable to telephone socket] [an Fernsprechdose direkt anschließbar]
direct control SWITCH **Direktsteuerung**
[elektron.Wählsystem]
direct control office (→direct SWITCH (→direktgesteuertes Vermittcontrol switching system) lungssystem)
direct control switching SWITCH **direktgesteuertes Vermittsystem lungssystem**
[electromechanical switch] [elektromechanisches Ver= direct switching system; direct con- mittlungssystem]
trol office; direct control system; di- = direktgesteuertes Wählsyrect pulsing system stem; Direktwahlsystem
↑ step-by-step switching system ↑ Schrittschaltsystem
direct control system (→direct SWITCH (→direktgesteuertes Vermittcontrol switching system) lungssystem)
direct convertion (→crash DATA PROC (→abrupter Übergang) conversion)
direct coupled logic MICROEL (→DCL)
(→DCL)
direct-coupled-transistor logic MICROEL (→DCTL)
(→DCTL)
direct coupling (→dc EL.TECH (→galvanische Kopplung) coupling)
direct current EL.TECH **Gleichstrom**
= dc current; dc (IEEE); d.c. (IEC); d- ≠ Wechselstrom
c; DC; unidirectional current; rectified current; dci
≠ alternating current
direct current amplifier CIRC.ENG **Gleichspannungsverstärker**
= dc amplifier = Gleichstromverstärker
direct current converter POWER SYS (→Gleichstromumrichter)
(→dc convertor)
direct-current generator POWER SYS **Gleichstromgenerator**
= dc generator ↑ Stromgenerator
↑ generator ↓ Lichtmaschine
↓ dynamo
direct-current inverter POWER SYS (→Wechselrichter)
(→inverter)
direct current level (→dc ELECTRON (→Gleichspannungspegel) level)
direct current measuring bridge INSTR (→Gleichstrom-Meßbrücke)
(→dc measuring bridge)
direct current motor (→dc POWER ENG (→Gleichstrommotor) motor)
direct current telegraphy TELEGR **Gleichstromtelegrafie**
= continuous current telegraphy; dc telegraphy
direct current voltage (→direct EL.TECH (→Gleichspannung) voltage)
direct current voltmeter (→dc INSTR (→Gleichspannungsmesser) tester)
direct data acquisition DATA PROC **direkte Datenerfassung**
= direct data entry; DDE = direkte Dateneingabe
direct data entry (→direct DATA PROC (→direkte Datenerfassung) data acquisition)
direct d.c. convertor POWER SYS **Gleichstromsteller**
(IEC) ↑ Gleichstromumrichter
= d.c. chopper convertor; dc chopper
direct dialing SWITCH (→Durchwahl)
(→in-dialing)
direct-dialing number SWITCH (→Durchwahl-Rufnummer)
(→in-dialing number)
direct dialling-in (BRI) SWITCH (→Durchwahl)
(→in-dialing)
direct digital control CONTROL **digitale Direktregelung**
= DDC = DDC
direct distance dialing (AM) SWITCH **Selbstwählferndienst**
= DDD (AM); subscriber trunk diall- = SWFD; Selbstwählferning (BRI); STD (BRI); nationwide kehr; Landesfernwahl
dialing
direct drive CONS.EL **Direktantrieb**
directed scan DATA PROC **Richtungssuche**
= directed search
directed search (→directed DATA PROC (→Richtungssuche) scan)
direct-fed EL.TECH **direktgespeist**

direct-fed aerial (→active ANT (→aktive Antenne) antenna)
direct-fed antenna (→active ANT (→aktive Antenne) antenna)
direct feed ANT **direkte Speisung**
directing force (→directive TECH (→Richtkraft) force)
direct-insert routine (→open DATA PROC (→offenes Unterprogramm) subroutine)
direct-insert subroutine DATA PROC (→offenes Unterprogramm)
(→open subroutine)
direct instruction DATA PROC (→Direktbefehl)
(→immediate instruction)
direct inward dialing (AM) SWITCH (→Durchwahl)
(→in-dialing)
directio finding station RADIO LOC **Funkpeilstelle**
= direction finder station
direction TECH **Richtung**
= orientation
direction 1 (→management 1) ECON (→Geschäftsführung)
direction 2 (→management 2) ECON (→Leitung 2)
directional TECH **gerichtet**
= vectored
directional (→directive) PHYS (→bündelnd)
directional aerial (→directional ANT (→Richtantenne) antenna)
directional antenna ANT **Richtantenne**
= directional aerial; directive antenna; = Richtstrahlantenne; Richtdirective antenna; beam antenna; beam strahler; bündelnde Antenaerial; beam; directive radiator ne
directional array ANT **Richtstrahlfeld**
= Richtantennensystem
directional arrow NETW.TH **Richtungspfeil**
≠ count arrow ≠ Zählpfeil
directional beam PHYS **Richtstrahl**
directional bridge INSTR (→Reflektometer)
(→reflectometer)
directional communications TELEC (→Verteilkommunikation)
(→distributive communication)
directional constant INSTR **Drehfederkonstante**
= torsional constant = Torsionsfederkonstante; Winkelrichtgröße
directional coupler MICROW **Richtkoppler**
directional coupler LINE TH **Richtungskoppler**
directional diagram ANT **Richtdiagramm**
[field strength distribution in a plane; [Feldstärkenverteilung in cut through the radiation pattern] einer Ebene; Schnitt durch
≈ radiation pattern die Strahlungscharakteri↓ sum diagram; difference diagram stik]
≈ Strahlungscharakteristik
↓ Summendiagramm; Differenzdiagramm
directional direction RADIO LOC **Vergleichspeilung**
finding
directional effect ANT (→Richtwirkung)
(→directivity 1)
directional filter RADIO REL **Richtungsweiche**
= branching filter; diplexer = Sende-Empfangs-Weiche;
≈ RF-combining circuit; polarization fil- Aufschaltzirkulator
ter ≈ RF-Anschaltung; Polarisationsweiche
directional filter (→duplexer RADIO (→Sende-Empfangs-Weiche) filter)
directional focussing PHYS **Richtungsfokussierung**
directional isolator MICROW **Richtungsisolator**
directional key TERM&PER **Richtungstaste**
= direction key; arrow key = Pfeiltaste
directional null ANT **Nullstelle**
= null; zero point = Nullwert; Richtungsnull
≠ lobe ≠ Keule
directional pattern (→radiation ANT (→Strahlungscharakteristik) pattern)
directional power meter INSTR **Durchgangsleistungsmesser**
= throughput power meter
directional radiation ANT **Richtstrahlung**
directional radio beacon RADIO NAV **Richtfunkbake**
directional reception (→directive RADIO (→Richtempfang) reception)
directional sensitivity PHYS **Richtungsempfindlichkeit**

directional solidification

directional solidification			gerichtetes Erstarren		
directional traffic		TELECONTR	**Richtungsverkehr**		
directional transmission		RADIO NAV	**Richtsendung**		
directional transmitter		RADIO NAV	**Richtsendeanlage**		
direction finder (→radio direction finder)		RADIO LOC	(→Funkpeiler)		
direction finder station (→directio finding station)		RADIO LOC	(→Funkpeilstelle)		
direction finding antenna		ANT	**Peilantenne**		
↓ frame antenna; rotary frame antenna; wullenwever antenna			↓ Rahmenantenne; Drehrahmenantenne; Wullenweverantenne		
direction focussing		PHYS	**richtungsfokussierend**		
direction key (→directional key)		TERM&PER	(→Richtungstaste)		
direction of propagation		PHYS	**Ausbreitungsrichtung**		
direction of rotation		PHYS	**Drehrichtung**		
direction of traction (→pulling direction)		MECH	(→Zugrichtung)		
direction of view (→line of sight)		TECH	(→Blickrichtung)		
directions for use (→use instruction 1)		TECH	(→Bedienungsanleitung)		
directive (n.)		DATA PROC	**Übersetzungsanweisung**		
= translation instruction					
directive (adj.)		PHYS	**bündelnd**		
= directional			= mit Richtwirkung		
≠ omnidirectional			≠ rundstrahlend		
directive (n.) (→guideline 1)		TECH	(→Richtlinie)		
directive aerial (→directional antenna)		ANT	(→Richtantenne)		
directive antenna (→directional antenna)		ANT	(→Richtantenne)		
directive coil		INSTR	**Richtmagnet**		
directive effect		PHYS	**Richteffekt**		
directive force		TECH	**Richtkraft**		
= directing force					
directive force spring		INSTR	**Richtkraft-Spiralfeder**		
directive gain		PHYS	**Bündelungsgewinn**		
directive gain (→gain)		ANT	(→Gewinn)		
directive radiator (→directional antenna)		ANT	(→Richtantenne)		
directive reception		RADIO	**Richtempfang**		
= directional reception			= gerichteter Empfang		
directivity 1		ANT	**Richtwirkung**		
= directional effect			= Bündelungsschärfe; Bündelungsgüte; Richtschärfe; Strahlschärfe; Strahlgüte		
≈ gain			≈ Gewinn		
directivity 2 (→gain)		ANT	(→Gewinn)		
directivity factor (→gain)		ANT	(→Gewinn)		
directivity verification standard		INSTR	**Richtschärfe-Prüfsatz**		
direct link set-up (→direct call set-up)		SWITCH	(→nicht versetzter Verbindungsaufbau)		
directly controlled system (→controlled system)		CONTROL	(→Regelstrecke)		
directly controlled variable (→controlled magnitude)		CONTROL	(→Regelgröße)		
directly heated cathode		ELECTRON	**direktgeheizte Kathode**		
			= direktgeheizte Katode		
direct mail		ECON	**Drucksachenwerbung**		
direct material (→production material)		MANUF	(→Fertigungsmaterial)		
direct memory access		DATA PROC	**direkter Speicherzugriff**		
= DMA; direct storage access; direct store access			= Direktspeicherzugriff; DMA		
direct memory access channel (→DMA channel)		DATA PROC	(→DMA-Kanal)		
direct mixer		HF	**Direktmischer**		
direct-moving		PHYS	**rechtläufig**		
direct numerical control		CONTROL	**numerische Direktsteuerung**		
[of a machine by a process computer, on-line]			[einer Maschine durch einen Prozeßrechner, on-line]		
= direct numeric control; DNC			= direkte numerische Steuerung; DNC-Steuerung		
direct numeric control (→direct numerical control)		CONTROL	(→numerische Direktsteuerung)		
director		ANT	**Wellenrichter**		
[parasitic antenna element in front of the main radiator, to modify the radiation pattern, e.g. of Yagi antennas]			[parasitäres Antennenelement vor dem Hauptstrahler, zur Beeinflussung der Richtcharakteristik, z.B. bei Yagi-Antennen]		
= director element			= Direktor		
≈ reflector			≈ Reflektor		
↑ parasitic element			↑ Parasitärelement		
director (→officer 1)		ECON	(→Leiter)		
director (→set point generator)		CONTROL	(→Sollwertgeber)		
director element (→director)		ANT	(→Wellenrichter)		
directory		DATA COMM	**Adreßbuch**		
			= Directory		
directory		DOC	**Verzeichnis**		
[a sequential list of items of a certain category]			[eine geordnete Auflistung von Einträgen einer bestimmten Kategorie]		
= survey			≈ Katalog		
≈ catalogue			↑ Liste		
↑ list			↓ Unterverzeichnis; Inhaltsverzeichnis; Register; Adreßbuch		
↓ index; register; table of contents					
directory		DATA PROC	**Verzeichnis**		
[list of files and programs with their names in a storage medium]			[Liste von Dateien und Programmen mit deren Name in einem Speicher]		
= catalog; catalogue			= Katalog		
≈ directory listing			≈ Inhaltsverzeichnis		
↓ main directory; sub-directory; disk directory; file directory			↓ Hauptverzeichnis; Nebenverzeichnis; Diskettenverzeichnis; Dateiverzeichnis		
directory (→telephone directory)		TELEC	(→Fernsprechverzeichnis)		
directory assistance		TELEPH	**Fernsprechauskunft**		
= directory enquiry service; directory enquiries			= Telefonauskunft; Telephonauskunft		
↑ special telephone service			↑ Fernsprechsonderdienst		
directory enquiries (→directory assistance)		TELEPH	(→Fernsprechauskunft)		
directory enquiry service (→directory assistance)		TELEPH	(→Fernsprechauskunft)		
directory listing		DATA PROC	**Inhaltsverzeichnis**		
[lists the files of a directory]			[listet die Dateien eines Dateiverzeichnisses auf]		
≈ directory			≈ Verzeichnis		
directory number (→subscriber number)		TELEC	(→Teilnehmerrufnummer)		
directory system		TELEC	**Directory-System**		
direct outward dialing		TELEPH	**Amtsberechtigung**		
[for extension user]			[für Nebenstellenteilnehmer]		
= DOD; non-restriction			= Vollamtsberechtigung		
direct plug connector		COMPON	**Direktsteckverbinder**		
= card edge connector			= direkter Steckverbinder		
direct pulsing system (→direct control switching system)		SWITCH	(→direktgesteuertes Vermittlungssystem)		
direct-radiator loudspeaker		EL.ACOUS	**trichterloser Lautsprecher**		
direct ray		RADIO PROP	**direkter Strahl**		
direct reading		INSTR	**Direktablesung**		
			= Direktanzeige		
direct recording		TERM&PER	**direkte Aufzeichnung**		
			[Faksimile]		
direct remote data processing (→direct teleprocessing)		DATA PROC	(→direkte Datenfernverarbeitung)		
directrix		MATH	**Leitschnitt**		
[conic section]			[Kegelschnitt]		
direct route		SWITCH	**Querleitung**		
[between two exchanges]			[direkte Leitung zwischen Vermittlungen]		
≈ primary route			= Ql; Querbündel; Direktweg		
			≈ Erstweg		
direct sorting		LING	**durchgehende Sortierung**		
[neglects spaces and commas]			[ignoriert Leerstellen und Kommas]		
direct station selection (→direct call)		TELEC	(→Direktruf)		
direct storage access (→direct memory access)		DATA PROC	(→direkter Speicherzugriff)		

direct store access (→direct DATA PROC (→direkter Speicherzugriff)
memory access)
direct switching system (→direct SWITCH (→direktgesteuertes Vermittcontrol switching system) lungssystem)
direct teleprocessing DATA PROC **direkte Datenfernverarbei-**
= on-line teleprocessing; direct remote **tung**
data processing = Online-Datenfernverarbei-
  tung
direct tension (→direct EL.TECH (→Gleichspannung)
voltage)
direct voltage EL.TECH **Gleichspannung**
= dc voltage; dc; d.c. voltage (IEC); d-c = Richtspannung
voltage; DC voltage; direct current ≠ Wechselspannung
voltage; rectified voltage; direct ten-
sion; dc tension; d-c tension; DC tension; rectified tension; dcv
≠ alternating voltage
direct wave RADIO PROP **Bodenwelle**
= ground wave (AM); surface wave = Grundwelle
dirt trap TECH **Schmutzfänger**
dirty bit DATA PROC **Speichermarke**
[flag bit to mark accomplished load-
ing]
disable (v.t.) ELECTRON **sperren**
= disenable (v.t.); inhibit (v.t.); block = unterbinden; deaktivieren;
(v.t.) Freigabe aufheben
≈ turn-off ≈ ausschalten
≠ enable ≠ freigeben
disable (n.) ELECTRON (→Abschaltung)
(→disabling)
disable (→disconnect) EL.TECH (→ausschalten)
disable bit (→inhibit bit) CODING (→Sperrbit)
disabled (→disconnected) SWITCH (→unbeschaltet)
disable pulse (→inhibit ELECTRON (→Sperrimpuls)
pulse)
disabler CIRC.ENG **Abschaltevorrichtung**
  = Ausschaltevorrichtung; Ab-
  schalter
disabling ELECTRON **Abschaltung**
= disable (n.); disconnection; disactiva- = Desaktivierung; Unwirk-
tion samschalten
≠ enabling ≈ Einschaltung
disabling key TECH **Sperrschlüssel**
disabling tone DATA COMM **Ausschalteton**
disactivation ELECTRON (→Abschaltung)
(→disabling)
disadvantage COLLOQ **Nachteil**
= detriment; adverse consequence;
prejudice; drawback
disassemble TECH **zerlegen**
= take apart; demount 2; dismount 2; ≈ abbauen; abnehmen; aus-
dissect bauen 2; demontieren
  ≠ zusammenbauen
disassembler DATA PROC **Disassemblierer**
[translates from machine code into [setzt von Maschinenspra-
machine-oriented language] che in maschinenorientier-
≠ assembler te Sprache um]
  = Disassembler
  ≈ Assemblierer
disassembly TECH **Zerlegung**
= dismount; dismantlement 2; dissection = Auseinanderbau
  ≈ Abbau
  ≠ Zusammenbau
disaster dump DATA PROC **Not-Speicherabzug**
disc (AM) MECH **Scheibe**
= disk; washer
disc (→disk) TERM&PER (→Platte)
disc access (→disk DATA PROC (→Plattenzugriff)
access)
disc access time (→disk access TERM&PER (→Spurzugriffszeit)
time)
discage antenna ANT **Discage-Antenne**
disc anode (→disk anode) ELECTRON (→Telleranode)
disc antenna ANT **Scheibenantenne**
= disk aerial (BRI)
discard (v.t.) TECH **verwerfen 1**
disc buffer (→disk DATA PROC (→Plattenpuffer)
buffer)
disc capacitor (→disk COMPON (→Scheibenkondensator)
capacitor)

disc cartridge (→disk TERM&PER (→Plattenkassette)
cartridge)
disc case (→disk jacket) TERM&PER (→Diskettenhülle)
disc cashing (→disk DATA PROC (→Plattenpufferung)
caching)
disc catalog (→disk DATA PROC (→Diskettenverzeichnis)
directory)
disc catalogue (→disk DATA PROC (→Diskettenverzeichnis)
directory)
disc controller (→disk DATA PROC (→Plattenlaufwerksteuerung)
controller)
disc copying (→disk DATA PROC (→Plattenkopieren)
copying)
disc directory (→disk DATA PROC (→Diskettenverzeichnis)
directory)
disc drive 1 (→disk drive 1) TERM&PER (→Plattenlaufwerk)
disc dump (→disk dump) DATA PROC (→Plattenspeicherabzug)
disc duplication (→disk DATA PROC (→Plattenkopieren)
copying)
disc envelope (→disk TERM&PER (→Diskettenhülle)
jacket)
discernible SCIE (→unterscheidbar)
(→differentiable)
discerning (→selective) TECH (→selektiv)
disc format (→disk TERM&PER (→Diskettenformat)
format)
disc formatting (→disk DATA PROC (→Plattenformatierung)
formatting)
discharge EL.TECH **entladen**
discharge PHYS **Entladung**
[flow of charges to compensate poten- [Ladungsfluß zum Aus-
tial differences between isolated con- gleich einer Potentialdiffe-
ductors] renz zwischen voneinander
  isolierten Leitern]
↓ gaseous discharge; breakdown
  ↓ Gasentladung; Durchschlag
discharge (→unload) TECH (→entladen)
discharge (→unloading) ECON (→Entladung)
discharge current PHYS **Entladestrom**
discharge lamp PHYS **Entladungslampe**
discharge of capacitor PHYS **Kondensatorentladung**
discharger COMPON **Schutzfunkenstrecke**
= voltage discharge gap; protective gap = Funkenstrecke
↑ high overvoltage protector ↑ Spannungsgrobsicherung
discharge space PHYS **Entladungsraum**
discharge tube (→glow COMPON (→Glimmlampe)
lamp)
discharge votage PHYS **Entladespannung**
discharging circuit CIRC.ENG **Entladekreis**
disc horn (→disk horn) ANT (→Scheibenhorn)
disc jacket (→disk TERM&PER (→Diskettenhülle)
jacket)
disclaimer ECON **Haftungsausschlußklausel**
discless (→diskless) DATA PROC (→plattenlos)
disc library (→disk DATA PROC (→Magnetplattenarchiv)
library)
disclose COLLOQ **offenbaren**
  = enthüllen
disclosure COLLOQ **Offenbarung**
disc memory (→magnetic disk TERM&PER (→Magnetplattenspeicher)
memory)
disc memory print (→disk DATA PROC (→Plattenspeicherabzug)
dump)
disc mirroring DATA PROC **Plattenspiegelung**
discone antenna ANT **Discone-Antenne**
↑ vertical antenna; biconical antenna = Diskone-Antenne; Schei-
  benkegelantenne
  ↑ Vertikalantenne;
  Doppelkonus-Antenne
disconnect 1 (v.t.) TELEC **abbauen**
[a link] [einer Verbindung]
disconnect 2 (v.t.) TELEC **abbrechen**
[a communication] [eines Gesprächs]
≈ break-off (v.t.) = abtrennen
disconnect (v.t.) EL.TECH **ausschalten**
= cut-off (v.t.); disable; deactivate; de- = abschalten; unwirksam
energize; turn-off (v.t.); power-down schalten
(v.t.); shut-down (v.t.); switch-out ≈ unterbrechen; sperren
(v.t.) [ELECTRON]
≈ interrupt; inhibit (v.t.)[ELECTRON] ≠ einschalten
≠ connect (v.t.)

disconnect

disconnect (n.) TELEPH **Trennen**
[a call without going on-hook] [Gespräch beenden ohne Hörer aufzulegen]
disconnect (→pinch off) MECH (→abkneifen)
disconnect (→release) SWITCH (→Auslösung)
disconnect (→separate) TECH (→trennen)
disconnectable ELECTRON **abschaltbar**
= defeatable
disconnected SWITCH **unbeschaltet**
= disabled; unobtainable
disconnected mode DATA PROC **Wartestellung**
= wait condition; wait state; camped-on condition = Wartezustand; Wartezyklus
disconnecting relay ELECTRON **Abschalterelais**
= overflux relay
disconnection 2 TELEC **Abbruch**
[of a call] [eines Gesprächs]
= Abtrennung; Trennung
disconnection 3 TELEC **Trennung**
≠ connection ≠ Verbindung
disconnection (→disabling) ELECTRON (→Abschaltung)
disconnection (→separation 2) MECH (→Abtrennung)
disconnection (→release) SWITCH (→Auslösung)
disconnection EL.TECH (→Unterbrechung)
(→interruption)
disconnection 1 (→connection SWITCH (→Verbindungsabbau)
tear-down)
disconnect signal (→clear-back SWITCH (→Schlußzeichen)
signal)
discontinue (→interrupt) EL.TECH (→unterbrechen)
discontinuity MATH **Unstetigkeit**
discontinuity (→step) ELECTRON (→Sprung)
discontinuity condition MATH **Unstetigkeitsbedingung**
discontinuity of media DATA PROC **Medienbruch**
= media crush
discontinuous MATH **unstetig**
= diskontinuierlich
discontinuous signaling SWITCH (→Impulszeichengabe)
(→impulse signaling)
disc operating system (→disk DATA PROC (→Platte-Betriebssystem)
operating system)
discount (n.) ECON **Preisnachlaß**
= allowance; price reduction; reduction; lowering; rebate = Nachlaß; Preisermäßigung; Ermäßigung; Rabatt; Skonto; Abschlag
↓ quantity discount; cash discount; loyality discount ↓ Mengenrabatt; Bahrzahlungsrabatt; Treuerabatt
discount rate ECON **Diskontsatz**
[interest rate charged from Central Bank to bussiness banks] [Zinssatz von Zentralbank zu Geschäftsbanken]
disc partition (→disk DATA PROC (→Plattenpartition)
partition)
disc plug (→disk plug) TECH (→Verschlußscheibe)
discrepancy COLLOQ **Unstimmigkeit**
= Diskrepanz
discrepancy switch TELECONTR **Quittungsschalter**
discrete MATH **diskret**
≠ continuous ≠ stetig
discrete COMPON **diskret**
discrete channel DATA COMM **diskreter Kanal**
discrete distribution MATH **diskrete Verteilung**
discrete electronic device COMPON **diskretes elektronisches Bauelement**
discrete Fourier transform INF.TECH **diskrete Fourier-Transformation**
= DFT
discrete in amplitude MODUL **amplitudendiskret**
discrete sentence intelligibility TELEPH (→Satzverständlichkeit)
(→phrase intelligibility)
discrete signal (→digital TELEC (→Digitalsignal)
signal)
discrete source INF **diskrete Quelle**
discrete sweep INSTR **diskrete Wobbelung**
discrete system SYS.TH **diskretes System**
≠ continuous system = diskontinuierliches System
≠ kontinuierliches System
discrete-time filter NETW.TH **zeitdiskretes Filter**
discrete words intelligibility TELEPH **Wortverständlichkeit**
= words articulation; words intelligibility ≈ Silbenverständlichkeit
≈ syllabic intelligibility

discretion COLLOQ **Ermessen**
[individual choice]
discretional matter COLLOQ **Ermessensfrage**
discretionary (→optional) ECON (→wahlweise)
discretionary hyphen (→soft DATA PROC (→Bedarfstrennstrich)
hyphen)
discriminable SCIE (→unterscheidbar)
(→differentiable)
discriminant MATH **Diskriminante**
discriminant analysis MATH **Diskriminanzanalyse**
discrimination RADIO LOC **Auflösungsvermögen**
discrimination digit SWITCH **Ausscheidungskennziffer**
= Ausscheidungsziffer
discriminator CIRC.ENG **Entscheider**
= decision circuit; decision element = Diskriminator
disc selector (→disk TERM&PER (→Plattenselektor)
selector)
disc shutter (→disk MICROW (→Scheibenblende)
shutter)
disc sleeve (→disk TERM&PER (→Diskettenhülle)
jacket)
disc storage (→magnetic disk TERM&PER (→Magnetplattenspeicher)
memory)
disc storage box (→disk DATA PROC (→Disketten-Box)
storage box)
disc thyristor (→disk MICROEL (→Scheibenthyristor)
thyristor)
disc track (→disk track) TERM&PER (→Plattenspur)
disc trimmer (→disk COMPON (→Scheibentrimmer)
trimmer)
disc unit (→magnetic disk TERM&PER (→Magnetplattenspeicher)
memory)
disc unit (→disk TERM&PER (→Plattenlaufwerk)
drive 1)
disc unit enclosure (→disk TERM&PER (→Magnetspeicheraufnahme)
unit enclosure)
discussion (→conference) ECON (→Besprechung)
disenable (v.t.) ELECTRON (→sperren)
(→disable)
disequilibrium TECH **Ungleichgewicht**
= inbalance
dish (→reflector 1) ANT (→Reflektor 1)
disintegrate (→pulverize) TECH (→zerstäuben)
disintegration TECH (→Zerstäubung)
(→pulverization)
dissipation factor measurement INSTR (→Verlustfaktormessung)
(→loss factor measurement)
disjoint (adj.) MATH **durchschnittsfremd**
[set theory] [Mengenlehre]
= disjunkt; elementfremd
disjunction (→OR ENG.LOG (→ODER-Verknüpfung)
operation)
disjunction circuit (→OR CIRC.ENG (→ODER-Glied)
gate)
disjunction element (→OR CIRC.ENG (→ODER-Glied)
gate)
disjunction gate (→OR CIRC.ENG (→ODER-Glied)
gate)
disk TERM&PER **Platte**
= disc; platter ↓ Magnetplatte; Schallplatte
↓ magnetic disk;
disk (→perforated screen) MICROEL (→Lochplatte)
disk (→disc) MECH (→Scheibe)
disk access DATA PROC **Plattenzugriff**
= disc access
disk access time TERM&PER **Spurzugriffszeit**
↑ positioning time ↑ Positionierungszeit
disk aerial (BRI) (→disc antenna) ANT (→Scheibenantenne)
disk anode ELECTRON **Telleranode**
= disc anode
disk buffer DATA PROC **Plattenpuffer**
= disc buffer
disk caching DATA PROC **Plattenpufferung**
[speeding-up access to fixed disk] [Beschleunigung des Festplattenzugriffs]
= disc cashing = Disk-Caching
disk capacitor COMPON **Scheibenkondensator**
= disc capacitor
disk cartridge TERM&PER **Plattenkassette**
= disc cartridge

disk case (→disk jacket) TERM&PER (→Diskettenhülle)
disk catalog (→disk DATA PROC (→Diskettenverzeichnis)
  directory)
disk catalogue (→disk DATA PROC (→Diskettenverzeichnis)
  directory)
disk clitch (→flange MICROW (→Flanschverbindung)
  joint)
disk controller DATA PROC **Plattenlaufwerksteuerung**
= disc controller
disk controller card TERM&PER **Plattenkontroller**
disk copying DATA PROC **Plattenkopieren**
= disk duplication; disc copying; disc
  duplication
disk crash (→head crash) TERM&PER (→Bauchlandung)
disk directory DATA PROC **Diskettenverzeichnis**
= disk catalog; disk catalogue; disc directory; disc catalog; disc catalogue
disk drive 1 TERM&PER **Plattenlaufwerk**
= disk unit; disc drive 1; disc unit ↑ Laufwerk
↑ drive ↓ Hartplattenlaufwerk; Dis-
↓ hard disk drive; floppy disk drive; kettenlaufwerk;
  fixed disk drive; removable disk drive Festplattenlaufwerk; Wechselplattenlaufwerk
disk drive 2 (→floppy disk TERM&PER (→Diskettenlaufwerk)
  drive)
disk dump DATA PROC **Plattenspeicherabzug**
= magnetic disk dump; disk memory = Magnetplattenspeicher-Ab-
  print; disc dump; magnetic disc zug
  dump; disc memory print
disk duplication (→disk DATA PROC (→Plattenkopieren)
  copying)
disk envelope (→disk TERM&PER (→Diskettenhülle)
  jacket)
diskette (→floppy disk) TERM&PER (→Diskette)
diskette drive (→floppy disk TERM&PER (→Diskettenlaufwerk)
  drive)
diskette maintenance TERM&PER **Disketten-Pflegegerät**
  equipment
diskette memory TERM&PER **Diskettenspeicher**
= diskette storage; floppy-disk mem- ↑ Magnetschichtspeicher
  ory; floppy-disk storage
↑ magnetic layer memory
diskette operating system DATA PROC **Disketten-Betriebssystem**
= floppy operation system; FOS
diskette storage (→diskette TERM&PER (→Diskettenspeicher)
  memory)
disk file (→magnetic disk DATA PROC (→Magnetplattendatei)
  file)
disk format TERM&PER **Diskettenformat**
= disc format
disk formatting DATA PROC **Plattenformatierung**
= disc formatting
disk horn ANT **Scheibenhorn**
= disc horn
disk index hole (→index TERM&PER (→Indexmarkierung)
  hole)
disk jacket TERM&PER **Diskettenhülle**
= disc jacket; floppy disk jacket; = Plattenhülle; Disketten-
  disc jacket; jacket; disk case; disc schutzhülle
  case; floppy disk case; floppy disc
  case; case; disk sleeve; disc sleeve;
  floppy disk sleeve; floppy disc sleeve;
  sleeve; disk envelope; disc envelope;
  floppy disk envelope; floppy disc envelope; envelope
diskless DATA PROC **plattenlos**
= discless
disk library DATA PROC **Magnetplattenarchiv**
= disc library
disk memory (→magnetic disk TERM&PER (→Magnetplattenspeicher)
  memory)
disk memory print (→disk DATA PROC (→Plattenspeicherabzug)
  dump)
disk operating system DATA PROC **Platte-Betriebssystem**
[supports the use of magnetic disk [unterstützt die Verwen-
  memories as external memory] dung von Magnetplatten
= DOS; disc operating system als externe Speicher]
↓ MS-DOS; PC-DOS = Plattenbetriebssystem;
  DOS
  ↓ MS-DOS; PC-DOS
disk pack (→magnetic disk TERM&PER (→Magnetplattenstapel)
  pack)
disk partition DATA PROC **Plattenpartition**
= disc partition
disk plug TECH **Verschlußscheibe**
= disc plug
disk-sealtube (→planar ELECTRON (→Scheibenröhre)
  tube)
disk-sealvalve (→planar ELECTRON (→Scheibenröhre)
  tube)
disk selector TERM&PER **Plattenselektor**
= disc selector
disk shutter MICROW **Scheibenblende**
= disc shutter
disk sleeve (→disk jacket) TERM&PER (→Diskettenhülle)
disk storage (→magnetic disk TERM&PER (→Magnetplattenspeicher)
  memory)
disk storage box DATA PROC **Disketten-Box**
= disc storage box = Diskettensarg
disk thyristor MICROEL **Scheibenthyristor**
= disc thyristor
disk track TERM&PER **Plattenspur**
= disc track ↓ Diskettenspur
↓ floppy-disk track
disk trimmer COMPON **Scheibentrimmer**
= disc trimmer
disk unit (→magnetic disk TERM&PER (→Magnetplattenspeicher)
  memory)
disk unit (→disk TERM&PER (→Plattenlaufwerk)
  drive 1)
disk unit enclosure TERM&PER **Magnetspeicheraufnahme**
= disc unit enclosure
dislocation MICROW **Versetzung**
dislocation (→lattice PHYS (→Gitterfehlstelle)
  vacancy)
dislocation density MICROEL **Versetzungsdichte**
dismantle TECH **abbauen**
= demount 1; dismount 1 = abmontieren; demontieren;
≈ dissasemble; detach; remove ausbauen 2
  ≈ zerlegen; lösen; entfernen
  ≠ aufbauen
dismantlement 1 TECH **Abbau**
= dismantling; removal = Demontage; Ausbau 2
≈ disassembly ≈ Zerlegung
  ≠ Zusammenbau
dismantlement 2 TECH (→Zerlegung)
  (→disassembly)
dismantling TECH (→Abbau)
  (→dismantlement 1)
dismember TECH **vereinzeln**
≈ fragment ≈ zerstückeln
dismount 1 (→dismantle) TECH (→abbauen)
dismount 2 (→disassemble) TECH (→zerlegen)
dismountable (→disassembly) TECH (→Zerlegung)
dismountable mast ANT **ausfahrbarer Mast**
disorder (n.) PHYS **Fehlordnung**
disorder (→lattice vacancy) PHYS (→Gitterfehlstelle)
disorderly close-down DATA PROC **vorwarnungsloser Zusammen-**
  bruch
disparate (→different) COLLOQ (→verschieden)
disparateness COLLOQ (→Verschiedenheit)
  (→difference)
dispatch (v.t.) DATA PROC **abfertigen**
[select for next task] [für nächsten Arbeits-
  schritt]
  = auswählen
dispatch (v.t.) TELEC **absetzen**
[a message] [eine Nachricht]
≈ despatch ≈ senden
≈ transmit
dispatch (→freight) ECON (→verfrachten)
dispatch (n. (→despatch) ECON (→Versand)
dispatch department ECON (→Spedition)
  (→forwarding)
dispatcher (→scheduler DATA PROC (→Abwickler)
  program)
dispatching business (→mail-order ECON (→Versandhaus)
  business)
dispatching discipline SWITCH **Abfertigungsdisziplin**
= service discipline

dispatching priority

dispatching priority DATA PROC **Abfertigungspriorität**		
= service priority		
dispatch note (→delivery note) ECON (→Lieferschein)		
dispatch system TERM&PER **Versandsystem**		
dispensation ELECTRON **Nachlieferung**		
[dispenser cathode] [Vorratskathode]		
dispense COLLOQ **austeilen**		
= deal out; administer [einem bestimmten Empfängerkreis]		
≈ distribute; partition (v.t.)		
↑ divide ≈ verteilen; aufteilen		
↑ teilen		
dispenser cathode ELECTRON **Vorratsbehälterkathode**		
= Vorratskathode		
dispersal gain (→carrier dispersal RADIO (→Trägerverwischungsgewinn)		
gain)		
dispersed data processing DATA PROC (→verteilte Datenverarbeitung)		
(→distributed data processing)		
dispersed intelligence DATA PROC (→verteilte Intelligenz)		
(→distributed intelligence)		
dispersing lens PHYS **Streuungslinse**		
dispersion PHYS **Dispersion**		
= chromatic dispersion = Farbenstreuung; Farbenzerstreuung		
dispersion INF **Abweichung**		
[from expected value] [vom Erwartungswert]		
dispersion (→alternative ENG.LOG (→Alternativverneinung)		
denial)		
dispersion coefficient (→leakage EL.TECH (→Streufaktor)		
factor)		
dispersion field (→stray PHYS (→Streufeld)		
field)		
dispersion-flattened fiber OPT.COMM **dispersionsabgeflachte Faser**		
dispersion-shifted OPT.COMM **dispersionsverschoben**		
dispersion-shifted fiber (AM) OPT.COMM **dispersionsoptimierte Faser**		
= dispersion-shifted fibre (BRI) = dispersionsverschobene Faser		
dispersion-shifted fibre (BRI) OPT.COMM (→dispersionsoptimierte Faser)		
(→dispersion-shifted fiber)		
displacable TECH **verschiebbar**		
= shiftable; offsetable = versetzbar		
displace TECH **verlagern**		
≈ shift; transpose ≈ verschieben; versetzen		
displace (→shift) COLLOQ (→verschieben 1)		
displacement DATA PROC **Distanz**		
= bias		
displacement TECH **Verlagerung**		
≈ shifting; transposition ≈ Verschiebung; Versetzung		
displacement PHYS **Verschiebung**		
= shift		
displacement (→displacement DATA PROC (→Distanzadresse)		
address)		
displacement address DATA PROC **Distanzadresse**		
= displacement; bias address; offset = Offset		
displacement current EL.TECH (→elektrischer Verschiebungsstrom)		
(→dielectric current)		
displacement device TECH **Verschiebeeinrichtung**		
displacement field PHYS **Verschiebungsfeld**		
displacement polarization EL.TECH **Verschiebungspolarisation**		
[erzwungene Ladungsverschiebung]		
↑ elektrische Polarisation		
displacement vector (→electric flux PHYS (→elektrische Flußdichte)		
density)		
display (n.) INSTR **Anzeige**		
= indication; monitoring; read-out (n.) ≈ Ablesung; Anzeiger		
≈ reading; indicator		
display 3 (n.) TERM&PER **Bildanzeige**		
= visual display = Schirmbilanzeige; Schirmbilddarstellung		
display 1 (n.) (→display TERM&PER (→Bildschirm 1)		
screen)		
display 2 (n.) (→display TERM&PER (→Sichtgerät)		
terminal)		
display (v.t.) (→indicate) INSTR (→anzeigen)		
display adapter DATA PROC **Auflösungsprüfadapter**		
display area TERM&PER **Bildfläche**		
display area (→display panel) INSTR (→Anzeigefeld)		
display attribute TERM&PER **Bildattribut**		
display background 1 DATA PROC **Bildhintergrund**		
= Bildschirmhintergrund		
display background 2 DATA PROC (→Bildschirmmaske)		
(→display mask)		
display background generator DATA PROC (→Maskenoperator)		
(→display mask generator)		
display build-up DATA PROC **Bildaufbau**		
display build-up time DATA PROC **Bildaufbauzeit**		
display build-up velocity DATA PROC **Bildaufbaugeschwindigkeit**		
display case (→show case) ECON (→Schaukasten)		
display character DATA PROC **Darstellungszeichen**		
display code TERM&PER **Bildschirmcode**		
= screen code		
display console (→data TERM&PER (→Datensichtgerät)		
display terminal)		
display controller TERM&PER **Bildschirmsteuergerät**		
display device (→display TERM&PER (→Sichtgerät)		
terminal)		
display element ELECTRON **Anzeigeelement**		
= annunciator		
display foreground DATA PROC **Bildvordergrund**		
= Bildschirmvordergrund		
display form (→display DATA PROC (→Bildschirmmaske)		
mask)		
display format DATA PROC **Darstellungsformat**		
= Bildschirmmaskenformat		
display formular (→display DATA PROC (→Bildschirmmaske)		
mask)		
display highlight (→highlight) DATA PROC (→Hervorhebung)		
displaying tube (→display ELECTRON (→Anzeigeröhre)		
tube)		
display lockout INSTR **Anzeigesperre**		
display marker (→marker) INSTR (→Marke)		
display mask DATA PROC **Bildschirmmaske**		
= screen mask; mask; display background 2; display form; screen form; screen formular; display formular = Maske 2; Bildschirmschablone, Schablone; Bildschirmformular; Formular; Formateinblendung		
↓ input mask ↓ Eingabemaske		
display mask generator DATA PROC **Maskenoperator**		
= screen format generator; display background generator		
display material (→exposition ECON (→Ausstellungsstück)		
specimen)		
display medium (→display DATA PROC (→Darstellungsunterlage)		
surface)		
display menu DATA PROC **Bildschirmmenü**		
display mode INSTR **Anzeigefunktion**		
= view state		
display mode (→screen TERM&PER (→Bildschirmdarstellung)		
mode)		
display on screen TELECONTR **Bildschirmanzeige**		
= video display		
display paging MOB.COMM **Anzeigefunkruf**		
↑ radio paging ↑ Funkruf		
display panel INSTR **Anzeigefeld**		
= display area		
display panel SYS.INST **Schaubild**		
[normally wall-mounted] [i.a. als Wandtafel]		
= mimic board		
display position TERM&PER **Bildschirmposition**		
= screen position		
display printout (→hard copy) TERM&PER (→Druckkopie)		
display processing unit DATA PROC (→Bildprozessor)		
(→display processor)		
display processor DATA PROC **Bildprozessor**		
= image processor; display processing unit; DPU		
display register DATA PROC **Bildregister**		
= image register = Anzeigeregister		
display resolution TERM&PER **Bildauflösung**		
= image resolution		
display scale (→indicating INSTR (→Anzeigeskala)		
scale)		
display screen TERM&PER **Bildschirm 1**		
= display 1 (n.); screen = Schirm		
≈ monitor; display terminal ≈ Monitor; Sichtgerät		
↑ luminescent screen ↑ Leuchtschirm		
↓ character display; raster scanned display; graphics screen; vector display ↓ Zeichenbildschirm; Rasterbildschirm; Graphikbildschirm; Vektorbildschirm		

display socket	COMPON	Display-Halterung	
display store [for screen content] = matrix store; matrix memory; frame storage; frame memory; frame buffer	TERM&PER	Bildspeicher [für Bildschirminhalt] = Bildschirmspeicher	
display surface = display medium	DATA PROC	Darstellungsunterlage	
display teleprinter	TERM&PER	Bildschirmfernschreiber	
display terminal [to display electronically or optically stored characters and graphics] = display unit; visual display device; display device; visual display unit; display 2 (n.) ≈ monitor; display screen ↓ data display terminal; microfilm reader	TERM&PER	Sichtgerät [zur Darstellung von elektronisch oder optisch gespeicherten Zeichen und Graphiken] = Sichtanzeigegerät; optisches Anzeigegerät; Bildschirm 2; Bildschirmeinheit; Bildschirmgerät ≈ Monitor; Bildschirm 1 ↓ Datensichtgerät; Mikrofilm-Lesegerät	
display tolerance	DATA PROC	Anzeigegenauigkeit	
display tube = indicator tube; displaying tube; indicating tube	ELECTRON	Anzeigeröhre = Indikatorröhre	
display type	DATA PROC	Anzeigemittel	
display typewriter ≈ text terminal equipment	TERM&PER	Bildschirm-Schreibmaschine ≈ Textendgerät	
display unit (→display terminal)	TERM&PER	(→Sichtgerät)	
display-update	TERM&PER	Anzeigeaktualisierung	
display window (→window)	DATA PROC	(→Fenster)	
display workstation	TERM&PER	Bildschirmarbeitsplatz	
disponibility (→availability)	QUAL	(→Verfügbarkeit)	
disposal = waste disposal ≠ supply ≈ recycling	TECH	Entsorgung ≠ Versorgung ≈ Rückgewinnung	
disproportionate	TECH	unverhältnismäßig	
dispute (n.) = controversy	ECON	Streit = Streitfall	
disregard (→neglect)	TECH	(→vernachlässigen)	
disruption (→breakdown)	PHYS	(→Durchschlag)	
disruption of insulation	PHYS	Isolationsdurchschlag	
disruptive discharge (→breakdown)	PHYS	(→Durchschlag)	
disruptive strength (→dielectric strength)	EL.TECH	(→Spannungsfestigkeit)	
disruptive voltage (→breakdown voltage)	PHYS	(→Durchbruchspannung)	
disruptive voltage (→breakdown voltage)	MICROEL	(→Durchbruchspannung)	
dissecate (→dry)	TECH	(→trocknen)	
dissecation ≈ drying	TECH	Entfeuchtung ≈ Trocknung	
dissect (→disassemble)	TECH	(→zerlegen)	
dissection (→disassembly)	TECH	(→Zerlegung)	
dissertation (→thesis)	SCIE	(→Doktorarbeit)	
dissimilar (→different)	COLLOQ	(→verschieden)	
dissimilarity	SCIE	Unähnlichkeit	
dissimlarity (→difference)	COLLOQ	(→Verschiedenheit)	
dissipated heat = dissipation 3 ≈ wasted heat	TECH	Verlustwärme ≈ Abwärme	
dissipation 2 (→dissipation power)	EL.TECH	(→Verlustleistung)	
dissipation 3 (→dissipated heat)	TECH	(→Verlustwärme)	
dissipation 1 (→heat transfer)	PHYS	(→Wärmeableitung)	
dissipation factor (→loss factor)	EL.TECH	(→Verlustfaktor)	
dissipation factor bridge (→loss factor bridge)	INSTR	(→Verlustfaktor-Meßbrücke)	
dissipation factor meter (→loss factor meter)	INSTR	(→Verlustfaktormesser)	
dissipation free (→loss-free)	EL.TECH	(→verlustfrei)	
dissipation heat (→dissipation power)	EL.TECH	(→Verlustleistung)	
dissipationless (→loss-free)	EL.TECH	(→verlustfrei)	
dissipation line	ANT	Schluckleitung	
dissipation power [active power converted into heat] = leakage power; dissipation 2; dissipation heat ≈ heat transfer [PHYS] ↑ active power	EL.TECH	Verlustleistung [in Wärme umgesetzte Wirkleistung] = Verlust 2 ≈ Wärmeableitung [PHYS] ↑ Wirkleistung	
dissipation resistor ≈ dummy load	ANT	Schluckwiderstand ≈ Abschlußwiderstand	
dissipative ending	ANT	Schluckende	
dissociated acceptor atom (→acceptor ion)	MICROEL	(→Akzeptor-Ion)	
dissociation	PHYS	Dissoziation	
dissociation (→connection tear-down)	SWITCH	(→Verbindungsabbau)	
dissociation energy [width of forbidden band]	PHYS	Dissoziationsenergie [Breite des verbotenen Bandes]	
dissociation voltage (→ionization voltage)	PHYS	(→Ionisierungsspannung)	
dissolution [process]	CHEM	Lösung 2 [Vorgang] = Auflösung	
dissolvable (→soluble)	CHEM	(→löslich)	
dissolve	PHYS	lösen = auflösen	
dissolve (v.t.) (→lap-dissolve)	TV	(→überblenden)	
dissolve (n.) (→lap-dissolve)	TV	(→Überblendung)	
dissymmetry (→unbalance)	NETW.TH	(→Unsymmetrie)	
dissymmetry (→assymmetry)	MATH	(→Asymmetrie)	
distance [in space or time domain] ≈ separation	PHYS	Entfernung [räumlich, zeitlich] = Distanz; Abstand	
distance (→path length)	PHYS	(→Weglänge)	
distance (→minimum code distance)	CODING	(→Code-Distanz)	
distance measurement	TECH	Entfernungsmessung	
distance measuring equipment = DME	RADIO NAV	Entfernungsmeßsystem = DME	
distance ring ≈ spacer ring	MECH	Abstandsring	
distant (→dx)	RADIO	(→dx)	
distant (→remote)	TECH	(→entfernt)	
distant exchange (→opposite exchange)	SWITCH	(→Gegenvermittlung)	
distant field (→far-field region)	ANT	(→Fernfeld)	
distant from ground	ANT	bodenfern	
distant-from-ground antenna = distant-to-ground aerial	ANT	bodenferne Antenne	
distant office (→opposite exchange)	SWITCH	(→Gegenvermittlung)	
distant signal	RAILW.SIGN	Vorsignal	
distant terminal = opposite terminal	TELEC	Gegenstelle	
distant-to-ground aerial (→distant-from-ground antenna)	ANT	(→bodenferne Antenne)	
distinctive (→characteristic)	TECH	(→charakteristisch)	
distinctive mark	TECH	Unterscheidungsmerkmal	
distinguish (→differ)	COLLOQ	(→unterscheiden) (v.r.)	
distinguishable (→differentiable)	SCIE	(→unterscheidbar)	
distort [twist out of normal position] ≈ deform	MECH	verdrehen ≈ drehen; verbiegen	
distorted ≠ undistorted	TELEC	verzerrt ≠ unverzerrt	
distortion ≈ tortion	MECH	Verdrehung ≈ Drehung	
distortion [optics]	PHYS	Verzeichnung [Optik; Verzerrung der Abbildung] = Distorsion	

distortion

distortion TELEC
[deviation from original signal form]

distortion analyzer INSTR
[measures total harmonic distortion]
= harmonic analyzer
↑ signal analyzer

distortion attenuation EL.TECH
(→harmonic distortion attenuation)

distortion bridge (→distortion INSTR
measuring bridge)

distortion factor (→harmonic EL.ACOUS
content)

distortion factor (→harmonic EL.TECH
content)

distortion factor meter INSTR
(→distortion measuring bridge)

distortionless line LINE TH
= zero-distortion line

distortion measurement INSTR

distortion measuring bridge INSTR
= distortion bridge; distortion factor meter; distortion meter; distortion measuring set

distortion measuring set INSTR
(→distortion measuring bridge)

distortion meter (→distortion INSTR
measuring bridge)

distortionless TELEC
= zero-distortion

distortive power (→harmonic NETW.TH
reactive power)

distress (→difficulty) COLLOQ

distress frecuency RADIO

distress message (→emergency TELEPH
call)

distress signal RADIO

distribute COLLOQ
≈ dispense; partition
↑ divide

distribute (→market) ECON

distributed MATH

distributed amplifier CIRC.ENG
(→cascade amplifier)

distributed array processor DATA PROC
(→array processor)

distributed capacitance LINE TH
= capacitance per unit length
↑ transmission-line constant

distributed coil (→divided COMPON
coil)

distributed computer system DATA PROC

distributed computing DATA PROC
(→distributed data processing)

distributed data base DATA PROC

distributed database management DATA PROC

distributed data processing DATA PROC
= DDP; distributed computing; dispersed data processing
≠ centralized data processing

distributed feedback laser OPTOEL
(→DFB laser)

distributed file system DATA PROC

distributed inductance LINE TH
= inductance per unit length
↑ transmission-line constant

distributed information processing INF.TECH

Verzerrung
[Abweichung vom Ursprungssignal]
= Verzeichnung

Verzerrungsanalysator
[mißt den Oberwellengehalt]
= Klirranalysator
↑ Signalanalysator

(→Klirrdämpfung)

(→Klirrfaktormeßbrücke)

(→Oberschwingungsgehalt)

(→Oberwellengehalt)

(→Klirrfaktormeßbrücke)

verzerrungsfreie Leitung
= verzerrungslose Leitung

Verzerrungsmessung

Klirrfaktormeßbrücke
= Klirrgradmeßbrücke; Verzerrungsmeßbrücke; Klirrfaktormesser; Klirrgradmesser; Verzerrungsmesser

(→Klirrfaktormeßbrücke)

(→Klirrfaktormeßbrücke)

verzerrungsfrei
= verzerrungslos
≈ verzerrungsarm

(→Oberwellenblindleistung)

(→Schwierigkeit)

Notruffrequenz
= Seenotfrequenz

(→Notruf)

Notsignal

verteilen
≈ austeilen; aufteilen
↑ teilen

(→vertreiben)

verteilt

(→Kaskadenverstärker)

(→Vektorrechner)

Kapazitätsbelag
[Kapazität pro Längeneinheit]
↑ Leitungskonstante

(→verteilte Spule)

verteiltes Rechnersystem
= verteiltes Computersystem

(→verteilte Datenverarbeitung)

verteilte Datenbank

verteilte Datenbasisverwaltung

verteilte Datenverarbeitung
= dezentrale Datenverarbeitung
≠ zentrale Datenverarbeitung

(→DFB-Laser)

verteiltes Dateiensystem

Induktivitätsbelag
[Induktivität pro Längeneinheit]
= Induktionsbelag
↑ Leitungskonstante

verteilte Informationsverarbeitung

distributed intelligence DATA PROC
= dispersed intelligence

distributed leakage LINE TH
= leakage per unit length; distributed leakance; leakance per unit length
↑ transmission-line constant

distributed leakance LINE TH
(→distributed leakage)

distributed network DATA COMM

distributed queen dual bus DATA COMM
= DQDB

distributed resistance LINE TH
(→resistance per unit length)

distributing box OUTS.PLANT
[distributor between primary and secondary cable]
= dividing box; terminal block; distribution terminal

distributing frame SYS.INST
(→distributor)

distribution MATH
↓ frequency distribution; probability distribution

distribution (→partitioning) TECH

distribution amplifier BROADC
[CATV]

distribution amplifier INSTR

distribution board (→mains EL.INST
multi-connector)

distribution cabinet OUTS.PLANT
[pedestal-mounted cabinet]
= cross connecting terminal; cross connecting distributor; cable distribution box

distribution cable OUTS.PLANT
[cable branching from main cable and from which drop wires lead to the subscriber's premises]
= branch cable; secondary cable

distribution cable POWER SYS
= distribution line; rising main

distribution cable OUTS.PLANT
[U.S. technique]

distribution closure OUTS.PLANT
(→distribution sleeve)

distribution communication TELEC
(→distributive communication)

distribution frame SYS.INST
(→distributor)

distribution function MATH
= cumulative distribution function
≈ probability function

distribution key ECON

distribution level (→cross-connect TELEC
level)

distribution line BROADC
[CATV]

distribution line POWER SYS
(→distribution cable)

distribution list DOC
(→distributor)

distribution network ECON
= sales network; dealership

distribution network 2 TELEC
(→subscriber network)

distribution network 1 TELEC
(→distributive network)

distribution panel POWER SYS

distribution service TELEC
= distributive service
≠ interactive service

distribution sleeve OUTS.PLANT
= distribution closure

verteilte Intelligenz

Ableitungsbelag
[Ableitung pro Längeneinheit]
= Leitwertsbelag
↑ Leitungskonstante

(→Ableitungsbelag)

verteiltes Netz

DQDB

(→Widerstandsbelag)

Kabelverzweiger
[Verteiler zwischen Primär- und Sekundärkabel]
= Endverzweiger; KVz

(→Verteiler)

Verteilung
↓ Häufigkeitsverteilung; Wahrscheinlichkeitsverteilung

(→Aufteilung)

Stammverstärkerstelle
[CATV]

Trennverstärker

(→Reihensteckdose)

Kabelverzweigergehäuse
[Gehäuse auf Sockel]
= KVz-Gehäuse; Kabelverzweigerschrank; KVz-Schrank; Verteilerkasten

Sekundärkabel
= Aufteilungskabel; Verzweigungskabel 1

Steigleitung

Verzweigungskabel 2
[US-Technik]

(→Aufteilungsmuffe)

(→Verteilkommunikation)

(→Verteiler)

Verteilungsfunktion
= Summenhäufigkeitsfunktion
≈ Wahrscheinlichkeitsfunktion

Verteilungsschlüssel
= Verteilerschlüssel

(→Rangierebene)

Stamm
[CATV]

(→Steigleitung)

(→Verteiler)

Vertriebsnetz

(→Teilnehmernetz 1)

(→Verteilnetz)

Verteilertafel

Verteildienst
≠ interaktiver Dienst

Aufteilungsmuffe

distribution strip	COMPON	**Verteilerleiste**	
= terminal strip		= Verteilerstreifen; Stützpunktleistung	
distribution system	TELEC	**Verteilsystem**	
distribution terminal	OUTS.PLANT	(→Kabelverzweiger)	
(→distributing box)			
distributive communication	TELEC	**Verteilkommunikation**	
= distribution communication; directional communications; broadcast communications		= gerichtete Kommunikation	
		≠ Dialogkommunikation	
≠ interactive communication			
distributive mode	DATA PROC	**Distributionsmethode**	
distributive network	TELEC	**Verteilnetz**	
= distribution network 1		= Distributionsnetz; distributives Netz	
≠ interactive network			
		≠ interaktives Netz	
distributive network	DATA COMM	**Distributionsnetz**	
distributive service (→distribution service)	TELEC	(→Verteildienst)	
distributive sort	DATA PROC	**Umgruppierung**	
distributor	DOC	**Verteiler**	
[of a written message]		[eines Schreibens]	
= distribution list; circulation list			
distributor	SYS.INST	**Verteiler**	
= distribution frame; distributing frame		= Verteilergestell	
↓ main distributing frame; patching distribution frame; power distributor		↓ Hauptverteiler; Rangierverteiler; Stromverteiler	
distributor	ECON	**Vertreiber**	
≈ seller		≈ Handelshaus	
		≈ Verkäufer	
distributor (→dealer)	ECON	(→Händler)	
distributor equipment plan	SYS.INST	**Verteilerbelegungsplan**	
distributor panel	EQUIP.ENG	**Verteilerfeld**	
district computer	TELECONTR	**Bezirksrechner**	
district exchange	TELEC	**Bezirksamt**	
= main district exchange			
district exchange	TELEGR	**Hauptamt**	
district network	TELEC	**Bezirksnetz**	
district traffic	TELEC	**Bezirksverkehr**	
≈ regional traffic		≈ Regionalverkehr	
disturbance (→disturbance variable)	CONTROL	(→Störgröße)	
disturbance (→interference)	TELEC	(→Störung)	
disturbance behaviour	CONTROL	(→Störverhalten)	
(→disturbance response)			
disturbance effect	ELECTRON	**Störeffekt**	
= hazard; spike			
disturbance elimination	ELECTRON	(→Entstörung)	
(→interference suppression)			
disturbance immunity	TELEC	(→Störfestigkeit)	
(→interference immunity)			
disturbance insensitive	TELEC	(→störungsunempfindlich)	
(→interference insensitive)			
disturbance level (→interference level)	TELEC	(→Störpegel)	
disturbance response	CONTROL	**Störverhalten**	
= disturbance behaviour			
disturbance suppression	ELECTRON	(→Entstörung)	
(→interference suppression)			
disturbance transfer function	CONTROL	**Störübertragungsfunktion**	
disturbance variable	CONTROL	**Störgröße**	
= perturbation variable; disturbance			
disturbance variable feed-forward	CONTROL	**Störgrößenaufschaltung**	
= perturbation variable feed-forward			
disturber (→interferer)	TELEC	(→Störer)	
disturbing current (→interfering current)	EL.TECH	(→Störstrom)	
disturbing noise (→interfering noise)	TELEC	(→Störgeräusch)	
disturbing pulse (→interfering pulse)	ELECTRON	(→Störimpuls)	
disturbing signal (→interfering signal)	TELEC	(→Störsignal)	
disturbing voltage (→interfering voltage)	TELEC	(→Störspannung)	
disturbing voltage content (→interfering voltage content)	TELEC	(→Fremdspannungsanteil)	
disturbing voltage suppression (→interference suppression)	TELEC	(→Störspannungsunterdrückung)	
ditch (→trench)	CIV.ENG	(→Graben)	
dithering 1	TERM&PER	**Farbmischung**	
[dots of different colors]		[Flecken verschiedener Farben]	
dithering 2	TERM&PER	**Rasterung 2**	
[dots of different size]		[Flecken unterschiedlicher Größe]	
diurnal (adj.)	TECH	**tageszeitlich**	
≠ nocturnal		[zur Tageszeit gehörend]	
		≠ nächtlich	
diurnal	ECON	**ganztägig**	
≠ semidiurnal		≠ halbtägig	
diurnal variation	TECH	**Tagesgang**	
diurnal wave	RADIO PROP	**Tagwelle**	
		= ganztägige Welle	
divalent (→bivalent)	MATH	(→zweiwertig 2)	
divergence	MATH	**Divergenz**	
[vector analysis]		[Vektorfeld]	
divergence (→deviation)	TECH	(→Abweichung)	
divergence loss	RADIO PROP	**Divergenzverlust**	
divergent	TECH	**abweichend**	
= deviating		= divergent	
divergent integral	MATH	**divergentes Integral**	
diverging lens	PHYS	**Zerstreuungslinse**	
= thinnest at the center		[in der Mitte am dünnsten]	
↓ biconcave lens; plane-concave lens; convexo-concave lens		↓ bikonkave Linse; plankonkave Linse; konvexkonkave Linse	
diverging wave	RADIO PROP	**divergierende Welle**	
diverse 2 (→various)	COLLOQ	(→vielfältig)	
diverse 1 (→different)	COLLOQ	(→verschieden)	
diversification	ECON	**Diversifikation**	
diversity (→diversity reception)	RADIO	(→Diversityempfang)	
diversity (→difference)	COLLOQ	(→Verschiedenheit)	
diversity receiver	RADIO	**Diversityempfänger**	
diversity reception	RADIO	**Diversityempfang**	
= diversity		= Mehrfachempfang; Diversity	
diversity spacing	RADIO	**Diversityabstand**	
divide	TECH	**teilen**	
= part; share		≈ trennen; spalten	
≈ separate; split		↓ aufteilen; verteilen; unterteilen; austeilen	
↓ partition (v.t.); distribute; subdivide; dispense			
divide	MATH	**dividieren**	
[- by]		= teilen	
divided coil	COMPON	**verteilte Spule**	
= distributed coil			
dividend 2	MATH	**Dividend 2**	
[number above the fraction line]		[Zahl über dem Bruchstrich]	
= numerator; teller		= Zähler	
≠ divisor 2		≠ Divisor 2	
dividend 1	MATH	**Dividend 1**	
[number divided by another]		[Zahl die durch andere geteilt wird]	
≠ divisor 1		≠ Divisor 1	
divider (→reducer)	CIRC.ENG	(→Untersetzer)	
divide reminder	MATH	**Divisionsrest**	
= reminder		= Restwert; Rest	
dividers	ENG.DRAW	**Stechzirkel**	
↑ compasses		= Teilzirkel	
		↑ Zirkel	
divide up (→partition)	TECH	(→aufteilen)	
dividing box (→distributing box)	OUTS.PLANT	(→Kabelverzweiger)	
dividing circuit (→reducer)	CIRC.ENG	(→Untersetzer)	
dividing plane (→interface)	TECH	(→Trennfläche)	
dividing surface (→interface)	TECH	(→Trennfläche)	
divisibility	MATH	**Teilbarkeit**	
divisible	MATH	**teilbar**	
division	MATH	**Division**	
= division operation		= Teilung; Dividieren	
division (→department)	ECON	(→Abteilung)	
division (→scale)	INSTR	(→Skala)	
division (→partition)	TECH	(→Teilung)	
division (→corporate division 1)	ECON	(→Unternehmensbereich)	

division

division (n.) (→branch) ECON (→Zweigwerk)
division bar (→fraction line) MATH (→Bruchstrich)
division check DATA PROC **Divisionsprobe**
division circuit (→reducer) CIRC.ENG (→Untersetzer)
division line (→fraction line) MATH (→Bruchstrich)
division operation (→Division) MATH (→Division)
division sign MATH **Divisionszeichen**
[symbol :] [Symbol :]
= division symbol = Teilungszeichen; Divis
≈ colon ≈ Doppelpunkt
division stroke (→fraction line) MATH (→Bruchstrich)
division stroke (→fraction line) MATH (→Bruchstrich)
division symbol (→division sign) MATH (→Divisionszeichen)
divisor 2 MATH **Divisor 2**
[number below the fraction line] [Zahl unter dem Bruchstrich]
= denominator = Nenner
≠ dividend 2 ≠ Dividend 2
↓ modulus 1 ↓ Modulus 1
divisor 1 MATH **Divisor 1**
[number dividing another number] [Zahl die eine andere teilt]
≠ dividend 1 = Teilungsfaktor
 ≠ Dividend 1
DL (→delete) DATA COMM (→Löschen)
D-layer RADIO PROP **D-Schicht**
= D-region
DLC (→digital loop carrier system) TRANS (→digitales Teilnehmermultiplexsystem)
DLE (→data link escape) DATA COMM (→Datenübertragungsumschaltung)
D lead (→tip2 wire) TELEPH (→D-Ader)
DL1FK beam antenna ANT **DL1FK-Antenne**
↑ rotary beam antenna ↑ Drehrichtstrahler
dm (→decimeter) PHYS (→Dezimeter)
DMA (→direct memory access) DATA PROC (→direkter Speicherzugriff)
DMA channel DATA PROC **DMA-Kanal**
= direct memory access channel
DME (→distance measuring equipment) RADIO NAV (→Entfernungsmeßsystem)
DML (→data manipulation language) DATA PROC (→Datenmanipulationssprache)
DMM (→digital multimeter) INSTR (→Digitalmultimeter)
DMOS MICROEL **DMOS**
= double diffused MOS [zweifach diffundierte MOS-Technologie]
D.M. quad (→multiple-twin quad) COMM.CABLE (→DM-Vierer)
DNC (→direct numerical control) CONTROL (→numerische Direktsteuerung)
do away with COLLOQ **abgehen von**
document (v.t) ECON **dokumentieren**
document (n.) DOC **Unterlage**
= record (n.)
document (n.) ECON **Urkunde**
= deed; instrument
document (n.) TERM&PER **Beleg**
[a data carrying form used in business] [in der Geschäftswelt verwendetes Datenträgerformular]
= voucher
↑ form ↑ Formular
↓ original document ↓ Originalbeleg
document (→record) OFFICE (→Akte)
document (→text file) DATA PROC (→Textdatei)
documentary channel BROADC **Studienprogramm**
document assembly (→document merge) DATA PROC **Belegmischung**
documentation (n.s.t.) DATA PROC **Dokumentation**
documentation aid DATA PROC **Dokumentationshilfe**
documentation department ECON **Schrifttum 1**
 [Organisation]
documentation filing department OFFICE **Registratur**
= filing department; registry [Organisationseinheit]

documentation system DATA PROC **Dokumentationssystem**
= data storage and retrieval system
document change (→update information) TECH (→Änderungsmitteilung)
document destroying device OFFICE **Aktenvernichter**
= shredding machine = Dokumentenvernichter; Aktenwolf; Reißwolf
documentfax TELEC **Dokumentenfax**
document feed TERM&PER **Belegzufuhr**
= document feeder; voucher feed; voucher feeder = Belegzuführung; Belegeinzug; Belegvorschub; Blattzufuhr; Blattzuführung; Blatteinzug; Blattvorschub; Belegtransport; Blatttransport; Belegeingabe; Blatteingabe
≈ form feed
↑ paper feed
 ≈ Formularzuführung
 ↑ Papierzufuhr
document feeder TERM&PER (→Belegzufuhr)
(→document feed)
document heading DOC **Unterlagenkopf**
document index ENG.DRAW **Zeichnungsverzeichnis**
= document survey
document merge DATA PROC **Belegmischung**
[into a common file] [in eine Datei]
= document assembly = Dokumentenmischung
documentor DATA PROC **Dokumentationsprogramm**
[a program]
document processing DATA PROC **Belegverarbeitung**
document reader TERM&PER **Belegleser**
↑ character reader ↑ Klarschriftleser
↓ optical document reader; magnetic document reader ↓ optischer Belegleser; Magnetschriftleser
documents (→source documents) OFFICE (→Schriftgut)
document sorter TERM&PER **Belegsortierer**
document storage DATA COMM **Dokumentenspeicher**
= DS
document survey ENG.DRAW (→Zeichnungsverzeichnis)
(→document index)
document telegraphy TELEGR **Dokumententelegrafie**
↑ facsimile telegraphy ↑ Faksimiletelegrafie
document type DOC **Unterlagenart**
DOD (→direct outward dialing) TELEPH (→Amtsberechtigung)
dog (n.) MECH **Klemmzapfen**
[fixing element of a setscrew] [Stellschrauben]
DO instruction (FORTRAN) DATA PROC (→Laufanweisung)
(→run statement)
Dolby strecher EL.ACOUS **Dolbisierung**
Dolby system EL.ACOUS **Dolby-Verfahren**
↑ noise suppression method = Dolby-System
 ↑ Rauschunterdrückungssystem
dollar exchange rate ECON **Dollarkurs**
dollar sign ECON **Dollarzeichen**
[symbol $] = Symbol $
= dollar symbol
dollar symbol (→dollar sign) ECON (→Dollarzeichen)
dolly (→camera dolly) TV (→Fahrstativ)
DO loop DATA PROC **DO-Schleife**
[FORTRAN] [FORTRAN]
Dolph-Chebychev distribution ANT **Dolph-Tschebycheff-Verteilung**
domain DATA PROC **Domäne**
= attribute domain = Attributwertebereich
domain PHYS **Kristallbereich**
[of a crystal] = Domäne
domain COLLOQ **Herrschaftsbereich**
domain (→validity range) MATH (→Gültigkeitsbereich)
domain boundary PHYS **Domänengrenzfläche**
domestic call (→inland toll call) TELEC (→Inlandsferngespräch)
domestic costumer (→private subscriber) TELEC (→Privatteilnehmer)
domestic distribution amplifier BROADC (→Hausverteilverstärker)
(→in-premises distribution amplifier)
domestic distribution system BROADC (→Hausverteilanlage)
(→in-premises distribution system)
domestic market (→home market) ECON (→Heimatmarkt)

domestic sales	ECON	**Inlandsvertrieb**		**doping factor** (→doping grade)	MICROEL	(→Dotierungsgrad)
[an activity or organization]				**doping grade**	MICROEL	**Dotierungsgrad**
domestic subscriber (→private subscriber)	TELEC	(→Privatteilnehmer)		= doping factor		= Dotierungsfaktor; Dop-Faktor
domestic tariff	ECON	**Inlandstarif**		**doping profile**	MICROEL	**Dotierungsprofil**
		= Binnentarif		= impurity concentration profile; impurity profile; defect profile		= Dotierprofil; Störstellenprofil; Fehlstellenprofil
domestic traffic	TELEC	**Inlandsverkehr**		**doping sequence**	MICROEL	**Dotierungsfolge**
domicile	ECON	**Sitz**				= Dotierfolge
		[einer Organisation]		**doping technique**	MICROEL	**Dotierungsverfahren**
dominant mode (→fundamental mode)	MICROW	(→Grundwelle)				= Dotierverfahren
dominant wave (→fundamental mode)	MICROW	(→Grundwelle)		**DOPOS** = doped polysilicon diffusion	MICROEL	**DOPOS**
donator (→donor)	PHYS	(→Donator)		**Doppler direction finder**	RADIO LOC	**Doppler-Peiler**
donator level (→donor level)	PHYS	(→Donatorniveau)		**Doppler effect**	PHYS	**Doppler-Effekt**
donator migration (→donor migration)	MICROEL	(→Donatorwanderung)		= doppler effect		
dongle	DATA PROC	**Kopierschutzschaltung**		**doppler effect** (→Doppler effect)	PHYS	(→Doppler-Effekt)
[a circuit or chip necessary to run a protected program]		[für den Betrieb der geschützen Software erforderlich]		**Doppler log**	RADIO NAV	**Doppler-Log**
↑ copy protection				**Doppler navigator**	RADIO NAV	**Doppler-Navigator**
donor	PHYS	**Donator**		**Doppler omnirange**	RADIO NAV	**Doppler-VOR**
= donator		[Elektronen abgebender Atom oder Störstelle]		**Doppler radar**	RADIO LOC	**Doppler-Radar**
↓ donor atom; donor ion		= Donor		**Doppler shift**	PHYS	**Dopplerverschiebung**
		↓ Donator-Atom; Donator-Ion		**DOR** (→optical disk)	TERM&PER	(→Bildplatte)
				dormant partner (→silent partner)	ECON	(→stiller Gesellschafter)
donor atom	PHYS	**Donator-Atom**		**dormer**	CIV.ENG	**Gaube**
↑ donor		↑ Donator		[vertical window on roof]		= Gaupe; Dachgaube
donor density	PHYS	**Donatorendichte**		**DOS** (→disk operating system)	DATA PROC	(→Platte-Betriebssystem)
donor exhaustion	PHYS	**Donatoren-Erschöpfung**				
donor ion	PHYS	**Donator-Ion**		**dosage**	TECH	**Dosierung**
= ionized donor atom		= ionisiertes Donator-Atom		= proportioning		
↑ donor		↑ Donator		**dose rate constant**	PHYS	**Dosisleistungskonstante**
donor level	PHYS	**Donatorniveau**		**DO statement** (FORTRAN) (→run statement)	DATA PROC	(→Laufanweisung)
= donator level				**dot** (v.t.) (AM)	ENG.DRAW	**stricheln**
donor migration	MICROEL	**Donatorwanderung**		[----]		[----]
= donator migration				= break		≈ punktieren
do-not-disturb (→station guarding)	TELEPH	(→Anrufschutz)		≈ punctuate		
do-nothing command (→blank instruction)	DATA PROC	(→Leeranweisung)		**dot**	LING	**Punkt 1**
				[symbol .]		[Symbol .]
do-nothing instruction (→blank instruction)	DATA PROC	(→Leeranweisung)		= period (AM); full stop (BRI); point 1		= Punktzeichen
do-nothing statement (→blank instruction)	DATA PROC	(→Leeranweisung)		↑ punctuation mark; end punctuation mark		↑ Satzzeichen; Satzschlußzeichen
DON'T TOUCH !	TECH	**NICHT BERÜHREN !**		**DOT**	TERM&PER	**DOT-Speicher**
[warning]		[Warnschild]		[domain tip]		
= NOLI ME TANGERE				**dot command**	DATA PROC	**Punktkommando**
donutron	MICROEL	**Doppelkäfig-Magnetron**				= separate Formatierungsanweisung
door 1	TECH	**Tür**		**dot cycle**	CODING	**Signalisierungszyklus**
door 2 (→flap)	TECH	(→Klappe)		**dot-dashed** (→chain-dotted)	ENG.DRAW	(→strichpunktiert)
door alarm	SIGN.ENG	**Türalarm**		**dot-dash line**	ENG.DRAW	**Strichpunktlinie**
doorknob transition	MICROW	**Türknauf-Übergang**				= strichpunktierte Linie
doorknob tube	ELECTRON	**Knopfröhre**		**dot frequency**	ELECTRON	**Punktfrequenz**
door lock (→drive lock)	TERM&PER	(→Laufwerksverriegelung)		= point frequency		
door monitoring system (→entrance monitoring system)	SIGN.ENG	(→Türanlage)		**dot-frequency diagram**	MATH	**Punktdiagramm**
				dot incidence angle	TV	**Bildpunkteinfallswinkel**
dopant	MICROEL	**Dotierungsmaterial**		**dot leaders** (→dotted line 1)	ENG.DRAW	(→punktierte Linie)
= doping agent; addition agent; dope additive		= Dotiermaterial; Dotiermittel		**dotless i**	LING	**türkisches I**
dope (v.t.)	MICROEL	**dotieren**				= I ohne Punkt; punktloses I
		= dopen		**dot matrix**	TERM&PER	**Punktmatrix**
dope additive (→dopant)	MICROEL	(→Dotierungsmaterial)				= Punktraster 1
doped	MICROEL	**dotiert**		**dot-matrix character**	TERM&PER	**Rasterdruckzeichen**
doped junction	MICROEL	**dotierte Schicht**		≠ fully formed character		≠ Volldruckzeichen
doped polysilicon diffusion (→DOPOS)	MICROEL	(→DOPOS)		**dot-matrix display**	TERM&PER	**Punktmatrix-Anzeige**
				dot-matrix printer	TERM&PER	**Rasterdrucker**
doped semiconductor	MICROEL	**dotierter Halbleiter**		= matrix printer		= Matrixdrucker; Mosaikdrucker
dope vector	DATA PROC	**Querverweisvektor**		↓ stylus printer		
doping	MICROEL	**Dotierung**				↓ Nadeldrucker
[introduction of impurities to change semiconductor characteristics]		[Verunreinigung zur Veränderung von Halbleitereigenschaften]		**dot-matrix printing**	TERM&PER	**Rasterdruckverfahren**
				= dot-matrix technique		= Punktraster-Druckverfahren; Matrix-Druckverfahren
		= Dotieren; Doping; Dopen		**dot-matrix technique** (→dot-matrix printing)	TERM&PER	(→Rasterdruckverfahren)
doping agent (→dopant)	MICROEL	(→Dotierungsmaterial)				
doping atom [MICROEL] (→foreign atom)	PHYS	(→Fremdatom)		**dot pattern**	TECH	**Punktmuster**
doping compensation	MICROEL	**Dotierungsausgleich**		**dot pitch**	TERM&PER	**Lochmaskenabstand**
doping density	MICROEL	**Dotierungsdichte**				= Pixelabstand

dot printing

dot printing	TERM&PER	**Punktdrucken**
dots per inch	TERM&PER	**Bildpunkte pro Zoll**
= d.p.i.; dpi		
dotted 1	ENG.DRAW	**gepunktet**
[....]		[...]
dotted 2 (AM)	ENG.DRAW	**gestrichelt**
[----]		[----]
= broken		
dotted line 2 (AM)	ENG.DRAW	**gestrichelte Linie**
[----]		[----]
= broken line		= gestrichtelte Kurve
dotted line 1	ENG.DRAW	**punktierte Linie**
[....]		[....]
= dot leaders; leaders		= Führungspunkte
double (v.t.) (→duplicate)	MATH	(→verdoppeln)
double (→twofold)	TECH	(→zweifach)
double-acting	TECH	**doppeltwirkend**
double-acting	TECH	**doppelwirkend**
= double-action		
double-action (→double-acting)	TECH	(→doppelwirkend)
double armature relay	COMPON	**Doppelankerrelais**
double-balanced mixer	HF	**doppeltabgestimmter Mischer**
double-balanced modulator	MODUL	**Doppelgegentaktmodulator**
double-base diode	MICROEL	(→Doppelbasisdiode)
(→unijunction transistor)		
double bazooka	ANT	**Doppelsperrtopf**
		= Doppelbazooka
double-beam cathode ray tube	ELECTRON	(→Zweistrahlröhre)
(→double-trace cathode ray tube)		
double-beam CRT	ELECTRON	(→Zweistrahlröhre)
(→double-trace cathode ray tube)		
double-beam tube	ELECTRON	(→Zweistrahlröhre)
(→double-trace cathode ray tube)		
double bend	MECH	**Doppelbiegung**
double bisquare antenna	ANT	**Doppel-Bisquare-Antenne**
double bond	CHEM	**Doppelbindung**
double bottom (→raised floor)	SYS.INST	(→Doppelboden)
double branch point	BROADC	**Zweifachabzweiger**
double-break contact	COMPON	**Öffner-Öffner**
		[Relais]
double-break-double-make contact	COMPON	**Zwillingswechsler**
		↑ Relaiskontakt
↑ relay contact		
double buffering	DATA PROC	**Doppelpufferspeicherung**
		= Doppelpufferung
double-button microphone	EL.ACOUS	**Doppelkapselmikrofon**
double-clad	ELECTRON	**doppelt kaschiert**
[PCB]		[Leiterplatte]
= double-sided		= beidseitig kaschiert
double click	TERM&PER	**Doppelklick**
[mouse]		[Maus]
double-coil mechanism	INSTR	**Doppelspulmeßwerk**
double command	TELECONTR	**Doppelbefehl**
double-conductor cord	EQUIP.ENG	**zweiadrige Schnur**
double connection	SWITCH	**Doppelverbindung**
double control	CONTROL	**Zweifachregelung**
= with two reference magnitudes and two controlled magnitudes		[zwei Führungs- und zwei Regelgrößen]
↑ multiple control		↑ Mehrfachregelung
double convertor (→two-way convertor)	POWER SYS	(→Umkehrstromrichter)
double current	TELEGR	**Doppelstrom**
= polar current (AM)		≠ Einfachstrom
≠ single current		
double-current keying	TELEGR	**Doppelstromtastung**
double current operation	TELEGR	(→Doppelstrombetrieb)
(→double current working)		
double-current signaling	TELEC	**Doppelstromsignalisierung**
double current transmission	TELEGR	(→Doppelstrombetrieb)
(→double current working)		
double current working	TELEGR	**Doppelstrombetrieb**
= double current transmission; double current operation		
double D beam	ANT	**Doppel-D-Beam**
double delta antenna	ANT	**Doppel-Delta-Antenne**
double delta loop	ANT	**Doppel-Delta-Loop**
= twin delta loop		
double density	TERM&PER	**doppelte Schreibdichte**
[floppy disk]		[Diskette]
= DD		= DD
double-density diskette	TERM&PER	(→DD-Diskette)
(→DD diskette)		
double diffused MOS	MICROEL	(→DMOS)
(→DMOS)		
double diode	COMPON	**Doppeldiode**
double-doublet antenna	ANT	**Doppel-Doublett-Antenne**
double echo (→reverberation)	ACOUS	(→Nachhall)
double-edged sword	COLLOQ	**zweischneidiges Schwert**
[fig.]		[fig.]
		= zweischneidiges Messer
double-ended queue	DATA PROC	(→Doppelendschlange)
(→deque 1)		
double eurocard size	ELECTRON	**Doppeleuropakarten-Format**
[PCB size of 91.9x65.4 sqin]		[Leiterplattenformat 233,4x166 mm^2]
		= Doppeleuropaformat; Doppeleuropakarte
double-face board	ELECTRON	(→doppelt kaschierte Leiterplatte)
(→double-face PCB)		
double-face PCB	ELECTRON	**doppelt kaschierte Leiterplatte**
= double-face board; two-sided PCB; two-sided board		= beidseitig kaschierte Leiterplatte
double feed	ANT	**Doppelspeisung**
double-focus tube	ELECTRON	**Doppelfokusröhre**
double-frame direction finder	RADIO LOC	**Doppelrahmenpeiler**
double-grid tube	ELECTRON	**Doppelgitterröhre**
= tetraode		
double-gun cathode ray tube	ELECTRON	(→Zweistrahlröhre)
(→double-trace cathode ray tube)		
double-gun CRT	ELECTRON	(→Zweistrahlröhre)
(→double-trace cathode ray tube)		
double-height	EQUIP.ENG	**zweizeilig**
= two-row		
double-height module	EQUIP.ENG	(→zweizeilige Baugruppe)
(→double-row module)		
double-height subassembly	EQUIP.ENG	(→zweizeilige Baugruppe)
(→double-row module)		
double helix	TECH	**Doppelwendel**
= double whip		
double helix antenna	ANT	**Doppelwendel-Antenne**
double hetereostructure	MICROEL	**Doppelhetereostruktur**
double hyphen	TYPOGR	**Doppelstrich**
[indicates a hyphen at the end of a line]		[kennzeichnet am Zeilenende einen Bindestrich]
≈ equal sign		≈ Gleichheitszeichen
↑ punctuation mark		↑ Satzzeichen
double image (→ghost image)	TV	(→Geisterbild)
double-image range finder	RADIO LOC	**Mischbild-Entfernungsmesser**
double integral	MATH	**Doppelintegral**
[integration over a piece of a planar surface]		[Integration über ein Stück einer Ebene]
≈ surface integral		≈ doppeltes Integral
↑ multiple integral		≈ Oberflächenintegral
		↑ mehrfaches Integral
double integrator	CIRC.ENG	**Zweifachintegrator**
[gives an output, which is proportional to the double integral]		[stellt eine Proportionalität zum Doppelintegral her]
		= Doppelintegrator
double layer	ELECTRON	**Doppelschicht**
double-layer screen	ELECTRON	**Doppelschichtschirm**
double-length arithmetic	DATA PROC	(→doppeltgenaue Arithmetik)
(→double-precision arithmetic)		
double line	ENG.DRAW	**Doppelstrich**
= double rule		= doppelte Linie
double line feed (→double line spacing)	TERM&PER	(→doppelter Zeilenabstand)
double line spacing	TERM&PER	**doppelter Zeilenabstand**
= double spacing; double line feed		= doppelter Zeilenvorschub
double make contact	COMPON	**Schließer-Schließer**
		[Relais]
double modulation	MODUL	**Zweistufenmodulation**
double operation	TECH	**Doppelbetrieb**
= duplex operation		
double pendulum	PHYS	**Doppelpendel**
double phantom	LINE TH	(→Achterleitung)
(→double-phantom circuit)		
double-phantom circuit	LINE TH	**Achterleitung**
= superphantom; double phantom		= Achter; Superphantom

double plate

double plate	PHYS	Doppelplatte	
double-point information	TELECONTR	Doppelmeldung	
double-polarized	PHYS	doppelt polarisiert	
≠ single-polarized		≠ einfach polarisiert	
double pole	OUTS.PLANT	Doppelgestänge	
double precision	DATA PROC	doppelte Genauigkeit	
double-precision arithmetic	DATA PROC	doppeltgenaue Arithmetik	
= double-length arithmetic			
double preselection	SWITCH	Doppelvorwahl	
double punch	TERM&PER	Doppellochung	
double quad element	ANT	Doppelquad-Element	
doubler (→doubler circuit)	CIRC.ENG	(→Verdopplerschaltung)	
double-ray cathode ray tube (→double-trace cathode ray tube)	ELECTRON	(→Zweistrahlröhre)	
double-ray CRT (→double-trace cathode ray tube)	ELECTRON	(→Zweistrahlröhre)	
doubler circuit	CIRC.ENG	Verdopplerschaltung	
= doubler			
↓ voltage doubler; frequency doubler		↓ Spannungsverdoppler; Frequenzverdoppler	
double reflecting aerial (→indirect-feed antenna)	ANT	(→indirektgespeiste Antenne)	
double reflecting antenna (→indirect-feed antenna)	ANT	(→indirektgespeiste Antenne)	
double refraction (→birefringence)	PHYS	(→Doppelbrechung)	
double refractive	PHYS	doppelbrechend	
double retaining pawl	TELEPH	Doppelsperrklinke	
double rhombic antenna	ANT	Doppelrhombus-Antenne	
double rhomboid antenna	ANT	Doppel-Rhomboid-Antenne	
double-ridge waveguide	MICROW	Doppel-Steghohlleiter	
double-row module	EQUIP.ENG	zweizeilige Baugruppe	
= double-row subassembly; double-height module; double-height subassembly			
double-row subassembly (→double-row module)	EQUIP.ENG	(→zweizeilige Baugruppe)	
double rule (→double line)	ENG.DRAW	(→Doppelstrich)	
double seizure	SWITCH	Kollision	
= glare			
double-sideband measurement	INSTR	Zweiseitenbandmessung	
double-sideband receiver	RADIO	Zweiseitenbandempfänger	
double sideband reception	RADIO	Zweiseitenbandempfang	
double-sideband reduced-carrier system	MODUL	Zweiseitenban-Übertragung mit gedämpftem Träger	
double-sideband signal	MODUL	Zweiseitenbandsignal	
double-sideband suppressed-carrier transmission	MODUL	Zweiseitenband-Übertragung mit unterdrücktem Träger	
double-sideband system	RADIO	Zweiseitenband-System	
= DSB system		= DSB-System	
double-sideband transmission	MODUL	Zweiseitenband-Übertragung	
double-sideband transmitter	RADIO	Zweiseitenbandsender	
double-sided (→double-clad)	ELECTRON	(→doppelt kaschiert)	
double-sided (→bilateral)	TECH	(→zweiseitig)	
double-sided disc (→DS disk)	TERM&PER	(→DS-Diskette)	
double-sided disk (→DS disk)	TERM&PER	(→DS-Diskette)	
double-sided diskette (→DS disk)	TERM&PER	(→DS-Diskette)	
double-sided rack	EQUIP.ENG	doppelseitiges Gestell	
double size	ENG.DRAW	Maßstab 2:1	
= scale 2:1			
double-slot antenna	ANT	Zweischlitzstrahler	
double snap-in terminal	COMPON	Doppelklemme	
double spacing (→double line spacing)	TERM&PER	(→doppelter Zeilenabstand)	
double-spiral winding (→bifilar winding)	COMPON	(→Bifilarwicklung)	
double-spot (→two-point)	EL.TECH	(→Zweipunkt-)	
double-spot tuning	ELECTRON	Zweipunktabstimmung	
= two-spot tuning			
double-star connection	POWER SYS	Doppelsternschaltung	
double star with interphase transformer (→double three-pulse mid-point circuit)	POWER ENG	(→Doppel-Dreipuls-Mittelpunkt-Schaltung)	
double-step multiplexer (→skip multiplexer)	TRANS	(→Doppelschrittmultiplexer)	
double-strength	TECH	doppelstark	
double stub	ANT	doppelte Stichleitung	
double superphantom (→quadruple-phantom circuit)	LINE TH	(→Sechzehnerleitung)	
doublet (→dipole)	ANT	(→Dipol) (n.m.)	
doublet (→diad)	DATA PROC	(→Dyade)	
double talk (→duplex telephony)	TELEPH	(→Gegensprechbetrieb)	
double talking (→duplex telephony)	TELEPH	(→Gegensprechbetrieb)	
doublet antenna (→dipole antenna)	ANT	(→Dipolantenne)	
double tariff	TELEC	Doppelgebühr	
		= Doppeltarif	
double-T dipole	ANT	Doppel-T-Dipol	
		= gefalteter Doppel-T-Dipol	
double tensor	MATH	Doppeltensor	
double-threaded	MECH	zweigängig [Gewinde]	
double three-pulse mid-point circuit	POWER ENG	Doppel-Dreipuls-Mittelpunkt-Schaltung	
= double star with interphase transformer		= Doppelstern mit Saugdrossel; Saugdrosselschaltung	
double-throw-contact (→changeover contact)	COMPON	(→Umschaltkontakt)	
double-tone coder (→dual-sound coder)	TV	(→Zwei-Ton-Coder)	
double-tone television transmission	TV	Zweiton-Fernsehübertragung	
		= Zweitonträger-Fernsehsystem	
double-trace cathode ray tube	ELECTRON	Zweistrahlröhre	
= double-trace CRT; double-ray cathode ray tube; double-ray CRT; double-beam CRT; double-gun cathode ray tube; double-gun CRT; double-beam tube		= Doppelstrahlröhre	
double-trace CRT (→double-trace cathode ray tube)	ELECTRON	(→Zweistrahlröhre)	
double track	TERM&PER	Doppelspur	
= dual track		[Magnetband]	
double-tuned (→two-circuit)	NETW.TH	(→zweikreisig)	
double-tuned amplifier	CIRC.ENG	Zweikreisverstärker	
double-tuned bandpass filter	NETW.TH	Zweikreisbandfilter	
double walled	TECH	doppelwandig	
double-way connection	POWER SYS	Zweiwegeschaltung	
= reversible connection			
double-way converter	POWER SYS	Zweiwegstromrichter	
= reversible converter			
double-way rectification	CIRC.ENG	Doppelweggleichrichtung	
= full-wave rectification		= Zweiweggleichrichtung; Vollweggleichrichtung	
double-way rectifier	CIRC.ENG	Doppelweggleichrichter	
= full-wave rectifier		= Zweiweggleichrichter; Vollweggleichrichter	
double-way thyristor	MICROEL	Doppelwegthyristor	
= full-wave thyristor		= Vollwegthyristor	
double whip (→double helix)	TECH	(→Doppelwendel)	
double winding (→bifilar winding)	COMPON	(→Bifilarwicklung)	
double-window fiber	OPT.COMM	Zweifensterfaser	
= double-window fibre (BRI)			
double-window fibre (BRI) (→double-window fiber)	OPT.COMM	(→Zweifensterfaser)	
double Wort	DATA PROC	Doppelwort	
double Yagi-Uda antenna	ANT	Doppel-Yagi-Uda-Antenne	
double zepp antenna	ANT	Doppelzepp	
double-zero key	TERM&PER	Doppelnulltaste	
doubling (→duplication)	MATH	(→Verdoppelung)	
doubling temperature	MICROEL	Verdoppelungstemperatur	
doubly-linked list	DATA PROC	Nachbarverkettungsliste	
dovetailed	TECH	schwalbenschwanzförmig	
dovetail (v.t.)	MECH	verschwalben	
≈ indent 1		≈ verzahnen 2	
dovetail (v.t.) (→indent)	COLLOQ	(→verzahnen)	
DOV modem	DATA COMM	DOV-Modem	
= data-above-voice modem			

dowel

dowel	TECH	Dübel	
dowel pin	MECH	Zylinderstift	
dowels for speaker cabinets	EL.ACOUS	Rahmendübel	
downcoming radiowave	RADIO PROP	(→Raumwelle)	
(→sky wave)			
downcoming wave (→sky wave)	RADIO PROP	(→Raumwelle)	
down conversion	HF	Abwärtsmischung	
= down mixing			
down-conversion mixer	HF	Abwärtsmischer	
downconverted	HF	abwärtsgemischt	
down converter	RADIO	Empfangsumsetzer	
down-count (n.)	TECH	Rückwärtszählung	
= regressive count			
down counter	CIRC.ENG	Rückwärtszähler	
= reverse counter; regressive counter; decrementer		= Abwärtszähler; Regressivzähler	
down-going (→inclined)	TECH	(→geneigt)	
down guy (→guy rope)	OUTS.PLANT	(→Abspannseil)	
down-lead (→drop cable)	ANT	(→Zuführungskabel)	
down-line loading	DATA COMM	Fernladen	
= downloading			
down-line processor	DATA COMM	Leitungsendprozessor	
down-link	SAT.COMM	Abwärtsstrecke	
= space-to-earth link		= Abwärtsrichtung	
download (v.t.)	DATA PROC	herunterladen	
[to transfer data to a smaller computer]		[Daten auf kleineren Rechner transferieren]	
≠ upload (v.t.)		≠ hinaufladen	
downloadable	DATA PROC	zuladbar	
[from large to smaller computer]		[von größerem zu kleinerem Rechner]	
≠ uploadable			
downloadable character set	DATA PROC	zuladbarer Zeichensatz	
[from CPU to a printer]		[von der Zentraleinheit zu einem Drucker]	
= downloadable font			
downloadable font	DATA PROC	(→zuladbarer Zeichensatz)	
(→downloadable character set)			
downloading (→down-line loading)	DATA COMM	(→Fernladen)	
down mixing (→down conversion)	HF	(→Abwärtsmischung)	
down payment	ECON	Anzahlung	
≈ prepayment		≈ Vorauszahlung	
downpour (n.)	METEOR	Regenguß	
= heavy shower			
down-scale (adj.)	INSTR	skalenabwärts	
down-scroll	DATA PROC	abwärtsrollen	
[a display content]		[Bilschirminhalt]	
downtime	QUAL	Ausfallzeit	
= down time; DT; idle time; fault time		= Ausfalldauer; Störungszeit; Totzeit	
≠ uptime		≠ Betriebszeit	
downtime	ECON	Stillstandszeit	
down time (→downtime)	QUAL	(→Ausfallzeit)	
downtime costs	ECON	Stillstandskosten	
= stillstand costs; idle costs			
down-time ratio (→non-availability)	QUAL	(→Nichtverfügbarkeit)	
downtown	ECON	Geschäftszentrum	
= city (BRI)			
downward (→retrograde)	TECH	(→rückläufig)	
downward arrow	TYPOGR	Pfeil nach unten	
downward compatibility	DATA PROC	Abwärtskompatibilität	
downward compatible	DATA PROC	abwärtskompatibel	
[compatible with preceding releases]		[mit Vorgängerversion kompatibel]	
		= rückwärtskompatibel	
downward modulation	MODUL	Negativmodulation	
downwards (adv.)	COLLOQ	abwärts (adv.)	
downward stroke	MECH	Abwärtshub	
downward trend	TECH	(→Abwärtstrend)	
dozen	MATH	Dutzend	
DP (→data processing)	INF.TECH	(→Datenverarbeitung)	
d.p.i. (→dots per inch)	TERM&PER	(→Bildpunkte pro Zoll)	
dpi (→dots per inch)	TERM&PER	(→Bildpunkte pro Zoll)	
DPLL	CIRC.ENG	DPLL	
= digital PLL; digitale phase-lock loop		= digitale PLL; digitaler Phasenregelkreis	
DP paper	TERM&PER	DV-Papier	
DPSK (→differential phase shift keying)	MODUL	(→Phasendifferenzmodulation)	
DPU (→display processor)	DATA PROC	(→Bildprozessor)	
DQ (→draft printing quality)	TERM&PER	(→Konzeptdruckqualität)	
DQDB (→distributed queen dual bus)	DATA COMM	(→DQDB)	
drab (adj.)	PHYS	gelblichgrau	
[ligt olive brown]			
draft 2 (n.)	TECH	Luftzug	
= draught 2(n.)(BRI); current of air		= Zug	
draft 1 (n.) (→design)	TECH	(→Entwurf)	
draft (v.t.) (→design 2)	TECH	(→konstruieren)	
draft (n.) (→drawing)	ENG.DRAW	(→Zeichnung)	
draft (→traction)	MECH	(→Zug)	
draft (v.t.) (→design 1)	TECH	(→entwerfen)	
drafting machine (NAM)	ENG.DRAW	Zeichenmaschine	
= draughting machine (BRI)			
draft mode	TERM&PER	Draft-Modus	
[option to frint faster but with degraded quality]		[Option schneller aber mit Qualitätsabstrichen zu drucken]	
draft printing quality	TERM&PER	Konzeptdruckqualität	
= draft quality; DQ		= EDV-Qualität; Entwurfqualität	
≠ letter quality		≠ Schönschrift	
draft quality (→draft printing quality)	TERM&PER	(→Konzeptdruckqualität)	
draft recommendation	TELEC	Empfehlungsentwurf	
draftsman	TECH	Konstrukteur	
= design engineer 2; designer 2; construction engineer		↓ Teilkonstrukteur	
↓ detailer			
draft text	LING	Textentwurf	
drag (v.t.)	DATA PROC	ausziehen	
[to move an image by moving a key-pressed mouse]		[eine Graphik mittels Maussteuerung bewegen]	
		= auseinanderziehen	
drag antenna	ANT	Schleppantenne	
= trailing antenna			
dragging	TECH	Mitführung	
dragging	DATA PROC	Ziehen	
[computer graphics]		[Computergraphik]	
drag-torque tachometer	INSTR	Wirbelstromtachometer	
drain (v.t.)	MICROEL	absaugen	
		= ausrοumen	
drain 3 (n.)	MICROEL	Senke	
[FET]		[FET]	
≠ source (n.)		= Drain 3; Abfluß; D-Pol	
		≈ Drainanschluß; Drainzone	
		≈ Quelle	
drain	INF	Senke	
= sink; destination		= Sinke	
≠ source		≠ Quelle	
		↓ Informationssenke; Nachrichtensenke	
drain 1 (→drain terminal)	MICROEL	(→Drainanschluß)	
drain 2 (→drain region)	MICROEL	(→Drainzone)	
drain (→load current)	EL.TECH	(→Laststrom)	
drainage	CIV.ENG	Entwässerung	
drainage coil	ELECTRON	Erdungsdrossel	
drain circuit	CIRC.ENG	Absaugschaltung	
drain current	MICROEL	Absaugstrom	
		= Ausräumstrom; Drainstrom	
drain current (→load current)	EL.TECH	(→Laststrom)	
drain electrode (→drain terminal)	MICROEL	(→Drainanschluß)	
draining transformer (→booster transformer)	EL.TECH	(→Saugtransformator)	
drain region	MICROEL	Drainzone	
= drain 2		= Drain 2	
drain-source breakdown voltage	MICROEL	Drain-Source-Durchbruchspannung	
drain-source voltage	MICROEL	Drain-Source-Spannung	
drain terminal	MICROEL	Drainanschluß	
[electrode for drawn current]		[Elektrode für gesteuerten Strom]	
= drain electrode; drain 1		= Drainelektrode; Drain 1	
drain valve	TECH	Ablaßventil	
= bleeder valve			

drain voltage	MICROEL	**Absaugspannung**		**drawing pin** (→thumb tack)	TECH	(→Reißzwecke)
		= Drainspannung		**drawing room**	ENG.DRAW	**Zeichensaal**
DRAM	MICROEL	**DRAM**		**drawing set**	ENG.DRAW	**Reißzeug**
= dynamic RAM; dynamic random access memory		= dynamisches RAM; dynamischer Direkzugriffsspeicher		= drawing instrument		
				drawing table	ENG.DRAW	**Zeichentisch**
≠ SRAM		≠ SRAM		≈ droawing board		≈ Zeichenbrett
↑ memory chip; random-access memory		↑ Speicherbaustein; Direktzugriffsspeicher		**drawing template**	ENG.DRAW	**Zeichenschablone**
				≈ ruler		= Schablone
						≈ Lineal; Kurvenlineal
draught 1 (n.) (BRI)	TECH	(→Entwurf)		**drawing triangle**	ENG.DRAW	(→Zeichendreieck)
(→design)				(→straightedge)		
draught 2(n.)(BRI)	TECH	(→Luftzug)		**draw tongs**	MECH	**Kniehebelklemme**
(→draft 2)				**draw to scale**	ENG.DRAW	**maßstäblich zeichnen**
draught (n.)(BRI)	ENG.DRAW	(→Zeichnung)		**DRCS**	DATA PROC	**DRCS-Zeichen**
(→drawing)				= dynamically redefinable character set		= frei definierbares Zeichen
draught (BRI) (→traction)	MECH	(→Zug)		**D-region** (→D-layer)	RADIO PROP	(→D-Schicht)
draughting machine (BRI)	ENG.DRAW	(→Zeichenmaschine)		**drier** (→dessicator)	TECH	(→Entfeuchter)
(→drafting machine)				**drift** (v.i.)	PHYS	**driften**
draughtsman (→drawer)	ENG.DRAW	(→technischer Zeichner)		**drift** (n.)	TECH	**Drift**
draw (v.t.)	ENG.DRAW	**zeichnen**		= runaway		= Trift; Abwanderung; Weglaufen
≈ sketch (v.t.); draft (v.t.)		≈ skizzieren; entwerfen		≈ slip (n.); deviation		
draw (v.t.)	METAL	**ziehen**				≈ Schlupf; Abweichung
↓ depth draw; iron (v.t.)		↓ tiefziehen; fließziehen		**drift** (n.)	PHYS	**Drift**
draw (→pull-in)	OUTS.PLANT	(→einziehen)		**drift** (v.i.) (→run away)	TECH	(→weglaufen)
draw (→pull)	COLLOQ	(→ziehen)		**drift** (n.) (→run-away)	TECH	(→Weglaufen)
DRAW (→WORM)	TERM&PER	(→WORM)		**drift characteristic** (→ageing rate)	QUAL	(→Alterungsrate)
drawback	ECON	**Rückzoll**				
= customs drawback		= Drawback; Zollzurückerstattung		**drift-compensated**	ELECTRON	**driftkompensiert**
				= drift-corrected		= driftkorrigiert
drawback (→disadvantage)	COLLOQ	(→Nachteil)		**drift compensation**	ELECTRON	**Driftkompensation**
draw cable (→winch cable)	OUTS.PLANT	(→Windenseil)		**drift-corrected** (→drift-compensated)	ELECTRON	(→driftkompensiert)
drawer (n.)	TECH	**Schublade**		**drift current**	PHYS	**Feldstrom**
[sliding box]		= Schubfach; Schubkasten		[caused by an electric field]		[durch ein elektrisches Feld erzeugt]
drawer	ENG.DRAW	**technischer Zeichner**		↑ charge carrier current		
= draughtsman; tracer		= Zeichner				= Driftstrom
drawer (→slide-in unit)	EQUIP.ENG	(→Einschub)				↑ Ladungsträgerstrom
drawer parts cabinet (→drawer storage cabinet)	TECH	(→Schubladenmagazin)		**drift distance**	PHYS	**Driftstrecke**
				drift factor	MICROEL	**Driftfaktor**
drawer storage cabinet	TECH	**Schubladenmagazin**		**drift field**	MICROEL	**Driftfeld**
[with many drawers]		[Kasten mit vielen Schubladen]		**drift field**	PHYS	**Driftfeld**
= drawer parts cabinet; storage cabinet; parts cabinet		= Kleinteilemagazin; Sortiment-Magazin; Magazin; Kleinteilebehälter		**drifting time** (→drift time)	PHYS	(→Driftzeit)
				drifting velocity (→drift velocity)	PHYS	(→Driftgeschwindigkeit)
≈ storage box		≈ Sortimentbox		**driftless**	ELECTRON	**driftlos**
drawing	ENG.DRAW	**Zeichnung**		**drift space**	MICROW	**Triftraum**
= draft (n.); draught (n.)(BRI)				[klystron]		[Klystron]
drawing	METAL	**Ziehen**				= Laufraum
↓ depth drawing; ironing		↓ Tiefziehen; Fließziehen		**drift speed** (→drift velocity)	PHYS	(→Driftgeschwindigkeit)
drawing (→traction)	MECH	(→Zug)		**drift stabilization**	CIRC.ENG	**Driftstabilisierung**
drawing archive	ENG.DRAW	**Zeichnungsmappe**		**drift time**	PHYS	**Driftzeit**
drawing board	ENG.DRAW	**Zeichenbrett**		= drifting time		
		= Reißbrett		**drift transistor**	MICROEL	**Drifttransistor**
drawing copier	OFFICE	**Zeichnungskopierer**		= graded-base transistor		
drawing die	METAL	**Ziehwerkzeug**		**drift tube**	MICROW	**Triftröhre**
drawing digitizing equipment	TERM&PER	**Zeichnungs-Digitalisierungsgerät**		= linear beam tube		↑ Laufzeitröhre
				↑ velocity-modulated tube		↓ Klystron
drawing element	DATA PROC	**Zeichenelement**		↓ klystron		
[most basic graphic element]		[kleinste graphische Darstellungseinheit]		**drift tunnel**	MICROEL	**Laufraumelektrode**
= graphics primitive; primitive 2; graphics element; primitive element		= graphisches Zeichenelement; graphisches Grundsymbol; Darstellungselement; Grundzeichenelement		**drift velocity**	PHYS	**Driftgeschwindigkeit**
				= drifting velocity; drift speed		= Wanderungsgeschwindigkeit
				drift voltage	PHYS	**Driftspannung**
				drill (v.t.)	MECH	**bohren**
				≈ bore (v.t.)		≈ aufbohren
				drill 1 (n.)	MECH	**Bohren**
drawing filing cabinet	ENG.DRAW	**Zeichnungsablage**		**drill 2** (n.)	MECH	**Bohrer 1**
drawing film	ENG.DRAW	**Zeichenfolie**		**drill chuck**	MECH	**Bohrkopf**
drawing ink	ENG.DRAW	**Tusche**				= Bohrfutter
= Chinese ink; Indian ink		≈ Tinte		**drilled**	MECH	**gebohrt**
↑ ink				= bored		= aufgebohrt
drawing instrument	ENG.DRAW	(→Reißzeug)		**drilled board**	ELECTRON	**Lochrasterplatte**
(→drawing set)				**drill gauge** (→drill jig)	MECH	(→Bohrlehre)
drawing lamp	ENG.DRAW	**Zeichenbrettleuchte**		**drill hole**	MECH	**Bohrloch**
drawing number	ENG.DRAW	**Zeichnungsnummer**		= borehole; boring; bore 1 (n.); hole		= Bohrung; Bohrzylinder
drawing office	ENG.DRAW	**Zeichenbüro**		**drilling gauge** (→drill jig)	MECH	(→Bohrlehre)
drawing paper	ENG.DRAW	**Zeichenpapier**		**drilling information**	MANUF	**Bohrdaten**
drawing pen	ENG.DRAW	**Reißfeder**		**drilling jig** (→drill jig)	MECH	(→Bohrlehre)
= ruling pen		= Ziehfeder; Zeichenfeder		**drill jig**	MECH	**Bohrlehre**
drawing pencil	ENG.DRAW	**Reißstift**		= drill gauge; drilling jig; drilling gauge		= Bohrschablone; Bohrvorrichtung

drill-jig bushing	MECH	**Bohrbuchse**
drill-out (v.t.)	MECH	**ausbohren**
= bore-out (v.t.)		
drill press	MECH	**Bohrmaschine**
= electric drill		= Bohrer 2
drill through	MECH	**durchbohren**
drip-proof	QUAL	**tropfwassergeschützt**
drive 1 (n.)	ELECTRON	**Ansteuerung**
= control; triggering		
drive (n.)	TECH	**Antrieb**
= transmission		
drive 2 (n.)	ELECTRON	**Aussteuerung**
≈ overload (n.)		≈ Überlastung
drive (→input signal)	ELECTRON	(→Eingangssignal)
drive (n.)	TERM&PER	**Laufwerk**
		= Antrieb
↓ disk drive; magnetic tape drive		↓ Plattenlaufwerk; Magnetbandantrieb
drive capability	MICROEL	**Treiberstärke**
drive circuit (→control circuit 1)	CIRC.ENG	(→Steuerkreis)
drive cord	INSTR	**Skalenseil**
drive designation	DATA PROC	**Laufwerksbezeichnung**
drive electronics	ELECTRON	**Treiberelektronik**
drive hole	TERM&PER	**Antriebsloch**
[of a diskette]		[einer Diskette]
= drive spindle hole		= Mittelloch
drive-in hole	MECH	**Mitnehmerloch**
drive letter	DATA PROC	**Laufwerkskennung**
drive lock	TERM&PER	**Laufwerksverriegelung**
= door lock		
drive logic (→control logic)	ELECTRON	(→Ansteuerlogik)
driven	ELECTRON	**fremdgesteuert**
driven element (→primary radiator)	ANT	(→Primärstrahler)
driven multivibrador (→one-shot multivibrator)	CIRC.ENG	(→stabile Kippschaltung)
driven transmitter	RADIO	**fremdgesteuerter Sender**
drive pulse	ELECTRON	**Ansteuerimpuls**
= driving pulse; control pulse		= Treiberimpuls
≈ trigger pulse; gate control pulse		≈ Triggerimpuls; Torimpuls
↑ control pulse		↑ Steuerimpuls
drive pulse (→control pulse)	ELECTRON	(→Steuerimpuls)
driver	TECH	**Schlüssel 2**
= wrench		
↑ tool		↑ Werkzeug
driver	CIRC.ENG	**Treiber**
[preamplifier driving final power amplifier stages]		[Vorverstärker zur Ansteuerung von Leistungsendstufen]
= driving stage; power driver; buffer		= Treiberstufe; Treiberverstärker; Leistungstreiber
↑ preamplifier		↑ Vorverstärker
driver	MICROEL	**Treiber**
[circuitry and software necessary to establish the sufficient signal levels when interfacing with microprocessors]		[Hardware und Software zur Herstellung der erforderlichen Signalleistung bei Anschluß von Mikroprozessoren]
driver (→driver software)	DATA PROC	(→Treiber)
drive range (→dynamic range)	ELECTRON	(→Dynamikbereich)
driver software	DATA PROC	**Treiber**
[program to control peripherals]		[Programm zur Anpassung von Hardware, z.B. Peripheriegeräte, oder von Software an eine bestimmte Anlage]
= driver; handler		
↓ hardware driver; software driver; printer driver; graphic driver		= Treibersoftware; Steuerungsprogramm; Hantierer; Handler
		↓ Hardwaretreiber; Softwaretreiber; Druckertreiber; Graphiktreiber
driver transistor	CIRC.ENG	**Treibertransistor**
= driving transistor		
driver tube	ELECTRON	**Treiberröhre**
= driver valve		
driver valve (→driver tube)	ELECTRON	(→Treiberröhre)
drive shaft	MECH	**Antriebswelle**
= driving shaft		
drive signal (→driving signal)	CIRC.ENG	(→Treibersignal)
drive spindle hole (→drive hole)	TERM&PER	(→Antriebsloch)
drive system engineering	POWER ENG	**Antriebstechnik**
drive transformer (→trigger transformer)	CIRC.ENG	(→Ansteuerübertrager)
drive voltage (→control voltage)	ELECTRON	(→Steuerspannung)
drive winding	COMPON	**Steuerwicklung**
= control winding		= Treiberwicklung
driving belt	MECH	**Treibriemen**
↓ V belt		= Antriebriemen
		↓ Keilriemen
driving circuit (→control circuit 1)	CIRC.ENG	(→Steuerkreis)
driving disk	MECH	**Mitnehmerscheibe**
driving key	MECH	**Mitnehmerkeil**
driving knob	TECH	**Antriebsknopf**
driving pawl	COMPON	**Stoßklinke**
driving pin	MECH	**Antriebsstift**
		= Mitnehmerstift
driving point impedance function	NETW.TH	**Zweipolfunktion**
↑ network function		↑ Netzwerkfunktion
driving power (→control power)	ELECTRON	(→Steuerleistung)
driving pulse (→drive pulse)	ELECTRON	(→Ansteuerimpuls)
driving shaft (→drive shaft)	MECH	(→Antriebswelle)
driving signal	CIRC.ENG	**Treibersignal**
= drive signal		
driving spindle	TERM&PER	**Bandantriebsspindel**
		= Antriebsspindel
driving stage (→driver)	CIRC.ENG	(→Treiber)
driving string	MECH	**Antriebsseil**
driving torque	MECH	**Antriebsmoment**
		= Antriebsdrehmoment
driving transistor (→driver transistor)	CIRC.ENG	(→Treibertransistor)
driving wheel (→pulley 3)	TECH	(→Antriebsscheibe)
drizzle	METEOR	**Nieselregen**
DRO (→destructive reading)	DATA PROC	(→löschendes Lesen)
droop (→ramp-off)	ELECTRON	(→Dachschräge)
droop (n.) (→proportional offset)	CONTROL	(→Proportionalabweichung)
DRO oscillator	MICROW	**DRO-Oszillator**
		= dielektrischer Oszillator
drop (n.)	TRANS	**Abzweig**
= derivation		
drop (n.)	TELEPH	**Klappe**
[indicator in a manual exchange]		
drop (n.) (→decay)	ELECTRON	(→Abfall)
drop (n.) (→release)	COMPON	(→Abfall)
drop (n.) (→descent)	TELEGR	(→Abfall)
drop (v.t.) (→branch-off)	TRANS	(→abzweigen)
drop (→drop cable)	OUTS.PLANT	(→Hauseinführungskabel)
drop (→decrease)	TECH	(→Verminderung)
drop (→voltage drop)	EL.TECH	(→Spannungsabfall)
drop-and-insert (→drop/insert)	TRANS	(→Abzweig und Wiederbelegung)
drop cable	OUTS.PLANT	**Hauseinführungskabel**
= drop; subscriber's drop		
drop cable	ANT	**Zuführungskabel**
= down-lead		= Zuleitung; Niederführung
drop channel operation	TRANS	**Abzweigbetrieb**
drop current	COMPON	**Abfallstrom**
[relay]		[Relais]
= release current		≈ Zündstrom
≈ trigger current		
drop dead halt (→abnormal end)	DATA PROC	(→Absturz)
drop dead halt (→dead halt)	DATA PROC	(→Blockierungsunterbrechung)
drop-down menu (→pull-down menu)	DATA PROC	(→Pull-down-Menü)

drop electrode		ELECTRON	Tropfenelektrode	drum printer		TERM&PER	Walzendrucker
drop exitation		COMPON	Abfallerregung	= barrel printer			= Typenwalzendrucker; Trommeldrucker
[relay]			[Relais]	↑ impact printer; type printer; line printer			↑ Anschlagdrucker; Typendrucker; Zeilendrucker
= drop power							
drop forging		METAL	Gesenkschmieden	drum scanner		TERM&PER	Trommelscanner
drop-in (n.) (→interfering signal)		ELECTRON	(→Störsignal)	drum sorting		DATA PROC	Magnettrommel-unterstützte Sortierung
drop in (→interfering signal)		TELEC	(→Störsignal)	drum storage (→ magnetic drum storage)		TERM&PER	(→Magnettrommelspeicher)
drop-in circulator		MICROW	Drop-in-Zirkulator	drum-type transceiver (→drum apparatus)		TELEGR	(→Trommelgerät)
drop-in pin		MECH	Rastbolzen				
drop/insert		TRANS	Abzweig und Wiederbelegung	dry (v.t.)		TECH	trocknen
= drop-and-insert; add-drop				= dissecate			
drop/insert capability		TRANS	Abzweigmöglichkeit	dry (adj.)		TECH	trocken
drop/insert multiplexer		TRANS	Abzweigmultiplexer	dry battery		POWER SYS	Trockenbatterie
= add/drop multiplexer; ADM							[Batterie von Trockenelementen]
drop/insert repeater		TRANS	Abzweig-Regenerator				
drop-out (v.i.)		COMPON	abfallen	dry cell		POWER SYS	Trockenelement
[relay]			[Relais]	[not rechargeable]			[nicht wiederaufladbar]
drop out (n.)		DATA COMM	Signalausfall	↑ Leclanché cell			↑ Leclanché-Element
			= Aussetzfehler	dry contact		COMPON	ungefritteter Kontakt
drop-out		TERM&PER	Signalausfall	dry contact (→ cold soldering point)		ELECTRON	(→kalte Lötstelle)
↑ magnetic tape error			↑ Magnetbandfehler				
dropout (→loss)		TELEC	(→Ausfall)	dryer (→ dessicator)		TECH	(→Entfeuchter)
drop-out (→ brief interruption)		TELEC	(→Kurzzeitunterbrechung)	dry etching		MICROEL	Trockenätzung
drop-out (→ line interruption)		TELEC	(→Leitungsunterbrechung)	dry-galvanize		METAL	trockenverzinken
				= sherardize			
dropout color (AM)		TERM&PER	Blindfarbe	drying (n.)		TECH	Trocknung
[detectable by human eye, but not by scanners, e.g. blue]			[vom menschlichen Auge, jedoch nicht von Beleglesern erkennbare Druckfarbe, z.B. Blau]	≈ dissecation			≈ Entfeuchtung
				drying cabinet		TECH	Trockenschrank
= dropout colour (BRI)				dry joint (→ cold soldering point)		ELECTRON	(→kalte Lötstelle)
dropout colour (BRI)		TERM&PER	(→Blindfarbe)				
(→dropout color)				dry plasma etching		MICROEL	Trockenplasmaätzung
drop-out value		COMPON	Abfallschwelle	dry rectifier		COMPON	Trockengleichrichter
[relay]			[Relais]	= metal rectifier; barrier-layer rectifier			= Sperrschicht-Gleichrichter
dropped channel		TELEC	Abzweigkanal				
			= abgezweigter Kanal	dry-reed contact		COMPON	Schutzgaskontakt
drop pen		ENG.DRAW	Nullenzirkel	↑ sealed contact			= trockener Zungenkontakt; Dry-Reed-Kontakt
= bow compass			↑ Zirkel				↑ Schutzrohrkontakt
dropping delay (→ release delay)		COMPON	(→Abfallverzögerung)	dry-reed relay		COMPON	Schutzgaskontakt-Relais
				↑ reed relay			= Dry-Reed-Relais
dropping resistor		EL.TECH	Vorschaltwiderstand				↑ Schutzrohrkontakt-Relais
= drop resistor			= Vorwiderstand	dry run (→ desk checking)		DATA PROC	(→Schreibtischtest)
≈ series resistance			≈ Reihenwiderstand				
drop power (→drop exitation)		COMPON	(→Abfallerregung)	dry running (→desk checking)		DATA PROC	(→Schreibtischtest)
drop ratio (→peak-to-valley ratio)		MICROEL	(→Höcker-Tal-Verhältnis)	dry toner		TERM&PER	Trockentoner
				= powder toner			= Pulvertoner
drop resistor (→ dropping resistor)		EL.TECH	(→Vorschaltwiderstand)	dry-type transformer		POWER ENG	Trockentransformator
				dry-up		TECH	austrocknen
drop sleeve		OUTS.PLANT	Abzweigmuffe	DS (→ digital signal)		TELEC	(→Digitalsignal)
= branching closure; branch-off sleeve				DS (→document storage)		DATA COMM	(→Dokumentenspeicher)
drop test		COMM.CABLE	Abtropfprüfung	DS1 (→basic rate)		TRANS	(→Basisrate)
drop test		QUAL	Fallprüfung	DSB system (→double-sideband system)		RADIO	(→Zweiseitenband-System)
drop time (→release time)		COMPON	(→Abfallzeit)				
drop-type switchboard		TELEPH	Klappenschrank	DS disc (→ DS disk)		TERM&PER	(→DS-Diskette)
drug-store terminal (AM)		TERM&PER	Apothekenterminal	DS disk		TERM&PER	DS-Diskette
= pharmacy terminal				["flippy" is sometimes intended as a DS disk used in a single-side drive]			= zweiseitig beschreibbare Diskette; zweiseitige Diskette
drum		TECH	Trommel				
[cylindrical object serving as container, or to wind something on]			[zylindrischer Gegenstand zur Aufnahme von etwas, oder zum Aufwickeln von etwas]	= double-sided disk; DS disc; double-sided disc; DS diskette; double-sided diskette; DS floppy disk; DS floppy disc;double-sided floppy disk; double-sided floppy disc; flippy disk; flippy disc; flippy diskette; flippy; two-sided disk; two-sided disc; two-sided diskette; two-sided floppy disk; two-sided floppy disc; 2-sided disk; 2-sided disc; 2-sided diskette; 2-sided floppy disk; 2-sided floppy disc			
≈ roll 2; roller							
			≈ Rolle 2; Walze				
drum		TERM&PER	Trommel				
drum (→cable drum)		COMM.CABLE	(→Kabeltrommel)				
drum apparatus		TELEGR	Trommelgerät				
= drum-type transceiver			= Trommeltransceiver				
↑ facsimile equipment			↑ Faksimilegerät				
drum armature		POWER ENG	Trommelanker				
= cylindrical armature				DS diskette (→ DS disk)		TERM&PER	(→DS-Diskette)
drum carrying trailer		OUTS.PLANT	Kabeltrommelanhänger	DSE (→data switching exchange)		DATA COMM	(→Datenvermittlungsstelle)
drum driving device		OUTS.PLANT	Abrollantrieb				
drum factor		TELEGR	Trommelfaktor	DS floppy disk (→ DS disk)		TERM&PER	(→DS-Diskette)
[facsimile; drum length to drum diameter]			[Faksimile; Trommellänge zu Trommeldurchmesser]	DS1 group		TELEC	DS1-Bündel
				[1.544 Mbit/s]			[1,544 Mbit/s]
drum plotter		TERM&PER	Trommelplotter	= T1 group			
			= Walzenplotter				

DS2 group

DS2 group [6.312 Mbit/s] = T2 group	TELEC	**DS2-Bündel** [6,312 Mbit/s]
DS3 group [44.736 Mbit/s] = T3 group	TELEC	**DS3-Bündel** [44,736 Mbit/s]
DS1 level [1.544 Mbit/s] = T1 level	TELEC	**DS1-Ebene** [1,544 Mbit/s]
DS2 level [6.312 Mbit/s] = T2 level	TELEC	**DS2-Ebene** [6,312 Mbit/s]
DS3 level [44.736 Mbit/s] = T3 level	TELEC	**DS3-Ebene** [44,736 Mbit/s]
DSR = device status report	DATA COMM	**Gerätestatusmeldung** [Code] = DSR
DSR [Data Set Ready; to receive]	DATA PROC	**Geräteempfangsbereitschaft**
DSR (→data set ready)	DATA PROC	(→Empfangsbereitschaft)
DSS (→decision support system)	DATA PROC	(→Entscheidungsvorbereitungssystem)
DSW [Device Status Word]	DATA PROC	**Gerätezustandswort**
DSW (→device status word)	DATA PROC	(→Gerätestatuswort)
DT (→downtime)	QUAL	(→Ausfallzeit)
DTE (→data terminal equipment)	DATA COMM	(→Datenendeinrichtung)
DTL = diode-transistor logic	MICROEL	**DTL** = Dioden-Transistor-Logik
DTLZ = diode-transistor logic with zener diodes	MICROEL	**DTLZ** = Dioden-Transistor-Logik mit Z-Dioden
DTMF (→pushbutton dialing)	SWITCH	(→Tastwahl)
D-to-A converter (→digital-to-analog converter)	CODING	(→D/A-Wandler)
DTP (→desktop publishing)	DATA PROC	(→Desktop Publishing)
DTR [Data Terminal Ready; to send]	DATA PROC	**Gerätesendebereitschaft** = Datenendgerät-Bereitschaftssignal
DTR (→non-availability)	QUAL	(→Nichtverfügbarkeit)
d-type flipflop (→delay element)	CIRC.ENG	(→Verzögerungsglied)
dual (→binary)	MATH	(→binär)
dual (→twofold)	TECH	(→zweifach)
dual amplitude modulation	MODUL	**Doppelamplitudenmodulation**
dual antenna filter	ANT	**Zweifach-Antennenweiche**
dual arithmetics (→binary arithmetics)	MATH	(→Binärarithmetik)
dual-band antenna (→two-band antenna)	ANT	(→Zweibandantenne)
dual-band short dipole = duoband short dipole	ANT	**Zweiband-Kurzdipol**
dual-beam oscilloscope [displays on the screen two measured magnitudes simultaneously] = dual-trace oscilloscope; dual-channel oscilloscope ↑ oscilloscope	INSTR	**Zweistrahloszilloskop** [stellt am Bildschirm zwei Meßgrößen gleichzeitig dar] = Zweistrahloszillograf; Zweikanaloszilloskop ↑ Oszilloskop
dual card (→dual punch card)	TERM&PER	(→Verbundlochkarte)
dual-carriage [printer]	TERM&PER	**zweibahnig** [Drucker]
dual-carriage way (→expressway)	CIV.ENG	(→Schnellstraße)
dual carrier system	TELEC	**Zweiträgersystem**
dual cassette deck = dual deck	CONS.EL	**Doppel-Cassettendeck**
dual channel board	TRANS	**Zwei-Kanal-Karte**
dual-channel cathode ray direction finder (→Watson-Watt direction finder)	RADIO LOC	(→Watson-Watt-Sichtfunkpeiler)
dual-channel controller [to read write simultaneously]	DATA PROC	**Zweikanalsteuerung** [zum simultanen Schreiben und Lesen]
dual-channel oscilloscope (→dual-beam oscilloscope)	INSTR	(→Zweistrahloszilloskop)
dual-channel pulse generator	INSTR	**Zweikanal-Pulsgenerator**
dual-channel signal analyzer	INSTR	**Zweikanal-Signalanalysator**
dual-channel spectrum analyzer	INSTR	**Zweikanal-Spektrumanalysator**
dual circuit	NETW.TH	**duale Schaltung**
dual code [place values ranged by powers of two] ↑ binary code	CODING	**Dualcode** [Stellenwerigkeiten nach Zweierpotenzen geordnet] = Bitcode ↑ Binärcode
dual computer system [real-time and batch computer accessing a common external memory] = twin computer system ↑ multi-computer system	DATA PROC	**Doppelsystem** [Realzeit- und Stapelsystem an einer gemeinsamen Datenbank] ↑ Mehrrechnersystem
dual contact (→twin contact)	COMPON	(→Doppelkontakt)
dual-contact spring	COMPON	**Doppelkontaktfeder**
dual-control	ELECTRON	**doppelgesteuert**
dual conversion receiver (→dual conversion superhet)	RADIO	(→Doppelsuperhet)
dual conversion superhet [with two IF] = dual-conversion superhet receiver; dual conversion receiver; dual converter ↑ heterodyne receiver	RADIO	**Doppelsuperhet** [mit zwei ZF] = Doppelsuper ↑ Überlagerungsempfänger
dual-conversion superhet receiver (→dual conversion superhet)	RADIO	(→Doppelsuperhet)
dual converter (→dual conversion superhet)	RADIO	(→Doppelsuperhet)
dual current	EL.TECH	**dualer Strom**
dual deck (→dual cassette deck)	CONS.EL	(→Doppel-Cassettendeck)
dual directional amplifier	INSTR	**Zweikanalverstärker**
dual directional coupler	INSTR	**Zweifach-Richtkoppler**
dual disk drive	TERM&PER	**Doppellaufwerk**
dual impedance	NETW.TH	**duale Impedanz** = dualer Widerstand
dual in line = DIL	COMPON	**Dual-in-line** = DIL; doppelreihig; Doppelreihen-
dual in-line connector (→DIL connector)	COMPON	(→DIL-Stecker)
dual-in-line package = DIP	COMPON	**DIL-Gehäuse** = DIP; Doppelreihenanschluß-Gehäuse
dual-in-line switch = DIL switch	COMPON	**DIL-Schalter** = Dual-in-line-Schalter; Doppelreihenanschluß-Schalter
dual intensity [regular and bold]	TERM&PER	**zweifache Intensität** [normal und fett]
duality	MATH	**Dualität**
duality invariant [factor for dual impedance]	NETW.TH	**Dualitätsinvariante** [Faktor für duale Impedanz]
dually polarized operation (→cochannel operation)	RADIO REL	(→Gleichkanalbetrieb)
dually polarized transmission (→cochannel operation)	RADIO REL	(→Gleichkanalbetrieb)
dual modulation	MODUL	**Doppelmodulation**
dual number (→binary number)		(→Binärzahl)
dual operational amplifier [two operational amplifiers in one case]	MICROEL	**Zweifach-Operationsverstärker** [zwei Operationsverstärker in einem Gehäuse]
dual output = twin output	EL.TECH	**Zweifachausgang** = Doppelausgang
dual-polarization aerial (→dual-polarization antenna)	ANT	(→doppelpolarisierte Antenne)
dual-polarization antenna = dual-polarization aerial	ANT	**doppelpolarisierte Antenne**
dual-pol operation (→cochannel operation)	RADIO REL	(→Gleichkanalbetrieb)
dual-pol transmission (→cochannel operation)	RADIO REL	(→Gleichkanalbetrieb)
dual port	MICROEL	**Dualport**
dual processor	DATA PROC	**Doppelprozessor**

dual punch card TERM&PER **Verbundlochkarte**
= dual card; dual-purpose card   = Verbundkarte
dual-purpose card (→dual TERM&PER (→Verbundlochkarte)
punch card)
dual quadripole (→dual NETW.TH (→dualer Vierpol)
two-port)
dual-sided disk drive TERM&PER **doppelseitiges Laufwerk**
dual-slope converter (→dual-slope INSTR (→Zweirampenumsetzer)
integrator)
dual-slope integrator INSTR **Zweirampenumsetzer**
= dual-slope converter   = Dual-slope-Umsetzer; Auf-ab-Integrator
dual-slope method INSTR **Zweirampenverfahren**
  = Dual-slope-Verfahren; Auf-ab-Verfahren
dual-sound coder TV **Zwei-Ton-Coder**
= double-tone coder
dual system (→duplex DATA PROC (→Duplexsystem)
computer system)
dual time base window INSTR **Doppelzeitbasis-Fenster**
dual-tone multifrequency dialing SWITCH (→Tastwahl)
(→pushbutton dialing)
dual-trace oscilloscope INSTR (→Zweistrahloszilloskop)
(→dual-beam oscilloscope)
dual track (→double TERM&PER (→Doppelspur)
track)
dual-track recorder EL.ACOUS **Zweispur-Tonbandgerät**
  = Doppelspur-Tonbandgerät
dual-transfer switch COMPON **Zweifach-Umschalter**
dual two-port NETW.TH **dualer Vierpol**
= dual quadripole   = widerstandsreziproker Vierpol
dual two-terminal (→dual NETW.TH (→dualer Zweipol)
two-terminal network)
dual two-terminal network NETW.TH **dualer Zweipol**
[impedance of two-terminal 1 is pro-   [die Impedanz des Zweipols 1 ist proportional zur
portional to admittance of two-termi-   Admittanz des Zweipols 2]
nal 2]
= dual two-terminal
dual voltage EL.TECH **duale Spannung**
duant electrometer INSTR **Duanten-Elektrometer**
dub (v.t.) CONS.EL **überspielen**
[to copy onto another tape]   [auf ein anderes Band kopieren]
= copy; re-record   = kopieren
dubb (n.) CONS.EL **Kopie**
= copy
dubbing CONS.EL **Überspielen**
= copying   = Überspielfunktion; Kopieren
dubbing (→sound mixing) EL.ACOUS (→Tonmischung)
dubbing lead CONS.EL **Überspielkabel**
= copying lead
duct RADIO PROP **Dukt**
  = Duct
duct OUTS.PLANT **Röhre**
= pipe
≈ single-duct conduit; cable conduit   ≈ Einrohrkanal; Kabelkanalzug
duct bank (→cable conduit) OUTS.PLANT (→Kabelkanal)
duct cable COMM.CABLE **Röhrenkabel**
= underground cable (AM); in-duct
cable
duct cleaner OUTS.PLANT **Rohrbürste**
duct laying OUTS.PLANT **Röhrenverlegung**
duct nest (→cable conduit) OUTS.PLANT (→Kabelkanal)
duct run (→cable conduit) OUTS.PLANT (→Kabelkanal)
due ECON **fällig**
= matured
due date ECON **Abgabetermin**
[of an offer]   [eines Angebots]
= bid closing date   = Abgabefrist
due date (→due time) ECON (→Fälligkeit)
dues (pl.t. →membership ECON (→Mitgliedsbeitrag)
subscription)
due time ECON **Fälligkeit**
= due date; expiry date; maturity date;   = Fälligkeitstag; Fälligkeitsdatum; Termin; Verfalldatum; Ablauffrist
term; expiration date
≈ appointed time [COLLOQ]; term   ≈ Laufzeit; Frist 2
↓ deadline   ↓ äußerster Termin

due time (→appointed time) COLLOQ (→Termin)
dull 1 (→mat) TECH (→matt)
dull 2 (→turbid) TECH (→trübe)
dull-bright TECH **mattglänzend**
dull emitter (→cold ELECTRON (→Kaltkathode)
cathode)
dull trading (→slow ECON (→schleppendes Geschäft)
business)
dumb (→detuned) ELECTRON (→verstimmt)
dumb aerial (→dumb antenna) ANT (→verstimmte Antenne)
dumb antenna ANT **verstimmte Antenne**
= dumb aerial
dumb terminal TERM&PER **nicht programmierbares Datensichtgerät**
[with no processing capability]
≠ intelligent terminal   = Einfachterminal
dummy (n.) DATA PROC **Platzhalter**
[provisional substitute for a not yet   [Provisorium für noch
worked out program module]   nicht detaillierte Strukturblöcke eines Programms]
dummy (adj.) TECH **Schein-**
  = Blind-; Leer-
dummy (n.) (→mock-up) TECH (→Attrappe)
dummy (→dummy signal) DATA PROC (→Füllsignal)
dummy address DATA PROC **Scheinadresse**
dummy aerial (→dummy antenna) ANT (→künstliche Antenne)
dummy antenna ANT **künstliche Antenne**
= dummy aerial; artificial antenna; arti-   = Kunstantenne; Antennennachbildung; Ersatzantenne; Blindantenne
ficial aerial; mute antenna; mute ae-
rial; phantom antenna; phantom
aerial   ≈ Abschlußwiderstand
≈ dummy load
dummy argument DATA PROC **Blindvariable**
  = Formalparameter
dummy bar TECH **Abdeckschiene**
  ≈ Abdeckleiste
  ↑ Schiene
dummy bit CODING **Leerbit**
= blank bit; null bit
dummy block DATA PROC **Leerblock**
dummy case EQUIP.ENG **Schlafgehäuse**
dummy command (→blank DATA PROC (→Leeranweisung)
instruction)
dummy corporation ECON **Scheingesellschaft**
= sham company
dummy cover MECH **Blindabdeckung**
= dummy plate   ↑ Abdeckung
dummy data DATA PROC **Leerdaten**
dummy element COMM.CABLE **Füllelement**
  = Blindelement; Trensen
dummy fuse COMPON **Blindsicherung**
dummy instruction (→blank DATA PROC (→Leeranweisung)
instruction)
dummy line DATA COMM **Leeranschluß**
dummy load NETW.TH **Blindlast**
≈ reactive load   = Ersatzlast; reaktive Last; Reaktanzlast
dummy load ANT **Abschlußwiderstand**
≈ dissipation resistor; dummy antenna   = Dummyload
  ≈ Schluckwiderstand; künstliche Antenne
dummy module EQUIP.ENG **Blindbaugruppe**
= blank module; dummy unit   = Leerbaugruppe
dummy plate (→dummy cover) MECH (→Blindabdeckung)
dummy section DATA PROC **Pseudoabschnitt**
↑ programm section   ↑ Programmabschnitt
dummy signal DATA PROC **Füllsignal**
= dummy   = Leersignal; Blindsignal
dummy sleeve OUTS.PLANT **Blindmuffe**
dummy statement (→blank DATA PROC (→Leeranweisung)
instruction)
dummy strip TECH **Abdeckleiste**
= cover strip   = Abdeckstreifen
  ≈ Abdeckschiene
  ↑ Leiste
dummy unit (→dummy EQUIP.ENG (→Blindbaugruppe)
module)
dummy variable DATA PROC **Scheinvariable**
  = Platzhaltevariable
dump (v.t.) (→output) DATA PROC (→ausgeben)
dump (n →printout) DATA PROC (→Ausdruck)

dumping

dumping	CIRC.ENG	**Bedämpfung**	
↑ protective circuit		↑ Schutzschaltung	
dun	ECON	**Mahnung**	
= reminder			
dunnage	ECON	**Füllmaterial**	
≈ packing material		= Zwischenpackmaterial	
		≈ Verpackungsmaterial	
duoband antenna (→two-band antenna)	ANT	(→Zweibandantenne)	
duo-bander (→two-band antenna)	ANT	(→Zweibandantenne)	
duoband short dipole (→dual-band short dipole)	ANT	(→Zweiband-Kurzdipol)	
duo-cone loudspeaker	EL.ACOUS	**Doppelkonus-Lautsprecher**	
duodecimal	MATH	**duodezimal**	
[with the basis of 12]		[mit der Basis 12]	
duodecimal system	MATH	**Duodezimalsystem**	
		= duodezimales System	
duolateral coil (→honeycomb coil)	COMPON	(→Honigwabenspule)	
duotricenary	MATH	**duotrizinär**	
duplet	PHYS	**Dublett**	
duplex (→duplex operation)	TELEC	(→Duplexbetrieb)	
duplex (→twofold)	TECH	(→zweifach)	
duplex computer (→duplex computer system)	DATA PROC	(→Duplexsystem)	
duplex computer system	DATA PROC	**Duplexsystem**	
[a second computer in hot-standby, executes batch processes in idle periods]		[ein zweiter Rechner als Heißersatz, löst nebenbei Stapelaufgaben]	
= duplex computer; dual system		↑ Mehrrechnersystem	
↑ multi-computer system			
duplexer 2	RADIO	**Antennenumschalter**	
= antenna change-over switch; TR switch		= Sende-Empfangs-Schalter; Duplexer; Antennenschalter; Sende-Empfangs-Umschalter; TR-Switch	
↓ T/R tube		↓ T/R-Röhre	
duplexer	TV	**TV/Ton-Weiche**	
		= Tonweiche	
duplexer 3 (→antenna duplexer)	RADIO	(→Sendeantennenweiche)	
duplexer 1 (→duplexer filter)	RADIO	(→Sende-Empfangs-Weiche)	
duplexer filter	RADIO	**Sende-Empfangs-Weiche**	
[filter to connect transmitters with receivers to a single antenna]		[Filter zum Anschluß von Sendern mit Empfängern an eine Antenne]	
= duplexer 1; directional filter; branching filter		= Richtungsweiche	
↑ diplexer filter		↑ Trennweiche	
duplexing	DATA PROC	**Hardware-Redundanz**	
		= Hardware-Ersatz	
duplex intercommunication system	TELEPH	**Gegensprechanlage**	
= talk-back; duplex intercom system		↑ Hausrufanlage	
↑ intercommunication system			
duplex intercom system (→duplex intercommunication system)	TELEPH	(→Gegensprechanlage)	
duplex mode (→duplex operation)	TELEC	(→Duplexbetrieb)	
duplex operation	TELEC	**Duplexbetrieb**	
[simultaneous transmission and reception]		[gleichzeitiges Senden und Empfangen]	
= duplex mode; duplex transmission; duplex; full duplex operation; full duplex mode; full duplex; fd; FD; fdx; FDX; bidirectional operaration; bidirectional mode		= Duplex; Gegenbetrieb; gleichzeiter Sende- und Empfangebetrieb; Zweiwegbetrieb; Zweiwegübertragung; zweiseitiger Betrieb; zweiseitige Übertragung	
≈ half duplex; semiduplex		≈ Halbduplexbetrieb	
↓ diplex operation; duplex telephony [TELEPH]; duplex telegraphy [TELEGR]		↓ Diplexbetrieb; Gegensprechbetrieb; Gegenschereibbetrieb	
duplex operation (→double operation)	TECH	(→Doppelbetrieb)	
duplex printing	TERM&PER	**beidseitiger Druck**	
duplex telegraphy	TELEGR	**Gegenschreibbetrieb**	
= full-duplex telegraphy		= Gegenschreibverkehr; Gegenschreiben	
↑ duplex operation [TELEC]		↑ Duplexbetrieb [TELEC]	
duplex telephony	TELEPH	**Gegensprechbetrieb**	
= double talking; double talk; full-duplex speech		= Gegensprechverkehr; Gegensprechen	
↑ duplex operation [TELEC]		↑ Duplexbetrieb [TELEC]	
duplex tool	TECH	**Zweifachwerkzeug**	
duplex transmission (→duplex operation)	TELEC	(→Duplexbetrieb)	
duplicate (v.t.)	DATA PROC	**duplizieren**	
[to copy data from a data carrier onto a similar one]		[Daten von einem Datenträger auf einen anderen gleichartigen überspielen]	
≈ copy (v.T.)		= doppeln	
		≈ kopieren	
duplicate (v.t.)	MATH	**verdoppeln**	
= double (v.t.)		↑ vervielfachen	
↑ multiplicate			
duplicate (n.)	DATA PROC	**Duplikat**	
duplicate (n.)	OFFICE	**Duplikat**	
		= Abschrift; Zweitschrift	
		≈ Pause; Faksimile	
duplicate (n.)	TECH	**Zweitausfertigung**	
		= Duplikat	
duplicated	EQUIP.ENG	**gedoppelt**	
= redundant		= redundant	
duplicating equipment (→copying machine)	OFFICE	(→Kopiergerät)	
duplicating programm	DATA PROC	**Duplizierprogramm**	
duplication	MATH	**Verdoppelung**	
= doubling		= Doppelung; Duplizierung	
↑ multiplication		↑ Vervielfachung	
duplication check	DATA PROC	**Doppelprüfung**	
[same result by separate runs]		[gleiches Ergebnis bei zwei Läufen]	
duplicator (→copying machine)	OFFICE	(→Kopiergerät)	
durability	TECH	**Dauerhaftigkeit**	
= endurance		= Haltbarkeit; Dauerfestigkeit; Beständigkeit	
≈ resistance; ruggedness		≈ Widerstandsfähigkeit; Robustheit	
durable	TECH	**dauerhaft**	
		= haltbar	
durable consumer good	ECON	**Gebrauchsgut**	
duration	PHYS	**Dauer**	
		= Zeitdauer; Länge 2	
duration charge	TELEC	**Zeitgebühr**	
= duration fee			
duration fee (→duration charge)	TELEC	(→Zeitgebühr)	
duration interference	RADIO	**Dauerstörung**	
duration of acceleration	PHYS	**Beschleunigungsdauer**	
duration of binary information	TELECONTR	**Meldungsdauer**	
duration of interruption (→off time)	TECH	(→Unterbrechungsdauer)	
duress alarm system (→emergency call system)	SIGN.ENG	(→Notrufanlage)	
dust (n.)	TECH	**Staub**	
dust (BRI) (→garbage)	COLLOQ	(→Müll)	
dust accumulation	TECH	**Verstaubung**	
		= Staubansammlung	
dust bin (BRI) (→garbage can)	COLLOQ	(→Mülleimer)	
dust cap (→dust cover)	TECH	(→Staubabdeckung)	
dust core	COMPON	**Pulverkern**	
= powder core; powdered iron core		= Massekern; Preßkern; Hochfrequenzeisenkern	
dust-core coil	COMPON	**Preßkernspule**	
= powder-core coil		= Pulverkernspule; Massekernspule	
dust cover	TECH	**Staubabdeckung**	
= dust cap		= Staubschutz; Staubkappe; Schutzumschlag	
≈ protective cover		≈ Abdeckhaube	
↑ cover		↑ Abdeckung	
dust-free	TECH	**staubfrei**	
dusting brush	TERM&PER	**Staubpinsel**	
dust jacket (→book jacket)	TYPOGR	(→Buchumschlag)	
dust lock	MANUF	**Staubschleuse**	
dust-proof	TECH	**staubdicht**	
= enclosed			

dust storm	METEOR	Staubsturm	
dust trap	TECH	Staubfänger	
DUT (→test specimen)	QUAL	(→Prüfling)	
duty (→load)	TECH	(→Belastung)	
duty (n.) (→operation 1)	TECH	(→Betrieb)	
duty (→charge 2)	ECON	(→Gebühr)	
duty (→tax)	ECON	(→Steuer)	
duty computer	CONTROL	Einsatzrechner	
duty cycle (→pulse duty factor)	ELECTRON	(→Tastgrad)	
duty factor (→pulse duty factor)	ELECTRON	(→Tastgrad)	
duty-free	ECON	zollfrei	
duty payment (→custom payment)	ECON	(→Verzollung)	
duty rate	TECH	nominelle Leistungsfähigkeit	
= rated performance		≈ Nennleistung	
duty-rated (→scheduled)	TECH	(→nominell)	
Dvorak keyboard	TERM&PER	Dvorak-Tastatur	
dwell (n.) (→dwell time)	TERM&PER	(→Verweilzeit)	
dwelling	CIV.ENG	Wohngebäude	
= residencial building		= Wohnhaus	
dwell phase (→rest condition)	TELEGR	(→Ruhezustand)	
dwell time [printer]	TERM&PER	Verweilzeit [Drucker]	
= dwell (n.)			
dwell time (→residence time)	DATA PROC	(→Verweilzeit)	
dwell time (→output pulse width)	CIRC.ENG	(→Verweilzeit)	
dwindling (→contraction)	TECH	(→Schrumpfung)	
D wire (→tip2 wire)	TELEPH	(→D-Ader)	
dx [amateur radiocommunications]	RADIO	dx [Funkamateurjorgon]	
= distant		= entfernt	
DX antenna	ANT	DX-Antenne	
DXer [short wave listener specialized in picking up fall-of radio stations]	RADIO	DX-er [auf Überreichweitenempfang spezialisierter Funkamateur]	
Dy (→dysprosium)	CHEM	(→Dysprosium)	
dyadic (→binary)	MATH	(→binär)	
dyadic logarithm (→logarithm to the base of 2)	MATH	(→dyadischer Logarithmus)	
dyadic operation (→binary operation)	ENG.LOG	(→binäre Funktion)	
dye	PHYS	Farbstoff	
= color pigment			
dye laser	OPTOEL	Farbstofflaser	
dyn (→dyne)	PHYS	(→Dyn)	
dynamic (adj.)	PHYS	dynamisch	
≠ static		≠ statisch	
dynamic address translation	DATA PROC	dynamische Adreßumsetzung	
= DAT		= DAM	
dynamic allocation [of reasources]	DATA PROC	dynamische Zuteilung [von Betriebsmitteln]	
dynamically redefinable character set (→DRCS)	DATA PROC	(→DRCS-Zeichen)	
dynamic behaviour	CONTROL	Dynamik	
= dynamic response		= dynamisches Verhalten	
dynamic buffer (→dynamic buffer store)	DATA PROC	(→dynamischer Pufferspeicher)	
dynamic buffer store	DATA PROC	dynamischer Pufferspeicher	
= dynamic buffer; elastic buffer store; elastic buffer		= dynamischer Puffer; elastischer Pufferspeicher; elastischer Puffer	
dynamic calibration	INSTR	dynamisches Einmessen	
		= dynamische Kalibrierung	
dynamic characteristics (→switching characteristics)	ELECTRON	(→Schaltverhalten)	
dynamic check (→operating test)	DATA PROC	(→Betriebsprüfung)	
dynamic coding (→dynamic programming)	DATA PROC	(→dynamische Programmierung)	
dynamic dissipation power	MICROEL	dynamische Verlustleistung	
dynamic dump	DATA PROC	dynamischer Speicherabzug	
dynamic effect	PHYS	Kraftwirkung	
dynamic equation	PHYS	Bewegungsgleichung	
dynamic error	INSTR	dynamischer Fehler	
dynamic headphone	EL.ACOUS	dynamischer Kopfhörer	
dynamicizer (→parallel-to-serial converter)	CIRC.ENG	(→Parallel-Seriell-Umsetzer)	
dynamic loudspeaker (→electrodynamic loudspeaker)	EL.ACOUS	(→elektrodynamischer Lautsprecher)	
dynamic memory [a very volatile memory, must be recharged or refreshed at frequent intervals]	DATA PROC	dynamischer Speicher [ein äußerst flüchtiger Speicher, muß in kurzen Abständen nachgeladen oder aufgefrischt werden]	
= dynamic storage; dynamic store			
↓ dynamic solid state memory		↓ dynamischer Halbleiterspeicher	
dynamic memory allocation (→dynamic storage allocation)	DATA PROC	(→dynamische Speicherzuteilung)	
dynamic microphone	EL.ACOUS	elektrodynamisches Mikrophon	
↓ moving-coil microphone; ribbon microphone		= dynamisches Mikrophon	
		↓ Tauchspulenmikrophon; Bändchenmikrophon	
dynamic multivibrator (→edge triggered flip-flop)	CIRC.ENG	(→dynamische Kippschaltung)	
dynamic mutual conductance	ELECTRON	Arbeitssteilheit	
dynamic noise immunity	ELECTRON	dynamische Störsicherheit	
dynamic parameter [generatd by a program and reused]	DATA PROC	dynamischer Parameter [von Programm erzeugt und weiterverwendet]	
dynamic programming	DATA PROC	dynamische Programmierung	
= dynamic coding			
dynamic program relocation	DATA PROC	dynamische Programmverschiebung	
= dynamic relocation; automatic program relocation; automatic relocation		= automatische Programmverschiebung; dynamische Programmversetzung; automatische Programmversetzung	
dynamic RAM (→DRAM)	MICROEL	(→DRAM)	
dynamic random access memory (→DRAM)	MICROEL	(→DRAM)	
dynamic range	ELECTRON	Dynamikbereich	
= drive range; control range; cutoff bias		= dynamischer Bereich; Aussteuerungsbereich; Aussteuerbereich	
≈ dynamics; overload point; regulation range		≈ Dynamik; Aussteuerungsgrenze; Regelbereich	
dynamic range (→volume range)	EL.ACOUS	(→Dynamikbereich)	
dynamic relocation (→dynamic program relocation)	DATA PROC	(→dynamische Programmverschiebung)	
dynamic resistance (→differential resistance)	MICROEL	(→differentieller Widerstand)	
dynamic response (→dynamic behaviour)	CONTROL	(→Dynamik)	
dynamics	MECH	Dynamik	
≠ statics		≠ Statik	
dynamic scheduling	DATA PROC	dynamische Disposition	
dynamics compression (→volume compression)	TRANS	(→Dynamikkompression)	
dynamic semiconductor memory (→dynamic solid state memory)	MICROEL	(→dynamischer Halbleiterspeicher)	
dynamic signal analyzer	INSTR	Dynamik-Signalanalysator	
dynamic solid state memory [a very volatile solid state memory, must be recharged or refreshed at short intervals]	MICROEL	dynamischer Halbleiterspeicher [äußerst flüchtiger Halbleiterspeicher, muß in kurzen Abständen nachgeladen oder aufgefrischt werden]	
= dynamic semiconductor memory		≈ flüchtiger Halbleiterspeicher	
≈ volatile solid state memory		↑ dynamischer Speicher	
↑ dynamic memory			
dynamic soubroutine [with case-by-case function]	DATA PROC	dynamisches Unterprogramm [mit fallweise definierter Aufgabe]	
dynamic storage (→dynamic memory)	DATA PROC	(→dynamischer Speicher)	
dynamic storage allocation	DATA PROC	dynamische Speicherzuteilung	
= dynamic memory allocation		= dynamische Speicherzuweisung	
dynamic store (→dynamic memory)	DATA PROC	(→dynamischer Speicher)	

dynamic test (→operating test)		DATA PROC	(→Betriebsprüfung)	earth (v.t.) (AM) = ground (v.t.)(BRI)		EL.TECH	erden
dynamic viscosity [SI unit: Pascal-second]		PHYS	dynamische Viskosität [SI-Einheit: Pascal-Sekunde]	earth (n.)		TECH	Erdreich = Erde
dynamo		POWER ENG	Lichtmaschine = Dynamo ↑ Gleichstromgenerator	earth (BRI) (→ground) earth (→chassis ground) earth bulge earth cable		EL.TECH ELECTRON RADIO PROP COMM.CABLE	(→Erde) (→Masse) Erdüberhöhung Erdkabel
dynamometer		INSTR	Dynamometer = elektrodynamisches Meßinstrument; Elektrodynamometer	= buried cable; underground cable (BRI); direct-buried cable ↑ below-ground cable			↑ unterirdisches Kabel
dynamometer dynamo sheet = electrical sheet		OUTS.PLANT METAL	Zugkraftgeber Dynamoblech	earth capacitance (BRI) (→capacitance to ground) earth conductivity		EL.TECH PHYS	(→Erdkapazität) Erdleitfähigkeit
dynamotor		POWER SYS	Gleichstrom-Gleichstrom-Einankerumformer = Dynamotor	earth conductor (→earth lead) earth connection = ground connection; earthing; grounding		EL.TECH EL.TECH	(→Erdungsleiter) Erdung = Erdverbindung; Masseverbindung
dynaquad dynatron dynatron oscillator dyne [unit for weight; = 1 g cm²/s] = dyn		MICROEL ELECTRON CIRC.ENG PHYS	Dynaquad Dynatron Dynatronoszillator Dyn [Einheit für Gewicht; = 1 g cm/s²] = dyn	earth connection pin = earth pin earth constants earth control center earth coverage (→global coverage)		EQUIP.ENG RADIO PROP ASTROPHYS SAT.COMM	Erdstift Bodenkonstante Bodenkontrollzentrum (→globale Bedeckung)
dynistor (→dynistor diode) dynistor diode = dynistor		MICROEL MICROEL	(→Dynistordiode) Dynistordiode = Dynistor	earth current earth current losses = ground losses earthed		EL.TECH ANT EL.TECH	Erdstrom Erdverluste geerdet
dynode		ELECTRON	Dynode = Prallelektrode; Prallanode; Sekundäremissionsanode	= grounded ≠ unearthed earthed neutral (→neutral conductor)		POWER ENG	≠ ungeerdet (→Nulleiter)
dysprosium = Dy dz (→metric quintal)		CHEM PHYS	Dysprosium = Dy (→Doppelzentner)	earth electrode = ground system ↓ depth earth electrode; ground rod; ground ribbon; lightning ground system; RF grounding system		EL.TECH	Erder ↓ Tiefenerder; Sternerder; Erdungsstab; Banderder; Blitzschutzerder; Hochfrequenzerder
E				earthenware earth exploration satellite service		TECH SAT.COMM	Steingut Erkundungsfunkdienst über Satelliten
E²CL (→EECL) E²PROM (→EEPROM) EA (→erase in area) EACH sign (→AT sign) EACH symbol (→AT sign) EAN code [European Article Number code] ↑ bar code		MICROEL MICROEL DATA COMM ECON ECON DATA PROC	(→EECL) (→EEPROM) (→Bereich löschen) (→Klammeraffe) (→Klammeraffe) EAN-Code [Europäischer Artikel-Nummer-Code] ↑ Balkemcode	earth fault (BRI) (→ground fault) earth-free (→isolated from earth) earth graph paper (→earth profile chart) earth green earth inductor earthing (→earth connection)		EL.TECH EL.TECH RADIO REL PHYS PHYS EL.TECH	(→Erdschluß) (→erdfrei) (→Schnittrahmen) grünerde Erdinduktor (→Erdung)
EAPROM [programmable with 12 V, erasable with UV light] = electrically alterable read-only memory; EAROM ≈ EEPROM ↑ read-only memory		MICROEL	EAPROM [programmierbar mit 12 V, löschbar mit UV-licht] = elektrisch änderbarer Festwertspeicher; EAROM ≈ EEPROM ↑ Festwertspeicher	earthing capacitor earthing clamp earthing resistance = ground resistance earthing switch = grounding switch earthing system		EL.TECH COMPON EL.TECH SYS.INST EL.TECH	Erdungskondensator Erdungsschelle Erdungswiderstand Erdungsschalter Erdungssystem
ear cushion ear inset Early effect		EL.ACOUS EL.ACOUS MICROEL	Ohrpolster Ohrmulde Basisbreitenmodulation = Early-Effekt	earth lead = ground lead; grounding lead; earth wire; ground wire; grounding wire; earth conductor; ground conductor; grounding conductor		EL.TECH	Erdungsleiter = Erdungsleitung; Erdleiter
Early equivalent network early failure = infant mortality		MICROEL QUAL	Early-Ersatzschaltung Frühausfall	earth leakage coil (→ground-fault neutralizer)		POWER SYS	(→Erdlöschspule)
early shift ≠ late shift ↑ shift		ECON	Frühschicht ≠ Spätschicht ↑ Schicht	earth leakage current (→short-circuit current to earth) earth loop		EL.TECH TELEC	(→Erdschlußstrom) Erdschleife
early warning ear microphone earning power (→profitability) earnings 1 ≈ profit; wage; salary; income		RADIO REL EL.ACOUS ECON ECON	Früherkennung Ohrmikrophon (→Wirtschaftlichkeit) Verdienst ≈ Gewinn; Lohn; Gehalt; Einkommen	= ground loop earth magnetism (→geomagnetism) earth pin (→earth connection pin)		PHYS EQUIP.ENG	(→Erdmagnetismus) (→Erdstift)
earnings 3 (→income) earnings 2 (→profit) EAROM (→EAPROM) earphone (→headphone) earphone socket (→headphone terminal)		ECON ECON MICROEL EL.ACOUS EQUIP.ENG	(→Einkommen) (→Gewinn) (→EAPROM) (→Kopfhörer) (→Kopfhöreranschluß)	earth potential = ground potential; zero potential ≈ neutral potential earth profile chart = earth graph paper; profile chart		EL.TECH RADIO REL	Erdpotential = Massepotential = Nullpotential Schnittrahmen = Geländeschnittkarte
ear piece = receiver cap ear response characteristic		TELEPH EL.ACOUS	Hörmuschel Ohrkurve	earthquake earth radius		GEOPHYS GEOPHYS	Erdbeben Erdradius

earth resistance meter	INSTR	**Erdungsmesser**
↑ ohmmeter		= Erdungswiderstandsmesser; Erdungsprüfer
		↑ Ohmmeter
earth return (→earth return circuit)	TELEC	(→Erdrückleitung)
earth return circuit	TELEC	**Erdrückleitung**
= ground return circuit; earth return; ground return		
earth return current	EL.TECH	**Erdrückstrom**
= ground return current		
earth rod (→ground rod)	EL.TECH	(→Erdungsstab)
earth segment	SAT.COMM	**Erdsegment**
earth's surface	GEOPHYS	**Erdoberfläche**
earth station	SAT.COMM	**Erdfunkstelle**
		= Erdfunkstelle; Bodenstation; Satellitenbodenstation
earth station	RADIO	**Landfunkstelle**
		= Bodenstation
		↓ Erdfunkstelle [SAT.COMM]
earth station antenna	ANT	**terrestrische Satellitenfunkantenne**
earth strip	ELECTRON	**Erdungsstreifen**
earth terminal	COMPON	**Erdklemme**
= ground terminal; grounding spanner		= Masseklemme
earth-to-space link (→up-link)	SAT.COMM	(→Aufwärtsstrecke)
earth wire (→earth lead)	EL.TECH	(→Erdungsleiter)
easiness	COLLOQ	**Leichtigkeit**
[low grade of difficulty]		[geringe Schwierigkeit]
easy	COLLOQ	**leicht**
≈ simple		[fig.]
		≈ einfach
easy control (→operator convenience)	TECH	(→Bedienungskomfort)
easy movable	TECH	**leichtbeweglich**
easy-to-maintain (→maintainable)	TECH	(→wartungsfreundlich)
easy-to-manufacture (adj.)	TECH	**fertigungsfreundlich**
= easy-to-produce		= fertigungsgünstig; fertigungsgerecht
easy-to-operate (adj.)	TECH	**betriebsfreundlich**
≈ simple-to-operate		= betriebsgerecht
easy-to-produce (→easy-to-manufacture)	TECH	(→fertigungsfreundlich)
easy-to-service (→maintainable)	TECH	(→wartungsfreundlich)
easy-to-use (adj.)	TECH	**gebrauchsfreundlich**
eaves (n.pl.t.)	CIV.ENG	**Traufe**
		= Dachtraufe
eavesdrop (v.i.)	TELEPH	**abhören**
[top listen secretly]		≈ mithören
≈ tap; bug		
≈ monitor		
eavesdropping device	SIGN.ENG	**Abhörvorrichtung**
= bugging device; surreptitious listening device		
EB (→erroned bit)	INF	(→fehlerbehaftetes Bit)
EBCD code	CODING	**EBCDI-Code**
[Extended Binary Coded Decimal Interchange code]		= EBCDI-Code
Ebers-Moll model	MICROEL	**Ebers-Moll-Modell**
EBR (→electron beam recording)	TERM&PER	(→Elektronenstrahlaufzeichnung)
eca silicon [MICROEL] (→germanium)	CHEM	(→Germanium)
eccentric	MATH	**exzentrisch**
eccentricity	MATH	**Exzentrizität**
[separation from cnter]		[Abstand vom Mittelpunkt]
ECCSL (→emitter-coupled logic)	MICROEL	(→emittergekoppelte Logik)
echelon antenna	ANT	**Echelon-Antenne**
echelon grating	PHYS	**Stufengitter**
echo	TELEC	**Echo** (n.)
		= Rückfluß
echo	ACOUS	**Echo**
[a sound reflection perceived with noticeable delay]		[mit deutlichem Verzug wahrgenommene Schallreflexion]
≈ reverberation		
↑ sound reflection		= Wiederhall
		≈ Nachhall
		↑ Schallreflexion
echo attenuation (→active return loss)	NETW.TH	(→Reflexionsdämpfung)
echo attenuation coefficient (→reflection coefficient)	NETW.TH	(→Reflexionsfaktor)
echo attenuation measuring set (→echo meter)	INSTR	(→Echometer)
echo canceller	TELEPH	**Echokompensator**
[operates with compensating signal]		[speist kompensierendes Gegensignal ein]
= echo compensator		≈ Echosperre
≈ echo suppressor		
echo check	DATA COMM	**Echoprüfung**
[by transmitting back]		[durch Rücksendung]
echo compensator (→echo canceller)	TELEPH	(→Echokompensator)
echo control equipment	TELEC	**Echounterdrückungsgerät**
↓ echo suppressor; echo canceller		↓ Echosperre; Echokompensator
echo disturbance	TELEC	**Echostörung**
echo loss (→active return loss)	NETW.TH	(→Reflexionsdämpfung)
echo meter	INSTR	**Echometer**
= echo attenuation measuring set		
echo phase	NETW.TH	**Echophase**
[imaginary part of return loss coefficient]		[Imaginärteil des Echodämpfungsmaßes]
		= Echowinkel
echoplexing	DATA COMM	**Spiegelung**
= echoplex mode		= Zurücksenden; Echoplex; Echo
echoplex mode (→echoplexing)	DATA COMM	(→Spiegelung)
echo principle	TELECONTR	**Echobetrieb**
= transmission with information feedback		
echo query (→reedback)	INSTR	(→Rückmeldung)
echosounder	RADIO NAV	**Echolotanlage**
echo sounding	RADIO NAV	**Echolot**
= sonic depth finder		= Echolotung
echo suppressor	TELEPH	**Echosperre**
[interrupts the echo path]		[unterbricht den Echopfad]
≈ echo canceller		= Rückflußsperre
		≈ Echokompensator
		↑ Echounterdrückungsgerät
echo transmission time	TELEC	**Echolaufzeit**
ECL	MIL.COMM	**ECL**
= electronic countermeasures		= elektronische Gegenmaßnahmen
ECL (→emitter-coupled logic)	MICROEL	(→emittergekoppelte Logik)
ECL device	MICROEL	**ECL-Baustein**
= ECL module		
ECL gate	MICROEL	**ECL-Gatter**
= ECL logic element		= ECL-Verknüpfungsglied
ECL logic element (→ECL gate)	MICROEL	(→ECL-Gatter)
ECL module (→ECL device)	MICROEL	(→ECL-Baustein)
ECMA data cassette	TERM&PER	**ECMA-Kassette**
economic (adj.)	ECON	**wirtschaftlich 1**
[related to economics]		[die Wirtschaft betreffend]
= fiscal (adj.)		
economic (→cost-effective 1)	ECON	(→wirtschaftlich 2)
economical analyst (→economist)	ECON	(→Wirtschaftler)
economical expert (→economist)	ECON	(→Wirtschaftler)
economic branch (→economic sector)	ECON	(→Wirtschaftszweig)
economic data	ECON	**Wirtschaftszahlen**
economic plan (→budget)	ECON	(→Etat)
economic planning (→budgeting)	ECON	(→Wirtschaftsplan)
economics	ECON	**Wirtschaftswissenschaften**
= economic sciences		= Betriebswirtschaftslehre
↓ business administration economics; political economics		↓ Volkswirtschaftslehre
economic sciences (→economics)	ECON	(→Wirtschaftswissenschaften)

English	Domain	German
economic sector	ECON	Wirtschaftszweig
= economic branch		
economist	ECON	Wirtschaftler
= economical analyst; economical expert		= Wirtschaftsfachmann; Wirtschaftsexperte
economize (→save)	ECON	(→sparen)
economy 1	ECON	Einsparung
[avoidance of costs]		
= saving		
economy 2	ECON	Wirtschaft
≈ business world		= Ökonomie
		≈ Geschäftswelt
economy circuit	CIRC.ENG	Sparschaltung
ECTL (→emitter-coupled logic)	MICROEL	(→emittergekoppelte Logik)
EDD (→removable-disk drive)	TERM&PER	(→Wechselplattenlaufwerk)
eddy (→curl)	PHYS	(→Wirbel)
eddy current	EL.TECH	Wirbelstrom
= Foucault current		= Foucaultscher Strom
eddy-current brake	PHYS	Wirbelstrombremse
eddy current damping	EL.TECH	Wirbelstromdämpfung
eddy current loss	EL.TECH	Wirbelstromverlust
edge (v.t.)	MECH	abkanten 3
		[mit Kante versehen]
edge (n.)	ELECTRON	Flanke
= slope; ramp; skirt		= Rampe 1
		↓ Impulsflanke
edge (n.)	MECH	Kante
edge (n.)	TYPOGR	Schnitt
[of the body of a book]		[Schnittfläche des Buchblocks]
edge (n.)	MATH	Kante
[theory of graphs; connecting line of nodes]		[Graphentheorie; Verbindungslinie zwischen Knoten]
≠ node		≠ Knoten
edge (→cutting edge)	METAL	(→Schneide)
edge board (→edge card)	ELECTRON	(→direktkontaktierte Leiterplatte)
edge card	ELECTRON	direktkontaktierte Leiterplatte
[PCB with contact strips]		
= edge board		= direktkontaktierte Platine
edge connector	COMPON	Randstecker
		= Winkelsteckleiste; Winkelstecker 2
edge controlled (→edge triggered)	ELECTRON	(→flankengesteuert)
edge cutter	TERM&PER	Führungsstreifen-Abtrenner
= edge trimmer		≈ Schlagschere
≈ burster		
edge finder	INSTR	Flankensuchfunktion
edge joint	MECH	Stirnstoß
edge-modulated pulse train	ELECTRON	flankenmodulierter Puls
edge notch	TERM&PER	Randkerbung
= margin notch; margin perforation		= Randlochung; Randperforation
edge-notched card	TERM&PER	(→Randlochkarte)
(→marginal punched card)		
edge-punched card	TERM&PER	(→Randlochkarte)
(→marginal punched card)		
edge-shaped obstruction	RADIO PROP	scharfkantiges Hindernis
edge sharpening	DATA PROC	Kantenkontrastierung
edge steepness	NETW.TH	Flankensteilheit
= rate of change		
edge triggered	ELECTRON	flankengesteuert
= edge controlled		= flankenmoduliert
edge triggered flip-flop	CIRC.ENG	dynamische Kippschaltung
= dynamic multivibrator		= flankengesteuerte Kippschaltung; dynamisches Flipflop
edge triggering	CIRC.ENG	Flankensteuerung
		[Digitalsteuerung]
edge trimmer (→edge cutter)	TERM&PER	(→Führungsstreifen-Abtrenner)
edge-wise	MECH	hochkant
= on end		
EDI (→electronic data interchange)	DATA COMM	(→elektronischer Datenaustausch)
E diagram (→E-plane pattern)	ANT	(→E-Diagramm)
edifice (→building)	CIV.ENG	(→Gebäude)
Edison cell	POWER SYS	Nickelakkumulator
edit	DATA PROC	editieren
[to check or modify data or texts]		[Daten oder Texte kontrollieren oder ändern]
		= aufbereiten
edit command	DATA PROC	Editierbefehl
edit controller	DATA COMM	Edit-Controller
editing	DATA PROC	Editieren
= revision editing		= Editing; Editierung; Aufbereitung
↓ printout editing		↓ Druckaufbereitung
editing (→revision editing)	DATA COMM	(→Redigieren)
editing capability	DATA PROC	Editierfähigkeit
editing copier	OFFICE	Editierkopierer
editing function	OFFICE	Editierfunktion
editing keyboard	TERM&PER	Editiertastatur
[videotex]		[Btx]
= information provider keyboard		
editing line	DATA PROC	Editierzeile
editing mark (→proofreaders' mark)	TYPOGR	(→Korrekturzeichen)
editing programm (→editor)	DATA PROC	(→Editor)
editing run	DATA PROC	Editierlauf
editing session	DATA PROC	Editiersitzung
editing term	DATA PROC	Editierterm
editing terminal	TELEC	Editierstation
[videotex]		[Btx]
= information provider terminal		
edition number (→issue number)	DOC	(→Ausgabestand)
edition printing	TYPOGR	Auflage 1
= press run		[Anzahl der gedruckten Exemplare]
edit key	DATA PROC	Editiertaste
edit mode	DATA PROC	Editionsmodus
editor	DATA PROC	Editor
[program to enter and modify data files interactively]		[Programm zur interaktiven Eingabe und Änderung von Dateien; editor = engl. „Herausgeber, Redakteur"]
= editing programm; editor program; edit programm; file editor		
↓ text editor		= Editierprogramm; Editor-Programm; Dateiaufbereiter; Dateiersteller
		↓ Text-Editor
editor program (→editor)	DATA PROC	(→Editor)
edit programm (→editor)	DATA PROC	(→Editor)
edit window	DATA PROC	Editierfenster
EDP (→electronic data processing)	INF.TECH	(→elektronische Datenverarbeitung)
EDPM (→data processing equipment)	DATA PROC	(→Datenverarbeitungsanlage)
EDS (→moving-disk memory)	TERM&PER	(→Wechselplattenspeicher)
educational television	BROADC	Bildungsfernsehen
= educational TV; instructional television; instructional TV		
educational TV (→educational television)	BROADC	(→Bildungsfernsehen)
EECL	MICROEL	EECL
= E²CL; emitter-emitter-coupled logic		= E²CL
EEL (→emitter-coupled logic)	MICROEL	(→emittergekoppelte Logik)
EEPROM	MICROEL	EEPROM
[erasable with higher voltage]		[mit erhöhter Spannung löschbar]
= E²PROM; electrically erasable programmable read-only memory; electrically erasable PROM		= E²PROM; elektrisch löschbarer programmierbarer Festwertspeicher
≈ EPROM; EAPROM		≈ EPROM; EAPROM
↑ EEROM		↑ EEROM
EEROM	MICROEL	EEROM
[Electrically Erasable Read-Only Memory]		= elektrisch löschbarer Festwertspeicher
= electrically erasable ROM		↓ EEPROM
↓ EEPROM		
effect (n.)	PHYS	Wirkung
≈ reaction		≈ Reaktion

effective TECH
= efficacious; effectual; operative
effective (→real) COLLOQ
effective address DATA PROC
(→absolute adress)
effective amplification NETW.TH
effective antenna height ANT
= effective height
≈ effective antenna length

effective antenna length ANT
≈ effective antenna height
effective antenna noise factor HF
(→effective antenna noise figure)
effective antenna noise figure HF
= effective antenna noise factor

effective aperture ANT
[of a receiving antenna]
= effective area; absorption area; capture area

effective area (→effective aperture) ANT
effective attenuation EL.TECH
effective attenuation constant NETW.TH
[attenuation constant under operational matching conditions; negative value of effective transfer constant]
= composite loss; composite attenuation; overall loss; working attenuation
≈ image attenuation constant
↑ attenuation constant; complex effective attenuation constant

effective attenuation factor NETW.TH
[attenuation factor under operational matching conditions at input and output; reciprocal of effective transmission coefficient]
= effective attenuation ratio; effective damping factor; effective damping coefficient; composite attenuation factor
≠ effective transmission coefficient
↑ attenuation factor

effective attenuation ratio NETW.TH
(→effective attenuation factor)
effective call TELEPH
= completed call
effective capacity ELECTRON
[electron tube]
effective chain parameter equation NETW.TH
effective chain parameter matrix NETW.TH
effective charge PHYS
effective cross-section PHYS
effective current (→active EL.TECH current)
effective current transmission coefficient NETW.TH
[current transmission coefficient under operational matching conditions]
= composite current transmission coefficient
↑ effective transmission coefficient; current transmission coefficient

effective damping coefficient NETW.TH
(→effective attenuation factor)
effective damping factor NETW.TH
(→effective attenuation factor)
effective distance RADIO PROP
[troposcatter]

effective-earth-radius factor RADIO PROP
(→K-factor)

wirksam
= wirkungsvoll; effektiv
(→tatsächlich)
(→absolute Adresse)

Wirkverstärkung
effektive Antennenhöhe
= effektive Höhe; wirksame Antennenhöhe; wirksame Höhe
≈ effektive Antennenlänge

wirksame Antennenlänge
≈ effektive Antennenhöhe
(→effektive Antennenrauschzahl)

effektive Antennenrauschzahl
= effektiver Antennenrauschfaktor

Antennenwirkfläche
[Apertur einer Empfangsantenne]
= Wirkfläche; Antennenabsorptionsfläche; Absoptionsfläche; effektive Fläche

Antennenwirkfläche
Wirkdämpfung
Betriebsdämpfungsmaß
[Dämpfungsmaß unter Betriebsbedingungen; Negativwert des Betriebsübertragungsmaßes]
= Betriebsdämpfung
≈ Wellendämpfungsmaß
↑ Dämpfungsmaß; komplexes Betriebsdämpfungsmaß

Betriebsdämpfungsfaktor
[Dämpfungsfaktor unter Betriebsbedingungen; Kehrwert des Betriebsübertragungsfaktors]
= Betriebsdämpfungskoeffizient; Betriebsdämpfungsfunktion
≠ Betriebsübertragungsfaktor
↑ Dämpfungsfaktor

(→Betriebsdämpfungsfaktor)

ausgeführtes Gespräch

Systemkapazität
[Elektronenröhre]
Betriebskettengleichung

Betriebskettenmatrix

wahre Ladung
Wirkungsquerschnitt
(→Wirkstrom)

Betriebs-Stromübertragungsfaktor
[Stromübertragungsfaktor unter Betriebsbedingungen]
↑ Betriebsübertragungsfaktor; Stromübertragungsfaktor

(→Betriebsdämpfungsfaktor)

(→Betriebsdämpfungsfaktor)

charakteristische Streuentfernung
[Überhorizontverbindung]
(→Krümmungsfaktor k)

effective echoing area RADIO LOC
(→radar cross section)
effective gain factor (→effective NETW.TH transmission coefficient)
effective height (→effective antenna ANT height)
effective instruction DATA PROC
[after modifications]
effective mass PHYS
[solid state physics]
effective matching loss NETW.TH
(→composite return loss)
effectiveness TECH
= efficaciousness; effectuality; operativity
≈ effectivity
effective parameter (→effective NETW.TH transmission parameter)
effective parameter theory NETW.TH
effective permeability PHYS

effective phase angle NETW.TH
[phase angle under operational matching conditions; negative value of effective phase angle factor]
= composite phase angle
≈ effective phase angle factor
↑ phase angle; complex effective attenuation constant

effective phase angle NETW.TH
factor
[phase angle factor under operational matching conditions; negative value of effective phase angle]
= effective phase constant; composite phase constant
≈ effective phase angle
↑ phase angle factor
effective phase constant NETW.TH
(→effective phase angle factor)
effective power TECH
effective power (→active EL.TECH power)
effective power match (→active EL.TECH power matching)
effective power matching EL.TECH
(→active power matching)
effective power meter (→active INSTR power meter)
effective radiated power ANT
= ERP

effective reflexion coefficient NETW.TH
(→composite return current coefficient)
effective resistance (→active NETW.TH resistance)
effective return current coefficient NETW.TH (→composite return current coefficient)
effective return loss NETW.TH
(→composite return loss)
effective sound pressure ACOUS
= sound pressure
effective transfer constant NETW.TH
[transfer constant under operational matching conditions; negative value of effective attenuation constant]
= composite gain
≈ image transfer constant
↑ transfer constant; complex effective transfer constant

effective transmission NETW.TH
coefficient
[transmission coefficient under operational matching conditions;

(→Radarquerschnitt)
(→Betriebsübertragungsfaktor)
(→effektive Antennenhöhe)

endgültiger Befehl

effektive Masse
[Halbleiterphysik]
(→Betriebsreflexionsdämpfungsmaß)

Wirksamkeit
= Effektivität
≈ Effizienz
(→Betriebsübertragungsparameter)

Betriebsparametertheorie
effektive Permeabilität
= Äquivalenzpermeabilität; gescherte Permeabilität
Betriebsdämpfungswinkel
[Dämpfungswinkel unter Betriebsbedingungen; Negativwert des Betriebsphasenmaßes]
≈ Betriebsphasenmaß
↑ Dämpfungswinkel; komplexes Betriebsdämunpfungsmaß

Betriebsphasenmaß
[Phasenmaß unter Betriebsbedingungen; Negativwert des Betriebsdämpfungswinkels]
= Betriebsübertragungswinkel
≈ Betriebsdämpfungswinkel
↑ Phasenmaß
(→Betriebsphasenmaß)

Nutzleistung
(→Wirkleistung)

(→Wirkleistungsanpassung)

(→Wirkleistungsanpassung)

(→Wirkleistungsmesser)

äquivalente Strahlungsleistung
= ERP
(→Betriebsreflexionsfaktor)

(→Wirkwiderstand)

(→Betriebsreflexionsfaktor)

(→Betriebsreflexionsdämpfungsmaß)
Schalldruck

Betriebsübertragungsmaß
[Übertragungsmaß unter Betriebsbedingungen; Negativwert des Betriebsdämpfungsmaßes]
= Betriebsverstärkungsmaßes
≈ Wellenübertragungsmaß
↑ Übertragungsmaß; komplexes Betriebsübetragungsmaß

Betriebsübertragungsfaktor
[Übertragungsfaktor unter Betriebsbedingungen; Kehrwert des Betriebs-

effective transmission factor

reciprocal of effective attenuation constant]
= effective transmission factor; effective gain factor; composite transmission coefficient; composite attenuation constant; composite gain factor
≠ effective attenuation constant
↑ transmission coefficient

effective transmission factor NETW.TH
(→effective transmission coefficient)
effective transmission NETW.TH
parameter
[response parameter of a quadripole under effective matching conditions at its input and output, existing under operational configuration]
= effective parameter; composite transmission parameter; composite parameter
≈ image parameter
↑ transmission parameter
↓ effective attenuation constant; effective transmission coefficient; complex effective transfer constant; complex effective attenuation constant; effective phase angle factor; effective phase angle

effective value PHYS
[root of time integral of squares, divided by t]
= root mean square; rms value; rms; RMS; heating value
≈ root sum square value [MATH]
effective voltage (→active EL.TECH voltage)
effective voltage transmission NETW.TH
coefficient
[voltage transmission coefficient under operational matching conditions]
= composite voltage transmission coefficient
↑ effective transfer coefficient; voltage transmission coefficient

effector DATA PROC
[peripheral with process control computer]
= actuator
effectual (→effective) TECH
effectuality (→effectiveness) TECH
efficacious (→effective) TECH
efficaciousness TECH
(→effectiveness)
efficiency PHYS
efficient ECON
= streamlined
effluence (→efflux) TECH
efflux TECH
= outflow; effluence
≠ influx
effort ECON
= expense; expenditure; erogation
≈ input
EFL (→emitter follower MICROEL logic)
EFS (→error-free second) TRANS
EGA TERM&PER
[Enhanced Graphics Adapter; graphics standard with 640x350 pixels and 16 out of 64 colors; with driver software also with 800x600 pixels]

dämpfungsfaktors]
= Betriebsverstärkungsfaktor; Betriebsübertragungsfuktion;
Betriebsübersetzungsverhältnis;
Betriebsübersetzung;
Betriebsübertragungsverhältnis
≠ betriebsdämpfungsfaktor
↑ Übertragungsfaktor
(→Betriebsübertragungsfaktor)
Betriebsübertragungsparameter
[Übertragungsparameter eines Vierpols mit seinen betrieblichen, effektiven Impedanzanpassungen an Ein- und Ausgang]
= Betriebsübertragungsgröße; Betriebsparameter
≈ Wellenparameter
↑ Übertragungsparameter
↓ Betriebsdämpfungsfaktor; Betriebsübertragungsfaktor; komplexes Betriebsübertragungsmaß; komplexes Betriebsdämpfungsmaß; Betriebsübertragungsmaß; Betriebsdämpfungsmaß; Betiebsphasenmaß: Betriebsdämpfungswinkel

Effektivwert
[Wurzel des Zeitintegrals der Quadrate, durch t]
≈ quadratisches Mittel [MATH]

(→Wirkspannung)
Betriebs-Spannungsübertragungsfaktor
[Spannungsübertragungsfaktor unter Betriebsverhältnissen]
↑ Betriebsübertragungsfaktor; Spannungsübertragungsfaktor

Effektor
[Peripheriegerät mit Prozeßrechner]
= Aktuator
(→wirksam)
(→Wirksamkeit)
(→wirksam)
(→Wirksamkeit)

Wirkungsgrad
= Nutzeffekt; Ergiebigkeit
rationell

(→Abfluss)
Abfluss
≠ Zufluß

Aufwand
≈ Einsatz

(→Emitterfolgerlogik)

(→fehlerfreie Sekunde)
EGA
[Graphikstandard mit 640x350 Punkten und 16 aus 64 Farben, mit Treibersoftware auch 800x600 Punkte]

EGA board TERM&PER
[Enhanced Graphics Adapter; controls EGA monitors]
↑ graphics board
EGA monitor TERM&PER
[resolution of 640x350 pixels]
↑ color monitor

egghead (AM) (→graduate) SCIE
ego-less programming DATA PROC

EHF (→millimetric waves) RADIO
eigenfrequency (→natural PHYS frequency)
eigenfunction (→characteristic MATH function)
eigenoscillation (→natural PHYS oscillation)
eigensolution MATH
eigenvalue MATH
= characteristic value
eigenvector MATH
= characteristic vector
eight-bit byte (→octet) CODING
eight-element dipole array ANT

eight-inch disk (→eight-inch TERM&PER floppy disk)
eight-inch drive TERM&PER

eight-inch floppy disk TERM&PER
= eight-inch disk; 8-in. floppy disk; 8-in. disk
eight-layer TECH
= eight-part
eight-part (→eight-layer) TECH
eight-track disk TERM&PER
eighty-column screen TERM&PER
einsteinium CHEM
= Es
EIRP (→equivalent isotropic RADIO radiated power)
EITHER-OR function (→OR ENG.LOG operation)
EITHER-OR operation (→OR ENG.LOG operation)
either-way (→two-way alternate) TELEC
either-way mode (→half duplex) TELEC
either-way operation (→half TELEC duplex)
either-way transmission (→half TELEC duplex)
ejaculatory word LING
(→interjection)
eject CONS.EL
= cassette ejector

eject (n.) (→ejector) TERM&PER
ejector TERM&PER
= eject (n.)

elapse (v.i.) PHYS
[time]
= pass (v.i.)
elastic (n.) CHEM
elastic (adj.) PHYS
elastic actuator (→elastic CONTROL control element)
elastic band TECH
elastic banding DATA PROC
= banding
elastic buffer (→dynamic DATA PROC buffer store)
elastic buffer store DATA PROC
(→dynamic buffer store)
elastic control element CONTROL
= elastic actuator

EGA-Karte
[steuert EGA-Monitoren]
= EGA-Adapter
↑ Graphikkarte
EGA-Monitor
[Auflösung von 640x350 Bildpunkten]
↑ Farbmonitor
(→Akademiker)
Kollektivprogrammierung
[mit verteilter Verantwortung]
(→Millimeter-Wellen)
(→Eigenfrequenz)

(→Eigenfunktion)

(→Eigenschwingung)

Eigenlösung
Eigenwert

Eigenvektor

(→Oktett)
Achterfeld
= Achterfeldantenne
(→Normaldiskette)

Normaldiskettenlaufwerk
= Acht-Zoll-Disketten-Laufwerk
Normaldiskette
= 8-Zoll-Diskette; Acht-Zoll-Diskette; Maxidiskette
achtlagig
= achtschichtig
(→achtlagig)
Acht-Spur-Platte
Achzig-Spalten-Bildschirm
Einsteinium
= Es
(→äquivalente isotrop abgestrahlte Leistung)
(→ODER-Verknüpfung)

(→ODER-Verknüpfung)

(→wechselseitig)
(→Halbduplexbetrieb)
(→Halbduplexbetrieb)

(→Halbduplexbetrieb)

(→Interjektion)

Kassettenausschub
= Cassettenausschub; Kassettenauswurf; Cassettenauswurf

(→Ausschubvorrihtung)
Ausschubvorrichtung
= Ausschub; Auswurfvorrichtung; Auswurf
ablaufen
[Zeit]
= verstreichen; vergehen
Elast
elastisch
(→Nachgebeglied)

Gummiband
Bildeinschnürung

(→dynamischer Pufferspeicher)
(→dynamischer Pufferspeicher)

Nachgebeglied
= DT1-Glied

elasticity	PHYS	**Elastizität**
elastic limit	PHYS	**Elastizitätsgrenze**
elastic wave	PHYS	**Elastizitätswelle**
elastomer	CHEM	**Elastomer**
elative form	LING	**Elativ**
[an absolute, rather than comparing, superlative]		[nicht vergleichender, sondern absoluter Superlativ; z.B. zumeist]
E-layer	RADIO PROP	**E-Schicht**
≈ E-region		= Heaviside-Schicht
elbow	MICROW	**Bogen**
= bond		
elbow	MECH	**Krümmer**
		= Winkelstück
elbow jack	COMPON	**Winkelbuchse**
≠ elbow plug		= Winkelkupplung
		≠ Winkelstecker
elbow plug	COMPON	**Winkelstecker 1**
≠ elbow jack		≠ Winkelbuchse
elbowroom (→margin)	COLLOQ	(→Spielraum)
EL display	COMPON	(→Elektrolumineszenzanzeige)
(→electroluminescence display)		
electomagnetic read-only memory	MICROEL	**elektromagnetischer Festwertspeicher**
= EROM		= EROM
electret	PHYS	**Elektret**
[permanently polarizable dielectric]		[permanent polarisierbares Dielektrikum]
electret condenser microphone	EL.ACOUS	**Elektret-Kondensator-Mikrofon**
electric	PHYS	**elektrisch**
= electrical		
electrical (→electric)	PHYS	(→elektrisch)
electrical apparatus	TECH	**Elektrogerät**
= electrical appliance		
electrical appliance (→electrical apparatus)	TECH	(→Elektrogerät)
electrical classification	COMM.CABLE	**Verseilkennzeichen**
electrical communication	INF.TECH	**elektrische Nachrichtenübertragung**
electrical components (→electro-material)	EL.INST	(→Elektromaterial)
electrical conductance (→electric conductivity)	PHYS	(→elektrischer Leitwert)
electrical conduction (→electric conduction)	PHYS	(→Elektrizitätsleitung)
electrical current	PHYS	**elektrischer Strom**
electrical energy	PHYS	**elektrische Energie**
electrical energy density	PHYS	**elektrische Energiedichte**
electrical engineer	TECH	**Elektroingenieur**
electrical engineering	TECH	**Elektrotechnik**
[science and technology of electical phenomena]		[Theorie und Technik elektrischer Vorgänge]
= electrotechnology; electrical technology		↑ Technik
↑ engineering		↓ Elektrizitätslehre; elektrische Energietechnik; Elektronik
↓ electricity 2; electrical power engineering; electronics		
electrical force	PHYS	**elektrische Kraft**
electrical fundamentals (→electricity 2)	PHYS	(→Elektrizitätslehre)
electrical image	PHYS	**Potentialbild**
electrical industry	TECH	**Elektroindustrie**
		= elektrotechnische Industrie
electrical installation	TECH	**Elektroinstallation**
electrical installation engineering	POWER SYS	**Installationstechnik**
electrical isolation (→dc decoupling)	EL.TECH	(→galvanische Trennung)
electrical length	ANT	**elektrische Länge**
electrical line (→electric line)	EL.INST	(→Elektrizitätsleitung)
electrically alterable read-only memory (→EAPROM)	MICROEL	(→EAPROM)
electrically asymmetric (→impedance unbalanced)	NETW.TH	(→widerstandsunsymmetrisch)
electrically erasable programmable read-only memory (→EEPROM)	MICROEL	(→EEPROM)
electrically erasable PROM (→EEPROM)	MICROEL	(→EEPROM)
electrically erasable ROM (→EEROM)	MICROEL	(→EEROM)
electrically insulated (→dc-insulated)	EL.TECH	(→galvanisch getrennt)
electrically long	NETW.TH	**elektrisch lang**
electrically operated switch (→electromagnetical switch)	COMPON	(→elektromagnetischer Schalter)
electrically short	NETW.TH	**elektrisch kurz**
electrically short dipole	ANT	**elektrisch kurzer Dipol**
electrically symmetric (→impedance-balanced)	NETW.TH	(→widerstandssymmetrisch)
electrically symmetric quadripole (→electrically symmetric two-port)	NETW.TH	(→widerstandssymmetrischer Vierpol)
electrically symmetric two-port	NETW.TH	**widerstandssymmetrischer Vierpol**
= electrically symmetric quadripole; symmetric two-port; symmetric quadripole		= symmetrischer Vierpol
electrical machine	POWER ENG	**Elektromaschine**
[converts into mechanical energy or viceversa]		[wandelt in elektrische Energie oder umgekehrt]
= electric machine		= elektrische Maschine
electrical machines engineering	POWER ENG	**Elektromaschinenbau**
↑ electrical power engineering		↑ elektrische Energietechnik
electrical measurement technique	INSTR	**elektrische Meßtechnik**
		= Elektromeßtechnik
electrical midpoint (→center tap)	EL.TECH	(→Mittelabgriff)
electrical network theory (→network theory)	EL.TECH	(→Netzwerktheorie)
electrical operating conditions	EL.TECH	**Anschlußbedingung**
electrical oscillation	EL.TECH	**elektrische Schwingung**
electrical potential	PHYS	**elektrisches Potential**
[SI unit: Volt]		[SI-Einheit: Volt]
≈ electric tension		= Coulomb-Einheit
		≈ elektrische Spannung
electrical-power cord (→power cable)	EQUIP.ENG	(→Netzanschlußkabel)
electrical power engineering	EL.TECH	**elektrische Energietechnik**
= power current engineering; power current technology; power engineering; heavy current engineering; heavy current technology		= Starkstromtechnik; Energietechnik
≠ low current engineering		≠ Schwachstromtechnik
↑ electrotechnology		↑ Elektrotechnik
↓ electrical machines engineering; industrial engineering; electric installation engineering		↓ Elektromaschinenbau; elektrische Anlagetechnik; elektrische Installationstechnik
electrical power line (→power line)	POWER ENG	(→Starkstromleitung)
electrical science (→electricity 2)	PHYS	(→Elektrizitätslehre)
electricals corporation	ECON	**Elektrokonzern**
electrical sheet (→dynamo sheet)	METAL	(→Dynamoblech)
electrical steel (IEC) (→soft iron)	METAL	(→Weicheisen)
electrical technology (→electrical engineering)	TECH	(→Elektrotechnik)
electrical utility (→electric utility)	ECON	(→Energieversorgungsunternehmen)
electric area flux	PHYS	**elektrischer Hüllenfluß**
electric authority (→electric utility)	ECON	(→Energieversorgungsunternehmen)
electric axis	PHYS	**elektrische Achse**
electric biasing (→pre-electrification)	PHYS	(→Vorelektrisierung)
electric boundary potential	PHYS	**elektrische Umlaufspannung**
		= elektrische Randspannung
electric bulb	COMPON	**Glühbirne**
electric bulb display	COMPON	**Glühlampenanzeige**
electric capacitance	PHYS	**elektrische Kapazität**
[SI unit: Farad]		[SI-Einheit: Farad]
= electric capacity; capacitance; C		= Kapazität; C
≈ capacitor [COMPON]		≈ Kondensator [COMPON]

electric capacity (→electric PHYS (→elektrische Kapazität)
capacitance)
electric charge PHYS **elektrische Ladung**
[SI unit: Coulomb] [SI-Einheit: Coulomb]
= Q = Elektrizitätsladung; Q
electric circuit NETW.TH (→Schaltkreis)
(→circuit)
electric circulation (→electric PHYS (→elektrische Durchflutung)
loading)
electric company (→electric ECON (→Energieversorgungsunter-
utility) nehmen)
electric conduction PHYS **Elektrizitätsleitung**
= electrical conduction
electric conductivity 1 PHYS **elektrische Leitfähigkeit**
[material constant of electrical con- [Materialfaktor des elektri-
ductance; SI unit: S/m] schen Leitwertes; SI-Ein-
≠ specific resistance heit: S/m]
 = elektrisches Leitvermögen
 ≠ spezifischer elektrischer
 Widerstand
electric conductivity 2 PHYS **elektrischer Leitwert**
[SI unit: Siemens] [SI-Einheit: Siemens]
= electrical conductance; conductance; = Leitwert; Konduktanz; G
G ≠ elektrischer Widerstand
≠ electrical resistance
electric conductor PHYS **Elektrizitätsleiter**
 = elektrischer Leiter; Strom-
 leiter
electric counter INSTR **Elektrizitätszähler**
electric current density PHYS **elektrische Stromdichte**
electric current strength EL.TECH (→Stromstärke)
(→current intensity)
electric dipole PHYS **elektrischer Dipol**
electric discharge PHYS **Elektrizitätsentladung**
electric displacement (→electric PHYS (→elektrische Flußdichte)
flux density)
electric displacement density PHYS (→elektrische Flußdichte)
(→electric flux density)
electric doublet (→elementary ANT (→Hertzscher Dipol)
electric dipole)
electric drill (→drill MECH (→Bohrmaschine)
press)
electric field PHYS **elektrisches Feld**
electric field line PHYS **elektrische Kraftlinie**
 = elektrische Feldlinie; E-Li-
 nie
electric field strength PHYS **elektrische Feldstärke**
electric flux 2 PHYS **elektrischer Fluß 2**
[surface integral of electric flux den- [Flächenintegral der elek-
sity; SI unit: Coulomb] trischen Flußdichte; SI-Ein-
≈ electric flux 1 heit: Coulomb]
 = Verschiebungsfluß
 ≈ elektrische Durchflutung
electric flux 1 (→electric PHYS (→elektrische Durchflutung)
loading)
electric flux density PHYS **elektrische Flußdichte**
[vector with magnitude equal to flux [Vektor dessen Betrag
per unit area, in direction of the elec- gleich dem Verschiebungs-
tric lines of force; SI unit: C/cm^2] fluß pro Flächeneinheit,
= dielectric displacement density; die- gleichgerichtet mit elektri-
lectric displacement; electric displace- schen Kraftlinien; SI-Ein-
ment; electric displacement density; heit: C/cm^2]
displacement vector; dielectric excita- = elektrische Verschiebungs-
tion dichte; Verschiebungsdich-
 te; Verschiebungsvektor;
 elektrische Verschiebung;
 dielektrische Erregung
electric generator POWER ENG (→Stromgenerator)
(→generator)
electric heat PHYS **Stromwärme**
electrician TECH **Elektriker**
 = Elektroinstallateur
electric installation POWER ENG **elektrische Installationstech-**
engineering **nik**
↑ electrical power engineering ↑ elektrische Energietechnik
electricity 1 PHYS **Elektrizität**
electricity 2 PHYS **Elektrizitätslehre**
= electrical fundamentals; electrical = Elektrophysik; theoreti-
science sche Elektrotechnik
≈ electrical engineering ≈ Elektrotechnik

↓ electrostatics; electrodynamis; magnet- ↓ Elektrostatik; Elektrodyna-
ism; network theory; line theory; high mik; Magnetismus; Netz-
frequency; microwaves wertheorie; Leitungstheo-
 rie; Hochfrequenztechnik;
 Mikrowellentechnik
electricity generation POWER SYS **Elektrizitätserzeugung**
 = Stromerzeugung
electricity supply (→mains) POWER ENG (→Starkstromnetz)
electric lens (→electron lens) ELECTRON (→Elektronenlinse)
electric line EL.INST **Elektrizitätsleitung**
= electrical line
electric loading PHYS **elektrische Durchflutung**
[surface integral of current density; [Flächenintegral der Strom-
current times number of turns; SI dichte; Strom mal Win-
unit: ampere] dungszahl; SI-Einheit:
= electric flux 1; electric circulation Ampere]
≈ magnetomotive force; electric flux 2 = elektrischer Fluß 1
 ≈ magnetomotorische Kraft;
 elektrischer Fluß 2
electric machine (→electrical POWER ENG (→Elektromaschine)
machine)
electric meter INSTR **Stromzähler**
electric network NETW.TH **Netzwerk**
[interconnection of several circuit ele- [Zusammenschaltung meh-
ments] rerer Bauelemente]
= network; system ↓ Schaltkreis; Zweipol;
↓ circuit; two-terminal network; two- Vierpol; Viertor
port network; four-port network
electric polarization PHYS **elektrische Polarisation**
[difference of flux density in dielec- [Differenz der Flußdichten
tric to vacuum] im Dielektrikum zum Va-
 kuum]
 ↓ Verschiebungspolarisation;
 Orientierungspolarisation
electric polarization PHYS (→Vorelektrisierung)
(→pre-electrification)
electric potential line PHYS **elektrische Potentiallinie**
 = elektrische Niveaulinie
electric power EL.TECH **elektrische Leistung**
= wattage
electric power plant 1 POWER SYS **Elektrizitätswerk**
electric power plant 2 POWER ENG **Starkstromanlage**
electric resistance PHYS **elektrischer Widerstand**
[voltage to current ratio; derived SI [elektrische Spannung zu
unit: ohm] elektrischem Strom; abge-
= resistance; R leitete SI-Einheit: Ohm]
 = Resistanz; R
electric space constant (→dielectric PHYS (→elektrische Feldkonstante)
constant 2)
electric susceptibility PHYS **elektrische Suszeptibilität**
[ratio of electric polarization in a [Verhältnis der elektri-
dielctric to electric flux density in va- schen Polarisation im Die-
cuum] lektrikum zur elektrischen
 Flußdichte im Vakuum]
electric typewriter TERM&PER **elektrische Schreibmaschine**
electric utility ECON **Energieversorgungsunterneh-**
= electric authority; electric company; **men**
electrical utility = EVU; Elektrizitätsunter-
 nehmen; Elektrizitätswerk
electric wind PHYS **elektrischer Wind**
electrifiable PHYS **elektrisierbar**
electrification PHYS **Elektrisierung**
electrify PHYS **elektrisieren**
electroacoustic (adj.) PHYS **elektroakustisch**
≈ acoustoelectric ≈ akustroelektrisch
electroacoustic effect PHYS **elektroakustischer Effekt**
electroacoustic index TELEPH **Empfangsübertragungsfaktor**
electroacoustic index EL.ACOUS **Übertragungsfaktor**
= acousto-electric index
electroacoustics PHYS **Elektroakustik**
electroacoustic transducer EL.ACOUS **Schallwandler**
= sound converter = elektroakustischer Wandler
↓ loudspeaker; microphone ↓ Lautsprecher; Mikrophon
electro-analysis PHYS **Elektroanalyse**
electro-chemical PHYS **elektrochemisch**
electrochemical equivalent CHEM **elektrochemisches Äquivalent**
electrochemical potential PHYS **elektrochemisches Potential**
electrochemical series PHYS **voltaische Spannungsreihe**
= contact series; electromotive series = elektrochemische Spannungsreihe

English	Domain	German
electrochemics	PHYS	Elektrochemie
= electro-chemistry		
electro-chemistry	PHYS	(→Elektrochemie)
(→electrochemics)		
electrochrome display	ELECTRON	elektrochrome Anzeige
electrode	PHYS	Elektrode
↓ anode; cathode		↓ Anode; Kathode
electrode	POWER SYS	Elektrode
= plate		[Batterie]
electrode characteristic	ELECTRON	Elektrodenkennlinie
[tube]		[Röhre]
electrode current	EL.TECH	Elektrodenstrom
electrode impedance	ELECTRON	Elektrodenimpedanz
electrodeless	ELECTRON	elektrodenlos
electrodeless tube	ELECTRON	elektrodenlose Röhre
electrode reactance	ELECTRON	Elektrodenreaktanz
electrode voltage	EL.TECH	Elektrodenspannung
electrodiochroism	PHYS	Elektropleochroismus
electrodynamic	EL.TECH	elektrodynamisch
electrodynamical measuring system	INSTR	dynamometrisches Meßwerk = elektrodynamisches Meßwerk
electrodynamical pick-up	INSTR	elektrodynamischer Meßfühler = Induktionsmeßfühler; Tauchspul-Meßfühler; generatorischer Meßfühler
electrodynamic loudspeaker	EL.ACOUS	elektrodynamischer Lautsprecher
= dynamic loudspeaker; moving-coil loudspeaker; coildriven louspeaker		↓ Druckkammerlautsprecher
↓ horn loudspeaker		
electrodynamic potential	PHYS	elektrodynamisches Potential
electrodynamic power meter	INSTR	elektrodynamischer Leistungsmesser
electrodynamics	EL.TECH	Elektrodynamik
↑ electrical fundamentals		↑ Elektrizitätslehre
electrodynamic transducer	EL.ACOUS	elektrodynamischer Wandler
↑ electroacoustic transducer		↑ Schallwandler
electrofax dry copier	OFFICE	Elektrofax-Trockenkopierer
electrofax wet copier	OFFICE	Elektrofax- Naßkopierer
electro furnace	TECH	Elektroofen
electro-galvanize	METAL	galvanisch verzinken
= galavanize		verzinken
electrographic printer	TERM&PER	(→elektrostatischer Drucker)
(→electrostatic printer)		
electrokinetic	PHYS	elektrokinetisch
electroluminescence	PHYS	Elektrolumineszenz
[induced by electric fields]		[durch elektrische Felder induziert]
electroluminescence display	COMPON	Elektrolumineszenzanzeige = EL-Anzeige; Lumineszenzanzeige
= electroluminescent display; EL display; electroluminescent panel; EL panel		
electroluminescent (adj.)	PHYS	elektrolumineszent
= electroluminescing		
electroluminescent display	COMPON	(→Elektrolumineszenzanzeige)
(→electroluminescence display)		
electroluminescent panel 1	COMPON	Lumineszenzplatte = Halbleiter-Flächenstrahler; Lumineszenzzelle; Leuchtkondensator; Elektrolumineszenzzelle
electroluminescent panel 2	COMPON	(→Elektrolumineszenzanzeige)
(→electroluminescence display)		
electroluminescing	PHYS	(→elektrolumineszent)
(→electroluminescent)		
electrolysis	PHYS	Elektrolyse
electrolyte	PHYS	Elektrolyt
electrolytic	PHYS	elektrolytisch
electrolytic capacitor	COMPON	Elektrolytkondensator = Elko
electrolytic conduction	PHYS	elektrolytische Leitung
electrolytic conductor	PHYS	Ionenleiter
electrolytic constant (→Faraday constant)	PHYS	(→Faraday-Konstante)
electrolytic copper	METAL	Elektrolytkupfer
electrolytic high-voltage capacitor	COMPON	Hochvolt-Elko
electrolytic potential	PHYS	elektrolytisches Potential
electrolytic recording [facsimile]	TERM&PER	elektrolytische Aufzeichnung [Faksimile]
electrolytic solution pressure	PHYS	elektrolytischer Lösungsdruck
electromagnet	EL.TECH	Elektromagnet
electromagnetic	PHYS	elektromagnetisch
electromagnetical switch	COMPON	elektromagnetischer Schalter
= electrically operated switch		
electromagnetic compatibility	ELECTRON	elektromagnetische Verträglichkeit
= EMC		= EMV
electromagnetic component	COMPON	elektromagnetisches Bauelement
electromagnetic delay line	COMPON	elektromagnetische Verzögerungsleitung
electromagnetic field	PHYS	elektromagnetisches Feld
electromagnetic force	PHYS	elektromagnetische Kraft
electromagnetic induction	PHYS	elektromagnetische Induktion
electromagnetic interference	RADIO	(→Störstrahlung)
(→unwanted emission)		
electromagnetic lens	ANT	Linse
electromagnetic memory	DATA PROC	elektromagnetischer Speicher
= electromagnetic storage; electromagnetic store		↓ Magnetkernspeicher; Magnetbandspeicher; Magnetblasenspeicher; Magnettrommelspeicher
↓ magnetic core memory; magnetic tape memory; magnetic bubble memory; magnetic drum memory		
electromagnetic moment	PHYS	elektromagnetisches Moment
electromagnetic pickup	EL.ACOUS	elektromagnetischer Tonabnehmer
= magnetic pickup		= Magnet-Tonabnehmersystem; Magnetsystem
electromagnetic radiation	PHYS	elektromagnetische Strahlung
= EMR		
electromagnetic spectrum	PHYS	elektromagnetisches Spektrum
electromagnetic storage	DATA PROC	(→elektromagnetischer Speicher)
(→electromagnetic memory)		
electromagnetic store	DATA PROC	(→elektromagnetischer Speicher)
(→electromagnetic memory)		
electromagnetic transducer	EL.ACOUS	Schallschnellewandler = elektromagnetischer Wandler
electromagnetism	PHYS	Elektromagnetismus
electro-material	EL.INST	Elektromaterial
= electrical components		
electromechanical	TECH	elektromechanisch
electromechanical converter	COMPON	(→elektromechanischer Wandler)
(→electromechanical transducer)		
electro-mechanical filter	COMPON	(→mechanisches Filter)
(→mechanical filter)		
electromechanical switching system	SWITCH	elektromechanisches Vermittlungssystem = elektromechanisches Wählsystem
electromechanical transducer	COMPON	elektromechanischer Wandler
= electromechanical converter		
electromechanic counter	INSTR	elektromechanischer Zähler
electromechanics	TECH	Elektromechanik
electromedicine	TECH	Elektromedizin
= medical engineering		= Medizintechnik; medizinische Technik
electrometer	INSTR	Elektrometer
electrometer bridge	INSTR	Elektrometerbrücke
electromigration	PHYS	Elektromigration
electromotive	PHYS	elektromotorisch
electromotive force	EL.TECH	(→Leerlaufspannung)
(→open-circuit voltage)		
electromotive series	PHYS	(→voltaische Spannungsreihe)
(→electrochemical series)		
electromotive voltage	EL.TECH	(→Leerlaufspannung)
(→open-circuit voltage)		
electromotor	POWER ENG	Elektromotor
electron	PHYS	Elektron
↑ elementary particle		↑ Elementarteilchen
electron accelerator	PHYS	Elektronenbeschleuniger
electron affinity	PHYS	Elektronenaffinität
electron avalanche	PHYS	Elektronenlawine
electron beam	ELECTRON	Elektronenstrahl
= electron jet		

English	Domain	German
electron beam deflection system	ELECTRON	**Elektronenstrahl-Ablenksystem**
= beam deflection system		= Strahlablenksystem
electron beam lithography	MICROEL	**Elektronenstrahl-Lithographie**
electron beam recording [directly on microfilm] = EBR	TERM&PER	**Elektronenstrahlaufzeichnung** = direkt auf Mikrofilm
electron-beam testing	MICROEL	**Elektronenstrahl-Testverfahren**
electron-beam writing	MICROEL	**Elektronenstrahl-Belichtung**
electron capture (→electron collection)	ELECTRON	(→Elektronenabsaugung)
electron cloud = electron sheath	ELECTRON	**Elektronenwolke**
electron collection = electron capture	ELECTRON	**Elektronenabsaugung**
electron collector ↑ collector	ELECTRON	**Elektronenkollektor** ↑ Kollektor
electron concentration ≈ electron density	PHYS	**Elektronenkonzentration** ≈ Elektronendichte
electron conduction (→n-type conduction)	PHYS	(→N-Leitung)
electron conductivity (→n-type conductivity)	PHYS	(→N-Leitfähigkeit)
electron coupling	ELECTRON	**Elektronenkopplung**
electron current = electron stream	PHYS	**Elektronenstrom**
electron density ≈ electron concentration	PHYS	**Elektronendichte** ≈ Elektronenkonzentration
electron detachment = electron removal	PHYS	**Elektronenablösung**
electron device (→electronic device)	COMPON	(→elektronisches Bauelement)
electron diffraction	PHYS	**Elektronenbeugung**
electron diffusion	PHYS	**Elektronendiffusion**
electron discharge	PHYS	**Elektronenentladung**
electron emission	PHYS	**Elektronenemission**
electron energy	PHYS	**Elektronenenergie**
electron gas	PHYS	**Elektronengas**
electron group [velocity-modulated tube]	ELECTRON	**Elektronenpaket** [Laufzeitröhren]
electron gun ↑ electron source [PHYS]	ELECTRON	**Elektronenkanone** ↑ Elektronenquelle [PHYS]
electronic	EL.TECH	**elektronisch**
electronic a.c. convertor (IEC) (→ac converter)	POWER ENG	(→Wechselstromumrichter)
electronically controlled	SWITCH	**elektronisch gesteuert**
electronically controlled switching system (→electronic switching system)	SWITCH	(→elektronisches Vermittlungssystem)
electronically programmable	DATA PROC	**elektronisch programmierbar**
electronical standard voltage source	INSTR	**elektronische Normalspannungsquelle** = Gleichspannungsnormal
electronic amplifier	CIRC.ENG	**elektronischer Verstärker**
electronic balance	INSTR	**elektronische Waage**
electronic banking (→telebanking)	TELEC	(→Telebanking)
electronic bulletin board (→bulletin board system)	DATA COMM	(→Schwarzes-Brett-System)
electronic calculator ≈ computer ≠ mechanical calculator ↑ calculator	DATA PROC	**Elektronenrechner** ≈ Computer ≠ mechanischer Rechner ↑ Rechner 1
electronic car aerial (→electronic car antenna)	ANT	(→elektronische Fahrzeugantenne)
electronic car antenna = electronic car aerial	ANT	**elektronische Fahrzeugantenne**
electronic charge	PHYS	**Elektronenladung**
electronic component (→electronic device)	COMPON	(→elektronisches Bauelement)
electronic computer system (→data processing equipment)	DATA PROC	(→Datenverarbeitungsanlage)
electronic control	CONTROL	**elektronische Steuerung**
electronic controller	CONTROL	**elektronischer Regler**
electronic counter [working with electronic circuitry]	INSTR	**elektronischer Zähler** [mit elektronischen Schaltungen aufgebaut]
electronic countermeasures (→ECL)	MIL.COMM	(→ECL)
electronic data interchange = EDI; electronic magazine	DATA COMM	**elektronischer Datenaustausch** = elektronisches Magazin
electronic data processing = EDP; automatic data processing; ADP ↑ data processing	INF.TECH	**elektronische Datenverarbeitung** = EDV; automatische Datenverarbeitung; ADV ↑ Datenverarbeitung
electronic data processing machine (→data processing equipment)	DATA PROC	(→Datenverarbeitungsanlage)
electronic data processing system (→data processing equipment)	DATA PROC	(→Datenverarbeitungsanlage)
electronic device = electron device; electronic component	COMPON	**elektronisches Bauelement**
electronic echo	EL.ACOUS	**Echogerät**
electronic engineer	TECH	**Elektronikingenieur**
electronic equipment	EQUIP.ENG	**elektronisches Gerät**
electronic equipment room	SYS.INST	**Geräteraum** = Gerätesaal
electronic excitation	PHYS	**Elektronenanregung**
electronic filing	DATA PROC	**elektronische Ablage**
electronic flash	ELECTRON	**Elektronenblitz**
electronic funds transfer	DATA COMM	**elektronische Banküberweisung**
electronic garbage	TECH	**Elektronik-Schrott**
electronic hash = chopper noise; chopper interference	ELECTRON	**Zerhackergeräusch**
electronic industry	ECON	**elektronische Industrie**
electronic journal	DATA PROC	**elektronisches Journal**
electronic lens (→electron lens)	ELECTRON	(→Elektronenlinse)
electronic letterbox (→electronic mailbox)	TELEC	(→elektronischer Briefkasten)
electronic load	INSTR	**elektronische Last**
electronic magazine (→electronic data interchange)	DATA COMM	**elektronischer Datenaustausch**
electronic mail = E-mail; email; electronic mail system; EMS; mailbox system; electronic text transfer; computer mail ↓ teletex; telefax; telebox	TELEC	**elektronischer Briefdienst** = elektronische Briefpost; elektronische Post; elektronische Textübermittlung; EMS; E-Mail; elektronisches Mitteilungssystem; EMS ↓ Teletex; Telefax; Telebox
electronic mailbox = electronic letterbox; telebox	TELEC	**elektronischer Briefkasten** = elektronisches Postfach; Telebox; Postfach
electronic mail system (→electronic mail)	TELEC	(→elektronischer Briefdienst)
electronic measuring equipment (→electronic measuring instrument)	INSTR	(→elektronisches Meßgerät)
electronic measuring instrument = electronic measuring equipment; electronic test equipment	INSTR	**elektronisches Meßgerät** = elektronisches Meßinstrument
electronic music	DATA PROC	**elektronische Musik**
electronic noise (→thermal noise)	TELEC	(→thermisches Rauschen)
electronic office = automated office	INF.TECH	**automatisiertes Büro**
electronic organizer	DATA PROC	**elektronisches Notizbuch**
electronic page composition (→electronic page make-up)	DATA PROC	(→elektronischer Seitenumbruch)
electronic page make-up [word processing] = electronic page composition; electronic pagination; electronic paste-up	DATA PROC	**elektronischer Seitenumbruch** [Textverarbeitung] = elektronische Seitenmontage
electronic pagination (→electronic page make-up)	DATA PROC	(→elektronischer Seitenumbruch)
electronic paintbrush [graphics program] = paintbrush	DATA PROC	**elektronischer Farbpinsel** [Graphikprogramm] = Farbpinsel; elektronischer Pinsel
electronic paste-up (→electronic page make-up)	DATA PROC	(→elektronischer Seitenumbruch)
electronic pen (→electronic pencil)	TERM&PER	(→elektronischer Bleistift)

electronic pencil	TERM&PER	elektronischer Bleistift
[hand-held device to input on a graphic tablet or screen]		[Handgerät zur Eingabe am Graphiktablett oder Bildschirm]
= electronic pen; electronic stylus		
↓ light pen		↓ Lichtgriffel
electronic point-of-sale	DATA COMM	Kassenterminalsystem
= EPOS		
electronic polarization	PHYS	Elektronenpolarisation
= electron polarization		
electronic power circuit	CIRC.ENG	Potenzierer
electronic publishing	DATA PROC	Electronic Publishing
[editing of printed matter on mainframe computers]		[Erstellen Druckerzeugnissen auf Großrechnern]
= EP; electronic technical publishing; ETP		≈ Desktop Publishing ↑ CAP 2
electronic relay (→transistor relay)	COMPON	(→Transistorrelais)
electronics	EL.TECH	Elektronik
[science and technology of devices working with flow of electrons through semiconductors, vacuum or gases; essentially: vacuum tubes, semiconductor devices and integrated circuits]		[Theorie und Technik von Vorrichtungen, die mit Elektronenfluß in Halbleitern, Vakuum oder Gasen funktionieren; im wesentlichen: Elektronenröhren, Halbleiterbauelemente und integrierte Schaltungen]
↓ consumer electronics; communication electronics; industrial electronics; microelectronics		≈ Schwachstromtechnik ↓ Konsumelektronik; Kommunikationselektronik; Industrieelektronik; Mikroelektronik
electronics amateur	ELECTRON	Hobby-Elektroniker
electronic scanning	RADIO LOC	elektronische Abtastung
= inertialess scanning		
electronics industry	ECON	Elektronikindustrie
electronic smog	EL.TECH	elektromagnetische Verseuchung
electronic spreadsheet	DATA PROC	(→Tabellenkalkulationsprogramm)
(→spreadsheet program)		
electronics technician	TECH	Elektroniker
electronic stencil	ENG.DRAW	Elektronikschablone
electronic storage	DATA PROC	elektronische Speicherung
electronic stylus (→electronic pencil)	TERM&PER	(→elektronischer Bleistift)
electronic switch (→semiconductor switch)		(→Halbleiterschalter)
electronic switching system	SWITCH	elektronisches Vermittlungssystem
= electronically controlled switching system		= elektronisch gesteuertes Vermittlungssystem; elektronisches Wählsystem; elektronisch gesteuertes Wählsystem
electronic technical publishing (→electronic publishing)	DATA PROC	(→Electronic Publishing)
electronic test equipment (→electronic measuring instrument)	INSTR	(→elektronisches Meßgerät)
electronic text transfer (→electronic mail)	TELEC	(→elektronischer Briefdienst)
electronic typesetting	TYPOGR	elektronischer Satz
↑ cold type		↑ Kaltsatz
electronic voltmeter	INSTR	elektronischer Spannungsmesser
electronic warefare	MIL.COMM	elektronische Kampfführung
= EW		= elektronische Kriegsführung
electron image	ELECTRON	Elektronenbild
electron image tube (→image converter tube)	ELECTRON	(→Bildwandlerröhre)
electron impact	PHYS	Elektronenstoß
electron impact spectroscopy	PHYS	Elektronenstoß-Spektrometrie
electron input	ELECTRON	Elektronenzufuhr
electron interferometry	PHYS	Elektroneninterferometrie
electron jet (→electron beam)	ELECTRON	(→Elektronenstrahl)
electron jump (→electron transition)	PHYS	(→Elektronenübergang)
electron lens	ELECTRON	Elektronenlinse
= electrostatic lens; aperture lens; electronic lens; electric lens		= elektronische Linse; elektrische Linse
electron mass	PHYS	Elektronenmasse
electron microscope	ELECTRON	Elektronenmikroskop
electron mirror	ELECTRON	Elektronenspiegel
electron mirror microscope	ELECTRON	Elektronen-Spiegelmikroskop
electron mobility	PHYS	Elektronenbeweglichkeit
electron multiplier	ELECTRON	Elektronenvervielfacher
= photomultiplier		= Sekundäremissionsvervielfacher; Sekundärelektronenvervielfacher; Photovervielfacher; Fotovervielfacher
↓ electron multiplier tube		↓ Elektronenvervielfachungsröhre
electron multiplier tube	ELECTRON	Elektronenvervielfachungsröhre
= photomultiplier tube		
↑ electron multiplier		↑ Elektronenvevielfacher
electron optical	PHYS	elektronenoptisch
= electro-optic		= elektrooptisch
electron optical image device	ELECTRON	elektronenoptisches Abbildungsgerät
electron optics	PHYS	Elektronenoptik
electron orbit	PHYS	Elektronenbahn
= electron trajectory		
electron physics	PHYS	Elektronenphysik
electron polarization (→electronic polarization)	PHYS	(→Elektronenpolarisation)
electron probe	ELECTRON	Elektronensonde
electron radius	PHYS	Elektronenradius
electron ray (→cathode ray)	PHYS	(→Kathodenstrahl)
electron removal (→electron detachment)	PHYS	(→Elektronenablösung)
electron scattering	PHYS	Elektronenstreuung
electron sheath (→electron cloud)	ELECTRON	(→Elektronenwolke)
electron shell	PHYS	Elektronenhülle
[the $2n^2$-limited number of allowed electron orbitals for a given principal quantum number n]		[die auf $2n^2$ begrenzte Zahl von Elektronenbahnen für eine gegebene Hauptquantenzahl n]
= shell		= Elektronenschale; Atomschale; Schale
≈ electron orbit		≈ Elektronenbahn
electron source	PHYS	Elektronenquelle
↓ electron gun [ELECTRON]		↓ Elektronenkanone [ELECTRON]
electron spin	PHYS	Elektronenspin
electron stream (→electron current)	PHYS	(→Elektronenstrom)
electron trajectory (→electron orbit)	PHYS	(→Elektronenbahn)
electron transition	PHYS	Elektronenübergang
= electron jump		
electron transit time	PHYS	Elektronenlaufzeit
electron tube	ELECTRON	Elektronenröhre
= thermoionic tube; tube; thermoionic valve (BRI); valve (BRI)		= Röhre ↓ Vakuumröhre; Gasentladungsröhre
↓ vacuum tube; gas discharge tube		
electron tube generator	ELECTRON	Elektronenröhrengenerator
electron vacancy (→hole)	PHYS	(→Loch)
electron-volt	PHYS	Elektronvolt
= eV		= eV; Elektronenvolt
electron wave	PHYS	Elektronenwelle
electro-optic (→electron optical)	PHYS	(→elektronenoptisch)
electro-optical effect	PHYS	elektrooptischer Effekt
electro-optical reconnaissance	OPT.COMM	elektrooptische Erkennung
= EOR		
electrooptic birefrigence (→Kerr effect)	PHYS	(→Kerr-Effekt)
electrophoresis	PHYS	Elektrophorese
		= Kathaphorese
electrophotographic	TECH	elektrofotografisch
		= elektrophotographisch
electro-photographic drawing copier	OFFICE	elektrofofographischer Zeichnungskopierer

electrophotographic recording [facsimile]	TERM&PER	elektrofotografische Aufzeichnung [Faksimile] = elektrophotographische Aufzeichnung
electroplate = galvanize; plate ↑ metallize	METAL	galvanisieren = plattieren 2; galvanisch plattieren; metallbeschichten ↑ metallisieren
electroplate (→plate)	METAL	(→plattieren 1)
electroplating ↑ plating	METAL	Elektroplattierung = galvanische Plattierung ↑ Plattierung
electropneumatic	TECH	elektropneumatisch
electrosensitive	TECH	elektrosensitiv
electrosensitive paper	TERM&PER	elektrosensitives Papier
electrosensitive printer ↑ non impact printer ↓ laser printer	TERM&PER	elektrosensitiver Drucker ↑ anschlagfreier Drucker ↓ Laserdrucker
electrosensitive recording [facsimile]	TERM&PER	elektrosensitive Aufzeichnung [Faksimile]
electrostatic	PHYS	elektrostatisch
electrostatic copying process = xerography	OFFICE	elektrostatisches Kopierverfahren = Xerographie
electrostatic discharge	PHYS	elektrostatische Entladung
electrostatic field	PHYS	elektrostatisches Feld
electrostatic generator = influence machine	PHYS	elektrostatischer Generator = Influenzmaschine
electrostatic induction	PHYS	Influenz
electrostatic lens (→electron lens)	ELECTRON	(→Elektronenlinse)
electrostatic loudspeaker	EL.ACOUS	elektrostatischer Lautsprecher
electrostatic measuring instrument	INSTR	elektrostatisches Meßinstrument
electrostatic measuring system	INSTR	elektrostatisches Meßwerk
electrostatic microphone (→capacitor microphone)	EL.ACOUS	(→Kondensatormikrophon)
electrostatic plotter	TERM&PER	elektrostatischer Plotter = Elektrostatik-Plotter
electrostatic potential	PHYS	elektrostatisches Potential
electrostatic printer = electrographic printer	TERM&PER	elektrostatischer Drucker
electrostatic recording [facsimile]	TERM&PER	elektrostatische Aufzeichnung [Faksimile]
electrostatics ↑ electrical fundamentals	PHYS	Elektrostatik ↑ Elektrizitätslehre
electrostatic screen	EL.TECH	elektrostatischer Schirm
electrostatic screening = electrostatic shield	EL.TECH	elektrostatische Abschirmung
electrostatic shield (→electrostatic screening)	EL.TECH	(→elektrostatische Abschirmung)
electrostatic storage	ELECTRON	elektrostatische Speicherung
electrostatic storage tube (→storage tube)	ELECTRON	(→Speicherröhre)
electrostatic transducer	EL.ACOUS	elektrostatischer Wandler
electrostriction	PHYS	Elektrostriktion
electrotator bidirectional array	ANT	Electrotator-Querstrahler
electrotechnology (→electrical engineering)	TECH	(→Elektrotechnik)
electrothermal (→electrothermic)	PHYS	(→elektrothermisch)
electrothermal printer [on electrosensitive paper]	TERM&PER	elektrothermischer Drucker [auf elektrosensitivem Papier]
electrothermal relay (→thermal relay)	COMPON	(→Thermorelais)
electrothermic = electrothermal	PHYS	elektrothermisch
electrotyping	TYPOGR	Elektrotypie
electrovalence (→heteropolar bond)	CHEM	(→heteropolare Bindung)
elegant (adj.)	TECH	elegant
element [of a thermocouple]	PHYS	Element
element	PHYS	Schenkel [eines Thermopaares]
element (→galvanic cell)	POWER SYS	(→galvanisches Element)
element (→transfer element)	CONTROL	(→Übertragungsglied)
elemental semiconductor	PHYS	Elementhalbleiter
elementary (adj.) ≈ essential	SCIE	elementar = grundlegend ≈ wesentlich
elementary access method (→basic access method)	DATA PROC	(→Basiszugriffsmethode)
elementary cable section (CCITT) (→regenerator spacing)	TRANS	(→Regeneratorfeldlänge)
elementary charge unit	PHYS	Elementarladung
elementary color = primary color	PHYS	Grundfarbe = Primärfarbe; Elementarfarbe
elementary diagram [a wiring diagram]	ELECTRON	Elementardiagramm [ein Verdrahtungsplan]
elementary dipole = current element ↓ elementary electric dipole; elementary megnatic dipole	ANT	Elementardipol = Elementarstrahler; Stromelement ↓ elektrischer Elementardipol; magnetischer Elementardipol
elementary electric dipole = Hertzian dipole; infinitesimal electric dipole; Hertzian doublet; electric doublet ↑ dipole	ANT	Hertzscher Dipol = elektrischer Elementardipol ≈ Kurzdipol ↑ Dipol
elementary event	MATH	Elementarereignis
elementary function	MATH	elementare Funktion
elementary magnet	PHYS	Elementarmagnet
elementary magnetic dipole = infinitesimal magnetic current element ↑ elementaey dipole	ANT	magnetischer Elementardipol = magnetisches Stromelement ↑ Elementardipol
elementary particle ≠ antiparticle	PHYS	Elementarteilchen ≠ Antiteilchen
elementary physics	PHYS	Elementarphysik
elementary quantum	PHYS	Elementarquantum
elementary school (→primary school)	SCIE	(→Grundschule)
elementary signal	TELEC	Elementarsignal
elementary wave	PHYS	Elementarwelle
element manager	TELEC	Elementsteuerung
elevated antenna	ANT	Hochantenne
elevation	ENG.DRAW	Aufriß
elevation	ANT	Elevation
elevation = raising ≠ recess	TECH	Erhöhung ≠ Vertiefung
elevation (→altitude)	GEOPHYS	(→Höhe)
elevation (→altitude)	ASTROPHYS	(→Höhe)
elevation adjustment	ANT	Höhenverstellung
elevation angle = take-off angle; horizon angle; look angle	ANT	Elevationswinkel = Höhenwinkel; Erhebungswinkel
elevation pattern (→vertical diagram)	ANT	(→Vertikaldiagramm)
elevator (NAM) = lift (BRI)	CIV.ENG	Aufzug = Lift
ELF (→etremely low frequency)	RADIO	(→Megameterwelle)
Elias theorem	INF	Theorem von Elias
eligible (adj.)	COLLOQ	auswählbar = wählbar
eliminate = eliminieren	COLLOQ	beseitigen
eliminate (→erase)	DATA PROC	(→löschen)
elimination	COLLOQ	Beseitigung = Eliminierung
elimination (→erasing)	ELECTRON	(→Löschung)
elliptically polarized	PHYS	elliptisch polarisiert
elliptically polarized wave	PHYS	elliptisch polarisierte Welle
elite [12 CPI type size]	TYPOGR	Elite [Schriftgröße von 12"]
ellipse ↑ conical section	MATH	Ellipse ↑ Kegelschnitt
ellipsis [obviously understandable word omission]	LING	Ellipse = Auslassung
ellipsis [symbol '; marks omission] ↓ apostrophe	TYPOGR	Auslassungszeichen [Symbol '] ↓ Apostroph
ellipsoid (n.)	MATH	Ellipsoid
ellipsoidal	MATH	ellipsoidförmig

ellipsoidal antenna	ANT	**Ellipsoidantenne**
		= Ellipsoidstrahler
ellipsoidal core antenna	ANT	**Ellipsoidkernantenne**
ellipsoid of revolution	MATH	**Rotationsellipsoid**
elliptic	MATH	**elliptisch**
= elliptical		
elliptical (→elliptic)	MATH	(→elliptisch)
elliptical waveguide	MICROW	**elliptischer Hohlleiter**
elliptic function	MATH	**elliptische Funktion**
elliptic integral	MATH	**elliptisches Integral**
ellipticity	MATH	**Elliptizität**
[major to minor axis ratio]		[große zu kleine Achse]
= axial ratio		= Achsenverhältnis
elliptic paraboloid	MATH	**elliptisches Paraboloid**
elliptic polarization	PHYS	**elliptische Polarisation**
Elmos's fire	PHYS	**Elmsfeuer**
ELOD	TERM&PER	**ELOD**
= erasable laser optical disk		= löschbare Bildplatte
		↑ Bildplatte
elongate (→strech)	PHYS	(→dehnen)
elongated	TYPOGR	**extraschmal**
elongated (→longish)	TECH	(→länglich)
elongation	PHYS	**Elongation**
		= Ausschlag
eloxadize	METAL	**eloxieren**
= anodize		
EL panel	COMPON	(→Elektrolumineszenzanzeige)
(→electroluminescence display)		
ELSE rule	DATA PROC	**Sonst-Regel**
		= ELSE-Regel
elucidation (→explanation)	LING	(→Erläuterung)
em	TYPOGR	**Geviert**
[dummy type or blank space with the width of an m]		[Leertype bzw. Leerstelle mit der Breite/Dicke einer m]
= em quad; em space		
≈ en		≈ Quadrat
		≈ Halbgeviert
EM (→end-of-medium)	DATA COMM	(→Ende der Aufzeichnung)
E-mail (→electronic mail)	TELEC	(→elektronischer Briefdienst)
email (→electronic mail)	TELEC	(→elektronischer Briefdienst)
emanation	PHYS	**Ausdunstung**
≈ evaporation		≈ Verdampfung
embargo	ECON	**Embargo**
embed	TECH	**einbetten**
= imbed		
embed (→nest)	DATA PROC	(→schachteln)
embedded coding	TELEC	**eingebettete Verschlüsselung**
embedded command	DATA PROC	**eingebettete Druckanweisung**
embedded computer (→control computer)	CONTROL	(→Steuerrechner)
embedded mixed crystal	PHYS	**Einlagerungs-Mischkristall**
embedding	TECH	**Einbettung**
embedding circuit	CIRC.ENG	**Umfeldschaltung**
embezzlement	DATA PROC	**Veruntreuungsmanipulation**
		= Buchführungsmanipulation
emblem (→badge)	TECH	(→Abzeichen)
emboss (v.t.)	METAL	**prägen**
↓ coin (v.t.)		↓ münzprägen
embossed	TYPOGR	**stark konturiert**
embossing	METAL	**Prägen**
embossing tool	TECH	**Stichel**
embossment	METAL	**Prägung**
EMC (→electromagnetic compatibility)	ELECTRON	(→elektromagnetische Verträglichkeit)
emendation (→improvement 1)	TECH	(→Verbesserung)
emerald (adj.) (→emerald green)	PHYS	(→smaragdgrün)
emerald green	PHYS	**smaragdgrün**
= emerald (adj.)		
emerge	COLLOQ	**aufkommen**
		[zum Vorschein kommen]
emergency battery	POWER SYS	**Notstrombatterie**
emergency call	TELEPH	**Notruf**
= distress message		
emergency call alarm equipment	TELEC	**Notrufmeldeeinrichtung**
emergency call equipment	TELEC	**Notrufeinrichtung**
emergency call number	TELEPH	**Notrufnummer**
emergency call service	TELEPH	**Notrufdienst**
emergency call system	SIGN.ENG	**Notrufanlage**
= duress alarm system		↑ Gefahrenmeldeanlage
↑ danger detection system		
emergency circuit	POWER SYS	**Notschaltung**
emergency operation	TECH	(→Notdienst)
(→emergency service)		
emergency plan (→contingency plan)	TECH	(→Notplan)
emergency power plant	POWER SYS	(→Notstromversorgung)
(→emergency power supply)		
emergency power supply	POWER SYS	**Notstromversorgung**
= stand-by power supply; mains failure supply; emergency power plant; backup power plant; stand-by set; stand-by power plant; auto mains failure system		= Notstromaggregat; Notstromanlage; Netzersatzanlage; Notstromzentrale
emergency reporting system (AM)	TELEC	(→Notrufsäule)
(→emergency telephone)		
emergency service	TECH	**Notdienst**
= emergency operation		= Notbetrieb
emergency shutdown	EL.TECH	**Notabschaltung**
		= Notausschaltung
emergency switch	EL.INST	**Notausschalter**
= panic switch; off emergency		
emergency telephone	TELEC	**Notrufsäule**
= emergency reporting system (AM); ERS (AM)		
emery paper	TECH	**Schmirgelpapier**
= abrasive paper; sand paper		= Schleifpapier; Sandpapier
↓ sandpaper		↓ Sandpapier
emf (→open-circuit voltage)	EL.TECH	(→Leerlaufspannung)
EMI (→unwanted emission)	RADIO	(→Störstrahlung)
EMI receiver	INSTR	**Störstrahlungsmesser**
		= EMI-Empfänger
emission	RADIO	**Aussendung**
= radiation		
emission (→irradiation)	PHYS	(→Abstrahlung)
emission current	ELECTRON	**Emissionsstrom**
emission microscope	ELECTRON	**Emissionsmikroskop**
emission resistance	PHYS	**Emissionswiderstand**
emission resistance (→emitter emission resistance)	MICROEL	(→Emitter-Emissionswiderstand)
emission threshold	MICROEL	**Emissionsschwelle**
emissivity (→emittance)	PHYS	(→Emissionsvermögen)
emit (→transmit)	RADIO	(→senden)
emit (→send)	TELEC	(→senden)
emit omnidirectionally	RADIO	**rundstrahlen**
emittance	PHYS	**Emissionsvermögen**
= emissivity		= Emissionsgrad; Emissionsfähigkeit
emitter 1	MICROEL	**Emitter 1**
[generatzes a carrier flux to the collector, controled by the basis]		[erzeugt einen von der Basis gesteuerten Ladungsträgerfluß zum Kollektor]
↑ transistor region		↑ Transistorzone
		↓ Emitterelektrode; Emitteranschluß; Emitterzone
emitter 2 (→emitter terminal)	MICROEL	(→Emitteranschluß)
emitter barrier (→emitter junction)	MICROEL	(→Emitter-Sperrschicht)
emitter barrier capacitance	MICROEL	(→Emittersperrschicht-Kapazität)
(→emitter junction capacitance)		
emitter base capacitance	MICROEL	**Emitter-Basis-Kapazität**
emitter-base cutoff current	MICROEL	**Emitter-Basis-Reststrom**
emitter-base junction	MICROEL	**Emitter-Basis-Diode**
(→emitter diode)		
emitter-base voltage	CIRC.ENG	**Emitter-Basis-Spannung**
emitter bulk resistance	MICROEL	(→Emitterbahnwiderstand)
(→emitter series resistance)		
emitter capacitance	MICROEL	**Emitter-Kapazität**
emitter conductance	MICROEL	**Emitterleitwert**
emitter connection (→emitter terminal)	MICROEL	(→Emitteranschluß)
emitter-coupled current steered logic (→emitter-coupled logic)	MICROEL	(→emittergekoppelte Logik)
emitter-coupled logic	MICROEL	**emittergekoppelte Logik**
= ECL; current-mode logic; CML; emitter-coupled transistor logic; ECTL; emitter-coupled current steered logic; ECCSL; emitter-emitter logic; EEL; current-switch logic; CSL		= emittergekoppelte Transistorlogik; ECL; CML; ECCSL; ECTL; EEL; CSL

emitter-coupled multivibrator CIRC.ENG **emittergekoppelte Kippschaltung**
= emittergekoppelter Multivibrator

emitter-coupled transistor logic MICROEL (→emittergekoppelte Logik)
(→emitter-coupled logic)

emitter current CIRC.ENG **Emitterstrom**

emitter depletion layer MICROEL (→Emitter-Sperrschicht)
(→emitter junction)

emitter diffusion capacitance MICROEL **Emitter-Diffusionskapazität**
emitter diffusion conductance MICROEL **Emitter-Diffusionsleitwert**
emitter diffusion resistance MICROEL **Emitter-Diffusionswiderstand**
emitter diode MICROEL **Emitter-Basis-Diode**
= emitter-base junction = Emitterdiode; Emitter-Basis-Strecke

emitter dip effect MICROEL **Emitter-Dip-Effekt**
= Emitter-Push-Effekt

emitter dissipation MICROEL **Emitter-Verlustleistung**
emitter dot MICROEL **Emitterpille**
emitter efficiency MICROEL **Emissions-Wirkungsgrad**
= Ermitter-Ergiebigkeit

emitter emission resistance MICROEL **Emitter-Emissionswiderstand**
= emission resistance

emitter-emitter-coupled logic MICROEL (→EECL)
(→EECL)

emitter-emitter logic MICROEL (→emittergekoppelte Logik)
(→emitter-coupled logic)

emitter follower logic MICROEL **Emitterfolgerlogik**
= EFL = EFL

emitter junction MICROEL **Emitter-Sperrschicht**
= emitter barrier; emitter depletion layer = Emitter-Grenzschicht

emitter junction capacitance MICROEL **Emittersperrschicht-Kapazität**
= emitter barrier capacitance

emitter series resistance MICROEL **Emitterbahnwiderstand**
= emitter bulk resistance

emitter series resistor MICROEL **Emitter-Vorwiderstand**
emitter terminal MICROEL **Emitteranschluß**
= emitter connection; emitter 2 = Emitter 2

emitter voltage CIRC.ENG **Emitterspannung**
emitter zone MICROEL **Emitterzone**
[emitter region plus junction] [Emitterbereich plus Sperrschicht]

emitting aerial (→transmitting ANT (→Sendeantenne)
antenna)

emitting antenna (→transmitting ANT (→Sendeantenne)
antenna)

emitting broadcast studio BROADC **Sendestudio**
emitting power (→transmit RADIO (→Sendeleistung)
power)

emittter resistance CIRC.ENG **Emitterwiderstand**
↓ emitter series resistance; emitter bulk resistance ↓ Emitter-Vorwiderstand; Emitter-Bahnwiderstand

emmentropic PHYS **normalsichtig**
[Optik]
= emmentropisch

emphasize LING **hervorheben**
= highlight (v.t.) [figurativ]
= betonen

emphasizing pronoun LING **verstärkendes Pronomen**
empirical data (→experimental SCIE (→Versuchsdaten)
data)

employ (→use) TECH (→benutzen)
employable (→serviceable) TECH (→brauchbar)
employed ECON **Arbeitnehmer**
↓ Angestellter; Arbeiter
= wage and salary earners (n.pl.t.)

employee ECON **Angestellter**
= clerk ↑ Arbeitnehmer

employees (→workforce) ECON (→Belegschaft)
employer ECON **Arbeitgeber**
employers' association ECON **Unternehmerverband**
= Arbeitgeberverband
≈ Industrieverband

employment (→engagement) ECON (→Einstellung)
empties ECON **Leergut**
empty TERM&PER **leer**
= void; blank = unbeschriftet

empty disk (→blank DATA PROC (→Leerplatte)
disk)

empty diskette (→blank DATA PROC (→Leerdiskette)
diskette)

empty list DATA PROC **Leerliste**
= void list = leere Liste; Nulliste

empty rack EQUIP.ENG **Leergestell**
empty set MATH **Leermenge**
= blank set; null set = Nullmenge

empty tape (→void tape) DATA PROC (→Leerband)
em quad (→em) TYPOGR (→Geviert)
EMR (→electromagnetic PHYS (→elektromagnetische Strahlung)
radiation)

EMS (→electronic mail) TELEC (→elektronischer Briefdienst)
E&M signaling SWITCH **E&M-Signalisierung**
em space (→em) TYPOGR (→Geviert)
emulate (v.t.) DATA PROC **emulieren**
[to have the same computing behaviour as another system] [sich datentechnisch wie ein anderes System verhalten; emulate = engl. „wetteifern"]
≈ simulate ≈ simulieren

emulation DATA PROC **Emulation**
[simulation of a different computer system, by hardware or software means] [Simulation eines anderen Computersystems, mit Hardware- oder Softwaremitteln]

emulation mode DATA PROC **Emulationsmodus**
emulator 1 DATA PROC **Emulator 1**
[program allowing to run a program written for another type of computer] [Programm zum Betreiben eines für anderen Computertyp geschriebenen Programms]
= Emulationsprogramm

emulator 2 DATA PROC **Emulator 2**
[hardware add-on translating the machine code of other computer type] [Hardware-Zusatz der den Maschinencode eines anderen Computertyps übersetzt]

emulsion CHEM **Emulsion**
en TYPOGR **Halbgeviert**
[dymmy type or blank space with half the width of an em, corresponds approximately to the width of an n] [Leertype bzw. Leerstelle mit der halben Breite/Dickte eines Gevierts, in etwa der des n]
= en quad; en space = Halbgeviertstärke
≈ em ≈ Geviert

enable (v.t.) ELECTRON **freigeben**
[a process by an electonic signal] [einen Vorgang durch elektrisches Signal]
≈ trigger ≈ auslösen
≠ disable ≠ sperren

enable (v.t.) DATA PROC (→auslösen)
(→trigger)

enable (→release) DATA PROC (→Freigabe)
enable (n.) (→enable ELECTRON (→Freigabeimpuls)
signal)

enable (n.) (→enabling) ELECTRON (→Einschaltung)
enable pulse (→enable ELECTRON (→Freigabeimpuls)
signal)

enable signal ELECTRON **Freigabeimpuls**
= enabling signal; enable pulse; enabling pulse; enable (n.); bus enable; strobe (n.) = Freigabesignal; Enable-Signal

enabling ELECTRON **Einschaltung**
= enable (n.); connection (NAM); connexion (BRI) = Freigabe
≈ activation ≈ Auslösung
≠ disabling ≠ Abschaltung

enabling pulse (→enable ELECTRON (→Freigabeimpuls)
signal)

enabling signal (→enable ELECTRON (→Freigabeimpuls)
signal)

enamel (v.t.; enameled or TECH **emaillieren**
enamelled)

enamel (n.) TECH **Email**
enamel coat TECH **Emailüberzug**
enameled cable COMM.CABLE **Lackkabel**
enameled copper wire COMM.CABLE **Kupferlackdraht**
= insulated copper wire

enameled wire (→varnished COMM.CABLE (→Lackdraht)
wire)

enantiomorphic CHEM (→enantiomorph)
(→enantiomorphous)
enantiomorphous (adj.) CHEM **enantiomorph**
[mirro-image structured] [mit spiegelbildlicher Struktur]
≈ enantiomorphic
enantiomorphy CHEM **Enantiomorphie**
enantiotropic CHEM **enantiotrop**
enantiotropy CHEM **Enantiotropie**
[ability to pass from one state into another] [von einer Zustandsform in andere überführbar]
enc (→annex) LING (→Anlage)
encapsulate TECH **verkapseln**
≈ inclose = kapseln
≈ einschließen
encapsulated TECH **gekapselt**
= capsulated; enclosed = verkappt
encapsulation TECH **Verkapselung**
= incapsulation = Kapselung; Verkappung
encipher (→cipher) TELEC (→verschlüsseln)
enciphrement INF.TECH (→Verschlüsselung)
(→encryption)
enclose (BRI) (→inclose) TECH (→einschließen)
enclosed (→encapsulated) TECH (→gekapselt)
enclosed (→dust-proof) TECH (→staubdicht)
enclose in brackets MATH **einklammern**
= bracket (v.t.)
enclosure (→annex) LING (→Anlage)
enclosure (→case) EQUIP.ENG (→Gehäuse)
enclosure (→case) COMPON (→Gehäuse)
encode (v.t.) INF **codieren**
= code (v.t.) = kodieren (Duden)
encode (→cipher) TELEC (→verschlüsseln)
encoded telegram POST **verschlüsseltes Telegramm**
encoder DATA PROC **Encoder**
[produces a combination of outputs from one input]
encoder (→coder) CIRC.ENG (→Codierer)
encoder disk COMPON **Codierscheibe**
= Codescheibe
encoding (→coding) INF (→Codierung)
encoding law (→coding law) CODING (→Coderegel)
encrypt (→cipher) TELEC (→verschlüsseln)
encrypted speech (→encrypted voice) TELEC (→verschlüsselte Sprache)
encrypted voice TELEC **verschlüsselte Sprache**
= encrypted speech; secure voice; secure speech; coded voice; coded speech ≠ natürliche Sprache
≠ plain language
encryption INF.TECH **Verschlüsselung**
= enciphrement; scrambling; coding 2 = Chiffrierung
≠ decryption ≠ Entschlüsselung
encryption equipment TELEC **Verschlüsselungsgerät**
= coder
encryption key INF.TECH **Verschlüsselungscode**
≠ decryption key ≠ Entschlüsselungscode
↑ key ↑ Schlüssel
encryption method (→cipher system) INF.TECH (→Verschlüsselungssystem)
encryption system (→cipher system) INF.TECH (→Verschlüsselungssystem)
encs (→annex) LING (→Anlage)
encyclopaedia LING (→Enzyklopädie)
(→encyclopedia)
encyclopedia LING **Enzyklopädie**
[alphabetic or thematic collection of knowledge of one or all branches of science] [alphabetische oder thematische Ordnung des Wissenstoffes eines oder aller Fachgebiete]
= encyclopaedia
↓ lexicon ↓ Lexikon
end (n.) TECH **Ende 1**
= extreme; final point = Extrem; Endpunkt
≈ side ≈ Seite
end-about shift (→end-around shift) DATA PROC (→Ringschieben)
end-around carry DATA PROC **Rückwärtsübertrag**
[from most to least significant place] [vom höchsten zum niedrigsten Stellenwert]
= Endübertrag

end-around shift DATA PROC **Ringschieben**
[bits falling outside of word are replaced by zeros] [außerhalb der Wortgrenzen fallende Bits werden durch Nullen ersetzt]
= end-about shift; circular shift; cyclic shift; logic shift; logical shift = logisches Verschieben; zyklische Stellenverschiebung; Stellenwertverschiebung; logischer Schiebebefehl
end-around shift register CIRC.ENG (→Ringschieberegister)
(→cyclic shift register)
end capacitance (→top load) ANT (→Endkapazität)
end cell POWER SYS **Zusatzzelle**
[communications battery] [Fernmeldebatterie]
end-consumer (→final consumer) ECON (→Endverbraucher)
end cutting nippers TECH **Rabitzzange**
end effect ANT **Endeffekt**
= Verkürzungseffekt
end exchange (→terminal exchange) SWITCH (→Endamt)
end-fed ANT **endgespeist**
end fed (→end feed) ANT (→Endspeisung)
end-fed antenna (→Zeppelin antenna) ANT (→Zeppelin-Antenne)
end feed ANT **Endspeisung**
= end fed
end-fire antenna ANT **Längsstrahler**
= end-fire array; end-fire array antenna = längsstrahlende Dipolanordnung
↓ meander antenna; fishbone antenna; Yagi-Uda antenna ↓ Mäanderantenne; Fischgrätenantenne; Yagi-Uda-Antenne
end-fire array (→end-fire antenna) ANT (→Längsstrahler)
end-fire array antenna (→end-fire antenna) ANT (→Längsstrahler)
endicon ELECTRON **Endicon**
end identifier (→end label) DATA PROC (→Endekennsatz)
ending TECH **Ende 2**
= end part [das letzte Teil]
= Endteil
ending COLLOQ **Beendigung**
[action]
end label DATA PROC **Endekennsatz**
= end identifier; end mark = Endekennung; Endemarke; Endezeichen; Endekriterium
endless TECH **endlos**
= continuous 2 ≈ kontinuierlich; stufenlos
≈ continuous 1; stepwise
endless cycle (→endless loop) DATA PROC (→Endlosschleife)
endless loop DATA PROC **Endlosschleife**
= endless cycle; infinite loop = endlose Schleife
endless phase shifter CIRC.ENG **Endlosphasenschieber**
endless-tape cassette TERM&PER (→Endloscassette)
(→continuous cassette)
end mark (→end label) DATA PROC (→Endekennsatz)
end message DATA COMM **Endemeldung**
end-of-address DATA COMM **Adreßende**
[code] [Code]
= EOA = EOA
end-of-block DATA PROC **Blockende**
= EOB
end-of-block signal DATA COMM **Blockendesignal**
end-of-charging indication POWER SYS **Ladeschlußanzeige**
end-of-charging voltage POWER SYS **Ladeschlußspannung**
[accumulator] [Akkumulator]
end of conversation SWITCH **Gesprächsende**
end-of-data DATA PROC (→Dateiende)
(→end-of-file)
end-of-document DATA PROC (→Dateiende)
(→end-of-file)
end-of-document marker DATA PROC (→Dateiendeblock)
(→end-of-file mark)
end-of-file DATA PROC **Dateiende**
= EOF; end-of-document; end-of-data; EOD = EOF; EOD

end-of-file mark

end-of-file mark	DATA PROC	**Dateiendeblock**
= end-of-document marker		
end of job	DATA PROC	**Auftragsende**
end-of-line	DATA COMM	**Datenzeilenende**
= EOL		= EOL
end-of-medium [code]	DATA COMM	**Ende der Aufzeichnung** [Code]
= EM		= EM
end of message	DATA COMM	**Nachrichtenende**
= EOM		
end of number	SWITCH	**Wahlende**
= EON; end of selection		
end-of-number signal	SWITCH	(→Wahlendzeichen)
(→end-of-selection signal)		
end-of-page halt	DATA PROC	**Seitenende-Halt**
end of pulsing	SWITCH	**Nummernende**
end-of-record	DATA PROC	**Datensatzende**
= EOR		= EOR
end-of-ribbon indication	OFFICE	**Farbbandende-Anzeige**
end-of-run routine	DATA PROC	**Laufbeendigungsroutine**
end of selection (→end of number)	SWITCH	(→Wahlende)
end-of-selection signal	SWITCH	**Wahlendzeichen**
= end-of-number signal		
end of signal	TELEC	**Signalende**
end-of-tape mark	TERM&PER	**Bandenmarke**
↑ tape mark		↑ Bandenmarke
↓ tape header; tape trailer		
end of test	DATA PROC	**Prüfende**
≠ start of test		≠ Prüfbeginn
end of text [code]	DATA COMM	**Textende** [Code]
= ETX; EOT; end-of-transmission		= ETX; EOT; Ende der Übertragung
end of transmission block [a code]	DATA COMM	**Ende des Datenübertragungs- blocks** [ein Code]
= ETB		= ETB
end-of-word character	DATA PROC	**Wortendezeichen**
endorsee [who receives all the rirhts by the endorsement]	ECON	**Indossatar** [auf den beim Indossament alle Rechte übergehen]
		= Indossat; Girat; Giratar
endorser (→bill endorser)	ECON	(→Wechselgirant)
end part (→ending)	TECH	(→Ende 2)
end-point [vector]	MATH	**Endpunkt** [Vektor]
end pole [open wire line]	OUTS.PLANT	**Endgestänge** [Freileitung]
= dead-end pole; stayed terminal pole; strutted terminal pole		= Endmast
end-product (→final product)	MANUF	(→Endprodukt)
end punctuation mark	LING	**Satzschlußzeichen**
↑ punctuation mark		↑ Satzzeichen
↓ dot; question mark; exclamation mark		↓ Punkt; Fragezeichen; Ausrufezeichen
endrow rack	SYS.INST	**Reihenendgestell**
≠ headrow rack		≠ Reihenanfangsstell
end-scale value	INSTR	**Skalenendwert**
= full-scale value		= Bereichsendwert
≈ full-scale		≈ Vollausschlag
end-side cover	EQUIP.ENG	**Endverkleidung**
end stroke	TYPOGR	**Abstrich**
≠ beginning stroke		≠ Aufstrich
end system	DATA COMM	**Endsystem**
end-to-end mode (→end-to-end signaling)	SWITCH	(→durchgehende Zeichengabe)
end-to-end signaling (AM)	SWITCH	**durchgehende Zeichengabe**
= end-to-end signalling (BRI); end-to-end mode		= durchgehende Signalisierung
end-to-end signalling (BRI) (→end-to-end signaling)	SWITCH	(→durchgehende Zeichengabe)
end-to-end test (→link test)	TELEC	(→Streckenmessung)
endurance (→durability)	TECH	(→Dauerhaftigkeit)
endurance limit	MECH	**Ermüdungsgrenze**
= fatigue limit; fatigue strength		= Ermüdungsfestigkeit; Dauerfestigkeit; Dauerbeanspruchungsgrenze
endurance test (→permanent test)	QUAL	(→Dauerprüfung)
endurance testing	QUAL	**Langzeitprüfung**
= long-time testing		
endure	TECH	**überstehen**
= stand		= überdauern
≈ resist		≈ widerstehen
end-user [anyone without computer training]	DATA PROC	**Endbenutzer** [im Sinne von DV-Laie]
end-user equipment (→user terminal)	TELEC	(→Teilnehmergerät)
end-user service	DATA PROC	**Endbenutzer-Service**
end-user system (→dialog system)	DATA PROC	(→Dialogsystem)
energize	ELECTRON	**erregen**
= excite		[Relais]
energized (→live)	EL.TECH	(→heiß)
energizing circuit	CIRC.ENG	**Erregerkreis**
energy [the capacity for doing work]	PHYS	**Energie** [die Fähigkeit Arbeit zu leisten]
energy balance	PHYS	**Energiebilanz**
energy band [energy levels smeared to a band]	PHYS	**Energieband** [zu einem Band verschmierte Energietermen]
↓ conduction band; valence band		↓ Leitungsband; Valenzband
energy band density	PHYS	**Energiebanddichte**
energy band diagram	PHYS	**Bändermodell**
= energy level diagram; energy band scheme; energy level scheme		= Energiebändermodell; Niveauschema; Termschema
energy band edge (→band edge)	PHYS	(→Bandkante)
energy band scheme (→energy band diagram)	PHYS	(→Bändermodell)
energy band structure	PHYS	**Bandstruktur**
energy cable (→power cable)	POWER ENG	(→Starkstromkabel)
energy consumption	TECH	**Energieverbrauch**
≈ expenditure of energy		≈ Energieaufwand
energy conversion	PHYS	**Energieumwandlung**
= energy transformation		= Energietransformation; Energiewandlung
energy convertor	PHYS	**Energiewandler**
energy demand (→energy requirement)	TECH	(→Energiebedarf)
energy density	PHYS	**Energiedichte**
energy dispersal	RADIO	**Energieverwischung**
energy distribution	TECH	**Energieverteilung**
energy dose [SI unit: Gray]	PHYS	**Energiedosis** [SI-Einheit: Gray]
		= absorbed dose
energy-economizing (→energy-saving)	PHYS	(→energiesparend)
energy economy (→energy savings)	TECH	(→Energieeinsparung)
energy equivalent	PHYS	**Energieäquivalent**
energy exchange	PHYS	**Energieaustausch**
energy flow (→energy flux)	PHYS	(→Energiestrom)
energy flux	PHYS	**Energiestrom**
= energy flow		= Energiefluß
≈ power		≈ Leistung
energy gap (→forbidden energy band)	PHYS	(→verbotenes Band)
energy generation	TECH	**Energieerzeugung**
energy input	TECH	**Energieaufnahme**
energy jump	PHYS	**Energiesprung**
energy level	PHYS	**Energieniveau**
≈ energy band		= Energieterm; Term
		≈ Energieband
energy level diagram (→energy band diagram)	PHYS	(→Bändermodell)
energy level scheme (→energy band diagram)	PHYS	(→Bändermodell)
energy loss	PHYS	**Energieverlust**
		≈ Energieverbrauch
energy quantum	PHYS	**Energiequant**
		= Energiequantum
energy release	PHYS	**Energieabgabe**
energy requirement	TECH	**Energiebedarf**
= energy demand		
energy -saving	PHYS	**energiesparend**
= energy-economizing		
energy savings	TECH	**Energieeinsparung**
= energy economy		

energy signal	TELEC	**Energiesignal**	
energy source	PHYS	**Energiequelle**	
energy state	PHYS	**Energiezustand**	
energy state density	PHYS	**Zustandsdichte**	
= state density		= Energieniveaudichte	
energy storage	PHYS	**Energiespeicher**	
energy supply	TECH	**Energieversorgung**	
= power generation and distribution			
energy supply (→energy conversion)	PHYS	(→Energiewandlung)	
energy system	POWER SYS	**Stromversorgungsanlage**	
= power supply system; power station; power generation plant		= Stromversorgung	
energy transfer	PHYS	**Energieübertragung**	
= power transmission			
energy transformation (→energy conversion)	PHYS	(→Energieumwandlung)	
enforceability	ECON	**Vollstreckbarkeit**	
enforceable	COLLOQ	**verbindlich**	
≈ mandatory		≈ vollstreckbar; einklagbar	
		≈ vorgeschrieben	
engaged (→busy)	SWITCH	(→belegt)	
engaged lamp (→busy lamp)	TERM&PER	(→Besetztlampe)	
engaged tone (BRI) (→busy tone)	TELEPH	(→Besetztton)	
engagement	ECON	**Einstellung**	
[of personnel]		[Personal]	
= recruitment; employment		= Anstellung	
engine	TECH	**Maschine**	
= machine		↓ Motor	
↓ motor			
engine (→motor)	TECH	(→Motor)	
engine (→processor 1)	DATA PROC	(→Prozessor 1)	
engineer	TECH	**Ingenieur**	
≈ technician		≈ Techniker	
		↓ Diplom-Ingenieur	
engineer (v.t.) (→develop)	TECH	(→entwickeln)	
engineering	TECH	**Technik 1**	
[procedures and equipment to shape processes and objects]		[Verfahren und Einrichtungen zur Gestaltung von Prozessen und Gegenständen]	
= technology 1; technique 2; technic 2; art			
≈ technology 2		≈ Technologie	
↓ electrical engineering; mechanical engineering; civil engineering; chemical engineering; manufacturing engineering		↓ Elektrotechnik; Maschinenbau; Bautechnik; Chemische Technik; Fertigungstechnik	
engineering (→lay-out)	TECH	(→Auslegung)	
engineering (→development)	TECH	(→Entwicklung)	
engineering change note	TECH	**Änderungsanweisung**	
= modification guideline		= Änderungsrichtlinie	
		≈ Änderungsmitteilung	
engineering design	EQUIP.ENG	**konstruktive Entwicklung**	
engineering drawing	TECH	**Technisches Zeichnen**	
= technical drawing		↓ Maschinenbauzeichnen	
↓ mechanical drawing			
engineering drawing	ENG.DRAW	(→Konstruktionszeichnung)	
(→design drawing)			
engineering logic	INF	**Schaltalgebra**	
[Boolean algebra applied to binary circuits]		[Anwendung der Booleschen Algebra auf binäre Schaltungen]	
= computer logic; switching algebra; switching theory; digital logic; circuit algebra; computer arithmetic		= Schaltungsalgebra; Schaltlogik; Kontaktalgebra	
≈ Boolean algebra [MATH]		≈ Boolesche Algebra [MATH]	
engineering model (→laboratory prototype)	TECH	(→Labormuster)	
engineering order wire (→service channel)	TELEC	(→Dienstkanal)	
engineering order wire equipment (BRI) (→service channel equipment)	TELEC	(→Dienstkanaleinrichtung)	
engineering sample	MICROEL	**Funktionsmuster**	
		= Entwicklungsmuster	
engineering society	ECON	**Ingenieursverband**	
engineering terms of delivery	TECH	**technische Lieferbedingungen**	
= technical terms of delivery			
engineering tool	DATA PROC	**Entwicklungswerkzeug**	
= development tool; designer tool; designer kit		= Software-Entwicklungswerkzeug; Programmentwicklungssystem; Programmentwicklungswerkzeug	
≈ development software		≈ Entwicklungssoftware	
↑ tools		↑ Tool	
engineering tool (→tool)	DATA PROC	(→Programmierwerkzeug)	
engineer order wire (BRI) (→service channel)	TELEC	(→Dienstkanal)	
engine output	TECH	**Motorleistung**	
engine trouble	TECH	**Maschinenschaden**	
English (BRI) (→type size 14 point)	TYPOGR	(→Schriftgröße 14 Punkt)	
English inch	PHYS	**englischer Zoll**	
[2.5399956 cm]		[2,5399956 cm]	
≈ standard inch		≈ Normalzoll	
↑ inch		↑ Zoll	
english red	PHYS	**caput mortuum**	
engrave (v.t.)	MECH	**gravieren**	
enhance (→enrich)	MICROEL	(→anreichern)	
enhance (→amplify)	ELECTRON	(→verstärken)	
enhanced keyboard	TERM&PER	**Multifunktionstastatur**	
[with separate numeric keypad and cursor pad]		[mit getrenntem Ziffern- und Cursorblock]	
		= MF-Tastatur	
enhancement (→enrichment)	MICROEL	(→Anreicherung)	
enhancement (→improvement 1)	TECH	(→Verbesserung)	
enhancement (→gain)	NETW.TH	(→Verstärkung 1)	
enhancement layer	MICROEL	**Anreicherungszone**	
= accumulation layer; carrier concentration layer		= Anreicherungsschicht	
enhancement mode	MICROEL	(→Anreicherungstyp)	
(→enhancement type)			
enhancement-mode FET	MICROEL	**Anreicherungs-Isolierschicht-Feldeffekttransistor**	
enhancement surface layer	MICROEL	**Anreicherungsrandschicht**	
enhancement transistor	MICROEL	**Anreicherungstransistor**	
enhancement type	MICROEL	**Anreicherungstyp**	
= enhancement mode		= Anreicherungssteuerung	
enhancer	TERM&PER	**Hervorhebungsmerkmal**	
enhancer	DATA PROC	**Verbesserungsprodukt**	
[hardware or software]		= Leistungssteigerungsprodukt	
enlarge	MECH	**erweitern**	
enlarge	TECH	**vergrößern**	
= amplify; increase; augment; blow-up		= erweitern	
≠ reduce 2		≠ verkleinern	
enlargement	TECH	**Vergrößerung**	
= magnification; amplification		≈ Zuwachs; Erweiterung	
≈ increment; expansion		≠ Verkleinerung	
≠ reduction			
enneode	ELECTRON	**Enneode**	
		= Nonode	
enormous sum	ECON	**Unsumme**	
E notation (→scientific notation)	MATH	(→wissenschaftliche Zahlenschreibweise)	
ENQ (→inquiry)	DATA COMM	(→Stationsaufforderung)	
ENQ (→inquiry character)	DATA PROC	(→Abfragezeichen)	
en quad (→en)	TYPOGR	(→Halbgeviert)	
enquire (v.i.) (BRI) (→inquire)	DATA PROC	(→abfragen)	
enquiring station (→inquiry station)	DATA COMM	(→Abfragestation)	
enquiry (BRI) (→inquiry)	DATA PROC	(→Abfrage)	
enquiry (n.) (→inquiry)	TELECONTR	(→Abfrage)	
enquiry (BRI) (→inquiry)	SWITCH	(→Abfragen)	
enquiry (BRI) (→inquiry)	DATA COMM	(→Stationsaufforderung)	
enquiry (BRI) (→inquiry)	ECON	(→Anfrage)	
enquiry character (→inquiry character)	DATA PROC	(→Abfragezeichen)	
enquiry key (→interrogation key)	TERM&PER	(→Abfragetaste)	
enquiry language (BRI) (→query language)	DATA PROC	(→Abfragesprache)	
enquiry/response	TELEC	(→Frage-/Antwort-)	
(→inquiry/response)			
enquiry signal (→interrogation signal)	TELECONTR	(→Abfragesignal)	
enquiry station (→inquiry station)	DATA COMM	(→Abfragestation)	

enquiry terminal TERM&PER (→Abfrageterminal)
(→interrogation terminal)
ENR (→excess noise ratio) HF (→Überschußrauschverhältnis)
enrich MICROEL **anreichern**
= enhance; accumulate
enrichment MICROEL **Anreicherung**
= enhancement; accumulation ≠ Verarmung
≠ depletion
en space (→en) TYPOGR (→Halbgeviert)
ensure (→secure) COLLOQ (→sicherstellen)
enter (→override) TELEPH (→aufschalten)
enter (→input) DATA PROC (→eingeben)
enter command (→command ELECTRON (→Befehlseingabe)
entry)
ENTER key TERM&PER **Eingabetaste**
[activates the read-in of the keyed in- [löst das Einlesen des ein-
formation or the execution of the getippten Textes oder die
keyed instruction; derived from the Ausführung des eingetipp-
carriage-return function] ten Befehls aus; von der
= entry key; RETURN key; data re- Wagenrücklauffunktion ab-
lease key; data transmit key geleitet]
≈ carriage return [TELEGR] = Freigabetaste; RETURN-
 Taste; Taste RETURN;
 ENTER-Taste; Taste EN-
 TER; Datenübertragungs-
 taste; Datenfreigabetaste
 ≈ Wagenrücklauftaste
 [TELEGR]
enterprise (→company) ECON (→Gesellschaft)
entertainment electronics ELECTRON **Unterhaltungselektronik**
= home entertainment ≈ Haushaltselektronik
≈ household electronics ↑ Konsumelektronik
↑ consumer electronics
enthalpy PHYS **Enthalpie**
entitlement ECON **Berechtigung**
= authorization; clearance
entitlement (→right of use) DATA PROC (→Benutzungsberechtigung)
entity SCIE **Objekt**
= object
entity DATA PROC **Subjekt**
[subject of database or file] = Titel
entrain TECH **mitschleppen**
entrainment TECH **Mitschleppung**
 = Mitnahme
entrance (→entrance link) TRANS (→Zubringerstrecke)
entrance cable OUTS.PLANT **Einführungskabel**
entrance conversation station TELEPH (→Türsprechstelle)
(→entrance telephone)
entrance fee TELEC **Einrichtungsgebühr**
= establishing charge = Bereitstellungsgebühr
entrance link TRANS **Zubringerstrecke**
= entrance; feeder link; feeder; spur = Zubringer 1; Zuführungs-
link; spur verbindung
↓ radio relay spur link ↓ Richtfunkzubringer
entrance monitoring system SIGN.ENG **Türanlage**
= door monitoring system; gate moni-
toring system
entrance telephone TELEPH **Türsprechstelle**
= entrance conversation station
entrance test QUAL **Eingangsprüfung**
= receiving inspection; incoming inspec- = Eingangsrevision; Waren-
tion eingangsprüfung
entrance wire OUTS.PLANT **Einführungsdraht**
entrepreneur ECON **Unternehmer**
= businessman; undertaker (BRI) [Eigentümer]
≈ manager = Geschäftsmann
entrepreneurial ECON **unternehmerisch**
entrepreneurship ECON **Unternehmerschaft**
 = Unternehmertum
entrepreneurship (→corporate ECON (→Unternehmensführung)
management)
entropy PHYS **Entropie**
["transformation content"; extent to [„Verwandlungsinhalt";
which energy can be converted to der in Arbeit wandelbare
work] Teil der Energie]
entropy (→information entropy) INF (→Informationsentropie)
entropy encoding (→statistical CODING (→statistische Codierung)
encoding)

entry LING **Eintrag**
[in a dictionary] [in einem Wörterbuch]
≈ headword = Eintragung; Wortstelle
 ≈ Stichwort
entry (→booking) ECON (→Buchung)
entry (→data input) DATA PROC (→Dateneingabe)
entry (→item) LING (→Position)
entry (→note) DOC (→Vermerk)
entry (→entry point) DATA PROC (→Einsprung)
entry condition DATA PROC **Einsprungbedingung**
= initial condition
entry control (→access SIGN.ENG (→Zutrittskontrolle)
control)
entry control system (→access SIGN.ENG (→Zutrittskontrollsystem)
control system)
entry control terminal TERM&PER (→Zutrittskontrollterminal)
(→access control terminal)
entry field DATA PROC **Eingabefeld**
entry instruction DATA PROC **Einsprungbefehl**
[the first executed]
entry-interruption key TERM&PER (→ESCAPE-Taste)
(→ESCAPE key)
entry key (→ENTER key) TERM&PER (→Eingabetaste)
entry keyboard TERM&PER **Eingabetastatur**
= input keyboard = Erfassungstastatur; Einga-
 betastenfeld
entry level DATA PROC **Einsteigerniveau**
 = Anfängerniveau
entry-level model DATA PROC **Einstiegsmodell**
 = Anfängermodell
entry-level user [DATA PROC] TECH (→Anfänger)
(→novice)
entry mode DATA PROC **Eingabemodus**
↓ insert mode; overtype mode ↓ Einfügemodus; Überschreibmodus
entry point DATA PROC **Einsprung**
[programm point which can be [über Sprungbefehl erreich-
reached by a jump instruction] bare Programmstelle]
= entry = Einsprungpunkt
entry program DATA PROC **Eingabeprogramm**
entry requirement TECH **Aufnahmebedingungen**
entry signal RAILW.SIGN **Einfahrtsignal**
entry terminal TERM&PER **Eingabeterminal**
= input terminal = Erfassungsterminal
entry time DATA PROC **Einsprungzeitpunkt**
entry to station RAILW.SIGN **Bahnhofseinfahrt**
entry word (dictionary) LING (→Stichwort)
(→headword)
enumerable MATH **aufzählbar**
enumerate MATH **aufzählen**
envelop (n.) (NAM) POST **Briefumschlag**
= envelope (BRI) = Umschlag
envelope DATA PROC **fehlergeschützte Bitgruppe**
[containing error protecting bits] = Enveloppe
= error-protected bit group
envelope MATH **Hüllkurve**
 = Enveloppe; Einhüllende
envelope (→harmonic) MODUL (→Oberwelle)
envelope (BRI) (→envelop) POST (→Briefumschlag)
envelope (→disk jacket) TERM&PER (→Diskettenhülle)
envelope delay (→group delay) TELEC (→Gruppenlaufzeit)
envelope delay distortion TELEC (→Gruppenlaufzeitverzer-
(→group delay distortion) rung)
envelope demodulator MODUL **Hüllkurvendemodulator**
= envelope detection
envelope detection (→envelope MODUL (→Hüllkurvendemodulator)
demodulator)
envelope feeder (BRI) TERM&PER (→Briefumschlag-Zufüh-
(→envelop feeder) rung)
envelope gain CONTROL **Modulationsverstärkung**
envelope surface MATH **Hüllfläche**
envelope velocity (→group EL.TECH (→Gruppengeschwindigkeit)
velocity)
envelop feeder (NAM) TERM&PER **Briefumschlag-Zuführung**
= envelope feeder (BRI)
enveloping machine OFFICE **Kuvertiermaschine**
 = Einlegemaschine
environment DATA PROC **Landschaft**
 = Umgebung; Umfeld

environment (→ambient environment)	TECH	(→Umwelt)		EPROM [erasable with UV light] = erasable programmable read-only memory; MOS PROM; ultraviolet light-erasable PROM ≈ EEPROM ↑ read-only memory	MICROEL	EPROM [durch UV-Licht löschbar] = änderbarer programmierbarer Festwertspeicher; löschbarer programmierbarer Festwertspeicher ≈ EEPROM ↑ Festwertspeicher
environmental climate = ambient climate	TECH	Umweltklima = Umgebungsklima				
environmental condition = ambient condition	TECH	Umweltbedingung = Umgebungsbedingung; Betriebsklima				
environmental division	DATA PROC	Maschinenteil [COBOL]		EPROM board	DATA PROC	EPROM-Karte
environmentally favorable = non-polluting	TECH	umweltfreundlich		EPROM eraser	TERM&PER	EPROM-Löschgerät
environmental stress	QUAL	umgebungsbedingte Beanspruchung		EPROM programmer	MICROEL	EPROM-Programmiergerät = EPROM-Ladegerät; EPROM-Brenner
environmental temperature = ambient temperature; surrounding temperature	TECH	Umgebungstemperatur = Raumtemperatur		epsilon (n.) [an insignificant quantity]	COLLOQ	Quäntchen
environmental test	QUAL	Umweltprüfung		equal (v.t.; equaled (AM); equalled (BRI)) = equate	COLLOQ	gleichstellen = gleichziehen; erreichen
environment-dependent (→environment-induced)	TECH	(→umgebungsbedingt)				
environment-induced = ambient-induced; environment-dependent	TECH	umgebungsbedingt = umgebungsabhängig		equal (adj.) ≠ unequal	MATH	gleich ≠ ungleich
				equal-access (adj.) = with equality of access	TELEC	gleichberechtigt
EOA (→end-of-address)	DATA COMM	(→Adreßende)		equal band mode = common band mode; co-band mode ≠ transposed band mode	TRANS	Gleichlageverfahren = Gleichlagebetrieb ≠ Getrenntlageverfahren
EOB (→end-of-block)	DATA PROC	(→Blockende)				
EOD (→end-of-file)	DATA PROC	(→Dateiende)				
E.& O.E. = errors and omissions excepted	ECON	Irrtum vorbehalten		equal channel mode (→cochannel mode)	TRANS	(→Gleichkanalverfahren)
EOF (→end-of-file)	DATA PROC	(→Dateiende)		equal channel transmission (→cochannel mode)	TRANS	(→Gleichkanalverfahren)
EOL (→end-of-line)	DATA COMM	(→Datenzeilenende)				
EOM (→end of message)	DATA COMM	(→Nachrichtenende)		equal in size	TECH	gleichgroß
EON (→end of number)	SWITCH	(→Wahlende)		equality ≠ inequality	MATH	Gleichheit ≠ Ungleichheit
EOR (→electro-optical reconnaissance)	OPT.COMM	(→elektrooptische Erkennung)		equality (→EXCLUSIVE-OR operation)	ENG.LOG	(→EXKLUSIV-ODER-Verknüpfung)
EOR (→end-of-record)	DATA PROC	(→Datensatzende)		equality function (→EXCLUSIVE-OR operation)	ENG.LOG	(→EXKLUSIV-ODER-Verknüpfung)
EOT (→end of text)	DATA COMM	(→Textende)				
EP (→electronic publishing)	DATA PROC	(→Electronic Publishing)		equality operation (→EXCLUSIVE-OR operation)	ENG.LOG	(→EXKLUSIV-ODER-Verknüpfung)
E pattern (→E-plane pattern)	ANT	(→E-Diagramm)		equalization	TRANS	Entzerrung
epen-tube process	MICROEL	Durchstömverfahren		equalization (→compensation)	TECH	(→Ausgleich)
ephemeral [of short duration] ≈ transient	SCIE	ephemer [von kurzem Bestand] ≈ vorübergehend		equalization (→smoothing)	EL.TECH	(→Glättung)
				equalization stage	CIRC.ENG	Entzerrungsstufe
				equalize = compensate; balance out	TECH	ausgleichen
ephemerides [tables of daily firmamental configuration]	ASTROPHYS	Ephemeriden [Tabellen der täglichen Gestirnkonfiguration]				
				equalize (→balancing)	NETW.TH	(→Abgleich)
				equalize (→smooth)	EL.TECH	(→glätten)
epi-base transistor	MICROEL	Epitaxial-Basistransistor = Epibasis-Transistor		equalizer [telephone set]	TELEPH	Ausgleichsschaltung [Fernsprechapparat]
EPIC process [epitaxial passivated integrated circuit process]	MICROEL	EPIC-Verfahren		equalizer	NETW.TH	Entzerrer
				equalizing capacitor = compensating capacitor	COMM.CABLE	Ausgleichkondensator
epicycloid	MATH	Epizykloid		equalizing charge [of an accumulator]	POWER SYS	Starkladung [eines Akkumulators]
episcope	OFFICE	Episkop				
epitaxial	MICROEL	epitaktisch		equalizing coil	CIRC.ENG	Ausgleichspule
epitaxial diffused-junction transistor	MICROEL	Epitaxial-Planartransistor = Epitaxie-Planartransistor		equalizing network 1 [of distortions] ≈ compensating network	CIRC.ENG	Entzerrernetzwerk = Entzerrungsnetzwerk ≈ Ausgleichsnetzwerk
epitaxial diode	MICROEL	Epitaxiediode				
epitaxial growth	MICROEL	Aufwachsen [Kristall]		equalizing network 2 (→correcting network)	CIRC.ENG	(→Ausgleichnetzwerk)
epitaxial growth method	MICROEL	Aufwachsverfahren = Epitaxieverfahren		equalizing pulse	TV	Ausgleichsimpuls = Trabant
epitaxial layer	MICROEL	Epitaxialschicht		equal lasting	INSTR	zeitsymmetrisch
epitaxial transistor	MICROEL	Epitaxialtransistor = epitaxialer Transistor		equally spaced (→equidistant)	MATH	(→äquidistant)
epitaxy	MICROEL	Epitaxie = Aufwachstechnik		equal sign [=] = equal symbol ↑ mathematical symbol	MATH	Gleichheitszeichen [=] ↑ mathematisches Zeichen
E plane	ANT	E-Ebene				
E-plane bend (→E-plane elbow)	MICROW	(→E-Bogen)				
E-plane elbow = E-plane bend; flatwise bend	MICROW	E-Bogen		equal symbol (→equal sign)	MATH	(→Gleichheitszeichen)
				equal-type ≈ homogeneous; similar	COLLOQ	gleichartig = artgleich ≈ homogen; ähnlich
E-plane pattern = E pattern; E diagram	ANT	E-Diagramm = E-Ebenen-Diagramm				
E-plane sectorial horn	ANT	E-Sektorhorn		equate	MATH	gleichsetzen
EPOS (→electronic point-of-sale)	DATA COMM	(→Kassenterminalsystem)		equate (→equal)	COLLOQ	(→gleichstellen)
				equation [equality of two mathematical expressions] = relation ≈ formula	MATH	Gleichung [Gleichsetzung zweier mathematischer Ausdrücke] ≈ Formel
epoxy	CHEM	Epoxyd				
epoxy casting resin	CHEM	Epoxidgießharz				
epoxy glass fiber board ↑ printed circuit board	ELECTRON	Epoxydglasfaserplatte ↑ Leiterplatte				
epoxy resin	CHEM	Epoxydharz		equiangular (→isogonal 1)	MATH	(→gleichwinklig)

equiangular spiral antenna (→logarithmic spiral antenna)	ANT	(→logarithmische Spiralantenne)	
equidistant	MATH	**äquidistant**	
= equally spaced		= abstandsgleich	
equidistant line	ENG.DRAW	**Äquidistante**	
		= Gleichabstandslinie	
equilateral	MATH	**gleichseitig**	
equilateral triangle	MATH	**gleichseitiges Dreieck**	
equilibrium	PHYS	**Gleichgewicht**	
equilibrium position	PHYS	**Gleichgewichtslage**	
		= Beharrungslage	
equilibrium state (→balanced state)	SCIE	(→Gleichgewichtszustand)	
equilibrium value (→final value)	TECH	(→Endwert)	
equip (v.t.)	TECH	**ausrüsten**	
		= ausstatten; einrichten	
equipartition (→uniform distribution)	MATH	(→Gleichverteilung)	
equiphase	PHYS	**Gleichphase**	
		≈ Phasengleichheit	
equipmemt configuration option	EQUIP.ENG	**Bestückungsvariante**	
= equipment variant		= Belegungsvariante	
equipmemt designer	TECH	**Geräteentwickler**	
equipment 1 (n.s.t.)	TECH	**Ausstattung**	
= fitting-out; outfit; features; furnishing; accoutrments		= Ausrüstung; Aufmachung	
equipment 2	TECH	**Gerät**	
[an implement to execute a process]		[Gegenstand zur Durchführung eines Prozesses]	
= rig (n.)			
≈ set of equipment; device		≈ Gerätschaften; Vorrichtung	
↓ apparatus		↓ Apparat	
equipment 3	TECH	**Geschirr**	
equipment (→set of equipment)	TECH	(→Gerätschaften) (pl.t.)	
equipment allocation	DATA PROC	**Gerätezuordnung**	
= device allocation; equipment assignment; device assignment			
equipment assignment (→equipment allocation)	DATA PROC	(→Gerätezuordnung)	
equipment case	EQUIP.ENG	**Gerätekoffer**	
equipment configuration plan (→allocation scheme)	EQUIP.ENG	(→Belegungsplan)	
equipment configuration (→equipping)	EQUIP.ENG	(→Gerätebestückung)	
equipment engineering	ELECTRON	**Gerätetechnik**	
≈ mechanical design		≈ Bauweise; Konstruktion; Geräteaufbau	
equipment expenditure	ELECTRON	**Geräteaufwand**	
equipment failure	EQUIP.ENG	**Gerätestörung**	
equipment failure alarm	EQUIP.ENG	**Gerätestörungsalarm**	
equipment fuse	EQUIP.ENG	**Gerätesicherung**	
equipment generation	EQUIP.ENG	**Gerätegeneration**	
equipment identification center	MOB.COMM	**Geräteidentifizierungszentrum**	
equipment implementation	ELECTRON	**Geräterealisierung**	
≈ hardware implementation [DATA PROC]		≈ Hardware-Realisierung [DATA PROC]	
equipment internal climate	QUAL	**Geräteinnenraumklima**	
equipment layout plan (→installation layout)	SYS.INST	(→Aufstellungsplan)	
equipment manual	TELEC	**Gerätehandbuch**	
		≈ Gebrauchsunterlage; Anwendungsrichtlinie	
equipment manufacturer	TECH	**Gerätehersteller**	
equipment mounting device	EQUIP.ENG	**Geräteaufnahme**	
≈ module frame		≈ Baugruppenrahmen	
equipment number	DOC	**Lagennummer**	
equipment number table	DOC	**Lagentabelle**	
equipment plan (→allocation scheme)	EQUIP.ENG	(→Belegungsplan)	
equipment redundancy	QUAL	**Geräteredundanz**	
equipment row (→rack row)	SYS.INST	(→Gestellreihe)	
equipment shelf (→module frame)	EQUIP.ENG	(→Baugruppenrahmen)	
equipment side	TELEC	**Geräteseite**	
≠ line side		≠ Leitungsseite	
equipment specifications	ELECTRON	**Gerätespezifikation**	
equipment suite (→rack row)	SYS.INST	(→Gestellreihe)	
equipment summary (→overall equipment list)	DOC	(→Geräteübersicht)	
equipment supplier	TECH	**Gerätelieferant**	
= equipment vendor			
equipment variant	EQUIP.ENG	(→Bestückungsvariante)	
(→equipmemt configuration option)			
equipment vendor (→equipment supplier)	TECH	(→Gerätelieferant)	
equipment wire	COMM.CABLE	**Schaltdraht**	
equiponderation	MATH	**Gleichbewertung**	
equipotent	MATH	**gleichmächtig**	
equipotential (n.)	PHYS	**Äquipotential**	
equipotential line	PHYS	**Äquipotentiallinie**	
= contour line			
equipotential surface	PHYS	**Äquipotentialfläche**	
		= Niveaufläche	
equipped	EQUIP.ENG	**bestückt**	
equipped PCB (→assembled PCB)	ELECTRON	(→bestückte Leiterplatte)	
equipping	EQUIP.ENG	**Gerätebestückung**	
≈ packaging; equipment configuration		= Bestückung	
equipping guide	EL.TECH	**Bestückungsanweisung**	
equiprobability	MATH	**Gleichwahrscheinlichkeit**	
equiprobability curve	MATH	**Gleichwahrscheinlichkeitskurve**	
equiprobable	MATH	**gleichwahrscheinlich**	
equisignal	RADIO NAV	**Leitstrahl**	
= ILS beam		= Equisignal	
equisignal line	RADIO NAV	**Leitstrahllinie**	
		= Equisignallinie	
equisignal lobing	RADIO NAV	**Leitstrahldrehung**	
equisignal localizer	RADIO LOC	**Leitstrahlortungsgerät**	
equisignal radio range beacon	RADIO NAV	**Leitstrahlbake**	
		= Equisignalbake	
equisignal sector	RADIO NAV	**Leitstrahlsektor**	
		= Equisignalsektor	
equisignal system	RADIO NAV	**Leitstrahlverfahren**	
equisignal zone	RADIO NAV	**Leitstrahlzone**	
		= Equisignalzone	
equity participation (→shareholding)	ECON	(→Kapitalbeteiligung)	
equity stock (→ordinary share)	ECON	(→Stammaktie)	
equivalence	ENG.LOG	**Äquivalenz**	
		= Entsprechung	
equivalence (→equivalence operation)	ENG.LOG	(→Äquivalenzverknüpfung)	
equivalence circuit (→equivalence gate)	CIRC.ENG	(→Äquivalenzglied)	
equivalence element (→equivalence gate)	CIRC.ENG	(→Äquivalenzglied)	
equivalence function (→equivalence operation)	ENG.LOG	(→Äquivalenzverknüpfung)	
equivalence gate	CIRC.ENG	**Äquivalenzglied**	
= equivalence element; equivalence circuit; EXNOR gate; EXNOR element; EXNOR circuit; identity gate; identity element; identity circuit		= Äquivalenzgatter; Äquivalenzelement; Äquivalenzschaltung; Äquivalenztor; EXKLUSIV-WEDER-NOCH-Glied; EXKLUSIV-WEDER-NOCH-Gatter; EXCLUSIV-WEDER-NOCH-Element; EXKLUSIV-WEDER-NOCH-Schaltung; EXKLUSIV-WEDER-NOCH-Tor; EXNOR-Glied; EXNOR-Gatter; EXNOR-Element; EX-NOR-Schaltung; EXNOR-Tor; Identitätsglied; Identitätsgatter; Identitätselement; Identitätsschaltung	
equivalence operation [output = 1 if A = B]	ENG.LOG	**Äquivalenzverknüpfung** [Ausgang = 1 wenn A = B]	
= equivalence function; equivalence; EXCLUSIVE-NOR operation; EX-		= Äquivalenzfunktion; Äquivalenz; EXKLUSIV-WE-	

CLUSIVE-NOR function; EXNOR operation; EXNOR function; EXNOR; identity operation; identity function; identity
↑ logic operation

equivalence relation ENG.LOG
(→equivalence operation)
equivalent (n.) ECON
equivalent (adj.) SCIE
≈ corresponding
equivalent (→of equal area) MATH
equivalent (→overall loss) TELEC
equivalent bit rate [DATA COMM] (→transmission rate) TELEC
equivalent capacitance NETW.TH

equivalent circuit NETW.TH
= equivalent circuit diagram; equivalent network; equivalent network diagram; replacement scheme

equivalent circuit diagram NETW.TH
(→equivalent circuit)
equivalent concentration PHYS
equivalent conductivity PHYS
equivalent dose PHYS
[SI unit: Sievert]
equivalent earth radius RADIO PROP
equivalent four-pole NETW.TH
(→equivalent two-port)
equivalent four wire carrier system TRANS
equivalent isotropic radiated power RADIO
= EIRP
equivalent line circuit NETW.TH
equivalent network NETW.TH
(→equivalent circuit)
equivalent network diagram NETW.TH
(→equivalent circuit)
equivalent noise bandwidth TELEC
equivalent noise conductance TELEC
equivalent of heath PHYS
= thermal equivalent
equivalent peak power TELEC
equivalent quadripole NETW.TH
(→equivalent two-port)
equivalent resistance EL.TECH
≈ loss resistance
equivalent sign MATH
= similar sign; proportional sign

equivalent solid angle ANT
equivalent source NETW.TH

equivalent thermal network MICROEL
equivalent time constant EL.TECH
equivalent T-network MICROEL
equivalent transistor circuit MICROEL
equivalent two-port NETW.TH
= equivalent quadripole; equivalent four-pole
equivalent two-terminal NETW.TH
(→equivalent two-terminal network)
equivalent two-terminal network NETW.TH
[with equal impedance]
= equivalent two-terminal

DER-NOCH-Verknüpfung; EXNOR-Verknüpfung; EXNOR-Funktion; Äquivalenzrelation; Identitätsverknüpfung; Identität
↑ logische Verknüpfung
(→Äquivalenzverknüpfung)

Gegenwert
äquivalent
= gleichwertig
≈ entsprechend
(→flächentreu)
(→Restdämpfung)
(→Übertragungsgeschwindigkeit)

Ersatzkapazität
[Rechengröße]
= Ersatzkondensator; Äquivalenzkapazität; Äquivalenzkondensator; Verlustkapazität; Verlustkondensator

Ersatzschaltbild
= Ersatzschaltung; Ersatzbild; äquivalente Schaltung; Äquivalenzschaltung; Ersatznetzwerk; äquivalentes Netzwerk; Äquivalenznetzwerk
(→Ersatzschaltbild)

Äquivalentkonzentration
Äquivalentleitfähigkeit
Äquivalentdosis
[SI-Einheit: Sievert]
äquivalenter Erdradius
(→Vierpolersatzschaltung)

Vierdrahtgetrenntlage-System

äquivalente isotrop abgestrahlte Leistung
= EIRP
Leitungsersatzschaltbild
(→Ersatzschaltbild)

(→Ersatzschaltbild)

äquivalente Rauschbandbreite
Rauschleitwerk
Wärmeäquivalent

äquivalente Spitzenleistung
(→Vierpolersatzschaltung)

Ersatzwiderstand
= Äquivalenzwiderstand
≈ Verlustwiderstand
Ähnlichzeichen
= Proportionalitätszeichen; Tilde
äquivalenter Raumwinkel
Ersatzstromquelle
= Ersatzgenerator
thermische Ersatzschaltung
Ersatzzeitkonstante
T-Ersatzschaltung
Transistor-Ersatzschaltung
Vierpolersatzschaltung
= äquivalenter Vierpol

(→äquivalenter Zweipol)

äquivalenter Zweipol
[gleicher Impedanz]

equivocality MATH
= ambiguity
equivocation INF
[the information lost on the transmission path]
Er (→erbium) CHEM
erasable DATA PROC
= clearable
erasable laser optical disk TERM&PER
(→ELOD)
erasable memory DATA PROC
= erasable storage; erasable store
↓ EPROM; REPROM

erasable optical memory TERM&PER
erasable programmable read-only memory (→EPROM) MICROEL
erasable storage (→erasable memory) DATA PROC
erasable store (→erasable memory) DATA PROC
erase (v.t.) DATA PROC
= cancel; destroy; clear; eliminate; remove; rub-out
↓ delete
erase (v.t.) ENG.DRAW
erase ELECTRON
= clear; delete; cancel; unmark; destroy; purge
≈ reset (v.t.); zero (v.t.)
≠ set
erase (→erasing) ELECTRON
erase character DATA PROC
= delete character; rub-out character
erase command (→delete statement) DATA PROC
erase head TERM&PER
= erasing head
≠ read head; write head
↑ magnetic head
erase in area DATA COMM
[code]
= EA
erase input CIRC.ENG
= delete input; cancel input; clear input
erase instruction (→delete statement) DATA PROC
erase job (→delete statement) DATA PROC
erase key (→backspace key) TERM&PER
eraser MICROEL
[for EPROM and similars]
≠ programmer
eraser ENG.DRAW
= rubber
erase signal (→reset pulse) ELECTRON
erase statement (→delete statement) DATA PROC
erasing ELECTRON
= erase; deletion; elimination
≈ zeroing; reset
erasing head (→erase head) TERM&PER
erasure character DATA PROC
= invalidation character
erasure of errores INF
erbium CHEM
= Er
erbium-doped OPT.COMM
erect TECH
= set-up

erection (→set up 1) TECH
erection kit (→kit) TECH
erg PHYS
[old unit for work; = 1 dyn · 1 cm]

ergodic MATH
ergodic theorem MATH
ergonometry ECON

Mehrdeutigkeit
= Vieldeutigkeit
↓ Zweideutigkeit
Äquivokation
[bei der Übertragung verlorengegangene Information]

(→Erbium)
löschbar

(→ELOD)

löschbarer Speicher
[überschreibbarer Festwertspeicher]
↓ EPROM; REPROM

löschbarer optischer Speicher
(→EPROM)

(→löschbarer Speicher)

(→löschbarer Speicher)

löschen
= entfernen
↓ ausfügen

radieren
löschen
= tilgen
≈ rücksetzen; nullen
≠ setzen

(→Löschung)
Löschzeichen

(→Löschanweisung)

Löschkopf
≠ Lesekopf; Schreibkopf 2
↑ Magnetkopf

Bereich löschen
[Code]
= EA
Löscheingang

(→Löschanweisung)

(→Löschanweisung)

(→Rücksetztaste 2)
Löschgerät
[für EPROM u. dgl]
≠ Programmiergerät
Radiergummi

(→Rücksetzimpuls)
(→Löschanweisung)

Löschung
= Löschen; Tilgung
≈ Nullung; Rücksetzen
(→Löschkopf)
Irrungszeichen
= Korrekturzeichen
Fehlerlöschung
Erbium
= Er
Erbium-dotiert
errichten
= aufbauen; aufstellen 1; montieren
(→Errichtung)
(→Baukasten)
Erg
[alte Einheit für Arbeit; = 1 Dyn · 1 cm]
ergodisch
Ergodentheorem
Ergonometrie

ergonomic 204

ergonomic (adj.)	ECON	ergonomisch	error-detecting	CODING	(→selbstprüfend)
ergonomics	SCIE	(→Ergonomie)	(→self-checking)		
(→ergonomy)			error-detecting code	CODING	selbstprüfender Code
ergonomy	ECON	Ergonomie	= error-detection code; error-checking		= fehlererkennender Code;
[science of optimum working conditions]		[Lehre von der optimalen Arbeitsbedingungen]	code; error-check code; self-checking code; self-check code		Fehlererkennungscode
= ergonomics		= Ergonomik; Arbeitswissenschaft	≈ error-correcting code		≈ fehlerkorrigierender Code
			error detection	INF	Fehlererkennung
erl (→erlang)	SWITCH	(→Erlang)	error detection (→fault detection)	TELECONTR	(→Fehlererfassung)
erlang	SWITCH	Erlang			
[traffic unit]		[Maß für Belegungsdichte]	error detection (→fault detection)	QUAL	(→Fehlererkennung)
= erl		= Erl	error-detection code	CODING	(→selbstprüfender Code)
Erlang distribution	MATH	Erlang-Verteilung	(→error-detecting code)		
erode	METAL	erodieren	error detector	CONTROL	Fehlerdetektor
erogation (→effort)	ECON	(→Aufwand)	error detector	INSTR	Fehlerdetektor
EROM (→electromagnetic read-only memory)	MICROEL	(→elektromagnetischer Festwertspeicher)	error diagnostics	DATA PROC	(→Diagnosemeldung)
			(→diagnostic message)		
erosion	METAL	Erodieren	erroed (→erroned)	INF.TECH	(→fehlerbehaftet)
		= Erosion	erroed bit (→erroned bit)	INF	(→fehlerbehaftetes Bit)
erosion printer	TERM&PER	Erosionsdrucker	erroed second (→erroned second)	TRANS	(→fehlerbehaftete Sekunde)
ERP (→effective radiated power)	ANT	(→äquivalente Strahlungsleistung)	error file	DATA PROC	Fehlerdatei
ERR (→excessive error rate)	TELEC	(→Fehlerhäufigkeitsüberschreitung)	= failure file; fault file		= Störungsdatei
			error flag	DATA PROC	Fehlerkennzeichen
erratic 2	TECH	anormal			[Fehler im Programm]
= abnormal		= von der Norm abweichend			↑ Merker
erratic 1 (→intermittent)	TECH	(→intermittierend)	error-free	TECH	einwandfrei
erratum (pl.-a)	TYPOGR	Druckfehler	= unobjectionable; error-free		= fehlerfrei
erroned	INF.TECH	fehlerbehaftet	error-free	INF	fehlerfrei
= errored		≠ fehlerfrei	= clean		
≠ error-free			error-free	TELEC	fehlerfrei
erroned bit	INF	fehlerbehaftetes Bit	≈ interference-free		≈ störungsfrei
= errored bit; EB		= gestörtes Bit	≠ erroned		≠ fehlerbehaftet
≈ bit error		≈ Bitfehler	error-free (→error-free)	TECH	(→einwandfrei)
erroned second	TRANS	fehlerbehaftete Sekunde	error-free (→fault-free)	TECH	(→fehlerfrei)
= errored second; ES		= gestörte Sekunde	error-free second	TRANS	fehlerfreie Sekunde
erroneous (→faulty)	QUAL	(→fehlerhaft)	= EFS		
erroneous decision	TECH	Fehlentscheidung	error function	MATH	Fehlerfunktion
= false decision			= aberration function		= Fehlerverteilungsfunktion
erroneous information (→misinformation)	TECH	(→Fehlinformation)	error guessing	DATA PROC	Fehlervermutung
			error handling	DATA PROC	Fehlerbehandlung
erroneous representation (→misrepresentation)	TECH	(→Fehldarstellung)	= error management; fault handling; fault management		= Störungsbehandlung
erroneuos identification (→misidentification)	TECH	(→Fehlerkennung)	error integral	MATH	Fehlerintegral
					= Gaußsches Integral
error	INF	Fehler	error interrupt 1	DATA PROC	Fehlerunterbrechung
error	INSTR	Fehler	error interrupt 2 (→error interrupt signal)	DATA PROC	(→Fehlerunterbrechungssignal)
		= Unsicherheit			
error	MATH	Fehler	error interrupt signal	DATA PROC	Fehlerunterbrechungssignal
error (→fault)	TECH	(→Fehler)	= error interrupt 2		
error (→deviation)	CONTROL	(→Regelabweichung)	error limit	INSTR	Fehlergrenze
error (→bug)	DATA PROC	(→Fehler)	error listing (→failure log)	DATA PROC	(→Störungsprotokoll)
error block number	TELECONTR	Fehlerblockzahl			
error burst	INF	Fehlerbündel	error log (→failure log)	DATA PROC	(→Störungsprotokoll)
		= plötzliche Fehlerhäufung			
error cause (→fault cause)	QUAL	(→Fehlerursache)	error logging (→failure log)	DATA PROC	(→Störungsprotokoll)
error-check code (→error-detecting code)	CODING	(→selbstprüfender Code)	error management (→error handling)	DATA PROC	(→Fehlerbehandlung)
error checking	DATA PROC	Fehlerprüfung	error measuring set	INSTR	Fehlermeßplatz
≈ error correction		Fehlerkorrektur	error message	DATA PROC	Fehlermeldung
error-checking code (→error-detecting code)	CODING	(→selbstprüfender Code)	[explanatory line]		[mit erklärendem Text]
			= diagnostic error message; diagnostics		= Fehlerhinweis
error code	CODING	Fehlercode	error message	DATA PROC	Fehlerhinweis
error condition	DATA PROC	Fehlerzustand	= diagnostics		
error control 1	DATA COMM	Fehlerkontrolle	error monitoring	DATA COMM	Fehlerüberwachung
≈ error monitoring		= Fehlerüberwachung	= error control 2		
error control 2 (→error monitoring)	DATA COMM	(→Fehlerüberwachung)	error printout (→failure log)	DATA PROC	(→Störungsprotokoll)
error corrected	TECH	fehlerbereinigt	error probability	DATA COMM	Fehlerwahrscheinlichkeit
error-correcting	CODING	fehlerkorrigierend	error propagation	MATH	Fehlerfortpflanzung
= self-correcting		= selbstkorrigierend	error-protected bit group (→envelope)	DATA PROC	(→fehlergeschützte Bitgruppe)
≈ self-checking		≈ selbstprüfend			
error-correcting code	CODING	fehlerkorrigierender Code	error protection	DATA COMM	Übertragungssicherung
= self-correcting code; correcting code		= selbstkorrigierender Code; Fehlerkorrekturcode	≈ error control		= Fehlerschutz
≈ error-detecting code		≈ selbstprüfender Code	error rate	INF	Fehlerrate
			[error per unit of time]		[Fehler pro Zeiteinheit]
error correction	CODING	Fehlerkorrektur	≈ error ratio		= Fehlerquote
= error recovery			error rate (→failure rate)	QUAL	(→Ausfallrate)
error detectability	TECH	Fehlererkennbarkeit			
		= Fehlernachweisbarkeit			

error ratio		INF	Fehlerquote	estimated value (→estimate)		MATH	(→Schätzwert)

error ratio | INF | Fehlerquote
[errored bits by total bits] | | [Fehler pro Gesamtbitzahl]
= bit error ratio; BER | | ≈ Fehlerrate
≈ error rate

error ratio | CONTROL | Regelfaktor
error recovery (→error correction) | CODING | (→Fehlerkorrektur)
error routine | DATA PROC | Fehlerroutine
errors and omissions excepted (→E.& O.E.) | ECON | (→Irrtum vorbehalten)
error source | TECH | Fehlerquelle
error tape | DATA PROC | Fehlerband
error trapping | DATA PROC | Fehlerabfangen
ERS (AM) (→emergency telephone) | TELEC | (→Notrufsäule)
ES (→erroned second) | TRANS | (→fehlerbehaftete Sekunde)
Es (→einsteinium) | CHEM | (→Einsteinium)
Esaki diode (→tunnel diode) | MICROEL | (→Tunneldiode)
ESC (→ESCAPE character 1) | DATA COMM | (→Codewechselzeichen)
ESC (→ESCAPE) | DATA PROC | (→Codeumschaltung)
escalation clause (→price escalation clause) | ECON | (→Preisformel)
escalation formula (→price escalation clause) | ECON | (→Preisformel)
escape (v.i.) | TECH | entweichen
ESCAPE (n.) | DATA PROC | Codeumschaltung
= ESC; code change | | [Code]
| | = ESC; Umschaltung; Codewechsel
ESCAPE character 1 | DATA COMM | Codewechselzeichen
= ESC | | = Codeumschaltezeichen
escape character 2 (→escape signal) | DATA COMM | (→Austrittssignal)
escape code | DATA COMM | Steuercode
escape instruction system | SIGN.ENG | Fluchtleitsystem
ESCAPE key | TERM&PER | ESCAPE-Taste
[activates the interruption of an instruction or program, or shift functions] | | [löst die Unterbrechung einer Befehlseingabe oder eines Programms aus, oder Umschaltefunktionen]
= ESC key; entry-interruption key | | = ESC-Taste; Codeumschalttaste; Codewechseltaste; Abbruchtaste; Eingabeunterbrechungs-Taste
↑ functional key | | ↑ Funktionstaste
escapement | TERM&PER | eingestellter Papiervorschub
= preset paper feed | |
escape sequence | DATA PROC | Escape-Sequenz
escape signal | DATA COMM | Austrittssignal
[indicates change to nonstandard transmission] | | = Escape-Folge
= escape character 2 | |
escape symbol | DATA PROC | Fluchtsymbol
ESC key (→ESCAPE key) | TERM&PER | (→ESCAPE-Taste)
escutcheon | TECH | Herstellerplakette
| | = Herstellerschild; Markenschild
ESD | COMPON | elektrostatisch gefährdet
[liable to electrostatic danger] | | = EGB
ESFI | MICROEL | ESFI-Technik
[epitaxial silicon film on insulator technique] | |
Es layer | RADIO PROP | Es-Schicht
establish (v.t.) | COLLOQ | feststellen
established | ECON | eingeführt
| | = etabliert
established (→usual) | COLLOQ | (→üblich)
establishing charge (→entrance fee) | TELEC | (→Einrichtungsgebühr)
estimate (v.t.) | ECON | veranschlagen
= rate; project | |
estimate (n.) | COLLOQ | Schätzung
= estimation; valuation; appraisal | | = Abschätzung
estimate (n.) | MATH | Schätzwert
= estimated value | |
estimate (n.) (→appraise) | COLLOQ | (→schätzen)
estimated 1 | COLLOQ | geschätzt
estimated 2 | COLLOQ | voraussichtlich

estimated value (→estimate) | MATH | (→Schätzwert)
estimated value (→appraised value) | ECON | (→Schätzwert)
estimation (→estimate) | COLLOQ | (→Schätzung)
estimation error | MATH | Schätzfehler
estimator | MATH | Schätzfunktion
ET (→exchange termination) | TELEC | (→Zellabschluß)
ETB (→end of transmission block) | DATA COMM | (→Ende des Datenübertragungsblocks)
etch | CHEM | ätzen
≈ corrode | | ≈ korrodieren
etch-back [PCB] | ELECTRON | rückätzen [Leiterplatten]
etch chemical | CHEM | Ätzmittel
= etching agent | |
etch-down (→etch-off) | CHEM | (→abätzen)
etched | CHEM | geätzt
etching | CHEM | Ätzung
≈ corrosion | | = Ätzen
| | ≈ Korrosion
etching agent (→etch chemical) | CHEM | (→Ätzmittel)
etching process | MICROEL | Ätzverfahren
= etching technique | | = Ätztechnik
etching technique (→etching process) | MICROEL | (→Ätzverfahren)
etch machine | MANUF | Ätzgerät
etch-off | CHEM | abätzen
= etch-down | |
↑ corrode | |
etch pit | MICROEL | Ätzgrube
etch-pit density | MICROEL | Ätzgrubendichte
etch polish | MICROEL | Ätzpolieren
| | = Politurätzen
etch resistant | CHEM | ätzbeständig
ethyl sulfate | CHEM | Äthylsulfat
ETP (→electronic publishing) | DATA PROC | (→Electronic Publishing)
etremely low frequency [100,000,000 m to 1,000,000 m] | RADIO | Megameterwelle [100.000.000 m bis 1.000.000 m]
= ELF | |
ETX (→end of text) | DATA COMM | (→Textende)
Eu (→europium) | CHEM | (→Europium)
Euclidian geometry | MATH | euklidische Geometrie
eureka (→constantan) | METAL | (→Konstantan)
eurocard | ELECTRON | Europakarte
eurocard size [PCB size 63x39.4 sqinch] | ELECTRON | Europakartenformat [Leiterplattenformat 160x100 mm^2]
= European standard size | | = Europaformat
≈ eurocard | | ≈ Europakarte
euro mains connector | EL.INST | Eurostecker
= mains connector European style | |
European A4 size | DOC | DIN-A4-Format
European standard size (→eurocard size) | ELECTRON | (→Europakartenformat)
European style [connector] | EL.INST | Europanorm [Stecker]
europium | CHEM | Europium
= Eu | | = Eu
eutectic (adj.) | METAL | eutektisch
eutectic die bonding | MICROEL | Eutectic-die-Bonding
eV (→electron-volt) | PHYS | (→Elektronvolt)
evaluate | TECH | auswerten
evaluation | TECH | Auswertung
evaluation circuit | CIRC.ENG | Auswerteschaltung
= evaluator; detection circuit; detector | | = Auswerter; Bewerteschaltung; Bewerter; Auswertevorrichtung; Detektor
evaluation function (→evaluator) | MATH | (→Auswertefunktion)
evaluation kit | MICROEL | Einarbeitungskit
= starter kit | |
evaluation module | MICROEL | Entwicklungsmodul
evaluation unit | TECH | Vorführgerät
= demonstration unit; trial equipment | |
evaluator | MATH | Auswertefunktion
= evaluation function | |
evaluator (→evaluation circuit) | CIRC.ENG | (→Auswerteschaltung)
evanesce (→decay) | PHYS | (→abklingen)

evanescent wave	PHYS	abklingende Welle
evaporate (→vaporize)	PHYS	(→verdampfen)
evaporation	PHYS	**Verdampfung**
[from liquid into gaseous state]		[von flüßigem in gasförmigen Zustand]
≈ sublimation; emanation		≈ Sublimation; Ausdunstung
↓ volatilization		↓ Verdunstung
evaporation point	PHYS	**Verdampfungspunkt**
= evaporation temperature; boiling point; b.p.; boiling temperature		= Verdampfungstemperatur; Siedepunkt; Siedetemperatur
		≈ Kondensationstemperatur
evaporation temperature	PHYS	(→Verdampfungspunkt)
(→evaporation point)		
even	MATH	**geradzahlig**
= even-numbered		= gerade 1
≠ uneven		≠ ungeradzahlig
even (→flat)	TECH	(→flach)
evencount	COLLOQ	**abgezählt**
evenness	MECH	**Ebenheit 2**
		= Gleichmäßigkeit
even-numbered (→even)	MATH	(→geradzahlig)
even numer	MATH	**gerade Zahl**
even parity (→parity)	CODING	(→Parität)
even spacing	TYPOGR	**fester Zeichenabstand**
event	MATH	**Ereignis**
event-by-event simulation	TELEC	**zeittreue Simulation**
event compatibility	TECH	**Ereigniskompatibilität**
event-compatible	TECH	**ereigniskompatibel**
event control	DATA PROC	**Ereignissteuerung**
event-controlled	DATA PROC	(→ereignisgesteuert)
(→event-driven)		
event-driven	DATA PROC	**ereignisgesteuert**
= event-controlled		
event indication	TELECONTR	**Ereignismeldung**
event mechanism	SWITCH	**Anreizmechanismus**
event number	SWITCH	**Anreiznummer**
event processing	DATA PROC	**Anreizverarbeitung**
event queue	DATA PROC	**Ereigniswarteschlange**
event register	DATA PROC	**Ereignisregister**
event synchronization	DATA PROC	**Ereignissynchronisation**
event table	SWITCH	**Anreiztabelle**
event triggering	ELECTRON	**Ereignistriggerung**
everyday 1	COLLOQ	**tagtäglich**
[without exception]		[täglich ohne Ausnahme]
≈ daily		≈ täglich
everyday 2 (adj.) (→general)	COLLOQ	(→gewöhnlich)
evolute	MATH	**Evolute**
evolutional trend	TECH	**Entwicklungstendenz**
= future trend		= Zukunftstendenz
evolutionary (→progressive 2)	TECH	(→fortschrittlich)
evolutionary step	TECH	**Entwicklungsschritt**
evolution capability	COLLOQ	**Entwicklungsfähigkeit**
EW (→electronic warefare)	MIL.COMM	(→elektronische Kampfführung)
E/W direction	RADIO REL	**A/B-Richtung**
[East/West]		
EW receiver	MIL.COMM	**EW-Empfänger**
exactness (→accuracy)	TECH	(→Genauigkeit)
examination (→check)	TECH	(→Prüfung)
examine (→check)	TECH	(→prüfen)
exceed (→transgres)	TECH	(→überschreiten)
exceeding probability	MATH	**Überschreitungswahrscheinlichkeit**
excelsior (AM) (→wood wool)	TECH	(→Holzwolle)
excentricity	MECH	**Mittenversatz**
≈ stroke 1		= Exzentrizität
		= Schlag
except function	ENG.LOG	(→EXKLUSIV-ODER-Verknüpfung)
(→EXCLUSIVE-OR operation)		
except operation	ENG.LOG	(→EXKLUSIV-ODER-Verknüpfung)
(→EXCLUSIVE-OR operation)		
exception	COLLOQ	**Ausnahme**
exceptional (→extraordinary)	COLLOQ	(→außerordentlich)
exception dictionary	DATA PROC	**Ausnahmenliste**
[list of exceptions to an algorithm]		[Auflistung von Ausnahmefällen eines Algorithmus]
		= Ausnahmeliste; Ausnahmenwörterbuch; Ausnahmelexikon
excerpt (→extract)	LING	(→Auszug)
excess	MATH	**Exzeß**
excess (n.) (→transgression)	TECH	(→Überschreitung)
excess bandwidth	CODING	**Zusatzbandrate**
excess current	MICROEL	**Überschußstrom**
[tunnel diode]		[Tunneldiode]
excess current	EL.TECH	(→Überstrom)
(→overcurrent)		
excess demand	ECON	**Übernachfrage**
excess electron	PHYS	**Überschußelektron**
excessive (→oversize)	TECH	(→übermäßig)
excessive error rate	TELEC	**Fehlerhäufigkeitsüberschreitung**
= ERR		
excessive production	ECON	**Überproduktion**
excess length	TECH	**Überlänge**
excess level	TELEC	**Überpegel**
≠ defect level		≠ Unterpegel
excess noise ratio	HF	**Überschußrauschverhältnis**
= ENR		
excess of electrons	PHYS	**Elektronenüberschuß**
excess population	PHYS	**Überbesetzung**
excess production	ECON	**Überschußproduktion**
= surplus production		
excess rainfall attenuation	RADIO PROP	**Regenzusatzdämpfung**
excess temperature	TECH	**Übertemperatur**
= overtemperature		
excess-three code	CODING	**Drei-Exzeß-Code**
[special binary coding of decimal digits, to ease arithmetic operations]		[spezielle binäre Darstellung von Dezimalziffern, zur Erleichterung arithmetischer Operationen]
= Stibitz code		= Stibitzcode; Exzess-3-Code; Dreier-Exzeß-Code
excess voltage	EL.TECH	(→Überspannung)
(→overvoltage)		
excess-voltage arrester	COMPON	(→Spannungssicherung)
(→overvoltage protector)		
excess-voltage cut-out	COMPON	(→Spannungssicherung)
(→overvoltage protector)		
excess voltage immunity	EL.TECH	(→Spannungsfestigkeit)
(→dielectric strength)		
excess-voltage protector	COMPON	(→Spannungssicherung)
(→overvoltage protector)		
excess weight	ECON	**Übergewicht**
excess zeros	CODING	**Nullenüberhang**
exchange (v.t.)	TECH	**auswechseln**
= substitute		≈ austauschen; vertauschen
≈ replace		≈ ersetzen
exchange 1 (n.)	ECON	**Börse**
≈ market (n.)		↓ Wertpapierbörse; Warenbörse
↓ stock exchange; commodity exchange		
exchange (n.)	TECH	**Vertauschung**
exchange	SWITCH	**Vermittlungsstelle**
= switching center (AM); switching centre (BRI); switching office; central office; switch		= Vermittlungszentrale; Vermittlungszentrum; Vermittlungsamt; Vermittlung
≈ switching system; switching equipment		≈ Vermittlungssystem; Vermittlungsanlage
↓ automatic exchange; local exchange; trunk exchange; transit exchange		↓ Wählvermittlungsstelle; Ortsvermittlungsstelle; Fernvermittlungsstelle; Durchgangsvermittlungsstelle
exchange (→replacement)	TECH	(→Austausch)
exchange (n.) (→barter business)	ECON	(→Tauschgeschäft)
exchangeability	TECH	**Auswechselbarkeit**
		= Austauschbarkeit 1
exchangeable	TECH	**auswechselbar**
≈ substitutable		= austauschbar; vertauschbar
≈ replaceable		≈ ersetzbar
exchangeable disk	TERM&PER	(→Wechselplatte)
(→removable disk)		
exchangeable disk drive	TERM&PER	**auswechselbares Laufwerk**
= exchangeable drive; exchangeable disk store; exchangeable disk storage; EDS		
exchangeable-disk drive	TERM&PER	(→Wechselplattenlaufwerk)
(→removable-disk drive)		

exchangeable-disk memory ($\rightarrow$moving-disk memory)	TERM&PER	($\rightarrow$Wechselplattenspeicher)	
exchangeable-disk store ($\rightarrow$moving-disk memory)	TERM&PER	($\rightarrow$Wechselplattenspeicher)	
exchangeable disk storage ($\rightarrow$exchangeable disk drive)	TERM&PER	($\rightarrow$auswechselbares Laufwerk)	
exchangeable disk storage ($\rightarrow$moving-disk memory)	TERM&PER	($\rightarrow$Wechselplattenspeicher)	
exchangeable disk store ($\rightarrow$exchangeable disk drive)	TERM&PER	($\rightarrow$auswechselbares Laufwerk)	
exchangeable drive ($\rightarrow$exchangeable disk drive)	TERM&PER	($\rightarrow$auswechselbares Laufwerk)	
exchange access network ($\rightarrow$subscriber network)	TELEC	($\rightarrow$Teilnehmernetz 1)	
exchange area = serving area; connecting range ↑ service area [TELEC]	SWITCH	**Anschlußbereich** = Vermittlungsbereich ↑ Versorgungsbereich [TELEC]	
exchange battery ($\rightarrow$central battery)	TELEPH	($\rightarrow$Zentralbatterie)	
exchange buffering	DATA PROC	**Austauschpufferung**	
exchange cabling = office cabling	SYS.INST	**Amtsverkabelung**	
exchange clock	TELEC	**Amtstakt**	
exchange code	SWITCH	**Kennzahl der Vermittlungsstelle**	
exchange configuration document ($\rightarrow$installation document)	TECH	($\rightarrow$Aufbauunterlage)	
exchange connection	TELEC	**Amtsanschluß**	
exchange copy = exchange print	DOC	**Austauschpause**	
exchange data ($\rightarrow$office data)	SWITCH	($\rightarrow$Amtsdaten)	
exchange energy	PHYS	**Austauschenergie**	
exchange environment = office environment	TELEC	**Amtsumgebung**	
exchange equipment = central office equipment; CO equipment; office terminal	TELEC	**Amtsausrüstung** = Amtsgerät	
exchange force	PHYS	**Austauschkraft**	
exchange installation technique [SWITCH] ($\rightarrow$office installation technique)	TELEC	($\rightarrow$Amtsbautechnik)	
exchange line ($\rightarrow$interoffice trunk)	TELEC	($\rightarrow$Amtsverbindungsleitung)	
exchange-located	TELEC	**vermittlungsseitig**	
exchange of goods	ECON	**Warenaustausch**	
exchange premises	TELEC	**Amtsgebäude**	
exchange print ($\rightarrow$exchange copy)	DOC	($\rightarrow$Austauschpause)	
exchange rate display	SIGN.ENG	**Kursanzeigegerät**	
exchange rate guarantee	ECON	**Wechselkursgarantie**	
exchange risk	ECON	**Wechselkursrisiko** = Kursrisiko	
exchange side [distributor]	SYS.INST	**Vermittlungsstellenseite** [Verteiler]	
exchange sort ($\rightarrow$bubble sorting)	DATA PROC	($\rightarrow$Bubble-Sortieren)	
exchange specific	SWITCH	**vermittlungsstellenspezifisch** = amtsspezifisch	
exchange termination [ISDN] = ET	TELEC	**Zellabschluß** [ISDN]	
excimer laser	OPTOEL	**Excimerlaser**	
excitation = stimulation; generation	PHYS	**Anregung** = Anfachung	
excitation	POWER SYS	**Erregung**	
excitation ($\rightarrow$call)	DATA PROC	($\rightarrow$Aufruf)	
excitation energy = excitation level; excited level ↓ activation energy	PHYS	**Anregungsenergie** = Anregungsniveau ↓ Aktivierungsenergie	
excitation level ($\rightarrow$excitation energy)	PHYS	($\rightarrow$Anregungsenergie)	
excitation potential	PHYS	**Anregungspotential**	
excitation state = excited state	PHYS	**Anregungszustand**	
excitation voltage ($\rightarrow$exciting voltage)	ELECTRON	($\rightarrow$Erregerspannung)	
excite = stimulate; generate	PHYS	**anregen**	
excite ($\rightarrow$call)	DATA PROC	($\rightarrow$aufrufen)	
excite ($\rightarrow$energize)	ELECTRON	($\rightarrow$erregen)	
excited level ($\rightarrow$excitation energy)	PHYS	($\rightarrow$Anregungsenergie)	
excited state ($\rightarrow$excitation state)	PHYS	($\rightarrow$Anregungszustand)	
exciter ($\rightarrow$primary radiator)	ANT	($\rightarrow$Primärstrahler)	
exciting coil = operating coil; relay coil	ELECTRON	**Erregerspule** = Relaisspule	
exciting voltage = excitation voltage	ELECTRON	**Erregerspannung**	
excitron ↑ mercury-vapour rectifier	POWER SYS	**Excitron** ↑ Quecksilberdampf-Gleichrichter	
excitron	MICROEL	**Exzitron**	
exclamation	LING	**Ausrufesatz**	
exclamation mark [!] ↑ punctuation mark; end punctuation mark	LING	**Ausrufezeichen** [!] = Ausrufungszeichen (ÖS, SWZ) ↑ Satzzeichen; Satzendzeichen	
exclude (v.t.)	COLLOQ	**auschließen**	
exclusion	TECH	**Ausschluß**	
exclusion ($\rightarrow$exclusion operation)	ENG.LOG	($\rightarrow$Inhibitionsverknüpfung)	
exclusion area	SAT.COMM	**Sperrgebiet**	
exclusion circuit ($\rightarrow$exclusion gate)	CIRC.ENG	($\rightarrow$Inhibitionsglied)	
exclusion element ($\rightarrow$exclusion gate)	CIRC.ENG	($\rightarrow$Inhibitionsglied)	
exclusion function ($\rightarrow$exclusion operation)	ENG.LOG	($\rightarrow$Inhibitionsverknüpfung)	
exclusion gate = exclusion element; exclusion circuit; NOT-IF-THEN gate; NOT-IF-THEN element; NOT-IF-THEN circuit ↑ logic gate	CIRC.ENG	**Inhibitionsglied** = Inhibitionsgatter; Inhibitionselement; Inhibitionsschaltung; Inhibitionstor ↑ Verknüpfungsglied	
exclusion operation [output = 1, if A = 1 and B = 0] = exclusion function; exclusion; NOT-IF-THEN operation; NOT-IF-THEN function; NOT-IF-THEN ↑ logic operation	ENG.LOG	**Inhibitionsverknüpfung** [Ausgang = 1, wenn A = 1 und B = 0] = Inhibitionsfunktion; Inhibition; NICHT-WENN-DANN-Verknüpfung; NICHT-WENN-DANN-Funktion ↑ logische Verknüpfung	
exclusion principle ($\rightarrow$Pauli principle)	PHYS	($\rightarrow$Pauli-Prinzip)	
exclusive ≠ inclusive	MATH	**ausschließend** = exklusiv ≠ einschließend	
exclusive exchange line ($\rightarrow$individual line)	TELEC	($\rightarrow$Einzelanschluß)	
EXCLUSIVE-NOR function ($\rightarrow$equivalence operation)	ENG.LOG	($\rightarrow$Äquivalenzverknüpfung)	
EXCLUSIVE-NOR operation ($\rightarrow$equivalence operation)	ENG.LOG	($\rightarrow$Äquivalenzverknüpfung)	
EXCLUSIVE OR ($\rightarrow$EXCLUSIVE-OR operation)	ENG.LOG	($\rightarrow$EXKLUSIV-ODER-Verknüpfung)	
EXCLUSIVE OR circuit ($\rightarrow$EXCLUSIVE OR gate)	CIRC.ENG	($\rightarrow$EXKLUSIV-ODER-Glied)	
EXCLUSIVE OR element ($\rightarrow$EXCLUSIVE OR gate)	CIRC.ENG	($\rightarrow$EXKLUSIV-ODER-Glied)	
EXCLUSIVE-OR function ($\rightarrow$EXCLUSIVE-OR operation)	ENG.LOG	($\rightarrow$EXKLUSIV-ODER-Verknüpfung)	
EXCLUSIVE OR gate = EXCLUSIVE OR element; EXCLUSIVE OR circuit; XOR gate; XOR element; XOR circuit; EXOR gate; EXOR element; EXOR circuit; anti-valence gate; antivalence element; antivalence circuit; anticoincidence gate; anticoincidence element; anticoincidence circuit; non-equivalence gate; non-equivalence element; non-equivalence circuit ↑ logic gate	CIRC.ENG	**EXKLUSIV-ODER-Glied** = EXKLUSIV-ODER-Gatter; EXKLUSIV-ODER-Element;EXKLUSIV-ODER-Schaltung; EXKLUSIV-ODER-Tor;AUSSCHLIESSLICHES-ODER-Glied; AUSSCHLIESSLICHES-ODER-Gatter; AUSSCHLIESSLICHES-ODER-Element; AUSSCHLIESSLICHES-	

EXCLUSIVE-OR operation   208

ODER-Schaltung; AUS-
SCHLIESSLICHES-
ODER-Tor; XOR-Glied;
XOR-Gatter; XOR-Element; XOR-Schaltung;
XOR-Tor; Antivalenzglied; Antivalenzgatter;Antivalenzelement; Antivalenzschaltung;
Antivalenztor;Antikoinzidenzglied; Antikoinzidenzgatter; Antikoinzidenzelement;
Antikoinzidenzschaltung; Antikoinzidenztor
↑ Verknüpfungsglied

EXCLUSIVE-OR operation ENG.LOG
[output = 1 if A different to B]
= EXCLUSIVE-OR function; EXCLUSIVE OR; XOR operation; XOR function; XOR; non-equivalence operation; non-equivalence function; non-equivalence; NEQ; inequivalence operation; inequivalence function; inequivalence; anticoincidence ; antivalence; OR-ELSE operation; OR-ELSE function; EXOR operation; EXOR function; destructive addition; equality function; equality operation; equality; except operation; except function; exjunction
≈ OR operation
↑ logic operation

EXKLUSIV-ODER-Verknüpfung
[Ausgang = 1 wenn A ungleich B]
= EXKLUSIV-ODER-Funktion; EXKLUSIV-ODER; AUSSCHLIESSLICHES-ODER-Verknüpfung; AUSSCHLIESSLICHES-ODER-Funktion; AUS-SCHLIESSLICHES ODER; XOR-Verknüpfung; XOR-Funktion; Antivalenz; Antikoinzidenz; ODER-ODER-Verknüpfung; ODER-ODER-Funktion; EXOR-Verknüpfung; EXOR-Funktion
≈ ODER-Verknüpfung
↑ logische Verknüpfung

exclusive sale franchise ECON
= sole selling rights
↑ franchise
exclusivity (→franchise) ECON
exclusivness (→franchise) ECON
excursion MODUL
= deviation; swing; shift; sweep
excursion PHYS
EXEC (→executable) DATA PROC
executable DATA PROC
= EXEC
executable statement DATA PROC
execute (v.t.) DATA PROC
[a program or process]
= run [v.t.]; perform; obey; carry out
execute command (→actuate TELECONTR message)
execute cycle (→execute DATA PROC phase)
execute mode DATA PROC
execute phase DATA PROC
= executing phase; execution phase; execute cycle; executing cycle; execution cycle; fetch-execute cycle; run phase; run cycle
≈ fetch-execute cycle; fetch cycle
execute signal DATA PROC
execute statement DATA PROC
executing cycle (→execute DATA PROC phase)
executing phase (→execute DATA PROC phase)
execution (→type) TECH
execution 2 (→program DATA PROC run)
execution 1 (→instruction execution)
execution cycle (→execute DATA PROC phase)
execution error DATA PROC
execution instruction DATA PROC
(→executive instruction)

Alleinverkaufsrecht
= Exklusivvertrieb
↑ Exklusivrecht
(→Exklusivrecht)
(→Exklusivrecht)
Hub

Schwingweg
(→ablauffähig)
ablauffähig
= lauffähig; ausführbar
ausführbare Anweisung
= ablauffähige Anweisung
ausführen
[ein Programm oder einen Prozeß]
= ablaufen lassen; fahren

(→Ausgabebefehl)

(→Ausführungsphase)

Ausführungszustand
Ausführungsphase
= Laufphase; Ausführungszyklus; Laufzyklus
≈ Abruf-/Ausführungsphase; Abrufzyklus

Ausführungssignal
Ausführungsanweisung
(→Ausführungsphase)

(→Ausführungsphase)

(→Ausführung 1)
(→Programmlauf)

(→Befehlsausführung)

(→Ausführungsphase)

Ausführungsfehler
(→Ausführungsbefehl)

execution phase (→execute DATA PROC phase)
execution time DATA PROC
= run time; turnaround (n.)
↓ instruction time; program execution time; job execution time

executive committee ECON
executive control program DATA PROC
(→control program)
executive director 1 ECON
[title]
executive instruction DATA PROC
= execution instruction
executive office OFFICE
executive override (→override) TELEPH
executive programm DATA PROC
(→executive routine)
executive regulation ECON
= regulations
executive routine DATA PROC
= executive programm; monitor
↑ control program
executive secretary OFFICE
= personal secretary
executive telephone TELEPH
executive telephone system TELEPH
exemption (n.) ECON
= release (n.)
exercise (n.) INSTR
[an equipment performs all its functions, to allow testing]

exerciser (n.) DATA PROC
= device tester
exhaust (v.t.) TECH
≈ ventilate
exhaust 2 (n.) (→output air) TECH
exhauster TECH
≈ ventilador
exhaustion region MICROEL
exhaustive (adj.) COLLOQ
= thorough
exhaustiveness COLLOQ
exhaust silencer POWER SYS
exhibit (n.) (→overview) LING
exhibit (n.) (→exposition ECON specimen)
exhibition (→exposition) ECON
exhibition hall ECON
exhibition stand (→booth) ECON
exhibitor ECON
exigent (→demanding) TECH
existence function DATA PROC
= system management function
existing COLLOQ
≈ available; in stock
exit (n.) PHYS
exit (n.) DATA PROC
[end of program execution]
exit signal RAILW.SIGN
exjunction ENG.LOG
(→EXCLUSIVE-OR operation)
EXNOR (→equivalence ENG.LOG operation)
EXNOR circuit (→equivalence CIRC.ENG gate)
EXNOR element CIRC.ENG
(→equivalence gate)
EXNOR function ENG.LOG
(→equivalence operation)
EXNOR gate (→equivalence CIRC.ENG gate)
EXNOR operation ENG.LOG
(→equivalence operation)
exorbitant ECON
= priceless
EXOR circuit (→EXCLUSIVE CIRC.ENG OR gate)

(→Ausführungsphase)
Ausführungszeit
= Ausführungsdauer; Laufzeit
↓ Befehlsausführungszeit; Programmausführungszeit; Auftragsausführungszeit

Zentralvorstand
(→Organisationsprogramm)

Direktor 1
[Titel]
Ausführungsbefehl

Chefraum
(→Aufschalten)
(→Hauptsteuerprogramm)

Vollzugsordnung
= Ausführungsbestimmung
Hauptsteuerprogramm
= Ablauteil; Monitor
↑ Organisationsprogramm
Chefsekretärin

Cheftelefon
Chef-Fernsprechanlage
Freistellung

Prüfablauf
[ein Gerät durläuft alle Betriebszustände, um geprüft werden zu können]

Prüfsystem

entlüften
≈ lüften
(→Abluft)
Entlüfter
≈ Ventilator
Erschöpfungsgebiet
ausführlich
= erschöpfend
Ausführlichkeit
Abgasschalldämpfer
(→Übersicht)
(→Ausstellungsstück)

(→Ausstellung)
Messehalle
(→Messestand)
Austeller
(→anspruchsvoll)
Existenzfunktion

vorhanden
≈ verfügbar; vorrätig
Austritt
Ausstieg
= Ausgang
Ausfahrtsignal
(→EXKLUSIV-ODER-Verknüpfung)
(→Äquivalenzverknüpfung)

(→Äquivalenzglied)

(→Äquivalenzglied)

(→Äquivalenzverknüpfung)

(→Äquivalenzglied)

(→Äquivalenzverknüpfung)

unbezahlbar
= exorbitant
(→EXKLUSIV-ODER-Glied)

exorciser (→exorciser) MICROEL (→Exorciser)
EXOR element CIRC.ENG (→EXKLUSIV-ODER-
(→EXCLUSIVE OR gate) Glied)
EXOR function ENG.LOG (→EXKLUSIV-ODER-Ver-
(→EXCLUSIVE-OR operation) knüpfung)
EXOR gate CIRC.ENG (→EXKLUSIV-ODER-
(→EXCLUSIVE OR gate) Glied)
EXOR operation ENG.LOG (→EXKLUSIV-ODER-Ver-
(→EXCLUSIVE-OR operation) knüpfung)
exorsiser MICROEL Exorciser
= exorciser
exothermic PHYS **exotherm**
expand METAL **aufweiten**
= aufdornen
expand TECH **ausbauen 1**
= upgrade = erweitern; aufrüsten
expand PHYS **ausdehnen**
= expandieren
≠ zusammenziehen
expand MATH **entwickeln**
[a formula] [einer Formel]
expandability (→expansion TECH (→Erweiterungsmöglichkeit)
capability)
expandable (→upgradable) TECH (→ausbaufähig)
expanded (→extended) TECH (→gedehnt)
expanded font TYPOGR **Breitschrift**
= wide font; expanded lettering; wide = Spreizschrift
lettering; expanded style: wide style
expanded lettering (→expanded TYPOGR (→Breitschrift)
font)
expanded memory DATA PROC **Erweiterungsspeicher**
[in PC terminology expanded [im PC-Jargon erfordern
memories require special software to Expansionsspeicher speziel-
be managed, extended memories le Hilfsprogramme zu ihrer
don't] Verwaltung, Extensions-
= expanded storage; expanded store; speicher hingegen nicht]
extended memory; extended storage; = Expansionsspeicher; Exten-
extended store sionsspeicher
≈ memory expansion ≈ Speichererweiterung
expanded storage DATA PROC (→Erweiterungsspeicher)
(→expanded memory)
expanded store (→expanded DATA PROC (→Erweiterungsspeicher)
memory)
expanded style: wide style TYPOGR (→Breitschrift)
(→expanded font)
expanded to full capacity (→fully TECH (→vollausgebaut)
expanded)
expander TELEC **Expander**
= Dehner
expanding METAL **Aufweiten**
= Aufdornen
expansion PHYS **Expansion**
[increase in volume] [Vergrößerung des Volu-
≠ compression mens]
= Ausdehnung 3
≠ Kompression
expansion MATH **Entwicklung**
[of a formula] [einer Formel]
expansion TECH **Erweiterung**
= extension; upgrading; growth en- = Ausbau 1
hancement ≈ Vergrößerung
≈ enlargement
expansion TELEC **Expandierung**
= Dehnung
expansion SWITCH **Expansion**
expansion (→stretch) PHYS (→Dehnung)
expansion board EQUIP.ENG **Erweiterungskarte**
= expansion card; expansion PCB; ex- = Erweiterungsplatine; Er-
pansion module; add-on board; add- weiterungssteckkarte; Zu-
on card; add-on PCB; add-on mo- satzkarte; Aufsteckkarte;
dule; add-in board; add-in card; Aufsteckplatine; Aufsteck-
add-in PCB; add-in module; addi- modul
tional board; additional card; addi- ≈ Huckepack-Baugruppe;
tional PCB; additional module; baby Steckbaugruppe
board; baby module; sub-board ≠ Hauptplatine
≈ piggy-pack board; plug-in module
≠ main board
expansion capability TECH **Erweiterungsmöglichkeit**
= growth capability; expandability; up- = Erweiterbarkeit; Ausbaufä-
gradability higkeit; Ausbaumöglichkeit

expansion card (→expansion EQUIP.ENG (→Erweiterungskarte)
board)
expansion chamber (→cloud PHYS (→Nebelkammer)
chamber)
expansion coefficient PHYS **Ausdehnungskoeffizient**
= Ausdehnungszahl
expansion equipment TECH **Ergänzungsausstattung**
≠ standard equipment ≠ Grundausstattung
expansion field MATH **Erweiterungskörper**
expansion gate CIRC.ENG **Expansionsstufe**
expansion instrument INSTR **Hitzdraht-Instrument**
= hot-wire instrument; hot-wire am- = Hitzdrahtstrommesser
meter
expansion interface DATA PROC **Erweiterungsschnittstelle**
expansion manual DOC **Erweiterungshandbuch**
expansion module EQUIP.ENG (→Erweiterungskarte)
(→expansion board)
expansion PCB (→expansion EQUIP.ENG (→Erweiterungskarte)
board)
expansion plan TECH **Ausbauplan**
expansion slot EQUIP.ENG **Erweiterungssteckplatz**
= peripheral slot [DATA PROC] ↑ Steckplatz
↑ slot
expansion stage SWITCH **Expansionsstufe**
expansion stage (→construction TECH (→Ausbaustufe)
stage)
expansion thermometer PHYS **Ausdehnungsthermometer**
expansion unit EQUIP.ENG **Ausbaueinheit**
= extension unit = Erweiterungseinheit
expectation INF **Erwartung**
expectation (→expected MATH (→Erwartungswert)
value)
expectation pattern INF **Erwartungsmuster**
expectation value INF **Erwartungswert**
= expected value; anticipation value;
anticipated value
expected value MATH **Erwartungswert**
= expectation
expected value (→expectation INF (→Erwartungswert)
value)
expedience (→expediency) TECH (→Zweckmäßigkeit)
expediency TECH **Zweckmäßigkeit**
= expedience ≈ Eignung
≈ suitability
expedient (n.) TECH **Notbehelf**
= makeshift; temporary solution; re- = Provisorium; Notlösung;
source Behelfslösung
expedient (adj.) TECH **zweckmäßig**
≈ suitable ≈ geeignet
expenditure (→effort) ECON (→Aufwand)
expenditure (→expense) ECON (→Ausgabe)
expenditure of energy TECH **Energieaufwand**
≈ energy consumption ≈ Energieverbrauch
expenditure of work ECON **Arbeitsaufwand**
= work effort
expense ECON **Ausgabe**
= expenditure; spending; outlay = Aufwendung; Unkosten
(pl.t.)
expense (→effort) ECON (→Aufwand)
expenses ECON **Spesen**
expensive ECON **teuer**
= dear; costly; pricey; high-priced = kostspielig
≈ aufwendig
experienced SCIE **sachkundig**
= skilled; versed = fachkundig; erfahren; be-
wandert; versiert
experiment (n.) PHYS **Versuch**
= Experiment
experiment SCIE **Experiment**
experiment (→trial) TECH (→Versuch)
experimental TECH **experimentell**
experimental board ELECTRON **Experimentierkarte**
= perforated sheet = Experimentierplatte; La-
borplatte
experimental data SCIE **Versuchsdaten**
= empirical data = empirische Daten; experi-
mentelle Daten; Beobach-
tungsmaterial
experimental device TECH **Versuchsvorrichtung**
experimental error PHYS **Versuchsfehler**

experimental game	ECON	**Planspiel**
experimental kit	ELECTRON	**Experimentierkasten**
= experimental system		
experimental link	TRANS	**Versuchsstrecke**
experimental mode (→test operation)	TECH	(→Probebetrieb)
experimental operation (→test operation)	TECH	(→Probebetrieb)
experimental physics	PHYS	**Experimentalphysik**
experimental prototype	TECH	**Versuchsmuster**
= prototype; test model		= Probemuster; Prototyp; Versuchsgerät; Funktionsmuster
experimental purpose	TECH	**Versuchszweck**
		= Experimentierzweck
experimental setup	TECH	**Versuchsanordnung**
experimental system	TECH	**Versuchssystem**
= pilot system; prototype system		= Pilotsystem; Prototypsystem
experimental system (→experimental kit)	ELECTRON	(→Experimentierkasten)
experimental transformator	ELECTRON	**Experimentiertrafo**
experimental value	TECH	**Erfahrungswert**
experimental value (→measured value)	PHYS	(→Meßwert)
expert (n.)	ECON	**Sachverständiger**
[person knowing a lot about a topic]		= Experte
≈ professional [COLLOQ]		≈ Fachmann [COLLOQ]
expertise (→technical knowledge)	SCIE	(→Fachkenntnis)
expert knowledge (→technical knowledge)	SCIE	(→Fachkenntnis)
expertness (→technical knowledge)	SCIE	(→Fachkenntnis)
expert opinion	TECH	**Gutachten**
= judgement		= Expertise
expert-support system	DATA PROC	**Experten-stützendes System**
expert system	DATA PROC	**Expertensystem**
[software with problem-solving expertise of a special field]		[das Fachwissen eines Spezialgebietes enthaltende Software]
= intelligent knowledge-based system; IKBS		= IKBS-System
↓ knowledge basis; inference module		↓ Wissensbasis; Inferenzmodul
expiration	ECON	**Verfall**
= expiry; maturity		= Ablauf; Erlöschen
expiration date (→due time)	ECON	(→Fälligkeit)
expire	ECON	**verfallen**
expire	COLLOQ	**verstreichen**
[term]		[Frist]
		= ablaufen
expired	ECON	**abgelaufen**
expired patent	ECON	**abgelaufenes Patent**
		= verfallenes Patent
expiry (→expiration)	ECON	(→Verfall)
expiry date (→due time)	ECON	(→Fälligkeit)
explanation	LING	**Erläuterung**
= elucidation; explanatory note; note		
explanatory note (→explanation)	LING	(→Erläuterung)
explicit address (→direct address)	DATA PROC	(→direkte Adresse)
explicit addressing (→direct addressing)	DATA PROC	(→direkte Adressierung)
exploded view	ENG.DRAW	**Explosionsdarstellung**
		= Explosionszeichnung
exploitation (→utilization)	TECH	(→Ausnutzung)
exploration (→hunting)	SWITCH	(→Absuchen)
exploration (→scanning)	ELECTRON	(→Abtastung)
explosion-proof	TECH	**explosionssicher**
explosion protection	TECH	**Explosionsschutz**
explosive (n.)	LING	**Explosivum**
[p,b,t,..]		[p,b,t,..]
= stop		= Explosivlaut; Verschlußlaut
explosive (adj.)	TECH	**explosionsgefährdet**
		= explosiv
exponent	MATH	**Exponent**
= power exponent		= Hochzahl; Potenzexponent
≠ basis		≠ Basis
↑ power		↑ Potenz

exponential	MATH	**exponentiell**
exponential antenna	ANT	**Exponentialantenne**
exponential distribution	MATH	**Exponentialverteilung**
exponential function	MATH	**Exponentialfunktion**
exponential horn	ANT	**Exponentialhorn**
= logarithmic horn		= Exponentialtrichter
exponential law	MATH	**Exponentialgesetz**
exponential line (→exponential transmission line)	LINE TH	(→Exponentialleitung)
exponentially distributed	MATH	**exponentialverteilt**
exponential smoothing	DATA PROC	**Exponentialglättung**
[heaviest weight]		[höchste Wichtung]
exponential transmission line	LINE TH	**Exponentialleitung**
= exponential line		
exponentiation	MATH	**Potenzierung**
export (v.t.)	DATA PROC	**überspielen**
[to hand over data stock to a foreign system]		[Datenbestände an ein Fremdsystem übergeben]
≠ import		= exportieren
		≠ einspielen
export credit insurance	ECON	**Exportkreditversicherung**
export declaration	ECON	**Ausfuhrerklärung**
export duty	ECON	**Ausfuhrzoll**
export licence (→export permit)	ECON	(→Ausfuhrgenehmigung)
export permit	ECON	**Ausfuhrgenehmigung**
= export licence		= Exportgenehmigung
export promotion	ECON	**Ausfuhrförderung**
		= Exportförderung
expose	PHYS	**belichten**
exposition	LING	**Ausführung**
		= Darlegung
exposition	ECON	**Ausstellung**
= exhibition; fair		= Messe
↓ trade show; industrial fair		↓ Handelsmesse; Industriemesse
exposition specimen	ECON	**Ausstellungsstück**
= display material; exhibit (n.)		= Exponat; Austellungsmaterial
exposure	PHYS	**Belichtung**
↓ underexposure; overexposure		↓ Unterbelichtung; Überbelichtung
exposure (→standard ion dose)	PHYS	(→Standardionendosis)
exposure meter	INSTR	**Belichtungsmesser**
exposure time	PHYS	**Belichtungszeit**
exposure unit	TECH	**Belichtungsgerät**
express (v.t.)	SCIE	**ausdrücken**
express air cargo	ECON	**Eilluftfracht**
express channel	TRANS	**Expresskanal**
[service channel]		[Dienstkanal]
express delivery	POST	**Eilzustellung**
expression	ENG.LOG	**Ausdruck**
[finite reasonable character sequence]		[sinnvolle endliche Zeichenfolge]
expression (→term)	MATH	(→Term)
expressway	CIV.ENG	**Schnellstraße**
= dual-carriage way		= Schnellverkehrstaße
↓ main highway; freeway		↓ Hauptstraße; Autobahn
extend (→prolong)	COLLOQ	(→verlängern)
extend (→relay)	TELEC	(→weiterleiten)
extend (v.t.) (→strech)	PHYS	(→dehnen)
extended	TYPOGR	**breit**
[font attribute]		[Schriftattribut]
= wide		
extended	TECH	**gedehnt**
= expanded		= erweitert
extended addressing	DATA PROC	**erweiterte Adressierung**
extended arithmetic	MATH	**erweiterte Arithmetik**
extended code	CODING	**erweiterter Code**
extended memory	DATA PROC	(→Erweiterungsspeicher)
(→expanded memory)		
extended partition	DATA PROC	**erweiterte Partition**
extended storage	DATA PROC	(→Erweiterungsspeicher)
(→expanded memory)		
extended store (→expanded memory)	DATA PROC	(→Erweiterungsspeicher)
extended temperature range	TECH	**erweiterter Temperaturbereich**

extender (→file extension)	DATA PROC	(→Dateikennung)
extender board (→board extender)	INSTR	(→Adapter)
extensibility	MECH	**Dehnbarkeit**
extensible	MECH	**dehnbar**
extensible language	DATA PROC	**erweiterbare Programmiersprache**
extension [increase of length] = lengthening; streching ≠ contraction ↑ strain	PHYS	**Ausdehnung 1** [Vergrößerung der Länge] = Dilatation; Streckung ≠ Kontraktion ↑ Dehnung
extension (→stretch)	PHYS	(→Dehnung)
extension (→expansion)	TECH	(→Erweiterung)
extension (→extension station)	SWITCH	(→Nebenstelle)
extension (→superstructure)	TECH	(→Überbau)
extension 2 (→extent)	TECH	(→Umfang)
extension 1 (→prolongation)	TECH	(→Verlängerung)
extension (→file extension)	DATA PROC	(→Dateikennung)
extensional mode (→longitudinal oscillation)	PHYS	(→Längsschwingung)
extension cable (→extension cord)	EQUIP.ENG	(→Verlängerungskabel)
extension cord = extension lead; extension cable	EQUIP.ENG	**Verlängerungskabel** = Verlängerungsschnur
extension lead (→extension cord)	EQUIP.ENG	(→Verlängerungskabel)
extension line	ENG.DRAW	**Maßhilfslinie**
extension line = PABX line	TELEPH	**Nebenanschlußleitung**
extension line = extension pad ↑ attenuator [NETW.TH]	TELEC	**Verlängerungsleitung** ↑ Dämpfungsglied [NETW.TH]
extension mast (→telescopic mast)	ANT	(→Teleskopmast)
extension number	SWITCH	**Nebenstellennummer**
extension pad (→extension line)	TELEC	(→Verlängerungsleitung)
extension register	DATA PROC	**Erweiterungsregister** = Zusatzregister
extension spindle = shaft extension	COMPON	**Verlängerungsachse**
extension station = extension	SWITCH	**Nebenstelle** = Nebenanschluß
extension station	TELEPH	**Zweitapparat**
extension telephone	TELEPH	**Nebenstellenfernsprecher**
extension unit (→expansion unit)	EQUIP.ENG	(→Ausbaueinheit)
extension user	SWITCH	**Nebenstellenteilnehmer**
extent (n.)	DATA PROC	**Bereich**
extent [figurative] = extension 2; scope ≈ range	TECH	**Umfang** [figurativ] = Ausmaß ≈ Reichweite
exterior premises mounting (→outdoor mounting)	EQUIP.ENG	(→Außenmontage)
exterior premises installation (→outdoor mounting)	EQUIP.ENG	(→Außenmontage)
external = extraneous; outside	TECH	**extern** = äußer; auswärtiger
external address	DATA PROC	**externe Adresse** = äußere Adresse
external blocking = all trunks busy; ATB ↑ blocking	SWITCH	**äußere Blockierung** ↑ Blockierung
external cable (→outside cable)	COMM.CABLE	(→Außenkabel)
external call	TELEPH	**Amtsgespräch** = Externgespräch
external clock = external clocking	ELECTRON	**externer Takt** = externe Taktzuführung
external clocking (→external clock)	ELECTRON	(→externer Takt)
external command (→transient command)	DATA PROC	(→transientes Kommando)
external data bank = information bank	DATA PROC	**externe Datenbank** = Informationsbank
external data file	DATA PROC	**Externdatei**
external data processing	DATA PROC	**externe Datenverarbeitung** [außer Haus]
external excitation = independent excitation	ELECTRON	**Fremderregung**
external installation (→outdoor mounting)	EQUIP.ENG	(→Außenmontage)
external interface	TELEC	**Externschnittstelle** = externe Schnittstelle
external interference ≈ interfering voltage ↑ interference	TELEC	**Fremdstörung** ≈ Störspannung ↑ Störung
external interrupt = external interrupt signal	DATA PROC	**externe Unterbrechung**
external interrupt signal (→external interrupt)	DATA PROC	(→externe Unterbrechung)
external label [identifying piece of paper stuck] = coder ≈ sticker	TECH	**Etikett** ≈ Aufkleber
externally commutated converter	POWER SYS	**fremdgeführter Stromrichter**
externally heated	ELECTRON	**fremdgeheizt**
external member	ENG.DRAW	**Außenteil**
external memory [not pertaining to the main memory] = external storage; external store; peripheral memory; peripheral storage; peripheral store; auxiliary memory; auxiliary store; backing memory 1; backing storage 1; backing store 1 ≈ mass memory ≠ main memory ↑ memory; peripheral equipment ↓ secundary memory; tertiary memory; back-up memory	DATA PROC	**Externspeicher** [nicht zur Zentraleinheit gehörend] = externer Speicher; äußerer Speicher; peripherer Speicher; Peripheriespeicher; Hilfsspeicher ≈ Hintergrundspeicher; Zubringerspeicher; Massenspeicher ≠ Hauptspeicher 1 ↑ Speicher; Peripheriegerät ↓ Sekundärspeicher; Tertiärspeicher; Sicherungsspeicher
external mounting (→outdoor mounting)	EQUIP.ENG	(→Außenmontage)
external noise	RADIO	**äußeres Rauschen**
external noise (→sky noise)	SAT.COMM	(→Himmelsrauschen)
external noise (→external noise voltage)	TELEC	(→Fremdspannung)
external noise voltage (→interfering voltage)	TELEC	(→Störspannung)
external plant (→local outside plant)	OUTS.PLANT	(→Ortsleitungsnetz)
external plant network (→outside plant network)	TELEC	(→Liniennetz 1)
external plant technique (→outside plant technique)	TELEC	(→Linientechnik)
external purchasing	ECON	**Zukauf**
external reference	DATA PROC	**Externverweis**
external sort	DATA PROC	**Externsortierung**
external storage (→external memory)	DATA PROC	(→Externspeicher)
external store (→external memory)	DATA PROC	(→Externspeicher)
external surface	MATH	**Außenfläche**
external symbol	DATA PROC	**Externsymbol** = externes Symbol
external symbol dictionary = external symbol file	DATA PROC	**Externsymbolverzeichnis**
external symbol file (→external symbol dictionary)	DATA PROC	(→Externsymbolverzeichnis)
external synchronization	TELEC	**Fremdsynchronisierung** = externe Synchronisierung
external thermal resistance [between case and ambient]	MICROEL	**äußerer Wärmewiderstand** [zwischen Gehäuse und Umgebung]
external thread	MECH	**Außengewinde**
external traffic	SWITCH	**Externverkehr**
extinction current	PHYS	**Löschstrom** [Gasentladung]
extinction fading	RADIO PROP	**Extinktionsschwund**
extinction ratio value	OPT.COMM	**Extinktionsverhältnis**

extinction voltage

extinction voltage	POWER SYS	**Löschspannung**	
extinguish	TECH	**auslöschen**	
extra (n.)	EQUIP.ENG	(→Zusatz)	
(→attachment)			
extra (adj) (→additional)	TECH	(→zusätzlich)	
extra bell	TERM&PER	**Zweitwecker**	
		= Zweitklingel	
extra bold	TYPOGR	**extrafett**	
= ultrabold; heavy		↑ Schriftattribut	
↑ font attribute			
extra charge (→surcharge)	ECON	(→Zuschlag 1)	
extracode (n.)	DATA PROC	**Hardware-Emulationsroutine**	
[routine emulating hardware]			
extract 1 (v.t.)	DATA PROC	**ausblenden**	
extract (v.t.)	TECH	**entnehmen**	
		= extrahieren; gewinnen; herausziehen	
extract (n.)	LING	**Auszug**	
= excerpt		≈ Aufriß	
≈ abstract			
extract (→decouple)	TELEC	(→auskoppeln)	
extract 2 (v.t.)	DATA PROC	(→abfragen)	
(→inquire)			
extract counter	TERM&PER	**Auswahlzähler**	
= extract tally counter		[Lochkarten]	
extracting tool	EQUIP.ENG	**Ziehwerkzeug**	
		= Baugruppen-Ziehwerkzeug	
extract instruction	DATA PROC	**Abfrageanweisung**	
extraction	DATA PROC	**Ausblendung**	
= blinding-out			
extraction	TECH	**Entnahme**	
		= Extraktion; Gewinnung	
extraction hook	EQUIP.ENG	**Ausziehhaken**	
extractor	ELECTRON	**Greifklemme**	
extractor (→mask)	DATA PROC	(→Maske 1)	
extract tally counter	TERM&PER	(→Auswahlzähler)	
(→extract counter)			
extract the root	MATH	**wurzelziehen**	
		= radizieren	
extra earphone	TERM&PER	**Zweithörer**	
[telephone set]		[Fernsprechapparat]	
extra-fine thread	MECH	**Sonderfeingewinde**	
extra-high vacuum	PHYS	**Höchstvakuum**	
extra light	TYPOGR	**extrafein**	
↑ font attribute		↑ Schriftattribut	
extra-long	TECH	**überlang**	
extra-long contact	COMPON	(→voreilender Kontakt)	
(→pre-mating contact)			
extraneous (→external)	TECH	(→extern)	
extraneous signal content	TELEC	(→Fremdspannungsanteil)	
(→interfering voltage content)			
extranuclear	PHYS	**kernfremd**	
		= außerhalb des Kernes	
extraordinary	COLLOQ	**außerordentlich**	
= exceptional		= außergewöhnlich; exzeptionell	
≈ above average		≈ überdurchschnittlich	
extrapolate	MATH	**extrapolieren**	
≠ interpolate		≠ interpolieren	
extrapolation	MATH	**Extrapolation**	
≠ interpolation		= Extrapolieren	
		≠ Interpolation	
extraterrestial noise	RADIO	**Weltraumrauschen**	
		= extraterrestrisches Rauschen	
extraterrestrial	ASTROPHYS	**extraterrestrisch**	
		= exterrestrisch; außerirdisch	
extremal (n.)	MATH	**Extremale**	
extreme (→end)	TECH	(→Ende 1)	
extreme error	INSTR	**Extremfehler**	
extremely	COLLOQ	**höchst**	
		= extrem	
extremely high frequency	RADIO	(→Millimeter-Wellen)	
(→millimetric waves)			
extremely sensitive	TECH	**höchstempfindlich**	
= supersensitive			
extreme value	MATH	**Extremum**	
↓ maximum value; minimum value		= Extremwert	
		≈ Grenzwert	
		↓ Größtwert; Kleinstwert	
extremize	MATH	**extremieren**	
↓ maximize; minimize		= extremate	
		↓ maximieren; minimieren	
extrinsic base resistance	MICROEL	**Basisbahnwiderstand**	
= internal base resistance; base bulk resistance		= Basiswiderstand	
extrinsic conduction	MICROEL	**Störstellenleitung**	
		= Fehlstellenleitung; Extrinsic-Leitung	
extrinsic conductor	MICROEL	**Störstellenleiter**	
= impurity conductor; defect conductor		= Fehlstellenleiter; Extrinsic-Leiter	
extrinsic photoelectric effect	PHYS	**äußerer Photoeffekt**	
		= äußerer Fotoeffekt; Maggi-Effekt	
extrinsic semiconductor	MICROEL	**Störstellenhalbleiter**	
= impurity semiconductor; defect semiconductor		= Fehlstellenhalbleiter; Extrinsic-Halbleiter	
extrude (v.t.)	METAL	**strangpressen**	
		= kaltspritzen	
extruded profile (→extruded structural shape)	METAL	(→Strangpreßprofil)	
extruded structural shape	METAL	**Strangpreßprofil**	
= extruded profile			
extruding shop	METAL	**Zieherei**	
extrusion	METAL	**Strangpressen**	
ex-work delivery time	ECON	**ab-Werk-Termin**	
ex works	ECON	**ab Werk**	
ex works price	ECON	**Werkpreis**	
		= Werkspreis (ÖS); Werkabgabepreis; Fabrikpreis	
eye	INSTR	**Auge**	
[of an oscillogram]		[Oszillogramm]	
= eye diagram; eye pattern			
eye	MECH	**Öse**	
= eyelet; lug		= Auge	
eye (→thimble)	TECH	(→Kausch)	
eyebolt	MECH	**Augenschraube**	
eye-catcher	DOC	**Blickfang**	
= bullet			
eye closure	INSTR	**Augenschließen**	
= closure			
eye diagram (→eye)	INSTR	(→Auge)	
eyelet	ELECTRON	**Anschlußauge**	
[PCB]		[Leiterplatte]	
eyelet (→soldering eyelet)	COMPON	(→Lötöse)	
eyelet (→eye)	MECH	(→Öse)	
eyelet terminal board	COMPON	**Lötösenleiste**	
eyeletting machine	MANUF	**Lötösenmaschine**	
eyelit (→soldering land)	ELECTRON	(→Lötauge)	
eyenut	MECH	**Augenmutter**	
eye pattern	MODUL	**Augendiagramm**	
		= Augenmuster	
eye pattern (→eye)	INSTR	(→Auge)	
eye-pattern display	INSTR	**Augenmusterdarstellung**	
eyepiece	PHYS	**Okular**	

F

f (→femto)	PHYS	(→Femto)	
f (→frequency)	PHYS	(→Frequenz)	
°F (→degree Fahrenheit)	PHYS	(→Grad Fahrenheit)	
F (→force)	PHYS	(→Kraft)	
F (→fluorine)	CHEM	(→Flour)	
F (→Farad)	PHYS	(→Farad)	
fabric (→workmanship 2)	TECH	(→Ausführungsqualität)	
fabricant (→manufacturer)	ECON	(→Hersteller)	
fabricating specification	TECH	(→Bauvorschrift)	
(→construction specification)			
fabrication (→manufacturing)	ECON	(→Fertigung)	
fabrication plan	MANUF	**Bauplan**	
fabrication specification	TECH	(→Bauvorschrift)	
(→construction specification)			
fabricator (→manufacturer)	ECON	(→Hersteller)	
fabric tape	POWER SYS	**Gewebeband**	
= woven tape			
Fabry-Perot interferometer	PHYS	**Fabry-Perot-Interferometer**	
Fabry-Perot laser	OPTOEL	**Fabry-Perot-Laser**	
= FP laser		= FP-Laser	

face (n.)	MATH	Seitenfläche	
FACE	MICROEL	FACE-Baustein	
= field-alterable control element			
face	MECH	Stirnfläche	
= front surface			
face [DATA PROC]	MATH	(→Polyeder)	
(→polyhedron)			
face (→typeface)	TYPOGR	(→Schriftbild)	
face bonding (→flip-chip bonding)	MICROEL	(→Flip-chip-Kontaktierung)	
face-centered	PHYS	flächenzentriert	
face-centered lattice	PHYS	flächenzentriertes Gitter	
face-down mode	TERM&PER	seitenrichtige Ablage	
face plan	EQUIP.ENG	Bestückungsplan	
= layout			
face plate (→front panel)	EQUIP.ENG	(→Frontabdeckung)	
face shear	PHYS	Flächenscherung	
face shear crystal	COMPON	Flächenscherungsschwinger [Quarz]	
face-shear mode	PHYS	Flächenscherungsschwingung	
= face-shear vibration		= Flächenschermode	
face-shear vibration (→face-shear mode)	PHYS	(→Flächenscherungsschwingung)	
facet (n.)	TECH	Facette	
[small surface]		[kleine Fläche]	
faceted	TECH	facettiert	
faceted surface	DATA PROC	Facettenoberfläche	
[computer graphics]		[Computergraphik]	
faceting	MICROEL	Facettierung	
face-up mode	TERM&PER	seitenverkehrte Ablage	
face value (→nominal value)	ECON	(→Nennwert)	
face value (→nominal amount)	ECON	(→Nominalbetrag)	
facility (→plant)	TECH	(→Anlage)	
facility (→feature)	TECH	(→Leistungsmerkmal)	
facility (→device)	TECH	(→Vorrichtung)	
facility request	DATA COMM	Leistungsmerkmalsanforderung	
		= Merkmalsanforderung	
facsimile	OFFICE	Faksimile	
[reproduction true in shape but not in scale]		[getreue, nicht maßstabgerechte Vervielfältigung]	
= facsimile copy		≈ Pause; Duplikat	
facsimile	TELEGR	Faksimileübertragung	
= facsimile transmission; fax 1		= Fax 1	
≈ telecopying		≈ Fernkopieren	
facsimile (→facsimile equipmemt)	TERM&PER	(→Faksimilegerät)	
facsimile (→facsimile message)	TELEC	(→Faksimilemitteilung)	
facsimile copy (→facsimile)	DOC	(→Faksimile)	
facsimile equipmemt	TERM&PER	Faksimilegerät	
= facsimile machine; facsimile transceiver; facsimile; fax equipment; fax machine; fax transceiver; fax 3		= Fax-Gerät; Bildtelegrafiegerät	
		≈ Fernkopierer	
facsimile machine (→facsimile equipmemt)	TERM&PER	(→Faksimilegerät)	
facsimile message	TELEC	Faksimilemitteilung	
= facsimile; fax 2		= Faksimilenachricht; Fax 2	
facsimile telegram	TELEC	Bildtelegramm [Fax]	
facsimile telegraphy	TELEGR	Faksimiletelegrafie	
↓ videotelegraphy; document telegraphy; meteorological telegraphy		↓ Bildtelegrafie; Dokumententelegrafie; Wetterkartentelegrafie	
facsimile transceiver	TERM&PER	(→Faksimilegerät)	
(→facsimile equipmemt)			
facsimile transmission (→facsimile)	TELGR	(→Faksimileübertragung)	
fact	COLLOQ	Tatsache	
= datum		= Fakt	
factor	MATH	Faktor	
[quantity to be multiblied with another]		[Größe mit der eine andere zu multiplizieren ist]	
↓ coefficient		↓ Koeffizient	
factor	SCIE	Faktor	
[important and influencing thing]		≈ Einflußgröße	
≈ parameter		≈ Parameter	
factor analysis	MATH	Faktorenanalyse	
		= Faktoranalyse	
factor group	MATH	Faktorgruppe	
factorial (n.)	MATH	Fakultät	
[symbol: !]		[Symbol: !]	
= factorial function			
factorial function	MATH	(→Fakultät)	
(→factorial)			
factoring	ECON	Factoring	
[sale of claims]		[Verkauf von Forderung]	
↑ financing		↑ Finanzierung	
factorization	MATH	Faktorenzerlegung	
factorize (v.t.)	MATH	in Faktoren zerlegen	
factor of merit	PHYS	Gütewert	
[characteristic of an energy store or resonant circuit]		[Kennzahl eines Energiespeichers oder Schwingkreises]	
= Q value; quality value; magnification factor; storage factor;		= Gütezahl	
≈ factor of quality [EL.TECH]		≈ Gütefaktor [EL.TECH]	
factor of merit (→factor of quality)	EL.TECH	(→Gütefaktor)	
factor of quality	EL.TECH	Gütefaktor	
[of an oscillating circuit]		[Schwingkreis]	
= factor Q; factor of merit; Q value; storage factor; figure of merit; quality factor; magnification factor		= Güte; Q-Wert	
		≠ Verlustwiderstand	
≠ loss resistance		↑ Gütewert [PHYS]	
↑ factor of merit [PHYS]			
factor Q	ANT	Antennengüte	
= Q		= Güte; Q-Faktor; Q	
factor Q (→factor of quality)	EL.TECH	(→Gütefaktor)	
factory	MANUF	Fabrik	
= works; manufacturing plant; plant		= Werk; Betrieb	
↑ production plant		↑ Produktionsanlage	
factory acceptance	TECH	Fabrikabnahme	
= factory inspection		= Werkabnahme; Werksabnahme (OES)	
factory acceptance tests	TECH	Fabrikabnahmemessungen	
		= Werkabnahmemessung; Werksabnahmemessungen (OES)	
factory automation	MANUF	(→Produktionsautomatisierung)	
(→production automation)			
factory cabling (→factory wiring)	EQUIP.ENG	(→Werkverdrahtung)	
factory inspection (→factory acceptance)	TECH	(→Fabrikabnahme)	
factory length	COMM.CABLE	Lieferlänge	
= production length		= Herstellungslänge; Fertigungslänge	
factory manager	MANUF	(→Fertigungsleiter)	
(→manufacturing manager)			
factory manager (→works manager)	ECON	(→Werkleiter)	
factory measurements	TELEC	Werkmessungen	
		= Werksmessungen; Fabrikmessungen	
factory network	TELEC	Fabriknetz	
factory-new (→brand-new)	ECON	(→fabrikneu)	
factory presetting (→factory setting)	EQUIP.ENG	(→Fabrikeinstellung)	
factory-programmable terminal (→smart terminal 2)	TERM&PER	(→herstellerprogrammierbares Datensichtgerät)	
factory-programmed option (→factory setting)	EQUIP.ENG	(→Fabrikeinstellung)	
factory setting	EQUIP.ENG	Fabrikeinstellung	
= factory presetting; factory-programmed option; manufacture setting; manufacture presetting		= Fabrikabgleich; Werkeinstellung; Werkseinstellung (OES); Werkabgleich; Werksabgleich (OES)	
factory specification	DOC	Werkunterlagen	
		= Werksunterlage	
factory-wired (→prewired)	EQUIP.ENG	(→vorverdrahtet)	
factory wiring	EQUIP.ENG	Werkverdrahtung	
= factory cabling		= Werksverdrahtung; Werkverkabelung	
≈ prewiring		≈ Vorverdrahtung	
factual (→real)	COLLOQ	(→tatsächlich)	
fade (→fading)	RADIO PROP	(→Schwund)	
fade (n.) (→lap-dissolve)	TV	(→Überblendung)	

fade 214

fade (n.) (→cross fade) EL.ACOUS (→Überblendung)
fade (v.i.) (→vanish) COLLOQ (→schwinden)
fade away (v.i.) (→fade out) ACOUS (→abklingen)
fade control CONS.EL **Überblendregler**
= fader; balance control = Fader
fade-in TV **einblenden**
fade margin (→fading RADIO (→Schwundreserve)
margin)
fade-out (v.t.) TV **abblenden**
fade out (v.i.) ACOUS **abklingen**
= fade away (v.i.)
fade out (v.i.) (→decay) PHYS (→abklingen)
fader (→fade control) CONS.EL (→Überblendregler)
fading RADIO PROP **Schwund**
= fade = Fading
fading area RADIO PROP **Schwundgebiet**
fading compensation RADIO **Schwundausgleich**
 = Fadingkompensation
fading depth RADIO PROP **Schwundtiefe**
 = Fadingtiefe
fading distribution RADIO PROP **Schwund-Überschreitungs-**
wahrscheinlichkeit
 = Fading-Überschreitungs-
wahrscheinlichkeit
fading margin RADIO **Schwundreserve**
= fade margin = Fadingreserve
fading-reducing aerial ANT (→schwundmindernde An-
(→anti-fading antenna) tenne)
fading-reducing antenna ANT (→schwundmindernde An-
(→anti-fading antenna) tenne)
fading response RADIO PROP **Schwundverhalten**
 = Fadingverhalten
fading simulator INSTR **Fadingsimulator**
fading type RADIO PROP **Schwundtyp**
 = Schwundart; Fadingtyp; Fa-
dingart
Fahrenheit temperature PHYS **Fahreneinheit-Temperatur**
[SI unit: degree Fahrenheit] [SI-Einheit: Grad Fahren-
heit]
fail (v.i.) QUAL **ausfallen**
fail (v.t.) COLLOQ **versäumen**
[to something] [etwas -]
fail (v.i.) (→crash) DATA PROC (→abstürzen)
failed call attempt SWITCH **Anruffehlversuch**
failing (→fatigue) QUAL (→Ermüdung)
fail-safe (adj.) TECH **selbstschützend**
= failsafe; self-protecting; autoprotec- = ausfallsicher
tive ≈ gefahrenlos
≈ secure
failsafe (→fail-safe) TECH (→selbstschützend)
fail-safe system TECH **Sicherheitssystem**
 = selbstschützendes System
fail-safe technique TECH **Sicherheitstechnik**
["security first"] [„Sicherheit über alles"]
 = Fail-safe-Technik
fail-soft system TECH **Betriebseinschränkungssy-**
stem
failure QUAL **Ausfall**
= outage; breakdown (n.); fallout (n.) ≈ Störung
failure MATH **Mißerfolg**
failure (→loss) TELEC (→Ausfall)
failure alarm TELEC **Fehleralarm**
failure criterion QUAL **Ausfallkriterium**
failure density QUAL **Ausfallhäufigkeitsdichte**
failure file (→error file) DATA PROC (→Fehlerdatei)
failure fraction QUAL **Ausfallsatz**
= cumulative failure frecuency
failure frequency QUAL **Ausfallhäufigkeit**
failure history file SWITCH **Fehlerdatei**
failure indication EQUIP.ENG **Fehleranzeige**
≈ failure report [TELECONTR] ≈ Fehlermeldung [TELE-
CONTR]
failure information (→failure TELECONTR (→Fehlermeldung)
report)
failure listing (→failure log) DATA PROC (→Störungsprotokoll)
failure log DATA PROC **Störungsprotokoll**
= failure logging; error log; error logg- = Fehlerprotokoll; Störungs-
ing; fault log; fault logging; failure list- auflistung; Fehlerauflistung
ing; error listing; fault listing; failure
printout; error printout; fault printout

failure logging (→failure log) DATA PROC (→Störungsprotokoll)
failure prediction QUAL **Fehlervoraussage**
failure printout (→failure DATA PROC (→Störungsprotokoll)
log)
failure quota (→failure rate) QUAL (→Ausfallrate)
failure rate QUAL **Ausfallrate**
[units: fit; kfit; % p.a.] [Einheiten: fit; kfit; % pro
= error rate; failure quota Jahr]
 = Fehlerrate; Fehlerhäufig-
keit; geschätzte Ausfallrate; Ausfallquote (DIN)
failure recovery QUAL **Fehlererholung**
 = Fehlerüberwindung
failure report TELECONTR **Fehlermeldung**
= fault report; defect report; malfunc- = Störungsmeldung
tion report; failure information; fault ≈ Fehleranzeige
information; defect information; mal- [EQUIP.ENG]
function information
≈ failure indication [EQUIP.ENG]
failure report QUAL **Fehlerbericht**
= trouble report
fair (→adequate) SCIE (→angemessen)
fair (→exposition) ECON (→Ausstellung)
fair copy OFFICE **Reinschrift**
= final copy
fair weather METEOR **Schönwetter**
≠ bad weather ≠ Schlechtwetter
fall 1 (n.) PHYS **Fall**
fall (n.) (→decay) ELECTRON (→Abfall)
fall 2 (n.) (→decay 1) PHYS (→Abklingvorgang)
fall (→decrease) TECH (→vermindern)
fall (→reduce 1) TECH (→vermindern)
fallacy SCIE **Trugschluß**
≈ wrong inference ≈ Fehlschluß
fall back (n.) SWITCH **Rückfall**
fall back (n.) DATA PROC **Bereitschaftsmaßnahme**
[instruction or procedure in fault situ-
ation]
fall-back circuit (→auxiliary CIRC.ENG (→Hilfskreis)
circuit)
fall back routine DATA PROC **Bereitschaftsroutine**
fall-back system (→stand-by DATA PROC (→Bereitschaftssystem)
system)
falling edge (→trailing edge) ELECTRON (→Abfallflanke)
falling inversion RADIO PROP **Absinkinversion**
falling pulse edge (→trailing ELECTRON (→Impulsabfallflanke)
pulse edge)
fall-off (n.) (→decay) ELECTRON (→Abfall)
fallout (n.) (→failure) QUAL (→Ausfall)
fall time MICROEL **Abfallzeit**
[transistor] [Transistor]
 = Abklingzeit
fall time (→decay time) ELECTRON (→Abfallzeit)
false alarm SIGN.ENG **Fehlalarm**
= nuisance alarm = Falschmeldung
false bottom (→raised floor) SYS.INST (→Doppelboden)
false count TECH **Falschzählung**
false decision (→erroneous TECH (→Fehlentscheidung)
decision)
false drop (→false DATA PROC (→Fehlsuche)
retrieval)
false error (→false error DATA PROC (→Fehlmeldung)
indication)
false error indication DATA PROC **Fehlmeldung**
= false error
false floor (→raised floor) SYS.INST (→Doppelboden)
false retrieval DATA PROC **Fehlsuche**
= false drop
FAM (→high-speed DATA PROC (→Schnellspeicher 1)
memory)
familiarization TECH **Einarbeitung**
[with a job] ≈ Lernphase
= secondment
≈ learning phase
familiarization time TECH **Einarbeitungszeit**
= secondment period
familiarize with a job TECH **einarbeiten**
family (→logic family) MICROEL (→Logikfamilie)
family of characteristics (→family MATH (→Kurvenschar)
of curves)

family of curves		MATH	Kurvenschar	Faraday cage	PHYS	Faraday-Käfig
= family of characteristics			= Kurvenlinienfeld	= screening cage; Faraday screen; Faraday shield		
FAMOST		MICROEL	**FAMOS-Transistor**	Faraday cell	OPTOEL	Faraday-Zelle
[Floating-gate Avalanche-injection Metal-Oxide Semiconductor Transistor; memory device loadable by avalanche injection, unloadable by ultraviolet radiation]			[durch Stoßinjektion ladbares und UV-Licht entladbares Speicherelement]	Faraday collector	ELECTRON	Faraday-Becher
				Faraday constant	PHYS	Faraday-Konstante
				= electrolytic constant		
				Faraday current	PHYS	Faraday-Strom
fan (v.t.)		TECH	blasen	Faraday disk	EL.TECH	Faraday-Scheibe
[to blow air for cooling]			[zur Kühlung]	Faraday effect	PHYS	Faraday-Effekt
fan (v.t.)		TECH	auffächern	[rotation of polarization plane by magnetic field]		[Drehung der Polarisationsebene durch Magnetfeld]
fan (n.)		MATH	Fächer (n.m.)			
[family of curves]			[Kurvenschar]	= Faraday rotation; non-reciprocal wave rotation; rotation		= Faraday-Drehung; Faraday-Rotation; Rotation
fan (n.)		TECH	Ventilator			
= ventilator; air circulator; cooling fan; blower			= Lüfter; Kühlgebläse; Kühlluftgebläse	Faraday rotation (→Faraday effect)	PHYS	(→Faraday-Effekt)
≈ exhaust			≈ Entlüfter	Faraday rotator	MICROW	Faraday-Rotator
↓ blast			↓ Gebläse	= non-reciprocal wave rotator		
fan antenna (→curtain antenna)		ANT	(→Vorhangantenne)	Faraday screen (→Faraday cage)	PHYS	(→Faraday-Käfig)
fan dipole		ANT	Fächerdipol	Faraday shield (→Faraday cage)	PHYS	(→Faraday-Käfig)
fanfold (adj.)		TERM&PER	zickzackgefaltet	Faraday switch	MICROW	Faraday-Schalter
[paper]			[Papier]	far end	TELEC	fernes Ende
= fanfolded; Z-folded; accordion-fold (adj.)			= leporellogefalzt	far-end actuated (→remotely operated)	TECH	(→fernbetätigt)
fanfold (n.) (→fan folding)		TERM&PER	(→Leporellofalzung)	far-end controlled (→remotely controlled)	TECH	(→ferngesteuert)
fanfolded (→fanfold)		TERM&PER	(→zickzackgefaltet)	far-end crosstalk	TRANS	Fernnebensprechen
fanfolded tape		TERM&PER	zusammengelegter Lochstreifen	= FEXT		= Gegensprechen
				far-end operated (→remotely operated)	TECH	(→fernbetätigt)
fanfold form		TERM&PER	Zickzackformular			
↑ continuous form			= Leporelloformular; Zickzackvordruck; Leporellovordruck	farewell	COLLOQ	Verabschiedung
				far-field (→far-field region)	ANT	(→Fernfeld)
				far-field diagram	ANT	Fernfelddiagramm
			↑ Endlosformular	far-field region	ANT	Fernfeld
fan folding		TERM&PER	Leporellofalzung	= far-field; distant field		= Frauenhoferregion
= leporello folding; fanfold (n.); accordion fold (n.); concertina fold (n.); Z folding; Z fold			= Zickzackfalzung	far-field strength	ANT	Fernfeldstärke
				farmline	OUTS.PLANT	Farmerleitung
				far point	PHYS	Fernpunkt
fanfold label		DATA PROC	Endlosetikette	far-reaching	COLLOQ	weitreichend
fanfold paper		TERM&PER	Leporellopapier	[figurative]		[figurativ]
= fanfold web			= Faltpapier; Zickzackpapier; zick-zack-gefalztes Endlospapier	= long-range; wide-ranging		
↑ continuous paper				far-to-near direction (→return direction)	TELEC	(→Rückrichtung)
			↑ Endlospapier	**FAS**	DATA COMM	**FAS**
fanfold web (→fanfold paper)		TERM&PER	(→Leporellopapier)	= flexible access system		
fan-in (n.) (→fan-in factor)		MICROEL	(→Eingangslastfaktor)	**FAS** (→frame alignment signal)	CODING	(→Rahmenkennungswort)
fan-in factor		MICROEL	Eingangslastfaktor			
[maximum number of inputs processably by a logic device]			[von einem Logikbaustein maximal verarbeitbare Anzahl von Eingängen]	fascia plate [DATA PROC] (→front panel)	EQUIP.ENG	(→Frontabdeckung)
= fan-in (n.)				fascicle (→volume)	TYPOGR	(→Band)
			= Eingangsfächerung; Eingangsfächer; Eingangslast; Fan-in	fashion (v.t.)	COLLOQ	gestalten
				= shape (v.t.)		
				≈ make		
fanlike		TECH	fächerförmig	fashion (→state)	TECH	(→Zustand)
			≈ fächerig	fast	TECH	schnell
fan marker		RADIO NAV	Fächermarkierungsbake	= quick; rapid; speedily; highspeed		= rasch; unverzüglich; geschwind
= fan-marker beakon				≈ quick acting		
fan-marker beacon (→fan marker)		RADIO NAV	(→Fächermarkierungsbake)			≈ schnellwirkend
				FAST	MICROEL	**FAST**
fanning		TECH	Auffächerung	[Fairchild advanced Schottky technology]		↑ Logikbaustein
			≈ Auffächern			
fanning comb (→fanning strip)		EQUIP.ENG	(→Verdrahtungskamm)	↑ logic device		
				fast access memory	DATA PROC	(→Schnellspeicher 1)
fanning strip		EQUIP.ENG	Verdrahtungskamm	(→high-speed memory)		
= fanning comb			= Drahtführungskamm	fast access storage	DATA PROC	(→Schnellspeicher 1)
Fano coding		CODING	Fano-Codierung	(→high-speed memory)		
fan-out (→fan-out factor)		MICROEL	(→Ausgangslastfaktor)	fast access store	DATA PROC	(→Schnellspeicher 1)
fan-out factor (n.)		MICROEL	Ausgangslastfaktor	(→high-speed memory)		
[quantity of circuits which can be connected to an output, or quantity of outputs a circuit can drive simultaneously]			[Anzahl Schaltung die an einen Ausgang gelegt werden können, oder Anzahl von Ausgängen die eine Schaltung gleichzeitig aussteuern kann]	fast acting (→quick acting)	TECH	(→schnellwirkend)
				fast-acting relay	ELECTRON	Schnellschaltrelais
				fast blowing fuse (→fast blowing melting fuse)	COMPON	(→flinke Schmelzsicherung)
= fan-out				fast blowing melting fuse	COMPON	flinke Schmelzsicherung
			= Ausgangsfächerung; Ausgangsverzweigung	= fast blowing fuse; quick reacting fuse		↑ Stromfeinsicherung
				↑ fine-wire fuse		
fan-type lock washer		MECH	Fächerscheibe	fast break (→rapid disconnection)	EL.TECH	(→Schnellabschaltung)
Farad		PHYS	**Farad**			
[SI unit for electric cpacity; = 1 C/V]			[SI-Einheit für elektrische Kapazität; = 1 C/V]	fast charge (→fast charging)	POWER SYS	(→Schnelladung)
= F			= F			

fast charging	POWER SYS	Schnelladung
= fast charge		
fast dial	TELEPH	schnelle Wählscheibe
fasten (v.t.)	TECH	befestigen
= fix		= festmachen; festsetzen; fixieren
≈ attach; hold		≈ anbringen; halten
fastener	COMPON	Montageteil
fastener (→mount)	TECH	(→Halterung)
fastener (→paper clip)	OFFICE	(→Büroklammer)
fastening 1 [process]	TECH	Befestigung 1 [Vorgang]
= fixing		= Festsetzung 1; Fixierung
fastening 2 (→mount)	TECH	(→Halterung)
fastening clamp	MECH	Befestigungsklemme
fastening screw	MECH	Befestigungsschraube
fastening thread	MECH	Befestigungsgewinde
fastest-growing	TECH	mit den höchsten Zuwachsraten
fast forward [cassette player]	CONS.EL	Vorlauf [Kassettengerät]
fast-forward key	CONS.EL	Vorlauftaste
fast-Fourier analyzer	INSTR	FFT-Analysator
fast Fourier transform	MATH	schnelle Fourier-Transformation
= FFT		≈ FFT
fast-growing	ECON	wachstumsträchtig
		= schnellwachsend
fast indication	INSTR	Schnellanzeige
fast line	DATA COMM	Hochgeschwindigkeitskanal
		= Hochgeschwindigkeits-Datenkanal; Hochgeschwindigkeitsleitung
fast memory (→high-speed memory)	DATA PROC	(→Schnellspeicher 1)
fast motion	TV	Zeitraffung
= quick motion		≠ Zeitlupe
≠ slow motion		
fast motion facility	TV	Zeitraffer
= quick motion facility		
fastness	COLLOQ	Schnelligkeit
= velocity; rapidity; quickness; speed		= Geschwindigkeit
fast printer (→high-speed printer)	TERM&PER	(→Schnelldrucker)
fast-release relay	COMPON	Schnellabschaltrelais
fast running	MECH	schnellaufend
fast select	DATA COMM	Einzelpaket
fast storage (→high-speed memory)	DATA PROC	(→Schnellspeicher 1)
fast store (→high-speed memory)	DATA PROC	(→Schnellspeicher 1)
fast tuning	ANT	Schnellabstimmung
fast-tuning device	ANT	Schnellabstimmgerät
fatal error	DATA PROC	fataler Fehler
[leads to collaps of program]		[führt zu Programmunterbrechung]
≠ nonfatal error		= gravierender Fehler; schwerer Fehler
		≠ nichtfataler Fehler
fat dipole	ANT	dicker Dipol
father file	DATA PROC	Vater-Datei
father tape	DATA PROC	Vater-Band
[generation principle]		[Generationsprinzip]
fathom	PHYS	Faden
[6 ft = 1.8288 m]		[1,8288 m]
fatigue (n.)	QUAL	Ermüdung
= wear; failing		≈ Verschleiß
≈ wearout (n.)		
fatigue fracture	MECH	Ermüdungsbruch
fatigue-free (→fatigue-proof)	TECH	(→ermüdungsfrei)
fatigue limit (→endurance limit)	MECH	(→Ermüdungsgrenze)
fatigue-proof	TECH	ermüdungsfrei
= fatigue-free		= ermüdungslos
fatigue strength (→endurance limit)	MECH	(→Ermüdungsgrenze)
faucet (AM) (→cock)	TECH	(→Hahn)
fault	TECH	Fehler
= defect; error; nonconformance		= Defekt
≈ malfunction; damage		≈ Funktionsstörung; Schaden
fault (→lack)	ECON	(→Mangel)
fault (→bug)	DATA PROC	(→Fehler)
fault cause	QUAL	Fehlerursache
= error cause		
fault-clear (v.t.)	TECH	entstören
fault clearance (→repair)	TECH	(→Instandsetzung)
fault-clearer	TECH	Entstörer
≈ trouble-shooter		
fault clearing (→repair)	TECH	(→Instandsetzung)
fault complaint service	TELEPH	Störungsannahme
= repair desk service		= Störungsannahmedienst
↑ special telephone service		↑ Fernsprechsonderdienst
fault condition	TECH	Störungszustand
fault coverage rate	MICROEL	Fehlerüberdeckung
= FCG		
fault current (→offset current)	EL.TECH	(→Fehlstrom)
fault detection	TELECONTR	Fehlererfassung
= error detection		= Fehlererkennung; Störungserfassung; Störungserkennung
fault detection	QUAL	Fehlererkennung
= error detection		= Störungserkennung; Fehlererfassung; Störungserfassung
fault detector	SIGN.ENG	Störungsmelder
fault diagnosis	QUAL	Fehlersuche
= trouble diagnosis; fault shooting; trouble shooting; fault localization; trouble localization; fault tracking; trouble tracking		= Fehlereingrenzung; Fehlerdiagnose; Fehlerlokalisierung; Fehlerortung; Fehlerortbestimmung; Störungssuche; Störungseingrenzung; Störungsdiagnose; Störungslokalisierung; Störungsortung; Störungsortbestimmung
fault file (→error file)	DATA PROC	(→Fehlerdatei)
fault-free	TECH	fehlerfrei
= faultless; unfaulted; error-free		= fehlerlos
≈ trouble-free; unobjectionable		≈ störungsfrei; einwandfrei
fault grade coverage	MICROEL	Fehlererfassungsgrad
fault handling (→error handling)	DATA PROC	(→Fehlerbehandlung)
fault information (→failure report)	TELECONTR	(→Fehlermeldung)
faulting	TECH	Verschmutzung
≈ contamination		≈ Verunreinigung
faultless (→fault-free)	TECH	(→fehlerfrei)
fault listing (→failure log)	DATA PROC	(→Störungsprotokoll)
fault localization (→fault diagnosis)	QUAL	(→Fehlersuche)
fault-localizing bridge	INSTR	Fehlerortungs-Meßbrücke
fault locating device (→fault locating equipment)	TRANS	(→Fehlerortungsgerät)
fault locating equipment	TRANS	Fehlerortungsgerät
= fault locating device		= Ortungsgerät; Fehlerortungsvorrichtung
fault locating loop	TRANS	Fehlerortungsschleife
= locating loop		= Ortungsschleife
fault locating measuring equipment	INSTR	Fehlerortmeßgerät
fault locating mode	TRANS	Fehlerortungsverfahren
		= Ortungsverfahren
fault-locating signal	TRANS	Fehlerortungssignal
= locating signal		= Ortungssignal
fault locating unit	TRANS	Fehlerortungseinschub
		= Ortungseinschub
fault location	TRANS	Fehlerortung
fault location program (→diagnostic program)	DATA PROC	(→Diagnoseprogramm)
fault log (→failure log)	DATA PROC	(→Störungsprotokoll)
fault logging (→failure log)	DATA PROC	(→Störungsprotokoll)
fault management	TELEC	Betriebsstörungsverwaltung
		= Betriebsstörungsmanagement
fault management (→error handling)	DATA PROC	(→Fehlerbehandlung)

fault power (→short-circuit power)	POWER ENG	(→Kurzschlußleistung)	
fault printout (→failure log)	DATA PROC	(→Störungsprotokoll)	
fault report (→failure report)	TELECONTR	(→Fehlermeldung)	
fault shooting (→fault diagnosis)	QUAL	(→Fehlersuche)	
fault simulation	MICROEL	**Fehlersimulation**	
fault time (→downtime)	QUAL	(→Ausfallzeit)	
fault tolerance	DATA PROC	**Fehlertoleranz**	
fault-tolerant	DATA PROC	**fehlertolerant**	
≈ fault-protected		≈ ausfallgeschützt	
fault-tolerant computer	DATA PROC	**fehlertoleranter Rechner**	
		= fehlertoleranter Computer	
fault-tolerant computing	DATA PROC	**fehlertolerante Datenverarbeitung**	
fault trace (n.)	DATA PROC	**Ablaufverfolgung**	
= trace (n.)		= Trace	
≈ fault diagnosis program		≈ Fehlerdiagnoseprogramm	
fault tracking (→fault diagnosis)	QUAL	(→Fehlersuche)	
fault tree	TECH	**Fehlerbaum**	
faulty	QUAL	**fehlerhaft**	
= defective; erroneous; flawed; out of order; O.O.O.; deficient		= mangelhaft; defekt; schadhaft	
		≈ beschädigt	
faulty design (→design mistake)	TECH	(→Konstruktionsfehler)	
faulty-line alarm	SWITCH	**Leitungsalarm**	
faulty polarization	EL.TECH	**Falschpolung**	
≈ polarity inversion		≈ Polaritätsumkehr	
faulty pulse	ELECTRON	**Fehlpuls**	
faulty selection	SWITCH	**Falschwahl**	
faulty soldering point (→cold soldering point)	ELECTRON	(→kalte Lötstelle)	
favourably oriented (→codirectional)	TECH	(→gleichsinnig)	
fax 1 (→facsimile)	TELEGR	(→Faksimileübertragung)	
fax 3 (→facsimile equipemnt)	TERM&PER	(→Faksimilegerät)	
fax 2 (→facsimile message)	TELEC	(→Faksimilemitteilung)	
fax equipment (→facsimile equipment)	TERM&PER	(→Faksimilegerät)	
fax machine (→facsimile equipment)	TERM&PER	(→Faksimilegerät)	
fax transceiver (→facsimile equipment)	TERM&PER	(→Faksimilegerät)	
F/B (→front-to-back ratio)	ANT	(→Rückstrahldämpfung)	
F'cap (→foolscap)	TYPOGR	(→Format 16x13 Zoll)	
FCB (→file control block)	DATA PROC	(→Dateisteuerblock)	
FCC	TELEC	**FCC**	
= Federal Communications Commission		[Fernmeldebehörde in USA]	
FCC (→font change character)	DATA PROC	(→Schriftartumschaltezeichen)	
FC capacitor	COMPON	**FC-Kondensator**	
FCFS	DATA COMM	**FCFS**	
[First Come First Served]		["wer zuerst kommt mahlt zuerst"]	
FCG (→fault coverage rate)	MICROEL	(→Fehlerüberdeckung)	
FCI	TERM&PER	**FCI**	
[unit for recording density of magnetic disk stores]		[Einheit für Beschreibungsdichte von Magnetplattenspeichern]	
= flix changes per inch		= Flußwechsel pro Zoll	
FCS (→block check sequence)	DATA COMM	(→Blockprüfzeichenfolge)	
FD (→floppy disk)	TERM&PER	(→Diskette)	
fd (→duplex operation)	TELEC	(→Duplexbetrieb)	
FD (→duplex operation)	TELEC	(→Duplexbetrieb)	
FD 4 antenna	ANT	**FD-4-Antenne**	
FDC (→floppy disk controller)	TERM&PER	(→Diskettensteuerung)	
fdc (→floppy disk controller)	TERM&PER	(→Diskettensteuerung)	
FDE (→frequency-domain equalizer)	NETW.TH	(→Frequenzbereichsentzerrer)	
F-distribution	MATH	**F-Verteilung**	
= variance-ratio distribution		= Fischer-Verteilung	
FDM (→frequency division multiplex)	TELEC	(→Frequenzmultiplex)	
FDMA (→frequency division multiple access)	TELEC	(→Frequenzvielfachzugriff)	
FDM carrier art (→FDM carrier transmission engineering)	TRANS	(→Trägerfrequenztechnik)	
FDM carrier frequency system = FDM carrier system	TRANS	**Trägerfrequenzsystem** = TF-System	
FDM carrier line (→FDM carrier line circuit)	TRANS	(→Trägerfrequenzleitung)	
FDM carrier line circuit = FDM carrier line	TRANS	**Trägerfrequenzleitung** = TF-Leitung	
FDM carrier link	TRANS	**TF-Verbindung**	
FDM carrier system (→FDM carrier frequency system)	TRANS	(→Trägerfrequenzsystem)	
FDM carrier transmission engineering	TRANS	**Trägerfrequenztechnik** = TF-Technik	
= FDM carrier transmission technique; FDM carrier art		≈ Frequenzmultiplex	
FDM carrier transmission technique (→FDM carrier transmission engineering)	TRANS	(→Trägerfrequenztechnik)	
FDM channel (→carrier channel)	TRANS	(→TF-Kanal)	
FDNR	CIRC.ENG	**FDNR**	
[Frequency-Dependent Negative Resistor; type of circuit]		[Schaltung mit frequenzabhängigem negativen Widerstand]	
FDX (→duplex operation)	TELEC	(→Duplexbetrieb)	
fdx (→duplex operation)	TELEC	(→Duplexbetrieb)	
Fe (→iron)	CHEM	(→Eisen)	
FE (→format effector)	DATA PROC	(→Formatsteuerung)	
feasible 1	TECH	**durchführbar**	
= viable; practicable; performable; realizable		= realisierbar; ausführbar; praktikabel; praktizierbar	
feasible 2 (→serviceable)	TECH	(→brauchbar)	
feasibility (→viability)	TECH	(→Durchführbarkeit)	
feasibility study	ECON	**Realisierbarkeitsstudie**	
		= Durchführbarkeitsstudie; Durchführbarkeitsuntersuchung	
feasible solution	MATH	**zulässige Lösung**	
= admissible solution; legal solution			
feathered key	MECH	**Paßfeder**	
= feather key; fitted key		≈ Keil	
feather key (→feathered key)	MECH	(→Paßfeder)	
feature 2 (n.)	TECH	**Leistungsmerkmal**	
= performance parameter; user facility; facility		= Leistungsparameter ↑ Merkmal	
↑ characteristic 1			
feature 1 (→characteristic 1)	TECH	(→Merkmal)	
feature extraction	DATA PROC	**Merkmalextraktion**	
features (→equipment 1)	TECH	(→Ausstattung)	
FEC (→forward error correction)	CODING	(→Vorwärtsfehlerkorrektur)	
FED (→field effect diode)	MICROEL	(→Feldeffektdiode)	
fed element (→primary radiator)	ANT	(→Primärstrahler)	
Federal Bank	ECON	**Bundesbank**	
Federal Communications Commission (→FCC)	TELEC	(→FCC)	
Federal Office	ECON	**Bundesamt**	
Federal Postal Administration	ECON	**Bundespost**	
Federal Railways	ECON	**Bundesbahn**	
federation (→association)	ECON	(→Verband)	
FEDS	TERM&PER	**Kombinations-Plattenspeicher**	
[Fixed and Exchangeable Disk Storage]			
fee	ECON	**Honorar**	
[charge for professional service]		[für professionelle Leistung]	
fee (a fixed charge)	ECON	(→Gebühr)	
(→charge 2)			
feeble current (→low current)	EL.TECH	(→Schwachstrom)	
feed (n.)	TERM&PER	**Vorschub**	
= advance (n.); throw (n.)		≈ Zeilenwechsel	
↓ line feed; reverse line feed; paper feed; tape feed 1; tape feed 2		↓ Zeilenvorschub; Zeilenrückschub; Papiervorschub; Bandvorschub; Streifenvorschub	

feed

feed (→antenna feeding)	ANT	(→Antennenspeisung)	
feed (n.) (→supply)	ELECTRON	(→Versorgung)	

feedback (n.)   CIRC.ENG   **Rückkopplung**
[part of output signal is fed to the input]
= feedback coupling; back coupling; reaction coupling; back-off
↓ positive feedback; negative feedback; voltage feedback; current feedback
[Rückführung eines Teils des Ausgangssignals auf den Eingang]
= Rückführung; Rückwirkung
↓ Mitkopplung; Gegenkopplung; Spannungsrückkopplung; Stromrückkopplung

feedback amount	CIRC.ENG	**Rückkopplungsgrad**
feedback amplifier	CIRC.ENG	**Rückkopplungsverstärker**
= reaction amplifier		
feedback coil (→reaction coil)	CIRC.ENG	(→**Rückkopplungsspule**)
feedback condition	CIRC.ENG	**Rückkopplungsbedingung**
feedback controller	CONTROL	**Rückführungsregler**
feedback coupling (→feedback)	CIRC.ENG	(→**Rückkopplung**)
feed-back distortion	TELEC	**Rückkoppelverzerrung**
feedback equation	CIRC.ENG	**Rückkopplungsgleichung**
feedback factor	CIRC.ENG	**Rückkopplungsfaktor**
feedback loss (→round-trip loss)	TELEPH	(→**Umlaufdämpfung**)
feedback network	CIRC.ENG	**Rückkopplungsnetzwerk**
feedback signal	CONTROL	**gemessene Regelgröße**

[input signal to the summing point; derived from the directly controlled variable]
≠ an den Eingang der Vergleichsstelle angepaßt

feedback transformer	CIRC.ENG	**Rückkopplungsübertrager**
feed cable	POWER SYS	**Zuführungskabel**
feed circuit	CIRC.ENG	**Speisekreis**
feed circuit (→feeding bridge)	TELEPH	(→**Speisebrücke**)
feed control	TERM&PER	**Vorschubsteuerung**
= advance control; carriage control		= Papiervorschubsteuerung
feed current (→supply current)	ELECTRON	(→**Speisestrom**)
feed device	TERM&PER	**Vorschubeinrichtung**
= feed mechanism		
feed direction (→transport direction)	MECH	(→**Transportrichtung**)
feeder	ANT	**Einspeisung**
≈ radiator		= Antenneneinspeisung
		≈ Strahler
feeder	DATA COMM	**Zubringer**
[signal carrying channel]		
feeder (AM) (→main cable)	OUTS.PLANT	(→**Hauptkabel**)
feeder (→feeder line)	ANT	(→**Speiseleitung**)
feeder (→entrance link)	TRANS	(→**Zubringerstrecke**)
feeder cable (AM) (→main cable)	OUTS.PLANT	(→**Hauptkabel**)
feeder line	ANT	**Speiseleitung**
= feeder		
feeder link (→entrance link)	TRANS	(→**Zubringerstrecke**)
feeder loop	OUTS.PLANT	**Primärkabel-Aderpaar**
feedforward (n.)	CONTROL	**Vorwärtsregelung**
		= Aufschaltung
feed forward converter	POWER ENG	**Durchflußwandler**
= forward DC converter		= Durchflußumrichter
feed gear	TERM&PER	**Vorschubgetriebe**
feed hole	TERM&PER	**Vorschubloch**
= sprocket hole; transport hole; center hole; centre hole		= Taktloch; Transportloch; Führungsloch
≠ information hole		≠ Informationsloch
feed holes	TERM&PER	**Transportlochung**
= sprocket holes; transport holes		= Vorschublochung; Führungslochung; Taktlochung
≠ code holes		≠ Informationslochung
feed hole scanning	TERM&PER	**Taktlochabschaltung**
feed horn	ANT	**Speisehorn**
= feedhorn		
feedhorn (→feed horn)	ANT	(→**Speisehorn**)
feed impedance	ANT	**Speisescheinwiderstand**
feeding	ELECTRON	**Einspeisung**
= injection		
feeding bridge	TELEPH	**Speisebrücke**
= supply bridge; feed circuit		
feeding loss	TELEPH	**Speisestromdämpfung**
feed instruction	DATA PROC	**Vorschubbefehl**
= advance instruction		
feed lever	TERM&PER	**Vorschubhebel**
feed line	POWER SYS	**Zuführungsleitung**
		= Zuführung
feed line loss	ANT	**Speiseleitungsverluste**
feedline with matching stub	ANT	**angezapfte Speiseleitung**
= stub feedline		
feed mechanism (→moving mechanism)	MECH	(→**Transporteinrichtung**)
feed mechanism (→feed device)	TERM&PER	(→**Vorschubeinrichtung**)
feed reel (→supply reel)	TERM&PER	(→**Abwickelspule**)
feed screw	TERM&PER	**Vorschubspindel**
feed-through (v.t.)	ELECTRON	**durchschleifen**
= loop-through (v.t.)		
feedthrough (n.) [PCB]	ELECTRON	**Durchkontaktierung** [Leiterplatte]
= through-plating; viahole		
feedthrough (n.)	CIRC.ENG	**Einstreuung**
feedthrough (→punch-through)	MICROEL	(→**Durchgriff**)
feedthrough capacitor	COMPON	**Durchführungskondensator**
= lead-through capacitor		= Duko
≈ suppression capacitor		≈ Funkentstörkondensator
feed-through input	TV	**Durchschleifeingang**
= loop-through input		
feed-through plug (→loop-through plug)	BROADC	(→**Durchschleifsteckdose**)
feedthrough termination	INSTR	**Durchführungsabschluß**
feed velocity	TERM&PER	**Vorschubgeschwindigkeit**
feed voltage (→supply voltage)	ELECTRON	(→**Speisespannung**)
feed wheel	TERM&PER	**Vorschubrad**
↓ sprocket wheel		↓ Stachelrad
feeed blanking	TELEPHON	**Speiselücke**
[short interruption]		[kurzzeitige Unterbrechung]
fee-metering signal (→metering signal)	SWITCH	(→**Zählzeichen**)
feep (n.) (→beep)	ACOUS	(→**Pieps**)
fee printing (→tax printing)	SWITCH	(→**Gebührenausdruck**)
FEFET	MICROEL	**Speicher-Feldeffekttransistor**
= ferroelectric field-effect transistor		= Speicher-FET; FEFET
fellow listener	TELEPH	**Mithörer**
felt	TECH	**Filz**
felt pad	TECH	**Filzpolster**
felt seal	MECH	**Filzdichtung**
felt-tip	OFFICE	**Filzstift**
FEM (→finite element method)	MATH	(→**Methode der endlichen Elemente**)
female (→jack)	COMPON	(→**Buchse**)
female cable connector (→cable jack)	COMPON	(→**Kabelbuchse**)
female connector (→jack)	COMPON	(→**Buchse**)
female die	MECH	**Matrize**
= bottom die		
female-female adapter	COMPON	**Doppelbuchse**
↑ in-series adapter		[beidseitig mit Buchse]
		↑ Kupplung 1
female plug (→jack)	COMPON	(→**Buchse**)
female terminal strip (→spring contact strip)	COMPON	(→**Federleiste**)
female-to-female adapter	COMPON	**Doppelkupplung** [Stecker]
↑ adapter 1		= Kupplungsstück
		↑ Übergangsstecker
femininie (n.)	LING	**Femininum**
		= weibliches Geschlecht
femto	PHYS	**Femto**
[ten to the power of minus fifteen]		[zehn hoch minus fünfzehn]
= f		= f
femtosecond	PHYS	**Femtosekunde**
= fs		= fs
fence	CIV.ENG	**Umzäunung**
= fencing		= Zaun
fence alarm	SIGN.ENG	**Zaunalarm**
fencing (→fence)	CIV.ENG	(→**Umzäunung**)

FEP (→front-end processor)	DATA COMM	(→Vorrechner)	**fetch-execute cycle** = fetch-run cycle; fetch-execute phase; fetch-run phase	DATA PROC	**Abruf-/Ausführungszyklus** = Abruf-/Ausführungsphase
Fermi degeneracy	PHYS	**Fermi-Entartung**	**fetch-execute cycle** (→fetch cycle)	DATA PROC	(→Abrufzyklus)
Fermi-Dirac distribution (→Fermi-Dirac function)	PHYS	(→Fermi-Dirac-Funktion)	**fetch-execute phase** (→fetch-execute cycle)	DATA PROC	(→Abruf-/Ausführungszyklus)
Fermi-Dirac function = Fermi-Dirac distribution; Fermi function; Fermi distribution	PHYS	**Fermi-Dirac-Funktion** = Fermi-Dirac-Verteilung; Fermi-Funktion; Fermi-Verteilung	**fetch instruction** (→instruction fetch)	DATA COMM	(→Befehlsabruf)
Fermi-Dirac gas	PHYS	**Fermi-Dirac-Gas**	**fetch out** (→output)	DATA PROC	(→ausgeben)
Fermi-Dirac statistics (→Fermi statistics)	PHYS	(→Fermi-Statistik)	**fetch phase** (→fetch cycle)	DATA PROC	(→Abrufzyklus)
Fermi distribution (→Fermi-Dirac function)	PHYS	(→Fermi-Dirac-Funktion)	**fetch protect**	DATA PROC	**Abrufsperre**
Fermi energy	PHYS	**Fermi-Energie**	**fetch-run cycle** (→fetch-execute cycle)	DATA PROC	(→Abruf-/Ausführungszyklus)
Fermi function (→Fermi-Dirac function)	PHYS	(→Fermi-Dirac-Funktion)	**fetch-run phase** (→fetch-execute cycle)	DATA PROC	(→Abruf-/Ausführungszyklus)
Fermi level	PHYS	**Fermi-Niveau** = Fermi-Kante	**fetch signal**	DATA PROC	**Abrufsignal**
Fermi potential	PHYS	**Fermi-Potential**	**FET differential amplifier**	CIRC.ENG	**FET-Differenzverstärker**
Fermi statistics = Fermi-Dirac statistics	PHYS	**Fermi-Statistik** = Fermi-Dirac-Statistik	**FET input**	CIRC.ENG	**FET-Eingang**
Fermi temperature	PHYS	**Fermi-Temperatur**	**FEXT** (→far-end crosstalk)	TRANS	(→Fernnebensprechen)
fermium = Fm	CHEM	**Fermium** = Fm	**FF** (→formular feed)	DATA COMM	(→Formularvorschub)
Ferraris motor	POWER SYS	**Ferraris-Motor**	**FF** (→form feed)	TERM&PER	(→Formularzuführung)
ferreed	COMPON	**Ferreed-Relais**	**FFT** (→fast Fourier transform)	MATH	(→schnelle Fourier-Transformation)
ferric oxide (→ferrite)	PHYS	(→Ferrit)	**fiber** (AM) = fibre (BRI, CCITT)	OPT.COMM	**Faser**
ferrimagnetism	PHYS	**Ferrimagnetismus**	**fiber** (AM) = fibre (BRI) ≈ thread	TECH	**Faser** = Fiber ≈ Faden
ferrite = ferric oxide	PHYS	**Ferrit**	**fiber buffer** (AM) = fibre buffer (BRI); fiber protection; fibre protection	OPT.COMM	**Faserschutz**
ferrite antenna ↓ magnetic rod antenna	ANT	**Ferritantenne** ↓ Ferritstabantenne	**fiber bundle** (→optical fiber bundle)	OPT.COMM	(→Faserbündel)
ferrite attenuator bead	ELECTRON	**Dämpfungsperle**	**fiber cable** (→optical fiber cable)	COMM.CABLE	(→Lichtwellenleiterkabel)
ferrite core ↑ magnetic core ↓ toroidal core	COMPON	**Ferritkern** ↑ Magnetkern ↓ Ringkern	**fiber cleaver** (→fiber cutting device)	OPT.COMM	(→Fasertrennvorrichtung)
ferrite cup core	COMPON	**Ferritschalenkern**	**fiber cutting device** (AM) = fibre cutting device (BRI); fiber cleaver	OPT.COMM	**Fasertrennvorrichtung**
ferrite rod	COMPON	**Ferritstab**	**fiber directional coupler** (AM) = fibre directional coupler (BRI)	OPT.COMM	**Faserrichtkoppler**
ferrite rod	ANT	**Antennenstab**	**fiber drawing** (AM) = fibre drawing (BRI)	OPT.COMM	**Faserziehen**
ferroelectric (n.)	PHYS	**Ferroelektrikum**	**fiber feeder** (→optical feeder loop)	TRANS	(→LWL-Transportsystem)
ferroelectric (adj.)	PHYS	**ferroelektrisch**	**fiber in the loop** = FITL ↓ fiber near the home; fiber to the home	TELEC	**Faser in der Teilnehmerleitung** ↓ Faser bis in Teilnehmernähe; Faser bis zum Teilnehmer
ferroelectric field-effect transistor (→FEFET)	MICROEL	(→Speicher-Feldeffekttransistor)			
ferroelectric hysteresis	PHYS	**ferroelektrische Hysterese**	**fiber line system** (→fiber optic system)	TRANS	(→Lichtwellenleitersystem)
ferroelectricity	PHYS	**Ferroelektrizität**	**fiber loop** = optical subscriber line; optical loop	TELEC	**Faser-Teilnehmerleitung**
ferromagnetic [strong, codirectional, remanent magnetization induced by external field]	PHYS	**ferromagnetisch** [starke, gleichsinnige, remanente Magnetisierung unter Einwirkung externen Magnetfelds]	**fiber near the home** = fiber to the curb; fibre to the kerb (BRI) ≈ fiber to the home	TELEC	**Faser bis in Teilnehmernähe** [engl."curb" = Bordstein] ≈ Faser bis zum Teilnehmer
ferromagnetic hysteresis	PHYS	**ferromagnetische Hysterese**	**fiber optic engineering** (→fiber optic technique)	TELEC	(→Lichtwellenleitertechnik)
ferromagnetic resonance	PHYS	**ferromagnetische Resonanz**	**fiber optic interferometer**	INSTR	**LWL-Interferometer**
ferromagnetism	PHYS	**Ferromagnetismus**	**fiber optic measuring technique** = optical fiber test technique	INSTR	**LWL-Meßtechnik**
ferrometer	INSTR	**Epstein-Apparat**			
ferrous coated tape (→magnetic tape)	ELECTRON	(→Magnetband)	**fiber optic repeater** = lightwave fiber repeater	TRANS	**LWL-Zwischenregenerator**
ferrous oxide ≈ rust (n.)	CHEM	**Eisenoxyd** ≈ Rost	**fiber optic system** = fiber line system; lightwave system	TRANS	**Lichtwellenleitersystem** = LWL-System; faseroptisches System
ferrule	COMPON	**Quetschhülse**	**fiber optic technique** = lightwave technique; fiber optic engineering; lightwave engineering	TELEC	**Lichtwellenleitertechnik** = Faseroptik
festoon (sl.) (→daisy-chain cable)	COMM.CABLE	(→Girlandenkabel)			
FET (→field effect transistor)	MICROEL	(→Feldeffekttransistor)	**fiber optic terminal** = lightwave fiber terminal (AM); lightwave terminal	TRANS	**LWL-Leitungsendgerät**
fetch (v.t.) [to locate in a memory and forward] ≈ retrieve	DATA PROC	**abrufen** [in einem Speicher finden und weiterleiten] ≈ wiedergewinnen	**fiber optic test equipment** (→lightwave test equipment)	INSTR	(→Lichtleiter-Meßgerät)
fetch (n.) = retrieval	DATA PROC	**Abruf** = Fetch			
fetch (n.) ≈ retrieval	DATA PROC	**Abruf** = Fetch ≈ Wiedergewinnung			
fetch (n.) (→instruction fetch)	DATA COMM	(→Befehlsabruf)			
fetch cycle ≈ fetch-execute cycle; fetch phase ≈ execute phase; instruction cycle	DATA PROC	**Abrufzyklus** = Abrufphase; Holzyklus; Holphase ≈ Ausfuhrungsphase; Befehlszyklus			

fiberotic

fiberotic (adj.) OPT.COMM **faseroptisch**
= optic = LWL-; fiberoptisch (SWZ); optisch
fiber protection (→fiber buffer) OPT.COMM (→**Faserschutz**)
fiber-reinforced board ELECTRON (→**glasfaserverstärkte Leiterplatte**)
(→fiber-reinforced PCB)
fiber-reinforced PCB ELECTRON **glasfaserverstärkte Leiterplatte**
= fiber-reinforced board
fiber ribbon (AM) TERM&PER **Faserband**
= fibre ribbon (BRI) = Faserfarbband
fiber-tip pen OFFICE **Faserschreiber**
fiber to the curb (→fiber near the home) TELEC (→**Faser bis in Teilnehmernähe**)
fiber to the home TELEC **Faser bis zum Teilnehmer**
= FITL; FTH ≈ Faser bis in Teilnehmernähe
≈ fiber to the curb
fibre (BRI, CCITT) OPT.COMM (→**Faser**)
(→fiber)
fibre (BRI) (→fiber) TECH (→**Faser**)
fibre buffer (BRI) (→fiber buffer) OPT.COMM (→**Faserschutz**)
fibre cutting device (BRI) OPT.COMM (→**Fasertrennvorrichtung**)
(→fiber cutting device)
fibre directional coupler (BRI) OPT.COMM (→**Faserrichtkoppler**)
(→fiber directional coupler)
fibre drawing (BRI) (→fiber drawing) OPT.COMM (→**Faserziehen**)
fibre protection (→fiber buffer) OPT.COMM (→**Faserschutz**)
fibre ribbon (BRI) (→fiber ribbon) TERM&PER (→**Faserband**)
fibre to the kerb TELEC (→**Faser bis in Teilnehmernähe**)
(→fiber near the home)
fiche (→microfiche) TERM&PER (→**Mikrofiche**)
Fick's diffusion law PHYS **Ficksches Diffusionsgesetz**
fictitious SCIE **fiktiv**
= fictive = scheinbar
fictitious address DATA PROC **fiktive Adresse**
= fictive address
fictive (→fictitious) SCIE (→**fiktiv**)
fictive address (→fictitious address) DATA PROC (→**fiktive Adresse**)
fidelity EL.ACOUS **Wiedergabetreue**
= reproduction quality = Wiedergabegüte; Wiedergabequalität; Originaltreue
fidelity (→accuracy) TECH (→**Genauigkeit**)
field TELEC **Feld**
field PHYS **Feld**
↓ field of force; gravitational field; electric field; magnetic field ↓ Kraftfeld; Gravitationsfeld; elektrisches Feld; magnetisches Feld
field TV **Halbbild**
[one half of a complete picture i.e. frame] [Teilbild des Zeilensprungverfahrens, besteht aus halbem Vollbild]
≠ frame = Halbraster
≠ Vollbild
↑ Teilbild
field MATH **Körper**
[set of elements obeying the field axioms] [Körperaxiome erfüllende Elementemenge]
↑ algebraic stucture ↑ algebraische Struktur
field (→data field) DATA PROC (→**Datenfeld 1**)
field (→visual field) PHYS (→**Gesichtsfeld**)
field (→subject field) LING (→**Fachgebiet**)
field (n.) (→field service) ECON (→**Außendienst**)
field-alterable TECH **am Einsatzort änderbar**
= feldänderbar
field-alterable control element MICROEL (→**FACE-Baustein**)
(→FACE)
field arrangement DATA PROC **Feldanordnung**
= fielding
field axiom MATH **Feldaxiom**
field-blanking interval (→vertical blanking interval) TV (→**Bildaustastlücke**)
field cable (→spiral-four cable 2) COMM.CABLE (→**Feldfernkabel**)
field coil EL.TECH **Feldspule**
field component PHYS **Feldkomponente**

field constant PHYS **Feldkonstante**
field current POWER ENG **Erregerstrom**
= Feldstrom
field current reversal POWER ENG **Erregerstromumkehr**
= Feldstromumkehr
field depth (→focus depth) PHYS (→**Schärfentiefe**)
field duration TV **Halbbilddauer**
≈ frame frquency ≈ Vollbilddauer
field effect PHYS **Feldeffekt**
field effect diode MICROEL **Feldeffektdiode**
= FED = FED
field effect transistor MICROEL **Feldeffekttransistor**
= FET = FET; Feldsteuerungstransistor
↑ unipolar transistor
↑ Unipolartransistor
field-effect varistor MICROEL **Feldeffektvaristor**
field emission PHYS **Feldemission**
field emission electron microscope ELECTRON **Feldelektronenmikroskop**
field energy PHYS **Feldenergie**
field engineer ECON **Außendienstingenieur**
field engineer (→costumer service engineer) TECH (→**Kundendienstingenieur**)
field equation PHYS **Feldgleichung**
field-expandable TECH **am Einsatzort erweiterbar**
= felderweiterbar
field expansion MATH **Körpererweiterung**
field experiment (→field test) TECH (→**Feldversuch**)
field flyback (AM) (→vertical flyback) TV (→**Vertikalrücklauf**)
field frequency TV **Teilbildfrequenz**
= vertical frequency = Vertikalfrequenz; V-Frequenz
= frame frequency
≈ Bildfolgefrequenz
field indicator DATA PROC **Feldkennzeichnung**
= field label
fielding (→field arrangement) DATA PROC (→**Feldanordnung**)
field installation TELEC **Feldmontage**
field intensity (→field strength) PHYS (→**Feldstärke**)
field label (→field indicator) DATA PROC (→**Feldkennzeichnung**)
field length DATA PROC **Feldlänge**
field magnet PHYS **Feldmagnet**
field mark (→field separator) DATA PROC (→**Feldtrennzeichen**)
field marker (→field separator) DATA PROC (→**Feldtrennzeichen**)
field marking TERM&PER **Feldmarkierung**
field number SWITCH **Feldnummer**
field of application TECH **Anwendungsfeld**
≈ applicability ≈ Anwendbarkeit
field office (→off-premises) ECON (→**Außenstelle**)
field of force PHYS **Kraftfeld**
= Kraftlinienfeld
field offset TV **Frequenzversatz**
field of knowledge SCIE **Wissensgebiet**
field of research SCIE **Forschungsgebiet**
field parameter PHYS **Feldgröße**
field pattern PHYS **Feldbild**
= Kraftlinienbild
field performance data TECH **Betriebsdaten**
= Betriebsergebnisse
field plate (→magnetoresistor) COMPON (→**Feldplatte**)
field-programmable TECH **am Einsatzort programmierbar**
= feldprogrammierbar
field-programmable device (→programmable logic array) MICROEL (→**programmierbare Logikanordnung**)
field-programmable logic array (→FPLA) MICROEL (→**FPLA**)
field-proved TECH **felderprobt**
= field-proven; field-tested
field-proven (→field-proved) TECH (→**felderprobt**)
field reactive power EL.TECH **Feldblindleistung**
[nichtsinusförmige Ströme]
field sensitivity EL.ACOUS **Feldübertragungsfaktor**

field separator	DATA PROC	**Feldtrennzeichen**		**filament cathode**	ELECTRON	**Fadenkathode**
= field marker; field mark		= Feldmarke		[tube]		[Röhre]
field service	ECON	**Außendienst**		= filamentary cathode		
= field (n.)				**filament circuit** (→filament heater circuit)	ELECTRON	(→Heizkreis)
field strength	PHYS	**Feldstärke**		**filament current** (→heater current)	ELECTRON	(→Heizstrom)
= field intensity; intensity						
field strength indicator	ANT	**Feldstärkeanzeigegerät**		**filament electrometer**	INSTR	**Fadenelektrometer**
field strength measurement	INSTR	**Feldstärkemessung**		= thread electrometer		
field strength meter	INSTR	**Feldstärkemeßgerät**		**filament heater circuit**	ELECTRON	**Heizkreis**
field strength pattern	ANT	**Feldstärkediagramm**		= filament circuit		
field survey (→site survey)	SYS.INST	(→Ortsbegehung)		**filament hum** (→cathode hum)	ELECTRON	(→Kathodenrauschen)
field technician (→customs service technician)	TECH	(→Kundendiensttechniker)		**filament pointer**	INSTR	**Fadenzeiger**
field telephone	TELEPH	**Feldfernsprecher**		↑ pointer		↑ Zeiger
		= Feldtelefon		**filament transformer**	ELECTRON	**Heiztransformator**
field test	TECH	**Feldversuch**		**filament transistor**	MICROEL	**Fadentransistor**
= field trial; field experiment		= Felderprobung; Einsatzerprobung		= filamentary transistor		
				file (v.t.)	OFFICE	**abheften**
field-tested (→field-proved)	TECH	(→felderprobt)				= einheften
field theory	PHYS	**Feldtheorie**		**file** (n.)	MECH	**Feile**
= theory of fields				↑ tool (n.)		↑ Werkzeug
field time constant	EL.TECH	**Feldkreiszeitkonstante**		**file** (n.)	OFFICE	**Aktenordner**
field trial (→field test)	TECH	(→Feldversuch)		[device to keep papers]		[zum Abheften von Papieren]
field-upgradable	TECH	**am Einsatzort aufrüstbar**				= Ordner; Hefter
field view	DATA PROC	**Kamerasichtfeld**		**file** (v.t.)	DATA PROC	**ablegen**
[computer graphics]		[Computergraphik]		[on external memory]		[extern speichern]
field weakening	PHYS	**Feldschwächung**		↑ store (v.t.)		= abspeichern
FIFO (→FIFO mode)	DATA PROC	(→FIFO-Modus)		↓ archive (v.t.)		↑ speichern
FIFO list (→push-up list)	DATA PROC	(→FIFO-Liste)				↓ archivieren
FIFO mode	DATA PROC	**FIFO-Modus**		**file** (→record)	OFFICE	(→Akte)
= FIFO; first-in-first-out mode		= FIFO; SILO-Modus; SILO		**file** (→data file)	DATA PROC	(→Datei)
≈ FIFO store		≠ LIFO-Modus		**file access**	DATA PROC	**Dateizugriff**
≠ LIFO mode				**file activity ratio** (→activity ratio)	DATA PROC	(→Bewegungsindex)
FIFO stack (→push-up storage)	DATA PROC	(→Silo-Speicher)		**file administration**	DATA PROC	**Dateiverwaltung**
FIFO store	DATA PROC	**FIFO-Speicher**		= file management		
= first-in-first-out store		= Silospeicher		**file attribute**	DATA PROC	**Dateiattribut**
fifth-generation computer	DATA PROC	**Rechner der fünften Generation**		**file backup**	DATA PROC	**Dateisicherung**
[VLSI]		[VLSI]		**file card**	OFFICE	**Karteikarte**
		= Computer der fünften Generation		= index card		
				file-card	DATA PROC	**Festplattenkarte**
figurative (→self-defining)	DATA PROC	(→selbstdefinierend)		[a fixed disk memory as plug-in module]		[Festplattenspeicher als Steckbaugruppe ausgeführt]
figure (n.)	DOC	**Bild**		= hard card		
= illustration 1; map (n.)		= Abbildung; Figur		**file cleanup**	DATA PROC	**Dateisäuberung**
figure (→illustration)	TYPOGR	(→Abbildung)		**file collating**	DATA PROC	**Dateienabgleich**
-**figure** (→-place)	MATH	(→-stellig)		= collating		= Abgleich
figure (→numeral)	MATH	(→numerisches Zeichen)		**file control block**	DATA PROC	**Dateisteuerblock**
figure-8 cable (→integral messenger cable)	COMM.CABLE	(→Tragseil-Luftkabel)		[list of active files]		[Liste der aktiven Dateien]
				= FCB		
figure-eight characteristic (→bilateral characteristic)	EL.ACOUS	(→Achtercharakteristik)		**file control program**	DATA PROC	**Dateiverwaltungsprogramm**
				= file management program		
figure-eight pattern	ANT	**Achtercharakteristik**		**file control system** (→file management system)	DATA PROC	(→Dateiverwaltungssystem)
= octagonal characteristic				**file conversion**	DATA PROC	**Dateiumsetzung**
figure-eight pattern (→bilateral characteristic)	EL.ACOUS	(→Achtercharakteristik)		**file copy**	OFFICE	**Aktenkopie**
				file creation	DATA PROC	**Dateierstellung**
figure of merit	TECH	**Gütezahl**		**filed**	MECH	**gefeilt**
≈ variable		≈ Kenngröße		**file deletion**	DATA PROC	**Dateilöschung**
figure of merit	HF	**Leistungszahl**		**file descriptor** (→file specification)	DATA PROC	(→Dateibezeichnung)
figure of merit (→factor of quality)	EL.TECH	(→Gütefaktor)		**file designation** (→file specification)	DATA PROC	(→Dateibezeichnung)
figures key (→numeric key)	TERM&PER	(→Zifferntaste)		**file directory**	DATA PROC	**Dateiverzeichnis**
figures keyboard (→numeric keypad)	TERM&PER	(→Zifferntastenblock)		**file editor** (→editor)	DATA PROC	(→Editor)
				file extension	DATA PROC	**Dateikennung**
figures shift	CODING	**Ziffernumschaltung**		[in MS-DOS a maximum of three characters to specify the type of file; detached from the filename by a separator]		[bei MS-DOS eine maximal dreistellige Kennzeichnung der Datei-Art, durch einen Separator vom Dateinamen getrennt]
[a code or a key]		[Code oder Taste]				
≈ letter shift		≈ Buchstabenumschaltung				
↑ letter-figure shift		↑ Buchstaben-Ziffern-Umschaltung				
figures shift character	CODING	**Ziffernumschaltzeichen**		= file name extension; filename extension; extension; extender; suffix		= Namenserweiterung; Dateinamen-Suffix; Suffix; Extender; Extension; Dateierweiterung
[telegraphic code]		[Fernschreibcode]		↑ file specification		
filament	ELECTRON	**Heizfaden**				↑ Dateibezeichnung
= heating wire		= Heizdraht; Glühfaden				
filamentary cathode	ELECTRON	(→Fadenkathode)		**file extent**	DATA PROC	**Dateibereich**
(→filament cathode)				[occupied storage area]		[belegte Speicherfläche]
filamentary transistor (→filament transistor)	MICROEL	(→Fadentransistor)				

file fragmentation	DATA PROC	Dateienfragmentierung	filing system [way to put documents in order]	OFFICE	Aktenordnung = Aktensystem; Ablagesystem
file gap	DATA PROC	Dateiendlücke			
file handling	DATA PROC	Dateihandhabung			
file handling routine	DATA PROC	Dateihandhabungsprogramm	filing system [software organizing files]	DATA PROC	Dateienorganisationssoftware
file hardness	MECH	Feilhärte			
file interrogation	DATA PROC	Dateiabfrage	fill (v.t.)	TECH	füllen
file label (→label 1)	DATA PROC	(→Etikett)	fill character	DATA COMM	(→Füllzeichen)
file layout	DATA PROC	Dateiaufbau = Dateianordnung	(→filler)		
			filled band	PHYS	vollbesetztes Band
file librarian (→data administrator)	DATA PROC	(→Datenbankbeauftragter)	filled orders (→turnover)	ECON	(→Umsatz)
			filler = time fill; padding signal; padding character; pad character; filling character; fill character; gap character	DATA COMM	Füllzeichen ≈ Leerzeichen
file locking	DATA PROC	Dateisperre			
file maintenance ≈ file processing	DATA PROC	Dateipflege			
file management (→file administration)	DATA PROC	(→Dateiverwaltung)			
			fillet = solder cup	MANUF	Lotkegel
file management program (→file control program)	DATA PROC	(→Dateiverwaltungsprogramm)	fill factor [solar panel]	POWER SYS	Füllfaktor [Solarzellen]
file management system = FMS; file control system	DATA PROC	Dateiverwaltungssystem	filling	DATA PROC	Füllen
			filling (→replenishment 1)	TECH	(→Auffüllung)
file manager = record manager	DATA PROC	Dateiverwalter	filling character (→filler)	DATA COMM	(→Füllzeichen)
file name [e.g. FILE in FILE.XXX] = filename ↑ file specification	DATA PROC	Dateiname [z.B. DATEI in DATEI.XXX] ↑ Dateibezeichnung	filling compound	COMM.CABLE	Füllmasse
			fill-in pulse	CODING	Füllimpuls
			fillister head	MECH	Linsenkopf [Schraube]
filename (→file name)	DATA PROC	(→Dateiname)	fillister-head screw	MECH	Linsenkopfschraube = Linsenschraube
filename extension (→file extension)	DATA PROC	(→Dateikennung)			
			fill up (→replenish 1)	TECH	(→auffüllen)
file name extension (→file extension)	DATA PROC	(→Dateikennung)	fill up again (→replenish 2)	TECH	(→nachfüllen)
			film (→layer)	TECH	(→Schicht)
file number (→reference number)	OFFICE	(→Aktenzeichen)	film capacitor	COMPON	Schichtkondensator
			film circuit	MICROEL	Schichtschaltung = Filmschaltung
file organization	DATA PROC	Dateiorganisation			
file processor	DATA PROC	Dateiprozessor	film copier	OFFICE	Filmkopiergerät
file protection	DATA PROC	Dateischutz	film hybrid technology	MICROEL	Filmhybridtechnik
file protection (→write protect)	TERM&PER	(→Schreibschutz)	film recorder ↑ display terminal	TERM&PER	Filmausgabegerät ↑ Sichtgerät
file protection ring (→write-enable ring)	TERM&PER	(→Schreibsicherungsring)	film resistor	COMPON	Schichtwiderstand = Filmwiderstand
file section	DATA PROC	Dateiabschnitt	filmsetting apparatus ≈ photo-composing equipment	TYPOGR	Lichtsetzgerät ≈ Fotosetzgerät
file separator = FS	DATA COMM	Hauptgruppentrennung [Code] = FS	filmsetting machine (→photocomposition system)	TYPOGR	Lichtsatzanlage
file server [manages the files of the users of a computer network] ≈ master computer [DATA PROC] ↑ server	DATA COMM	Datenserver [verwaltet zentral die Dateien der Teilnehmer eines Rechnerverbundes] = Datenbankserver; Dateiserver; Dateizugriffssteuerung; Speicherzugriffssteuerung; Fileserver ≈ Leitrechner [DATA PROC] ↑ Server	film technology ↓ thick-film technology; thin-film technology	MICROEL	Filmtechnik = Schichttechnik ↓ Dickfilmtechnik; Dünnfilmtechnik
			film trimmer	COMPON	Schichttrimmer ↑ Trimmerpotentiometer
			film video recording	TV	FAZ
			FILO mode (→LIFO mode)	DATA PROC	(→LIFO-Modus)
			FILO stack (→push-down storage)	DATA PROC	(→Kellerspeicher)
			filter (v.t.) = filtrate	EL.TECH	filtern ≈ sieben
file specification [composed in MS-DOS of file name and file extension] = file designation; file descriptor	DATA PROC	Dateibezeichnung [besteht bei MS-DOS aus Dateiname und Dateikennung]	filter (v.t.)	PHYS	filtern
			filter (v.t.) ≈ sieve	TECH	filtern ≈ sieben
			filter (n.) = selective circuit	NETW.TH	Filter = Siebschaltung
file storage (→file store)	DATA PROC	(→Dateispeicher)	filter (n.)	PHYS	Filter
file store = file storage	DATA PROC	Dateispeicher	filter (n.) ≈ sieve (n.)	TECH	Filter ≈ Sieb
file structure	DATA PROC	Dateistruktur	filter capacitor	POWER SYS	Ladekondensator
file system (→database)	DATA PROC	(→Datenbank) (pl.-en)	filter capacitor	COMPON	Siebkondensator
file transfer (→data file transfer)	DATA PROC	(→Dateitransfer)	filter case	COMPON	Filtergehäuse
			filter coupling	BROADC	Bandfilterkopplung
filing [documents put in order] ↓ record filing	OFFICE	Ablage [geordnete Aufbewahrung von Dokumenten] = Archivierung ↓ Aktenablage	filter crossbar	TV	Filterkreuzschiene
			filter crystal = resonant crystal	COMPON	Filterquarz
			filter edge	NETW.TH	Filterkante
filing cabinet	OFFICE	Aktenschrank = Registraturschrank	filtering 1	NETW.TH	Siebung = Filterung
filing cabinet (→card-index cabinet)	OFFICE	(→Karteikasten)	filtering 2 (→filtering means)	NETW.TH	(→Siebmittel)
filing department (→documentation filing department)	OFFICE	(→Registratur)	filtering (→filtration)	TECH	(→Filterung)
			filtering means = filtering 2	NETW.TH	Siebmittel
filing shelf	OFFICE	Registraturregal = Aktenregal			

filtering module	EQUIP.ENG	**Siebbaugruppe**	final test	MANUF	**Endprüfung**
filter kit	ELECTRON	**Filterbausatz**	= final inspection		= Endmessung; Endtest
= coil set					≈ Systemprüfung
filter program	DATA PROC	**Filterprogramm**	final value	TECH	**Endwert**
[deletes or modifies selected characters]		[entfernt oder verändert bestimmte Zeichen]	= ultimate value; equilibrium value		= Endgröße
			finance	ECON	**finanzieren**
filter section	NETW.TH	**Siebglied**	finances		**Finanzen** (pl.t.)
filter slope	NETW.TH	**Filterflanke**	financial costs	ECON	**Finanzierungskosten**
filter synthesis	NETW.TH	**Filtersynthese**	financial institution	ECON	**Geldinstitut**
filter theory	NETW.TH	**Filtertheorie**	↓ bank		↓ Bank
		= Siebschaltungstheorie	financial recovery	ECON	**Sanierung**
filtrate (→filter)	EL.TECH	(→filtern)	financial statement (→balance sheets)	ECON	(→Bilanz)
filtration	TECH	**Filterung**			
= filtering		= Filtrierung	financial statement (AM)	ECON	**Jahresabschluß**
≈ sifting		≈ Siebung	= annual accounts (BRI)		= Abschluß
fin (→rib)	MECH	(→Rippe)			≈ Bilanz; Gewinn- und Verlustrechnung; Geschäftsbericht
final (adj.)	TECH	**abschließend**			
= conclusive					
final acceptance	TECH	**endgültige Abnahme**	financing	ECON	**Finanzierung**
final accounting	ECON	**Endabrechnung**	↓ leasing; factoring		↓ Leasing; Factoring
final amplifier (→power amplifier)	CIRC.ENG	(→Leistungsverstärker)	fin antenna	ANT	**Flossenantenne**
			find (v.t.)	TECH	**finden**
final anode	ELECTRON	**Endanode**			= wiederfinden
final assembly	MANUF	**Endmontage**	find-and-replace	DATA PROC	**Such-und-Einsetzfunktion**
final capacity (→maximum capacity)	TECH	(→Endausbau)	= search-and-replace		
			finder	DATA PROC	**Suchprogramm**
final character	DATA COMM	**Schlußzeichen**	finder	SWITCH	**Suchwähler**
= terminator		= Abschlußzeichen; Schlußsignal	[a switch searching for call wishes and connecting to appropriate switching stage]		[sucht Leitungen nach Neubelegungen ab und verbindet mit den gewünschten Vermittlungseinrichtungen]
final check	QUAL	**Endkontrolle**			
final considerations	DOC	**Schlußbemerkungen**	↓ line finder; zone finder; trunk finder		↓ Anrufsucher;
		= Schlußbetrachtung			
final consumer	ECON	**Endverbraucher**	fine (→sophisticated)	TECH	(→hochwertig)
= ultimate consumer; end-consumer		≈ Endabnehmer	fine (→penalty)	ECON	(→Strafe)
final control element	CONTROL	**Stellglied**	fine adjustment	ELECTRON	**Feinabgleich**
[generates from a weaker actuating signal the manipulated variable]		[erzeugt aus schwächerem Signal die Stellgröße]	= fine tuning; mop-up		= Feinabstimmung; Lupe [RADIO]
			fine adjustment	TECH	**Justierung**
= correcting element; correcting unit; actuator 1; actuating element; actuating unit; forward controlling element		↓ Aktuator	≈ adjustment; positioning		= Feineinstellung; Präzisionseinstellung; Justage
					≈ Einstellung; Positionierung
↓ actuator 2			fine adjustment drive	INSTR	**Präzisionsantrieb**
final copy (→fair copy)	OFFICE	(→Reinschrift)	fine arts	COLLOQ	**Bildende Künste**
final costumer	ECON	**Endabnehmer**	↓ plastic arts; painting; graphic arts; architecture; arts and crafts		↓ Plastik; Malerei; Graphik; Baukunst; Kunstgewerbe
= ultimate buyer; ultimate taker		= Endkunde			
≈ final consumer		≈ Endverbraucher	fine azimuth adjustment	ANT	**Seitenfeineinstellung**
final design	DATA PROC	**Feinentwurf**	fine-grained	TECH	**feinkörnig**
final deviation	CONTROL	**Endabweichung**	= finely grained; finely granular		
final equipment (→maximum capacity)	TECH	(→Endausbau)	fine-grind	MECH	**ziehschleifen**
			finely grained	TECH	(→feinkörnig)
final group selector	SWITCH	**Leitungsgruppenwähler**	(→fine-grained)		
= final group switch		= LGW	finely granular	TECH	(→feinkörnig)
↑ line selector		↑ Leitungswähler	(→fine-grained)		
final group switch (→final group selector)	SWITCH	(→Leitungsgruppenwähler)	fine-meshed	TECH	**engmaschig**
			= close-meshed		= feinmaschig
final inspection (→final test)	MANUF	(→Endprüfung)	fineness	TECH	**Feinheit**
final meeting	ECON	**Abschlußbesprechung**	fine overvoltage protection	ELECTRON	**Überspannungsfeinschutz**
final operation	MANUF	**Endarbeitsgang**			
final point (→end)	TECH	(→Ende 1)	fine overvoltage protector	COMPON	**Spannungsfeinsicherung**
final product	MANUF	**Endprodukt**	[protects against overvoltages under 2,000 V]		[schützt für Spannungen unter 2.000 V]
= finished good; end-product; finished product		= Fertigprodukt; Enderzeugnis; Fertigerzeugnis			
			↓ gas-discharge protector; carbon		↓ Gasentladungsableiter; Kohlespannungsableiter
final report	ECON	**Abschlußbericht**			
		= Schlußbericht	fine-polish	MECH	**feinpolieren**
final result	TECH	**Endergebnis**	fine print	TYPOGR	**Kleindruck**
final route (→last-choice route)	SWITCH	(→Letztweg)			= Kleingedrucktes
			fine protecting device (→fine protection)	COMPON	(→Feinsicherung 1)
final selector	SWITCH	**Leitungswähler**			
[connects in step-by-step switching systems the desired user line]		[schaltet in Direktwahlsystemen den gewünschten Teilnehmer an]	fine protection	CIRC.ENG	**Feinschutz**
			fine protection	COMPON	**Feinsicherung 1**
= line selector; primary lineswitch; lineswitch		= LW	= fine protecting device		[schützt vor kleinen Überspannungen oder -strömen]
					↓ Stromfeinsicherung
final stage	CIRC.ENG	**Endstufe**	fine radiolocation	RADIO LOC	**Feinortung**
= output stage		= Ausgangsstufe	fine sheet	METAL	**Feinblech**
≈ final amplifier		≈ Endverstärker	↑ sheet metal		↑ Blech
final stage tube (→output tube)	ELECTRON	(→Endröhre)	fine spark gap	COMPON	**Feinfunkenstrecke**
			finess of scanning (→scanning density)	ELECTRON	(→Rasterfeinheit)
final stage valve (→output tube)	ELECTRON	(→Endröhre)			
			fine station tuning	CONS.EL	**Sender-Feinabstimmung**
final stroke	COLLOQ	**Schlußstrich**			
[fig.]		[fig.]			

fine structure

English	Domain	German
fine structure	TECH	**Feinstruktur**
fine thread	MECH	**Feingewinde**
fine-tune (v.t.)	ELECTRON	**feinabstimmen**
fine tuning (→fine adjustment)	ELECTRON	(→Feinabgleich)
fine-tuning control	ELECTRON	**Feinregler**
fine-wire fuse [protects against currents of below 1 A] ↑ overcurrent protector ↓ cartridge fuse; slow-blowing melting fuse	COMPON	**Stromfeinsicherung** [schützt vor Strömen unter 1 A] = Feinsicherung 2 ↑ Stromsicherung ↓ Glasrohr-Feinsicherung; träge Feinsicherung
finger hole	TERM&PER	**Fingerloch**
finger stop	TERM&PER	**Fingeranschlag**
finger wheel (→dialing disk)	TERM&PER	(→Wählscheibe)
finish (n.)	MECH	**Feinbearbeitung** = Fertigbearbeitung
finish (v.t.)	MECH	**feinbearbeiten** = schlichten; fertigbearbeiten
finish (→polish)	TECH	(→polieren)
finish 2 (n.) (→workmanship 2)	TECH	(→Ausführungsqualität)
finish-bore (v.t.)	MECH	**fertigbohren**
finished	MECH	**bearbeitet** = fertigbearbeitet
finished	TYPOGR	**druckfertig**
finished good (→final product)	MANUF	(→Endprodukt)
finished part	MANUF	**Fertigteil**
finished product (→final product)	MANUF	(→Endprodukt)
finishes	METAL	**Fertigzeug**
finish-grind (v.t.)	MECH	**feinschleifen**
finish-grind (n.) [surface grade]	MECH	**Feinschliff** [Oberflächengüte] ↑ Schliff
finishing	MANUF	**Endfertigung**
finishing draw [deep drawing]	METAL	**Fertigzug 2** [Tiefziehen]
finishing pass	METAL	**Fertigzug 1** [Reduzierziehen]
finish-machine	MECH	**maschinell feinbearbeiten**
finish machining	MECH	**maschinelle Feinbearbeitung**
finish-polishing	MECH	**Feinpolieren**
finite ≠ infinite	MATH	**endlich** = finit ≠ unendlich
finite element method = FEM	MATH	**Methode der endlichen Elemente**
finite group	MATH	**endliche Gruppe**
finite-impulse-response system (→FIR system)	NETW.TH	(→FIR-System)
finit elements method	DATA PROC	**Finite-Elemente-Methode**
finite number	MATH	**endliche Zahl**
finned (→ribbed)	TECH	(→gerippt)
finning	MECH	**Berippung**
fire alarm (→fire detector)	SIGN.ENG	(→Feuermelder)
fire alarming (→fire signaling)	SIGN.ENG	(→Brandschutz)
fire alarm system	SIGN.ENG	**Feuermeldeeinrichtung** = Feuermeldeanlage; Feuermeldesystem; Brandmeldeeinrichtung; Brandmeldeanlage; Brandmeldesystem
fire button [joystick] = fire knob	TERM&PER	**Feuerknopf** [Steuerknüppel] = Reaktionsknopf
Fire code	CODING	**Fire-Code**
fire detector = fire alarm ≈ flame alarm system ↓ smoke detector; heat detector; flame detector	SIGN.ENG	**Feuermelder** = Brandmelder ≈ Flammenmelder ↓ Rauchmelder; Wärmemelder; Flammenmelder
fire-inhibiting (→flame-retardant)	TECH	(→flammwidrig)
fire knob (→fire button)	TERM&PER	(→Feuerknopf)
fire point	TECH	**Brennpunkt**
fire prevention	TECH	**Brandverhütung** = Feuerverhütung
fire-proof = flameproof 2; refractory ≈ fire-resistant; flameproof 1	TECH	**feuerfest** = feuerbeständig; flammsicher ≈ feuerhemmend; schlagwettergeschützt
fire-reporting telephone	TELEPH	**Brandmeldetelefon**
fire-resistant (→flame-retardant)	TECH	(→flammwidrig)
fire signaling = fire alarming	SIGN.ENG	**Brandschutz** = Brandmeldung; Feuermeldung
fire station	SIGN.ENG	**Feuerwache**
firing (→triggering)	ELECTRON	(→Auslösung)
firing angle [static power converter]	POWER SYS	**Zündwinkel** [Stromrichter] = Steuerwinkel
firing current (→trigger current)	ELECTRON	(→Zündstrom)
firing device (→trigger 1)	ELECTRON	(→Auslöser)
firing voltage (→trigger voltage)	ELECTRON	(→Zündspannung)
firm (→company)	ECON	(→Gesellschaft)
firmkey	INSTR	**Festtaste**
firmware [software or data stored indelibly on ROM's by the system manufacturer] ↑ software	DATA PROC	**Firmware** [auf Festwertspeichern vom Systemhersteller gespeicherte Software und Daten] = festgespeichertes Standardprogramm ↑ Software
first-angle projection	ENG.DRAW	**europäische Projektion**
first breakdown	MICROEL	**Primärdurchbruch** = erster Durchbruch
first-choice route (→primary route)	SWITCH	(→Erstweg)
first connection	TELEC	**Neuanschluß** = Neuanschließung
first cut (→first-order approximation)	TECH	(→erste Näherung)
first degree (BRI) (→master degree)	SCIE	(→Diplom)
first detector (→mixer)	HF	(→Mischer)
first floor (NAM) = ground floor	CIV.ENG	**Erdgeschoß**
first-generation computer [valve-based, beginning fifties]	DATA PROC	**Rechner der ersten Generation** [mit Röhren, Anfang 50er-Jahre] = Computer der ersten Generation
first-generation image (→master copy)	DOC	(→Stammkopie)
first harmonic (→fundamental wave)	PHYS	(→Grundschwingung)
first-in-first-out mode (→FIFO mode)	DATA PROC	(→FIFO-Modus)
first-in-first-out store (→FIFO store)	DATA PROC	(→FIFO-Speicher)
first-in-last-out mode (→LIFO mode)	DATA PROC	(→LIFO-Modus)
first-level address (→direct address)	DATA PROC	(→direkte Adresse)
first line indent	TYPOGR	**Erstzeileneinzug**
first loading section [pupin-coil loading]	OUTS.PLANT	**Anlauffeld** [Pupinisierung]
first option (→right of preemption)	ECON	(→Vorkaufsrecht)
first-order approximation = first-order treatment; first cut	TECH	**erste Näherung**
first-order predicate logic	DATA PROC	**Prädikatenlogik ersten Ranges**
first-order treatment (→first-order approximation)	TECH	(→erste Näherung)
first route (→primary route)	SWITCH	(→Erstweg)

first schools (BRI) (→primary school) SCIE (→Grundschule)
first shell PHYS 1.Schale
= shell K = K-Schale
first supplier (→main supplier) ECON (→Hauptlieferant)
FIR system NETW.TH FIR-System
= finite-impulse-response system
fiscal (adj.) (→economic) ECON (→wirtschaftlich 1)
fiscal law ECON Steuerrecht
= tax law
fish-bite protected COMM.CABLE fischbißgeschützt
fishbone antenna ANT Fischgrätenantenne
= christmastree antenna; Xmastree antenna = Fischgrätantenne; Fischbeinantenne; logarithmisch-periodische V-Antenne; LPV-Antenne; Tannenbaumantenne
↑ end-fire antenna; travelling wave antenna ↑ Längsstrahler; Wandelwellenantenne; Dipolwand
fisheye lens PHYS Fischaugenlinse
fishing-rod beam antenna ANT Angelruten-Beam
fission PHYS Spaltung
↓ nuclear fission ↓ Kernspaltung
fissure (→crack) TECH (→Sprung)
fist TYPOGR Hinweiszeichen
[a fist as indicating symbol] [eine Hand mit ausgestrecktem Zeigefinger]
= index
fit (n.) ENG.DRAW Passung
fit (n.) MECH Sitz
  [Passung]
fit (n.) MATH Anpassung
fit QUAL fit
[failure in time; 10 to the power of minus 9 failures per hour] [10 hoch minus 9 Ausfälle pro Stunde]
FITL (→fiber to the home) TELEC (→Faser bis zum Teilnehmer)
fitted bolt MECH Paßbolzen
= reamed bolt
fitted key (→feathered key) MECH (→Paßfeder)
fitter (→mechanic) TECH (→Monteur)
fitting TECH Armatur
fitting (→sleeve) TECH (→Muffe)
fitting (→connecting piece) MECH (→Verbindungsstück)
fitting (→accessory) TECH (→Zubehör)
fitting (→curve fitting) MATH (→Kurvenermittlung)
fitting-out (→equipment 1) TECH (→Ausstattung)
five-arm transformer EL.TECH Fünfschenkeltransformator
five-by-five RADIO beidseitig guter Empfang
= bidirectional good transmission
five-circuit ... NETW.TH Fünfkreis...
five-layer TECH fünflagig
≈ five-part = fünfschichtig
  = fünfteilig
five-part TECH fünfteilig
↑ multisectional ↑ mehrteilig
five-unit alphabet TELEGR Fünferalphabet
[with 5 information pulses] [mit 5 Informationsschritten]
= five-unit code = Fünfschrittcode; Fünf-Schritt-Code; 5-Schritt-Code; 5-er-Code
↓ telegraph code ↓ Fernschreibalphabet
five-unit code (→five-unit alphabet) TELEGR (→Fünferalphabet)
five-wire TELEC fünfdrähtig
  = fünfadrig
five-wire element COMM.CABLE Fünfer
fix (→fasten) TECH (→befestigen)
fixed TELEC festgeschaltet
= dedicated ≈ gemietet [ECON]
≈ leased [ECON]
fixed TECH festeingestellt
= fixly adjusted; fixly set-up (adj.) ≠ veränderbar
≠ alterable
fixed (→stationary) TECH (→ortsfest)
fixed DATA PROC (→speicherresident)
  (→memory-resident)
fixed (→preset) COLLOQ (→vorgegeben)
fixed Adcock antenna ANT Fest-Adcock-Antenne
≠ rotary Adcock antenna ≠ Dreh-Adcock-Antenne

fixed antenna ANT Festantenne
fixed area DATA PROC Festbereich
fixed bit rate TELEC Festbitrate
fixed bit rate mode TELEC Festbitratenbetrieb
= constant bit rate mode
fixed capacitor COMPON Festkondensator
fixed carbon film resistor COMPON Kohleschicht-Festwiderstand
fixed circuit (→fixed line) TELEC (→Standleitung)
fixed command control CONTROL Festwertregelung
= constant-value control; set-value control = Konstantwertregelung
fixed connection (→fixed line) TELEC (→Standleitung)
fixed connection fee TELEC Festverbindungsgebühr
fixed cycle DATA PROC Festzyklus
fixed-cycle operation DATA PROC Festzyklusbetrieb
fixed data (→static data) DATA PROC (→feste Daten)
fixed decimal point (→fixed point) DATA PROC (→Festkomma) (im Englischen verwendet man Dezimal-„Punkte" statt -„Kommas")
fixed decimal point calculation MATH Festkommarechnung
= fixed point computation = Fixkommarechnung; Festpunktrechnung; Fixpunktrechnung
fixed disc (→fixed disk) TERM&PER (→Festplatte)
fixed disk TERM&PER Festplatte
[not removable from disk driver] [aus dem Laufwerk nicht ausbaubar]
= fixed disc = Fixplatte
≈ harddisk ≈ Hartplatte
≠ removable disk ≠ Wechselplatte
↓ Winchester disk ↓ Winchester-Platte
fixed-disk drive TERM&PER Festplattenlaufwerk 1
[for not removable disks] [für nicht auswechselbare Platten]
= hard-disk drive = Festplattenantrieb
≠ removable-disk drive ≈ Hartplattenlaufwerk
↑ disk drive ≠ Wechselplattenlaufwerk
  ↑ Plattenlaufwerk
fixed-disk memory TERM&PER Festplattenspeicher
[disks cannot be removed from disk drive] [mit Laufwerk fest montierte, d.h. nicht auswechselbare Platten]
= fixed-disk storage; fixed-disk store ≈ Hartplattenspeicher; Festplattenspeicher [DATA PROC]
≈ hard-disk memory; hard-disk memory [DATA PROC]
≠ moving-disk memory ≠ Wechselplattenspeicher
↑ magnetic disk memory ↑ Magnetplattenspeicher
↓ Winchester-disk memory ↓ Winchester-PLattenspeicher
fixed-disk storage (→fixed-disk memory) TERM&PER (→Festplattenspeicher)
fixed-disk store (→fixed-disk memory) TERM&PER (→Festplattenspeicher)
fixed echo TELEC Dauerecho
fixed film resistor COMPON Schichtfestwiderstand
fixed format DATA PROC festes Format
≠ free format ≠ freies Format
↓ sector format [TERM&PER] ↓ Sektorformat [TERM&PER]
fixed frame (→fixed image) TV (→Festbild)
fixed-frequency monitor TERM&PER Festfrequenz-Monitor
[works only with one graphics standard] [arbeitet nur mit einem Graphikstandard]
≠ multi-frequency monitor ≠ Autosync-Monitor
fixed-head-disk drive TERM&PER Festkopf-Plattenlaufwerk
≠ movable-head-disk drive = Festkopflaufwerk
  ≠ Bewegtkopf-Plattenlaufwerk
fixed image TV Festbild
= fixed frame
fixed-image communication TELEC Festbildkommunikation
= non-motion video; freeze-frame system; slow-scan system
fixed information length DATA PROC feste Informationslänge
≈ fixed word length ≈ feste Wortlänge
fixed-length record DATA PROC Datensatz fester Länge
= fixed-size record

fixed-length word

fixed-length word	DATA PROC	**Festwort**	
		= Wort fester Länge	
fixed line	TELEC	**Standleitung**	
= nailed line; dedicated line; tie line; point-to-point line; designated line; permanent line; fixed connection; nailed connection; dedicated connection; point-to-point connection 2; fixed circuit; nailed circuit; dedicated circuit; tie circuit; point-to-point circuit; designated circuit; permanent circuit		= Standverbindung; Festverbindung; Punkt-zu-Punkt-Verbindung 2; festgeschaltete Verbindung; permanente Verbindung; überlassene Verbindung; überlassene Leitung	
≈ leased line		≈ Mietleitung	
≠ switched line		≠ Wählleitung	
fixed-lines network (→fixed network)	TELEC	(→Festnetz)	
fixed memory (→fixed storage)	TERM&PER	(→Festspeicher)	
fixed metal-film resistor	COMPON	**Metallfilm-Festwiderstand**	
↑ metal-film resistor		= Metallschicht-Festwiderstand	
		↑ Metallfilmwiderstand	
fixed metal-oxide-film resistor	COMPON	**Metalloxidschicht-Festwiderstand**	
= metal-oxide-film resistor			
↑ metal oxide resistor		↑ Metalloxidschicht-Widerstand	
fixed network	TELEC	**Festnetz**	
= fixed-lines network		= festgeschaltetes Netz; Festleitungsnetz; Standnetz; Standleitungsnetz	
≠ switched network		≠ Wählnetz	
fixed-network operator	TELEC	**Festnetzbetreiber**	
fixed numbering (→uniform numbering)	SWITCH	(→Feststellennumerierung)	
fixed point	MATH	**Festkomma** (im Englischen verwenden man Dezimal-„Punkte" statt -„Kommas")	
[Continental Europeans use decimal "commas" instead of "points"]			
= fixed decimal point		= Festpunkt; Fixkomma; Fixpunkt	
≠ floating decimal point		≠ Gleitkomma	
fixed point addition	MATH	**Festkommaaddition**	
		= Festpunktaddition	
fixed-point arithmetic	MATH	**Festkomma-Arithmetik**	
		= Festpunktarithmetik	
fixed point computation (→fixed decimal point calculation)	MATH	(→Festkommarechnung)	
fixed point constant	MATH	**Festkommakonstante**	
		= Festpunktkonstante	
fixed point division	MATH	**Festkommadivision**	
		= Festpunktdivision	
fixed point exponent	MATH	**Festkommaexponent**	
		= Festpunktexponent	
fixed point multiplication	MATH	**Festkommamultiplikation**	
		= Festpunktmultiplikation	
fixed-point notation	MATH	**Festkommadarstellung**	
(→fixed-point representation)			
fixed point number	MATH	**Festpunktzahl**	
≠ floating point number		= Festpunktzahl	
		≠ Gleitkommazahl	
fixed point operation	MATH	**Festkommaoperation**	
		= Festpunktoperation	
fixed point processor	DATA PROC	**Festkommaprozessor**	
		= Festpunktprozessor	
fixed point register	DATA PROC	**Festkommaregister**	
		= Festpunktregister	
fixed-point representation	MATH	**Festkommadarstellung**	
= fixed-point notation		= Festpunktschreibweise; Festpunktdarstellung; Festkommanotierung; Festpunktnotierung	
fixed point routine	DATA PROC	**Festkommaroutine**	
		= Festpunktroutine	
fixed point subtraction	MATH	**Festkommasubtraktion**	
		= Festpunktsubtraktion	
fixed-program (adj.)	DATA PROC	**festprogrammiert**	
fixed program	DATA PROC	**Festprogramm**	
= hardwired program; wired program		= festverdrahtetes Programm; verdrahtetes Programm	
≠ stored program		≠ Speicherprogramm	
fixed-program computer	DATA PROC	(→festprogrammierter Computer)	
(→hardwired-program computer)			
fixed programming	DATA PROC	**Festprogrammierung**	
≠ stored programming		= nicht variierbar Speicherprogrammierung	
		≠ Speicherprogrammierung	
fixed radio service	TELEC	**fester Funkdienst**	
fixed radio station	RADIO	**Funkfeststation**	
fixed rate (→flat-rate tariff)	ECON	(→Pauschaltarif)	
fixed resistance	COMPON	**Festwiderstand**	
fixed routing	SWITCH	**feste Verkehrslenkung**	
fixed satellite service	TELEC	**fester Funkdienst über Satelliten**	
fixed service	TELEC	**fester Dienst**	
fixed sine	INSTR	**Festfrequenz-Sinussignal**	
fixed-size record	DATA PROC	(→Datensatz fester Länge)	
(→fixed-length record)			
fixed spacing	TERM&PER	**Festzeichenabstand**	
fixed star	ASTROPHYS	**Fixstern**	
fixed storage	TERM&PER	**Festspeicher**	
= fixed memory			
fixed subscriber	TELEC	**Festteilnehmer**	
≠ mobile telephone subscriber		≠ Mobilfunkteilnehmer	
fixed time (→appointed time)	COLLOQ	(→Termin)	
fixed time call	TELEPH	**Festzeitgespräch**	
fixed value	MATH	**Festwert**	
fixed vertical loop antenna	ANT	**Festrahmenantenne**	
fixed wire-wound resistor	COMPON	**Drahtfestwiderstand**	
fixed word length	DATA PROC	**feste Wortlänge**	
fixed-word-length arithmetics	DATA PROC	**Festwortarithmetik**	
≠ character arithmetics		≠ Zeichenarithmetik	
fixing	MECH	**Fixierung**	
fixing (→fastening 1)	TECH	(→Befestigung 1)	
fixing element (→mount)	TECH	(→Halterung)	
fixing pin	MECH	**Fixierstift**	
≈ alignment pin		≈ Führungsstift	
fixing strip	MECH	**Befestigungsstreifen**	
fixing varnish	TECH	**Sicherungslack**	
fixly adjusted (→fixed)	TECH	(→festeingestellt)	
fixly set-up (adj.) (→fixed)	TECH	(→festeingestellt)	
fixly syntonized (→fixly tuned)	CIRC.ENG	(→festabgestimmt)	
fixly tuned	CIRC.ENG	**festabgestimmt**	
= fixly syntonized			
fixture (→clamping device)	MECH	(→Spannelement)	
FLAD	COMPON	**FLAD**	
= fluorescence activated display			
flag	DATA PROC	**Merker**	
[indicator for special conditions]		[Zeichen zur Signalisierung bestimmter Zustände]	
= indicator flag; flag bit; condition bit 2; status bit; functional bit		= Kennzeichen; Hinweiszeichen; Flag; Kennzeichenbit; Zustandsbit; Markierung 1; Statusbit; funktionales Bit	
↓ error flag		↓ Fehlerkennzeichen	
flag bit (→flag)	DATA PROC	(→Merker)	
flag code	DATA COMM	**Hinweiscode**	
flag event	DATA PROC	**Hinweisereignis**	
flag function	TERM&PER	**Merkerfunktion**	
flagging	DATA PROC	**Hinweiszeichensetzung**	
= marking 3		= Kennzeichensezung; Markierung 3	
flag memory	TERM&PER	**Merkerspeicher**	
flagpole antenna (→dielectric rod antenna)	ANT	(→Stabantenne)	
flag register	DATA PROC	**Flag-Register**	
flag sequence (→framing bit sequence)	DATA COMM	(→Synchronisierbitmuster)	
flagship product (→pilot product)	ECON	(→Leitprodukt)	
flake (n.)	TECH	**Haarriß**	
= hairline crack			
flake off (v.i.)	TECH	**abblättern**	
= peel off		≈ abschälen	
		≈ ablösen	
flame detector	SIGN.ENG	**Flammenmelder**	
≈ fire detector		≈ Feuermelder	

flameproof 1	TECH	schlagwettergeschützt
≈ fireproof		≈ feuerfest
flameproof 2 (→fire-proof)	TECH	(→feuerfest)
flame resistance	TECH	**Flammwidrigkeit**
≈ incombustibility		= Unbrennbarkeit
flame-retardant	TECH	**flammwidrig**
= fire-inhibiting; fire-resistant		= feuerhemmend
≈ incombustible; fire-proof		= unbrennbar; feuersicher
flame-retardant cable	COMM.CABLE	**feuerhemmendes Kabel**
flame welding	METAL	**Flammenschweißen**
flaming point	TECH	**Flammpunkt**
flammability (→inflammability)	TECH	(→Entflammbarkeit)
flammable (→inflammable)	TECH	(→entflammbar)
flange (v.t.)	MECH	bördeln
flange (n.)	MECH	**Flansch**
flange coupling (→flange joint)	MICROW	(→Flanschverbindung)
flanged	MECH	gebördelt
flanged connector (→flange joint)	MICROW	(→Flanschverbindung)
flanged panel plug	COMPON	**Flanschstecker**
		≠ Einbaustecker mit Zentralbefestigung
		↑ Einbaustecker
flanged panel socket	COMPON	**Flanschbuchse**
		≠ Einbaubuchse mit Zentralbefestigung
		↑ Einbaubuchse
flange joint	MICROW	**Flanschverbindung**
= flange coupling; flanged connector; disk clitch		= Flanschkopplung; Flanschkupplung; Scheibenkupplung
flange mounting	COMPON	**Flanschbefestigung**
≠ circular mounting		[von Einbausteckverbindern]
		≠ Zentralbefestigung
flange sleeve	OUTS.PLANT	**Flanschmuffe**
flange-type bearing	MECH	**Flanschlager**
flanging device	MECH	**Bördelgerät**
= flanging tool		= Bördelwerkzeug
flanging tool (→flanging device)	MECH	(→Bördelgerät)
flank	MECH	**Gewindeflanke**
		= Flanke
flap	TECH	**Klappe**
= door 2		≈ Deckel
≈ lid		
flap (v.t.) (→impact)	MECH	(→aufschlagen)
flapper	TECH	**Prallplatte**
flash (n.)	TECH	**Lichtblitz**
flash (v.i.) (→flicker)	TECH	(→blinken)
flash (n.) (→overlay)	DATA PROC	(→Überlagerung)
flash butt-welding (→flash welding)	METAL	(→Abschmelzschweißung)
flash EPROM	MICROEL	**Flash-EPROM**
[partially alterable several times]		[nur teilweise mehrmals beschreibbar]
↑ memory chip		↑ Speicherbaustein
flash harden	METAL	teilhärten
flash heat	TECH	teilerhitzen
flashing	TERM&PER	blinkend
= blinking; flushing		↑ Bildschirmdarstellung
↑ screen mode		
flashing character	DATA PROC	**Flackerzeichen**
[with flashing intensity]		= Blinkzeichen; flackerndes Zeichen; blinkendes Zeichen
flashing indication	EQUIP.ENG	**Blinkanzeige**
= flickering indication; pulsing indication		= pulsierende Anzeige
flashing light system	SIGN.ENG	**Blinklichtanlage**
flashing signal	TECH	**Flackerzeichen**
flash interval	AERON	**Blitzabstand**
[light]		[Feuer]
flash lamp	PHYS	**Blitzlichtlampe**
flash-over (→spark-over)	PHYS	(→Funkenüberschlag)
flash test	EL.TECH	**Stoßprüfung**
[insulation test]		[Isolationsprüfung]
flash tube	ELECTRON	**Blitzröhre**
↑ ion tube; cold-cathode tube		↑ Ionenröhre; Kaltkathodenröhre
flash welding	METAL	**Abschmelzschweißung**
= flash butt-welding		
flask (→bottle)	TECH	(→Flasche)
flat (adj.)	TECH	**flach**
[relatively smooth]		[ohne größere Welligkeit]
= even; planar; plane; plain (obs.)		= plan; eben
		≈ glatt
flat (→frequency-flat)	ELECTRON	(→frequenzgerade)
flat aluminium (→flat-bar aluminum)	METAL	(→Flachaluminium)
flat band	MICROEL	**Flachband**
flat bar	METAL	**Flachstange**
flat-bar aluminum	METAL	**Flachaluminium**
= flat aluminium		
flat-bar brass	METAL	**Flachmessing**
= flat brass		
flat bar copper	METAL	**Flachkupfer**
= flat copper		
flat-bar iron	METAL	**Flacheisen**
= flat iron		
flat-bar steel	METAL	**Flachstahl**
= flat steel		
flat bed feed	TERM&PER	**Flachbettzufuhr**
flatbed plotter	TERM&PER	**Flachbettplotter**
[plots over a flat surface]		[zeichnet auf flacher Unterlage]
= desk plotter		= Tischplotter
flatbed scanner	TERM&PER	**Flachbettscanner**
[scans from flat surface]		[tastet flach liegende Vorlagen ab]
		= Flachbett-Abtastsystem
flatbed scanner (→flatbed scanning system)	TELEGR	(→Flachbett-Abtastsystem)
flatbed scanning system	TELEGR	**Flachbett-Abtastsystem**
= flatbed scanner		[Faksimile]
		= Flachbettscanner
flat brass (→flat-bar brass)	METAL	(→Flachmessing)
flat cable (→ribbon cable)	COMM.CABLE	(→Bandkabel)
flat coat (v.t.)	TECH	grundieren
flat coat (n.)	TECH	**Grundierung**
flat coil	COMPON	**Flachspule**
= pancake coil; slab coil		
flat-coil instrument	INSTR	**Flachspulinstrument**
flat color	TECH	**Grundierfarbe**
flat construction	EQUIP.ENG	**Flachbauweise**
flat copper (→flat bar copper)	METAL	(→Flachkupfer)
flat display (→flat screen)	TERM&PER	(→Flachbildschirm)
flat fading	RADIO PROP	**Flachschwund**
= amplitude fading		= flacher Schwund
flat file	DATA PROC	zweidimensionale Datei
flat file	OFFICE	**Flachdatei**
flat filter	NETW.TH	**flaches Filter**
flat head	MECH	**Senkkopf**
		[Schraube]
flat-head screw	MECH	**Senkkopfschraube**
flat iron (→flat-bar iron)	METAL	(→Flacheisen)
flat key	MECH	**geradstirnige Flachpaßfeder**
flat line	COMM.CABLE	**Flachleitung**
flat line (→matched feeder)	ANT	(→angepaßte Speiseleitung)
flat loudspeaker	EL.ACOUS	**Flachlautsprecher**
= pancake loudspeaker		
flat module	EQUIP.ENG	**Flachbaugruppe**
↑ module		↑ Baugruppe
flatness (→planeness)	MECH	(→Ebenheit 1)
flatness (→levelness)	MECH	(→Ebenheit 3)
flatness [INSTR] (→frequency response)	NETW.TH	(→Frequenzgang)
flatness quality	MECH	**Ebenheitsqualität**
flat-nose pliers	TECH	**Flachzange 2**
≈ combination pliers		[klein]
↑ pliers		= Klemmzange
		≈ Kombinationszange
		↑ Zange 2
flat-nose tongs	TECH	**Flachzange 1**
↑ tongs		[groß]
		↑ Zange 1

flat pack

flat pack [with horizontal leads]	MICROEL	**Flachgehäuse** [mit horizontalen Anschlüssen] = Flat-Pack	
flat-panel display (→flat screen)	TERM&PER	(→Flachbildschirm)	
flat-panel screen (→flat screen)	TERM&PER	(→Flachbildschirm)	
flat plate (→grid plate)	POWER SYS	(→Gitterplatte)	
flat plate (→planar array)	ANT	(→ebene Gruppe)	
flat point	MECH	**Kegelkuppe** [Gewindestift]	
flat rate (→lump sum)	ECON	(→Pauschalbetrag)	
flat-rate tariff = fixed rate	ECON	**Pauschaltarif** = Pauschalgebühr	
flat reed contact	COMPON	**Flach-reed-Kontakt**	
flat relay	COMPON	**Flachrelais** ≈ Flachankerrelais	
flat scanner [facsimile]	TELEGR	**Flachabtaster** [Faksimile]	
flat screen = flat-panel screen; flat display; flat-panel display ↓ plasma display; LCD	TERM&PER	**Flachbildschirm** ≈ flacher Bildschirm ↓ Plasmabildschirm; Flüssigkristall-Bildschirm	
flat spring	MECH	**Flachmaterialfeder**	
flat steel (→flat-bar steel)	METAL	(→Flachstahl)	
flatten	METAL	**breitschlagen**	
flattened	TECH	**abgeflacht**	
flat-top antenna (→aperture antenna)	ANT	(→Aperturantenne)	
flatwise	MECH	**flachkantig**	
flatwise bend (→E-plane elbow)	MICROW	(→E-Bogen)	
flaw (→crack)	TECH	(→Sprung)	
flaw (→windflaw)	METEOR	(→Windstoß)	
flawed (→faulty)	QUAL	(→fehlerhaft)	
F-layer = F-region	RADIO PROP	**F-Schicht**	
flection = flexion; bending	PHYS	**Biegung**	
fleeting information ≈ transient information	TELECONTR	**Kurzzeitmeldung** ≈ Wischermeldung	
flesh colored	PHYS	**fleischfarben**	
flex (→stranded wire)	COMM.CABLE	(→Litzendraht)	
flexibility ≈ pliability	MECH	**Biegsamkeit** = Flexibilität ≈ Geschmeidigkeit	
flexibility [fig.]	COLLOQ	**Flexibilität** [fig.]	
flexible ≈ pliable	TECH	**biegsam** = biegbar; flexibel = plastisch; geschmeidig	
flexible (→stranded wire)	COMM.CABLE	(→Litzendraht)	
flexible access system (→FAS)	DATA COMM	(→FAS)	
flexible board (→flexible PCB)	ELECTRON	(→flexible Leiterplatte)	
flexible cable	COMM.CABLE	**flexibles Kabel**	
flexible disc (→floppy disk)	TERM&PER	(→Diskette)	
flexible disk (→floppy disk)	TERM&PER	(→Diskette)	
flexible network acess system = FNAS	TELEC	**flexibles Teilnehmerzugangssystem**	
flexible PCB = flexible board	ELECTRON	**flexible Leiterplatte**	
flexible shaft	MECH	**biegsame Welle**	
flexible-tube thermometer	INSTR	**Federthermometer**	
flexible twin cable	COMM.CABLE	**Zwillingslitze**	
flexible waveguide	MICROW	**flexibler Hohlleiter**	
flexible working time ≈ flexy time; floating time	ECON	**Gleitzeit**	
flexion (→flection)	PHYS	(→Biegung)	
flexion enforcement [the property to enforce a flexion]	LING	**Rektion** [Fähigkeit eine Biegung zu bestimmen]	
flex-twist section	MICROW	**Flex/Twist-Stück**	
flexural mode	PHYS	**Biegungsschwingung**	
flexural wave (→flexure mode)	PHYS	(→Biegeschwingung)	
flexure (→bending strength)	PHYS	(→Biegefestigkeit)	
flexure mode = flexural wave	PHYS	**Biegeschwingung**	
flexy time (→flexible working time)	ECON	(→Gleitzeit)	
flicker (v.i.) = flash (v.i.); blink (v.i.)	TECH	**blinken** = flackern; flimmern	
flicker (→flickering)	TECH	(→Flimmern)	
flicker effect	PHYS	**Funkeleffekt**	
flicker-free	TECH	**flimmerfrei**	
flicker-free image	TERM&PER	**flimmerfreies Bild**	
flicker frequency	TECH	**Flackerfrequenz** = Flackertakt	
flickering = flicker	TECH	**Flimmern** = Flickern; Flackern	
flickering indication (→flashing indication)	EQUIP.ENG	(→Blinkanzeige)	
flicker noise (→semiconductor noise)	TELEC	(→Funkelrauschen)	
flickery	TECH	**flimmernd** = flackernd	
flight (n.)	AERON	**Flug**	
flight analyzer (→flight log)	RADIO NAV	(→Flugwegschreiber)	
flight computer (→onboard computer)	DATA PROC	(→Bordcomputer)	
flight log = flight analyzer	RADIO NAV	**Flugwegschreiber**	
flight path = flight track; flight trayectory	AERON	**Flugbahn** = Flugweg; Flugspur ≈ Kurs	
flight path angle	RADIO NAV	**Flugwegwinkel**	
flight-path computer	RADIO NAV	**Flugwegrechner**	
flight simulator	RADIO NAV	**Flugsimulator**	
flight target	RADIO LOC	**Flugziel**	
flight target data	RADIO LOC	**Flugzieldaten**	
flight track (→flight path)	AERON	(→Flugbahn)	
flight traffic (→air traffic)	AERON	(→Luftverkehr)	
flight trayectory (→flight path)	AERON	(→Flugbahn)	
flimsy paper (→carbon paper)	OFFICE	(→Kohlepapier)	
flint glass	PHYS	**Flintglas**	
flip chip 1 [chip with terminals on its back, to apply on hybrid film circuits]	MICROEL	**Flip-chip** [Halbleiterchip mit Anschlüssen auf der Rückseite, zur Aufbringung auf hybriden Schichtschaltungen] ≈ Chip	
flip-chip 2 (→flip-chip bonding)	MICROEL	(→Flip-chip-Kontaktierung)	
flip-chip bonding [leadless bonding of chips with substrate] = flip-chip 2; face bonding	MICROEL	**Flip-chip-Kontaktierung** [drahtlose Kontaktierung zwischen Chip und Substrat] = Kopfüber-Kontaktierung	
flip-flop [with two stable output conditions] = bistable flip-flop; bistable multivibrator; bistable trigger circuit ↑ one-shot multivibrator	CIRC.ENG	**bistabile Kippschaltung** [mit zwei stabilen Ausgangszuständen] = Flipflop; bistabiler Multivibrator ↑ stabile Kippschaltung ↓ JK-Flipflop	
flip-flop chain	CIRC.ENG	**Kippstufenkette**	
fliptop case [hinged]	EQUIP.ENG	**Klappgehäuse** [mit Scharnieren] = Fliptop	
flix changes per inch (→FCI)	TERM&PER	(→FCI)	
float (n.) = task interval	DATA PROC	**Task-Intervall**	
float (→buffering)	POWER SYS	(→Pufferbetrieb)	
float-charge (v.t.) ≈ trickle-charge	POWER SYS	**puffern**	
float charging [accumulator] = floating charge; compensating charge	POWER SYS	**Pufferung** [Akkumulator] = Erhaltungsladung	

floating	EL.TECH	**schwebend**
[not connected to earth or other potential]		[nicht mit Erd- oder anderem Potential verbunden]
≈ power-off		≈ mit Schwebepotential
		≈ stromlos
floating	TECH	**schwimmend**
floating accent	TERM&PER	**fliegender Akzent**
[handled independently from vocals]		[unabhängig vom Umlaut behandelt]
= flying accent		
floating address	DATA PROC	**relative Adresse**
[related to another address, e.g. to the program starting point]		[auf andere Adresse, z.B. Programmanfang, bezogen]
= relative address		≈ indirekte Adresse
≈ indirect address		≠ echte Adresse
≠ real address		
floating battery	POWER SYS	**Pufferbatterie**
floating channel addressing	SWITCH	**veränderliche Kanaladressierung**
floating charge (→float charging)	POWER SYS	(→Pufferung)
floating contact	COMPON	**schwimmender Kontakt**
floating decimal point	MATH	**Gleitkomma**
[Continental Europeans use decimal "commas" instead of "points"]		[im Englischen verwendet man Dezimal-„Punkte" statt -„Kommas"]
= floating point		= Fließkomma; Gleitpunkt; Fließpunkt
≠ fixed decimal point		≠ Festkomma
floating fate	MICROEL	**schwebender Steueranschluß**
		= schwebendes Gate
floating head (→flying head)	TERM&PER	(→schwimmender Magnetkopf)
floating point (→floating decimal point)	MATH	(→Gleitkomma)
floating point addition	MATH	**Gleitkomma-Addition**
		= Gleitpunktaddition; Fließkomma-Addition; Fließpunktaddition
floating-point arithmetic	MATH	**Gleitkomma-Arithmetik**
		= Gleitpunktarithmetik; Fließkomma-Arithmetik; Fließpunktarithmetik
floating point calculation	MATH	**Gleitkommarechnung**
		= Gleitpunktrechnung; Fließkommarechnung; Fließpunktrechnung
floating point constant	MATH	**Gleitkommakonstante**
		= Gleitpunktkonstante; Fließkommakonstante; Fließpunktkonstante
floating point division	MATH	**Gleitkommadivision**
		= Gleitpunktdivision; Fließkommadivision; Fließpunktdivision
floating point exponent	MATH	**Gleitpunktexponent**
= characteristic (n.)		= Gleitkommaexponent
floating point multiplication	MATH	**Gleitkommamultiplikation**
		= Gleitpunktmultiplikation; Fließkommamultiplikation; Fließpunktmultiplikation
floating point number	MATH	**Gleitkommazahl**
≠ fixed-point number		= Gleitpunktzahl; Fließkommazahl; Fließpunktzahl
		≠ Festkommazahl
floating-point-number exponent (→biased exponent)	MATH	(→Gleitkommaexponent)
floating point operation	MATH	**Gleitkommaoperation**
= FLOP		= Gleitpunktoperation; Fließkommaoperation; Fließpunktoperation; FLOP
floating point operations per second (→flops)	DATA PROC	(→Flops)
floating point overflow	MATH	**Gleitkommaüberlauf**
		= Gleitpunktüberlauf; Fließkommaüberlauf; Fließpunktüberlauf
floating point processor	DATA PROC	**Gleitpunktprozessor**
= FPP		= Gleitkommaprozessor; Fließpunktprozessor; Fließkommaprozessor; FPP
floating point register	DATA PROC	**Gleitkommaregister**
= floating register; FPR		= Gleitpunktregister; Fließkommaregister; Fließpunktregister; FPR
floating point representation	MATH	**Gleitkommadarstellung**
		= Gleitpunktdarstellung; Fließkommadarstellung; Fließpunktdarstellung; Gleikommaschreibweise; Gleitpunktschreibweise; Fließkommaschreibweise; Fließpunktschreibweise
floating point routine	DATA PROC	**Gleitkommaroutine**
		= Gleitpunktroutine; Fließkommaroutine; Fließpunktroutine
floating point subtraction	MATH	**Gleitkommasubtraktion**
= floating subtract		= Gleitpunktsubtraktion; Fließkommasubtraktion; Fließpunktsubtraktion
floating point underflow	MATH	**Gleitkommaunterlauf**
		= Gleitpunktunterlauf; Fließkommaunterlauf; Fließpunktunterlauf
floating potential	EL.TECH	**Schwebepotential**
floating register (→floating point register)	DATA PROC	(→Gleitkommaregister)
floating subtract (→floating point subtraction)	MATH	(→Gleitkommasubtraktion)
floating time (→flexible working time)	ECON	(→Gleitzeit)
floating voltage	CIRC.ENG	**Leerlaufgleichspannung**
		= Schwebespannung
floating-zone process	MICROEL	**Schwebezonenverfahren**
float relocation	DATA PROC	**Relativ-Absolut-Adressenwandlung**
[conversion of relative into absolute addresses]		
float voltage (→compensation voltage)	POWER SYS	(→Erhaltungsladespannung)
float zone process	MICROEL	**tiegelfreies Schmelzverfahren**
floodlight	TECH	**Flutlicht**
floor (n.)	CIV.ENG	**Sohle**
[of a channel or manhole]		[Boden eines Kanals oder Schachts]
= bottom		
floor	CIV.ENG	**Fußboden**
		= Boden
floor covering	SYS.INST	**Bodenbelag**
floor installation	TECH	**Bodenaufstellung**
= ground mounting		
floor load (→floor loading)	CIV.ENG	(→Bodenbelastung)
floor loading	CIV.ENG	**Bodenbelastung**
= floor load		
floor loading capability	CIV.ENG	**Bodenbelastbarkeit**
		= Bodentragfähigkeit
floor mounting	EQUIP.ENG	**Bodenbefestigung**
floor mounting	SYS.INST	**Reihenaufbau**
≠ wall mounting		≠ Wandaufbau
floor plan (→block layout)	MICROEL	(→Block Layout)
floor plate	SYS.INST	**Bodenplatte**
= base plate		
floor space	TECH	**Stellfläche**
= footprint; shelf place [EQUIP.ENG]		= Stellplatz
≈ space requirement		≈ Raumbedarf
floor stand	EQUIP.ENG	**Standfuß**
[to install a desk model on the floor]		[um ein Tischgerät auf dem Boden aufzustellen]
↑ stand		↑ Fuß
floor switch (→foot switch)	COMPON	(→Fußschalter)
floor turtle	TERM&PER	**Schildkröte**
= turtle		= Igel
↑ rollover device		↑ Abrollgerät
floor-type cabinet	EQUIP.ENG	**Standgehäuse**
		= Standschrank
FLOP (→floating point operation)	MATH	(→Gleitkommaoperation)
floppy (→floppy disk)	TERM&PER	(→Diskette)
floppy cartridge disc (→floppy disk)	TERM&PER	(→Diskette)

floppy cartridge disk (→floppy disk)	TERM&PER	(→Diskette)	
floppy disc (→floppy disk)	TERM&PER	(→Diskette)	
floppy disc case (→disk jacket)	TERM&PER	(→Diskettenhülle)	
floppy disc envelope (→disk jacket)	TERM&PER	(→Diskettenhülle)	
floppy disc jacket (→disk jacket)	TERM&PER	(→Diskettenhülle)	
floppy disc sleeve (→disk jacket)	TERM&PER	(→Diskettenhülle)	
floppy disk	TERM&PER	**Diskette**	

= floppy disc; floppy cartridge disk; floppy cartridge disc; floppy; FD; diskette; flexible disk; flexible disc
≠ hard disk
↑ magnetic disk
↓ 8-in. floppy disk; 5 1/4-in. floppy disk; 3 1/4-in. floppy disk; 3-in. floppy disk; SD floppy disk; DD floppy disk; HD floppy disk

= Magnetdiskette; flexible Magnetplatte; Floppydisk; Weichplatte; Speicherfolie
≠ Hartplatte
↑ Magnetplatte
↓ Normaldiskette; Minidiskette; Mikrodiskette; Kompaktdiskette; SD-Diskette; DD-Diskette; HD-Diskette

floppy disk case (→disk jacket)	TERM&PER	(→Diskettenhülle)	
floppy disk controller	TERM&PER	**Diskettensteuerung**	
= fdc; FDC			
floppy disk drive	TERM&PER	**Diskettenlaufwerk**	
= disk drive 2; diskette drive; floppy drive; floppy disk unit		= Floppy-Laufwerk; Floppydisk-Laufwerk	
floppy disk envelope (→disk jacket)	TERM&PER	(→Diskettenhülle)	
floppy disk jacket (→disk jacket)	TERM&PER	(→Diskettenhülle)	
floppy-disk memory (→diskette memory)	TERM&PER	(→Diskettenspeicher)	
floppy disk sleeve (→disk jacket)	TERM&PER	(→Diskettenhülle)	
floppy-disk storage (→diskette memory)	TERM&PER	(→Diskettenspeicher)	
floppy disk unit (→floppy disk drive)	TERM&PER	(→Diskettenlaufwerk)	
floppy drive (→floppy disk drive)	TERM&PER	(→Diskettenlaufwerk)	
floppy loader	TERM&PER	**Floppy-Lader**	
floppy operation system (→diskette operating system)	DATA PROC	(→Disketten-Betriebssystem)	
floppy streamer	DATA PROC	**Floppy-streamer**	
[controlled by floppy drive]		[von Diskettenlaufwerk gesteuertes Bandlaufwerk]	
floppy tape (→streamer)	TERM&PER	(→Streamer-Magnetbandgerät)	
flops	DATA PROC	**Flops**	
= floating point operations per second; FLOP/s		[eine Gleitpunktoperation pro Sekunde]	
FLOP/s (→flops)	DATA PROC	(→Flops)	
flour spar (→fluorite)	CHEM	(→Flußspat)	
flow (v.i.)	MECH	**fließen**	
[liquid, gas]		[Flüssigkeit, Gas]	
		= strömen	
flow (v.i.)	PHYS	**strömen**	
		= fließen	
flow 2 (n.)	PHYS	**Durchfluß**	
		= Durchflußstärke	
		≈ Durchflußmenge von Gasen oder Flüssigkeiten pro Zeiteinheit	
		↓ Massendurchfluß; Volumendurchfluß	
flow 1 (n.)	PHYS	**Strömung**	
		= Fluß 2	
flowchart (→timing diagram)	TECH	(→Ablaufdiagramm)	
flowcharter	DATA PROC	**Flußdiagramm-Programm**	
flowchart symbol	DATA PROC	**Flußdiagrammsymbol**	
flowchart template	DATA PROC	**Flußdiagrammschablone**	
flow control	DATA COMM	**Flußregelung**	
↑ congestion control		= Flußsteuerung	
		↑ Überlastabwehr	
flow diagram (→timing diagram)	TECH	(→Ablaufdiagramm)	

flow direction	DATA PROC	**Ablaufrichtung**	
[order of occurrences]			
flow equation	PHYS	**Strömungsgleichung**	
flowline	DATA PROC	**Flußlinie**	
[indicates direction in data flowcharts]		[zeigt in Datenflußplänen die Richtung an]	
		= Ablauflinie	
flow of goods	ECON	**Warenfluß**	
		= Güterfluß	
flow of information (→information stream)	INF	(→Informationsfluß)	
flow over (v.i.) (→overspill)	TECH	(→überlaufen)	
flow production	MANUF	**Fließfertigung**	
flow resistance	PHYS	**Strömungswiderstand**	
flow sheet (→timing diagram)	TECH	(→Ablaufdiagramm)	
flow simulation	DATA PROC	**Strömungssimulation**	
flow soldering	MANUF	**Schwallbadlötung**	
= wave soldering		= Wellenbadlötung; Fließlöten	
flow-soldering bath	MANUF	**Schwallbad**	
= wave-soldering bath		= Wellenlötbad	
↑ solder bath		↑ Lötbad	
flow text	TYPOGR	**Fließtext**	
[body of a text, with uniform font]		[Haupttext mit einheitlicher Schriftgestaltung]	
fluctuate	TECH	**schwanken**	
[oscilate slowly]		[langsam schwingen]	
≈ sway		≈ schwingen; variieren	
≈ oscillate; variate			
fluctuating	TECH	**schwankend**	
fluctuation	TECH	**Schwankung**	
[a slow oscillation]		[langsame Schwingung]	
≈ sway		≈ Schwingung; Variierung	
≈ oscillation; variation			
fluctuation margin	TECH	**Schwankungsbreite**	
flue gas	TECH	**Rauchgas**	
[waste gas with soot]		[Abgas mit Ruß]	
fluid (adj.) (→liquid)	PHYS	(→flüssig)	
fluid (n.) (→liquid)	PHYS	(→Flüssigkeit)	
fluid crystal (→liquid crystal)	PHYS	(→Flüssigkristall)	
fluidic gate	DATA PROC	**fluidisches Schaltelement**	
[fluidics]		[Fluidik]	
= hydraulic gate; pneumatic gate		= hydraulisches Schaltelement; pneumatisches Schaltelement	
fluidics	DATA PROC	**Fluidik**	
[realization of logic connections with hydraulic devices]		[Realisierung logischer Verknüpfungen mit hydraulischen Vorrichtungen]	
fluid laser	OPTOEL	**Flüssigkeitslaser**	
fluorescence	PHYS	**Fluoreszenz**	
[characteristic radiation externally exited, stops instantaneously with the exitation]		[durch Erregungsenergie mit anderer Wellenlänge erzeugte Eigenstrahlung, erlischt sofort mit der Erregung]	
≈ phosphorescence		≈ Phosphoreszenz	
↑ luminescence		↑ Lumineszenz	
fluorescence activated display (→FLAD)	COMPON	(→FLAD)	
fluorescence radiation	PHYS	**Fluoreszenzstrahlung**	
fluorescent display tube	ELECTRON	**Vakuumfluoreszenzanzeige**	
fluorescent lamp	EL.INST	**Leuchtstofflampe**	
fluorescent screen (→luminescent screen)	ELECTRON	(→Leuchtschirm)	
fluoride	CHEM	**Fluorid**	
fluorine	CHEM	**Flour**	
= F		= F	
↑ halogen		↑ Halogen	
fluorite	CHEM	**Flußspat**	
= flour spar			
flush (v.t.)	DATA PROC	**räumen**	
[to erase content of e.g. a storage, queue or file]		[Speicherplatz, Schlange, Datei u. dgl.]	
		= freimachen	
flush (adj.)	TECH	**bündig**	
		= fluchtend	
flush (→set flush)	TYPOGR	(→bündig)	

flush-disc antenna	ANT	Flush-disc-Antenne	FMS (→file management system)	DATA PROC	(→Dateiverwaltungssystem)
flushing (→flashing)	TERM&PER	(→blinkend)	FM threshold	RADIO REL	FM-Schwelle
flush left (adj.)	TECH	linksbündig			[Analog-Richtfunk]
flush left (→left-justified)	TYPOGR	(→linksbündig)			= Mindestempfangspegel
flush-mounted antenna	ANT	Flush-Antenne	FNAS (→flexible network acess system)	TELEC	(→flexibles Teilnehmerzugangssystem)
flush mount speaker	EL.ACOUS	Einbaulautsprecher [Autoradio]	foam (→foam materials)	TECH	(→Schaumstoff)
flush right	TECH	rechtsbündig	foam materials	TECH	Schaumstoff
flush right (→right-aligned)	TYPOGR	(→rechtsbündig)	= foam		
flutter (n.) [fast fluctuations of amplitude]	EL.ACOUS	Trillern [schnelle Amplitudenschwankungen]	foam rubber ≈ Schwammgummi; Schaumstoff	TECH	Schaumgummi ≈ Schwammgummi
flutter (n.) [unwanted speed fluctuations] ≈ wow	TERM&PER	Gleichlaufschwankung = Gleichlaufstörung	foam rubber speaker front focal aperture = diaphragm aperture; aperura	EL.ACOUS PHYS	Schaumfront Blendenöffnung
flutter echo (→multiple echo)	TELEC	(→Mehrfachecho)	focal distance focal line	PHYS PHYS	Brennweite Brennlinie
flutter fading	RADIO PROP	Flatterschwund = Flatterfading	focal plane focal plane antenna	PHYS ANT	Brennebene Focal-plane-Antenne
flux (n.) [electric, magnetic]	PHYS	Fluß 1 [elektrisch, magnetisch]	focal point = focus (pl.-es, foci)	PHYS	Brennpunkt = Fokus
flux 1 (→soldering flux)	METAL	(→Flußmittel 1)			≈ Brennfleck
flux 2 (→welding flux)	METAL	(→Flußmittel 2)	focal ratio	PHYS	Blendenzahl
flux-controlled	EL.TECH	durchflutungsgesteuert	focal spot	PHYS	Brennfleck
flux density	PHYS	Kraftflußdichte = Flußdichte; Kraftliniendichte	= luminescent spot focus (focused, focussed; focusing, focusing)	PHYS	≈ Brennpunkt fokussieren [Optik]
flux dispersion (→flux leakage)	PHYS	(→Kraftlinienzerstreuung)			= scharfeinstellen
fluxing (→welding flux)	METAL	(→Flußmittel 2)	focus (pl.-es, foci) (→focal point)	PHYS	(→Brennpunkt)
flux leakage = flux dispersion	PHYS	Kraftlinienzerstreuung = Flußsteuerung	focus depth = field depth; in-depth definition	PHYS	Schärfentiefe = Tiefenschärfe; Fokustiefe
fluxless	METAL	flußmittelfrei	focus-fed	ANT	fokusgespeist
flux line	PHYS	Flußlinie	focusing		Fokussierung
flux linkage	PHYS	Kraftlinienverkettung = Flußverkettung	= focussing		= Bündelung; Strahlungskonzentrierung
flux meter ≈ creeping galvanometer ↑ moving-coil galvanometer	INSTR	Fluxmeter ≈ Kriechgalvanometer ↑ Drehspulgalvanometer	focusing anode = focussing anode focusing coil = focussing coil	ELECTRON ELECTRON	Fokussierungsanode = Bündelungsanode Fokussierspule = Fokussierungsspule; Abbildungsspule
flux of heat (→thermal flux)	PHYS	(→Wärmefluß)	focusing electrode	ELECTRON	Fokussierelektrode
flux path	PHYS	Kraftlinienweg = Flußweg	= focussing electrode		= Fokussierungselektrode; Bündelungselektrode
flyback = line flyback	ELECTRON	Zeilenrücklauf	focusing fading = focussing fading	RADIO PROP	Fokussierungsschwund
flyback (→reverse action)	ELECTRON	(→Rücklauf)	focusing lens (→converging lens)	PHYS	(→Sammellinse)
flyback converter	POWER SYS	Sperrwandler = Sperrwandlernetzteil; Sperrumrichter	focussing (→focusing) focussing anode (→focusing anode)	PHYS ELECTRON	(→Fokussierung) (→Fokussierungsanode)
flyback transformer (→line scan transformer)	TV	(→Horizontal-Ablenktransformator)	focussing coil (→focusing coil)	ELECTRON	(→Fokussierspule)
flyer (→leaflet)	DOC	(→Prospekt)	focussing device	ELECTRON	Fokussiereinrichtung
flying = airborne	TECH	fliegend	focussing electrode (→focusing electrode)	ELECTRON	(→Fokussierelektrode)
flying accent (→floating accent)	TERM&PER	(→fliegender Akzent)	focussing fading (→focusing fading)	RADIO PROP	(→Fokussierungsschwund)
flying clock [Standard Time]	INSTR	mobile Uhr [Zeitnormal]	Foerster probe fog	INSTR METEOR	Förstersonde Nebel
flying head = floating head	TERM&PER	schwimmender Magnetkopf = schwebender Magnetkopf	= mist fog	TV	Schleier
flying spot = optical spot; optical dot ≈ optical pointer	ELECTRON	Lichtpunkt ≈ Lichtmarke	foil fold (v.t.)	TECH TECH	Folie falten = einklappen
flying-spot scanner	TERM&PER	Lichtpunktabtaster	fold (n.)	MECH	Faltung
flying-spot scanning	TERM&PER	Lichtpunktabtastung = Flying-spot-Abtastung	fold ≈ chamfer; bevel	MECH.ENG	abkanten 2 [Kanten umbiegen]
flywheel	MECH	Schwungrad			= falten
flywheel moment	PHYS	Schwungmoment			= abfasen; abschrägen
FM (→frequency modulation)	MODUL	(→Frequenzmodulation)	foldback current limiting	POWER SYS	Kurzschlußstrom-Rückregelung
Fm (→fermium)	CHEM	(→Fermium)	= cutback current limiting; foldback		
FM broadcast	BROADC	UKW-Sendung = UKW-Programm	mode; cutback mode foldback mode (→foldback	POWER SYS	(→Kurzschlußstrom-Rückre-
FM broadcast transmitter	BROADC	FM-Rundfunksender	current limiting)		gelung)
FM code (→two-frequency recording)	DATA PROC	(→Wechseltaktschrift)	folded ≈ beveled	MECH	abgekantet ≈ abgeschrägt
FM-CW-Radar (→frequency modulated continuous wave radar)	RADIO LOC	(→frequenzmodulierter Dauerstrichradar)	folded folded (→tapered) folded antenna	TECH MECH ANT	gefaltet (→abgeschrägt) Faltantenne
FMFB (→frequency modulation feedback)	MODUL	(→Frequenz-Gegenkopplungs-Demodulation)			

English	Domain	German
folded dipole	ANT	**Faltdipol**
= folded dipole antenna; skirt dipole		= Faltdipolantenne; Schleifendipol
folded dipole antenna (→folded dipole)	ANT	(→Faltdipol)
folded ground plane	ANT	**Falt-Groundplane-Antenne**
folded grouping	SWITCH	**Umklappgruppierung**
folded monopole antenna	ANT	**Faltmonopol**
		= Faltunipol
folded T antenna	ANT	**gefaltete T-Antenne**
folded-top antenna	ANT	**Sperrtopfantenne**
= detuning-sleeve antenna; sleeve dipole		
folder	DATA PROC	**Ordner**
[file subdirectory]		[Datei-Unterverzeichnis]
folder (→leaflet)	DOC	(→Prospekt)
folder (→letter-file)	OFFICE	(→Schnellhefter)
folder top (→bazooka)	ANT	(→Viertelwellen-Sperrtopf)
folding (→convolution)	MATH	(→Faltung)
folding (→hinged)	MECH	(→schwenkbar)
folding cabinet feet	EQUIP.ENG	**einklappbarer Aufstellfuß**
folding machine	OFFICE	**Falzmaschine**
		= Faltmaschine
fold-in handle (→clasp handle)	TECH	(→Klappgriff)
foldover	MODUL	**Spektrumüberlappung**
fold-over (→convolution)	MATH	(→Faltung)
fold-over (→ghost image)	TV	(→Geisterbild)
foldover distortion (→aliasing)	MODUL	(→Rückfaltung)
folio (→page number)	TYPOGR	(→Seitenzahl 2)
follow-up (v.t.)	TECH	**nachrücken**
follow-up (→pursuit)	COLLOQ	(→verfolgen)
follow-up control	CONTROL	**Folgeregelung**
= tracking control		= Nachlaufregelung
≈ servomechanism		≈ Servomechanismus
follow-up document	DOC	**Folgeunterlage**
follow-up file (→appointment book)	OFFICE	(→Terminkalender)
follow-up portfolio	OFFICE	**Terminmappe**
		= Wiedervorlagemappe
font (n.)	TYPOGR	**Schriftzeichensatz**
[assortment of types of same style and size]		[Satz von Schriftzeichen gleicher Schriftart und Größe]
= fount (BRI); type font; character font; graphic character set		= Schriftzeichenvorrat; Schriftzeichenrepertoire; Schriftsatz; Zeichensatz [TERM&PER]; Font [TERM&PER]
↓ printer font [TERM&PER]; display font [TERM&PER]		
		≈ Schriftart
		↓ Druckfont [TERM&PER]; Bildschirmfont [TERM&PER]
font attribute (→typeface design)	TYPOGR	(→Schriftschnitt)
font card (→font cartridge)	TERM&PER	(→Schriftartkassette)
font cartridge	TERM&PER	**Schriftartkassette**
[for printer]		[für Drucker]
= font card		= Font-Kassette; Font-Karte
font change	TERM&PER	**Schriftartwechsel**
font change character	DATA PROC	**Schriftartumschaltezeichen**
= FCC		
font design (→typeface design)	TYPOGR	(→Schriftschnitt)
font disk	TERM&PER	**Zeichensatz-Diskette**
= font diskette		= Font-Diskette
font diskette (→font disk)	TERM&PER	(→Zeichensatz-Diskette)
font editor	DATA PROC	**Zeichensatz-Editor**
		= Font-Editor
font family (→type family)	TYPOGR	(→Schriftfamilie)
font file	TERM&PER	**Schriftdatei**
font height (→type height)	TYPOGR	(→Schrifthöhe)
font recognition (→character recognition)	TERM&PER	(→Klarschriftlesen)
font selection	TERM&PER	**Schriftartwahl**
font size (→type size)	TYPOGR	(→Schriftgröße)
font style (→type style)	TYPOGR	(→Schriftart)
food waste (→garbage)	COLLOQ	(→Müll)
foolproof	COLLOQ	**idiotensicher**
foolscap	TYPOGR	**Format 16x13 Zoll**
[paper size 16x13"]		
= foo's cap; F'cap		
foo's cap (→foolscap)	TYPOGR	(→Format 16x13 Zoll)
foot (pl. feet)	PHYS	**Fuß**
[= 12 inches; = 0.3048 m]		[angloamerikanisches Längenmaß; = 0,3048 m]
foot	TYPOGR	**Fuß**
foot (→stand)	EQUIP.ENG	(→Fuß)
footer	TYPOGR	**Fußzeile**
[repetitive footnote]		[sich wiederholende Fußnote]
= footing		≈ Fußnote
≈ footnote		
footing (→footer)	TYPOGR	(→Fußzeile)
foot line	TYPOGR	**Fußlinie**
= footline; foot rule		
footline (→foot line)	TYPOGR	(→Fußlinie)
footnote	TYPOGR	**Fußnote**
≈ footer		≈ Fußzeile
footnote reference mark	TYPOGR	**Fußnotenzeichen**
footprint (→illumination spot)	SAT.COMM	(→Ausleuchtzone)
footprint (→floor space)	TECH	(→Stellfläche)
foot rail	EQUIP.ENG	**Fußschiene**
foot rail alarm device	SIGN.ENG	**Fußschienenalarmsystem**
foot rule (→foot line)	TYPOGR	(→Fußlinie)
foot switch	COMPON	**Fußschalter**
= floor switch; pedal switch		
for bench use (→for laboratory use)	INSTR	(→für Laboreinsatz)
forbid (v.t.; forbade-forbidden)	COLLOQ	**verbieten**
= prohibit; interdict; inhibit		= untersagen
forbidden band (→forbidden energy band)	PHYS	(→verbotenes Band)
forbidden energy band	PHYS	**verbotenes Band**
= forbidden band; energy gap		= verbotenes Energieband; verbotene Zone; Energielücke; Energiezwischenband
force (v.t.)	TECH	**erzwingen**
= constrain; compel		
force (n.)	PHYS	**Kraft**
[product of mass and acceleration; SI unit: Newton]		[Masse mal Beschleunigung; SI-Einheit: Newton]
= F		= F
↓ weight		↓ Gewicht
forced commutation	POWER SYS	**Zwangskommutierung**
[static power converter]		[Stromrichter]
forced-commutation converter (→self-commutated converter)	POWER SYS	(→selbstgeführter Stromrichter)
forced convection	PHYS	**erzwungene Konvektion**
force density	PHYS	**Kraftdicke**
force density per unit length	POWER ENG	**Kraftbelag**
forced justification	TERM&PER	**Zwangsausschließen**
[printer]		[Drucker]
= compulsory justification		
forced oscillation	PHYS	**erzwungene Schwingung**
forced-read	DATA PROC	**Lesezwang**
forced release	SWITCH	**Trennen**
		[das Auslösen einer bestehenden Verbindung um eine bevorrechtigte Verbindung herzustellen]
forced termination (→abnormal end)	DATA PROC	(→Absturz)
force due to gravity (→weight)	PHYS	(→Gewicht)
forced ventilation	TECH	**Zwangsbelüftung**
		= Zwangslüftung; Fremdbelüftung; Fremdlüftung
force fit (→interference fit)	ENG.DRAW	(→Preßpassung)
force open (→burst open)	TECH	(→aufbrechen)
force vector	PHYS	**Kraftvektor**
force-ventilate	TECH	**zwangsbelüften**
forcibly disconnect (→lock-out)	SWITCH	(→Abwerfen)
forecastable (→predictable)	SCIE	(→vorhersehbar)
foreground (n.)	TECH	**Vordergrund**
≠ background (n.)		≠ Hintergrund

foreground color TERM&PER **Vordergrundfarbe**
foregrounding (→foreground DATA PROC (→Vordergrundverarbeitung)
processing)
foreground job (→priority DATA PROC (→Prioritätsprogramm)
program)
foreground processing DATA PROC **Vordergrundverarbeitung**
[of high-priority programs] [der Programme höchster
= foregrounding Priorität]
≠ background processing ≠ Hintergrundverarbeitung
foreground program DATA PROC (→Prioritätsprogramm)
(→priority program)
foreground task (→priority DATA PROC (→Prioritätsprogramm)
program)
foreign (adj.) ECON **ausländisch**
= outlandish
foreign atom PHYS **Fremdatom**
= impurity atom; doping atom = Störatom [MICOEL]
[MICOEL] ↑ Störstelle [MICOEL]
↑ imperfection [MICOEL]
foreign commerce (AM) ECON (→Außenhandel)
(→foreign trade)
foreign data file DATA PROC **Fremddatei**
= foreign file; foreign dataset
foreign dataset (→foreign DATA PROC (→Fremddatei)
data file)
foreign exchange clause ECON (→Währungsklausel)
(→currency clause)
foreign file (→foreign data DATA PROC (→Fremddatei)
file)
foreign language secretary OFFICE **Fremdsprachensekretärin**
foreign notes and coins ECON **Sorten**
[Zahlungsmittel in fremder
Währung]
= Devisen 2
↑ Devisen 1
foreign trade ECON **Außenhandel**
= foreign commerce (AM)
foreign voltage (→interfering TELEC (→Störspannung)
voltage)
foreman ECON **Vorarbeiter**
= lead hand (AM); superintendent [in Industriebetrieb]
= Werkmeister; Meister 2;
Faktor [TYPOGR]
foremarker (→outer RADIO NAV (→Voreinflugzeichen)
marker)
fore-scattering (→forward PHYS (→Vorwärtsstreuung)
scattering)
foreshorten ENG.DRAW **verkürzen**
foreshortened view ENG.DRAW **verkürzte Ansicht**
foreshortening ENG.DRAW **Verkürzung**
foresight TECH **Vorausschau**
forest (n.) DATA PROC **Datenstrukturwald**
[interconnected data structure
trees]
forestall TECH **zuvorkommen**
= anticipate = vorwegnehmen
forfeiture (→loss) ECON (→Verlust)
forge METAL **schmieden**
forgeability METAL **Schmiedbarkeit**
= malleability
forgeable MECH **schmiedbar**
= malleable ≈ plastisch [TECH]
≈ plastic [TECH]
forged ECON **gefälscht**
forged METAL **geschmiedet**
= wrought
forgery ECON **Fälschung**
forge welding METAL **Feuerschweißen**
= hammer welding ≈ Preßschweißen
≈ pressure welding
forging 1 METAL **Schmieden**
≈ pressure welding = Schmiedung
≈ Preßschweißen
forging 2 METAL **Schmiedestück**
FOR instruction (ALGOL) DATA PROC (→Laufanweisung)
(→run statement)
fork (→branching) NETW.TH (→Verzweigung)
fork truck TECH **Gabelstapler**
for laboratory use INSTR **für Laboreinsatz**
= for bench use; laboratory-bench

FOR list DATA PROC **Laufliste**
form (n.) TECH **Form**
form (n.) TERM&PER **Formular**
= preprinted form; formular = Vordruck
↓ continuous form; single form; docu- ↓ Endlosformular; Einzelfor-
ment; coding form [DATA PROC]; mular; Beleg;
test plan [DATA PROC]; display Programmvordruck [DA-
form TA PROC]; Testvordruck
[DATA PROC]; Bild-
schirmformular
form (→standard form) OFFICE (→Formular)
form advance (→form feed) TERM&PER (→Formularzuführung)
formal education (→formation) SCIE (→Bildung)
formalism SCIE **Formalismus**
formal parameter DATA PROC **Platzhaltesymbol**
= formaler Parameter
formant (→characteristic ACOUS (→Formant)
frequency)
format (v.t.) DATA PROC **formatieren**
[to structure a magnetic data carrier] [einen magnetischen Da-
= initialize 3 tenträger strukturieren]
= initialisieren 3
format 1 (n.) DATA PROC **Format**
[way of arranging data] [Art der Anordnung von
Daten]
format TECH **Format**
≈ dimensions ≈ Dimensionen
format 2 (→instruction DATA PROC (→Befehlssyntax)
syntax)
format buffer (→vertical TERM&PER (→Formularformatspeicher)
format buffer)
format check DATA PROC **Formatkontrolle**
format control character DATA COMM **Formatsteuerzeichen**
format effector DATA PROC **Formatsteuerung**
= FE [Code]
= FE
formation SCIE **Bildung**
= formal education
formation (→twisting) COMM.CABLE (→Verseilung)
format label DATA PROC **Formatetikett**
= Format-Kennsatz
format manager DATA PROC **Formatverwalter**
format mode (→form DATA PROC (→Maskenbetrieb)
mode)
format packed CODING **gepackte Formatdarstellung**
format storage DATA PROC **Formularspeicherung**
formatted DATA PROC **formatiert**
≠ unformatted = formatgebunden
≠ unformatiert
formatted capacity DATA PROC **Nennkapazität**
[of a memory] [eines Speichers]
= Nettospeicherkapazität;
Nettokapazität
formatted display DATA PROC **formatierte Bildschirmanzeige**
formatted input DATA PROC **formatierte Eingabe**
formatter DATA PROC **Formatierer**
formatting DATA PROC **Formatierung**
= initialization 2 = Formatieren; Initialisie-
rung 2
form drive (→form feed) TERM&PER (→Formularzuführung)
formed TECH **geformt**
= shaped
former (→template) TECH (→Schablone)
form factor EL.TECH **Formfaktor**
form factor TECH **Länge-/Breite-Verhältnis**
= Formfaktor
form feed TERM&PER **Formularzuführung**
= form feeding; form feeder; FF; form = Formularzufuhr; Formular-
transport; form advance; form drive transport; Formulareinga-
≈ document feed be; Formulareinzug;
↑ paper feed Formularvorschub; Vor-
↓ continuous form feeding; single form druckzuführung; Vordruck-
feeding zufuhr; Vordrucktransport;
Vordruckeingabe; Vor-
druckeinzug; Vordruckvor-
schub
≈ Belegzufuhr
↑ Papiervorschub
↓ Endlosformularzuführung;
Einzelformularzuführung

form feeder (→form feed)	TERM&PER	(→Formularzuführung)	
form feeding (→form feed)	TERM&PER	(→Formularzuführung)	
form fidelity	TECH	**Formtreue**	
= contour fidelity			
form flash	DATA PROC	**Formulareinblendung**	
= form overlay			
form handling equipment	TERM&PER	**Formularhantierungsgerät**	
↓ decollator		↓ Trennmaschine	
forming [electrolysis]	PHYS	**Formierung** [Elektrolyse]	
forming board	MANUF	**Kabelformbrett**	
= lacing board			
form letter [word processing]	DATA PROC	**Standardbrief** [Textverarbeitung]	
= standard letter			
form letter (→boilerplate letter)	OFFICE	(→Standardbrief)	
form-letter program	DATA PROC	**Standardbriefprogramm**	
= mail-merge program		= Serienbriefprogramm	
form manager	DATA PROC	**Formular-Manager**	
		= Formular-Verwalter	
form mask	DATA PROC	**Formularmaske**	
form mode [data can be entered in a dispayed form]	DATA PROC	**Maskenbetrieb** [mit eingeblendeter Formularmaske]	
= format mode			
form overlay (→form flash)	DATA PROC	(→Formulareinblendung)	
form-permanent (→rigid 2)	TECH	(→formbeständig)	
form printer	TERM&PER	**Formulardrucker**	
form release	TERM&PER	**Papierlöser**	
forms ejection (→forms stacker)	TERM&PER	(→Formularstapler)	
forms hopper (→paper tray)	TERM&PER	(→Papiernachfüllmagazin)	
form signal	RAILW.SIGN	**Formsignal**	
forms layout gage (→line gauge)	TYPOGR	(→Zeilenmaß)	
forms stacker	TERM&PER	**Formularstapler**	
= forms ejection		↑ Papierstapler; Ablagefach	
↑ paper stacker; stacker			
form stack (→paper stack)	TERM&PER	(→Papierstapel)	
form tolerance (→tolerance of form)	ENG.DRAW	(→Formtoleranz)	
form transport (→form feed)	TERM&PER	(→Formularzuführung)	
former	COLLOQ	**ehemalig**	
≈ past		= vormalig = vergangen	
formula (pl.-as; -ae) [symbols or digits representing a mathematical relation] ≈ equation	MATH	**Formel** [Symbole oder Ziffern zur Darstellung eines mathematischen Zusammenhangs] ≈ Gleichung	
formula portability [spreadsheet calculation]	DATA PROC	**Formelübertragbarkeit** [Tabellenkalkulation]	
formular (→display mask)	TERM&PER	(→Bildschirmmaske)	
formular feed [code] = FF	DATA COMM	**Formularvorschub** [Code] = FF	
FOR-NEXT loop [repetition till a limit condition]	DATA PROC	**FOR-NEXT-Schleife** [wiederholt eine Operation bis zu einer Endbedingung]	
FOR statement (ALGOL) (→run statement)	DATA PROC	(→Laufanweisung)	
forthcoming = coming	TECH	**kommend** [fig.] = im Kommen befindlich	
for the attention of (→care of)	OFFICE	(→zu Händen)	
fortitude (→constancy)	COLLOQ	(→Konstanz)	
fortnight (BRI) ["fourteen nights"] = two weeks (AM)	COLLOQ	**zwei Wochen** = vierzehn Tage	
fortnightly	COLLOQ	**vierzehntägig**	
FORTRAN [FORmula TRANslator] ↑ high-level programming language	DATA PROC	**FORTRAN** ↑ problemorientierte Programmiersprache	
fortuitous (→irregular)	TECH	(→unregelmäßig)	
fortuitous (→random)	TECH	(→zufällig)	
fortuitous distortion	DATA COMM	**unregelmäßige Verzerrung**	
fortuitousness (→irregularity)	TECH	(→Unregelmäßigkeit)	
forward (n.) [of a carry on a new account]	ECON	**Vortrag** [eines Überlaufs auf ein neues Konto]	
forward (adv.)	COLLOQ	**vorwärts** (adv.)	
forward (n.) (→forward wind)	TERM&PER	(→Aufspulung)	
forward (v.t.) (→send)	ECON	(→übersenden 1)	
forward (→relay)	TELEC	(→weiterleiten)	
forward-acting controller	CONTROL	**Vorwärtsregler**	
= forward-acting regulator			
forward-acting regulator (→forward-acting controller)	CONTROL	(→Vorwärtsregler)	
forward AGC (→forward automatic gain control)	MICROEL	(→Aufwärtsregelung)	
forward automatic gain control	MICROEL	**Aufwärtsregelung**	
= forward AGC; forward control			
forward blocking voltage	ELECTRON	**Vorwärtsblockierspannung**	
forward chain (→forward chaining)	DATA PROC	(→Vorwärtskettung)	
forward chaining	DATA PROC	**Vorwärtskettung**	
= forward chain; forward concatenation		= Vorwärtsverkettung	
forward channel (→main channel)	TELEC	(→Hauptkanal)	
forward characteristic	ELECTRON	**Vorwärtskennlinie**	
= forward voltage-current characteristic		= Vorwärts-Durchlaßkennlinie; Durchlaßkennlinie	
forward clearance (→forward release)	SWITCH	(→Vorwärtsauslösung)	
forward concatenation (→forward chaining)	DATA PROC	(→Vorwärtskettung)	
forward control (→forward automatic gain control)	MICROEL	(→Aufwärtsregelung)	
forward controlling element (→final control element)	CONTROL	(→Stellglied)	
forward current	ELECTRON	**Durchlaßstrom**	
= conducting state current; on-state current ≠ reverse current		= Vorwärtsstrom; Flußstrom ≠ Sperrstrom	
forward current transfer ratio	ELECTRON	**Gleichstromverhältnis**	
forward current transfer ratio	NETW.TH	**Kurzschluß-Übertragungsfaktor**	
forward cutoff region	MICROEL	**Vorwärtssperrbereich**	
forward DC converter (→feed forward converter)	POWER ENG	(→Durchflußwandler)	
forward direction ≠ reverse direction	ELECTRON	**Durchlaßrichtung** = Vorwärtsrichtung; Schaltrichtung ≠ Sperrichtung	
forward direction	SWITCH	**Vorwärtsrichtung**	
forward error correction = FEC	CODING	**Vorwärtsfehlerkorrektur** = FEC	
forwarding = dispatch department	ECON	**Spedition** = Versandabteilung	
forwarding (→despatch)	ECON	(→Versand)	
forwarding charges (→freight 1)	ECON	(→Frachtkosten)	
forward jump	DATA PROC	**Vorwärtssprung**	
forward loss	MICROEL	**Durchlaßverlust**	
forward-propagation ionospheric scatter (→ionospheric scatter)	RADIO PROP	(→ionosphärische Streuausbreitung)	
forward recovery time	MICROEL	**Durchlaßverzögerung** = Durchlaßerholungszeit; Vorwärtserholungszeit	
forward release = forward clearance; clear forward ↑ release	SWITCH	**Vorwärtsauslösung** = Vorwärtsauslösen ↑ Auslösung	
forward resistance = on-state resistance	MICROEL	**Durchlaßwiderstand** = Vorwärtswiderstand; Gleichstromwiderstand vorwärts	
forward scan (→forward scanning)	ELECTRON	(→Vorwärtsabtastung)	
forward scanning = forward scan	ELECTRON	**Vorwärtsabtastung**	

forward scatter (→forward scattering)		PHYS	(→Vorwärtsstreuung)	four-layer		TECH	vierlagig = vierschichtig
forward scattering = fore-scattering; forward scatter; forward stray		PHYS	Vorwärtsstreuung	four-part ↑ multisectional		TECH	vierteilig ↑ mehrteilig
forward-scattering cross section (→radar cross section)		RADIO LOC	(→Radarquerschnitt)	four-party line with selective ringing ↑ party line		TELEC	Viereranschluß ↑ Gemeinschaftsanschluß
forward signal ↑ switching signal		SWITCH	Vorwärtszeichen = Vorwärtskennzeichen ↑ Kennzeichen	four-phase modulator		CIRC.ENG	Vierphasenmodulator
				four-phase operation		MICROEL	Vierphasenbetrieb [MOS]
forward stray (→forward scattering)		PHYS	(→Vorwärtsstreuung)	four-phase technique [dynamic shift register]		CIRC.ENG	Vierphasentechnik [dynamisches Schieberegister]
forward stroke		MECH	Vorwärtshub				
forward test		MICROEL	Durchgangsprüfung [Diode]	four-place = four-figure ↑ of many places		MATH	vierstellig ↑ vielstellig
forward transconductance [FET]		CIRC.ENG	Gate-Steilheit [FET]	four-plus-one address		DATA PROC	Vier-plus-Eins-Adresse
forward transfer (→forward transfer signal)		SWITCH	(→Eintretezeichen)	four-point probe method		INSTR	Vierspitzenmethode
				four-pole (→two-port network)		NETW.TH	(→Vierpol)
forward transfer signal = forward transfer		SWITCH	Eintretezeichen	fourpole (→two-port network)		NETW.TH	(→Vierpol)
forward voltage		MICROEL	Vorwärtsspannung = Durchlaßspannung	four-pole analysis (→two-port network analysis)		NETW.TH	(→Vierpoltheorie)
forward voltage-current characteristic (→forward characteristic)		ELECTRON	(→Vorwärtskennlinie)	four-pole connection (→two-port connection)		NETW.TH	(→Vierpolzusammenschaltung)
				four-pole determinant (→network determinant)		NETW.TH	(→Vierpoldeterminante)
forward wind (n.) = forward (n.)		TERM&PER	Aufspulung [Band]	four-pole dissection (→two-port dissection)		NETW.TH	(→Vierpolzerlegung)
FOS (→diskette operating system)		DATA PROC	(→Disketten-Betriebssystem)	four-pole equation (→network equation)		NETW.TH	(→Vierpolgleichung)
Foster reactance theorem = reactance theorem		NETW.TH	Reaktanztheorem = Reaktanzsatz; Theorem von Foster; Fostersches Theorem	four-pole matrix (→network matrix)		NETW.TH	(→Vierpolmatrix)
				four-pole network (→two-port network)		NETW.TH	(→Vierpol)
FOT		HF	FOT [fréquence optimale de trafique]	fourpole network (→two-port network)		NETW.TH	(→Vierpol)
				four-pole parameter (→network parameter)		NETW.TH	(→Vierpolparameter)
Foucault current (→eddy current)		EL.TECH	(→Wirbelstrom)	four-pole transformation (→two-port network transformation)		NETW.TH	(→Vierpoltransformation)
foul (v.t.) ≈ contaminate		TECH	verschmutzen ≈ verunreinigen	four-pole wave (→two-port wave)		NETW.TH	(→Vierpolwelle)
found (v.t.) = cast		METAL	gießen	four-port (→four-port network)		NETW.TH	(→Viertor)
foundation		CIV.ENG	Fundament	four-port circulator		MICROW	Vierarmzirkulator
foundation bolt = anchor bolt; holding-down bolt		MECH	Ankerschraube	four-port network = four-port ↑ network		NETW.TH	Viertor = Achtpol ↑ Netzwerk
foundry		METAL	Gießerei	four-quadrant drive (→four-quadrant operation)		POWER SYS	(→Vierquadrantenbetrieb)
fount (BRI) (→font)		TYPOGR	(→Schriftzeichensatz)				
fountain-pen = pen		OFFICE	Füllfederhalter = Füller; Füllfeder	four-quadrant multiplier		CIRC.ENG	Vierquadranten-Multiplizierer
four-address instruction = 4-address instruction ↓ three-plus-one address		DATA PROC	Vieradreßbefehl ↓ Drei-plus-Eins-Befehl	four-quadrant operation = four-quadrant drive		POWER SYS	Vierquadrantenbetrieb
				four-quad series array		ANT	Vier-quad-Serien-Antenne
four band T antenna		ANT	Vierband-T-Antenne	four-row		EQUIP.ENG	vierzeilig
four-band Windom antenna		ANT	Vierband-Windom-Antenne	four-step ↑ multi-step		TECH	vierstufig ↑ mehrstufig
four-channel digitizing oscilloscope		INSTR	Vierkanal-Digitaloszilloskop	four-terminal equation (→network equation)		NETW.TH	(→Vierpolgleichung)
four-channel tape		TERM&PER	Vierspurlochstreifen	four-terminal network (→two-port network)		NETW.TH	(→Vierpol)
four-circuit		NETW.TH	Vierkreis- = vierkreisig	fourth-generation computer [LSI; since 1972]		DATA PROC	Rechner der vierten Generation [LSI; ab ca. 1972] = Computer der vierten Generation
four-element array		ANT	Vier-Element-Gruppenantenne				
four-figure (→four-place)		MATH	(→vierstellig)	four-to-two-wire (→termination hybrid)		TELEC	(→Gabelschaltung)
fourfold = quadruple; quad		MATH	vierfach	forty-track disk = 40-track disk; fourty-track diskette; 40-track diskette		TERM&PER	Vierzig-Spuren-Diskette = 40-Spuren-Diskette
four-frequency duplex telegraphy		HF	Vierfrequenz-Duplex-Telegraphie				
four-half-dipole antenna		ANT	Viererfeldantenne	forty-track diskette (→fourty-track disk)		TERM&PER	(→Vierzig-Spuren-Diskette)
Fourier analysis = harmonic analysis		MATH	Fourier-Analyse = harmonische Analyse; Fourier-Zerlegung	four-valued = quadrivalent		MATH	vierwertig
Fourier analyzer ↑ signal analyzer		INSTR	Fourier-Analysator ↑ Signalanalysator	four-way box (→four-way system)		EL.ACOUS	(→Vierwegsystem)
Fourier series		MATH	Fourier-Reihe = trigonometrische Reihe				
Fourier transform = Fourier transformation		MATH	Fourier-Transformation				
Fourier transformation (→Fourier transform)		MATH	(→Fourier-Transformation)				

four-way system	EL.ACOUS	**Vierwegsystem**	
= four-way box		= Vierwegebox	
four-wheel drive	TECH	**Allradantrieb**	
four-wire ...	TELEC	**vieradrig**	
		= vierdrähtig; Vierdraht-	
four-wire amplifier	TRANS	**Vierdrahtverstärker**	
= four-wire repeater		↓ Vierdraht-Endverstärker; Vierdraht-Zwischenverstärker	
↓ four-wire terminal repeater; four-wire intermediate repeater			
four-wire bus	DATA COMM	**Vierdrahtbus**	
four-wire carrier system	TRANS	(→Vierdrahtsystem)	
(→four-wire system)			
four-wire connection (→four-wire link)	TELEC	(→Vierdrahtverbindung)	
four-wire feeder	ANT	**Vierdraht-Speiseleitung**	
four-wire line	TELEC	**Vierdrahtleitung**	
four-wire link	TELEC	**Vierdrahtverbindung**	
= four-wire connection			
four-wire operation	TELEC	**Vierdrahtbetrieb**	
four-wire polar current	TELEGR	**Vierdrahtdoppelstrom**	
four-wire repeater (→four-wire amplifier)	TRANS	(→Vierdrahtverstärker)	
four-wire switching	SWITCH	**Vierdraht-Durchschaltung**	
		= Vierdraht-Vermittlung	
four-wire system	TRANS	**Vierdrahtsystem**	
= four-wire carrier system			
four-wire terminal repeater	TRANS	**Vierdraht-Endverstärker**	
fox (→fox message)	TELEGR	(→Fox-Code)	
fox message	TELEGR	**Fox-Code**	
[test message containing all alpha-numerics of a teletype keyboard, e.g. "The quick brown fox jumped over the lazy's dog back 1234567890"]		[alle Zeichen der Fernschreibtastatur enthaltender Prüftext]	
= fox			
FPIS (→ionospheric scatter)	RADIO PROP	(→ionosphärische Streuausbreitung)	
FPLA	MICROEL	**FPLA**	
= field-programmable logic array		= frei programmierbare logische Anordnung	
FP laser (→Fabry-Perot laser)	OPTOEL	(→Fabry-Perot-Laser)	
FPP (→floating point processor)	DATA PROC	(→Gleitpunktprozessor)	
FPR (→floating point register)	DATA PROC	(→Gleitkommaregister)	
Fr (→francium)	CHEM	(→Francium)	
fraction	COLLOQ	**Bruchteil**	
= fractional part			
fraction (→ratio)	MATH	(→Bruch)	
fractional (adj.)	MATH	**gebrochen**	
≠ integer		≠ ganzzahlig	
fractional (n.) (→fractional number)	MATH	(→Bruchzahl)	
fractional digit	MATH	**Nachkommastelle**	
fractional exponent	MATH	**Bruchexponent**	
fractionally spaced filter	NETW.TH	**Fractionally-Spaced-Filter**	
fractional number	MATH	**Bruchzahl**	
= fractional (n.)		≈ Bruch; Quotient 2; gebrochene Zahl	
≈ fraction; quotient			
fractional part (→fraction)	COLLOQ	(→Bruchteil)	
fractional power	MATH	**Bruchpotenz**	
fractionation (→subdivision)	TECH	(→Unterteilung)	
fraction bar (→fraction line)	MATH	(→Bruchstrich)	
fraction detector	SIGN.ENG	**Bruchmelder**	
fraction line	MATH	**Bruchstrich**	
= fraction bar; fraction stroke; fraction symbol; division line; division bar; division stroke; division symbol			
fraction stroke (→fraction line)	MATH	(→Bruchstrich)	
fraction symbol (→fraction line)	MATH	(→Bruchstrich)	
fragile	TECH	**zerbrechlich**	
= frangible; breakable; frail		≈ unempfindlich; spröde	
≈ brittle		≠ unzerbrechlich	
≠ unbreakable			
fragility	TECH	**Zerbrechlichkeit**	
= frangibility			
fragility (→sensitivity)	NETW.TH	(→Empfindlichkeit)	
fragment (v.t.)	TECH	**zerstückeln**	
= fragmentize; cleave		≈ vereinzeln	
≈ dismember			
fragmentation	TECH	**Zerstückelung**	
= cleavage			
fragmentize (→fragment)	TECH	(→zerstückeln)	
frail (→fragile)	TECH	(→zerbrechlich)	
frame	TERM&PER	**Bandsprosse**	
[place demand of a character on a magnetic tape]		[Platz den ein Zeichen auf einem Magnetband einnimmt]	
frame	MECH	**Rahmen**	
frame	CODING	**Rahmen**	
= pulse frame		= Pulsrahmen	
frame	TV	**Vollbild**	
↓ subframe; field		↓ Teilbild; Halbbild	
frame	DOC	**Kästchen**	
= box			
frame (→data transmission block)	DATA COMM	(→Datenübertragungsblock)	
frame (→case)	TECH	(→Gehäuse)	
frame (→rack)	EQUIP.ENG	(→Gestell)	
frame (→page frame)	DATA PROC	(→Seite)	
frame address	CODING	**Rahmenadresse**	
frame agreement	ECON	**Rahmenabkommen**	
		= Rahmenvereinbarung	
frame alignment	CODING	**Rahmengleichlauf**	
= frame synchronization; frame synchronism		= Rahmensynchronismus; Rahmensynchronisierung; Pulsrahmengleichlauf; Rahmensynchronisation	
frame alignment loss	CODING	**Rahmengleichlaufstörung**	
= frame synchronization loss			
frame alignment signal	CODING	**Rahmenkennwort**	
= FAS; frame alignment word		= FAS; Rahmenkennungssignal; Synchronwort; Rahmenkennwort	
frame alignment word (→frame alignment signal)	CODING	(→Rahmenkennungswort)	
frame analyzer	INSTR	**Rahmenanalysator**	
frame antenna	ANT	**Rahmenantenne**	
= loop antenna; loop aerial; coil antenna; coil aereal		= Ringantenne 2; Schleifenantenne	
↑ direction finding antenna		↑ Peilantenne	
frame bit	CODING	**Rahmenbit**	
frame blanking (→frame suppression)	TV	(→Bildaustastung)	
frame buffer (→display store)	TERM&PER	(→Bildspeicher)	
frame buffer (→graphics memory)	DATA PROC	(→Graphikspeicher)	
frame buffer (→refresh memory)	DATA PROC	(→Bildwiederholspeicher)	
frame buffer store	TRANS	**Rahmenpufferspeicher**	
frame check sequence (→block check sequence)	DATA COMM	(→Blockprüfzeichenfolge)	
frame clock	CODING	**Rahmentakt**	
		= Rahmensynchrontakt; Pulsrahmentakt; Pulsrahmensynchrontakt	
frame clock pulse (→frame synchronization pulse)	CODING	(→Rahmensynchrimpuls)	
frame colour	TERM&PER	**Rahmenfarbe**	
frame connector (→connector strip)	COMPON	(→Steckerleiste)	
frame duration	TV	**Bilddauer**	
		= Vollbilddauer	
frame equipping (→module frame packaging)	EQUIP.ENG	(→Baugruppenträgerbestückung)	
frame flyback (→vertical flyback)	TV	(→Vertikalrücklauf)	
frame frequency	TV	**Bildfolgefrequenz**	
= picture frequency; video frequency; frame repetition rate		= Bildfrequenz 2; Bildwiederholfrequenz; Bildwiederholrate; Vollbildfrequenz	
≈ field frequency		≈ Teilbildfrequenz	
frame frequency	CODING	**Rahmenfrequenz**	
		= Pulsrahmenfrequenz	
frame generation	CODING	**Rahmenbildung**	
= framing		= Rahmenerzeugung	

frame grabber		DATA PROC	**Bildfangschaltung**	**franking machine**	OFFICE	**Frankiermaschine**
= video grabber; grabber				**Franklin antenna**	ANT	(→Marconi-Franklin-Antenne)
frame grabbing		DATA PROC	**Bildeinfangung**	(→Marconi-Franklin antenna)		
= video grabbing; grabbing				**fraud** (v.t.)	ECON	**betrügen**
frame ground (→chassis ground)		ELECTRON	(→Masse)	**fraud** (n.)	ECON	**Betrug**
				Fraunhofer diffraction	PHYS	**Fraunhofersche Beugung**
frame grounding		EQUIP.ENG	**Gehäuseerde**	**freak**	DATA PROC	**Freak**
= chassis grounding				[an amateur dealing intensively with computing]		[der sich hobbymäßig sehr intensiv mit Computertechnik befaßt]
frame layout (→bay face)		TELEC	(→Gestellansichtsplan)	= computernik		
frame length		CODING	**Rahmenlänge**	≈ computer professional		≈ Computerfachmann
= frame period			= Rahmenperiode; Pulsrahmenlänge; Pulsrahmenperiode	**freak failure** (→random failure)	QUAL	(→Zufallsausfall)
				free 1 (adj.)	COLLOQ	**frei** (adj.)
frame mark bit		CODING	**Rahmenkennungsbit**	≈ available; unrestricted		≈ verfügbar; unbeschränkt
= frame marking bit			= Rahmenmarkierungsbit	**free** (→free-hand)	TECH	(→freihändig)
frame marking		CODING	**Rahmenkennung**	**free 2** (adj.) (→voluntary)	COLLOQ	(→freiwillig)
			= Rahmenmarkierung; Pulsrahmenkennung; Pulsrahmenmarkierung	**free call**	TELEPH	**gebührenfreier Anruf**
						= gebührenfreies Gespräch
frame marking bit (→frame mark bit)		CODING	(→Rahmenkennungsbit)	**free-code** (→without charge)	TELEC	(→gebührenfrei)
				freedom	TECH	**Freiheit**
frame memory (→display store)		TERM&PER	(→Bildspeicher)	**freedom from blocking**	SWITCH	**Blockierungsfreiheit**
				freedom from wear	QUAL	**Verschleißfreiheit**
frame organization (→frame structure)		CODING	(→Rahmenstruktur)	**free electron** (→conduction electron)	PHYS	(→Leitungselektron)
frame packaging (→module frame packaging)		EQUIP.ENG	(→Baugruppenträgerbestückung)	**free fares**	ECON	**Nulltarif**
				free format	DATA PROC	**freies Format**
frame pattern (→frame structure)		CODING	(→Rahmenstruktur)	≠ fixed format		≠ festes Format
				free-form graphic	DATA PROC	**Freiformgraphik**
frame period (→frame length)		CODING	(→Rahmenlänge)	[a graphic not limited to geometrical figures]		[nicht auf geometrische Figuren beschränkte Graphik]
framer dissolution		CODING	**Rahmenabbau**			
			= Pulsrahmenabbau	**free-form surface**	DATA PROC	**Freiformfläche**
frame reject		DATA COMM	**Blockrückweisung**			= Fläche höherer Ordnung
frame repetition rate (→frame frequency)		TV	(→Bildfolgefrequenz)			↓ Spline-Fläche; Bezier-Fläche
frame rotation		TV	**Bilddrehung**	**free-hand** (adj.)	TECH	**freihändig**
= image rotation				= freehand; free; hands off		
frame slip		CODING	**Rahmenschlupf**	**freehand** (→free-hand)	TECH	(→freihändig)
= controlled slip			= Pulsrahmenschlupf	**freehand drawing**	ENG.DRAW	**Bildfangschaltug**
frame storage (→display store)		TERM&PER	(→Bildspeicher)	= freehand illustration; freehand sketch		= Freihandillustration; Freihandskizze
frame structure		CODING	**Rahmenstruktur**	**freehand illustration** (→freehand drawing)	ENG.DRAW	(→Bildfangschaltug)
= frame organization; frame pattern; framing pattern			= Rahmenaufbau; Rahmenmuster; Pulsrahmenstruktur; Pulsrahmenaufbau; Pulsrahmenmuster			
				freehand sketch (→freehand drawing)	ENG.DRAW	(→Bildfangschaltug)
frame suppression		TV	**Bildaustastung**	**free information entropy**	INF	**freie Informationsentropie**
= frame blanking; vertical blanking; vertical interval				**freelance**	ECON	**freiberuflich**
frame synchronism (→frame alignment)		CODING	(→Rahmengleichlauf)	= self-employed		
				freelancer	ECON	**freier Mitarbeiter**
frame synchronization (→frame alignment)		CODING	(→Rahmengleichlauf)	**free line** (→idle state)	SWITCH	(→Ruhezustand)
				freely (adv.)	COLLOQ	**frei** (adv.)
frame synchronization loss (→frame alignment loss)		CODING	(→Rahmengleichlaufstörung)	**free of charge**	ECON	**kostenlos**
				= gratis; without costs		= gratis; kostenfrei; unentgeltlich
frame synchronization pulse		CODING	**Rahmensynchronimpuls**	**free of charge** (→without charge)	TELEC	(→gebührenfrei)
= frame clock pulse			= Rahmentaktpuls	**free of interest**	ECON	**zinslos**
framework		TECH	**Gerüst**	**free of scales**	TECH	**zunderfrei**
≈ skeleton			≈ Gerippe	**freephone** (→handsfree talking)	TELEPH	(→Freisprechen)
framework		SCIE	**Gerüst**			
[fig.]			[fig.]	**free plug** (BRI) (→cable plug)	COMPON	(→Kabelstecker)
= basic structure			= Grundstruktur			
framing		TECH	**Umrandung**	**free radiowave** (→sky wave)	RADIO PROP	(→Raumwelle)
			= Umrahmung			
framing (→frame generation)		CODING	(→Rahmenbildung)	**free run** (→free running)	TECH	(→Freilauf)
				free running (n.)	TECH	**Freilauf**
framing bit sequence		DATA COMM	**Synchronisierbitmuster**	= free run; free wheeling; holdover		
= flag sequence			[grenzt Beginn und Ende eines Rahmens ab]	**free-running** (adj.)	TECH	**freilaufend**
				= free-wheeling; overrunning		
framing pattern (→frame structure)		CODING	(→Rahmenstruktur)	**free-running**	ELECTRON	**freischwingend**
				free-running (→self-oscillating)	ELECTRON	(→selbstschwingend)
franchise		ECON	**Exklusivrecht**			
= exclusivness; exclusivity			↓ Alleinverkaufsrecht	**free-running capability** (→holdover capability)	ELECTRON	(→Freilaufvermögen)
↓ exclusive sale franchise						
franchised dealer		ECON	**Vertragshändler**	**free-running mode**	ELECTRON	**Freilaufbetrieb**
francium		CHEM	**Francium**	= holdover mode		
= Fr			= Fr	**free-running multivibrator** (→astable multivibrator)	CIRC.ENG	(→astabile Kippschaltung)
frangibility (→fragility)		TECH	(→Zerbrechlichkeit)			
frangible (→fragile)		TECH	(→zerbrechlich)	**free-running oscillator**	CIRC.ENG	**freischwingender Oszillator**

free sample	ECON	**Werbemuster**
free socket (BRI) (→cable jack)	COMPON	(→Kabelbuchse)
free space	PHYS	**Freiraum**
free-space attenuation	RADIO PROP	**Freiraumdämpfung**
= free-space loss; basic transmission loss		= Grundübertragungsdämpfung
free-space impedance	RADIO PROP	**Feldwellenwiderstand**
= intrinsic impedance		
free-space loss	SAT.COMM	**Freiraumdämpfung**
= spreading loss		
free-space loss (→free-space attenuation)	RADIO PROP	(→Freiraumdämpfung)
free-space permeability	PHYS	**Induktionskonstante**
free-space propagation	RADIO PROP	**Freiraumausbreitung**
= free-space transmission		
free-space transmission (→free-space propagation)	RADIO PROP	(→Freiraumausbreitung)
free-standing (→self-supporting)	TECH	(→selbsttragend)
free vibration	PHYS	**freie Schwingung**
freeware (→public-domain software)	DATA PROC	(→Public-domain-Software)
free wave (→sky wave)	RADIO PROP	(→Raumwelle)
freeway	CIV.ENG	**Autobahn**
[with limited access, separated tracks and no intersections]		[kreuzungsfrei, mit beschränktem Zutritt und abgegrenzten Fahrbahnen]
= motorway (BRI); superhighway; autobahn		↑ Schnellstraße
↑ expressway		
free-wheeling (→free-running)	TECH	(→freilaufend)
free wheeling (→free running)	TECH	(→Freilauf)
free-wheeling diode	ELECTRON	**Freilaufdiode**
		= Löschdiode
		↑ Schutzdiode
free-wheeling path	CIRC.ENG	**Freilaufzweig**
free-wheeling rectifier	POWER SYS	**Nullanode**
= free-wheeling valve		= Nullventil; Freilaufventil
free-wheeling thyristor	POWER SYS	**Freilaufthyristor**
free-wheeling valve (→free-wheeling rectifier)	POWER SYS	(→Nullanode)
freeze (v.i.)	PHYS	**gefrieren**
freeze-frame system (→fixed-image communication)	TELEC	(→Festbildkommunikation)
freezing point	PHYS	**Gefrierpunkt**
freezing process (→frost 1)	PHYS	(→Gefrierung)
F-region (→F-layer)	RADIO PROP	(→F-Schicht)
freight (v.t.)	ECON	**verfrachten**
= dispatch; send off		= versenden
↓ ship (v.t.)		↓ verschiffen
freight 2 (n.)	ECON	**Fracht 1**
= cargo; carriage		= Frachtgut; Frachtladung
freight 1 (n.)	ECON	**Frachtkosten**
= carriage; freight charges; freightage (AM); transport costs; transportation costs; forwarding charges		= Fracht 2; Transportkosten
≈ freight rate		≈ Frachtarif
freightage (AM) (→freight 1)	ECON	(→Frachtkosten)
freight charges (→freight 1)	ECON	(→Frachtkosten)
French curve	ENG.DRAW	**Kurvenlineal**
= irregular curve; curve templet		= Kurvenschablone
≈ drawing templet		≈ Zeichenschablone
french keyboard	TERM&PER	**französische Tastatur**
= AZERTY keyboard		= AZERTY-Tastatur
Frenkel defect	PHYS	**Frenkel-Fehlstelle**
		= Frenkel-Defekt
frequency	PHYS	**Frequenz**
[oscillations in a unit of time; SI unit: Hertz]		[Schwingungszahl pro Zeiteinheit; SI-Maßeinheit: Hertz]
= f		= f
frequency	MATH	**Häufigkeit**
frequency accuracy	ELECTRON	**Frequenzgenauigkeit**
frequency administration	RADIO	**Frequenzverwaltung**
frequency agile	MIL.COMM	**frequenzagil**
frequency-agile signal simulator	INSTR	**frequenzagiler Signalsimulator**
frequency agility	MIL.COMM	**Frequenzagilität**
frequency allocation scheme [FDM]	MODUL	**Frequenzschema** [Trägerfrequenztechnik] = Frequenzplan
frequency analysis	TELEC	**Frequenzanalyse** [Signaltheorie]
= wave analysis		
≠ frecuency synthesis		≠ Frequenzsynthese
frequency analyzer (→spectrum analyzer)	INSTR	(→Spektrumanalysator)
frequency assignment	RADIO	**Frequenzzuteilung** = Frequenzzuweisung
frequency band	PHYS	**Frequenzband**
frequency band	RADIO	**Frequenzband**
= frequency range; waveband; wavelength band; band; wave range; radio frequency band ; RF band; radio frequency range 2; RF range 2; range		= Frequenzbereich; Wellenbereich; Band; Radiofrequenzbereich 2; RF-Bereich 2; RF-Band; Bereich
frequency band analysis	INSTR	**Frequenzbandanalyse**
frequency bar chart (→histogram)	MATH	(→Histogramm)
frequency bridge (→frequency measuring bridge)	INSTR	(→Frequenzmeßbrücke)
frequency calibration	INSTR	**Frequenzeichung**
frequency change	TELEC	**Frequenzwechsel**
frequency changer (→frequency converter)	CIRC.ENG	(→Frequenzumsetzer)
frequency characteristic	TELEC	**Frequenzkennlinie**
= Bode diagram		
frequency clock	DATA PROC	**Netzuhr**
frequency-code modulation	TELEGR	**Frequenzcodemodulation**
frequency compensation	CIRC.ENG	**Frequenzkompensation** = Frequenzausgleich
frequency constancy (→frequency stability)	PHYS	(→Frequenzstabilität)
frequency control	CIRC.ENG	**Frequenznachsteuerung**
= frequency tuning		= Frequenznachstellung; Frequenznachstimmung
frequency control	POWER SYS	**Frequenzregelung**
		↑ Netzregelung
frequency control pilot [FDM]	TRANS	**Frequenzvergleichspilot** [TF-Technik]
frequency conversion	MODUL	**Frequenzumsetzung**
= frequency translation		
frequency converter	CIRC.ENG	**Frequenzumsetzer**
= frequency translator; frequency changer		= Frequenzumrichter; Frequenzwandler; Frequenzumformer
frequency converter	POWER SYS	**Frequenzumrichter**
= frequency convertor		↑ Stromrichter; Umrichter
↑ static power converter; voltage system converter		
frequency convertor (→frequency converter)	POWER SYS	(→Frequenzumrichter)
frequency counter	INSTR	**Frequenzzähler**
= frequency meter		= Frequenzmesser
frequency coverage	INSTR	**Frequenzabdeckung** = Frequenzbereich
frequency curve	INSTR	**Frequenzablauf**
frequency curve	MATH	**Häufigkeitskurve**
frequency decade	INSTR	**Frequenzdekade**
frequency departure (→frequency offset)	EL.TECH	(→Frequenzversatz)
frequency dependence	PHYS	**Frequenzabhängigkeit**
frequency dependence (→frequency response)	NETW.TH	(→Frequenzgang)
frequency dependent	PHYS	**frequenzabhängig**
frequency deviation (→frequency shift)	MODUL	(→Frequenzhub)
frequency deviation (→frequency offset)	EL.TECH	(→Frequenzversatz)
frequency diplexer	RADIO REL	**Frequenzweiche**
frequency discriminator	MODUL	**Frequenzdiskriminator**
frequency distribution	MATH	**Häufigkeitsverteilung**
↑ distribution		↑ Verteilung
frequency diversity	RADIO	**Frequenzdiversity**
frequency divider	CIRC.ENG	**Frequenzteiler**
↑ divider		= Frequenzdividierer; Frequenzuntersetzer
		↑ Teiler
frequency-division mode	TRANS	**Frequenzgetrenntlage-Verfahren**

frequency division multiple access	TELEC	Frequenzvielfachzugriff	
= FDMA		= Vielfachzugriff in der Frequenzebene; FDMA	
frequency division multiplex	TELEC	Frequenzmultiplex	
= FDM		= Frequenzvielfach	
		≈ Trägerfrequenztechnik	
frequency domain	TELEC	Frequenzbereich	
≠ time domain		≠ Zeitbereich	
frequency domain analysis	TELEC	Frequenzbereichdarstellung	
frequency-domain equalizer	NETW.TH	Frequenzbereichsentzerrer	
= FDE			
frequency doubler	CIRC.ENG	Frequenzverdoppler	
↑ frequency multiplier; doubler circuit		= Frequenzdoppler	
		↑ Frequenzvervielfacher; Verdopplerschaltung	
frequency drift	ELECTRON	Frequenzabwanderung	
= frequency sliding		= Frequenzinkonstanz	
frequency economy	RADIO	Frequenzökonomie	
		↓ Frequenzbandökonomie	
frequency entrainment	ELECTRON	Frequenzmitnahme	
frequency error	PHYS	Frequenzfehler	
≈ frequency offset		= Frequenzunsicherheit	
		≈ Frequenzversatz	
frequency extension	INSTR	Frequenzbereichserweiterung	
frequency -fixing	ELECTRON	frequenzbestimmend	
frequency-flat (adj.)	ELECTRON	frequenzgerade	
= flat			
frequency fluctuation	PHYS	Frequenzschwankung	
frequency frogging	TRANS	Bandtausch	
[FDM]		[TF-Technik]	
= frogging		= Frequenztausch	
frequency function	MATH	Häufigkeitsfunktion	
		= Frequenzfunktion	
frequency generator	INSTR	Frequenzgenerator	
frequency hop (→frequency jump)	PHYS	(→Frequenzsprung)	
frequency hopping	RADIO	Frequenzsprungverfahren	
frequency-independent	PHYS	frequenzunabhängig	
frequency indication	INSTR	Frequenzanzeige	
frequency-indipendent antenna	ANT	frequenzunabhängige Antenne	
↑ broadband antenna		↑ Breitbandantenne	
↓ logarithmically periodic antenna; spiral antenna		↓ logarithmisch-periodische Antenne; Spiralantenne	
frequency interlace	TELEC	Frequenzverschachtelung	
frequency interlace (→frequency interlacing)	TV	(→Spektralverkämmung)	
frequency interlacing	TV	Spektralverkämmung	
= frequency interlace; frequency interleaving			
frequency interleaving (→frequency interlacing)	TV	(→Spektralverkämmung)	
frequency jump	PHYS	Frequenzsprung	
= frequency hop			
frequency keying (→frequency shift keying)	MODUL	(→Frequenzumtastung)	
frequency linearity	ELECTRON	Frequenzlinearität	
frequency locking	ELECTRON	Frequenzeinrastung	
frequency management	RADIO	Funkverwaltung	
		≈ Funküberwachung	
frequency mark	INSTR	Frequenzmarke	
frequency measurement	INSTR	Frequenzmessung	
frequency measuring bridge	INSTR	Frequenzmeßbrücke	
= frequency bridge		= Frequenzbrücke	
frequency meter (→frequency counter)	INSTR	(→Frequenzzähler)	
frequency method	INSTR	Frequenzverfahren	
frequency modulated	MODUL	frequenzmoduliert	
frequency modulated continuous wave radar	RADIO LOC	frequenzmodulierter Dauerstrichradar	
= FM-CW-Radar		= FM-CW-Radar	
frequency modulation	MODUL	Frequenzmodulation	
= FM		= FM	
		↑ Winkelmodulation	
frequency modulation feedback	MODUL	Frequenz-Gegenkopplungs-Demodulation	
= FMFB		= FMFB	
frequency-modulation with a noise carrier	MODUL	Rauschfrequenz-Modulation	
frequency modulator	MODUL	Frequenzmodulator	
frequency monitor	RADIO	Frequenzmonitor	
frequency multiplication (→harmonic generation)	EL.TECH	(→Frequenzvervielfachung)	
frequency multiplier	CIRC.ENG	Frequenzvervielfacher	
↑ multiplier		↑ Vervielfacher	
frequency normalization	NETW.TH	Frequenznormierung	
frequency offset	EL.TECH	Frequenzversatz	
= frequency deviation; frequency departure; frequency shift		= Frequenzablage; Frequenzverschiebung; Frequenzabweichung; Frequenzverwerfung	
≈ frequency error; frequency recovery offset		≈ Frequenzfehler; Frequenzrückgewinnungsfehler	
frequency pattern	RADIO	(→Radiofrequenzraster)	
(→radio-frequency pattern)			
frequency planning	RADIO	Frequenzplanung	
frequency polygon		Häufigkeitspolygon [MATH]	
frequency profiling	INSTR	Frequenzformung	
frequency range (→frequency band)	RADIO	(→Frequenzband)	
frequency ratio (→relative frequency)	MATH	(→relative Häufigkeit)	
frequency recovery offset	TELEC	Frequenzrückgewinnungsfehler	
≈ frequency offset		≈ Frequenzversatz	
frequency response	NETW.TH	Frequenzgang	
= frequency dependence; level flatness [INSTR]; flatness [INSTR]		= Frequenzabhängigkeit	
frequency response of amplitude	TELEC	Amplitudenfrequenzgang	
= amplitude response		= Amplitudengang	
frequency-response synthesis	INSTR	Frequenzgangsynthese	
frequency reuse	RADIO	Frequenzwiederbenutzung	
↓ cochannel operation [RADIO REL]		= Frequenzwiederverwendung; Frequenzmehrfachnutzung	
		↓ Gleichkanalbetrieb [RADIO REL]	
frequency reuse (→cochannel operation)	RADIO REL	(→Gleichkanalbetrieb)	
frequency-selective voltmeter (→selective voltmeter)	INSTR	(→Selektivspannungsmesser)	
frequency-sensitive	PHYS	frequenzempfindlich	
frequency sensitivity	PHYS	Frequenzempfindlichkeit	
frequency separating filter	TELEGR	Frequenzweiche	
frequency set	RADIO	Frequenzvorrat	
frequency settability	INSTR	Frequenzeinstellbarkeit	
frequency setting	ELECTRON	Frequenzeinstellung	
frequency sharing	RADIO	Frequenzbandmitbenutzung	
frequency shift	MODUL	Frequenzhub	
= frequency swing; frequency deviation; frequency sweep			
frequency shift (→frequency offset)	EL.TECH	(→Frequenzversatz)	
frequency-shifted	EL.TECH	frequenzversetzt	
frequency shift keying	MODUL	Frequenzumtastung	
= FSK; frequency shift modulation; carrier shift; frequency keying		= Frequenztastung; FSK; Trägerverschiebung	
frequency shift modulation (→frequency shift keying)	MODUL	(→Frequenzumtastung)	
frequency sliding (→frequency drift)	ELECTRON	(→Frequenzabwanderung)	
frequency spacing	RADIO	Frequenzabstand	
frequency span (→sweep span)	ELECTRON	(→Wobbelbandbreite)	
frequency spectrum	PHYS	Frequenzspektrum	
frequency stability	PHYS	Frequenzstabilität	
= frequency constancy		= Frequenzkonstanz	
frequency-stabilize	PHYS	frequenzstabilisieren	
frequency standard	INSTR	Frequenznormal	
		= Frequenzstandard	
frequency step	MODUL	Frequenzschritt	
		= Frequenzstufe	
frequency-stepped sweep	INSTR	frequenzgestufte Wobbelung	
frequency sweep	INSTR	Frequenzwobbelung	
frequency sweep (→frequency shift)	MODUL	(→Frequenzhub)	
frequency swing (→frequency shift)	MODUL	(→Frequenzhub)

frequency synthesis	TELEC	Frequenzsynthese [Signaltheorie]	front coupler	OPTOEL	Stirnflächenkoppler
= wave synthesis			front cover (→cover sheet)	TYPOGR	(→Deckblatt)
≠ frequency analysis		≠ Frequenzanalyse	front cover (→front panel)	EQUIP.ENG	(→Frontabdeckung)
frequency synthesizer	CIRC.ENG	Frequenzsynthesizer	front-end (adj.)	TELEC	vorgezogen
= synthesizer		= Synthesizer			= vorgelagert; Vorfeld-
frequency table		Häufigkeitstabelle [MATH]	front-end (→front-end processor)	DATA COMM	(→Vorrechner)
frequency thyristor	COMPON	Frequenzthyristor	front-end computer	DATA COMM	(→Vorrechner)
		= T-Thyristor	(→front-end processor)		
frequency-time-domain transformation	TELEC	Frequenz-Zeit-Transformation	front-end development	TECH	Vorfeldentwicklung
frequency tolerance	PHYS	Frequenztoleranz	= advanced development		
frequency-to-voltage converter	COMPON	Frequenz-Spannungs-Wandler	front-end equipment [equipment between subscriber and switch, to economize lines]	TELEC	Vorfeldeinrichtung [Einrichtung zwischen Teilnehmer und Vermittlung, zur Einsparung von Leitungen]
frequency transformation	NETW.TH	Frequenztransformation	↓ remote equipment; pair-saving device		
frequency translation (→frequency conversion)	MODUL	(→Frequenzumsetzung)	↓ stand-alone concentrator; remote concentrator		↓ Wählsterneinrichtung; Konzentrator
frequency translator (→frequency converter)	CIRC.ENG	(→Frequenzumsetzer)	front-end processor [performs central processing tasks in a computer network and controls host computers]	DATA COMM	Vorrechner [nimmt in einem Mehrrechnersystem zentrale Abwicklungsaufgaben wahr und steuert Verarbeitungsrechner aus]
frequency transposition (→transposition)	SAT.COMM	(→Transponierung)			
frequency tripler	CIRC.ENG	Frequenzverdreifacher			
↑ frequency multiplier		↑ Frequnzvervielfacher	= FEP; front-end computer; front-end		
frequency tuning (→frequency control)	CIRC.ENG	(→Frequenznachsteuerung)	≈ preprocessor 1 [DATA PROC]; slave computer		≈ Vorverarbeitungsrechner [DATA PROC]; Nebenrechner
frequency vernier	INSTR	Frequenzfeineinstellung	≠ host computer		
frequent (adj.)	COLLOQ	häufig (adj.)	↓ communications computer; communications front-end processor		≠ Verarbeitungsrechner
frequently (adv.)	COLLOQ	häufig (adv.)			↓ Kommunikationsrechner; Datenübertragungsvorrechner
= often; oftentimes; offtimes					
Fresnel diffraction	PHYS	Fresnelsche Beugung			
Fresnel ellipsoid	RADIO PROP	Fresnelellipsoid			
Fresnel lens antenna	ANT	Fresnel-Linsen-Antennen	front-end processor (→terminal computer)	DATA COMM	(→Datenstationsrechner)
Fresnel zone	RADIO PROP	Fresnelzone			
friction (v.t.)	PHYS	reiben	front-end processor (→preprocessor 1)	DATA PROC	(→Vorverarbeitungsrechner)
friction (n.)	PHYS	Reibung	front-end technique	OUTS.PLANT	Vorfeldtechnik
↓ frictional grip; sliding friction; rolling friction		= Friktion ↓ Haftreibung; Gleitreibung; Rollreibung	front-facing connector	EQUIP.ENG	Frontstecker
			front feed	TERM&PER	Fronteinzug
frictional electrical machine	PHYS	Reibungselektrisiermaschine	front handle	EQUIP.ENG	Frontgriff
frictional grip	PHYS	Haftreibung	= front-panel handle		
friction clutch	MECH	Reibungskupplung	frontispiece (→cover 2)	TYPOGR	(→Titelbild)
friction coefficient	PHYS	Reibungszahl	front lobe (→mayor lobe)	ANT	(→Hauptkeule)
= friction factor		= Reibungskoeffizient	front loops	MICROEL	Stirnschleifen
friction damping	PHYS	Reibungsdämpfung	front mounting	EQUIP.ENG	Frontmontage
friction electricity (→static electricity)	PHYS	(→Reibungselektrizität)	≠ rear mounting		≠ Rückseitenmontage
			front page (→title page)	TYPOGR	(→Titelseite)
friction factor (→friction coefficient)	PHYS	(→Reibungszahl)	front panel	EQUIP.ENG	Frontabdeckung
			= front cover; front plate; face plate; fascia plate [DATA PROC]		= Frontplatte; Frontblende ≠ rückseitige Abdeckung
friction feed	TERM&PER	Friktionsantrieb	≠ rear panel		↑ Abdeckung
≠ tractor feed		≠ Raupenvorschub	↑ panel		↓ Baugruppenabdeckung
↑ paper feed		↑ Papiervorschub	↓ module front panel		
friction-locked	TECH	kraftschlüßig	front-panel connection	EQUIP.ENG	Frontplattenanschluß
= solid			front-panel handle (→front handle)	EQUIP.ENG	(→Frontgriff)
friction welding	METAL	Reibschweißen			
fringe area	SCIE	Randgebiet	front part	TECH	Vorderteil
fritting (→wetting 1)	COMPON	(→Frittung)	≠ rear part		≠ Hinterteil
fritting resistance (→wetting resistance)	COMPON	(→Frittwiderstand)	front plate (→front panel)	EQUIP.ENG	(→Frontabdeckung)
fritting voltage (→wetting voltage)	COMPON	(→Frittspannung)	front porch	TV	vordere Schwarzschulter
			front scanning	ELECTRON	Vorderabtastung
frob (v.t.) [to fiddle with a picking device]	DATA PROC	herumfahren [mit einem Eingabegerät]	front side	TECH	Vorderseite
			= front		= Frontseite
frogging (→frequency frogging)	TRANS	(→Bandtausch)	≠ rear side		≠ Rückseite
			front surface (→face)	MECH	(→Stirnfläche)
frogging repeater	TRANS	Zweidraht-Zwischenverstärker mit Bandtausch	front-to-back ratio	ANT	Rückstrahldämpfung
= low-high repeater			= F/B; front-to-rear ratio; side lobe level		= Vor-Rück-Verhältnis; Antennenrückdämpfung; Rückdämpfung
FROM 1 [Factory ROM; can be reprogrammed only in factory]	MICROEL	FROM 1 [nur im Werk veränderbarer Festwertspeicher]	≈ side -lobe level		≈ Nebenkeulen-Richtfaktor
↑ ROM		↑ Festwertspeicher	front-to-rear ratio (→front-to-back ratio)	ANT	(→Rückstrahldämpfung)
FROM 2 (→fusible-link PROM)	MICROEL	(→Fusible-link-PROM)			
			front-to-side ratio	ANT	Vor-Seit-Verhältnis
front (→frontal)	TECH	(→frontseitig)	front view	SYS.INST	Ansichtsplan
front (→front side)	TECH	(→Vorderseite)			= AP
frontal (adj.)	TECH	frontseitig	front view	ENG.DRAW	Vorderansicht
= front;		= frontal; vorderer			= Hauptansicht
≠ rear		≠ rückseitig	frost 2 (n.)	METEOR	Frosttemperatur
frontal connection	EQUIP.ENG	Frontanschluß	= frost temperature		
frontal mounting	EQUIP.ENG	Fronteinbau			
frontal wave	PHYS	Frontalwelle			

frost 1 (n.) PHYS **Gefrierung**
= freezing process = Gefriervorgang
frost 3 METEOR **Reif**
= frozen fog [gefrorener Niederschlag
↓ hoarfrost aus Luftfeuchte]
 ↓ Rauhreif
frosted glass TECH **Mattglas**
= milky glass; opalescent glass; opal = Milchglas
  glass
frosted plate PHYS **Mattscheibe**
= ground glass
frost temperature METEOR (→Frosttemperatur)
(→frost 2)
frown (v.t.) COLLOQ **verpönen**
= frown upon
frown upon (→frown) COLLOQ (→verpönen)
frozen fog (→frost 3) METEOR (→Reif)
fruit (n.) RADIO LOC **nichtsynchrone Empfangsstö-**
[a grapefruit-shaped radar display] **rung**
frustum (pl. -ums, -a) MATH **Stumpf**
[cone-shaped solid cut parallel to [parallel zur Grundfläche
base] abgeschnittener konusför-
 miger Körper]
 = Kegelstumpf
fry (v.t.) ELECTRON **durchbrennen**
frying (→contact noise) EL.ACOUS (→Kratzgeräusch)
frying (BRI) (→transmitter TELEPH (→Mikrophongeräusch)
  noise)
frying noise (→contact EL.ACOUS (→Kratzgeräusch)
  noise)
FS (→file separator) DATA COMM (→Hauptgruppentrennung)
fs (→femtosecond) PHYS (→Femtosekunde)
FSK (→frequency shift MODUL (→Frequenzumtastung)
  keying)
FSK analyzer INSTR **Telegrafieanalysator**
 = Telegraphieanalysator
FTH (→fiber to the home) TELEC (→Faser bis zum Teilnehmer)
fuel (→combustible) TECH (→Brennstoff)
fuel cell (→hydrogen fuel POWER SYS (→Brennstoffzelle)
  cell)
fuel filter TECH **Kraftstoffilter**
fuel service tank (→fuel TECH (→Brennstoffbehälter)
  tank)
fuel tank TECH **Brennstoffbehälter**
= fuel service tank = Treibstoffbehälter; Kraft-
 stoffbehälter; Brennstoff-
 tank; Treibstofftank;
 Kraftstofftank
full (adj.) TECH **voll** (adj.)
= plenished = gefüllt
full accessibility SWITCH **volle Erreichbarkeit**
= full availability = volle Verfügbarkeit
full adder CIRC.ENG **Volladdierer**
[considers carries] [berücksichtigt Überträge]
= three input adder ≠ Halbaddierer
≠ half adder
full area coverage (→global TELEC (→Flächendeckung)
  coverage)
full automation TECH **Vollautomatisierung**
full availability (→full SWITCH (→volle Erreichbarkeit)
  accessibility)
full-bore version TECH **voll ausgebaute Version**
= maximum-configuration version; ≠ abgemagerte Version
  fully expanded version
≠ crippled version
full capacity stage (→maximum TECH (→Endausbau)
  capacity)
full-cost pricing ECON **Vollkostenrechnung**
 = Vollkostenkalkulation
full costs ECON **Vollkosten**
full-coverage (adj.) TELEC **flächendeckend**
= global-coverage ≈ landesweit
full coverage (→global TELEC (→Flächendeckung)
  coverage)
full custom MICROEL **voll kundesspezifisch**
full-custom design MICROEL **voll kundenspezifischer Ent-**
 wurf
full custom IC MICROEL **Vollkundenschaltung**
≈ application-specific IC = voll kundenspezifischer
↑ custom IC Schaltkreis; voll kundenspezifische integrierte Schal-
 tung; voll kundenspezifischer IC; Full-custom
 ≈ anwendungsspezifische
 integrierte Schaltung
 ↑ kundenspezifische integrierte Schaltung
full discharge POWER SYS **Vollentladung**
[accumulator] [Akkumulator]
full duplex (→duplex TELEC (→Duplexbetrieb)
  operation)
full duplex data communi- DATA COMM (→beidseitige Datenüber-
cation (→two-way simultaneous com- mittlung)
  munication)
full duplex mode (→duplex TELEC (→Duplexbetrieb)
  operation)
full duplex operation (→duplex TELEC (→Duplexbetrieb)
  operation)
full-duplex speech (→duplex TELEPH (→Gegensprechbetrieb)
  telephony)
full-duplex telegraphy (→duplex TELEGR (→Gegenschreibbetrieb)
  telegraphy)
full duty (→full load) TECH (→Vollast)
full echo suppressor TELEPH **Vollechosperre**
full expansion (→maximum TECH (→Endausbau)
  capacity)
full-frame display DATA PROC **Vollformatanzeige**
= full-screen display = Vollbild; Vollschirmdarstel-
 lung
full insulation (→solid COMM.CABLE (→Vollisolierung)
  insulation)
full-line catalog (NAM) TECH **Gesamtkatalog**
= full-line catalogue (BRI); general
  catalog (NAM); general catalogue
  (BRI)
full-line catalogue (BRI) TECH (→Gesamtkatalog)
(→full-line catalog)
full load TECH **Vollast**
= full duty
full-motion picture (→moved TELEC (→Bewegtbild)
  picture)
full-motion video (→moving TELEC (→Bewegtbild-Kommunika-
  picture communication) tion)
full output POWER ENG **Vollaussteuerung**
full-page DOC **ganzseitig**
full-page display DATA PROC **Ganzseitendarstellung**
$[21 \times 28 \text{ cm}^2]$ $[21 \times 28 \text{ cm}^2]$
full-page graphic TERM&PER **Ganzseitengraphik**
full-production status (→maturity MANUF (→Serienreife)
  for series production)
full-range tunable INSTR **durchstimmbar**
full rate TELEC **volle Gebühr**
full scale INSTR **Vollausschlag**
≈ end-scale value = Gesamtmeßbereich
 ≈ Skalenendwert
full-scale sensitivity INSTR **Endempfindlichkeit**
full scale setting INSTR **Skalenendwert-Einstellung**
 = Bereichsendwert-Einstel-
 lung
full-scale value (→end-scale INSTR (→Skalenendwert)
  value)
full-screen display DATA PROC (→Vollformatanzeige)
(→full-frame display)
full-screen editing DATA PROC **Vollschirmeditierung**
full screen monitor TERM&PER **Ganzseiten-Bildschirm**
[for a full page] [z.B. für DIN A4]
= full-size display = Ganzseitenmonitor
full section ENG.DRAW **Halbschnitt**
full size ENG.DRAW **Maßstab 1:1**
= scale 1:1 = natürlicher Maßstab
full-size display (→full screen TERM&PER (→Ganzseiten-Bildschirm)
  monitor)
full-slice technology MICROEL **Vollscheibentechnik**
full-span sweep ELECTRON **Totalwobbelung**
full stop (BRI) (→dot) LING (→Punkt 1)
full subtracter CIRC.ENG **Vollsubtrahierer**
= full subtractor ≠ Halbsubtrahierer
≠ half subtracter
full subtractor (→full CIRC.ENG (→Vollsubtrahierer)
  subtracter)

full-text searching	DATA PROC	**Ganztextsuche**		**functional simulation**	MICROEL	**Funktionssimulation**
full-time worker	ECON	**Vollarbeitskraft**		**functional specification**	DOC	(→Funktionsbeschreibung)
full-wave dipole	ANT	**Ganzwellendipol**		(→functional description)		
= lambda dipole		= Lambdadipol		**functional stress**	QUAL	**funktionsbedingte Beanspruchung**
↑ geometrically thick antenna		↑ geometrisch dicke Antenne				
full-wave loop	ANT	**Ganzwellenschleife**		**functional test** (→function test)	TECH	(→Funktionsprüfung)
full-wave rectification	CIRC.ENG	(→Doppelweggleichrichtung)		**functional unit**	TECH	**Funktionseinheit**
(→double-way rectification)				= module		= Modul
full-wave rectifier	CIRC.ENG	(→Doppelweggleichrichter)		**function code**	DATA PROC	**Funktionscode**
(→double-way rectifier)				= function digit		
full-wave thyristor	MICROEL	(→Doppelwegthyristor)		**function digit** (→function code)	DATA PROC	(→Funktionscode)
(→double-way thyristor)						
full-wave zepp antenna (→full-wave Zeppelin antenna)	ANT	(→Ganzwellen-Zeppelinantenne)		**function expansion**	TECH	**Funktionserweiterung**
				function generator	CIRC.ENG	**Funktionsgenerator**
full-wave Zeppelin antenna	ANT	**Ganzwellen-Zeppelinantenne**				= Funktionsgeber; Function Generator
= full-wave zepp antenna		= Ganzwellen-Zepp		**function generator**	INSTR	(→Signalformgenerator)
fully (adv.)	COLLOQ	**völlig** (adv.)		(→waveform generator)		
= totally		= gänzlich		**function key**	TERM&PER	**Funktionstaste**
fully automatic	TECH	**vollautomatisch**		[assignable by program to special function; generally labeled F1, F2,..]		[per Programm bestimmten Funktionen zuordenbar; i.a. mit F1, F2 usf. bezeichnet]
fully decoded selection	DATA PROC	**Auswahl mit vollständiger Decodierung**				
fully electronic	ELECTRON	**vollelektronisch**		= programmable function key; control key 2; soft key		= Steuertaste 2; Programmsteuertaste; Softkey
= all-electronic				≈ control key 1; release key		
fully equipped	EQUIP.ENG	**vollbestückt**		↓ command key		≈ Steuerungstaste; Auslösetaste
fully expanded	TECH	**vollausgebaut**				↓ Kommandotaste
= expanded to full capacity				**function keyboard**	TERM&PER	**Funktionstastenblock**
fully expanded version (→full-bore version)	TECH	(→voll ausgebaute Version)		= functional keyboard		= Funktionstastatur
				function macro	MICROEL	**Funktionsmakro**
fully formed character	TERM&PER	**Volldruckzeichen**		= soft macro		
≠ dot-matrix character		≠ Rasterdruckzeichen		**function network**	CIRC.ENG	**Funktionsnetzwerk**
fully modular	EQUIP.ENG	**vollmodular**		**function of complex variables**	MATH	**komplexwertige Funktion**
fully restricted	TELEPH	**nichtamtsberechtigt**				= Funktion komplexer Variable
[allows only internal calls]		[kann nur intern, nicht mit dem Amt, verbunden werden]		**function-oriented**	TECH	**funktionsorientiert**
				function-related (adj.)	TECH	**funktional**
fully transistorized	ELECTRON	(→transistorisiert)		= functional 1		[Funktion betreffend]
(→transistorized)						= funktionsbezogen
function (v.i.)	TECH	**funktionieren**		**function routine** (→function subprogram)	DATA PROC	(→Funktionsunterprogramm)
= operate; work (v.i.)		= arbeiten				
function (n.)	TECH	**Funktion**		**function sharing**	SWITCH	**Funktionsteilung**
≈ task		≈ Aufgabe				= Funktionsverbund
function (n.)	MATH	**Funktion**		**function simulator**	MICROEL	**Funktionssimulator**
function (→task)	SWITCH	(→Aufgabe)		**function subprogram**	DATA PROC	**Funktionsunterprogramm**
function (→operation)	ENG.LOG	(→Verknüpfung)		= function routine		= Funktionsroutine
functional (n.)	MATH	**Funktional**		**function table**	DATA PROC	**Funktionstafel**
functional 2 (adj.)	TECH	**funktionell**		**function test**	TECH	**Funktionsprüfung**
[performing a function]		[Funktion erfüllend]		= functional test; performance test		= Funktionstest
functional 3	TECH	**funktionsbedingt**		**function test module** (→test board)	EQUIP.ENG	(→Testbaugruppe)
[conditioned by a function]						
functional 1	TECH	(→funktional)		**functor**	ENG.LOG	**Funktor**
(→function-related)				[symbol for truth function]		[Symbolzeichen für Wahrheitsfunktion]
functional bit (→flag)	DATA PROC	(→Merker)				
functional block	MICROEL	**Funktionsblock**		**fundamenatl oscillation**	PHYS	(→Grundschwingung)
functional byte (→status byte)	DATA PROC	(→Zustandsbyte)		(→fundamental wave)		
				fundamental cell (→unit cube)	PHYS	(→Elementarwürfel)
functional description	DOC	**Funktionsbeschreibung**		**fundamental characteristic**	TECH	(→Grundeigenschaft)
= functional specification		= Funktionenbeschreibung		(→fundamental property)		
functional determinant	MATH	**Funktionaldeterminante**		**fundamental crystal**	COMPON	**Grundschwingungsquarz**
functional diagram (→block diagram)	TELEC	(→Blockdiagramm)		**fundamental frequency**	PHYS	**Grundfrequenz**
				= basic frequency		≈ Grundschwingung
functional drawing	ENG.DRAW	**Funktionszeichnung**		≈ fundamental oscillation		≠ Oberfrequenz
functional flow	TECH	**Funktionsablauf**		≠ harmonic frequency		
functionality (→capability)	TECH	(→Leistungsfähigkeit)		**fundamental frequency**	ACOUS	(→Formant)
functional keyboard	TERM&PER	(→Funktionstastenblock)		(→characteristic frequency)		
(→function keyboard)				**fundamental mode**	MICROW	**Grundwelle**
functionally essential	TECH	**funktionswichtig**		= fundamental wave; dominant mode; dominant wave		= Haupttyp; Hauptwelle; Basismodus
functional mock-up	TECH	**Versuchsmuster**				
(→experimental prototype)				**fundamental model** (→basic model)	SCIE	(→Grundmodell)
functional parameter information	TELECONTR	**Funktionsmeldung**				
		= Betriebsmeldung; Funktionsüberwachungsmeldung; Kennmeldung		**fundamental power**	NETW.TH	**Grundschwingungsleistung**
				fundamental property	TECH	**Grundeigenschaft**
functional principle	TECH	**Funktionsprinzip**		= fundamental characteristic; basic property; basic characteristic		= Hauptcharakteristik; Grundcharakteristik
		≈ Funktionsweise				
functional programming language	DATA PROC	**funktionale Programmiersprache**		**fundamental reactive power**	NETW.TH	**Grundschwingungs-Blindleistung**
= functional language		= funktionale Sprache				[Komponente der Blindleistung]
functional sample	TECH	**funktionstüchtiges Muster**				

fundamental research	SCIE	**Grundlagenforschung**	
= basic research		≈ Grundsatzentwicklung [TECH]	
≈ basic R&D [TECH]			
fundamentals (→principles)	SCIE	(→Grundlagen)	
fundamental service (→basic service)	TELEC	(→Grunddienst)	
fundamental tone	EL.ACOUS	**Grundton**	
fundamental unit	PHYS	**Grundgröße**	
fundamental unit (→primitive 1)	DATA PROC	(→Grundelement)	
fundamental wave	PHYS	**Grundschwingung**	
= first harmonic; fundamenatl oscillation; main oscillation; basic oscillation		= Grundharmonische; Grundwelle	
≈ fundamental frequency		≈ Grundfrequenz	
≠ harmonic wave		≠ Oberwelle	
fundamental wave (→fundamental mode)	MICROW	(→Grundwelle)	
fundamental wave content	PHYS	**Grundschwingungsgehalt**	
funds (pl.t.)	ECON	**Mittel**	
= assets		= Geldmittel	
↑ resources		↑ Ressourcen	
funnel	TECH	**Trichter**	
= horn			
funnel antenna (→horn antenna)	ANT	(→Hornantenne)	
funnel mouth	TECH	**Trichteröffnung**	
= horn mouth			
funware	DATA PROC	**Unterhaltungssoftware**	
= game software; game programs		= Funware	
furnace	MANUF	**Lötofen**	
furnace (→stove)	TECH	(→Ofen)	
furnish	ECON	**ausliefern**	
= deliver		= liefern	
furnishing (→equipment 1)	TECH	(→Ausstattung)	
further development	TECH	**Weiterentwicklung**	
further education	SCIE	**Weiterbildung**	
= continuing education		≈ Schulung	
≈ training			
further processing	MANUF	**Weiterverarbeitung**	
= supplementary processing		= Weiterveredelung	
fuse (v.t.)		**durchschmelzen**	
fuse (→fuse-link)	EL.INST	(→Sicherungseinsatz)	
fuse (→melting fuse)	COMPON	(→Schmelzsicherung)	
fuse-base	COMPON	**Sicherungsunterteil** [Sockel der Stromgrobsicherung, mit Anschlußklemmen]	
fuse carrier	COMPON	**Sicherungseinsatzträger**	
fuse cartridge (→cartridge fuse)	COMPON	(→Glasrohr-Feinsicherung)	
fused fiber splice (→fusion splice)	OPT.COMM	(→thermischer Spleiß)	
fuse-disconnector (IEC) (→circuit breaker)	COMPON	(→Schutzschalter)	
fuse element (IEC) (→fusible wire)	COMPON	(→Schmelzdraht)	
fuseholder	COMPON	**Sicherungshalter**	
= fuse-holder			
fuse-holder (→fuseholder)	COMPON	(→Sicherungshalter)	
fuse-isolator (IEC) (→circuit breaker)	COMPON	(→Schutzschalter)	
fuse-link	EL.INST	**Sicherungseinsatz** [der austauschbare Porzellankörper einer Stromgrobsicherung]	
= fuse-unit; fuse			
fuse panel	EQUIP.ENG	**Sicherungsfeld**	
fuse panel	POWER SYS	**Sicherungsschiene**	
fuse strip	COMPON	**Sicherungsstreifen**	
fuse-switch	COMPON	**Sicherungslastschalter**	
fuse-unit (→fuse-link)	EL.INST	(→Sicherungseinsatz)	
fuse wire (→fusible wire)	COMPON	(→Schmelzdraht)	
fusible	TECH	**schmelzbar**	
fusible cut-out (→melting fuse)	COMPON	(→Schmelzsicherung)	
fusible link	MICROEL	**Sicherungselement**	
fusible-link PROM	MICROEL	**Fusible-link-PROM**	
= fusible ROM; FROM 2		= FROM 2	
fusible part (→fusible wire)	COMPON	(→Schmelzdraht)	
fusible ROM (→fusible-link PROM)	MICROEL	(→Fusible-link-PROM)	
fusible wire	COMPON	**Schmelzdraht**	
= fuse wire; fuse element (IEC); fusible part		[Schmelzsicherung]	
fusing point (→melting point)	PHYS	(→Schmelzpunkt)	
fusion	METAL	**Verschmelzung**	
fusion heat	PHYS	**Schmelzwärme**	
= melting heat			
fusion splice	OPT.COMM	**thermischer Spleiß**	
= fused fiber splice		≠ mechanischer Spleiß	
≠ mechanical splice			
fusion welding	METAL	**Schmelzschweißung**	
future-oriented	TECH	**zukunftsorientiert**	
≈ trend setting; progressive; modern		= zukunftsgerichtet	
		≈ zukunftsweisend; fortschrittlich; modern	
future-proof	TECH	**zukunftssicher**	
		= zukunftsreich; zukunftsträchtig	
future tense	LING	**Futur**	
		= Zukunft	
		↓ erstes Futur; zweites Futur	
future trend (→evolutional trend)	TECH	(→Entwicklungstendenz)	
fuze (→melting fuse)	COMPON	(→Schmelzsicherung)	
fuzziness (→uncertainty)	SCIE	(→Unschärfe)	
fuzzy (→blurred)	TV	(→unscharf)	
fuzzy (→jaggy)	TYPOGR	(→unscharf)	
fuzzy algorithm	DATA PROC	**unscharfer Algorithmus**	
		= qualitativer Algorithmus	
fuzzy logic	DATA PROC	**Qualitativaussagenlogik**	
[a mathematical treatment of uncertaqin or qualitative statements]		[mathematische Behandlung ungewisser oder qualitativer Aussagen]	
= fuzzy theory			
		= unscharfe Logik; Fuzzy Logik	
fuzzy theory (→fuzzy logic)	DATA PROC	(→Qualitativaussagenlogik)	

G

G	PHYS	**G**	
[10E9; in computing = 2E30 = 1,073,741,824]		[10E9; in der Datentechnik = 2E30 = 1.073.741.824]	
= giga; thousand mega		= Giga-	
G (→weight)	PHYS	(→Gewicht)	
g (→gram)	PHYS	(→Gramm)	
g (→acceleration of gravity)	PHYS	(→Fallbeschleunigung)	
G (→electric conductivity)	PHYS	(→elektrischer Leitwert)	
G (→Gauss)	PHYS	(→Gauß)	
Ga (→gallium)	CHEM	(→Gallium)	
GaAs (→gallium arsenide)	CHEM	(→Gallium-Arsenid)	
gable (n.)	CIV.ENG	**Giebel**	
		= Dachgiebel	
gadget (→device)	TECH	(→Vorrichtung)	
gadolonium	CHEM	**Gadolinium**	
= Gd		= Gd	
gage (→gauging)	INSTR	(→Eichung)	
gage (→gauge)	MECH	(→Lehre)	
gage (→gauge)	MECH	(→prüfen)	
gage block	MECH	**Endmaß**	
gage kit (→gauge kit)	MECH	(→Lehrensatz)	
gage plug (→plug gauge)	MECH	(→Lehrdorn)	
gage ring (→gauge ring)	MECH	(→Paßring)	
gaging	MECH	**Prüfung** [mit Lehren]	
gaging (→gauging)	INSTR	(→Eichung)	
gain	ANT	**Gewinn** [Verhältnis der effektiven zur isotrop verteilten Strahlstärke]	
[relation of effective to isotropic distributed radiation intensity]			
= antenna gain; directive gain; directivity factor; directivity 2		= effektiver Gewinn; Antennengewinn; Richtfaktor; Strahlungsgewinn	
≈ directivity		≈ Richtwirkung	
gain 1 (n.)	NETW.TH	**Verstärkung 1** [Verhältnis einer Größe am Ausgang zu dessen Wert am Eingang eines	
[output to input ratio of a magnitude passing a fourpole, if this qoutient is greater than one]			

gain 2 | 244

= amplification; enhancement; gain factor; gain constant; amplification factor
≠ attenuation
↑ transmission coefficient
↓ current gain; voltage gain

gain 2 (n.) NETW.TH
[logarithm of gain factor]

gain (→profit) ECON
gain adjustment CIRC.ENG
gain-bandwidth product MICROEL
[parameter for HF characteristics of transistors]
= transition frequency
↑ ß-cutoff
gain-bandwith product ELECTRON

gain constant (→gain) NETW.TH
gain-controlled laser diode OPTOEL
= GLD
gain estimate ANT
gain factor (→transmission coefficient) NETW.TH
gain factor (→gain) NETW.TH
gainless TELEC
gain-to-noise-temperature ratio SAT.COMM
= G/T
gain transistor CIRC.ENG
galactic noise (→cosmic noise) RADIO
galvanize METAL
(→electro-galvanize)
galaxy ASTROPHYS
galena CHEM
gallery (→cable subway) OUTS.PLANT
galley (→proof) TYPOGR
galley proof (→proof) TYPOGR
gallium CHEM
= Ga
gallium antimonide CHEM
= GaSb
gallium arsenide CHEM
= GaAs
gallium phosphide CHEM
= GaP
Galois group MATH
galvanic PHYS
galvanic bath TECH
galvanic battery POWER SYS
(→battery)
galvanic cell POWER SYS
[combination of three types of conductors, one of them an electrolyte]
= voltaic cell; cell; voltaic couple; element
≈ battery
↓ accumulator
galvanic contact MICROEL

galvanic contact (→metallic contact) EL.TECH
galvanic separation transformer COMPON
(→isolating transformer)
galvanic tension PHYS

galvanization METAL

galvanize (→electroplate) METAL
galvanized METAL
galvanized steel METAL
galvanomagnetic PHYS
galvanometer INSTR
galvanoplastics METAL

Vierpols, wenn dieser Quotient größer als eins ist]
= Gewinn; Verstärkungsfaktor
≠ Dämpfung
↑ Übertragungsfaktor
↓ Stromverstärkung; Spannungsverstärkung; Leistungsverstärkung
Verstärkungsmaß
[Logarithmus des Verstärkungsfaktors]
(→Gewinn)
Verstärkungseinstellung
Transitfrequenz
[HF-Kenngröße von Transistoren]
= ß1-Grenzfrequenz
↑ Beta-Grenzfrequenz
Verstärkung-Bandbreite-Produkt
= Bandbreitenprodukt
(→Verstärkung 1)
gewinngeführter Laser
= GLD
Gewinnabschätzung
(→Übertragungsfaktor)

(→Verstärkung 1)
nichtverstärkend
Empfangsempfindlichkeit
= G/T
Hubtransistor
(→kosmisches Rauschen)

(→galvanisch verzinken)

Galaxie
Bleiglanz
(→begehbarer Kabelkanal)

(→Korrekturfahne)
(→Korrekturfahne)
Gallium
= Ga
Gallium-Antimonid
= GaSb
Gallium-Arsenid
= GaAs
Gallium-Phosphid
= GaP
Galoissche Gruppe
galvanisch
elektrolytisches Bad
(→Batterie)

galvanisches Element
[Kombination dreier Leiter, davon ein Elektrolyt]
≈ Element
≈ Batterie
↓ Akkumulator; Primärelement
ohmscher Kontakt
= sperrschichtfreier Kontakt
(→galvanischer Kontakt)

(→Trennübertrager)

galvanische Spannung
= elektrochemische Spannung
Galvanisierung
= Plattierung 2
(→galvanisieren)
verzinkt
verzinkter Stahl
galvanomagnetisch
Galvanometer
Galvanoplastik

game (n.) COLLOQ **Spiel**
= Unterhaltungsspiel
game paddle (→paddle) TERM&PER (→Drehregler)
game port DATA PROC **Gameport**
[for connection of a joystick] [Anschluß für Joystick]
game programs DATA PROC (→Unterhaltungssoftware)
(→funware)
game software DATA PROC (→Unterhaltungssoftware)
(→funware)
games theory (→theory of games) MATH (→Spieltheorie)
gamma distribution MATH **Gammaverteilung**
gamma function MATH **Gammafunktion**
= Eulersches Integral zweiter Gattung
gamma matching ANT **Gamma-Anpassung**
gamma radiation PHYS **Gammastrahlung**
gamut TERM&PER **Farbpalette**
= color range (AM); colour range (BRI)
GAN DATA COMM **GAN**
[Global Area Network; a data communication network over satellite] [ein Datennetz über Satelliten]
gang (→assortment) ECON (→Sortiment)
ganged COMPON **gleichlaufend**
ganged potentiometer COMPON **Mehrfachpotentiometer**
ganged switch COMPON **Paketschalter**
= Mehrfachschalter
ganging COMPON **Gleichlauf**
Gantt chart DATA PROC **Gantt-Diagramm**
gap (n.) CODING **Lücke**
= Nennmarkierung
gap (n.) EL.TECH **Spalte**
[Magnetkreis]
= Trennstrecke
gap (n.) TECH **Spalt**
[narrow, long separation between two solid objects] [schmale, längliche Trennung zweier Gegenstände]
≈ interstice; crack = Ritze; Lücke; Trennstrecke
↑ opening ≈ Zwischenraum; Sprung
↑ Öffnung
gap (→band gap) PHYS (→Bandabstand)
gap (→magnetic gap) PHYS (→Magnetluftspalt)
gap (→interblock gap) TERM&PER (→Blockzwischenraum)
GaP (→gallium phosphide) CHEM (→Gallium-Phosphid)
gap character CODING **Lückenzeichen**
gap character (→filler) DATA COMM (→Füllzeichen)
gap clock CODING **Lückentakt**
gap detector TERM&PER **Blocklückendetektor**
gap digit (→gap element) CODING (→Füllelement)
gap element CODING **Füllelement**
= gap digit
gap-filler BROADC **Fernsehfüllsender**
gapless TECH **spaltlos**
= lückenlos
gap loss TERM&PER **Spaltdämpfung**
[due to misalignment of read/write head] [durch Fehljustierung des Magnetkopfes]
garbage COLLOQ **Müll**
= refuse; dust (BRI); sweepings; food waste ≈ Abfall; Unrat; Schrott; Schutt
garbage DATA PROC **Datensalat**
[useless or erroneous data] [unbrauchbare oder fehlerhafte Daten]
= junk; garbled data; gospel; gibberish = Datenabfall; Schunt; unbrauchbares Ergebnis; wertlose Daten
garbage can (AM) COLLOQ **Mülleimer**
= garbage pail (AM); dust bin (BRI) = Abfalleimer
garbage collection DATA PROC **Speicherbereinigung**
garbage collection ECON **Müllabfuhr**
= scavenging (BRI)
garbage pail (AM) (→garbage can) COLLOQ (→Mülleimer)
garble (→mutilate) TELEC (→verstümmeln)
garbled (→interfered) RADIO (→gestört)
garbled data (→garbage) DATA PROC (→Datensalat)
garbling RADIO LOC **Schlüsselverwirrung**
garnet CHEM **Granat**
↑ silicate ↑ Silikat

gas	PHYS	Gas	
≈ vapor		= gasförmiger Stoff	
		≈ Dampf	
gas arrester (→gas-discharge protector)	COMPON	(→Gasentladungsableiter)	
gas arrester (→gas-discharge protector)	COMPON	(→Gasentladungsableiter)	
gas attenuation	RADIO PROP	Gasbabsorptionsdämpfung	
GaSb (→gallium antimonide)	CHEM	(→Gallium-Antimonid)	
gas barrier	MICROW	Gassperre	
gas cutting	METAL	Gasschneiden	
gas cylinder	TECH	Druckluftflasche	
gas discharge (→gaseous discharge)	PHYS	(→Gasentladung)	
gas discharge display (→plasma display 1)	TERM&PER	(→Plasmaanzeige)	
gas-discharge protector	COMPON	Gasentladungsableiter	
= gas arrester; gas arrester; gas-filled arrester; gas-tube surge arrester		= Gasableiter	
		↑ Spannungsfeinsicherung	
gas discharge rectifier	ELECTRON	Gasentladungsgleichrichter	
↑ arc discharge tube		↑ Bogenentladungsröhre	
gas discharge relay	COMPON	Glimmrelaisröhre	
↑ glow discharge tube		= Glimmschaltröhre; Glimmrelais	
		↑ Glimmentladungsröhre	
gas-discharge tube	ELECTRON	Gasentladungsröhre	
= gas tube		↑ Elektronenröhre; Ionenröhre	
↑ electronic tube; ionic tube			
↓ glow-discharge tube		↓ Glimmentladungsröhre; Bogenentladungsröhre	
gas display panel (→plasma display 1)	TERM&PER	(→Plasmaanzeige)	
gaseous	PHYS	gasförmig	
gaseous deposition	OPT.COMM	Gasphasenabscheidung	
gaseous discharge	PHYS	Gasentladung	
= gas discharge		↑ Entladung	
↑ discharge		↓ Kanalentladung; Glimmentladung; Bogenentladung	
↓ glow discharge; arc discharge			
gaseous epitaxy	MICROEL	Gasphasenepitaxie	
gaseous phase	PHYS	Gasphase	
gas etching	MICROEL	Gasätzung	
gas-filled	TECH	gasgefüllt	
gas-filled arrester	COMPON	(→Gasentladungsableiter)	
(→gas-discharge protector)			
gasket (for fixed parts)	MECH	(→Dichtung)	
(→seal)			
gas laser	OPTOEL	Gaslaser	
gasoline (AM)	TECH	Benzin	
= petrol (BRI)			
gas panel (→plasma display 1)	TERM&PER	(→Plasmaanzeige)	
gas plasma display (→plasma display 1)	TERM&PER	(→Plasmaanzeige)	
gas ratio	ELECTRON	Vakuumfaktor	
gas scrubbing	PHYS	Gasreinigung	
gas soldering iron	ELECTRON	Gaslötstift	
gas thermometer	PHYS	Gasthermometer	
gastight	TECH	gasdicht	
= sealed		↓ luftdicht	
gas tube (→gas-discharge tube)	ELECTRON	(→Gasentladungsröhre)	
gas-tube surge arrester (→gas-discharge protector)	COMPON	(→Gasentladungsableiter)	
gas welding	METAL	Gasschweißung	
gate (v.t.)	ELECTRON	auftasten	
gate (v.t.) [pulses]	ELECTRON	durchschalten [Impulstechnik]	
gate (n.)	MICROW	Tor	
gate (→logic gate)	CIRC.ENG	(→Verknüpfungsglied)	
gate (→gate terminal)	MICROEL	(→Steueranschluß)	
gate (v.t.) (→link)	ENG.LOG	(→verknüpfen)	
gate array	MICROEL	Gate Array	
[application-specific "wireable" standard array of gates]		[anwendungsspezifisch „verdrahtete" Standard-Anordnung von Gattern]	
= uncommited logic array; ULA; logic array		= ULA; Gatter-Anordnung; Logikanordnung	
≈ logic chip		↓ Logikbaustein	
↑ ASIC		↑ ASIC	
gate array master	MICROEL	Gate-array-Master	
= master		= vorgefertigter Halbleiterbaustein	
gate characteristic [thyristor]	MICROEL	Zündkennlinie [Thyristor]	
gate circuit (→logic gate)	CIRC.ENG	(→Verknüpfungsglied)	
gate control [FET]	CIRC.ENG	Gate-Steuerung [FET]	
gate-controlled rise time	MICROEL	Durchschaltezeit [Thyristor]	
gate control pulse	ELECTRON	Torimpuls	
= gate pulse; gating pulse		≈ Ansteuerimpuls	
↓ drive pulse		↑ Steuerimpuls	
↑ control impulse			
gate current	MICROEL	Steuerstrom	
gate delay	MICROEL	Gatterlaufzeit	
= gate propagation delay			
gate electrode (→gate terminal)	MICROEL	(→Steueranschluß)	
gate equivalent (→gate function)	MICROEL	(→Gatterfunktion)	
gate floating	MICROEL	Floating-gate-Struktur	
gate function	MICROEL	Gatterfunktion	
= gate equivalent			
gate injection MOS (→GIMOS)	MICROEL	(→GIMOS-Technik)	
gate length	MICROEL	Gate-Länge	
gate-level simulator	MICROEL	Gatterebenen-Simulator	
gate monitoring system (→entrance monitoring system)	SIGN.ENG	(→Türanlage)	
gate noise	MICROEL	Gate-Rauschen	
gate oxide	MICROEL	Gate-Oxid	
gate propagation delay (→gate delay)	MICROEL	(→Gatterlaufzeit)	
gate protection	CIRC.ENG	Gate-Schutz	
gate pulse (→gate control pulse)	ELECTRON	(→Torimpuls)	
gate region [FET]	MICROEL	Gate-Zone [FET]	
gate terminal	MICROEL	Steueranschluß	
[control electrode of a transistor, thyristor etc.]		[Steuerelektrode eines Transistors, Thyristors u.dgl.]	
= control-current terminal; gate electrode; gate		= Gate-Anschluß; Gate-Elektrode; Gate; Steuerelektrode; Tor; G-Pol	
gate turn-off current	MICROEL	Löschstrom	
gate-turn-off thyristor	MICROEL	Abschaltthyristor	
= GTO thyristor; GTO		= GTO-Thyristor; GTO	
gate turn-off voltage	MICROEL	Löschspannung	
gate turn-on time [thyristor]	MICROEL	Zündzeit [Thyristor]	
		= Zündverzugszeit	
gate voltage	MICROEL	Torspannung	
		= Gate-Spannung	
gateway (n.)	DATA COMM	Überleiteinrichtung	
[converts protocols to connect different data networks]		[setzt, zur Verbindung unterschiedlicher Datennetze, deren Datenprotokolle um]	
≈ interworking unit; network node		= Netzkoppler; Gateway	
↑ internetworking processor		≈ Netzanpassungsgerät; Netzknoten	
		↑ Verbindungsrechner	
gateway	TELEC	Netzübergang	
≈ interworking unit [DATA COMM]		= Gateway	
		≈ Netzanpassungsgerät [DATA COMM]	
gateway center (→gateway exchange)	SWITCH	(→Kopfvermittlungsstelle)	
gateway exchange	SWITCH	Kopfvermittlungsstelle	
= gateway center; gateway office (AM)		= Kopfvermittlung; Kopfamt	
↓ ionternational gateway exchange		↓ Auslands-Kopfvermittlungsstelle	
gateway office (AM) (→gateway exchange)	SWITCH	(→Kopfvermittlungsstelle)	
gateway switch	DATA COMM	Durchgangsvermittlung	
gather (→collect)	TECH	(→einsammeln)	
gather write	DATA PROC	Sammelschreiben	
[of separate records as one data block]		[getrennte Dtensätze in einem Datenblock]	

gating

gating	MICROEL	Torsteuerung
gating period	MICROEL	Torsteuerungszeit
gating pulse (→gate control pulse)	ELECTRON	(→Torimpuls)
gating valve (→blocking valve)	TECH	(→Sperrventil)
gauge (v.t.)	MECH	prüfen
= gage		
gauge (n.)	MECH	Lehre
= gage; jig; appliance 1		= Eichmaß
↓ measuring gage; checking appliance; assembly appliance		≈ Schablone
		↓ Meßlehre; Prüflehre; Montagelehre
gauge (v.t.) (→calibrate)	INSTR	(→eichen)
gauge (→gauging)	INSTR	(→Eichung)
gauge condition	INSTR	Eichbedingung
gauge kit	MECH	Lehrensatz
= gage kit		
gauge plug (→plug gauge)	MECH	(→Lehrdorn)
gauge ring	MECH	Paßring
= gage ring		≈ Lehrring
gauging	INSTR	Eichung
= gaging; calibration; standardizing; gauge; gage		= Eichen; Einmessung; Einmessen; Kalibrierung
≈ traceability		≈ Nachweisbarkeit
gauging office	INSTR	Eichamt
Gauss	PHYS	Gauß
[unit for magnetic flux density; = 0,0001 T]		[Einheit für magnetische Flußdichte; = 0,0001 T]
= G		= G
Gauss distribution (→normal distribution)	MATH	(→Normalverteilung)
Gaussian filter	NETW.TH	Gauß-Filter
Gaussian hump	MATH	Gaußsche Glocke
Gaussian method of square of error	MATH	Gaußsche Fehlerquadratmethode
Gauss type lowpass	NETW.TH	Gaußscher Tiefpaß
gauze	TECH	Gaze
GB (→gigabyte)	DATA PROC	(→Gigabyte)
Gbit (→gigabit)	TELEC	(→Gigabit)
GByte (→gigabyte)	DATA PROC	(→Gigabyte)
GCR (→group-coded recording)	DATA PROC	(→gruppencodierte Aufzeichnung)
Gd (→gadolonium)	CHEM	(→Gadolinium)
Ge (→germanium)	CHEM	(→Germanium)
gear 1 (v.t.)	MECH	verzahnen 1
[to connect by toothed surfaces]		[Ineinandergreifen von Verzahnungen]
= indent 2		
gear (n.)	MECH	Zahnrad
= toothed wheel; cogwheel		↓ Ritzel; Stirnrad; Schneckenrad; Kegelrad
↓ pinion; spur gear; worm wheel; bevel gear		
gear 2 (→indent 1)	MECH	(→verzahnen 2)
gear (v.t.) (→indent)	COLLOQ	(→verzahnen)
gear coupling	MECH	Zahnkopplung
gear hobbing machine	MECH	Wälzfräsmaschine
gearing 1	MECH	Getriebe
gearing 2	MECH	Verzahnung
= indentation 2		
gearing	COLLOQ	Verzahnung
[figuratively]		[figurativ]
≈ indentation; interleaving		≈ Verschachtelung; Verquikkung; Verschränkung
gear ratio	MECH	Übersetzung
= ratio		
Geiger counter	PHYS	Geigerzähler
Geiger-Müller counter	PHYS	Geiger-Müller-Zähler
gelatinized	TECH	gelierend
gender	LING	Genus (n.n.)
↓ masculine; feminine; neuter		[pl.-era]
		= grammatisches Geschlecht; Geschlecht
		↓ Maskulinum; Femininum; Neutrum
gender changer	COMPON	Stecker-Buchsen-Übergang
gender changer (→in-series adapter)	COMPON	(→Kupplung 1)
general (adj.)	COLLOQ	gewöhnlich
= ordinary; common; everyday 2 (adj.)		= gängig; alltäglich
≠ special		≠ speziell
general (→public)	ECON	(→öffentlich)
general agency	ECON	Generalvertretung
general catalog (NAM) (→full-line catalog)	TECH	(→Gesamtkatalog)
general catalogue (BRI) (→full-line catalog)	TECH	(→Gesamtkatalog)
general description	DOC	Rahmenbeschreibung
general expenses (→indirect costs)	ECON	(→Gemeinkosten)
general instruction (→macro instruction 2)	DATA PROC	(→Makroanweisung)
general interrogation command	TELECONTR	Generalabfragebefehl
		= Gruppenabfragebefehl
generality	DATA PROC	Anwendungsflexibilität
generalization	SCIE	Verallgemeinerung
		= Generalisierung
generalize	SCIE	verallgemeinern
generalized cell	MICROEL	allgemeine Zelle
generalized routine	DATA PROC	Mehrzweckroutine
generally valid	SCIE	allgemeingültig
= universally valid		
general office (→head office)	ECON	(→Zentralverwaltung)
general premises administration	ECON	Standortverwaltung
general-purpose	TECH	(→Mehrzweck-)
(→multipurpose)		
general-purpose (→universal)	TECH	(→universell)
general-purpose computer	DATA PROC	Universalrechner
[universally programable]		[universell programmierbar]
= universal computer; GPC; all-purpose computer		= Universalcomputer; universeller Rechner; universeller Computer
≈ mainframe		≈ Großrechner
≠ special-purpose computer		≠ Spezialrechner
general-purpose grabber	COMPON	Universalklemme
general-purpose interface bus (→IEC bus)	DATA COMM	(→IEC-Bus)
general-purpose lamp	EL.INST	Allgebrauchslampe
general-purpose programm	DATA PROC	Universalprogramm
		= Mehrzweckprogramm
general-purpose register (→general register)	DATA PROC	(→allgemeines Register)
general-.purpose saw	TECH	Universalsäge
		= Multisäge
general-purpose video terminal	TERM&PER	Mehrzweck-Datensichtgerät
general register	DATA PROC	allgemeines Register
= general-purpose register		= Mehrzweckregister
≈ working register; register memory		≈ Arbeitsregister; Registerspeicher
general services	ECON	Betriebsunterhaltung
= operation support		
general specification	ECON	Rahmenpflichtenheft
		= Rahmenspezifikation
general switched telephone network (→public switched telephone network)	TELEC	(→öffentliches Fernsprechwählnetz)
general telegraph exchange service (→gentex)	TELEC	(→Gentex)
general tolerance	ENG.DRAW	Freimaßtoleranz
general validity	SCIE	Allgemeingültigkeit
= universal validity		= allgemeine Gültigkeit
generate	TECH	erzeugen
		= erstellen; generieren
generate (→excite)	PHYS	(→anregen)
generate a message	SWITCH	eine Meldung erstellen
= build-up a message		
generated address (→synthetic address)	DATA PROC	(→synthetische Adresse)
generated error	DATA PROC	generierter Fehler
generated in the transmitter	RADIO	sendereigen
generating function	MATH	Erzeugende
generating set	POWER SYS	Stromaggregat
= genset		= Aggregat
generating station (→power station)	POWER ENG	(→Kraftwerk)
generation 2	TECH	Generation
↓ system generation; computer generation		≈ Entwicklungsstand
		↓ Systemgeneration; Rechnergeneration

generation 1	TECH	**Erzeugung**
= production		= Erstellung; Generierung
generation (→excitation)	PHYS	(→Anregung)
generation current	MICROEL	**Erzeugungsstrom**
		= Generationsstrom
generation principle	DATA PROC	**Generationen-Prinzip**
[backup by maintaining a grandfather, father and son copy of files]		[Dateiensicherung durch Führen von Großvater-, Vater- und Sohn-Kopien]
= grandfather-father-son concept; grandfather file; ancestral file		= Generations-Prinzip; Großvater-Vater-Sohn-Prinzip
generation rate	MICROEL	**Erzeugungsrate**
		= Generationsrate
generator	ELECTRON	**Generator**
		= Erzeuger
generator	POWER ENG	**Stromgenerator**
[converts mechanical into electrical energy]		[wandelt mechanische in elektrische Energie]
= electric generator		= elektrischer Generator; Generator
↓ ac generator; dc generator		↓ Wechselstromgenerator; Gleichstromgenerator
generator	INSTR	**Generator**
= source		= Quelle
generator (→program generator)	DATA PROC	(→Programmgenerator)
generator program	DATA PROC	(→Programmgenerator)
(→program generator)		
generic (n.)		**Generikum**
[compatible with the software or the hardware line of a vendor]		[mit einer Software- oder Hardware-Linie eines Herstellers kompatibel]
generic 2 (adj.)	DATA PROC	**auswählbar**
= selectable		
≈ variable		
generic 1 (adj.)	DATA PROC	**herstellerkompatibel**
[compatible with a product family]		[mit Produkten eines Herstellers]
generic term	LING	**Oberbegriff**
≠ derivative term		= Hyperonym; Supernym
		≠ Unterbegriff
genitive	LING	**Genitiv**
		= Genetiv; Wesfall; 2. Fall; zweiter Fall
genlock	TERM&PER	**Genlock**
[equipment to mix graphics with video]		[Gerät zum Mischen von Computer- und Videobildern]
		= Bildmischer
genset (→generating set)	POWER SYS	(→Stromaggregat)
gentex	TELEC	**Gentex**
[USA]		[Telegrammwähldienst in USA]
= general telegraph exchange service		
gently rolling terrain	RADIO PROP	**leichtwelliges Gelände**
= slightly ondulating terrain		
genuine (adj.)	COLLOQ	**authentisch**
= authentic		
genuine (→unadulterated)	TECH	(→unverfälscht)
genuine accessory	ECON	**Originalzubehör**
geocentric	PHYS	**geozentrisch**
geocoding	DATA PROC	**geographische Darstellung**
geographical map	GEOPHYS	**Landkarte**
= map		= Karte
geomagnetic	PHYS	**erdmagnetisch**
geomagnetic field	PHYS	**Erdmagnetfeld**
geomagnetism	PHYS	**Erdmagnetismus**
= earth magnetism		
geometric (→geometrical)	MATH	(→geometrisch)
geometrical	MATH	**geometrisch**
= geometric		
geometrically thick antenna	ANT	**geometrisch dicke Antenne**
↓ cage antenna; cage dipole; discone antenna; biconical antenna; full-wave dipole; cup antenna		↓ Breitbandreuse; Reusendipol; Disconeantenne; Doppelkonusantenne; Ganzwellendipol; Kelchantenne
geometrical mean (→geometric mean)	MATH	(→geometrisches Mittel)
geometrical progression (→geometric progression)	MATH	(→geometrische Reihe)
geometrical representation	MATH	**geometrische Darstellung**
geometrical series (→geometric progression)	MATH	(→geometrische Reihe)
geometric distortion	TV	**Geometriefehler**
geometric distribution	MATH	(→geometrische Verteilung)
(→spatial distribution)		
geometric horizon	RADIO PROP	**geometrischer Horizont**
geometric mean	MATH	**geometrisches Mittel**
[n-th root of the product A1 times ... An]		[n-te Wurzel des Produktes A1 mal ... An]
= geometrical mean		= mittlere Proportionale
geometric progression	MATH	**geometrische Reihe**
= geometric progression; geometric series; geometrical series		= geometrische Progression
geometric series (→geometric progression)	MATH	(→geometrische Reihe)
geometry	MATH	**Geometrie**
		= Raumlehre
geophysics	PHYS	**Geophysik**
geostationary	SAT.COMM	**geostationär**
= geosynchronous		= geosynchron
geostationary orbit	ASTROPHYS	(→Synchronlaufbahn)
(→synchronous orbit)		
geostationary satellite	SAT.COMM	(→Synchronsatellit)
(→synchronous satellite)		
geosynchronous	SAT.COMM	(→geostationär)
(→geostationary)		
geosynchronous satellite	SAT.COMM	(→Synchronsatellit)
(→synchronous satellite)		
German double s	LING	**scharfes S**
[ß]		[ß]
		= Eszett (OES)
German horsepower	PHYS	(→Pferdestärke)
(→horsepower)		
germanium	CHEM	**Germanium**
= Ge; eca silicon [MICROEL]		= Ge; Eca-Silizium [MICROEL]
German sectional view	ENG.DRAW	**deutsche Schnittdarstellung**
German silver	METAL	**Neusilber**
= nickel-silver		
German-standard keyboard	TERM&PER	**DIN-Tastatur**
= QWERTZ keyboard		= QWERTZ-Tastatur; deutsche Schreibmaschinentastatur
German three-wire plug	EL.INST	(→Schukostecker)
(→Schuko plug)		
German three-wire socket	EL.INST	(→Schukosteckdose)
(→Schuko socket)		
gerund	LING	**Gerundium**
[infinitive with adverbial qualities to express generalization or continuance; e.g. working]		[gebeugter Infinitiv zum Ausdruck einer Verallgemeinerung oder Dauerhaftigkeit; z.B. arbeitend]
≈ gerundive; participle		≈ Gerundiv; Partizip
gerundive	LING	**Gerundiv**
[future passive participle expressing necessity, e.g. in Latin delenda]		[Notwendigkeit ausdrückendes Partizip des Passivs des Futur; z.B. lat. delenda]
≈ gerund		= Gerundivum
		≈ Gerundium
get (v.t.) (→load)	DATA PROC	(→laden)
GET instruction (→GET statement)	DATA PROC	(→Holanweisung)
GET statement	DATA PROC	**Holanweisung**
= GET instruction		
get stuck (→stick)	TECH	(→klemmen 2)
getter	PHYS	**Getter**
		= Fangstoff
gettering	PHYS	**Getterung**
GFLOP (→gigaflop)	DATA PROC	(→Gigaflop)
Gflop (→gigaflop)	DATA PROC	(→Gigaflop)
ghost (→ghost image)	TV	(→Geisterbild)
ghost cursor	DATA PROC	**Zweitkursor**
= second cursor		
ghost hyphen (→soft hyphen)	DATA PROC	(→Bedarfstrennstrich)
ghost image	TV	**Geisterbild**
= double image; ghost; fold-over; multipath effect		
GHz (→gigahertz)	PHYS	(→Gigahertz)

Giacoletto equivalent circuit	MICROEL	Giacoletto-Ersatzschaltung	glaze (v.t.)	TECH	glasieren
gibberish (→garbage)	DATA PROC	(→Datensalat)	glaze	TECH	Lasur
gib-head key	MECH	Nasenkeil	[transparent coating]		[durchsichtige Farbe]
giga (→G)	PHYS	(→G)			= Glasur
gigabit	TELEC	Gigabit			≈ Lack; Firnis
[10E9 bits]		[10E9 Bit]	GLD (→gain-controlled laser diode)	OPTOEL	(→gewinngeführter Laser)
= Gbit		= Gbit	glich (→interfering pulse)	ELECTRON	(→Störimpuls)
gigabyte	DATA PROC	Gigabyte	glide path	RADIO NAV	Gleitweg
[10E30 = 1,073,741,824 bytes]		[10E30 = 1.073.741.824 Bytes]			= Gleitebene
= GB; GByte		= GByte	glide path transmitter (→glide slope)	RADIO NAV	(→Gleitwegsender)
gigaflop	DATA PROC	Gigaflop	glide slope	RADIO NAV	Gleitwegsender
[10E9 FLOPs]		[10E9 Fließkommaoperationen]	= GS; glide path transmitter		
= Gflop; GFLOP			gliding-surface bearing	MECH	Gleitlager
gigahertz	PHYS	Gigahertz	glimmer (v.i.)	TECH	glimmern
[10E9 hertz]		[10E9 Hertz]			= schimmern
= GHz		= GHz	G line (→surface wave transmission line)	LINE TH	(→Oberflächenwellenleitung)
GIGO	DATA PROC	GIGO	glitter (v.i.)	TECH	glitzern
[Garbage-in-Garbage-Out]		["so unbrauchbare Daten man eingibt, so unbrauchbare Ergebnisse kommen heraus"]	= sparkle (v.i.)		= funkeln
			glitter (n.)	TECH	Glitzern
gimbal (→gimbal ring)	TECH	(→Kompaßring)			= Glitzer (obs.); Funkeln
gimbal ring	TECH	Kompaßring	Glixon code	CODING	Glixon-Code
= gimbal			global	DATA PROC	allgemeingültig
GIMOS	MICROEL	GIMOS-Technik	[covering everything]		= durchgängig; global; umfassend
= gate injection MOS			≠ local		
gin pole	ANT	Hilfsmast			≠ singulär
girder 2	CIV.ENG	Profilträger	global (→overall)	TECH	(→Gesamt-)
= joist		↑ Träger	global address	DATA COMM	Generaladresse
↑ support			global beam	SAT.COMM	Globalausleuchtung
girder 1	CIV.ENG	Tragebalken	global beam antenna	SAT.COMM	Globalstrahlantenne
		↑ Träger			= Globalantenne
Giro system (BRI) (→postal check service)	ECON	(→Postgiro)	global character (→wild card)	DATA PROC	(→Stellvertreterzeichen)
gitter	ELECTRON	Gitter	global communications (→global telecommunications)	TELEC	(→weltweite Telekommunikation)
[vacuum tube]		[Elektronenröhre]	global coverage	SAT.COMM	globale Bedeckung
give-away price	ECON	Schleuderpreis	= earth coverage		
given (→preset)	COLLOQ	(→vorgegeben)	global coverage	TELEC	Flächendeckung
give notice	ECON	kündigen	= full coverage; full area coverage		
give on lease (v.t.)	ECON	verpachten	global-coverage (→full-coverage)	TELEC	(→flächendeckend)
[to grant use of real estate for a fixed time and pay]		[gegen Entgelt befristet zur Nutzung überlassen]	global exchange	DATA PROC	Globalaustausch
= lease 1		≈ vermieten	[word processing; e.g. of a word throughout a text]		[Textverarbeitung; z.B. eines Wortes im ganzen Text]
≈ take on rent		≠ pachten			
≠ take on lease			global operation	DATA PROC	durchgängige Operation
give on rent	ECON	mieten	global tab erasing (→global tabulator erasing)	TERM&PER	(→Tabulator-Gesamtlöscher)
[to take into temporary use for a fixed pay]		[gegen Entgelt vorübergehend in Gebrauch nehmen]	global tabulator erasing	TERM&PER	Tabulator-Gesamtlöscher
= rent 1; hire 2		≈ entleihen; pachten	= global tab erasing		= Tab-Gesamtlöscher
≈ lend; take on lease		≠ vermieten	global telecommunications	TELEC	weltweite Telekommunikation
≠ take on rent			= global communications		
GKS (→graphical kernel system)	DATA PROC	(→graphische Kernroutinen)	global variable	DATA PROC	globale Variable
glare (v.t.)	TECH	blenden	[valid for and accesible by the whole program]		[für das gesamte Programm gültig und zugänglich]
glare (n.)	TERM&PER	Bilschirmreflexion	≠ local variable		
glare (n.)	TECH	Blendung			≠ lokale Variable
glare (→double seizure)	SWITCH	(→Kollision)	globe (→sphere)	MATH	(→Kugel)
glas-fiber reinforced synthetic	TECH	Glasfaserkunststoff	globe (→ball)	TECH	(→Kugel)
			globe valve	MECH	Kugelventil
glass	TECH	Glas	gloss (n.)	TECH	Hochglanz
glassbreak vibration detector	SIGN.ENG	Scheibenbruchmelder	= highlight (n.)		
		= Glasbruchmelder	gloss (n.)	LING	Kurzerläuterung
glass bulb	TECH	Glaskolben	glossary	LING	Begriffserklärung
glass capacitor	COMPON	Glaskondensator	[collection of terms of a subject field, with short explanations]		[Sammlung der Begriffe eines Fachgebiets, mit Kurzerläuterungen]
glasses (→spectacles)	TECH	(→Brille)			
glass fiber (AM)	OPT.COMM	Glasfaser			= Begriffssammlung; Glossar
= glass fibre (BRI); lightwave fiber		= Lichtleiterfaser	glossy (adj.)	TECH	hochglänzend
≈ optical waveguide		≈ Lichtwellenleiter	glow (v.i.)	TECH	glühen
glass fibre (BRI) (→glass fiber)	OPT.COMM	(→Glasfaser)	glow discharge	PHYS	Glimmentladung
glass insulator	OUTS.PLANT	Glasisolator	= luminous discharge		↑ Gasentladung
		[Freileitung]	↑ gaseous discharge		
glassivation	TECH	Glasierung	glow discharge diode	ELECTRON	Glimmschaltdiode
glass semiconductor	MICROEL	Glashalbleiter	glow discharge tube	ELECTRON	Glimmentladungsröhre
glass thermometer	INSTR	Glasthermometer	↑ gas discharge tube		= Glimmröhre
glass wool	TECH	Glaswolle			↑ Gasentladungsröhre
glassy (→vitreous)	TECH	(→gläsern)	glow-discharge voltage regulator (→voltage regulator tube)	ELECTRON	(→Stabilisatorröhre)
G layer	RADIO PROP	G-Schicht			
= G region					

glow display tube	COMPON	**Glimmanzeigeröhre**	
= glow indicating tube			
glow indicating tube (→glow display tube)	COMPON	(→Glimmanzeigeröhre)	
glow lamp	COMPON	**Glimmlampe**	
= glow tube; discharge tube		↓ Neonlampe	
↓ neon lamp			
glow light	PHYS	**Glimmlicht**	
= blue glow			
glow rectifier	ELECTRON	**Glimmgleichrichter**	
glow stabilizing tube	ELECTRON	**Glimmstabilisatorröhre**	
glow tube (→glow lamp)	COMPON	(→Glimmlampe)	
glue (→adhesive)	TECH	(→Kleber)	
glue logic	MICROEL	**Randlogik**	
gluing machine	OFFICE	**Verleimgerät**	
glycerine	CHEM	**Glyzerin**	
GND (→ground)	EL.TECH	(→Erde)	
gnomon (n.)	DATA PROC	**Gnomon**	
[aid to represent 3 in 2 dimensions]		[Behelf zur Darstellung von 3 auf 2 Dimensionen]	
gnomon (n.)	PHYS	**Gnomon**	
= sundial style		= Sonnenuhrstab	
gnomon (n.)	MATH	**Gnomon**	
[remainder after removal of similar partial parallelogram]		[Rest nach Auschneiden eines ähnlichen Teilparallelogramms]	
goal (→scope)	COLLOQ	(→Zielsetzung)	
go-and-return test (→loop measurement)	TELEC	(→Schleifenmessung)	
go direction	TELEC	**Hin-Richtung**	
= near-to-far direction		≠ Rück-Richtung	
≠ return direction		↑ Übertragungsrichtung	
↑ transmission direction			
goggles (→safety goggles)	TECH	(→Schutzbrille)	
Golay cell	INSTR	**Golay-Zelle**	
gold	CHEM	**Gold**	
= Au		= Au	
goldbeam	MICROEL	**Goldbahn**	
gold bonded	MICROEL	**goldkontaktiert**	
gold-bonded diode	MICROEL	**Golddrahtdiode**	
gold-clad	METAL	**goldplattieren**	
gold-cladded (→gold-plated)	METAL	(→goldplattiert)	
gold cladding	METAL	**Vergoldung**	
= gold plating 2		↑ Galvanisierung	
↑ galvanization			
gold contact	COMPON	**Goldkontakt**	
= gold-plated contact			
gold doping	MICROEL	**Golddotierung**	
golden 1	PHYS	**golden 1**	
[of gold]		[aus Gold]	
golden 2	PHYS	**goldfarben**	
[with the color of gold]		= goldfarbig; golden 2	
golden yellow (→canary yellow)	PHYS	(→goldgelb)	
gold-plate	METAL	**vergolden**	
↑ electroplate		↑ galvanisieren	
gold-plated	METAL	**goldplattiert**	
= gold-cladded		↑ vergoldet	
gold-plated contact (→gold contact)	COMPON	(→Goldkontakt)	
gold plating 1	METAL	**Goldplattierung**	
↑ gold cladding		↑ Vergoldung	
gold plating 2 (→gold cladding)	METAL	(→Vergoldung)	
golf ball (→print ball)	TERM&PER	(→Kugelkopf)	
golf-ball printer (→type-ball printer)	TERM&PER	(→Kugelkopfdrucker)	
gommet (→sleeve)	TECH	(→Hülse)	
gon	MATH	**Gon**	
[angular measuring unit; complete angle/400]		[Winkelmaß, 1 Vollwinkel/400]	
= new degree		= Neugrad	
gong (→single-stroke bell)	EL.ACOUS	(→Einschlagwecker)	
goniometer (→protractor)	INSTR	(→Winkelmaß)	
goniometer (→protractor)	ENG.DRAW	(→Winkelmaß)	
goniometric direction finder	RADIO LOC	**Goniometerpeiler**	
goniospectrophotometry	PHYS	**Goniospektrophotometrie**	
good (n.)	ECON	**Ware**	
= commodity; merchandise; commercial product; produce		= Handelsartikel; Handelsware; Handelsgut; Gut	
goods import (→import)	ECON	(→Einfuhr)	
goods inward department	ECON	(→Warenannahme)	
(→goods receiving department)			
goods marking equipment	TERM&PER	**Warenauszeichnungsgerät**	
goods receiving department	ECON	**Warenannahme**	
= goods inward department		[Organisation]	
go off-hook (→lift)	TELEPH	(→abheben)	
go on-hook	TELEPH	**auflegen**	
[handset]		[den Hörer]	
		= einhängen	
goose neck	ANT	**Bischofsstab**	
[a form of antenna feeder]		[Speisesystem für Parabolantenne]	
gopher (AM)	COLLOQ	(→Nagetier)	
(→rodent)			
gopher attack (→rodent attack)	TECH	(→Nagetierfraß)	
gopher-protected cable (AM)	COMM.CABLE	**nagetiergeschütztes Kabel**	
= rodent-protected cable			
gopher protection (AM) (→rodent protection)	COMM.CABLE	(→Nagetierschutz)	
GOS (→grade of service)	SWITCH	(→Verkehrsgüte)	
gospel (→garbage)	DATA PROC	(→Datensalat)	
gothic (→sans serif)	TYPOGR	(→serifenlos)	
Gothic (→black letter)	TYPOGR	(→gotische Schrift)	
go-to statement (→jump instruction)	DATA PROC	(→Sprungbefehl)	
go up (→rise)	ECON	(→steigen)	
governmental (→state-)	ECON	(→staatlich)	
governmental carrier (→governmental telecommunications carrier)	TELEC	(→staatliche Fernmeldeverwaltung)	
governmental communication carrier (→governmental telecommunications carrier)	TELEC	(→staatliche Fernmeldeverwaltung)	
governmental telecommunications carrier	TELEC	**staatliche Fernmeldeverwaltung**	
= governmental communication carrier; governmental carrier; state-run telecommunication carrier; state-run communication carrier; state-run carrier		↑ Fernmeldeverwaltung	
government contract (→public order)	ECON	(→Staatsauftrag)	
government contractor	ECON	**Staatslieferant**	
governor (→controller)	CONTROL	(→Regler)	
GPC (→general-purpose computer)	DATA PROC	(→Universalrechner)	
GPIB bus (→IEC bus)	DATA COMM	(→IEC-Bus)	
GPR (→general register)	DATA PROC	(→allgemeines Register)	
grab (v.t.)	COLLOQ	**schnappen** (v.r.) [fig.]	
grabber (→terminal)	COMPON	(→Anschlußklemme)	
grabber (→frame grabber)	DATA PROC	(→Bildfangschaltung)	
grabbing (→frame grabbing)	DATA PROC	(→Bildeinfangung)	
graceful degradation	DATA PROC	**Teilausfall**	
[limited operation still possible]		[eingeschränkter Betrieb weiterhin möglich]	
gradation (→grading)	TECH	(→Staffelung)	
gradation distortion	TV	**Gradationsfehler**	
= gradation error			
gradation error (→gradation distortion)	TV	(→Gradationsfehler)	
grade	TECH	**Gütegrad**	
≈ type; quality		≈ Ausführung; Qualität	
grade (quality) (→sort)	QUAL	**Güteklasse**	
graded	TECH	(→sortieren)	
= graduated; staggered	COLLOQ	**gestaffelt**	
graded-base transistor (→drift transistor)	MICROEL	(→Drifttransistor)	
graded index	OPT.COMM	**Gradientenindex**	
graded junction (→progressive junction)	MICROEL	(→allmählicher Übergang)	
graded multiple	SWITCH	**Teilvielfachfeld**	
= partial multiple			
grade of expansion	TELEC	**Ausbaugrad**	

grade of service SWITCH
[service quality depending from the dimensioning of the system]
= GOS; grade of switching performance; service quality
↓ loss; delay probability

grade of service (→operating quality) TELEC

grade-of-service monitoring SWITCH
grade of switching performance SWITCH (→grade of service)
gradient MATH
gradient fiber (AM) OPT.COMM
= gradient fibre (BRI,CCITT)
≈ multimode fiber
gradient fibre (BRI,CCITT) OPT.COMM (→gradient fiber)
gradient microphone EL.ACOUS
grading SWITCH
[interconnection scheme to economize crosspoints, with limited accessibility to outlets]
= mixing
↓ skipping; slipping

grading TECH
= graduation; gradation
grading group SWITCH
[all inlets have access to the same outlets]
= subgroup
≈ incoming group
gradually (→stepwise) TECH
graduate (v.t.) TECH
graduate (n.) SCIE
[holding academic degree]
= egghead (AM)
graduated (→graded) COLLOQ
graduated engineer (→academic engineer) SCIE
graduated interest ECON
graduated scale INSTR
graduation 1 INSTR
↑ scale
graduation ENG.DRAW
graduation 2 (→scale) INSTR
graduation (→grading) TECH
graduation mark INSTR
Graetz connection (→rectifier bridge circuit) CIRC.ENG
grain boundary PHYS
[crystals]
grain noise ELECTRON
grain-oriented METAL
gram PHYS
[= 0.001 kg]
= gramme; g
gram-atomic weight PHYS
grammar LING
[rules for pronunciation, words and sentence formation]
↑ linguistics
↓ phonetics; morphology; syntax

grammar school (BRI) SCIE
= high school (AM)
grammatical error (→syntax error) DATA PROC
grammatical mistake DATA PROC
(→syntax error)
gramme (→gram) PHYS
gramophone (trademark in U.S.A.) (→record player) CONS.EL
grandfather cycle DATA PROC
grandfathered (→obsolete) TECH

Verkehrsgüte
[durch die Bemessung der Anlage bedingte Qualität der Verkehrsabwicklung]
= Vermittlungsgüte; Dienstgüte
↓ Verlust; Wartewahrscheinlichkeit

(→Betriebsgüte)

Verkehrsgüteüberwachung
(→Verkehrsgüte)

Gradient
[Vektorrechnung]
Gradientenfaser
≈ Mehrmodenfaser

(→Gradientenfaser)

Gradientenmikrophon
Mischung
[Koppelpunkte sparendes Verdrahtungsprinzip, mit Einschränkung der Erreichbarkeit]
↓ Staffeln; Übergreifen; Verschränken

Staffelung
= Abstufung
Staffelgruppe
[alle Eingänge haben zu denselben Ausgängen Zugang]
≈ Zubringerteilgruppe
(→schrittweise)
abstufen
Akademiker
= Diplomierter; Graduierter; Hochschulabsolvent
(→gestaffelt)
(→Hochschulingenieur)

Staffelzins
Strichskale
Gradeinteilung
↑ Skaleneinteilung
Maßteilung
(→Skala)
(→Staffelung)
Skalenmarke
(→Graetz-Schaltung)

Korngrenze
[Kristall]
Kornrauschen
kornorientiert
Gramm
[= 0,001 kg]
= g
Grammatom
Grammatik
[Regeln der Aussprache, Wortformen und Satzaufbauten]
= Sprachlehre
↑ Sprachwissenschaft
↓ Phonetik; Morphologie; Syntax

Gymnasium

(→Syntaxfehler)

(→Syntaxfehler)

(→Gramm)
(→Plattenspieler)

Großvaterzyklus
(→veraltet)

grandfather-father-son concept (→generation principle) DATA PROC
grandfather file (→generation principle) DATA PROC
grandfather tape DATA PROC
[generations principle]
grand total MATH
= sum total; total amount; total
graniness TECH
= granularity
grant (n.) (→allocation) TECH
granted patent ECON
grant time (→allow deferment) ECON
granularity DATA PROC
[size of memory segments]
granularity (→graniness) TECH
granularity (→switching layer) TRANS
graph NETW.TH
graph DOC
= chart
graph (→characteristic 3) TECH
graph (→diagram 2) ECON
graphic DOC
= graphic representation; chart

graphic (→graphical) DATA PROC
graphic adapter DATA PROC

graphical (adj.) DATA PROC
= graphic
graphical character (→graphic character) LING
graphical data DATA PROC
= graphic data
graphical integration MATH
graphical kernel system DATA PROC
= GKS

graphically (adv.) DATA PROC
graphical processing DATA PROC
graphical symbol CIRC.ENG

graphical tool DOC
graphical user interface DATA PROC
= GUI; graphic user surface
graphic board (→digitizing tablet) TERM&PER
graphic capability TERM&PER
graphic character LING
= graphical character
≈ type [TYPOGR]
↑ sign
↓ letter
graphic character DATA PROC
= graphics character
graphic character set TYPOGR
(→font)
graphic CRT display TERM&PER
(→graphic display)
graphic cursor DATA PROC
graphic data (→graphical data) DATA PROC
graphic data base DATA PROC

graphic digitizer (→digitizing tablet) TERM&PER
graphic display TERM&PER
= graphic display screen; graphic CRT display
graphic display mode DATA PROC
= graphic mode; graphics mode
graphic display screen TERM&PER
(→graphic display)
graphic environment DATA PROC
graphic equalizer CONS.EL

(→Generationen-Prinzip)

(→Generationen-Prinzip)

Großvaterband
[Generationen-Prinzip]
Gesamtsumme
= Gesamtwert
Körnigkeit

(→Zuordnung)
erteiltes Patent
(→stunden)

Granularität

(→Körnigkeit)
(→Schaltebene)

Graph
Kurvenblatt

(→Kennlinie)
(→Schaubild)
Graphik
= Grafik; graphische Darstellung; grafische Darstellung

(→graphisch) (adj.)
Graphikadapter
= Grafikadapter
graphisch (adj.)

(→Schriftzeichen)

Graphikdaten

graphische Integration
graphische Kernroutinen
= graphisches Kernsystem; GKS

graphisch (adv.)
graphische Datenverarbeitung
Schaltzeichen
= Schaltsymbol
graphisches Hilfsmittel
graphische Benutzeroberfläche

(→Digitalisiertablett)

Graphikfähigkeit
Schriftzeichen
≈ Drucktype [TYPOGR]
↑ Zeichen
↓ Buchstabe

Graphikzeichen
= Grafikzeichen
(→Schriftzeichensatz)

(→graphischer Bildschirm)

Graphikcursor
(→Graphikdaten)

Graphikdatenbank
= Grafikdatenbank
(→Digitalisiertablett)

graphischer Bildschirm

Graphikmodus
= Grafikmodus
(→graphischer Bildschirm)

graphische Umgebung
Graphic-Equalizer
= Grafik-Equalizer; graphischer Equalizer; grafischer Equalizer

graphic hardware	DATA PROC	**Graphik-Hardware** = Grafik-Software	**graphics printer** [printer able to mimic, with degraded quality, a plotter] = high-resolution printer; printer-plotter ↓ plotter ↑ display terminal	TERM&PER	**Graphikdrucker** [Drucker der, mit geringer Auflösung, ein Zeichengerät simulieren kann] = Grafikdrucker; hochauflösender Drucker ≈ Plotter ↑ Sichtgerät
graphic input	DATA PROC	**graphische Eingabe**			
graphicist	TYPOGR	**Graphiker**			
graphic language = graphic programming language	DATA PROC	**Graphiksprache** = graphische Programmiersprache			
graphic limits	DATA PROC	**graphische Begrenzungslinie**	**graphics processor** [supports the main processor for graphic programs] = pixel processor	DATA PROC	**Graphikprozessor** [entlastet den Hauptprozessor bei Graphikprogrammen] = Grafikprozessor; Bildschirmprozessor; Bildpunktprozessor; Pixelprozessor
graphic mode (→ graphic display mode)	DATA PROC	(→ Graphikmodus)			
graphic plotter (→ plotter)	TERM&PER	(→ Plotter)			
graphic programming	DATA PROC	**Graphikprogrammierung** = Grafikprogrammierung			
graphic programming language (→ graphic language)	DATA PROC	(→ Graphiksprache)			
graphic rendition	TERM&PER	**Darstellungsart**	**graphics program** ≈ graphics software	DATA PROC	**Graphikprogramm** = Grafikprogramm ≈ Graphik-Software
graphic representation (→ graphic)	DOC	(→ Graphik)			
graphic resolution [maximum pixel density]	TERM&PER	**Graphikauflösung** [maximale Bildpunktdichte] = graphische Auflösung	**graphics screen** = graphics display; graphics monitor ↑ display terminal ↓ raster screen; vector display	TERM&PER	**Graphikbildschirm** = Grafikbildschirm; Graphikmonitor; Grafikmonitor ↑ Sichtgerät ↓ Rasterbildschirm; Vektorbildschirm
graphics board 1 [controls the video display of characters and/or graphics] = graphics card; video board; video card ↓ monochrome graphics board; MGA board; Hercules board; color graphics board; CGA board; AGA board; EGA board; VGA board; PGA board	TERM&PER	**Graphikkarte** [steuert die Bildschirmdarstellung von Zeichen und/oder Graphiken] = Grafikkarte; Graphikadapterkarte; Grafikadapterkarte; Videokarte ↓ Monochrom-Graphikkarte; MGA-Karte; Hercules-Karte; Farbgraphikkarte; CGA-Karte; AGA-Karte; EGA-Karte; VGA-Karte; PGA-Karte			
			graphics standard ↓ MGA; Hercules; CGA; EGA; VGA; PGA	TERM&PER	**Graphikstandard** ↓ MGA; Hercules; CGA; EGA; VGA; PGA
			graphics tablet (→ digitizing tablet)	TERM&PER	(→ Digitalisiertablett)
			graphics terminal (→ graphics output hardware)	TERM&PER	(→ Graphikausgabegerät)
			graphics VDU	DATA PROC	**Graphikstation** = graphisches Datensichtgerät
graphics board 2 (→ digitizing tablet)	TERM&PER	(→ Digitalisiertablett)			
graphics card (→ graphics board)	TERM&PER	(→ Graphikkarte)			
graphics character (→ graphic character)	DATA PROC	(→ Graphikzeichen)	**graphic symbol** **graphic user surface** (→ graphical user interface)	DOC DATA PROC	**graphisches Symbol** (→ graphische Benutzeroberfläche)
graphics digitizer (→ digitizing tablet)	TERM&PER	(→ Digitalisiertablett)	**graphic workstation** ≈ graphics output hardware	DATA PROC	**graphisches Arbeitsplatzsystem** = Graphik-Arbeitsstation; graphischer Arbeitsplatz; graphische Arbeitsstation; Grafik-Workstation ≈ Graphikausgabegerät
graphics display	INSTR	**Graphikanzeige**			
graphics display (→ graphics screen)	TERM&PER	(→ Graphikbildschirm)			
graphics element (→ drawing element)	DATA PROC	(→ Zeichenelement)			
graphics input device (→ graphics input hardware)	TERM&PER	(→ Graphikeingabegerät)	**graphite** **graph plotter** (→ X-Y plotter)	CHEM TERM&PER	**Graphit** (→ Koordinatenschreiber)
graphics input hardware = graphics input device ↓ digitizing tablet; light pen	TERM&PER	**Graphikeingabegerät** = Grafikeingabegerät ↓ Digitalisiertablett; Lichtgriffel	**graph plotter** (→ plotter) **graph tablet** (→ digitizing tablet)	TERM&PER TERM&PER	(→ Plotter) (→ Digitalisiertablett)
			graph theory	MATH	**Graphentheorie**
			grasping grip	TECH	**Umfassungsgriff**
graphics integration [word processing]	DATA PROC	**Graphikeinbindung** [Textverarbeitung]	**grate** **graticule** = time frame	TECH INSTR	**Gitter** **Zeitraster**
graphics library	DATA PROC	**Graphikbibliothek**	**graticule** (→ grid)	TECH	(→ Raster)
graphics light pen	TERM&PER	**Graphikstift** = Graphikgriffel	**graticule line** = grid line	TECH	**Rasterlinie**
graphics memory = video memory; frame buffer; bit map 1	DATA PROC	**Graphikspeicher** = Grafikspeicher; Videospeicher	**graticule point** = grid point	TECH	**Rasterpunkt**
			grating	PHYS	**Gitter 2** [Optik]
graphics mode (→ graphic display mode)	DATA PROC	(→ Graphikmodus)	**grating** (→ grid)	SYS.INST	(→ Rost)
graphics monitor (→ graphics screen)	TERM&PER	(→ Graphikbildschirm)	**grating constant** (→ lattice constant)	PHYS	(→ Gitterkonstante)
graphic software ≈ graphic programm	DATA PROC	**Graphiksoftware** = Grafiksoftware ≈ Graphikprogramm	**gratis** (→ free of charge) **grave accent** [à] = accent grave ↑ diacritic mark; accent	ECON LING	(→ kostenlos) **Gravis** [à] = Accent grave ↑ diakritisches Zeichen; Akzent
graphics output device (→ graphics output hardware)	TERM&PER	(→ Graphikausgabegerät)			
graphics output hardware = graphics output device; graphics terminal ≈ graphic workstation ↓ plotter; graphics printer	TERM&PER	**Graphikausgabegerät** = Grafikausgabegerät; Graphikterminal; Grafikterminal ≈ graphisches Arbeitsplatzsystem ↓ Plotter; Graphikdrucker			
			gravel	CIV.ENG	**Kies**
			gravitation = gravity	PHYS	**Gravitation** = Erdanziehung
			gravitational constant [attraction of two masses of 1 g at distance of 1 cm] ≈ acceleration of gravity	PHYS	**Gravitationskonstante** [Anziehungskraft zweier Massen von 1 g im abstand von 1 cm] ≈ Fallbeschleunigung
graphics pad (→ digitizing tablet)	TERM&PER	(→ Digitalisiertablett)			
graphics primitive (→ drawing element)	DATA PROC	(→ Zeichenelement)			

gravitational field

gravitational field PHYS **Gravitationsfeld**
= gravity field = Schwerefeld; Erdanziehungsfeld

gravitational force PHYS **Gravitationskraft**
= gravity force = Schwerkraft; Erdanziehungskraft; Anziehungskraft der Erde
≈ weight ≈ Gewicht

gravitational potential PHYS **Gravitationspotential**

gravitational wave PHYS **Gravitationswelle**
 = Schwerewelle

gravity (→gravitation) PHYS (→Gravitation)
gravity field (→gravitational field) PHYS (→Gravitationsfeld)
gravity force (→gravitational force) PHYS (→Gravitationskraft)

gray (adj.) (NAM) PHYS **grau**
= grey (BRI) ≈ aschgrau

Gray PHYS **Gray**
[SI unit for energy dosis] [SI-Einheit für Energiedosis]
= Gy = Gy

gray area COLLOQ **Grauzone**
[fig.] [fig.]

gray cast iron METAL **Grauguß**

Gray code CODING **Gray-Code**
[a binary code with only one bit changing between sequential numbers] [binärer Code bei dem sich nur ein Bit von Wert zu Wert ändert]
= cyclic binary code; reflected binary code; reflected binary unit-distance code = zyklisch permutierter binärer Code; reflektierter binärer Code; zyklischer binärer Code
↑ cyclic code ↑ zyklischer Code

gray heat PHYS **Grauglut**

gray level PHYS **Graustufe**
= gray shade; gray tone; gray scale; half tone; screen [TV] = Grauwert; Grauton; Grautönung; Graustufung; Halbton

gray scale (→gray wedge) TV (→Graukeil)
gray scale (→gray level) PHYS (→Graustufe)
gray shade (→gray level) PHYS (→Graustufe)
gray tone (→gray level) PHYS (→Graustufe)

gray-tone transmission TELEGR **Grauwertübertragung**
= half-tone transmission

gray wedge TV **Graukeil**
= gray scale = Gradationskeil; Grautreppe

grazing TECH **streifend**
grazing sight RADIO PROP **streifende Sicht**
gread leak resistor ELECTRON **Gitterableitwiderstand**

grease (v.t.) MECH **einfetten**
≈ lubricate = fetten
 ≈ schmieren

grease (n.) MECH **Fett**
[thick lubricant] [dickflüssiges Schmiermittel]
↑ lubricant ↑ Schmiermittel

grease cup MECH **Fettbüchse**
greased MECH **gefettet**
grease lubrication MECH **Fettschmierung**
 ↑ Schmierung

great circle direction HF **Großkreisrichtung**
great circle distance (→greater circle distance) MATH (→Großkreisabstand)

greater circle MATH **Großkreis**
greater circle distance MATH **Großkreisabstand**
= great circle distance = Großkreisentfernung

greater or equal MATH **größer-gleich**
greater-or-equal sign MATH **Größer-gleich-Zeichen**
[symbol ≥] [Symbol ≥]
= greater-or-equal symbol

greater-or-equal symbol MATH (→Größer-gleich-Zeichen)
(→greater-or-equal sign)

greater relation MATH **Größer-Beziehung**
 = Größer-Relation

greater than MATH **größer als**
greater-than sign MATH **Größerzeichen**
[symbol >] [Symbol >]
= greater-than symbol; more-than sign; more-than symbol = Größer-als-Zeichen
↑ mathematical symbol ↑ mathematisches Zeichen

greater-than symbol MATH (→Größerzeichen)
(→greater-than sign)

greatest measure TECH **Höchstmaß**
 = Größtmaß

Great Primer (BRI) (→type size 18 point) TYPOGR (→Schriftgröße 18 Punkt)

grecian antenna (→Bruce antenna) ANT (→Bruce-Antenne)

green PHYS **grün**
greenfield site TECH **grüne Wiese**
[fig.] [fig.]

greenish PHYS **grünlich**
= greeny

green monitor TERM&PER **Grünmonitor**
= green phosphor monitor

green phosphor monitor TERM&PER (→Grünmonitor)
(→green monitor)

greeny (→greenish) PHYS (→grünlich)
G region (→G layer) RADIO PROP (→G-Schicht)

Gregorian reflector antenna ANT **Gregory-Antenne**
 = Gregorianische Antenne

Gregory system ANT **Gregory-System**
gremlin (n.) DATA PROC **Aussetzer**
[unexplained fault] [Fehler ungeklärter Ursache]

Gremmelmaier process MICROEL **Gremmelmaier-Verfahren**
grey (BRI) (→gray) PHYS (→grau)

grid SYS.INST **Rost**
= runway; shelf; grating; rack 2 ↓ Kabelrost
↓ cable rack

grid TECH **Raster**
= graticule = Gitterraster; Rasterfeld
≈ coordinate field [MATH] ≈ Koordinatenfeld [MATH]

grid antenna ANT **Grid-Antenne**
grid array ANT **Gittergruppe**
grid bias (→grid polarization voltage) ELECTRON (→Gittervorspannung)

grid capacitance ELECTRON **Gitterkapazität**
grid control ELECTRON **Gittersteuerung**
grid current ELECTRON **Gitterstrom**

gridding DATA PROC **Rastermodus**
[coincidence of all endpoints with grid point] [alle Endpunkte decken sich mit Rasterpunkten]

grid-dip meter INSTR **Grid-Dip-Meter**
grid dipper (→resonance frequency meter) INSTR (→Resonanzfrequenzmesser)

grid emission ELECTRON **Gitteremission**
grid leak ELECTRON **Gitterableitung**
grid leakage current PHYS **Gitterfehlstrom**
grid leak capacitor ELECTRON **Gitterableitkondensator**
grid-leak detector CIRC.ENG (→Audion)
(→regenerative detector)

grid line (→graticule line) TECH (→Rasterlinie)
grid network TELEC **Gitternetz**

grid plate POWER SYS **Gitterplatte**
= pasted plate; flat plate [Akkumulator]
 = OGi-Platte

grid-plate capacitance ELECTRON **Gitter-Anoden-Kapazität**
grid-plate transconductance ELECTRON (→Steilheit)
(→transconductance)

grid point (→graticule point) TECH (→Rasterpunkt)

grid polarization voltage ELECTRON **Gittervorspannung**
= grid bias ↑ Gitterspannung
↑ grid voltage

grid reflector ANT **Gitterreflektor**
grid-separation circuit CIRC.ENG (→Gitterbasisschaltung)
(→grounded-grid circuit)

gridsheet (→coordinate paper) ENG.DRAW (→Koordinatenpapier)

grid test pattern TV **Gittermuster**
grid voltage ELECTRON **Gitterspannung**
[electron tube] [Elektronenröhre]
↓ grid polarization voltage ↓ Gittervorspannung

grind (v.t.) TECH **schleifen**
≈ polish ≈ polieren; schärfen

grind (n.) MECH **Schliff**
[surface treatment] [Oberflächenbehandlung]
= grinding = Schleifen
↓ finish grind; rough grind ↓ Feinschliff; Grobschliff

grinding (→grind)	MECH	(→Schliff)
grinding machine	MECH	Schleifmaschine
grip 1 (n.) [action] ≡ manipulation 1	TECH	Handgriff 2 [Tätigkeit]
grip 2 (n.) = clutch ≈ access	TECH	Zugriff ≈ Zugang
grip 3 (n.)	TECH	Seilklemme
grip 4 (n.) = claw (n.) ≈ jaw	TECH	Greifer ≈ Klaue
grip (v.t.) = gripe (v.t.) ≈ jam	TECH	greifen ≈ klemmen 1
gripe (v.t.) (→grip)	TECH	(→greifen)
gripper arm	TECH	Greifarm
gripping	TECH	griffig
grooming (→traffic sorting)	TELEC	(→Dienstetrennung)
groove (v.t.)	MECH	nuten
groove 3 (n.) [a narrow channel to fit with a counterpiece] ≈ notch	TECH	Nut (n.f.; pl.-en) [längliche Vertiefung zur Aufnahme eines Gegenstücks] = Nute ≈ Kerbe ≠ Feder (Holz)
groove 1 (n.) [long narrow depression in hard material] ≈ groove 2 (n.)	TECH	Rille [lange schmale Vertiefung in hartem Material] = Riefe ≈ Furche
groove 2 (n.) [a linear V-shaped depression] ≈ groove 1	TECH	Furche [linienmäßige keilförmige Vertiefung] ≈ Rille
groove [phonogram record] ↓ monophonic groove; stereofonic groove	EL.ACOUS	Rille [Schallplatte] ↓ Mono-Rille; Stereo-Rille
groove (→joint 3)	MECH	(→Fuge)
grooved (→riveted)	METAL	(→genietet)
grooved pin	MECH	Kerbstift
groove speed	EL.ACOUS	Rillengeschwindigkeit
groove weld	METAL	Fugenhaht [Schweißen]
gross	ECON	brutto
gross bit rate (→transmission rate)	TELEC	(→Übertragungsgeschwindigkeit)
gross sales	ECON	Bruttoumsatz = Rohumsatz
gross timer (→long timer)	CIRC.ENG	(→Langzeitglied)
gross weight	ECON	Bruttogewicht = Rohgewicht
groumet = rubber	COMPON	Durchführungstülle
ground (n.) ↓ soil; floor	TECH	Boden 1 ↓ Erdboden; Fußboden
ground (adj.)	MECH	geschliffen
ground (AM) = earth (BRI); GND ≈ neutral conductor; protective earth conductor	EL.TECH	Erde = Masse ≈ Nulleiter; Schutzerdungsleiter
ground (v.t.)(BRI) (→earth)	EL.TECH	(→erden)
ground (→chassis ground)	ELECTRON	(→Masse)
ground anchorage	CIV.ENG	Bodenverankerung
ground antenna = near-to-ground antenna ↓ Beverage antenna	ANT	Erdantenne 1 = erdnahe Antenne ↓ Beverage-Antenne
ground antenne	ANT	Bodenantenne
ground-based duct (→surface duct)	RADIO PROP	(→erdnaher Dukt)
ground bus = ground distributor ↑ busbar	SYS.INST	Erdsammelschiene = Erdschiene; Erdverteilschiene; Erdungsschiene ↑ Sammelschiene
ground clamp	COMPON	Erdungsklemme
ground communications equipment	SAT.COMM	Bodengerät
ground conductor (→earth lead)	EL.TECH	(→Erdungsleiter)
ground connection (→earth connection)	EL.TECH	(→Erdung)
ground distance [distance between two points of equal elevation] ≠ slant distance	RADIONAV	Bodenentfernung [Entfernung zweier Punkte gleicher Elevation] ≠ Geradeausentfernung
ground distributor (→ground bus)	SYS.INST	(→Erdsammelschiene)
grounded (→earthed)	EL.TECH	(→geerdet)
grounded-anode circuit = cathode-coupled circuit; cathode follower	CIRC.ENG	Anoden-Basis-Schaltung = Kathodenfolger; Kathodenverstärker
grounded-base circuit = inter-base circuit; grounded-base connection; inter-base connection	CIRC.ENG	Zwischen-Basis-Schaltung
grounded base circuit (→common base connection)	CIRC.ENG	(→Basisschaltung)
grounded-base connection (→common base connection)	CIRC.ENG	(→Basisschaltung)
grounded-base connection (→grounded-base circuit)	CIRC.ENG	(→Zwischen-Basis-Schaltung)
grounded cathode amplifier (→cathode base circuit)	CIRC.ENG	(→Kathodenbasisschaltung)
grounded cathode circuit (→cathode base circuit)	CIRC.ENG	(→Kathodenbasisschaltung)
grounded emitter circuit (→common-emitter connection)	CIRC.ENG	(→Emitterschaltung)
grounded-grid amplifier = ground-grid amplifier	CIRC.ENG	Gitterbasisverstärker
grounded-grid circuit = current follower; grid-separation circuit	CIRC.ENG	Gitterbasisschaltung
ground effect	ANT	Bodeneffekt
ground fault (AM) = earth fault (BRI); contact to earth; short-circuit to earth; short-circuit to ground; accidental ground; ground leak; shorting to earth; short to earth; short to ground;	EL.TECH	Erdschluß = Erdfehler; Masseschluß
ground-fault neutralizer = Petresen coil; earth leakage coil	POWER SYS	Erdlöschspule = Petersen-Spule; Erdschlußlöschspule
ground floor (→first floor)	CIV.ENG	(→Erdgeschoß)
ground glass (→frosted plate)	PHYS	(→Mattscheibe)
ground-grid amplifier (→grounded-grid amplifier)	CIRC.ENG	(→Gitterbasisverstärker)
grounding (→earth connection)	EL.TECH	(→Erdung)
grounding conductor = ground lead	EQUIP.ENG	Erdleitung = Masseleitung
grounding conductor (→earth lead)	EL.TECH	(→Erdungsleiter)
grounding contact (→protective contact)	EL.INST	(→Schutzkontakt)
grounding key	TELEPH	Erdtaste
grounding kit	ANT	Erdungsmuffe
grounding lead (→earth lead)	EL.TECH	(→Erdungsleiter)
grounding spanner (→earth terminal)	COMPON	(→Erdklemme)
grounding strap	COMPON	Erdungslasche
grounding switch (→earthing switch)	SYS.INST	(→Erdungsschalter)
grounding washer	COMPON	Erdungsscheibe
grounding wire (→earth lead)	EL.TECH	(→Erdungsleiter)
ground lead (→grounding conductor)	EQUIP.ENG	(→Erdleitung)
ground lead (→earth lead)	EL.TECH	(→Erdungsleiter)
ground leak (→ground fault)	EL.TECH	(→Erdschluß)
ground loop (→earth loop)	TELEC	(→Erdschleife)
ground losses (→earth current losses)	ANT	(→Erdverluste)
ground-moded ≠ overmoded	MICROW	grundmodiert ≠ Übermodiert

ground mounted

ground mounted	TECH	am Boden aufgestellt	
		= für Bodenaufstellung	
ground mounting (→floor installation)	TECH	(→Bodenaufstellung)	
ground plane (→counterpoise)	ANT	(→Gegengewicht)	
ground plane antenna	ANT	Groundplane-Antenne	
↑ vertical antenna; monopole antenna		↑ Vertikalantenne; Monopol	
ground potential (→earth potential)	EL.TECH	(→Erdpotential)	
ground reflection	RADIO PROP	Bodenreflexion	
ground-reflection multiplier	ANT	Multiplikationsfaktor	
ground resistance (→earthing resistance)	EL.TECH	(→Erdungswiderstand)	
ground return (→earth return circuit)	TELEC	(→Erdrückleitung)	
ground return circuit (→earth return circuit)	TELEC	(→Erdrückleitung)	
ground return current (→earth return current)	EL.TECH	(→Erdrückstrom)	
ground ribbon	SYS.INST	Banderder	
ground rod	EL.TECH	Erdungsstab	
= earth rod		= Erdstab; Erdungsstange; Staberder; Rohrerder	
↑ earth electrode		↑ Erder	
ground signal	TELEC	Erdimpuls	
ground signaling	TELEC	Erdimpulsgabe	
ground state	PHYS	Grundniveau	
		= Grundzustand	
ground system	ANT	Radialnetz	
		= Erdnetz	
ground system (→earth electrode)	EL.TECH	(→Erder)	
ground terminal (→earth terminal)	COMPON	(→Erdklemme)	
ground wave (AM) (→direct wave)	RADIO PROP	(→Bodenwelle)	
ground-wave attenuation	RADIO PROP	Bodenwellendämpfung	
ground wire	POWER ENG	Erdseil	
= screen wire; guard wire		= Blitzschutzseil; Schutzleiter 2; Schutzdraht	
ground wire	ANT	Erdleitung	
		= Erddraht	
ground wire (→earth lead)	EL.TECH	(→Erdungsleiter)	
group (v.t.)	TECH	gruppieren	
= cluster			
group (n.)	TECH	Gruppe	
= cluster		= Komplex	
≈ array; system		≈ Anordnung; System	
group (n.)	MATH	Gruppe	
≈ cluster		≈ Klumpen	
group 1	TRANS	Gruppe 1	
[FDM]		[TF-Technik]	
↓ primary group; secundary group		↓ Primärgruppe; Sekundärgruppr	
group (→trunk group)	SWITCH	(→Leitungsbündel)	
group 2 (→primary group)	TRANS	(→Primärgruppe)	
group (→grouping)	TECH	(→Gruppierung)	
group address	DATA COMM	Gruppenadresse	
group address code	DATA COMM	Gruppenkennzeichenwahl	
group calling	TELEPH	Gruppenruf	
group change	SWITCH	Gruppenwechsel	
group clock generator	SWITCH	Gruppentaktgenerator	
group code	CODING	Gruppencode	
group-coded recording	DATA PROC	gruppencodierte Aufzeichnung	
= GCR		= GCR-Aufzeichnung; GCR-Codierung; Gruppencodierung	
↑ recording mode		↑ Aufzeichnungsmethode	
group communication system	DATA COMM	Gruppenkommunikationssystem	
group-controlled	SWITCH	gruppengesteuert	
[electromechanic switching system]		[elektromechanische Wählsysteme]	
group-controlled switching system	SWITCH	gruppengesteuertes Vermittlungssystem	
[electromechanical switching system]		[elektromechanisches Vermittlungssystem]	
		= gruppengesteuertes Wählsystem	
group delay	TELEC	Gruppenlaufzeit
= envelope delay; group delay time		= Signallaufzeit
group delay distortion	TELEC	Gruppenlaufzeitverzerrung
= envelope delay distortion		
group delay measuring set	INSTR	Gruppenlaufzeit-Meßgerät
group delay time (→group delay)	TELEC	(→Gruppenlaufzeit)
grouped-time mode	TELEC	(→Zeitgetrenntlage-Verfahren)
(→time-division mode)		
grouped time operation (→burst operation)	TRANS	(→Burstverfahren)
grouped-time operation (→time-division mode)	TELEC	(→Zeitgetrenntlage-Verfahren)
group frame	SYS.INST	Gruppenrahmen
group information	TELECONTR	(→Sammelmeldung)
(→collective indication)		
grouping (n.)	TECH	Gruppierung
= group; cluster; clustering		
grouping	SWITCH	Bündelung
[of lines with equal traffic relation]		[Zusammenfassung von Leitungen gleicher Verkehrsbeziehung]
= trunking		
grouping	MATH	Klasseneinteilung
[statistics]		[Statistik]
		= Klassifizierung
grouping (→clustering)	DATA PROC	(→Gruppierung)
grouping mark	DATA PROC	Gruppenmarke
= group mark; group marker		
grouping of cables	POWER ENG	Häufung
group interconnection	SWITCH	Gruppenverbindung
group interconnection plan	SWITCH	Gruppierungsübersicht
		= Gruppenverbindungsplan
group leader	ECON	Gruppenführer
group mark (→grouping mark)	DATA PROC	(→Gruppenmarke)
group marker (→grouping mark)	DATA PROC	(→Gruppenmarke)
group modulator	TRANS	Primärgruppenumsetzer
[FDM]		[TF-Technik]
		= PGU
group of trunks (→trunk group)	SWITCH	(→Leitungsbündel)
group pilot	TRANS	Gruppenpilot
[FDM]		[TF-Technik]
↓ primary grpou pilot; supergroup pilot		↓ Primärgruppenpilot; Sekundärgruppenpilot
group poll	DATA PROC	Gruppenabfrage
group processor	SWITCH	Gruppenprozessor
group selection	SWITCH	Gruppenwahl
		[Richtungswahl]
group-selection stage	SWITCH	Richtungswahlstufe
		= Gruppenwahlstufe; Verteilstufe
group selector	SWITCH	Gruppenwähler
[selects the direction in step-by-step systems, directly controlled by digits of the called number]		[Richtungswahl in Direktwahlsystemen, durch Rufnummerstellen direkt gesteuert]
= group switch; group switching unit		= Richtungskoppelfeld; Gruppenkoppler
group separator	DATA PROC	Gruppentrennung
[code]		[Code]
= GS		= GS
group statements	ECON	Konzernabschluß
group switch (→group selector)	SWITCH	(→Gruppenwähler)
group switching unit (→group selector)	SWITCH	(→Gruppenwähler)
group theory	MATH	Gruppentheorie
[algebra]		[Algebra]
group velocity	EL.TECH	Gruppengeschwindigkeit
= envelope velocity		
group velocity of sound (→sound particle velocity)	ACOUS	(→Schallschnelle)
grow (→pull)	MICROEL	(→ziehen)
grow back	MICROEL	Grow Back
growing (→pulling)	MICROEL	(→Ziehen)
grown diode	MICROEL	gezogene Diode
growth	ECON	Wachstum
growth (→increase 1)	TECH	(→Zunahme)

growth capability (→expansion capability)	TECH	(→Erweiterungsmöglichkeit)	
growth enhancement (→expansion)	TECH	(→Erweiterung)	
growth figure	OUTS.PLANT	Wachstumskennzahl	
growth from melt	MICROEL	Schmelzzüchtung	
growth impulse	ECON	Wachstumsimpuls	
growth industry	ECON	Wachstumsindustrie	
growth market	ECON	Wachstumsmarkt	
growth promoting	ECON	wachstumsfördernd	
growth rate	ECON	Wachstumsrate	
growth rate	TECH	Zuwachsrate	
growth time (→rise time)	ELECTRON	(→Anstiegzeit)	
GS (→glide slope)	RADIO NAV	(→Gleitwegsender)	
GS (→group separator)	DATA PROC	(→Gruppentrennung)	
G/T (→gain-to-noise-temperature ratio)	SAT.COMM	(→Empfangsempfindlichkeit)	
G time (→mean Greewich time)	PHYS	(→Weltzeit)	
GTO (→gate-turn-off thyristor)	MICROEL	(→Abschaltthyristor)	
GTO thyristor (→gate-turn-off thyristor)	MICROEL	(→Abschaltthyristor)	
guarantee (v.t.) = warrant	ECON	garantieren	
guarantee 2 (→guaranty 2)	ECON	(→Bürgschaft)	
guarantee 1 (→guaranty 1)	ECON	(→Garantie)	
guarantee (→secure)	COLLOQ	(→sicherstellen)	
guaranty 2 = guarantee 2; surety	ECON	Bürgschaft	
guaranty 1 = guarantee 1; warranty	ECON	Garantie = Gewährleistung; Zusicherung	
guaranty (→security 1)	ECON	(→Kaution)	
guard (→tip wire)	TELEPH	(→C-Ader)	
guard (n.) (→protection)	TECH	(→Schutz)	
guard action (→speech protection)	TELEC	(→Sprachschutz)	
guard band	RADIO	Schutzband = Schutzfrequenzband	
guard bit	DATA PROC	Schutzbit	
guard cap (→protecting cap)	TECH	(→Schutzkappe)	
guard delay	SWITCH	Sperrzeit	
guard digit	DATA COMM	Schutzziffer	
guarded protected area	DATA PROC	überwachter Bereich	
guard electrode	INSTR	Schutzelektrode	
guarding (→speech protection)	TELEC	(→Sprachschutz)	
guard protection systems	SIGN.ENG	Wächterschutzsystem	
guard-ring capacitor	COMPON	Schutzringkondensator	
guard sector = protection sector	RADIO	Schutzsektor	
guard separation	RADIO	Schutzabstand	
guard signal	DATA PROC	Ausblendsignal	
guard supervision system	SIGN.ENG	Wächterkontrollsystem	
guard time (→blocking time)	ELECTRON	(→Sperrzeit)	
guard wire (→tip wire)	TELEPH	(→C-Ader)	
guard wire (→ground wire)	POWER ENG	(→Erdseil)	
guest computer (→slave computer)	DATA COMM	(→Nebenrechner)	
guest-host effect	OPTOEL	Gast-Wirt-Effekt	
GUI (→graphical user interface)	DATA PROC	(→graphische Benutzeroberfläche)	
guidance ≈ control	TECH	Lenkung ≈ Steuerung	
guide (n.)	MECH	Führung	
guide	DOC	Leitfaden	
guide (→steer)	TECH	(→lenken)	
guide bar [at the beginning and end of a bar code]	DATA PROC	Trennungszeichen [am Anfang und Ende eines Strichcode]	
guide bolt	MECH	Führungsbolzen	
guide bushing	MECH	Führungsbuchse	
guided mode (→core mode)	OPT.COMM	(→Kernmodus)	
guided probe	INSTR	rechnergeführter Tastkopf	
guided wave (→transverse electromagnetic wave)	LINE TH	(→transversale elektromagnetische Welle)	
guide hole	MECH	Führungsloch	
guideline 1 = directive (n.)	TECH	Richtlinie = Leitlinie; Richtschnur	
guideline	TYPOGR	Zeilenstrich	
guideline 2 (→reference value 1)	TECH	(→Richtwert)	
guide pin = aligning plug; spigot ≈ fixing pin	MECH	Führungsstift = Paß-Stift ≈ Fixierstift	
guide pulley	MECH	Führungsscheibe	
guide pulley	OUTS.PLANT	Kabelführungsrolle	
guide rail	MECH	Führungsschiene	
guide ring	MECH	Führungsring	
guide rod	MECH	Führungsstab	
guide roller	MECH	Führungsrolle	
guide sleeve	MECH	Führungshülse	
guide slot	MECH	Führungsnut	
guide strip	MECH	Führungsleiste	
guide way	MECH	Führungsbahn	
guiding dimension	TECH	Richtmaß	
guiding price (→target price)	ECON	(→Richtpreis)	
guiding value (→reference value 1)	TECH	(→Richtwert)	
guitar microphone	EL.ACOUS	Gitarre-Mikrofon	
gulp (n.) [group of bytes]	DATA PROC	Bytegruppe	
gummed	TECH	gummiert	
gummed label (→sticker)	TECH	(→Aufkleber)	
gun current	ELECTRON	Elektronenkanonenstrom	
Gunn diode	MICROEL	Gunndiode	
Gunn effect	PHYS	Gunn-Effekt	
Gunn oscillator	MICROW	Gunnoszillator	
gust (→windflaw)	METEOR	(→Windstoß)	
gutta-percha	TECH	Guttapercha	
gutter (n.)	TECH	Rinne	
gutter (n.) [inside margin of facing pages or margin between columns] ≈ gutter margin	TYPOGR	Zwischenschlag [unbedruckter Innenrand zwischen gegenüberliegenden Seiten oder Spalten] = Bundsteg = Bund	
gutter margin [inside margin of a page] ≈ gutter	TYPOGR	Bund [innerer Rand einer Seite] ≈ Zwischenschlag	
gutter rule	TYPOGR	Spaltentrennlinie	
guy (n.)	OUTS.PLANT	Abspannung	
↓ guy rope		↓ Abspannseil	
guy (v.t.) (→brace 1)	TECH	(→verspannen)	
guy anchor (→stay)	OUTS.PLANT	(→Anker)	
guyed [mast, tower] ≠ self-supporting	CIV.ENG	abgespannt [Mast, Turm] ≠ freistehend	
guyed antenna mast = guyed antenna tower	ANT	abgespannter Antennenmast = abgespannter Antennenturm	
guyed antenna tower (→guyed antenna mast)	ANT	(→abgespannter Antennenmast)	
guyed mast = guyed tower;	CIV.ENG	abgespannter Mast = abgespannter Turm	
guyed tower (→guyed mast)	CIV.ENG	(→abgespannter Mast)	
guy insulator [open wire line]	OUTS.PLANT	Abspannisolator [Freileitung]	
guy rope = stay; stay rope; stay wire; anchor guy; down guy; guy wire ↑ guy	OUTS.PLANT	Abspannseil = Ankerdraht; Pardune (f.); Pardun (n.) ↑ Abspannung	
guy wire (→guy rope)	OUTS.PLANT	(→Abspannseil)	
Gy (→Gray)	PHYS	(→Gray)	
gypsum	CHEM	Gips (m.)	
gyrator	NETW.TH	Gyrator = Dualübersetzer	
gyrator	MICROW	Gyrator	
gyrofrequency (→angular frequency)	PHYS	(→Kreisfrequenz)	
gyromagnetic effect	PHYS	gyromagnetischer Effekt	
gyromagnetic ratio = gyromagnetic relation	PHYS	gyromagnetisches Verhältnis	
gyromagnetic relation (→gyromagnetic ratio)	PHYS	(→gyromagnetisches Verhältnis)	
gyroscope	PHYS	Kreisel	

H

h (→hour) PHYS (→Stunde)
H (→hydrogen) CHEM (→Wasserstoff)
H (→high level) MICROEL (→Hochpegelzustand)
H (→Henry) PHYS (→Henry)
ha (→hectare) PHYS (→Hektar)
Ha (→hahnium) CHEM (→Hahnium)
habituation COLLOQ **Gewöhnung**
hacker 1 DATA PROC **Hacker 1**
  [somebody trying to learn computing by trial and error] [jemand der durch Probieren Computerei lernen will]
hacker 2 DATA PROC **Hacker 2**
  [person breaking into computer systems via data communication] [jemand der per Datenübertragung arglistig in fremde Datenbanken eindringt]
  ≈ cracker; wizard ≈ Cracker; Wizard
H Adcock antenna ANT **H-Adcock-Antenne**
hafnium CHEM **Hafnium**
  = Hf = Hf
haft (n.) (→handle 1) TECH (→Handgriff 1)
hahnium CHEM **Hahnium**
  = Ha = Ha
hairline (adj.) TYPOGR **besonders dünn**
  [especially thin typeface] = zart
  ≈ light ↑ Schriftschnitt
  ↑ type design
hairline DATA PROC **Hairline**
  [the finest stroke width of a graphics program] [dünnste Strichstärke eines Graphikprogramms]
hairline crack (→flake) TECH (→Haarriß)
hairpin- (→U-shaped) TECH (→U-förmig)
hairpin loop LINE TH **Haarnadelschleife**
half (pl. -ves) MATH **Hälfte**
half (pl. -ves) TECH **Hälfte**
  [of a piece] [halber Gegenstand]
half-adder CIRC.ENG **Halbaddierer**
  [neglects carries] [vernachlässigt Überträge]
  = one-digit adder; two-input adder = Halbadder; einstelliges Addierwerk
  ≠ full adder ≠ Volladdierer
half bridge CIRC.ENG **Halbbrücke**
half-byte (→tetrad) CODING (→Tetrade)
half-carry MATH **Halbübertrag**
half dipole ANT **Halbdipol**
half duplex TELEC **Halbduplexbetrieb**
  [communication possible in either direction, but not simultaneously] [Übertragung in beiden Richtungen möglich, jedoch nicht gleichzeitig]
  = half-duplex operation; half-duplex transmission; half-duplex mode; either-way operation; either-way transmission; either-way mode; two-way alternate operation; two-way alternate transmission; two-way alternate mode; simplex operation [TELEPH] = Halbduplex; Wechselverkehr; wechselseitiger Betrieb; wechselseitige Übertragung; absatzweise Sende- und Empfangsbetrieb
  ≈ semi-duplex ≈ Semiduplexbetrieb
half-duplex data communication (→two-way alternate communication) DATA COMM (→wechselseitige Datenübermittlung)
half-duplex mode (→half duplex) TELEC (→Halbduplexbetrieb)
half-duplex operation (→half duplex) TELEC (→Halbduplexbetrieb)
half-duplex transmission (→half duplex) TELEC (→Halbduplexbetrieb)
half echo suppressor TELEPH **Halbechosperre**
half frame CODING **Halbrahmen**
half height (→half value) MATH (→Halbwert)
half-life PHYS **Halbwertszeit**
  = half-life period; half-time period [Kernzerfall]
half-life period (→half-life) PHYS (→Halbwertszeit)
half-mat TYPOGR **halbmatt**
  = half-matt; half-matte; semi-mat; semi-matt; semi-matte
half-matt (→half-mat) PHYS (→halbmatt)
half-matte (→half-mat) PHYS (→halbmatt)
half of earth surface (→hemisphere) GEOPHYS (→Hemisphäre)

half period PHYS **Halbperiode**
  ≈ half wave ≈ Halbwelle
half-plane MATH **Halbebene**
half power PHYS **Halbwertsleistung**
half-power angle ANT **Halbwertswinkel**
half-power beamwidth ANT **Halbwertsbreite**
  = antenna beamwidth; half power width
half power width (→half-power beamwidth) ANT (→Halbwertsbreite)
half-round MECH **halbrund**
halfround-head screw (→round-head screw) MECH (→Halbrundkopfschraube)
half-section NETW.TH **Halbglied**
half section ENG.DRAW **halb Schnitt – halb Ansicht**
half-shell TECH **Halbschale**
half-shell sleeve OUTS.PLANT **Halbschalenmuffe**
half size ENG.DRAW **Maßstab 1:2**
  = scale 1:2
half sloper (→quarter-wave half sloper antenna) ANT (→Halb-Sloper)
half-space MATH **Halbraum**
half-square antenna ANT **Half-square-Antenne**
half subtracter CIRC.ENG **Halbsubtrahierer**
  = one-digit subtracter; half subtractor ≠ Vollsubtrahierer
  ≠ full subtracter
half subtractor (→half subtracter) CIRC.ENG (→Halbsubtrahierer)
half-time period (→half-life) PHYS (→Halbwertszeit)
half title TYPOGR **Respektblatt**
  [sheet preceding the titel page, void or with short title] [Blatt vor Titelseite eines Buches, leer oder mit Kurztitel]
  ≈ title page = Respekt; Schmutztitel; Schmutzblatt
  ≈ Titelblatt
halftone ACOUS **Halbton**
halftone (→continuous tone) TYPOGR (→Halbton)
half tone (→gray level) PHYS (→Graustufe)
half-tone master TYPOGR **Halbtonvorlage**
half-tone transmission (→gray-tone transmission) TELEGR (→Grauwertübertragung)
halftoning DATA PROC **Punktschattierung**
half value MATH **Halbwert**
  = half height
half wave PHYS **Halbwelle**
  ≈ half period; alternation ≈ Halbperiode; Wechsel
half-wave antenna ANT **Halbwellenantenne**
half-wave balun ANT **Balun-Leitung**
half-wave circuit EL.TECH **Einwegschaltung**
half wave dipole ANT **Halbwellendipol**
  = Lambdahalbedipol
half-wave end-fed antenna ANT **Halbwellen-Zeppelinantenne**
  = half-wave Zeppelin antenna ≈ Halbwellen-Zepp
half wave folded dipole ANT **Halbwellenfaltdipol**
half wavelength PHYS **Halbwellenlänge**
half-wavelength section NETW.TH (→Lambda-Halbe-Transformator)
  (→half-wavelength transformer)
half-wavelength transformer NETW.TH **Lambda-Halbe-Transformator**
  = half-wavelength section; half-wave transformer; half-wave section; half-wave matching line = Lambda-Halbe-Anpassungsglied; Lambda-Halbe-Leitunggalbwellen-Anpassungsglied; Halbwellentransformator; Halbwellenleitung; Halbwellen-Anpaßleitung
half-wave loop antenna ANT **Halo-Antenne**
  = halo antenna ≈ Ringdipol
  ≈ circular dipole ↑ Faltdipol
  ↑ folded dipole
half-wave matching line NETW.TH (→Lambda-Halbe-Transformator)
  (→half-wavelength transformer)
half-wave power supply POWER SYS **Halbwellenstromversorgung**
half-wave radiator ANT **Halbwellenstrahler**
  ↓ half-wave dipole ↓ Halbwellendipol
half-wave rectifier POWER SYS **Einweggleichrichter**
  = single-way rectifier
half-wave section NETW.TH (→Lambda-Halbe-Transformator)
  (→half-wavelength transformer)

half wave transformer	EL.TECH	Halbwellentransformator	hand drive (→manual drive)	TECH	(→Handantrieb)
half-wave transformer	NETW.TH	(→Lambda-Halbe-Transformator)	hand-held ...	EQUIP.ENG	Hand-...
(→half-wavelength transformer)			hand-held (→portable)	TECH	(→tragbar)
half-wave Zeppelin antenna	ANT	(→Halbwellen-Zeppelinantenne)	hand-held computer	DATA PROC	(→Taschencomputer)
(→half-wave end-fed antenna)			(→pocket computer)		
half width	PHYS	Halbwertsbreite	handheld mobile telephone	MOB.COMM	Handfunktelefon
halfword	DATA PROC	Halbwort	↑ cellular phone		↑ Zellulartelefon
half-yearly (→semiannual)	COLLOQ	(→halbjährlich)	handheld radiotelephone	RADIO	(→Handfunksprechgerät)
hall 1	CIV.ENG	Halle	(→walkie-talkie)		
[a large building consisting of one large room, or such a large room for public or economic activities]		[großes Gebäude aus einem großen Raum, oder ein solcher großer Raum für öffentliche oder kommerzielle Zwecke]	hand-held reader	TERM&PER	Handleser
			= hand reader; hand-held scanner; hand scanner; manual reader; manual scanner		= Handlesegerät; Handscanner
≈ hall 2			≠ stationary document reader		≠ stationärer Belegleser
		≈ Saalbau; Saal	↑ document reader		↑ Belegleser
hall 2	CIV.ENG	Saal	↓ code pen; scanning pistol		↓ Lesestift; Lesepistole
[a large room in a building]		[großer Raum in einem Gebäude]	hand-held scanner	TERM&PER	(→Handleser)
≈ hall 1			(→hand-held reader)		
↑ room 1		≈ Halle	hand-held terminal	EQUIP.ENG	Handbediengerät
		↑ Raum	handheld transceiver	RADIO	(→Handfunksprechgerät)
Hall constant	PHYS	Hall-Konstante	(→walkie-talkie)		
		= Hall-Koeffizient	handing over (→delivery)	ECON	(→Lieferung)
Hall effect	PHYS	Hall-Effekt	handing-over	TELEC	(→Weitersendung)
Hall effect sensor	COMPON	Hall-Magnetfeldsensor	handi-talkie (→walkie-talkie)	RADIO	(→Handfunksprechgerät)
Hall-effect vane switch	COMPON	Hall-Magnetgabelschranke	handle (v.t.)	ECON	abwickeln 3
= Hall magnet barrier		= Magnetschraube	[a matter]		[einen Vorgang]
Hall generator	COMPON	Hall-Generator	handle (v.t.)	SWITCH	abwickeln 2
		= Hall-Sonde; Hall-Element; Hall-Multiplikator	[traffic]		[Verkehr]
			handle 1 (n.)	TECH	Handgriff 1
Hall magnet barrier	COMPON	(→Hall-Magnetgabelschranke)	[device to be grasped by hand]		[Vorrichtung zum Anfassen]
(→Hall-effect vane switch)			= haft (n.); helve		= Griff
Hall mobility	PHYS	Hall-Beweglichkeit	↓ carrying handle; clasp handle; stick (n.); handle 2; knob		↓ Tragegriff; Klappgriff; Stiel; Henkel; Knopf
Hall voltage	PHYS	Hall-Spannung			
halo (pl.-os, -oes)	PHYS	Lichthof	handle 2 (n.)	TECH	Henkel
≈ corona		= Halo (n.n., pl.-onen)	↑ handle 1		↑ Handgriff 1
		≈ Hof	handlebar	TECH	Lenkstange
halo antenna (→half-wave loop antenna)	ANT	(→Halo-Antenne)			= Lenker
			handle cover	EQUIP.ENG	Griffblende
halogen	CHEM	Halogen	= handle strip		≈ Frontabdeckung
halt (n.)	DATA PROC	Halt	≈ face plate		
= halt instruction; breakpoint instruction		= Haltanweisung; Stopp	handler (→driver software)	DATA PROC	(→Treiber)
halt (v.t.) (→suspend)	ELECTRON	(→anhalten)	handle recess (→recessed grip)	TECH	(→Griffmulde)
halt command	DATA PROC	Abbruchbefehl	handle strip (→handle cover)	EQUIP.ENG	(→Griffblende)
= kill job		= Abbruchanweisung	handling	DATA PROC	Abwicklung
halt condition (→hold)	DATA PROC	(→Haltezustand)	= servicing		
halt instruction (→halt)	DATA PROC	(→Halt)	handling	TECH	Handhabung
haltpoint (→breakpoint)	DATA PROC	(→Unterbrechungspunkt)	≈ manipulation 2		= Bedienung; Hantierung; Umgang
halve (v.t.)	TECH	halbieren	≈ operation		≈ Betrieb
= bisect		= zweiteilen	handling	OFFICE	Bearbeitung
		≈ trennen	= transaction handling		= Vorgangsbearbeitung
ham (→radio amateur)	RADIO	(→Funkamateur)	handling (→transshipment)	ECON	(→Warenumschlag)
ham antenna	ANT	Amateurfunkantenne	handling charge	ECON	Bearbeitungsgebühr
Hamilton operator (→vector operator)	MATH	(→Nablaoperator)	handling noise	EL.ACOUS	Körperschall
			hand microphone	EL.ACOUS	Handmikrophon
hammer (v.t.)	METAL	hämmern	= close-talking micriphone		
hammer (n.)	TECH	Hammer	hand-off 1 (n.)	MOB.COMM	Gesprächsumschaltung
↑ tool		↑ Werkzeug	= hand-over (n.)		= Verbindungsumschaltung; Hand-off
hammer	INSTR	Hammer	↓ intracell hand-off; intercell hand-off		
[for modal analysis]		[für Modenanalyse]			↓ Kanalwechsel; Zellenwechsel
hammer (→bell hammer)	COMPON	(→Glockenköppel)			
hammered	METAL	gehämmert	hand-off 2 (→intracell hand-off)	MOB.COMM	(→Kanalwechsel)
hammer-effect varnish	TECH	Hammerschlaglack			
hammer welding (→forge welding)	METAL	(→Feuerschweißen)	hand-off receiver	RADIO	Absetzempfänger
			hand operation (→manual operation)	TECH	(→manueller Betrieb)
Hamming code	CODING	Hammingcode			
Hamming distance	CODING	Hammingdistanz	hand over	COLLOQ	übergeben
= signal distance		= Hammingabstand	hand-over (n.)	MOB.COMM	(→Gesprächsumschaltung)
		≈ Stellendistanz	(→hand-off 1)		
handbill	COLLOQ	Flugblatt	hand-portable (→portable)	TECH	(→tragbar)
handbook	TECH	Handbuch	handpunch (→card puncher)	TERM&PER	(→Kartenlocher)
= manual; reference manual					
hand calculator (→pocket calculator)	DATA PROC	(→Taschenrechner)	hand reader (→hand-held reader)	TERM&PER	(→Handleser)
hand dictating equipment	OFFICE	(→Handdiktiergerät)	hand saw (→ripsaw)	TECH	(→Handsäge)
(→hand dictating set)			hand scanner (→hand-held reader)	TERM&PER	(→Handleser)
hand dictating set	OFFICE	Handdiktiergerät			
= hand dictating equipment		= Reisediktiergerät; Taschendiktiergerät			

handset	TELEPH	**Handapparat**
		= Hörer; Fernhörer; Mikrotelefon
handset cord	TELEPH	**Hörerschnur**
= telephone cord; telephone cable		= Handapparateschnur; Anschlußschnur; Fernsprecher-Anschlußschnur; Telefon-Anschlußschnur
↑ connecting cord [EQUIP.ENG]		↑ Anschlußschnur [EQUIP.ENG]
handsfree dialling	MOB.COMM	**Freihandswahl**
handsfree equipment	TELEPH	(→Freisprechtelefon)
(→handsfree telephone)		
handsfree operation	TELEPH	(→Freisprechen)
(→handsfree talking)		
handsfree talking	TELEPH	**Freisprechen**
= handsfree operation; freephone		[ohne den Hörer abzunehmen]
		= Freihandsprechen
handsfree telephone	TELEPH	**Freisprechtelefon**
= handsfree equipment		= Freisprechtelephon; Freisprechanlage; Freisprechgerät; Freispracheinrichtung
handshake (n.)	DATA COMM	(→Quittungsaustausch)
(→handshaking)		
handshake mode	DATA COMM	(→Quittungsaustausch)
(→handshaking)		
handshaking	DATA COMM	**Quittungsaustausch**
[standardized procedure to establish and check a communication]		[Ritual zum Aufbau und Prüfen einer Verbindung]
= handshake (n.); handshake mode		= Quittungsbetrieb; einleitender Signalisierungsaustausch; Handshake-Betrieb
≈ protocoll		≈ Protokoll
hands off (→free-hand)	TECH	(→freihändig)
hands-on (n.)	TECH	**Handanlegen**
hands-on training	TECH	**Schulung am Gerät**
hand tool	TECH	**Handwerkzeug**
hand turner	TERM&PER	**Handspuler**
[for cassettes]		[Cassette]
hand twist (→twisted joint)	OUTS.PLANT	(→Würgeverbindung)
handwriting 1	LING	**Handschrift 1**
≈ manuscript		≈ Manuskript
handwriting 2	LING	**Handschrift 2**
[freehand written letters]		[freihändig erzeugte Schrift]
= handwritten lettering		= Schreibschrift
handwriting reader	TERM&PER	**Handschriftenleser**
= handwritten document reader		= Handschriftleser
handwritten	LING	**handschriftlich**
handwritten document reader	TERM&PER	(→Handschriftenleser)
(→handwriting reader)		
handwritten lettering	LING	(→Handschrift 2)
(→handwriting 2)		
handy (adj.)	TECH	**handlich**
hanging indent (→hanging indentation)	TYPOGR	(→hängender Einzug)
hanging indentation	TYPOGR	**hängender Einzug**
[of a line block with respect to the begin of the chapter or the first line]		[Einzug eines Zeilenblocks gegenüber dem Absatzanfang oder der ersten Zeile]
= hanging indent; hanging indention		
hanging indention (→hanging indentation)	TYPOGR	(→hängender Einzug)
hangover time	ELECTRON	**Nachwirkzeit**
		≈ Haltezeit
hangup (n.)	DATA PROC	**nichtprogrammierter Schleifenstop**
		= Hänger
haptic (adj.)	SCIE	**taktil**
[related to the sense of touch]		[den Tastsinn betreffend]
hard (adj.)	ELECTRON	**Hochvakuum-**
≠ gas-discharge (adj.)		≠ Gasentladungs-
hard	TECH	**hart**
hard-aluminum	METAL	**Duraluminium**
hard blank	DATA PROC	**harte Leerstelle**
[adjoining words are not separated by line breaks]		[angrenzende Wörter werden beim Zeilenumbruch nicht getrennt]
hardbound	TYPOGR	**mit hartem Einband**
hard card (→file-card)	DATA PROC	(→Festplattenkarte)
hard chrome plating	METAL	**Hartverchromung**
hard-clip area	DATA PROC	**Zeichenfläche**
hard-contact printing	TERM&PER	**Hartanschlagdruck**
hard copper	METAL	**Hartkupfer**
hard copy	TERM&PER	**Druckkopie**
[printed copy of a screen content]		[Ausdruck eines Bildschirminhalts]
= display printout; screen printout; screen dump		= Hardcopy; Bildschirmausdruck
↑ printout		≈ Druckerausdruck
		↑ Ausdruck
hardcopy printer	TERM&PER	**Hardcopygerät**
[a printer annexed to a display, to print screen contents]		[einem zum Ausdrucken von Bildschirminhalten vorgesehener Drucker]
		= Kopiendrucker 1
hard-copy terminal (→printer terminal)	DATA COMM	(→Terminaldrucker)
hard disc (→hard disk)	TERM&PER	(→Hartplatte)
hard disc controller	DATA PROC	**Festplattencontroller**
[controls data transfer with main memory]		[steuert Datentransfer mit dem Hauptspeicher]
hard disk	TERM&PER	**Hartplatte**
[rigid disk]		[aus steifem Material]
= hard disc; rigid disk; rigid disc; HD		= starre Magnetplatte; Hard Disk; HD
≈ fixed disk		≈ Festplatte
≠ floppy disk		≠ Diskette
↑ magnetic disk		↑ Magnetplatte
hard-disk controller card	TERM&PER	**Festplatten-Steuerkarte**
		= Festplatten-Kontrollkarte
hard-disk drive	TERM&PER	**Hartplattenlaufwerk**
[in German PC terminology the equivalent to "fixed-disk drive" is preferred]		[im deutschen PC-Jargon bevorzugt man den Begriff „Festplattenlaufwerk"]
= hard-disk unit		= Hartplatteneinheit; Festplattenlaufwerk 2 (PC)
≈ fixed-disk drive		≈ Festplattenlaufwerk 1
↑ disk drive		↑ Plattenlaufwerk
hard-disk drive lamp	DATA PROC	**Festplattenlämpchen**
hard disk memory	DATA PROC	**Festplattenspeicher**
[substitutive expression used in PC computing for "external, non-volatile, high-capacity random-access memory", as PC's employ mostly hard magnetic disks for this function; the German equivalent "fixed disk memory" refers to the fact that non-removable-disk drives are generally used]		[im PC-Jargon angewandter stellvertretender Ausdruck für „externer nichtflüchtiger Direktzugriffsspeicher großer Kapazität", da in PC's diese Funktion zumeist mit Festplattenspeichern realisiert wird; die englischsprachige Bezeichnung bezieht sich auf den Umstand, daß Festplattenspeicher mit Hartplatten ausgestattet sind]
= hard disk storage ; hard disk store		≈ Festplattenspeicher [TERM&PER]
≈ hard disk memory [TERM&PER]		
hard disk storage (→hard disk memory)	DATA PROC	(→Festplattenspeicher)
hard disk store (→hard disk memory)	DATA PROC	(→Festplattenspeicher)
hard-disk unit (→hard-disk drive)	TERM&PER	(→Hartplattenlaufwerk)
harden	METAL	**härten**
hardened	METAL	**gehärtet**
hardening	METAL	**Härtung**
		= Verhärtung
hard error	DATA PROC	**hardwarebedingter Fehler**
≈ hardware fault		≈ Hardwarefehler
hard failure (→hardware fault)	DATA PROC	(→Hardwarefehler)
hard glass	CHEM	**Hartglas**
hard gold plating	METAL	**Hartvergoldung**
hardhole	TERM&PER	**Verstärkungsring**
[floppy disk]		[Diskette]
		= Hubring
hard hyphen	DATA PROC	**echter Trennstrich**
[must be printed in any case]		[muß immer gedruckt werden]
≠ soft hyphen		≠ Bedarfstrennstrich

hard keying	MODUL	**Harttastung**	
= click			
hard magnetic (→magnetically hard)	PHYS	(→hartmagnetisch)	
hardness	TECH	**Härte**	
hard-of-hearing telephone (→deaf-aid telephone)	TELEPH	(→Fernsprechapparat mit Hörverstärker)	
hard printout (→printout)	DATA PROC	(→Ausdruck)	
hard rubber	CHEM	**Hartgummi**	
hard sector	TERM&PER	**Hartsektor**	
hard sectored	TERM&PER	**hartsektoriert** [Diskette]	
hard-sectored disc (→hard-sectored diskette)	TERM&PER	(→hartsektorierte Diskette)	
hard-sectored diskette [beginning of every sector marked by a hole]	TERM&PER	**hartsektorierte Diskette** [Anfang jedes Sektors durch ein Loch markiert]	
= hard-sectored floppy disk; hard-sectored disc			
hard-sectored floppy disk (→hard-sectored diskette)	TERM&PER	(→hartsektorierte Diskette)	
hard sectoring [diskette]	TERM&PER	**Hartsektorierung** [Diskette]	
hard-solder (v.t.)	METAL	**hartlöten**	
= braze 2		↓ messinghartlöten	
hard solder (n.)	METAL	**Hartlot**	
= higher-melting-point solder		↓ Messinghartlot	
↓ brazing solder			
hard-soldered	METAL	**hartgelötet**	
hard soldering	METAL	**Hartlötung**	
↓ brazing		↓ Messinghartlötung	
hardware	DATA PROC	**Hardware**	
[the material components of a computer]		[die materiellen Bestandteile eines Computers]	
≠ software		= Gerätschaften (pl.t.)	
↓ CPU; peripheral equipment		≠ Software	
		↓ Zentraleinheit; Peripheriegerät	
hardware accelerator	DATA PROC	**Hardware-Beschleuniger**	
hardware clock	DATA PROC	**Hardware-Uhr**	
hardware compatibility	DATA PROC	**Hardware-Kompatibilität**	
hardware configuration	DATA PROC	**Hardware-Konfiguration**	
hardware-dependent	DATA PROC	**hardwareabhängig**	
= machine-dependent		= maschinenabhängig	
hardware description language	MICROEL	**Hardware-Beschreibungssprache**	
= HDL			
hardware developer	DATA PROC	**Hardware-Entwickler**	
hardware driver	DATA PROC	**Hardware-Treiber**	
hardware failure (→hardware fault)	DATA PROC	(→Hardwarefehler)	
hardware fault	DATA PROC	**Hardware-Fehler**	
= hardware failure; hard failure; machine error		≈ hardwarebedingter Fehler ↑ Fehler	
≈ hard error			
↑ bug			
hardware identification	DATA PROC	**Hardware-Kennung**	
hardware-independent	DATA PROC	**hardwareunabhängig**	
= machine-independent		= maschinenunabhängig	
hardware interlocking	DATA PROC	**Hardware-Verbund**	
hardware interrupt	DATA PROC	**Hardware-Interrupt**	
hardware key	DATA PROC	**Hardware-Schlüssel**	
hardware resources	DATA PROC	**Hardware-Ressourcen**	
		= Hardware-Möglichkeit	
hardware security	DATA PROC	**Hardware-Absicherung**	
hardware specialist	DATA PROC	**Hardware-Spezialist**	
hardware technology	DATA PROC	**Gerätetechnik**	
hard-wearing (→resistant)	TECH	(→widerstandsfähig)	
hardwired [program]	DATA PROC	**festverdrahtet** [Programm]	
= wired-in; wired		= verdrahtet	
hardwired logic	DATA PROC	**festverdrahtete Logik**	
hardwired program (→fixed program)	DATA PROC	(→Festprogramm)	
hardwired-program computer	DATA PROC	**festprogrammierter Computer**	
= wired program computer; fixed-program computer		= Festprogrammcomputer	
harmful	TECH	**schädlich**	
= detrimental; mischievous		↓ giftig	
↓ toxic			
harmful agent (→polluant)	TECH	(→Schadstoff)	
harmful gas	QUAL	**Schadgas**	
harmonic (n.)	ACOUS	**Oberschwingung**	
= overtone		= Oberton; Harmonische; harmonische Schwingung	
harmonic (n.)	MODUL	**Oberwelle**	
= envelope			
harmonic (→harmonic wave)	PHYS	(→Oberwelle)	
harmonic analysis (→Fourier analysis)	MATH	(→Fourier-Analyse)	
harmonic analyzer (→distortion analyzer)	INSTR	(→Verzerrungsanalysator)	
harmonic antenna	ANT	**Oberwellenantenne**	
harmonic content	EL.ACOUS	**Oberschwingungsgehalt**	
= distortion factor; total harmonic distortion; THD		= Oberwellengehalt; Klirrfaktor; Klirrgrad	
harmonic content	EL.TECH	**Oberwellengehalt**	
= harmonics content; ripple content; k-rating; distortion factor		= Oberschwingungsgehalt; Klirrfaktor; Klirrfaktorkoeffizient	
harmonic dipole	ANT	**Oberwellendipol**	
harmonic distortion 1	EL.TECH	**Klirren**	
harmonic distortion	TELEGR	**harmonische Verzerrung**	
harmonic distortion 2 (→harmonic distortion attenuation)	EL.TECH	(→Klirrdämpfung)	
harmonic distortion attenuation [logarithm of harmonic content]	EL.TECH	**Klirrdämpfung** [Logarithmus des Klirrfaktors]	
= distortion attenuation; harmonic distortion 2			
harmonic filter	NETW.TH	**Oberwellenfilter**	
= harmonics filter			
harmonic filtering	NETW.TH	**Oberwellenfilterung**	
= harmonics filtering			
harmonic frequency	PHYS	**Oberfrequenz**	
≈ harmonic wave		= harmonische Frequenz	
≠ fundamental frequency		≈ Oberschwingung	
		≠ Grundfrequenz	
harmonic generation	EL.TECH	**Frequenzvervielfachung**	
= harmonics generation; frequency multiplication			
harmonic mean	MATH	**harmonisches Mittel**	
[n divided by the sum of 1/A1 to 1/An]		[n durch die Summe von 1/A1 bis 1/An]	
harmonic mixer	HF	**harmonischer Mischer**	
= harmonics mixer			
harmonic mode crystal (→overtone crystal)	COMPON	(→Oberschwingungsquarz)	
harmonic oscillator (→sine-wave generator)	CIRC.ENG	(→Sinusgenerator)	
harmonic power	EL.TECH	**Oberwellenleistung**	
		= Oberschwingungsleistung	
harmonic reactive power	NETW.TH	**Oberwellenblindleistung**	
= distortive power		[Komponente der Blindleistung]	
		= Verzerrungsleistung; Oberschwingungsblindleistung	
harmonics content (→harmonic content)	EL.TECH	(→Oberwellengehalt)	
harmonics filter (→harmonic filter)	NETW.TH	(→Oberwellenfilter)	
harmonics filtering (→harmonic filtering)	NETW.TH	(→Oberwellenfilterung)	
harmonics generation (→harmonic generation)	EL.TECH	(→Frequenzvervielfachung)	
harmonics generator	CIRC.ENG	**Oberwellengenerator**	
harmonics mixer (→harmonic mixer)	HF	(→harmonischer Mischer)	
harmonics oscillator (→sine-wave generator)	CIRC.ENG	(→Sinusgenerator)	
harmonic wave (n.)	PHYS	**Oberwelle**	
= harmonic		= Oberschwingung; Harmonische	
≈ harmonic frequency		≈ Oberfrequenz	
≠ fundamental wave		≠ Grundwelle	
harness wiring	EQUIP.ENG	**Stammverdrahtung**	

harness wiring plan

harness wiring plan (→cable forming plan)	DOC	(→Kabelformplan)
harp antenna (→curtain antenna)	ANT	(→Vorhangantenne)
Hartley oscillator	CIRC.ENG	**Hartley-Oszillator**
↑ LC oscillator		= Hartley-Schaltung
		↑ LC-Oszillator
Hartley oscillator circuit	CIRC.ENG	**induktive Dreipunktschaltung**
hash (n.) (→snow)	TV	(→Schnee)
hash code (→hash coding)	DATA PROC	(→Hash-Codierung)
hash coding	DATA PROC	**Hash-Codierung**
[physical store address is derived from an ordering argument inherent to the data set]		[physikalische Adresse wird aus einem dem Datensatz inhärenten Ordnungsbegriff abgeleitet; hash = engl. Hackfleisch]
= hash code; hashing; key-to-address		= Hash-Code; Faltung
hash file	DATA PROC	**Hash-Datei**
[with hashing]		[mit Hash-Codierung]
hashing (→hash coding)	DATA PROC	(→Hash-Codierung)
hash total (→running digital sum)	DATA PROC	(→Prüfsumme)
hatch (v.t.)	ENG.DRAW	**schraffieren**
[to mark with fine, closely spaced lines]		≈ riffeln; schattieren
≈ checker; shade		
hatched	ENG.DRAW	**schraffiert**
= section-lined		≈ schattiert
≈ shaded		
hatching	ENG.DRAW	**Schraffur**
= section lining		= Schraffierung
≈ checker; shading		≈ Riffelung; Schattierung
↓ cross-hatching		↓ Kreuzschraffierung
haulage rope	TECH	**Schleppseil**
↑ traction rope		↑ Zugseil
haver-	INSTR	**90° phasenverschoben**
haverfunction	INSTR	**90° phasenverschobene Funktion**
haversine	INSTR	**90° phasenverschobener Sinus**
[a 90° phase-shifted sine]		
havertriangle	INSTR	**90° phasenverschobenes Dreieck**
Hay bridge	INSTR	**Hay-Brücke**
Hayes modem	DATA COMM	**Hayes-Modem**
hazard (→risk)	ECON	(→Risiko)
hazard (→disturbance effect)	ELECTRON	(→Störeffekt)
hazard bacon	RADIO NAV	**Warnungsbake**
= warning beacon		
hazard bonus	ECON	**Gefahrenzulage**
= danger money		
hazard notice (→danger notice)	TECH	(→Warntafel)
hazard paint (→warning paint)	TECH	(→Warnbemalung)
HB9CV array	ANT	**HB9CV-Gruppenstrahler**
HB9 multiband delta loop antenna	ANT	**HB9-Multiband-delta-loop-Antenne**
HBT	MICROEL	**HBT**
= hetero-bipolar transistor		= Hetero-Bipolar-Transistor
HCDR (→high-capacity digital radio link)	TRANS	(→Breitband-Digitalrichtfunk-Verbindung)
HCMOS	MICROEL	**HCMOS**
= high-performance CMOS		
HCR (→high-capacity radio)	TRANS	(→Breitbandrichtfunk)
HD (→hard disk)	TERM&PER	(→Hartplatte)
HD (→high-density)	TERM&PER	(→HD)
HDB3	CODING	**HDB3**
= high density bipolar of order 3		
HD disk (→HD diskette)	TERM&PER	(→HD-Diskette)
HD diskette	TERM&PER	**HD-Diskette**
[9,640 bpi, 96 tpi]		[3.795 Bit/cm, 38 Spuren/cm]
= high-density diskette; HD disk		= High-Density-Diskette
HDDR (→high density digital magnetic recording)	TERM&PER	(→HDDR)
H diagram (→H-plane pattern)	ANT	(→H-Diagramm)
HDL (→hardware description language)	MICROEL	(→Hardware-Beschreibungssprache)
HDLC (→high-level data link control)	DATA COMM	(→HDLC-Verfahren)
HDLC procedure	DATA COMM	**HDLC-Prozedur**
↑ transmission procedure		↑ Übertragungsprozedur
HDR (→header label)	DATA PROC	(→Dateianfangs-Etikett)
HDTV (→high-definition TV)	TV	(→hochauflösendes Fernsehen)
He (→helium)	CHEM	(→Helium)
head (n.)	TERM&PER	**Kopf**
↓ write head; magnetic head		↓ Schreibkopf; Magnetkopf
head	TYPOGR	**oberer Papierrand**
[portion of page above first line]		
head (→test probe)	INSTR	(→Prüfspitze)
head (→phonograph pickup)	EL.ACOUS	(→Tonabnehmer)
head (→heading)	LING	(→Überschrift)
head (→headword)	LING	(→Stichwort)
head alignment	TERM&PER	**Magnetkopfjustierung**
headband	EL.ACOUS	**Kopfhörerbügel**
[headphone]		= Kopfband
head cleaner (→tape head cleaner)	EL.ACOUS	(→Tonkopfreiniger)
head cleaning disk	TERM&PER	**Reinigungsdiskette**
= cleaning diskette		
headcount	ECON	**Kopfzahl**
≈ workforce		≈ Personalbestand
head crash	TERM&PER	**Bauchlandung**
[collision of read/write head with recording surface]		[Schreib-Lesekopf berührt die Plattenoberfläche]
= disk crash		
head distance	TERM&PER	**Flughöhe**
[separation of magnetic head from disk]		[Abstand Magnetkopf von Platte]
= head gap; air gap		= Luftspalt
head drum (→tape drum)	TERM&PER	(→Bandführungstrommel)
head-end	BROADC	**Empfangsstelle**
head-end	DATA COMM	**Kopfstelle**
header	MICROEL	**Chip-Halterung**
header	TELEC	**Anfangsblock**
[SONET; STM; ATM]		[SONET, STM, ATM]
		= Steuerungsblock; Header
header	DATA PROC	**Anfangsblock**
header (→header line)	DATA PROC	(→Kopfzeile)
header (→message header)	DATA COMM	(→Nachrichtenkopf)
header (→heading)	LING	(→Überschrift)
header (→header label)	DATA PROC	(→Dateianfangs-Etikett)
header (→cell header)	TELEC	(→Zellenkopf)
header bit	DATA COMM	**Kopfbit**
header block (→header label)	DATA PROC	(→Dateianfangs-Etikett)
header card	DATA PROC	**Anfangskarte**
		= Vorlaufkarte
header label	DATA PROC	**Dateianfangs-Etikett**
[identifies beginning of file]		[kennzeichnet den Dateianfang]
= header; prefix; preamble; beginning of file; BOF; header block; leader label; leader; HDR		= Dateianfangs-Kennsatz; Dateikennsatz; Anfangsetikett; Anfangskennsatz; Header-Etikett; Header-Kennsatz; Dateivorsatz; Vorsatz; Dateivorspann; Vorspann; Präfix; Präambel
≠ trailer label		≠ Dateiend-Etikett
↑ label		↑ Etikett
header line	DATA PROC	**Kopfzeile**
= heading line; header; heading		
header line (→headline)	TYPOGR	(→Überschriftzeile)
header statement	DATA PROC	**Anfangsanweisung**
head gap (→head distance)	TERM&PER	(→Flughöhe)
head guy	OUTS.PLANT	**Längsabspannung**
[in line with the pole line]		= Linienanker
		↑ Abspannseil
head inductance	TERM&PER	**Kopfinduktivität**
heading	METAL	**Kopfanstauchen**
heading	LING	**Überschrift**
[of a text]		[Inhaltskennzeichnung über einem Text]
= title; header; caption 1; head; chapter heading		= Titel; Titelkopf; Titelzeile; Kapitelüberschrift
≈ headline; headwort		≈ Überschriftzeile; Stichwort

heading (→header line)	DATA PROC	(→Kopfzeile)	
heading (→message header)	DATA COMM	(→Nachrichtenkopf)	
heading line (→header line)	DATA PROC	(→Kopfzeile)	
heading line (→headline)	TYPOGR	(→Überschriftzeile)	
heading marker	RADIO NAV	Zielmarkierung	
headless setscrew	MECH	Madenschraube	
headline	TYPOGR	Überschriftzeile	
= header line; heading line		= Titelzeile; Kopfzeile	
≈ heading		≈ Überschrift	
headline	LING	Schlagzeile	
[at the head of a newspaper]		[auffällige, erste Überschrift einer Zeitung]	
= catchline; banner headline (AM)			
head office	ECON	Zentralverwaltung	
= central office; principal office; general office		= Zentrale	
		≈ Stammhaus	
≈ parent company			
head of form	DATA PROC	Formularkopf	
= HOF		= Formularanfang	
head-on collision (→call collision)	DATA COMM	(→Verbindungszusammenstoß)	
headphone	EL. ACOUS	Kopfhörer	
= headset; earphone; head receiver;		= Hörer	
≈ headset [TELEPH]		= Kopfsprechhörer [TELEPH]	
headphone output	CIRC. ENG	Hörerausgang	
headphone terminal	EQUIP. ENG	Kopfhöreranschluß	
= earphone socket			
head positioning	TERM&PER	Kopfpositionierung	
head pressure	TERM&PER	Bandandruck	
headquarters	ECON	Hauptverwaltung	
≈ administrative headquartes		≈ Zentralverwaltung	
head receiver (→headphone)	EL. ACOUS	(→Kopfhörer)	
headrow rack	SYS. INST	Reihenanfanggestell	
≠ endrow rack		≠ Reihenendgestell	
head screw	MECH	Kopfschraube	
= headscrew; cap screw			
headscrew (→head screw)	MECH	(→Kopfschraube)	
head separation	TERM&PER	Kopfabstand	
headset	TELEPH	Kopfsprechhörer	
≈ headphone [EL. ACOUS]		= Sprechgarnitur; Abfragegarnitur	
		≈ Kopfhörer [EL. ACOUS]	
headset (→headphone)	EL. ACOUS	(→Kopfhörer)	
head setting equipment	OFFICE	Titelsatzgerät	
head sort (→tree sort)	DATA PROC	(→Baumsortierung)	
head switching	TERM&PER	Schreib-Lesekopf-Umschaltung	
head switching circuit	TV	Kopfumschalter	
		[Videorecorder]	
head tip projection	TV	Kopfüberstand	
head-up display	RADIO NAV	Blickfeld-Darstellungsgerät	
= HUD			
head wheel	TERM&PER	Bandandruckrolle	
head window	TERM&PER	Schreib-Lese-Öffnung	
[floppy disk]		[Diskette]	
= read-write window		= Schreib-Lese-Fenster; Kopffenster	
headword	LING	Stichwort	
= head; entry word (dictionary); reference entry		= Schlagwort; Lemma (in Lexikas)	
≈ entry; theme		≈ Eintrag; Thema	
heap	DATA PROC	Halde	
[temporarily borrowed memory location]		[vorübergehend ausgeliehene Speicherplätze]	
hearing aid	EL. ACOUS	Hörgerät	
heat (v.t)	PHYS	erwärmen	
= warm; heat-up		= erhitzen	
		≈ aufwärmen	
heat (n.)	PHYS	Wärme	
= warmth [TECH]; hotness [TECH]		= Hitze [TECH]	
≈ heat quantity		≈ Wärmemenge	
heatable	TECH	beheizbar	
		= heizbar	
heat absorption	PHYS	Wärmeaufnahme	
heat accumulation (→heat concentration)	PHYS	(→Wärmestau)	
heat baffle (→heat shield)	TECH	(→Wärmeschild)	
heat balance	PHYS	Wärmebilanz	
= thermal balance		= Wärmehaushalt; thermische Bilanz; thermische Bilanz	
heat capacity	PHYS	Wärmekapazität	
= thermal capacity		= thermische Kapazität	
heat-coil fuse	COMPON	Hitzdrahtsicherung	
[for telecommunication equipment]		[Fernmeldesicherung]	
≈ resolderable fuse; reversible fuse		≈ Rücklotsicherung; Umkehrauslöser	
heat concentration	PHYS	Wärmestau	
= heat accumulation			
heat conducting	PHYS	wärmeleitend	
= thermal conducting			
heat conduction (→thermal conduction)	PHYS	(→Wärmeleitung)	
heat conductivity	PHYS	Wärmeleitfähigkeit	
[heat quantity flowing through 1 cm^2 x 1 cm at 1 degree of temperature difference]		[bei 1 Grad Temperaturdifferenz durch 1 cm^2 x 1 cm durchfließende Wärmemenge]	
= thermal conductivity		= Wärmeleitzahl; thermische Leitfähigkeit; spezifisches Wärmeleitvermögen	
heat conductor (→thermal conductor)	PHYS	(→Wärmeleiter)	
heat content (→thermal content)	PHYS	(→Wärmeinhalt)	
heat convection	PHYS	Wärmekonvektion	
= thermal convection			
heat detector	SIGN. ENG	Wärmemelder	
↑ fire detector		↑ Feuermelder	
heat dissipation (→heat transfer)	PHYS	(→Wärmeableitung)	
heat-energy density	PHYS	Wärmeenergiedichte	
heater	TECH	Erhitzer	
= warmer		= Erwärmer	
≈ heat exchanger		≈ Wärmeaustauscher	
heater battery	ELECTRON	Heizbatterie	
heater current	ELECTRON	Heizstrom	
= filament current			
heater voltage	ELECTRON	Heizspannung	
heat exchange	PHYS	Wärmeausgleich	
≈ heat balance		= Wärmeaustausch	
		≈ Wärmebilanz	
heat exchanger	TECH	Wärmeaustauscher	
↓ refrigerator; heater		= Wärmetauscher	
		↓ Kühler; Erhitzer	
heat flow (→thermal flux)	PHYS	(→Wärmefluß)	
heat-flow density	PHYS	Wärmestromdichte	
heat flux (→thermal flux)	PHYS	(→Wärmefluß)	
heating	PHYS	Erwärmung	
		= Erhitzung	
heating	TECH	Heizung	
heating characteristic	MICROEL	Erwärmungskenngröße	
heating coil	ELECTRON	Heizwicklung	
heating time	ELECTRON	Anheizzeit	
= warm-up period		= Aufheizzeit	
heating-up (→preheating)	TECH	(→Vorwärmung)	
heating-up time constant	MICROEL	Anheiz-Zeitkonstante	
		= Aufheiz-Zeitkonstante	
heating value (→effective value)	PHYS	(→Effektivwert)	
heating wire (→filament)	ELECTRON	(→Heizfaden)	
heat insulation (→thermal insulation)	PHYS	(→Wärmeisolierung)	
heat pipe	COMPON	Wärmeleitrohr	
[coooling device with a liquid evaporating/condensing in its interior]		[Kühlkörper in dessen Inneren eine Flüssigkeit durch Verdampfen/Kondensieren Wärme transportiert]	
		= Heat-pipe-Wärmeableiter	
heat quantity	PHYS	Wärmemenge	
[SI unit: Joule]		[SI-Einheit: Joule]	
= Q		= Q	
≈ heat		≈ Wärme	
heat radiation (→thermal radiation)	PHYS	(→Wärmestrahlung)	
heat resistance (→thermal stability)	TECH	(→Wärmebeständigkeit)	

heat-resistant		TECH	**hitzebeständig**			= Hektometerwellen; MW; LF
= resistant to heat			= hitzefest; wärmebeständig; wärmefest	**heel effect**	ELECTRON	**Anodenschatten**
heat resistivity (→thermal stability)		TECH	(→**Wärmebeständigkeit**)	**Hefner candle**	PHYS	**Hefnerkerze**
				[SI unit for luminous density; = 0,903 cd]		[Maßeinheit für Lichtstärke; = 0,903 cd]
heat-sensitive paper		TERM&PER	(→**Thermopapier**)	= H K		= H K
(→temperature-sensitive paper)				**height**	MATH	**Höhe**
heat shield		TECH	**Wärmeschild**	[geometry; perpendicular distance]		[Geometrie; vertikaler Abstand]
= heat baffle			= Hitzeschild	= altitude		
heat-shrinkable cap		OUTS.PLANT	**Schrumpfkappe**	↑ dimension		↑ Dimension
heat-shrinkable closure		OUTS.PLANT	(→**Schrumpfmuffe**)	**height-adjustable** (→adjustable in height)	TECH	(→**höhenverstellbar**)
(→shrinkage sleeve)						
heat-shrinkable sleeves wrap		OUT.PLANT	**Schrumpfmanschette**	**height adjustment**	TECH	**Höheneinstellung**
heat-shrinkable tube		OUTS.PLANT	(→**Schrumpfschlauch**)			= Höhenverstellung
(→shrinkage tube)				**height-position indicator**	RADIO NAV	**Höhe- und Ortsgeber**
heat-shrinking		TECH	(→**wärmeschrumpfend**)	= HPI		
(→thermocontractible)				**height-range indicator**	RADIO NAV	**Höhe- und Entfernungsgeber**
heat sink		COMPON	**Wärmesenke**	= HRI		
			= Wärmeableiter; Kühlkörper; Kühler	**helical**	TECH	**schraubenförmig**
				= spiral; corkscrew; helicoid; helicoidal		= spiralförmig
heat sink compound		ELECTRON	**Wärmeleitpaste**	**helical antenna** (→helix antenna)	ANT	(→**Wendelantenne**)
= thermal compound						
heat source		PHYS	**Wärmequelle**	**helical conductor**	ELECTRON	**Wendelleitung**
= thermal source				= spiral conductor		
heat transfer		PHYS	**Wärmeableitung**	**helical filament**	ELECTRON	**Wendel**
= heat dissipation; dissipation 1; thermal removal			= Wärmeabfuhr; Wärmeübergang; Wärmeverlust; Wärmeabgabe	↑ filament		↑ Glühfaden
≈ thermal conduction; dissipation power [EL.TECH]			≈ Wärmeleitung; Verlustleistung [EL.TECH]	**helical filament**	ELECTRON	**Heizspirale**
				helical gear	MECH	**Schraubenrad**
						[Zahnrad]
heat-transfer coefficient		PHYS	**Wärmeübergangskoeffizient**	**helical recording**	TV	**Schrägaufzeichnung**
heat-transfer resistance		PHYS	**Wärmeübergangswiderstand**			= Schrägspuraufzeichnung
heat transmission coefficient		PHYS	**Wärmedurchgangswiderstand**	**helical scanning**	ELECTRON	**Schraubenlinien-Abtastung**
heat transport (→thermal conduction)		PHYS	(→**Wärmeleitung**)	**helical spring**	MECH	**Spiralfeder**
				= spiral spring		= Schraubenfeder; Wendelfeder
heat-treat		METAL	**warmbehandeln**			
			= vergüten	**helical waveguide**	MICROW	**Wendelhohlleiter**
heat treatment		TECH	**Warmbehandlung**			= Wendelwellenleiter
			= Wärmebehandlung; Vergütung [METAL]	**helicoid** (→helical)	TECH	(→**schraubenförmig**)
				helicoidal (→helical)	TECH	(→**schraubenförmig**)
heat-up (→heat)		PHYS	(→**erwärmen**)	**helicoidal field**	PHYS	**schraubenförmiges Feld**
heavily damaged		TECH	**schwerbeschädigt**	= corkscrew field		
Heaviside-Campbell inductance bridge		INSTR	**Heaviside-Campbell-Induktivitätsbrücke**	**helicoidal motion** (→screw motion)	MECH	(→**Schraubenbewegung**)
heavy (→extra bold)		TYPOGR	(→**extrafett**)	**helicoidal potentiometer**	COMPON	**Wendelpotentiometer**
heavy current		POWER SYS	**Starkstrom**			= Helipot; Mehrgangspotentiometer
= power current			= Kraftstrom			
≠ low current			≠ Schwachstrom	**helicopter**	AERON	**Hubschrauber**
heavy current engineering (→electrical power engineering)		EL.TECH	(→**elektrische Energietechnik**)			= Helikopter
				helicopter antenna	ANT	**Hubschrauberantenne**
heavy current technology (→electrical power engineering)		EL.TECH	(→**elektrische Energietechnik**)	**heliogravure** (→photogravure)	TYPOGR	(→**Heliographie**)
				heliostationary	ASTROPHYS	**sonnenstationär**
heavy-duty.. (→high-performance ...)		TECH	(→**Hochleistungs-**)	**heliotrope**	PHYS	**lila**
heavy-duty (→rugged)		TECH	(→**robust**)	= lilac		= erika
heavy engineering		MECH	**Schwermaschinenbau**	**helium**	CHEM	**Helium**
heavy-lift helicopter		AERON	**Schwerlasthubschrauber**	= He		= He
heavy lorry (BRI) (→heavy motor truck)		TECH	(→**Schwerlastwagen**)	**helix**	MATH	**Schraubenlinie**
				= spiral		= Spirale; Wendel
heavy metal		CHEM	**Schwermetall**	**helix angle**	MECH	**Steigungswinkel**
heavy motor truck		TECH	**Schwerlastwagen**			[Schraubenlinie]
= heavy lorry (BRI)			= Schwerlaster	**helix antenna**	ANT	**Wendelantenne**
heavy route (→high capacity link)		TRANS	(→**Breitbandstrecke**)	= helical antenna; spiral antenna; corkskrew antenna		= Spulenantenne; Schraubenantenne; Helixantenne; Korkenzieherantenne; Spiralantenne
heavy shower (→downpour)		METEOR	(→**Regenguß**)	≈ frequency-indipendent antenna		
heavy time (→heavy-traffic period)		TELEC	(→**verkehrsstarke Zeit**)	↑ travelling-wave antenna		
				↓ zigzag antenna		↑ Wanderwellenantenne; frequenzunabhängige Antenne
heavy-traffic period		TELEC	**verkehrsstarke Zeit**			↓ Zickzackantenne
= heavy time; busy time; peak-traffic time			= verkehrsreiche Zeit; Spitzenzeit; Stoßzeit	**helix filter**	NETW.TH	**Helix-Filter**
			≈ Hauptverkehrszeit	**"hello!"**	TELEPH	**„Hallo!"**
heavy water		CHEM	**Schwerwasser**	**Hell printer**	TELEGR	**Hell-Schreiber**
hectare		PHYS	**Hektar**	**Helmhotz equivalent-source theorem**	NETW.TH	**Helmhotzscher Satz**
[unit for land areas; = 100 a = 100 m²100 m = 10,000 m²]			[Maß für Grundstücksflächen; = 100 AR = 100 m²100 m = 10.000 m²]			= Satz von der Zweipolquelle
= ha			= ha	↓ Thevenin's theorem; Norton's theorem		↓ Theorem von Thevenin; Theorem von Norton
hectometric waves		RADIO	**Mittelwellen**	**Helmhotz resonator**	EL.ACOUS	**Helmholtz-Resonator**
[1 km- 100 m; 300 kHz-3 MHz]			[1 km-100 m; 300 kHz-3 MHz]	**helpful** (→useful)	TECH	(→**nützlich**)
= medium waves; low frequency						

English	Domain	German
HELP function [user guidance]	DATA PROC	Hilfe-Funktion [Bedienerführung] = HELP-Funktion
HELP key	TERM&PER	Hilfe-Taste = HELP-Taste
helpware = software literature	DATA PROC	Software-Literatur
helve (→handle 1)	TECH	(→Handgriff 1)
helvetica [a sans-serif type style]	TYPOGR	Helvetica [serifenlose Schriftart]
hemisphere	MATH	Halbkugel = Hemisphäre
hemisphere = half of earth surface	GEOPHYS	Hemisphäre = Erdhalbkugel; Erdhälfte
hemisphere [celestial]	ASTROPHYS	Hemisphäre = Himmelshalbkugel
hemispheric = hemispherical	MATH	halbkugelförmig = hemisphärisch
hemispherical (→hemispheric)	MATH	(→halbkugelförmig)
hemispherical beam = hemispheric beam	SAT.COMM	hemisphärische Ausleuchtung = hemisphärischer Strahl
hemispheric beam (→hemispherical beam)	SAT.COMM	(→hemisphärische Ausleuchtung)
hemp	TECH	Hanf
hemp rope	TECH	Hanfseil
HEMT [High Electron Mobility Transistor]	MICROEL	HEMT
Henry [SI unit for inductance and permeance; = 1 Wb/A] = H	PHYS	Henry [SI-Einheit für Induktivität und Permeanz; = 1 Wb/A] = H
heptagon	MATH	Siebeneck = Heptagon ↑ Vieleck
heptagonal ↑ polygonal	MATH	siebeneckig ↑ vieleckig
heptahedral	MATH	siebenflächig
heptahedron (pl. -ons, -a)	MATH	Heptaeder
heptode	ELECTRON	Heptode
Hercules [Hercules Graphics Adapter; graphics standard with 720x348 pixels, 50 Hz, monochrome] = HGA	TERM&PER	Hercules [Graphikstandard mit 720x348 Bildpunkten, 50 Hz, monochrom] = HGA
Hercules board = HGA board ↑ MGA board; monochrome graphics board; graphics board	TERM&PER	Hercules-Karte = HGA-Karte; Hercules-Adapter; HGA-Adapter ↑ MGA-Karte; Monochrom-Graphikkarte; Graphikkarte
hereinafter (→stated below)	OFFICE	(→nachstehend)
hermetic (→airtight)	TECH	(→luftdicht)
hertz (→cycles per second)	PHYS	(→Hertz)
Hertz effect ↑ photo effect	PHYS	Hertz-Effekt ↑ Photoeffekt
Hertzian dipole (→elementary electric dipole)	ANT	(→Hertzscher Dipol)
Hertzian doublet (→elementary electric dipole)	ANT	(→Hertzscher Dipol)
Hertzian oscillator	PHYS	Hertzscher Oszillator
Hertz rate (→mains frequency)	POWER SYS	(→Netzfrequenz)
hesitation (→arrest)	TECH	(→Hemmung)
hetero-bipolar transistor (→HBT)	MICROEL	(→HBT)
heterodiode	MICROEL	Heterodiode
heterodyne (v.t.)	HF	überlagern
heterodyne (n.) (→heterodyne receiver)	RADIO	(→Überlagerungsempfänger)
heterodyne detection (→beat reception)	HF	(→Überlagerungsempfang)
heterodyne detector (→heterodyne receiver)	RADIO	(→Überlagerungsempfänger)
heterodyne frequency = beat frequency	INSTR	Schwebungsfrequenz
heterodyne frequency meter = heterodyne wave meter	INSTR	Überlagerungs-Frequenzmesser = Schwebungsfrequenzmesser; Heterodyn-Frequenzmesser
heterodyne oscillator	INSTR	Schwebungssender
heterodyne principle (→heterodyning)	HF	(→Überlagerung)
heterodyne receiver = superheterodyne receiver; superhet; heterodyne detector; heterodyne (n.) ↓ single conversion superhet; double conversion superhet	RADIO	Überlagerungsempfänger = Superheterodyn-Empfänger; Superhet; Heterodynempfänger; Überlagerer ↓ Einfachsuperhet; Doppelsuperhet
heterodyne reception (→beat reception)	HF	(→Überlagerungsempfang)
heterodyne wave meter (→heterodyne frequency meter)	INSTR	(→Überlagerungs-Frequenzmesser)
heterodyning [linear summing of a receive frequency f with a carrier frequency c, in order to rationalize amplification, executing her at the beat frequency f-c, called intermediate frequency] = beat (n.); superposition; heterodyne principle	HF	Überlagerung [lineare Summierung einer Empfangsfrequenz f mit einer Trägerfrequenz t , i.a. um die Verstärkung bei der tieferen Schwebungsfrequenz f-t ,genannt Zwischenfrequenz, kostengünstiger durchführen zu können] = Überlagerungsprinzip
heteroepitaxy	MICROEL	Heteroepitaxie
heterogeneous	SCIE	heterogen
heterogeneous multiplexer = heterogeneous multiplexor	DATA COMM	gemischter Multiplexer = heterogener Multiplexer
heterogeneous multiplexing	DATA COMM	Mischmultiplexierung
heterogeneous multiplexor (→heterogeneous multiplexer)	DATA COMM	(→gemischter Multiplexer)
heterojunction	MICROEL	Heteroübergang
heteropolar	PHYS	heteropolar
heteropolar bond = ionic bond; electrovalence	CHEM	heteropolare Bindung = Ionenbindung
heterostructure	MICROEL	Heterostruktur
heuristic (n.) [theory and practise of methodical problem solution]	SCIE	Heuristik [Theorie und Praxis methodischer Problemlösung]
heuristic (adj.) [relative to a methodical problem solution]	SCIE	heuristisch [eine methodische Problemlösung betreffend]
heuristic (adj.) [learning from experience]	DATA PROC	heuristisch [aus Erfahrung lernend]
Heusler alloy	METAL	Heuslersche Legierung
hex (→hexadecimal)	MATH	(→hexadezimal)
hexadecimal [with base 16] = hex; sexadecimal	MATH	hexadezimal [mit Grundzahl 16] = sedezimal
hexadecimal code = hex code	DATA PROC	Hexadezimalcode = Hex-Code
hexadecimal digit = sexadecimal digit	MATH	Hexadezimalziffer = Sedezimalziffer; hexadezimale Ziffer; sedezimale Ziffer
hexadecimal notation (→hexadecimal number system)	MATH	(→Hexadezimalsystem)
hexadecimal number = sexadecimal number	MATH	Hexadezimalzahl = Sedezimalzahl
hexadecimal number system = hexadecimal notation; hexadecimal system; sexadecimal number system; sexadecimal notation; sexadecimal system ↑ denominational number system	MATH	Hexadezimalsystem [Zahlendarstellung mit 16 Zifferncodes] = hexadezimales Zahlensystem; hexadezimale Darstellung; Hexazimaldarstellung; Sedezimalsystem; sedezimales Zahlensystem; sedezimale Darstellung; Sedezimaldarstellung ↑ Stellenwertsystem
hexadecimal pad = hex pad; Sexadecimal pad	TERM&PER	Hexadezimaltastatur = Sedezimaltastatur
hexadecimal point = sexadecimal point	MATH	Hexadezimalpunkt = Sedezimalpunkt
hexadecimal system (→hexadecimal number system)	MATH	(→Hexadezimalsystem)
hexagon ↑ polygon	MATH	Sechseck = Hexagon ↑ Vieleck

hexagonal

hexagonal	MATH	sechseckig
		= hexagonal
hexagon bar iron	METAL	Sechskantstahl
= hexagon iron		
hexagon bolt	MECH	Sechskantkopfschraube 2 [mit Mutter]
hexagon driver	MECH	Stiftschlüssel
hexagon head	MECH	Sechskantkopf [Schrauben]
hexagon head screw	MECH	Sechskantkopfschraube 1 [ohne Mutter]
hexagon iron (→hexagon bar iron)	METAL	(→Sechskantstahl)
hexagon nut	MECH	Sechskantmutter
hexagon socket	MECH	Innensechskant
hexahedral	MATH	sechsflächig
hexahedron (pl.-ons) (→cube 2)	MATH	(→Würfel)
hexavalent	MATH	hexavalent
hex code (→hexadecimal code)	DATA PROC	(→Hexadezimalcode)
hexode	ELECTRON	Hexode
hex pad (→hexadecimal pad)	TERM&PER	(→Hexadezimaltastatur)
HF (→high-frequency)	EL.TECH	(→hochfrequent)
HF (→high frequency)	EL.TECH	(→Hochfrequenz)
HF (→short waves)	RADIO	(→Kurzwelle)
Hf (→hafnium)	CHEM	(→Hafnium)
HF amplifier	HF	HF-Verstärker
= high-frequency amplifier		= Hochfrequenzverstärker
≈ RF amplifier		≈ RF-Verstärker
HF cable (→high-frequency cable)	COMM.CABLE	(→Trägerfrequenzkabel)
HF choke coil	COMPON	Hochfrequenzdrossel
HF engineering (→high-frequency engineering)	EL.TECH	(→Hochfrequenztechnik)
HF heating	POWER ENG	HF-Heizung
= high-frequency heating		= Hochfrequenzheizung
HF line	HF	HF-Leitung
= high frequency line		= Hochfrequenzleitung
HF probe (→RF probe)	INSTR	(→HF-Tastkopf)
HF-proof (→RF radiation proof)	EL.TECH	(→HF-dicht)
HF propagation forecast (→radiopropagation forecast)	RADIO	(→Funkprognose)
HF-radiation proof (→RF radiation proof)	EL.TECH	(→HF-dicht)
HF radio (→high frequency radio)	RADIO	(→Kurzwellenfunk)
HF resistance	EL.TECH	HF-Widerstand
= high-frequency resistance		= Hochfrequenzwiderstand
HF resistance bridge	INSTR	HF-Widerstandsmeßbrücke
= high-frequency resistance bridge; RF resistance bridge; radio frequency bridge		= Hochfrequenz-Widerstandsbrücke
HF transceiver	HF	HF-Transceiver
= high-frequency transceiver; transceiver		= Transceiver
HF transistor	COMPON	HF-Transistor
= high-frequency transistor		= Hochfrequenztransistor
HF wattmeter	INSTR	HF-Wattmeter
Hg (→mercury)	CHEM	(→Quecksilber)
HGA (→Hercules)	TERM&PER	(→Hercules)
HGA board (→Hercules board)	TERM&PER	(→Hercules-Karte)
hidden	TECH	verborgen
		= versteckt
hidden defect	ECON	verborgener Mangel
hidden file	DATA PROC	versteckte Datei
hidden line	DATA PROC	verdeckte Linie
hidden-line removal	DATA PROC	Entfernen verdeckter Linien
hidden object	DATA PROC	verdeckter Gegenstand
hidden outline	ENG.DRAW	verdeckte Kante
hidden surface	DATA PROC	verdeckte Fläche
hide (v.t.)	COLLOQ	verbergen
= conceal		= verstecken
hide (→cover 2)	TECH	(→decken)
hiding material (→covering material)	TECH	(→Deckmittel)
hiding power (→covering power)	TECH	(→Deckfähigkeit)

hierarchical database	DATA PROC	hierarchische Datenbank
hierarchical directory	DATA PROC	hierarchisches Verzeichnis
hierarchical modell	DATA PROC	hierarchisches Modell
hierarchical network	TELEC	hierarchisches Netz
= despotic network		= zwangssynchronisiertes Netz
hierarchical order	TRANS	Hierarchiestufe
≈ translation stage		= Hierarchieebene; Multiplexstufe; Stufe
		≈ Umsetzerstufe
hierarchical route	SWITCH	Kennzahlweg
≈ final route		= Letztweg
hierarchical structure (→tree topology)	TELEC	(→Baumstruktur)
hierarchy	TELEC	Hierarchie
↓ multiplex hierarchy; network hierarchy		↓ Multiplexhierarchie; Netzhierarchie
HIFAM (→high fidelity amplitude modulation)	BROADC	(→Breitband-Amplitudenmodulation)
HiFi (→high fidelity)	EL.ACOUS	(→hohe Wiedergabetreue)
high (adj.)	TECH	hoch
≠ deep		≠ tief
high (→high level)	MICROEL	(→Hochpegelzustand)
high address	DATA PROC	höherwertige Adresse
= higher-order address		
high-angle radiation (→steep radiation)	ANT	(→Steilstrahlung)
high-angle radiator	ANT	Steilstrahler
≈ steep radiator		
high-bit-rate (adj.)	TELEC	hochbitratig
high button head	MECH	Hochrundkopf
high button head rivet	MECH	Hochrundkopfniet
high byte	CODING	höherwertiges Byte
high-capacity.. (→high-performance ...)	TECH	(→Hochleistungs-)
high-capacity cable	COMM.CABLE	hochpaariges Kabel
= large-size cable		
high-capacity digital radio link	TRANS	Breitband-Digitalrichtfunk-Verbindung
= HCDR; digital high-capacity line-of-sight radiolink; digital high capacity LOS radiolink; digital broadband digital radiolink; digital broadband LOS link; digital broadband radiolink; digital broadband LOS		
high-capacity line-of-sight radio (→high-capacity radio)	TRANS	(→Breitbandrichtfunk)
high capacity link	TRANS	Breitbandstrecke
= high capacity route; heavy route; broadband link; broadband route; wideband link; wideband route		
high-capacity LOS (→high-capacity radio)	TRANS	(→Breitbandrichtfunk)
high-capacity microwave (→high-capacity radio)	TRANS	(→Breitbandrichtfunk)
high-capacity radio	TRANS	Breitbandrichtfunk
= HCR; high-capacity radio relay; high-capacity line-of-sight radio; high-capacity LOS; high-capacity microwave; broadband radio-relay; broadband radio; broadband line-of-sight radio; broadband LOS; broadband microwave; high-density radio; high-density radio relay; high-density line-of-sight radio; high-density LOS; high-density microwave		= Vielkanal-Richtfunk
		≈ Weitverkehrsrichtfunk
≈ backbone microwave		
high-capacity radio relay (→high-capacity radio)	TRANS	(→Breitbandrichtfunk)
high capacity route (→high capacity link)	TRANS	(→Breitbandstrecke)
high-capacity system	TRANS	Breitbandsystem
= broadband system; wideband system; high density carrier; higher-order transmission system		= Vielkanalsystem; Übertragungssystem höherer Ordnung
≈ long-haul system		≈ Weitverkehrssystem
≠ low-capacity system		≠ Kleinkanalsystem
high-conductivity copper	METAL	Kupfer für Leitzwecke
high-contrast	TV	kontrastreich

high-current fuse	COMPON	**Stromgrobsicherung**	
↑ overcurrent protector		[schützt vor Strömen über 1 A]	
		= Edisonsicherung	
		↑ Stromsicherung	
high-def (→high-resolution)	TERM&PER	(→hochauflösend)	
high-definition (adj.)	TV	**hochauflösend**	
= with high definition			
high-definition (adj.)	TERM&PER	(→hochauflösend)	
(→high-resolution)			
high-definition television	TV	(→hochauflösendes Fernsehen)	
(→high-definition TV)			
high-definition TV	TV	**hochauflösendes Fernsehen**	
= HDTV; high-definition television		= HDTV; Hochzeilenfernsehen	
high-density	TERM&PER	**HD**	
[diskette]		[Diskette]	
= HD			
high density bipolar of order 3	CODING	(→HDB3)	
(→HDB3)			
high density carrier	TRANS	(→Breitbandsystem)	
(→high-capacity system)			
high density digital magnetic recording	TERM&PER	**HDDR**	
= HDDR			
high-density diskette (→HD diskette)	TERM&PER	(→HD-Diskette)	
high-density drive	TERM&PER	**High-Density-Laufwerk**	
		= HD-Laufwerk	
high-density line-of-sight radio	TRANS	(→Breitbandrichtfunk)	
(→high-capacity radio)			
high-density LOS	TRANS	(→Breitbandrichtfunk)	
(→high-capacity radio)			
high-density microwave	TRANS	(→Breitbandrichtfunk)	
(→high-capacity radio)			
high-density radio	TRANS	(→Breitbandrichtfunk)	
(→high-capacity radio)			
high-density radio relay	TRANS	(→Breitbandrichtfunk)	
(→high-capacity radio)			
high-doped (adj.)	MICROEL	**hochdotiert**	
high-end (adj.)	ECON	**der oberen Preisklasse**	
high-energy	PHYS	**energiereich**	
higher-bit-rate	TELEC	**höherbitratig**	
higher degree (→master degree)	SCIE	(→Diplom)	
higher-level ...	MODUL	**höherstufig**	
higher-level (→higher-order)	CODING	(→höherwertig)	
higher mathematics	MATH	**höhere Mathematik**	
higher-melting-point solder	METAL	(→Hartlot)	
(→hard solder)			
higher-order (adj.)	CODING	**höherwertig**	
= higher significant; higher-level		= signifikanter	
higher-order address (→high address)	DATA PROC	(→höherwertige Adresse)	
higher-order transmission system	TRANS	(→Breitbandsystem)	
(→high-capacity system)			
higher significant	CODING	(→höherwertig)	
(→higher-order)			
highest-order (→most significant)	CODING	(→höchstwertig)	
high fidelity	EL.ACOUS	**hohe Wiedergabetreue**	
= HiFi		= Hifi	
high fidelity amplitude modulation	BROADC	**Breitband-Amplitudenmodulation**	
[CATV]		[CATV]	
= HIFAM			
high frequency (n.)	EL.TECH	**Hochfrequenz**	
[in a narrow sense: 20 kHz to 100 MHz; in a broader sense: 20 kHz to 100 GHz then including the range of super high frequencies; the quasi-synonym "radiofrequency" is often preferred in English, but its use is restricted to radio-related topics in German]		[im engeren Sinne: 20 kHz bis 100 MHz; im weiteren Sinne: 20 kHz bis 100 GHz, dann also den Bereich der Höchstfrequenzen mit einschließend; in der Funktechnik – im Englischen oft auch generell – wird die Bezeichnung „Radiofrequenz" bevorzugt]	
= HF		≈ Radiofrequenz; Höchstfrequenz	
≈ radiofrequency; super high frequency		= HF	
high-frequency (adj.)	EL.TECH	**hochfrequent**	
= HF			
high frequency	POWER ENG	**Hochfrequenz**	
[above 10 kHz]		[ab 10 kHz]	
high frequency (→short waves)	RADIO	(→Kurzwelle)	
high-frequency amplifier (→HF amplifier)	HF	(→HF-Verstärker)	
high-frequency cable	COMM.CABLE	**Trägerfrequenzkabel**	
= HF cable		= TF-Kabel	
high-frequency engineering	EL.TECH	**Hochfrequenztechnik**	
= HF engineering		= HF-Technik	
↑ electrical fundamentals		↑ Elektrizitätslehre	
high-frequency heating	POWER ENG	(→HF-Heizung)	
(→HF heating)			
high frequency line (→HF line)	HF	(→HF-Leitung)	
high-frequency probe (→RF probe)	INSTR	(→HF-Tastkopf)	
high frequency radio	RADIO	**Kurzwellenfunk**	
= HF radio		= KW-Funk; HF-Funk	
high-frequency resistance	EL.TECH	(→HF-Widerstand)	
(→HF resistance)			
high-frequency resistance bridge	INSTR	(→HF-Widerstandsmeßbrücke)	
(→HF resistance bridge)			
high-frequency transceiver (→HF transceiver)	HF	(→HF-Transceiver)	
high-frequency transistor (→HF transistor)	COMPON	(→HF-Transistor)	
high-grade (→sophisticated)	TECH	(→hochwertig)	
high-grade steel	METAL	**Edelstahl**	
high-impedance (adj.)	EL.TECH	**hochohmig**	
= high ohmic			
high-impedance status	CIRC.ENG	**logisch neutraler Zustand**	
high-intensity obstruction light	AERON	**Hochleistungshindernisfeuer**	
high level	MICROEL	**Hochpegelzustand**	
[a positive potential by positive logic; neutral potential with negative logic]		[positives Potential bei positiver Logik; Nullpotential bei negativer Logik]	
= high signal; H level; H signal; H; logical high; signal high; high		= Hochpegelsignal; Hochpegel; H-Zustand; H-Pegel; H-Signal;High-Zustand; Signal-high-Pegel; High-Signal; H	
≠ low level		≠ Tiefpegelzustand	
↑ pulse level		↑ Impulspegel	
high-level data link control	DATA COMM	**HDLC-Verfahren**	
= HDLC		= codeunabhängiges Steuerungsverfahren	
high-level language	DATA PROC	(→problemorientierte Programmiersprache)	
(→high-level programming language)			
high-level logic	MICROEL	**Hochpegel-Logik**	
= HLL		= HLL; störsichere Logik	
≠ low-level logic		≠ Tiefpegellogik	
high-level modulation	MODUL	**Hochpegel-Modulation**	
		= Endstufenmodulation	
high-level programming language	DATA PROC	**problemorientierte Programmiersprache**	
[language optimized for specific types of problems, without direct correspondence to machine commands; the languages from the third generation onwards; examples: ADA; ALGOL; APL; BASIC; C; COBOL; COMAL; CORAL; FORTH; FORTRAN; LISP; LOGO; PASCAL; PL/1; POP-2; PROLOG language]		[am Aufgabenbereich orientierte Sprache, ohne direkte Entsprechung zu Maschinenbefehlen; die Sprachen ab der 3. Generation; z.B.: ADA; ALGOL; APL; BASIC; C; COBOL; COMAL; CORAL; FORTH; FORTRAN; LISP; LOGO; PASCAL; PL/1; POP-2; PROLOG Programmiersprache]	
= high-level language; HLL; high-order language; HOL; problem-oriented programming language; problem-orientated programming language; POL		= problemorientierte Sprache; höhere Programmiersprache; Hochpegel-Programiiersprache; Hochpegelsprache; Hochsprache	
≠ low-level programming language; machine-oriented programming		≠ niedere Programmiersprache; maschinenorientierte	
↑ procedure-oriented programming language; symbolic programming language		↑ prozedurorientierte Programmiersprache; symbolische Programmiersprache	
↓ compile-level language; interpreter language			

high-level selection   266

↓ Kompilersprache; Interpretersprache; Anfängersprache
high-level selection (→high-level signaling)　TRANS　(→Hochpegelsignalisierung)
high level signal (→large signal)　ELECTRON　(→Großsignal)
high-level signaling [FDM]　TRANS　**Hochpegelsignalisierung** [TF-Technik]
= high-level selection　　= Hochpegelwahl
highlight (v.t.)　DATA PROC　**hervorheben**
[on a display]　　auf einem Bildschirm]
= mark　　= markieren
highlight (n.)　DATA PROC　**Hervorhebung**
[emphasis of an an object on a display, e.g. by bold type]　　[am Bildschirm, z.B. durch Fettschrift]
= highlightning; display highlight; marking 2　　= Markierung 2
highlight (adj.)　TERM&PER　**hell**
↑ screen mode　　↑ Bildschirmdarstellung
highlight (n.) (→gloss)　TECH　(→Hochglanz)
highlight (v.t.) (→emphasize)　LING　(→hervorheben)
highlightning (→highlight)　DATA PROC　(→Hervorhebung)
high limit　ENG.DRAW　**Größtmaß**
= maximum dimension
high limit　TECH　**Oberwert**
high-loss　EL.TECH　**verlustreich**
≈ lossy　　≈ verlustbehaftet
≠ low-loss　　≠ verlustarm
highly developed　ECON　**hochentwickelt**
highly diluted　TECH　**hochverdünnt**
highly integrated (→large-scale integrated)　MICROEL　(→hochintegriert)
highly-integrated circuit (→large-scale integrated circuit)　MICROEL　(→hochintegrierte Schaltung)
highly integrated IC (→large-scale integrated circuit)　MICROEL　(→hochintegrierte Schaltung)
highly qualified　TECH　**hochqualifiziert**
highly sensitive　TECH　**hochempfindlich**
= high-sensitive
highly specialized　ECON　**hochspezialisiert**
highly-stable　TECH　**hochstabil**
high-noise-immunity logic (→HNIL)　MICROEL　(→HNIL)
high ohmic (→high-impedance)　EL.TECH　(→hochohmig)
high-order language (→high-level programming language)　DATA PROC　(→problemorientierte Programmiersprache)
high-order zero　DATA PROC　**führende Null**
[zeros in front of the highest ranking digit, e.g. in 007]　　[vor der höchstwertigen Ziffer stehende Null, z.B. in 007]
= leading zero
high overvoltage arrester (→high overvoltage protector)　COMPON　(→Spannungsgrobsicherung)
high overvoltage protector　COMPON　**Spannungsgrobsicherung**
[protects against overvoltages above 2,000 V]　　[schützt vor Spannungen über 2.000 V]
= high overvoltage arrester　　↓ Schutzfunkenstrecke; Hörnerableiter
high pass (→high pass filter)　NETW.TH　(→Hochpaßfilter)
high pass filter　NETW.TH　**Hochpaßfilter**
= high pass　　= Hochpaß
high-pass/low-pass filter (→separating filter)　NETW.TH　(→Weiche)
high-performance ...　TECH　**Hochleistungs-**
= high-capacity..; heavy-duty..; large-power-; performance-　　[mit hervorragenden Eigenschaften]
high-performance antenna　ANT　**Hochleistungs-Antenne** [hervorragender Eigenschaften]
high-performance chip　MICROEL　**Hochleistungschip** [hevorragender Eigenschaften]
high-performance CMOS (→HCMOS)　MICROEL　(→HCMOS)
high-performance computer　DATA PROC　**Hochleistungsrechner**
= Hochleistungscomputer

high-performance equipment　TECH　**Hochleistungsgerät**
high permittivity　COMPON　**HDK**
= hohe Dielektrizitätskonstante
high-permittivity capacitor　COMPON　**HDKK-Kondensator**
high persistence　PHYS　**große Nachleuchtdauer**
high power attenuator　INSTR　**Hochleistungsabschwächer**
= Hochleistungsdämpfungsglied; Leistungs-Dämpfungsglied
high-power sensor　INSTR　**Hochleistungsmeßkopf**
high-power speaker　EL.ACOUS　**Großlautsprecher**
high-power transformer　POWER SYS　**Umspanner**
↑ transformator [EL.TECH]　　[Transformator für große Leistungen]
　　↑ Transformator [EL.TECH]
high pressure　PHYS　**Hochdruck**
high-pressure fan (→blast)　TECH　(→Gebläse)
high-pressure jet　TERM&PER　**Hochdruck**
　　[hoher Druck der Tinte]
high-pressure mercury vapour lamp　EL.INST　**Quecksilberdampf-Hochdrucklampe**
high-pressure sodium vapour lamp　EL.INST　**Natriumdampf-Hochdrucklampe**
high-priced (→expensive)　ECON　(→teuer)
high-priority programm (→priority program)　DATA PROC　(→Prioritätsprogramm)
high-quality ... (→quality ...)　TECH　(→Qualitäts-)
high-quality paper　TERM&PER　**Qualitätspapier**
high-rate　TELEC　**hochratig**
high-res (→high-resolution)　TERM&PER　(→hochauflösend)
high-res graphics (→high-resolution graphics)　DATA PROC　(→hochauflösende Graphik)
high-resistance direction (→reverse direction)　ELECTRON　(→Sperrichtung)
high-resistance meter　INSTR　**Hochwiderstandsmesser**
high-resolution (adj.)　TERM&PER　**hochauflösend**
= high-res; high-definition (adj.); high-def　　≠ niedrigauflösend
≠ low-resolution
high-resolution graphics　DATA PROC　**hochauflösende Graphik**
= high-res graphics; HRG
high-resolution printer (→graphics printer)　TERM&PER　(→Graphikdrucker)
high school (AM) (→grammar school)　SCIE　(→Gymnasium)
high schools (BRI)　SCIE　**Mittelschule Oberstufe**
high-sensitive (→highly sensitive)　TECH　(→hochempfindlich)
high-sensitivity power sensor　INSTR　**Hochempfindlichkeitsmeßkopf**
high signal (→high level)　MICROEL　(→Hochpegelzustand)
high spec (→high specification)　TECH　(→detaillierte Spezifikation)
high specification　TECH　**detaillierte Spezifikation**
= high spec
high speed (n.)　TECH　**Hochgeschwindigkeit**
high-speed (adj.)　DATA PROC　**schnell**
≈ quick-access　　≈ zugriffszeitfrei
highspeed (→fast)　TECH　(→schnell)
high-speed carry　DATA PROC　**Hochgeschwindigkeitsübertrag**
high-speed digital stream　TELEC　**Hochgeschwindigkeits-Bitstrom**
high-speed dubbing　CONS.EL　**Schnellkopiervorrichtung**
= Schnellüberspielfunktion
high-speed facsimile　TELEGR　**Schnellfax**
high-speed memory　DATA PROC　**Schnellspeicher 1**
[additionally to the main memory, a memory with extremely short access times]　　[in Ergänzung zum Arbeitsspeicher, ein Speicher extrem kurzer Zugriffszeit, meist geringer Kapazität]
= HSM; high-speed storage; high-speed store; quick-access memory; quick-access storage; quick-access store; very fast memory; very fast storage; very fast store; zero-access memory; zero-access storage; zero-access store; fast access memory; FAM; fast access storage; fast access store; fast memory; fast storage; fast store; immediate access store; IAS　　= Schnellzugriffsspeicher; Sofortzugriffsspeicher

high-speed printer = fast printer ≈ line printer	TERM&PER	Schnelldrucker = Schnellschreiber ≈ Zeilendrucker
high-speed relay	COMPON	Schnellrelais
high-speed shutter	TECH	Hochgeschwindigkeitsverschluß [Optik]
high-speed storage (→high-speed memory)	DATA PROC	(→Schnellspeicher 1)
high-speed store (→high-speed memory)	DATA PROC	(→Schnellspeicher 1)
high speed transistor-transistor logic (→HSTTL)	MICROEL	(→HSTTL)
high storage [upper address range]	DATA PROC	oberer Speicherbereich
high-strength	TECH	hochfest
hight dependent	PHYS	höhenabhängig
high-tech (→high technology)	TECH	(→Hochtechnologie)
high technology = high-tech	TECH	Hochtechnologie
high-temperature superconductor	PHYS	Hochtemperatur-Supraleiter
high-tensile	TECH	hochzugfest
high tension (→high voltage)	POWER ENG	(→Hochspannung)
high-tension battery (→plate battery)	POWER SYS	(→Anodenbatterie)
high-tension instrument (→high-voltage instrument)	INSTR	(→Hochspannungs-Meßinstrument)
high-tension line ↑ power line	POWER ENG	Hochspannungsleitung ↑ Starkstromleitung
high-tension measurement	INSTR	Hochspannungs-Meßtechnik
high-tension probe = high-voltage probe	INSTR	Hochspannungs-Meßkopf = Hochspannungstastkopf
high-tension protection	POWER ENG	Hochspannungsschutz
high threshold logic (→HTL)	MICROEL	(→HTL)
high-usage route (→high-usage trunk)	SWITCH	(→Querweg)
high-usage trunk (AM) = high-usage route	SWITCH	Querweg
high vacuum	PHYS	Hochvakuum
high vacuum tube	ELECTRON	Hochvakuumröhre
high voltage [above 1kV (650 V in GB)] = HV; high tension; HT; supervoltage	POWER ENG	Hochspannung [über 1 kV (650 V in GBr)]
high-voltage battery	POWER SYS	Hochspannungsbatterie
high-voltage cable (→power cable)	POWER ENG	(→Starkstromkabel)
high-voltage convertor	POWER ENG	Hochvoltumrichter
high-voltage engineering	POWER ENG	Hochspannungstechnik
high-voltage FET	MICROEL	Hochspannungs-FET
high-voltage installation	POWER ENG	Hochspannungsanlage
high-voltage instrument = high-tension instrument	INSTR	Hochspannungs-Meßinstrument = Hochspannungsinstrument
high-voltage power supply	EQUIP.ENG	Hochspannungs-Stromversorgung
high-voltage probe (→high-tension probe)	INSTR	(→Hochspannungs-Meßkopf)
high-voltage rectifier	POWER ENG	Hochspannungsgleichrichter
high-voltage resistor	COMPON	Hochohmwiderstand
high-voltage transformer	COMPON	Hochohmübertrager
high-voltage transmission line ↑ transmission line	POWER ENG	Hochspannungs-Freileitung ↑ Starkstrom-Freileitung
high-volume bell	ACOUS	Starktonglocke
highway [path multiply used by time division] = multiple lead	SWITCH	Multiplexleitung [im Zeitvielfach] = Zeitmultiplexleitung; Vielfachleitung; Highway
highway 2 (→address bus)	DATA PROC	(→Adreßbus)
highway 1 (BRI) (→bus)	DATA PROC	(→Bus) (pl. Busse)
highway (→data bus)	DATA COMM	(→Datenbus)
highway (→main highway)	CIV.ENG	(→Hauptverkehrsstraße)
highway communication system	SIGN.ENG	Autobahn-Notruftechnik
highway control center	SIGN.ENG	Autobahnmeisterei
highway emergency telephone	SIGN.ENG	Autobahn-Notrufsäule
Hilbert transformation	MATH	Hilbert-Transformation
hill-and-dale recording [phonographic record]	EL.ACOUS	Tiefenschrift [Schallplatte]
hilly terrain	GEOPHYS	hügeliges Gelände
hinge (n.)	MECH	Scharnier
hinge (→hinge point)	MECH	(→Scharniergelenk)
hinged = swiveling; swivelling; tilting; pivoting; folding ≈ steerable	MECH	schwenkbar = klappbar; aufklappbar ≈ lenkbar
hinged-armature magnet = clapper armature magnet	COMPON	Klappankermagnet
hinged-armature relay = clapper armature relay	COMPON	Klappankerrelais
hinged clamp	TECH	Klappschelle
hinged handle (→clasp handle)	TECH	(→Klappgriff)
hinge pin	MECH	Scharnierbolzen
hinge point = hinge	MECH	Scharniergelenk
hint (→cue)	COLLOQ	(→Tip)
hints for the operator (→notes for the operator)	TECH	(→Bedienungshinweise)
hire (n.) (→rental tariff)	ECON	(→Mietgebühr)
hire 1 (→hire out)	ECON	(→vermieten)
hire 2 (→give on rent)	ECON	(→mieten)
hire out [to let for temporary use for a fixed sum] = rent 2 (AM); hire 1 ≈ borrow	ECON	vermieten [vorübergehend gegen Entgelt überlassen] ≈ verleihen ≠ mieten
hire-purchase (→installment purchase)	ECON	(→Ratenkauf)
hiss (n.) (→hissing)	ACOUS	(→Zischgeräusch)
hissing = hiss (n.); sibilance	ACOUS	Zischgeräusch = Zischen
histogram [a bar chart with bar widths proportional to class intervals] = frequency bar chart ↑ bar chart	MATH	Histogramm [Balkendiagramm dessen Balkenbreite proportional, zum Häufigkeitsintervall ist] = Staffelbild; Treppendiagramm ↑ Balkendiagramm
history file	DATA PROC	Ereignisspeicher
hit (v.t.) = beat; strike ≈ impact	TECH	schlagen ≈ aufschlagen
hit (n.) [successful search]	DATA PROC	Treffer [bei Datensuche]
hit 1 (n.) (→stroke 1)	MECH	(→Schlag)
hit 2 (n.) (→slip)	MECH	(→Schlupf)
hit (→interfering pulse)	ELECTRON	(→Störimpuls)
hitless [switchover]	TRANS	stoßfrei [Umschaltung] = schlupffrei; schlupflos; unterbrechungsfrei
hitless (→bumpless)	TECH	(→stoßfrei)
hitless switching	TRANS	schlupflose Umschaltung = schlupflose Steuerung; schlupflose Ersatzschaltung
hit second	TELEC	Störsekunde
H&J (→hyphenation and justification)	TYPOGR	(→Silbentrennung und Zeilenausschluß)
H K (→Hefner candle)	PHYS	(→Hefnerkerze)
H level (→high level)	MICROEL	(→Hochpegelzustand)
HLL (→high-level logic)	MICROEL	(→Hochpegel-Logik)
HLL (→high-level programming language)	DATA PROC	(→problemorientierte Programmiersprache)
HLS [color definition by Hue, Lightness and Saturation]	TERM&PER	HLS [Farbdefinitionsverfahren auf der Basis Farbe, Helligkeit und Sättigung]
h-matrix	MICROEL	H-Matrix
HMI (→man-machine interface)	DATA PROC	(→Mensch-Maschine-Schnittstelle)
HMI (→human-machine interaction)	DATA PROC	(→Mensch-Maschine-Dialog)
HMOS [High performance Metal-Oxide Semiconductor]	MICROEL	HMOS
HMOS technology	MICROEL	HMOS-Technik

HNIL	MICROEL	HNIL
= high-noise-immunity logic		
H-noise margin	ELECTRON	H-Störabstand
Ho (→holmium)	CHEM	(→Holmium)
hoar (→hoarfrost)	METEOR	(→Rauhreif)
hoarfrost	METEOR	Rauhreif
[frost with ice crystals]		[Reif mit sichtbar großen Eiskristallen]
= hoar		↑ Reif
↑ frost 3		
hob (v.t.)	MECH	abwälzfräsen
hobby computer	DATA PROC	Hobbycomputer
≈ home computer		= Hobbyrechner
↑ microcomputer		≈ Heimcomputer
		↑ Mikrocomputer
hodograph	MATH	Hodograph
HOF (→head of form)	DATA PROC	(→Formularkopf)
hoist (n.)	MECH	Hebevorrichtung
= hoisting gear; hoisting equipment; lifting apparatus		= Hebezeug
hoisting equipment (→hoist)	MECH	(→Hebevorrichtung)
hoisting gear (→hoist)	MECH	(→Hebevorrichtung)
hoisting rope	TECH	Hebeseil
= hoist line		↑ Zugseil
↑ traction rope		
hoist line (→hoisting rope)	TECH	(→Hebeseil)
hoist rope	ANT	bewegliches Gut
		[bewegbares Seil]
		= laufendes Gut
HOL (→high-level programming language)	DATA PROC	(→problemorientierte Programmiersprache)
hold (v.t.)	SWITCH	besetzthalten
hold (n.)	TV	Bildfang
= image retention		
hold (n.)	DATA PROC	Halten
= hold state; halt condition		= Haltezustand; Hold-Zustand
hold (n.)	SWITCH	Hold-Zustand
= hold state		
hold (→queing)	SWITCH	(→Warten)
hold circuit	CIRC.ENG	Halteglied
= keep-alive circuit		= Halteschaltung
hold current (→holding current)	ELECTRON	(→Haltestrom)
hold-down reel	TERM&PER	Bandspulenhalterung
holder	MECH	Halter
≈ support		= Träger
		≈ Stütze
holder (→case)	COMPON	(→Gehäuse)
holder (→mount)	TECH	(→Halterung)
holding (→seizure)	SWITCH	(→Belegung)
holding (→holding company)	ECON	(→Holdinggesellschaft)
holding appliance	TECH	Haltevorrichtung
holding circuit	ELECTRON	Haltestromkreis
holding coil	ELECTRON	Haltespule
holding company	ECON	Holdinggesellschaft
= holding		= Dachgesellschaft
≈ parent company		≈ Stammhaus
holding current	ELECTRON	Haltestrom
= hold current		
holding-down bolt (→foundation bolt)	MECH	(→Ankerschraube)
holding helix	COMM.CABLE	Haltewendel
holding loop	DATA PROC	Halteschleife
≈ waiting loop		≈ Warteschleife
holding piece (→mount)	TECH	(→Halterung)
holding time	SWITCH	Belegungsdauer
= busy time		= Belegungszeit
holding winding	ELECTRON	Haltewicklung
holdoff (→delay)	ELECTRON	(→Verzögerungszeit)
holdover (→free running)	TECH	(→Freilauf)
holdover capability	ELECTRON	Freilaufvermögen
= free-running capability		
holdover mode (→free-running mode)	ELECTRON	(→Freilaufbetrieb)
hold range	CIRC.ENG	Haltebereich
hold state (→hold)	DATA PROC	(→Halten)
hold state (→hold)	SWITCH	(→Hold-Zustand)
"hold the line"	TELEPH	„bleiben Sie am Apparat"
hold time	ELECTRON	Haltezeit
		≈ Nachwirkzeit
hold time	MICROEL	Übernahmezeit
hold together	TECH	zusammenhalten
holdup (n.)	POWER SYS	Reichweite
[time of power supply of an UPS]		[einer Ersatzstromversorgung]
		= Versorgungsreichweite
holdup alarm device	SIGN.ENG	Überfallmelder
= assault alarm device; raider alarm device		
holdup alarm system	SIGN.ENG	Überfallmeldesystem
= assault alarm system		
holdup button	SIGN.ENG	Überfallalarmknopf
= assault button		
hole (n.)	ELECTRON	Lochung
[PCB]		[Leiterplatte]
hole 1	TECH	Loch 1
= opening		[offene Stelle]
hole 2	TECH	Loch 2
= pit; depression		[Vertiefung]
hole	PHYS	Loch
[in valence band]		= Defektelektron; Elektronenlücke; Elektronenfehlstelle; Mangelelektron
= defect electron; electron vacancy		
hole (→drill hole)	MECH	(→Bohrloch)
hole-and-slot-anode magnetron	MICROW	Zylinder-Spalt-Magnetron
		= Schlitz-und-Loch-Magnetron
hole center distance (→hole pitch)	MECH	(→Lochabstand)
hole concentration	PHYS	Löcherkonzentration
		= Defektelektronen-Konzentration; Mangelelektronen-Konzentration
hole conduction	PHYS	Löcherleitung
= defect conduction; p-type conduction; impurity conduction		= Defektelektronenleitung; P-Leitung; Defektleitung; Mangelelektronenleitung; Mangelleitung; Störleitung
hole current	PHYS	Löcherstrom
		= Defektelektronenstrom; Mangelelektronenstrom; P-Strom
hole density	PHYS	Löcherdichte
		= Defektelektronendichte
hole etching	MICROEL	Lochätzen
hole gas	PHYS	Löchergas
		= Defektelektronengas
hole metal plate	MECH	Lochblende
hole mobility	PHYS	Löcherbeweglichkeit
		= Defektelektronen-Beweglichkeit; Mangelelektronen-Beweglichkeit
hole pattern	TERM&PER	Lochkombination
hole pitch	MECH	Lochabstand
= hole center distance		
hole row	TERM&PER	Lochreihe
= row		↓ Sprosse 1; Lochspur
↓ row frame; channel		
hole semiconductor	PHYS	Löcherhalbleiter
= p-type semiconductor		= P-Halbleiter; P-Typ-Halbleiter; Fehlstellenhalbleiter; P-Material
hole site	TERM&PER	Lochstelle
= punch position		
hole spacing	TERM&PER	Lochteilung
hole-storage effect	MICROEL	Trägerstaueffekt
= carrier jam effect		
hole-type conductivity	PHYS	Löcherleitfähigkeit
= p-type conductivity		= Defektelektronen-Leitfähigkeit; P-Leitfähigkeit; Mangelelektronen-Leitfähigkeit
holiday	ECON	Feiertag
≈ Sunday		= Sonntag
≠ working day		≠ Werktag
holiday (BRI) (→vacation)	ECON	(→Urlaub)
holiday shutdown (→works holidays)	ECON	(→Werkferien)
Hollerith card	TERM&PER	Hollerith-Lochkarte
Hollerith code	CODING	Hollerith-Code

hollow	TECH	hohl
≈ concave [MATH]; recessed		≈ konkav [MATH]; vertieft
≠ raised		≠ erhaben
hollow (→cavity)	TECH	(→Hohlraum)
hollow anode	ELECTRON	Hohlanode
hollow cathode	ELECTRON	Hohlkathode
= concave cathode		= Hohlspiegelkathode
hollow-conical antenna	ANT	Hohlkegelantenne
hollow cylinder	MATH	Hohlzylinder
hollow cylindrical antenna	ANT	Hohlzylinderantenne
hollow leading rope	ANT	Hohlseil
hollow punch	MECH	Locheisen
hollow rivet	MECH	Hohlniet
holmium	CHEM	Holmium
= Ho		= Ho
hologram	OPTOEL	Hologramm
↓ surface hologram; volume hologram		↓ Flächenhologramm; Volumenhologramm
holographic image (→volume hologram)	OPTOEL	(→Volumenhologramm)
holographic interferometry	TECH	holografische Interferometrie
holographic memory	TERM&PER	holographischer Speicher
= holographic store; holographic storage 2		= Hologrammspeicher
↑ optical memory		↑ optischer Speicher
holographic storage 1	TERM&PER	Hologrammspeicherung
		= holographische Speicherung
holographic storage 2 (→holographic memory)	TERM&PER	(→holographischer Speicher)
holographic store (→holographic memory)	TERM&PER	(→holographischer Speicher)
holography	OPTOEL	Holographie
holomorph function	MATH	holomorphe Funktion
		= differenzierbare Funktion; monogene Funktion
home (v.i.)	ELECTRON	rücklaufen
home (→normal position)	TECH	(→Normalstellung)
home (→cursor home position)	TERM&PER	(→Schreibmarken-Normalstellung)
home appliance	TECH	Haushaltsgerät
= household appliance		
home banking (→telebanking)	TELEC	(→Telebanking)
homebrew (adj.)	TECH	selbstgebastelt
= homemade		= selbstgemacht
home computer	DATA PROC	Heimcomputer
[designed for home uses]		[für Anwendungen im Privathaushalt]
≈ hobby computer		≈ Heimrechner; Home-Computer
↑ microcomputer		≈ Hobbycomputer
		↑ Mikrocomputer
home contact	SWITCH	Ruhekontakt
= rest contact		
home entertainment (→entertainment electronics)	ELECTRON	(→Unterhaltungselektronik)
home file (→home location register)	MOB.COMM	(→Heimatdatei)
home-grown software	DATA PROC	selbstgeschriebene Software
HOME key	TERM&PER	HOME-Taste
		= POS1-Taste
home location register	MOB.COMM	Heimatdatei
= home file		≠ Besucherdatei
≠ visitor location register		
homemade (→homebrew)	TECH	(→selbstgebastelt)
home market	ECON	Heimatmarkt
= domestic market		= Heimmarkt; Binnenmarkt
home network	MOB.COMM	Heimatnetz
home position	DATA PROC	Ausgangsstellung
		= Ausgangsposition; Ruheposition; Home Position
home position	SWITCH	Ruhelage
= normal position; rest position		= Ruheposition; Normallage
home position (→normal position)	TECH	(→Normalstellung)
home receiver	BROADC	Heimempfänger
home record	DATA PROC	Ausgangssatz
home signal	RAIL W.SIGN	Hauptsignal
homing	RADIO NAV	Zielanflug
		= Zielflug; Zielansteuerung; Heimflug
homing (→reverse action)	ELECTRON	(→Rücklauf)
homing device	RADIO NAV	Zielanfluggerät
homing receiver	RADIO NAV	Zielanflugempfänger
		= Zielflugempfänger
homocentric	PHYS	homozentrisch
homochronous	TELEC	homochron
[with constant phase difference at significant instants]		[mit fester Phasendifferenz der Kennzeitpunkte]
		≈ synchron
homodyne detection (→homodyne reception)	HF	(→Homodynempfang)
homodyne reception	HF	Homodynempfang
= homodyne detection; straight reception; straight detection		= Direktempfang; Geradeausempfang
homoepitaxy	MICROEL	Homöoepitaxie
homogeneous	SCIE	homogen
≠ inhomogeneous		≠ inhomogen
homogeneous field	PHYS	homogenes Feld
homogeneous grading	SWITCH	homogene Mischung
homogeneous line	LINE TH	homogene Leitung
homogeneous multiplexer	DATA COMM	homogener Multiplexer
= homogeneous multiplexor		
homogeneous multiplexing	DATA COMM	homogene Multiplexierung
[of equal protocol and rate signals]		
homogeneous multiplexor (→homogeneous multiplexer)	DATA COMM	(→homogener Multiplexer)
homograph	LING	Homonym
[word written in the same form but with different meaning]		[gleichgeschriebenes Wort unterschiedlicher Bedeutung]
= homonym 2		≈ Homophon
homologation (→type approval)	TECH	(→Typenabnahme)
homologation test	QUAL	Typprüfung
= type acceptance test; type approval test; conformance test		= Typmusterprüfung; Typenmusterprüfung; Typenprüfung; Zulassungsprüfung; Homologationsprüfung; Konformitätsprüfung; Konformitätstest
homomorphic	MATH	homomorph
homomorphism	MATH	Homomorphie
		= Homomorphismus
homonym 2 (→homograph)	LING	(→Homonym)
homonym 1 (→homophone)	LING	(→Homophon)
homophone	LING	Homophon
[word with same pronounciation but different spelling]		[Wort gleicher Aussprache aber unterschiedlicher Schreibweise]
= homonym 1		≈ Homonym
homotope	MATH	homotop
hone (v.t.)	METAL	honen
[to work a surface with extreme precision]		[eine Oberfläche extrem fein bearbeiten]
honed	METAL	gehont
[polished to extremly high precision]		[extrem fein geglättet]
honeycomb	EL.ACOUS	Honeycomb
		= Bienenwabe
honeycomb coil	COMPON	Honigwabenspule
= duolateral coil		≈ Kreuzwickelspule
honeycomb cooling radiator	TECH	Wabenkühler
= honeycomb radiator		
honeycomb radiator (→honeycomb cooling radiator)	TECH	(→Wabenkühler)
hood (n.)	TECH	Haube
= cap		= Kappe
↑ cover		↑ Abdeckung
↓ protective hood		↓ Schutzhaube
hood (→cover 1)	TYPOGR	(→Kappe)
hook (v.i.)	MECH	einhaken
hook (n.)	TECH	Haken
hook (→cradle)	TELEPH	(→Gabel)
hook bolt	MECH	Hakenkopfschraube
= clip bolt		= Hakenschraube
hooked	TECH	hakenförmig
hooks	OUTS.PLANT	Steigeisen
= lineman's climbers; climbing irons		
hookswitch	TELEPH	Gabelumschalter
= switchhook; cradle switch		= Hakenumschalter
≈ cradle 1		≈ Gabel

hookswitch contact

hookswitch contact	TELEPH	Gabelumschaltekontakt
hook tip adapter	INSTR	Hakenspitzenadapter
hook transistor	MICROEL	Vierschicht-Transistor
		= Hook-Koolektor-Transistor; Hook-Transistor; Hakentransistor
hooter (→horn)	EL.ACOUS	(→Hupe)
hooter signal	EL.ACOUS	Hupe
↑ audible signal		[Signal]
		↑ Schallsignal
hop	RADIO PROP	Funkfeld
= radiolink hop		↓ Richtfunkfunkfeld
↓ microwave hop		
hop interference	RADIO REL	Funkfeldbeeinflussung
Hopkinson coefficient	EL.TECH	(→Streufaktor)
(→leakage factor)		
hop length	RADIO PROP	Funkfeldlänge
= path length		
hopper (→vessel)	TECH	(→Behälter)
hopper (→card hopper)	TERM&PER	(→Kartenstapler)
hop rate	INSTR	Sprungrate
horizon	GEOPHYS	Horizont
		= Gesichtskreis
horizon angle (→elevation angle)	ANT	(→Elevationswinkel)
horizontal	MATH	waagrecht
≠ vertical		= horizontal
		≠ senkrecht
horizontal amplifier (→X amplifier)	ELECTRON	(→Horizontalverstärker)
horizontal bar 2 (→hyphen 2)	LING	(→Bindestrich)
horizontal bar 1 (→hyphen 1)	LING	(→Trennungsstrich)
horizontal blanking	TV	Zeilenaustastung
= line blanking; line suppression		= Zeilenunterdrückung; Teilbildaustastung
horizontal blanking interval	TV	Zeilenaustastlücke
= line blanking interval		↑ Austastlücke
↑ blanking interval		
horizontal blanking pulse	TV	Horizontalaustastimpuls
horizontal blanking time	TV	Horizontalaustastzeit
horizontal centering (→horizontal centering control)	TV	(→Horizontalregelung)
horizontal centering control	TV	Horizontalregelung
= horizontal centering		
horizontal check (→horizontal check sum)	CODING	(→Quersummenprüfung)
horizontal check (→horizontal parity check)	DATA COMM	(→Längsparitätsprüfung)
horizontal check parity (→horizontal parity)	CODING	(→Längsparität)
horizontal check sum	CODING	Quersummenprüfung
= horizontal check		
horizontal check sum (→horizontal parity check)	DATA COMM	(→Längsparitätsprüfung)
horizontal construction practice	EQUIP.ENG	Horizontalbauweise
= horizontal design; horizontal equipment practice		= Längsaufbau; horizontale Bauweise; quergestreifte Bauweise
		≠ Vertikalbauweise
horizontal deflection	TV	Horizontalablenkung
= horizontal sweep		
horizontal design (→horizontal construction practice)	EQUIP.ENG	(→Horizontalbauweise)
horizontal diagram	ANT	Horizontaldiagramm
= horizontal pattern; horizontal radiation pattern		≈ E-Diagramm
≈ E pattern		
horizontal dipole	ANT	Horizontaldipol
= horizontal half-wave pattern		
horizontal drive (→horizontal drive control)	ELECTRON	(→Horizontalsteuerung)
horizontal drive control [thyratron]	ELECTRON	Horizontalsteuerung [Thyratron]
= horizontal drive		
horizontal equipment practice (→horizontal construction practice)	EQUIP.ENG	(→Horizontalbauweise)
horizontal focus	TV	Horizontalfokussierung
horizontal frequency (→line frequency)	TV	(→Horizontalfrequenz)

horizontal gain	ELECTRON	Horizontalverstärkung
= X gain		= X-Verstärkung
horizontal half-wave pattern (→horizontal dipole)	ANT	(→Horizontaldipol)
horizontal oscillator (→line frequency oscillator)	TV	(→Horizontalfrequenzoszillator)
horizontal output transformer (→line scan transformer)	TV	(→Horizontal-Ablenktransformator)
horizontal parity	CODING	Längsparität
[longitudinal check bit inserted after a block]		[blockweise eingefügtes Längsparitätsbit]
= longitudinal parity; horizontal check parity		= Horizontalparität; Quersummenparität
↓ block parity		↓ Blockparität
horizontal parity character	DATA COMM	Längsparitätszeichen
[inserted character of horizontal parity bits]		[aus Längsparitätsbits bestehendes eingefügtes Zeichen]
= horizontal parity check character; longitudinal parity check character; longitudinal parity check character; longitudinal redundancy check character; LRC character		= LRC-Zeichen
		↓ Blockprüfzeichen
↓ block check character		
horizontal parity check	DATA COMM	Längsparitätsprüfung
= horizontal redundancy check; HRC; horizontal check sum; horizontal check; longitudinal parity check; longitudinal redundancy check; LRC; longitudinal check; longitudinal check sum; longitudinal check		= Längsparitätskontrolle; Längsprüfung; Längskontrolle; Längsprüfverfahren; Horizontal-Paritätsprüfung; Horizontal-Paritätskontrolle; Horizontalprüfung; Horizontalkontrolle; Horizontalprüfverfahren; Quersummenprüfung; Quersummenkontrolle; Querprüfung; Querkontrolle; Longitudinalprüfung
↓ block check; casting-out-nines		↓ Blockprüfung; Neunerprüfung
horizontal parity check character (→horizontal parity character)	DATA COMM	(→Längsparitätszeichen)
horizontal pattern (→horizontal diagram)	ANT	(→Horizontaldiagramm)
horizontal polarization	PHYS	Horizontalpolarisation
		= horizontale Polarisation
horizontal print density (→character density)	TYPOGR	(→Zeichendichte)
horizontal radiation pattern (→horizontal diagram)	ANT	(→Horizontaldiagramm)
horizontal redundancy check (→horizontal parity check)	DATA COMM	(→Längsparitätsprüfung)
horizontal resolution	TV	Horizontalauflösung
horizontal scanning (→line scanning)	ELECTRON	(→Zeilenabtastung)
horizontal scrolling (→side-scrolling)	DATA PROC	(→seitliches Rollen)
horizontal size control	TV	Zeilenbreitenregler
horizontal sum	MATH	Quersumme
horizontal sweep (→horizontal deflection)	TV	(→Horizontalablenkung)
horizontal sweep transformer (→line scan transformer)	TV	(→Horizontal-Ablenktransformator)
horizontal synchronization (→line synchronization)	TV	(→Zeilensynchronisierung)
horizontal synchronizing pulse	TV	Horizontal-Synchronimpuls
		= H-Synchronimpuls; H-Impuls
horizontal tabulation	DATA PROC	Horizontal-Tabulator
= HT		= HT
horizontal wraparound [of a cursor]	DATA PROC	Cursorrücksprung
horn	EL.ACOUS	Hupe
= buzzer; hooter; klaxon		[Vorrichtung]
horn (→horn radiator)	ANT	(→Hornstrahler)
horn (→funnel)	TECH	(→Trichter)
horn aerial (→horn antenna)	ANT	(→Hornantenne)
horn antenna	ANT	Hornantenne
= horn aerial; funnel antenna		= Trichterantenne
horn feed	ANT	Hornspeisung

270

horn lousspeaker	EL.ACOUS	**Druckkammerlautsprecher**	**hot-forge**	METAL	**warmschmieden**
= horn-type loudspeaker; horn speaker; horn-type speaker		= Hornlautsprecher; Trichterlautsprecher	**hot forging**	METAL	**Warmschmiedung** = Warmschmieden
↑ electrodynamic loudspeaker		↑ elektrodynamischer Lautsprecher	**hot-galvanize** = hot-dip galvanize	METAL	**feuerverzinken** ↑ verzinken
horn mouth (→funnel mouth)	TECH	(→Trichteröffnung)	↑ zink-coat		
horn-parabolic antenna	ANT	(→Hornparabolantenne)	**hot-galvanizing**	METAL	**Feuerverzinkung**
(→horn-reflector antenna)			**hot line** (→automatic connection)	TELEPH	(→automatischer Verbindungsaufbau)
horn-parabolic feed	ANT	**Hornparabolspeisung**	**hotness** [TECH] (→heat)	PHYS	(→Wärme)
horn-parabolic radiator	ANT	**Hornparabolstrahler**	**hot-press** (v.t.)	METAL	**warmpressen**
horn radiator	ANT	**Hornstrahler**	**hot pressing**	METAL	**Warmpressen**
= horn		= Horn; Trichterstrahler	**hot-roll** (v.t.)	METAL	**warmwalzen**
↑ waveguide radiator		≈ Hornantenne	**hot spot**	MICROEL	**Überhitzungspunkt**
↓ pyramidal horn; sectorial horn; diagonal horn; conical horn; Potter horn; corrugated horn		↑ Hohlleiterstrahler ↓ Pyramidenhorn; Sektorhorn; Diagonalhorn; Konushorn; Potterhorn; Rillenhorn	**hot standby**	RADIO REL	**Geräteersatz** = Hot-Stand-By
			hot stand-by	QUAL	**Heißersatz**
horn-reflector antenna	ANT	**Hornparabolantenne**	≠ cold stand-by		= Geräteersatz
= horn-parabolic antenna		= Hornreflektorantenne			≠ Kaltersatz
horn spark gap	ANT	**Hörnerblitzableiter**	**hot-tinned** = tin-coated	METAL	**feuerverzinnt**
		= Hörnerfunkenstrecke	**hot-wire ammeter** (→expansion instrument)	INSTR	(→Hitzdraht-Instrument)
horn speaker (→horn lousspeaker)	EL.ACOUS	(→Druckkammerlautsprecher)	**hot-wire expansion** (→hot-wire movement)	INSTR	(→Hitzdrahtmeßwerk)
horn tweeter	EL.ACOUS	**Hochtonhorn**	**hot-wire instrument** (→expansion instrument)	INSTR	(→Hitzdraht-Instrument)
horn-type loudspeaker (→horn lousspeaker)	EL.ACOUS	(→Druckkammerlautsprecher)	**hot wire microphone**	EL.ACOUS	**Hitzdrahtmikrophon**
horn-type speaker (→horn lousspeaker)	EL.ACOUS	(→Druckkammerlautsprecher)	**hot-wire movement** = hot-wire expansion	INSTR	**Hitzdrahtmeßwerk**
horsepower	PHYS	**Pferdestärke**	**hot-work**	METAL	**warmverformen**
[= 0.735498759 kW]		[= 0,735498759 kW]	**hot working**	METAL	**Warmverformung**
= German horsepower; PS; HP		= PS	**hot zone** [where hyphenation might occur]	TYPOGR	**Zeilenausgang** [Bereich der Wortrennungen]
horsepower-hour	PHYS	**PS-Stunde**	= justification range		
= PS h		= PS h			= Zeilenendezone; Randzone; Ausschließbereich; Ausschließzone
hose (n.)	TECH	**Schlauch**			
[a flexible tube]		[biegsames Rohr]	**hour**	PHYS	**Stunde**
= tube 2		≈ Rohr	[= 60 min]		[= 60 min]
host (n.)	COLLOQ	**Gastgeber**	= h		= h
host (→host computer)	DATA COMM	(→Hauptrechner)	**hourly**	COLLOQ	**stündlich**
host-cluster mode	DATA COMM	**Host-Cluster-Betrieb**	**hourly output**	MANUF	**Stundenleistung**
host computer	DATA COMM	**Hauptrechner**	**24-hours satellite** (→synchronous satellite)	SAT.COMM	(→Synchronsatellit)
[the main computer in a computer network]		[der zentrale Rechner in einem Mehrrechnersystem]	**house** (v.t.) = accomodate	EQUIP.ENG	**unterbringen**
= host processor; host; main computer; central computer; master computer; master processor; master; main computer; main processor; central computer; central processor		= Zentralrechner; Leitrechner; Verarbeitungsrechner; Dienstleistungsrechner; Steuerrechner; Host-Rechner; Host; Master	**house** (→company)	ECON	(→Gesellschaft)
			house cable (→indoor cable)	COMM.CABLE	(→Innenraumkabel)
			household	ECON	**Haushalt**
			household ac (→mains)	POWER ENG	(→Starkstromnetz)
≈ front-end computer; mainframe computer [DATA PROC]		≈ Vorrechner; Großrechner [DATA PROC];	**household appliance** (→home appliance)	TECH	(→Haushaltsgerät)
≠ slave computer		≠ Nebenrechner	**household electronics**	ELECTRON	**Haushaltselektronik**
host computer (→central computer)	DATA PROC	(→Zentralrechner)	≈ entertainment electronics ↑ consumer electronics		≈ Unterhaltungselektronik ↑ Konsumelektronik
host crystal	PHYS	**Grundkristall**	**household television**	TV	**Haushaltsfernsehen**
		= Wirtskristall	**housekeeping**	DATA PROC	**Systemverwaltung**
host file	MOB.COMM	**Fremddatei**	= system management; system administration; bookkeeping		= Systemorganisation; Organisation
hostile takeover	ECON	**feindliche Übernahme**	**housekeeping instruction**	DATA PROC	**Organisationsbefehl**
= unfriendly takeover			= overhead instruction		
host language	DATA PROC	**Gastsprache**	**housekeeping operation** (→housekeeping sequence)	DATA PROC	(→Organisationsablauf)
host material (→substrate)	MICROEL	(→Substrat)	**housekeeping program** = housekeeping routine; management program; management routine	DATA PROC	**Verwaltungsroutine** = Verwaltungsprogramm
host processor (→host computer)	DATA COMM	(→Hauptrechner)	**housekeeping register** (→control program)	DATA PROC	(→Organisationsprogramm)
host program	DATA PROC	**Wirtsprogramm**	**housekeeping routine** (→housekeeping program)	DATA PROC	(→Verwaltungsroutine)
hot	PHYS	**heiß**			
≈ worm		≈ warm	**housekeeping sequence**	DATA PROC	**Organisationsablauf**
hot (→live)	EL.TECH	(→heiß)	= housekeeping sequency; housekeeping operation; overhead operation; bookkeeping operation		= Verwaltungsablauf; organisatorische Operation
hot air	TECH	**Heißluft**			
hot-barrier diode (→Schottky barrier diode)	MICROEL	(→Schottky-Diode)	**housekeeping sequency** (→housekeeping sequence)	DATA PROC	(→Organisationsablauf)
hot cathode (→thermoionic cathode)	ELECTRON	(→Glühkathode)	**house style**	TYPOGR	**Verlagsstil**
hot-dip galvanize (→hot-galvanize)	METAL	(→feuerverzinken)	**house style**	ECON	**Firmendesign**
hot electrode (→incandescent electrode)	ELECTRON	(→Glühelektrode)			
hot-electron transistor	MICROEL	**ballistischer Transistor**			
hotel telephone	TELEPH	**Hotelfernsprecher**			
hot end	ELECTRON	**heißes Ende**			

housetop terrace

housetop terrace	CIV.ENG	Dachterrasse
		= Terrassendach
		≈ Flachdach
house wiring	EQUIP.ENG	Gehäuseverdrahtung
housing (→case)	TECH	(→Gehäuse)
housing (→case)	EQUIP.ENG	(→Gehäuse)
howler (n.)	COLLOQ	dummer Fehler
[trivial but serious mistake]		= Schnitzer
howler	SWITCH	Daueraushängeschnarre
[buzzer indicating continuous off-hook of a subscriber]		[weist auf einen Teilnehmer hin der nicht richtig aufgehängt hat]
Hoyer bridge	INSTR	Hoyer-Brücke
Hoyt balancing network	TELEPH	Hoyt-Nachbildung
= special balancing network		
HP (→horsepower)	PHYS	(→Pferdestärke)
h parameter	MICROEL	h-Parameter
= hybrid parameter		= Hybridparameter; Kleinsignalparameter; h-Vierpol-Parameter
		↑ Transistorkenngröße
H pattern (→H-plane pattern)	ANT	(→H-Diagramm)
HPI (→height-position indicator)	RADIO NAV	(→Höhe- und Ortsgeber)
hp-IB bus (→IEC bus)	DATA COMM	(→IEC-Bus)
H plane	ANT	H-Ebene
90° H-plane elbow	MICROW	H-Bogen 90°
H-plane pattern	ANT	H-Diagramm
= H pattern; H diagram		= H-Ebenen-Diagramm
HRC (→horizontal parity check)	DATA COMM	(→Längsparitätsprüfung)
HRG (→high-resolution graphics)	DATA PROC	(→hochauflösende Graphik)
HRI (→height-range indicator)	RADIO NAV	(→Höhe- und Entfernungsgeber)
H shaped sectorial horn	ANT	H-Sektorhorn
H signal (→high level)	MICROEL	(→Hochpegelzustand)
HSM (→high-speed memory)	DATA PROC	(→Schnellspeicher 1)
HSTTL	MICROEL	HSTTL
= high speed transistor-transistor logic; HTTL		= HTTL
HT (→high voltage)	POWER ENG	(→Hochspannung)
HT (→horizontal tabulation)	DATA PROC	(→Horizontal-Tabulator)
HTL	MICROEL	HTL
= high threshold logic		
HTTL (→HSTTL)	MICROEL	(→HSTTL)
hub (n.)	MECH	Nabe
[central piece of a revolving object]		[Mittelstück eines drehbaren Gegenstandes]
= key		
↓ wheel hub		↓ Radnabe
hub (n.)	TERM&PER	Plattenmitte
[central part of a disk]		= Plattenzentrum
hub (n.) (→network node)	DATA COMM	(→Netzknoten)
hubbing	TELEC	Verkehrskonzentrierung
= consolidation		
hub unit	TELECONTR	Unterzentrale
= sub-central		≈ Nebenzentrale
≈ stand-by central		
HUD (→head-up display)	RADIO NAV	(→Blickfeld-Darstellungsgerät)
hue	PHYS	Farbton
= color hue (AM); colour hue (BRI); color tone; colour tone; tone; tint		= Farbtönung; Farbwert
		= Farbstärke; Farbnuance
≈ color intensity; tint 2		↓ Dunkeltönung; Helltönung
↓ shade; tint 1		
hue control	TV	Farbtonregler
≈ color intensity control		= Farbstärkeregler; Phasensteuerung
Huffman code	CODING	Huffman-Code
hula-hoop antenna (→DDRR antenna)	ANT	(→DDRR-Antenne)
hum	TELEC	Brumm
[audible noise due to residual power supply frequency]		[hörbares Geräusch durch Netzspannungsreste]
= hum noise; supply noise; residual hum; residual power supply voltage		= Stromversorgungsgeräusch; Brummspannung; Restbrumm
≈ residual ripple [EL.TECH]		≈ Restwelligkeit [EL.TECH]
human body detection	SIGN.ENG	Personenerfassung
human-computer interface (→man-machine interface)	DATA PROC	(→Mensch-Maschine-Schnittstelle)
human error (→human failure)	COLLOQ	(→menschliches Versagen)
human factors engineering	INF.TECH	Anthropotechnik
human failure	COLLOQ	menschliches Versagen
= human error		
human language	LING	menschliche Sprache
human-machine interaction	DATA PROC	Mensch-Maschine-Dialog
= HMI		
human-machine interface (→man-machine interface)	DATA PROC	(→Mensch-Maschine-Schnittstelle)
hum bucking coil (→hum suppression coil)	BROADC	(→VF-Drosselspule)
hum-free (adj.)	TELEC	brummfrei
humid	PHYS	feucht
= moist; damp; dank		= naß
≈ wet		
humidity	PHYS	Feuchtigkeit
= dampness; moisture		= Feuchte
humidity barrier	TECH	Feuchteschutz
≈ water blocking		= Feuchtigkeitssperre
		≈ Nässeschutz
humidity class	QUAL	Feuchteklasse
humidity sensor	COMPON	Feuchtesensor
= dampness sensor		
hum interference	TELEC	Brummeinstreuung
hum modulation	TELEC	Brummodulation
hum noise (→hum)	TELEC	(→Brumm)
hump control computer	RAIL.W.SIGN	Ablaufsteuerrechner
hum rejection	TELEC	Brummunterdrückung
hum suppression coil [CATV]	BROADC	VF-Drosselspule [CATV]
= hum bucking coil		
hundredweight 1	PHYS	hundert Pounds
[100 pounds]		
hundred-weight 2 (→metric hundredweight)	PHYS	(→Zentner)
hunt group (→PBX line group)	SWITCH	(→Sammelanschluß)
hunting	SWITCH	Absuchen
[serching for seizure]		[feststellen ob belegt]
= exploration		= Abtastung
hunting	CONTROL	Regelschwingung
		= Pendelschwingung; Pendelung
hunting (→scrolling)	TERM&PER	(→Bildverschiebung)
hunting (→search)	DATA PROC	(→Suche)
hunting contact (→multiple contact)	COMPON	(→Mehrfachkontakt)
hunting group number (→line group number)	SWITCH	(→Sammelrufnummer)
hunting sequence	SWITCH	Absuchsequenz
Hurwitz criterion	NETW.TH	Hurwitz-Kriterium
Hurwitz polynomial	NETW.TH	Hurwitz-Polynom
hut	CIV.ENG	Häuschen
Huygen's principle	PHYS	Huygens-Fresnel-Prinzip
HV (→high voltage)	POWER ENG	(→Hochspannung)
HVD (→hyperabrupt varactor diode)	MICROW	(→Hyperabrupt-Varakterdiode)
hybrid (adj.)	TECH	hybrid
hybrid (adj.)	MICROEL	hybrid
≠ monobrid		≠ monobrid
hybrid (→termination hybrid)	TELEC	(→Gabelschaltung)
hybrid amplifier	TRANS	Gabelverstärker
= termination amplifier		
hybrid attenuation	TELEC	Gabeldämpfung
hybrid circuit (→termination hybrid)	TELEC	(→Gabelschaltung)
hybrid circuit (→hybrid IC)	MICROEL	(→Hybridschaltung)
hybrid coil (→hybrid transformer)	TELEC	(→Gabelübertrager)
hybrid computer	DATA PROC	Hybridrechner
[analog and digital]		[digital und analog]
hybrid computer system	DATA PROC	hybrides Rechnersystem
= hybrid system		= hybrides System
hybrid coupler (→three-dB coupler)	MICROW	(→Drei-dB-Koppler)

hybrid divider	MICROW	Hybridkoppler	hyperbolic logarithm (→natural logarithm)	MATH	(→natürlicher Logarithmus)
hybrid IC	MICROW	Hybridschaltung	hyperbolic navigation system	RADIO NAV	Hyperbel-Navigationsverfahren
= hybrid integrated circuit; hybrid circuit			hyperbolic orbit	PHYS	Hyperbelbahn
hybrid integrated circuit (→hybrid IC)	MICROEL	(→Hybridschaltung)	hyperbolic paraboloid	MATH	hyperbolisches Paraboloid
hybrid light barrier	COMPON	Gabellichtschranke	hyperbolic secant	MATH	Secans hyperbolicus
hybrid mode	MICROW	Hybridwelle	= sch		= sch; Sekans hyperbolicus; Hyperbelsekans
hybrid network	DATA COMM	hybrides Netz	↑ hyperbolic function		↑ Hyperbelfunktion
		= gemischtes Netz	hyperbolic sine	MATH	Sinus hyperbolicus
hybrid parameter (→h parameter)	MICROEL	(→h-Parameter)	= sh ↑ hyperbolic function		= sh; Sinus hyperbolicus; Hyperbelsinus
hybrid system (→hybrid computer system)	DATA PROC	(→hybrides Rechnersystem)			↑ Hyperbelfunktion
hybrid T (→magic T)	MICROW	(→magisches T)	hyperbolic spiral	MATH	hyperbolische Spirale
hybrid technology	MICROEL	Hybridtechnologie	hyperbolic tangent	MATH	Tangens hyperbolicus
		= hybride Technik	= tanh; th		= Tangens hyperbolicus; Hyperbeltangens; tanh; th
hybrid tee (→magic T)	MICROW	(→magisches T)			↑ Hyperbeltangens
hybrid transformer	TELEC	Gabelübertrager	hyperboloid	MATH	Hyperboloid
= hybrid coil		= Brückenübertrager	hypergeometric	MATH	hypergeometrisch
hybrid twin quad antenna	ANT	Hybrid-Doppelquad-Antenne	hypergeometric distribution	MATH	hypergeometrische Verteilung
hybrid value	EL.TECH	Mischgröße	hyperspace	MATH	Überraum
		[Gleich- plus Wechselanteil]	hypervapotron	ELECTRON	Hypervapotron
hydrated	CHEM	hydratiert	[vapour cooling method for power tubes]		[Dampfkühlungsverfahren für Leistungsröhren]
[containing chemically bonded water]		[chemisch gebundenes Wasser enthaltend]	≈ vapotron; supervapotron		≈ Vapotron; Supervapotron
≈ hydrous		≈ wässrig	hyphen 2	LING	Bindestrich
hydraulic	PHYS	hydraulisch	[symbol – ; to compound words, in German also to mark omitted word elements]		[Symbol- ; verbindet zusammengehörige Wörter oder ersetzt ausgelassene Wortteile]
hydraulic energy	TECH	hydraulische Energie			
hydraulic engineering	CIV.ENG	Wassertechnik			
≈ hydraulics [SCIE]		≈ Hydraulik [SCIE]	= horizontal bar 2; short hyphen		= Divis [TYPOGR]
hydraulic gate (→fluidic gate)	DATA PROC	(→fluidisches Schaltelement)	≈ hyphen 1; dash		≈ Trennungsstrich; Gedankenstrich
hydroacoustics	ACOUS	Hydroakustik	hyphen 1	LING	Trennungsstrich
hydrochloric acid	CHEM	Salzsäure	[symbol -; to separate syllables or compound words]		[Symbol -; zur Trennung von Silben oder zusammengesetzter Wörter]
[aqueous solution of hydron chloride]		[wäßrige Lösung von Chlorwasserstoff]			
		= Chlorwasserstoffsäure	= horizontal bar 1		= Trennungszeichen; Trennstrich; Teilungszeichen; Divis
hydrodynamics	PHYS	Hydrodynamik	≈ hyphen 2		
hydroelectric power plant	TECH	Wasserkraftwerk	↑ punctuation mark		
hydrogen	CHEM	Wasserstoff			≈ Bindestrich
= H		[H]			↑ Satzzeichen
		= Hydrogen; Hydrogenium	hyphenated (adj.)	LING	mit Tennstrichen geschrieben
hydrogen-anneal	METAL	wasserstoffglühen			= in Trennstrich-Schreibweise
hydrogen chloride	CHEM	Chlorwasserstoff	hyphenation	DATA PROC	automatische Trennhilfe
[HCl]		[HCl]	[word processing]		[Textverarbeitung]
hydrogen electrode	PHYS	Wasserstoffelektrode			= Hyphenation
hydrogen fuel cell	POWER SYS	Brennstoffzelle	hyphenation	LING	Silbentrennung
= fuel cell					= Trennung
hydrogen-iron resistance	PHYS	Eisen-Wasserstoff-Widerstand	hyphenation algorithm	DATA PROC	Silbentrennungsalgorithmus
= barretter		= Barretter	[word processing]		[Textverarbeitung]
hydrous	CHEM	wasserhaltig			= Trennungsalgorithmus
≈ aqueous; hydrated		≈ wässrig; hydratiert	hyphenation and justification	TYPOGR	Silbentrennung und Zeilenausschluß
Hygens source	ANT	Hygens-Strahler	= H&J		
		= Hygensche Quelle	hyphenation error	LING	Trennungsfehler
hygrometer	INSTR	Hygrometer			= Trennfehler
= moisture meter		= Feuchtemesser	hyphenation program	DATA PROC	Silbentrennungsprogramm
hygroscopic	PHYS	wasseranziehend	[word processing]		[Textverarbeitung]
		≈ hygroskopisch	= hyphenation routine		= Trennprogramm; Silbentrennroutine; Trennungsroutine
hyl	PHYS	Hyl			
[= 9.806650 g]		[= 9,806650 g]			
		= hyl	hyphenation routine	DATA PROC	(→Silbentrennungsprogramm)
hyperabrupt varactor diode	MICROW	Hyperabrupt-Varakterdiode	(→hyphenation program)		
= HVD			hypoid	MATH	Hypoid
hyperbola	MATH	Hyperbel	hyponym (→derivative term)	LING	(→Unterbegriff)
↑ conical section		↑ Kegelschnitt	hypotenuse	MATH	Hypotenuse
hyperbolic	MATH	hyperbolisch	= hypothenuse		
hyperbolic cosecant	MATH	Cosecans hyperbolicus	hypothenuse (→hypotenuse)	MATH	(→Hypotenuse)
= csch		= csch; Kosekans hyperbolicus; Hyperbelkosekans	hypothesis (→supposition)	SCIE	(→Annahme)
hyperbolic cosine	MATH	Cosinus hyperbolicus	hypothetical digital reference circuit	TELEC	hypothetischer digitaler Bezugskreis
= ch		= ch; Kosinus hyperbolikus; Hyperbelkosinus	hypothetical reference circuit	TELEC	hypothetischer Bezugskreis
↑ hyperbolic function		↑ Hyperbelfunktion	hypsogram (→level chart)	TELEC	(→Pegeldiagramm)
hyperbolic cotangent	MATH	Cotangens hyperbolicus	hypsometer (→level meter)	INSTR	(→Pegelmesser)
= cth		= cth; Kotagens hyperbbbolikus; Hyperbelkotangens; Hyperbelcotangens	hysteresis	PHYS	Hysterese
					= Hysteresis

hysteresis curve

hysteresis curve (→hysteresis loop)	PHYS	(→Hystereseschleife)	
hysteresis loop	PHYS	**Hystereseschleife**	
= hysteresis curve		= Hysteresekurve	
hysteresis loss (→hysteretic loss)	PHYS	(→Hystereseverlust)	
hysteretic loss	PHYS	**Hystereseverlust**	
= hysteresis loss		= Hysteresewärme	
Hz (→cycles per second)	PHYS	(→Hertz)	

I

I (→luminous flux)	PHYS	(→Lichtstrom)	
I (→current intensity)	EL.TECH	(→Stromstärke)	
I²L (→inegrated injection logic)	MICROEL	(→IIL)	
IA5 (→intenational alphabet no.5)	TELEGR	(→internationales Alphabet Nr.5)	
IAM (→intermediate access memory)	DATA PROC	(→Speicher mittlerer Zugriffsgeschwindigkeit)	
"I am pulling through"	TELEPH	„ich verbinde"	
= "I'll put you through"			
IAR (→instruction counter)	DATA PROC	(→Befehlszähler)	
IAS (→high-speed memory)	DATA PROC	(→Schnellspeicher 1)	
IBG (→interblock gap)	TERM&PER	(→Blockzwischenraum)	
IBM-compatible	DATA PROC	**IBM-kompatibel**	
IBM format card [187.3 x 82,5 mm, 80 columns, 12 lines]	TERM&PER	**IBM-Lochkarte** [187,3 x 82,5 mm, 80 Spalten, 12 Zeilen]	
IC (pl. ICs) (→integrated circuit)	MICROEL	(→integrierte Schaltung)	
icand (→multiplicand)	MATH	(→Multiplikand)	
IC bus	MICROEL	**IC-Bus**	
= inter-IC bus		= Inter-IC-Bus	
IC chip	MICROEL	**IC-Baustein**	
↑ chip		= IC-Chip	
		↑ Chip	
IC design	MICROEL	**IC-Entwicklung**	
		= IC-Entwurf	
ICE (→in-circuit emulator)	MICROEL	(→Echtzeit-Testadapter)	
ice blue	PHYS	**hellblau**	
= light blue			
ice green	PHYS	**eisgrün**	
ice load	OUTS.PLANT	**Eislast**	
ice-up	TECH	**vereisen**	
IC extractor	ELECTRON	**IC-Greifklemme**	
icing-up	TECH	**Vereisung**	
IC inserter	ELECTRON	**IC-Montagewerkzeug**	
icon [small on-screen symbol] = ikon; pictograph ↓ cursor; turtle	DATA PROC	**Piktogramm** [am Bildschirm gezeigtes Symbol] = Bildzeichen; Ikon; Icon; ikonisches Zeichen; Bildschirmsymbol ↓ Cursor; Schildkröte	
iconoscope	ELECTRON	**Ikonoskop**	
I controller [integral action] = integral controller ↑ continuous-action controller	CONTROL	**I-Regler** [integral wirkend] = Integralregler; integraler Regler ↑ stetiger Regler	
icosahedron (pl.-ons, -a) ↑ polyhedron	MATH	**Ikosaeder** = Zwanzigflächner ↑ Polyeder	
IC socket	MICROEL	**IC-Fassung**	
ID (→user identification)	DATA PROC	(→Anwenderkennung)	
id (→keyword 1)	DATA PROC	(→Paßwort)	
id (→code number)	DATA PROC	(→Schlüsselwort)	
IDC (→insulation piercing connection)	COMPON	(→Schneidklemmverbindung)	
ID card (→identification card)	TERM&PER	(→ID-Karte)	
ID code (→identification code)	DATA PROC	(→Kennungscode)	
ID digits (→identity number)	DATA PROC	(→Kennungsnummer)	
IDDN video telephone	TERM&PER	**ISDN-Bildtelefon** = ISDN-Bildfernsprecher	
ideal filter	NETW.TH	**ideales Filter**	
ideal sampling = ideal scanning; ideal scan	ELECTRON	**punktförmige Abtastung** = ideale Abtastung	
ideal scan (→ideal sampling)	ELECTRON	(→punktförmige Abtastung)	
ideal scanning (→ideal sampling)	ELECTRON	(→punktförmige Abtastung)	
ideal transformer	EL.TECH	**idealer Transformator** = idealer Übertrager [COMPON]	
idempotent (n.)	MATH	**Idempotenz**	
identical (adj.)	COLLOQ	**identisch**	
identical mapping [set theory]	MATH	**identische Abbildung** [Mengenlehre]	
identical sign [symbol ≡]	MATH	**Identitätszeichen** [Symbol ≡] = Kongruenzzeichen	
identifiability (→detectability)	TECH	(→Erkennbarkeit)	
identifiable (→recognizable)	TECH	(→erkennbar)	
identification = state estimation; parameter identification	CONTROL	**Identifikation** = Parameterschätzung; Parameteridentifikation; Zustandsschätzung	
identification	TELEC	**Kennung** = Identifizierung; Identifikation	
identification = marking; lettering; labeling ≈ description	TECH	**Kennzeichnung** ≈ Bezeichnung	
identification (→detection)	INF.TECH	(→Erkennung)	
identification beacon	RADIO NAV	**Kennbake**	
identification card = ID card	TERM&PER	**ID-Karte**	
identification card (USA) (→identity card)	ECON	(→Personalausweis)	
identification character (→station identification)	DATA COMM	(→Stationskennung)	
identification code = ID code ≈ identification number	DATA PROC	**Kennungscode** = Kenncode ≈ Kennungsnummer	
identification digit	SWITCH	**Kennziffer**	
identification division ↑ COBOL	DATA PROC	**Erkennungsteil** ↑ COBOL	
identification field	DATA PROC	**Kennungsfeld** = Identifikationsfeld	
identification frequency = characteristic frequency	TRANS	**Kennfrequenz**	
identification frequency locating	TRANS	**Kennfrequenzortung**	
identification friend or foe = IFF	MIL.COMM	**Freund-Feind-Erkennung** = IFF	
identification mark (→characteristic 1)	TECH	(→Merkmal)	
identification number	TECH	**Kennzahl 1** = Kennummer	
identification number (→code number)	SWITCH	(→Kennzahl)	
identification please = who are you?	TELEGR	**wer da ?**	
identification system	TERM&PER	**Identifikationssystem**	
identifier [COBOL]	DATA PROC	**Bezeichner** [COBOL] = Kennzeichner 2	
identifier = track descriptor ≈ key	TERM&PER	**Spurenkennzblock** [Magnetplattenspeicher, freie Formatierung] ≈ Schlüsselfeld	
identifier (→station identification)	DATA COMM	(→Stationskennung)	
identifier check (→label check)	DATA PROC	(→Etikettprüfung)	
identifier generator (→answerback generator)	DATA COMM	(→Kennungsgeber)	
identifier group (→label group)	DATA PROC	(→Etikettfolge)	

identifier storage	DATA COMM	(→Kennungsspeicher)	
(→answerback code storage)			
identifier word (→keyword 1)	DATA PROC	(→Paßwort)	
identify	TECH	**kennzeichnen** (v.t.)	
= mark (v.t.); label(v.t.); letter (v.t.)			
identify (→detect)	INF.TECH	(→erkennen)	
identifying	SWITCH	**Identifizieren**	
identifying character (→label	DATA PROC	(→Etikettenkennzeichen)	
identifier)			
identifying code (→station	DATA COMM	(→Stationskennung)	
identification)			
identity	COLLOQ	**Identität**	
identity (→equivalence	ENG.LOG	(→Äquivalenzverknüpfung)	
operation)			
identity burst	DATA COMM	**Kennbitfolge**	
identity card	ECON	**Personalausweis**	
= identification card (USA)		= amtlicher Ausweis	
↑ card			
identity card system	TERM&PER	**Ausweissystem**	
identity circuit (→equivalence	CIRC.ENG	(→Äquivalenzglied)	
gate)			
identity digits (→identity	DATA PROC	(→Kennungsnummer)	
number)			
identity element (→equivalence	CIRC.ENG	(→Äquivalenzglied)	
gate)			
identity function (→equivalence	ENG.LOG	(→Äquivalenzverknüpfung)	
operation)			
identity gate (→equivalence	CIRC.ENG	(→Äquivalenzglied)	
gate)			
identity number	DATA PROC	**Kennungsnummer**	
= ID number; identity digits; ID digits		= Kennummer; Kennziffern	
≈ identity code		≈ Kennungscode	
identity operation	ENG.LOG	(→Äquivalenzverknüpfung)	
(→equivalence operation)			
identity papers	ECON	**Personalpapiere**	
IDF (→intermediate distribution	SYS.INST	(→Zwischenverteiler)	
frame)			
idiom	LING	**Redewendung**	
		= Redensart	
idiostatic circuit	PHYS	**idiostatische Schaltung**	
idle	SWITCH	**unbelegt**	
≠ busy		= frei	
		≠ belegt	
idle (→unutilized)	TECH	(→ungenutzt)	
idle accessibility	SWITCH	**Leerlauf-Erreichbarkeit**	
= idle availability			
idle availability (→idle	SWITCH	(→Leerlauf-Erreichbarkeit)	
accessibility)			
idle channel noise	TELEC	**Leerkanalgeräusch**	
idle character (→blank	DATA COMM	(→Leerzeichen)	
character)			
idle character (→space	TELEGR	(→Zwischenraumzeichen)	
character)			
idle code	TELEC	**Leerlaufcode**	
idle condition (→idle state)	SWITCH	(→Ruhezustand)	
idle condition (→rest condition)	TELEGR	(→Ruhezustand)	
idle costs (→downtime costs)	ECON	(→Stillstandskosten)	
idle current	EL.TECH	**Blindstrom**	
= reactive current; reactance current			
idle line	TELEC	**freie Leitung**	
idle motion	MECH	**Leerlauf**	
idle noise (→basic noise)	TELEC	(→Grundgeräusch)	
idler pulley	MECH	**Spannrolle**	
= tension roller			
idle state	SWITCH	**Ruhezustand**	
= idle condition; free line		= Freizustand	
≠ busy state		≠ Besetztzustand	
idle task	DATA PROC	**Ruheschleife**	
idle time	TECH	**Leerlaufzeit**	
idle time (→downtime)	QUAL	(→Ausfallzeit)	
idle tone	TELEPH	**Freiton**	
[indicates that subscriber is beeing called; e.g. 1 s (425 Hz) and 3 s pause]		[der gewählte Anschluß ist frei und wird gerufen; z.B. 1 s lang (mit 425 Hz) und 4 s Pause]	
= ringing tone; calling tone			
≈ dial tone		= Freizeichen; Rufton; Signalton	
≠ busy tone		≈ Wählton	
↑ audible signal		≠ Besetztton	
		↑ Hörton	
IDN (→integrated digital	DATA COMM	(→integriertes digitales Netz)	
network)			
ID number (→identity	DATA PROC	(→Kennungsnummer)	
number)			
IDP (→integrated data	DATA PROC	(→integrierte Datenverarbeitung)	
processing)			
IEC	EL.TECH	**IEC**	
[international organization for the uniformization of national electro-technical standards; in French "Commission Electrotechnique Internationale]		[internationale Organisation zur Vereinheitlichung nationaler elektrotechnischer Normen; franz. Commission Electrotechnique Internationale]	
= International Electrotechnical Commission; CEI		= CEI	
IEC bus	DATA COMM	**IEC-Bus**	
= IEEE-488 bus; hp-IB bus; GPIB bus; general-purpose interface bus		= IEEE-488-Bus; HP-IB-Bus; GPIB-Bus	
IEE	EL.TECH	**IEE**	
[of UK]		[Verein der Elektrotechniker Großbritanniens]	
= Institution of Eelectrical Engineers			
IEEE	EL.TECH	**IEEE**	
[USA]		[Verein der Elektrotechniker und Elektroniker der USA; im Englischen mit „ai triple i" ausgesprochen]	
= Institute of Elecrical and Electronic Engineers			
IEEE-488 bus (→IEC bus)	DATA COMM	(→IEC-Bus)	
I element	CONTROL	**I-Glied**	
		= Integrierglied; Integralglied; integrales Übertragungsglied	
ier (n.) (→multiplier)	MATH	(→Multiplikator)	
IF (→intermediate	RADIO	(→Zwischenfrequenz)	
frequency)			
IFAC	ELECTRON	**IFAC**	
= International Federation of Automatic Control			
IF amplifier	RADIO	**ZF-Verstärker**	
		= Zwischenfrequenzverstärker	
IF derivation	RADIO	**ZF-Auskopplung**	
= intermediate frequency derivation		= Zwischenfrequenzauskopplung	
IFF (→identification friend or foe)	MIL.COMM	(→Freund-Feind-Erkennung)	
IF filter	RADIO	**ZF-Filter**	
= intermediate frequency filter		= Zwischenfrequenzfilter	
IF instruction (→IF statement)	DATA PROC	(→WENN-Anweisung)	
IF interconnection	RADIO	**ZF-Durchschaltung**	
= intermediate frequency interconnection		= Zwischenfrequenzdurchschaltung	
IF rejection	MODUL	**ZF-Unterdrückung**	
= IF suppression			
IFS (→ionospheric scatter)	RADIO PROP	(→ionosphärische Streuausbreitung)	
IF statement	DATA PROC	**WENN-Anweisung**	
= IF instruction		= IF-Anweisung	
IF suppression (→IF rejection)	MODUL	(→ZF-Unterdrückung)	
IF THEN (→implication operation)	ENG.LOG	(→Implikationsverknüpfung)	
IF-THEN circuit (→implication gate)	CIRC.ENG	(→Implikationsglied)	
IF-THEN element (→implication gate)	CIRC.ENG	(→Implikationsglied)	
IF-THEN-ELSE instruction (→IF-THEN-ELSE statement)	DATA PROC	(→WENN-DANN-SONST-Anweisung)	
IF-THEN-ELSE statement	DATA PROC	**WENN-DANN-SONST-Anweisung**	
= IF-THEN-ELSE instruction		= IF-THEN-ELSE-Anweisung	
IF-THEN function (→implication operation)	ENG.LOG	(→Implikationsverknüpfung)	
IF-THEN gate (→implication gate)	CIRC.ENG	(→Implikationsglied)	
IF-THEN instruction (→IF-THEN statement)	DATA PROC	(→WENN-DANN-Anweisung)	
IF-THEN operation (→implication operation)	ENG.LOG	(→Implikationsverknüpfung)	
IF-THEN statement	DATA PROC	**WENN-DANN-Anweisung**	
= IF-THEN instruction		= IF-THEN-Anweisung	

IF transformer

IF transformer = intermediate frequency transformer	RADIO	**ZF-Übertrager** = Zwischenfrequenz-Übertrager
IGFET (→insulated gate field-effect transistor)	MICROEL	(→IGFET)
ignition (→triggering)	ELECTRON	(→Auslösung)
ignition expansion time [final phase of the ignition process of a thyristor]	MICROEL	**Zündausbreitungszeit** [Endphase der Zündung von Thyristoren]
ignition mechanism [thyristor]	MICROEL	**Zündmechanismus** [Thyristor]
ignition point	CHEM	**Entzündungspunkt**
ignition threshold [thyristor]	MICROEL	**Zündschwelle** [Thyristor]
ignition transformer	ELECTRON	**Zündtrafo**
ignitor (→trigger 1)	ELECTRON	(→Auslöser)
ignitron	POWER SYS	**Ignitron** = Zündstiftröhre ↑ steuerbarer Quecksilber-Gleichrichter
ignitron rectifier	POWER SYS	**Ignitrongleichrichter**
ignore	COLLOQ	**übergehen** [fig.] = ignorieren
ignore (→skip)	DATA PROC	(→überspringen)
ignore character = cancel character; invalid character	DATA COMM	**Ungültigkeitszeichen**
IIL (→inegrated injection logic)	MICROEL	(→IIL)
IIR (→infinite impulse response)	NETW. TH	(→IIR)
IKBS (→expert system)	DATA PROC	(→Expertensystem)
ikon (→icon)	DATA PROC	(→Piktogramm)
i layer (→intrinsic layer)	MICROEL	(→Intrinsic-Schicht)
ILD (→injection laser diode)	OPTOEL	(→Injektionslaserdiode)
illegal (adj.)	DATA PROC	**unzulässig**
illegal copy (→pirated copy)	DATA PROC	(→Raubkopie)
illegal instruction	DATA PROC	**unzulässiger Befehl**
illegal operation	DATA PROC	**unzulässige Operation**
illicit listener (→unlicensed listener)	BROADC	(→Schwarzhörer)
illicit transmitter (→unlicensed transmitter)	RADIO	(→Schwarzsender)
illicit TV viewer (→unlicensed TV viewer)	BROADC	(→Schwarzfernseher)
illigal (→unlawful)	ECON	(→unerlaubt)
illiterate	LING	**Analphabet**
"I'll put you through" (→"I am pulling through")	TELEPH	(→"ich verbinde")
illuminance (→illumination)	PHYS	(→Beleuchtungsstärke)
illuminate	TECH	**beleuchten**
illuminated push button switch	COMPON	**Leuchtdruckschalter**
illuminating beam	SAT. COMM	**Ausleuchtbündel**
illumination	TECH	**Beleuchtung**
illumination [luminous flux per unit area; SI unit: lux] = illuminance	PHYS	**Beleuchtungsstärke** [Lichtstrom pro Flächeneinheit; SI-Einheit: Lux]
illumination (→aperture illumination)	ANT	(→Ausleuchtung)
illumination cone	TECH	**Beleuchtungskegel**
illumination engineering	TECH	**Leuchttechnik**
illumination spot = spot; footprint	SAT. COMM	**Ausleuchtzone** = Ausleuchtgebiet
illustrate	DOC	**bebildern** = illustrieren
illustrate ≈ clarify	COLLOQ	**veranschaulichen** = illustrieren ≈ verdeutlichen
illustrated data sheet	DOC	**Bildblatt**
illustration = figure; image	TYPOGR	**Abbildung** = Bild; Figur
illustration	DOC	**Bebilderung** = Illustration
illustration ≈ clarification	COLLOQ	**Veranschaulichung** = Illustration ≈ Verdeutlichung
illustration 1 (→figure)	DOC	(→Bild)
ILS (→instrument landing system)	RADIO NAV	(→Instrumentenlandesystem)
ILS beam (→equisignal)	RADIO NAV	(→Leitstrahl)
"I'm afraid we were cut off"	TELEPH	**„wir wurden leider unterbrochen"**
imag (→imaginary part)	MATH	(→Imaginärteil)
image (n.)	DATA PROC	**Abbildung** = Abbild
image (n.) ≈ reflection	MATH	**Spiegelbild** ≈ Reflexion
image (→illustration)	TYPOGR	(→Abbildung)
image analysis	INF. TECH	**Bildanalyse**
image antenna	ANT	**Antennenspiegelbild**
image area [area covered by the text and the figures of a print page] = type area	TYPOGR	**Satzspiegel** [von Text und Abbildungen eingenommene Fläche einer Druckseite] = Spiegel
image attenuation = two-port attenuation ≈ complex image transfer constant	NETW. TH	**Wellendämpfung** = Vierpoldämpfung ≈ komplexes Wellendämpfungsmaß
image attenuation coefficient (→image attenuation constant)	NETW. TH	(→Wellendämpfungsmaß)
image attenuation constant [attenuation constant with both sides impedance-matched; negative value of the image transfer constant] = image attenuation coefficient; two-port attenuation constant; two-port attenuation coefficient ≈ image transfer constant ↑ attenuation constant	NETW. TH	**Wellendämpfungsmaß** [Dämpfungsmaß bei beiseitiger Anpassung; Negativwert des Wellenübertragungsmaßes] = Vierpoldämpfungsmaß ≈ Wellenübertragungsmaß ↑ Dämpfungsmaß; komplexes Wellendämpfungsmaß
image attenuation factor [attenuation factor with both sides impedance-matched] = two-port attenuation factor ↑ attenuation factor	NETW. TH	**Wellendämpfungsfaktor** [Dämpfungsfaktor bei beidseitiger Anpassung] ↑ Dämpfungsfaktor
image-audio separating filter	TV	**Bildtonweiche**
image-centering control	TV	**Bildmitteneinstellung**
image coding	SIGN. TH	**Bildcodierung** [Signalverarbeitung]
image communication ↑ telecommunications ↓ television; videotelephony	TELEC	**Bildkommunikation** ↑ Telekommunikation ↓ Fernsehen; Bildfernsprechen
image converter (→picture converter)	ELECTRON	(→Bildwandler)
image converter tube = electron image tube	ELECTRON	**Bildwandlerröhre** = Elektronenbildwandler
image current transfer coefficient [current transmission coefficient with impedance matching on both sides] = image current transfer factor ↑ image transmission coefficient; current transmission coefficient	NETW. TH	**Wellenstromübertragungsfaktor** [Stromübertragungsfaktor bei beidseitiger Anpassung] ↑ Wellenübertragungsfaktor; Stromübertragungsfaktor
image current transfer factor (→image current transfer coefficient)	NETW. TH	(→Wellenstromübertragungsfaktor)
image curvature	PHYS	**Bildwölbung**
image curvature aberration	PHYS	**Bildfeldwölbung**
image degradation	TERM&PER	**Bildverschlechterung**
image distance	PHYS	**Bildweite**
image dot [facsimile]	TELEGR	**Bildpunkt** [Faksimile]
image dot frequency	TELEGR	**Bildpunktfrequenz** [Faksimile]
image enhancement	DATA PROC	**Bildverstärkung**
image frequency [facsimile]	TELEGR	**Bildfrequenz** [Faksimile]
image frequency	MODUL	**Spiegelfrequenz**
image frequency rejection = image rejection; image rejection ratio	MODUL	**Spiegelselektion** = Spiegelfrequenzabstand; Spiegelfrequenzsicherheit
image gain coefficient (→image transfer constant)	NETW. TH	(→Wellenübertragungsmaß)
image gain constant (→image transfer constant)	NETW. TH	(→Wellenübertragungsmaß)

image gain factor (→image transfer constant) NETW.TH (→Wellenübertragungsmaß)
image generation (→rendering) DATA PROC (→Bildaufbereitung)
image impedance EL.TECH allgemeiner Kennwiderstand
image intensifier tube ELECTRON **Bildverstärkerröhre**
= Bildverstärker
image matching NETW.TH **Wellenanpassung**
[termination of a two-port with its input and output impedance respectively] [beidseitiger Abschluß eines Vierpols mit seinem Eingangs- bzw. Ausgangswiderstand]
↑ matching 1 = Spiegelanpassung
↑ Anpassung 1
image modulation (→vision modulation) TV (→Bildmodulation)
image orthicon ELECTRON **Superorthikon**
= superorthicon = Zwischenbild-Orthikon
image parameter NETW.TH **Wellenparameter**
[parameter of two-ports impedance-matched on both sides] [Parameter von beidseitig mit ihren Wellenwiderständen abgeschlossenen Vierpolen]
image parameter filter NETW.TH **Wellenparameterfilter**
image parameter theory NETW.TH **Wellenparametertheorie**
image pattern recognition INF.TECH **Bildmustererkennung**
image phase angle NETW.TH **Wellendämpfungswinkel**
[phase angle with impedance matching on both sides, negative value of the image phase angle factor] [Dämpfungswinkel bei beidseitiger Anpassung, Negativwert des Wellenphasenmaßes]
≈ image phase angle factor = Wellenwinkel
↑ phase angle ≈ Wellenphasenmaßes
↑ Dämpfungswinkel
image phase angle factor NETW.TH **Wellenphasenmaß**
[phase angle factor with impedance-matching on both sides, negative value of the image phase angle] [Phasenmaß bei beidseitiger Wellenanpassung, Negativwert des Wellendämpfungswinkels]
≈ image phase angle ≈ Wellendämpfungswinkel
↑ phase angle factor ↑ Phasenmaß
image phase coefficient NETW.TH (→Vierpolwinkelmaß)
(→image phase constant)
image phase constant NETW.TH **Vierpolwinkelmaß**
= image phase coefficient
image precision TV **Bildschärfe**
image presentation TERM&PER **Bildeindruck**
image processing INF.TECH **Bildverarbeitung**
= picture processing
image processor (→display processor) DATA PROC (→Bildprozessor)
image reconstruction SIGN.TH **Bildrekonstruktion**
image regeneration TERM&PER (→Bildwiederholung)
(→refresh)
image regeneration store DATA PROC (→Bildwiederholspeicher)
(→refresh memory)
image register (→display register) DATA PROC (→Bildregister)
image rejection (→image frequency rejection) MODUL (→Spiegelselektion)
image rejection ratio (→image frequency rejection) MODUL (→Spiegelselektion)
image resistance EL.TECH **Kennwiderstand**
image resolution (→display resolution) TERM&PER (→Bildauflösung)
image response RADIO **Spiegelfrequenz-Empfindlichkeit**
image retention (→hold) TV (→Bildfang)
image return coefficient NETW.TH **Wellenreflexionsfaktor**
image rotation (→frame rotation) TV (→Bilddrehung)
image scanner TERM&PER **Bildabtaster**
= video scanner; scanner; scanner camera = Bildsensor
image sensor TV **Bildsensor**
↓ Festkörperbildsensor
image signal MODUL **Spiegelsignal**
image splitting TV **Bildteilung**
image stability TERM&PER **Flimmerfreiheit**
= Bildstabilität

image storage space DATA PROC **Bilspeicherplatz**
image surface PHYS **Bildschale**
image synthesis DATA PROC (→Bildsynthese)
(→imaging)
image tranmission coefficient NETW.TH (→Wellenübertragungsmaß)
(→image transfer constant)
image transfer coefficient NETW.TH (→Wellenübertragungsmaß)
(→image transfer constant)
image transfer constant NETW.TH **Wellenübertragungsmaß**
[transfer constant with impedance matching on both sides, negative value of the image attenuetion constant] [Übertragungsmaß bei beidseitiger Anpassung, Negativwert des Wellendämpfungsmaßes]
= image transfer coefficient; image transfer factor; image transmission constant; image tranmission coefficient; image transmission factor; image gain constant; image gain coefficient; image gain factor; two-port transmission constant; two-port transmission coefficient; two-port transmission factor = Vierpolübertragungsmaß; Wellenübertragungsfaktor; Vierpolübertragungsfaktor
≈ Wellendämpfungsmaß
↑ Übertragungsmaß; komplexes Wellenübertragungsmaß
≈ image attenuation constant
↑ transfer constant; complex image transfer constant
image transfer factor (→image transfer constant) NETW.TH (→Wellenübertragungsmaß)
image transmission constant (→image transfer constant) NETW.TH (→Wellenübertragungsmaß)
image transmission factor (→image transfer constant) NETW.TH (→Wellenübertragungsmaß)
image voltage transmission coefficient NETW.TH **Wellenspannungsübertragungsfaktor**
[voltage transmission coefficient with impedance matching on both sides] [Spannungsübertragungsfaktor bei beidseitiger Anpassung]
= image voltage transmission factor
↑ image transmission coefficient; voltage transmission coefficient ↑ Wellenübertragungsfaktor; Spannungsübertragungsfaktor
image voltage transmission factor (→image voltage transmission coefficient) NETW.TH (→Wellenspannungsübertragungsfaktor)
imaginary MATH **imaginär**
≠ real ≠ real
imaginary component MATH (→Imaginärteil)
(→imaginary part)
imaginary number MATH **imaginäre Zahl**
≠ real number = rein imaginäre Zahl
≠ reelle Zahl
imaginary part MATH **Imaginärteil**
= imaginary component; imag
imaginary unit MATH **imaginäre Einheit**
[square root of −1] [Quadratwurzel von −1]
imaging DATA PROC **Bildsynthese**
= image synthesis
imbalance COLLOQ **Mißverhältnis**
imbalance (→unbalance) NETW.TH (→Unsymmetrie)
imbed (→embed) TECH (→einbetten)
IMF (→International Monetary Fund) ECON (→Weltwährungsfond)
imitate TECH **nachbilden**
= copy (v.t.) = nachahmen; imitieren
imitation (→copy) TECH (→Nachbildung)
immeasurable COLLOQ **unmeßbar**
= immensurable ≈ unzählbar; immens
≈ innumerable; immense
immediate (→instant-) TECH (→Sofort-)
immediate access DATA PROC **Sofortzugriff**
= instantaneous access = Schnellzugriff; Instantanzugriff
immediate access memory (→main memory 1) DATA PROC (→Hauptspeicher 1)
immediate access store (→high-speed memory) DATA PROC (→Schnellspeicher 1)
immediate address DATA PROC **unmittelbare Adresse**
[contains the value of the operand] [enthält den gesuchten Operandenwert direkt]
= literal address = Literaladresse
≈ direct address ≈ direkte Adresse

immediate addressing   DATA PROC   **unmittelbare Adressierung**
[containing also the operand value]   [enthält auch den Operandenwert]
= literal addressing   = Literaladressierung
≈ direct addressing   ≈ direkte Adressierung

immediate answer   DATA COMM   **Sofortantwort**

immediate instruction   DATA PROC   **Direktbefehl**
[with integrated operand]   [enthält bereits den Operanden]
= direct instruction; literal instruction; immediate mode command   = Sofortbefehl

immediate measure   COLLOQ   **Sofortmaßnahme**

immediate mode   DATA PROC   **Sofortausführung**
[immediate execution of commands]   [von Befehlen]

immediate mode command   DATA PROC   (→Direktbefehl)
(→immediate instruction)

immediate operand   DATA PROC   **Direktoperand**
[fetched simultaneously with the instruction]   [gleichzeitig mit der Anweisung abgerufen]

immediate processing   DATA PROC   (→Geradewohl-Verarbeitung)
(→in-line processing)

immediate ringing   SWITCH   **Sofortruf**
= splash ringing

immediate tracing   SWITCH   **Sofortfangen**

immense number   COLLOQ   **Unzahl**

immensurable   COLLOQ   (→unmeßbar)
(→immeasurable)

immersion lens   ELECTRON   (→Beschleunigungslinse)
(→acceleration lens)

immitance (→input quantity)   NETW.TH   (→Eingangsgröße)

immitance converter   NETW.TH   (→Impedanzkonverter)
(→impedance converter)

immovable   TECH   **unbeweglich**
≈ rigid   = unbewegbar
   ≈ starr

immune (→insentitive)   TECH   (→unempfindlich)

immunity (→strength)   TECH   (→Festigkeit)

IMOS   MICROEL   **IMOS-Technik**
[ion-implanted metal-oxide semiconductor technology]

impact (v.t.)   MECH   **aufschlagen**
≈ flap (v.t.)   ≈ aufprallen; prallen
≈ hit   ≈ schlagen

impact (n.)   MECH   **Aufschlag**
   = Aufprall
   ↑ Schlag

impact   TERM&PER   **Anschlag**
[of a type]   [einer Drucktype]
≈ touch

impact (→shock)   PHYS   (→Stoß)

impact avalanche transit diode   MICROEL   **Impatt-Diode**
= impatt diode; avalanche transit time diode   = NPIP-Diode; Lawinenlaufzeitdiode
↑ avalanche photodiode   ↑ Lawinenlaufzeitdiode

impact bending strength   MECH   **Schlagbiegefestigkeit**

impact control (→penetration control)   TERM&PER   (→Anschlagregler)

impact cover (→protective cover)   TECH   (→Schutzabdeckung)

impact excitation   PHYS   **Stoßerregung**

impact ionization   PHYS   **Stoßionisierung**
= collision ionization   = Stoßionisation

impact molding (→cold forging)   METAL   (→Fließpressen)

impact parameter   PHYS   **Stoßparameter**

impact printer   TERM&PER   **Anschlagdrucker**
= mechanical printer   = mechanischer Drucker; Impaktdrucker
↓ type wheel printer; stylus printer; drum printer; chain printer; belt printer   ↓ Typenraddrucker; Nadeldrucker; Walzendrucker; Kettendrucker; Banddrucker

impact-proof (→impact-resistant)   TECH   (→schlagfest)

impact resistance (→shock resistance)   TECH   (→Schlagfestigkeit)

impact-resistant   TECH   **schlagfest**
= shock-resistant; impact-proof; shock-proof   = stoßfest
≈ knock-resistant [QUAL]   ≈ klopffest [QUAL]

impact sensation   TERM&PER   **Anschlaggefühl**

impact test   QUAL   **Schlagprüfung**

impact test (→shock test)   QUAL   (→Stoßprüfung)

impair (v.t.)   TECH   **beeinträchtigen**
≈ affect; limit (v.t.)   ≈ beeinflussen; beschränken

impair (→decrease)   TECH   (→vermindern)

impaired (→damaged)   TECH   (→beschädigt)

impairment   TECH   **Beeinträchtigung**
≈ damage (n.); limitation; reduction 1   ≈ Schaden; Einschränkung; Verminderung

impatt diode (→impact avalanche transit diode)   MICROEL   (→Impatt-Diode)

impedance   NETW.TH   **komplexer Scheinwiderstand**
[reciprocal of admittance; vector sum of resistance and reactance]   [komplexe Spannung zu komplexem Strom; Realteil = Wirkwiderstand, Imaginärteil = Blindwiderstand]
= variational resistance; ac resistance; apparent resistance; Z; vector impedance   = elektische Impedanz; Impedanz; komplexer Widerstand; Scheinwiderstand 1; Wechselstromwiderstand; Z

impedance analyzer   INSTR   **Impedanzanalysator**

impedance balance   NETW.TH   (→Symmetrie)
(→balance)

impedance-balanced   NETW.TH   **widerstandssymmetrisch**
= electrically symmetric   = torsymmetrisch

impedance chart (→transmission line chart)   LINE TH   (→Leitungsdiagramm)

impedance converter   NETW.TH   **Impedanzkonverter**
= immitance converter; converter   = Immitanzkonverter; Konverter

impedance discontinuity   NETW.TH   **Stoßstelle**
= impedance jump

impedance equation   NETW.TH   **Widerstandsgleichung**
↑ two-port equation   ↑ Vierpolgleichung

impedance inverter   NETW.TH   **Impedanzinverter**
   = Inveter

impedance jump (→impedance discontinuity)   NETW.TH   (→Stoßstelle)

impedance line transformer   NETW.TH   **Transformationsgleichung**

impedance matching (→matching 1)   NETW.TH   (→Anpassung 1)

impedance matching section   NETW.TH   **Transformationsglied**
   = Anpaßglied; Impedanzanpassungsglied

impedance matrix   NETW.TH   **Widerstandsmatrix**
= Z matrix   = Impedanzmatrix; Z-Matrix

impedance measuring   INSTR   **Scheinwiderstandsmessung**
   = Impedanzmessung

impedance measuring bridge   INSTR   **Impedanzmeßbrücke**
= impedance test set   = Scheinwiderstands-Meßbrücke; Impedanzmeßplatz
↓ LCR meter; impedance analyzer   ↓ LCR-Messer; Impedanzanalysator

impedance normalizing   NETW.TH   **Impedanznormung**

impedance parameter   NETW.TH   **Widerstandsparameter**

impedance representation   NETW.TH   **Widerstandsform**
[of fourpole equations]   [Vierpolgleichungen]

impedance test set (→impedance measuring bridge)   INSTR   (→Impedanzmeßbrücke)

impedance transformation   NETW.TH   **Impedanzwandlung**
   = Impedanztransformation

impedance transformer   NETW.TH   **Impedanzwandler**
   = Impedanztransformator

impedance unbalance (→unbalance)   NETW.TH   (→Unsymmetrie)

impedance unbalanced   NETW.TH   **widerstandsunsymmetrisch**
= electrically asymmetric   = torunsymmetrisch

impediment   LING   **Sprachfehler**
[of speech articulation]

impediment   COLLOQ   **Hinderungsgrund**

impeller   MECH   **Flügelrad**

impenetrability   TECH   **Undurchdringlichkeit**
≈ impermeability   = Undurchdringbarkeit
   ≈ Undurchlässigkeit

impenetrable   TECH   **undurchdringbar**
≈ impermeable   = undurchdringlich
   ≈ undurchlässig

imperative (n.) (→imperative mood)   LING   (→Imperativ)

imperative mood	LING	**Imperativ**	
= imperative (n.)		= Befehlsform; Aufforderungsform	
↑ verbal mode		↑ Modus	
imperative sentence	LING	**Aufforderungssatz**	
imperfect	PHYS	**fehlgeordnet** [Kristall]	
imperfect (→simple past)	LING	(→Imperfekt)	
imperfect bunch (→imperfect trunk group)	SWITCH	(→unvollständiges Bündel)	
imperfect crystal	PHYS	**fehlgeordnetes Kristall**	
imperfection	MICROEL	**Störstelle**	
= impurity; defect		= Fehlstelle	
↓ lattice vacanvy; foreign atom		↓ Gitterfehlstelle; Fremdatom	
imperfect trunk group	SWITCH	**unvollständiges Bündel**	
= imperfect bunch			
impermeability	TECH	**Undurchlässigkeit**	
= imperviousness		≈ Undichtigkeit; Undurchdringlichkeit	
≈ leak; impenetrability			
≠ permeability		≠ Durchlässigkeit	
impermeable	TECH	**undurchlässig**	
= impervious		= dicht 2; undurchdringbar	
≈ tight; impenetrable		≠ durchlässig	
≠ permeable			
impersonal verb	LING	**Impersonale**	
		= unpersönliches Zeitwort	
impervious (→impermeable)	TECH	(→undurchlässig)	
imperviousness (→impermeability)	TECH	(→Undurchlässigkeit)	
impinge [ray]	PHYS	**einfallen** [Strahl]	
= incide			
impingement (→incidence)	PHYS	(→Einfall)	
impinging (→incident)	PHYS	(→einfallend)	
impinging beam (→incident beam)	PHYS	(→einfallender Strahl)	
impinging ray (→incident beam)	PHYS	(→einfallender Strahl)	
implantation	MICROEL	**Implantation**	
implement (v.t.)	DATA PROC	**implementieren**	
[to bring a software design into an executable form, by translation or complementation]		[einen Programmentwurf ablauffähig machen, durch Übersetzung oder Ergänzung]	
≈ program (v.t.)		= realisieren	
		≈ programmieren	
implement (n.) (→tool)	TECH	(→Werkzeug)	
implement (→realize)	TECH	(→realizieren)	
implementation	DATA PROC	**Implementierung** [Inbetriebnahme von Geräten oder Programmen]	
[to put hardware and software operative]		= Realisierung; Verwirklichung	
implementation	TECH	(→Realisierung)	
(→realization)			
implementation language	DATA PROC	**Implementierungssprache**	
= system programming language		↓ Sprache C	
implementation strategy	ECON	**Einführungsstrategie**	
implementing regulations	ECON	**Durchführungsbestimmung**	
implements	TECH	**Utensilien**	
implication (→implication operation)	ENG.LOG	(→Implikationsverknüpfung)	
implication circuit (→implication gate)	CIRC.ENG	(→Implikationsglied)	
implication element (→implication gate)	CIRC.ENG	(→Implikationsglied)	
implication function (→implication operation)	ENG.LOG	(→Implikationsverknüpfung)	
implication gate	CIRC.ENG	**Implikationsglied**	
= implication element; implication circuit; IF-THEN gate; IF-THEN element; IF-THEN circuit		= Implikationsgatter; Implikationselement; Implikationsschaltung; Implikationstor; WENN-DANN-Glied; WENN-DANN-Gatter; WENN-DANN-Element; WENN-DANN-Schaltung	
↑ logic gate		↑ Verknüpfungsglied	
implication operation	ENG.LOG	**Implikationsverknüpfung**	
[output = 0 only if A = 1 and B = 0]		[Ausgang = 0 nur wenn A = 1 und B = 0]	
= implication function; implication; inclusion operation; inclusion function; inclusion; IF-THEN operation; IF-THEN function; IF THEN		= Implikationsfunktion; Implikation; WENN-DANN-Verknüpfung; WENN-DANN-Funktion; WENN-DANN	
↑ logic operation		↑ logische Verknüpfung	
implied address (→indirect address)	DATA PROC	(→indirekte Adresse)	
implied addressing (→indirect addressing)	DATA PROC	(→indirekte Adressierung)	
implosion	TECH	**Implosion**	
import (v.t.)	DATA PROC	**einspielen** [Datenbestände einer Fremddatei übernehmen]	
[to take over data stock of a foreign system]		= importieren; übernehmen	
≈ retrieve		≠ wiedergewinnen	
≠ export		≠ überspielen	
import (n.)	ECON	**Einfuhr**	
= goods import; out-of-country purchase		= Import; Wareneinfuhr	
importation	ECON	**Importvorgang**	
		= Einfuhrvorgang	
import certificate	ECON	**Einfuhrbescheinigung**	
		= Einfuhrbestätigung; Importbescheinigung; Importbestätigung	
import documents	ECON	**Importpapiere**	
		= Einfuhrpapiere	
impossible event	MATH	**unmögliches Ereignis**	
impossible figure [computer graphics]	DATA PROC	**unmögliche Figur** [Computergraphik]	
impracticable	TECH	**nichtausführbar**	
= not executable			
imprecise (→inaccurate)	TECH	(→ungenau)	
impreciseness (→inaccuracy)	TECH	(→Ungenauigkeit)	
imprecision (→inaccuracy)	TECH	(→Ungenauigkeit)	
impregnate	TECH	**imprägnieren**	
= soak (v.t.)		= tränken	
impregnation	TECH	**Imprägnierung**	
= soak (n.); soakage		≈ Tränkung	
impress [current, voltage]	EL.TECH	**einprägen** [Strom, Spannung]	
impressed [current]	EL.TECH	**eingeprägt** [Strom]	
impressed current	EL.TECH	**eingeprägter Strom**	
impressed current	NETW.TH	**Urstrom**	
		= Quellenstrom	
impressed voltage	EL.TECH	**eingeprägte Spannung**	
imprimatur [license to print]	TYPOGR	**Imprimatur** (n.n., OES: n.f.)	
		= impr.; imp.; Druckerlaubnis; Druckgenehmigung	
imprint (v.t.)	TECH	**aufdrucken**	
imprint (n.)	TYPOGR	**Impressum**	
imprinted stamp	ECON	**Wertstempel**	
↑ stamp		↑ Wertzeichen	
improper	COLLOQ	**unsachgemäß**	
= inexpert; inadequate			
improper integral	MATH	**uneigentliches Integral**	
improve	TECH	**veredeln**	
= refine		= vergüten	
≈ process		≈ verarbeiten	
improved T antenna	ANT	**optimierte T-Antenne**	
improvement 1	TECH	**Verbesserung**	
= enhancement; emendation		= Aktualisierung	
improvement 2	TECH	**Veredelung**	
= refinement		= Verfeinerung; Vergütung	
≈ processing		≈ Verarbeitung	
improvement proposal	TECH	**Verbesserungsvorschlag**	
improvised	COLLOQ	**provisorisch**	
= provisional; tentative		= behelfsmäßig; Behelfs-; vorläufig	
impulse [non-repetitive surge]	ELECTRON	**Impuls 1** [kurzzeitiger aperiodischer Vorgang]	
≈ pulse; pulse train		= Stromstoß	
		≈ Impuls 2; Puls	

impulse behaviour (→pulse response)	ELECTRON	(→Impulsantwort)	**impulse tail** (→pulse tail)	ELECTRON	(→Impulsschwanz)
impulse circuit (→pulse circuit)	CIRC.ENG	(→Impulsschaltung)	**impulse telegraphy**	TELEGR	**Impulstelegraphie** = Impulstelegrafie
impulse command (→pulse command)	TELECONTR	(→Impulsbefehl)	**impulse top** (→pulse top)	ELECTRON	(→Impulsdach)
impulse counter (→pulse counter)	CIRC.ENG	(→Impulszähler)	**impulse transformer** (→pulse transformer)	ELECTRON	(→Impulsübertrager)
impulse counting (→pulse metering)	ELECTRON	(→Impulszählung)	**impulse voltage** (→surge voltage)	EL.TECH	(→Stoßspannung)
impulse crest (→pulse peak)	ELECTRON	(→Impulsspitze)	**impulsing relay** (→impulse relay)	COMPON	(→Impulsrelais)
impulse delay (→pulse delay)	ELECTRON	(→Impulsverzögerung)	**impulsive** = impulse-shaped; pulse-shaped ≈ pulsed	ELECTRON	**impulsförmig** ≈ gepulst
impulse duration (→pulse duration)	ELECTRON	(→Impulsdauer)	**impulsive force** [force times impact duration]	PHYS	**Stoßkraft** [Kraft mal Stoßdauer]
impulse duration jitter (→pulse duration jitter)	ELECTRON	(→Impulsdauerflattern)	**impulsive noise** = impulse noise; pulse noise; click ≈ glitch [ELECTRON]	TELEC	**Impulsgeräusch** = kurzzeitiges Geräusch ≈ Störimpuls [ELECTRON]
impulse echo meter = pulse echometer	INSTR	**Impulsechometer**	**impurity** = contamination	PHYS	**Verunreinigung**
impulse edge (→pulse edge)	ELECTRON	(→Impulsflanke)	**impurity** (→imperfection)	MICROEL	(→Störstelle)
impulse energy (→pulse energy)	ELECTRON	(→Impulsenergie)	**impurity addition**	MICROEL	**Fremdatomzusatz**
impulse excitation (→pulse excitation)	ELECTRON	(→Impulserregung)	**impurity atom** (→foreign atom)	PHYS	(→Fremdatom)
impulse exciter (→impulse generator)	CIRC.ENG	(→Impulsgenerator 1)	**impurity band**	PHYS	**Störband**
			impurity band conduction	PHYS	**Störbandleitung**
impulse form: impulse shape (→pulse form)	ELECTRON	(→Impulsform)	**impurity concentration** (→impurity density)	MICROEL	(→Störstellendichte)
impulse function = pulse function	MATH	**Impulsfunktion** = Stoßfunktion	**impurity concentration profile** (→doping profile)	MICROEL	(→Dotierungsprofil)
impulse gate	CIRC.ENG	**Impulsgatter**	**impurity conduction** (→hole conduction)	PHYS	(→Löcherleitung)
impulse generator [produces single surges] = impulse exciter; impulse sender	CIRC.ENG	**Impulsgenerator 1** [erzeugt Einzelimpulse] = Impulserzeuger	**impurity conductor** (→extrinsic conductor)	MICROEL	(→Störstellenleiter)
impulse hit (→spike)	ELECTRON	(→Zacken)	**impurity density** = impurity concentration; defect density; defect concentration	MICROEL	**Störstellendichte** = Fehlerdichte; Fehlstellendichte
impulse intensity (→pulse intensity)	ELECTRON	(→Impulsintensität)	**impurity exhaustion** = defect exhaustion	MICROEL	**Störstellenerschöpfung** = Fehlstellenerschöpfung
impulse localization	TRANS	**Impulsortung**	**impurity-film resistance**	MICROEL	**Fremdschichtwiderstand**
impulse magnetization	PHYS	**Impulsmagnetisierung** = Pulsmagnetisierung	**impurity generation** = defect generation	MICROEL	**Störstellenerzeugung** = Fehlstellenerzeugung
impulse meter (→pulse multiplier)	ELECTRON	(→Impulsmultiplikator)	**impurity mobility** = defect mobility	MICROEL	**Störstellenbeweglichkeit** = Fehlstellenbeweglichkeit
impulse noise (→impulsive noise)	TELEC	(→Impulsgeräusch)	**impurity profile** (→doping profile)	MICROEL	(→Dotierungsprofil)
impulse pattern ≈ pulse train	ELECTRON	**Impulsmuster** ≈ Puls	**impurity semiconductor** (→extrinsic semiconductor)	MICROEL	(→Störstellenhalbleiter)
impulse peak (→pulse peak)	ELECTRON	(→Impulsspitze)	**imref level** (→quasi-Fermi level)	PHYS	(→Quasi-Fermi-Niveau)
impulse permeability	PHYS	**Impulspermeabilität**	**IMS** (→information management system)	DATA PROC	(→Informationsverwaltungssystem)
impulse profiling (→pulse shaping)	CIRC.ENG	(→Impulsformung)	**In** (→indium)	CHEM	(→Indium)
impulse reflectometer = time-domain reflectometer	INSTR	**Impulsreflektometer** = Time-Domain-Reflektometer	**in** (→inch)	PHYS	(→Zoll)
			IN (→intelligent network)	TELEC	(→intelligentes Netz)
impulse regeneration (→pulse regeneration)	ELECTRON	(→Impulsregenerierung)	**inaccessibility**	COLLOQ	**Unzugänglichkeit**
impulse relay = impulsing relay; pulse relay	COMPON	**Impulsrelais**	**inaccessible**	COLLOQ	**unzugänglich**
impulse response	MATH	**Gewichtsfunktion**	**inaccuracy** = inexactitude; inexactness; imprecision; impreciseness	TECH	**Ungenauigkeit**
impulse response (→pulse response)	ELECTRON	(→Impulsantwort)	**inaccurate** = inexact; imprecise ≈ incorrect	TECH	**ungenau** = unpräzise; unpräzis ≈ fälschlich
impulse sender (→impulse generator)	CIRC.ENG	(→Impulsgenerator 1)	**inactive** = not working; not running	DATA PROC	**inaktiv**
impulse separation (→pulse separation)	ELECTRON	(→Impulspause)	**inactive character**	DATA COMM	**nichtaktives Zeichen** = inaktives Zeichen
impulse sequence (→pulse train)	ELECTRON	(→Puls)	**inactive costs**	ECON	**ruhende Kosten**
impulse-shaped (→impulsive)	ELECTRON	(→impulsförmig)	**inactive partner** (→silent partner)	ECON	(→stiller Gesellschafter)
impulse shaper (→pulse shaper)	CIRC.ENG	(→Impulsformer)	**inactive window**	DATA PROC	**inaktives Fenster**
impulse shaping (→pulse shaping)	CIRC.ENG	(→Impulsformung)	**inadequacy** ≈ insufficiency	COLLOQ	**Unangemessenheit** ≈ Unzulänglichkeit
impulse signaling = discontinuous signaling ≈ impulse dialing	SWITCH	**Impulszeichengabe** = Impulssignalisierung ≈ Impulswahl	**inadequate** (→unfit)	TECH	(→ungeeignet)
			inadequate (→improper)	COLLOQ	(→unsachgemäß)
			inadmissibility	COLLOQ	**Unzulässigkeit**
			inadmissible	COLLOQ	**unzulässig**
impulse slope (→pulse edge)	ELECTRON	(→Impulsflanke)	**InAs** (→indium arsenide)	CHEM	(→Indium-Arsenid)
impulse sound-level meter	INSTR	**Impulsschallpegelmesser**	**inbalance** (→disequilibrium)	TECH	(→Ungleichgewicht)

in-band energy	CIRC.ENG	Inbandenergie
inband signaling	TRANS	Inbandsignalisierung
= in-band signaling; speech-plus signaling		= Tonwahl; Inbandzeichengabe
in-band signaling (→inband signaling)	TRANS	(→Inbandsignalisierung)
inband telegraphy	TELEGR	Inbandtelegraphie
= speech-plus telegraphy		
inbound traffic (→incoming traffic)	SWITCH	(→kommender Verkehr)
inbuild (→built-in)	TECH	(→eingebaut)
in bulk	COLLOQ	in rauhen Mengen
incandescent cathode	ELECTRON	(→Glühkathode)
(→thermoionic cathode)		
incandescent electrode	ELECTRON	Glühelektrode
= hot electrode		
incandescent lamp	EL.INST	Glühlampe
= bulb		
incapsulation (→encapsulation)	TECH	(→Verkapselung)
inception (→begin)	TECH	(→Beginn)
inceptor (→beginner)	ECON	(→Berufsanfänger)
incessant (→continuous 1)	TECH	(→kontinuierlich)
inch	PHYS	Zoll
[0.083 ft = 25.40 mm]		[25,40 mm]
= symbol " ; in		= Symbol „ ; in
↓ standard inch; english inch		↑ Längenmaß
		↓ Normalzoll; englischer Zoll
in charge of (→responsible for)	ECON	(→zuständig)
inches-per-second	TERM&PER	Zoll pro Sekunde
[tape speed]		[Bandgeschwindigkeit]
= IPS; ips		= IPS; ips
inch graduation	TECH	Zollteilung
inch mark (→inch sign)	TYPOGR	(→Zoll-Symbol)
inch sign	TYPOGR	Zoll-Symbol
["]		["]
= inch symbol; inch mark		= Zoll-Zeichen
inch symbol (→inch sign)	TYPOGR	(→Zoll-Symbol)
incide (→impinge)	PHYS	(→einfallen)
incidence	PHYS	Einfall
[ray]		[Strahl]
= impingement		
incidence	MATH	Inzidenz
incidence matrix	DATA PROC	(→Verbindungsmatrix)
(→connection matrix)		
incident	PHYS	einfallend
= impinging		
incidental (→random)	TECH	(→zufällig)
incidental (→spurious)	INF.TECH	(→unerwünscht)
incident beam	PHYS	einfallender Strahl
= incident ray; impinging beam; impinging ray		= Einfallstrahl
incident ray (→incident beam)	PHYS	(→einfallender Strahl)
incipient (→beginning)	TECH	(→beginnend) (adj.)
in-circuit emulator	MICROEL	Echtzeit-Testadapter
= ICE		= ICE
in-circuit tester	INSTR	Schaltkreisprüfgerät
= circuit tester		= In-circuit-tester
inclination	MECH	Neigung
[deviation from horizontal]		[Abweichung von der Horizontalen]
= slope; tilt		= Schräglage
↓ descent; rise		↑ Schräge
		↓ Gefälle; Anstieg
inclination	PHYS	Inklination
inclination (→slope)	MATH	(→Neigung)
inclination angle (→bevel)	MECH	(→Schrägungswinkel)
incline	MECH	schrägstellen
= tilt		= neigen; kippen 2
inclined	TECH	geneigt
[deviating from horizontal]		[von Horizontale abweichend]
= sloping; tilt; down-going; negative-going		= schräg abfallend; abfallend; schief 2
≈ skew		≈ schief 1
inclined (→oblique)	TYPOGR	(→schräg)
inclined plane	MECH	schiefe Ebene
inclose (v.t.) (NAM)	TECH	einschließen
= enclose (BRI)		≈ verkapseln
≈ encapsulate		
inclose (→annex)	LING	(→Anlage)
inclosure (→annex)	LING	(→Anlage)
include (→contain)	COLLOQ	(→enthalten)
included in the price	DATA PROC	(→im Preis inbegriffen)
(→bundled)		
inclusion	MATH	Inklusion
inclusion (→implication operation)	ENG.LOG	(→Implikationsverknüpfung)
inclusion function (→implication operation)	ENG.LOG	(→Implikationsverknüpfung)
inclusion operation (→implication operation)	ENG.LOG	(→Implikationsverknüpfung)
inclusive (adj.)	COLLOQ	einschließlich
		= inklusiv
INCLUSIVE-OR circuit (→OR gate)	CIRC.ENG	(→ODER-Glied)
INCLUSIVE-OR element (→OR gate)	CIRC.ENG	(→ODER-Glied)
INCLUSIVE-OR function (→OR operation)	ENG.LOG	(→ODER-Verknüpfung)
INCLUSIVE-OR gate (→OR gate)	CIRC.ENG	(→ODER-Glied)
INCLUSIVE-OR operation (→OR operation)	ENG.LOG	(→ODER-Verknüpfung)
incoherent	PHYS	inkohärent
incoherent sampling (→random sampling)	INSTR	(→inkohärente Abtastung)
incoherent scanning (→random sampling)	INSTR	(→inkohärente Abtastung)
incombustibility	TECH	Unbrennbarkeit
= noncombustibility		≈ Flammwidrigkeit
≈ flame resistance		
incombustible	TECH	unbrennbar
≈ flame-retardant		= unverbrennbar
		≈ flammwidrig
income	ECON	Einkommen
= earnings; revenues; yield		= Einkünfte; Einnahmen; Bezüge; Verdienst
incoming	SWITCH	kommend
≠ outgoing		= ankommend
		≠ gehend
incoming call	SWITCH	ankommender Anruf
= receiving call		= ankommendes Gespräch
incoming call barred with advise	DATA COMM	Anschlußsperre mit Hinweisgabe
incoming calls telephone	TELEPH	anrufbares Telefon
incoming circuit	SWITCH	ankommender Satz
↑ junctor		↑ Verbindungssatz
incoming communication (→incoming message)	DATA COMM	(→ankommende Nachricht)
incoming group	SWITCH	Zubringerteilgruppe
≈ grading group		≈ Staffelgruppe
incoming highway	SWITCH	Zubringermultiplexleitung
[an incoming line common to several links]		[eine mehreren Verbindungen gemeinsame Zubringerleitung]
↑ offering line		= kommende Abnehmerleitung; Zeitmultiplex-Zubringerleitung
		↑ Zubringerleitung
incoming inspection (→entrance test)	QUAL	(→Eingangsprüfung)
incoming line (→offering line)	SWITCH	(→Zubringerleitung)
incoming message	DATA COMM	ankommende Nachricht
= incoming communication		
incoming point	COLLOQ	Anlaufstelle
incoming post	OFFICE	ankommende Post
		= Eingangspost; Posteingang
incoming seizure	SWITCH	kommende Belegung
incoming selector	SWITCH	Eingangswähler
incoming signal	TELEC	ankommendes Signal
incoming signal (→input signal)	ELECTRON	(→Eingangssignal)
incoming tolerance	QUAL	Anlieferungstoleranz
incoming traffic	SWITCH	kommender Verkehr
= inbound traffic		= ankommender Verkehr
≈ offered traffic		≈ Endverkehr; Verkehrsangebot
incoming trunk (→offering line)	SWITCH	(→Zubringerleitung)

incoming trunk circuit	SWITCH	kommender Leitungssatz		incrementally 1 (→scaled)	TECH	(→stufenweise 1)
incoming trunk group	SWITCH	Zubringerbündel		incrementally 2 (→stepwise)	TECH	(→schrittweise)
incommensurable	MATH	inkommensurabel		incremental permeability	EL.TECH	Überlagerungspermeabilität
[not divisible by common divisor]		[ohne gemeinsamen Teiler]		[with bias polarization]		[bei Vorpolarisierung]
3 1/2 in. compact floppy disk	TERM&PER	(→Mikrodiskette)		incremental plotter	TERM&PER	Inkrementalplotter
(→3 1/2 in. floppy disk)				[operates with increment data]		[arbeitet mit Zuwachsdaten]
incompatible	COLLOQ	unvereinbar		incremental recorder	TERM&PER	Schrittrecorder
= inconsistent; incongruent		= inkompatibel		incremental resistance	MICROEL	(→differentieller Widerstand)
incompatible	TECH	unverträglich		(→differential resistance)		
		= inkompatibel		incremental spacing	TERM&PER	Mikroabstand
incompatible (→mutually exclusive)	MATH	(→einander ausschließend)		(→microspacing)		
incomplete	COLLOQ	unvollständig		incremental-step converter	CIRC.ENG	Stufenkompensationsumsetzer
≈ partial		≈ teilweise		= voltage comparison encoder		= Stufenumsetzer; Stufenkompensator; Stufenverschlußler
incomplete connection	SWITCH	unvollständige Verbindung				
= incomplete link						
incomplete dialing	SWITCH	unvollständige Wahl				
incomplete link (→incomplete connection)	SWITCH	(→unvollständige Verbindung)		incrementation parameter	DATA PROC	Schrittweitenparameter
				[of a loop]		[einer Schleife]
incompletness	COLLOQ	Unvollständigkeit		increment control	CIRC.ENG	Schrittweitensteuerung
incongruent (→incompatible)	COLLOQ	(→unvereinbar)		increment stage (→construction stage)	TECH	(→Ausbaustufe)
inconsistent (→incompatible)	COLLOQ	(→unvereinbar)				
inconstancy (→variability)	TECH	(→Veränderlichkeit)		increment time	MICROEL	Fortschaltzeit
inconstant (→variable)	TECH	(→veränderlich)		[memory]		[Speicher]
incontestable	COLLOQ	unanfechtbar		incumbent (n.)	ECON	Stellungsinhaber
= irrefutable; incontrovertible						= Amtsinhaber
incontrovertible (→incontestable)	COLLOQ	(→unanfechtbar)		incumbent (adj.)	ECON	amtierend
				indebtedness	ECON	Verschuldung
incorporate (→integrate)	TECH	(→integrieren)		≈ debts		≈ Schulden
incorporated (→built-in)	TECH	(→eingebaut)		indecomposable	SCIE	unzerlegbar
incorporation	ECON	(→Aktiengesellschaft)		= indivisible		= unteilbar; untrennbar
(→corporation 2)				indefinite integral	MATH	unbestimmtes Integral
incorrect (adj.)	TECH	fälschlich		indefinite pronoun	LING	Indefinitpronomen
≈ inaccurate; faulty		= unrichtig				= unbestimmtes Pronomen; unbestimmtes Fürwort
		≈ ungenau; fehlerhaft				
incorrect access	DATA PROC	fälschlicher Zugriff		indelible	OFFICE	untilgbar
in correct position	TECH	lagerichtig		[ink]		[Tinte u. dgl.]
increase 1 (n.)	TECH	Zunahme		= inextinguishable		
= rise; increment; growth; accretion		= Zuwachs; Inkrement		indemnification	ECON	(→Schadenersatz)
≠ decrease		≠ Abnahme		(→compensation)		
increase 2 (n.)	TECH	Zuschlag		indent (v.t.)	TYPOGR	einrücken
↓ safety increase		↓ Sicherheitszuschlag				= einziehen
increase (v.i.) (→rise)	ECON	(→steigen)		indent 1 (v.t.)	MECH	verzahnen 2
increase (→enlarge)	TECH	(→vergrößern)		[to provide with gearing notches]		[mit Zähnen versehen]
increased stress	QUAL	erhöhte Beanspruchung		= gear 2		≈ verschwalben
increase in bandwidth	MODUL	Bandbreitenerhöhung		≈ dovetail		
increase the price	ECON	verteuern		indent (v.t.)	COLLOQ	verzahnen
≠ cheapen		≠ verbilligen		[fig.]		[fig.]
increment (v.t.)	DATA PROC	inkrementieren		= gear (v.t.); cog; dovetail (v.t.)		= verquicken
≠ decrement		= erhöhen				≈ verschachteln
		≠ vermindern		indent (n.)	TYPOGR	Einzug
increment (v.i.)	TECH	weiterrücken		[inward shift of lines; a separation from left margin of lines, or of end of line from right margin]		[Einrücken von Zeilen; ein Abstand vom linken Zeilenrand, oder vonZeilenende zu rechtem Rand]
= progress (v.i.)						
increment (n.)	DATA PROC	Schrittweite		= indentation; indention		
[of a loop]		[einer Schleife]		↓ hanging indent; paragraph indentation; first-line indent		= Einrücken; Einrückung
= step rate; step size		= Inkrement				↓ hängender Einzug; Asatzeinzug; Erstzeileneinzug
increment (→advance)	ELECTRON	(→Fortschaltung)				
increment (→increase 1)	TECH	(→Zunahme)		indent 2 (→gear)	MECH	(→verzahnen 1)
incremental	ELECTRON	fortschaltend		indentation (→notch)	TECH	(→Kerbe)
		= inkrementierend; vorrückend		indentation (→indent)	TYPOGR	(→Einzug)
				indentation 2 (→gearing 2)	MECH	(→Verzahnung)
incremental (→differential)	MATH	(→differentiell)		indented	TYPOGR	eingerückt
incremental angle-position encoder	COMPON	Winkelschrittgeber		indented cross	MECH	Kreuzschlitz
				= cross recess		
incremental compiler	DATA PROC	Inkrementalcompiler		indention (→indent)	TYPOGR	(→Einzug)
incremental computer	DATA PROC	Inkrementalrechner		independent	MATH	unabhängig
[stores differences to start value]		[speichert Veränderungen zum Anfangswert]		independent	TECH	unabhängig
				= standalone; autarkic; autarchic		= selbstständig; autark
		= digitaler Integrator		independent event	MATH	unabhängiges Ereignis
incremental conductance	MICROEL	(→differentieller Leitwert)		independent excitation	ELECTRON	(→Fremderregung)
(→differential conductance)				(→external excitation)		
incremental costs (→marginal costs)	ECON	(→Grenzkosten)		independently operating	TECH	alleinbetriebsfähig
				≈ automatic		≈ automatisch
incremental counter	DATA PROC	Schrittzähler		independent sideband system	BROADC	(→ISB-System)
incremental data	DATA PROC	Inkrementdaten		(→ISB system)		
[differences to start value]		[den Unterschied zum Ausgangswert betreffend]		independent variable	MATH	unabhängige Variable
				in-depth (adj.)	COLLOQ	tiefgründig
		= inkrementelle Daten		[fig.]		[fig.]
incremental dimension	CONTROL	Kettenmaß				= tiefschürfend
incrementally 2 (→stepwise)	TECH	(→schrittweise)				

in-depth definition (→focus depth)	PHYS	(→Schärfentiefe)	
indeterminacy (→uncertainty)	SCIE	(→Unschärfe)	
indeterminate (→undefined)	SCIE	(→unbestimmt)	
index	MATH	Index	
index (→subject index)	LING	(→Sachwortverzeichnis)	
index (→fist)	TYPOGR	(→Hinweiszeichen)	
index card (→file card)	OFFICE	(→Karteikarte)	
index-controlled laser	OPTOEL	indexgeführter Laser [Laserdiode mit eingebautem Wellenleiter]	
indexed = subscripted	MATH	indexiert ["indiziert" als Synonym zwar gebräuchlich aber nicht korrekt, da von „Indiz" = Anzeichen abgeleitet] = indiziert	
indexed access method [works with an index list, where the physical store site is indicated for every data record] ≈ indexed sequential access method	DATA PROC	Indexzugriff [funktioniert mit einer Indexliste, die für jeden Datensatz den physikalischen Speicherplatz angibt] ≈ idiziert sequentieller Zugriff	
indexed address	DATA PROC	indexierte Adresse	
indexed addressing	DATA PROC	indexierte Adressierung = indizierte Adressierung	
indexed file	DATA PROC	indexierte Datei = indizierte Datei	
indexed instruction	DATA PROC	indexierte Anweisung	
indexed-sequential	MATH	index-sequentiell	
indexed sequential access method [combination of indexed access and sequential access] = index-sequential access mode; ISAM	DATA PROC	indexiert-sequentieller Zugriff [Kombination von Indexzugriff und sequentiellem Zugriff] = index-sequentieller Zugriff; indexiert-sequentielle Zugriffsmethode; index-sequentielle Zugriffsmethode; ISZM; ISAM; index-sequentieller Dateizugriff; indiziert-sequentieller Zugriff	
indexed variable = subscripted variable	DATA PROC	indexierte Variable = indizierte Variable	
indexer = indexing programm	DATA PROC	Indexierprogramm	
index file	DATA PROC	Indexdatei	
index gap (→index hole)	TERM&PER	(→Indexmarkierung)	
index generation [word processing]	DATA PROC	Indexgenerierung [Textbearbeitung]	
index hole [hole on floppy disk, for sectoring] = index gap; indexing hole; disk index hole	TERM&PER	Indexmarkierung [Loch auf Diskette, zur Sektorierung] = Indexloch; Indexmarke; Sektor-Indexloch	
indexing	SCIE	Indexierung	
indexing hole (→index hole)	TERM&PER	(→Indexmarkierung)	
indexing programm (→indexer)	DATA PROC	(→Indexierprogramm)	
index letter	MATH	Indexbuchstabe	
index matching fluid	OPT.COMM	Immersionsflüssigkeit	
index number	MATH	Indexzahl	
index of cooperation [facsimile]	TELEGR	Arbeitsmodul [Faxsimile]	
index page [videotext]	TELEC	Indexseite [Bildschirmtext]	
index profile	OPT.COMM	Indexprofil	
index profile (→refractive index distribution)	OPT.COMM	(→Brechzahlprofil)	
index register = IR	DATA PROC	Indexregister = IR	
index-sequential access mode (→indexed sequential access method)	DATA PROC	(→indexiert-sequentieller Zugriff)	
index-sequential storage	DATA PROC	index-sequentielle Speicherung	
index tube	ELECTRON	Indexröhre	
index value (→set value)	CONTROL	(→Sollwert)	
index value word = index word	DATA PROC	Indexwort	
index word (→index value word)	DATA PROC	(→Indexwort)	
in-dialing = direct inward dialing (AM); DID (AM); direct dialling-in (BRI); direct dialing	SWITCH	Durchwahl	
in-dialing number = direct-dialing number	SWITCH	Durchwahl-Rufnummer	
Indian ink (→drawing ink)	ENG.DRAW	(→Tusche)	
indian red	PHYS	indisch rot	
indicate = display (v.t.)	INSTR	anzeigen	
indicated angle (→adjusted angle)	TECH	(→Einstellwinkel)	
indicating device (→indicator)	ELECTRON	(→Anzeigevorrichtung)	
indicating instrument	INSTR	Anzeigeinstrument	
indicating range	INSTR	Anzeigebereich	
indicating scale = display scale	INSTR	Anzeigeskala	
indicating telegram	DATA COMM	Meldetelegramm	
indicating tube (→display tube)	ELECTRON	(→Anzeigeröhre)	
indication (→display)	INSTR	(→Anzeige)	
indication (→message)	TELECONTR	(→Meldung)	
indication amplifier	INSTR	Anzeigeverstärker	
indication error	INSTR	Anzeigefehler	
indicative (→indicative mood)	LING	(→Indikativ)	
indicative mood = indicative ↑ mood	LING	Indikativ = Wirklichkeitsform ↑ Modus	
indicator = indicating device	ELECTRON	Anzeigevorrichtung = Anzeigegerät ≈ Auswertevorrichtung	
indicator = annunciator ≈ display	INSTR	Anzeiger = Indikator ≈ Anzeige	
indicator (→indicator lamp)	COMPON	(→Anzeigelampe)	
indicator bit	SWITCH	Indikatorbit	
indicator contact	EQUIP.ENG	Meldekontakt	
indicator flag (→flag)	DATA PROC	(→Merker)	
indicator lamp = signal lamp; indicator light; indicator	COMPON	Anzeigelampe = Signallampe; Signalleuchte ≈ Lampenanzeige	
indicator light (→indicator lamp)	COMPON	(→Anzeigelampe)	
indicator plate [melting fuse]	COMPON	Kennmeldeplättchen [Schmelzsicherung]	
indicator tube (→display tube)	ELECTRON	(→Anzeigeröhre)	
indirect (adj.)	COLLOQ	indirekt	
indirect address [indicates memory location, where the proper address can be found] = deferred address; implied address; inherent address ≈ relative address ≠ direct address	DATA PROC	indirekte Adresse [gibt Speicherplatz an, wo die eigentliche Adresse zu finden ist] = mittelbare Adresse; implizite Adresse; inhärente Adresse ≈ relative Adresse ≠ direkte Adresse	
indirect addressing [indication of address for proper address] = indirect addressing mode; deferred addressing; deferred addressing mode; multilevel addressing; multilevel addressing mode; implied addressing; inherent addressing ≠ direct addressing	DATA PROC	indirekte Adressierung [Angabe der Adresse für die eigentliche Adresse] = mittelbare Adressierung; implizite Adressierung; inhärente Adressierung ≠ direkte Adressierung	
indirect addressing mode (→indirect addressing)	DATA PROC	(→indirekte Adressierung)	
indirect control	SWITCH	indirekte Steuerung	
indirect-controlled	SWITCH	indirekt gesteuert	

indirect control switching system SWITCH
[switch networks are controlled after temporary store and processing of dial pulses]
↓ register system; switching system with stored program; crossbar switching system

indirect costs ECON
= overhead costs; oncosts; burden (AM); general expenses

indirect data acquisition DATA PROC
(→indirect data recording)

indirect data processing DATA PROC
indirect data recording DATA PROC
= indirect data acquisition

indirect-feed aerial (→indirect-feed antenna) ANT

indirect-feed antenna ANT
= double reflecting antenna; indirect-feed aerial; double reflecting aerial

indirect heating ELECTRON
indirectly heated cathode ELECTRON
indirect plug connector COMPON
indirect programming DATA PROC
indirect radiowave (→sky wave) RADIO PROP
indirect recording TERM&PER
[facsimile]
indirect signal RADIO PROP
indirect speech LING
indirect wave (→sky wave) RADIO PROP
8-in. disk (→eight-inch floppy disk) TERM&PER
3 1/2 in. diskette (→3 1/2 in. floppy disk) TERM&PER
5 1/4 in. diskette (→5 1/4 in. floppy disk) TERM&PER
indispensable COLLOQ
indium CHEM
= In
indium antimonide CHEM
= InSb
indium arsenide CHEM
= InAs
indium phosphide CHEM
= InP
indium tin oxide CHEM
≠ ITO
individual board (→individual PCB) ELECTRON
individual communication TELEC
(→interactive communication)
individual distortion TELEGR
individual drive MECH
= single drive
individual image TV
individual indication TELECONTR
= individual report; single-point information
individual jack TELEPH
≠ jack strip
individual joint OPT.COMM
individual line TELEC
= exclusive exchange line; single line
individual lock TECH
individual PCB ELECTRON
= individual board
individual port MICROW
= separate port
individual report TELECONTR
(→individual indication)

indirekt gesteuertes Vermittlungssystem
[Ansteuerung der Koppelanordnungen erfolgt nach Zwischenspeicherung und Auswertung der Rufnummer]
= indirekt gesteuertes Wählsystem
↓ Registerwählsystem; speicherprogrammiertes Vermittlungssystem; Koordinatenschaltersystem; Maschinenwählsystem

Gemeinkosten

(→indirekte Datenerfassung)

indirekte Datenverarbeitung
indirekte Datenerfassung

(→indirektgespeiste Antenne)

indirektgespeiste Antenne
= doppelreflektierende Antenne

indirekte Heizung
indirekt geheizte Kathode
= indirekt geheizte Katode
indirekter Steckverbinder
indirekte Programmierung
(→Raumwelle)

indirekte Aufzeichnung
[Faksimile]
Umwegsignal
= indirektes Signal
indirekte Rede
(→Raumwelle)
(→Normaldiskette)

(→Mikrodiskette)

(→Minidiskette)

unabkömmlich
Indium
= In
Indium-Antimonid
= InSb
Indium-Arsenid
= InAs
Indium-Phosphid
= InP
Indium-Zinn-Oxyd
= ITO
(→Einzelleiterplatte)

(→Dialogkommunikation)

individuelle Verzerrung
Einzelantrieb

Einzelbild
Einzelmeldung

Einzelklinke
≠ Klinkenstreifen
Einfachspleiß
Einzelanschluß

Einzelverriegelung
Einzelleiterplatte

Separatzugang

(→Einzelmeldung)

individual test program SWITCH **Einzelprüfprogramm**
individual trunk SWITCH **individuelle Abnehmerleitung**
indivisible SCIE (→unzerlegbar)
(→indecomposable)
indoor aerial (→indoor antenna) ANT (→Innenantenne)
indoor antenna ANT **Innenantenne**
= indoor aerial; room antenna; room aerial; inside antenna; inside aerial = Zimmerantenne
indoor box EQUIP.ENG **Innenraumgehäuse**
indoor cable COMM.CABLE **Innenraumkabel**
= house cable; internal cable = Innenkabel
indoor climate (→room climate) QUAL (→Innenraumklima)
indoor installation EQUIP.ENG (→Innenmontage)
(→indoor mounting)
indoor laying COMM.CABLE **Innenverlegung**
= In-Haus-Verlegung
indoor mounting EQUIP.ENG **Innenmontage**
= indoor installation = Innenraummontage; Innenaufbau; Innenraumaufbau
≠ outdoor mounting
↓ office installation ≠ Außenmontage
↓ Stationsaufbau
indoors TECH **Innenraum**
induce PHYS **induzieren**
induced charge PHYS **induzierte Ladung**
induced current (→induction current) PHYS (→Induktionsstrom)
induced failure QUAL **induzierter Fehler**
induced interference TELEC **Induktionsstörung**
induced magnetism PHYS **induzierter Magnetismus**
induced power noise (→mains hum) TELEC (→Netzbrumm)
induced voltage PHYS **Induktionsspannung**
= induzierte Spannung
inductance PHYS **Induktivität**
[SI unit: Henry] [SI-Einheit: Henry]
= inductivity ↓ Selbstinduktivität
inductance (→inductive reactance) NETW.TH (→induktiver Widerstand)
inductance coil PHYS **Induktionsspule**
= induction coil; Ruhmkorff coil = Funkeninduktor; Induktionsrolle
inductance measurement INSTR **Induktivitätsmessung**
inductance measuring bridge INSTR **Induktionsmessbrücke**
inductance meter INSTR **Induktivitätsmesser**
inductance per unit length LINE TH (→Induktivitätsbelag)
(→distributed inductance)
in-duct cable (→duct cable) COMM.CABLE (→Röhrenkabel)
induction SCIE **Induktion**
[logic] [Logik]
≠ deduction ≠ Deduktion
induction (→influence) PHYS (→Beeinflussung)
induction coil (→inductance coil) PHYS (→Induktionsspule)
induction coil EL.TECH (→Induktor)
(→inductor)
induction current PHYS **Induktionsstrom**
= induced current
induction flux (→magnetic flux) PHYS (→magnetischer Induktionsfluß)
induction furnace TECH **Induktionsofen**
induction-harden METAL **induktionshärten**
induction hardening METAL **Induktionshärtung**
induction instrument INSTR **Induktionsinstrument**
= Wendelfeldinstrument; Induktionsmeßwerk
induction machine POWER SYS **Induktionsmaschine**
= asynchronous machine; induction motor; asynchronous motor = Asynchronmaschine
induction motor POWER ENG (→Asynchronmotor)
(→asynchronous machine)
induction motor (→induction machine) POWER SYS (→Induktionsmaschine)
inductive PHYS **induktiv**
inductive compensation CIRC.ENG (→induktive Kompensation)
(→inductive neutralization)
inductive conclusion (→inductive reasoning) SCIE (→Induktionsschluß)

inductive coupling	EL.TECH	**induktive Kopplung**
= magnetic coupling		
inductive heating	PHYS	**induktive Erwärmung**
inductive inference (→inductive reasoning)	SCIE	(→Induktionsschluß)
inductive link	HF	**Koppelschleife**
inductive load	EL.TECH	**induktive Last**
inductive loading (→coil-loading)	COMM.CABLE	(→Bespulung)
inductive neutralization	CIRC.ENG	**induktive Kompensation**
= inductive compensation		= induktive Neutralisierung
inductive pick-up	INSTR	**induktiver Meßfühler**
		= induktive Meßsonde
inductive potentiometer	COMPON	**Spulenpotentiometer**
= ipot		= induktives Potentiometer
inductive reactance [ΩL]	NETW.TH	**induktiver Widerstand** [ΩL]
= inductance; inductive resistance		= induktiver Blindwiderstand; Induktanz
inductive reasoning	SCIE	**Induktionsschluß**
= inductive conclusion; inductive inference		= induktiver Schluß
inductive resistance (→inductive reactance)	NETW.TH	(→induktiver Widerstand)
inductive selection	SWITCH	**Induktivwahl**
inductive train control system	RAILW.SIGN	**Zugbeeinflussungssystem**
inductivity (→inductance)	PHYS	(→Induktivität)
inductor	EL.TECH	**Induktor**
= induction coil		= Induktionsspule
inductor 1 [device used because of its inductance (IEC)]	COMPON	**induktives Bauelement** = Induktor
↓ coil; transformator		↓ Spule; Übertrager
inductor 2 (→choke)	COMPON	(→Drossel)
inductor generator (→magneto generator)	TELEPH	(→Kurbelinduktor)
inductorless filter	NETW.TH	**spulenloses Filter**
≠ coil filter		≠ Spulenfilter
in due course	COLLOQ	**zur gegebenen Zeit**
industrial	ECON	**industriell**
industrial electronics	ELECTRON	**Industrieelektronik**
		= industrielle Elektronik
industrial engineering	POWER ENG	**elektrische Anlagentechnik**
↑ electrical power engineering		↑ elektrische Energietechnik
industrial engineering [branch]	TECH	**Fertigungstechnik** [Fachgebiet]
= production engineering; manufacturing engineering 1		= Produktionstechnik
industrial equipment	TECH	**Fertigungsgerät**
industrial espionage	ECON	**Werkspionage**
industrial exposition (→industrial fair)	ECON	(→Industriemesse)
industrial fair	ECON	**Industriemesse**
= industrial exposition		= Industrieausstellung
industrial power system	POWER SYS	**Industrienetz**
industrial roboter [programmable manipulator]	TECH	**Industrieroboter** [programmierbares Handhabungsgerät]
industrial robotics	CONTROL	**Industrierobotertechnik**
		= Industrierobotik
industrial standard	ECON	**Industrienorm**
industrial TV	TV	**Industriefernsehen**
↑ closed circuit TV		↑ nichtöffentliches Fernsehen
industry	ECON	**Industrie**
industry (→business)	ECON	(→Gewerbe)
ineffective	TECH	**unwirksam**
= inoperative; void		= betriebsunfähig; arbeitsunfähig
ineffectiveness	TECH	**Unwirksamkeit**
inegrated injection logic	MICROEL	**IIL**
= IIL; I²L; merged-transistor logic; MTL		= I²L; MTL; integrierte Injektionslogik
inelastic	TECH	**unelastisch**
inelasticity	PHYS	**Unelastizität**
inequal (→unequal)	MATH	(→ungleich)
inequality	MATH	**Ungleichung**
= unequality; unevenness		= Ungleichheit
inequivalence (→EXCLUSIVE-OR operation)	ENG.LOG	(→EXKLUSIV-ODER-Verknüpfung)
inequivalence function (→EXCLUSIVE-OR operation)	ENG.LOG	(→EXKLUSIV-ODER-Verknüpfung)
inequivalence operation (→EXCLUSIVE-OR operation)	ENG.LOG	(→EXKLUSIV-ODER-Verknüpfung)
inert	PHYS	**träge**
= slow-acting		
inert gase (→noble gas)	CHEM	(→Edelgas)
inertia	PHYS	**Trägheit**
		= Massenträgheit
inertia-free	PHYS	**trägheitslos**
= inertialess		= trägheitsfrei
inertialess (→inertia-free)	PHYS	(→trägheitslos)
inertialess scanning (→electronic scanning)	RADIO LOC	(→elektronische Abtastung)
inertial force	PHYS	**Trägheitskraft**
inertial guidance	TECH	**Trägheitslenkung**
inertial navigation	RADIO NAV	**Trägheitsnavigation**
inertial system	PHYS	**Inertialsystem**
inertial torque (→moment of inertia)	PHYS	(→Trägheitsmoment)
inexact (→inaccurate)	TECH	(→ungenau)
inexactitude (→inaccuracy)	TECH	(→Ungenauigkeit)
inexcactness (→inaccuracy)	TECH	(→Ungenauigkeit)
inexhaustible	TECH	**unerschöpflich**
inexpert (→improper)	COLLOQ	(→unsachgemäß)
inextinguishable (→indelible)	OFFICE	(→untilgbar)
infant mortality (→early failure)	QUAL	(→Frühausfall)
infer (→derive)	SCIE	(→ableiten)
inference (→logical inference)	DATA PROC	(→Inferenz)
inference (→conclusion)	SCIE	(→Schlußfolgerung)
inference control	DATA PROC	**Datenschutzprüfung**
inference engine (→logical inference program)	DATA PROC	(→Inferenzprogramm)
inference machine (→logical inference program)	DATA PROC	(→Inferenzprogramm)
inference module (→logical inference program)	DATA PROC	(→Inferenzprogramm)
inference program (→logical inference program)	DATA PROC	(→Inferenzprogramm)
inferior (→low-quality)	ECON	(→minderwertig)
inferior (→below value)	TECH	(→unterwertig)
inferior (→subscript)	TYPOGR	(→Tiefstellung)
inferior figure (→subscript)	TYPOGR	(→Tiefstellung)
inferior quality (→low-quality)	ECON	(→minderwertig)
inferior quality (→poor quality)	TECH	(→schlechte Qualität)
in-fiber component (→lightwave component)	OPT.COMM	(→Lichtleiterbauteil)
infinite ≠ finite	MATH	**unendlich** ≠ endlich
infinite (→infinitive mood)	LING	(→Infinitiv)
infinite impulse response = IIR	NETW.TH	**IIR**
infinite loop (→endless loop)	DATA PROC	(→Endlosschleife)
infinitely variable (→continuously adjustable)	ELECTRON	(→stufenlos einstellbar)
infinite number [symbol ∞]	MATH	**unendliche Zahl** [Symbol ∞]
infinitesimal	MATH	**infinitesimal**
infinitesimal electric dipole (→elementary electric dipole)	ANT	(→Hertzscher Dipol)
infinitesimal magnetic current element (→elementary magnetic dipole)	ANT	(→magnetischer Elementardipol)
infinitive mood (n.) [e.g.: to be]	LING	**Infinitiv** [z.B.: hören]
= infinite		= Nennform; Grundform des Verbs
↑ mood		↑ Modus
infinity	MATH	**Unendlich** (n.)
		= Unendlichkeit
infinity sign [symbol ∞]	MATH	**Unendlichzeichen** [Symbol ∞]
= infinity symbol		

infinity symbol

infinity symbol (→infinity sign)	MATH	(→Unendlichzeichen)
infix notation	MATH	**Infixschreibweise** = Infixnotation; Infixdarstellung
inflammability = flammability	TECH	**Entflammbarkeit** = Entzündbarkeit
inflammable = flammable ≈ combustible	TECH	**entflammbar** = entzündbar; entzündlich; feuergefährlich ≈ brennbar
inflection = inflexion ↑ morphology ↓ declination; conjugation	LING	**Flexion** = Beugung ↑ Morphologie ↓ Deklination; Konjugation
inflection point [point of a curve where the concavity changes its orientation]	MATH	**Wendepunkt** [Kurvenpunkt in dem die Konkavität Seite wechselt]
inflexion (→inflection)	LING	(→Flexion)
3 1/2 in. floppy disk = microfloppy disk; micro-diskette; 3 1/2 in. compact floppy disk; 3 1/2 in. diskette ↑ floppy disk	TERM&PER	**Mikrodiskette** = 3 1/4-Zoll-Diskette ↑ Diskette
5 1/4 in. floppy disk = mini-floppy disk; minifloppy; mini-disk; 5 1/4 in. diskette ↑ floppy disk	TERM&PER	**Minidiskette** = Minidisk; Flippy; 5 1/4-Zoll-Diskette ↑ Diskette
8-in. floppy disk (→eight-inch floppy disk)	TERM&PER	(→Normaldiskette)
inflow (v.i.) ≠ outflow	TECH	**zufließen** ≠ abfließen
inflow (→influx)	TECH	(→Zufluß)
influence (v.i.)	PHYS	**beeinflussen**
influence (n.) = induction	PHYS	**Beeinflussung** [elektrostatisch, elektromagnetisch] = Influenz; Induktion 1
influence	INSTR	**Einfluß**
influence (→affect)	TECH	(→beinflussen)
influence machine (→electrostatic generator)	PHYS	(→elektrostatischer Generator)
influx = inflow ≠ efflux; outflow	TECH	**Zufluß** = Zustrom; Zulauf ≠ Abfluß; Ausfluß
informal ≈ non-binding	ECON	**informell** = formlos ≈ unverbindlich
informatics (n.plt.) [science of information processing] = computer science; computing science (BRI) ≈ information theory; telematics; computer engineering ↑ information technology ↓ applied informatics	INF.TECH	**Informatik** [Wissenschaft der Datenverarbeitung] = Kerninformatik = Informationstheorie; Telematik; Informationsverarbeitung 2; Computerelektronik ↑ Nachrichtentechnik ↓ angewandte Informatik
information = notification	ECON	**Benachrichtigung**
information [content of a message, reduces the uncertainty of the receiver] ≈ message	INF	**Information** [Inhalt einer Nachricht, verringert Unsicherheiten des Empfängers] ≈ Nachricht
information access control = information flow control	DATA PROC	**Informationszugangskontrolle**
information age	SCIE	**Informationszeitalter**
information and communication technology (→information technology)	EL.TECH	(→Nachrichtentechnik)
information bank (→external data bank)	DATA PROC	(→externe Datenbank)
information bar	TELEC	**Informationsschiene**
information bearer (→information carrier)	DATA PROC	(→Informationsträger)
information bit	DATA COMM	**Informationsbit**
information bit (→information pulse)	TELEGR	(→Informationsschritt)
information booklet	DOC	**Informationsbrochüre**
information bureau	ECON	**Informationbureau**
information carrier = information bearer ≈ data medium	DATA PROC	**Informationsträger** ≈ Datenträger
information channel (→data channel)	DATA PROC	(→Datenkanal)
information channel (→communication channel)	INF	(→Übertragungskanal)
information content	INF	**Informationsgehalt** = Informationsinhalt
information drain (→information sink)	INF	(→Informationssenke)
information element (→information pulse)	TELEGR	(→Informationsschritt)
information entropy = entropy; mean information content	INF	**Informationsentropie** = Negentropie; mittlerer Informationsgehalt; durchschnittlicher Informationseingabe; mittlerer Informationsgehalt; Informationsbelag
information exchange = information interchange ↑ communication	TELEC	**Informationsaustausch** ↑ Kommunikation
information field	DATA COMM	**Daten-/Textfeld**
information flow (→message flow)	DATA COMM	(→Nachrichtenfluß)
information flow control (→information access control)	DATA PROC	(→Informationszugangskontrolle)
information incidence	TELECONTR	**Meldungsaufkommen**
information input	DATA PROC	**Informationszufluß** = Informationseingabe
information interchange (→information exchange)	TELEC	(→Informationsaustausch)
information junk	INF.TECH	**Informationsmüll**
information length [storage space for an information category of a data base, e.g. "profession"] ≈ word length	DATA PROC	**Informationslänge** [Speicherplatz für eine Informationskategorie einer Datenbank, z.B. „Beruf"] ≈ Wortlänge
information line [across a screen]	INF.TECH	**Informationszeile** [auf einem Bildschirm]
information loss	INF	**Informationsverlust**
information management system = IMS	DATA PROC	**Informationsverwaltungssystem** = IMS
information memory ≈ message memory	INF	**Informationsspeicher** ≈ Nachrichtenspeicher
information network	DATA COMM	**Informationsnetz**
information output	DATA PROC	**Informationsausgabe**
information parameter	INF	**Informationsparameter**
information processing 2 [human processing of information] ≈ message processing; data processing	INF.TECH	**Informationsverarbeitung 2** [Verarbeitung von Informationen durch den Menschen] ≈ Nachrichtenverarbeitung; Datenverarbeitung
information processing 1 (→data processing)	INF.TECH	(→Datenverarbeitung)
information processing center (→computing center)	DATA PROC	(→Rechenzentrum)
information processing curriculum (→data processing curriculum)	DATA PROC	(→DV-Fachausbildung)
information processing machine (→information processor)	DATA PROC	(→informationsverarbeitende Maschine)
information processor = information processing machine ↓ computer	DATA PROC	**informationsverarbeitende Maschine** ↓ Computer
information provider [videotext] = IP	DATA COMM	**Informationsanbieter** [Btx]
information provider = ip	INF.TECH	**Informationslieferant**
information provider keyboard (→editing keyboard)	TERM&PER	(→Editiertastatur)
information provider terminal (→editing terminal)	TELEC	(→Editierstation)

286

information pulse	TELEGR	**Informationsschritt**	
= information signal; information element; information bit; information unit		= Kombinationsschritt; Nutzschritt; Informationselement; Informationszeichen; Nutzelement; Nutzzeichen	
≠ start pulse; stop pulse		≠ Anlaufschritt; Sperrschritt	
↑ unit interval		↑ Zeichenelement	
information quantity	INF	**Informationsmenge**	
= information volume		≈ Nachrichtenmenge	
≈ message volume			
information rate (→information stream)	INF	(→Informationsfluß)	
information recording	TERM&PER	**Informationsaufzeichnung**	
information retrieval	DATA COMM	**Informationsabruf**	
= IR			
information retrieval	DATA PROC	**Informationswiedergewinnung**	
information securing	INF	**Informationssicherung**	
information security (→data integrity)	DATA PROC	(→Datensicherheit)	
information separator	DATA COMM	**Informationstrennung** [Code]	
= IS		= IS	
information service (→call information service)	DATA COMM	(→Auskunftdienst)	
information signal	INF	**Informationszeichen**	
information signal (→information pulse)	TELEGR	(→Informationsschritt)	
information sink	INF	**Informationssenke**	
= information drain		= Informationssinke	
≈ message sink		≈ Nachrichtensenke	
information source	INF	**Informationsquelle**	
≈ message source		≈ Nachrichtenquelle	
information storage	DATA PROC	**Informationsspeicherung**	
information storage and retrieval	DATA PROC	**Informationsspeicherung und Wiedergewinnung**	
information stream	INF	**Informationsfluß**	
= information rate; flow of information		≈ Nachrichtenfluß	
≈ message flow			
information system	DATA PROC	**Informationssystem**	
information technique	EL.TECH	(→Nachrichtentechnik)	
(→information technology)			
information technology	EL.TECH	**Nachrichtentechnik**	
[engineering of processing and transfer of informations by electronic means]		[Technik der elektronischen Verarbeitung und Übermittlung von Informationen]	
= IT; information technique; information and communication technology		= elektrische Nachrichtentechnik; Informationstechnik; Informationstechnologie; Informationselektronik; Informations- und Kommunikationstechnologie; IKT	
≈ applied informatics; telematics		≈ angewandte Informatik; Telematik	
↓ computing; telecommunication engineering		↓ Datentechnik; Telekommunikationstechnik	
information theory	INF.TECH	**Informationstheorie**	
≈ informatics		≈ Informatik	
information time slot	CODING	**Informationszeitschlitz**	
information track	TERM&PER	**Informationsspur**	
= bit track; logical track		≈ Lesespur	
≈ reading track		↑ Spur	
↑ track			
information transfer	INF.TECH	**Informationsübermittlung**	
= information transmission		= Informationsübertragung; Informationstransfer	
≈ message transfer [DATA COMM]		≈ Kommunikation; Nachrichtentransfer [DATA COMM]	
information transfer (→message transfer)	DATA COMM	(→Nachrichtenübermittlung)	
information transfer channel (→communication channel)	INF	(→Übertragungskanal)	
information transmission (→information transfer)	INF.TECH	(→Informationsübermittlung)	
information unit	DATA PROC	**Informationseinheit**	
[set of characters or data treated as a unit in logic or physical processes]		[bei logischen oder physikalischen Vorgängen als Einheit behandelter Satz von Zeichen oder Daten]	
↓ bit; Byte; data record; data field; data block		↓ Bit; Byte; Datensatz; Datenfeld; Datenblock	
information unit (→information pulse)	TELEGR	(→Informationsschritt)	
information user [videotex]	DATA COMM	**Informationsbenutzer** [Btx]	
information user	INF	**Informationsempfänger**	
≈ message receiver		= Informationsverbraucher	
		≈ Nachrichtenempfänger	
information utility (→computer utility)	DATA PROC	**DV-Dienstleistungsbetrieb**	
information volume (→information quantity)	INF	(→Informationsmenge)	
informative offer	ECON	**Informativangebot**	
≈ budgetary offer		≈ Budgetangebot	
informative set	DOC	**Informationssatz**	
infrared	PHYS	**infrarot**	
= IR		= IR	
infrared detector	COMPON	**Infrarotdetektor** [Sensor]	
= infrared sensor			
infrared emitter	PHYS	**Infrarotstrahler**	
infrared emitter	COMPON	**Infrarotstrahler**	
infrared head-phone	EL.ACOUS	**Infrarot-Tonempfänger**	
infrared light barrier	SIGN.ENG	**Infrarot-Lichtschranke**	
infrared light emitting diode (→IRED)	MICROEL	(→IRED)	
infrared radiation	PHYS	**Infrarotstrahlung**	
[wavelength greater than 0.8 μ]		[Wellenbereich größer 0,8μ]	
= ultrared radiation		= infrarote Strahlung; Ultrarotstrahlung; ultrarote Strahlung	
infrared remote control	TERM&PER	**Infrarotfernbedienung**	
infrared sensor (→infrared detector)	COMPON	(→Infrarotdetektor)	
infrared transmission	ELECTRON	**Infrarotübertragung**	
infrasonic	PHYS	**Infraschall-...**	
infrasound	ACOUS	**Infraschall**	
[oscillations below 16 Hz]		[Schwingungen unter 16 Hz]	
= subsound		= Unterschall	
≠ ultrasound		≠ Ultraschall	
↑ sound 1		↑ Schall 1	
infrastructure	TECH	**Infrastruktur**	
infringe (v.t.)	COLLOQ	**verletzen**	
[fig.; a rule, law]		[fig.; eine Regel, Gesetz]	
		= verstoßen	
infringement	COLLOQ	**Verletzung**	
[fig.]		[fig.]	
		= Verstoß	
ingenious	TECH	**ausgeklügelt**	
= tricky		≈ raffiniert	
≈ sophisticated			
ingot	METAL	**Rohblock**	
ingot steel (→mild steel)	METAL	(→Flußstahl)	
ingress	TECH	**Eindringen**	
		[z.B. Wasser]	
inherent	TECH	**Eigen-**	
= intrinsic			
inherent address (→indirect address)	DATA PROC	(→indirekte Adresse)	
inherent addressing (→indirect addressing)	DATA PROC	(→indirekte Adressierung)	
inherent error	MATH	**mitgeschleppter Fehler**	
= propagated error		= mitlaufender Fehler; inhärenter Fehler; übernommener Fehler	
inherent fault	QUAL	**anhaftender Mangel**	
= inherent weakness			
inherent jitter	ELECTRON	**Eigenjitter**	
inherent weakness (→inherent fault)	QUAL	(→anhaftender Mangel)	
inhibit (v.t.) (→disable)	ELECTRON	(→sperren)	
inhibit (→inhibition)	ELECTRON	(→Sperrung)	
inhibit (→forbid)	COLLOQ	(→verbieten)	

inhibit bit = disable bit	CODING	**Sperrbit**
inhibit flipflop	CIRC.ENG	**Sperrkippstufe**
inhibit gate	CIRC.ENG	**Sperrgatter** = Inhibit-Gatter
inhibiting condition (→blocking state)	ELECTRON	(→Sperrzustand)
inhibiting input (→inhibit input)	CIRC.ENG	(→Sperreingang)
inhibit input = inhibiting input	CIRC.ENG	**Sperreingang** = Inhibit-Eingang
inhibition = inhibit; interlock; retention ≈ suppression	ELECTRON	**Sperrung** = Halten ≈ Unterdrückung
inhibit pulse = disable pulse	ELECTRON	**Sperrimpuls**
inhibit switch (→lockout switch)	DATA PROC	(→Sperrschalter)
inhibit wire [magnetic core memory]	DATA PROC	**Inhibitdraht** [Kernspeicher] = Sperrdraht
inhomogeneity	SCIE	**Inhomogenität**
inhomogeneous	SCIE	**inhomogen**
in-house = inhouse; corporate	ECON	**firmenintern** = betriebsintern; innerbetrieblich; hausintern
inhouse (→in-house)	ECON	(→firmenintern)
in-house development	TECH	**Eigenentwicklung**
in-house network (→local area network)	DATA COMM	(→lokales Netz)
in-house system [corporate videotex]	DATA COMM	**In-house-System** [betriebsinternes Btx]
in-house testing (→alpha testing)	DATA PROC	(→firmeninterner Feldversuch)
INIC	NETW.TH	**stromumkehrender Negativ-Impedanzkonverter**
initial (v.t.) [sign with initials]	OFFICE	**abzeichnen** [mit Namenskurzzeichen]
initial (n.) = initial letter ≈ capital letter	TYPOGR	**Anfangsbuchstabe** = Initialbuchstabe; Initiale (OES); Initial ≈ Großbuchstabe
initial (adj.) = outgoing ≈ beginning	TECH	**anfänglich** [am Anfang vorhanden] = Ausgangs- ≈ beginnend
initial address (→start address)	DATA PROC	(→Startadresse)
initial alignment control	SWITCH	**Synchronisationssteuerung**
initial armature force [relay]	COMPON	**Ankervorspannung** [Relais]
initial condition = starting condition	MATH	**Anfangsbedingung**
initial condition (→entry condition)	DATA PROC	(→Einsprungbedingung)
initial configuration (→initial equipment)	EQUIP.ENG	(→Erstausbau)
initial costs = starting costs	ECON	**Anlaufkosten**
initial deviation = starting deviation	CONTROL	**Anfangsabweichung**
initial equipment = initial configuration; initial stage of construction	EQUIP.ENG	**Erstausbau**
initial error	DATA PROC	**Ausgangsfehler** = Anfangsfehler
initialing (→beginning)	TECH	(→beginnend) (adj.)
initial instruction	DATA PROC	**Startbefehl**
initialization 1 [to put parameters to initial value] = initializing; presetting; job initialization; job preparation; preparation ≈ start; initiation	DATA PROC	**Initialisierung 1** [Parameteranfangswerte einstellen] = Vorbereitung ≈ Anlauf; Einleitung
initialization 2 (→formatting)	DATA PROC	(→Formatierung)
initialization file	DATA PROC	**Initialisierungsdatei**
initialization logic	MICROEL	**Initialisierungssteuerung**
initialization mode	DATA COMM	**Vorbereitungsphase**
initialization time (→set-up time)	DATA PROC	(→Vorbereitungszeit)
initialize 1 [to set parameters to starting values, to establish start conditions] = preset	DATA PROC	**initialisieren 1** [Parameter auf Anfangswerte stellen, Startzustand herstellen]
initialize 3 (→format)	DATA PROC	(→formatieren)
initialize 2 (→preformat)	DATA PROC	(→vorformatieren)
initializing (→initialization)	DATA PROC	(→Initialisierung 1)
initial letter (→initial)	TYPOGR	(→Anfangsbuchstabe)
initialling (→beginning)	TECH	(→beginnend) (adj.)
initial magnetization curve [ferromagnetism] = normal magnetization curve; virgin curve; neutral curve; rise path	PHYS	**Neukurve** [Ferromagnetismus] = jungfräuliche Kurve
initial period [e.g. the first three minutes]	TELEPH	**erstes Zeitintervall** [z.B. drei Minuten]
initial period charge	TELEC	**Mindestgebühr** [Gebühr für erstes Zeitintervall]
initial permeability	PHYS	**Anfangspermeabilität**
initial point [of a vector]	MATH	**Anfangspunkt** [Vektor]
initial position = starting posicion	MECH	**Anfangsstellung** = Ausgangsstellung
initial program loader [programm to load the operational program] = IPL; bootstrap loader ≈ bootstrap program	DATA PROC	**Urlader** [Programm zur Ladung des Organisationsprogramms] = Urladeprogramm; Bootstrap-Lader ≈ Ureingabeprogramm
initial program loading (→bootstrapping)	DATA PROC	(→Urladen)
initial resistance	ELECTRON	**Anfangswiderstand** = Kaltwiderstand
initial sensitivity	INSTR	**Anfangsempfindlichkeit**
initial stage of construction (→initial equipment)	EQUIP.ENG	(→Erstausbau)
initial standard profile (→default profile)	TELEC	(→Ausgangsprofil)
initial start	DATA PROC	**Urstart**
initial state = original state; normal position	CIRC.ENG	**Grundstellung**
initial state (→initial status)	ENG.LOG	(→Anfangszustand)
initial status = initial state	ENG.LOG	**Anfangszustand**
initial value = basic value	DATA PROC	**Anfangswert**
initial zero time	TELEC	**Ausgangszeitwert**
initiate (→trigger)	DATA PROC	(→auslösen)
initiation (n.) ≈ start; initialization	DATA PROC	**Einleitung** = Anstoß; Auslösung ≈ Anlauf; Initialisierung
initiator (→job initiator)	DATA PROC	(→Vorbereiter)
inject (→apply)	EL.TECH	(→anlegen)
injected power (→launched power)	OPT.COMM	(→eingespeiste Leistung)
injection	CODING	**Einblendung**
injection	MICROEL	**Injektion**
injection [set theory]	MATH	**Injektion** [Mengenlehre]
injection (→feeding)	ELECTRON	(→Einspeisung)
injection current	MICROEL	**Injektionsstrom**
injection efficiency	MICROEL	**Injektionswirkungsgrad** = Emitterwirkungsgrad; Emissionswirkungsgrad
injection laser diode = ILD	OPTOEL	**Injektionslaserdiode** = ILD
injection logic	MICROEL	**Injektionslogik**
injection-molded [plastics]	TECH	**spritzgegossen** [Kunststoff]
injection molding [plastics]	TECH	**Spritzguß** [Kunststoff]
injector [binistor]	MICROEL	**Injektor** [Binistor]
injury (→damage)	TECH	(→Schaden)
ink ≈ drawing ink	ENG.DRAW	**Tinte** ≈ Tusche

ink cartridge	TERM&PER	**Tintenpatrone**				
inked ribbon	OFFICE	**Farbband**				
= ink ribbon; printer ribbon; ribbon				inner marker beacon	RADIO NAV	**Hauptanflugbake**
inking	DATA PROC	**Freihandzeichnen**		inner partition (→ partition)	TECH	(→**Zwischenwand**)
[computer graphics; freehand drawing with the cursor]		[mit der Schreibmarke] = Inking		inner program loop (→ inner loop)	DATA PROC	(→**verschachtelte Programmschleife**)
inking roller	TERM&PER	**Farbwalze**		inner sheath	COMM.CABLE	**Innenmantel**
ink-jet oscillograph	INSTR	**Flüssigkeitsstrahlschreiber**		= inner cladding		
= ink-jet recorder; ink-vapor recorder		= Strahlschreiber; Tinten-Schnellschreiber; Tintenschreiber; Flüssigkeitsstrahl-Oszillograph; Flüssigkeitsstrahl-Oszillograf		innovation	TECH	**Innovation** = Neuerung; Erneuerung 2
↑ oscillograph				innovation consultancy	ECON	**Innovationsberatung**
				innovation cycle	TECH	**Innovationszyklus**
				innovative	TECH	**innovativ**
		↑ Oszillograph		≈ progressive 2		≈ fortschrittlich
ink-jet printer	TERM&PER	**Tintendrucker**		innovative drive	TECH	**Innovationsanstoß** = Innovationsanstoß; Innovationsimpuls
↓ non impact printer		= Tintenstrahldrucker; Tintenspritzdrucker; Farbstrahldrucker; Tintenstrahlschreiber				
				innumerable	COLLOQ	**zahllos**
				= numberless; countless; myriad (adj.)		= unzählbar
		↑ anschlagfreier Drucker		≈ immense; immeasurable		≈ immens; unmeßbar
ink-jet printhead	TERM&PER	**Tintenkopf**		in one go	COLLOQ	**in einem Zug**
ink-jet printing	TERM&PER	**Tintendruck**		in on-stage (→ conducting)	ELECTRON	(→**durchlässig**)
= ink printing				in open circuit	EL.TECH	**im Leerlauf**
ink-jet printing mechanism	TERM&PER	**Tintendruckwerk**		= open-circuited		
ink-jet recorder (→ ink-jet oscillograph)	INSTR	(→**Flüssigkeitsstrahlschreiber**)		in operation (→ on-line)	TELEC	(→**in Betrieb**)
				inoperative (→ ineffective)	TECH	(→**unwirksam**)
ink-jet recording	TERM&PER	**Farbstrahldruckverfahren**		in opposition	PHYS	**gegenphasig**
ink pad	OFFICE	**Stempelkissen**		= quadrature-phase		≈ gleichphasig
ink pen	TERM&PER	**Tintengriffel**		≠ in-phase		
ink point marking	MICROEL	**Inken**		InP (→ indium phosphide)	CHEM	(→**Indium-Phosphid**)
ink printing (→ ink-jet printing)	TERM&PER	(→**Tintendruck**)		in-phase	PHYS	**gleichphasig**
				= cophasal; common-mode		= phasengleich
ink recording	TERM&PER	**Tintenschrift**		≠ in opposition		≠ gegenphasig
ink ribbon (→ inked ribbon)	OFFICE	(→**Farbband**)		in-premises distribution amplifier	BROADC	**Hausverteilverstärker**
ink roller	TERM&PER	**Farbrolle**		= domestic distribution amplifier		
ink-vapor recorder (→ ink-jet oscillograph)	INSTR	(→**Flüssigkeitsstrahlschreiber**)		in-premises distribution system	BROADC	**Hausverteilanlage**
				= domestic distribution system		
inland toll call (AM)	TELEC	**Inlandsferngespräch**		in preparation	COLLOQ	**in Vorbereitung**
= inland trunk call (BRI); national long-distance call; domestic call				= in the works (AM)		
				input (v.t.)	DATA PROC	**eingeben**
inland trunk call (BRI) (→ inland toll call)	TELEC	(→**Inlandsferngespräch**)		= enter; post (v.t.); introduce ≈ read-in (v.t.); key-in; insert; load (v.t.)		≈ einlesen; eintasten; einfügen; laden
inlay (n.)	TECH	**Einlage**		≠ output (v.t)		
= insert				input (n.)	EL.TECH	**Eingang**
inlay (→ inlay mode)	TV	(→**Inlay-Verfahren**)		= i/p; I/P; inlet		≠ Ausgang
inlay card	TERM&PER	**Beschriftungskarte**		≠ output (n.)		
inlay mode	TV	**Inlay-Verfahren**		input (n.)	ECON	**Einsatz**
= inlay				= deployment		[Personal-, Material- ...]
inlet (→ input)	EL.TECH	(→**Eingang**)				≈ Aufwand
inlet temperature	TECH	**Eintrittstemperatur**		input (→ data input)	DATA PROC	(→**Dateneingabe**)
inlet valve	TECH	**Einlaßventil**		input acknowledgment	DATA PROC	**Eingabequittung**
in-line (n.)	MICROEL	**In-line**		input admittance	NETW.TH	**Eingangsleitwert**
[pin arrangement in row(s)]		= Inline; Reihenaschluß				= Eingangsadmittanz; Eingangsscheinleitwert
= inline		↓ DIL				
↓ DIL				input admittance, output shorted (→ short-circuit input admittance)	NETW.TH	(→**Kurzschluß-Eingangsadmittanz**)
in-line (adj.)	TECH	**hintereinander**				
inline- (→ outlined)	TYPOGR	(→**konturiert**)				
inline (→ in-line)	MICROEL	(→**In-line**)		input air	TECH	**Zuluft**
in-line coding (→ straight-line programming)	DATA PROC	(→**gestreckte Programmierung**)		≠ output air		≠ Abluft
				input alphabet	DATA PROC	**Eingangsalphabet**
in-line processing	DATA PROC	**Geradewohl-Verarbeitung**		input amplifier	CIRC.ENG	**Eingangsverstärker**
[data are processed when and as they appear]		[Daten werden so wie sie anfallen sofort verarbeitet]		input amplitude	ELECTRON	**Eingangsamplitude**
				input area	DATA PROC	**Eingabebereich**
= demand processing; immediate processing; instant processing		= unmittelbare Verarbeitung; Sofortverarbeitung		[main memory]		[Arbeitsspeicher]
				≠ input section		
≈ real-time operation		≈ Echtzeitbetrieb		input balance	NETW.TH	**Eingangssymmetrie**
in-line programming (→ straight-line programming)	DATA PROC	(→**gestreckte Programmierung**)		= input symmetry		
				input bias input	CIRC.ENG	**Eingangsruhestrom**
in loop-back condition	TELEC	**in Schleife**		[operational amplifier]		[Operationsverstärker]
= looped				input-bound (→ input-limited)	DATA PROC	(→**eingabebegrenzt**)
inner cladding (→ inner sheath)	COMM.CABLE	(→**Innenmantel**)				
				input buffer	DATA PROC	**Eingabepuffer**
inner conductor	COMM.CABLE	**Innenleiter**		input buffer register	DATA PROC	**Eingabepufferregister**
= center conductor; centre conductor		= Mittelleiter		input capacitance	NETW.TH	**Eingangskapazität**
inner loop	DATA PROC	**verschachtelte Programmschleife**		input cell	MICROEL	**Eingangszelle**
[loop inside a loop]				input characteristic	ELECTRON	**Eingangskennlinie**
= nested loop; inner program loop; nested program loop; nesting loop		[Schleife in einer Schleife] = verschachtelte Schleife; geschachtelte Programm-				[Thyristor]
				input characteristic impedance	NETW.TH	**Eingangswellenwiderstand**

input circuit CIRC.ENG **Eingangsschaltung**
input command (→input DATA PROC (→Eingabebefehl)
   instruction)
input condition ELECTRON **Eingangszustand**
= input state
input control DATA PROC **Eingabesteuerung**
input controller DATA PROC **Eingebewerk**
  [controls the data input to the CPU] [steuert den Datentransfer
  = input processor zur Zentraleinheit]
  ↑ input/output controller = Eingabeprozessor
   ↑ Ein-Ausgabe-Werk
input current EL.TECH **Eingangsstrom**
input cut-off frequency MICROEL **Eingangs-Grenzfrequenz**
input data DATA PROC **Eingabedaten**
   = Eingangsdaten
input data carrier (→input DATA PROC (→Eingabemedium)
   medium)
input data medium (→input DATA PROC (→Eingabemedium)
   medium)
input device (→input TERM&PER (→Eingabegerät)
   equipment)
input disable DATA PROC **Eingabesperre**
input drift CIRC.ENG **Eingangsdrift**
input equipment TERM&PER **Eingabegerät**
  = input device; input unit = Eingabevorrichtung; Eingabeeinheit
input filter NETW.TH **Eingangsfilter**
input impedance NETW.TH **Eingangsimpedanz**
input impulse (→unitary ELECTRON (→Einheitsimpuls)
   impulse)
input information SWITCH **Eingabeinformation**
  = order
input instruction DATA PROC **Eingabebefehl**
  = input statement; input command
input job stream DATA PROC **Auftragseingabefluß**
  = input stream; job stream ≈ Eingabefluß
  ≈ job queue ≈ Auftragsschlange
input keyboard (→entry TERM&PER (→Eingabetastatur)
   keyboard)
input lead TERM&PER **Eingabeanschluß**
input level NETW.TH **Eingangspegel**
input-limited DATA PROC **eingabebegrenzt**
  = input-bound
input line terminating NETW.TH **Eingangs-Abschlußwiderstand**
   impedance
input load EL.TECH **Eingangslast**
   = Eingangsbelastung
input magazine DATA PROC **Eingabemagazin**
  [reserved memory location] [reservierter Speicherplatz]
  ↑ magazine ↑ Magazin
input magazine (→paper tray) TERM&PER (→Papiernachfüllmagazin)
input manager DATA PROC **Eingabe-Manager**
   = Eingabe-Verwalter; Input-Manager; Input-Verwalter
input mask DATA PROC **Eingabemaske**
  = data entry mask = Maske 3; Erfassungsmaske
  ↑ display mask ↑ Bildschirmmaske
input medium DATA PROC **Eingabemedium**
  = input data medium; input data carrier = Eingabedatenträger
input memory DATA PROC **Eingabespeicher**
  = input storage; input store
input mode DATA PROC **Eingabebetrieb**
  = input operation
input offset voltage CIRC.ENG **Eingangs-Nullspannung**
  [operational amplifier] [Operationsverstärker]
   = Eingangs-Offsetspannung;
   Eingangs-Fehlspannung
input operation (→input DATA PROC (→Eingabebetrieb)
   mode)
input/output ELECTRON **Ein-/Ausgabe**
  = I/O = E/A; Eingabe/Ausgabe
input/output-bound DATA PROC **Ein-Ausgabe-bedingt**
  = I/O-bound; input/output-limited; I/O-limited = E/A-gebunden; Ein-Ausgabe-gebunden
input/output buffer DATA PROC **Ein-/Ausgabe-Puffer**
  = I/O buffer = E/A-Puffer; Ein-/Ausgabe-Pufferspeicher; E/A-Pufferspeicher; Eingabe-Ausgabe-Puffer; Eingabe-Ausgabe-Pufferspeicher

input/output bus DATA PROC **Ein-/Ausgabe-Bus**
  = I/O bus = Eingabe-Ausgabe-Bus; E/A-Bus
input/output cell MICROEL **Ein-/Ausgabe-Zelle**
  = I/O cell = E/A-Zelle; I/O-Zelle; Eingabe-Ausgabe-Zelle
input/output channel DATA PROC **Ein-/Ausgabe-Kanal**
  = I/O channel = Eingabe-Ausgabe-Kanal; E/A-Kanal
input/output control CIRC.ENG **Ein-/Ausgabe-Steuerung**
  = I/O control = E/A-Steuerung; Eingabe-Ausgabe-Steuerung
input/output controller DATA PROC **Ein-/Ausgabe-Werk**
  [part of CPU controlling the data exchange with peripherals or external memories] [Teil der Zentraleinheit der den Datenaustausch mit Peripheriegeräten oder externen Speichern steuert]
  = I/O controller; input/output control system; I/O control system; IOCS; input/output system; I/O system; input/output processor; I/O processor; IOP; interface processor = E/A-Werk; Ein-/Ausgabe-Prozessor; E/A-Prozessor; Ein-Ausgabe-Steuerung; E/A-Steuerung; Schnittstellenprozessor; Ein-/Ausgabe-System; E/A-System; EAS; IOCS; IOP; Eingabe-Ausgabe-Werk; Eingabe-Ausgabe-Prozessor; Eingabe-Ausgabe-Steuerung; Eingabe-Ausgabe-System
  ≈ BIOS; ROS ≈ BIOS; ROS
  ↑ central processing unit ↑ Zentraleinheit
  ↓ input controller; output controller ↓ Eingabewerk; Ausgabewerk
input/output control DATA PROC **Ein-/Ausgabe-Kontrollprogramm**
   program
  = I/O control program; input/output executive; I/O executive = E/A-Kontrollprogramm; Eingabe-Ausgabe-Kontrollprogramm
input/output control system DATA PROC (→Ein-/Ausgabe-Werk)
   (→input/output controller)
input/output device TERM&PER **Ein-Ausgabe-Gerät**
  = I/O device = E/A-Gerät
  ↑ peripheral equipment ↑ Peripheriegerät
input/output device TERM&PER **Ein-/Ausgabe-Vorrichtung**
  [permits both] [erlaubt beides]
  = I/O device = Ein-/Ausgabe-Gerät; E/A-Vorrichtung; E/A-Gerät; Eingabe-Ausgabe-Vorrichtung; Eingabe-Ausgabe-Gerät
  ↑ peripheral equipment ↑ Peripheriegerät
input/output driver DATA PROC **Ein-/Ausgabe-Treiber**
  = I/O driver = E/A-Treiber
input/output executive DATA PROC (→Ein-/Ausgabe-Kontrollprogramm)
   (→input/output control program)
input/output file DATA PROC **Ein-/Ausgabe-Datei**
  = I/O file = E/A-Datei; Eingabe-Ausgabe-Datei
input-output instruction DATA PROC **Ein-/Ausgabe-Befehl**
  = I/O instruction = E/A-Befehl; Eingabe-Ausgabe-Befehl
input/output interface DATA PROC **Ein-/Ausgabe-Schnittstelle**
  = I/O interface = E/A-Schnittstelle; Eingabe-Ausgabe-Schnittstelle
input/output interrupt DATA PROC **Ein-/Ausgabe-Unterbrechungssignal**
  = I/O interrupt = E/A-Unterbrechungssignal; Ein-Ausgabe-Interrupt; Eingabe-/Ausgabe-Interrupt; E/A-Interrupt
input/output library DATA PROC **Ein-/Ausgabe-Programmbibliothek**
  = I/O library; input/output program library; I/O program library = E/A-Programmbibliothek; Eingabe-Ausgabe-Programmbibliothek
input/output-limited DATA PROC (→Ein-Ausgabe-bedingt)
   (→input/output-bound)
input/output module TELECONTR **Eingabe-Ausgabe-Baugruppe**
  = I/O module = Ein-/Ausgabe-Baugruppe; E/A-Baugruppe

English	Domain	German
input/output port = I/O port	CIRC.ENG	**Ein-/Ausgabe-Port** = E/A-Port; Eingabe-Ausgabe-Port
input/output processor (→input/output controller)	DATA PROC	(→Ein-/Ausgabe-Werk)
input/output program library (→input/output library)	DATA PROC	(→Ein-/Ausgabe-Programmbibliothek)
input/output referencing = I/O referencing	DATA PROC	**Ein-/Ausgabegerät-Bezugnahme** = E/A-Gerät-Bezugnahme
input/output register = I/O register	DATA PROC	**Ein-/Ausgabe-Register** = E/A-Register; I/O-Register; Eingabe-Ausgabe-Register
input/output request [from a CPU] = I/O request; IORQ	DATA PROC	**Ein-/Ausgabe-Anforderung** [seitens einer Zentraleinheit] = E/A-Anforderung; Eingabe-Ausgabe-Anforderung
input/output status word = I/O status word	DATA PROC	**Ein-/Ausgabe-Statuswort** = E/A-Statuswort; Eingabe-Ausgabe-Statuswort
input/output symbol = I/O symbol	DATA PROC	**Ein-Ausgabe-Symbol** = E/A-Symbol; Eingabe-Ausgabe-Symbol
input/output system (→input/output controller)	DATA PROC	(→Ein-/Ausgabe-Werk)
input port	NETW.TH	**Eingangstor**
input power	EL.TECH	**Eingangsleistung**
input processor (→input controller)	DATA PROC	(→Eingebewerk)
input program (→input routine)	DATA PROC	(→Einleseroutine)
input protection circuit	MICROEL	**Eingangsschutzschaltung**
input quantity = immittance	NETW.TH	**Eingangsgröße** = Eingangsfunktion; Immitanz
input record	DATA PROC	**Eingabesatz**
input register	DATA PROC	**Eingaberegister**
input resonator [klystron]	MICROW	**Eingangsresonator** [Klystron]
input routine = read-in routine; read-in program; readin program; input program	DATA PROC	**Einleseroutine** = Eingaberoutine; Einleseprogramm
input section (→input area)	DATA PROC	(→Eingabebereich)
input sensitivity	ELECTRON	**Eingangsempfindlichkeit**
input signal = incoming signal; drive	ELECTRON	**Eingangssignal**
input-signal delay	TELECONTR	**Erkennungszeit**
input stage	CIRC.ENG	**Eingangsstufe** = Anfangsstufe
input state (→input condition)	ELECTRON	(→Eingangszustand)
input statement (→input instruction)	DATA PROC	(→Eingabebefehl)
input station	DATA PROC	**Eingabestation**
input storage (→input memory)	DATA PROC	(→Eingabespeicher)
input store (→input memory)	DATA PROC	(→Eingabespeicher)
input stream (→input job stream)	DATA PROC	(→Auftragseingabefluß)
input swing	CIRC.ENG	**Eingangshub**
input symmetry (→input balance)	NETW.TH	(→Eingangssymmetrie)
input terminal (→entry terminal)	TERM&PER	(→Eingabeterminal)
input time constant	MICROEL	**Eingangs-Zeitkonstante**
inputting (→data input)	DATA PROC	(→Dateneingabe)
input unit (→input equipment)	TERM&PER	(→Eingabegerät)
input voltage	EL.TECH	**Eingangsspannung**
input-work queue	DATA PROC	**Arbeitsanfallschlange**
inquire (v.i.) (NAM) = enquire (v.i.) (BRI); query (v.i.); request (v.i.); extract 2 (v.t.); poll (v.t.)	DATA PROC	**abfragen** = auslesen; wiedergewinnen
inquiring station (→inquiry station)	DATA COMM	(→Abfragestation)
inquiry (n.) (NAM) = enquiry (BRI); query; request; interrogation	DATA PROC	**Abfrage** = Anfrage
inquiry (n.) = enquiry (n.); query (n.); interrogation; scan (n.) ↓ poll	TELECONTR	**Abfrage** ↓ zyklische Abfrage
inquiry (n.) (NAM) = enquiry (BRI); query; interrogation	SWITCH	**Abfragen**
inquiry (NAM) [code] = enquiry (BRI); ENQ; query; interrogation; interrogating	DATA COMM	**Stationsaufforderung** [Code] = ENQ
inquiry (AM) = enquiry (BRI)	ECON	**Anfrage**
inquiry (→call-back)	TELEPH	(→Rückfrage)
inquiry (→investigation)	TECH	(→Untersuchung)
inquiry (→census)	MATH	(→Erhebung)
inquiry character = enquiry character; ENQ; query character	DATA PROC	**Abfragezeichen**
inquiry information ≠ spontaneous information	TELECONTR	**Abfragemeldung** ≠ Spontanmeldung
inquiry key (→interrogation key)	TERM&PER	(→Abfragetaste)
inquiry language (NAM) (→query language)	DATA PROC	(→Abfragesprache)
inquiry mode	DATA PROC	**Abfragemodus**
inquiry position	TELEPH	**Rückfrageplatz**
inquiry processing = Anfrageverarbeitung	DATA PROC	**Abfrageverarbeitung**
inquiry/response = enquiry/response; query/response	TELEC	**Frage-/Antwort-**
inquiry signal (→interrogation signal)	TELECONTR	(→Abfragesignal)
inquiry station = enquiry station; query station; inquiring station; enquiring station; retrieval terminal	DATA COMM	**Abfragestation** = abfragende Station
inquiry terminal (→interrogation terminal)	TERM&PER	(→Abfrageterminal)
in round figures (adv.) = roundly	COLLOQ	**rund** (adv.) [fig.] = in runden Zahlen
inrush current (→switch-on peak)	EL.TECH	(→Einschaltstromspitze)
InSb (→indium antimonide)	CHEM	(→Indium-Antimonid)
inscription	MOB.COMM	**Einbuchen**
insects-proof (n.)	QUAL	**Insektensicherheit**
insects-proof (adj.)	QUAL	**insektensicher**
insensitive to light = light-insensitive	PHYS	**lichtunempfindlich**
insensitive = immune	TECH	**unempfindlich**
in-series adapter = within-series adapter; intra-series adapter; gender changer ↑ adapter 2 ↓ male-male adapter; female-female adapter	COMPON	**Kupplung 1** [Übergang für Anschlüsse derselben Steckerfamilie] = Übergangsverbinder 2 ↑ Übergangsstecker ↓ Doppelstecker; Doppelkupplung
insert (v.t.)	TELEC	**einfügen**
insert (v.t.) (→nest)	DATA PROC	(→schachteln)
insert (→plug-in)	ELECTRON	(→stecken)
insert (→inlay)	TECH	(→Einlage)
insert character (→insertion character)	DATA PROC	(→Einfügungszeichen)
insert command = insertion command; insert instruction	DATA PROC	**Einfügungsbefehl** = Einfügungskommando; Einfügekommando
insert/delete key (→backspace key)	TERM&PER	(→Rücksetztaste 2)
inserted subroutine = eingefügte Subroutine	DATA PROC	**eingefügte Unterroutine**
inserter	ELECTRON	**Montagewerkzeug**
insert instruction (→insert command)	DATA PROC	(→Einfügungsbefehl)
insertion	DATA PROC	**Einfügung** = Einfügen
insertion	MECH	**Einführung** = Steckung
insertion (→placement)	MANUF	(→Bestückung)
insertion amplifier	TV	**Einfügungsverstärker**

insertion attenuation NETW.TH (→Einfügungsdämpfung)
(→insertion loss)
insertion character DATA PROC **Einfügungszeichen**
= insert character = Einfügezeichen; Einfügungscharakter
insertion command (→insert DATA PROC (→Einfügungsbefehl)
command)
insertion force COMPON **Steckkraft**
= Einsteckkraft
insertion gain NETW.TH **Einfügungsgewinn**
insertion loss NETW.TH **Einfügungsdämpfung**
= insertion attenuation = Einfügungsverlust
insertion machine MANUF **Automatenbestückung**
[Leiterplattenbestückung]
insertion mode (→sifting) DATA PROC (→Einfügungsbetrieb)
insertion of components MANUF **Leiterplattenbestückung**
= printed board assembly; PCB assembly; module mounting = Baugruppenbestückung
insertion signal generator TV **Prüfzeilen-Signalgenerator**
insertion unit TRANS **Einfügungsgerät**
INSERT key TERM&PER **Einfügetaste**
[activates or disactivates an insert mode] [eröffnet einen Einfügemodus, bzw. hebt ihn wieder auf]
↑ functional key = Taste EINFG; EINFG-Taste; INSERT-Taste; Taste INSERT
↑ Funktionstaste
insert mode DATA PROC **Einfügemodus**
[allows insertion of characters by shifting of existing text] [erlaubt Einfügen von Zeichen mit Verschiebung des vorhandenen Textes]
≠ overtype mode = Einfügungsmodus; Insert-Modus
↑ entry mode ≠ Überschreibmodus
↑ Eingabemodus
insert position (→board EQUIP.ENG (→Baugruppenposition)
position)
in-service monitoring TRANS **In-Betrieb-Überwachung**
= ISM = In-Service-Monitoring; ISM
inset EQUIP.ENG **Geräteeinsatz**
= unit; subrack = Einsatz
= module frame = Baugruppenrahmen
inset mounting device EQUIP.ENG **Einsatzaufnahme**
inset variant (→variant of EQUIP.ENG (→Einsatzvariante)
subrack)
inset wiring (→intra-shelf EQUIP.ENG (→Einsatzverdrahtung)
wiring)
inside TECH **Innenseite**
= internal side
inside aerial (→indoor ANT (→Innenantenne)
antenna)
inside antenna (→indoor ANT (→Innenantenne)
antenna)
inside call transfer TELEPH **Interngesprächsweitergabe**
inside cover TYPOGR **Deckelinnenseite**
inside diameter TECH **Innendurchmesser**
inside dimension MECH **Innenmaß**
inside margin TECH **Innenrand**
inside wall TECH **Innenwand**
insolation (→solar radiation) PHYS (→Sonneneinstrahlung)
insoluble CHEM **unlöslich**
inspeccion mirror ELECTRON **Reparaturspiegel**
inspect (→check) TECH (→prüfen)
inspection QUAL **Prüfung**
= test (n.); measurement; check; checking = Test
= Erprobung; Messung
inspection (→check) TECH (→Prüfung)
inspection by attributes QUAL **attributive Prüfung**
inspection by variables QUAL **Variablen-Prüfung**
inspection level QUAL **Prüfniveau**
inspection stamp ECON **Prüfstempel**
instability TECH **Instabilität**
= unstableness; lability = Labilität; Unstabilität
≈ variability ≈ Variabilität
instable (→unstable) TECH (→instabil)
instable state CIRC.ENG **instabiler Zustand**
install (v.t.) TECH **installieren**
≈ mount; assemble; set-up = einrichten
≈ montieren; zusammenbauen

install (v.t.) DATA PROC **installieren**
[to customize a program to a specific computer configuration] [ein Programm auf die Gegebenheiten einer spezifischen Computereinlage einstellen]
installation ELECTRON **Anlage**
= plant
installation TECH **Installierung**
≈ mounting; assembly; set-up 1 (n.) = Einrichtung; Installation 2
≈ Montage; Zusammenbau; Errichtung
installation DATA PROC **Installation**
= setup = Installierung; Inbetriebnahme
installation 1 (→plant) TECH (→Anlage)
installation (→mounting 2) TECH (→Aufstellung 3)
installation and commissioning SYS.INST (→Montage und Einschaltung)
[TRANS] (→installation and cutover)
installation and cutover SYS.INST **Montage und Einschaltung**
[SWITCH]
= installation and commissioning [TRANS]
installation cable COMM.CABLE **Installationskabel**
installation disk DATA PROC **Installationsdiskette**
= set-up disk = Einrichtungsdiskette
installation document TECH **Aufbauunterlage**
= exchange configuration document = Ausführungsunterlage
installation drawing ENG.DRAW **Montagezeichnung**
≈ assembly drawing ≈ Aufbauzeichnung
≈ Zusammenbbauzeichnung
installation expenses TECH **Montageaufwand**
installation instructions 1 TECH **Aufstellanleitung**
installation instructions 2 TECH **Einbauanleitung**
= mounting instructions = Einbauanweisung
installation layout SYS.INST **Aufstellungsplan**
= layout; layout plan; equipment layout plan = AP
installation manager SYS.INST **Montageleiter**
installation manual SYS.INST **Montagehandbuch**
= Aufbauhandbuch
↑ Projektunterlagen
installation material TECH **Montagematerial**
= Installationsmaterial
installation overhead costs ECON **Montage-Betriebskosten**
= MBK
installation pitch (→mounting EQUIP.ENG (→Einbauteilung)
pitch)
installation planning SYS.INST **Aufbauplanung**
= Amtsbauplanung; Montageplanung
installation program (→setup DATA PROC (→Installationsprogramm)
program)
installation specification SYS.INST **Montagevorschrift**
= mounting instructions = Aufbauvorschrift; Montageanweisung
≈ assembly instructions ≈ Bauvorschrift; Zusammenbauvorschrift
installation staff SYS.INST **Montagepersonal**
installation toolbox TECH **Montagekoffer**
↑ toolbox ↑ Werkzeugkoffer
installation wire TECH **Installationsdraht**
installment (NAM) ECON **Abzahlungsrate**
= instalment (BRI) = Rate; Teilzahlung 2
installment payment ECON **Ratenzahlung**
= instalment payment; payment on deferred terms; payment on account = Abschlagzahlung; Teilzahlung 1
installment purchase ECON **Ratenkauf**
= hire-purchase = Abzahlungskauf
instalment (BRI) (→installment) ECON (→Abzahlungsrate)
instalment payment (→installment ECON (→Ratenzahlung)
payment)
instant- TECH **Sofort-**
= immediate
instant PHYS **Zeitpunkt**
= moment = Moment
↑ time ↑ Zeit
instant (→instantaneous) TECH (→sofortig)
instantaneous (adj.) TECH **sofortig**
= instant = instantan; augenblicklich
≈ temporary ≈ vorübergehend

instantaneous access DATA PROC (→Sofortzugriff)
 (→immediate access)
instantaneous measurand TELECONTR **Momentanwert**
instantaneous power NETW.TH **Augenblicksleistung**
instantaneous release EL.TECH **Schnellauslösung**
instantaneous value (→instant PHYS (→Augenblickswert)
 value)
instantaneous value converter INSTR **Momentanwertumsetzer**
instantaneous-value store CIRC.ENG **Momentanwertspeicher**
 ↓ sample-hold circuit ↓ Abtast-Halte-Glied
instant dialing (→abbreviated SWITCH (→Kurzwahl)
 dialing)
instant of failure QUAL **Ausfallzeitpunkt**
instant print DATA PROC **Instantandruck**
instant processing (→in-line DATA PROC (→Geradewohl-Verarbei-
 processing) tung)
instant value PHYS **Augenblickswert**
 = instantaneous value; momentary = Momentanwert
 value; actual value ↑ Zeitwert
 ↑ time value
in-station test (→station SYS.INST (→Stationsmessung)
 test)
INST/DEL key (→backspace TERM.&PER (→Rücksetztaste 2)
 key)
in steps (→stepwise) TECH (→schrittweise)
in step with (→synchronized) TECH (→synchronisiert)
in step with (→synchronous) TELEC (→synchron)
Institute of Electrical and EL.TECH (→IEEE)
 Electronic Engineers (→IEEE)
institution ECON **Anstalt**
 = Institution
institutional ad (→institutional ECON (→Repräsentativwerbung)
 advertising)
institutional advertising ECON **Repräsentativwerbung**
 = institutional ad = institutionelle Werbung
Institution of Electrical EL.TECH (→IEE)
 Engineers (→IEE)
in stock ECON **vorrätig**
 ≈ available; existent = verfügbar; vorhanden
instruct (v.t.) DATA PROC **anweisen**
instruction TECH **Anleitung**
 = Anweisung
instruction DATA PROC **Befehl**
 [elementary instruction of a program [elementarste Anweisungs-
 in machine code; consisting of ad- einheit eines Programms in
 dress and operator part] Maschinencode; besteht
 = command 1 ("loosely"); statement 2 aus Adreßteil und Opera-
 (mach.orient. languages); micro code tionsteil]
 ≈ microprogram; program step = elementarer Befehl; Ele-
 ↑ statement; program mentaroperation;
 ↓ machine instruction; one-address in- Mikrobefehl; Mikroinstruk-
 struction; multi-address instruction; tion; Mikrocode; Anwei-
 input instruction; output instruction; sung 2 (masch.orient. Spra-
 memory instruction; transfer instruc- chen); Instruktion;
 tion; logic instruction; address compu- Operation; Kommando 1
 tation instruction; jump instruction; ≈ Mikroprogramm; Pro-
 internal instruction; external instruc- grammschritt
 tion; command 2 ↑ Anweisung; Programm
 ↓ Maschinenbefehl;
 Ein-Adreß-Befehl; Mehr-
 Adreß-Befehl; Eingabebe-
 fehl; Ausgabebefehl; Speicherbefehl; Transferbefehl;
 logischer Befehl; Adreß-
 rechnungsbefehl; Sprungbefehl; interner Befehl; externer Befehl
instruction SCIE **Lehre**
 = tuition 1 = Unterricht
instruction address DATA PROC **Befehlsadresse**
 = address location
instruction address register DATA PROC (→Befehlszähler)
 (→instruction counter)
instructional dictating system OFFICE **Diktierlehranlage**
instructional phonotyping system OFFICE **Phonotypier-Lehranlage**
instructional television BROADC (→Bildungsfernsehen)
 (→educational television)
instructional TV (→educational BROADC (→Bildungsfernsehen)
 television)
instruction area DATA PROC **Befehlsbereich**

instruction block DATA PROC **Befehlsblock**
 = command block = Kommandoblock
instruction cache DATA PROC **Befehls-Cache-Speicher**
 = command cache = Kommando-Cache-Spei-
  cher
instruction catena DATA PROC (→Befehlskette)
 (→instruction chain)
instruction chain DATA PROC **Befehlskette**
 = command chain; instruction catena; = Kommandokette
 command catena
instruction chaining DATA PROC **Befehlsverkettung**
 = command chaining = Befehlsverkettung; Kom-
  mandoverkettung
instruction character DATA PROC **Befehlszeichen**
 = command character; instruction sig- = Befehlssignal; Kommando-
 nal; command signal zahl
instruction code (→operating DATA PROC (→Befehlsschlüssel)
 code)
instruction code (→operation DATA PROC (→Operationscode)
 code)
instruction counter DATA PROC **Befehlszähler**
 [register within the CPU, indicating [Register im Steuerwerk
 the address of the next instruction to der Zentraleinheit, proto-
 be executed] kolliert die Adresse des
 = address counter; program counter; in- nächsten auszuführenden
 struction address register; IAR; loca- Befehls]
 tion counter; command counter = Befehlszählregister; BZR;
  Befehlsadreßregister; Pro-
  grammzähler 1; Adreßzäh-
  ler; Kommandozähler
instruction cycle DATA PROC **Befehlszyklus**
 [time to fetch, interprete and execute [Dauer für Lesen, Interpre-
 an instruction] tieren und Ausführen ei-
 = operation cycle; command cycle; in- nes Befehls]
 struction phase = Operationszyklus; Kom-
 ≈ fetch cycle mandozyklus
  ≈ Abrufzyklus
instruction cycle time DATA PROC **Befehlszykluszeit**
 = operation cycle time; command cycle = Operationszykluszeit; Kom-
 time mandozykluszeit
 ↑ cycle time ↑ Zykluszeit
instruction decoder DATA PROC **Befehlsdecodierer**
 [part of control unit in CPU] [Teil des Steuerwerks in
 = command decoder der Zentraleinheit]
  = Kommandodecodierer; Be-
  fehlsentschlüssler; Kom-
  mandoentschlüssler
instruction decoding DATA PROC **Befehlsentschlüsselung**
 = command decoding = Befehlsdecodierung; Kom-
  mandodecodierung
instruction execution DATA PROC **Befehlsausführung**
 = execution 1 = Ausführung
instruction fetch DATA COMM **Befehlsabruf**
 = fetch instruction; fetch (n.) = Holphase; Abrufbefehl
instruction file DATA PROC **Befehlsdatei**
 = command file = Kommandodatei
instruction format DATA PROC **Befehlsformat**
 = command format = Kommandoformat
instruction length DATA PROC **Befehlslänge**
 = command length = Kommandolänge
instruction line DATA PROC **Befehlszeile**
 [sequential unit of a program] [Gliederungseinheit eines
 = program line; command line Programms]
  = Programmzeile; Komman-
  dozeile
instruction list DATA PROC **Befehlsliste**
 = statement list; command list = Kommandoliste
 ≈ statement sequence ≈ Anweisungsfolge
instruction lookahead DATA PROC (→Vorabbefehlsaufnahme)
 (→prefetching)
instruction loop DATA PROC **Befehlschleife**
instruction normalization DATA PROC **Befehlsnormalisierung**
 = command normalization; statement = Kommandonormalisierung
 normalization
instruction phase DATA PROC (→Befehlszyklus)
 (→instruction cycle)
instruction pipelining DATA PROC (→Fließband-Verarbeitung)
 (→pipeline processing)
instruction prefetch DATA PROC **Befehlsvorauslesen**
 = command prefetch = Kommandovorauslesen

instruction processing	DATA PROC	**Befehlsverarbeitung**	**instrument transformer**	COMPON	(→Meßwandler)
= command processing		= Kommandoverarbeitung	(→measuring transducer)		
instruction processing time	DATA PROC	**Befehlsverarbeitungszeit**	**insufficiency**	COLLOQ	**Unzulänglichkeit**
		= Kommandoverarbeitungszeit	≈ inadequacy		≈ Unangemessenheit
= command processing time			**insufficient**	COLLOQ	**unzulänglich**
instruction processor	DATA PROC	**Befehlsprozessor**	≈ inadequate		= unzureichend
= command processor		= Kommandoprozessor			≈ unangemessen
instruction register	DATA PROC	**Befehlsregister**	**insulant**	PHYS	**Isolierstoff**
[part of CPU, for intermediate storage of instructions]		[Teil des Steuerwerks der Zentraleinheit, zur Zwischenspeicherung von Befehlen]	= insulating material		
			insulate	PHYS	**isolieren**
			insulated	EL.TECH	**berührungssicher**
= IR; command register; order register; control register			**insulated copper wire**	COMM.CABLE	(→Kupferlackdraht)
			(→enameled copper wire)		
≈ control field; statement register		= Instruktionsregister; Kommandoregister; Steuerregister	**insulated gate field-effect transistor**	MICROEL	**IGFET**
					= Isolierschicht-Feldeffekttransistor
		≈ Kontrollfeld; Anweisungsregister	= IGFET		
			insulated terminal	INSTR	**Polklemme**
instruction reject	DATA PROC	**Befehlsrückweisung**	**insulated wire**	EL.TECH	**Isolierdraht**
= command reject		= Kommandorückweisung	**insulating bead**	COMPON	**Isolierperle**
instruction repertoire	DATA PROC	(→Befehlsvorrat)	≈ grommet		
(→instruction set)			**insulating bush**	COMPON	**Isolierbuchse**
instruction retry	DATA PROC	**Befehlswiederholung**			= Isoliernippel
= command retry		= Kommandowiederholung	**insulating bushing**	COMM.CABLE	(→Isolierhülse)
instruction sequence	DATA PROC	**Befehlsfolge**	(→insulating covering)		
= command string; command sequence		≈ Anweisungsfolge (masch.-or. Spr.)	**insulating clamp**	COMPON	**Isolierklemme**
			insulating cover	EQUIP.ENG	**Isolierabdeckung**
≈ statement sequence (mash.-or. lang)			**insulating covering**	COMM.CABLE	**Isolierhülse**
instruction set	DATA PROC	**Befehlsvorrat**	= insulating bushing		= Isolierhülle
= instruction repertoire; command set; command repertoire; repertoire		= Befehlssatz; Befehlsrepertoire; Kommandovorrat; Instruktionssatz	**insulating fault**	PHYS	**Isolationsfehler**
			insulating foil (→insulating sheet)	EQUIP.ENG	(→Isolierfolie)
instructions for use (→use instruction 1)	TECH	(→Bedienungsanleitung)	**insulating helix**	COMM.CABLE	**Isolierwendel**
			insulating housing	EQUIP.ENG	**Isoliergehäuse**
instruction signal	DATA PROC	(→Befehlszeichen)	**insulating layer**	MICROEL	**Isolierschicht**
(→instruction character)			**insulating material**	PHYS	(→Isolierstoff)
instructions section	DATA PROC	(→Anweisungsteil)	(→insulant)		
(→statements section)			**insulating paper**	EL.TECH	**Isolierpapier**
instruction staticizing	DATA PROC	**Befehlsübernahme**	**insulating plate**	TECH	**Isolierplatte**
= command staticizing; staticizing			**insulating sheet**	EQUIP.ENG	**Isolierfolie**
instruction structure	DATA PROC	**Befehlsaufbau**	= insulating foil		
instruction syntax	DATA PROC	**Befehlssyntax**	**insulating sheet**	COMM.CABLE	**Isolierschlauch**
= format 2			**insulating tape**	EL.INST	**Isolierband**
instruction telegram	TELECONTR	**Befehlstelegramm**	**insulating transformer**	COMPON	(→Trennübertrager)
instruction time	DATA PROC	**Befehlsausführungsdauer**	(→isolating transformer)		
instruction type	DATA PROC	**Befehlsart**	**insulating varnish**	TECH	**Isolierlack**
instruction wait list	DATA PROC	**Befehlswarteliste**	**insulating washer**	TECH	**Isolierscheibe**
= command wait list		= Kommandowarteliste	**insulating wax**	TECH	**Isolierwachs**
instruction word	DATA PROC	**Befehlswort**	**insulation**	PHYS	**Isolation**
= command word		= Kommandowort			= Isolierung
instructor	ECON	**Lehrer**	**insulation** (→isolation)	EL.TECH	(→Trennung)
		≈ Ausbilder	**insulation by oxidated porous silicon process** (→IPOS process)	MICROEL	(→IPOS-Verfahren)
instrument (n.)	INSTR	**Instrument**			
↓ measuring instrument		↓ Meßgerät	**insulation cut-back**	EL.TECH	**Abmantelung**
instrument (→document)	ECON	(→Urkunde)	[of a cable]		[eines Kabels]
instrument (→tool)	TECH	(→Werkzeug)	**insulation displacement connection** (→insulation piercing connection)	COMPON	(→Schneidklemmverbindung)
instrumental (adj.)	LING	**instrumental**			
instrumental input	DATA PROC	**maschinelle Eingabe**			
instrumental sentence	LING	**Instrumentalsatz**	**insulation piercing connection**	COMPON	**Schneidklemmverbindung**
↑ causal sentence		= Satz des Mittels oder Werkzeugs			
			= insulation displacement connection; IDC		
		↑ Kausalsatz			
instrumentation	TECH	**Instrumentierung**	**insulation resistance** (→leak resistance)	LINE TH	(→Isolationswiderstand)
instrumentation correction (→calibration factor)	INSTR	(→Kalibrierfaktor)			
			insulation test	INSTR	**Isolationsprüfung**
instrumentation meter	INSTR	**Einbauinstrument**	**insulation tester**	INSTR	**Isolationsmesser**
= panel meter		= Einbaumeßgerät; Panelmeter	↑ ohmmeter		= Isolationsprüfer
					↑ Ohmmeter
instrumentation tool (→measuring set)	INSTR	(→Meßplatz)	**insulation transformer** (→isolating transformer)	COMPON	(→Trennübertrager)
instrument calibration	INSTR	**Meßgeräteeichung**	**insulator**	PHYS	**Isolator**
instrument grade	QUAL	**Meßgeräte-Güteklasse**	= non-conductor		= Nichtleiter
instrument landing system	RADIO NAV	**Instrumentenlandesystem**	≠ conductor		≠ Leiter
= ILS; blind landing system		= ILS-System; ILS; Blindlandesystem	**insulator**	OUTS.PLANT	**Isolierkörper**
					[Freileitung]
instrument navigation	RADIO NAV	**Instrumentennavigation**			= Isolator
= blind navigation; blind flying		= Blindnavigation; Blindflug			↓ Abspannisolator
instrument panel	INSTR	**Instrumentenfeld**	**insulator pin**	OUTS.PLANT	**Isolatorstütze**
instrument range (→measurement range)	INSTR	(→Messbereich)	**insurance**	ECON	**Versicherung**
			= assurance		
Instrument Society of America (→ISA)	ELECTRON	(→ISA)	**insurance certificate**	ECON	**Versicherungszertifikat**

insurance policy	ECON	**Versicherungspolice**	
= policy		= Police; Versicherungsurkunde	
insurance premium	ECON	**Versicherungsprämie**	
= premium		= Versicherungsgebühr; Prämie	
insurance underwriter (→insurer)	ECON	(→Versicherer)	
insure	ECON	**versichern**	
= assure			
insured (n.)	ECON	**Versicherter**	
= assured		= Versicherungsnehmer	
≠ insurer		≠ Versicherer	
insured value	ECON	**Versicherungswert**	
= assured value			
insurer	ECON	**Versicherer**	
= assurer; assuror; insurance underwriter; underwriter		= Versicherungsträger	
		≠ Versicherter	
≠ insured			
in tabular form (→tabular)	DOC	(→tabellarisch)	
intact (→undamaged)	ECON	(→unversehrt)	
intaglio (→intaglio printing)	TYPOGR	(→Tiefdruck)	
intaglio printing	TYPOGR	**Tiefdruck**	
[printing parts are below the non printing parts]		[druckende Teile liegen tiefer als die nichtdruckenden]	
= intaglio		≠ Hochdruck	
≠ letterpress		↑ Druckverfahren	
↑ printing process		↓ Heliographie; Rotationstiefdruck	
↓ photogravure; rotogravure			
integer (n.)	MATH	**Ganzzahl**	
= integer number; whole number		= ganze Zahl; Integer	
integer (→integral)	MATH	(→ganzzahlig)	
integer BASIC	DATA PROC	**Integer-BASIC**	
integer mathematics	MATH	**ganzzahlige Arithmetik**	
integer multiple	MATH	**ganzzahliges Vielfache**	
integer number (→integer)	MATH	(→Ganzzahl)	
integer part	DATA PROC	**ganzzahliger Teil**	
integer symbol (→integral sign)	MATH	(→Integralzeichen)	
integer value	MATH	**ganzzahliger Wert**	
integer variable	DATA PROC	**Integralvariable**	
integral (n.)	MATH	**Integral 1**	
≠ derivative		≠ Ableitung	
↓ undefined integral; defined integral; line integral; surface integral; volume integral; time integral		↓ unbestimmtes Integral; bestimmtes Integral; Kurvenintegral; Doppelintegral; dreifaches Integral; Oberflächenintegral; Zeitintegral	
integral (adj.)	MATH	**ganzzahlig**	
= integer		= integral	
≠ fractional		≠ gebrochen	
integral action	CONTROL	**I-Verhalten**	
integral calculus	MATH	**Integralrechnung**	
↑ analysis; calculus		↑ Analysis; Infinitesimalrechnung	
integral communications control unit	DATA COMM	**integrierte Datenübertragungseinheit**	
= integral controller			
integral control	CONTROL	**Integralregelung**	
integral controller (→I controller)	CONTROL	(→I-Regler)	
integral controller (→integral communications control unit)	DATA COMM	(→integrierte Datenübertragungseinheit)	
integral cosine	MATH	**Integralkosinus**	
		= Integralcosinus	
integral equation	MATH	**Integralgleichung**	
integral form	MATH	**Integralform**	
integral messenger cable	COMM.CABLE	**Tragseil-Luftkabel**	
= figure-8 cable; catenary wire aerial cable		[mit eingebautem Tragseil]	
↑ self-supporting cable		↑ selbsttragendes Luftkabel; Tragseilkabel	
integral sign	MATH	**Integralzeichen**	
= integer symbol		= Integral 2	
↑ mathematical symbol		↑ mathematisches Zeichen	
integral sinus	MATH	**Integralsinus**	
integral solution (→complete solution)	DATA PROC	(→Komplettlösung)	
integral transform	MATH	**Integraltransformation**	
integrand	MATH	**Integrand**	
integrate	MATH	**integrieren**	
integrate	TECH	**integrieren**	
= incorporate		= einbinden; einfügen; zusammenfügen; inkoporieren; eingliedern	
≈ mount		≈ montieren	
integrated (→built-in)	TECH	(→eingebaut)	
integrated aerial (→active antenna)	ANT	(→aktive Antenne)	
integrated antenna (→active antenna)	ANT	(→aktive Antenne)	
integrated capacitor	MICROEL	**integrierter Kondensator**	
integrated cash register	TERM&PER	**Kassenverbundsystem**	
integrated circuit	MICROEL	**integrierte Schaltung**	
[complete circuits on a piece of silicon]		[komplette Schaltungen auf kleinem Siliziumplättchen]	
= IC (pl. ICs)		= integrierter Schaltkreis; integrierte Halbleiterschaltung; IC (pl. ICs); IS	
≈ chip		≈ Chip	
↓ monolithic IC; SSI; MSI; LSI; VLSI		↓ Monolithschaltung; Kleinintegration; Mittelintegration; Großintegration; Größtintegration	
integrated data processing	DATA PROC	**integrierte Datenverarbeitung**	
= IDP			
integrated digital network	DATA COMM	**integriertes digitales Netz**	
= integrated text and data network; IDN		= integriertes Text-und Datennetz; integriertes Digitalnetz; IDN	
integrated diode	MICROEL	**integrierte Diode**	
integrated junction capacitor	MICROEL	**intergrierter Sperrschichtkondensator**	
integrated measurand	TELECONTR	**Integralwert**	
integrated optical circuit	OPTOEL	**integrierte optische Schaltung**	
integrated optics	OPTOEL	**integrierte Optik**	
integrated package	DATA PROC	(→integriertes Softwarepaket)	
(→integrated software package)			
integrated radome antenna	ANT	**Radant-Antenne**	
integrated resistor	MICROEL	**integrierter Widerstand**	
integrated Schottky logic (→ISL)	MICROEL	(→ISL-Technik)	
integrated-services	TELEC	**dienstintegriert**	
integrated-services digital network	TELEC	**dienstintegriertes Digitalnetz**	
= ISDN		= ISDN; dienstintegriertes digitales Netz	
integrated software	DATA PROC	**integrierte Software**	
integrated software package	DATA PROC	**integriertes Softwarepaket**	
= integrated package		= integriertes Paket	
integrated text and data network (→integrated digital network)	DATA COMM	(→integriertes digitales Netz)	
integrating circuit (→integrating network)	CIRC.ENG	(→Integrierglied)	
integrating converter	INSTR	**integrierender Umsetzer**	
integrating device with time lag	CONTROL	**IT-Glied**	
integrating network	CIRC.ENG	**Integrierglied**	
[pulse shaping network]		[impulsformendes lineares Netzwerk]	
= integrating circuit; integrator		= Integrationsglied; Integrierschaltung; Integrationsschaltung; Integrator; Integrierer; integrierendes Netzwerk	
≈ adder		≈ Addierer	
≠ differentiating network		≠ Differenzierglied	
integrating network	DATA COMM	**Integrationsnetz**	
integration	TECH	**Integration**	
		= Integrierung; Einbindung; Einbinden; Zusammenfügung; Zusammenfügen	
integration	MATH	**Integration**	
integration	MICROEL	**Integration**	
= integration scale; level of integration		= Integrationsgrad	
↓ small-scale integration; medium-scale integration; large-scale integration; very-large-scale integration; ultra-large-scale integration		↓ Kleinstintegration; mittlerer Integrationsgrad; Großintegration; Größtintegration; Ultrahöchstintegration	

integration 296

integration (→system integration)	DATA PROC	(→Systemintegration)	
integration interval = region of integration	MATH	**Integrationsintervall**	
integration limit	MATH	**Integrationsgrenze**	
integration path	MATH	**Integrationsweg**	
integration scale (→integration)	MICROEL	(→Integration)	
integration surface	MATH	**Integrationsfläche**	
integration time	ELECTRON	**Integrierzeit**	
integrator (→integrating network)	CIRC.ENG	(→Integrierglied)	
integrity	TECH	**Integrität** = Unversehrtheit; Unverfälschtheit	
integrity [state of no corruption]	DATA PROC	**Integrität** = Unverfälschtheit	
integrration 2 (→software integration)	DATA PROC	(→Software-Integration)	
intelligent concentrator	DATA COMM	**intelligenter Konzentrator**	
intelligent keyboard [significance of keys varying with the type of task] = alternate keyboard; programmable keyboard	TERM&PER	**intelligente Tastatur** [die Bedeutung der Tasten variiert je nach Aufgabe] = alternative Tastatur; programmierbare Tastatur	
intelligent knowledge-based system (→expert system)	DATA PROC	(→Expertensystem)	
intelligent multiplexer (→cross-connect multiplexer)	TRANS	(→Verteilmultiplexer)	
intelligent network = IN	TELEC	**intelligentes Netz**	
intelligent programing language	DATA PROC	**intelligente Programmiersprache**	
intelligent sensor	COMPON	**intelligenter Sensor**	
intelligent spacer (→automatic hyphenation)	DATA PROC	(→automatische Silbentrennung)	
intelligent terminal [terminal which permits a programmable processing of introduced data before transmission, or of received data before output] = programmable terminal; smart terminal ≠ dumb terminal	TERM&PER	**programmierbare Datenstation** [Endgerät welches eine programmierbare Verarbeitung eingegebener Daten vor deren Übertragung, oder empfangener Daten vor deren Ausgabe erlaubt] = programmierbares Terminal; intelligente Datenstation; intelligentes Terminal; programmierbares Datensichtgerät ≠ unintelligente Datenstation	
intelligent video display terminal	TERM&PER	**anwenderprogrammierbares Datensichtgerät**	
intelligibility = articulation; intelligibleness	TELEC	**Verständlichkeit**	
intelligible ≠ unintelligible	TELEC	**verständlich** ≠ unverständlich	
intelligible crosstalk = uninverted crosstalk	TELEC	**verständliches Nebensprechen**	
intelligibleness (→intelligibility)	TELEC	(→Verständlichkeit)	
intenational alphabet no.5 [contains 7 information bits and one parity bit] = IA5; ISO 8-bit code	TELEGR	**internationales Alphabet Nr.5** [enthält 7 Informationsbits u. ein Paritätsbit] = IA5; ISO-8-Bit-Code	
intensifier	TECH	**Verstärker**	
intensifying electrode (→post-accelerating electrode)	ELECTRON	(→Nachbeschleunigungselektrode)	
intensity (→field strength)	PHYS	(→Feldstärke)	
intensity (→radiance)	PHYS	(→Strahlungsdichte 1)	
intensity dot	INSTR	**Intensitätspunkt**	
intensity level	PHYS	**Intensitätspegel**	
intensity modulation	OPT.COMM	**Intensitätsmodulation**	
intensive user ≠ occasional user	TECH	**Vielbenutzer** ≠ Wenigbenutzer	
interact (→interwork)	TECH	(→zusammenwirken)	
interaction = mutual action; reciprocal action	PHYS	**Wechselwirkung**	
interaction	TECH	**Zusammenspiel** = Zusammenwirken	
interaction (→intervention)	TECH	(→Eingriff)	
interaction gap	MICROW	**Koppelspalt**	
interaction impedance	MICROW	**Koppelimpedanz**	
interaction loss	EL.TECH	**Wechselwirkungs-Dämpfungsmaß**	
interaction space	MICROW	**Koppelraum**	
interactive = conversational	DATA PROC	**Dialog-** = Dialogverkehrs-; interaktiv; dialogorientiert; dialogfähig; konversational	
interactive communication = dialog communication; individual communication ≠ distributive communication	TELEC	**Dialogkommunikation** = Individualkommunikation ≠ Verteilkommunikation	
interactive communication [DATA COMM] (→interactive processing)	DATA PROC	(→Dialogbetrieb)	
interactive computer	DATA PROC	**Dialogrechner** = interaktiver Rechner; Dialogcomputer; interaktiver Computer	
interactive computer graphics	DATA PROC	**interaktive grafische Datenverarbeitung**	
interactive data processing (→interactive processing)	DATA PROC	(→Dialogbetrieb)	
interactive dialog system (→voice recognizing dialog system)	DATA PROC	(→sprachverstehendes Dialogsystem)	
interactive display terminal	TERM&PER	**dialogfähige Bildschirmstation**	
interactive enquiry (→interactive inquiry)	DATA PROC	(→Dialogabfrage)	
interactive graphics	DATA PROC	**Dialoggraphik** = interaktive Graphik; interaktive graphische Datenverarbeitung	
interactive information system (→conversational information system)	DATA PROC	(→passives Informationssystem)	
interactive inquiry = interactive enquiry; interactive query	DATA PROC	**Dialogabfrage** = interaktive Abfrage	
interactive keyboard	TERM&PER	**interaktive Tastatur**	
interactive mode (→interactive processing)	DATA PROC	(→Dialogbetrieb)	
interactive network ≠ distributive network	TELEC	**interaktives Netz** = Interaktionsnetz ≠ Verteilnetz	
interactive operation (→user interface)	DATA PROC	(→Benutzeroberfläche)	
interactive processing [question-answer dialog] = interactive data processing; interactive mode; conversational processing; conversational mode; transaction processing; transaction mode; interactive communication [DATA COMM]; conversational interaction [DATA COMM]; dialog processing; dialog data processing >; dialog mode; dialog ≠ batch processing	DATA PROC	**Dialogbetrieb** [Frage-Antwort-Spiel] = Dialogdatenverarbeitung; Dialogverarbeitung; Dialogverkehr [DATA COMM]; Dialog ≠ Stapelverarbeitung	
interactive program (NAM) (→dialog program)	DATA PROC	(→Dialogprogramm)	
interactive programme (BRI) (→dialog program)	DATA PROC	(→Dialogprogramm)	
interactive query (→interactive inquiry)	DATA PROC	(→Dialogabfrage)	
interactive routine (→dialog program)	DATA PROC	(→Dialogprogramm)	
interactive service ≠ distribution service	TELEC	**interaktiver Dienst** = Dialogdienst ≠ Verteildienst	
interactive system (→dialog system)	DATA PROC	(→Dialogsystem)	
interactive terminal = conversational terminal; dialog station; conversational device ↑ terminal station	DATA COMM	**Dialogstation** = Dialoggerät ↑ Datenstation	
interactive videography (→interactive videotex)	TELEC	(→Bildschirmtext)	

interactive videotex TELEC
[interactive system to call-up informations via telephone line, and display them on TV receiver]
= videotex; interactive videotex service; interactive videography; viewdata service
≈ teletext
↓ PRESTEL (Great Britain)

interactive videotex service TELEC
(→interactive videotex)
inter-base circuit (→common CIRC.ENG base connection)
inter-base circuit CIRC.ENG
(→grounded-base circuit)
inter-base connection CIRC.ENG
(→common base connection)
inter-base connection CIRC.ENG
(→grounded-base circuit)
interblock gap TERM&PER
[magnetic storage]
= IBG; record gap; block gap; inter-record gap; gap
inter-call pause SWITCH
intercapital network (→toll TELEC network)
intercapital route (→long-haul TRANS route)
intercapital system (→long-haul TRANS system)
intercarrier noise TV
intercarrier system TV
intercell hand-off MOB.COMM
↑ hand-off
intercell wiring MICROEL
intercept SWITCH
= lock-out
interceptability RADIO LOC
intercept probability RADIO LOC

intercept reach RADIO LOC
interchange TELEC

interchangeability TECH
interchangeable TECH
interchange circuit DATA COMM
(→interface circuit)
interchange point (→delivery TELEC point)
intercharacter spacing TYPOGR
(→character spacing)
intercity trunk (AM) (→toll TELEC trunk)
intercommunication system TELEPH
= intercom system
↓ simplex intercommunication system; duplex intercommunication system
intercompany ECON
intercompany business ECON
(→corporate business)
intercom system TELEPH
(→intercommunication system)
interconnect EL.TECH
interconnect EL.TECH

interconnect (n.) (→wiring) MICROEL
interconnect channel (→wiring MICROEL channel)
interconnecting EL.TECH
(→interconnection)

Bildschirmtext
[Zweiweg-Kommunikationssystem zum Abruf von Informationen über Fernsprechleitung und Wiedergabe an Fernsehempfänger]
= Btx; Videotex; Bildschirmtextdienst
≈ Teletext
↓ PRESTEL (Großbritannien)

(→Bildschirmtext)

(→Basisschaltung)
(→Zwischen-Basis-Schaltung)
(→Basisschaltung)
(→Zwischen-Basis-Schaltung)
Blockzwischenraum
[Magnetspeicher]
= Blocklücke; Satzlücke; Lücke; Kluft
≈ Start-Stop-Lücke
Wahlbereinigungspause
= Wahlberuhigungspause
(→Fernnetz)

(→Ferntrasse)

(→Weitverkehrssystem)

Intercarrierbrumm
Differenzträgerverfahren
= Intercarrierverfahren
Zellenwechsel
↑ Gesprächsübergabe
Interzellenverdrahtung
abfangen

Entdeckungsfähigkeit
= Erfassungsfähigkeit
Entdeckungswahrscheinlichkeit
= Erfassungswahrscheinlichkeit
Entdeckungsreichweite
= Erfassungsreichweite
austauschen
[Infomationen]
Austauschbarkeit 2
[untereinander]
untereinander austauschbar
(→Schnittstellenleitung)

(→Übergabepunkt)

(→Zeichenabstand)

(→Fernleitung)

Hausrufanlage
↓ Gegensprechanlage; Wechselsprechanlage

zwischenbetrieblich
(→Verbundgeschäft)

(→Hausrufanlage)

durchverbinden
zusammenschalten
= untereinander verbinden
(→Verdrahtung)
(→Verdrahtungskanal)

(→Zusammenschaltung)

interconnecting cable EQUIP.ENG
≈ connecting cable
interconnecting density DATA COMM
interconnection SWITCH
[trunk exchange]
interconnection POWER ENG
interconnection EL.TECH
= interconnecting; strapping; connection
≈ connection
interconnection (→wiring) MICROEL
interconnect layer (→wiring MICROEL layer)
interconnect level (→wiring MICROEL level)
intercontinental call TELEPH
↑ international call

interdependence SCIE

interdependent SCIE
interdevice (adj.) ELECTRON
interdialing SWITCH
= interdialling
interdialling SWITCH
(→interdialing)
interdict (→forbid) COLLOQ
interdigital coupler MICROW
interdigital filter MICROW
= strip line filter

interdigital interval (→interdigit SWITCH time)
interdigital magnetron MICROW
interdigital pause (→interdigit SWITCH time)
interdigital time (→interdigit SWITCH time)
interdigit interval (→interdigit SWITCH time)
interdigit pause (→interdigit SWITCH time)
interdigit time SWITCH
= interdigit interval; interdigital time; interdigital interval; interdigital pause; interdigital pause
interdisciplinary SCIE

interdot TV
interest ECON
interest (→shareholding) ECON
interest-free ECON
interest rate ECON

interest rate level ECON
interest return ECON

interexchange carrier (→trunk TELEC carrier)
interexchange carrier TRANS
(→transmission system)
interexchange circuit (→interofficeTELEC trunk)
interexchange line (→interoffice TELEC trunk)
interexchange link (→interoffice TELEC trunk)
interexchange trunk (→interoffice TELEC trunk)
interface (n.) TECH
= dividing surface; dividing plane; separating surface; separating plane
interface ELECTRON

interface adapter ELECTRON
= interface pod

Verbindungskabel
≈ Anschlußkabel
Verknüpfungsdichte
Durchgang
[Fernamt]
Netzkupplung
Zusammenschaltung
= Verbindung; Durchverbindung
≈ Anschluß
(→Verdrahtung)
(→Verdrahtungslage)

(→Verdrahtungsebene)

Überseegespräch
= interkontinentales Gespräch
↑ Auslandsgespräch
Interdependenz
= wechlelseitige Abhängigkeit
wechselseitig abhängig
zwischen Geräten
Zwischenwahl

(→Zwischenwahl)

(→verbieten)
Interdigitalkoppler
Interdigitalfilter
= Striplinefilter
≈ Kammfilter
(→Zeichenpause)

Doppelkamm-Magnetron
(→Zeichenpause)

(→Zeichenpause)

(→Zeichenpause)

(→Zeichenpause)

Zeichenpause
= Wählpause; Zwischenwahlzeit

bereichsüberschreitend
= interdisziplinär; überbereichlich
Zwischenpunkt
Zins
(→Kapitalbeteiligung)
unverzinslich
Zinssatz
= Zinsfuß
Zinsniveau
Verzinsung 2
[Gewinn aus Zinsen]
(→Fernnetzbetreiber)

(→Übertragungssystem)

(→Amtsverbindungsleitung)

(→Amtsverbindungsleitung)

(→Amtsverbindungsleitung)

(→Amtsverbindungsleitung)

Trennfläche
= Trennebene

Schnittstelle
= Übergabestelle; Nahtstelle; Interface
Schnittstellenanpassung
= Schnittstellenadapter

interface board (→interface module) EQUIP.ENG (→Schnittstellenbaugruppe)
interface cable EQUIP.ENG **Schnittstellenkabel**
interface card (→interface module) EQUIP.ENG (→Schnittstellenbaugruppe)
interface circuit DATA COMM **Schnittstellenleitung**
= interchange circuit
interface comparator DATA COMM **Schnittstellenvergleicher**
= Schnittstellenkomparator
interface condition ELECTRON **Schnittstellenbedingung**
interface control DATA COMM **Schnittstellensteuerung**
interface control line DATA COMM **Schnittstellensteuerleitung**
interface converter DATA COMM **Schnittstellenwandler**
= Schnittstellenumsetzer; Schnittstellenkonverter; Interface-Konverter
interface description DOC **Schnittstellenbeschreibung**
interface driver DATA COMM **Schnittstellentreiber**
interface equipment DATA COMM **Schnittstellengerät**
interface expander DATA COMM (→Schnittstellenverfielfacher)
(→interface multiplier)
interface module EQUIP.ENG **Schnittstellenbaugruppe**
= interface board; interface card
= Schnittstellenkarte; Anschlußbaugruppe; Anschlußkarte
interface multiplier DATA COMM **Schnittstellenverfielfacher**
= interface expander
= Schnittstellenmultiplikator
interface plan ELECTRON **Schnittstellenplan**
interface plug connector COMPON **Schnittstellen-Steckverbinder**
interface pod (→interface adapter) ELECTRON (→Schnittstellenanpassung)
interface procedure DATA COMM **Schnittstellenverfahren**
= Schnittstellenprozedur
interface processor DATA PROC (→Ein-/Ausgabe-Werk)
(→input/output controller)
interface protocol DATA COMM **Schnittstellenprotokoll**
interface requirements ELECTRON (→Schnittstellenspezifikation)
(→interface specification)
interface routine DATA PROC **Schnittstellenroutine**
interface simulator DATA COMM **Schnittstellensimulator**
interface specification ELECTRON **Schnittstellenspezifikation**
= interface requirements
= Schnittstellenbedingungen
interface standard ELECTRON **Schnittstellennorm**
interface test DATA PROC **Schnittstellentest**
interface unit EQUIP.ENG **Schnittstelleneinheit**
interfere EL.TECH **beeinflussen**
interfered RADIO **gestört**
= garbled
interference EL.TECH **Beeinflussung**
≈ noise [TELEC] [Störung]
≈ Geräusch [TELEC]
interference ENG.DRAW **Durchdringung**
≠ play (n.) ≠ Spielraum
interference RADIO PROP **Fremdstörung**
= Einstreuung
interference PHYS **Interferenz**
interference TELEC **Störung**
[unwanted impairment of wanted signal] [unerwünschte Beinträchtigung des Nutzsignals]
= disturbance = Signalstörung; Störeinfluß; Interferenz
≈ interfering signal ≈ Störsignal; Beeinlussung
↓ noise; harmonic distortion; intermodulation; crosstalk; selective interference ↓ Rauschen; Klirren; Intermodulation; Nebensprechen; Selektivstörung
interference (→jamming) MIL.COMM (→Störung)
interference analysis TELEC **Interferenzanalyse**
= Störanalyse
interference cancellation PHYS **Interferenzauslöschung**
interference canceller RADIO REL **Interferenzunterdrücker**
interference current EL.TECH (→Störstrom)
(→interfering current)
interference effect TELEC **Störwirkung**
= Interferenzwirkung
interference elimination ELECTRON (→Entstörung)
(→interference suppression)
interference factor TELEC **Störfaktor**
= Interferenzfaktor

interference fading RADIO PROP (→Interferenzschwund)
(→multipath fading)
interference field PHYS **Störfeld**
interference filter PHYS **Interferenzfilter**
interference fit ENG.DRAW **Preßpassung**
= force fit = Preßsitz
↑ class of fit ↑ Passungsklasse
interference-free TELEC **störungsfrei**
≈ error-free = ungestört
≈ fehlerfrei; entstört
interference fringe PHYS **Interferenzstreifen**
interference immunity TELEC **Störfestigkeit**
= disturbance immunity = Störsicherheit; Störungsunempfindlichkeit
≠ susceptibility
↓ noise immunity ≠ Störempfindlichkeit
↓ Rauschfestigkeit
interference insensitive TELEC **störungsunempfindlich**
= disturbance insensitive; perturbance insensitive
interference level TELEC **Störpegel**
= interfering level; disturbance level; noise floor ≈ Grundgeräusch
↓ Rauschpegel
≈ basic noise
↓ noise level
interference limiter ELECTRON (→Störspannungsbegrenzer)
(→transitient limiter)
interference margin TELEC (→Störabstand)
(→signal-to-interference ratio)
interference microscope PHYS **Interferenzmikroskop**
interference pattern PHYS **Interferenzmuster**
interference power TELEC **Störleistung**
↓ noise power ↓ Rauschleistung
interference pulse ELECTRON (→Störimpuls)
(→interfering pulse)
interference reduction factor RADIO **Interferenz-Unterdrückungsfaktor**
= IRF
interference rejection TELEC (→Störspannungsunterdrückung)
(→interference suppression)
interference sensibility TELEC **Störempfindlichkeit**
= susceptibility = Beeinflußbarkeit; Suszeptibilität
≈ immunity to interference
≠ interference immunity ≠ Störsicherheit
↓ noise sensibility ↓ Rauschempfindlichkeit
interference sensitive TELEC **störungsempfindlich**
interference signal ELECTRON (→Störsignal)
(→interfering signal)
interference source TELEC **Störquelle**
↓ noise source ↓ Rauschquelle
interference spectroscopy PHYS **Interferenzspektroskopie**
interference suppression ELECTRON **Entstörung**
= interference elimination; disturbance suppression; disturbance elimination; anti-interference = Störschutz
↓ Funkentstörung
interference suppression TELEC **Störspannungsunterdrückung**
= interference rejection; disturbing voltage suppression = Störunterdrückung; Störsignalunterdrückung
↓ noise suppression ↓ Rauschspannungsunterdrückung
interference suppression capacitor (→anti-interference capacitor) CIRC.ENG (→Entstörungskondensator)
interference suppression component COMPON **Funkenstörungsbauelement**
interference suppression filter CIRC.ENG **Entstörfilter**
= interference trap; noise filter; noise trap; noise killer = Entstörnetzwerk; Rauschfilter; Netzentstörfilter; Störschutzfilter; Entstörungsglied
interference suppressor RADIO (→Rauschsperre)
(→squelch circuit)
interference threshold CIRC.ENG **Störschwelle**
interference trap CIRC.ENG (→Entstörfilter)
(→interference suppression filter)
interference voltage (→interfering voltage) TELEC (→Störspannung)
interferer TELEC **Störer**
= disturber
interfering carrier MODUL **Störträger**
interfering current EL.TECH **Störstrom**
= disturbing current; parasitic current; interference current = Fremdstrom

interfering field strength RADIO **Störfeldstärke**
interfering frequency TELEC **Störfrequenz**
interfering hop RADIO PROP **Störfunkfeld**
interfering level (→interference TELEC (→Störpegel)
level)
interfering mode MICROW **Störwelle**
= unwanted mode = Störmode
interfering noise TELEC **Störgeräusch**
= disturbing noise
interfering pulse ELECTRON **Störimpuls**
= interference pulse; disturbing pulse; = impulsförmiges Störsignal;
noise peak; glich; transient disturb- Störspannungsspitze; Störspitze;
ance; perturbation pulse; hit; tran- Spannungsspitze;
sient 2 Geräuschspitze; Fehlerursache
≈ impulsive noise
≈ Impulsgeräusch; Aussetzer
interfering signal ELECTRON **Störsignal**
= interference signal; parasitic signal;
perturbation signal; drop-in (n.)
interfering signal TELEC **Störsignal**
= spurious signal; disturbing signal; ↓ Rauschsignal
drop in
↓ noise signal
interfering transmitter MIL.COMM (→Störsender)
(→jamming transmitter)
interfering voltage TELEC **Störspannung**
= interference voltage; disturbing volt- = Fremdspannung
age; foreign voltage; conducted inter- ↓ Rauschspannung
ference; conducted emission
↓ noise voltage
interfering voltage content TELEC **Fremdspannungsanteil**
= extraneous signal content; disturbing = Fremdsignalanteil; Störspannungsanteil
voltage content
interfering wave RADIO **Störwelle**
interferometer PHYS **Interferometer**
[Optik]
interferometer RADIO LOC **Interferometer**
interferometer antenna ANT **Interferometer-Antenne**
interferometer homing RADIO NAV **Interferometeranflug**
interferometry TECH **Interferometrie**
interframe prediction TV **Bild-zu-Bild-Prädiktion**
= Interframe-Prädiktion
inter-group (adj.) ECON **bereichsüberschreitend**
= bereichsübergreifend
inter-group sale (→corporate ECON (→Verbundgeschäft)
business)
inter-IC bus (→IC bus) MICROEL (→IC-Bus)
interim TECH **zwischenzeitlich**
= intermediate 2
interim bill ECON **Zwischenrechnung**
interim reply OFFICE **Zwischenbescheid**
interior hole surface TECH **Lochwandung**
interjection LING **Interjektion**
= ejaculatory word = Ausrufewort
interlace (v.t.) TECH **verflechten**
= lace ≈ verschachteln; verzahnen;
≈ interleave; gear; cross verschränken
interlaced antenna ANT **verschachtelte Antenne**
interlaced cubical quad antenna ANT **verschachtelte Cubical-quad-Antenne**
interlaced mode TERM&PER **Halbbildbetrieb**
interlaced mode (→interlacing) TV (→Zeilensprungverfahren)
interlaced scanning (→interlacing) TV (→Zeilensprungverfahren)
interlacement TECH **Verflechtung**
interlacing TV **Zeilensprungverfahren**
= interlaced scanning; line jump [Übertragung in zwei Halbbildern]
method; interlaced mode
= Zwischenzeilenverfahren
inter-lattice gap MICROEL **Zwischengitter-Haftstelle**
interleave TECH **verschachteln**
= nest = ineinanderschachteln;
≈ gear (v.t.); cross (v.t.); interlace schachteln
≈ verzahnen; verschränken;
verflechten
interleaved TECH **verschachtelt**
= nested = ineinandergeschachtelt; geschachtelt
= interlaced
≈ verflochten
interleaved operation RADIO REL (→kreuzpolare Nachbarkanalbelegung)
(→interleaved pattern)

interleaved pattern RADIO REL **kreuzpolare Nachbarkanalbelegung**
= interleaved operation
≠ cochannel pattern = kreuzpolare Belegung;
Kreuzpolarbelegung; kreuzpolarer Nachbarkanalbetrieb; Kreuzpolarbetrieb
≠ Gleichkanalbelegung
interleaved pattern RADIO **versetztes Raster**
interleaved pattern 2 (→offset RADIO REL (→Zwischenraster)
pattern)
interleaved winding EL.TECH **Scheibenwicklung**
[transformer] [Transformator]
= sandwich coil winding = Kammerwicklung
≠ layer winding ≠ Zylinderwicklung
interleave factor DATA PROC **Versetzungsfaktor**
[number of disk revolutions for com- [Anzahl der Plattenumdrehungen zum vollständigen
plete reading of a sector] Lesen eines Sektors]
= Interleave-Faktor
interleaving TECH **Verschachtelung**
= interlocking = Ineinanderschachtelung;
≈ indentation 2; crossing; interlacing Schachtelung; Verkämmung
≈ Verzahnung; Verschränkung; Verflechtung
interleaving (→nesting) DATA PROC (→Verschachtelung)
interline flicker TV **Zeilenflimmern**
inter-line leading (→lead) TYPOGR (→Durchschuß)
interlink (→concatenate) TECH (→verketten)
interlinked enterprises ECON **verflochtene Unternehmen**
= interlocked enterprises
interlinking TECH (→Verkettung)
(→concatenation)
interlock (n.) DATA PROC **Verriegelung**
= barring; lock; lockout 1 = Sperre
interlock CONTROL **Verriegelung**
[conditioning of processes] [Konditionierung von Vorgängen]
interlock (→inhibition) ELECTRON (→Sperrung)
interlock (→lock 2) TECH (→verriegeln)
interlock (→locking 2) TECH (→Verriegelung)
interlock code DATA COMM **Verknüpfungscode**
interlock diagram RADIO **Blockierungsplan**
interlocked enterprises ECON (→verflochtene Unternehmen)
(→interlinked enterprises)
interlocked network DATA PROC (→Mehrrechnersystem)
(→multi-computer system)
interlocked processing DATA PROC **verriegelte Verarbeitung**
interlocking (→interleaving) TECH (→Verschachtelung)
interlocking circuit CIRC.ENG **Verriegelungsschaltung**
interlude (n.) DATA PROC **Vorprogramm**
[small start routine] [kleines Startprogramm]
= preliminary housekeeping
intermediary ECON **Zwischenhändler**
= middleman; jobber [AM]; com- ≈ Wiederverkäufer
mission agent
≈ reseller
intermediary (→mediator) ECON (→Vermittler)
intermediate 1 TECH **dazwischenliegend**
intermediate 2 (→interim) TECH (→zwischenzeitlich)
intermediate access DATA PROC **Speicher mittlerer Zugriffsgeschwindigkeit**
memory
[with intermediate access time] = mittelschneller Speicher
= IAM; intermediate access storage; intermediate access store
intermediate access storage DATA PROC (→Speicher mittlerer Zugriffsgeschwindigkeit)
(→intermediate access memory)
intermediate access store DATA PROC (→Speicher mittlerer Zugriffsgeschwindigkeit)
(→intermediate access memory)
intermediate amplifier TRANS (→Zwischenverstärker)
(→intermediate repeater)
intermediate branch OUTS.PLANT **Zwischenabzug**
intermediate character DATA PROC **Zwischenzeichen**
intermediate code DATA PROC **Zwischencode**
intermediate derivation TRANS **Unterwegsabzweig**
= way-side derivation; intermediate = Unterwegsausstieg
drop-out
intermediate distribution SYS.INST **Zwischenverteiler**
frame
= IDF; trunk distribution frame; TDF

intermediate drop-out (→intermediate derivation)	TRANS	(→Unterwegsabzweig)	
intermediate echo suppressor	TELEPH	**Unterwegsechosperre**	
intermediate exchange = intermediate office	SWITCH	**Zwischenamt**	
intermediate file	DATA PROC	**Zwischendatei**	
intermediate frequency = IF	RADIO	**Zwischenfrequenz** = ZF	
intermediate frequency derivation (→IF derivation)	RADIO	(→ZF-Auskopplung)	
intermediate frequency filter (→IF filter)	RADIO	(→ZF-Filter)	
intermediate frequency interconnection (→IF interconnection)	RADIO	(→ZF-Durchschaltung)	
intermediate frequency transformer (→IF transformer)	RADIO	(→ZF-Übertrager)	
intermediate image = intrafield	TV	**Zwischenbild** = Innerbild	
intermediate language [stepwise program translation]	DATA PROC	**Zwischensprache** [bei stufenweiser Programmübersetzung]	
intermediate layer ≈ interface	TECH	**Zwischenschicht** ≈ Zwischenlage	
intermediate light	AERON	**Zwischenfeuer**	
intermediate memory [part of a sectioned main memory, where input/output operations and floating or fixed point operations are performed] = intermediate storage 2; intermediate store	DATA PROC	**Zwischenspeicher 1** [bei unterteiltem Hauptspeicher der Teil, in dem die Ein- und Ausgabeoperationen, sowie die Festpunkt- oder Gleitpunktoperationen durchgeführt werden] = Schnellspeicher 2	
intermediate node (→transit node)	DATA COMM	(→Durchgangsknoten)	
intermediate office (→intermediate exchange)	SWITCH	(→Zwischenamt)	
intermediate part	TECH	**Mittelteil**	
intermediate regenerator	TRANS	**Zwischenregenerator**	
intermediate register (→buffer register)	DATA PROC	(→Zwischenregister)	
intermediate repeater = repeater 2; intermediate amplifier ≈ intermediate regenerator ≠ terminal repeater ↑ line repeater	TRANS	**Zwischenverstärker** ≈ Zwischenregenerator ≠ Endverstärker ↑ Leitungsverstärker	
intermediate result = provisional result	TECH	**Zwischenergebnis** = provisorisches Ergebnis	
intermediate stage (→interstage)	TECH	(→Zwischenstufe)	
intermediate station = repeater station ↓ radio relay repeater station [RADIO REL]; line repeater station; regenerator station	TRANS	**Zwischenstelle** ↓ Relaisstelle [RADIO REL]; Leitungsverstärkerstelle; Regeneratorstelle	
intermediate storage = temporary storage	ECON	**Zwischenlagerung**	
intermediate storage 1	DATA PROC	**Zwischenspeicherung**	
intermediate storage 2 (→intermediate memory)	DATA PROC	(→Zwischenspeicher 1)	
intermediate store (→intermediate memory)	DATA PROC	(→Zwischenspeicher 1)	
intermediate subcarrier	MODUL	**Zwischenhilfsträger**	
intermediate switch (→cable switch)	COMPON	(→Zwischenschalter)	
intermediate synchronization	DATA COMM	**Nachsynchronisierung**	
intermediate system (→transit system)	DATA COMM	(→Transitsystem)	
intermediate total (→subtotal)	MATH	(→Zwischensumme)	
intermediate trade ≈ resale	ECON	**Zwischenhandel** ≈ Wiederverkauf	
intermeshed network = meshed network	TELEC	**Maschennetz** = vermaschtes Netz	
intermeshing = meshing ≈ networking	TELEC	**Vermaschung** ≈ Vernetzung	
intermeshing	NETW.TH	**Vermaschung**	
intermetallic	PHYS	**intermetallisch**	
intermingle (v.t.) (→intermix)	TECH	(→vermischen)	
intermingling (→intermixing)	TECH	(→Vermischung)	
intermittent = sporadic; erratic 1	TECH	**intermittierend** = aussetzend; sporadisch; unstetig	
intermittent contact (→loose contact)	ELECTRON	(→Wackelkontakt)	
intermittent DC flow	POWER SYS	**Lückbetrieb**	
intermittent duty (→intermittent operation)	TECH	(→intermittierender Betrieb)	
intermittent fault = sporadic fault	QUAL	**intermittierender Fehler** = sporadischer Fehler	
intermittent operation = intermittent duty	TECH	**intermittierender Betrieb** = Aussetzbetrieb	
intermix (v.t.) = intermingle (v.t.) ≈ mix (v.t.)	TECH	**vermischen** ≈ mischen	
intermixing = intermingling	TECH	**Vermischung** ≈ Mischung	
intermix printing	TYPOGR	**Gemischtdruck**	
intermodulation	MODUL	**Intermodulation** = Kombinationsschwingung	
intermodulation distortion	TRANS	**Intermodulationsverzerrung**	
intermodulation factor	MODUL	**Intermodulationsfaktor** = Differenztonfaktor	
intermodulation noise	TELEC	**Intermodulationsgeräusch** = Intermodulationsrauschen	
intermodulation noise measurement = noise-in-slot measurement	INSTR	**Rauschklirrmessung**	
intermodulation product	MODUL	**Intermodulationsprodukt**	
intermolecular	CHEM	**zwischenmolekular** = intermolekular	
internal	TECH	**hausintern**	
internal base point	MICROEL	**innerer Basispunkt**	
internal base resistance (→extrinsic base resistance)	MICROEL	(→Basisbahnwiderstand)	
internal blocking = matching loss	SWITCH	**interne Blockierung**	
internal cable (→indoor cable)	COMM.CABLE	(→Innenraumkabel)	
internal cabling	EQUIP.ENG	**Internverkabelung**	
internal character code	DATA PROC	**interner Zeichencode**	
internal clock	DATA PROC	**interne Uhr**	
internal connection = intraconnection	ELECTRON	**Innenverbindung**	
internal data processing [im Hause]	DATA PROC	**interne Datenverarbeitung**	
internal data representation	DATA PROC	**interne Datendarstellungsweise**	
internal emitter point	MICROEL	**innerer Emitterpunkt**	
internal financing	ECON	**Innenfinanzierung**	
internal font ≠ soft font	DATA PROC	**residenter Zeichensatz** = fester Zeichensatz ≠ Soft-Font	
internal format	DATA PROC	**internes Format**	
internal interface	TELEC	**Internschnittstelle** = interne Schnittstelle	
internal language	DATA PROC	**interne Sprache**	
internal member	ENG.DRAW	**Innenteil**	
internal memory (→main memory 1)	DATA PROC	(→Hauptspeicher 1)	
internal reporting	ECON	**internes Berichtswesen**	
internal resistance (→intrinsic resistance)	NETW.TH	(→Innenwiderstand)	
internal side (→inside)	TECH	(→Innenseite)	
internal sort [using main memory only]	DATA PROC	**Internsortierung** [im Hauptspeicher ablaufend]	
internal storage (→main memory 1)	DATA PROC	(→Hauptspeicher 1)	
internal store (→main memory 1)	DATA PROC	(→Hauptspeicher 1)	
internal thermal resistance [between junction and case]	MICROEL	**innerer Wärmewiderstand** [zwischen Sperrschicht und Gehäuse] = Wärmeinnenwiderstand	
internal thread	MECH	**Innengewinde**	
internal traffic	SWITCH	**Internverkehr** = interner Verkehr	

international ECON
= transnational
≈ cross-border
international access code SWITCH
(→international code)
international call TELEPH
↓ intercontinental call
international candle PHYS
[unit for luminous intensity; = 1,019 cd]
international code SWITCH
[regional identity code + country code, e.g. 44 for Great Britain and Northern Ireland; not appliable when dialing North American subscribers from outside]
= international access code
≈ country prefix
international economy (→world ECON economy)
International Electrotechnical EL.TECH **Commission** (→IEC)
International Federation of ELECTRON **Automatic Control** (→IFAC)
international gateway SWITCH **exchange**
↑ gateway exchange

internationaL market (→world ECON market)
International Monetary Fund ECON
= IMF
international number SWITCH
[national number + international code]
International Organization for TECH **Standardization** (→ISO)
international prefix SWITCH
[digits to access to international gateway exchange]
international reply coupon RADIO
[amateur radio]
= IRC
International Telecommunication TELEC **Union** (→ITU)
international telegraph alphabet TELEGR **no.2**
[a 5-bit code with start and stop element; the stop element length of Baudot code (2) used in USA is 1.42 instead of 1.5]
= ITA code no.2; ITA no.2; international telegraph code no.2; Baudot code no.2
≈ Baudot code (1)
↑ telegraph code
international telegraph code no.2 TELEGR
(→international telegraph alphabet no.2)
international trade (→world ECON trade)
international traffic SWITCH

internetworking (→networking) TELEC
internetworking processor DATA COMM
↓ gateway; bridge processor
interoffice circuit (→interoffice TELEC trunk)
interoffice line (→interoffice TELEC trunk)
interoffice link (→interoffice TELEC trunk)

international
= zwischenstaatlich; transnational
≈ grenzüberschreitend
(→internationale Kennzahl)

Auslandsgespräch
↓ Überseegespräch
internationale Kerze
[Maßeinheit für Lichtstärke; = 1,019 cd]
internationale Kennzahl
[Weltnummerierungszone + Landeskennzahl, z.B. 49 für BRD]
≈ Landesvorwahl

(→Weltwirtschaft)

(→IEC)

(→IFAC)

Auslands-Kopfvermittlungsstelle
= internationales Fernamt; Auslandsvermittlungsstelle
↑ Kopfvermittlungsstelle

(→Weltmarkt)

Weltwährungsfond
= Internationaler Währungsfond
internationale Rufnummer
[nationale Rufnummer + internationale Kennzahl]
= internationale Nummer
(→ISO)

internationale Verkehrsausscheidungszahl
[im Netz der DBP: OO]
= Auslandsvorwählnummer
internationales Antwortschein
[Amateurfunk]

(→UIT)

internationales Telegraphenalphabet Nr.2
[ein Fünf-Schritt-Code mit Anlauf – u. Sperrschritt; der in USA gebräuchliche Baudot-Code (2) hat Sperrschrittlänge 1,42 statt 1,5]
= internationales Telegrafenalfabet Nr.2; ITA Nr.2; CCITT Nr.2; Baudot-Code (2)
≈ Baudot-Code (1)
↑ Fernschreibcode
(→internationales Telegraphenalphabet Nr.2)

(→Welthandel)

grenzüberschreitender Verkehr
= internationaler Verkehr
(→Vernetzung)
Verbindungsrechner
↓ Überleiteinrichtung; Bridge
(→Amtsverbindungsleitung)

(→Amtsverbindungsleitung)

(→Amtsverbindungsleitung)

interoffice trunk TELEC
[line between exchanges]
= interexchange trunk; interswitch trunk; trunk 2; interoffice line; interexchange line; interswitch line; junction line; exchange line; trunk line 2; interoffice link; interexchange link; interswitch link; interoffice circuit; interexchange circuit; interswitch circuit
↓ local trunk
interoperability TECH
(→compatibility)
interoperate (→interwork) TECH
inter-paragraph leading TYPOGR
interphase PHYS
interphase EL.TECH
interphase transformer POWER ENG
interphone TELEPH

interpolate MATH
≠ extrapolate
interpolation MATH
≠ extrapolation
interpolation stage INSTR
interpolator CIRC.ENG
interpole POWER SYS
interpret DATA PROC
= recognize
interpretation DATA PROC
↑ program translation
interpretative programm DATA PROC
(→interpreter 2)
interprete LING
[translate in oral form]
↑ translate
interpreter LING
↑ translator
interpreter 2 DATA PROC
[adapts a basic program for a similar special task]
= interpretative programm
≈ programm generator

interpreter 1 DATA PROC
[translates source language into machine code statement by statement executing them immediately]
= language interpreter
≠ compiler

interpreter language DATA PROC
[language subject to translation into machine code by interpreter]
= interpreter 1
≠ compiler language
↑ high-level programming language

interpretive programing DATA PROC

interpretive routine DATA PROC
inter-record gap (→interblock TERM&PER gap)
interregister signal (→register SWITCH signal)
interrelated SCIE

interrelation SCIE

interrogating (→inquiry) DATA COMM
interrogation (→inquiry) DATA PROC
interrogation (→inquiry) TELECONTR
interrogation (→inquiry) SWITCH
interrogation (→inquiry) DATA COMM

Amtsverbindungsleitung
[Verbindung zwischen Vermittlungen]
= Verbindungsleitung; AVL; Zwischenamtsleitung
↓ Ortsverbindungsleitung

(→Kompatibilität)

(→zusammenwirken)
Absatzabstand
Phasengrenzschicht
Zwischenphase
Saugdrossel
Haustelefon
= Haustelephon; Hausfernsprechapparat
interpolieren
≠ extrapolieren
Interpolation
≠ Extrapolation
Interpolationsstufe
Interpolator
Wendepol
interpretieren

Interpretierung
↑ Programmübersetzung
(→Interpretierer 2)

dolmetschen
[mündlich übersetzen]
↑ übersetzen
Dolmetscher
↑ Übersetzer
Interpretierer 2
[adaptiert ein Grundprogramm für eine ähnlich gelagerte spezielle Aufgabe]
= Interpretierprogramm; Interpreter 2; interpretierendes Programm
≈ Programmgenerator
Interpretierer 1
[übersetzt ein Quellprogramm befehlsweise in Maschinensprache und führt jeden Befehl sofort aus]
= Interpreter
≠ Kompilierer

Interpretersprache
[Programmiersprache die per Iterpretierer in Maschinensprache übersetzt werden muß]
= Interpretierersprache; Sprachinterpretierer
≠ Kompilersprache
↑ problemorientierte Programmiersprache
interpretierendes Programmieren
Interpretierroutine
(→Blockzwischenraum)

(→Registerzeichen)

interreliert
= wechselbezogen; gegenseitig bezogen
Interrelation
= Wechselbeziehung
(→Stationsaufforderung)
(→Abfrage)
(→Abfrage)
(→Abfragen)
(→Stationsaufforderung)

interrogation cycle   TELECONTR   **Abfragezyklus**
interrogation key   TERM&PER   **Abfragetaste**
= enquiry key; inquiry key; answering key
interrogation language   DATA PROC   (→Abfragesprache)
(→query language)
interrogation path   RADIO LOC   **Abfrageweg**
interrogation point (→question   LING   (→Fragezeichen)
mark)
interrogation signal   TELECONTR   **Abfragesignal**
= query signal; inquiry signal; enquiry signal; polling signal   = Abrufsignal; Abfragebefehl
interrogation terminal   TERM&PER   **Abfrageterminal**
= enquiry terminal; inquiry terminal
interrogative adverb   LING   **Interrogativadverb**
[e.g. where, when]   [z.B. wo, wann]
= Frageumstandswort
interrogative pronoun   LING   **Interrogativpronomen**
[e.g. who, what]   [z.B. wer, was]
= question word   = Fragefürwort; Fragewort
interrogator   RADIO NAV   **Abfragesender**
= Interrogator; Abfragegerät
interrupt (v.t.)   EL.TECH   **unterbrechen**
= discontinue; break (v.t.); defeat; shut-off (v.t.)   ≈ ausschalten
≈ disconnect
interrupt (n.)   EL.TECH   (→Unterbrechung)
(→interruption)
interrupt 1(n.) (→program   DATA PROC   (→Programmunterbrechung)
interrupt)
interrupt 2 (n.) (→interrupt   DATA PROC   (→Unterbrechungskennzeichen)
signal)
interrupt acknowledge   DATA COMM   **Unterbrechungs-Quittierungssignal**
interrupt bus   DATA PROC   **Unterbrechungsbus**
= daisy chain interrupt   = Interrupt-Bus
interrupt control   DATA COMM   **Unterbrechungssteuerung**
= interruption control   = Interrupt-Steuerung
interrupt disable   DATA PROC   **Unterbrechungsdesaktivierung**
interrupt-driven   DATA PROC   **Interrupt-intensiv**
[making extensive use of interrupts]
interrupt enable   DATA PROC   **Unterbrechungsaktivierung**
interrupter (→break   COMPON   (→Ruhekontakt)
contact)
interrupt handling   DATA PROC   **Unterbrechungsbehandlung**
= interrupt servicing   = Interrupt-Behandlung
interruption   EL.TECH   **Unterbrechung**
= interrupt (n.); disconnection; blackout; cut-out; break (n.)   = Abschaltung; Unterbruch (SWZ)
interruption (→loss of   TELEC   (→Dienstunterbrechung)
service)
interruption (→program   DATA PROC   (→Programmunterbrechung)
interrupt)
interruption (→interrupt   DATA PROC   (→Unterbrechungskennzeichen)
signal)
interruption control   DATA COMM   (→Unterbrechungssteuerung)
(→interrupt control)
interruption-free power supply POWER SYS   (→unterbrechungsfreie Stromversorgung)
(→uninterruptable power supply)
interruption line (→interrupt   DATA PROC   (→Unterbrechungsleitung)
line)
interruption mask   DATA PROC   (→Unterbrechungsmaske)
(→interrupt mask)
interruption rule   DATA COMM   (→Unterbrechungsprozedur)
(→interrupt procedure)
interruption state   DATA PROC   (→Unterbrechungszustand)
(→interrupt state)
interruption technique   DATA PROC   (→Unterbrechungstechnik)
(→interrupt technique)
interruption time (→off time)   TECH   (→Unterbrechungsdauer)
interruption vector   DATA PROC   (→Unterbrechungsvektor)
(→interrupt vector)
interrupt level   SWITCH   **Dringlichkeit**
interrupt level (→interrupt   DATA PROC   (→Unterbrechungspriorität)
priority)
interrupt line   DATA PROC   **Unterbrechungsleitung**
= interruption line   = Interrupt-Leitung
interrupt mask   DATA PROC   **Unterbrechungsmaske**
= interruption mask   = Interrupt-Maske

interrupt mask register   DATA PROC   **Unterbrechungsmaskenregister**
[register which permits hinder or execute programmed interruptions]   [Register mit dem man programmgesteuerte Unterbrechungen unterbinden oder zulassen kann]
= Interrupt-Masken-Register
interrupt package   DATA PROC   (→Unterbrechungsregister)
(→interrupt register)
interrupt priority   DATA PROC   **Unterbrechungspriorität**
= interrupt level   = Interrupt-Priorität
interrupt procedure   DATA COMM   **Unterbrechungsprozedur**
= interruption rule   = Unterbrechungsregel; Interrupt-Prozedur; Interrupt-Regel
interrupt program   DATA PROC   **Unterbrechungsprogramm**
= interrupt routine   = Unterbrechungsroutine; Interrupt-Programm; Interrupt-Routine
interrupt register   DATA PROC   **Unterbrechungsregister**
[register to store interrupt requests]   [Register zur Speicherung von Unterbrechungsanforderungen]
= interrupt package   = Interrupt-Register
interrupt request   DATA PROC   **Unterbrechungsanforderung**
= IRQ   = Interrupt-Anforderung; Unterbrechungsaufforderung; Interrupt-Aufforderung
interrupt routine (→interrupt   DATA PROC   (→Unterbrechungsprogramm)
program)
interrupt servicing (→interrupt   DATA PROC   (→Unterbrechungsbehandlung)
handling)
interrupt signal   DATA PROC   **Unterbrechungskennzeichen**
= interrupt 2 (n.); interruption   = Unterbrechungssignal; Interrupt-Kennzeichen; Interrupt-Signal
interrupt stacking   DATA PROC   **Unterbrechungsstapelung**
interrupt state   DATA PROC   **Unterbrechungszustand**
= interruption state   = Interrupt-Zustand
interrupt suppression   DATA PROC   (→Unterbrechungsaufhebung)
(→lockout 2)
interrupt technique   DATA PROC   **Unterbrechungstechnik**
= interruption technique   = Interrupt-Technik
interrupt vector   DATA PROC   **Unterbrechungsvektor**
= interruption vector   = Interrupt-Vektor
intersatellite service   SAT.COMM   **Intersatellitenfunkdienst**
= ISL
intersect (v.t.)   MATH   **schneiden**
[e.g. a curve]   [z.B. eine Kurve]
= durchkreuzen
intersect (→cross 1)   TECH   (→kreuzen)
intersection   CIV.ENG   **Straßenkreuzung**
= Kreuzung
intersection   TECH   **Überschneidung**
≈ overlapping   ≈ Überlappung
intersection (→AND   ENG.LOG   (→UND-Verknüpfung)
operation)
intersectional bearing   RADIO LOC   **Schnittpeilung**
intersection line   MATH   **Schnittlinie**
intersection point   MATH   **Schnittpunkt**
inter-series adapter   COMPON   **Adapter 2**
= between-series adapter; cross-series adapter   [stellt den Übergang zwischen unterschiedlichen Steckerfamilien her]
↑ adapter 1   ↑ Übergangsstecker
inter-shelf cabling   SYS.INST   **Gestellreihenverkabelung**
interspace (→interstice)   TECH   (→Zwischenraum)
interstage   TECH   **Zwischenstufe**
= intermediate stage   = Zwischenzustand
interstage circuit (→interstage   CIRC.ENG   (→Stufenkopplungsnetzwerk)
network)
interstage network   CIRC.ENG   **Stufenkopplungsnetzwerk**
= interstage circuit
interstice   TECH   **Zwischenraum**
= interspace; space interval   = Zwickel
≈ separation; gap   ≈ Abstand; Spalt
interstice (→lattice   PHYS   (→Gitterzwischenplatz)
interstitial)
intersticial   COMM.CABLE   **Zwickel-**

intersticial atom		PHYS	Zwischengitteratom	intra-shelf wiring		EQUIP.ENG	Einsatzverdrahtung

intersticial atom	PHYS	Zwischengitteratom
interstitial (n.)	TECH	Zwischenplatz
interstitial impurity	PHYS	Zwischengitterverunreinigung
interswitch circuit (→interoffice trunk)	TELEC	(→Amtsverbindungsleitung)
interswitch line (→interoffice trunk)	TELEC	(→Amtsverbindungsleitung)
interswitch link (→interoffice trunk)	TELEC	(→Amtsverbindungsleitung)
interswitch trunk (→interoffice trunk)	TELEC	(→Amtsverbindungsleitung)
inter-symbol interference	CODING	Intersymbolstörung
= ISI		= Intersymbolinterferenz; Impulsnebensprechen; Zwischenzeicheninterferenz
intertrain pause (→pulse separation)	ELECTRON	(→Impulspause)
interturn (v.i.)	TECH	ineinanderdrehen
interval	PHYS	Intervall
↓ time interval		= Abschnitt
		↓ Zeitintervall
interval (→scale)	INSTR	(→Skala)
interval estimation	MATH	Intervallschätzung
interval signal	BROADC	Pausenzeichen
interval timer	DATA PROC	Intervallzeitgeber
		= Intervallzeitregister; Intervalltaktgeber
intervention	TECH	Eingriff
= interaction; adjustment		= Eingreifen
interview [application]	ECON	Vorstellung [Bewerbung]
interwinding capacity (→winding capacitance)	COMPON	(→Wicklungskapazität)
interword spacing [word processing]	TYPOGR	Wortzwischenraum [Textbearbeitung]
= space between words		= Worttrenner
interwork	TECH	zusammenwirken
= interoperate; interact; cooperate		= zusammenarbeiten; kooperieren; interagieren
interworking unit	DATA COMM	Netzanpassungsgerät
≈ gateway		≈ Überleiteinrichtung
interzone call (AM)	TELEPH	Nahgespräch
[call between zones of a large metropolitan area]		≈ Ortsgespräch
= toll call 2 (BRI)		≠ Ferngespräch
≠ toll call 1 (AM)		
in the works (AM) (→in preparation)	COLLOQ	(→in Vorbereitung)
intilialization mode	DATA COMM	Vorbereitungsbetrieb
= preparation mode		= Initialisierungsbetrieb
intonation	LING	Intonation
		= Stimmführung
intraband telegraph channel	TELEC	Einlagerungskanal
		= Zwischenlagerungskanal
intraband telegraphy	TELEGR	Einlagerungstelegrafie
intracell hand-off	MOB.COMM	Kanalwechsel
[to another radio channel within the same cell]		[innerhalb einer Funkzelle]
= hand-off 2		↑ Gesprächsübergabe
intracell wiring	MICROEL	Intrazellenverdrahtung
intracity trunk (AM) (→local trunk)	TELEC	(→Ortsverbindungsleitung)
intraconnection (→internal connection)	ELECTRON	(→Innenverbindung)
intra-exchange circuit	SWITCH	Internsatz
↑ junctor		= interner Satz
		↑ Verbindungssatz
intra-exchange traffic	SWITCH	amtsinterner Verkehr
= own-exchange traffic		
intrafield (→intermediate image)	TV	(→Zwischenbild)
intrafield prediction	TV	Zwischenbild-Prädiktion
		= Innerbild-Prädiktion; Intrafield-Prädiktion
intramolecular	CHEM	innermolekular
		= intramolekular
in transit (→on the way)	COLLOQ	(→unterwegs)
intransitive (adj.)	LING	intransitiv
		= nichtzielend
intra-series adapter (→in-series adapter)	COMPON	(→Kupplung 1)
intra-shelf wiring	EQUIP.ENG	Einsatzverdrahtung
= shelf wiring; inset wiring; unit wiring		
intra-system	TECH	(→systemeigen)
(→system-inherent)		
in-tray	OFFICE	Eingangskorb
intricacy (→complexity)	TECH	(→Komplexität)
intricate (→complicated)	COLLOQ	(→kompliziert)
intrinsic (adj.)	MICROEL	eigenleitend
= self-conducting; i-type		= i-leitend
intrinsic (→inherent)	TECH	(→Eigen-)
intrinsically safe telephone (→mine telephone)	TERM&PER	(→Grubentelefon)
intrinsic attenuation	EL.TECH	Eigendämpfung
intrinsic conduction	PHYS	Eigenleitung
intrinsic conduction density	PHYS	Eigenleitungsdichte
intrinsic conductivity [without foreign atoms]	PHYS	Eigenleitfähigkeit [ohne Fremdatome]
		= Intrinsic-Leitfähigkeit
intrinsic conductivity range	PHYS	Eigenleitungsbereich
intrinsic conductor	PHYS	Eigenleiter
intrinsic consumption	ELECTRON	Eigenverbrauch
intrinsic energy	PHYS	Eigenenergie
intrinsic error	DATA COMM	Ursprungsfehler
intrinsic error	INSTR	Eigenfehler
intrinsic free space impedance	RADIO PROP	Freiraumimpedanz
intrinsic frequency	MECH	Schwingfrequenz
≈ vibration frequency		= Eigenfrequenz
		≈ Schwingungsfrequenz
intrinsic frequency (→natural frequency)	PHYS	(→Eigenfrequenz)
intrinsic harmonic distortion	ELECTRON	Eigenklirren
intrinsic impedance (→free-space impedance)	RADIO PROP	(→Feldwellenwiderstand)
intrinsic layer	MICROEL	Intrinsic-Schicht
= intrinsic region; intrinsic zone; i layer; i region; i zone		= i-Schicht; i-Zone; Eigenleitfähigkeitsschicht
intrinsic mobility	PHYS	Eigenbeweglichkeit
intrinsic mode (→natural oscillation)	PHYS	(→Eigenschwingung)
intrinsic noise (→basic noise)	TELEC	(→Grundgeräusch)
intrinsic number	MICROEL	Intrinsic-Zahl
= inversion density		= Inversionsdichte; Inversionskonzentration
intrinsic oscillation (→natural oscillation)	PHYS	(→Eigenschwingung)
intrinsic photoelectric effect	PHYS	innerer Photoeffekt
		= innerer Fotoeffekt; Photo-Volumeneffekt
intrinsic radiation	PHYS	Eigenstrahlung
= normal radiation		
intrinsic region (→intrinsic layer)	MICROEL	(→Intrinsic-Schicht)
intrinsic resistance	NETW.TH	Innenwiderstand
= internal resistance; source impedance		= Quellenwiderstand
≠ load resistance		≠ Lastwiderstand
intrinsic resonance	PHYS	Eigenresonanz
= natural resonance		
intrinsic semiconductor	MICROEL	Eigenhalbleiter
[intrinsic conductivity much higher than extrinsic conductivity]		[so rein, daß die Störstellenleitfähigkeit wesentlich kleiner als die Eigenleifähigkeit]
		= Intrinsic-Halbleiter; eigenleitender Halbleiter
intrinsic transistor	MICROEL	innerer Transistor
		= Intrinsic-Transistor
intrinsic vibration (→natural oscillation)	PHYS	(→Eigenschwingung)
intrinsic zone (→intrinsic layer)	MICROEL	(→Intrinsic-Schicht)
introduce (→input)	DATA PROC	(→eingeben)
introducer	DATA COMM	Einleitungszeichen
introduction	LING	Einleitung
≈ rationale		≈ Begründung
introductory clause	LING	Satzauftakt
introductory text (→rationale)	LING	(→Begründung)

intrude

intrude (→override)	TELEPH	(→aufschalten)	
intruder alarm system (→intrusion detection system)	SIGN.ENG	(→Einbruchmeldeanlage)	
intruder detection system (→intrusion detection system)	SIGN.ENG	(→Einbruchmeldeanlage)	
intrusion	SIGN.ENG	**Intrusion** [unbefugter Eintritt]	
intrusion (→override)	TELEPH	(→Aufschalten)	
intrusion alarm (→burglar alarm)	SIGN.ENG	(→Einbruchalarm)	
intrusion alarm system (→intrusion detection system)	SIGN.ENG	(→Einbruchmeldeanlage)	
intrusion detection	SIGN.ENG	**Einbruchsicherung** = Intrusionssicherung	
intrusion detection system	SIGN.ENG	**Einbruchmeldeanlage** = Einbruchmeldesystem; Intrusionsmeldeanlage ↑ Gefahrenmeldeanlage	
= intrusion detection system; intruder detection system; intruder alarm system; burglar alarm system ↑ danger detection system			
intrusion guard	TELEPH	**Innenverbindung sperren**	
invalid (adj.) = void	ECON	**ungültig**	
invalid (→cancel)	DATA COMM	(→ungültig)	
invalidation character (→erasure character)	DATA PROC	(→Irrungszeichen)	
invalid character (→cancel character)	TERM&PER	(→ungültiges Zeichen)	
invalid character (→ignore character)	DATA COMM	(→Ungültigkeitszeichen)	
invariable (adj.) = constant ≈ continuous; time-invariant ≠ variable	TECH	**unveränderlich** = konstant; gleichbleibend ≈ stetig; zeitinvariant ≠ veränderlich	
invariance	SCIE	**Invarianz** = Unveränderlichkeit	
invariant (n.) ≈ constant	MATH	**Invariante** = Konstante	
invariant (adj.)	SCIE	**invariant** = unveränderlich	
invent (v.i.)	TECH	**erfinden**	
invention	TECH	**Erfindung**	
inventor	TECH	**Erfinder**	
inventory ↓ stocktaking	ECON	**Inventur** = Bestandsaufnahme ↓ Lagerbestandsaufnahme	
inventory (→stock)	ECON	(→Vorrat)	
inventory data [data on actual inventory] ≈ master data	DATA PROC	**Bestandsdaten** [Daten über augenblicklichen Bestand] ≈ Stammdaten	
inventory file	DATA PROC	**Bestandsdatei**	
inventory management	ECON	**Bestandsverwaltung**	
inventory record keeping (→accounting management)	TELEC	(→Bestandsverwaltung)	
inventory tape	DATA PROC	**Bestandsband**	
inverse (n.) ≈ Gegensatz	SCIE	**Gegenteil**	
inverse (→reverse)	TECH	(→gegensinnig)	
inverse action	ELECTRON	**inverser Betrieb**	
inverse action (→inverse operation)	MICROEL	(→Inversbetrieb)	
inverse bias (→reverse voltage)	ELECTRON	(→Sperrspannung)	
inverse bias test (→reverse bias test)	MICROEL	(→Sperrprüfung)	
inverse characteristic (→reverse characteristic)	ELECTRON	(→Rückwärtskennlinie)	
inverse common base	CIRC.ENG	**Basisschaltung rückwärts**	
inverse common collector circuit (→inverse common collector connection)	CIRC.ENG	(→Kollektorschaltung rückwärts)	
inverse common collector connection = inverse common collector circuit	CIRC.ENG	**Kollektorschaltung rückwärts**	
inverse common emitter (→inverse common emitter connection)	CIRC.ENG	(→Emitterschaltung rückwärts)	
inverse common emitter circuit (→inverse common emitter connection)	CIRC.ENG	(→Emitterschaltung rückwärts)	
inverse common emitter connection = inverse common emitter circuit; inverse common emitter	CIRC.ENG	**Emitterschaltung rückwärts**	
inverse converter (→inverse transducer)	COMPON	(→Rückwandler)	
inverse correlation	MATH	**negative Korrelation**	
inverse current (→return current)	EL.TECH	(→Rückstrom)	
inverse current (→reverse current)	ELECTRON	(→Sperrstrom)	
inverse current (→cutoff current 1)	MICROEL	(→Sperrstrom)	
inverse dc resistance (→reverse dc resistance)	ELECTRON	(→Sperrwiderstand)	
inverse direction (→reverse direction)	ELECTRON	(→Sperrichtung)	
inverse function 1	MATH	**Umkehrfunktion 1** = inverse Funktion 1	
inverse function 2 (→inverse mapping)	MATH	(→inverse Abbildung)	
inverse integrator ≈ summing integrator	CIRC.ENG	**Umkehrintegrator** ≈ Summenintegrator	
inverse mapping [set theory] = inverse function 2	MATH	**inverse Abbildung** [Mengenlehre] = Umkehrabbildung; inverse Funktion 2; Umkehrfunktion 2	
inverse mixing	HF	**Rückmischung**	
inverse operation = inverse action	MICROEL	**Inversbetrieb** [Transistor] = inverser Betrieb	
inverse photoelectric effect	PHYS	**inverser Photoeffekt**	
inverse polarity (→polarity inversion)	EL.TECH	(→Polaritätsumkehr)	
inverse power loss (→reverse power loss)	MICROEL	(→Sperrverlustleistung)	
inverse reading (→reverse reading)	DATA PROC	(→Rückwärtslesen)	
inverse recovery current (→reverse recovery current)	MICROEL	(→Sperrverzögerungsstrom)	
inverse recovery time (→reverse recovery time)	MICROEL	(→Sperrverzögerungszeit)	
inverse region	MICROEL	**inverser Bereich**	
inverse-saturation current = reverse-saturation current	MICROEL	**Sperrsättigungsstrom**	
inverse search (→reverse reading)	DATA PROC	(→Rückwärtslesen)	
inverse SWR (→inverse voltage standing wave ratio)	LINE TH	(→Anpassungsfaktor)	
inverse transducer = inverse converter	COMPON	**Rückwandler** = Rückumwandler; Rückumsetzer	
inverse transform	MATH	**Rücktransformation**	
inverse video (→reverse video)	TERM&PER	(→invertierte Darstellung)	
inverse voltage (→reverse voltage)	EL.TECH	(→Gegenspannung)	
inverse voltage (→reverse voltage)	ELECTRON	(→Sperrspannung)	
inverse voltage protection (→reverse voltage protection)	POWER SYS	(→Gegenspannungsschutz)	
inverse voltage standing wave ratio [voltage minimum to voltage maximum] = inverse SWR ≈ reflection coefficient [NETW.TH] ≠ voltage standing wave ratio	LINE TH	**Anpassungsfaktor** [Spannungsminima zu Spannungsmaxima] ≈ Reflexionsfaktor [NETW.TH] ≠ Welligkeitsfaktor	
inversion	MATH	**Inversion**	
inversion [transition from n to p]	MICROEL	**Inversion** [Übergang von n zu p]	
inversion [from DC into DC] ≠ rectification	EL.TECH	**Wechselrichtung** [von Gleichstrom in Wechselstrom] ≠ Gleichrichtung	
inversion (→NOT operation)	ENG.LOG	(→NICHT-Verknüpfung)	
inversion (→temperature inversion)	METEOR	(→Temperaturinversion)	

inversion (→reversion)	TECH	(→Umkehrung)	investigator (→researcher)	SCIE	(→Forscher)
inversion area	MICROEL	Inversionsfläche	investment	ECON	Investition
inversion by advection	METEOR	Advektionsinversion	invisible	TECH	unsichtbar
inversion chip	MICROEL	Inverter-Chip	invisible antenna	ANT	unsichtbare Antenne
inversion density (→intrinsic number)	MICROEL	(→Intrinsic-Zahl)	invitation	TELEC	Aufforderung
			invitation (→request)	TELECONTR	(→Aufforderung)
inversion function (→NOT operation)	ENG.LOG	(→NICHT-Verknüpfung)	invitation to bid (AM) (→invitation to tender)	ECON	(→Ausschreibung)
inversion layer = inversion region	MICROEL	Inversionsschicht = Inversionsgebiet	invitation to clear (→clear request)	DATA COMM	=Auslöseanforderung
inversion layer	METEOR	Inversionsschicht	invitation to send	TELEC	Sendeaufforderung
inversion of distribution [laser]	PHYS	Besetzungsumkehr [Laser]	= ITS invitation to tender	ECON	Ausschreibung
inversion operation (→NOT operation)	ENG.LOG	(→NICHT-Verknüpfung)	= invitation to bid (AM); call for tenders; call for bids (AM); request for quotation; RFQ; request for tender; request of offer; request for proposal; solicitation; tender 2		= Angebotsaufforderung; Angebotseinholung; Submission
inversion region (→inversion layer)	MICROEL	(→Inversionsschicht)			
inversion theorem (→reciprocity theorem)	ANT	(→Umkehrsatz)			
			invitatio to break	DATA PROC	Anhaltanzeige
inversive point [curve changes abruptly its sense, the tangents coincide]	MATH	Rückkehrpunkt [Kurve ändert sprunghaft ihre Richtung, die Kurventangenten decken sich dabei]	invocate (→call)	DATA PROC	(→aufrufen)
			invocation (→call)	DATA PROC	(→Aufruf)
			invoice (v.t.)	ECON	verrechnen
			= charge		= fakturieren
= cusp		= Umkehrpunkt	invoice (n.) [shows what has to be payed]	ECON	Rechnung [weist den zu zahlenden Betrag aus]
≈ peak; break point		≈ Scheitel; Knickpunkt			
↑ singularity		↑ Singularität	= account (n.); bill (n.)		= Faktura; Faktur
invert (v.t.)	TECH	invertieren	invoice figure	ECON	Rechnungsbetrag
= reverse (v.t.); turn-over (v.t.)		= umkehren	invoicing	ECON	Inrechnungstellung
invert (v.t.)	CIRC.ENG	invertieren			= Fakturierung
= revert (v.t.); negate		= negieren; umkehren	invoicing machine	OFFICE	Fakturiermaschine
inverted (→reverse)	TECH	(→gegensinnig)	invoke (v.t.) (→call)	DATA PROC	(→aufrufen)
inverted comma [symbol ']	LING	Hochkomma [Symbol ']	involuntary	COLLOQ	unfreiwillig
			involute	MATH	Evolvente
= quote		≈ Anführungszeichen			= Involute
≈ quote mark			involute gearing	MECH	Evolventenverzahnung
inverted commas (→quotation mark)	TYPOGR	(→Anführungszeichen)	inward WATS [arrangement whereby a called subscriber takes over the charge of outside callers]	TELEC	ankommende Ferngesprächspauschale [der Angerufene übernimmt gegen eine Monatspauschale die Gebühr der ankommenden Gespräche]
inverted commas (→quote mark)	LING	(→Anführung)			
inverted crosstalk (→unintelligible crosstalk)	TELEC	(→unverständliches Nebensprechen)			
inverted file	DATA PROC	invertierte Datei			
inverted group selection	SWITCH	invertierte Gruppenwahl			
inverted image	TV	Kehrbild	in writing = written form	ECON	Schriftform
inverted matrix	MATH	inverse Matrix			
inverted position (→reverse frequency position)	TRANS	(→Kehrlage)	I/O (→input/output)	ELECTRON	(→Ein-/Ausgabe)
			I/O-bound (→input/output-bound)	DATA PROC	(→Ein-Ausgabe-bedingt)
inverted question mark	LING	umgekehrtes Fragezeichen			
inverted rectifier (→inverter)	POWER SYS	(→Wechselrichter)	I/O buffer (→input/output buffer)	DATA PROC	(→Ein-/Ausgabe-Puffer)
			I/O bus (→input/output bus)	DATA PROC	(→Ein-/Ausgabe-Bus)
inverted speech (→inverted voice)	TELEC	(→invertierte Sprache)	I/O cell (→input/output cell)	MICROEL	(→Ein-/Ausgabe-Zelle)
inverted T match	ANT	umgekehrte T-Anpassung			
inverted voice	TELEC	invertierte Sprache	I/O channel (→input/output channel)	DATA PROC	(→Ein-/Ausgabe-Kanal)
= inverted speech					
inverter [converts DC into AC]	POWER SYS	Wechselrichter [wandelt Gleichstrom in Wechselstrom]	I/O control (→input/output control)	CIRC.ENG	(→Ein-/Ausgabe-Steuerung)
			I/O controller (→input/output controller)	DATA PROC	(→Ein-/Ausgabe-Werk)
= direct-current inverter; DC inverter; current inverter; DC-AC converter; ondulator; vibrator (AM); inverted rectifier		= Inverter	I/O control program (→input/output control program)	DATA PROC	(→Ein-/Ausgabe-Kontrollprogramm)
		↑ Umrichter	I/O control system (→input/output controller)	DATA PROC	(→Ein-/Ausgabe-Werk)
↑ voltage system converter					
inverter 2 (→NOT gate)	CIRC.ENG	(→NICHT-Glied)	IOCS (→input/output controller)	DATA PROC	(→Ein-/Ausgabe-Werk)
inverter 1 (→inverting circuit)	CIRC.ENG	(→Umkehrschaltung)			
			I/O device (→input/output device)	TERM&PER	(→Ein-Ausgabe-Gerät)
inverter stability limit	POWER SYS	Wechselrichter-Trittgrenze = Trittgrenze; Wechselrichter-Kippgrenze	I/O device (→input/output device)	TERM&PER	(→Ein-/Ausgabe-Vorrichtung)
			iodine	CHEM	Jod
invertible (→reversible)	PHYS	(→umkehrbar)	= J		= J
inverting circuit	CIRC.ENG	Umkehrschaltung			
= inverter 1		↓ Phasenumkehrschaltung; NICHT-Glied	I/O driver (→input/output driver)	DATA PROC	(→Ein-/Ausgabe-Treiber)
↓ phase inverter; NOT gate			I/O executive (→input/output control program)	DATA PROC	(→Ein-/Ausgabe-Kontrollprogramm)
inverting input	CIRC.ENG	invertierender Eingang			
inverting stage (→reversal stage)	CIRC.ENG	(→Umkehrstufe)	I/O file (→input/output file)	DATA PROC	(→Ein-/Ausgabe-Datei)
inverting transformer	POWER SYS	Invertiertransformator			
investigation	TECH	Untersuchung	I/O instruction (→input-output instruction)	DATA PROC	(→Ein-/Ausgabe-Befehl)
= inquiry					
≈ examination		≈ Prüfung			

I/O interface

Term	Domain	German
I/O interface (→input/output interface)	DATA PROC	(→Ein-/Ausgabe-Schnittstelle)
I/O interrupt (→input/output interrupt)	DATA PROC	(→Ein-/Ausgabe-Unterbrechungssignal)
I/O library (→input/output library)	DATA PROC	(→Ein-/Ausgabe-Programmbibliothek)
I/O-limited (→input/output-bound)	DATA PROC	(→Ein-Ausgabe-bedingt)
I/O module (→input/output module)	TELECONTR	(→Eingabe-Ausgabe-Baugruppe)
ion	PHYS	**Ion**
↓ cation; anion		↓ Kation; Anion
ion avalanche	PHYS	**Ionenlawine**
ion beam	PHYS	**Ionenstrahl**
ion beam sputtering	ELECTRON	**Ionenstrahlzerstäubung**
ion burn (→ion spot)	ELECTRON	(→Ionenfleck)
ion charge	PHYS	**Ionenladung**
ion cloud	PHYS	**Ionenwolke**
ion conduction	PHYS	**Ionenleitung**
ion crystal	PHYS	**Ionenkristall**
ion dose	PHYS	**Ionendosis**
ion drift (→ion migration)	PHYS	(→Ionenwanderung)
ion etching	MICROEL	**Ionenätzung**
ion exchange	PHYS	**Ionenaustausch**
ion gun (→ion source)	PHYS	(→Ionenquelle)
ionic bond (→heteropolar bond)	CHEM	(→heteropolare Bindung)
ionic fire detector	SIGN.ENG	**Ionisations-Feuermelder**
ionic focussing	PHYS	**Ionenfokussierung**
ionic mobility	PHYS	**Ionenbeweglichkeit**
= ion mobility		
ionic semiconductor	PHYS	**Ionenhalbleiter**
ion implantation	MICROEL	**Ionenimplantation**
ionization	PHYS	**Ionisierung**
		= Ionisation
ionization chamber	PHYS	**Ionisationskammer**
ionization current	PHYS	**Ionisationsstrom**
		= Ionisierungsstrom
ionization energy	PHYS	**Ionisierungsenergie**
≈ work function		= Ionisationsenergie; Ionisierungsarbeit; Ionisationsarbeit
↑ activation energy		≈ Ablösearbeit
		↑ Aktivierungsenergie
ionization level	MICROEL	**Außenraum-Niveau**
= zero-energy level		= Außenniveau
ionization potential	PHYS	**Ionisierungspotential**
ionization rate	PHYS	**Ionisationsrate**
ionization voltage	PHYS	**Ionisierungsspannung**
= dissociation voltage		= Dissoziationsspannung
ionized	PHYS	**ionisiert**
ionized acceptor atom (→acceptor ion)	MICROEL	(→Akzeptor-Ion)
ionized donor atom (→donor ion)	PHYS	(→Donator-Ion)
ionized layer	GEOPHYS	**ionisierte Schicht**
ionizing	PHYS	**ionisierend**
ion lattice	PHYS	**Ionengitter**
ion loudspeaker	EL.ACOUS	**Ionenlautsprecher**
= ionophone		= Ionophon-Lautsprecher
ion magnetron	MICROW	**Ionenmagnetron**
ion migration	PHYS	**Ionenwanderung**
= ion drift; ion moving		
ion mobility (→ionic mobility)	PHYS	(→Ionenbeweglichkeit)
ion moving (→ion migration)	PHYS	(→Ionenwanderung)
ionogram	RADIO PROP	**Ionogramm**
ionophone (→ion loudspeaker)	EL.ACOUS	(→Ionenlautsprecher)
ionosphere	GEOPHYS	**Ionosphäre**
= ionospheric layer		= Ionosphärenschicht
ionospheric	GEOPHYS	**ionosphärisch**
ionospheric forward scatter (→ionospheric scatter)	RADIO PROP	(→ionosphärische Streuausbreitung)
ionospheric layer (→ionosphere)	GEOPHYS	(→Ionosphäre)
ionospheric radiowave (→sky wave)	RADIO PROP	(→Raumwelle)
ionospheric reflection	RADIO PROP	**Ionosphärenreflexion**
ionospheric scatter	RADIO PROP	**ionosphärische Streuausbreitung**
= ionospheric forward scatter; IFS; forward-propagation ionospheric scatter; FPIS		
ionospheric storm	RADIO PROP	**Ionosphärensturm**
ionospheric wave (→sky wave)	RADIO PROP	(→Raumwelle)
ion pair	PHYS	**Ionenpaar**
ion plating	METAL	**Ionenplattierung**
ion polarisation	PHYS	**Ionenpolarisation**
ion pump	PHYS	**Ionenpumpe**
ion repeller	ELECTRON	**Ionenreflektor**
ion source	PHYS	**Ionenquelle**
= ion gun		= Ionenkanone
ion spot	ELECTRON	**Ionenfleck**
= ion burn		[Kathodenstrahlröhre]
ion trap	PHYS	**Ionenfalle**
ion tube	ELECTRON	**Ionenröhre**
↓ gas-discgarge tube; flash tube; thyratron; ignitron		↓ Gasentladungsröhre; Blitzröhre; Thyratron; Ignitron
ion yield	PHYS	**Ionenausbeute**
IOP (→input/output controller)	DATA PROC	(→Ein-/Ausgabe-Werk)
I/O port (→input/output port)	CIRC.ENG	(→Ein-Ausgabe-Port)
I/O processor (→input/output controller)	DATA PROC	(→Ein-/Ausgabe-Werk)
I/O program library (→input/output library)	DATA PROC	(→Ein-/Ausgabe-Programmbibliothek)
I/O referencing (→input/output referencing)	DATA PROC	(→Ein-/Ausgabegerät-Bezugnahme)
I/O register (→input/output register)	DATA PROC	(→Ein-/Ausgabe-Register)
I/O request (→input/output request)	DATA PROC	(→Ein-/Ausgabe-Anforderung)
IORQ (→input/output request)	DATA PROC	(→Ein-/Ausgabe-Anforderung)
I/O status word (→input/output status word)	DATA PROC	(→Ein-/Ausgabe-Statuswort)
I/O symbol (→input/output symbol)	DATA PROC	(→Ein-Ausgabe-Symbol)
I/O system (→input/output controller)	DATA PROC	(→Ein-/Ausgabe-Werk)
IOU (→promissory note)	ECON	(→Eigenwechsel)
I/P (→input)	EL.TECH	(→Eingang)
i/p (→input)	EL.TECH	(→Eingang)
IP (→information provider)	DATA COMM	(→Informationsanbieter)
ip (→information provider)	INF.TECH	(→Informationslieferant)
IPL (→bootstrapping)	DATA PROC	(→Urladen)
IPL (→initial program loader)	DATA PROC	(→Urlader)
IPOS process	MICROEL	**IPOS-Verfahren**
= insulation by oxidated porous silicon process		
ipot (→inductive potentiometer)	COMPON	(→Spulenpotentiometer)
IPS (→inches-per-second)	TERM&PER	(→Zoll pro Sekunde)
ips (→inches-per-second)	TERM&PER	(→Zoll pro Sekunde)
I/Q gain	MODUL	**I/Q-Verstärkung**
I/Q gain unbalance	MODUL	**I/Q-Verstärkungsassymetrie**
IR (→instruction register)	DATA PROC	(→Befehlsregister)
IR (→index register)	DATA PROC	(→Indexregister)
IR (→information retrieval)	DATA COMM	(→Informationsabruf)
IR (→infrared)	PHYS	(→infrarot)
IRC (→international reply coupon)	RADIO	(→internationales Antwortschein)
IRED	MICROEL	**IRED**
= infrared light emitting diode		
i region (→intrinsic layer)	MICROEL	(→Intrinsic-Schicht)
IRF (→interference reduction factor)	RADIO	(→Interferenz-Unterdrückungsfaktor)
iris-in (→dim)	PHYS	(→abblenden)
iris-out	PHYS	**aufblenden**
		[Blende öffnen]

iron	CHEM	**Eisen**	
= Fe		= Fe	
iron (v.t.) (→smooth)	TECH	(→glätten)	
iron accumulator	POWER SYS	**Stahlakkumulator**	
iron band	METAL	**Bandeisen**	
= iron tape; iron hoop			
iron casting	METAL	**Eisenguß**	
ironclad	METAL	**eisenbeschichtet**	
iron core	EL.TECH	**Eisenkern**	
iron-core coil	EL.TECH	**Eisenkernspule**	
iron detector	TECH	**Eisensuchgerät**	
iron-free	TECH	**eisenfrei**	
= ironless; air-core [EL.TECH]		= eisenlos	
iron-free booster	CIRC.ENG	**eisenlose Endstufe**	
iron hoop (→iron band)	METAL	(→Bandeisen)	
ironing	METAL	**Fließziehen**	
ironless (→iron-free)	TECH	(→eisenfrei)	
iron loss (→core loss)	EL.TECH	(→Eisenverlust)	
iron-needle instrument	INSTR	**Eisennadel-Instrument**	
iron sheet	METAL	**Eisenblech**	
= sheet iron		= Blecheisen	
iron soldering	METAL	**Kolbenlöten**	
iron strip	METAL	**Eisenband**	
iron tape (→iron band)	METAL	(→Bandeisen)	
IRQ (→interrupt request)	DATA PROC	(→Unterbrechungsanforderung)	
irradiance	PHYS	**Bestrahlungsstärke**	
irradiance (→spurious irradiation)	ELECTRON	(→Einstrahlung)	
irradiate	PHYS	**ausstrahlen**	
↑ radiate		= abstrahlen	
		↑ strahlen	
irradiation	PHYS	**Abstrahlung**	
= emission		= Ausstrahlung; Emission	
↑ radiation		↑ Strahlung	
irradiation (→spurious irradiation)	ELECTRON	(→Einstrahlung)	
irrational number	MATH	**Irrationalzahl**	
[not expressable as ratio of integers]		[nicht als Bruch ganzer Zahlen ausdrückbar]	
= surd		≠ Rationalzahl	
≠ rational number			
irredeemable	ECON	**untilgbar**	
irredeemable (→non-terminable)	ECON	(→unkündbar)	
irrefutable (→incontestable)	COLLOQ	(→unanfechtbar)	
irregular	TECH	**unregelmäßig**	
= fortuitous		≈ zufällig	
≈ random			
irregular curve (→French curve)	ENG.DRAW	(→Kurvenlineal)	
irregularity	TECH	**Unregelmäßigkeit**	
= fortuitousness			
irregular terrain attenuation (→diffraction loss)	RADIO PROP	(→Beugungsdämpfung)	
irrelevance	INF	**Irrelevanz**	
[interfering "not relevant" information, summing to the useful information along the transmission path]		[unterwegs einstreuende „nicht zur Sache gehörende" Störinformation]	
= prevarication		= Störinformationsentropie; Belanglosigkeit	
irreparable	TECH	**irreparabel**	
≈ irretrievable		= nicht reparierbar; nicht behebbar	
irreparable error (→permanent error)	DATA PROC	(→irreparabler Fehler)	
irretrievable	DATA PROC	**unwiederbringlich**	
= unrecoverable		= unwiedergewinnbar	
irretrievable (→uncollectable)	ECON	(→uneinbringlich)	
irreversible	SCIE	**irreversibel**	
= uninvertible		= nicht umkehrbar; nicht invertierbar	
irreversible transducer	EL.ACOUS	**irreversibler Wandler**	
irrevocable	ECON	**unwiderruflich**	
irrevocable letter of credit	ECON	**unwiderrufliches Akkreditiv**	
		= unwiderrufliches L/C	
irrotational	MATH	**wirbelfrei**	
= swirl-free; potential (adj.)		= rotationsfrei	
irrotational field	PHYS	**wirbelfreies Feld**	
		[mit Rotation gleich Null]	
irrotational flow	PHYS	**Potentialströmung**	
		= rotationsfreie Strömung	
irrotationality	MATH	**Wirbelfreiheit**	
IS (→information separator)	DATA COMM	(→Informationstrennung)	
ISA	ELECTRON	**ISA**	
= Instrument Society of America			
ISAM (→indexed sequential access method)	DATA PROC	(→indexiert-sequentieller Zugriff)	
ISB system	BROADC	**ISB-System**	
= independent sideband system			
ISDN (→integrated-services digital network)	TELEC	(→dienstintegriertes Digitalnetz)	
ISDN basic access	TELEC	**ISDN-Basisanschluß**	
[offers two 64 kbit/s payload channels ("channels B") and a 16 kbit/s service channel ("channel D")]		[bietet zwei 64 kbit/s-Nutzkanäle („B-Kanäle") und einen 16-kbit/s-Signalkanal („D-Kanal")]	
= basic access; BA		= Basisanschluß	
≈ basic channel		≈ Basiskanal	
ISDN capability	TELEC	**ISDN-Fähigkeit**	
= ISDN functionality			
ISDN functionality (→ISDN capability)	TELEC	(→ISDN-Fähigkeit)	
ISDN plug (→Western plug)	TELEC	(→ISDN-Stecker)	
ISDN primary access	TELEC	**ISDN-Primäranschluß**	
[gives 30 channels 64 kbit/s over a 2 Mbit/s group, resp. 24 channels over 1.4 Mbit/s in the USA]		[bietet über ein 2-Mbit/s-Bündel 30 Kanäle 64-kbit/s, bzw. 24 Kanäle über 1,4 Mbit/s in den USA]	
= primary access		= Primäranschluß	
ISDN switch	SWITCH	**ISDN-Vermittlung**	
ISDN telephone	TERM&PER	**ISDN-Fernsprecher**	
		= ISDN-Telefon	
ISDN terminal	TERM&PER	**ISDN-Endgerät**	
isentropic	PHYS	**isentropisch**	
[with constant isentropy]		[bei gleicher Entropie]	
≈ adiabatisch		≈ isentrop	
		≈ adiabatisch	
ISFET	MICROEL	**ISFET**	
[ion-selective field effect transistor]			
ISI (→inter-symbol interference)	CODING	(→Intersymbolstörung)	
ISL	MICROEL	**ISL-Technik**	
= integrated Schottky logic			
island	MICROEL	**Insel**	
island effect	ELECTRON	**Inselbildung**	
[tube]		[Röhre]	
ISM (→in-service monitoring)	TRANS	(→In-Betrieb-Überwachung)	
ISO	TECH	**ISO**	
[union of all national standardization committees]		= internationaler Normenausschuß, Zusammenschluß aller Normungsausschüsse	
= International Organization for Standardization			
isobar (n.)	PHYS	**Isobare**	
[line connecting points of equal pressure]		[Verbindungslinie von Punkten gleichen Drucks]	
isobaric 1 (adj.)	PHYS	**isobar 1**	
[of equal pressure]		[gleichen Drucks]	
isobaric 2	PHYS	**isobar 2**	
[having same number of neutrons]		[mit gleicher Anzahl von Neutronen]	
		= isoton	
ISO 6-bit code	TELEGR	**ISO-6-Bit-Code**	
ISO 7-bit code	TELEGR	**ISO-7-Bit-Code**	
ISO 8-bit code (→intenational alphabet no.5)	TELEGR	(→internationales Alphabet Nr.5)	
isochorous	PHYS	**isochor**	
[of constant volume]		[konstanten Volumens]	
isochromatic	PHYS	**isochrom**	
[equal response to all colours]		[für alle Farben gleich empfindlich]	
		= isochromatisch	
isochrone (→isochronous)	PHYS	(→isochron)	
isochronous	PHYS	**isochron**	
[of same duration]		[gleich lang dauernd]	
= isochrone		≈ synchron	
≈ synchronous			
isochronous distortion	TELEGR	**Isochron-Verzerrung**	
isochronous signal	DATA COMM	**Isochronsignal**	

isoclinal

isoclinal (→isoclinic line)	GEOPHYS	(→Isokline)	
isoclinic line	GEOPHYS	**Isokline**	
[line connecting points of equal magnetic inclination]		[Verbindungslinie zwischen Punkten gleicher magnetischer Inklination]	
= isoclinal			
ISO code	CODING	**ISO-Code**	
[for 6,7 or 8 bits]		[für 6,7 oder 8 Bits]	
isodyn	PHYS	**Isodyne**	
[line connecting points of equal force]		[Verbindungslinie zwischen Punkten gleicher Kraft]	
isodynamic (→isodynamic line)	GEOPHYS	(→Isodyname)	
isodynamic line	GEOPHYS	**Isodyname**	
[line connecting points of equal magnetic intensity]		[Verbindungslinie zwischen Punkten gleicher magnetischer Stärke]	
= isodynamic			
isoelectric	PHYS	**potentialgleich**	
isoelectronic	PHYS	**isoelektronisch**	
[with the same number of valence electrons]		[mit gleicher Anzahl von Valenzelektronen]	
isogon	MATH	**Isogon**	
[polygon having equal angles]		= regelmäßiges Vieleck	
isogonal 1	MATH	**gleichwinklig**	
= isogonic; equiangular			
isogonal 2 (→angle-preserving)	MATH	(→winkeltreu)	
isogonal (→isogonic line)	GEOPHYS	(→Isogone)	
isogonal antenna	ANT	**Gleichwinkelantenne**	
isogonic (→isogonal 1)	MATH	(→gleichwinklig)	
isogonic (→isogonic line)	GEOPHYS	(→Isogone)	
isogonic line	GEOPHYS	**Isogone**	
[line connecting points of equal magnetic declination]		[Verbindungslinie von Orten gleicher magnetischer Deklination]	
= isogonic; isogonal			
isogram	GEOPHYS	**Isolinie**	
[line connecting points of equal value]		[Verbindungslinie zwischen Punkten gleichen Wertes]	
= isometric line			
isohel	GEOPHYS	**Isohelie**	
[connection of point with equal duration of sunshine]		[Verbindung von Punkten gleich langer Sonneneinstrahlung]	
isohyet	GEOPHYS	**Isohyete**	
[line connecting points of equal rainfall]		[Verbindungslinie zwischen Orten gleicher Niederschlagsmenge]	
= isohyetal line			
isohyetal line (→isohyet)	GEOPHYS	(→Isohyete)	
isohypse	GEOPHYS	**Isohypse**	
[line connecting points of equal hight above sea level]		[Verbindungslinie von Orten gleicher Meereshöhe]	
= level line		= Höhenlinie	
isohypse voltage (→peak-point voltage)	MICROEL	(→Höckerspannung)	
isokeraunic level	GEOPHYS	**isokeraunischer Pegel**	
[days with thunderstorms per year]		[Gewittertage pro Jahr]	
isolate	TECH	**absondern**	
= segregate		≈ trennen	
isolate	QUAL	**lokalisieren**	
= pinpoint		[Fehler]	
		= eingrenzen	
isolated	TECH	**vereinzelt**	
≈ single; individual		≈ einzelner; individueller	
isolated	PHYS	**isoliert**	
[thermally, electrically, magnetically]		[thermisch, elektrisch, magnetisch]	
isolated from earth	EL.TECH	**erdfrei**	
= earth-free		≠ geerdet	
≠ earthed			
isolated region	MICROEL	**Isolationsinsel**	
isolating capacitor	CIRC.ENG	(→Entkopplungskondensator)	
(→decoupling capacitor)			
isolating diffusion (→isolation process)	MICROEL	(→Isolationsdiffusion)	
isolating ring	MICROEL	**Isolationskranz**	
isolating spark gap	EL.TECH	**Trennfunkenstrecke**	
isolating transformer	COMPON	**Trenntübertrager**	
= insulating transformer; isolation transformer; insulation transformer; galvanic separation transformer; decoupling transformer		= Trenntransformator	
isolation	EL.TECH	**Trennung**	
= insulation; separation		≈ Entkopplung	
≈ decoupling			
isolation between aerials (→isolation between antennas)	ANT	(→Antennenentkopplung)	
isolation between antennas	ANT	**Antennenentkopplung**	
= isolation between aerials			
isolation capacitor (→decoupling capacitor)	CIRC.ENG	(→Entkopplungskondensator)	
isolation process	MICROEL	**Isolationsdiffusion**	
= isolating diffusion			
isolation relay	POWER SYS	**Trennrelais**	
= connect/disconnect relay			
isolation stub (→decoupling stub)	ANT	(→Entkopplungs-Stumpf)	
isolation transformer (→isolating transformer)	COMPON	(→Trennübertrager)	
isolation well	MICROEL	**Isolationswanne**	
isolator	NETW.TH	**Isolator**	
isolator	MICROW	**Richtungsleitung**	
		= Isolator; Einwegleitung; Entkoppler	
isolator electronics	ELECTRON	**Isolatorelektronik**	
ISO layer model (→ISO reference model)	DATA COMM	(→ISO-Referenzmodell)	
isomeric	CHEM	**isomer**	
[of equal composition but with different properties]		[aus gleichen Bestandteilen aber unterschiedlicher Eigenschaften]	
isomerism	CHEM	**Isomerie**	
isometric line (→isogram)	GEOPHYS	(→Isolinie)	
isometrism	GEOPHYS	**Isometrie**	
		= Längengleichheit; Längentreue	
isomorph	PHYS	**isomorph**	
[of equal shape or structure]		[gleicher Gestalt oder Struktur]	
isomorph	MATH	**isomorph**	
[of identical structure]		[strukturell völlig gleich]	
isomorphism	PHYS	**Mischkristallbildung**	
		≈ Isomorphie	
isomorphism	MATH	**Isomorphie**	
		= Isomorphismus	
isoperimetric	MATH	**isoperimetrisch**	
[of equal circumference]		[gleichen Umfangs]	
isoplanar	MATH	**isoplanar**	
[of same plane]		[der gleichen Ebene]	
isoplanar technology	MICROEL	**Isoplanarverfahren**	
		= LOCOS-Verfahren	
isopycnic	METEOR	**Isopykne**	
[line connecting points of equal atmospheric density]		[Verbindungslinie von Punkten gleicher Luftdichte]	
ISO reference model	DATA COMM	**ISO-Referenzmodell**	
= ISO layer model; OSI model		= ISO/OSI-Modell; OSI-Modell; ISO-Schichtenmodell; ISO-Datenübertragungsmodell	
↑ layer model			
		↑ Schichtenmodell	
isosceles	MATH	**gleichschenklig**	
isosceles triangle	MATH	**gleichschenkliges Dreieck**	
isoseismal line	GEOPHYS	**Isoseiste**	
[line connecting points of equal earthquake]		[Verbindungslinie von Punkten gleicher Erdbebenstärke]	
isotherm	PHYS	**isotherm**	
= isothermic			
isotherm (→isothermal line)	GEOPHYS	(→Isotherme)	
isotherm (→isothermal line)	PHYS	(→Isotherme)	
isothermal (n.) (→isothermal line)	PHYS	(→Isotherme)	
isothermal line	GEOPHYS	**Isotherme**	
[between points of equal temperature]		[Verbindungslinie zwischen Orten gleicher Temperatur]	
= isotherm			
isothermal line	PHYS	**Isotherme**	
[line of equal temperature]		[Linie gleicher Temperatur]	
= isotherm; isothermal (n.)			
isothermic (→isotherm)	PHYS	(→isotherm)	

isotope	PHYS	**Isotop**
isotopic	PHYS	**isotopisch**
[with equal atomic number but different mass number]		[mit gleicher Atomnummer aber unterschiedlicher Massenzahl]
isotrop (→isotropic radiator)	ANT	(→Kugelstrahler)
isotropic	PHYS	**isotrop**
[equal properties in all directions]		[nach allen Richtungen gleiche Eigenschaften]
≠ anisotropic		≠ anisotrop
isotropic antenna (→isotropic radiator)	ANT	(→Kugelstrahler)
isotropic etching	MICROEL	**isotropes Ätzen**
isotropic pattern	ANT	**Kugelcharakteristik**
isotropic radiator	ANT	**Kugelstrahler**
[equal radiation characteristic in all directions]		[in allen Richtungen gleiche Strahlungscharakteristik]
= isotropic antenna; spherical antenna; isotrop		= Isotopstrahler; isotroper Strahler; Isotropantenne; Kugelantenne
≈ omnidirectional antenna		≈ Rundstrahlantenne
isotropy	PHYS	**Isotropie**
issue	DOC	**Ausgabe**
= revision		= Auflage
issue (v.t.) (→output)	DATA PROC	(→ausgeben)
issue number	DOC	**Ausgabestand**
= revision level; edition number; release number		
IT (→information technology)	EL.TECH	(→Nachrichtentechnik)
ITA code no.2 (→international telegraph alphabet no.2)	TELEGR	(→internationales Telegraphenalphabet Nr.2)
italic (adj.)	TYPOGR	**kursiv**
[slightly inclined to the right]		[etwas nach rechts geneigt]
= cursive		
↑ typeface design		↑ Schriftschnitt
italic (n.) (→italic face)	TYPOGR	(→Kursivschrift)
italic face	TYPOGR	**Kursivschrift**
= italic (n.); cursive face; italic mode		= Kursive; Italique
italicize	TYPOGR	**kursiv legen**
italic mode (→italic face)	TYPOGR	(→Kursivschrift)
ITA no.2 (→international telegraph alphabet no.2)	TELEGR	(→internationales Telegraphenalphabet Nr.2)
item	TECH	**Position**
= entry		= Pos.; Item; Einzelposten; Posten 2; Gegenstand; Artikel
item (→piece)	ECON	(→Stück)
itemized (→detailed)	DOC	(→detailliert)
itemized billing (→toll ticketing)	SWITCH	(→Einzelgebührenerfassung)
itemized breakdown	COLLOQ	**Aufschlüsselung**
= breakdown		[in Einzelpositionen]
item number	DOC	**Positionsnummer**
		= Gegenstandsnummer
iterate (→repeat)	DATA PROC	(→wiederholen)
iteration	MATH	**Iteration**
iteration algorithm	MATH	**Iterationsalgorithmus**
= iterative algorithm		= iterativer Algorithmus
iterative	MATH	**iterativ**
iterative algorithm (→iteration algorithm)	MATH	(→Iterationsalgorithmus)
iterative attenuation constant	NETW.TH	**Kettendämpfung**
= attenuation constant		= Kettendämpfungsfaktor
iterative coder	CIRC.ENG	**Wägecodierer**
iterative coding	CODING	**Wägecodierung**
iterative impedance	NETW.TH	**Kettenwiderstand**
= iterative resistance		
iterative method (→iterative procedure)	MATH	(→Iterationsverfahren)
iterative phase constant	NETW.TH	**Kettenwinkelmaß**
iterative procedure	MATH	**Iterationsverfahren**
= iterative method		
iterative propagation constant	NETW.TH	**Kettenübertragungsmaß**
= propagation coefficient		
iterative resistance (→iterative impedance)	NETW.TH	(→Kettenwiderstand)
iterative run	DATA PROC	**Iterationslauf**
ITS (→invitation to send)	TELEC	(→Sendeaufforderung)

ITU	TELEC	**UIT**
[head organisation uf UNO for telecommunications standardization]		[Dachorganisation der Normenausschüsse der Vereinten Nationen für das Fernmeldewesen]
= International Telecommunication Union; UIT		= ITU
i-type (→intrinsic)	MICROEL	(→eigenleitend)
i zone (→intrinsic layer)	MICROEL	(→Intrinsic-Schicht)

J

J (→Joule)	PHYS	(→Joule)
J (→iodine)	CHEM	(→Jod)
jack	COMPON	**Buchse**
[female part of a connecting device]		[weiblicher Teil der Steckverbindung]
= socket (BRI); plug socket; female plug; receptacle; female connector; female		= Steckerbuchse; Mutterteil; Mutter; Kupplung 2; Kuppler
≠ plug		≠ Stecker
		↓ Netzsteckdose
jack	TELEPH	**Klinke**
[special jack of telephony, for repetitive multipolar connexions]		[Spezialbuchse der Fernsprechtechnik, zur wiederholten Herstellung mehrpoliger Kontakte]
= switchboard jack		
≠ jack connector		= Klinkenbuchse; Steckbuchse
		≠ Klinkenstecker
		↓ Klinkenkupplung; Klinken-Einbaubuchse
jack (US slang) (→money)	ECON	(→Geld)
jack connector	TELEPH	**Klinkenstecker**
[male counterpart of jack]		[Spezialstecker der Fernsprechtechnik, Gegenstück zur Klinke]
= switchboard plug; telephone plug; phone plug		= Stöpsel
≠ jack		≠ Klinke
jacket (→coat)	TECH	(→Mantel)
jacket (→protective casing)	TECH	(→Schutzhülle 1)
jacket (→disk jacket)	TERM&PER	(→Diskettenhülle)
jacketed	COMM.CABLE	**ummantelt**
jack field (→jack panel)	TELEPH	(→Klinkenfeld)
jack panel	TELEPH	**Klinkenfeld**
= jack field; concentrator panel		
jack strip	TELEPH	**Klinkenstreifen**
≠ individual jack		≠ Einzelklinke
jagged (→jaggy)	TYPOGR	(→unscharf)
jaggies (n.pl.t.)	ELECTRON	**Verzackung**
[accumulation of spikes in a signal]		[einer Signalform]
jaggy	TYPOGR	**unscharf**
= jagged; fuzzy		
jam (v.t.)	TECH	**klemmen 1** (v.t.)
= clamp (v.t.)		= einklemmen
≈ grip (v.t.); crush (v.t.)		= greifen; quetschen
jam (v.t.)	MECH	**kontern**
jam (n.)	TECH	**Stau**
= jamming		≈ Blockierung; Quetschung
≈ blocking; crush (n.)		
jammer (→jamming transmitter)	MIL.COMM	(→Störsender)
jamming	TERM&PER	**Papierstau**
jamming	MIL.COMM	**Störung**
= interference		
jamming (→jam)	TECH	(→Stau)
jamming transmitter	MIL.COMM	**Störsender**
= jammer; interfering transmitter		= Störer
jam nut	MECH	**Kontermutter**
Janet system	RADIO	**Janet-System**
J antenna	ANT	**J-Antenne**
= J pole antenna		
jar (→concuss)	TECH	(→erschüttern)
jargon	LING	**Jargon**
= argot		= Sondersprache; Soziolekt
≈ technical language		≈ Fachsprache
jargon (→parlance)	LING	(→Fachjargon)
jaw (n.)	TECH	**Klaue**
≈ grip (n.)		= Backe
		≈ Greifer

jaw

jaw (→chuck jaw) MECH (→Spannbake)
jaw clutch MECH **Klauenkupplung**
JCL (→job control language) DATA PROC (→Auftragssprache)
Jeckles-Jordan flipflop (→JK flipflop) CIRC.ENG (→JK-Flipflop)
Jefferson map RADIO NAV **Jefferson-Karte**
jelly COMM.CABLE **Vaseline**
= petroleum jelly; petrolatum
jelly-filled cable COMM.CABLE **gefülltes Kabel**
jerky (adj.) TECH **ruckartig**
≈ discontinuous
= sprungartig
≈ unstetig
jet (→squirt) TECH (→spritzen)
JFET MICROEL **JFET**
= junction FET; junction field effect transistor
= Sperrschicht-FET; Sperrschicht-Feldeffekttransistor
Jg,Kg flipflop (→storage circuit) CIRC.ENG (→Speicherglied)
jig (→gauge) MECH (→Lehre)
jig (→device) TECH (→Vorrichtung)
jitter (n.) ELECTRON **Jitter**
[unwanted, small, rapid aberration in time or size]
↓ phase jitter
[unerwünschte, kleine, schnelle Schwankungen in Zeit oder Amplitude]
= Zittern; Flattern; Signalschwankung
↓ Phasenschwankung
jitter-disturbed ELECTRON **jittergestört**
= jitter-interfered
jitter-free ELECTRON **jitterfrei**
jitter hit ELECTRON **Jitter-Sprung**
jitter immunity ELECTRON **Jitterverträglichkeit**
jitter-interfered ELECTRON (→jittergestört)
(→jitter-disturbed)
jitter peak ELECTRON **Jitter-Spitzenwert**
jitter spectrum ELECTRON **Jitter-Spektrum**
jitter suppression ELECTRON **Jitterunterdrückung**
jitter suppression factor ELECTRON **Jitterunterdrückungsfaktor**
jitter test set INSTR **Jitter-Meßplatz**
JK flipflop CIRC.ENG **JK-Flipflop**
= Jeckles-Jordan flipflop
↑ bistabile Kippschaltung
job (n.) DATA PROC **Auftrag**
[sequence of specific tasks to solve a problem, may constitute a program or chain of programs]
↓ task
[Sequenz spezifischer Teilaufgaben zur Lösung einer Aufgabe; kann ein Programm oder eine Programmkette bilden]
= Arbeitsauftrag; Aufgabe; Job
≈ Anweisung; Befehl
↓ Teilaufgabe
job ECON **Stelle**
= post
= Arbeitsstelle
job (→task) TECH (→Aufgabe)
job accounting DATA PROC **Auftragsabrechnung**
job account number ECON **Verrechnungskonto**
job advertisement ECON **Stellenausschreibung**
job assignment DATA PROC **Auftragsanweisung**
jobber [AM] (→intermediary) ECON (→Zwischenhändler)
job composition TYPOGR **Akzidenzsatz**
job control (→job management) DATA PROC (→Auftragsverwaltung)
job control language DATA PROC **Auftragssprache**
= JCL
↑ command control language
= Jobkontrollsprache; Job-Betriebssprache
↑ Kommandosprache
job-dependent TECH (→anwendungsspezifisch)
(→application-specific)
job description (→position description) ECON (→Arbeitsplatzbeschreibung)
job execution DATA PROC **Auftragsdurchführung**
job file DATA PROC **Auftragsdatei**
[file of jobs to be processed]
[Datei der auszuführenden Aufträge]
job handling (→order processing) DATA PROC (→Auftragsabwicklung)
job initialization DATA PROC (→Initialisierung 1)
(→initialization)

job initiator DATA PROC **Vorbereiter**
= initiator
= Initialisierer
job level DATA PROC **Auftragsebene**
= task level
job macro DATA PROC **Auftragsmakro**
job management DATA PROC **Auftragsverwaltung**
[automatic processing of program strings]
= job control
≈ task management
↑ control program
[automatische Verarbeitung von Programmketten]
= Aufgabensteuerung; Job Management
↑ Organisationsprogramm
job number DATA PROC **Auftragsnummer**
job order MANUF **Fertigungsauftrag**
= manufacturing order; production order
= Fertigungsorder
job-oriented DATA PROC (→anwendungsorientiert)
(→application-oriented)
job-orintated DATA PROC (→anwendungsorientiert)
(→application-oriented)
job preparation DATA PROC (→Initialisierung 1)
(→initialization)
job priority DATA PROC **Auftragspriorität**
= Aufgabenpriorität
job processing (→order processing) DATA PROC (→Auftragsabwicklung)
job queue DATA PROC **Auftragsschlange**
= job string
≈ input job stream
= Aufgabenschlange
≈ Auftragseingabefluß
job scheduler DATA PROC **Auftragsdisponent**
job scheduling DATA PROC **Auftragsdisposition**
= Aufgabendisposition
job sequence TECH **Arbeitsablauf**
= workflow
job-specific TECH (→anwendungsspezifisch)
(→application-specific)
job statement DATA PROC **Auftraganweisung**
= Aufgabenanweisung
job step DATA PROC **Auftragsschritt**
= step
= Aufgabenschritt; Step
job stream (→input job stream) DATA PROC (→Auftragseingabefluß)
job string (→job queue) DATA PROC (→Auftragsschlange)
job submission DATA PROC **Auftragserteilung**
job-to-job transition DATA PROC **Auftragsvorbereitung**
job turnaround DATA PROC **Auftragsausführungszeit**
↑ turnaround (n.)
↑ Ausführungszeit
job work (→unit production) MANUF (→Einzelanfertigung)
jog (n.) CONS.EL **manueller Suchlauf**
[video recorder]
[Videorecorder]
jog (n.) MATH **Kurvenknick**
jog (→bend) MECH (→Knick)
jogging equipment OFFICE **Glattstoßgerät**
Johnson counter CIRC.ENG **Johnson-Zähler**
join (v.t.) TECH **verbinden**
join (n.) (→joint 3) MECH (→Fuge)
joint 3 (n.) MECH **Fuge**
= groove; join (n.)
= Trennfuge
joint 4 (n.) MECH **Gelenk**
↓ hinge joint; knuckle joint; universal joint; ball-and-socket joint
↓ Scharniergelenk; Gabelgelenk; Kardangelenk; Kugelgelenk
joint 2 (n.) MECH **Stoßstelle**
[a mechanical interface]
= Stoß
joint 1 (n.) MECH **Verbindung**
[a linking device]
= Verbinder
joint (v.t.) (→splice) COMM.CABLE (→spleißen)
joint denial (→NOR operation) ENG.LOG (→NOR-Verknüpfung)
joint design TECH **Gemeinschaftsentwicklung**
jointing chamber (BRI) OUTS.PLANT (→Kabelschacht)
(→cable chamber)
jointing closure (→jointing sleeve) OUTS.PLANT (→Verbindungsmuffe)
jointing sleeve OUTS.PLANT **Verbindungsmuffe**
= splicing sleeve; jointing closure; straight sleeve
= Spleißmuffe
jointing strip OUTS.PLANT **Verbindungsleiste**
jointing technique (→splicing technique) OPT.COMM (→Spleißtechnik)

joint rate	ECON	**Verbundtarif**	
		= Einheitstarif	
joint venture	ECON	**Gemeinschaftsunternehmen**	
≈ associated company		= Joint Venture	
		≈ Beteiligungsgesellschaft	
joist (→girder 2)	CIV.ENG	(→Profilträger)	
joist (→support)	CIV.ENG	(→Träger)	
jolt (v.t.)	PHYS	**stoßen**	
= shock (v.t.)		≈ schütteln	
≈ shake			
jolt (n.) (→shock)	PHYS	(→Stoß)	
jolting table	QUAL	**Schütteltisch**	
Josephson effect	PHYS	**Josephson-Effekt**	
Josephson junction circuit	COMPON	**Josephson-Element**	
Joule	PHYS	**Joule**	
[SI unit for work, energy and heat quantity; = 1 N m = 1 W s]		[SI-Einheit für Arbeit, Energie und Wärmemenge; = 1 N m = 1 W s]	
= J		= J	
Joule effect (→Joule-Thomson effect)	PHYS	(→Joule-Effekt)	
Joule's heat	PHYS	**Joulesche Wärme**	
Joule's law	PHYS	**Joulesches Gesetz**	
Joule-Thomson effect	PHYS	**Joule-Effekt**	
= Joule effect			
journal	MECH	**Lagerzapfen**	
[of a gliding-surface bearing]		[eines Gleitlagers]	
journal	INF.TECH	**Journal**	
[complete record for control purposes]		[eine vollständige Aufzeichnung für Kontrollzwecke]	
		= Protokoll	
journal file	DATA PROC	**Journaldatei**	
		= Protokolldatei	
journal file	SWITCH	**Rufdatenaufzeichnung**	
journal printer	TERM&PER	**Journaldrucker**	
journal reader	TERM&PER	**Journalstreifenleser**	
= journal scanner			
journal roll	TERM&PER	**Journalrolle**	
journal scanner (→journal reader)	TERM&PER	(→Journalstreifenleser)	
journeyman	ECON	**Geselle**	
		≈ Lehrling	
joystick	TERM&PER	**Steuerknüppel**	
		= Joystick; Steuerhebel	
J pole antenna (→J antenna)	ANT	(→J-Antenne)	
judder (n.)	ELECTRON	**Verwacklung**	
		≈ Verwackeln	
judgement (→expert opinion)	TECH	(→Gutachten)	
judgement (BRI) (→judgment)	ECON	(→Urteil)	
judgment (NAM)	ECON	**Urteil**	
= judgement (BRI)			
Julian number	DATA PROC	**julianische Datumsangabe**	
jumbo chip	MICROEL	**Jumbochip**	
[making use of a whole wafer]		[belegt eine ganze Siliziumscheibe]	
jumbogroup	TRANS	**Jumbogruppe**	
[FDM, 3600 channels]		[TF-Technik, 3600-Kanal-Bündel nach US-Norm]	
jump (→step)	ELECTRON	(→Sprung)	
jump (→program jump)	DATA PROC	(→Programmsprung)	
jump acknowledgment	DATA PROC	**Sprungquittung**	
≈ branch acknowledgment		≈ Vertweigungsquittung	
jump address (→transfer address)	DATA PROC	(→Verzweigungsadresse)	
jump condition	DATA PROC	**Sprungbedingung**	
= branch condition		= Verzweigungsbedingung	
≈ jump instruction; programm branch		↓ Sprungbefehl; Programmsprung	
jump destination	DATA PROC	**Sprungziel**	
jumper	ELECTRON	**Jumper**	
[flexible wire connection on a PCB]		[flexibler Leiterplattenverbinder]	
≈ wire link		≈ Drahtbrücke	
jumper (v.t.)	EQUIP.ENG	(→rangieren)	
(→cross-connect)			
jumper (→cross connect)	SYS.INST	(→Rangierleitung)	
jumper (n.) (→jumpering)	SYS.INST	(→Rangierung)	
jumper cable (→cross connect)	SYS.INST	(→Rangierleitung)	
jumper field (→patching distribution frame)	SYS.INST	(→Rangierverteiler)	
jumpering	SYS.INST	**Rangierung**	
= jumper (n.); patching; cross-connecting			
jumpering panel	EQUIP.ENG	**Rangierfeld**	
= cross-connecting panel			
jumper line (→cross connect)	SYS.INST	(→Rangierleitung)	
jumper plug (→plug link)	ELECTRON	(→Brückenstecker)	
jumper-selectable 1	ELECTRON	**mit Drahtbrücke einstellbar**	
jumper-selectable 2	ELECTRON	**mit Steckbrücke einstellbar**	
[by a pluggable jumper]			
jumper wire (→wire link)	ELECTRON	(→Drahtbrücke)	
jumping cam	MECH	**Nockenscheibe**	
= cam disk 2			
jumping wire (→strap)	SYS.INST	(→Rangierdraht)	
jump instruction	DATA PROC	**Sprungbefehl**	
= jump statement; go-to statement; branch statement; branch instruction		= Verzweigungsbefehl	
≈ programm jump; jump condition		≈ Programmsprung; Sprungbedingungen	
jump-on	MECH	**aufspringen**	
[spring]		[Feder]	
jump operation	DATA PROC	**Sprungoperation**	
jump routine	DATA PROC	**Sprungroutine**	
jump statement (→jump instruction)	DATA PROC	(→Sprungbefehl)	
jump to subroutine	DATA PROC	**Unterprogrammsprung**	
↑ program jump		↑ Sprungbefehl	
junction	MICROEL	**Übergang**	
[interface or region between semiconductors of different type, doping or degree of doping]		[Fläche oder Zone zwischen Halbleitern von unterschiedlichem Typ, Dotierung oder Dotierungsgrad]	
= semiconductor junction; transition region; transition zone; transition		= Zonenübergang; Störstellenübergang; Kontaktstelle; Übergangszone; Übergangsgebiet 1; Übergangsbereich 1	
↓ depletion layer		↓ Sperrschicht	
junction	MICROW	**Verzweiger**	
		= Verzweigung	
junction	DATA PROC	**Zusammenführung**	
junction (→cable splice)	COMM.CABLE	(→Kabelspleiß)	
junction box	OUTS.PLANT	**Kabelkasten**	
junction box	INSTR	**Verzweigungsstück**	
junction cable	OUTS.PLANT	**Ortsverbindungskabel**	
junction capacitor	COMPON	**Sperrschichtkondensator**	
= barrier-layer capacitor		= RKN-Kondensator; Keramikkondensator Typ 3	
junction capacity	MICROEL	**Sperrschichtkapazität**	
= depletion layer capacity			
junction center (→secondary exchange)	SWITCH	(→Knotenvermittlungsstelle)	
junction circuit (BRI) (→local trunk)	TELEC	(→Ortsverbindungsleitung)	
junction depth	MICROEL	(→Sperrschichtdicke)	
(→depletion-layer thickness)			
junction device	MICROEL	**Sperrschicht-Bauelement**	
↓ junction transistor; junction diode		↓ Sperrschicht-Transistor; Sperrschicht-Diode	
junction diode	MICROEL	**Sperrschicht-Diode**	
≈ junction varactor		= Flächendiode	
↑ junction device		↑ Sperrschichtvaraktor	
↓ tunnel diode		↑ Sperrschicht-Element	
		↓ Tunneldiode	
junction FET (→JFET)	MICROEL	(→JFET)	
junction field effect transistor (→JFET)	MICROEL	(→JFET)	
junction insulation	MICROEL	**Sperrschichtisolation**	
junction line (→interoffice trunk)	TELEC	(→Amtsverbindungsleitung)	
junction line (→connecting line)	MATH	(→Verbindungslinie)	
junction photo detector	COMPON	**Sperrschicht-Photoempfänger**	
junction pole	OUTS.PLANT	**Verzweigungsmast**	
junction rectifier	MICROEL	**Flächengleichrichter**	
		= Sperrschichtgleichrichter	
junction temperature	MICROEL	**Sperrschichttemperatur**	
junction thickness	MICROEL	(→Sperrschichtdicke)	
(→depletion-layer thickness)			

junction transistor	MICROEL	Flächentransistor	
≠ point-contact transistor		≠ Spitzentransistor	
↓ alloy-junction transistor; diffusion transistor		↓ Legierungstransistor; Diffusionstransistor	
junction transistor	MICROEL	Sperrschicht-Transistor	
↑ junction device		↑ Sperrschicht-Bauelemnt	
junction varactor	MICROEL	Sperrschichtvaraktor	
≈ junction diode		≈ Sperrschichtdiode	
↑ varactor diode		↑ Varaktordiode	
junction width (→depletion-layer thickness)	MICROEL	(→Sperrschichtdicke)	

junctor SWITCH Verbindungssatz
[temporarily assigned to a call, interconnects to switching matrices and other internal functional units of a switch]
[für die Dauer der Verbindung zugeordnet, stellt den Anschluß an Koppelfelder oder andere interne Funktionseinheiten her]
= connector circuit
↑ circuit
↓ outgoing circuit; incoming circuit; intraexchange circuit; special-services circuit
↓ Satz
↓ abgehender Satz; ankommender Satz; Internsatz; Sonderdienstsatz

junk COLLOQ Schrott
≈ scrap; garbage; litter; rubbish
= Altmaterial
≈ Abfall; Müll; Unrat; Schutt

junk (→garbage) DATA PROC (→Datensalat)
justification DATA PROC Bündigkeit
justification TYPOGR Ausschluß 2
[to adjust a line by spacings and hyphenation to a given length]
[durch Zwischenräume und Silbentrennung eine Zeile auf vorgegebene Länge bringen]
= Ausschließen; Ausrichtung

justification (→pulse stuffing) CODING (→Stopfen)
justification bit (→stuffing bit) CODING (→Stopfbit)
justification digit (→stuffing bit) CODING (→Stopfbit)
justification font TYPOGR Ausschluß 1
[nonprinting types]
[nichtdruckende Typen für Zwischenräume]
justification frame CODING Stopfrahmen
= stuffing frame
justification jitter CODING Stopfjitter
justification mode (→stuffing mode) CODING (→Stopfverfahren)
justification program DATA PROC Ausschließprogramm
[word processing]
[Textverarbeitung]
= justification routine
= Ausschließroutine
justification range (→hot zone) TYPOGR (→Zeilenausgang)
justification routine (→justification program) DATA PROC (→Ausschließprogramm)
justification service bit CODING Stopfinformationsbit
= justification service digit
= Stopfsteuerbit; Füllinformation
≈ Stopfbit
justification service digit (→justification service bit) CODING (→Stopfinformationsbit)
justified composition (→justified typesetting) TYPOGR (→Blocksatz)
justified print (→justified typesetting) TYPOGR (→Blocksatz)
justified setting (→justified typesetting) TYPOGR (→Blocksatz)
justified type (→justified typesetting) TYPOGR (→Blocksatz)
justified typesetting TYPOGR Blocksatz
[print format with equal long lines, flush on both sides]
[Druckbild mit beidseitig bündigen, gleich langen Zeilen]
= justified setting; justified type; justified print; justified composition; automatic formatting [DATA PROC]
= ausgeschlossener Satz
≠ unjustified typesetting
≠ Flattersatz
justify TYPOGR ausschließen
[to bring a line to a given length by spacings and hyphenation]
[durch Zwischenräume und Silbentrennung eine Zeile auf vorgegebene Länge bringen]
= ausrichten

justify (→pulse stuff) CODING (→stopfen)
justifying digit (→stuffing bit) CODING (→Stopfbit)
jute TECH Jute
jute serving COMM.CABLE Juteumwicklung
juxtaposition SCIE Juxtaposition
≈ comparison [COLLOQ]
= Nebeneinanderstellung; Zusammenrückung
≈ Vergleich [COLLOQ]

K

K (→potassium) CHEM (→Kalium)
K (→Kelvin) PHYS (→Kelvin)
Kalman filtering CONTROL Kalman-Filterung
K antenna factor (→antenna factor) ANT (→Antennenfaktor)
Karnaugh map ENG.LOG Karnaugh-Veitch-Diagramm
= KV-Diagramm; Karnaugh-Veitch-Tafel; KV-Tafel; Karnaugh-Plan
kbit (→kilobit) INF (→Kilobit)
kc (→kilo character) DATA PROC (→tausend Zeichen)
keen 2 (→pungent) TECH (→stechend)
keen 1 (→sharp) MECH (→scharf)
keen (→cutting) MECH (→schneidend)
keep-alive circuit (→hold circuit) CIRC.ENG (→Halteglied)
keep-alive electrode ELECTRON Halteanode
keep dry ! ECON trocken aufbewahren !
keeper of minutes OFFICE Protokollführer
= Schriftführer
keeping with requirement TECH bedarfsgerecht
= suited to demand
keep-out area TECH verbotene Zone
keep pace COLLOQ schritthalten
keep up (→maintain) COLLOQ (→aufrechterhalten)
kell factor TV Kellfaktor
Kelvin PHYS Kelvin
[SI unit for thermodynamic temperature]
[SI-Einheit für thermodynamische Temperatur]
= K
= K
≈ Grad Celsius
Kelvin-Bridge (→Thomson bridge) INSTR (→Thomson-Meßbrücke)
Kelvin probe set INSTR Kelvin-Klemme
Kelvin-Varley divider CIRC.ENG Kelvin-Varley-Teiler
Kennedy key MECH Diagonalpaßfeder
Kennelly-Heavyside layer GEOPHYS Kennely-Heavyside-Schicht
kerb (BRI) (→curb) CIV.ENG (→Bordstein)
kern (v.t.) TYPOGR unterschneiden
[to reduce the separation of specific character combinations for visual reasons]
[aus optischen Gründen den Buchstabenabstand bestimmter Buchstabenkombinationen verringern]
kern (n.) TYPOGR Überhang
[part of type proyecting beyond body]
[über den Typenkörper hervorragendes Relief]
kernel (n.) (→nucleus) DATA PROC (→Nukleus)
kernel MATH Integralkern
kernel (→core) TECH (→Kern)
kernel (→kernel routines) DATA PROC (→Kernroutinen)
kernel routines DATA PROC Kernroutinen
[set of basic commands]
= kernel system; kernel
kernel system (→kernel routines) DATA PROC (→Kernroutinen)
kerning TYPOGR Unterschneidung
[reduction of space between letters]
[Abstände zwischen Buchstaben reduzieren]
kerosene CHEM Petroleum 1
[Brennstoff]
Kerr cell PHYS Kerrzelle
Kerr effect PHYS Kerr-Effekt
= electrooptic birefrigence
= elektrooptische Doppelbrechung
key (v.t.) TERM&PER eintasten
[to enter by keyboard]
[über Tastatur eingeben]
≈ enter (v.t.)
≈ eingeben

key (v.t.)		MODUL	tasten		
= keyboard (v.t.)			= umtasten		
key (n.)		TELEGR	Geber		
			= Taster		
key (n.)		DATA PROC	Kennbegriff		
[identifying part of a record]			[identifizierender Teil eines Datensatzes]		
= search word; search key; control field 2			= Suchargument; Suchkriterium; Suchbegriff; Schlüsselwort 1		
≈ sorting key; key			≈ Sortierschlüssel; Kennbegriff		
key (n.)		TECH	Schlüssel 1		
[to a lock]			[für ein Schloß]		
key (n.)		TERM&PER	Taste		
[element of a keyboard]			[Bedienelement einer Tastatur]		
= digital (n.); button			= Drucktaste; Druckknopf		
≈ key [COMPON]			≈ Taste [COMPON]		
↓ alphanumeric key; alphabetical key; numeric key; function key; ENTER key; ESCAPE key; SHIFT key; SHIFT LOCK key; ERASE key; dead key			↓ Datentaste; Buchstabentaste; Zifferntaste; Funktionstaste; Eingabetaste; ESCAPE-Taste; Umschalttaste; Umschaltfeststelltaste; Löschtaste; Tottaste		
key		COMPON	Taste		
≈ key [TERM&PER]			≈ Taste [TERM&PER]		
↓ pushbutton; touch key; sensor key; toggle key; toggle switch; pushbutton key; magnetic pushbutton key; luminous key; non-locking key			↓ Drucktaste; Tipptaste; Berührungstaste; Kipptaste; Kippschalter; Tastschalter; Magnettaste; Leuchttaste; Taste		
key (→hub)		MECH	(→Nabe)		
key (→key field)		DATA PROC	(→Schlüsselfeld)		
key (→cipher key)		INF.TECH	(→Schlüssel)		
keyboard (n.)		TERM&PER	Tastatur		
[an array of keys]			[Anordnung von Tasten]		
= key panel; pushbutton panel			= Tastenfeld		
≈ keypad 2			≈ Tastaturfeld		
↓ keypad; control keyboard; computer keyboard; teletypewriter keyboard; telephone keypad; single-key keyboard; membrane keyboard; alphabetical keyboard; numeric keyboard; alphanumeric keyboard; US-standard keyboard; German-standard keyboard; Dvorak keyboard; Maltron kexboard			↓ Kleintastatur; Bedientastatur; Datentastatur; Fernschreibtastatur; Fernsprechertastentastatur; Einzeltasten-Tastatur; Folientastatur; Buchstabentastatur; Zifferntastatur; alphanumerische Tastatur; amerikanische Tastatur; DIN-Tastatur; Dvorak-Tastatur; Maltron-Tastatur		
keyboard (v.t.) (→key)		MODUL	(→tasten)		
keyboard calling (→pushbutton dialing)		SWITCH	(→Tastwahl)		
keyboard code		DATA PROC	Tastencode		
keyboard coder		TERM&PER	Tastaturcodierer		
= keyboard encoder					
keyboard connection		TERM&PER	Tastaturanschluß		
keyboard console		TERM&PER	Tastaturgehäuse		
			= Tastaturkonsole		
keyboard-controlled		DATA PROC	tastaturgesteuert		
= keyboard-driven; key-controlled; key-driven			≈ tastengesteuert		
keyboard controller		DATA PROC	Tastaturkontroller		
keyboard cover		TERM&PER	Tastaturabdeckung		
keyboard data entry device		TERM&PER	(→Tastatur-Dateneingabegerät)		
(→key data entry device)					
keyboard dialing (→pushbutton dialing)		SWITCH	(→Tastwahl)		
keyboard dialling (→pushbutton dialing)		SWITCH	(→Tastwahl)		
keyboard-driven		DATA PROC	(→tastaturgesteuert)		
(→keyboard-controlled)					
keyboard driver		TERM&PER	Tastaturtreiber		
			= Tastenfeldtreiber		
keyboard encoder		TERM&PER	(→Tastaturcodierer)		
(→keyboard coder)					
keyboard entry		DATA PROC	Tastatureingabe		
= keyboard input; keyboarding; key entry			= Tasteneingabe		
keyboard entry station		TERM&PER	(→Tastaturterminal)		
(→keyboard terminal)					
keyboarding (→keying)		TERM&PER	(→Eintastung)		
keyboarding (→keyboard entry)		DATA PROC	(→Tastatureingabe)		
keyboarding error (→keying error)		TERM&PER	(→Eingabefehler)		
keyboarding mistake		TERM&PER	(→Eingabefehler)		
(→keying error)					
keyboard input (→keyboard entry)		DATA PROC	(→Tastatureingabe)		
keyboard layout		TERM&PER	Tastaturbelegung		
			= Tastenbelegung		
keyboard lock		TERM&PER	Tastatursperre		
≈ keylock			≈ Tastensperre		
keyboard overlay (→key overlay)		TERM&PER	(→Funktionstasten-Beschriftungsstreifen)		
keyboard perforator (→card puncher)		TERM&PER	(→Kartenlocher)		
keyboard punch (→card puncher)		TERM&PER	(→Kartenlocher)		
keyboard selection		SWITCH	(→Tastwahl)		
(→pushbutton dialing)					
keyboard send/receive		TERM&PER	(→Tastaturterminal)		
(→keyboard terminal)					
keyboard strip (→key overlay)		TERM&PER	(→Funktionstasten-Beschriftungsstreifen)		
keyboard switch (→pushbutton key)		COMPON	(→Tastschalter)		
keyboard template		TERM&PER	Tastaturschablone		
keyboard terminal		TERM&PER	Tastaturterminal		
= keyboard entry station; key station; keyboard send/receive; KSR			= Tastatureingabestation		
keyboard-to-disk system		TERM&PER	Magnetplatten-Eintastsystem		
keyboard-to-disk unit		TERM&PER	Magnetplatten-Eintastgerät		
keyboard-to-tape system		TERM&PER	Magnetband-Eintastsystem		
keyboard-to-tape unit		TERM&PER	Magnetband-Eintastgerät		
keyboard transmission		TELEGR	Handsendung		
key body		COMPON	Tastenkörper		
key bounce		TERM&PER	Tastenprellung		
key button (→keytop)		TERM&PER	(→Tastenknopf)		
key cap		COMPON	Tastenkuppe		
			= Tastenkopf		
key coding switch		COMPON	Tast-Codierschalter		
			↑ Codierschalter		
key connector		COMPON	Schlüsselstecker		
key-controlled		DATA PROC	(→tastaturgesteuert)		
(→keyboard-controlled)					
key data		DATA PROC	Ordnungsdaten		
≈ ordering argument			≈ Ordnungsbegriff		
key data entry device		TERM&PER	Tastatur-Dateneingabegerät		
= keyboard data entry device					
key day		ECON	Stichtag		
= cut-off day					
key depression		TERM&PER	Tastendruck		
= keystroke			= Tastenanschlag		
key disk		DATA PROC	Schlüsseldiskette		
key-driven		DATA PROC	(→tastaturgesteuert)		
(→keyboard-controlled)					
key drop		TERM&PER	Tastenhub		
keyed control		CONTROL	Tastregelung		
= sampled data control			= Abtastregelung; getastete Regelung; Taststeuerung; getastete Steuerung		
key entry (→keyboard entry)		DATA PROC	(→Tastatureingabe)		
key field		DATA PROC	Schlüsselfeld		
[magnetic disk]			[bei Magnetplattenspeichern freier Formatierung]		
= key					
≈ identifier			≈ Spurenkennblock		
key function		DATA PROC	Tastenfunktion		
			= Tastaturfunktion		
key industry		TECH	Schlüsselindustrie		
keying		TERM&PER	Eintastung		
= keyboarding			= Eintasten		
			≈ Eingabe		
keying		MODUL	Tastung		
keying error		TERM&PER	Eingabefehler		
= keying mistake; keyboarding error; keyboarding mistake			= Eintastfehler; Tippfehler [OFFICE]; Fehleintastung		

keying mistake (→keying error) — TERM&PER — (→Eingabefehler)
keying ratio (→pulse duty ratio) — ELECTRON — (→Tastverhältnis)
key interrogation program — DATA PROC — **Tastenabfrageprogramm**
key lever — TERM&PER — **Tasthebel**
keylock — TECH — **Sperrschloß**
keylock — TERM&PER — **Tastensperre**
≈ keyboard lock — ≈ Tastatursperre
key number — TECH — **Eckwert**
key number (→keyword 1) — DATA PROC — (→Paßwort)
key on flat (→plain taper key) — MECH — (→Flachkeil)
key-operated switch (→keyswitch 1) — COMPON — (→Schlüsselschalter)
key overlay — TERM&PER — **Funktionstasten-Beschriftungsstreifen**
[exchangeable strip describing the key functions for a specific software] [auswechselbarer Streifen mit Angabe der für eine Software gültigen Tastaturbelegung]
= keyboard overlay; key strip; keyboard strip
= Tastaturbelegungsstreifen
keypad 1 — TERM&PER — **Kleintastatur**
= small keyboard
keypad 2 — TERM&PER — **Tastaturfeld**
[section of a keyboard] [Bereich einer Tastatur]
= key panel; pushbotton panel = Tastenfeld; Tastenblock; Tastaturblock; Block 2; Tastaturbereich
≈ keyboard ≈ Tastatur
↓ alphanumeric pad; cursor pad; numeric keypad ↓ alphanumerisches Tastaturfeld; Cursorblock; Ziffernblock
keypad 2 (→cursor pad) — TERM&PER — (→Schreibmarkenblock)
key panel (→keyboard) — TERM&PER — (→Tastatur)
key panel (→keypad 2) — TERM&PER — (→Tastaturfeld)
key pressure — TERM&PER — **Tastenkraft**
keypunch (→card puncher) — TERM&PER — (→Kartenlocher)
keyseat — MECH — **Wellennut**
key shift — TERM&PER — **Tastenumschaltung**
key signal — CODING — **Schlüsselzeichen**
key station (→keyboard terminal) — TERM&PER — (→Tastaturterminal)
keystone distortion (→trapezium distortion) — TV — (→Trapezfehler)
key strip — COMPON — **Tastenstreifen**
key strip (→key overlay) — TERM&PER — (→Funktionstasten-Beschriftungsstreifen)
keystroke (→key depression) — TERM&PER — (→Tastendruck)
keystroke verification — DATA PROC — **Tastenanschlagkontrolle**
keyswitch 1 — COMPON — **Schlüsselschalter**
= key switch 1; key-operated switch
key switch 2 (→pushbutton key) — COMPON — (→Tastschalter)
key switch 1 (→keyswitch 1) — COMPON — (→Schlüsselschalter)
key system — TELEPH — **Reihenanlage**
= Key-Anlage
key technology — TECH — **Schlüsseltechnologie**
= Schrittmachertechnologie
key-to-address (→hash coding) — DATA PROC — (→Hash-Codierung)
keytop — TERM&PER — **Tastenknopf**
= key button
key-verifying unit (→punched card verifier) — TERM&PER — (→Kartenprüfer)
keyway — MECH — **Nabennut**
= Keilnut
keyword 1 — DATA PROC — **Paßwort**
= password; right-of-access code; user identification; user id; identifier word; id; code number; key number
= Berechtigungszeichen; Berechtigungsschlüssel; Berechtigungscode; Kennwort; Kennungswort; Schlüsselwort 2; Codewort; Benutzeridentifikation; Benutzerkennung
keyword — LING — **Schlagwort**
= slogan; buzzword; catchphrase
keyword 2 (→descriptor) — DATA PROC — (→Beschreiber)

key-word-in-context (→KWIC) — DATA PROC — (→zusammenhangbezogenes Schlüsselwort)
keyword macro — DATA PROC — **Schlüsselwort-Makro**
K-factor — RADIO PROP — **Krümmungsfaktor k**
= effective-earth-radius factor
K factor (→antenna factor) — ANT — (→Antennenfaktor)
kg (→kilogram) — PHYS — (→Kilogramm)
kHz (→kilocycle/s) — PHYS — (→Kilohertz)
Kichhoff's loop law — EL.TECH — **zweites Kirchhoffsches Gesetz**
= Maschenregel
kill (→abort) — DATA PROC — (→abbrechen)
kill job (→halt command) — DATA PROC — (→Abbruchbefehl)
kilobit — INF — **Kilobit**
= kbit = kbit
kilobyte — DATA PROC — **Kilobyte**
[2E10 = 1,024 bytes] [zwei hoch zehn = 1.024 Bytes]
= KByte
kilo character — DATA PROC — **tausend Zeichen**
= kc = Kilozeichen
kilocycle/s — PHYS — **Kilohertz**
= kilohertz; kHz = kHz
kilogram — PHYS — **Kilogramm**
[SI unit for mass] [SI-Basiseinheit für Masse]
= kg = kg
kilohertz (→kilocycle/s) — PHYS — (→Kilohertz)
kilometer (NAM) — PHYS — **Kilometer**
[1,000 m] [1.000 m]
= kilometre (BRI); km = km
kilometers per hour — TECH — **Stundenkilometer**
= kmph; km/h = Kilometer pro Stunde; km/h
≈ mph ≈ Meilenkilometer
kilometre (BRI) (→kilometer) — PHYS — (→Kilometer)
kilometric waves — RADIO — **Langwellen**
[10km-1km; 30kHz-300kHz] [10km-1km; 30kHz-300kHz]
= medium frequencies; MF; long waves; LW = LW; Kilometerwellen; MF
kilopond — PHYS — **Kilopond**
[unit for weight; = 9.80665 N] [Einheit für Gewicht; = 9,80665 N]
= kp = kp
kilopond-meter — PHYS — **Kilopondmeter**
= kp m = kp m
kilowatt-hour — PHYS — **Kilowattstunde**
[= 3.6 Megajoule] [= 3,6 Megajoule]
= kWh = kWh
kiloword — DATA PROC — **Kiloworte**
[1,026 words] [1.024 Worte]
= kW = kW
kind — TECH — **Sorte**
= sort n.; variety
kind of business (→branch 1) — ECON — (→Branche)
kind of transport (→transportation mode) — ECON — (→Transportart)
kinematic model — DATA PROC — **Kinematikmodell**
kinematics — PHYS — **Kinematik**
[theory of time-dependence of movements] [Theorie der Bewegungen in Abängigkeit der Zeit]
≈ kinetics ≈ Kinetik
kinematic viscosity — PHYS — **kinematische Viskosität**
[SI unit: m^2/s] [SI-Einheit: m^2/s]
kinetic energy — PHYS — **kinetische Energie**
= Energie der Bewegung; aktuelle Energie; Wucht
KIPS — DATA PROC — **KIPS**
[Kilo Instructions Per Second; measure of computer power] [1000 Anweisungen pro Sekunde; Maß für Computerleistung]
Kirchhoff's laws — EL.TECH — **Kirchhoffsches Gesetz**
= Kirchhoff's rules = Kirchhoffsche Sätze; Kirchhoffsche Regeln
Kirchhoff's node law — EL.TECH — **erstes Kirchhoffsches Gesetz**
= Knotenregel
Kirchhoff's rules (→Kirchhoff's laws) — EL.TECH — (→Kirchhoffsches Gesetz)
kit (n.) — ELECTRON — **Bausatz**
= Teilesatz; Kit
≈ mounting kit [TECH] ≈ Montagesatz [TECH]

kit (n.)		TECH	**Baukasten**	**Krarup cable**	COMM.CABLE	(→Krarup-Kabel)
= erection kit				(→continuously loaded cable)		
kite antenna		ANT	**Drachenantenne**	**Krarup loading**	COMM.CABLE	(→Krarupisierung)
klaxon (→horn)		EL.ACOUS	(→Hupe)	(→continuous loading)		
kludge		DATA PROC	**Flickschusterei**	**k-rating** (→harmonic content)	EL.TECH	(→Oberwellengehalt)
[provisional correction of deficient software]			[behelfsmäßige Korrektur einer mangelhaften Software]	**Kraus formula**	ANT	**Kraus-Formel**
= kluge; makeshift				**KS capacitor**	COMPON	**KS-Kondensator**
kluge (→kludge)		DATA PROC	(→Flickschusterei)	↑ polystyrene capacitor		↑ Polystyrolkondensator
klystron		MICROW	**Klystron**	**KSR** (→keyboard terminal)	TERM&PER	(→Tastaturterminal)
≠ linear-beam tube			↑ Triftröhre	**KT capacitor**	COMPON	**KT-Kondensator**
↓ reflex klystron			↓ Reflexklystron	**K-type carrier system**	TRANS	**Z12-System**
km (→kilometer)		PHYS	(→Kilometer)	[FDM carrier system]		[TF-Technik]
km/h (→kilometers per hour)		TECH	(→Stundenkilometer)	**Ku** (→Kurchatovium)	CHEM	(→Kurtschatowium)
kmph (→kilometers per hour)		TECH	(→Stundenkilometer)	**Kurchatovium**	CHEM	**Kurtschatowium**
knee		MATH	**Stufe**	= Ku		= Ku
[curve]			[Kurve]	**kW** (→kiloword)	DATA PROC	(→Kiloworte)
= step (n.)				**kWh** (→kilowatt-hour)	PHYS	(→Kilowattstunde)
knee (→bend)		MECH	(→Knick)	**KWIC**	DATA PROC	**zusammenhangbezogenes Schlüsselwort**
knee voltage		MICROEL	**Kniespannung**	= key-word-in-context		
= collector saturation voltage			= Kollektor-Restspannung			
knife		TECH	**Messer**	**L**		
↑ cutting tool			↑ Schneidewerkzeug			
knife-blade connector		COMPON	**Messersteckverbinder**	**l** (→length)	PHYS	(→Länge 1)
knife-blade contact		COMPON	**Messerkontakt**	**L** (→luminance)	PHYS	(→Leuchtdichte)
knife-edge diffraction		RADIO PROP	**Messerkantenbeugung**	**L** (→low level)	MICROEL	(→Tiefpegelzustand)
= knife-edge effect				**L** (NAM) (→liter)	PHYS	(→Liter)
knife-edge effect		RADIO PROP	(→Messerkantenbeugung)	**L** (→Lambert)	PHYS	(→Lambert)
(→knife-edge diffraction)				**La** (→lanthanum)	CHEM	(→Lanthan)
knob		TECH	**Knopf**	**lab** (→laboratory)	SCIE	(→Laboratorium)
↑ handle			↑ Handgriff 1	**lab bench** (→laboratory bench)	TECH	(→Labortisch)
knob		COMPON	**Knopf**	**label 1** (n.)	DATA PROC	**Etikett**
↓ rotary knob			= Bedienungsknopf; Bedienknopf; Geräteknopf	[data block to mark and protect files on magnetic layer memories]		[Datenblock auf Magnetschichtspeichern zur Markierung und Sicherung von Dateien]
			↓ Drehknopf	= label record; file label		
knob (→rotary knob)		COMPON	(→Drehknopf)	↓ volume label; header label; trailer label; section label; checkpoint label; format label; standard label; user label; tape label		= Kennsatz; Marke; Dateietikett
knob dial		COMPON	**Knopfskala**			≈ Vorspann
knock-in bolt		MECH	**Einschlagschraube**			↓ Datenträger-Etikett; Dateianfangs-Etikett; Dateiend-Etikett; Abschnitts-Etikett; Fixpunkt-Etikett; Format-Etikett; Standardetikett; Benutzeretikett; Bandetikett
knockresistant (→antiknock)		QUAL	(→klopffest)			
knot (n.)		TECH	**Knoten**			
≈ node [MATH]			[in einer Schleife]			
			≈ Knoten [MATH]			
know how (→technical knowledge)		SCIE	(→Fachkenntnis)			
know-how agreement		ECON	**Know-how-Vertrag**	**label** (n.)	TECH	**Kennzeichnungsschild**
			≈ Lizenzvertrag	**label 2** (n.)	DATA PROC	**Sprungmarke**
know-how bearer		SCIE	**Wissensträger**	= name (n.)		
knowledge base		DATA PROC	**Wissensbasis**	**label** (n.) (→designation label)	TECH	(→Beschriftungsschild)
[data on facts and procedures of a speciality]			[Daten über Fakten und Verfahren eines Fachgebietes]	**label** (v.t.) (→identify)	TECH	(→kennzeichnen) (v.t.)
↑ expert system			↑ Expertensystem	**label 3** (→address)	DATA PROC	(→Adresse)
knowledge engineer		DATA PROC	**Wissensingenieur**	**label check**	DATA PROC	**Etikettprüfung**
knowledge engineering		DATA PROC	**Wissenstechnik**	= tag check; identifier check		= Kennsatzprüfung
knowledge industry		DATA PROC	**Wissensindustrie**	**labeled** (AM)	TECH	**gekennzeichnet**
knowledge of foreign language		LING	(→Sprachkenntnisse)	= labelled (BRI); marked		= markiert
(→linguistic attainments)				**label field**	DATA PROC	**Etikettenfeld**
knowledge processing		DATA PROC	(→Wissensbearbeitung)	**label group**	DATA PROC	**Etikettfolge**
(→knowledge work)				= tag group; identifier group		= Kennsatzgruppe
knowledge representation		DATA PROC	**Wissensrepräsentation**	**label identifier**	DATA PROC	**Etikettenkennzeichen**
knowledge work		DATA PROC	**Wissensbearbeitung**	= tag; identifying character		= Kennsatzname; Identifizierungskennzeichen
= knowledge processing			= Wissensverarbeitung			
knuckle (→bend)		MECH	(→Knick)	**labeling** (AM)	TECH	**Etikettierung**
knuckle joint		MECH	**Kniegelenk**	= labelling (BRI)		
			= Gabelgelenk	**labeling** (→lettering)	EQUIP.ENG	(→Beschriftung)
knuckle pin		MECH	**Gelenkbolzen**	**labeling** (→identification)	TECH	(→Kennzeichnung)
knuckle thread		MECH	**Rundgewinde**	**labeling document** (→lettering plan)	TECH	(→Beschriftungsplan)
knuckle yoke		MECH	**Gelenkgabel**			
knurle (n.)		MECH	**Rändel**	**labeling machine**	TERM&PER	**Etikettiermaschine**
knurled		MECH	**gerändelt**	**labeling plan** (→lettering plan)	TECH	(→Beschriftungsplan)
knurled disk		MECH	**Rändelscheibe**			
= thumbwheel				**labeling space** (→lettering space)	EQUIP.ENG	(→Beschriftungsfeld)
knurled nut		MECH	**Rändelmutter**			
knurled screw		MECH	**Rändelschraube**	**labeling strip** (→designation strip)	EQUIP.ENG	(→Beschriftungsstreifen)
kovar		METAL	**Kovar**			
kp (→kilopond)		PHYS	(→Kilopond)	**labeling system** (→marking system)	SIGN.ENG	(→Auszeichnungssystem)
KP capacitor		COMPON	**KP-Kondensator**			
= polypropylene capacitor						
kp m (→kilopond-meter)		PHYS	(→Kilopondmeter)			
Kr (→crypton)		CHEM	(→Krypton)			

labelled

labelled (BRI) (→labeled)	TECH	(→gekennzeichnet)
labelling (BRI) (→labeling)	TECH	(→Etikettierung)
label print program	DATA PROC	**Etikettendruckprogramm**
label record (→label 1)	DATA PROC	(→Etikett)
labile (→unstable)	TECH	(→instabil)
lability (→instability)	TECH	(→Instabilität)
labor	ECON	**Arbeitskräfte**
laboratory	SCIE	**Laboratorium**
= lab		= Labor
laboratory application (→bench application)	ELECTRON	(→Laboranwendung)
laboratory bench	TECH	**Labortisch**
= lab bench; bench		
laboratory-bench (→for laboratory use)	INSTR	(→für Laboreinsatz)
laboratory engineer	TECH	**Laboringenieur**
laboratory probe	INSTR	**Labormeßfühler**
laboratory prototype	TECH	**Labormuster**
= engineering model		= Entwicklungsmuster
laboratory technician	SCIE	**Laborant**
		≈ Laborantin
labor force (→workforce)	ECON	(→Belegschaft)
labor-intensive	ECON	**personalintensiv**
labor union (AM) (→union)	ECON	(→Gewerkschaft)
labour force	ECON	**Arbeitskraft**
= manpower; workforce		
labour market	ECON	**Arbeitsmarkt**
labour union (BRI) (→union)	ECON	(→Gewerkschaft)
lace (v.t.)	TECH	**schnüren**
≈ strangulate		≈ einschnüren
lace (→cord)	TECH	(→Schnur)
lace (→interlace)	TECH	(→verflechten)
lace, a cable	SYS.INST	**binden, ein Kabel**
lacing	TERM&PER	**Vollochung**
lacing board (→forming board)	MANUF	(→Kabelformbrett)
lack	ECON	**Mangel**
= fault		= Fehlen
≈ shortage		≈ Knappheit
lack of space	TECH	**Raummangel**
= space restriction		
lacquer (n.)	CHEM	**Lack**
= varnish 2 (n.)		≈ Firnis
≈ varnish 1 (n.)		
lacquer (n.)	TECH	**Siegellack**
lacquered	TECH	**lackiert**
= varnished		
lacquer-free	TECH	**lackfrei**
lacquer seal	TECH	**Lacksicherung**
lacquer stamp	TECH	**Lackstempel**
ladder	TECH	**Leiter**
		[Steiggerät]
ladder antenna	ANT	**Leiterantenne**
ladder network	NETW.TH	**Kettenschaltung**
		= Abzweigschaltung; Leiternetzwerk
ladder-type filter	NETW.TH	**Siebkette**
lade (→load)	TECH	(→beladen)
lag (v.i.)	TECH	**nacheilen**
lag (n.) (→time-lag)	TECH	(→Nacheilung)
lag (n.) (→offset)	TECH	(→Versatz)
lag (→delay)	ELECTRON	(→Verzögerungszeit)
lag angle (→trailing angle)	PHYS	(→Nacheilwinkel)
lagging (→trailing)	TECH	(→nacheilend)
lagging angle (→trailing angle)	PHYS	(→Nacheilwinkel)
lake	TECH	**Farblack**
lambda dipole (→full-wave dipole)	ANT	(→Ganzwellendipol)
Lambert	PHYS	**Lambert**
[unit for brightness; 1/π sb]		[Maßeinheit für Leuchtdichte; 1/π sb]
= L		= L
Lamb shift	PHYS	**Lambsche-Verschiebung**
laminar	PHYS	**laminar**
≠ turbulent		≠ turbulent
laminar box	MANUF	**Laminarbox**
laminar flow	PHYS	**laminare Strömung**
		= Laminarströmung
laminar layer	PHYS	**laminare Schicht**
		= Laminarschicht
laminate	TECH	**schichten**
		= lamellieren
laminated (→stratified)	TECH	(→geschichtet)
laminated aerial (→laminated antenna)	ANT	(→geschichtete Antenne)
laminated antenna	ANT	**geschichtete Antenne**
= laminated aerial		
laminated core	EL.TECH	**Schnittbandkern**
= tape-wound core; tape-wounded core; tape core; core stack; core assembly		= Blechpaket; Bandkern; bandumwickelter Kern
laminated paper	ELECTRON	**Hartpapier**
= bakelized paper; phenolic paper		
lamination	METAL	**Walzen**
= rolling		
lamp	COMPON	**Lampe**
lamp cap	COMPON	**Lampenkappe**
lamp extractor	COMPON	**Lampenzieher**
lamp holder (→lamp socket)	COMPON	(→Lampenfassung)
lamp indication	EQUIP.ENG	**Lampenanzeige**
≈ indication lamp		≈ Anzeigelampe
lamp key (→luminous key)	COMPON	(→Leuchttaste)
lamp panel	EQUIP.ENG	**Lampenfeld**
lamp socket	COMPON	**Lampenfassung**
= lamp holder; clamp		
lamp strip	COMPON	**Lampenstreifen**
LAN (→local area network)	DATA COMM	(→lokales Netz)
lance (→pierce)	METAL	(→lochen)
lance pointer	INSTR	**Lanzenzeiger**
↑ pointer		↑ Zeiger
lancing	METAL	**Einstechen 1**
lancing (→louvering)	METAL	(→Durchreißen)
land	MICROEL	**Kontaktfleck**
land (→soldering land)	ELECTRON	(→Lötauge)
landed hole	ELECTRON	**Lötaugenloch**
[PCB]		[Leiterplatte]
Landé factor	PHYS	**Landé-Faktor**
landing	OUTS.PLANT	**Anlandung**
[sea cable]		[Seekabel]
landing	AERON	**Landung**
landing aid	RADIO NAV	**Landungsanweisung**
landing beacon (→approach radiobacon)	RADIO NAV	(→Landebake)
landing beam beacon	RADIO NAV	**Einflugzeichensender**
landing direction	RADIO NAV	**Anfluggrundlinie**
landing path	AERON	**Landeweg**
= approach path		= Anflugweg
landing point	OUTS.PLANT	**Anlandepunkt**
[sea cable]		[Seekabel]
		= Seekabellandepunkt
landing system	RADIO NAV	**Landesystem**
landing zone	TERM&PER	**Landezone**
[blank track to park a magnetic head]		[Leerspur zum Parken eines Manetkopfes]
landless	MICROEL	**anschlußlos**
land mobile	TELEC	**Landfunk**
land pattern	ELECTRON	**Lötaugenbild**
[PCB]		[Leiterplatte]
= pad pattern		
land radio service	TELEC	**Landfunkdienst**
landscape (→landscape format)	TYPOGR	(→Querformat)
landscape format	TYPOGR	**Querformat**
= landscape		≠ Hochformat
≠ portrait format		
landscape printing	TERM&PER	**Querdruck**
land transportation vehicle	TECH	**Landfahrzeug**
language	LING	**Sprache**
[an agreed set of vowels and characters to transmit information]		[für Informationsübermittlung vereinbarter Satz von Lauten und Symbolen]
≈ national language; foreign language; dialect		= Sprachsystem
		≈ Landessprache; Fremdsprache; Dialekt
language (→programming language)	DATA PROC	(→Programmiersprache)
language C	DATA PROC	**Sprache C**
[derived from UNIX, for PC's]		[für PC's aus UNIX abgeleitet]
↑ high-level programming language		= C
		↑ problemorientierte Programmiersprache

language code	DOC	Sprachkennzeichen	large-area (adj.)	TECH	großflächig
language department	ECON	Sprachendienst	↑ large		= ausgedehnt
[for interpreter services and translations]		[Organisationseinheit für Übersetzungs- und Dolmetscherdienste]			↑ groß
			large area office (→open-plan office)	OFFICE	(→Großraumbüro)
↓ translations department		↓ Übersetzungsbüro	large base direction finder	RADIO LOC	(→Großbasispeiler)
language description	DATA PROC	Sprachbeschreibungssprache	(→wide-aperture direction finder)		
language		↑ Metasprache	large-base principle	RADIO LOC	(→Großbasisverfahren)
= LDL			(→wide-aperture principle)		
↑ metalanguage			large-capacity	COMM.CABLE	hochpaarig
language editor	DATA PROC	Spracheneditor	large color screen	TERM&PER	Farbgroßbildschirm
language interpreter	DATA PROC	(→Interpretierer 1)	large computer (→mainframe computer)	DATA PROC	(→Großrechner)
(→interpreter 1)					
language level	DATA PROC	Sprachebene	large format	TECH	Großformat
language processor	DATA PROC	(→Übersetzer)	large-format drafting plotter	TERM&PER	Großformat-Plotter
(→translator)					
language prompt	DATA PROC	Aufforderungstext	large-format scanner	TERM&PER	Großformat-Scanner
↑ promt		↑ Bereitmeldung	large-power- (→high-performance ...)	TECH	(→Hochleistungs-)
language rule	DATA PROC	Sprachregel			
≈ syntax		≈ Syntax	large quantity (→bulk)	COLLOQ	(→Unmenge)
language subset	DATA PROC	Teilsprache	large scale ...	TECH	Massen-
language translation	DATA PROC	Sprachübersetzung	large-scale application	TECH	Massenanwendung
language translator	DATA PROC	(→Übersetzer)	large-scale changeover	SWITCH	Massenumschaltung
(→translator)			= bulk changeover		
L antenna	ANT	L-Antenne	large scale computer	DATA PROC	(→Großrechner)
lanthanum	CHEM	Lanthan	(→mainframe computer)		
= La		= La	large-scale IC (→large-scale integrated circuit)	MICROEL	(→hochintegrierte Schaltung)
lap (v.t.)	MECH	läppen			
lap 1 (n.)	TECH	Überlappungsgrad	large-scale integrated	MICROEL	hochintegriert
[the amount overlapping]			= highly integrated		
lap 2 (→overlapping)	TECH	(→Überlappung)	large-scale integrated circuit	MICROEL	hochintegrierte Schaltung
LAP (→link access procedure)	DATA COMM	(→Leitungszugangsverfahren)	= LSIC; highly-integrated circuit; large-scale IC; highly integrated IC		= LSIC; hochintegrierter Schaltkreis; gruppenintegrierte Schaltung; gruppenintegrierter Schaltkreis
lap computer (→laptop computer)	DATA PROC	(→Aktentaschencomputer)			
lap-dissolve (v.t.)	TV	überblenden	large-scale integration (→LSI)	MICROEL	(→Großintegration)
[gradual transition from one picture to another]		[gradueller Übergang von einem Bild in ein anderes]	large-scale manufacturing	MANUF	Großserienfertigung
			large-scale production (→mass production)	MANUF	(→Massenproduktion)
= dissolve (v.t.); cross-fade		↑ mischen			
↑ mix			large screen	TERM&PER	Großbildschirm
lap-dissolve (n.)	TV	Überblendung	large-screen terminal	TERM&PER	Großbildschirm-Terminal
= dissolve (n.); cross fade (n.); fade (n.)			large signal	ELECTRON	Großsignal
lap dissolve shutter	TV	Überblendeinrichtung	= high level signal		
lapel microphone	EL.ACOUS	Knopflochmikrophon	large-signal amplification	CIRC.ENG	Großsignalverstärkung
= button microphone			large-signal amplifier	CIRC.ENG	Großsignalverstärker
lapheld (→laptop computer)	DATA PROC	(→Aktentaschencomputer)	large-size cable (→high-capacity cable)	COMM.CABLE	(→hochpaariges Kabel)
lapheld computer (→laptop computer)	DATA PROC	(→Aktentaschencomputer)	large user	TELEC	Großteilnehmer
			≈ business subscriber		≈ Geschäftsteilnehmer
Laplace differential equation	MATH	Laplace-Differentialgleichung	Larmor frequency	PHYS	Larmorfrequenz
= Laplace's equation		= Laplace-Potentialgleichung	lase-beam saw	MICROEL	Lichtsäge
Laplace transform	MATH	Laplace-Transformation	laser	OPTOEL	Laser
laplacian	MATH	Laplace-Operator	[light amplification by stimulated emission of radiation]		= Quanten-Generator (DDR); optischer Maser
		[Vektorrechnung]			
		= Laplacescher Operator	laser altimeter	INSTR	Laser-Höhenmesser
lapped	MECH	geläppt	laser amplifier	OPTOEL	Laserverstärker
lapping	MECH	Läppen	laser anemometer	INSTR	Laseranemometer
lapping (→overlap)	POWER SYS	(→Überlappung)	laser calibration	MANUF	Laserabgleich
lapping (→serving)	COMM.CABLE	(→Umwicklung)	= laser trimming		= Lasertrimmen
lapping machine	COMM.CABLE	Umwicklungsmaschine	laser calorimeter	INSTR	Laser-Kalorimeter
lap-size computer (→laptop computer)	DATA PROC	(→Aktentaschencomputer)	laser chirping	OPTOEL	Laserchirpen
			laser cut-off	TRANS	Laserabschaltung
laptop (→laptop computer)	DATA PROC	(→Aktentaschencomputer)	laser diode	OPTOEL	Laserdiode
laptop computer		Aktentaschencomputer	laser diode source	INSTR	Laserdiodenquelle
[a complete PC with the size of a briefcase; lap = person's knees when sitting]		[kompletter Personalcomputer in Aktentaschenformat; lap (Engl.) = Schoß]	= LD source		
			laser display	TERM&PER	Laserbildschirm
					= Laser-Bildplatte
		= Laptop-Computer; Portable-Computer; Laptop; Lapheld Computer; Lapheld	laser drill	TECH	Laserbohrer
= lap-size computer; laptop; lapheld computer; laphold; lap computer; briefcase computer			laser facsimile equipment	TERM&PER	Laser-Faksimilegerät
			= laser fax		= Laserfax
≈ hand-held computer		≈ Taschencomputer	laser fax (→laser facsimile equipment)	TERM&PER	(→Laser-Faksimilegerät)
↑ personal computer; portable computer		↑ Personalcomputer; tragbarer Computer			
			laser gyroscope	INSTR	Lasergyroskop
LARAM	MICROEL	LARAM	laser head	INSTR	Laserkopf
= line-addressable random-access memory			laser imager (→laser typesetter)	TERM&PER	(→Laserbelichter)
			laser oscillator	OPTOEL	Laseroszillator
large (adj.)	TECH	groß	laser position transducer	INSTR	Laser-Positionsgebersystem
↓ voluminous; large-area; long; high		↓ voluminös; großflächig; lang; hoch			

laser printer

laser printer	TERM&PER	**Laserdrucker**	
↑ non impact printer; electrophotographic printer		↑ anschlagfreier Drucker; elektrofotografischer Drukker	
laser storage	TERM&PER	**Laserspeicher**	
laser storage disk (→optical disk)	TERM&PER	(→Bildplatte)	
laser trimming (→laser calibration)	MANUF	(→Laserabgleich)	
laser typesetter	TERM&PER	**Laserbelichter**	
[photocomposition]		[Lichtsatz]	
= laser imager		= Lasersatzanlage	
lash (v.t.)	TECH	**laschen**	
		= festbinden	
lashing (→strapping)	TECH	(→Laschung)	
lashing wire	OUTS.PLANT	**Laschdraht**	
last-choice route	SWITCH	**Letztweg**	
= final route		≈ Kennzahlweg	
≈ hierarquical route			
last-choice trunk group	SWITCH	**Letztwegbündel**	
last-in-first-out mode (→LIFO mode)	DATA PROC	(→LIFO-Modus)	
last look mode	TELEPH	**Last-look-Verfahren**	
last number redial (→redialing)	TELEPH	(→Wahlwiederholung)	
last year (→preceding year)	COLLOQ	(→Vorjahr)	
latch (v.t.)	TECH	**einklinken**	
latch (v.i.)	ELECTRON	**einrasten**	
[to set a state]		= einklinken; verklinken; schalten auf	
latch (n.)	CIRC.ENG	**Latch**	
[type of trigger circuit]		[„Klinke"; eine Art Kippschaltung]	
latch (n.)	DATA PROC	**Signalspeicher**	
[a special buffer memory]		= Auffang-Flipflop	
latch (→address latch)	SWITCH	(→Adressenspeicher)	
latch (→catch)	MECH	(→Schnapper)	
latching current	MICROEL	**Einraststrom**	
[thyristor]		[Thyristor]	
latching mechanism	TECH	**Verklinkung**	
latching relay (→remanent relay)	COMPON	(→Haftrelais)	
latching solenoid	COMPON	**Einklinkspule**	
		= Verklinkspule; Einrastspule	
latch-up effect	MICROEL	**Latch-up-Effekt**	
[undesired holding of semiconductor devices]		[unerwünschtes Klemmen von Halbleiterbausteinen]	
		= Einklink-Effekt	
late (→belated)	TECH	(→verspätet)	
late failure	QUAL	**Spätausfall**	
≈ ageing failure		≈ Ermüdungsausfall	
≠ early failure		≠ Frühausfall	
latency	SCIE	**Latenz**	
latency (→latency time)	TERM&PER	(→Latenzzeit)	
latency (→response time)	TECH	(→Reaktionszeit)	
latency time	TERM&PER	**Latenzzeit**	
= latency		↑ Zugriffszeit	
↑ access time			
latent	SCIE	**latent**	
lateral (adj.)	TECH	**seitlich**	
lateral adjustment	MECH	**Seitenverstellung**	
lateral correction magnet	TV	**Lateralmagnet**	
		= Blau-Lateralmagnet; Blau-Schiebemagnet	
lateral cover (→side cover)	EQUIP.ENG	(→Seitenabdeckung)	
lateral dimension	PHYS	**Querabmessung**	
≈ width		≈ Breite	
lateral guide	MECH	**Seitenführung**	
lateral rail	EQUIP.ENG	**Seitenschiene**	
lateral sound	EL.ACOUS	**Seitenschall**	
lateral tower	ANT	**Außenmast**	
lateral transistor	MICROEL	**Lateral-Transistor**	
late shift	ECON	**Spätschicht**	
≈ night shift		≈ Nachtschicht	
≠ early shift		≠ Frühschicht	
↑ shift		↑ Schicht	
lathe 2 (n.)	TECH	**Drehbank**	
lathe 1 (n.)	TECH	**Drehen**	
		[spanabhebende Bearbeitung]	
latin square	MATH	**lateinisches Quadrat**	
latitude	GEOPHYS	**geographische Breite**	
lattice	PHYS	**Gitter 1**	
[solid state]		[Festkörper]	
lattice	TECH	**Linienraster**	
[regular array of lines]			
lattice binding (→lattice bond)	PHYS	(→Gitterbindung)	
lattice bond	PHYS	**Gitterbindung**	
= lattice binding			
lattice conductivity	PHYS	**Gitterleitfähigkeit**	
[solid state]		[Festkörper]	
lattice constant	PHYS	**Gitterkonstante**	
[solid state]		[Festkörper]	
= lattice parameter; grating constant			
lattice defect (→lattice imperfection)	PHYS	(→Gitterfehler)	
lattice defect (→lattice vacancy)	PHYS	(→Gitterfehlstelle)	
lattice design	MATH	**Gitterplan**	
lattice dislocation (→lattice imperfection)	PHYS	(→Gitterfehler)	
lattice disorder (→lattice vacancy)	PHYS	(→Gitterfehlstelle)	
lattice distance	PHYS	**Gitterabstand**	
[solid state]		[Festkörper]	
lattice electron	PHYS	**Gitterelektron**	
lattice energy band	PHYS	**Gitterenergieband**	
lattice equivalent form	NETW.TH	**Differentialbrückenschaltung**	
≈ differential filter		≈ äquivalente Brückenschaltung	
		≈ Differentialfilter	
lattice filter	NETW.TH	**Brückenfilter**	
lattice filter (→all-pass filter)	NETW.TH	(→Allpaß)	
lattice imperfection (→crystal imperfection)	PHYS	(→Kristallbaufehler)	
lattice interstitial	PHYS	**Gitterzwischenplatz**	
= interstice		= Zwischengitterplatz	
lattice mast	CIV.ENG	**Gittermast**	
= lattice work mast; derrick-style mast; lattice pylon		≈ Gitterturm	
≈ lattice tower		↓ Stahlgittermast	
↓ steel lattice mast			
lattice parameter (→lattice constant)	PHYS	(→Gitterkonstante)	
lattice plane	PHYS	**Netzebene**	
= net plane			
lattice pylon (→lattice mast)	CIV.ENG	(→Gittermast)	
lattice section	NETW.TH	**X-Schaltung**	
↑ attenuator		= X-Glied	
		↑ Dämpfungsglied	
lattice structure	PHYS	**Gitterstruktur**	
= reticular structure			
lattice tower	CIV.ENG	**Gitterturm**	
≈ lattice mast		≈ Gittermast	
↓ steel lattice tower		↓ Stahlgitterturm	
lattice vacancy	PHYS	**Gitterfehlstelle**	
= vacancy; lattice void; void; lattice dislocation; dislocation; lattice disorder; disorder; lattice defect; defect		= Gitterlücke; Lücke; Gitterleerstelle; Leerstelle	
↑ lattice imperfection; imperfection [MICROEL]		↑ Kristallbaufehler; Störstelle [MICROEL]	
lattice vibration	PHYS	**Gitterschwingung**	
lattice void (→lattice vacancy)	PHYS	(→Gitterfehlstelle)	
lattice work mast (→lattice mast)	CIV.ENG	(→Gittermast)	
Laue pattern	PHYS	**Laue-Diagramm**	
launched power	OPT.COMM	**eingespeiste Leistung**	
= injected power			
launching fiber	OPT.COMM	**Vorlauffaser**	
Laurent transform	MATH	**Laurent-Transformation**	
		= zweiseitige Z-Transformation	
lava	PHYS	**lava**	
[color]		[Farbe]	
lavalier clip-on microphone (→clip-on microphone)	EL.ACOUS	(→Ansteckmikrofon)	

law	SCIE	Gesetz	
= rule		= Gesetzmäßigkeit	
law of gravity	PHYS	Gravitationsgesetz	
law of induction	PHYS	Induktionsgesetz	
law of large numbers	MATH	Gesetz der großen Zahlen	
law of magnetic flux	PHYS	Durchflutungsgesetz	
law of mass action	PHYS	Massenwirkungsgesetz	
Lawrence tube (→chromatron tube)	ELECTRON	(→Gitterablenkröhre)	
lawrencium	CHEM	Lawrencium	
= Lr		= Lr	
law suit	ECON	Prozeß	
= legal action; litigation		= Gerichtsverfahren; Rechtsstreit	
lawyer (AM)	ECON	Rechtsanwalt	
= solicitor (BRI); barrister; attorney 1; attorney-at-law (AM); barrister-at-law (BRI); legal adviser		= Anwalt; Rechtsberater	
lay (v.t.)	OUTS.PLANT	verlegen	
≈ bury		= legen	
		≈ eingraben	
lay (n.)	ENG.DRAW	Oberflächenzeichnung	
lay (n.)	COMM.CABLE	Schlaglänge	
[distance of complete twist]		= Schlag; Kabelschritt	
layer	COMPON	Lage	
[PCB]		[Leiterplatte]	
layer	COMM.CABLE	Lage	
layer	TECH	Schicht	
= film		= Film	
≈ coating		≈ Beschichtung	
layer	COMPON	Schicht	
[coil]		[Spule]	
layer	RADIO PROP	Schicht	
layer (→protocol layer)	DATA COMM	(→Protokollschicht)	
layered (→stratified)	TECH	(→geschichtet)	
layered structure	TECH	Mehrschichtenaufbau	
= sandwich structure; sandwich		= Mehrschichtstruktur; Sandwich-Struktur; Schichtstruktur	
layering	DATA PROC	Ausschnittbetrachtung	
[graphics program]		[Graphikprogramm]	
layer link	DATA COMM	Leitungssteuerung	
		[Protokollschicht]	
layer management	DATA COMM	Schichtenmanagement	
layer model	DATA COMM	Schichtenmodell	
= reference model		= Datenübertragungsmodell	
↓ ISO reference model		↓ ISO-Referenzmodell	
layer thickness	PHYS	Schichtdicke	
layer-thickness measurement	INSTR	Schichtdickenmessung	
layer twist	COMM.CABLE	Lagendrall	
layer twisting	COMM.CABLE	Lagenverseilung	
layer winding	EL.TECH	Zylinderwicklung	
[transformer]		[Transformator]	
≠ interleaved winding		= Lagenwicklung	
		≠ Scheibenwicklung	
laying (→cable laying)	OUTS.PLANT	(→Kabelverlegung)	
laying depth	OUTS.PLANT	Verlegetiefe	
– burying depth		= Verlegungstiefe; Legetiefe	
≈ covering depth		≈ Überdeckung	
laying length	OUTS.PLANT	Verlegelänge	
		= Legelänge	
laying ship (→cable ship)	TELEC	(→Kabelschiff)	
laying technique	OUTS.PLANT	Verlegetechnik	
		= Legetechnik	
layman (→layperson)	TECH	(→Laie)	
lay-out	TECH	Auslegung	
= design 2; make (n.); engineering		= Bauart	
= equipment 1		= Ausstattung	
layout	MICROEL	Strukturentwurf	
		= Layout	
layout (→configuration)	TECH	(→Aufbau 2)	
layout (→installation layout)	SYS.INST	(→Aufstellungsplan)	
layout (→face plan)	EQUIP.ENG	(→Bestückungsplan)	
layout (→project)	TECH	(→Plan)	
layout foil	ELECTRON	Layout-Folie	
layout list	SYS.INST	Aufstellungsliste	
layout plan (→installation layout)	SYS.INST	(→Aufstellungsplan)	
layout plan (→survey plan)	TECH	(→Übersichtsplan)	
layout sheet	ENG.DRAW	Entwurfspapier	
layperson	TECH	Laie	
= layman		= Nichtfachmann	
LB (→local battery)	TELEPH	(→Ortsbatterie)	
L/C (→letter of credit)	ECON	(→Akkreditiv)	
LCB (→line control block)	DATA COMM	(→Leitungssteuerungsblock)	
LCD (→LCD display)	COMPON	(→Flüssigkristallanzeige)	
LCD display	COMPON	Flüssigkristallanzeige	
= LCD; liquid crystal display		= LCD-Anzeige; LCD	
LCD screen (→liquid-crystal display monitor)	TERM&PER	(→Flüssigkristall-Bildschirm)	
LCDTL	MICROEL	LCDTL	
= load-compensated diode-transistor logic			
LC filter	NETW.TH	LC-Filter	
		= Reaktanzfilter	
LC filtering	NETW.TH	LC-Siebung	
LC generator	CIRC.ENG	LC-Generator	
= LC oscillator		= LC-Oszillator	
LC measuring generator	INSTR	LC-Meßgenerator	
		= LC-Signalgenerator	
LCN (→local communication network)	DATA COMM	(→lokales Kommunikationssystem)	
LC network	NETW.TH	LC-Zweipol	
LC oscillator (→LC generator)	CIRC.ENG	(→LC-Generator)	
LCP (→call control procedure)	SWITCH	(→Verbindungssteuerungsverfahren)	
LCR measuring bridge	INSTR	LCR-Meßbrücke	
= LCR meter		= LCR-Meßgerät	
↑ impedance measuring bridge		↑ Impedanz-Meßbrücke	
LCR meter (→LCR measuring bridge)	INSTR	(→LCR-Meßbrücke)	
LC section	NETW.TH	LC-Glied	
LCZ meter	INSTR	LCZ-Messer	
LDL (→language description language)	DATA PROC	(→Sprachbeschreibungssprache)	
LDR (→photoresistor)	COMPON	(→Photowiderstand)	
LD source (→laser diode source)	INSTR	(→Laserdiodenquelle)	
leach	TECH	auslaugen	
lead (n.)	COMPON	Anschlußdraht	
= component lead; wire lead; connecting lead; contact		= Anschlußleiter; Anschluß; Zuleitung	
lead (n.)	MECH	Steigung	
[thread]		[Gewinde]	
lead (n.)	COLLOQ	Vorsprung	
= advance (n.)		[figurativ]	
lead (n.)	ELECTRON	Zuleitung	
		= Anschlußleitung	
lead (n.)	TYPOGR	Durchschuß	
[thin metal strip to separate lines; separation between upper and lower border of characters]		[Metallstreifen zur Zeilentrennung; Zeilenabstand von Kegelunterkante zu Kegeloberkante]	
= inter-line leading; leading		= Zeilendurchschuß	
≈ line spacing [TERM&PER]		= Zeilenabstand [TERM&PER]	
lead	CHEM	Blei	
= Pb		= Pb	
lead	TECH	Lot	
[vertical reference]		[vertikal]	
lead (→wire)	COMM.CABLE	(→Ader)	
lead (→connecting wire)	TELEC	(→Anschlußdraht)	
lead (n.) (→time lead)	TECH	(→Voreilung)	
lead acid accumulator (→lead acid cell)	POWER SYS	(→Bleiakkumulator)	
lead acid cell	POWER SYS	Bleiakkumulator	
= lead acid accumulator		= Bleibatterie	
lead adapter	POWER SYS	Netzteiladapter	
lead bundle (→wire bundle)	COMM.CABLE	(→Aderbündel)	
lead cable joint (→lead sleeve)	OUTS.PLANT	(→Bleimuffe)	
lead-calcium battery	POWER SYS	Blei-Kalzium-Akku	
lead capacitance	ELECTRON	Zuleitungskapazität	
lead-coat (v.t.)	METAL	verbleien	
lead coating	METAL	Verbleiung	
lead-covered cable	COMM.CABLE	Bleikabel	
= lead-sheathed cable			

lead cycle	DATA PROC	**Lesezyklus**
leaded (adj.)	MICROEL	**mit Anschluß**
leader	ENG.DRAW	**Bezugslinie 2**
[reference line]		[amerikan.Linienart]
leader 1 (→tape leader 1)	TERM&PER	(→Bandführung)
leader (→leader tape)	BROADC	(→Vorspannband)
leader 2 (→tape leader 2)	TERM&PER	(→Vorspannband)
leader (→header label)	DATA PROC	(→Dateianfangs-Etikett)
leader (→leader character)	TYPOGR	(→Füllzeichen)
leader character	TYPOGR	**Füllzeichen**
= leader		↓ Führungspunkt
↓ leading dot		
leader label (→header label)	DATA PROC	(→Dateianfangs-Etikett)
leaders (→dotted line 1)	ENG.DRAW	(→punktierte Linie)
leader tape	BROADC	**Vorspannband**
[movie; videocassette]		[Film; Videocassette]
= leader		= Startband
lead hand (AM) (→foreman)	ECON	(→Vorarbeiter)
lead-in	ANT	**Einführung**
lead-in (→connecting wire)	TELEC	(→Anschlußdraht)
lead-in (→antenna feeding line)	ANT	(→Antennenspeiseleitung)
lead inductance	ELECTRON	**Zuleitungsinduktivität**
[parasitic inductance]		[parasitäre Induktivität]
leading	ECON	**federführend**
leading	TECH	**voreilend**
≠ trailing		≠ nacheilend
leading (→lead)	TYPOGR	(→Durchschuß)
leading angle (→lead-phase angle)	PHYS	(→Voreilwinkel)
leading costumer (→trend setting costumer)	ECON	(→Leitkunde)
leading dot	TYPOGR	**Führungspunkt**
leading edge (→rising edge)	ELECTRON	(→Anstiegflanke)
leading-edge overshoot (→preshoot)	ELECTRON	(→Vorschwingung)
leading-edge technology (→top technology)	TECH	(→Spitzentechnologie)
leading figure	COLLOQ	**Leitfigur**
leading-in conductor	ELECTRON	**Durchführung**
leading-in tube	OUTS.PLANT	**Kabeleinführstutzen**
leading pulse edge (→rising pulse edge)	ELECTRON	(→Impulsanstiegflanke)
lead-in groove (→lead-in spiral)	EL.ACOUS	(→Randrille)
leading transition time (→rise time)	ELECTRON	(→Anstiegzeit)
leading videotext page	TELEC	**Leitseite**
		[Btx]
leading zero (→high-order zero)	DATA PROC	(→führende Null)
lead-in insulator	COMPON	**Durchführungsisolator**
= lead-through insulator		
lead-in spiral	EL.ACOUS	**Randrille**
[phonogram record]		[Schallplatte]
= lead-in groove		
leadless	MICROEL	**ohne Anschlußstift**
lead-out groove (→lead-out spiral)	CONS.EL	(→Auslaufrille)
lead-out spiral	CONS.EL	**Auslaufrille**
[phonogram record]		[Schallplatte]
= lead-out groove; throw-out		
lead pair (→wire pair)	COMM.CABLE	(→Aderpaar)
lead-phase angle	PHYS	**Voreilwinkel**
= leading angle		= voreilender Winkel
lead resistance	ELECTRON	**Zuleitungswiderstand**
lead salt laser	OPTOEL	**Bleisalzlaser**
lead screw	MECH	**Leitspindel**
lead sheat (→cover sheet)	TYPOGR	(→Deckblatt)
lead sheath	COMM.CABLE	**Bleimantel**
lead-sheathed cable (→lead-covered cable)	COMM.CABLE	(→Bleikabel)
lead sleeve	OUTS.PLANT	**Bleimuffe**
= lead cable joint		
lead solder	METALL	**Bleilot**
lead-through capacitor (→feedthrough capacitor)	COMPON	(→Durchführungskondensator)
lead-through insulator (→lead-in insulator)	COMPON	(→Durchführungsisolator)
lead time	TECH	**Vorlaufzeit**
≈ time lead		≈ Vorlauf
leaf	MATH	**Blatt**
[theory of graphs; final node]		[Graphentheorie; Knoten ohne Nachfolger]
↑ node		↑ Knoten
leaf (→leaf node)	DATA PROC	(→Astknoten)
leaf electroscope	PHYS	**Blattelektroskop**
leaf green	PHYS	**laubgrün**
leaflet	DOC	**Prospekt**
[printed matter on single sheet, often folded and illustrated]		[kleines, meist bebildertes Faltblatt mit Werbeinformation]
= folder; prospectus; flyer		≈ Datenblatt; Kennblatt; Broschüre; Faltprospekt
≈ data sheet; brochure		
leaf node	DATA PROC	**Astknoten**
[tree sort]		[Baumsortierung]
≈ leaf; root node		
leaf spring	MECH	**Blattfeder**
= plate spring		
leaf spring block	MECH	**Blattfederpaket**
leak (n.)	TECH	**Undichtigkeit**
≈ leakage; impermeability		= Leck
≠ tightness		≈ Leckverlust; Undurchlässigkeit
		≠ Dichtigkeit
leak (adj.)	TECH	**undicht**
≈ leaky		= leck
≈ impermeable		undurchlässig
≠ tight		≠ dicht 2
leak (v.i.)	TECH	**lecken**
leakage	LINE TH	**Ableitung**
= leakance		= Betriebsableitung
≈ distributed leakage		≈ Ableitungsbelag
leakage	TECH	**Leckverlust**
≈ leak		≈ Undichtigkeit
leakage coefficient (→leakage factor)	EL.TECH	(→Streufaktor)
leakage conductance	PHYS	**Leckleitwert**
leakage current	ELECTRON	**Reststrom**
= residual current		
leakage current (→leak current)	EL.TECH	(→Leckstrom)
leakage current (→cutoff current 2)	MICROEL	(→Reststrom)
leakage-current noise	MICROEL	**Leckstromrauschen**
[MOS transistor]		[MOS-Transistor]
leakage factor	EL.TECH	**Streufaktor**
[transformer]		[Übertrager]
= leakage coefficient; Hopkinson coefficient; dispersion coefficient		= Streugrad; Streuzahl; Streuziffer
leakage field (→stray field)	PHYS	(→Streufeld)
leakage inductance (→stray inductance)	PHYS	(→Streuinduktivität)
leakage per unit length (→distributed leakage)	LINE TH	(→Ableitungsbelag)
leakage power (→dissipation power)	EL.TECH	(→Verlustleistung)
leakage protection	EL.TECH	**Isolationsfehlerschutz**
leakage resistance (→leck resistance)	EL.TECH	(→Ableitungswiderstand)
leakance	NETW.TH	**Streuleitwert**
leakance (→leakage)	LINE TH	(→Ableitung)
leakance per unit length (→distributed leakage)	LINE TH	(→Ableitungsbelag)
leak current	EL.TECH	**Leckstrom**
= leakage current		≈ Streustrom; Sperrstrom [ELECTRON]
≈ stray current; reverse current [ELECTRON]		↓ Kriechstrom
↓ tracking current		
leak loss	EL.TECH	**Ableitverlust**
leakproof	TECH	**lecksicher**
= leaktight		= auslaufsicher
≈ spillproof		≈ dicht 1
leakproofness	TECH	**Lecksicherheit**
≈ tightness		≈ Dichtigkeit
leak resistance	LINE TH	**Isolationswiderstand**
= insulation resistance		

leaktight (→leakproof)	TECH	(→lecksicher)	
leaky (→leak)	TECH	(→undicht)	
leaky feeder	MOB.COMM	**Leckleitung**	
leaky mode	OPT.COMM	**Leckwelle**	
leaky-pipe antenna (→slot antenna)	ANT	(→Schlitzstrahler)	
leaky-wave antenna	ANT	**Leckwellenantenne**	
leaky waveguide	MICROW	**Schlitzhohlleiter**	
leapfrog filter	NETW.TH	**Leapfrog-Filter**	
leapfrog test [memory testing by random skipping]	DATA PROC	**Bocksprungtest** [Speichertest durch wahlloses Springen]	
learnable = trainable	SCIE	**erlernbar**	
learning algorithm	DATA PROC	**Lernalgorithmus**	
learning curve	SCIE	**Erfahrungskurve** = Lernkurve	
learning machine	INF	**Lernautomat** = lernender Automat	
learning matrix	INF	**Lernmatrix**	
learning step ≈ teaching step	SCIE	**Lernschritt** ≈ Lehrschritt	
lease	ECON	**Pacht**	
lease 1 (→give on lease)	ECON	(→verpachten)	
lease 2 (→take on lease)	ECON	(→pachten)	
leased circuit (→leased line)	TELEC	(→Mietleitung)	
leased-circuit data network = data network with fixed connection; direct-call network	DATA COMM	**Direktrufnetz**	
leased line = private line; leased circuit; private circuit (BRI) ≈ fixed line	TELEC	**Mietleitung** = gemietete Standverbindung ≈ Standleitung	
leasing 2 ↑ financing	ECON	**Leasing** [Vermieten von Anlagegegenständen mit Serviceleistung] ↑ Finanzierung	
leasing 1 = letting	ECON	**Verpachten**	
least costs	ECON	**Minimalkosten**	
least significant = lowest order	CODING	**niederwertigst** = niedrigstwertig; wertniedrigst	
least significant bit = LSB	CODING	**niedrigstwertiges Bit** = LSB	
least significant character = LSC	CODING	**niederwertigstes Zeichen**	
least significant digit = LSD	MATH	**niedrigstwertige Stelle** = LSD	
least square method	MATH	**Fehlerquadratmethode**	
leather back	TYPOGR	**Lederrücken**	
leave (AM) (→vacation)	ECON	(→Urlaub)	
Lecher line (→two-wire line)	LINE TH	(→Lecher-Leitung)	
Lecher wires (→two-wire line)	LINE TH	(→Lecher-Leitung)	
leck resistance = bleeder resistance; leakage resistance	EL.TECH	**Ableitungswiderstand** = Ableitwiderstand; Leckwiderstand	
Leclanché cell = Leclanché element ↑ primary cell ↓ dry cell	POWER SYS	**Leclanché-Element** ↑ Primärelement ↓ Trockenelement	
Leclanché element (→Leclanché cell)	POWER SYS	(→Leclanché-Element)	
lecture ≈ talk ≈ speech	SCIE	**Vortrag** ≈ Vorlesung ≈ Rede	
lecturer	SCIE	**Vortragender** ≈ Redner	
lecture room	OFFICE	**Vortragsraum**	
LED (→light emitting diode)	MICROEL	(→Lumineszenzdiode)	
LED clip	COMPON	**LED-Fassung**	
LED display	COMPON	**LED-Anzeige**	
LED printer	TERM&PER	**LED-Drucker**	
LED source	INSTR	**LED-Quelle**	
leeway (→margin)	COLLOQ	(→Spielraum)	
left aligned = left justified; left flush; left-hand justified	TYPOGR	**linksbündig**	
left blank [of a character]	TYPOGR	**Vorbreite** [Leerraum rechts eines Buchstabens]	
left bracket [(= open bracket	MATH	**Klammer auf** [(	
left flush (→left aligned)	TYPOGR	(→linksbündig)	
left-hand (adj.) [thread]	MECH	**linkssteigend** [Gewinde]	
left-hand (→counterclockweise)	TECH	(→linksdrehend)	
left-hand helix	MECH	**Linksschraube**	
left-hand justified (→left aligned)	TYPOGR	(→linksbündig)	
left-hand rule	EL.TECH	**Linke-Hand-Regel**	
left-hand thread	MECH	**Linksgewinde**	
left-hand twist	PHYS	**Linksdrall**	
left-in station	TELEPH	**vorübergehend abgeschalteter Fernsprechanschluß** = vorübergehend abgeschalteter Telephonanschluß	
left-justified	CODING	**linksbündig**	
left justified (→left aligned)	TYPOGR	(→linksbündig)	
left shift	CODING	**Linksverschiebung**	
leftward arrow	TYPOGR	**Pfeil nach links**	
leg = arm	MATH	**Schenkel**	
legal (→permissible)	TECH	(→zulässig)	
legal (→valid)	DATA PROC	(→gültig)	
legal action (→law suit)	ECON	(→Prozeß)	
legal adviser (→lawyer)	ECON	(→Rechtsanwalt)	
legal claim	ECON	**Rechtsanspruch**	
legal holiday	ECON	**gesetzlicher Feiertag**	
legalization (→validation)	ECON	(→Legalisierung)	
legally valid	ECON	**rechtsgültig**	
legal provision	ECON	**Rechtsvorschrift**	
legal solution (→feasible solution)	MATH	(→zulässige Lösung)	
legal tender	ECON	**gesetzliches Zahlungsmittel**	
legend [explanatory comment to symbols] ≈ caption	LING	**Zeichenerklärung** = Legende ≈ Bildlegende	
Legendre filter	NETW.TH	**Legendre-Filter**	
legitimacy (→validity)	DATA PROC	(→Zulässigkeit)	
lightweight (→light)	PHYS	(→leicht)	
lemon yellow	PHYS	**zitronengelb** = zitrongelb	
Lenard window	PHYS	**Lenard-Fenster**	
lend 1 [to give for temporary use] = loan ≠ borrow	ECON	**verleihen** [vorübergehend zur Verfügung stellen] = ausleihen 2; borgen 1; leihen 1 ≠ entleihen	
lend 2 (v.t.) [to take for temporary use] = loan (v.t.) ≠ borrow	ECON	**entleihen** [vorübergehend in Besitz nehmen] = ausleihen 1; borgen 2; leihen 2 ≠ verleihen	
lender	ECON	**Kreditgeber**	
lender 1 [who gives] ≈ creditor ≠ borrower	ECON	**Verleiher** = Leiher; Borger 1 ≈ Gläubiger ≠ Entleiher	
length [spacial; SI unit: meter] = l ≈ longitudinal dimension ↑ dimension	PHYS	**Länge 1** [räumlich; SI-Einheit: Meter] = l ≈ Längsabmessung ↑ Dimension	
length dependence	PHYS	**Längenabhängigkeit**	
length element (→unit interval)	TELEGR	(→Zeichenelement)	
lengthen [pulse]	ELECTRON	**dehnen** [Impuls]	
lengthen (→prolong)	COLLOQ	(→verlängern)	

lengthening

lengthening (→extension)	PHYS	(→Ausdehnung 1)
lengthening (→stretch)	PHYS	(→Dehnung)
lengthening (→prolongation)	TECH	(→Verlängerung)
lengthening bar	TECH	**Verlängerungsstange**
lengthening sleeve	TECH	**Verlängerungshülse**
length indicator	SWITCH	**Längenindikator**
		= Längenkennung; Längenkennzeichnung
length-preserving	TECH	**längentreu**
length-to-diameter ratio	TECH	**Schlankheitsgrad**
lens	PHYS	**Linse**
↓ spherical lens; aspherical lens		↓ sphärische Linse; asphärische Linse
lens antenna	ANT	**Linsenantenne**
= delay lens		= Verzögerungslinse
lens effect	OPTOEL	**Linsen-Effekt**
lens equation	PHYS	**Linsenformel**
		= Linsengleichung
lentil head (→oval head)	MECH	(→Linsensenkkopf)
lentil-head screw (→oval-head screw)	MECH	(→Linsensenkkopfschraube)
Lenz's law	EL.TECH	**Lenzsche Regel**
leporello folding (→fan folding)	TERM&PER	(→Leporellofalzung)
lessee (→tenant)	ECON	(→Mieter)
lessen (→decrease)	TECH	(→vermindern)
lessen (→reduce 1)	TECH	(→vermindern)
less or equal	MATH	**kleiner-gleich**
less-or-equal sign [symbol ≦]	MATH	**Kleiner-gleich-Zeichen** [Symbol ≦]
= less-or-equal symbol		
less-or-equal symbol (→less-or-equal sign)	MATH	(→Kleiner-gleich-Zeichen)
less than	MATH	**kleiner als**
less-than sign [symbol <]	MATH	**Kleinerzeichen** [Symbol <]
= less-than symbol		= Kleiner-als-Zeichen
less-than symbol (→less-than sign)	MATH	(→Kleinerzeichen)
let-go current	EL.TECH	**Loslaßstrom**
let-go threshold	EL.TECH	**Loslaßschwelle**
letter	LING	**Brief**
		= Schreiben 2
letter	LING	**Buchstabe**
[a graphic character symbolizing a sound or a sound combination]		[einen Laut oder eine Lautverbindung symbolisierendes Schriftzeichen]
= alphabetic character		= alphabetisches Zeichen; Alphazeichen
↑ graphic character		↑ Schriftzeichen
letter	ECON	**Vermieter**
≠ lessee		≠ Mieter
letter (→type)	TYPOGR	(→Drucktype)
letter (v.t.) (→identify)	TECH	(→kennzeichnen) (v.t.)
letter box	POST	**Briefkasten**
letter closing device	OFFICE	**Briefschließgerät**
		= Briefschließmaschine
letter family (→type family)	TYPOGR	(→Schriftfamilie)
letter-figure shift	TERM&PER	**Buchstaben-Ziffern-Umschaltung**
↓ letter shift; figure shift		= Bu-Zi-Umschaltung
		↓ Buchstabenumschaltung; Ziffernumschaltung
letter-file	OFFICE	**Schnellhefter**
= folder		
letter height (→type height)	TYPOGR	(→Schrifthöhe)
lettering	EQUIP.ENG	**Beschriftung**
= labeling; marking; designation		= Auszeichnung
lettering (→identification)	TECH	(→Kennzeichnung)
lettering attribute (→typeface design)	TYPOGR	(→Schriftschnitt)
lettering equipment	OFFICE	**Beschriftungsgerät**
lettering field	TERM&PER	**Schriftfeld**
= title field		
lettering label (→designation label)	TECH	(→Beschriftungsschild)
lettering plan	TECH	**Beschriftungsplan**
= labeling plan; labeling document; designation plan; marking plan		= Beschriftungsunterlage; Auszeichnungsplan; Auszeichnungsunterlage
lettering size (→type size)	TYPOGR	(→Schriftgröße)
lettering space	EQUIP.ENG	**Beschriftungsfeld**
= designation space; marking space; labeling space		= Auszeichnungsfeld
lettering stencil	ENG.DRAW	**Schriftschablone**
lettering strip (→designation strip)	EQUIP.ENG	(→Beschriftungsstreifen)
lettering type (→typeface design)	TYPOGR	(→Schriftschnitt)
letter of credit	ECON	**Akkreditiv**
= L/C		= Dokumentenakkreditiv; L/C
letter of intent	ECON	**Absichtserklärung**
= LOI		
letter opening device	OFFICE	**Brieföffnergerät**
		= Brieföffnermaschine
letterpress	TYPOGR	**Hochdruck**
[printing parts hihger than non printing ones]		[druckende Teile liegen höher als nicht druckende]
= letterprint; relief printing		≠ Tiefduck
≠ intaglio printing		↑ Druckverfahren
↑ printing process		↓ Typendruck
↓ type printing		
letterprint (→letterpress)	TYPOGR	(→Hochdruck)
letter print quality (→letter quality)	TERM&PER	(→Schönschrift)
letter quality	TERM&PER	**Schönschrift**
[quality of a daisy-weel typewriter]		[Qualität einer Typenradschreibmaschine]
= LQ; correspondence quality; letter print quality; calligraphy		= Schönschriftqualität; Schöndruckqualität; Korrespondenzqualität; Briefqualität
≠ draft quality		
		≠ Konzeptdruckqualität
letter-quality printer	TERM&PER	**Schönschreibdrucker**
= correspondence quality printer		= Schönschriftdrucker; Korrespondenzdrucker
lettersapce (→character spacing)	TYPOGR	(→Zeichenabstand)
letter scale	OFFICE	**Briefwaage**
letter service	POST	**Briefdienst**
letter shift	TELEGR	**Buchstabenumschaltung**
[key or code]		[Code oder Taste]
= shift letter		= Buchstabenwechsel
≈ figures shift		≈ Ziffernumschaltung
↑ letter-figure shift		↑ Buchstaben-Ziffern-Umschaltung
letterspace (v.t.)	TYPOGR	**spatiieren**
[to increase the intercharacter spacing for visual reasons]		[aus optischen Gründen den Buchstabenabstand erhöhen]
= space		= spationieren; sperren
letterspacing (→character spacing)	TYPOGR	(→Zeichenabstand)
letter type (→type style)	TYPOGR	(→Schriftart)
letting (→leasing 1)	ECON	(→Verpachten)
level	TECH	**Pegel**
		= Niveau
level	TELEC	**Pegel**
level (n.) (→plane)	MATH	(→Ebene)
level (→plane)	DATA PROC	(→Ebene)
level (→protocol layer)	DATA COMM	(→Protokollschicht)
level adjustment (→level equalization)	TELEC	(→Pegelausgleich)
level adjustment (→level controller)	EL.ACOUS	(→Pegelregler)
level chart	TELEC	**Pegeldiagramm**
= level diagram; hypsogram		
level compensation (→level equalization)	TELEC	(→Pegelausgleich)
level control (→level regulation)	ELECTRON	(→Pegelregelung)
level controller	EL.ACOUS	**Pegelregler**
= level regulator; volume control; volume regulator; level adjustment		= Pegelregelung; Pegelsteller; Lautstärkeregler; Lautstärkeregelung; Laustärkesteller
↓ fader		↓ Überblendregler
level converter	CIRC.ENG	**Pegelumsetzer**

level crossing (→rail crossing)	RAILW.SIGN	(→Bahnübergang)		lexicographical sorting [considers space lower as comma, comma lower as a]	LING	lexikographische Sortierung [wertet Leerstelle kleiner als Komma, Komma kleiner als a]
level curve	INSTR	Pegelkurve				
level dependence	TELEC	Pegelabhängigkeit				
level diagram (→level chart)	TELEC	(→Pegeldiagramm)		lexicographic code	CODING	lexikografischer Code
level difference (→relative level)	TELEC	(→relativer Pegel)		lexicon (pl.: -a, -ons) [alphabetically arranged encyclopedia]	LING	Lexikon 1 (pl.: -a oder -en) [alphabetisch geordnete Enzyklopädie] ↓ Konversationslexikon
level difference meter	INSTR	Pegeldifferenzmesser				
leveled	CIRC.ENG	pegelgeregelt				
level equalization = level compensation; level adjustment	TELEC	Pegelausgleich = Einpegelung		lexicon [language with definition of all terms]	DATA PROC	Lexikon
level flatness [INSTR] (→frequency response)	NETW.TH	(→Frequenzgang)		LF (→voice frequency)	TELEC	(→Niederfrequenz)
				LF (→line feed)	TERM&PER	(→Zeilenschalter)
level fluctuation = level variation	TELEC	Pegeländerung = Pegelschwankung		LF (→line feed)	TERM&PER	(→Zeilenvorschub)
				LF (→line feed)	DATA COM	(→Zeilenvorschub)
level generator [for FDM measurements] = signal generator 2; level source ↑ transmission measuring set	INSTR	Pegelsender = Pegelgenerator ↑ TF-Pegelmeßplatz		LF transmitter	RADIO	Langwellensender
				L group (NAM) (→basic group)	TRANS	(→Grund-Primärgruppe)
				Li (→lithium)	CHEM	(→Lithium)
level indicator (→level meter)	INSTR	(→Pegelmesser)		liabilities [balance] = accounts payable	ECON	Verbindlichkeiten [Bilanz] = Kreditoren
leveling (→level regulation)	ELECTRON	(→Pegelregelung)				
level jump ≈ level step	TELEC	Pegelsprung ≈ Pegelstufe		liability (→security)	ECON	(→Haftung)
				liability (→susceptibility)	QUAL	(→Störanfälligkeit)
level line (→isohypse)	GEOPHYS	(→Isohypse)		liability reserves (→provision 2)	ECON	(→Rückstellung)
levelling [curve]	MATH	abflachen [Kurvenverlauf]				
				liable to recourse	ECON	regreßpflichtig
levelling	TECH	Einebnung		librarian	SCIE	Bibliothekar
level measurement	INSTR	Pegelmessung		librarian (→library program)	DATA PROC	(→Bibliotheksprogramm)
level measuring program	SWITCH	Pegelmeßprogramm				
level meter [FDM tests] = level indicator; hypsometer ↑ transmission measuring set	INSTR	Pegelmesser = Pegelmeßgerät; Pegelempfänger ↑ TF-Pegelmeßplatz		library	DATA PROC	Bibliothek
				library administration = library maintenance; library management	DATA PROC	Bibliothekverwaltung = Bibliothekmanagement
				library maintenance (→library administration)	DATA PROC	(→Bibliothekverwaltung)
levelness = flatness	MECH	Ebenheit 3 = Waagrechtigkeit				
level of coverage	TELEC	Flächendeckungsgrad		library management (→library administration)	DATA PROC	(→Bibliothekverwaltung)
level of integration (→integration)	MICROEL	(→Integration)		library manager [a program]	DATA PROC	Bibliotheksverwalter [ein Pròrgramm]
level range	TELEC	Pegelbereich				
level recorder	INSTR	Pegelschreiber		library program = librarian	DATA PROC	Bibliotheksprogramm
level regulation = leveling; level control	ELECTRON	Pegelregelung				
				library routine	DATA PROC	Bibliotheksroutine
level regulator	TELEC	Pegelregler		library track [on magnetic disc or tape for data on content]	TERM&PER	Hinweisspur [auf Magnetplatten oder -bänder mit Inhaltshinweisen]
level regulator (→level controller)	EL.ACOUS	(→Pegelregler)				
level-sensitive	ELECTRON	pegelempfindlich				
level shift	CIRC.ENG	Potentialverschiebung = Pegelverschiebung		Libraw-Craig code = switched tailring counter code	CODING	Libraw-Craig-Code
level sonde	MICROW	Pegelsonde		licence (→permission)	ECON	(→Genehmigung)
level source (→level generator)	INSTR	(→Pegelsender)		licence (BRI) (→license)	ECON	(→Lizenz)
level stabilizing amplifier	TRANS	Regelverstärker		licence (→admission)	ECON	(→Zulassung)
level standard	TELEC	Pegelplan		licence agreement = licence agreement	ECON	Lizenzvertrag = Lizenzabkommen
level tracer	INSTR	Pegelbildempfänger = Pegelbildgerät				
				licence agreement (→licence agreement)	ECON	(→Lizenzvertrag)
level variation (→level fluctuation)	TELEC	(→Pegeländerung)		licence fee = royalty	ECON	Lizenzgebühr
lever	MECH	Hebel				
leverage	MECH	Hebelwirkung		licence holder	ECON	Lizenzinhaber
lever force = lever power	MECH	Hebelkraft		license (NAM) = licence (BRI)	ECON	Lizenz
lever key (→toggle switch)	COMPON	(→Kippschalter)		licensed (adj.) = admitted ↓ officially licensed	ECON	zugelassen = lizensiert ↓ amtlich zugelassen
lever power (→lever force)	MECH	(→Hebelkraft)				
lever type decade resistance (→crank resistance)	INSTR	(→Kurbelwiderstand)				
				licensee	ECON	Lizenznehmer
levorotatory ≠ dextrorotatory	PHYS	linksdrehend = lävogyr ≠ dextrogyr		license granter (→licenser)	ECON	(→Lizenzgeber)
				licenser = licensor; license granter	ECON	Lizenzgeber
levorotatory (→counterclockwise)	TECH	(→linksdrehend)				
				licensor (→licenser)	ECON	(→Lizenzgeber)
levy (→charge 2)	ECON	(→Gebühr)		lid ≈ flap ↑ cover	TECH	Deckel ≈ Klappe ↑ Abdeckung
levy (→allocation)	ECON	(→Umlage)				
Lewis key	MECH	Tangentialpaßfeder				
lexeme [word not composed by other words, basic unit of a vocabulary; a morpheme designing by itself an object or topic]	LING	Lexem (n.n.) [nicht zusammengesetzes Wort, Grundeinheit des Wortschatzes; Morphem das für sich einen Gegenstand oder Sachverhalt darstellt]		LIDAR [light detecting and ranging]	OPTOEL	LIDAR
				Lie group	MATH	Liesche Gruppe
				life (→life time)	QUAL	(→Lebensdauer)
				life cycle	TECH	Lebenszyklus
				life-cycle costing	ECON	Lebensdauerkalkulation
				life expectation = lifetime expectancy	QUAL	Lebenserwartung

life test		QUAL	**Lebensdauerprüfung**	**light emitting diode**		MICROEL	**Lumineszenzdiode**
			= Lebensdauertest	= LED			= Leuchtdiode; lichtemittierende Diode; LED; Elektrolumineszenzdiode
life time		QUAL	**Lebensdauer**	≈ photodiode			
[till the end of operability]			[bis zum Ende der Funktionsfähigkeit]				≈ Photodiode
= operation life; life				**lightfast**		TECH	**lichtecht**
≈ utilizatio time; useful life			= Betriebslebensdauer	[color]			[Farbe]
			≈ Nutzungsdauer; Gebrauchslebensdauer	**light flesh tint**		PHYS	**fleischfarbe hell**
				light flux (→luminous flux)		PHYS	(→Lichtstrom)
life-time distribution		QUAL	**Lebensdauerverteilung**	**light fork coupler**		OPTOEL	**Lichtgabelkoppler**
lifetime expectancy (→life expectation)		QUAL	(→Lebenserwartung)	= optoelectronic fork coupled device			= Gabellichtkoppler
life utility (→useful life)		QUAL	(→Betriebsbrauchbarkeitsdauer)	**light frequency**		PHYS	**Lichtfrequenz**
				= light wavelength			≈ Lichtwellenlänge
life zero		INSTR	**lebender Nullpunkt**	**light-frequency converter**		OPTOEL	**Lichtfrequenzwandler**
LIFO mode		DATA PROC	**LIFO-Modus**	**light green** (→pale green)		PHYS	(→hellgrün)
= last-in-first-out mode; FILO mode; first-in-last-out mode			[Auslesen des zuletzt Eingeschriebenen]	**light grey**		PHYS	**mittelgrau**
			= FILO-Modus	**lighthouse**		SIGN.ENG	**Leuchtturm**
≠ FIFO mode			≠ FIFO-Modus	**lighthouse tube** (→planar tube)		ELECTRON	(→Scheibenröhre)
LIFO stack (→push-down storage)		DATA PROC	(→Kellerspeicher)	**light-hydraulic effect**		PHYS	**lichthydraulischer Effekt**
				lighting engineering		POWER ENG	**Lichttechnik**
lift (v.t.)		TELEPH	**abheben**				= Beleuchtungstechnik
[handset]			[Handapparat]	**light-insensitive** (→insensitive to light)		PHYS	(→lichtunempfindlich)
= go off-hook			= aushängen				
≠ go on-hook			≠ auflegen	**light italic**		TYPOGR	**zartes Kursiv**
lift (→black-level set-up)		TV	(→Schwarzabhebung)	↑ typeface design			= zart kursiv
lift (BRI) (→elevator)		CIV.ENG	(→Aufzug)				↑ Schriftschnitt
lifter		TERM&PER	**Abstandhalter**	**light loading**		OUTS.PLANT	**leichte Bespulung**
= spacer			= Abstandstück	**light metal**		METAL	**Leichtmetall**
lifting apparatus (→hoist)		MECH	(→Hebevorrichtung)	**light modulator**		OPTOEL	**Lichtmodulator**
lifting equipment		TECH	**Hubgerät**	**lightness 2**		PHYS	**Leichtigkeit**
lift-off force		COMPON	**Abhebekraft**	[low weight]			
[relay]			[Relais]	**lightness** (→luminance)		PHYS	(→Leuchtdichte)
lift-out ribbon		EQUIP.ENG	**Entnahmeschlaufe**	**lightning**		AERON	**Befeuerung**
[e.g. in a battery compartment]			[z.B. in Batteriefach]	= aircraft warning lightning			[an Hindernissen, Masten]
ligature		TYPOGR	**Ligatur**	↓ obstruction lightning			↓ Hindernisbefeuerung
[union of two letters in one type, e.g. Æ]			[Verbindung zweier Buchstaben auf einer Type, z.B. Æ]	**lightning**		METEOR	**Blitz**
				≈ thunderstrike			≈ Blitzschlag
light (n.)		PHYS	**Licht**	↑ thunderstroke			↑ atmosphärische Entladung
light (n.)		AERON	**Feuer**	**lightning arrester** (→lightning protection)		EL.TECH	(→Blitzschutz)
↓ obstruction light			↓ Hindernisfeuer	**lightning arrester** (→overvoltage protector)		COMPON	(→Spannungssicherung)
light (adj.)		PHYS	**leicht**				
[weight]			[Gewicht]	**lightning call**		TELEPH	**Blitzgespräch**
= leightweight				**lightning current**		EL.TECH	**Bliztstrom**
light (adj.)		TYPOGR	**mager**	**lightning earthing electrode**		EL.TECH	**Blitzschutzerder**
≈ hairline			= zart; dünn				= Blitzerder
↑ typeface design			≈ besonders dünn	**lightning protection**		EL.TECH	**Blitzschutz**
			↑ Schriftschnitt	= lightning arrester			= Blitzschutzautomat
light 2 (adj.) (→bright)		PHYS	(→hell)	↑ overvoltage protection			↑ Überspannungsschutz
light-absorptive		PHYS	**lichtschluckend**	**lightning protection device**		EQUIP.ENG	**Blitzschutzeinrichtung**
light barrier		SIGN.ENG	**Lichtschranke**				
			↑ Strahlenschranke	**lightning protector** (→overvoltage protector)		COMPON	(→Spannungssicherung)
light beam		PHYS	**Lichtstrahl**				
= light ray; luminous ray			↑ Strahl	**lightning resistance** (→surge resistance)		QUAL	(→Blitzfestigkeit)
↑ ray							
light-beam instrument		INSTR	**Lichtstrahlinstrument**	**lightning rod**		CIV.ENG	**Blitzableiter**
light-beam oscillograph		INSTR	**Lichtstrahl-Oszillograph**	**light path**		PHYS	**Lichtweg**
			= Lichtstrahl-Oszillograf; Lichtpunkt-Linienschreiber; Lichtpunktschreiber	**light pen**		TERM&PER	**Lichtgriffel**
				[detects pixels on a screen]			[tastet Bildelemente am Bildschirm ab]
light blue (→ice blue)		PHYS	(→hellblau)	↑ electronic pencil			= Lichtstift; Lichtschreiber; Auswahlstift; Lightpen
light box		TECH	**Schachtel**				
≈ box; case			[leichter Kleinbehälter mit Deckel oder Klappe]				↑ elektronischer Bleistift
↓ cardboard box			≈ Kasten; Gehäuse	**light propagation**		PHYS	**Lichtausbreitung**
			↓ Pappschachtel	**light quantity**		PHYS	**Lichtmenge**
light brown (→pale brown)		PHYS	(→hellbraun)	**light radiation**		PHYS	**Lichtstrahlung**
light button		DATA PROC	**Lichttaste**	**light ray** (→light beam)		PHYS	(→Lichtstrahl)
light chopper		TELEGR	**Lichtsäge**	**light red** (→pale red)		PHYS	(→hellrot)
			= Lochscheibe	**light regulator** (→dimmer)		EL.INST	(→Helligkeitsregler)
light coupling		OPT.COMM	**Lichteinkopplung**	**light route** (→low-capacity link)		TRANS	(→Kleinkanalstrecke)
light current (→low current)		EL.TECH	(→Schwachstrom)				
light-current engineering (→low-current engineering)		EL.TECH	(→Schwachstromtechnik)	**light scattering**		PHYS	**Lichtstreuung**
				↑ scattering			↑ Streuung
light/dark current ratio		COMPON	**Hell-Dunkelstrom-Verhältnis**	**light sensitive** (→photosensitive)		PHYS	(→lichtempfindlich)
light-dependent resistor (→photoresistor)		COMPON	(→Photowiderstand)	**light signal**		RAILW.SIGN	**Lichtsignal**
				light simulation		TECH	**Lichtsimulation**
				light source		PHYS	**Lichtquelle**
				= luminous source			

light speed (→light velocity)	PHYS	(→Lichtgeschwindigkeit)	
light switch	EL.INST	Lichtschalter	
light-tight	TECH	lichtdicht	
light trap	PHYS	Lichtfalle	
light velocity	PHYS	Lichtgeschwindigkeit	
= light speed			
light wave	PHYS	Lichtwelle	
= lightwave; optical wave			
lightwave (→light wave)	PHYS	(→Lichtwelle)	
lightwave cable (→optical fiber cable)	COMM.CABLE	(→Lichtwellenleiterkabel)	
lightwave component	OPT.COMM	Lichtleiterbauteil	
= in-fiber component		= LWL-Bauteil; Lichtwellenleiter-Bauteil; Lichtwellenkomponente	
lightwave component analyzer	INSTR	Lichtleiterbauteile-Analysator	
		= Lichtwellenkomponenten-Analysator	
lightwave couple element (→optical coupling device)	OPTOEL	(→Optokoppler)	
lightwave coupler (→optical coupling device)	OPTOEL	(→Optokoppler)	
lightwave engineering (→fiber optic technique)	TELEC	(→Lichtwellenleitertechnik)	
lightwave fiber (→glass fiber)	OPT.COMM	(→Glasfaser)	
lightwave fiber repeater (→fiber optic repeater)	TRANS	(→LWL-Zwischenregenerator)	
lightwave fiber terminal (AM) (→fiber optic terminal)	TRANS	(→LWL-Leitungsendgerät)	
light wavelength (→light frequency)	PHYS	(→Lichtfrequenz)	
lightwave receiver	INSTR	Lichtwellenempfänger	
= optical receiver		= LWL-Empfänger; optischer Empfänger	
lightwave signal analyzer (→optical signal analysator)	INSTR	(→optischer Signalanalysator)	
lightwave system (→fiber optic system)	TRANS	(→Lichtwellenleitersystem)	
lightwave technique (→fiber optic technique)	TELEC	(→Lichtwellenleitertechnik)	
lightwave terminal (→fiber optic terminal)	TRANS	(→LWL-Leitungsendgerät)	
lightwave test equipment	INSTR	Lichtleiter-Meßgerät	
= fiber optic test equipment		= LWL-Meßgerät	
lightweight construction	TECH	Leichtbauweise	
Likasiewicz notation	MATH	umgekehrte polnische Schreibweise	
[algebraic notation without parentheses, operators acting towards the left]		[klammerlose algebraische Schreibweise, die Operatoren wirken nach links]	
		= umgekehrte polnische Notation; Lukasiewicz-Schreibweise; Lukasiewicz-Notation	
likelihood	MATH	Mutmaßlichkeit	
≈ probability		≈ Wahrscheinlichkeit	
likelihood equation	MATH	Mutmaßlichkeitsgleichung	
		= Likelihood-Gleichung	
likelihood function	MATH	Mutmaßlichkeitsfunktion	
		= Likelihood-Funktion	
likelihood ratio test	MATH	Mutmaßlichkeits-Quotiententest	
		= Likelihood-Quotiententest	
lilac (→heliotrope)	PHYS	(→lila)	
limb (→margin)	TECH	(→Rand)	
limit (v.t.)	TECH	begrenzen	
= bound; confine		= limitieren	
≈ enclose			
limit (→boundary)	TECH	(→Grenze)	
limit (→limit value)	MATH	(→Grenzwert)	
limit angle (→limiting angle)	TECH	(→Grenzwinkel)	
limitation	TECH	Begrenzung	
= bound		= Limitierung	
≈ restriction		≈ Beschränkung	
limitation (→clipping)	CIRC.ENG	(→Begrenzung)	
limitation (→restriction)	TECH	(→Beschränkung)	
limitation (→bound)	DATA PROC	(→Beschränkung)	
limit channel (→outboard channel)	RADIO	(→Randkanal)	
limit check	DATA PROC	Bereichsprüfung	
limit dimension	ENG.DRAW	Grenzmaß	
limited (→restricted)	TECH	(→eingeschränkt)	
limited accessibility	SWITCH	begrenzte Erreichbarkeit	
= limited availability		= begrenzte Verfügbarkeit	
limited availability (→limited accessibility)	SWITCH	(→begrenzte Erreichbarkeit)	
limited distance modem transmission (→baseband transmission)	DATA COMM	(→Basisbandtechnik)	
limited in time	COLLOQ	befristed	
limited partnership	ECON	Kommanditgesellschaft	
↑ company		= KG	
		↑ Gesellschaft	
limited saturation device	MICROEL	LSD	
= LSD			
limiter (→limiter circuit)	CIRC.ENG	(→Begrenzerschaltung)	
limiter characteristic	ELECTRON	Begrenzerkennlinie	
limiter circuit	CIRC.ENG	Begrenzerschaltung	
= limiter; clipper circuit; clipping circuit; clipper		= Begrenzer; Begrenzerkreis	
≈ amplitude limiter circuit		↓ Amplitudenbegrenzerschaltung	
limiter diode	CIRC.ENG	Begrenzerdiode	
≈ clamping diode 1		≈ Kappdiode	
limit frequency	NETW.TH	Eckfrequenz	
= corner frequency; coutoff frequency		= Randfrequenz; Knickfrequenz	
≠ center frequency		≠ Mittenfrequenz	
limit gage	MECH	Grenzlehre	
limit indicator	INSTR	Grenzwertmelder	
limiting (→restrictive)	TECH	(→einschränkend)	
limiting angle	TECH	Grenzwinkel	
= limit angle; critical angle			
limiting coil (→current limiting coil)	CIRC.ENG	(→Begrenzungsdrossel)	
limiting condition (→boundary condition)	MATH	(→Randbedingung)	
limiting frequency (→cutoff frequency)	EL.TECH	(→Grenzfrequenz)	
limiting quality (→rejectable quality level)	QUAL	(→Rückweisgrenze)	
limiting stage	CIRC.ENG	Begrenzerstufe	
limiting value	COMPON	Grenzwert	
limiting voltage	MICROEL	Grenzspannung	
limitless (→unrestricted)	COLLOQ	(→unbeschrαnkt)	
limit measuring bridge (→tolerance measuring bridge)	INSTR	(→Toleranzmeßbrücke)	
limit of audibility	ACOUS	Hörgrenze	
[at about 20 kHz]		[bei ca. 20 kHz]	
limit of liability	ECON	Haftungsgrenze	
limit of resolution	PHYS	Auflösungsgrenze	
≈ resolution 2		≈ Auflösungsvermögen	
limit of singularity	MICROW	Eindeutigkeitsgrenze	
limit setting	CONTROL	Grenzwerteinstellung	
limit temperature	PHYS	Grenztemperatur	
↓ maximum temperature; minimum temperature		↓ Höchsttemperatur; Tiefsttemperatur	
limit theorem	MATH	Grenzwertsatz	
limit theorem of Laplace	MATH	Laplacescher Grenzwertsatz	
limit value	MATH	Grenzwert	
= limit		↓ Höchstwert; Tiefstwert	
Lindenblad antenna	ANT	Lindenblad-Antenne	
line (n.)	LINE TH	Leitung	
line (n.)	TYPOGR	Zeile	
= row		↓ Druckzeile; Leerzeile	
↓ print line; blank line			
line (n.)	ELECTRON	Zeile	
= scanning line		= Abtastzeile	
line	TELEC	Leitung	
= link; circuit			
line	ENG.DRAW	Linie	
↓ guideline		= Strich	
		↓ Zeilenstrich	
line	TECH	Leine	
[a slender cord]		[dünnes Seil]	
≈ cord; rope; cable		≈ Seil; Strick; Tau; Kabel	

line (→rack row)	SYS.INST	(→Gestellreihe)	
line (→trunk 1)	TELEC	(→Fernmeldelinie)	
line (→assortment)	ECON	(→Sortiment)	
line (AM) (→queue)	COLLOQ	(→Schlange)	
line access	DATA COMM	Leitungszugang	
line-addressable random-access memory (→LARAM)	MICROEL	(→LARAM)	
line advance (→line spacing)	TERM&PER	(→Zeilenabstand)	
line air traffic	AERON	**Linienflugverkehr**	
↑ air traffic		= Linienluftverkehr ↑ Luftverkehr	
lineal polarization	PHYS	**Linearpolarisation** = lineare Polarisation	
line amplifier (→line repeater)	TRANS	(→Leitungsverstärker)	
linear = straight-line (adj.)	MATH	**linear**	
linear algebra	MATH	**lineare Algebra** = Linearalgebra	
linear amplifier	CIRC.ENG	**linearer Verstärker** = Linearverstärker	
linear antenna ↓ half-wave dipole; full-wave dipole; long-wire antenna	ANT	**Linearantenne** = gestreckte Antenne ↓ Halbwellendipol; Ganzwellendipol; Langdrahtantenne	
linear antenna array (→linear array)	ANT	(→lineare Gruppe)	
linear array = linear antenna array; linear array antenna	ANT	**lineare Gruppe** = lineare Gruppenantenne; Lineargruppe	
linear array antenna (→linear array)	ANT	(→lineare Gruppe)	
linear beam tube (→drift tube)	MICROW	(→Triftröhre)	
linear characteristic	ELECTRON	**lineare Kennlinie**	
linear coding (→linear programming)	DATA PROC	(→lineare Programmierung)	
linear control area	CONTROL	**betragslineare Regelfläche**	
linear controller	CONTROL	**linearer Regler**	
linear correlation	MATH	**lineare Korrelation**	
linear distortion	TELEC	**lineare Verzerrung**	
linear expansion	PHYS	**Längenausdehnung**	
linear expansion coefficient	PHYS	**Längenausdehnungskoeffizient**	
linear flow	PHYS	**lineare Strömung** = Linearströmung	
linear IC (→linear integrated circuit)	MICROEL	(→integrierte Analogschaltung)	
linear integrated circuit = linear IC	MICROEL	**integrierte Analogschaltung** = integrierte Linearschaltung; lineares IC	
linear interpolation	MATH	**Linearinterpolation** = lineare Interpolation	
linearity 1 = straightness	MATH	**Geradlinigkeit** = Geradläufigkeit; Geradheit	
linearity 2	MATH	**Linearität**	
linearity control	TV	**Linearitätsregelung**	
linearity error	INSTR	**Linearitätsfehler**	
linearization	ELECTRON	**Linearisierung**	
linearization circuit	CIRC.ENG	**Linearisierungsschaltung**	
linearize	ELECTRON	**linearisieren**	
linear list (→sequential list)	DATA PROC	(→sequentielle Liste)	
linear modulation	MODUL	**lineare Modulation**	
linear motor	POWER ENG	**Linearmotor**	
linear network	NETW.TH	**lineares Netzwerk** = lineares Netz	
linear path	CONTROL	**Strecke**	
linear path control	CONTROL	**Streckensteuerung** ↑ numerische Steuerung	
linear polarization	RADIO	**Linearpolarisation**	
linear polarized	PHYS	**linear polarisiert**	
linear polarized wave	PHYS	**linear polarisierte Welle**	
linear predictive coding = LPC	CODING	**LPC-Codierung**	
linear program (→sequential program)	DATA PROC	(→lineares Programm)	
linear programming [for optimization of a system with linear functions]	DATA PROC	**lineare Programmierung** [zur Optimierung eines Systems über lineare Zusammenhänge; Operations Research] = lineare Optimierung; lineare Planungsrechnung; LP ↓ Simplexmethode; Distributionsmethode; Floodsche Zurechnungstechnik	
= linear coding; LP ↓ simplex mode; distributive move			
linear pulse technique	ELECTRON	**lineare Impulstechnik**	
linear quadripole (→linear two-port)	NETW.TH	(→linearer Vierpol)	
linear regression	MATH	**lineare Regression**	
linear regulator	POWER ENG	**Linearregler**	
linear resistance (→ohmic resistance)	EL.TECH	(→ohmscher Widerstand)	
linear resistance (→ohmic resistor)	COMPON	(→ohmscher Widerstand)	
linear resistor (→ohmic resistor)	COMPON	(→ohmscher Widerstand)	
linear search = sequential search	DATA PROC	**lineare Suchmethode** = sequentielle Suchmethode	
linear selection	DATA PROC	**lineare Auswahl**	
linear structure	DATA PROC	**lineare Struktur**	
linear sweep	INSTR	**Linearwobbelung** = lineare Wobbelung	
linear system	SYS.TH	**lineares System**	
line art [without half tones]	TYPOGR	**Strichvorlage** [ohne Graustufungen]	
linear transfer element	CONTROL	**lineares Übertragungsglied**	
linear two-port = linear quadripole	NETW.TH	**linearer Vierpol**	
linear two-terminal (→linear two-terminal network)	NETW.TH	(→linearer Zweipol)	
linear two-terminal network [I proportional to U] = linear two-terminal	NETW.TH	**linearer Zweipol** [I proportional zu U]	
linear wave [propagates only in one direction]	PHYS	**lineare Welle** [breitet sich nur in eine Richtung aus] = Linearwelle	
line-at-a-time printer (→line printer)	TERM&PER	(→Zeilendrucker)	
line attenuation	LINE TH	**Leitungsdämpfung**	
line-balancing network = balancing network; artificial network; simulation network; artificial line; balancing circuit; simulation circuit ≈ line build-out	TELEC	**Leitungsnachbildung** = Nachbildung ≈ Leitungsverlängerung	
line battery (→telgraph battery)	POWER SYS	(→Telegrafenbatterie)	
line bit rate	TRANS	**Leitungstakt** = Leitungsbitrate	
line blanking (→horizontal blanking)	TV	(→Zeilenaustastung)	
line blanking interval (→horizontal blanking interval)	TV	(→Zeilenaustastlücke)	
line booster	RADIO	**Kabelverstärker**	
line-bound (→conducted)	TELEC	(→leitergebunden)	
line break = word wrapping; wrapping; word wrap ↑ makeup	TYPOGR	**Zeilenumbruch** ↑ Umbruch	
line break (→line interruption)	TELEC	(→Leitungsunterbrechung)	
line broadcasting (→wired broadcasting)	BROADC	(→Kabelrundfunk)	
line broadening	PHYS	**Linienverbreiterung**	
line buffer	DATA COMM	**Leitungspuffer**	
line build-out ≈ line-balancing network	TRANS	**Leitungsverlängerung** ≈ Leitungsnachbildung	
line busy tone (AM) (→busy tone)	TELEPH	(→Besetztton)	
line cancellation (→line deletion)	DATA PROC	(→Zeilenlöschung)	
line change (→new line)	TERM&PER	(→Zeilenwechsel)	
line chart	DOC	**Liniendiagramm**	
line chart (→transmission line chart)	LINE TH	(→Leitungsdiagramm)	
line check ≈ line measuring	SWITCH	**Leitungsprüfung** ≈ Leitungsmessung	
line circuit (→subscriber-line circuit)	SWITCH	(→Teilnehmersatz)	
line code = transmission code	CODING	**Leitungscode** = Übertragungscode	

line communications (→line transmission)	TRANS	(→Leitungsübertragung)
line-commutated	POWER SYS	netzgeführt
line-commutated converter	POWER SYS	netzgeführter Stromrichter
line concentrator 1 (→trunk concentrator 1)	SWITCH	(→Verkehrskonzentrator)
line concentrator 2 (→stand-alone concentrator)	SWITCH	(→Wählsterneinrichtung)
line configuration (→path configuration)	TRANS	(→Streckenkonfiguration)
line connecting equipment (→line terminating unit)	DATA COMM	(→Leitungsabschlußeinrichtung)
line connection unit	SWITCH	Leitungsanschaltesatz
line control block = LCB	DATA COMM	Leitungssteuerungsblock
line control unit	TELECONTR	Kanalgerät
line-coupling balanced (→reciprocal)	NETW.TH	(→übertragungssymmetrisch)
line current = telegraph current	TELEGR	Fernschreibstrom = Telegrafierstrom; Linienstrom
line current (→mains current)	POWER ENG	(→Netzstrom)
line deletion [word processing] ≈ line erase; line cancellation	DATA PROC	Zeilenlöschung [Textverarbeitung] = Zeilenlöschen
line detector (→metal detector)	OUTS.PLANT	Metallsuchgerät
line display = line layout	TERM&PER	Zeilenanordnung
line distance (→line spacing)	TERM&PER	(→Zeilenabstand)
line diversity (→path diversity)	TELEC	(→Mehrwegeführung)
lined paper = ruled paper	OFFICE	liniiertes Papier = Linienpapier
line drawing	ENG.DRAW	Strichzeichnung
line driver	CIRC.ENG	Leitungstreiber
line driver [magnetic core memory]	DATA PROC	Zeilentreiber [Magnetkernspeicher]
line duration = line time	TV	Zeilendauer = Zeilenzeit
line editor [allows to edit a line at a time]	DATA PROC	Zeileneditor [erlaubt zeilenweises Editieren] = Zeilentexteditor
line end (→circuit end)	TELEC	(→Leitungsende)
line ending	DATA PROC	Zeilenende
line engineering	OUTS.PLANT	Leitungsbau
line equalizer	TRANS	Leitungsentzerrer
line equipment ↓ line terminating equipment; line repeater	TRANS	Leitungsgerät = Leitungsausrüstung ↓ Leitungsendgerät; Leitungsverstärker
line erase (→line deletion)	DATA PROC	(→Zeilenlöschung)
line feed = LF; line spacer ≈ line space regulator ↓ line space lever	TERM&PER	Zeilenschalter = Zeilentransportschalter ≈ Zeileneinsteller ↓ Zeilenschalthebel
line feed ≈ linefeed; LF ≈ new line; vertical spacing ≠ reverse line feed ↑ feed	TERM&PER	Zeilenvorschub = Zeilentransport ≈ Zeilenwechsel; Zeilenabstand ≠ Zeilenrückschub ↑ Vorschub
line feed [code] = LF	DATA COMM	Zeilenvorschub [Code] = LF
linefeed (→line feed)	TERM&PER	(→Zeilenvorschub)
line feedthrough	CIRC.ENG	Netzspannungseinstreuung
line filter = surge protector; transient suppressor	CIRC.ENG	Netzfilter
line finder [searches for subscriber lines wishing to initiate a call] = switch finder; line switch; call finder ≈ preselector ↑ finder	SWITCH	Anrufsucher [sucht in elektromechanischen Vermittlungssystemen die Teilnehmerleitungen nach Neubelegungen ab] ≈ Vorwähler ↑ Suchwähler
line flicker	TV	Teilbildflimmern = Zwischenzeilenflimmern
line flyback (→reverse action)	ELECTRON	(→Rücklauf)
line flyback (→flyback)	ELECTRON	(→Zeilenrücklauf)
line folding (→line format)	TYPOGR	(→Zeilenformat)
line format = line folding ≈ page format	TYPOGR	Zeilenformat ≈ Seitenformat
line forward equalizer (→predistorter)	TELEC	(→Vorverzerrer)
line frequency = horizontal frequency	TV	Horizontalfrequenz = H-Frequenz; Zeilenfrequenz; Abtastfrequenz
line frequency	TELEGR	Sendefrequenz
line frequency oscillator = line sweep oscillator; horizontal oscillator	TV	Horizontalfrequenzoszillator = Zeilenfrequenzoszillator; Zeilenoszillator; Zeilenkipposzillator
line fuse (→main fuse)	EL.INST	(→Hauptsicherung)
line gauge [special ruler used in typography] = forms layout gage	TYPOGR	Zeilenmaß [spezielles Lineal des Druckwesens] = Zeilenlineal; Typometer
line generator	DATA PROC	Liniengenerator
line group (→PBX line group)	SWITCH	(→Sammelanschluß)
line group number = hunting group number	SWITCH	Sammelrufnummer ≈ Sammelanschluß [TELEC]
line height [measuring unit: lines per vertical inch]	TERM&PER	Zeilenhöhe
line identification (→call line identification)	DATA COMM	(→Anschlußkennung)
line incidence angle	TV	Zeileneinfallswinkel
line input	DATA COMM	Zeileneingabe
line insertion	DATA PROC	Zeileneinfügung [Textverarbeitung]
line integral [integration over a segment of a planar or spacial curve] ≈ undefined integral ↓ circulation	MATH	Kurvenintegral [Integration über ein Stück einer ebenen oder räumlichen Kurve] = Linienintegral ≈ bestimmtes Integral ↓ Umlaufintegral
line interface	TELEC	Leitungsschnittstelle
line interface unit (→line terminal equipment)	TRANS	(→Leitungsendgerät)
line interruption = line break; drop-out	TELEC	Leitungsunterbrechung
line jump (→line skip)	TV	(→Zeilensprung)
line jump method (→interlacing)	TV	(→Zeilensprungverfahren)
line layout (→line display)	TERM&PER	(→Zeilenanordnung)
line length [number of characters per line]	DATA PROC	Zeilenlänge [Zeichen pro Zeile]
line loop (→subscriber line)	TELEC	(→Teilnehmerleitung)
line loss	LINE TH	Leitungsverlust
lineman [working on pole routes]	OUTS.PLANT	Leitungsaufseher [Betrieb und Wartung von Freileitungen]
lineman's climbers (→hooks)	OUTS.PLANT	(→Steigeisen)
line margin	TYPOGR	Zeilenrand
line measuring ≈ line check	SWITCH	Leitungsmessung ≈ Leitungsprüfung
line mode (→on-line operation)	DATA COMM	(→Leitungsbetrieb)
line module	SWITCH	Leitungsmodul
line network = line plant ↑ outside plant network	TELEC	Leitungsnetz [Netz der leitergebundenen Verbindungswege] ↑ Liniennetz
line noise = circuit noise; random circuit noise	TELEC	Leitungsgeräusch = Leitungsrauschen
line number	SWITCH	Leitungsnummer
line number	TYPOGR	Zeilennummer

line number		TV	**Zeilenzahl**	**line receiver**	CIRC.ENG	**Line-Receiver**
line number		DATA PROC	**Zeilennummer**	**line recorder**	INSTR	**Linienschreiber**
line of code		DATA PROC	(→Programmzeile)	**line rectifier**	POWER SYS	**Netzgleichrichter**

line number TV **Zeilenzahl**
line number DATA PROC **Zeilennummer**
line of code DATA PROC (→Programmzeile)
  (→instruction line)
line of curvature MATH **Krümmungslinie**
line of flux (→line of PHYS (→Kraftlinie)
  force)
line of flux and force (→line of PHYS (→Kraftlinie)
  force)
line of force PHYS **Kraftlinie**
  = line of flux; line of flux and force; = Feldlinie; Stromlinie
  streamline ≈ Potentiallinie
  ≈ potential line
line offset TV **Zeilenoffset**
  = Zeilenversatz 2
line of sight TECH **Blickrichtung**
  = direction of view
line of sight (→line-of-sight RADIO (→Richtfunk)
  radio)
line-of-sight communication TRANS (→Richtfunkübertragung)
  (→radio relay transmission)
line-of-sight connection RADIO PROP **Sichtverbindung**
line-of-sight distance (→slant RADIO PROP (→Luftlinienentfernung)
  distance)
line-of-sight distance (→visibility PHYS (→Sichtbereich)
  range)
line-of-sight microwave link RADIO REL (→Richtfunkstrecke)
  (→line-of-sight radiolink)
line-of-sight RADIO PROP **freie Ausbreitung**
  propagation
  = LOS
line-of-sight radio RADIO **Richtfunk**
  = line of sight; LOS; microwave radio ≈ Richtfunkübertragung
  relay; radio relay; microwave radio-
  link
  ≈ radio relay transmission
line-of-sight radiolink RADIO REL **Richtfunkstrecke**
  = LOS radiolink; line-of-sight micro- = Richtfunkverbindung
  wave link; microwave radiolink;
  radio relay link; microwave link
  ↓ radioenlace de visibilidad directa
line-of-sight radio system (→radio TRANS (→Richtfunksystem)
  relay system)
line-of-sight route (→radio relay TRANS (→Richtfunktrasse)
  route)
line output stage TV **Zeilenendstufe**
line pairing TV **Zeilenpaarigkeit**
line path MICROEL **Zeilenleitung**
  = line track ≠ Spaltenleitung
  ≠ column path
line path TELEC **Grundleitung**
line path OUTS.PLANT **Leitungsweg**
  ↓ cable path
line pilot TRANS **Leitungspilot**
line plan EL.TECH **Leitungsplan**
line plant (→line network) TELEC (→Leitungsnetz)
line plot DATA PROC **Liniengraphik**
line polling DATA COMM **Leitungsabruf**
  = circuit polling
line-polling operation DATA COMM **Leitungsabrufbetrieb**
  = circuit-polling operation; polled-line
  operation; polled-circuit operation
line port MICROEL **Zeileneingang**
  ≠ column port ≠ Spalteneingang
line-powered (→remotely TELEC (→ferngespeist)
  fed)
line powering (→remote power TRANS (→Fernspeisung)
  feeding)
line powering voltage (→remote TRANS (→Fernspeisespannung)
  feeding voltage)
line printer TERM&PER **Zeilendrucker**
  [prints a line at once] [druckt eine Zeile gleichzei-
  = line-at-a-time printer tig]
  ≈ high-speed printer ≈ Schnelldrucker
  ≠ page printer; character printer ≠ Zeichendrucker; Seiten-
  ↑ parallel printer drucker
  ↑ Paralleldrucker
line printing TERM&PER **Zeilendruck**
line reactor POWER SYS **Netzdrossel**
  = Drosselspule

line receiver CIRC.ENG **Line-Receiver**
line recorder INSTR **Linienschreiber**
line rectifier POWER SYS **Netzgleichrichter**
  = power rectifier
line repeater TRANS **Leitungsverstärker**
  = line amplifier; repeater 1; amplifier = Verstärker
  ≈ regenerator ≈ Regenerator
  ↑ line equipment ↑ Leitungsgerät
  ↓ terminal repeater; intermediate re- ↓ Endverstärker; Zwischen-
  peater verstärker
line repeater station TRANS **Leitungsverstärkerstelle**
  ↑ repeater station = Verstärkerstelle;
  Verstärkeramt
  ↑ Zwischenstelle
line resonator (→coaxial cavity MICROW (→Topfkreis)
  resonator)
line reversal TELEPH **Hartumpolung**
line routing (→line tracing) TELEC (→Leitungsführung)
line scanner SWITCH **Teilnehmerabtaster**
line scanning ELECTRON **Zeilenabtastung**
  = horizontal scanning = Horizontalabtastung
line scan transformer TV **Horizontal-Ablenktransfor-**
  = line transformer; flyback transfor- **mator**
  mer; horizontal sweep transformer; = Zeilentransformator; Zei-
  horizontal output transformer lentrafo; Zeilenübertrager;
  Horizontal-Ausgangstrans-
  formator; Zeilenablenk-
  transformator
line scratches (BRI) (→contact EL.ACOUS (→Kratzgeräusch)
  noise)
line section TRANS **Streckenabschnitt**
  = path section = Leitungsabschnitt
  ≈ Teilstrecke
line seizure by home station SWITCH **Eigenbelegung**
line seizure by other SWITCH **Fremdbelegung**
  station
line selection register SWITCH **Leitungswahlregister**
line selection stage SWITCH **Leitungswahlstufe**
  = Punktwahlstufe; Anrufsu-
  cher-Wahlstufe; Mischwahl-
  stufe
line selector (→final selector) SWITCH (→Leitungswähler)
line separating filter TRANS **Leitungsweiche**
line separation (→line TERM&PER (→Zeilenabstand)
  spacing)
line-serially TERM&PER **zeilenweise**
line side TELEC **Leitungsseite**
  ≠ equipment side ≠ Geräteseite
line signal SWITCH **Leitungszeichen**
  ↑ switching signal [für leitungsindividuelle Si-
  gnalisierung]
  ↑ Kennzeichen
line signaling SWITCH **Leitungssignalisierung**
  = Leitungszeichengabe
line skew ELECTRON **Zeilenschräglauf**
line skip TV **Zeilensprung**
  = line jump
line skip suppression TELEGR **Zeilensprungunterdrückung**
  [videotex] [Btx]
line source ANT **Linienquelle**
line space lever TERM&PER **Zeilenschalthebel**
  [line feed of a typewriter] [Zeilenschalter einer
  Schreibmaschine]
line spacer (→line feed) TERM&PER (→Zeilenschalter)
line space regulator TERM&PER **Zeileneinsteller**
  [to regulate the vertical spacing] [zur Regulierung des Zeile-
  ≈ line feed nabbstands]
  = Zeilensteller
  = Zeilenschalter
line spacing TERM&PER **Zeilenabstand**
  [measuring unit: LPI / lines per inch] [Maßeinheit: LPI / Zeilen
  = line separation; line distance; line ad- pro Zoll]
  vance; scanning separation; scanning = Zeilenzwischenraum
  pitch; vertical spacing; vertical separ- = Zeilenvorschub; Durch-
  ation; vertical distance; vertical ad- schuß [TYPOGR]
  vance
  ≈ line feed; leading [TYPOGR]
line spectrum PHYS **Linienspektrum**
  = diskontinuierliches Spek-
  trum

line speed		DATA COMM	Leitungsgeschwindigkeit	line trunk group	SWITCH	Anschlußgruppe

line speed DATA COMM **Leitungsgeschwindigkeit**
= channel speed
lines per inch TERM&PER **Zeilen pro Zoll**
= LPI
lines per minute TERM&PER **Zeilen pro Minute**
= LPM
line splitting PHYS **Linienaufspaltung**
line start (→beginning of TYPOGR (→Zeilenanfang)
line)
line supervision SWITCH **Leitungsüberwachung**
line supplement (→trunk SWITCH (→Leitungssatz)
circuit)
line suppression (→horizontal TV (→Zeilenaustastung)
blanking)
line surge ELECTRON **Netzstoß**
line sweep oscillator (→line TV (→Horizontalfrequenzoszilla-
frequency oscillator) tor)
line switch (→line finder) SWITCH (→Anrufsucher)
lineswitch (→final selector) SWITCH (→Leitungswähler)
line switching (→circuit SWITCH (→Durchschaltevermittlung)
switching)
line synchronization TV **Zeilensynchronisierung**
= horizontal synchronization = Horizontalsynchronisierung
line synchronizing pulse TV **Zeilensynchronimpuls**
= Zeilenimpuls
line systems TRANS **Leitungssystem**
[uses physical means as copper [überträgt auf verlegten
cables, open wire or optical cables] Leitern, wie Kupferkabeln, Freileitungen oder
LWL-Kabeln]
line terminal amplifier TRANS **Leitungsendverstärker**
= Kabelendverstärker
line terminal equipment TRANS **Leitungsendgerät**
= line terminating equipment; line in- = Leitungsendeinrichtung;
terface unit; terminal equipment; line LE
terminating unit; LTU ↑ Leitungsgerät
↑ line equipment
line terminating equipment DATA COMM (→Leitungsabschlußeinrich-
(→line terminating unit) tung)
line terminating equipment TRANS (→Leitungsendgerät)
(→line terminal equipment)
line terminating unit DATA COMM **Leitungsabschlußeinrichtung**
= LTU; line terminating equipment; = Leitungsabschlußeinheit;
line connecting equipment; circuit ter- LTU
minating equipment
line terminating unit (→line TRANS (→Leitungsendgerät)
terminal equipment)
line termination LINE TH **Leitungsabschluß**
line termination TELEC **Leitungsabschluß**
[ISDN] [ISDN]
= LT
line theory EL.TECH **Leitungstheorie**
↑ electrical fundamentals ↑ Elektrizitätslehre
line thickness ENG.DRAW **Strichdicke**
line thyristor MICROEL **Netzthyristor**
line time (→line duration) TV (→Zeilendauer)
line tracing TELEC **Leitungsführung**
= line tracking; line routing = Leitungsverlauf
line track (→line path) MICROEL (→Zeilenleitung)
line tracking (→line tracing) TELEC (→Leitungsführung)
line transformer TELEC **Leitungsübertrager**
line transformer (→line scan TV (→Horizontal-Ablenktrans-
transformer) formator)
line transformer (→mains POWER ENG (→Netztransformator)
transformer)
line transmission TRANS **Leitungsübertragung**
[on physical lines as copper cables, [Übertragung auf verlegten
open-wire lines or optical cables] Leitern wie Kupferkabeln,
= line communications Freileitungen oder LWL-
Kabeln]
= leitergebundene Übertragung
line transmission technique TRANS **leitergebundene Übertra-**
[on metallic or optical lines] **gungstechnik**
↓ metallic-line transmission technique [auf metallischen oder optischen Leitern]
= Leitungsübertragungstech-
nik
↓ drahtgebundene Übertra-
gungstechnik

line trunk group SWITCH **Anschlußgruppe**
line-type network TELEC **linienförmiges Netz**
= Liniennetz 2
line-up (n.) TRANS **Einmessung**
≈ commissioning = Einpegelung
≈ Einschaltung
line-up (→commission) TELEC (→einschalten)
line-up engineer TELEC (→Einschaltingenieur)
(→commissioning engineer)
line voltage POWER ENG **Netzspannung**
= AC line voltage; mains voltage = Leitungsspannung
line width PHYS **Linienbreite**
line wire OUTS.PLANT **Leitungsdraht**
line working (→on-line DATA COMM (→Leitungsbetrieb)
operation)
linguistic attainments LING **Sprachkenntnisse**
= knowledge of foreign language
linguistics SCIE **Sprachwissenschaft**
= philology = Linguistik; Philologie
↓ grammar ↓ Grammatik
lining TECH **Auskleidung**
link (v.t.) ENG.LOG **verknüpfen**
= gate (v.t.); combine
link (n.) TECH **Bindeglied**
= linking element
link (n.) MECH **Gelenkglied**
= Glied
link (n.) TRANS **Strecke**
= path; transmission link = Leitung; Übertragungsstrek-
≈ route ke; Übertragungsleitung
≈ Trasse
link SWITCH **Zwischenleitung**
[connects switching matrices] [verbindet Koppelvielfache]
link (v.t.) (→attach) TECH (→anbringen)
link (v.t.) (→connect 1) EL.TECH (→anschließen)
link (v.t.) DATA PROC (→binden)
(→link-edit)
link (→strap) ELECTRON (→Brücke)
link (→line) TELEC (→Leitung)
link (→connection) TELEC (→Verbindung)
link (→concatenate) TECH (→verketten)
link access procedure DATA COMM **Leitungszugangsverfahren**
= LAP
link address (→reference DATA PROC (→Verweisadresse)
address)
linkage (→concatenation) TECH (→Verkettung)
linkage (→chaining) DATA PROC (→Kettung)
linkage address (→reference DATA PROC (→Verweisadresse)
address)
linkage editing DATA PROC **Programmbindung**
linkage editor DATA PROC **Binderprogramm**
[assembles programm modules to [fügt Programmodule zu
programs] Programmen zusammen]
= linker = Binder; Modulbinder; Lin-
kage-Editor; Linker
linkage editor listing DATA PROC **Binderprotokoll**
linkage flux PHYS **Verkettungsfluß**
= magnetic loading = magnetische Durchflutung
linkage software DATA PROC **Binder-Software**
linkage specification DATA PROC **Bindevorschrift**
link analysis (→path TRANS (→Streckenberechnung)
calculation)
link-attached [DATA COMM] TELEC (→abgesetzt)
(→remote)
link attenuation (→path TRANS (→Streckendämpfung)
attenuation)
link budget analysis SAT.COMM **Leistungsbilanz**
link-by-link TELEC **abschnittsweise**
link-by-link mode (→link-by-link SWITCH (→abschnittsweise Zeichen-
signaling) gabe)
link-by-link signalling (AM) SWITCH **abschnittsweise Zeichengabe**
= link-by-link signalling (BRI); link-by- = abschnittsweise Signalisie-
link mode rung
link-by-link signalling (BRI) SWITCH (→abschnittsweise Zeichen-
(→link-by-link signaling) gabe)
link cable OUTS.PLANT **Querkabel**
↑ local cable ↑ Ortskabel
link calculation (→path TRANS (→Streckenberechnung)
calculation)
link charge (→connection fee) TELEC (→Verbindungsgebühr)

link connector (→plug link) ELECTRON (→Brückenstecker)
link control (→data transmission control) DATA COMM (→Datenübertragungssteuerung 1)
link control (→call control) SWITCH (→Verbindungssteuerung)
link control character (→call control character) SWITCH (→Verbindungssteuerungszeichen)
link control procedure (→call control procedure) SWITCH (→Verbindungssteuerungsverfahren)
link control signal (→call control character) SWITCH (→Verbindungssteuerungszeichen)
link converter POWER SYS Zwischenkreisumrichter
link data (→call data) SWITCH (→Verbindungsdaten)
link design engineer (→path design engineer) TRANS (→Streckenplaner)
link design engineering (→path design engineering) TRANS (→Streckenplanung)
linked file (→chained file) DATA PROC (→verkettete Datei)
link-edit DATA PROC binden
= link (v.t.); bind (v.t.) = verknüpfen
linked list (→chained list) DATA PROC (→verkettete Liste)
linked numbering (→closed numbering) SWITCH (→verdeckte Numerierung)
linked subroutine (→closed subroutine) DATA PROC (→geschlossenes Unterprogramm)
link engineering (→path design engineering) TRANS (→Streckenplanung)
linker (→linkage editor) DATA PROC (→Binderprogramm)
linker (→program linker) DATA PROC (→Programmbinder)
link establishment (→connection set-up) SWITCH (→Verbindungsaufbau)
link failure (→call failure) SWITCH (→Verbindungsstörung)
link fee (→connection fee) TELEC (→Verbindungsgebühr)
link identifier (→call identifier) DATA COMM (→Verbindungskennung)
linking (→concatenation) TECH (→Verkettung)
linking (→chaining) DATA PROC (→Kettung)
linking element (→link) TECH (→Bindeglied)
linking loader DATA PROC Bindelader
= link loader
link layer DATA COMM Sicherungsschicht
[2nd layer of OSI; defines parameters relevant for error protection] [2. Schicht im ISO-Schichtenmodell; legt Parameter für Fehlersicherung fest]
= data link layer
link loader (→linking loader) DATA PROC (→Bindelader)
link logging (→call logging) DATA COMM (→Verbindungsprotokollierung)
link register (→call register) SWITCH (→Verbindungsspeicher)
link request (→call request) SWITCH (→Verbindungsanforderung)
link sectioning SWITCH Leitungsauftrennung
link sequence (→connection sequence) SWITCH (→Verbindungsablauf)
link set-up (→connection set-up) SWITCH (→Verbindungsaufbau)
link set-up time (→call set-up time) DATA COMM (→Verbindungsaufbaudauer)
link state (→call state) SWITCH (→Verbindungszustand)
link system DATA PROC Linksystem
[each file entry contains the address of its successor] [jeder Dateieintrag enthält die Adresse seines Nachfolgers]
link system SWITCH Linksystem
[array of switching matrices, intermeshed by intermediate links, suited for conjugate selection] [durch Zwischenleitungen vermaschte Anordnung von Koppelvielfachen, geeignet für bedingte Wegesuche]
= conjugate selection system; conditional selection system = Zwischenleitungsanordnung; Linkanordnung
link tear-down (→connection tear-down) SWITCH (→Verbindungsabbau)

link test TELEC Streckenmessung
= end-to-end test; section test = Punkt-zu-Punkt-Messung; Abschnittsmessung
link trials DATA PROC Verknüpfungstest
[interworking of program modules] [des Zusammenspiels von Programmteilen]
Linvill model MICROEL Linvill-Ersatzschaltung
LIPS DATA PROC LIPS
[measuring unit for expert systems] [Maßeinheit für Expertensysteme]
= logical inferences per second = logische Verknüpfungen pro Sekunde
liquefaction PHYS Verflüssigung
≈ melting ≈ Schmelzen
liquid (n.) PHYS Flüssigkeit
= fluid (n.)
liquid (n.) LING Liquidum
[l,r,...] [l,r,...]
= Fließlaut; Dauerlaut
liquid (adj.) PHYS flüssig
= fluid (adj.)
liquidation (→bankruptcy) ECON (→Konkurs)
liquid crystal PHYS Flüssigkristall
= fluid crystal
liquid crystal display (→LCD display) COMPON (→Flüssigkristallanzeige)
liquid-crystal display monitor TERM&PER Flüssigkristall-Bildschirm
= liquid-crystal screen; LCD screen = LCD-Bildschirm
↑ flat screen ↑ Flachbildschirm
liquid-crystal screen TERM&PER (→Flüssigkristall-Bildschirm)
(→liquid-crystal display monitor)
liquidity ECON Liquidität
liquid jet TECH Flüssigkeitsstrahl
liquid phase epitaxy MICROEL Flüssigphasenepitaxie
liquid semiconductor PHYS flüssiger Halbleiter
liquid state PHYS Flüssigphase
liquidus line PHYS Liquidus-Kurve
liquify PHYS verflüssigen
≈ melt ≈ schmelzen
LISP DATA PROC LISP
[List Processing] ↑ höhere Programmiersprache
↑ high-level programing language
Lissajus figure (→Lissajus pattern) PHYS (→Lissajus-Figur)
Lissajus pattern PHYS Lissajus-Figur
= Lissajus figure
list (v.t.) DATA PROC auflisten
list (n.) DATA PROC Liste
[ALGOL, FORTRAN] = Meldung
= report (COBOL)
list DOC Liste
= record; listing; statement = Auflistung; Aufstellung
≈ catalogue ≈ Katalog
↓ directory ↓ Verzeichnis
list code CODING Listencode
listener COLLOQ Hörer
[Person]
= Zuhörer
listener (→radio broadcast listener) BROADC (→Rundfunkhörer)
listener (→listening master station) TELECONTR (→mithörende Zentrale)
listening (→monitoring) TELEPH (→Mithören)
listening-in (→monitoring) TELEPH (→Mithören)
listening jack TELEPH Mithörklinke
listening key TELEPH Mithörtaste
listening master station TELECONTR mithörende Zentrale
= listener
listening switch TELEPH Mithörschalter
list format (→list layout) DOC (→Listenbild)
list generator (→report generator) DATA PROC (→Listengenerator)
list header DOC Listenkopf
listing DATA PROC Protokoll
= logging; log (n.); summary = Auflistung
≈ printout ≈ Ausdruck
↓ translator listing ↓ Übersetzungsprotokoll
listing (→list) DOC (→Liste)
listing paper (→continuous paper) TERM&PER (→Endlospapier)

list layout	DOC	**Listenbild**
= list format		
list of acronyms and abbreviations	DOC	**Abkürzungsverzeichnis**
list processing	DATA PROC	**Listenverarbeitung**
list processing language [LISP; PROLOG]	DATA PROC	**Listenverarbeitungssprache** [LISP; PROLOG]
liter [1 cubic decimeter] = litre; L (NAM)	PHYS	**Liter** [1 Kubikdezimeter] = l; L ↑ Volumeneinheit
literal (n.) (→literal constant)	DATA PROC	(→Literalkonstante)
literal address (→immediate address)	DATA PROC	(→unmittelbare Adresse)
literal addressing (→immediate addressing)	DATA PROC	(→unmittelbare Adressierung)
literal constant [operand of instruction, containing actual information] = literal (n.); constant (n.)	DATA PROC	**Literalkonstante** [Operand eines Befehls, enthält direkt die Nutzinformation] = Literal; Befehlskonstante; Konstante ≠ Variable
literal instruction (→immediate instruction)	DATA PROC	(→Direktbefehl)
literature (→references)	LING	(→Literatur)
lithium = Li	CHEM	**Lithium** = Li
lithium battery	POWER SYS	**Lithium-Batterie**
lithium-manganese battery	POWER SYS	**Lithium-Mangan-Batterie**
lithium niobate	CHEM	**Lithiumniobat**
lithography [a smooth surface is treated to absorb ink on the printing parts and to repell it on the non-printing ones] ↑ planographic printing	TYPOGR	**Lithographie** [eine glatte Fläche wird so präpariert, daß die druckenden Teile Tinte aufnehmend und die nichtdruckenden Teile abweisend sind] ↑ Flachdruck
litigation (→law suit)	ECON	(→Prozeß)
litography	MICROEL	**Lithografie**
litre (→liter)	PHYS	(→Liter)
litz wire (→stranded wire)	COMM.CABLE	(→Litzendraht)
Liupanow stability	CONTROL	**Lijapunow-Stabilität**
live (adj.) = alive; energized; power-on; hot ≠ dead ↓ current-carrying; voltage-carrying	EL.TECH	**heiß** = aktiv ≠ kalt ↓ stromführend; spannungsführend
live (adj.) = alive	BROADC	**direktübertragen**
live broadcast	BROADC	**Direktübertragung** = Direktsendung; Livesendung
live data	DATA PROC	**heiße Daten**
live insertion [interruptionless replacement of boards]	EQUIP.ENG	**unterbrechungsweise Auswechslung**
liveware [users and operators] = computing staff	DATA PROC	**DVA-Personal** [Anwender und Bedienpersonal] = Liveware
living (→lodging)	ECON	(→Verpflegung)
LLL (→low-level logic)	MICROEL	(→Tiefpegellogik)
LLL (→low-level programming language)	DATA PROC	(→niedere Programmiersprache)
lm (→lumen)	PHYS	(→Lumen)
L match	ANT	**L-Anpassung** = Haarnadelanpassung; Induktor-Match
L matching ↑ beta matching	NETW.TH	**L-Anpassung** ↑ Beta-Anpassung
ln (→natural logarithm)	MATH	(→natürlicher Logarithmus)
L-network (→L-section)	NETW.TH	(→L-Schaltung)
L-noise margin	ELECTRON	**L-Störabstand**
load (v.t.) = lade	TECH	**beladen** = belasten
load (n.) = duty ≈ operation	TECH	**Belastung** = Last; Beladung; Ladung ≈ Betrieb
load (n.) ≈ power load ≈ power absorber	EL.TECH	**Last** = Belastung ≈ Verbraucher; Stromverbraucher
load (n.) ≈ weight	MECH	**Last** = Gewicht
load (→stress)	QUAL	(→Beanspruchung)
load (v.t.) [from external to main memory] = get (v.t.) ≈ input (v.t.)	DATA PROC	**laden** [von Externspeicher auf Hauptspeicher] ≈ eingeben
load (→cargo)	ECON	(→Ladung)
load (→load resistance)	NETW.TH	(→Lastwiderstand)
load (→pupinize)	COMM.CABLE	(→bespulen)
loadable (→transient)	DATA PROC	(→transient)
load address	DATA PROC	**Ladeadresse**
load admittance	EL.TECH	**Belastungsleitwert**
load alternation	MECH	**Lastwechsel** ≈ Lastumkehrung
load-and-go (→self-triggering)	DATA PROC	(→selbststartend)
load-and-run (→self-triggering)	DATA PROC	(→selbststartend)
load at deflection [spring]	MECH	**Auslenkbelastung** [Feder]
load capability (→load capacity)	TECH	(→Belastbarkeit)
load capacitor	EL.TECH	**Lastkondensator**
load capacity = loading capacity; power rating; rating; load capability; load-carrying ability ↓ nominal load ability	TECH	**Belastbarkeit** = Tragfähigkeit ↓ Nennbelastbarkeit
load capacity (→overload point)	ELECTRON	(→Aussteuerungsgrenze)
load capacity table (→ampacity table)	POWER ENG	(→Strombelastbarkeitstabelle)
load card	DATA PROC	**Ladekarte**
load-carrying ability (→load capacity)	TECH	(→Belastbarkeit)
load carrying ability (→load carrying capacity)	DATA PROC	(→Belastbarkeit)
load carrying capacity = load carrying ability	DATA PROC	**Belastbarkeit**
load characteristic (→loading characteristic)	TECH	(→Belastungskennlinie)
load circuit	EL.TECH	**Lastkreis** = Belastungskreis
load commutated	POWER SYS	**lastgeführt**
load commutated converter	POWER SYS	**lastgeführter Stromrichter**
load-compensated diode-transistor logic (→LCDTL)	MICROEL	(→LCDTL)
load control ↑ congestion control	POWER ENG	**Leistungsregelung** ↑ Netzregelung
load control	DATA COMM	**Lastregelung** ↑ Überlastabwehr
load current = drain current; drain	EL.TECH	**Laststrom**
load curve (→loading characteristic)	TECH	(→Belastungskennlinie)
load-dependent	TELEC	**belegungsabhängig**
load-dependent psophometric noise	TELEC	**belegungsabhängiges Geräusch**
load diagram (→loading characteristic)	TECH	(→Belastungskennlinie)
load dispatching station	POWER ENG	**Lastverteiler**
load distribution (→load sharing)	SWITCH	(→Lastverteilung)
loaded (→coil-loaded)	COMM.CABLE	(→bespult)
loaded antenna	ANT	**belastete Antenne**
loaded cable = pupinized cable	OUTS.PLANT	**bespultes Kabel** = pupinisiertes Kabel
loaded line = pupinized line	OUTS.PLANT	**Pupinleitung** = bespulte Leitung
loaded noise = noise with load; noise with tone ≈ weighted noise with load; signal-to-noise with load ratio	TELEC	**Rauschen bei Belastung** = Rauschen mit Belastung; belastetes Rauschen ≈ Signal-Geräusch-Abstand bei Belastung

loaded VF circuit	OUTS.PLANT	**bespulte NF-Leitung**	
= loaded VF link			
loaded VF link (→ loaded VF circuit)	OUTS.PLANT	(→ bespulte NF-Leitung)	
load effect (→ load regulation)	POWER SYS	(→ Lastregelung)	
load enable	DATA PROC	**Ladefreigabe**	
loader (→ loader programm)	DATA PROC	(→ Programmlader)	
loader programm	DATA PROC	**Programmlader**	
[loads into the main memory]		[Programm zum Laden in den Hauptspeicher]	
= loading routine; loader routine; loader		= Ladeprogramm; Laderoutine; Lader	
↓ absolute program loader; relative program loader; initial program loader		↓ Absolutlader; Relativlader; Urlader	
loader routine (→ loader programm)	DATA PROC	(→ Programmlader)	
load factor	EL.TECH	**Lastfaktor**	
load-frequency control	POWER ENG	**Leistungs-Frequenz-Regelung**	
loading (→ despatch)	ECON	(→ Versand)	
loading capacity (→ load capacity)	TECH	(→ Belastbarkeit)	
loading characteristic	TECH	**Belastungskennlinie**	
= load characteristic; load diagram; load curve; load line		= Belastungskurve; Belastungscharakteristik; Belastungsdiagramm	
loading characteristic [accumulator]	POWER SYS	**Ladecharakteristik** [Akkumulator]	
		= Ladekennlinie	
loading coil	CIRC.ENG	**Lastspule**	
loading coil	COMM.CABLE	**Pupinspule**	
= Pupin coil			
loading coil	ANT	**Belastungsspule**	
		= Ladespule	
loading-coil case	OUTS.PLANT	**Pupinspulenkasten**	
loading coil inductance	COMM.CABLE	**Spuleninduktivität**	
loading corotron	TERM&PER	**Ladecoroton**	
↑ lase printer		↑ Lasedrucker	
loading guide (→ loading scheme)	OUTS.PLANT	(→ Bespulungsplan)	
loading limit	TECH	**Belastungsgrenze**	
loading point	OUTS.PLANT	**Spulenpunkt**	
loading ramp	CIV.ENG	**Laderampe**	
		= Rampe	
loading range	TECH	**Belastungsbereich**	
loading resistor (→ load resistor)	EL.TECH	(→ Belastungswiderstand)	
loading resistor (→ load resistance)	NETW.TH	(→ Lastwiderstand)	
loading routine (→ loader programm)	DATA PROC	(→ Programmlader)	
loading scheme	OUTS.PLANT	**Bespulungsplan**	
= loading guide		= Pupinisierungsplan	
loading section (→ coil section)	OUTS.PLANT	(→ Spulenfeldlänge)	
load instruction	DATA PROC	**Ladeanweisung**	
		= Ladebefehl; Ladeinstruktion	
load interlocking (→ load sharing)	DATA PROC	(→ Lastverbund)	
load-invariant	TELEC	**belegungsunabhängig**	
load-invariant noise (→ basic noise)	TELEC	(→ Grundgeräusch)	
load line	POWER SYS	**Verbraucherleitung**	
load line (→ loading characteristic)	TECH	(→ Belastungskennlinie)	
load matching	NETW.TH	**Lastanpassung**	
load measurement	SWITCH	**Belastungsmessung**	
load module	DATA PROC	**Lademodul**	
load per unit area (→ area load)	CIV.ENG	(→ Flächenbelastung)	
load point [tape]	TERM&PER	**Anfangsmarke** [Magnetband]	
load pulling	CIRC.ENG	**Loadpulling**	
[variation of impedance matching in order to minimize ditortions]		[Variation der Scheinwiderstandsanpassung zur Verringerung der Nichtlinearitäten]	
load pulse	ELECTRON	**Ladeimpuls**	
load rate	MECH	**Federsteife**	
[excursion gradient of spring force]		[Kraftzunahme mit Auslenkung]	
		≈ Federspannung; Federkraft	
load reduction	EL.TECH	**Entlastung**	
load regulation	POWER SYS	**Lastregelung**	
= load effect			
load relay	COMPON	**Lastrelais**	
load resistance	NETW.TH	**Lastwiderstand**	
= load resistor; loading resistor; load		= Arbeitswiderstand; Außenwiderstand; äußerer Widerstand; äußere Last; Last	
≈ terminating resistance		≈ Abschlußwiderstand	
≠ intrinsic resistance		≠ Innenwiderstand	
load resistor	EL.TECH	**Belastungswiderstand**	
= loading resistor			
load resistor (→ load resistance)	NETW.TH	(→ Lastwiderstand)	
load reversion	MECH	**Lastumkehrung**	
		≈ Lastwechsel	
load sensitivity	TECH	**Belastungsempfindlichkeit**	
load sharing	SWITCH	**Lastverteilung**	
= load distribution		= Lastverbund; Lastteilung; Belastungsteilung	
load sharing	DATA PROC	**Lastverbund**	
= load interlocking		↑ Mehrrechnersystem	
↑ multi-computer system			
load simulation	TECH	**Belastungssimulation**	
load transfer switch [accumulator]	POWER SYS	**Ladeumschaltung** [Akkumulator]	
load voltage	EL.TECH	**Lastspannung**	
loan 1 (n.)	ECON	**Darlehen**	
loan (→ credit 2)	ECON	(→ Kredit)	
loan (→ lend 1)	ECON	(→ verleihen)	
loan (v.t.) (→ lend 2)	ECON	(→ entleihen)	
lobby (n.)	ECON	**Pforte**	
[entrance to corporate premises]		[z.B. eines Firmengeländes]	
lobe	ANT	**Keule**	
		= Zipfel; Lappen	
lobe shoulder	ANT	**Keulenschulter**	
lobe switching	ANT	**Keulenumschaltung**	
LOC (→ localizer)	RADIO NAV	(→ Landekurssender)	
LOC (→ instruction line)	DATA PROC	(→ Befehlszeile)	
local (adj.)	COLLOQ	**örtlich**	
local (adj.)	DATA PROC	**singulär**	
[only valid in a section]		[nur in bestimmtem Abschnitt gültig]	
≠ global		= lokal	
		≠ allgemeingültig	
local area (→ local exchange area)	SWITCH	(→ Ortsanschlußbereich)	
local area network	DATA COMM	**lokales Netz**	
[a locally restricted private data network]		[ein räumlich begrenztes privates Datennetz]	
= LAN; in-house network		= örtliches Netz; LAN; Grundstücksnetz; lokales Netzwerk; Standortnetz	
≈ corporate network [TELEC]		≈ Firmennetz [TELEC]	
local batch processing	DATA PROC	**Ortsstapelverarbeitung**	
≠ remote batch processing		≠ Stapelfernverarbeitung	
↑ batch processing		↑ Stapelverarbeitung	
local battery	TELEPH	**Ortsbatterie**	
= LB		= OB	
≠ central battery		≠ Zentralbatterie	
local battery telephone	TELEPH	**OB-Fernsprecher**	
		= Ortsbatterie-Fernsprecher	
local cable (→ local junction cable)	COMM.CABLE	(→ Ortskabel)	
local call	SWITCH	**Ortsgespräch**	
		[zwischen Anschlüssen gleicher Vorwählnummer]	
		= Amtsanruf	
		≈ Nahgespräch	
local call rate (→ local call tariff)	SWITCH	(→ Ortsgebühr)	
local call tariff	SWITCH	**Ortsgebühr**	
= local call rate; local charge			
local carrier	TELEC	**Ortsnetzbetreiber**	
= local exchange carrier		↑ Betriebsgesellschaft	
↑ operating company			

local central office (AM) SWITCH
[central office where subscribers are connected directly]
= local exchange (BRI); telephone central office; telephone exchange (BRI)
≈ end office
↑ local office

local charge (→local call tariff) SWITCH

local charging area (→local exchange area) SWITCH

local circuit (→local trunk) TELEC

local civil time PHYS
= local time; standard time

local communication network DATA COMM
= LCN

local connection SWITCH
local content ECON
local declaration DATA PROC
local dose PHYS
local exchange (BRI) (→local central office) SWITCH

local exchange area SWITCH
= local service area; local charging area; telephone exchange (AM); local area; short-haul area
≈ multi-exchange area; local network

local exchange carrier (→local carrier) TELEC

local hardening METAL
local intelligence TERM&PER
localizer RADIO NAV
= LOC
localizer (→locator) TECH
local junction cable COMM.CABLE
= local cable
↓ transmission cable'

local junction circuit (BRI) TELEC
(→local trunk)

local line TELEC
[a line interconnecting subscriber terminals with local exchanges or between local exchanges]
↓ local trunk

local line plant (→local network) TELEC

local loop (→subscriber line) TELEC

local-loop operation TELEGR
= local mode

locally added carrier TELEGR
(→supplementary-added carrrier)

local memory DATA PROC
= local storage; local store

local mode DATA PROC

local mode (→local-loop operation) TELEGR

local network TELEC
[consists of subscriber-premises equipment, local outside line plant and local switching]
= local line plant; local plant; metropolitan network
≈ local outside plant
↑ telecommunications network
↓ local outside line plant

local network code (→area code) SWITCH

local network planner TELEC
= local planner

local node DATA COMM
local oscillator MICROW
local outside line plant OUTS.PLANT
(→local outside plant)

Teilnehmervermittlungsstelle
[Ortsvermittlungsstelle an die Teilnehmer direkt angeschlossen sind]
↑ Ortsvermittlungsstelle

(→Ortsgebühr)

(→Ortsanschlußbereich)

(→Ortsverbindungsleitung)
Ortszeit

lokales Kommunikationssystem
= lokales Kommunikationsnetz; LCN

Nahanschluß
lokale Wertschöpfung
lokale Vereinbarung
Ortsdosis
(→Teilnehmervermittlungsstelle)

Ortsanschlußbereich
= Ortznetzbereich; Ortsgebührbereich; Ortsverkehrsbereich; Nahbereich; Nahverkehrsbereich
≈ Ortsnetz

(→Ortsnetzbetreiber)

örtliche Härtung
lokale Intelligenz
Landekurssender
= Ansteuerungssender
(→Suchgerät)
Ortskabel
↓ Übertragungskabel

(→Ortsverbindungsleitung)

Ortslinie
[Fernmeldelinie für Fernsprechortsverkehr, Teilnehmeranschluß]
↑ Fernmeldelinie
↓ Ortsanschlußleitung; Ortsverbindungsleitung

(→Ortsnetz)

(→Teilnehmerleitung)

Lokalbetrieb

(→Zusatzträger)

Lokalspeicher

Lokalbetrieb
≈ Offline-Betrieb
(→Lokalbetrieb)

Ortsnetz
[besteht aus Teilnehmereinrichtungen, Ortsleitungsnetz und Ortsvermittlungsstellen]
= Ortsleitungsnetz
↑ Fernmeldenetz
↓ Ortsleitungsnetz

(→Ortskennzahl)

Ortsnetzplaner

Ortsvermittlung
Lokaloszillator
(→Ortsleitungsnetz)

local outside plant OUTS.PLANT
[all out-of-doors equipment between customer premises ond local office]
= local outside line plant; local plant; outside plant; external plant
↑ local network

local oxidation MICROEL

local oxidation of silicon and sapphire technology (→LOSOS technology) MICROEL

local planner (→local network planner) TELEC

local plant (→local outside plant) OUTS.PLANT

local plant (→local network) TELEC

local service area (→local exchange area) SWITCH

local shift DATA COMM
= locking shift

local stability CONTROL
local station BROADC
local storage (→local memory) DATA PROC

local store (→local memory) DATA PROC

local subscriber TELEC

local subscriber link TELEC
[from subscriber to exchange of the same local network]
↑ subscriber line

local switch (→local switching center) SWITCH

local switching SWITCH
[process]

local switching center SWITCH
= local switch; urban switch; local switching office
≈ local trunk exchange
↑ exchange (n.)
↓ local central office; local tandem exchange

local switching office (→local switching center) SWITCH

local tandem access office SWITCH
(→local tandem exchange)

local tandem exchange SWITCH
= local transit exchange; local tandem access office

local television link BROADC
= local TV link

local time (→local civil time) PHYS

local traffic TELEC
≈ short-distance traffic
≠ toll traffic

local transit exchange (→local tandem exchange) SWITCH

local trunk (AM) TELEC
= local junction circuit (BRI); junction circuit (BRI); local circuit; intracity trunk (AM); interoffice trunk
↑ trunk; interoffice trunk

local trunk exchange SWITCH

local TV link (→local television link) BROADC

local variable DATA PROC
[valid for and accesible only by certain program sections]
≠ global variable

Ortsleitungsnetz
[alle Einrichtungen zum Verbinden des Teilnehmerapparates mit der Ortsvermittlung]
↑ Ortsnetz

Oxydwand-Isolation
= Oxydwall-Isolation

(→LOSOS-Technik)

(→Ortsnetzplaner)

(→Ortsleitungsnetz)

(→Ortsnetz)
(→Ortsanschlußbereich)

sperrende Umschaltung

lokale Stabilität
Ortssender
(→Lokalspeicher)

(→Lokalspeicher)

Ortsteilnehmer
= lokaler Teilnehmer

Ortsanschlußleitung
[von Vermittlung zu Teilnehmer im gleichen Ortsnetz]
= Ortsanschlußlinie
↑ Teilnehmerleitung

(→Ortsvermittlungsstelle)

Ortsvermittlung
[Vorgang]

Ortsvermittlungsstelle
[Vermittlungsstelle mit Ortsnetzfunktionen]
≈ Ortsamt; Ortsvermittlung 2
≈ Endvermittlung
↑ Vermittlungsstelle
↓ Teilnehmervermittlungsstelle;
Orts-Durchgangsvermittlungsstelle

(→Ortsvermittlungsstelle)

(→Orts-Durchgangsvermittlungsstelle)

Orts-Durchgangsvermittlungsstelle
= Ortsknotenamt (DBP); Gruppenvermittlungsstelle (DBP); Ortsdurchgangsvermittlung
↑ Ortsvermittlungsstelle

Fernseh-Ortsleitung
= TV-Ortsleitung
(→Ortszeit)
Ortsverkehr
≈ Nahverkehr
≠ Fernverkehr

(→Orts-Durchgangsvermittlungsstelle)

Ortsverbindungsleitung
= OVL; Ortsverbindung; Ortskreis
↑ Verbindungsleitung; Amtsverbindungsleitung

Endvermittlungsstelle
≈ Ortsvermittlungsstelle

(→Fernseh-Ortsleitung)

lokale Variable
[nur für bestimmte Programmteile gültig und zugänglich]
≠ globale Variable

local zoning	SWITCH	lokale Verzonung
locating	TECH	Ortung
locating loop (→fault locating loop)	TRANS	(→Fehlerortungsschleife)
locating signal (→fault-locating signal)	TRANS	(→Fehlerortungssignal)
location	COMPON	Einbauplatz
location (→position 1)	TECH	(→Lage)
location (→memory location)	DATA PROC	(→Speicherstelle)
location (→site)	ECON	(→Standort)
location advantage	ECON	Standortvorteil
locational requirement (→positioning requirement)	TECH	(→Lagevorschrift)
locational tolerance (→position tolerance)	ENG.DRAW	(→Lagetoleranz)
location change	ECON	Standortwechsel
= resiting		= Standortverlegung
location counter (→instruction counter)	DATA PROC	(→Befehlszähler)
location hole	MECH	Paßloch
location plan	TECH	Lageplan
location screw driver	ELECTRON	Festhalte-Schraubendreher
locator	TECH	Suchgerät
= localizer		
locator (→sound locator)	ACOUS	(→Schallortungsgerät)
lock 2 (v.t.)	TECH	verriegeln
= interlock		= sperren
lock 1 (v.i.)	TECH	rasten
		= einrasten
lock 1 (n.)	TECH	Schloß
lock 2 (n.)	TECH	Rast
lock (→arrest)	MECH	(→arretieren)
lock (→blocking)	ELECTRON	(→Sperre)
lock (→interlock)	DATA PROC	(→Verriegelung)
lock 3 (n.) (→seal)	TECH	(→Verschluß)
lockable	TECH	abschließbar
		= absperrbar
lockable potentiometer	COMPON	Raster-Potentiometer
lock code	DATA PROC	Sperrcode
locked disk	TERM&PER	Diskette mit Schreibschutz
= write-protected disk		
locked-up keyboard	DATA PROC	gesperrte Tastatur
lock error	CIRC.ENG	Rastfehler
lock-in amplifier	INSTR	Lock-in-Verstärker
locking 1 (n.)	TECH	Einrasten
locking 2 (n.)	TECH	Verriegelung
= interlock		
locking (adj.)	TECH	rastend
		= einrastend; mit Rast
locking clip	COMPON	Rastklinke
locking cog	MECH	Arretierzahn
locking device (→arrest)	TECH	(→Hemmung)
locking plate	TECH	Sicherungsblech
locking plunger	INSTR	Abstimmkolben
locking pushbutton key	COMPON	rastender Tastschalter
		= Tastschalter mit Rast
locking relay (→remanent relay)	COMPON	(→Haftrelais)
locking screw (→lock screw)	MECH	(→Feststellschraube)
locking shift (→local shift)	DATA COMM	(→sperrende Umschaltung)
locking wheel	MECH	Arretierrad
lock-in range (→capture range)	ELECTRON	(→Fangbereich)
lockless platen	TERM&PER	Stechwalze [Schreibmaschine]
lock nut	MECH	Sicherungsmutter
		= Gegenmutter
lock onto	ELECTRON	aufsynchronisieren, sich (v.r.)
lockout (v.t.) [prevent others from sending]	DATA COMM	sperren [andere am Senden hindern]
lockout 2 (n.)	DATA PROC	Unterbrechungsaufhebung
= interrupt suppression		
lock-out	SWITCH	Abwerfen
= forcibly disconnect		
lock-out (→intercept)	SWITCH	(→abfangen)
lockout 1 (→interlock)	DATA PROC	(→Verriegelung)
lockout switch	DATA PROC	Sperrschalter
= inhibit switch		
lock screw	MECH	Feststellschraube
= locking screw		
lock-up (n.)	DATA PROC	Totalsperre
lock washer	MECH	Sicherungsscheibe
= spring washer; snap ring; spring lock washer; retainer; retaining washer		= Benzingscheibe; Federring; Federscheibe
locus	MATH	geometrischer Ort
locus	NETW.TH	Ortskurve
↓ transmission line chart [LINE TH]		↓ Leitungsdiagramm [LINE TH]
locus recorder	INSTR	Ortskurvenschreiber
lodging	ECON	Verpflegung
= living		
lodging	ECON	Unterkunft
= accomodation		
loft antenna	ANT	Dachbodenantenne
log (v.t.)	DATA PROC	protokollieren
= print-out (v.t.)		= ausdrucken
log (n.) (→block)	TECH	(→Klotz)
log (→logarithmic)	MATH	(→logarithmisch)
log (n.) (→listing)	DATA PROC	(→Protokoll)
log (v.t.) (→register)	INSTR	(→registrieren)
logarithm	MATH	Logarithmus
[power to which a basis must be raised to produce a given number]		[Zahl, mit der die Basis potenziert, den Numerus ergibt]
↓ natural logarithm; common logarithm		↓ natürlicher Logarithmus; dekadischer Logarithmus
logarithmic	MATH	logarithmisch
= log		
logarithmically periodic aerial (→logarithmically periodic antenna)	ANT	(→logarithmisch-periodische Antenne)
logarithmically periodic antenna	ANT	logarithmisch-periodische Antenne
= log-periodic antenna; logarithmically periodic aerial; log-periodic aerial; coaxial-fed log-periodic antenna		= LPC-Antenne
↑ frequency-independent antenna		↑ frequenzunabhängige Antenne
logarithmic amplifier	CIRC.ENG	logarithmischer Verstärker
		= Logarithmierverstärker; Logarithmierer
logarithmic decrement	PHYS	logarithmisches Dekrement
logarithmic graph paper	DOC	Logarithmierpapier
= logarithmic paper		= Logarithmenpapier
logarithmic horn (→exponential horn)	ANT	(→Exponentialhorn)
logarithmic-normal distribution	MATH	Lognormalverteilung
logarithmic number	MATH	Numerus
[number resulting from raising the base by the logarithm]		[Zahl die sich aus der Potenzierung einer Basis mit einem Loraithmus ergibt]
logarithmico-normal	MATH	logarithmisch-normal
= log-normal		
logarithmic paper (→logarithmic graph paper)	DOC	(→Logarithmierpapier)
logarithmic-periodic	MATH	logarithmisch-periodisch
= log-periodic		
logarithmic spiral	MATH	logarithmische Spirale
logarithmic spiral antenna	ANT	logarithmische Spiralantenne
= logspiral antenna; equiangular spiral antenna		= gleichwinklige Spiralantenne
logarithmic system	MATH	Logarithmensystem
logarithm to the base of 2	MATH	dyadischer Logarithmus
= dyadic logarithm		= Logarithmus zur Basis 2; dualer Logarithmus; Logarithmus dualis
logarithm to the base of 10 (→common logarithm)	MATH	(→dekadischer Logarithmus)
logarithm to the basis of e (→natural logarithm)	MATH	(→natürlicher Logarithmus)
logatom	TELEPH	Logatom
[artificial word for intelligibility tests]		[Kunstwort für Verständlichkeitstests]
logger	INSTR	Meßwertdrucker
= printing recorder		= Mitschreiber
logging (→listing)	DATA PROC	(→Protokoll)
logging (→printout)	DATA PROC	(→Ausdruck)

logic gate

logging file	DATA PROC	**Logdatei**	
		= Logbuch	
logging-in (→log-on)	DATA PROC	(→Anmeldung)	
logging-off (→log-off)	DATA PROC	(→Abmeldung)	
logging-on (→log-on)	DATA PROC	(→Anmeldung)	
logic (n.)	SCIE	**Logik**	
[science of formal reasoning]		[Lehre des folgerichtigen Denkens]	
logic (→logic circuit)	CIRC.ENG	(→Logikschaltung)	
logical add (→OR operation)	ENG.LOG	(→ODER-Verknüpfung)	
logical addition (→OR operation)	ENG.LOG	(→ODER-Verknüpfung)	
logical address (→virtual address)	DATA PROC	(→virtuelle Adresse)	
logical analysis (→logic analysis)	DATA PROC	(→logische Analyse)	
logical analyzer	DATA PROC	**logischer Analysator**	
logical chart (→logic diagram)	CIRC.ENG	(→Signalflußplan)	
logical decision	DATA PROC	**logische Entscheidung**	
= logic decision			
logical design (→logic design)	DATA PROC	(→Logikentwurf)	
logical diagram (→logic diagram)	CIRC.ENG	(→Signalflußplan)	
logical drive (→partition)	DATA PROC	(→Partition)	
logical element (→logic gate)	CIRC.ENG	(→Verknüpfungsglied)	
logical error	DATA PROC	**Logikfehler**	
= logic error		= logischer Fehler	
logical file (→logic file)	DATA PROC	(→logische Datei)	
logical flowchart (→logic flowchart)	DATA PROC	(→logischer Flußplan)	
logical flow diagram (→logic flowchart)	DATA PROC	(→logischer Flußplan)	
logical function (→logical operation)	ENG.LOG	(→logische Verknüpfung)	
logical gain	INF	**logischer Hub**	
= logic swing			
logic algebra (→Boolean algebra)	MATH	(→Boolesche Algebra)	
logical high (→high level)	MICROEL	(→Hochpegelzustand)	
logical inference	DATA PROC	**Inferenz**	
[conclusion from given facts]		[Schlußfolgerung auf der Basis vorliegender Fakten]	
= inference		= logische Inferenz; Schlußfolgerung; Ableitung	
logical inference program	DATA PROC	**Inferenzprogramm**	
[set of algorithms of an expert system]		[Algorithmen eines Expertensystems]	
= inference program; inference module; inference engine; inference machine		= Inferenzmodul; Problemlöser; Inferenzmechanismus; Schlußfolgerungsprogramm	
logical inferences per second (→LIPS)	DATA PROC	(→LIPS)	
logical instruction (→logic instruction)	DATA PROC	(→logischer Befehl)	
logical low (→low level)	MICROEL	(→Tiefpegelzustand)	
logical map (→logic flowchart)	DATA PROC	(→logischer Flußplan)	
logical multiply (→AND operation)	ENG.LOG	(→UND-Verknüpfung)	
logical network (→combinatorial circuit)	CIRC.ENG	(→Schaltnetz)	
logical operation	ENG.LOG	**logische Verknüpfung**	
= logic operation; logical function; logic function		= logische Operation	
↓ AND operation; OR operation; exclusive-OR operation; NOT operation; NAND operation; NOR operation; exclusion operation; implication operation		↓ UND-Verknüpfung; ODER-Verknüpfung; Exklusiv-ODER-Verknüpfung; NICHT-Verknüpfung; NAND-Verknüpfung; NOR-Verknüpfung; Inhibitionsverknüpfung; Implikationsverknüpfung	
logical operator	DATA PROC	**logischer Operator**	
↓ relational operator; Boolean operator		↓ Vergleichsoperator; Verknüpfungsoperator	
logical product (→AND operation)	ENG.LOG	(→UND-Verknüpfung)	
logical programming	DATA PROC	**logische Programmierung**	
= logic programming			
logical record (→data record)	DATA PROC	(→Datensatz)	
logical representation	ENG.LOG	**logische Darstellung**	
= logic representation			
logical shift (→end-around shift)	DATA PROC	(→Ringschieben)	
logical sum (→OR operation)	ENG.LOG	(→ODER-Verknüpfung)	
logical swing	MICROEL	**logischer Hub**	
		[Differenz H-Pegel zu L-Pegel]	
logical symbol	ENG.LOG	**Logiksymbol**	
= logic symbol		= logisches Symbol	
logical track (→information track)	TERM&PER	(→Informationsspur)	
logical unit name	DATA PROC	**logischer Gerätename**	
logical unit number	DATA PROC	**logische Gerätenummer**	
logical value	ENG.LOG	**logischer Wert**	
= logic value			
logic analysis	DATA PROC	**logische Analyse**	
= logical analysis		= Logikanalyse	
logic analyzer	INSTR	**Logikanalysator**	
[tests complex logic circuits]		[testet komplexe Logikschaltungen]	
= logic state analyzer		= logischer Analysator	
logic array (→gate array)	MICROEL	(→Gate Array)	
logic board	DATA PROC	**Logikkarte**	
= logic card			
logic bomb	DATA PROC	**Logikbombe**	
[intentional coding of a program crash]		[mutwillige Programmierung eines Programmabsturzes]	
logic card (→logic board)	DATA PROC	(→Logikkarte)	
logic chart (→logic diagram)	CIRC.ENG	(→Signalflußplan)	
logic chip	MICROEL	**Logikbaustein**	
= logic device; logic unit			
logic circuit	CIRC.ENG	**Logikschaltung**	
[executes complex logical operations]		[realisiert komplexe Operationen der Schaltalgebra]	
= logic; decision-making network		= logische Schaltung; Logik	
≈ logic gate		≈ Verknüpfungsglied	
logic clip	INSTR	**Logikclip**	
logic comparator	INSTR	**Logikkomparator**	
logic decision (→logical decision)	DATA PROC	(→logische Entscheidung)	
logic design	DATA PROC	**Logikentwurf**	
= logical design		= logischer Entwurf	
logic device (→logic chip)	MICROEL	(→Logikbaustein)	
logic diagram	CIRC.ENG	**Signalflußplan**	
= logic chart; logical diagram; logical chart		= logisches Schaltbild	
logic element (→logic gate)	CIRC.ENG	(→Verknüpfungsglied)	
logic error (→logical error)	DATA PROC	(→Logikfehler)	
logic family	MICROEL	**Logikfamilie**	
= family		= Schaltkreisfamilie; Schaltungsfamilie	
logic file	DATA PROC	**logische Datei**	
= logical file			
logic flowchart	DATA PROC	**logischer Flußplan**	
= logical flowchart; logic map; logical map; logic flow diagram; logical flow diagram		= logisches Flußdiagramm; Logikflußplan	
logic flow diagram (→logic flowchart)	DATA PROC	(→logischer Flußplan)	
logic function	INF	**Logikfunktion**	
logic function (→logical operation)	ENG.LOG	(→logische Verknüpfung)	
logic gate	CIRC.ENG	**Verknüpfungsglied**	
[performs basic logical operations]		[realisiert Grundoperationen der Schaltalgebra]	
= gate; logic element; logical element; switching element; gate circuit		= Gatterschaltung; Gatter; Torschaltung; Tor; Ver-	
≈ logic circuit			

logic generator 336

↓ AND gate; OR gate; exclusive OR gate; NOT gate; NAND gate; NOR gate

knüpfungsschaltung; logische Grundschaltung; Logikelement; Logikgatter; verknüpfendes Schaltungsglied; Schaltglied; Glied; Gate
≈ Logikschaltung
↓ UND-Glied; ODER-Glied; Exklusiv-ODER-Glied; NICHT-Glied; NAND-Glied; NOR-Glied

logic generator INSTR **Logic Generator**
logic instruction DATA PROC **logischer Befehl**
= logical instruction
≠ arithmetic instruction
↓ comparing instruction; Boolean instruction

= logische Anweisung; Verknüpfungsbefehl; Verknüpfungsanweisung; Logikbefehl; Logikanweisung
≠ arithmetischer Befehl
↓ Vergleichsbefehl; Boole'scher Befehl

logic level MICROEL **Logikpegel**
≈ logic state [ENG.LOG]
≈ logischer Pegel
≈ logischer Zustand [ENG.LOG]

logic map (→logic flowchart) DATA PROC (→logischer Flußplan)
logic operation (→logical operation) ENG.LOG (→logische Verknüpfung)
logic pattern triggering INSTR **Signalmustertriggerung**
logic probe (→logic tester) INSTR (→Logiktester)
logic probe indicator INSTR **Logiktastkopf-Anzeige**
logic procedure INF **Loppgikprozedur**
logic programming (→logical programming) DATA PROC (→logische Programmierung)
logic representation (→logical representation) ENG.LOG (→logische Darstellung)
logic-seeking (adj.) TERM&PER **druckwegoptimierend**
logic-seek printing TERM&PER **Druckwegoptimierung**
logic sequence ENG.LOG **logischer Ablauf**
logic shift (→end-around shift) DATA PROC (→Ringschieben)
logic simulation MICROEL **Logiksimulation**
logic simulator DATA PROC **Logiksimulator**
logic state ENG.LOG **logischer Zustand**
≈ logic level [MICROEL]
= Logikzustand
= Logikpegel [MICROEL]
logic state analyzer (→logic analyzer) INSTR (→Logikanalysator)
logic swing (→logical gain) INF (→logischer Hub)
logic symbol (→logical symbol) ENG.LOG (→Logiksymbol)
logic synthesis MICROEL **Logigsynthese**
logic system INF **Logiksystem**
logic tester INSTR **Logiktester**
[to detect logic levels]
= logic probe
≈ test probe
[zur Feststellung logischer Pegel]
= Logik-Probe; Logiktastkopf
≈ Prüfspitze
logic unit (→logic chip) MICROEL (→Logikbaustein)
logic value (→logical value) ENG.LOG (→logischer Wert)
log-in (v.t.) (→log-on) DATA PROC (→anmelden)
log-in (→log-on) DATA PROC (→Anmeldung)
log-in mode (→log-on procedure) DATA PROC (→Anmeldeverfahren)
log-in name DATA PROC **Kennungsname**
log-in procedure (→log-on procedure) DATA PROC (→Anmeldeverfahren)
log measuring technique INSTR **Protokollmeßtechnik**
log-normal MATH (→logarithmisch-normal)
(→logarithmico-normal)
log-normal distribution MATH **logarithmische Normalverteilung**
logo (→logotype) ECON (→Logotype)
log-off (v.t) DATA PROC **abmelden**
[to finish a session]
= log off; logoff
≠ log-on
[eine Sitzung beenden]
≠ anmelden

log-off (n.) DATA PROC **Abmeldung**
[termination of a session]
= logoff; log-out; logout; logging-off
≈ XOFF [DATA COMM]
≠ log-in
[Beendigung einer Sitzung]
= Schließungsprozedur; Log-off; Logoff
≈ XOFF [DATA COMM]
≠ Anmeldung
logoff (→log-off) DATA PROC (→abmelden)
log off (→log-off) DATA PROC (→abmelden)
logoff (→log-off) DATA PROC (→Abmeldung)
log-off procedure DATA COMM **Abmeldeverfahren**
log-off request DATA COMM **Abmeldungsanforderung**
log-on (v.t.) DATA PROC **anmelden**
= log-in (v.t.)
log-on (n.) DATA PROC **Anmeldung**
[to start a session]
= log-in; log-on; logon; logging-in; logging-on
≈ XON [DATA COMM]
≠ log-off
[Beginn einer Sitzung]
= Eröffnungsprozedur; Log-in; Log-on; Logon
≈ XON [DATA COMM]
≠ Abmeldung
logon (→log-on) DATA PROC (→Anmeldung)
log-on (→log-on) DATA PROC (→Anmeldung)
log-on mode (→log-on procedure) DATA PROC (→Anmeldeverfahren)
log-on procedure DATA PROC **Anmeldeverfahren**
= log-on mode; log-in procedure; log-in mode
= Anmeldemodus
log-on request DATA COMM **Anmeldeanforderung**
logotom articulation (→syllabic intelligibility) TELEPH (→Silbenverständlichkeit)
logotype ECON **Logotype**
= logo
↓ corporate logo; trademark
= Signet; Logo
↓ Firmenzeichen; Markenzeichen
logout (→log-off) DATA PROC (→Abmeldung)
log-out (→log-off) DATA PROC (→Abmeldung)
log-periodic MATH (→logarithmisch-periodisch)
(→logarithmic-periodic)
log-periodic aerial ANT (→logarithmisch-periodische Antenne)
(→logarithmically periodic antenna)
log-periodic antenna ANT (→logarithmisch-periodische Antenne)
(→logarithmically periodic antenna)
log-periodic dipole antenna ANT **logarithmisch-periodische Dipolantenne**
= LPD-Antenne
log-periodic folded dipole antenna ANT (→logarithmitisch-periodische Faltdipol-Antenne)
(→LPFD antenna)
log-periodic ladder antenna ANT **logarithmisch-periodische Leiterantenne**
= LPL-Antenne
log-periodic unipole ANT **logarithmisch-periodische Unipolantenne**
= LP-unipole
= LPU-Antenne
log-periodic Yagi antenna ANT **logarithmisch-periodische Yagi-Antenne**
= LPY-Antenne
logspiral antenna (→logarithmic spiral antenna) ANT (→logarithmische Spiralantenne)
log-Yag array ANT **Log-Yag-Antenne**
= LPDA Yagi antenna
LOI (→letter of intent) ECON (→Absichtserklärung)
long PHYS **lang**
long-awaited COLLOQ **langerwartet**
long-axis plotting TERM&PER **Längsachsenplotten**
long distance barred SWITCH **Fernsperre**
long-distance call (→toll call 1) TELEPH (→Ferngespräch)
long-distance call fee TELEC **Ferngebühr**
= trunk charge; long-distance charge; toll
long-distance-call-fee meter SWITCH **Ferngebührenzähler**
long-distance carrier (→trunk carrier) TELEC (→Fernnetzbetreiber)
long-distance charge TELEC (→Ferngebühr)
(→long-distance call fee)
long-distance coin telephone TERM&PER **Fernwahl-Münzfernsprecher**
long-distance communications TELEC (→Fernverkehr)
(→toll traffic)
long-distance connection TELEC **Fernverbindung**

long-distance data transmission (→remote data transmission)	DATA COMM	(→Datenfernübertragung)
long-distance dialing	SWITCH	**Fernwahl**
= trunk dialing		
long-distance exchange (→trunk exchange)	SWITCH	(→Fernvermittlungsstelle)
long-distance line (→toll trunk)	TELEC	(→Fernleitung)
long-distance line-of-sight (→backbone microwave)	RADIO REL	(→Weitverkehrsrichtfunk)
long-distance LOS (→backbone microwave)	RADIO REL	(→Weitverkehrsrichtfunk)
long-distance microwave (→backbone microwave)	RADIO REL	(→Weitverkehrsrichtfunk)
long-distance network (→toll network)	TELEC	(→Fernnetz)
long-distance propagation	RADIO PROP	**Fernausbreitung**
long-distance radio relay (→backbone microwave)	RADIO REL	(→Weitverkehrsrichtfunk)
long-distance route (→long-haul route)	TRANS	(→Ferntrasse)
long-distance scatter	RADIO PROP	**Entfernungsstreuung**
long-distance subscriber	TELEGR	**Fernteilnehmer**
long-distance system (→long-haul system)	TRANS	(→Weitverkehrssystem)
long-distance traffic (→toll traffic)	TELEC	(→Fernverkehr)
long duration pulse	TELECONTR	**Langimpuls**
long-duration signal element	MODUL	**Langschritt**
long echo	TELEPH	**Langzeitecho**
long-haul carrier (→trunk carrier)	TELEC	(→Fernnetzbetreiber)
long-haul communications (→toll traffic)	TELEC	(→Fernverkehr)
long-haul communications engineering	TELEC	**Weitverkehrstechnik**
= long-haul transmission technique; long-range transmission engineering; transmission trunking engineering		↑ Nachrichtenübertragungstechnik
↑ communications transmission engineering		
long-haul line-of-sight (→backbone microwave)	RADIO REL	(→Weitverkehrsrichtfunk)
long-haul LOS (→backbone microwave)	RADIO REL	(→Weitverkehrsrichtfunk)
long-haul microwave (→backbone microwave)	RADIO REL	(→Weitverkehrsrichtfunk)
long-haul network (→toll network)	TELEC	(→Fernnetz)
long-haul radio relay (→backbone microwave)	RADIO REL	(→Weitverkehrsrichtfunk)
long-haul route	TRANS	**Ferntrasse**
= long-range route; long-distance route; intercapital route; toll route (AM); trunking route		= Weitverkehrstrasse
≈ backbone route		≈ Haupttrasse
long-haul system	TRANS	**Weitverkehrssystem**
= toll transmission system (AM); transmission trunking system; trunking system; backbone system; long-distance system; long-range system; intercapital system		= Fernverkehrssystem; Fernnetzsystem
≈ high-capacity system		≈ Breitbandsystem
long-haul traffic (→toll traffic)	TELEC	(→Fernverkehr)
long-haul transmission equipment	TRANS	**Weitverkehrsgerät**
= transmission trunking equipment		
long-haul transmission technique (→long-haul communications engineering)	TELEC	(→Weitverkehrstechnik)
longish	TECH	**länglich**
= oblong; elongated		
longitude	GEOPHYS	**Länge**
longitude	GEOPHYS	**geographische Länge**
longitudinal	TECH	**Längs-**
		= Longitudinal-
longitudinal aberration	PHYS	**Längsaberration**
longitudinal attenuation	TELEC	**Längsdämpfung**
longitudinal axis	MATH	**Längsachse**
longitudinal check (→horizontal parity check)	DATA COMM	(→Längsparitätsprüfung)
longitudinal check (→horizontal parity check)	DATA COMM	(→Längsparitätsprüfung)
longitudinal check sum (→horizontal parity check)	DATA COMM	(→Längsparitätsprüfung)
longitudinal controller	POWER SYS	**Längsregler**
longitudinal fin	MECH	**Längsrippe**
= longitudinal rib		
longitudinally stable	TECH	**längsstabil**
longitudinally symmetric	MATH	**längssymmetrisch**
longitudinally symmetric two-port	NETW.TH	**längssymmetrischer Vierpol**
= longitudinally symmtric quadripole		
longitudinally symmtric quadripole (→longitudinally symmetric two-port)	NETW.TH	(→längssymmetrischer Vierpol)
longitudinal mode	PHYS	**Longitudinalmode**
longitudinal motion	MECH	**Längsbewegung**
longitudinal oscillation	PHYS	**Längsschwingung**
= extensional mode		
longitudinal parity (→horizontal parity)	CODING	(→Längsparität)
longitudinal parity character (→horizontal parity character)	DATA COMM	(→Längsparitätszeichen)
longitudinal parity check (→block check)	CODING	(→Blockprüfung)
longitudinal parity check (→horizontal parity check)	DATA COMM	(→Längsparitätsprüfung)
longitudinal parity check character (→horizontal parity character)	DATA COMM	(→Längsparitätszeichen)
longitudinal play	MECH	**Längenspielraum**
longitudinal recording	TV	**Längsspurverfahren**
longitudinal recording	TERM&PER	**Longitudinalaufzeichnung**
≠ transversal recording		≠ Transversalaufzeichnung
longitudinal redundancy check (→block check)	CODING	(→Blockprüfung)
longitudinal redundancy check (→horizontal parity check)	DATA COMM	(→Längsparitätsprüfung)
longitudinal redundancy check character (→horizontal parity character)	DATA COMM	(→Längsparitätszeichen)
longitudinal rib (→longitudinal fin)	MECH	(→Längsrippe)
longitudinal section	NETW.TH	**Längsglied**
longitudinal section	ENG.DRAW	**Längsschnitt**
longitudinal strength (→tensile strength 1)	MECH	(→Zugfestigkeit)
longitudinal support (→boom)	ANT	(→Längsträger)
longitudinal symmetry	MATH	**Längssymmetrie**
longitudinal tightness	COMM.CABLE	**Längsdichtigkeit**
longitudinal voltage	EL.TECH	**Längsspannung**
longitudinal wave	PHYS	**Longitudinalwelle**
		= longitudinale Welle; Längswelle
long-line effect	CIRC.ENG	**Long-Line-Effekt**
long-line effect	RADIO PROP	**Long-Line-Effekt**
		= Langleitungseffekt
long name	LING	**Langform** [eines Namens]
long-nose pliers	TECH	**Langschnabelzange**
long play (→long playing record)	CONS.EL	(→Langspielplatte)
long playing record	CONS.EL	**Langspielplatte**
= long play		↑ Spielplatte
Long Primer (BRI) (→type size 10 point)	TYPOGR	(→Schriftgröße 10 Punkt)
long-range (→far-reaching)	COLLOQ	(→weitreichend)
long-range carrier (→trunk carrier)	TELEC	(→Fernnetzbetreiber)
long-range communications (→toll traffic)	TELEC	(→Fernverkehr)
long-range force	PHYS	**Fernkraft**
long-range interference suppression	RADIO	**Fernentstörung**
long-range interferometer	RADIO LOC	**Langstreckeninterferometer**
long-range navigation	RADIO NAV	**Weitbereichsnavigation**
= LORAN		= Langstreckennavigation; LORAN

long-range network (→toll TELEC (→Fernnetz)
network)
long-range radar RADIO LOC **Weitbereichsradar**
≈ LORAN = Weitstreckenradar
 ≈ LORAN
long-range route (→long-haul TRANS (→Ferntrasse)
route)
long-range system (→long-haul TRANS (→Weitverkehrssystem)
system)
long-range traffic (→toll TELEC (→Fernverkehr)
traffic)
long-range transmission engi- TELEC (→Weitverkehrstechnik)
neering (→long-haul communicat-
ions engineering)
long-term ... TECH **Langzeit-**
= long-time; longtime
long-term analysis TECH **Langzeituntersuchung**
long-term behaviour TECH **Langzeitverhalten**
long-term constancy TECH **Langzeitkonstanz**
long-term fading RADIO PROP **Langzeitschwund**
long-term plan ECON **langfristiger Plan**
long-term stability TECH **Langzeitstabilität**
long-term storage DATA PROC **Langzeitspeicher**
[maintains after powering-off the PC, [bleibt auch nach Ausschal-
e.g. a hard disk memory] ten des PC erhalten, z.B.
≈ external memory ein Festplattenspeicher]
 = Langzeitgedächtnis
 ≈ Externspeicher
long-term storage DATA PROC **Langzeitspeicherung**
[process] [Vorgang]
long-term tendency (→long-term TECH (→Langzeittrend)
trend)
long-term trend TECH **Langzeittrend**
= long-term tendency
long-time (→long-term ...) TECH (→Langzeit-)
longtime (→long-term ...) TECH (→Langzeit-)
long-time drift TECH **Langzeitdrift**
long-time interference RADIO **Langzeitstörung**
long timer CIRC.ENG **Langzeitglied**
= gross timer = Langzeitgeber
long-time testing (→endurance QUAL (→Langzeitprüfung)
testing)
long-wave ... MATH **langwellig**
long waves (→kilometric RADIO (→Langwellen)
waves)
long-wire aerial (→long-wire ANT (→Langdrahtantenne)
antenna)
long-wire antenna ANT **Langdrahtantenne**
= long-wire aerial ↑ Wanderwellenantenne
↑ travelling-wave antenna
long Yagi antenna ANT **Lang-Yagi-Antenne**
look-ahead DATA PROC **Vorgriff**
= lookahead
lookahead (→look-ahead) DATA PROC (→Vorgriff)
look-alike (n.) (→copy) TECH (→Nachbildung)
look-alike program (n.) DATA PROC **Doppelgängerprogramm**
[program imitating another]
look angle (→elevation angle) ANT (→Elevationswinkel)
lookup table DOC **Nachschlagtabelle**
look-up table DATA PROC **Verweistabelle**
[collection of stored results] [Ergebnissammlung]
= LUT = Nachschlagtabelle
loop (n.) NETW.TH **Schleife**
[closed path with no branch passing a [geschlossener Stromweg,
node more than once] deren Zweige die Knoten
↑ mesh nur einmal berühren]
 = Schlaufe (SWZ)
 ↑ Masche
loop (n.) TELEC **Schleife**
= loop-back; round-trip
loop (→program loop) DATA PROC (→Programmschleife)
loop (→subscriber line) TELEC (→Teilnehmerleitung)
loop aerial (→frame antenna) ANT (→Rahmenantenne)
loop analysis (→mesh NETW.TH (→Schleifenanalyse)
analysis)
loop antenna (→frame antenna) ANT (→Rahmenantenne)
loopback (v.t.) SWITCH **spiegeln**
loopback (n.) PHYS **Rückschluß**
loopback (→test loop) TELEC (→Prüfschleife)
loop-back (→loop) TELEC (→Schleife)

loopback (→loop TELEC (→Schleifenmessung)
measurement)
loopback code TELEC **Prüfschleifencode**
loop-back operation SWITCH **Spiegelbetrieb**
loopback register SWITCH **Spiegelregister**
loop body DATA PROC **Schleifenkern**
loop bus (→ring bus) DATA COMM (→Ringbus)
loop cable (→ring cable) DATA COMM (→Ringkabel)
loop cable connection DATA COMM **Ringleitungsanschluß**
loop carrier (→subscriber loop TRANS (→Teilnehmermultiplexsy-
carrier) stem)
loop carrier system (→subscriber TRANS (→Teilnehmermultiplexsy-
loop carrier) stem)
loop check DATA COMM **Rückwärtsübertragungs-Kon-**
 trolle
loop closed TELEC **Schleifenschluß**
= loop closure
loop closure (→loop closed) TELEC (→Schleifenschluß)
loop closureswitch TELEC **Schleifenschlußschalter**
= loop switch; loop connector = Schleifenschalter
loop code (→cyclic DATA PROC (→zyklische Programmie-
programming) rung)
loop coding (→cyclic DATA PROC (→zyklische Programmie-
programming) rung)
loop connector (→loop TELEC (→Schleifenschlußschalter)
closureswitch)
loop counter DATA PROC **Schleifenzähler**
= cycle counter
loop current EL.TECH **Schleifenstrom**
loop dialing SWITCH **Schleifenwahl**
loop direction finder RADIO LOC **Rahmenpeiler**
= DF-loop
loop-disconnect dial (→rotary TERM&PER (→Nummernschalter)
dial switch)
loop-disconnect dialing SWITCH **Schleifenimpulswahl**
loop-disconnect signaling SWITCH (→Impulswahl)
(→pulse dialing)
looped (→in loop-back TELEC (→in Schleife)
condition)
loop gain CIRC.ENG **Schleifenverstärkung**
 = Ringverstärkung; Kreis-
 stärkung; Umlaufverstär-
 kung
loop gain TELEPH **Umlaufverstärkung**
[sum of all gain contributions of a [Summe aller Verstärkun-
four-wire loop, without discount for gen eines Vierdrahtkreises,
attenuation contributions of passive ohne Einrechnung der
elements] dämpfenden Elemente]
= round-trip gain = Schleifenverstärkung
loop holding CIRC.ENG **schleifenhaltend**
loophole (n.) DATA PROC **Schlupfloch**
looping DATA PROC **Schleifendurchlauf**
loopless DATA PROC **schleifenlos**
 = schleifenfrei
loop limit DATA PROC **Schleifenbegrenzung**
loop measurement TELEC **Schleifenmessung**
= loopback; go-and-return test
loop oscillograph INSTR **Schleifenoszillograph**
 = Schleifenschwinger-Oszillo-
 graph
loop program DATA PROC **Schleifenprogramm**
loop resistance EL.TECH **Schleifenwiderstand**
loop signaling SWITCH **Schleifensignalisierung**
 = Schleifengabe
loop stick antenna (→magnetic rod ANT (→Ferritstabantenne)
antenna)
loop switch (→loop TELEC (→Schleifenschlußschalter)
closureswitch)
loop-through (v.t.) ELECTRON (→durchschleifen)
(→feed-through)
loop-through input (→feed-through TV (→Durchschleifeingang)
input)
loop-through plug BROADC **Durchschleifsteckdose**
= feed-through plug
loop topology (→ring bus) DATA COMM (→Ringbus)
loose (→unmounted) TECH (→lose)
loose (v.t.) (→undo) TECH (→lösen)
loose contact ELECTRON **Wackelkontakt**
= intermittent contact = intermittierender Kontakt
looseness ENG.DRAW **Passungsspiel**

English	Domain	German
loosening	TECH	**Lockerung**
loose packaged structure	OPT.COMM	**loser Aufbau**
= loose-tube construction; loose-tube structure		
loose-sheet binding system	OFFICE	**Lose-Blatt-Bindesystem**
loosest fit	ENG.DRAW	**weitester Sitz**
loose-tube construction	OPT.COMM	(→loser Aufbau)
(→loose packaged structure)		
loose-tube structure (→loose packaged structure)	OPT.COMM	(→loser Aufbau)
LORAN (→long-range navigation)	RADIO NAV	(→Weitbereichsnavigation)
Lorentz force	PHYS	**Lorentz-Kraft**
lorry (BrE) (→truck)	TECH	(→Lastkraftwagen)
LOS (→line-of-sight propagation)	RADIO PROP	(→freie Ausbreitung)
LOS (→line-of-sight radio)	RADIO	(→Richtfunk)
Loschmidt number 1	CHEM	**Loschmidtsche Zahl 1**
[number of atoms per qcm standard gas]		[Anzahl Atome pro qcm Normalgas]
= Avogadro number 2		= Avogadro-Konstante 2
Loschmidt number 2 (→Avogadro constant 1)	CHEM	(→Avogadro-Konstante 1)
LOS communication (→radio relay transmission)	TRANS	(→Richtfunkübertragung)
LOS equipment (→radio relay equipment)	RADIO REL	(→Richtfunkgerät)
losing bussiness	ECON	**Verlustgeschäft**
LOSOS technology	MICROEL	**LOSOS-Technik**
= local oxidation of silicon and sapphire technology		
LOS radio antenna (→radio link antenna)	ANT	(→Richtfunkantenne)
LOS radiolink (→line-of-sight radiolink)	RADIO REL	(→Richtfunkstrecke)
LOS relay station (→radio relay station)	RADIO REL	(→Richtfunkstation)
LOS route (→radio relay route)	TRANS	(→Richtfunktrasse)
loss	TELEC	**Ausfall**
= dropout; failure; break down		[eines Signals]
loss	SWITCH	**Verlust**
[figure of grade of service for loss systems]		[Maß der Verkehrsgüte von Verlustsystemen]
↑ grade of service		↑ Verkehrsgüte
loss	ECON	**Verlust**
= forfeiture		
loss	EL.TECH	**Verlust 1**
loss (→attenuation)	LINE TH	(→Dämpfung)
loss (→contraction)	TECH	(→Schrumpfung)
loss angle	EL.TECH	**Verlustwinkel**
[complementary angle to 90° between current and voltage]		[Komplementärwinkel zu 90° zwischen Strom und Spannung; Winkel um den der Gesamtstrom dem Ladestrom nacheilt]
≈ loss factor		≈ Verlustfaktor
loss factor	EL.TECH	**Verlustfaktor**
[active power to absolute value of reactive power; tangens of loss angle]		[Wirkleistung zu Betrag der Blindleistung; Tangens des Verlustwinkels]
= dissipation factor; dielectric loss factor; dielectric dissipation factor		= dielektrischer Verlustfaktor; Tangens Delta
≈ loss angle		≈ Verlustwinkel
loss factor bridge	INSTR	**Verlustfaktor-Meßbrücke**
= dissipation factor bridge		= Tangens-Delta-Meßbrücke
loss factor measurement	INSTR	**Verlustfaktormessung**
= disipation factor measurement		= Tangens-Delta-Messung
loss factor meter	INSTR	**Verlustfaktormesser**
= dissipation factor meter		= Tangens-Delta-Messer
loss-free	EL.TECH	**verlustfrei**
= lossless; zero-loss; dissipation free; dissipationless; zero-dissipation		≈ verlustlos
≈ low-loss		≈ verlustarm
≠ lossy		≠ verlustbehaftet
loss-free network	NETW.TH	**verlustfreies Netzwerk**
loss-free quadripole (→lossless two-port)	NETW.TH	(→verlustfreier Vierpol)
loss-free two-port (→lossless two-port)	NETW.TH	(→verlustfreier Vierpol)
loss in path-antenna gain	ANT	(→Antennenkopplungsverlust)
(→aperture-to-medium coupling loss)		
lossless (→loss-free)	EL.TECH	(→verlustfrei)
lossless line (→zero-loss line)	LINE TH	(→verlustlose Leitung)
lossless quadripole (→lossless two-port)	NETW.TH	(→verlustfreier Vierpol)
lossless two-port	NETW.TH	**verlustfreier Vierpol**
= lossless quadripole; loss-free two-port; loss-free quadripole		= verlustloser Vierpol
loss of service	TELEC	**Dienstunterbrechung**
= interruption		
loss of sound	BROADC	**Tonausfall**
loss of synchronism	TECH	**Gleichlaufstörung**
= synchronisation error; synchronisation trouble		= Synchronisationsausfall; Gleichlaufausfall; Desynchronisation
loss probability	SWITCH	**Verlustwahrscheinlichkeit**
= lost-call probability		
loss resistance	EL.TECH	**Verlustwiderstand**
≈ equivalent resistance		≈ Ersatzwiderstand
≠ Q factor		≠ Gütefaktor
loss resistance (→antenna loss resistance)	ANT	(→Antennenverlustwiderstand)
loss system	SWITCH	**Verlustsystem**
[system dimensioned for economic reasons with limited resources, so that a prefixed percentage of simultaneous bids cannot be served]		[System welches mit begrenztem Aufwand dimensioniert ist, sodaß ein vorgegebener Prozentsatz an gleichzeitig auftretenden Verbindungswünschen nicht erfüllt wird]
loss time	ELECTRON	**Verlustzeit**
loss traffic (→lost traffic)	SWITCH	(→Verlustverkehr)
lossy	EL.TECH	**verlustbehaftet**
≈ high-loss		≈ verlustreich
≠ loss-free		≠ verlustfrei
lossy line	LINE TH	**verlustreiche Leitung**
lost	SWITCH	**abgewiesen**
lost call	SWITCH	**zurückgewiesener Verbindungswunsch**
		= abgewiesene Belegung
lost-call probability (→loss probability)	SWITCH	(→Verlustwahrscheinlichkeit)
lost calls (→lost traffic)	SWITCH	(→Verlustverkehr)
lost traffic	SWITCH	**Verlustverkehr**
= loss traffic; lost calls		= verlorener Verkehr; Verlustbelegung
lot 1 (n.)	ECON	**Los**
= batch		↓ Lieferlos
↓ delivery lot		
lot (n.)	QUAL	**Los**
		= Prüflos
lot 2 (n.)	ECON	**Posten 1**
= parcel (n.)		[Lieferung]
lot size	QUAL	**Losgröße**
= batch size		= Losumfang
loud	ACOUS	**laut**
loudness	ACOUS	**Lautheit**
		= Lautstärke
loudness level (→volume of sound)	ACOUS	(→Lautstärkepegel)
loudness rating	TELEPH	**Lautheitsgrad**
loudspeaker	EL.ACOUS	**Lautsprecher**
= speaker		↑ Schallwandler
↑ electroacoustic transducer		
loudspeaker baffle 1	EL.ACOUS	**Lautsprecher-Abdeckung**
= speaker baffle		
loudspeaker baffle 2	EL.ACOUS	**Lautsprecherwand**
loudspeaker box	EL.ACOUS	**Lautsprecherbox**
= speaker box; loudspeaker cabinet 2		
loudspeaker cabinet 1	EL.ACOUS	**Lautsprecher-Leergehäuse**
		= Lautsprecherchassis
loudspeaker cabinet 2 (→loudspeaker box)	EL.ACOUS	(→Lautsprecherbox)
loudspeaker capacitor	EL.ACOUS	**Lautsprecherkondensator**
loudspeaker connector	EL.ACOUS	**Lautsprecher-Stecker**
= connector speaker		
loudspeaker protector	EL.ACOUS	**Lautsprecher-Schutzschalter**
= spaeker protector		

loudspeaker set	EL.ACOUS	Lautsprecher-Set		**lower part**	TECH	**Unterteil**
= speaker set				= base part		≠ Oberteil
loudspeaker system	EL.ACOUS	**Lautsprechersystem**		≠ upper part		
= speaker system				**lower sideband**	MODUL	**unteres Seitenband**
loudspeaker telephone	TELEPH	(→Lauthörtelefon)		**lower significant**	DATA PROC	(→niederwertig)
(→loudspeaker telephone set)				(→low-order)		
loudspeaker telephone set	TELEPH	**Lauthörtelefon**		**lowest level**	COLLOQ	**Tiefstand**
= loudspeaker telephone		= Lauthörfernsprecher; Laut-		= low point		= Tiefpunkt
= hands-free telephone		hörapparat; Lauthörgerät		**lowest order** (→least	CODING	(→niederwertigst)
loudspeaker textile	EL.ACOUS	**Lautsprecher-Bespannungs-**		significant)		
= speaker textile		**gewebe**		**lowest usable frequency**	RADIO	**untere Grenzfrequenz**
		= Lautsprecherseide		= LUF		
loudspeaker volume control	EL.ACOUS	(→L-Regler)		**low frequency** (n.)	POWER SYS	**Niederfrequenz**
(→L-pad)				[up to 150 Hz]		[bis 150 Hz]
loupe (→magnifying glass)	PHYS	(→Lupe)		**low-frequency** (adj.)	TELEC	**niederfrequent**
loupe scanner	TERM&PER	**Abtastlupe**		**low frequency** (→hectometric	RADIO	(→Mittelwellen)
louver	EL.ACOUS	**Lautsprecherschirm**		waves)		
louvering	METAL	**Durchreißen**		**low frequency** (→voice	TELEC	(→Niederfrequenz)
= slitting; lancing		[Stanzvorgang, Einschnei-		frequency)		
		den und Biegen]		**low-frequency ferrite**	METAL	**NF-Ferrit**
low	ACOUS	**tief**		**low frequency measuring**	INSTR	**NF-Meßtechnik**
low-aperture direction	RADIO LOC	**Kleinbasispeiler**		**technique**		
finder				**low frequency signaling** (→voice	TELEC	(→Tonfrequenz-Signalisie-
= small-base direction finder				frequency signaling)		rung)
low band (→lower band)	RADIO	(→Unterband)		**low-high repeater** (→frogging	TRANS	(→Zweidraht-Zwischen-
low-band system	TV	**Lowband-Verfahren**		repeater)		verstärker mit Band-
low-bed trailer	TECH	**Tieflader**				tausch)
low byte	CODING	**niederwertiges Byte**		**low-hum** (adj.)	TELEC	**brummarm**
low-capacitance (adj.)	EL.TECH	**kapazitätsarm**		**low-impedance**	EL.TECH	**niederohmig**
= anticapacitance (adj.)				≠ high-impedance		≠ hochohmig
low-capacity link	TRANS	**Kleinkanalstrecke**		**low insulation**	EL.TECH	**schlechte Isolation**
= low-capacity route; light route		= Schmalbandstrecke		**low level**	MICROEL	**Tiefpegelzustand**
low-capacity route (→low-	TRANS	(→Kleinkanalstrecke)		= low signal; L-signal; L; logical low		= Tiefpegelsignal; Tiefpegel;
capacity link)				↑ pulse level		L-Zustand; L-Pegel; L-Si-
low-capacity system	TRANS	**Kleinkanalsystem**				gnal; Low-Zustand; Low-
= small-band system; narrow-band sys-		= Schmalbandsystem				Pegel; Low-Signal; L
tem; low-density carrier		≠ Breitbandsystem				≠ Hochpegelzustand
≠ high-capacity system						↑ Impulspegel
low current	EL.TECH	**Schwachstrom**		**low-level language**	DATA PROC	(→niedere Programmierspra-
= feeble current; light current; weak		≠ Starkstrom		(→low-level programming lan-		che)
current				guage)		
≠ heavy current				**low-level logic**	MICROEL	**Tiefpegellogik**
low-current engineering	EL.TECH	**Schwachstromtechnik**		= LLL		= LLL
= weak-current enginnering; light-cur-		≈ Nachrichtentechnik		≠ high-level logic		≠ Hochpegellogik
rent engineering		≠ Starkstromtechnik		**low-level modulation**	MODUL	**Tiefpegel-Modulation**
≠ electrical power engineering						= Vorstufenmodulation
low definition (→low	TERM&PER	(→niedrige Auflösung)		**low-level programming**	DATA PROC	**niedere Programmiersprache**
resolution)				**language**		[mehr auf Verarbeitungsge-
low-density carrier	TRANS	(→Kleinkanalsystem)		[aiming at velocity of processing		schwindigkeit als auf
(→low-capacity system)				rather than of human learning]		schnelle Erlernbarkeit aus-
low-distortion (adj.)	EL.TECH	**klirrarm**		= low-level language; LLL		gerichtet]
low-distortion	TELEC	**verzerrungsarm**		≠ high-level programming language		= Tiefsprache
≈ distortionless		≈ verzerrungsfrei		↓ machine language; assembler lan-		≠ problemorientierte
low-drift (adj.)	ELECTRON	**driftarm**		guage		Programmiersprache
low-duty-cycle signal	TELEC	**Signal mit geringem Tastver-**				↓ Maschinensprache; As-
		hältnis				semblersprache
low-end (adj.)	ECON	**der unteren Preisklasse**		**low level selction** (→low level	TRANS	(→Tiefpegelsignalisierung)
low-energy	PHYS	**energiearm**		signalling)		
lower (v.t.)	TECH	**senken**		**low level signalling**	TRANS	**Tiefpegelsignalisierung**
lower band	RADIO	**Unterband**		= low level selction		= Tiefpegelwahl
= low band		≠ Oberband		**low limit**	ENG.DRAW	**Kleinstmaß**
≠ upper band				= minimum dimension		
lower band limit	RADIO	**Bandanfang**		**low limit**	TECH	**Unterwert**
		≠ Bandende 2		**low loss ...**	TELEC	**dämpfungsarm**
		↑ Bandende 1		**low loss**	EL.TECH	**verlustarm**
lower case letter (→small	TYPOGR	(→Kleinbuchstabe)		[adj.]		≠ verlustreich
character)				≈ lossy		≈ verlustfrei
lower case printing	TYPOGR	**Kleinschreibung**		≈ loss-free		
lower end (→base)	ANT	(→Fußpunkt)		**low-loss circular waveguide**	\MICROW	**Hohlkabel**
lowering	COLLOQ	**Senkung**		**low noise**	TELEC	**geräuscharm**
= reduction; cut; curtailment		= Verminderung; Reduzie-		≈ noiseless		= rauscharm; lärmarm
		rung				≈ geräuschlos
lowering (→discount)	ECON	(→Preisnachlaß)		**low-order** (adj.)	DATA PROC	**niederwertig**
lower mode	TERM&PER	**Grundebene**		= lower significant		
[the lower symbols of a double as-		[es gelten die unteren Zei-		**lowpass**	NETW.TH	**Tiefpaß**
signed keyboard are valid]		chen einer doppelt beleg-		= lowpass filter		= Tiefpaßfilter
≈ primary occupation		ten Tastatur]		**lowpass cutoff frequency**	NETW.TH	**Tiefpaßgrenzfrequenz**
≠ upper mode		≈ primäre Belegung		**lowpass filter** (→lowpass)	NETW.TH	(→Tiefpaß)
		≠ Umschaltebene		**low point** (→lowest level)	COLLOQ	(→Tiefstand)

low-power diode-transistor logic = LPDTL	MICROEL	**LPDTL**	
low-power laser	OPTOEL	**Kleinleistungslaser**	
low-power relay	COMPON	**Schwachstromrelais**	
low-power Schottky (→LPS)	MICROEL	(→LPS)	
low-power Schottky transistor-transistor logic = LSTTL; LPTTL	MICROEL	**Low-Power-Schottky-TTL** = LSTTL; LPTTL	
low-power transistor-transistor logic (→LTTL)	MICROEL	(→LTTL)	
low pressure ≠ high pressure	TECH	**Niederdruck** ≠ Hochdruck	
low priority program (→background program)	DATA PROC	(→Hintergrundprogramm1)	
low-quality = inferior quality; substandard; inferior	ECON	**minderwertig**	
low-radiation (adj.)	RADIO	**abstrahlarm** = strahlungsarm	
low-res (→low-resolution)	TERM&PER	(→niedrigauflösend)	
low-res graphics (→low-resolution graphics)	DATA PROC	(→niedrigauflösende Graphik)	
low resolution (n.) = low definition	TERM&PER	**niedrige Auflösung**	
low-resolution (adj.) = low-res ≠ high-resolution	TERM&PER	**niedrigauflösend** ≠ hochauflösend	
low-resolution graphics = low-res graphics	DATA PROC	**niedrigauflösende Graphik**	
low signal (→low level)	MICROEL	(→Tiefpegelzustand)	
low-speed	TECH	**langsamlaufend**	
low-speed data	TELEC	**langsame Daten**	
low-speed logic with high noise immunity (→LSL)	MICROEL	(→LSL)	
low-tape alarm contact	TELEGR	**Papiervoralarmkontakt**	
low-temperature brittle	TECH	**kaltbrüchig**	
low tension (→low voltage)	POWER ENG	(→Niederspannung)	
low voltage [under 250 V] = low tension	POWER ENG	**Niederspannung** [unter 250 V]	
low-voltage cable	POWER ENG	**Schwachstromkabel**	
low-voltage switchgear	POWER ENG	**Niederspannungsschaltgerät**	
lozenge = diamond	DATA PROC	**Rhombuszeichen** = Raute; Suppenstern	
LP (→linear programming)	DATA PROC	(→lineare Programmierung)	
L-pad = loudspeaker volume control	EL.ACOUS	**L-Regler** = Lautsprecher-Leistungsregler; Lautsprecher-Pegelregler	
LPC (→linear predictive coding)	CODING	(→LPC-Codierung)	
LPDA Yagi antenna (→log-Yag array)	ANT	(→Log-Yag-Antenne)	
LPDTL (→low-power diode-transistor logic)	MICROEL	(→LPDTL)	
LPFD antenna = log-periodic folded dipole antenna	ANT	**logarithmitisch-periodische Faltdipol-Antenne** = LPFD-Antenne	
LPI (→lines per inch)	TERM&PER	(→Zeilen pro Zoll)	
LPM (→lines per minute)	TERM&PER	(→Zeilen pro Minute)	
LPS = low-power Schottky	MICROEL	**LPS**	
LPTTL (→low-power Schottky transistor-transistor logic)	MICROEL	(→Low-Power-Schottky-TTL)	
LP-unipole (→log-periodic unipole)	ANT	(→logarithmisch-periodische Unipolantenne)	
LQ (→rejectable quality level)	QUAL	(→Rückweisgrenze)	
LQ (→letter quality)	TERM&PER	(→Schönschrift)	
Lr (→lawrencium)	CHEM	(→Lawrencium)	
LRC (→block check)	CODING	(→Blockprüfung)	
LRC (→horizontal parity check)	DATA COMM	(→Längsparitätsprüfung)	
LRC character (→horizontal parity character)	DATA COMM	(→Längsparitätszeichen)	
LSB (→least significant bit)	CODING	(→niedrigstwertiges Bit)	
LSC (→least significant character)	CODING	(→niederwertigstes Zeichen)	
LSD (→least significant digit)	MATH	(→niedrigstwertige Stelle)	
LSD (→limited saturation device)	MICROEL	(→LSD)	
L-section = L-network ↑ attenuator	NETW.TH	**L-Schaltung** = L-Glied; Spannungsteiler ↑ Dämpfungsglied	
L-shaped hexagon key ↑ Allen wrench	MECH	**Inbusschlüssel** [Innensechskantschlüssel der Firma Bauer und Schnaurte] ↑ Innensechskantschlüssel	
L shell (→second shell)	PHYS	(→2.Schale)	
LSI [500 to 10,000 components per IC's] = large-scale integration	MICROEL	**Großintegration** [500 bis 10.000 Komponenten pro IC] = LSI; hoher Integrationsgrad; Gruppenintegration	
LSIC (→large-scale integrated circuit)	MICROEL	(→hochintegrierte Schaltung)	
L-signal (→low level)	MICROEL	(→Tiefpegelzustand)	
LSI test system	INSTR	**LSI-Prüfsystem**	
LSL = low-speed logic with high noise immunity	MICROEL	**LSL**	
LSTTL (→low-power Schottky transistor-transistor logic)	MICROEL	(→Low-Power-Schottky-TTL)	
LT (→line termination)	TELEC	(→Leitungsabschluß)	
Ltd. (→company with limited liability)	ECON	(→GmbH)	
LTTL = low-power transistor-transistor logic	MICROEL	**LTTL**	
LTU (→line terminating unit)	DATA COMM	(→Leitungsabschlußeinrichtung)	
LTU (→line terminal equipment)	TRANS	(→Leitungsendgerät)	
Lu (→lutecium)	CHEM	(→Lutetium)	
lube-oil (→lubricating oil)	MECH	(→Schmieröl)	
lube-oil filter	MECH	**Schmierölfilter**	
lube-oil pressure controller	TECH	**Öldruckwächter**	
lubricant (n.) ↓ grease; lubricating oil	MECH	**Schmiermittel** = Gleitmittel ↓ Fett; Schmieröl	
lubricate ↓ grease; oil	MECH	**schmieren** ↓ fetten; ölen	
lubricated magnetic disk (→Winchester disk)	DATA PROC	(→Winchesterplatte)	
lubricating grease	TECH	**Gleitfett**	
lubricating oil = lube-oil	MECH	**Schmieröl**	
lubrication	MECH	**Schmierung** ↓ Fettschmierung; Ölschmierung	
lucrativeness (→profitability)	ECON	(→Wirtschaftlichkeit)	
Luenburg observer	CONTROL	**Luenberg-Beobachter**	
LUF (→lowest usable frequency)	RADIO	(→untere Grenzfrequenz)	
lug (n.)	MECH	**Lappen** = Ansatz 1; Ohr; Nase	
lug	COMPON	**Fahne** = tag	
lug (n.) (→strap)	TECH	(→Lasche)	
lug (→eye)	MECH	(→Öse)	
luggable (→transportable)	TECH	(→transportierbar)	
lumber (→wood)	TECH	(→Holz)	
lumen [unit for luminous flux; = 1 cd sr] = lm	PHYS	**Lumen** [SI-Einheit für Lichtstrom; = 1 cd sr] = lm	
luminance [SI unit: candela/m²] = photometric brightness; brightness; luminosity; L; lightness	PHYS	**Leuchtdichte** [SI-Einheit: Candela/m²] = photometrische Helligkeit; Helligkeit; Flächenhelligkeit; L	
luminance (→brightness)	TV	(→Helligkeit)	
luminance content	SIGN.TH	**Leuchtdichteinformation**	
luminance signal	TV	**Leuchtdichtesignal** = Helligkeitssignal; Bildinhaltssignal; Luminanzsignal; Y-Signal	

luminance-to-chrominance delay   342

luminance-to-chrominance delay   TV   **Luminanz/Chrominanz-Verzögerung**
luminator crystal   PHYS   **Leuchtquarz**
= luminous crystal
luminescence   PHYS   **Lumineszenz**
[externally excited radiation]   [angeregte Eigenstrahlung]
↓ fluorescence; phosphorescence; photoluminescence; thermoluminescence; chemoluminescence; electroluminescence; radioluminescence; bioluminescence; triboluminescence; catodoluminescence   ↓ Fluoreszenz; Phosphoreszenz; Photolumineszenz; Thermolumineszenz; Chemolumineszenz; Elektrolumineszenz; Radiolumineszenz; Biolumineszenz; Tribolumineszenz; Kathodolumineszenz

luminescent material   PHYS   **Leuchtstoff**
= phosphor (n.)
luminescent screen   ELECTRON   **Leuchtschirm**
= luminous screen; fluorescent screen   = Fluoreszenzschirm
↓ display screen [TERM&PER]; video screen [TV]   ↓ Bildschirm [TERM&PER]; Bildschirm [TV]
luminescent spot (→focal spot)   PHYS   (→**Brennfleck**)
luminophor   PHYS   **Luminophor**
luminosity (→brightness)   TV   (→**Helligkeit**)
luminosity (→luminance)   PHYS   (→**Leuchtdichte**)
luminosity flicker   TV   **Helligkeitsflimmern**
luminous crystal (→luminator crystal)   PHYS   (→**Leuchtquarz**)
luminous discharge (→glow discharge)   PHYS   (→**Glimmentladung**)
luminous efficiency   PHYS   **Lichtausbeute**
luminous flux   PHYS   **Lichtstrom**
[SI unit: lumen]   [SI-Einheit: Lumen]
= light flux; I   = I
luminous intensity   PHYS   **Lichtstärke**
[luminous flux per solid angle; SI unit: candela]   [Lichtstrom pro Raumwinkel; SI-Einheit: Candela]
luminous key   COMPON   **Leuchttaste**
= lamp key   = Lampentaste
luminous pointer   ELECTRON   **Lichtmarke**
= optical pointer   = Leuchtmarke; optische Marke
≈ flying spot   ≈ Lichtpunkt
luminous-pointer galvanometer   INSTR   **Lichtmarken-Galvanometer**
luminous power (→optical power)   PHYS   (→**Lichtleistung**)
luminous ray (→light beam)   PHYS   (→**Lichtstrahl**)
luminous screen   ELECTRON   (→**Leuchtschirm**)
(→luminescent screen)
luminous slit   PHYS   **Lichtspalt**
luminous source (→light source)   PHYS   (→**Lichtquelle**)
luminous spot   PHYS   **Lichtpunkt**
lumious board   EQUIP.ENG   **Leuchtschrifttafel**
lump amount (→lump sum)   ECON   (→**Pauschalbetrag**)
lumped circuit element   NETW.TH   **konzentriertes Schaltkreiselement**
= lumped device
lumped device (→lumped circuit element)   NETW.TH   (→**konzentriertes Schaltkreiselement**)
lumped loading   COMM.CABLE   (→**Bespulung**)
(→coil-loading)
lump sum   ECON   **Pauschalbetrag**
= lump amount; flat rate   = Pauschale; Preispauschale
Luneberg lens antenna   ANT   **Luneberg-Linsenantenne**
lusterless (→mat)   TECH   (→matt)
lustrous (→bright)   TECH   (→glänzend)
LUT (→look-up table)   DATA PROC   (→Verweistabelle)
lute (v.t.)   TECH   **verkitten**
≈ seal (v.t.)   ≈ dichten
lute (→sealing material)   TECH   (→**Dichtungsmasse**)
lutecium   CHEM   **Lutetium**
= Lu; lutetium   = Lu
lutetium (→lutecium)   CHEM   (→Lutetium)
Luxemburg effect   RADIO   **Luxemburg-Effekt**
LW (→kilometric waves)   RADIO   (→Langwellen)
lye   CHEM   **Lauge**
= causting solution
Lyman series   PHYS   **Lyman-Serie**

M

M (→torque)   PHYS   (→Drehmoment)
m (→mass)   PHYS   (→Masse)
m (→meter)   PHYS   (→Meter) (n.m. oder n.n.)
μ (→micrometer)   PHYS   (→Mikrometer)
m² (→square meter)   PHYS   (→Quadratmeter)
M²FM code   DATA PROC   **M²FM-Code**
machinability   MECH   **maschinelle Bearbeitbarkeit**
= Zerspanbarkeit
machinable   MECH   **zerspanbar**
= maschinell bearbeitbar
machine (v.t.) (→tool)   MECH   (→bearbeiten)
machine (→engine)   TECH   (→Maschine)
machine- (→mechanical)   TECH   (→maschinell)
machine (→computer)   DATA PROC   (→Computer)
machine address (→absolute address)   DATA PROC   (→absolute Adresse)
machine addressing (→absolute addressing)   DATA PROC   (→absolute Adressierung)
machine-aided   DATA PROC   **maschinenunterstützt**
≈ automatic   ≈ automatisch
machine-aided coding (→machine-aided programming)   DATA PROC   (→automatische Programmierung)
machine-aided programming   DATA PROC   **automatische Programmierung**
= automatic programming; computer-aided programing; machine-aided coding; automatic coding; computer-aided coding; computer-aided software engineering; computer assisted software engineering; CASE   = maschinenunterstützte Programmierung; rechnergestützte Programmierung; computergestützte Programmierung; CASE
machine code 1   DATA PROC   **Maschinencode 1**
[machine-specific instruction code, controls the computer without further translation]   [maschinenspezifischer Befehlscode, steuert den Computer ohne weitere Übersetzung]
= machine format 1; native code; computer code; absolute code; direct code; one-level code; specific code   = maschineninterner Code; anlageninterner Code; interner Code; Interncode; Maschinenprogrammcode; Absolutcode; Direktcode
≈ machine language; object code   ≈ Maschinensprache; Objektcode
machine code 2 (→machine language)   DATA PROC   (→Maschinensprache)
machine code instruction (→machine instruction)   DATA PROC   (→Maschinenanweisung)
machine control electronic   ELECTRON   **Maschinensteuerungselektronik**
machine cycle   DATA PROC   **Maschinenzyklus**
[time interval in which a CPU executes a unit operation]   [einer Zentraleinheit zur Durchführung einer Operation]
machine data acquisition   DATA PROC   **Betriebsdatenerfassung von Maschinen**
= BDM
machine-dependent (→computer-oriented)   DATA PROC   (→maschinenorientiert)
machine-dependent (→hardware-dependent)   DATA PROC   (→hardwareabhängig)
machine-dependent program   DATA PROC   **maschinenorientiertes Programm**
[written in machine code or assembler]   [in Maschinensprache oder Assemblierer geschrieben]
machine element   MECH   **Maschinenelement**
machine equation   DATA PROC   **Rechnergleichung**
[an analog computer has been designed for]   [für die ein Analogrechner ausgelegt wurde]
= machine function   = Maschinegleichung; Maschinengleichung; Rechnerfunktion; Maschinefunktion; Maschinenfunktion
machine error (→hardware fault)   DATA PROC   (→Hardwarefehler)
machine format 1 (→machine code 1)   DATA PROC   (→Maschinencode 1)
machine format 2 (→machine language)   DATA PROC   (→Maschinensprache)

machine function (→machine DATA PROC (→Rechnergleichung)
equation)
machine-independent DATA PROC (→hardwareunabhängig)
(→hardware-independent)
machine instruction DATA PROC **Maschinenanweisung**
[directly executable by the computer] [vom Computer direkt ausführbar]
= machine code instruction; computer instruction = Maschinenbefehl; Maschineninstruktion
↑ instruction ↑ Befehl
machine intelligence DATA PROC **Maschinenintelligenz**
machine language **Maschinensprache**
[computer-specific programming language, consists of instructions formulated in machine code] [maschinenspezifische Programmiersprache, besteht aus in Maschinencode formulierten Befehlen]
= computer language; machine format 2; machine code 2 = Computersprache; Rechnersprache; Maschinencode 2
≈ machine-oriented programming language; machine code 1 ≈ maschinenorientierte Programmiersprache; Maschinencode 1; Zielsprache
≠ symbolic program language
↑ object language; programming language ≠ symbolische Programmiersprache
↑ Zielsprache; Programmiersprache
machine operation DATA PROC (→Rechneroperation)
(→computer operation 1)
machine-oriented DATA PROC (→maschinenorientiert)
(→computer-oriented)
machine-oriented language DATA PROC (→maschinenorientierte Programmiersprache)
(→machine-oriented programming language)
machine-oriented programming language DATA PROC **maschinenorientierte Programmiersprache**
[programming language matched to the machine language, i.e.computer-specific; the languages of seccond generation] [an die Maschinensprache angelehnte, rechnerspezifische Programmiersprache; die Sprachen der 2. Generation]
= machine-oriented language; computer-oriented programming language; computer-oriented language; computer-dependent programming language; computer-dependent language = maschinenorientierte Sprache; maschinennahe Programmiersprache; maschinennahe Sprache
≈ machine language ≈ Maschinensprache
≠ problem-oriented programming language ≠ problemorientierte Maschinensprache
↑ symbolic programming language; procedure-oriented programming language ↑ symbolische Programmiersprache; prozedurorientierte Programmiersprache
↓ assembler language ↓ Assemblersprache
machine point DATA PROC **Maschinenkomma**
= Maschinenpunkt
machine programm DATA PROC **Maschinenprogramm**
[program written in or translated into machine language] [in Maschinensprache geschriebenes oder in die Maschinensprache übersetztes Programm]
↓ absolute program; object program ↓ Absolutprogramm; Objektprogramm
machine-readable TERM&PER **maschinenlesbar**
= computer-readable; machine-sensible; computer-sensible = computerlesbar; rechnerlesbar
machine-readable character TERM&PER **maschinenlesbares Zeichen**
= computer-readable character = rechnerlesbares Zeichen; computerlesbares Zeichen
machine-readable document TERM&PER **Maschinenbeleg**
= machine-sensible document = maschinenlesbarer Beleg
machine room (→computer SYS.INST (→Rechnerraum)
room)
machine run (→program DATA PROC (→Programmlauf)
run)
machinery TECH **Maschinerie**
machine-sensible TERM&PER (→maschinenlesbar)
(→machine-readable)
machine-sensible document TERM&PER (→Maschinenbeleg)
(→machine-readable document)
machine test DATA PROC **Maschinentest**
[of software] [von Software]
machine time (→computer DATA PROC (→Maschinenzeit)
time)
machine tool MECH **Werkzeugmaschine**
= tool
machine word (→word) DATA PROC (→Wort)
machine word length DATA PROC **Maschinenwortlänge**
= computer word length
machining 1 MECH **maschinelle Bearbeitung**
machining 2 (→cutting MECH (→spanabhebende Bearbeitung)
shaping)
macro (n.) MICROEL **Funktioneinheit**
= Makro
macro 2 (→macro DATA PROC (→Makroanweisung)
instruction 2)
macro 3 (→macro call) DATA PROC (→Makroaufruf)
macro 1 (→macro DATA PROC (→Makrobefehl)
instruction 1)
macro assembler DATA PROC **Makroassembler**
= macro assembly program = Makroassemblierer
macro assembly program DATA PROC (→Makroassembler)
(→macro assembler)
macrobending OPT.COMM **Makrobiegung**
macro call DATA PROC **Makroaufruf**
= macro 3 = Makro 3
macrocell MICROEL **Makrozelle**
macro code (→macro DATA PROC (→Makrobefehl)
instruction 1)
macro coding DATA PROC **Makrocodierung**
macro command (→macro DATA PROC (→Makrobefehl)
instruction 1)
macro definition DATA PROC **Makrodefinition**
macro directory DATA PROC **Makroverzeichnis**
macroelement DATA PROC **Makroelement**
macro expansion (→macro DATA PROC (→Makroumwandlung)
generation)
macro generation DATA PROC **Makroumwandlung**
= macro expansion; macrogeneration
macrogeneration (→macro DATA PROC (→Makroumwandlung)
generation)
macro instruction 2 DATA PROC **Makroanweisung**
[standardized sequence of instructions] [standardisierte Befehlsfolge]
= macro statement; macroinstruction 2; macro 2; general instruction = Makroinstruktion; Makro 2
macro instruction 1 DATA PROC **Makrobefehl**
[represents several instruction in the machine code] [entspricht mehreren Befehlen in der Maschinensprache]
= macroinstruction 1; macro code; macro command; macro 1 = Makrocode; Globalbefehl; Makro 1
macroinstruction 2 (→macro DATA PROC (→Makroanweisung)
instruction 2)
macroinstruction 1 (→macro DATA PROC (→Makrobefehl)
instruction 1)
macro-instruction memory DATA PROC **Makrobefehlsspeicher**
= macro-instruction storage; macro-instruction store
macro-instruction storage DATA PROC (→Makrobefehlsspeicher)
(→macro-instruction memory)
macro-instruction store DATA PROC (→Makrobefehlsspeicher)
(→macro-instruction memory)
macro language DATA PROC **Makrosprache**
macro library DATA PROC **Makrobibliothek**
macron LING **Makron**
= Längezeichen
macro operand DATA PROC **Makrooperand**
macroplasma PHYS **Makroplasma**
macro processing DATA PROC **Makroübersetzung**
= macroprocessing
macroprocessing (→macro DATA PROC (→Makroübersetzung)
processing)
macro programming DATA PROC **Makroprogrammierung**
= macroprogramming
macroprogramming (→macro DATA PROC (→Makroprogrammierung)
programming)
macro recorder DATA PROC **Makroaufzeichner**
[aloows storage and call of macros by combination of keys] [erlaubt Speicherung und Abruf von Makros über Tastenkombinationen]
macroscopic PHYS **makroskopisch**

English	Domain	German
macroscopic instruction (→visual inspection)	QUAL	(→Sichtprüfung)
macro statement (→macro instruction 2)	DATA PROC	(→Makroanweisung)
MAD (→magnetic anomaly detector)	INSTR	(→Magnetometer)
madistor [operated at low temperature, characteristic controlled by external magnetic field] ↓ magnetodiode; madistor transistor	MICROEL	**Madistor** [bei Tieftemperaturen betrieben, Kennlinie durch externes Magnetfeld gesteuert] ↓ Magnet-Diode; Madistor-Transistor
madistor-transistor ↑ madistor	MICROEL	**Madistor-Transistor** ↑ Madistor
MADT = micro-alloy diffused-base transistor	MICROEL	**MADT-Transistor** = MADT; Mikro-alloy-Diffusionstransistor
mag (→magnetic)	PHYS	(→magnetisch)
magazine [reserved memory location] ↓ input magazine; output magazine	DATA PROC	**Magazin** [reservierter Speicherplatz] ↓ Eingabemagazin; Ausgabemagazin
magazine [publication on a weekly or longer basis] = review ≈ newspaper ↓ weekly paper; monthly magazine	COLLOQ	**Zeitschrift** [wöchentlich bis mehrmals jährlich erscheinende, geheftete Druckschrift] ≈ Zeitung ↓ Wochenzeitung; Monatszeitschrift
magazine	TECH	**Magazin**
magenta [deep purplish red]	PHYS	**Magenta** = dunkles Karminrot
magic eye	COMPON	**magisches Auge**
magic number	MATH	**magische Zahl**
magic T = magic tee; hybrid T; hybrid tee ↑ parallel-T network	MICROW	**magisches T** = kompensiertes magisches T; Doppel-T-Verzweigung
magic tee (→magic T)	MICROW	(→magisches T)
magnesium = Mg	CHEM	**Magnesium** = Mg
magnet ↓ permanent magnet; electromagnet [EL.TECH]	PHYS	**Magnet** ↓ Dauermagnet; Elektromagnet [EL.TECH]
magnet armature ≈ magnet yoke	PHYS	**Magnetanker** ≈ Magnetjoch
magnet coil = solenoid	COMPON	**Magnetspule** = Solenoid
magnetic (adj.) = mag	PHYS	**magnetisch**
magnetic accounting card	TERM&PER	**Magnetkontokarte**
magnetically hard = hard magnetic	PHYS	**hartmagnetisch** [hohe Remanenz und Koerzitivkraft] = magnetisch hart
magnetically soft [with high permeability and low coercive force] = soft magnetic	PHYS	**weichmagnetisch** [hohe Permeabilität und geringe Koerzitivkraft] = magnetisch weich
magnetically soft iron (IEC) (→soft iron)	METAL	(→Weicheisen)
magnetically soft metal (→soft metal)	METAL	(→Weichmetall)
magnetic amplification	PHYS	**magnetische Verstärkung**
magnetic amplifier (→transductor amplifier)	CIRC.ENG	(→Transduktorverstärker)
magnetic anomaly detector = MAD; magnetometer	INSTR	**Magnetometer**
magnetic axis	PHYS	**magnetische Achse**
magnetic biasing (→premagnetization)	PHYS	(→Vormagnetisierung)
magnetic board = speedboard; white board	OFFICE	**Magnettafel** = Hafttafel
magnetic boundary potential (→magnetomotive force)	PHYS	(→magnetische Umlaufspannung)
magnetic bubble = bubble	PHYS	**Magnetblase** = magnetische Blase; Magnetdomäne; Domäne; Bubble
magnetic bubble memory = bubble memory ↑ magnetic layer memory	TERM&PER	**Magnetblasenspeicher** = Blasenspeicher; Magnetdomänenspeicher; Domänentransportspeicher ↑ Magnetschichtspeicher
magnetic card ≈ smart card	TERM&PER	**Magnetkarte** ≈ Chipkarte
magnetic card authorization	TERM&PER	**Magnetkartenberechtigung**
magnetic card reader	TERM&PER	**Magnetkartenleser**
magnetic card storage ≈ magnetic strip meory	TERM&PER	**Magnetkartenspeicher** ≈ Magnetstreifenspeicher
magnetic card telephone	TERM&PER	**Magnetkartentelefon** = Magnetkartenfernsprecher
magnetic cell	MICROEL	**magnetisches Speicherelement**
magnetic character = magnetic ink character	TERM&PER	**Magnetschriftzeichen**
magnetic character printer = magnetic ink character printer; magnetic font printer	TERM&PER	**Magnetschriftdrucker**
magnetic character reader [reads visual characters impressed with magnetic ink] = magnetic ink character reader; MICR; magnetic document reader; magnetic ink scanner; magnetic font reader ↑ character reader; document reader	TERM&PER	**Magnetschriftleser** [liest mit magnetisierter Tinte gedruckte Klarschriftzeichen] = Magnetschrift-Lesegerät; Magnetschriftbelegleser; magnetischer Belegleser ↑ Klarschriftleser; Belegleser
magnetic character recognition = magnetic ink character recognition; MICR	TERM&PER	**Magnetschrift-Zeichenerkennung** = Magnetschrifterkennung
magnetic circuit	PHYS	**Magnetkreis** = magnetischer Kreis; Eisenkreis
magnetic clutch	POWER ENG	**Magnetkupplung**
magnetic compass	INSTR	**Magnetkompass**
magnetic conductance (→permeance)	PHYS	(→Permeanz)
magnetic core ↓ ferrite core	COMPON	**Magnetkern** ↓ Ferritkern
magnetic core antenna	ANT	**Magnetkernantenne**
magnetic core measuring system	INSTR	**Kernmagnetmeßwerk**
magnetic core memory = magnetic core storage; magnetic core store; core memory; core storage; core store ↑ magnetic memory	TERM&PER	**Magnetkernspeicher** = Kernspeicher; Ferritkernspeicher; Ringkernspeicher ↑ Magnetspeicher
magnetic core storage (→magnetic core memory)	TERM&PER	(→Magnetkernspeicher)
magnetic core store (→magnetic core memory)	TERM&PER	(→Magnetkernspeicher)
magnetic coupling (→inductive coupling)	EL.TECH	(→induktive Kopplung)
magnetic cut-out	POWER ENG	**magnetischer Auslöser** = Magnetschütz
magnetic declination (→declination)	PHYS	(→Mißweisung)
magnetic dipole	PHYS	**magnetischer Dipol**
magnetic dipole moment	PHYS	**magnetisches Dipolmoment**
magnetic disc device (→magnetic disk memory)	TERM&PER	(→Magnetplattenspeicher)
magnetic disc dump (→disk dump)	DATA PROC	(→Plattenspeicherabzug)
magnetic disc memory (→magnetic disk memory)	TERM&PER	(→Magnetplattenspeicher)
magnetic disc recorder system (→MDR system)	TV	(→MDR-SYSTEM)
magnetic disc storage (→magnetic disk memory)	TERM&PER	(→Magnetplattenspeicher)
magnetic disc unit (→magnetic disk memory)	TERM&PER	(→Magnetplattenspeicher)
magnetic disk ↓ hard disk; floppy disk	TERM&PER	**Magnetplatte** ↓ Hartplatte; Diskette
magnetic disk cartridge	TERM&PER	**Magnetplattenkassette**
magnetic disk device (→magnetic disk memory)	TERM&PER	(→Magnetplattenspeicher)

magnetic disk dump (→ disk DATA PROC (→Plattenspeicherabzug)
dump)
magnetic disk file DATA PROC **Magnetplattendatei**
= disk file = Plattendatei
magnetic disk maintenance TERM&PER **Magnetband-Pflegegerät**
equipment
magnetic disk memory TERM&PER **Magnetplattenspeicher**
= magnetic disc memory; disk memory; = Plattenspeicher; Magnet-
disc memory; magnetic disk storage; scheibenspeicher; Magnet-
magnetic disc storage; disk storage; platteneinheit;
disc storage; magnetic disk device; Magnetplattengerät
magnetic disc device; magnetic disk ↑ Magnetschichtspeicher
unit; magnetic disc unit; disk unit; ↓ Festplattenspeicher; Wech-
disc unit selplattenspeicher
↑ magnetic layer memory
↓ fixed-disk memory; moving-disk
memory
magnetic disk pack TERM&PER **Magnetplattenstapel**
= disk pack = Plattenstapel; Plattenturm;
Plattensatz
magnetic disk storage TERM&PER (→Magnetplattenspeicher)
(→magnetic disk memory)
magnetic disk unit TERM&PER (→Magnetplattenspeicher)
(→magnetic disk memory)
magnetic document reader TERM&PER (→Magnetschriftleser)
(→magnetic character reader)
magnetic document sorter TERM&PER **Magnetschriftsortierer**
= magnetic ink document sorter; mag- = Magnetschriftbeleg-Sortie-
netic font character reader rer
magnetic domain PHYS **Magnetisierungsbereich**
magnetic drum TERM&PER **Magnettrommel**
magnetic drum storage TERM&PER **Magnettrommelspeicher**
= drum storage = Trommelspeicher
magnetic energy PHYS **magnetische Energie**
magnetic energy density PHYS **magnetische Energiedichte**
magnetic field PHYS **Magnetfeld**
= magnetisches Feld
magnetic field probe INSTR **Magnetfeldfühler**
↓ magnistor [MICROEL] = Magnetfühler; Magnetfeld-
sonde
↓ Magnistor [MICROEL]
magnetic field strength PHYS **magnetische Feldstärke**
[SI unit: ampere/meter] [SI-Einheit: Ampere/Me-
= magnetizing force ter]
= magnetische Erregung
magnetic film memory TERM&PER (→Magnetschichtspeicher)
(→magnetic layer memory)
magnetic film storage TERM&PER (→Magnetschichtspeicher)
(→magnetic layer memory)
magnetic film store TERM&PER (→Magnetschichtspeicher)
(→magnetic layer memory)
magnetic fluctuation noise PHYS (→Barkhausen-Effekt)
(→Barkhausen effect)
magnetic flux 1 PHYS **magnetischer Induktionsfluß**
[surface integral of magnetic flux den- [Flächenintegral der ma-
sity; SI unit: Weber] gnetischen Flußdichte; SI-
= magnetic induction flux; induction Einheit: Weber]
flux = magnetischer Fluß; Magnet-
fluß; Kraftfluß
magnetic flux 2 (→magnetic flux PHYS (→magnetische Flußdichte)
density)
magnetic flux density PHYS **magnetische Flußdichte**
[magnetic flux per cross-sectional [magnetischer Fluß pro
area; derived SI unit: Tesla] Querschnittsfläche; abgelei-
= magnetic flux 2; magnetic induction tete SI-Einheit: Tesla]
= magnetische Kraftflußdich-
te; Induktion 2;
magnetische Induktion
magnetic flux quantum PHYS **magnetisches Flußquant**
[Konstante]
magnetic font TERM&PER **Magnetschrift**
[optical character with magnetic ink] [Klarschrift mit magneti-
= magnetic ink font scher Farbe]
↓ analog magnetic font; digital mag- ↓ Analogschrift; Digitalschrift
netic font
magnetic font character reader TERM&PER (→Magnetschriftsortierer)
(→magnetic document sorter)
magnetic font document TERM&PER (→Magnetschriftbeleg)
(→magnetic ink document)

magnetic font printer TERM&PER (→Magnetschriftdrucker)
(→magnetic character printer)
magnetic font reader TERM&PER (→Magnetschriftleser)
(→magnetic character reader)
magnetic force PHYS **magnetische Kraft**
= Magnetkraft
magnetic gap PHYS **Magnetluftspalt**
= air gap; pole gap; gap = Magnetspalt; Luftspalt;
Spalt; Interferricum
magnetic head TERM&PER **Magnetkopf**
≈ sound head [EL.ACOUS] ≈ Tonkopf [EL.ACOUS]
↓ read/write head; read head; write ↓ Schreib-Lese-Kopf; Lese-
head; magnetic tape head kopf; Schreibkopf; Magnetbandkopf
magnetic induction (→magnetic PHYS (→magnetische Flußdichte)
flux density)
magnetic induction flux PHYS (→magnetischer Induktions-
(→magnetic flux) fluß)
magnetic ink TERM&PER **Magnettinte**
= magnetische Tinte
magnetic ink character TERM&PER (→Magnetschriftzeichen)
(→magnetic character)
magnetic ink character printer TERM&PER (→Magnetschriftdrucker)
(→magnetic character printer)
magnetic ink character reader TERM&PER (→Magnetschriftleser)
(→magnetic character reader)
magnetic ink character recog- TERM&PER (→Magnetschrift-Zeichener-
nition (→magnetic character recogni- kennung)
tion)
magnetic ink document TERM&PER **Magnetschriftbeleg**
= magnetic font document
magnetic ink document sorter TERM&PER (→Magnetschriftsortierer)
(→magnetic document sorter)
magnetic ink font (→magnetic TERM&PER (→Magnetschrift)
font)
magnetic ink scanner TERM&PER (→Magnetschriftleser)
(→magnetic character reader)
magnetic iron (→ soft iron) METAL (→Weicheisen)
magnetic layer memory TERM&PER **Magnetschichtspeicher**
[information stored in a magnetic [Speicherung der Informa-
layer] tion in magnetisierbarer
= magnetic layer storage; magnetic Schicht]
layer store; magnetic film memory; ↑ Magnetspeicher
magnetic film storage; magnetic film ↓ Magnetbandspeicher; Ma-
store gnetplattenspeicher; Dis-
↑ magnetic memory kettenspeicher
↓ magnetic tape memory; magnetic
disk memory; diskette memory
magnetic layer storage TERM&PER (→Magnetschichtspeicher)
(→magnetic layer memory)
magnetic layer store TERM&PER (→Magnetschichtspeicher)
(→magnetic layer memory)
magnetic lens PHYS **magnetische Linse**
= Magnetlinse
magnetic loading (→linkage PHYS (→Verkettungsfluß)
flux)
magnetic material PHYS **Magnetmaterial**
= magnetic medium
magnetic medium (→magnetic PHYS (→Magnetmaterial)
material)
magnetic memory TERM&PER **Magnetspeicher**
= magnetic storage = magnetischer Speicher
↓ magnetic layer memory; magnetic ↓ Magnetschichtspeicher;
core memory Magnetkernspeicher
magnetic metal (→ soft METAL (→Weichmetall)
metal)
magnetic microphone EL.ACOUS **Magnetmikrophon**
magnetic mirror ELECTRON **magnetischer Spiegel**
magnetic moment PHYS **magnetisches Moment**
magnetic needle PHYS **Magnetnadel**
magnetic particle TERM&PER **Magnetteilchen**
magnetic pickup EL.ACOUS (→elektromagnetischer Ton-
(→electromagnetic pickup) abnehmer)
magnetic polarization 1 PHYS **magnetische Polarisation**
magnetic polarization 2 PHYS (→Vormagnetisierung)
(→premagnetization)
magnetic pole PHYS **Magnetpol**
= magnetischer Pol
magnetic potential PHYS **magnetostatisches Potential**

magnetic potential difference PHYS (→magnetische Umlaufspannung)
(→magnetomotive force)

magnetic printer TERM&PER (→magnetographischer Drucker)
(→magnetographic printer)

magnetic pushbutton key COMPON **Magnettaste**

magnetic quantum number PHYS **magnetische Quantenzahl**

magnetic read head TERM&PER (→Lesekopf)
(→reading head)

magnetic reading head TERM&PER (→Lesekopf)
(→reading head)

magnetic recording ELECTRON **Magnetaufzeichnung**
↓ tape recording; sound tape recording [EL.ACOUS]; video tape recording [TV]
= MAZ
↓ Magnetbandaufzeichnung; Tonbandaufzeichnung [EL.ACOUS]; Magnetband-Fernsehaufzeichnung [TV]

magnetic reluctance EL.TECH **magnetischer Widerstand**
[ratio of magnetomotive force to magnetic flux]
[magnetische Kraft zu magnetischem Fluß]
= reluctance; magnetic resistance 1
= Reluktanz

magnetic resistance (→magnetic resistor) COMPON (→magnetfeldabhängiger Widerstand)

magnetic resistance 1 EL.TECH (→magnetischer Widerstand)
(→magnetic reluctance)

magnetic resistivity PHYS (→Reluktivität)
(→reluctivity)

magnetic resistor COMPON **magnetfeldabhängiger Widerstand**
[sensor]
= magnetic resistance
[Sensor]

magnetic resonance PHYS **magnetische Resonanz**

magnetic reversal PHYS **Ummagnetisierung**

magnetic reversal loss PHYS **Ummagnetisierungsverlust**

magnetic rod antenna ANT **Ferritstabantenne**
= loop stick antenna
↑ ferrite antenna
↑ Ferritantenne

magnetic screen EL.TECH **magnetischer Schirm**

magnetic screening EL.TECH **magnetische Abschirmung**
= magnetic shield

magnetic shield (→magnetic screening) EL.TECH (→magnetische Abschirmung)

magnetic space constant PHYS (→magnetische Feldkonstante)
(→coefficient of self-inductance)

magnetic storage (→magnetic memory) TERM&PER (→Magnetspeicher)

magnetic storage core COMPON (→Speicherkern)
(→storage core)

magnetic storm GEOPHYS **magnetisches Gewitter**

magnetic strip TERM&PER **Magnetstreifen**
= magstripe

magnetic strip card TERM&PER **Magnetstreifenkarte**

magnetic strip memory TERM&PER **Magnetstreifenspeicher**
= magstripe memory
≈ magnetic card storage
≈ Magnetkartenspeicher

magnetic strip reader TERM&PER **Magnetstreifenleser**

magnetic susceptibility PHYS **magnetische Suszeptibilität**

magnetic switch (→contactor) POWER ENG (→Schütz)

magnetic tape ELECTRON **Magnetband**
= recording tape; ferrous coated tape; megtape
↓ Datenband [TERM&PER]; Tonband [EL.ACOUS]; Videoband [TV]
↓ data tape [TERM&PER]; sound tape [EL.ACOUS]; video tape [TV]

magnetic tape cartridge TERM&PER (→Magnetbandkassette)
(→magnetic tape cassette)

magnetic tape cartridge system TERM&PER **Magnetbandkassetten-Gerät**

magnetic tape cassette TERM&PER **Magnetbandkassette**
= magnetic tape cartridge; cassette; tape cassette; tapecartridge
= Kassette; Cassette
↑ Einschubkassette
↑ cartridge
↓ audio cassette; data cassette; recorder cassette
↓ Tonbandkassette; Datenkassette; Recorderkassette

magnetic tape cassette drive DATA PROC **Magnetbandkassetten-Laufwerk**
= cartridge tape drive

magnetic tape cassette recorder (→cassette tape recorder) TERM&PER (→Kassettenrecorder)

magnetic tape coil system TERM&PER **Spulen-Magnetbandgerät**

magnetic tape command DATA PROC (→Magnetbandbefehl)
(→magnetic tape instruction)

magnetic tape controller TERM&PER **Magnetbandsteuerung**

magnetic tape deck TERM&PER (→Magnetbandgerät)
(→magnetic tape device)

magnetic tape density TERM&PER **Magnetband-Aufzeichnungsdichte**

magnetic tape device TERM&PER **Magnetbandgerät**
[drive mechanism and magnetic head]
[Laufwerk und Magnetkopf]
= tape device; magnetic tape deck; tape deck; magnetic tape unit; tape unit; magnetic tape station; tape station; magnetic tape handler; tape handler
= Bandgerät; Magnetbandeinheit
≈ Tonbandgerät [EL.ACOUS]
≈ tape recorder [EL.ACOUS]; drive
↓ magnetic tape drive; streamer; magnetic tape coil system; magnetic tape cartridge system
↓ Magnetbandlaufwerk; Streamer-Magnetbandgerät; Spulen-Magnetbandgerät; Kassetten-Magnetbandgerät

magnetic tape drive TERM&PER **Magnetbandantrieb**
= magnetic tape transport; tape drive; tape transport; magnetic tape drive servo; tape drive servo
= Magnetbandtransport; Bandantrieb; Bandtransport; Magnetbandlaufwerk; Bandlaufwerk
↑ drive
↑ Laufwerk

magnetic tape drive servo TERM&PER (→Magnetbandantrieb)
(→magnetic tape drive)

magnetic tape error TERM&PER **Magnetbandfehler**
= tape error
↓ drop-out; drop-in
↓ Signalausfall; Störsignal

magnetic tape file DATA PROC **Magnetbanddatei**
= tape file
= Banddatei

magnetic tape handler TERM&PER (→Magnetbandgerät)
(→magnetic tape device)

magnetic tape head TERM&PER **Magnetbandkopf**
↑ magnetic head
↑ Magnetkopf

magnetic tape instruction DATA PROC **Magnetbandbefehl**
= magnetic tape command

magnetic tape memory TERM&PER **Magnetbandspeicher**
= tape memory; magnetic tape storage; tape storage
= Bandspeicher
↑ magnetic layer memory
↑ Magnetschichtspeicher
↓ streamer; cassette tape memory
↓ Streamer; Magnetband-Kassettenspeicher

magnetic tape reader TERM&PER **Magnetbandleser**
↓ cassette tape reader

magnetic tape recorder TERM&PER (→Bandaufzeichnungsgerät)
(→tape recorder)

magnetic tape recording ELECTRON **Magnetbandaufzeichnung**
= tape recording
= Bandaufzeichnung
↑ magnetic recording
↑ Magnetaufzeichnung
↓ sound tape recording [EL.ACOUS]; video tape recording [TV]
↓ Tonbandaufzeichnung [EL.ACOUS]; Fernsehaufzeichnung [TV]

magnetic tape reel TERM&PER **Magnetbandrolle**
[spool with tape]
[mit aufgespultem Band]
= magnetic tape spool
= Magnetbandspule
↑ tape reel; reel
↑ Bandrolle; Rolle

magnetic tape sorting DATA PROC **magnetbandunterstütztes Sortieren**
= Magnetbandsortierung

magnetic tape spool TERM&PER **Magnetbandspule**
[a flanged spool without tape]
[leere Spule]
= spool frame
= Spulenkörper
≈ magnetic tape reel
≈ Magnetbandrolle
↑ tape spool; spool
↑ Bandspule; Spule

magnetic tape station TERM&PER (→Magnetbandgerät)
(→magnetic tape device)

magnetic tape storage TERM&PER (→Magnetbandspeicher)
(→magnetic tape memory)

magnetic tape transport TERM&PER (→Magnetbandantrieb)
(→magnetic tape drive)

magnetic tape unit TERM&PER (→Magnetbandgerät)
(→magnetic tape device)

magnetic track TERM&PER **Magnetspur**

magnetic type library (→tape library) DATA PROC (→Magnetbandarchiv)

magnetic voltmeter INSTR **magnetischer Spannungsmesser**

magnetic wire PHYS **Magnetdraht**

magnetic wire memory PHYS **Magnetdrahtspeicher**
= plated wire memory; magnetic wire storage; plated wire storage
= Drahtspeicher

magnetic wire storage	TERM&PER	(→Magnetdrahtspeicher)	**magnetorestrictive store**	TERM&PER	(→magnetorestriktiver Speicher)
(→magnetic wire memory)			(→magnetorestrictive memory)		
magnetism	PHYS	**Magnetismus**	**magnetosensitive**	PHYS	**magnetempfindlich**
magnetite	PHYS	**Magneteisenstein**			= magnetfeldabhängig
		= Magnetit	**magnetostatic**	PHYS	**magnetostatisch**
magnetizable	PHYS	**magnetisierbar**	**magnetostatic field**	PHYS	**magnetostatisches Feld**
magnetization	PHYS	**Magnetisierung**	**magnetostatics**	PHYS	**Magnetostatik**
magnetization curve	PHYS	**Magnetisierungskurve**	**magnetostriction**	PHYS	**Magnetostriktion**
magnetization heat	PHYS	**Magnetisierungswärme**	**magnetostrictive hysteresis**	PHYS	**magnetostriktive Hysterese**
magnetization intensity	PHYS	**Magnetisierungsstärke**	**magnetostrictive**	EL.ACOUS	**magnetostriktiver Wandler**
magnetization reversal	PHYS	**Magnetisierungswechsel**	transducer		↑ Schallwandler
magnetize	PHYS	**magnetisieren**	↑ electroacoustic transducer		
magnetizing force (→magnetic field	PHYS	(→magnetische Feldstärke)	**magnetron**	MICROW	**Magnetron**
strength)					= Magnetfeldröhre
magneto (→magneto	TELEPH	(→Kurbelinduktor)			↑ Wanderfeldröhre
generator)			**magnet steel**	METAL	**Magnetstahl**
magneto-crystaline anisotropy	PHYS	**magnetische Kristallanisotropie**	**magnet yoke**	EL.TECH	**Magnetjoch**
			= yoke		= Joch
magnetodiode	MICROEL	**Magnetdiode**	= armature		= Anker
↑ madistor		↑ Madistor	↓ transformer yoke		↓ Transformatorjoch
magnetodynamic (adj.)	PHYS	**magnetodynamisch**	**magnification**	PHYS	**Vergrößerung**
magnetodynamics	PHYS	**Magnetodynamik**	[optics]		[Optik]
magnetoelastic	PHYS	**magnetoelastisch**	**magnification** (→enlargement)	TECH	(→Vergrößerung)
magnetofluid	EL.ACOUS	**Magnetofluid**	**magnification factor** (→factor	EL.TECH	(→Gütefaktor)
magneto generator	TELEPH	**Kurbelinduktor**	of quality)		
= magneto-inductor; magneto; inductor		= Induktormaschine	**magnification factor** (→factor of	PHYS	(→Gütewert)
generator			merit)		
magnetographic printer	TERM&PER	**magnetographischer Drucker**	**magnifier lamp**	ELECTRON	**Lupenleuchte**
= magnetic printer		= magnetografischer Drucker; Magnetdrucker	**magnify**	PHYS	**vergrößern**
			[optics]		[Optik]
magnetohydronynamics	PHYS	**Magnetohydrodynamik**	**magnifying glass**	PHYS	**Lupe**
magneto-inductor (→magneto	TELEPH	(→Kurbelinduktor)	= loupe		[Optik]
generator)					= Vergrößerungsglas
magnetomechanic transducer	COMPON	**magnetomechanischer Wandler**	**magnistor**	MICROEL	**Magnistor**
			↑ magnetic field probe [INSTR]		↑ Magnetfeldfühler [INSTR]
magnetometer (→magnetic	INSTR	(→Magnetometer)	**magnitude** (→absolute value)	MATH	(→Absolutwert)
anomaly detector)			**magnitude** (→amplitude)	PHYS	(→Amplitude)
magnetomotive force	PHYS	**magnetische Umlaufspannung**	**magnitude accuracy**	INSTR	**Amplitudengenauigkeit**
[SI unit: Ampere; line integral of		[SI-Einheit: Ampere; Linienintegral der magnetischen	**magnitude comparator**	CIRC.ENG	**Wertevergleicher**
magnetic field strength]		Feldstärke]	= value comparator		= Größenvergleicher
= MMF; magnetic boundary potential;		= magnetomotorische Kraft;	**magnitude of admittance**	NETW.TH	**Betrag des Scheinleitwerts**
magnetic potential difference		MMK; magnetische Spannung	= absolute value of admittance		= Scheinleitwert 2; Betrag des komplexen Leitwerts
≈ electric loading		≈ elektrische Durchflutung	**magnitude of impedance**	NETW.TH	**Betrag des Scheinwiderstandes**
			= absolute value of impedance		
magnetomotoric memory	TERM&PER	**magnetomotorischer Speicher**			= Scheinwiderstand 2
[requires mechanical movement for		[erfordert zum Speichern	**magstripe** (→magnetic	TERM&PER	(→Magnetstreifen)
storage and retrieval]		und Lesen eine Bewegung]	strip)		
= magnetomotoric storage		↓ Magnettrommelspeicher;	**magstripe memory**	TERM&PER	(→Magnetstreifenspeicher)
↓ magnetic drum memory; magnetic		Magnetbandspeicher	(→magnetic strip memory)		
tape memory			**mail**	OFFICE	**Post**
magnetomotoric storage	TERM&PER	(→magnetomotorischer Speicher)	[written communication]		[schriftliche Nachricht]
(→magnetomotoric memory)			≈ correspondence; letter		≈ Korrespondenz; Brief
magneton	PHYS	**Magneton**	**mail** (→postal administration)	POST	(→Postverwaltung)
[unit for magnetic momentum]		[Einheit für magnetisches Moment]	**mailbox**	TELEC	**Mitteilungsspeicher**
			mailbox	DATA PROC	**Briefbox**
magnetooptical effect	PHYS	**magnetooptischer Effekt**	**mailbox system** (→electronic	TELEC	(→elektronischer Briefdienst)
magneto-optical memory	TERM&PER	(→magnetooptischer Speicher)	mail)		
(→magneto-optic memory)			**mail circular**	POST	**Postwurfsendung**
magneto-optical storage	TERM&PER	(→magnetooptischer Speicher)	**mail dispatch service**	POST	**Postversanddienst**
(→magneto-optic memory)			**mail handling**	OFFICE	**Postbearbeitung**
magneto-optic memory	TERM&PER	**magnetooptischer Speicher**	**mailing list program**	DATA PROC	**Adressetikettenprogramm**
= magneto-optic storage; magneto-optic store; magneto-optical memory;		[mit Laserstrahl]	**mailman** (AM) (→postman)	POST	(→Postbote)
magneto-optical storage			**mail merge**	DATA PROC	**Serienbriefherstellung**
magnetooptics	PHYS	**Magnetooptik**	= mail-merging		= Adress-Serienbrief-Mischfunktion
magneto-optic storage	TERM&PER	(→magnetooptischer Speicher)	**mail-merge program**	DATA PROC	(→Standardbriefprogramm)
(→magneto-optic memory)			(→form-letter program)		
magneto-optic store	TERM&PER	(→magnetooptischer Speicher)	**mail-merging** (→mail	DATA PROC	(→Serienbriefherstellung)
(→magneto-optic memory)			merge)		
magnetoresistor	COMPON	**Feldplatte**	**mail-order business**	ECON	**Versandhaus**
= field plate			= dispatching business		= Versandhandel
magnetoresistor	COMPON	**Feldplattenpotentiometer**	**mail room**	OFFICE	**Poststelle**
potentiometer			**mail service**	ECON	**Postdienst**
magnetorestrictive memory	TERM&PER	**magnetorestriktiver Speicher**	= postal service		= Postwesen
= magnetorestrictive storage; magnetorestrictive store		↑ Laufzeitspeicher	≈ postal administration		
↑ delay-line memory			**mail service employee**	POST	**Postangestellter**
magnetorestrictive storage	TERM&PER	(→magnetorestriktiver Speicher)			= Postler
(→magnetorestrictive memory)			**main address** (→base	DATA PROC	(→Grundadresse)
			address)		

348 main application

main application TECH **Einsatzschwerpunkt**
= main use
= Anwendungsschwerpunkt; Hauptanwendung

main axis (→ principal axis) MATH (→Hauptachse)
main axis transformation MATH **Hauptachsentransformation**
main board DATA PROC **Hauptplatine**
[main PCB of a microcomputer, containing processor chips and connectors for other boards]
[zentrale Leiterplatte eines Mikrokomputers, mit Prozessor-Bausteinen und Stecker zum Anschluß weiterer Platinen]
= main card; mother board; mother card; system board
= Grundplatine; Hauptleiterplatte; Hauptkarte; Grundleiterplatte
≠ expansion board
≠ Erweiterungskarte

main building ECON **Hauptgebäude**
main cable (BRI) OUTS.PLANT **Hauptkabel**
[from exchange to first branching point]
[vom Vermittlungsamt zum ersten Vetrzweigungspunkt]
= primary cable; feeder cable (AM); feeder (AM)
= Primärkabel (USA); Zuleitungskabel
≈ branch feeder cable
≈ Verteilkabel

main card (→ main board) DATA PROC (→Hauptplatine)
main carrier MODUL **Hauptträger**
≠ subcarrier ≠ Hilfsträger
main catalog (→ root directory) DATA PROC (→Hauptverzeichnis)
main channel TELEC **Hauptkanal**
= forward channel ≠ Rückkanal
≠ backward channel
main clause LING **Hauptsatz**
= main sentence = Matrixsatz
main clock ELECTRON **Haupttakt**
= master clock = Haupttaktgeber
main clock (→reference clock) TELEC (→Bezugstaktgeber)
main computer (→ host computer) DATA COMM (→Hauptrechner)
main computer (→ host computer) DATA COMM (→Hauptrechner)
main contractor ECON **Generalunternehmer**
= Hauptunternehmer
main control loop CONTROL **Hauptregelkreis**
main corridor SYS.INST **Hauptgang**
[between rack rows] [zwischen Gestellreihen]
main current (→principal current) MICROEL (→Hauptstrom)
main diagonal NETW.TH **Hauptdiagonale**
main diagonal element NETW.TH **Hauptdiagonalelement**
main directory DATA PROC (→Hauptverzeichnis)
(→root directory)
main distribution frame TELEC **Hauptverteiler**
= MDF; master distributor; main distributor
= HVT
main distributor (→main distribution frame) TELEC (→Hauptverteiler)
main district exchange (→district exchange) TELEC (→Bezirksamt)
main entrance signal RADIO NAV **Haupteinflugzeichen**
= middle marker = Halbwegbake
main entry (→basic entry) LING (→Grundeintrag)
main file (BRI) (→ master file 2) DATA PROC (→Bestandsaufnahme)
main flux PHYS **Hauptfluß**
[magnetism] [Magnetismus]
mainframe INSTR **Grundgerät**
mainframe (→ mainframe computer) DATA PROC (→Großrechner)
mainframe (→ basic equipment) EQUIP.ENG (→Grundgerät)
mainframe computer DATA PROC **Großrechner**
= mainframe; large scale computer; large computer; number cruncher
= Großcomputer; Mainframe; Host
≈ general-purpose computer; supercomputer
= Universalrechner; Größtrechner
mainframe manufacturer DATA PROC **Großrechnerhersteller**
mainframe memory DATA PROC **Hauptspeicher 3**
[Speicher eines Großrechners]
= Großrechnerspeicher

main function TECH **Hauptfunktion**
= chief function = wichtigste Funktion
main fuse EL.INST **Hauptsicherung**
= line fuse
main highway CIV.ENG **Hauptverkehrsstraße**
= highway ≈ Schnellverkehrsstraße
≈ dual-carriage way
main key TERM&PER **Haupttaste**
main line TELEPH **Hauptanschluß**
= main station line; main station
= Hauptstelle; Hauptanschlußleitung
main line TELEC **Hauptleitung**
[line between switching center and front-end equipment]
[Leitung zwischen Vermittlung und Vorfeldeinrichtung]
≈ branch line
≈ Zweigleitung
main line transportation RAILW.SIGN **Fernverkehr**
main lobe (→ mayor lobe) ANT (→Hauptkeule)
main memory 1 DATA PROC **Hauptspeicher 1**
[addressable storage of CPU, where data and programs are stored for the execution of its tasks, with direct acess from the processor and input/output unit; generally implemented with RAM chips, therefore often named RAM in microcomputer terminology]
[adressierbarer Speicher für die Daten und Programme, welche die Zentraleinheit für die Erfüllung ihrer Aufgaben benötigt; Prozessor sowie Ein-Ausgabewerk haben darauf direkten Zugriff; meist mit RAM-Bausteinen realisiert, daher im Fachjargon der Mikrocomputertechnik vielfach als RAM bezeichnet]
= main storage 1; central memory; central storage; prime memory; prime storage; core memory; core storage; internal memory; internal storage; internal store; RAM 2; immediate access memory
= Arbeitsspeicher 1; Zentralspeicher; Primärspeicher; Speicherwerk; Speichersystem; Internspeicher; interner Speicher; innerer Speicher; RAM 2
≠ external memory
≠ externer Speicher
↑ central processing unit 1
↑ Zentraleinheit 1
↓ Hauptspeicher 2; Arbeitsspeicher 2; Ergänzungsspeicher

main memory 2 DATA PROC **Hauptspeicher 2**
[in a structured main memory 1 that part, where resident programs are stored]
[bei Untergliederung eines Hauptspeichers 1 der Teil, in dem arbeitsspeicherresidente Programme gespeichert werden]
= main storage 2; main store 2
= Arbeitsspeicher 2
main memory address DATA PROC **Hauptspeicheradresse**
= main storage address; central memory address; primary storage address
= Arbeitsspeicheradresse; Zentralspeicheradresse; Primärspeicheradresse; Speicherwerkadresse
main memory allocation DATA PROC **Arbeitsspeicherzuweisung**
= main memory assignment
= Hauptpeicherzuweisung
↑ memory management
↑ Speicherverwaltung
main memory assignment DATA PROC (→Arbeitsspeicherzuweisung)
(→main memory allocation)
main memory resident DATA PROC (→speicherresident)
(→memory-resident)
main-memory-resident command (→ resident command) DATA PROC (→residentes Kommando)
main-memory-resident program (→ resident programm) DATA PROC (→residentes Programm)
main-memory-resident software (→ resident software) DATA PROC (→residente Software)
main menu DATA PROC **Hauptmenü**
[list of primary options] [erste Ebene der Auswahlmöglichkeiten eines Menüs]
= master menu
main oscillation (→fundamental wave) PHYS (→Grundschwingung)
main pattern RADIO REL **Hauptraster**
main processor MICROEL **Hauptprozessor**
= master processor
main processor (→host computer) DATA COMM (→Hauptrechner)
main program DATA PROC **Hauptprogramm**
≈ background program ≈ Hintergrundprogramm
≠ subroutine ≠ Unterprogramm

main quantum number PHYS **Hauptquantenzahl**
= principal quantum number ↑ Quantenzahl
↑ quantum number
main-radiation direction ANT (→Hauptstrahlrichtung)
(→boresight)
main receiver RADIO REL **Grundempfänger**
≠ diversity receiver ≠ Diversityempfänger
main receiving direction ANT **Hauptempfangsrichtung**
main reflector ANT **Hauptreflektor**
≠ subreflector ≠ Nebenreflektor
main routine DATA PROC **Hauptroutine**
mains POWER ENG **Starkstromnetz**
= mains electricity supply; electricity = Stromversorgungsnetz;
supply; commercial power; commer- Wechselstromnetz; Strom-
cial supply; commercial line; house- versorgung; öffentliches
hold ac; power supply Netz; Hauptnetz; Netz
mains-borne interference EL.TECH **Netzstörung**
 [vom Netz kommend]
 = Netzinterferenz
mains cable (→power cable) EQUIP.ENG (→Netzanschlußkabel)
mains connection (→ac power POWER SYS (→Netzanschluß)
 supply)
mains connector European style EL.INST (→Eurostecker)
(→euro mains connector)
mains control POWER ENG **Netzregelung**
= source regulation ↓ Frequenzregelung; Lei-
↓ frequency control; load control; load- stungsregelung; Leistungs-
 frequency control Frequenzregelung
mains current POWER ENG **Netzstrom**
= line current = Leitungsstrom
mains distribution rack POWER ENG **Netzschaltfeld**
= cubicle = Netzfeld
mains electricity supply POWER ENG (→Starkstromnetz)
(→mains)
main sentence (→main clause) LING (→Hauptsatz)
mains failure POWER SYS **Netzausfall**
= mains interruption = Netzspannungsausfall
↑ power failure ↑ Stromausfall
mains failure supply POWER SYS (→Notstromversorgung)
(→emergency power supply)
mains feeder POWER SYS **Netzanschlußleitung**
mains frequency POWER SYS **Netzfrequenz**
= ac line frequency; commercial-line
 frequency; power frequency; Hertz
 rate
main shell PHYS **Hauptschale**
mains hum TELEC **Netzbrumm**
= mains noise; power-line hum; power- = Netzgeräusch; Netzspan-
 line noise; induced power noise nungsbrumm
mains interruption (→mains POWER SYS (→Netzausfall)
 failure)
mains lead (→power EQUIP.ENG (→Netzanschlußkabel)
 cable)
mains line antenna ANT **Netzantenne**
mains multi-connector EL.INST **Reihensteckdose**
= distribution board [mehrere Steckdosen auf einer Leiste]
mains noise (→mains hum) TELEC (→Netzbrumm)
mains operated (→mains EL.TECH (→netzgespeist)
 powered)
mains overload filter EL.INST **Überspannungsfilter**
mains plug EL.INST **Netzstecker**
≠ mains socket ≠ Netzsteckdose
 ↑ Steckverbindung
mains powered EL.TECH **netzgespeist**
= mains operated = netzbetrieben; netzversorgt
mains power line POWER ENG **Netzleitung**
≠ AC power line
mains power supply EQUIP.ENG **Netzstromversorgungsgerät**
= ac power supply; ac power source; = Netzstromversorgung;
 power pack Netzversorgung; Netzan-
 schlußgerät; Netzgerät;
 Netzteil; Netzspannungs-
 quelle; Wechselstromver-
 sorgung
mains socket EL.INST **Netzsteckdose**
= power socket; utility socket; plug con- = Netzdose; Steckdose; An-
 tact; outlet; socket schlußdose; Dose
≠ mains plug ≠ Netzstecker
 ↑ Steckverbindung

mains switch EL.INST **Netzschalter**
= power switch
main station TELECONTR **Zentrale**
= master station; control station; con- = Leitzentrale; Leitstation;
 trol center (AM); control centr (BRI) Leitstelle; Leitzentrum;
≠ tributary station Kontrollstelle; Kontrollzen-
 trum; Fernwirkzentrale;
 Fernwirk-Zentralstation
 ≠ Unterstation
main station (→main line) TELEPH (→Hauptanschluß)
main station line (→main TELEPH (→Hauptanschluß)
 line)
main storage 1 (→main DATA PROC (→Hauptspeicher 1)
 memory 1)
main storage 2 (→main DATA PROC (→Hauptspeicher 2)
 memory 2)
main storage address (→main DATA PROC (→Hauptspeicheradresse)
 memory address)
main store 1 (→main DATA PROC (→Hauptspeicher 1)
 memory 1)
main store 2 (→main DATA PROC (→Hauptspeicher 2)
 memory 2)
mains transformer POWER ENG **Netztransformator**
= line transformer; power transformer; = Netzanschlußtransforma-
 supply transformer tor; Netztrafo
main studio BROADC **Hauptstudio**
= central studio
main supplier ECON **Hauptlieferant**
= first supplier
mains voltage (→line POWER ENG (→Netzspannung)
 voltage)
main switch (→master EL.INST (→Hauptschalter)
 switch)
maintain COLLOQ **aufrechterhalten**
= keep up; retain = beibehalten
≈ ensure ≈ sicherstellen
maintain TECH **warten**
≈ operate = unterhalten; instandhalten
↓ test; repair = betreiben
 ↓ prüfen; instandsetzen
maintainability TECH **Wartungsfreundlichkeit**
≈ serviceability; user friendliness = Servicefreundlichkeit;
 Wartbarkeit; Instandsetz-
 barkeit
 ≈ Bedienerfreundlichkeit; Benutzerfreundlichkeit
maintainable TECH **wartungsfreundlich**
= easy-to-service; easy-to-maintain = wartungsgerecht; wartbar
≈ serviceable; user-friendly ≈ bedienerfreundlich; benut-
 zerfreundlich
maintained TECH **gewartet**
≈ staffed ≈ bemannt
maintenace instruction TECH **Wartungsanleitung**
= servicing instructions
maintenance TECH **Wartung**
= servicing; service = Unterhaltung; Instandhal-
↓ repair tung
 ↓ Instandsetzung
maintenance alarm TELEC **Wartungsalarm**
maintenance center TECH **Wartungszentrum**
 = Unterhaltungszentrum; Un-
 terhaltungsstelle; Enstö-
 rungsstelle
maintenance channel TELEC **Wartungskanal**
maintenance channel unit TELEC **Wartungskanaleinheit**
maintenance console EQUIP.ENG (→Wartungsfeld)
(→maintenance panel)
maintenance contract ECON **Wartungsvertrag**
= servicing contract
maintenance control panel EQUIP.ENG (→Wartungsfeld)
(→maintenance panel)
maintenance costs ECON **Wartungskosten**
= servicing costs = Unterhaltungskosten
maintenance dependent TECH **wartungsbedingt**
 = unterhaltungsbedingt
maintenance engineer TECH **Wartungsingenieur**
= servicing engineer
maintenance equipment TERM&PER **Pflegegerät**
maintenance handbook DOC (→Wartungshandbuch)
(→maintenance manual)

maintenance instruction TECH **Wartungsvorschrift**
= servicing instruction; maintenance standards
maintenance interval TECH **Wartungsfrist**
= servicing interval = Wartungsintervall
maintenance manual DOC **Wartungshandbuch**
= maintenance handbook
maintenance medium TECH **Pflegemittel**
maintenance mode TECH **Wartungsbetrieb**
maintenance-oriented TECH **wartungsorientiert**
maintenance panel EQUIP.ENG **Wartungsfeld**
= servicing panel; maintenance control panel; maintenance console
maintenance personnel TECH **Wartungspersonal**
= maintenance staff; service personnel ≈ Betriebspersonal
≈ operating personnel
maintenance processor DATA PROC **Wartungsprozessor**
= service processor = Serviceprozessor
maintenance programm TECH **Wartungsprogramm**
= maintenance routine; maintenance schedule; servicing program; servicing routine; servicing schedule = Wartungszeitplan
maintenance programmer DATA PROC **Wartungsprogrammierer**
= Änderungsprogrammierer
maintenance routine TECH (→Wartungsprogramm)
(→maintenance programm)
maintenance schedule TECH (→Wartungsprogramm)
(→maintenance programm)
maintenance service ECON **Wartungsdienst**
≈ after-sales service = Wartungsservice; Wartungsdienstleistung
≈ Kundendienst
maintenance staff (→maintenance TECH (→Wartungspersonal)
personnel)
maintenance standards TECH (→Wartungsvorschrift)
(→maintenance instruction)
maintenance technician ECON (→Wartungstechniker)
(→serviceman 2)
maintenance test DATA PROC (→Betriebsprüfung)
(→operating test)·
maintenance works TECH **Wartungsarbeiten**
= servicing works
main terminal MICROEL **Hauptstromanschluß**
= principal terminal
main text (→current text) TYPOGR (→Lauftext)
main transmitter (→master RADIO NAV (→Hauptsender)
transmitter)
main tuning (→coarse ELECTRON (→Grobabgleich)
tuning)
main use (→main application) TECH (→Einsatzschwerpunkt)
main verb LING **Vollverb**
≠ auxiliary verb ≠ Hilfsverb
Majorana effect PHYS **Majorana-Effekt**
majorant MATH **Majorante**
[referential series whoes elements are not less] [Vergleichsreihe dessen Glieder nicht kleiner sind]
≠ minorant = Oberreihe
≠ Minorante
major cycle DATA PROC **Hauptzyklus**
major graduation INSTR **Hauptskalenteilung**
majority SCIE **Majorität**
= Mehrheit
majority bit CODING **Stopf-Mehrheitsbit**
majority carrier (→majority charge PHYS (→Majoritätsträger)
carrier)
majority carrier current PHYS **Majoritätsträgerstrom**
majority carrier density PHYS **Majoritätsträgerdichte**
majority charge carrier PHYS **Majoritätsträger**
= majority carrier = Majoritätsladungsträger
majority circuit (→majority CIRC.ENG (→Majoritätsgatter)
gate)
majority decision MATH **Majoritätsentscheidung**
majority function MATH **Majoritätsfunktion**
majority gate CIRC.ENG **Majoritätsgatter**
= majority circuit = Majoritätsschaltung; Majoritätsglied
majority logic MATH **Majoritätslogik**
major sorting key DATA PROC **Hauptsortierschlüssel**
= Hauptsortierungsschlüssel
majuscule (→capital character) TYPOGR (→Großbuchstabe)

make COMPON **schließen**
[contact] [Kontakt]
= close
make (→type) TECH (→Ausführung 1)
make (n.) (→lay-out) TECH (→Auslegung)
make (s.) (→product) ECON (→Produkt)
make-and-break contact ELECTRON **Ruhe-und-Arbeitskontakt**
make a phone call TELEPH (→telefonieren)
(→telephone)
make-before-breake contact COMPON **Folgeumschaltekontakt**
≠ breake-before-make contact [„Schließen vor Öffnen"]
↑ relay contact = unterbrechungsloser Umschaltekontakt;
Folgewechsler; Folge-Arbeits-Ruhekontakt;
Schleppkontakt; Folgekontakt Schließen-vor-Öffnen
≠ Folgekontakt Öffnen-vor-Schließen
↑ Relaiskontakt
make-break contact COMPON **Folgekontakt**
[relay] [Relais]
= sequence contact; trailing contact
make contact COMPON **Arbeitskontakt**
[closed when operated, open when disactivated] [geschlossen bei Betätigung, offen in Ruhelage]
= normally open contact; N/O contact; NOC = Schließerkontakt; Schließer
≠ break contact ≠ Ruhekontakt
↑ relay contact ↑ Relaiskontakt
make-make contact COMPON **Zwillingsschließer**
↑ relay contact = Zwillingsarbeitskontakt
↑ Relaiskontakt
makeshift (→kludge) DATA PROC (→Flickschusterei)
makeshift (→expedient) TECH (→Notbehelf)
make sure (→secure) COLLOQ (→sicherstellen)
make time COMPON **Schließzeit**
[contact] [Kontakt]
make to wastepaper TYPOGR **makulieren**
= einstampfen
makeup TYPOGR **Umbruch**
[to arrange into columns and pages] [Einteilung in Zeilen, Spalten und Seiten]
≈ typesetting = Umbrechen
↓ line break; page makeup ≈ Satz
↓ Zeilenumbruch; Seitenumbruch
making (→manufacturing) ECON (→Fertigung)
maladjustment ELECTRON (→Fehleinstellung)
(→misadjustment)
malalignment ELECTRON (→Fehlabstimmung)
(→misalignment)
male (→plug) COMPON (→Stecker)
male connector (→plug) COMPON (→Stecker)
male-male adapter COMPON **Doppelstecker**
= biplug [beidseitig mit Stecker]
↑ in-series adapter ↑ Übergangsstecker
male short (→short-circuit COMPON (→Kurzschlußstecker)
plug)
malfunction (v.i.) TECH **versagen**
= Fehlverhalten aufweisen
malfunction (n.) TECH **Funktionsstörung**
= trouble; malfunctioning; maloperation = Fehlfunktion; Fehlverhalten; Versagen; Störung
≈ fault; failure ≈ Fehler; Ausfall
malfunction effect TECH **Störungsauswirkung**
malfunction indication TELECONTR **Störungsanzeige**
= trouble indication
malfunction information TELECONTR (→Fehlermeldung)
(→failure report)
malfunctioning (adj.) TECH **schlecht funktionierend**
malfunctioning (n.) TECH (→Funktionsstörung)
(→malfunction)
malfunction report (→failure TELECONTR (→Fehlermeldung)
report)
malicious call TELEPH **belästigender Anruf**
malicious call identification SWITCH (→Fangen)
(→call tracing)
malicious call tracing (→call SWITCH (→Fangen)
tracing)

malicious call tracing data	SWITCH	Fangdaten				≈ vielfältig; vielfach 1; verschieden
malicious call tracing device	SWITCH	Fangvorrichtung				
		= Beobachtungseinrichtung	manifold (n.)		MATH	Mannigfaltigkeit
malleability (→forjeability)	METAL	(→Schmiedbarkeit)	manifoldness		COLLOQ	Mannigfaltigkeit
malleabilize	METAL	glühfrischen	= multifariousness			≈ Vielfalt; Vielfältigkeit; Verschiedenartigkeit
		[Wärmebehandlung]	≈ variety; variousness; dofference			
malleable (→forgeable)	MECH	(→schmiedbar)	manipulate (→tamper)		TECH	(→manipulieren)
malleablized	METAL	getempert	manipulate (→process)		TECH	(→verarbeiten)
mallet	TECH	Holzhammer	manipulated variable		CONTROL	(→Stellgröße)
maloperation (→malfunction)	TECH	(→Funktionsstörung)	(→control variable)			
malposition	TECH	Fehlstellung	manipulation 1 (→grip 1)		TECH	(→Handgriff 2)
= wrong position			manipulation 2 (→handling)		TECH	(→Handhabung)
maltese cross antenna	ANT	Malteserkreuz-Antenne	manipulation variable		CONTROL	(→Stellgröße)
≈ square loop antenna		≈ Quadratantenne	(→control variable)			
Maltron keyboard	TERM&PER	Maltron-Tastatur	manipulator		TECH	Manipulator
manage	COLLOQ	bewältigen				= Handhabungsgerät; Ferngreifer
[difficulties, problems]		[Schwierigkeit]				
= solve		= meistern	man-machine		DATA PROC	Mensch-Maschine
manage (→administer)	ECON	(→verwalten)	man-machine interface		DATA PROC	Mensch-Maschine-Schnittstelle
manageable	COLLOQ	bewältigbar	= human-machine interface; man-computer interface; human-computer interface; HMI			
managed multiplexer	TRANS	(→Verteilmultiplexer)				
(→cross-connect multiplexer)						
management 1	ECON	Geschäftsführung	man-machine interpreter		DATA PROC	MML-Umsetzer
[institution]		[Institution]	man-machine language		DATA PROC	Mensch-Maschine-Sprache
= direction 1		= Management; Direktion 1; Leitung 1	= MML			= Bedienersprache; MML
≈ head office			man-made noise		RADIO	Industriestörungen
management 2	ECON	Leitung 2				= technische Störstrahlung; Störung durch elektrische Maschinen und Anlagen
[function]		[Funktion]				
= direction 2		= Führung; Direktion 2				
management 3	ECON	(→Verwaltung 1)	man-month		ECON	Mann-Monat
(→administration 1)			manned		TECH	bemannt
management graphics	DATA PROC	(→Geschäftsgraphik)	[vehicle]			[Fahrzeug, Flugzeug]
(→business graphics)			manned (→staffed)		TELEC	(→bemannt)
management information system	DATA PROC	Management-Informationssystem	manner (→mode)		COLLOQ	(→Weise)
			manoeuvre (→maneuver)		TECH	(→Manöver)
= MIS			manometer		INSTR	Manometer
management library	DATA PROC	Verwaltungsbibliothek	= pressure gage			= Druckmesser; Druckmeßgerät
management processor	COMPON	Leitprozessor				
management program	DATA PROC	(→Verwaltungsroutine)	manpower (→labour force)		ECON	(→Arbeitskraft)
(→housekeeping program)			mantissa		MATH	Mantisse
management routine	DATA PROC	(→Verwaltungsroutine)	manual		TECH	manuell
(→housekeeping program)			= by hand			= handgeführt; handbetrieben; handbedient
manager (→director)	ECON	(→Direktor)				
manager (→officer 1)	ECON	(→Leiter)	manual (→handbook)		TECH	(→Handbuch)
manager 2 (→administrator)	ECON	(→Verwalter)	manual calling (→manual dialing)		DATA COMM	(→manuelles Wählen)
man-computer interface	DATA PROC	(→Mensch-Maschine-Schnittstelle)				
(→man-machine interface)			manual control		CONTROL	Handsteuerung
mandate (→power)	ECON	(→Vollmacht)				= Handleitstelle
mandatory	COLLOQ	vorgeschrieben	manual dialing		DATA COMM	manuelles Wählen
= obligatory; compulsory		= obligatorisch; zwingend	= manual calling			= manuelle Wahl
≈ enforceable			manual direction finder		RADIO LOC	Handpeiler
mandatory leave	ECON	Fenstertag	manual drive		TECH	Handantrieb
man-day	ECON	Mann-Tag	= hand drive			
mandrel	MECH	Dorn	manual exchange		SWITCH	Handvermittlung 1
= pin			= switchboard (AM)			[Vorrichtung]
mandrel	OUTS.PLANT	Kabelziehkopf	manual input		DATA PROC	manuelle Eingabe
maneuver	TECH	Manöver	manual input medium		TERM&PER	handgeführtes Eingabemittel
[controlled displacement]		[sorgfältige Bewegung]	manual insertion		MANUF	Handbestückung
= manoeuvre			[PCB]			[Leiterplattenfertigung]
manganese	CHEM	Mangan	manual layout		MICROEL	Hand-Layout
= Mn		= Mn	manual measurement		INSTR	Handmessung
manganese dioxide	CHEM	Braunstein	manual network		DATA COMM	Handvermittlungsnetz
		= Mangandioxyd; Prolusit	manual operation		TECH	manueller Betrieb
Mangin's lens	PHYS	Mangin-Linse	= hand operation			= Handbetrieb; Handbetätigung; Handbedienung
		= Spiegellinse				
manhole	OUTS.PLANT	Mannloch	manual operation		DATA PROC	manuelle Verarbeitung
≈ chamber		≈ Schacht	manual reader (→hand-held reader)		TERM&PER	(→Handleser)
manhole cover	OUTS.PLANT	Schachtdeckel				
= cover slab (of concrete)		= Schachtabdeckung	manual scanner (→hand-held reader)		TERM&PER	(→Handleser)
manhole floor	OUTS.PLANT	Schachtsohle				
manhole guard	OUTS.PLANT	Absperrgestell	manual soldering		MANUF	Handlöten
manhole opening	OUTS.PLANT	Einstiegöffnung	manual switching		SWITCH	Handvermittlung 2
man-hour	ECON	Mann-Stunde				[Vorgang]
manifold (n.)	COLLOQ	Vielfach	manual switching system		SWITCH	(→Handvermittlungssystem)
= multiple (n.)			(→operator service system)			
manifold 2 (adj) (→multiplle)	COLLOQ	(→vielfach)	manual tuning		ELECTRON	Handabstimmung
manifold 1 (adj.)	COLLOQ	mannigfaltig	manufacture (v.t.)			fertigen
[with great variety and quantity]		[vielzählig und verschiedenartig zugleich]	= produce (v.t.)			[ECON]
= multifold 1; multifarious		= vielerlei	≈ process			= herstellen; produzieren
≈ various; multiple; different						≈ verarbeiten

manufacture

manufacture (→manufacturing) ECON (→Fertigung)
manufacture (s.) (→product) ECON (→Produkt)
manufacture presetting EQUIP.ENG (→Fabrikeinstellung)
 (→factory setting)
manufacturer ECON **Hersteller**
 = producer; fabricant; fabricator = Fabrikant; Produzent; Erzeuger
 ≈ vendor 1
 ≈ Lieferant
manufacturers logo ECON **Herstellerzeichen**
 = Herstellermonogramm; Herstellerlogo
manufacturer's nameplate EQUIP.ENG **Firmenschild**
 = name plate
manufacturer's software DATA PROC **Herstellersoftware**
 [programming aids] [Programmierhilfen]
manufacture setting EQUIP.ENG (→Fabrikeinstellung)
 (→factory setting)
manufacture to order (→unit MANUF (→Einzelanfertigung)
 production)
manufacturing ECON **Fertigung**
 = manufacture; production 2; fabrication; making = Produktion 2; Herstellung; Fabrikation
 ↑ production 1 ↑ Erzeugung
manufacturing aid (→production MANUF (→Fertigungsmittel)
 means)
manufacturing automation MANUF (→Produktionsautomatisierung)
 (→production automation)
manufacturing capacity MANUF (→Fertigungskapazität)
 (→production capacity)
manufacturing chain MANUF (→Fertigungskette)
 (→production chain)
manufacturing control MANUF (→Fertigungslenkung)
 (→production control)
manufacturing costs (→production ECON (→Herstellkosten)
 costs)
manufacturing data MANUF (→Betriebsdaten)
 (→production data)
manufacturing defect QUAL **Fertigungsfehler**
 = Herstellungsfehler
manufacturing documents MANUF (→Fertigungsunterlage)
 (→production documents)
manufacturing engineer MANUF (→Fertigungsingenieur)
 (→production engineer)
manufacturing engineering 2 MANUF **Fertigungsplanung**
 = production planning; routing = Fepla; Produktionsplanung
manufacturing engineering 1 TECH (→Fertigungstechnik)
 (→industrial engineering)
manufacturing facility MANUF (→Fertigungseinrichtung)
 (→production facility)
manufacturing flow MANUF (→Fertigungsablauf)
 (→production flow)
manufacturing line (→assembly MANUF (→Fließband)
 line)
manufacturing loss (→production MANUF (→Produktionsausfall)
 loss)
manufacturing lot MANUF **Fertigungslos**
 = Herstellungslos
manufacturing manager MANUF **Fertigungsleiter**
 = production manager; works manager; works superintendent; factory manager = Werkleiter; Werksleiter (OES); Fabrikleiter
manufacturing means MANUF (→Fertigungsmittel)
 (→production means)
manufacturing method MANUF **Fertigungsverfahren**
 = production method = Herstellungsverfahren; Produktionsverfahren
 ≈ production process ≈ Fertigungsprozeß; Fertigungsablauf
manufacturing order (→job MANUF (→Fertigungsauftrag)
 order)
manufacturing plant MANUF **Werkanlage**
 = Werksanlage (OES); Fabrikanlage
manufacturing plant MANUF (→Produktionsanlage)
 (→production plant)
manufacturing process MANUF (→Fertigungsprozeß)
 (→produccional process)
manufacturing program MANUF (→Fertigungsprogramm)
 (→production program)

manufacturing range MANUF (→Fertigungsspektrum)
 (→production range)
manufacturing schedule MANUF **Fertigungsplan**
 = production schedule
manufacturing scheduling MANUF (→Zeitwirtschaft)
 (→time management)
manufacturing secret ECON **Fertigungsgeheimnis**
manufacturing sequence MANUF (→Fertigungsablauf)
 (→production flow)
manufacturing site MANUF **Fertigungsstandort**
manufacturing specification TECH (→Bauvorschrift)
 (→construction specification)
manufacturing stage MANUF (→Fertigungsstufe)
 (→production stage)
manufacturing start MANUF (→Fertigungsaufnahme)
 (→production start)
manufacturing technology MANUF **Fertigungstechnik**
 = production technique = Produktionstechnik
 ≈ production method ≈ Fertigungsverfahren
manufacturing time MANUF **Fertigungszeit**
manufacturing under licence ECON **Nachbau**
manufacturing year MANUF **Baujahr**
manuscript LING **Manuskript**
manuscript holder (→original TERM&PER (→Konzepthalter)
 holder)
man-year ECON **Mann-Jahr**
 = work-year
many-valued (→ambiguous) MATH (→mehrdeutig)
map (n.) (→mapping) MATH (→Abbildung)
map (n.) (→figure) DOC (→Bild)
map (→geographical map) GEOPHYS (→Landkarte)
map (→memory map 1) DATA PROC (→Speicherabbild)
mapped (→memory-mapped) DATA PROC (→speicherkonform)
mapper TELEC **Digitalhierarchie-Umsetzer**
 = digital hierarchy converter = Mapper
mapping MATH **Abbildung**
 [set theory] [Mengenlehre]
 ≈ map (n.) ≈ Funktion; Operation; Verknüpfung; Operator; Funktional; Morphismus; Funktor
 ≈ function; operation; operator; functional; functor
mapping TELEC **Digitalhierarchieumsetzung**
 [e.g. from 140 to 155 Mbit/s] [z.B. von 140 auf 155 Mbit/s]
 = digital hierarchy conversion
mapping DATA PROC **Umsetzung**
mapping (→memory-mapped DATA PROC (→speicherkonforme Bildschirmanzeige)
 video)
mapping PROM DATA PROC **Abbildungs-PROM**
MAR (→memory address DATA PROC (→Speicheradressregister)
 register)
marble CHEM **Marmor**
Marconi antenna ANT **Marconi-Antenne**
 ↑ monopole; vertical antenna ↑ Monopol; Vertikalantenne
Marconi-Franklin antenna ANT **Marconi-Franklin-Antenne**
 = Franklin antenna = Franklin-Antenne
margin (n.) COLLOQ **Spielraum**
 [fig.] [fig.]
 = leeway; elbowroom; allowance
margin (n.) TELEGR **Spielraum**
margin TECH **Rand**
 = border; limb ≈ Grenze
 ≈ limit
margin 1 TYPOGR **Seitenrand**
 [unprinted margin of a page] = Rand; Papierrand
margin ECON **Spanne**
 = spread = Marge
 ≈ profit ≈ Gewinn
 ↓ Vertriebsspanne; Gewinnspanne
margin 2 (→margin distance) TYPOGR (→Randabstand)
margin adjust (→margin TYPOGR (→Randausgleich)
 compensation)
margin adjustment (→margin TYPOGR (→Randausgleich)
 compensation)
marginal breakdown MICROEL **Randdurchbruch**
marginal checking ELECTRON **Grenzwertprüfung**
marginal contribution ECON **Deckungsbeitrag**

marginal costs	ECON	Grenzkosten	marker amplitude	INSTR	Markenamplitude
= incremental costs		= Marginalkosten	marker beacon	RADIO NAV	Markierungsfunkfeuer
marginal distribution	MATH	Randverteilung	= marker		= Markierungsbake; Marker;
marginalia (n.pl.t.)	TYPOGR	Marginalie	↑ beacon		Einflugzeichen
= marginal notes		[Randeintrag außerhalb der Spalte]			↑ Bake
margin alignment (→margin compensation)	TYPOGR	(→Randausgleich)	marker bit	DATA PROC	Markierbit
			marker frequency	INSTR	Markenfrequenz
			marker function	INSTR	Markenfunktion
marginal punched card	TERM&PER	Randlochkarte	marker position	INSTR	Markenposition
= edge-notched card; edge-punched card; border-punched card; margin-notched card; tape card		= Kerblochkarte			= Markenstellung
			marker setting	INSTR	Markeneinstellung
					= Markensetzung
marginal resolution [optics]	PHYS	Randschärfe [Optik]	marker sweep	INSTR	Markenwobbelung
			marker value	INSTR	Markenwert
margin compensation	TYPOGR	Randausgleich	market (v.t.)	ECON	vertreiben
= margin adjustment; margin adjust; margin justification; margin alignment			= commercialize; distribute		= verkaufen 2; vermarkten
			market	ECON	Markt
margin distance	TYPOGR	Randabstand	market (n.) (→exchange 1)	ECON	(→Börse)
= margin 2			marketable life	ECON	Vermarktungslebensdauer
margin justification (→margin compensation)	TYPOGR	(→Randausgleich)	market agreement	ECON	Marktabsprache
					= Marktabrede
margin notch (→edge notch)	TERM&PER	(→Randkerbung)	market analysis	ECON	Marktanalyse
margin-notched card (→marginal punched card)	TERM&PER	(→Randlochkarte)	≈ marketing analysis		= Marktuntersuchung
					≈ Absatzanalyse
margin of error [acceptable number of errors]	DATA PROC	Fehlertoleranzgrenze [zulässige Fehlerzahl]	market controlling (→market dominating)	ECON	(→marktbeherrschend)
			market dominating	ECON	marktbeherrschend
margin-perforated	TERM&PER	randgelocht	= market controlling		
margin-perforated paper	TERM&PER	randgelochtes Papier	market-driven	ECON	marktgesteuert
margin perforation (→edge notch)	TERM&PER	(→Randkerbung)	market fragmentation	ECON	Marktzersplitterung
			market growth	ECON	Marktwachstum
margin release [permits the transgression of a fixed margin]	TERM&PER	Randlöser [ermöglicht die Überschreitung eines fixierten Randes]	marketing	ECON	Marketing
			≈ sales; sale		= Vermarktung
					≈ Vertrieb; Verkauf 1
		= Randauslöser	marketing (→sales department)	ECON	(→Vertriebsabteilung)
margin stop	TERM&PER	Randsteller	marketing department (→sales department)	ECON	(→Vertriebsabteilung)
Maria maluca antenna [triband TV receive antenna]	ANT	Maria-maluca-Antenne [Dreiband-TV-Empfangsantenne]			
			marketing experience	ECON	Vertriebserfahrung
marine control equipment	TECH	Schiffssteuergerät	marketing research (→market research)	ECON	(→Marktforschung)
marine electronics	ELECTRON	Schiffselektronik			
marine engineering (→shipbuilding industry)	TECH	(→Schiffbauindustrie)	market investigation (→market research)	ECON	(→Marktforschung)
			market licence (→market qualification)	ECON	(→Marktzulassung)
marine signaling equipment	TECH	Schiffssignalgerät			
maritime mobile satellite service	SAT.COMM	beweglicher Seefunkdienst über Satelliten	market qualification	ECON	Marktzulassung
			= market licence		
maritime radio	RADIO	Seefunk	market research	ECON	Marktforschung
maritime radio call	TELEC	Seefunkgespräch	= marketing research; market investigation		
maritime radionavigation service	RADIO NAV	Seenavigationsfunkdienst			
			marking	SWITCH	Markieren
maritime radio service	RADIO	Seefunkdienst	marking	AERON	Markierung
maritime radio station	RADIO	Seefunkstelle	≈ lightning		≈ Befeuerung
maritime transport	ECON	Seetransport	marking (→lettering)	EQUIP.ENG	(→Beschriftung)
= sea transport			marking (→identification)	TECH	(→Kennzeichnung)
mark (v.t.)	ECON	markieren	marking 2 (→highlight)	DATA PROC	(→Hervorhebung)
[a merchandise]		[eine Ware]	marking 3 (→flagging)	DATA PROC	(→Hinweiszeichensetzung)
		= signieren	marking current	TELEGR	Arbeitsstrom
mark	MICROEL	Marke	= open-circuit current		= Zeichenstrom
mark (→name)	TECH	(→bezeichnen)	≠ spacing current		≠ Ruhestrom
mark (v.t.) (→identify)	TECH	(→kennzeichnen) (v.t.)	marking frequency	TELEGR	Zeichenfrequenz
mark (→characteristic 1)	TECH	(→Merkmal)			= Stromschrittfrequenz
mark (→marking pulse)	TELEGR	(→Stromschritt)	marking label (→designation label)	TECH	(→Beschriftungsschild)
mark (v.t.) (→dial)	TELEC	(→wählen)			
mark (→sign)	LING	(→Zeichen)	marking of goods	ECON	Warenauszeichnung
mark (→highlight)	DATA PROC	(→hervorheben)			= Warenkennzeichnung
mark (→one)	CODING	(→Eins)	marking pen	TERM&PER	Markierstift
mark detection (→mark sensing)	TELEGR	(→Zeichentastung)			= markierender Stift
marked (→labeled)	TECH	(→gekennzeichnet)	marking plan (→lettering plan)	TECH	(→Beschriftungsplan)
marker	INSTR	Marke	marking post (→buried cable marker)	OUTS.PLANT	(→Kabelmerkstein)
= display marker		= Anzeigenmarke; Anzeigenmarkierung			
			marking pulse	TELEGR	Stromschritt
marker	SWITCH	Markierer	= mark pulse; mark polarity; mark; current condition; current polarity; polarity; current pulse		= Zeichenschritt 2; Zeichenlage; Zeichenpolarität; Zeichen; Telegrafierschritt
[central control device for pathfinding and throughconnection]		[zentrale Steuereinrichtung zur Wegsuche und Durchschaltung]			
			≈ signal condition A; signal condition Z; spacing current		≈ Startpolarität; Stoppolarität; Zeichenstrom
marker	RADIO	Eichpunktgeber			
marker (→tape mark)	TERM&PER	(→Bandmarke)	≠ spacing pulse		≠ Pausenschritt
marker (→buried cable marker)	OUTS.PLANT	(→Kabelmerkstein)	↑ unit interval; signal condition [DATA COMM]		↑ Zeichenelement; Zeichenlage [DATA COMM]
marker (→marker beacon)	RADIO NAV	(→Markierungsfunkfeuer)			
marker (→cursor)	TERM&PER	(→Schreibmarke)			

marking space (→lettering space)	EQUIP.ENG	(→Beschriftungsfeld)
marking spring [crossbar switch]	SWITCH	**Markierfeder** [Kreuzschienenschalter]
marking strip (→designation strip)	EQUIP.ENG	(→Beschriftungsstreifen)
marking system = labeling system	SIGN.ENG	**Auszeichnungssystem** = Markierungssystem; Beschriftungssystem
Markow chain	INF	**Markow-Kette**
mark polarity (→marking pulse)	TELEGR	(→Stromschritt)
mark pulse (→marking pulse)	TELEGR	(→Stromschritt)
mark reader [reads strokes or other marks on documents called mark sheets] = mark sensing device; mark sheet reader; optical mark reader ↑ document reader	TERM&PER	**Markierungsleser** [liest Striche oder andere Markierungen auf Markierungsbelegen] = optischer Markierungsleser ↑ Belegleser
mark sense card [with sense fields to put machine-readable marks]	TERM&PER	**Zeichenlochkarte** [mit Feldern zum Anbringen maschinenlesbarer Markierungen]
mark sense character	TERM&PER	**Markierungszeichen**
mark sensing = mark detection	TELEGR	**Zeichentastung** = Zeichenabfühlung
mark sensing	TERM&PER	**Markierungslesen** = Markierungsabtastung
mark sensing device (→mark reader)	TERM&PER	(→Markierungsleser)
mark sheet [with boxes to be marked]	TERM&PER	**Markierungsbeleg** [zum Anstreichen von Kästchen] = Strichmarkierungsbeleg
mark sheet reader (→mark reader)	TERM&PER	(→Markierungsleser)
mark-to-space ratio	TELEGR	**Zeichen-Pausen-Verhältnis** = Puls-Pausen-Verhältnis; Impulstastverhältnis
martensite = martensitic steel	METAL	**martensitischer Stahl**
martensitic steel (→martensite)	METAL	(→martensitischer Stahl)
m-ary	MATH	**m-wertig**
masculine (n.) = masculine gender	LING	**Maskulinum 2** = männliches Geschlecht
masculine gender (→masculine)	LING	(→Maskulinum 2)
masculine substantive (n.)	LING	**Maskulinum 1** = männliches Substantiv
maser [microwave amplification by stimulated emission of radiation]	MICROW	**Maser** = Molekularverstärker
MASFET = metal alumina silicon field-effect transistor	MICROEL	**MASFET**
masher	TECH	**Stampfer**
mask (v.t.) ≈ cover 1 (v.t.)	TECH	**verkleiden** ≈ zudecken
mask (n.) ≈ template	TECH	**Maske** ≈ Schablone
mask (n.) [software to filter character patterns] = extractor; picture (COBOL)	DATA PROC	**Maske 1** [Software zum Ausblenden bestimmter Zeichenfolgen]
mask (→mask-off)	ELECTRON	(→ausblenden)
mask (→display mask)	DATA PROC	(→Bildschirmmaske)
mask (→diffusion mask)	MICROEL	(→Diffusionsmaske)
mask (n.) (→mask word)	DATA PROC	(→Maskenwort)
maskable	DATA PROC	**maskierbar**
maskable interrupt	DATA PROC	**maskierbare Unterbrechung** = maskierbares Interrupt; maskenunterdrückbare Unterbrechung
mask alignment	MICROEL	**Maskenjustierung**
mask-based	DATA PROC	**maskengesteuert**
mask-based user guidance	DATA PROC	**maskengesteuerte Bedienerführung**
mask bit	DATA PROC	**Maskenbit**
mask byte	DATA PROC	**Maskenbyte**
mask design	MICROEL	**Maskenentwurf**
masked	MICROEL	**maskenprogrammiert**
masked read-only memory = masked ROM	MICROEL	**maskenprogrammierter Festwertspeicher**
masked ROM (→masked read-only memory)	MICROEL	(→maskenprogrammierter Festwertspeicher)
mask handling	DATA PROC	**Maskenhantierung**
masking	MICROEL	**Maskierung**
masking = trim ≈ cover (n.)	TECH	**Verkleidung** ≈ Abdeckung
masking layer	MICROEL	**Maskierungsschicht**
masking-off (→blanking)	ELECTRON	(→Austastung)
masking-out (→blanking)	ELECTRON	(→Austastung)
masking procedure	MICROEL	**Maskierverfahren**
masking step	MICROEL	**Maskierschritt**
mask making	MICROEL	**Maskenherstellung** = Maskenerstellung
mask-off = mask-out; mask; blind-out	ELECTRON	**ausblenden** [Impuls]
mask-out (→mask-off)	ELECTRON	(→ausblenden)
mask programming	MICROEL	**Anwenderprogrammierung** [Mikroprozessor] = Maskenprogrammierung
mask selection	DATA PROC	**Maskenauswahl**
mask set	MICROEL	**Maskensatz**
mask word = mask (n.)	DATA PROC	**Maskenwort**
Mason formula	MICROEL	**Mason-Formel**
masonry	CIV.ENG	**Mauerwerk**
mass [SI unit: kilogram] = m ≈ weight	PHYS	**Masse** [SI-Einheit: Kilogramm] = m ≈ Gewicht
massage (v.t.) (→process)	DATA PROC	(→verarbeiten)
mass communication medium	ECON	**Massenmedium**
mass defect	PHYS	**Massendefekt**
mass density = density 2	PHYS	**Massendichte** = Dichte 2
mass memory = bulk memory; bulk storage; storage memory; mass storage ≈ magnetic disk memory ↑ external memory	DATA PROC	**Massenspeicher** = Großraumspeicher; Magnetplattenspeicher ↑ externer Speicher
mass number (→atomic mass number)	PHYS	(→Massenzahl)
mass point	PHYS	**Massenpunkt**
mass production = large-scale production	MANUF	**Massenproduktion** = Massenfertigung; Massenherstellung
mass-production good	MANUF	**Massengut**
mass spectrograph = mass spectrometer	PHYS	**Massenspektrograph**
mass spectrometer (→mass spectrograph)	PHYS	(→Massenspektrograph)
mass storage (→mass memory)	DATA PROC	(→Massenspeicher)
mass storage system	DATA PROC	**Massenspeichersystem**
mass transportation	RAILW.SIGN	**Nahverkehr**
mast [high pole or structure to mount sails, flags, lines, illuminations, tents etc.] ≈ tower; pillar ↓ mastil	CIV.ENG	**Mast** [hochragende Stange oder pfeilerähnliche Struktur, aus Holz, Metall oder Beton, zur Befestigung von Segeln, Fahnen, Leitungen, Beleuchtungen, Zeltplanen u.dgl.] ≈ Turm; Pfeiler
mast = pole ↓ fixture; antenna mast	OUTS.PLANT	**Mast** ↓ Freileitungsmast; Antennenmast
master (→master craftsman)	ECON	(→Handwerksmeister)
master (→gate array master)	MICROEL	(→Gate-array-Master)
master (→host computer)	DATA COMM	(→Hauptrechner)
master cashbox	TERM&PER	**Leitkasse**
master clear	DATA PROC	**Hauptlöschschalter**
master clock [for Standard Time]	INSTR	**Hauptuhr** [für Standardzeit]

master clock (→reference clock)	TELEC	(→Bezugstaktgeber)
master clock (→main clock)	ELECTRON	(→Haupttakt)
master clock (→timing generator)	CIRC.ENG	(→Taktgeber)
master clock pulse	TELEC	Haupttakt = Mastertakt
master computer [main computer in a multiprocessor system]	DATA PROC	Leitrechner [Hauptrechner in einem Mehrrechnersystem]
master computer (→host computer)	DATA COMM	(→Hauptrechner)
master condition	TECH	Rahmenbedingung
master control	CONTROL	Führungssteuerung
master copy = first-generation image	DOC	Stammkopie = Hauptkopie
master craftsman = master	ECON	Handwerksmeister = Meister 1
master data [infrequently or never altered data] ≈ inventory data ≠ variable data	DATA PROC	Stammdaten [selten oder überhaupt nicht sich ändernde Daten] ≈ Bestandsdaten ≠ variable Daten
master degree = master's degree; first degree (BRI); higher degree	SCIE	Diplom [akademischer Grad]
master disk [with masters of programs]	TERM&PER	Masterplatte [mit Masterfassungen von Programmen]
master diskette (→master floppy disk)	TERM&PER	(→Masterdiskette)
master distributor (→main distribution frame)	TELEC	(→Hauptverteiler)
master drawing	DOC	Reinzeichnung
master exchange (→parent exchange)	SWITCH	(→Mutteramt)
master file 2 = main file (BRI)	DATA PROC	Bestandsaufnahme
master file 1 ≠ transaction file	DATA PROC	Stammdatei = Hauptdatei
master floppy disk = master diskette	TERM&PER	Masterdiskette
master gage	TECH	Urlehre
master gear	MECH	Lehrzahnrad
mastergroup 1 [FDM, 600 channel group by US standard]	TRANS	Mastergruppe [TF-Technik, 600-Kanal-Gruppe nach US-Norm]
mastergroup 2 [FDM] = teriary group	TRANS	Tertiärgruppe [TF-Technik] = TG
mastergroup modulator	TRANS	Tertiärgruppenumsetzer = TGU
master index	DATA PROC	Hauptindex
master mask	MICROEL	Muttermaske
master menu (→main menu)	DATA PROC	(→Hauptmenü)
master office (→parent exchange)	SWITCH	(→Mutteramt)
master oscillator	CIRC.ENG	Hauptoszillator
master print	TERM&PER	Druckoriginal
master processor (→main processor)	MICROEL	(→Hauptprozessor)
master processor (→host computer)	DATA COMM	(→Hauptrechner)
master program (→control program)	DATA PROC	(→Organisationsprogramm)
master record	DATA PROC	Stammsatz
master's degree (→master degree)	SCIE	(→Diplom)
master-slave computer system	DATA PROC	Master-slave-Rechnersystem
master-slave flipflop	CIRC.ENG	Master-slave-Flipflop
master-slave JK flipflop	CIRC.ENG	Master-slave-JK-Flipflop
master-slave synchronization	TELEC	despotische Synchronisierung = Master-slave-Synchronisierung
master slice IC	MICROEL	Zellenbaustein = Master-slice-Baustein
master-standby (→operation-standby)	TECH	(→Betrieb-Ersatz)
master station	TELEGR	Hauptstelle
master station (→master transmitter)	BROADC	(→Bezugssender)
master station (→transmitting terminal)	DATA COMM	(→Sendestation)
master station (→main station)	TELECONTR	(→Zentrale)
master station location	TELECONTR	Zentralstelle [Ort mit Fernwirkzentrale]
master switch	POWER SYS	Generalschalter = Generalausschalter
master switch = main switch	EL.INST	Hauptschalter
master tape 2	DATA PROC	System-Urband = Master-Band
master tape 1	DATA PROC	Stammband
master telephone system = transmission reference system	TELEPH	Ureichkreis
master telephone transmission reference system	TELEPH	Fernsprech-Ureichkreis
master thesis	SCIE	Diplomarbeit
master transmitter = master station	BROADC	Bezugssender = Muttersender
master transmitter = main transmitter	RADIO NAV	Hauptsender
mast fundament ≈ tower fundament	CIV.ENG	Mastfundament ≈ Turmfundament
mat (n.)	TECH	Matte
mat (adj.) = matt; matte; lusterless; dull 1 ≈ turbid	TECH	matt = glanzlos ≈ trübe
mat (v.t.)	TECH	mattieren
mat black	TECH	mattschwarz
match (v.t.)	NETW.TH	anpassen
match (v.t.)	DATA PROC	auf Gleichheit prüfen = gleichheitsprüfen
match 1 (→matching 1)	NETW.TH	(→Anpassung 1)
matchbox (→tuner)	ANT	(→Anpaßgerät)
match code	DATA PROC	Match-Code = Abgleichcode
matched = with matched load; terminated by matched load	EL.TECH	angepaßt = angepaßt abgeschlossen
matched	TECH	angepaßt [verändert]
matched dipole	ANT	direkt angepaßter Dipol
matched feeder = flat line	ANT	angepaßte Speiseleitung = Flat-line
matched filter	NETW.TH	Wurzel-Nyquist-Filter = angepaßtes Filter
matched line ↑ terminated line	LINE TH	angepaßte Leitung ↑ abgeschlossene Leitung
matched response	NETW.TH	angepaßter Frequenzgang
matching (adj.) = mating	TECH	paarig
matching 1 [optimization of energy or signal transfer between networks, by adapting the impedances] = impedance matching; match 1; adaptation 1 ↓ power matching; overmatching; undermatching; image matching; resonance matching; active power matching	NETW.TH	Anpassung 1 [Optimierung der Energie- oder Signalübertragung zwischen Netzwerken, durch Anpassung der Scheinwiderstände] = Scheinwiderstandsanpassung; Impedanzanpassung; Wirkleistungsanpassung ↓ Leistungsanpassung; Spannungsanpassung; Stromanpassung; Wellenanpassung; Resonanzanpassung
matching	DATA PROC	Gleichheitsprüfung = Paarigkeitsvergleich
matching 2 (→power matching)	NETW.TH	(→Leistungsanpassung)
matching attenuation (→active return loss)	NETW.TH	(→Reflexionsdämpfung)
matching circuit = adapter	CIRC.ENG	Anpassungsschaltung = Anpassung
matching loss (→internal blocking)	SWITCH	(→interne Blockierung)
matching loss (→active return loss)	NETW.TH	(→Reflexionsdämpfung)
matching pad = matching section	NETW.TH	Anpassungsglied

matching section

matching section (→matching pad)	NETW.TH	(→Anpassungsglied)	
matching stub	ANT	**Abstimmleitung**	
matching transformer	EL.TECH	**Anpassungsübertrager**	
		= Anpassungstransformator; Anpassungstrafo	
matching unit (→tuner)	ANT	(→Anpaßgerät)	
match maker	ANT	**Matchmaker**	
[to measure active resistance]		[für Wirkwiderstandsmessungen]	
mate (v.t.)	MECH	**paaren**	
[parts]		[Paßteile]	
material	TECH	**Werkstoff**	
≈ raw material		= Material	
		≈ Rohstoff	
material bill (→parts list)	MANUF	(→Stückliste)	
material consumption	ECON	**Materialverbrauch**	
= material usage		≈ Materialeinsatz	
≈ material deployment			
material defect	QUAL	**Materialfehler**	
material deployment	ECON	**Materialeinsatz**	
≈ material consumption		≈ Materialverbrauch	
material dispersion	OPT.COMM	**Materialdispersion**	
material fatigue	QUAL	**Materialermüdung**	
material flow	MANUF	**Materialfluß**	
material flow (→material migration)	PHYS	(→Materialwanderung)	
materialization (→realization)	TECH	(→Realisierung)	
materialize (→realize)	TECH	(→realizieren)	
material list (→parts list)	MANUF	(→Stückliste)	
material migration	PHYS	**Materialwanderung**	
= material flow		= Materialfluß	
material requirement	MANUF	**Materialanforderung**	
materials administration	MANUF	**Materialwirtschaft**	
materials analysis	TECH	**Materialanalyse**	
≈ material testing		≈ Materialprüfung	
materials requisition	MANUF	**Materialentnahme**	
material store	MANUF	**Materiallager**	
material testing	TECH	**Materialprüfung**	
≈ material analysis		≈ Werkstoffprüfung	
		≈ Materialanalyse	
material usage (→material consumption)	ECON	(→Materialverbrauch)	
math capability	INSTR	**Rechenfunktion**	
mathematical	MATH	**mathematisch**	
mathematical expression	MATH	**mathematischer Ausdruck**	
mathematical logic	MATH	**mathematische Logik**	
mathematical physics	PHYS	**mathematische Physik**	
mathematical software	DATA PROC	**mathematische Software**	
mathematical symbol	MATH	**mathematisches Zeichen**	
↓ plus sign; minus sign; plus-minus sign; multiply sign; division sign; equal sign; identical sign; approximately-equal sign; greater-than sign; greater-or-equal sign; lees-than sign; less-or-equal sign; integral sign; root symbol; infinity sign; sum sign; product sign		↑ Zeichen ↓ Pluszeichen; Minuszeichen; Plus-Minus-Zeichen; Multiplikationszeichen; Divisionszeichen; Gleichheitszeichen; Identitätszeichen; Ungefähr-gleich-Zeichen; Größerzeichen; Größer-gleich-Zeichen; Wesentlicgrößer-Zeichen; Kleinerzeichen; Kleiner-gleich-Zeichen; Wesentlich-kleiner-Zeichen; Integralzeichen; Wurzelzeichen; Unendlichzeichen; Summenzeichen; Produktzeichen	
mathematical tool	SCIE	**mathematisches Hilfsmittel**	
mathematician	SCIE	**Mathematiker**	
mathematics	SCIE	**Mathematik**	
mathematics coprocessor (→maths coprocessor)	DATA PROC	(→mathematischer Koprozessor)	
maths chip (→maths coprocessor)	DATA PROC	(→mathematischer Koprozessor)	
maths coprocessor	DATA PROC	**mathematischer Koprozessor**	
= maths chip; mathematics coprocessor; number cruncher		= mathematischer Koprozessor; mathematischer Hilfsprozessor; Numerikprozessor; Arithmetik-Prozessor	
mating (n.)	MECH	**Paarung**	
[of parts]		[von Paßteilen]	
mating (adj.)	TECH	**passend**	
mating (→matching)	TECH	(→paarig)	
mating component (→counterpart)	TECH	(→Gegenstück)	
mating connector	COMPON	**Gegenstecker**	
mating part	MECH	**Paßteil**	
		= Paßstück	
mating part (→counterpart)	TECH	(→Gegenstück)	
matrix (pl.-ces; -xes)	MATH	**Matrix** (pl.-izen)	
matrix (pl.-ces;-xes)	SWITCH	**Matrix**	
matrix (→switching matrix)	SWITCH	(→Koppelvielfach)	
matrix (→array 1)	DATA PROC	(→Datenfeld 2)	
matrix array	ANT	**Matrixantenne**	
matrix calculus	MATH	**Matrizenrechnung**	
matrix character	TERM&PER	**Matrixzeichen**	
matrix circuit	CIRC.ENG	**Matrixschaltkreis**	
matrix decoder	CIRC.ENG	**Matrixdecodierer**	
matrix element	MATH	**Matrixelement**	
matrix equation	MATH	**Matrizengleichung**	
matrix function	MATH	**Matrixfunktion**	
matrix memory	DATA PROC	**Matrixspeicher**	
= matrix storage			
matrix memory (→display store)	TERM&PER	(→Bildspeicher)	
matrix notation (→matrix representation)	MATH	(→Matrixdarstellung)	
matrix-oriented	DATA PROC	**rasterorientiert**	
matrix printer (→dot-matrix printer)	TERM&PER	(→Rasterdrucker)	
matrix representation	MATH	**Matrixdarstellung**	
= matrix notation		= Matrizenschreibweise	
matrix screen (→raster screen)	TERM&PER	(→Rasterbildschirm)	
matrix storage (→matrix memory)	DATA PROC	(→Matrixspeicher)	
matrix store (→display store)	TERM&PER	(→Bildspeicher)	
matrix switching relay (→switching relay)	SWITCH	(→Koppelrelais)	
matrix switch path	SWITCH	**Koppelfeldweg**	
matrix transformation	MATH	**Matrixtransformation**	
matrix wire	SWITCH	**Koppelader**	
matt (→mat)	TECH	(→matt)	
matte (→mat)	TECH	(→matt)	
matter	OFFICE	**Vorgang**	
= topic; transaction		= Bearbeitungsvorgang	
≈ file		≈ Akte	
matter	PHYS	**Materie**	
≈ substance		≈ Substanz	
matter wave (→de Broglie wave)	PHYS	(→Materialwelle)	
mattress wiring	EQUIP.ENG	**Mattenverdrahtung**	
mature (adj.)	TECH	**erprobt**	
		= ausgereift; eingelaufen	
mature (adj.)	COLLOQ	**ausgereift**	
[fig.]		[fig.]	
matured (→due)	ECON	(→fällig)	
mature technology	TECH	**ausgereifte Technik**	
		= ausgereifte Technologie	
maturity (→expiration)	ECON	(→Verfall)	
maturity date (→due time)	ECON	(→Fälligkeit)	
maturity for series production	MANUF	**Serienreife**	
= full-production status			
MAVAR (→parametric amplifier)	CIRC.ENG	(→parametrischer Verstärker)	
maximable	SCIE	**maximierbar**	
maximal deviation	CONTROL	**Maximalabweichung**	
= peak deviation; crest deviation			
maximally flat filter	NETW.TH	**Potenzfilter**	
= Butterworth filter		= Butterworth-Filter; maximal flaches Filter	
maximization problem	SCIE	**Maximierungsproblem**	
maximize	MATH	**maximieren**	
↑ extremize		↑ extremieren	
maximum ...	COLLOQ	**Höchst-**	
		= Maximal-	
maximum (adj.) (→maximum permissible)	TECH	(→maximal zulässig)	

maximum amount	ECON	**Höchstbetrag**
		= Maximalbetrag
maximum capacity	TECH	**Endausbau**
= maximum configuration; ultimate configuration; final equipment; full expansion; final capacity; full capacity stage		= Maximalausbau; Vollausbau; Höchstausbau; Endausbaustufe
maximum clearance	ENG.DRAW	**Größtspiel**
maximum configuration ($\rightarrow$ maximum capacity)	TECH	($\rightarrow$ Endausbau)
maximum-configuration version ($\rightarrow$ full-bore version)	TECH	($\rightarrow$ voll ausgebaute Version)
maximum diameter	MECH	**Größtdurchmesser**
maximum dimension ($\rightarrow$ high limit)	ENG.DRAW	($\rightarrow$ Größtmaß)
maximum duty ($\rightarrow$ peak load)	TECH	($\rightarrow$ Belastungsspitze)
maximum forward blocking voltage	ELECTRON	**Nullkippspannung**
maximum likelihood method	MATH	**Maximum-likelihood-Methode**
		= Methode der maximalen Mutmaßlichkeit
maximum limited stress	QUAL	**Grenzbeanspruchung**
= tolerated stress		
maximum load ($\rightarrow$ peak load)	TECH	($\rightarrow$ Belastungsspitze)
maximum modulating frequency	TV	**Bildfrequenz 1**
maximum oscillation frequency	ELECTRON	**Schwinggrenzfrequenz**
		= Schwinggrenze
maximum overshoot	ELECTRON	**Überschwingweite**
maximum permissible	TECH	**maximal zulässig**
= maximum (adj.)		
maximum principle	CONTROL	**Maximumprinzip**
maximum-radiation direction ($\rightarrow$ boresight)	ANT	($\rightarrow$ Hauptstrahlrichtung)
maximum radio direction finder	RADIO LOC	**Maximumpeiler**
		= Maximum-Funkpeiler
maximum steering	RADIO LOC	**Maximumpeilung**
$\neq$ null steering		$\neq$ Nullpeilung
maximum temperature	PHYS	**Höchsttemperatur**
		= Maximaltemperatur
		$\uparrow$ Grenztemperatur
maximum usable frequency [HF communications]	RADIO	**obere Grenzfrequenz** [HF-Funk]
= maximum useful frequency; MUF		= MUF
maximum useful frequency ($\rightarrow$ maximum usable frequency)	RADIO	($\rightarrow$ obere Grenzfrequenz)
maximum value	MATH	**Größtwert**
= peak value		= Maximalwert; Maximum; Höchstwert; Spitzenwert; Scheitelwert
$\uparrow$ extreme value		$\uparrow$ Grenzwert; Extremum
maximum-value limiter ($\rightarrow$ peak-value limiter)	ELECTRON	($\rightarrow$ Höchstwertbegrenzer)
maxterm	ENG.LOG	**Volldisjunktion**
$\uparrow$ term		= Maxterm
		$\uparrow$ Term
Maxwell [unit for magnetic flux]	PHYS	**Maxwell** [Einheit für magnetischen Fluß]
Maxwell-Boltzmann distribution	PHYS	**Maxwell-Boltzmann-Verteilung**
Maxwell-Boltzmann statistics	PHYS	**Maxwell-Boltzmann-Statistik**
Maxwell bridge	INSTR	**Maxwell-Brücke**
= Maxwell inductance bridge		= Maxwellsche Induktivitätsbrücke; LLRR-Brücke
Maxwell DC commutator bridge	INSTR	**Maxwellsche Kommutatorbrücke**
Maxwellian distribution ($\rightarrow$ Maxwellian velocity distribution)	PHYS	($\rightarrow$ Maxwellsche Geschwindigkeitsverteilung)
Maxwellian velocity distribution	PHYS	**Maxwellsche Geschwindigkeitsverteilung**
= Maxwellian distribution		
Maxwell inductance bridge ($\rightarrow$ Maxwell bridge)	INSTR	($\rightarrow$ Maxwell-Brücke)
Maxwell mutual inductance bridge	INSTR	**Maxwellsche Gegeninduktivitätsbrücke**
Maxwell's equations	PHYS	**Maxwell-Gleichungen**
		= Maxwellsche Gleichungen
Maxwell's law	PHYS	**Maxwellsche Relation**
Maxwell-Wien bridge	INSTR	**Maxwell-Wien-Brücke**
		= LRRC-Brücke
Maxwellian theory	PHYS	**Maxwellsche Theorie**
mayor alarm ($\rightarrow$ urgent alarm)	EQUIP.ENG	($\rightarrow$ dringender Alarm)
mayor lobe	ANT	**Hauptkeule**
= main lobe; front lobe		= Hauptstrahlbereich; Hauptstrahlungskeule; Hauptmaximum
$\neq$ side lobe		$\neq$ Nebenzipfel
MBE ($\rightarrow$ molecular beam epitaxy)	MICROEL	($\rightarrow$ Molekularstrahlepitaxie)
Mbit ($\rightarrow$ megabit)	TELEC	($\rightarrow$ Megabit)
Mbyte ($\rightarrow$ megabyte)	INF	($\rightarrow$ Megabyte)
Md ($\rightarrow$ mendelevium)	CHEM	($\rightarrow$ Mendelevium)
MDA [Monochrome Display Adapter; monochrome, not graphicable video board]	TERM&PER	**MDA** = monochrome, nicht graphikfähige Videokarte
m-derived filter section	NETW.TH	**versteilertes Siebglied**
m-derived section	NETW.TH	**Versteilerungsglied**
		= Zobel-Glied; m-Glied; Wagner-Glied
MDF ($\rightarrow$ main distribution frame)	TELEC	($\rightarrow$ Hauptverteiler)
MDR ($\rightarrow$ memory data register)	DATA PROC	($\rightarrow$ Speicherdatenregister)
MDR system	TV	**MDR-SYSTEM**
= magnetic disc recorder system		
MDT ($\rightarrow$ mean down time)	QUAL	($\rightarrow$ mittlere Ausfalldauer)
mealy automat	ENG.LOG	**endlicher Automat**
mean (adj.)	MATH	**durchschnittlich**
= average (adj.)		= mittlerer
mean ($\rightarrow$ average)	MATH	($\rightarrow$ Mittel)
mean access time	DATA PROC	**mittlere Zugriffszeit**
mean call time	SWITCH	**mittlere Belegungsdauer**
mean consumption	TECH	**Durchschnittsverbrauch**
		= mittlerer Verbrauch; durchschnittlicher Verbrauch
mean date rate	DATA COMM	**mittlere Übertragungsgeschwindigkeit**
= average transmission rate		
meander	TECH	**Mäander**
[sequence of tightly contiguous windings]		[dicht aneinanderliegende Windungen]
$\approx$ coil		$\approx$ Schlange
meander-line antenna	ANT	**Mäanderantenne**
$\approx$ Bruce antenna; Sterba array		$\approx$ Bruce-Antenne; Sterba-Antenne
mean down time	QUAL	**mittlere Ausfalldauer**
= MDT		= MDT
$\approx$ MTTR		$\approx$ MTTR
meandrous	TECH	**mäanderförmig**
mean free path	PHYS	**mittlere freie Weglänge**
mean Greewich time	PHYS	**Weltzeit**
= G time; Zulu time		
mean information content ($\rightarrow$ information entropy)	INF	($\rightarrow$ Informationsentropie)
mean interconnecting number	SWITCH	**Mischungsverhältnis**
mean life (IEC)	QUAL	**mittlere Lebensdauer**
[sum of all life times divided by quantity]		[Summe aller Lebensdauer durch Anzahl]
mean surface	MECH	**Mitteloberfläche**
		[Oberflächengüte]
mean time between failures	QUAL	**mittlerer Ausfallabstand**
= MTBF		= MTBF; mittlerer Fehlerabstand; mittlere ausfallfreie Zeit
mean time to repair ($\rightarrow$ MTTR)	QUAL	($\rightarrow$ MTTR)
mean value ($\rightarrow$ average)	MATH	($\rightarrow$ Mittel)
mearurement	PHYS	**Messung**
= measuring; test [INSTR]		= Messen
$\approx$ test [QUAL]		$\approx$ Prüfung [QUAL]
measurable	PHYS	**messbar**
measurable quantity ($\rightarrow$ measurand)	PHYS	($\rightarrow$ Meßgröße)
measurable variable ($\rightarrow$ measurand)	PHYS	($\rightarrow$ Meßgröße)
measurand	PHYS	**Meßgröße**
= quantity under test; measured quantity; measurable variable; measurable quantity		= Beobachtungsgröße
		= Meßwert
measurand transducer	INSTR	**Meßumformer**
= transmitter		= Meßgrößenumformer; Meßwertumformer

measure

measure (v.t.)	PHYS	messen	
≈ test (v.t.)		≈ prüfen	
measure (n.)	PHYS	Maß	
= size (n.)			
measure (n.)	COLLOQ	Maßnahme	
≈ precaution		≈ Vorkehrung	
measured	MECH	abgemessen	
measured curve	PHYS	Meßkurve	
measured-data processing	INSTR	(→Meßwertverarbeitung)	
(→measured-value processing)			
measured quantity	PHYS	(→Meßgröße)	
(→measurand)			
measured value	PHYS	Meßwert	
= experimental value; value		= Wert	
≈ measurand; measurement result		≈ Meßgröße: Meßergebnis	
measured value (→test result)	INSTR	(→Meßwert)	
measured-value processing	INSTR	Meßwertverarbeitung	
= measured-data processing; test-value processing; test result processing			
measurement (→inspection)	QUAL	(→Prüfung)	
measurement (→test)	INSTR	(→Messung)	
measurement accuracy	INSTR	(→Meßgenauigkeit)	
(→measuring precision)			
measurement amplifier	INSTR	Meßverstärker	
= measuring amplifier			
measurement concept	INSTR	Prüfkonzept	
measurement current	INSTR	Meßstrom	
= current through unknown			
measurement data	INSTR	Meßdaten	
		= Meßwerte	
measurement device (→test arrangement)	INSTR	(→Meßaufbau)	
measurement equipment	INSTR	Meßapparatur	
≈ measurement instrument; test equipment		≈ Meßgerät; Prüfapparatur	
measurement gear	TECH	Meßzeug	
measurement method	INSTR	(→Meßverfahren)	
(→measurement procedure)			
measurement of pH value	CHEM	pH-Wert-Messung	
measurement practice	INSTR	(→Meßverfahren)	
(→measurement procedure)			
measurement probability	INSTR	Meßstochastik	
measurement procedure	INSTR	Meßverfahren	
= measurement practice; measurement technique; measurement method; measuring technique; measuring practice; measuring procedure; measuring method		= Meßmethode; Meßtechnik	
		≈ Prüfverfahren [QUAL]	
≈ test procedure [QUAL]			
measurement range	INSTR	Messbereich	
= measuring range; instrument range			
measurement requirement	TELEC	Meßauftrag	
= test requirement			
measurement requirement	INSTR	Meßanforderung	
measurement result	PHYS	Meßergebnis	
≈ test result; measured value		≈ Prüfergebnis; Meßwert	
measurement result (→test result)	INSTR	(→Meßwert)	
measurement setup (→test arrangement)	INSTR	(→Meßaufbau)	
measurements record	INSTR	Meßprotokoll	
= test record			
measurement system	PHYS	Maßsystem	
= system of units			
measurement task	INSTR	Meßaufgabe	
measurement technique	INSTR	(→Meßverfahren)	
(→measurement procedure)			
measurement throughput	INSTR	Meßdurchsatz	
measurement time	INSTR	Meßzeit	
≈ test time		≈ Prüfzeit	
measurement uncertainty	INSTR	(→Meßunsicherheit)	
(→measuring inaccuracy)			
measurement unit	PHYS	Maßeinheit	
		= Meßeinheit	
measure of capacity (→volumetric measure)	PHYS	(→Raummaß)	
measure of volume (→volumetric measure)	PHYS	(→Raummaß)	
measuring (→mearurement)	PHYS	(→Messung)	
measuring access point (→test point)	ELECTRON	(→Meßpunkt)	
measuring adapter	INSTR	Meßvorsatz	
= test adapter; test fixture		= Meßadapter; Prüfadapter; Testadapter	
measuring amplifier	INSTR	(→Meßverstärker)	
(→measurement amplifier)			
measuring appliance (→measuring gauge)	MECH	(→Meßlehre)	
measuring automat (→automatic measuring equipment)	INSTR	(→Meßautomat)	
measuring branch	CIRC.ENG	Meßzweig	
measuring bridge	INSTR	Messbrücke	
measuring capacitance	INSTR	Meßkondensator	
measuring chain	INSTR	Meßkette	
measuring chamber	INSTR	Meßraum	
≈ test room [QUAL]		≈ Prüfraum [QUAL]	
measuring chopper amplifier	INSTR	Zerhackermeßverstärker	
measuring circuit	CIRC.ENG	Meßschaltung	
measuring climate	QUAL	Meßklima	
measuring coil	INSTR	Meßspule	
measuring console	SWITCH	Meßplatz	
measuring current source	INSTR	Meßstromquelle	
		= Meßspannungsquelle	
measuring device	INSTR	Meßvorrichtung 2	
≈ testing device		≈ Meßgerät; Meßinstrument; Meßapparat; Prüfvorrichtung	
measuring device (→measuring set)	INSTR	(→Meßplatz)	
measuring device (→testing device)	INSTR	(→Prüfvorrichtung)	
measuring diode	ELECTRON	Meßdiode	
measuring director	INSTR	Meßdirektor	
		= steuernde Meßeinrichtung	
measuring electrode	INSTR	Meßelektrode	
= test electrode			
measuring equipment	INSTR	(→Meßplatz)	
(→measuring set)			
measuring error	PHYS	Meßfehler	
measuring facility (→measuring set)	INSTR	(→Meßplatz)	
measuring frequency	INSTR	Meßfrequenz	
≈ test frequency		≈ Prüffrequenz	
measuring function	INSTR	Meßfunktion	
= personality			
measuring gage (→measuring gauge)	MECH	(→Meßlehre)	
measuring gauge (n.)	MECH	Meßlehre	
= measuring gage; measuring appliance		≈ Prüflehre	
≈ checking appliance			
measuring generator (→signal generator 1)	INSTR	(→Universal-Meßsender)	
measuring head (→test probe)	INSTR	(→Prüfspitze)	
measuring inaccuracy	INSTR	Meßunsicherheit	
= measurement uncertainty		≈ Meßtoleranz	
≈ measuring tolerance			
measuring instrument	INSTR	Meßgerät	
= meter		= Meßinstrument; Messer	
≈ measurement equipment; measuring set; test instrument; test equipment; measuring set		≈ Meßapparatur; Meßvorrichtung; Prüfgerät; Prüfapparatur; Meßplatz	
↑ instrument		↑ Instrument	
measuring lead (→measuring line)	INSTR	(→Meßleitung)	
measuring line	INSTR	Meßleitung	
= test lead wire; test lead; measuring lead; test line		= Meßzuleitung; Prüfleitung; Prüfschnur; Meßschnur	
measuring loop	ELECTRON	Meßschleife	
≈ test loop		≈ Prüfschleife	
measuring mandrel	MECH	Meßdorn	
measuring method	INSTR	(→Meßverfahren)	
(→measurement procedure)			
measuring microphone (→standard microphone)		(→Meßmikrophon)	
measuring panel	EQUIP.ENG	Meßfeld	
= metering panel			
measuring point (→test point)	ELECTRON	(→Meßpunkt)	
measuring potentiometer	INSTR	Meßpotentiometer	

measuring practice		INSTR	(→Meßverfahren)		
(→measurement procedure)					
measuring precision		INSTR	**Meßgenauigkeit**		
= measurement accuracy; meter accuracy					
measuring principle		INSTR	**Meßprinzip**		
measuring probe (→test probe)		INSTR	(→Prüfspitze)		
measuring procedure		INSTR	**Meßverfahren**		
(→measurement procedure)					
measuring range (→measurement range)		INSTR	(→Messbereich)		
measuring range selector		INSTR	**Messbereichswahlschalter**		
measuring receiver (→test receiver)		INSTR	(→Meßempfänger)		
measuring rectifier		INSTR	**Meßgleichrichter**		
measuring resistance		COMPON	(→Normalwiderstand)		
(→precision resistor)					
measuring resistor (→precision resistor)		COMPON	(→Normalwiderstand)		
measuring responder		INSTR	**Meßresponder**		
= controlled measuring device			= gesteuerte Meßeinrichtung		
measuring result (→test result)		INSTR	(→Meßwert)		
measuring scale		INSTR	**Meßskala**		
measuring set		INSTR	**Meßplatz**		
[equipment with several measurement functions]			[Gerät mit mehreren zusammenwirkenden Meßfunktionen]		
= test position; measuring equipment; measuring facility; measuring device; instrumentation tool			= Meßvorrichtung 1; Meßeinrichtung		
≈ measuring instrument; test set; testing device; test tools; test arrangement			≈ Meßgerät; Prüfplatz; Prüfeinrichtung; Meßaufbau		
measuring signal (→test signal)		ELECTRON	(→Testsignal)		
measuring step		INSTR	**Meßschritt**		
measuring system		INSTR	**Meßwerk**		
measuring tape		TECH	**Maßband**		
			= Bandmaß		
measuring technique		INSTR	(→Meßverfahren)		
(→measurement procedure)					
measuring template		TECH	**Meßschablone**		
measuring thermistor		COMPON	**Meßheißleiter**		
			= Meßthermistor		
measuring tool		TECH	**Meßmittel**		
measuring transducer		COMPON	**Meßwandler**		
= instrument transformer			= Meßtransformator		
			≈ Wandler		
measuring transmitter		INSTR	**Meßsender 2**		
			= Hochfrequenz-Signalgenerator; Labormeßsender		
measuring voltage		INSTR	**Meßspannung**		
≈ test voltage [ELECTRON]			≈ Prüfspannung [ELECTRON]		
measuring watch tower		CONTROL	**Meßwarte**		
mechanic (n.)		TECH	**Monteur**		
= fitter; assembler			= Mechaniker		
mechanical (adj.)		TECH	**maschinell**		
≈ machine-			≈ mechanisch		
≈ automatic			≈ automatisch		
mechanical		PHYS	**mechanisch**		
mechanical characteristic		TECH	**konstruktives Merkmal**		
= mechanical feature; constructional characteristic; constructional feature; design characteristic					
mechanical component		EQUIP.ENG	**Konstruktionsteil**		
= structural part					
mechanical converter		POWER ENG	**Maschinenumformer**		
mechanical design		EQUIP.ENG	**Konstruktion**		
= construction; packaging; design			= konstruktiver Aufbau; Geräteaufbau		
≈ construction practice			≈ Bauweise; Aufbau 2		
mechanical drawing		ENG.DRAW	**Maschinenbauzeichnen**		
↑ engineering drawing 1			↑ technisches Zeichen		
mechanical energy		PHYS	**mechanische Energie**		
mechanical engineer		TECH	**Maschinenbauingenieur**		
mechanical engineering		TECH	**Maschinenbau**		
↑ engineering			↑ Technik		
mechanical equivalent of light		PHYS	**Lichtäquivalent**		
mechanical feature (→mechanical characteristic)		TECH	(→konstruktives Merkmal)		
mechanical filter		COMPON	**mechanisches Filter**		
= electro-mechanical filter			= elektromechanisches Filter		
mechanical impedance		MECH	**Standwert**		
[acting force to velocity]			[einwirkende Kraft zu Geschwindigkeit]		
			= mechanische Impedanz		
mechanical latching relay		COMPON	**Verklinkrelais**		
mechanically tunable		INSTR	**mechanisch abstimmbar**		
mechanical oscillation		MECH	(→mechanische Schwingung)		
(→vibration)					
mechanical pliers		TECH	**Mechanikerzange**		
mechanical printer (→impact printer)		TERM&PER	(→Anschlagdrucker)		
mechanical relay		COMPON	**mechanisches Relais**		
mechanical resistance		MECH	**mechanische Resistanz**		
[real component of mechanical impedance]			[Realteil des Standwertes]		
mechanical splice		OPT.COMM	**mechanischer Spleiß**		
≠ fused splice			= Klebespleiß		
			≠ thermischer Spleiß		
mechanical stress		MECH	**Spannung**		
[SI unit: Pascal]			[SI-Einheit: Pascal]		
= stress; tension; σ; strain			= mechanische Spannung; σ		
≈ pressure			≈ Zug; Druck		
mechanical translation		DATA PROC	**maschinelle Übersetzung**		
mechanical workshop		TECH	**mechanische Werkstatt**		
mechanics		PHYS	**Mechanik**		
mechanism		TECH	**Mechanismus**		
			= Werk		
mechanization		TECH	**Mechanisierung**		
media (n.pl.t.) (→communication media)		TELEC	(→Kommunikationsmedien) (n.pl.t.)		
media crush (→discontinuity of media)		DATA PROC	(→Medienbruch)		
media diversity		TRANS	**Mehrmedienführung**		
median 2 (n.)		MATH	**Medianwert**		
[stays in the center of a growing sequence of samples]			[in einer steigenden Reihung in der Mitte liegend]		
= medium value			= Zentralwert		
			≈ Mittelwert		
median 1 (n.)		MATH	**Seitenhalbierende**		
			= Mediane		
median life		QUAL	**zentrale Lebensdauer**		
[till 50 % have failed]			[Ausfall von 50 %]		
median offset		ELECTRON	**Mittelwert-Offset**		
media protection		TRANS	**Medienersatz**		
media specialist		DATA PROC	**Speichermedienspezialist**		
mediate		ECON	**vermitteln**		
mediation		ECON	**Vermittlung**		
mediation function		TELEC	**Vermittlungsfunktion**		
mediation device		TELEC	**Verbindungsglied**		
[TMN]			[TMN]		
mediator		ECON	**Vermittler**		
= intermediary					
medical engineering		TECH	(→Elektromedizin)		
(→electromedicine)					
mediocre		COLLOQ	**mittelmäßig**		
[of moderate quality]			[abwertend]		
≈ second grade; ordinary			≈ zweitklassig; gewöhnlich		
medium (adj.)		TYPOGR	**normal**		
= regular; normal; standard			↑ Schriftattribut		
↑ font attribute					
medium (→storage medium)		DATA PROC	(→Speichermedium)		
medium base directional antenna		ANT	**Mittelbasis-Richtstrahlantenne**		
medium-capacity system		TRANS	**Mittelbandsystem**		
medium fit		MECH	**Laufsitz**		
medium force fit		MECH	**Festsitz**		
medium frequencies		RADIO	(→Langwellen)		
(→kilometric waves)					
medium-frequency convertor		POWER ENG	**Mittelfrequenzumrichter**		
medium-heavy loading		OUTS.PLANT	**mittelschwere Bespulung**		
medium-high tension		POWER ENG	(→Mittelspannung)		
(→medium-high voltage)					
medium-high voltage		POWER ENG	**Mittelspannung**		
[250 to 1.000 V]			[250 bis 1.000 V]		
= medium-high tension; medium voltage; medium tension					

medium lettering TYPOGR **Mittelschrift**
medium-level management ECON **mittlerer Führungskreis**
= middle-level management; middle management
= mittleres Management; mittlere Führungsschicht
medium of exchange (→payment means) ECON (→Zahlungsmittel)
medium-range navigation RADIO NAV **Mittelstreckennavigation**
medium scale integration (→MSI) MICROEL (→Mittelintegration)
medium size TECH **mittlere Größe**
medium-sized TECH **mittelgroß**
medium-speed TECH **mittelschnell**
medium tension (→medium-high voltage) POWER ENG (→Mittelspannung)
medium-term ECON **mittelfristig**
medium-term plan ECON **mittelfristiger Plan**
medium value (→median 2) MATH (→Medianwert)
medium voltage (→medium-high voltage) POWER ENG (→Mittelspannung)
medium warm grey PHYS **dunkelgrau**
medium-wave ferrite METAL **MW-Ferrit**
medium waves (→hectometric waves) RADIO (→Mittelwellen)
meet (v.t.) ECON **tagen**
= sit (v.i.)
meet (→concur) TECH (→zusammentreffen)
meeting ECON **Tagung**
= congress
≈ conference
= Kongreß; Treffen
≈ Konferenz; Besprechung
meeting (→conference) ECON (→Besprechung)
meeting (→session) ECON (→Sitzung)
meeting place ECON **Tagungsort**
= venue
= Veranstaltungsort; Kongreßort
meet-me-conference call TELEC **Verabredungskonferenz**
= meet-me-conference connection; meet-me-conference link
meet-me-conference connection (→meet-me-conference call) TELEC (→Verabredungskonferenz)
meet-me-conference link (→meet-me-conference call) TELEC (→Verabredungskonferenz)
meg (v.t.) INSTR **hohe Widerstände messen**
megabit TELEC **Megabit**
[1,000,000 bit]
= Mbit
[1.000.000 Bit]
= Mbit
megabit DATA PROC **Megabit**
[1,048,575 bit]
[1.048.575 Bit]
= Mbit
megabyte INF **Megabyte**
[1,024 kbyte = 1.048.576 byte]
= Mbyte
[1.024 kByte = 1.048.576 Byte]
= MByte
megaflops DATA PROC **Megaflops**
[one million floating point operations per second; measure of computing power]
= MFLOPS
[eine Million Gleitpunktoperationen pro Sekunde; Maß für Computerleistung]
= MFLOPS
megahertz PHYS **Megahertz**
= MHz
= MHz
megaphone EL.ACOUS **Megaphon**
megapond PHYS **Megapond**
[= 9,806.650 N]
= Mp
[= 9.806,650 N]
= Mp
megawatt PHYS **Megawatt**
= MW
= MW
megtape (→magnetic tape) ELECTRON (→Magnetband)
Meissner oscillator CIRC.ENG **Meißner-Oszillator**
↑ LC oscillator
↑ LC-Oszillator
melt (v.t.) PHYS **schmelzen**
≈ liquify
≈ verflüssigen
melt (n.) MICROEL **Schmelze**
meltback MICROEL **Rückschmelzen**
melt-back technique (→remelt process) MICROEL (→Nachschmelzverfahren)
meltback transistor MICROEL **Rückschmelztransistor**
= Schmelzperlen-Transistor
melting PHYS **Schmelzen**
≈ liquefaction
= Schmelze
≈ Verflüssigung

melting fuse COMPON **Schmelzsicherung**
= fuse; safety fuse; fuze; fusible cut-out
↑ overcurrent protector
[schützt durch Abschmelzen eines Schmelzleiters]
= mittelträge Feinsicherung
↑ Stromsicherung
melting heat (→fusion heat) PHYS (→Schmelzwärme)
melting point PHYS **Schmelzpunkt**
= fusing point
≈ melting temperature
≈ Schmelztemperatur
melt-quench transistor MICROEL **Rückschmelz-Abschreck-Transistor**
melt zone MICROEL **Schmelzzone**
member price ECON **Mitgliedspreis**
= Mitgliederpreis
membership subscription ECON **Mitgliedsbeitrag**
= dues (pl.t.; NAM)
membrane TECH **Membran**
= diaphragm; septum
= Membrane; Scheidewand
membrane (→diaphragm) EL.ACOUS (→Membrane)
membrane keyboard TERM&PER **Folientastatur**
= pressure-sensitive keyboard
≈ touch-sensitive keyboard
≠ single-key keyboard
≈ Berührungstastatur
≠ Einzeltasten-Tastatur
memo (→memorandum) OFFICE (→Notiz)
memo pad (→note pad) OFFICE (→Notizblock)
memorandum OFFICE **Notiz**
[informal written record]
= memo; note
[informelle schriftliche Aufzeichnung]
= Aktenvermerk
memory DATA PROC **Speicher**
[device to store data and retrieve them on request; "storage" often has the connotation of external, higher capacity and slower access than "memory" and expresses also the process of storing]
= storage 2; store (BRI)
↓ main memory; external memory; random access memory; sequential access memory; volatile memory; non-volatile memory; solid state memory; electromagnetic memory
[Vorrichtung zum Festhalten von Daten, die bei Bedarf wieder abgerufen werden können]
↓ Hauptspeicher 1; externer Speicher; Direktzugriffsspeicher; sequentieller Speicher; flüchtiger Speicher; nichtflüchtiger Speicher; Halbleiterspeicher; elektromagnetischer Speicher
memory access DATA PROC **Speicherzugriff**
= storage access; store access
memory access time DATA PROC **Speicherzugriffszeit**
= storage access time; store access time
memory address DATA PROC **Speicheradresse**
= storage address; store address
memory address register DATA PROC **Speicheradressregister**
= MAR; storage address register; store address register
= MAR
memory allocation (→memory management) DATA PROC (→Speicherverwaltung)
memory area DATA PROC **Speicherbereich**
= storage area; store area; memory sector; storage sector; store sector; memory zone; storage zone; store zone; bucket
= Speicherfeld; Speicherzone
memory area protection DATA PROC **Speicherbereichsschutz**
= storage area protection
memory array MICROEL **Speicherfeld**
memory backup capacitor DATA PROC **Speicherkondensator**
memory bank DATA PROC **Speicherbank**
= storage bank; store bank
memory bar TERM&PER **Speicherbalken**
memory board DATA PROC **Speicherkarte**
= memory card
memory bus DATA PROC **Speicherbus**
memory byte DATA PROC **Speicherbyte**
= storage byte; store byte
memory capacity DATA PROC **Speicherkapazität**
= storage capacity; store capacity; memory space 1; storage space 1; store space 1
= Aufnahmekapazität
memory card (→memory board) DATA PROC (→Speicherkarte)
memory chip MICROEL **Speicherbaustein**
= memory device; memory component; storage chip; storage device; storage
= Speicherchip; Speicherbauelement

component; store chip; store device; store component
↑ chip
↓ read-only memory; PROM; EEPROM; EPROM; FROM 1; FROM 2; read-write memory; random access memory; DRAM; SRAM
memory component (→memory chip) MICROEL
memory contents DATA PROC **Speicherinhalt**
= storage contents; store contents
memory control DATA PROC **Speichersteuerung**
= storage control; store control
memory cycle DATA PROC **Speicherzyklus**
= storage cycle; store cycle
memory-cycle counter DATA PROC **Speicherzyklus-Zähler**
= storage-cycle counter; store-cycle counter
memory cycle request DATA PROC **Speicherzyklus-Anforderung**
= storage-cycle request; store-cycle request
memory cycle time DATA PROC **Speicherzykluszeit**
↑ cycle time ↑ Zykluszeit
memory data DATA PROC **Speicherdaten**
= storage data; store data
memory data register DATA PROC **Speicherdatenregister**
= MDR; storage data register; store data register = MDR
memory device (→memory chip) MICROEL (→Speicherbaustein)
memory dump DATA PROC **Speicherabzug**
[output of memory content on printer or display] [Ausgabe des Speicherinhalts auf Drucker oder Bildschirm]
= storage dump; store dump; post-mortem dump = Speicherauszug
≈ memory map 1 ≈ Speicherabbild
↑ printout ↑ Ausdruck
↓ main-memory dump; magnetic tape memory dump; disk memory dump ↓ Arbeitsspeicherabzug; Magnetbandauszug; Plattenabzug
memory effect OPTOEL **Speichereffekt**
memory efficiency (→storage efficiency) DATA PROC (→Speicherausnutzung)
memory element DATA PROC **Speicherelement**
[for one bit] [zur Speicherung eines Bits]
= storage element; store element ≈ Speicherzelle; Speicherglied [CIRC.ENG]
memory-equipped typewriter (→memory typewriter) OFFICE (→Speicherschreibmaschine)
memory expansion DATA PROC **Speichererweiterung**
= storage expansion; store expansion; memory extension; storage extension; store extension ≈ Speicherexpansion
≈ Erweiterungsspeicher
≈ expanded memory
memory expansion board DATA PROC **Speichererweiterungskarte**
memory extension DATA PROC (→Speichererweiterung)
(→memory expansion)
memory ferrite METAL **Speicher-Ferrite**
↑ square-loop ferrite ↑ Rechteckferrit
memory flipflop (→storage circuit) CIRC.ENG (→Speicherglied)
memory hierarchy DATA PROC **Speicherhierarchie**
= storage hierarchy; store hierarchy
memory instruction DATA PROC **Speicherbefehl**
= storage instruction; store instruction ≠ Registerbefehl
≠ register instruction
memory key TERM&PER **Speichertaste**
memory list DATA PROC **Speicherliste**
= memory table; storage list; store list
memory location DATA PROC **Speicherstelle**
= storage location; store location; location; memory space 2; storage space 2; store space 2 = Speicherplatz; Speicherwort 2
≈ Speicherzelle
≈ storage cell
memory management DATA PROC **Speicherverwaltung**
= storage management; store management; memory allocation; storage allocation; store allocation = Speicherzuweisung
≈ Speicherbelegung
↓ Arbeitsspeicherzuweisung
≈ memory occupancy
↓ main memory assignment

memory management unit DATA PROC **Speicherverwaltungseinheit**
= MNU
memory map 2 DATA PROC **Adressenumsetzer**
memory map 1 DATA PROC **Speicherabbild**
[correspondence of the memory location of an information to its location in a display] [Entsprechung der Speicheranordnung einer Information zu ihrer Anordnung in einer Darstellung]
= storage map; store map; map ≈ Speicherabzug
≈ memory dump
memory-mapped DATA PROC **speicherkonform**
= storage-mapped; store-mapped; mapped = speicherabbildgetreu
≈ bit-mapped ≈ einzelbitweise
memory-mapped screen DATA PROC (→speicherkonforme Bildschirmanzeige)
(→memory-mapped video)
memory-mapped video DATA PROC **speicherkonforme Bildschirmanzeige**
= memory-mapped screen; bit-mapped video; bit-mapped screen; bit-mapped graphics; bit-mapping; mapping = speicherkonforme Darstellung;
speicherabbildgetreue Bildschirmanzeige
memory matrix (→storage matrix) COMPON (→Speichermatrix)
memory medium (→storage medium) DATA PROC (→Speichermedium)
memory modification DATA PROC **Speicheränderung**
= storage modification; store modification = Speichermodifizierung
memory module EQUIP.ENG **Speicherbaugruppe**
= storage module; store module
memory occupancy DATA PROC **Speicherbelegung**
= storage occupancy; store occupancy ≈ Speicherverwaltung
≈ memory management
memory organization DATA PROC **Speicherorganisation**
= storage organization; store organization
memory partitioning DATA PROC (→Speicherplatzaufteilung)
(→partitioning)
memory protection DATA PROC **Speicherschutz**
= storage protection; store protection = Speicherabsicherung
memory refresh DATA PROC **Speicherauffrischung**
memory refresh cycle DATA PROC **Speicherauffrischzyklus**
memory register DATA PROC **Speicherregister**
[part of the main memory] [Teil des Hauptspeichers]
= storage register; store register
memory request DATA PROC **Speicheranforderung**
= storage request; store request
memory requirement DATA PROC (→Speicherplatzbedarf)
(→memory space requirement)
memory-resident DATA PROC **speicherresident**
[constantly in the main memory, therefore always and directly callable] [ständig im Arbeitsspeicher, daher immer und direkt aktivierbar]
= main memory resident; storage-resident; store-resident; resident; built-in; fixed = arbeitsspeicherresident; hauptspeicherresident; resident
≠ transient ≠ transient
memory-resident command DATA PROC (→residentes Kommando)
(→resident command)
memory-resident program DATA PROC (→residentes Programm)
(→resident programm)
memory-resident software DATA PROC (→residente Software)
(→resident software)
memory scan (→memory scanning) DATA PROC (→Speicherabtastung)
memory scanning DATA PROC **Speicherabtastung**
= memory scan; storage scanning; storage scan; store scanning; store scan
memory sector (→memory area) DATA PROC (→Speicherbereich)
memory sense amplifier MICROEL (→Leseverstärker)
(→sense amplifier)
memory size DATA PROC **Speichergröße**
= storage size
memory sniffing DATA PROC **Speicherdauerüberwachung**
memory space 1 (→memory capacity) DATA PROC (→Speicherkapazität)
memory space 2 (→memory location) DATA PROC (→Speicherstelle)

memory space requirement DATA PROC **Speicherplatzbedarf**
= memory requirement; storage place = Speicherbedarf
requirement; storage requirement;
store requirement
memory structure DATA PROC **Speicherstruktur**
= storage structure; store structure = Speichereinteilung
memory table (→memory DATA PROC (→Speicherliste)
list)
memory transistor MICROEL **Speichertransistor**
memory typewriter OFFICE **Speicherschreibmaschine**
= memory-equipped typewriter = electronische Schreibmaschine
memory unit DATA PROC **Speichereinheit**
= storage unit; store unit ≈ Speicherelement; Speicherzelle
≈ memory element; memory cell
↓ bank ↓ Bank
memory utilization DATA PROC (→Speichernutzung)
(→storage utilization)
memory word DATA PROC **Speicherwort 1**
= storage word
memory zone (→memory DATA PROC (→Speicherbereich)
area)
mendelevium CHEM **Mendelevium**
= Md = Md
menu DATA PROC **Menü**
[list of options presented on a display] [am Bildschirm dem Benutzer angebotene Auswahlliste]
↓ main menu; sequential menu; submenu
 = Auswahlliste
 ↓ Hauptmenü; Folgemenü; Untermenü
menu bar DATA PROC **Menüleiste**
 = Menübalken
menu display DATA PROC **Menümaske**
menu-driven DATA PROC **menügesteuert**
≠ command-driven = menügeführt; selbsterklärend
 ≠ kommandogeteuert
menu driven program DATA PROC **menügeführtes Programm**
≠ command-driven program = menügesteuertes Programm; selbsterklärendes Programm
 ≠ kommandogesteuertes Programm
menu-driven software DATA PROC **menügesteuerte Software**
menu feature (→menu item) DATA PROC (→Menüpunkt)
menu form DATA PROC **Menüblatt**
menu generator DATA PROC **Menügenerator**
[for menu displays] = Menümaskengenerator
menu item DATA PROC **Menüpunkt**
= menu feature
menu level DATA PROC **Menüebene**
menu logic (→menu mode) DATA PROC (→Menütechnik)
menu manager DATA PROC **Menü-Manager**
 = Menü-Verwalter
menu mode DATA PROC **Menütechnik**
= menu prompt; menu logic = Menüsteuerung; Menüführung
↑ user guidance
 ↑ Bedienerführung
menu option DATA PROC **Auswahlposition**
= option [Menu]
menu program DATA PROC **Menüprogramm**
menu prompt (→menu mode) DATA PROC (→Menütechnik)
menu tree DATA PROC **Menübaum**
merchandise (→good) ECON (→Ware)
merchandise security system SIGN.ENG **Artikelsicherungssystem**
merchant (→dealer) ECON (→Händler)
mercury CHEM **Quecksilber**
= Hg = Hg
mercury contact COMPON **Quecksilberkontakt**
= wetted contact 2 = nasser Zungenkontakt
 ↑ Schutzrohrkontakt
mercury lamp (→mercury-vapour PHYS (→Quecksilberdampflampe)
lamp)
mercury memory (→mercury ELECTRON (→Quecksilberspeicher)
storage)
mercury relay COMPON **Quecksilber-Relais**
= mercury-wetted-contact relay; mercury-wetted relay = Quecksilberfilm-Relais
 ↑ Schutzrohrkontakt-Relais
↑ reed relay

mercury storage ELECTRON **Quecksilberspeicher**
= mercury memory
mercury switch COMPON (→Quecksilberschalter)
(→mercury-wetted switch)
mercury vapor (AM) PHYS **Quecksilberdampf**
= mercury vapour (BRI)
mercury vapour (BRI) (→mercury PHYS (→Quecksilberdampf)
vapor)
mercury-vapour lamp PHYS **Quecksilberdampflampe**
= mercury lamp = Quecksilberlampe
mercury-vapour rectifier PHYS **Quecksilberdampfgleichrichter**
mercury-wetted-contact relay COMPON (→Quecksilber-Relais)
(→mercury relay)
mercury-wetted relay COMPON (→Quecksilber-Relais)
(→mercury relay)
mercury-wetted switch COMPON **Quecksilberschalter**
= mercury switch
merge DATA PROC **mischen**
[to compare data and combine them in a sorted manner] [Daten vergleichen u. sortiert zusammenfügen]
= collate; coalesce; reassemble = zusammenlegen; abgleichen; vereinigen
merged-transistor logic MICROEL (→IIL)
(→inegrated injection logic)
merge program DATA PROC **Mischprogramm**
merger (n.) ECON **Unternehmenszusammenschluß**
 = Zusammenschluß; Fusion
merger (→unification) COLLOQ (→Vereinigung)
merge sorting DATA PROC **Mischsortieren**
[sorting procedure with presorting and merging runs] [Sortiervefahren mit Vorsortierungen und anschließenden Mischläufen]
= collating sort
≠ polyphase sorting = Mischsortierung
↑ sorting ≠ Polyphasensortieren
 ↑ Sortieren
merge sorting algorithm DATA PROC **Mischsortieralgorithmus**
merging DATA PROC **Mischen**
= collation; collating = Mischung; Zusammenlegung
merging run DATA PROC **Mischlauf**
[sorting] [Sortieren]
meridian GEOPHYS **Meridian**
mesa diode MICROEL **Mesadiode**
mesa transistor MICROEL **Mesatransistor**
↑ bipolar transistor; diffusion transistor ↑ Bipolartransistor; Diffusionstransistor
MESFET MICROEL **MESFET**
= metal semiconductor field-effect transistor
mesh (v.t.) TECH **ineinandergreifen**
mesh (v.t.) TELEC **vernetzen**
= network (v.t.)
mesh (n.) NETW.TH **Masche**
[a loop witch contains no further loops] [keine weiteren Schleifen enthaltende Schleife]
 ↑ Schleife
mesh (n.) TECH **Masche**
mesh (→picture element) INF.TECH (→Bildpunkt)
mesh analysis NETW.TH **Schleifenanalyse**
= loop analysis = Maschenanalyse
meshed TELEC **vernetzt**
= networked
meshed network (→intermeshed TELEC (→Maschennetz)
network)
mesh equation EL.TECH **Maschengleichung**
meshing TECH **Vernetzung**
meshing (→intermeshing) TELEC (→Vermaschung)
mesh size TECH **Maschenweite**
mesh topolgy TELEC **Maschenstruktur**
 = vermaschte Topologie
mesial power level OPT.COMM **mesialer Leistungspegel**
[50% amplitude level] [50% Amplitudenpegel]
mesochronous TELEC **mesochron**
[with equal average distance between significant instants] [mit gleichem Mittelwert der Kennzeitpunkt-Abständen]
= synchronous
 ≈ synchron
mesochronous network TELEC **mesochrones Netz**

meson	PHYS	Meson
mesosphere	GEOPHYS	Mesosphäre
message (v.t.)	COLLOQ	mitteilen
message	TELECONTR	Meldung
= indication; report; monitored binary information		
message	INF	Nachricht
[signal acting on sense organs, conveying information]		[auf Sinnesorgane einwirkendes, Informationen übermittelndes, Signal]
= communication		= Mitteilung
≈ information		≈ Information
message accounting (→tax metering)	SWITCH	(→Gebührenzählung)
message block	DATA COMM	Nachrichtenblock
message buffer	SWITCH	Nachrichtenverteiler
message channel	SWITCH	Nachrichtenkanal
message code	DATA COMM	Mitteilungscode
message content	INF	Nachrichteninhalt
message delay	DATA COMM	Mitteilungsverzug
		= Mitteilungsverzögerung; Mitteilungslaufzeit; Nachrichtenverzug; Nachrichtenverzögerung; Nachrichtenlaufzeit
message discrimination	SWITCH	Meldungsunterscheidung
message distribution	SWITCH	Meldungsverteilung
message distributor	SWITCH	Meldungsverteiler
message drain	INF	Nachrichtensenke
= message sink		≈ Nachrichtensinke
≈ information sink		≈ Informationssenke
message element	INF	Nachrichtenelement
message flow	DATA COMM	Nachrichtenfluß
= information flow		≠ Datenübertragungsgeschwindigkeit
message format (→message structure)	DATA COMM	(→Nachrichtenaufbau)
message handler	SWITCH	Nachrichtenzuteiler
message handling (→messaging)	DATA COMM	(→Mitteilungsübermittlung)
message handling system	DATA COMM	Mitteilungsübermittlungssystem
= MHS; computer-based message system; CBMS		= Mitteilungssystem
message header	DATA COMM	Nachrichtenkopf
= header; heading		= Nachrichtenvorsatz; Kopf
message-line subscription	TELEC	Einzelgebührenanschluß
message memory	INF	Nachrichtenspeicher
= message store		= Informationsspeicher
≈ information memory		
message output queue	DATA PROC	Ausgabewarteschlange
message processing	INF.TECH	Nachrichtenverarbeitung
≈ informatics; information processing		≈ Informatik; Informationsverarbeitung
↑ communications		↑ Kommunikation
↓ data processing		↓ Datenverarbeitung
message queing	DATA PROC	Meldungswarteschlangen-Betrieb
message queue	DATA COMM	Meldungswarteschlange
message receiver	INF	Nachrichtenempfänger
message retrieval	DATA PROC	Nachrichtenwiedergewinnung
message routing	SWITCH	Meldungslenkung
message service (→announcement service)	TELEPH	(→Ansagedienst)
message sink (→message drain)	INF	(→Nachrichtensenke)
message source	INF	Nachrichtenquelle
= originator [DATA COMM]		≈ Informationsquelle
≈ information source		
message storage switching (→store-and-forward switching)	DATA COMM	(→Speichervermittlung)
message store (→message memory)	INF	(→Nachrichtenspeicher)
message storing system (→message switching)	SWITCH	(→Sendungsvermittlung)
message structure	DATA COMM	Nachrichtenaufbau
= message format		= Nachrichtenformat
message switching	SWITCH	Sendungsvermittlung
[with block-by-block transmission]		[mit blockweiser Übermittlung]
= message storing system		= Nachrichtenvermittlung
↑ store-and-forward switching		↑ Speichervermittlung
message throughput	INF	Nachrichtendurchsatz
message transfer	DATA COMM	Nachrichtenübermittlung
= information transfer		= Nachrichtentransfer; Mitteilungstransfer; Informationstransfer
message transfer agent entity	DATA COMM	Instanz des Transport-Systemteils
= MTAE		
message transfer part	SWITCH	Kennzeichentransfer-Teil
message volume	INF	Nachrichtenmenge
≈ information volume		≈ Informationsmenge
messaging	DATA COMM	Mitteilungsübermittlung
= message handling		= Mitteilungsverarbeitung
messenger	POST	Telegrammbote
messenger	COLLOQ	Bote
messenger (→messenger wire)	OUTS.PLANT	(→Tragseil)
messenger cable (→suspension guy)	CIV.ENG	(→Tragseil)
messenger strand (→messenger wire)	OUTS.PLANT	(→Tragseil)
messenger wire	OUTS.PLANT	Tragseil
= messenger strand; messenger; catenary wire; suspension strand		= Spannseil; Spanndraht
messenger wire (→suspension guy)	CIV.ENG	(→Tragseil)
metacharacter	DATA PROC	Metazeichen
metacompiler	DATA PROC	Metakompilierung
metal	CHEM	Metall
metal alumina silicon field-effect transistor (→MASFET)	MICROEL	(→MASFET)
metalanguage	INF	Metasprache
= meta language		
meta language (→metalanguage)	INF	(→Metasprache)
metal armouring	COMM.CABLE	Metallbewehrung
metal band resistor	COMPON	Metallbandwiderstand
metal case	EQUIP.ENG	Metallgehäuse
metal casting	METAL	Metallgießen
metal ceramic	CHEM	Metallkeramik
metal character (→type)	TYPOGR	(→Drucktype)
metal-clad	METAL	metallbeschichtet
≈ metallized		= metallgekapselt
		= metallisiert
metal construction	TECH	Metallbau
metal container (→can)	TECH	(→Kanister)
metal cover	TECH	Metallschutzkappe
metal detector	OUTS.PLANT	Metallsuchgerät
= line detector		= Metalldetektor
metal-enclosed relay	COMPON	Stahlrelais
= steel-sealed relay		
metal film	METAL	Metallschicht
= metallization 2; metalization 2		= Metallbelag; Metallisierung 2; Metallfilm
metal film resistor (→metallic film resistor)	COMPON	(→Metallfilmwiderstand)
metal-film strain gauge	INSTR	Metallfilm-Dehnmeßstreifen
metal foil	TECH	Metallfolie
metal-free	TECH	metallfrei
metal-free cable	COMM.CABLE	metallfreies Kabel
= dielectric cable		≠ Metalleiterkabel
≠ metallic cable		
metal gate technology	MICROEL	Metall-gate-Technik
metal gauze	TECH	Drahtgitter
metal halide lamp	EL.INST	Halogen-Metalldampflampe
metal-insulator semiconductor	MICROEL	MIS
= MIS		
metal-insulator semiconductor FET (→MISFET)	MICROEL	(→MISFET)
metal interface transistor	MICROEL	Metallzwischenschicht-Transistor
= MI transistor		= MI-Transistor
metalization 1 (→metallization 1)	METAL	(→Metallisierung 1)
metalization 2 (→metal film)	METAL	(→Metallschicht)
metalize (→metallize)	METAL	(→metallisieren)
metal lattice	PHYS	Metallgitter
[solid state physics]		[Festkörperphysik]
metallic	PHYS	metallisch

metallic atom	CHEM	**Metallatom**	metal spraying	TECH	**Spritzmetallisierung**
metallic bond	CHEM	**Metallbindung**	metal strip	METAL	**Metallband**
		= metallische Bindung	metal tape	TERM&PER	**Metallschichtband**
metallic cable	COMM.CABLE	**Metalleiterkabel**	metal tape	CONS.EL	**Reineisenband**
≠ metal-free cable		≠ metallfreies Kabel	↑ sound tape		↑ Tonband
metallic contact	EL.TECH	**galvanischer Kontakt**	metal thick oxide semiconductor (→MTOS)	MICROEL	(→MTOS)
= galvanic contact		≈ ohmscher Kontakt			
≈ ohmic contact			metamagnetism	PHYS	**Metamagnetismus**
metallic film resistor	COMPON	**Metallfilmwiderstand**	meta-meta language	DATA PROC	**Meta-Metasprache**
= metal film resistor		= Metallschichtwiderstand	metamouse	TERM&PER	**Metamaus**
metallic isolation (→dc decoupling)	EL.TECH	(→galvanische Trennung)	metastability	CIRC.ENG	**Metastabilität**
			metastable	PHYS	**metastabil**
metallic-line transmission technique	TRANS	**drahtgebundene Übertragungstechnik**	metastable energy level	PHYS	**metastabiles Energieniveau**
[on metallic lines]		[mit metallischen Leitern]	meteor burst link (→meteoric scatter link)	RADIO	(→Meteorscatterverbindung)
↑ line transmission technique		↑ leitergebundene Übertragungstechnik	meteor burst scatter (→meteor scatter)	RADIO PROP	(→Meteorscatter)
metallic luster	TECH	**Metallglanz**	meteoric scatter (→meteor scatter)	RADIO PROP	(→Meteorscatter)
metallic package	COMPON	**Metallgehäuse**			
metallic work function (→work function)	PHYS	(→Ablösearbeit)	meteoric scatter link	RADIO	**Meteorscatterverbindung**
			= meteor burst link		
metalliozed plastic-film capacitor	COMPON	**MK-Kondensator**	meteorological	METEOR	**metereologisch**
		= metallisierter Kunststoff-lien-Kondensator	meteorological facsimile equipment	TERM&PER	**Wetterfaxapparat**
					= Wetterkartengerät
metallization 1	METAL	**Metallisierung 1**	meteorological radar	RADIO LOC	**Wetterradar**
= metalation 1; metallizing		↓ Plattierung; Galvanisierung	= weather radar		
↓ plating; galvanization			meteorological satellite	SAT.COMM	**Wettersatellit**
metallization 2 (→metal film)	METAL	(→Metallschicht)			= meteorologischer Satellit
metallize	METAL	**metallisieren**	meteorological satellite service	SAT.COMM	**Wetterfunkdienst über Satelliten**
= metalize		↓ plattieren; galvanisieren			
↓ plate; galvanize					= Wettersatellitendienst
metallized capacitor	COMPON	(→Metallpapierkondensator)	meteorological service	METEOR	**Wetterdienst**
(→metallized paper capacitor)					= metereologischer Dienst
metallized paper	TECH	**Metallpapier**	meteorological telegraphy	TELEGR	**Wetterkartentelegrafie**
metallized paper capacitor	COMPON	**Metallpapierkondensator**	↑ facsimile telegraphy		↑ Faksimiletelegrafie
= metallized capacitor		= MP-Kondensator	meteorology	SCIE	**Meteorologie**
metallized-paper capacitor with plastic film dielectric	COMPON	**MPK-Kondensator**			= Wetterkunde
			meteor scatter	RADIO PROP	**Meteorscatter**
metallized paper recording	INSTR	**Metallpapierschrift**	= meteoric scatter; meteor burst scatter		
metallized polycarbonate capacitor	COMPON	**MKC-Kondensator**	meter (NAM)	PHYS	**Meter** (n.m. oder n.n.)
= MKC capacitor			[SI unit for physical length]		[SI-Basiseinheit für räumliche Länge]
metallized polyethylene thereptalate capacitor (→MKT capacitor)	COMPON	(→MKT-Kondensator)	= metre (BRI); m		
					= m
metallizing (→metallization 1)	METAL	(→Metallisierung 1)	meter (→measuring instrument)	INSTR	(→Meßgerät)
metallography	METAL	**Metallographie**	meter (→counter)	INSTR	(→Zähler)
metalloid 1 (→non-metal)	CHEM	(→Metalloid)	meter (→counter)	CIRC.ENG	(→Zähler)
metall-organic vapor-phase epitaxy (→MOVPE)	MICROEL	(→MOVPE)	meter accuracy (→measuring precision)	INSTR	(→Meßgenauigkeit)
			meter bundle	COMM.CABLE	**Zählbündel**
metallurgical plant (→smelting works)	METAL	(→Hüttenwerk)	= counting bundle		
			meter constant (→counter constant)	INSTR	(→Zählerkonstante)
metal migration	PHYS	**Metallwanderung**			
↑ material migration		↑ Materialwanderung	metered measurand (→counted measurand)	TELECONTR	(→Zählwert)
metal money (→coin)	ECON	(→Münzgeld)			
metal-nitride oxide semiconductor (→MNOS)	MICROEL	(→MNOS)	metering (→tax metering)	SWITCH	(→Gebührenzählung)
			metering (→time measurement)	INSTR	(→Zeitmessung)
metal-nitride-oxide-semiconductor FET (→MNOFET)	MICROEL	(→MNOSFET)	metering failure	SWITCH	**Zählstörung**
			= metering fault		
metal-nitride-semiconductor FET (→MNSFET)	MICROEL	(→MNSFET)	metering fault (→metering failure)	SWITCH	(→Zählstörung)
metal oxide	CHEM	**Metalloxid**	metering key	INSTR	**Zähltaste**
metal-oxide-film resistor (→fixed metal-oxide-film resistor)	COMPON	(→Metalloxidschicht-Festwiderstand)	= counter key; meter key		
			metering noise	TELEPH	**Zählgeräusch**
metal-oxide semiconductor (→MOS)	MICROEL	(→Metalloxid-Halbleiter)	metering panel (→measuring panel)	EQUIP.ENG	(→Meßfeld)
metal-oxide semiconductor field-effect transistor (→MOSFET)	MICROEL	(→MOSFET)	metering pulse	SWITCH	**Zählimpuls**
			= meter pulse		= Takt
metal-oxide semiconductor tetrode	MICROEL	**MIS-Tetrode**	≈ metering signal		≈ Zählzeichen
			metering pulse [TELEC] (→time pulse)	ELECTRON	(→Zeitsteuertakt)
metal rectifier (→dry rectifier)	COMPON	(→Trockengleichrichter)			
			metering pulse generator	TELEC	**Impulstaktgeber**
metal reel	TERM&PER	**Metall-Leerspule**	metering pulse generator	SWITCH	**Zählimpulsgeber**
metal-resistance strain gauge	INSTR	**Metalldehnmeßstreifen**	= meter pulse generator		= ZIG; Zählimpulsgenerator; Zählspannungsgenerator
metal screen	EL.TECH	**Metallschirm**			
metal-semiconductor contact	MICROEL	**Metall-Halbleiter-Kontakt**	metering signal	SWITCH	**Zählzeichen**
metal semiconductor FET (→MSFET)	MICROEL	(→MSFET)	= meter signal; fee-metering signal; tele-tax signal		= Gebührenzählzeichen
metal semiconductor field-effect transistor (→MESFET)	MICROEL	(→MESFET)	≈ metering pulse; switching pulse		≈ Zählimpuls; Kennzeichen

metering zone	SWITCH	**Gebührenzone**
= tariff zone; charging area; charging zone		= Tarifzone
meter key (→metering key)	INSTR	(→Zähltaste)
meter per second (NAM)	PHYS	**Meter durch Sekunde**
[SI unit for velocity]		[SI-Einheit für Geschwindigkeit]
= metre per second (BRI); m/s		= m/s
meter pulse (→metering pulse)	SWITCH	(→Zählimpuls)
meter pulse generator	SWITCH	(→Zählimpulsgeber)
(→metering pulse generator)		
meter reading 1 (→counter reading 1)	INSTR	(→Zählerablesung)
meter reading 2 (→counter reading 2)	INSTR	(→Zählerstand)
meter reading day (→counter reading date)	ECON	(→Zählerablestag)
meter signal (→metering signal)	SWITCH	(→Zählzeichen)
meter wire	COMM.CABLE	**Zählader**
= counting element		
method	SCIE	**Methode**
= approach		= Lösungsweg
≈ procedure		≈ Verfahren
method (→operating mode)	TECH	(→Betriebsart)
method of determination	PHYS	**Bestimmungsmethode**
method of electric images	EL.TECH	**Spiegelbildmethode**
method of operation (→operating mode)	TECH	(→Betriebsart)
methodology	SCIE	**Methodologie**
metre (BRI) (→meter)	PHYS	(→Meter) (n.m. oder n.n.)
metre per second (BRI) (→meter per second)	PHYS	(→Meter durch Sekunde)
metric	PHYS	**metrisch**
= metrical		
metrical (→metric)	PHYS	(→metrisch)
metric carat	PHYS	**metrisches Karat**
[= 0.2 g]		[= 0,2 g]
= ct		= Kt
metric hundredweight	PHYS	**Zentner**
[metric unit for mass; = 50 kg]		[Maßeinheit für Masse; = 50 kg]
= hundredweight 2; hundred-weight 2; hundredweights 2		= ztr
metric pound	PHYS	**Pfund**
[= 0.5 kg]		[Maßeinheit für Masse; = 0,5 kg]
metric quintal	PHYS	**Doppelzentner**
[measuring unit for mass; = 100 kg]		[Maßeinheit für Masse; = 100 kg]
= quintal; dz		= dz
metric system	PHYS	**metrisches Maßsystem**
metric ton	PHYS	**Tonne**
[unit for mass; = 1,000 kg]		[Einheit für Masse; = 1.000 kg]
= ton; t		= t
≈ metric ton-force		≈ Tonne-Kraft
metric ton-force	PHYS	**Tonne-Kraft**
[1000 kp]		[1000 kp]
= ton-force; t*		= t*
≈ metric ton		≈ Tonne
metric waves	RADIO	**Ultrakurzwellen**
[10 m-1 m; 30 MHz-300 MHz]		[10 m-1 m; 30 MHz-300 MHz]
= very high frequency; VHF		= UKW; Meter-Wellen; VHF
metrological	INSTR	**meßtechnisch**
metrology grade	QUAL	**Metrologie-Güteklasse**
		= Laborqualität
metropolitan area	ECON	**Ballungsgebiet**
= conurbation		= Ballungszentrum; Großstadtbereich
≈ urban area		≈ Stadtbereich
metropolitan network (→local network)	TELEC	(→Ortsnetz)
metropolitan paging service	MOB.COMM	**Stadtfunkrufdienst**
		= Cityruf (BRD)
MF (→kilometric waves)	RADIO	(→Langwellen)
MFC (→multifrequency code)	CODING	(→Mehrfrequenzcode)
MFC dialing	SWITCH	**MFC-Wahl**
= multi-frequency-code dialing; multi-frequency dialing		= Mehrfrequenzwahl
MFC signaling	SWITCH	**MFC-Signalisierung**
= multi-frequency signaling		= Mehrfrequenzsignalisierung
MF/HF receiver	RADIO	**Grenzwellenempfänger**
MFLOPS (→megaflops)	DATA PROC	(→Megaflops)
MFM code (→modified FM code)	DATA PROC	(→modifizierte Wechseltaktschrift)
Mg (→magnesium)	CHEM	(→Magnesium)
MGA	TERM&PER	**MGA**
[Monochrome Graphics Adapter; monochrome graphics standard with 740x348 pixels]		[monochromer Graphikstandard mit 740x348 Bildpunkten]
		↓ Hercules
MGA board	TERM&PER	**MGA-Karte**
= monochrome graphics adapter		= MGA-Adapter
↑ graphics board		↑ Graphikkarte
↓ Hercules board		↓ Hercules-Karte
MHO (→siemens)	PHYS	(→Siemens)
MHS (→message handling system)	DATA COMM	(→Mitteilungsübermittlungssystem)
MHz (→megahertz)	PHYS	(→Megahertz)
mi (→mile)	PHYS	(→Meile)
MIC (→microwave integrated circuit)	MICROEL	(→integrierter Mikrowellenschaltkreis)
mica	CHEM	**Glimmer**
mica capacitor	COMPON	**Glimmer-Kondensator**
mica-foil	CHEM	**Mikafolium**
mickey-match	ANT	**Mickeymatch**
[primitive reflectometer]		[behelfsmäßiger Reflektometer]
MICR (→magnetic character reader)	TERM&PER	(→Magnetschriftleser)
MICR (→magnetic character recognition)	TERM&PER	(→Magnetschrift-Zeichenerkennung)
micro (→microcomputer)	MICROEL	(→Mikrocomputer)
micro (→microcomputer)	DATA PROC	(→Mikrocomputer)
microacoustics	MICROEL	**Mikroakustik**
microaddress converter	DATA PROC	**Mikroadreßwandler**
microaddress register	SWITCH	**Mikroadreßregister**
micro-alloy diffused-base transistor (→MADT)	MICROEL	(→MADT-Transistor)
micro-alloy transistor	MICROEL	**Mikro-alloy-Transistor**
microammeter	INSTR	**Mikroampermeter**
microampere	PHYS	**Mikroampere**
microbeam	PHYS	**Feinstrahl**
microbending	OPT.COMM	**Mikrokrümmung**
		= Mikrobiegung
microchip (→chip)	MICROEL	(→Chip)
microchip card	TERM&PER	**Mikrochipkarte**
microchip card coder	TERM&PER	**Mikrochipkartendecodierer**
		= Mikrochipkartendekodierer
microchip card reader	TERM&PER	**Mikrochipkartenleser**
microchip card terminal	TERM&PER	**Mikrochipkarten-Terminal**
microcircuit	MICROEL	**Mikroschaltung**
microclimate	QUAL	**Mikroklima**
micro code	DATA PROC	(→Befehl)
(→instruction)		
microcoding	DATA PROC	**Mikrocodierung**
microcomputer	MICROEL	**Mikrocomputer**
[a chip with full computer functionality with CPU]		[als Computer funktionsfähige integrierte Schaltung]
= single-chip computer; microprocessor system; microprocessor 2; micro		= Mikroprozessor 2; Ein-Chip-Computer; Mikrorechner 1; Mikroprozessorsystem; Mikro
microcomputer	DATA PROC	**Mikrocomputer**
[a computer with a microprocessor as CPU]		[ein Computer dessen Zentraleinheit ein Mikroprozessor ist]
= micro		
↓ personal computer; home computer; hobby computer		= Mikrorechner; Mikroprozessorsystem
		↓ Personal Computer; Heimcomputer; Hobbycomputer
microcomputer chip	DATA PROC	(→Ein-Chip-Mikrocomputer)
(→single-chip microcomputer)		
microcontroller	DATA PROC	**Mikrocontroller**
[a microprocessor for specific control functions]		[ein Mikroprozessor für spezifische Steueraufgaben]
microcrack	TECH	**Mikroriß**
microcrystal	PHYS	**Mikrokristall**

microcycle		MICROEL	Mikrozyklus	microphony (→microphonic effect)		ELECTRON	(→Mikrophonieeffekt)

microcycle MICROEL Mikrozyklus
micro-diskette (→3 1/2 in. floppy disk) TERM&PER (→Mikrodiskette)
microelectronics ELECTRON Mikroelektronik
↓ semiconductor technology; film technology; optoelectronics; microacoustics; quantum microelectronics; insulator electronics; neuristor electronics
≈ Miniaturelektronik
↓ Halbleitertechnik; Filmtechnik; Optoelektronik; Mikroakustik; Quantenmikroelektronik; Isolatorelektronik; Neuristorelektronik
microfiche TERM&PER Microfiche
[sheet of many mirofilms]
= fiche
↑ microfilm
[Milkrofilm mit Anreihung von Mikrokopien]
= Microfiche
↑ Mikrofilm
microfiche reader TERM&PER Mikrofiche-Leser
microfiche store TERM&PER Mikrofiche-Speicher
microfilm TERM&PER Mikrofilm
↓ microfiche ↓ Mikrofiche
microfilm cabinet OFFICE Mikrofilmschrank
microfilm hand viewers OFFICE Mikrofilm-Leselupe
= Mikrofilm-Handlesegerät
microfilm planetary camera OFFICE Mikrofilm-Schrittschaltkamera
microfilm plotter TERM&PER Mikrofilm-Plotter
microfilm projector OFFICE Mikrofilm-Projektor
microfilm reader TERM&PER Mikrofilm-Lesegerät
microfilm rotary camera OFFICE Mikrofilm-Durchlaufkamera
microfilm shredder OFFICE Mikrofilmvernichter
microfilm step-and-repeat camera OFFICE Mikroplanfilm-Kamera
microfilm storage TERM&PER Mikrofilm-Speicher
microflame PHYS Mikroflamme
microfloppy disk (→3 1/2 in. floppy disk) TERM&PER (→Mikrodiskette)
microform TERM&PER Mikroform
↓ microfiche; microfilm ↓ Mokrofiche; Mikrofilm
microform memory (→microform storage) TERM&PER (→Mikroformspeicher)
microform storage TERM&PER Mikroformspeicher
= microform memory; microform store
microform store (→microform storage) TERM&PER (→Mikroformspeicher)
microjustification DATA PROC Mikroausschluß
microline TERM&PER Mikrozeile
micrologic DATA PROC Mikrologik
micromanipulator TECH Mikromanipulator
micromatch ANT Mikromatch
micrometer (NAM) [0.000 001 m] PHYS Mikrometer [0,000 001 m]
= micrometre (BRI); μ = μ
micrometer caliper MECH Mikrometerschraube
= Mikrometer
micrometer screw MECH Feinmeßschraube
micrometre (BRI) (→micrometer) PHYS (→Mikrometer)
micro-miniaturization TECH Mikrominiaturisierung
microoperation DATA PROC Mikrooperation
microphone EL.ACOUS Mikrophon
↑ electroacoustic transducer
= Mikrofon
≈ Schallempfänger
↑ Schallwandler
microphone TELEPH Mikrophon
= transmitter = Mikrofon
microphone capsule (→transmission capsule) TELEPH (→Sprechkapsel)
microphone inset (→transmission capsule) TELEPH (→Sprechkapsel)
microphone noise (→transmitter noise) TELEPH (→Mikrophongeräusch)
microphone system EL.ACOUS Mikrophonanlage
microphonic effect ELECTRON Mikrophonieeffekt
= microphony; microphonics; microphonism
[unerwünschte Änderung elektrischer Eigenschaften durch Erschütterungen]
= Mikrophonie
microphonics (→microphonic effect) ELECTRON (→Mikrophonieeffekt)
microphonism (→microphonic effect) ELECTRON (→Mikrophonieeffekt)

microphony (→microphonic effect) ELECTRON (→Mikrophonieeffekt)
microphysical PHYS mikrophysikalisch
micro-plasma PHYS Mikroplasma
microprocessor 1 MICROEL Mikroprozessor 1
[an IC with the functions of a CPU] [integrierte Schaltung mit den Funktionen einer Zentraleinheit]
≈ microcomputer
= Mikrorechner
≈ Mikrocomputer
microprocessor 2 (→microcomputer) MICROEL (→Mikrocomputer)
microprocessor system (→microcomputer) MICROEL (→Mikrocomputer)
microprocessor technique DATA PROC Mikroprozessortechnik
microprogram DATA PROC Mikroprogramm
[instruction element, composed of a sequence of micro-instructions] [Befehlselement, bestehend aus einer Folge von Mikrobefehlen]
microprogram control DATA PROC Mikroprogammsteuerung
microprogrammability DATA PROC Mikroprogrammierbarkeit
microprogrammable DATA PROC mikroprogrammierbar
microprogrammed DATA PROC mikroprogammiert
microprogram memory (→microprogram storage) DATA PROC (→Mikroprogrammspeicher)
microprogramming DATA PROC Mikroprogrammierung
[programming of the elementary instructions] [Programmierung der Elementaroperationen]
microprogram storage DATA PROC Mikroprogrammspeicher
= microprogram memory; microprogram store
microprogram store (→microprogram storage) DATA PROC (→Mikroprogrammspeicher)
micro-scanner TERM&PER Mikro-Scanner
microscope PHYS Mikroskop
microscopic PHYS mikroskopisch
microsecond PHYS Mikrosekunde
= μs; us = μs
micro sign PHYS Mikrozeichen
micro source (→microwave generator) MICROW (→Mikrowellengenerator)
microspacing TERM&PER Mikroabstand
= incremental spacing
microstop SWITCH Mikrostop
microstrip MICROW Mikrostreifen
microstrip antenna ANT Mikrostrip-Antenne
= printed-circuit antenna = Platinen-Antenne; Mikrostreifenleitungs-Antenne
microstrip array ANT Mikrostrip-Gruppe
= Mikrostreifengruppe
microstrip dipole ANT Mikrostrip-Dipol
microstrip line MICROW Mikrostreifenleiter
= Mikrostripleitung
microstructure PHYS Mikrostruktur
microswitch COMPON Mikroschalter
microswitch relay COMPON Mikroschalter-Relais
microsynchronization SWITCH Mikrosynchronbetrieb
microvoltmeter INSTR Mikrovoltmeter
microwave (→super high frequency) EL.TECH (→Höchstfrequenz)
microwave antenna 1 ANT Mikrowellenantenne
= SHF antenna ↓ Radarantenne; Richtfunkantenne
↓ radar antenna; radiolink antenna
microwave antenna 2 (→radio link antenna) ANT (→Richtfunkantenne)
microwave carrier supply MICROW Mikrowellen-Trägerversorgung
microwave counter INSTR Mikrowellenzähler
= microwave frequency counter = Mikrowellen-Frequenzzähler
microwave diode MICROEL Mikrowellendiode
microwave engineering EL.TECH Mikrowellentechnik
= microwave technique = Höchstfrequenztechnik
microwave entrance link (→radio relay spur link) TRANS (→Richtfunkzubringer)
microwave frequency counter (→microwave counter) INSTR (→Mikrowellenzähler)
microwave generator MICROW Mikrowellengenerator
= micro source = Höchstfrequenzgenerator

microwave generator	INSTR	(→Mikrowellengenerator)	
(→microwave source)			
microwave hybrid	MICROW	**Hohlleitergabel**	
microwave integrated circuit	MICROEL	**integrierter Mikrowellenschaltkreis**	
= MIC		= MIC	
microwave isolator	MICROW	**Hohlleiterisolator**	
microwave landing system	RADIO NAV	**MLS**	
= MLS		= Mikrowellen-Landesystem	
microwave limiter	MICROW	**Mikrowellenbegrenzer**	
microwave line	TELEC	**Richtfunklinie**	
[a multichannel connection via a common LOS]		↑ Fernmeldelinie	
= microwave transmission line			
↑ trunk 1			
microwave link	RADIO REL	(→Richtfunkstrecke)	
(→line-of-sight radiolink)			
microwave measurement	INSTR	**Mikrowellen-Meßtechnik**	
		= Höchstfrequenz-Meßtechnik	
microwave mixer	MICROEL	**Mikrowellenmischer**	
microwave modulation analyzer	INSTR	**Mikrowellen-Modulationsanalysator**	
↑ signal analizer		↑ Signalanalysator	
microwave network (→radio relay network)	TRANS	(→Richtfunknetz)	
microwave network analyzer	INSTR	**Mikrowellen-Netzwerkanalysator**	
microwave network planning	RADIO REL	(→Richtfunk-Netzplanung)	
(→radio-relay-network planning)			
microwave oscillator	MICROW	**Mikrowellenoszillator**	
		= Höchstfrequenzoszillator	
microwave-oscillator diode	MICROEL	**Mikrowellen-Oszillatordiode**	
microwave power meter	INSTR	**Mikrowellen-Leistungsmesser**	
microwave power transistor	MICROEL	**Mikrowellen-Leistungstransistor**	
microwave preamplifier	INSTR	**Mikrowellen-Vorverstärker**	
microwave radiolink	RADIO	(→Richtfunk)	
(→line-of-sight radio)			
microwave radiolink	RADIO REL	(→Richtfunkstrecke)	
(→line-of-sight radiolink)			
microwave radio measuring technique	INSTR	**Richtfunkmeßtechnik**	
microwave radio network	TRANS	(→Richtfunknetz)	
(→radio relay network)			
microwave radio noise and interference test set	INSTR	**Richtfunk-Rausch-Stör-Meßplatz**	
microwave radio relay	RADIO	(→Richtfunk)	
(→line-of-sight radio)			
microwave repeater (→radio relay repeater station)	RADIO REL	(→Relaisstelle)	
microwave repeater station	RADIO REL	(→Relaisstelle)	
(→radio relay repeater station)			
microwave route (→radio relay route)	TRANS	(→Richtfunktrasse)	
microwaves (→centimetric waves)	RADIO	(→Zentimeterwellen)	
microwave semiconductor	MICROEL	**Mikrowellenhalbleiter**	
microwave sensor	INSTR	**Mikrowellensensor**	
microwave source	INSTR	**Mikrowellengenerator**	
= microwave generator			
microwave spectrometer	INSTR	**Mikrowellenspektrometer**	
microwave spectrum analyzer	INSTR	**Mikrowellen-Spektrumanalysator**	
microwave spur (→radio relay spur link)	TRANS	(→Richtfunkzubringer)	
microwave spur link (→radio relay spur link)	TRANS	(→Richtfunkzubringer)	
microwave system (→radio relay system)	TRANS	(→Richtfunksystem)	
microwave technique	EL.TECH	(→Mikrowellentechnik)	
(→microwave engineering)			
microwave test equipment	INSTR	**Mikrowellen-Meßgerät**	
microwave tower	OUTS.PLANT	**Richtfunkturm**	
= radio relay tower		↑ Fernmeldeturm	
↑ communication tower			
microwave transistor	MICROW	**Mikrowellentransistor**	
microwave transmission line	TELEC	(→Richtfunklinie)	
(→microwave line)			
microwave trunk analyzer	INSTR	**Richtfunkanalysator**	
microwave tube	ELECTRON	**Mikrowellen-Elektronenröhre**	
= microwave valve		= Mikrowellenröhre; Höchstwellenröhre	
microwave valve	ELECTRON	(→Mikrowellen-Elektronenröhre)	
(→microwave tube)			
midband 1	RADIO	**Mittelband**	
≈ band center		≈ Bandmitte	
midband 2 (→band center)	RADIO	(→Bandmitte)	
midband frequency	NETW.TH	**Bandmittenfrequenz**	
= band center frequency			
midband frequency (→band center frequency)	RADIO	(→Bandmittenfrequenz)	
midband telegraph system	TELEGR	**Mittelfrequenztelegraphie**	
		= MT	
middle dot	MATH	**Mittelpunkt 2**	
		= Punkt in der Mitte	
middle-level management	ECON	(→mittlerer Führungskreis)	
(→medium-level management)			
middleman (→intermediary)	ECON	(→Zwischenhändler)	
middle management	ECON	(→mittlerer Führungskreis)	
(→medium-level management)			
middle manager	ECON	**mittlere Führungskraft**	
middle marker (→main entrance signal)	RADIO NAV	(→Haupteinflugzeichen)	
middle position	TECH	**Mittelstellung**	
middle school (BRI)	SCIE	**Mittelschule**	
middleware	DATA PROC	**Middleware**	
[software and hardware to run a program on other computer types]		[Hardware und Software zum Betreiben eines Programms auf anderem Computermodell]	
↓ emulator; simulator program		↓ Emulator; Simulierer	
middling	TECH	**mittelmäßig**	
mid frequency	TELEC	**Mittenfrequenz**	
= center frequency (AM); centre frequency (BRI)			
mid frequency	POWER SYS	**Mittelfrequenz**	
[from 150 Hz to 10 kHz]		[von 150 Hz bis 10 kHz]	
midget (→miniature)	TECH	(→Kleinst-)	
midget set	RADIO	**Kleinstempfänger**	
midgettape (→midget tape recorder)	EL.ACOUS	(→Taschentonbandgerät)	
midget tape recorder	EL.ACOUS	**Taschentonbandgerät**	
= midgettape			
MIDI	ELECTRON	**MIDI**	
[musical instrument digital interface]			
midplane (→central plane)	ENG.DRAW	(→Mittelebene)	
midpoint	MATH	**Mittenwert**	
≈ average		≈ Mittel	
midpoint connection	POWER SYS	**Mittelpunktschaltung**	
mid-range	EL.ACOUS	**Mitteltonlautsprecher**	
		= Mitteltöner	
mid-scale	INSTR	**Skalenmitte**	
mid-span compatibility	TRANS	**Luftübergabe** [fig.]	
mid-wire	POWER SYS	**Mittelleiter**	
[conductor of a dc system]		[Leiter eines Gleichstromnetzes]	
↓ neutral conductor		↓ Nulleiter	
mignon	POWER SYS	**Mignon**	
[a standard accumulatoe cell 14x50 mm^2]		[Akkumulator mit ca. 14x50 mm^2]	
		= Mignonzelle	
mil	MECH	**Tausendstelzoll**	
[1/1000 inch]			
MIL (→military standard)	MIL.COMM	(→Militärnorm)	
mil (→mile)	PHYS	(→Meile)	
mild steel	METAL	**Flußstahl**	
[produced in liquid phase]		[im Flüssigzustand erzeugt]	
= ingot steel		= Flußeisen	
mile	PHYS	**Meile**	
[5 280 ft = 1 609,344 m]		[1 609,344 m]	
= mi; mil		= mi	
≈ nautical mile		≈ Seemeile	
miles per hour (→mph)	TECH	(→Meilen pro Stunde)	
milestone	COLLOQ	**Meilenstein**	
[figurative]		[figurativ]	
military communications	TELEC	**Militärkommunikation**	

military electronics

military electronics (→defense electronics)		ELECTRON	(→Verteidigungselektronik)
military network	TELEC		**Militärnetz**
↑ private network			= militärisches Netz
			↑ Privatnetz
military personnel (→servicemen)	ECON		(→Militärpersonal)
military standard	MIL.COMM		**Militärnorm**
= MIL			= MIL-Standard
military technology	TECH		**Militärtechnik**
= defense technology (NAM); defence technology (BRI)			= Wehrtechnik
milky glass (→frosted glass)	TECH		(→Mattglas)
mill (v.t.)	METAL		**fräsen**
milled 1	MECH		**gefräst**
milled 2	MECH		**gekerbt**
Miller capacitance	ELECTRON		**Miller-Kapazität**
Miller compensation	ELECTRON		**Miller-Kompensation**
Miller effect	ELECTRON		**Miller-Effekt**
Miller index	MICROEL		**Miller-Index**
Miller indices	PHYS		**Millersche Indizes**
			= Flächenindizes
Miller integrator	CIRC.ENG		**Miller-Integrator**
= bootstrap generator			= Bootstrap-Generator
↑ sawtooth generator			↑ Sägezahn-Generator
millimeter (NAM)	PHYS		**Millimeter**
[0.001 m]			[0,001 m]
= millimetre (BRI); mm			= mm
millimeter measurement (→millimeter wave measurement)	INSTR		(→Millimeterwellen-Messung)
millimeter mercury	PHYS		**Millimeter Quecksilbersäule**
[= 133,322 Pascal]			[= 133,322 Pascal]
= mm Hg			= mm Hg
millimeter paper	DOC		**Millimeterpapier**
= rectilinear graph paper; rectilinear paper			
millimeter wave measurement	INSTR		**Millimeterwellen-Messung**
= millimeter measurement			
millimetre (BRI) (→millimeter)	PHYS		(→Millimeter)
millimetric waves	RADIO		**Millimeter-Wellen**
[0.01 m-0.001 m; 30 GHz-300 GHz]			[0,01 m-0,001 m; 30 GHz-300 GHz]
= extremely high frequency; EHF			= EHF
millimicrosecond (→nanosecond)	PHYS		(→Nanosekunde)
milling	METAL		**Fräsen**
			= Fräsung
milling machine	METAL		**Fräsmaschine**
milliohmmeter	INSTR		**Milliohmmeter**
million	MATH		**Million**
			= Mio.; Mill.
million (BRI) (→billion)	MATH		(→Milliarde)
millisecond	PHYS		**Millisekunde**
= ms			= ms; Tausendstelsekunde
millivoltmeter	INSTR		**Millivoltmeter**
MIMD processor	DATA PROC		**MIMD-Prozessor**
[multiple instruction, multiple data]			[mehrere Datenfelder werden gleichzeitig, nach individuellen Anweisungen gleichzeitig bearbeitet]
↑ array processor			↑ Feldrechner
mimic board (→display panel)	SYS.INST		(→Schaubild)
mimic diagram	TELECONTR		**Blindschaltfeld**
min (→minute)	PHYS		(→Minute)
mineral oil (→petroleum)	TECH		(→Erdöl)
mine telephone	TERM&PER		**Grubentelefon**
= intrinsically safe telephone			
mini (n.) (→minicomputer)	DATA PROC		(→Minicomputer)
mini-AT-style case	DATA PROC		**Baby-AT-Gehäuse**
			= Babyzeile
miniature (adj.)	TECH		**Kleinst-**
= midget; subminiature			≈ Mindest-
≈ minimum			
miniature accumulator	POWER SYS		**Kleinakkumulator**
miniature electric drill	TECH		**Kleinstbohrmaschine**
miniature electronics	ELECTRON		**Miniaturelektronik**
≈ microelectronics			≈ Mikroelektronik
miniature lavalier clip-on microphone	EL.ACOUS		**Miniatur-Ansteckmikrofon**
miniature pushbutton	COMPON		**Miniatur-Tastenschalter**
miniature rotary switch	COMPON		**Kleindrehschalter**
miniature speaker	EL.ACOUS		**Kleinlautsprecher**
miniature switch (→miniature toggle switch)	COMPON		(→Miniatur-Kippschalter)
miniature toggle switch	COMPON		**Miniatur-Kippschalter**
= miniature switch			
miniaturization	ELECTRON		**Miniaturisierung**
miniaturization electronics	ELECTRON		**Miniaturisierungstechnik**
minibeam	OPT.COMM		**Minibündel**
minibeam	ANT		**Minibeam**
↑ directional antenna			↑ Richtantenne
minicartridge	TERM&PER		(→Minikassette)
(→minicassette)			
minicassette	TERM&PER		**Minikassette**
= minicartridge			
minicomputer	DATA PROC		**Minicomputer**
[between a microcomputer and a mainframe computer]			[zwischen Mikrocomputer und Großrechner]
= mini (n.)			= Minirechner; Mini
≈ microcomputer; small computer			≈ Mikrocomputer; Kleinrechner
minidisk (→5 1/4 in. floppy disk)	TERM&PER		(→Minidiskette)
mini-distributor	SYS.INST		**Miniverteiler**
minifloppy (→5 1/4 in. floppy disk)	TERM&PER		(→Minidiskette)
mini-floppy disk (→5 1/4 in. floppy disk)	TERM&PER		(→Minidiskette)
minimable	SCIE		**minimierbar**
minimal-phase ...	NETW.TH		**minimalphasig**
minimax (v.t.)	TECH		**Maximum minimieren**
[to minimize the maximum]			
minimax (n.)	MATH		**Maximum-Minimierung**
minimization	MATH		**Minimierung**
minimization problem	SCIE		**Minimierungsproblem**
minimize	MATH		**minimieren**
↑ extremize			↑ extremieren
minimum (adj.)	TECH		**Mindest-**
			= Minimal-
minimum amount	ECON		**Mindestbetrag**
minimum capacity	TECH		**Minimalausbau**
= minimum equipment			= Mindestausbau; Mindestbestückung
minimum clearance (→minimum distance)	TECH		(→Mindestabstand)
minimum clearance (→tightest fit)	ENG.DRAW		(→Kleinstsitz)
minimum code distance	CODING		**Code-Distanz**
[smallest Hamming distance]			[kleinste aller Hamming-Distanzen]
= minimum-code distance; distance			= Distanz; Abstand
minimum-code distance (→minimum code distance)	CODING		(→Code-Distanz)
minimum deviation	CONTROL		**Minimalabweichung**
minimum diameter	ENG.DRAW		**Kleinstdurchmesser**
minimum dimension (→low limit)	ENG.DRAW		(→Kleinstmaß)
minimum discernable sensitivity	INSTR		**minimale Empfindlichkeit**
minimum distance	TECH		**Mindestabstand**
= minimum clearance			= Minimalabstand; minimaler Abstand
minimum-distance code	CODING		**Mindest-Abstands-Code**
[code maintaining a given minimum Hamming distance]			[Code der einen vorgegebenen Hammingabstand immer einhält]
minimum equipment (→minimum capacity)	TECH		(→Minimalausbau)
minimum-noise tuning (→noise tuning)	HF		(→Rauschabstimmung)
minimum-phase network	NETW.TH		**Minimalphasen-Vierpol**
			= Mindestphasen-Vierpol; allpassfreier Vierpol
minimum purchase	ECON		**Mindestabnahme**
minimum radio direction finder	RADIO LOC		**Minimumpeiler**
			= Minimum-Funkpeiler
minimum-runtime coding (→minimum-runtime programming)	DATA PROC		(→zeitoptimale Programmierung)

minimum-runtime programming	DATA PROC	**zeitoptimale Programmierung**	**mirror** (n.)		PHYS	**Spiegel**
= minimum-runtime coding			**mirror** (→ reflector)		MICROW	(→ Reflektor)
minimum steering	RADIO LOC	**Minimumpeilung**	**mirror disk**		DATA PROC	**Spiegelplatte**
		= Minimum-Funkpeilung	**mirrored**		TECH	**gespiegelt**
minimum-store coding	DATA PROC	(→ speicherplatzoptimale Programmierung)	≈ reflected			≈ reflektiert
(→ minimum-store programming)			**mirror-finish** (v.t.)		TECH	**hochglanzpolieren**
minimum-store programming	DATA PROC	**speicherplatzoptimale Programmierung**	**mirror finish** (n.)		MECH	**Hochglanzpolitur**
= minimum-store coding			**mirror galvanometer**		INSTR	**Spiegelgalvanometer**
minimum temperature	PHYS	**Tiefsttemperatur**	**mirroring**		TECH	**Spiegelung**
↑ limit temperature		= Minimaltemperatur	**mirror plane**		MATH	**Spiegelebene**
		↑ Grenztemperatur	**mirror point**		MATH	**Spiegelpunkt**
minimum value	MATH	**Kleinstwert**	= reflection point			= Reflexionspunkt
↑ extreme value		= Minimalwert; Minimum; Mindestwert	**mirror positioning system**		TERM&PER	**Spiegelpositioniersystem**
			mirror screw		MECH	**Spiegelschraube**
		↑ Extremum	**mirror test**		RADIO PROP	**Spiegelung**
minimum-value limiter	CIRC.ENG	**Minimalwertbegrenzer**	[hop survey]			[Funkfeld-Survey]
Minion (BRI) (→ type size 7 point)	TYPOGR	(→ Schriftgröße 7 Punkt)	**mirror wave**		MODUL	**Spiegelwelle**
miniprobe	INSTR	**Miniaturtastkopf**	**mirror-wave discrimination**		MODUL	**Spiegelwellenselektion**
minister of finance	ECON	**Finanzminister**				= Spiegelwellenunterdrückung
= treasury secretary (AM)			**mirror wheel**		MECH	**Spiegelrad**
minium	METAL	**Mennige**	**MIS** (→ metal-insulator semiconductor)		MICROEL	(→ MIS)
= red lead						
mini vice (BRI) (→ mini vise)	MECH	(→ Mini-Schraubstock)	**MIS** (→ management information system)		DATA PROC	(→ Management-Informationssystem)
mini vise	MECH	**Mini-Schraubstock**				
= mini vice (BRI)			**misadjustment**		ELECTRON	**Fehleinstellung**
minor alarm (→ non-urgent alarm)	EQUIP.ENG	(→ nicht dringender Alarm)	= maladjustment			= Einstellfehler
			≈ misalignment			≈ Fehlabgleich
minorant	MATH	**Minorante**	**misalignment**		ELECTRON	**Fehlabstimmung**
[a referential series whoes elemnts are not greater]		[Vergleichsreihe deren Glieder nicht größer sind]	= malalignment			= Fehlabgleich
			≈ misadjustment			≈ Fehleinstellung
≠ minorant		= Unterreihe	**misalignment**		TECH	**Fluchtungsfehler**
		≠ Majorante	= angular misalignmemt			≈ Versatz
minor graduation	INSTR	**Nebenskalenteilung**	≈ offset			
minor hysteresis loop	PHYS	**innere Hystereseschleife**	**miscalculate**		ECON	**verkalkulieren** (v.r.)
minority carrier	MICROEL	**Minoritätsträger**				= verrechnen (v.r.)
= minority charge carrier		= Minoritätsladungsträger	**mischievous** (→ harmful)		TECH	(→ schädlich)
minority carrier current	PHYS	**Minoritätsträgerstrom**	**MISFET**		MICROEL	**MISFET**
minority carrier density	PHYS	**Minoritätsträgerdichte**	= metal-insulator semiconductor FET			
minority charge carrier (→ minority carrier)	MICROEL	(→ Minoritätsträger)	**misfire** (→ misfiring)		TECH	(→ Fehlzündung)
			misfiring		TECH	**Fehlzündung**
minor lobe (→ side lobe)	ANT	(→ Nebenzipfel)	= misfire			
minor sorting key	DATA PROC	**Hilfssortierschlüssel**	**misidentification**		TECH	**Fehlerkennung**
		= Hilfssortierungsschlüssel	= erroneuos identification			= Fehlidentifizierung
minterm	ENG.LOG	**Vollkonjunktion**	**misinformation**		TECH	**Fehlinformation**
↑ term		= Minterm	= erroneous information			= Desinformation; Falschinformation
		↑ Term				
minuend	MATH	**Minuend**	**mismatch**		EL.TECH	**Fehlanpassung**
[number to reduced by the subtrahend]		[um den Subtrahenden zu verringernde Zahl]	**mismatch**		INSTR	**Fehlanpassungsglied**
			mismatched		EL.TECH	**fehlabgeschlossen**
≠ subtrahend		≠ Subtrahend				= fehlangepaßt
minus	MATH	**minus**	**mismatch factor** (→ reflection coefficient)		NETW.TH	(→ Reflexionsfaktor)
minuscule (n.) (→ small character)	TYPOGR	(→ Kleinbuchstabe)				
			misplug		TELEC	**falschstöpseln**
minus flag	DATA PROC	**Minusanzeige**	**mispunching** (→ punching error)		TERM&PER	(→ Fehllochung)
minus potential	PHYS	**Minuspotential**				
		= negatives Potential	**misrepresentation**		TECH	**Fehldarstellung**
minus sign	MATH	**Minuszeichen**	= erroneous representation			
[symbol -]		[Symbol -]	**misroute**		SWITCH	**fehlleiten**
		= Wenigerzeichen	**mist** (→ fog)		METEOR	(→ Nebel)
minute	PHYS	**Minute**	**mistake**		TECH	**Fehlgriff**
[= 60 s]		[= 60 s]	**misuse**		QUAL	**unzulässige Beanspruchung**
= min; '		= min; '	**misuse protection**		DATA PROC	**Mißbrauchsschutz**
↑ unit of time		↑ Zeitmaß	**MIS varactor**		MICROEL	**MIS-Varaktor**
minute	MATH	**Minute**	[metal-insulator semiconductor varactor]			= MIS-Varactor
[1°/60]		[1°/60]				↑ Varaktor
= symbol '		= Symbol '	↑ varactor			
		↑ Gradmaß	**miter**		TECH	**Gehrung**
minutes	ECON	**Protokoll**				[schräger Zuschnitt]
↓ minutes of meeting		↓ Besprechungsprotokoll	**mitigation technique** (→ countermeasure)		TECH	(→ Gegenmaßnahme)
minutes (→ record)	OFFICE	(→ Niederschrift)				
minutes of meeting	ECON	**Besprechungsprotokoll**	**MI transistor** (→ metal interface transistor)		MICROEL	(→ Metallzwischenschicht-Transistor)
		= Sitzungsprotokoll				
		≈ Verhandlungsprotokoll	**mix** (v.t.)		TECH	**mischen**
minutes of negotiation	ECON	**Verhandlungsprotokoll**	**mix** (v.t.)		TV	**mischen**
≈ minutes of meeting		≈ Besprechungsprotokoll	[to combine voice, music, noise and picture]			[Sprache, Musik, Geräusche und Bild aufeinander abstimmen]
MIPS	DATA PROC	**MIPS**				
[mega instructions per second]		[Millionen Anweisungen pro Sekunde]	↓ lap-dissolve			↓ überblenden

mix (n.) DATA PROC **Mix**
[a set of instructions to evaluate the performance of computers] [Mischung von Befehlen zur Ermittlung der Leistungsfähigkeit von Computern]
mix (n.) TECH **Mischung**
mix (n.) TV **Mischen**
mixed color PHYS **Mischfarbe**
mixed communications TELEC (→Mischkommunikation)
(→combined communications)
mixed crystal PHYS **Mischkristall**
mixed lattice PHYS **Mischgitter**
mixed load DATA COMM **Mischlast**
[call + data] [Ruf + Daten]
mixed loading RADIO **Mischbelegung**
mixed network TELEC **Verbundnetz**
mixed traffic DATA COMM **gemischter Verkehr**
= batch-conversational mode
mixer TV **Mischer**
 = Bildmischer
mixer HF **Mischer**
= first detector = Mischerschaltung; Frequenzumsetzer; Modulator
mixer diode MICROEL **Mischdiode**
mixer stage HF **Mischstufe**
= mixing stage
mixer transconductance HF **Mischsteilheit**
 = Konversionssteilheit
mixer tube ELECTRON **Mischröhre**
= mixer valve; mixing tube; mixing valve
mixer valve (→mixer tube) ELECTRON (→Mischröhre)
mixing HF **Mischung**
[frequency conversion with filtering of useful products] [Frequenzumsetzung mit Ausfilterung des interessierenden Mischproduktes]
= conversion
↓ up-conversion; down-conversion ↓ Aufwärtsmischung; Abwärtsmischung
mixing (→grading) SWITCH (→Mischung)
mixing amplifier CIRC.ENG **Mischverstärker**
 [Summierung niederfrequenter Signale]
mixing console BROADC **Mischpult**
= audio mixer
mixing oscillator HF **Mischoszillator**
 = Überlagerungsoszillator
mixing stage (→mixer stage) HF (→Mischstufe)
mixing tube (→mixer tube) ELECTRON (→Mischröhre)
mixing valve TECH **Mischventil**
= blending valve
mixing valve (→mixer tube) ELECTRON (→Mischröhre)
mixture CHEM **Gemisch**
≈ composition ≈ Zusammensetzung
mixture frequency HF **Mischfrequenz**
MKC capacitor (→metallized polycarbonate capacitor) COMPON (→MKC-Kondensator)
MKP capacitor COMPON **MKP-Kondensator**
MKS capacitor COMPON **MKS-Kondensator**
MKT capacitor COMPON **MKT-Kondensator**
= metallized polyethylene thereptalate capacitor
MKU capacitor COMPON **MKU-Kondensator**
 = Lackfolien-Kondensator; Metallackkondensator
MLS (→microwave landing system) RADIO NAV (→MLS)
mm (→millimeter) PHYS (→Millimeter)
MMF (→magnetomotive force) PHYS (→magnetische Umlaufspannung)
mm Hg (→millimeter mercury) PHYS (→Millimeter Quecksilbersäule)
MML (→man-machine language) DATA PROC (→Mensch-Maschine-Sprache)
Mn (→manganese) CHEM (→Mangan)
mnemonic DATA PROC **mnemotechnisch**
= mnemotechnic = mnemonisch; merkfähig
mnemonic code CODING **mnemonischer Code**

mnemonic language DATA PROC **mnemotechnische Sprache**
mnemotechnic DATA PROC (→mnemotechnisch)
(→mnemonic)
mnemotechnic address DATA PROC (→symbolische Adresse)
(→symbolic address)
MNOFET MICROEL **MNOSFET**
= metal-nitride-oxide-semiconductor FET
MNOS MICROEL **MNOS**
= metal-nitride oxide semiconductor
MNSFET MICROEL **MNSFET**
= metal-nitride-semiconductor FET
MNU (→memory management unit) DATA PROC (→Speicherverwaltungseinheit)
Mo (→molybdenum) CHEM (→Molybdän)
mobile (adj.) TECH **beweglich**
= moveable; movable = ortsveränderlich
≈ transporteable ≈ transportierbar
≠ stationary ≠ ortsfest
↓ portable; passable ↓ tragbar; fahrbar
mobile (→non-steady) TECH (→nichtstationär)
mobile aerial 2 (→steerable antenna) ANT (→einstellbare Antenne)
mobile antenna 1 ANT **mobile Antenne**
 = Mobilantenne
mobile antenna 2 (→steerable antenna) ANT (→einstellbare Antenne)
mobile cellular phone (→car phone) MOB.COMM (→Autotelefon)
mobile radiocommunications TELEC **Mobilfunk**
↓ radiotelephony; radio paging; trunking ↓ Funktelefonie; Funkruf; Bündelfunk
mobile radio service TELEC **beweglicher Funkdienst**
mobile radio station RADIO **ortsveränderliche Funkstelle**
mobile satellite service TELEC **beweglicher Funkdienst über Satelliten**
mobile service TELEC **beweglicher Dienst**
mobile service switching center (→mobile switching center) MOB.COMM (→Funkvermittlungsstelle)
mobile station MOB.COMM (→Funktelefon)
(→radiotelephone)
mobile switching center MOB.COMM **Funkvermittlungsstelle**
= mobile service switching center = Funkkonzentrator; FuKo; Mobilvermittlungsstelle; Mobilkommunikations-Vermittlungsstelle
mobile telephone connection TELEC **Funktelefonanschluß**
mobile telephone number TELEC **Funktelefonnummer**
mobile telephone service MOB.COMM (→Funktelefonie)
(→radiotelephony)
mobile telephone subscriber TELEC **Mobilfunkteilnehmer**
≠ fixed subscriber = beweglicher Teilnehmer
 ≠ Festteilnehmer
mobile telephony exchange MOB.COMM **Überleiteinrichtung**
 = ÜLE
mobile telephony network MOB.COMM **Mobilfunknetz**
= cellular network = Mobiltelefonnetz; Mobiltelephonnetz
 ≈ Zellularnetz
mobile TV link BROADC **mobile Fernsehverbindung**
 = mobile TV-Verbindung
mobility PHYS **Beweglichkeit**
 = Mobilität
mock-up (n.) TECH **Attrappe**
[a full-scale imitation] [formgetreue Nachbildung in Originalgröße]
= dummy (n.) = Blindmuster
↑ model (n.) ↑ Modell
MOCVD MICROEL **MOCVD**
[Metal Organic Chemical Vapour Deposition]
modal (→modal verb) LING (→Modalverb)
modal analysis INSTR **Modalanalyse**
modal auxiliary LING **modales Hilfsverb**
modal dipersion OPT.COMM **Modendispersion**
modal noise OPT.COMM **Modenrauschen**
= speckle noise
modal sentence LING **Modalsatz**
↑ conjunctional sentence = Satz der Art und Weise
 ↑ Konjunktionalsatz

modal verb		LING	**Modalverb**	**mode scrambler**	MICROW	**Modenmischer**
= modal				= mode mixer		= Modenscrambler
mode (n.)		COLLOQ	**Weise**	**mode selection key**	TERM&PER	**Moduswähltatste**
= manner				**mode shift**	MICROW	**Modenwechsel**
mode (n.)		MATH	**häufigster Wert**	**mode suppressor**	MICROW	**Modensperre**
			= Modalwert	**mode transducer** (→mode transformer)	MICROW	(→**Modenwandler**)
mode (→operating mode)		TECH	(→**Betriebsart**)			
mode (→class of operation)		ELECTRON	(→**Betriebsart**)	**mode transformer**	MICROW	**Modenwandler**
mode (→waveform)		PHYS	(→**Schwingungsform**)	= mode transducer; mode changer; transducer		= Wellenformwandler; Wellentypwandler
mode (→waveguide mode)		MICROW	(→**Wellentyp**)			
mode changer (→mode transformer)		MICROW	(→**Modenwandler**)	**MODFET**	MICROEL	**MODFET**
				= modulation-doped FET		= modulationsdotierter Feldeffekttransistor
mode-dependent		TECH	**betriebsartabhängig**			
mode filter		MICROW	**Modenfilter**	**modification**	DATA PROC	**Änderung**
= waveguide filter			= Wellentypfilter; Wellenfilter; Hohlleiterfilter			= Modifikation
				modification (→remodelling)	TECH	(→**Umbau**)
mode-independent		TECH	**betriebsartunabhängig**	**modification block**	DATA PROC	**Änderungsblock**
mode jump		MICROW	**Modensprung**	= change block		= Modifikationsblock
			= Schwingungsartsprung	**modification entry**	DATA PROC	**Änderungseintrag**
model (n.)		MATH	**Modell**	= change entry		= Modifikationseintrag
model 1 (n.)		TECH	**Modell**	**modification guideline**	TECH	(→**Änderungsanweisung**)
[scaled reproduction or representation of an object]			[maßstabgerechte Ausführung oder Darstellung eines Gegenstands]	(→engineering change note)		
				modification instruction	TECH	**Umbauanweisung**
≈ model 2; copy (n.); model 3				**modification log**	DATA PROC	**Änderungsprotokoll**
↓ mock-up; type model			≈ Muster; Nachbildung; Bauform	= change log		= Modifikationsprotokoll
				modification loop	DATA PROC	**Änderungsschleife**
			↓ Attrappe; Ausführungsmuster	= change loop		
				modification mode	DATA PROC	**Änderungsmodus**
model 2 (n.)		TECH	**Muster 1**	= change mode		= Modifikationsmodus
[form or model for imitation]			[als Vorlage dienende Form oder Modell]	**modification parameter**	DATA PROC	**Änderungsparameter**
= pattern 1				= change parameter		= Modifikationsparameter
↓ pattern 2			≈ Modell	**modification record**	DATA PROC	(→**Änderungssatz**)
			↓ Schnittmuster	(→addition record)		
model 3 (→type)		TECH	(→**Ausführung 1**)	**modification run**	DATA PROC	**Änderungslauf**
model (→type)		ECON	(→**Typ**)	= change run		= Modifikationslauf
model (n.) (→copy 3)		OFFICE	(→**Vorlage 1**)	**modifications tape** (→change tape)	DATA PROC	(→**Änderungsband**)
model A antenna		ANT	**Modell-A-Antenne**			
modelability		TECH	**Modellierbarkeit**	**modification tape** (→change tape)	DATA PROC	(→**Änderungsband**)
modelable		TECH	**modellierbar**			
model antenna		ANT	**Antennenmodell**	**modified FM code**	DATA PROC	**modifizierte Wechseltaktschrift**
model B antenna		ANT	**Modell-B-Antenne**	= modified two-frequency recording; MFM code		
model C antenna		ANT	**Modell-C-Antenne**			= MFM-Code
model D antenna		ANT	**Modell-D-Antenne**	**modified two-frequency recording** (→modified FM code)	DATA PROC	(→**modifizierte Wechseltaktschrift**)
modeling		TECH	**Modellierung**			
model number (→code number)		TECH	(→**Sachnummer**)	**modify** (→change)	TECH	(→**ändern**)
model of protocol layers		DATA COMM	**Architekturmodell**	**moding**	ELECTRON	**Frequenzspringen**
model shop		MANUF	**Musterbau**	[frequency hopping of an oscillator]		[Oszillator]
			[Organisationseinheit]	**modul**	COMPON	**Modul**
model simulation		TECH	**Modellsimulierung**	**modulability**	MODUL	**Modulierbarkeit**
			= Modellsimulation	**modular**	EQUIP.ENG	**modular**
modem (→data modem)		DATA COMM	(→**Datenmodem**)			= bausteinartig
modem (→modem equipmemt)		RADIO REL	(→**Modulationsgerät**)	**modular coding**	DATA PROC	**modulare Programmierung**
				= modular programming		
modem attachment		DATA COMM	**Modemanschluß**	**modular construction**	EQUIP.ENG	(→**Modularbauweise**)
modem eliminator		DATA COMM	**Modem-Eliminator**	(→modular design)		
modem equipmemt		RADIO REL	**Modulationsgerät**	**modular design**	EQUIP.ENG	**Modularbauweise**
[modulator + demodulator]			[Modulator + Demodulator]	= modular construction		= Modulartechnik; Modulbauweise; modulare Bauweise; modulare Technik; modularer Aufbau; Baukastenprinzip
= modem			= Modulationseinrichtung; Modem	≈ modularity [TECH]		
modem filter		TRANS	**Modemfilter**			≈ Modularität [TECH]
mode mixer (→mode scrambler)		MICROW	(→**Modenmischer**)	**modularity**	TECH	**Modularität**
				modular programming (→modular coding)	DATA PROC	(→**modulare Programmierung**)
modem rack (→modulation rack)		RADIO REL	(→**Modulationsgestell**)			
mode of application		TECH	**Anwendungsart**	**modular splice**	OPT.COMM	**Modulspleiß**
= mode of use			= Einsatzart	**modular system**	DATA PROC	**Modularsystem**
mode of calculation		MATH	**Berechnungsart**	↓ modular hardware system; modular software system		= modulares System
= calculation method; calculus			= Rechnungsart			↓ modulares Hardwaresystem; modulares Softwaresystem
mode of operation		DATA COMM	**Betriebsverfahren**			
mode of operation (→operating mode)		TECH	(→**Betriebsart**)	**modulating frequency**	MODUL	**Modulationsfrequenz**
				= modulation frequency		[Frequenz des modulierenden Signals]
mode of propagation		PHYS	**Ausbreitungsform**	≈ modulating wave		
mode of use (→mode of application)		TECH	(→**Anwendungsart**)			≈ Modulationsschwingung
modern (adj.)		TECH	**modern**	**modulating wave**	MODUL	**Modulationsschwingung**
= up-to-date; state-of-the-art; timely			= neuzeitlich; zeitgemäß	= modulation wave		= modulierende Schwingung; Signalschwingung
≈ progressive; future-oriented			≈ fortschrittlich; zukunftsorientiert	≈ modulating frequency		
				≠ carrier wave		≈ Modulationsfrequenz
modernization		TECH	**Modernisierung**			≠ Trägerschwingung

modulation

modulation TELEC
[modification of parameters of a carrier signal with a useful signal]
↓ wave-carrier modulation; amplitude modulation; angle modulation; frequency modulation; phase modulation; pulse modulation; pulse amplitude modulation; pulse time modulation; pulse duration modulation; pulse position modulation; pulse frequency modulation

Modulation
[Veränderung der Parameter eines Trägersignals mittels eines Nutzsignals]
↓ Schwingungsmodulation; Amplitudenmodulation; Winkelmodulation; Frequenzmodulation; Phasenmodulation; Pulsmodulation; Pulsamplitudenmodulation; Pulszeitmodulation; Pulsdauermodulation; Pulslagenmodulation; Pulsfrequenzmodulation

modulation amplifier RADIO **Modulationsverstärker**
= aktiver Modulator

modulation analyzer INSTR **Modulationsanalysator**
[indicates characteristic ratios of AM, FM and PM]
[mißt Kennwerte von Amplituden-, Frequenz- und Phasenmodulation]

modulation capability MODUL **Aussteuerbarkeit**
[limit to unacceptable distortion]

modulation carrier (→carrier wave) MODUL (→Trägerschwingung)

modulation characteristic MODUL **Modulationskennlinie**

modulation depth MODUL **Modulationsgrad**
= modulation factor; modulation percentage; modulation ratio; degree of modulation
= Modulationstiefe; Modulationsfaktor

modulation distortion TELEC **Modulationsverzerrung**
modulation distortion degree MODUL **Modulationsklirrgrad**
modulation domain MODUL **Modulationsbereich**
modulation-doped FET MICROEL (→MODFET)
(→MODFET)
modulation envelope MODUL **Modulationshüllkurve**
modulation factor TELEC **Modulationsfaktor**
[nonlinearity index]
[Nichtlinearitätsmaß]
modulation factor (→modulation depth) MODUL (→Modulationsgrad)
modulation frequency MODUL (→Modulationsfrequenz)
(→modulating frequency)
modulation index MODUL **Modulationsindex**
[FM]
[FM]
modulation-matched coding CODING (→unterlegte Codierung)
(→underlaying coding)
modulation method MODUL **Modulationsverfahren**
modulation noise MODUL **Modulationsgeräusch**
modulation noise EL.ACOUS **Modulationsrauschen**
modulation-on-pulse ELECTRON **impulsüberlagerte Modulation**
modulation percentage MODUL (→Modulationsgrad)
(→modulation depth)
modulation plan TRANS **Modulationsplan**
[TF-Technik]
modulation rack RADIO REL **Modulationsgestell**
= modem rack
modulation rate MODUL **Modulationsrate**
modulation rate [TELEGR] TELEC (→Schrittgeschwindigkeit)
(→telegraph speed)
modulation ratio (→modulation depth) MODUL (→Modulationsgrad)
modulation slope MODUL **Modulationssteilheit**
modulation source INSTR **Modulationsquelle**
modulation stage MODUL **Modulationsstufe**
modulation transfer function MODUL **Modulations-Übertragungsfunktion**
modulation voltage MODUL **Modulationsspannung**
modulation wave (→modulating wave) MODUL (→Modulationsschwingung)
modulator CIRC.ENG **Modulator**
modulator DATA COMM **Modulator**
[converts data signals from A/D]
[ein A/D-Wandler von Datensignalen]
modulator circuit CIRC.ENG **Modulatorschaltung**
modulator-demodulator DATA COMM (→Datenmodem)
(→data modem)
module EQUIP.ENG **Baugruppe**
= mounted board; board; subassembly; assembly; assy; construction unit
= Modul; Kassette
↓ flat module; plug-in module
↓ Flachbaugruppe; Steckbaugruppe

module COMPON **Baustein**
= device
≈ chip [MICROEL]; component
≈ Chip [MICROEL]; Bauelement
module (→functional unit) TECH (→Funktionseinheit)
module (→program module) DATA PROC (→Programmmodul)
module address SWITCH **Baugruppenadresse**
= subassembly address
module case COMPON **Modulgehäuse**
module completeness EQUIP.ENG **Baugruppenvollständigkeit**
= module integrity
module frame EQUIP.ENG **Baugruppenrahmen**
= mounting shelf; shelf; subrack; equipment shelf; subassembly frame; subassembly shelf; card cage; card chassis; card frame; mounting chassis; mounting frame; printed circuit shelf; PCB case; printed circuit frame; PCB frame; printed circuit shelf; PCB shelf; printed circuit chassis; PCB chassis
= Baugruppenträger; Chassis; Einbaurahmen; Einschubgehäuse; Einschubrahmen; Leiterplattenaufnahme; Leiterplattenrahmen; Leiterplattengehäuse; Platinengehäuse
≈ inset; mounting device
≈ Geräteeinsatz; Geräteaufnahme
module frame equipping EQUIP.ENG (→Baugruppenträgerbestückung)
(→module frame packaging)
module frame layout EQUIP.ENG **Baugruppenrahmenbelegung**
module frame packaging EQUIP.ENG **Baugruppenträgerbestückung**
= frame packaging; shelf packaging; subrack packaging; module packaging; module frame equipping; frame equipping; shelf equipping; subrack equipping
= Baugruppenträgerbestückung; Baugruppenbestückung; Geräteeinsatz-Bestückung; Einsatzbestückung
module-frame wiring EQUIP.ENG **Rahmenverdrahtung**
module front panel EQUIP.ENG **Baugruppenabdeckung**
= subassembly panel
= Baugruppenblende
↑ front panel
↑ Frontabdeckung
module guide bar EQUIP.ENG **Baugruppenführung**
= subassembly guide bar; board guide bar
module integrity (→module completeness) EQUIP.ENG (→Baugruppenvollständigkeit)
module labeling (AM) EQUIP.ENG **Baugruppenbeschriftung**
= module labelling (BRI); subassembly labeling
module labelling (BRI) EQUIP.ENG (→Baugruppenbeschriftung)
(→module labeling)
module mounting (→insertion of components) MANUF (→Leiterplattenbestückung)
module packaging (→module frame packaging) EQUIP.ENG (→Baugruppenträgerbestückung)
module position (→board position) EQUIP.ENG (→Baugruppenposition)
module replacement EQUIP.ENG **Baugruppentausch**
= subassembly replacement
module variant (→board variant) EQUIP.ENG (→Baugruppenvariante)
modulo (prep.) MATH **modulo**
[with respect to a modulus]
[auf ein Modul bezogen]
= mod
modulo addressing DATA PROC **Moduloadressierung**
modulo arithmetic MATH **Modulo-Arithmetik**
= Restwertverfahren
modulo-n counter CIRC.ENG **Modulo-N-Zähler**
[resets after every n units and counts the modulo n units]
[beginnt nach zehn Einheiten von vorne und addiert die Modulo-n-Einheiten]
modulus PHYS **Modul** (n.m.pl.-n)
= Materialkonstante
modulus 1 MATH **Modul 1** (n.m.pl.-n)
[divisor giving the same remainder for a set of dividends]
[Divisor der für einen Satz von Dividenden denselben Restwert ergibt]
modulus 2 MATH **Modul 2**
[factor to convert logarithmic systems]
[Umrechnungsfaktor zwischen Logarithmiersystemen]
modulus of elasticity PHYS **Elastizitätsmodul**
modus converter DATA COMM **Moduskonverter**
Mohs scale PHYS **Ritzhärteskala**
= Mohssche Härteskala
moiré TV **Moiré**
= moiré effect
= Moire-Effekt

monochromatic

moiré effect (→moiré) TV (→Moiré)
moist (→humid) PHYS (→feucht)
moisture (→humidity) PHYS (→Feuchtigkeit)
moisture-barrier cable COMM.CABLE (→Schichtenmantelkabel)
 (→PAL-sheath cable)
moisture meter (→hygrometer) INSTR (→Hygrometer)
moisture-repellant TECH feuchtedicht
 ≈ moisture-resistant ≈ feuchteunempfindlich
moisture-resistant TECH feuchteunempfindlich
 = feuchtigkeitsfest
 ≈ feuchtedicht
mol (→mole) PHYS (→Mol)
molar gas constant PHYS molare Gaskonstante
molar mass PHYS Molarmasse
 = molare Masse
mold (n.) (NAM) METAL Gußform
 = mould (BRI) = Form
molded (NAM) TECH gegossen
 = moulded (BRI) [Kunststoff]
molding strip (→ornamental TECH (→Zierleiste)
 strip)
mole PHYS Mol
 [SI unit for quantity of substance] [SI-Einheit für Stoffmenge]
 = mol = mol
molecular CHEM molekular
molecular beam PHYS Molekularstrahl
 = molecular ray
molecular beam epitaxy MICROEL Molekularstrahlepitaxie
 = MBE = MBE
molecular bond CHEM Molekülbindung
molecular electronics ELECTRON Molerkularelektronik
molecular field (→Weiss PHYS (→Weissches Feld)
 field)
molecular magnet PHYS Molekularmagnet
molecular motion (→molecular PHYS (→Molekularbewegung)
 movement)
molecular movement PHYS Molekularbewegung
 = molecular motion
molecular ray (→molecular PHYS (→Molekularstrahl)
 beam)
molecular resonance PHYS Molekularresonanz
molecular spectrum PHYS Molekülspektrum
molecular weight PHYS Molekulargewicht
molecule PHYS Molekül
 = Molekel
molten solder MANUF Lotbad
 = Lötbad
molybdenum CHEM Molybdän
 = Mo = Mo
moment (→instant) PHYS (→Zeitpunkt)
momentary (→temporary) TECH (→vorübergehend)
momentary contact (→wiping COMPON (→Wischkontakt)
 contact)
momentary contact switch COMPON (→Taster)
 (→non-locking key)
momentary value (→instant PHYS (→Augenblickswert)
 value)
moment generating function MATH momenterzeugende Funktion
moment of force (→torque) PHYS (→Drehmoment)
moment of inertia PHYS Trägheitsmoment
 = inertial torque = Drehmasse
moment of momentum (→angular PHYS (→Drehimpuls)
 momentum)
momentum PHYS Impuls
 [mass x velocity] [Masse x Geschwindigkeit]
 = Bewegungsgröße
momentum MATH Moment
 [vectorial calculus] [Vektorrechnung]
monadic DATA PROC monadisch
 [with only one operand] [mit einem Operanden]
 = unary
monaural ACOUS monoaural
 = monoaural = einohrig
 = monophonic [EL.ACOUS] = monophon [EL.ACOUS]
 ≠ binaural ≠ zweiohrig
money ECON Geld
 = jack (US slang) ↑ Zahlungsmittel
 ↑ payment means ↓ Zentralbankgeld; Giralgeld;
 ↓ central bank money; bank money; de- Quasigeld
 mand deposit

money order (→payment order) ECON (→Zahlungsanweisung)
money-order form ECON Zahlkarte
money testing equipment OFFICE Geldprüfgerät
 = cash testing equipment
monimatch ANT Monimatch
 [for VSWR monitoring] [zur Dauerüberwachung
 des Stehwellenverhältnis-
 ses]
monitor (v.t.) TELEPH mithören
 ≈ eavesdrop ≈ abhören
monitor (n.) BROADC Kontrollempfänger
 = Monitor
monitor (n.) TERM&PER Monitor
 = picture monitor; picture and wave- [Sichtgerät für Kontroll-
 form monitor zwecke]
 ≈ display terminal ≈ Bildkontrollempfänger;
 Bildmonitor; Kontrollemp-
 fänger
 ≈ Sichtgerät
monitor CONTROL Monitor
 [control of thresholds] [Grenzwertüberwachung]
monitor (→executive DATA PROC (→Hauptsteuerprogramm)
 routine)
monitor (v.t.) (→tracer) DATA PROC (→Überwacher)
monitor (v.t.) (→supervise) TECH (→überwachen)
monitor diode ELECTRON Monitordiode
monitored TECH überwacht
 = supervized ≠ unüberwacht
 ≠ unmonitored
monitored binary information TELECONTR (→Meldung)
 (→message)
monitored information TELECONTR Überwachungsinformation
monitoring MIL.COMM Abhören
monitoring TELEPH Mithören
 = listening; listening-in
monitoring (→display) INSTR (→Anzeige)
monitoring (→supervision) TECH (→Überwachung)
monitoring channel TELEC (→Überwachungskanal)
 (→supervisory channel)
monitoring circuit (→control CIRC.ENG (→Überwachungsschaltung)
 circuit 2)
monitoring demodulator RADIO Überwachungsdemodulator
monitoring device EQUIP.ENG (→Überwachungsvorrich-
 (→supervisory device) tung)
monitoring direction TELECONTR Überwachungsrichtung
monitoring function EQUIP.ENG Überwachungsfunktion
monitoring point ELECTRON Überwachungspunkt
 ≈ test point ≈ Meßpunkt
monitoring printing TELEGR Kontrolldruck
monitoring receiver RADIO Überwachungsempfänger
 [radio monitoring] [Funküberwachung]
 = surveillance receiver
monitoring relay (→sensor COMPON (→Überwachungsrelais)
 relay)
monitoring terminal TERM&PER Mitlesemaschine
monitoring threshold ELECTRON (→Überwachungsschwelle)
 (→supervision threshold)
monitor man (→sound EL.ACOUS (→Toningenieur)
 engineer)
monitor tube ELECTRON Monitorröhre
 ↑ cathode ray tube; picture tube ↑ Kathodenstrahlröhre; Bildwiedergaberöhre
monkey wrench TECH Universalschraubenschlüssel
mono (n.) POWER SYS Monozelle
 [battery of approx. size 13x24"] [Zellengröße ca. 33x61
 mm^2]
monoaural (→monaural) ACOUS (→monoaural)
monoband antenna (→single-range ANT (→Einbandantenne)
 antenna)
monoband dipole ANT Einbanddipol
 = Monoband-Dipol
monobander (→single-range ANT (→Einbandantenne)
 antenna)
monobrid MICROEL monobrid
 ≠ hybrid ≠ hybrid
monobrid technology MICROEL Monobridtechnik
 ≠ hybrid technology ≠ Hybridtechnik
monochromatic (→monochrome) PHYS (→monochrom)
monochromatic (→black and TV (→Schwarz-Weiß)
 white)

monochromatic light	PHYS	monochromatisches Licht	= monostable trigger circuit; monoflop; univibrator		gen stabilen Zustand am Ausgang zurück]
monochromatic radiation	PHYS	monochromatische Strahlung	↑ one-shot multivibrator		= Monoflop; Univibrator; Monovibrator
monochrome	PHYS	monochrom			↑ stabile Kippstufe
= monochromatic		= monochromatisch; einfarbig	monostable trigger circuit	CIRC.ENG	(→monostabile Kippstufe)
monochrome display	ELECTRON	(→Schwarz-Weiß-Bildröhre)	(→monostable multivibrator)		
(→black-and-white tube)			monostatic cross section	RADIO LOC	monostatische Reflexionsflä-
monochrome graphics adapter	TERM&PER	(→MGA-Karte)	[scatters back to sending antenna]		che
(→MGA board)					[strahlt zur Antenne zu-
monochrome graphics board	TERM&PER	Monochrom-Graphikkarte			rück]
↑ graphics board		↑ Graphikkarte	monotone (adj.)	MATH	monoton
↓ MGA board; Hercules board		↓ MGA-Karte; Hercules-Kar-	[steadily increasing or steadily de-		[immer wachsend oder im-
		te	creasing]		mer fallend]
monochrome monitor	TERM&PER	monochromer Monitor	= monotonic		= echt monoton; eigentlich
monochrome plotter	TERM&PER	Einfarb-Plotter			monoton
		= Monochrom-Plotter; mono-			↓ monoton wachsend; monot-
		chromatischer Plotter			on fallend
monochrome printer	TERM&PER	Schwarz-Weiß-Drucker			
		= monochromer Drucker	monotonic (→monotone)	MATH	(→monoton)
monochrome scanner	TERM&PER	Schwarz-/Weiß-Scanner	monotonicity (→monotony)	MATH	(→Monotonie)
monochrome tube	ELECTRON	(→Schwarz-Weiß-Bildröhre)	monotony	ELECTRON	Monotonie
(→black-and-white tube)			monotony	MATH	Monotonie
monoclinic	PHYS	monoklin	= monotonicity		
		[Kristall]	Monte-Carlo method	MATH	Monte-Carlo-Methode
monocrystal (→single	PHYS	(→Einkristall)	= random-walk method		
crystal)			month	PHYS	Monat
monoflop (→monostable	CIRC.ENG	(→monostabile Kippstufe)	monthly (n.) (→monthly	COLLOQ	(→Monatszeitschrift)
multivibrator)			magazine)		
monograph	LING	Fachbuch	monthly access fee	TELEC	Monatsgebühr
≈ textbook		≈ Monographie	= monthly charge		= monatliche Grundgebühr
		≈ Lehrbuch	monthly charge (→monthly access	TELEC	(→Monatsgebühr)
monolayer PBC	ELECTRON	Einebenen-Leiterplatte	fee)		
monolithic	MICROEL	monolitisch	monthly magazine	COLLOQ	Monatszeitschrift
monolithic filter	MICROEL	monolithisches Filter	= monthly (n.)		= Monatsschrift; Monatsheft
monolithic integrated	MICROEL	monolithische integrierte	↑ magazine		↑ Zeitschrift
circuit		Schaltung	monthly wage	ECON	Monatslohn
= semiconductor block technology;		= monolithischer Schaltkreis;	≈ monthly salary		≈ Monatsgehalt
monolothic IC		Halbleiterblocktechnik	↑ wage		↑ Lohn
		(DDR)	mood	LING	Modus
monolithic layer-built capacitor	COMPON	(→Vielschichtkondensator)	↓ indicative mood; subjunctive mood;		= Aussageart
(→multilayer capacitor)			imperative mood		↓ Indikativ; Konjunktiv; Im-
monolothic IC (→monolithic	MICROEL	(→monolithische integrierte			perativ
integrated circuit)		Schaltung)	mop-up (→fine	ELECTRON	(→Feinabgleich)
monomial (n.)	MATH	Monom	adjustment)		
[a single-term expression]		[eingliedriger mathemati-	mop-up network	CIRC.ENG	Feinabgleich-Netzwerk
		scher Ausdruck]	more data	DATA COMM	Folgepaket
		= Monomon	more-than sign (→greater-than	MATH	(→Größerzeichen)
monomial (adj.)	MATH	monomisch	sign)		
[composed by a single term]		[aus nur einem Glied beste-	more-than symbol (→greater-than	MATH	(→Größerzeichen)
		hend]	sign)		
		= mononomisch	Morgan connection	POWER SYS	Morgan-Schaltung
monomode fiber	OPT.COMM	(→Einmodenfaser)	= Troeger connection		≈ Tröger-Schaltung
(→single-mode fiber)			Morgan's theorems	ENG.LOG	Theorem von De Morgan
monomode optical waveguide	OPT.COMM	(→Einmodenfaser)			= De Morgansche Theoreme
(→single-mode fiber)			morpheme	LING	Morphem (n.n.)
monomode range	MICROW	Eindeutigkeitsbereich	[smallest meaning-bearing word ele-		[kleinstes bedeutungtragen-
		= Einwelligkeitsbereich	ment, as "car" or "s" in "cars"; typi-		des Wortelement, wie "er",
monophonic groove	EL.ACOUS	Mono-Rille	cally 10,000 per language]		„bau" oder „en" in „erbau-
monophonic transmission	TELEC	monophone Übertragung			en"; typisch 10.000 pro
≠ stereophonic transmission		≠ Stereoübertragung			Sprache]
monopole	ANT	Monopol			= Phonemfolge
= monopole antenna; monopole aereal;		= Monopolantenne; Unipol;			≈ Morph
unipole		Unipolantenne; einpolige	morphology	LING	Morphologie
≈ vertical antenna		Antenne	[rules for word formation]		[Regeln der Formenbil-
≠ dipole		≈ Vertikalantenne	↑ grammar		dung von Wörtern]
↓ Marconi antenna; folded monopole		≠ Dipol	↓ declination; conjugation		= Formenlehre
		↓ Marconi-Antenne;			↑ Grammatik
		Faltmonopol			↓ Deklination; Konjugation
monopole (→single-pole)	EL.TECH	(→einpolig)	Morse code	CODING	Morse-Code
monopole (→unipolar)	PHYS	(→unipolar)	= Morse telegraph code		= Morse-Alphabet
monopole aereal (→monopole)	ANT	(→Monopol)	Morse dash	TELEGR	Morsestrich
monopole antenna (→monopole)	ANT	(→Monopol)	Morse key	TELEGR	Morsetaste
monopulse antenna	ANT	Monopuls-Antenne	Morse telegraph code (→Morse	CODING	(→Morse-Code)
monopulse mode	RADIO LOC	Monopulse-Verfahren	code)		
monoscope	ELECTRON	Monoskop	Morse telegraphy	TELEGR	Morse-Telegraphie
monospaced font	TYPOGR	(→Konstantschrift)	mortgage (→pledge)	ECON	(→Pfand)
(→constant-width font)			mortise (v.t.)	TECH	verzapfen
monostable	TECH	monostabil	MOS	MICROEL	Metalloxid-Halbleiter
monostable multivibrator	CIRC.ENG	monostabile Kippstufe	= metal-oxide semiconductor		= MOS
[returns to its sole stable output con-		[kippt aus einem quasistabi-	mosaic electrode	ELECTRON	Mosaikelektrode
dition]		len Zustand in seinen einzi-			

mosaic panel	TERM&PER	**Mosaikschaubild**
MOS capacitor	MICROEL	**MOS-Kondensator**
MOSFET	MICROEL	**MOSFET**
= MOS transistor; metal-oxide-semi-conductor field-effect transistor		= MOS-Transistor
↑ unipolar transistor		↑ Unipolartransistor
MOS PROM (→EPROM)	MICROEL	(→EPROM)
moss [color]	PHYS	**moosfarben** = moos
MOS transistor (→MOSFET)	MICROEL	(→MOSFET)
most significant	CODING	**höchstwertig**
= highest-order		= werthöchst
most significant bit	CODING	**höchstwertiges Bit**
= MSB		= werthöchstes Bit; MSB
most significant byte	CODING	**höchstwertiges Byte**
		= werthöchstes Byte
mother board [DATA PROC] (→backplane)	EQUIP.ENG	(→Rückwandleiterplatte)
mother board (→main board)	DATA PROC	(→Hauptplatine)
mother card (→main board)	DATA PROC	(→Hauptplatine)
motion	PHYS	**Bewegung**
= movement		
motion artifact	INF.TECH	**Bewegungsfehler**
motion blur [computer graphics]	DATA PROC	**Bewegungsunschärfe** [Computergraphik]
motion compensated	SIGN.TH	**bewegungskompensiert** [Signalverarbeitung]
motion detector (→movement detector)	SIGN.ENG	(→Bewegungsmelder)
motionless	TECH	**bewegungslos**
motion planning	CONTROL	**Bewegungsprogrammierung**
motion sensor (→movement detector)	SIGN.ENG	(→Bewegungsmelder)
motion transmission	MECH	**Bewegungsübertragung**
= transmission		
motion video (→moving picture communication)	TELEC	(→Bewegtbild-Kommunikation)
motor	TECH	**Motor**
= engine		↑ Maschine
↑ machine		↓ Verbrennungsmotor; Elektromotor
↓ combustion machine; electric machine		
motorboat (v.t.)	EL.ACOUS	**blubbern**
motor capacitor (→motor starting capacitor)	POWER SYS	(→Motorkondensator)
motor convertor (IEC)	POWER SYS	**Kaskadenumformer** (IEC)
motor-generator set	POWER SYS	**Motorgenerator**
= motor genset		↑ Umformer 1
↓ diesel generating set		↓ Dieselaggregat
motor genset (→motor-generator set)	POWER SYS	(→Motorgenerator)
motor speed	TECH	**Motordrehzahl**
motor starting capacitor	POWER SYS	**Motorkondensator**
= motor capacitor		
motor vehicle	TECH	**Kraftfahrzeug**
= automobile		= Kraftwagen; Kfz
motorway (BRI) (→freeway)	CIV.ENG	(→Autobahn)
motor winch	OUTS.PLANT	**Kabelkraftwinde**
		= Spillwinde
mould (BRI) (→mold)	METAL	(→Gußform)
moulded (BRI) (→molded)	TECH	(→gegossen)
mount (v.t.)	TECH	**montieren**
≈ assemble (v.t.); install (v.t.)		= einbauen
		≈ zusammenbauen; installieren
mount (n.) [device]	TECH	**Halterung** [Vorrichtung]
= fixing element; fastening 2; fastener; holding piece; holder		= Befestigung 2; Fixierung; Befestigungselement
≈ support		≈ Stütze
mount (→test probe)	INSTR	(→Prüfspitze)
mounted [component inserted on a socket]	ELECTRON	**gesockelt** [auf Sockel aufgestecktes Bauteil]
mounted board (→module)	EQUIP.ENG	(→Baugruppe)
mounting 1 (n.)	TECH	**Einbau**
mounting 3	TECH	**Montage**
≈ installation; assembly		≈ Installation; Zusammenbau
mounting 2	TECH	**Aufstellung 3**
= installation; setting-up		= Aufstellen
↓ table mounting; wall mounting; floor installation		↓ Tischaufstellung; Wandaufstellung; Bodenaufstellung
mounting base	TECH	**Befestigungssockel**
mounting bracket	EQUIP.ENG	**Montagewinkel**
mounting chassis (→module frame)	EQUIP.ENG	(→Baugruppenrahmen)
mounting device	EQUIP.ENG	**Aufnahme**
mounting dimension	MECH	**Montagemaß**
mounting frame	MANUF	**Montagerahmen**
≈ assembly appliance		≈ Montagelehre
mounting frame (→module frame)	EQUIP.ENG	(→Baugruppenrahmen)
mounting height	EQUIP.ENG	**Einbauhöhe**
		= Bauhöhe
mounting instructions (→installation instructions 2)	TECH	(→Einbauanleitung)
mounting instructions (→installation specification)	SYS.INST	(→Montagevorschrift)
mounting jack	COMPON	**Einbaubuchse**
= panel socket (BRI); chassis receptacle		= Gerätebuchse; Gerätedose; Einbaukupplung; Gehäusekuppler;
		≠ Kabelbuchse; Einbaustecker
		↓ Flanschbuchse; Einbaubuchse mit Zentralbefestigung
mounting jack	TELEPH	**Klinken-Einbaubuchse**
		≈ Klinkenkupplung
		↑ Klinke
mounting kit	TECH	**Montagesatz**
≈ kit [ELECTRON]		≈ Bausatz [ELECTRON]
mounting level	EQUIP.ENG	**Einbauebene**
mounting location (→mounting place)	EQUIP.ENG	(→Einbauplatz)
mounting pitch	EQUIP.ENG	**Einbauteilung**
= installation pitch		
mounting place	EQUIP.ENG	**Einbauplatz**
= mounting position 2; mounting location; slot		= Einbauort; Einbaulage 2
		≈ Einbaustellung
↓ plug-in place		↓ Steckplatz
mounting place code (→slot code)	EQUIP.ENG	(→Einbaukennung)
mounting plate (→base plate)	MECH	(→Grundplatte)
mounting plug	COMPON	**Einbaustecker**
= panel plug		= Gerätestecker
		≠ Einbaubuchse; Kabelbuchse
		↓ Flanschstecker
mounting position 1	EQUIP.ENG	**Einbaustellung**
		= Einbaulage 1
		≈ Einbauplatz
mounting position 2 (→mounting place)	EQUIP.ENG	(→Einbauplatz)
mounting position code (→slot code)	EQUIP.ENG	(→Einbaukennung)
mounting shelf (→module frame)	EQUIP.ENG	(→Baugruppenrahmen)
mounting socket	COMPON	**Einbausteckdose**
mounting strap	TECH	**Spannband**
mounting unit	EQUIP.ENG	**Einbaueinheit**
mounting width	COMPON	**Baubreite**
mount on	TECH	**aufmontieren**
mouse (→control mouse)	TERM&PER	(→Maus) (pl.Mäuse)
mouse-based (→mouse-controlled)	DATA PROC	(→mausgesteuert)
mouse button	TERM&PER	**Mausknopf**
		= Maustaste
mouse click	TERM&PER	**Mausklick**
mouse connector	DATA PROC	**Mausanschluß**
mouse-controlled	DATA PROC	**mausgesteuert**
= mouse-driven; mouse-based		= mausgestützt
mouse-driven (→mouse-controlled)	DATA PROC	(→mausgesteuert)
mouse driver	DATA PROC	**Maustreiber**
mouse pad	TERM&PER	**Mausunterlage**
		= Mauspad

mouse pointer	DATA PROC	Mauszeiger
mouse rest	TERM&PER	Mausgarage
mouthpiece of microphone	EL.ACOUS	Mikrophonbecher
m-out-of-n code	CODING	m-aus-n-Code
movable (→mobile)	TECH	(→beweglich)
movable disk relay	COMPON	Scheibenrelais
movable-head disk unit [one head for several disks] ≠ fixed-head disk unit	TERM&PER	Bewegtkopf-Plattenlaufwerk [ein Kopf für mehrere Platten] ≠ Festkopf-Plattenlaufwerk
moveable (→mobile)	TECH	(→beweglich)
moved image (→moved picture)	TELEC	(→Bewegtbild)
moved picture = moving picture; moved image; moving image; full-motion picture	TELEC	Bewegtbild
moved picture communication (→moving picture communication)	TELEC	(→Bewegtbild-Kommunikation)
movement (→motion)	PHYS	(→Bewegung)
movement data (→transaction data)	DATA PROC	(→Bewegungsdaten)
movement detector = movement sensor; motion sensor; motion detector	SIGN.ENG	Bewegungsmelder
movement file (→transaction file)	DATA PROC	(→Bewegungsdatei)
movement sensor (→movement detector)	SIGN.ENG	(→Bewegungsmelder)
movement simulation	DATA PROC	Bewegungssimulation
movie-sound recording	EL.ACOUS	Filmtonaufnahme
moving	ECON	Umzug = Standortwechsel
moving average	DATA PROC	gleitende Mittelwertbildung
moving-coil galvanometer ↓ creeping galvanometer	INSTR	Drehspulgalvanometer ↓ Kriechgalvanometer
moving-coil instrument = moving-coil meter	INSTR	Drehspulinstrument
moving-coil loudspeaker (→electrodynamic loudspeaker)	EL.ACOUS	(→elektrodynamischer Lautsprecher)
moving-coil mechanism	INSTR	Drehspulmeßwerk
moving-coil meter (→moving-coil instrument)	INSTR	(→Drehspulinstrument)
moving-coil microphone ↑ dynamic microphone	EL.ACOUS	Tauchspulmikrophon = Tastspulmikrophon ↑ dynamisches Mikrophon
moving-coil pickup	EL.ACOUS	elektrodynamischer Tonabnehmer
moving-coil ratiometer	INSTR	Drehspulquotientenmesser
moving-coil relay	COMPON	Drehspulrelais
moving-disk memory [disk or disk stack can be removed] = moving-disk storage; removable-disk memory; removable-disk storage; exchangeable disk memory; exchangeable disk storage; exchangeable disk stor; EDS ≈ exchangeable disk drive; hard disk memory ≠ fixed-disk memory ↑ disk memory	TERM&PER	Wechselplattenspeicher [Platte oder Plattenstapel kann ausgewechselt werden] = Wechsel-Magnetplattenspeicher ≈ Wechselplattenlaufwerk; Hartplattenspeicher ≠ Festplattenspeicher ↑ Magnetplattenspeicher
moving-disk storage (→moving-disk memory)	TERM&PER	(→Wechselplattenspeicher)
moving element (→pointer)	INSTR	(→Zeiger)
moving image (→moved picture)	TELEC	(→Bewegtbild)
moving-iron instrument	INSTR	Dreheiseninstrument
moving junction	MICROW	Drehkupplung
moving magnet (→rotary magnet)	EL.TECH	(→Drehmagnet)
moving-magnet galvanometer ↓ vibration galvanometer	INSTR	Drehmagnetgalvanometer ↓ Vibrationsgalvanometer
moving-magnet mechanism	INSTR	Drehmagnet-Meßwerk
moving-magnet ratiometer	INSTR	Drehmagnet-Quotientenmesser
moving-magnet vibration galvanometer	INSTR	Drehmagnet-Vibrationsgalvanometer
moving mechanism = feed mechanism; advance mechanism	MECH	Transporteinrichtung
moving-needle galvanometer	INSTR	Nadelgalvanometer
moving picture (→moved picture)	TELEC	(→Bewegtbild)
moving picture communication = moved picture communication; full-motion video; motion video	TELEC	Bewegtbild-Kommunikation
moving target indication [radar] = MTI	RADIO LOC	Festzeichenunterdrückung [Radar] = Festzeichenlöschung; MTI; Festzielunterdrückung
MOVPE = metall-organic vapor-phase epitaxy	MICROEL	MOVPE = metallorganisch-chemische Gasphasenepitaxie
Mp (→megapond)	PHYS	(→Megapond)
mph = miles per hour ≈ kilometers per hour	TECH	Meilen pro Stunde ≈ Stundenkilometer
MQW laser diode	OPTOEL	MQW-Laserdiode
MRF (→multipath transmission factor)	RADIO PROP	(→Mehrwege-Reduktionsfaktor)
µs (→microsecond)	PHYS	(→Mikrosekunde)
m/s (→meter per second)	PHYS	(→Meter durch Sekunde)
ms (→millisecond)	PHYS	(→Millisekunde)
MSB (→most significant bit)	CODING	(→höchstwertiges Bit)
MSFET = metal semiconductor FET	MICROEL	MSFET
M shell (→third shell)	PHYS	(→3.Schale)
MSI [10 to 500 components per IC] = medium scale integration	MICROEL	Mittelintegration [10 bis 500 Komponenten pro IC] = MSI; mittlerer Integrationsgrad
MTAE (→message transfer agent entity)	DATA COMM	(→Instanz des Transport-Systemteils)
MTBF (→mean time between failures)	QUAL	(→mittlerer Ausfallabstand)
MTI (→moving target indication)	RADIO LOC	(→Festzeichenunterdrückung)
MTL (→inegrated injection logic)	MICROEL	(→IIL)
MTOS = metal thick oxide semiconductor	MICROEL	MTOS
MTTR = mean time to repair ≈ mean down time	QUAL	MTTR = mittlere Instandsetzungsdauer; mittlere Reparaturdauer ≈ mittlere Ausfalldauer
much-greater sign [symbol: ≫] = much-greater symbol	MATH	Wesentlich-größer-Zeichen [Symbol: ≫]
much-greater symbol (→much-greater sign)	MATH	(→Wesentlich-größer-Zeichen)
much-less sign [symbol: ≪] = much-less symbol	MATH	Wesentlich-kleiner Zeichen [Symbol: ≪]
much-less symbol (→much-less sign)	MATH	(→Wesentlich-kleiner Zeichen)
MUF (→maximum usable frequency)	RADIO	(→obere Grenzfrequenz)
muffler (→deadener)	ACOUS	(→Schalldämpfer)
muldex [multiplexer + demultiplexer]	TRANS	Muldex [Multiplexer + Demultiplexer]
muldex slide-unit	TRANS	Muldex-Einschub
muldipol antenna	ANT	Muldipol-Antenne
multee two-band antenna	ANT	Multee-Zweibandantenne
multi-access computer	DATA PROC	Vielfachzugriff-Computer
multi-access line (→party line 2)	TELEC	(→Gemeinschaftsanschluß)
multiaddress call (→conference call)	TELEC	(→Konferenzverbindung)
multi-address calling (→multi-address message)	DATA COMM	(→Rundschreiben)
multiaddress circuit (→conference call)	TELEC	(→Konferenzverbindung)
multi-address instruction = multiple address instruction ≠ single-address instruction ↓ two-address instruction; three-address instruction	DATA PROC	Mehr-Adreß-Befehl ≠ Ein-Adreß-Befehl ↓ Zwei-Adreß-Befehl; Drei-Adreß-Befehl
multi-address machine	DATA PROC	Mehr-Adreß-Maschine

multi-address message		DATA COMM	**Rundschreiben**	**multicomponent** (→composite)	TECH	(→zusammengesetzt)
[message is sent to several addressees indicated by the sender]			[Nachricht wird an mehrere, vom Absender genannte, Empfänger gesandt]	**multi-computer system**	DATA PROC	**Mehrrechnersystem**
				[interaction of various computers for a common task]		[Kopplung mehrerer Rechner zu einer gemeinsamen Aufgabe]
= multi-address calling; multiple-address message; broadcasting; broadcast			= Rundsenden	= computer network; interlocked network		= Rechnerverbund; Computerverbund; Rechnerkopplung; Funktionsverbund; Verbundsystem; Rechnernetz; Verbundschaltung
multiband antenna		ANT	**Mehrbandantenne**	≈ multi-processor system; network [DATA COMM]		
			= Allbandantenne; Multibandantenne	↓ load sharing; data sharing; front-end computer system; dual computer system; parallel computer system; duplex computer system		≈ Mehrprozessorsystem; Netzwerk [DATA COMM]
multiband beam antenna		ANT	**Mehrbandrichtstrahler**			
multiband delta loop antenna		ANT	**Mehrband-delta-loop-Antenne**			↓ Lastverbund; Datenverbund; Vorrechnersystem; Doppelsystem; Parallelsystem; Duplexsystem
multiband element		ANT	**Mehrbandelement**			
multiband filter		NETW.TH	**Mehrbandfilter**			
multiband groundplane antenna		ANT	**Mehrband-groundplane-Antenne**	**multiconductor**	COMM.CABLE	**mehradrig**
				= multicore; multiwire		= vielpolig
multiband operation		ANT	**Mehrbandbetrieb**	**multi-connection endpoint identifier**	DATA COMM	**Zielpunktkennung**
multiband trap antenna		ANT	**Allband-Trap-Antenne**			
multiband tuner		ANT	**Allbandanpaßgerät**	**multicontact connector**	COMPON	**Vielpolstecker**
[for all radio amateur bands]			[für alle Amateurfunkbänder]	**multicore**	COMM.CABLE	(→mehradrig)
				(→multiconductor)		
multiband vertical antenna		ANT	**Mehrleiter-Groundplane**	**multi-core cable**	POWER ENG	**Mehraderleitung**
multiband Zeppelin antenna		ANT	**Allband-Zeppelin-Antenne**	**multicoupler** (→receiving multicoupler)	RADIO	(→Trennverstärker)
multi-beam antenna		ANT	**Mehrbündelantenne**			
			= Mehrkeulenantenne	**multi-destination**	TELEGR	**Sammelschaltung**
multibranched		TECH	**mehrverzweigt**	**multidimensional**	MATH	**mehrdimensional**
			= mehrzweigig; vielzweigig; vielverzweigt	= multivariate		= vieldimensional
				multidinous (→numerous)	COLLOQ	(→zahlreich)
multiburst signal		TV	**Multiburst**	**multi-domain network**	DATA COMM	**Mehrdomänennetz**
multibus (adj.)		DATA PROC	**mit mehreren Bussen**	**multidrop**	DATA COMM	**Mehrpunkt**
multibyte instruction		DATA PROC	(→**Mehrwortbefehl**)			= Knotenpunkt
(→multiword instruction)				**multidrop line**	DATA COMM	**Gruppenverbindung**
multi-carrier system		TELEC	**Mehrbetreibersystem**	**multi-duct conduit**	OUTS.PLANT	**Mehrröhrenkanal**
multicathode tube		ELECTRON	**Vielkathodenröhre**	↑ cable conduit		↑ Kabelkanal
multicavity klystron		MICROW	**Vielkammerklystron**	**multielectrode tube**	ELECTRON	**Mehrgitterröhre**
			= Vielkreisklystron; Mehrkammerklystron	= multigrid tube; multielectrode valve; multigrid valve		= Mehrelektrodenröhre
multicavity magnetron		MICROW	(→**Wanderfeldmagnetron**)	**multielectrode valve**	ELECTRON	(→**Mehrgitterröhre**)
(→travelling-wave magnetron)				(→multielectrode tube)		
multicavity v.m. tube		MICROEL	**Mehrkreis-Triftröhre**	**multi-emitter transistor**	MICROEL	**Mehremitter-Transistor**
multicell horn		ANT	**Facettenhorn**			= Multiemitter-Transistor
multicellular loudspeaker		EL.ACOUS	**Vielzellenlautsprecher**	**multi-endpoint connection**	DATA COMM	(→**Mehrpunktleitung**)
multichannel ...		TELEC	**Mehrkanal-**	(→multipoint circuit)		
			= Vielkanal-	**multi-equipment system**	DATA PROC	**Geräteverbund**
multi-channel analyzer		INSTR	**Vielkanalanalysator**	**multifarious** (→manifold 1)	COLLOQ	(→mannigfaltig)
multichannel circuit (→carrier link line)		TRANS	(→**TF-Grundleitung**)	**multifariousness** (→manifoldness)	COLLOQ	(→Mannigfaltigkeit)
multichannel connection		TELEC	(→**Mehrkanalverbindung**)	**multi-fee metering**	SWITCH	**Mehrfachgebührenzählung**
(→multichannel link)				[differentiated metering for own-exchange calls]		[differenzierte Gebührenzählung für Gespräche innerhalb der eigenen Vermittlung]
multichannel link		TELEC	**Mehrkanalverbindung**			
= multichannel connection			= Vielkanalverbindung	= multimetering		
multichannel loudspeaker		EL.ACOUS	(→**Lautsprecherkombination**)			= Mehrfachzählung
(→composite loudspeaker)				**multifiber buffer tube**	OPT.COMM	**Bündelader**
multichannel switch		DATA PROC	**Mehrkanalschalter**	**multifold 1** (→manifold 1)	COLLOQ	(→mannigfaltig)
multichannel system		TELEC	**Mehrkanalsystem**	**multifold 2** (adj.)	COLLOQ	(→vielfach)
			= Vielkanalsystem	(→multiple)		
multichannel transmission		TELEC	**Mehrkanalübertragung**	**multifont printer**	TERM&PER	**Mehrschriftendrucker**
= channelization 2			= Vielkanalübertragung	**multi-font reader**	TERM&PER	**Mehrschriftenleser**
multichannel transmitter		BROADC	**Mehrkanal-Rundfunksender**	= multi-font scanner		= Multifontleser
multi-channel transmitter		RADIO	**Vielkanalsender**	≈ omni-font reader		≈ Allschriftenleser
multi-chip hybrid technology		MICROEL	**Multichip-Hybridtechnik**	↑ character reader		↑ Klarschriftenleser
				multi-font scanner	TERM&PER	(→**Mehrschriftenleser**)
multi-chip technology		MICROEL	**Multichiptechnik**	(→multi-font reader)		
[several chips in a case]			[mehrere Chips in einem Gehäuse]	**multiform** (adj.)	COLLOQ	**verschiedenförmig**
				= multishaped		= verschiedengestaltet
multi-chute device		TERM&PER	**Mehrschachteinzug**	**multiframe**	CODING	**Mehrfachrahmen**
multi-circuit (adj.)		NETW.TH	**mehrkreisig**	= superframe		= Überrahmen; Mehrfachpulsrahmen
multi-circuit filter		NETW.TH	**mehrkreisiges Filter**			
multi-collector transistor		MICROEL	**Mehrkollektor-Transistor**	**multifrequency code**	CODING	**Mehrfrequenzcode**
multicolored printing		TYPOGR	(→**Mehrfarbendruck**)	= MFC		= MFC
(→multicolor print)				**multi-frequency-code dialing** (→MFC dialing)	SWITCH	(→MFC-Wahl)
multicolor print		TYPOGR	**Mehrfarbendruck**			
= multicolour print; multicolored printing			↓ Zweifarbendruck; Dreifarbendruck	**multi-frequency dialing** (→MFC dialing)	SWITCH	(→MFC-Wahl)
↓ two-color print; three-color print				**multi-frequency display**	TERM&PER	(→Autosync-Monitor)
multi-coloured		TECH	**mehrfarbig**	(→multi-frequency monitor)		
= coloured				**multifrequency modulation**	MODUL	**Mehrfrequenzmodulation**
multicolour print (→multicolor print)		TYPOGR	(→**Mehrfarbendruck**)			
multi-column (adj.)		TYPOGR	**mehrspaltig**			

multi-frequency monitor TERM&PER
[adapts itself automatically to the type of graphic board]
= multiscan monitor; autosync monitor; multisync monitor; multi-frequency display; multiscan display; autosync display
≠ fixed-frequency monitor

Autosync-Monitor
[stellt sich automatisch auf die Graphikkarten-Variante ein]
= Multisync-Monitor; Multiscan-Monitor; Autosync-Bildschirm; Multisync-Bildschirm; Multiscan-Bildschirm
≠ Festfrequenz-Monitor

multi-frequency signaling SWITCH
(→MFC signaling)
(→MFC-Signalisierung)

multifrequency transmitter RADIO **Mehrfrequenzsender**
multifunctional board DATA PROC **Multifunktionskarte**
multifunctional chip card TERM&PER **Multifunktions-Chipkarte**
multifunctionality TECH **Multifunktionalität**
multifunctional terminal TERM&PER **Mehrdienste-Endgerät**
= Multifunktions-Endgerät; Multifunktionsterminal

multifunction network TELEC (→Mehrdienstenetz)
(→multipurpose network)

multi-functions peripherie DATA COMM **Multifunktionsperipherie**

multifunction synthesizer INSTR **Multifunktions-Synthesizer**
multifunction workstation TERM&PER **Mehrfunktionsplatz**
multigrid tube ELECTRON (→Mehrgitterröhre)
(→multielectrode tube)

multigrid valve ELECTRON (→Mehrgitterröhre)
(→multielectrode tube)

multi-hole coupler MICROEL **Mehrlochkoppler**
multi-hop link RADIO **Mehrsprungverbindung**
[Kurzwellenverbindung]

multi-input (→multiple input) ELECTRON (→Mehrfacheingang)

multi-job operation DATA PROC **Mehrauftragbetrieb**
= Vielfachauftragbetrieb

multi-job processing DATA PROC **Mehrauftragverarbeitung**
multilayer (n.) TECH **Mehrschicht**
multilayer (adj.) TECH **mehrlagig**
= multi-part; multilayered
= mehrschichtig; vielschichtig

multilayer (→multilayer wiring) EQUIP.ENG (→Mehrlagenverdrahtung)

multilayer capacitor COMPON **Vielschichtkondensator**
= monolithic layer-built capacitor
= Stapelkondensator

multilayered (→multilayer) TECH (→mehrlagig)
multilayer metallization MICROEL **Mehrlagenmetallisierung**
multilayer PCB (→multilayer printed circuit board) ELECTRON (→Mehrlagenleiterplatte)

multilayer printed circuit board ELECTRON **Mehrlagenleiterplatte**
= multilayer PCB; multiple board; multiple PCB
= Mehrlagenplatte; Mehrschichtleiterplatte; Mehrebenenleiterplatte; Mehrfachleiterplatte; Mehrfachverdrahtungsplatte; Multilayer

multilayer technique MICROEL **Mehrlagentechnik**
= Multilayer-Technik; Mehrschichtverfahren

multilayer wiring EQUIP.ENG **Mehrlagenverdrahtung**
= multilayer
= Mehrebenenverdrahtung; Multilayer

multileaf (adj.) MATH **mehrblättrig**
multi-level (adj.) TECH **mehrstufig**
≠ single-level
= vielstufig
≠ einstufig

multilevel (→value-discrete) INF (→wertdiskret)
multilevel addressing DATA PROC (→indirekte Adressierung)
(→indirect addressing)

multilevel addressing mode DATA PROC (→indirekte Adressierung)
(→indirect addressing)

multilevel code (→multivalid code) CODING (→mehrwertiger Code)

multilevel modulation MODUL **Mehramplitudenmodulation**
multi-level modulation MODUL **Mehrstufenmodulation 1**
[mit mehrstufigen Kennzuständen]

multi-line (adj.) TYPOGR **mehrzeilig**
multiline (→multiple circuits) DATA COMM (→Paket-Reihenanschluß)

multilinear algebra MATH **multilineare Algebra**

multilingual LING **mehrsprachig**
multilink DATA COMM **Mehrfach-Übermittlungsabschnitt**
multilinked list DATA PROC **mehrfach verknüpfte Liste**
multilinking DATA PROC **Mehrfachverknüpfung**
multiloop (adj.) CONTROL **mehrschleifig**
multiple (adj.) (→manifold) COLLOQ (→vielfach)
multimedia mail TELEC **Mitteilungssystem für Text und Sprache**
≈ voice annotation
≈ Sprachanmerkung

multimeter INSTR **Universalmeßgerät**
= mutipurpose instrument; multirange instrument; multispan instrument
= Vielfachinstrument; Mehrbereichsinstrument; Vielbereichsinstrument; Multimeter; Vielfachmeßgerät

multimetering (→multi-fee metering) SWITCH (→Mehrfachgebührenzählung)

multi-mode fiber OPT.COMM **Mehrmodenfaser**
= multimode fiber
= Multimode-Faser
↓ Gradientenfaser; Stufenindexfaser

multimode fiber (→multi-mode fiber) OPT.COMM (→Mehrmodenfaser)

multimode laser OPTOEL **Multimodenlaser**
multimode waveguide MICROW **Multimodewellenleiter**
multinomial distribution MATH **Multinomialverteilung**
multi-octave amplifier MICROW **Multioktav-Verstärker**
multi-octave band MICROW **Multioktavenband**
multi-output (→multiple output) ELECTRON (→Mehrfachausgang)

multi-page TYPOGR **mehrseitig**
multipair cable COMM.CABLE **vielpaariges Kabel**
= mehrpaariges Kabel

multi-part (→multilayer) TECH (→mehrlagig)
multipartite (→multisectional) TECH (→mehrteilig)
multipart paper TERM&PER **mehrlagiges Papier**
multiparty line (→party line 2) TELEC (→Gemeinschaftsanschluß)
multipass (n.) DATA PROC (→Mehrfachdurchgang)
(→multipassing)

multipassing DATA PROC **Mehrfachdurchgang**
= multipass (n.)

multipass sort DATA PROC **Mehrschrittsortierung**
multipath ... TELEC **Mehrwege-**
multipath coupling loss ANT (→Antennenkopplungsverlust)
(→aperture-to-medium coupling loss)

multipath effect (→ghost image) TV (→Geisterbild)
multipath fading RADIO PROP **Interferenzschwund**
= selective fading; interference fading
= Mehrwegeschwund; Selektivschwund; selektiver Schwund; dispersiver Schwund

multipath propagation RADIO PROP **Mehrwegeausbreitung**
= multipath transmission

multipath reception RADIO **Mehrwegeempfang**
multipath routing (→path diversity) TELEC (→Mehrwegeführung)

multipath sorting DATA PROC **Mehrwegesortieren**
↑ sorting
↑ Sortieren

multipath transmission RADIO PROP (→Mehrwegeausbreitung)
(→multipath propagation)

multipath transmission (→path diversity) TELEC (→Mehrwegeführung)

multipath transmission factor RADIO PROP **Mehrwege-Reduktionsfaktor**
[shortwave transmission]
= MRF
[Kurzwellenübertragung]
= MRF-Wert

multiphase (→polyphase) POWER ENG (→mehrphasig)
multiphase current POWER ENG (→Mehrphasenstrom)
(→polyphase current)

multiphase modulation MODUL **Multiphasenmodulation**
multiphase system POWER SYS (→Mehrphasensystem)
(→polyphase system)

multiple MECH **mehrgängig**
[thread]
[Gewinde]

multiple (adj.) COLLOQ **vielfach 1**
= manifold 2; multifold 2
≈ mannigfaltig
= manifold 1

multiple (n.) (→manifold) COLLOQ (→Vielfach)
multiple (→multiple field) SWITCH (→Vielfachfeld)

English	Domain	German
multiple access	TELEC	**Vielfachzugriff**
		= Mehrfachzugriff
multiple address	TELECONTR	**Mehradresse**
multiple address instruction	DATA PROC	(→Mehr-Adreß-Befehl)
(→multi-address instruction)		
multiple-address message	DATA COMM	(→Rundschreiben)
(→multi-address message)		
multiple antenna	ANT	**Mehrfachantenne**
multiple-bit error	CODING	**Mehrbitfehler**
multiple board (→multilayer printed circuit board)	ELECTRON	(→Mehrlagenleiterplatte)
multiple branch	CIRC.ENG	**Mehrfachverzweigung**
multiple-byte error	CODING	**Mehrbytefehler**
multiple-byte instruction	DATA PROC	**Mehrbytebefehl**
= variable-length instruction		
multiple chain (→multiple chaining)	DATA PROC	(→Mehrfachkettung)
multiple chaining	DATA PROC	**Mehrfachkettung**
= multiple chain; multiple concatenation		= Mehrfachverkettung
multiple circuits	DATA COMM	**Paket-Reihenanschluß**
= multiline		
multiple collision	PHYS	**Vielfachstoß**
multiple commutating	POWER SYS	**Mehrfachkommutierung**
multiple concatenation (→multiple chaining)	DATA PROC	(→Mehrfachkettung)
multiple connection	COMPON	**Mehrfachsteckverbindung**
= multi-way connection		= Vielfachsteckverbindung
multiple connection	DATA COMM	**Mehrfachverbindung**
multiple connection	TELEC	**Mehrgeräteanschluß**
multiple connector	COMPON	**Mehrfachsteckverbinder**
[pair of multiple plugs and multiple jacks]		= Vielfachsteckverbinder
↓ multiple plug; multiple jack		↓ Mehrfachstecker; Mehrfachbuchse
multiple contact	COMPON	**Mehrfachkontakt**
= hunting contact		
multiple control	CONTROL	**Mehrfachregelung**
[with several reference and controlled magnitudes]		[mit mehreren Führungs- und Regelgrößen]
↓ double control		↓ Zweifachregelung
multiple crosstalk	TELEC	**Mehrfachnebensprechen**
multiple die	TECH	**Mehrfachwerkzeug**
multiple diode	MICROEL	**Mehrfachdiode**
multiple dipole (→multiple dipole antenna)	ANT	(→Mehrfachdipol)
multiple dipole antenna	ANT	**Mehrfachdipol**
= multiple dipole		
multiple-disc clutch	MECH	**Mehrscheibenkupplung**
multiple echo	TELEC	**Mehrfachecho**
= flutter echo		
multiple exploitation	TECH	**Mehrfachnutzung**
= multiple use		
multiple fault	DATA PROC	**Mehrfachfehler**
multiple field	SWITCH	**Vielfachfeld**
= multiple		
multiple frequency keying	MODUL	**Mehrfrequenztastung**
multiple-helix potentiometer	COMPON	**Mehrfachwendelpotentiometer**
multiple input	ELECTRON	**Mehrfacheingang**
= multi-input		= mehrfacher Eingang
multiple integral	MATH	**mehrfaches Integral**
↓ double integral; triple integral		= Mehrfachintegral
		↓ Doppelintegral; dreifaches Integral
multiple jack	COMPON	**Mehrfachbuchse**
= multi-way jack		= Vielfachbuchse
≠ multiple plug		≠ Mehrfachstecker
↑ multiple connector		
multiple jack	TELEPH	**Vielfachklinke**
multiple-joint main sleeve	OUTS.PLANT	**Verzweigungsmuffe**
multiple lead (→highway)	SWITCH	(→Multiplexleitung)
multiple line	TELEC	**Mehrfachleitung**
		= Mehrleitersystem
multiple matching	ANT	**Mehrfachanpassung**
multiple modulation 1	MODUL	**Mehrfachmodulation**
[FDM]		[TF-Technik]
multiple modulation 2	MODUL	**Mehrstufenmodulation 2**
[modulation of premodulated signals]		[Modulation vormodulierter Signale]
		↓ Zweistufenmodulation
multiple output	ELECTRON	**Mehrfachausgang**
= multi-output		= mehrfacher Ausgang
multiple-pass printing	TERM&PER	**Mehrfachdruck**
[to reach print quality]		[zur Erzielung eines besonderen Druckbildes]
= overstriking; overwriting; multiple striking		
multiple PCB (→multilayer printed circuit board)	ELECTRON	(→Mehrlagenleiterplatte)
multiple plug	COMPON	**Mehrfachstecker**
[male connector with several contacts]		[stellt mehrere Kontakte gleichzeitig her; männlicher Teil]
= multi-way plug		= Vielfachstecker
≠ multiple jack		≠ Mehrfachbuchse
↑ multiple connector		↑ Mehrfachsteckverbinder
multiple printing	TERM&PER	**Mehrfachdrucken**
multiple processing	DATA PROC	**Simultanarbeit 2**
[simultaneous processing under different aspects]		[gleichzeitiges Bearbeiten eines Vorgangs nach verschiedenen Gesichtspunkten]
≈ integrated data processing		= simultane Verarbeitung
		≈ intergrierte Datenverarbeitung
multiple reception	TELEC	**Vielfachempfang**
multiple recording	TERM&PER	**Mehrfachbeschriftung**
multiple reflection	TELEC	**Mehrfachumlauf**
multiple reflection	PHYS	**Mehrfachreflexion**
		= Vielfachreflexion; Mehrfachspiegelung; Vielfachspiegelung
multiple rhombic antenna	ANT	**Rhombus-Gruppenantenne**
↓ rhombic line; rhombic row		↓ Rhombus-Linie; Rhombus-Reihe
multiple rotatable condensator	COMPON	**Mehrfachdrehkondensator**
multiple routing (→path diversity)	TELEC	(→Mehrwegeführung)
multiple sampling inspection	QUAL	**Mehrfach-Stichprobenprüfung**
multiple scattering	PHYS	**Mehrfachstreuung**
= plural scattering		= Vielfachstreuung
multiple selection	DATA PROC	**Mehrfachauswahl**
		= mehrfache Alternative
multiple-slot antenna	ANT	**Mehrschlitzstrahler**
multiple sorting	DATA PROC	**Mehrdateiensortierung**
multiple splice	OPT.COMM	**Mehrfachspleiß**
multiple striking	TERM&PER	(→Mehrfachdruck)
(→multiple-pass printing)		
multiple subtracter	CIRC.ENG	**Mehrfachsubtrahierer**
multiple teeing	OUTS.PLANT	**Parallelschalten**
= bridged taps		
multiple thread	MECH	**mehrgängiges Gewinde**
multiple-tuned antenna 1	ANT	**Mehrfachresonanz-Antenne**
≈ multiple-tuned antenna 2		= mehrfach abgestimmte Antenne; mehrfach geerdete Antenne
		≈ Alexanderson-Antenne
multiple-tuned antenna 2	ANT	(→Alexanderson-Antenne)
(→Alexanderson antenna)		
multiple tuning	ANT	**Vielfachabstimmung**
multiple twin formation	COMM.CABLE	**DM-Verseilung**
		= Dieselhorst-Martin-Verseilung
multiple-twin quad	COMM.CABLE	**DM-Vierer**
= D.M. quad		= Dieselhorst-Martin-Vierer
↑ quad; stranding element		↑ Viererseil; Verseilelement
multiple unit steerable array (→MUSA)	ANT	(→MUSA-Antenne)
multiple-unit tube	ELECTRON	**Verbundröhre**
= multitube		
multiple use (→multiplex)	TELEC	(→Multiplex)
multiple use (→multiple exploitation)	TECH	(→Mehrfachnutzung)
multiple-valued logic (→multivalid logic)	MATH	(→mehrwertige Logik)
multiple-way loudspeaker	EL.ACOUS	**Mehrwege-Lautsprecher**
multiple-wire antenna	ANT	**Mehrdrahtantenne**
= multiwire antenna		

multiplex

multiplex (v.t.) TELEC **multiplexieren**
≈ bundle ≈ multiplexen
 ≈ bündeln
multiplex TELEC **Multiplex**
= multiplexing; multiple use = Multiplexierung; Mehrfachnutzung 2; Mehrfachausnutzung

multiplex channel DATA PROC **Multiplexkanal**
[data channel for simultaneous interconnect a computer with several peripheral equipment] [Datenkanal zur gleichzeitigen Anbindung mehrerer Peripheriegeräte an einen Computer]
= multiplexer channel = MPX-Kanal
≠ selector channel ≠ Selektorkanal
↓ block multiplex channel; byte multiplex channel ↓ Blockmultiplexkanal; Bytemultiplexkanal

multiplex convention (→multiplex rule) TELEC (→Multiplexvorschrift)

multiplexed TELEC **gemultiplext**
 = multiplexiert

multiplex equipment (→multiplexer) TRANS (→Multiplexgerät)

multiplexer TRANS **Multiplexgerät**
= multiplex equipment; multiplexor (BRI) = Multiplexer

multiplexer (→multiplexor) DATA PROC (→Multiplexer)
multiplexer channel (→multiplex channel) DATA PROC (→Multiplexkanal)
multiplex hierarchy TRANS **Multiplexhierarchie**
multiplex highway SWITCH **Multiplexschiene**
multiplexing (→multiplex) TELEC (→Multiplex)
multiplexing (→multiplex operation) TELEC (→Multiplexbetrieb)
multiplexing controller DATA PROC **multiplexierende Steuerung**
multiplex link DATA COMM **Mehrfachanschluß**
multiplex mode (→multiplex operation) TELEC (→Multiplexbetrieb)
multiplex mode (→concurrent processing) DATA PROC (→verzahnte Verarbeitung)
multiplex operation TELEC **Multiplexbetrieb**
= multiplex mode; multiplexing = Multiplexverfahren

multiplexor DATA PROC **Multiplexer**
[device allowing to share a computer channel] [Vorrichtung zur Mehrfachnutzung eines Datenkanals]
= multiplexer

multiplexor (BRI) (→multiplexer) TRANS (→Multiplexgerät)

multiplex rule TELEC **Multiplexvorschrift**
= multiplex convention

multiplex scheme TRANS **Multiplexschema**
[FDM] [TF-Technik]
 = Multiplexplan

multiplex telegraphy TELEGR **Mehrfach-Telegrafie**
multiplicand (n.) MATH **Multiplikand**
[number multiplied by the multiplier] [Zahl die mit dem Multiplikator multipliziert werden soll]
= icand
 = erster Faktor

multiplicand register DATA PROC **Multiplikandenregister**
[a register of the arithmetic and logic unit of the CPU, to execute multiplications] [ein Register des Rechenwerks in der Zentraleinheit, zur Asführung von Multiplikationen]
↑ arithmetic and logic unit ↑ Rechenwerk

multiplicate MATH **vervielfachen**
= multiply 2 ≈ multiplizieren
≈ multiply 1 ↓ verdoppeln; verdreifachen; vervierfachen; verfünffachen; versechsfachen; verzehnfachen
↓ duplicate; triplicate; quadruplicate; quintuplicate; sextuplicate; decuple

multiplicate (→multiply) MATH (→multiplizieren)
multiplication MATH **Multiplikation**
≈ product ≈ Malnehmen; Vervielfachung; Vervielfachen
↓ duplication; triplication; quadruplication; quintuplication; sextuplication; multiplication by ten ≈ Produkt
 ↓ Verdoppelung; Verdreifachung; Vervierfachung; Verfünffachung; Versechsfachung; Verzehnfachung

multiplication TECH **Vermehrung**

multiplication 1 MATH **Vervielfachung**
≈ multiplication 2 ≈ Multiplikation
↓ duplication; triplication; quadruplication; quintuplication; sextuplication; multiplication by ten ↓ Verdoppelung; Verdreifachung; Vervierfachung; Vefünffachung; Versechsfachung; Verzehnfachung

multiplication by ten MATH **Verzehnfachung**
↑ multiplication ↑ Vervielfachung
multiplication factor MATH **Vervielfachungsfaktor**
= multiplying factor; multiplier factor = Multiplikationsfaktor
multiplicative MATH **multiplikativ**
multiplicative mixing HF **multiplikative Mischung**
multiplicator noise INSTR **Multiplikatorrauschen**
multiplier MATH **Multiplikator**
[number by which the multiplicand is multiplied] [Zahl mit der der Multiplikand multipliziert wird]
= ier (n.) = zweiter Faktor

multiplier CIRC.ENG **Vervielfacher**
= multiplier circuit; multiplier chain; multiplying circuit; multiplying chain = Vervielfacherschaltung; Multiplizierer; Multiplikationsschaltung
↑ scaler ↑ Festwertmultiplikator
↓ frequency multiplier ↓ Frequenzvervielfacher

multiplier (→multiplier resistor) INSTR (→Vorwiderstand)

multiplier chain (→multiplier) CIRC.ENG (→Vervielfacher)
multiplier circuit (→multiplier) CIRC.ENG (→Vervielfacher)
multiplier diode COMPON **Vervielfacherdiode**
multiplier element ENG.LOG **Multiplizierglied**
multiplier factor (→multiplication factor) MATH (→Vervielfachungsfaktor)

multiplier-quotient register DATA PROC **Multiplikator-Quotienten-Register**
[an auxiliary register of the arithmetic and logic unit of a CPU, to store temporarily multiplication and division results] [eim Hilfsregister des Rechenwerks der Zentraleinheit, zur Zwischenspeicherung von Produkten oder Quotienten]
↑ arithmetic and logic unit ↑ Rechenwerk

multiplier resistor INSTR **Vorwiderstand**
= multiplier
multiplier stage (→multiplying stage) CIRC.ENG (→Vervielfachungsstufe)
multiply MATH **multiplizieren**
= multiplicate = malnehmen; vervielfachen
multiply ELECTRON **vielfachschalten**
multiply 2 (→multiplicate) MATH (→vervielfachen)
multiplying chain (→multiplier) CIRC.ENG (→Vervielfacher)
multiplying circuit (→multiplier) CIRC.ENG (→Vervielfacher)
multiplying factor (→multiplication factor) MATH (→Vervielfachungsfaktor)
multiplying stage CIRC.ENG **Vervielfachungsstufe**
= multiplier stage
multiply sign MATH **Multiplikationszeichen**
[x or a dot on half hight] [x oder ein Punkt auf halber Höhe]
= multiply symbol = Malzeichen

multiply symbol (→multiply sign) MATH (→Multiplikationszeichen)
multipoint ... TECH **Mehrpunkt-**
multipoint (→multipoint connection) TRANS (→Mehrpunktverbindung)
multipoint circuit DATA COMM **Mehrpunktleitung**
= multipoint connection; multi-endpoint connection = Mehrpunktschaltung; Mehrpunktverbindung; Knotenverbindung

multipoint connection TRANS **Mehrpunktverbindung**
= multipoint; party line = Gruppenverbindung; Multipointverbindung; Mehrpunktverbindung; Liniennetz; Kettennetz

multipoint connection (→multipoint circuit) DATA COMM (→Mehrpunktleitung)
multipoint controller CONTROL **Mehrpunktregler**
multipoint network DATA COMM **Mehrpunktnetz**
 = Knotennetz; Multipointnetz
multipoint operation DATA COMM **Mehrpunktbetrieb**
 = Multipoint-Betrieb
multi-point traffic TELECONTR **Gemeinschaftsverkehr**

multipolar ELECTRON
= multi-pole
multipole ANT
↓ tripole; quadripole
multi-pole (→multipolar) ELECTRON
multi-pole switch COMPON
multiport NETW.TH
multiport modem DATA COMM
multi-position system DATA PROC
(→multi-user system)
multiprocessing DATA PROC
[parallel processing of several tasks by several CPU]
≈ multiprogramming
↑ parallel processing

multi-processor system DATA PROC
[a computer with various processors under common control]
≠ single-processor system
↓ vector procesor

multiprogramming DATA PROC
[simultaneous or interleaved processing of several programs by the same CPU]
= multitasking; concurrent processing; concurrent mode; parallel processing; parallel mode; parallel working; parallel operation
≈ multiprocessing; multi-user system
≠ single programming; serial processing
↓ simultaneous processing; concurrent processing

multipurpose (adj.) TECH
= general-purpose
≈ universal
multipurpose computer DATA PROC
multipurpose network TELEC
= multifunction network; multiservice network
multi-purpose terminal TERM&PER
≠ applications terminal
multirange instrument INSTR
(→multimeter)
multi-reed contact COMPON
multiscan display TERM&PER
(→multi-frequency monitor)
multiscan monitor TERM&PER
(→multi-frequency monitor)
multisectional TECH
= multipartite
↓ two-part; three-part; four-part; five-part
multisection RC network NETW.TH
multisegment magnetron MICROW
(→travelling-wave magnetron)
multi-server architecture DATA COMM
multiservice network TELEC
(→multipurpose network)
multiservice operation TELEC
multishaped (→multiform) COLLOQ
multispan instrument INSTR
(→multimeter)
multistage CIRC.ENG
= polystage
≠ single-stage
multistage amplifier (→cascade CIRC.ENG amplifier)

vielpolig
= mehrpolig
Multipole
↓ Tripol; Quadrupol
(→vielpolig)
mehrpoliger Schalter
Mehrtor
Mehrkanalmodem
(→Mehrplatzsystem)

Mehrprozessorbetrieb
[gleichzeitige Bearbeitung mehrerer Aufgaben durch mehrere Zentraleinheiten]
≈ Mehrprogrammverarbeitung
↑ Parallelverarbeitung

Mehrprozessorsystem
[Rechner mit mehreren zentral gesteuerten Prozessoren]
= Multiprozessorsystem
≠ Einprozessorsystem
↓ Zweiprozessorsystem; Vektorrechner

Mehrprogrammbetrieb
[gleichzeitige oder verschachtelte Verarbeitung mehrerer Programme durch eine Zentraleinheit]
= Mehrprogrammverarbeitung; Multiprogrammverarbeitung; Multiprogramming; Parallelverarbeitung; Parallelbetrieb; Paralleldatenverarbeitung; Multitasking
≈ Mehrprozessorbetrieb
≠ Einprogrammbetrieb; Serienverarbeitung
↓ Simultanbetrieb; Multiplexbetrieb; Teilnehmersystem

Mehrzweck-
= Vielzweck-
≈ Universal-
Mehrzweckrechner
Mehrdienstenetz
= Mehrfachnutzungsnetz

Mehrzweckendgerät
≠ Spezialendgerät
(→Universalmeßgerät)

Multireed-Kontakt
(→Autosync-Monitor)

(→Autosync-Monitor)

mehrteilig
= vielteilig
↓ zweiteilig; dreiteilig; vierteilig; fünfteilig
RC-Kette
(→Wanderfeldmagnetron)

Multi-Server-Architektur
(→Mehrdienstenetz)

Mehrfachnutzung 1
= Vielfachnutzung; Mehrdienstbetrieb
(→verschiedenförmig)
(→Universalmeßgerät)

mehrstufig
= vielstufig
≠ einstufig
(→Kaskadenverstärker)

multi-stage sampling MATH
multi-standard reception TV
multi-state (adj.) CIRC.ENG
multi-station system DATA PROC
(→multi-user system)
multisync monitor TERM&PER
(→multi-frequency monitor)
multitasking DATA PROC
(→multiprogramming)
multiterm (→polynomial) MATH
multiterminal microcomputer DATA PROC
multiterminal minicomputer DATA PROC
multi-terminal operation TERM&PER
(→multi-terminal service)
multi-terminal service TERM&PER
= multi-terminal operation
multithreading DATA PROC
(→thread 2)
multi-throw switch COMPON
multitube (→multiple-unit tube) ELECTRON
multi-tubular plate POWER SYS

multiturn dial COMPON
multi-twin COMM.CABLE
multiuse capability DATA PROC

multi-user-band radio MOB.COMM
(→trunking)
multi-user computer DATA PROC
(→multi-user system)
multi-user mode DATA PROC
= multiusing; shared logic
↓ time-sharing operation; multi-user system
multi-user system DATA PROC
[allows simultaneous work on the same computer and programm system]
= multi-user computer; multi-station system; multi-position system
≈ time-sharing operation
≠ single-user system
↑ multi-user mode

multiusing (→multi-user DATA PROC mode)
multivalent TECH
= polyvalent; multivalid
multivalid (→multivalent) TECH
multivalid code CODING
= multilevel code
multivalid logic MATH
= multiple-valued logic
multivariate CONTROL
multivariate (→multidimensional) MATH
multivariate normal distribution MATH
multivendor show ECON
multi-vendor system DATA PROC
[assembled with system components of several manufacturers, or compatibel with systems of other manufacturers]

multivibrator 1 CIRC.ENG
[circuit with discrete output signals depending not only from the instant value at the input]
= sweep circuit
≈ trigger circuit
↓ one-shot multivibrator; flip-flop

Mehrstufen-Stichprobenverfahren
Mehrnormenempfang
Mehrzustands-
(→Mehrplatzsystem)

(→Autosync-Monitor)

(→Mehrprogrammbetrieb)

(→polynomisch)
Mehrplatz-Mikrocomputer
= Mehrplatz-Mikrorechner
Mehrplatz-Minicomputer
= Mehrplatz-Minirechner
(→Mehrplatzbedienung)

Mehrplatzbedienung

(→Teilprozeß)

Mehrstufenschalter
(→Verbundröhre)

Panzerplatte
[Akkumulator]
Präzisionsdrehknopf
vielpaarig
Mehrbenutzbarkeit
= Multi-user-Fähigkeit
(→Bündelfunk)

(→Mehrplatzsystem)

Mehrbenutzerbetrieb
↓ Teilnehmerbetrieb; Mehrplatzsystem
Mehrplatzsystem
[erlaubt gleichzeitiges Arbeiten am selben Rechner und Programmsystem]
= Mehrbenutzersystem 2; Mehrplatzrechner; Vielfachzugriffssystem; Gemeinschaftsrechner; Multi-user-System
≈ Teilnehmerbetrieb
≠ Einplatzsystem
↑ Mehrbenutzerbetrieb

(→Mehrbenutzerbetrieb)

mehrwertig
= vielwertig
(→mehrwertig)
mehrwertiger Code
= mehrstufiger Code
mehrwertige Logik

Mehrgrößen-
(→mehrdimensional)
mehrdimensionale Normalverteilung
Multivendor-Schau
Multivendorsystem
[aus Geräten verschiedener Hersteller zusammengefügt, oder mit Systemen anderer Hersteller kompatibel]

Kippschaltung
[Schaltung mit diskreten Ausgangssignalen, die nicht nur vom Augenblickswert am Eingang abhängen]
= Multivibrator; Kippstufe; Kippglied; Kippanordnung
≈ Triggerschaltung
↓ JK-Flipflop; bistabile Kippschaltung

multivibrator 2 (→astable multivibrator)		CIRC.ENG	(→astabile Kippschaltung)	mutual forward impedance	NETW.TH	Kernwiderstand vorwärts
multiviewport		TERM&PER	Mehrfachbildschirm	mutual impedance	NETW.TH	Kernwiderstand

multivibrator 2 (→astable multivibrator) CIRC.ENG (→astabile Kippschaltung)
multiviewport TERM&PER Mehrfachbildschirm
multi-volume file DATA PROC Mehrträgerdatei
multi-waveform generator INSTR Multi-Signalform-Generator
multi-way connection (→multiple connection) COMPON (→Mehrfachsteckverbindung)
multi-way jack (→multiple jack) COMPON (→Mehrfachbuchse)
multi-way plug (→multiple plug) COMPON (→Mehrfachstecker)
multi-window DATA PROC Multisichtfenster
multi-window display DATA PROC Mehrfensteranzeige
= multi-window technique = Mehrfenstertechnik
multi-window editor DATA PROC Mehrsichtfenster-Editor
multi-window technique (→multi-window display) DATA PROC (→Mehrfensteranzeige)
multiwire (→multiconductor) COMM.CABLE (→mehradrig)
multiwire antenna (→multiple-wire antenna) ANT (→Mehrdrahtantenne)
multiwire element ANT Mehrdraht-Element
multiwire technique ELECTRON Multiwire-Technik
[PCB with wires instead of cladding] [Leiterplatte mit Drähten statt Kupferkaschierung]
multiwire-triatic antenna (→triangle antenna) ANT (→Dreiecksantenne)
multiword instruction DATA PROC Mehrwortbefehl
= multibyte instruction
mu metal METAL Mumetall
≈ Permalloy
municipal administration ECON Stadtverwaltung
= Gemeindeverwaltung
muon PHYS Myon
= μ-Meson
MUSA ANT MUSA-Antenne
= multiple unit steerable array
musical language DATA PROC Musikprogrammiersprache
musician's microphone EL.ACOUS Musikermikrophon
music note DOC Notenzeichen
music on hold TELEPH Musik bei Warten
music power CONS.EL Musikleistung
≈ nominal power = Impuls-Verstärkerleistung
≈ Nennleistung
music processor DATA PROC Musiksynthesizer
mutated vowel (→umlaut 1) LING (→Umlaut)
mute aerial (→dummy antenna) ANT (→künstliche Antenne)
mute antenna (→dummy antenna) ANT (→künstliche Antenne)
mutilate TELEC verstümmeln
= garble = verfälschen
mutilated data DATA PROC verstümmelte Daten
mutilation TELEC Verstümmelung
= Verfälschung
muting CONS.EL Stummschaltung
[disables as long as reception signal is insufficient] [unterdrückt bei unzureichendem Empfangssignal]
= Rauschsperre; Muting
muting button CONS.EL Ton-Stopptaste
mutipurpose instrument (→multimeter) INSTR (→Universalmeßgerät)
MUTOS DATA PROC MUTOS
[Multi User Task Operating System] ↑ Betriebssystem
↑ operating system
mutual COLLOQ gegenseitig
= wechselseitig
mutual action (→interaction) PHYS (→Wechselwirkung)
mutual admittance NETW.TH Kernleitwert
[current at output to voltage at input] [Strom am Ausgang zu Spannung am Eingang]
= transfer admittance; coupling admittance = Übertragungsleitwert; Koppelleitwert; Übergangsleitwert
↑ transmission coefficient ↑ Übertragungsfaktor
mutual admittance MICROEL Kurzschlußsteilheit
mutual backward impedance NETW.TH Kernwiderstand rückwärts
mutual capacitance LINE TH Betriebskapazität
mutual capacitance EL.TECH Gegenkapazität
mutual conductance (→transconductance) ELECTRON (→Steilheit)
mutual coupling PHYS Strahlungskopplung

mutual forward impedance NETW.TH Kernwiderstand vorwärts
mutual impedance NETW.TH Kernwiderstand
[voltage at output to current at input] [Spannung am Ausgang zu Strom am Eingang]
= transfer impedance; coupling impedance = Übertragungswiderstand; Kopplungswiderstand; Übergangswiderstand
↑ transmission coefficient ↑ Übertragungsfaktor
mutual inductance EL.TECH Gegeninduktivität
= mutual inductivity
mutual inductance coefficient EL.TECH Gegeninduktivitäts-Koeffizient
mutual-inductance Heaviside bridge INSTR Heaviside-Gegeninduktivitätsbrücke
mutual induction EL.TECH Gegeninduktion
= gegenseitige Induktion
mutual inductivity (→mutual inductance) EL.TECH (→Gegeninduktivität)
mutual information (→transinformation) INF (→Transinformation)
mutually exclusive MATH einander ausschließend
= incompatible = gegenseitig ausschließend
mutual synchronization TELEC gegenseitige Synchronisierung
MW (→megawatt) PHYS (→Megawatt)
myopic PHYS kurzsichtig
myriad (adj.) (→innumerable) COLLOQ (→zahllos)
myriametric waves RADIO Myriameter-Wellen
[30 km-10 km; 10 kHz-30 kHz] [30 km-10 km; 10 kHz-30 kHz]
= very low frequencies; VLF = Längstwellen; VLF

N

N (→nitrogen) CHEM (→Stickstoff)
Na (→sodium) CHEM (→Natrium)
nail (v.t.) TECH nageln
nail (n.) TECH Nagel
↓ tack ↓ Zwecke
nail bed MANUF Nagelbrett
nailed MECH genagelt
nailed circuit (→fixed line) TELEC (→Standleitung)
nailed connection (→fixed line) TELEC (→Standleitung)
nailed line (→fixed line) TELEC (→Standleitung)
nailed-up connection SWITCH semipermanente Durchschaltung
nailhead bonding MICROEL Nagelkopfbondierung
= ball bonding = Nagelkopfschweißen; Nagelkopfkontaktieren; Nagelkopfverfahren; Ball-Bonden
naive user [DATA PROC] (→novice) TECH (→Anfänger)
NAK (→negative acknowledge) DATA COMM (→negative Rückmeldung)
NAM) (→membership subscription) ECON (→Mitgliedsbeitrag)
name (v.t.) TECH bezeichnen
= describe; mark = benennen
name (n.) (→description) TECH (→Bezeichnung)
name (n.) (→label 2) DATA PROC (→Sprungmarke)
name key TERM&PER Namentaste
name plate (→manufacturer's nameplate) EQUIP.ENG (→Firmenschild)
nameplate (→type label) EQUIP.ENG (→Typenschild)
name server DATA COMM Namenserver
NAND circuit (→NAND gate) CIRC.ENG (→NAND-Glied)
NAND element (→NAND gate) CIRC.ENG (→NAND-Glied)
NAND function (→NAND operation) ENG.LOG (→NAND-Verknüpfung)
NAND gate CIRC.ENG NAND-Glied
= NAND element; NAND circuit; nonconjunction gate; nonconjunction element; nonconjunction circuit; Sheffer gate; Sheffer element; Sheffer circuit = NAND-Gatter; NAND-Element; NAND-Schaltung; NAND-Tor; NICHT-UND-Glied; NICHT-UND-Gatter;
↑ logic gate

		NICHT-UND-Element; NICHT-UND-Schaltung; NICHT-UND-Tor; Sheffer-Gatter; Sheffer-Element; Sheffer-Schaltung; Sheffer-Tor ↑ Verknüpfungsglied	
NAND operation ENG.LOG [output = 0 only if simultaneously A = 1 and B = 1] = NAND function; nonconjunction; Sheffer operation; Sheffer function ≈ alternative denial ↑ logic operation		**NAND-Verknüpfung** [Ausgang nur dann = 0 wenn gleichzeitig A = 1 und B = 1] = NAND-Funktion; NICHT-UND-Verknüpfung; NICHT-UND-Funktion; Sheffer-Verknüpfung; Sheffer-Funktion ↑ logische Verknüpfung	
nanocomputer DATA PROC [processing in nanoseconds]		**Nanosekunden-Computer**	
nanometer (NAM) PHYS = nanometre (BRI); nm		**Nanometer** [0,000 001 m] = nm	
nanometre (BRI) (→nanometer) PHYS		(→Nanometer)	
nano programm DATA PROC		**Nanoprogramm**	
nanosecond PHYS = ns; millimicrosecond		**Nanosekunde** = ns; Millimikrosekunde	
naples yellow PHYS ≈ ocher		**neapelgelb** ≈ ocker	
narrative (n.) ECON		**freier Text**	
narrowband (adj.) TELEC		**schmalbandig**	
narrowband TELEC = narrow band ≠ broadband		**Schmalband** ≠ Breitband	
narrow band (→narrowband) TELEC		(→Schmalband)	
narrowband antenna ANT		**Schmalbandantenne**	
narrowband detection INSTR		**Schmalbanddetektion**	
narrowband feeding ANT		**Schmalbandspeisung**	
narrowband filter NETW.TH		**Schmalbandfilter**	
narrowband noise TELEC		**Schmalbandrauschen** = farbiges Rauschen	
narrow-band sweep INSTR		**Schmalbandwobbeln**	
narrow-band system TRANS (→low-capacity system)		(→Kleinkanalsystem)	
narrowband television TV		**Schmalbandfernsehen**	
narrow dimension MECH = small face ≠ broad dimension		**Schmalseite** ≠ Breitseite	
narrowing (→taper) MATH		(→Verjüngung)	
n-ary MATH		**n-stellig**	
nasal (n.) LING		**Nasal** = Nasallaut; Nasenlaut	
NASC (USA) TECH = National Aircraft Standard		**NASC** = amerikanische Flugzeugnorm	
National Aircraft Standard TECH (→NASC)		(→NASC)	
national coarse thread MECH = NC		**amerikanisches Grobgewinde**	
national extra fine thread MECH = NEF		**amerikanisches Sonderfeingewinde**	
national fine thread MECH = NF		**amerikanisches Feingewinde**	
nationalization 1 ECON		**Verstaatlichung**	
nationalize ECON		**verstaatlichen**	
national long-distance call TELEC (→inland toll call)		(→Inlandsferngespräch)	
national network TELEC		**Landesnetz**	
national number SWITCH [suscriber number + trunk code]		**nationale Rufnummer** [Teilnehmerrufnummer + Ortskennzahl]	
national screw-thread standards MECH		**amerikanische Schraubengewindeform**	
national special thread MECH = NS		**amerikanisches Spezialgewinde**	
national straight pipe thread MECH		**amerikanisches zylindrisches Rohrgewinde**	
national thread MECH		**amerikanisches Gewinde**	
nationwide TELEC = country-wide; with global coverage		**landesweit** ≈ flächendeckend	
nationwide dialing (→direct distance dialing) SWITCH		(→Selbstwählferndienst)	
native (n.) COLLOQ		**Einheimischer**	
native (adj.) COLLOQ		**einheimisch**	
native code (→machine code 1) DATA PROC		(→Maschinencode 1)	
native compiler DATA PROC		**maschinenspezifischer Kompilierer**	
native speaker LING		**Muttersprachler**	
natural color TECH		**Naturfarbe** = natürliche Farben	
natural cooling TECH		**natürliche Kühlung**	
natural frequency PHYS = intrinsic frequency; eigenfrequency		**Eigenfrequenz** = Eigenschwingungsfrequenz	
natural-language (adj.) INF.TECH = plain-language		**natürlichsprachlich**	
natural logarithm MATH = ln; hyperbolic logarithm; neperian logarithm; logarithm to the basis of e		**natürlicher Logarithmus** = ln; Neperscher Logarithmus; Logarithmus zur Basis e; hyperbolischer Logarithmus; Logarithmus naturalis	
natural magnet PHYS		**natürlicher Magnet**	
natural mode (→natural oscillation) PHYS		(→Eigenschwingung)	
natural mold finish METAL		**preßblank**	
natural number (→cardinal number) MATH		(→Kardinalzahl)	
natural oscillation PHYS = intrinsic oscillation; natural vibration; intrinsic vibration; normal vibration; natural mode; intrinsic mode; normal mode; eigenoscillation		**Eigenschwingung** ≈ Eigenresonanz	
natural resonance (→intrinsic resonance) PHYS		(→Eigenresonanz)	
natural vibration (→natural oscillation) PHYS		(→Eigenschwingung)	
nautical RADIO NAV		**nautisch**	
nautical mile PHYS [6,076.115 ft or 1,852 m by international standard] ≈ Admiralty mile; mile		**Seemeile** [nach internationaler Normung: 1.852 m] = nautische Meile; sm ≈ englische Seemeile; Meile	
nautical point PHYS [11° 15']		**nautischer Strich** [11° 15']	
nave (→wheel hub) MECH		(→Radnabe)	
navigation radar RADIO LOC		**Schiffsradar**	
navigation satellite SAT.COMM		**Navigationssatellit**	
navigator AERON		**Navigator**	
Nb (→niobium) CHEM		(→Niob)	
NC (→national coarse thread) MECH		(→amerikanisches Grobgewinde)	
NC (→numerical control) CONTROL		(→numerische Steuerung)	
N/C contact (→break contact) COMPON		(→Ruhekontakt)	
n-channel field-effect transistor (→NFET) MICROEL		(→NFET)	
n-channel metal-oxide semiconductor (→NMOS technology) MICROEL		(→NMOS-Technik)	
n-channel MOSFET MICROEL		**N-Kanal-MOSFET** = N-Kanal-MOS-Transistor	
n-channel transistor MICROEL		**N-Kanal-Transistor**	
NC machine tool TECH = numerically controlled machine tool		**numerisch gesteuerte Werkzeugmaschine**	
n-conducting PHYS = n-type		**N-leitend**	
NCP (→pressure-sensitive paper) TERM&PER		(→druckempfindliches Papier)	
NC postprocessor (→postprocessor) DATA PROC		(→Postprozessor)	
NC programming MANUF		**NC-Programmierung**	
Nd (→neodymium) CHEM		(→Neodym)	
NDB (→omnidirectional radiobeacon) RADIO NAV		(→Rundstrahlbake)	
n doped MICROEL		**N-dotiert**	
NDRM (→non-destructive readout memory) DATA PROC		(→nichtlöschbarer Lesespeicher)	
Ne (→neon) CHEM		(→Neon)	

near

near	COLLOQ	**nah**	
= close; close-in; proximal; proximate		= nahegelegen; naheliegend	
near band (→adjacent band)	RADIO	(→Nachbarband)	
near-channel interference (→adjacent-channel interference)	MODUL	(→Nachbarkanalbeeinflussung)	
near-channel operation (→adjacent-channel operation)	RADIO	(→Nachbarkanalbetrieb)	
near-channel selection (→adjacent-channel selection)	MODUL	(→Nachbarkanalentkopplung)	
near-channel selectivity (→adjacent-channel selection)	MODUL	(→Nachbarkanalentkopplung)	
near echo	TELEC	**Nahecho**	
near end	TELEC	**nahes Ende**	
near-end crosstalk = NEXT	TELEC	**Nahnebensprechen**	
near-end operated	TELEC	**nahgesteuert**	
nearest match (→best fit)	MATH	(→beste Anpassung)	
near-field diagram = near-field pattern	ANT	**Nahfelddiagramm**	
near-field fading	RADIO PROP	**Nahschwund** = Nahfeldschwund	
near-field pattern (→near-field diagram)	ANT	(→Nahfelddiagramm)	
near-field region = proximity zone	RADIO PROP	**Nahfeld** = Nahfeldbereich	
near-letter quality (→NLQ printing)	TERM&PER	(→Beinahe-Schönschrift)	
nearness (→proximity)	TECH	(→Nähe)	
near-singing	CIRC.ENG	**Pfeifneigung**	
near-to-far direction (→go direction)	TELEC	(→Hin-Richtung)	
near-to-ground antenna (→ground antenna)	ANT	(→Erdantenne 1)	
near zone (AM) = toll area (BRI)	TELEPH	**Nahzone** = Nahgesprächsbereich; Nahgesprächszone	
nebulum	ASTROPHYS	**Nebel**	
neck	METAL	**Einstich**	
neck	MECH	**Hals**	
necking down (→strangulation)	MECH	(→Einschnürung)	
need for action	COLLOQ	**Handlungsbedarf**	
needle	TECH	**Nadel**	
needle = stylus	EL.ACOUS	**Nadel** = Abtastnadel	
needle (→pointer)	INSTR	(→Zeiger)	
needle chatter	EL.ACOUS	**Nadelflattern**	
needle counter [for radioactivity]	INSTR	**Spitzenzähler** [für Radioaktivität]	
needle electrometer	INSTR	**Nadelelektrometer**	
needle file	TECH	**Schlüsselfeile**	
needle pressure	EL.ACOUS	**Nadeldruck**	
needle scratch	EL.ACOUS	**Nadelkratzen**	
needle throw (→pointer throw)	INSTR	(→Zeigerausschlag)	
Neél temperature	PHYS	**Neél-Temperatur**	
NEF (→national extra fine thread)	MECH	(→amerikanisches Sonderfeingewinde)	
negate (v.t.)	ENG.LOG	**negieren**	
negate (→invert)	CIRC.ENG	(→invertieren)	
negation	LING	**Negation** = Verneinung	
negative (adj.) [screen mode] = reverse	TERM&PER	**negativ** [Bildschirmdarstellung] = invertiert ↑ Bildschirmdarstellung	
negative (adj.) ≠ positive	MATH	**negativ** ≠ positiv	
negative (adj.) ≠ affirmative	LING	**verneinend** ≠ bejahend	
negative = at negative pole ≠ positive	PHYS	**negativ** = an Minuspol ≠ positiv	
negative acknowledge [code] = NAK	DATA COMM	**negative Rückmeldung** [Code] = NAK	
negative acknowledge mode	TELECONTR	**Rückfragebetrieb**	
negative acknowledgment	DATA COMM	**Schlechtquittung** = negative Quittung	
negative allowance (→tightest fit)	ENG.DRAW	(→größte Durchdringung)	
negative amplitude modulation	TV	**Negativ-Amplitudenmodulation**	
negative boosting transformer (→booster transformer)	EL.TECH	(→Saugtransformator)	
negative conductor (→negative wire)	EL.TECH	(→Minusader)	
negative current feedback [part of output current is fed in opposition to the input] ↑ negative feedback	CIRC.ENG	**Stromgegenkopplung** [Teil des Ausgangsstromes wird gegenphasig an den Eingang gelegt] ↑ Gegenkopplung	
negative differential resistance [characteristic of decreasing current with increasing voltage]	ELECTRON	**negativ differentieller Widerstand** [Kennlinie mit abnehmendem Strom bei zunehmender Spannung]	
negative distortion (→pincushion distortion)	PHYS	(→kissenförmige Verzeichnung)	
negative edge (→trailing edge)	ELECTRON	(→Abfallflanke)	
negative electrode (→cathode)	PHYS	(→Kathode)	
negative feedback = reverse feedback; degenerative feedback ↑ feedback	CIRC.ENG	**Gegenkopplung** = negative Rückkopplung ↑ Rückkopplung	
negative feedback amplifier	CIRC.ENG	**gegengekoppelter Verstärker**	
negative feedback network	CIRC.ENG	**Gegenkopplungsnetzwerk**	
negative frequency modulation	TV	**Negativ-Frequenzmodulation**	
negative-going (→inclined)	TECH	(→geneigt)	
negative-going edge (→trailing edge)	ELECTRON	(→Abfallflanke)	
negative gradient = antigradient	MATH	**negativer Gradient** = Antigradient	
negative gyrator	NETW.TH	**Negativgyrator**	
negative impedance booster (→NLT amplifier)	TRANS	(→NLT-Verstärker)	
negative impedance converter = NIC ↑ impedance converter	NETW.TH	**Negativ-Impedanzkonverter** = Negativübersetzer; Negativ-Impedanzwandler ↑ Impedanzkonverter	
negative impedance repeater (→NLT amplifier)	TRANS	(→NLT-Verstärker)	
negative justification = negative stuffing	CODING	**negatives Stopfen** = Negativstopfen	
negative line (→negative wire)	EL.TECH	(→Minusader)	
negative line amplifier (→NLT amplifier)	TRANS	(→NLT-Verstärker)	
negative logic [positive or high voltage for state 0] = negative true logic ≠ positive logic	CIRC.ENG	**negative Logik** [positive oder hohe Spannung für Zustand 0] ≠ positive Logik	
negatively coupled	CIRC.ENG	**gegengekoppelt**	
negative message	DATA COMM	**Schlechtmeldung** = negative Meldung	
negative pressure = vacuum 2 ≈ vacuum 1	PHYS	**Unterdruck** ≈ Vakuum	
negative pulse edge (→trailing pulse edge)	ELECTRON	(→Impulsabfallflanke)	
negative repeater (→NLT amplifier)	TRANS	(→NLT-Verstärker)	
negative resist	MICROEL	**Negativlack**	
negative resistance	EL.TECH	**negativer Widerstand**	
negative sentence	LING	**Verneinungssatz** = verneinter Satz	
negative stuffing (→negative justification)	CODING	(→negatives Stopfen)	
negative temperature coefficient thermistor (→NTC thermistor)	COMPON	(→Heißleiter)	
negative temperature coefficient resistor (→NTC thermistor)	COMPON	(→Heißleiter)	
negative true logic (→negative logic)	CIRC.ENG	(→negative Logik)	
negative voltage feedback [part of output voltage is fed in counterphase to the input] ≠ positive voltage feedback ↑ voltage feedback; negative feedback	CIRC.ENG	**Spannungsgegenkopplung** [ein Teil der Ausgangsspannung wird gegenphasig zum Eingang rückgekoppelt]	

		≠	Spannungsmitkopplung
		↑	Spannungsrückkopplung; Gegenkopplung
negative wire	EL.TECH		**Minusader**
= negative conductor; negative line		=	Minusleiter
negator (→NOT gate)	CIRC.ENG		(→NICHT-Glied)
neglect	TECH		**vernachlässigen**
= disregard			
negligible	MATH		**vernachlässigbar**
negotiate	ECON		**verhandeln**
= bargain			
negotiation	ECON		**Verhandlung**
= bargaining			
negotiation round	ECON		**Verhandlungsrunde**
negotiation skill	ECON		**Verhandlungsgeschick**
neighborhood	COLLOQ		**Nachbarschaft**
= neighbourhood		=	Umgebung
≈ proximity		≈	Nähe
neighboring atom	PHYS		**Nachbaratom**
neighboured (→adjacent)	TECH		(→benachbart)
neighbourhood	COLLOQ		(→Nachbarschaft)
(→neighborhood)			
nematic	PHYS		**nematisch**
NEMP strength	ELECTRON		**NEMP-Festigkeit**
[nuclear electromagnetic pulse]			
neodymium	CHEM		**Neodym**
= Nd		=	Nd
neon	CHEM		**Neon**
= Ne		=	Ne
neon indicator	COMPON		**Neon-Signalleuchte**
NEP (→noise equivalent power)	TELEC		(→äquivalente Rauschleistung)
neper	TELEC		**Neper**
[8.686 dB]			[8,686 dB]
neperian logarithm (→natural logarithm)	MATH		(→natürlicher Logarithmus)
neptunium	CHEM		**Neptunium**
= Np		=	Np
NEQ (→EXCLUSIVE-OR operation)	ENG.LOG		(→EXKLUSIV-ODER-Verknüpfung)
nerd (→computer amateur)	DATA PROC		(→Computer-Amateur)
Nernst-Einstein relation	PHYS		**Nernstsche Gleichung**
Nernst voltage	PHYS		**Nernstspannung**
nest (v.t.)	DATA PROC		**schachteln**
= embed; insert (v.t.)		=	einfügen
nest (→interleave)	TECH		(→verschachteln)
nested (→interleaved)	TECH		(→verschachtelt)
nested block (→nested program block)	DATA PROC		(→verschachtelter Programmblock)
nested loop (→inner loop)	DATA PROC		(→verschachtelte Programmschleife)
nested program	DATA PROC		**verschachteltes Unterprogramm**
= nesting program; nested routine; nesting routine		=	geschachteltes Unterprogramm; verschachtelte Routine; geschachtelte Routine
nested program loop (→inner loop)	DATA PROC		(→verschachtelte Programmschleife)
nested programm block	DATA PROC		**verschachtelter Programmblock**
= nested block; nesting program block; nesting block		=	geschachtelter Programmblock
nested routine (→nested program)	DATA PROC		(→verschachteltes Unterprogramm)
nesting	DATA PROC		**Verschachtelung**
= interleaving		=	Verzahnung; Nesting
nesting block (→nested programm block)	DATA PROC		(→verschachtelter Programmblock)
nesting loop (→inner loop)	DATA PROC		(→verschachtelte Programmschleife)
nesting program (→nested program)	DATA PROC		(→verschachteltes Unterprogramm)
nesting program block (→nested programm block)	DATA PROC		(→verschachtelter Programmblock)
nesting routine (→nested program)	DATA PROC		(→verschachteltes Unterprogramm)
net	ECON		**netto**
net amount	ECON		**Nettobetrag**
netlist	MICROEL		**Netzliste**
net loss (→overall loss)	TELEC		(→Restdämpfung)
net plane (→lattice plane)	PHYS		(→Netzebene)
netting	TECH		**Geflecht**
net weight	ECON		**Nettogewicht**
		=	Reingewicht
network	TELEC		**Netz**
network	DATA COMM		**Netzwerk**
≈ multi-computer system		≈	Netz
		≈	Mehrrechnersystem
network (→electric network)	NETW.TH		(→Netzwerk)
network (v.t.) (→mesh)	TELEC		(→vernetzen)
network access	TELEC		**Teilnehmerzugang**
		=	Netzzugang; Netznutzzugang
network access system	TELEC		**Teilnehmerzugangssystem**
≈ suscriber loop carrier		≈	Netzzugangssystem
		≈	Teilnehmermultiplexsystem
network address	DATA COMM		**Netzadresse**
↑ subscriber number [TELEC]		≈	Rufnummer
		↑	Teilnehmerrufnummer [TELEC]
network analysator	INSTR		**Netzwerkanalysator**
[measures response of linear networks in the frequency domain]			[mißt Frequenzgänge von linearen Netzwerken]
= network analyzer			
network analysis	NETW.TH		**Netzwerkanalyse**
↑ network theory		↑	Netzwerktheorie
		↓	Knotenanalyse; Schleifenanalyse
network analysis (→network planning technique)	ECON		(→Netzplantechnik)
network analyzer (→network analysator)	INSTR		(→Netzwerkanalysator)
network architecture	TELEC		**Netzarchitektur**
= network structure; network topology; network configuration		=	Netzstruktur; Netztopologie; Netzform; Netzaufbau; Netzkonfiguration
network autarchy	TELEC		**Netzhoheit**
network board	DATA PROC		**Netzwerkkarte**
			[zum Anschluß an ein Computernetz]
network carrier	TELEC		**Netzanbieter**
= network provider			
network carrier (→operating company)	TELEC		(→Betriebsgesellschaft)
network chart (→network plan)	ECON		(→Netzplan)
network compatibility	TELEC		**Netzkompatibilität**
network component	TELEC		**Netzkomponente**
network configuration (→network architecture)	TELEC		(→Netzarchitektur)
network congestion	TELEC		**Netzüberlastung**
network control	TELEC		**Netzsteuerung**
≈ network management		≈	Netzverwaltung
network control center	TELEC		**Netzkontrollzentrum**
network control information	DATA COMM		**Netzsteuerungsinformation**
network control program	DATA COMM		**Netzsteuerprogramm**
network conversion	NETW.TH		**Netzwerkumwandlung**
= network transformation			
network core	OUTS.PLANT		**Kernbereich**
network database	DATA PROC		**Netzwerk-Datenbank**
≠ relational database		≠	relationale Datenbank
network determinant	NETW.TH		**Vierpoldeterminante**
= two-port determinant; four-pole determinant			
networked (→meshed)	TELEC		(→vernetzt)
network element	TELEC		**Netzelement**
network element controller	TELEC		**Netzelementsteuerung**
network element management	TELEC		**Netzelementverwaltung**
network equation	NETW.TH		**Vierpolgleichung**
= two-port equation; four-terminal equation; four-pole equation			
network expansion	TELEC		**Netzausbau**
network failure	DATA COMM		**Netzstörung**
network function	TELEC		**Netzwerkfunktion**
network identification signal	DATA COMM		**Netzkennung**
= network identify; network identification utilities		=	Netzkennzahl

network identification utilities 386

network identification util- DATA COMM (→Netzkennung)
ities (→network identification
signal)
network identify (→network DATA COMM (→Netzkennung)
identification signal)
network information DATA COMM **Netzinformation**
networking TELEC **Vernetzung**
= internetworking; network intercon- ≈ Vermaschung
nection
≈ intermeshing
networking flexibility TELEC **Netzflexibilität**
network integration DATA COMM **Netzwerkverbund**
network interconnection TELEC (→Vernetzung)
(→networking)
network layer DATA COMM **Vermittlungsschicht**
[3rd layer of OSI; decides parameters [3. Schicht im ISO-Schich-
relevant for the establishment of con- tenmodell; legt die für den
nections] Verbindungsaufbau rele-
 vanten Daten fest]
 = Netzwerkschicht; Netz-
 steuerung; Netzwerk-
 ebene
network layout DOC **Netzplan**
network level TELEC **Netzebene**
network management TELEC **Netzverwaltung**
≈ network control = Netzführung; Netzmanage-
 ment
 ≈ Netzsteuerung
network manager TELEC **Netzverwalter**
network matrix NETW.TH **Vierpolmatrix**
= two-port matrix; four-pole matrix
network message (→service DATA COMM (→Dienstsignal)
signal)
network monopoly TELEC **Netzmonopol**
network node (n.) DATA COMM **Netzknoten**
= node; hub (n.) = Konzentrator
≈ gateway ≈ Überleiteinrichtung
network node TELEC **Netzknoten**
= node = Knotenpunkt; Knoten
network-node computer DATA COMM **Netzknotenrechner**
= network node processor; network = Knotenrechner
node controller; remote communica-
tion computer; remote front-end processor
network node controller DATA COMM (→Netzknotenrechner)
(→network-node computer)
network node interface TELEC **Netzknotenschnittstelle**
network node processor DATA COMM (→Netzknotenrechner)
(→network-node computer)
network occupancy TELEC **Netzbelegung**
network operating company TELEC (→Betriebsgesellschaft)
(→operating company)
network operating system DATA COMM **Netzwerkbetriebssystem**
= NOS = NOS
network operations center TELEC **Netzleitzentrale**
network operator (→operating TELEC (→Betriebsgesellschaft)
company)
network-oriented DATA PROC **netzwerk-orientiert**
network parameter NETW.TH **Vierpolparameter**
= two-port parameter; four-pole pa- = Vierpolkoeffizient
rameter
network periphery TELEC **Netzausläufer**
 = Netzperipherie
network plan ECON **Netzplan**
= network chart [Netzplantechnik]
network planning technique ECON **Netzplantechnik**
= NPT; network analysis
network provider (→operating TELEC (→Betriebsgesellschaft)
company)
network reconfiguration TELEC **Netzumrangierung**
network resilience TELEC **Netzelastizität**
network resilient TELEC **netzelastisch**
network simulation INSTR **Netznachbildung**
network structure NETW.TH **Netzwerkstruktur**
network structure (→network TELEC (→Netzarchitektur)
architecture)
network structure (→network NETW.TH (→Netzwerktopologie)
topology)
network synchronization TELEC **Netzsynchronisation**
= network timing

network synthesis EL.TECH **Netzwerksynthese**
≠ network analysis = Schaltungssynthese
↑ network theory ≠ Netzwerkanalyse
 ↑ Netzwerktheorie
network termination TELEC **Netzabschlußeinrichtung**
= NT [ISDN]
 = NT-Gerät; Netzabschluß
network theory EL.TECH **Netzwerktheorie**
= electrical network theory ↑ Elektrizitätslehre
↑ electrical fundamentals ↓ Netzwerkanalyse; Netz-
↓ network analysis; network synthesis werksynthese
network timing (→network TELEC (→Netzsynchronisation)
synchronization)
network topology NETW.TH **Netzwerktopologie**
= network structure; topology = Netzstruktur
network topology (→network TELEC (→Netzarchitektur)
architecture)
network transformation NETW.TH (→Netzwerkumwandlung)
(→network conversion)
network unit TELEC **Netzeinheit**
network user TELEC **Netzbenutzer**
≠ network operator = Netzanwender
 ≠ Netzbetreiber
network user access DATA COMM (→Teilnehmeradresse)
(→transport address)
network user identification DATA COMM (→Teilnehmerkennung)
(→subscriber identification)
network utility DATA COMM **Netzmerkmal**
network-wide TELEC **netzweit**
neural network DATA PROC **Neuronennetz**
neuristor MICROEL **Neuristor**
neuristor electronics MICROEL **Neuristorelektronik**
neurocomputing DATA PROC **Neurocomputing**
neuronal computer DATA PROC **Neurorechner**
 = Neuro-Computer
neuter (n.) LING **Neutrum**
= neuter gender = sächliches Substantiv
neuter gender (→neuter) LING (→Neutrum)
neutral TECH **neutral**
neutral POWER SYS **Neutralleiter**
[conductor of ac systems] [Wechselstromnetz]
neutral conductor (AC) POWER ENG **Nulleiter**
[mid-wire with protective function] [Mittelleiter mit Schutz-
= return conductor (DC); third wire; funktion]
zero conductor; earthed neutral = Mittelpunktsleiter;
≈ protective earth conductor; earth Sternpunktleiter
↑ mid-wire ≈ Schutzerdungsleiter; Erde
 ↑ Mitteleiter
neutral current (AM) (→single TELEGR (→Einfachstrom)
current)
neutral curve (→initial PHYS (→Neukurve)
magnetization curve)
neutralization CIRC.ENG **Neutralisation**
[of an output-to-input feedback] [der Rückkopplung Ausgang auf Eingang]
 ↑ Kompensationsmaßnahme
neutralization network CIRC.ENG **Neutralisationsnetzwerk**
neutralize TECH **neutralisieren**
neutralizing capacitor CIRC.ENG (→Entkopplungskondensa-
(→decoupling capacitor) tor)
neutral potential EL.TECH **Nullpotential**
≈ earth potential ≈ Erdpotential
neutral solution CHEM **neutrale Lösung**
neutral track TERM&PER **Zwischenspur**
neutral zone PHYS **neutrale Zone**
neutron PHYS **Neutron**
neutron diffraction PHYS **Neutronenbeugung**
neutron flux PHYS **Neutronenfluß**
new (adj.) COLLOQ **neu**
≈ novel; modern ≈ neuartig; modern
new degree (→gon) MATH (→Gon)
new design TECH **Neukonstruktion**
new developed TECH **neuentwickelt**
new line TYPOGR **neue Zeile**
new line TERM&PER **Zeilenwechsel**
[carriage return + line feed] [Wagenrücklauf + Zeilen-
= line change vorschub]
 = Zeilenumschaltung
new minute PHYS **Neuminute**
[0.01 gon] [0,01 gon]

news agency		COLLOQ	Presseagentur	nine's complement		DATA PROC	(→Neunerkomplement)
new second		PHYS	Neusekunde	(→complement to nine)			
[0.0001 gon]			[0,0001 gon]	niobite		CHEM	Niobit
new services		TELEC	neue Dienste	niobium		CHEM	Niob
newsletter (→circular)		ECON	(→Rundschreiben)	= Nb			= Nb
newspaper		COLLOQ	Zeitung	nipper pliers		TECH	Beißzange
[unstapled publication edited daily or in other short period]			[täglich oder in sonstigen kurzen Abständen erscheinende, nicht geheftete Druckschrift]	= carpenter's pincers; pincers ≈ nippers ↑ tongs			= Kneifzange ↑ Zange 1
≈ magazine			≈ Zeitschrift	nipple		MECH	Nippel
↓ daily paper; weekly paper			↓ Tageszeitung; Wochenzeitung				≈ Pimpel
				nipple stud		MECH	Pimpel
newspaper article		COLLOQ	Zeitungsartikel	= stud			≈ Nippel
= press item			= Zeitungsnotiz	nit		PHYS	Nit
new start (→restart)		DATA PROC	(→Neustart)	[unit for brightness; = 1 cd/m^2]			[Maßeinheit für Leuchtdichte; = 1 cd/m^2]
newvicon		TV	Sperrschicht-Vidikon	= nt			= nt
			= Newvicon	nitrate		CHEM	Nitrat
NEXT (→near-end crosstalk)		TELEC	(→Nahnebensprechen)	nitric acid		CHEM	Salpetersäure
next line		DATA COMM	Folgezeile	nitriding		METAL	Nitrierhärtung
next page (→continuation page)		TYPOGR	(→Folgeseite)	nitrocellulose laquer		CHEM	Nitrolack
				nitrogen		CHEM	Stickstoff
NF (→national fine thread)		MECH	(→amerikanisches Feingewinde)	= N			= N
				nixie (→Nixie tube)		ELECTRON	(→Nixie-Röhre)
NFET		MICROEL	NFET	Nixie tube		ELECTRON	Nixie-Röhre
= n-channel field-effect transistor				[Burroughs Corp. USA]			[Burroughs Corp. USA]
n-gate thyristor		MICROEL	anodenseitig steuerbarer Thyristor	= nixie			
				NL (→nil)		DATA COMM	(→Null)
Ni (→nickel)		CHEM	(→Nickel)	n-level (→value-discrete)		INF	(→wertdiskret)
nibble (→tetrad)		CODING	(→Tetrade)	n-level signal		CODING	wertdiskretes Signal
NIC (→negative impedance converter)		NETW.TH	(→Negativ-Impedanzkonverter)	NLQ printing = near-letter quality		TERM&PER	Beinahe-Schönschrift = NLQ-Schrift; Nahezu-Korrespondenzqualität; Fast-Brief-Qualität
Nichols diagram		CONTROL	Nichols-Diagramm				
nick (v.t.)		TECH	zwicken	NLT amplifier		TRANS	NLT-Verstärker
			= kneifen	= negative impedance booster; negative impedance repeater; negative line amplifier			
nickel		CHEM	Nickel				
= Ni			= Ni				
nickel bronze		METAL	Nickelbronze	nm (→nanometer)		PHYS	(→Nanometer)
nickel-cadmium battery		POWER SYS	Nickel-Cadmium-Batterie	NMI (→non-maskable interrupt)		DATA PROC	(→nichtmaskierbare Unterbrechung)
			= NiCd-Batterie				
nickel-cadmium cell		POWER SYS	Nickel-Cadmium-Zelle	NMOS technology		MICROEL	NMOS-Technik
			= NiCd-Zelle	= n-channel metal-oxide semiconductor			
			↑ Stahlakkumulator				
nickel-iron battery		POWER SYS	Nickel-Eisen-Batterie	N + N system (→two-wire carrier system)		TRANS	(→Zweidraht-Getrenntlage-System)
			= NiFe-Batterie				
nickel-iron cell		POWER SYS	Nickel-Eisen-Zelle	No (→nobelium)		CHEM	(→Nobelium)
			= NiFe-Zelle	no-address (→addressless)		DATA PROC	(→adressenlos)
			↑ Stahlakkumulator	nobelium		CHEM	Nobelium
nickel-plate		METAL	vernickeln	= No			= No
↑ galvanize			↑ galvanisieren	noble gas		CHEM	Edelgas
nickel plating		METAL	Vernickelung	= inert gase; rare gase			
nickel-silver (→german silver)		METAL	(→Neusilber)	noble-gas cell		PHYS	Edelgaszelle
nickel steel		METAL	Nickelstahl	noble metal		CHEM	Edelmetall
NIGFET		MICROEL	NIGFET	= precious metal			
= non-isolated-gate field effect transistor				no-break power supply (→uninterruptable power supply)		POWER SYS	(→unterbrechungsfreie Stromversorgung)
night-alarm switch		SIGN.ENG	Nachtschalter	NOC (→make contact)		COMPON	(→Arbeitskontakt)
night answer		SWITCH	Nachtabfrage	no-charge (→without charge)		TELEC	(→gebührenfrei)
night answer connection		SWITCH	Nachtschaltung	N/O contact (→make contact)		COMPON	(→Arbeitskontakt)
night answering number		SWITCH	Nachtrufnummer	nocturnal (adj.)		COLLOQ	nächtlich
night effect		RADIO LOC	Nachteffekt	= nightly			≠ tageszeitlich [TECH]
[bearing errors due to nocturnal sky waves]			[Peilstörung durch nächtliche Raumwellen]	≠ diurnal			
				nocturnal tariff		TELEC	Nachttarif
nightly (→nocturnal)		COLLOQ	(→nächtlich)	= overnight rate			= Nachtgebühr
night shift		MANUF	Nachtschicht	≠ daytime tariff			≠ Taggebühr
≈ late shift			≈ Spätschicht	no-current condition (→spacing pulse)		TELEGR	(→Pausenschritt)
↑ shift			↑ Schicht				
ni-junction		MICROEL	NI-Übergang	nodal equation		EL.TECH	Knotengleichung
nil		DATA COMM	Null	nodal operation		RADIO REL	Knotenbetrieb
[code]			[Code]	nodal plane		PHYS	Knotenebene
= NL			= NUL	nodal point		EL.TECH	Knotenpunkt
nil		MATH	Nullzeichen	= node			
			= Null	node (n.)		PHYS	Schwingungsknoten
nil (→zero)		MATH	(→Null)	= oscillation node			= Knoten
niladic (adj.)		DATA PROC	operandenlos	≠ antinode			≠ Schwingungsbauch
[without specified operand]				node (n.)		MATH	Knoten
nil pointer		DATA PROC	Listenendezeige	[theory of graphs]			[Graphentheorie]
nine-layer		TECH	neunlagig	= edge			= Knotenpunkt
= nine-part			= neunschichtig	↓ root 3; leaf			≠ Kante
nine-part (→nine-layer)		TECH	(→neunlagig)				↓ Wurzel 3; Blatt

node		NETW.TH	**Knoten**	noiselessness	TECH	**Geräuschlosigkeit**
			= Verzweigungspunkt	noise level	TELEC	**Geräuschpegel**
node (→network node)		DATA COMM	(→Netzknoten)	noise level measurement	INSTR	**Geräuschpegelmessung**
node (→nodal point)		EL.TECH	(→Knotenpunkt)	noise level meter	INSTR	**Geräuschmesser**
node (→network node)		TELEC	(→Netzknoten)	= noise meter		= Geräuschpegelmesser
node analysis		NETW.TH	**Knotenanalyse**	noise limiter	CIRC.ENG	**Störbegrenzer**
no feed-through		SYS.INST	**fädelfrei**	noise margin (→signal-to-noise ratio)	TELEC	(→Rauschabstand)
[office cable]			[Amtskabel]			
NO function (→NOT operation)		ENG.LOG	(→NICHT-Verknüpfung)	noise margin (→noise threshold)	CIRC.ENG	(→Rauschschwelle)
no-hardened		METAL	**ungehärtet**	noise matching	HF	**Rauschimpedanzanpassung**
noise		TELEC	**Rauschen**			= Rauschanpassung
[a random interference, with instant values following a statistical law]			[zeitlich unregelmäßige Störung, deren Augenblickswerte statistisch verteilt sind]	noise measurement	TELEC	**Rauschmessung**
				noise meter (→noise level meter)	INSTR	(→Geräuschmesser)
↑ interference			↑ Störung	noise peak (→interfering pulse)	ELECTRON	(→Störimpuls)
↓ weighted noise			↓ Geräusch			
noise		ACOUS	**Geräusch**	noise power	TELEC	**Rauschleistung**
= audible noise			= Hörgeräusch; Krach	↑ interference power		↑ Störleistung
noise-absorbing cover		TERM&PER	**Schallschluckhaube**	noise-power density	TELEC	**Rauschleistungsdichte**
= acoustic hood; soundproof cover; sound-insulating cover; acoustical sound enclosure			= Geräuschhaube; Schallschluckglocke	noise power ratio	TELEC	**Geräuschleistungsverhältnis**
				= NPR		= Störleistungsverhältnis
				noise reception	ACOUS	**Geräuschimmission**
						= Schallimmission
noise-absorptive (→sound-absorptive)		ACOUS	(→schallschluckend)	noise resistance	INSTR	**Rauschwiderstand**
				noise response	TELEC	**Rauschverhalten**
noise amplitude		TELEC	**Rauschamplitude**	↑ interference response		↑ Störverhalten
noise analysis		TELEC	**Rauschanalyse**	noise sensitive	TELEC	**geräuschempfindlich**
			= Geräuschanalyse	≈ interference sensitive		≈ rauschempfindlich
noise and interference test set		INSTR	**Rausch-Stör-Meßplatz**			≈ störempfindlichkeit
				noise sensitivity	TELEC	**Rauschempfindlichkeit**
noise-attenuating (→sound-insulating)		ACOUS	(→schalldämmend)	≠ noise immunity		= Geräuschempfindlichkeit
				↑ interference sensitivity		≠ Rauschfestigkeit
noise averaging		INSTR	**Rauschmittelung**			↑ Störempfindlichkeit
noise bandwidth		TELEC	**Rauschbandbreite**	noise sideband	TELEC	**Rausch-Seitenband**
noise blanking		ELECTRON	**Störaustastung**	noise signal	TELEC	**Rauschsignal**
noise bridge		INSTR	**Rauschbrücke**	↑ interfering signal		= Geräuschsignal
[measures antenna impedance]			[mißt Antennenimpedanz]			↑ Störsignal
noise cancellation		EL.ACOUS	**Störschall-Unterdrückung**	noise source	TECH	**Geräuschquelle**
noise carrier		MODUL	**Rauschträger**	noise source (→noise generator)	INSTR	(→Rauschgenerator)
noise choke		COMPON	**Entstördrossel**	noise spectrum	TELEC	**Rauschspektrum**
noise contribution		TELEC	**Geräuschbeitrag**	noise standard temperature	TELEC	**Rausch-Bezugstemperatur**
noise current		TELEC	**Rauschstrom**	noise suppression	TELEC	**Rauschunterdrückung**
↑ interfering current			↑ Störspannung	↑ interference suppression		↑ Störunterdrückung
noise diode		ELECTRON	**Rauschdiode**	noise suppression (→noise suppression system)	EL.ACOUS	(→Rauschunterdrückungssystem)
noise equivalent power		TELEC	**äquivalente Rauschleistung**			
= NEP				noise suppression method (→noise suppression system)	EL.ACOUS	(→Rauschunterdrückungssystem)
noise factor (→noise figure)		TELEC	(→Rauschzahl)			
				noise suppression system	EL.ACOUS	**Rauschunterdrückungssystem**
noise-factor measurement		INSTR	**Rauschzahlmessung**	= noise suppression method; noise suppression		= Rauschunterdrückungsverfahren
			= Rauschfaktormessung			
noise figure		ANT	**Grenzempfindlichkeit**	↓ Dolby system		↓ Dolby-Verfahren
noise figure		TELEC	**Rauschzahl**	noise temperature	TELEC	**Rauschtemperatur**
= noise factor			= Rauschmaß; Rauschfaktor	noise threshold	CIRC.ENG	**Rauschschwelle**
noise figure meter		INSTR	**Rauschfaktormesser**	= noise margin		
			= Rauschfaktor-Meßgerät	noise trap (→interference suppression filter)	CIRC.ENG	(→Entstörfilter)
noise filter (→interference suppression filter)		CIRC.ENG	(→Entstörfilter)			
				noise tuning	HF	**Rauschabstimmung**
noise floor (→interference level)		TELEC	(→Störpegel)	= minimum-noise tuning		= Rauschminimum-Abstimmung
noise four-pole (→noise two-port)		NETW.TH	(→Rauschvierpol)	noise two-port	NETW.TH	**Rauschvierpol**
				= noise four-pole		
noise generator		INSTR	**Rauschgenerator**	noise voltage	TELEC	**Rauschspannung**
= noise source			= Rauschquelle	↑ interfering voltage		↑ Störspannung
noise immunity		TELEC	**Rauschfestigkeit**	noise-voltage source	EL.TECH	**Rauschspannungsquelle**
≠ noise sensitivity			= Geräuschfestigkeit; Rauschunempfindlichkeit; Geräuschunempfindlichkeit	noise weighting	TELEC	**Rauschbewertung**
↑ interference immunity				noise weighting filter	TELEC	**Rauschbewertungsfilter**
				= psophometric filter		= psophometrisches Filter
			≠ Rauschempfindlichkeit	noise with load (→loaded noise)	TELEC	(→Rauschen bei Belastung)
			↑ Störfestigkeit			
noise-in-slot measurement (→intermodulation noise measurement)		INSTR	(→Rauschklirrmessung)	noise with tone (→loaded noise)	TELEC	(→Rauschen bei Belastung)
				noisy	TECH	**geräuschvoll**
noise killer (→interference suppression filter)		CIRC.ENG	(→Entstörfilter)	≠ noiseless		= rauschig
						≠ geräuschlos
noiseless		TELEC	**geräuschfrei**	NOLI ME TANGERE (→DON'T TOUCH!)	TECH	(→NICHT BERÜHREN!)
= quiet						
noiseless		TECH	**geräuschlos**	no-load ac	EL.TECH	**Leerlaufwechselspannung**
= quiet			= leise	no-load current (→open-circuit current)	EL.TECH	(→Leerlaufstrom)
≈ low-noise			≈ lärmarm			

no-load mode (→no-load operation)	EL.TECH	(→Leerlaufbetrieb)		**no-name** (n.) [compatible product of unknown producer]	DATA PROC	**Noname** [markenloses kompatibles Produkt]
no-load operation = no-load mode	EL.TECH	**Leerlaufbetrieb**		≈ clone		= No-name ≈ Clone
no-load operation (→open-circuit operation)	EL.TECH	(→Leerlauf)		**non-associated mode** (→non-associated signaling)	SWITCH	(→nichtassoziierte Zeichengabe)
no-load voltage (→open-circuit voltage)	EL.TECH	(→Leerlaufspannung)		**non-associated signaling** (AM) = non-associated signalling (BRI); non-associated mode	SWITCH	**nichtassoziierte Zeichengabe** = nichtassoziierte Signalisierung
N.O.L. top loaded antenna	ANT	**N.O.L.-Schirmantenne**		**non-associated signalling** (BRI) (→non-associated signaling)	SWITCH	(→nichtassoziierte Zeichengabe)
nomenclature (→terminology)	LING	(→Terminologie)		**non-availability** = down-time ratio; DTR	QUAL	**Nichtverfügbarkeit**
nominal (→scheduled)	TECH	(→nominell)		**non-binding** = without obligation; not obligatory; without engagement ≈ informal	ECON	**unverbindlich** = freibleibend ≈ informell
nominal amount = face value ≈ nominal value	ECON	**Nominalbetrag** = Nennbetrag ≈ Nominalwert				
nominal black signal (→artificial black signal)	TELEGR	(→künstliches Schwarz)		**non-blocking**	SWITCH	**blockierungsfrei**
nominal cross-section = rated cross-section	TECH	**Nennquerschnitt**		**non-branched** = unbranched	DATA PROC	**unverzweigt**
nominal current = rated current	EL.TECH	**Nennstrom** = Sollstrom		**non-carbon paper** (→pressure-sensitive paper)	TERM&PER	(→druckempfindliches Papier)
nominal diameter	MECH	**Nenndurchmesser**		**non-circular**	TECH	**unrund**
nominal dimension (→nominal size)	ENG.DRAW	(→Nennmaß)				= nicht kreisrund
nominal frequency = characteristic frequency	EL.TECH	**Nennfrequenz** = Sollfrequenz		**noncircularity**	TECH	**Unrundheit**
nominal illuminance	EL.INST	**Nennbeleuchtungsstärke**		**non-coded** (→uncoded)	INF	(→uncodiert)
nominal level	TELEC	**Nennpegel** = Sollpegel		**noncoded information** = bit mapped information	DATA PROC	**nichtcodierte Information** = uncodierte Information
nominal load (→nominal load capacity)	TECH	(→Nennbelastbarkeit)		**non-coherent** = noncoherent	PHYS	**nichtkohärent**
nominal load capacity = nominal load; rated load; design rating ≈ load capacity	TECH	**Nennbelastbarkeit** = Nennbelastung; Nennlast; Bauleistung ≈ Belastbarkeit		**noncoherent** (→non-coherent)	PHYS	(→nichtkohärent)
				noncombustibility (→incombustibility)	TECH	(→Unbrennbarkeit)
nominal power = rated power; power rating; wattage rating	TECH	**Nennleistung** = Solleistung		**non-competitive**	ECON	**ausschreibungslos**
				non-conducting ≈ insulating	PHYS	**nichtleitend** ≈ isolierend
nominal power = rated power ≈ music power	CONS.EL	**Nennleistung** = Sinusleistung; Sinusdauertonleistung ≈ Musikleistung		**nonconducting hole** [PCB]	ELECTRON	**Freilochung** [Leiterplatte]
				non-conductor (→insulator)	PHYS	(→Isolator)
nominal size = nominal dimension ≈ prescribed dimension	ENG.DRAW	**Nennmaß** = Nenngröße ≈ Sollmaß		**nonconformance** (→fault)	TECH	(→Fehler)
				non-conformance (→non-fulfilment)	ECON	(→Nichterfüllung)
nominal size	ENG.DRAW	**Passungsnennmaß**		**non-conformity** (→unconformity)	TECH	(→Nichtübereinstimmung)
nominal speed = rated speed	MECH	**Nenndrehzahl**		**nonconjunction** (→NAND operation)	ENG.LOG	(→NAND-Verknüpfung)
nominal temperature ≈ operating temperature	TECH	**Nenntemperatur** ≈ Betriebstemperatur		**nonconjunction circuit** (→NAND gate)	CIRC.ENG	(→NAND-Glied)
nominal value = rated value; desired value; reference value; rating 1	TECH	**Nennwert** = Sollwert		**nonconjunction element** (→NAND gate)	CIRC.ENG	(→NAND-Glied)
				nonconjunction gate (→NAND gate)	CIRC.ENG	(→NAND-Glied)
nominal value = face value; par value; stated value	ECON	**Nennwert** = Nominalwert		**noncontact** (adj.) (→contactless)	PHYS	(→berührungslos)
nominal voltage [accumulator]	POWER SYS	**Leerlaufspannung** [Akkumulator] = Ruhespannung		**non-corroding** = corrosion-proof; corrosion-protected; non-corrosive; corrosion-resistant ≈ stainless; acid-resistant	TECH	**korrosionsfest** = korrosionsbeständig; korrosionsgeschützt ≈ rostfrei; säurebeständig
nominal voltage = rated voltage; voltage rating	EL.TECH	**Nennspannung** = Sollspannung		**noncorrosive**	CHEM	**nichtkorrodierend**
nominal white signal (→artificial white signal)	TELEGR	(→künstliches Weiß)		**non-corrosive** (→non-corroding)	TECH	(→korrosionsfest)
nominative	LING	**Nominativ** = Werfall; 1. Fall; erster Fall		**non-cutting**	TECH	**nichtschneidend**
				non-cutting	MECH	**spanlos**
nomogram = nomograph	NETW.TH	**Nomogramm**		**non-cutting shaping** = working	MECH	**spanlose Bearbeitung**
nomograph (→nomogram)	NETW.TH	(→Nomogramm)		**non-cutting shaping**	MECH	**spanlose Formung**
nomography	NETW.TH	**Nomographie**		**non-dedicated server**	DATA COMM	**nichtdedizierter Server**
nonacceptance = rejection	SWITCH	**Abweisung**		**non-delivery**	ECON	**Nichtlieferung**
				non-delivery indication	DATA COMM	**Nichtübergabe-Anzeige**
non-additive	PHYS	**nicht additiv**		**non-destructive**	QUAL	**zerstörungsfrei**
nonaddressable memory (→shaded memory)	DATA PROC	(→Ergänzungsspeicher)		**non-destructive reading**	DATA PROC	**nichtlöschendes Lesen**
nonaddressable storage (→shaded memory)	DATA PROC	(→Ergänzungsspeicher)		**non-destructive readout memory** = NDRM	DATA PROC	**nichtlöschbarer Lesespeicher**
nonagon ↑ polygon	MATH	**Neuneck** = Nonagon ↑ Vieleck		**non-destructive test**	QUAL	**zerstörungsfreie Prüfung**
				non-directional ≈ omnidirectional	TECH	**ungerichtet** ≈ allseitig gerichtet
nonagonal	MATH	**neuneckig**		**nondirectional aerial** (→omnidirectional antenna)	ANT	(→Rundstrahlantenne)

nondirectional antenna ANT (→Rundstrahlantenne)
(→omnidirectional antenna)
non-directional beacon RADIO NAV (→Rundstrahlbake)
(→omnidirectional radiobeacon)
non-directional microphone EL.ACOUS (→Kugelmikrophon)
(→omnidirectional microphone)
non-directional radiobeacon RADIO NAV (→Rundstrahlbake)
(→omnidirectional radiobeacon)
nondirective aerial ANT (→Rundstrahlantenne)
(→omnidirectional antenna)
nondirective antenna ANT (→Rundstrahlantenne)
(→omnidirectional antenna)
nondisjunction (→NOR ENG.LOG (→NOR-Verknüpfung)
operation)
nondisjunction circuit (→NOR CIRC.ENG (→NOR-Glied)
gate)
nondisjunction element CIRC.ENG (→NOR-Glied)
(→NOR gate)
nondisjunction gate (→NOR CIRC.ENG (→NOR-Glied)
gate)
non-documentary OFFICE **beleglos**
= nichtdokumentär
nonelectrical PHYS **nichtelektrisch**
non-equivalence ENG.LOG (→EXKLUSIV-ODER-Ver-
(→EXCLUSIVE-OR operation) knüpfung)
non-equivalence circuit CIRC.ENG (→EXKLUSIV-ODER-
(→EXCLUSIVE OR gate) Glied)
non-equivalence element CIRC.ENG (→EXKLUSIV-ODER-
(→EXCLUSIVE OR gate) Glied)
non-equivalence function ENG.LOG (→EXKLUSIV-ODER-Ver-
(→EXCLUSIVE-OR operation) knüpfung)
non-equivalence gate CIRC.ENG (→EXKLUSIV-ODER-
(→EXCLUSIVE OR gate) Glied)
non-equivalence operation ENG.LOG (→EXKLUSIV-ODER-Ver-
(→EXCLUSIVE-OR operation) knüpfung)
non-erasable memory 1 DATA PROC **nichtlöschbarer Speicher 1**
[content cannot be changed nor er- [Inhalt kann weder geän-
ased] dert noch gelöscht werden]
= non-erasable storage 1; non-erasable ≈ Dauerspeicher 3; Perma-
store 1; permanent memory 3; perma- nentspeicher 3
nent storage 3 ≈ Festwertspeicher
≈ read-only memory
non-erasable storage 1 DATA PROC (→nichtlöschbarer Spei-
(→non-erasable memory 1) cher 1)
non-erasable store 1 DATA PROC (→nichtlöschbarer Spei-
(→non-erasable memory 1) cher 1)
non-euclidian MATH **nichteuklidisch**
nonexecutable statement DATA PROC **nichtausführbare Anweisung**
nonfatal error DATA PROC **nichtfataler Fehler**
[program continues, although not cor- [Programm läuft weiter,
rectly] wennauch mit Einschrän-
≠ fatal error kungen]
 ≠ fataler Fehler
non-ferrous metal METAL **NE-Metall**
= Nicht-Eisen-Metall; Bunt-
metall
nonformatted DATA PROC (→formatfrei)
(→unformatted)
nonformatted input DATA PROC **formatfreie Eingabe**
= unformatted input
non-fulfilment ECON **Nichterfüllung**
= non-performance; non-conform-
ance
non-fused earthed conductor POWER ENG (→Schutzerdungsleiter)
(→protective earth conductor)
nonglare TECH **spiegelfrei**
nongraphic character TERM&PER **nichtgraphisches Zeichen**
nonharmonic (→anharmonic) PHYS (→nichtharmonisch)
non-hierarchical network DATA COMM **nicht-hierarchisches Netz**
= democratic network ≈ demokratisches Netz
non-impact (adj.) TERM&PER **anschlagfrei**
[printer] [Drucker]
= nonimpact = anschlaglos; nichtmecha-
 nisch
nonimpact (→non-impact) TERM&PER (→anschlagfrei)
non-impact printer TERM&PER **anschlagfreier Drucker**
↓ ink-jet printer; thermal printer; elec- = anschlagloser Drucker;
trophotographic printer; laser printer nichtmechanischer Drucker
 ↓ Tintendrucker; Thermo-
 drucker; elektrofotografi-
 scher Drucker; Laserdruk-
 ker
noninductive winding (→bifilar COMPON (→Bifilarwicklung)
winding)
non-insulated (→bare) COMM.CABLE (→blank)
non-interlaced mode TERM&PER **Vollbildbetrieb**
non-intrusive monitoring TELECONTR **unterbrechungsfreie Überwachung**
non-invertible TECH **nichtumkehrbar**
non-isolated-gate field effect MICROEL (→NIGFET)
transistor (→NIGFET)
non-licensed transmitter RADIO **Schwarzsender**
(→unlicensed transmitter)
nonlinear MATH **nichtlinear**
= non-linear ≈ krummlinig
≈ curvilinear
non-linear (→nonlinear) MATH (→nichtlinear)
nonlinear distortion TV **Linearitätsfehler**
nonlinear distortion TELEC **nichtlineare Verzerrung**
nonlinearity MATH **Nichtlinearität**
nonlinear optics PHYS **nichtlineare Optik**
nonlinear programming DATA PROC **nichtlineare Programmierung**
non-linear pulse technique ELECTRON **nichtlineare Impulstechnik**
nonlinear resistance EL.TECH **nichtlinearer Widerstand**
[physical magnitude] [physikalische Größe]
≠ ohmic resistance ≠ ohmscher Widerstand
nonlinear resistor COMPON **nichtlinearer Widerstand**
[component] [Bauelement]
≠ ohmic resistor ≠ ohmscher Widerstand
↓ termistor; varistor ↓ Thermistor; Varistor
nonlinear scale INSTR **nichtlineare Skala**
non-linear transfer element CONTROL **nichtlineares Übertragungsglied**
non-linked code SWITC **offene Kennzahl**
[exchange code separated from sub- [zusätzlich zu Teilnehmer-
scriber's number] nummer ist die Kennzahl
= open code der Vermittlung zu wählen]
non-linked numbering (→open SWITCH (→offene Numerierung)
numbering)
non-load dc EL.TECH **Leerlaufgleichspannung**
nonloaded (→unloaded) COMM.CABLE (→unbespult)
nonloaded cable COMM.CABLE (→unbespultes Kabel)
(→unloaded cable)
non-locking COMPON **nichtrastend**
= ohne Rast
non-locking key COMPON **Taster**
[contact closing while depressed] [nur solange geschlossen
= momentary contact switch bleibender Kontakt, wie er
≠ switch betätigt wird]
↑ key ≠ Schalter
 ↑ Taste
non-locking pushbutton key TERM&PER **nichtrastender Tastschalter**
 = Tastschalter ohne Rast
nonlocking toggle switch COMPON **Kipptaste**
≈ toggle switch ≈ Kippschalter
non-magnetic PHYS **nichtmagnetisch**
= nonmagnetic = unmagnetisch
nonmagnetic (→non-magnetic) PHYS (→nichtmagnetisch)
non-marking pen TERM&PER **nicht markierender Stift**
non-maskable interrupt DATA PROC **nichtmaskierbare Unterbre-
= NMI chung**
 = nichtunterdrückbare Unterbrechung; nichtmaskierba-
 res Interrupt
non-metal (n.) CHEM **Metalloid**
= metalloid 1 ≈ Nichtmetall
≈ metalloid 2
non metallic TECH **nichtmetallisch**
non-monetary capital ECON **Realkapital**
non-motion video (→fixed-image TELEC (→Festbildkommunikation)
communication)
nonnegative integer (→cardinal MATH (→Kardinalzahl)
number)
nonnumeric INF **nichtnumerisch**
nonnumeric programming DATA PROC **nichtnumerische Programmierung**
nonoverlapping processing DATA PROC **nichtüberlappende Verarbeitung**
Nonpareil (BRI) (→type size 6 TYPOGR (→Schriftgröße 6 Punkt)
point)

non-performance	ECON	(→Nichterfüllung)	non-trigger	ELECTRON	nichtzündend
(→non-fulfilment)			non-uniform	TECH	ungleichförmig
nonpermanent (→temporary)	TECH	(→vorübergehend)			= ungleichmäßig
non-polarized relay	COMPON	neutrales Relais	non-uniform coding	CODING	nichtgleichmäßige Codierung
non-polluting (→environmentally favorable)	TECH	(→umweltfreundlich)	non-uniform quantification	CODING	nichtgleichförmige Quantisierung
non-preemtive priority	SWITCH	nichtunterbrechende Priorität			= nichtgleichmäßige Quantisierung
nonprinting	DATA PROC	nichtdruckend	non-unique mapping	MATH	(→Korrespondenz 2)
nonprocedural query language	DATA PROC	nicht verfahrensorientierte Abfragesprache	(→correspondence 2) non-urgent alarm	EQUIP.ENG	nicht dringender Alarm
non-proprietary	TECH	herstellerneutral	= deferred alarm; minor alarm		
non-pulse period (→pulse separation)	ELECTRON	(→Impulspause)	nonverbal communication (→non-voice communication)	TELEC	(→Nicht-Sprache-Kommunikation)
non-reactive	PHYS	induktionsfrei	nonverbal service (→non-voice	TELEC	(→Nicht-Sprache-Kommuni-
nonreactive	CIRC.ENG	rückwirkungsfrei	communication)		kation)
= reaction-free		= rückkopplungsfrei	non-voice communication	TELEC	Nicht-Sprache-Kommunika-
nonreactive amplifier	CIRC.ENG	(→rückwirkungsfreier Verstärker)	= nonverbal communication; non-telephone communication; non-voice ser-		tion = Nichtfernsprechdienst;
(→reaction-free amplifier)			vice; nonverbal service;		Nichttelefondienst; Nicht-
nonreciprocal	MATH	nichtreziprok	non-telephone service		Sprache-Dienst
= non-reciprocal			≠ voice communication		≠ Sprachkommunikation
non-reciprocal	MATH	(→nichtreziprok)			↓ Datenkommunikation
(→nonreciprocal)			non-voice service (→non-voice	TELEC	(→Nicht-Sprache-Kommuni-
non-reciprocal wave rotation	PHYS	(→Faraday-Effekt)	communication)		kation)
(→Faraday effect)			non volatile	DATA PROC	nichtflüchtig
non-reciprocal wave rotator	MICROW	(→Faraday-Rotator)	non-volatile	TECH	nichtflüchtig
(→Faraday rotator)			≠ volatile		≠ flüchtig
nonrecurrent (→one-time)	COLLOQ	(→einmalig)	non-volatile memory	DATA PROC	nichtflüchtiger Speicher
non-reflecting	PHYS	reflexionsfrei	[maintains content even after removing operational power]		[bewahrt Speicherinhalt auch nach Abschalten der
= reflectionless; non-reflective		= nichtreflektierend	= non-volatile storage; non-volatile		Betriebsspannung]
non-reflective	PHYS	reflexionsfrei	store; permanent memory; perma-		= Permanentspeicher 1; Dau-
(→non-reflecting)			nent storage; permanent store		erspeicher 1
nonreflecting ink	TERM&PER	nichtreflektierende Farbe	≈ non-erasable memory 1		≈ nichtlöschbarer Speicher
= read ink			≠ volatile memory		≠ flüchtiger Speicher
non-regular (→unplanned)	COLLOQ	(→unplanmäßig)	↓ ROM; non-volatile solid state memory		↓ Festwertspeicher; nichtflüchtiger Halbleiterspei-
non-relocatable programm	DATA PROC	unverschiebliches Programm			cher
non-resident	DATA PROC	nichtresident	non-volatile RAM	DATA PROC	nichtflüchtiger Direkzugriffs-
non-resident program	DATA PROC	nichtresidentes Programm	= NV-RAM		speicher
nonresonant aerial (→aperiodic antenna)	ANT	(→aperiodische Antenne)	nonvolatile semiconductor memory (→nonvolatile solid state	MICROEL	= NV-RAM (→nichtflüchtiger Halbleiter-
nonresonant antenna (→aperiodic antenna)	ANT	(→aperiodische Antenne)	memory) nonvolatile solid state	MICROEL	speicher) nichtflüchtiger Halbleiterspei-
nonresonant long wire antenna	ANT	abgeschlossene Langdrahtantenne	memory [maintains its content when power is		cher [erhält den Inhalt auch ohne Stromversorgung]
non-restricted	TELEPH	vollamtsberechtigt	turned off]		
[allowed to dial long-distance calls]		[darf Ferngespräche wählen]	= nonvolatile semiconductor memory ↑ nonvolatile memory		↑ nichtflüchtiger Speicher
≠ outward restricted		= amtsberechtigt	non-volatile storage	DATA PROC	(→nichtflüchtiger Speicher)
		≠ halbamtsberechtigt	(→non-volatile memory)		
non-restriction (→direct outward dialing)	TELEPH	(→Amtsberechtigung)	non-volatile store (→non-volatile memory)	DATA PROC	(→nichtflüchtiger Speicher)
nonretractable screw	MECH	Einwegschraube	no-op command (→blank	DATA PROC	(→Leeranweisung)
[resists removal]		[kann nicht mehr gelöst	instruction)		
= one-way screw		werden]	NO operation (→NOT	ENG.LOG	(→NICHT-Verknüpfung)
non-return-to-zero code (→NRZ code)	CODING	(→NRZ-Code)	operation) no-operation instruction	DATA PROC	(→Leeranweisung)
non-return-to-zero recording	DATA PROC	(→Wechselschrift)	(→blank instruction)		
(→NRZ recording)			no-op instruction (→blank	DATA PROC	(→Leeranweisung)
non-rusting (→rustless)	CHEM	(→rostfrei)	instruction)		
nonsaturated logic	MICROEL	ungesättigte Logik	no-op statement (→blank	DATA PROC	(→Leeranweisung)
non-scrap manufacturing	MANUF	abfallose Fertigung	instruction)		
nonsealed relay	COMPON	offenes Relais	noose (→sling)	TECH	(→Schlinge)
≠ hermetically sealed relay			NOP instruction (→blank	DATA PROC	(→Leeranweisung)
non-self-maintained discharge	PHYS	unselbständige Entladung	instruction)		
nonsinusoidal	MATH	nichtsinusförmig	norator	NETW.TH	Norator
= nonsinusoidal			NOR circuit (→NOR gate)	CIRC.ENG	(→NOR-Glied)
nonsinusoidal	MATH	(→nichtsinusförmig)	NORD antenna	ANT	NORD-Antenne
(→non-sinusoidal)			Nordberg key	MECH	Rundpaßfeder
non-skid	TECH	rutschfest	no release	SWITCH	keine Auslösung
= slip-resistant					= keine Trennung
non-standard	TECH	nicht normgerecht	NOR element (→NOR gate)	CIRC.ENG	(→NOR-Glied)
non-stationary (→non-steady)	TECH	(→nichtstationär)	NOR function (→NOR	ENG.LOG	(→NOR-Verknüpfung)
non-steady	TECH	nichtstationär	operation)		
= non-stationary; unsteady; mobile		= mobil	NOR gate	CIRC.ENG	NOR-Glied
nonswitched	TELEC	nichtvermittelt	= NOR element; NOR circuit; nondis-		= NOR-Gatter; NOR-Element; NOR-Schaltung;
non-telephone communication (→non-voice communication)	TELEC	(→Nicht-Sprache-Kommunikation)	junction gate; nondisjunction ele-		
non-telephone service (→non-voice communication)	TELEC	(→Nicht-Sprache-Kommunikation)			
non-terminable	ECON	unkündbar			
= irredeemable					

norm

ment; nondisjunction circuit; Pierce gate; Pierce element; Pierce circuit ↑ logic gate		NOR-Tor; WEDER-NOCH-Glied; WEDER-NOCH-Gatter; WEDER-NOCH-Element; WEDER-NOCH-Schaltung; WEDER-NOCH-Tor; Pierce-Gatter; Pierce-Element; Pierce-Schaltung; Pierce-Tor	
norm (→standard)	TECH	(→Norm)	
normal (→medium)	TYPOGR	(→normal)	
normal (n.) (→perpendicular)	MATH	(→Senkrechte)	
normal cassette	TERM&PER	**Normalkassette**	
normal charging time [accumulator]	POWER SYS	**Normalladezeit** [Akkumulator]	
normal contact (→break contact)	COMPON	(→Ruhekontakt)	
normal crystal	PHYS	**Normalkristall**	
normal disconnected mode	DATA COMM	**abhängiger Wartezustand**	
normal dispersion = normale Dispersion	PHYS	**Normaldispersion**	
normal distribution = Gauss distribution	MATH	**Normalverteilung** = Gauß-Verteilung	
normal equation	MATH	**Normalgleichung**	
normal hysteresis loop	PHYS	**normale Hystereseschleife**	
normalization (→standardization)	TECH	(→Standardisierung)	
normalize = normalisieren	METAL	**normalglühen**	
normalize	NETW.TH	**normieren**	
normalize = standardize ≈ uniform	TECH	**standardisieren** = normieren; normen ≈ vereinheitlichen	
normalized equation	MATH	**normierte Gleichung**	
normalized frequency	NETW.TH	**normierte Frequenz**	
normalizing = stress-relieving anneal	METAL	**Normalisieren; Entspannungsglühen**	
normalizing [simplification of formulas] ≠ denormalizing ↓ frequency normalizing; impedance normalizing; scaling; time normalizing	NETW.TH	**Normierung** [Vereinfachung von Formeln] ≠ Entnormierung ↓ Frequenznormierung; Impedanznormierung; Skalierung; Zeitnormierung	
normalizing frequency [fo in f/fo]	NETW.TH	**Normierungsfrequenz** [fo in f/fo]	
normally closed contact (→break contact)	COMPON	(→Ruhekontakt)	
normally open contact (→make contact)	COMPON	(→Arbeitskontakt)	
normal magnetization curve [magnetization] = commutation curve	PHYS	**Kommutierungskurve** [Ferromagnetismus, Magnetisierungskurve]	
normal magnetization curve (→initial magnetization curve)	PHYS	(→Neukurve)	
normal mode (→natural oscillation)	PHYS	(→Eigenschwingung)	
normal offset	BROADC	**Normaloffset**	
normal operation	TECH	**Normalbetrieb**	
normal paper = standard paper; plain paper	TERM&PER	**Normalpapier** = normales Papier	
normal position = preferred position; standard position; home position; home	TECH	**Normalstellung** = Normallage; Nennlage; Regelstellung; Regellage; Vorzugsstellung; Vorzugslage	
normal position (→initial state)	CIRC.ENG	(→Grundstellung)	
normal position (→home position)	SWITCH	(→Ruhelage)	
normal price = standard price	ECON	**Normalpreis**	
normal propagation	RADIO PROP	**Normalausbreitung**	
normal radiation (→intrinsic radiation)	PHYS	(→Eigenstrahlung)	
normal rate (→normal tariff)	TELEC	(→Normaltarif)	
normal rating	TECH	**Dauerleistung**	
normal route (→primary route)	SWITCH	(→Erstweg)	
normal spectrum	PHYS	**Normalspektrum**	
normal style (→normal type)	TERM&PER	(→Normalschrift)	
normal subscriber (→ordinary subscriber)	TELEC	(→Normalteilnehmer)	
normal tariff = normal rate	TELEC	**Normaltarif**	
normal type = normal style	TERM&PER	**Normalschrift**	
normal velocity = normale Geschwindigkeit	PHYS	**Normalgeschwindigkeit**	
normal-velocity surface	PHYS	**Normalenfläche**	
normal vibration (→natural oscillation)	PHYS	(→Eigenschwingung)	
NOR operation [output = 1 if simultaneously A = 0 and B = 0, output = 0 if at least one input = 1] = NOR function; nondisjunction; Pierce operation; Pierce function; joint denial ↑ logic operation	ENG.LOG	**NOR-Verknüpfung** [Ausgang = 1 wenn gleichzeitig A = 0 und B = 0, Ausgang = 0 wenn mindestens ein Eingang = 1] = NOR-Funktion; WEDER-NOCH-Verknüpfung; WEDER-NOCH-Funktion; Pierce-Verknüpfung; Pierce-Funktion ↑ logische Verknüpfung	
North American hierarchy = Bell hierarchy	TELEC	**nordamerikanische Hierarchie**	
northern light = aurora borealis	GEOPHYS	**Nordlicht**	
north pole	GEOPHYS	**Nordpol**	
Norton amplifier	CIRC.ENG	**Norton-Verstärker**	
Norton factor	DATA PROC	**Norton-Faktor**	
Norton's theorem ↑ Helmholtz equivalent-source theorem	NETW.TH	**Theorem von Norton** = Nortonsches Theorem ↑ Helmholtzscher Satz	
Norton transformation	NETW.TH	**Norton-Transformation**	
NOS (→network operating system)	DATA COMM	(→Netzwerkbetriebssystem)	
NOSFER	TELEPH	**NOSFER** [von CCITT genormter Eichkreis für subjektive Messungen der Bezugsdämpfung]	
no-station address	DATA COMM	**Sperradresse**	
not accepted	DATA COMM	**abgelehnt**	
notary	ECON	**Notar**	
notation = representation	MATH	**Schreibweise** = Darstellung; Notation	
notation (→representation)	DOC	(→Darstellung)	
notation (→number system)	MATH	(→Zahlensystem)	
notch (n.) [an angular short cut] = indentation ≈ cleft; gap; split; groove 3; dent; incision	TECH	**Kerbe** [kurze, V-förmige Vertiefung] = Kerb (n.m. pl.-e); Falz (Holz); Dalle; Einkerbung ≈ Schlitz; Spalt; Spalte; Nut; Delle; Einschnitt	
notch	RADIO PROP	**Dämpfungsmaximum**	
notch antenna	ANT	**Kerbantenne** = Nutantenne; Notch-Antenne	
notch diplexer	TV	**Filterweiche**	
notched	TECH	**ausgeklinkt**	
notched filter [INSTR] (→bandstop filter)	NETW.TH	(→Bandsperrfilter)	
notched taper pin	MECH	**Kegelkerbstift**	
notch filter (→twin-T-filter)	NETW.TH	(→Doppel-T-Filter)	
notch filter (→blocking filter)	NETW.TH	(→Sperrfilter)	
notch frequency	RADIO REL	**Notch-Frequenz**	
notch frequency (→stop frequency)	NETW.TH	(→Sperrfrequenz)	
notching	METALL	**Ausklinken** [Stanzen]	
notching relais [activated by a given number of impulses]	COMPON	**Stromstoßrelais** = Fortschaltrelais	
NOT circuit (→NOT gate)	CIRC.ENG	(→NICHT-Glied)	
note (n.) = entry	DOC	**Vermerk** = Hinweis	

note (→explanation)	LING	(→Erläuterung)
note (→memorandum)	OFFICE	(→Notiz)
note (→promissory note)	ECON	(→Eigenwechsel)
note bit	DATA COMM	Merkbit
notebook	DOC	Merkbuch
		= Notizbuch
notebook computer	DATA PROC	Notizbuchcomputer
		= Note-book-Computer
NOT element (→NOT gate)	CIRC.ENG	(→NICHT-Glied)
note pad	OFFICE	Notizblock
= memo pad; pad; copy block; scribbling block; block; scratch pad; sketch pad		= Block; Schreibblock; Schmierblock
note pad (→scratch pad memory)	DATA PROC	(→Notizblockspeicher)
not equal (→unequal)	MATH	(→ungleich)
not-equal sign	MATH	Ungleichzeichen
not-erasable memory 2 (→read-only memory)	DATA PROC	(→Festwertspeicher)
notes for the operator	TECH	Bedienungshinweise
= hints for the operator		≈ Bedienungsanleitung; Betriebsvorschrift
≈ use instruction; operating instruction		
not executable (→impracticable)	TECH	(→nichtausführbar)
NOT function (→NOT operation)	ENG.LOG	(→NICHT-Verknüpfung)
NOT gate	CIRC.ENG	NICHT-Glied
= NOT element; NOT circuit; inverter 2; negator		= NICHT-Gatter; NICHT-Element; NICHT-Schaltung; NICHT-Tor; Negationsglied; Negationsgatter; Negationselement; Negationsschaltung; Negatorglied; Negatorgatter; Negatorelement; Negatorschaltung; Negator; Inverter-Glied; Inverter-Gatter; Inverter-Element; Inverter-Schaltung; Inverter; Invertierer; NEIN-Glied; NEIN-Gatter; NEIN-Element; NEIN-Schaltung
↑ logic gate		↑ Verknüpfungsglied
notice	ECON	Kündigung
noticeably	COLLOQ	spürbar
≈ clearly		≈ merkbar
		≈ eindeutig
notice board (→bulletin board)	COLLOQ	(→Anzeigetafel)
notification (→information)	ECON	(→Benachrichtigung)
NOT-IF-THEN (→exclusion operation)	ENG.LOG	(→Inhibitionsverknüpfung)
NOT-IF-THEN circuit (→exclusion gate)	CIRC.ENG	(→Inhibitionsglied)
NOT-IF-THEN element (→exclusion gate)	CIRC.ENG	(→Inhibitionsglied)
NOT-IF-THEN function (→exclusion operation)	ENG.LOG	(→Inhibitionsverknüpfung)
NOT-IF-THEN gate (→exclusion gate)	CIRC.ENG	(→Inhibitionsglied)
NOT-IF-THEN operation (→exclusion operation)	ENG.LOG	(→Inhibitionsverknüpfung)
not included in the price (→unbundled)	DATA PROC	(→nicht im Preis einbegriffen)
not obligatory (→non-binding)	ECON	(→unverbindlich)
not obtainable	DATA COMM	nicht erreichbar
NOT operation	ENG.LOG	NICHT-Verknüpfung
[output = 1 if input = 0, and viceversa]		[Ausgang = 1 wenn Eingang = 0, und umgekehrt]
= NOT function; NO operation; NO function; inversion operation; inversion function; inversion		= NICHT-Funktion; Negationsverknüpfung; Negationsfunktion; Negation; Invertierung
↑ logic operation		↑ logische Verknüpfung
not running (→inactive)	DATA PROC	(→inaktiv)
not transferable	ECON	unübertragbar
= unassignable		
not working (→inactive)	DATA PROC	(→inaktiv)
nought (→zero)	MATH	(→Null)
noun 2	LING	Nomen 2
[declinable word]		[deklinierbares Wort]
↓ noun 1; pronoun; adjective		↓ Substantiv; Pronom; Adjektiv
noun 1 (→substantive)	LING	(→Substantiv)
no-value sample	POST	Muster ohne Wert
= pattern-post		
novel (adj.)	COLLOQ	neuartig
= unprecedented		= beispiellos
≈ new; innovative		≈ neu; innovativ
novelty	COLLOQ	Neuheit
		= Neuartigkeit; Novum
novice	TECH	Anfänger
= beginner; entry-level user [DATA PROC]; naive user [DATA PROC]		= Einsteiger [DATA PROC]
novice licence	RADIO	Anfängerlizenz
nozzle	TECH	Düse
nozzle (→stem)	TECH	(→Stutzen)
nozzle-drawing (→burring)	METAL	(→Düsenziehen)
Np (→neptunium)	CHEM	(→Neptunium)
NPIN transistor	MICROEL	NPIN-Transistor
np junction	PHYS	NP-Übergang
n-plus-one instruction	DATA PROC	N-plus-Eins-Adresse
npn transistor	MICROEL	NPN-Transistor
n-port	NETW.TH	n-Tor
NPR (→noise power ratio)	TELEC	(→Geräuschleistungsverhältnis)
NPT (→network planning technique)	ECON	(→Netzplantechnik)
NRZ code	CODING	NRZ-Code
= non-return-to-zero code		
NRZ recording	DATA PROC	Wechselschrift
= non-return-to-zero recording		= NRZ-Schrift; Non-return-to-zero-Schrift
↑ recording mode		↑ Aufzeichnungsverfahren
ns (→nanosecond)	PHYS	(→Nanosekunde)
NS (→national special thread)	MECH	(≈ amerikanisches Spezialgewinde)
NT (→network termination)	TELEC	(→Netzabschlußeinrichtung)
nt (→nit)	PHYS	(→Nit)
NTC thermistor	COMPON	Heißleiter
= negative temperature coefficient thermistor; negative temperature coefficient resistor; sensistor		= NTC-Widerstand; Kaltwiderstand; TN-Halbleiterwiderstand (DDR); Thernewid; Sensistor
		↑ Thermistor
n-th root	MATH	n-te Wurzel
NTSC	TV	NTSC
[National Television Systems Committee]		[US-Norm, mit 60 Hz, 525 Zeilen]
n-type (→n-conducting)	PHYS	(→N-leitend)
N type carrier system	TRANS	Vierdrahtgleichlage-System
N2 type carrier system	TRANS	V12-System
N3 type carrier system	TRANS	V24-System
n-type conduction	PHYS	N-Leitung
= electron conduction		= Überschußleitung; Elektronenleitung
n-type conductivity	PHYS	N-Leitfähigkeit
= electron conductivity		= Elektronenleitfähigkeit
n-type conductor	PHYS	N-Leiter
n-type semiconductor	PHYS	N-Halbleiter
		= N-Typ-Halbleiter; N-Material; Überschuß-Halbleiter
NUA (→transport address)	DATA COMM	(→Teilnehmeradresse)
nuclear installation (→nuclear plant)	POWER ENG	(→Kernanlage)
nuclear magnetic resonance	PHYS	Kernresonanz
nuclear magneton	PHYS	Kernmagneton
[constant]		[Konstante]
nuclear mass	PHYS	Kernmasse
nuclear number (→atomic mass number)	PHYS	(→Massenzahl)
nuclear physics	PHYS	Kernphysik
≈ atomic physics		≈ Atomphysik
nuclear plant	POWER ENG	Kernanlage
= nuclear installation		
nuclear power station	POWER ENG	Kernkraftanlage
nuclear reactor	POWER SYS	Nuklearreaktor
= reactor		= Reaktor

nuclear research	PHYS	Kernforschung
nuclear-resonance fluorescence	PHYS	Kernresonanzfluoreszenz
nuclear spin	PHYS	Kernspin
nucleon	PHYS	Nukleon
nucleon number (→atomic mass number)	PHYS	(→Massenzahl)
nucleus	DATA PROC	Nukleus
[part of the executive routine permanently resident in the main memory]		[dauernd im Arbeitsspeicher residenter Teil des Organisationsprogramms]
≠ transient program area		≠ Übergangsbereich
nucleus (→atomic core)	PHYS	(→Atomkern)
nucleus (→core)	TECH	(→Kern)
NUI (→subscriber identification)	DATA COMM	(→Teilnehmerkennung)
nuisance alarm (→false alarm)	SIGN.ENG	(→Fehlalarm)
null (→zero)	MATH	(→Null)
null (→zero point)	MATH	(→Nullstelle)
null (→directional null)	ANT	(→Nullstelle)
nullator	NETW.TH	Nullator
≈ nullor		≈ Nullor
null bit (→dummy bit)	CODING	(→Leerbit)
null command (→blank instruction)	DATA PROC	(→Leeranweisung)
null cycle	DATA PROC	Nullzyklus
[run time without introducing new data]		[Programmlaufzeit mit unveränderten Daten]
null detector	CIRC.ENG	Nulldetektor
null detector (→zero instrument)	INSTR	(→Nullinstrument)
nullify (→override)	TECH	(→lahmlegen)
null indicator (→zero instrument)	INSTR	(→Nullinstrument)
nulling (→zero balancing)	ELECTRON	(→Nullabgleich)
null instruction (→blank instruction)	DATA PROC	(→Leeranweisung)
null matrix	MATH	Nullmatrix
null method (→zero method)	INSTR	(→Nullmethode)
null modem	DATA COMM	Nullmodem
nullode	ELECTRON	Nullode
nullor	NETW.TH	Nullor
[nullator + norator]		[Nullator + Norator]
null set (→empty set)	MATH	(→Leermenge)
null statement (→blank instruction)	DATA PROC	(→Leeranweisung)
null steering	RADIO LOC	Nullpeilung
≠ maximum steering		≠ Maximumpeilung
null-steering antenna	ANT	Nullpeilantenne
null string	DATA PROC	Leerkette
= blank string		
null suppression	DATA PROC	Nullzeichenunterdrückung
number	SWITCH	Nummer
= code 2		≈ Kennzahl
≈ code number		↓ Vorwahlnummer; Teilnehmernummer
↓ prefix plus code number; subscriber number		
number	MATH	Zahl 1
[a set related to the unit one, represented by a symbol, like a cipher, or a combination of symbols]		[durch Zeichen, wie Ziffern, oder Zeichenkombinationen dargestellte, auf die Grundeinheit Eins bezogene Zählmenge]
≈ digit		= Nummer
		≈ Ziffer
number	LING	Numerus
		= Zahl
number base (→base 1)	MATH	(→Basis)
number checking equipment	TERM&PER	Nummernprüfgerät
number cruncher	DATA PROC	(→Großrechner)
(→mainframe computer)		
number cruncher (→maths coprocessor)	DATA PROC	(→mathematischer Koprozessor)
numbered clause (→paragraph sign 1)	TYPOGR	(→Paragraphzeichen)
number format	DATA PROC	Zahlenformat
number identification (→call number identification)	TELEC	(→Rufnummeridentifizierung)
numbering	SWITCH	Numerierung
numbering	TECH	Zählfolge
		= Numerierung; Benummerung
numbering area	SWITCH	Numerierungsbereich
numbering plan	SWITCH	Numerierungsplan
= numbering scheme		
numbering scheme (→numbering plan)	SWITCH	(→Numerierungsplan)
numberless (→innumerable)	COLLOQ	(→zahllos)
number of ampere turns	PHYS	Amperewindungszahl
= ampere turns 2		
number of defects	QUAL	Ausschußzahl
number of impulsive noise	ELECTRON	Störimpulshäufigkeit
number of pages	TYPOGR	Seitenzahl 1
		[Gesamtzahl der Seiten]
number of parts (→quantity)	ECON	(→Stückzahl)
number of seizures (→peg)	SWITCH	(→Belegungszahl)
number of threads	MECH	Gangzahl
		[Gewinde]
number of turns	EL.TECH	Windungszahl
number plate	TERM&PER	Ziffernring
[rotary dial selector]		[Nummernschalter]
= dial 2		= Ziffernblatt
number preselection	TECH	Ziffernvorwahl
= number presetting		
number presetting (→number preselection)	TECH	(→Ziffernvorwahl)
number protection code	CODING	Ziffernsicherungscode
		= Zahlensicherungscode
number representation	DATA PROC	Zahlendarstellung
number sequence	MATH	Zahlenfolge
[infinite sequence of numbers arranged by a determined law]		[unendliche Menge in einer bestimmten Reihenfolge angeordneter Zahlen]
number symbol (→numeral)	MATH	(→numerisches Zeichen)
number system	MATH	Zahlensystem
= counting system; notation; numeric system		= Zahlendarstellung
↓ denominational number system; decimal number system; binary number system		↓ Stellenwertsystem; Dezimalsystem; Dualsystem
"number unobtainable"	TELEPH	„kein Anschluß unter dieser Nummer"
numeral	LING	Numerale (n.n.)
↓ cardinal number; ordinal number		= Zahlenwort; Zahlwort
		↓ Grundzahl; Ordnungszahl
numeral	MATH	numerisches Zeichen
[element of a set of characters to represent numbers]		[Element eines Zeichenvorrats zur Darstellung von Zahlen]
= number symbol; numeric character; figure; digit; cipher		= Zahlzeichen; Zahlensymbol; Chiffre; Ziffer; Zahl 2
≈ number		≈ Zahl 1
↑ sign		↑ Zeichen
numeralization	CODING	Zifferndarstellung
[of characters]		[von Buchstaben]
		= Numeralisierung
numerator (→dividend 2)	MATH	(→Dividend 2)
numeric (n.)	MATH	Numerik
numeric (adj.)	MATH	numerisch
= numerical		= zahlenmäßig
numeric	CODING	numerisch
= numerical		≈ alphanumerisch
≈ alphanumerical		≠ alphabetisch
≠ alphabetic		
numerical (→numeric)	MATH	(→numerisch)
numerical (→numeric)	CODING	(→numerisch)
numerical aperture	PHYS	numerische Apertur
numerical calculation	MATH	numerische Berechnung
numerical code	CODING	Nummernschlüssel
= numeric code		= Nummerncode; numerischer Code; Zifferncode
numerical constant	PHYS	Zahlenkonstante
		= numerische Konstante
numerical control	CONTROL	numerische Steuerung
= numeric control; NC		= NC-Steuerung; NC-Technik; numerische Maschinensteuerung
↓ numerical machine tool control		↓ numerische Werkzeugmaschinensteuerung

numerical data (→numeric DATA PROC (→numerische Daten)
data)
numerical display (→numeric INSTR (→Ziffernanzeige)
display)
numerical distance ANT **numerische Entfernung**
numerical indicator (→numeric INSTR (→Ziffernanzeige)
display)
numerical indicator tube ELECTRON (→Ziffernanzeigeröhre)
(→numeric display tube)
numerically controlled machine TECH (→numerisch gesteuerte
tool (→NC machine tool) Werkzeugmaschine)
numerical machine tool CONTROL **numerische Werkzeugmaschi-**
control **nensteuerung**
↑ numeric control ↑ numerische Steuerung
numerical mathematics MATH **numerische Mathematik**
numerical progression (→numeric MATH (→numerische Reihe)
progression)
numerical series (→numeric MATH (→numerische Reihe)
progression)
numerical simulation DATA PROC (→Simulationstechnik)
(→simulation technique)
numerical term MATH **Zahlenterm**
↑ term ↑ Term
numerical value MATH **Zahlenwert**
[result of a mathematical operation] [Ergebnis einer mathemati-
= value schen Operation]
  = numerischer Wert; Wert
numerical-value equation PHYS **Zahlenwertgleichung**
≠ dimensional equation ≠ Größengleichung
numeric character (→numeral) MATH (→numerisches Zeichen)
numeric code (→numerical code) CODING (→Nummernschlüssel)
numeric control (→numerical CONTROL (→numerische Steuerung)
control)
numeric data DATA PROC **numerische Daten**
= numerical data
numeric display INSTR **Ziffernanzeige**
= numerical display; digital display; = numerische Anzeige; digita-
numeric indicator; numerical indica- le Anzeige; Digitalanzeige
tor; digital indicator; numeric rea-
dout; digital readout
numeric display tube ELECTRON **Ziffernanzeigeröhre**
= digital display tube; numerical indica- = Ziffernröhre
tor tube
numeric indicator (→numeric INSTR (→Ziffernanzeige)
display)
numeric key TERM&PER **Zifferntaste**
= figures key ↑ Taste
↑ key
numeric keyboard (→numeric TERM&PER (→Zifferntastenblock)
keypad)
numeric keypad TERM&PER **Zifferntastenblock**
[special keypad or sector of a key- [Spezialtastatur oder Tasta-
board, containing numeric keys, oper- turbereich, mit Ziffernta-
ator keys etc.] sten, Operatortasten u. dgl.]
= numeric keyboard; figures keyboard; = Zifferntastatur; Ziffern-
figures keypad; ten-key pad; adding block; Zehnertastatur; Zeh-
machine keypad nerblock; Rechenblock;
  Blocktastatur; numerisches
  Tastenfeld; numerische Tastatur; numerischer Tastenblock
  ↑ Tastatur
numeric place (→digit DATA PROC (→Ziffernstelle)
position)
numeric position (→digit DATA PROC (→Ziffernstelle)
position)
numeric printer TERM&PER **Zifferndrucker**
[prints only digits] [druckt nur Ziffern]
numeric progression MATH **numerische Reihe**
= numerical progression; numeric = numerische Progression;
series; numerical series numerische Folge
numeric readout (→numeric INSTR (→Ziffernanzeige)
display)
numeric series (→numeric MATH (→numerische Reihe)
progression)
numeric string DATA PROC **Ziffernfolge**
numeric system (→number MATH (→Zahlensystem)
system)
numerous COLLOQ **zahlreich**
= multidinous = vielzählig

NUMLOCK (→NUMLOCK TERM&PER (→NUM-Taste)
key)
NUMLOCK key TERM&PER **NUM-Taste**
[changes from primary to secondary [schaltet zwischen primärer
allocation of numeric keyboard] und sekundärer Belegung
= NUMLOCK des Ziffernblocks um]
  = Taste NUM; NUMLOCK-
  Taste; NUMLOCK
Nusselt number PHYS **Nußelt-Zahl**
nut (n.) MECH **Mutter**
≠ bolt (n.) ≠ Schraube 2
nutate PHYS **kreiseln**
NV-RAM (→non-volatile DATA PROC (→nichtflüchtiger Direkzu-
RAM) griffsspeicher)
n-well MICROEL **N-Wanne**
n-well technology MICROEL **N-Wannen-Technik**
nybble (→tetrad) CODING (→Tetrade)
Nyquist characteristic (→Nyquist MODUL (→Nyquist-Charakteristik)
slope)
Nyquist criterion CONTROL **Nyquist-Kriterium**
Nyquist frequency INF **Nyquist-Frequenz**
Nyquist measuring modulator INSTR **Nyquist-Meßmodulator**
Nyquist noise TELEC **Nyquist-Rauschen**
Nyquist slope MODUL **Nyquist-Charakteristik**
= Nyquist characteristic = Nyquist-Flanke

O

Ω (→ohm) PHYS (→Ohm)
Ω (→solid angle) MATH (→Raumwinkel)
O (→oxygen) CHEM (→Sauerstoff)
oakum TECH **Werg**
[fiber impregnated with tar] [Dichtungsmaterial aus ge-
  teerten Fasern]
obey (→execute) DATA PROC (→ausführen)
object LING **Objekt**
  = Satzergänzung
object (→entity) SCIE (→Objekt)
object code (→object DATA PROC (→Zielsprache)
language)
object computer DATA PROC **Objektprogramm-Computer**
[used to run an object program] [zum Ablaufen eines Ob-
  jektprogramms]
objection ECON **Einspruch**
= protest; appeal
objection (→claim 2) ECON (→Beanstandung)
objective PHYS **Objektiv**
[single lens or system of lenses to [Linse oder System von
map an object on a photographic film] Linsen zur Abbildung ei-
≈ lens nes Objektes auf einenpho-
  tographischen Film]
  ≈ minse
objective (→scope) COLLOQ (→Zielsetzung)
objective variable CONTROL **Hilfsregelgröße**
= auxiliary controlled variable
object language DATA PROC **Zielsprache**
[language into which a source pro- [Sprache in die ein Quell-
gram is translated, generally the ma- programm übersetzt wird,
chine code] i.a. die Maschinensprache]
= object code; target language; target = Objektcode
code ≠ Quellsprache
≠ source language ↓ Maschinensprache
↓ machine language
object language program DATA PROC (→Objektprogramm)
(→object program)
object module DATA PROC **Objektmodul**
object-oriented DATA PROC **objektorientiert**
= object-related ≠ pixelorientiert
≠ pixel oriented
object-oriented language DATA PROC **objektorientierte Sprache**
object program DATA PROC **Objektprogramm**
[source program translated into ob- [in Zielsprache/Maschinen-
ject/machine code] sprache übersetztes Quell-
= object language program; target pro- programm]
gram; translated program; secondary = Zielprogramm; übersetztes
program Programm; Sekundärpro-
≈ absolute program gramm
≠ source program ≈ Absolutprogramm
↑ machine program ≠ Quellprogramm
  ↑ Maschinenprogramm

object protection	SIGN.ENG	**Objektschutz**
= spot protection; point protection		
object-related	DATA PROC	(→objektorientiert)
(→object-oriented)		
object-related database	DATA PROC	**objektorientierte Datenbank**
object-related programming	DATA PROC	**objektorientierte Programmierung**
object safeguarding	SIGN.ENG	**Objektsicherung**
obligation	ECON	**Verpflichtung**
= commitment		
obligatory (→mandatory)	COLLOQ	(→vorgeschrieben)
oblige	ECON	**verpflichten**
obliging (→binding)	ECON	(→verbindlich)
obligor (→debtor)	ECON	(→Schuldner)
oblique	TYPOGR	**schräg**
[typeface]		[Schrift]
= inclined; slant		= schräggelegt
oblique (→slant)	TECH	(→schräg)
oblique angle	MATH	**schiefer Winkel**
oblique incidence	PHYS	**schräger Einfall**
		= Schrägeinfall; schiefer Einfall
oblique line (→slash)	LING	(→Schrägstrich)
oblique plane	MECH	**schräge Ebene**
oblique wire antenna	ANT	**Schrägdrahtantenne**
oblong (→longish)	TECH	(→länglich)
oblong antenna	ANT	**Oblong-Antenne**
oblong hole	MECH	**Langloch**
obscure (adj.)	COLLOQ	**völlig unverständlich**
observation	SCIE	**Beobachtung**
observation (→check)	TECH	(→Prüfung)
observation error	SCIE	**Beobachtungsfehler**
obsolescence	TECH	**Veralterung**
		= Überalterung
obsolete	TECH	**veraltet**
= outmoded; outdated; out-of-date; ageing; grandfathered		= überholt; obsolete
obstacle light (→obstruction light)	AERON	(→Hindernisfeuer)
obstruct (→sreen)	RADIO PROP	(→abschirmen)
obstruct (→clog)	TECH	(→verstopfen)
obstruction	RADIO PROP	(→Sichtbehinderung)
(→screening)		
obstruction fading	RADIO PROP	**Beugungsschwund**
= diffraction fading; power fading		= Behinderungsschwund; Hindernisschwund
obstruction light	AERON	**Hindernisfeuer**
= obstacle light		
obstruction loss	RADIO PROP	(→Beugungsdämpfung)
(→diffraction loss)		
obstruction of line-of-sight	RADIO PROP	(→Sichtbehinderung)
(→screening)		
obstruction paint (→warning paint)	TECH	(→Warnbemalung)
obtain (→achieve)	TECH	(→erzielen)
obtainment (→procurement)	ECON	(→Beschaffung)
obtuse (→obtuse angled)	MATH	(→stumpfwinklig)
obtuse angle	MATH	**stumpfer Winkel**
obtuse angled	MATH	**stumpfwinklig**
= obtuse		
≠ acute angled		≠ spitzwinklig
obtuse-angled triangle (→obtuse triangle)	MATH	(→stumpfwinkliges Dreieck)
obtuse triangle	MATH	**stumpfwinkliges Dreieck**
= obtuse-angled triangle		
O.B. van (→outside broadcast van)	BROADC	(→Übertragungswagen)
OC (→operation code)	DATA PROC	(→Operationscode)
occasional student (BRI)	SCIE	**Gasthörer**
occasional user	TECH	**Wenigbenutzer**
= casual user		= gelegentlicher Benutzer; gelegentlicher Anwender
≠ intensive user		≠ Vielbenutzer
occupancy	TELEC	**Belegung**
= usage		
occupancy level	SWITCH	**Füllgrad**
occupation	TECH	**Belegung**
≈ allocation; assignment		≈ Zuteilung; Festlegung
occupation (→seizure)	SWITCH	(→Belegung)
occupation (→activity)	ECON	(→Tätigkeit)
occupational education (→training)	TECH	(→Schulung)
occupation probability	PHYS	**Besetzungswahrscheinlichkeit**
occupation state (→busy state)	SWITCH	(→Besetztzustand)
occupy	DATA PROC	**belegen**
= seize		
ocean freight (AM) (→sea freight)	ECON	(→Seefracht)
ocher	PHYS	**ocker**
= ochre		= ockerfarben; ockerfarbig; gelbbraun; ockerbraun
≈ naples yellow; ocher yellow		≈ neapelgelb; ockerbraun
ocher yellow	PHYS	**ockergelb**
= brownish yellow		= bräunlichgelb
≈ ocher; naples yellow		≈ ocker; neapelgelb
ochre (→ocher)	PHYS	(→ocker)
OCR (→optical character recognition)	TERM&PER	(→optische Zeichenerkennung)
OCR constant coders	TERM&PER	**OCR-Konstantencodierer**
		= OCR-Konstantenkodierer
OCR document reader	TERM&PER	**OCR-Schriftleser**
OCR font (→OCR letter)	TERM&PER	(→OCR-Schrift)
OCR letter	TERM&PER	**OCR-Schrift**
[letter alphabet standardized internationally to be scannable by optical readers]		[international genormte, durch optische Leser erfaßbare Schrift]
= OCR font; optical character recognition letter		↓ OCR-A-Schrift; OCR-B-Schrift
↓ OCR letter type A; OCR letter type B		
OCR post-coders	TERM&PER	**OCR-Nachcodierer**
		= OCR-Nachkodierer
OCR precoders	TERM&PER	**OCR-Vorcodierer**
		= OCR-Vorkodierer
OCR reader (→optical character reader)	TERM&PER	(→optischer Leser)
OCR terminal (→optical character reader)	TERM&PER	(→optischer Leser)
octagon	MATH	**Achteck**
↑ polygon		= Oktagon
		↑ Vieleck
octagonal	MATH	**achteckig**
octagonal characteristic (→bilateral characteristic)	EL.ACOUS	(→Achtercharakteristik)
octagonal characteristic (→figure-eight pattern)	ANT	(→Achtercharakteristik)
octagon antenna	ANT	**Achteckantenne**
octahedral	MATH	**achtflächig**
octahedron (pl.-ons,-a)	MATH	**Oktaeder**
↑ polyhedron		= Achtflächner
↓ polygon		↑ Polyeder
octal (adj.)	MATH	**oktal**
octal digit	MATH	**Oktalziffer**
octal encoder	CODING	**Oktalcodierer**
octal notation (→octal number system)	MATH	(→Oktalsystem)
octal number	MATH	**Oktalzahl**
= octal numeral		
octal number system	MATH	**Oktalsystem**
[number system to the base 8]		[Zahlensystem zur Basis 8]
= octal notation		= oktales Zahlensystem
		↑ Stellenwertssystem
octal numeral (→octal number)	MATH	(→Oktalzahl)
octal point	MATH	**Oktalpunkt**
octant	MATH	**Oktant**
[1/8 of circle area]		[1/8 Kreisfläche]
octavalent	TECH	**achtwertig**
octave	MATH	**Oktave**
octave amplifier	MICROW	**Oktav-Verstärker**
octave analysis	INSTR	**Oktavenanalyse**
octave band	MICROW	**Oktavband**
octave filter	NETW.TH	**Oktavbandpass**
octet	CODING	**Oktett**
[group of eight digits or bits handled as unit]		[als Einheit behandelte Gruppe von acht Ziffern oder Bits]
= eight-bit byte		= Acht-Bit-Byte
↑ byte		↑ Byte

octet		MATH	Oktett		
octet-by-octet		TECH	oktettweise		
octode		ELECTRON	Oktode		
ocurrence probability		MATH	Ereigniswahrscheinlichkeit		
odd (→uneven)		MATH	(→ungeradzahlig)		
odd-harmonics crystal oscillator		CIRC.ENG	Quarz-Oberton-Oszillator		
odd number (→uneven number)		MATH	(→ungerade Zahl)		
odd parity [odd number of binary ones] = unparity; uneven parity ≠ parity		CODING	ungerade Parität [Ungeradzahligkeit binärer Einser] = Imparität ≠ Parität		
odd parity check = unparity check; odd parity control; unparity control ≠ parity check		CODING	Imparitätskontrolle ≠ Paritätskontrolle		
odd parity control (→odd parity check)		CODING	(→Imparitätskontrolle)		
odograph		RADIO NAV	Kursschreiber		
odometer		OUTS.PLANT	Meßrad = Odometer		
OEIC (→opto-electronic integrated circuit)		OPTOEL	(→OEIC-Baustein)		
OEM = original equipment manufacturer; original equipment supplier		ECON	OEM-Lieferant		
OEM device		EQUIP.ENG	Fremdgerät		
OEM equipment		ECON	OEM-Gerät		
OEM product ≈ purchased part		ECON	OEM-Produkt ≈ Fremdfabrikat		
Oersted [old unit for magnetic field strength; = $10/4\pi$ A/m]		PHYS	Oersted [alte Einheit für magnetische Feldstärke; = $10/4\pi$ A/cm]		
of equal area = equivalent		MATH	flächentreu		
off ≠ on		EL.TECH	aus [ausgeschaltet] ≠ ein		
off-center		MECH	außermittig		
off-center fed antenna (→Windom antenna)		ANT	(→Windom-Antenne)		
off-delay (→release delay)		COMPON	(→Abfallverzögerung)		
off emergency (→emergency switch)		EL.INST	(→Notausschalter)		
offer (v.t.)		SWITCH	anbieten		
offer (v.t.)		ECON	anbieten		
offer (n.) = tender offer; tender 1; proposal; quotation; bid 2 (AM) ↓ quotation; bidn 1 (n.)		ECON	Angebot = Offerte ↓ Preisangebot; Preisangebot		
offered call (→call attempt)		SWITCH	(→Wählversuch)		
offered traffic ≈ incoming traffic		SWITCH	Verkehrsangebot = Angebot ≈ kommender Verkehr		
offeree [receiver of an offer]		ECON	Angebotsempfänger		
offering		SWITCH	Anbieten		
offering (→override)		TELEPH	(→Aufschalten)		
offering channel ≈ offering line		SWITCH	Zubringerkanal ≈ Zubringerleitung		
offering line [conveys traffic to a switching matrix] = incoming line; offering trunk; incoming trunk ≠ outgoing line ↓ incoming highway		SWITCH	Zubringerleitung [führt einer Koppeleinrichtung Verkehr zu] ≈ Zubringerkanal; Zubringerbündel ≠ Abnehmerleitung ↓ Zubringer-Multiplexleitung		
offering signal [of an operator]		TELEPH	Aufschaltezeichen [durch eine Dienststelle der Fernmeldeverwaltung] = Aufschalteton		
offering trunk (→offering line)		SWITCH	(→Zubringerleitung)		
offeror (→bidder)		ECON	(→Anbieter)		
off-hook [handset]		TELEPH	abgehoben [Handapparat] = im Aushängezustand		
off-hook condition = off-hook state		TELEPH	Aushängezustand		
off-hook signal (AM) = answer signal (BRI); seizure signal		SWITCH	Beginnzeichen = Belegzeichen		
off-hook state (→off-hook condition)		TELEPH	(→Aushängezustand)		
office [institution] = agency		ECON	Amt [Institution] = Dienststelle		
office (→bureau)		OFFICE	(→Büro)		
office (→central office)		SWITCH	(→Fernsprechvermittlungsstelle)		
office automation		INF.TECH	Büroautomatisierung = Büroautomation; Bürotik		
office building		CIV.ENG	Bürogebäude		
office cabinet		OFFICE	Büroschrank		
office cable ≈ connecting cable [COMM.CABLE]		SYS.INST	Amtskabel = Stationskabel ≈ Schaltkabel [COMM.CABLE]		
office cabling (→exchange cabling)		SYS.INST	(→Amtsverkabelung)		
office chair		OFFICE	Bürostuhl		
office communication (→office communications)		INF.TECH	(→Bürokommunikation)		
office communications = office communication		INF.TECH	Bürokommunikation		
office communication system		DATA COMM	Bürokommunikationssystem = K-I-System; Kommunikations- und Informationsverarbeitungssystem		
office computer		DATA PROC	Bürocomputer		
office data = exchange data		SWITCH	Amtsdaten		
office dictating equipment (→office dictating set)		OFFICE	(→Bürodiktiergerät)		
office dictating machine (→office dictating set)		OFFICE	(→Bürodiktiergerät)		
office dictating set = office dictating machine; office dictating equipment		OFFICE	Bürodiktiergerät		
office environment (→exchange environment)		TELEC	(→Amtsumgebung)		
office equipment = office machine; business equipment; business machine; bureau equipment; bureau machine		TERM&PER	Bürogerät = Büromaschine		
office equipment [procedures and equipment] = business equipment ↑ office systems		OFFICE	Bürotechnik [Verfahren und Geräte] ↑ Bürowirtschaft		
office hours ↑ working time		ECON	Dienstzeit = Dienststunden ↑ Arbeitszeit		
office information system		INF.TECH	Büroinformationssystem		
office installation = station installation 1 ≠ outdoor installation		SYS.INST	Stationsaufbau = Amtsaufbau; Stationsbau ≠ Freiluftaufbau		
office installation (→office installation technique)		TELEC	(→Amtsbautechnik)		
office installation technique = station installation technique; exchange installation technique [SWITCH]; office installation; station installation		TELEC	Amtsbautechnik = Stationsbautechnik; Amtsbau; Stationsbau		
office machine (→office equipment)		TERM&PER	(→Bürogerät)		
office management ↑ office systems		OFFICE	Büroorganisation ↑ Bürowirtschaft		
office organization and equipment (→office systems)		TECH	(→Bürowirtschaft)		
officer 1 [person holding a position of authority in an organization] = director; manager		ECON	Leiter = Direktor; Geschäftsführer		
officer 2 (→official)		ECON	(→Beamter)		
office routine		OFFICE	Büroablauf		
office selector		SWITCH	Amtswähler		
office software = bureau software		DATA PROC	Bürosoftware		

office systems

office systems	TECH	**Bürowirtschaft**
= office organization and equipment		↓ Bürotechnik; Büroorganisation
↓ office equipment; office organization		
office terminal (→exchange equipment)	TELEC	(→Amtsausrüstung)
office typewriter (→bureau typewriter)	OFFICE	(→Büroschreibmaschine)
office workstation	TERM&PER	**Büro-Arbeitsplatzsystem**
official (n.)	ECON	**Beamter**
= officer 2; civil servant		↓ Staatsbeamter; Stadtbeamter
↓ government officer; municipal officer		
official (adj.)	ECON	**amtlich**
		= offiziell
official call (→service call)	TELEPH	(→Dienstgespräch)
officially licensed	ECON	**amtlich zugelassen**
official network	TELEC	**Behördennetz**
↑ private network		↑ Privatnetz
official print	DOC	**offizielle Pause**
off-line	DATA PROC	**offline**
[not in direct communication with CPU]		[mit der Zentraleinheit nicht direkt verbunden]
= offline		= abgetrennt
≠ on-line		≠ online
offline (→off-line)	DATA PROC	(→offline)
off-line data processing	DATA PROC	**Offline-Datenverarbeitung**
= off-line processing		= Offline-Verarbeitung
off-line maintenance	DATA PROC	**Offline-Wartung**
off-line mode (→off-line operation)	DATA PROC	(→Offline-Betrieb)
off-line operation	DATA PROC	**Offline-Betrieb**
= off-line mode; off-line processing		[getrennt von einer DVA betrieben]
		= rechnerunabhängiger Betrieb
		≈ Lokalbetrieb
off-line peripheral equipment	DATA PROC	**Offline-Peripheriegerät**
off-line processing (→off-line operation)	DATA PROC	(→Offline-Betrieb)
off-line processing (→off-line data processing)	DATA PROC	(→Offline-Datenverarbeitung)
off-line teleprocessing	DATA PROC	**indirekte Datenfernverarbeitung**
		= Off-line-Datenfernverarbeitung
off-load (v.t.)	DATA PROC	**entlasten**
off loss	ELECTRON	**Sperrverlust**
off-page connector	DATA PROC	**Seitenübergangsstelle**
off period (→blocking time)	ELECTRON	(→Sperrzeit)
off-position	TECH	**Ausschaltstellung**
off-premises	ECON	**Außenstelle**
= field office		[an anderem Standort]
≈ branch office		≈ Filiale
off-premises (→remote)	TELEC	(→abgesetzt)
off-premises station	TELEC	**abgesetzte Nebenstelle**
off punch (→punching error)	TERM&PER	(→Fehllochung)
offset (n.)	ELECTRON	**Offset**
offset (n.)	TECH	**Versatz**
= lag (n.)		= Ablage
≈ misalignment		= Fluchtungsfehler
offset (n.)	MICROW	**Versatz**
offset (→displacement address)	DATA PROC	(→Distanzadresse)
offset (→deviation)	CONTROL	(→Regelabweichung)
offset (→offset printing)	TYPOGR	(→Offsetdruck)
offsetable (→displacable)	TECH	(→verschiebbar)
offset antenna	ANT	**Offset-Antenne**
offset copier	OFFICE	**Offsetkopierer**
offset current	EL.TECH	**Fehlstrom**
= fault current		= Fehlerstrom
offset current [operarional amplifier]	CIRC.ENG	**Offset-Strom** [Operationsverstärker]
offset diode	CIRC.ENG	**Offsetdiode**
		= Potentialverschiebediode
offset measurement (→relative measurement)	INSTR	(→Relativmessung)
offset paper	OFFICE	**Offsetpapier**
offset parabolic antenna	ANT	**Offset-Parabolantenne**
= offset paraboloidal reflector antenna		
offset parabolic reflector	ANT	**Offset-Parabolspiegel**
offset paraboloidal reflector antenna	ANT	(→Offset-Parabolantenne)
(→offset parabolic antenna)		
offset pattern	RADIO REL	**Zwischenraster**
= interleaved pattern 2		
offset printer	OFFICE	**Offsetdrucker**
offset printing	TYPOGR	**Offsetdruck**
[the ink is first applied to a rubber-blanked surface, and than transferred to paper]		[die Tinte wird zuerst auf eine mit Gummi überzogene Fläche aufgebracht, und von dort aufs Papier]
= offset		
↑ planographic printing		↑ Flachdruck
offset quaternary phase shift keying (→OQPSK)	MODUL	(→OQPSK)
offset voltage	EL.TECH	**Fehlspannung**
		= Offset-Spannung; Nullspannung
offset voltage [operational amplifier]	CIRC.ENG	**Offset-Spannung** [Operationsverstärker]
		= Nullspannung
off-site paging	MOB.COMM	**öffentlicher Funkruf**
↑ radio paging		
off-state (→reverse characteristic)	ELECTRON	(→Rückwärtskennlinie)
off-the-shelf	ECON	**ab Lager**
		= sofort lieferbar
off-the-shelf business	ECON	**Seriengeschäft**
off-the-shelf instrument	INSTR	**Komplettgerät**
off-the-shelf model (→serial model)	TECH	(→Seriengerät)
off time	TECH	**Unterbrechungsdauer**
= interruption time; duration of interruption		
off time (→blocking time)	ELECTRON	(→Sperrzeit)
offtimes (→frequently)	COLLOQ	(→häufig) (adv.)
off transistor [power transistor]	MICROEL	**Abschaltezeit** [Leistungstransistor]
of many places	MATH	**mehrstellig**
↓ two-place; three-place; four-place		= vielstellig
		↓ zweistellig; dreistellig; vierstellig
oft-cited (→often-cited)	COLLOQ	(→vielzitiert)
often (→frequently)	COLLOQ	(→häufig) (adv.)
often-cited	COLLOQ	**vielzitiert**
= oft-cited		
oftentimes (→frequently)	COLLOQ	(→häufig) (adv.)
ogonek	DOC	**Ogonek**
ohm	PHYS	**Ohm**
[SI unit for electric resistance, reactance, impedance and characteristic impedance]		[SI-Einheit für elektrischen Widerstand, Blindwiderstand, Scheinwiderstand und Wellenwiderstand]
= Ω		= Ω
ohmic	PHYS	**ohmsch**
ohmic contact	EL.TECH	**ohmscher Kontakt**
[voltage drop proportional to current]		[hat dem Strom proportionalen Spannungsabfall]
≈ metallic contact		≈ galvanischer Kontakt
ohmic drop (→ohmic voltage drop)	EL.TECH	(→ohmscher Spannungsabfall)
ohmic loss	EL.TECH	**ohmscher Verlust**
[power dissipated in an active resistance]		[in einem Wirkwiderstand umgesetzte Verlustwärme]
= wattful loss		
ohmic resistance	EL.TECH	**ohmscher Widerstand**
[physical magnitude, linear relation between current and voltage]		[physikalische Größe, linearer Abhängigkeit des Stroms von der Spannung]
= linear resistance		= linearer Widerstand
≠ nonlinear resistance		≠ nichtlinearer Widerstand
ohmic resistance (→ohmic resistor)	COMPON	(→ohmscher Widerstand)
ohmic resistor [component]	COMPON	**ohmscher Widerstand** [Bauteil]
= linear resistor; ohmic resistance; linear resistance		= linearer Widerstand; Linearwiderstand
≠ nonlinear resistor		≠ nichtlinearer Widerstand
↑ resistor		↑ Widerstand

ohmic voltage drop		EL.TECH	ohmscher Spannungsabfall	omnidirectional operation	RADIO	Rundstrahlbetrieb
= ohmic drop				omnidirectional radiobeacon	RADIO NAV	Rundstrahlbake
ohmmeter		INSTR	Ohmmeter	= non-directional radiobeacon; non-directional beacon; NDB		= Kreisfunkfeuer; ungerichtetes Funkfeuer
[instrument exclusively designed for resistance measurements]			[nur für Widerstandsmessungen ausgelegtes Instrument]			
↑ resistance meter; insulation tester; earth resistance meter			↑ Widerstandsmesser	omnidirectional range	RADIO NAV	(→Drehfunkfeuer)
↓ teraohmmeter;			↓ Teraohmmeter; Isolationsmesser; Erdungsmesser	(→rotating radiobeacon)		
				omnidirectional wideband antenna	ANT	(→Breitbandrundstrahler)
ohm sign		PHYS	Ohm-Zeichen	(→omnidirectional broadband antenna)		
[Ω]			[Ω]			
Ohm's law		PHYS	Ohmsches Gesetz	omni-font reader	TERM&PER	Allschriftleser
oil (v.t.)		TECH	ölen	= omni-font scanner		= universeller Schriftenleser
↑ lubrication			↑ schmieren	≈ multi-font reader		≈ Mehrschriftleser
oil		CHEM	Öl	↑ optical character reader		↑ optischer Leser
oilcloth (→oil-cloth lining)		TECH	(→Wachstuch)	omni-font scanner	TERM&PER	(→Allschriftleser)
oil-cloth lining		TECH	Wachstuch	(→omni-font reader)		
= oilcloth; American cloth (BRI)				omnipresent (→ubiquitous)	COLLOQ	(→allgegenwärtig)
oil cooling		TECH	Ölkühlung	omnirange (→rotating radiobeacon)	RADIO NAV	(→Drehfunkfeuer)
oiled paper		TECH	Ölpapier			
oil-filled transformer		POWER ENG	ölgefüllter Transformator	omnirange (→omnidirectional)	TECH	(→allseitig gerichtet)
			= Öltransformator	on	EL.TECH	ein
oil hardening		METAL	Ölhärtung	≠ off		≠ aus
oil hole		MECH	Schmierbohrung	ONA (→open network architecture)	DATA COMM	(→offene Netzarchitektur)
oil level		TECH	Ölstand			
oil lubrication		TECH	Ölschmierung	on application (→on request)	ECON	(→auf Anfrage)
			↑ Schmierung	on a revolving basis	ECON	(→revolvierend)
oil of vitriol (→sulphuric acid)		CHEM	(→Schwefelsäure)	(→revolving)		
				onboard antenna	ANT	Bordantenne
oil paint		TECH	Ölfarbe	onboard computer	DATA PROC	Bordcomputer
oil retainer		TECH	Öldichtung	= airborne computer (aircraft); flight computer; seaborne computer (vessel)		= Bordrechner
OLC (→optical loop carrier system)		TELEC	(→optisches Teilnehmermultiplexsystem)			
				onboard instrument	RADIO NAV	Bordinstrument
old degree (→degree)		MATH	(→Grad)	onboard power supply	RADIO NAV	Bordnetz
Old Face		TYPOGR	Alter Schnitt	on both sides (→bilateral)	TECH	(→zweiseitig)
[font type]			[Schriftart]	on call	COLLOQ	auf Abruf
olive (→olivegreen)		PHYS	(→olivgrün)	oncosts (→indirect costs)	ECON	(→Gemeinkosten)
olive drab (→olive gray)		PHYS	(→olivgrau)	on due time (→on time)	ECON	(→fristgerecht)
olive gray		PHYS	olivgrau	ondulation (→corrugation)	MECH	(→Wellung)
= olive drab				ondulator (→inverter)	POWER SYS	(→Wechselrichter)
olivegreen		PHYS	olivgrün	one	CODING	Eins
= olive			= oliv	= mark		
O&M (→operation and maintenance)		TELEC	(→Betrieb und Wartung)	one-address computer	DATA PROC	(→Ein-Adreß-Computer)
				(→single-address computer)		
O&M center		TELEC	O&M-Wartung	one-address instruction	DATA PROC	Ein-Adreß-Befehl
omega match		ANT	Omega-Anpassung	= single-address instruction		≠ Mehr-Adreß-Befehl
omission		COLLOQ	Unterlassung	≠ multi-address instruction		
≈ failure			= Versäumnis; Weglassung, Auslassung	one-address machine	DATA PROC	(→Ein-Adreß-Computer)
			≈ Fehler	(→single-address computer)		
omit		COLLOQ	weglassen	one-at-a-time mode	DATA PROC	(→sequentieller Betrieb)
			= auslassen	(→sequential operation)		
omniband doublet antenna		ANT	Allband-Dipol	one-bit error	CODING	Ein-Bit-Fehler
omniband Z tuner		ANT	Allband-Z-Tuner	one-cable separator	TV	Einkabelweiche
omnidirectional		TECH	allseitig gerichtet	one-chip computer	DATA PROC	(→Ein-Chip-Mikrocomputer)
= omnirange			≈ ungerichtet	(→single-chip microcomputer)		
≠ non-directional				one-chip equipment	EQUIP.ENG	Ein-Chip-Gerät
omnidirectional		PHYS	rundstrahlend	one-digit adder	CIRC.ENG	(→Halbaddierer)
≠ directive			≠ bündelnd	(→half-adder)		
omnidirectional aerial		ANT	(→Rundstrahlantenne)	one-digit subtracter (→half subtracter)	CIRC.ENG	(→Halbsubtrahierer)
(→omnidirectional antenna)						
omnidirectional antenna		ANT	Rundstrahlantenne	one-dimensional	MATH	eindimensional
[no directivity in a plane, but not necessarily so in the orthogonal]			[fehlende Richtwirkung in einer Ebene, nicht unbedingt auch in der orthogonalen]	= unidimensional; univariate		
				one-element	MATH	Eins-Element
				one-element (→one-part)	TECH	(→einteilig)
= omnidirectional aerial; nondirectional antenna; nondirectional aerial; nondirective antenna; nondirective aerial			= Rundstrahler; allseitige Antenne; flächendeckende Antenne	one-figure (→one-place)	MATH	(→einstellig)
				one-finger activation	TERM&PER	Ein-Finger-Betätigung
				one-layer	TECH	einlagig
				≈ one-part		= einschichtig
≈ isotropic radiator			≈ Kugelstrahler			≈ einteilig
omnidirectional broadband antenna		ANT	Breitbandrundstrahler	one-level (adj.)	TECH	einstufig
				= single-step		≠ mehrstufig
= omnidirectional wideband antenna				≠ multi-level		
				one-level code (→machine code 1)	DATA PROC	(→Maschinencode 1)
omnidirectional characteristic		EL.ACOUS	Kugelcharakteristik			
omnidirectional double quad radiator		ANT	Doppelquad-Rundstrahler	one-milliwatt generator	INSTR	(→Normalgenerator 1)
				(→standard generator)		
omnidirectional microphone		EL.ACOUS	Kugelmikrophon	on end (→edge-wise)	MECH	(→hochkant)
= non-directional microphone			= Allrichtungsmikrophon	one-out-of-ten code	CODING	Ein-aus-Zehn-Code

one-part TECH **einteilig**		
= one-element		≈ einlagig
≈ one-layer		≠ mehrteilig
≠ multisectional		
one-part paper TERM&PER **einlagiges Papier**		
one-phase (→single-phase) PHYS (→einphasig)		
one-place (adj.) MATH **einstellig**		
= one-figure		≈ unär
= unary		
one-pole plug COMPON **einpoliger Stecker**		
one-port (→two-terminal) NETW.TH (→Zweipol)		
one's complement DATA PROC (→Einerkomplement)		
(→complement to one)		
one-shot multivibrator CIRC.ENG **stabile Kippschaltung**		
[reaches a stable output condition after each input trigger signal]		[durch ein Eingangssignal wird ein stabiler Zustand am Ausgang ausgelöst]
= start-stop multivibrator; driven multivibrador		= fremdgesteuerte Kippschaltung; fremdgesteuerter Multivibrator
↓ monostable multivibrator; flip-flop		↓ monostabile Kippschaltung; bistabile Kippschaltung
one-sided disc (→SS disk) TERM&PER (→SS-Diskette)		
one-sided disk (→SS disk) TERM&PER (→SS-Diskette)		
one-sided diskette (→SS disk) TERM&PER (→SS-Diskette)		
one-third octave filter NETW.TH **Terzbandpaß**		
one-time COLLOQ **einmalig**		
= nonrecurrent		
one-valued (→unambiguous) MATH (→eindeutig)		
one-way TELEC **einfachgerichtet**		
= unidirectional		≈ einseitig gerichtet; einseitig; Einweg-
≠ both-way		≈ simplex
		≠ doppeltgerichtet
one-way communication DATA COMM **einseitige Datenübermittlung**		
= simplex data communication		
one-way line ELECTRON **Einwegleitung**		
one-way operation TELEC (→Simplexbetrieb)		
(→simplex)		
one-way polar reversible circuit CIRC.ENG **Wendeschaltung**		
one-way receptable TECH **Einwegbehälter**		
one-way screw (→nonretractable screw) MECH (→Einwegschraube)		
one-way videotex TELEC (→Teletext)		
(→teletext)		
one-wire feeding ANT **Eindrahtspeisung**		
on first try COLLOQ **auf Anhieb**		
ongoing TECH **laufend 1**		
= current; undergoing; under way		= in Gang befindlich
on-hook (n.) TELEPH **aufgelegt**		
[handset]		[Hörer]
		= eingehängt
on-hook (n.) TELEPH **Auflegen**		
		= Einhängen
on-hook condition TELEPH **Einhängezustand**		
on-hook pulse TELEPH **Einhängepuls**		
on-hook signal TELEPH **Höreraufegesignal**		
on-line TELEC **in Betrieb**		
= operating; working; in operation		
on-line DATA PROC **online**		
[in direct communication with CPU]		[mit Zentraleinheit direkt verbunden]
= online		
≠ off-line		≠ offline
online (→on-line) DATA PROC (→online)		
on-line backup DATA PROC **Online-Reserve**		
on-line data transmission DATA COMM **Datendirektübertragung**		
on-line devices (→on-line peripherals) DATA PROC (→Online-Peripherie)		
on-line dialog DATA COMM **Online-Dialog**		
on-line maintenance DATA PROC **Online-Wartung**		
on-line mode (→on-line operation) DATA PROC (→Leitungsbetrieb)		
on-line mode (→on-line operation) DATA PROC (→Online-Betrieb)		
on-line operation DATA COMM **Leitungsbetrieb**		
= line working; on-line mode; line mode		
on-line operation DATA PROC **Online-Betrieb**		
= on-line mode		
on-line operation CONTROL **Online-Kopplung**		
		[Prozeßrechner]
on-line peripherals DATA PROC **Online-Peripherie**		
= on-line devices		
on-line processing DATA PROC **Online-Verarbeitung**		
on-line software DATA PROC **Online-Software**		
on-line teleprocessing DATA PROC (→direkte Datenfernverarbeitung)		
(→direct teleprocessing)		
on-line teller transaction DATA PROC **Online-Kassenverkehr**		
on-line user DATA PROC **Teilhaber**		
[user common programs and files]		[benutzt gemeinsame Programme und Dateien]
on local (→stand-alone) DATA PROC (→allein operierend)		
on-off EQUIP.ENG **ein-aus**		
on-off (→two-point) EL.TECH (→Zweipunkt-)		
on-off action (→two-position control) CONTROL (→Zweipunktregelung)		
on-off control (→two-position control) CONTROL (→Zweipunktregelung)		
on-off controller CONTROL **Zweipunktregler**		
[has only two output states]		[hat nur zwei Ausgangszustände]
= snap-acting controller		
on-off keying (→single tone operation) TELEGR (→Einfachtonbetrieb)		
on period ELECTRON **Einschaltzeit**		
= activation period; turn-on time		
on-pulse modulation MODUL **pulsüberlagerte Modulation**		
on request ECON **auf Anfrage**		
= on application		
on-request service RADIO NAV **angeforderte Verbindung**		
= OR service		
on schedule (→on time) ECON (→fristgerecht)		
onset TECH **Einsetzen**		
[fig.]		[fig.]
≈ begin (n.)		≈ Beginn
onset antenna ANT **Onset-Antenne**		
on site TECH **vor Ort**		
on-site paging MOB.COMM **privater Funkruf**		
↑ radio paging		= Grundstückspersonenruf
on-state MICROEL **Durchlaßzustand**		
on-state current (→forward current) ELECTRON (→Durchlaßstrom)		
on-state resistance (→forward resistance) MICROEL (→Durchlaßwiderstand)		
on-the-fly printing TERM&PER **fliegender Druck**		
on-the-job training (→training on the job) ECON (→Ausbildung am Arbeitsplatz)		
on the leading edge COLLOQ **an der Spitze**		
[fig.]		[fig.]
on the transmit side (adj.) TELEC **sendeseitig**		
= transmit-; send-side-		
on the way COLLOQ **unterwegs**		
= in transit		
on time ECON **fristgerecht**		
= timely; on schedule; on due time; punctual		= fristgemäß; termingerecht; termingemäß; pünktlich
O.O.O. (→faulty) QUAL (→fehlerhaft)		
opacity PHYS **Undurchsichtigkeit**		
≠ transparency		≠ Durchsichtigkeit
opalescent glass (→frosted glass) TECH (→Mattglas)		
opal glass (→frosted glass) TECH (→Mattglas)		
op-amp (→operational amplifier) CIRC.ENG (→Operationsverstärker)		
opaque PHYS **undurchsichtig**		
≠ transparent		≠ durchsichtig
op code (→operating code) DATA PROC (→Befehlsschlüssel)		
opcode (→operation code) DATA PROC (→Operationscode)		
op-code (→operation code) DATA PROC (→Operationscode)		
open (v.t.) DATA PROC **eröffnen**		
open (n.) INSTR **Leerlaufabschluß**		
open (adj.) TECH **offen**		
open-air (→outdoor) TECH (→im Freien)		

open air climate (→outdoor climate) QUAL (→Außenraumklima)
open antenna (→outdoor antenna) ANT (→Außenantenne)
open bracket (→left bracket) MATH (→Klammer auf)
open circuit (→open-circuit operation) EL.TECH (→Leerlauf)
open-circuit admittance NETW.TH **Leerlaufadmittanz**
open-circuit attenuation NETW.TH **Leerlaufdämpfung**
open-circuit current EL.TECH **Leerlaufstrom**
= open-loop current; no-load current
open-circuit current (→marking current) TELEGR (→Arbeitsstrom)
open-circuited (→in open circuit) EL.TECH (→im Leerlauf)
open-circuit gain factor CIRC.ENG **Leerlauf-Verstärkungsfaktor**
open-circuit impedance NETW.TH **Leerlaufimpedanz**
= Leerlaufscheinwiderstand; Leerlaufwiderstand
open-circuit input admittance NETW.TH **Leerlauf-Eingangsadmittanz**
= Leerlauf-Eingangsleitwert
open-circuit input impedance NETW.TH **Leerlauf-Eingangsimpedanz**
open-circuit loss EL.TECH **Leerlauf-Verlust**
open-circuit operation EL.TECH **Leerlauf**
= open circuit; no-load operation
= offener Stromkreis
open-circuit operation TELEGR (→Arbeitsstrombetrieb)
(→open-circuit working)
open-circuit output admittance NETW.TH **Leerlauf-Ausgangsadmittanz**
= Leerlauf-Ausgangsleitwert
open-circuit output impedance NETW.TH **Leerlauf-Ausgangsimpedanz**
= Leerlauf-Ausgangswiderstand
open-circuit reverse voltage transfer ratio MICROEL **Leerlauf-Spannungsrückwirkung**
open-circuit signaling (→tone-off idle) TRANS (→Arbeitsstromverfahren)
open-circuit transmission factor NETW.TH **Leerlauf-Übertragungsfaktor**
open-circuit voltage EL.TECH **Leerlaufspannung**
= no-load voltage; electromotive voltage; electromotive force; emf
= Quellenspannung; Urspannung; elektromotorische Kraft; EMK; eingeprägte Kraft
open-circuit voltage gain CIRC.ENG **Leerlauf-Spannungsverstärkung**
= open-loop gain
= Leerlaufverstärkung
open-circuit working TELEGR **Arbeitsstrombetrieb**
= open-circuit operation
≠ Ruhestrombetrieb
open code (→non-linked code) SWITC (→offene Kennzahl)
open collector MICROEL **offener Kollektor**
open-collector output MICROEL **Open-Kollektor-Ausgang**
open-ended (→upgradable) TECH (→ausbaufähig)
opening COLLOQ **Öffnung**
[act of opening] [Vorgang des Öffnens]
opening TECH **Öffnung**
[feature] [offene Stelle]
≈ hole; interstice; cutout; slot; gap
≈ Loch; Zwischenraum; Durchbruch; Schlitz; Spalt
opening (→hole 1) TECH (→Loch 1)
opening (→wall opening) CIV.ENG (→Wanddurchbruch)
opening hours ECON **Öffnungszeiten**
opening of tender ECON **Angebotseröffnung**
= tender opening
opening routine (→open routine 1) DATA PROC (→Eröffnungsroutine)
opening screen DATA PROC **Eröffnungsbild**
open listening TELEPH **Lauthören**
open loop CONTROL **offener Regelkreis**
= Steuerkette
open-loop control CONTROL **Steuerung**
[action on a control quantity as a function of a reference magnitude, without any feedback]
[Beeinflussung einer Steuergröße nur durch eine Führungsgröße, ohne Rückkopplung]
= control
≈ regulation
= Vorwärtssteuerung; rückführungslose Steuerung
≈ Regelung
open-loop current EL.TECH (→Leerlaufstrom)
(→open-circuit current)

open-loop gain (→open-circuit voltage gain) CIRC.ENG (→Leerlauf-Spannungsverstärkung)
open-loop mode DATA PROC **offen prozeßgekoppelter Betrieb**
[process control] [Prozeßsteuerung]
= open-loop operation
= Open-loop-Betrieb
open-loop operation DATA PROC (→offen prozeßgekoppelter Betrieb)
(→open-loop mode)
open network DATA COMM **offenes Netz**
open network architecture DATA COMM **offene Netzarchitektur**
= ONA
open numbering SWITCH **offene Numerierung**
= non-linked numbering; prefix numbering
open oscillatory circuit (→tank circuit) CIRC.ENG (→offener Schwingungskreis)
open-plan office OFFICE **Großraumbüro**
= large area office
open routine 1 DATA PROC **Eröffnungsroutine**
= opening routine
open routine 2 (→open subroutine) DATA PROC (→offenes Unterprogramm)
open sentence DATA PROC **Aussageform**
open shop operation DATA PROC **Open-shop-Betrieb**
[user can operate the computer] [Anwender hat Zugang zur DVA]
≠ closed shop operation
≠ Closed-shop-Betrieb
open-site (→outdoor) TECH (→im Freien)
open start pulse (→start pulse) TELEGR (→Anlaufschritt)
open statement DATA PROC **Eröffnungsanweisung**
open subroutine DATA PROC **offenes Unterprogramm**
= direct-insert subroutine; open routine 2; direct-insert routine
= offene Routine
open system (→open system interconnection) DATA COMM (→offenes Kommunikationssystem)
open system interconnection DATA COMM **offenes Kommunikationssystem**
= OSI; open system
= OSI; offener Computerverbund; offener Rechnerverbund; offenes Verbundnetz; offenes Computersystem; offenes Rechnersystem; offenes System; Kommunikation offener Systeme; offene Systemverbindung;
open-wire carrier equipment TRANS **Freileitungssystem**
open-wire line OUTS.PLANT **Freileitung**
= overhead line
= Freileitungslinie; oberirdische Linie
open-wire pole OUTS.PLANT **Freileitungsmast**
= Stange
open-wire trunk OUTS.PLANT **Fernmelde-Freileitung**
operability TECH **Betriebsfähigkeit**
= operating capacity
≈ serviceability; user friendliness
= Funktionsfähigkeit; Funktionstüchtigkeit
≈ Brauchbarkeit; Benutzerfreundlichkeit
operability (→ready status) DATA COMM (→Betriebsbereitschaft)
operand DATA PROC **Operand**
[information operated upon] [Information die Gegenstand eines Befehles ist]
operand MATH **Operand**
[parameter operated upon] [Gegenstand einer Operation]
= Rechengröße
operand address DATA PROC **Operandenadresse**
operand field (→address field) DATA PROC (→Adreßfeld)
operand part DATA PROC **Operandenteil**
[part of command indicating address] [Adreßteil eines Befehls]
↓ operand address; jump address
↓ Operandenadresse; Sprungadresse
operate COMPON **anziehen**
= pull-in; respond
[Relais]
≈ ansprechen 1
operate 2 TECH **betreiben**
= run (v.t.)
≈ betätigen; bedienen
≈ actuate

operate 1

operate 1		TECH	**bedienen**		
= service (v.t.); run			= betreiben		
≈ actuate			≈ betätigen		
operate 3 (→function)		TECH	(→funktionieren)		
operate level (→response level)		ELECTRON	(→Ansprechpegel)		
operating (adj.)		TECH	**Betriebs-**		
= working; service					
operating (→on-line)		TELEC	(→in Betrieb)		
operating (→working)		TECH	(→funktionstüchtig)		
operating capacity (→operability)		TECH	(→Betriebsfähigkeit)		
operating characteristic 2		TECH	**Arbeitskennlinie**		
operating characteristics 1		TECH	**Betriebsdaten**		
			= Operations-Charakteristik		
operating code (→operation code)		DATA PROC	(→Operationscode)		
operating coil (→exciting coil)		ELECTRON	(→Erregerspule)		
operating company		TELEC	**Betriebsgesellschaft**		
= network operating company; network operator; network carrier; telecom operator			= Betreibergesellschaft; Betreiber; Netzbetreibergesellschaft; Netzbetreiber; Netzträger		
≈ telecommunications carrier; network operating company; service provider			≈ Fernmeldegesellschaft; Konzessionär; Diensteanbieter; Betriebsgesellschaft		
≠ network user			≠ Netzbenutzer		
↓ trunk carrier; local carrier			↓ Fernnetzbetreiber; Ortsnetzbetreiber		
operating condition		TECH	**Betriebsbedingung**		
			= Einsatzbedingung; Betriebsverhältnisse (pl.t.)		
operating control system		SIGN.ENG	**Einsatzleitsystem**		
operating costs		ECON	**Betriebskosten**		
= working costs; running charge			≈ Betriebsausgaben		
≈ working expenses; operating expenses					
operating current		EL.TECH	**Betriebsstrom**		
= working current; service current			≈ Arbeitsstrom		
operating cycle		TECH	**Arbeitszyklus**		
operating desk (→control desk)		TECH	(→Kontrollpult)		
operating device		TERM&PER	**Bedienelement**		
= operator device			= Bedienungsgerät		
operating documentation		TECH	**Gebrauchsunterlagen**		
= operating manual; user's manual; run manual [DATA PROC]			= Bedienhandbuch; Bedienungshandbuch		
≈ equipment manual; application instruction; user documentation			≈ Gerätehandbuch; Anwendungsrichtlinie; Benutzerdokumentation		
operating expenses		ECON	**Betriebsausgaben**		
≈ operating costs			≈ Betriebskosten		
operating failure		TECH	**Betriebsausfall**		
operating feature		TECH	**Einsatzmerkmal**		
			= Betriebsmerkmal		
operating funds		ECON	**Betriebsmittel**		
operating guide (→use instruction 1)		TECH	(→Bedienungsanleitung)		
operating hours		TECH	**Betriebsstunden**		
operating hours counter		INSTR	**Betriebsstundenzähler**		
operating instruction 1		TECH	**Betriebsvorschrift**		
= operating rule; practice; practise			= Betriebsanweisung; Arbeitsanweisung		
≈ use instruction; notes for the operator			≈ Bedienungsanleitung; Bedienungshinweise		
operating instructions 2 (→use instruction 1)		TECH	(→Bedienungsanleitung)		
operating level		TELEC	**Betriebspegel**		
operating level [CATV]		BROADC	**Systempegel** [CATV]		
			= Betriebspegel		
operating manual (→operating documentation)		TECH	(→Gebrauchsunterlagen)		
operating method (→operating mode)		TECH	(→Betriebsart)		
operating mode		TECH	**Betriebsart**		
= mode of operation; operation mode; working mode; working principle; service mode; mode; operating method; method of operation; action; method			= Betriebsweise; Arbeitsweise; Wirkungsweise; Betriebsverfahren; Arbeitsverfahren; Modus		
≈ response; operation			≈ Verhalten; Betrieb		
operating noise		TECH	**Betriebsgeräusch**		
operating overheads		ECON	**Betriebsgemeinkosten**		
operating panel		EQUIP.ENG	**Bedienfeld**		
= operator control panel; control panel			= Bedienungsfeld		
			≈ Bedienungspult		
operating personnel		TECH	**Betriebspersonal**		
≈ maintenance personnel			≈ Wartungspersonal		
operating place (→operating site)		TECH	(→Einsatzort)		
operating point		TECH	**Arbeitspunkt**		
= working point					
operating point		ELECTRON	**Arbeitspunkt**		
= bias point; working point					
operating point adjustment		ELECTRON	**Arbeitspunkteinstellung**		
= bias point adjustment			= Arbeitspunktfestlegung		
operating position		TELEC	**Arbeitsplatz**		
operating position		TECH	**Arbeitsstellung**		
operating program		DATA PROC	**Betriebsprogramm**		
= running program					
operating quality		TELEC	**Betriebsgüte**		
= quality of service; grade of service					
operating range		TECH	**Betriebsbereich**		
= operating region; working range; working region			= Arbeitsbereich		
operating region (→operating range)		TECH	(→Betriebsbereich)		
operating revenues		ECON	**Betriebseinnahmen**		
			= Betriebseinkünfte		
operating room		TECH	**Betriebsraum**		
= workroom			≈ Arbeitsraum		
operating rule (→operating instruction 1)		TECH	(→Betriebsvorschrift)		
operating site		TECH	**Einsatzort**		
= operating place			≈ Anwendungsort		
≈ application site					
operating speed		TECH	**Arbeitsgeschwindigkeit**		
operating speed		MECH	**Betriebsdrehzahl**		
operating speed (→plot speed)		TERM&PER	(→Zeichengeschwindigkeit)		
operating state (→operational state)		DATA PROC	(→Betriebszustand)		
operating status (→operational state)		DATA PROC	(→Betriebszustand)		
operating supplies		MANUF	**Betriebsmittel**		
= supplies			= Betriebsstoff		
operating system		DATA PROC	**Betriebssystem**		
[part of system software, to coordinate hardware capabilities with user programs]			[Teil der Systemsoftware welcher Anwendersoftware mit den Hardwaremöglichkeiten der Anlage kompatibilisiert]		
= OS			= Bedienungssystem; BS; OS		
↑ system software			↑ Systemsoftware		
↓ executive control program; tape operating system; disk operating system; time-sharing operating system; basic operating system			↓ Organisationsprogramm; Band-Betriebssystem; Platten-Betriebssystem; Teilnehmer-Betriebssystem; Basis-Betriebssystem		
operating system residence		DATA PROC	**Betriebssystem-Residenz**		
[external memory for seldom used parts of the executive routine]			[Externspeicher für sporadisch notwendige Teile des Organisationsprogramms]		
operating telephone company (NAM) (→telephone administration)		TELEPH	(→Fernsprechverwaltung)		
operating temperature		TECH	**Betriebstemperatur**		
= working temperature			= Arbeitstemperatur		
≈ nominal temperature			≈ Nenntemperatur		
operating test		DATA PROC	**Betriebsprüfung**		
= maintenance test; dynamic test; dynamic check					
operating time		TECH	**Betriebszeit**		
operating voltage		EL.TECH	**Betriebsspannung**		
= working voltage; service voltage			= Arbeitsspannung		
operation		MANUF	**Arbeitsgang**		
operation 1		TECH	**Betrieb**		
= running; service; working; use (n.); duty (n.)			= Bedienung; Betreiben; Einsatz		
≈ operating mode; load; handling			≈ Betriebsart; Belastung; Handhabung		
operation		MATH	**Operation**		

operation	ENG.LOG	**Verknüpfung**
= connective (n.); function		= Funktion
operation 2	TECH	**Vorgang**
= process		= Operation; Prozeß 1
≈ procedure		≈ Verfahren
operation (→class of operation)	ELECTRON	(→Betriebsart)
operational	TECH	**betriebsfähig**
= serviceable		= funktionstüchtig
≈ working		
operational amplifier	CIRC.ENG	**Operationsverstärker**
[dc amplifier designed to perform a specific computing or transfer function]		[für spezifische Übertragungs- oder Rechenfunktion ausgelegter Gleichspannungsverstärker]
= op-amp		= Rechenverstärker
↓ integrating amplifier; summing integrator		↓ Integrationsverstärker; Summenintegrator
operational efficiency (→profitability)	ECON	(→Wirtschaftlichkeit)
operational empairment (→operational restriction)	TECH	(→betriebliche Einschränkung)
operational entity	ECON	**Meisterei**
operational experience	TECH	**Betriebserfahrung**
operational reliability (→service reliability)	QUAL	(→Betriebszuverlässigkeit)
operational requirement	TECH	**betriebliche Anforderung**
operational research (→operations research)	ECON	(→Unternehmensforschung)
operational restriction	TECH	**betriebliche Einschränkung**
= operational empairment		
operational result	ECON	**Wirtschaftsergebnis**
		= WE
operational state	DATA PROC	**Betriebszustand**
= operational status; operating state; operating status		
operational status (→operational state)	DATA PROC	(→Betriebszustand)
operation and maintenance	TELEC	**Betrieb und Wartung**
= O&M		= Bedienung und Wartung; O&M
operation building	TELEC	**Betriebsgebäude**
operation center	TELEC	**Betriebszentrum**
		≈ O&M-Zentrum
operation code	DATA PROC	**Operationscode**
[part of machine-code instruction defining the type of operation]		[Teil des Befehlscodes der die Operationsart definiert]
= op-code; opcode; OC; instruction code; operation part; operating code		= OC; Op-Code; Operationsschlüssel; Befehlscode; Operationsteil
operation concession	ECON	**Betriebskonzession**
operation control	DATA PROC	**Operationssteuerung**
[in the CPU]		[Steuerung der Befehlsausführung, im Steuerwerk der Zentraleinheit]
operation cycle (→instruction cycle)	DATA PROC	(→Befehlszyklus)
operation cycle time (→instruction cycle time)	DATA PROC	(→Befehlszykluszeit)
operation depending stress	QUAL	**betriebsbedingte Beanspruchung**
operation life (→life time)	QUAL	(→Lebensdauer)
operation mode	ENG.LOG	**Verknüpfungsart**
operation mode (→operating mode)	TECH	(→Betriebsart)
operation mode (→class of operation)	ELECTRON	(→Betriebsart)
operation mode (→processing mode)	DATA PROC	(→Betriebsart)
operation part (→operation code)	DATA PROC	(→Operationscode)
operation path	DATA PROC	**Operationspfad**
operation ratio (→availability)	QUAL	(→Verfügbarkeit)
operation register	DATA PROC	**Operationsregister**
operations controller	DATA PROC	**Operationen-Steuerung**
operation set	DATA PROC	**Operationsvorrat**
operations research (AM)	ECON	**Unternehmensforschung**
= operational research		= Planungsforschung
operations system	DATA PROC	**Betriebsführungssystem**
operation-standby	TECH	**Betrieb-Ersatz**
= master-standby; regular-standby		
operation supervision	TECH	**Betriebsüberwachung**
operation support (→general services)	ECON	(→Betriebsunterhaltung)
operation time	DATA PROC	**Operationszeit**
= command execution time		
operation tolerance	QUAL	**Betriebstoleranz**
operative (→effective)	TECH	(→wirksam)
operative memory	DATA PROC	**Operativspeicher**
operativity (→effectiveness)	TECH	(→Wirksamkeit)
operator	TELEPH	**Beamtin**
[in Europe generally female]		[Hinweis: in USA ist in der Fernvermittlung vielfach männliches Personal tätig]
= central (n.); attendant		
↓ toll operator		= Mädchen vom Amt
		↓ Fernbeamtin
operator	TECH	**Bedienungsperson**
= attendant		= Bedienperson; Betreiber
≈ user		≈ Anwender
operator	MATH	**Operator**
operator 1	DATA PROC	**Operator 1**
[part of address, defining type of operation]		[ALGOL, COBOL]
		≠ Befehlsteil, die Operationsart kennzeichnend
operator (→radio operator)	RADIO	(→Funker)
operator (→call handling)	TELEPH	(→Vermittlung)
operator 2 (→computer operator)	DATA PROC	(→Rechnerbediener)
operator-assisted	SWITCH	**handvermittelt**
		= mit Platzbeteiligung
operator-assited call	TELEPH	**handvermitteltes Gespräch**
operator-controlled	TECH	**bedienergesteuert**
operator control panel (→operating panel)	EQUIP.ENG	(→Bedienfeld)
operator convenience	TECH	**Bedienungskomfort**
= easy control; convenience; convenience feature		
operator device (→operating device)	TERM&PER	(→Bedienelement)
operator error	TECH	**Bedienungsfehler**
		= Fehlbedienung
operator guidance (→user interface)	DATA PROC	(→Benutzeroberfläche)
operator identification code	DATA PROC	**Bedienercode**
operator position (→operator's position)	TECH	(→Bedienplatz)
operator prompting (→user interface)	DATA PROC	(→Benutzeroberfläche)
operator recall	SWITCH	**Platzherbeiruf**
= attendant recall		= Bedienerherbeiruf; Schrankherbeiruf; Eintreteaufforderung
operator's console	SWITCH	**Vermittlungsplatz**
= console; operator's position; switching position; switchboard; supervisor position		= Vermittlungsschrank; Aufsichtsplatz
operator's console (→control desk)	TECH	(→Kontrollpult)
operator's console typewriter	TERM&PER	**Bedienungsblattschreiber**
operator's delay	SWITCH	**Warten auf Abfragen**
operator's desk (→control desk)	TECH	(→Kontrollpult)
operator's board (→control desk)	TECH	(→Kontrollpult)
operator's position	TECH	**Bedienplatz**
= operator's terminal; attendant position; attendant's terminal		= Bedienungsplatz; Bedienstation; Betriebsplatz
↓ control desk		↓ Kontrollpult
operator service	TELEC	**handvermittelter Dienst**
operator service system	SWITCH	**Handvermittlungssystem**
= OSS; manual switching system		
operator's position (→operator's console)	SWITCH	(→Vermittlungsplatz)
operator's terminal (→operator's position)	TECH	(→Bedienplatz)
ophtalmometer	PHYS	**Ophtalmometer**
[for optical measurement of distance]		[zur optischen Abstandsmessung]

opinion (→statement) ECON (→Stellungnahme)
opponent color PHYS **Gegenfarbe**
opposing field (→opposite field) PHYS (→Gegenfeld)
opposite (adj.) TECH **gegenüberliegend**
opposite direction (→return direction) TELEC (→Rückrichtung)
opposite exchange SWITCH **Gegenvermittlung**
= distant exchange; opposite office; distant office
= Gegenvermittlungsstelle; Gegenamt
opposite field PHYS **Gegenfeld**
= opposing field; counter field
opposite office (→opposite exchange) SWITCH (→Gegenvermittlung)
opposite side TECH **Gegenseite**
= other side
= gegenüberliegende Seite
opposite side (→solder side) ELECTRON (→Lötseite)
opposite terminal (→distant terminal) TELEC (→Gegenstelle)
opposition PHYS **Gegenphase**
opposition (→opposition to patent) ECON (→Patenteinspruch)
opposition (→contraposition) SCIE (→Kontraposition)
opposition to patent ECON **Patenteinspruch**
= opposition
= Einspruch
O-4PSK (→OQPSK) MODUL (→OQPSK)
optic (→fiberotic) OPT.COMM (→faseroptisch)
optical PHYS **optisch**
≈ visible
≈ sichtbar
optical attenuation (→optical loss) OPT.COMM (→optische Dämpfung)
optical attenuator OPT.COMM **optisches Dämpfungsglied**
= optischer Abschwächer
optical axis PHYS **Binormale**
= visual axis
= Sehachse
optical cable (→optical fiber cable) COMM.CABLE (→Lichtwellenleiterkabel)
optical character TERM&PER **Klarschrift**
[character detectable by human eye]
[vom menschlichen Auge lesbare Schrift, wie Druck- oder Handschrift]
= plain text
≈ plain writing
≠ machine-readable character
= Klartext
≠ Maschinenschrift
optical character display TERM&PER **Klarschriftanzeige**
optical character paper roll TERM&PER **Klarschrift-Papierrolle**
optical character reader TERM&PER **optischer Leser**
[reads standardized characters]
[liest genormte optische Schriftzeichen]
= OCR reader; optical reader; OCR terminal; optical character scanner; optical scanner; visual scanner
= OCR-Leser; OCR-Lesestation; optisches Lesegerät; optischer Codierzeilenleser
↑ character reader
↑ Klarschriftleser
↓ omni-font reader; multi-font reader; strip reader; page reader; document reader
↓ Allschriftenleser; Mehrschriftenleser; Streifenleser; Blattleser; Beleglesser
optical character recognition TERM&PER **optische Zeichenerkennung**
= OCR
= OCR
↑ character recognition
↑ Klarschriftlesen
optical character recognition letter (→OCR letter) TERM&PER (→OCR-Schrift)
optical character scanner (→optical character reader) TERM&PER (→optischer Leser)
optical communications TELEC **optische Nachrichtentechnik**
optical communications transmission TELEC **optische Nachrichtenübertragung**
= optical-fiber transmission; optical telecomunications
optical computer DATA PROC **optischer Computer**
optical coupled isolator (→optical coupling device) OPTOEL (→Optokoppler)
optical coupling device OPTOEL **Optokoppler**
= optocoupler; optical coupled isolator; opto-isolator; lightwave coupler; lightwave couple element
= Optoisolator; Lichtkoppler; Lichtwellenkoppler; LWL-Koppler; optischer Koppler
optical crosstalk OPT.COMM **optisches Nebensprechen**

optical data processing DATA PROC **optische Datenverarbeitung**
optical disc (→optical disk) TERM&PER (→Bildplatte)
optical disk TERM&PER **Bildplatte**
[stores data by laser-burned hole strings]
[speichert Daten durch mit Laser gebrannte Lochsequenzen]
= optical storage disk; video disk; optical disc; digital optical reading; DOR; laser storage disk
= optische Speicherplatte; Laserplatte; optische Platte; Videoplatte
↓ CD-ROM; WORM; TMO; PCR
↑ optischer Speicher
↓ CD-ROM; WORM; TMO; PCR
optical dispersion PHYS **optische Dispersion**
optical document reader TERM&PER **optischer Belegleser**
↑ document reader
↑ Belegleser
optical dot (→flying spot) ELECTRON (→Lichtpunkt)
optical feeder loop TRANS **LWL-Transportsystem**
= fiber feeder
optical fiber bundle (AM) OPT.COMM **Faserbündel**
= optical fibre bundle (BRI); fiber bundle
optical fiber cable COMM.CABLE **Lichtwellenleiterkabel**
= lightwave cable; optical cable; fiber cable
= LWL-Kabel; Lichtleiterkabel; Glasfaserkabel; optisches Kabel; faseroptisches Kabel
optical fiber splicing device OPT.COMM **LWL-Spleißgerät**
= Lichtwellenleiter-Spleißgerät
optical fiber test technique INSTR (→LWL-Meßtechnik)
(→fiber optic measuring technique)
optical-fiber transmission TELEC (→optische Nachrichtenübertragung)
(→optical communications transmission)
optical fibre bundle (BRI) OPT.COMM (→Faserbündel)
(→optical fiber bundle)
optical filter PHYS **optisches Filter**
optical head INSTR **optischer Meßkopf**
optical horizon RADIO PROP **optischer Horizont**
optical line-of-sight RADIO PROP **Sichtbegrenzungslinie**
optical link TELEC **optische Leitung**
optical loop (→fiber loop) TELEC (→Faser-Teilnehmerleitung)
optical loop carrier system TELEC **optisches Teilnehmermultiplexsystem**
= OLC
optical loss OPT.COMM **optische Dämpfung**
= optical attenuation
optical mark reader (→mark reader) TERM&PER (→Markierungsleser)
optical memory TERM&PER **optischer Speicher**
= optical storage; optical store
↓ Laserplatte; magnetooptischer Speicher; fotografischer Speicher; Mikrofilm; holographischer Speicher
↓ optical disk; magneto-optical memory; photographic memory; microfilm; holographic memory
optical page reader TERM&PER **optischer Seitenleser**
optical pointer (→cursor) TERM&PER (→Schreibmarke)
optical pointer (→luminous pointer) ELECTRON (→Lichtmarke)
optical power PHYS **Lichtleistung**
= luminous power
≈ light efficiency
optical power budget OPT.COMM **optische Leistungsbilanz**
optical power meter INSTR **optischer Pegelmesser**
= optischer Leistungsmesser
optical power source INSTR **optischer Pegelsender**
optical power splitter OPTOEL **optischer Leistungsteiler**
optical pumping OPTOEL **optisches Pumpen**
optical pumping circuit OPTOEL **Pumpkreis**
optical range (→visibility range) PHYS (→Sichtbereich)
optical reader (→optical character reader) TERM&PER (→optischer Leser)
optical receiver (→lightwave receiver) INSTR (→Lichtwellenempfänger)
optical recording TERM&PER **optische Speicherung**
optical relay COMPON **optisches Relais**
optical resonator OPTOEL **optischer Resonator**
optical scanner (→optical character reader) TERM&PER (→optischer Leser)

optical scanning	TERM&PER	optische Abtastung	optionally	ECON	bei Bedarf	
		= photoelektrische Abtastung			= optional	
optical sensor	COMPON	Lichtsensor	optional user facility (→optional facility)	TECH	(→wahlfreies Leistungsmerkmal)	
		= Lichtfühler; optischer Sensor	option bit (→check bit)	CODING	(→Prüfbit)	
optical signal analysator	INSTR	optischer Signalanalysator	option icon [menu]	DATA PROC	Auswahlbild [Piktogramm der Menütechnik]	
= lightwave signal analyzer						
optical signal generator (→optical signal source)	INSTR	(→optischer Signalgenerator)	option key	TERM&PER	Umschalttaste 2	
			options display [menu]	DATA PROC	Auswahlfeld [Menütechnik]	
optical signal source	INSTR	optischer Signalgenerator				
= optical signal generator		= optische Signalquelle	option selector (→dialing switch)	EQUIP.ENG	(→Wählschalter)	
optical spot (→flying spot)	ELECTRON	(→Lichtpunkt)				
			optoacoustical effect	PHYS	optoakustischer Effekt	
optical storage (→optical memory)	TERM&PER	(→optischer Speicher)			= photoakustischer Effekt	
			optocoupler (→optical coupling device)	OPTOEL	(→Optokoppler)	
optical storage disk (→optical disk)	TERM&PER	(→Bildplatte)				
			optoelectronic	MICROEL	optoelektronisch	
optical store (→optical memory)	TERM&PER	(→optischer Speicher)	= opto-electronic			
			opto-electronic (→optoelectronic)	MICROEL	(→optoelektronisch)	
optical subscriber line (→fiber loop)	TELEC	(→Faser-Teilnehmerleitung)				
			optoelectronic amplifier	OPTOEL	optoelektronischer Verstärker	
optical switch (→opto-switch)	OPTOEL	(→Optoschalter)	optoelectronic component	COMPON	optoelektronisches Bauelement	
optical telecommunications (→optical communications transmission)	TELEC	(→optische Nachrichtenübertragung)	= optoelectronic device			
			optoelectronic converter	COMPON	(→optoelektronischer Wandler)	
optical time domain reflectometer	INSTR	Rückstreumeßplatz	(→optoelectronic transducer)			
= OTDR		= Reflektometer; LWL-Rückstreumeßgerät; optisches Rückstreumeßgerät	optoelectronic device (→optoelectronic component)	COMPON	(→optoelektronisches Bauelement)	
			optoelectronic fork coupled device (→light fork coupler)	OPTOEL	(→Lichtgabelkoppler)	
optical video disk	CONS.EL	optische Bildplatte				
optical wave (→light wave)	PHYS	(→Lichtwelle)	opto-electronic integrated circuit	OPTOEL	OEIC-Baustein	
optical waveguide	OPT.COMM	Lichtwellenleiter	= OEIC			
= OWG lightguide		= optischer Wellenleiter; Lichtleiter; faseroptischer Wellenleiter; faseroptischer Leiter	optoelectronics	MICROEL	Optoelektronik	
≈ glass fiber			= optronics		= Optronik	
			optoelectronic transducer	COMPON	optoelektronischer Wandler	
		≈ Glasfaser	= optoelectronic converter			
optics	PHYS	Optik	optogalvanic effect	PHYS	optogalvanischer Effekt	
optimal code	CODING	redundanzsparender Code			= optovoltaischer Effekt	
		= Optimalcode	opto-isolator (→optical coupling device)	OPTOEL	(→Optokoppler)	
optimal control	CONTROL	optimale Steuerung				
optimization	TECH	Optimierung	opto-switch	OPTOEL	Optoschalter	
optimization algorithm (→optimizing algorithm)	MATH	(→Optimierungsalgorithmus)	= optical switch		= optischer Schalter	
			optronics (→optoelectronics)	MICROEL	(→Optoelektronik)	
optimization of application (→optimization of use)	TECH	(→Einsatzoptimierung)				
			OQPSK	MODUL	OQPSK	
optimization of use	TECH	Einsatzoptimierung	= offset quaternary phase shift keying; O-4PSK		= O-4PSK	
= optimization of application		= Anwendungsoptimierung				
optimization problem	SCIE	Optimierungsproblem	orange	PHYS	orange	
optimize	TECH	optimieren	orbit	PHYS	Umlaufbahn	
optimizing algorithm	MATH	Optimierungsalgorithmus	↑ trayectory		↑ Bahn	
= optimization algorithm		= optimierender Algorithmus	orbital electron	PHYS	Hüllenelektron	
optimizing compiler	DATA PROC	optimierender Kompilierer	orbital position	SAT.COMM	Orbitposition	
		= optimierender Kompiler; optimierender Compiler			= Umlaufbahnposition	
			orbital spacing	SAT.COMM	Orbitabstand	
optimum magnitude	CONTROL	Betragsoptimum	orbit coordination	SAT.COMM	Orbitkoordinierung	
		[Gütekriterium für Regelverhalten]	orbit utilization	SAT.COMM	Orbitnutzung	
			OR circuit (→OR gate)	CIRC.ENG	(→ODER-Glied)	
optimum matched filter [correlation electronics]	SIGN.TH	Optimalfilter [Korrelationselektronik]	order (n.)	ECON	Auftrag	
			order (→purchase order)	ECON	(→Bestellung)	
= analog matched filter		= optimales Filter	order (→input information)	SWITCH	(→Eingabeinformation)	
optimum-path (→path-optimized)	TECH	(→wegoptimiert)	order (→sequence)	COLLOQ	(→Reihenfolge)	
			order (→ordinance)	ECON	(→Verordnung)	
optimum shaped antenna	ANT	Vogelschwingenantenne	order (→significance)	INF	(→Wertigkeit)	
optimum-time (adj.)	TECH	zeitoptimal	order (→state)	TECH	(→Zustand)	
optimum-time (→time-optimized)	TECH	(→zeitoptimiert)	order (→arrangement)	MATH	(→Anordnung)	
			order acknowledgment (→order confirmation)	ECON	Auftragsbestätigung	
optimum working frequency	HF	günstigste Verkehrsfrequenz				
= OWF			order confirmation	ECON	Auftragsbestätigung	
option	ECON	Auswahlposition	= order acknowledgment			
≈ equipping option [EQUIP.ENG]; variant		= Option	ordered	MATH	geordnet	
		≈ Bestückungsvariante [EQUIP.ENG]; Variante; Wahlmöglichkeit	ordered set	MATH	geordnete Menge	
			order entry	DATA COMM	Auftragserfassung	
			order handling (→order processing)	DATA PROC	(→Auftragsabwicklung)	
option (→menu option)	DATA PROC	(→Auswahlposition)				
optional (adj.)	ECON	wahlweise	ordering argument	DATA PROC	Ordnungsbegriff	
= discretionary; alternative (adj.)		= wahlfrei; wählfrei; optional; nach Bedarf	= defining argument; sequence argument		= Schlüssel	
			≈ key data		≈ Ordnungsdaten	
optional facility	TECH	wahlfreies Leistungsmerkmal	ordering guide	ECON	Bestellanweisung	
= optional user facility						

ordering information		ECON	**Bestellangaben**			
			= Bestellinformation			tionsgatter; Adjunktionsschaltung; Adjunktionstor;
order of magnitude		MATH	**Größenordnung**			Alternator; Alternativ-Gatter; Mischglied; Mischgat-
= scale						ter; Mischschaltung; Mischelement
order of priority (→rule of precedence)		MATH	(→Rangfolge)			↑ Verknüpfungsglied
order of rank (→rule of precedence)		MATH	(→Rangfolge)	**orgware**	DATA PROC	**Orgware**
order of succession (→sequence)		COLLOQ	(→Reihenfolge)	[organizational measures]		[organisatorische Maßnahmen]
order processing center		MANUF	**Abwicklungszentrum**	**orientation**	PHYS	**Orientierung**
order processing		DATA PROC	**Auftragsabwicklung**	[crystal]		[Kristall]
= order handling; task processing; task handling; job processing; job handling				**orientation** (→alignment)	ANT	(→Ausrichtung)
				orientation (→direction)	TECH	(→Richtung)
				orientational polarization 2	PHYS	**Gitterpolarisation**
order processing		ECON	**Auftragsabwicklung**	[solid state]		[Festkörper]
order processing pool		MANUF	**Auftragsabwicklungszentrum**	**orientational polarization 1**	PHYS	**Orientierungspolarisation**
			= AZ	[by external field]		[Dipoldrehung durch äußeres Feld]
order register (→instruction register)		DATA PROC	(→Befehlsregister)			↑ elektrische Polarisation
order statistics		MATH	**Anordnungsstatistik**	**origin**	ECON	**Ursprung**
order wire		TELEC	**Dienstleitung**	= sourcing		
ordinal number		MATH	**Ordinalzahl**	**origin**	SWITCH	**Ursprung**
[designating the sequence, e.g. "first"]			[die Reihenfolge kennzeichnend, z.B. „erster"]	**original**	OFFICE	**Original**
≈ cardinal number			= Ordnungszahl	≠ copy 2		≈ Vorlage
			≈ Kardinalzahl			≠ Pause
ordinance		ECON	**Verordnung**	**original** (→original text)	LING	(→Urschrift)
= order			≈ Bestimmung	**original data** (→primary data)	DATA PROC	(→Primärdaten)
≈ regulation				**original document**	DATA PROC	**Originalbeleg**
ordinary (→regular)		TECH	(→regelmäßig)	= original voucher; source document; source voucher		= Urbeleg; Ursprungsbeleg
ordinary (→general)		COLLOQ	(→gewöhnlich)			↑ Beleg
ordinary dividend		ECON	**Stammdividende**	↑ document		
ordinary share		ECON	**Stammaktie**	**original drawing**	ENG.DRAW	**Originalzeichnung**
= common stock (AM); equity stock				**original equipment manufacturer**	ECON	(→OEM-Lieferant)
				(→OEM)		
ordinary share capital		ECON	**Stammkapital**	**original equipment supplier**	ECON	(→OEM-Lieferant)
ordinary subscriber		TELEC	**Normalteilnehmer**	(→OEM)		
= normal subscriber				**original holder**	TERM&PER	**Konzepthalter**
ordinate		MATH	**Ordinate**	= manuscript holder		= Vorlagenhalter; Manuskripthalter
= Y axis			= Y-Achse			
↑ coordinate			↑ Koordinate	**original stacker**	TERM&PER	**Konzeptablage**
ordnance 1		ECON	**Nachschub**			= Vorlagenablage
[militärisch]			[militärisch]	**original state** (→initial state)	CIRC.ENG	(→Grundstellung)
≈ supply (n.)			≈ Versorgung			
ordnance 2 (→ordnance department)		ECON	(→Zeugamt)	**original text**	LING	**Urschrift**
				= original		
ordnance department		ECON	**Zeugamt**	**original voucher** (→original document)	DATA PROC	(→Originalbeleg)
[for military supplies]			= militärische Beschaffungsbehörde			
= ordnance 2				**originate**	TELEPH	**erzeugen**
ore		METAL	**Erz**	[a call]		[einen Anruf]
OR element (→OR gate)		CIRC.ENG	(→ODER-Glied)	= send		≈ anrufen
OR-ELSE function		ENG.LOG	(→EXKLUSIV-ODER-Verknüpfung)	≈ call (v.t.)		
(→EXCLUSIVE-OR operation)				**originate** (→cause)	TECH	(→verursachen)
OR-ELSE operation		ENG.LOG	(→EXKLUSIV-ODER-Verknüpfung)	**originating exchange**	SWITCH	**Ursprungsvermittlung**
(→EXCLUSIVE-OR operation)				= originating switching center		= Ursprungsvermittlungsstelle; Ursprungsamt
OREM		TELEPH	**OBDM**	≠ terminating exchange		≠ Zielvermittlung
[objective reference equivalent measurement]			[objektiver Bezugsdämpfungsmeßplatz]	**originating point**	TELEC	**Ursprungspunkt**
				originating register	SWITCH	**Leitregister**
OR function (→OR operation)		ENG.LOG	(→ODER-Verknüpfung)	**originating station** (→calling station)	DATA COMM	(→rufende Station)
organic semiconductor		PHYS	**organischer Halbleiter**	**originating subscriber**	TELEC	**Ursprungsteilnehmer**
organizational instruction		DATA PROC	**organisatorischer Befehl**	**originating switching center**	SWITCH	(→Ursprungsvermittlung)
organization channel		MOB.COMM	**Organisationskanal**	(→originating exchange)		
organization chart		ECON	**Organisationsplan**	**originating traffic**	SWITCH	**Ursprungsverkehr**
			= Orgplan			≈ abgehender Verkehr
organizer		ECON	**Veranstalter**	**originator** [DATA COMM] (→message source)	INF	(→Nachrichtenquelle)
			= Organisator			
OR gate		CIRC.ENG	**ODER-Glied**	**originator** (→causer)	TECH	(→Verursacher)
= OR element; OR circuit; disjunction gate; disjunction element; disjunction circuit; INCLUSIVE-OR gate; INCLUSIVE-OR element; INCLUSIVE-OR circuit; adjunction gate; adjunction element; adjunction circuit			= ODER-Gatter; ODER-Schaltung; ODER-Tor; Disjunktionsglied; Disjunktionsgatter; Disjunktionsschaltung; Disjunktionstor; INKLUSIV-ODER-Glied; INKLUSIV-ODER-Gatter; INKLUSIV-ODER-Schaltung; INKLUSIV-ODER-Tor; Adjunktionsglied; Adjunk-	**origin code**	SWITCH	**Ursprungskennung**
				ornamental strip	TECH	**Zierleiste**
				= molding strip		↑ Leiste
				↑ strip 2		
				orography	GEOPHYS	**Orographie**
				[description of terrain formation]		[Beschreibung der Geländeformation]
				= topography of terrain; terrain formation		= Geländeformation; Geländetopographie
↑ logic gate				≈ topography		≈ Topographie

OR operation ENG.LOG
[output = 1 if at least one input = 1]
= OR function; INCLUSIVE-OR operation; INCLUSIVE-OR function; disjunction; logical addition; logical add; adjunction; logical sum; alternation; EITHER-OR operation; EITHER-OR function
≈ EXCLUSIVE OR operation
↑ logic operation

orphan TYPOGR
[first line of a paragraph remaining in the preceding page or column]
= orphan line
≈ widow

orphan line (→orphan) TYPOGR
OR service (→on-request RADIO NAV service)
orthicon ELECTRON
↑ camera tube
orthogonal 2 MATH
[function]
orthogonal 1 (→right-angled) MATH
orthogonality 2 MATH
[of a function]
orthogonality 1 MATH
[the quality of right angle]
orthogonal polarisation PHYS
orthogonal projection ENG.DRAW
(→orthographic projection)
orthogonal system DATA PROC
orthographic projection ENG.DRAW
= orthogonal projection
orthography LING
= spelling
orthoscopic PHYS
OS (→operating system) DATA PROC
Os (→osmium) CHEM
oscillate EL.TECH
oscillate MECH
= vibrate; undulate
↓ swing; fluctuate

oscillating circuit NETW.TH
= oscillatory circuit
↓ resonant circuit
oscillating circuit (→ring-back POWER SYS circuit)
oscillating discharge PHYS
oscillating frequency EL.TECH
oscillating frequency (→vibration MECH frequency)
oscillating function MATH
oscillating sorting DATA PROC
↑ sorting
oscillating threshold CIRC.ENG

oscillation PHYS
[periodical process in the time or space domain]
≈ wave
↓ vibration
oscillation (→vibration) MECH
oscillation amplitude PHYS
oscillation antinode (→vibration MECH antinode)
oscillation conditions CIRC.ENG

oscillation frequency EL.TECH

ODER-Verknüpfung
[Ausgang = 1 wenn mindestens ein Eingang = 1]
= ODER-Funktion; INKLUSIVES ODER; EINSCHLIESSLICHES ODER; logische Addition; Disjunktion; Alternative; Adjunktion
≈ EXKLUSIV-ODER-Verknüpfung
↑ logische Verknüpfung

Schusterjunge
[erste Zeile eines Abschnitts, die auf der vorangehenden Seite oder Spalte verblieben ist]
= Schusterbube; alleinstehende Zeile
≈ Überhangzeile

(→Schusterjunge)
(→angeforderte Verbindung)

Orthicon
↑ Bildaufnahmeröhre
orthogonal 2
[Funktion]
(→rechtwinklig)
Orthogonalität 2
[einer Funktion]
Rechtwinkligkeit
= Orthogonalität 1
orthogonale Polarisation
= Orthogonalpolarisation
(→rechtwinklige Projektion)

Orthogonal-System
rechtwinklige Projektion

Rechtschreibung
= Orthographie
≈ Schreibweise
orthoskopisch
(→Betriebssystem)
(→Osmium)
schwingen
= oszillieren
schwingen
= vibrieren
↓ pendeln; schaukeln; schwanken

Schwingkreis
= Schwingungskreis
↓ Resonanzkreis
(→Umschwingkreis)

oszillierende Entladung
Schwingungsfrequenz
= Schwingungszahl; Oszillationsfrequenz
(→Schwingungsfrequenz)

oszillierende Funktion
oszillierendes Sortieren
↑ Sortieren
Schwingeinsatzpunkt
= Schwingungsschwelle
Schwingung
[zeitlich oder räumlich periodischer Vorgang]
≈ Welle
↓ mechanische Schwingung
(→mechanische Schwingung)
Schwingungsamplitude
(→Schwingungsbauch)

Schwingungsbedingungen
[Oszillator]
Schwingfrequenz
= Eigenfrequenz

oscillation generation CIRC.ENG
oscillation mode (→waveform) PHYS
oscillation node (→node) PHYS
oscillation node (→vibration MECH node)
oscillator CIRC.ENG
[circuit to generate peridic signals]
↓ sine wave generator; square wave generator; sawtooth generator

oscillator frequency ELECTRON
oscillatory circuit (→oscillating NETW.TH circuit)
oscillogram INSTR
[display on oscilloscope]
oscillograph INSTR
[instrument recording rapidly varying quantities]
≈ oscilloscope
↓ cathode ray oscillograph; ink jet recorder; light-beam oscillograph; sampling oscillograph

oscillograph tube ELECTRON
(→osciloscope tube)
oscilloscope INSTR
[instrument visualizing on a display rapidly varying quantities]
= scope
≈ oscillograph
↓ cathode ray oscilloscope; sampling oscilloscope; dual-channel oscilloscope; logic analyzer

oscilopscope tube ELECTRON
= oscillograph tube
↑ picture tube
OSI (→open system DATA COMM interconnection)
OSI model (→ISO reference DATA COMM model)
osmium CHEM
= Os
OSS (→operator service SWITCH system)
OTA CIRC.ENG
= transconductance operational amplifier
OTC (NAM) TELEPH
(→telephone administration)
OTDR (→optical time domain INSTR reflectometer)
other side (→opposite side) TECH
O type carrier system TRANS
[U.S. standard]
ourside circle ENG.DRAW
ouside conductor POWER ENG
(→phase)
outage (→failure) QUAL
outband (adj.) TRANS

outband signaling TRANS
= outband signalling; out-of-band signaling (AM)

outband signalling (→outband TRANS signaling)
outboard channel RADIO
= limit channel
outbound (→outgoing) TELEC
outbound bearing (→back RADIO NAV bearing)

Schwingungserzeugung
(→Schwingungsform)
(→Schwingungsknoten)
(→Schwingungsknoten)

Oszillator
[Schaltung zur Erzeugung periodischer Signale]
↓ Schwingungserzeuger
↓ Sinusgenerator; Rechteckgenerator; Sägezahngenerator

Oszillatorfrequenz
(→Schwingkreis)

Oszillogramm
[Darstellung auf Oszilloskop]
Oszillograph
[Meßgerät zum Registrieren schneller Größen]
= Oszillograf; Schwingungsschreiber
≈ Oszilloskop
↓ Elektronenstrahl-Oszillograph; Flüssigkeitsstrahlschreiber; Lichtstrahl-Oszillograph; Abtastoszillograph

(→Oszilloskopröhre)
Oszilloskop
[auf Bildschirm schnelle Größen darstellendes Meßgerät]
= Elektronenstrahl-Oszillograph 2
≈ Oszillograph
↓ Elektronenstrahl-Oszilloskop; Abtast-Oszilloskop; Zweikanaloszilloskop; Logikanalysator;

Oszilloskopröhre
= Oszillographenröhre
↑ Bildwiedergaberöhre
(→offenes Kommunikationssystem)
(→ISO-Referenzmodell)

Osmium
= Os
(→Handvermittlungssystem)

Transkonduktanzverstärker
= OTA-Verstärker

(→Fernsprechverwaltung)

(→Rückstreumeßplatz)

(→Gegenseite)
Z16F-System
[nach US-Norm]
Kopfkreis
(→Außenleiter)

(→Ausfall)
Außerband-
= Outband-
Außerbandsignalisierung
= systemeigene Wahl; Außerbandwahl; Außerbandzeichengabe

(→Außerbandsignalisierung)

Randkanal
= Eckkanal
(→abgehend)
(→Abflugpeilung)

outbound traffic (→outgoing SWITCH (→gehender Verkehr)
traffic)
outdated (→obsolete) TECH (→veraltet)
outdent (v.t.) TYPOGR **ausrücken**
out-diffusion MICROEL **Ausdiffusion**
outdoor TECH **im Freien**
= open-air; open-site = Freigelände
outdoor- (adj.) TECH (→wettergeschützt)
(→weather-protected)
outdoor antenna ANT **Außenantenne**
= outside antenna; open antenna
outdoor apparatus TECH **Freiluftapparatur**
outdoor bell TERM&PER **Außenwecker**
= weatherproof bell = schlagwettergeschützter
 Wecker
outdoor cabinet EQUIP.ENG (→Wettergehäuse)
(→weatherproof housing)
outdoor cable (→outside COMM.CABLE (→Außenkabel)
cable)
outdoor case (→weatherproof EQUIP.ENG (→Wettergehäuse)
housing)
outdoor climate QUAL **Außenraumklima**
= open air climate; outdoor environ- = Freiraumklima; Freiluftkli-
ment; outdoor service environ- ma
ment
outdoor enclosure EQUIP.ENG (→Wettergehäuse)
(→weatherproof housing)
outdoor environment (→outdoor QUAL (→Außenraumklima)
climate)
outdoor housing EQUIP.ENG (→Wettergehäuse)
(→weatherproof housing)
outdoor installation EQUIP.ENG **Freiluftaufbau**
(→outdoor mounting) (→Außenmontage)
 ≠ Stationsasaufbau
outdoor monitoring system SIGN.ENG **Freilandüberwachungsanlage**
outdoor mounting EQUIP.ENG **Außenmontage**
= external mounting; exterior-premises = Freiraummontage; Freiluft-
mounting; external installation; out- montage; Außenaufbau;
door installation; exterior premises Freiraumaufbau; Freiluft-
installation aufbau
≠ indoor mounting ≠ Innenmontage
outdoor service environment QUAL (→Außenraumklima)
(→outdoor climate)
outdoor telephone TELEPH **Freilufttelefon**
 = Außentelefon
outdoor temperature QUAL **Außentemperatur**
outer conductor LINE TH **Außenleiter**
[coaxial pair] [Koaxialpaar]
outer mark beakon (→outer RADIO NAV (→Voreinflugzeichen)
marker)
outer marker RADIO NAV **Voreinflugzeichen**
= outer mark beakon; foremarker = Voreinflugbake
outer-shell electron (→bonding PHYS (→Valenzelektron)
electron)
outfit (→equipment 1) TECH (→Ausstattung)
outflow (→efflux) TECH (→Abfluss)
outflow valve TECH **Auslaufventil**
outgoing TELEC **abgehend**
[call, link] [Verbindung]
= outbound = gehend
≠ incoming ≠ ankommend
outgoing (→initial) TECH (→anfänglich)
outgoing air (→output air) TECH (→Abluft)
outgoing call SWITCH **gehendes Gespräch**
outgoing call blocking TELEPH **Ausgangssperre**
outgoing channel SWITCH **Abnehmerkanal**
= serving channel ≈ Abnehmerleitung
outgoing circuit SWITCH **abgehender Satz**
↑ junction ↑ Verbindungssatz
outgoing counter SWITCH **gehender Zähler**
outgoing fraction defective QUAL (→mittlere Auslieferqualität)
(→average outgoing quality)
outgoing highway SWITCH **Abnehmer-Multiplexleitung**
= serving highway = gehende Multiplexleitung;
 Zeitmultiplex-Abnehmer-
 leitung
 ↑ Abnehmerleitung
outgoing line SWITCH **Abnehmerleitung**
[conveys traffic from a switching ma- [führt Verkehr von Koppel-
trix] einrichtungen ab]

= outgoing trunk ≈ Abnehmerkanal; Abneh-
≠ offering line merbündel
 ≠ Zubringerleitung
 ↓ Abnehmer-Multiplex-
 leitung
outgoing post OFFICE **abgehende Post**
 = Ausgangspost; Postausgang
outgoing seizure SWITCH **gehende Belegung**
outgoing signal TELEC **abgehendes Signal**
≈ transmitted signal = Ausgangssignal
 = Sendesignal
outgoing signal distortion TELEGR **Sendeverzerrung**
outgoing subscriber (→calling SWITCH (→rufender Teilnehmer)
subscriber)
outgoing traffic SWITCH **gehender Verkehr**
= outbound traffic = abgehender Verkehr
 ≈ Ursprungsverkehr
outgoing trunk (→outgoing SWITCH (→Abnehmerleitung)
line)
outgoing trunk circuit SWITCH **abgehender Leitungssatz**
 = gehender Leitungssatz
outgoing trunk group SWITCH **Abnehmerbündel**
= service trunk group ≈ Abnehmerleitung
outlandish (→foreign) ECON (→ausländisch)
outlay (→expense) ECON (→Ausgabe)
outlet TECH **Auslass**
outlet (→output) EL.TECH (→Ausgang)
outlet (→mains socket) EL.INST (→Netzsteckdose)
outlet temperature TECH **Austrittstemperatur**
outlier TECH **Ausreißer**
[something away from main part] [fig.]
outline (n.) ENG.DRAW **Kante**
↓ break ↓ Bruchkante
outline (n.) TECH **Umriß**
= contour; profile = Kontur; Profil
≈ silhouette ≈ Silhouette
outline LING **Umriß**
[of a text] [Text]
outline (n.) (→design) TECH (→Entwurf)
outline- (→outlined) TYPOGR (→konturiert)
outline character (→outline TYPOGR (→Konturschrift)
type)
outlined (adj.) TYPOGR **konturiert**
[formed by contoures] [aus Konturen gebildet]
= outline-; inline-
outlined letter (→outline TYPOGR (→Konturschrift)
type)
outline font DATA PROC **Outline-Font**
[geometrically defined font for laser [geometrisch definierter
printer, scalable] Zeichensatz für Laserdruk-
 ker, skalierbar]
 = vektororientierter Zeichensatz
outline font (→outline type) TYPOGR (→Konturschrift)
outline information DOC **Übersichtsinformation**
outliner DATA PROC **Gliederungsfunktion**
[word processing] [Textverarbeitung]
outline terms ECON **Rahmenbestimmungen**
outline type TYPOGR **Konturschrift**
[with characters designed only by [nur mit Umrißlinien ge-
contour lines] zeichnete Buchstaben]
= outline typefont; outline font; outline = Konturbuchstabe
character; outlined letter
outline typefont (→outline TYPOGR (→Konturschrift)
type)
outlook COLLOQ **Vorschau**
outmode (v.t.) (→become TECH (→veralten)
outdated)
outmoded (→obsolete) TECH (→veraltet)
outnumber (v.t.) COLLOQ **zahlenmäßig übertreffen**
out-of-band radiation RADIO **Außerbandstrahlung**
 = Randaussendung
out-of-band signaling (AM) TRANS (→Außerbandsignalisierung)
(→outband signaling)
out-of-country purchase ECON (→Einfuhr)
(→import)
out-of-date (→obsolete) TECH (→veraltet)
out of focus (→blurred) TV (→unscharf)
out of order TECH **außer Betrieb**
= out of service = gestört; gesperrt

out of order (→faulty)	QUAL	(→fehlerhaft)	
out-of-parity	CODING	**fehlerhafte Parität**	
out of phase	PHYS	**phasenverschoben**	
= phase-shifted; phase-displayed			
out of scale	ENG.DRAW	**nicht maßstabgerecht**	
out of service (→out of order)	TECH	(→außer Betrieb)	
out of shape (→warped)	TECH	(→windschief)	
outpulsing	SWITCH	**Adreßinformationsaustausch**	
output (n.)	MANUF	**Ausstoß**	
		= Ausbringung	
output	EL.TECH	**Ausgang**	
= outlet		≠ Eingang	
≠ input			
output (n.)	DATA PROC	**Ausgabe**	
= sending		≈ Einlesen	
≈ read-out (n.)		≠ Eingabe	
≠ input (n.)		↓ Datenausgabe	
↓ data output			
output (v.t)	DATA PROC	**ausgeben**	
= dump (v.t.); issue (v.t.); fetch out		≈ auslesen	
≈ read-out		≠ eingeben	
≠ input (v.t.)			
output acknowledgment	SWITCH	**Ausgabequittung**	
output address register	DATA PROC	**Ausgabeadreßregister**	
output admittance	NETW.TH	**Ausgangsadmittanz**	
		= Ausgangs-Scheinleitwert	
output admittance, input shorted (→short-circuit output admittance)	NETW.TH	(→Kurzschluß-Ausgangsadmittanz)	
output air	TECH	**Abluft**	
= exhaust 2 (n.); outgoing air		≠ Zuluft	
≠ input air			
output alphabet	DATA PROC	**Ausgangsalphabet**	
output amplifier	CIRC.ENG	**Ausgangsverstärker**	
output amplitude	EL.TECH	**Ausgangsamplitude**	
output attenuator	CIRC.ENG	**Ausgangsteiler**	
output block	DATA PROC	**Ausgabeblock**	
output buffer	DATA PROC	**Ausgabepufferspeicher**	
output capacitance	EL.TECH	**Ausgangskapazität**	
output cell	MICROEL	**Ausgangszelle**	
output channel	DATA PROC	**Ausgabekanal**	
output characteristic	MICROEL	**Ausgangskennlinie**	
output characteristic impedance	LINE TH	**Ausgangswellenwiderstand**	
output circuit	CIRC.ENG	**Ausgangsschaltung**	
		= Ausgangskreis	
output condition	ELECTRON	**Ausgangszustand**	
output controller	DATA PROC	**Ausgabewerk**	
[controls data output of CPU]		[steuert die Datenausgabe der Zentraleinheit]	
↑ input/output controller		= Ausgabeprozessor	
		↑ Ein-Ausgabe-Werk	
output current	EL.TECH	**Ausgangsstrom**	
output cutoff frequency	MICROEL	**Ausgangskennfrequenz**	
output data	DATA PROC	**Ausgabedaten**	
output date	DATA PROC	**Ausgabedatum**	
output destination	SWITCH	**Ausgabeziel**	
output device (→output unit)	TERM&PER	(→Ausgabeeinheit)	
output disk	TERM&PER	**Ausgabeplatte**	
output driver	CIRC.ENG	**Ausgangstreiber**	
output enable input	MICROEL	**Ausgangsfreigabe-Eingang**	
output file	DATA PROC	**Ausgabedatei**	
output filter	CIRC.ENG	**Ausgangsfilter**	
output format	DATA PROC	**Ausgabeformat**	
output gap (→catcher space)	MICROW	(→Auskoppelraum)	
output gate	MICROEL	**Ausgabegatter**	
output impedance	NETW.TH	**Ausgangsimpedanz**	
output jack	ELECTRON	**Ausgangsbuchse**	
output level	TELEC	**Ausgangspegel**	
output loading capability	ELECTRON	**Ausgangsbelastbarkeit**	
output magazine	DATA PROC	**Ausgabemagazin**	
[reserved memory location]		[reservierter Speicherplatz]	
↑ magazine		↑ Magazin	
output oscillation	CIRC.ENG	**Ausgangsschwingung**	
output port	NETW.TH	**Ausgangstor**	
output power	EL.TECH	**Ausgangsleistung**	
output program	DATA PROC	**Ausgabeprogramm**	
= output writer			
output pulse width	CIRC.ENG	**Verweilzeit**	
[gate]		[Kippstufe]	
= dwell time			
output quantity	PHYS	**Ausgangsgröße**	
output repeater	SWITCH	**Ausgangsübertrager**	
output signal	ELECTRON	**Ausgangssignal**	
output stacker	TERM&PER	**Ablagefach**	
[for documents or punched cards]		[für Belege oder Lochkarten]	
= stacker		= Papierablage; Stapler	
↓ card stacker; forms stacker		↓ Lochkartenstapler; Formularstapler	
output stage (→final stage)	CIRC.ENG	(→Endstufe)	
output teleprinter	TERM&PER	**Ausgabedrucker**	
		[Fernschreibtechnik]	
output terminal	EQUIP.ENG	**Ausgangsklemme**	
output time constant	MICROEL	**Ausgangs-Zeitkonstante**	
output transformer	CIRC.ENG	**Ausgangsübertrager**	
output tube	ELECTRON	**Endröhre**	
= final stage tube; output valve; final stage valve		= Endverstärkerröhre	
output unit	TERM&PER	**Ausgabeeinheit**	
= output device		= Ausgabegerät	
↑ peripheral equipment		↑ Peripheriegerät	
output valve (→output tube)	ELECTRON	(→Endröhre)	
output voltage	EL.TECH	**Ausgangsspannung**	
output voltage swing	ELECTRON	**Ausgangsspannungshub**	
output writer (→output program)	DATA PROC	(→Ausgabeprogramm)	
outside (n.)	TECH	**Außenseite**	
≠ inside		≠ Innenseite	
outside (adj.)	TECH	**auswärtig**	
		= äußerer	
outside (→external)	TECH	(→extern)	
outside antenna (→outdoor antenna)	ANT	(→Außenantenne)	
outside broadcast	BROADC	**Außenübertragung**	
outside broadcast transmission	BROADC	**Zubringerübertragung**	
outside broadcast transmitter	BROADC	**Reportagesender**	
outside broadcast van	BROADC	**Übertragungswagen**	
= O.B. van		= Ü-Wagen	
↓ television van		↓ Fernsehwagen	
outside cable	COMM.CABLE	**Außenkabel**	
= outdoor cable; external cable			
outside consultation	TELEPH	**Externrückfrage**	
outside diameter	MATH	**Außendurchmesser**	
outside diameter	MECH	**Kopfkreisdurchmesser**	
[cogwheel]		[Zahnrad]	
outside dimension	ENG.DRAW	**Außenmaß**	
outside plant (→local outside plant)	OUTS.PLANT	(→Ortsleitungsnetz)	
outside plant network	TELEC	**Liniennetz 1**	
= external plant network; communication lines network		[Gesamtheit der Kabel-, Freileitungs- oder Funklinien]	
↓ line plant; microwave link plant		↓ Leitungsnetz; Richtfunknetz	
outside plant technique	TELEC	**Linientechnik**	
= external plant technique; communication line technique			
outside supplier	ECON	**Fremdlieferant**	
outside television broadcast link	BROADC	**Fernsehzubringerleitung**	
= outside TV broadcast link		[von Außenreportage zu Studio]	
		= TV-Zubringerleitung; Fernseh-Außenübertragungsleitung	
outside termination	CIRC.ENG	**Außenbeschaltung**	
outside TV broadcast link	BROADC	(→Fernsehzubringerleitung)	
(→outside television broadcast link)			
outsize (n.)	ECON	**Übergröße**	
outskirt (→suburb)	ECON	(→Vorort)	
outslot signaling	TELEC	**Outslot-Signalisierung**	
outsort (v.t.)	TECH	**aussondern**	
outsorting (→rejection)	DATA PROC	(→Aussteuerung)	
outsourcing	ECON	**Außer-Haus-Beschaffung**	
		= Fremdbezug	
outstanding	COLLOQ	**unerledigt**	

Term	Domain	Translation
outstanding (→unpaid)	ECON	(→unbezahlt)
outstation	DATA COMM	Außenstation
outstation (→tributary station)	TELECONTR	(→Unterstation)
outstation location	TELECONTR	Unterstelle [Ort mit Unterstation]
out-tray	OFFICE	Ausgangskorb
outward-restricted [cannot place outward calls directly] = toll-restricted ≠ non-restricted	TELEPH	halbamtsberechtigt [kann das Amt nicht direkt anwählen] ≠ vollamtsberechtigt
outward WATS [the arrangement holds for outgoing calls]	TELEC	abgehende Ferngesprächspauschale [gilt für abgehende Gespräche]
oval (adj.)	TECH	oval
oval head = lentil head	MECH	Linsensenkkopf [Schrauben]
oval-head screw = lentil-head screw	MECH	Linsensenkkopfschraube
ovecurrent protection ↑ overload protection	EL.TECH	Überstromschutz ↑ Überlastschutz
oven	ELECTRON	thermostatgeregelt
overall = total; global	TECH	Gesamt-
overall amplification	NETW.TH	Betriebsverstärkung
overall bit rate (→transmission rate)	TELEC	(→Übertragungsgeschwindigkeit)
overall delay	TELEC	Gesamtverzögerung
overall dimension	ENG.DRAW	Gesamtmaß
overall equipment list = equipment summary	DOC	Geräteübersicht
overall equivalent (→overall loss)	TELEC	(→Restdämpfung)
overall loss = net loss; overall equivalent; zero-insertion loss; equivalent	TELEC	Restdämpfung = Restdämpfungsmaß
overall loss (→effective attenuation constant)	NETW.TH	(→Betriebsdämpfungsmaß)
overall-performance test (→system test)	DATA PROC	(→Systemtest)
overall sensitivity	INSTR	Gesamtempfindlichkeit
overall transmission index	TELEPH	Gesamtübertragungsfaktor
overbuild (→overengineer)	TECH	(→überdimensionieren)
overburden (→overload)	TECH	(→überlasten)
overcapacity	ECON	Überkapazität
overcapitalization	ECON	Überkapitalisierung
over-car antenna	ANT	Dachantenne 2 [Autoantenne]
overcharge (→overload)	TECH	(→überlasten)
overcharge (n.) (→overload)	TECH	(→Überlastung)
overcharging	ECON	Überteuerung ↑ Verteuerung
overcharging protection	POWER SYS	Überladeschutz
overcompensate	TECH	überkompensieren = überausgleichen
overcompensation	TECH	Überkompensation = Überausgleich
overcorrect (v.t.)	TECH	überkorrigieren
overcorrection	TECH	Überkorrektur
overcritical	TECH	überkritisch
overcurrent = excess current	EL.TECH	Überstrom
overcurrent cut-out = overcurrent release	POWER SYS	Überstromauslöser
overcurrent protecting device (→overcurrent protector)	COMPON	(→Stromsicherung)
overcurrent protection switch = overload protection switch ↑ overcurrent protector	COMPON	Überstromschutzschalter ↑ Stromsicherung
overcurrent protector [protects from too high currents] = overcurrent protecting device ↓ fine-wire fuse; high-current fuse; overcurrent protection switch; melting fuse	COMPON	Stromsicherung [schützt vor unzulässig hohen Strömen] = Überstromschutzorgan; Sicherung 2 ↑ Sicherung 1 ↓ Stromgrobsicherung; Stromfeinsicherung; Schmelzsicherung; Überstromschutzschalter
overcurrent release (→overcurrent cut-out)	POWER SYS	(→Überstromauslöser)
overdamp (v.t.)	PHYS	überdämpfen
overdamped	PHYS	überdämpft
overdamping	PHYS	Überdämpfung = überkritische Dämpfung
overdesign (v.t.) (→overengineer)	TECH	(→überdimensionieren)
over-design (n.) (→over-engineering)	TECH	(→Überdimensionierung)
overdimension (v.t.) (→overengineer)	TECH	(→überdimensionieren)
overdraftm credit	ECON	Überziehungskredit
overdrive = overload	ELECTRON	übersteuern
overdrive-resistant = overload-resistant	ELECTRON	übersteuerungsfest
overdriving = overload; overstress; overpower ≈ overmodulation [MODUL]; saturation	ELECTRON	Übersteuerung = Überlastung; Übererregung ≈ Übermodulation [MODUL]; Sättigung
overdriving capability (→overload capability)	ELECTRON	(→Übersteuerbarkeit)
overdue	ECON	überfällig
overdue payment	ECON	rückständige Zahlung = überfällige Zahlung
overengineer (v.t.) = overdesign (v.t.); oversize (v.t.); overbuild; overdimension (v.t.)	TECH	überdimensionieren = überzüchten
over-engineering = over-design (n.); over-sizing	TECH	Überdimensionierung = Überzüchtung
overequip	EQUIP.ENG	überbestücken
overequipment (→over-equipping)	EQUIP.ENG	(→Überbestückung)
over-equipping = overequipment	EQUIP.ENG	Überbestückung
overexcitation	CIRC.ENG	Übererregung
overexposure	PHYS	Überbelichtung
overflow (n.) [result digits of an arithmetic operation, which exceed the storage capacity of the result register] ≈ carry [MATH] ≠ underflow (n.)	DATA PROC	Überlauf [im Ergebnisregister nicht mehr unterbringbare Ergebnisstellen einer arithmetischen Operation] = Bereichsüberschreitung ≈ Übertrag [MATH] ≠ Unterlauf
overflow (n.) = overrun (n.); spillover (n.)	ELECTRON	Überlauf
overflow 1 [traffic exceeding the capacity of a switching facility] = spillover	SWITCH	Überlauf [die Kapazität einer Vermittlungseinrichtung überschreitender Verkehr]
overflow 2 (→overflow traffic)	SWITCH	(→Überlaufverkehr)
overflow 1 (→carry)	MATH	(→Übertrag)
overflow 2 (→carry digit)	MATH	(→Übertragziffer)
overflow bit (→carry bit)	DATA PROC	(→Übertragbit)
overflow digit (→carry digit)	MATH	(→Übertragziffer)
overflow flag (→carry bit)	DATA PROC	(→Übertragbit)
overflow operation	SWITCH	Überlaufbetrieb
overflow program [to process overflows]	DATA PROC	Überlaufprogramm [zur Verarbeitung von Überläufen]
overflow register [registers overflows]	DATA PROC	Überlaufregister [registriert aufgetretene Überläufe]
overflow traffic = overflow 2; overshoot traffic; overshoot 2	SWITCH	Überlaufverkehr = überlaufender Verkehr; überfließender Verkehr
overflow valve	TECH	Überlaufventil
overflux relay (→disconnecting relay)	ELECTRON	(→Abschalterelais)
overfrequency	TECH	Überfrequenz
overglassing = protective film; protective layer	MICROEL	Schutzschicht
overhang (n.)	TECH	Ausladung = Überhang; Überstehen

overhaul (n.)		TECH	**Überarbeitung**	**overload** (→overdriving)	ELECTRON	(→Übersteuerung)
= refit (n.); rework (n.)			= Überholung	**overload capability**	ELECTRON	**Übersteuerbarkeit**
overhaul (v.t.) (→rework)		TECH	(→überarbeiten)	= overload tolerance; overdriving capability		
overhead (→above ground)		TECH	(→oberirdisch)	**overload coupling**	MECH	**Überlastkupplung**
overhead bit		TRANS	**Zusatzbit**	**overload level**	ELECTRON	**Überlastungsgrenze**
overhead cable grid (→planar cable grid)		SYS.INST	(→Flächenkabelrost)	= overload margin; burn-out level		
overhead capacity		TRANS	**Zusatzkapazität**	**overload margin** (→overload level)	ELECTRON	(→Überlastungsgrenze)
= auxiliary capacity						
overhead channel		TRANS	**Zusatzkanal**	**overload point**	ELECTRON	**Aussteuerungsgrenze**
overhead costs (→indirect costs)		ECON	(→Gemeinkosten)	= load capacity		≈ Aussteuerungsbereich
overhead instruction		DATA PROC	**Organisationsbefehl**	**overload protection**	EL.TECH	**Überlastschutz**
(→housekeeping instruction)				↓ overcurrent protection; overvoltage protection		↓ Überstromschutz; Überspannungsschutz
overhead line		POWER ENG	**Fahrleitung**	**overload protection device**	EL.TECH	**Überlastschutzorgan**
= contact wire; traction line; caternary				**overload protection switch**	COMPON	**Überstromschutzschalter**
overhead line (→open-wire line)		OUTS.PLANT	(→Freileitung)	(→overcurrent protection switch)		
overhead operation		DATA PROC	**Organisationsablauf**	**overload relay**	COMPON	**Überstromrelais**
(→housekeeping sequence)				**overload-resistant**	ELECTRON	**übersteuerungsfest**
overhead power line		POWER ENG	**Starkstrom-Freileitung**	(→overdrive-resistant)		
(→transmission line)				**overload spectrum**	ELECTRON	**Überlastspektrum**
overhead projector		OFFICE	**Folienprojektor**	**overload tolerance**	ELECTRON	(→Übersteuerbarkeit)
			= Tageslichtprojektor	(→overload capability)		
overhead transparency		OFFICE	(→Transparentfolie)	**overmatch** (→overmatching)	NETW.TH	(→Spannungsanpassung)
(→transparency)				**overmatching**	NETW.TH	**Spannungsanpassung**
overheating		TECH	**Überhitzung**	[load resistance very larger than intrinsic resistance]		[Außenwiderstand viel größer als Innenwiderstand]
over-land path		RADIO PROP	**Landstrecke**	= overmatch		= Überanpassung
overlap (v.t.)		TECH	**überlappen**	≠ undermatching		≠ Stromanpassung
≈ intersect; overlay			= teilweise überdecken	↑ matching 1		↑ Anpassung 1
			≈ überschneiden; überlagern	**overmoded**	MICROW	**übermodiert**
overlap (n.)		DATA PROC	**Überlappung**	≠ basic-moded		≠ grundmodiert
[simultaneous execution of sequential operations, mostly related to input/output]			[simultane Ausführung von an sich sequentiellen Vorgängen, meist im Zusammenhang mit Ein- und Ausgabeoperationen]	**overmodulate**	MODUL	**übermodulieren**
						= übersteuern
				overmodulated	MODUL	**übermoduliert**
= pipelining						= übersteuert
overlap (n.)		POWER SYS	**Überlappung**	**overmodulation**	MODUL	**Übermodulation**
[line commutated converter]			[netzgeführte Stromrichter]			= Übersteuerung
overlap (→overlapping)		TECH	(→Überlappung)	**overmodulation capability**	MODUL	**Übermodulierbarkeit**
overlap distortion (→aliasing)		MODUL	(→Rückfaltung)			= Übersteuerbarkeit
overlap in time		TECH	**zeitlich überlappen**	**overnight rate** (→nocturnal tariff)	TELEC	(→Nachttarif)
overlapping		TECH	**Überlappung**	**overpayment**	ECON	**Überzahlung**
= overlap; lap 2			= Überfaltung; Überdeckung	**overpower** (→overdriving)	ELECTRON	(→Übersteuerung)
≈ intersection			≈ Überschneidung	**overpressure**	PHYS	**Überdruck**
overlapping		MODUL	**Überlappung**	≈ high pressure		≈ Hochdruck
			= Überfaltung	**overpressure valve**	TECH	**Überdruckventil**
overlapping		COMPON	**Überwicklung**	= relief valve		
[of turns]				**overprint** (v.t.)	TERM&PER	**überdrucken**
overlapping (→overlap)		POWER SYS	(→Überlappung)			= überschreiben
overlapping (→window overlapping)		DATA PROC	(→Fensterüberlappung)	**overprint** (→overwrite)	DATA PROC	(→überschreiben)
				overprinted character	TERM&PER	**Doppelzeichen**
overlap time		POWER SYS	**Überlappungszeit**	**overprinting**	TERM&PER	**Doppeldruckverfahren**
[line commutated converter]			[netzgeführte Stromrichter]			= Doppeldruck
overlay (v.t.)		TECH	**überdecken**	**overprinting** (→overwriting)	DATA PROC	(→Überschreiben)
≈ cover (v.t.); superimpose; overlap			≈ bedecken; überlagern; überlappen	**overprint lock** (→overwrite lock)	DATA PROC	(→Überschreibsperre)
overlay (n.)		DATA PROC	**Überlagerung**	**overpunch**	TERM&PER	**Überlochung**
= flash (n.)			= Overlay; Einblendung	= overpunching		
≈ swapping			≈ Swapping	**overpunching** (→overpunch)	TERM&PER	(→Überlochung)
overlay		INSTR	**Transparentschablone**	**overradiation**	ANT	**Vorbeistrahlung**
overlay (→paging)		DATA PROC	(→Seitenwechsel)	= spillover (n.)		= Überstrahlung
overlay network		TELEC	**Überlagerungsnetz**	**overrange** (n.) (→transgression)	TECH	(→Bereichsüberschreitung)
			= Overlay-Netz	**overranging**	INSTR	**Bereichsüberschreitung**
overlay segment		DATA PROC	**Überlagerungssegment**	**overreach**	RADIO PROP	**Überreichweite**
[program segment loaded into the main memory only on demand]			[Programmsegment das nur bedarfsweise in den Hauptspeicher geladen wird]	**overreach interference**	RADIO REL	**Fernstörung**
						= Überreichweitenstörung
↑ program segment			↑ Programmsegment	**override** (v.t.)	TELEPH	**aufschalten**
overlay transistor		MICROEL	**Overlay-Transistor**	= intrude; enter; break-in		
overline (v.t.)		LING	**überstreichen**	**override** (v.t.)	TECH	**lahmlegen**
≠ underline			≠ unterstreichen	= annul (v.t.); nullify		= außerkraftsetzen; unwirksam machen; ausschalten
overload (v.t.)		TECH	**überlasten**			
= overcharge; overburden			= überladen	**override** (v.t.)	ELECTRON	**von außen eingreifen**
≈ overstress			≈ überbeanspruchen	**override** (n.)	TELEPH	**Aufschalten**
overload (n.)		TECH	**Überlastung**	[authorized station can break-in into a busy station and request a priority call]		[Möglichkeit sich auf ein Gespräch aufzuschalten und einen Gesprächswunsch zu äußern]
= overcharge (n.)			= Überlast; Überbelastung; Überladung			
≈ overstress (n.); congestion				= executive override; offering; intrusion; break-in		
			≈ Überbeanspruchung			≈ Anklopfen
overload (→overdrive)		ELECTRON	(→übersteuern)	≈ call waiting		

override security	TELEPH	**Aufschalteschutz**
overrun (→overrun error)	DATA PROC	(→Überlauffehler)
overrun (v.t.) (→transgres)	TECH	(→überschreiten)
overrun (n.) (→transgression)	TECH	(→Überschreitung)
overrun (n.) (→overflow)	ELECTRON	(→Überlauf)
overrun error	DATA PROC	**Überlauffehler**
[loss of data because the receiver cannot follow the transmission rate]		[Verlust von Daten weil der Empfänger der Senedegeschwindigkeit nicht folgen kann]
= overrun		
≈ data loss		≈ Datenverlust
overrunning (→free-running)	TECH	(→freilaufend)
overrunning clutch	MECH	**Freilaufkupplung**
oversampling	CODING	**Überabtastung**
overseas	ECON	**Übersee**
overseas market	ECON	**Überseemarkt**
oversheath	COMM.CABLE	**Schutzhülle**
overshoot (v.i.)	ELECTRON	**überschwingen**
overshoot (n.)	ELECTRON	**Überschwingung**
↑ transient		↑ Übergangsvorgang; Überschwinger
overshoot 2 (→overflow traffic)	SWITCH	(→Überlaufverkehr)
overshoot frequency (→ringing frequency)	ELECTRON	(→Überschwingfrequenz)
overshoot traffic (→overflow traffic)	SWITCH	(→Überlaufverkehr)
oversize (adj.)	TECH	**übermäßig**
= excessive		= übergroß
oversize (v.t.) (→overengineer)	TECH	(→überdimensionieren)
over-sizing (→over-engineering)	TECH	(→Überdimensionierung)
overspill (v.t.)	TECH	**überlaufen**
= flow over (v.i.)		= überquellen
overstep (→transgres)	TECH	(→überschreiten)
overstepping (→transgression)	TECH	(→Überschreitung)
overstraining	COLLOQ	**Überforderung**
overstress (v.t.)	TECH	**überbeanspruchen**
≈ overload		= strapazieren
		= überlasten
overstress (n.)	TECH	**Überbeanspruchung**
≈ overload		≈ Überlastung
overstress (→overdriving)	ELECTRON	(→Übersteuerung)
overstrike (→overwrite)	DATA PROC	(→überschreiben)
overstriking (→overwriting)	DATA PROC	(→Überschreiben)
overstriking (→multiple-pass printing)	TERM&PER	(→Mehrfachdruck)
overtemperature (→excess temperature)	TECH	(→Übertemperatur)
overtemperature protection	TECH	**Übertemperaturschutz**
overtime	ECON	**Überstunde**
overtime charge	TELEC	**Gebühr für überschießende Minute**
overtone (→harmonic)	ACOUS	(→Oberschwingung)
overtone crystal	COMPON	**Oberschwingungsquarz**
= harmonic mode crystal		= Oberwellenschwinger; Obertonquarz
overtype mode	DATA PROC	**Überschreibmodus**
[permits substitution of characters by new ones]		[erlaubt Ersatz vorhandener Zeichen durch neue]
≠ insert mode		≠ Einfügemodus
↑ entry mode		↑ Eingabemodus
overview (v.t.)	COLLOQ	**einen Überblick geben**
= to give a bird's eye view		= überblicken
overview (n.)	LING	**Übersicht**
= survey (n.); exhibit (n.)		= Überblick
≈ summary; outline		≈ Zusammenfassung; Umriß
overvoltage	EL.TECH	**Überspannung**
= excess voltage		≠ Unterspannung
≠ undervoltage		
overvoltage arrester (→overvoltage protector)	COMPON	(→Spannungssicherung)
overvoltage crowbar (→crowbar)	POWER SYS	(→Eingangskurzschluß)
overvoltage cut-out (→overvoltage protector)	COMPON	(→Spannungssicherung)
overvoltage limiter (→overvoltage protector)	COMPON	(→Spannungssicherung)
overvoltage protection	EL.TECH	**Überspannungsschutz**
= surge protection		↑ Überlastschutz
↑ overload protection		↓ Blitzschutz; Eingangskurzschluß [POWER SYS]
↓ lightning protection; crowbar [POWER SYS]		
overvoltage protector	COMPON	**Spannungssicherung**
[protects against unadmissibly high voltages]		[schützt vor unzulässig hohen Spannungen]
= lightning protector; overvoltage arrester; surge arrester; lightning arrester; arrester; arrestor; surge-voltage protector; excess-voltage protector; excess-voltage arrester; excess-voltage cut-out; overvoltage cut-out; overvoltage limiter; surge diverter		= Überspannungsableiter; Spannungsableiter; Blitzableiter; Ableiter; Überspannungsbegrenzer
↑ protector		↑ Sicherung
↓ high overvoltage protector; fine overvoltage protector		↓ Spannungsgrobsicherung; Spannungsfeinsicherung
overvoltage relay	COMPON	**Überspannungsrelais**
over-water path	RADIO PROP	**Überwasserstrecke**
		= Seestrecke
overwrite	DATA PROC	**überschreiben**
= overprint; overstrike		↑ schreiben
↑ write		
overwrite lock	DATA PROC	**Überschreibsperre**
= overprint lock		
overwriting	DATA PROC	**Überschreiben**
[to substitute stored data by others]		[gespeicherte Daten durch andere ersetzen]
= overprinting; overstriking		
overwriting (→multiple-pass printing)	TERM&PER	(→Mehrfachdruck)
ovevoltage factor	CIRC.ENG	**Überspannungsfaktor**
[load rating of componenets]		[Sicherheitsfaktor zur Dimensionierung von Bauteilen]
ovonic device	MICROEL	**Ovonik-Bauelement**
ovonik memory	MICROEL	**Ovonik-Speicher**
OWF (→optimum working frequency)	HF	(→günstigste Verkehrsfrequenz)
OWG lightguide (→optical waveguide)	OPT.COMM	(→Lichtwellenleiter)
own consumption	ECON	**Selbstverbrauch**
owner	ECON	**Eigentümer**
= proprietor		
ownership (→propriety)	ECON	(→Eigentum)
own-exchange traffic (→intra-exchange traffic)	SWITCH	(→amtsinterner Verkehr)
oxidate (→oxidize)	CHEM	(→oxydieren)
oxidation	CHEM	**Oxydation**
oxide	CHEM	**Oxyd**
oxide cathode	ELECTRON	**Oxydkathode**
= oxide coated cathode		
oxide coated cathode (→oxide cathode)	ELECTRON	(→Oxydkathode)
oxide film	CHEM	**Oxydschicht**
oxide isolation	MICROEL	**Oxydisolation**
oxide masking	MICROEL	**Oxydmaskierung**
oxide wall	MICROEL	**Oxydwall**
oxidize	CHEM	**oxydieren**
= oxidate		
oxygen	CHEM	**Sauerstoff**
= O		= Oxygenium; O
oxygen cutter	METAL	**Schneidbrenner**
oxygen cutting	METAL	**Schneidbrennen**
		= Brennschneiden
oxyhydrogen voltmeter	INSTR	**Knallgasvoltameter**
ozalid print	DOC	**Oxalidpause**
ozone	CHEM	**Ozon**
ozonosphere	GEOPHYS	**Ozonosphäre**
		= Ozonschicht

P

P (→phosphorus)	CHEM	(→Phosphor)
p (→point 1)	TYPOGR	(→Punkt)
p (→pond)	PHYS	(→Pond)
P (→Poise)	PHYS	(→Poise)
Pa (→Pascal)	PHYS	(→Pascal)

Term	Domain	Translation
Pa (→protactinium)	CHEM	(→Protactinium)
PABX (→private automatic branch exchange)	SWITCH	(→automatische Nebenstellenanlage)
PABX line (→extension line)	TELEPH	(→Nebenanschlußleitung)
pace making function	COLLOQ	Schrittmacherfunktion
pacing	DATA COMM	Datenflußdosierung = Pacing
pack (v.t.) [to code two decimal digits in one byte] ≠ unpack	DATA PROC	packen [zwei Dezimalziffern in einem Byte darstellen] ≠ entpacken
pack (v.t.)	TECH	verpacken = packen
pack (→compress)	DATA PROC	(→verdichten)
package (→case)	COMPON	(→Gehäuse)
package (→software package)	DATA PROC	(→Software-Paket)
packaged software (→software package)	DATA PROC	(→Software-Paket)
package style = package type	COMPON	Gehäuseform
package type (→package style)	COMPON	(→Gehäuseform)
packaging = assembly	MICROEL	Montage
packaging (→equipping)	EQUIP.ENG	(→Gerätebestückung)
packaging (→mechanical design)	EQUIP.ENG	(→Konstruktion)
packaging (→packing 1)	TECH	(→Verpackung)
packaging density = component density; packing density; density ≈ space requirement [TECH]	EQUIP.ENG	Packungsdichte = Bauelementedichte; Baudichte; Kompaktheit ≈ Raumbedarf [TECH]
packaging density (→recording density)	DATA PROC	(→Speicherdichte)
packaging material (→packing material)	TECH	(→Verpackungsmaterial)
packaging standard (→construction standard)	EQUIP.ENG	(→Baunorm)
packaging structure (→construction practice)	EQUIP.ENG	(→Bauweise)
packaging system (→construction practice)	EQUIP.ENG	(→Bauweise)
packaging technique (→construction practice)	EQUIP.ENG	(→Bauweise)
packed	DATA COMM	gepackt
packet (→data packet)	DATA COMM	(→Datenpaket)
packet assembling/ dissambling = PAD	DATA COMM	Paketieren/Depaketieren = PAD
packet assembling process	DATA COMM	Paketiervorgang
packet assembly	DATA COMM	Paketierung = Paketieren
packet assembly/disassembly facility = PAD facility	DATA COMM	Paketierer/Depaketierer = PAD-Einrichtung
packet data	DATA PROC	gepackte Daten
packet decimal	DATA PROC	gepackte Dezimale
packet disassembly	DATA COMM	Depaketierung
packet format	DATA COMM	Paketformat
packet interleaved	DATA COMM	paketweise verschachtelt
packet length	DATA COMM	Paketlänge
packet level	DATA COMM	Paketebene
packet mode (→packet oriented)	DATA COMM	(→paketorientiert)
packet oriented = packet mode	DATA COMM	paketorientiert
packet retransmission	DATA COMM	Paketwiederholung
packet sequence	DATA COMM	Paketfolge
packet sequence numbering	DATA COMM	Paketnumerierung
packet sequencing	DATA COMM	Paketreihung
packet-switched	SWITCH	paketvermittelt
packet-switched network ↑ store-and-forward network	DATA COMM	Paketvermittlungsnetz = paketvermitteltes Netz ↑ speichervermitteltes Netz
packet-switched public data network = PSPDN	TELEC	paketvermitteltes öffentliches Datennetz
packet switching [store-and-forward switching by sub-packets] ↑ store-and-forwardswitching	SWITCH	Paketvermittlung [Speichervermittlung in Teilpaketen] = Datenpaketvermittlung ↑ Speichervermittlung
packet switching technique	SWITCH	Paketvermittlungstechnik
packet-type identifier	DATA COMM	Pakettypkennzeichen
packetization delay [ATM]	TELEC	Paketierungsverzug [ATM]
packing 1 = packaging; wrapping	TECH	Verpackung = Umhüllung
packing (for moving parts) (→seal)	MECH	(→Dichtung)
packing (→stratification)	TECH	(→Schichtung)
packing density (→packaging density)	EQUIP.ENG	(→Packungsdichte)
packing density (→recording density)	DATA PROC	(→Speicherdichte)
packing engineering	TECH	Verpackungstechnik
packing instruction	TECH	Verpackungsanweisung
packing list = Verpackungsliste	ECON	Packliste
packing material = packaging material ≈ dunnage	TECH	Verpackungsmaterial = Packmaterial ≈ Füllmaterial
packing specification	TECH	Verpackungsvorschrift
pad 2	MICROEL	Füller
pad ↓ transistor pad	COMPON	Isolierscheibe
pad	TECH	Polster
pad 3 [base of a chip]	MICROEL	Unterbau [Chip]
pad 1 (→terminal pad)	MICROEL	(→Anschlußfleck)
pad (→pad character)	DATA PROC	(→Ausfüllzeichen)
pad (→attenuator)	NETW.TH	(→Dämpfungsglied)
pad (→soldering land)	ELECTRON	(→Lötauge)
pad (→note pad)	OFFICE	(→Notizblock)
PAD (→packet assembling/dissambling)	DATA COMM	(→Paketieren/Depaketieren)
pad (v.t.) (→pulse stuff)	CODING	(→stopfen)
pad character = pad	DATA PROC	Ausfüllzeichen = Auffüllzeichen
pad character (→filler)	DATA COMM	(→Füllzeichen)
PAD clearing	DATA COMM	PAD-Auslösung
PAD command	DATA COMM	PAD-Befehl
padding	DATA PROC	Auffüllen
padding bit (→stuffing bit)	CODING	(→Stopfbit)
padding character (→filler)	DATA PROC	(→Füllzeichen)
padding signal (→filler)	DATA COMM	(→Füllzeichen)
paddle [peripheral for computer games, with two knobs] = game paddle	TERM&PER	Drehregler [Bediengerät für Computerspiele; mit zwei Knöpfen] = Paddle
PAD facility (→packet assembly/disassembly facility)	DATA COMM	(→Paketierer/Depaketierer)
PAD identification	DATA COMM	PAD-Kennung
pad master [PCB]	MANUF	Lötaugenschablone [Leiterplatten] = Lötaugenvorlage
PAD message	DATA COMM	PAD-Mitteilung
pad pattern (→land pattern)	ELECTRON	(→Lötaugenbild)
pad recall	DATA COMM	PAD-Rückruf
pad signal (→filler)	DATA COMM	(→Füllzeichen)
PAD technique (→alloy diffusion technique)	MICROEL	(→AD-Technik)
PAD transistor = post-alloy diffused transistor	MICROEL	PAD-Transistor
page [one of the faces of a sheet] ≈ sheet	TYPOGR	Seite [eine der beiden Flächen eines Blattes] ≈ Blatt
page (→page frame)	DATA PROC	(→Seite)
page assembly (→page makeup)	TYPOGR	(→Seitenumbruch)
page-at-a-time printer (→page printer 2)	TERM&PER	(→Seitendrucker)
page break (→page makeup)	TYPOGR	(→Seitenumbruch)

page change	TERM&PER	**Seitenvorschub**	
= page skip		= Seitenwechsel; Umblättern	
≈ paging			
page composition (→page makeup)	TYPOGR	(→Seitenumbruch)	
page description language	DATA PROC	**Seitenbeschreibungssprache**	
PAGE-DOWN key	TERM&PER	**BILD-ABWÄRTS-Taste**	
		= PAGE-DOWN-Taste	
page fault	DATA PROC	**Seitenfehler**	
pagefax (→pressfax)	TELEGR	(→Pressefax)	
page foot	TYPOGR	**Seitenfuß**	
page format	TYPOGR	**Seitenformat**	
≈ line format		≈ Zeilenformat	
page formatting	DATA PROC	**Seitenformatierung**	
page frame	DATA PROC	**Seite**	
[subdivision of main memory or program]		[Unterteilung des Hauptspeichers oder eines Programms]	
= frame; page		= Kachel; Rahmen	
page header	TYPOGR	**Seitenkopf**	
= page heading		= Seitenüberschrift	
≈ catchword			
page heading (→page header)	TYPOGR	(→Seitenkopf)	
page home line	TERM&PER	**Seitenanfang**	
PAGE key	TERM&PER	**Bild-Taste**	
page makeup	TYPOGR	**Seitenumbruch**	
[way of distribution on a page]		[Art der Verteilung auf eine Druckseite]	
= page composition; page break; page assembly; area composition [DATA PROC]		= Seitenmontage	
↑ makeup (n.)		↑ Umbruch	
page montage program	DATA PROC	**Seitenmontageprogramm**	
page number	TYPOGR	**Seitenzahl 2**	
= pagination 2; folio		[eine Seite kennzeichnende Zahl]	
page numbering	TYPOGR	**Seitennumerierung**	
= pagination 1		= Paginierung	
page printer 2	TERM&PER	**Seitendrucker**	
[prints a page at once]		[druckt eine Seite auf einmal]	
= page-at-a-time printer			
≠ line printer; character printer		≠ Zeilendrucker; Zeichendrucker	
page printer 1	TERM&PER	**Blattschreiber**	
[prints on pages]		[druckt auf Blättern]	
= page teleprinter; console typewriter		= Blattdrucker; Blattfernschreiber	
page reader	TERM&PER	**Blattleser**	
= page scanner		= Seitenleser	
↑ character reader		↑ Klarschriftleser	
page scanner (→page reader)	TERM&PER	(→Blattleser)	
page scrolling (→paging)	TERM&PER	(→Bildschirmblättern)	
page skip (→page change)	TERM&PER	(→Seitenvorschub)	
pages per hour	TERM&PER	**Seiten pro Stunden**	
= pph			
page teleprinter (→page printer 1)	TERM&PER	(→Blattschreiber)	
page terminator	TERM&PER	**Seitenende**	
page turning (→paging)	TERM&PER	(→Bildschirmblättern)	
PAGE-UP key	TERM&PER	**BILD-AUFWÄRTS-Taste**	
		= PAGE-UP-Taste	
pagination 1 (→page numbering)	TYPOGR	(→Seitennumerierung)	
pagination 2 (→page number)	TYPOGR	(→Seitenzahl 2)	
paging	TERM&PER	**Bildschirmblättern**	
[to replace full pages or display contents]		[Wechseln ganzer Seiten oder Bildschirminhalte]	
= page turning; page scrolling		= Blättern	
≈ scrolling		≈ Bildverschiebung	
paging	DATA PROC	**Seitenwechsel**	
[alternation of programs between storages]		[Hin- und Herladen von Programmen zwischen Speichern]	
= swapping; overlay		= Seitenüberlagerung; Überlagerung; Paging	
paging (→radio paging)	MOB.COMM	(→Funkruf)	
paging system	TELEC	**Personenrufanlage**	
paint (n.)	TECH	**Anstrich**	
= paint coat		= Anstrichfarbe; Farbüberzug	
≈ coating		≈ Beschichtung; Schutzschicht	
paint (→varnish color)	TECH	(→Lackfarbe)	
paintbrush (→electronic paintbrush)	DATA PROC	(→elektronischer Farbpinsel)	
paint coat (→paint)	TECH	(→Anstrich)	
pain threshold	ACOUS	**Schmerzschwelle**	
paint-on process	MICROEL	**Film-Verfahren**	
pair (→wire pair)	COMM.CABLE	(→Aderpaar)	
pair annihilation	PHYS	**Paarvernichtung**	
pair binding	PHYS	**Paarbindung**	
pair cable (→paired cable)	COMM.CABLE	(→Paarkabel)	
pair creation (→pair production)	PHYS	(→Paarbildung)	
paired	COMM.CABLE	**paarverseilt**	
paired cable	COMM.CABLE	**Paarkabel**	
= pair cable		= paarverseiltes Kabel; paarig verseiltes Kabel; Doppelladerkabel	
pair formation (→pair production)	PHYS	(→Paarbildung)	
pair forming coefficient	PHYS	**Paarbildungskoeffizient**	
pair gain system (→subscriber loop carrier)	TRANS	(→Teilnehmermultiplexsystem)	
pair generation	DATA PROC	**Paarbildung**	
= pairing			
pair generation (→pair production)	PHYS	(→Paarbildung)	
pairing	COMM.CABLE	**Paarverseilung**	
= twisting by pairs			
pairing (→pair production)	PHYS	(→Paarbildung)	
pairing (→pair generation)	DATA PROC	(→Paarbildung)	
pairing machine	COMM.CABLE	**Paarverseilmaschine**	
↑ stranding machine		↑ Verseilmaschine	
pair of compasses (→compasses)	ENG.DRAW	(→Zirkel)	
pair production	PHYS	**Paarbildung**	
= pair creation; pair formation; pair generation; pairing		= Paarerzeugung; Paargenerierung	
pair resistor	COMPON	**gepaarter Widerstand**	
[selected to minimum difference]		[auf Minimalabweichung ausgesucht]	
pair-saving device (→front-end equipment)	TELEC	(→Vorfeldeinrichtung)	
paket-mode operation	DATA COMM	**Paketbetrieb**	
		= Paketmodus	
PAL	TV	**PAL**	
[phase alternation line]			
PAL (→programmable array logic)	MICROEL	(→PAL)	
palatal sound	LING	**Palatal**	
		= Gaumenlaut	
pale brown	PHYS	**hellbraun**	
= light brown			
pale green	PHYS	**hellgrün**	
= light green			
pale grey	PHYS	**hellgrau**	
pale red	PHYS	**hellrot**	
= light red		≈ rosa	
pale yellow	PHYS	**hellgelb**	
palladium	CHEM	**Palladium**	
= Pd		= Pd	
palladium plate	METAL	**palladinieren**	
palladium plating	METAL	**Palladinierung**	
palmtop (→slate PC)	DATA PROC	(→Slate-PC)	
palmtop PC (→slate PC)	DATA PROC	(→Slate-PC)	
PAL sheath	COMM.CABLE	**Schichtenmantel**	
PAL-sheath cable	COMM.CABLE	**Schichtenmantelkabel**	
= moisture-barrier cable			
PAM (→pulse amplitude modulation)	MODUL	(→Pulsamplitudenmodulation)	
pamphlet (→publication)	DOC	(→Druckschrift)	
pancake coil (→flat coil)	COMPON	(→Flachspule)	
pancake loudspeaker (→flat loudspeaker)	EL.ACOUS	(→Flachlautsprecher)	
panel (→slide-in unit)	EQUIP.ENG	(→Einschub)	
panel (→console)	EQUIP.ENG	(→Pult)	
panel cabinet (→control cubicle)	POWER SYS	(→Schaltschrank)	
panel door	EQUIP.ENG	**Abdeckklappe**	

panel instrument	INSTR	**Schalttafelinstrument**	
= switchboard instrument			
panel light socket	COMPON	**Einbaufassung**	
		[für Signallampen]	
panel meter (→instrumentation meter)	INSTR	(→Einbauinstrument)	
panel model	TERM&PER	**Einbaumodell**	
panel plug (→mounting plug)	COMPON	(→Einbaustecker)	
panel socket (BRI) (→mounting jack)	COMPON	(→Einbaubuchse)	
pan head [thread]	MECH	**Zylinderkopf** [Schraube] = Flachkegelkopf	
pan-head rivet	MECH	**Flachkegelkopfniet**	
pan-head screw (AM) = cheese-head screw (BRI)	MECH	**Zylinderkopfschraube** = Zylinderschraube	
panic switch (→emergency switch)	EL.INST	(→Notausschalter)	
panning	TV	**Kameraschwenk** = Schwenk	
panning [horizontal displacement on a display]	DATA PROC	**Schwenk** [auf Bildschirm] = Schwenken	
panora adapter (→panoramic adapter)	INSTR	(→Panorama-Adapter)	
panoramic adapter = panora adapter	INSTR	**Panorama-Adapter**	
panoramic display	SIGN.ENG	**Panoramatafel**	
panoramic radar = surveillance radar; search radar	RADIO LOC	**Panoramaradar** = Rundsichtradar; Suchradar; Aufklärungsradar	
paper	TECH	**Papier**	
paper advance (→paper feed)	TERM&PER	(→Papiervorschub)	
paper batch (→paper stack)	TERM&PER	(→Papierstapel)	
paper bill (→paper currency)	ECON	(→Banknote)	
paperboard (→cardboard)	TECH	(→Pappe)	
paper capacitor	COMPON	**Papierkondensator**	
paper clip [bent piece of metallic or plastic wire] = clip; paper fastener; fastener ≈ staple [TECH]	OFFICE	**Büroklammer** [gebogenes Stück aus Draht oder Kunststoff] = Heftklammer 2 ≈ Heftklammer 1 [TECH]	
paper covered (→paper insulated)	COMM.CABLE	(→papierisoliert)	
paper currency = paper money; paper bill; bill	ECON	**Banknote** = Geldschein; Papiergeld	
paper cutting machine (→cutting machine)	TERM&PER	(→Schneidemaschine)	
paper deflector	TERM&PER	**Papierableiter**	
paper drilling machine	OFFICE	**Papierbohrmaschine**	
paper drive (→paper feed)	TERM&PER	(→Papiervorschub)	
paper eyeletting machine	OFFICE	**Papierösmaschine**	
paper fastener (→paper clip)	OFFICE	(→Büroklammer)	
paper feed = paper transport; paper advance; paper drive ≈ paper throw ↓ form feed; document feed; friction feed; tractor feed	TERM&PER	**Papiervorschub** = Papiertransport; Papierzufuhr; Papiereingabe; Papiereinzug ≈ Papierauswurf ↑ Vorschub ↓ Formularzufuhr; Belegzufuhr; Blattzufuhr; Friktionsantrieb; Raupenvorschub	
paper guide	TERM&PER	**Papierführung**	
paper input tray (→single-sheet feed)	TERM&PER	(→Einzelblattzuführung)	
paper insulated = paper covered	COMM.CABLE	**papierisoliert**	
paper insulated cable	COMM.CABLE	**papierisoliertes Kabel** = Papierkabel	
paper mail	POST	**Briefpost**	
paper money (→paper currency)	ECON	(→Banknote)	
paper park function	TERM&PER	**Papierparkfunktion**	
paper pressure rod	TERM&PER	**Papierandrückbügel** = Papierhalter	
paper pressure roll = pressure roll; paper roll 2	TERM&PER	**Papierandrückrolle** = Papierhalterrolle	
paper rewinder	OFFICE	**Papierumroller**	
paper roll 1 = roll	TERM&PER	**Papierrolle** = Rolle	
paper roll 2 (→paper pressure roll)	TERM&PER	(→Papierandrückrolle)	
paper sheet	OFFICE	**Papierbogen**	
paper size	OFFICE	**Papierformat**	
paper slew (AM) (→paper throw)	TERM&PER	(→Papierauswurf)	
paper slip box	OFFICE	**Zettelkasten**	
paper stack = form stack; paper batch	TERM&PER	**Papierstapel**	
paper stacker ↓ form stacker	TERM&PER	**Papierstapler** ↓ Formularstapler	
paper stamping machine (→stamping machine)	OFFICE	(→Stanzmaschine)	
paper supply	TERM&PER	**Papiervorrat**	
paper tape	TECH	**Papierband**	
paper tape punch (→tape punch)	TERM&PER	(→Lochstreifenstanzer)	
paper taper reader (→punched tape reader)	TERM&PER	(→Lochstreifenleser)	
paper tensioner	TERM&PER	**Papierstraffer**	
paper throw = paper slew (AM) ≈ paper feed	TERM&PER	**Papierauswurf** = schneller Papiervorschub ≈ Papiervorschub	
paper tractor = tractor ≈ pinfeed platen	TERM&PER	**Vorschubraupe** = Transportraupe; Formulartraktor ≈ Stachelrad	
paper transport (→paper feed)	TERM&PER	(→Papiervorschub)	
paper tray = tray; input magazine; forms hopper	TERM&PER	**Papiernachfüllmagazin** = Eingabemagazin; Papierschacht; Zuführungsschacht; Schacht; Papierkassette;
paper web	TERM&PER	**Papierbahn**	
paper wrapping	COMM.CABLE	**Papierumwicklung**	
parabeam antenna	ANT	**Parabeam**	
parabola ↑ conical section	MATH	**Parabel** ↑ Kegelschnitt	
parabolic	MATH	**parabolisch** = parabelförmig	
parabolic antenna = parabolic reflector antenna; paraboloidal antenna	ANT	**Parabolantenne** = Rotationsparabolantenne; Parabolreflektorantenne	
parabolic cylinder antenna	ANT	**Zylinderparabolantenne**	
parabolic cylindrical	MATH	**parabolzylindrisch**	
parabolic grid antenna	ANT	**Gitterparabolantenne**	
parabolic mirror	PHYS	**Parabolspiegel**	
parabolic multiplier [analog computer]	DATA PROC	**Parabelmultiplikator** [Analogrechner] = Parabelmultiplizierer	
parabolic reflector = paraboloidal reflector	ANT	**Parabolspiegel** = Parabolreflektor; Rotationsparabolspiegel; Paraboloidalspiegel; Paraboloidalreflektor	
parabolic reflector antenna (→parabolic antenna)	ANT	(→Parabolantenne)	
paraboloid [all intersections are parabolas, parabolas and ellipses, or parabolas and hyperbolas] ↓ paraboloid of revolution; elliptic paraboloid; hyperbolic paraboloid	MATH	**Paraboloid** [alle Schnitte sind Parabeln, Parabeln und Ellipsen oder Parabeln und Hyperbeln] ↓ Rotationsparaboloid; elliptisches Paraboloid; hyperbolisches Paraboloid	
paraboloidal antenna (→parabolic antenna)	ANT	(→Parabolantenne)	
paraboloidal reflector (→parabolic reflector)	ANT	(→Parabolspiegel)	
paraboloid of revolution	MATH	**Rotationsparaboloid** ↑ Paraboloid	
paradigm	SCIE	**Musterbeispiel**	
parafil rope	ANT	**Parafil-Seil**	

paragraph 1 LING
[continuation of a text with a new and offset line]
≈ section; article; paragraph 2

paragraph 2 LING
[a section of a text identified by a section clause and a number]
≈ paragraph 1; section

paragraph indent (→paragraph indentation) TYPOGR

paragraph indentation TYPOGR
= paragraph indention; paragraph indent

paragraph indention TYPOGR
(→paragraph indentation)

paragraph marker DATA PROC
[word processing]

paragraph sign 1 TYPOGR
= paragraph symbol 1; section clause; section mark 1; section sign; pilcrow; numbered clause
≈ paragraph sign 2

paragraph sign 2 TYPOGR
= paragraph symbol 2; section mark 2
≈ paragraph sign 1
↑ punctuation mark

paragraph symbol 1 TYPOGR
(→paragraph sign 1)

paragraph symbol 2 TYPOGR
(→paragraph sign 2)

parallactic PHYS
parallax PHYS
parallax error PHYS
parallax-free PHYS
= anti-parallax

parallel (n.) MATH
parallel (adj.) MATH
≈ codirectional [TECH]
≠ anti-parallel

parallel (adj) TECH
≠ sequential

parallel (→ anti-parallel) MATH
parallel (v.t.) (→connect in parallel) EL.TECH

parallel adder CIRC.ENG
= parallel carry counter

parallel addition DATA PROC
parallel branch NETW.TH
parallel branching TRANS
parallel bus DATA COMM

parallel by character INF.TECH
(→character-parallel)

parallel cable COMM.CABLE
parallel capacitance EL.TECH
= shunt capacitance

parallel carry counter CIRC.ENG
(→parallel adder)

parallel code CODING
parallel computer DATA PROC
(→simultaneous computer)

parallel connection EL.TECH
= paralleling
≈ shunt
≠ series connection

parallel data DATA PROC

parallel display ELECTRON
parallelepiped MATH
[a prism with two parallel parallelograms as ground faces; a solid formed by six parallelograms]
↑ prism; polyhedron
↓ rectangular prism; cube

Absatz
[Fortsetzung eines Textes mit einer abgesetzten neuen Zeile]
≈ Abschnitt; Paragraph

Paragraph
[mit dem Paragraphenzeichen und einer Zahl gekennzeichneter Abschnitt eines Textes]
≈ Absatz; Abschnitt

(→Absatzeinzug)

Absatzeinzug

(→Absatzeinzug)

Absatzmarke
[Textverarbeitung]

Paragraphzeichen
= Paragraphenzeichen
≈ amerikanisches Paragraphenzeichen

amerikanisches Paragraphenzeichen
= Abschnittszeichen
≈ Paragraphenzeichen
↑ Satzzeichen

(→Paragraphenzeichen)

(→amerikanisches Paragraphenzeichen)

parallaktisch
Parallaxe
Parallaxefehler
parallaxfrei

Parallele
parallel
≈ gleichsinnig [TECH]
≠ antiparallel

parallel
= gleichzeitig
≠ sequentiell

(→antiparallel)
(→parallelschalten)

Paralleladdierwerk
= Parallel-Carry-Zähler

Paralleladdition
Parallelzweig
Parallelabzweig
Parallelbus
= paralleler Bus

(→zeichenparallel)

Beilaufkabel
Parallelkapazität

(→Paralleladdierwerk)

Parallelcode
(→Parallelrechner)

Parallelschaltung
≈ Nebenschluß
≠ Reihenschaltung

parallele Daten
= Paralleldaten

Parallelanzeige
Parallelepiped
[ein Prisma mit parallelen Parallelogrammen als Grundflächen; ein von sechs Parallelogrammen gebildeter Körper]

↑ Prisma; Polyeder
↓ Quader; Würfel

parallel equivalent circuit NETW.TH **Parallelersatzschaltung**
parallel inductance EL.TECH **Parallelinduktivität**
= shunt inductance

paralleling (→ parallel connection) EL.TECH (→Parallelschaltung)

parallel-in/parallel-out register DATA PROC **PIPO-Register**
= PIPO register

parallel input/output DATA PROC **parallele Ein-/Ausgabe**
= parallel I/O; PIO

parallel-in/serial-out register DATA PROC **PISO-Register**
= PISO register

parallel interface DATA PROC **parallele Schnittstelle**
parallel interface modem DATA COMM **Parallelschnittstellen-Modem**
parallel inverter POWER SYS **Parallelwechselrichter**
parallel I/O (→parallel input/output) DATA PROC (→parallele Ein-/Ausgabe)

parallelism MATH **Parallelität**
parallel memory (→parallel storage) DATA PROC (→Parallelspeicher)

parallel mode DATA PROC (→Mehrprogrammbetrieb)
(→multiprogramming)

parallelogram MATH **Parallelogramm**
[quadrangle with equal and parallel opposite sides]
↑ quadrangle
↓ rhomboid; rhombus; rectangle; quadrate

[Viereck mit gegenüberliegenden gleichen und parallelen Seiten]
↑ Viereck
↓ Rhomboid; Rhombus; Rechteck; Quadrat

parallel operation RADIO REL **Parallelbetrieb**
parallel operation DATA PROC (→Mehrprogrammbetrieb)
(→multiprogramming)

parallel plates lens ANT **Parallelplattenlinse**
parallel polarization PHYS **parallele Polarisation**
= Parallelpolarisation

parallel printer TERM&PER **Paralleldrucker**
[prints several characters simultaneously]
≠ serial printer
↓ line printer; page printer

[druckt mehrere Zeichen gleichzeitig]
≠ Serialdrucker
↓ Zeilendrucker; Seitendrucker

parallel processing DATA PROC (→Mehrprogrammbetrieb)
(→multiprogramming)

parallel resistance EL.TECH **Parallelwiderstand**
= shunt resistance
≠ cross resistance
= parallelgeschalteter Widerstand
≠ Querwiderstand

parallel resonance NETW.TH **Parallelresonanz**
= current resonance
= Stromresonanz

parallel resonance ANT **Spannungsresonanz**
[feeding in the voltage antinode]
[Anregung im Spannungsbauch]

parallel-resonant circuit NETW.TH **Parallelschwingkreis**
= rejector circuit; rejector; anti-resonant circuit; blocking circuit; wave trap
= Stromresonanzkreis; Parallelresonanzkreis; Sperrkreis

parallel ruler ENG.DRAW **Parallelführung**
parallel run (→parallel running) DATA PROC (→Parallellauf)

parallel running DATA PROC **Parallellauf**
[to process a task with several different modes for security reasons]
= parallel run

[eine Aufgabe aus Sicherheitsgründen mit verschiedenen Verfahren bearbeiten]

parallel-search memory DATA PROC (→Assoziativspeicher)
(→associative storage)

parallel-search storage DATA PROC (→Assoziativspeicher)
(→associative storage)

parallel-serial conversion CODING **Parallel-Seriell-Umsetzung**
parallel-serial transfer DATA COMM **Parallel-Serien-Übertragung**
parallel-series connection NETW.TH **Parallel-/Reihenschaltung**
↑ Vierpolzusammenschaltung
parallel-series matrix NETW.TH **Parallel-/Reihenmatrix**
parallel stabilization voltage POWER SYS **Parallelstabilisierung**
parallel stabilizing circuit CIRC.ENG **Parallelstabilisierungsschaltung**

parallel station TELEPH **Parallelapparat**

English	Domain	German
parallel storage	DATA PROC	Parallelspeicher
[access time not depending on order of storage]		≠ sequentieller Speicher
= parallel memory		
≠ sequential access memory		
parallel-to-serial converter	CIRC.ENG	Parallel-Seriell-Umsetzer
= serializer; dynamicizer		= Parallel-Seriell-Wandler; Parallel-Serien-Wandler; Parallel-Serien-Umsetzer
parallel-to-serial register	DATA PROC	Parallel-Serien-Register
parallel transfer (→parallel transmission)	TELEC	(→Parallelübertragung)
parallel transmission	TELEC	Parallelübertragung
[all code elements of a character are transmitted simultaneously]		[alle Codeelemente eines Zeichens werden gleichzeitig übertragen]
= parallel transfer		= Parallelübergabe; parallele Übertragung
≠ serial transmission		≠ Serienübertragung
parallel winding	COMPON	Parallelwicklung
parallel winding (→shunt winding)	POWER ENG	(→Nebenschlußwicklung)
parallel-wire line (→two-wire line)	LINE TH	(→Lecher-Leitung)
parallel working (→multiprogramming)	DATA PROC	(→Mehrprogrammbetrieb)
paralysis circuit (→blocking circuit)	CIRC.ENG	(→Sperrschaltung)
paralysis time (→blocking time)	ELECTRON	(→Sperrzeit)
paramagnetic	PHYS	paramagnetisch
paramagnetism	PHYS	Paramagnetismus
[weak magnetic induction, with equal sense of inducing field and vanishing with him]		[schwache magnetische Induktion, gleichsinnig mit induzierendem Feld, erlischt bei dessen Wegfall]
parameter	MATH	Parameter
parameter	DATA PROC	Parameter
[variable assuming properties of constant for a specific task]		[aufgabenspezifisch festgeschriebene Variable]
≈ variable; argument		≈ Variable; Argument
parameter area	DATA PROC	Parameterbereich
parameter block	DATA PROC	Parameterblock
		= Parametersatz
parameter card (→control card)	DATA PROC	(→Steuerlochkarte)
parameter delimiter	DATA PROC	Parameterbegrenzer
parameter hand-off	DATA PROC	Parameterübergabe
parameter identification (→identification)	CONTROL	(→Identifikation)
parameter listing	DATA PROC	Parameterprotokollierung
= parameter logging		
parameter logging (→parameter listing)	DATA PROC	(→Parameterprotokollierung)
parameter setting	DATA PROC	Parametrierung
parameter string	DATA PROC	Parameterfolge
parameter sub-string	DATA PROC	Parameterteilfolge
parameter test	MICROEL	Parametertest
parameter transformation	MATH	Parametertransformation
parameter variation mode	DATA PROC	Variatenprinzip
[computer graphics]		[Computergraphik]
parameter word	DATA PROC	Parameterwort
parameterizable	DATA PROC	parametrierbar
parametric	ELECTRON	parametrisch
parametric (→parametric subroutine)	DATA PROC	(→parametrisches Unterprogramm)
parametric amplifier	CIRC.ENG	parametrischer Verstärker
= reactance amplifier; MAVAR		= Reaktionsverstärker; Reaktanzverstärker
parametric circuit	CIRC.ENG	parametrischer Schaltkreis
parametric converter	ELECTRON	parametrischer Umsetzer
parametric diode	MICROEL	parametrische Diode
parametric oscillator	ELECTRON	parametrischer Oszillator
parametric programming	DATA PROC	parametrisches Programmieren
parametric resonance	PHYS	parametrische Resonanz
parametric subroutine	DATA PROC	parametrisches Unterprogramm
= parametric		= Parameterunterprogramm
parametrization	DATA PROC	Parametersetzung
[definition of parameters]		= Parametrisierung
paramid (→pyramidal circuit)	ENG.LOG	(→Pyramidenschaltung)
paraphase amplifier (→push-pull amplifier)	CIRC.ENG	(→Gegentaktverstärker)
parasitic	EL.TECH	parasitär
parasitic (→spurious)	INF.TECH	(→unerwünscht)
parasitically excited	ANT	strahlungsgekoppelt
parasitic antenna (→reflector 1)	ANT	(→Reflektor 1)
parasitic capacitance	EL.TECH	Parasitärkapazität
		= parasitäre Kapazität
parasitic current (→interfering current)	EL.TECH	(→Störstrom)
parasitic deflection	INSTR	Störausschlag
parasitic effect (→spurious oscillation)	PHYS	(→Nebenschwingung)
parasitic element	ANT	Parasitärelement
≠ fed element		= Parasitärstrahler; parasitäres Element; parasitär erregtes Element; passiver Strahler
↓ director element; reflector element		≠ gespeistes Element
		↓ Direktor; Reflektor
parasitic multiband element	ANT	parasitäres Mehrbandelement
parasitic oscillation (→spurious oscillation)	PHYS	(→Nebenschwingung)
parasitic signal (→interfering signal)	ELECTRON	(→Störsignal)
parasitic transistor (→substrate transistor)	MICROEL	(→Substrat-Transistor)
paraxial	PHYS	paraxial
parcel	POST	Postpaket
		= Paket
parcel (n.) (→lot 2)	ECON	(→Posten 1)
parcel service	POST	Paketdienst
PARD (→ac component)	EL.TECH	(→Wechselstromkomponente)
parent (→parent record)	DATA PROC	(→Stammdatensatz)
parent company	ECON	Stammhaus
= parent enterprise		= Muttergesellschaft; Stammfirma
≈ head office; holding company		≈ Zentralverwaltung; Holdinggesellschaft
≠ subsidiary		≠ Tochtergesellschaft
parent enterprise (→parent company)	ECON	(→Stammhaus)
parent exchange	SWITCH	Mutteramt
= master exchange; parent office; master office		= Hauptamt
parenthesis (pl.-es) [symbol ( )]	MATH	runde Klammer [Symbol ( )]
= curves; brackets 3 (n.)		↑ Klammer
↑ bracket		
parent office (→parent exchange)	SWITCH	(→Mutteramt)
parent population	MATH	Grundgesamtheit
parent record	DATA PROC	Stammdatensatz
= parent		≠ abgeleiteter Datensatz
≠ child record		
parity	CODING	Parität
[even number of binary ones]		[Geradzahligkeit binärer Einser]
= even parity		= gerade Parität
≠ unparity		≠ Imparität
parity bit	CODING	Paritätsbit
= P bit		= Paritätsschritt; Paritybit
↑ check bit		↑ Prüfbit
parity check	CODING	Paritätskontrolle
= parity control; parity checking		= Paritätsprüfung; Paritätsüberwachung; Vergleichskontrolle
≠ unparity check		≠ Imparitätskontrolle
↓ horizontal parity check; vertical parity check		↓ Horizontal-Paritätskontrolle; Vertikal-Paritätskontrolle
parity checking (→parity check)	CODING	(→Paritätskontrolle)
parity conservation	CODING	Paritätserhaltung

parity control (→parity check) CODING (→Paritätskontrolle)
parity error CODING **Paritätsfehler**
parity generator CODING **Paritätsgenerator**
parity network CIRC.ENG **Paritätsnetz**
parity track TERM&PER **Paritätsbitspur**
= vertical parity bit track = Querparitätsbit-Spur
parity violation CODING **Paritätsverletzung**
= bit parity violation
parking (magnetic head) TERM&PER **Parken** [Magnetkopf]
parking (→call parking) TELEPH (→Parken)
parking lot CIV.ENG **Parkplatz** [AM]
= parking place
parking orbit SAT.COMM **Parkumlaufbahn**
parking place (→parking lot) CIV.ENG (→Parkplatz)
parking track TERM&PER **Parkspur**
parlance LING **Fachjargon**
= jargon ≈ Fachsprache
≈ technical language
parralel-fed (→shunt-fed) ANT (→nebenschlußgespeist)
parser DATA PROC **Analysealgorithmus**
[a routine to analyze syntax] [Unterprogramm zur syntaktischen Analyse]
 = lexikalischer Analysator; Analysierer; Parser
Parseval's relation TELEC **Theorem von Parseval**
 = Parsevalsches Theorem
parsing 1 DATA PROC **syntaktische Analyse**
[to dismember statements into syntactic units] [Anweisungen in syntaktische Bestandteile auflösen]
 = Parsing 1
parsing 2 DATA PROC **lexikalische Analyse**
[to brak down character strings into processable units] [eine Zeichenfolge in leichter verarbeitbare Teile zerlegen]
 = Parsing 2
part (n.) TECH **Teil**
≈ portion; member; constituent ≈ Anteil; Glied; Bestandteil
part (→piece) ECON (→Stück)
part (→divide) TECH (→teilen)
part (n.) (→segment) SCIE (→Segment)
part amount ECON **Teilbetrag**
≈ tranche ≈ Tranche
part failure rate QUAL **Teilausfallrate**
partial (n.) ACOUS **Teilton**
partial acceptance ECON **Teilabnahme**
partial carry MATH **Teilzehnerübertrag**
= partial transfer = Teilüberhang
partial conjunction ENG.LOG **partielle Konjunktion**
partial consignment (→partial delivery) ECON (→Teillieferung)
partial cover (→cover shortage) ECON (→Unterdeckung)
partial current EL.TECH **Teilstrom**
partial damage ECON **Teilschaden**
≠ total damage ≠ Totalschaden
partial delivery ECON **Teillieferung**
= partial consignment = Teilsendung; Teilübergabe
≈ advance delivery ≈ Vorauslieferung
partial differential equation MATH **partielle Differentialgleichung**
partial dispersion PHYS **Partialdispersion**
 = partielle Dispersion
partial equipping TECH **Teilbestückung**
≈ underequipping ≈ Unterbestückung
partial failure QUAL **Änderungsausfall**
 = Parameterausfall; Teilausfall
partial fraction MATH **Partialbruch**
 = Teilungsbruch
partial fraction arrangement NETW.TH **Partialbruchschaltung**
partial line DATA PROC **Teilzeile**
partially allocated SWITCH **teilbeschaltet**
partially common trunk SWITCH **teilweise gemeinsame Abnehmerleitung**
partially equipped TECH **teilbestückt**
≈ partially expanded; underequipped ≈ teilaugebaut; unterbestückt
partially expanded EQUIP.ENG **teilausgebaut**
≈ partially equipped ≈ teilbestückt
partially perforated tape (→chadless tape) TERM&PER (→Schuppenlochstreifen)
partial multiple (→graded multiple) SWITCH (→Teilvielfachfeld)
partial oscillation PHYS **Partialschwingung**
≈ suboscillation = Teilschwingung
partial power supply POWER SYS **Teilnetzstromversorgung**
partial pressure PHYS **Partialdruck**
partial problem (→subproblem) SCIE (→Teilproblem)
partial removal NETW.TH **Teilabbau**
partial response CIRC.ENG **Teilerregung**
 = Partial-response
partial-response code CODING **Partial-response-Code**
partial set (→subset) MATH (→Teilmenge)
partial sum MATH **Partialsumme**
 = Teilsumme
partial tension (→partial voltage) EL.TECH (→Teilspannung)
partial transfer (→partial carry) MATH (→Teilzehnerübertrag)
partial vacuum PHYS **Vorvakuum**
partial voltage EL.TECH **Teilspannung**
= partial tension
partial wave PHYS **Teilwelle**
≈ subwave = Partialwelle
partial word DATA PROC (→Teilwort)
(→partword)
participation COLLOQ **Teilnahme**
= attendance (of persons)
participative network DATA COMM **Partizipationsnetz**
participle LING **Partizip**
[verbal form with adjective function, e.g. listening, listened] [deklinierbare Verbalform, z.B. hörender, gehörter]
≈ gerund = Mittelwort
↓ participle of present; participle of past ≈ Gerundium
 ↓ Partizip I; Partizip II
particle PHYS **Teilchen**
= corpuscle = Partikel; Korpuskel
↓ elementary particle ↓ Elementarteilchen
particle accelerator PHYS **Teilchenbeschleuniger**
particle displacement ACOUS **Schallauschlag**
particle radiation (→corpuscular radiation) PHYS (→Korpuskularstrahlung)
particle wave (→de Broglie wave) PHYS (→Materialwelle)
particular (→characteristic) TECH (→charakteristisch)
partition (v.t.) TECH **aufteilen**
= divide up; break down = einteilen; zerteilen
≈ subdivide; distribute; dispense; fragment ≈ unterteilen; verteilen; austeilen; zerstückeln
↑ divide ↑ teilen
partition (n.) TECH **Teilung**
= division ≈ Trennung; Unterteilung
= separation; subdivision
partition (n.) SWITCH **Unterteilung**
partition (n.) TECH **Zwischenwand**
= inner partition
partition CIV.ENG **Abteilung**
[of a room] [eines Raumes]
partition 1 DATA PROC **Partition**
[a segregated memory area, treated as independent store] [wie ein getrennter Speicher behandelt]
= logical drive = logisches Laufwerk; Speicherplatzabtrennung
partition 2 (→program segment) DATA PROC (→Programmsegment)
partitioning TECH **Aufteilung**
= distribution; assignment; apportionment = Verteilung; Einteilung
≈ subdivision; division; classification; assignment ≈ Unterteilung; Teilung; Klassifizierung; Zuordnung
partitioning DATA PROC **Speicherplatzaufteilung**
= memory partitioning; storage partitioning; store partitioning = Partitionierung; Speicherplatzuntergliederung; Untergliederung
partition noise MICROEL **Stromverteilungsrauschen**
partition wall TECH **Trennwand**
partner (→associate) ECON (→Teilhaber)

partnership (→company with limited liability)	ECON	(→GmbH)	**passive antenna**	ANT	**strahlungsgekoppelte Antenne**
			= passive aerial		= passive Antenne
part number (→code number)	TECH	(→Sachnummer)	**passive bus connection**	DATA COMM	**passive Sternanschaltung**
part of sentence	LING	**Satzglied**	= branched star connection		
↓ subject; verb 2; object; adverbial		= Satzteil	**passive component**	COMPON	**passives Bauelement**
		↓ Subjekt; Prädikat; Objekt; adverbiale Bestimmung	= passive device		
part of speech (→word class)	LING	(→Wortart)	**passive decoding**	RADIO LOC	**passive Decodierung**
parts (→parts and pieces)	MANUF	(→Satzteile)	**passive device** (→passive component)	COMPON	(→passives Bauelement)
parts and pieces	MANUF	**Satzteile**	**passive gate**	CIRC.ENG	**passives Gatter**
= parts; components			**passive information system**	DATA PROC	(→passives Informationssystem)
parts cabinet (→drawer storage cabinet)	TECH	(→Schubladenmagazin)	(→conversational information system)		
parts list	MANUF	**Stückliste**	**passive jamming**	MIL.COMM	**passive Störung**
= variety parts list; variety plan; material list; bill of materials; material bill; stock list ≈ component list		= Sammelkarte; Materialliste; Schaltteilliste; Teileliste ≈ Bauteileübersicht	**passive network**	NETW.TH	**passives Netzwerk**
			passive network	CIRC.ENG	**passiver Schaltkreis**
			passive quadripole (→passive two-port)	NETW.TH	(→passiver Vierpol)
parts manufacture	MANUF	**Vorfertigung**	**passive radiator** (→reflector 1)	ANT	(→Reflektor 1)
= prefabrication; preproduction			**passive reflector**	ANT	**Umlenkspiegel**
parts programming	CONTROL	**Teileprogrammierung**	= reflector 3		= Umlenkanordnung; passiver Reflektor
[for numerical control]		[von Steuerprogrammen für Werkzeugmaschinen]	≈ passive repeater [RADIO RELAY]		≈ passive Relaisstelle [RADIO RELAY]
parts-tester	INSTR	**Bauelemente-Testgerät**	**passive repeater**	RADIO REL	**passive Relaisstelle**
= components tester		= Bauteile-Tester	= passive repeater system		= passiver Repeater
partword	DATA PROC	**Teilwort**	≈ passive reflector [ANT]		≈ passiver Reflektor [ANT]
= partial word			**passive repeater system**	RADIO REL	(→passive Relaisstelle)
party (→suscriber)	TELEC	(→Teilnehmer)	(→passive repeater)		
party line	TELEC	**Gemeinschaftsanschluß**	**passive satellite**	SAT.COMM	**passiver Satellit**
[serves several users over a single pair]		[mehrere Teilnehmer über ein Leiterpaar]	**passive station**	TELECONTR	**Wartestation**
= multiparty line; multi-access line; shared-service line (BRI)		= Gemeinschaftsleitung; Gesellschaftsanschluß; Mehrfachanschluß; Omnibusleitung	[station which is active only by poll]		[Station die nur auf Abfrage agiert]
↓ two-party line with selective ringing; two-party omnibus line			↑ tributary station		↑ Unterstation
			passive two-port	NETW.TH	**passiver Vierpol**
		↓ Zweieranschluß; Doppelanschluß	= passive quadripole		
party line (→multipoint connection)	TRANS	(→Mehrpunktverbindung)	**passive two-terminal** (→passive two-terminal network)	NETW.TH	(→passiver Zweipol)
party-line equipment	TELEPH	**Gemeinschaftseinrichtung**	**passive two-terminal network**	NETW.TH	**passiver Zweipol**
party-line mode	DATA COMM	**Linienverkehr**	= passive two-terminal		
		= Partyline-Verkehr; Multipoint-Verkehr	**passive voice**	LING	**Passiv**
par value (→nominal value)	ECON	(→Nennwert)	[e.g. I am listened]		[z.B. ich werde gehört]
Pascal	PHYS	**Pascal**	= passive (n.)		= Passivform; Leideform
[SI unit for pressure; = 1 N/m^2]		[SI-Einheit für Druck; = 1 N/m^2]	≠ active voice		≠ Aktiv
= Pa		= Pa	**password** (→keyword 1)	DATA PROC	(→Paßwort)
PASCAL	DATA PROC	**PASCAL**	**past** (n.) (→past tense)	LING	(→Vergangenheit)
↑ high-level programming language		≈ problemorientierte Programmiersprache	**paste** (n.)	TECH	**Paste**
			pasted plate (→grid plate)	POWER SYS	(→Gitterplatte)
pass (n.)	DATA PROC	**Durchlauf**	**past participle**	LING	**Partizip II**
pass (n.) (→sweep)	ELECTRON	(→Wobbeldurchgang)	[e.g. listened]		[z.B. gehörter]
pass (v.i.) (→elapse)	PHYS	(→ablaufen)	↑ participle		= Partizip der Vergangenheit; Mittelwort der Vergangenheit; Partizip der Vorzeitigkeit
pass a message	COLLOQ	**ausrichten**			
passable	TECH	**fahrbar**			
↑ mobile		↑ beweglich			↑ Partizip
passage	LING	**Passage**	**past perfect**	LING	**Plusquamperfekt**
[relevant portion of text]		[relevante Textstelle]	[e.g. I had listened, I had been listened]		[z.B. ich hatte gehört, ich war gehört worden]
pass band	NETW.TH	**Durchlaßbereich**			= 3.Vergangenheit; dritte Vergangenheit; vollendete Vergangenheit; Vorvergangenheit
= pass-band; passband		≠ Sperrbereich			
≠ stop band					
pass-band (→pass band)	NETW.TH	(→Durchlaßbereich)	**past perfect progressive**	LING	**Verlaufsform des Plusquamperfekts**
passband (→pass band)	NETW.TH	(→Durchlaßbereich)	[e.g. I had been working]		
passband attenuation	NETW.TH	**Durchlaßdämpfung**	**past progressive**	LING	**Verlaufsform des Präteritums**
pass direction	PHYS	**Durchlaßrichtung**	[e.g. I was working]		[im Englischen zum Ausdrücken einer in der Vergangenheit gerade ablaufenden Handlung verwendet]
passengers carrier	ECON	**Verkehrsunternehmen**			
↑ carrier		↑ Transportunternehmen 2			
pass-fail testing	QUAL	**Gut-Schlecht-Prüfung**			
passing contact (→wiping contact)	COMPON	(→Wischkontakt)			= Verlaufsform der Vergangenheit
passing light	PHYS	**Durchlicht**	**past record** (→reference)	ECON	(→Referenz)
passivation	MICROEL	**Passivierung**	**past tense**	LING	**Vergangenheit**
passivation (→surface passivation)	METAL	(→Oberflächenpassivierung)	= past (n.)		↓ Imperfekt; Perfekt; Plusquamperfekt
passive	EL.TECH	**passiv**	↓ simple past; present perfect; past perfect		
		≈ stromlos	**pasty**	TECH	**pastös**
passive (n.) (→passive voice)	LING	(→Passiv)	**patch** (v.t.)	DATA PROC	**korrigieren**
passive aerial (→passive antenna)	ANT	(→strahlungsgekoppelte Antenne)			

patch 1 (n.)	DATA PROC	**Direktkorrektur**
= patching		
patch 3 (n.)	DATA PROC	**Korrekturprogramm**
		= Korrekturroutine
patch 2 (n.)	DATA PROC	**Teilfläche**
[computer graphics]		[Computergraphik]
patch 4 (→correction instruction)	DATA PROC	(→Korrekturanweisung)
patch board (→patch panel)	EQUIP.ENG	(→Schaltfeld)
patch-board control	DATA PROC	**Stecktafelsteuerung**
= plug-board control		= Schalttafelsteuerung
patch card	DATA PROC	**Korrekturkarte**
		= Patch-Karte
patch cord	ELECTRON	**Steckschnur**
= patching cord; cord		= Verbindungsleitung
patch cord	INSTR	**Adapterkabel**
patching (→patch 1)	DATA PROC	(→Direktkorrektur)
patching (→jumpering)	SYS.INST	(→Rangierung)
patching cord (→patch cord)	ELECTRON	(→Steckschnur)
patching distribution frame	SYS.INST	**Rangierverteiler** ↑ Verteiler
= cross-connection field; jumper field ↑ distributor		
patching wire (→strap)	SYS.INST	(→Rangierdraht)
patch panel	EQUIP.ENG	**Schaltfeld**
= patch board; plugboard; plugtable		= Schalttafel; Stecktafel [DATA PROC]
patchwork	COLLOQ	**Flickwerk**
patent (v.t.)	ECON	**patentieren**
patent	ECON	**Patent**
patentable	ECON	**patentfähig**
		= patentierbar; schutzfähig
patent action	ECON	**Patentklage**
patent agent (→patent attorney)	ECON	(→Patentanwalt)
patent application	ECON	**Patentanmeldung**
patent attorney	ECON	**Patentanwalt**
= patent agent		
patent case	ECON	**Patentstreit**
patent claim	ECON	**Patentanspruch**
patented	ECON	**patentiert**
≈ proprietary 1		= patentrechtlich geschützt ≈ firmeneigen
patentee	ECON	**Patentinhaber**
= patent holder		
patent grant (→patent issue)	ECON	(→Patenterteilung)
patent holder (→patentee)	ECON	(→Patentinhaber)
patent issue	ECON	**Patenterteilung**
= patent grant		
patent letter	ECON	**Patentschrift**
patent office	ECON	**Patentamt**
patent pending	ECON	**Patentierung läuft**
patent refusal	ECON	**Patentversagung**
patent specification	ECON	**Patentbeschreibung**
patent term	ECON	**Patentlaufzeit**
path	SCIE	**Pfad**
		= Weg
path	SWITCH	**Weg**
= route		
path	DATA PROC	**Pfad 1**
[sequence of events or instructions]		[Abfolge von Ereignissen oder Befehlen]
↓ access path; search path		↓ Zugriffspfad; Suchpfad
path	MATH	**Wegstrecke**
path (→trayectory)	PHYS	(→Bahn)
path (→data transmission path)	DATA COMM	(→Datenübertragungsweg)
path (→communication path)	TELEC	(→Nachrichtenweg)
path (→link)	TRANS	(→Strecke)
path (→path length)	PHYS	(→Weglänge)
path analysis (→path calculation)	TRANS	(→Streckenberechnung)
path attenuation	TRANS	**Streckendämpfung**
= link attenuation		
path calculation	TRANS	**Streckenberechnung**
= link calculation; path analysis; link analysis		
path clearance	RADIO PROP	**Hindernisfreiheit**
= clearance		
path conductor (→track conductor)	ELECTRON	(→Leiterbahn)
path configuration	TRANS	**Streckenkonfiguration**
= line configuration		
path control	DATA COMM	**Pfadsteuerung**
path design engineer	TRANS	**Streckenplaner**
= link design engineer		
path design engineering	TRANS	**Streckenplanung**
= path engineering; link design engineering; link engineering		
path difference	PHYS	**Wegdifferenz**
path disconnection	SWITCH	**Wegeabbau**
= route disconnection		
path diversity	TELEC	**Mehrwegeführung**
= multiple routing; multipath routing; multipath transmission; line diversity; route diversity		= Mehrwegeübertragung; Streckenersatz; Linienersatz
≠ single-path routing		≠ Einwegeführung
path engineering (→path design engineering)	TRANS	(→Streckenplanung)
path-finding section	SWITCH	**Wegesuchabschnitt**
= route-finding section		
path interrogation	SWITCH	**Wegeabfragen**
= route interrogation		
path irregularity (→terrain irregularity)	RADIO PROP	(→Geländeunregelmäßigkeit)
path length	PHYS	**Weglänge**
= path; distance		= Weg
path length (→hop length)	RADIO PROP	(→Funkfeldlänge)
path line (→trayectory)	PHYS	(→Bahn)
path loss	RADIO PROP	**Funkfelddämpfung**
path management	TELEC	**Wegeverwaltung**
≈ transmission management		≈ Übertragungsverwaltung
path name	DATA PROC	**Suchwegbezeichnung**
[in DOS generally the indication of drive, directory and file]		[in DOS i.a. Bezeichnung des Laufwerks, Dateiverzeichnis und Datei]
path-optimized	TECH	**wegoptimiert**
= optimum-path		
path overhead	TELEC	**Path-overhead**
[SDH/SONET; contro information additional to payload]		[SDH/SONET] = POH; Pfadzusatz
= POH		
path profile	RADIO PROP	**Geländeschnitt**
≈ terrain profile		
path section (→line section)	TRANS	(→Streckenabschnitt)
path setting (→path setup)	SWITCH	(→Wegedurchschaltung)
path setup	SWITCH	**Wegedurchschaltung**
= path setting; route setup; route setting		= Wegeinstellung
path survey	SYS.INST	**Streckenbegehung**
↑ site survey		= Streckenbesichtigung; Strecken-Survey ↑ Ortsbegehung
path type	RADIO PROP	**Funkfeldtyp**
pattern 2 (n.)	TECH	**Muster 2**
[a regular decorative design]		[regelmäßige Verzierung]
≈ raster		≈ Raster
pattern 3 (n.)	TECH	**Schnittmuster**
[form used to cut by shape]		
pattern	MATH	**Muster**
pattern	RADIO	**Raster**
= raster		
pattern 1 (→model 2)	TECH	(→Muster 1)
pattern analysis	SIGN.TH	**Musteranalyse**
[signal processing]		[Signalverarbeitung]
patterned line (→decorative line)	TYPOGR	(→Zierlinie)
pattern factor (→radiation factor)	ANT	(→Strahlungsmaß)
pattern generator	DATA PROC	**Flächenmustergenerator**
		= Pattern Generator
pattern generator	INSTR	**Bitmustergenerator**
= test pattern generator; pattern source; test pattern source		= Prüfmustergenerator; Mustergenerator; Pattern-Generator
pattern generator (→test pattern generator)	TV	(→Bildmustergenerator)
pattern generator tape	MICROEL	**Maskensteuerband**

pattern maker	TECH	**Modelltischler**	**payor** (→ payer)	ECON	(→ Zahler)
pattern-post (→ no-value sample)	POST	(→ Muster ohne Wert)	**pay phone** (AM) (→ coin telephone)	TELEPH	(→ Münzfernsprecher)
pattern range (→ test range)	ANT	(→ Antennenmeßfeld)	**paystation** (AM) (→ coin telephone)	TELEPH	(→ Münzfernsprecher)
pattern recognition [signal processing]	INF.TECH	**Mustererkennung** [Signalverarbeitung]	**pay television** (→ pay TV)	BROADC	(→ Abonnementsfernsehen)
pattern source (→ pattern generator)	INSTR	(→ Bitmustergenerator)	**pay TV** = pay television; subscription TV; subscription television	BROADC	**Abonnementsfernsehen**
pattern structure	RADIO	**Rasterarchitektur**	**Pb** (→ lead)	CHEM	(→ Blei)
Pauli principle = exclusion principle; Pauli exclusion principle	PHYS	**Pauli-Prinzip** = Pauli-Verbot; Ausschließungsprinzip	**P bit** (→ parity bit)	CODING	(→ Paritätsbit)
			PBX (→ private branch exchange)	SWITCH	(→ Nebenstellenanlage)
pause = break ≈ interruption; suspension	COLLOQ	**Pause** ≈ Unterbrechung ↓ Ruhepause	**PBX group** (→ PBX line group)	SWITCH	(→ Sammelanschluß)
pause (→ spacing pulse)	TELEGR	(→ Pausenschritt)	**PBX hunting group** (→ PBX line group)	SWITCH	(→ Sammelanschluß)
pause key [stops a display output]	TERM&PER	**PAUSE-Taste** [hält Bildschirmausgabe an] = Taste PAUSE = Taste PAUSE/UNTEBR	**PBX line group** = line group; PBX hunting group; hunt group; PBX group	SWITCH	**Sammelanschluß** = Sammelanschlußbündel; Sammelgruppe ≈ Sammelrufnummer
pave the way [fig.]	COLLOQ	**den Weg bereiten** [fig.]	**PC** (→ personal computer)	DATA PROC	(→ Personal-Computer)
pawl (n.)	MECH	**Klinke** ≈ Schnapper; Anschlag	**PC/AT** (→ AT computer)	DATA PROC	(→ AT-Computer)
pawl (n.) [controls movement of electromechanical switch] ↓ stepping pawl; holding pawl	SWITCH	**Klinke** [steuert Bewegungen eines elektromechanischen Wählers] ↓ Schaltklinke; Sperrklinke (→ Sperrad)	**PCAT** (→ AT computer)	DATA PROC	(→ AT-Computer)
			PCB (→ printed circuit board)	ELECTRON	(→ Leiterplatte)
			PCB artwork (→ artwork master)	MANUF	(→ Druckvorlage)
pawl wheel (→ rached wheel)	MECH	(→ Sperrad)	**PCB artwork creation** = printed circuit drafting; PCB artwork generation; PCB artwork production; PCB design; PCB drafting and design; routing	ELECTRON	**Leiterplattenentflechtung** = Leiterplattenauflösung; Strukturentflechtung; Entflechtung
pawn (v.t.) = pledge	ECON	**verpfänden**			
pawn (→ pledge)	ECON	(→ Pfand)			
pawning = pledge ≈ mortgage	ECON	**Verpfändung** ≈ Hypothek	**PCB artwork generation** (→ PCB artwork creation)	ELECTRON	(→ Leiterplattenentflechtung)
Pawsey balun	ANT	**Pawsey-Symmetrierglied**	**PCB artwork production** (→ PCB artwork creation)	ELECTRON	(→ Leiterplattenentflechtung)
pay (v.t.)	ECON	**zahlen** = bezahlen; einlösen	**PCB assembly** (→ insertion of components)	MANUF	(→ Leiterplattenbestückung)
pay (→ payment)	ECON	(→ Bezahlung)	**PCB base** = printed circuit base	ELECTRON	**Leiterplattenträger**
pay (→ wage)	ECON	(→ Lohn)	**PCB case** (→ module frame)	EQUIP.ENG	(→ Baugruppenrahmen)
payable	ECON	**zahlbar**			
pay duty (→ clear)	ECON	(→ verzollen)	**PCB chassis** (→ module frame)	EQUIP.ENG	(→ Baugruppenrahmen)
payee (→ remittee)	ECON	(→ Geldempfänger)			
payer = payor ≠ payee	ECON	**Zahler** = Bezahler ≠ Zahlungsempfänger	**PCB design** (→ PCB artwork creation)	ELECTRON	(→ Leiterplattenentflechtung)
pay interests	ECON	**verzinsen**	**PCB drafting and design** (→ PCB artwork creation)	ELECTRON	(→ Leiterplattenentflechtung)
payload	ECON	**Nutzlast**			
payload (→ useful information)	CODING	(→ Nutzinformation)	**PCB edge contact**	COMPON	**gedruckter Randkontakt**
payload bit (→ useful bit)	CODING	(→ Nutzbit)	**PCB frame** (→ module frame)	EQUIP.ENG	(→ Baugruppenrahmen)
payload cell [ATM]	TELEC	**Nutzzelle** [ATM]	**PCB manufacture** (→ PCB production)	MANUF	(→ Leiterplattenfertigung)
payload channel (→ basic channel)	TELEC	(→ Basiskanal)	**PCB production** = PCB manufacture	MANUF	**Leiterplattenfertigung**
payload field [ATM]	TELEC	**Nutzinformationsfeld** [ATM]	**PCB relay**	COMPON	**Kartenrelais**
payload rate	TELEC	**Nutzbitrate**	**PCB shelf** (→ module frame)	EQUIP.ENG	(→ Baugruppenrahmen)
payment = pay; remuneration ≈ wage; salary; remittance	ECON	**Bezahlung** = Zahlung ≈ Lohn; Gehalt; Überweisung	**PCB test system** (→ board test system)	INSTR	(→ Leiterplatten-Prüfsystem)
			PCB track (→ track conductor)	ELECTRON	(→ Leiterbahn)
payment date (→ payment term)	ECON	(→ Zahlungsfrist)	**PC fax board**	DATA PROC	**PC-Faxkarte**
payment demand	ECON	**Zahlungsaufforderung**	**p-channel isolated gate FET** (→ PIGFET)	MICROEL	(→ PIGFET)
payment means = medium of exchange; tender 2 ↓ money; check	ECON	**Zahlungsmittel** ↓ Geld; Scheck	**p-channel MOSFET**	MICROEL	**P-Kanal-MOSFET** = P-Kanal-MOS-Transistor
payment of interests	ECON	**Verzinsung 1** [Zahlung von Zinsen]	**PCI** = programmable communication interface	MICROEL	**PCI**
payment of the balance	ECON	**Restzahlung**	**PCM** (→ plug-compatible manufacturer)	EQUIP.ENG	(→ steckerkompatibler Hersteller)
payment on account (→ installment payment)	ECON	(→ Ratenzahlung)			
payment on deferred terms (→ installment payment)	ECON	(→ Ratenzahlung)	**PCM** (→ pulse code modulation)	CODING	(→ Pulscodemodulation)
payment order = money order	ECON	**Zahlungsanweisung**	**PCM** (→ process-control monitor)	MICROEL	(→ Teststruktur)
payment term = payment date	ECON	**Zahlungsfrist** = Zahlungsziel	**PCM highway**	SWITCH	**PCM-Leitung**
paynes grey	PHYS	**paynesgrau**	**PCM insertion unit**	TRANS	**PCM-Einfügung**
payoff (→ advantage)	COLLOQ	(→ Vorteil)	**PCM-instrumentation recorder**	EL.ACOUS	**PCM-Bandspeicher**

PCM measuring technique	INSTR	PCM-Meßtechnik	
PCM multiplex equipment	TRANS	PCM-Multiplexgerät	
= PCM multiplexer		= PCM-Multiplexer	
PCM multiplexer (→PCM multiplex equipment)	TRANS	(→PCM-Multiplexgerät)	
PCM programm channel system	TRANS	PCM-Tonkanalsystem	
PCM switch	SWITCH	PCM-Stufe	
PCM switching	SWITCH	PCM-Vermittlung	
PCM system		PCM-System	
PCM/TDM error measuring set	INSTR	PCM/TDM-Fehlermeßplatz	
PCM technique	TELEC	PCM-Technik	
PCM telemetering	TELECONTR	PCM-Telemetrie	
PCM transmission	TELEC	PCM-Übertragung	
PCM transmission system	TRANS	PCM-Übertragungssystem	
PCM word	TELEC	Codewort	
PCN (→personal communications network)	MOB.COMM	(→Netz für persönliche Kommuikation)	
P-controller	CONTROL	Proportionalregler	
[proportional action]		[proportional wirkend]	
= proportional controler		= P-Regler	
↑ continuous-action controller		↑ stetiger Regler	
PCR (→phase change recording)	TERM&PER	(→Phasenänderungsverfahren)	
PC/XT (→XT computer)	DATA PROC	(→XT-Computer)	
PCXT (→XT computer)	DATA PROC	(→XT-Computer)	
Pd (→palladium)	CHEM	(→Palladium)	
PD controller	CONTROL	PD-Regler	
PDM (→pulse duration modulation)	MODUL	(→Pulsdauermodulation)	
PDV bus	DATA COMM	PDV-Bus	
peak 1 (n.)	MATH	Höcker	
[in a curve]		[in Kurve]	
peak 2 (n.)	MATH	Spitze 1	
[of a solid]		[eines Körper]	
peak 3 (n.)	MATH	Scheitel 2	
[point of a curve where the curvature reaches a minimum or maximum]		[Kurvenpunkt in dem die Krümmung ein Maximum oder Minimum erreicht]	
= crest		= Scheitelpunkt 2; Spitze 2	
≈ cusp		≈ Rückkehrpunkt	
peak ammeter (→peak-current meter)	INSTR	(→Scheitelstrommesser)	
peak amplitude	EL.TECH	Spitzenamplitude	
peak clipper (→amplitude limiter circuit)	CIRC.ENG	(→Amplitudenbegrenzerschaltung)	
peak clipping (→peak limiting)	ELECTRON	(→Spitzenbegrenzung)	
peak consumption	ECON	Spitzenverbrauch	
peak current	EL.TECH	Scheitelstrom	
= crest current		= Spitzenstrom	
peak-current meter	INSTR	Scheitelstrommesser	
= peak ammeter		= Spitzenstrommesser	
↑ peak-value meter		↑ Scheitelwertmesser	
peak demand	ECON	Spitzenbedarf	
		= Spitzennachfrage	
peak deviation (→maximal deviation)	CONTROL	(→Maximalabweichung)	
peak duty (→peak load)	TECH	(→Belastungsspitze)	
peak envelope power	RADIO	Spitzenleistung	
peak factor (→crest factor)	EL.TECH	(→Spitzenfaktor)	
peak forward voltage	ELECTRON	Spitzenvorwärtsspannung	
		= Spitzendurchlaßspannung	
peak frequency	NETW.TH	Höckerfrequenz	
peak inverse voltage	ELECTRON	Spitzensperrspannung	
= peak reverse voltage; crest inverse voltage; crest reverse voltage			
peak limiter (→amplitude limiter circuit)	CIRC.ENG	(→Amplitudenbegrenzerschaltung)	
peak limiting	ELECTRON	Spitzenbegrenzung	
= peak clipping			
peak load	TECH	Belastungsspitze	
= maximum load; peak duty; maximum duty		= Belastungsmaximum; Höchstlast; Höchstbelastung; Maximalbelastung	
peak load	EL.TECH	Spitzenlast	
		= Spitzenbelastung	
peak output	TECH	Höchstleistung	
peak point	MATH	Gipfelpunkt	
↑ surface point		= elliptischer Flächenpunkt	
		↑ Flächenpunkt	
peak point	MICROEL	Höckerpunkt	
[tunnel diode]		[Tunneldiode]	
peak-point current	MICROEL	Höckerstrom	
[tunnel diode]		[Tunneldiode]	
peak-point voltage	MICROEL	Höckerspannung	
[tunnel diode]		[Tunneldiode]	
= isohypse voltage			
peak power	EL.TECH	Spitzenleistung	
peak power meter	INSTR	Spitzenleistungsmesser	
peak power point	ANT	Strahlungsmaximum	
peak power sensor	INSTR	Spitzenleistungs-Meßkopf	
peak reverse voltage (→peak inverse voltage)	ELECTRON	(→Spitzensperrspannung)	
peak search function	INSTR	Spitzenwert-Suchfunktion	
peak-to-average ratio (→crest factor)	EL.TECH	(→Spitzenfaktor)	
peak-to-peak	EL.TECH	Spitze-Spitze	
= pk-pk			
peak-to-valley height (→roughness height)	MECH	(→Rauhigkeitshöhe)	
peak-to-valley ratio	MICROEL	Höcker-Tal-Verhältnis	
= drop ratio			
peak traffic	TELEC	Spitzenverkehr	
peak-traffic time (→heavy-traffic period)	TELEC	(→verkehrsstarke Zeit)	
peak-type rectifier	CIRC.ENG	Spitzenwertgleichrichter	
		= Spitzengleichrichter	
peak value (→maximum value)	MATH	(→Größtwert)	
peak-value limiter	ELECTRON	Höchstwertbegrenzer	
= maximum-value limiter		= Maximalwertbegrenzer	
peak-value meter	INSTR	Scheitelwertmesser	
↓ peak voltmeter; peak-current meter		↓ Scheitelspannungsmesser; Scheitelstrommesser	
peak voltage	EL.TECH	Spitzenspannung	
		= Scheitelspannung	
peak voltmeter	INSTR	Scheitelspannungsmesser	
= crest voltmeter		= Spitzenspannungsmesser	
↑ peak-value meter		↑ Scheitelwertmesser	
PEARL	DATA PROC	PEARL	
["Process and Experiment Automation Realtime Language"; a programming language for automation tasks, derived from BASIC]		[aus BASIC abgeleitete Programmiersprache für Automatisierungsaufgaben]	
pearl (→bead)	TECH	(→Kügelchen)	
Pearl (BRI) (→type size 5 point)	TYPOGR	(→Schriftgröße 5 Punkt)	
pebble gray	PHYS	kieselgrau	
PE cellular insulation	COMM.CABLE	Zell-PE-Isolierung	
pecularity (→special feature)	TECH	(→Besonderheit)	
peculiar (→characteristic)	TECH	(→charakteristisch)	
pedal switch (→foot switch)	COMPON	(→Fußschalter)	
pedestal	TECH	Sockel	
= rest; support		= Piedestal; Bock; Postament	
		≈ Ständer	
pedestal copier	OFFICE	Standkopierer	
pedestal-mounted	EQUIP.ENG	auf Sockel aufgestellt	
		= zur Aufstellung auf Sockel	
pedestal mounting	TECH	Sockelmontage	
peek (n.)	DATA PROC	Direktleseanweisung	
[instruction allowing to look at any store place]			
peeling	METAL	Abschälung	
[unwanted detachment of plated metal]		[unerwünschte Ablösung einer Metallisierung]	
peel off (→flake off)	TECH	(→abblättern)	
peel-off strength	METAL	Abschälkraft	
peen (v.t.)	MECH	vernieten 1	
		[mit Hammerfinne]	
peen end	METAL	Nietzapfen	
peg	SWITCH	Belegungszahl	
= number of seizures			
peg count	SWITCH	Belegungszählung	
peg count	QUAL	Stichprobenzählung	
pel (→picture element)	TV	(→Abtastwert)	
P element	CONTROL	P-Glied	
		= Proportionalglied; proportionales Übertragungsglied	
Peltier effect	PHYS	Peltier-Effekt	
↑ thermoelectric effect		↑ thermoelektrischer Effekt	

pen		TERM&PER	**Stift**	**percussion drilling**		TECH	**Schlagbohren**

pen		TERM&PER	**Stift**
↓ color pen			↓ Farbstift
pen (→fountain-pen)		OFFICE	(→Füllfederhalter)
penalty		TECH	**Mehraufwand**
penalty		ECON	**Strafe**
= fine			
pencil		OFFICE	**Bleistift**
			= Stift
			≈ Farbstift
pencil (→ray bunch)		PHYS	(→Strahlenbüschel)
pencil beam		PHYS	**Schmalbündel**
pencil-beam antenna		ANT	**Bleistiftkeulen-Antenne**
pencil galvanometer		INSTR	**Stiftgalvanometer**
pencil plotter		TERM&PER	**Stiftplotter**
= pen-on-paper plotter; pen plotter			= Zeichenstiftplotter; Bleistiftplotter
↑ plotter			↑ Plotter
pending patent		ECON	**angemeldetes Patent**
= applied-for patent			
pendulum		PHYS	**Pendel**
penetrate		PHYS	**durchdringen**
= permeate; pervade			
penetration		TERM&PER	**Anschlagstärke**
= print intensity; print force			
penetration		PHYS	**Durchdringung**
penetration control		TERM&PER	**Anschlagregler**
= impact control			
penetration depth		TECH	**Eindringtiefe**
= skin depth			= Eindringungstiefe
penetration factor		MICROEL	(→Durchgriff)
(→punch-through)			
penetration voltage		MICROEL	(→Durchgreifspannung)
(→punch-through voltage)			
pen force		TERM&PER	**Auflagekraft**
[plotter]			[Plotter]
pen lift		TERM&PER	**Stifthub**
pen-on-paper plotter		TERM&PER	(→Stiftplotter)
(→pencil plotter)			
pen plotter (→pencil plotter)		TERM&PER	(→Stiftplotter)
pentagon		MATH	**Fünfeck**
			= Pentagon
pentagonal		MATH	**fünfeckig**
pentagondodecahedron		MATH	**Pentagondodekaeder**
(pl.-ons, -a)			= Zwölfflächner
[with twelve faces]			↑ Polyeder
↑ polyhedron			
pentagon head		MECH	**Fünfkant**
pentagon nut		MECH	**Fünfkantmutter**
pentahedral		MATH	**fünfflächig**
pentahedron (pl.-ons,-a)		MATH	**Pentaeder**
[solid of five faces]			= Fünfflach; Fünfflächner
↑ polyhedron			↑ Polyeder
pentode		ELECTRON	**Pentode**
↑ screen-grid tube			= Fünfgitterröhre
			↑ Schirmgitterröhre
penumbra		PHYS	**Halbschatten**
perceivability		COLLOQ	**Wahrnehmbarkeit**
= perceptibility			
perceivable		COLLOQ	**wahrnehmbar**
= perceptible			≈ beobachtbar
≈ observable			
perceive		COLLOQ	**wahrnehmen**
= register			= registrieren
≈ observe			≈ beobachten
percent (adj.)		TECH	**prozentual**
percent		MATH	**Prozent**
[% ; p.c.]			[% ; v.H.; p.c.]
= per centum			= von Hundert
percentage error		INSTR	**prozentualer Fehler**
percent sign		ECON	**Prozentzeichen**
[%]			[%]
= percent symbol			
percent symbol (→percent sign)		ECON	(→Prozentzeichen)
per centum (→percent)		MATH	(→Prozent)
perceptibility (→perceivability)		COLLOQ	(→Wahrnehmbarkeit)
perceptible (→perceivable)		COLLOQ	(→wahrnehmbar)
perception		COLLOQ	**Wahrnehmung**
= registration			= Beobachtung
≈ observation			
percussion drilling		TECH	**Schlagbohren**
per diem (AM) (→daily allowance)		ECON	(→Tagegeld)
perfect bunch (→perfect trunk group)		SWITCH	(→vollständiges Bündel)
perfect code		CODING	**vollständiger Code**
perfect trunk group		SWITCH	**vollständiges Bündel**
= perfect bunch			
perforate (→punch)		TERM&PER	(→lochen)
perforate (→pierce)		METAL	(→lochen)
perforated board		OFFICE	**Lochtafel**
perforated cathode		ELECTRON	**Lochkathode**
perforated screen		MICROEL	**Lochplatte**
= disk			
perforated screen lens		ANT	**Lochplattenlinse**
perforated sheet (→experimental board)		ELECTRON	(→Experimentierkarte)
perforated tape		TERM&PER	**Lochstreifen**
= punched tape 1			
↑ tape			
perforated-tape equipment (→perforated-tape teleprinter)		TELEGR	(→Lochstreifengerät)
perforated tape reader (→punched tape reader)		TERM&PER	(→Lochstreifenleser)
perforated-tape teleprinter		TELEGR	**Lochstreifengerät**
= perforated-tape equipment			[Fernschreiber mit Lochstreifenzusatz]
perforating (→piercing)		METAL	(→Lochen)
perforating pin		TERM&PER	**Lochernagel**
perforation		TERM&PER	**Lochung**
= punching; perfs (pl.t.)			= Lochen; Perforation
perforation (→cutout)		TECH	(→Durchbruch)
perform (→execute)		DATA PROC	(→ausführen)
performability (→viability)		TECH	(→Durchführbarkeit)
performable (→feasable)		TECH	(→durchführbar)
performance		TECH	**Leistung**
performance- (→high-performance ...)		TECH	(→Hochleistungs-)
performance (→capability)		TECH	(→Leistungsfähigkeit)
performance figure		TECH	**Kennzahl 2**
performance monitor		DATA PROC	**Leistungsüberwachungsprogramm**
[a program]			
performance monitoring		TELEC	**Leistungsverwaltung**
			= Leistungsmanagement
performance parameter (→feature)		TECH	(→Leistungsmerkmal)
performance range		TECH	**Funktionsumfang**
performance specification		TECH	**Leistungsbeschreibung**
performance test (→function test)		TECH	(→Funktionsprüfung)
performance universal counter		INSTR	**Hochleistungsuniversalzähler**
performing arts		COLLOQ	**Darstellende Kunst**
↓ dramatic art; dance			↓ Schauspielkunst; Tanzkunst
PERFORM instruction (COBOL) (→run statement)		DATA PROC	(→Laufanweisung)
PERFORM statement (COBOL) (→run statement)		DATA PROC	(→Laufanweisung)
perfory (n.) (→pin-feed edge)		TERM&PER	(→Führungsstreifen)
perfs (pl.t.) (→perforation)		TERM&PER	(→Lochung)
perigee		ASTROPHYS	**Perigäum**
[orbital point nearest to earth]			[Bahnpunkt geringster Entfernung zur Erde]
≠ apogee			≠ Apogäum
↑ apsis			↑ Apside
perihelion		ASTROPHYS	**Perihel**
[orbital point nearest to sun]			[Bahnpunkt geringster Entfernung zur Sonne]
↑ apsis			= Perihelium
			↑ Apside
perimeter		MATH	**Umfang**
[lenth of an closed boundary line]			[Länge einer geschlossenen Begrenzung]
= circumference 2			
↓ circumference 1			↓ Kreisumfang
period 1		PHYS	**Periode**
= cycle			≈ Zeitabschnitt
period		TELEPH	**Zeitintervall**
period (AM) (→dot)		LING	(→Punkt 1)
period 2 (→sentence)		LING	(→Satz)

period 2 (→time interval)	PHYS	(→Zeitintervall)	
period duration (→period length)	MATH	(→Periodenlänge)	
periodic [regularly at equal intervals] = at intervals ≈ recurrent; cyclic	SCIE	**periodisch** [regelmäßig in gleichen Abständen] = wiederholend; zyklisch	
periodic aerial (→periodic antenna)	ANT	(→abgestimmte Antenne)	
periodical line displacement	TV	**Zeilenversatz 1** = Bauchtanz	
periodic and random deviation (→ac component)	EL.TECH	(→Wechselstromkomponente)	
periodic antenna = tuned antenna; periodic aerial; tuned aerial	ANT	**abgestimmte Antenne** = periodische Antenne	
periodicity	SCIE	**Periodizität**	
periodic law (→periodic system)	CHEM	(→Periodensystem)	
periodic noise	TELEC	**periodisches Rauschen** = periodischer Störer	
periodic pulse (→time pulse)	ELECTRON	(→Zeitsteuertakt)	
periodic-pulse metering (→time-pulse metering)	SWITCH	(→Zeitimpulszählung)	
periodic ringing	SWITCH	**Weiterruf**	
periodic sampling (→repetitive sampling)	ELECTRON	(→periodische Abtastung)	
periodic signal (→repetitive signal)	INSTR	(→periodisches Signal)	
periodic system = periodic law ≈ periodic table	CHEM	**Periodensystem** = Periodensystem der Elemente ≈ Periodentafel der Elemente	
periodic table	CHEM	**Periodentafel**	
period length = period duration; cycle	MATH	**Periodenlänge** = Periodendauer; Schwingungsdauer	
period of depreciation	ECON	**Abschreibungsdauer** = Abschreibungszeitdauer	
period of limitation (→prescription)	ECON	(→Verjährung)	
period of time (→term)	COLLOQ	(→Frist 1)	
period of validity	ECON	**Geltungsdauer**	
peripheral (n.) (→peripheral equipment)	DATA PROC	(→Peripheriegerät)	
peripheral adapter	TERM&PER	**Peripherie-Anschlußeinheit**	
peripheral cell	MICROEL	**Randzelle** = Peripheriezelle	
peripheral device (→peripheral equipment)	DATA PROC	(→Peripheriegerät)	
peripheral driver (→device driver)	DATA PROC	(→Peripheriegerätetreiber)	
peripheral equipment [equipment for input, output or storage of data] = peripheral unit; peripheral device; peripheral (n.) ↓ input device; output device; input/output device; external memory	DATA PROC	**Peripheriegerät** [Gerät für Eingabe, Ausgabe oder Speicherung von Daten] = periphere Einheit; peripheres Gerät; Anschlußgerät; Peripherie ↓ Eingabegerät; Ausgabegerät; Ein-/Ausgabegerät; externer Speicher	
peripheral intelligence (→device intelligence)	DATA PROC	(→Geräteintelligenz)	
peripheral interface adapter (→PIA)	MICROEL	(→PIA)	
peripheral memory (→external memory)	DATA PROC	(→Externspeicher)	
peripheral mode	OPT.COMM	**Mantelmode**	
peripheral slot [DATA PROC] (→expansion slot)	EQUIP.ENG	(→Erweiterungssteckplatz)	
peripherals priority (→device priority)	DATA PROC	(→Gerätepriorität)	
peripheral storage (→external memory)	DATA PROC	(→Externspeicher)	
peripheral store (→external memory)	DATA PROC	(→Externspeicher)	
peripheral unit (→peripheral equipment)	DATA PROC	(→Peripheriegerät)	
peripheral units	SWITCH	**vermittlungstechnische Peripherie**	
periphery	SCIE	**Peripherie** ≈ Umgebung; Umfeld	
periscope antenna	ANT	**Periskopantenne**	
permalloy	METAL	**Permalloy** ≈ Mumetall	
permanence [relation of magnetic flux density and magnetic field strength; equal to remanence when characteristic is linear] ≈ residual induction	PHYS	**magnetische Permanenz** [Relation von magnetischer Flußdichte zu magnetischer Feldstärke; bei Linearität gleich mit remanter Induktion] = Permanenz ≈ remanente Induktion	
permanent charge (→continuous load)	TECH	(→Dauerbelastung)	
permanent circuit (→fixed line)	TELEC	(→Standleitung)	
permanent connection 1	TELEC	**starre Durchschaltung**	
permanent copy	TERM&PER	**Dauerkopie**	
permanent employee	ECON	**Festangestellter**	
permanent error	QUAL	**stetiger Fehler**	
permanent error = irreparable error	DATA PROC	**irreparabler Fehler**	
permanent line (→fixed line)	TELEC	(→Standleitung)	
permanent load (→continuous load)	TECH	(→Dauerbelastung)	
permanently active	DATA COMM	**daueraktiv**	
permanent magnet	PHYS	**Dauermagnet** = Permanentmagnet; permanenter Magnet	
permanent magnetization	PHYS	**permanente Magnetisierung**	
permanent memory 2 (→read-only memory)	DATA PROC	(→Festwertspeicher)	
permanent memory 1 (→non-volatile memory)	DATA PROC	(→nichtflüchtiger Speicher)	
permanent mold	METAL	**Dauerformguß**	
permanent noise	TELEC	**Dauergeräusch**	
permanent staff	ECON	**Stammpersonal**	
permanent storage 2 (→read-only memory)	DATA PROC	(→Festwertspeicher)	
permanent storage 1 (→non-volatile memory)	DATA PROC	(→nichtflüchtiger Speicher)	
permanent store 2 (→read-only memory)	DATA PROC	(→Festwertspeicher)	
permanent store 1 (→non-volatile memory)	DATA PROC	(→nichtflüchtiger Speicher)	
permanent supervision	ELECTRON	**Dauerüberwachung**	
permanent test = continuous test; endurance test	QUAL	**Dauerprüfung** = Dauertest	
permanent virtual circuit = permanent virtual connection; permanent virtual link	DATA COMM	**feste virtuelle Verbindung**	
permanent virtual connection (→permanent virtual circuit)	DATA COMM	(→feste virtuelle Verbindung)	
permanent virtual link (→permanent virtual circuit)	DATA COMM	(→feste virtuelle Verbindung)	
permeability 1	PHYS	**Durchlässigkeit 1**	
permeability 2 [ratio of magnetic flux density to magnetic field strength] ≠ reluctivity	PHYS	**Permeabilität** [magn. Flußdichte zu magn. Feldstärke] = Magnetisierungskonstante; magnetische Leitfähigkeit ≠ Reluktivität	
permeability index (→relative permeability)	PHYS	(→Permeabilitätszahl)	
permeability of vacuum (→coefficient of self-inductance)	PHYS	(→magnetische Feldkonstante)	
permeable = pervious ≠ impermeable	TECH	**durchlässig** ≠ undurchlässig	
permeance [SI unit: Henry] = magnetic conductance	PHYS	**Permeanz** [SI-Einheit: Henry] = magnetischer Leitwert	
permeate (→penetrate)	PHYS	(→durchdringen)	
permendur	METAL	**Permendur**	
perminvar	METAL	**Perminvar**	
permissible = safe; tolerated; legal ≈ permissive	TECH	**zulässig**	

permissible (→valid)		DATA PROC	(→zulässig)	**pertinent** (→relevant)		SCIE	(→relevant)

permissible (→valid) — DATA PROC — (→zulässig)
permission — ECON — **Genehmigung**
= licence
≈ entitlement
= Erlaubnis
≈ Berechtigung
permittivity (→dielectric constant 1) — PHYS — (→Dielektrizitätskonstante)
permittivity (→dielectric constant 2) — PHYS — (→elektrische Feldkonstante)
permittivity of vacuum (→dielectric constant 2) — PHYS — (→elektrische Feldkonstante)
permittivity ratio (→relative permittivity) — PHYS — (→Dielektrizitätszahl)
permutation — MATH — **Permutation**
[any of different arrangement alternatives for a given set of items]
[jede der Anordnungsmöglichkeiten einer betrachteten Anzahl von Elementen]
permutation group — MATH — **Permutationsgruppe**
permute — MATH — **permutieren**
perpendicular (n.) — MATH — **Senkrechte**
= vertical (n.); normal (n.)
≠ horizontal (n.)
= Normale
≠ Waagrechte
perpendicular (→vertical) — MATH — (→senkrecht)
persistence — PHYS — **Nachleuchtdauer**
persistence characteristic — PHYS — **Nachleuchtcharakteristik**
= decay characteristic
persistency check — DATA PROC — **Fehlerbestätigung**
persistent — TECH — **anhaltend**
= andauernd; ständig
persistent command — TELECONTR — **Dauerbefehl**
persistent information — TELECONTR — **Dauermeldung**
persistent screen — TERM&PER — **nachleuchtender Bildschirm**
= afterglow screen
personal communications — TELEC — **Personalkommunikation**
personal communications network — TELEC — **Personalkommunikationsnetz**
= PCN
= Netz für persönliche Kommunikation; PCN
personal computer — DATA PROC — **Personal-Computer**
= PC; personal microcomputer
≈ workstation computer
↑ microcomputer
= PC
≈ Arbeitsplatzrechner
↑ Mikrocomputer
personal data acquisition — DATA PROC — **Personaldatenerfassung**
= PDE
personal data sheet — ECON — **Kurzlebenslauf**
personal dose — PHYS — **Personendosis**
personal form — LING — **Personalform**
= finite Form
personality (→measuring function) — INSTR — (→Meßfunktion)
personalize — TECH — **personalisieren**
[to adapt to a specific user]
≈ customize
[für einen bestimmten Benutzer anpassen]
≈ kundenspezifisch anpassen
personal microcomputer (→personal computer) — DATA PROC — (→Personal-Computer)
personal pronoun — LING — **Personalpronomen**
[e.g. he]
[z.B. er]
= persönliches Fürwort
personal record (→resumé) — ECON — (→Lebenslauf)
personal secretary (→executive secretary) — OFFICE — (→Chefsekretärin)
personal speech recognition — INF.TECH — (→Stimmerkennung)
(→personal voice recognition)
personal voice recognition — INF.TECH — **Stimmerkennung**
[recognition of the voice of a specific person]
= personal speech recognition
↑ speech recognition
[Erkennung der Stimme einer bestimmten Person]
↑ Spracherkennung
person in charge — OFFICE — **Sachbearbeiter**
personnel — ECON — **Personal**
= staff
≈ Arbeiter und Angestellten
personnel (→workforce) — ECON — (→Belegschaft)
personnel training — ECON — **Personalschulung**
person-to-person calling — TELEPH — **Voranmeldungsgespräch**
perspective — MATH — **Perspektive**
per-thousand sign — ECON — **Promillzeichen**
[‰]
= per-thousand symbol
[‰]
per-thousand symbol — ECON — (→Promillzeichen)
(→per-thousand sign)
pertinax — ELECTRON — **Pertinax**

pertinent (→relevant) — SCIE — (→relevant)
perturbance insensitive — TELEC — (→störungsunempfindlich)
(→interference insensitive)
perturbance variable — CONTROL — (→Störgröße)
(→disturbance variable)
perturbance variable feed-forward (→disturbance variable feed-forward) — CONTROL — (→Störgrößenaufschaltung)
perturbation — PHYS — **Störung**
perturbation-free — TECH — (→störungsfrei)
(→trouble-free)
perturbation function — MATH — **Störfunktion**
perturbation pulse — ELECTRON — (→Störimpuls)
(→interfering pulse)
perturbation sensitive — TECH — **störungsempfindlich**
perturbation signal — ELECTRON — (→Störsignal)
(→interfering signal)
pervade (→penetrate) — PHYS — (→durchdringen)
perveance — ELECTRON — **Perveanz**
[performance parameter of electronic tubes]
[Parameter für Elektronenröhren]
pervious (→permeable) — TECH — (→durchlässig)
Petresen coil (→ground-fault neutralizer) — POWER SYS — (→Erdlöschspule)
Petri net — DATA PROC — **Petri-Netz**
petrol (BRI) (→gasoline) — TECH — (→Benzin)
petrolatum (→jelly) — COMM.CABLE — (→Vaseline)
petroleum — TECH — **Erdöl**
= crude oil; mineral oil
= Mineralöl; Rohöl; Petroleum 2
petroleum jelly — COMM.CABLE — (→Vaseline)
(→jelly)
pF (→picofarad) — PHYS — (→Pikofarad)
PFM (→pulse frequency modulation) — MODUL — (→Pulsfrequenzmodulation)
PGA — TERM&PER — **PGA**
[Professional Graphics Adapter; graphics standard with 640x840 pixels]
[Graphikstandard mit 640x840 Bildpunkten]
PGA board — TERM&PER — **PGA-Karte**
↑ graphics board
= PGA-Adapter
↑ Graphikkarte
p-gate thyristor — MICROEL — **kathodenseitig steuerbarer Thyristor**
pH (→pH value) — CHEM — (→pH-Wert)
phantom (→phantom circuit) — COMM.CABLE — (→Phantomkreis)
phantom aerial (→dummy antenna) — ANT — (→künstliche Antenne)
phantom antenna (→dummy antenna) — ANT — (→künstliche Antenne)
phantom circuit — COMM.CABLE — **Phantomkreis**
= phantom; side circuit
≠ physical circuit
= Viererleitung; Phantom
≠ Stammleitung
phantom circuit loading — COMM.CABLE — (→Viererbespulung)
(→phantom loading)
phantom-circuit loading coil — OUTS.PLANT — **Phantom-Pupinspule**
phantom coil — OUTS.PLANT — **Phantomübertrager**
= side-circuit coil
= Fernleitungsübertrager
phantoming — COMM.CABLE — **Phantombildung**
phantom loading — COMM.CABLE — **Viererbespulung**
= phantom circuit loading
= Phantombespulung; Phantomkreisbespulung
phantom telegraphy — TELEGR — **Vierertelegraphie**
= VT
phantom-to-phantom circuit — COMM.CABLE — (→Übersprechen)
(→side-to-side crosstalk)
pharmacy terminal — TERM&PER — (→Apothekenterminal)
(→drug-store terminal)
phase (v.t.) — PHYS — **einphasen**
= in Phase bringen
phase — PHYS — **Phase**
phase — POWER ENG — **Außenleiter**
[ac system]
= phase conductor; ouside conductor
[Wechselstromnetz]
phase accuracy — EL.TECH — **Phasengenauigkeit**
phase advance (→phase lead) — ELECTRON — (→Phasenvoreilung)
phase angle — PHYS — **Phasenwinkel**

phase angle

phase angle (→ phase shift) PHYS (→ Phasenverschiebung)
phase angle control POWER SYS **Phasenanschnittsteuerung**
= Anschnittsteuerung
phase angle factor NETW.TH **Phasenmaß**
[imaginary part of the complex transfer constant; negative value of thr phase angle]
[Imaginärteil des komplexen Übertragungsmaßes; Negativwert des Dämpfungswinkels]
≈ phase angle
↓ image phase angle factor; effective phase angle factor
= Übertragungswinkel
≈ Dämpfungswinkel
↓ Wellenphasenmaß; Betriebsphasenmaß
phase center ANT **Phasenzentrum**
phase-change coefficient LINE TH (→ Phasenkonstante)
(→ phase constant)
phase changer (→ phase shifter) MICROW (→ Phasenschieber)
phase change recording TERM&PER **Phasenänderungsverfahren**
[optical disk]
= PCR
[optische Speicherplatte]
phase coefficient (→ phase constant) LINE TH (→ Phasenkonstante)
phase commutation POWER SYS **Phasenlöschung**
[Wechselrichter]
phase comparator CIRC.ENG **Phasenvergleichsschaltung**
= Phasenvergleicher; Phasenkomparator
phase compensation EL.TECH **Phasenausgleich**
= phase correction
= Phasenkompensation
phase compensator (→ phase equalizer) CIRC.ENG (→ Phasenentzerrer)
phase condition CIRC.ENG **Phasenbedingung**
[of an oszillator]
[Oszillator]
phase conductor (→ phase) POWER SYS (→ Außenleiter)
phase constant LINE TH **Phasenkonstante**
[imaginary part of the propagation coefficient]
[Imaginärteil der Fortpflanzungskonstante]
= phase-change coefficient; phase coefficient
= Phasenmaß; Phasenbelag; Phasenkoeffizient; Winkelkonstante
phase-continuous ELECTRON **phasenstetig**
= phasenkontinuierlich
phase-continuous sweep INSTR **phasenstetige Wobbelung**
phase contrast PHYS **Phasenkontrast**
phase control CIRC.ENG **Phasenanschnitt**
phase control ELECTRON **Phasenregelung**
= phase regulation
= Phasensteuerung
phase control TV **Phasensteuerung**
≈ hue control
≈ Farbtonregelung
phase-controlled antenna ANT **phasengesteurte Antenne**
phase converter POWER SYS **Phasenumformer**
= Phasenwandler
phase correction (→ phase compensation) EL.TECH (→ Phasenausgleich)
phase current POWER SYS **Phasenstrom**
= Außenleiterstrom
phased (→ scaled) TECH (→ stufenweise 1)
phased (→ stepwise) TECH (→ schrittweise)
phase delay (→ phase lag) PHYS (→ Phasenverzögerung)
phase delay time EL.TECH **Phasenlaufzeit**
phase detector (→ phase discriminator) CIRC.ENG (→ Phasendiskriminator)
phase deviation MODUL **Phasenhub**
= phase swing
phase deviation (→ phase shift) PHYS (→ Phasenverschiebung)
phased expansion (→ scaled expansion) TECH (→ stufenweise Erweiterung)
phase diagram (→ state diagram) DATA PROC (→ Zustandsdiagramm)
phase difference PHYS **Phasendifferenz**
≈ phase shift
= Phasenunterschied
≈ Phasenverschiebung
phase discriminator CIRC.ENG **Phasendiskriminator**
= phase detector
= Phasendetektor; Riegerschaltung
phase displacement (→ phase shift) PHYS (→ Phasenverschiebung)
90° phase displacement circuit NETW.TH **Mummelschaltung**
phase-displayed (→ out of phase) PHYS (→ phasenverschoben)

phase distortion TELEC **Phasenverzerrung**
= phase-frequency distortion
≈ Phasenverschiebung [PHYS]
≈ phase shift [PHYS]
phased monopole antenna ANT **gephaste Monomodeantenne**
phased transition TECH **gleitender Übergang**
phase encoding DATA PROC **Richtungstaktschrift**
= two-phase recording
= PE-Schrift; Zweiphasenschrift; Phase-encoding-Schrift
↑ recording mode
↑ Aufzeichnungsverfahren
phase equality PHYS **Phasengleichheit**
= Gleichphase
phase equalizer CIRC.ENG **Phasenentzerrer**
= phase compensator
phase filter (→ all-pass filter) NETW.TH (→ Allpaß)
phase focusing PHYS **Phasenfokussierung**
= phase focussing
phase focussing (→ phase focusing) PHYS (→ Phasenfokussierung)
phase-frequency distortion (→ phase distortion) TELEC (→ Phasenverzerrung)
phase grouping PHYS **Phasengruppierung**
phase hit (→ phase jump) EL.TECH (→ Phasensprung)
phase-insensitive EL.TECH **phasenunempfindlich**
phase inversion (→ phase reversal) PHYS (→ Phasenumkehrung)
phase inverter CIRC.ENG **Phasenumkehrschaltung**
= phase inverter circuit; phase inverter stage
= Phaseninverter; Phasenumkehrstufe
phase inverter circuit (→ phase inverter) CIRC.ENG (→ Phasenumkehrschaltung)
phase inverter stage (→ phase inverter) CIRC.ENG (→ Phasenumkehrschaltung)
phase inverter tube ELECTRON **Kathodynschaltung**
= Katodynschaltung; Phasenumkehrröhre
phase inverting amplifier CIRC.ENG **Phasenumkehrverstärker**
= Umkehrverstärker; invertierender Verstärker
phase jitter ELECTRON **Phasenschwankung**
[fast phase fluctuations]
= Phasenjitter; Phasenzittern
≈ wander
≈ Wander
↑ jitter
↑ Jitter
phase jump EL.TECH **Phasensprung**
= phase hit; phase transition
= Phasenübergang; Phasenänderung
≈ phase shift
≈ Phasenverschiebung
phase keyed MODUL **phasengetastet**
phase keying circuit CIRC.ENG **Phasenumtaster**
= Phasensprungschalter; Phasenschalter
phase lag PHYS **Phasenverzögerung**
= phase delay; phase retardation
= Phasennacheilung
phase-lag compensation CONTROL **Phase-lag-Kompensation**
phase lattice PHYS **Phasengitter**
phase lead ELECTRON **Phasenvoreilung**
= phase advance
phase-lead compensation CONTROL **Phase-lead-Kompensation**
phase linearity ELECTRON **Phasenlinearität**
phase-locked ELECTRON **phasenstarr**
= phasengerastet; phasensynchron; phasensynchronisiert
phase locked loop CIRC.ENG **Phasenregelkreis**
= PLL
= Phasenregelschleife; PLL
phase-locked oscillator CIRC.ENG **phasenstarrer Oszillator**
= PLO
= phasensynchronisierter Oszillator
phase locking CIRC.ENG **Phasenrastung**
phase margin CONTROL **Phasenreserve**
= Phasenrand; Phasenvorrat
phase meter INSTR **Phasenwinkelmesser**
= Phasenmesser
phase modifier (→ phase shifter) POWER ENG (→ Phasenschieber)
phase-modulated ELECTRON **phasenmoduliert**
phase modulation MODUL **Phasenmodulation**
= PM
↑ Winkelmodulation
↑ angle modulation
↓ Pulsphasenmodulation
↓ pulse phase modulation

phase noise		TELEC	Phasenrauschen	phase voltage	POWER SYS	Phasenspannung
phase pattern		ANT	Phasencharakteristik			= Sternspannung
			= Phasendiagramm	phase wire	POWER SYS	Phasenseil
phase perturbation		TELEC	Phasenstörung	phasing	TELEGR	Einphasen
phase plan		DATA PROC	Phasenplan	phasing line	ANT	Phasenleitung
phase plot		INSTR	Phasendiagramm	phasor (→rotating phasor)	EL.TECH	(→Wechselstromzeiger)
phase position (→phase relationship)		PHYS	(→Phasenlage)	phasor power (→complex power 1)	EL.TECH	(→komplexe Wechselleistung)
phase quadrature		MATH	90°-Phasenverschiebung	phenolic paper (→laminated paper)	ELECTRON	(→Hartpapier)
phase-pure (→correctly phased)		EL.TECH	(→phasenrein)	phenolic resin moulding	CHEM	Phenolharzpreßstoff
phase recorder		INSTR	Phasenschreiber	phenomenon	SCIE	Erscheinung
phase refractive index		PHYS	Gruppenbrechzahl			= Phänomen
phase regulation (→phase control)		ELECTRON	(→Phasenregelung)	Philbert transformer	COMPON	Philbert-Transformator
				Phillips screw (→cross-recessed screw)	MECH	(→Kreuzschlitzschraube)
phase regulator		CIRC.ENG	Phasenregler	philology (→linguistics)	SCIE	(→Sprachwissenschaft)
phase relation		EL.TECH	Phasenbeziehung	phon	ACOUS	Phon
			= Phasenrelation	phone (v.i.) (→telephone)	TELEPH	(→telefonieren)
phase relationship = phase position		PHYS	Phasenlage	phone book (→telephone directory)	TELEC	(→Fernsprechverzeichnis)
phase resonance		NETW.TH	Phasenresonanz	phone box (→telephone booth)	TELEPH	(→Fernsprechzelle)
phase response		EL.TECH	Phasengang	phone call record	OFFICE	Telefonnotiz
phase retardation (→phase lag)		PHYS	(→Phasenverzögerung)	phonecard	TERM&PER	Telefonkarte
phase reversal = phase inversion		PHYS	Phasenumkehrung = Phaseninversion	= taxcard; telecard ↓ telephone chip card; magnetic stripe card		↓ Telefon-Chipkarte; Magnetstreifenkarte; Hologrammkarte
phase reversal keying (→2 PSK)		MODUL	(→2 PSK)	phone hood (→telephone hood)	TELEPH	(→Telefonhaube)
phase rotator = rotator; wave rotator		MICROW	Phasendreher = Rotator	phoneme [smallest significance-carrying sound element; typically 50 per language] ↓ vowel; consonant; prosodeme	LING	Phonem [kleinste bedeutungsunterscheidende Lauteinheit; typisch 50 pro Sprache] ↓ Vokal; Konsonant; Prosodem
phase selective measuring rectifier		INSTR	phasenabhängiger Meßgleichrichter = gesteuerter Meßgleichrichter; phasenselektiver Meßgleichrichter; getasteter Meßgleichrichter; phasensynchroner Meßgleichrichter			
				phone plug (→jack connector)	TELEPH	(→Klinkenstecker)
phase-sensitive		EL.TECH	phasenempfindlich	phonetic alphabet	TELEC	Buchstabieralphabet
phase-sequence commutation		POWER SYS	Phasenfolgelöschung			= Buchstabierliste
phase shift		PHYS	Phasenverschiebung	phonetics [rules for pronunciation] ↑ grammar	LING	Phonetik [Regeln der Aussprache] = Lautlehre ↑ Grammatik
= phase deviation; phase displacement; phase angle; phase slip ≈ phase difference; phase jump; phase distortion [TELEC]			= Phasendrehung; Phasenfehler; Phasenabweichung; Phasenschlupf ≈ Phasendifferenz; Phasensprung; Phasenverzerrung [TELEC]			
				phono cartridge (→phonograph pickup)	EL.ACOUS	(→Tonabnehmer)
				phonogram record = phonograph record; record ↓ long-play record	CONS.EL	Schallplatte = Platte ↓ Langspielplatte
phase-shifted (→out of phase)		PHYS	(→phasenverschoben)	phonograph (→record player)	CONS.EL	(→Plattenspieler)
phase shifter		NETW.TH	Phasenkette	phonographic recording	CONS.EL	Schallplattenaufnahme
phase shifter = phase modifier; compensator; condenser		POWER ENG	Phasenschieber			= Plattenaufnahme
				phonograph pickup = phono pickup; pickup; pickup head; head; phono reproducer; reproducer; phono cartridge; cartridge	EL.ACOUS	Tonabnehmer = Schalldose
phase shifter = phase changer		MICROW	Phasenschieber			
phase shift keying = PSK		MODUL	Phasenumtastung = PSK	phonograph record (→phonogram record)	CONS.EL	(→Schallplatte)
phase shift oscillator		CIRC.ENG	Phasenkettenoszillator	phono jack = phono socket (BRI)	COMPON	Tonabnehmerbuchse
phase-shift transformer		EL.TECH	Phasendrehtrafo	phonon	PHYS	Phonon
phase slip (→phase shift)		PHYS	(→Phasenverschiebung)			= Schwingungsquantum
phase space		MATH	Phasenraum	phono pickup (→phonograph pickup)	EL.ACOUS	(→Tonabnehmer)
phase-space diagram		PHYS	Phasenraumdiagramm			
phase-space representation		EL.TECH	Phasenraumdarstellung	phono plug	COMPON	Tonabnehmerstecker
phase spectrum		TELEC	Phasenwinkelspektrum	phono reproducer (→phonograph pickup)	EL.ACOUS	(→Tonabnehmer)
			= Phasenspektrum	phono socket (BRI) (→phono jack)	COMPON	(→Tonabnehmerbuchse)
phase speed = phase velocity		PHYS	Phasengeschwindigkeit			
phase splitter		CIRC.ENG	Phasentrenner	phonotypist 1 [female]	OFFICE	Phonotypistin
phase spreading		PHYS	Phasenverschmierung	phonotypist 2 [male]	OFFICE	Phonotypist
phase swing (→phase deviation)		MODUL	(→Phasenhub)	phontyping [typing of recorded texts]	OFFICE	Phonotypie [Schreiben nach Tonträgeraufzeichnung]
phase synchronization		EL.TECH	Phasensynchronisierung			
phase tracking		ELECTRON	Phasengleichlauf			
phase transformer		POWER SYS	Drehtransformator	phosphate coating	METAL	Phosphatierung
phase transformer		ANT	Phasentransformator	phosphor (n.) (→luminescent material)	PHYS	(→Leuchtstoff)
phase transition (→phase jump)		EL.TECH	(→Phasensprung)			
phase velocity (→phase speed)		PHYS	(→Phasengeschwindigkeit)			

phosphor-bronze	METAL	**Phosphorbronze**
phosphor dot [color tube]	ELECTRON	**Leuchtstoffpunkt** [Farbbildröhre]
phosphorescence [luminescence of relatively long afterglow] ≈ fluorescence ↑ luminescence	PHYS	**Phosphoreszenz** [Luminszenz relativ langer Nachleuchtdauer] ≈ Fluoreszenz ↑ Luminszenz
phosphor triple	TERM&PER	**Farbtripel**
phosphorus = P	CHEM	**Phosphor** = P
photistor (→phototransistor)	MICROEL	(→Phototransistor)
photoarray	MICROEL	**Photo-Array** [Anordnung von Photobauteilen] = Foto-Array
photocapacitive effect	PHYS	**photokapazitiver Effekt** = fotokapazitiver Effekt
photo cathode	PHYS	**Photokathode** = Fotokathode
photocell [semiconductor device with a light-sensitive characteristic] = photoelectric cell; photovoltaic cell; photosensitive cell ≈ photovoltaic cell; phototube	MICROEL	**Photozelle** [Halbleiterbauelement mit lichtempfindlicher Kennlinie] = Fotozelle; photoelektrische Zelle; fotoelektrische Zelle; lichtelektrische Zelle; Lichtzelle; photoelektronisches Bauelement ≈ Photoelement; Photodiode
photochemics	CHEM	**Photochemie** = Fotochemie
photochopper	COMPON	**Photozerhacker** = Lichtzerhacker; Photochopper; Fotozerhacker; Fotochopper
photocomp device (→photocomposition system)	TYPOGR	(→Lichtsatzanlage)
photocomp machine (→photocomposition system)	TYPOGR	(→Lichtsatzanlage)
photo-composing equipment (→phototypesetting equipment)	TYPOGR	(→Fotosetzanlage)
photocomposition (→phototypesetting)	TYPOGR	(→Fotosatz)
photocomposition device (→photocomposition system)	TYPOGR	(→Lichtsatzanlage)
photocomposition machine (→photocomposition system)	TYPOGR	(→Lichtsatzanlage)
photocomposition system = photocomposition device; photocomposition machine; photocomp system; photocomp device; photocomp machine; filmsetting machine	TYPOGR	**Lichtsatzanlage** = Lichtsetzanlage
photocomp system (→photocomposition system)	TYPOGR	(→Lichtsatzanlage)
photoconduction	PHYS	**Photoleitung** = Fotoleitung
photoconductive cell (→photoresistor)	COMPON	(→Photowiderstand)
photo conductor	PHYS	**Photoleiter**
photocopy	OFFICE	**Fotokopie** = Photokopie; Ablichtung ≈ Lichtpause ↑ Pause
photocurrent = photoelectric current	PHYS	**Photostrom** = Fotostrom; photoelektrischer Strom; fotoelektrischer Strom
photo-Darlington transistor	MICROEL	**Photo-Darlington-Transistor** = Foto-Darlington-Transistor
photodetection	PHYS	**Photodetektion** = Fotodetektion
photodetector	COMPON	**Photodetektor** = Fotodetektor
photo detector	PHYS	**Lichtempfänger**
photodielectric effect	PHYS	**photodielektrischer Effekt** = fotodielektrischer Effekt
photodiode ≈ light emitting diode	MICROEL	**Photodiode** = Fotodiode ≈ Lumineszenzdiode
photo effect = photoelectric effect	PHYS	**Photoeffekt** = Fotoeffekt; lichtelektrischer Effekt; photoelektrischer Effekt; fotoelektrischer Effekt
photoelectric	PHYS	**lichtelektrisch** = photoelektrisch; fotoelektrisch
photoelectric cell (→photocell)	MICROEL	(→Photozelle)
photoelectric current (→photocurrent)	PHYS	(→Photostrom)
photoelectric effect (→photo effect)	PHYS	(→Photoeffekt)
photoelectric emission	PHYS	**photoelektrische Emission** = fotoelektrische Emission
photoelectric voltage (→photovoltage)	PHYS	(→Photospannung)
photoelectric yield	PHYS	**photoelektrische Ausbeute**
photoelectron	PHYS	**Photoelektron** = Fotoelektron; Leuchtelektron
photoelement (→photovoltaic cell)	MICROEL	(→Photoelement)
photoemission	PHYS	**Photoemission** = Fotoemission
photo etching	CHEM	**Photoätzverfahren** = Fotoätzverfahren
photogalvanic effect	OPTOEL	**Dember-Effekt** = photogalvanischer Effekt; fotogalvanischer Effekt
photographic emulsion	CHEM	**photografische Emulsion** = fotographische Emulsion
photographic-grade line	TRANS	**Bildleitung**
photography	TECH	**Fotografie** = Photographie
photogravure [intaglio printing with plates produced by photoetching] = heliogravure ↑ intaglio printing	TYPOGR	**Heliographie** [Tiefdruck mit in Photoätztechnik hergestellten Platten] = Heliogravüre ↑ Tiefdruck
photolithographic	MICROEL	**fotolithografisch** = photolithographisch
photolithography	MICROEL	**Fotolithografie** = Photolithographie; optische Lithografie
photoluminescence [induced by optic radiation] ↑ luminescence ↓ fluorescence; phosphorescence	PHYS	**Photolumineszenz** [durch optische Strahlen angeregt] = Fotolumineszenz ↑ Lumineszenz ↓ Fluoreszenz; Phosphoreszenz
photomagnetic	PHYS	**fotomagnetisch** = photomagnetisch
photomask	MICROEL	**Fotomaske** = Photomaske
photometric brightness (→luminance)	PHYS	(→Leuchtdichte)
photometry	PHYS	**Photometrie** = Fotometrie
photomultiplier (→electron multiplier)	ELECTRON	(→Elektronenvervielfacher)
photomultiplier tube (→electron multiplier tube)	ELECTRON	(→Elektronenvervielfachungsröhre)
photon	PHYS	**Photon** = Lichtquant; Foton
photon absorption	PHYS	**Photonenabsorption** = Fotonenabsorption
photon dose equivalent	PHYS	**Photonen-Äquivalentdosis**
photon energy	PHYS	**Photonenenergie** = Fotonenenergie
photon gas	PHYS	**Photonengas** = Fotonengas
photonic cell (→photovoltaic cell)	MICROEL	(→Photoelement)
photo-optic memory	TERM&PER	**photooptischer Speicher**
photo-pattern generation	MICROEL	**Lichtmaskenverfahren**
photoplotter	TERM&PER	**Lichtzeichenmaschine** = Fotoplotter

photopolymer	CHEM	Photopolymer	physical atmosphere	PHYS	physische Atmosphäre
photorefractive effect	PHYS	photorefraktiver Effekt	[= 101,325 pascal]		[= 101.325 Pascal]
		= fotorefraktiver Effekt	= atm		= atm
photorelay	COMPON	Lichtrelais	physical circuit	LINE TH	Stammleitung
		= Photorelais; Fotorelais	= physical line; real circuit; side circuit		= Stammkreis; Stamm
photoresist	MICROEL	Photolack	≠ phantom circuit		≠ Phantomkreis
↓ positive resist; negative resist		= Fotolack; Photoabdeckung; Fotoabdeckung	physical constant	PHYS	Konstante der Physik
		↓ Positivlack; Negativlack	physical layer	DATA COMM	Bitübertragungsschicht
photoresistor	COMPON	Photowiderstand	[1st layer of OSI; defines bit rate, power, transmission medium etc.]		[1.Schicht im ISO-Schichtentenmodell; legt Übertragungsgeschwindigkeit, Sendeleistung, Übertragungsmedium etc. fest]
= photoconductive cell; light-dependent resistor; LDR		= Fotowiderstand; Photovaristor; Fotovaristor; Photowiderstandszelle; Fotowiderstandszelle			= physikalische Schicht; physikalische Ebene
photo semiconductor	MICROEL	Photohalbleiter	physical line (→physical circuit)	LINE TH	(→Stammleitung)
		= Fotohalbleiter	physical protocol	DATA COMM	Bitübertragungsprotokoll
photosensitive	PHYS	lichtempfindlich	physical record (→data block)	DATA PROC	(→Datenblock)
= light sensitive		= photoempfindlich; fotoempfindlich	physical scientist	SCIE	Physiker
photosensitive cell	MICROEL	(→Photozelle)	physical unit	TECH	Baueinheit
(→photocell)			physics	SCIE	Physik
photosensitive paper	TERM&PER	Photopapier	PIA	MICROEL	PIA
		= Fotopapier; lichtempfindliches Papier	= peripheral interface adapter		
phototelegraph apparatus	TERM&PER	Bildtelegrafiegerät 1	pica 1	TYPOGR	Pica 1
		= Telebildgerät	[typefont with size of 12 points]		[Schriftart mit Schriftgröße 12 Point]
phototelegraph position	TELEGR	Bildvermittlungsplatz	pica 2	TYPOGR	Pica 2
phototelegraphy	TELEGR	(→Bildtelegrafie)	[typoghraphic measuring unit, 1/6 inch, aprox. 4,2 inch]		[typographische Maßeinheit, 1/6 Zoll, ca. 4,2 mm]
(→videotelegraphy)			≈ point		≈ Punkt
photothyristor	MICROEL	Photothyristor	Pica (BRI) (→type size 12 point)	TYPOGR	(→Schriftgröße 12 Punkt)
		= Fotothyristor	pi character (→special character)	DATA PROC	(→Sonderzeichen)
phototransistor	MICROEL	Phototransistor	pick-up level (→response level)	ELECTRON	(→Ansprechpegel)
= photistor		= Fototransistor; Optotransistor; Photistor	pick-and-place robot	MANUF	Bestückungsrobot
phototropic	PHYS	phototrop	picker	TERM&PER	Identifiziergerät
[changing when exposed to light]		[sich unter Lichteinwirkung verändernd]	picking device	TERM&PER	Bildschirmeingabegerät
		= fototrop	[to enter data on a screen]		↓ Maus; Lichtgriffel; Steuerknüppel
phototropy	PHYS	Phototropie	↓ mouse; light pen; joystick		
		= Fototropie	pickle (v.t.)	CHEM	abbeizen
phototube	ELECTRON	Vakuumphotozelle	= scour (v.t.)		= beizen
		= Hochvakuumzelle	↑ corrode		
phototypesetter	TYPOGR	(→Fotosetzanlage)	pickled	TECH	gebeizt
(→phototypesetting equipment)					[Reinigung]
phototypesetting	TYPOGR	Fotosatz	pick time	TERM&PER	Ansprechzeit
= photocomposition		= Photosatz; Lichtsatz	[repeat function of a keyboard]		[Wiederholfunktion einer Tastatur]
↑ cold type		↑ Kaltsatz	pick-up (→ test probe)	INSTR	(→Prüfspitze)
phototypesetting equipment	TYPOGR	Fotosetzanlage	pickup (→phonograph pickup)	EL.ACOUS	(→Tonabnehmer)
= phototypesetter; photo-composing equipment		= Photosetzanlage; Fotosetzgerät; Photosetzgerät	pickup arm	EL.ACOUS	Tonabnehmerarm
≈ filmsetting equipment		≈ Lichtsatzgerät	= tone arm; swinging arm		= Tonarm
photovoltage	PHYS	Photospannung	pickup arm cable	EL.ACOUS	Tonarmleitung
= photoelectric voltage		= Fotospannung	pick-up coil	ELECTRON	Sondenspule
photovoltaic cell	MICROEL	Photoelement	pick-up current (→response current)	ELECTRON	(→Ansprechstrom)
[converts radiation into electricity, without external operation voltage]		[wandelt Strahlung in Elektrizität, ohne Anlegen einer Betriebsspannung]	pick-up delay (→response delay)	ELECTRON	(→Ansprechverzögerung)
= photoelement; photonic cell		= Fotoelement; Photo-Spannungszelle; Photo-Sperrschichtzelle	pick-up electrode (→collector)	ELECTRON	(→Kollektor)
≈ photodiode; photocell		≈ Photodiode; Photozelle	pickup head (→phonograph pickup)	EL.ACOUS	(→Tonabnehmer)
photovoltaic cell (→photocell)	MICROEL	(→Photozelle)	pick-up magnet	EL.ACOUS	Abnahmemagnetkopf
photovoltaic power supply	POWER SYS	(→Solarstromversorgung)	≈ read head [TERM&PER]		= Lesekopf [TERM&PER]
(→solar power supply)			↑ magnetic head [TERM&PER]		↑ Magnetkopf [TERM&PER]
photovoltaics	PHYS	Photovoltaik	pickup recess (→recessed grip)	TECH	(→Griffmulde)
		= Fotovoltaik	pick-up threshold (→response threshold)	ELECTRON	(→Ansprechschwelle)
phrase intelligibility	TELEPH	Satzverständlichkeit			
= discrete sentence intelligibility			pick-up time (→response time)	ELECTRON	(→Ansprechzeit)
phraseology	LING	Phraseologie			
[rules of typical word combinations of a language]		[Regeln typischer Wortverbindungen und Redensarten einer Sprache]	pick-up value (→response value)	ELECTRON	(→Ansprechwert)
≈ syntax		≈ Syntax	pick-up voltage (→response voltage)	ELECTRON	(→Ansprechspannung)
pH value	CHEM	pH-Wert			
= pH		= pH			
physical	PHYS	physikalisch			
physical address (→absolute address)	DATA PROC	(→absolute Adresse)			
physical addressing (→absolute addressing)	DATA PROC	(→absolute Adressierung)			

picoammeter	INSTR	Picoampermeter
picocomputer [working in picoseconds]	DATA PROC	Pikosekunden-Computer
picofarad [10E-12 Farad] = pF	PHYS	Pikofarad [10E-12 Farad] = pF
pi connection (→delta section)	NETW.TH	(→Pi-Schaltung)
PI controller [proportional-integral action] ↑ continuous-action controller	CONTROL	PI-Regler [proportional-integral wirkend] ↑ stetiger Regler
picosecond [10E-12 seconds] = ps; psec	PHYS	Pikosekunde [10E-12 Sekunden] = ps; psec
pictograph (→icon)	DATA PROC	(→Piktogramm)
picture (→TV picture)	TV	(→Fernsehbild)
picture (COBOL) (→mask)	DATA PROC	(→Maske 1)
picture and waveform monitor	TERM&PER	(→Monitor)
picture black	TERM&PER	Bildschwarz [Fax]
picture converter = image converter	ELECTRON	Bildwandler
picture dot (→picture element)	INF.TECH	(→Bildpunkt)
picture element = pel ≈ picture element [INF.TECH]	TV	Abtastwert ≈ Bildpunkt [INF.TECH]
picture element [smallest addressable and reproducable picture element] = picture dot; point; mesh; scanning point; pixel ≈ picture element [TV]	INF.TECH	Bildpunkt [kleinstes adressierbares und darstellbares Bildelement] = Bildelement; Pixel ≈ Abtastwert [TV]
picture evaluation	SIGN.TH	Bildauswertung
picture frequency (→frame frequency)	TV	(→Bildfolgefrequenz)
picture graph	DOC	Symbolsäulendiagramm
picture information	TELEC	Bildinhalt
picture monitor (→monitor)	TERM&PER	(→Monitor)
picture processing (→image processing)	INF.TECH	(→Bildverarbeitung)
picture reproduction	TV	Bildwiedergabe
picture signal (→video signal)	TV	(→Videosignal)
picture switching	SWITCH	Bildvermittlung
picture synchronization	TV	Bildsynchronisierung
picture synthesis = Bildsynthese	TV	Bildzusammenstellung
picture telegraphy (→videotelegraphy)	TELEGR	(→Bildtelegrafie)
picture tone	TV	Bildträgerfrequenz
picture transmission (AM) (→videotelegraphy)	TELEC	(→Bildtelegrafie)
picture transmitter	TV	Bildsender
picture tube ↑ cathode ray tube ↓ TV picture tube; monitor tube; oscilloscope tube; radar tube	ELECTRON	Bildwiedergaberöhre = Bildröhre; Elektronenstrahl-Bildröhre; Kathodenstrahl-Bildröhre; CRT-Bildschirm; Bildwiedergabe-Elektronenröhre ↑ Kathodenstrahlröhre ↓ Fernsehbildröhre; Monitorröhre; Oszilloskopröhre; Radarröhre
PID controller	CONTROL	PID-Regler
piece = item; part	ECON	Stück
piece of data (→datum)	DATA PROC	(→Datum) (pl. -en)
piecewise	TECH	stückweise
pie chart	MATH	Kreisdiagramm = Kreisgraphik; Tortendiagramm; Tortengraphik; Kuchendiagramm; Kuchengraphik
pierce (v.t.) = perforate; lance ↑ stamp	METAL	lochen = durchstechen ↑ stanzen
Pierce circuit (→NOR gate)	CIRC.ENG	(→NOR-Glied)
Pierce element (→NOR gate)	CIRC.ENG	(→NOR-Glied)
Pierce function (→NOR operation)	ENG.LOG	(→NOR-Verknüpfung)
Pierce gate (→NOR gate)	CIRC.ENG	(→NOR-Glied)
Pierce operation (→NOR operation)	ENG.LOG	(→NOR-Verknüpfung)
Pierce oscillator ↑ crystal oscillator	CIRC.ENG	Pierce-Oszillator ↑ Quarzoszillator
piercing = perforating ↑ stamping	METAL	Lochen
pie recorder = circular chart recorder	INSTR	Kreisblatt-Schreiber
piezo buzzer	EL.ACOUS	Piezosummer
piezodiode	MICROEL	Piezodiode
piezoelectric	PHYS	piezoelektrisch
piezoelectric bell = Piezotonruf	TERM&PER	elektronischer Tonruf
piezoelectric ceramic	PHYS	Piezokeramik
piezoelectric component = piezoelectric device	COMPON	piezoelektrisches Bauelement
piezoelectric converter (→piezoelectric transducer)	COMPON	(→piezoelektrischer Wandler)
piezoelectric crystal	PHYS	Piezokristall
piezoelectric device (→piezoelectric component)	COMPON	(→piezoelektrisches Bauelement)
piezoelectric effect = Piezoeffekt	PHYS	piezoelektrischer Effekt
piezoelectricity	PHYS	Piezoelektrizität
piezoelectric loudspeaker = crystal loudspeaker; piezo speaker	EL.ACOUS	piezoelektrischer Lautsprecher = Kristallautsprecher; Piezolautsprecher
piezoelectric microphone (→crystal microphone)	EL.ACOUS	(→Kristallmikrophon)
piezoelectric pick-up	INSTR	piezoelektrischer Meßfühler
piezoelectric resonator	COMPON	piezoelektrischer Resonator
piezoelectric transducer = piezoelectric converter	COMPON	piezoelektrischer Wandler
piezoelectric transistor	COMPON	Piezotransistor
piezo horn tweeter	EL.ACOUS	Piezo-Hochtonhorn
piezomagnetic pick-up	INSTR	piezomagnetischer Meßfühler
piezo speaker (→piezoelectric loudspeaker)	EL.ACOUS	(→piezoelektrischer Lautsprecher)
piezo tweeter	EL.ACOUS	Piezo-Hochtonlautsprecher
PIGFET = p-channel isolated gate FET	MICROEL	PIGFET
piggy-pack board [pluggable onto a mother board] = expansion board	EQUIP.ENG	Huckepack-Baugruppe [auf eine Hauptplatine aufsteckbar] ≈ Erweiterungsplatine
pig iron	METAL	Roheisen
pigtail = pigtail fiber	OPT.COMM	Anschlußfaser = Anschlußlichtleiter; Pigtail
pigtail	POWER ENG	Bürstenlitze
pigtail fiber (→pigtail)	OPT.COMM	(→Anschlußfaser)
pilcrow (→paragraph sign 1)	TYPOGR	(→Paragraphzeichen)
pile (v.t.) = pile up; staple (v.t.) ≈ accumulate; stack (v.t.)	TECH	stapeln = aufstapeln ≈ anhäufen; häufen
pile (n.) = staple ≈ stack	TECH	Stapel ≈ Haufen; Schub
pile [ferrite core memory]	COMPON	Stapelblock [Ferritkernspeicher]
pile up (→pile)	TECH	(→stapeln)
piling (→stacking)	COLLOQ	(→Stapelung)
pillar	MICROEL	Kontakthöcker
pillar [an upright support for a superstructure] ≈ column; pylon; pilaster; mast	CIV.ENG	Pfeiler [freistehende Stütze eines Bauwerks, meist rechteckigen Querschnitts] ≈ Säule; Mast
pill-box aerial (→pill-box antenna)	ANT	(→Tortenschachtelantenne)
pill-box antenna = pill-box aerial ↑ segment antenna	ANT	Tortenschachtelantenne = Pillbox-Antenne ↑ Segmentantenne

pilot [FDM]		TRANS	**Pilot** [TF-Technik]		
pilot (→aircraft pilot)		AERON	(→Flugzeugpilot)		
pilot blocking filter (→pilot suppression filter)		TRANS	(→Pilotsperre)		
pilot channel		TELEGR	**Pilotkanal**		
pilot control (→pilot supervision)		TRANS	(→Pilotüberwachung)		
pilot detector (→pilot receiver)		TRANS	(→Pilotempfänger)		
pilot extraction		TRANS	**Pilotauskopplung**		
pilot failure		TRANS	**Pilotausfall**		
= pilot loss			= Pilotunterbrechung		
pilot frequency		TRANS	**Pilotfrequenz**		
pilot frequency technique		TRANS	**Pilottechnik**		
= pilot technique					
pilot generator		TRANS	**Pilotgenerator**		
pilot injection		TRANS	**Piloteinspeisung**		
pilot lamp		EQUIP.ENG	**Kontrollampe**		
= check lamp			= Kontrolleuchte		
pilot lamp		ANT	**Glühlampenindikator**		
pilot level		TRANS	**Pilotpegel**		
pilot loss (→pilot failure)		TRANS	(→Pilotausfall)		
pilot PCB		ELECTRON	**Versuchsleiterplatte**		
= prototype PCB					
pilot plant		TECH	**Versuchsanlage**		
pilot product		ECON	**Leitprodukt**		
= flagship product					
pilot project		TECH	**Pilotprojekt**		
pilot receipt confirmation [R2]		SWITCH	**Pilotrückmeldung** [R2]		
pilot receiver		TRANS	**Pilotempfänger**		
= pilot detector					
pilot regulating range		TRANS	**Pilotregelbereich**		
pilot regulation		TRANS	**Pilotregelung**		
pilot series (→test series)		MANUF	(→Vorserie)		
pilot supervision		TRANS	**Pilotüberwachung**		
= pilot control					
pilot suppression filter		TRANS	**Pilotsperre**		
= pilot blocking filter					
pilot system (→experimental system)		TECH	(→Versuchssystem)		
pilot technique (→pilot frequency technique)		TRANS	(→Pilottechnik)		
pilot tone		EL.ACOUS	**Pilotton**		
pilot-tone method		BROADC	**Pilotton-Verfahren** [UKW]		
pin		TECH	**Stift**		
↓ centering pin			↓ Zentrierstift		
pin [connector] ↑ terminal ↓ contact pin; terminal pin; solder pin; wrap pin		COMPON	**Stift** [Steckverbindung] ↓ Kontaktstift; Anschlußstift; Lötstift; Wrapstift		
pin (→bolt)		MECH	(→Bolzen)		
pin (→mandrel)		MECH	(→Dorn)		
pin arrangement		COMPON	**Stiftanordnung**		
= pin disposition					
pin assignment		EQUIP.ENG	**Stiftbelegung**		
			= Steckerbelegung		
pin cable lug		COMPON	**Stiftkabelschuh**		
pincers		TECH	**Pinzette**		
= tweezers					
pincers (→nipper pliers)		TECH	(→Beißzange)		
pinch-in effect		MICROEL	**Einschnürungseffekt**		
pinch off (v.t.)		MECH	**abkneifen**		
= disconnect			= abklemmen		
pinch-off (n.)		MICROEL	**Kanalabschnürung**		
pinch-off current		MICROEL	**Abschnürstrom**		
			= Pinch-off-Strom		
pinch-off effect		MICROEL	**Abschnüreffekt**		
			= Pinch-off-Effekt		
pinch-off voltage		MICROEL	**Abschnürspannung**		
			= Pinch-off-Spannung		
pinch resistor		MICROEL	**Pinch-Widerstand**		
pinch roller		TERM&PER	**Andruckrolle**		
pin compatibility		ELECTRON	**Steckerkompatibilität**		
= plug compatibility			= Anschlußkompatibilität; Pin-Kompatibilität		
pin compatible (→plug compatible)		EQUIP.ENG	(→steckkompatibel)		
pin contact		COMPON	**Stiftkontakt**		
pin contact strip		COMPON	**Stiftleiste**		
= pin strip			↑ Kontaktleiste		
↑ contact strip					
pincushion distortion		PHYS	**kissenförmige Verzeichnung** [Optik]		
= negative distortion			= Kissenverzeichnung		
pin depth		COMPON	**Stifteindringtiefe**		
PIN diode (→PIN photodiode)		MICROEL	(→PIN-Photodiode)		
PIN diode rectifier		CIRC.ENG	**PIN-Gleichrichter**		
pin disposition (→pin arrangement)		COMPON	(→Stiftanordnung)		
pi network (→delta section)		NETW.TH	(→Pi-Schaltung)		
pin-feed edge		TERM&PER	**Führungsstreifen**		
= sprockered margin; perfory (n.)			= perforierter Randstreifen		
pinfeed platen		TERM&PER	**Stachelrad**		
= sprocket wheel			≈ Vorschubtraktor		
≈ paper tractor			↑ Vorschubrad		
↑ feed wheel					
pin-for-pin compatible (→plug compatible)		EQUIP.ENG	(→steckkompatibel)		
pingpong (v.t.)		DATA PROC	**umherschalten**		
pinhole		MICROEL	**Pinhole**		
pinhole plate		PHYS	**Lochblende**		
pinion		MECH	**Ritzel**		
↑ gear			↑ Zahnrad		
pinlimited		MICROEL	**pinlimitiert**		
pin pattern		COMPON	**Stiftraster**		
PIN photodiode		MICROEL	**PIN-Photodiode**		
= PIN diode			= PIN-Diode		
pinpoint (v.t.) (→delimit)		TECH	(→eingrenzen)		
pinpoint (→isolate)		QUAL	(→lokalisieren)		
pin strip (→pin contact strip)		COMPON	(→Stiftleiste)		
pin wheel		MECH	**Stiftrad**		
PIO (→parallel input/output)		DATA PROC	(→parallele Ein-/Ausgabe)		
pip (n.)		TELEPH	**Tuten**		
pip [e.g. for calibration purposes] = blip		RADIO LOC	**Bildschirmmarkierung** [z.B. zu Eichzwecken]		
pipe (→duct)		OUTS.PLANT	(→Röhre)		
pipe (→tube)		TECH	(→Rohr)		
pipe burst		TECH	**Rohrbruch**		
pipe clamp		TECH	**Rohrschelle**		
piped heat		TECH	**Fernheizung**		
			= Fernwärme		
pipeline mode (→pipeline processing)		DATA PROC	(→Fließband-Verarbeitung)		
pipeline processing [beginning an instruction while still processing the previous one] = pipeline mode; instruction pipelining		DATA PROC	**Fließband-Verarbeitung** [Beginn der Ausführung eines Befehls während der Vorgänger noch bearbeitet wird] = Pipeline-Verarbeitung; Pipeline-Verfahren		
pipelining (→overlap)		DATA PROC	(→Überlappung)		
pipe thread		MECH	**Rohrgewinde**		
pipe tongs (→pipe wrench)		TECH	(→Rohrzange)		
pipe wall		TECH	**Rohrwandung**		
pipe wrench		TECH	**Rohrzange**		
= pipe tongs			= Wasserrohrzange; Pumpenzange		
piping diagram		TECH	**Rohrnetzplan**		
PIPO register (→parallel-in/parallel-out register)		DATA PROC	(→PIPO-Register)		
piracy		DATA PROC	**Piraterie**		
pirated copy		DATA PROC	**Raubkopie**		
= illegal copy; bootleg					
pirated edition		DOC	**Raubdruck**		
pirate transmitting station (→unlicensed transmitter)		RADIO	(→Schwarzsender)		
pi section (→delta section)		NETW.TH	(→Pi-Schaltung)		
PISO register (→parallel-in/serial-out register)		DATA PROC	(→PISO-Register)		
piston		MECH	**Kolben**		
piston diaphragm		EL.ACOUS	**Kolbenmembran**		

pit (→hole 2)	TECH	(→Loch 2)	
pitch (v.i.)	TECH	**stampfen**	
[angular oscillation along a lateral axis]		[um Querachse schwingen]	
pitch (n.)	MECH	**Ganghöhe**	
[thread]		[Gewinde]	
pitch	ENG.DRAW	**Teilungsmaß**	
		= Teilung	
pitch	ACOUS	**Tonhöhe**	
pitch	TERM&PER	**Teilung**	
≈ characters per inch		≈ Zeichen pro Zoll	
pitch 1 (→timbre)	ACOUS	(→Klangfarbe)	
pitch 2 (→standard tuning tone)	ACOUS	(→Stimmton)	
pitch (→character spacing)	TYPOGR	(→Zeichenabstand)	
pitch axis	TECH	**Stampfachse**	
		= Nickachse	
pitch circle	ENG.DRAW	**Teilkreis**	
pitch circle of gear	ENG.DRAW	**Zahnradteilkreis**	
pitch diameter 1	MECH	**Flankendurchmesser**	
pitch diameter 2	MECH	**Teilkreisdurchmesser**	
[gear]		[Zahnrad]	
pivot	MECH	**Drehzapfen**	
		= Zapfen	
pivot (→swivel)	MECH	(→schwenken)	
pivoting (→hinged)	MECH	(→schwenkbar)	
pivoting mechanism	MECH	**Schwenkmechanismus**	
= swiveling mechanism		= Schwenkvorrichtung	
≈ swing mechanism		≈ Drehvorrichtung	
pivot joint	MECH	**Drehzapfengelenk**	
pixel (→picture element)	INF.TECH	(→Bildpunkt)	
pixel array (→bit map 2)	TERM&PER	(→Pixelmuster)	
pixel map (→bit map 2)	TERM&PER	(→Pixelmuster)	
pixel memory	TERM&PER	**Bildpunktspeicher**	
pixel-oriented	DATA PROC	**pixelorientiert**	
≠ object-oriented		≠ objektorientiert	
pixel processor (→graphics processor)	DATA PROC	(→Graphikprozessor)	
Pixie tube	ELECTRON	**Pixie-Röhre**	
pk-pk (→peak-to-peak)	EL.TECH	(→Spitze-Spitze)	
PLA (→programmable logic array)	MICROEL	(→programmierbare Logikanordnung)	
-place	MATH	**-stellig**	
		= -figure	
place (n.) (→position 1)	TECH	(→Lage)	
place (→position)	DATA PROC	(→Stelle)	
placement	MANUF	**Bestückung**	
= insertion			
placement	ELECTRON	**Plazierung**	
[of components on PCB or chip]		[von Bauelementen auf Leiterplatte oder Chip]	
place of destination	ECON	**Bestimmungsort**	
= point of destination; destination		= Zielort	
place of work	TECH	**Arbeitsplatz**	
place value	CODING	(→Stellenwert)	
(→significance)			
plain	MECH	**geradstirnig**	
plain (obs.) (→flat)	TECH	(→flach)	
plain (obs.) (→planar)	MATH	(→planar)	
plain angle (→plane angle)	MATH	(→ebener Winkel)	
plain connector	MICROW	**Flachflanschverbinder**	
plain language	INF.TECH	**natürliche Sprache**	
[uncoded voice or text]		[unverschlüsselte Stimme oder Text]	
↓ plaintext		↓ Klartext	
plain-language (→natural-language)	INF.TECH	(→natürlichsprachlich)	
plain-old-telephone service (→POT service)	TELEC	(→konventioneller Fernsprechdienst)	
plain paper (→normal paper)	TERM&PER	(→Normalpapier)	
plain paper desk copier	OFFICE	**Normalpapier-Tischkopierer**	
plain paper pedestal copier	OFFICE	**Normalpapier-Standkopierer**	
plain taper key	MECH	**Flachkeil**	
= key on flat			
plaintext (n.)	INF.TECH	**Klartext**	
= plain writing; uncoded text; clear text		= unverschlüsselter Text	
≈ optical character [TERM&PER]		≈ Klarschrift [TERM&PER]	
≠ ciphertext		≠ verschlüsselter Text	
↑ plain language		↑ natürliche Sprache	
plain text (→optical character)	TERM&PER	(→Klarschrift)	
plain-text document	DATA PROC	**Klarschriftbeleg**	
plain washer	MECH	**Beilagscheibe**	
= washer		= Unterlegscheibe	
plain writing (→plaintext)	INF.TECH	(→Klartext)	
plan (v.t.)	ECON	**planen**	
plan (n.)	ECON	**Plan**	
= schedule; project		= Vorhaben; Projekt	
plan (→plan view)	ENG.DRAW	(→Grundriß)	
plan (→project)	TECH	(→Plan)	
planar	MATH	**planar**	
= plane; plain (obs.)		= eben	
planar (→flat)	TECH	(→flach)	
planar array	ANT	**ebene Gruppe**	
= flat plate		= planare Antenne; planare Gruppenantenne	
↓ circular array		↓ Kreisgruppenantenne	
planar cable grid	SYS.INST	**Flächenkabelrost**	
= overhead cable grid; cable grid; planar cable shelf		= Flächenrost; Rostmatte; Mattenkabelrost	
planar cable shelf (→planar cable grid)	SYS.INST	(→Flächenkabelrost)	
planar diode	MICROEL	**Planardiode**	
planar dipole array	ANT	**Dipolfeld**	
= dipole curtain array; dipole array		= Dipolwand; Dipolgruppe	
↓ curtain antenna		↓ Vorhangantenne	
planar doped	MICROEL	**planar dotiert**	
planar emitter	MICROEL	**Flächenemitter**	
planar spiral antenna	ANT	**ebene Spiralantenne**	
planar technique (→planar technology)	MICROEL	(→Planartechnik)	
planar technology	MICROEL	**Planartechnik**	
= planar technique			
planar transistor	MICROEL	**Planartransistor**	
↑ bipolar transistor		↑ Bipolartransistor	
planar tube	ELECTRON	**Scheibenröhre**	
= planar valve; disk-sealtube; disk-sealvalve; lighthouse tube		= Scheibentriode	
planar valve (→planar tube)	ELECTRON	(→Scheibenröhre)	
plan cabinet	ENG.DRAW	**Planschrank**	
Planck's constant	PHYS	**Plancksches Wirkungsquantum**	
[h = 6.626x10E-34 Js]		[h = 6,626x10E-34 Js]	
= Planck's radiation constant		= Wirkungsquantum; Planck-Konstante; Plancksche Strahlungskonstante	
Planck's law	PHYS	**Plancksches Gesetz**	
Planck's radiation constant (→Planck's constant)	PHYS	(→Plancksches Wirkungsquantum)	
plane (v.t.)	TECH	**hobeln**	
= shape (v.t.)			
plane (n.)	MATH	**Ebene**	
= level (n.)			
plane (n.)	TECH	**Hobel**	
plane (n.)	ENG.DRAW	**Projektionsebene**	
plane	DATA PROC	**Ebene**	
= level		= Stufe	
plane (→flat)	TECH	(→flach)	
plane (→planar)	MATH	(→planar)	
plane angle	MATH	**ebener Winkel**	
= plain angle			
plane antenna (→aperture antenna)	ANT	(→Aperturantenne)	
plane-concave	PHYS	**plankonkav**	
[optics]		[Optik]	
plane mirror	PHYS	**ebener Spiegel**	
planeness	MECH	**Ebenheit 1**	
= flatness		= Flachheit; Planheit	
plane of incidence	PHYS	**Einfallsebene**	
plane of vibration	PHYS	**Schwingungsebene**	
plane old telephone service	TELEC	**konventioneller Fernsprechdienst**	
= POT			
plane-parallel (adj.)	MATH	**planparallel**	
planetary gear (→planetary gear train)	MECH	(→Planetengetriebe)	
planetary gear train	MECH	**Planetengetriebe**	
= planetary gear		= Planetenradtrieb	

plane wave	PHYS	ebene Welle	plastic-encapsulated	COMPON	kunststoffgekapselt
planform (→plan view)	ENG.DRAW	(→Grundriß)	plastic-film capacitor	COMPON	**Kunststoffolien-Kondensator**
planimeter	MATH	Planimeter	= plastic-foil capacitor		= kunststoffkondensator; Metallfolienkondensator; Metall-Kunststoff-Kondensator; KC-Kondensator; Belagfolienkondensator
planish (v.t.)	METAL	flachstanzen			
planish (v.t.)	MECH	planieren			
planishing	METAL	Flachstanzen			
planned targed	ECON	Plansoll			
planning	ECON	Planung	plastic-foil capacitor	COMPON	(→Kunststoffolien-Kondensator)
planning guide	TECH	Planungsleitfaden	(→plastic-film capacitor)		
planning manual	TECH	Planungshandbuch	plastic housing	COMPON	**Kunststoffgehäuse**
planning phase	TECH	Planungsphase	= plastic package		= Plastikgehäuse
planographic printing	TYPOGR	**Flachdruck**	plastic-insulant cable	COMM.CABLE	kunststoffisoliertes Kabel
[printing and non-printing parts on the same level]		[druckende und nichtdruckende Teile in einer Ebene]	= plastic cable 2		= Kunststoffkabel 2
			plastic-insulated	COMM.CABLE	kunststoffisoliert
= planography		↑ Druckverfahren	plastic insulation	COMM.CABLE	**Kunststoffisolierung**
↑ printing process		↓ Lithographie; Offsetdruck	plasticity	TECH	**Formbarkeit**
↓ lithography; offset printing			≈ workability		≈ Verformbarkeit
planography (→planographic printing)	TYPOGR	(→Flachdruck)	plastic material (→synthetic material)	CHEM	(→Kunststoff)
plan position indicator	RADIO LOC	**Plan-Position-Indicator**	plastic package (→plastic housing)	COMPON	(→Kunststoffgehäuse)
= PPI		= PPI	plastic part	TECH	**Kunststoffteil**
plansheet (→worksheet)	OFFICE	(→Arbeitsblatt)	plastic picture (→plastic effect)	TV	(→Plastikeffekt)
plant	TECH	**Anlage**	plastic processing	TECH	**Kunststoffverarbeitung**
= works; installation 1; facility		[Einrichtung] = Installation 1	plastics chemistry	CHEM	**Kunststoffchemie**
plant (→production plant)	MANUF	(→Produktionsanlage)	plastic sheath	COMM.CABLE	**Kunststoffmantel**
plant (→installation)	ELECTRON	(→Anlage)	plastic-sheathed cable	COMM.CABLE	**Kunststoffmantel-Kabel**
plant (→controlled system)	CONTROL	(→Regelstrecke)	= plastic cable 1		= Kunststoffkabel 1
			plastic sleeve	OUTS.PLANT	**Kunststoffmuffe**
plant computer	DATA PROC	**Betriebsrechner**			= Plastikmuffe
≠ process computer		= Dispositionsrechner; Verwaltungsrechner ≠ Prozeßrechner	plastic technology	TECH	**Kunststofftechnik**
			plate (v.t.)	METAL	**plattieren 1**
plant construction	TECH	**Anlagenbau**	[mechanically, electrically or chemically]		[mechanisch, galvanisch oder chemisch]
Plante cell (→Plante plate)	POWER SYS	(→Großoberflächenplatte)	↑ clad; metallize		↑ metallisieren; kaschieren
Plante plate	POWER SYS	**Großoberflächenplatte**	↓ electroplate		↓ galvanisieren
= Plante cell		[Akkumulator]	plate (n.)	METAL	**Grobblech**
plant TV	TV	**Betriebsfernsehen**	plate (n.)	TECH	**Platte**
↑ closed circuit TV		↑ nichtöffentliches Fernsehen			= Platine
plan view	ENG.DRAW	**Grundriß**	plate	POWER SYS	**Platte**
= plan; planform			[accumulator]		[Akkumulator]
plasma	PHYS	**Plasma**	plate (AM) (→anode)	ELECTRON	(→Anode)
plasma beam	PHYS	**Plasmastrahl**	plate (AM) (→anode)	POWER SYS	(→Anode)
plasma display 1	TERM&PER	**Plasmaanzeige**	plate (AM) (→anode)	COMPON	(→Anode)
= plasma display panel; plasma panel; gas display panel; gas panel; gas plasma display; gas discharge display		= Gasplasma-Anzeige; Gasentladungsanzeige	plate (→electrode)	POWER SYS	(→Elektrode)
			plate (→electroplate)	METAL	(→galvanisieren)
			plate battery	POWER SYS	**Anodenbatterie**
plasma display 2 (→plasma screen)	TERM&PER	(→Plasma-Bildschirm)	= anode battery; high-tension battery; B-battery (NAM)		↑ Trockenbatterie
plasma display panel (→plasma display 1)	TERM&PER	(→Plasmaanzeige)	plate capacitor	COMPON	**Plattenkondensator**
			= plate condenser		
plasma etching	MICROEL	**Plasmaätzen**	plate condenser (→plate capacitor)	COMPON	(→Plattenkondensator)
plasma panel (→plasma display 1)	TERM&PER	(→Plasmaanzeige)	plate copier	OFFICE	**Plattenkopiergerät**
			plate current (→anode current)	ELECTRON	(→Anodenstrom)
plasma physics	PHYS	**Plasmaphysik**			
plasma screen	TERM&PER	**Plasma-Bildschirm**	plate deck	COMPON	**Plattenpaket**
= plasma display 2		= Plasma-Paneel; Plasma-Monitor; Gasplasma-Bildschirm; Gasplasma-Paneel; Gasplasma-Monitor ↑ Flachbildschirm			[Kondensator]
↑ flat screen			plate dissipation	ELECTRON	**Anodenverlustleistung**
			plated-through	ELECTRON	(→durchkontaktiert)
			(→through-contacted)		
plasma sputtering	MICROEL	**Plasmazerstäubung**	plated-through board	ELECTRON	(→durchkontaktierte Leiterplatte)
plasma video terminal	TERM&PER	**Plasma-Bildschirmterminal**	(→through-hole plated PCB)		
plastic (adj.)	TECH	**plastisch**	plated-through PCB	ELECTRON	(→durchkontaktierte Leiterplatte)
≈ forgeable [METAL]		= formbar ≈ schmiedbar [METAL]; biegsam	(→through-hole plated PCB)		
			plated wire memory	TERM&PER	(→Magnetdrahtspeicher)
			(→magnetic wire memory)		
plastic (n.) (→synthetic material)	CHEM	(→Kunststoff)	plated wire storage	TERM&PER	(→Magnetdrahtspeicher)
plastic cable 2	COMM.CABLE	(→kunststoffisoliertes Kabel)	(→magnetic wire memory)		
(→plastic-insulant cable)			platen	OFFICE	**Schreibwalze**
plastic cable 1	COMM.CABLE	(→Kunststoffmantel-Kabel)			= Walze
(→plastic-sheathed cable)			platen detent	OFFICE	**Zeilenraste**
plastic card coder	TERM&PER	**Plastikkartencodierer**	platen knob	OFFICE	**Walzendrehknopf**
		= Plastikkartencodierer	[typewriter]		[Schreibmaschine]
plastic card reader	TERM&PER	**Plastikkartenleser**	plate spring (→leaf spring)	MECH	(→Blattfeder)
plastic card terminal	TERM&PER	**Plastikkartenterminal**	plate tank circuit	ELECTRON	**Anodenschwingkreis**
plastic-cladded	TECH	kunststoffkaschiert	plate touch	POWER SYS	**Plattenschluß**
plastic-coated	TECH	kunststoffbeschichtet			[Akkumulator]
plastic effect	TV	**Plastikeffekt**	plate voltage (→anode voltage)	ELECTRON	(→Anodenspannung)
= plastic picture					

plate-voltage/current characteristic	ELECTRON	Anodenkennlinie	
platform = scaffold 2	CIV.ENG	Plattform ≈ Gerüst	
plating [mechanically, electrically or chemically] ↓ electroplating; cladding	METAL	Plattierung 1 [mechanisch, galvanisch oder chemisch] ↓ Elektroplattierung; mechanische Plattierung	
platinize	METAL	platinieren	
platinum = Pt	CHEM	Platin = Pt	
platter (→disk)	TERM&PER	(→Platte)	
plausibility = reasonableness ≈ validity	DATA PROC	Plausibilität ≈ Zulässigkeit	
plausibility check = reasonableness check ≈ validity check	DATA PROC	Plausibilitätsprüfung = Plausibilitätskontrolle; Plausibilitätstest ≈ Zulässigkeitsprüfung	
play (n.) = clearance; slackness ≠ interference	ENG.DRAW	Spielraum = Spiel ≠ Durchdringung	
playback (v.t.) ≈ rewind	TERM&PER	rückspielen [an den Bandanfang] ≈ rückspulen	
playback (n.) ≈ rewinding	TERM&PER	Rückspielen = Wiederabspielen ≈ Rückspulung	
playback	BROADC	Playback	
play key	CONS.EL	Wiedergabetaste	
PLC (→power-line cycle)	EL.TECH	(→Netzspannungsperiode)	
PLD (→programmable logic array)	MICROEL	(→programmierbare Logikanordnung)	
pleading	ECON	Schriftsatz	
"please hold the line"	TELEPH	„bitte warten"	
pledge = pawn; mortgage	ECON	Pfand	
pledge (→pawn)	ECON	(→verpfänden)	
pledge (→pawning)	ECON	(→Verpfändung)	
plenary assembly	ECON	Vollversammlung	
plenished (→full)	TECH	(→voll) (adj.)	
pleochroic	PHYS	pleochroisch	
pleochroism [direction-dependent colour decomposition]	PHYS	Pleochroismus [Zerlegung in verschiedenen Farben in unterschiedlichen Richtungen]	
plesiochronous [nominally synchronous, but within established tolerances] ≈ synchronous	TELEC	plesiochron [nominell synchron, jedoch innerhalb vorgegebener Toleranzen] ≈ synchron	
plesiochronous hierarchy	TELEC	Plesiochron-Hierarchie	
plesiochronous mode (→plesiochronous operation)	TELEC	(→plesiochroner Betrieb)	
plesiochronous network	TELEC	plesiochrones Netz	
plesiochronous operation = plesiochronous mode	TELEC	plesiochroner Betrieb	
plexiglas	TECH	Plexiglas	
pliability = suppleness ≈ flexibility	TECH	Geschmeidigkeit ≈ Biegsamkeit	
pliable = supple ≈ flexible	TECH	geschmeidig ≈ biegsam	
pliers ↓ combination pliers; side cutting pliers; adjusting pliers; flat-nose pliers	TECH	Zange 2 [klein] ↓ Kombinationszange; Seitenschneider; Justierzange; Flachzange 2	
pliers ammeter [with a hinged iron core] = clamp ammeter	INSTR	Zangenstrommeter [mit aufklappbarem Eisenkern] = Zangenamperemeter	
pliers meter = clamp meter; clamp-on meter	INSTR	Zangenmeßgerät	
pliers power meter [with a hinged iron core] = clamp power meter	INSTR	Zangenleistungsmesser [mit aufklappbarem Eisenkern] = Zangenwattmeter	
PLL (→phase locked loop)	CIRC.ENG	(→Phasenregelkreis)	
PLL demodulator	CIRC.ENG	PLL-Demodulator	
PLO (→phase-locked oscillator)	CIRC.ENG	(→phasenstarrer Oszillator)	
plot (v.t) = represent graphically	DOC	graphisch darstellen = grafisch darstellen	
plot (n.) ≈ diagram	DATA PROC	graphische Darstellung ≈ Diagramm	
plot speed = plotting speed; operating speed	TERM&PER	Zeichengeschwindigkeit	
plotter [digitally driven drawing machine, generally working with pencils] = graphic plotter; graph plotter; digital plotter ≈ graphics printer ↑ graphics output hardware ↓ flat bed plotter; drum plotter; X-Y plotter; incremental plotter; pencil plotter; thermal plotter; electrostatic plotter; photoplotter	TERM&PER	Plotter [digital gesteuertes Zeichengerät, meist mit Stiften arbeitend] = Zeichengerät; Digitalplotter; Kurvenschreiber ≈ Graphikdrucker ↑ Graphikausgabegerät ↓ Flachbettplotter; Trommelplotter; Koordinatenschreiber; Inkrementalplotter; Stiftplotter; Thermoplotter; elektrostatischer Plotter; Lichtzeichenmaschine	
plotter film	TERM&PER	Plotterfilm	
plotter pen	TERM&PER	Zeichenstift	
plotter roll	TERM&PER	Plotterfolie	
plotting head	TERM&PER	Schreibkopf 3 [Plotter]	
plotting plate	ENG.DRAW	Zeichenplatte	
plotting speed (→plot speed)	TERM&PER	(→Zeichengeschwindigkeit)	
plug (v.t.)	TELEPH	stöpseln	
plug (n.) [male part of a connecting device] = male connector; male ≠ jack	COMPON	Stecker [männlicher Teil der Steckverbindung] ≠ Buchse ↓ Klinkenstecker [TELEPH]; Netzstecker [EL.INST]	
plug (n.) ≠ jack	TELEPH	Stöpsel ≠ Klinke	
plug (n.)	TECH	Verschlußstopfen	
plugboard (→patch panel)	EQUIP.ENG	(→Schaltfeld)	
plug-board control (→patch-board control)	DATA PROC	(→Stecktafelsteuerung)	
plug body	COMPON	Steckerkörper	
plug charger [for rechargable batteries, accumulators]	POWER SYS	Steckerlader [für ladbare Batterien, Akkumulatoren]	
plug clamp	COMPON	Schnellspannklemme	
plug compatibility (→pin compatibility)	ELECTRON	(→Steckerkompatibilität)	
plug compatible = pin compatible; pin-for-pin compatible ≈ interface compatible	EQUIP.ENG	steckkompatibel = steckerkompatibel; pinkompatibel; anschlußkompatibel; sockelkompatibel ≈ schnittstellenkompatibel	
plug-compatible manufacturer = PCM	EQUIP.ENG	steckerkompatibler Hersteller = PCM	
plug connection (→plug-in connection)	COMPON	(→Steckverbindung)	
plug connector [pair of male and female connector] = connector	COMPON	Steckverbinder [Pärchen von Stecker und Buchse]	
plug contact	COMPON	Steckkontakt	
plug contact (→mains socket)	EL.INST	(→Netzsteckdose)	
plug fuse	COMPON	Schraubsicherung [schraubbare Stromsicherung]	
pluggable = plug-in type; connectorized ↓ repluggable	ELECTRON	steckbar ↓ umsteckbar	
pluggable cable (→plug-in cable)	EQUIP.ENG	(→Steckkabel)	
plug gage (→plug gauge)	MECH	(→Lehrdorn)	
plug gauge = plug gage; gauge plug; gage plug	MECH	Lehrdorn	
plugging (→plugging break)	POWER ENG	(→Gegenstrombremsung)	
plugging break = plugging	POWER ENG	Gegenstrombremsung	

plug-in (v.t.)		ELECTRON	**stecken**	**pneumatic** (adj.)	TECH	**pneumatisch**
= insert				**pneumatically activated**	TECH	(→druckluftbetätigt)
plug-in (→slide-in unit)		EQUIP.ENG	(→Einschub)	(→pneumatically operated)		
plug-in board (→plug-in module)		EQUIP.ENG	(→Steckbaugruppe)	**pneumatically actuated** (→pneumatically operated)	TECH	(→druckluftbetätigt)
plug-in cable		EQUIP.ENG	**Steckkabel**	**pneumatically controlled**	TECH	(→druckluftbetätigt)
= connector-ended cable; connectorized cable; pluggable cable;			= Steckerkabel ≈ vorgefertigtes Kabel	(→pneumatically operated) **pneumatically operated**	TECH	**druckluftbetätigt**
≈ preformed cable				= pneumatically activated; pneumatically actuated; pneumatically controlled; air-operated; air-activated; air-actuated; air-controlled; pressure-operated; pressure-activated; pressure-actuated; pressure-controlled		= pneumatisch betätigt; druckluftgesteuert; pneumatisch gesteuert
plug-in card (→plug-in module)		EQUIP.ENG	(→Steckbaugruppe)			
plug-in connection		COMPON	**Steckverbindung**			
= plug connection						
plug-in filter		INSTR	**Einschubfilter**			
plug-in joint		OPT.COMM	**Fügeverbindung**	**pneumatic control**	TECH	**pneumatische Steuerung**
plug-in module		EQUIP.ENG	**Steckbaugruppe**	= air control		= Pneumatiksteuerung; Drucklufsteuerung
= plug-in board; plug-in card			= Einschubmodul; Steckmodul; Steckkarte	**pneumatic damping**	TECH	**Luftdämpfung**
≈ expansion module			≈ Erweiterungskarte	= air damping; air-friction damping		= pneumatische Dämpfung
↑ module			↑ Baugruppe	**pneumatic dispatch** (→pneumatic post)	POST	(→Rohrpost)
plug-in option		INSTR	**Einschuboption**	**pneumatic gate** (→fluidic gate)	DATA PROC	(→fluidisches Schaltelement)
plug-in place		EQUIP.ENG	**Steckplatz**			
↑ mounting place			↑ Einbauplatz	**pneumatic post**	POST	**Rohrpost**
↓ expansion slot			↓ Erweiterungssteckplatz	= pneumatic dispatch; pneumatic tube system		
plug-in position (→board position)		EQUIP.ENG	(→Baugruppenposition)	**pneumatic tube system** (→pneumatic post)	POST	(→Rohrpost)
plug-in power supply		EQUIP.ENG	**Steckernetzgerät** = Steckernetzteil	**pneumatic valve**	TECH	**Druckluftventil**
plug-in type (→pluggable)		ELECTRON	(→steckbar)	**pnin transistor**	MICROEL	**PNIN-Transistor**
plug link		ELECTRON	**Brückenstecker**	**pn junction**	PHYS	**PN-Übergang**
= link connector; jumper plug; U plug			= Steckerbrücke	**pn photo effect**	PHYS	**Sperrschicht-Photoeffekt**
≈ wire link			≈ Drahtbrücke			= photovoltaischer Effekt; fotovoltaischer Effekt; PN-Photoeffekt
↑ strap			↑ Brücke			
plug locking		EQUIP.ENG	**Steckerverriegelung**	**pnpn diode**	MICROEL	**Vierschichtdiode**
plug pin (→connector pin)		COMPON	(→Steckerstift)	≈ trigger diode [CIRC.ENG]; binistor		= PNPN-Diode ≈ Triggerdiode [CIRC.ENG]; Binistor
plug resistance box		PHYS	**Stöpselwiderstand**			
plug socket (→jack)		COMPON	(→Buchse)	**pnpn structure**	MICROEL	**PNPN-Struktur**
plugtable (→patch panel)		EQUIP.ENG	(→Schaltfeld)	**pnpn transistor**	MICROEL	**PNPN-Transistor**
plug with grounding contact (→Schuko plug)		EL.INST	(→Schukostecker)	**pnp transistor**	MICROEL	**PNP-Transistor**
				Po (→polonium)	CHEM	(→Polonium)
plug with protective ground (→Schuko plug)		EL.INST	(→Schukostecker)	**poaching** [unlicensed search of data]	DATA PROC	**unbefugte Datensuche**
plumb (v.t.)		CIV.ENG	**loten**	**POB** (→post office box)	POST	(→Postfach)
[to adjust the vertical]			[die Senkrechte einstellen]	**POB process** [push-out base]	MICROEL	**POB-Methode** = POB-Verfahren; POB-Prozess
plumb (adj.)		TECH	**lotrecht**			
≈ vertical [MATH]			≈ senkrecht [MATH]	**POB technique** (→alloy diffusion technique)	MICROEL	(→AD-Technik)
plump (v.t.)		TECH	**verplomben**			
≈ seal 2 (v.t.)			≈ versiegeln	**pocket**	TECH	**Fach**
plumber		TECH	**Installateur**	= stacker		[zur Ablage]
plumbicon		ELECTRON	**Plumbikon**	**pocket book**	TYPOGR	**Taschenbuch**
↑ camera tube			= Plumbicon; Leddicon ↑ Bildaufnahmeröhre	≈ paperback		
				pocket calculator	DATA PROC	**Taschenrechner**
plunger-coil controller		CONTROL	**Tauchspulenregler**	= hand calculator		≈ Taschencomputer
plunger relay		COMPON	**Hubrelais**	≈ pocket computer		
plunger-type armature		COMPON	**Tauchanker**	**pocket computer**	DATA PROC	**Taschencomputer**
plural		LING	**Plural**	= hand-held computer		= Handheld-Computer; Pocket-Computer
			= Mehrzahl	≈ pocket calculator; laptop computer		≈ Taschenrechner; Aktentaschencomputer
plural scattering (→multiple scattering)		PHYS	(→Mehrfachstreuung)	**pocket meter**	INSTR	**Taschen-Meßgerät**
plus		MATH	**plus**	**pocket size**	TECH	**Taschenformat**
= and			= und	**pocket telephone**	MOB.COMM	**Taschentelefon**
plus-minus sign		MATH	**Plus-minus-Zeichen**	**pod** (→test probe)	INSTR	(→Prüfspitze)
[symbol ±]			[Symbol ±]	**POH** (→path overhead)	TELEC	(→Path Overhead)
= plus-minus symbol				**Poincaré sphere**	NETW.TH	**Poincaré-Kugel**
↑ mathematical symbol						= Polarisationskugel
plus-minus symbol (→plus-minus sign)		MATH	(→Plus-minus-Zeichen)	**point** (n.) [to separate decimal fractions; observation: in German speaking countries the punctuation mark used therefor is the comma, consequently the correct translation is "Komma"]	MATH	**Komma** (n.n.; pl.-s oder -tas) [zur Abtrennung von Dezimalstellen; Hinweis: im angelsächsischem Sprachraum verwendet man dafür den Punkt, daher mit „point" zu übersetzen]
plus potential		PHYS	**Pluspotential**			
			= positives Potential			
plus sign		MATH	**Plus-Zeichen**			
[+]			[+]			
= plus symbol			≈ Und-Zeichen			
≈ ampersand						
plus symbol (→plus sign)		MATH	(→Plus-Zeichen)			
plutonium		CHEM	**Plutonium**	= decimal point		= Dezimalkomma; Dezimalpunkt; Radixpunkt
= Pu			= Pu			
plywood		TECH	**Sperrholz**			
PM (→phase modulation)		MODUL	(→Phasenmodulation)			
Pm (→promethium)		CHEM	(→Promethium)			
p-MOS technology		MICROEL	**PMOS-Technik**			

point

point (n.)   LING   **Punkt 2**
[part of a text]   [Textteil]
point   MECH   **Kuppe**
[of a screw or bolt]   [einer Schraube]
≠ screw head   ≠ Schraubenkopf
point   TYPOGR   **Punkt**
[typographic unit, used especially to indicate character hights; 0.376 065 mm]   [typographische Maßeinheit, insbesonders für Angabe von Schriftgrößen verwendet; 0,376 065 mm]
= printer's point; p   = typographischer Punkt; p
↓ anglo-american point; Didot point   ≈ Pica 1; Cicero
    ↓ angelsächsischer Punkt; kontinentaleuropäischer Punkt
point (→picture element)   INF.TECH   (→Bildpunkt)
point 1 (→dot)   LING   (→Punkt 1)
point (v.t.)   ENG.DRAW   (→punktieren)
(→punctuate)
point 1 (→punctuation mark)   LING   (→Satzzeichen)
point brightness   PHYS   **Punkthelligkeit**
point cathode   ELECTRON   **Punktkathode**
point charge   PHYS   **Punktladung**
    = punktförmige Ladung
point contact   MICROEL   **Spitzenkontakt**
    = Punktkontakt
point contact diode   MICROEL   **Spitzendiode**
    = Spitzenkontaktdiode
point-contact rectifier   MICROEL   **Spitzengleichrichter**
    = Punktgleichrichter
point-contact transistor   MICROEL   **Spitzentransistor**
≠ junction transistor   = Punktkontakttransistor
↑ bipolar transistor   ≠ Flächentransistor
    ↑ Bipolartransistor
point dipole   ANT   **Punktbündel-Dipol**
point discharge   PHYS   **Spitzenentladung**
pointed   TECH   **spitz**
≈ pungent   = spitzig
≠ blunt 1   ≈ stechend
    ≠ stumpf 1
point electrode   ELECTRON   **Spitzenelektrode**
pointer   MICROEL   **Hinweisadresse**
[microprocessor]   [Mikroprozessor]
    = Absolutzeiger; Zeiger
pointer   TECH   **Marke**
pointer   INSTR   **Zeiger**
= needle; moving element   = Nadel
↓ lance pointer; rod pointer; blade-type pointer; filament pointer   ↓ Lanzenzeiger; Stabzeiger; Messerzeiger; Fadenzeiger
pointer   DATA PROC   **Zeiger**
[an identifier indicating the location of an item of data]   [ein Hinweis auf die Adresse einer bestimmten Information]
    = Absolutzeiger; Hinweisadresse; Pointer
pointer   TRANS   **Zeiger**
[SDH/SONET; defines the phase lag between the begin of payload information and frame]   [SDH/SONET; gibt die Phasenbeziehung zwischen Beginn der Nutzinformation und Rahmen an]
    = Pointer
pointer (→cursor)   TERM&PER   (→Schreibmarke)
pointer (→chain field)   DATA PROC   (→Kettfeld)
pointer throw   INSTR   **Zeigerausschlag**
= needle throw   = Nadelausschlag
point estimation   MATH   **Punktschätzung**
point frequency (→dot frequency)   ELECTRON   (→Punktfrequenz)
point-group marking   SWITCH   **Punkt-Bündel-Markierung**
point indication   DATA PROC   **Punktbeschreibung**
pointing loss   SAT.COMM   **Mißweisungsdämpfung**
point of destination (→place of destination)   ECON   (→Bestimmungsort)
point of intersection   MATH   **Doppelpunkt**
↑ singularity   ↑ Singularität
point of presence (→delivery point)   TELEC   (→Übergabepunkt)
point of rotation   PHYS   **Drehpunkt**
point-of-sale   TERM&PER   **Kassenplatz**
= POS   = Kasse
point-of-sale terminal   TERM&PER   **Kassenterminal**
= POS terminal   = POS-Kasse
point of termination (→delivery point)   TELEC   (→Übergabepunkt)
point protection (→object protection)   SIGN.ENG   (→Objektschutz)
point recorder   TERM&PER   **Punktdrucker**
    = Punktschreiber
point set (→set of points)   MATH   (→Punktmenge)
point set curve   ENG.DRAW   **Segmentkurve**
point shifting   DATA PROC   **Kommaverschiebung**
point source   PHYS   **Punktquelle**
    = punktförmige Quelle; Punktstrahler; punktförmiger Strahler
point-to-multipoint   TELEC   **Punkt zu Mehrpunkt**
point-to-multipoint connection   TELEC   **Punkt-zu-Mehrpunkt-Verbindung**
point-to-multipoint operation   TELEC   **Punkt-zu-Mehrpunkt-Betrieb**
= point-to-multipoint service; broadcast operation; broadcast
point-to-multipoint service   TELEC   (→Punkt-zu-Mehrpunkt-Betrieb)
(→point-to-multipoint operation)
point-to-point   TECH   **punktweise**
    = Punkt-zu-Punkt
point-to-point   TELEC   **Punkt zu Punkt**
point-to-point circuit (→fixed line)   TELEC   (→Standleitung)
point-to-point connection 1   TELEC   **Punkt-zu-Punkt-Verbindung 1**
≠ point-to-multipoint connection
    = Zweipunktverbindung; Einzelverbindung
    ≠ Punkt-zu-Mehrpunkt-Verbindung
point-to-point connection 2   TELEC   (→Standleitung)
(→fixed line)
point-to-point line (→fixed line)   TELEC   (→Standleitung)
point-to-point measurement   INSTR   **punktweises Messen**
= spot measurement; spot analysis   = Punktmessen
≠ sweep measurement   ≠ Wobbelmessung
point-to-point operation   TELEC   **Punkt-zu-Punkt-Betrieb**
= point-to-point service
point-to-point position control   CONTROL   **Punktsteuerung**
    ↑ numerische Steuerung
point-to-point service   TELEC   (→Punkt-zu-Punkt-Betrieb)
(→point-to-point operation)
point-to-point traffic   TELECONTR   **End-End-Verkehr**
    = Punkt-zu-Punkt-Verkehr
Poise   PHYS   **Poise**
[measuring unit for dynamic viscosity; = 1 g/cm · s]   [Maßeinheit für dynamische Viskosität; = 1 g/cm · s]
= P   = P
poisoning (→contamination)   TECH   (→Verunreinigung)
Poisson distribution   MATH   **Poisson-Verteilung**
Poisson process   TELEC   **Poisson-Prozeß**
Poisson's equation   PHYS   **Poisson-Gleichung**
    = Poissonsche Differentialgleichung
poke (n.)   DATA PROC   **Direktspeicheranweisung**
[instruction allowing to place in any store location]
poke (v.t.) (→read-in)   DATA PROC   (→Einlesen)
POL (→procedure-oriented programming language)   DATA PROC   (→prozedurorientierte Programmiersprache)
POL (→high-level programming language)   DATA PROC   (→problemorientierte Programmiersprache)
polar (→polarized)   PHYS   (→polarisiert)
polar chart (→polar diagram)   EL.TECH   (→Polardiagramm)
polar circuit   NETW.TH   **Polarkreis**
polar coordinate   MATH   **Polarkoordinate**
[bidimensional]   [zweidimensional]
= spherical coordinate   = polare Koordinate; Kugelkoordinate; sphärische Kooordinate
polar current (AM) (→double current)   TELEGR   (→Doppelstrom)
polar diagram   EL.TECH   **Polardiagramm**
= polar chart   = polares Schaubild

polar distance	MATH	Polabstand
polar intensity	PHYS	Polstärke
= pole strength		
polar inversion (→polar reversal)	EL.TECH	(→Umpolung)
polarity	PHYS	Polarität
polarity (→marking pulse)	TELEGR	(→Stromschritt)
polarity correlator	CIRC.ENG	Polaritätskorrelator
polarity integrity	TELEC	Polaritätstreue
polarity inversion	EL.TECH	Polaritätsumkehr
= inverse polarity; reverse connection; reverse polarity		= Verpolung
≈ faulty polarization		≈ Falschpolung
polarity inversion relay	POWER SYS	Polaritätsumschalterelais
polarity reset	ELECTRON	Rückpolung
polarizability	PHYS	Polarisierbarkeit
polarization	PHYS	Polarisation
		= Polarisierung
polarization (→biasing)	PHYS	(→Vorpolung)
polarization diplexer	RADIO REL	(→Polarisationsweiche)
(→polarization filter)		
polarization diversity	RADIO	Polarisationsdiversity
polarization efficiency	ANT	Polarisationswirkungsgrad
polarization ellipse	ANT	Polarisationsellipse
polarization fading	RADIO PROP	Polarisationsschwund
polarization filter	PHYS	Polarisationsfilter
polarization filter	RADIO REL	Polarisationsweiche
= polarization diplexer		≈ Richtungsweiche
≈ directional filter		
polarization/frequency diplexer	RADIO REL	Polarisation-/Frequenz-Weiche
= quadruplexer		
≈ system diplexer		≈ system diplexer
polarization match	ANT	Polarisationsanpassung
polarization mismatch	ANT	Polarisations-Fehlanpassung
polarization pattern	ANT	Polaristionsverteilung
polarization plane	PHYS	Polarisationsebene
polarize	PHYS	polarisieren
polarize (→premagnetize)	PHYS	(→vormagnetisieren)
polarized	PHYS	polarisiert
= polar		
polarized connector	COMPON	gepolter Stecker
[avoids false plugging]		[verhindert Fehlsteckung]
polarized light	PHYS	polarisiertes Licht
polarized relay	COMPON	polarisiertes Relais
[maintains the contact position after interruption of activation current]		[Schaltstellung bleibt auch nach Unterbrechung des Erregerstromes]
↓ remanent relay		↓ Haftrelais
polarizer	MICROW	Polarisationsfilter
		= Polarisator
polarizing current	PHYS	Polarisationsstrom
polarizing filter	TERM&PER	Polarisationsfilter
polarizing potential	PHYS	Polarisationsspannung
polarizing slot	COMPON	Polarisierungsschlitz
		[gepolter Stecker]
polar light	GEOPHYS	Polarlicht
polar mount	ANT	Polarmounthalterung
polar reversal	EL.TECH	Umpolung
= pole reversal; polar inversion; pole inversion; poling		= Polwendung
pole	PHYS	Pol
≈ electrode		≈ Elektrode
pole	NETW.TH	Pol
pole	GEOPHYS	Pol
pole (→mast)	OUTS.PLANT	(→Mast)
pole arm (→crossarm)	MECH	(→Querträger)
pole climbers	OUTS.PLANT	Steigestütze
= pole steps		
pole detector	INSTR	Polaritätsdetektor
[indicates the sense of direct current]		[zeigt die Richtung des Gleichstroms an]
		= Polaritätsanzeiger
pole effect	PHYS	Pol-Effekt
pole extension	OUTS.PLANT	Mastverlängerung
pole fender	OUTS.PLANT	Prellpfahl
pole gap (→magnetic gap)	PHYS	(→Magnetluftspalt)
pole hole	OUTS.PLANT	Stangenloch
pole inversion (→polar reversal)	EL.TECH	(→Umpolung)

pole line	OUTS.PLANT	Mastlinie
		= Gestänge; Stangenlinie
pole-mounted	EQUIP.ENG	an Mast befestigt
		= für Mastbefestigung
pole-mounted container	OUTS.PLANT	Mastbehälter
pole mounting	EQUIP.ENG	Mastmontage
pole piece (→pole shoe)	EL.TECH	(→Polschuh)
pole point	NETW.TH	Polstelle
pole Q	NETW.TH	Polgüte
pole removal	NETW.TH	Polabspaltung
↓ total removal		↓ Vollabbau
pole reversal (→polar reversal)	EL.TECH	(→Umpolung)
pole shifter	POWER ENG	Polumschalter
pole shoe	EL.TECH	Polschuh
= pole piece		
pole socket	OUTS.PLANT	Stangenfuß
= butt end		
pole steps (→pole climbers)	OUTS.PLANT	(→Steigestütze)
pole strength (→polar intensity)	PHYS	(→Polstärke)
pole-zero configuration	NETW.TH	PN-Plan
police (n.)	ECON	Polizei
police call device	TERM&PER	Polizeirufeinrichtung
police radar	RADIO LOC	Polizeiradar
police telegraph equipment	TERM&PER	Polizei-Bildtelegraphiegerät
policing (→cell stream policing)	TELEC	(→Zellstromkontrolle)
policy (→insurance policy)	ECON	(→Versicherungspolice)
poling (→polar reversal)	EL.TECH	(→Umpolung)
polish (v.t.)	TECH	polieren
= finish		≈ schleifen
≈ grind		
polish (n.)	TECH	Polieren
polish notation	MATH	polnische Schreibweise
[algebraic notation without parantheses, with operand acting to the right]		[klammerlose algebraische Schreibweise, bei der die Operatoren nach rechts wirken]
= postfix notation		= polnische Notation; klammerfreie Schreibweise; Postfixnotation
political economy	ECON	Volkswirtschaftslehre
↑ economics		↑ Wirtschaftswissenschaften
poll (v.t.)	TELECONTR	zyklisch abfragen
↑ inquiry cyclically		= pollen
		↑ abfragen
poll (n.)	DATA COMM	Abruf
[invitation of a central towards peripherals to send]		[Sendeaufforderung durch eine Zentrale]
= polling 1		= Sendeabruf; Aufruf; Sendeaufruf; Polling
poll	ECON	Meinungsumfrage
poll (v.t.) (→request)	DATA COMM	(→abrufen)
poll (v.t.) (→inquire)	DATA PROC	(→abfragen)
polled-circuit operation	DATA COMM	(→Leitungsabrufbetrieb)
(→line-polling operation)		
polled-line operation	DATA COMM	(→Leitungsabrufbetrieb)
(→line-polling operation)		
polled station	TELECONTR	zyklisch abgefragte Station
		= gepollte Station
polling	TELECONTR	zyklische Abfrage
		= Sendeaufruf
polling 1 (→poll)	DATA COMM	(→Abruf)
polling 2 (→polling mode)	DATA COMM	(→Abrufbetrieb)
polling mode	DATA COMM	Abrufbetrieb
= selecting mode; polling 2		= Aufrufbetrieb; Aufrufverfahren; Anrufbetrieb
≠ request mode		≠ Anforderungsbetrieb
polling mode	TELECONTR	Aufrufbetrieb
[transmission is activated by interrogation commands]		[Übermittlung wird durch Abrufbefehl ausgelöst]
= transmission on demand		= Abfragebetrieb
polling signal	DATA COMM	Datenabrufsignal
		= Abrufsignal
polling signal	TELECONTR	(→Abfragesignal)
(→interrogation signal)		
polling station	TELEC	Abfragestation
= answering station		= Abfragestelle; abfragende Station
polling station	TELECONTR	zyklisch abfragende Station
		= pollende Station

polling time		TELECONTR	Abfragezeit	pool (n.)	DATA PROC	Pool
			= Pollingzeit	pooler	DATA PROC	Datenaufbereiter
polluant (n.)		TECH	Schadstoff	[device to preprocess key entry data]		
= toxid agent; harmful agent				pooling	DATA PROC	Pooling
pollute (→contaminate)		TECH	(→verunreinigen)	pool management	DATA PROC	Poolverwaltung
pollution (→pollution of environment)		TECH	(→Umweltverschmutzung)	pool organization	DATA PROC	Poolorganisation
pollution (→contamination)		TECH	(→Verunreinigung)	poor conductor	PHYS	schlechter Leiter
pollution control		TECH	Umweltschutz	poor quality	TECH	schlechte Qualität
pollution of environment		TECH	Umweltverschmutzung	= inferior quality		= minderwertige Qualität; unterwertige Qualität
= pollution			= Umweltbelastung			
polonium		CHEM	Polonium	poor sale	ECON	schlechter Absatz
= Po			= Po	pop (v.t.)	DATA PROC	abheben
polyacetal		CHEM	Polyacetal	[retrieve from top of stack]		[den obersten Speicherinhalt]
polyamide		CHEM	Polyamid	= pull (v.t.)		
↓ nylon			↓ Nylon	≠ push (v.t.)		≠ draufegen
polycarbonate		CHEM	Polycarbonat	popcorn noise (AM)	TELEC	(→Funkelrauschen)
polycristalline structure		PHYS	polykristalline Struktur	(→semiconductor noise)		
polycrystal		PHYS	Polykristall	pop instruction	DATA PROC	Abhebungsanweisung
polycrystalline		PHYS	polykristallin	= pull instruction		
polyester		CHEM	Polyester	populated board (→assembled PCB)	ELECTRON	(→bestückte Leiterplatte)
polyester capacitor		COMPON	Polyester-Kondensator			
polyester resin		CHEM	Polyesterharz	population	MATH	Gesamtheit
polyethylene		CHEM	Polyäthylen	= totality		[Statistik]
polyethylene disk		DATA COMM	Polyäthylenscheibe	pop-up antenna (→telescopic antenna)	ANT	(→Teleskopantenne)
polygon		MATH	Vieleck			
			= Polygon	pop-up menu	DATA PROC	Pop-up-Menü
polygonal		MATH	vieleckig	[can be displayed with a mouse from bottom upward]		[wird mit Maus vom unteren Bildrand nach oben aufgerollt]
			= polygonal; mehreckig			
polyhedron (pl.-ons,-a)		MATH	Polyeder	≠ pull-down menu		≠ Pull-down-Menü
[solid of plane faces]			[von Ebenen begrenzter Körper]	↑ selection menu		↑ Auswahlmenü
= face [DATA PROC]				porcelain	TECH	Porzellan
↓ prism; parallelepiped; rectangular prism; cube			= Vielflächner	porcelain insulator	OUTS.PLANT	Porzellanisolator
			↓ Prisma; Parallelepiped; Quader; Würfel	porcentage	TECH	Prozentsatz
				porch	TV	Schwarzschulter
polyimide		CHEM	Polyimid	[step in the TV signal, at 75% modulation, before and behind the sync pulse]		[Stufe im Fernsehsignal, bei 75% Modulation, vor und hinter dem Synchronimpuls]
polymer		CHEM	polymer			
polymerize		CHEM	aushärten			
			[Harz]	≈ black level		≈ Austastschulter; Schwarztreppe
			= härten	↓ front porch; back porch		
polymeter		METEOR	Polymeter			≈ Schwarzpegel; Austastimpuls
[hygrometer with thermometer]			[Hygro- u. Thermometer]			
polynomial (n.)		MATH	Polynom			↓ vordere Schwarzschulter; hintere Schwarzschulter
			= mehrgliedriger Ausdruck			
polynomial (adj.)		MATH	polynomisch	porosity	TECH	Porosität
= multiterm			= vielgliedrig; mehrgliedrig	porous	TECH	porös
polyphase		POWER ENG	mehrphasig	port (v.t.)	DATA PROC	portieren
= multiphase				[to execute a program on a different computer]		[ein Programm auf einem unterschiedlichen Computer ablaufen lassen]
polyphase current		POWER ENG	Mehrphasenstrom			
= multiphase current			↓ Drehstrom			
↓ three-phase current				port (n.)	DATA PROC	Port
polyphase machine		POWER ENG	Drehfeldmaschine	[circuit or program matching data with an interface]		[Schaltung oder Programm zur Anpassung von Daten an eine Schnittstelle]
polyphase sorting		DATA PROC	Polyphasensortieren			
[sorting procedure employing several scratch tapes]			[Sortierverfahren unter Zuhilfenahme mehrerer Arbeitsbänder]			
				port (n.)	NETW.TH	Tor
≠ merge sorting			≠ Mischsortieren	[related pair of terminals of a network]		[zusammenhängiges Klemmenpaar eines Netzwerkes]
polyphase system		POWER SYS	Mehrphasensystem			
= multiphase system				port	COMPON	Anschlußbuchse
polypropylene		CHEM	Polypropylen	portability	TECH	Tragbarkeit
polypropylene capacitor		COMPON	Polypropylen-Kondensator	≈ transportability		≈ Transportierbarkeit
↓ KS capacitor			↓ KS-Kondensator	portability	DATA PROC	Übertragbarkeit
polypropylene capacitor (→KP capacitor)		COMPON	(→KP-Kondensator)	[feature of a program, to be loadable on different computers; "ability to be connected to different ports"]		[Eigenschaft von Software, auf unterschiedlichen Computern einsatzfähig zu sein]
polyscope		INSTR	Polyskop			
polysilicon		MICROEL	Polysilizium			
polystage (→multistage)		CIRC.ENG	(→mehrstufig)	= software portability; software transportability		= Software-Übertragbarkeit; Portabilität; Austauschbarkeit
polystyrene		CHEM	Polystyrol			
polystyrene capacitor		COMPON	Polystyrolkondensator			
polyurethane		CHEM	Polyurethan	portable	TECH	tragbar
polyvalence		MATH	Mehrwertigkeit	≈ hand-held; hand-portable		≈ transportierbar; beweglich
polyvalent (→multivalent)		TECH	(→mehrwertig)	≈ transportable; mobile		
polyvinyl		CHEM	Polyvinyl	portable (→portable equipment)	EQUIP.ENG	(→tragbares Gerät)
pompeian red		PHYS	pompejanischrot			
pond		PHYS	Pond	portable antenna	ANT	Portable-Antenne
[unit for weight; = 0,009806650 N]			[Einheit für Gewicht; = 0,009806650 N]			[für tragbares Gerät]
= p			= p; Gewichtsgramm (obs.)	portable case	EQUIP.ENG	Koffer
				≈ case		= Traggehäuse
ponder (→trade-off)		COLLOQ	(→abwägen)	portable computer	DATA PROC	tragbarer Computer
ponderation (→trade-off)		COLLOQ	(→Abwägung)	↓ laptop computer		= tragbarer Rechner
						↓ Aktentaschenkomputer

portable equipment	EQUIP.ENG	tragbares Gerät	
= portable set; portable			
portable program	DATA PROC	portierbares Programm	
[executable on compatible computers]		[auf kompatiblen Rechnern ablauffähig]	
portable radio receiver	CONS.EL	Kofferadio	
= portable receiver			
portable receiver (→portable radio receiver)	CONS.EL	(→Kofferadio)	
portable set (→portable equipment)	EQUIP.ENG	(→tragbares Gerät)	
portable software	DATA COMM	Portable-Software	
≈ portability		≈ Übertragbarkeit	
portable typewriter	TERM&PER	Reiseschreibmaschine	
portait printing	TERM&PER	Hochformatdruck	
port extension	TERM&PER	Port-Erweiterung	
portmanteau word	LING	Schachtelwort	
[word formed by merging in unusual way different terms, e.g. smog]		↑ Kunstwort	
= blend			
↑ artificial word			
port match	NETW.TH	Anschlußanpassung	
port radar	RADIO LOC	Hafenradar	
portrait (→portrait format)	TYPOGR	(→Hochformat)	
portrait format	TYPOGR	Hochformat	
= portrait		≠ Querformat	
≠ landscape format			
port-specific	TELEC	portindividuell	
POS (→point-of-sale)	TERM&PER	(→Kassenplatz)	
position (v.t.)	TERM&PER	positionieren	
[a plotter, read-write head]		[einen Plotter, Schreib-/Lesekopf]	
		= verstellen	
position 2 (vertical or horizontal)	TECH	Stellung	
		≈ Stelle	
position (v.t.)	TECH	positionieren	
≈ adjust; adjust precisely		≈ einstellen; justieren	
position 1 (n.)	TECH	Lage	
= location; place (n.); attitude		= Position; Stelle; Ort	
≈ spot		≈ Stellung	
position	SWITCH	Anschlußlage	
position	DATA PROC	Stelle	
= place		[Platz für ein Zeichen in einem Datenträger, oder Lage eines Zeichens in einer Datenfolge]	
↓ digit position; alphabetic position; alphanumeric position			
		↓ Ziffernstelle; Buchstabenstelle; alphanumerische Stelle	
-position	TECH	-teilig	
position	TRANS	Übertragungslage	
[FDM]		[TF-Technik]	
↓ regular frequency position; reverse frequency position		↓ Regellage; Kehrlage	
position (→site)	TELEC	(→Standort)	
position accuracy	TERM&PER	Abstandsgenauigkeit	
		[Plotter]	
positional addressing	DATA PROC	Stelleadressierung	
positional notation	MATH	Stellenwertsystem	
[number system, where each digit position of a number has another value]		[Zahlendarstellungssystem bei dem jeder Stelle innerhalb einer Zahl ein anderer Wert zugeordnet ist]	
= positional representation; denominational number system; denominational notation		= Stellenschreibweise; polyadisches Zahlensystem; B-adisches Zahlensystem	
↑ number system		↑ Zahlensystem	
↓ radix notation; binary number system; octal number system; decimal number system; hexadecimal number system		↓ Radixschreibweise; Dualsystem; Oktalsystem; Dezimalsystem; Sedezimalsystem	
positional representation (→positional notation)	MATH	(→Stellenwertsystem)	
position control	CONTROL	Lageregelung	
		= Positionsregelung	
position description	ECON	Arbeitsplatzbeschreibung	
≈ job description		≈ Tätigkeitsbeschreibung	
position fixing	RADIO NAV	Standortbestimmung	
position indicator (→cursor)	TERM&PER	(→Schreibmarke)	
positioning	TECH	Positionierung	
≈ adjustment		≈ Verstellung	
		≈ Einstellung	
positioning appliance	MECH	Aufnahmelehre	
positioning requirement	TECH	Lagevorschrift	
= locational requirement			
positioning speed	TERM&PER	Positioniergeschwindigkeit	
[of a plotter, read-write head]		[eines Plotters, Schreib-/Lesekopfes]	
		= Positionierungsgeschwindigkeit	
positioning time	TERM&PER	Positionierzeit	
[plotter, read-write head]		[Plotter, Schreib-/Lesekopf]	
= seek time		= Positionierungszeit	
↑ access time		↑ Zugriffszeit	
↓ disk access time		↓ Spurzugriffszeit	
position number	SYS.INST	Platznummer	
position sensor	COMPON	Positionssensor	
position tolerance	ENG.DRAW	Lagetoleranz	
= locational tolerance			
positive (n.)	LING	Positiv	
[uninflected form adjective]		[Grundstufe des Adjektivs]	
		= Grundstufe	
positive (adj.)	MATH	positiv	
≠ negative		≠ negativ	
positive (adj.)	PHYS	positiv	
= at positive pole		= an Pluspol	
≠ negative		≠ negative	
positive (→positive pole)	EL.TECH	(→Pluspol)	
positive acknowledgment	DATA COMM	Gutmeldung	
		= Gutquittung	
positive allowance	ENG.DRAW	gewünschtes Kleinstspiel	
positive conductor	EL.TECH	Plusleiter	
		= positiver Leiter	
positive correlation	MATH	positive Korrelation	
positive current feedback	CIRC.ENG	Strommitkopplung	
[part of output current is fed in-phase to the input]		[Teil des Ausgangsstroms wird gleichphasig an den Eingang gelegt]	
↑ positive feedback		↑ Mitkopplung	
positive distortion (→barrel distortion)	PHYS	(tonnenförmige Verzeichnung)	
positive electrode (→anode)	PHYS	(→Anode)	
positive feedback	CIRC.ENG	Mitkopplung	
[part of output signal is fed in-phase to the input]		[Teil des Ausgangssignals wird gleichphasig an den Eingang gelegt]	
↑ feedback		= positive Rückkopplung	
↓ positive voltage feedback; positive current feedback		↑ Rückkopplung	
		↓ Spannungsmitkopplung; Strommitkopplung	
positive justification	CODING	positives Stopfen	
= positive stuffing		= Positivstopfen	
positive logic	CIRC.ENG	positive Logik	
[negative or low tension for logical state 0]		[negative oder niedrige Spannung für logischen Zustand 0]	
= active high data; positive true logic			
positive pole	EL.TECH	Pluspol	
= positive		= positiver Pol	
positive pulse edge (→rising pulse edge)	ELECTRON	(→Impulsanstiegflanke)	
positive ray (→canal ray)	ELECTRON	(→Kanalstrahl)	
positive resist	MICROEL	Positivlack	
positive sentence	LING	Aussagesatz	
positive stuffing (→positive justification)	CODING	(→positives Stopfen)	
positive temperature coefficient thermistor (→PTC thermistor)	COMPON	(→Kaltleiter)	
positive true logic (→positive logic)	CIRC.ENG	(→positive Logik)	
positive type	TERM&PER	Positivschrift	
[dark characters on bright background]		[dunkle Schrift auf hellem Untergrund]	
positive voltage feedback	CIRC.ENG	Spannungsmitkopplung	
[part of output voltage is fed in phase to the input]		[Teil der Ausgangsspannung wird gleichphasig auf den Eingang rückgekoppelt]	
≠ negative voltage feedback		= Spannungsgegenkopplung	
↑ voltage feedback; positive feedback		↑ Spannungsrückkopplung; Mitkopplung	

positron		PHYS	Positron	post-mortem program	DATA PROC	Post-mortem-Programm
possesive pronoun [e.g. my]		LING	Possesivpronomen [z.B. mein] = besitzanzeigendes Fürwort; Possessivum	[generates an outprint of the main memory after occurrence of a fault] = post-mortem routine ↑ memory dump program		[erzeugt ein Protokoll des Hauptspeicherinhalts nach einem Fehler] ↑ Speicherabzugprogramm
post (n.) [vertical support, generally wooden]		CIV.ENG	Pfosten [senkrechte Stütze, i.a. aus Holz] ≈ Mast	post-mortem routine (→post-mortem program)	DATA PROC	(→Post-mortem-Programm)
post (v.t.) (→book)		ECON	(→buchen)	post office	POST	Postamt
post (v.t.) (→input)		DATA PROC	(→eingeben)	post office box	POST	Postfach
post (→postal administration)		POST	(→Postverwaltung)	= POB post office counter service	POST	= Postschließfach Postschalterdienst
post (→job)		ECON	(→Stelle)	postpay coin telephone (AM)	TELEPH	Münzfernsprecher mit Nach-
post-accelerating electrode = intensifying electrode		ELECTRON	Nachbeschleunigungselektrode	= postpay paystation (AM) postpay paystation (AM) (→postpay coin telephone)	TELEPH	zahlung (→Münzfernsprecher mit Nachzahlung)
post-acceleration		ELECTRON	Nachbeschleunigung	postpone	COLLOQ	vertagen
postage = postal rate		POST	Postgebühr = Porto	= adjourn; put off; defere = delay; suspend		= verschieben 2 = verzögern; einstellen
postage		POST	Frankierung	postponement	COLLOQ	Vertagung
postage stamp 2 (n.) [adhesive stamp] ≈ adhesive stamp [ECON] ↑ postage stamp 1		POST	Briefmarke ≈ Wertmarke [ECON] ↑ Postwertzeichen	= adjournment postponement (→deferment) postponement of appointed time	ECON ECON	= Verschiebung 2; Verlegung 2 = Stundung Terminverlängerung = Fristverlängerung
postage stamp 1 [adhesive or impressed] ≈ stamp [ECON] ↓ postage stamp 2		POST	Postwertzeichen = Wertzeichen ≈ Wertzeichen [ECON] ↓ Briefmarke	post-processing ↓ call data post-processing post-processing (→post-production)	SWITCH BROADC	Nachverarbeitung ↓ Rufdatennachbearbeitung (→Nachbearbeitung)
postage stamp printer		OFFICE	Wertzeichendrucker = Postwertzeichendrucker	postprocessor [complementing programm to drive	DATA PROC	Postprozessor [Zusatzprogramm zur
postal address = address		POST	Postanschrift = Anschrift; Postadresse; Adresse	an external machine] = NC postprocessor		Steuerung einer Maschine] = Nachübersetzer; Nachverarbeiter; NC-Postprozessor
postal administration = mail; post; record carrier ≈ mail service		POST	Postverwaltung = gelbe Post; Post 1 ≈ Postdienst ↑ Post- und Fernmeldeverwaltung	post-production = post-processing ↓ post-sound-tracking postscript (→supplement) postscript laser printer	BROADC LING TERM&PER	Nachbearbeitung = Nachverarbeitung ↓ Nachvertonung (→Nachtrag) Postscript-Laserdrucker
postal check service = Giro system (BRI)		ECON	Postgiro = Postscheckdienst	postscript printer post-selection (→suffix	TERM&PER SWITCH	Postscript-Drucker (→Nachwahl)
post-alloy diffused transistor (→PAD transistor)		MICROEL	(→PAD-Transistor)	dialing) Posttechnisches Zentralamt [postal engineering center of DBP]	POST	Posttechnisches Zentralamt [Institution der DBP]
post-alloy diffusion technique (→alloy diffusion technique)		MICROEL	(→AD-Technik)	post, telegraph and telephone administration	ECON	Post- und Fernmeldeverwaltung
postal modem [installed by the common carrier]		DATA COMM	Post-Modem [von der Fernmeldeverwaltung installiert]	= PTT ↓ postal administration; telecommunications administration		↓ Postverwaltung; Fernmeldeverwaltung
postal rate (→postage)		POST	(→Postgebühr)	post-transformation [computer graphics; automatic correc-	DATA PROC	Nachtransformation [Computergraphik; auto-
postal service (→mail service)		ECON	(→Postdienst)	tion of contour lines, dimensions, hatching etc.]		matische Korrektur von Konturlinien, Bemaßung,
postal wrapper		POST	Streifband	≈ streching		Schraffur etc.]
postamble (→trailer label)		DATA PROC	(→Dateiend-Etikett)			≈ Stretching
postcedent (n.)		MATH	Hinterglied	post-trigger delay	INSTR	Nachtriggerverzögerung
postcode (BRI) = zip code (AM)		POST	Postleitzahl	post-triggering pot	INSTR TECH	Nachtriggerung Tiegel
post-connected		EL.TECH	nachgeschaltet			↓ Schmelztiegel
post-dialing delay [of indirect-control switching systems]		SWITCH	Rufverzug [bei indirekt gesteuerten Vermittlungssystemen]	POT (→plane old telephone service) pot (→potentiometer)	TELEC COMPON	(→konventioneller Fernsprechdienst) (→Regelwiderstand)
post edit (→post-editing)		DATA PROC	(→Nacheditierung)	pot (→pot core) POT (→delivery point)	COMPON TELEC	(→Schalenkern) (→Übergabepunkt)
post-editing = post edit		DATA PROC	Nacheditierung	potassium = K	CHEM	Kalium = K
post-equalization		TRANS	Nachentzerrung	potassium chloride	CHEM	Kaliumchlorid
poster		ECON	Plakat	= sylvine		= Sylvin
poste restante		POST	postlagernd	potassium niobate	CHEM	Kaliumniobat
POS terminal (→point-of-sale terminal)		TERM&PER	(→Kassenterminal)	potassium tantalate pot coil (→pot-core coil)	CHEM COMPON	Kaliumtantalat (→Schalenkernspule)
postfix notation (→polish notation)		MATH	(→polnische Schreibweise)	pot core = pot; cup core	COMPON	Schalenkern
post-graduate course		SCIE	Aufbaustudium	pot-core coil	COMPON	Schalenkernspule
posting (→booking)		ECON	(→Buchung)	= pot coil; cup-core coil		
post-injection voltage		MICROEL	Post-Injektion-Spannung	pot-core transformer	COMPON	Schalenkernübertrager
postman = mailman (AM)		POST	Postbote	potential potential (adj.)	PHYS MATH	Potential (→wirbelfrei)
postmark		POST	Poststempel	(→irrotational)		
post-mortem dump (→memory dump)		DATA PROC	(→Speicherabzug)	potential analogy	NETW.TH	Potentialanalogie

potential barrier	PHYS	**Potentialberg**
= potential hill		= Potentialbarriere; Potentialwall; potentialgebirge
≠ potential well		≠ Potentialmulde
potential compensation (→potential equalization)	PHYS	(→Potentialausgleich)
potential connection	EL.TECH	**Potentialverbindung**
potential converter	EL.TECH	**Potentialwandler**
potential difference	PHYS	**Potentialdifferenz**
↓ voltage [EL.TECH]		↓ Spannung [EL.TECH]
potential distribution	PHYS	**Potentialverteilung**
potential divider (→voltage divider)	EL.TECH	(→Spannungsteiler)
potential energy	PHYS	**potentielle Energie**
= static energy; rest energy		= Potentialenergie; statische Energie; Energie der Lage; Ruheenergie; virtuelle Energie
potential equalization	PHYS	**Potentialausgleich**
= potential compensation		
potential equation	PHYS	**Potentialgleichung**
potential field	PHYS	**Potentialfeld**
potential gradient	PHYS	**Potentialgradient**
potential hill (→potential barrier)	PHYS	(→Potentialberg)
potential insulation (→potential separation)	EL.TECH	(→Potentialtrennung)
potential layer	MICROEL	**Potentiallage**
potential line	PHYS	**Potentiallinie**
≈ line of force		≈ Niveaulinie
		≈ Kraftlinie
potential profile	MICROEL	**Potentialprofil**
		= Spannungsprofil; Spannungsverlauf
potential separation	EL.TECH	**Potentialtrennung**
= potential insulation		
potential sink	PHYS	**Potentialsenke**
potential slope	PHYS	**Potentialgefälle**
potential surface	PHYS	**Potentialfläche**
potential threshold	PHYS	**Potentialschwelle**
potential transformer	LINE TH	**Potentialtransformator**
[balncing plus 1:4 transformer loop]		[Symmetrier- plus Transformationsschleife 1:4]
≈ balun		≈ Symmetrierglied
potential transformer (→voltage converter)	INSTR	(→Spannungswandler)
potential trough (→potential well)	PHYS	(→Potentialmulde)
potential well	PHYS	**Potentialmulde**
= square well; potential trough		= Potentialtopf; Potentialtrog
≠ potential barrier		≠ Potentialberg
potentiometer	COMPON	**Regelwiderstand**
[adjustable resistor with three leads; adjustment is by wiping contact]		[einstellbarer Widerstand mit 3 Anschlüssen, über Schleifenkontakt einstellbar]
= pot		
↑ adjustable resistor		= Potentiometer; stellbarer Widerstand
↓ trimming potentiometer; rotatable resistor		↑ veränderbarer Widerstand
		↓ Trimmwiderstand; Drehwiderstand; Schieberegler
potentiometer circuit	INSTR	**Potentiometerverfahren**
potentiometer circuit (→potentiometer connection)	PHYS	(→Potentiometerschaltung)
potentiometer connection	PHYS	**Potentiometerschaltung**
= potentiometer circuit		
potentiostate	INSTR	**Potentiostat**
pothead	POWER SYS	**Endverschluß**
pothead (→terminal)	OUTS.PLANT	(→Endverschluß)
POT line (→telephone connection 1)	TELEC	(→Fernsprechanschluß)
POTS (→POT service)	TELEC	(→konventioneller Fernsprechdienst)
POT service	TELEC	**konventioneller Fernsprechdienst**
= POTS; plain-old-telephone service; conventional telephonic service; voice service		= Sprachdienst
↑ telephonic service; basic service		↑ Fernsprechdienst; Grunddienst
Potter horn (→Potter horn antenna)	ANT	(→Potter-Horn-Antenne)
Potter horn antenna	ANT	**Potter-Horn-Antenne**
= Potter horn		= Potter-Horn
↑ horn radiator		↑ Hornstrahler
pound 1	PHYS	**Pound 1**
[of the avoirdupois system; = 16 ounces; = 0.4536 kg]		[im Sechzehn-Unzen-System; = 0,4536 kg]
= pound (ar); lb; lb (ar) avoirdupois pound		= angelsächsisches Pfund
pound 2	PHYS	**Pound 2**
[of the troy system; = 1.2153 kg]		[im Troy-System; = 1,2153 kg]
= pound (tr); lb (tr); troy pound		= Troy-Pfund
pound sign (→pound symbol)	ECON	(→Pfund-Zeichen)
pound symbol [£]	ECON	**Pfund-Zeichen** [£]
= pound sign		= Pfund-Symbol
powder (n.)	TECH	**Pulver**
≈ dust		≈ Staub
powder (v.t.) (→pulverize)	TECH	(→zerstäuben)
powder cathode	ELECTRON	**Sinterkathode**
powder core (→dust core)	COMPON	(→Pulverkern)
powder-core coil (→dust-core coil)	COMPON	(→Preßkernspule)
powdered iron core (→dust core)	COMPON	(→Pulverkern)
powder electroluminescence	PHYS	**Pulverelektrolumineszenz**
powder metal	METAL	**Pulvermetall**
= cermet		= Sintermetall
powder metallurgy	METAL	**Pulvermetallurgie**
powder metal part	METAL	**Sinterteil**
powder toner (→dry toner)	TERM&PER	(→Trockentoner)
power (v.t.)	MATH	**potenzieren**
= raise to the power of		↓ quadrieren
↓ square		
power (n.)	PHYS	**Leistung**
[work per unit time; SI unit: Watt]		[Arbeit pro Zeiteinheit; SI-Einheit: Watt]
≈ energy flow		≈ Energiestrom
power (n.)	MATH	**Potenz**
[mutiple multiplication of a number with itself]		[mehrfache Multiplikation einer Zahl mit sich selbst]
↓ base; exponent; decade		↓ Basis; Exponent; Zehnerpotenz
power	ECON	**Vollmacht**
= mandate		= Bevollmächtigung
↓ subdelegated power		↓ Untervollmacht
power absorber	EL.TECH	**Stromverbraucher**
≈ load		≈ Last
power alarm	TELEC	**Stromversorgungsalarm**
power amplification (→power gain)	NETW.TH	(→Leistungsverstärkung)
power amplifier	CIRC.ENG	**Leistungsverstärker**
= final amplifier		= Leistungsendverstärker; Endverstärker; Leistungsendstufe; Leistungsendverstärkerstufe; Endverstärkerstufe
≈ final stage; output amplifier		
		≈ Endstufe; Ausgangsverstärker
power attenuation factor	NETW.TH	**Leistungsdämpfungsfaktor**
[power at input to power at output]		[Leistung am Eingang zu Leistung am Ausgang]
		= Leistungsdämpfungsmaß
power bandwidth	EL.ACOUS	**Leistungsbandbreite**
[hifi]		[HiFi]
power bus	POWER SYS	**Potentialschiene**
power cable	EQUIP.ENG	**Netzanschlußkabel**
= mains cable; mains lead; AC-line-supply cable; electrical-power cord		= Netzkabel; Netzleitung; Netzschnur; Stromversorgungskabel; Versorgungskabel
↑ connecting cable		
↓ mains cable		↑ Anschlußkabel
		↓ Netzkabel
power cable	POWER ENG	**Starkstromkabel**
= high-voltage cable; energy cable; power transmission cable		= Stromkabel; Energiekabel
power capacitor	COMPON	**Starkstromkondensator**
power combiner	RADIO	**Leistungssummierer**
		= Leistungskombinator

power consumption

power consumption (→power input)	EL.TECH	(→Leistungsaufnahme)
power contactor	POWER ENG	**Leistungsschütz**
power cross	EL.TECH	**Netzberührung**
power current (→heavy current)	POWER SYS	(→Starkstrom)
power current engineering (→electrical power engineering)	EL.TECH	(→elektrische Energietechnik)
power current technology (→electrical power engineering)	EL.TECH	(→elektrische Energietechnik)
power cut [intentional]	POWER SYS	**Stromabschaltung** [beabsichtigte]
= power-off; power dump		≈ Stromausfall
≈ power failure		↑ Stromunterbrechung
↑ power outage		
power decrease (→power drop)	PHYS	(→Leistungsabfall)
power demand	EL.TECH	**Leistungsbedarf**
= required power		≈ Leistungsaufnahme; Strombedarf; Stromaufnahme
≈ power consumption; current demand; current consumption		
power density	PHYS	**Leistungsdichte 1**
power density spectrum	TELEC	**Leistungsspektralfunktion**
power derating (→back-off)	ELECTRON	(→Untersteuerung)
power diode	MICROEL	**Leistungsdiode**
power dissipation curve	MICROEL	**Leistungshyperbel**
power distribution panel	POWER SYS	**Stromverteilertafel**
= power switchboard		
power distributor	POWER SYS	**Stromverteiler**
power divider	ANT	**Leistungsteiler**
		= Lastverteiler
power divider (→power splitter)	RADIO	(→Leistungsverteiler)
power divider (→power splitter)	MICROW	(→Leistungsteiler)
power-down (n.)	DATA PROC	**Abschaltevorgang**
= power-down process; power-off (n.); power-off process		= Abschaltvorgang; Abschalten; Ausschaltevorgang; Ausschaltvorgang; Ausschalten
power-down (v.t.) (→disconnect)	EL.TECH	(→ausschalten)
power-down process (→power-down)	DATA PROC	(→Abschaltevorgang)
power drain (→power input)	EL.TECH	(→Leistungsaufnahme)
power driver (→driver)	CIRC.ENG	(→Treiber)
power drop	PHYS	**Leistungsabfall**
= power decrease		
power dump (→power cut)	POWER SYS	(→Stromabschaltung)
power electronics	ELECTRON	**Energieelektronik**
↑ electronics		= Leistungselektronik
		↑ Elektronik
power element	ELECTRON	**Leistungsglied**
power engineering (→electrical power engineering)	EL.TECH	(→elektrische Energietechnik)
power equipment (→power supply)	EQUIP.ENG	(→Stromversorgungsgerät)
power exponent (→exponent)	MATH	(→Exponent)
power factor	EL.TECH	**Leistungsfaktor** [Kosinus des Phasenwinkels zwischen Spannung und Strom]
power fading (→obstruction fading)	RADIO PROP	(→Beugungsschwund)
power fail	DATA PROC	**Wiedereinschaltautomatik**
= automatic restart		
power failure (→black-out)	POWER SYS	(→Spannungsausfall)
power failure protection	DATA PROC	**Netzausfallsicherung**
power flatness	INSTR	**Leistungspegelgang**
power flow	POWER SYS	**Lastfluß**
power flow	PHYS	**Leistungsfluß**
power-flow computer [for power sypply]	DATA PROC	**Lastflußrechner** [Starkstromnetz]
power flux density (→Poynting's vector)	PHYS	(→Poynting-Vektor)
power frequency (→mains frequency)	POWER SYS	(→Netzfrequenz)
power frequency bridge (→Schering bridge)	INSTR	(→Schering-Brücke)
powerful	TECH	**leistungsfähig**
≈ efficient		= leistungsstark
power gain	NETW.TH	**Leistungsverstärkung**
= power amplification		↑ Verstärkung
power generating plant (→power station)	POWER ENG	(→Kraftwerk)
power generation	PHYS	**Energiegewinnung**
power generation and distribution (→energy supply)	TECH	(→Energieversorgung)
power generation plant (→energy system)	POWER SYS	(→Stromversorgungsanlage)
power indicator	EL.ACOUS	**Leistungsindikator**
power influence	POWER ENG	**Starkstrombeinflussung**
↓ mains interference		
power input	EL.TECH	**Leistungsaufnahme**
= absorbed power; power consumption; power drain; wattage		= Stromverbrauch ≈ Stomaufnahme; Stromverbrauch; Leistungsbedarf; Strombedarf
≈ current input; current consumption; power demand; current demand		
power jack (→power supply jack)	EQUIP.ENG	(→Stromversorgungsbuchse)
power level	TELEC	**Leistungspegel**
≠ voltage level		≠ Spannungspegel
power line	POWER SYS	**Netzzuleitung**
power line	POWER ENG	**Starkstromleitung**
= electrical power line		↓ Hochspannungsleitung
↓ high-tension line		
powerline carrier transmission (→carrier transmission over power lines)	TRANS	(→TFH-Technik)
power-line cycle	EL.TECH	**Netzspannungsperiode**
= PLC		
power-line hum (→mains hum)	TELEC	(→Netzbrumm)
power-line noise (→mains hum)	TELEC	(→Netzbrumm)
power line tower	POWER ENG	**Hochspannungsmast**
power load (→load)	EL.TECH	(→Last)
power loss	TECH	**Leistungsverlust**
power loudspeaker	EL.ACOUS	**Kraftlautsprecher**
power match (→power matching)	NETW.TH	(→Leistungsanpassung)
power matching	NETW.TH	**Leistungsanpassung**
[load resistance equal to intrinsic resistance; maximum energy transfer to load, but with equal internal dissipation of the energy source]		[Innenwiderstand gleich Außenwiderstand, maximale Leistungsübertragung auf den Verbraucher, jedoch mit gleich großer Innenverlustleistung der Quelle]
= power match; matching 2; adaptation 2		= Anpassung 2
↑ matching 1		↑ Anpassung 1
power measurement	INSTR	**Leistungsmessung**
power meter	INSTR	**Leistungsmesser**
= wattmeter		= Wattmeter; Leistungsmeßgerät
		↓ Wirkleistungsmesser; Blindleistungsmesser; Scheinleistungsmesser
power-off (n.) (→power-down)	DATA PROC	(→Abschaltevorgang)
power-off (→power cut)	POWER SYS	(→Stromabschaltung)
power-off (→dead)	EL.TECH	(→kalt)
power-off process (→power-down)	DATA PROC	(→Abschaltevorgang)
power of three (→cube 1)	MATH	(→Kubikzahl)
power-on (→live)	EL.TECH	(→heiß)
power oscillator	CIRC.ENG	**Leistungsoszillator**
power outage	POWER SYS	**Stromunterbrechung**
↓ powerfailure; power cut		↓ Stromausfall; Stromabschaltung
power output	BROADC	**Nutzausgangsleistung**
power pack (→mains power supply)	EQUIP.ENG	(→Netzstromversorgungsgerät)
power plant (→power station)	POWER ENG	(→Kraftwerk)
power rating (→load capacity)	TECH	(→Belastbarkeit)
power rating (→nominal power)	TECH	(→Nennleistung)
power-ratio meter	INSTR	**Leistungsfaktormesser**
		= Kosinus-Phi-Messer

power rectifier	MICROEL	**Leistungsgleichrichter**	
power rectifier (→line rectifier)	POWER SYS	(→Netzgleichrichter)	
power relay	COMPON	**Leistungsrelais**	
power restoration trailer	POWER SYS	**Notstromanhänger**	
power saving	EL.TECH	**stromsparend**	
power semiconductor	MICROEL	**Leistungshalbleiter**	
power sensor	INSTR	**Leistungsmeßkopf**	
power series	MATH	**Potenzreihe**	
power series expansion	MATH	**Potenzreihenentwicklung**	
power set	MATH	**Potenzmenge**	
power socket (→mains socket)	EL.INST	(→Netzsteckdose)	
power source	EL.TECH	**Leistungsquelle**	
power source (→power supply)	EQUIP.ENG	(→Stromversorgungsgerät)	
power spectrum	PHYS	**Leistungsspektrum**	
power splitter	RADIO	**Leistungsverteiler**	
= power divider		= Leistungsteiler	
power splitter	MICROW	**Leistungsteiler**	
= power divider			
power station	POWER ENG	**Kraftwerk**	
= power plant; power generating plant; generating station		↓ Wasserkraftwerk; Kohlekraftwerk; Atomkraftwerk	
power station (→energy system)	POWER SYS	(→Stromversorgungsanlage)	
power supply	CIRC.ENG	**Stromversorgung**	
= current supply			
power supply	EQUIP.ENG	**Stromversorgungsgerät**	
= power source; power equipment		= Stromversorgung	
power supply (→mains)	POWER ENG	(→Starkstromnetz)	
power supply bar (→current bus)	POWER ENG	(→Stromschiene)	
power supply bus (→current bus)	POWER ENG	(→Stromschiene)	
power supply jack	EQUIP.ENG	**Stromversorgungsbuchse**	
= power jack		= Fremdstrombuchse; Strombuchse	
power supply rejection	ELECTRON	**Netzunterdrückungsfaktor**	
		= Netzunterdrückung	
power supply system	POWER SYS	**Stromnetz**	
power supply system (→energy system)	POWER SYS	(→Stromversorgungsanlage)	
power supply transformer	POWER SYS	**Stromversorgungstransformator**	
power sweep	INSTR	**Leistungswobbelung**	
power switch	EQUIP.ENG	**Stromversorgungsschalter**	
power switch (→mains switch)	EL.INST	(→Netzschalter)	
power switch (→circuit breaker)	POWER ENG	(→Trennschalter)	
power switchboard (→power distribution panel)	POWER SYS	(→Stromverteilertafel)	
power switchgear	POWER ENG	**Starkstrom-Schaltgerät**	
power system room	SYS.INST	**Stromversorgungsraum**	
power tetrode	ELECTRON	**Leistungstetrode**	
↑ power tube		= Sendetetrode	
		↑ Leistungsröhre	
power thyristor	MICROEL	**Leistungsthyristor**	
power transfer	ANT	**Leistungsübertragung**	
power transformer	COMPON	**Leistungsübertrager**	
power transformer	POWER ENG	**Leistungstransformator**	
power transformer (→mains transformer)	POWER ENG	(→Netztransformator)	
power transistor	MICROEL	**Leistungstransistor**	
		= End-Transistor	
power transmission	MECH	**Kraftübertragung**	
power transmission (→energy transfer)	PHYS	(→Energieübertragung)	
power transmission cable (→power cable)	POWER ENG	(→Starkstromkabel)	
power transmission coefficient (→power transmission factor)	NETW.TH	(→Leistungsübertragungsfaktor)	
power transmission factor [power output/input ratio]	NETW.TH	**Leistungsübertragungsfaktor** [Leistung am Ausgang zu Leistung am Eingang]	
= power transmission coefficient			
≠ power attenuation factor		≠ Leistungsdämpfungsfaktor	
↓ image power transmission factor; effective power transmission factor		↓ Wellenleistungsübertragungsfaktor; Betriebsleistungsübertragungsfaktor	
power triode	ELECTRON	**Leistungstriode**	
↑ power tube		↑ Leistungsröhre	
power tube	ELECTRON	**Leistungsröhre**	
= power valve			
power-type relay	COMPON	**Starkstromrelais**	
power-up (v.t.) (→connect 2)	EL.TECH	(→einschalten)	
power-up (v.t.) (→start-up)	DATA PROC	(→hochfahren)	
power-up (n.) (→start-up)	DATA PROC	(→Hochfahren)	
power-up diagnosis (→self test)	DATA PROC	(→Eigentest)	
power-up test (→self test)	DATA PROC	(→Eigentest)	
power valve (→power tube)	ELECTRON	(→Leistungsröhre)	
Poynting's vector	PHYS	**Poynting-Vektor**	
= power flux density		= Strahlungsvektor; Ausbreitungsvektor; Leistungsflußdichte; Strahlungsdichte 2; Leistungsdichte 2	
pph (→pages per hour)	TERM&PER	(→Seiten pro Stunden)	
PPI [microprocessors] = programmable peripheral interface	DATA PROC	**PPI** [Mikroprozessor]	
PPI (→plan position indicator)	RADIO LOC	(→Plan-Position-Indicator)	
PPM (→pulse position modulation)	MODUL	(→Pulslagemodulation)	
Pr (→praseodymium)	CHEM	(→Praseodym)	
practicability (→viability)	TECH	(→Durchführbarkeit)	
practicable (→feasible 1)	TECH	(→durchführbar)	
practical	TECH	**praxisbezogen**	
practical equivalent transistor network	MICROEL	**praktische Transistorersatzschaltung**	
practicality (→viability)	TECH	(→Durchführbarkeit)	
practical training	ECON	**Praktikum**	
practice (v.t.) = practise (v.t.)	COLLOQ	**praktizieren**	
practice (n.) = practise (n.)	TECH	**Praxis**	
practice (→operating instruction 1)	TECH	(→Betriebsvorschrift)	
practice (→usage)	COLLOQ	(→Gepflogenheit)	
practice (→technology 2)	TECH	(→Technologie)	
practise (→operating instruction 1)	TECH	(→Betriebsvorschrift)	
practise (n.) (→practice)	TECH	(→Praxis)	
practise (v.t.) (→practice)	COLLOQ	(→praktizieren)	
Prandtl number	PHYS	**Prandtl-Zahl**	
praseodymium = Pr	CHEM	**Praseodym** = Pr	
Pratt & Whitney key	MECH	**Flachpaßfeder**	
PRBS generator	INSTR	**Generator für pseudozufällige Daten**	
preadjust (v.t.) (→preset)	CIRC.ENG	(→voreinstellen)	
preamble (→header label)	DATA PROC	(→Dateianfangs-Etikett)	
preamplifier	CIRC.ENG	**Vorverstärker**	
↓ driver		↓ Treiber	
pre-attenuation	TELEC	**Vordämpfung**	
preavis	TELEPH	**Voranmeldung**	
precalculate (→precompute)	TECH	(→vorausrechnen)	
precalculation (→precomputation)	TECH	(→Vorausberechnung)	
precanned (adj.) (→canned)	TECH	(→konfektioniert)	
precanned routines (→standard software)	DATA PROC	(→Standard-Software)	
precaution	COLLOQ	**Vorkehrung**	
≈ arrangement		≈ Vorsichtsmaßnahme; Vorsicht	
≈ measure		≈ Maßnahme	
precaution (→prevention)	TECH	(→Vorbeugung)	
precautionary (→preventive)	TECH	(→vorbeugend)	
precautionary measure	COLLOQ	**Vorsorgemaßnahme**	
≈ protective measure		≈ Schutzmaßnahme	
precedence	COLLOQ	**Vorrang**	
= priority; privilege		= Vortritt; Vorrecht; Priorität; Bevorrechtigung	
≈ preference; urgency		≈ Bevorzugung; Dringlichkeit	

preceding line 444

preceding line	TERM&PER	**Vorherzeile**
preceding month	COLLOQ	**Vormonat**
= previous month		
preceding year	COLLOQ	**Vorjahr**
= last year		
precession	PHYS	**Präzession**
precheck (v.t.)	MANUF	**vorprüfen**
= pretest (v.t.)		= vortesten; vorabprüfen; vorabtesten
		≈ vormessen
prechecking (→pretesting)	MANUF	(→Vorprüfung)
precious metal (→noble metal)	CHEM	(→Edelmetall)
precipitation (→deposition)	TECH	(→Ablagerung)
precipitation (→atmospheric precipitation)	METEOR	(→atmosphärischer Niederschlag)
precipitation attenuation	RADIO PROP	**Niederschlagsdämpfung**
↓ rain attenuation		↓ Regendämpfung
precipitous	TECH	**abschüßig**
= steep		= steil
precise value	TECH	**Genauwert**
		= exakter Wert
precision (→accuracy)	TECH	(→Genauigkeit)
precision approach radar	RADIO NAV	**Präzisions-Anflug-Radar**
= APR		= PAR-Gerät
precision audio amplifier	INSTR	**Präzisions-NF-Verstärker**
precision balance	TELEC	**Präzisionsnachbildung**
precision casting	METAL	**Genauigkeitsguß**
precision engineering	MECH	**Feinmechanik**
precision etching	ELECTRON	**Feinätztechnik**
precision instrument	INSTR	**Präzisionsinstrument**
precision load	INSTR	**Präzisionslastwiderstand**
precision mains power supply	ELECTRON	**Präzisionsnetzgerät**
precision measuring generator	INSTR	**Präzisionsmeßgenerator**
precision plotter	TERM&PER	**Präzisionszeichenmaschine**
precision potentiometer	CONTROL	**Präzisionspotentiometer**
precision rectifier	INSTR	**Präzisionsgleichrichter**
precision resistance 1 (→standard resistor 1)	INSTR	(→Eichwiderstand)
precision resistance (→precision resistor)	COMPON	(→Normalwiderstand)
precision resistor	COMPON	**Normalwiderstand**
[commercial resistor with maximum precision and stability]		[Verbrauchswiderstand höchster Ansprüche]
= standard resistor 2; precision resistance; standard resistance 2; measuring resistor; measuring resistance		= Widerstandsnormal; Meßwiderstand; Präzisionswiderstand
≈ standard resistance 1 [INSTR]		≈ Eichwiderstand [INSTR]
precision resistor decade	INSTR	**Präzisions-Widerstandsdekade**
precision short	INSTR	**Präzisionskurzschluß**
precision stamp	MECH	**Feinstanzen**
precision sweep	INSTR	**Präzisionswobbelung**
precompiler	DATA PROC	**Vorkompilierer**
[program to convert non-standard program parts into a standard program language]		[Programm zur Übersetzung nicht genormter Programmteile in eine standardisierte Programmiersprache]
= preprocessor 1		= Precompiler
↑ preprocessor 2		↑ Preprozessor
precomputation	TECH	**Vorausberechnung**
= precalculation		= Vorkalkulation
precompute (v.t.)	TECH	**vorausberechnen**
= precalculate		= vorkalkulieren
precondition	COLLOQ	(→Voraussetzung)
(→prerequisite)		
precondition input	CIRC.ENG	**Vorbereitungseingang**
[multivibrator]		[Kippschaltung]
= preparatory input		
preconnect	EL.TECH	**vorschalten**
preconnected	EL.TECH	**vorgeschaltet**
≈ series-connected		≈ seriengeschaltet
preconnected fiber	OPT.COMM	**Vorschaltfaser**
precoupler	TELEC	**Vorkoppler**
precure (v.t.)	METAL	**vorhärten**
precuring	METAL	**Vorhärten**
predecessor	COLLOQ	**Vorgänger**
pre-decision	COLLOQ	**Vorentscheidung**
predefined (→preset)	COLLOQ	(→vorgegeben)
predefinition	DATA PROC	(→Vorgabe)
(→default)		
predetection	ELECTRON	**Vorausnachweis**
predetermination	COLLOQ	**Vorausbestimmung**
predetermined (→preset)	COLLOQ	(→vorgegeben)
predicate	LING	**Prädikat**
= verb 2		[Satzteil der über das Subjekt aussagt]
		= Satzaussage
predicate logic	MATH	**Prädikatenlogik**
predicate logic language	MATH	**prädikatenlogische Sprache**
predicative (n.)	LING	**Prädikativ**
		= Prädikativum; Prädikatsnomen; Gleichsetzungsnominativ; Artergänzung
predictable	SCIE	**vorhersehbar**
= predictive; forecastable		= prädiktiv
prediction	TELEC	**Prädiktion**
[signal processing]		[Signalverarbeitung]
		= Vorhersage
prediction	MATH	**Vorhersage**
		= Voraussage; Prädiktion
prediction algorithm	MATH	**Vorhersagealgorithm**
= predictive algorithm		= Prädiktionsalgorithmus
prediction error	TELEC	**Prädiktionsfehler**
		= Vorhersagefehler
prediction model	SCIE	**Vorhersagemodell**
= predictive model		= prädiktives Modell
predictive	TECH	**vorraussagend**
predictive (→predictable)	SCIE	(→vorhersehbar)
predictive algorithm (→prediction algorithm)	MATH	(→Vorhersagealgorithmus)
predictive coding	CODING	**prädiktive Codierung**
predictive model (→prediction model)	SCIE	(→Vorhersagemodell)
predischarge	PHYS	**Vorentladung**
prediscriminador	CIRC.ENG	**Vorentscheider**
predistorter	RADIO REL	**Vorverzerrer**
		= Predistorter
predistorter	TELEC	**Vorverzerrer**
= line forward equalizer		≈ Vorentzerrer
≈ pre-equalizer		
predistortion	TELEC	(→Vorverzerrung)
(→pre-emphasis)		
pre-drilled rail	MECH	**Lochschiene**
pre-electrification	PHYS	**Vorelektrisierung**
= electric biasing; electric polarization		↑ Vorpolung
↑ biasing		
preeminence	COLLOQ	**Vorrangstellung**
pre-emphasis	TELEC	**Vorverzerrung**
= predistortion		= Preemphasis
≠ de-emphasis		≈ Vorentzerrung
		≠ Rückentzerrung
pre-emphasis (→treble correction)	EL.ACOUS	(→Höhenanhebung)
preemption (→right of preemption)	ECON	(→Vorkaufsrecht)
preemptive priority	SWITCH	**unterbrechende Priorität**
		= absolute Priorität
pre-equalization	TELEC	**Vorentzerrung**
≈ pre-emphasis		≈ Vorverzerrung
pre-equalizer	TELEC	**Vorentzerrer**
≈ predistorter		≈ Vorverzerrer
prefabricate	MANUF	**vorfertigen**
prefabricated	MANUF	**vorgefertigt**
prefabricated cable	EQUIP.ENG	(→vorgefertigtes Kabel)
(→preformed cable)		
prefabricated part	MANUF	**vorgefertigtes Teil**
prefabrication (→parts manufacture)	MANUF	(→Vorfertigung)
preference state	CIRC.ENG	**Vorzugslage**
[multivibrator]		[Kippschaltung]
= preferred position		
preferred direction	PHYS	**Vorzugsrichtung**
= preferred orientation		= Vorzugsorientierung
preferred fits	ENG.DRAW	**Vorzugspassungen**
preferred nominal dimensions	ENG.DRAW	**Vorzugsnennmaße**

preferred orientation (→preferred direction)	PHYS	(→Vorzugsrichtung)	
preferred parameter	TECH	**Vorzugsparameter**	
preferred part	COMPON	**Vorzugstype**	
= preferred type			
preferred position (→normal position)	TECH	(→Normalstellung)	
preferred position (→preference state)	CIRC.ENG	(→Vorzugslage)	
preferred quantity	DATA PROC	**Vorzugsgröße**	
= preferred variable		= Vorzugsvariable	
preferred range	TECH	**Vorzugsbereich**	
preferred tolerances	ENG.DRAW	**Vorzugstoleranzen**	
preferred type (→preferred part)	COMPON	(→Vorzugstype)	
preferred value	TECH	**Vorzugswert**	
preferred variable (→preferred quantity)	DATA PROC	(→Vorzugsgröße)	
prefetch	DATA PROC	**Vorauslesen**	
prefetching	DATA PROC	**Vorabbefehlsaufnahme**	
= instruction lookahead		= Befehlsvorverarbeitung	
prefetch register	DATA PROC	**Vorabbefehlsregister**	
		= Pre-fetch-Register	
prefilter (n.)	NETW.TH	**Vorfilter**	
prefinance	ECON	**vorfinanzieren**	
pre-financing	ECON	**Vorfinanzierung**	
prefix	SWITCH	**Verkehrsausscheidungszahl**	
[one or two digits to access to the trunk exchange or gateway exchanche]		[ein oder zwei Ziffern zum Ansteuern einer nationalen oder internationalen Fernvermittlung]	
= access code		= Verkehrsausscheidungsziffer; Zugangskennzahl; Zugangszahl; Zugangskennziffer; Zugangsziffer	
prefix	PHYS	**Vorsatz**	
[for decimal multiples or fractions]		[für dezimale Vielfache oder Bruchteile]	
prefix	LING	**Präfix**	
↑ affix		= Vorsilbe	
		↑ Affix	
prefix	RADIO	**Landeskenner**	
[amateur radio]		[Amateurfunk]	
prefix (→header label)	DATA PROC	(→Dateianfangs-Etikett)	
prefix analysis	SWITCH	**Präfixauswertung**	
prefix numbering (→open numbering)	SWITCH	(→offene Numerierung)	
prefix plus code number	SWITCH	**Vorwahlnummer**	
[sequence of digits to be dialled before the subscriber number]		[Ziffernfolge die vor der Teilnehmerrufnummer gewählt werden muß, d.h. Verkehrsausscheidungszahl + Kennzahl; z.B. 089 für München]	
≈ prefix		= Vorwählnummer; Vorwahl 2	
		= Kennzahl	
preform (n.)	MANUF	**Rohling**	
= unmachined part; slug		= Rohteil; Ausgangsteil; Vorform	
preform	OPT.COMM	**Vorform**	
= boule			
preformat (v.t.)	DATA PROC	**vorformatieren**	
= initialize 2			
preformatting	DATA PROC	**Vorformatierung**	
		= Vorformatieren	
preformed cable	EQUIP.ENG	**vorgefertigtes Kabel**	
= prefabricated cable		= fabrikkonfektioniertes Kabel; konfektioniertes Kabel	
≈ plug-in cable		≈ Steckkabel	
preformulated	DOC	**vorformuliert**	
[standard text]		[Normtext]	
= boilerplate-			
p-register	DATA PROC	**P-Register**	
[indicates location of present instruction]		[zeigt Speicherplatz der laufenden Anweisung an]	
pregroup	TRANS	**Vorgruppe**	
[FDM]		[TF-Technik]	
preheat (v.t.) (→warm-up)	TECH	(→anwärmen)	
preheating	TECH	**Vorwärmung**	
= heating-up		= Vorerhitzung; Anheizung; Aufheizung	
preheating time	TECH	**Vorheizzeit**	
≈ warm-up time		≈ Aufwärmzeit	
preheat resistor	ELECTRON	**Vorheizwiderstand**	
preimpulse	ELECTRON	**Vorimpuls**	
preintegration	SYS.INST	**Vorintegration**	
		= Vormontage	
prejudice (→disadvantage)	COLLOQ	(→Nachteil)	
preliminary	COLLOQ	**Vorab-**	
= advance-			
preliminary adjustment	ELECTRON	**Vorabgleich**	
preliminary contract	ECON	**Vorvertrag**	
preliminary design	DATA PROC	**Grobentwurf**	
preliminary discussion (→preliminary meeting)	ECON	(→Vorbesprechung)	
preliminary draft	DOC	**Vorentwurf**	
= rough draft		= Grobentwurf	
preliminary drawing	ENG.DRAW	**Zeichnungsentwurf**	
preliminary housekeeping (→interlude)	DATA PROC	(→Vorprogramm)	
preliminary information	TECH	**Vorabinformation**	
= advance information			
preliminary meeting	ECON	**Vorbesprechung**	
≈ preliminary discussion			
preliminary product (→pre-product)	MANUF	(→Vorerzeugnis)	
preliminary standard	TECH	**Vornorm**	
preliminary work	COLLOQ	**Vorarbeit**	
pre-machine (→pre-tool)	MECH	(→vorbearbeiten)	
pre-machined	METAL	**geschruppt**	
premagnetization	PHYS	**Vormagnetisierung**	
= magnetic polarization 2; magnetic biasing; presaturation		≈ magnetische Polarisierung	
		↑ Vorpolung	
≈ magnetic polarization 1			
↑ biasing			
premagnetize	PHYS	**vormagnetisieren**	
= polarize			
pre-mating contact	COMPON	**voreilender Kontakt**	
= protuding contact; extra-long contact			
premature	COLLOQ	**vorzeitig**	
↑ untimely		≈ voreilig	
		↑ unzeitig	
premature disconnection	SWITCH	(→vorzeitige Auslösung)	
(→premature release)			
premature release	SWITCH	**vorzeitige Auslösung**	
= premature disconnection		= vorzeitige Trennung	
premise (→prerequisite)	COLLOQ	(→Voraussetzung)	
premises	CIV.ENG	**Räumlichkeiten**	
[building or part of it]		≈ Gebäude	
= conveniencies			
≈ building			
premises	ECON	**Gelände**	
premises (→site)	TELEC	(→Standort)	
premises protection system	SIGN.ENG	(→Geländeüberwachungsanlage)	
(→premises surveillance system)			
premises surveillance system	SIGN.ENG	**Geländeüberwachungsanlage**	
= premises protection system			
premium (→insurance premium)	ECON	(→Versicherungsprämie)	
premodulation	MODUL	**Vormodulation**	
pre-oscillation current	EL.TECH	(→Anschwingstrom)	
(→starting current)			
prepaid	POST	**franko**	
prepaid card (→prepaid phonecard)	TERM&PER	(→Guthabenkarte)	
prepaid phonecard	TERM&PER	**Guthabenkarte**	
= prepaid card; debit card; cash-card		[beim Kauf vorausbezahlt]	
↑ chip phonecard		= Kaufkarte; Debit-Karte	
		↑ Telefon-Chipkarte	
preparation (→treatment)	TECH	(→Behandlung)	
preparation (→initialization)	DATA PROC	(→Initialisierung 1)	
preparation mode (→intilialization mode)	DATA COMM	(→Vorbereitungsbetrieb)	
preparatory input (→precondition input)	CIRC.ENG	(→Vorbereitungseingang)	
preparatory programm	DATA PROC	**vorbereitendes Programm**	
preparatory signal	SWITCH	**Vorbereitungszeichen**	

prepay coin telephone (AM) TELEPH
= prepay paystation (AM)
prepayment ECON
= advance payment; advance
≈ down payment
prepay paystation (AM) TELEPH
(→prepay coin telephone)
preposition LING
[specifies spacial or temporal relation; e.g. in, after]
prepositional adverb LING
prepositional object LING
preprint (n.) TYPOGR
preprinted form (→form) TERM&PER
preprocessing DATA PROC
preprocessor 1 DATA PROC
[checks or compresses data before entering them into a mainframe]
= front-end processor
≈ front-end processor [DATA COMM]
preprocessor 2 DATA PROC
[a program to prepare or generate data in adequate form]
↓ precompiler
pre-product MANUF
= preliminary product
preproduction (→parts MANUF
manufacture)
preproduction series (→test MANUF
series)
prerecorded message TELEC
prerecorded tape TERM&PER
prerecorded text element DATA PROC
(→boilerplate)
prerequisite COLLOQ
= condition; premise; precondition
≈ supposition
presaturation PHYS
(→premagnetization)
prescaler INSTR
prescribed dimension ENG.DRAW
= set dimension
≈ nominal size
prescription 2 ECON
= period of limitation
prescription 1 ECON
= regulatory standard
≈ regulation
prescriptive (→superannuated) ECON
preselectable TECH
= presettable
preselect command TELECONTR
preselection SWITCH
preselection TECH
= presetting
preselection filter NETW.TH
preselection stage SWITCH
(→preselector)
preselector HF
preselector SWITCH
[concentration stage in electromechanical switches]
= preselection stage
≈ line finder
present ECON
= unveil
presentation TECH
≈ demostration

Münzfernsprecher mit Vorauszahlung
Vorauszahlung
= Vorwegzahlung; Vorschuß; Vorleistung 2
≈ Anzahlung
(→Münzfernsprecher mit Vorauszahlung)
Präposition
[bezeichnet örtliche oder zeitliche Verhältnisse; z.B. in, nach]
= Verhältniswort
Pronominaladverb
= Umstandsfürwort
Präpositionalobjekt
Vorabdruck
(→Formular)
Vorverarbeitung
Vorverarbeitungsrechner
[prüft oder verdichtet Daten vor Eingabe in einen Großrechner]
= Vorrechner; Satellit
≈ Vorrechner [DATA COMM]
Preprozessor
[ein Programm zur Aufbereitung oder Generierung von Daten in adequater Form]
= Vorübersetzer; Vorverarbeiter; Vorprozessor
↓ Vorkompilierer
Vorerzeugnis
= Vorprodukt
(→Vorfertigung)
(→Vorserie)
Bandansage
bespieltes Band
(→Normtext)
Voraussetzung
= Prämisse; Vorbedingung
≈ Annahme
(→Vormagnetisierung)
Prescaler
[digitaler Frequenzteiler]
Sollmaß
≈ Nennmaß
Verjährung
Vorschrift
≈ Regelung
(→verjährt)
voreinstellbar
Vorbefehl
Vorwahl 1
Vorwahl
Vorselektionsfilter
(→Vorwähler)
Selektor
Vorwähler
[Konzentrationsstufe in elektromechanischen Wählern]
= Vorwahlstufe
≈ Anrufsucher
vorstellen
≈ präsentieren
Präsentation
≈ Vorführung

presentation DATA PROC
(→representation)
presentation (→submission) OFFICE
presentation graphics DATA PROC
≈ business graphics
presentation image DATA COMM
presentation-image DATA COMM
definition
presentation layer DATA COMM
[6th layer os OSI; defines format, code]
presentation protocol DATA COMM
presentation service DATA COMM
present participle LING
[e.g. listening]
↑ participle
present perfect LING
[e.g. I have heard; I have been heard]
present perfect progressive LING
[e.g. I have been working]
present progressive LING
[e.g. I am working]
present tense LING
[e.g. I listen]
↓ simple present; present progressive
present value ECON
= current value; present worth (AM)
present worth (AM) (→present ECON
value)
preservation ECON
≈ maintenance
preset (v.t.) CIRC.ENG
= preadjust (v.t.)
≈ set (v.t.)
preset (adj.) COLLOQ
= default; predetermined; predefined; fixed; given
preset (v.t.) (→set) CIRC.ENG
preset (n.) CIRC.ENG
(→presetting)
preset (→initialize 1) DATA PROC
preset counter (→presettable CIRC.ENG
counter)
preset paper feed TERM&PER
(→escapement)
presettable (→preselectable) TECH
presettable counter CIRC.ENG
= preset counter
presetting CIRC.ENG
= preset (n.)
presetting (→initialization) DATA PROC
presetting (→preselection) TECH
preshaping filter NETW.TH
preshoot (n.) ELECTRON
= leading-edge overshoot
↑ transient
presidency (→chairmanship) ECON
president (→chairman) ECON

(→Darstellung)
(→Vorlage 2)
Präsentationsgrafik
= Präsentationsgraphik
≈ Geschäftsgrafik
Begriffsvorrat
Begriffsvorrat-Syntax
Darstellungsschicht
[6.Schicht im ISO-Schichtenmodell]
= Darstellungsebene; Darstellung; Präsentationsebene
Darstellungsprotokoll
= Präsentationsprotokoll
Darstellungsdienst
= Präsentationsdienst
Partizip I
[z.B. hörender]
= Partizip des Präsens; Mittelwort der Gegenwart
↑ Partizip
Perfekt
[z.B. ich habe gehört, ich wurde gehört; im Englischen zur Aussage daß oder ob etwas geschehen ist]
= 2. Vergangenheit; zweite Vergangenheit; vollendete Gegenwart; Vorgegenwart
Verlaufsform des Perfekts
Verlaufsform des Präsens
[im Englischen übliche Tempusform zum Ausdruck eines gerade Geschehenden]
= Verlaufsform der Gegenwart
Präsens
[z.B. ich höre]
= Gegenwart
↓ Verlaufsform des Präsens
Barwert
= Zeitwert
(→Barwert)
Erhaltung
≈ Wartung
voreinstellen
≈ vorbereiten
≈ setzen (v.t.)
vorgegeben
= vorbestimmt; vordefiniert; fest
(→setzen)
(→Voreinstellung)
(→initialisieren 1)
(→Vorwählzähler)
(→eingestellter Papiervorschub)
(→voreinstellbar)
Vorwählzähler
Voreinstellung
= Vorwahl
(→Initialisierung 1)
(→Vorwahl)
Spektrumformungsfilter
Vorschwingung
= Vorschwinger; vorlaufender Überschwinger
↑ Übergangsvorgang
(→Vorsitz)
(→Vorsitzender)

presort (v.t.)	DATA PROC	vorsortieren	
press (n.)	COLLOQ	Presse	
press (n.)	METAL	Presse	
		= Druckpresse	
press button (→pushbutton)	COMPON	(→Druckknopf)	
press cable lug (→crimp cable lug)	COMPON	(→Quetschkabelschuh)	
press casting	METAL	Preßgießen	
press connector (→crimp cable lug)	COMPON	(→Quetschkabelschuh)	
pressed	TECH	gepreßt	
pressed glass	TECH	Preßglas	
pressfax	TELEGR	Pressefax	
= pagefax			
press-fit technique (→press-in technique)	COMPON	(→Einpreßverfahren)	
press-fit terminal	COMPON	Einpreßstift	
press-in	MECH	einpressen	
pressing	METAL	Pressen	
pressing die	METAL	Preßform	
pressing screw	MECH	Druckschraube	
= thumbscrew			
press-in nut	MECH	Einpreßmutter	
press-in technique	COMPON	Einpreßverfahren	
= press-fit technique			
press item (→newspaper article)	COLLOQ	(→Zeitungsartikel)	
press-on felt	TERM&PER	Andruckfilz	
press radio	RADIO	Pressefunk	
press release	ECON	Pressemitteilung	
		= Presseinformation	
press report	COLLOQ	Pressemeldung	
press run (→edition printing)	TYPOGR	(→Auflage 1)	
press sleeve	COMPON	Preßhülse	
press-stud connector	POWER SYS	Druckknopfkontakt [Batterie-Zubehör]	
= snap-on connector			
press terminal (→crimp cable lug)	COMPON	(→Quetschkabelschuh)	
pressure	PHYS	Druck	
[force per unit area; SI unit: Pascal]		[Kraft durch Fläche; SI-Einheit: Pascal]	
≈ mechanical stress		≈ mechanische Spannung	
		↓ Überdruck; Unterdruck	
pressure-activated (→pneumatically operated)	TECH	(→druckluftbetätigt)	
pressure-actuated (→pneumatically operated)	TECH	(→druckluftbetätigt)	
pressure air (→compressed air)	PHYS	(→Druckluft)	
pressure alarm system	SIGN.PROC	Druckmeldesystem	
pressure angle	MECH	Eingriffswinkel [Zahnrad]	
pressure balance	PHYS	Druckausgleich	
= pressure compensation		= Druckkompensation	
pressure-cast	METAL	preßgießen	
pressure clamp	EL.INST	Klemme	
≈ tweezer; clip (n.)		≈ Klemmleiste	
pressure compensation (→pressure balance)	PHYS	(→Druckausgleich)	
pressure connection (→crimp connection)	COMPON	(→Quetschverbindung)	
pressure-controlled (→pneumatically operated)	TECH	(→druckluftbetätigt)	
pressure controller	POWER ENG	Druckwächter	
pressure-dependent	TECH	druckabhängig	
pressure die-cast (adj.)	METAL	spritzgegossen	
pressure die-casting	METAL	Druckguß	
= die-casting; die-cast		= Spritzguß	
pressure drop	PHYS	Druckabfall	
pressure gage (→manometer)	INSTR	(→Manometer)	
pressure-measuring transducer	COMPON	Druckmeßumformer	
= pressure transducer			
pressure microphone	EL.ACOUS	Schalldruckmikrophon	
pressure monitoring	OUTS.PLANT	Druckluftüberwachung	
pressure-operated (→pneumatically operated)	TECH	(→druckluftbetätigt)	
pressure probe	INSTR	Druckmeßsonde	
pressure rise	PHYS	Druckanstieg	
pressure roll (→paper pressure roll)	TERM&PER	(→Papieranddrückrolle)	

pressure-sensitive (adj.)	TECH	druckempfindlich	
≈ touch-sensitive		≈ berührungsempfindlich	
pressure-sensitive keyboard (→membrane keyboard)	TERM&PER	(→Folientastatur)	
pressure-sensitive paper	TERM&PER	druckempfindliches Papier	
= action paper; non-carbon paper; NCP		= Non-Karbon-Papier; Durchschreibpapier; durchschreibendes Papier; Aktionspapier	
pressure-sensitive pen	TERM&PER	druckempfindlicher Lesestift	
pressure sensitivity	EL.ACOUS	Druckübertragungsfaktor	
pressure sensitivity	TECH	Druckempfindlichkeit	
pressure sensor	COMPON	Drucksensor	
		= Druckfühler	
pressure spring (→compression spring)	MECH	(→Druckfeder)	
pressure tight	TECH	druckdicht	
pressure transducer	COMPON	(→Druckmeßumformer)	
(→pressure-measuring transducer)			
pressure transmitter	INSTR	Durchflußmeßumformer	
pressure vessel	TECH	Druckbehälter	
pressure wave (→compressional wave)	PHYS	(→Druckwelle 1)	
pressure welding	METAL	Preßschweißen	
≈ forge welding		≈ Feuerschweißung	
pressure window	MICROW	Druckfenster	
prestrained	MECH	vorverformt	
prestressed	MECH	vorgespannt	
prestressed concrete	CIV.ENG	Spannbeton	
prestressed concrete tower	CIV.ENG	Spannbetonturm	
prestudy (n.)	TECH	Vorstudie	
presupposition	SCIE	Prämisse	
preterit (→simple past)	LING	(→Imperfekt)	
preterite (→simple past)	LING	(→Imperfekt)	
pretest (v.t.) (→precheck)	MANUF	(→vorprüfen)	
pretesting	MANUF	Vorprüfung	
= prechecking			
pre-tool (v.t.)	MECH	vorbearbeiten	
= pre-machine			
pre-transformer	POWER SYS	Vorschalttransformator	
pre-trigger	INSTR	Vortriggerung	
pre-trigger delay	INSTR	Vortriggerverzögerung	
pre-trigger display	INSTR	Pre-trigger-Anzeige	
prevarication (→irrelevance)	INF	(→Irrelevanz)	
prevention	TECH	Vorbeugung	
= precaution		= Vorsorge	
preventive (n.)	TECH	Schutzmittel	
preventive (adj.)	TECH	vorbeugend	
= precautionary		= vorsorglich	
preventive maintenance	TECH	vorsorgliche Wartung	
		= vorbeugende Wartung	
preview function [word processing]	DATA PROC	Preview-Funktion [Textverarbeitung]	
		= Seitenübersicht	
previous month (→preceding month)	COLLOQ	(→Vormonat)	
previous quarter	ECON	Vorquartal	
preview box	DATA PROC	Vorschaubox	
preview function	DATA PROC	Seitenansichtsfunktion	
		= Vorschaufunktion; Layout-Darstellungsfunktion	
prewired	EQUIP.ENG	vorverdrahtet	
≈ factory-wired		≈ vorverkabelt	
prewiring	EQUIP.ENG	Vorverdrahtung	
≈ factory wiring		≈ Werkverdrahtung	
price (n.)	ECON	Preis	
↓ sales price; transfer price		↓ Verkaufspreis; Verrechnungspreis	
price collapse	ECON	Preisverfall	
		= Preissturz	
price escalation clause	ECON	Preisformel	
= escalation clause; price escalation formula; escalation formula; price revision formula		= Preisgleitklausel; Preisanpassungsklausel; PGK; Gleitklausel	
price escalation formula (→price escalation clause)	ECON	(→Preisformel)	
price excess	ECON	Überpreis	
≈ excessive price		≈ überhöhter Preis	
price ex factory	ECON	Fabrikabgabepreis	
= price ex works		= Fabrikpreis	

price ex works (→price ex factory) ECON (→Fabrikabgabepreis)
price formation (→pricing) ECON (→Preisbildung)
price increase ECON **Teuerung**
= rising prices = Verteuerung
≠ price reduction ≠ Verbilligung
priceless (→exorbitant) ECON (→unbezahlbar)
price list ECON **Preisliste**
price-performance payoff ECON **Preis-Leistungs-Verhältnis**
price reduction (→discount) ECON (→Preisnachlaß)
price reduction (→price increase) ECON (→Verteuerung)
price revision formula (→price escalation clause) ECON (→Preisformel)
pricey (→expensive) ECON (→teuer)
pricing ECON **Preisbildung**
= price formation = Preisstellung
primary (n.) (→primary print) DATA PROC (→Primärausdruck)
primary ac POWER SYS **Primärwechselstrom**
primary access TELEC **Primärratenanschluß**
[1,984 kbit/s] [1,984 kbit/s]
primary access (→ISDN primary access) TELEC (→ISDN-Primäranschluß)
primary battery POWER SYS **Elementen-Batterie**
primary block (→digroup) TELEC (→Primärmultiplexbündel)
primary cable (→main cable) OUTS.PLANT (→Hauptkabel)
primary cell (→primary galvanic cell) POWER SYS (→Primärelement)
primary coating OPT.COMM **Primärbeschichtung**
primary code (→source code) CODING (→Primärcode)
primary coding CODING **Quellencodierung**
[coding within the source] [erfolgt in der Signalquelle]
= source coding; source encoding ≠ Kanalcodierung
≠ channel coding
primary color (→elementary color) PHYS (→Grundfarbe)
primary commodity (→raw material) TECH (→Rohstoff)
primary control SWITCH **Leitsteuerung**
primary data DATA PROC **Primärdaten**
= original data
primary data acquisition DATA COMM **Primärdatenerfassung**
= primary data entry = Primäreingabe
primary data entry DATA COMM (→Primärdatenerfassung)
(→primary data acquisition)
primary detector (→transducer) INSTR (→Meßaufnehmer)
primary electron PHYS **Primärelektron**
primary element (→transducer) INSTR (→Meßaufnehmer)
primary energy TECH **Primärenergie**
primary exchange SWITCH **Hauptvermittlungsstelle**
= HVSt
primary frequency standard INSTR **Primärfrequenznormal**
[does not require any other reference for calibration] [benötigt keine Eichnormal]
= Primärfrequenzstandard
primary fuse COMPON **Grobsicherung**
primary galvanic cell POWER SYS **Primärelement**
= primary cell [sich irreversibel entladendes Element]
= primäres Element
↑ galvanisches Element
primary group TRANS **Primärgruppe**
[FDM] [TF-Technik]
= channel group (AM); group 2 = Gruppe 2
≈ basic group ≈ Grund-Primärgruppe
primary key DATA PROC **Primärschlüssel**
primary lineswitch (→final selector) SWITCH (→Leitungswähler)
primary multiplex access TELEC **Primarmultiplexanschluß**
[ISDN]
= PMXA
primary multiplex analyzer INSTR **Primärmultiplex-Analysator**
primary multiplexer TELEC **Primärmultiplexer**
= channel bank (AM) = erste Umsetzerstufe
↓ FDM channel modulation; PCM ↓ Kanalumsetzung; PCM

primary multiplex group TELEC (→Primärmultiplexbündel)
(→digroup)
primary occupation TERM&PER **primäre Belegung**
[of numeric keyboard, when NUM-LOCK key is activatet; digits can be entered] [des Ziffernblocks; bei eingeschalteter NUM-Taste; es können Ziffern eingegeben werden]
≈ lower mode ≈ Grundebene
primary power (→prime power) POWER SYS (→Primärstromversorgung)
primary print DATA PROC **Primärausdruck**
= primary (n.)
primary radar RADIO LOC **Primärradar**
= passive Rückstrahlortung
primary radiator ANT **Primärstrahler**
= active radiator; exciter; fed element; driven element = Erregerstrahler; Erreger; aktiver Strahler; gespeistes aktives Element; gespeistes Element
≠ parasitic element ≠ Parasitärstrahler
primary rate TRANS **Primärrate**
[1,544 kbit/s] [USA, 1,544 kbit/s für 23 + 1 PCM-Kanäle]
= DS1; T1
primary route SWITCH **Erstweg**
= first-choice route; first route; normal route = erster Leitweg; Regelweg
≈ direct route ≈ Direktweg
primary school SCIE **Grundschule**
= elementary school; first schools (BRI) = Volksschule; Elementarschule; Primarschule
↑ school ↑ Lehranstalt
primary shift MANUF **Normalschicht**
primary signal TV **Primär-Farbartsignal**
≈ chrominance signal ≈ Chrominanzsignal
primary standard PHYS **Urnormal**
primary storage address DATA PROC (→Hauptspeicheradresse)
(→main memory address)
primary track DATA PROC **erste Spur**
= track 0 = Spur 0
primary winding EL.TECH **Primärwicklung**
↑ transformer winding ↑ Transformatorwicklung
primary word LING **Grundwort**
prime MATH **teilerfremd**
[having no factor except itself and 1] [mit keinem gemeinsamen Teiler außer 1]
= relativ prim
prime memory (→main memory 1) DATA PROC (→Hauptspeicher 1)
prime mover TECH **Kraftmaschine**
prime number MATH **Primzahl**
[cardinal number divisible only by 1 or itself] [nur durch 1 oder sich selbst teilbare Kardinalzahl]
prime power POWER SYS **Primärstromversorgung**
= primary power = Primärenergie
= station power supply ≈ Amtsstromversorgung
prime storage (→main memory 1) DATA PROC (→Hauptspeicher 1)
prime time BROADC **Hauptsendezeit**
primitive 1 (n.) DATA PROC **Grundelement**
[lowest unit of programming language] [kleinste Befehlseinheit auf tiefstem Sprachniveau]
= basic unit; fundamental unit
primitive TECH **primitiv**
= rudimental; unsophisticated = simpel; schlicht
≈ provisional ≈ behelfsmäßig
≠ sophisticated ≠ raffiniert
primitive 2 (→drawing element) DATA PROC (→Zeichenelement)
primitive element (→drawing element) DATA PROC (→Zeichenelement)
principal axis MATH **Hauptachse**
= main axis
principal benefit TECH **Hauptvorteil**
principal current MICROEL **Hauptstrom**
= main current
principally (→basically) COLLOQ (→grundsätzlich)
principal office (→head office) ECON (→Zentralverwaltung)

principal quantum number (→main	PHYS	(→Hauptquantenzahl)
quantum number)
principal terminal (→main	MICROEL	(→Hauptstromanschluß)
terminal)
principle	COLLOQ	**Grundsatz**
		= Prinzip
principle of duality	MATH	**Dualitätsprinzip**
principle of operation	TECH	**Arbeitssystem**
principles	SCIE	**Grundlagen**
= fundamentals; basics
print (v.t.)	TERM&PER	**drucken**
print (v.t.)	OFFICE	**pausen**
= copy (v.t.)		= kopieren
print (n.)	TERM&PER	**Druck**
= printing		= Drucklegung
print (→copy 2)	OFFICE	(→Pause)
print (→typesetting)	TYPOGR	(→Satz)
printable position (→print	TERM&PER	(→Druckstelle)
position)
print ball	TERM&PER	**Kugelkopf**
= golf ball		= Kugelschreibkopf
↑ print head; type ball; type head; type		↑ Schreibkopf 1; Typenträger
carrier
print bar (→type bar)	OFFICE	(→Typenhebel)
print buffer	DATA PROC	**Druckpuffer**
= printing buffer; printer buffer		= Druckerpuffer
print chain	TERM&PER	**Druckkette**
[of a chain printer]		[eines Kettendruckers]
= type chain		= Typenkette
↑ type carrier		↑ Typenträger
print chart	DATA PROC	**Ausdruck-Spezifikation**
print command (→print	DATA PROC	(→Druckbefehl)
instruction)
print control character	DATA PROC	**Drucksteuerzeichen**
print controller	TERM&PER	**Drucksteuergerät**
print density (→printing	TERM&PER	(→Druckdichte)
density)
printed board (→printed	ELECTRON	(→Leiterplatte)
circuit board)
printed board assembly	MANUF	(→Leiterplattenbestückung)
(→insertion of components)
printed circuit (→printed	ELECTRON	(→Leiterplatte)
circuit board)
printed-circuit antenna	ANT	(→Mikrostrip-Antenne)
(→microstrip antenna)
printed circuit base (→PCB	ELECTRON	(→Leiterplattenträger)
base)
printed circuit board	ELECTRON	**Leiterplatte**
= PCB; printed wiring board; PWB;		= gedruckte Schaltung; Schal-
printed circuit; board; printed circuit;		tungsplatine; Printplatte;
circuit board		Platine; Karte
printed circuit chassis	EQUIP.ENG	(→Baugruppenrahmen)
(→module frame)
printed circuit drafting	ELECTRON	(→Leiterplattenentflechtung)
(→PCB artwork creation)
printed circuit frame	EQUIP.ENG	(→Baugruppenrahmen)
(→module frame)
printed circuit shelf	EQUIP.ENG	(→Baugruppenrahmen)
(→module frame)
printed circuit shelf	EQUIP.ENG	(→Baugruppenrahmen)
(→module frame)
printed matter	POST	**Drucksache**
printed tape	TERM&PER	**Druckstreifen**
printed wire (→track	ELECTRON	(→Leiterbahn)
conductor)
printed wiring	ELECTRON	**gedruckte Verdrahtung**
printed wiring board	ELECTRON	(→Leiterplatte)
(→printed circuit board)
print element (→print	TERM&PER	(→Druckkopf)
head)
printer	TERM&PER	**Drucker**
↓ type printer; dot-matrix printer; im-		↓ Typendrucker; Rasterdruk-
pact printer; non impact printer		ker; Anschlagdrucker; an-
		schlagfreier Drucker
printer buffer (→print	DATA PROC	(→Druckpuffer)
buffer)
printer cable	TERM&PER	**Druckerkabel**
printer chassis	TERM&PER	**Druckergestell**
printer connection (→printer	EQUIP.ENG	(→Druckeranschluß)
output)
printer control command	TERM&PER	**Druckersteuerbefehl**
printer controller	TERM&PER	**Druckersteuergerät**
printer emulation	DATA PROC	**Druckeremulation**
printer font	TERM&PER	**Druckerzeichensatz**
↑ font		= Druckerfont; Druckfont
		↑ Zeichensatz
printer listing	DATA COMM	**Druckerprotokoll**
≈ computer listing		≈ Rechnerprotokoll
printer output	EQUIP.ENG	**Druckeranschluß**
= printer connection; printout connec-		= Druckerausgang
tion
printer output	DATA PROC	(→Ausdruck)
(→printout)
printer-plotter (→graphics	TERM&PER	(→Graphikdrucker)
printer)
printer ribbon (→inked	OFFICE	(→Farbband)
ribbon)
print error (→write	TERM&PER	(→Druckfehler)
error)
printer server (→print	DATA PROC	(→Druckserver)
server)
printer shaft	TERM&PER	**Druckerwelle**
printer's ink	TYPOGR	**Druckerschwärze**
≈ toner		= Schwärze
		≈ Toner
printer's point (→point 1)	TYPOGR	(→Punkt)
printer stand	TERM&PER	**Druckerständer**
printer terminal	DATA COMM	**Terminaldrucker**
[terminal station with printing func-		[Datenstation mit Drucker-
tion]		funktion]
= printing terminal; hard-copy termi-		= Druckerstation;
nal; printing station terminal		Druckwerkterminal;
↑ terminal station; data terminal		Schreibstation
		≈ Drucker
		↑ Datenstation
print force	TERM&PER	(→Anschlagstärke)
(→penetration)
print format	TERM&PER	**Druckbild**
= printing format; printout format		= Druckformat
print formatter	DATA PROC	**Druckformatierer**
print generator	DATA PROC	**Druckgenerator**
		= Printgenerator
print hammer	TERM&PER	**Druckhammer**
		= Typenhammer
print head	TERM&PER	**Druckkopf**
= printhead; type head; typehead;		= Schreibkopf 1
write head; writehead; writing		↑ Typenträger
head; print element; write ele-		↓ Kugelkopf; Typenkopf
ment
↑ type carrier
↓ print ball; type pallet
printhead (→print head)	TERM&PER	(→Druckkopf)
printing	TERM&PER	**Drucken**
printing (→print)		(→Druck)
printing buffer (→print	DATA PROC	(→Druckpuffer)
buffer)
printing character	TERM&PER	**Druckbuchstabe**
printing density	TERM&PER	**Druckdichte**
= print density		= Schreibdichte
printing element	TERM&PER	**Druckelement**
↓ metallic needle; ink ejector		↓ Metallnadel; Tintendüse
printing format (→print	TERM&PER	(→Druckbild)
format)
printing mechanism	TERM&PER	**Druckwerk**
		= Druckmechanismus
printing performance (→print	TERM&PER	(→Druckleistung)
speed)
printing process	TYPOGR	**Druckverfahren**
↓ letterpress; intaglio printing; plano-		↓ Hochdruck; Tiefdruck;
graphic printing		Flachdruck
printing program (→print	DATA PROC	(→Druckprogramm)
program)
printing punch	TERM&PER	**Schreiblocher**
[makes both at the same time]		[bedruckt gleichzeitig beim
		Lochen]
printing recorder (→logger)	INSTR	(→Meßwertdrucker)
printing speed (→print	TERM&PER	(→Schreibleistung)
rate)
printing station terminal	TERM&PER	**Druckwerkterminal**
= printing terminal

printing station terminal 450

printing station terminal	DATA COMM	(→Terminaldrucker)	
(→printer terminal)			
printing terminal (→printer terminal)	DATA COMM	(→Terminaldrucker)	
printing terminal (→printing station terminal)	TERM&PER	(→Druckwerkterminal)	
print inhibit	DATA PROC	**Drucksperre**	
		= Schreibsperre	
print instruction	DATA PROC	**Druckbefehl**	
= print command		= Druckanweisung; Druckkommando	
print intensity	TERM&PER	(→Anschlagstärke)	
(→penetration)			
print letter	TYPOGR	**Druckschrift**	
[font type]		[Schriftart]	
print line	TYPOGR	**Druckzeile**	
print manager	DATA PROC	**Druckverwalter**	
print mask	DATA PROC	**Druckmaske**	
printout	DATA PROC	**Ausdruck**	
= printer output; hard printout; dump (n; AM); logging; recording		= Druckerausdruck; Ausdruck; Druckerausgabe; Protokollierung; Aufzeichnung	
≈ hard copy; printer listing; listing		≈ Druckkopie; Druckerprotokoll; Auflistung	
↓ hardcopy		↓ Hardcopy	
print-out (v.t.) (→log)	DATA PROC	(→protokollieren)	
printout connection (→printer output)	EQUIP.ENG	(→Druckeranschluß)	
printout editing	DATA PROC	**Druckaufbereitung**	
↑ editing		↑ Editieren	
printout format (→print format)	TERM&PER	(→Druckbild)	
print position	TERM&PER	**Druckstelle**	
= printable position		= bedruckbare Stelle	
print preparation program	DATA PROC	**Druckaufbereitungsprogramm**	
print program	DATA PROC	**Druckprogramm**	
= printing program			
print quality	TERM&PER	**Druckqualität**	
↓ draft print quality; near-letter quality; letter quality		↓ Konzeptdruckqualität; Nahezu-Korrespondenzqualität; Korrespondenzqualität	
print queue	DATA PROC	**Druckschlange**	
print rate	TERM&PER	**Schreibleistung**	
= printing speed			
print reader	TERM&PER	**Druckschriftleser**	
print run	DATA PROC	**Drucklauf**	
PRINT-SCREEN key	TERM&PER	**PRINTSCREEN-Taste**	
[actives the printout of actual screen content]		[veranlasst den Druck des aktuellen Bildschirminhaltes]	
		= DRUCK-Taste	
print server	DATA PROC	**Druckserver**	
[manages the print jobs of a computer network]		[verwaltet die Druckaufträge eines Rechnernetzes]	
= printer server		↑ Sever	
↑ server			
print speed	TERM&PER	**Druckleistung**	
[units of measure: LPS = lines per second, CPS = characters per second]		[Maßeinheiten: LPS = Zeilen pro Sekunde, Z/sec oder CPS = Zeichen pro Sekunde]	
= printing performance			
print wheel (→typewheel)	TERM&PER	(→Typenrad)	
print wheel-printer (→typew-heel printer)	TERM&PER	(→Typenraddrucker)	
prior charges	ECON	**Vorlasten**	
prioritirazing	TELEC	(→Prioritätensetzung)	
(→prioritization)			
prioritization	TELEC	**Prioritätensetzung**	
= prioritirazing			
priority (→precedence)	COLLOQ	(→Vorrang)	
priority (→significance)	INF	(→Wertigkeit)	
priority call	SWITCH	**Vorrangsverbindung**	
		= Prioritätsverbindung	
priority class (→priority level)	DATA PROC	(→Prioritätsebene)	
priority command	TELECONTR	**Schnellbefehl**	
priority control	DATA PROC	**Prioritätssteuerung**	
		= Vorrangssteuerung	
priority encoder	DATA PROC	**Prioritätscodierer**	
priority interrupt	DATA PROC	**Prioritätsunterbrechung**	
		= Vorrangsunterbrechung	
priority level	DATA PROC	**Prioritätsebene**	
= priority class			
priority logic	DATA PROC	**Prioritätslogik**	
		= Vorrangslogik	
priority processing	DATA PROC	**Prioritätsverarbeitung**	
		= Vorrangverarbeitung	
priority program	DATA PROC	**Prioritätsprogramm**	
= high-priority programm; foreground program; foreground job; foreground task		= Vordergrundsprogramm	
		≠ Hintergrundsprogramm	
≠ background program			
priority return information	TELECONTR	**Schnellrückmeldung**	
priority state information	TELECONTR	**Schnellmeldung**	
prior-ranking	COLLOQ	**vorrangig**	
= with priority		= vorrechtlich; prioritätisch	
pripheral electron (→bonding electron)	PHYS	(→Valenzelektron)	
prism	MATH	**Prisma**	
[parralel polygons as ground faces, side faces by parallelograms]		[parallele Vielecke als Grundflächen und Parallelogramme als Seitenflächen]	
↑ polyhedrons		↑ Polyeder	
↓ parallelepiped; rectangular rism; cube		↓ Parallelepiped; Quader; Würfel	
prism antenna	ANT	**Reusenantenne**	
= pyramid antenna		≈ Reuse	
≈ cage antenna		≈ Käfigantenne	
prismatic	MATH	**prismatisch**	
Pritchard equivalent circuit	MICROEL	**Pritchard-Ersatzschaltung**	
privacy	INF.TECH	**Geheimhaltung**	
= secrecy		≈ Vertraulichkeit; Fernmeldegeheimnis	
≈ confidentiality; communications confidentiality			
Privacy Act (U.S.A.)	ECON	(→Datenschutzgesetz)	
(→Data Protection Act)			
private address	POST	**Privatanschrift**	
private automatic branch exchange	SWITCH	**automatische Nebenstellenanlage**	
[providing access to the public network]		[mit Anschluß am öffentlichen Netz]	
= PABX			
≈ private automatic exchange			
private branch exchange	SWITCH	**Nebenstellenanlage**	
= PBX; branch exchange		= Teilnehmerzentrale	
private branch net (→private branch network)	TELEC	(→Nebenstellennetz)	
private branch network	TELEC	**Nebenstellennetz**	
= private branch net			
private call	TELEPH	**Privatgespräch**	
private circuit (BRI) (→leased line)	TELEC	(→Mietleitung)	
private company	ECON	**kleine Aktiengesellschaft**	
[by british law, with limited liability]		[nach britischem Recht, mit beschränkter Haftung]	
private correspondence	POST	**Privatkorrespondenz**	
private costumer	ECON	**Privatkunde**	
private costumer (→private subscriber)	TELEC	(→Privatteilnehmer)	
private line (→leased line)	TELEC	(→Mietleitung)	
private network	TELEC	**Privatnetz**	
= dedicated network		= Sondernetz; Kundennetz; nichtöffentliches Netz; privates Netz; Geschäftsnetz; Teilnehmernetz 2	
≈ closed user network			
≠ public network			
↓ corporate network; official network; military network; user network 2		≈ geschlossenes Teilnehmernetz	
		≠ öffentliches Netz	
		↓ Firmennetz; Behördennetz; Militärnetz	
private subscriber	TELEC	**Privatteilnehmer**	
= private costumer; residential subscriber; residential costumer; domestic subscriber; domestic costumer		≠ Geschäftsteilnehmer	
≠ business subscriber			
private switching	SWITCH	**private Vermittlungstechnik**	

privatization	ECON	Privatisierung	problem program	DATA PROC	Problemprogramm	
privatize	ECON	privarisieren	problem solution	MATH	Problemlösung	
privilege	ECON	Vergünstigung	problem solving	DATA PROC	Problemlösung	
= benefit			procedural	DATA PROC	verfahrensorientiert	

privatization · ECON · Privatisierung
privatize · ECON · privarisieren
privilege · ECON · Vergünstigung
= benefit
privilege (→precedence) · COLLOQ · (→Vorrang)
privileged instruction · DATA PROC · privilegierter Befehl
  [exclusive for the executive routine] · [für das Organisationsprogramm reserviert]
PRK (→2 PSK) · MODUL · (→2 PSK)
probabilistic (→stochastic) · MATH · (→stochastisch)
probabilistic model · SCIE · Wahrscheinlichkeitsmodell
probability · MATH · Wahrscheinlichkeit
≈ likelihood · ≈ Mutmaßlichkeit
probability calculus · MATH · Wahrscheinlichkeitsrechnung
probability curve · MATH · Wahrscheinlichkeitskurve
probability density · PHYS · Dichtefunktion
  = Aufenthaltwahrscheinlichkeit
probability density function · MATH · Wahrscheinlichkeitsdichte
  = Verteilungsdichte
probability distribution · MATH · Wahrscheinlichkeitsverteilung
↑ distribution · ↑ Verteilung
probability function · MATH · Wahrscheinlichkeitsfunktion
  [dependence of probability for intervals of random variable] · [Verlauf der Wahrscheinlichkeiten pro Intervall der Zufallsvariable]
≈ distribution function · ≈ Verteilungsfunktion
probability mass · MATH · Wahrscheinlichkeitsmasse
probability paper · MATH · Wahrscheinlichkeitspapier
  = Wahrscheinlichkeitsnetz
probability theory · MATH · Wahrscheinlichkeitstheorie
≈ statistics · ≈ Statistik
probability wave · PHYS · Wahrscheinlichkeitswelle
probable · MATH · wahrscheinlich
probation period (→probatory period) · ECON · (→Probezeit)
probatory period · ECON · Probezeit
= probation period
probe (v.t.) · COLLOQ · ergründen
probe (n.) · MICROW · Sonde
= sonde
probe (→check) · TECH · (→prüfen)
probe (→test probe) · INSTR · (→Prüfspitze)
probe accessory · INSTR · Tastkopfzubehör
probe barrel insulator · COMPON · Röllchenisolator
probe card · MICROEL · Nadelkarte
probe coil · ELECTRON · Suchspule
probe coil (→search coil) · ELECTRON · (→Suchspule)
probe multiplexer · INSTR · Tastkopf-Multiplexer
probe power · INSTR · Tastkopfversorgung
probe tip (→test probe) · INSTR · (→Prüfspitze)
problem analysis · DATA PROC · Problemanalyse
  = Aufgabenanalyse
problem definition · DATA PROC · Aufgabenstellung
= problem determination; problem formulation
  = Problembestimmung; Problemdefinition; Problemformulierung; Aufgabenstellung; Aufgabenbestimmung; Aufgabendefinition; Aufgabenformulierung
problem description · DATA PROC · Problembeschreibung
problem determination · DATA PROC · (→Problemstellung)
(→problem definition)
problem formulation · DATA PROC · (→Problemstellung)
(→problem definition)
problem-free · TECH · (→problemlos)
(→straightforward)
problem language · DATA PROC · Problemsprache
problem-oriented program- · DATA PROC · (→problemorientierte Programmiersprache)
ming language (→high-level programming language)
problem-oriented · DATA PROC · problemorientiert
= problem-orientated · = problembezogen; problemnah
problem-orientated · DATA PROC · (→problemorientiert)
(→problem-oriented)
problem-oriented programming language (→high-level programming language) · DATA PROC · (→problemorientierte Programmiersprache)

problem program · DATA PROC · Problemprogramm
problem solution · MATH · Problemlösung
problem solving · DATA PROC · Problemlösung
procedural · DATA PROC · verfahrensorientiert
= prozedural
procedural language · DATA PROC · (→prozedurorientierte Programmiersprache)
(→procedure-oriented programming language)
procedural programming · DATA PROC · (→prozedurorientierte Programmiersprache)
language (→procedure-oriented programming language)
procedure · DATA PROC · Prozedur
[a routine of a problem-oriented language, recallable by proper name] · [mit Eigennamen abrufbare Routine einer problemorientierten Programmiersprache]
≈ subroutine; routine · ≈ Unterprogramm; Routine
↓ code procedure · ↓ Codeprozedur
procedure · TECH · Verfahren
= process (n.) · = Prozeß 2
≈ method; sequence of operation; operation 2 · ≈ Methode; Technik 2; Ablauf; Vorgang
procedure declaration · DATA PROC · Prozedurvereinbarung
(ALGOL)
= declarative (COBOL)
procedure diversity · DATA COMM · Prozedurwandlung
procedure division · DATA PROC · Prozedurteil
↑ COBOL program · ↑ COBOL-Programm
procedure error · DATA PROC · Ablauffehler
procedure identifier · DATA PROC · Prozedurname
(ALGOL)
= procedure name (COBOL)
procedure name (COBOL) · DATA PROC · (→Prozedurname)
(→procedure identifier)
procedure-oriented language · DATA PROC · (→prozedurorientierte Programmiersprache)
(→procedure-oriented programming language)
procedure-oriented programming language · DATA PROC · prozedurorientierte Programmiersprache
[language where the instructions must follow the sequence of the process; the languages of generation 1 to 3] · [Sprache bei der die Befehlssequenz dem Ablauf entsprechen muß; die Sprachen der 1. bis 3. Generation]
= procedure-oriented language; POL; procedural programming language; procedural language · = prozedurale Programmiersprache; prozedurorientierte Sprache; prozedurale Sprache; verfahrensorientierte Programmiersprache; verfahrensorientierte Sprache
≈ machine language · ≈ Maschinensprache
≠ symbol-oriented programming language · ≠ symbolorientierte Programmiersprache
↓ machine-oriented progamming language; problem-oriented programming language · ↓ maschinenorientierte Programmiersprache; problemorientierte Programmiersprache
procedure statement · DATA PROC · Prozeduranweisung
proceedings 1 (pl.t.) · ECON · Verfahren
[judicial] · [gerichtlich]
= case · = Prozess
proceedings 2 (n.plt.) · ECON · (→Tagungsbericht)
(→transactions)
proceeds (→revenue) · ECON · (→Erlös)
proceed-to-dial signal (→dial tone) · TELEPH · (→Wählton)
proceed-to-select protocol · DATA COMM · Aufforderungsprotokoll
proceed-to-send signal · DATA COMM · Sendefreigabe
process (v.t.) · ECON · abwickeln 1
[an order] · [einen Auftrag]
process (v.t.) · SWITCH · abwickeln 1
[a call] · [eines Gesprächs]
process (v.t.) · TECH · verarbeiten
= manipulate · ≈ fertigen; veredeln; bearbeiten [MECH]
≈ manufacture; improve; tool (v.t.) [MECH]
process (v.t.) · DATA PROC · verarbeiten
= massage (v.t.) · = bearbeiten; abarbeiten
≈ execute · ≈ ausführen

process

process (n.)   DATA PROC   **Prozeß**
[process computer]   [Prozeßrechner]
= task
process (n.) (→procedure)   TECH   (→Verfahren)
process (→operation 2)   TECH   (→Vorgang)
process automation   CONTROL   **Prozeßautomatisierung**
process chart (→timing   TECH   (→Ablaufdiagramm)
diagram)
process computer (→process   DATA PROC   (→Prozeßrechner)
control computer)
process control   CONTROL   **Prozeßsteuerung**
≈ process data processing   = Prozeßleitung; Prozeßführung
    ≈ Prozeßdatenverarbeitung
process control computer   DATA PROC   **Prozeßrechner**
[a computer dedicated to control processes, with specialized interfacing, instruction set and environmental conditions]   [für Prozeßsteuerungen spezialisierter Rechner, für spezielle Schnittstellen, Befehlsvorräte und Umgebungsbedingungen optimiert]
= process computer   ≈ Prozeßleitrechner
≈ robot [CONTROL]   ≈ Roboter [CONTROL]
≠ business computer; plant computer   ≠ Geschäftscomputer; Betriebsrechner
↓ traffic control computer   ↓ Verkehrsrechner
process control engineering   CONTROL   **Prozeßleittechnik**
process-control monitor   MICROEL   **Teststruktur**
= PCM; test structure   = Prüfstruktur
process control peripheral unit   DATA PROC   **Prozessorperipherie**
process data   CONTROL   **Prozeßdaten**
process data processing   CONTROL   **Prozeßdatenverarbeitung**
≈ process control   ≈ Prozeßsteuerung
process data transmission   CONTROL   **Prozeßdatenübertragung**
process engineering   TECH   **Verfahrenstechnik**
processible   DATA PROC   **verarbeitbar**
processing   TECH   **Verarbeitung 1**
≈ manufacturing; improvement; work [MECH]; workmanship 2   = Fertigung; Veredelung; Bearbeitung [MECH]; Verarbeitungsqualität
↓ finish 1 (n.)   ↓ Fertigbearbeitung
processing   DATA PROC   **Verarbeitung**
≈ execution; computation   = Bearbeitung; Abarbeitung; Berechnung
    ≈ Ausführung
processing (→work)   MECH   (→Bearbeitung)
processing method   MECH   **Bearbeitungsmethode**
processing mode   DATA PROC   **Betriebsart**
= operation mode
processing module   DATA PROC   **Verarbeitungsmodul**
processing plant   MANUF   **Verarbeitungsanlage**
processing speed   MICROEL   **Durchsatz**
[memory]   [Speicher]
processing state   DATA PROC   **Verarbeitungszustand**
processing symbol   DATA PROC   **Verarbeitungssymbol**
processing time   DATA PROC   **Verarbeitungszeit**
processing traffic   ECON   **Veredelungsverkehr**
processing unit   SWITCH   **Verarbeitungseinheit**
processing unit (→central processing unit 1)   DATA PROC   (→Zentraleinheit 1)
processing width   DATA PROC   **Verarbeitungsbreite**
[number of bits processed simultaneously]   [Anzahl simultan verarbeiteter Bits]
process interface equipment   TELECONTR   **Prozeßperipherie**
process management   CONTROL   **Prozeßverwaltung**
processor 2   DATA PROC   **Prozessor 2**
[program to translate into a language of of numeric control]   [Programm zur Übersetzung in eine Werkzeugmaschinen steuernde Sprache]
↑ translator   ↑ Übersetzer
↓ procesor 2
processor 1   DATA PROC   **Prozessor 1**
[generic term for arithmetic-logic unit plus control unit; in microprocessor terminology synonymous to CPU]   [Oberbegriff zu Steuerwerk plus Rechenwerk, ohne Hauptspeicher; in der Mikroprozessortechnik Synonym zu Zentralprozessor]
= central processing unit 2; CPU 2; basic processing unit 2; BPU; engine
↓ arithmetic-logic unit; control unit   = Verarbeitungsprozessor;

Zentralprozessor 2; CPU 2; Zentraleinheit 2
↓ Rechenwerk; Steuerwerk;
processor-bound   DATA PROC   **prozessorgebunden**
= computer-bound   = prozessorbedingt; rechnergebunden
≈ computer-oriented   ≈ maschinenorientiert
processor chip   MICROEL   **Prozessorchip**
processor chip card (→chip card)   TERM&PER   (→Chip-Karte)
processor development module   MICROEL   **Prozessorentwicklungsmodul**
processor-intensive   DATA PROC   **prozessorintensiv**
processor interrupt facility   MICROEL   **Prozessorunterbrechbarkeit**
processor performance   DATA PROC   **Prozessorleistung**
[unit: MIPS / Mega-Instructions Per Second]   [Maßeinheit: MIPS / Millionen Befehle pro Sekunde]
process patent   ECON   **Verfahrenspatent**
process quantity   CONTROL   **Prozeßgröße**
= process variable   = Prozeßvariable
process simulation   TECH   **Prozeßsimulation**
    = Prozeßsimulierung
process time   MANUF   **Durchlaufzeit**
process variable (→process quantity)   CONTROL   (→Prozeßgröße)
procurement   ECON   **Beschaffung**
= acquisition; purchase; buying; obtainment   = Bezug; Anschaffung
produce (v.t.)   (→fertigen)
(→manufacture)
produce (→good)   ECON   (→Ware)
producer (→manufacturer)   ECON   (→Hersteller)
producer's liability   ECON   **Produzentenhaftung**
product   ECON   **Produkt**
= manufacture (s.); make (s.)   = Fabrikat; Erzeugnis
product   MATH   **Produkt**
[result of multiplication]   [Ergebnis der Multiplikation]
≈ multiplication   ≈ Multiplikation
product advertising   ECON   **Produktwerbung**
= product publicity
product audit   QUAL   **Produkt-Audit**
product brochure (→product information)   TECH   (→Produktschrift)
product catalogue (→product line summary)   ECON   (→Produktübersicht)
product demodulator   CIRC.ENG   **Phasendemodulator**
    = Koinzidenzdemodulator; Produktdemodulator; Quadraturdemodulator; Phasendrehdemodulator
    ≈ Phi-Detektor
product documentation   DOC   **Produktunterlage**
product family   ECON   **Produktfamilie**
product generation   TECH   **Produktgeneration**
product improvement   ECON   **Produktverbesserung**
product information   TECH   **Produktschrift**
= product brochure   = Produktbroschüre
↓ leaflet   ↓ Kennblatt
product introduction   ECON   **Produkteinführung**
production 1   ECON   **Erzeugung**
↓ manufacturing   = Produktion 1
    ≈ Herstellung
    ↓ Fertigung
production 2   ECON   (→Fertigung)
(→manufacturing)
production (→generation 1)   TECH   (→Erzeugung)
production aid (→production means)   MANUF   (→Fertigungsmittel)
productional process   MANUF   **Fertigungsprozeß**
= manufacturing process   = Herstellungsprozeß; Produktionsprozeß; Fertigungsvorgang
≈ manufacturing method; production flow   ≈ Fertigungsverfahren; Fertigungsablauf
production automation   MANUF   **Produktionsautomatisierung**
= factory automation; manufacturing automation   = Herstellungsautomatisierung; Fabrikationsautomatisierung

production bottleneck	MANUF	**Fertigungsengpaß**	production sequence	MANUF	(→Fertigungsablauf)
		= Produktionsendpaß	(→ production flow)		
production capacity	MANUF	**Fertigungskapazität**	production shortfall	MANUF	(→Produktionsausfall)
= manufacturing capacity		= Produktionskapazität	(→ production loss)		
production chain	MANUF	**Fertigungskette**	production stage	MANUF	**Fertigungsstufe**
= manufacturing chain		= Herstellungskette	= manufacturing stage		≈ Fertigungsschritt
production control	MANUF	**Fertigungslenkung**	≈ production step		
= manufacturing control		= Produktionslenkung; Fertigungssteuerung; Produktionssteuerung	production start	MANUF	**Fertigungsaufnahme**
≈ work preparation			= manufacturing start		= Produktionsaufnahme
			production step	MANUF	**Fertigungsschritt**
		≈ Arbeitsvorbereitung	≈ production stage		≈ Fertigungsstufe
production control engineering	CONTROL	**Produktionsleittechnik**	production supervision	MANUF	**Fertigungsüberwachung**
					= Produktionsüberwachung
production control post	MANUF	**Fertigungsleitstand**	production technique	MANUF	(→Fertigungstechnik)
production costs	ECON	**Herstellkosten**	(→ manufacturing technology)		
= manufacturing costs		= Herstellungskosten; HL; Fertigungskosten; Produktionskosten	production test area (→ test department)	MANUF	(→Prüffeld)
			production test floor (→ test department)	MANUF	(→Prüffeld)
production data	MANUF	**Betriebsdaten**			
= manufacturing data			productivity	ECON	**Produktivität**
production data acquisition	DATA PROC	**Betriebsdatenerfassung**	productization	ECON	**Industrialisierung** [eines Produkts]
		= BDE	product knowledge	TECH	**Produktkenntnisse**
production data management	MANUF	**Produktionsdatenmanagement**	product line	ECON	**Produktspektrum**
			= product range; range		= Produktpalette; Spektrum
production data processing	DATA PROC	**Betriebsdatenverarbeitung**	≈ assortment		≈ Sortiment
		= BDV	↓ delivery range; production range		↓ Lieferspektrum; Fertigungsspektrum
production documents	MANUF	**Fertigungsunterlage**			
= manufacturing documents		= FU; Bauunterlage	product line summary	ECON	**Produktübersicht**
production engineer	MANUF	**Fertigungsingenieur**	= product catalogue		
= manufacturing engineer			product list	ECON	**Warenverzeichnis**
production engineering	TECH	(→Fertigungstechnik)			= Warenliste; Produktverzeichnis; Produktliste
(→ industrial engineering)					
production facility	MANUF	**Fertigungseinrichtung**	product manager	ECON	**Produktverantwortlicher**
= manufacturing facility		= Produktionseinrichtung			= Produktmanager
production flow	MANUF	**Fertigungsablauf**	product modulator	CIRC.ENG	**Produktmodulator**
= manufacturing flow; production sequence; manufacturing sequence		≈ Fertigungsprozeß	product planning	ECON	**Produktplanung**
			product publicity (→ product advertising)	ECON	(→Produktwerbung)
≈ produccional process					
production grade	QUAL	**Produktionsgüteklasse**	product range (→ product line)	ECON	(→Produktspektrum)
production length	COMM.CABLE	(→Lieferlänge)			
(→ factory length)			product sign [capital letter of greek pi]	MATH	**Produktzeichen** [Großbuchstabe griechisch Pi]
production line (→ assembly line)	MANUF	(→Fließband)			
			= product symbol		
production loss	MANUF	**Produktionsausfall**	product specification	ECON	**Produktbeschreibung**
= production shortfall; manufacturing loss		= Fertigungsausfall			= Produktspezifikation
			product symbol (→ product sign)	MATH	(→Produktzeichen)
production manager	MANUF	(→Fertigungsleiter)			
(→manufacturing manager)			profession	ECON	**Beruf**
production master (→ artwork master)	MANUF	(→Druckvorlage)	= vocation		≈ Beschäftigung
			professional	SCIE	**Fachmann** (pl.-leute)
production material	MANUF	**Fertigungsmaterial**	= specialist		= Spezialist
= direct material		≈ Materialaufwand	≈ expert		≈ Fachpersonal; Sachverständiger [ECON]
production means	MANUF	**Fertigungsmittel**			
= production aid; manufacturing means; manufacturing aid		= Produktionsmittel	professional edication	ECON	**Fachausbildung**
			= professional training; curriculum		
production method (→manufacturing method)	MANUF	(→Fertigungsverfahren)	professional education	ECON	**Berufsausbildung**
			= vocational training		
production number	MANUF	**Fabriknummer**	professional journal	DOC	**Fachzeitschrift**
production order (→ job order)	MANUF	(→Fertigungsauftrag)	= technical journal		= technische Zeitschrift
production planning	MANUF	(→Fertigungsplanung)	professional training (→ professional education)	ECON	(→Fachausbildung)
(→manufacturing engineering 2)					
production plant	MANUF	**Produktionsanlage**	professional use	TECH	**professionelle Anwendung**
= manufacturing plant; plant		= Fertigungsanlage; Fertigungsstätte; Produktionsstätte; Werkanlage; Werksanlage (OES); Fabrikanlage; Fabrikationsstätte	profil	DATA COMM	**Profil**
↑ factory			profile	MECH	**Profil**
			= structural shape		
			profile (→ outline)	TECH	(→Umriß)
			profile chart (→ earth profile chart)	RADIO REL	(→Schnittrahmen)
		↑ Fabrik			
production program	MANUF	**Fertigungsprogramm**	profile meter	INSTR	**Profil-Einbauinstrument**
= manufacturing program			profile milling	METAL	**Formfräsen**
production range	MANUF	**Fertigungsspektrum**	profile section	ENG.DRAW	**Profilschnitt**
= manufacturing range		= Produktionsspektrum	profiling (→ shaping)	TECH	(→Formung)
production run	DATA PROC	**Produktivlauf**	profil selection	DATA COMM	**Profilauswahl**
		= Betriebslauf	profit (n.)	ECON	**Gewinn**
production schedule	MANUF	(→Fertigungsplan)	= gain; earnings 2; yield; return		= Profit; Ertrag
(→manufacturing schedule)			≈ earnings 1		≈ Verdienst
production scrap	MANUF	**Mehrverbrauch**	profit (v.t.) (→yield)	ECON	(→einbringen)
= additional consumption					

profitability ECON **Wirtschaftlichkeit**
= profitableness; cost effectiveness; operational efficiency; lucrativeness; earning power
= Rentabilität; Ertragskraft; Lukrativität; Ertragsfähigkeit; Ertragspotential; Gewinnträchtigkeit
≈ return on investment; earning performance
≈ Rendite auf investiertes Kapital; Ertragslage

profitable ECON **rentabel**
≈ economic; cheap
≈ wirtschaftlich 2; preiswert
profitableness (→profitability) ECON (→Wirtschaftlichkeit)
profound (→deep) TECH (→tief)
program (v.t.) MICROEL **schießen**
[write data into a EPROM] [von Daten in einen EPROM]
= write; blast (v.t.); blow (v.t.); burn-in (v.t.) = brennen

program (v.t.) DATA PROC **programmieren**
≈ implement (v.t.) ≈ implementieren
program (NAM) (n.) DATA PROC **Programm**
[a sequence of computer instructions to solve a task] [eine Folge von Computerbefehlen zur Lösung einer Aufgabe]
= programme (BRI); computer program = Rechnerprogramm; Computerprogramm
≈ routine; software ≈ Routine; Software
↓ machine program; user program; system program; setup program ↓ Maschinenprogramm; Anwenderprogramm; Systemprogramm; Installationsprogramm

program (n.) COLLOQ **Programm**
[NAM]
= programme (BRI)
≈ schedule ≈ Ablaufplan

program (NAM) (→broadcast program) BROADC (→Rundfunkprogramm)
program analyzer DATA PROC **Programmanalysator**
program architecture DATA PROC **Programmarchitektur**
= program structure = Programmstruktur
program bank DATA PROC **Programmbank**
≈ program library ≈ Programmbibliothek
program branch DATA PROC **Programmzweig**
[splitting of program sequence, depending from intermediate results] [Konditionierung des Programmablaufs durch ein Zwischenergebnis]
≈ branch ≈ Programmverzweigung; Verzweigung
≈ jump instruction; program jump; program switch ≈ Sprungbefehl; Programmsprung; Programmschalter

program button TERM&PER **Programmtaste**
program card DATA PROC **Programmkarte**
= solid-state cartridge
program chaining DATA PROC **Programmverkettung**
program channel (→sound channel) BROADC (→Tonkanal)
program chapter (→chapter) DATA PROC (→Programmkapitel)
program circuit TELEC **Rundfunkleitung**
= programme circuit (BRI)
program coding (→programming) DATA PROC (→Programmierung)
program contribution BROADC **Programmbeitrag**
≈ program exchange ≈ Programmaustausch
program counter (→instruction counter) DATA PROC (→Befehlszähler)
program crash DATA PROC **Programmabsturz**
↑ abnormal end = Programmabbruch
↓ dead halt ↑ Absturz
↓ Blockierungsunterbrechung
program crash (→abnormal end) DATA PROC (→Absturz)
program description DATA PROC **Programmbeschreibung**
program design aid (→tool) DATA PROC (→Programmierwerkzeug)
program development cycle DATA PROC **Programmierungsstufe**
program disk DATA PROC **Programmdiskette**
program distribution BROADC **Programmverteilung**
program distribution line (→TV program distribution line) BROADC (→Fernsehverteilleitung)
program documentation DATA PROC **Programmdokumentation**
program edition (→program release) DATA PROC (→Programmversion)
program editor DATA PROC **Programmeditor**

program error DATA PROC **Programmfehler**
≈ programming error; software error; software fault = Programmierfehler; Software-Fehler
↑ bug ↑ Fehler
program exchange BROADC **Programmaustausch**
≈ program contribution ≈ Programmbeitrag
program exchange line BROADC **Austauschleitung**
[for national or international program exchange] [für nationalen oder internationalen Programmaustausch]
↓ TV program exchange line ↓ Fernsehaustauschleitung

program execution DATA PROC (→Programmlauf)
(→program run)
program execution time DATA PROC **Programmlaufzeit**
= program run time = Programmausführungszeit
↑ execution time ↑ Ausführungszeit
program family DATA PROC **Programmfamilie**
= suite
program file DATA PROC **Programmdatei**
program flow DATA PROC **Programmablauf**
≈ program run ≈ Programmlauf
program flowchart DATA PROC **Programmablaufplan**
≈ programming flowchart ≈ PAP; Programmschema
program generator DATA PROC **Programmgenerator**
[adapts a basic program to specific cases] [paßt ein Grundprogramm an Spezialfälle an]
= generator program; generator = Generatorprogramm; Generator; generierendes Programm; erzeugendes Programm
≈ interpreter 2
↓ sort-merge generator
≈ Interpretierer 2
↓ Sortier-Misch-Generator

program header DATA PROC **Programmanfangsblock**
program ID (→program identification) DATA PROC (→Programmkennzeichnung)
program identification DATA PROC **Programmkennzeichnung**
= program ID
programing librarian DATA PROC **Beauftragter für Programmierhilfen**
program interrupt DATA PROC **Programmunterbrechung**
[for higher-priority tasks] [wegen vorrangiger Aufgabe]
= interrupt 1(n.); interruption; trapping = Unterbrechung; Interrupt
program jump DATA PROC **Programmsprung**
[continuation of program with non subsequent operation] [Fortsetzung des Programms mit nicht unmittelbar folgender Operation]
≈ jump = Sprung
≈ program branch ≈ Programmzweig
program language (→programming language) DATA PROC (→Programmiersprache)
program level counter DATA PROC **Programmstufenzähler**
= Programmzähler 2
program library DATA PROC **Programmbibliothek**
≈ program bank ≈ Programmbank
program library management DATA PROC **Programmbibliotheks-Verwaltung**
program line (→instruction line) DATA PROC (→Befehlszeile)
program linkage DATA PROC **Programmverknüpfung**
≈ linkage editor ≈ Binderprogramm
program linker DATA PROC **Programmbinder**
= linker
program listing (→source listing) DATA PROC (→Programmliste)
program loop DATA PROC **Programmschleife**
= loop; cyclic program; cyclic code = Schleife; zyklisches Programm; zyklischer Code
programmability ELECTRON **Programmierbarkeit**
programmable ELECTRON **programmierbar**
programmable array logic MICROEL **PAL**
= PAL
programmable attenuator MICROW **programmierbares Dämpfungsglied**
programmable clock CIRC.ENG **programmierbarer Taktgeber**
[with settable clock rate]
programmable communication interface (→PCI) MICROEL (→PCI)
programmable function key TERM&PER **programmierbare Funktionstaste**

programmable function key TERM&PER (→Funktionstaste)
(→function key)
programmable keyboard TERM&PER (→intelligente Tastatur)
(→intelligent keyboard)
programmable logic array MICROEL **programmierbare Logikan-**
[IC permanently programmable by **ordnung**
braking matrix points] [durch Trennen von Matrix-
= PLA; programmable logic device; punkten
PLD; field-programmable device festprogrammierbarer IC]
 = PLA; PLD; programmier-
 barer Logikbaustein
programmable logic device MICROEL (→programmierbare Logi-
(→programmable logic array) kanordnung)
programmable memory DATA PROC (→veränderbarer Speicher)
(→alterable memory)
programmable multiplexer TRANS (→Verteilmultiplexer)
(→cross-connect multiplexer)
programmable peripheral DATA PROC (→PPI)
interface (→PPI)
programmable read-only MICROEL (→PROM)
memory (→PROM)
programmable storage DATA PROC (→veränderbarer Speicher)
(→alterable memory)
programmable store DATA PROC (→veränderbarer Speicher)
(→alterable memory)
programmable terminal TERM&PER (→programmierbare Daten-
(→intelligent terminal) station)
program maintenance DATA PROC **Programmpflege**
 = Programmwartung
program manager DATA PROC **Programmverwalter**
programme (BRI) (→program) COLLOQ (→Programm)
programme (BRI) (→broadcast BROADC (→Rundfunkprogramm)
program)
programme (BRI) DATA PROC (→Programm)
(→program)
programme circuit (BRI) TELEC (→Rundfunkleitung)
(→program circuit)
programmed barring DATA PROC (→programmierte Verriege-
(→programmed interlock) lung)
programmed check DATA PROC **programmierte Prüfung**
programmed interlock DATA PROC **programmierte Verriegelung**
= programmed barring
programmed label DATA PROC **programmierter Etikettenaus-**
 druck
programme listing (→source DATA PROC (→Programmliste)
listing)
program memory DATA PROC **Programmspeicher**
[sector of main memory reserved for [für Programme reservier-
programs] ter Teil des Hauptspei-
= program storage chers]
 ≈ Hauptspeicher
programmer DATA PROC **Programmierer**
= software programmer; software de- = Programmentwickler; Soft-
veloper ware-Entwickler; Software-
↓ system programmer; application pro- Programmierer
grammer ↓ Systemprogrammierer; An-
 wendungsprogrammierer
programmer board (→PROM MICROEL (→PROM-Programmierge-
programmer) rät)
programme transmission (BRI) BROADC (→Rundfunkübertragung)
(→program transmission)
programming DATA PROC **Programmierung**
= program coding; coding 3 = Programmieren
programming aid (→tool) DATA PROC (→Programmierwerkzeug)
programming button CONS.EL **Speichertaste**
programming convention DATA PROC **Programmkonvention**
= convention = Programmierungskonven-
 tion
programming device MICROEL **Programmiergerät**
[to program PROM's and similars] [zum Schießen von
= burner PROM's u. dgl]
≠ eraser ≠ Löschgerät
programming effort DATA PROC **Programmieraufwand**
= programming expenditure
programming error DATA PROC (→Programmfehler)
(→program error)
programming expenditure DATA PROC (→Programmieraufwand)
(→programming effort)
programming flowchart DATA PROC (→Programmablaufplan)
(→program flowchart)

programming language DATA PROC **Programmiersprache**
[artificial language to formulate in- [Kunstsprache zur Formu-
structions] lierung von Anweisungen]
= program language; language = Sprache
↓ machine language; symbolic pro- ↓ Maschinensprache; symbo-
gramming language lische Programmiersprache
programming mode DATA COMM **Programmierbetrieb**
programming product DATA PROC (→Software-Produkt)
(→software product)
programming system DATA PROC **Programmiersystem**
programming team DATA PROC **Programmierteam**
program modification DATA PROC **Programmodifikation**
program module DATA PROC **Programmodul**
[part of a program which can be [als Einheit verarbeitbarer
handled as unit, can be assembled to Teil eines Programms, mit
programs by a linkage editor] Binderprogramm zu einem
= module; program unit Programm zusammenfüg-
 bar]
 = Programmbaustein; Modul
program package (→software DATA PROC (→Software-Paket)
package)
program release DATA PROC **Programmversion**
= program edition; program version = Programmausgabe
↑ software release ↑ Software-Version
program run DATA PROC **Programmlauf**
= computer run; run; program execu- = Rechnerlauf; Computer-
tion; execution 2; machine run lauf; Lauf; Programmaus-
≈ programm flow führung
↓ search run ↓ Programmablauf
 ↓ Suchlauf
program run (→sequence of TECH (→Ablauf)
operations)
program run time (→program DATA PROC (→Programmlaufzeit)
execution time)
program segment DATA PROC **Programmsegment**
[fractionation of a program intro- [für Hauptspeicher-Einspa-
duced to economize main memory] rung definierte Teilung ei-
= partition 2 nes Programms]
↓ root segment; overlay segment = Segment; Programmbereich
 ↓ arbeitsspeicherresidentes
 Programmsegment; Überla-
 gerungssegment
program site CONS.EL **Programmspeicherplatz**
 = Programmplatz
program specification DATA PROC **Programmspezifikation**
program stack (→stack DATA PROC (→Stapelspeicher)
storage)
program status word DATA PROC **Programmzustandswort**
 = Programmstatuswort
program step DATA PROC **Programmschritt**
≈ instruction = Operation; Verarbeitungs-
 schritt; Rechenschritt
 ≈ Befehl
program stop DATA PROC **Programmstopp**
program storage (→program DATA PROC (→Programmspeicher)
memory)
program structure (→program DATA PROC (→Programmarchitektur)
architecture)
program switch DATA PROC **Programmschalter**
[conditioning of a program brach to [Konditionierung einer Pro-
the content of a determined memory grammverzweigung an ei-
place] nen bestimmten
= switch Speicherinhalt]
≈ program branch = Programmweiche; Schalter;
 Weiche
 ≈ Programmzweig
program syntax (→coding DATA PROC (→Programmierungssyntax)
syntax)
program tape DATA PROC **Programmstreifen**
[punched tape] [Lochstreifen]
program test DATA PROC **Programmtest**
= test; program testing; testing = Test
program testing (→program DATA PROC (→Programmtest)
test)
program translation DATA PROC **Programmübersetzung**
[from a programming language into [von einer Programmier-
another or into machine language] sprache in andere oder in
↓ compilation; interpretation Maschinensprache]
 ↓ Kompilierung; Interpretie-
 rung

program transmision BROADC **Rundfunkübertragung**
= programme transmission (BRI); ↓ Fernsehübertragung; Hörfunkübertragung
radio program transmission
↓ TV transmission; sound program transmission

program unit (→program module) DATA PROC (→Programmodul)

program update DATA PROC **Programmaktualisierung**
= update (n.)

program upgrade DATA PROC **Programmerweiterung**
= upgrade (n.)

program version (→program release) DATA PROC (→Programmversion)

progress (n.) TECH **Fortschritt**
= advance (n.) ≈ Errungenschaft

progress (v.i.) (→increment) TECH (→weiterrücken)

progression (→series) MATH (→Reihe)

progressive 1 TECH **fortschreitend**

progressive 2 TECH **fortschrittlich**
= advanced; evolutionary = fortgeschritten; hochentwickelt
≈ future-oriented; modern; innovative
≈ zukunftsorientiert; modern; innovativ

progressive conference TELEC **Einberufer-Konferenz**

progressive counter (→up-counter) CIRC.ENG (→Vorwärtszähler)

progressive form LING **Verlaufsform**
[expresses action or state in progress at the time of speaking or spoken on] = Aspekt; Aktionsart

progressive grading SWITCH **Staffel**
= progressive Mischung

progressive junction MICROEL **allmählicher Übergang**
= graded junction = Übergangsschicht

progressive path finding SWITCH **schrittweise Wegsuche**

progress report (→activity report) ECON (→Tätigkeitsbericht)

prohibit (→forbid) COLLOQ (→verbieten)

project (v.t.) TECH **planen**
≈ design (v.t.) ≈ entwerfen

project (n.) TECH **Plan**
= layout; plan

project (v.t.) (→protrude) TECH (→herausragen)

project (→plan) ECON (→Plan)

project (→estimate) ECON (→veranschlagen)

project-bound (→project-linked) ECON (→projektgebunden)

project description ECON **Projektbeschreibung**

project execution ECON **Projektausführung**

project implementation (→proyect implementation) ECON (→Projektrealisierung)

projecting (adj.) MECH **vorspringend**
= protrusible

projecting (n.) (→projection) ENG.DRAW (→Projektion)

projection (fig.) SCIE **Projektion** [fig.]

projection ENG.DRAW **Projektion**
= projecting (n.)

projection MECH **Vorsprung**
= Warze

projection transparency (→transparency) OFFICE (→Transparentfolie)

project-linked ECON **projektgebunden**
= project-bound

project manager ECON **Projektführer**
= project team leader = Projektmanager

project mangement ECON **Projektleitung**
= Projektmanagement

projector TECH **Projektor**
↓ overhead projector = Projektionsapparat
↓ Folienprojektor

project pace ECON **Durchführungsgeschwindigkeit**

project plan ECON **Projektplan**

project site (→works) TECH (→Baustelle)

project-specific ECON **projektspezifisch**

project sponsor ECON **Projektträger**

project team leader (→project manager) ECON (→Projektführer)

PROLOG DATA PROC **PROLOG**
[programming in logic; a symbol-oriented programming language] [eine symbolorientierte Programmsprache]

prolong (v.t.) COLLOQ **verlängern**
= prolongate; lengthen; extend ≠ verkürzen
≠ shorten

prolongate (→prolong) COLLOQ (→verlängern)

prolongation TECH **Verlängerung**
= extension 1; lengthening ≠ Verkürzung
≠ shortening

PROM MICROEL **PROM**
[not erasable and therefore programmable only once, with difference to EPROM and EEPROM; once programmed, it is a ROM] [nicht löschbar, daher nur einmal programmierbar, zum Unterschied zu EPROM und EEPROM; wird nach seiner Programmierung zum Festwertspeicher]
= programmable read-only memory
= programmierbarer Festwertspeicher

PROM burner (→PROM programmer) MICROEL (→PROM-Programmiergerät)

pro memoria item ECON **Erinnerungsposten**

promethium CHEM **Promethium**
= Pm = Pm

prominence (→protuberance) TECH (→Protuberanz)

promise ECON **Zusage**
= commitment

promissory note ECON **Eigenwechsel**
["I owe you"] ["gegen diesen Wechsel zahle ich"]
= IOU; note; certificate of indebtness = Schuldschein; Schuldverschreibung; Solawechsel; gezogener Wechsel; trockener Wechsel
↑ bill of exchange ↑ Wechsel

promotion purpose ECON **Werbezweck**

PROM programmer MICROEL **PROM-Programmiergerät**
= PROM burner; programmer board

prompt (v.t.) DATA PROC **auffordern**
= anfordern; veranlassen

prompt (n.) DATA PROC **Bereitmeldung**
[indication that computer is ready to accept entries; in DOS generally] [Anzeige daß Rechner für nächste Eingabe bereit ist; in DOS]
= prompting 1; cue = Bereitschaftsanzeige; Aufforderung; Auffordern; Anforderung; Anfordern; Prompt; Bereitschaftszustand;
↓ prompt character; language promt ↓ Aufforderungszeichen; Aufforderungstext

prompt alarm (→urgent alarm) EQUIP.ENG (→dringender Alarm)

prompter DATA PROC **Wecker**

prompter period DATA PROC **Weckertakt**
= prompter timing

prompter timing (→prompter period) DATA PROC (→Weckertakt)

prompting 2 (→user interface) DATA PROC (→Benutzeroberfläche)

prompting 1 (→prompt) DATA PROC (→Bereitmeldung)

prompt instant DATA PROC **Weckerzeitpunkt**
= prompt time

prompt time (→prompt instant) DATA PROC (→Weckerzeitpunkt)

promt character DATA PROC **Anforderungszeichen**
[character indicating readiness of the computer for entries, e.g. blinking cursor or > sign] [Hinweiszeichen daß der Computer für eine Eingabe bereit ist, z.B. durch blinkende Schreibmarke oder > Zeichen]
↑ prompt = Bereitschaftszeichen
↑ Bereitmeldung

pronoun LING **Pronomen**
= Fürwort

pronounce LING **aussprechen**

pronounced [fig.] COLLOQ **ausgeprägt** [fig.]
= salient (fig.)

pronunciation LING **Aussprache**

proof (n.) TYPOGR **Korrekturfahne**
= galley proof; galley = Korrekturabzug; Korrekturbogen; Probedruck; Fahne

proof (n.) (→proof copy)	TYPOGR	(→Andruck)	
proof copy	TYPOGR	**Andruck**	
= proof impression; proof (n.)		[Probedruck]	
proof copy (→test print)	TERM&PER	(→Probedruck)	
proof-finding program	DATA PROC	**beweisführendes Programm**	
proof impression (→proof copy)	TYPOGR	(→Andruck)	
proofing (→protection)	TECH	(→Schutz)	
proofread (v.t.)	TYPOGR	**korrekturlesen**	
proofreaders' mark	TYPOGR	**Korrekturzeichen**	
= correction sign; editing mark ↑ sign		↑ Zeichen	
proofreading	TYPOGR	**Korrekturlesung**	
		= Korrekturlesen	
proof sum (→control total)	ECON	(→Kontrollsumme)	
prooftest (v.t.)	QUAL	**erproben**	
= test (v.t.); trial (v.t.)			
prooftesting	QUAL	**Erprobung**	
= trial; tryout			
prooftesting center	QUAL	**Erprobungsstelle**	
proof total (→control total)	ECON	(→Kontrollsumme)	
propaganda (→advertising)	ECON	(→Werbung)	
propagated error (→inherent error)	MATH	(→mitgeschleppter Fehler)	
propagation	PHYS	**Ausbreitung**	
		= Fortpflanzung	
propagation coefficient	LINE TH	**Fortpflanzungskonstante**	
= propagation constant		= Fortpflanzungsmaß; Übertragungskonstante; Übertragungsmaß; Ausbreitungskonstante; Ausbreitungskoeffizient; Ausbreitungszahl	
propagation coefficient	NETW.TH	(→Kettenübertragungsmaß)	
(→iterative propagation constant)			
propagation constant	LINE TH	(→Fortpflanzungskonstante)	
(→propagation coefficient)			
propagation delay	COMPON	**Stufenverzögerung**	
[of digital devices]		[Reaktionszeit digitaler Bausteine]	
propagation delay (→propagation time)	PHYS	(→Laufzeit)	
propagation loss	RADIO PROP	**Ausbreitungsdämpfung**	
propagation mode (→waveguide mode)	MICROW	(→Wellentyp)	
propagation problem	RADIO PROP	**Ausbreitungsproblem**	
propagation speed (→speed of propagation)	PHYS	(→Fortpflanzungsgeschwindigkeit)	
propagation test	RADIO PROP	**Ausbreitungsmessung**	
propagation time	PHYS	**Laufzeit**	
= transit time; travel time; runtime; propagation delay; delay		≈ Verzug	
propagation-time compensation	TELEC	**Laufzeitausgleich**	
≈ delay equalization		≈ Laufzeitentzerrung	
propagation velocity (→speed of propagation)	PHYS	(→Fortpflanzungsgeschwindigkeit)	
properly (→correctly)	COLLOQ	(→korrekt) (adv.)	
proper motion	PHYS	**Eigenbewegung**	
property (→characteristic 1)	TECH	(→Merkmal)	
proportion 2	MATH	**Verhältnisgleichung**	
[equality of two ratios; a:b = c:d]		[Gleichheit zweier Verhältnisse; a:b = c:d]	
		= Proportion 2	
proportion 1 (→ratio)	MATH	(→Bruch)	
proportional	MATH	**proportional**	
proportional controler	CONTROL	(→Proportionalregler)	
(→P-controller)			
proportional elastic limit	MECH	**Proportionalitätsgrenze**	
proportionality	MATH	**Proportionalität**	
proportionally spaced font	TYPOGR	**Proportionalschrift**	
[spacing varying with character width]		[Schriftart variiert mit der Buchstabenbreite]	
= proportionally spaced print; proportionally spaced printing		= Proportionaldruck	
≠ constant-width font		≠ Konstantschrift	
proportionally spaced print	TYPOGR	(→Proportionalschrift)	
(→proportionally spaced font)			
proportionally spaced printing	TYPOGR	(→Proportionalschrift)	
(→proportionally spaced font)			
proportional offset	CONTROL	**Proportionalabweichung**	
= droop (n.)		= P-Abweichung	
proportional part (→quota)	TECH	(→Anteil)	
proportional sign (→equivalent sign)	MATH	(→Ähnlichzeichen)	
proportional spacing	TYPOGR	**proportionaler Zeichenabstand**	
proportioning (→dosage)	TECH	(→Dosierung)	
proposal	COLLOQ	**Vorschlag**	
≈ suggestion		≈ Anregung	
proposal (→offer)	ECON	(→Angebot)	
proposal form	ECON	**Angebotsformular**	
proposition (n.)	ENG.LOG	**Aussage**	
propositional calculus	MATH	**Aussagenlogik**	
↓ Boolean algebra		= Aussagenkalkül	
		↓ Boolesche Algebra	
propositional function	ENG.LOG	**Aussageform**	
propositional variable	ENG.LOG	**Aussagevariable**	
proprietary 1	ECON	**firmeneigen**	
= corporate (adj.)		= herstellerindividuell; herstellerspezifisch	
≈ patented		≈ patentiert	
proprietary software	DATA PROC	**urheberrechtlich geschützte Software**	
≠ public-domain software		≠ Public-domain-Software	
proprietor (→owner)	ECON	(→Eigentümer)	
propriety	ECON	**Eigentum**	
= ownership			
prop-word	LING	**Stützwort**	
prosodeme	LING	**Prosodem**	
[differenciation of pitch or accentuation in pronouncing vowels or consonants]		[Tonhöhen- oder Betonungsunterschied von Vokalen oder Konsonanten]	
↑ phoneme		↑ Phonem	
prospects	COLLOQ	**Zukunftsaussichten**	
		= Perspektiven; Aussichten	
prospectus (→leaflet)	DOC	(→Prospekt)	
protactinium	CHEM	**Protactinium**	
= Pa		= Pa	
protectable	TECH	**schützbar**	
protected	TELEC	**gesichert**	
= secured		= geschützt	
protected	TELEC	**ersatzgeschützt**	
≠ unprotected		= geschützt	
		≠ ungeschützt	
protected against splashing water (→splash-proof)	TECH	(→spritzwasserfest)	
protected direction	RADIO	**Schutzrichtung**	
protected field	TERM&PER	**geschütztes Feld**	
protected memory area	DATA PROC	**Speicherschutzbereich**	
= protected storage area; protected store area; protected memory sector; protected storage sector; protected store sector		= Speicherschutzsektor	
protected memory sector (→protected memory area)	DATA PROC	(→Speicherschutzbereich)	
protected storage area (→protected memory area)	DATA PROC	(→Speicherschutzbereich)	
protected storage sector (→protected memory area)	DATA PROC	(→Speicherschutzbereich)	
protected store area (→protected memory area)	DATA PROC	(→Speicherschutzbereich)	
protected store sector (→protected memory area)	DATA PROC	(→Speicherschutzbereich)	
protecting cap	TECH	**Schutzkappe**	
= protective cap; guard cap		≈ Schutzabdeckung; Schutzgehäuse	
≈ cover			
protecting device (→protector)	COMPON	(→Sicherung 1)	
protecting diode (→protective diode)	ELECTRON	(→Schutzdiode)	
protecting frame (→protective frame)	TECH	(→Schutzrahmen)	
protection	ECON	**Absicherung**	
[against risks]		[Risiko]	
protection	TECH	**Schutz**	
= guard (n.); proofing			
protection channel (→stand-by channel)	TRANS	(→Ersatzkanal)	
protection circuit	CIRC.ENG	**Schutzschaltung**	
= protective circuit		[schützende Schaltung]	
protection current	POWER SYS	**Schutzstrom**	

protection device TECH **Schutzvorrichtung**
= safety device = Sicherheitsvorrichtung
protection earthing POWER ENG (→Schutzerdung)
(→protective earthing)
protection frame (→protective TECH (→Schutzrahmen)
frame)
protection potential POWER ENG **Schutzpotential**
protection sector (→guard RADIO (→Schutzsektor)
sector)
protection switching TRANS **Ersatzschaltung**
= changeover to standby = Schutzschaltung
protection switching device TRANS (→Schutzschalteinrichtung)
(→protection switching equipment)
protection switching TRANS **Schutzschalteinrichtung**
equipment = Ersatzschalteinrichtung;
= protection switching system; protec- Umschalteinrichtung;
tion switching device; automatic Schutzschaltgerät; Ersatz-
protection switch; APS schaltgerät; Umschaltgerät;
protection switching system TRANS (→Schutzschalteinrichtung)
(→protection switching equipment)
protection switching technique TRANS **Schutzschalttechnik**
= Ersatzschalttechnik; Um-
schaltetechnik
protection switching threshold TRANS **Umschaltschwelle**
= switching threshold; switchover thre- = Umschaltpunkt; Schutz-
shold; changeover threshold schaltschwelle; Ersatz-
schaltschwelle;
Schaltschwelle
protection tube TECH **Schutzrohr**
protective anode POWER SYS **Schutzanode**
protective cap (→protecting TECH (→Schutzkappe)
cap)
protective case TECH **Schutzgehäuse**
≈ hood; protective casing; protective ≈ Schutzabdeckung; Schutz-
cap hülle 1; Schutzkappe
protective casing TECH **Schutzhülle 1**
[of stiff material] [aus steifem Material]
= jacket ≈ Schutzabdeckung; Schutz-
≈ hood; protective case gehäuse
↑ Hülle
protective circuit (→protection CIRC.ENG (→Schutzschaltung)
circuit)
protective coating TECH **Schutzschicht**
= protective layer; resist = Abdeckmittel
≈ protective paint ≈ Schutzanstrich
protective conductor POWER ENG (→ungeerdeter Schutzleiter)
(→unearthed protective conductor)
protective contact EL.INST **Schutzkontakt**
= grounding contact = Schuko
protective cover TECH **Schutzabdeckung**
= impact cover = Abdeckhaube
≈ protective case; protective casing; ≈ Schutzgehäuse; Schutzhülle
protective cap 1; Schutzhaube; Staub-
↑ cover schutz
↑ Abdeckung
protective cover OFFICE **Schutzhülle**
protective covering TECH **Schutzverkleidung**
protective diode ELECTRON **Schutzdiode**
= protecting diode ↓ Freilaufdiode; Kappdiode
↓ free-wheeling diode; clamping diode
protective earth POWER SYS **Schutzerde**
= protector ground; safety earth ≈ Betriebserde
protective earth conductor POWER ENG **Schutzerdungsleiter**
= protective ground conductor; safety = Schutzerder; geerdeter
earth conductor; non-fused earthed Schutzleiter; Schutzleiter 1
conductor ≈ Nulleiter; Erde
≈ neutral conductor; earth
protective earthing POWER ENG **Schutzerdung**
= protection earthing; protective
grounding; safety earthing
protective film MICROEL (→Schutzschicht)
(→overglassing)
protective frame TECH **Schutzrahmen**
= protecting frame; protection frame
protective gap (→discharger) COMPON (→Schutzfunkenstrecke)
protective gas TECH **Schutzgas**
protective gas capacitor COMPON **Schutzgaskondensator**
= Preßgaskondensator
protective ground conductor POWER ENG (→Schutzerdungsleiter)
(→protective earth conductor)

protective grounding POWER ENG (→Schutzerdung)
(→protective earthing)
protective lacquer TECH **Schutzlack**
protective layer (→protective TECH (→Schutzschicht)
coating)
protective layer MICROEL (→Schutzschicht)
(→overglassing)
protective measure COLLOQ **Schutzmaßnahme**
≈ security measure; precautionary ≈ Sicherheitsmaßnahme; Vor-
measure sorgemaßnahme
protective paint TECH **Schutzanstrich**
= protective paint coating ≈ Schutzschicht
≈ protective coating
protective paint coating TECH (→Schutzanstrich)
(→protective paint)
protective resistance CIRC.ENG **Schutzwiderstand**
protective strip OUTS.PLANT **Schutzleiste**
protective wrapping TECH **Schutzhülle 2**
protective wrapping (→coat) TECH (→Mantel)
protector COMPON **Sicherung 1**
[against electric harm] [Schutz vor elektrischer Be-
= protecting device; cut-out schädigung]
↓ Spannungssicherung;
Stromsicherung
protector OUTS.PLANT **Protektor**
[underwater cable] [Unterwasserkabel]
protector ground POWER SYS (→Schutzerde)
(→protective earth)
protest 2 (n.) ECON **Rekurs**
[against a decision] = Einspruch
protest COLLOQ **Protest**
protest (→objection) ECON (→Einspruch)
protocol (n.) DATA COMM **Protokoll**
[procedure for data transfer] [Verfahrensvorschrift für
≈ handshaking Übermittlung]
≈ Quittungsaustausch
protocol analysator INSTR **Protokollanalysator**
= protocol analyzer
protocol analyzer (→protocol INSTR (→Protokollanalysator)
analysator)
protocol architecture DATA COMM **Protokollarchitektur**
protocol converter DATA COMM **Protokollwandler**
= Protokollkonverter
protocol layer DATA COMM **Protokollschicht**
[OSI] [OSI-Modell]
= layer; protocol stack; level = Protokollebene; Schicht;
Ebene
protocol of acceptance ECON (→Abnahmeprotokoll)
(→acceptance certificate)
protocol stack (→protocol DATA COMM (→Protokollschicht)
layer)
protocol tester INSTR **Protokolltester**
proton PHYS **Proton**
proton-implanted laser OPTOEL **protonenimplantierter Laser**
prototype (→experimental TECH (→Versuchsmuster)
prototype)
prototype (→type model) TECH (→Ausführungsmuster)
prototype PCB (→pilot ELECTRON (→Versuchsleiterplatte)
PCB)
prototype slice (→prototype MICROEL (→Prototypenwafer)
wafer)
protocol stack DATA COMM **Protokollprofil**
prototype statement DATA PROC **Musteranweisung**
prototype system (→experimental TECH (→Versuchssystem)
system)
prototype wafer MICROEL **Prototypenwafer**
= prototype slice
protract COLLOQ **hinausziehen**
[to prolong deliberately in time] [mutwillig zeitlich verlän-
≈ prolong gern]
≈ verlängern
protractor INSTR **Winkelmaß**
= goniometer = Winkelmesser; Goniometer
≈ strightedge ≈ Zeichendreieck
protractor ENG.DRAW **Winkelmaß**
= goniometer = Winkelmesser; Goniometer
≈ straightedge ≈ Zeichendreieck
protrude TECH **herausragen**
= project (v.t.)
protrusible (→projecting) MECH (→vorspringend)

protuberance TECH **Protuberanz**
= bulge (n.); prominence
≈ projection
protuding contact (→pre-mating COMPON (→voreilender Kontakt)
contact)
proved (→checked) TECH (→geprüft)
proven (→checked) TECH (→geprüft)
provide (→supply) ELECTRON (→versorgen)
provision 2 (n.) ECON **Rückstellung**
= liability reserves; reserves ; accrual
provision 3 (→stockpiling) ECON (→Bevorratung)
provision 1 (→supply) ECON (→Versorgung)
provision 4 (→stock) ECON (→Vorrat)
provisional (→improvised) COLLOQ (→provisorisch)
provisional acceptance TECH **provisorische Abnahme**
= vorläufige Abnahme
provisional antenna ANT **Behelfsantenne**
≈ auxiliary antenna ≈ Hilfsantenne
provisional result (→intermediate TECH (→Zwischenergebnis)
result)
provisioning security QUAL **Bevorratungssicherheit**
= Vorsorge-Prozentwert
proximal (→near) COLLOQ (→nah)
proximate (→near) COLLOQ (→nah)
proximity TECH **Nähe**
= nearness; closeness = Näherung
≈ neighborhood; approximation ≈ Nachbarschaft; Annäherung
proximity alarm system SIGN.ENG **Annäherungsmeldesystem**
= capacitance alarm system
proximity area OUTS.PLANT **Näherungsbereich**
[approximation of open-wire communications to power lines] [Näherung von Fernmeldefreileitung an Starkstromfreileitung]
proximity detector SIGN.ENG **Annäherungsmelder**
= capacitance detector
proximity effect ANT **Proximity-Effekt**
proximity effect EL.TECH **Stromverdrängungseffekt**
proximity switch SIGN.ENG **Annäherungsschalter**
= Näherungsschalter
proximity zone (→near-field RADIO PROP (→Nahfeld)
region)
proyect documentation TECH **Projektunterlagen**
≈ station manual ≈ Stationsuntrelagen; Montagehandbuch
proyect engineering TECH **Projektplanung**
= proyect planning = Projektierung
proyect implementation ECON **Projektrealisierung**
= project implementation
proyect planning (→proyect TECH (→Projektplanung)
engineering)
proyect scheduling ECON **Projektplanung 2**
[zeitlich]
prussian blue PHYS **pariserblau**
PS (→horsepower) PHYS (→Pferdestärke)
ps (→picosecond) PHYS (→Pikosekunde)
psec (→picosecond) PHYS (→Pikosekunde)
pseudo address DATA PROC **Pseudoadresse**
[entspricht keiner Adresse im Hauptspeicher]
pseudocode DATA PROC **Pseudocode**
= pseudolanguage; quasilanguage
pseudo-decimal digit CODING **Pseudodezimale**
↓ pseudo-tetrad [Bitkombination eines Binärcodes, der keine Dezimalziffer entspricht]
↓ Pseudotetrade
pseudo instruction (→pseudo DATA PROC (→Pseudobefehl)
statement)
pseudolanguage DATA PROC (→Pseudocode)
(→pseudocode)
pseudomorphic PHYS **pseudomorph**
pseudooperation DATA PROC **Pseudobetrieb**
pseudo-quaternary CODING **pseudoquaternär**
pseudo-random CODING **pseudostatistisch**
= pseudozufällig
pseudo-random number MATH **Pseudozufallszahl**
pseudoscopic PHYS **pseudoskopisch**
pseudo-sporadic SCIE **pseudosporadisch**
pseudo-sporadic fault QUAL **pseudosporadischer Fehler**

pseudo statement DATA PROC **Pseudobefehl**
= pseudo instruction [nicht interpretierbarer Befehl, aus Irrtum oder absichtlich]
= Pseudoanweisung; Pseudoinstruktion
pseudo-ternary CODING **pseudoternär**
pseudoternary signal CODING **pseutoternäres Signal**
pseudotetrade CODING **Pseudotetrade**
[one of the six bit combinations, whom doesn't correspond any decimal digit, in a four-bit code] [eine der sechs Bitkombinationen, denen im vierstelligen Binärcode keine Dezimalziffer entspricht]
↑ tetrade; pseudo-decimal digit ↑ Tetrade; Pseudodezimale
pseudo-variable DATA PROC **pseudovariabel**
pseufraction MATH **Scheinbruch**
PS h (→horsepower-hour) PHYS (→PS-Stunde)
2 PSK MODUL **2 PSK**
= PRK; two-level phase shift keying; phase reversal keying = PRK
PSK (→phase shift keying) MODUL (→Phasenumtastung)
4PSK (→quaternary phase shift MODUL (→QPSK)
keying)
PSK modulator MODUL **PSK-Modulator**
psn diode MICROEL **PSN-Diode**
psophometer INSTR **Geräuschspannungsmesser**
= Psophometer
psophometric TELEPH **psophometrisch**
[simulating sensitivity of ear] [die Empfindlichkeit des Ohres simulierend]
psophometric filter (→noise TELEC (→Rauschbewertungsfilter)
weighting filter)
psophometric noise (→weighted TELEC (→Geräusch)
noise)
psophometric noise voltage TELEC **Geräuschspannung**
= weighted noise voltage
psophometric weighting TELEC **psophometrische Bewertung**
PSPDN (→packet-switched TELEC (→paketvermitteltes öffentliches Datennetz)
public data network)
Pt (→platinum) CHEM (→Platin)
PTC thermistor COMPON **Kaltleiter**
= positive temperature coefficient thermistor = PTC-Widerstand; TP-Halbleiterwiderstand (DDR)
↑ Thermistor
PTM (→pulse-time MODUL (→Pulszeitmodulation)
modulation)
PTT (→post, telegraph and ECON (→Post- und Fernmeldeverwaltung)
telephone administration)
PTT network TELEC **postalisches Netz**
≈ public network ≈ öffentliches Netz
p-type conduction (→hole PHYS (→Löcherleitung)
conduction)
p-type conductivity (→hole-type PHYS (→Löcherleitfähigkeit)
conductivity)
p-type conductor PHYS **P-Leiter**
= Mangelelektronenleiter; Defektelektronenleiter; Löcherleiter
p-type doping PHYS **P-Dotierung**
p-type semiconductor (→hole PHYS (→Löcherhalbleiter)
semiconductor)
Pu (→plutonium) CHEM (→Plutonium)
public (n.) ECON **Öffentlichkeit**
public (adj.) ECON **öffentlich**
= general
public address system EL.ACOUS **Rundrufanlage**
publication DOC **Druckschrift**
= pamphlet = Veröffentlichung
publication language DATA PROC **Verlagssprache**
[für Verlage geeignete Programmiersprache]
publication microfilming TERM&PER **Schrifttumsverfilmung**
publications department ECON **Druckschriftenverwaltung**
= Druckschriftenlager
public bond (→public loan) ECON (→Staatsanleihe)
public call station TELEC **öffentliche Sprechstelle**
public communication carrier TELEC (→Fernmeldeverwaltung)
(→telecommunication administration)

public communications

public communications	TELEC	**öffentliche Telekommunikation**
= public telecommunications; common carrier communications		= öffentliches Nachrichtenwesen
public communications common carrier (→telecommunication administration)	TELEC	(→Fernmeldeverwaltung)
public communications engineering	TELEC	**öffentliche Nachrichtentechnik**
= public telecommunications engineering; common carrier communications engineering		
public domain	COLLOQ	**Allgemeingut**
public domain (→public-domain software)	DATA PROC	(→Public-domain-Software)
public-domain software	DATA PROC	**Public-domain-Software**
[freed for unrestricted copying, modification or passing-on]		[darf frei kopiert, modifiziert und weitergegeben werden]
= freeware; public domain		= Freeware
≈ shareware		≈ Shareware
≠ proprietary software		≠ urheberrechtlich geschützte Software
publicity	ECON	**Publizität**
		≈ Werbung
publicity (→advertising)	ECON	(→Werbung)
publicity agency (→advertising agency)	ECON	(→Werbefirma)
publicity campaign	ECON	**Werbekampagne**
publicity costs (→advertising costs)	ECON	(→Werbekosten)
publicity department (→advertising department)	ECON	(→Werbeabteilung)
publicity expenditures (→advertising costs)	ECON	(→Werbekosten)
public limited company (BRI) (→corporation 2)	ECON	(→Aktiengesellschaft)
public loan	ECON	**Staatsanleihe**
= public bond		
public network	TELEC	**öffentliches Netz**
= common carrier network		≈ postalisches Netz
≈ PTT network		≠ Privatnetz
≠ private network		
public order	ECON	**Staatsauftrag**
= government contract		
public relations	ECON	**Öffentlichkeitsarbeit**
public switched network	TELEC	**öffentliches Wählnetz**
public switched telephone network	TELEC	**öffentliches Fernsprechwählnetz**
= general switched telephone network		
public switching	SWITCH	**öffentliche Vermittlungstechnik**
= central-office switching		
public telecommunications (→public communications)	TELEC	(→öffentliche Telekommunikation)
public telecommunications engineering (→public communications engineering)	TELEC	(→öffentliche Nachrichtentechnik)
public telephone station	TELEC	**öffentliche Fernsprechstelle**
public tender	ECON	**öffentliche Ausschreibung**
pull (v.t.)	COLLOQ	**ziehen**
= draw		≈ schleppen
≈ drag		≠ schieben
≠ push		
pull (v.t.)	MICROEL	**ziehen**
[a crystal]		[Kristalle]
= grow		
pull (→pull-in)	OUTS.PLANT	(→einziehen)
pull (→traction)	MECH	(→Zug)
pull (v.t.) (→pop)	DATA PROC	(→abheben)
pull-down (n.)	DATA PROC	**Pull-down**
[of cursor, releasing mouse button at desired field]		[Herunterziehen des Cursors und Loslassen der Maustaste beim gewünschten Feld]
pull-down menu	DATA PROC	**Pull-down-Menü**
[can be roll up, by mouse movement, from top to bottom]		[rollt sich, durch Mausbewegung, von der angeklickten Menüleiste nach unten auf]
= drop-down menu		= Drop-down-Menu
≠ pop-up menu		≠ Pop-up-Menü
↑ selection menu		↑ Auswahlmenü
pull-down resistor	CIRC.ENG	**Pull-down-Widerstand**
pulley 3	TECH	**Antriebsscheibe**
= driving wheel		↓ Seilscheibe
↓ sheave		
pulley 1	TECH	**Umlenkrolle**
= turning swivel		
pulley 2 (→tackle)	TECH	(→Flaschenzug)
pulley block	TECH	**Seilrolle**
pull-in	OUTS.PLANT	**einziehen**
= pull; draw		[Kabel]
≈ lay		≈ ziehen
		≈ verlegen
pull-in (→operate)	COMPON	(→anziehen)
pull-in current (→response current)	ELECTRON	(→Ansprechstrom)
pull-in delay (→response delay)	ELECTRON	(→Ansprechverzögerung)
pulling	MICROEL	**Ziehen**
[of crystals]		[Kristall]
= growing		
pulling direction	OUTS.PLANT	**Ziehrichtung**
pulling direction	MECH	**Zugrichtung**
= direction of traction		
pulling eyer (→cable grip)	OUTS.PLANT	(→Ziehstrumpf)
pulling force	MECH	**Zugkraft**
≈ pull		≈ Zug
pulling-in line (→winch cable)	OUTS.PLANT	(→Windenseil)
pulling length	OUTS.PLANT	**Einziehlänge**
= pull-in length		= Kabeleinziehlänge
pulling operation	OUTS.PLANT	**Einziehvorgang**
pulling speed	OUTS.PLANT	**Einziehgeschwindigkeit**
pulling technique	OUTS.PLANT	**Einziehtechnik**
pulling test (→tension test)	MECH	(→Zugprüfung)
pulling tower	OPT.COMM	**Ziehturm**
pulling tractor	TERM&PER	**Zugtraktor**
↑ tractor feed		↑ Raupenvorschub
pull-in length (→pulling length)	OUTS.PLANT	(→Einziehlänge)
pull-in level (→response level)	ELECTRON	(→Ansprechpegel)
pull-in power (→response excitation)	COMPON	(→Ansprecherregung)
pull-in range (→tuning range)	CIRC.ENG	(→Ziehbereich)
pull-in sensitivity (→responsivity)	ELECTRON	(→Ansprechempfindlichkeit)
pull instruction (→pop instruction)	DATA PROC	(→Abhebungsanweisung)
pull-in threshold (→response threshold)	ELECTRON	(→Ansprechschwelle)
pull-in time (→response time)	ELECTRON	(→Ansprechzeit)
pull-in value (→response value)	ELECTRON	(→Ansprechwert)
pull-in voltage (→response voltage)	ELECTRON	(→Ansprechspannung)
pull-lift	OUTS.PLANT	**Hebezug**
pull of demand	ECON	**Nachfragesog**
pull-off strength	MECH	**Abreißfestigkeit**
pull-through scanner	TERM&PER	**Durchzugleser**
= slot reader		
pull-up resistor	CIRC.ENG	**Endwiderstand**
		= Pull-in-Widerstand
pulp (→cellulose)	TECH	(→Zellstoff)
pulp-insulated	COMM.CABLE	**zellstoffisoliert**
pulp insulated cable	COMM.CABLE	**zellstoffisoliertes Kabel**
pulp insulation	COMM.CABLE	**Zellstoffisolierung**
pulpit	TECH	**Pult**
[desk type construction with inclined platform]		[tischartige Konstruktion mit schräggestelltem Aufsatz]
≈ console; desk		≈ Konsole; Tisch
pulsate	TECH	**pulsieren**
pulsating field (→alternating field)	PHYS	(→Wechselfeld)
pulsating power (→pulsed power)	ELECTRON	(→Impulsleistung)

pulsation (→angular frequency)   PHYS   (→Kreisfrequenz)
pulse (v.i.)   ELECTRON   **takten**
pulse (n.)   ELECTRON   **Impuls 2**
[element of a periodic sequence]   [Element eines Pulses]
≈ impulse   ≈ Impuls 1
pulse amplifier   CIRC.ENG   **Impulsverstärker**
   = Pulsverstärker
pulse amplitude   ELECTRON   **Impulsamplitude**
≈ pulse train amplitude   = Impulshöhe
   ≈ Pulsamplitude
pulse amplitude modulation   MODUL   **Pulsamplitudenmodulation**
= PAM   = PAM
pulse behaviour (→pulse response)   ELECTRON   (→Impulsantwort)
pulse blanking (→strobing)   ELECTRON   (→Impulsausblendung)
pulse broadening   ELECTRON   **Impulsverbreiterung**
= pulse lengthening   = Impulsverlängerung; Impulsdehnung
≈ pulse train spreading   ≈ Pulsverbreiterung
pulse capacitor   COMPON   **Impulskondensator**
pulse carrier   ELECTRON   **Trägerpuls**
pulse characteristic   ELECTRON   **Pulskennlinie**
   = Impulskennlinie
pulse characterization   INSTR   **Impulsbewertung**
pulse circuit   CIRC.ENG   **Impulsschaltung**
= impulse circuit
pulse clock   SIGN.ENG   **Impulsuhr**
pulse clock (→telegraph speed)   TELEC   (→Schrittgeschwindigkeit)
pulse code   CODING   **Pulscode**
pulse-code modulated   CODING   **pulsecodemoduliert**
pulse code modulation   CODING   **Pulscodemodulation**
= PCM   = PCM
pulse command   TELECONTR   **Impulsbefehl**
= impulse command
pulse-compression filter   CIRC.ENG   **Pulskompressionsfilter**
pulse counter   CIRC.ENG   **Impulszähler**
= impulse counter
pulse crest (→pulse peak)   ELECTRON   (→Impulsspitze)
pulsed   ELECTRON   **gepulst**
≈ impulsive   ≈ impulsförmig
pulse decay time (→decay time)   ELECTRON   (→Abfallzeit)
pulse delay   ELECTRON   **Impulsverzögerung**
= impulse delay   = Impulsverzug; Pulsverzögerung
pulse delay time   ELCTRON   **Impulsverzögerungszeit**
pulse-delay-time jitter   ELECTRON   **Impulsverzögerungszeit-Jitter**
   = Pulsverzögerungszeit-Jitter
pulse dialing   SWITCH   **Impulswahl**
= loop-disconnect signaling   ≈ Impulssignalisierung
≈ pulse signaling
pulse dialing transmitter   SWITCH   **Impulswahlsender**
pulse-digit spacing (→pulse separation)   ELECTRON   (→Impulspause)
pulse distortion   ELECTRON   **Impulsverzerrung**
pulse divider   CIRC.ENG   **Taktteiler**
pulsed laser   OPTOEL   **Impulslaser**
≠ continuous-wave laser   ≠ Dauerstrichlaser
pulsed operation   ELECTRON   **Pulsbetrieb**
≈ pulse operation   ≈ Impulsbetrieb
pulsed oscillator   MICROW   **Impulsoszillator**
pulsed power   ELECTRON   **Impulsleistung**
= pulsating power   = Pulsleistung
≠ continuos-wave power   ≠ Dauerstrichleistung
pulsed radar   RADIO LOC   **Pulsradar**
   = Impulsradar
pulse droop (→ramp-off)   ELECTRON   (→Dachschräge)
pulsed sweep   INSTR   **gepulstes Wobbeln**
pulse duration   ELECTRON   **Impulsdauer**
= pulse length; pulse width 1; impulse duration   = Impulslänge; Impulsbreite; Impulsweite; Pulsdauer 2
≈ pulse-train duration; pulse width 2; pulse interval   ≈ Pulsdauer 1; Pulsweite; Impulsperiodendauer
pulse duration jitter   ELECTRON   **Impulsdauerflattern**
= impulse duration jitter   = Impulsdauer-Jitter
pulse duration modulation   MODUL   **Pulsdauermodulation**
= PDM; pulse length modulation; pulse width modulation   = PDM; Pulsbreitenmodulation; Pulsweitenmodulation; PWMPulslängenmodulation; PLM
↑ pulse modulation; pulse duration modulation   ↑ Pulsmodulation; Pulszeitmodulation
pulse-duration modulator   MODUL   **Pulsdauermodulator**
   = Pulslängenmodulator
pulse duty cycle (→pulse duty factor)   ELECTRON   (→Tastgrad)
pulse duty factor   ELECTRON   **Tastgrad**
[pulse duration to pulse interval]   [Impulsdauer zu Impulsperiodendauer]
= pulse duty cycle; duty cycle; duty factor; reciprocal pulse duty ratio   ≠ Tastverhältnis
≠ pulse duty ratio
pulse duty ratio   ELECTRON   **Tastverhältnis**
[pulse interval to pulse duration]   [Pulsperiodendauer zu Impulsdauer]
= keying ratio   = Stromstoßverhältnis; Schaltverhältnis; relative Einschaltdauer
≠ pulse duty factor   ≠ Tastgrad
pulse echometer (→impulse echo meter)   INSTR   (→Impulsechometer)
pulse edge   ELECTRON   **Impulsflanke**
= pulse slope; impulse edge; impulse slope   ↓ Impulsanstiegflanke; Impulsabfallflanke
pulse energy   ELECTRON   **Impulsenergie**
= impulse energy
pulse excitation   ELECTRON   **Impulserregung**
= impulse excitation
pulse fault location   TRANS   **Impulsfehlerortung**
   = Pulsfehlerortung
pulse form   ELECTRON   **Impulsform**
= pulse shape; impulse form: impulse shape   = Pulsform
pulse-forming network   CIRC.ENG   **impulsformendes Netzwerk**
pulse frame (→frame)   CODING   (→Rahmen)
pulse frequency (→pulse repetition rate)   ELECTRON   (→Pulsrate)
pulse frequency modulation   MODUL   **Pulsfrequenzmodulation**
= PFM   = PFM
↑ pulse time modulation   ↑ Pulszeitmodulation
pulse function   PHYS   **Dirac-Puls**
= Dirac delta function; delta function; Dirac delta pulse; Dirac pulse function   = Einheitsimpulsfunktion
pulse function (→impulse function)   MATH   (→Impulsfunktion)
pulse/function generator   INSTR   **Puls-/Funktionsgenerator**
pulse gate   CIRC.ENG   **Zeilentor**
[allows passage of pulses only during definite time intervals]   [läßt Impulse nur während definierter Zeitintervalle durch]
pulse generator   CIRC.ENG   **Pulsgenerator**
[of repetitive pulses]   [erzeugt periodische Folge von Impulsen]
= pulser   = Impulsgenerator 2
pulse generator   INSTR   **Impulsgenerator**
= pulse source   = Pulsgenerator; Impulsquelle; Pulsquelle
pulse intensity   ELECTRON   **Impulsintensität**
= impulse intensity
pulse interleaving   ELECTRON   **Impulsverflechtung**
pulse interval   ELECTRON   **Impulsperiodendauer**
[pulse duration + pulse separation]   [Impulsdauer + Impulspause]
= pulse spacing   = Pulsperiodendauer; Taktintervall; Impulsabstand 2
pulse length (→pulse duration)   ELECTRON   (→Impulsdauer)
pulse lengthening (→pulse broadening)   ELECTRON   (→Impulsverbreiterung)
pulse length modulation (→pulse duration modulation)   MODUL   (→Pulsdauermodulation)
pulse level   MICROEL   **Impulspegel**
↓ high level; low level   ↓ Hochpegel; Tiefpegel

pulse metering

pulse metering	ELECTRON	**Impulszählung**
= impulse counting		
pulse modulation	MODUL	**Pulsmodulation**
≠ sinus carrier modulation		≠ Schwingungsmodulation
↓ pulse amplitude modulation; pulse time modulation; pulse duration modulation		↓ Pulsamplitudenmodulation; Pulszeitmodulation; Pulsdauermodulation; Pulslagenmodulation
pulse modulator	MODUL	**Pulsmodulator**
pulse motor (→stepper motor)	POWER ENG	(→Schrittmotor)
pulse multiplier	ELECTRON	**Impulsmultiplikator**
= impulse meter		
pulse noise (→impulsive noise)	TELEC	(→Impulsgeräusch)
pulse number check	TELECONTR	**Synchronschlußkontrolle**
pulse operation	ELECTRON	**Impulsbetrieb**
≈ pulsed operation		≈ Pulsbetrieb
pulse overlapping	ELECTRON	**Impulsüberlagerung**
pulse pattern generator	INSTR	**Pulsmustergenerator**
pulse peak	ELECTRON	**Impulsspitze**
= pulse crest; impulse peak; impulse crest		= Impulskuppe
≈ pulse top		≈ Impulsdach
pulse peak power	ELECTRON	**Impulsspitzenleistung**
pulse phase modulation (→pulse position modulation)	MODUL	(→Pulslagemodulation)
pulse position modulation	MODUL	**Pulslagemodulation**
= PPM; pulse phase modulation;		= PLM; Pulsphasenmodulation; PPM; Pulspositionsmodulation
↑ pulse time modulation		↑ Pulszeitmodulation
pulse profiling (→pulse shaping)	CIRC.ENG	(→Impulsformung)
pulser (→pulse generator)	CIRC.ENG	(→Pulsgenerator)
pulse rate (→pulse repetition rate)	ELECTRON	(→Pulsrate)
pulse rate meter	INSTR	**Pulsfrequenzmesser**
		= Impulsfrequenzmesser; Pulsratenmesser; Impulsratenmesser
pulse recurrency (→pulse train)	ELECTRON	(→Puls)
pulse recurrency frequency (→pulse repetition rate)	ELECTRON	(→Pulsrate)
pulse regeneration	ELECTRON	**Impulsregenerierung**
= impulse regeneration		= Impulserneuerung; Pulsregenerierung; Pulserneuerung
pulse relay (→impulse relay)	COMPON	(→Impulsrelais)
pulse repetition (→pulse train)	ELECTRON	(→Puls)
pulse repetition frequency (→pulse repetition rate)	ELECTRON	(→Pulsrate)
pulse repetition rate	ELECTRON	**Pulsrate**
= pulse rate; pulse repetition frequency; pulse frequency; pulse recurrency frequency		= Impulsrate; Pulsfrequenz; Impulsfolgefrequenz; Pulsfolgefrequenz
≈ clock frequency		≈ Taktfrequenz
pulse response	ELECTRON	**Impulsantwort**
= impulse response; pulse behaviour; impulse behaviour		= Pulsantwort; Impulssprungverhalten; Pulssprungverhalten; Stoßantwort
pulse response measurement	INSTR	**Impulsantwortmessung**
= pulse response testing		
pulse response testing (→pulse response measurement)	INSTR	(→Impulsantwortmessung)
pulse separation	ELECTRON	**Impulspause**
[separation of consecutive pulses]		= Pulspause; Impulsabstand 1; Skew
= intertrain pause; non-pulse period; impulse separation; skew 1 (n.); pulse-digit spacing		
pulse sequence (→pulse train)	ELECTRON	(→Puls)
pulse shape (→pulse form)	ELECTRON	(→Impulsform)
pulse-shaped (→impulsive)	ELECTRON	(→impulsförmig)
pulse shaper	CIRC.ENG	**Impulsformer**
= impulse shaper		= Pulsformer
pulse shaping	CIRC.ENG	**Impulsformung**
= impulse shaping; pulse profiling; impulse profiling		= Pulsformung
pulse signal	SWITCH	**Impulskennzeichen**
pulse signaling	SWITC	**Impulskennzeichenverfahren**
pulse slope (→pulse edge)	ELECTRON	(→Impulsflanke)
pulse source (→pulse generator)	INSTR	(→Impulsgenerator)
pulse spacing (→pulse interval)	ELECTRON	(→Impulsperiodendauer)
pulse spreading (→pulse-train spreading)	ELECTRON	(→Pulsverbreiterung)
pulse storage time (→carrier storage time)	MICROEL	(→Speicherzeit)
pulse string (→pulse train)	ELECTRON	(→Puls)
pulse stuff (v.t.)	CODING	**stopfen**
= justify; pad (v.t.)		
pulse stuffing	CODING	**Stopfen**
[change of data rate from inherent rate in order to grant synchronization]		[Anpassen der Übertragungsgeschwindigkeit zwecks Gleichlauf]
= justification; stuffing		= Impulsstopfen
pulse synchronisation	TELEGR	**Schrittsynchronisierung**
[by synchronizing information inherent ot every pulse]		[durch jedem Schritt inhärente Gleichlaufinformation]
		= Schrittsynchronisation
pulse tail	ELECTRON	**Impulsschwanz**
= impulse tail; tail		= Flankenabfall
≈ trailing pulse edge		≈ Impulsabfallflanke
pulse technique	ELECTRON	**Impulstechnik**
[generation, processing and transfer of pulse-shaped signals]		[Erzeugung, Verarbeitung und Übertragung impulsförmiger Signale]
≠ sinusoidal signal technique		= Pulstechnik
		≠ Sinustechnik
		↓ lineare Impulstechnik; nichtlineare Impulstechnuk
pulse telegram	TELECONTR	**Impulstelegramm**
pulse tilt	TV	**Dachschräge**
= tilt		
pulse time meter	INSTR	**Impulsmesser**
pulse-time modulated	MODUL	**pulszeitmoduliert**
		= pulswinkelmoduliert
pulse-time modulation	MODUL	**Pulszeitmodulation**
= PTM		= Pulswinkelmodulation
↓ pulse-phase modulation; pulse-frequency modulation; pulse-duration modulation		↓ Pulsphasenmodulation; Pulsfrequenzmodulation; Pulsdauermodulation
pulse top	ELECTRON	**Impulsdach**
= impulse top; topline		= Pulsdach
≈ pulse peak		≈ Impulsspitze
pulse train	ELECTRON	**Puls**
[periodic or quasiperiodic seqence of uniform pulses]		[periodische oder quasiperiodische Folge gleichgeformter Impulse]
= pulse sequence; pulse repetition; pulse string; pulse recurrency; impulse sequence		= Pulsfolge; Impulszug; Impulsfolge; Impulsreihe; Impulsserie; Pulsstrom
≈ impulse; pulse; impulse pattern		≈ Impuls 1; Impuls 2; Impulsmuster
pulse-train duration	ELECTRON	**Pulsdauer 1**
≈ pulse duration		≈ Impulsdauer
pulse-train spreading	ELECTRON	**Pulsverbreiterung**
= pulse spreading		≈ Impulsverbreiterung
≈ pulse broadening		
pulse transformer	ELECTRON	**Impulsübertrager**
[a broadband transformer]		[ein breitbandiger Übertrager]
= impulse transformer		
pulse transformer	POWER SYS	**Zündübertrager**
[static power converter]		[Stromrichter]
pulse transmission system	TELEC	**Pulsübertragungssystem**
pulse transmitter	RADIO	**Impulssender**
		[Funkstörmessung]
pulse width 2	ELECTRON	**Pulsweite**
≈ pulse duration		= Pulsbreite
		≈ Impulsdauer

pulse width 1 (→pulse duration)		ELECTRON		**punched tape reader** = peforated tape reader; paper tape reader; tape reader; reperforator 2 ↑ taper	TERM&PER	**Lochstreifenleser** = Streifenleser 1; Lochstreifenabtaster ↑ Lochstreifengerät
pulse-width modulated		MODUL	**pulsweitenmoduliert** = pulsbreitenmoduliert; pulsdauermoduliert			
				punched ticket	TERM&PER	**Lochticket**
pulse-width-modulated inverter		POWER SYS	**Pulswechselrichter**	**punching**	METAL	**Freischneiden** [Stanzen]
pulse width modulation (→pulse duration modulation)		MODUL	(→Pulsdauermodulation)	**punching** (→perforation) **punching die** (→punch die) **punching error** = mispunching; off punch	TERM&PER MECH TERM&PER	(→Lochung) (→Stanzwerkzeug) **Fehllochung** = Lochungsfehler; Falschlochung
pulse-width recording (→two-frequency recording)		DATA PROC	(→Wechseltaktschrift)			
pulsing indication (→flashing indication)		EQUIP.ENG	(→Blinkanzeige)	**punch position** (→hole site) **punch station** (→punched card station)	TERM&PER TERM&PER	(→Lochstelle) (→Lochstation)
pulverization = sputtering; atomization; disintegration ≈ disintegration		TECH	**Zerstäubung** = Pulverisierung	**punch-through** = reach-through; penetration factor; feedthrough	MICROEL	**Durchgriff**
pulverize = powder (v.t.); spray 2 (v.t.); sputter (v.t.); atomize; disintegrate		TECH	**zerstäuben** = pulverisieren	**punch-through** [figure of merit of a tube, relation of plate-cathode capacity to grid-cathode capacity]	ELECTRON	**Durchgriff** [Kenngröße einer Elektronenröhre, die Barkhausen-Formel setzt sie mit den Kenngrößen Steilheit und Innenwiderstand in Beziehung]
pump (v.t.)		TECH	**pumpen**			
pump (n.)		TECH	**Pumpe**			
pumping		OPTOEL	**Pumpen**			
pumping energy		OPTOEL	**Pumpenergie**			
pumping frequency		OPTOEL	**Pumpfrequenz**			
pumping oscillator		OPTOEL	**Pumposzillator**			
pumping power		OPTOEL	**Pumpleistung**	**punch-through breakdown** (→breakdown)	PHYS	(→Durchschlag)
pumping source		OPTOEL	**Pumpquelle** = Pumpgenerator	**punch-through effect** **punch-through voltage** = penetration voltage; reach-through voltage	MICROEL MICROEL	**Durchgreifeffekt** **Durchgreifspannung** = Durchreichspannung; Sperrschicht-Berührungsspannung
punch (v.t.) = perforate		TERM&PER	**lochen** = stanzen; ablochen			
punch (n.)		OFFICE	**Locher**			
punch (→cut free)		METAL	(→freischneiden)	**punctual** (→on time)	ECON	(→fristgerecht)
punch (→punch die)		MECH	(→Stanzwerkzeug)	**punctuate** [.....] ≈ point (v.t.) ≈ dot (AM)	ENG.DRAW	**punktieren** [.....] ≈ stricheln
punch block		TERM&PER	**Stanzeinrichtung**			
punch card (→punched card)		TERM&PER	(→Lochkarte)			
punch card machine (→card puncher)		TERM&PER	(→Kartenlocher)	**punctuation** ≈ punctuation mark	LING	**Zeichensetzung** = Interpunktion = Satzzeichen
punch clock (→time clock)		SIGN.ENG	(→Stechuhr)	**punctuation character** (→punctuation mark)	LING	(→Satzzeichen)
punch die = stamping die; punch; punching die		MECH	**Stanzwerkzeug** = Stanzstempel	**punctuation mark** [mark to structure texts and clarify meanings] = punctuation symbol; punctuation character; point 1 ≈ diacritic; punctuation ↑ sign ↓ point 2; comma; colon; semicolon; exclamation point; interrogation mark; hyphen; end punctuation mark	LING	**Satzzeichen** [Zeichen zur Gliederung von Texten und Ergänzung des Ausdrucks] = Interpunktionszeichen ≈ diakritisches Zeichen; Zeichensetzung ↑ Zeichen ↓ Punkt; Komma; Doppelpunkt; Semikolon; Ausrufezeichen; Fragezeichen; Trennungsstrich; Satzschlußzeichen
punched card = punch card ↑ card		TERM&PER	**Lochkarte** ↑ Karte			
punched card collator = card collator; collator		TERM&PER	**Lochkartenmischer** = Kartenmischer			
punched card deck = card deck; deck		TERM&PER	**Lochkartenstapel** = Kartenstapel			
punched card equipment (→punched card machine)		TERM&PER	(→Lochkartenmaschine)			
punched card format		TERM&PER	**Lochkartenformat**			
punched card interpreter (→punched card scanner)		TERM&PER	(→Lochkartenleser)			
punched card machine = punched card equipment		TERM&PER	**Lochkartenmaschine** = Lochkartengerät	**punctuation rule**	LING	**Interpunktionsregel**
punched card reader (→punched card scanner)		TERM&PER	(→Lochkartenleser)	**punctuation symbol** (→punctuation mark)	LING	(→Satzzeichen)
punched card reproducer = card reproducer; reproducer		TERM&PER	**Lochkartendoppler** = Kartendoppler	**puncture** (→breakdown) **puncture voltage** (→breakdown voltage)	PHYS PHYS	(→Durchschlag) (→Durchbruchspannung)
punched card row = card row		TERM&PER	**Lochkartenzeile** = Kartenzeile	**pungent** ≈ keen 2 ≈ pointed	TECH	**stechend** ≈ spitz
punched card scanner = card scanner; punched card reader; card reader; punched card interpreter; card interpreter		TERM&PER	**Lochkartenleser** = Kartenleser; Lochkartenabtaster; Kartenabtaster	**Pupin coil** (→loading coil)	COMM.CABLE	(→Pupinspule)
				Pupin-coil loading (→coil-loading)	COMM.CABLE	(→Bespulung)
punched card station = punch station		TERM&PER	**Lochstation**	**pupinize** = coil-load; load	COMM.CABLE	**bespulen** = pupinisieren
punched card tabulator (→tabulation machine)		TERM&PER	(→Tabelliermaschine)	**pupinized cable** (→loaded cable)	OUTS.PLANT	(→bespultes Kabel)
punched card verifier = card verifier; verifier; key-verifying unit		TERM&PER	**Kartenprüfer** = Lochkartenprüfer; Prüflocher	**pupinized line** (→loaded line)	OUTS.PLANT	(→Pupinleitung)
punched hole		TERM&PER	**Stanzloch**	**purchase 1** (n.)	ECON	**Kauf** = Erwerb; Einkauf
punched label		TERM&PER	**Lochetikette**			
punched tape 2 [a punched tape with mayor width]		TERM&PER	**Lochband** [ein breiter Lochstreifen] ↑ Lochstreifen	**purchase** [activity]	ECON	**Einkauf 2** [Tätigkeit]
				purchase (→procurement)	ECON	(→Beschaffung)
punched tape 1 (→perforated tape)		TERM&PER	(→Lochstreifen)	**purchase 2** (→purchased part)	ECON	(→Fremdfabrikat)

purchased part

purchased part (n.)	ECON	**Fremdfabrikat**
= third-party product; purchase 2		= Fremdhandelsware; Fremdprodukt; Handelsware; Kaufteil; Zukaufteil; Mitvertriebsprodukt
≈ OEM product		
		≈ OEM-Produkt
purchase order	ECON	**Bestellung**
= order		
purchaser	ECON	**Käufer**
= buyer; vendee		= Abnehmer
≈ costumer; consumer		≈ Kunde; Verbraucher
purchasing department	ECON	**Einkaufsabteilung**
		= Einkauf 2
pure aluminium	METAL	**Reinaluminium**
pure-chance (→random)	TECH	(→zufällig)
pure-chance traffic	SWITCH	**regelloser Verkehr**
= random traffic		
pure iron	METAL	**Reineisen**
purely imaginary	MATH	**rein imaginär**
pure procedure	DATA PROC	**reine Prozedur**
		= Reinprozedur
purge (→erase)	ELECTRON	(→löschen)
purge air	TECH	**Spülluft**
purity coil	TV	**Farbreinheitsspule**
purity degree	TECH	**Reinheitsgrad**
purple	PHYS	**purpur**
purposive sentence	LING	**Finalsatz**
↑ causal sentence		= Zwecksatz
		↑ Kausalsatz
purr (n.) (→rattle)	ACOUS	(→Schnarren)
purring (→buzz tone)	TELEPH	(→Summton)
pursuit	COLLOQ	**verfolgen**
= follow-up		= nachfassen
≈ follow		≈ folgen
purveyor (→supplier 1)	ECON	(→Lieferant)
pusbutton receiver	SWITCH	**Tastwahlempfänger**
push (n.)	MECH	**Schub 1**
		= Vortrieb
pushbotton panel (→keypad 2)	TERM&PER	(→Tastaturfeld)
pushbutton	COMPON	**Druckknopf**
= press button		
pushbutton calling (→pushbutton dialing)	SWITCH	(→Tastwahl)
pushbutton dialing	SWITCH	**Tastwahl**
= push-button dialling; pushbutton selection; pushbutton calling; keyboard dialing; keyboard dialling; keyboard selection; keyboard calling; touch dialing; touch dialling; touch selection; touch calling; dual-tone multifrequency dialling; DTMF		= Tastenwahl; Tastaturwahl
pushbutton dialing block (→pushbutton set module)	TERM&PER	(→Tastwahlblock)
push-button dialling (→pushbutton dialing)	SWITCH	(→Tastwahl)
pushbutton key	COMPON	**Tastschalter**
= pushbutton switch; keyboard switch		= Tastenschalter; Druckstastenschalter; Drucktaster
≈ pushbutton		≈ Drucktaste
↓ locking pushbutton key; non-locking pushbutton kex		↓ rastender Tastschalter; nichtrastender Tastenschalter
pushbutton panel (→keyboard)	TERM&PER	(→Tastatur)
pushbutton selection (→pushbutton dialing)	SWITCH	(→Tastwahl)
pushbutton set module	TERM&PER	**Tastwahlblock**
= pushbutton dialing block		
pushbutton station (→touch-tone telephone)	TELEPH	(→Tastwahltelefon)
pushbutton subscriber	SWITCH	**Tastwahlteilnehmer**
pushbutton switch (→pushbutton key)	COMPON	(→Tastschalter)
pushbutton telephon (→touch-tone telephone)	TELEPH	(→Tastwahltelefon)
push-down list (→push-down storage)	DATA PROC	(→Kellerspeicher)
push-down memory (→push-down storage)	DATA PROC	(→Kellerspeicher)
push-down stack (→push-down storage)	DATA PROC	(→Kellerspeicher)
push-down storage	DATA PROC	**Kellerspeicher**
[a stack storage working in the LIFO mode. i.e. the last entry is the first retrieved]		[nach dem LIFO-Prinzip arbeitender Stapelspeicher, d.h. ausgelagert wird zuerst der zuletzt eingelagerte Eintrag]
= push-down store; push-down memory; push-down list; push-down stack; LIFO stack; FILO stack; cellar		= LIFO-Stapelspeicher; FILO-Stapelspeicher
≈ push-pop stack		≈ LIFO-Adressregister
≠ push-up storage		≠ Silospeicher
↑ stack storage		↑ Stapelspeicher
push-down store (→push-down storage)	DATA PROC	(→Kellerspeicher)
pusher tug	TECH	**Schubschlepper**
pushing tractor	TERM&PER	**Schubtraktor**
= push tractor		↑ Raupenvorschub
↑ tractor feed		
push-on connector (→crimp cable lug)	COMPON	(→Quetschkabelschuh)
push-out base technique (→alloy diffusion technique)	MICROEL	(→AD-Technik)
push-pop stack	DATA PROC	**LIFO-Adressregister**
[LIFO register storing address registers]		≈ Kellerspeicher
≈ push-down store		
push-pull	CIRC.ENG	**Gegentakt**
push-pull amplifier	CIRC.ENG	**Gegentaktverstärker**
= balanced amplifier; paraphase amplifier		= Push-pull-Verstärker
push-pull arrangement (→push-pull circuit)	CIRC.ENG	(→Gegentaktschaltung)
push-pull circuit	CIRC.ENG	**Gegentaktschaltung**
= push-pull arrangement		
push-pull collector circuit	CIRC.ENG	**Gegentaktkollektorschaltung**
push-pull complementary collector circuit	CIRC.ENG	**Gegentakt-Komplementärkollektorschaltung**
push-pull input	MICROEL	**Gegentakteingang**
push-pull microphone (→differential microphone)	EL.ACOUS	(→Doppelkohlemikrofon)
push-pull mixer (→balanced mixer)	MICROW	(→Gegentaktmischer)
push-pull modulator (→balanced modulator)	CIRC.ENG	(→Gegentaktmodulator)
push-pull operation	CIRC.ENG	**Gegentaktbetrieb**
push-pull oscillator	CIRC.ENG	**Gegentaktoszillator**
push-pull output	MICROEL	**Gegentaktausgang**
push-pull rectifier	POWER SYS	**Gegentaktgleichrichter**
		= Mittelpunktgleichrichter
push-pull stage	CIRC.ENG	**Gegentaktstufe**
push-pull transformer	COMPON	**Gegentakt-Übertrager**
push-pull voltage transformer	CIRC.ENG	**Gegentaktspannungswandler**
push tractor (→pushing tractor)	TERM&PER	(→Schubtraktor)
push-up list	DATA PROC	**FIFO-Liste**
= FIFO list		
push-up storage	DATA PROC	**Silo-Speicher**
[stack memory working with the FIFO mode, i.e. entries are retrieved in the sequence input]		[Stapelspeicher nach dem FIFO-Prinzip, d.h. Abruf in der Reihenfolge der Eingabe]
= FIFO stack; FIFO store; first-in-first-out store		= FIFO-Stapelspeicher; FIFO-Speicher
≠ push-down storage		≠ Kellerspeicher
put before (fig.)	COLLOQ	**vorlegen** [fig.]
put off (→postpone)	COLLOQ	(→vertagen)
put right (→amend)	TECH	(→berichtigen)
put through to	TELEPH	**durchverbinden**
putty (v.t.)	TECH	**spachteln**
PWB (→printed circuit board)	ELECTRON	(→Leiterplatte)
pylon antenna (→tower antenna)	ANT	(→Mastantenne)
pyramid	MATH	**Pyramide**
[the base is a polygon, the faces are triangles with a common vertex]		[die Grundfläche ist ein Polyeder, die Seitenflächen sind Dreiecke mit gemeinsamen Scheitel]
↑ polyhedron		↑ Polyeder

pyramidal antenna	ANT	Drahtpyramide
pyramidal circuit	ENG.LOG	Pyramidenschaltung
= paramid		= Schaltpyramide
		↑ Reihenparallelschaltung
pyramidal horn	ANT	Pyramidenhorn
= square-end horn		↑ Hornstrahler
↑ horn radiator		
pyramidal horn antenna	ANT	Pyramidenhorn-Antenne
pyramid antenna (→prism antenna)	ANT	(→Reusenantenne)
pyroelectric detector	INSTR	pyroelektrischer Detektor
		= Pyrodetektor
pyroelectric effect	PHYS	pyroelektrischer Effekt
pyroelectricity	PHYS	Pyroelektrizität
pyrolisis [chemical dissociation by high temperature]	CHEM	Pyrolyse [chemische Zersetzung durch hohe Temperatur]
pyrometry	PHYS	Pyrometrie

Q

q (→area)	MATH	(→Fläche)
Q (→electric charge)	PHYS	(→elektrische Ladung)
Q (→heat quantity)	PHYS	(→Wärmemenge)
Q (→factor Q)	ANT	(→Antennengüte)
QAM (→quadrature amplitude modulation)	MODUL	(→Quadratur-Amplitudenmodulation)
QAM (→quadrature amplitude modulation)	TV	(→Quadraturmodulation)
Q band	RADIO	Q-Band
Q bit (→qualifier bit)	DATA COMM	(→Unterscheidungsbit)
Q external	HF	Lastgüte
Q factor	PHYS	Kreisgüte
QH beam antenna (→quick heading beam antenna)	ANT	(→Quick-heading-beam-Antenne)
QIL	MICROEL	QIL
= quad in line		
Q loaded	CIRC.ENG	Gesamtgüte
Q match (→quarter-wave transformer)	NETW.TH	(→Lambda-Viertel-Transformator)
Q meter	INSTR	Gütefaktormeßgerät
= quality factor meter		= Q-Messer
Q multiplier	CIRC.ENG	Q-Multiplier
QPSK (→quaternary phase shift keying)	MODUL	(→QPSK)
QPSK modulator	CIRC.ENG	Vierphasen-Umtastmodulator
		= QPSK-Modulator
QTH locator map [amateur radio]	RADIO	QTH-Kenner-Karte [Amateurfunk]
		= QTH-Locator-Karte
quad (v.t.)	TYPOGR	ausbringen [durch Vergrößern der Wortzwischenräume die Zeilenlänge oder Zeilenzahl erhöhen]
quad (n.)	COMM.CABLE	Viererseil
↑ standing element		= Vierer
↓ star quad; multiple-twin quad		↑ Verseilelement
		↓ Sternvierer; DM-Vierer
quad (→quadrilateral)	MATH	(→Viereck)
quad (→fourfold)	MATH	(→vierfach)
quad antenna (→cubical quad antenna)	ANT	(→Cubical-quad-Antenne)
quad-bit ...	INF	Vierbit-
quadded cable	COMM.CABLE	Viererkabel
↓ star-quad cable; multiple-twin cable		↓ Sternviererkabel; DM-Vierer-Kabel
quad density [magnetic disk]	TERM&PER	Vierfachdichte [Magnetplatte]
quad formation	COMM.CABLE	Sternverseilung
↑ twisting		= Sternviererverseilung
		↑ Verseilung
quad in line (→QIL)	MICROEL	(→QIL)
quad-in-line package (→QUIL package)	MICROEL	(→QUIL-Gehäuse)
quad loop (→cubical quad antenna)	ANT	(→Cubical-quad-Antenne)
quad output	EL.TECH	Vierfachausgang
quad plane antenna	ANT	Quad-plane-Antenne
quadrangle (→quadrilateral)	MATH	(→Viereck)
quadrangular (→quadrilateral)	MATH	(→viereckig)
quadrant	MATH	Quadrant
quadrant antenna (→angular dipole)	ANT	(→Winkeldipol)
quadrant electrometer	PHYS	Quadrantenelektrometer
quadrant multiplier	COMPON	Quadrantenmultiplizierer
quadrant symmetry	NETW.TH	Quadrantensymmetrie
quadrate (n.) (→square)	MATH	(→Quadrat)
quadratic (→square)	MATH	(→quadratisch)
quadratic control area	CONTROL	quadratische Regelfläche
quadrature	MATH	Quadratur
quadrature amplitude modulation	MODUL	Quadratur-Amplitudenmodulation
= QAM		= QAM
quadrature amplitude modulation [NTSC, PAL]	TV	Quadraturmodulation [NTSC, PAL]
= QUAM; QAM		= QUAM; QAM
quadrature component	EL.TECH	Quadraturkomponente
quadrature component (→reactive component)	NETW.TH	(→Blindanteil)
quadrature distortion	MODUL	Quadraturverzerrung
quadrature error	MODUL	Quadraturfehler
quadrature modulation	MODUL	Quadraturmodulation
quadrature-phase (→in opposition)	PHYS	(→gegenphasig)
quadrature phase shift keying (→quaternary phase shift keying)	MODUL	(→QPSK)
quadrilateral (n.)	MATH	Viereck
= quadrangle; quad		↑ Vieleck
↑ polygon		↓ Parallelogram; Trapez
↓ parallelogram; trapezium		
quadrilateral (adj.)	MATH	viereckig
= quadrangular; tetragonal		= tetragonal
↑ polygonal		↑ vieleckig
quadripartite	COLLOQ	vierlateral
↑ multilateral		↑ multilateral
quadriphase (→quaternary phase shift keying)	MODUL	(→QPSK)
quadriphase keying (→quaternary phase shift keying)	MODUL	(→QPSK)
quadripole	ANT	Quadripol
↑ multipole		↑ Multipol
quadripole (→two-port network)	NETW.TH	(→Vierpol)
quadripole analysis (→two-port network analysis)	NETW.TH	(→Vierpoltheorie)
quadripole connection (→two-port connection)	NETW.TH	(→Vierpolzusammenschaltung)
quadripole dissection (→two-port dissection)	NETW.TH	(→Vierpolzerlegung)
quadripole transformation (→two-port network transformation)	NETW.TH	(→Vierpoltransformation)
quadrivalent (→four-valued)	MATH	(→vierwertig)
quadrophony	EL.ACOUS	Quadrophonie
quadruple (→quadruplicate)	MATH	(→vervierfachen)
quadruple (→fourfold)	MATH	(→vierfach)
quadruple diversity	RADIO	Vierfachdiversity
quadruple phantom	LINE TH	(→Sechzehnerleitung)
(→quadruple-phantom circuit)		
quadruple-phantom circuit	LINE TH	Sechzehnerleitung
= quadruple phantom; double super-phantom		
quadruplexer (→polarization/frequency diplexer)	RADIO REL	(→Polarisations-/Frequenz-Weiche)
quadruplex operation	TELEGR	Doppelgegenschreiben
quadruplex recording (→transversal recording)	TV	(→Querspurverfahren)
quadruplicate	MATH	vervierfachen
= quadruple		↑ vervielfachen
↑ multiply		
quadruplication	MATH	Vervierfachung
↑ multiplication		↑ Vervielfachung
quadrupole	PHYS	Quadrupol
quadrupole slot antenna		Vierschlitzstrahler
quad transistor	MICROEL	Vierfachtransistor
quagi antenna	ANT	Quagi-Antenne
		= Quasi-Yagi-Antenne

qualification (→suitability)	TECH	(→Eignung)
qualification test	QUAL	**Qualifikationstest**
qualified component	QUAL	**zugelassenes Bauteil**
		= qualifiziertes Bauteil; homologiertes Bauteil
qualifier bit	DATA COMM	**Unterscheidungsbit**
= Q bit		= Q-Bit
qualitative	SCIE	**qualitativ**
≠ quantitative		≠ quantitativ
quality	TECH	**Qualität**
≈ grade; property		= Güte
		≈ Gütegrad; Eigenschaft
quality ...	TECH	**Qualitäts-**
= high-quality ...		
quality assurance	QUAL	**Qualitätssicherung**
≈ quality control; quality test; quality surveillance		= Gütesicherung
		≈ Qualitätskontrolle; Qualitätsprüfung; Qualitätsüberwachung
quality assurance engineer	QUAL	**Qualitätssicherungsingenieur**
quality assurance manager	QUAL	**Qualitätsbeauftragter**
quality audit	QUAL	**Qualitäts-Audit**
quality characteristic	QUAL	**Qualitätsmerkmal**
		= Qualitätsparameter
quality check (→quality test)	QUAL	(→Qualitätsprüfung)
quality control	QUAL	**Qualitätskontrolle**
≈ cuality test; quality assurance; quality surveillance		≈ Qualitätsprüfung; Qualitätssicherung; Qualitätsüberwachung
quality defect	QUAL	**Qualitätsmangel**
quality degradation (→quality impairment)	QUAL	(→Qualitätseinbuße)
quality deterioration (→quality impairment)	QUAL	(→Qualitätseinbuße)
quality factor (→factor of quality)	EL.TECH	(→Gütefaktor)
quality factor meter (→Q meter)	INSTR	(→Gütefaktormeßgerät)
quality impairment	QUAL	**Qualitätseinbuße**
= quality degradation; quality deterioration		= Qualitätsminderung; Qualitätsverschlechterung
quality inspection (→quality test)	QUAL	(→Qualitätsprüfung)
quality of service (→operating quality)	TELEC	(→Betriebsgüte)
quality-of-service processor	SWITCH	**Betriebsgüte**
quality requirement	TECH	**Qualitätsanforderung**
quality rules	QUAL	**Qualitätsordnung**
quality supervision (→quality surveillance)	QUAL	(→Qualitätsüberwachung)
quality surveillance	QUAL	**Qualitätsüberwachung**
= quality supervision		≈ Qualitätskontrolle; Qualitätsprüfung
≈ quality control; quality test		
quality test	QUAL	**Qualitätsprüfung**
= quality check; quality inspection		= Gütetest; Qualitätstest; Güteprüfung
≈ quality control; quality assurance; quality surveillance; acceptance inspection		≈ Qualitätskontrolle; Qualitätssicherung; Qualitätsüberwachung; Abnahmeprüfung
quality value (→factor of merit)	PHYS	(→Gütewert)
QUAM (→quadrature amplitude modulation)	TV	(→Quadraturmodulation)
quantifier	LING	**Mengenangabe**
[e.g. some, lots of]		[z.B. etwas, kein]
quantitative	SCIE	**quantitativ**
≠ qualitative		≠ qualitativ
quantity	MATH	**Größe**
= dimension		= Dimension
quantity	ECON	**Stückzahl**
= number of parts		
quantity	SCIE	**Menge**
= amount (n.)		= Quantität
quantity discount	ECON	**Mengenrabatt**
= quantity rebate; volume discount		
quantity equation (→dimensional equation)	PHYS	(→Größengleichung)
quantity of electricity	EL.TECH	**Elektrizitätsmenge**
quantity rebate (→quantity discount)	ECON	(→Mengenrabatt)
quantity structure	ECON	**Mengengerüst**
quantity under test (→measurand)	PHYS	(→Meßgröße)
quantity value	PHYS	**Größenwert**
quantization	CODING	**Quantisierung**
= quantizing		= Quantelung; Quantisieren
≈ digitization [INF.TECH]		≈ Digitalisierung [INF.TECH]
quantization cycle	CODING	**Quantisierungszyklus**
quantization distortion	TELEC	**Quantisierungsverzerrung**
= quantizing distortion		≈ Quantisierungsgeräusch
quantization error	CODING	**Quantisierungsfehler**
= quantizing error		
quantization interval	CODING	**Quantisierungsintervall**
= quantizing interval		
quantization noise	TELEC	**Quantisierungsgeräusch**
= quantizing noise		= Quantisierungsrauschen
≈ quantization distortion		≈ Quantisierungsverzerrung
quantization step	CODING	**Quantisierungsschritt**
= quantizing step; quantization value; quantizing value		= Quantisierungsstufe; Quantisierungswert; Quantisierungseinheit
quantization value (→quantization step)	CODING	(→Quantisierungsschritt)
quantize	CODING	**quantisieren**
[to divide a continuous range of a variable into discrete intervals]		[einen kontinuierlichen Variablenbereich in diskrete Intervalle unterteilen]
≈ digitize [INF.TECH]		≈ digitalisieren [INF.TECH]
quantizing (→quantization)	CODING	(→Quantisierung)
quantizing distortion (→quantization distortion)	TELEC	(→Quantisierungsverzerrung)
quantizing error (→quantization error)	CODING	(→Quantisierungsfehler)
quantizing interval (→quantization interval)	CODING	(→Quantisierungsintervall)
quantizing noise (→quantization noise)	TELEC	(→Quantisierungsgeräusch)
quantizing step (→quantization step)	CODING	(→Quantisierungsschritt)
quantizing value (→quantization step)	CODING	(→Quantisierungsschritt)
quantor	MATH	**Quantor**
quantum (pl.-a)	PHYS	**Quant** (pl.-en)
quantum condition	PHYS	**Quantenbedingung**
quantum efficiency	PHYS	**Quantenausbeute**
		= Quantenwirkungsgrad
quantum electrodynamics	PHYS	**Quantenelektrodynamik**
quantum electronics	PHYS	**Quantenelektronik**
quantum energy	PHYS	**Quantenenergie**
quantum jump	PHYS	**Quantensprung**
= quantum leap; quantum transition		= Quantenübergang
quantum leap (→quantum jump)	PHYS	(→Quantensprung)
quantum level	PHYS	**Quantenzustand**
quantum limited	PHYS	**quantenbegrenzt**
quantum mechanics	PHYS	**Quantenmechanik**
≈ quantum theory		≈ Quantentheorie
quantum noise	OPTOEL	**Quantenrauschen**
quantum number	PHYS	**Quantenzahl**
↓ main quantum number; angular momentum quantum number; spin quantum number		↓ Hauptquantenzahl; Nebenquantenzahl; Spinquantenzahl
quantum optics	PHYS	**Quantenoptik**
quantum physics	PHYS	**Quantenphysik**
quantum statistics	PHYS	**Quantenstatistik**
quantum theory	PHYS	**Quantentheorie**
≈ quantum mechanics		≈ Quantenmechanik
quantum transition (→quantum jump)	PHYS	(→Quantensprung)
quantum well channel	MICROEL	**Quantentopf-Elektronenkanal**
quarter (n.)	ECON	**Quartal**
[three month]		[drei Monate]
= trimester		= Trimester; Vierteljahr
quarter	MATH	**Viertel**
[1/4]		[1/4]
quarterly	ECON	**vierteljährlich**
= trimestral; trimestrial		= quartalsweise

quarter size = scale 1:4	ENG.DRAW	**Maßstab 1:4**
quarter turn	TECH	**Vierteldrehung**
quarter wave ≈ quarter period	PHYS	**Viertelwelle** ≈ Viertelperiode
quarter-wave antenna	ANT	**Viertelwellenantenne**
quarter-wave counterpoise	ANT	**Viertelwellen-Gegengewicht**
quarter-wave dipole	ANT	**Viertelwellendipol**
quarter-wave half sloper antenna = half sloper	ANT	**Halb-Sloper**
quarter-wave line (→quarter-wave transformer)	NETW.TH	(→Lambda-Viertel-Transformator)
quarter-wave section (→quarter-wave transformer)	NETW.TH	(→Lambda-Viertel-Transformator)
quarter-wave transformer = quarter-wave section; quarter-wave line; Q match	NETW.TH	**Lambda-Viertel-Transformator** = Viertelwellen-Anpassungsglied; Viertelwellentransformator; Viertelwellenleitung; Lambda-Viertel-Leitung; Q-Anpassung
quartile	MATH	**Quartil**
quarto [cut-four paper size]	TYPOGR	**Viertelbogen**
quartz = crystal 2 ≈ silica	CHEM	**Quarz** ↑ Siliziumdioxyd
quartz (→quartz resonator)	COMPON	(→Schwinquarz)
quartz clock	CIRC.ENG	**Quarzuhr**
quartz-controlled measuring generator	INSTR	**quarzgesteuerter Meßgenerator**
quartz crystal	PHYS	**Quarzkristall**
quartz generator = crystal generator ≈ crystal oscillator	CIRC.ENG	**Quarzgenerator** ≈ Quarzoszillator
quartz oscillator = crystal oscillator; crystal-controlled oscillator	CIRC.ENG	**Quarzoszillator** = quarzgesteuerter Oszillator; Steuerquarz ≈ Quarzgenerator
quartzose (adj.)	CHEM	**quarzhaltig** = quarzhältig (OES); quarzig
quartz pressure probe	INSTR	**Quarz-Druckmeßsonde**
quartz resonator = crystal resonator; crystal unit; quartz; crystal	COMPON	**Schwinquarz**
quartz standard (→crystal standard)	CIRC.ENG	(→Quarznormal)
quasi-associated signalling (BRI) (→quasi-associated signaling)	SWITCH	(→quasi-assoziierte Zeichengabe)
quasi-associated mode (→quasi-associated signaling)	SWITCH	(→quasi-assoziierte Zeichengabe)
quasi-associated signaling (AM) = quasi-associated signalling (BRI); quasi-associated mode	SWITCH	**quasi-assoziierte Zeichengabe** = quasi-assoziierte Signalisierung
quasi-complementary circuit	CIRC.ENG	**Quasikomplementärschaltung**
quasi-Fermi level = imref level	PHYS	**Quasi-Fermi-Niveau** = Imref
quasifree	PHYS	**quasifrei**
quasilanguage (→pseudocode)	DATA PROC	(→Pseudocode)
quasilinear	MATH	**quasilinear**
quasilogarithmic	MATH	**quasilogarithmisch**
quasilogarithmic coding	CODING	**quasilogarithmische Codierung**
quasi optical sight	RADIO PROP	**quasioptische Sicht**
quasiparticle	PHYS	**Quasiteilchen**
quasi-peak [radio interference measurement]	INSTR	**Quasi-Spitzenwert** [Störstrahlungsmessung]
quasi-peak adapter	INSTR	**Quasi-Spitzenwert-Adapter**
quasi-peak detection	TELEC	**Quasi-Spitzenwert-Gleichrichtung**
quasiperiodic = almost periodic	PHYS	**quasiperiodisch**
quasistatic	PHYS	**quasistatisch**
quasistationary = quasisteady	PHYS	**quasistationär**
quasistationary field	PHYS	**quasistationäres Feld**
quasisteady (→quasistationary)	PHYS	(→quasistationär)
quasi-synonym	LING	**Quasisynonym** ≈ Nebenbegriff
quasi-ternary keying	MODUL	**quasiternäre Tastung**
quaternary	MATH	**quaternär**
quaternary phase shift keying = quadrature phase shift keying; QPSK; 4PSK; quadriphase keying; quadriphase	MODUL	**QPSK** = 4PSK
queing = waiting; hold	SWITCH	**Warten** [Zwischenspeicherung]
quench (v.t.)	METAL	**abschrecken**
quench (v.t.)	PHYS	**löschen**
quench (→spark extinction)	PHYS	(→Funkenlöschung)
quenched gap (→quenched spark-gap)	PHYS	(→Löschfunkenstrecke)
quenched spark	PHYS	**Löschfunke**
quenched spark-gap = quenched gap	PHYS	**Löschfunkenstrecke** = Funkenlöschstrecke
quenching	METAL	**Abschreckung**
quenching [discharge]	PHYS	**Löschung** [Gasentladung]
quenching (→spark extinction)	PHYS	(→Funkenlöschung)
quenching circuit	POWER SYS	**Löschschaltung**
quench tube	ELECTRON	**Löschröhre** = Quenchröhre
query (→inquiry)	DATA PROC	(→Abfrage)
query (n.) (→inquiry)	TELECONTR	(→Abfrage)
query (→inquiry)	SWITCH	(→Abfragen)
query (→inquiry)	DATA COMM	(→Stationsaufforderung)
query (v.i.) (→inquire)	DATA PROC	(→abfragen)
query character (→inquiry character)	DATA PROC	(→Abfragezeichen)
query language [for direct user access to files] = enquiry language (BRI); inquiry language (NAM); retrieval language; interrogation language	DATA PROC	**Abfragesprache** [für direkten Benutzerzugriff zu Dateien] = Dialogsprache; eigenständige Datenbanksprache; Manipulationssprache; Suchsprache
query/response (→inquiry/response)	TELEC	(→Frage-/Antwort-)
query signal (→interrogation signal)	TELECONTR	(→Abfragesignal)
query station (→inquiry station)	DATA COMM	(→Abfragestation)
question-answer mode	TECH	**Frage-Antwort-Spiel**
questionary = questionnaire	DOC	**Fragebogen**
question mark [?] = interrogation point ↑ punctuation mark; end punctuation mark	LING	**Fragezeichen** [?] ↑ Satzzeichen; Satzschlußzeichen
questionnaire (→questionary)	DOC	(→Fragebogen)
question tag	LING	**Frageanhängsel**
question word (→interrogative pronoun)	LING	(→Interrogativpronomen)
queue (n.) = line (AM); cue	COLLOQ	**Schlange** ≈ Reihe
queue (→waiting queue)	SWITCH	(→Warteschlange)
queued access mode	DATA PROC	**Warteschlangen-Zugriffsmethode**
queueing system (→queuing system)	SWITCH	(→Wartesystem)
queue management	SWITCH	**Warteschlangenverwaltung**
queue manager	DATA PROC	**Warteschlangenverwalter**
queue organisation	SWITCH	**Warteschlangenorganisation**
queuing delay	SWITCH	**Warteschlangenverzugszeit** = Wartezeit
queuing operation	SWITCH	**Warteschlangenbetrieb** = Wartebetrieb
queuing probability (→delay probability)	SWITCH	(→Wartewahrscheinlichkeit)
queuing system = queueing system; call queuing system	SWITCH	**Wartesystem**
queuing theory ↑ probability theory	MATH	**Warteschlangentheorie** ↑ Wahrscheinlichkeitstheorie

quick (→fast)	TECH	(→schnell)	
quick-access	DATA PROC	zugriffszeitfrei	
≈ highspeed		≈ schnell	
quick-access memory	DATA PROC	(→Schnellspeicher 1)	
(→high-speed memory)			
quick-access storage	DATA PROC	(→Schnellspeicher 1)	
(→high-speed memory)			
quick-access store	DATA PROC	(→Schnellspeicher 1)	
(→high-speed memory)			
quick acting	TECH	schnellwirkend	
= fast acting; quick-reacting		≈ schnell	
≠ fast			
quick-action	ELECTRON	(→verzögerungsfrei)	
(→undelayed)			
quick-action switch	COMPON	Momentanschalter	
= quick-rupture switch		= Momentanschalter	
quick-break (→quick-breake fuse)	COMPON	(→Schnellsicherung)	
quick-breake fuse	COMPON	Schnellsicherung	
= quick-break		= Momentunterbrecher	
quick heading beam antenna	ANT	Quick-heading-beam-Antenne	
= QH beam antenna		≈ QH-Beam	
quick motion (→fast motion)	TV	(→Zeitraffung)	
quick motion facility (→fast motion facility)	TV	(→Zeitraffer)	
quickness (→fastness)	COLLOQ	(→Schnelligkeit)	
quick-operating (→undelayed)	ELECTRON	(→verzögerungsfrei)	
quick-reacting (→quick acting)	TECH	(→schnellwirkend)	
quick reacting fuse (→fast blowing melting fuse)	COMPON	(→flinke Schmelzsicherung)	
quick-response (→undelayed)	ELECTRON	(→verzögerungsfrei)	
quick-rupture switch (→quick-action switch)	COMPON	(→Momentanschalter)	
quicksort (v.t.)	DATA PROC	schnellsortieren	
quicksort (n.)	DATA PROC	Schnellsortierung	
quicksort algorithm	DATA PROC	Schnellsortieralgorithmus	
quiescent carrier (→suppressed carrier)	MODUL	(→unterdrückter Träger)	
quiescent current	ELECTRON	Ruhestrom	
= rest current; steady current; closed-circuit current		≠ Arbeitsstrom	
≠ working current			
quiescent phase (→rest condition)	TELEGR	(→Ruhezustand)	
quiescent point	MICROEL	Ruhepunkt	
quiescent potential	ELECTRON	Ruhepotential	
quiescent power dissipation (→static power dissipation)	MICROEL	(→statische Verlustleistung)	
quiescent state	ELECTRON	Ruhezustand	
= closed-circuit state			
quiescent voltage	ELECTRON	Ruhespannung	
= reset voltage; steady voltage			
quiet (→noiseless)	TELEC	(→geräuschfrei)	
quiet (→noiseless)	TECH	(→geräuschlos)	
quietized (→sound-insulating)	ACOUS	(→schalldämmend)	
quiet mode	TERM&PER	Leisebetrieb	
= quiet operation			
quiet operation (→quiet mode)	TERM&PER	(→Leisebetrieb)	
QUIL package	MICROEL	QUIL-Gehäuse	
= quad-in-line package			
quinary	MATH	quinär	
[with the basis of 5, or composed of 5 elements]		[mit der Basis 5 oder, aus 5 Elementen bestehend]	
quincunx	TV	Quincunx	
quincunx scanning pattern	TV	Quincunx-Abtastraster	
quintal (→metric quintal)	PHYS	(→Doppelzentner)	
quintuple (adj.)	MATH	fünffach	
quintuple (→quintuplicate)	MATH	(→verfünffachen)	
quintuplicate	MATH	verfünffachen	
= quintuple		↑ vervielfachen	
↑ multiplicate			
quintuplication	MATH	Verfünfachung	
quirl (n.) (→conical-scan tracking)	RADIO LOC	(→Quirlen)	

quit (→terminate)	DATA PROC	(→beenden)	
Q unloaded	ELECTRON	Leerlaufgüte	
quorum	ECON	Beschlußfähigkeit	
quota	TECH	Anteil	
= proportional part		= Quote	
		≈ Teil	
quota	ECON	Quote	
quotation (→offer)	ECON	(→Angebot)	
quotation mark	TYPOGR	Anführungszeichen	
[symbol "]		[Symbol „]	
= inverted commas		= Anführungsstrich; Gänsefüßchen	
≈ quote			
↑ punctuation mark		≈ Hochkomma	
↓ quote mark; unquote mark		↑ Satzzeichen	
		↓ Anführung; Abführung	
quote	DATA PROC	Anführungszeichen	
[special character of programming languages]		[Sonderzeichen von Programmiersprachen]	
quote (→inverted comma)	LING	(→Hochkomma)	
quote mark	LING	Anführung	
= quote sign; quotes; inverted commas		= öffnendes Anführungszeichen	
≠ unquote mark		≠ Abführung	
↑ quotation mark		↑ Anführungszeichen	
quotes (→quote mark)	LING	(→Anführung)	
quote sign (→quote mark)	LING	(→Anführung)	
quotidian (→daily 1)	COLLOQ	(→täglich) (adj.)	
quotient	MATH	Quotient 2	
[number resulting from division]		[Ergebnis einer Division]	
≈ ratio; fractional number; division		= Teilzahl; Teilungsverhältnis	
		≈ Bruch; Bruchzahl; Division	
quotient field	MATH	Quotientenkörper	
Q value (→factor of quality)	EL.TECH	(→Gütefaktor)	
Q value (→factor of merit)	PHYS	(→Gütewert)	
QWERTY keyboard	TERM&PER	(→amerikanische Tastatur)	
(→US-standard keyboard)			
QWERTZ keyboard	TERM&PER	(→DIN-Tastatur)	
(→german-standard keyboard)			

R

R (→electric resistance)	PHYS	(→elektrischer Widerstand)	
Ra (→radium)	CHEM	(→Radium)	
race condition	DATA PROC	Wettlaufsituation	
[whichs instruction finishes first]			
rached wheel	MECH	Sperrad	
= pawl wheel; ratched wheel			
rack	EQUIP.ENG	Gestell	
= bay; frame		= Bucht	
rack	TECH	Regal	
= stand		= Gestell	
rack (→rack panel)	EQUIP.ENG	(→Gestellrahmen)	
rack (IEC) (→cable rack)	SYS.INST	(→Kabelrost)	
rack 2 (→grid)	SYS.INST	(→Rost)	
rack and stack case	INSTR	Stapel-/Einbau-Gehäuse	
rack arrangement (→rack equipment)	EQUIP.ENG	(→Gestellbelegung)	
rack cabinet	EQUIP.ENG	Gestellschrank	
rack cable	EQUIP.ENG	Gestellkabel	
rack construction	EQUIP.ENG	Gestellaufbau	
= rack mounting			
rack design	EQUIP.ENG	Gestellbauweise	
rack equipment	EQUIP.ENG	Gestellbelegung	
= rack profile; rack arrangement		= Gestellbestückung	
rack equipment list	DOC	Belegungsliste	
rack extension	EQUIP.ENG	Gestellaufsatz	
rack flange kit (→rack-mount kit)	INSTR	(→Gestelleinbausatz)	
rack foot	EQUIP.ENG	Gestellfuß	
rack frame (→rack panel)	EQUIP.ENG	(→Gestellrahmen)	
rack line (→rack row)	SYS.INST	(→Gestellreihe)	
rack mountable	EQUIP.ENG	rahmenmontierbar	
rack-mountable shelf unit	EQUIP.ENG	Gestelleinschubrahmen	
rack mounting (→rack construction)	EQUIP.ENG	(→Gestellaufbau)	
rack-mount kit	INSTR	Gestelleinbausatz	
= rack flange kit			

rack-mount option	EQUIP.ENG	**Gestelleinbau-Option**	
rack panel	EQUIP.ENG	**Gestellrahmen**	
= rack frame; rack			
rack profile (→rack equipment)	EQUIP.ENG	(→Gestellbelegung)	
rack row	SYS.INST	**Gestellreihe**	
= equipment row; row; rack suite; equipment suite; suite; rack line; bay line; line		= Gestellzeile; Reihe; Zeile	
rack row arrangement (→rack suite construction)	SYS.INST	(→Gestellreihenaufbau)	
rack suite (→rack row)	SYS.INST	(→Gestellreihe)	
rack suite construction	SYS.INST	**Gestellreihenaufbau**	
= rack row arrangement			
rack top	EQUIP.ENG	**Gestelloberteil**	
		= Gestellkopf	
rack-type cabinet	EQUIP.ENG	**Schrankgestell**	
rack wiring	EQUIP.ENG	**Gestellverdrahtung**	
= shelf cabling		= Gestellverkabelung	
rad (→radian)	MATH	(→Radiant)	
radar	RADIO LOC	**Radar**	
= radiodetection and ranging		= Rückstrahlortung	
↑ radio location		↑ Funkortung	
radar antenna	ANT	**Radarantenne**	
radar beacon (→secondary surveillance radar)	RADIO LOC	(→Sekundärradar)	
radar cross section	RADIO LOC	**Radarquerschnitt**	
= RCS; effective echoing area; back-scattering cross section; forward-scattering cross section; bistatic-scattering cross section			
radar equation	RADIO LOC	**Radargleichung**	
= radar range equation; range equation			
radar horizon	RADIO LOC	**Radarhorizont**	
radar measurement	INSTR	**Radarmessung**	
radar modulator	RADIO LOC	**Radarmodulator**	
		≈ Ladeschaltung	
radar picture tube	ELECTRON	**Radarbildröhre**	
radar radome		**Radarhaube**	
radar range	RADIO NAV	**Radarreichweite**	
radar range equation (→radar equation)	RADIO LOC	(→Radargleichung)	
radar screen	RADIO LOC	**Radarbildschirm**	
radar screen picture	RADIO LOC	**Radarschirmbild**	
radar shadow	RADIO LOC	**Radarschatten**	
radar transmitter	RADIO LOC	**Radarsender**	
radar tube	ELECTRON	**Radarröhre**	
↑ picture tube		↑ Bildwiedergaberöhre	
radial	MATH	**radial**	
≠ axial		≠ axial	
radial	TECH	**strahlenförmig**	
≈ stelliform		≈ sternförmig	
radial 3 (→counterpoise)	ANT	(→Gegengewicht)	
radial 1 (→radial wire 1)	ANT	(→Radial 1) (n.n.)	
radial 2 (→radial wire 2)	ANT	(→Radial 2) (n.n.)	
radial beam	ELECTRON	**Radialstrahl**	
		= radialer Strahl	
radial bearing	MECH	**Radiallager**	
radial earth electrode	ANT	**Radialerder**	
		[vom Antennenfuß radial verlegt]	
radial field	PHYS	**radiales Feld**	
		= Radialfeld	
radial force	PHYS	**Radialkraft**	
		= radiale Kraft	
radial frequency (→angular frequency)	PHYS	(→Kreisfrequenz)	
radial frequency deviation (→angular frequency shift)	MODUL	(→Kreisfrequenzhub)	
radial frequency sweep (→angular frequency shift)	MODUL	(→Kreisfrequenzhub)	
radial frequency swing (→angular frequency shift)	MODUL	(→Kreisfrequenzhub)	
radial ground system	ANT	**Strahlenerder**	
radial lead	COMPON	**Radialanschluß**	
		= radialer Anschlußdraht	
radial length (→arc length)	MATH	(→Bogenlänge)	
radial load	MECH	**Radiallast**	
		= radiale Last	
radially symmetric	MATH	**radialsymmetrisch**	
radial mode (→radial oscillation)	PHYS	(→Radialschwingung)	
radial network (→star network)	TELEC	(→Sternnetz)	
radial oscillation	PHYS	**Radialschwingung**	
= radial mode			
radial play	MECH	**Radialspiel**	
radial wire 1	ANT	**Radial 1** (n.n.)	
[buried grounding wires]		[eingegrabene Erdleiter]	
= radial 1			
radial wire 2	ANT	**Radial 2** (n.n.)	
[wire of a counterpoise]		[Leiter eines Gegengewichts]	
= radial 2			
radian	MATH	**Radiant**	
[complementary SI unit for plane angles]		[ergänzende SI-Einheit für ebene Winkel]	
= rad		= rad; Radian	
		↑ Bogenmaß	
radiance	PHYS	**Strahlungsdichte 1**	
[radiation energy per solid angle]		[Strahlungsleistung pro Raumwinkel]	
= radiation density; radiant intensity; intensity		= Strahldichte; Flächenintensität; Intensität; Leistungsbedeckung	
radian frequency (→angular frequency)	PHYS	(→Kreisfrequenz)	
radian frequency deviation (→angular frequency shift)	MODUL	(→Kreisfrequenzhub)	
radian frequency sweep (→angular frequency shift)	MODUL	(→Kreisfrequenzhub)	
radian frequency swing (→angular frequency shift)	MODUL	(→Kreisfrequenzhub)	
radian measure	MATH	**Bogenmaß**	
[measure of plane angle by the ratio of arc length to radius; SI unit: radian]		[Maßgröße für ebene Winkel, Bogenlänge durch Radius; SI-Einheit: Radiant]	
≈ arc length		≈ Bogenlänge	
↑ angular dimension		↑ Winkelmaß	
↓ radian		↓ Radiant	
radian per second	PHYS	**Radiant durch Sekunde**	
[SI unit for angular velocity]		[SI-Einheit für Rotationsgeschwindigkeit]	
= rad/s		= rad/s	
radiant energy	PHYS	**Strahlungsenergie**	
= radiation energy			
radiant exposure	PHYS	**Bestrahlung**	
radiant flux density	PHYS	**Strahlungsflußdichte**	
radiant intensity (→radiance)	PHYS	(→Strahlungsdichte 1)	
radiate	PHYS	**strahlen**	
↓ irradiate		↓ ausstrahlen	
radiated emission (→unwanted emission)	RADIO	(→Störstrahlung)	
radiated intensity (→beam power)	ANT	(→Strahlstärke)	
radiated interference (→unwanted emission)	RADIO	(→Störstrahlung)	
radiated interference power	RADIO	**Störstrahlleistung**	
radiated power (→beam power)	ANT	(→Strahlstärke)	
radiated susceptibility	TELEC	**strahlungsgebundene Störfestigkeit**	
= RS			
radiatiant flux	PHYS	**Strahlungsfluß**	
= radiation flux			
radiating cable	HF	**abstahlendes Kabel**	
radiating element	ANT	**Einzelstrahler**	
= antenna element		= Einzelelement; Antennenelement; Strahlungselement; Strahlelement	
↑ radiator		↑ Strahler	
radiation	PHYS	**Strahlung**	
↓ irradiation		↓ Ausstrahlung	
radiation (→emission)	RADIO	(→Aussendung)	
radiation angle	PHYS	**Strahlungswinkel**	
= beam angle; angle of radiation		= Strahlwinkel	
radiation barrier	SIGN.ENG	**Strahlenschranke**	
↓ light barrier		↓ Lichtschranke	
radiation beam (→ray)	PHYS	(→Strahl)	
radiation characteristic	PHYS	**Abstrahlcharakteristik**	
radiation chart (→radiation pattern)	ANT	(→Strahlungscharakteristik)	

radiation cone	ANT	Strahlungskegel
radiation constant	PHYS	Strahlungskonstante
radiation damping	PHYS	Strahlungsdämpfung
radiation density (→radiance)	PHYS	(→Strahlungsdichte 1)
radiation detector	INSTR	Strahlungsdetektor
= Strahlungsempfänger		
radiation efficiency	ANT	Strahlungswirkungsgrad
= Antennenwirkungsgrad		
radiation energy (→radiant energy)	PHYS	(→Strahlungsenergie 1)
radiation factor	ANT	Strahlungsmaß
= pattern factor		
radiation field	PHYS	Strahlungsfeld
radiation flux (→radiatiant flux)	PHYS	(→Strahlungsfluß)
radiation-free (→radiationless)	PHYS	(→strahlungsfrei)
radiation hardening	MICROEL	Strahlungsabsicherung
radiation hazard	PHYS	Strahlengefährdung
radiation intensity	PHYS	Strahlungsintensität
= Strahlstärke		
radiation inversion	RADIO PROP	Strahlungsinversion
radiationless	PHYS	strahlungsfrei
= radiation-free		
radiation lobe	PHYS	Strahlungskeule
= Strahlungslappen		
radiation loss	ANT	Abstrahlungsverlust
radiation loss	PHYS	Strahlungsverlust
radiation measuring instrument	INSTR	Strahlungsmeßinstrument
radiation path	RADIO PROP	Strahlungsweg
radiation pattern	ANT	Strahlungscharakteristik
[spacial distribution of a radiation parameter]		[räumliche Verteilung eines Strahlungsparameters]
= directional pattern; antenna pattern; radiation chart		= Richtcharakteristik; Richtstrahlcharakteristik; Antennencharakteristik; Strahlungsdiagramm; Antennendiagramm
≈ directional diagram		≈ Richtdiagramm
radiation pattern envelope	ANT	Winkeldämpfungs-Hüllkurve
= RPE		= RPE
radiation physics	PHYS	Strahlenphysik
radiation pressure	PHYS	Strahlungsdruck
radiation protection	PHYS	Strahlenschutz
radiation pyrometer	INSTR	Strahlungspyrometer
radiation quantum	PHYS	Strahlungsquant
radiation resistance	ANT	Strahlungswiderstand
↑ Antennenwiderstand		
radiation source (→radiator)	PHYS	(→Strahler)
radiation sphere	ANT	Strahlungskugel
radiator	PHYS	Strahler
= radiation source		= Strahlenquelle; Strahlungsquelle
radiator	ANT	Strahler
≈ antenna		≈ Antenne
↓ radiating element		↓ Einzelstrahler
radiator aperture	ANT	Strahleröffnung
radical 1 (→radical expression)	MATH	(→Wurzelausdruck)
radical 2 (→root symbol)	MATH	(→Wurzelzeichen)
radical expression	MATH	Wurzelausdruck
= radical 1		
radical law	MATH	Wurzelgesetz
radical sign (→root symbol)	MATH	(→Wurzelzeichen)
radicand	MATH	Radikand
[quantity under root symbol]		[Zahl deren Wurzel berechnet werden soll]
radio	TELEC	Funk
= wireless (BRI)		
radioactive chain (→radioactive decay series)	PHYS	(→Zerfallsfolge)
radioactive decay series	PHYS	Zerfallsfolge
= radioactive chain		
radioactivity	PHYS	Radioaktivität
radio advertising	BROADC	Rundfunkwerbung
↓ TV advertizing		↓ Fernsehwerbung; Hörfunkwerbung
radio altimeter	RADIO NAV	Funkhöhenmesser
radio amateur	RADIO	Funkamateur
= amateur radio operator; amateur; ham		= Amateurfunker
		≈ Radiobastler
radio amateur band	RADIO	Amateurfunkband
		= Funkamateurband
radio amateur installation	RADIO	Amateurfunkanlage
		= Funkamateuranlage
radio amateur licence	RADIO	Amateurfunklizenz
		= Funkamateurlizenz
radio amateur receiver	RADIO	Amateurfunkempfänger
		= Funkamateurempfänger
radio amateur transmitter	RADIO	Amateurfunksender
		= Funkamateursender
radioastronomy	ASTROPHYS	Radioastronomie
radio atmosphere	GEOPHYS	Radioatmosphäre
radio audience	BROADC	Rundfunkhörerschaft
		↑ Rundfunkteilnehmer
radio balloon	RADIO	Radiosonde
= radio sonde		
radio base station	MOB.COMM	Funk-Basisstation
= base station		= Basisstation; ortsfeste Funkstelle
radio beacon	RADIO NAV	Funkfeuer
= radiophare; radioguiding; range		= Radiobake
radio beam	RADIO	Funkstrahl
= radio ray beam		
radio bearing (→radio direction finding)	RADIO	(→Funkpeilung)
radio broadcast (→broadcast transmission)	BROADC	(→Rundfunksendung 1)
radio broadcasting	RADIO	Rundfunk
= broadcast; broadcasting		↓ Hörfunk; Fernsehrundfunk
↓ sound broadcasting; television broadcasting		
radio broadcasting (→sound broadcasting)	BROADC	(→Hörfunk)
radio broadcasting receiver	BROADC	Hörfunkempfänger
= radio receiver		= Hörfunkgerät
radio broadcast listener	BROADC	Rundfunkhörer
= listener		= Radiohörer; Hörer
≈ radio broadcast service user		≈ Hörfunkteilnehmer
radio broadcast user (→sound broadcast user)	BROADC	(→Hörfunkteilnehmer)
radio button	DATA PROC	Radio-button
radio car	RADIO	Funkmeßwagen
		= Funkwagen
radio cell	MOB.COMM	Funkzelle
= cell		= Zelle; Funkzone
radio cell boundary	MOB.COMM	Funkzellgrenze
= cell boundary; cell limit		= Zellgrenze; Zellengrenze; Funkzonengrenze
radio channel (→radiofrequency channel)	RADIO	(→Funkkanal)
radiochannel frame	RADIO REL	Richtfunkrahmen
radio circuit (→radio link)	RADIO	(→Funkverbindung)
radiocode test set	INSTR	Radiocode-Test-Set
radio communication (→radio link)	RADIO	(→Funkverbindung)
radio communications	TELEC	Funkwesen
radio communications (→radio traffic)	RADIO	(→Funkverkehr)
radiocommunication service	TELEC	Funkdienst
= radio service		
radiocommunication service monitor (→radiocommunication tester)	INSTR	(→Funkmeßplatz)
radio communications network	RADIO	Funknetz
radiocommunications subscriber	TELEC	Funkteilnehmer
radiocommunication tester	INSTR	Funkmeßplatz
= transceiver test equipment; radiocommunication service monitor; CMS		= Funkgeräte-Meßplatz
radio continuity pilot	RADIO REL	Funkpilot
radio control (→radio telecontrol)	RADIO	(→Funkfernsteuerung 1)
radiodetection and ranging (→radar)	RADIO LOC	(→Radar)
radio detector	HF	Radiodetektor
radiodetermination-satellite service (→radiolocation-satellite service)	SAT.COMM	(→Ortungsfunkdienst)
radio direction finder	RADIO LOC	Funkpeiler
= direction finder; DF		= Peiler; Peilempfänger; Funkerfassungspeiler
↓ automatic direction finder		

radio direction finding RADIO **Funkpeilung**
= radio bearing ↑ Peilung
radio disciplines RADIO **Funktechniken**
 [Spezialzweige der Funktechnik]
radio disturbance (→radio RADIO (→**Funkstörung**)
 interference)
radioelectric coverage RADIO **Funkversorgung**
radioelectric wave (→radio RADIO (→**Funkwelle**)
 wave)
radio engineering TELEC **Funktechnik**
radio equipment RADIO **Funkgerät**
= radio set ↓ Transceiver; Funkempfänger; Funksender
radio equipment RADIO REL (→**Funkgerät**)
 (→transceiver)
radio exchange HF **Funkfernamt**
radio fee (→radio licence) BROADC (→**Rundfunkgebühr**)
radiofrequencies spectrum RADIO **Funkwellenspektrum**
= radio spectrum = Funkspektrum
radio frequency RADIO **Radiofrequenz**
 [from 10 kHz to 100 GHz; in contrast to English, this term is not used in German outside radio related field as a synonym to "high frequency"] [von 10 kHz bis 100 GHz; im Englischen wird „radio frequency" gelegentlich auch außerhalb der Funktechnik als Synonym zu „high frequency" benutzt]
= RF = Funkfrequenz; RF
≈ high frequency [TELEC] ≈ Hochfrequenz [TELEC]
radio frequency band RADIO **Frequenzband**
 (→frequency band)
radio frequency bridge (→HF INSTR (→**HF-Widerstandsmeßbrücke**)
 resistance bridge)
radio-frequency cable COMM.CABLE **RF-Kabel**
= RF cable
radiofrequency channel RADIO **Funkkanal**
= RF channel; radio frequency channel; radio channel = Radiofrequenzkanal; RF-Kanal
radio frequency channel RADIO (→**Funkkanal**)
 (→radiofrequency channel)
radio frequency coder TERM&PER **Radiofrequenzcodierer**
= RF coder = Radiofrequenzkodierer; RF-Codierer
radio frequency detector TERM&PER **Radiofrequenzdetektor**
= RF detector = RF-Detektor
radiofrequency interference RADIO (→**Funkstörung**)
 (→radio interference)
radio-frequency pattern RADIO **Radiofrequenzraster**
= RF pattern; radio-frequency raster; RF raster; frequency pattern = RF-Raster; Frequenzraster
≈ channel configuration ≈ Kanalraster
radio-frequency range 1 RADIO **Radiofrequenzbereich 1**
 [from 10 kHz to 100 GHz] [von 10 kHz bis 100 GHz]
= RF range 1 = RF-Bereich 1
≈ high-frequency range ≈ Hochfrequenzbereich
radio frequency range 2 RADIO (→**Frequenzband**)
 (→frequency band)
radio-frequency raster RADIO (→**Radiofrequenzraster**)
 (→radio-frequency pattern)
radio frequency spectroscopy PHYS **Hochfrequenz-Spektroskopie**
radio frequency voltmeter (→RF INSTR (→**HF-Spannungsmesser**)
 voltmeter)
radiogoniometer RADIO LOC **Radiogoniometer**
radiogram TELEC **Funktelegramm**
= radio-telegram; wireless message 2 = Radiotelegramm
radioguiding (→radio RADIO NAV (→**Funkfeuer**)
 beacon)
radio horizon RADIO PROP **Radiohorizont**
 = Funkhorizont; Radiosichtweite
radio installation RADIO **Funkanlage**
= wireless plant
radio interference RADIO **Funkstörung**
= radio disturbance; radiofrequency interference; RFI = HF-Störung
radio interference level RADIO **Funkstörgrad**
radio interference measuring receiver INSTR **Funkstörmeßempfänger**
 = Störmeßempfänger
radio interference measuring technique INSTR **Funkstörmeßtechnik**

radio interference power RADIO **Funkstörleistung**
radio interference suppression RADIO **Funkentstörung**
 ↑ Entstörung
radio interferometer ANT **Radiointerferometer**
 [array of antennas with extremely high angular resolution, for radioastronomic measurements] [Antennenanordnung höchster Winkelauflösung für radioastronomische Messungen]
radio in the loop RADIO **Teilnehmerfunk**
= RITL
radio licence BROADC **Rundfunkgebühr**
= radio fee
radio licence collecting center BROADC **Gebühreneinzugszentrale**
radio line-of-sight RADIO PROP **Radiosicht**
radio link RADIO **Funkverbindung**
= radiolink; radio circuit; radio communication ↓ Richtfunkverbindung
radiolink (→radio link) RADIO (→**Funkverbindung**)
radio link antenna ANT **Richtfunkantenne**
 [for line-of-sight radio links] ↑ Mikrowellenantenne
= radio-relay antenna; LOS radio antenna; microwave antenna 2
↑ microwave antenna 1
radiolink calculation RADIO REL **Richtfunkstreckenberechnung**
radiolink equipment (→radio RADIO REL (→**Richtfunkgerät**)
 relay equipment)
radiolink hop (→hop) RADIO PROP (→**Funkfeld**)
radiolink network (→radio relay TRANS (→**Richtfunknetz**)
 network)
radio location RADIO **Funkortung**
= radiolocation; radiotetermination = Ortungsfunk
↓ radar ↓ Radar
radiolocation (→radio RADIO (→**Funkortung**)
 location)
radiolocation-satellite service SAT.COMM **Ortungsfunkdienst**
= radiodetermination-satellite service
radioluminiscence PHYS **Radiolumineszenz**
 [induced by ionizing radiation] [durch ionisierende Strahlung induziert]
radio magnetic indicator RADIO LOC **Radiokompaßrose**
= RMI = Kompaßrose; RMI
radio-mesh RADIO NAV **Funkgitter-Navigationssystem**
= radio-web = Radio-mailles
radio message 1 TELEC **Funkspruch**
= wireless message 1
radiometer INSTR **Radiometer**
 = Strahlungsmesser; Strahlenmesser
radio model control RADIO **Funkfernsteuerung 2**
 [Modellbau]
radio monitoring RADIO **Abhören**
radiomonitoring RADIO **Funküberwachung**
 = Funkerfassung; Erfassung; Abhören
 ≈ Funkverwaltung
radio monitoring recorder RADIO LOC **Frequenzbandschreiber**
radio monitoring service RADIO **Funkkontrollmeßdienst**
radio navigation RADIO **Funknavigation**
 = Radionavigation
radio navigation aid RADIO NAV **Funknavigationshilfe**
radionavigation-satellite system SAT.COMM **Navigationsdienst über Satelliten**
radionavigation service RADIO NAV **Navigationsfunkdienst**
radio noise RADIO **Funkgeräusch**
radio noise field strength RADIO **Funkstörfeldstärke**
radio noise voltage RADIO **Funkstörspannung**
radionuclide PHYS **Radionuklid**
radio operator RADIO **Funker**
= operator
radio paging MOB.COMM **Funkruf**
 [unidireccional transmission of messages by radio] [einseitig gerichtete Nachrichtenübermittlung über Funk]
= radiopaging; paging; beeper = Personenruf; Personensuchsystem
↓ voice paging; tone-only paging; display paging; off-site paging; on-site paging ↓ Sprechfunkruf; Tonfunkruf; Anzeigefunkruf; öffentlicher Funkruf; privater Funkruf

radiopaging

radiopaging (→radio paging)	MOB.COMM	(→Funkruf)
radio paging service	TELEC	**Funkrufdienst**
radiopaque	PHYS	**strahlenundurchlässig**
radio path (→ray path)	RADIO PROP	(→Strahlenweg)
radiophare (→radio beacon)	RADIO NAV	(→Funkfeuer)
radiophone set (→radiotelephone)	RADIO	(→Funksprechgerät)
radiophony (→sound broadcasting)	BROADC	(→Hörfunk)
radio program transmission (→program transmision)	BROADC	(→Rundfunkübertragung)
radiopropagation (→radio wave propagation)	RADIO PROP	(→Funkwellenausbreitung)
radiopropagation forecast = HF propagation forecast	RADIO	**Funkprognose** = Funkwetter
radiopropagation path	RADIO PROP	**Funkweg**
radio range (→radio reach)		(→Funkreichweite)
radio range beacon = RNG	RADIO NAV	**Kursfunkbake** = Kursfunkfeuer
radio range leg	RADIO NAV	**Bakenleitstrahl**
radio ray beam (→radio beam)	RADIO	(→Funkstrahl)
radio reach = radio range	RADIO	**Funkreichweite**
radio receiver ↑ radio equipment	RADIO	**Funkempfänger** = Funkempfangsgerät ↑ Funkgerät
radio receiver (→radio broadcasting receiver)	BROADC	(→Hörfunkempfänger)
radio receiving (→radio reception)	RADIO	(→Funkempfang)
radio receiving station	RADIO	**Funkempfangsstation**
radio reception = radio receiving	RADIO	**Funkempfang**
radio recorder	CONS.EL	**Radiorecorder**
Radio Regulations [UIT]	RADIO	**Funk-Vollzugsordnung** [UIT]
radio relay (→line-of-sight radio)	RADIO	(→Richtfunk)
radio-relay antenna (→radio link antenna)	ANT	(→Richtfunkantenne)
radio relay entrance link (→radio relay spur link)	TRANS	(→Richtfunkzubringer)
radio relay equipment = LOS equipment; radiolink equipment	RADIO REL	**Richtfunkgerät**
radio relay link (→line-of-sight radiolink)	RADIO REL	(→Richtfunkstrecke)
radio relay network = microwave radio network; microwave network; radiolink network	TRANS	**Richtfunknetz**
radio-relay-network planning = microwave network planning	RADIO REL	**Richtfunk-Netzplanung**
radio relay repeater station = microwave repeater station; repeater station; microwave repeater; relay site; repeater ↑ radio relay station; intermediate station [TRANS]	RADIO REL	**Relaisstelle** = Repeaterstelle; Repeater; Zwischenstelle ↑ Richtfunkstation; Zwischenstelle [TRANS]
radio relay route = line-of-sight route; LOS route; microwave route	TRANS	**Richtfunktrasse**
radio relay spur (→radio relay spur link)	TRANS	(→Richtfunkzubringer)
radio relay spur link = radio relay spur; radio relay entrance link; microwave spur link; microwave spur; microwave entrance link ↑ spur link	TRANS	**Richtfunkzubringer** ↑ Zubringerstrecke
radio relay station = relay; booster station	RADIO	**Relaisfunkstelle**
radio relay station = LOS relay station ↓ radio-relay repeater station	RADIO REL	**Richtfunkstation** ↓ Relaisstelle
radio relay station (→retransmitter)	BROADC	(→Ballsender)
radio relay system (CCIR) = line-of-sight radio system; microwave system	TRANS	**Richtfunksystem**
radio relay tower (→microwave tower)	OUTS.PLANT	(→Richtfunkturm)
radio relay transmission = line-of-sight communication; LOS communication ≈ microwave radio relay	TRANS	**Richtfunkübertragung** ≈ Richtfunk
radio room [ship]	RADIO LOC	**Funkraum** [Schiff]
radiosensitive	PHYS	**strahlenempfindlich**
radio service (→radiocommunication service)	TELEC	(→Funkdienst)
radio set (→radio equipment)	RADIO	(→Funkgerät)
radio shadow (→dead spot)	RADIO PROP	(→Funkschatten)
radio silence = silent period	RADIO	**Funkstille**
radio sonde (→radio balloon)	RADIO	(→Radiosonde)
radio spectrum (→radiofrequencies spectrum)	RADIO	(→Funkwellenspektrum)
radio station = radiostation	RADIO	**Funkstation** = Funkstelle
radiostation (→radio station)	RADIO	(→Funkstation)
radio telecontrol = radio control	RADIO	**Funkfernsteuerung 1** = Funksteuerung
radio-telegram (→radiogram)	TELEC	(→Funktelegramm)
radiotelegraphy = radio teleprinting; radio teletyping; teleprinting over radio; TOR	TELEGR	**Funktelegrafie** = Funkfernschreiben; Radiotelegrafie
radiotelephone = radiophone set; radiotelephony terminal ≈ radiotelephone [MOB.COMM] ↓ walkie-talkie	RADIO	**Funksprechgerät** = Sprechfunkgerät; Funksprechstelle ≈ Funktelefon [MOB.COMM] ↓ Handfunksprechgerät
radiotelephone = mobile station ≈ radiotelephone [RADIO] ↓ cellular phone; car phone; mobile telephone	MOB.COMM	**Funktelefon** = Mobilstation ≈ Funksprechgerät [RADIO] ↓ Zellulartelefon; Autotelefon; Mobiltelefon
radiotelephonic measuring set	INSTR	**Funksprechmeßplatz**
radiotelephonic traffic	TELEC	**Sprechfunkverkehr**
radiotelephony = RT ≈ radiotelephony [MOB.COMM]	TELEC	**Funksprechwesen** = Radiotelephonie; Radiotelefonie; Sprechfunk ≈ Funktelefonie [MOB.COMM]
radiotelephony [bidirectional] = mobile telephone service ↑ mobile radiocommunications	MOB.COMM	**Funktelefonie** [zweiseitig gerichtet] = Funksprechen; Funkfernsprechen; Sprechfunk; Funksprechdienst ↑ Mobilfunk
radiotelephony terminal (→radiotelephone)	RADIO	(→Funksprechgerät)
radiotelephony test equipment = RT test equipment	INSTR	**Sprechfunkgeräte-Meßplatz**
radio teleprinting (→radiotelegraphy)	TELEGR	(→Funktelegrafie)
radio telescope	RADIO	**Radioteleskop**
radioteletype traffic	RADIO	**Funkfernschreib-Verkehr**
radioteletype transmitter	RADIO	**Funkfernschreibsender** = RTTY-Sender
radio teletyping (→radiotelegraphy)	TELEGR	(→Funktelegrafie)
radiotetermination (→radiolocation)	RADIO LOC	(→Ortungsfunk)
radiotheodolite	RADIO	**Radiotheodolit**
radio tinker ≈ radio amateur	RADIO	**Radiobastler** = Radioamateur ≈ Funkamateur
radio traffic = radio communications	RADIO	**Funkverkehr** = Funksprechverkehr
radiotransmission	RADIO	**Funkübertragung**
radio transmitter = radiotransmitter	RADIO	**Funksender**
radiotransmitter (→radio transmitter)	RADIO	(→Funksender)
radiovision (→television broadcasting)	BROADC	(→Fernsehrundfunk)

radio wave RADIO	**Funkwelle**	
= radioelectric wave	= Radiowelle	
radio wave propagation RADIO PROP	**Funkwellenausbreitung**	
= radiopropagation	= Wellenausbreitung; Funkausbreitung	
radio-web (→radio-mesh) RADIO NAV	(→Funkgitter-Navigationssystem)	
radio window RADIO PROP	**Radiofenster**	
= window		
radium CHEM	**Radium**	
= Ra	= Ra	
radius (pl.-ii, -iuses) MATH	**Radius** (pl.-ien)	
	= Halbmesser	
radius gage MECH	**Radiuslehre**	
radix (→base 1) MATH	(→Basis)	
radix complement DATA PROC	**Basiskomplement**	
[the complement are related to the number base]	[bezogen auf die Basis des Zahlensystems]	
= true complement	= B-Komplement	
↓ complement to two; complement to ten	↓ Zweierkomplement; Zehnerkomplement	
radix-minus-one complement DATA PROC	**Basis-minus-Eins-Komplement**	
[the complements are related to the number base reduced by one]	[die Komplemente werden auf die um Eins reduzierte Basis des Zahlensystems gebildet]	
= diminished radix complement		
↓ complement to one; complement to nine	= B-minus-Eins-Komplement	
	↓ Einerkomplement; Neunerkomplement	
radix notation MATH	**Radixschreibweise**	
[the value of a digit on position n is digit times base to the power of n]	[Wert der Ziffer in Stelle n ist Ziffer mal n-te Potenz der Basis]	
= base notation; radix representation		
↑ positional notation	↑ Stellenwertsystem	
radix point MATH	**Basiskomma**	
	= Radixkomma	
↓ decimal point; binary point; octal point	↓ Dezimalkomma; Binärkomma; Oktalkomma	
radix representation (→radix notation) MATH	(→Radixschreibweise)	
radix sorting (→digital sorting) DATA PROC	(→Digitalsortierung)	
radix sorting algorithm DATA PROC	**Radixkomma-Sortieralgorithmus**	
radom ANT	**Radom**	
["radar dome"]	= Schutzhaube; Antennenkuppel	
radom loss ANT	**Radomdämpfung**	
radon CHEM	**Radon**	
= Rn	= Rn	
rad/s (→radian per second) PHYS	(→Radiant durch Sekunde)	
ragged composition (→ragged typesetting) TYPOGR	(→Flattersatz)	
ragged left (→ragged-left typesetting) TYPOGR	(→rechtsbündiger Flattersatz)	
ragged-left composition TYPOGR (→ragged-left typesetting)	(→rechtsbündiger Flattersatz)	
ragged-left print (→ragged-left typesetting) TYPOGR	(→rechtsbündiger Flattersatz)	
ragged-left setting (→ragged-left typesetting) TYPOGR	(→rechtsbündiger Flattersatz)	
ragged-left type (→ragged-left typesetting) TYPOGR	(→rechtsbündiger Flattersatz)	
ragged-left typesetting TYPOGR	**rechtsbündiger Flattersatz**	
= ragged-left setting; ragged-left type; ragged-left print; ragged-left composition	= linksseitiger Flattersatz	
ragged print (→ragged typesetting) TYPOGR	(→Flattersatz)	
ragged-right (→ragged-right typesetting) TYPOGR	(→linksbündiger Flattersatz)	
ragged-right composition (→ragged-right typesetting) TYPOGR	(→linksbündiger Flattersatz)	
ragged-right print (→ragged-right typesetting) TYPOGR	(→linksbündiger Flattersatz)	
ragged-right setting (→ragged-right typesetting) TYPOGR	(→linksbündiger Flattersatz)	
ragged-right type (→ragged-right typesetting) TYPOGR	(→linksbündiger Flattersatz)	
ragged-right typesetting TYPOGR	**linksbündiger Flattersatz**	
= ragged-right setting; ragged-right type; ragged-right print; ragged-right composition; ragged-right typesetting	= rechtsseitiger Flattersatz	
ragged setting (→ragged typesetting) TYPOGR	(→Flattersatz)	
ragged type (→ragged typesetting) TYPOGR	(→Flattersatz)	
ragged typesetting TYPOGR	**Flattersatz**	
[print format flush only on one side]	[nur einseitig bündiges Druckbild]	
= ragged setting; ragged type; ragged composition; unjustified typesetting; unjustified setting; unjustified type; ragged print; unjustified print; unjustified composition	= Rauhsatz	
	≠ Blocksatz	
≈ ragged typesetting with hyphenation		
≠ justified typesetting		
ragged typesetting with hyphenation TYPOGR	**Rauhsatz**	
[ragged print moderated by hyphenation]	[durch Silbentrennung abgeschwächter Flattersatz]	
= unjustified typesetting with hyphenation	≈ Flattersatz	
≈ ragged typesetting	↑ Satz	
↑ typesetting		
raider alarm device (→holdup alarm device) SIGN.ENG	(→Überfallmelder)	
rail METAL	**Schiene**	
rail car (→wagon) ECON	(→Waggon)	
rail contact RAILW.SIGN	**Gleiskontakt**	
rail crossing RAILW.SIGN	**Bahnübergang**	
= level crossing	= Niveauübergang	
rail gutter antenna ANT	**Dachrinnenantenne**	
railing CIV.ENG	**Geländer**	
railroad 1 (AM) TECH	**Eisenbahn**	
= railway 1 (BRI)		
railroad 2 (AM) (→railroad track 2) TECH	(→Eisenbahnlinie)	
railroad station (AM) (→railway station) TECH	(→Bahnhof)	
railroad telephone TELEPH	**Bahnfernsprecher**	
railroad track 2 (AM) TECH [fig.]	**Eisenbahnlinie**	
= railroad 2 (AM); railway track 2 (BRI); railway 2 (BRI)		
railway 1 (BRI) TECH (→railroad 1)	(→Eisenbahn)	
railway 2 (BRI) (→railroad track 2) TECH	(→Eisenbahnlinie)	
railway radio communications RADIO	**Zugfunk**	
railway signalling SIGN.ENG	**Eisenbahnsignaltechnik**	
railway station TECH	**Bahnhof**	
= railroad station (AM)		
railway track 1 TECH	**Bahngleis**	
	= Gleis	
railway track 2 (BRI) (→railroad track 2) TECH	(→Eisenbahnlinie)	
rain (→rainfall) METEOR	(→Regenfall)	
rain attenuation RADIO PROP	**Regendämpfung**	
= rainfall loss	↑ Niederschlagsdämpfung	
↑ precipitation attenuation		
rain barrel hollowness TELEPH	**Hohlklang**	
rain cell METEOR	**Regenzelle**	
raindrop METEOR	**Regentropfen**	
rainfall METEOR	**Regenfall**	
= rain shower; rain	= Regenschauer; Regen	
↑ atmospheric precipitation	↑ atmosphärischer Niederschlag	
rainfall climatic region GEOPHYS	**Regenklimazone**	
= rainfall region		
rainfall loss (→rain attenuation) RADIO PROP	(→Regendämpfung)	
rainfall region (→rainfall climatic region) GEOPHYS	(→Regenklimazone)	
rain intensity METEOR	**Regenintensität**	
rain shower (→rainfall) METEOR	(→Regenfall)	
raised TECH	**erhaben**	
≈ convex [MATH]	= erhöht	
≠ recessed; hollow	≈ gewölbt; konvex [MATH]	
	≠ vertieft; hohl	

raised floor

raised floor	SYS.INST	**Doppelboden**	
= raised flooring; false floor; false bottom; double bottom; computer floor; computer flooring		= doppelter Boden; Kriechboden; Montageboden	
raised flooring (→raised floor)	SYS.INST	(→Doppelboden)	
raise to the power of (→power)	MATH	(→potenzieren)	
raising (→elevation)	TECH	(→Erhöhung)	
RAM 1 (→random-access memory)	DATA PROC	(→Direktzugriffsspeicher)	
RAM 2 (→main memory 1)	DATA PROC	(→Hauptspeicher 1)	
Raman effect	PHYS	**Raman-Effekt**	
ramark (→secondary surveillance radar)	RADIO LOC	(→Sekundärradar)	
RAM board	DATA PROC	**RAM-Karte**	
[expands main memory]		[erweitert Hauptspeicher]	
= RAM cartridge		↑ Speichererweiterungskarte	
↑ memory expansion board			
RAM cartridge (→RAM board)	DATA PROC	(→RAM-Karte)	
RAM disk	DATA PROC	**virtuelles Laufwerk**	
[section of main memory made behaving like a disk drive by special software]		[Bereich der Hauptspeichers der sich durch Software wie ein Laufwerkspeicher verhält]	
= RAMDISK; virtual drive		= RAM-Disk; RAMDISK	
↓ virtual floppy disk		↓ virtuelle Diskette	
RAMDISK (→RAM disk)	DATA PROC	(→virtuelles Laufwerk)	
Ramey amplifier	CIRC.ENG	**Ramey-Verstärker**	
↑ voltage driving transducer amplifier		= magnetischer Rücklaufverstärker	
		↑ spannungsteuernder Magnetverstärker	
RAM floppy disk	TERM&PER	**RAM-Diskette**	
		= RAM-Floppy; Pseudo-Floppy	
ramp (→edge)	ELECTRON	(→Flanke)	
ramp (→sawtooth waveform)	ELECTRON	(→Sägezahnkurve)	
ramp characteristics	ELECTRON	**Sägezahn-Kenndaten**	
		= Rampen-Kenndaten	
ramp function	ELECTRON	**Rampenfunktion**	
ramp-off (n.)	ELECTRON	**Dachschräge**	
[inclination of pulse top]		[Impuls]	
= pulse droop; droop			
ramp response	ELECTRON	**Rampenantwort**	
ramp signal (→sawtooth signal)	TELEC	(→Sägezahnsignal)	
ramp sweep	INSTR	**Rampenwobbelung**	
ramp time	ELECTRON	**Rampenzeit**	
↓ rise time; decay time		↓ Anstiegzeit; Abfallzeit	
ramp-up	ELECTRON	**stufenweiser Anstieg**	
ramp voltage (→sawtooth voltage)	ELECTRON	(→Sägezahnspannung)	
RAM test	DATA PROC	**RAM-Test**	
random	TECH	**zufällig**	
= aleatory; casual; incidental; fortuitous; accidental; pure-chance		= wahllos; beiläufig	
≈ stochastic [MATH]; statistic [MATH]; irregular		≈ stochastisch [MATH]; statistisch [MATH]; unregelmäßig	
random (→statistical)	MATH	(→statistisch)	
random access	DATA COMM	**stochastischer Zugang**	
random access (→direct access)	DATA PROC	(→Direktzugriff)	
random-access memory	DATA PROC	**Direktzugriffsspeicher**	
[a direct access read-write memory into which information can be entered or called up whenever necessary; intensively used for main memories, therefore the term RAM is used as synonym for main memory in the PC world]		[Schreib-Lese-Speicher mit direktem Zugriff, kann beliebig oft beschrieben und gelesen werden; u.a. für die Realisierung von Hauptspeichern verwendet, in der PC-Welt wird deshalb RAM auch als Synonym für Hauptspeicher verwendet]	
= RAM 1		= RAM-Speicher; Randomspeicher; RAM 1	
≠ sequential access memory		≠ sequentieller Speicher	
↑ read-write memory		↑ Schreib-Lese-Speicher	
↓ SRAM; DRAM		↓ SRAM; DRAM	
random circuit noise (→line noise)	TELEC	(→Leitungsgeräusch)	
random current (→boundary current)	MICROEL	(→Randströmung)	
random distribution	MATH	**statistische Verteilung**	
random error	TECH	**Zufallsfehler**	
= stochastic error		= stochastischer Fehler	
random error (→statistical error)	MATH	(→statistischer Fehler)	
random event (→random result)	MATH	(→Zufallsergebnis)	
random event generator (→random sequence generator)	DATA PROC	(→Zufallsgenerator)	
random failure	QUAL	**Zufallsausfall**	
= freak failure		≠ systematischer Ausfall	
≠ systematic failure			
random graphic	DATA PROC	**Zufallsgraphik**	
randomly distributed	MATH	**statistisch verteilt**	
= statistically distributed			
random number	MATH	**Zufallszahl**	
= stochastic number		= stochastische Zahl; Zufallsgröße; stochastische Größe	
random number generator (→random sequence generator)	DATA PROC	(→Zufallsgenerator)	
random numbering	DATA PROC	**systemlose Numerierung**	
random process	MATH	**Zufallsprozeß**	
= stochastic process		= stochastischer Prozess	
random processing	DATA PROC	(→wahlfreie Verarbeitung)	
(→direct-acces processing)			
random result	MATH	**Zufallsergebnis**	
[statistics]		[Statistik]	
= stochastic result; random event; stochastic event		= stochastisches Ergebnis	
random sampling	INSTR	**inkohärente Abtastung**	
= incoherent sampling; random scanning; incoherent scanning		= zufällige Abtastung; inkohärentes Sampling; zufälliges Sampling; Random-Sampling	
random sampling	QUAL	**Stichprobenkontrolle**	
= sampling inspection		= Stichprobenprüfung	
random-scan display (→vector display)	TERM&PER	(→Vektorbildschirm)	
random scanning (→random sampling)	INSTR	(→inkohärente Abtastung)	
random sequence	MATH	**Zufallsfolge**	
[statistics]		[Statistik]	
random sequence generator	DATA PROC	**Zufallsgenerator**	
= random event generator; REG; random number generator; RNG; stochastic event generator		= Zufallszahlengenerator; stochastischer Generator	
random signal	INF	**Zufallssignal**	
= stochastic signal		= nichtdeterministisches Signal; stochastisches Signal	
random storage	DATA PROC	**wahlfreie Speicherung**	
random traffic (→pure-chance traffic)	SWITCH	(→regelloser Verkehr)	
random variable	MATH	**Zufallsvariable**	
= stochastic variable; variate		= stochastische Variable	
random-walk method (→Monte-Carlo method)	MATH	(→Monte-Carlo-Methode)	
random wire	ANT	**Zufallsdraht**	
range 1 (n.)	MATH	**Bereich**	
range 1	TECH	**Bereich**	
= region			
range	COLLOQ	**Reichweite**	
[figurative]		[figurativ]	
≈ reach		≈ Umfang	
≈ extent			
range (→frequency band)	RADIO	(→Frequenzband)	
range (→radio beacon)	RADIO NAV	(→Funkfeuer)	
range (→product line)	ECON	(→Produktspektrum)	
range 2 (→reach)	TECH	(→Reichweite)	
range 2 (→variation range)	MATH	(→Spannweite)	
range calibrator	INSTR	**Bereichskalibrator**	
range check	DATA PROC	**Spannweitenüberprüfung**	
range equation (→radar equation)	RADIO LOC	(→Radargleichung)	
range finder	RADIO NAV	**Entfernungsmesser**	
range of application (→application field)	TECH	(→Anwendungsbereich)	

range of values (→set of values)		MATH	(→Wertevorrat)	rate (→estimate)	ECON	(→veranschlagen)
range recognizer		INSTR	Bereichserkenner	rate action (→derivative action)	CONTROL	(→Vorhalt)
range setting		INSTR	Bereichswahl	rate counter (→tax meter)	SWITCH	(→Gebührenzähler)
= ranging				rated (→scheduled)	TECH	(→nominell)
range switch		INSTR	Bereichsumschalter	rated cross-section (→nominal cross-section)	TECH	(→Nennquerschnitt)
ranging (→range setting)		INSTR	(→Bereichswahl)	rated current (→nominal current)	EL.TECH	(→Nennstrom)
rank (v.t.)		DATA PROC	ordnen [in der Rangfolge]	rated lightning impulse withstand voltage	POWER ENG	Nenn-Steh-Blitzstoßspannung
rank (n.)		SWITCH	Rang	rated load (→nominal load capacity)	TECH	(→Nennbelastbarkeit)
rank correlation		MATH	Rangkorrelation	rated performance (→duty rate)	TECH	(→nominelle Leistungsfähigkeit)
rapid (→fast)		TECH	(→schnell)			
rapid decollation set		TERM&PER	Schnelltrennsatz			
rapid disconnection		EL.TECH	Schnellabschaltung	rated power (→nominal power)	TECH	(→Nennleistung)
= fast break			= schnelle Abschaltung	rated power (→nominal power)	CONS.EL	(→Nennleistung)
rapidity (→fastness)		COLLOQ	(→Schnelligkeit)			
rapid set square		ENG.DRAW	Schnellzeichendreieck	rated power-frequency withstand voltage	POWER ENG	Nenn-Steh-Wechselspannung
rare earth		CHEM	seltene Erde			
rarefaction		TECH	Verdünnung 3 [Gas]	rated reliability	QUAL	Nennzuverlässigkeit
				rated speed (→nominal speed)	MECH	(→Nenndrehzahl)
rarefy (v.t.)		TECH	verdünnen 3 [Gas]	rated switching impulse withstand voltage	POWER ENG	Nenn-Steh-Schaltstoßspannung
= rarify						
rare gase (→noble gas)		CHEM	(→Edelgas)	rated torque	MECH	Nenndrehmoment
rarify (→rarefy)		TECH	(→verdünnen 3)	rated value (→nominal value)	TECH	(→Nennwert)
RAS facility		DATA PROC	RAS-Einrichtung	rated voltage (→nominal voltage)	EL.TECH	(→Nennspannung)
[improving reliability, availability and serviceability]						
raster		TERM&PER	Raster	rate effect	MICROEL	Geschwindigkeitseffekt
[graphics]			[Graphik]	rate grown	MICROEL	gezogen
↓ dot matrix			↓ Punktraster	rate-grown transistor	MICROEL	Wachstumstransistor
raster (→pattern)		RADIO	(→Raster)	rate growth process	MICROEL	Stufenziehverfahren
raster code		TERM&PER	Mosaik-Code	rate meter (→tax meter)	SWITCH	(→Gebührenzähler)
[dot matrix to reproduce a character, e.g. 5x7]			[Punktraster zur Darstellung eines Zeichens, z.B. 5x7]	rate of change (→edge steepness)	NETW.TH	(→Flankensteilheit)
				rate of voltage rise	MICROEL	Spannungssteilheit
			= Raster-Code; Matrix-Code	rate printing (→tax printing)	SWITCH	(→Gebührenausdruck)
raster coordinate		DATA PROC	Raster-Koordinate			
raster discrimination		ELECTRON	(→Rasterfeinheit)	rate setting	MANUF	Zeitvorgabe
(→scanning density)				rate time (→derivative time)	CONTROL	(→Vorhaltezeit)
raster display (→raster screen)		TERM&PER	(→Rasterbildschirm)	rating 3 (→load capacity)	TECH	(→Belastbarkeit)
raster fill		TERM&PER	Rasterauffüllung	rating 2 (→dimensioning)	TECH	(→Bemessung)
raster graphics		DATA PROC	Rastergraphik	rating 1 (→nominal value)	TECH	(→Nennwert)
[computer graphics]			[Computergraphik]	rating plate	TECH	Leistungsschild
= bit-mapped graphic			= Rastergrafik; Pixelgrafik; Bit-map-Graphik	ratio	MATH	Bruch
≠ vector graphics			≠ Vektorgraphik	[mathematical expression formed by dividend and divisor]		[aus Dividend und Divisor bestehender mathematischer Ausdruck]
raster graphics (→raster scan mode)		TERM&PER	(→Rasterverfahren)	= fraction; proportion 1		= Quotient 1; Verhältnis; Proportion 1
raster graphics screen (→raster screen)		TERM&PER	(→Rasterbildschirm)	≈ quotient; fractional number; division; ratio; proportion 2		≈ Quotient 2; Bruchzahl; Division; Verhältniszahl; Verhältnisgleichung; Relation
raster image processor		DATA PROC	Raster-image-Prozessor			
[composition computer]			[Satzrechner] = RIP			
raster plotter		TERM&PER	Rasterplotter	ratio	ECON	Schlüssel
raster scan (→raster scanning)		TERM&PER	(→Rasterabtastung)	ratio 2	MATH	Verhältniszahl
				ratio (→gear ratio)	MECH	(→Übersetzung)
raster scan mode		TERM&PER	Rasterverfahren	ratio accuracy	INSTR	Verhältnisgenauigkeit
[generation of display line by line]			[zeilenweiser Bildaufbau]	ratio arm (→bridge arm)	NETW.TH	(→Brückenzweig)
= raster graphics			= Punktrasterung	ratio controller	CONTROL	Verhältnisregler
raster scanned screen		TERM&PER	(→Rasterbildschirm)	ratio detector	CIRC.ENG	Verhältnisdiskriminator
(→raster screen)				[FM demodulator]		[FM-Demodulator]
raster scanner		TERM&PER	Raster-Abtaster			= Verhältnisdemodulator; Verhältnisgleichrichter; Ratiodetektor
			= Raster-Scanner			
raster scanning		TERM&PER	Rasterabtastung	ratio measurement	INSTR	Verhältnismessung
= raster scan						= Relationsmessung
raster scan screen (→raster screen)		TERM&PER	(→Rasterbildschirm)	ratio measuring system (→ratio meter)	INSTR	(→Quotientenmesser)
raster screen		TERM&PER	Rasterbildschirm	ratio meter	INSTR	Quotientenmesser
[maps on the basis of a point matrix]			[bildet aus einem Punktraster ab]	= ratio measuring system		= Verhältnismesser; Quotientenmeßwerk
= raster scanned screen; raster scan screen; raster graphics screen; raster display; bit map screen; matrix screen			= Matrixbildschirm ≠ Vektorbildschirm ↑ Sichtgerät; Graphikbildschirm	rationale = introductory text ≈ introduction	LING	Begründung = Themenstellung; einleitender Text ≈ Einleitung
≠ vector display ↑ display terminal				rationalization = streamlining	ECON	Rationalisieren
rate (n.)		SCIE	Rate	rationalize	ECON	rationalisieren
rate (v.t.) (→dimension)		TECH	(→bemessen)	= streamline (v.t.)		
rate (proportional to measured service) (→charge 2)		ECON	(→Gebühr)			
rate (→unit rate)		ECON	(→Satz)			

rational number	MATH	Rationalzahl	RC integrator	CIRC.ENG	RC-Integrierglied
≠ irrational number		≠ Irrationalzahl	RC low-pass	NETW.TH	RC-Tiefpaß
ratio of quantities	PHYS	Verhältnisgröße	RC measuring generator	INSTR	RC-Meßgenerator
ratio resistor	INSTR	Verhältniswiderstand	RC network	NETW.TH	RC-Zweipol
rat's nest	DATA PROC	Nagetierbau	RC oscillator	CIRC.ENG	RC-Oszillator

rat's nest [display of computer-determined conductor tracks] — [Darstellung der rechnerbestimmten Leiterbahnen]

RCS (→radar cross section) RADIO LOC (→Radarquerschnitt)

rat-tail file (→round file)	MECH	(→Rundfeile)	RC section (→RC element)	NETW.TH	(→RC-Glied)
rattle (v.i.)	TECH	rasseln	RCTL	MICROEL	RCTL
rattle (n.)	ACOUS	Schnarren	= resistor-capacitor-transistor logic		
= purr (n.)			R&D (→research and development)	ECON	(→Forschung und Entwicklung)
raw	TECH	roh	RDS (→running digital sum)	DATA PROC	(→Prüfsumme)

raw = unwrought; unmachined; unfinished — = unbearbeitet; unfertig; halbfertig ≈ rauh

			re [commercial letter; "regarding"] = subject	OFFICE	Betreff [Geschäftsbrief] = Betr.
raw data (→source data)	DATA PROC	(→Ursprungsdaten)	Re (→rhenium)	CHEM	(→Rhenium)
raw material	TECH	Rohstoff	reach (n.)	TECH	Reichweite
= primary commodity; commodity		= Rohmaterial ≈ Grundstoff	= range 2		
raw materials processing	MANUF	Rohstoffverarbeitung	reach (→achieve)	TECH	(→erzielen)
raw materials store	MANUF	Rohstofflager	reach (→range)	COLLOQ	(→Reichweite)
raw product	TECH	Rohprodukt	reach-through	MICROEL	(→Durchgriff)
ray	PHYS	Strahl	(→punch-through)		

ray [perpendicular to a wave front, geometric concept, no physical magnitude] — [Normale zur Wellenfläche; geometrische Definition, keine physikalische Größe]

reach-through voltage (→punch-through voltage)	MICROEL	(→Durchgreifspannung)		
reactance	NETW.TH	Blindwiderstand		

reactance [imaginary part of impedance] = reactive impedance; reactive resistance; X — [Imaginärteil des Scheinwiderstandes] = Reaktanz; imaginärer Widerstand; X

= radiation beam; beam
≈ ray bundle; ray bunch
↓ acoustical ray [ACOUS]; light beam; X ray

= Strahlenbündel; Strahlenbüschel
↓ Schallstrahl [ACOUS], Wärmestrahl; Lichtstrahl, Röntgenstrahl

			reactance amplifier (→parametric amplifier)	CIRC.ENG	(→parametrischer Verstärker)
			reactance chart	HF	Hochfrequenztapete = HF-Tapete
ray bunch	PHYS	Strahlenbüschel	reactance current (→idle current)	EL.TECH	(→Blindstrom)

ray bunch [a bunch of waves diverging from a source, e.g. of a spherical wave] — [Bündel von einem Erregungszentrum divergierender Strahlen, z.B. einer Kugelwelle]

= bunch; ray pencil; pencil
≈ ray bundle; ray

≈ Strahlenbündel; Strahl

			reactance modulator	MODUL	Reaktanzmodulator
			reactance output (→reactive power)	EL.TECH	(→Blindleistung)
ray bundle	PHYS	Strahlenbündel	reactance theorem (→Foster reactance theorem)	NETW.TH	(→Reaktanztheorem)

ray bundle [bundle of parallel rays of a plane wave] — [Bündel paralleler Strahlen einer ebenen Welle]

= bundle
≈ ray bunch; beam
↑ ray

≈ Strahlenbüschel; Strahl

			reactance tube = reactance valve	ELECTRON	Reaktanzröhre = Blindröhre
ray casting (→ray tracing)	DATA PROC	(→Strahlverfolgung)	reactance-tube circuit	CIRC.ENG	Reaktanzröhren-Schaltung
Rayleigh scattering	PHYS	Rayleigh-Streuung	reactance valve (→reactance tube)	ELECTRON	(→Reaktanzröhre)
ray path = radio path	RADIO PROP	Strahlenweg	reaction	CHEM	Reaktion
			reaction	PHYS	Rückwirkung
ray pencil (→ray bunch)	PHYS	(→Strahlenbüschel)	= retroaction		= Reaktion
ray surface	PHYS	Strahlenfläche	↓ recoil		↓ Rückstoß
ray tracing	DATA PROC	Strahlverfolgung	reaction amplifier (→feedback amplifier)	CIRC.ENG	(→Rückkopplungsverstärker)

ray tracing [computer graphics; method for solid modelling] — [Computergraphik; Verfahren zur Darstellung von Festkörperbildern] ≈ Bildberechnung

= ray casting

			reaction coil = tickler; feedback coil	CIRC.ENG	Rückkopplungsspule = Stromstärkeregler
			reaction coupling (→feedback)	CIRC.ENG	(→Rückkopplung)
ray trajectory	PHYS	Strahlverlauf			
ray velocity	PHYS	Strahlengeschwindigkeit	reaction force = restoring force	MECH	Rückstellkraft
Rb (→rubidium)	CHEM	(→Rubidium)	reaction-free (→nonreactive)	CIRC.ENG	(→rückwirkungsfrei)
RB code = relocatable binary code	CODING	RB-Code	reaction-free amplifier = nonreactive amplifier	CIRC.ENG	rückwirkungsfreier Verstärker
RBOC (→Regional Bell Operating Company)	TELEC	(→regionale Bell-Betreibergesellschaft)	reaction suppressor (→anti-feedback device)	ELECTRON	(→Rückkopplungssperre)
RBW (→resolution bandwidth)	INSTR	(→Auflösungsbandbreite)	reaction time (→response time)	TECH	(→Reaktionszeit)
RC amplifier (→RC-coupled amplifier)	CIRC.ENG	(→RC-Verstärker)	reactive component = quadrature component	NETW.TH	Blindanteil = Blindkomponente
RC circuit	NETW.TH	RC-Schaltung	reactive current (→idle current)	EL.TECH	(→Blindstrom)
RC-coupled amplifier = RC amplifier; resistance-coupled amplifier	CIRC.ENG	RC-Verstärker = Widerstandsverstärker	reactive factor	NETW.TH	Blindfaktor
			reactive field ≈ near-field region	ANT	Reaktanzfeld ≈ Nahfeldbereich
RC coupling	NETW.TH	RC-Kopplung = Widerstand-Kondensator-Kopplung	reactive four-terminal network	NETW.TH	Reaktanzvierpol
			reactive impedance (→reactance)	NETW.TH	(→Blindwiderstand)
RC device (→RC element)	NETW.TH	(→RC-Glied)	reactive load (→dummy load)	NETW.TH	(→Blindlast)
RC differentiating element	NETW.TH	RC-Differenzglied	reactive power = reactance output	EL.TECH	Blindleistung [U(eff) x I(eff) x sin] = Reaktanzleistung
RC element = RC section; RC device	NETW.TH	RC-Glied			
RC filter	NETW.TH	RC-Filter			
RC generator	CIRC.ENG	RC-Generator			
RC high-pass	NETW.TH	RC-Hochpaß			

reactive power meter	INSTR	**Blindleistungsmesser**	
= reactive volt-ampere meter; varmeter		↑ Leistungsmesser	
reactive resistance	NETW.TH	(→Blindwiderstand)	
(→reactance)			
reactive torque	PHYS	**Gegendrehmoment**	
reactive two-terminal function	NETW.TH	**Reaktanz-Zweipolfunktion**	
		= Reaktanzfunktion	
reactive two-terminal network	NETW.TH	**Reaktanzzweipol**	
reactive voltage	NETW.TH	**Blindspannung**	
reactive volt-ampere meter	INSTR	(→Blindleistungsmesser)	
(→reactive power meter)			
reactive watt	PHYS	**Blindwatt**	
[1 W reactive power]		[1 W Blindleistung]	
= bW		= bW	
reactor (→choke)	COMPON	(→Drossel)	
reactor (→nuclear reactor)	POWER SYS	(→Nuklearreaktor)	
read (v.t.)	TERM&PER	**lesen**	
read (v.t.)	DATA PROC	**lesen**	
≠ write		≠ schreiben	
↓ read-in (v.t.); read-out (v.t.)		↓ einlesen; auslesen	
read (n.)	DATA PROC	**Lesen**	
= reading (n.)		= Einlesen; Auslesen	
↓ read-in; read-out			
read (→read-off)	INSTR	(→ablesen)	
readability	TERM&PER	**Lesbarkeit**	
readability	INSTR	**Ablesbarkeit**	
		= Lesbarkeit	
readable	INSTR	**ablesbar**	
readable	TERM&PER	**lesbar**	
read after write	DATA PROC	**prüflesen**	
		= kontrollesen	
read-after-write (→read control)	OFFICE	(→Kontrollesen)	
read clock	CIRC.ENG	**Auslesetakt**	
read control	OFFICE	**Kontrollesen**	
= read-after-write			
readdress	DATA PROC	**neuadressieren**	
reader (→scanner)	ELECTRON	(→Abtaster)	
reader (→reading device)	TERM&PER	(→Lesegerät)	
reader pistol (→scanning pistol)	TERM&PER	(→Lesepistole)	
read error	TERM&PER	**Lesefehler**	
= transient read			
reader sorter	TERM&PER	**Sortierleser**	
read head (→reading head)	TERM&PER	(→Lesekopf)	
read-in (n.)	DATA PROC	**Einlesen**	
= poke (v.t.)		≈ Eingabe	
= input (n.)		≠ Auslesen	
≠ read-out (n.)		↑ Lesen	
↑ read (n.)			
read-in (v.t.)	DATA PROC	**einlesen**	
[from external to internal memory]		[von externen auf internen Speicher]	
≈ output (v.t)		≈ eingeben	
≠ read-out (v.t.)		≠ auslesen	
↑ read (v.t.)		↑ lesen	
read-in address	CIRC.ENG	**Einleseadresse**	
reading	INSTR	**Ablesung**	
= readout		≈ Anzeige	
≈ indication			
reading	TERM&PER	**Lesen**	
reading (n.) (→read)	DATA PROC	(→Lesen)	
reading accuracy	INSTR	**Ablesegenauigkeit**	
= readout accuracy		= Ablesegenauigkeit	
≈ indication accuracy		≈ Anzeigegenauigkeit	
reading beam	ELECTRON	**Lesestrahl**	
reading device	TERM&PER	**Lesegerät**	
= reader; scanner		= Leser; Scanner	
reading head	TERM&PER	**Lesekopf**	
= read head; magnetic reading head; magnetic read head		= Magnetlesekopf	
		≠ Schreibkopf 2; Löschkopf	
≠ write head; erase head		↑ Magnetkopf	
↑ magnetic head			
reading rate	TERM&PER	**Lesegeschwindigkeit**	
= reading speed			
reading rate	INSTR	**Meßrate**	
reading speed (→reading rate)	TERM&PER	(→Lesegeschwindigkeit)	
reading track	TERM&PER	**Lesespur**	
≈ information track		≈ Informationsspur	
reading wand (→code pen)	TERM&PER	(→Lesestift)	
read ink (→nonreflective ink)	TERM&PER	(→nichtreflektierende Farbe)	
readin program (→input routine)	DATA PROC	(→Einleseroutine)	
read-in program (→input routine)	DATA PROC	(→Einleseroutine)	
read-in routine (→input routine)	DATA PROC	(→Einleseroutine)	
readjust	TECH	**nachregeln**	
		= nachjustieren; nacheinstellen; neuabgleichen	
readjustment	ELECTRON	**Nachregelung**	
		= Neuabgleich; Neueinstellung	
read line	DATA COMM	**Leseleitung**	
		[Bus]	
read line (→read wire)	TERM&PER	(→Lesedraht)	
read memory	DATA PROC	**Lesespeicher**	
read-off	INSTR	**ablesen**	
= read			
read-only file	DATA PROC	**Nur-Lese-Daten**	
= RO file			
read-only memory	DATA PROC	**Festwertspeicher**	
[contents only alterable with special devices]		[Inhalt nur durch Spezialeinrichtungen änderbar]	
= ROM; ROM memory; ROM storage; RO memory; RO storage; RO store; read-only storage; not-erasable memory 2; permanent memory 2; permanent storage 2; permanent store 2		= Festspeicher; Nur-Lese-Speicher; ROM-Speicher; ROM; ausschließlich lesbarer Speicher; Totspeicher; Permanentspeicher 2; festverdrahteter Speicher; nichtlöschbarer Speicher 2; Dauerspeicher 2	
≈ non-volatile memory		≈ nichtflüchtiger Speicher	
↑ memory chip		↑ Speicherbaustein	
↓ PROM; EEPROM; EPROM; FROM		↓ PROM; EEPROM; EPROM; FROM	
read-only-memory control	CONTROL	**Festwertsteuerung**	
read-only optical store (→CD-ROM)	TERM&PER	(→CD-ROM)	
read-only storage (→read-only memory)	DATA PROC	(→Festwertspeicher)	
read operation	DATA PROC	**Lesevorgang**	
read-out (v.t.)	DATA PROC	**auslesen**	
[from internal to external memory]		[von internem auf externen Speicher]	
= retrieve; destage		= ausspeichern	
≈ fetch out (v.t.)		≈ ausgeben	
≠ write-in (v.t.)		≠ einschreiben	
		↑ lesen	
read-out (n.)	DATA PROC	**Auslesen**	
= readout (n.)		= Ausspeichern	
≈ output (n.)		≈ Ausgabe	
≠ read-in (n.)		≠ Einlesen	
↑ read		↑ Lesen	
readout (→reading)	INSTR	(→Ablesung)	
read-out (n.) (→display)	INSTR	(→Anzeige)	
readout (n.) (→read-out)	DATA PROC	(→Auslesen)	
readout accuracy (→reading accuracy)	INSTR	(→Ablesegenauigkeit)	
read-out speed	DATA PROC	**Auslesegeschwindigkeit**	
read register	DATA PROC	**Leseregister**	
read signal	ELECTRON	**Lesesignal**	
= sense signal			
read winding	TERM&PER	**Lesewicklung**	
= sense winding			
read wire	TERM&PER	**Lesedraht**	
[magnetic core memory]		[Magnetkernspeicher]	
= sense wire; read line; sense line		≈ Schreib-Lese-Draht	
= read-write wire		≠ Schreibdraht	
≠ write wire			
read-write ...	TERM&PER	**Schreib-Lese-**	
= R/W; write-read ...		= Lese-Schreib-	
read-write head	TERM&PER	**Schreib-Lese-Kopf**	
= combined head		= Kombikopf	
↑ magnetic head		↑ Magnetkopf	

read-write memory DATA PROC **Schreib-Lese-Speicher**
= RWM; R/W memory = RWM-Speicher
↑ memory chip ≈ Direktzugriffsspeicher
↓ random access memory ↑ Speicherbaustein
↓ Direktzugriffspeicher

read-write speed (→read-write TERM&PER (→Schreib-Lese-Geschwin-
velocity) digkeit)

read-write velocity TERM&PER **Schreib-Lese-Geschwindig-**
= read-write speed; R/W velocity; R/W **keit**
speed

read-write window (→head TERM&PER (→Schreib-Lese-Öffnung)
window)

read-write wire TERM&PER **Schreib-Lese-Draht**
[magnetic core memory] [Magnetkernspeicher]

ready DATA PROC **bereit**
= betriebsklar

ready condition DATA COMM **Bereitzustand**
= ready state

ready criterion SWITCH **Bereitschaftskriterium**

ready for execution DATA PROC **ablaufbereit**

ready for receiving DATA COMM (→Empfangsbereitschaft)
(→ready-to-receive state)

ready for sending DATA COMM (→sendebereit)
(→ready-to-transmit)

ready state (→ready DATA COMM (→Bereitzustand)
condition)

ready status DATA COMM **Betriebsbereitschaft**
= operability

ready-to-operate DATA COMM **betriebsbereit**

ready to receive DATA COMM **empfangsbereit**
= receive ready; RR

ready-to-receive state DATA COMM **Empfangsbereitschaft**
= ready for receiving

ready-to-send state DATA COMM (→Sendebereitschaft)
(→ready-to-transmit state)

ready-to-transmit (adj.) DATA COMM **sendebereit**
= ready for sending

ready-to-transmit state DATA COMM **Sendebereitschaft**
= ready-to-send state

real MATH **reell**
≠ imaginary ≠ imaginär

real COLLOQ **tatsächlich**
= factual; effective; actual; realized = effektiv; faktisch

real DATA PROC **real**
≠ virtual ≠ virtuell

real address (→absolute DATA PROC (→absolute Adresse)
address)

real addressing (→absolute DATA PROC (→absolute Adressierung)
addressing)

real circuit (→physical LINE TH (→Stammleitung)
circuit)

real component (→real part) MATH (→Realteil)

realizability (→viability) TECH (→Durchführbarkeit)

realizable (→feasable) TECH (→durchführbar)

realization TECH **Realisierung**
= materialization; implementation = Konkretisierung; Verwirklichung

realization expenditure TECH **Realisierungsaufwand**

realize ECON **verflüssigen**

realize TECH **realisieren**
= materialize; implement = konkretisieren; verwirklichen; implementieren

realized gain ANT **tatsächlicher Gewinn**

real memory DATA PROC **realer Speicher**
= real storage; real store = Realspeicher
≠ virtual memory ≠ virtueller Speicher

real number MATH **reelle Zahl**
≠ imaginary number ≠ imaginäre Zahl

real part MATH **Realteil**
= real component = Realkomponente

real pole NETW.TH **reeller Pol**

real power (→active power) EL.TECH (→Wirkleistung)

real power meter (→active power INSTR (→Wirkleistungsmesser)
meter)

real storage (→real DATA PROC (→realer Speicher)
memory)

real store (→real DATA PROC (→realer Speicher)
memory)

real time INF.TECH **Echtzeit**
= Realzeit; Istzeit; Real-time

real time analysis INF.TECH **Echtzeitanalyse**

real-time animation DATA PROC **Echtzeit-Animation**
[graphics] [Graphik]

real time clock DATA PROC **Realzeituhr**

real-time clock TECH **Echtzeituhr**

real-time data acquisition DATA PROC **Echtzeit-Datenerfassung**
= source-data acquisition; source-data = quellorientierte Datenerfassung; Realzeiterfassung;
entry; source-data automation Quelldateneingabe

real-time image generation DATA PROC **Echtzeit-Bildgenerierung**

real-time language (→real- DATA PROC (→Realzeit-Programmiersprache)
time program language)

real time operating system DATA PROC **Echtzeit-Betriebssystem**
= real-time OS = Realzeit-Betriebssystem

real time operation DATA PROC **Echtzeitbetrieb**
[fast enough to keep pace with under- [hinreichend schnell, um
going process] mit laufendem Vorgang
= real time processing Schritt zu halten]
≈ in-line processing = Echtzeitverarbeitung; Echtzeitverfahren; Realzeitbetrieb; Realzeitverfahren; Realzeitverarbeitung; schritthaltende Verarbeitung; Real-time-Betrieb; Real-time-Verarbeitung
≈ Geradewohl-Verarbeitung

real-time OS (→real time DATA PROC (→Echtzeit-Betriebssystem)
operating system)

real time processing (→real DATA PROC (→Echtzeitbetrieb)
time operation)

real time program DATA PROC **Echtzeitprogramm**

real-time program language DATA PROC **Realzeit-Programmiersprache**
= real-time language

real-time simulation TECH **Echtzeitsimulierung**
= Echtzeitsimulation; Realzeitsimulierung; Realzeitsimulation

real-time simulator TECH **Echtzeitsimulator**
= Realzeitsimulator

real traffic SWITCH **wahrer Verkehr**

real value ECON **Sachwert**

real variable DATA PROC **Realvariable**
= reale Variable

real zero point NETW.TH **reelle Nullstelle**

ream (v.t.) MECH **reiben**
[to smooth and widen an opening] [ein Loch glätten oder erweitern]

reamed bolt (→fitted bolt) MECH (→Paßbolzen)

reamer MECH **Reibahle**
[rotating tool to finish holes] [Werkzeug zum Glätten von Löchern, durch Drehbewegungen]

reapeter TRANS **Repeater**
↓ intermediate amplifier; intermediate ↓ Zwischenverstärker; Zwischenregenerator
regenerator

rear (adj.) TECH **rückseitig**
= backward; revertive = rückwärtig
≈ back (adj.) ≠ frontseitig
≠ front (adj.)

rear 2 (→rear side) TECH (→Rückseite)

rear 1 (→rear part) TECH (→Hinterteil)

rear cover TECH **rückseitige Abdeckung**
= rear panel; back plate = Rückwand
≠ front cover ≠ Frontabdeckung
↑ cover ↑ Abdeckung

rear cover (→back plate) EQUIP.ENG (→Rückwand)

rear input terminal (→rear EQUIP.ENG (→rückseitiger Anschluß)
panel connection)

rear mounting EQUIP.ENG **Rückseitenmontage**
≠ front mounting ≠ Frontmontage

rear panel (→rear cover) TECH (→rückseitige Abdeckung)

rearpanel (→back plate) EQUIP.ENG (→Rückwand)

rear panel (→back EQUIP.ENG (→Rückwand)
plate)

rearpanel board EQUIP.ENG (→Rückwandleiterplatte)
(→backplane)

rearpanel cable connector COMPON (→Rückwandkabelstecker)
(→backplane cable connector)

rear panel connection EQUIP.ENG **rückseitiger Anschluß**
= rear input terminal = Rückwandanschluß

rearpanel wiring (→backplane	EQUIP.ENG	(→Rückwandverdrahtung)	
wiring)			
rear part		TECH	Hinterteil
= back 1 (n.); rear 1			≈ Rückseite
≈ rear side			≠ Vorderteil
≠ front part			
rearrange		DATA PROC	reorganisieren
[a memory content]			[eines Speicherinhalts]
= reorganize			
rearrange (→remodel)		TECH	(→umbauen)
rearrangement		TECH	Umordnung
≈ remodelling			≈ Umbau
rearrangement (→remodelling)		TECH	(→Umbau)
rear scan (→rear		ELECTRON	(→Rückwärtsabtastung)
scanning)			
rear scanning		ELECTRON	Rückwärtsabtastung
= rear scan; reverse scanning; reverse			
scan; backward scanning; backward			
scan			
rear side		TECH	Rückseite
= back 2; rear 2			= Hinterseite
≈ rear part			≈ Hinterteil
			≠ Vorderseite
rear side (→verso)		TYPOGR	(→Rückseite)
rear view		ENG.DRAW	Rückansicht
reasonable		SCIE	sinnvoll
reasonableness		DATA PROC	(→Plausibilität)
(→plausibility)			
reasonableness check		DATA PROC	(→Plausibilitätsprüfung)
(→plausibility check)			
reasoning (→conclusion)		SCIE	(→Schlußfolgerung)
reassemble (→merge)		DATA PROC	(→mischen)
reassembling		DATA COMM	Vereinigung
rebate (→discount)		ECON	(→Preisnachlaß)
reboot (v.t.)		DATA PROC	(→neustarten)
(→restart)			
rebooting (→restart)		DATA PROC	(→Neustart)
rebound		TECH	abprallen
rebound (v.i.) (→bounce)		MECH	(→prellen)
rebound (n.) (→bounce)		MECH	(→Prellen)
rebroadcast transmitter		BROADC	(→Fernsehumsetzer)
(→television transposer)			
recalculate		MATH	nachrechnen
= check a calculation			
recalculation		DATA PROC	Nachrechnen
[spreadsheet analysis]			[Tabellenkalkulation]
recalibrate		INSTR	nacheichen
recalibration		INSTR	Nacheichung
recalibration period		INSTR	Eichperiode
recall (n.)		SWITCH	Wiederanruf
			= Wiederherbeiruf
recall (n.)		COLLOQ	Abruf
recall (→call back 2)		TELEPH	(→rückfragen)
recall (→call-back)		TELEPH	(→Rückfrage)
recall ratio		SCIE	Behaltensquote
recall signal (→reringing		SWITCH	(→Nachrufzeichen)
signal)			
receipt		ECON	Quittung
= voucher 2			= Beleg; Zahlungsbeleg;
↑ voucher 1			Rechnungsbeleg; Abrech-
			nungsbeleg; Kassenzettel;
			Empfangsbestätigung
			↑ Beleg
receipt (→acknowledgment)		TELEC	(→Quittierung)
receipt confirmation		TELEC	(→Quittierung)
(→acknowledgment)			
receipted bill		ECON	quittierte Rechnung
receipt signal (→acknowledgment		TELEC	(→Quittungszeichen)
signal)			
receivable interest (→debtor		ECON	(→Sollzins)
interest)			
receive		TELEC	empfangen
receive (→answer)		TELEPH	(→antworten)
receive acknowledgment		DATA COMM	Empfangsquittung
receive antenna		ANT	Empfangsantenne
= receiving antenna; receive-only an-			
tenna			
receive circuit		CIRC.ENG	Empfangsschaltung
			= Empfängerschaltung
receive clock		TELEC	Empfangstakt
receive control		DATA COMM	Empfangssteuerung
			= Empfängersteuerung
received data		DATA COMM	Empfangsdaten
received power		RADIO	Empfangsleistung
received signal		TELEC	Empfangssignal
receive input		TRANS	Empfangseingang
= reception input			= F1an (DBP)
receive level		TELEC	Empfangspegel
= reception level			= Eingangspegel
receive memory		DATA COMM	Empfangsspeicher
receive mode		DATA COMM	Empfangsbetrieb
receive module		EQUIP.ENG	Empfangsbaugruppe
			= Empfängerbaugruppe
receive not ready		DATA COMM	nicht empfangsbereit
= RNR			
receive-only (adj)		TERM&PER	nur empfangend
			= Nur-Empfangs-
receive-only antenna (→receive		ANT	(→Empfangsantenne)
antenna)			
receive-only terminal		TERM&PER	Nur-Empfangs-Endgerät
= RO terminal			= Nur-Empfangs-Terminal
receive operating margin		TELEGR	Empfangsspielraum
[distortion]			[Verzerrung]
= receiver margin			
receive output		TRANS	Empfangsausgang
= reception output			= F2ab (DBP)
receive part (→receive		EQUIP.ENG	(→Empfangsteil)
section)			
receiver		TELEC	Empfänger
= receiving equipment			= Empfangsgerät; Empfang-
			sapparatur
receiver (→receiver inset)		TELEPH	(→Telefonkapsel)
receiver cap (→ear piece)		TELEPH	(→Hörmuschel)
receiver capsule (→receiver		TELEPH	(→Telefonkapsel)
inset)			
receive ready (→ready to		DATA COMM	(→empfangsbereit)
receive)			
receive register		DATA COMM	Empfangsregister
receive relay		TELEGR	Empfgangsrelais
receiver horizon		RADIO PROP	Radiohorizont des Empfän-
			gers
			= Funkhorizont des Empfän-
			gers
receiver inset		TELEPH	Telefonkapsel
= receiver capsule; receiver			= Telephonkapsel; Hörkap-
			sel; Telefon 2; Telephon 2
receiver magnet (→selector		TELEGR	(→Empfangsmagnet)
magnet)			
receiver margin (→receive		TELEGR	(→Empfangsspielraum)
operating margin)			
receiver noise temperature		SAT.COMM	Empfängerrauschtemperatur
receiver rest (→cradle)		TELEPH	(→Gabel)
receiver shaft		TERM&PER	Empfängerwelle
			[Telegraf]
receiver-signal-element		DATA COMM	Empfangsschrittakt
timing			
receiver station		RADIO	Empfangsstation
receive section		EQUIP.ENG	Empfangsteil
= receive part			
receive sequence number		DATA COMM	Empfangsfolgenummer
receive telegram		DATA COMM	Empfangstelegramm
receiving antenna (→receive		ANT	(→Empfangsantenne)
antenna)			
receiving band		RADIO	Empfangsband
receiving call (→incoming		SWITCH	(→ankommender Anruf)
call)			
receiving equipment		TELEC	(→Empfänger)
(→receiver)			
receiving inspection (→entrance		QUAL	(→Eingangsprüfung)
test)			
receiving multicoupler		RADIO	Trennverstärker
[to connect several receivers with one			[zum Anschluß mehrerer
antenna]			Empfänger an eine Anten-
= multicoupler			ne]
receiving plant		TELEC	Empfangsanlage
receiving reference		TELEPH	Empfangsbezugsdämpfung
equivalent			
receiving side		TELEC	Empfangsseite
receiving tube		ELECTRON	Empfängerröhre
= receiving valve			= Empfangsröhre

receiving valve (→receiving ELECTRON (→Empfängerröhre)
tube)
receptacle TECH **Gefäß**
≈ vessel ≈ Behälter
receptacle COMPON **Steckhülse**
receptacle (→vessel) TECH (→Behälter)
receptacle (→jack) COMPON (→Buchse)
reception TELEC **Empfang**
reception central CIRC.ENG **Empfangszentrale**
 [PCM]
reception chain TELEC **Empfangskette**
reception confirmation signal DATA COMM (→Anrufbestätigung)
 (→call confirmation signal)
reception input (→receive TRANS (→Empfangseingang)
input)
reception interruption TELEC **Empfangsunterbrechung**
reception level (→receive TELEC (→Empfangspegel)
level)
reception output (→receive TRANS (→Empfangsausgang)
output)
reception room OFFICE **Empfangsraum**
recess (n.) TECH **Vertiefung**
= slot 2 = Mulde
≈ cavity ≈ Hohlraum
≠ elevation ≠ Erhöhung
↓ recessed grip ↓ Griffmulde
recessed TECH **vertieft**
≠ raised ≠ erhaben
recessed collar head-screw MECH **Halsschraube**
recessed grip TECH **Griffmulde**
= pickup recess; handle recess = Grifftasche; Griffschale
↑ recess ↑ Vertiefung
recessed handle TECH **Schalengriff**
recession ECON **Rezession**
recession TECH **Einbuchtung**
≈ strangulation ≈ Einschnürung
recess turning METAL **Einstechen 2**
 [Drehen]
rechargable POWER SYS **wiederaufladbar**
 [accumulator] [Akkumulator]
recharge (v.t.) POWER SYS **nachladen**
recharge (n.) POWER SYS **Nachladung**
rechargeable battery POWER SYS **wiederaufladbare Batterie**
re-check (v.t.) TECH **nachprüfen**
= review; revise; re-examine; audit [nochmals prüfen]
≈ check; verify = überprüfen
 = prüfen; bestätigen
re-check (n.) TECH **Nachprüfung**
= revision; re-examination; audit (n.) = Überprüfung
≈ check = Prüfung
recherche DATA PROC **Recherche**
recipient (→consignee) ECON (→Lieferungsempfänger)
reciprocal (n.) MATH **Kehrwert**
= reciprocal value = reziproker Wert
reciprocal (adj.) MATH **reziprok**
reciprocal (adj.) NETW.TH **übertragungssymmetrisch**
= line-coupling balanced = kopplungssymmetrisch;
 kernsymmetrisch; reziprok
reciprocal action PHYS (→Wechselwirkung)
 (→interaction)
reciprocal integration MATH **Kehrwertintegration**
reciprocal lattice PHYS **reziprokes Gitter**
reciprocal length PHYS **reziproke Länge**
reciprocal meter PHYS **reziprokes Meter**
≈ diopter ≈ Dioptrie
reciprocal network NETW.TH **duales Netzwerk**
reciprocal pulse duty ratio ELECTRON (→Tastgrad)
 (→pulse duty factor)
reciprocal quadripole NETW.TH (→kernsymmetrischer Vier-
 (→reciprocal two-port) pol)
reciprocal second PHYS **reziproke Sekunde**
 [SI unit for speed of revolution turns; [SI-Einheit für Drehzahl;
 = one turn per second] = eine Umdrehung pro Se-
 = 1/s kunde]
 = 1/s
reciprocal two-port NETW.TH **kernsymmetrischer Vierpol**
= reciprocal quadripole = übertragungssymmetri-
 scher Vierpol; kopplungs-
 symmetrischer Vierpol;
 reziproker Vierpol

reciprocal value MATH (→Kehrwert)
 (→reciprocal)
reciprocating motion MECH **Hin- und Herbewegung**
reciprocity MATH **Reziprozität**
 = Umkehrbarkeit 2
reciprocity theorem NETW.TH **Reziprozitätstheorem**
reciprocity theorem ANT **Umkehrsatz**
= inversion theorem = Reziprozitätssatz; Reziprozitätstheorem
recirculating current POWER SYS (→Umschwingstrom)
 (→ring-back current)
recirculating delay (→round-trip TELEPH (→Umlaufzeit)
delay)
recirculating memory DATA PROC (→Umlaufspeicher)
 (→circulating memory)
recirculating memory TERM&PER (→Umlaufspeicher)
 (→rotating memory)
recirculating storage DATA PROC (→Umlaufspeicher)
 (→circulating memory)
recirculating storage TERM&PER (→Umlaufspeicher)
 (→rotating memory)
recirculating store DATA PROC (→Umlaufspeicher)
 (→circulating memory)
recirculating store (→rotating TERM&PER (→Umlaufspeicher)
memory)
recirculation (→cycle) TECH (→Umlauf)
reckon (v.t.) ECON **ermitteln**
 [an amount] [einen Betrag]
≈ calculate [MATH] ≈ rechnen [MATH]
recode CODING **umcodieren**
recognition (→detection) INF.TECH (→Erkennung)
recognition logic (→detection DATA PROC (→Erkennungslogik)
logic)
recognizable TECH **erkennbar**
= detectable; identifiable = nachweisbar; identifizierbar
recognize (→detect) INF.TECH (→erkennen)
recognize (→interpret) DATA PROC (→interpretieren)
recoil (n.) PHYS **Rückstoß**
↑ reaction ↑ Rückwirkung
recoil TECH **zurückspringen**
 = zurückprallen
recoil electron PHYS **Rückstoßelektron**
recoil radiation PHYS **Rückstoßstrahlung**
recombination PHYS **Rekombination**
 = Wiedervereinigung
recombination center PHYS **Rekombinationszentrum**
= deathnium center; deathnium = Deathnium
recombination coefficient PHYS **Rekombinationskoeffizient**
recombination radiation PHYS **Rekombinationsstrahlung**
recombination rate PHYS **Rekombinationsrate**
recombination velocity PHYS **Rekombinationsgeschwindigkeit**
recommencement COLLOQ (→Wiederaufnahme)
 (→resumption)
recommendation TELEC **Empfehlung**
 [CCITT, CCIR]
recompatibility TV **Rekompatibilität**
= reverse compatibility
recompile DATA PROC **rekompilieren**
recondition (→repair) TECH (→instandsetzen)
recondition (→rework) TECH (→überarbeiten)
reconditioning (→repair) TECH (→Instandsetzung)
reconfiguration DATA PROC **Rekonfiguration**
 [Ersatz fehlerhafter Hardware-Teile]
reconfiguration TELEC **Umrangierung**
reconfigure (v.t.) DATA PROC **rekonfigurieren**
reconnaisance INF.TECH (→Erkennung)
 (→detection)
reconnection TELEC **Wiederanschließung**
reconstitute TECH **zurückbilden**
reconstructable SCIE **rekonstruierbar**
reconstruction DATA PROC **Rekonstruktion**
reconstruction filter CODING **Rekonstruktionsfilter**
reconversion TECH **Zurückwandlung**
= retransformation = Rücktransformation
reconvert (v.t.) TECH **zurückwandeln**
= retransform (v.t.) = rücktransformieren
record (v.t.) DATA PROC **erfassen**
 = registrieren

record (n.) OFFICE **Akte**
[collection of documents about a [Sammlung von Schriftstük-
topic] ken einer Angelegenheit]
= file; document
record (n.) OFFICE **Niederschrift**
= minutes ≈ Protokoll
record (→ report) LING (→ Bericht)
record (→ list) DOC (→ Liste)
record (v.t.) (→ register) INSTR (→ registrieren)
record (→ data record) DATA PROC (→ Datensatz)
record (→ phonogram CONS.EL (→ Schallplatte)
record)
record (n.) (→ document) DOC (→ Unterlage)
record accessories CONS.EL **Schallplattenzubehör**
 = Plattenzubehör
record address DATA PROC **Satzadresse**
record archive BROADC **Schallplattenarchiv**
 = Plattenarchiv
record block (→ data DATA PROC (→ Datenblock)
block)
record carrier (→ postal POST (→ Postverwaltung)
administration)
record changer CONS.EL **Plattenwechsler**
 [Plattenspieler mit Wechs-
 ler]
 = Schallplattenwechsler
record chart INSTR **Registrierstreifen**
record circuit TELEC **Meldeleitung**
= recording circuit; recording trunk
record company CONS.EL **Plattenfirma**
 = Plattenfirma
record cover CONS.EL **Schallplattenhülle**
[the outer protection; illustrated] [äußerer Schutz; bedruckt]
= cover = Plattenhülle; Plattencover;
≈ record pocket Cover
 ≈ Schallplattentasche
record density (→ recording DATA PROC (→ Speicherdichte)
density)
recorded announcement service TELEPH **automatische Ansage**
recorder BROADC **Aufnahmegerät**
recorder (→ recording INSTR (→ Registriergerät)
instrument)
recorder cassette TERM&PER **Recorderkassette**
↑ magnetic tape cassette ↑ Magnetbandkassette
record file (→ data file) DATA PROC (→ Datei)
record format DATA PROC **Satzformat**
record gap (→ interblock TERM&PER (→ Blockzwischenraum)
gap)
record group (→ record DATA PROC (→ Satzgruppe)
set)
record head (→ write TERM&PER (→ Schreibkopf 2)
head)
record identifier (→ record DATA PROC (→ Satzkennung)
label)
record industry CONS.EL **Schallplattenindustrie**
 = Plattenindustrie
recording BROADC **Aufnahme**
recording TERM&PER **Aufzeichnung**
recording DATA PROC **Erfassung**
≈ logging ≈ Protokollierung
recording INSTR **Registrierung**
 = Aufzeichnung
recording (→ printout) DATA PROC (→ Ausdruck)
recording beam ELECTRON **Schreibstrahl**
= writing beam
recording circuit (→ record TELEC (→ Meldeleitung)
circuit)
recording density DATA PROC **Speicherdichte**
[number or data which can be stored [Anzahl der pro geometri-
per geometrical unit] schen Einheit speicherba-
= record density; packing density; bit ren Daten]
density; character density; packaging = Aufzeichnungsdichte;
density; density Schreibdichte; Zeichendich-
 te; Bitdichte; Packungsdich-
 te
recording equipment TERM&PER **Aufzeichnungsgerät**
recording error (→ write TERM&PER (→ Druckfehler)
error)
recording head (→ recording EL.ACOUS (→ Aufzeichnungsmagnet-
magnetic head) kopf)

recording head (→ write head) TERM&PER (→ Schreibkopf 2)
recording instrument INSTR **Registriergerät**
= recorder = Registrierinstrument;
 Schreiber
recording magnetic head EL.ACOUS **Aufzeichnungsmagnetkopf**
= recording head = Aufnahmemagnetkopf
≈ write head [TERM&PER] ≈ Schreibkopf
↑ magnetic head [TERM&PER] [TERM&PER]
 ↑ Magnetkopf
 [TERM&PER]
recording method TERM&PER (→ Aufzeichnungsverfahren)
(→ recording mode)
recording mode TERM&PER **Aufzeichnungsverfahren**
[on magnetic layer memories] [auf Magnetschichtspei-
= recording method chern]
↓ NRZ recording; two-frequency recor- = Aufzeichnungsmethode
ding; phase encoding; group-coded ↓ Wechselschrift; Wechsel-
recording taktschrift; Richtungstakt-
  schrift; gruppencodierte
  Aufzeichnung
recording output CIRC.ENG **Registrierausgang**
recording speed (→ write TERM&PER (→ Schreibgeschwindigkeit)
velocity)
recording system TERM&PER (→ Aufzeichnungssystem)
(→ transceiver)
recording tape (→ magnetic ELECTRON (→ Magnetband)
tape)
recording tool EL.ACOUS **Aufnahmewandler**
recording track (→ track) TERM&PER (→ Spur)
recording trunk (→ record TELEC (→ Meldeleitung)
circuit)
record label DATA PROC **Satzkennung**
= record identifier
record layout (→ record DATA PROC (→ Satzaufbau)
structure)
record length DATA PROC **Satzlänge**
record locking DATA PROC **Datensatzsperre**
[modifiable only by authorized user] [nur durch bestimmte Anwender veränderbar]
record manager (→ file DATA PROC (→ Dateiverwalter)
manager)
record mark DATA PROC **Satzmarke**
record number DATA PROC **Satznummer**
record of detailed measures TECH **Aufmaß**
 [Feststellung von Maßen
 im einzelnen]
record paper INSTR **Registrierpapier**
record player CONS.EL **Plattenspieler**
= phonograph; gramophone (trade- = Schallplattenspieler
mark in U.S.A.) ≈ Plattenwechsler
≈ turntable
record pocket CONS.EL **Schallplattentasche**
[direct protection of the disk] [unmittelbarer Schutz der
≈ record cover Platte]
 = Plattentasche
 ≈ Schallplattenhülle
record segment DATA PROC **Satzsegment**
record separator DATA COMM **Untergruppentrennung**
= RS [Code]
 = RS
record set DATA PROC **Satzgruppe**
= record group
records filing OFFICE **Aktenablage**
record structure DATA PROC **Satzaufbau**
= record layout = Satzstruktur
recourse ECON **Regress**
recover (→ restore) TECH (→ wiederherstellen)
recoverable TECH **behebbar**
recoverable error DATA PROC **behebbarer Fehler**
recovered charge MICROEL **Sperrverzugsladung**
recover financially ECON **sanieren**
recovery ELECTRON **Rückgewinnung**
[of a signal] [Signal]
= retrieval
recovery ECON **Wiederaufschwung**
 = Erholung
recovery 1 DATA PROC **Wiederherstellung**
[of lost data] [verlorener Daten]
 = Behebung; Recovery
recovery 2 (→ restart) DATA PROC (→ Neustart)

recovery 3 (→retrieval) DATA PROC (→Wiedergewinnung)
recovery (→restoration) TECH (→Wiederherstellung)
recovery phase MICROEL **Erholungsphase**
recovery procedure DATA COMM **Rückstellprozedur**
= Wiederherstellungsprozedur
recovery program SWITCH **Anlaufprogramm**
recovery signal (→start signal) SWITCH (→Anlaufsignal)
recovery time CIRC.ENG **Erholungszeit**
= restabilization time = Erholzeit; Stabilisierungszeit
recovery time DATA PROC **Holzeit**
recreational software DATA PROC (→Spielprogramm)
(→computer game program)
recruitment (→engagement) ECON (→Einstellung)
recrystallization PHYS **Rekristallisation**
rectangle MATH **Rechteck**
[rectangular parallelogram, with equal opposite sides] [rechtwinkliges Parallelogramm, mit gegenüberliegenden gleichen Winkeln]
↑ parallelogram ↑ Parallelogramm
↓ square (n.) ↓ Quadrat
rectangular 2 MATH **rechteckig 2**
[relative to rectangles] [ein Rechteck betreffend]
≈ right-angled ≈ rechtwinklig
rectangular 1 (→right-angled) MATH (→rechtwinklig)
rectangular coordenate MATH (→kartesische Koordinate)
(→cartesian coordenate)
rectangular coordinate system MATH (→kartesisches Koordinatensystem)
(→cartesian coordinate system)
rectangular failure ELECTRON **Rechteckfehler**
rectangular-hysteresis ferrite METAL (→Rechteck-Ferrit)
(→square-loop ferrite)
rectangular hysteresis loop PHYS **rechteckige Hystereseschleife**
rectangular impedance chart LINE TH (→Schmidt-Buschbeck-Diagramm)
(→rectangular transmission line chart)
rectangular prism MATH **Quader**
[a parallelepiped with two parallel rectangles as ground faces; a solid formed by six rectangles] [gerades Parallelepiped mit parallelen Rechtecken als Grundflächen; ein von sechs Rechtecken gebildeter Körper]
↑ parallelepiped; prism; polyhedron ↑ Parallelepiped; Prisma; Polyeder
↓ cube ↓ Würfel
rectangular pulse (→square pulse) ELECTRON (→Rechteckimpuls)
rectangular repetition rate ELECTRON (→Rechteckwelle)
(→square-wave reversals)
rectangular sine wave generator CIRC.ENG **Sinus-Rechteckgenerator**
= sine/square source
rectangular transmission line chart LINE TH **Schmidt-Buschbeck-Diagramm**
= rectangular impedance chart; bipolar transmission line chart; bipolar impedance chart = Buschbeck-Diagramm
↑ transmission line chart ↑ Leitungsdiagramm
rectangular wave EL.TECH **Rechteckschwingung**
↑ relaxation oscillation ↑ Kippschwingung
rectangular wave ELECTRON (→Rechteckwelle)
(→square-wave reversals)
rectangular-wave generator CIRC.ENG (→Rechteckgenerator)
(→square-wave generator)
rectangular waveguide MICROW **Rechteckhohlleiter**
= rechteckförmiger Hohlleiter
rectangular waveguide branch MICROW **Rechteckhohlleiterarm**
rectangular-wave measuring generator (→square-wave measuring generator) INSTR (→Rechteckwellen-Meßgenerator)
rectification EL.TECH **Gleichrichtung**
[from AC into DC] [von Wechselstrom in Gleichstrom]
≠ inversion ≠ Wechselrichtung
rectification (→amendment) LING (→Berichtigung)
rectified current EL.TECH **Richtstrom**
≈ dc current = gleichgerichteter Strom
≈ Gleichstrom
rectified voltage EL.TECH **Richtspannung**
= rectified tension = gleichgerichtete Spannung
≈ direct voltage ≈ Gleichspannung
rectifier EL.TECH **Gleichrichter**
= current changer
rectifier POWER SYS **Gleichrichter**
[AC into DC] [Wechselstrom in Gleichstrom]
↑ static power converter ↑ Stromrichter
rectifier (→demodulator) MODUL (→Demodulator)
rectifier bridge circuit CIRC.ENG **Graetz-Schaltung**
= Graetz connection
rectifier circuit CIRC.ENG **Gleichrichterschaltung**
= rectifier connection; rectifying circuit
rectifier connection (→rectifier circuit) CIRC.ENG (→Gleichrichterschaltung)
rectifier diode MICROEL **Gleichrichterdiode**
≠ signal diode = Kristallgleichrichter
↑ diode ≠ Signaldiode
↑ Halbleiterdidoe
rectifier element MICROEL **Richtleiter**
[with polarization-dependent resistance] [mit polarisationsabhängigem Widerstand]
rectifier instrument INSTR **Gleichrichtermeßgerät**
rectifier stack POWER SYS **Stapelgleichrichter**
rectifier tube ELECTRON **Gleichrichterröhre**
= rectifier valve; detector tube; detector valve; valve detector = Röhrendiode
rectifier valve (→rectifier tube) ELECTRON (→Gleichrichterröhre)
rectify EL.TECH **gleichrichten**
rectifying circuit (→rectifier circuit) CIRC.ENG (→Gleichrichterschaltung)
rectifying transition MICROEL **Gleichrichterübergang**
rectilinear MATH **geradlinig**
= straight-lined; straight; colinear = gerade 2; geradläufig
rectilinear graph paper DOC (→Millimeterpapier)
(→millimeter paper)
rectilinear paper (→millimeter paper) DOC (→Millimeterpapier)
recurrent TECH **wiederkehrend**
[returning from time to time] = rekursiv; rekurrent
≈ repetitive; cyclic ≈ zyklisch
recurrent code (→chain code) CODING (→Kettencode)
recurrent four-terminal NETW.TH (→Kettenleiter)
(→recurrent network)
recurrent network NETW.TH **Kettenleiter**
= recurrent two-pole; recurrent four-terminal = Vierpolkette
recurrent two-pole (→recurrent network) NETW.TH (→Kettenleiter)
recursion DATA PROC **Rekursion**
[continued repetition] [mehrfache Wiederholung]
recursive MATH **rekursiv**
recursive algorithm DATA PROC **rekursiver Algorithmus**
recursive call (→recursive subroutine) DATA PROC (→rekursives Unterprogramm)
recursive code CODING **rekursiver Code**
recursive filter (→recursive-type digital filter) NETW.TH (→rekursives Digitalfilter)
recursive function MATH **rekursive Funktion**
recursive procedure DATA PROC **rekursive Prozedur**
recursive subroutine DATA PROC **rekursives Unterprogramm**
= recursive call
recursive-type digital filter NETW.TH **rekursives Digitalfilter**
= recursive filter = rekursives Filter
≈ Transversalfilter
recycling TECH **Rückgewinnung**
[from junk] [aus Altmaterial]
≈ disposal ≈ Entsorgung
red PHYS **rot**
reddish PHYS **rötlich**
redeem ECON **tilgen**
[to eliminate by payment] [durch Zurückbezahlen aufheben]
≈ amortize; repay = abzahlen
≈ amortisieren; zurückbezahlen
redeem (→repurchase) ECON (→zurückkaufen)
redeemable (→repayable) ECON (→rückzahlbar)

redefinable	DATA PROC	redefinierbar = neudefinierbar	reduction	MECH	Untersetzung
redefine	DATA PROC	redefinieren = neu definieren	reduction = diminution; retrenchment; cutback 1 ≈ decrease ≠ enlargement	TECH	Verkleinerung = Abbau ≈ Verminderung ≠ Vergrößerung
redelivery (→return delivery)	ECON	(→Rücklieferung)			
redemption ≈ amortization; repayment	ECON	Tilgung ≈ Amortisation; Zurückzahlung	reduction [optics]	PHYS	Verkleinerung [Optik]
redemption (→refund)	ECON	(→Rückerstattung)	reduction (→cut)	ECON	(→Kürzung)
redemption (→withdrawal)	ECON	(→Widerruf 2)	reduction (→discount)	ECON	(→Preisnachlaß)
redesign	MICROEL	Redesign	reduction (→lowering)	COLLOQ	(→Senkung)
red-green-blue monitor (→RGB monitor)	TERM&PER	(→RGB-Monitor)	reduction (→data compression)	DATA PROC	(→Datenverdichtung)
red-green-blue signal (→RGB signal)	TV	(→RGB-Signal)	reduction factor [interference of power lines into communication lines] = derating factor	LINE TH	Reduktionsfaktor [Starkstrombeeinflussung von Fernmeldeleitungen]
red heat	TECH	Rotglut			
red-hot	TECH	glutrot = rotglühend			
redial (n.) (→redialing)	TELEPH	(→Wahlwiederholung)	reduction factor	TELEPH	Reduzierungsfaktor
redialing = repeated call attempt	SWITCH	Anrufwiederholung	reduction factor	ANT	Verkürzungsfaktor
			reduction gearing = Reduktionsgetriebe	MANUF	Vorgelege
redialing = last number redial; redial (n.) ≈ saved number redial	TELEPH	Wahlwiederholung = Wiederwahl	reduction of charge (→allowance of charge)	ECON	(→Gebührenermäßigung)
			reduction of cross section ≈ tapering	TECH	Querschnittsverjüngung
redirect = reroute	TELEC	umleiten	reduction scale	ENG.DRAW	Verkleinerungsmaßstab
redirected call = redirected connection; redirected link	SWITCH	umgeleitete Verbindung	redundancy	INF	Redundanz = Weitschweifigkeit
redirected connection (→redirected call)	SWITCH	(→umgeleitete Verbindung)	redundancy ≈ operation/stand-by	QUAL	Redundanz ≈ Betrieb/Ersatz
redirected link (→redirected call)	SWITCH	(→umgeleitete Verbindung)	redundancy bit = redundant bit	DATA COMM	Redundanzbit
redirection = rerouting	TELEC	Umleitung = Umleiten; Umlenkung; Umwegführung	redundancy check	INF	Redundanzprüfung = Redundanzkontrolle
			redundancy check (→block control)	DATA COMM	(→Blocksicherung)
redirection address (→rerouting address)	DATA COMM	Umleitadresse (→Umleitadresse)	redundancy compression	SIGN.TH	Bildkompression = Redundanzreduzierung
redistribute	TECH	umverteilen	redundancy reduction	INF	Redundanzminderung = Redundanzreduktion
red lead (→minium)	METAL	(→Mennige)			
red shift	PHYS	Rotverschiebung	redundant	INF	redundant = weitschweifig
reduce 1 (v.t.) [the intensity of something] = diminish; lessen; decline; retrench; fall ≈ decrease; reduce 2; impair ≠ augment	TECH	vermindern [in der Intensität abschwächen] = mindern; verringern; reduzieren; zurückgehen (fig.) ≈ abnehmen; verkleinern; beeinträchtigen ≠ verstärken	redundant (→duplicated)	EQUIP.ENG	(→gedoppelt)
			redundant bit (→redundancy bit)	DATA COMM	(→Redundanzbit)
			redundant character	INF	redundantes Zeichen = selbstprüfendes Zeichen
			redundant code	CODING	redundanter Code = Sicherheitscode
reduce 2 (v.t.) [the size of something] ≈ reduce 1 ≠ enlarge	TECH	verkleinern [die Größe] ≈ vermindern ≠ vergrößern	redundant quaternary	CODING	redundant quaternär
			redundant ternary	CODING	redundant ternär
			red violet	PHYS	rotviolett
reduce [optics]	PHYS	verkleinern [Optik]	reed = tongue	TECH	Zunge
reduce (→cut)	ECON	(→kürzen)	reed (→contact reed)	COMPON	(→Kontaktzunge)
reduced instruction set computer = RISC ≠ complex instruction set computer	DATA PROC	RISC-Computer [mit eingeschränktem Befehlsvorrat] ≠ CISC-Computer	reedback = echo query	INSTR	Rückmeldung
			reed capsule	COMPON	Schutzrohr [Reed-Relais]
reduced-rate (→cheapened)	ECON	(→verbilligt)	reed contact (→sealed contact)	COMPON	(→Schutzrohrkontakt)
reduced tariff (→reduced tax)	TELEC	(→Billigtarif)	reed relay [contacts protected by gas or mercury] ↓ gas-protected relay; mercury relay; remanent relay	COMPON	Schutzrohrkontakt-Relais [Kontakte mit Gas oder Quecksilber geschützt] = Reed-Relais; Schutzkontakt-Relais; Herkon-Relais ↓ Schutzgaskontakt-Relais; Quecksilber-Relais; Haftrelais
reduced tariff (→allowance of charge)	ECON	(→Gebührenermäßigung)			
reduced tax = reduced tariff	TELEC	Billigtarif			
reducer	MECH	Reduzierstück			
reducer = reducing circuit; dividing circuit; division circuit; divider ↑ scaler ↓ frequency divider	CIRC.ENG	Untersetzer = Teiler; Teilerschaltung; Dividierer; Verkleinerer; Verkleinerungsschaltung ↑ Festwertmultiplikator ↓ Frequenzteiler			
			reed switch	COMPON	Reed-Schalter
			reed-type frequency meter = vibrating-reed frequency meter; vibrating-reed instrument	INSTR	Zungenfrequenzmesser = Vibrationsfrequenzmesser
			reel (v.t.)	TECH	aufrollen
reducing	METAL	Verengen = Reduzieren	reel (n.) [device on which something is wound] = spool 1; bobbin	TECH	Spule 1 [Körper auf den man etwas aufwickelt] = Spulenkörper; Aufwickelrolle; Rolle 3 ≈ Rolle 1
reducing circuit (→reducer)	CIRC.ENG	(→Untersetzer)			
reducing valve	TECH	Reduzierventil			
reduction	NETW.TH	Drosselung			

reel

reel (n.) TERM&PER
[device, mostly flanged, on which tape can be wound]
= coil
↓ magnetic tape reel; peforated tape reel

reel (n.) TECH
[cylindrical device to reel]
≈ windlass

reel (→cable drum) COMM.CABLE
reel 2 (→roll 1) TECH
reemission PHYS
re-enlargement DOC
re-enlarger OFFICE
reenter DATA PROC

reenterable DATA PROC
(→reentrant)
reenterable code DATA PROC
(→reentrant code)
reenterable program DATA PROC
[runnable simultaneously in various application programs]
= sharable program; reentrant program
≠ serially reusable program
↑ reusable program

reentrant DATA PROC
[can be activated simultaneously by several user programs]
= reenterable; sharable
≠ serially reusable

reentrant code DATA PROC
= reenterable code

reentrant program DATA PROC
(→reenterable program)
reentrant subroutine DATA PROC

reentry DATA PROC
[from subroutine to main program]
= return jump; return (n.)

reestablish (→restore) TECH
reestablishment TECH
(→restoration)
re-examination (→re-check) TECH
re-examine (→re-check) TECH
reexport ECON
reference LING
= referencing
≈ references
reference ECON
= past record
reference OFFICE
[in commercial correspondence]

reference COLLOQ

reference address DATA PROC
[to link separately stored but correlated data]
= linkage address; link address; continuation address; chaining address

reference address (→base address) DATA PROC

reference antenna ANT
= reference radiator

Spule
= Spulenkörper; Aufwickelkörper; Bandspule; Leerspule
≈ Rolle
↓ Magnetbandspule

Haspel
[zylinderförmige Vorrichtung zum Aufwickeln]
≈ Winde

(→Kabeltrommel)
(→Rolle 1)
Reemission
Rückvergrößerung
Rückvergrößerer
wiedereingeben
= erneut eingeben
(→ablaufinvariant)

(→ablaufinvariant codiertes Programm)
ablaufinvariables Programm
[in mehreren Anwenderprogrammen gleichzeitig anwendbar]
≠ seriell mehrfach aufrufbares Programm
↑ mehrfach aufrufbares Programm

ablaufinvariant
[in mehreren Anwenderprogrammen gleichzeitig aktivierbar]
= wiedereintrittsinvariant; parallel wiederverwendbar; ablaufinvariabel
≠ seriell wiederverwendbar

ablaufinvariant codiertes Programm
= wiedereintrittsinvarianter Code

(→ablaufinvariables Programm)
ablaufinvariantes Unterprogramm
Rücksprung
[von Unterprogramm in Hauptprogramm]
= Rückkehrsprung; Rückkehr; Return

(→wiederherstellen)
(→Wiederherstellung)

(→Nachprüfung)
(→nachprüfen)
Wiederausfuhr
= Reexport
Literaturhinweis
= Bezug; Bezugnahme
≈ Literatur
Referenz
≈ Empfehlungsschreiben
Zeichen
[Bezugnahme in Geschäftsbrief]

Bezugnahme
= Bezug; Referenz; Verweis
Verweisadresse
[zum Verketten getrennt gespeicherter, jedoch korrelierter Daten]
= Folgeadresse; Verkettungsadresse; Kettungsadresse; Verknüpfungsadresse
(→Grundadresse)

Bezugsantenne
= Vergleichsantenne; Referenzantenne
≈ Meßantenne

reference attenuation TELEC
(→reference equivalent)
reference bit DATA COMM
reference book LING
↓ encyclopedia; lexicon; dictionary

reference boresight ANT
[direction to which the antenna is aligned]
≈ boresight

reference circuit TELEC

reference circuit (→standard transmission line) TELEPH
reference clock TELEC
= master clock; main clock
reference clock signal TELEC
reference color (AM) TV
= reference colour (BRI)
reference colour (BRI) (→reference color) TV
reference condition (→set condition) TECH
reference coupling EL.TECH
reference dimension DOC

reference diode (→voltage reference diode) MICROEL
reference dipole ANT
↑ standard antenna

reference earth EL.TECH
reference edge TERM&PER
[document evaluator]
= aligning edge
reference electrode PHYS

reference entry (→headword) LING
reference equivalent TELEC
[attenuation on stardard system NOSFER to achieve equal volume of sound]
= reference attenuation
reference equivalent measuring equipment TELEPH
reference frequency EL.TECH
reference frequency (→standard frequency) INSTR
reference inductor INSTR
reference input (→reference magnitude) CONTROL
reference input signal CONTROL
(→reference magnitude)
reference installation ECON
reference level (→relative level) TELEC
reference level accuracy INSTR
reference line (→datum line) ENG.DRAW
reference list ECON
reference magnitude TECH
= reference quantity
reference magnitude CONTROL
[a magnitude preset as reference, whom the controlled magnitude has to follow]
= reference variable; reference input; reference input signal
≠ controlled magnitude
reference manual (→handbook) TECH
reference method (→comparison method) INSTR
reference model (→layer model) DATA COMM
reference node NETW.TH

reference number OFFICE
= file number

(→Bezugsdämpfung)

Referenzbit
Nachschlagewerk
↓ Enzyklopädie; Lexikon; Wörterbuch

Zielrichtung
[Richtung in die ausgerichtet wird]
= Seelenachse
≈ Hauptstrahlrichtung
Bezugskreis
= Bezugsverbindung
(→Eichleitung)

Bezugstaktgeber
= Referenztaktgeber
Bezugstakt
Bezugsfarbe

(→Bezugsfarbe)

(→Sollzustand)

Referenzkopplung
Informationsmaß
[tech. Zeichnen]
(→Referenzdiode)

Normdipol
= Referenzdipol; Bezugsdipol
↑ Meßantenne

Bezugserde
Bezugskante
[Dokumentleser]

Bezugselektrode
[phys. Chemie]
(→Stichwort)
Bezugsdämpfung
[im Ureichkreis NOSFER erforderliche Dämpfung für gleiche Laustärke]

Bezugsdämpfungsmeßplatz

Bezugsfrequenz
(→Normalfrequenz)

Bezugsinduktivität
(→Führungsgröße)

(→Führungsgröße)

Referenzanlage
(→relativer Pegel)

Referenzpegelgenauigkeit
(→Bezugslinie 1)

Referenzliste
Bezugsgröße

Führungsgröße
[eine vorgegebene Größe, der die Regelgröße folgen soll]
≠ Regelgröße

(→Handbuch)
(→Vergleichsmethode)

(→Schichtenmodell)

Bezugsknoten
= Referenzknoten
Aktenzeichen

reference oscillator	CIRC.ENG	**Referenzoszillator**
reference parameter	DATA PROC	**Bezugsparameter**
reference point	TECH	**Bezugspunkt**
		= Führungspunkt; Referenzpunkt
reference potential	PHYS	**Bezugspotential**
reference quantity (→reference magnitude)	TECH	(→Bezugsgröße)
reference radiator (→reference antenna)	ANT	(→Bezugsantenne)
references	LING	**Literatur**
= literature		= Bibliographie; Schrifttum
reference signal	TELEC	**Bezugssignal**
		= Referenzsignal
reference stimuli	TV	**Primärvalenz**
reference surface	OPTOEL	**Referenzoberfläche**
reference symbol [arrow]	TYPOGR	**Verweiszeichen** [Pfeil]
reference tape	TERM&PER	**Bezugsband**
reference temperature	PHYS	**Bezugstemperatur**
reference tone	TELEC	**Normalton**
= reftone		
reference transfer function	CONTROL	**Führungsübertragungsfunktion**
reference value 1	TECH	**Richtwert**
= guiding value; guideline 2		
reference value 2	TECH	**Vergleichswert**
= comparison value		
reference value 3 (→nominal value)	TECH	(→Nennwert)
reference value (→set value)	CONTROL	(→Sollwert)
reference value (→bias)	EL.TECH	(→Bezugswert)
reference variable (→reference magnitude)	CONTROL	(→Führungsgröße)
reference voltage	ELECTRON	**Vergleichsspannung**
		= Referenzspannung
reference voltage generator	CIRC.ENG	**Referenzspannungsquelle**
reference voltage tube	ELECTRON	**Vergleichsspannungsröhre**
reference wave	PHYS	**Bezugswelle**
		= Referenzwelle
reference white level	TV	**Weißpegel**
= white level; white signal		[10 % der Maximalamplitude]
		= Weißwert
referencial address (→base address)	DATA PROC	(→Grundadresse)
referencing (→reference)	DOC	(→Bezug)
refine (→improve)		(→veredeln)
refinement	DATA PROC	**Verfeinerung**
refinement (→improvement 2)	TECH	(→Veredelung)
refinery	TECH	**Raffinerie**
refit (v.t.) (→rework)	TECH	(→überarbeiten)
refit (n.) (→overhaul)	TECH	(→Überarbeitung)
reflect	PHYS	**reflektieren**
		= zurückstrahlen; zurückwerfen
reflect	ENG.DRAW	**spiegeln**
reflectance (→reflectivity)	PHYS	(→Reflexionsvermögen)
reflectance ink	TERM&PER	**Reflexionsfarbe**
reflected	TECH	**spiegelbildlich**
≈ symmetrical		≈ symmetrisch
reflected binary code (→Gray code)	CODING	(→Gray-Code)
reflected binary unit-distance code (→Gray code)	CODING	(→Gray-Code)
reflected code (→cyclic code)	CODING	(→zyklischer Code)
reflected copy	DATA COMM	**Reflected-copy**
reflected lettering	TYPOGR	**Spiegelschrift**
reflecting curtain	ANT	**Reflektorwand**
reflecting goniometer	RADIO LOC	**Reflexionsgoniometer**
reflecting layer	PHYS	**Reflexionsschicht**
reflecting power (→reflectivity)	PHYS	(→Reflexionsvermögen)
reflecting surface (→area of reflection)	RADIO PROP	(→Reflexionsfläche)
reflection (AM)	PHYS	**Reflexion**
= reflexion (BRI)		
↓ diffuse reflection		↓ Streureflexion
reflection coefficient [reflected to injected wave]	NETW.TH	**Reflexionsfaktor** [reflektierende zu eingespeister Wellengröße]
= return current coefficient; mismatch factor; echo attenuation coefficient; reflection factor		= Reflexionskoeffizient; Echofaktor; Echoübertragungsfaktor; Stoßfaktor; Anpassungsfaktor; Anpassungskoeffizient
≈ inverse voltage SWR		≈ Anpassungsfaktor; Welligkeitsfaktor
		↓ Betriebsreflexionsfaktor
reflection coefficient	PHYS	**Reflexionsgrad**
reflection coefficient measuring bridge	INSTR	**Reflexionsfaktor-Meßbrücke**
reflection detour	RADIO PROP	**Reflexionsumweg**
reflection factor (→reflection coefficient)	NETW.TH	(→Reflexionsfaktor)
reflection grating	PHYS	**Reflexionsgitter**
reflection hologram	PHYS	**Reflexionshologramm**
reflectionless (→non-reflecting)	PHYS	(→reflexionsfrei)
reflection light barrier	SIGN.ENG	**Reflexionslichtschranke**
= reflex light barrier; reflex optical sensor		= Reflexlichtschranke
reflection loss	PHYS	**Reflexionsverlust**
reflection loss 1 (→return loss coefficient)	NETW.TH	(→Echodämpfungsmaß)
reflection loss 2 (→active return loss)	NETW.TH	(→Reflexionsdämpfung)
reflection loss coefficient (→return loss coefficient)	NETW.TH	(→Echodämpfungsmaß)
reflection mixer	OPTOEL	**Reflexionsmischer**
reflection point (→mirror point)	MATH	(→Spiegelpunkt)
reflectivity	PHYS	**Reflexionsvermögen**
= reflecting power; reflectance		= Reflexionseigenschaft
reflectometer	INSTR	**Reflektometer**
= directional bridge		= Reflexionsmesser; Reflektometerbrücke
reflector 1	ANT	**Reflektor 1**
= dish; secondary radiator; passive radiator; parasitic antenna		= Spiegel; Sekundärstrahler
reflector	MICROW	**Reflektor**
= mirror		= Spiegel
reflector	PHYS	**Reflektor**
		= Rückstrahler
reflector 3 (→passive reflector)	ANT	(→Umlenkspiegel)
reflector 2 (→reflector element)	ANT	(→Reflektor 2)
reflector antenna	ANT	**Spiegelantenne**
		= Reflektorantenne
reflector element [linear directive antenna]	ANT	**Reflektor 2** [lineare Richtantenne]
= reflector 2		= Reflektorelement
↑ parasitic element		↑ Parasitärstrahler
reflector scatterer	ANT	**Rückstrahler**
reflector voltage [klystron]	ELECTRON	**Reflektorspannung** [Klystron]
reflexion (BRI) (→reflection)	PHYS	(→Reflexion)
reflexive	LING	**reflexiv**
		= rückbezüglich
reflexive pronoun [e.g. myself]	LING	**Reflexivpronomen** [z.B: mich]
		= rückbezügliches Fürwort
reflex klystron	MICROW	**Reflexklystron**
reflex light barrier (→reflection light barrier)	SIGN.ENG	(→Reflexionslichtschranke)
reflex optical sensor (→reflection light barrier)	SIGN.ENG	(→Reflexionslichtschranke)
reflow	TECH	**Rückfluß**
reflow soldering	MICROEL	**Aufschmelzlöten**
		= Reflow-Löten
reform (v.t.) (→rework)	TECH	(→überarbeiten)
reformat (v.t.)	DATA PROC	**umformatieren**
refracting (→refractive)	PHYS	(→lichtbrechend)
refracting power	PHYS	**Brechkraft**
refraction [desviation of a wave from stright path, caused by inhomogenities of the medium]	PHYS	**Brechung** [Abweichung einer Welle von der Geradlinigkeit, verursacht durch Inhomogenitäten des Mediums]
≈ diffraction; scattering		

refraction index

↓ superrefraction [RADIO PROP]; subrefraction [RADIO PROP]
≈ Beugung; Streueung
↓ Superrefraktion [RADIO PROP]; Subrefraktion [RADIO PROP]

refraction index (→refractive index) PHYS (→Brechungsindex)
refraction law PHYS **Brechungsgesetz**
refractive PHYS **lichtbrechend**
= refracting = brechend
refractive angle (→angle of refraction) PHYS (→Brechungswinkel)
refractive fading RADIO PROP **Brechungsschwund**
refractive index PHYS **Brechungsindex**
= refraction index = Brechzahl
refractive index distribution OPT.COMM **Brechzahlprofil**
= index profile
refractivity RADIO PROP **Brechwert**
refractivity conditions RADIO PROP **Brechungsverhältnisse**
refractivity gradient RADIO PROP **Delta-N-Wert**
refractivity N RADIO PROP **N-Einheit**
refractometer PHYS **Refraktometer**
refractory (→fire-proof) TECH (→feuerfest)
reframing CODING **Rahmenwiederherstellung**
refresh (v.t.) DATA PROC **auffrischen**
  [a memory content] [einen Speicherinhalt]
  = wiederauffrischen
refresh (n.) TERM&PER **Bildwiederholung**
= image regeneration = Refresh; Bildauffrischung; Bildregenerierung
refresh (n.) DATA PROC (→Auffrischen)
(→refreshing)
refresh circuit CIRC.ENG **Auffrisch-Schaltung**
refresh cycle DATA PROC **Auffrischzyklus**
= refreshing cycle = Refresh-Zyklus;
  Auffrischrate
  ↓ Bildwiederholzyklus
refresh display (→refresh screen) TERM&PER (→Wiederholbildschirm)
refresher DATA PROC **Auffrischer**
  [a dynamic store] [dynamischer Speicher]
refresh indicator DATA PROC **Wiederholanzeige**
  = Refresh-Anzeige
refreshing DATA PROC **Auffrischen**
  [of a dynamic store] [eines dynamischen Speichers]
  = refresh (n.) = Auffrischung; wiederholtes Einschreiben
refreshing cycle (→refresh cycle) DATA PROC (→Auffrischzyklus)
refresh memory DATA PROC **Bildwiederholspeicher**
= image regeneration store; refresh storage; frame buffer; video memory; video RAM; VRAM = Auffrischspeicher; Video-Speicher; Video-RAM; VRAM
≈ graphics memory ≈ Graphikspeicher
refresh pulse ELECTRON **Auffrischimpuls**
= Refresh-Impuls
refresh rate TERM&PER **Auffrischrate**
≈ refresh cycle = Auffrischfrequenz
  ≈ Auffrischzyklus
  ↓ Bildwiederholfrequenz
refresh screen TERM&PER **Wiederholbildschirm**
= refresh display = Wiederholschirm; Refresh-Bildschirm; Bildwiederholschirm
↓ raster display ↓ Rasterbildschirm
refresh storage (→refresh memory) DATA PROC (→Bildwiederholspeicher)
refrigerant TECH **Kühlmittel**
= coolant; cooling medium; cooling agent ↓ Kühlflüssigkeit; Kühlwasser; Kühlluft
↓ cooling liquid; cooling water; cooling air
refrigerate (→cool) TECH (→kühlen)
refrigeration (→cooling) TECH (→Kühlung)
refrigerator TECH **Kühler**
= cooler; cooling device = Kühlvorrichtung
≈ heat exchanger ≈ Wärmeaustauscher
reftone (→reference tone) TELEC (→Normalton)
refund (v.t.) (→repay) ECON (→zurückbezahlen)

refund (n.) ECON **Rückerstattung**
= reimbursement; redemption; repayment = Rückzahlung; Rückvergütung; Kostenerstatung
refundable (→repayable) ECON (→rückzahlbar)
refurbish TECH **aufpolieren**
= spruce up ≈ erneuern
≈ renovate
refurbishment TECH **Aufpolierung**
≈ renewal ≈ Erneuerung 1
refuse (n.) QUAL **Ausschuß**
= rejects (pl.t.)
≈ waste (n.)
refuse (→garbage) COLLOQ (→Müll)
refuse (→deny) COLLOQ (→verweigern)
refuse a patent ECON **ein Patent versagen**
refused call TELEPH **abgelehntes Gespräch**
reg (→register) OFFICE (→Register)
REG (→random sequence generator) DATA PROC (→Zufallsgenerator)
regenerability ELECTRON **Regenerierbarkeit**
regenerate ELECTRON **regenerieren**
regeneration TRANS **Regenerierung**
  = Regeneration
regeneration ELECTRON **Regenerierung**
regeneration site (→regenerator station) TRANS (→Regeneratorstelle)
regenerative detector CIRC.ENG **Audion**
= grid-leak detector = Regenerativempfänger
regenerative memory DATA PROC **regenerativer Speicher**
= regenerative storage; regenerative store = Regenerativspeicher
  ↑ Laufzeitspeicher
regenerative repeater TELEGR **Fernschreibentzerrer**
regenerative repeater TRANS **Regenerator**
= regenerator = Regenerationsverstärker
  ≈ Leitungsverstärker; Zwischenverstärker
regenerative storage DATA PROC (→regenerativer Speicher)
(→regenerative memory)
regenerative store DATA PROC (→regenerativer Speicher)
(→regenerative memory)
regenerator (→regenerative repeater) TRANS (→Regenerator)
regeneratorless cable TELEC (→repeaterloses Kabel)
(→repeaterless cable)
regenerator section TRANS **Regeneratorfeld**
  [of digital systems] [von Digitalsystemen]
≈ repeater section ≈ Verstärkerfeld
↑ section ↑ Feld
regenerator spacing TRANS **Regeneratorfeldlänge**
  [of digital systems] [von Digitalsystemen]
= regenerator span; elementary cable section (CCITT) = Regeneratorabstand
  ≈ repeater spacing 2
↑ section length ↑ Feldlänge
regenerator span (→regenerator spacing) TRANS (→Regeneratorfeldlänge)
regenerator station TRANS **Regeneratorstelle**
= regeneration site = Regenerationspunkt
↑ repeater station ↑ Zwischenstelle
regime (→operation mode) TECH (→Betriebsart)
region COLLOQ **Gebiet**
= zone = Region; Zone
region (→range 1) TECH (→Bereich)
region (→zone) MICROEL (→Zone)
regional air route AERON **Regionalflugverbindung**
regional beam SAT.COMM **Regionalbündel**
regional beam antenna SAT.COMM **Regionalstrahlantenne**
  = Zonenantenne
Regional Bell Operating Company TELEC **regionale Bell-Betreibergesellschaft**
= RBOC; Baby Bell
regional exchange SWITCH **Zentralvermittlungsstelle**
  [represents, e.g. in the network of Deutsche Bundespost, the highest switching level] [höchste Netzebene der Landesfernwahl der DBP]
  = ZVSt
regional identity code SWITCH **Weltnumerierungszone**
  [first digit in front of the country code, to select the continent. e.g. 1 for North America or 4 for Central and Northern Europe] [erste Ziffer der internationalen Kennzahl, kennzeichnet den Weltkontinent, z.B. 1 für Nordamerika, 4 für Zentral- und Nordeuropa]
↑ code number ↑ Kennzahl

regional illumination spot SAT.COMM
= regional spot
regional office (→branch office) ECON
regional spot (→regional SAT.COMM
illumination spot)
regional traffic TELEC
region of integration MATH
(→integration interval)
region sequence (→zone MICROEL
sequence)
registartion TYPOGR
[exact matching of front with rear
side]
register (v.t.) INSTR
= record (v.t.); log (v.t.)
register (n.) DATA PROC
[memory with short access time, for
the temporary storage of small
amount of datas]
↓ instruction register; index register; interrupt register

register (n.) OFFICE
[a generally alphanumeric directory]
= reg
↑ directory
register 2 SWITCH
register 1 (→signaling SWITCH
circuit)
register (→perceive) COLLOQ
register addressing DATA PROC
register-controlled call set-up SWITCH
= register-controlled connection set-up;
register-controlled link set-up
register-controlled connection SWITCH
set-up (→register-controlled call
set-up)
register-controlled link set-up SWITCH
(→register-controlled call set-up)
registered design ECON
= registered utility model (AM); utility model
registered letter (→registered POST
mail)
registered mail (AM) POST
= registered post (BRI); registered letter
registered office ECON
≈ corporate headquarter
registered post (BRI) (→registered POST
mail)
registered sign ECON
registered trademark ECON
registered-trademark sign ECON
registered utility model (AM) ECON
(→registered design)
register instruction DATA PROC
≠ memory instruction
register memory DATA PROC
= register store; register storage
≈ general register
register pair DATA PROC
register select DATA PROC
register signal SWITCH
= interregister signal
register storage (→register DATA PROC
memory)
register store (→register DATA PROC
memory)
register system SWITCH
↑ indirect-control switching system

regionale Ausleuchtzone
(→Filiale)
(→regionale Ausleuchtzone)

Regionalverkehr
(→Integrationsintervall)

(→Zonenfolge)

Register
[das genaue Aufeinanderpassen von Vorder- und Rückseite]
= Registerhaltung
registrieren
= aufzeichnen
≈ Speichern
Register (n.n.)
[Speicher kurzer Zugriffszeit, für vorübergehende Speicherung geringer Datenmengen]
↓ Befehlsregister; Indexregister; Unterbrechungsregister; Basisadreßregister; Akkumulator
Register
[ein meist alphanumerisch geordnetes Verzeichnis]
↑ Verzeichnis
Datensatz
(→Signalisierungssatz)

(→wahrnehmen)
Registeradressierung
teilversetzter Verbindungsaufbau
= teilversetzte Verbindungsherstellung

(→teilversetzter Verbindungsaufbau)

(→teilversetzter Verbindungsaufbau)
Gebrauchsmuster

Einschreiben
[Brief]

Firmensitz
≈ Firmenzentrale
(→Einschreiben)

Gebrauchsmusterzeichen
eingetragenes Warenzeichen
Markenschutzzeichen
(→Gebrauchsmuster)

Registerbefehl
= Registeranweisung
≠ Speicherbefehl
Registerspeicher
≈ allgemeines Register

Registerpaar
Registerauswahl
Registerzeichen

(→Registerspeicher)

(→Registerspeicher)

Registerwählsystem
[Steuern der Koppeleinrichtungen über Register,

registrated user DATA PROC
registration SWITCH
registration (→storage) ELECTRON
registration (→perception) COLLOQ
registration accepted SWITCH
registration completion SWITCH
registration fee ECON
registration of trademark ECON
registration request SWITCH
registry (→documentation filing OFFICE
department)
regress (v.i.) (→decrement) TECH
regression MATH

regression analysis DATA PROC
regression coefficient MATH
regressive (→retrograde) TECH
regressive count (→down-count) TECH
regressive counter (→down CIRC.ENG
counter)
regroup TECH
= reschedule
regrouping TECH
= rescheduling
regulable TECH
≈ controllable
regular TECH
= routine; ordinary
≈ scheduled; usual
regular PHYS
[crystal]
= cubic
regular (→medium) TYPOGR
regular (→usual) COLLOQ
regular costumer ECON
regular frequency position TRANS
= regular position
↑ transmission position
regularity TECH
regular line TELEC
regular mode LING
≠ contracted mode
regular polyhedron MATH
[polyhedron whoes faces are all identical]
↓ tetrahedron; cube; octahedron; dodecahedron; icosahedron

regular position (→regular TRANS
frequency position)
regular reflection PHYS

regular span OUTS.PLANT
regular-standby TECH
(→operation-standby)
regulate CONTROL
≈ control

regulating amplifier CIRC.ENG
= automatic gain control amplifier; AGC amplifier
regulating characteristic CONTROL
regulating choke (→regulating EL.TECH
inductor)
regulating inductor EL.TECH
= regulating choke
regulating valve TECH
= controlling valve
regulation ECON
= settlement; rule
≈ prescription
regulation (→closed-loop CONTROL
control)

in dem die Rufnummern zwischengespeichert und ausgewertet werden]
= Registersystem
↑ indirekt gesteuertes Vermittlungssystem

registrierter Benutzer
Registrierung
(→Speicherung)
(→Wahrnehmung)
Registrierungsannahme
Registrierungsvollzug
Teilnahmegebühr
Warenzeicheneintragung
Registrierungsaufforderung
(→Registratur)

(→zurückrücken)
Regression
≠ Progression
Regressionsanalyse
Regressionskoeffizient
(→rückläufig)
(→Rückwärtszählung)
(→Rückwärtszähler)

umschichten
= umgruppieren
Umschichtung
= Umgruppierung
regulierbar
= regelbar 2
≈ steuerbar
regelmäßig
= routinemäßig
≈ planmäßig; üblich
regulär
[Kristall]
= kubisch
(→normal)
(→üblich)
Stammkunde
Regellage
↑ Übertragungslage

Regelmäßigkeit
Betriebsleitung
Lentoform
≠ Allegroform
reguläres Polyeder
[Polyeder deren Flächen alle gleich sind]
↓ Tetraeder; Würfel; Oktaeder; Dodekaeder; Ikosaeder

(→Regellage)

Spiegelung
= reguläre Reflexion
Regelspannweite
(→Betrieb-Ersatz)

regeln
= regulieren
≈ steuern
Regelverstärker
= Reglerverstärker

Regelkennlinie
(→Regeldrossel)

Regeldrossel
= regulating choke
Regelventil
= Steuerventil; Stellventil
Regelung
= Regel
≈ Vorschrift
(→Regelung)

regulation circuit	CIRC.ENG	**Regelschaltung**
regulation loop	CIRC.ENG	**Regelspannungsschleife**
regulation range	CONTROL	**Regelbereich**
= control range		= Regelumfang
≈ dynamic range [ELECTRON]		≈ Dynamikbereich [ELECTRON]
regulation	ECON	**Bestimmung**
≈ ordinance		= Regelung
↓ executive ordinance		= Verordnung
		↓ Ausführungsbestimmungen
regulations (→executive regulation)	ECON	(→Vollzugsordnung)
regulator (→controller)	CONTROL	(→Regler)
regulator tube	ELECTRON	**Regelröhre**
= variable mu tube		
regulatory authority	ECON	**Überwachungsbehörde**
		= Kontrollbehörde
regulatory standard (→prescription)	ECON	(→Vorschrift)
rehash	DATA PROC	**Rehash**
reimburse (→repay)	ECON	(→zürückbezahlen)
reimbursement (→refund)	ECON	(→Rückerstattung)
reimbursment (→compensation)	ECON	(→Schadenersatz)
Reinartz antenna (→Reinartz radar antenna)	ANT	(→Reinartz-Radarantenne)
Reinartz loop (→Reinartz radar antenna)	ANT	(→Reinartz-Radarantenne)
Reinartz radar antenna	ANT	**Reinartz-Radarantenne**
= Reinartz antenna; Reinartz loop		= Reinartz-Antenne
reinforce	TECH	**verstärken**
= strengthen		≈ versteifen
≈ stiffen		
reinforced concrete	CIV.ENG	**Stahlbeton**
= steel concrete		
reinforced concrete tower	CIV.ENG	**Stahlbetonturm**
= steel concrete tower		
reinforcement	MECH	**Verstärkung**
= strengthening		≈ Versteifung
≈ stiffening		
reinforcement post	OUTS.PLANT	**Stützpfahl**
= reinforcement stub		
reinforcement stub (→reinforcement post)	OUTS.PLANT	(→Stützpfahl)
reinforcing fin (→reinforcing rib)	MECH	(→Verstärkungsrippe)
reinforcing rib	MECH	**Verstärkungsrippe**
= reinforcing fin		
reinforcing spring	MECH	**Überfeder**
reinsert	TRANS	**Wiederbelegung**
= reuse		
reinsurance	ECON	**Rückversicherung**
reinsurance company	ECON	**Rückversicherungsgesellschaft**
reinvest	ECON	**reinvestieren**
reissue	DATA PROC	**Neuausgabe**
reject	TECH	**zurückweisen**
≈ outsort		aussondern
reject (→suppress)	ELECTRON	(→unterdrücken)
reject (→request for repeat)	DATA COMM	(→Wiederholungsaufforderung)
rejectable quality level	QUAL	**Rückweisgrenze**
= RQL; limiting quality; LQ		
rejection	DATA PROC	**Aussteuerung**
[punched cards]		[Lochkarten]
= outsorting		= Zurückweisung
rejection	QUAL	**Rückweisung**
		= Ablehnung
rejection (→nonacceptance)	SWITCH	(→Abweisung)
rejection 2 (→stop-band attenuation)	NETW.TH	(→Sperrdämpfung)
rejection 1 (→blocking)	NETW.TH	(→Sperrung)
rejection (→suppression)	ELECTRON	(→Unterdrückung)
rejection band (→stop band)	NETW.TH	(→Sperrbereich)
rejection number	QUAL	**Rückweiszahl**
		= Schlechtzahl
rejection region	QUAL	**Ablehnungsbereich**
rejector (→parallel-resonant circuit)	NETW.TH	(→Parallelschwingkreis)
rejector circuit (→parallel-resonant circuit)	NETW.TH	(→Parallelschwingkreis)
reject pocket	TERM&PER	**Fehlerfach**
[sorting device, reading device]		[Sortiergerät, Lesegerät]
= reject stacker		= Restfach
rejects (pl.t.) (→refuse)	QUAL	(→Ausschuß)
reject stacker (→reject pocket)	TERM&PER	(→Fehlerfach)
related (→connected)	TECH	(→zusammenhängend)
related term	LING	**Nebenbegriff**
≈ quasisynonym		≈ Quasisynonym
relation	MATH	**Relation**
[table with lines of equal type]		[Tabelle mit gleichartigen Zeilen]
relation (→equation)	MATH	(→Gleichung)
relational	MATH	**relational**
relational database	DATA PROC	**relationale Datenbank**
[permits to link data of different files by key words]		[erlaubt Verknüpfung von Daten unterschiedlicher Dateien mittels Schlüsselbegriffe]
		≠ Netzwerk-Datenbank
relational expression	DATA PROC	**Vergleichsausdruck**
relational operator	DATA PROC	**Vergleichsoperator**
[symbols ≧; ≦]		[Symbole ≧; ≦]
= comparator		↑ logischer Operator
↑ logic operator		
relational structure	DATA PROC	**relationale Datenstruktur**
relationship	ECON	**Verbindung**
= connection (NAM); connexion (BRI)		= Beziehung
relative	COLLOQ	**relativ**
≈ correspondent; pertinent		≈ entsprechend; betreffend
relative address (→floating address)	DATA PROC	(→relative Adresse)
relative addressing	DATA PROC	**relative Adressierung**
= base addressing		≈ relative Codierung
≈ relative coding		
relative bandwidth	NETW.TH	**relative Bandbreite**
		= normierte Bandbreite
relative clause	LING	**Relativsatz**
		= Bezugswortsatz; Bezugssatz
relative coding (→relative programming)	DATA PROC	(→relative Programmierung)
relative dielectric constant (→relative permittivity)	PHYS	(→Dielektrizitätszahl)
relative error	INSTR	**relativer Fehler**
relative frequency	MATH	**relative Häufigkeit**
= frequency ratio		
relative humidity	PHYS	**relative Feuchtigkeit**
		= relative Feuchte
relative hysteresis coefficient	PHYS	**relativer Hysteresebeiwert**
relative level	TELEC	**relativer Pegel**
= reference level; level difference		= Bezugspegel; Referenzpegel; Pegeldifferenz
relative measurement	INSTR	**Relativmessung**
= offset measurement		= Offsetmessung
relative motion	PHYS	**Relativbewegung**
= relative movement		
relative movement (→relative motion)	PHYS	(→Relativbewegung)
relative permeability	PHYS	**Permeabilitätszahl**
= permeability index		= relative Permeabilität
relative permittivity	PHYS	**Dielektrizitätszahl**
= permittivity ratio; relative dielectric constant		= relative Dielektrizitätskonstante; Elektrisierungszahl; Permittivitätszahl; relative Permittivität
		↑ Dielektrizitätskonstante
relative power level	TELEC	**relativer Leistungspegel**
relative programming	DATA PROC	**relative Programmierung**
[machine instructions with relative addressing]		[Maschinenbefehle mit relativer Adressierung]
= relative coding		= relative Codierung
relative pronoun	LING	**Relativpronomen**
[e.g. who, which, that]		[z.B. welcher]
		= bezügliches Fürwort; Relativum
relative redundancy	INF	**relative Redundanz**
relatives	MATH	**Meßzahlen**
		[Statistik]
relative voltage level	TELEC	**relativer Spannungspegel**
relativistic	PHYS	**relativistisch**

English	Domain	German
relativity	PHYS	Relativität
relativity theory	PHYS	Relativitätstheorie
relaxation	PHYS	Relaxation
[process toward steady state]		[Vorgang zur Erreichung des Gleichgewichtzustands]
≈ building-up transient		≈ Einschwingvorgang
relaxational dispersion	PHYS	Relaxationsdispersion
relaxation oscillation	EL.TECH	Kippschwingung
↓ sawtooth wave; square wave		= Relaxationsschwingung; ↓ Sägezahnschwingung; Rechtecksignal
relaxation oscillator	CIRC.ENG	Relaxationsoszillator
↓ toggle generator		↓ Kippgenerator
relaxation time	PHYS	Relaxationszeit
[to reach steady state]		[Zeit zur Erreichung des Gleichgewichtzustands]
relay (v.t.)	TELEC	weiterleiten
= extend; forward; route onward; transfer		= weitergeben
relay (n.)	COMPON	Relais
[device for electromagnetic activation of contacts]		[Bauteil zur elektromagnetischen Betätigung von Kontakten]
relay (→radio relay station)	RADIO	(→Relaisfunkstelle)
relay amplifier	CIRC.ENG	Relaisverstärker
relay broadcast	BROADC	Ballsendung
relay coil (→exciting coil)	ELECTRON	(→Erregerspule)
relay contact	COMPON	Relaiskontakt
↓ make contact; break contact; break-break contact; make-make contact; double-break-double-make contact; make-before-break contact		↓ Arbeitskontakt; Ruhekontakt; Zwillingsöffner; Zwillingsschließer; Zwillingswechsler; Folgeumschaltkontakt
relay controller	CONTROL	Relaisregler
relay correlator	CIRC.ENG	Relaiskorrelator
relay demodulator (→repeater demodulator)	RADIO REL	(→Zwischenstellendemodulator)
relay driver	ELECTRON	Relaistreiber
relay extractor	ELECTRON	Relaiszieher
relay group	COMPON	Relaissatz
= relay set		
relay interrupter	TELEGR	Relaisunterbrecher
relay list	SWITCH	Relaisübersicht
		= Relaisspiegel
relay modem (→repeater modem)	RADIO REL	(→Zwischenstellenmodem)
relay modulator (→repeater modulator)	RADIO REL	(→Zwischenstellenmodulator)
relay multiplicator	CIRC.ENG	Relaismultiplikator
relay receiver	BROADC	Ballempfänger
= retransmission receiver		
relay reception	BROADC	Ballempfang
relay repeater	SWITCH	Relaisübetragung
		= Übertragung
relay set (→relay group)	COMPON	(→Relaissatz)
relay site (→radio relay repeater station)	RADIO REL	(→Relaisstelle)
relay station	RADIO REL	Zwischenstelle
= repeater station		= Relaisstelle
relay storage	DATA PROC	Relaisspeicher
relay system (→transit system)	DATA COMM	(→Transitsystem)
releasable (→detachable)	MECH	(→abnehmbar)
release (v.t.)	SWITCH	auslösen
[of a connection, to pass from busy to idle state]		[einer Verbindung, von Belegt- in Freizustand überführen]
= clear; clear down		
↓ release back		↓ rückauslösen
release (v.t.)	TELEC	freigeben
release (n.)	COMPON	Abfall
[relay, return to rest position]		[Relais, Rückkehr in die Ruhelage]
= drop (n.)		
release (n.)	DOC	Freigabe
[of a documentation]		[einer Dokumentation]
release (n.)	DATA PROC	Freigabe
[of software]		[von Software]
= enable		
≈ release version		≈ Software-Version
release	SWITCH	Auslösung
= clearing forward; clearance; disconnection; disconnect		= Auslösen; Trennung; Trennen
↓ forward release; backward release		↓ Vorwärtsauslösung; Rückwärtsauslösung
release (v.t.)	DATA PROC	(→auslösen)
(→trigger)		
release (n.) (→cut-out)	POWER SYS	(→Auslöser)
release (→approval)	QUAL	(→Freigabe)
release (n.) (→exemption)	ECON	(→Freistellung)
release (n.)	ELECTRON	(→Auslösung)
(→triggering)		
release (→release version)	DATA PROC	(→Software-Version)
release (→trip)	POWER SYS	(→auslösen)
release (→trip)	POWER SYS	(→Auslösung)
release back	SWITCH	rückauslösen
↑ release		↑ auslösen
release button	ELECTRON	Auslöseknopf
release current (→drop current)	COMPON	(→Abfallstrom)
release current (→blowing current)	COMPON	(→Auslösestrom)
release delay	COMPON	Abfallverzögerung
[relay]		[Relais]
= off-delay; dropping delay		
release guard	SWITCH	Auslösequittung
release-guard signal	SWITCH	Auslösequittungszeichen
release key	TERM&PER	Auslösetaste
= action key		≈ Funktionstaste
≈ function key		
release number	DATA PROC	Versionsnummer
= version number		
release number (→issue number)	DOC	(→Ausgabestand)
release signal	SWITCH	Auslösezeichen
= clear forward signal		
release signal (→trip signal)	POWER SYS	(→Auslösesignal)
release time	COMPON	Abfallzeit
[relay]		[Relais]
= drop time		
release time	SWITCH	Freiwerdezeit
release time (→decay time)	ELECTRON	(→Abfallzeit)
release version	DATA PROC	Software-Version
= release; software version; version		= Version; Software-Fassung; Fassung
↓ program release		≈ Freigabe
		↓ Programmversion
relevant	SCIE	relevant
= pertinent		= sachbezogen
reliability	QUAL	Zuverlässigkeit
= dependability 1		
reliability function	QUAL	Überlebenswahrscheinlichkeit
		= Zuverlässigkeitsfunktion
reliability in operation (→service reliability)	QUAL	(→Betriebszuverlässigkeit)
reliability prediction	QUAL	Zuverlässigkeitsprognose
reliability test	QUAL	Zuverlässigkeitstest
		= Zuverlässigkeitsprüfung
reliable	QUAL	zuverlässig
= dependable		
reliable transfer server	DATA COMM	gesicherter Transfer-Server
= RTS		
relief (v.t.)	TECH	entlasten
relief (n.)	TECH	Entlastung
relief printing	TYPOGR	(→Hochdruck)
(→letterpress)		
relief valve (→overpressure valve)	TECH	(→Überdruckventil)
relinquishment	ECON	(→Verzicht)
(→renunciation)		
reload 2 (v.t.)	DATA PROC	nachladen
reload (v.t.)	ECON	umladen
= tranship (v.t.)		
reload 1 (v.t.)	DATA PROC	(→neustarten)
(→restart)		
reloading	ECON	Umladung
= transshipment		

reloading (→restart) DATA PROC (→Neustart)
relocatability DATA PROC **Relativierbarkeit**
= Verschiebbarkeit; Versetzbarkeit; Umsetzbarkeit
relocatable DATA PROC **relativierbar**
[executable on any memory region] [auf beliebegem Speicherplatz betreibbar]
= verschiebbar; versetzbar; umsetzbar
relocatable address DATA PROC **relativierbare Adresse**
= verschiebbare Adresse; versetzbare Adresse; umsetzbare Adresse
relocatable binary code (→RB code) CODING (→RB-Code)
relocatable loader DATA PROC **Relativlader**
[program to load programs into the main memory, with freely selectable load address] [Programm zum Laden von Programmen in de Arbeitsspeicher, mit frei wählbarer Ladeadresse]
= relocating loader
≠ absolute loader ≠ Absolutlader
relocatable program DATA PROC **Relativprogramm**
[loadable in any region of main memory] [an jeder Stelle des Hauptspeichers ladbar]
= self-relocating program = relativierbares Programm
relocate (→shift) DATA PROC (→verschieben)
relocating loader DATA PROC **Relativlader**
(→relocatable loader)
relocation COLLOQ **Verschiebung 1**
[räumlich]
= Verlegung 1
relocation (→shifting) DATA PROC (→Verschiebung)
relocation information DATA PROC **Relativierungsinformation**
= Verschiebungsinformation; Versetzungsinformation; Umsetzungsinformation
reluctance (→magnetic reluctance) EL.TECH (→magnetischer Widerstand)
reluctivity PHYS **Reluktivität**
[reciprocal of permeability] [Kehrwert der Permeabilität]
= specific magnetic resistance; magnetic resistivity = spezifischer magnetischer Widerstand
≠ permeability ≠ Permeabilität
rem PHYS **Rem**
[unit for equivalent dose] [Maßeinheit für Äquivalentdosis]
re-machine (→re-tool) MECH (→nachbearbeiten)
remagnification TECH **Rückvergrößerung**
remagnify TECH **rückvergrößern**
remainder (→remaining stock) ECON (→Restbestand)
remaining amount ECON **Restbetrag**
= residual amount
remaining error rate (→residual error rate) CODING (→Restfehlerrate)
remaining stock ECON **Restbestand**
= remainder
remains (→residue) TECH (→Rückstand)
remanence (→residual magnetism) PHYS (→Restmagnetisierung)
remanence ratio PHYS **Remanenzverhältnis**
remanent induction PHYS **Remanenzinduktion**
remanent magnetism (→residual magnetism) PHYS (→Restmagnetisierung)
remanent magnetization (IEC) (→residual magnetism) PHYS (→Restmagnetisierung)
remanent permeability PHYS **remanente Permeabilität**
= Remanenzpermeabilität
remanent polarization PHYS **Remanenzpolarisation**
remanent relay COMPON **Haftrelais**
[works with a permanent relay] [funktioniert mit einem Dauermagnet]
= latching relay; locking relay = Remanenzrelais; Halterelais; Einklinkrelais; Verklinkrelais; Einrastrelais
↑ reed relay ↑ Schutzrohrkontakt-Relais
remanent voltage PHYS **Remanenzspannung**
remedy (→countermeasure) TECH (→Gegenmaßnahme)
remelt process MICROEL **Nachschmelzverfahren**
= melt-back technique

reminder (→divide reminder) MATH (→Divisionsrest)
reminder (→dun) ECON (→Mahnung)
reminder lamp EQUIP.ENG **Erinnerungslampe**
remit (→transfer) ECON (→überweisen)
remittance ECON **Überweisung**
[of money] = Transfer; Geldüberweisung
= transfer 2 ≈ Zahlung
remittance (→short-circuit reverse-transfer admittance) MICROEL (→Kurzschluß-Rückwärtssteilheit)
remittee ECON **Geldempfänger**
[receiver of money]
= payee
remodel TECH **umbauen**
= rearrange = umgestalten; umrangieren; umordnen; umdisponieren
≈ change ≈ ändern
remodelling TECH **Umbau**
= modification; rearrangement = Modifikation; Umgestaltung
remote TELEC **abgesetzt**
[costumer, station, office, equipment] [Teilnehmer, Amt, Gerät]
= detached; off-premises; link-attached [DATA COMM] = entfernt; abgelegen
remote TECH **entfernt**
= distant = entlegen
= detached ≈ abgesetzt
remote (→tele...) INF.TECH (→Fern...)
remote access DATA PROC **Fernzugriff**
remote action technique TELEC (→Fernwirktechnik)
(→telecontrol engineering)
remote alarm TELECONTR **Fernmeldung**
remote alarm reception TRANS **Fernalarmempfang**
= RMT
remote antenna (→remotely controlled antenna) ANT (→Motorantenne)
remote area DATA COMM **Fernbereich**
remote batch (→remote batch processing) DATA COMM (→Stapelfernverarbeitung)
remote batch computing DATA COMM (→Stapelfernverarbeitung)
(→remote batch processing)
remote batch processing DATA COMM **Stapelfernverarbeitung**
= remote batch computing; remote batch = Fernstapelverarbeitung; Remote-batch-Processing
≠ local batch processing ≠ Ortsstapelverarbeitung
↑ batch processing ↑ Stapelverarbeitung
remote command TELECONTR (→Fernsteuern)
(→telecommand)
remote communication computer DATA COMM (→Netzknotenrechner)
(→network-node computer)
remote computing (→remote data processing) DATA PROC (→Datenfernverarbeitung)
remote concentrator SWITCH **Konzentrator 2**
[a concentrating device of a switch, dislocated to the subscriber end for pair saving purposes] [eine zur Einsparung von Anschlußleitungen, am Teilnehmende abgesetzte Konzentrationsstufe einer Vermittlung]
≈ stand-alone concentrator = Leitungskonzentrator 2; Verkehrskonzentrator; abgesetzterKonzentrator
↑ front-end equipment ≈ Wählsterneinrichtung
↑ Vorfeldeinrichtung
remote connection DATA COMM **Fernanschluß**
remote control CONS.EL **Fernbedienung**
remote control 1 CONTROL **Fernsteuerung 1**
[process] [Vorgang]
remote control 2 CONTROL **Fernsteuerung 2**
[device] [Vorrichtung]
remote control (→telecontrol command) TELECONTR (→Fernwirkbefehl)
remote control (→remote sensing) INSTR (→Fernfühlung)
remote control feature TECH **Fernsteuerbarkeit**
remote control switch TELECONTR **Fernschalter**
remote control unit TELECONTR **Fernschaltgerät**
[subscriber equipment for the connection to a telex or data network, pereforms all functions to establish or dissolve a connection] [Teilnehmergerät zum Anschluß an das Fernschreib- oder Datennetz, erfüllt alle zum Aufbau und Lösen der Verbindungen erforderlichen Funktionen]
= signalling unit

remote data processing	DATA PROC	Datenfernverarbeitung	
= teleprocessing; remote computing; remote processing		= Fernverarbeitung; Teleprocessing	
		≈ Datenübertagung; Datenübertragung; Datenkommunikation	
remote data processing system	DATA PROC	Datenfernverarbeitungssystem	
remote data transmission	DATA COMM	Datenfernübertragung	
= long-distance data transmission		= Fernübertragung	
		≈ Datenübertragung; Datenfernverarbeitung; Datenkommunikation; Datenübermittlung	
remote enquiry	TERM&PER	Fernabfrage	
= remote inquiry			
remote equipment (→front-end equipment)	TELEC	(→Vorfeldeinrichtung)	
remote error sensing (→remote sensing)	INSTR	(→Fernfühlung)	
remote exchange	SWITCH	fremdes Vermittlungsamt	
remote feeding (→remote power feeding)	TRANS	(→Fernspeisung)	
remote feeding current	TRANS	Fernspeisestrom	
= remote power feeding current			
remote feeding section	TRANS	Fernspeiseabschnitt	
remote feeding unit	TRANS	Fernspeiseeinsatz	
		= Fernspeisegerät; Fernspeisestromversorgung	
remote feeding voltage	TRANS	Fernspeisespannung	
= remote power feeding voltage; line powering voltage			
remote front-end processor (→network-node computer)	DATA COMM	(→Netzknotenrechner)	
remote indication	TELECONTR	Fernanzeige	
remote indication (→teleindication)	TELECONTR	(→Fernanzeigen)	
remote inquiry (→remote enquiry)	TERM&PER	(→Fernabfrage)	
remote interlock	ELECTRON	Fernverriegelung	
remote job entry	DATA PROC	Auftragsferneingabe	
= RJE		= Auftagsfernverarbeitung; Jobferneingabe; Jobfernverarbeitung	
remote loader control	DATA COMM	Fernladesteuerung	
remote location	TELECONTR	Fernlokalisierung	
		= Fernortung	
remote loop	TELEC	ferne Schleife	
remotely actuated (→remotely operated)	TECH	(→fernbetätigt)	
remotely configure	DATA PROC	fernkonfigurieren	
		= ferneinstellen	
remotely controlled	TECH	ferngesteuert	
= far-end controlled		= fernbetätigt	
≈ remotely actuated			
remotely controlled antenna	ANT	Motorantenne	
= remote antenna		≈ fernbediente Antenne	
remotely fed	TELEC	ferngespeist	
= line-powered			
remotely operated	TECH	fernbetätigt	
= far-end operated; remotely actuated; far-end actuated		= fernbetrieben	
		≈ ferngesteuert	
≈ remotely controlled			
remote measuring	TELECONTR	(→Fernmessen)	
(→telemetering)			
remote meintenance	DATA PROC	(→Fernwartung)	
(→teleservice)			
remote off-control	TELECONTR	Fernabschaltung	
remote operation	TELECONTR	Fernbetrieb	
remote power feeding	TRANS	Fernspeisung	
= remote feeding; line powering			
remote power feeding current (→remote feeding current)	TRANS	(→Fernspeisestrom)	
remote power feeding voltage (→remote feeding voltage)	TRANS	(→Fernspeisespannung)	
remote printer	TERM&PER	Ferndrucker	
[for batch terminals]		[in Stapelstationen]	
≈ teleprinter		≈ Fernschreiber	
remote processing (→remote data processing)	DATA PROC	(→Datenfernverarbeitung)	
remote programming (→remote sensing)	INSTR	(→Fernfühlung)	
remote ring	DATA COMM	Fernring	
remote sensing	INSTR	Fernfühlung	
= remote error sensing; remote control; remote programming		= Fernfehlerfühlung; Fernsteuerung; Fernprogrammierung	
remote setting	TECH	Ferneinstellung	
remote site	TELEC	abgesetzter Standort	
remote station	TELEC	fernes Amt	
remote subscriber	TELEC	abgesetzter Teilnehmer	
remote supervision (→telemonitoring)	TELECONTR	(→Fernüberwachen)	
remote supervisory (→telemonitoring)	TELECONTR	(→Fernüberwachen)	
remote terminal	DATA COMM	abgesetztes Endgerät	
remote transmission procedure	DATA COMM	Fernübertragungsprozedur	
removable (→detachable)	MECH	(→abnehmbar)	
removable disk	TERM&PER	Wechselplatte	
= exchangeable disk		[auswechselbare]	
≠ fixed disk		≠ Festplatte	
removable-disk drive	TERM&PER	Wechselplattenlaufwerk	
= exchangeable-disk drive; EDD		= Wechselplattenantrieb; auswechselbares Plattenlaufwerk	
≈ moving-disk memory			
≠ fixed-disk drive			
		≈ Wechselplattenspeicher	
		≠ Festplattenlaufwerk 1	
removable-disk memory (→moving-disk memory)	TERM&PER	(→Wechselplattenspeicher)	
removable-disk storage (→moving-disk memory)	TERM&PER	(→Wechselplattenspeicher)	
removable medium (→removable storage medium)	TERM&PER	(→Wechselspeichermedium)	
removable storage medium	TERM&PER	Wechselspeichermedium	
= removable medium			
removal	COLLOQ	Wegnahme	
		= Beseitigung	
removal (→dismantlement 1)	TECH	(→Abbau)	
removal from service	TECH	Außerbetriebnahme	
remove	MECH	ablösen	
		≈ abblättern	
remove	SWITCH	aushängen	
[from queue]		[aus Warteschlange]	
= unlink			
remove	TECH	entfernen	
≈ dismantle; detach		≈ abbauen; lösen	
remove (→erase)	DATA PROC	(→löschen)	
removeable	TECH	entfernbar	
removed section	ENG.DRAW	Profilschnitt außerhalb der Ansicht	
remove from service (→take out of service)	TECH	(→außerbetriebsetzen)	
remove rust	METAL	entrosten	
remove the sheet	EL.TECH	abmanteln	
[of a cable]		[Kabel]	
= cut back the insulation			
remunerate	ECON	vergüten	
		≈ erstatten	
remuneration	ECON	Vergütung	
≈ refund (n.)		≈ Rückerstattung	
remuneration (→payment)	ECON	(→Bezahlung)	
rename	TECH	umbenennen	
		= neu benennen	
rendering	DATA PROC	Bildaufbereitung	
[computer graphics]		[Computergraphik]	
= image generation		= Bildgenerierung; Bilderzeugung; künstlerische Aufbereitung	
rending rate (→debtor interest)	ECON	(→Sollzins)	
rendition (→reproduction)	TERM&PER	(→Wiedergabe)	
rendition characteristic (→reproducing characteristic)	TERM&PER	(→Wiedergabecharakteristik)	
renewal	TECH	Erneuerung 1	
[of an object]		[eines Gegenstandes]	
≈ renovation		≈ Aufpolierung	
≈ refurbishment			
renewal function (→replacement function)	MATH	(→Erneuerungsfunktion)	

renewal process (→replacement process) QUAL (→Erneuerungsprozess)
renovation (→renewal) TECH (→Erneuerung 1)
rent (n.) TECH **Riß 2**
[separation of parts] [Trennung der Teile]
rent 2 (AM) (→hire out) ECON (→vermieten)
rent 1 (→give on rent) ECON (→mieten)
rent (n.) (→rental 1) ECON (→Vermietung)
rental 1 ECON **Vermietung**
= rent (n.)
rental (a fixed periodical pecunary burden) (→charge 2) ECON (→Gebühr)
rental 2 (→rental tariff) ECON (→Mietgebühr)
rental charge (→basic rental) TELEC (→Grundgebühr)
rental charge (→rental tariff) ECON (→Mietgebühr)
rental tariff ECON **Mietgebühr**
= rental charge; rental 2; hire (n.) = Miete
renumber TECH **umnumerieren**
= umbeziffern; neunumerieren
renunciation ECON **Verzicht**
= waiver; relinquishment
reorder (→repeat order) ECON (→Nachbestellung)
reorder point MANUF **Nachbestellungsschwelle**
reorganization ECON **Umorganisation**
reorganize (→rearrange) DATA PROC (→reorganisieren)
repack TECH **umpacken**
repair (v.t.) TECH **instandsetzen**
= recondition = reparieren; wiederinstandsetzen; ausbessern
≈ restore; rework
↑ maintain ≈ wiederherstellen; überarbeiten
↑ warten
repair (n.) TECH **Instandsetzung**
= fault clearance; fault clearing; reconditioning; debugging = Wiederinstandsetzung; Reparatur; Störungsbeseitigung; Entstörung; Ausbesserung; Fehlerbehebung; Fehlerbeseitigung
≈ corrective maintenance [QUAL]; restoration
≈ korrigierende Wartung [QUAL]; Wiederherstellung
repair desk (→complaints desk) TELEC (→Störungsannahmeplatz)
repair desk service (→fault complaint service) TELEPH (→Störungsannahme)
repair instruction TECH **Reparaturanweisung**
= Instandsetzungsanleitung
repair order ECON **Reparaturauftrag**
repair service TECH **Reparaturdienst**
repay ECON **tilgen**
= redeem = abzahlen
≈ amortize ≈ amortisieren
repay ECON **zürückbezahlen**
= reimburse; refund = erstatten; rückerstatten; zurückerstatten; abzahlen
≈ amortize; redeem ≈ amortisieren; tilgen; vergüten
repayable ECON **rückzahlbar**
= refundable; redeemable
repayment (→refund) ECON (→Rückerstattung)
repayment (→redemption) ECON (→Tilgung)
repeat (v.t.) DATA PROC **wiederholen**
= rerun; retry; iterate
repeat (n.) DATA COMM **Wiederholung**
= retry (n.)
repeat (→rerun 1) DATA PROC (→Wiederholungslauf)
repeatability TERM&PER **Wiederholgenauigkeit**
[plotter] [Plotter]
repeatability INSTR **Wiederholgenauigkeit**
[width of accuracy window under identical conditions] [Breite des Streubereichs unter konstanten Bedingungen]
repeatability (→reproducibility) QUAL (→Wiederholbarkeit)
repeat-action key (→run-out key) TERM&PER (→Dauertaste)
repeat counter DATA PROC **Wiederholzähler**
repeat decimal number MATH **periodischer Dezimalbruch**
[e.g. 0.3333..] [z.B. 0,3333..]

repeated call attempt (→redialing) SWITCH (→Anrufwiederholung)
repeater 1 (→line repeater) TRANS (→Leitungsverstärker)
repeater (→radio relay repeater station) RADIO REL (→Relaisstelle)
repeater 2 (→intermediate repeater) TRANS (→Zwischenverstärker)
repeater demodulator RADIO REL **Zwischenstellendemodulator**
= relay demodulator
repeater ferrite METAL **Übertrager-Ferrit**
repeaterless cable TELEC **repeaterloses Kabel**
= regeneratorless cable = regeneratorloses Kabel
↑ daisy-chain cable ↓ Girlandenkabel
repeater modem RADIO REL **Zwischenstellenmodem**
= relay modem
repeater modulator RADIO REL **Zwischenstellenmodulator**
= relay modulator
repeater section TRANS **Verstärkerfeld**
[of analog systems, by CCITT definition a cable section plus one repeater] [von Analogsystemen, nach CCITT-Definition ein Kabelabschnitt einschließlich eines Leitungsverstärkers]
≈ regenerator section
↑ section ≈ Regeneratorfeld
↑ Feld
repeater spacing TRANS **Verstärkerfeldlänge**
[analog systems] [Analogtechnik]
≈ regenerator spacing = Repeaterabstand
↑ section length ≈ Regeneratorfeldlänge
↑ Feldlänge
repeater station (→radio relay repeater station) RADIO REL (→Relaisstelle)
repeater station (→relay station) RADIO REL (→Zwischenstelle)
repeater station (→intermediate station) TRANS (→Zwischenstelle)
repeat feature (→repeat function) TERM&PER (→Wiederholfunktion)
repeat function TERM&PER **Wiederholfunktion**
[keyboard feature] [Tastatur]
= repeat feature; auto-repeat
repeating coil OUTS.PLANT **Trennübertrager**
repeat key TERM&PER **Wiederholtaste**
≈ run-out key = Wiederholungstaste; REPEAT-Taste
≈ Dauertaste
repeat order ECON **Nachbestellung**
= reorder = Nachorder
repeat specification (FORTRAN) (→replicator) DATA PROC (→Wiederholangabe)
reperforator TELEGR **Empfangslocher**
reperforator 1 TERM&PER **Lochstreifendoppler**
↑ taper = Streifendoppler; Lochstreifenübertrager
↑ Lochstreifengerät
reperforator 2 (→punched tape reader) TERM&PER (→Lochstreifenleser)
repertoire COLLOQ **Vorrat**
= set [figurativ]
repertoire (→instruction set) DATA PROC (→Befehlsvorrat)
repertoire of values (→set of values) MATH (→Wertevorrat)
repertory dialer TERM&PER **Namentaster**
repertory grid DATA PROC **Konstruktgitter-Verfahren**
[experts system] [Expertensysteme]
repetition frequency ELECTRON **Wiederholungsgeschwindigkeit**
= repetition rate
= Wiederholungsfrequenz; Folgefrequenz
repetition instruction (→rerun instruction) DATA PROC (→Wiederholbefehl)
repetition of transmission (→retransmission 1) TELEC (→Übertragungswiederholung)
repetition program (→rerun program) DATA PROC (→Wiederholprogramm)
repetition rate (→repetition frequency) ELECTRON (→Wiederholungsgeschwindigkeit)
repetition routine (→rerun program) DATA PROC (→Wiederholprogramm)

repetitive		TECH	wiederholt	representation (→notation)		MATH	(→Schreibweise)
= continual 2			≈ periodisch; kontinuierlich	representation mode		DOC	Darstellungsweise
≈ periodic; continuos				representative		ECON	Vertreter
repetitive addressing		DATA PROC	Wiederholungsadressierung	= sales agent; commercial agent			= Repräsentant; Verkaufs- agent
repetitive error		DATA PROC	Wiederholungsfehler	≈ representation			≈ Vertretung
			= wiederholter Fehler	representative (→deputy)		ECON	(→Stellvertreter)
repetitive sampling		ELECTRON	periodische Abtastung	represent graphically (→plot)		DOC	(→graphisch darstellen)
= periodic sampling				reprint (n.)		TYPOGR	Nachdruck
repetitive signal		INSTR	periodisches Signal				= Neudruck
= periodic signal; repetitive waveform				reproduce		TERM&PER	vervielfältigen
repetitive waveform (→repetitive signal)		INSTR	(→periodisches Signal)	reproducer (→punched card reproducer)		TERM&PER	(→Lochkartendoppler)
replace (v.t.)		TECH	ersetzen	reproducer (→phonograph pickup)		EL.ACOUS	(→Tonabnehmer)
= supersede			≈ auswechseln	reproducibility		TECH	Reproduzierbarkeit
≈ exchange				reproducibility		QUAL	Wiederholbarkeit
replaceable		TECH	ersetzbar	= repeatability			
≈ exchangeable			≈ auswechselbar	reproducing characteristic		TERM&PER	Wiedergabecharakteristik
replaceable unit		EQUIP.ENG	Austauscheinheit	= rendition characteristic; restitution characteristic			
replacement		TECH	Austausch	reproduction		TERM&PER	Wiedergabe
= exchange			= Auswechslung; Ersatz 1	= rendition; restitution			= Reproduktion
replacement (→reposition)		ECON	(→Wiederbeschaffung)	≈ fidelity			≈ Widergabetreue
replacement function		MATH	Erneuerungsfunktion	reproduction (→copy 2)		OFFICE	(→Pause)
= renewal function				reproduction quality (→fidelity)		EL.ACOUS	(→Wiedergabetreue)
replacement part (→spare part)		TECH	(→Ersatzteil)				
replacement process		QUAL	Erneuerungsprozess	reprogram (v.t.)		DATA PROC	umprogrammieren
= renewal process				reprograming		DATA PROC	Umprogrammierung
replacement scheme		NETW.TH	(→Ersatzschaltbild)	reprogrammable		MICROEL	wiederprogrammierbar
(→equivalent circuit)							= mehrfach programmierbar
replenish 1		TECH	auffüllen	reprogrammable read-only memory (→REPROM)		MICROEL	(→REPROM)
= fill up			↓ wiederauffüllen	reprographics		OFFICE	Reprographie
↓ replenish 2				REPROM		MICROEL	REPROM
replenish 2		TECH	nachfüllen	= reprogrammable read-only memory			= mehrfach programmierba- rer Festwertspeicher
= fill up again			= wiederauffüllen				
↓ replenish 1			↑ auffüllen	repulsion		PHYS	Abstoßung
replenishment 1		TECH	Auffüllung	repurchase		ECON	zurückkaufen
= filling			= Füllung	= redeem; buy back			
↓ replenishment 2			↓ Wiederauffüllung	repurchasing		ECON	Wiederkauf
replenishment 2		TECH	Wiederauffüllung	reputable (→well-known)		COLLOQ	(→namhaft)
[repetitive action of filling up again]			= Wiederaufstockung (fig.)	reputation		ECON	Ruf
↑ replenishment 1			↑ Auffüllung	= standing			
replicate gate		TERM&PER	Kopiertor	request (v.t.)		DATA COMM	abrufen
[magnetic bubble memory]			[Magnetblasenspeicher]	= call up; poll (v.t.)			
replicator (ALGOL)		DATA PROC	Wiederholangabe	request (v.t.)		SWITCH	anfordern
= repeat specification (FORTRAN)				request (n.)		SWITCH	Anforderung
repluggable		ELECTRON	umsteckbar	≈ call attempt			= Abrufen
↑ pluggable			↑ steckbar				≈ Wählversuch
reply (→answer)		TECH	(→Antwort)	request (n.)		TELECONTR	Aufforderung
report (n.)		LING	Bericht	= invitation			
= record; account				request (→inquiry)		DATA PROC	(→Abfrage)
report (COBOL) (→list)		DATA PROC	(→Liste)	request (→application)		ECON	(→Antrag)
report (→message)		TELECONTR	(→Meldung)	request (→call request)		SWITCH	(→Verbindungsanforderung)
reporter microphone		EL.ACOUS	Reporter-Mikrophon	request (v.i.) (→inquire)		DATA PROC	(→abfragen)
report file		DATA PROC	Listendatei	request communication		TELEC	Abrufkommunikation
report generator		DATA PROC	Listengenerator	requester		DATA COMM	Requester
= report program generator; RPG; re- port writer; list generator			= Listenprogrammgenerator; LPG; RPG	= client			≠ Server
report of the supervisory board		ECON	Aufsichtsratbericht	≠ server			
report program generator (→report generator)		DATA PROC	(→Listengenerator)	request for connection (→request for service)		SWITCH	(→Vermittlungswunsch)
report programm		DATA PROC	Listenprogramm	request for information		TELECONTR	Aufruf
[program for individual list printouts]			[Programm für individuelle Listenausdrucke]	request for proposal (→invitation to tender)		ECON	(→Ausschreibung)
report to		COLLOQ	unterstellt sein	request for quotation (→invitation to tender)		ECON	(→Ausschreibung)
report writer (→report generator)		DATA PROC	(→Listengenerator)	request for repeat		DATA COMM	Wiederholungsaufforderung
reposition		ECON	Wiederbeschaffung	= reject			
= replacement			= Wiederanschaffung; Neube- schaffung	request for service		SWITCH	Vermittlungswunsch
reposition time		ECON	Wiederbeschaffungszeit	= request for connection; bid			
represent		LING	darstellen	request for tender (→invitation to tender)		ECON	(→Ausschreibung)
= depict				request handling		DATA PROC	Anforderungsbearbeitung
represent		ECON	vertreten	= request servicing			
representation		DOC	Darstellung	request indicator		DATA COMM	Anforderungskennzeichnung
= notation				= request signal			
representation		DATA PROC	Darstellung	request mode		DATA COMM	Anforderungsbetrieb
= presentation				= contention mode			= Konkurrenzbetrieb
representation		ECON	Vertretung	≠ polling mode			≠ Abrufbetrieb
≈ representative			= Repräsentation				
			≈ Vertreter				
representation		ENG.DRAW	Darstellung				

request of offer (→ invitation to tender) ECON (→ Ausschreibung)
request servicing (→ request handling) DATA PROC (→ Anforderungsbearbeitung)
request signal (→ request indicator) DATA COMM (→ Anforderungskennzeichnung)
required power (→ power demand) EL.TECH (→ Leistungsbedarf)
required service RADIO NAV dauernd erforderliche Verbindung
= R service
requirement TECH **Anforderung**
= Erfordernis
requirement (→ claim 3) ECON (→ Forderung)
requirements analysis DATA PROC **Anforderungsstudie**
[software production] [Software-Erstellung]
requirements list (→ requirements specification) TECH (→ Anforderungskatalog)
requirements specification TECH **Anforderungskatalog**
= requirements list
requiring habituation (adj.) COLLOQ gewöhnungsbedürftig
requisition (→ call) ECON (→ Abruf)
reradiation RADIO **Rückstrahlung**
re-record (→ dub) CONS.EL (→ überspielen)
rereel (v.t.) (→ rewind) ELECTRON (→ zurückspulen)
rering SWITCH **Nachrufen**
reringing (→ reringing signal) SWITCH (→ Nachrufzeichen)
reringing signal SWITCH **Nachrufzeichen**
= reringing; recall signal
reroll (v.t.) (→ rewind) ELECTRON (→ zurückspulen)
reroute (→ redirect) TELEC (→ umleiten)
rerouting (→ redirection) TELEC (→ Umleitung)
rerouting address DATA COMM **Umleitadresse**
= redirection address
rerun 1 (n.) DATA PROC **Wiederholungslauf**
= repeat; retry = Wiederholung
rerun 2 (→ restart) DATA PROC (→ Neustart)
rerun (→ repeat) DATA PROC (→ wiederholen)
rerun instruction DATA PROC **Wiederholbefehl**
= repetition instruction = Wiederholungsbefehl
rerun point (→ checkpoint) DATA PROC (→ Fixpunkt)
rerun program DATA PROC **Wiederholprogramm**
= rerun routine; repetition program; repetition routine; rollback program; rollback routine = Wiederholungsprogramm; Wiederholungsroutine
rerun routine (→ rerun program) DATA PROC (→ Wiederholprogramm)
resale ECON **Wiederverkauf**
≈ intermediate trade ≈ Zwischenhandel
reschedule (→ regroup) TECH (→ umschichten)
rescheduling (→ regrouping) TECH (→ Umschichtung)
rescind (→ cancel) ECON (→ rückgängigmachen)
rescission ECON **Rücktritt 1**
[of a contract] [aus Vertrag]
rescission by the byer ECON **Wandelung**
[Rücktritt seitens des Käufers]
= Wandlung
rescue bacon RADIO LOC **Rettungsbake**
research SCIE **Forschung**
research activity SCIE **Forschungstätigkeit**
research and development ECON **Forschung und Entwicklung**
= R&D = F&E
research assistant SCIE **Forschungsassistent**
research center SCIE **Forschungszentrum**
researcher SCIE **Forscher**
= investigator ≈ Entwickler
≈ developer
research institute SCIE **Forschungsinstitut**
= research station ≈ Forschungsanstalt
≈ Versuchsanstalt
research lab (→ research laboratory) SCIE (→ Forschungslabor)
research laboratory SCIE **Forschungslabor**
= research lab
research station (→ research institute) SCIE (→ Forschungsinstitut)
research target SCIE **Forschungsziel**
reseat pressure TECH **Schließdruck**

reseize (v.t.) DATA PROC **Maßstab ändern**
≈ scale (v.t.) ≈ skalieren
resell ECON **wiederverkaufen**
reseller ECON **Wiederverkäufer**
= retailer ≈ Zwischenhändler
≈ intermediary
reservation ECON **Vorbehalt**
reservation (→ booking) ECON (→ Buchung)
reservation terminal TERM&PER (→ Buchungsplatz)
(→ booking terminal)
reserve (→ stand-by) TELEC (→ Ersatz)
reserve (n.) (→ stock) ECON (→ Vorrat)
reserve accumulator DATA PROC **Hilfsakkumulator**
reserve capacity TELEC **Ersatzkapazität**
reserved DATA PROC **reserviert**
[e.g. memory locations] [z.B. Speicherplätze]
= dedicated
reserved character DATA PROC **reserviertes Zeichen**
reserved word DATA PROC **reserviertes Wort**
reserves (→ provision 2) ECON (→ Rückstellung)
reset (v.t.) DATA PROC **rücksetzen**
[to restore the original state] [den Ausgangszustand wiederherstellen]
≈ zero (v.t.); erase (v.t.) [DATA PROC]; abort (v.t.) [DATA PROC] = rückstellen; rückschalten; zurücksetzen
≠ set (v.t.) ≈ nullen; löschen [DATA PROC]; abbrechen [DATA PROC]
≠ setzen
reset (n.) CIRC.ENG **Rücksetzen**
= resetting = Rücksetzung; Rückstellen; Rückstellung; Rückschalten; rücksetzen; Zurücksetzung
≈ zeroing; erasing [DATA PROC]
≈ Nullung; Löschung [DATA PROC]
reset (→ resetting) TECH (→ Rückstellung)
reset button (→ reset key 1) TERM&PER (→ Rücksetztaste 1)
reset confirmation DATA COMM **Rücksetzbestätigung**
reset control circuit CONTROL **Rückstellungsregelungskreis**
reset key EQUIP.ENG **Rücksetztaste**
≈ cancel key; correction key = Rückstelltaste
≈ Löschtaste; Korrekturtaste
reset key 1 TERM&PER **Rücksetztaste 1**
[resets the computer to start condition] [setzt den Computer in den Startzustand zurück]
= reset button = RESET-Taste
reset key 2 (→ backspace key) TERM&PER (→ Rücksetztaste 2)
reset line ELECTRON **Rücksetzleitung**
= Rückstelleitung
reset protocol DATA COMM **Rücksetzprotokoll**
reset pulse ELECTRON **Rücksetzimpuls**
= clear pulse; erase signal = Rückstellimpuls; Nullungsimpuls; Löschimpuls
reset relay (→ restoring relay) ELECTRON (→ Rückstellrelais)
reset signal CIRC.ENG **Rücksetzsignal**
reset switch EQUIP.ENG **Rücksetzschalter**
= resetting switch; restoring switch
reset terminal ELECTRON **Rücksetzeingang**
reset time CONTROL **Nachstellzeit**
resetting ELECTRON **Rückmagnetisierung**
[magnetic core] [Magnetspeicher]
resetting TECH **Rückstellung**
= reset
resetting (→ reset) CIRC.ENG (→ Rücksetzen)
resetting cause DATA COMM **Rücksetzgrund**
resetting switch (→ reset switch) EQUIP.ENG (→ Rücksetzschalter)
reset voltage (→ quiescent voltage) ELECTRON (→ Ruhespannung)
reshape TECH **umformen**
≈ rework ≈ umarbeiten
reship (→ return) ECON (→ zurücksenden)
residence telephone TELEPH **Privatanschluß**
≠ business telephone ≠ Geschäftsanschluß
residence time DATA PROC **Verweilzeit**
[storage] [Speicher]
= retention time; dwell time

residencial building (→dwelling) CIV.ENG (→Wohngebäude)
resident DATA PROC (→speicherresident)
(→memory-resident)
resident command DATA PROC **residentes Kommando**
= memory-resident command; main-memory-resident command; build-in command
residential costumer (→private TELEC (→Privatteilnehmer)
subscriber)
residential subscriber (→private TELEC (→Privatteilnehmer)
subscriber)
resident programm DATA PROC **residentes Programm**
= memory-resident program; main-memory-resident program
 = speicherresidentes Programm; hauptspeicherresidentes Programm; arbeitsspeicherresidentes Programm
resident software DATA PROC **residente Software**
= memory-resident software; main-memory-resident software
 = speicherresidente Software; hauptspeicherresidente Software; arbeitsspeicherresidente Software
residual amount (→remaining ECON (→Restbetrag)
amount)
residual BER (→residual bit error TELEC (→Grundfehlerquote)
rate)
residual BER (→background bit CODING (→Grundbitfehlerrate)
error rate)
residual bit error rate TELEC **Grundfehlerquote**
= residual BER
residual bit error rate CODING (→Grundbitfehlerrate)
(→background bit error rate)
residual-bulk resistance MICROEL (→Sättigungswiderstand)
(→saturation resistance)
residual current ELECTRON **Anlaufstrom**
[electron tubes] [Röhren]
residual current (→leakage ELECTRON (→Reststrom)
current)
residual echo TELEPH **Restecho**
residual echo level TELEPH **Restechopegel**
residual error MATH **Restfehler**
= uncorrected error
residual error probability MATH **Restfehlerwahrscheinlichkeit**
residual error rate CODING **Restfehlerrate**
= remaining error rate
residual hum (→hum) TELEC (→Brumm)
residual induction PHYS **remanente Induktion**
[ferromagnetism: value of residual magnetism]
 [Ferromagnetismus: Wert der Restmagnetisierung]
 = Remanenzflußdichte; Remanenz 2
residual interference (→residual INSTR (→Reststörsignal)
spurious)
residual magnetism PHYS **Restmagnetisierung**
[a ferroelectric effect]
= residual magnetization; remanent magnetism; remanent magnetization (IEC); remanence
≈ permanence
 [Effekt in ferromagnetischen Materialien]
 = Remanenz 1; Restmagnetismus; Remanenzmagnetisierung; remanente Magnetisierung
 ≈ Permanenz
residual magnetization (→residual PHYS (→Restmagnetisierung)
magnetism)
residual modulation MODUL **Trägerrauschpegel**
residual noise INSTR **Restrauschen**
residual phase noise ELECTRON **Restphasenrauschen**
residual power supply voltage TELEC (→Brumm)
(→hum)
residual ripple EL.TECH **Restwelligkeit**
≈ hum [TELEC]
↑ ripple
 = Brumm [TELEC]
 ↑ Welligkeit
residual short-circuit MICROEL **Kurzschluß-Reststrom**
current
= cutoff collector current, base and emitter shorted
residual spurious INSTR **Reststörsignal**
= residual interference = Restinterferenz
residual voltage ELECTRON **Restspannung**
residue TECH **Rückstand**
= remains

resignation ECON **Rücktritt 2**
[from an office] [von einem Amt]
resilience MECH **Federung**
 = Nachgiebigkeit; Federungsarbeit
resilient MECH **federnd**
resin 1 CHEM **Harz**
resin (→soldering flux) METAL (→Flußmittel 1)
resin 2 (→rosin) CHEM (→Kolophonium)
resin core METAL **Flußmittelseele**
resin-cored solder ELECTRON **Röhrenlötzinn**
resin-free CHEM **harzfrei**
resinous CHEM **harzhaltig**
resist (v.t.) TECH **widerstehen**
≈ endure ≈ überstehen
resist (→protective coating) TECH (→Schutzschicht)
resistance PHYS **Widerstand**
[ratio of a motive force to the flow generated by it]
↓ electric resistance; thermal resistance; flow resistance
 [Verhältnis einer treibenden Kraft zum von ihr erzeugten Fluß]
 ↓ elektrischer Widerstand; Wärmewiderstand; Strömungswiderstand
resistance TECH **Widerstandsfähigkeit**
= resistibility; resistivity
≈ ruggedness; durability; strength
 = Strapazierfähigkeit
 ≈ Robustheit; Dauerhaftigkeit; Festigkeit
resistance (→resistor) COMPON (→Widerstand)
resistance (→electric PHYS (→elektrischer Widerstand)
resistance)
resistance (→active NETW.TH (→Wirkwiderstand)
resistance)
resistance bridge (→Wheatstone INSTR (→Wheatstonebrücke)
bridge)
resistance coefficient PHYS (→Temperaturkoeffizient
(→temperature coefficient of resistance) des Widerstandes)
resistance-coupled amplifier CIRC.ENG (→RC-Verstärker)
(→RC-coupled amplifier)
resistance coupling (→dc EL.TECH (→galvanische Kopplung)
coupling)
resistance decade INSTR **Widerstandsdekade**
 = Stufenwiderstand
resistance drop (→voltage EL.TECH (→Spannungsabfall)
drop)
resistance lattice (→resistance NETW.TH (→Widerstandnetzwerk)
network)
resistanceless EL.TECH **widerstandsfrei**
≠ resistive
 = widerstandslos
 ≠ widerstandsbehaftet
resistance-loaded EL.TECH (→widerstandsbehaftet)
(→resistive)
resistance measuring bridge INSTR **Widerstandsmeßbrücke**
resistance metal METAL **Widerstandsmetall**
[with high electric resistance] [mit hohem elektrischen Widerstand]
resistance meter INSTR **Widerstandsmesser**
↓ ohmmeter
 = Wirkwiderstandsmesser
 ↓ Ohmmeter
resistance network NETW.TH **Widerstandnetzwerk**
= resistance lattice = Widerstandsnetz
resistance pad NETW.TH (→Dämpfungsglied)
(→attenuator)
resistance per square (→surface PHYS (→Oberflächenwiderstand)
impedance)
resistance per unit length LINE TH **Widerstandsbelag**
= distributed resistance
↑ transmission line constant
 ↑ Leitungskonstante
resistance thermometer INSTR **Widerstandsthermometer**
resistance to wear TECH **Verschleißfestigkeit**
resistance welding METAL **Widerstandsschweißen**
 = Widerstandschweißen
resistance wire (→resistor COMPON (→Widerstandsdraht)
wire)
resistant TECH **widerstandsfähig**
= resistive; resistible; strong; hard-wearing
≈ rugged; durable
 = strapazierfähig; stark
 ≈ robust; dauerhaft
resistant to heat TECH (→hitzebeständig)
(→heat-resistant)

resistibility (→resistance) TECH (→Widerstandsfähigkeit)
resistible (→resistant) TECH (→widerstandsfähig)
resistive EL.TECH **widerstandsbehaftet**
= resistance-loaded
≠ resistanceless ≠ widerstandslos
resistive (→resistant) TECH (→widerstandsfähig)
resistive pickup INSTR **Widerstandsmeßfühler**
= ohmscher Meßfühler
resistivity (→specific electric resistance) PHYS (→spezifischer elektrischer Widerstand)
resistivity (→resistance) TECH (→Widerstandsfähigkeit)
resistor COMPON **Widerstand**
= resistance ↑ Bauelement
↑ component ↓ Festwiderstand; veränder-
↓ fixed resistor; adjustable resistor; barer Widerstand;
ohmic resistor; nonlinear resistor; ohmscher Widerstand;
film resistor; composition resistor; nichtlinearer Widerstand;
wire resistor Schichtwiderstand; Massewiderstand; Drahtwiderstand

resistor-capacitor-transistor logic (→RCTL) MICROEL (→RCTL)
resistor-transistor logic (→RTL) MICROEL (→RTL)
resistor wire COMPON **Widerstandsdraht**
= resistance wire
resistron ELECTRON **Resistron**
↑ camera tube ↑ Bildaufnahmeröhre
resiting (→location change) ECON (→Standortwechsel)
resizing DATA PROC **Maßstabänderung**
≈ scaling ≈ Skalierung
resolderable ELECTRON **rücklötbar**
resolderable fuse COMPON **Rücklötsicherung**
= resoldering fuse ≈ Hitzdrahtsicherung; Umkehrauslöser
↑ telephone service fuse
↓ heat-coil fuse; reversible fuse ↓ Fernmeldesicherung
resoldering fuse (→resolderable fuse) COMPON (→Rücklötsicherung)
resolution 1 PHYS **Auflösung**
= definition ↓ Abbildungsschärfe
↓ sharpness
resolution TERM&PER **Auflösung**
= darstellbare Punkte
resolution 2 PHYS **Auflösungsvermögen**
= resolution capacity ≈ Auflösungsgrenze
↓ limit of resolution
resolution (→scale resolution) INSTR (→Skalenauflösung)
resolution bandwidth INSTR **Auflösungsbandbreite**
= RBW ≈ Auflösebandbreite
resolution capacity (→resolution 2) PHYS (→Auflösungsvermögen)
resolution chart (→test pattern) TV (→Testbild)
resolution of forces PHYS **Kräftezerlegung**
resolution pattern (→test pattern) TV (→Testbild)
resolution pattern generator (→test pattern generator) TV (→Bildmustergenerator)
resolver (→rotary resolver) INSTR (→Drehmelder)
resonance PHYS **Resonanz**
= Mitschwingen
resonance (→antenna resonance) ANT (→Antennenresonanz)
resonance absorption PHYS **Resonanzabsorption**
resonance amplifier (→tuned amplifier) CIRC.ENG (→Resonanzverstärker)
resonance band PHYS **Resonanzband**
resonance capacitor INSTR **Resonanzkondensator**
resonance curve (→resonating curve) PHYS (→Resonanzkurve)
resonance frequency (→resonant frequency) PHYS (→Resonanzfrequenz)
resonance frequency meter INSTR **Resonanzfrequenzmesser**
[antenna measurements] [Messung der Güte von Antennen]
= dip meter; grid dipper ≈ Wellenmesser; Dipmeter
resonance insulator MICROW **Resonanzisolator**
resonance line PHYS **Resonanzlinie**

resonance matching NETW.TH **Resonanzanpassung**
[selective impedance matching using response of resonant circuits] [schmalbandige Impedanzanpassung durch Ausnutzung der Eigenschaften von Schwingkreisen]
↑ matching 1 ↑ Anpassung 1
resonance mode PHYS **Resonanzform**
= resonant mode
resonance mode INSTR **Resonanzverfahren**
resonance sharpness NETW.TH **Resonanzschärfe**
resonant antenna ANT **Resonanzantenne**
resonant breaker loop ANT **Resonanzunterbrecher**
resonant cavity (→coaxial cavity resonator) MICROW (→Topfkreis)
resonant circuit NETW.TH **Resonanzkreis**
= resonating circuit = Resonanzschaltung
↑ oscillating circuit ↑ Schwingkreis
resonant circuit converter POWER SYS **Schwingkreisumrichter**
resonant circuit inverter POWER SYS **Schwingkreiswechselrichter**
resonant crystal (→filter crystal) COMPON (→Filterquarz)
resonant frequency PHYS **Resonanzfrequenz**
= resonance frequency
resonant impedance NETW.TH **Resonanzwiderstand**
resonant length ANT **Resonanzlänge**
resonant mode (→resonance mode) PHYS (→Resonanzform)
resonant rhombic antenna ANT **resonante Rhombusantenne**
resonant-ring filter MICROW **Resonant-ring-Filter**
↑ directional coupler ↑ Richtkoppler
resonant shunt NETW.TH **Resonanznebenschluß**
resonate PHYS **mitschwingen**
resonating circuit (→resonant circuit) NETW.TH (→Resonanzkreis)
resonating curve PHYS **Resonanzkurve**
= resonance curve
resonator PHYS **Resonator**
resonator ACOUS **Resonanzkörper**
= Schallkörper; Resonanzkasten; Schallkasten
resource (→expedient) TECH (→Notbehelf)
resource allocation DATA PROC **Betriebsmittelzuteilung**
= Betriebsmittelzuweisung
resource file DATA PROC **Betriebsmitteldatei**
resource leveling DATA PROC **Betriebsmittelausgleich**
resource management DATA PROC **Betriebsmittelverwaltung**
resource management SWITCH (→Betriebsmittelverwaltung)
(→auxiliary resource manager)
resources DATA PROC **Betriebsmittel**
resources ECON **Ressourcen**
↓ funds = Mittel
↓ Geldmittel
resource sharing DATA PROC **Betriebsmittelteilung**
= shared resource
respond (→operate) COMPON (→anziehen)
responder RADIO NAV **Antwortsender**
responder SWITCH **Responder**
[for test equipment] [für Prüfeinrichtungen]
↑ signaling converter ↑ Kennzeichenumsetzer
responder beacon RADIO NAV (→Antwortbake)
(→transponder)
responding current ELECTRON (→Ansprechstrom)
(→response current)
responding dc voltage (→dc spark-over voltage) COMPON (→Ansprechgleichspannung)
responding delay (→response delay) ELECTRON (→Ansprechverzögerung)
responding excitation COMPON (→Ansprecherregung)
(→response excitation)
responding level (→response level) ELECTRON (→Ansprechpegel)
responding sensitivity ELECTRON (→Ansprechempfindlichkeit)
(→responsitivity)
responding threshold ELECTRON (→Ansprechschwelle)
(→response threshold)
responding time (→response time) ELECTRON (→Ansprechzeit)
responding value (→response value) ELECTRON (→Ansprechwert)

responding voltage	ELECTRON	(→Ansprechspannung)
(→response voltage)		
response	ELECTRON	Antwort
= response action; behaviour (n.)		= Ansprechverhalten
response	PHYS	Gang
↓ thermal response; frequency response		↓ Temperaturgang; Frequenzgang
response	TECH	Verhalten
= response action; behaviour		= Antwort; Ansprechverhalten
≈ operating mode		≈ Betriebsart
response action	ELECTRON	(→Antwort)
(→response)		
response action (→response)	TECH	(→Verhalten)
response current	ELECTRON	Ansprechstrom
= responding current; pull-in current; pick-up current		= Anzugstrom
response dc (→dc spark-over voltage)	COMPON	(→Ansprechgleichspannung)
response delay	ELECTRON	Ansprechverzögerung
= responding delay; slow operation; pull-in delay; pick-up delay		= Anzugverzögerung
response excitation	COMPON	Ansprecherregung
= responding excitation; pull-in power		= Anzugerregung
response frame	TELEC	Antwortseite
[videotex]		[Btx]
= user action frame		= Dialogseite
response level	ELECTRON	Ansprechpegel
= responding level; operate level; pull-in level; pich-up level		= Anzugspegel
response message	DATA COMM	Antwortnachricht
		= Antwortblock
response sensitivity	ELECTRON	(→Ansprechempfindlichkeit)
(→responsitivity)		
response signal (→backward signal)	SWITCH	(→Rückwärtszeichen)
response telegram	TELECONTR	Antworttelegramm
response threshold	ELECTRON	Ansprechschwelle
= responding threshold; pull-in threshold; pick-up threshold		= Ansprechgrenze; Anzugsschwelle; Anzugsgrenze
response time	ELECTRON	Ansprechzeit
= responding time; pull-in time; pick-up time		= Anzugzeit (Relais); Antwortzeit
response time	INSTR	Beruhigungszeit
response time	TECH	Reaktionszeit
= reaction time; latency; start-up time		= Antwortzeit; Anlaufdauer
response value	ELECTRON	Ansprechwert
= responding value; pull-in value; pick-up value		= Anzugwert
response voltage	ELECTRON	Ansprechspannung
= responding voltage; pull-in voltage; pick-up voltage		= Anzugsspannung
responsibility	ECON	Verantwortlichkeit
≈ competence		≈ Zuständigkeit
responsible	ECON	verantwortlich
≈ competent		≈ zuständig
responsible for	ECON	zuständig
= in charge of		= verantwortlich
≈ competent		≈ kompetent
responsitivity	ELECTRON	Ansprechempfindlichkeit
= responsivness; response sensitivity; responding sensitivity; pull-in sensitivity		= Anzugsempfindlichkeit (Relais)
responsive (adj.)	COLLOQ	reaktionsschnell
responsivness (→responsitivity)	ELECTRON	(→Ansprechempfindlichkeit)
respool (v.t.) (→rewind)	ELECTRON	(→zurückspulen)
rest (n.)	MECH	Ablage
rest (→pedestal)	TECH	(→Sockel)
restabilization time (→recovery time)	CIRC.ENG	(→Erholungszeit)
rest against (v.i.)	TECH	aufsitzen
restart (v.t.)	DATA PROC	neustarten
= reload 1 (v.t.); reboot (v.t.)		= neuladen; rebooten
restart (n.)	DATA PROC	Neustart
[reloading of the main memory]		[Neuladen des Arbeitsspeichers]
= recovery 2; rerun 2; reloading; rebooting; new start		= Wiederanlauf; Restart
restart cause	DATA PROC	Wiederanlaufgrund
		= Restartgrund
restart indication	DATA PROC	Wiederanlaufanzeige
		= Restartanzeige
restart point (→checkpoint)	DATA PROC	(→Fixpunkt)
restart program (→restart routine)	DATA PROC	(→Wiederanlaufroutine)
restart-proof	DATA COMM	restartsicher
= restart-protected		
restart-protected (→restart-proof)	DATA COMM	(→restartsicher)
restart request	DATA PROC	Wiederanlaufanforderung
		= Restartforderung
restart routine	DATA PROC	Wiederanlaufroutine
= restart program		= Wiederanlaufprogramm; Restartroutine; Restartprogramm
≈ checkpoint routine		≈ Fixpunktroutine
rest condition	TELEGR	Ruhezustand
= idle condition; dwell phase; quiescent phase		= Freizustand
rest contact (→home contact)	SWITCH	(→Ruhekontakt)
rest current (→quiescent current)	ELECTRON	(→Ruhestrom)
rest energy (→potential energy)	PHYS	(→potentielle Energie)
resting contact (→break contact)	COMPON	(→Ruhekontakt)
restitution (→reproduction)	TERM&PER	(→Wiedergabe)
restitution characteristic (→reproducing characteristic)	TERM&PER	(→Wiedergabecharakteristik)
rest mass	PHYS	Ruhemasse
rest noise	MODUL	Ruhegeräusch
restorability	QUAL	Instandsetzbarkeit
restoral	TRANS	Zurückschaltung
[protection switching]		[Ersatzschaltung]
restoration	TECH	Wiederherstellung
= reestablishment; recovery		≈ Instandsetzung
≈ repair		
restore (v.t.)	DATA PROC	umspeichern
restore (v.t.)	TECH	wiederherstellen
= reestablish; recover		= rückstellen
≈ repair		≈ instandsetzen
RESTORE key	TERM&PER	RESTORE-Taste
restoring	DATA PROC	Umspeicherung
		= Umspeichern
restoring force (→reaction force)	MECH	(→Rückstellkraft)
restoring relay	ELECTRON	Rückstellrelais
= reset relay		
restoring rod	MECH	Rückzugstange
restoring spring	MECH	Rückzugfeder
restoring switch (→reset switch)	EQUIP.ENG	(→Rücksetzschalter)
restoring torque	TECH	Rückstellmoment
		= Rücksetzmoment
rest position (→home position)	SWITCH	(→Ruhelage)
restraint (→restriction)	TECH	(→Beschränkung)
restrict (→constrain)	TECH	(→einschränken)
restricted	TECH	eingeschränkt
= limited		= beschränkt
restricted access	DATA PROC	beschränkter Zugriff
		= beschränkter Zugang
restriction	TECH	Beschränkung
= restraint; limitation; bound		= Einschränkung; Begrenzung; Limitierung; Drosselung (fig.)
≈ impairment		≈ Beeinträchtigung
restriction	SWITCH	Sperre
= barring		
restrictive	TECH	einschränkend
= limiting		= beschränkend; restriktiv
≈ impairing		≈ beinträchtigend
restructuralization	SCIE	Umstrukturierung
restructuralize (→restructure)	SCIE	(→umstrukturieren)
restructure	SCIE	umstrukturieren
= restructuralize		
resubmission	ECON	Wiedervorlage
result	MATH	Ergebnis
result	ECON	Ergebnis
≈ yield; revenue		≈ Ertrag; Erlös

resultant

resultant (n.)	MATH	**Resultante**
result digit	DATA PROC	**Ergebnisstelle**
result message	SWITCH	**Ergebnismeldung**
result register	SWITCH	**Ergebnisregister**
resumé	ECON	**Lebenslauf**
= curriculum vitae; personal record		= beruflicher Werdegang
resumé (→summary)	LING	(→Zusammenfassung)
resume (→summary)	LING	(→Zusammenfassung)
resumption	COLLOQ	**Wiederaufnahme**
= recommencement		= Wiederbeginn
resurgence	TECH	**Wiederanstieg**
		= Wiederanwachsen
retail (→retail trade)	ECON	(→Einzelhandel)
retailer	ECON	**Einzelhändler**
≠ wholesale dealer		≠ Großhändler
↑ dealer		↑ Händler
retailer (→reseller)	ECON	(→Wiederverkäufer)
retail trade	ECON	**Einzelhandel**
= retail		= Kleinhandel
retain (→maintain)	COLLOQ	(→aufrechterhalten)
retained image	ELECTRON	**Nachwirkungsbild**
retainer (→lock washer)	MECH	(→Sicherungsscheibe)
retaining clip	MECH	**Halteklammer**
retaining pawl	MECH	**Sperrklinke**
retaining ring	MECH	**Sicherungsring**
retaining washer (→lock washer)	MECH	(→Sicherungsscheibe)
retardation (→breaking 1)	MECH	(→Bremsung)
retardation (→dead time)	CONTROL	(→Totzeit)
retardation radiation (→bremsstrahlung)	PHYS	(→Bremsstrahlung)
retardation spectrum	PHYS	**Bremsspektrum**
= bremsspectrum		
retard coil (→choke)	COMPON	(→Drossel)
retarded contact	COMPON	**nacheilender Kontakt**
retarding electrode	ELECTRON	**Bremselektrode**
= decaying electrode		= Verzögerungselektrode
retarding field	PHYS	**Bremsfeld**
= break-field		
retarding-field tube	ELECTRON	**Bremsfeldröhre**
= break-field tube		
retarding force	PHYS	**Bremskraft**
retarding potential	PHYS	**Bremspotential**
retarding thermistor	POWER ENG	**Bremswiderstand**
retention	ECON	**Zurückbehaltung**
		= Einbehaltung
retention (→inhibition)	ELECTRON	(→Sperrung)
retention force	MECH	**Haltekraft**
retention spring	MECH	**Haltefeder**
retention time	ELECTRON	**Speicherzeit**
= storage time		
retention time (→residence time)	DATA PROC	(→Verweilzeit)
retest signal (→test sequence)	DATA COMM	(→Prüffolge)
reticle	MICROEL	**Zwischenmaske**
		= Reticle
reticular structure (→lattice structure)	PHYS	(→Gitterstruktur)
reticule (→cross-line)	INSTR	(→Fadenkreuz)
retina	PHYS	**Netzhaut**
re-tool (v.t.)	MECH	**nachbearbeiten**
= re-machine		
retourned good (→return delivery)	ECON	(→Rücklieferung)
retrace (→reverse action)	ELECTRON	(→Rücklauf)
retract	TECH	**einziehen**
≈ sink		≈ versenken
retract (→withdraw)	TECH	(→zurückziehen)
retractable	TECH	**einziehbar**
= retractile		≈ versenkbar
≈ sinkable		
retractable antenna	ANT	**Versenkantenne**
retractile (→retractable)	TECH	(→einziehbar)
retractile cord	TERM&PER	**Spiralschnur**
= spiral cord; tinsel cord; coil cord		= dehnbare Handapparateschnur; Litzenschnur
≈ tinsel conductor [COMM.CABLE]		≈ Lahnlitze [COMM.CABLE]
retrain (v.t.)	ECON	**umschulen**
retraining	ECON	**Umschulung**
retransform (v.t.) (→reconvert)	TECH	(→zurückwandeln)
retransformation (→reconversion)	TECH	(→Zurückwandlung)
retransmission 1	TELEC	**Übertragungswiederholung**
= repetition of transmission; retry		= nochmalige Übertragung; erneute Übertragung
≈ retransmission 2		≈ Weitersendung
retransmission 2 (→delivery)	TELEC	(→Weitersendung)
retransmission buffer	SWITCH	**Wiederholspeicher**
retransmission receiver (→relay receiver)	BROADC	(→Ballempfänger)
retransmitter	BROADC	**Ballsender**
= radio relay station		
retrench (→decrease)	TECH	(→vermindern)
retrench (→reduce 1)	TECH	(→vermindern)
retrenchment (→reduction)	TECH	(→Verkleinerung)
retrieval	DATA PROC	**Wiedergewinnung**
= recovery 3; retrieving		= Wiederauffinden; Wiederabrufen; Retrieval
≈ fetch		≈ Abruf
retrieval (→recovery)	ELECTRON	(→Rückgewinnung)
retrieval language (→query language)	DATA PROC	(→Abfragesprache)
retrieval process (→search process)	DATA PROC	(→Suchvorgang)
retrieval specification (→search specification)	DATA PROC	(→Suchvorschrift)
retrieval strategy (→search strategy)	DATA PROC	(→Suchstrategie)
retrieval terminal (→inquiry station)	DATA COMM	(→Abfragestation)
retrieve	TECH	**zurückgewinnen**
		≈ wiedergewinnen
retrieve	DATA PROC	**wiedergewinnen**
[of data from a data stock]		[von Daten aus einem Datenbestand]
≈ import; inquire; read-out		≈ einspielen; abfragen; auslesen
retrieve (→read-out)	DATA PROC	(→auslesen)
retrieving (→retrieval)	DATA PROC	(→Wiedergewinnung)
retrigger	ELECTRON	**umtriggern**
retroaction (→reaction)	PHYS	(→Rückwirkung)
retroactive	ECON	**rückwirkend**
retrofit	EQUIP.ENG	**nachrüsten**
= add-on		= nachbestücken
retrofit (→convert)	TECH	(→umrüsten)
retrofit kit	EQUIP.ENG	**Nachrüstsatz**
= add-on kit		
retrofitting	EQUIP.ENG	**Nachrüstung**
retrofitting (→conversion)	TECH	(→Umrüstung)
retrograde	TECH	**rückläufig**
= declining; downward; regressive		= regressiv
≈ backward 2; countermoving		≈ rückwärtig; gegenläufig
≠ forward		≠ vorwärts
retrospect	SCIE	**Rückblick**
retrospective	SCIE	**rückblickend**
		= retrospektiv
retry (→retransmission 1)	TELEC	(→Übertragungswiederholung)
retry (→repeat)	DATA PROC	(→wiederholen)
retry (→rerun 1)	DATA PROC	(→Wiederholungslauf)
retry (n.) (→repeat)	DATA COMM	(→Wiederholung)
retune	EL.TECH	**umstimmen**
return (v.t.)	ECON	**zurücksenden**
= send back; reship		≈ zurückschicken
return (v.i.)	ELECTRON	**zurückspringen**
return (n.)	ECON	**Rendite**
= yield (n.)		≈ Erlös
≈ revenue		
return (n.)	TELEPH	**Rücklauf**
[dial]		[Nummernscheibe]
return (→profit)	ECON	(→Gewinn)
return (→returned good)	ECON	(→Retoure)
return (n.) (→reentry)	DATA PROC	(→Rücksprung)
return address	DATA PROC	**Rücksprungadresse**
		= Rückkehradresse; Rückadresse; Absprungadresse

return circuit (→return wire)	EL.TECH	(→Rückleiter)		**revenue stamp**	ECON	**Stempelmarke**
return conductor (DC) (→neutral conductor)	POWER ENG	(→Nulleiter)		**reverberation** [overlapping and fading-out sound reflection]	ACOUS	**Nachhall** [überlagerte und abklingende Schallreflexion]
return criterion	DATA PROC	**Rückkriterium**		= double echo		≈ Echo
return current	EL.TECH	**Rückstrom**		≈ echo		↑ Schallreflexion
= inverse current; reverse current				↑ sound reflection		
return current coefficient (→reflection coefficient)	NETW.TH	(→Reflexionsfaktor)		**reverberation amplifier** (→reverberator)	EL.ACOUS	(→Nachhallgerät)
return delivery	ECON	**Rücklieferung**		**reverberation chamber** (→reverberation room)	ACOUS	(→Nachhallraum)
= redelivery; returned good		≈ Retoure		**reverberation room**	ACOUS	**Nachhallraum**
return direction	TELEC	**Rückrichtung**		= reverberation chamber		= Nachhallkammer
= opposite direction; far-to-near direction		= Gegenrichtung		≠ anechoic room		≠ schalltoter Raum
		≠ Hin-Richtung		**reverberation time**	ACOUS	**Nachhallzeit**
≠ go direction		↑ Übertragungsrichtung		**reverberator**	EL.ACOUS	**Nachhallgerät**
↑ transmission direction				= reverberation amplifier		
returned good	ECON	**Retoure**		**reversal**	PHYS	**Wechsel**
= return				[between states]		[zwischen Zuständen]
return freight	ECON	**Rückfracht**		= alternation		
return information	TELECONTR	**Rückmeldung**		**reversal**	TELEC	**Wechselzeichen**
return instruction	DATA PROC	**Rücksprungbefehl**		= alternation		= Wechsel
		= Rückkehrbefehl; Absprungbefehl; Return-Befehl		**reversal** (→current reversal)	EL.TECH	(→Stromumkehr)
				reversal (→reversion)	TECH	(→Umkehrung)
return jump (→reentry)	DATA PROC	(→Rücksprung)		**reversal margin**	INSTR	**Umkehrspanne**
RETURN key (→ENTER key)	TERM&PER	(→Eingabetaste)		**reversals** (→square-wave reversals)	ELECTRON	(→Rechteckwelle)
RETURN key (→carriage return key)	TELEGR	(→Wagenrücklauftaste)		**reversal stage**	CIRC.ENG	**Umkehrstufe**
				= inverting stage		≈ Umkehrschaltung
return line (→return wire)	EL.TECH	(→Rückleiter)		≈ inverting circuit		
				reversals transmitter [telegraphy]	RADIO	**Wechselsender** [Telegrafie]
return loss 1 (→return loss coefficient)	NETW.TH	(→Echodämpfungsmaß)		**reversals transmitter** (→single current test transmitter)	TELEGR	(→Einfachstromwechselsender)
return loss 2 (→active return loss)	NETW.TH	(→Reflexionsdämpfung)		**reverse** (n.)	TYPOGR	**Umkehrung**
return loss coefficient	NETW.TH	**Echodämpfungsmaß**		= reverse representation		= invertierte Darstellung
[log.amplitude ratio of reflected to incident wave]		[log.Maß des komplexen Reflexionsfaktors; Realteil = Reflexionsdämpfung, Imaginärteil = Echophase]		**reverse** (n.)	TECH	**Kehrseite**
= return loss 1; reflection loss coefficient; reflection loss 1		= Echomaß; Rückflußdämpfungsmaß; Stoßdämpfungsmaß				≈ Revers
				reverse (adj.)	TECH	**gegensinnig**
				= reverse-acting; inverse; inverted; contradirectional		= umgekehrt; rückwärtsgerichtet; entgegengesetzt; entgegengerichtet; invers
				≈ anti-parallel [MATH]		≈ antiparallel [MATH]
return postage	POST	**Rückporto**		≠ codirectional		≠ gleichsinnig
return spring	TECH	**Rückführungsfeder**		**reverse**	TERM&PER	**Rücklauf**
		= Rückholfeder				= Reverse
return to normal [of a switch]	SWITCH	**Rücklauf** [Wähler]		**reverse** (→negative)	TERM&PER	(→negativ)
return to service	TECH	**Wiederinbetriebnahme**		**reverse** (n.) (→reverse printing)	TERM&PER	(→invertierter Druck)
return-to-zero	CODING	**RZ**		**reverse** (v.t.) (→invert)	TECH	(→invertieren)
= RZ				**reverse** (n.) (→reversion)	TECH	(→Umkehrung)
return trace (→reverse action)	ELECTRON	(→Rücklauf)		**reverse-acting** (→reverse)	TECH	(→gegensinnig)
				reverse action [of a beam]	ELECTRON	**Rücklauf** [Elektronenstrahl; Bildpunkt]
return warranty	ECON	**Rückgabegarantie**		= retrace; return trace; beam return; flyback; homing; line flyback		= Zeilenrücklauf
return wire	EL.TECH	**Rückleiter**		≠ trace		≠ Hinlauf
= return line; return circuit		= Rückleitung		**reverse AGC** (→reverse automatic gain control)	CONTROL	(→Abwärtsregelung)
reusable	DATA PROC	**mehrfach abrufbar**				
≈ reentrant		≈ eintrittsinvariant		**reverse automatic gain control**	CONTROL	**Abwärtsregelung**
		≈ ablaufinvariant		= reverse AGC		
reusable	TECH	**wiederverwendbar**		**reverse bias** (→reverse voltage)	ELECTRON	(→Sperrspannung)
		≈ wiederverwertbar; wiederbenutzbar		**reverse biased**	ELECTRON	**in Sperrichtung betrieben**
reusable program [can be reused by several user without reloading]	DATA PROC	**eintrittsinvariantes Programm** [ohne Nachladen von mehreren Benutzern wieder abrufbar]		**reverse bias test** [diode]	MICROEL	**Sperrprüfung** [Diode]
≈ reentrant program		≈ ablaufinvariantes Programm		= inverse bias test		
				reverse blocking (n.)	MICROEL	**Rückwärtssperrung**
reusable routine	DATA PROC	**mehrfach aufrufbare Routine**		**reverse-blocking** (adj.)	ELECTRON	**rückwärts sperrend**
reuse (n.)	RADIO	**Doppelnutzung**		**reverse blocking state**	MICROEL	**Rückwärtssperrzustand**
reuse (n.)	TECH	**Wiederverwendung**		**reverse blocking thyristor diode**	MICROEL	**rückwärts sperrende Thyristordiode**
		= Wiederverwertung; Wiederbenutzung		**reverse blocking triode thyristor**	MICROEL	**rückwärts sperrende Thyristortriode**
reuse (→reinsert)	TRANS	(→Wiederbelegung)		**reverse channel** (→backward channel)	TELEC	(→Rückkanal)
revenue	ECON	**Erlös**				
= proceeds		= Einnahme		**reverse characteristic**	ELECTRON	**Rückwärtskennlinie**
≈ return (n.)		≈ Ertrag; Ergebnis; Rendite		= inverse characteristic; blocking characteristic; off-state		= Rückwärts-Durchlaßkennlinie; Sperrkennlinie
revenue (→turnover)	ECON	(→Umsatz)				
revenue loss	ECON	**Einnahmeausfall**				
revenues (→income)	ECON	(→Einkommen)				

reverse charging TELEC	Gebührenübernahme
[charge taken over by the called subscriber]	[durch Gerufenen]
	= Gebührenumkehr
	≈ R-Gespräch
reverse coducting triode thyristor MICROEL	rückwärts leitende Thyristortriode
reverse compatibility TV	(→Rekompatibilität)
(→recompatibility)	
reverse conducting ELECTRON	rückwärtsleitend
reverse conducting thyristor diode MICROEL	rückwärts leitende Thyristordiode
= reverse polarity thyristor diode	
reverse connection (→polarity inversion) EL.TECH	(→Polaritätsumkehr)
reverse-connect protection EL.TECH	Verpolschutz
reverse counter (→down counter) CIRC.ENG	(→Rückwärtszähler)
reverse current MICROEL	Rückwärtsstrom
= inverse current; cutoff current 1	= negativer Sperrstrom; Sperrstrom; Rückstrom
reverse current ELECTRON	Sperrstrom
= inverse current	= Rückwärtsstrom; Rückstrom
≈ leak current	≈ Leckstrom
≠ conducting-state current	≠ Durchlaßstrom
reverse current (→return current) EL.TECH	(→Rückstrom)
reverse current transfer ratio MICROEL	Kurzschluß-Rückwärts-Stromverstärkung
reversed [font] TYPOGR	negativ [Schrift]
reversed action tweezers TECH	Kreuzpinzette [öffnet sich beim Drehen]
reverse dc resistance ELECTRON	Sperrwiderstand
= inverse dc resistance	= Gleichstromwiderstand rückwärts
reversed image (→reverse image) TV	(→Negativbild)
reverse diode COMPON	Rückspeisediode
	= Blindleistungsdiode
reverse direction ELECTRON	Sperrichtung
= inverse direction; backward direction; back direction; high-resistance direction	= Rückwärtsrichtung
	≠ Durchlaßrichtung
≠ forward direction	
reversed quadripole (→reversed two-port) NETW.TH	(→umgekehrter Vierpol)
reversed two-port NETW.TH	umgekehrter Vierpol
= reversed quadripole	
reverse engineering TECH	Umkehrtechnik
reverse feedback (→negative feedback) CIRC.ENG	(→Gegenkopplung)
reverse feeding CIRC.ENG	Gegenspeisung
reverse frequency position TRANS	Kehrlage
= inverted position	↑ Übertragungslage
↑ transmission position	
reverse function DATA PROC	Folgeumkehrfunktion
reverse gate voltage MICROEL	Rückwärtssteuerspannung
reverse grid current ELECTRON	positiver Gitterstrom
reverse grid voltage ELECTRON	Gittergegenspannung
reverse image TV	Negativbild
= reversed isolation image	
reverse isolation MICROEL	Umkehrspannung
[breakdown voltage of a diode]	[Durchbruchspannung einer Diode]
= turn-over voltage	
≈ breakdown voltage	≈ Durchbruchspannung
reverse line feed TERM&PER	Zeilenrücklauf
≠ line feed	≠ Zeilenvorschub
↑ feed	↑ Vorschub
reverse magnetization EL.TECH	Gegenmagnetisierung
reverse motion TV	Wiedergabe rückwärts
reverse pn junction MICROEL	Rückwärtssperrbereich
reverse polarity (→polarity inversion) EL.TECH	(→Polaritätsumkehr)
reverse polarity thyristor diode MICROEL	(→rückwärts leitende Thyristordiode)
(→reverse conducting thyristor diode)	
reverse power loss MICROEL	Sperrverlustleistung
[in rectifier, due to reverse current]	[Verlustleistung in Gleichrichter, durch Sperrstrom]
= inverse power loss	
reverse power protection EL.INST	Rückleiterschutz
reverse power protection MICROW	Rückleistungsschutz
reverse printing TERM&PER	invertierter Druck
= reverse (n.)	= Umkehrung; Reverse
reverser (→card reverser) TERM&PER	(→Kartenwender)
reverse reading DATA PROC	Rückwärtslesen
= inverse reading; reverse search; inverse search	= Rückwärtssuchen
reverse recovery current MICROEL	Sperrverzögerungsstrom
= inverse recovery current	
reverse recovery time MICROEL	Sperrverzögerungszeit
= inverse recovery time; backward recovery time	= Sperrerholzeit
reverse representation (→reverse) TYPOGR	(→Umkehrung)
reverse-saturation current MICROEL	(→Sperrsättigungsstrom)
(→inverse-saturation current)	
reverse scan (→rear scanning) ELECTRON	(→Rückwärtsabtastung)
reverse scanning (→rear scanning) ELECTRON	(→Rückwärtsabtastung)
reverse-scroll (→scroll down) TERM&PER	(→zurückrollen)
reverse scrolling (→scrolling down) TERM&PER	(→Zürückrollen)
reverse search (→reverse reading) DATA PROC	(→Rückwärtslesen)
reverse slant (→backslash) TYPOGR	(→verkehrter Schrägstrich)
reverse solidus (→backslash) TYPOGR	(→verkehrter Schrägstrich)
reverse transfer admittance NETW.TH	Rückwärtssteilheit
	= Rückwärtsleitwert; Rückwärtsadmittanz
reverse-transfer admittance MICROEL	(→Kurzschluß-Rückwärtssteilheit)
(→short-circuit reverse-transfer admittance)	
reverse transfer capacitance MICROEL	Rückwirkungskapazität
reverse transfer impedance NETW.TH	Rückwirkungswiderstand
	= Rückwirkungsimpedanz
reverse transfer inductance NETW.TH	Rückwirkungsinduktivität
reverse type TERM&PER	Negativschrift
[white characters on dark background]	[helle Schrift auf dunklem Hintergrund]
reverse video TERM&PER	invertierte Darstellung
= inverse video	= Invers-Darstellung; Revers-Darstellung
≠ positive type	≠ positive Schrift
reverse voltage EL.TECH	Gegenspannung
= inverse voltage	
reverse voltage MICROEL	Rückwärtsspannung
	= negative Sperrspannung
reverse voltage ELECTRON	Sperrspannung
= inverse voltage; blocking voltage; reverse bias; inverse bias; blocking bias; cutoff voltage	= Rückwärtsspannung; negative Sperrspannung
	≠ Durchlaßspannung
≠ conduction voltage	
reverse voltage protection POWER SYS	Gegenspannungsschutz
= inverse voltage protection	
reverse voltage transfer NETW.TH	Spannungsrückwirkung
[from output to input of a four-terminal network]	[von Ausgangstor eines Vierpols auf Eingangstor]
reverse voltage transfer ratio characteristic MICROEL	Spannungsrückwirkungs-Kennlinie
	= Rückwirkungskennlinie
reversibility MATH	Reversibilität
≠ irreversibility	= Umkehrbarkeit 1
	≠ Irreversibilität
reversible PHYS	umkehrbar
= invertible	= reversibel; invertierbar
reversible connection POWER SYS	(→Zweiwegeschaltung)
(→double-way connection)	
reversible converter POWER SYS	(→Zweiwegstromrichter)
(→double-way converter)	
reversible convertor POWER SYS	(→Umkehrstromrichter)
(→two-way convertor)	
reversible counter (→up-down counter) CIRC.ENG	(→Vor-Rückwärts-Zähler)
reversible drive (→reversing drive) POWER SYS	(→Umkehrantrieb)

reversible fuse		COMPON	**Umkehrauslöser**	**rework** (n.) (→overhaul)	TECH	(→Überarbeitung)
↑ telephone service fuse			↑ Fernmeldesicherung	**rewrite** (→rewrite-in)	DATA PROC	(→wiedereinschreiben)
reversible permeability		PHYS	**reversible Permeabilität**	**rewrite-in** (v.t.)	DATA PROC	**wiedereinschreiben**
reversible transducer		EL.ACOUS	**reversibler Wandler**	[to erase and reset]		= neueinschreiben
reversible two-port		NETW.TH	**umkehrbarer Vierpol**	= rewrite		
reversing drive		POWER SYS	**Umkehrantrieb**	**Reynolds number**	PHYS	**Reynolds-Zahl**
= reversible drive				**RF** (→radio frequency)	RADIO	(→Radiofrequenz)
reversing entry		ECON	**Rückbuchung**	**RF band** (→frequency band)	RADIO	(→Frequenzband)
reversing reactor		POWER SYS	**Umschwingdrossel**	**RF band utilization**	RADIO	**RF-Bandnutzung**
= ring-back reactor				**RF blocking transformer**	ANT	**HF-Trenntransformator**
reversion		TECH	**Umkehrung**	= braid breaker		
= reversal; reverse (n.); inversion				**RF cable** (→radio-frequency cable)	COMM.CABLE	(→RF-Kabel)
revert (v.t.) (→invert)		CIRC.ENG	(→invertieren)	**RF channel** (→radiofrequency channel)	RADIO	(→Funkkanal)
revertible quadripole (→revertible two-port)		NETW.TH	(→leistungssymmetrischer Vierpol)	**RF channel identification**	RADIO REL	**Kanalkennung**
revertible two-port		NETW.TH	**leistungssymmetrischer Vierpol**	**RF choke**	EL.TECH	**HF-Drossel**
[with equal performance in both directions]			[mit gleichem Verhalten in beiden Richtungen]	**RF coder** (→radio frequency coder)	TERM&PER	(→Radiofrequenzcodierer)
= revertible quadripole				**RF combining circuit**	RADIO REL	**RF-Anschaltung**
revertive (→rear)		TECH	(→rückseitig)	≈ directional filter		= RF-Kanal-Aufschaltung; RF-Anschlußbaugruppe
revertive barring (→revertive blocking)		ELECTRON	(→rückwärtige Sperre)			≈ Richtungsweiche
revertive blocking		ELECTRON	**rückwärtige Sperre**	**RF connector**	COMPON	**HF-Stecker**
= backward blocking; revertive barring; backward barring				**RF decoupling**	RADIO	**Radiofrequenzentkopplung**
review (→re-check)		TECH	(→nachprüfen)			= RF-Entkopplung
review (→magazine)		COLLOQ	(→Zeitschrift)	**RF detection probe** (→RF probe)	INSTR	(→HF-Tastkopf)
revise (→re-check)		TECH	(→nachprüfen)	**RF detector** (→radio frequency detector)	TERM&PER	(→Radiofrequenzdetektor)
revision (→issue)		DOC	(→Ausgabe)	**RF detector probe**	INSTR	**HF-Detektortastkopf**
revision (→re-check)		TECH	(→Nachprüfung)	**RF earthing electrode**	EL.TECH	**Hochfrequenzerder**
revision editing		DATA COMM	**Redigieren**			= HF-Erder
= editing				**RFI** (→radio interference)	RADIO	(→Funkstörung)
revision editing (→editing)		DATA PROC	(→Editieren)	**RF modulator**	MODUL	**HF-Modulator**
revision level (→issue number)		DOC	(→Ausgabestand)	**RF network analyzer**	INSTR	**HF-Netzwerkanalysator**
revocation		ECON	**Widerruf 1**	**RF oscillator**	CIRC.ENG	**RF-Oszillator**
[of a power]			[einer Vollmacht]	**RF pattern** (→radio-frequency pattern)	RADIO	(→Radiofrequenzraster)
revocation (→withdrawal)		ECON	(→Widerruf 2)			
revoke		ECON	**widerrufen**	**RF pip**	INSTR	**HF-Spitze**
≈ cancel			≈ rückgängigmachen			= Hochfrequenzspitze
revolution		PHYS	**Umdrehung**	**RF power transistor**	MICROW	**RF-Leistungstransistor**
[motion about a center or axis]			[Bewegung um einen Punkt oder Achse]	**RF preselector**	INSTR	**HF-Preselector**
↑ twist			↑ Drehung	**RF probe**	INSTR	**HF-Tastkopf**
revolutional turns speed (→rotational speed)		PHYS	(→Drehzahl)	= HF probe; high-frequency probe; RF detection probe		
revolution counter		INSTR	**Drehzahlmesser**	**RF-proof** (→RF radiation proof)	EL.TECH	(→HF-dicht)
= speed counter			= Umdrehungsmesser; Tourenzähler	**RFQ** (→invitation to tender)	ECON	(→Ausschreibung)
revolutions per minute		PHYS	**Umdrehungen pro Minute**	**RF radiation proof**	EL.TECH	**HF-dicht**
[old unit for speed of revolution turns]			[alte Maßeinheit für Drehzahl]	= RF-proof; HF-radiation proof; HF-proof		
= rpm; r.p.m.; r/m			= U/min	**RF range 2** (→frequency band)	RADIO	(→Frequenzband)
revolutions per second		PHYS	**Umdrehungen pro Sekunde**	**RF range 1** (→radio-frequency range 1)	RADIO	(→Radiofrequenzbereich 1)
[old unit for spedd of revolution turns]			[alte Maßeinheit für Drehzahl]	**RF raster** (→radio-frequency pattern)	RADIO	(→Radiofrequenzraster)
= rps; r.p.s.; r/s			= U/s; r/s			
revolution transducer		TERM&PER	**Drehzahlaufnehmer**	**RF resistance bridge** (→HF resistance bridge)	INSTR	(→HF-Widerstandsmeßbrücke)
revolve (→rotate)		PHYS	(→rotieren)	**RF shielding**	EL.TECH	**HF-Abschirmung**
revolved section		ENG.DRAW	**Profilschnitt innerhalb der Ansicht**	**RF switch**	RADIO REL	**RF-Umschalter**
revolved view		ENG.DRAW	**gedrehte Ansicht**	**RF transformer**	RADIO	**RF-Übertrager**
revolving		ECON	**revolvierend**	**RF unit**	MICROW	**RF-Einschub**
= on a revolving basis; rolling				**RF voltmeter**	INSTR	**HF-Spannungsmesser**
revolving field (→rotating field)		PHYS	(→Drehfeld)	= radio frequency voltmeter		
revolving mechanism (→swing mechanism)		TECH	(→Drehvorrichtung)	**RG** (→ring ground)	SYS.INST	(→Ring-Erdung)
rewind (v.t.)		ELECTRON	**zurückspulen**	**RG** (→ringing generator)	TELEC	(→Rufgenerator)
= rereel (v.t.); respool (v.t.); reroll (v.t.)			= umspulen	**RGB camera**	TV	**RGB-Kamera**
rewind (v.t.)		TERM&PER	**rückspulen**			= Rot-Grün-Blau-Kamera
≈ playback			≈ rückspielen; umspulen	**RGBI mobitor**	TERM&PER	**RGBI-Monitor**
rewinder		TECH	**Umspulvorrichtung**	↑ color monitor		↑ Farbmonitor
			= Umspuler	**RGB monitor**	TERM&PER	**RGB-Monitor**
rewinding		TERM&PER	**Rückspulung**	[with separate electron guns for Red, Green and Blue]		[mit je einer Elektronenkanone für Rot, Grün und Blau]
≈ playback (n.)			≈ Rückspielen			
rework (v.t.)		TECH	**überarbeiten**	= red-green-blue monitor		
[eliminate defects]			[Fehler beseitigen]	↑ color monitor		↑ Farbmonitor
= recondition; overhaul (v.t.); refit (v.t.); reform (v.t.)			≈ überholen; umarbeiten ≈ überprüfen; umformen; instandsetzen	**RGB signal**	TV	**RGB-Signal**
≈ revise; reshape; repair				= red-green-blue signal		= Rot-Grün-Blau-Signal
				RGBY camera	TV	**RGBY-Kamera**

Rh   502

Rh (→rhodium)   CHEM   (→Rhodium)
rhenium   CHEM   **Rhenium**
= Re   = Re
rheostat   PHYS   **Regelwiderstand**
= variable resistor   = Schiebewiderstand
rhodium   CHEM   **Rhodium**
= Rh   = Rh
rhodium-plate   METAL   **rhodinieren**
rhombic   MATH   **rhombisch**
≈ rhomboid   = rautenförmig
       ≈ rhomboid
rhombic antenna   ANT   **Rhombusantenne**
= diamond antenna; diamond-shaped antenna   = Rautenantenne
↑ travelling-wave antenna; long-wire antenna   ↑ Wanderwellenantenne; Langdrahtantenne
↓ rhomboid antenna   ↓ Rhomboid-Antenne
rhombic line   ANT   **Rhombus-Linie**
rhombic row   ANT   **Rhombus-Reihe**
rhombiquad   ANT   **Rhombiquad**
[for amateur radio]   [für Amateurfunk]
↑ multiband antenna   ↑ Allbandantenne
rhombohedron (pl.-ons, -a)   MATH   **Rhomboeder**
[delimited by six regular rhombuses]   [von sechs gleichen Rhomben begrenzt]
↑ parallelepiped   ↑ Parallelepiped
rhomboid   MATH   **Rhomboid**
[oblique parallelogramm, with unequal adjacent sides]   [schiefwinkliges Parallelogramm, mit paarweise ungleichen Seiten]
↑ parallelogram   ↑ Parallelogramm
rhomboid antenna   ANT   **Rhomboid-Antenne**
↑ rhombic antenna   ↑ Rhombusantenne
rhombus (pl.-uses, -i)   MATH   **Rhombus** (pl.-en)
[oblique, equilateral parallelogram]   [schiefwinkliges, gleichschenkliges Parallelogramm]
↑ parallelogram   = Raute
       ↑ Parallelogramm
rho-theta navigation   RADIO NAV   **Rho-Theta-Navigation**
= rho-theta system   = Rho-Theta-Verfahren
rho-theta system (→rho-theta navigation)   RADIO NAV   (→Rho-Theta-Navigation)
rhytm   COLLOQ   **Rhythmus**
rhythmic (→rhythmical)   COLLOQ   (→rhythmisch)
rhythmical   COLLOQ   **rhythmisch**
= rhythmic
rib   MECH   **Rippe**
= fin
ribbed   TECH   **gerippt**
= finned
ribbon (→inked ribbon)   OFFICE   (→Farbband)
ribbon (→tape)   TECH   (→Band)
ribbon advance facility   OFFICE   **Farbbandantrieb**
= ribbon feed   = Farbbandtransport
ribbon cable   COMM.CABLE   **Bandkabel**
= strip cable; flat cable; twin lead; balanced feeder cable   = Flachkabel; Bandleitung; Twin-lead-Kabel
ribbon cartridge   OFFICE   **Farbbandkassette**
ribbon feed (→ribbon advance facility)   OFFICE   (→Farbbandantrieb)
ribbon lifter   OFFICE   **Farbbandgabel**
ribbon microphone   EL.ACOUS   **Bändchenmikrofon**
       = Bändchenmikrophon
ribbon reverse   OFFICE   **Farbbandumkehr**
ribbon shift   OFFICE   **Farbbandumschaltung**
       = Farbzoneneinsteller
ribbon shift black   TERM&PER   **Schwarzschreibung**
ribbon shift red   TERM&PER   **Rotschreibung**
ribbon spool   OFFICE   **Farbbandrolle**
       = Farbbandspule
ribbon tweeter   EL.ACOUS   **Bändchen-Hochtöner**
rickwork   CIV.ENG   **Ziegel 1**
       [gebrannter Ton]
rider   OFFICE   **Reiter**
[for file cards]   [für Karteikarten]
= tab
ridged waveguide (→ridge waveguide)   MICROW   (→Steghohlleiter)

ridge waveguide   MICROW   **Steghohlleiter**
= ridged waveguide
rift (→crack)   TECH   (→Sprung)
rig (n.) (→equipment 2)   TECH   (→Gerät)
right (→claim 1)   ECON   (→Anspruch)
right (→correct)   TECH   (→genau)
right-aligned   TYPOGR   **rechtsbündig**
= right-justified; right flush; flush right; right-hand justified
right aligned (→right justified)   CODING   (→rechtsbündig)
right angle   MATH   **rechter Winkel**
       = rechtwinkliger Winkel
right-angle curve (→square loop)   PHYS   (→Rechteckschleife)
right-angled   MATH   **rechtwinklig**
= right angular; rectangular 1; orthogonal 1   = orthogonal 1; winkelrecht; rechteckig 1
≈ perpendicular   ≈ senkrecht
right-angled triangle (→right triangle)   MATH   (→rechtwinkliges Dreieck)
right angular (→right-angled)   MATH   (→rechtwinklig)
right blank   TYPOGR   **Nachbreite**
       [Leerraum rechts des Zeichens]
right bracket   MATH   **Klammer zu**
[symbol)]   [Symbol)]
= close bracket
right flush (→right-aligned)   TYPOGR   (→rechtsbündig)
right-hand ... (→clockwise)   TECH   (→rechtsdrehend)
right handed system   MATH   **Rechtssystem**
right-hand helix   MATH   **Rechtsschraube**
right-hand justified (→right-aligned)   TYPOGR   (→rechtsbündig)
right-hand lay (→right-hand twist)   PHYS   (→Rechtsdrall)
right-hand rule (→three-fingers rule)   EL.TECH   (→Dreifingerregel)
right-hand screw rule (→three-fingers rule)   EL.TECH   (→Dreifingerregel)
right-hand thread   MECH   **Rechtsgewinde**
right-hand twist   PHYS   **Rechtsdrall**
= right-hand lay
right justified   CODING   **rechtsbündig**
= right aligned
right-justified (→right-aligned)   TYPOGR   (→rechtsbündig)
right-of-access code (→keyword 1)   DATA PROC   (→Paßwort)
right of preemption   ECON   **Vorkaufsrecht**
= preemption; first option
right of use   DATA PROC   **Benutzungsberechtigung**
= entitlement
right of way   OUTS.PLANT   **Wegerecht**
= wayleave   = Verlegerecht
right-of-way company (AM)   TELEC   **privater Netzbetreiber mit eigenen Verlegerechten**
[utility company with own telecommunication network]   [Elektrizitätsversorgungsunternehmen; Eisenbahn]
= ROW
right triangle   MATH   **rechtwinkliges Dreieck**
= right-angled triangle
rightward arrow   TYPOGR   **Pfeil nach rechts**
rigid 2 (adj.)   TECH   **formbeständig**
= form-permanent   ≈ starr
≈ stiff
rigid 1 (→stiff)   TECH   (→starr)
rigid body   MECH   **Starrkörper**
       = starrer Körper
rigid cable   COMM.CABLE   **Vollmantelkabel**
rigid coaxial cable   LINE TH   **Rohrleitung**
rigid disc (→hard disk)   TERM&PER   (→Hartplatte)
rigid disk (→hard disk)   TERM&PER   (→Hartplatte)
rigid foam plastic   CHEM   **Hartschaum**
rigidity   MECH   **Starrheit**
= stiffness   = Steifheit
rigidity 1 (→stiffness)   TECH   (→Starrheit)
rigidity modulus (→shear modulus)   MECH   (→Schubmodul)

rigid stay (→terminal pole)	OUTS.PLANT	(→Abspannmast)
rim	MECH	**Kranz**
		↓ Radkranz
ring (v.i.)	TELEC	**rufen**
= call		
ring (n.) [one of the contacts of the telephone jack connector, connected to ring wire] ≈ tip; sleeve; ground	TELEPH	**Ring** [einer der Pole des Klinkensteckers, mit B-Ader verbunden] ≈ Spitze; Hals; Masse
ring 3 (n.) [set of elements with defined addition and multiplication] ↑ algebraic structure	MATH	**Ring 3** [Menge in der Addition und Multiplikation definiert sind] ↑ algebraische Struktur
ring (→ring wire)	TELEPH	(→B-Ader)
ring 2 (→annulus)	MATH	(→Kreisring)
ring 1 (n.) (→torus)	MATH	(→Torus) (pl.-i)
ring (v.i.) (→telephone)	TELEPH	(→telefonieren)
ring (n.) (→circular list)	DATA PROC	(→Ringliste)
ring acknowledge	DATA COMM	**Ringquittung**
ring armature	PHYS	**Ringanker**
ring-around circuit (→ring-back circuit)	POWER SYS	(→Umschwingkreis)
ring-around current (→ring-back current)	POWER SYS	(→Umschwingstrom)
ring array (→circular array antenna)	ANT	(→Kreisgruppenantenne)
ring back (→call back 2)	TELEPH	(→rückfragen)
ring back (→call-back)	TELEPH	(→Rückfrage)
ringback (→call-back)	TELEPH	(→Rückfrage)
ring-back circuit = ring-around circuit; oscillating circuit	POWER SYS	**Umschwingkreis** = Umschwingzweig
ring-back current = ring-around current; recirculating current	POWER SYS	**Umschwingstrom**
ring-back reactor (→reversing reactor)	POWER SYS	(→Umschwingdrossel)
ring balance	INSTR	**Ringwaage**
ring binder	DOC	**Ringmappe**
ring blocking circuit	TELEC	**Rufsperrschaltung**
ring book	OFFICE	**Ringbuch**
ring bus = loop bus; ring topology; loop topology ≈ ring network	DATA COMM	**Ringbus** = Ringtopologie ≈ Ringnetz
ring cable = loop cable	DATA COMM	**Ringkabel** = Ringleitung ≈ Ringbus
ring connection	DATA COMM	**Ringanschluß**
ring counter ↑ counter	CIRC.ENG	**Ringzähler** ↑ Zähler
ring current = circulating current	POWER ENG	**Kreisstrom**
ring-current reactor	POWER ENG	**Kreisstromdrossel**
ringdown signal (→ringing signal)	TELEC	(→Rufsignal)
ringer (→bell)	TERM&PER	(→Klingel)
ringer (→ringing converter)	TELEC	(→Rufumsetzer)
ring ground = RG	SYS.INST	**Ring-Erdung**
ringing	TELEC	**Rufen** = Ruf
ringing condition = ringing state; calling condition; calling state	SWITCH	**Rufzustand**
ringing conversion	TELEC	**Rufumsetzung**
ringing converter = ringer	TELEC	**Rufumsetzer**
ringing current (→dialing current)	TELEC	(→Rufstrom)
ringing device	TERM&PER	**Ruforgan**
ringing frequency = calling frequency	TELEC	**Rufstromfrequenz**
ringing frequency = overshoot frequency; singing frequency	ELECTRON	**Überschwingfrequenz**
ringing generator = RG; ringing supply	TELEC	**Rufgenerator** = Rufstromgenerator
ringing key	TERM&PER	**Ruftaste**
ringing machine	TELEC	**Rufmaschine**
ringing queue	SWITCH	**Rufwarteschlange**
ringing relay	TERM&PER	**Rufanschaltrelais**
ringing repeater	TELEPH	**Rufübertrager**
ringing signal = ringdown signal; call signal; calling signal; subscriber ringing signal	TELEC	**Rufsignal** = Rufzeichen; Ruf; Teilnehmerruf; Anrufsignal; Anrufzeichen
ringing state (→ringing condition)	SWITCH	(→Rufzustand)
ringing supply (→ringing generator)	TELEC	(→Rufgenerator)
ringing time	TELEC	**Rufzeit**
ringing tone (→idle tone)	TELEPH	(→Freiton)
ringing tone connection	TELEC	**Rufanschaltung** = Rufstromanschaltung
ring mixer	HF	**Ringmischer**
ring modulator = ring-type modulator	MODUL	**Ringmodulator** = Sternmodulator
ring n. (→call)	TELEPH	(→Anruf)
ring network = ring-type network; ring-structure network ≈ annular bus [DATA COMM]	TELEC	**Ringnetz** = ringförmiges Netz ≈ Ringbus [DATA COMM]
ring oscillator	CIRC.ENG	**Ringoszillator**
ring radiator	ANT	**Ringantenne 1**
ring radiator antenna	ANT	**Ringstrahler**
ring resonator	OPTOEL	**Ringresonator**
ring-shaped (→annular)	TECH	(→ringförmig)
ring shift register (→cyclic shift register)	CIRC.ENG	(→Ringschieberegister)
ring-structure network (→ring network)	TELEC	(→Ringnetz)
ring topology (→ring bus)	DATA COMM	(→Ringbus)
ring tripping	TELEC	**Rufabschaltung** = Rufstromabschaltung
ring-type modulator (→ring modulator)	MODUL	(→Ringmodulator)
ring-type network (→ring network)	TELEC	(→Ringnetz)
ring winding	COMPON	**Ringwicklung**
ring wire = ring; B wire; B-lead	TELEPH	**B-Ader**
ripple [alternating content to continuous value] ↓ voltage ripple; current ripple; residual ripple	EL.TECH	**Welligkeit** [Wechselanteil zu Gleichwert] ↓ Spannungswelligkeit; Stromwelligkeit; Restwelligkeit
ripple [voltage at bump frequency to voltage at center frequency]	NETW.TH	**Welligkeit** [Spannungswert bei Hökkerfrequenz zu Spannungswert in Bandmitte]
ripple band	NETW.TH	**Welligkeitsbereich**
ripple blanking	INSTR	**Nullstellenausblendung**
ripple blanking (→zero compression)	DATA PROC	(→Nullunterdrückung)
ripple carry = ripple-through carry	DATA PROC	**Schnellübertrag** = Ripple-Übertrag
ripple carry counter	MICROEL	**Ripple-carry-Zähler**
ripple content [rms of superimposed ac to no-load dc] ≈ superimposed ac ↑ ripple	EL.TECH	**Spannungswelligkeit** [Effektivwert überlagerter Wechselspannungen zur Leerlaufgleichspannung] = überlagerte Welligkeit ≈ überlagerte Wechselspannung ↑ Welligkeit
ripple content (→harmonic content)	EL.TECH	(→Oberwellengehalt)
ripple control	TELECONTR	**Rundsteuerung**
ripple control receiver	SIGN.ENG	**Rundsteuerempfänger**
ripple control system [transmission of HF pulses over the electric supply network, for control purposes]	TELECONTR	**Rundsteueranlage** [Übertragung von HF-Impulsen über das Stromversorgungsnetz zu Steuerzwecken]
ripple filter	CIRC.ENG	**Welligkeitsfilter**
ripple sort (→bubble sorting)	DATA PROC	(→Bubble-Sortieren)

ripple-through carry ($\rightarrow$ripple DATA PROC carry) ($\rightarrow$Schnellübertrag)
ripple voltage ($\rightarrow$superimposed ac) EL.TECH ($\rightarrow$überlagerte Wechselspannung)
ripsaw TECH **Handsäge**
= hand saw
↑ saw ↑ Säge
RISC ($\rightarrow$reduced instruction set computer) DATA PROC ($\rightarrow$RISC-Computer)
rise (v.t.) ECON **steigen**
= go up; increase (v.i.); ascend
rise (n.) ELECTRON **Anstieg**
= buildup ≠ Abfall
≠ fall
rise ($\rightarrow$increase 1) TECH ($\rightarrow$Zunahme)
rise path ($\rightarrow$initial magnetization curve) PHYS ($\rightarrow$Neukurve)
riser SYS.INST **Steigschacht**
[vertically running duct]
rise response CONTROL **Anstiegverhalten**
↑ transient response ↑ Übergangsverhalten
rise time CONTROL **Anregelzeit**
rise time ELECTRON **Anstiegzeit**
[between 10% and 90% amplitude points] [zwischen 10% und 90% Amplitude]
= build-up time; growth time; leading transition time; transition time; slew time = Steigzeit; Flankenanstiegzeit
≠ decay time ≠ Abfallzeit
↑ ramp time ↑ Rampenzeit
rising edge ELECTRON **Anstiegflanke**
= leading edge [eines Signals, Impulses]
≠ trailing edge = Vorderflanke; steigende Flanke; vordere Flanke
↓ rising pulse edge
 ≠ Abfallflanke
 ↓ Impulsanstiegflanke
rising main ($\rightarrow$distribution cable) POWER SYS ($\rightarrow$Steigleitung)
rising prices ($\rightarrow$price increase) ECON ($\rightarrow$Teuerung)
rising pulse edge ELECTRON **Impulsanstiegflanke**
= leading pulse edge; positive pulse edge = Impulsvorderflanke; steigende Impulsflanke; vordere Impulsflanke; Pulsvorderflanke; Pulsanstiegsflanke; steigende Pulsflanke
≠ trailing pulse edge
 ≠ Impulsabfallflanke
risk ECON **Risiko**
= hazard
RITL ($\rightarrow$radio in the loop) RADIO ($\rightarrow$Teilnehmerfunk)
rival ($\rightarrow$competitor) ECON ($\rightarrow$Mitbewerber)
rival product ECON **Konkurrenzprodukt**
= Konkurrenzerzeugnis
rivet (v.t.) MECH **vernieten 3**
[mit Niet]
= annieten; nieten
rivet (n.) METAL **Niet**
riveted METAL **genietet**
= grooved ↓ meißelgenietet
riveter ($\rightarrow$riveting machine) METAL ($\rightarrow$Nietmaschine)
rivet head METAL **Nietkopf**
riveting METAL **Nieten**
 ↓ Meißelnieten
riveting machine METAL **Nietmaschine**
= riveter
riveting nut MECH **Annietmutter**
riveting pin METAL **Nietstift**
rivet joint METAL **Nietverbindung**
RJE ($\rightarrow$remote job entry) DATA PROC ($\rightarrow$Auftragsferneingabe)
RLCM network NETW.TH **RLCÜ-Zweipol**
RLC network NETW.TH **RLC-Schaltung**
RL network NETW.TH **RL-Zweipol**
= RL-Schaltung
r/m ($\rightarrow$revolutions per minute) PHYS ($\rightarrow$Umdrehungen pro Minute)
RMI ($\rightarrow$radio magnetic indicator) RADIO LOC ($\rightarrow$Radiokompaßrose)
R/MOS process MICROEL **R/MOS-Verfahren**
[refractory-metal gate metal-oxide semiconductor process]

RMS ($\rightarrow$effective value) PHYS ($\rightarrow$Effektivwert)
rms ($\rightarrow$effective value) PHYS ($\rightarrow$Effektivwert)
rms converter COMPON **RMS-Wandler**
rms current ($\rightarrow$active current) EL.TECH ($\rightarrow$Wirkstrom)
rms jitter ELECTRON **effektive Jitteramplitude**
rms measurement INSTR **Effektivwertmeßung**
rms rectifier POWER SYS **Effektivwertgleichrichter**
= quadratischer Gleichrichter
rms value ($\rightarrow$effective value) PHYS ($\rightarrow$Effektivwert)
rms voltage ($\rightarrow$active voltage) EL.TECH ($\rightarrow$Wirkspannung)
Rn ($\rightarrow$radon) CHEM ($\rightarrow$Radon)
RNG ($\rightarrow$radio range beacon) RADIO NAV ($\rightarrow$Kursfunkbake)
RNG ($\rightarrow$random sequence generator) DATA PROC ($\rightarrow$Zufallsgenerator)
RNR ($\rightarrow$receive not ready) DATA COMM ($\rightarrow$nicht empfangsbereit)
road traffic control ($\rightarrow$traffic control) SIGN.ENG ($\rightarrow$Verkehrsregelung)
road traffic control computer SIGN.ENG ($\rightarrow$Verkehrsrechner)
($\rightarrow$traffic control computer)
road-traffic control engineering SIGN.ENG **Straßenverkehrs-Signaltechnik**
road traffic engineering TECH **Straßenverkehrstechnik**
= traffic control engineering
road user SIGN.ENG **Verkehrsteilnehmer**
roam (v.t.) TERM&PER **schwenken**
[move a display window] [ein Bildschirmfenster]
roamer MOB.COMM **Gastteilnehmer**
[user outside his home service area] [aus einem fremden Versorgungsgebiet]
= roaming subscriber
roaming MOB.COMM **Gesprächsübergabe**
roaming agreement MOB.COMM **Gesprächsübergabeabkommen**
roaming subscriber MOB.COMM ($\rightarrow$Gastteilnehmer)
($\rightarrow$roamer)
roar (v.i.) ACOUS **dröhnen**
= boom (v.i.)
robot CONTROL **Roboter**
≈ process control computer [DATA PROC]; automaton [INF.TECH] = Handhabungsgerät; Handhabungsautomat; Handhabungseinrichtung
 ≈ Prozeßrechner [DATA PROC]; Automat [INF.TECH]
robot-control language DATA PROC **Robotsteuersprache**
robotics CONTROL **Robotik**
= Robotertechnik
robotic workcell MANUF **Roboterarbeitsplatz**
robotization TECH **Robotisierung**
≈ automation ≈ Automatisierung
robotized manufacturing MANUF ($\rightarrow$Roboterfertigung)
($\rightarrow$robot manufacturing)
robotized production ($\rightarrow$robot manufacturing) MANUF ($\rightarrow$Roboterfertigung)
robot manufacturing MANUF **Roboterfertigung**
[equipped with r.] [mit R. ausgerüstete F.]
= robot production; robotized manufacturing; robotized production = automatisierte Fertigung
≈ automated manufacturing
robot production ($\rightarrow$robot manufacturing) MANUF ($\rightarrow$Roboterfertigung)
robust ($\rightarrow$rugged) TECH ($\rightarrow$robust)
robustness ($\rightarrow$ruggedness) TECH ($\rightarrow$Robustheit)
rock ($\rightarrow$swing) MECH ($\rightarrow$schaukeln)
rocker ($\rightarrow$rocker arm) TECH ($\rightarrow$Schwinghebel)
rocker arm TECH **Schwinghebel**
= rocker = Wippe; Kipphebel
rocker switch COMPON **Wippenschalter**
= rocking switch = Wippschalter
rocking motion ($\rightarrow$rotational oscillation) PHYS ($\rightarrow$Drehschwingung)
rocking switch ($\rightarrow$rocker switch) COMPON ($\rightarrow$Wippenschalter)
rod MECH **Rundstab**
= round bar = Rundstange
 ↑ Stab
rod ($\rightarrow$stick) COMPON ($\rightarrow$Stäbchen)
rod ($\rightarrow$wire rod) METAL ($\rightarrow$Walzdraht)
rod antenna ($\rightarrow$dielectric rod antenna) ANT ($\rightarrow$Stabantenne)

rod chart	MATH	**Stabdiagramm**
[with non touching rods]		[Darstellung mit nicht aneinander anstoßenden Stäbe]
↑ bar chart		↑ Balkendiagramm
rod core balun	ANT	**Stabkern-Balun**
rodent	COLLOQ	**Nagetier**
= gopher (AM)		
rodent attack	TECH	**Nagetierfraß**
= gopher attack		
rodent-protected cable	COMM.CABLE	(→nagetiergeschütztes Kabel)
(→gopher-protected cable)		
rodent protection	COMM.CABLE	**Nagetierschutz**
= gopher protection (AM)		
rod memory	DATA PROC	**Stäbchenspeicher**
rod pointer	INSTR	**Stabzeiger**
↑ pointer		↑ Zeiger
rod reflector	ANT	**Stabreflektor**
RO file (→read-only file)	DATA PROC	(→Nur-Lese-Daten)
roll (v.i.)	TECH	**schlingern**
[slow swing on longitudinal axis]		[langsames um Längsachse schwanken]
		= rollen
roll (n.)	TERM&PER	**Rolle**
[reel with windings]		[Spulenkörper mit Wicklung]
↓ magnetic tape roll; peforated tape roll		↓ Magnetbandrolle; Lochstreifenrolle
roll 1 (n.)	TECH	**Rolle 1**
[something rolled into shape of cylinder or ball]		[etwas walzenförmig zusammengerolltes]
≈ reel 2		= Spule 2
≈ coil		
roll 2	TECH	**Rolle 2**
[something performing a rolling action]		[Kugel, Walze, Rad, Scheibe oder sonstiger Körper, auf dem etwas rollt oder gleitet]
≈ drum; roller		≈ Trommel; Walze
↓ sheave		↓ Seilscheibe
roll (→ paper roll 1)	TERM&PER	(→Papierrolle)
roll (v.t.) (→swing)	MECH	(→schaukeln)
roll (→scroll)	TERM&PER	(→rollen)
roll axis	TECH	**Schlingerachse**
		= Rollachse
rollback program (→rerun program)	DATA PROC	(→Wiederholprogramm)
rollback routine (→rerun program)	DATA PROC	(→Wiederholprogramm)
rolled	METAL	**gewalzt**
rolled steel	METAL	**Walzstahl**
rolled transposition (→transposition system)	OUTS.PLANT	(→Kreuzungsschema)
roller	TECH	**Walze**
[revolving cylindrical body, used to transmit force, move or to press something]		[länglicher Körper mit kreisförmigem Querschnitt, zur Kraftübertragung, Fortbewegung oder Glättung]
≈ drum; roll 2		≈ Trommel; Rolle
roller bearing	MECH	**Rollenlager**
roller-burnishing	TECH	**Polierrollen**
roller chain	MECH	**Gelenkkette**
roll-front cabinet	TECH	**Rollschrank**
rolling	TERM&PER	**verikaler Bilddurchlauf**
rolling (→scrolling)	TERM&PER	(→Bildverschiebung)
rolling (→revolving)	ECON	(→revolvierend)
rolling (→lamination)	METAL	(→Walzen)
rolling-contact bearing	MECH	**Wälzlager**
rolling cut	TV	**Rollschnitt**
rolling down (→scroll down)	TERM&PER	(→zurückrollen)
rolling down (→scrolling down)	TERM&PER	(→Zurückrollen)
rolling friction	PHYS	**Rollreibung**
↑ friction		= rollende Reibung
		↑ Reibung
rolling mill	METAL	**Walzwerk**
rolling stock	RAILW.SIGN	**Triebfahrzeug**
rolling up (→scrolling up)	TERM&PER	(→Vorrollen)
roll-in / roll-out	DATA PROC	**Ein-Aus-Speicher**
		= Roll-in/Roll-out
ROLL key (→SCROLL key)	TERM&PER	(→Taste ROLLEN)
roll-off (n.)	NETW.TH	**Flankenabfall**
[a smooth variation of frequency response]		= Abfall; Roll-off; Frequenzgangabsenkung; Absenkung
roll-off factor	NETW.TH	**Roll-off-Faktor**
rollover (n.) (→scrolling)	TERM&PER	(→Bildverschiebung)
rollover (n.) (→rollover buffer)	DATA PROC	(→Überlaufpuffer)
rollover buffer	DATA PROC	**Überlaufpuffer**
= rollover (n.)		
rollover device	TERM&PER	**Abrollgerät**
↓ mouse; turtle		↓ Maus; Schildkröte
rollpaper	TERM&PER	**Rollenpapier**
= continuous rollpaper		↑ Endlospapier
= continuous paper		
rollpaper form	TERM&PER	**Rollenformular**
↑ continuous form		= Rollenvordruck
		↑ Endlosformular
roll-up (→scroll-up)	TERM&PER	(→vorrollen)
ROLL-UP key	TERM&PER	**ROLL-UP-Taste**
ROM (→read-only memory)	DATA PROC	(→Festwertspeicher)
roman (n.)	TYPOGR	**Roman**
[font type]		[Schriftart]
= roman type		= Antiqua; lateinische Schrifttart
roman type (→roman)	TYPOGR	(→Roman)
ROM cartridge	TERM&PER	**Festspeicherkassette**
= solid-state cartridge		= Festspeichercassette; ROM-Kassette
RO memory (→read-only memory)	DATA PROC	(→Festwertspeicher)
ROM memory (→read-only memory)	DATA PROC	(→Festwertspeicher)
ROM programming system	TERM&PER	**ROM-Programmierer**
ROM storage (→read-only memory)	DATA PROC	(→Festwertspeicher)
roof antenna	ANT	**Dachantenne 1**
= top antenna		
roof cover	ANT	**Dachabdeckblech**
roofed over	CIV.ENG	**überdacht**
room 1	CIV.ENG	**Raum**
[partitioned space inside a building]		[abgetrennter Raum im Inneren eines Gebäudes]
↓ chamber; room 2; hall		↓ Kammer; Zimmer; Saal
room 2	CIV.ENG	**Zimmer**
[a room of an apartment]		
↑ room 1		↑ Raum
room acoustics (→acoustics of room)	ACOUS	(→Raumakustik)
room aerial (→indoor antenna)	ANT	(→Innenantenne)
room antenna (→indoor antenna)	ANT	(→Innenantenne)
room climate	QUAL	**Innenraumklima**
= indoor climate		
room monitoring system	SIGN.ENG	**Raumüberwachungsanlage**
room noise	ACOUS	**Raumgeräusch**
= ambient noise		
room shielding	EL.TECH	**Raumabschirmung**
root	MECH	**Fußkreis**
[gear]		[Zahnrad]
root 1	MATH	**Wurzel 1**
↓ square root; cube root		↓ Quadratwurzel; Kubikwurzel
root 3	MATH	**Wurzel 3**
[theory of graphs; node without antecessor]		[Graphentheorie; Knoten ohne Vorgänger]
↑ node		= Root
		↑ Knoten
root circle	ENG.DRAW	**Fußkreis**
root directory	DATA PROC	**Hauptverzeichnis**
[the top directory]		[oberstes Dateiverzeichnis]
= main directory; main catalog		= Haupt-Dateiverzeichnis; Haupt-Arbeitsbereich; Hauptinhaltsverzeichnis; Wurzelverzeichnis; Hauptkatalog

root element (→rooter circuit)		CIRC.ENG	(→Radizierer)	**rotary joint**	MECH	**Drehgelenk** = Rotationsgelenk
rooter circuit = root element		CIRC.ENG	**Radizierer** = Radizierelement	**rotary knob** = knob	COMPON	**Drehknopf** ↑ Knopf
root index		MATH	**Wurzelindex**	**rotary magnet** = moving magnet	EL.TECH	**Drehmagnet**
root locus		CONTROL	**Wurzelortskurve** = Wurzelort	**rotary magnetism**	PHYS	**Rotationsmagnetismus**
root locus		MATH	**Wurzelort**	**rotary motion** (→rotary movement)	PHYS	(→Drehbewegung)
root mean square (→effective value)		PHYS	(→Effektivwert)	**rotary movement** = rotational motion; rotary motion; rotational movement; rotation; ≈ circular movement	PHYS	**Drehbewegung** ≈ Kreisbewegung
root-mean-square indication		INSTR	**Effektivwertanzeige**			
root node (→leaf node)		DATA PROC	(→Astknoten)			
root of unit		MATH	**Einheitswurzel**	**rotary phase shifter**	EL.TECH	**Drehphasenschieber**
root operator		MATH	**Wurzeloperator**	**rotary printing** [paper runs in between two rolls, one of them carrying the printing form] ↓ rotogravure	TYPOGR	**Rotationsdruck** [Papier läuft zwischen zwei Walzen, von denen eine die Druckform trägt] ↓ Rotationstiefdruck
root segment [permanently in the main memory] ≠ overlay segment		DATA PROC	**arbeitsspeicherresidentes Programmsegment** [ständig im Arbeitsspeicher stehend] = Rumpfteil; Rumpfsegment ≠ Überlagerungssegment			
				rotary resolver = resolver; synchrogenerator; synchro	INSTR	**Drehmelder** = Synchro; Resolver
root segment		MATH	**Wurzelsegment**	**rotary rheostat** (→crank resistance)	INSTR	(→Kurbelwiderstand)
root-sum-square value [square root] ≈ effective value [PHYS]		MATH	**quadratisches Mittel** [Wurzel der durch n geteilten Summe A1² plus ... An²] ≈ Effektivwert [PHYS]			
				rotary selector = rotary switch ↑ selector	SWITCH	**Drehwähler** ↑ Wähler
root symbol [symbol √] = radical sign; radical 2		MATH	**Wurzelzeichen** [Symbol √]	**rotary selector** (→rotary switch)	COMPON	(→Drehschalter)
				rotary selector switch	COMPON	**Stufendrehschalter**
rope [wires or fibers twisted together] ≈ cord; line; rope ↓ cable		TECH	**Seil** [aus Fäden oder Drähten gedreht] ≈ Schnur; Leine; Strick ↓ Hanfseil; Tau	**rotary switch** = turn switch; rotary selector ↓ rotary selector switch	COMPON	**Drehschalter** = Drehwähler ↓ Stufendrehschalter
				rotary switch (→rotary selector)	SWITCH	(→Drehwähler)
				rotary symmetry	MATH	**Drehsymmetrie**
rope		ANT	**Gut** [Seil oder Draht eines Antennenaufbaus]	**rotary table** (→turntable)	TECH	(→Drehtisch)
				rotatable = rotary; versatile	MECH	**drehbar**
rope sheave		TECH	**Seilroller**	**rotatable capacitor** = rotatable condenser ↑ variable capacitor	COMPON	**Drehkondensator** = Regelkondensator
rope traction = conveyor trip		MECH	**Seilzug**			
rope wire		TECH	**Seildraht**	**rotatable condenser** (→rotatable capacitor)	COMPON	(→Drehkondensator)
rose madder		PHYS	**krapprosa**			
Rose metal ↑ solder		METAL	**Rose-Metall** ↑ Lot	**rotatable resistor** ↑ variable resistor	COMPON	**Drehwiderstand** ↑ Regelwiderstand
Rosen's theorem		NETW.TH	**Theorem von Rosen** = Rosensches Theorem	**rotate** (v.i.) [to move about an axis or center] = revolve; turn 1 ≈ circulate; turn 2	PHYS	**rotieren** [sich um einen Punkt oder eine Achse bewegen] ≈ kreisen; drehen
rosette		MECH	**Rosette**			
rosewood [color]		PHYS	**rosenholz** [Farbe]	**rotate** [a graphics on the display]	DATA PROC	**rotieren** [eine Graphik oder Schrift drehen]
rosin = resin 2 ↑ resin 1		CHEM	**Kolophonium** ↑ Harz			
				rotating = rotational	TECH	**Dreh-**
RO storage (→read-only memory)		DATA PROC	(→Festwertspeicher)	**rotating beacon**	RADIO NAV	**Drehbake**
RO store (→read-only memory)		DATA PROC	(→Festwertspeicher)	**rotating direction finder**	RADIO LOC	**Umlaufpeiler**
				rotating field = revolving field	PHYS	**Drehfeld**
rot (→curl)		MATH	(→Rotation)	**rotating field antenna**	ANT	**Drehfeldantenne**
rotary (→rotatable)		MECH	(→drehbar)	**rotating field feed**	ANT	**Drehfeldspeisung**
rotary Adcock antenna		ANT	**Dreh-Adcock-Antenne**	**rotating field instrument**	INSTR	**Drehfeldinstrument**
rotary antenna ↓ rotary beam antenna		ANT	**Drehantenne** = drehbare Antenne ↓ Drehrichtstrahler	**rotating frame**	MECH	**schwenkbarer Rahmen**
				rotating frame antenna = rotating loop antenna ↑ direction finding antenna	ANT	**Drehrahmenantenne** = Drehrahmenpeiler ↑ Peilantenne
rotary arm		MECH	**Dreharm**			
rotary-beam antenna ≠ fixed beam antenna ↑ rotary antenna		ANT	**Drehrichtungsstrahler** = Drehrichtstrahler ≠ fester Richtstrahler ↑ Drehantenne	**rotating-frame direction finder**	RADIO LOC	**Drehrahmenpeiler**
				rotating loop antenna (→rotating frame antenna)	ANT	(→Drehrahmenantenne)
rotary convertor (IEC)		POWER SYS	**Einankerumformer**	**rotating memory** = rotating storage; rotating store; recirculating memory; recirculating storage; recirculating store	TERM&PER	**Umlaufspeicher** = umlaufender Speicher
rotary current (→three-phase current)		POWER ENG	(→Drehstrom)			
rotary dial (→dial pulsing)		SWITCH	(→Nummernschalterwahl)			
rotary dial selection (→dial pulsing)		SWITCH	(→Nummernschalterwahl)	**rotating phasor** [phasor representation] = phasor; sinor; rotating vector	EL.TECH	**Wechselstromzeiger** [Zeigerdiagramm] = Zeiger; Radiusvektor ≈ Vektor
rotary dial switch = dial switch; dial; loop-disconnect dial ≈ dialling disk; number plate		TERM&PER	**Nummernschalter** = Drehnummernschalter ≈ Wählscheibe; Zifferring			
				rotating radiobeacon = omnidirectional range; omnirange	RADIO NAV	**Drehfunkfeuer**
rotary drive		MECH	**Drehantrieb**			
rotary field design		POWER ENG	**Drehfeldbauart**	**rotating storage** (→rotating memory)	TERM&PER	(→Umlaufspeicher)
rotary filing stand		OFFICE	**Registraturdrehständer**			

rotating store (→rotating memory)		TERM&PER	(→Umlaufspeicher)	
rotating vector (→rotating phasor)		EL.TECH	(→Wechselstromzeiger)	
rotation (→rotary movement)		PHYS	(→Drehbewegung)	
rotation (→twist 3)		PHYS	(→Drehung)	
rotation (→Faraday effect)		PHYS	(→Faraday-Effekt)	
rotational (→rotating)		TECH	(→Dreh-)	
rotational delay		TERM&PER	**Drehverzug**	
= Drehwartezeit				
rotational distortion		PHYS	**Bildzerdrehung**	
rotational field		MATH	**Wirbelfeld**	
= solenoidal field				
rotational frequency (→rotational speed)		PHYS	(→Drehzahl)	
rotationally symmetric (→rotational-symmetric)		MATH	(→rotationssymmetrisch)	
rotational motion (→rotary movement)		PHYS	(→Drehbewegung)	
rotational movement (→rotary movement)		PHYS	(→Drehbewegung)	
rotational oscillation		PHYS	**Drehschwingung**	
= rocking motion				
rotational spectrum		PHYS	**Rotationsspektrum**	
rotational speed		PHYS	**Drehzahl**	
[SI unit: reciprocal second]			[SI-Einheit: reziproke Sekunde]	
= revolutional turns speed; speed 1; rotational frequency			≈ Umdrehungen pro Minute; Winkelgeschwindigkeit	
≈ revolutions per minute; angular velocity				
rotational surface distribution		PHYS	**Flächenwirbel**	
rotational-symmetric		MATH	**rotationssymmetrisch**	
= rotationally symmetric				
rotation axis (→spin axis)		PHYS	(→Drehachse)	
rotation circulator		MICROW	**Faraday-Zirkulator**	
= wave rotation circulator				
rotation vector		MATH	**Drehvektor**	
rotator (→phase rotator)		MICROW	(→Phasendreher)	
rotatory dispersion		PHYS	**Rotationsdispersion**	
RO terminal (→receive-only terminal)		TERM&PER	(→Nur-Empfangs-Endgerät)	
rotor		POWER SYS	**Läufer**	
[the moving part of an electric engine]			[beweglicher Teil eines Elektromotors]	
≠ stator			= Rotor	
			≠ Ständer	
rough (v.t.)		MECH	**schruppen**	
[to shape roughly in a preliminary way]			[grob vorbearbeiten]	
rough (adj.)		MECH	**rauh**	
≈ raw			≈ roh; grob	
rough (→approximate)		MATH	(→näherungsweise)	
rough adjustment (→coarse adjustment)		ELECTRON	(→Grobeinstellung)	
rough calculation		MATH	**Überschlagsrechnung**	
≈ rough estimate; approximation			= grobe Rechnung	
			= grobe Schätzung; Näherung	
rough draft (→preliminary draft)		DOC	(→Vorentwurf)	
roughen		MECH	**aufrauhen**	
rough estimate		MATH	**grobe Schätzung**	
≈ rough calculation			≈ Überschlagsrechnung	
rough formula		PHYS	**Faustformel**	
= rule of thumb			= Daumenregel; Faustregel	
rough grind		MECH	**Grobschliff**	
[surface grade]			[Oberflächengüte]	
rough-machine		MECH	**maschinell schruppen**	
roughness		MECH	**Rauhigkeit**	
roughness		RADIO PROP	**Welligkeit**	
[path profile]			[Geländeprofil]	
			= Rauhigkeit	
≈ rugosity				
roughness height		MECH	**Rauhigkeitshöhe**	
[surface grade]			[Oberflächengüte]	
= peak-to-valley height			= Rauhtiefe	
roughness width		MECH	**Rauhigkeitsbreite**	
[surface grade]			[Oberflächengüte]	
roughness-width cutoff		ENG.DRAW	**Bezugsstrecke**	
			[Oberflächengüte]	
rough terrain		RADIO PROP	**rauhes Gelände**	
			= welliges Gelände	

round (v.t.)		MATH	**runden**	
≈ truncate			≈ abstreichen	
↓ round-up; round-off			↓ aufrunden; abrunden	
round (adj.)		MATH	**rund**	
≈ circular; spheric; ring-shaped			≈ kreisförmig; sphärisch; ringförmig	
round (n.) (→rounding)		MATH	(→Rundung)	
roundabout way (→detour)		COLLOQ	(→Umweg)	
round aluminum (→round-bar aluminum)		METAL	(→Rundaluminium)	
round bar (→rod)		MECH	(→Rundstab)	
round-bar aluminum		METAL	**Rundaluminium**	
= round aluminum				
round-bar brass		METAL	**Rundmessing**	
round-bar bronze		METAL	**Rundbronze**	
= round bronze				
round-bar copper (→round conductor)		EL.TECH	(→Rundleiter)	
round-bar iron		METAL	**Rundstahl**	
= round iron			≈ Walzdraht	
≈ wire rod				
round bronze (→round-bar bronze)		METAL	(→Rundbronze)	
round cable		COMM.CABLE	**Rundkabel**	
round chart (→circular chart)		NETW.TH	(→Kreisblatt)	
round chart diagram (→circular chart diagram)		NETW.TH	(→Kreisblattdiagramm)	
round-coil measuring system		INSTR	**Rundspulmeßwerk**	
round conductor		EL.TECH	**Rundleiter**	
= round-bar copper			= Rundkupfer	
round-down (→round-off)		MATH	(→abrunden)	
rounded		MECH	**abgerundet**	
			= gerundet	
rounded		MATH	**abgerundet 2**	
↓ rounded-off; rounded-up			[zu kleinerem oder größerem runden Wert]	
			↓ abgerundet 1; aufgerundet	
round-edge obstruction		RADIO PROP	**rundes Hindernis**	
rounded-off		MATH	**abgerundet 1**	
			[zu kleinerem runden Wert]	
			≠ aufgerundet	
round file		MECH	**Rundfeile**	
= rat-tail file				
round head		MECH	**Halbrundkopf 2**	
			[Schraube]	
round-head screw		MECH	**Halbrundkopfschraube**	
= halfround-head screw; button-headed screw				
round-head-wood screw		MECH	**Halbrundholzschraube**	
rounding		MATH	**Rundung**	
= round (n.)			↓ Abrundung; Aufrundung	
↓ rounding-off; rounding-up				
rounding error		MATH	**Rundungsfehler**	
rounding off		MATH	**Abrundung 1**	
= roundoff			[nach oben oder unten]	
			↓ Abrundung 2; Aufrundung	
rounding up		MATH	**Aufrundung**	
round iron (→round-bar iron)		METAL	(→Rundstahl)	
roundly (→in round figures)		COLLOQ	(→rund) (adv.)	
roundness		TECH	**Rundheit**	
round-off (v.t.)		MECH	**abkanten 1**	
[sharp edges]			[scharfe Kanten beseitige]	
= blunt (v.t.)			= Kanten brechen; kantenbrechen	
round-off		MATH	**abrunden**	
[to turn into the next lowest round number]			[zum nächstkleineren ganzzahligen Wert]	
= round-down			= kaufmännisch abrunden	
≈ truncate			≈ abstreichen	
≠ round-up			≠ aufrunden	
↑ round 1			↑ runden	
roundoff (→rounding off)		MATH	(→Abrundung 1)	
round-off error		MATH	**Abrundungsfehler**	
round package		MICROEL	**Rundgehäuse**	
round relay		COMPON	**Rundrelais**	
round-the-earth		TECH	**erdumspannend**	
round-top counter-sunk rivet		MECH	**Linsenkopfniet**	
round-trip (→loop)		TELEC	(→Schleife)	

round-trip attenuation	TELEPH	(→Umlaufdämpfung)	
(→round-trip loss)			
round-trip delay	TELEPH	**Umlaufzeit**	
= recirculating delay			
round-trip gain (→loop gain)	TELEPH	(→Umlaufverstärkung)	
round-trip loss	TELEPH	**Umlaufdämpfung**	
[total loss of a four-wire loop, including the two trans-hybrid losses, not considering active elements]		[Gesamtdämpfung eines Vierdrahtkreises, einschließlich der zwei Gabelübergangsdämpfungen, ohne Berücksichtigung verstärkender Elemente]	
= feedback loss; round-trip attenuation			
round-up (v.t.)	MATH	**aufrunden**	
≠ round-off		[zu nächsthöherem ganzzahligen Wert]	
↑ round 1		= absolut aufrunden	
		≠ abrunden 1	
		↑ runden	
route (n.)	SWITCH	**Leitweg**	
[most favorable route]		[günstigster Weg]	
= routing			
route (n.)	TRANS	**Trasse**	
= transmission route		= Übertragungsstrecke	
≈ link		≈ Strecke	
		↓ Ferntrasse	
route (→course 2)	TECH	(→Kurs)	
route (→path)	SWITCH	(→Weg)	
route (→course 1)	TECH	(→Kurs)	
route administration (→alternate routing)	SWITCH	(→Leitweglenkung)	
route alarm	SWITCH	**Streckenalarm**	
route disconnection (→path disconnection)	SWITCH	(→Wegeabbau)	
route discriminating digit	SWITCH	**Richtungsausscheidungsziffer**	
route diversity (→path diversity)	TELEC	(→Mehrwegeführung)	
route-finding section (→path-finding section)	SWITCH	(→Wegesuchabschnitt)	
route interrogation (→path interrogation)	SWITCH	(→Wegeabfragen)	
route list	SWITCH	**Leitwegliste**	
route of line (→route tracing)	TELEC	(→Streckenführung)	
route onward (→relay)	TELEC	(→weiterleiten)	
router	DATA COMM	**Router**	
route searching (→traffic route searching)	SWITCH	(→Verkehrswegsuche)	
route selection	TRANS	**Trassenwahl**	
route selection (→alternate routing)	SWITCH	(→Leitweglenkung)	
route setting (→path setup)	SWITCH	(→Wegedurchschaltung)	
route setup (→path setup)	SWITCH	(→Wegedurchschaltung)	
route store	SWITCH	**Leitwegspeicher**	
route tracing	TELEC	**Streckenführung**	
= routing; tracing; tracking; route of line		= Trassenführung [TRANS]; Linienführung; Streckenverlauf; Wegeführung	
routine	DATA PROC	**Routine**	
[small program for recurrent tasks; primarily used in assemler languages]		[kleines Programm für häufig vorkommende Aufgaben; primär im Zusammenhang mit Assemblersprachen verwendet]	
≈ subroutine; procedure; macro call; program section			
↑ program			
↓ input routine; output routine; print routine		≈ Unterprogramm; Prozedur; Makroaufruf; Programmteil	
		↑ Programm	
		↓ Eingaberoutine; Ausgaberoutine; Druckroutine	
routine (→regular)	TECH	(→regelmäßig)	
routine measurement	TECH	**Routinemessung**	
≈ routine test		≈ Routineprüfung	
routine test	TECH	**Routineprüfung**	
≈ routine measurement		≈ Routinemessung	
routing	SWITCH	**Lenkung**	
routing (→manufacturing engineering 2)	MANUF	(→Fertigungsplanung)	
routing (→tracking)	TECH	(→Führung)	
routing (→PCB artwork creation)	ELECTRON	(→Leiterplattenentflechtung)	
routing (→route)	SWITCH	(→Leitweg)	
routing (→alternate routing)	SWITCH	(→Leitweglenkung)	
routing (→route tracing)	TELEC	(→Streckenführung)	
routing data	SWITCH	**Leitwegangaben**	
		= Leitweglenkungsdaten	
routing engineer	MANUF	**Fertigungsplaner**	
routing map	SWITCH	**Leitwegplan**	
routing selector	SWITCH	**Umsteuerwähler**	
row	COLLOQ	**Reihe**	
≈ line; series; sequence		≈ Zeile; Serie; Reihenfolge	
↑ array		↑ Anordnung	
row	TERM&PER	**Sprosse 2**	
[transversal set of recording positions which can be assembled on a multitrack magnetic tape]		[quer zur Spurrichtung bildbarer Satz von Speicherplätzen eines mehrspurigen Magnetbandes]	
≈ track		≈ Spur	
row (→rack row)	SYS.INST	(→Gestellreihe)	
row (→hole row)	TERM&PER	(→Lochreihe)	
ROW (→right-of-way company)	TELEC	(→privater Netzbetreiber mit eigenen Verlegerechten)	
row (→line)	TYPOGR	(→Zeile)	
row frame	TERM&PER	**Sprosse 1**	
[holes in transversal array to the punched tape]		[Lochreihe quer zur Lochstreifenrichtung]	
↑ hole row		↑ Lochreihe	
row matrix	MATH	**Zeilenmatrix**	
row scanning	ELECTRON	**Reihenabtastung**	
royalty (→licence fee)	ECON	(→Lizenzgebühr)	
RPE (→radiation pattern envelope)	ANT	(→Winkeldämpfungs-Hüllkurve)	
RPG (→report generator)	DATA PROC	(→Listengenerator)	
r.p.m. (→revolutions per minute)	PHYS	(→Umdrehungen pro Minute)	
rpm (→revolutions per minute)	PHYS	(→Umdrehungen pro Minute)	
r.p.s. (→revolutions per second)	PHYS	(→Umdrehungen pro Sekunde)	
rps (→revolutions per second)	PHYS	(→Umdrehungen pro Sekunde)	
RQL (→rejectable quality level)	QUAL	(→Rückweisgrenze)	
RR (→ready to receive)	DATA COMM	(→empfangsbereit)	
RS (→record separator)	DATA COMM	(→Untergruppentrennung)	
RS (→radiated susceptibility)	TELEC	(→strahlungsgebundene Störfestigkeit)	
r/s (→revolutions per second)	PHYS	(→Umdrehungen pro Sekunde)	
R service (→required service)	RADIO NAV	(→dauernd erforderliche Verbindung)	
RS flipflop	CIRC.ENG	**RS-Flipflop**	
R2 signaling	SWITCH	**R2-Signalisierung**	
RT (→radiotelephony)	TELEC	(→Funksprechwesen)	
RTL	MICROEL	**RTL**	
= resistor-transistor logic		= Widerstand-Transistor-Logik	
RTS (→reliable transfer server)	DATA COMM	(→gesicherter Transfer-Server)	
RT test equipment (→radiotelephony test equipment)	INSTR	(→Sprechfunkgeräte-Meßplatz)	
Ru (→ruthenium)	CHEM	(→Ruthenium)	
rub (v.t.)	MECH	**scheuern**	
= scrub (v.t.)		= schrubben	
rubber	CHEM	**Gummi**	
rubber (→groumet)	COMPON	(→Durchführungstülle)	
rubber (→eraser)	ENG.DRAW	(→Radiergummi)	
rubber banding	DATA PROC	**Gummibandverfahren**	
[computer graphics; the possibility to dislocate an element and its connections aper cursor]		[Computergraphik; Möglichkeit ein Teil samt Verbindungen per Cursor zu verlagern]	
		= Dehnlinienverfahren	
rubber cement	TECH	**Alleskleber**	
↑ adhesive		↑ Kleber	
rubber-covered cable	EL.INST	**Gummikabel**	
= rubber-insulated cable			
rubber-covered wire	EL.INST	**Gummidraht**	

rubber effect [computer graphics]	DATA PROC	**Gummieffekt** [Computergraphik]
rubber feet	EQUIP.ENG	**Gummifuß**
rubber-insulated cable (→rubber-covered cable)	EL.INST	(→Gummikabel)
rubber mat ↑ antistatic mat	ELECTRON	**Vollgummi-Gittermatte** ↑ Antistatik-Matte
rubber seal	TECH	**Gummidichtung**
rubbish = rubble ≈ scrap; garbage; litter; junk	COLLOQ	**Schutt** ≈ Abfall; Müll; Unrat; Schrott
rubbish (→scrap)	TECH	(→Abfall)
rubble (→rubbish)	COLLOQ	(→Schutt)
rubidium = Rb	CHEM	**Rubidium** = Rb
rubidium frequency standard = rubidium vapor frequency standard; rubidium vapor standard; rubidium gas cell frequency standard	INSTR	**Rubidium-Frequenzstandard** = Rubidium-Frequenznormal; Rubidiumnormal
rubidium gas cell frequency standard (→rubidium frequency standard)	INSTR	(→Rubidium-Frequenzstandard)
rubidium magnetometer	INSTR	**Rubidium-Magnetometer**
rubidium oscillator	CIRC.ENG	**Rubidium-Oszillator**
rubidium vapor frequency standard (→rubidium frequency standard)	INSTR	(→Rubidium-Frequenzstandard)
rubidium vapor standard (→rubidium frequency standard)	INSTR	(→Rubidium-Frequenzstandard)
rub-out (→erase)	DATA PROC	(→löschen)
rub-out character (→erase character)	DATA PROC	(→Löschzeichen)
rub-out command (→delete statement)	DATA PROC	(→Löschanweisung)
rub-out instruction (→delete statement)	DATA PROC	(→Löschanweisung)
rub-out job (→delete statement)	DATA PROC	(→Löschanweisung)
rub-out statement (→delete statement)	DATA PROC	(→Löschanweisung)
ruby	CHEM	**Rubin**
ruby laser	OPTOEL	**Rubinlaser**
Rüdenberg equation	ANT	**Rüdenbergsche Gleichung**
rudimental (→primitive)	TECH	(→primitiv)
rugged = robust; heavy-duty ≈ resistant; stable; durable	TECH	**robust** = kräftig; hochbelastbar ≈ widerstandsfähig; stabil; dauerhaft
ruggedness = robustness ≈ resistance; durability	TECH	**Robustheit** ≈ Widerstandsfähigkeit; Dauerhaftigkeit
ruggerized	TECH	**widerstandsfähiger** = auf höhere Belastungen ausgelegt; für höhere Beanspruchungen
rugosity (→roughness)	RADIO PROP	(→Welligkeit)
Ruhmkorff coil (→inductance coil)	PHYS	(→Induktionsspule)
rule (n.)	COLLOQ	**Regel**
rule (→law)	SCIE	(→Gesetz)
rule (→regulation)	ECON	(→Regelung)
ruled paper (→lined paper)	OFFICE	(→liniiertes Papier)
rule of precedence = order of priority; order of rank	MATH	**Rangfolge**
rule of thumb (→rough formula)	PHYS	(→Faustformel)
ruler ≈ straightedge	ENG.DRAW	**Lineal** ≈ Zeichendreieck
ruler (→ruler line)	DATA PROC	(→Zeilenlineal)
ruler line [line on top edge of display] = ruler	DATA PROC	**Zeilenlineal** [Anzeige am oberen Bildschirmrand]
ruling pen (→drawing pen)	ENG.DRAW	(→Reißfeder)
rumble [phonogram record]	EL.ACOUS	**Rumpelgeräusch** [Schallplatte]
rumble filter	CONS.EL	**Rumpelfilter**
rump-down	ELECTRON	**stufenweiser Abfall**
run (v.i.)	TECH	**verlaufen**
run (n.) [of an engine]	MECH	**Lauf** [Maschine]
run [v.t.] (→execute)	DATA PROC	(→ausführen)
run (v.t.) (→operate 2)	TECH	(→betreiben)
run (→program run)	DATA PROC	(→Programmlauf)
run (→operate)	TECH	(→bedienen)
runaround [to fit text around a graphic] = run around	TYPOGR	**Formsatz** [Umrandung einer Graphik mit Text]
run around (→runaround)	TYPOGR	(→Formsatz)
run away (v.i.) [fig.] = drift (v.i.)	TECH	**weglaufen** [fig.] = wegdriften; abwandern
run-away (n.) [fig.] = drift (n.) ≈ slip (n.); deviation	TECH	**Weglaufen** [fig.] = Wegdriften; Drift; Trift; Abwanderung ≈ Schlupf; Abweichung
runaway (→drift)	TECH	(→Drift)
run cycle (→execute phase)	DATA PROC	(→Ausführungsphase)
run instruction (→run statement)	DATA PROC	(→Laufanweisung)
RUN key	TERM&PER	**RUN-Taste**
run-length coding [facsimile; coding of the number of consecutive identical picture elements]	TERM&PER	**Lauflängencodierung** [Faksimile; Codierung der Anzahl aufeinanderfolgender gleicher Bildpunkte]
run manual [DATA PROC] (→operating documentation)	TECH	(→Gebrauchsunterlagen)
running (→operation 1)	TECH	(→Betrieb)
running (→active)	DATA PROC	(→aktiv)
running (→active)	TECH	(→aktiv)
running charge (→operating costs)	ECON	(→Betriebskosten)
running digital sum = RDS; check sum; checksum; CKSM; check total; hash total	DATA PROC	**Prüfsumme** = Kontrollsumme; Checksumme
running headline [varying with page content]	TYPOGR	**Kolumnentitel** [mit dem Seiteninhalt variierend]
running of cables (→cable run)	SYS.INST	(→Kabelführung)
running program (→operating program)	DATA PROC	(→Betriebsprogramm)
running wave (→travelling wave)	PHYS	(→fortschreitende Welle)
run-off (→sequence of operations)	TECH	(→Ablauf)
runout (n.) [termination of a mechanical run]	TECH	**Auslauf 2** [Beendigung eines mechanischen Laufs]
run-out key = repeat-action key ≈ repeat key	TERM&PER	**Dauertaste** = Dauerfunktionstaste; Dauerauslösetaste ≈ Wiederholtaste
run-out model	TECH	**Auslaufmodell**
run phase (→execute phase)	DATA PROC	(→Ausführungsphase)
run statement = run instruction; FOR statement (ALGOL); FOR instruction (ALGOL); DO statement (FORTRAN); DO instruction (FORTRAN); PERFORM statement (COBOL); PERFORM instruction (COBOL)	DATA PROC	**Laufanweisung** = FOR-Anweisung (ALGOL); Schleifenanweisung (FORTRAN); DO-Befehl (FORTRAN); DO-Anweisung (FORTRAN); PERFORM-Befehl (COBOL); PERFORM-Anweisung (COBOL)
RUN/STOP key	TERM&PER	**RUN/STOP-Taste**
run time (→execution time)	DATA PROC	(→Ausführungszeit)
run time (→turnaround time)	DATA COMM	(→Durchlaufzeit)
runtime (→propagation time)	PHYS	(→Laufzeit)
runway ≈ taxiway; apron ↑ airstrip	AERON	**Start-und-Landebahn** = Piste; Runway ≈ Rollbahn; Vorfeld ↑ Rollfeld ↓ Landestreifen
runway (→grid)	SYS.INST	(→Rost)
rupture joint	MECH	**Sollbruchstelle**
rural exchange	SWITCH	**Landzentrale** = Ruralzentrale

rural subscriber		TELEC	**Ruralteilnehmer**	safety labeling		TECH	**Sicherheitshinweis**

rural subscriber　TELEC　**Ruralteilnehmer**
　= Überlandteilnehmer
rural telephony　TELEC　**Ruraltelephonie**
rust (v.i.)　METAL　**verrosten**
rust (n.)　CHEM　**Rost**
　≈ ferrous oxide　≈ Eisenoxyd
rust (adj.)　PHYS　**rostfarben**
　[colour]　= rost
rustless　CHEM　**rostfrei**
　= non-rusting; rustproof; stainless　≈ nichtrostend; rostsicher
　　≈ korrosionsfest; rostbeständig
rustproof (→rustless)　CHEM　(→rostfrei)
rust protection　TECH　**Rostschutz**
rust resisting　CHEM　**rostbeständig**
　≈ stainless　≈ rostfrei
rusty　CHEM　**rostend**
　↑ oxidizable　↑ oxydierend
ruthenium　CHEM　**Ruthenium**
　= Ru　= Ru
R/W (→read-write ...)　TERM&PER　(→Schreib-Lese-)
RWM (→read-write memory)　DATA PROC　(→Schreib-Lese-Speicher)
R/W memory (→read-write memory)　DATA PROC　(→Schreib-Lese-Speicher)
R/W speed (→read-write velocity)　TERM&PER　(→Schreib-Lese-Geschwindigkeit)
R/W velocity (→read-write velocity)　TERM&PER　(→Schreib-Lese-Geschwindigkeit)
Rydberg constant　PHYS　**Rydberg-Konstante**
Rydberg frequency　PHYS　**Rydberg-Frequenz**
RZ (→return-to-zero)　CODING　(→RZ)

S

1/s (→cycles per second)　PHYS　(→Hertz)
s (→second)　PHYS　(→Sekunde)
σ (→mechanical stress)　MECH　(→Spannung)
S (→sulphur)　CHEM　(→Schwefel)
1/s (→reciprocal second)　PHYS　(→reziproke Sekunde)
Sa (→samarium)　CHEM　(→Samarium)
sack (→bag)　TECH　(→Sack)
saddle　OUTS.PLANT　**Stangenkappe**
saddle key　MECH　**Hohlkeil**
saddle point　MATH　**Sattelpunkt**
　↑ surface point　= hyperbolischer Punkt
　　↑ Flächenpunkt
safe (→permissible)　TECH　(→zulässig)
safeguard (n.) (→precaution)　COLLOQ　(→Vorkehrung)
safeguard (v.t.) (→secure)　COLLOQ　(→sicherstellen)
safeguarding ...　TECH　**Sicherheits-**
　= security ...　= Sicherungs-
safeguarding　TECH　**Sicherung**
　= securing
safeguarding of future　ECON　**Zukunftssicherung**
safekeeping (→deposit 1)　ECON　(→Verwahrung)
safe load　TECH　**zulässige Belastung**
　　= zulässige Last
safe operating area (→SOAR)　MICROEL　(→SOAR)
safety (→security)　COLLOQ　(→Sicherheit)
safety belt　TECH　**Sicherheitsgürtel**
safety class　POWER SYS　**Schutzklasse**
safety code (AM) (→safety rule)　TECH　(→Sicherheitsvorschrift)
safety connector　EL.INST　**Schutzstecker**
safety device (→protection device)　TECH　(→Schutzvorrichtung)
safety earth (→protective earth)　POWER SYS　(→Schutzerde)
safety earth conductor (→protective earth conductor)　POWER ENG　(→Schutzerdungsleiter)
safety earthing (→protective earthing)　POWER ENG　(→Schutzerdung)
safety factor (→safety margin)　TECH　(→Sicherheitsfaktor)
safety fuse (→melting fuse)　COMPON　(→Schmelzsicherung)
safety goggles　TECH　**Schutzbrille**
　= goggles
safety key　EL.TECH　**Sicherheitsschalter**

safety labeling　TECH　**Sicherheitshinweis**
safety margin　TECH　**Sicherheitsfaktor**
　= safety factor　= Sicherheitszahl
safety measure (→security measure)　TECH　(→Sicherheitsvorkehrung)
safety radio service　RADIO　**Sicherheitsfunkdienst**
　= safety service
safety regulation (→safety rule)　TECH　(→Sicherheitsvorschrift)
safety representative　MANUF　**Sicherheitsbeauftragter**
safety requirement　TECH　**Sicherheitserfordernis**
safety rule　TECH　**Sicherheitsvorschrift**
　= safety code (AM); safety regulation; safety standard　= Sicherheitsbestimmung; Sicherheitsnorm
safety service (→safety radio service)　RADIO　(→Sicherheitsfunkdienst)
safety shutdown　TECH　**Sicherheitsabschaltung**
safety standard (→safety rule)　TECH　(→Sicherheitsvorschrift)
safety stock　ECON　**Sicherheitsbestand**
safety switch (→circuit breaker)　COMPON　(→Schutzschalter)
sag (n.)　TECH　**Durchhang**
sag control　OUTS.PLANT　**Durchhangsprüfung**
sag gauge　OUTS.PLANT　**Winkelhaken**
　[sag control]　[Durchhangskontrolle]
salable　ECON　**verkäuflich**
　= vendible; vendable　≈ veräußerbar
　≈ alienable
salami technique　COLLOQ　**Salamitaktik**
salaried employee　ECON　**Gehaltsempfänger**
　= salary-earner　≈ Lohnempfänger
salary　ECON　**Gehalt**
　[a regular fixed pay for work]　≈ Bezahlung; Lohn
　= compensation　↓ Monatsgehalt; Jahresgehalt
　≈ wage
　↓ monthly salary; annual salary
salary-earner (→salaried employee)　ECON　(→Gehaltsempfänger)
sale (v.t.)　ECON　**verkaufen**
　= commercialize　≈ veräußern
　≈ alienate
sale 1 (n.)　ECON　**Verkauf 1**
　[the activity of selling]　[Tätigkeit]
　= selling　= Absatz; Vertrieb 1; Akquisition 1
　≈ marketing; alienation
　　≈ Veräußerung; Vermarktung
sales (AM) (→turnover)　ECON　(→Umsatz)
sales 2 (→sales department)　ECON　(→Vertriebsabteilung)
sales activity (→sale)　ECON　(→Verkauf)
sales agent (→representative)　ECON　(→Vertreter)
sales and marketing (→sales department)　ECON　(→Vertriebsabteilung)
sales and marketing department (→sales department)　ECON　(→Vertriebsabteilung)
sales and support office (→branch office)　ECON　(→Filiale)
sales department　ECON　**Vertriebsabteilung**
　[organizational unit]　[Organisationseinheit]
　= sales and marketing department; sales and marketing; sales 2; marketing department; marketing　= Vertrieb 2; Verkaufsabteilung; Verkauf 2; Akquisition 2
　≈ marketing department　≈ Marketingabteilung
sales letter　ECON　**Werbebrief**
salesman (→seller)　ECON　(→Verkäufer)
sales network (→distribution network)　ECON　(→Vertriebsnetz)
salesperson (→seller)　ECON　(→Verkäufer)
sales price　ECON　**Verkaufspreis**
　= selling price　↑ Preis
　↑ price
sales region　ECON　**Vertriebsgebiet**
　= sales territory　= Verkaufsgebiet
sales territory (→sales region)　ECON　(→Vertriebsgebiet)
sales volume (AM) (→turnover)　ECON　(→Umsatz)
salient (fig.) (→pronounced)　COLLOQ　(→ausgeprägt)
salt (n.)　CHEM　**Salz**

salt spray	QUAL	Salznebel	
salutation (→salutation clause)	ECON	(→Begrüßungsformel)	
salutation clause	ECON	Begrüßungsformel	
[business letter]		[Geschäftsbrief]	
= salutation			
SAM (→sequential access)	DATA PROC	(→sequentieller Zugriff)	
samarium	CHEM	Samarium	
= Sa		= Sm	
SAMNOS technology	MICROEL	SAMNOS-Technologie	
[self-aligned metal-nitride-oxide-silicon technology]			
SAMOS transistor	MICROEL	SAMOS-Transistor	
[stacked-gate avalanche-injection metal-oxide semiconductor transistor]			
sample (v.t.)	ELECTRON	abtasten	
= scan (v.t.)		≈ abfühlen	
≈ sense (v.t.)			
sample (n.)	MATH	Stichprobe	
= spot check		= Probe	
sample (n.) (→sampling value)	ELECTRON	(→Abtastwert)	
sample (n.) (→type model)	TECH	(→Ausführungsmuster)	
sample-and-hold circuit	CODING	Abtast-Halte-Schaltung	
= sample-hold unit		= Abtast-Halte-Glied; Abtast-und-Halte-Schaltung	
↑ instantaneous-value store		↑ Momentanwertspeicher	
sample correlation	MATH	Stichprobenkorrelation	
sample data	DATA PROC	Stichprobendaten	
sampled data control (→keyed control)	CONTROL	(→Tastregelung)	
sampled-data controller	CONTROL	Abtastregler	
		≠ kontinuierlicher Regler	
sampled-data filter	NETW.TH	Abtastfilter	
sampled-data period	ELECTRON	(→Abtastperiode)	
(→scanning period)			
sampled-data time	ELECTRON	(→Abtastperiode)	
(→scanning period)			
sample distribution	QUAL	Stichprobenverteilung	
sampled value (→sampling value)	ELECTRON	(→Abtastwert)	
sample-hold unit	CODING	(→Abtast-Halte-Schaltung)	
(→sample-and-hold circuit)			
sampler (→scanner)	ELECTRON	(→Abtaster)	
samples fair	ECON	Mustermesse	
sample size	MATH	Stichprobenumfang	
sample strew	MATH	Exemplarstreuung	
sample value	MATH	Stichprobenwert	
sampling	QUAL	Probeentnahme	
sampling	CODING	Abtastung	
		= Probeentnahme	
sampling	MATH	Stichprobenentnahme	
sampling (→scanning)	ELECTRON	(→Abtastung)	
sampling circuit	CIRC.ENG	Abtastschaltung	
[digital]		[digital]	
= sensing circuit			
sampling cycle	MODUL	Abtastintervall	
= sampling interval; scanning interval; scan interval		= Abtastzyklus; Impulsabstand	
sampling frequency	MODUL	Abtastfrequenz	
= sampling rate; scanning frequency; scan frequency; digitizing rate		= Abtastrate; Abtastfolge; Abfragefrequenz; Digitalisierungsrate	
sampling function	MODUL	Abtastfunktion	
sampling inspection (→random sampling)	QUAL	(→Stichprobenkontrolle)	
sampling instant	ELECTRON	Abtastzeitpunkt	
= scanning instant; scan instant			
sampling instruction	QUAL	Stichprobenanweisung	
sampling interval (→sampling cycle)	MODUL	(→Abtastintervall)	
sampling lattice	SIGN.TH	Abtastraster	
sampling method	ELECTRON	Abtastverfahren	
		= Sampling-Verfahren	
sampling oscillograph	INSTR	Abtastoszillograph	
		= Abtastoszillograf; Sampling-Oszillograph; Sampling-Oszillograf	
sampling oscilloscope	INSTR	Abtastoszilloskop	
		= Sampling-Oszilloskop	
sampling plan	QUAL	Stichprobenplan	
= sampling scheme		= Stichprobensystem	
sampling point	SIGN.TH	Abtastpunkt	
sampling rate (→sampling frequency)	MODUL	(→Abtastfrequenz)	
sampling scheme (→sampling plan)	QUAL	(→Stichprobenplan)	
sampling signal	ELECTRON	Abtastsignal	
= scanning signal			
sampling spectrum	MODUL	Abtastspektrum	
sampling switch	CIRC.ENG	Abtastschalter	
= scanning switch; scan switch			
sampling technique	ELECTRON	Abtasttechnik	
= scanning circuit; scan circuit			
sampling theorem	INF	Abtasttheorem	
sampling time	ELECTRON	Abtastdauer	
sampling unit	MATH	Auswahleinheit	
sampling value	ELECTRON	Abtastwert	
= sampled value; sample (n.)		= Abtastinformation; Abtastprobe	
sampling voltmeter	INSTR	Sampling-Spannungsmesser	
		= Sampling-Voltmeter	
sand-cast alloy	METAL	Sandgußlegierung	
sand casting	METAL	Sandgießen	
sand-glass	DATA PROC	Sanduhr	
[generally indicates waiting state]		[zeigt i.a. Wartezustand an]	
↑ cursor		↑ Schreibmarke	
sand paper (→emery paper)	TECH	(→Schmirgelpapier)	
sandwich (→layered structure)	TECH	(→Mehrschichtaufbau)	
sandwich coil winding	EL.TECH	(→Scheibenwicklung)	
(→interleaved winding)			
sandwich structure	MICROEL	Sandwich-Struktur	
		= Schichtstruktur; Mehrschichtstruktur	
sandwich structure (→layered structure)	TECH	(→Mehrschichtaufbau)	
sanguine	PHYS	rötel	
sanserif (→sans serif)	TYPOGR	(→serifenlos)	
sans serif	TYPOGR	serifenlos	
= sanserif; gothic		≈ Grotesk	
sans serif	TYPOGR	Grotesk (n.f.)	
= sansserif		[serifenlose, gleichmäßig starke Antiquaschrift]	
↑ font type		= Groteskschrift	
		↑ Schriftart	
sansserif (→sans serif)	TYPOGR	(→Grotesk) (n.f.)	
sap green	PHYS	saftgrün	
sapphire	CHEM	Saphir	
satellite	ASTROPHYS	Satellit	
satellite broadcast	BROADC	Satellitenrundfunk	
satellite channel renting	TELEC	Satellitenkanalvermietung	
= satellite frequency renting		= Satellitenfrequenzvermietung	
satellite communication	TELEC	Satelliten-Telekommunikation	
		= Satelliten-Fernmeldewesen	
satellite communication hop	SAT.COMM	Satellitenabschnitt	
= satellite hop; satellite section			
satellite computer	DATA PROC	Zubringeranlage	
= satellite system		= Satellitensystem	
satellite frequency renting	TELEC	(→Satellitenkanalvermietung)	
(→satellite channel renting)			
satellite hop (→satellite communication hop)	SAT.COMM	(→Satellitenabschnitt)	
satellite link	SAT.COMM	Satellitenverbindung	
satellite navigation	RADIO NAV	Satellitennavigation	
satellite news gathering	SAT.COMM	Pressefunkdienst über Satelliten	
= SNG			
satellite processor (→slave computer)	DATA COMM	(→Nebenrechner)	
satellite radio	RADIO	Satellitenfunk	
satellite ranging	SAT.COMM	Satellitenpositionierung	
satellite reception unit	BROADC	Satelliten-Empfangsanlage	
satellite section (→satellite communication hop)	SAT.COMM	(→Satellitenabschnitt)	
satellite service	TELEC	Satellitenleistung	
satellite system (→satellite computer)	DATA PROC	(→Zubringeranlage)	
satellite town	ECON	Trabantenstadt	

satellite transmission	TELEC	Satellitenübertragung
satellite TV (→television by satellite)	BROADC	(→Satelliten-Fernsehen)
saticon (→saticon tube)	TV	(→Saticon-Röhre)
saticon tube	TV	Saticon-Röhre
= saticon		= Saticon
↑ camera tube		↑ Bildaufnahmeröhre
satin paper	TERM&PER	satiniertes Papier
SATO process	MICROEL	SATO-Verfahren
[self-aligned thick-oxide process]		
saturate	PHYS	sättigen
saturated logic circuit	MICROEL	gesättigte Logik
		= Sättigungslogik
		≈ TTL
saturated solution	CHEM	gesättigte Lösung
saturated vapor	PHYS	gesättigter Dampf
		= Sattdampf
saturation	PHYS	Sättigung
saturation	ELECTRON	Sättigung
≈ overdriving		≈ Übersteuerung
saturation (→color intensity)	PHYS	(→Farbstärke)
saturation choke (→saturation reactor)	POWER SYS	(→Sättigungsdrossel)
saturation current	PHYS	Sättigungsstrom
saturation current (→cutoff current 2)	MICROEL	(→Reststrom)
saturation factor	MICROEL	Übersteuerungsfaktor
saturation fall time	MICROEL	Sättigungs-Abfallzeit
saturation humidity	PHYS	Sättigungsfeuchte
≈ dew point		≈ Taupunkt
saturation hysteresis loop	PHYS	äußere Hystereseschleife
saturation induction	PHYS	Sättigungsinduktion
saturation magnetization	PHYS	Sättigungsmagnetisierung
saturation point	PHYS	Sättigungspunkt
saturation potential	PHYS	Sättigungspotential
saturation pressure	PHYS	Sättigungsdruck
saturation range (→saturation region)	ELECTRON	(→Sättigungsbereich)
saturation reactor	POWER SYS	Sättigungsdrossel
= saturation choke		
saturation region	ELECTRON	Sättigungsbereich
= saturation range		= Übersteuerungsbereich; Sättigungsgebiet
saturation region	PHYS	Sättigungsgebiet
saturation resistance	MICROEL	Sättigungswiderstand
= residual-bulk resistance		= Restwiderstand
saturation rise time	MICROEL	Sättigungs-Anstiegzeit
saturation routing	SWITCH	Verkehrlenkung mit Zielsuche
saturation state	PHYS	Sättigungszustand
saturation voltage	PHYS	Sättigungsspannung
saturation voltage	MICROEL	Sättigungsspannung
		= Restspannung
sausage dipole (→cage dipole)	ANT	(→Reusendipol)
save (v.t.)	ECON	sparen
= economize		
save (v.t.)	DATA PROC	sichern
= backup (v.t.)		= sicherstellen; saven
≈ store		≈ speichern
save area (→saving area)	DATA PROC	(→Sicherungsbereich)
saved number redial	TELEPH	Merken
save/recall capability	INSTR	Speicher-/Ladefunktion
saving	DATA PROC	Sicherung
= backing-up; back-up		
saving (→economy 1)	ECON	(→Einsparung)
saving area	DATA PROC	Sicherungsbereich
= save area; backup area; backing-up area		
saw	TECH	Säge
↓ circular saw; ripsaw		↓ Kreissäge; Handsäge
SAW (→surface-acoustic wave)	PHYS	(→akustische Oberflächenwelle)
sawdust	TECH	Sägemehl
		= Sägespäne
saw file	TECH	Sägefeile
SAW filter (→surface-acoustic-wave filter)	COMPON	(→Oberflächenwellenfilter)
sawtooth	TECH	Sägezahn
sawtooth converter	CIRC.ENG	Sägezahn-Umsetzer
= single-slope converter		= Sägezahn-Verschlüssler
sawtooth current	ELECTRON	Sägezahnstrom
sawtooth curve (→sawtooth waveform)	ELECTRON	(→Sägezahnkurve)
saw-toothed (→sawtooth-shaped)	TECH	(→sägezahnförmig)
saw-toothed wave (→sawtooth wave)	ELECTRON	(→Sägezahnschwingung)
sawtooth generator	CIRC.ENG	Sägezahn-Generator
↑ oscillator		↑ Oszillator
↓ Miller integrator; transitron		↓ Miller-Integrator; Transitron
sawtooth pulse (→sawtooth wave)	ELECTRON	(→Sägezahnschwingung)
sawtooth-shaped	TECH	sägezahnförmig
= saw-toothed		
sawtooth signal	TELEC	Sägezahnsignal
= ramp signal		
sawtooth voltage	ELECTRON	Sägezahnspannung
= ramp voltage		
sawtooth voltage method	CODING	Sägezahnmethode
sawtooth wave	ELECTRON	Sägezahnschwingung
= saw-toothed wave; sawtooth pulse; triangular wave; triangle pulse		= Sägezahnwelle; Dreieckschwingung; Dreieckwelle; Dreieckswelle; Dreieckspannung; Dreieckimpuls
≈ sawtooth waveform		≈ Sägezahnkurve
↑ relaxation oscillation		↑ Kippschwingung
sawtooth waveform	ELECTRON	Sägezahnkurve
= sawtooth curve; ramp		= Sägezahnwellenform; Rampe 2
≈ sawtooth wave		≈ Sägezahnschwingung
sb (→stilb)	PHYS	(→Stilb)
Sb (→antimony)	CHEM	(→Antimon)
S band	RADIO	S-Band
[1,500 to 5,200 MHz]		[1.500 MHz bis 5.200 MHz]
SBC (→single-board computer)	DATA PROC	(→Einkartenrechner)
SBC (→small business computer)	DATA PROC	(→kleiner Geschäftscomputer)
SBC technology	MICROEL	SBC-Technik
[standard buried collector technology]		
Sc (→scandium)	CHEM	(→Scandium)
scaffold 1	CIV.ENG	Gerüst
		≈ Plattform
scaffold 2 (→platform)	CIV.ENG	(→Plattform)
scalar (n.)	MATH	Skalar
[magnitude which can be characterized by numbers alone]		[nur durch Zahlen charakterisierte Größe]
= scalar quantity		= skalare Größe
≠ vector		≠ Vektor
scalar (adj.)	MATH	skalar
≠ vectorial		≠ vektoriell
scalar field	PHYS	Skalarfeld
scalar integral	MATH	skalares Integral
scalar measurement	INSTR	skalare Messung
[of amplitude only]		[nur Amplitude]
≠ scalar measurement		≠ Vektormessung
scalar network analyzer	INSTR	skalarer Netzwerkanalysator
scalar potential	PHYS	skalares Potential
scalar processor	DATA PROC	Skalarrechner
≠ vector processor		= Skalarprozessor
		≠ Vektorrechner
scalar product	MATH	Skalarprodukt
		= skalares Produkt
scalar quantity (→scalar)	MATH	(→Skalar)
scalar triple product	MATH	Spatprodukt
scale (v.t.)	DATA PROC	skalieren
[continuously enlarge or reduce characters or graphics]		[kontinuierliches Vergrößern oder Verkleinern von Zeichensätzen oder Graphiken]
↓ scale-up; scale-down		↓ vergrößern; verkleinern
scale (n.)	ENG.DRAW	Maßstab
scale (n.)	INSTR	Skala
= graduation 2; scale marks; scale division; division; scale interval; interval		= Skalenteilung; Skalenteil; Skale; Maßeinteilung
↓ graduation 1		↓ Gradeinteilung

scale (→order of magnitude)	MATH	(→Größenordnung)	scan (v.t.) (→sample)	ELECTRON	(→abtasten)
scale 1:1 (→full size)	ENG.DRAW	(→Maßstab 1:1)	scan (n.) (→scanning)	SWITCH	(→Abtasten)
scale 1:2 (→half size)	ENG.DRAW	(→Maßstab 1:2)	scan (n.) (→scanning)	ELECTRON	(→Abtastung)
scale 1:4 (→quarter size)	ENG.DRAW	(→Maßstab 1:4)	scan (n.) (→scanning)	RADIO LOC	(→Abtastung)
scale 2:1 (→double size)	ENG.DRAW	(→Maßstab 2:1)	scan area	TERM&PER	**Abtastfläche**
scale constant	INSTR	**Skalenkonstante**	scan area (→scan sector)	ANT	= Abtastbereich
scale correction	INSTR	**Skalenkorrektur**	scan area (→scanning area)	INF	= Abtastgebiet
scaled (adj.)	ENG.DRAW	**maßstäblich**	scan circuit (→sampling technique)	ELECTRON	(→Abtasttechnik)
[represented conforming a given scale]		[in einem angegebenen Maßstab dargestellt]	scan density (→scanning density)	ELECTRON	(→Rasterfeinheit)
≈ true to scale		≈ Maßstabgerecht	scandium	CHEM	**Scandium**
scaled	TECH	**stufenweise 1**	= Sc		= Sc
[time domain]		[zeitlich]	scan frequency (→sampling frequency)	MODUL	(→Abtastfrequenz)
= phased; incrementally 1			scan instant (→sampling instant)	ELECTRON	(→Abtastzeitpunkt)
scaled	INSTR	**untersetzt**	scan interval (→sampling cycle)	MODUL	(→Abtastintervall)
scaled (→stepwise)	TECH	(→schrittweise)	scan line	TERM&PER	**Abtastlinie**
scaled expansion	TECH	**stufenweise Erweiterung**	scan line (→scanning line)	TV	(→Abtastzeile)
= phased expansion			scannable	ELECTRON	**abtastbar**
scale division (→scale)	INSTR	(→Skala)	scanner	ELECTRON	**Abtaster**
scale-down	DATA PROC	**verkleinern**	= sampler; reader; scanning device		= Abtasteinrichtung
[computer graphics]		[Computergraphik]	scanner	DATA PROC	**Eingabe-Multiplexer**
↑ scale		↑ skalieren			= Scanner
scale factor	INSTR	**Skalenwert**	scanner	RADIO LOC	**Scanner**
= scale value		= Skalenfaktor; Skalierung	[swiweling mechanism of a radar antenna]		[Schwenkmechanismus einer Radarantenne]
scale fidelity	INSTR	**Skalentreue**			
		= Skalengenauigkeit; Skalierungsgenauigkeit; Anzeigegenauigkeit	scanner (→reading device)	TERM&PER	(→Lesegerät)
			scanner (→image scanner)	TERM&PER	(→Bildabtaster)
scale interval (→scale)	INSTR	(→Skala)	scanner camera (→image scanner)	TERM&PER	(→Bildabtaster)
scale length	INSTR	**Skalenlänge**	scanner head (→test probe)	INSTR	(→Prüfspitze)
scale length converter	INSTR	**Skalenstreckenumsetzer**	scanner probe (→test probe)	INSTR	(→Prüfspitze)
scale loupe	INSTR	**Skalenlupe**	scanning	SWITCH	**Abtasten**
scale marks (→scale)	INSTR	(→Skala)	= scan (n.)		
scalene	MATH	**ungleichseitig**	scanning	ELECTRON	**Abtastung**
scalene triangle	MATH	**ungleichseitiges Dreieck**	= scan (n.); sampling; exploration; scansion; sensing		= Abtasten; Abfühlen
scale numbering	INSTR	**Skalenbezifferung**	↓ front scanning		↓ Vorderabtastung
scaler	CIRC.ENG	**Festwertmultiplikator**	scanning	RADIO LOC	**Abtastung**
= scaling circuit		↓ Teiler; Vervielfacher	= scan (n.)		[Radar]
↓ reducer; multiplier					= Absuchen
scale range	INSTR	**Skalenbereich**	scanning	TV	**Bildabtastung**
= scale span					= Bildzerlegung
scale resolution	INSTR	**Skalenauflösung**	scanning area	INF	**Abtastgebiet**
= resolution			= scan area		
scale span (→scale range)	INSTR	(→Skalenbereich)	scanning circuit (→sampling technique)	ELECTRON	(→Abtasttechnik)
scale switching	INSTR	**Skalierungsumschaltung**	scanning density	TELEGR	**Linienzahl**
scale-up	DATA PROC	**vergrößern**	[facsimile]		[Faksimile]
[computer graphics]		[Computergraphik]	scanning density	ELECTRON	**Rasterfeinheit**
↑ scale		↑ skalieren	= scan density; raster discrimination; fineness of scanning		
scale value (→scale factor)	INSTR	(→Skalenwert)	scanning device	ELECTRON	(→Abtaster)
scaling (adj.)	TECH	**zundernd**	(→scanner)		
scaling	NETW.TH	**Skalierung**	scanning electron microscope	ELECTRON	**Rasterelektronenmikroskop**
[normalization of frequency and impedance]		[Normierung von Frequenz und Impedanz]			= Rastermikroskop
↑ normalizing		= Doppelnormierung	scanning frequency (→sampling frequency)	MODUL	(→Abtastfrequenz)
		↑ Normierung	scanning instant (→sampling instant)	ELECTRON	(→Abtastzeitpunkt)
scaling	MICROEL	**Skalierung**	scanning interval (→sampling cycle)	MODUL	(→Abtastintervall)
scaling	DATA PROC	**Skalierung**	scanning line	TV	**Abtastzeile**
[computer graphics]		[Computergraphik]	= scan line		= Bildzeile
↓ zooming		↓ Maßstabsänderung	scanning line (→line)	ELECTRON	(→Zeile)
↓ scaling-up; scaling-down		↓ Vergrößerung; Verkleinerung	scanning period	ELECTRON	**Abtastperiode**
scaling circuit (→scaler)	CIRC.ENG	(→Festwertmultiplikator)	= scan period; sampled-data period; sampled-data time		= Abtastzeit
scaling constant (→scaling factor)	DATA PROC	(→Skalierungsfaktor)	scanning pistol	TERM&PER	**Lesepistole**
scaling-down factor	INSTR	**Untersetzungsfaktor**	= reader pistol		↑ Handleser; Belegleser
= scaling factor			↑ hand-held reader; document reader		
scaling factor	DATA PROC	**Skalierungsfaktor**	scanning pitch (→line spacing)	TERM&PER	(→Zeilenabstand)
[computer graphics]		[Computergraphik]			
= scaling constant			scanning point (→picture element)	INF.TECH	(→Bildpunkt)
scaling factor (→scaling-down factor)	INSTR	(→Untersetzungsfaktor)	scanning process	ELECTRON	**Abtastvorgang**
scaling resistor	CIRC.ENG	**Teilerwiderstand**	= scan process		
scallop (n.)	TECH	**Ausbogung**			
[cyclic discontinuity of a border]					
scallop coastal cable (→daisy-chain cable)	COMM.CABLE	(→Girlandenkabel)			
scan (v.t.) (→search)	DATA PROC	(→suchen)			
scan (n.) (→inquiry)	TELECONTR	(→Abfrage)			

scanning processor	CIRC.ENG	**Scanprozessor**	
scanning separation (→line spacing)	TERM&PER	(→Zeilenabstand)	
scanning signal (→sampling signal)	ELECTRON	(→Abtastsignal)	
scanning spot	ELECTRON	**Abtastfleck** = Schreibfleck	
scanning switch (→sampling switch)	CIRC.ENG	(→Abtastschalter)	
scanning terminal	TELEC	**Abtaststation** [Btx]	
scanning tube	ELECTRON	**Bildabtaströhre**	
scanning tunneling microscope	ELECTRON	**Rastertunnelmikroskop**	
scan path (→test bus)	MICROEL	(→Prüfbus)	
scan period (→scanning period)	ELECTRON	(→Abtastperiode)	
scan process (→scanning process)	ELECTRON	(→Abtastvorgang)	
scan sector = scan area	ANT	**Abtastbereich**	
scan sequence	ELECTRON	**Abtastfolge**	
scansion (→scanning)	ELECTRON	(→Abtastung)	
scan switch (→sampling switch)	CIRC.ENG	(→Abtastschalter)	
scan tuning (→station search)	CONS.EL	(→Sendersuchlauf)	
scan width (→sweep range)	ELECTRON	(→Wobbelbereich)	
scarcely populated	ECON	**dünnbesiedelt**	
scarcity (→shortage)	ECON	(→Knappheit)	
SCART connector	COMPON	**SCART-Steckverbinder**	
scatter (v.i.) [values]	MATH	**streuen** [Werte]	
scatter (v.t.)	PHYS	**streuen**	
scatter (→scattering)	PHYS	(→Streuung)	
scatter angle = angular distance	RADIO PROP	**Streustrahlwinkel**	
scatter coefficient	PHYS	**Streukoeffizient**	
scatter diagram	PHYS	**Streudiagramm**	
scattered beam = scattered ray	PHYS	**Streustrahl**	
scattered loading (→scatter load)	DATA PROC	(→gestreutes Laden)	
scattered radiation = stray radiation ≈ scattering ↓ forward scattering; backscattering; light scattering	PHYS	**Streustrahlung** ≈ Streuung ↓ Vorwärtsstreuung; Rückwärtsstreuung; Lichtstreuung	
scattered ray (→scattered beam)	PHYS	(→Streustrahl)	
scatter fading	RADIO PROP	**Streuschwund**	
scatter hop	RADIO PROP	**Streustrahlfunkfeld**	
scattering = scatter; stray ≈ scattered radiation	PHYS	**Streuung** ≈ Streustrahlung	
scattering [of values]	MATH	**Streuung 3** [von Werten]	
scattering angle	PHYS	**Streuwinkel**	
scattering cone	PHYS	**Streukegel**	
scattering cross section	PHYS	**Streuquerschnitt**	
scattering cross section	RADIO LOC	**Reflexionsfäche**	
scattering effect	PHYS	**Streu-Effekt**	
scattering equation	MATH	**Streugleichung**	
scattering layer	PHYS	**Streuschicht**	
scattering loss	PHYS	**Streuverlust**	
scattering matrix	MATH	**Streumatrix** = Scattering-Matrix	
scattering parameter = s-parameter	NETW.TH	**Streuparameter** = S-Parameter	
scattering value	MATH	**Streuwert** = streuender Wert	
scatter link (→troposcatter radio link)	RADIO REL	(→Streustrahl-Richtfunkverbindung)	
scatter load [into non-contiguous memory locations] = scattered loading	DATA PROC	**gestreutes Laden** [in nicht benachbarte Speicherplätze]	
scatter propagation = spread	RADIO PROP	**Streuausbreitung**	
scavenge (→clean)	TECH	(→säubern)	
scavenged (→clean)	TECH	(→sauber)	
scavenging (BRI) (→garbage collection)	ECON	(→Müllabfuhr)	
scenario [simulation of event sequences]	DATA PROC	**Szenarium** [Simulation komplexer Situation] = Szenario; Szenar	
scenario technique	DATA PROC	**Szenariotechnik**	
SC filter (→switched capacitor filter)	CIRC.ENG	(→Schalterfilter)	
sch (→hyperbolic secant)	MATH	(→Secans hyperbolicus)	
schedule (→timing diagram)	TECH	(→Ablaufdiagramm)	
schedule (→plan)	ECON	(→Plan)	
schedule (→table)	DOC	(→Tabelle)	
schedule (→time schedule)	TECH	(→Zeitplan)	
scheduled (adj.) = duty-rated; rated; nominal; desired	TECH	**nominell** = Nenn- = Soll-	
scheduled = according to plan ≈ regular	TECH	**planmäßig** ≈ regelmäßig	
scheduled = set ≈ nominal	TECH	**Soll-** ≈ Nenn-	
scheduled conception	DATA PROC	**Soll-Konzept**	
scheduled/effective comparation	CONTROL	**Soll-/Ist-Vergleich**	
scheduled value (→set value)	CONTROL	(→Sollwert)	
schedule flight	AERON	**Linienflug**	
scheduler [person programming and assigning]	ECON	**Disponent**	
scheduler (→sequence control)	DATA PROC	(→Ablaufsteuerung)	
scheduler (→scheduler program)	DATA PROC	(→Abwickler)	
scheduler program [assigns the time slots] = scheduler; dispatcher	DATA PROC	**Abwickler** [weist die Zeitscheiben zu] = Scheduler; Dispatcher	
scheduling [of material flow]	ECON	**Disposition** [Materialfluß]	
scheduling (→sequence control)	DATA PROC	(→Ablaufsteuerung)	
schematic	SCIE	**schematisch**	
schematic capture	MICROEL	**Schaltkreiseingabe**	
schematic circuit diagram (→circuit diagram)	ELECTRON	(→Stromlaufplan)	
schematic diagram = Prinzipbild	TECH	**Prinzipdarstellung**	
schematic diagram (→circuit diagram)	ELECTRON	(→Stromlaufplan)	
schematic symbol	CIRC.ENG	**Schaltbildsymbol**	
scheme	SCIE	**Schema**	
Schering bridge = power frequency bridge	INSTR	**Schering-Brücke**	
Schmitt trigger ≈ squaring circuit	CIRC.ENG	**Schmitt-Trigger** = Versteiler ≈ Rechteckformer	
scholar [student receiving grant-in-aid]	SCIE	**Stipendiat**	
scholarship = bursary	SCIE	**Stipendium**	
school ↓ primary school; secundary school; tertiary school	SCIE	**Lehranstalt** = Schule ↓ Grundschule; Oberschule; Hochschule	
school (→train)	TECH	(→schulen)	
schooling	SCIE	**Schulbildung**	
schooling (→training)	TECH	(→Schulung)	
school leaving certificate	SCIE	**Schulabgangszeugnis**	
Schottky barrier	MICROEL	**Schottky-Übergang**	
Schottky barrier diode = Schottky diode; hot-barrier diode	MICROEL	**Schottky-Diode** = Metall-Halbleiter-Diode; Hot-Carrier-Diode	
Schottky defect	PHYS	**Schottky-Fehlstelle** = Schottky-Defekt	
Schottky diode (→Schottky barrier diode)	MICROEL	(→Schottky-Diode)	
Schottky effect	PHYS	**Schottky-Effekt**	

Schottky equation	MICROEL	Schottky-Gleichung
Schottky photodiode	MICROEL	Schottky-Fotodiode
Schottky transistor	MICROEL	Schottky-Transistor
Schottky TTL	MICROEL	Schottky-TTL
Schuko cable socket	EL.INST	Schuko-Kupplung
Schuko plug	EL.INST	Schukostecker
= German three-wire plug; three-wire plug; plug with protective ground; plug with grounding contact		
Schuko socket	EL.INST	Schukosteckdose
= German three-wire socket; three-wire socket; socket with protective contact; socket with grounding contact		= Schukodose ≈ Schuko-Kupplung
≈ Schuko cable socket		
science	SCIE	Wissenschaft
scientific	SCIE	wissenschaftlich
scientific computer	DATA PROC	wissenschaftlicher Computer = wissenschaftlicher Rechner
scientific notation	MATH	wissenschaftliche Zahlenschreibweise
= E notation		≈ wissenschaftliche Notation
scientist	SCIE	Wissenschaftler ≈ Forscher
scimitar antenna	ANT	Sichelantenne
= cornucopia antenna		= Scimitarantenne; Türkensäbelantenne
scimitar cut	COMPON	Sichelplattenschnitt
= cornucopia shape		
scintillation	PHYS	Szintillation
scintillation counter	INSTR	(→Szintillationsdetektor)
(→scintillation detector)		
scintillation detector	INSTR	Szintillationsdetektor
= scintillation counter		= Szintillationszähler
scintillation fading	RADIO PROP	Szintillationsschwund
scissoring	DATA PROC	Beschneidung
scissors (→shears)	TECH	(→Schere)
scissors bonding (→stitch bonding)	MICROEL	(→Stichkontaktierung)
scleroscope hardness	MECH	Rückprallhärte
scope	DATA PROC	Gültigkeitsbereich
scope	COLLOQ	Zielsetzung
= goal; objective		= Objektiv
scope (→scope of supply)	ECON	(→Lieferumfang)
scope (→oscilloscope)	INSTR	(→Oszilloskop)
scope (→extent)	TECH	(→Umfang)
scope of application (→scope of validity)	COLLOQ	(→Geltungsbereich)
scope of supply	ECON	Lieferumfang
= scope		≈ Leistungsumfang
≈ scope of work		
scope of validity	COLLOQ	Geltungsbereich
= scope of application		
scope of work	ECON	Leistungsumfang
		= Projektumfang = Lieferumfang
scotchtape (→adhesive tape)	TECH	(→Klebeband)
scotch tape (→self-adhesive tape)	TECH	(→Selbstklebeband)
scour (v.t.) (→pickle)	CHEM	(→abbeizen)
SCPC (→single-carrier per channel)	MODUL	(→Einkanalträger)
SCPC (→single-carrier-per-channel multiple access)	SAT.COMM	(→SCPC-Mehrfachzugriff)
SCR (→thyristor)	MICROEL	(→Thyristor)
scramble (→cipher)	TELEC	(→verschlüsseln)
scramble (v.t.)	CODING	verwürfeln = verscrambeln
scrambler	CIRC.ENG	Verwürfler
= scrambling circuit		= Verwürfelungsschaltung; Scrambler
scrambling	CODING	Verwürfelung = Verscrambelung
scrambling (→encryption)	INF.TECH	(→Verschlüsselung)
scrambling circuit (→scrambler)	CIRC.ENG	(→Verwürfler)
scrap (v.t.)	TECH	verschrotten
scrap (n.)	DATA PROC	Papierkorb
[temporary storage for deleted texts]		[Pufferspeicher für gelöschte Texte]
scrap	TECH	Abfall
= trash; rubbish		= Ausschuß
≈ garbage; litter; junk; rubbish		≈ Müll; Unrat; Schrott; Schutt
scrapbook	DATA PROC	Sammelalbum
[store for frequently used text elements]		[Speicher für häufig benutzte Textelemente]
scrape	TECH	schaben
scrap iron	METAL	Alteisen
scrapping	TECH	Verschrottung
scrap value	ECON	Schrottwert
scratch (v.t.)	TECH	ritzen = einritzen; kratzen
scratch (n.)	TECH	Kratzer = Ritz; Schramme
scratch (→surface noise)	EL.ACOUS	(→Abspielgeräusch)
scratch file	DATA PROC	ungeschützte Datei
[a temporary file]		≈ Arbeitsdatei
≈ work file		
scratch hardness	PHYS	Ritzhärte
scratch noise (→surface noise)	EL.ACOUS	(→Abspielgeräusch)
scratch pad (→auxiliary register)	DATA PROC	(→Hilfsregister)
scratch pad (→note pad)	OFFICE	(→Notizblock)
scratch pad memory	DATA PROC	Notizblockspeicher
[small fast store for temporarily needed data]		[kleiner schneller Zwischenspeicher]
= note pad; backing memory; backing store; backing storage; sketch pad		= Zwischenspeicher 2; Notizblock; Hilfsspeicher; Arbeitsspeicher 4; Notepad; Scratch-Pad-Speicher
≈ scratch pad		≈ Hilfsregister
scratch paper	OFFICE	Schmierpapier
scratch tape	DATA PROC	Arbeitsband
[sorting]		[Sortieren]
screen 1 (n.)	EL.TECH	Schirm
screen (v.t.) (→screen-off)	EL.TECH	(→abschirmen)
screen (→display screen)	TERM&PER	(→Bildschirm 1)
screen (→video screen)	TV	(→Bildschirm)
screen (→cable screen)	COMM.CABLE	(→Kabelschirm)
screen 2 (n.) (→screening wall)	EL.TECH	(→Schirmwand)
screen [TV] (→gray level)	PHYS	(→Graustufe)
screen (v.t.)	RADIO PROP	abschirmen
= obstruct		= abschatten
screenage (→screening)	EL.TECH	(→Abschirmung)
screen attribute	DATA PROC	Bildschirmattribut
[flickering, bold, ..]		[blinkend, fett, ..]
screen-based	DATA PROC	bildschirmorientiert
= screen-oriented		
screen camera	TERM&PER	Bildschirmkamera
screen character (→screen font)	TERM&PER	(→Bildschirmschrift)
screen code (→display code)	TERM&PER	(→Bildschirmcode)
screen colour	TERM&PER	Bildschirmfarbe
↓ foreground colour; background colour		↓ Vordergrundfarbe; Hintergrundfarbe
screen diagonal	TERM&PER	Bildschirmdiagonale
screen dump	DATA PROC	Bildschirmabzug
screen dump (→hard copy)	TERM&PER	(→Druckkopie)
screened	EL.TECH	abgeschirmt
= shielded		
screened aerial (→screened antenna)	ANT	(→abgeschirmte Antenne)
screened antenna	ANT	abgeschirmte Antenne
= shielded antenna; screened aerial; shielded aerial		
screened cable (→shielded cable)	COMM.CABLE	(→geschirmtes Kabel)
screen editor	DATA PROC	Bildschirm-Editor
		= seitenorientierter Editor
screened pair (→shielded pair)	COMM.CABLE	(→geschirmtes Adernpaar)
screen font	TERM&PER	Bildschirmschrift
= screen character		= Bildschirm-Zeichensatz; Bildschirmfont
≈ soft font		≈ Softfont
screen form (→display mask)	DATA PROC	(→Bildschirmmaske)

screen format

screen format (→screen size)	TERM&PER	(→Bildschirmformat)
screen format generator (→display mask generator)	DATA PROC	(→Maskenoperator)
screen form generator = screen formular generator	TERM&PER	**Bildschirmformular-Generator**
screen formular (→display mask)	DATA PROC	(→Bildschirmmaske)
screen formular generator (→screen form generator)	TERM&PER	(→Bildschirmformular-Generator)
screenful	TERM&PER	**Bildschirminformation** = Bildschirminhalt
screen function	DATA PROC	**Bildschirmfunktion**
screen generator	DATA PROC	**Bildschirmgenerator**
screen grid	ELECTRON	**Schirmgitter**
screen grid current	ELECTRON	**Schirmgitterstrom**
screen grid modulation	MODUL	**Schirmgittermodulation**
screen grid tube ↓ tetrode; pentode	ELECTRON	**Schirmgitterröhre** ↓ Tetrode; Pentode
screen grid voltage	ELECTRON	**Schirmgitterspannung**
screen image editor	DATA PROC	**Rasterbild-Editor**
screening = shielding; shield; screenage	EL.TECH	**Abschirmung** = Schirmung
screening = obstruction of line-of-sight; obstruction; shielding ≠ clearance	RADIO PROP	**Sichtbehinderung** = Abschirmung; Abschattung ≠ Sichtfreiheit
screening [shadowing of graphics]	TERM&PER	**Rasterung 1** [Schattierung von Graphiken]
screening cage (→Faraday cage)	PHYS	(→Faraday-Käfig)
screening conductor	OUTS.PLANT	**Schirmleiter**
screening cover	ELECTRON	**Abschirmkappe**
screening foil	ELECTRON	**Abschirmfolie**
screening inspection = sort check	QUAL	**Sortierprüfung** = Auseleseprüfung
screening wall = screen 2 (n.)	EL.TECH	**Schirmwand**
screen language	DATA PROC	**Bildschirmsprache**
screen management	DATA PROC	**Bildschirmgestaltung**
screen mask (→display mask)	DATA PROC	(→Bildschirmmaske)
screen mode = visual display mode; display mode ↓ highlight; negative; flushing	TERM&PER	**Bildschirmdarstellung** ↓ hell; negativ; blinkend
screen-off (v.t.) = screen (v.t.); shield (v.t.)	EL.TECH	**abschirmen** = schirmen
screen-oriented (→screen-based)	DATA PROC	(→bildschirmorientiert)
screen persistence (→afterglow)	TV	(→Nachleuchten)
screen position (→display position)	TERM&PER	(→Bildschirmposition)
screen printing (→silk-screen printing)	TYPOGR	(→Siebdruck)
screen printout (→hard copy)	TERM&PER	(→Druckkopie)
screen radiation	TERM&PER	**Bildschirmstrahlung**
screen resolution	TERM&PER	**Bildschirmauflösung**
screen saver [a software utility]	DATA PROC	**Bildschirmabschalter** [eine Software-Utility]
screen size = screen format	TERM&PER	**Bildschirmformat** = Bildschirmgröße
screen splitting (→split screen)	DATA PROC	(→geteilter Bildschirm)
screen storage material	TERM&PER	**Speicherleuchtstoff**
screen turtle (→turtle)	DATA PROC	(→Schildkröte)
screen update	DATA PROC	**Bildschirmaktualisierung**
screen window (→window)	DATA PROC	(→Fenster)
screen wire (→ground wire)	POWER ENG	(→Erdseil)
screw (v.t.) [without nut]	MECH	**schrauben 1** [ohne Mutter] = verschrauben 1
screw (n.) ↓ cap screw; setscrew	MECH	**Schraube 1** [ohne Mutter verwendbar] ↓ Kopfschraube; Stellschraube
screw connection (→screwed joint)	MECH	(→Schraubverbindung 1)
screw core [magnetic coil]	COMPON	**Schraubkern** [Spule]
screw coupling	COMPON	**Steckschraubverbindung** [Koxialstecker]
screwdriver	MECH	**Schraubendreher** (DIN) = Schraubenzieher
screwed	MECH	**geschraubt 1** [ohne Mutter]
screwed connection (→screwed joint)	MECH	(→Schraubverbindung 1)
screwed joint = screw joint; screwed connection; screw connection	MECH	**Schraubverbindung 1** [ohne Mutter]
screw head = bolt head ≠ point	MECH	**Schraubenkopf** = Kopf ≈ Kopffläche ≠ Kuppe
screw joint (→screwed joint)	MECH	(→Schraubverbindung 1)
screw lock	MECH	**Schraubverschluß**
screw-locked sleeve	OUTS.PLANT	**Schraubmuffe**
screw locking device = screw retainer	MECH	**Schraubensicherung**
screw motion = helicoidal motion	MECH	**Schraubenbewegung**
screw on (v.t.)	MECH	**anschrauben**
screw-on (adj.)	MECH	**anschraubbar**
screw-out (→unscrew)	MECH	(→herausdrehen)
screw plug	MECH	**Gewindestopfen**
screw retainer (→screw locking device)	MECH	(→Schraubensicherung)
screw terminal (→solderless lug)	MECH	(→Schraubklemme)
screw thread (n.) = thread 1 (n.)	MECH	**Gewinde** = Schraubgewinde
screw thread	MECH	**Schraubengewinde**
scribbling block (→note pad)	OFFICE	(→Notizblock)
scribe line	MICROEL	**Ritzrahmen**
scriber	MICROEL	**Ritzgerät**
script [printed lettering imitating handwritten lettering]	TYPOGR	**Schreibschrift** [Handschrift imitierende Schriftart] = Script
script ≈ document	LING	**Schriftstück** ≈ Schrift 2 ≈ Dokument
scroll (v.t.) [continuous shift of a display content, vertical or horizontal] = roll ≈ page (v.t.) ↓ scroll up; scroll down; side-scroll	TERM&PER	**rollen** [kontinuierliches Bewegen einen Bildschirminhalts, vertikal oder horizontal] = verschieben; scrollen ≈ blättern ↓ vorrollen; zurückrollen; seitlich rollen
scroll bar	DATA PROC	**Rollbalken** = Bildlauflinie
scroll down [a screen display] = reverse-scroll; rolling down ↑ scroll	TERM&PER	**zurückrollen** [Bildschirminhalt] = nach unten rollen; rückwärtsrollen ↑ rollen
scrolling [shifting of display contents by characters or lines] = rolling; hunting; rollover ≈ paging ↓ scrolling up; scrolling down	TERM&PER	**Bildverschiebung** [zeilen- oder zeichenweises Verschieben von Bildschirminhalten] = Bilddurchlauf; Bildlauf; Verschiebung; Rollen; Abrollen ≈ Bildschirmblättern ↓ Vorrollen; Zurückrollen
scrolling down = reverse scrolling; rolling down ↑ scrolling	TERM&PER	**Zürckrollen** = Rückwärtsrollen ↑ Bildverschiebung
scrolling up = rolling up ↑ scrolling	TERM&PER	**Vorrollen** [Zeilen am Bildschirm] = Aufwärtsrollen ↑ Bildverschiebung
SCROLL key = ROLL key; SCROLL-LOCK key; SCROLL-LOCK ↑ functional key	TERM&PER	**Taste ROLLEN** = ROLLEN-Taste; SCROLL-LOCK-Taste ↑ Funktionstaste

SCROLL-LOCK	TERM&PER	(→Taste ROLLEN)		sealing wax	TECH	**Siegelwachs**
(→SCROLL key)				seam (v.t.)	METAL	falzen
SCROLL-LOCK key	TERM&PER	(→Taste ROLLEN)		seam (n.)	METAL	**Falz**
(→SCROLL key)				seam (n.)	METAL	**Naht**
scroll-up (v.t.)	TERM&PER	**vorrollen**		seamed (adj.)	MECH	gefalzt
[lines on a display]		[von Zeilen auf dem Bildschirm]		seaming	METAL	**Falzen**
= roll-up				seamless	METAL	nahtlos
↑ scroll		= aufwärtsrollen; nach oben rollen		seamless [fig.]	COLLOQ	nahtlos [fig.]
		↑ rollen		seamless tube	METAL	**nahtloses Rohr**
scrub (v.t.) (→rub)	MECH	(→scheuern)		seam welding	METAL	**Nahtschweißung**
SCT (→surface-charge transistor)	MICROEL	(→Oberflächenladungstransistor)		= continuous welding		= Wiederstandsnahtschweißen
SD disk (→SD diskette)	TERM&PER	(→SD-Diskette)		sea-proof packing	ECON	**Seeverpackung**
SD diskette	TERM&PER	**SD-Diskette**		= seaworthy packing		= seemäßige Verpackung; Überseeverpackung
[2.768 bpi, 48 tpi]		[1.090 Bit/cm, 19 Spuren/cm]		search (v.t.)	DATA PROC	**suchen**
= single-density diskette; SD disk		= Single-density-Diskette		= scan (v.t.)		= absuchen
SDE (→submission and delivery entity)	DATA COMM	(→Sende- und Empfangsinstanz)		search (n.)	DATA PROC	**Suche**
				= hunting		
SDH (→synchronous digital hierarchy)	TELEC	(→Synchronhierarchie)		search (→search process)	DATA PROC	(→Suchvorgang)
				search aid	DATA PROC	**Suchhilfe**
SDLC procedure	DATA COMM	**SDLC-Prozedur**		search algorithm	DATA PROC	**Suchalgorithmus**
↑ transmission procedure		↑ Übertragungsprozedur		= searching algorithm		
Se (→selenium)	CHEM	(→Selen)		search-and-replace (→find-and-replace)	DATA PROC	(→Such-und-Einsetzfunktion)
seaborne computer (vessel) (→onboard computer)	DATA PROC	(→Bordcomputer)		search coil	ELECTRON	**Suchspule**
				= probe coil		
sea cable (→submarine cable)	COMM.CABLE	(→Seekabel)		search command (→search instruction)	DATA PROC	(→Suchanweisung)
seacable (→submarine cable)	COMM.CABLE	(→Seekabel)		search cycle	DATA PROC	**Suchzyklus**
				= searching cycle		
sea freight	ECON	**Seefracht**		search entry	DATA PROC	**Sucheintrag**
= ocean freight (AM)				search information	DATA PROC	**Suchinformation**
sea green	PHYS	**meergrün**		searching algorithm (→search algorithm)	DATA PROC	(→Suchalgorithmus)
seal 1 (v.t.)	TECH	**abdichten**				
[to secure against leakage]		= dichten		searching cycle (→search cycle)	DATA PROC	(→Suchzyklus)
seal 2 (v.t.)	TECH	**versiegeln**				
[to put a distinctive seal]		≈ plombieren		searching time (→access time)	DATA PROC	(→Zugriffszeit)
≈ plumb						
seal 2 (n.)	TECH	**Siegel**		search instruction	DATA PROC	**Suchanweisung**
		≈ Plombe		= search command		= Suchbefehl; Suchkommando
seal 3 (n.)	TECH	**Verschluß**				
≈ shutter; lock (n.)				search jammer	MIL.COMM	**Suchstörsender**
seal 1 (n.)	TECH	**Dichtung**		search key (→key)	DATA PROC	(→Kennbegriff)
= gasket (for fixed parts); packing (for moving parts); sealing		= Abdichtung		search memory (→associative storage)	DATA PROC	(→Assoziativspeicher)
seal (→sealing body)	TECH	(→Dichtungskörper)		search method (→search strategy)	DATA PROC	(→Suchstrategie)
sealed (→gastight)	TECH	(→gasdicht)				
sealed contact	COMPON	**Schutzrohrkontakt**		search path	MATH	**Suchweg**
= reed contact		= geschützter Kontakt; Reed-Kontakt; Zungenkontakt; Herkon		search process	DATA PROC	**Suchvorgang**
↓ gas-protected contact				= retrieval process; search		≈ Wiedergewinnung
		↓ Schutzgaskontakt; Quecksilberkontakt		≈ retrieval		
				search radar (→panoramic radar)	RADIO LOC	(→Panoramaradar)
sea level	GEOPHYS	**Meeresspiegel**				
≈ altitude		[Mittelwert zwischen Tidehochwasser und Tideniedrigwasser]		search run	DATA PROC	**Suchlauf**
				↑ program run		↑ Programmlauf
		= Meereshöhe; Normalpegel Null; Normalnull; NN; N.N.		search software	DATA PROC	**Suchsoftware**
				search specification	DATA PROC	**Suchvorschrift**
		≈ Höhenlage		= retrieval specification		
sealing (n.)	TECH	**Versiegelung**		search strategy	DATA PROC	**Suchstrategie**
sealing (adj.)	TECH	**abdichtend**		= retrieval strategy; search method		= Suchlogik; Suchverfahren
		= dichtend		search string	DATA PROC	**Suchkette**
sealing (→seal)	MECH	(→Dichtung)		search time (→access time)	DATA PROC	(→Zugriffszeit)
sealing body	TECH	**Dichtungskörper**		search tone	INSTR	**Suchton**
= seal				search-tone method	INSTR	**Suchtonverfahren**
sealing compound (→sealing material)	TECH	(→Dichtungsmasse)		= sound analysis method		= Suchtonanalyse; Suchtonmethode
sealing cord (→closure sealing cord)	TECH	(→Dichtungsschnur)		search v.	DATA PROC	**durchsuchen**
				[in a file]		[Datei]
sealing device (→capping system)	TECH	(→Abdichtungsvorrichtung)		search word (→key)	DATA PROC	(→Kennbegriff)
				sea transport (→maritime transport)	ECON	(→Seetransport)
sealing material	TECH	**Dichtungsmasse**				
= sealing compound; sealing medium; lute		= Dichtungsmaterial; Dichtungskitt; Abdichtmasse; Ausgießmasse; Vergußmasse		seaworthy packing (→sea-proof packing)	ECON	(→Seeverpackung)
				sec (→secant)	MATH	(→Sekans)
				SECAM	TV	**SECAM**
sealing medium (→sealing material)	TECH	(→Dichtungsmasse)				[Abkürzung des Französichen „séquentiel à mémoire"]
sealing ring	TECH	**Dichtungsring**				

SECAM decoder	TV	SECAM-Decoder
SECAM system	TV	SECAM-System
secant (n.) [hypotenuse to ancathete] = sec ↑ trigonometric function	MATH	Sekans [Hypotenuse zu Ankathete] = sec ↑ trigonometrische Funktion
seccionable (→separable)	TECH	(→trennbar)
second [SI unit for time, !/3,600 of an hour] = s ↑ time measure	PHYS	Sekunde [SI-Basiseinheit für Zeit, 1/3.600 einer Stunde] = s ↑ Zeitmaß
second [1/3,600 of a degree] ↑ angular measure	MATH	Sekunde [1/3.600 eines Grades] = " ↑ Gradmaß
secondary AC	POWER SYS	Sekundärwechselstrom
secondary action (→secondary effect)	TECH	(→Nebenwirkung)
secondary alarm (→sequence alarm)	EQUIP.ENG	(→Folgealarm)
secondary allocation (→upper mode)	TERM&PER	(→Umschaltebene)
secondary battery = storage battery ↑ battery	POWER SYS	Sammelbatterie [Batterie von Akkumulatoren] ↑ Batterie
secondary cable (→distribution cable)	OUTS.PLANT	(→Sekundärkabel)
secondary card [punched card]	TERM&PER	Zweitkarte [Lochkarte]
secondary cell (→accumulator)	POWER SYS	(→Akkumulator)
secondary coating	OPT.COMM	Sekundärbeschichtung
secondary color	PHYS	Sekundärfarbe
secondary corridor	SYS.INST	Nebengang
secondary data acquisition = secondary data entry	DATA PROC	Sekundärdatenerfassung = Sekundäreingabe
secondary data entry (→secondary data acquisition)	DATA PROC	(→Sekundärdatenerfassung)
secondary defect (→secondary failure)	QUAL	(→Folgeausfall)
secondary diagonal	NETW.TH	Nebendiagonale
secondary diagonal element	NETW.TH	Nebendiagonalelement
secondary digital carrier	SWITCH	Sekundärmultiplexleitung
secondary effect = secondary action	TECH	Nebenwirkung
secondary electron	PHYS	Sekundärelektron
secondary emission	PHYS	Sekundäremission = Sekundärelektronenemission
secondary exchange = junction center; sector exchange ↑ trunk exchange	SWITCH	Knotenvermittlungsstelle = Knotenamt ↑ Fernvermittlungsstelle
secondary failure = dependent failure; sequence failure; secondary defect	QUAL	Folgeausfall
secondary frame	CODING	Sekundärrahmen
secondary frequency standard [requires calibration at intervals]	INSTR	Sekundärfrequenznormal [erfordert periode Nacheichung] = Sekundärfrequenzstandard
secondary key	DATA PROC	Sekundärschlüssel
secondary line switch	SWITCH	Mischwähler
secondary lobe (→side lobe)	ANT	(→Nebenzipfel)
secondary main lobe (→side lobe)	ANT	(→Nebenzipfel)
secondary maximum	MATH	Nebenmaximum
secondary memory [external memory with on-line access] = secondary storage; ancillary memory; ancillary storage; ancillary store ≠ primary storage; tertiary storage ↑ external memory	DATA PROC	Sekundärspeicher [externer, online zugreifbarer Speicher] ≠ Primärspeicher; Tertiärspeicher ↑ Externspeicher
secondary multiplexer	TRANS	Sekundärmultiplexer
secondary peak (→side lobe)	ANT	(→Nebenzipfel)
secondary program (→object program)	DATA PROC	(→Objektprogramm)
secondary quantum number = angular momentum quantum number ↑ quantum number	PHYS	Nebenquantenzahl = Bahndrehimpuls-Quantenzahl; azimutale Quantenzahl ↑ Quantenzahl
secondary radar (→secondary surveillance radar)	RADIO LOC	(→Sekundärradar)
secondary radiation	PHYS	Sekundärstrahlung
secondary radiator (→reflector 1)	ANT	(→Reflektor 1)
secondary rainbow	METEOR	Nebenregenbogen
secondary regulator	CONTROL	Nachregler
secondary route	TELEC	Zweitweg
secondary school ↑ school ↓ college; highschool	SCIE	Oberschule = Sekundarschule ↑ Schule ↓ Hauptschule; Realschule; Gymnasium
secondary signal branch	CIRC.ENG	Signalnebenzweig
secondary standard	INSTR	Gebrauchsnormal = Sekundärnormal
secondary station (→tributary station)	TELECONTR	(→Unterstation)
secondary storage (→secondary memory)	DATA PROC	(→Sekundärspeicher)
secondary studio	BROADC	Nebenstudio
secondary surveillance radar = SSR; secondary radar; radar beacon; ramark	RADIO LOC	Sekundärradar = Radar-Bake; Transponder; aktive Rückstrahlortung; SSR
secondary-to-main-studio line	BROADC	Nebenstudioleitung [DBP/ARD, von Nebenstudio zu Hauptstudio]
secondary tone (→sidetone 2)	TELEPH	(→Nebenton)
secondary transmitter	RADIO NAV	Nebensender
secondary voltage	POWER SYS	Sekundärspannung = Verbraucherspannung
secondary winding ↑ transformer winding	EL.TECH	Sekundärwicklung ↑ Transformatorwicklung
second breakdown	MICROEL	Sekundärdurchbruch = zweiter Durchbruch
second-choice route	SWITCH	Zweitweg
second color = second colour	TERM&PER	Zweitfarbe
second-color control	TERM&PER	Zweitfarbensteuerung
second colour (→second color)	TERM&PER	(→Zweitfarbe)
second cursor (→ghost cursor)	DATA PROC	(→Zweitkursor)
second-generation computer [transistorized, approx. 1955 to 1965]	DATA PROC	Rechner der zweiten Generation [transistorisiert; ca. 1955 bis 1965] = Computer der zweiten Generation
second-grade ≈ second-order; mediocre	COLLOQ	zweitklassig ≈ zweitrangig; mittelmäßig
secondment (→familiarization)	TECH	(→Einarbeitung)
secondment period (→familiarization time)	TECH	(→Einarbeitungszeit)
second-order ≈ second-grade	COLLOQ	zweitrangig ≈ zweitklassig
second set	BROADC	Zweitgerät
second shell = L shell	PHYS	2.Schale = L-Schale
second source = alternative supplier	ECON	Zweitlieferant = Zweithersteller; Second-source
second telephone	TELEPH	Zweittelephon
second to the last	COLLOQ	vorletzter = zweitletzter
secrecy (→privacy)	INF.TECH	(→Geheimhaltung)
secret ≈ confidential	OFFICE	geheim ≈ vertraulich
secretarial unit	TELEPH	Vorzimmeranlage
secretariat	OFFICE	Sekretariat
secretary 1 [female]	OFFICE	Sekretärin
secretary 2 [male]	OFFICE	Sekretär

secretary telephone	TELEPH	**Sekretärtelefon**	
section (n.)	TELEC	**Abschnitt**	
[of a link]		[einer Verbindung]	
section	LING	**Abschnitt**	
[contextual part of a text]		[zusammenhängender Teil eines Textes]	
≈ paragraph; article		= Kapitel; Passus	
		≈ Absatz; Paragraph	
section	TRANS	**Feld**	
↓ repeater section		↓ Verstärkerfeld; Regeneratorfeld	
section	NETW.TH	**Glied**	
section	MATH	**Schnitt 2**	
↓ cross section		[darstellende Geometrie]	
		↓ Querschnitt	
section	ENG.DRAW	**Schnitt**	
section (→department)	ECON	(→Abteilung)	
section (→circuit)	NETW.TH	(→Schaltkreis)	
sectional cut	ENG.DRAW	**Teilungsschnitt**	
[e.g. of a exposition sample]		[z.B. eines Anschauungsmusters]	
sectional drawing	ENG.DRAW	**Schnittzeichnung**	
sectionalized	TECH	**unterteilt**	
sectional view	ENG.DRAW	**Schnittdarstellung**	
		= Schnittansicht	
section clause (→paragraph sign 1)	TYPOGR	(→Paragraphzeichen)	
section label	DATA PROC	**Abschnittsetikett**	
[identifies tape section marks]		[kennzeichnet Abschnittsmarken]	
↑ label		= Abschnittskennsatz	
		↑ Etikett	
section length	TRANS	**Feldlänge**	
↓ repeater spacing; regenerator spacing		↓ Verstärkerfeldlänge; Regeneratorfeldlänge	
section-lined (→hatched)	ENG.DRAW	(→schraffiert)	
section lining (→hatching)	ENG.DRAW	(→Schraffur)	
section mark 1 (→paragraph sign 1)	TYPOGR	(→Paragraphzeichen)	
section mark 2 (→paragraph sign 2)	TYPOGR	(→amerikanisches Paragraphenzeichen)	
section mark (→tape section mark)	TERM&PER	(→Bandabschnittsmarke)	
section sign (→paragraph sign 1)	TYPOGR	(→Paragraphzeichen)	
section test (→link test)	TELEC	(→Streckenmessung)	
sector	MATH	**Sektor**	
[area delimited by an arc and corresponding radii]		[Fläche zwischen Kreisbogen und den begrenzenden Radien]	
		= Kreisausschnitt; Kreissektor	
sector	TERM&PER	**Sektor**	
[of magnetic disk]		[Magnetplatte]	
		= Spurblock; Block 1	
sector cell	MOB.COMM	**Sektorzelle**	
sector chart	DOC	**Sektorschaubild**	
sector control	CIRC.ENG	**Abschnittssteuerung**	
[power control]		[Leistungssteuerung]	
		= Sektorsteuerung	
		≠ Phasenanschnittsteuerung	
sector exchange (→secondary exchange)	SWITCH	(→Knotenvermittlungsstelle)	
sector format	TERM&PER	**Sektorformat**	
[track formatting procedure for magnetic disk memories]		[Spureneinteilungsverfahren bei Magnetplattenspeicher]	
↑ fixed format [DATA PROC]		≠ freies Format	
		↑ festes Format [DATA PROC]	
sectorial antenna	ANT	**Sektorantenne**	
sectorial horn	ANT	**Sektor-Horn**	
↑ horn radiator		= Mehrzellenhorn	
		↑ Hornstrahler	
sectoring	TECH	**Sektorbildung**	
sectoring	DATA PROC	**Sektorierung**	
[diskette]		[Diskette]	
↓ hard sectoring; soft sectoring		↓ Hartsektorierung; Weichsektorierung	
sector management	TERM&PER	**Sektorenverwaltung**	
[magnetic store]		[Magnetspeicher]	
sector scanning	ANT	**Sektorabtastung**	
secure (v.t.)	COLLOQ	**sicherstellen**	
= safeguard (v.t.); guarantee; ensure; make sure		= sichern	
≈ maintain		≈ aufrechterhalten	
secure (adj.)	TECH	**gefahrenlos**	
≈ fail-safe		= sicher	
		= selbstschützend	
secure (→arrest)	MECH	(→arretieren)	
secured (→protected)	TELEC	(→gesichert)	
secure kernel	DATA PROC	**Sicherheitskern**	
secure speech (→encrypted speech)	TELEC	(→verschlüsselte Sprache)	
secure voice (→encrypted voice)	TELEC	(→verschlüsselte Sprache)	
securing (→safeguarding)	TECH	(→Sicherung)	
security	ECON	**Haftung**	
= liability		≈ Sicherheit	
security 1	ECON	**Kaution**	
= guaranty			
security	COLLOQ	**Sicherheit**	
= safety			
security 2	ECON	**Wertpapier**	
[a document]		= Geldmarktpapier; Effekten	
security ... (→safeguarding ...)	TECH	(→Sicherheits-)	
security and alarm engineering	ELECTRON	(→Sicherheitstechnik)	
(→security engineering)			
security control	DATA PROC	**Befugniskontrolle**	
= authorization control			
security control center	SIGN.ENG	**Sicherheitszentrale**	
security engineering	ELECTRON	**Sicherheitstechnik**	
= security and alarm engineering		= Sicherheitselektronik; Sicherungstechnik	
security file (→backup file)	DATA PROC	(→Sicherungsdatei)	
security glass	TECH	**Sicherheitsglas**	
		= Sicherungsglas	
security management	TELEC	**Sicherheitsmanagement**	
		= Sicherheitsverwaltung	
security mean (→security measure)	TECH	(→Sicherheitsvorkehrung)	
security measure	TECH	**Sicherheitsvorkehrung**	
= security mean; safety measure		= Sicherheitsmaßnahme	
≈ protective measure; precautionary measure		≈ Schutzmaßnahme; Vorsorgemaßnahme	
security monitor	SIGN.ENG	(→Alarmierungsfeld)	
(→annunciator)			
security program	DATA PROC	(→Sicherungsprogramm)	
(→sefeguarding program)			
security test lead	INSTR	**Sicherheitslaborbuchse**	
see	DOC	**siehe**	
Seebeck effect	PHYS	**Seebeck-Effekt**	
↑ thermoelectric effect		↑ thermoelektrischer Effekt	
seed (n.)	DATA PROC	**Startparameter**	
[of a random sequence generator]		[eines Zufallsgenerators]	
seed crystal	MICROEL	**Kristallkeim**	
		= Keimling; Sämling; Impfkristall	
seed single-crystal	MICROEL	**Impf-Einkristall**	
seek (n.)	DATA PROC	**Positionierung**	
seek (→station search)	CONS.EL	(→Sendersuchlauf)	
seek time (→positioning time)	TERM&PER	(→Positionierzeit)	
sefeguarding program	DATA PROC	**Sicherungsprogramm**	
= security program		= Sicherheitsprogramm	
Seger cone	INSTR	**Segerkegel**	
[temperature measurement]		[Temperaturmessgerät]	
segment (v.t.)	SCIE	**segmentieren**	
segment (n.)	SCIE	**Segment**	
= part (n.)		= Teil; Ausschnitt	
segment	MATH	**Segment**	
[area between chord and arc]		[Fläche zwischen Kreisbogen und Sehne]	
		= Kreisabschnitt	
segment antenna	ANT	**Segmentantenne**	
↑ cylindrical parabol antenna		↑ Zylinderparabol-Antenne	
↓ cheese antenna; pill-box antenna		↓ Käseantenne; Tortenschachtelantenne	
segmentation	MATH	**Segmentierung**	
segmented	MATH	**segmentiert**	
segmented bar chart	DOC	**segmentiertes Säulendiagramm**	

segmented characteristic

segmented characteristic	CODING	Segmentkennlinie
= segmented law		
segmented encoding law	CODING	segmentierte Codierungskennlinie
segmented law (→segmented characteristic)	CODING	(→Segmentkennlinie)
segregate (→isolate)	TECH	(→absondern)
seizability	SWITCH	Belegungsfähigkeit
seize	SWITCH	belegen
= busy (v.t.)		
seize (→occupy)	DATA PROC	(→belegen)
seize (v.t.) (→dimension)	TECH	(→bemessen)
seizing (→seizure)	SWITCH	(→Belegung)
seizure	SWITCH	Belegung
= occupation; holding; seizing		= Belegungsvorgang; Belegen
seizure acknowledgement (→seizure acknowledgment)	SWITCH	(→Belegungsquittung)
seizure acknowledgment	SWITCH	Belegungsquittung
= seizure acknowledgement		
seizure command	DATA PROC	Belegungsbefehl
seizure indication	SWITCH	Belegungsanzeige
seizure signal (→off-hook signal)	SWITCH	(→Beginnzeichen)
select (v.t.)	SCIE	selektieren
		= aussondern
select (→dial)	TELEC	(→wählen)
selectable (→ generic 2)	DATA PROC	(→auswählbar)
selectance (→selectivity)	NETW.TH	(→Selektion)
selecting mode (→polling mode)	DATA COMM	(→Abrufbetrieb)
selection	SWITCH	Auswählen
selection	DATA PROC	Auswahl
selection	SCIE	Selektion
		= Selektierung; Auswahl
selection (→selectivity)	NETW.TH	(→Selektion)
selection key	DATA PROC	Auswahlbegriff
selection logic	SWITCH	Auswahllogik
selection matrix [memory]	DATA PROC	Selektionsmatrix [Speicher]
selection menu	DATA PROC	Auswahlmenü
↑ Menü		↑ Menü
↓ pull-down menu; pop-up menu		↓ Pull-down-Menü; Pop-up-Menü
selection rule	PHYS	Auswahlregel
selection signal	TELEPH	Wählzeichen
= dial train		= Wählzeichenfolge
↓ dial pulse; key pulse		≈ Wählton
selection sort	DATA PROC	Auswahlsortierung
selection stage	SWITCH	Wahlstufe
↑ switching stage		↑ Koppelstufe
selection structure	DATA PROC	Selektionsstruktur
= decision structure		
selection tree	DATA PROC	Suchbaum
selective	EL.TECH	selektiv
≠ broadband		= frequenzselektiv; trennscharf
		≠ breitbandig
selective	TECH	selektiv
= discerning		
selective amplifier	CIRC.ENG	Schmalbandverstärker
		= Selektivverstärker
selective call	SWITCH	Selektivruf
≠ multiaddress call		= Einzelruf
		≠ Sammelruf
selective circuit (→filter)	NETW.TH	(→Filter)
selective command	TELECONTR	Anwahlbefehl
selective fading (→multipath fading)	RADIO PROP	(→Interferenzschwund)
selective interference	TELEC	Selektivstörer
		= Selektivstörung
selective jamming	MIL.COMM	Selektivstörung
selective level measurement	INSTR	selektive Pegelmessung
		= Einzelpegelmessung
selective level meter	INSTR	selektiver Pegelmesser
[a selective voltmeter specific for FDM transmission]		[für TF-Technik spezialisierter Selktivspannungsmesser]
= SLM; carrier frequency voltmeter		= Trägerfrequenz-Voltmeter; TF-Voltmeter
↑ selective voltmeter		↑ Selektivspannungsmesser
selective photo effect	PHYS	selektiver Photoeffekt
		= selektiver Fotoeffekt; spektraler selektiver Effekt; Vektoreffekt
selective voltmeter	INSTR	Selektivspannungsmesser
[not swept amplitude and frequency measurements, within a tuned bandslot]		[ungewobbelte Messung von Amplitude und Frequenz innerhalb eines abstimmbaren Bandschlitzes]
= frequency-selective voltmeter; wave analyzer		= selektiver Spannungsmesser; frequenzabhängiger Voltmeter; Wellenanalysator
↑ signal analyzer		↑ Signalanalysator
↓ selective level meter		↓ selektiver Pegelmesser
selectivity	NETW.TH	Selektion
= selection; selectance; clearness of tuning		= Selektivität; Trennschärfe; Abstimmschärfe
selector	DATA PROC	Selektor
selector	SWITCH	Wähler
[functional unit composed of electromechanical switching networks]		[aus elektromechanischen Koppelanordnungen gebildete Funktionseinheit]
= switch		= Schalter
↓ rotary selector; two-motion selector; step-by-step selector; crossbar selector		↓ Drehwähler; Hebdrehwähler; Schrittschaltwähler; Kreuzschienenschalter
selector bar	SWITCH	Wählschiene
selector channel	DATA PROC	Selektorkanal
[high-speed data channel connecting a computer to a single peripheral equipment]		[Datenkanal hoher Transfergeschwindigkeit zwischen Computer und einem Peripheriegerät]
≈ block multiplex channel		≈ Blockmultiplexkanal
≠ multiplexer channel		≠ Multiplexkanal
selector hum	TELEPH	Wählergeräusch
selector magnet	TELEGR	Empfangsmagnet
= receiver magnet		
selector plug	TELEPH	Wählklinke
selector pulse (→dial pulse)	SWITCH	(→Wählimpuls)
selector-repeater	SWITCH	Mitlaufwähler
selector subchannel	DATA PROC	Selektorunterkanal
selector switch	CIRC.ENG	Auswahlschalter
selenide	CHEM	Selenid
selenium	CHEM	Selen
= Se		= Se
selenium cell	PHYS	Selenzelle
selenium electronics	CIRC.ENG	Seleniumelektronik
selenium rectifier	COMPON	Selengleichrichter
		= Selenventil
self-acting (→automatic)	TECH	(→automatisch)
self-actuated (→automatic)	TECH	(→automatisch)
self-adapting	TECH	selbstanpassend
= adaptive		= selbsteinstellend; adaptiv
≈ automatic		≈ automatisch
self-adapting program	DATA PROC	selbstanpassendes Programm
		= Selbstanpaßprogramm
self-adhesive	TECH	selbstklebend
self-adhesive label	TECH	Selbstklebeschild
self-adhesive tape	TECH	Selbstklebeband
= scotch tape		↑ Klebeband
↑ adhesive tape		
self-adjoint (adj.)	MATH	selbstadjungiert
self-adjoint operator	MATH	selbstadjungierter Operator
self-adjusting	ELECTRON	selbstabgleichend
= self-balancing; automatic balancing		
self-aligned	TECH	selbstjustierend
self-aligning bearing	MECH	Pendellager
self-aligning gate	MICROEL	selbstjustierender Steueranschluß
		= selbstjustierendes Gate
self-balancing (→self-adjusting)	ELECTRON	(→selbstabgleichend)
self-bias	ELECTRON	Selbstvorspannung
self-blocking oscillator (→blocking oscillator)	CIRC.ENG	(→Sperrschwinger)
self-calibrating	ELECTRON	selbsteichend
= self-gauging; auto-zeroing		

self-calibration	ELECTRON	**Selbsteichung**	
= self-gauging; auto-zeroing		= laufende Eichung	
self-capacitance	PHYS	**Eigenkapazität**	
self-centering connector	COMPON	**selbstzentrierender Stecker**	
self check (→self test)	DATA PROC	(→Eigentest)	
self-check code	CODING	(→selbstprüfender Code)	
(→error-detecting code)			
self-checking	CODING	**selbstprüfend**	
= error-detecting; self-validating		≈ selbstkorrigierend	
≈ error-correcting			
self-checking code	CODING	(→selbstprüfender Code)	
(→error-detecting code)			
self-checking number	DATA PROC	**selbstprüfende Zahl**	
self-clocked	ELECTRON	**selbsttaktend**	
= with internal clock			
self-clocked modem	DATA COMM	**selbsttaktendes Modem**	
self-commutated	POWER SYS	**selbstgeführt**	
self-commutated converter	POWER SYS	**selbstgeführter Stromrichter**	
= forced-commutation converter		= zwangskommutierter Stromrichter	
self-compiling	DATA PROC	**selbstkompilierend**	
self-complementing	DATA PROC	**selbstergänzend**	
self-conducting (→intrinsic)	MICROEL	(→eigenleitend)	
self-contained (→built-in)	TECH	(→eingebaut)	
self-contained construction	EQUIP.ENG	**geschlossener Aufbau**	
self-controlling	CONTROL	**selbststeuernd**	
self-correcting code	CODING	(→fehlerkorrigierender Code)	
(→error-correcting code)			
self-correlation	MATH	(→Eigenkorrelation)	
(→autocorrelation)			
self-corrosion	CHEM	**Eigenkorrosion**	
= autocorrosion		= Autokorrosion	
self-damping	PHYS	**Selbstdämpfung**	
self-defining	DATA PROC	**selbstdefinierend**	
= figurative		= figurativ	
self diagnosis (→self test)	DATA PROC	(→Eigentest)	
self-diagnostic (adj.)	DATA PROC	**selbstdiagnostisch**	
		= eigendiagnostisch	
self-dialing	TELEC	(→selbstwählend)	
(→auto-dialing)			
self-discharge (v.i.)	PHYS	**selbstentladen**	
self-discharge (n.)	PHYS	**Selbstentladung**	
self-employed (→freelance)	ECON	(→freiberuflich)	
self-energized (→self-powered)	TECH	(→eigenangetrieben)	
self-excitation	PHYS	**Selbsterregung**	
= self-oscillation			
self-excited	PHYS	**selbsterregt**	
self-exciting	PHYS	**selbsterregend**	
self-explanatory	SCIE	**selbsterklärend**	
self-focussing (n.)	PHYS	**Selbstfokussierung**	
self-focussing (adj.)	PHYS	**selbstfokussierend**	
self-gauging	ELECTRON	(→selbsteichend)	
(→self-calibrating)			
self-gauging	ELECTRON	(→Selbsteichung)	
(→self-calibration)			
self-hardening	METAL	**selbsthärtend**	
self-healing (n.)	COMPON	**Selbstheilung**	
self-healing (adj.)	COMPON	**selbstheilend**	
[capacitor]		[Kondensator]	
self-healing network	TELEC	**selbstheilendes Netzwerk**	
self-heating	TECH	**Eigenerwärmung**	
self-impedance	EL.TECH	**Eigenimpedanz**	
self-imposed (→voluntary)	COLLOQ	(→freiwillig)	
self-inductance	EL.TECH	**Selbstinduktivität**	
		= Selbstinduktions-Koeffizient; Eigeninduktivität	
self-inductance	EL.TECH	(→Selbstinduktion)	
(→self-induction)			
self-induction	EL.TECH	**Selbstinduktion**	
= self-inductance			
self-interrupting	ELECTRON	**selbstunterbrechend**	
= autointerrupting			
self-interruption	ELECTRON	**Selbstunterbrechung**	
= autointerruption			
self-learning	DATA PROC	**selbstlernend**	
self-learning computer	DATA PROC	**selbstlernender Computer**	
		= selbstlernender Rechner	
self-limiting	TECH	**selbstbegrenzend**	
self-loading	DATA PROC	**selbstladend**	
= bootstrap			
self-loading programm	DATA PROC	**selbstladendes Programm**	
self-locking	TECH	**selbstsperrend**	
		= selbsthaltend	
self-lubricating (adj.)	MECH	**selbstschmierend**	
self-luminous	PHYS	**selbstleuchtend**	
self-made	ELECTRON	**Selbstbau**	
		[Basteln]	
self-magnetic	PHYS	**eigenmagnetisch**	
self-maintained discharge	PHYS	**selbstständige Entladung**	
self-maintained gaseous discharge	PHYS	(→selbststständige Gasentladung)	
(→self-sustaining gaseous discharge)			
self-monitoring (adj.)	ELECTRON	**selbstüberwachend**	
self-monitoring	ELECTRON	(→Selbstüberwachung)	
(→self-supervision)			
self-noise (→basic noise)	TELEC	(→Grundgeräusch)	
self-operated (→automatic)	TECH	(→automatisch)	
self-operating (→automatic)	TECH	(→automatisch)	
self-optimizing	TECH	**selbstoptimierend**	
self-organizing	DATA PROC	**selbstorganisierend**	
self-organizing programm	DATA PROC	**selbstorganisierendes Programm**	
self-oscillating	ELECTRON	**selbstschwingend**	
= free-running			
self-oscillation	PHYS	(→Selbsterregung)	
(→self-excitation)			
self-powered	TECH	**eigenangetrieben**	
= self-energized		= selbstangetrieben	
self-powered	EQUIP.ENG	(→batteriebetrieben)	
(→battery-operated)			
self-protecting (→fail-safe)	TECH	(→selbstschützend)	
self-pulsing	OPTOEL	**Selbstpulsation**	
[laser]		[Laser]	
self-quenched detector	ELECTRON	**Pendel-Rückkopplungsaudion**	
self-reading	ELECTRON	**selbstregistrierend**	
self-relocating program	DATA PROC	(→Relativprogramm)	
(→relocatable program)			
self-reproducing	TECH	**selbstreproduzierend**	
		= eigenreproduzierend	
self-resetting	ELECTRON	**selbstrücksetzend**	
		= selbstrückstellend	
self-resetting loop	DATA PROC	**selbstrücksetzende Schleife**	
self-saturating	ELECTRON	**selbstsättigend**	
self-scanning	MICROEL	**selbstabtastend**	
self-sealing	TECH	**selbstdichtend**	
		= selbstabdichtend	
self-starting	CIRC.ENG	**selbststartend**	
		= selbstanlaufend	
self-starting	DATA PROC	(→selbststartend)	
(→self-triggering)			
self-structuring	DATA PROC	**selbststrukturierend**	
self-sufficiency	ECON	**Selbstversorgung**	
self-supervision	ELECTRON	**Selbstüberwachung**	
= self-monitoring		= Eigenüberwachung; Mitlaufüberwachung	
self-supporting	TECH	**selbsttragend**	
= unsupported; free-standing		= freistehend	
self-supporting	COMM.CABLE	**selbsttragung**	
		= freitragend	
self-supporting antenna mast	ANT	**freistehender Antennenmast**	
self-supporting cable	COMM.CABLE	**selbsttragendes Kabel**	
↓ integral-messenger cable		↓ Tragseil-Luftkabel	
self-supporting mast	CIV.ENG	**freistehender Mast**	
= self-supporting tower		= selbsttragender Mast; freistehender Turm; selbsttragender Turm	
self-supporting tower	CIV.ENG	(→freistehender Mast)	
(→self-supporting mast)			
self-sustaining gaseous discharge	PHYS	**selbstständige Gasentladung**	
= self-maintained gaseous discharge			
self-synchronism	ELECTRON	**Selbstsynchronisierung**	
self-synchronized	ELECTRON	**selbstsynchronisierend**	
self test	DATA PROC	**Eigentest**	
= self diagnosis; self check; power-up test; power-up diagnosis		= Eigendiagnose; Eigenprüfung; Eigenkontrolle; Selbsttest; Selbstdiagnose; Selbstprüfung; Selbstkontrolle; Einschalttest; Einschaltdiagnose; Einschaltprüfung;	

self-test	TECH	Selbstprüfung	
self-triggering	DATA PROC	selbststartend	
= self-starting; load-and-go; load-and-run		= selbstanlaufend; laden-und-ausführen	
self-validating	CODING	(→selbstprüfend)	
(→self-checking)			
sell by auction	ECON	versteigern	
= auction (v.t.)			
seller	ECON	Verkäufer	
= vendor 1; salesman; vender; salesperson		≈ Vertreiber; Händler	
≈ distributor; dealer			
seller (→dealer)	ECON	(→Händler)	
selling (→sale)	ECON	(→Verkauf 1)	
selling price (→sales price)	ECON	(→Verkaufspreis)	
semantics	LING	Semantik	
[science of meaning of words]		= Wortbedeutungslehre; Bedeutungslehre; Semasiologie	
= semasiology			
semaphore	DATA PROC	Semaphor	
[synchronization primitive]		[Synchronisierungsbefehl]	
semasiology (→semantics)	LING	(→Semantik)	
semestral (→semiannual)	COLLOQ	(→halbjährlich)	
semestrial (→semiannual)	COLLOQ	(→halbjährlich)	
semi-amplitude	MATH	Halbamplitude	
semiannual	COLLOQ	halbjährlich	
= half-yearly; semestral; semestrial			
semiautomatic	TECH	halbautomatisch	
= semi-automatic		= halbselbsttätig	
semi-automatic	TECH	(→halbautomatisch)	
(→semiautomatic)			
semiaxis	MATH	Halbachse	
semibold	TYPOGR	halbfett	
= semi bold; demi		↑ Schriftschnitt	
↑ typeface design			
semi bold (→semibold)	TYPOGR	(→halbfett)	
semi bold italic	TYPOGR	halbfett kursiv	
↑ typeface design		↑ Schriftschnitt	
semicap	MICROEL	Semicap	
semicircle	MATH	Halbkreis	
semicircular	MATH	halbkreisförmig	
		= halbkreisartig	
semicolon	LING	Semikolon	
[;]		[;]	
↑ punctuation mark		= Strichpunkt	
		↑ Satzzeichen	
semiconducting	PHYS	halbleitend	
semiconductor	PHYS	Halbleiter	
semiconductor	MICROEL	(→Halbleiter-)	
(→solid-state)			
semiconductor amplifier	CIRC.ENG	(→Halbleiterverstärker)	
(→solid state amplifier)			
semiconductor block technology	MICROEL	(→monolithische integrierte Schaltung)	
(→monolithic integrated circuit)			
semiconductor chip	MICROEL	(→Chip)	
(→chip)			
semiconductor component	MICROEL	(→Halbleiterbauelement)	
(→semiconductor device)			
semiconductor detector	MICROEL	Halbleiterdetektor	
semiconductor device	MICROEL	Halbleiterbauelement	
= semiconductor component; solid state device; solid-state component		= Halbleiterbaustein; Halbleiterschaltung; Festkörperbauelement; Festkörperbaustein; Festkörperschaltung	
≈ chip		≈ Chip	
semiconductor die (→chip)	MICROEL	(→Chip)	
semiconductor diode	MICROEL	(→Diode)	
(→diode)			
semiconductor electronics	ELECTRON	Halbleiterelektronik	
= solid state electronics			
semiconductor film technology	MICROEL	Halbleiterfilmtechnik	
↓ SOS		↓ SOS	
semiconductor gate	MICROEL	Halbleiterventil	
↑ semiconductor diode		= Halbleiter-Stromrichterventil	
↓ thyristor		↑ Diode	
		↓ Thyristor	
semiconductor junction	MICROEL	(→Übergang)	
(→junction)			

semiconductor laser	OPTOEL	Halbleiterlaser	
semiconductor material	MICROEL	Halbleiterwerkstoff	
semiconductor measurements	INSTR	Halbleitermeßtechnik	
= semiconductor testing			
semiconductor memory (→solid state memory)	MICROEL	(→Halbleiterspeicher)	
semiconductor memory chip	MICROEL	(→Halbleiterspeicherchip)	
(→solid state memory chip)			
semiconductor noise	TELEC	Funkelrauschen	
= flicker noise; popcorn noise (AM)		= Halbleiterrauschen; Flikkerrauschen	
semiconductor oscillator	MICROW	Halbleiteroszillator	
semiconductor parameter analyzer	INSTR	Halbleiterparameter-Analysator	
semiconductor parameter test set	INSTR	(→Halbleiter-Tester)	
(→semiconductor tester)			
semiconductor physics	PHYS	Halbleiterphysik	
≈ solid state physics		≈ Festkörperphysik	
semiconductor pressure sensor	INSTR	Halbleiter-Drucksensor	
semiconductor production	MICROEL	Halbleiterfertigung	
semiconductor protection	COMPON	Halbleitersicherung	
		[Schutz von Halbleitern]	
semiconductor rectifier diode	MICROEL	Halbleiter-Gleichrichterdiode	
semiconductor region	PHYS	Halbleiterzone	
semiconductor relay	COMPON	(→Transistorrelais)	
(→transistor relay)			
semiconductor resistance strain gage	INSTR	Halbleiter-Dehnmeßstreifen	
= semiconductor resistance strain gauge			
semiconductor resistance strain gauge (→semiconductor resistance strain gage)	INSTR	(→Halbleiter-Dehnmeßstreifen)	
semiconductor storage (→solid state memory)	MICROEL	(→Halbleiterspeicher)	
semiconductor switch		Halbleiterschalter	
= electronic switch		[MICROEL]	
		= kontaktloser Schalter; elektronischer Schalter	
semiconductor technique	MICROEL	(→Halbleitertechnik)	
(→semiconductor technology)			
semiconductor technology	MICROEL	Halbleitertechnik	
= semiconductor technique		= Halbleitertechnologie	
semiconductor tester	INSTR	Halbleiter-Tester	
= semiconductor parameter test set		= Halbleiterparameter-Prüfsystem	
↓ transistor tester			
semiconductor testing	INSTR	(→Halbleitermeßtechnik)	
(→semiconductor measurements)			
semiconductor trap	PHYS	Halbleiterhaftstelle	
		= Elektronenhaftstelle	
semicustom (→semicustom IC)	MICROEL	(→Semikundenschaltung)	
semicustom IC	MICROEL	Semikundenschaltung	
= semicustom		= halbkundenspezifische Schaltung; semi-kundenspezifische integrierte Schaltung	
semidiurnal	ECON	halbtägig	
≠ diurnal		≠ ganztägig	
semi-duplex	TELEC	Semiduplexbetrieb	
[one end is duplex, the other beeing simplex]		[von einem Ende ist Duplexbetrieb, vom anderen nur Simplexbetrieb möglich]	
≈ half-duplex		≈ Halbduplexbetrieb	
semidynamic	TECH	semidynamisch	
		= halbdynamisch	
semiempirical	TECH	halbempirisch	
semi-finished	MECH	teilweise bearbeitet	
		= halbbearbeitet	
semi-finished (n.)	MANUF	(→halbfertiges Erzeugnis)	
(→semi-finished good)			
semi-finished	MANUF	(→halbfertig)	
(→semimanufactured)			
semi-finished good	MANUF	halbfertiges Erzeugnis	
= semi-finished (n.); semimanufactured good; stock		= H-Erzeugnis; Halbzeug; Halbfabrikat	
		≈ unfertiges Erzeugnis	
semigraphic	DATA PROC	Halbgraphik	
[computer graphics]		[Computergraphik]	
		= Semigraphik	

semi-group	MATH	Halbgruppe
semi-hard	TECH	halbhart
semilogarithmic	MATH	semilogarithmisch
		= halblogarithmisch
semimanufactured	MANUF	halbfertig
= semi-finished		= halbbearbeitet; teilweise bearbeitet
semimanufactured good (→semi-finished good)	MANUF	(→halbfertiges Erzeugnis)
semi-mat (→half-mat)	PHYS	(→halbmatt)
semi-matt (→half-mat)	PHYS	(→halbmatt)
semi-matte (→half-mat)	PHYS	(→halbmatt)
semimetal	CHEM	Halbmetall
seminar	ECON	Seminar
≈ symposium; congress		≈ Symposium; Kongress
seminar for management personnel	ECON	Führungskräfteseminar
semiotics	LING	Semiologie
= theory of signs		= Semiotik; Zeichentheorie
semipermanent	TECH	semipermanent
≈ continuous		≈ dauerhaft
semipermanent data	DATA PROC	semipermanente Daten
semipermanent link	TELEC	semipermanente Verbindung
semipermanent memory	DATA PROC	semipermanenter Speicher
= semipermanent storage; semipermanent store		= Semipermanentspeicher; Semipermanenzspeicher
≈ read-only memory		≈ Festwertspeicher
semipermanent storage (→semipermanent memory)	DATA PROC	(→semipermanenter Speicher)
semipermanent store (→semipermanent memory)	DATA PROC	(→semipermanenter Speicher)
semipermeable	TECH	halbdurchlässig
semiprofessional	ECON	halbprofessionell
		= semiprofessionell
semirandom	MATH	semistochastisch
semistable	TECH	semistabil
		= halbstabil
semistatic	TECH	semistatisch
		= halbstatisch
semi-trailer truck (AM)	TECH	Sattelschlepper
= articulated lorry (BRI)		
semitransparent	PHYS	teildurchsichtig
[optics]		[Optik]
		= teildurchlässig; halbdurchsichtig; halbdurchlässig
senary	MATH	senär
[based-on or characterized by six]		= auf die Zahl 6 bezogen
send (v.t.)	TELEC	senden
= emit; transmit 2		≈ übertragen; absetzen
≈ transmit 1; dispatch		
send (v.t.)	ECON	übersenden 1
[merchandise]		[Ware]
= forward (v.t.); consign; ship (v.t.)		
send (→transmit)	RADIO	(→senden)
send (→originate)	TELEPH	(→erzeugen)
send back (→return)	ECON	(→zurücksenden)
send control	DATA COMM	Sendesteuerung
sender	INF.TECH	Absender
= addresser		= Adressant
sender (→transmitter)	RADIO	(→Sender)
sender identification	TELEC	Absenderkennung
send frequency (→transmit frequency)	RADIO	(→Sendefrequenz)
sending (→output)	DATA PROC	(→Ausgabe)
sending aerial (→transmitting antenna)	ANT	(→Sendeantenne)
sending antenna (→transmitting antenna)	ANT	(→Sendeantenne)
sending equipment (→transmit equipment)	TELEC	(→Sendegerät)
sending field	DATA PROC	Sendefeld
sending-off (→despatch)	ECON	(→Versand)
sending reference equivalent	TELEPH	Sendebezugsdämpfung
= transmission reference equivalent		= SBZ
send key	TERM&PER	Sendetaste
send level (→transmission level)	TRANS	(→Sendepegel)
send mixer	RADIO	Sendemischer
send mode (→transmit mode)	TELEC	(→Sendebetrieb)
send off (→freight)	ECON	(→verfrachten)
send/receive switch	TERM&PER	Sende-Empfangs-Umschalter
= transceiver switch; transmit/receive switch		
send register	DATA COMM	Senderegister
send sequence number	DATA COMM	Sendefolgenummer
		= Sendelaufnummer
send-side ... (→on the transmit side)	TELEC	(→sendeseitig)
send signal (→transmitted signal)	TELEC	(→Sendesignal)
sendytron	ELECTRON	Sendytron
		= Senditron
seniority	ECON	Betriebszugehörigkeit
sensation	SCIE	Sinnesempfindung
sensation of pain	ACOUS	Schmerzempfindung
sense (v.t.)	ELECTRON	abfühlen
≈ scan (v.t.)		≈ abtasten
sense (n.)	SCIE	Sinn
[perceiving faculty]		[Wahrnehmungsfähigkeit]
sense (n.)	PHYS	Sinn
[one of two contrary directions]		= Bewegungssinn; Bewegungsrichtung
sense amplifier	MICROEL	Leseverstärker
= memory sense amplifier		
sense byte	DATA PROC	Fehlerbyte
sense finding	RADIO LOC	Seitenbestimmung
sense indicator	NETW.TH	Strompfeil
sense lead	ELECTRON	Fühlerleitung
sense line (→read wire)	TERM&PER	(→Lesedraht)
sense signal (→read signal)	ELECTRON	(→Lesesignal)
sense switch	DATA PROC	Abfühlschalter
[can be interrogated by a program]		[per Programm abfragbar]
sense winding (→read winding)	TERM&PER	(→Lesewicklung)
sense wire (→read wire)	TERM&PER	(→Lesedraht)
sensing (→scanning)	ELECTRON	(→Abtastung)
sensing circuit (→sampling circuit)	CIRC.ENG	(→Abtastschaltung)
sensing device (→test probe)	INSTR	(→Prüfspitze)
sensing element (→test probe)	INSTR	(→Prüfspitze)
sensing head (→test probe)	INSTR	(→Prüfspitze)
sensing pin (→digitizing pen)	TERM&PER	(→Digitalisierstift)
sensing threshold (→supervision threshold)	ELECTRON	(→Überwachungsschwelle)
sensistor (→NTC thermistor)	COMPON	(→Heißleiter)
sensitive	TECH	empfindlich
≈ susceptible; fragile		= sensibel; sensitiv
		≈ störanfällig; zerbrechlich
sensitive data	DATA PROC	schutzwürdige Daten
sensitiveness (→sensitivity)	TECH	(→Empfindlichkeit)
sensitivity	TECH	Empfindlichkeit
= sensitiveness		= Sensibilität
≈ fragility		≈ Zerbrechlichkeit
sensitivity	NETW.TH	Empfindlichkeit
= fragility		= Sensibilität
sensitivity analysis	CIRC.ENG	Empfindlichkeitsanalyse
		= Sensibilitätsanalyse
sensitivity control	ELECTRON	Empfindlichkeitsregelung
sensitivity factor	ELECTRON	Empfindlichkeitsfaktor
		= Sensibilitätsfaktor
sensitization	PHYS	Sensibilisierung
sensor	COMPON	Fühler
[detection device]		= Sensor
≈ transducer [PHYS]		≈ Wandler [PHYS]
sensor	SIGN.ENG	Melder
= detector		= Detektor
sensor (→test probe)	INSTR	(→Prüfspitze)
sensor array processing	INF.TECH	Sensorfeldverarbeitung
sensor-controlled	ELECTRON	sensorgesteuert
sensor key (→touch key)	COMPON	(→Berührungstaste)
sensor module	INSTR	Sensormodul
sensor relay	COMPON	Überwachungsrelais
= monitoring relay; supervision relay		
sensor switch	COMPON	Berührungsschalter
		= Sensorschalter

sensor technique

sensor technique	INSTR	**Sensortechnik**
		= Sensorik
sensory key (→touch key)	COMPON	(→Berührungstaste)
sentence	LING	**Satz**
[group of words forming a self-contained speech, concluding with end punctuation]		[eine selbständige Aussage beinhaltende Wortgruppe, mit einem Satzschlußzeichen endend]
= period 2		
sentence	CODING	**Satz**
[sequence of words considered as a unit]		[als Einheit betrachtete Folge von Wörtern]
separable	TECH	**trennbar**
= seccionable		= separierbar
separata	DOC	**Sonderbeilage**
[of a magazine]		[einer Zeitschrift]
separate	TECH	**trennen**
= disconnect		= abtrennen
≈ divide; segregate		≈ teilen; absondern
separate	EL.TECH	**trennen**
≈ decouple		= entkoppeln
separate cover	OFFICE	**getrennte Post**
separated	TECH	**getrennt**
≈ detached		= gesondert
		≈ abgesetzt
separate port (→individual port)	MICROW	(→Separatzugang)
separating branch instruction	DATA PROC	**Abgrenzungssprungbefehl**
separating filter	NETW.TH	**Weiche**
= high-pass/low-pass filter; decoupling filter		= Weichenfilter
separating plane (→interface)	TECH	(→Trennfläche)
separating surface (→interface)	TECH	(→Trennfläche)
separation 1	TECH	**Abstand**
= spacing 2		≈ Entfernung; Zwischenraum; Spalt
≈ distance; interstice; gap		
separation 2	MECH	**Abtrennung**
= disconnection		= Trennung
≈ division; segregation		≈ Teilung; Absonderung
separation	TERM&PER	**Vereinzelung**
[of paper]		[von Papier]
separation (→isolation)	EL.TECH	(→Trennung)
separation line	TYPOGR	**Trennungslinie**
		= Trennlinie
separator	DATA PROC	**Trennzeichen**
[to delimit variable-length data fields; in MS-DOS a period]		[zur Abgrenzung variabler Datenfelder; bei MS-DOS ein Punkt]
= tag; delimiter symbol; delimiter		= Trennsymbol; Abgrenzungszeichen; Begrenzungszeichen; Abgrenzungssymbol; Abgrenzer; Begrenzungssymbol; Begrenzer; Separator
separator	TERM&PER	**Separator**
[of copying paper sets]		[trennt Durchschreibsätze]
separator (→trap)	TECH	(→Abscheider)
sepia	PHYS	**sepiabraun**
septenary	MATH	**septenär**
[related to number 7]		[auf die Zahl 7 bezogen]
septum (→membrane)	TECH	(→Membran)
sequence (v.t.)	SCIE	**sequentialisieren**
sequence	COLLOQ	**Reihenfolge**
= order of succession; order		= Folge 1; Sequenz; Abfolge
≈ row		≈ Reihe
sequence	SCIE	**Sequenz**
[a specific order]		[eine vorgegebene Reihenfolge]
= sequency; succession		
≈ series		≈ Reihenfolge; Serie
sequence (→string 1)	DATA PROC	(→String 1)
sequence action relay (→two-step relay)	COMPON	(→Zweistufenrelais)
sequence alarm	EQUIP.ENG	**Folgealarm**
= secondary alarm		= Sekundäralarm
sequence argument (→ordering argument)	DATA PROC	(→Ordnungsbegriff)
sequence check	CODING	**Folgeprüfung**
sequence circuit (→sequencial circuit)	MICROEL	(→Folgeschaltung)
sequence contact (→make-break contact)	COMPON	(→Folgekontakt)
sequence control	DATA PROC	**Ablaufsteuerung**
= scheduling; scheduler; sequencer		≈ Ablaufteil
≈ executive routine		
sequence control	CONTROL	**Folgesteuerung**
= cascade control		= Ablaufsteuerung; sequentielle Steuerung; Kaskadenregelung; Kaskadensteuerung
sequence-controlled	CONTROL	**folgegesteuert**
sequence control memory (→sequence control store)	DATA COMM	(→Einstellspeicher)
sequence control storage (→sequence control store)	DATA COMM	(→Einstellspeicher)
sequence control store	DATA COMM	**Einstellspeicher**
= sequence control storage; sequence control memory		
sequence error	MATH	**Folgefehler**
sequence failure (→secondary failure)	QUAL	(→Folgeausfall)
sequence manufacture (→series production)	MANUF	(→Serienfertigung)
sequence number (→sequential number)	MATH	(→laufende Zahl)
sequence of operations	TECH	**Ablauf**
= run-off; program run		≈ Prozedur
≈ procedure		
sequence planning	TECH	**Ablaufplanung**
= sequence scheduling		
sequence program	DATA PROC	**Ablaufprogramm**
sequencer	MANUF	**Umgurter**
[bets components in their mounting sequence]		[gurtet Bauteile in ihrer Montagefolge]
		= Sequenzer
sequencer (→sequence control)	DATA PROC	(→Ablaufsteuerung)
sequence scheduling (→sequence planning)	TECH	(→Ablaufplanung)
sequencing	SCIE	**Sequentialisierung**
		= Sequentialisieren
sequency (→sequence)	SCIE	(→Sequenz)
sequency mode	INSTR	**Sequenzbetriebsart**
sequentail lobing	RADIO LOC	**Hauptkeulenüberlappung**
sequential	TECH	**sequentiell**
[in a specified sequence]		[in einer vorgegebenen Reihenfolge]
≈ serial		≈ seriell
sequential (→consecutive)	TECH	(→aufeinanderfolgend)
sequential access	DATA PROC	**sequentieller Zugriff**
[sequential reading of all data, till the procured ones are found]		[es werden sämtliche Daten nacheinander überlesen, bis die gesuchten gefunden werden]
= sequential access method; SAM; serial access		= serieller Zugriff; Linearzugriff; Reihenfolgezugriff
≠ direct access		≠ Direktzugriff
sequential access memory	DATA PROC	**sequentieller Speicher**
= sequential access storage; sequential acces store		[mit seriellem Zugriff]
≈ random access memory		= Sequentialspeicher; Serienspeicher
↓ delay-line memory; magnetic tape memory		≠ Direktzugriffsspeicher; Parallelspeicher
		↓ Laufzeitspeicher; Magnetbandspeicher
sequential access method (→sequential access)	DATA PROC	(→sequentieller Zugriff)
sequential access storage (→sequential access memory)	DATA PROC	(→sequentieller Speicher)
sequential acces store (→sequential access memory)	DATA PROC	(→sequentieller Speicher)
sequential algorithm	DATA PROC	**sequentieller Algorithmus**
sequential analysis	MATH	**Sequenzanalyse**
		= sequentielle Analyse
sequential circuit	MICROEL	**Folgeschaltung**
= sequence circuit		
sequential computer	DATA PROC	**Linearrechner**
[executes only one instruction at a time]		[führt immer nur einen Befehl aus]
≠ parallel computer		= von-Neumann-Rechner; se-

sequential damage ECON
= contiguous data structure
sequential data structure DATA PROC
sequential file DATA PROC
sequential function MATH
↓ slant function
sequential instruction DATA PROC
= continuation instruction
sequential list DATA PROC
[stored in contiguous locations]
= dense list; linear list
sequential logic CIRC.ENG
[output depends on previous inputs]
= sequential logic system
↑ logic circuit
sequential logic system CIRC.ENG
(→sequential logic)
sequential menu DATA PROC
sequential mode TELEC
(→serial transmission)
sequential number MATH
= sequence number; consecutive number; serial number
sequential operation DATA PROC
= one-at-a-time mode
sequential operation (→sequential TELEC mode)
sequential operator DATA PROC
sequential processing DATA PROC
≠ direct access processing
sequential program DATA PROC
[without branches or loops]
= linear program
sequential sampling INSTR
= coherent sampling; sequential scanning; coherent scanning
sequential scanning (→sequential INSTR sampling)
sequential search (→linear DATA PROC search)
sequential storage DATA PROC
sequential transmission (→serial TELEC transmission)
serial TECH
[one after other]
≈ sequential
≠ parallel
serial access (→sequential DATA PROC access)
serial adder CIRC.ENG

serial bus DATA PROC
serial carry DATA PROC
serial-carry counter MICROEL

serial data DATA PROC

serial data driver INSTR

serial data processing DATA PROC
(→serial processing)
serial dc power feeding TRANS
serial-in/parallel-out DATA PROC
= SIPO
serial-in/serial-out DATA PROC
= SISO

quentieller Computer; sequentieller Rechner
≠ Parallelrechner
Folgeschaden
sequentielle Datenstruktur

sequentielle Datei
Sequenzfunktion
= sequentielle Funktion
↓ Slant-Funktion
Folgebefehl

sequentielle Liste
[in benachbarten Speicherplätzen gespeichert]
= lineare Liste
sequentielle Logik
[Ausgang von vorangegangenen Eingangszuständen abhängig]
= Schaltwerk
↑ Logikschaltung
(→sequentielle Logik)

Folgemenü
(→Serienübertragung)

laufende Zahl
= sequentielle Zahl; Folgezahl; laufende Nummer; sequentielle Nummer; Folgenummer; Laufnummer

sequentieller Betrieb

(→Sequenztechnik)

Folgeoperator
sequentielle Verarbeitung
≠ wahlfreie Verarbeitung
lineares Programm
[ohne Verzweigungen oder Schleifen]
= sequentielles Programm
kohärente Abtastung
= sequentielles Sampling

(→kohärente Abtastung)

(→lineare Suchmethode)

sequentielle Speicherung
(→Serienübertragung)

seriell
[hintereinander und nicht gleichzeitig]
= reihenmäßig; nacheinender
≈ sequentiell
≠ parallel
(→sequentieller Zugriff)

Serienaddierer
= Serienadder; Serienaddierwerk
serieller Bus
Serienübertrag
Serienübertragzähler
= Serial-carry-Zähler
serielle Daten
= Serielldaten
Serielldatentreiber
= Serial-data-Driver
(→Serienverarbeitung)

Gleichstromreihenspeisung
SIPO

SISO

serial interface DATA PROC
serializability DATA PROC
serialize CODING

serializer CIRC.ENG
(→parallel-to-serial converter)
serial letter DATA PROC
serially reusable program DATA PROC
[can be runned sequentially for various application programs, without necessity to reload into the main memory]
≠ reenterable program
↑ reusable program

serial mode (→serial TELEC transmission)
serial mode (→serial DATA PROC processing)
serial model TECH
= off-the-shelf model
≈ standard version
≠ prototype; special make
serial number MANUF

serial number (→sequential MATH number)
serial numbering SWITCH
= consecutive numbering
serial operation (→serial DATA PROC processing)
serial port DATA PROC
serial printer TERM&PER
[prints character-by-character]
≠ parallel printer
↓ dot-matrix printer; ink-jet printer; type wheel printer

serial processing DATA PROC
[tasks are processed in sequence, without simultaneity or temporal interleaving]
= serial mode; serial data processing; serial operation
≠ multiprogramming

serial scanning ELECTRON
serial transmission TELEC
[sequential transmission of the code elements of a character]
= sequential transmission; serial mode; sequential mode
≠ parallel transmission

serial voltage regulator CIRC.ENG
serial voltage stabilization POWER SYS
series (n. pl.-ies) SCIE
= batch
≈ sequence
series MATH
= progression
↑ array
↓ Fourier series; power series

serielle Schnittstelle
seriell belegbar
in serielle Form bringen
= serialisieren
(→Parallel-Seriell-Umsetzer)

Serienbrief
seriell mehrfach aufrufbares Programm
[ohne Neuladen in den Arbeitsspeicher, hintereinander in mehreren Anwenderprogrammen einsetzbar]
= seriell wiederverwendbares Programm
≠ ablaufinvariantes Programm
↑ mehrfach aufrufbares Programm
(→Serienübertragung)

(→Serienverarbeitung)

Seriengerät
≈ Grundbauform
≠ Prototyp; Sonderanfertigung
Seriennummer
= Fertigungsnummer
(→laufende Zahl)

fortlaufende Numerierung

(→Serienverarbeitung)

serieller Anschluß
Serialdrucker
[druckt Zeichen nach Zeichen]
= Seriendrucker; Serielldrucker; Zeichendrucker
≠ Paralleldrucker
↓ Rasterdrucker; Tintendrucker; Typenraddrucker
Serienverarbeitung
[Aufgaben werden nacheinander bearbeitet, ohne Simultanität oder zeitliche Verschachtelung]
= serielle Verarbeitung; Seriellverarbeitung; Serialverarbeitung; Serialbetrieb; Serienverarbeitung; Serienbetrieb; serielle Arbeitsweise; serielle Datenverarbeitung; serielle Verarbeitung; Seriellbetrieb; serieller Betrieb
≠ Mehrprogrammbetrieb
Serienabtastung
Serienübertragung
[sequenzielle Übertragung der Codeelemente eines Zeichens]
= serielle Übertragung; Serienprinzip; Sequenztechnik
≠ Parallelübertragung
Serienregler
Serienstabilisierung
Serie
≈ Sequenz

Reihe
= Progression; Folge
↓ Fourier-Reihe; Potenzreihe

series capacitor	EL.TECH	**Reihenkondensator** = Serienkondensator	serigraphic template	MANUF	**Siebdruckschablone**
series circuit (→series connection)	EL.TECH	(→Reihenschaltung)	serigraphy (→silk-screen printing)	TYPOGR	(→Siebdruck)
series code	CODING	**Reihencode**	serrated [pulse shape]	ELECTRON	**eingesägt** [Impulsform]
series-connected ≈ preconnected	EL.TECH	**seriengeschaltet** = in Serie geschaltet = vorgeschaltet	serrations	MECH	**Kerbverzahnung**
			serration shaft	MECH	**Kerbzahnwelle**
series connection = series circuit; cascade connection; cascade circuit; cascade; cascading; tandem circuit; tandem connection; tandem ≠ parallel connection	EL.TECH	**Reihenschaltung** = Serienschaltung; Hintereinanderschaltung; Kaskadenschaltung; Kaskadenverbindung; Kaskade; Kettenschaltung ≠ Parallelschaltung	server (n.) [computer or peripheral performing a central function in a computer network] ≈ master computer ≠ requester; workstation ↓ file server; print server; communication server; dedicated server	DATA COMM	**Server** [in einem Rechnerverbund zentrale Funktionen ausübender Rechner oder Peripheriegerät] ≈ Hauptrechner ≠ Requester; Workstation ↓ Datenserver; Druckserver; Kommunikationsserver; dedizierter Server
series equivalent circuit	NETW.TH	**Reihenersatzschaltung**			
series expansion	MATH	**Reihenentwicklung**			
series-fed	ANT	**seriengespeist**	server	SWITCH	**Bedieneinheit**
series-fed aerial (→series-fed antenna)	ANT	(→fußpunktgespeiste Antenne)	service = service category	TELEC	**Dienst**
series-fed antenna = series-fed aerial	ANT	**fußpunktgespeiste Antenne**	service	ECON	**Dienstleistung**
			service (→support)	ECON	(→Betreuung)
series feed	EL.TECH	**Serienspeisung**	service (→operation 1)	TECH	(→Betrieb)
series feed (→base feed)	ANT	(→Fußpunktspeisung)	service (→operating)	TECH	(→Betriebs-)
series loading	EL.TECH	**Serienbelastung**	service (→after-sales service)	ECON	(→Kundendienst)
series motor (→series-wound motor)	POWER SYS	(→Reihenschlußmotor)	service (→coverage)	TELEC	(→Versorgung)
			service (→maintenance)	TECH	(→Wartung)
series-parallel connection	NETW.TH	**Reihen-Parallel-Schaltung** = Reihen-Parallel-Form	service (v.t.) (→operate)	TECH	(→bedienen)
series-parallel conversion	CODING	**Serien-Parallel-Umsetzung** = Serien-Parallel-Wandlung; Reihen-Parallel-Wandlung; Serienumsetzung; Serienwandlung	serviceability (→serviceableness)	TECH	(→Brauchbarkeit)
			serviceable = utilizable; usable; employable; feasible 2 ≈ useful; suited	TECH	**brauchbar** = verwendbar; verwertbar; benutzbar; nutzbar; dienlich; benutzerfreundlich ≈ nützlich; geeignet
series-parallel converter = staticizer	CIRC.ENG	**Serien-Parallel-Umsetzer** = Serien-Parallel-Wandler; Serienwandler; Serienumsetzer; Reihen-Parallel-Wandler	serviceable (→user-friendly)	TECH	(→benutzerfreundlich)
			serviceable (→operational)	TECH	(→betriebsfähig)
			serviceableness = serviceability; usability ≈ usefulness; usability; suitability; operability; user friendliness	TECH	**Brauchbarkeit** = Verwendbarkeit; Benutzbarkeit; Betriebsfreundlichkeit ≈ Nützlichkeit; Nutzbarkeit; Eignung; Betriebsfähigkeit; Benutzerfreundlichkeit
series-parallel matrix	NETW.TH	**Reihen-Parallel-Matrix**
series-parallel winding	ELECTRON	**Reihen-Parallel-Wicklung**
series production = sequence manufacture	MANUF	**Serienfertigung** = Serienproduktion; Serienherstellung
series resistance = series resistor	NETW.TH	**Längswiderstand** = Reihenwiderstand; Serienwiderstand			
			service accessories	ELECTRON	**Werkstattzubehör**
series resistance = series resistor ≈ dropping resistor	EL.TECH	**Reihenwiderstand** = Serienwiderstand ≈ Vorschaltwiderstand	service access point	DATA COMM	**Dienstzugangspunkt**
			service alarm	EQUIP.ENG	**Betriebsalarm**
			service area = servicing area; coverage area; catchment area ↓ broadcast service area [BROADC]; exchange area [SWITCH]; management domain [DATA COMM]	TELEC	**Versorgungsbereich** = Versorgungsgebiet; Einzugsbereich; Zuständigkeitsbereich ↓ Rundfunk-Versorgungsbereich [BROADC]; Anschlußbereich [SWITCH]
series resistor (→series resistance)	NETW.TH	(→Längswiderstand)
series resistor (→series resistance)	EL.TECH	(→Reihenwiderstand)
series resonance = voltage resonance	NETW.TH	**Serienresonanz** = Spannungsresonanz
series resonance [feeding in the current antinode]	ANT	**Stromresonanz** [Anregung im Strombauch]
			service assortment	ELECTRON	**Service-Sortiment**
series-resonance inverter	POWER SYS	**Reihenschwingkreis-Wechselrichter**	service bit	TRANS	**Service-Bit** = Dienstbit
			service bureau (→computer utility)	DATA PROC	(→DV-Dienstleistungsbetrieb)
series-resonant circuit = acceptor circuit; acceptor	NETW.TH	**Reihenschwingkreis** = Serienresonanzkreis; Serienschwingkreis; Saugkreis	service call = official call	TELEPH	**Dienstgespräch**
series-section transformer [impedance transformation]	ANT	**Serien-Sektion-Anpassung** [Impedanztransformation]	service capability	TELEC	**Dienstleistungsfähigkeit**
			service category (→service)	TELEC	(→Dienst)
series spectrum	PHYS	**Serienspektrum**	service category (→user group)	DATA COMM	(→Teilnehmerbetriebsklasse)
series T	MICROW	**Serienverzweigung**	service center = service centre (BRI)	TECH	**Bedienungszentrum**
series winding ≈ bank winding	ELECTRON	**Reihenwicklung** ≈ verschachtelte Wicklung	service centre (BRI) (→service center)	TECH	(→Bedienungszentrum)
series-wound motor = series motor	POWER SYS	**Reihenschlußmotor** = Hauptschlußmotor	service channel = engineering order wire; engineer order wire (BRI)	TELEC	**Dienstkanal**
serif [decorative transversal short lines at the extremes of letters] = ceriph	TYPOGR	**Serife** (n.f. pl.-n) [dekorative Querstriche an den Enden von Buchstaben] = Schraffe	service channel equipment = engineering order wire equipment (BRI)	TELEC	**Dienstkanaleinrichtung**
			service computer	SWITCH	**Bedienrechner** = Bedienungsrechner
serif font = ceriph font; serif font style; ceriph font style	TYPOGR	**Serifenschrift** = serifenhaltige Schrift	service computer center	DATA PROC	**Service-Rechnezentrum** = Service-Computerzentrum
serif font style (→serif font)	TYPOGR	(→Serifenschrift)			

service contract	ECON	**Servicevertrag**	
		= Wartungsvertrag	
service control point	TELEC	**Dienstesteuerungspunkt**	
service current (→operating current)	EL.TECH	(→Betriebsstrom)	
service demand	TELEC	**Dienstnachfrage**	
service demand (→service request)	DATA COMM	(→Dienstanforderung)	
service discipline (→dispatching discipline)	SWITCH	(→Abfertigungsdisziplin)	
service handset	MOB.COMM	**Bedienhörer**	
service identification signal	DATA COMM	**Dienstkennzeichen**	
service indicator	DATA COMM	**Dienstkennung**	
service instruction	ECON	**Dienstvorschrift**	
service instrument	TELEC	**Betriebsmeßgerät**	
		= Betriebsinstrument	
service integration	TELEC	**Dienstintegration**	
service intercept	SWITCH	**Gesprächsumleitung zur Beamtin**	
service interruption	TECH	**Betriebsunterbrechung**	
service interruption (→breakdown 2)	TECH	(→Betriebsstörung)	
service line connection	TELEC	**Dienstanschluß**	
serviceman 2	ECON	**Wartungstechniker**	
= service technician; maintenance technician		= Servicetechniker	
serviceman 1	ECON	**Militärangehöriger**	
service management	TELEC	**Betriebsverwaltung**	
		= Betriebsmanagement	
servicemen (pl.t.)	ECON	**Militärpersonal**	
= military personnel			
service mode (→operating mode)	TECH	(→Betriebsart)	
service multiplicity	TELEC	**Dienstevielfalt**	
= service variety			
service offer	TELEC	**Dienstangebot**	
service personnel (→maintenance personnel)	TECH	(→Wartungspersonal)	
service primitive	DATA COMM	**Dienstprimitiv**	
		= Dienst-Stammelement	
service priority (→dispatching priority)	DATA PROC	(→Abfertigungspriorität)	
service processor (→maintenance processor)	DATA PROC	(→Wartungsprozessor)	
service program (→utility program)	DATA PROC	(→Dienstprogramm)	
service provider	TELEC	**Diensterbringer**	
≈ operating company		= Diensteanbieter	
		≈ Netzbetreibergesellschaft	
service provider	ECON	**Dienstleistungsunternehmen**	
		≈ Versorgungsunternehmen	
service quality (→grade of service)	SWITCH	(→Verkehrsgüte)	
servicer (→computer utility)	DATA PROC	(→DV-Dienstleistungsbetrieb)	
service ready	DATA COMM	**dienstbereit**	
service reliability	QUAL	**Betriebszuverlässigkeit**	
= operational reliability; dependability 2; reliability in operation		= Betriebssicherheit	
service request	DATA COMM	**Dienstanforderung**	
= service demand			
service routine (→utility program)	DATA PROC	(→Dienstprogramm)	
service signal	DATA COMM	**Dienstsignal**	
= call progress signal; network message		= Netzmeldung	
service-specific	TELEC	**dienstespezifisch**	
service technician (→serviceman 2)	ECON	(→Wartungstechniker)	
service telephone	O&M	**Bedienfernsprecher**	
		= Bedienungsfernsprecher	
service traffic	DATA COMM	**Dienstverkehr**	
service trunk group (→outgoing trunk group)	SWITCH	(→Abnehmerbündel)	
service user	DATA COMM	**Dienstteilnehmer**	
		= Dienstbenutzer	
service variety (→service multiplicity)	TELEC	(→Dienstevielfalt)	
service voltage (→operating voltage)	EL.TECH	(→Betriebsspannung)	
service word	TRANS	**Meldewort**	
= SW		= SW	
servicing (→handling)	DATA PROC	(→Abwicklung)	
servicing (→maintenance)	TECH	(→Wartung)	
servicing area (→service area)	TELEC	(→Versorgungsbereich)	
servicing contract (→maintenance contract)	ECON	(→Wartungsvertrag)	
servicing costs (→maintenance costs)	ECON	(→Wartungskosten)	
servicing engineer (→maintenance engineer)	TECH	(→Wartungsingenieur)	
servicing instruction (→maintenance instruction)	TECH	(→Wartungsvorschrift)	
servicing instructions (→maintenace instruction)	TECH	(→Wartungsanleitung)	
servicing interval (→maintenance interval)	TECH	(→Wartungsfrist)	
servicing panel (→maintenance panel)	EQUIP.ENG	(→Wartungsfeld)	
servicing program (→maintenance program)	TECH	(→Wartungsprogramm)	
servicing routine (→maintenance program)	TECH	(→Wartungsprogramm)	
servicing schedule (→maintenance program)	TECH	(→Wartungsprogramm)	
servicing works (→maintenance works)	TECH	(→Wartungsarbeiten)	
serving (n.)	SWITCH	**Bedienen**	
serving	COMM.CABLE	**Umwicklung**	
= wrapping; lapping		= Bewicklung	
serving area (→exchange area)	SWITCH	(→Anschlußbereich)	
serving channel (→outgoing channel)	SWITCH	(→Abnehmerkanal)	
serving highway (→outgoing highway)	SWITCH	(→Abnehmer-Multiplexleitung)	
servo (→servomechanism)	CONTROL	(→Servomechanismus)	
servo (→servomotor)	TECH	(→Stellmotor)	
servo amplifier	CONTROL	**Servoverstärker**	
servo drive	CONTROL	**Stellantrieb**	
= adjusting drive			
servomechanism	CONTROL	**Servomechanismus**	
= servo		= Stellmechanismus	
≈ follow-up control		≈ Folgeregelung	
servo mode	TERM&PER	**Servotechnik**	
[disk memory]		[Magnetplattenspeicher]	
≠ stepping mode		≠ Schritttechnik	
servomotor	TECH	**Stellmotor**	
= servo		= Servomotor	
servo multiplier	COMPON	**Servomultiplikator**	
servosystem	CONTROL	**Servosystem**	
session	ECON	**Sitzung**	
= meeting; conference; sitting		≈ Versammlung; Tagung; Besprechung; Konferenz	
session	DATA PROC	**Sitzung**	
		= Session	
session	DATA COMM	**Sitzung**	
= communication session			
session layer	DATA COMM	**Kommunikationssteuerungsschicht**	
[5th layer in the OSI model; refers to appointments to establish logical links between end-users]		[5.Schicht im OSI-Referenzmodell; bertrifft Festlegungen zur Herstellung logischer End-zu-End-Verbindungen]	
= session service		= Kommunikationssteuerungsebene	
session manager	DATA PROC	**Session-Manager**	
session service (→session layer)	DATA COMM	(→Kommunikationssteuerungsschicht)	
set (v.t.)	CIRC.ENG	**setzen**	
= preset (v.t.)		≠ rücksetzen; löschen	
≠ reset (v.t.); erase			
set 1 (v.t.)	DATA PROC	**setzen**	
[to give a value to a variable]		[einer Variable einen Wert geben]	
		= einstellen 1	

set | | | | 528

set (n.)		MATH	**Menge**	**setting appliance** (→setting gauge)		MECH	(→Paßlehre)

set (n.) MATH **Menge**
 [collection of distinguishable items to a unit] [Zusammenfassung wohl-unterschiedener Objekte zu einem Ganzen]
 ↓ subset; cut set; set of points; union of sets = Satz
 ↓ Teilmenge; Schnittmenge; Punktmenge; Vereinigungsmenge

set (n.) TECH **Satz**
 [group of related pieces] [Gruppe von zusammengehörigen Teilen]
 = Garnitur

set TERM&PER **Gerät**
 = subset; apparatus = Apparat
 ↓ Fernsprechapparat

set (→repertoire) COLLOQ (→Vorrat)
set (→scheduled) TECH (→Soll-)
set 2 (v.t.) (→configure) DATA PROC (→konfigurieren)
setback COLLOQ **Rückschlag**
 ≈ Mißerfolg
set-change fee TELEC **Änderungsgebühr**
 = Auswechslungsgebühr
set condition TECH **Sollzustand**
 = reference condition; desired condition
set dimension (→prescribed dimension) ENG.DRAW (→Sollmaß)
set flush TYPOGR **bündig**
 [even with a margin] [mit einem Rand fluchtend]
 = flush ↓ linksbündig; rechtsbündig
 ↓ flush left; flush right
set free TECH **freimachen**
 = freisetzen
set length TECH **Sollänge**
set of apparatus TECH **Apparatur**
 ≈ set of equipment [Satz zusammengehöriger Apparate]
 ≈ Gerätschaften
set of drawings ENG.DRAW **Zeichnungssatz**
set of equipment TECH **Gerätschaften** (pl.t.)
 [set of related equipment] [Satz zusammengehöriger Geräte]
 = equipment = Gerät 2 (s.t.)
 ≈ apparatus ≈ Apparatur
set-off (adj) (→detached) TECH (→abgesetzt)
set of points MATH **Punktmenge**
 = point set
set of rules ECON **Regelwerk**
set of values MATH **Wertevorrat**
 = range of values; repertoire of values
set path CONTROL **Sollbahn**
set point (→set value) CONTROL (→Sollwert)
set point adjustment CONTROL **Sollwerteinstellung**
set-point correction CONTROL **Sollwertkorrektur**
set point generator CONTROL **Sollwertgeber**
 = director = Sollwerteinsteller
set-point potentiometer CONTROL **Sollwertpotentiometer**
set-point signal ELECTRON **Sollwertsignal**
set position CONTROL **Sollstellung**
set pressure TECH **Solldruck**
set range CONTROL **Sollbereich**
setscrew MECH **Stellschraube**
 = setting screw; adjusting screw = Gewindeschraube; Abgleichschraube; Einstellschraube
set size TYPOGR **Laufweite**
set square ENG.DRAW (→Zeichendreieck)
 (→straightedge)
settability (→adjustability) TECH (→Einstellbarkeit)
settability (→ajustment accuracy) ELECTRON (→Einstellgenauigkeit)
settable DATA PROC (→einstellbar)
 (→configurable)
settable (→adjustable) TECH (→einstellbar)
set theory MATH **Mengenlehre**
setting SWITCH **Einstellen**
setting (→adjustment) ELECTRON (→Einstellung)
setting (→typesetting) TYPOGR (→Satz)
setting accuracy (→ajustment accuracy) ELECTRON (→Einstellgenauigkeit)
setting angle (→adjusted angle) TECH (→Einstellwinkel)

setting appliance (→setting gauge) MECH (→Paßlehre)
setting error (→set-up error) INSTR (→Einstellfehler)
setting gauge MECH **Paßlehre**
 = adjusting gage; setting appliance; adjusting appliance = Einstellehre
setting knob EQUIP.ENG **Einstellknopf**
 = adjusting knob
setting mark INSTR **Einstellmarke**
 = adjustment mark
setting range (→adjustment range) TECH (→Einstellbereich)
setting rule CONTROL **Einstellregel**
setting screw (→setscrew) MECH (→Stellschraube)
setting time TECH **Einstellzeit**
 = adjusting time
setting time (→call set-up time) DATA COMM (→Verbindungsaufbaudauer)
setting up 1 (→set up 1) TECH (→Errichtung)
setting-up 2 (→mounting 2) TECH (→Aufstellung 3)
setting-up time (→call set-up time) DATA COMM (→Verbindungsaufbaudauer)
setting value (→adjusted value) ELECTRON (→Einstellwert)
settlement (→regulation) ECON (→Regelung)
settlement (→charging) ECON (→Verrechnung)
settling characteristics (→transient response) ELECTRON (→Übergangsverhalten)
settling error CONTROL **Ausregelfehler**
settling period (→build-up time) ELECTRON (→Einschwingzeit)
settling process (→transient 1) ELECTRON (→Übergangsvorgang)
settling time CONTROL **Ausregelzeit**
settling time (→build-up time) ELECTRON (→Einschwingzeit)
set up 1 (n.) TECH **Errichtung**
 = setting up; erection = Aufbau 1; Montage; Aufstellung 1; Aufstellen
set-up TELEC **Aufbau**
 [of a connection] [einer Verbindung]
 = build-up
set-up (→adjust) TECH (→einstellen)
setup (→adjustment) ELECTRON (→Einstellung)
set-up (→erect) TECH (→errichten)
setup (→installation) DATA PROC (→Installation)
set up a call SWITCH **eine Verbindung aufbauen**
 = set up a connection; set up a link ≈ durchschalten [TELEC]
 ≈ through connect [TELEC]
set up a connection (→set up a call) SWITCH (→eine Verbindung aufbauen)
set up a link (→set up a call) SWITCH (→eine Verbindung aufbauen)
set-up disk (→installation disk) DATA PROC (→Installationsdiskette)
set-up error INSTR **Einstellfehler**
 = setting error
set-up menu (→configuration menu) DATA PROC (→Konfigurationsmenu)
setup program DATA PROC **Installationsprogramm**
 = installation program
set-up time DATA PROC **Vorbereitungszeit**
 = initialization time = Rüstzeit; Einrichtzeit; Aufsetzzeit; Inizialisierungdzeit
set-up time ELECTRON **Vorbereitungszeit**
 = setup time
set-up time (→call set-up time) DATA COMM (→Verbindungsaufbaudauer)
setup time (→set-up time) ELECTRON (→Vorbereitungszeit)
set value CONTROL **Sollwert**
 = index value; reference value; scheduled value; desired value; set point
set-value control (→fixed command control) CONTROL (→Festwertregelung)
set velocity TECH **Sollgeschwindigkeit**
set width TECH **Sollbreite**
set width (→character width) TYPOGR (→Dickte)
set wire (→write wire) TERM&PER (→Schreibdraht)
seven-layer TECH **siebenschichtig**
 = seven-part = siebenlagig
seven-part (→seven-layer) TECH (→siebenschichtig)

seven-segment display	COMPON	Siebensegmentanzeige
severe	TECH	streng
severe (→stringend)	COLLOQ	(→streng)
severely errored second	TRANS	stark gestörte Sekunde
severe operating condition	TECH	rauhe Betriebsbedingung
sewage (→waste-water)	TECH	(→Abwasser)
sewage clarification plant	TECH	Abwasserkläranlage
sexadecimal (→hexadecimal)	MATH	(→hexadezimal)
sexadecimal digit (→hexadecimal digit)	MATH	(→Hexadezimalziffer)
sexadecimal notation (→hexadecimal number system)	MATH	(→Hexadezimalsystem)
sexadecimal number (→hexadecimal number)	MATH	(→Hexadezimalzahl)
sexadecimal number system (→hexadecimal number system)	MATH	(→Hexadezimalsystem)
Sexadecimal pad (→hexadecimal pad)	TERM&PER	(→Hexadezimaltastatur)
sexadecimal point (→hexadecimal point)	MATH	(→Hexadezimalpunkt)
sexadecimal system (→hexadecimal number system)	MATH	(→Hexadezimalsystem)
sextant [goniometer to measure altitude of celestial bodies]	ASTROPHYS	Sextant [Winkelmeßinstrument zur Bestimmung der Höhe eines Gestirns]
sextillon (AM) [ten to the power of twenty one] = thousand trillions (BRI)	MATH	Trillarde [zehn hoch einundzwanzig]
sextuple (→sixfold)	COLLOQ	(→sechsfach)
sextuple (→sextuplicate)	MATH	(→versechsfachen)
sextuplicate = sextuple ↑ multiplicate	MATH	versechsfachen ↑ vervielfachen
S.F.E.R.T.	TELEPH	S.F.E.R.T. [europäischer Ureichkreis zur subjektiven Messung der Bezugsdämpfung – Système Fondamental Européen de Référence pour la Transmission Téléphonique]
sgd. (→signed)	ECON	(→gezeichnet)
sh (→hyperbolic sine)	MATH	(→Sinus hyperbolicus)
shackle = split cable grip	OUTS.PLANT	Kabelziehschlauch
shackle [chaining element closed by a bolt] = Bride	TECH	Schäkel [mit Bolzen schließbares Verbindungselement]
shade (v.t.) ≈ hatch (v.t.)	ENG.DRAW	schattieren ≈ schraffieren
shade (n.) = color shade (NAM); colour shade (BRI) ↑ hue	PHYS	dunkelgetönte Farbe = dunkler Farbton; Dunkeltönung ↑ Farbton
shade (→shadow)	COLLOQ	(→Schatten)
shaded (→shadow printing)	TYPOGR	(→Schattendruck)
shaded memory [sector of main memory 1 for data of quick access, not addressable by the program] = nonaddressable memory; shaded storage; nonaddressable storage ↑ main memory 1	DATA PROC	Ergänzungsspeicher [Teil des Hauptspeichers 1 für Daten geringer Zugriffszeit, vom Programm nicht abrufbar] ↑ Hauptspeicher 1
shaded printing (→shadow printing)	TYPOGR	(→Schattendruck)
shaded storage (→shaded memory)	DATA PROC	(→Ergänzungsspeicher)
shading	TV	Bildabschattung
shading = shadowing ≈ hatching	ENG.DRAW	Schattierung ≈ Schraffur
shading mast [to dampen radiation in the direction of transmitters with equal frequency]	ANT	Ausblendemast [zur Dämpfung der Austrahlung in Richtung gleichfrequenter Rundfunksender]
shadow = shade	COLLOQ	Schatten
shadow (→dead spot)	RADIO PROP	(→Funkschatten)
shadow (→shadow printing)	TYPOGR	(→Schattendruck)
shadowing (→shading)	ENG.DRAW	(→Schattierung)
shadow mask = Schattenmaske	TV	Lochmaske
shadow mask tube = Maskenröhre; Deltaröhre	TV	Lochmaskenröhre
shadow printing = shaded printing; shaded; shadow	TYPOGR	Schattendruck [versetzter Doppelschrift] = Schattenschrift
shadow region	RADIO PROP	Schattengebiet
shaft ↓ drive shaft; capstan [TERM&PER]	MECH	Welle ↓ Antriebswelle; Capstan [TERM&PER]
shaft extension (→extension spindle)	COMPON	(→Verlängerungsachse)
shake (v.t.) ≈ concuss	TECH	schütteln = rütteln ≈ erschüttern
shake (n.) ≈ concussion	TECH	Schütteln ≈ Erschütterung
shake-proof	MECH	schüttelfest
shake test	QUAL	Schüttelprüfung
sham company (→dummy corporation)	ECON	(→Scheingesellschaft)
shank	MECH	Schaft
Shannon-Fano code	CODING	Shannon-Fano-Code
shape (v.t.) (→plane)	TECH	(→hobeln)
shape [COLLOQ] (→state)	TECH	(→Zustand)
shape (v.t.) (→fashion)	COLLOQ	(→gestalten)
shape accuracy	MECH	Formgenauigkeit
shape-accurate (→true to shape)	MECH	(→formgerecht)
shaped (→formed)	TECH	(→geformt)
shape factor	TECH	Formparameter
shape factor measurement	INSTR	Formfaktormessung
shaper = shaping circuit ≈ equalizer	CIRC.ENG	Formungsschaltung = Former ≈ Entzerrer
shape ratio [ratio of circumference to area]	MICROEL	Umfang-Flächen-Verhältnis
shaping = profiling	TECH	Formung
shaping circuit (→shaper)	CIRC.ENG	(→Formungsschaltung)
shaping filter	TELEC	Formungsfilter
shaping treatment = shaping work	MECH	formgebende Bearbeitung
shaping work (→shaping treatment)	MECH	(→formgebende Bearbeitung)
sharable (→reentrant)	DATA PROC	(→ablaufinvariant)
sharable program (→reenterable program)	DATA PROC	(→ablaufinvariables Programm)
share = stock (AM)	ECON	Aktie
share (→divide)	TECH	(→teilen)
shared file	DATA PROC	Gemeinschaftsdatei
shared logic (→multi-user mode)	DATA PROC	(→Mehrbenutzerbetrieb)
shared-logic LAN	DATA COMM	zentralisiertes lokales Netz
shared resource (→resource sharing)	DATA PROC	(→Betriebsmittelteilung)
shared-resource LAN	DATA COMM	verteiltes lokales Netz
shared-service line (BRI) (→party line 2)	TELEC	(→Gemeinschaftsanschluß)
shareholder (BR) (→stockholder)	ECON	(→Aktionär)
shareholding = equity participation; interest ↓ stockholding	ECON	Kapitalbeteiligung ↓ Aktienbeteiligung
shareware [can be tried freely, with the claim of remuneration in case of regular use; cannot be modified or passed on] ≈ public-domain software	DATA PROC	Shareware [frei ausprobierbar, mit der Auflage eines Entgelts im Falle der regelmäßigen Nutzung; darf nicht verändert oder freigegeben werden] = Prüfprogramm 2 ≈ Public-domain-Software
share wave	MECH	Scherungswelle = Schubwelle

sharp [easily cutting] = keen 1 ≈ pointed ≠ blunt 2 ↑ cutting ↓ sharp-edged	MECH	**scharf** [gut schneidend] ≈ spitzig ≠ stumpf 2 ↑ schneidend ↓ scharfkantig	
sharp-edged ↑ sharp	MECH	**scharfkantig** ↑ scharf	
sharpen	MECH	**schärfen** ≈ schleifen	
sharpening	DATA PROC	**Kontrastierung**	
sharpness [optics] ↓ resolution	PHYS	**Schärfe** [Optik] = Abbildungsschärfe ↓ Auflösung	
sharpness [cutting property]	MECH	**Schärfe** [schneidendes Merkmal]	
shave hook	OUTS.PLANT	**Bleifeile**	
shaving	METAL	**Nachschneiden**	
shear (v.t.)	MECH	**scheren** = abscheren	
shear (n.) = shearing	MECH	**Scherung** = Abscherung; Schub 2	
shear force (→shearing force)	MECH	(→Scherungskraft)	
shear fracture	MECH	**Scherungsbruch**	
shearing (→shear)	MECH	(→Scherung)	
shearing (→shearing movement)	MECH	(→Scherungsbewegung)	
shearing deformation	MECH	**Schubverformung**	
shearing force = shear force	MECH	**Scherungskraft** = Schubkraft	
shearing load	MECH	**Scherbelastung**	
shearing movement = shearing	MECH	**Scherungsbewegung**	
shearing pin	MECH	**Scherstift**	
shearing strain	MECH	**Abscherungsbeanspruchung**	
shearing strength (→shear strength)	MECH	(→Scherfestigkeit)	
shearing stress (→shear stress)	MECH	(→Scherspannung)	
shearing viscosity = shear viscosity	MECH	**Scherungsviskosität** = Schubviskosität	
shear mode = shear oscillation	MECH	**Scherungsschwingung**	
shear modulus = rigidity modulus	MECH	**Schubmodul**	
shear oscillation (→shear mode)	MECH	(→Scherungsschwingung)	
shears = scissors	TECH	**Schere**	
shear stiffness	MECH	**Scherungsfestigkeit**	
shear strength = shearing strength	MECH	**Scherfestigkeit**	
shear stress = shearing stress	MECH	**Scherspannung** = Schubspannung	
shear stress line (→shear trajectory)	MECH	(→Scherungslinie)	
shear trajectory = shear stress line	MECH	**Scherungslinie** = Schublinie	
shear viscosity (→shearing viscosity)	MECH	(→Scherungsviskosität)	
shear zone	MECH	**Schubgebiet**	
sheath (→cable sheath)	COMM.CABLE	(→Kabelmantel)	
sheathed line	ANT	**Schlauchleitung**	
sheath-kilometers	TELEC	**Kabel-Kilometer**	
sheave [a grooved pulley] ↑ pulley	MECH	**Seilscheibe** [Rolle mit Rillen] = Seilrad ↑ Antriebsscheibe	
sheet [a piece of paper, usually rectangular] ≈ page	OFFICE	**Blatt** [ein, meist rechteckiges, Stück Papier] = Seite; Druckbogen [TYPOGR] ↓ Buchseite; Heftseite; Zeitungsseite	
sheet	TECH	**Lamelle**	
sheet (→sheet metal)	METAL	(→Blech)	
sheet aluminium	METAL	**Aluminiumblech**	
sheet antenna (→aperture antenna)	ANT	(→Aperturantenne)	
sheet bar (→sheet-bar iron)	METAL	(→Breiteisen)	
sheet-bar iron = sheet bar	METAL	**Breiteisen**	
sheet brass	METAL	**Messingblech**	
sheet copper	METAL	**Kupferblech**	
sheet current	LINE TH	**Mantelwelle**	
sheet feed (→single-sheet feed)	TERM&PER	(→Einzelblattzuführung)	
sheet feeder (→single-sheet feed)	TERM&PER	(→Einzelblattzuführung)	
sheet iron (→iron sheet)	METAL	(→Eisenblech)	
sheet metal = sheet ↓ thin sheet	METAL	**Blech** ↓ Feinblech	
sheet-metal cutting tool	MECH	**Blechschere**	
sheet-metal gage	MECH	**Blechlehre**	
sheet-metal nibbler	MECH	**Blechschneider**	
sheet-metal spring	MECH	**Federblech**	
sheet resistance	PHYS	**Schichtwiderstand**	
sheet zinc	METAL	**Zinkblech**	
Sheffer circuit (→NAND gate)	CIRC.ENG	(→NAND-Glied)	
Sheffer element (→NAND gate)	CIRC.ENG	(→NAND-Glied)	
Sheffer function (→NAND operation)	ENG.LOG	(→NAND-Verknüpfung)	
Sheffer gate (→NAND gate)	CIRC.ENG	(→NAND-Glied)	
Sheffer operation (→NAND operation)	ENG.LOG	(→NAND-Verknüpfung)	
shelf (→module frame)	EQUIP.ENG	(→Baugruppenrahmen)	
shelf (→grid)	SYS.INST	(→Rost)	
shelf cabling (→rack wiring)	EQUIP.ENG	(→Gestellverdrahtung)	
shelf equipping (→module frame packaging)	EQUIP.ENG	(→Baugruppenträgerbestückung)	
shelf life (→storage life)	TECH	(→Lagerzeit)	
shelf packaging (→module frame packaging)	EQUIP.ENG	(→Baugruppenträgerbestückung)	
shelf place [EQUIP.ENG] (→floor space)	TECH	(→Stellfläche)	
shelf wiring (→intra-shelf wiring)	EQUIP.ENG	(→Einsatzverdrahtung)	
shell [gliding-surface bearing]	MECH	**Hülse** [Gleitlager]	
shell 1 = bowl ≈ sleeve	TECH	**Schale** ≈ Hülse	
shell (→user interface)	DATA PROC	(→Benutzeroberfläche)	
shell (→electron shell)	PHYS	(→Elektronenhülle)	
shell 2 (→sleeve)	TECH	(→Hülse)	
shellac	CHEM	**Schellack**	
shell antenna	ANT	**Muschelantenne**	
shell K (→first shell)	PHYS	(→1.Schale)	
shell model	PHYS	**Schalenmodell**	
shell type transformer	PHYS	**Manteltransformator**	
shelter = container; custom building (AM); building (AM)	SYS.INST	**Shelter** = Container	
sherardize (→dry-galvanize)	METAL	(→trockenverzinken)	
SHF (→centimetric waves)	RADIO	(→Zentimeterwellen)	
SHF (→super high frequency)	EL.TECH	(→Höchstfrequenz)	
SHF antenna (→microwave antenna 1)	ANT	(→Mikrowellenantenne)	
shield (v.t.) (→screen-off)	EL.TECH	(→abschirmen)	
shield (→screening)	EL.TECH	(→Abschirmung)	
shield (n.) (→shroud)	ANT	(→Kragen)	
shielded	PHYS	**geschirmt**	
shielded (→screened)	EL.TECH	(→abgeschirmt)	
shielded aerial (→screened antenna)	ANT	(→abgeschirmte Antenne)	
shielded antenna (→screened antenna)	ANT	(→abgeschirmte Antenne)	
shielded cable = screened cable	COMM.CABLE	**geschirmtes Kabel**	
shielded galvanometer	INSTR	**Panzergalvanometer**	

shielded loop antenna	ANT	abgeschirmte Rahmenantenne
		= abgeschirmte Schleifenantenne
shielded pair	COMM.CABLE	geschirmtes Adernpaar
= screened pair		= geschirmtes Paar
shielded wire	COMM.CABLE	Schirmdraht
= shield wire		
shield factor	EL.TECH	Abschirmfaktor
= shielding factor		= Schirmfaktor
shielding	MECH	Abschirmung
shielding (→screening)	EL.TECH	(→Abschirmung)
shielding (→screening)	RADIO PROP	(→Sichtbehinderung)
shielding cover	EQUIP.ENG	Schirmgehäuse
= shielding enclosure		
shielding enclosure	EQUIP.ENG	(→Schirmgehäuse)
(→shielding cover)		
shielding factor (→shield factor)	EL.TECH	(→Abschirmfaktor)
shield wire (→shielded wire)	COMM.CABLE	(→Schirmdraht)
shift (v.t.)	COLLOQ	verschieben 1
= displace		[räumlich]
shift (v.t.)	DATA PROC	verschieben
[e.g. data in a register]		[z.B. Daten in einem Register]
= relocate		= stellenversetzen
shift	MANUF	Schicht
↓ primary shift; night shift		↓ Arbeitsschicht
		↓ Normalschicht; Nachtschicht
shift (→case shift)	TERM&PER	(→Groß-Klein-Umschaltung)
shift (→excursion)	MODUL	(→Hub)
shift (→displacement)	PHYS	(→Verschiebung)
shiftable (→displacable)	TECH	(→verschiebbar)
shift-click (v.t.)	DATA PROC	umschaltklicken
shift clock	CIRC.ENG	Schiebetakt
shift counter	CIRC.ENG	Schiebezähler
shift-in	DATA COMM	Rückschaltung
= SI		[Code]
		= SI
shifting	DATA PROC	Verschiebung
= relocation		[von Registerinhalten]
shift instruction	DATA PROC	Schiebebefehl
[shifting of a stored bit sequence by a number of positions]		[Verschieben einer gespeicherten Bitfolge um eine befohlene Stellenzahl]
		= Verschiebebefehl
SHIFT key	TERM&PER	Umschalttaste 1
[while pressed, upper-case symbols are sent]		[solange gedrückt gehalten, werden Großbuchstaben bzw. die oberen Tastenzeichen ausgegeben]
= case shift key		= Umschalter; Umschaltetaste; SHIFT-Taste; Taste SHIFT
≈ SHIFT LOCK key		≈ Umschaltfeststelltaste
shift letter (→letter shift)	TELEGR	(→Buchstabenumschaltung)
SHIFT LOCK (→SHIFT LOCK key)	TERM&PER	(→Umschaltfeststelltaste)
SHIFT-LOCK (→SHIFT LOCK key)	TERM&PER	(→Umschaltfeststelltaste)
SHIFT LOCK key	TERM&PER	Umschaltfeststelltaste
[after activation, capital letters are sent]		[nach Betätigung werden Großbuchstaben ausgegeben]
= SHIFT LOCK; capitals lock key; CAPS-LOCK; SHIFT-LOCK		= Großschreibtaste; CAPS-LOCK-Taste; SHIFT-LOCK-Taste; Feststelltaste; Umschaltfeststeller; Feststeller
≈ shift key		≈ Umschalttaste
shift of operating point	ELECTRON	Arbeitspunktverschiebung
= shift of working point		
shift of working point (→shift of operating point)	ELECTRON	(→Arbeitspunktverschiebung)
shift-out	DATA COMM	Dauerumschaltung
[a code]		[ein Code]
= SO		= SO
shift register	DATA PROC	Schieberegister
shift work	ECON	Schichtarbeit
shilling sign (→shilling symbol)	ECON	(→Shilling-Zeichen)
shilling symbol	ECON	Shilling-Zeichen
= shilling sign		
shining (→bright)	TECH	(→glänzend)
ship 1 (v.t.)	ECON	verschiffen
↑ freight		↑ verfrachten
ship (v.t.) (→send)	ECON	(→übersenden 1)
shipboard antenna	ANT	Schiffsantenne
shipbuilding (→shipbuilding industry)	TECH	(→Schiffbauindustrie)
shipbuilding engineering	TECH	Schiffsbau
shipbuilding industry	TECH	Schiffbauindustrie
= shipbuilding; marine engineering		
shipment 2 (→despatch)	ECON	(→Versand)
shipment 1 (→shipping 1)	ECON	(→Verschiffung 1)
shipper (→consigner)	ECON	(→Absender)
shipping 1	ECON	Verschiffung 1
= shipment 1		↑ Verfrachtung
↑ dispatch		
shipping 2 (→despatch)	ECON	(→Versand)
shipping date	ECON	Versanddatum
= date of despatch		
shipping documents	ECON	Versandpapiere
= shipping papers; waybill		= Transportpapiere; Verschiffungspapiere; Verladepapiere; Versanddokumente; Transportdokumente; Verschiffungsdokumente; Verladedokumente; Warenbegleitpapiere
↓ waybill		↓ Frachtbrief
shipping papers (→shipping documents)	ECON	(→Versandpapiere)
shipping spacer (→transport lock)	TECH	(→Transportsicherung)
shipping weight	ECON	Versandgewicht
shock (n.)	PHYS	Stoß
[a short force action]		[kurzfristige Kraftwirkung]
= impact; jolt (n.); collision		≈ Erschütterung [TECH]
≈ concussion [TECH]		↓ Zusammenstoß [MECH]
↓ collision [MECH]		
shock (v.t.) (→jolt)	PHYS	(→stoßen)
shock absorber	MECH	Stoßdämpfer
		= Schwingmetall
shock absorption	MECH	Stoßdämpfung
shock front	PHYS	Stoßfront
Shockley diode	MICROEL	Shockley-Diode
shock mounted frame	EQUIP.ENG	Schwingrahmen
		[Transportschutz]
shock-proof	ELECTRON	hochspannungsfest
		= hochspannungssicher
shock-proof (→impact-resistant)	TECH	(→schlagfest)
shock resistance	TECH	Schlagfestigkeit
= impact resistance		= Stoßfestigkeit
shock-resistant (→impact-resistant)	TECH	(→schlagfest)
shock test	QUAL	Stoßprüfung
= impact test		= Schockprüfung; Erschütterungsprüfung
shock wave	PHYS	Stoßwelle
shop (AM) (→workshop)	TECH	(→Werkstatt)
shop scale	TERM&PER	Ladenwaage
shore-based radar	RADIO LOC	Küstenradar
shore-based station (→coastal radio station)	RADIO NAV	(→Küstenfunkstelle)
shoreline	GEOPHYS	Küstenlinie
= coastal line		
short	INSTR	Kurzschlußnebenschluß
short (v.t.)	EL.TECH	(→kurzschließen)
(→short-circuit)		
short (n.)	EL.TECH	(→Kurzschluß)
(→short-circuit)		
shortage	ECON	Knappheit
= tightness; stringency; scarcity		= Verknappung
≈ lack		≈ Mangel
shortage (→shortfall)	ECON	(→Fehlmenge)
short antenna	ANT	kurze Antenne

short backfire antenna

short backfire antenna	ANT	kurze Backfire-Antenne
		= Short-backfire-Antenne
short-cicuit output impedance	NETW.TH	Kurzschluß-Ausgangsimpedanz
		= kurzschluß-Ausgangswiderstand
short-circuit (n.)	EL.TECH	Kurzschluß
= short (n.)		
short-circuit (v.t.)	EL.TECH	kurzschließen
= short (v.t.)		
short-circuit current	EL.TECH	Kurzschlußstrom
		= Einströmung; Urstrom
short-circuit current to earth	EL.TECH	Erdschlußstrom
= earth leakage current		
short-circuited winding	EL.TECH	Kurzschlußwicklung
= slug		= Kurzschlußwindung
short-circuit forward current transfer	MICROEL	Stromsteuerkennlinie
short-circuit forward current transfer ratio (→small-signal short-circuit forward transfer ratio)	MICROEL	(→Kurzschluß-Stromverstärkung)
short-circuit forward transfer admittance (→short-circuit transconductance)	MICROEL	(→Transmittanz)
short-circuit impedance	NETW.TH	Kurzschlußimpedanz
		= Kurzschlußwiderstand
short-circuit input admittance	NETW.TH	Kurzschluß-Eingangsadmittanz
= input admittance, output shorted		= Kurzschluß-Eingangsleitwert
short-circuit input impedance	NETW.TH	Kurzschluß-Eingangsimpedanz
		= Kurzschluß-Eingangawiderstand
short-circuit loss	NETW.TH	Kurzschlußdämpfung
short-circuit output admittance	NETW.TH	Kurzschluß-Ausgangsadmittanz
= output admittance, input shorted		= Kurzschluß-Ausgangsleitwert
short-circuit plug	COMPON	Kurzschlußstecker
= male short		
short-circuit power	POWER ENG	Kurzschlußleistung
= fault power		
short-circuit proof	EL.TECH	kurzschlußfest
= short-circuit protected		= kurzschlußsicher
short-circuit protected	EL.TECH	(→kurzschlußfest)
(→short-circuit proof)		
short-circuit rating	EL.TECH	Kurzschlußbelastbarkeit
short-circuit reverse-transfer admittance	MICROEL	Kurzschluß-Rückwärtssteilheit
= reverse-transfer admittance; remittance		[Transistorkenngröße]
↑ transistor parameter		= Remittanz; Kernleitwert rückwärts; Kurzschluß-Übertragungsadmittanz
short-circuit time	EL.TECH	Kurzschlußdauer
short-circuit to earth (→ground fault)	EL.TECH	(→Erdschluß)
short-circuit to ground (→ground fault)	EL.TECH	(→Erdschluß)
short-circuit transadmittance	NETW.TH	Kurzschluß-Transadmittanz
short-circuit transconductance	MICROEL	Transmittanz
= short-circuit forward transfer admittance		= Kernleitwert vorwärts; Kurzschluß-Vorwärtssteilheit; Vorwärtssteilheit
↑ transistor parameter		↑ Transistorkenngröße
short code (→abbreviated number)	SWITCH	(→Kurzrufnummer)
short code dialing (→abbreviated dialing)	SWITCH	(→Kurzwahl)
short cord	TECH	Strick
≈ line; rope; cable		[kurzes Seil]
↑ cord		≈ Leine; Tau; Kabel
		↑ Seil
short dipole	ANT	verkürzter Dipol
		= Kurzdipol
short-distance LOS	RADIO REL	(→Nahbereichsrichtfunk)
(→short-distance microwave)		
short-distance microwave	RADIO REL	Nahbereichsrichtfunk
= short-range microwave; short-range radio relay; short-distance LOS		= Nahverkehrsrichtfunk
short-distance modem	DATA COMM	Kurzdistanzmodem
short-distance traffic	TELEC	Nahverkehr
= short-haul traffic		≈ Ortsverkehr
≈ local traffic		≠ Fernverkehr
≠ toll traffic		
short duration pulse	TELEGR	Kurzimpuls
short-duration signal element	TELEGR	Kurzschritt
shorten 1	COLLOQ	verkürzen 1
[in space and figuratively]		[räumlich und figurativ]
≠ prolong		≠ verlängern
↑ reduce		
shorten 2	COLLOQ	verkürzen 2
[in time]		[zeitlich]
shortening	COLLOQ	Verkürzung
≠ prolongation		≠ Verlängerung
shortfall	ECON	Fehlmenge
= shortage ; deficiency		= Fehlbestand; Defizit
short-form catalog (NAM)	TECH	Kurzkatalog
= short-form catalogue (BRI)		
short-form catalogue (BRI)	TECH	(→Kurzkatalog)
(→short-form catalog)		
shorthand typist 1	OFFICE	Stenotypistin
[female]		
shorthand typist 2	OFFICE	Stenotypist
[male]		
shorthand writing	OFFICE	(→Stenographie)
(→stenography)		
short-haul area (→local exchange area)	SWITCH	(→Ortsanschlußbereich)
short-haul traffic	TELEC	(→Nahverkehr)
(→short-distance traffic)		
short hyphen (→hyphen 2)	LING	(→Bindestrich)
shorting to earth (→ground fault)	EL.TECH	(→Erdschluß)
short instruction	DATA PROC	Kurzbefehl
short interference	TELEC	Kurzzeitstörung
shortlived	COLLOQ	kurzlebig
short memorandum	OFFICE	Kurznotiz
short message	TELECONTR	Kurzmeldung
		= Kurztelegramm
short monopole	ANT	verkürzter Monopol
short path	SWITCH	Kurzweg
short precision (→single precision)	DATA PROC	(→Einfachgenauigkeit)
short-range microwave (→short-distance microwave)	RADIO REL	(→Nahbereichsrichtfunk)
short-range order	PHYS	Nahordnung
short-range radar	RADIO NAV	Nahbereichsradar
short-range radio relay (→short-distance microwave)	RADIO REL	(→Nahbereichsrichtfunk)
short spec (→abbreviated specification)	TECH	(→Kurzspezifikation)
short style antenna	ANT	Kurzstabantenne
short supply	ECON	Unterangebot 1
		[Mangel an Angebot]
short take-off and landing	RADIO NAV	Kurzstart und -landung
= STOL		= STOL
short-term fading	RADIO PROP	Kurzzeitschwund
short-term memory (→short-term storage 1)	DATA PROC	(→Kurzzeitspeicher)
short-term plan	ECON	kurzfristiger Plan
short-term storage 1	DATA PROC	Kurzzeitspeicher
= short-term memory; short-term store		= Kurzzeitgedächtnis
short-term storage 2	DATA PROC	Kurzzeitspeicherung
short-term store (→short-term storage 1)	DATA PROC	(→Kurzzeitspeicher)
short test	MANUF	Kurzprüfung
short timer	CIRC.ENG	Kurzzeitglied
short to earth (→ground fault)	EL.TECH	(→Erdschluß)
short to ground (→ground fault)	EL.TECH	(→Erdschluß)
short-wave	MATH	kurzwellig
[adj.]		
short-wave antenna	ANT	Kurzwellenantenne
= SW antenna		= KW-Antenne
short-wave connection (→short-wave link)	TELEC	(→Kurzwellenverbindung)
short-wave ferrite	METAL	KW-Ferrit
short-wave fine tuning	RADIO	KW-Lupe

short-wave link	TELEC	Kurzwellenverbindung
= SW link; short-wave connection; SW connection		= KW-Verbindung
short waves [100m-10m; 3MHz-30MHz]	RADIO	Kurzwelle [100m-10m; 3MHz-30MHz]
= high frequency; HF; decametric waves		= Dekameterwelle; KW; HF
short-wave transmitter	RADIO	Kurzwellensender
= SW transmitter		= KW-Sender
short weight (→underweight)	ECON	(→Untergewicht)
shot effect (→shot noise)	TELEC	(→Schrotrauschen)
shot noise	TELEC	Schrotrauschen
= shot effect		= Schroteffekt; Prasselstörung; Prasseln
shot welding	METAL	Schußschweißung
shoulder	MECH	Schulter
		= Bund 2; Ansatz 2
shoulder screw	MECH	Ansatzschraube
show case	ECON	Schaukasten
= display case		
showmaster	BROADC	Moderator
show piece	ECON	Schaustück
show room	ECON	Ausstellungsraum
shred (v.t.)	TECH	zerschnitzeln
= snip (v.t.); chip (v.t.)		= schnitzeln; schnippeln; schnipseln; schnipsen
shred (→chip)	TECH	(→Schnitzel)
shredder	OFFICE	Vernichter
shredding machine (→document destroying device)	OFFICE	(→Aktenvernichter)
shrink (n.)	MECH	Verzug
= warp		= Schrumpfung
shrink (→contract)	TECH	(→schrumpfen)
shrinkage (→contraction)	TECH	(→Schrumpfung)
shrinkage sleeve	OUTS.PLANT	Schrumpfmuffe
= heat-shrinkable closure		
shrinkage tube	OUTS.PLANT	Schrumpfschlauch
= heat-shrinkable tube		= Wärmeschrumpfschlauch
shrink fit	MECH	Schrumpfsitz
shrinking force (→contracting force)	MECH	(→Schrumpfkraft)
shrink on	MECH	aufschrumpfen
shrivel varnish	CHEM	Schrumpflack
shroud	ANT	Kragen
= shield (n.)		= Antennenkragen; Randblende
shrouded plug	INSTR	Sicherheitslaborstecker
shunt (n.)	EL.TECH	Nebenschluß
= current shunt		= Stromnebenschluß
≈ parallel connection		≈ Parallelschaltung
shunt	INSTR	Nebenwiderstand
shunt (v.t.) (→connect in parallel)	EL.TECH	(→parallelschalten)
shunt capacitance (→parallel capacitance)	EL.TECH	(→Parallelkapazität)
shunt circuit	EL.TECH	Nebenschlußschaltung
		= Nebenstromkreis
shunt current (→cross current)	PHYS	(→Querstrom)
shunted	EL.TECH	nebengeschlossen
		= parallelgeschaltet
shunt excitation	POWER ENG	Nebenschlußerregung
shunt exciting (→shunt feeding)	ANT	(→Nebenschlußspeisung)
shunt-fed	ANT	nebenschlußgespeist
= parralel-fed		= parallelgespeist
shunt-fed aerial (→shunt-fed antenna)	ANT	(→parallelgespeiste Antenne)
shunt-fed antenna	ANT	parallelgespeiste Antenne
= shunt-fed aerial		
shunt-fed system	ANT	Anzapfspeisung
		= Kurzschlußspeisung
shunt feed	EL.TECH	Parallelspeisung
shunt feed (→top feed)	ANT	(→Obenspeisung)
shunt feeding	ANT	Nebenschlußspeisung
= shunt exciting		
shunt field	POWER ENG	Nebenschlußfeld
shunt inductance (→parallel inductance)	EL.TECH	(→Parallelinduktivität)
shunt machine (→shunt-wound machine)	POWER ENG	(→Nebenschlußmaschine)
shunt relay	COMPON	Nebenschlußrelais
shunt resistance (→parallel resistance)	EL.TECH	(→Parallelwiderstand)
shunt resistance (→cross resistance)	PHYS	(→Querwiderstand)
shunt resistor	POWER ENG	Nebenschlußwiderstand
		= Nebenwiderstand; Shunt
shunt winding	POWER ENG	Nebenschlußwicklung
= parallel winding		
shunt-wound generator	POWER ENG	Nebenschlußgenerator
		↑ Gleichstrom-Nebenschlußmaschine
shunt-wound machine	POWER ENG	Nebenschlußmaschine
= shunt machine		↑ Gleichstrommaschine
shut (v.t.)	TECH	verschließen
shut-down (n.)	ECON	Stillegung
= closure		
shut-down (v.t.) (→disconnect)	EL.TECH	(→ausschalten)
shut-down (v.t.) (→close-down)	ECON	(→stillegen)
shut-off (v.t.) (→interrupt)	EL.TECH	(→unterbrechen)
shutter (→seal)	TECH	(→Verschluß)
shuttle (n.) [video recorder]	CONS.EL	Pendelsuchlauf [Videorecorder]
shuttle armature (→two-pole armature)	EL.TECH	(→Doppel-T-Anker)
SI (→shift-in)	DATA COMM	(→Rückschaltung)
Si (→silicon)	CHEM	(→Silizium)
SI base unit [one of the seven basic units defined within the Inzernational System of Units]	PHYS	SI-Basiseinheit [eine der im International System of Units definierte Basisgrößen: m, kg, A, K, mol, cd]
= base SI unit		
sibilance (→hissing)	ACOUS	(→Zischgeräusch)
sibilant	LING	Zischlaut
		= Sibilant
side [polygon]	MATH	Seite [Vieleck]
sideband	MODUL	Seitenband
= side band		
side band (→sideband)	MODUL	(→Seitenband)
sideband fading	HF	Seitenbandfading
sideband filter	MODUL	Seitenbandfilter
sideband frequency	MODUL	Seitenbandfrequenz
sideband suppression	MODUL	Seitenbandunterdrückung
sideband theory	MODUL	Seitenbandtheorie
sidebuilding	CIV.ENG	Anbau
= annex (n.)		= Nebengebäude
side circuit (→phantom circuit)	COMM.CABLE	(→Phantomkreis)
side circuit (→physical circuit)	LINE TH	(→Stammleitung)
side-circuit coil (→phantom coil)	OUTS.PLANT	(→Phantomübertrager)
side circuit loading	COMM.CABLE	Stammkreispupinisierung
side cover	EQUIP.ENG	Seitenabdeckung
= lateral cover		
side cutting pliers	TECH	Seitenschneider
= wire cutter; cutter		= Drahtschneider
1-sided disk (→SS disk)	TERM&PER	(→SS-Diskette)
1-sided diskette (→SS disk)	TERM&PER	(→SS-Diskette)
side effect	DATA PROC	Seiteneffekt
		= Nebenwirkung
side elevation (→side view)	ENG.DRAW	(→Seitenansicht)
side event	ECON	Rahmenveranstaltung
		= Begleitveranstaltung
side frequency	MODUL	Seitenfrequenz
side guy [crossing the direction of line]	OUTS.PLANT	Querabspannseil = Querabspannung
↑ guy rope		↑ Abspannseil
side handle	EQUIP.ENG	Seitengriff
side lobe	ANT	Nebenzipfel
= sidelobe; minor lobe; secondary lobe; secondary main lobe; secondary peak		= Nebenkeule; Nebenlappen; Nebenmaximum
≠ mayor lobe		≠ Hauptkeule
↓ back lobe		↓ Rückwärtskeule

sidelobe

sidelobe (→side lobe)	ANT	(→Nebenzipfel)
side lobe attenuation	ANT	**Nebenzipfeldämpfung**
= side lobe suppression		= Nebenkeulendämpfung; Nebenzipfelunterdrükkung; Nebenkeulenunterdrückung
side-lobe level	ANT	**Nebenkeulen-Richtfaktor**
≈ side-lobe level		≈ Rückstrahldämpfung
side lobe level (→front-to-back ratio)	ANT	(→Rückstrahldämpfung)
side lobe suppression (→side lobe attenuation)	ANT	(→Nebenzipfeldämpfung)
side one (→component side)	ELECTRON	(→Bauelementeseite)
side panel (→side section)	SYS.INST	(→Seitenteil)
side-rolling (→side-scrolling)	DATA PROC	(→seitliches Rollen)
side-scroll (v.t.)	DATA PROC	**seitlich rollen**
= creep (v.t.)		↑ rollen
↑ scroll		
side-scrolling	DATA PROC	**seitliches Rollen**
= side-rolling; creeping; horizontal scrolling		
↑ scrolling		
side section	SYS.INST	**Seitenteil**
= side panel		
side street	TECH	**Anliegerstraße**
side strut	ANT	**Schwenkarm**
sidetone 2	TELEPH	**Nebenton**
= secondary tone		
sidetone 1	TELEPH	**Rückhören**
[acoustic feedback from microphone to receiver capsule]		[Rückkopplung von Sprechkapsel auf Hörkapsel]
= side tone; telephone sidetone		
side tone (→sidetone 1)	TELEPH	(→Rückhören)
sidetone attenuation	TELEPH	**Rückhördämpfung**
sidetone reference equivalent	TELEPH	**Rückhörbezugsdämpfung**
		= RBD
side-to-phantom crosstalk	COMM.CABLE	**Mitsprechen**
[crosstalk between side and phantom circuits or viceversa]		[Nebensprechen von Phantomkreis zu Stammleitung oder umgekehrt]
≠ side-to-side crosstalk		≠ Übersprechen
↑ crosstalk [TELEC]		↑ Nebensprechen [TELEC]
side-to-side coupling	COMM.CABLE	**Übersprechkopplung**
= cross coupling		= Kreuzkopplung
side-to-side crosstalk	COMM.CABLE	**Übersprechen**
[crosstalk between side circuits or between phantom circuits]		[Nebensprechen zwischen Stammleitungen oder zwischen Phantomkreisen]
= phantom-to-phantom circuit		≠ Mitsprechen
≠ side-to-phantom crosstalk		↑ Nebensprechen [TELEC]
↑ crosstalk [TELEC]		
side two (→solder side)	ELECTRON	(→Lötseite)
side view	ENG.DRAW	**Seitenansicht**
= side elevation		= Seitenriß
siemens	PHYS	**Siemens**
[SI unit for electric conductivity, susceptance and admittance; = 1/Ω]		[SI-Einheit für elektrischen Leitwert, Blindleitwert und Scheinleitwert; = 1/Ω]
= MHO		
sienna	PHYS	**siena**
		= rotbraun
sieve (v.t.)	TECH	**sieben**
= sift (f.t.)		≈ filtern
≈ filter (v.t.)		
sieve (n.)	TECH	**Sieb**
Sievert	PHYS	**Sievert**
[SI unit for equivalent dose]		[SI-Einheit für Äquivalentdosis]
= Sv		= Sv
sift (v.t.)	DATA PROC	**herauspflücken**
[to extract data from a large amount]		[von Daten aus einem großen Bestand]
sift (f.t.) (→sieve)	TECH	(→sieben)
sifting	DATA PROC	**Einfügungsbetrieb**
= insertion mode		
sight (n.)	PHYS	**Sicht**
sight draft	ECON	**Sichttratte**
sight-glass	TECH	**Sichtglas**
sign (v.t.)	OFFICE	**unterschreiben**
		= unterzeichnen
		≈ firmieren
sign	MATH	**Vorzeichen**
= algebraic sign; signum		= Signum
sign	LING	**Zeichen**
[an agreed graphical symbol]		[vereinbartes graphischer Symbol]
= mark; symbol		= Symbol
↓ character; punctuation mark; diacritic mark; numeral; mathematical symbol; proofreaders' marks		↓ Schriftzeichen; diakritisches Zeichen; Satzzeichen; numerisches Zeichen; mathematisches Zeichen; Korrekturzeichen
sign (→corporate logo)	ECON	(→Firmenzeichen)
sign (→characteristic 1)	TECH	(→Merkmal)
signal (n.)	INF	**Signal**
[state or change of physical magnitude representing information]		[Zustand oder Änderung physikalischer Größe zur Darstellung von Information]
≈ character		= Informationssignal; Zeichenträger
		≈ Zeichen
signal (→switching signal)	SWITCH	(→Kennzeichen)
signal amplifier	CIRC.ENG	**Signalverstärker**
signal amplitude	ELECTRON	**Signalamplitude**
signal analyzer	INSTR	**Signalanalysator**
[instrument for frequency-domain measurements]		[Gerät für Messungen im Frequenzbereich]
↓ spectrum analyzer; Fourier analyzer; selective voltmeter; distortion analyzer; modulation analyzer		↓ Spektrumanalysator; Fourier-Analysator; Selektivspannungsmesser; Verzerrungsanalysator; Modulationsanalysator
signal box (→switch tower)	RAILW.SIGN	(→Stellwerk)
signal branching	CIRC.ENG	**Signalverzweiger**
signal breakdown	TELEC	**Signalverlust**
		= Signaleinbruch
signal breakup	ELECTRON	**Signalverfälschung**
= breakup		
signal buffer	CIRC.ENG	**Zeichenpuffer**
signal cabin (BRI) (→switch tower)	RAILW.SIGN	(→Stellwerk)
signal channel	TELEC	**Signalkanal**
[ISDN]		[ISDN]
= channel D		= D-Kanal
signal condition	ELECTRON	**Signalzustand**
= signal state		
signal condition	DATA COMM	**Zeichenlage**
= condition; signal state; state		= Signalzustand
↓ signal condition A; signal condition B		↓ Trennstrom; Zeichenstrom
signal condition A	DATA COMM	**Startpolarität**
[condition of start signal]		[Kennzustand des Startschrittes]
= condition A; start polarity		= A-Zustand; Signalzustand EIN
≈ marking pulse; pause pulse		≈ Stromschritt; Pausenschritt
signal conditioning	ELECTRON	**Signalaufbereitung**
≈ signal processing; signal shaping		= Signalunformung
		≈ Signalverarbeitung; Signalformung
signal condition Z	DATA COMM	**Stoppolarität**
[condition of stop signal]		[Kennzustand des Stoppschrittes]
= condition Z; stop polarity		= Z-Zustand; Signalzustand AUS
≈ marking pulse; pause pulse		≈ Stromschritt; Pausenschritt
signal constellation	MODUL	**Signalkonstellation**
= constellation		= Konstellation
signal contact	EQUIP.ENG	**Signalkontakt**
signal conversion	TELEC	**Signalumsetzung**
		= Zeichenumsetzung; Signalkonversion; Zeichenkonversion
signal converter	CIRC.ENG	**Signalumsetzer**
signal converter storage tube	ELECTRON	**Signalspeicherröhre**
signal delay	TELEC	**Signalverzögerung**
signal detection	TELEC	**Signalentscheidung**
signal digit (→dialing bit)	DATA COMM	(→Wählbit)
signal diode	MICROEL	**Signaldiode**
≠ rectifier diode		≠ Gleichrichterdiode
↑ diode		↑ Halbleiterdiode

signal distance (→ Hamming distance)	CODING	(→ Hammingdistanz)	
signal distortion	TELEC	**Signalverzerrung** = Zeichenverzerrung	
signal distributor	TELEC	**Signalverteiler**	
signal edge	ELECTRON	**Signalflanke**	
signal element (→ unit interval)	TELEGR	(→ Zeichenelement)	
signal element error probability = signalling error rate	TELEGR	**Schrittfehlerwahrscheinlichkeit**	
signal element length (→ unit interval)	TELEGR	(→ Zeichenelement)	
signal-element timing	CODING	**Schrittakt**	
signal energy	INF	**Signalenergie**	
signal engineering = Meldetechnik	ELECTRON	**Signal- und Sicherungstechnik**	
signal enhancement	RADIO PROP	**Signalüberhöhung**	
signal follower (→ signal tracer)	INSTR	(→ Signalverfolger)	
signal form = signal shape; waveform; waveshape ≈ waveform [PHYS]	ELECTRON	**Signalform** = Signalverlauf; Wellenform; Wellenverlauf ≈ Wellenform [PHYS]	
signal generator 1 [generates a modulable signal with variable frequency and level] = measuring generator; test generator	INSTR	**Universal-Meßsender** [erzeugt modulierbares Signal mit einstellbarem Pegel und Frequenz] = Meßgenerator; Signalgenerator	
signal generator	TELEGR	**Zeichengeber** = Zeichengenerator	
signal generator 2 (→ level generator)	INSTR	(→ Pegelsender)	
signal generator (→ signaling transmitter)	ELECTRON	(→ Signalgeber)	
signal generator (→ signal source)	INSTR	(→ Signalgenerator)	
signal ground (→ common return)	EL. TECH	(→ Betriebserde)	
signal-harmonics ratio	MODUL	**Oberwellenabstand**	
signal high (→ high level)	MICROEL	(→ Hochpegelzustand)	
signal imitation	INF	**Zeichenimitation**	
signal information field · (→ signaling information field)	SWITCH	(→ Zeichen-Informationsfeld)	
signaling (AM) = signalling (BRI)	TELEC	**Zeichengabe** = Signalisierung; Signalübertragung	
signaling button (→ signaling key)	TERM&PER	(→ Signaltaste)	
signaling channel (AM) = signalling channel (BRI); signalling link (BRI)	SWITCH	**Zeichenkanal** = Zeichengabekanal; Signalisierungskanal	
signaling circuit [functional unit assigned temporarily to a call with signaling functions] = digit receiver; register 1 ↑ circuit ↓ digit input circuit; digit output circuit	SWITCH	**Signalisierungssatz** [für Signalisierungsaufgaben einer Verbindung zeitweilig zugeordnete Funktionseinheit] = Wahlsatz; Register; Empfangssatz ↑ Satz ↓ Wahlaufnahmesatz; Wahlsendesatz	
signaling control 2	SWITCH	**Leitungstechnik**	
signaling control 1 (AM) = signalling control (BRI); signaling management (AM); signalling management (BRI)	SWITCH	**Zeichengabesteuerung** = Signalisierungssteuerung	
signaling control link (AM) = signalling control link (BRI)	SWITCH	**Zeichengabeabschnitt** = Signalisierungsabschnitt	
signaling conversion [PCM]	TRANS	**Kennzeichenumsetzung** [PCM]	
signaling converter [PCM]	TRANS	**Kennzeichenumsetzer** [PCM] = KZU; Signalisierungsumsetzer	
signaling distortion	TELEC	**Wählzeichenverzerrung**	
signaling engineering	ELECTRON	**elektronische Sicherungstechnik** = Sicherungselektronik	
signaling frame	CODING	**Signalrahmen**	
signaling frequency	ELECTRON	**Schrittfrequenz**	
signaling information field (AM) = signalling information field (BRI); signal information field	SWITCH	**Zeichen-Informationsfeld**	
signaling key = signaling button	TERM&PER	**Signaltaste**	
signaling management (AM) (→ signaling control 1)	SWITCH	(→ Zeichengabesteuerung)	
signaling management processor (AM) = signalling management processor	SWITCH	**Zeichengabe-Leitprozessor** = Signalisieruns-Leitprozessor	
signaling measurement set (→ signaling test set)	INSTR	(→ Signalisierungsmeßplatz)	
signaling point (AM) = signalling point (BRI)	SWITCH	**Zeichengabepunkt** = Signalisierungspunkt	
signaling protocol	SWITCH	**Signalisierungsprotokoll**	
signaling rate (→ telegraph speed)	TELEC	(→ Schrittgeschwindigkeit)	
signaling rate generator (→ baud rate generator)	DATA COMM	(→ Baudratengenerator)	
signaling system (AM) = signalling system (BRI)	SWITCH	**Zeichengabeverfahren** = Zeichengabesystem; Signalisierungsverfahren; Signalisierungssystem	
signaling test set = signaling measurement set	INSTR	**Signalisierungsmeßplatz** = Signalisierungsprüfplatz	
signaling transfer point (AM) = signalling transfer point (BRI); signal transfer point	SWITCH	**Zeichengabe-Transferpunkt** = Signalisierungs-Transferpunkt	
signaling transmitter = signal generator; tone generator	ELECTRON	**Signalgeber** = Signalgenerator	
signal interception system	RADIO LOC	**Erfassungs-und Peilanlage**	
signal interval (→ unit interval)	TELEGR	(→ Zeichenelement)	
signal inversion	INF	**Signalumkehr**	
signal lamp (→ indicator lamp)	COMPON	(→ Anzeigelampe)	
signal lead	TELEC	**Signalader**	
signal level	TV	**Kanalpegel**	
signal level	TRANS	**Signalpegel**	
signal level swing [difference of high to low level]	MICROEL	**Signalhub** [Differenz von H- zu L-Zustand]	
signal line ≈ control bus	DATA PROC	**Signalleitung** ≈ Steuerbus	
signalling (BRI) (→ signaling)	TELEC	(→ Zeichengabe)	
signalling channel (BRI) (→ signaling channel)	SWITCH	(→ Zeichenkanal)	
signalling control (BRI) (→ signaling control 1)	SWITCH	(→ Zeichengabesteuerung)	
signalling control link (BRI) (→ signaling control link)	SWITCH	(→ Zeichengabeabschnitt)	
signalling error rate (→ signal element error probability)	TELEGR	(→ Schrittfehlerwahrscheinlichkeit)	
signalling information field (BRI) (→ signaling information field)	SWITCH	(→ Zeichen-Informationsfeld)	
signalling link (BRI) (→ signaling channel)	SWITCH	(→ Zeichenkanal)	
signalling management (BRI) (→ signaling control 1)	SWITCH	(→ Zeichengabesteuerung)	
signalling management processor (→ signaling management processor)	SWITCH	(→ Zeichengabe-Leitprozessor)	
signalling message	DATA COMM	**Kennzeichenblock**	
signalling point (BRI) (→ signaling point)	SWITCH	(→ Zeichengabepunkt)	
signalling system (BRI) (→ signaling system)	SWITCH	(→ Zeichengabeverfahren)	
signalling transfer point (BRI) (→ signaling transfer point)	SWITCH	(→ Zeichengabe-Transferpunkt)	
signalling unit (→ remote control unit)	DATA COMM	(→ Fernschaltgerät)	
signal low = signal low level	MICROEL	**Signal-Tiefpegel** = Signal-low-Pegel	
signal low level (→ signal low)	MICROEL	(→ Signal-Tiefpegel)	
signal mark (→ switching signal)	SWITCH	(→ Kennzeichen)	
signal mark generator	SWITCH	**Kennzeichengenerator**	

signal matching ELECTRON Signalanpassung
signal multiplexer (→character INF.TECH (→Zeichenmultiplexer)
  multiplexer)
signal network SWITCH Signalwegenetz
signal panel EQUIP.ENG Signalfeld
signal parameter INF Signalparameter
signal path ELECTRON Signalpfad
  = Signalweg
signal pattern recognizer INSTR Signalmustererkenner
signal probe ELECTRON Signalprobe
signal processing TELEC Signalverarbeitung
≈ signal conditioning [ELECTRON] ≈ Signalaufbereitung [ELECTRON]
signal processing antenna ANT Signalprozessor-Antenne
signal processor MICROEL Signalprozessor
signal quality TELEC Signalgüte
signal receiver TRANS Signalempfänger
  = tone detector
signal reconstruction ELECTRON (→Signalrekonstruktion)
  (→waveform reconstruction)
signal repertoire INF Signalvorrat
  = signal set
signal representation ELECTRON Signaldarstellung
signal selector CIRC.ENG Signalwähler
signal set (→signal INF (→Signalvorrat)
  repertoire)
signal shape (→signal ELECTRON (→Signalform)
  form)
signal shaper CIRC.ENG Signalformer
signal shaping ELECTRON Signalformung
≈ signal conditioning ≈ Signalaufbereitung
signal simulation system INSTR Signalsimulationssystem
signal source INSTR Signalgenerator
  = signal generator = Signalquelle
signal space TELEC Signalraum
signal state (→signal ELECTRON (→Signalzustand)
  condition)
signal state (→signal DATA COMM (→Zeichenlage)
  condition)
signal strength meter (→S INSTR (→S-Meter)
  meter)
signal string DATA COMM Signalfolge
signal-to-cross-modulation ratio MODUL Kreuzmodulationsabstand
  = signal-to-x-modulation ratio
signal-to-crosstalk ratio TELEC Grundwert des Nebensprechens
signal-to-hum ratio TELEC Brummabstand
signal-to-interence ratio TELEC Störabstand
  = S/I ratio; interference margin ↓ Rauschabstand
  ↓ signal-to-noise ratio
signal-to-interference ratio TELEC Störabstand
  = S/I ratio; interference margin ↓ Rauschabstand
  ↓ signal-to-noise ratio
signal-to-intermodulation EL.ACOUS Intermodulationsabstand
  ratio
signal-to-intermodulation ratio TV Moirédämpfung
signal tone (→idle tone) TELEPH (→Freiton)
signal-to-noise ratio TELEC Rauschabstand
  = S/N ratio; S/N; noise margin; SNR = Störabstand; Signal-Geräusch-Abstand
  ≈ signal-to-psophometric-noise ratio ≈ Geräuschabstand
signal-to-noise-with-load ratio TELEC Rauschabstand bei Belastung
  = signal-to-noise-with-tone ratio; S/N = Störabstand bei Belastung;
    with load ratio Signal-Geräusch-Abstand
    bei Belastung
signal-to-noise-with-tone ratio TELEC (→Rauschabstand bei Belastung)
  (→signal-to-noise-with-load ratio)
signal-to-pilot ratio TRANS Pilotabstand
  [FDM] [TF-Technik]
signal-to-psophometric-noise TELEC Geräuschabstand
  ratio ≈ Rauschabstand
≈ signal-to-noise ratio
signal-to-x-modulation ratio MODUL (→Kreuzmodulationsabstand)
  (→signal-to-cross-modulation ratio)
signal tracer INSTR Signalverfolger
  = signal follower
signal track INSTR Signalgleichlauf
≈ center frequency tracking ≈ Mittenfrequenznachlauf
signal track (→center frequency INSTR (→Mittenfrequenznachführung)
  tracking)

signal transfer point SWITCH (→Zeichengabe-Transferpunkt)
  (→signaling transfer point)
signal unit SWITCH Signaleinheit
  = Zeicheneinheit
signal velocity PHYS Signalgeschwindigkeit
signal voltage ELECTRON Signalspannung
signal wiring list DOC Signalverbindungsliste
signatory (→signer) OFFICE (→Unterzeichner)
signature RADIO REL Signatur
signature MICROEL Signatur
signature OFFICE Unterschrift
= signing (act) = Unterzeichnung
signature analysis INSTR Signaturanalyse
= signature diagnosis = Signaturdiagnose
signature analyzer INSTR Signatur-Analysator
signature diagnosis (→signature INSTR (→Signaturanalyse)
  analysis)
signature portfolio OFFICE Unterschriftsmappe
signature reader TERM&PER Unterschriftenleser
= signature verification system = Unterschriften-Verifikationsgerät
signature register MICROEL Signaturregister
signature verification system TERM&PER (→Unterschriftenleser)
  (→signature reader)
sign bit CODING Vorzeichenbit
sign digit CODING Vorzeichenziffer
signed MATH vorzeichenbehaftet
  = mit Vorzeichen
signed ECON gezeichnet
  [corrspondence] [Brief]
= sgd. = gez.
signer OFFICE Unterzeichner
= signatory = Unterzeichnete; Signatar
sign extension DATA PROC Vorzeichenerweiterung
sign flag DATA PROC Vorzeichenmerker
significance CODING Stellenwert
= place value = Stellenwertigkeit; Signifikanz
significance INF Wertigkeit
= signification; order; priority; weight = Wichtung; Priorität; Signifikanz
significance level MATH Signifikanzzahl
  [statistics] [Statistik]
  = Signifikanzstufe; Signifikanzniveau
significance test MATH Signifikanztest
significant SCIE signifikant
  = bezeichnend; erheblich; bedeutungsvoll
significant condition TELEGR Kennzustand
= significant state
significant instant TELEGR Kennzeitpunkt
  = Schritteinsatz
significant interval (→unit TELEGR (→Zeichenelement)
  interval)
significant state (→significant TELEGR (→Kennzustand)
  condition)
signification (→significance) INF (→Wertigkeit)
signing (act) (→signature) OFFICE (→Unterschrift)
signing-on (→sign-on DATA PROC (→Bereitschaftsmeldung)
  message)
sign inversion MATH Vorzeichenumkehr
sign-on (→sign-on DATA PROC (→Bereitschaftsmeldung)
  message)
sign-on message DATA PROC Bereitschaftsmeldung
= signing-on; sign-on [Mikroprozessor]
sign position DATA PROC Vorzeichenstelle
signum (→sign) MATH (→Vorzeichen)
silane CHEM Silan
silane pyrolisis CHEM Sylan-Pyrolyse
silent discharge PHYS stille Entladung
silent partner (AM) ECON stiller Gesellschafter
= dormant partner; sleeping partner; inactive partner = stiller Teilhaber
silent period (→radio RADIO (→Funkstille)
  silence)
silent reversal TELEC Weichumpolung
≈ soft keying [MODUL] ≈ Weichtastung [MODUL]
silent zone (→dead RADIO PROP (→tote Zone)
  zone)

silica		CHEM **Siliziumdioxyd**
= silicon dioxide		= Siliziumdioxid; Siliciumdioxid
↓ quartz		↓ Quarz
silica fiber	OPT.COMM	**Quarzfaser**
silica gel	CHEM	**Silikagel**
silica glass	CHEM	**Quarzglas**
		= Kieselglas
		↑ Siliziumdioxyd
silicide	CHEM	**Silizid**
silicon carbide	CHEM	**Siliziumkarbid**
silicon	CHEM	**Silizium**
= Si		= Silicium; Si
silicon chip (→chip)	MICROEL	(→Chip)
silicon controlled rectifier	MICROEL	(→Thyristor)
(→thyristor)		
silicon crystal	PHYS	**Siliziumkristall**
silicon die (→chip)	MICROEL	(→Chip)
silicon diode	COMPON	**Siliziumdiode**
silicon dioxide (→silica)	CHEM	(→Siliziumdioxyd)
silicone	CHEM	**Silikon**
		= Silicon
silicon foundry	MICROEL	**Chip-Hersteller**
= foundry		
silicon-gate technology	MICROEL	**Siliziumgattertechnologie**
		= Silizium-Steuerelektronen-Technologie; Silizium-gate-Technologie
silicon-gate transistor	MICROEL	**Silizium-gate-Transistor**
silicon island	MICROEL	**Siliziuminsel**
silicon layer	MICROEL	**Siliziumschicht**
silicon nitride	CHEM	**Siliziumnitrid**
silicon nitride passivation	MICROEL	**Siliziumnitrid-Passivierung**
silicon slice (→wafer crystal)	MICROEL	(→Kristallscheibe)
silicon solar cell	MICROEL	**Silizium-Solarzelle**
silicon transistor	MICROEL	**Siliziumtransistor**
silicon wafer (→wafer crystal)	MICROEL	(→Kristallscheibe)
silk-screen printing	TYPOGR	**Siebdruck**
= screen printing; serigraphy		
silver	CHEM	**Silber**
= Ag		= Ag; Argentum
silver (v.t.) (→silver-plate)	METAL	(→versilbern)
silver-clad (→silver-plate)	METAL	(→silberplattieren)
silver-clad (v.t.) (→silver-plate)	METAL	(→versilbern)
silver-cladded (→silver-plated)	METAL	(→versilbert)
silver cladding (→silver plating)	METAL	(→Versilberung)
silver contact	COMPON	**Silberkontakt**
= silver-plated contact		
silvered (→silver-plated)	METAL	(→versilbert)
silvering (→silver plating)	METAL	(→Versilberung)
silverly	METAL	**silbrig**
= silvery		≈ silbern; versilbert
≈ silver (adj.); silver-plated		
silver-plate (v.t.)	METAL	**versilbern**
= silver-clad (v.t.); silver (v.t.)		= silberplattieren; silberkontaktieren [COMPON]
↑ galvanize		↑ galvanisieren
silver-plated	METAL	**versilbert**
= silver-cladded; silvered		= silberplattiert
		≈ silberkontaktiert [COMPON]
silver-plated contact (→silver contact)	COMPON	(→Silberkontakt)
silver plating	METAL	**Versilberung**
= silver cladding; silvering		= Silberplattierung
		≈ Silberkontaktierung [COMPON]
silver-solder	METAL	**silberlöten**
silver solder	METAL	**Silberlot**
silver soldering	METAL	**Silberlötung**
silvery (→silverly)	METAL	(→silbrig)
SIMD processor	DATA PROC	**SIMD-Prozessor**
[single instruction, multiple data]		[bearbeitet mit einer Programmanweisung mehrere Datenfelder gleichzeitig]
↑ array processor		↑ Feldrechner
similar	COLLOQ	**ähnlich**
≈ comparable; equal-type		≈ vergleichbar; gleichartig
similarity	COLLOQ	**Ähnlichkeit**
≈ comparibility		≈ Vergleichbarkeit; Artgleichheit
similarity	MATH	**Ähnlichkeit**
		= geometrische Verwandschaft
similarity theorem	MATH	**Ähnlichkeitssatz**
similarity transform (→similarity transformation)	MATH	(→Ähnlichkeitstransformation)
similarity transformation	MATH	**Ähnlichkeitstransformation**
= similarity transform		
similar sign (→equivalent sign)	MATH	(→Ähnlichzeichen)
SIMOS	MICROEL	**SIMOS**
[stacked gate injection MOS]		
simple	TECH	**einfach**
≈ elementary; simplified		≈ elementar; vereinfacht
simple-acting (adj.)	TECH	**einfachwirkend**
= simple-action (adj.); single-acting; single-action (adj.)		
simple-action (adj.)	TECH	(→einfachwirkend)
(→simple-acting)		
simple buffering	DATA PROC	**Einfachpufferung**
simple chain (→simple chaining)	DATA PROC	(→Einfachkettung)
simple chaining	DATA PROC	**Einfachkettung**
= simple chain; simple concatenation		= Einfachverkettung
simple concatenation (→simple chaining)	DATA PROC	(→Einfachkettung)
simple past	LING	**Imperfekt**
[e.g. I heard, I was heard]		[z.B. ich hörte, ich wurde gehört]
≈ preterite; preterit; imperfect		= Präteritum; 1. Vergangenheit; erste Vergangenheit
↑ past tense		↑ Vergangenheit
simple-to-operate (→easy-to-operate)	TECH	(→betriebsfreundlich)
simplex	TELEC	**Simplexbetrieb**
[communication in one direction only]		[Übermittlung in nur einer Richtung]
= simplex operation; simplex transmission; simplex mode; one-way operation		= Richtungsbetrieb; Einzelbetrieb; einseitiger Betrieb; einseitige Übertragung
		↓ Sendebetrieb; Empfangsbetrieb
simplex data communication (→one-way communication)	DATA COMM	(→einseitige Datenübermittlung)
simplex intercommunication system	TELEPH	**Wechselsprechanlage**
= simplex mode intercom; talk-through facility; two-way telephone system		↑ Haussprechanlage
↑ intercommunication system		
simplex mode (→simplex)	TELEC	(→Simplexbetrieb)
simplex mode intercom (→simplex intercommunication system)	TELEPH	(→Wechselsprechanlage)
simplex operation	TELEPH	**Wechselsprechen**
simplex operation [TELEPH]	TELEC	(→Halbduplexbetrieb)
(→half duplex)		
simplex operation (→simplex)	TELEC	(→Simplexbetrieb)
simplex transmission (→simplex)	TELEC	(→Simplexbetrieb)
simplification	TECH	**Vereinfachung**
≈ unification		≈ Vereinheitlichung
≠ complication		≠ Komplizierung
simplified calculus (→approximation calculus)	MATH	(→Näherungsrechnung)
simplified drawing	ENG.DRAW	**Maßbild**
simplified equation (→approximation equation)	MATH	(→Näherungsgleichung)
simplified formula (→approximation formula)	MATH	(→Näherungsformel)
simplify	COLLOQ	**vereinfachen**
≠ complicate		≠ komplizieren
simulate	TELEC	**nachbilden**
simulate	DATA PROC	**simulieren**
[to imitate a real situation or a machine]		[einen reellen Vorgang oder eine Maschine nachahmen]
≈ emulate		≈ emulieren

simulation

simulation	DATA PROC	**Simulation**
		= Nachbilden
simulation (→copy)	TECH	(→Nachbildung)
simulation circuit	TELEC	(→Leitungsnachbildung)
(→line-balancing network)		
simulation device	MICROEL	**Bauelemente-Simulation**
simulation network	CIRC.ENG	**Nachbildung**
= balancing circuit		
simulation network	TELEC	(→Leitungsnachbildung)
(→line-balancing network)		
simulation program	DATA PROC	**Simulierer**
[allows to run a program written in one machine language on another type of computer]		[Programm welches den Ablauf eines Programm einer Maschinensprache auf einem anderen Anlagenmodell ermöglicht]
↑ middleware		= Simulationsprogramm
		↑ Middleware
simulation programming language	DATA PROC	**Simulations-Programmiersprache**
simulation software	DATA PROC	**Simulationssoftware**
simulation technique	DATA PROC	**Simulationstechnik**
= numerical simulation; computer-aided simulation		= numerische Simulation
simulator	DATA PROC	**Simulator**
		[Programm zur Nachbildung eines Vorgangs]
simultaneity	COLLOQ	**Gleichzeitigkeit**
≈ contemporaneity		= Simultaneität; Simultanität
simultaneous	COLLOQ	**gleichzeitig**
≈ concurrent 1		= simultan
		≈ zusammenfallend
simultaneous computer	DATA PROC	**Parallelrechner**
= parallel computer		= Parallelcomputer; Nicht-von-Neumann-Rechner
simultaneous data acquisition	DATA PROC	**simultane Datenerfassung**
simultaneous data processing	DATA PROC	(→Simultanverarbeitung)
(→simultaneous processing)		
simultaneous input/output	DATA PROC	**simultane Ein-/Ausgabe**
simultaneous mode	DATA PROC	(→Simultanverarbeitung)
(→simultaneous processing)		
simultaneous operation	DATA PROC	**Simultanbetrieb 1**
[of various peripherals by one computer]		[gleichzeitiges Aussteuern mehrerer Peripheriegeräte durch einen Rechner]
≈ multiprogramming		= Simultanarbeit 1
		≈ Mehrprogrammbetrieb
simultaneous processing	DATA PROC	**Simultanverarbeitung**
[performance of several tasks at the same instant]		[simultane Bearbeitung mehrerer Programme]
= simultaneous mode; simultaneous data processing		= Simultanbetrieb; simultane Verarbeitung; simultane Datenverarbeitung
≠ concurrent processing		≠ verzahnte Verarbeitung
↑ parallel processing		↑ Parallelbetrieb
simultaneous transmitter	TELEGR	**Schrittsender**
sin² pulse (→sine-squared pulse)	ELECTRON	(→sin²-Impuls)
Sincerely (AM) (→Very sincerely yours)	ECON	(→Mit freundlichen Grüßen Ihr)
Sincerely yours (AM)	ECON	**Hochachtungsvoll**
[neutral complimentary close]		[neutraler Briefschluß]
= Yours faithfully (BRI); Yours very truly (BRI); Yours truly (BRI); Very truly yours (AM)		= Mit vorzüglicher Hochachtung; Mit freundlichen Grüßen
≈ Yours sincerely		≈ Mit freundlichen Grüßen Ihr
sinc function	MATH	**Si-Funktion**
		= sinx/x
sine	MATH	**Sinus**
↑ trigonometric function		↑ trigonometrische Funktion
sine curve	MATH	**Sinuskurve**
sine die	ECON	**auf unbestimmte Zeit**
sine function	MATH	**Sinusfunktion**
= sinusoidal function		
sine-function network	NETW.TH	**Sinusfunktionsnetzwerk**
sine generator (→sine-wave generator)	CIRC.ENG	(→Sinusgenerator)
sine mode	ELECTRON	**Sinusbetrieb**
sine-shaped (→sinusoidal)	MATH	(→sinusförmig)
sine source (→sine-wave generator)	CIRC.ENG	(→Sinusgenerator)
sine-squared pulse	ELECTRON	**sin²-Impuls**
= sin² pulse		= Glockenimpuls
sine/square source	CIRC.ENG	(→Sinus-Rechteckgenerator)
(→rectangular sine wave generator)		
sine-wave (→sinusoidal)	MATH	(→sinusförmig)
sinewave (→sinusoidal wave)	PHYS	(→Sinuswelle)
sinewave characteristics	ELECTRON	**Sinus-Kenndaten**
		= Sinus-Eigenschaften
sine-wave generator	CIRC.ENG	**Sinusgenerator**
= harmonic oscillator; harmonics oscillator; sine source; sine generator		= Sinusoszillator; Sinusquelle
		↑ Oszillator
↑ oscillator		
sing	CIRC.ENG	**pfeifen**
singing	CIRC.ENG	**Pfeifen**
singing frequency	CIRC.ENG	**Pfeiffrequenz**
= singing point frequency		
singing frequency (→ringing frequency)	ELECTRON	(→Überschwingfrequenz)
singing interference	TELEC	**Pfeifstörung**
singing margin	CIRC.ENG	**Pfeifabstand**
		= Pfeifsicherheit
singing point	CIRC.ENG	**Pfeifpunkt**
singing point frequency	CIRC.ENG	(→Pfeiffrequenz)
(→singing frequency)		
singing-point method	INSTR	**Pfeifpunktverfahren**
single (→unprotected)	TELEC	(→ungeschützt)
single acquisition terminal	DATA COMM	**Einzelerfassungsplatz**
single-acting (→simple-acting)	TECH	(→einfachwirkend)
single-action (adj.)	TECH	(→einfachwirkend)
(→simple-acting)		
single-address computer	DATA PROC	**Ein-Adreß-Computer**
= one-address computer; single-address machine; one-address machine		= Ein-Adreß-Maschine
single-address instruction	DATA PROC	(→Ein-Adreß-Befehl)
(→one-address instruction)		
single-address machine	DATA PROC	(→Ein-Adreß-Computer)
(→single-address computer)		
single antenna system	BROADC	**Einzelantennenanlage**
[for one household]		[für einen Haushalt]
		= EA
single-band antenna (→single-range antenna)	ANT	(→Einbandantenne)
single bin cut-sheet feeder	TERM&PER	(→Einzelblattzuführung)
(→single-sheet feed)		
single board	ELECTRON	**Einfachverdrahtungsplatte**
[PCB]		[Leiterplatte]
single-board	EQUIP.ENG	(→Einkarten-)
(→single-module)		
single-board computer	DATA PROC	**Einkartenrechner**
= SBC; board computer		= Einplatinenrechner; Einkartencomputer; Einplatinencomputer
single-board equipment	EQUIP.ENG	(→Einplattengerät)
(→single-board unit)		
single-board unit	EQUIP.ENG	**Einplattengerät**
= single-board equipment		
single branch point	BROADC	**Einfachabzweiger**
single-carriage (adj.)	TERM&PER	**einbahnig**
[printer]		[Drucker]
single-carrier per channel	MODUL	**Einkanalträger**
= SCPC		= SCPC
single-carrier-per-channel multiple access	SAT.COMM	**SCPC-Mehrfachzugriff**
= SCPC		
single-cavity magnetron	MICROW	**Einkammer-Magnetron**
single-cavity v.m. tube	MICROW	**Einkreis-Triftröhre**
single-channel ...	TELEC	**Einkanal-**
		= Einzelkanal-
single-channel antenna	ANT	**Einkanalantenne**
single-channel codec	TRANS	**Einkanalcodec**
single-channel direction finder	RADIO LOC	**Einkanalfunkpeiler**
		= Einkanalpeiler
single-channel modulation	MODUL	**Einzelkanalmodulation**
single-channel technique	DATA COMM	**Einkanatechnik**
single-channel transmitter	RADIO	**Einkanal-Sender**
single-chip computer	DATA PROC	(→Ein-Chip-Mikrocomputer)
(→single-chip microcomputer)		

single-chip computer MICROEL (→Mikrocomputer)
(→microcomputer)
single-chip microcomputer DATA PROC **Ein-Chip-Mikrocomputer**
= single-chip computer; computer-on-a-chip; microcomputer chip; one-chip computer
= Ein-Chip-Computer; Ein-Chip-Rechner
single-chip processor MICROEL **Ein-Chip-Prozessor**
single-chip solution CIRC.ENG **Ein-Chip-Lösung**
single-chip technology MICROEL **Monochiptechnik**
single-chute feed device TERM&PER **Einschachteinzug**
single-clear tab TERM&PER (→Tabulator-Einzellöscher)
(→single-clear tabulator)
single-clear tabulator TERM&PER **Tabulator-Einzellöscher**
= single-clear tab = Tab-Einzellöscher
single-clock cell MICROEL **Eintaktzelle**
single command TELECONTR **Einzelbefehl**
single-conductor cord EQUIP.ENG **einadrige Schnur**
single control CONTROL **einfachregelung**
single controller CONTROL **Einfachregler**
single conversion superhet RADIO **Einfachsuperhet**
[with one IF] [mit einer ZF]
↑ heterodyne receiver ↑ Überlagerungsempfänger
single-core (→single-wired) EL.TECH (→einadrig)
single-core cable POWER ENG **Einleiterkabel**
single crystal PHYS **Einkristall**
= monocrystal
single current TELEGR **Einfachstrom**
= neutral current (AM) ≠ Doppelstrom
≠ double current
single-current keying TELEGR **Einfachstromtastung**
single current test transmitter TELEGR **Einfachstromwechselsender**
= reversals transmitter
single current transmission TELEGR (→Einfachstrombetrieb)
(→single current working)
single current working TELEGR **Einfachstrombetrieb**
= single current transmission
single density TERM&PER **einfache Schreibdichte**
[Diskette]
single-density diskette (→SD TERM&PER (→SD-Diskette)
diskette)
single-disk clutch MECH **Einscheibenkupplung**
single document TERM&PER **Einzelbeleg**
= single voucher ↑ Einzelformular; Beleg
↑ single form; document
single-document advance TERM&PER (→Einzelbelegzuführung)
(→single-document feed)
single-document feed TERM&PER **Einzelbelegzuführung**
= single-voucher feed; single-document transport; single-voucher transport; single-document advance; single-voucher advance
= Einzelbelegtransport; Einzelbelegeingabe; Einzelbelegeinzug; Einzelbelegvorschub
↑ single-form feed ↑ Einzelformularzuführung
single-document transport TERM&PER (→Einzelbelegzuführung)
(→single-document feed)
single drive (→individual MECH (→Einzelantrieb)
drive)
single-duct conduit OUTS.PLANT **Einrohrkanal**
≈ cable conduit ≈ Einröhrenkanal
≠ multiple-duct conduit ≈ Kabelkanalzug
≠ Mehrrohrkanal
single-ended DC converter POWER ENG (→Eintaktgleichspannungswandler)
(→single-phase DC converter)
single-ended forward converter POWER ENG (→Eintaktdurchflußwandler)
verter (→single-phase feed-forward converter)
single-fiber cord OPT.COMM **Ein-Faser-Innenkabel**
single form TERM&PER **Einzelformular**
= single-sheet form = Einzelvordruck
≠ continuous form ≠ Endlosformular
↑ form ↑ Formular
↓ single document ↓ Einzelbeleg
single-form advance TERM&PER (→Einzelformularzuführung)
(→single-form feed)
single-form drive TERM&PER (→Einzelformularzuführung)
(→single-form feed)
single-form feed TERM&PER **Einzelformularzuführung**
= single-form transport; single-form feeding; single-form advance; single-form drive
= Einzelvordruckzuführung; Einzelformulartransport; Einzelvordrucktransport;
≈ single-sheet feed
≠ continuous form feed
↑ form feed
↓ single-document feed
Einzelformulareingabe; Einzelvordruckeingabe; Einzelformulareinzug; Einzelvordruckeinzug; Einzelformularvorschub; Einzelvordruckvorschub
≈ Einzelblattzuführung
≠ Endlosformularzuführung
↑ Formularzuführung
↓ Einzelbelegzuführung
single-form feeding TERM&PER (→Einzelformularzuführung)
(→single-form feed)
single-form transport TERM&PER (→Einzelformularzuführung)
(→single-form feed)
single-height module EQUIP.ENG (→einzeilige Baugruppe)
(→single-row module)
single-height subassembly EQUIP.ENG (→einzeilige Baugruppe)
(→single-row module)
single-in-line package MICROEL **SIP-Gehäuse**
= SIP
single-input mixing (→additive HF (→additive Mischung)
mixing)
single-key keyboard TERM&PER **Einzeltasten-Tastatur**
≠ membrane keyboard ≠ Folientastatur
single-key set-up INSTR **Einknopfbedienung**
single-length arithmetic DATA PROC (→einfachgenaue Arithmetik)
(→single-precision arithmetic)
single lens ELECTRON **Einzellinse**
single line (→individual line) TELEC (→Einzelanschluß)
single line feed (→single line TERM&PER (→einfacher Zeilenabstand)
spacing)
single line spacing TERM&PER **einfacher Zeilenabstand**
= single line feed = einfacher Zeilenvorschub
single link DATA COMM **Einfach-Übermittlungsabschnitt**
single metering SWITCH **Einfachzählung**
= unit-fee metering
single-mode fiber OPT.COMM **Einmodenfaser**
= single-mode optical waveguide; singlemode fibre (BRI); monomode fiber; monomode optical waveguide
= Monomode-Faser; Singlemode-Faser; SM-Faser
singlemode fibre (BRI) OPT.COMM (→Einmodenfaser)
(→single-mode fiber)
single-mode optical waveguide OPT.COMM (→Einmodenfaser)
(→single-mode fiber)
single-module EQUIP.ENG **Einkarten-**
= single-board
single-page data print TERM&PER **Einzelblattdatendruck**
single-page processing TERM&PER **Einzelblattverarbeitung**
single part (→component TECH (→Einzelteil)
part)
single-part production (→unit MANUF (→Einzelanfertigung)
production)
single-path routing TELEC **Einwegführung**
≠ path diversity ≠ Mehrwegeführung
single pattern recognition INSTR **Einzelsignal-Mustererkennung**
single-petticoat insulator OUTS.PLANT **Einzelglocke**
= single shed insulator [Freileitung]
single-phase PHYS **einphasig**
= one-phase
single-phase bridge CIRC.ENG **Einphasenbrückenschaltung**
single-phase clocked system MICROEL **Einphasentaktsystem**
single-phase DC converter POWER ENG **Eintaktgleichspannungswandler**
= single-ended DC converter
single-phase feed-forward converter POWER ENG **Eintaktdurchflußwandler**
= single-ended forward converter
single-phase midpoint POWER SYS (→Zweipuls-Mittelpunktschaltung)
connection (→two-pulse midpoint connection)
single-phase transformer POWER ENG **Einphasentransformator**
single-point information TELECONTR (→Einzelmeldung)
(→individual indication)
single-polarization aerial ANT (→einfach polarisierte Antenne)
(→single-polarization antenna)
single-polarization antenna ANT **einfach polarisierte Antenne**
= single-polarization aerial
single-polarized PHYS **einfach polarisiert**
≠ double-polarized ≠ doppelt polarisiert

single-pole

English	Domain	German
single-pole	EL.TECH	**einpolig**
= monopole; unipolar		
single-pole double throw switch	MICROEL	**einpoliger Umschalter**
= SPDT		
single precision	DATA PROC	**Einfachgenauigkeit**
= short precision		= einfache Genauigkeit
single-precision arithmetic	DATA PROC	**einfachgenaue Arithmetik**
= single-length arithmetic		
single-programming	DATA PROC	**Einprogrammbetrieb**
= single tasking		= Einprogrammverarbeitung
≠ multi-programming		≠ Mehrprogrammbetrieb
single-purpose (adj.)	TECH	**Einzweck-**
single-purpose register	DATA PROC	**Einzweckregister**
single-purpose terminal	TERM&PER	(→Spezialendgerät)
(→applications terminal)		
single-quadrant multiplier	COMPON	**Einquadranten-Multiplikator**
single quenching	EL.TECH	**Einzellöschung**
single quotation mark	LING	(→Apostroph)
(→apostrophe)		
single-range antenna	ANT	**Einbandantenne**
= single-band antenna; monoband antenna; monobander		= Monobandantenne
single rapid decollation set	TERM&PER	**Einzeltrennsatz**
single-row ...	EQUIP.ENG	**einzeilig**
single-row module	EQUIP.ENG	**einzeilige Baugruppe**
= single-row subassembly; single-height module; single-height subassembly		
single-row subassembly	EQUIP.ENG	(→einzeilige Baugruppe)
(→single-row module)		
single-service network	TELEC	**Monodienstnetz**
single shed insulator	OUTS.PLANT	(→Einzelglocke)
(→single-petticoat insulator)		
single-sheet feed	TERM&PER	**Einzelblattzuführung**
= single-sheet feeding; single-sheet transport; bill feed; cut sheet feeding; single bin cut-sheet feeder; paper input tray; single-sheet insertion; sheet feeder; sheet feed		= Einzelblattzufuhr; Einzelblattransport; Einzelblatteingabe; Einzelblatteinzug; Einzelblattanlage; Einzelblattschacht; Einzelschacht
≈ single-form feed		≈ Einzelformularzuführung
↑ paper feed		↑ Papiervorschub
↓ single-form feed; cut-sheet feeder		↓ Einzelformularzuführung; Einzelblattschacht
single-sheet feeding	TERM&PER	(→Einzelblattzuführung)
(→single-sheet feed)		
single-sheet form (→single form)	TERM&PER	(→Einzelformular)
single-sheet insertion	TERM&PER	(→Einzelblattzuführung)
(→single-sheet feed)		
single-sheet reader	TERM&PER	**Einzelblattbelegleser**
single-sheet transport	TERM&PER	(→Einzelblattzuführung)
(→single-sheet feed)		
single shot	CODING	**Einzelfehler**
single-shot analysis	INSTR	**Single-shot-Analyse**
single sideband	MODUL	**Einseitenband**
= SSB		
single-sideband measurement	INSTR	**Einseitenbandmessung**
single-sideband mixer	MODUL	**Einseitenbandmischer**
single sideband modulation	MODUL	**Einseitenband-Modulation**
single sideband modulator	MODUL	**Einseitenbandumsetzer**
single-sideband system	RADIO	**Einseitenbandsystem**
= SSB system		= SSB-System
single-sideband transmission	TELEC	**Einseitenbandübertragung**
single-sided	TECH	**einseitig**
= unilateral		
single-sided disc (→SS disk)	TERM&PER	(→SS-Diskette)
single-sided disk (→SS disk)	TERM&PER	(→SS-Diskette)
single-sided diskette (→SS disk)	TERM&PER	(→SS-Diskette)
single-sided drive	TERM&PER	**einseitiges Laufwerk**
single-sided floppy disc (→SS disk)	TERM&PER	(→SS-Diskette)
single-sided floppy disk (→SS disk)	TERM&PER	(→SS-Diskette)
single-slope converter (→sawtooth converter)	CIRC.ENG	(→Sägezahn-Umsetzer)
single-slot antenna	ANT	**Einschlitzstrahler**
single-span system (→daisy-chain cable)	COMM.CABLE	(→Girlandenkabel)
single-stage connecting network (→single-stage switching network)	SWITCH	(→einstufige Koppelanordnung)
single-stage switching network	SWITCH	**einstufige Koppelanordnung**
= single-stage connecting network		
single station	TERM&PER	**Einzelstation**
single step (n.)	DATA PROC	**Einzelschritt**
single-step (→one-level)	TECH	(→einstufig)
single-step control (→step-by-step control)	SWITCH	(→Einzelschrittsteuerung)
single-step mode	DATA PROC	**Einzelschrittmodus**
single-step operation	DATA PROC	**Einzelschrittverarbeitung**
single-stroke bell	EL.ACOUS	**Einschlagwecker**
= gong		= Gong
single tasking (→single-programming)	DATA PROC	(→Einprogrammbetrieb)
single-terminal cash register	TERM&PER	**Einplatz-Registrierkasse**
single-terminal microcomputer	DATA PROC	**Einplatz-Mikrocomputer**
		= Einplatz-Mikrorechner
single-terminal minicomputer	DATA PROC	**Einplatz-Minicomputer**
		= Einplatz-Minirechner
single-terminal operation (→single-terminal service)	TERM&PER	(→Einzelplatzbedienung)
single-terminal service	TERM&PER	**Einzelplatzbedienung**
= single-terminal operation		
single-terminal system (→single-user system)	DATA PROC	(→Einplatzsystem)
single thread	MECH	**eingängiges Gewinde**
singleton	MATH	**Einermenge**
single tone operation	TELEGR	**Einfachtonbetrieb**
= on-off keying		
single-track recording	TERM&PER	**Einspurtechnik**
		[Magnetspeicher]
single-user system	DATA PROC	**Einplatzsystem**
= single-terminal system; stand-alone system		[nur für einen Benutzer]
≠ multi-user system		≠ Mehrplatzsystem
single voucher (→single document)	TERM&PER	(→Einzelbeleg)
single-voucher advance (→single-document feed)	TERM&PER	(→Einzelbelegzuführung)
single-voucher feed (→single-document feed)	TERM&PER	(→Einzelbelegzuführung)
single-voucher transport (→single-document feed)	TERM&PER	(→Einzelbelegzuführung)
single-way rectifier (→half-wave rectifier)	POWER SYS	(→Einweggleichrichter)
single-wire (→surface wave transmission line)	LINE TH	(→Oberflächenwellenleitung)
single-wire antenna	ANT	**Eindrahtantenne**
		= Einleiterantenne
single-wired	EL.TECH	**einadrig**
= unifilar; single-core		= eindrähtig
single-wire line	OUTS.PLANT	**Eindrahtleitung**
single-word instruction	DATA PROC	**Ein-Wort-Befehl**
single-word recognition system	TERM&PER	**Einzelwort-Erkennungssystem**
singular (adj.)	MATH	**singulär**
singular	LING	**Singular**
		= Einzahl
singular integral	MATH	**singuläres Integral**
singularity	MATH	**Singularität**
= singular point		= singulärer Punkt; singuläre Stelle
↓ point of intersection; isolated point; inversive point; break point; asymptotic point		↑ ausgezeichneter Punkt
		↓ Doppelpunkt; isolierter Punkt; Rückkehrpunkt; Knickpunkt; asymptotischer Punkt
singular point (→singularity)	MATH	(→Singularität)
singular signal	INF	**singuläres Signal**
singular surface	MATH	**Unstetigkeitsfläche**
sink (v.t.)	TECH	**versenken**
sink (→drain)	INF	(→Senke)
sinor (→rotating phasor)	EL.TECH	(→Wechselstromzeiger)
sinter (v.t.)	METAL	**sintern**
sintering	METAL	**Sinterung**
		= Sintern

sintonize		HF	**abstimmen**	**size** (v.t.)	MECH	**kalibrieren**
[to optimaze a frequency setting]			[eine Frequenzeinstellung optimieren]	[to give exact size]		[auf exaktes Maß bringen]
= tune				**size** (n.)	TECH	**Größe**
sintonized (→tuned)		CIRC.ENG	(→abgestimmt)	**size** (n.) (→measure)	PHYS	(→Maß)
sinuous		TECH	**geschlängelt**	**size consistency** (→size permanency)	TECH	(→Maßbeständigkeit)
= tortuous; winding				**size permanency**	TECH	**Maßbeständigkeit**
sinuous (→sinusoidal)		MATH	(→sinusförmig)	= accuracy to size; size consistency		≈ Maßhaltigkeit
sinusoidal		MATH	**sinusförmig**	**sizing**	METAL	**Feinziehen**
= sine-wave; sine-shaped; sinuous			= Sinus-	**sizing**	MECH	**Kalibrierung**
sinusoidal frequency shift keying		MODUL	**SFSK**	**sizing** (→dimensioning)	TECH	(→Bemessung)
= sinusoidal FSK				**sizing die**	MECH	**Kalibrierwerkzeug**
sinusoidal FSK (→sinusoidal frequency shift keying)		MODUL	(→SFSK)	**skeletal code** (→code skeleton)	DATA PROC	(→Programmgerippe)
sinusoidal function (→sine function)		MATH	(→Sinusfunktion)	**skeleton**	TECH	**Gerippe**
				≈ framework		≈ Gerüst
sinusoidal oscillation		PHYS	**Sinusschwingung**	**skeleton slot antenna**	ANT	**Skelettschlitz-Antenne**
			= harmonische Schwingung 2; Sinusvorgang			= Skelettschlitzstrahler
				skelp	METAL	**Röhrenstreifen**
sinusoidal signal technique		ELECTRON	**Sinustechnik**	**sketch** (v.t.)	ENG.DRAW	**skizzieren**
[generation, processing and transfer of sinusoidal signals]			[Erzeugung, Verarbeitung und Übertragung von sinusförmigen Signalen]	≈ draft (v.t.); draw		≈ entwerfen; zeichnen
				sketch (n.)	ENG.DRAW	**Skizze**
≠ pulse technique			≠ Impulstechnik	≈ draft (n.); schematic representation		≈ Entwurf; schematische Darstellung
sinusoidal wave		PHYS	**Sinuswelle**	**sketch pad** (→note pad)	OFFICE	(→Notizblock)
= sinewave				**sketch pad** (→scratch pad memory)	DATA PROC	(→Notizblockspeicher)
SIP (→single-in-line package)		MICROEL	(→SIP-Gehäuse)	**skew** (n.)	TERM&PER	**Bitversatz**
SIPO		DATA PROC	(→SIPO)	[magnetic tape]		[Magnetband]
(→serial-in/parallel-out)				**skew 3** (n.)	ELECTRON	**Schrägverzerrung**
S/I ratio (→signal-to-interence ratio)		TELEC	(→Störabstand)	**skew** (n.)	TECH	**Schräglauf**
				skew 1 (n.)	ELECTRON	**Zeitversatz**
siren (pl.-ns)		EL.ACOUS	**Sirene**	[difference in propagation time]		= Laufzeitunterschied
SISO		DATA PROC	(→SISO)			
(→serial-in/serial-out)				**skew** (adj.)	MATH	**windschief**
SIS transmission (→sound-in-syncs transmission)		TV	(→SIS-Übertragung)	[not in a same plane]		[nicht in einer Ebene liegend]
sit (v.i.) (→meet)		ECON	(→tagen)	**skew** (adj.)	TECH	**schief 1**
site		TELEC	**Standort**	[deviating from vertical]		[von vertikaler Richtung abweichend]
= premises; position			= Gelände	≈ warped; inclined		
site		ECON	**Standort**	↑ slant		≈ windschief; geneigt
= location; ubication			= Stelle			↑ schräg
≈ premises			≈ Gelände	**skew 1** (n.) (→pulse separation)	ELECTRON	(→Impulspause)
			↓ Fertigungsstandort			
site determination		ECON	**Standortbestimmung**	**skew** (→tape skew)	TERM&PER	(→Bandschräglauf)
site engineer (→works manager)		TECH	(→Bauleiter)	**skew-angled**	MATH	**schiefwinklig**
site selection		ECON	**Standortwahl**	**skew-angled parallel coordinates**	MATH	**schiefwinklige Parallelkoordinaten**
site shielding		RADIO PROP	**Geländeabschirmung**			
site survey		SYS.INST	**Ortsbegehung**	**skew flipflop**	CIRC.ENG	**Skew-Flipflop**
= field survey; site visit; survey; visit			= Ortsbesichtigung; Begehung; Besichtigung; Survey; Ortsauskundung	**skewness**	MATH	**Schiefe**
↓ station survey; path survey				≈ assymmetry		≈ Asymmetrie
				skew symmetric	MATH	**schiefsymmetrisch**
			↓ Stationsbegehung; Streckenbegehung	**skiatron** (→dark-trace tube)	ELECTRON	(→Dunkelschriftröhre)
site test (→station test)		SYS.INST	(→Stationsmessung)	**skilful** (BRI) (→skillful)	TECH	(→kunstfertig)
site visit (→site survey)		SYS.INST	(→Ortsbegehung)	**skill**	TECH	**Kunstfertigkeit**
sitting (→session)		ECON	(→Sitzung)	**skilled** (→experienced)	SCIE	(→sachkundig)
SI unit		PHYS	**SI-Einheit**	**skilled crafts**	TECH	**Handwerk**
[International System of Units]			[internationales Einheitssystem]	**skilled personnel** (→specialized personnel)	ECON	(→Fachpersonal)
Si vidicon		ELECTRON	**Si-Vidikon**	**skilled worker**	ECON	**Facharbeiter**
six-channel ...		TRANS	**Sechskanal-**	= specialized worker		
six-circuit ...		NETW.TH	**Sechskreis-**	**skillful** (NAM)	TECH	**kunstfertig**
six-circuit filter		NETW.TH	**Sechskreisfilter**	= skilful (BRI)		
sixfold		COLLOQ	**sechsfach**	**skin depth**	RADIO PROP	**Eindringtiefe**
= sextuple				**skin depth** (→penetration depth)	TECH	(→Eindringtiefe)
six-layer ...		TECH	**sechslagig**	**skin effect**	EL.TECH	**Skineffekt**
= six-part			= sechsschichtig			= Hauteffekt
six-part (→six-layer)		TECH	(→sechslagig)	**skinning tool** (→wire-end stripper)	EL.TECH	(→Abisolierzange)
six-pulse bridge		POWER ENG	**Sechspuls-Brückenschaltung**	**skip** (v.t.)	DATA PROC	**überspringen**
≈ three-phase bridge			≈ Drehstrom-Brückenschaltung	[e.g. an instruction]		[z.B. eine Anweisung]
				= ignore		≈ überlesen; übergehen; ignorieren
six-pulse mid-point circuit		POWER ENG	**Sechspuls-Mittelpunkt-Schaltung**			
≈ double-star connection			≈ Doppelsternschaltung	**skip** (v.t.)	TERM&PER	**überspringen**
six-shooter broadside antenna		ANT	**Six-shooter-Querstrahler**	[e.g. a punched hole]		[z.B. eine Lochung]
six-sigma principle		QUAL	**Sechs-Sigma-Konzept**	**skip** (→step)	ELECTRON	(→Sprung)
sixteen-bit chip		MICROEL	**Sechzehn-Bit-Chip**	**skip command** (→blank instruction)	DATA PROC	(→Leeranweisung)
sixteen-element dipole antenna		ANT	**Sechzehnerfeld**			
sixth		COLLOQ	**Sechstel**			

skip distance RADIO PROP **Sprungentfernung**
[HF, shortest return point to earth] [Kurzwellenverbindung, kürzester Rückkehrpunkt zur Erde]

skip instruction (→blank instruction) DATA PROC (→Leeranweisung)

skip keying ELECTRON **Impulsfolge-Frequenzteilung**

skip multiplexer TRANS **Doppelschrittmultiplexer**
= double-step multiplexer = Direktmultiplexer

skipping SWITCH **Übergreifen**
[interconnection of equal numbered choices of nonadjacent groups] [Zusammenschalten gleichnummerierter Ausgänge nicht benachbarter Teilgruppen]
↑ grading ↑ Mischen

skip statement (→blank instruction) DATA PROC (→Leeranweisung)

skirt (→edge) ELECTRON (→Flanke)
skirt dipole (→folded dipole) ANT (→Faltdipol)
skirting board EQUIP.ENG **Sockelleiste**
= skirting plate; bumper strip = Sockel; Sockelblech
skirting plate (→skirting board) EQUIP.ENG (→Sockelleiste)
sky noise SAT.COMM **Himmelsrauschen**
= external noise
sky radiowave (→sky wave) RADIO PROP (→Raumwelle)
sky wave RADIO PROP **Raumwelle**
= ionospheric wave; indirect wave; atmospheric radiowave; downcoming wave; space wave; free wave; sky radiowave; ionospheric radiowave; indirect radiowave; downcoming radiowave; space radiowave; free radiowave = Ionosphärenwelle

skywave reflector ANT **Steilstrahlungsreflektor**
slab (n.) DATA PROC **Wortelement**
slab METAL **Bramme**
slabbing mill (→blooming mill) METAL (→Vorwalzwerk)
slab coil (→flat coil) COMPON (→Flachspule)
slack MECH **schlaff**
≈ loose ≈ lose
slack (→unmounted) TECH (→lose)
slack (n.) (→slip) DATA COMM (→Schlupf)
slack hours (→slack traffic period) TELEC (→verkehrsschwache Zeit)
slackness (→play) ENG.DRAW (→Spielraum)
slackness loop EQUIP.ENG **Bewegungsschleife**
[Kabel]
slack period (→slack traffic period) TELEC (→verkehrsschwache Zeit)
slack traffic period TELEC **verkehrsschwache Zeit**
= slack period; slack hours = verkehrsarme Zeit
slant (n.) TECH **Schräge**
[Abweichung von horizontaler oder vertikaler Sollrichtung]
↓ Neigung
slant (adj.) TECH **schräg**
[deviating from nominal angle] [von vertikaler oder horizontaler Sollrichtung abweichend]
= oblique; aslant
≈ diagonal; cross ≈ schräggestellt
↓ skew ≈ diagonal; quer
≈ geneigt; schief
slant (→oblique) TYPOGR (→schräg)
slant (→slash) LING (→Schrägstrich)
slant distance RADIO PROP **Luftlinienentfernung**
= line-of-sight distance
slant distance RADIO NAV **Geradeausentfernung**
[direct ddistance between two point of different elevation] [direkte Entfernung zweier Punkte unterschiedlicher Elevation]
≠ ground distance ≠ Bodenentfernung
slant function MATH **Slant-Funktion**
↑ sequential function ↑ Sequenzfunktion
slanting TECH **schräglaufend**
slash LING **Schrägstrich**
[/] [/]
≈ virgule; slant; solidus; oblique line; cross-line; bar; cross stroke ≈ Querstrich; Querbalken
≠ backslash ≠ verkehrter Schrägstrich

slashed zero (→barred zero) DATA PROC (→durchgestrichene Null)
slate PHYS **schiefer**
[color] [Farbe]
slate PC DATA PROC **Slate-PC**
[without keypad, input by sensor screen] [tastenloser PC mit Eingabe über Sensorbildschirm; slate = engl. „Schiefertafel"]
= palmtop PC; palmtop = Palm-Top
slave (→subordinate) TECH (→untergeordnet)
slave computer DATA COMM **Nebenrechner**
= satellite processor; guest computer = Satellitenrechner; Nebencomputer; Satellitencomputer; Nebenprozessor
≈ front-end processor ≈ Vorrechner
≠ host computer ≠ Hauptrechner
slave processor DATA PROC **untergeordneter Prozessor**
slave station (→tributary station) TELECONTR (→Unterstation)
SLC (→subscriber loop carrier) TRANS (→Teilnehmermultiplexsystem)
sleeping partner (→silent partner) ECON (→stiller Gesellschafter)
sleep mode INSTR **Ruhezustand**
sleet METEOR **Graupel** (n.n.)
[frozen rain] [gefrohrener Regentropfen]
≈ Hagelkorn
sleet melting ANT **Antennenheizung**
sleeve MECH **Buchse**
= bushing
sleeve TELEPH **Hals**
[one of the contacts of telrphone jack connector, connected to the sleeve wire] [einer der Pole des Klinkensteckers, mit C-Ader verbunden]
≈ ring; tip; ground ≈ Ring; Spitze; Masse
sleeve TECH **Hülse**
[tubular casing] [röhrenförmige Hülle]
= can; shell 2; gommet = Tülle
≈ jacket; shell 1 ≈ Schutzhülle; Schale
sleeve TECH **Muffe**
= coupling; union; fitting
sleeve OUTS.PLANT **Muffe**
sleeve MICROW **Symmetrietopf**
sleeve (→disk jacket) TERM&PER (→Diskettenhülle)
sleeve antenna ANT **Koaxialdipol**
= sleeve dipole antenna = Koaxialdipolantenne; Vertikaldipol; Dipolantenne mit koaxialem Schirm; Sleeve-Antenne; koaxiale Monopolantenne
sleeve antenna array ANT **Mehrfachkoaxialdipol**
sleeve bearing MECH **Radialgleitlager**
sleeve dipole (→folded-top antenna) ANT (→Sperrtopfantenne)
sleeve dipole antenna (→sleeve antenna) ANT (→Koaxialdipol)
sleeve nut MECH **Überwurfmutter**
sleeve terminal COMPON **Hülsenklemme**
sleeve wire (→tip wire) TELEPH (→C-Ader)
slewing rate (→slew rate 1) ELECTRON (→Anstiegsgeschwindigkeit)
slewing speed INSTR **Ablenkgeschwindigkeit**
[speed of a device searching for information or scanning]
= deflection speed
slew rate 1 ELECTRON **Anstiegsgeschwindigkeit**
= slewing rate
slew rate 2 ELECTRON **Ausgangsspannungs-Schwankung**
slew time (→rise time) ELECTRON (→Anstiegzeit)
SLIC (→subscriber-line circuit) SWITCH (→Teilnehmersatz)
slice (→wafer crystal) MICROEL (→Kristallscheibe)
slide (v.t.) TECH **gleiten**
= slip
slide (n.) EQUIP.ENG **Laufschiene**
slide caliper (→slide gauge) MECH (→Schiebelehre)
slide gauge MECH **Schiebelehre**
= slide caliper; vernier rule; vernier caliper = Schublehre

slide-in coupling		COMPON	Einschubverbindung [Koaxialstecker]	slope regulator = tilt regulator	TRANS	Neigungsregler = Schräglagenregler
slide-in unit = panel; plug-in; drawer		EQUIP.ENG	Einschub ≈ Baugruppe	slope time	ELECTRON	Flankenzeit
slide rule		MATH	Rechenschieber	sloping (→inclined)	TECH	(→geneigt)
slide switch (→wiper switch)		COMPON	(→Schiebeschalter)	sloping arrow	TYPOGR	schräger Pfeil = Schrägpfeil
slide wire = wiping wire		EL.TECH	Schleifdraht	slot [a long, narrow opening or cut]	TECH	Schlitz [längliche, schmale Öffnung]
sliding contact		COMPON	Gleitkontakt	= slit		
sliding friction = slipping friction		PHYS	Gleitreibung = gleitende Reibung	≈ incision; split; gap; interstice; notch (n.)		≈ Einschnitt; Spalte; Spalt; Zwischenraum; Kerbe
sliding load		INSTR	Schiebelast	↑ opening		↑ Öffnung
sliding mismatch		INSTR	verschiebbares Fehlanpassungsglied	slot (→mounting place)	EQUIP.ENG	(→Einbauplatz)
sliding nut		MECH	Gleitmutter	slot (→coin slot)	TERM&PER	(→Münzeinwurfschlitz)
slightly ondulating terrain (→gently rolling terrain)		RADIO PROP	(→leichtwelliges Gelände)	slot 2 (→recess)	TECH	(→Vertiefung)
				slot antenna = slotted antenna; leaky-pipe antenna; slot radiator; slotted tubular antenna; slotted tube antenna	ANT	Schlitzstrahler = Schlitzantenne; Spaltantenne; Schlitzrohrstrahler; Rohrschlitzstrahler
slim jim antenna ["slim J-type integrated matching stub"]		ANT	Slim-jim-Antenne ↑ J-Antenne			
slim line cutter		TECH	Slim-line-Seitenschneider	slot code = mounting place code; mounting position code	EQUIP.ENG	Einbaukennung = Einbaulagennummer
slim line style		DATA PROC	Slim-line-Bauweise [halbe Bauhöhe]	slot coupling	MICROW	Schlitzkopplung
slim rack		EQUIP.ENG	Schmalgestell	slot filter [INSTR] (→bandstop filter)	NETW.TH	(→Bandsperrfilter)
sling [contractible looped device]		TECH	Schlinge [zusammenziehbare Verknüpfung]	slotline (→slotted line)	MICROW	(→Schlitzleitung)
≈ noose			≈ Schleife	slot machine	TECH	Münzautomat
≈ loop				slot radiator (→slot antenna)	ANT	(→Schlitzstrahler)
slinky dipole		ANT	Slinky-Dipol	slot reader (→pull-through scanner)	TERM&PER	(→Durchzugleser)
slip (v.t.)		TECH	rutschen = schlüpfen	slotted antenna (→slot antenna)	ANT	(→Schlitzstrahler)
slip (n.) = slippage; hit 2 (n.) ≈ drift		TECH	Schlupf = Rutsch ≈ Drift	slotted core cable	COMM.CABLE	Kammerkabel
				slotted head screw	MECH	Schlitzkopfschraube
slip (n.) = slack (n.)		DATA COMM	Schlupf	slotted line = slotline	MICROW	Schlitzleitung
slip (n.)		OFFICE	Zettel	slotted line	ANT	Meßleitung
slip (→slide)		TECH	(→gleiten)	slotted nut	MECH	Schlitzmutter
slip contact (→wiping contact)		COMPON	(→Wischkontakt)	slotted tube antenna (→slot antenna)	ANT	(→Schlitzstrahler)
slip measurement		INSTR	Schlupfmessung	slotted tubular antenna (→slot antenna)	ANT	(→Schlitzstrahler)
slippage (→slip)		TECH	(→Schlupf)			
slippery (adj.) = slippy		TECH	schlüpfrig	slow (n.)	RAILW.SIGN	Langsamfahrt
				slow (adj.)	TECH	langsam
slipping [interconnection of differently numbered choices of nonadjacent groups] ↑ grading		SWITCH	Verschränken [Zusammenschalten unterschiedlich nummerierter Ausgänge nichtbenachbarter Teilgruppen] ↑ Mischung	slow-acting (→inert)	PHYS	(→träge)
				slow-action (→slow-reacting)		(→träge)
				slow blowing fuse = delayed-action fuse; time-delay fuse	COMPON	träge Schmelzsicherung = Feinsicherung mit Zeitverzögerung
slipping friction (→sliding friction)		PHYS	(→Gleitreibung)	slow-blowing melting fuse ↑ fine-wire fuse	COMPON	träge Feinsicherung ↑ Stromfeinsicherung
slip printer		TERM&PER	Kassenbelegdrucker	slow business = dull trading	ECON	schleppendes Geschäft
slippy (→slippery)		TECH	(→schlüpfrig)			
slip resistance		TECH	Rutschfestigkeit	slow-down (n.) ↑ time scaling	TECH	Zeitdehnung ↑ Zeitmaßstabsänderung
slip-resistant (→non-skid)		TECH	(→rutschfest)	slow-down (→decelerate)	TECH	(→verlangsamen)
slit (adj.)		MECH	geschlitzt	slow evaporation (→volatilization)	PHYS	(→Verdunstung)
slit (→slot)		TECH	(→Schlitz)			
slit pin (→cotter pin)		MECH	(→Splint)	slow motion ≠ fast motion	TV	Zeitlupe ≠ Zeitraffung
slitting (→louvering)		METAL	(→Durchreißen)			
slit-tube balun = split-sheath balun		ANT	Schlitzbalun	slow motion facility	TV	Zeitlupengerät = Slow-motion-Gerät
SLM (→selective level meter)		INSTR	(→selektiver Pegelmesser)	slow motion reverse	TV	Zeitlupe rückwärts
slogan (→keyword)		LING	(→Schlagwort)	slow-operating (→slow-reacting)	TECH	(→träge)
slope [of a curve] = inclination ≈ steepness		MATH	Neigung [Kurvenverlauf] = Steigung ≈ Steiheit	slow operation (→response delay)	ELECTRON	(→Ansprechverzögerung)
				slow-reacting = slow-response; slow-action; slow-operating; delayed-action; sluggish	TECH	träge = langsamwirkend
slope (→edge)		ELECTRON	(→Flanke)			
slope (→inclination)		MECH	(→Neigung)			
sloped case		EQUIP.ENG	Pultgehäuse	slow relay (→slow-releasing relay)	COMPON	(→langsam abfallendes Relais)
slope detector		CIRC.ENG	Flankendiskriminator ≈ Resonanzkreisumformer			
slope distortion		MODUL	Steilheitsverzerrung	slow-releasing relay = slow relay	COMPON	langsam abfallendes Relais
slope drive		MECH	Seiltrieb			
slope equalizer = tilt equalizer		TRANS	Neigungsentzerrer = Schräglagenentzerrer	slow-response (→slow-reacting)	TECH	(→träge)
slope limiter		CIRC.ENG	Steilheitsbegrenzer	slow-scan system (→fixed-image communication)	TELEC	(→Festbildkommunikation)
sloper dipole		ANT	Sloper = schräggestellter Halbwellendipol	slow-scan TV = SSTV	TV	Slow-scan-TV = SSTV

SLT MICROEL **SLT**
 [solid logic technology] [Speicherelement in Dickschichttechnik]
slug (n.) TYPOGR **Typenstein**
slug (n.) TYPOGR **Zeilentype**
 [a line cast as one type] [Drucktype für eine ganze Zeile]
slug (→ tuning screw) MICROW (→ Abgleichschraube)
slug (→ token) TERM&PER (→ Einwurfmünze)
slug (→ short-circuited winding) EL.TECH (→ Kurzschlußwicklung)
slug (→ preform) MANUF (→ Rohling)
sluggisch (→ slow-reacting) TECH (→ träge)
sluggish (→ viscous) TECH (→ dickflüssig)
sluggishness (→ viscosity) PHYS (→ Zähflüssigkeit)
slug tuning ELECTRON **Magnetkernabstimmung**
slur (v.t.) COLLOQ **überspielen**
  = darüberhinweggehen
small-band system (→ low-capacity system) TRANS (→ Kleinkanalsystem)
small-base diode MICROEL **Schmalbasisdiode**
small-base direction finder RADIO LOC **Kleinbasispeiler**
 (→ low-aperture direction finder)
small business ECON **Kleinunternehmen**
small business computer DATA PROC **kleiner Geschäftscomputer**
 = SBC
small business system DATA PROC **mittlere Datentechnik**
small cap (→ small capitals) TYPOGR (→ Kapitälchen)
small capitals TYPOGR **Kapitälchen**
 [capitals with the size of small characters] [auf die Größe von Kleinbuchstaben verkleinerte Großbuchstaben]
 = small cap
small case EQUIP.ENG **Kleingehäuse**
small character TYPOGR **Kleinbuchstabe**
 = small letter; lower case letter; minuscule (n.) = Minuskel; Gemeine
small computer DATA PROC **Kleinrechner**
 ≈ minicomputer; compact computer = Kleincomputer
  ≈ Minicomputer; Kompaktrechner
small DF (→ small direction finder) RADIO LOC (→ Kleinpeiler)
small direction finder RADIO LOC **Kleinpeiler**
 = small DF
small exchange SWITCH **Kleinvermittlung**
  = Kleinzentrale
small face (→ narrow dimension) MECH (→ Schmalseite)
small keyboard (→ keypad 1) TERM&PER (→ Kleintastatur)
small letter (→ small character) TYPOGR (→ Kleinbuchstabe)
Small Pica (BRI) (→ type size 11 point) TYPOGR (→ Schriftgröße 11 Punkt)
small plate MECH **Plättchen**
small punched card TERM&PER **Kleinlochkarte**
small relay COMPON **Kleinrelais**
small scale integration (→ SSI) MICROEL (→ Kleinstintegration)
small-scale production MANUF **Kleinserienfertigung**
small series MANUF **Miniserie**
small signal ELECTRON **Kleinsignal**
small-signal amplification ELECTRON **Kleinsignalverstärkung**
small-signal amplification factor ELECTRON **Kleinsignal-Stromverstärkungsfaktor**
small-signal amplifier CIRC.ENG **Kleinsignalverstärker**
small-signal analysis ELECTRON **Kleinsignalanalyse**
small-signal driving ELECTRON **Kleinsignalsteuerung**
small-signal equivalent circuit ELECTRON **Kleinsignalmodell**
small-signal permittivity ELECTRON **Kleinsignal-Permittivität**
 [ferroelectric] [Ferroelektrikum]
small-signal resistance POWER SYS **Kleinsignalwiderstand**
  [Halbleitergleichrichter]
small-signal response ELECTRON **Kleinsignalverhalten**
small-signal short-circuit forward transfer ratio MICROEL **Kurzschluß-Stromverstärkung**
 = short-circuit forward current transfer ratio; current transfer ratio; current gain, output shorted = Kleinsignal-Kurzschlußstromverstärkung; Stromverstärkungsfaktor

small-signal transadmittance ELECTRON **Kleinsignal-Transadmittanz**
small-signal transistor MICROEL **Kleinsignaltransistor**
  = Vorstufentransistor
small tube COMM.CABLE **Kleintube**
↑ coaxial tube [1,2/4,4 mm]
  = Zwergtube
  ↑ Koaxialtube
smart card (→ chip card) TERM&PER (→ Chip-Karte)
smart terminal 2 TERM&PER **herstellerprogrammierbares Datensichtgerät**
 = factory-programmable terminal
smart terminal (→ intelligent terminal) TERM&PER (→ programmierbare Datenstation)
SMD (→ surface mounted device) COMPON (→ SMD-Bauelement)
SMD placement machine MANUF **SMD-Bestückungsautomat**
 = SMT placement machine = SMT-Bestückungsautomat
SMD technology ELECTRON **SMD-Technik**
 = SMT technology; surface mounted technology = SMT-Technik
  ≈ Oberflächenmontage
 ≈ surface mounting
SMD test probe INSTR **SMD-Tastkopf**
 = surface mount device test probe
smear (n.) (→ blur) TERM&PER (→ Unschärfe)
smearing TV **Nachzieheffekt**
smear-proof (→ smudge-proof) TECH (→ wischfest)
smelter TECH **Schmelzofen**
↑ stove ↑ Ofen
smelting works METAL **Hüttenwerk**
 = metallurgical plant
S meter INSTR **S-Meter**
 [HF] [HF]
 = signal strength meter
SMIF box MICROEL **SMIF-Box**
 [Standard Mechanical InterFace]
Smith chart LINE TH **Smith-Diagramm**
↑ transmission line chart = Reflexionsfaktorkarte
  ↑ Leitungsdiagramm
smoke detector SIGN.ENG **Rauchmelder**
 = smoke sensor = Rauchfühler
↑ fire detector ↑ Feuermelder
smoke meter SIGN.ENG **Rauchmesser**
smoke sensor (→ smoke detector) SIGN.ENG (→ Rauchmelder)
smoke test DATA PROC **Feuerprobe**
smolder (v.i.) TECH **schwelen**
 [to burn without flame] [flammenlos verbrennen]
 = smoulder
smolder (v.i.) TECH **glimmen**
 [to burn without flame] [schwach brennen]
 = smoulder
smolder (n.) TECH **Schwelbrand**
 = smoulder
smooth (v.t.) TECH **glätten**
 = iron (v.t.) = bügeln
  ≈ ebnen
smooth (v.t.) EL.TECH **glätten**
 = equalize = abflachen
smooth 2 (adj.) TECH **sanft**
 ≈ soft ≈ weich
smooth (→ stepless) TECH (→ stufenlos)
smoothed TECH **geglättet**
smoothed curve MATH **geglättete Kurve**
smoothed motion TV **ruckfreie Bewegung**
smooth file TECH **Schlichtfeile**
smoothing EL.TECH **Glättung**
 = equalization = Ebnung; Abflachung
smoothing choke EL.TECH **Glättungsdrossel**
 = smoothing reactor
smoothing choke POWER SYS **Speicherdrossel**
  ≈ Glättungsinduktivität
smoothing circuit CIRC.ENG **Glättungskreis**
  = Glättungsschaltung; Abflachungsschaltung
smoothing factor EL.TECH **Glättungsfaktor**
smoothing filter CIRC.ENG **Glättungsfilter**
  = Abflachungsfilter
smoothing reactor (→ smoothing choke) EL.TECH (→ Glättungsdrossel)
smoothing resistor CIRC.ENG **Glättungswiderstand**
smooth-machine (v.t.) MECH **maschinell schlichten**

smoothness		TECH	Glätte	soft copy		DATA PROC	Softcopy

English	Domain	German
smoothness	TECH	Glätte
smoulder (→smolder)	TECH	(→schwelen)
smoulder (→smolder)	TECH	(→Schwelbrand)
smoulder (→smolder)	TECH	(→glimmen)
SMT placement machine	MANUF	(→SMD-Bestückungsautomat)
(→SMD placement machine)		
SMT technology (→SMD technology)	ELECTRON	(→SMD-Technik)
smudge-proof [paint]	TECH	wischfest [Farbe]
= smear-proof		
S/N (→signal-to-noise ratio)	TELEC	(→Rauschabstand)
Sn (→tin)	CHEM	(→Zinn)
snap-acting controller	CONTROL	(→Zweipunktregler)
(→on-off controller)		
snap-in	MECH	einschnappen
snap-in connection	COMPON	Klemmverbindung
snap-in termination	COMPON	Klemmenanschluß
snap-in wiring	ELECTRON	Snap-in-Verdrahtung
snap-off diode	MICROEL	Snap-off-Diode
↑ charge-storage diode		= Abreißdiode
		↑ Speicherschaltdiode
snap-on cable lug	SYS.INST	Federkabelschuh
snap-on connector	POWER SYS	(→Druckknopfkontakt)
(→press-stud connector)		
snap-on coupling	COMPON	Steckrastverbindung [Koxialstecker]
snap ring (→lock washer)	MECH	(→Sicherungsscheibe)
snap switch (→toggle switch)	COMPON	(→Kippschalter)
SNG (→satellite news gathering)	SAT.COMM	(→Pressefunkdienst über Satelliten)
snip (→chip)	TECH	(→Schnitzel)
snip (v.t.) (→shred)	TECH	(→zerschnitzeln)
snipe nose pliers (→adjusting pliers)	ELECTRON	(→Justierzange)
snow (n.)	METEOR	Schnee
snow (n.) [visual static]	TV	Schnee
= hash (n.)		
SNR (→signal-to-noise ratio)	TELEC	(→Rauschabstand)
S/N ratio (→signal-to-noise ratio)	TELEC	(→Rauschabstand)
S/N with load ratio (→signal-to-noise-with-load ratio)	TELEC	(→Rauschabstand bei Belastung)
Snyder dipole	ANT	Snyder-Dipol
SO (→shift-out)	DATA COMM	(→Dauerumschaltung)
soak (n.) (→impregnation)	TECH	(→Imprägnierung)
soak (v.t.) (→impregnate)	TECH	(→imprägnieren)
soakage (→impregnation)	TECH	(→Imprägnierung)
soaked	TECH	getränkt
soaked in oil	TECH	ölgetränkt
SOAR	MICROEL	SOAR
= safe operating area		= sicherer Arbeitsbereich
society (→company)	ECON	(→Gesellschaft)
socket	COMPON	Fassung
= base		= Sockel
		↓ Röhrenfassung; Lampenfassung
socket (BRI) (→jack)	COMPON	(→Buchse)
socket (→mains socket)	EL.INST	(→Netzsteckdose)
socket driver	MECH	Steckschraubenschlüssel
= socket wrench		= Steckschlüssel
socket driver set	MECH	Steckschlüsselsatz
		= Steckschlüsselsortiment
socket-head cap screw	MECH	Inbusschraube
		↑ Innensechskantschraube
socket with grounding contact (→Schuko socket)	EL.INST	(→Schukosteckdose)
socket with protective contact (→Schuko socket)	EL.INST	(→Schukosteckdose)
socket wrench (→socket driver)	MECH	(→Steckschraubenschlüssel)
SOD [silicon on diamond technology]	MICROEL	SOD-Technik
soda lye	CHEM	Natronlauge
soder-eyelet ring	COMPON	Lötösenring
sodium	CHEM	Natrium
= Na		= Na
soft annealing	METAL	Weichglühen
soft copy [output on a screen]	DATA PROC	Softcopy [Ausgabe am Bildschirm]
≠ hard copy		= flüchtige Anzeige; flüchtiges Bild
		≠ Hardcopy
soft current limit	ELECTRON	weicher Stromgrenzwert
soft decision	CODING	Weichentscheidung
		= weiche Entscheidung
soft-doped zone	MICROEL	s-Zone
= s zone		
softener	CHEM	Weichmacher
soft failure	DATA PROC	weicher Fehler
soft font [data file with fonts]	DATA PROC	Soft-font [Datei für Zeichensätze]
≠ internal font		= ladbarer Zeichensatz
		≠ fester Zeichensatz
soft hyphen [printed only when word must be splitted at end of line]	DATA PROC	Bedarfstrennstrich [nur bei Worttrennung am Zeilenende gedruckt]
= ghost hyphen; discretionary hyphen		= Trennfuge
≠ hard hyphen		≠ echter Trennstrich
soft iron	METAL	Weicheisen
= magnetically soft iron (IEC); electrical steel (IEC); magnetic iron		
soft-iron instrument	INSTR	Weicheiseninstrument
soft key (→function key)	TERM&PER	(→Funktionstaste)
soft keyboard [represented on a display]	DATA PROC	Bildschirmtastatur
soft keyboard	TERM&PER	programmierbare Tastatur
= programmable keyboard		
soft keying	MODUL	Weichtastung
≈ silent reversal [TELEC]		≈ Weichumpolung [TELEC]
soft macro (→function macro)	MICROEL	(→Funktionsmakro)
soft magnetic (→magnetically soft)	PHYS	(→weichmagnetisch)
soft metal	METAL	Weichmetall
= magnetically soft metal; magnetic metal		
soft pad 1 (→touch key)	COMPON	(→Berührungstaste)
soft rubber	CHEM	Weichgummi
soft sector	TERM&PER	Weichsektor
soft-sectored diskette [one hole marks the beginn of the diskette]	DATA PROC	weichsektorierte Diskette [ein Loch markiert den Anfang der Diskette]
soft sectoring [diskette]	TERM&PER	Weichsektorierung [Diskette]
		= Softsektorierung
soft solder (v.t.)	METAL	weichlöten
soft solder (n.)	METAL	Weichlot
soft-soldered	METAL	weichgelötet
soft soldering	METAL	Weichlötung
soft voltage limit	ELECTRON	weicher Spannungsgrenzwert
software [general term for the immaterial components which run a computer, i.e. for programs and data]	DATA PROC	Software [Überbegriff für die immateriellen Bestandteile eines Computers, d.h. für Programme und Daten]
≈ programming		= immaterielle Ware; nicht materielle Bestandteile
≠ hardware		≈ Programmierung
↓ system software; application software; firmware		≠ Hardware
		↓ Systemsoftware; Anwendersoftware 1; Firmware
software base	DATA PROC	Softwarebasis
software broker	DATA PROC	Software-Händler
software calibration	INSTR	Software-Kalibrierung
software company	DATA PROC	Software-Haus
= software house		
software compatibility	DATA PROC	Software-Kompatibilität
software configuration manager	DATA PROC	Software-Konfigurationsmanagement
software conversion	DATA PROC	Software-Umstellung
software developer (→programmer)	DATA PROC	(→Programmierer)
software development	DATA PROC	Software-Entwicklung
≈ programming		≈ Programmierung
software development environment	DATA PROC	Software-Entwicklungsumgebung
software documentation	DATA PROC	Software-Dokumentation

English	Domain	German
software driver	DATA PROC	Software-Treiber
[adapts software to the computer]		[paßt Software an die Anlage an]
software duplication	DATA PROC	Software-Duplizierung
software encryption	DATA PROC	Software-Verschlüsselung
software engineering	DATA PROC	Software-Engineering
software ergonometry	DATA PROC	Software-Ergonometrie
software error (→program error)	DATA PROC	(→Programmfehler)
software fault (→program error)	DATA PROC	(→Programmfehler)
software flexibility	DATA PROC	Software-Flexibilität
software house (→software company)	DATA PROC	(→Software-Haus)
software integration	DATA PROC	Software-Integration
= integrration 2		= Integration 2
software librarian	DATA PROC	Software-Verwalter
software licence	DATA PROC	Software-Lizenz
software literature (→helpware)	DATA PROC	(→Software-Literatur)
software maintenance	DATA PROC	Software-Pflege
		= Software-Wartung
software monitor	DATA PROC	Prüfsoftware
software nucleus	SWITCH	Sockel
software pac (→software package)	DATA PROC	(→Software-Paket)
software package	DATA PROC	Software-Paket
= program package; package; software pac; packaged software		= Programmpaket
software piracy	DATA PROC	Software-Piraterie
software portability (→portability)	DATA PROC	(→Übertragbarkeit)
software product	DATA PROC	Software-Produkt
= programming product		
software programmer (→programmer)	DATA PROC	(→Programmierer)
software project management	DATA PROC	Software-Projektmanagement
software protection (→software security)	DATA PROC	(→Software-Schutz)
software publisher	DATA PROC	Software-Verlag
software resources	DATA PROC	Software-Möglichkeiten
software security	DATA PROC	Software-Schutz
= software protection		= Software-Sicherheit
software service	DATA PROC	Software-Dienstleistung
software specialist	DATA PROC	Software-Spezialist
software support	DATA PROC	Software-Unterstützung
software technology	DATA PROC	Software-Technologie
software tool	DATA PROC	Software-Werkzeug
software transportability (→portability)	DATA PROC	(→Übertragbarkeit)
software update	DATA PROC	Software-Aktualisierung
software version (→release version)	DATA PROC	(→Software-Version)
software virus (→virus)	DATA PROC	(→Virus)
SOH (→start of heading)	DATA COMM	(→Anfang des Kopfes)
soil conductivity	EL.TECH	bodenleitfähigkeit
soiling resistance	TECH	Schmutzunempfindlichkeit
soil mechanics	CIV.ENG	Bodenmechanik
solar battery	POWER SYS	Solarbatterie
		= Sonnenbatterie
solar cell	POWER SYS	Solarzelle
solar energy	POWER SYS	Sonnenenergie
solar module (→solar panel)	POWER SYS	(→Solarpaneel)
solar panel	POWER SYS	Solarpaneel
= solar module		= Solarmodul
solar power supply	POWER SYS	Solarstromversorgung
= photovoltaic power supply		
solar radiation	PHYS	Sonneneinstrahlung
= insolation		
solar time	PHYS	Sonnenzeit
solder (n.)	METAL	Lot
↓ tin solder; soft solder; hard solder; brazing solder		= Lotmittel; Lötmetall ↓ Lotzinn; Weichlot; Hartlot; Messinghartlot
solder (v.t.)	METAL	löten
[union of parts by wetting them with lower melting solder]		[Verbindung durch Benetzung der Teile mit niedriger schmelzendem Lot]
≈ weld (v.t.)		≈ schweißen
solderability	METAL	Lötbarkeit
		= Lötfähigkeit
solder aggregation	MANUF	Lotanhäufung
solder and terminal block	EL.INST	Lötklemmleiste
solder bath	MANUF	Lötbad
[PCB]		[Leiterplatten]
= solder pot		≈ Lotbad
↓ flow-soldering bath		↓ Schwallbad
solder bath wave	MANUF	Lötbadwelle
[PCB]		[Leiterplatten]
solder braid	MANUF	Lötlitze
solder bridge	MANUF	Lotbrücke
solder coating	METAL	Lotüberzug
solder contact	COMPON	Lötkontakt
solder cup (→fillet)	MANUF	(→Lotkegel)
soldered	METAL	gelötet
		≈ geschweißt
soldered connection	ELECTRON	Lötanschluß
soldered connection	METAL	Lötverbindung
= soldered joint; soldering joint		
soldered joint (→soldering point)	ELECTRON	(→Lötstelle)
soldered joint (→soldered connection)	METAL	(→Lötverbindung)
solder-eyelet tail	COMPON	Lötösenfahne
solder fuse	COMPON	Lötsicherung
solder-in	ELECTRON	einlöten
soldering	METAL	Löten
		= Lötung
soldering accessories	ELECTRON	Lötzubehör
soldering aid (→soldering tool)	METAL	(→Löthilfe)
soldering behaviour	METAL	Lötverhalten
soldering defect	MANUF	Lötfehler
soldering eye (→soldering eyelet)	COMPON	(→Lötöse)
soldering eyelet	COMPON	Lötöse
= soldering eye; eyelet		↑ Lötanschlußpunkt
↑ solder terminal		
soldering fluid (→soldering varnish)	METAL	(→Lötlack)
soldering flux	METAL	Flußmittel 1
= flux 1; resin		[Löten]
soldering grease	METAL	Lötfett
soldering gun	ELECTRON	Lötpistole
		= Schnellöter
soldering iron	METAL	Lötkolben
		= Lötgerät; Löter
soldering iron holder	ELECTRON	Lötkolbenständer
soldering joint (→soldered connection)	METAL	(→Lötverbindung)
soldering land	ELECTRON	Lötauge
[PCB]		[Leiterplatte]
= land; soldering pad; pad; terminal pad; eyelet		
soldering list	MANUF	Lötliste
soldering lug	COMPON	Lötfahne
= soldering tag 2; soldering tail		= Lötanschlußfahne
↑ solder terminal		↑ Lötanschlußpunkt
soldering lug	EL.INST	Lötschuh
= solder lug		
soldering pad (→soldering land)	ELECTRON	(→Lötauge)
soldering paste	MANUF	Lötpaste
= solder paste		
soldering pencil	ELECTRON	Feinlötkolben
		= Lötpencil
soldering period	METAL	Lötzeit
soldering pin (→solder pin)	COMPON	(→Lötstift)
soldering plate	ELECTRON	Lötösenplatte
soldering point	ELECTRON	Lötstelle
= soldering spot; solder point; solder joint; soldered joint		= Lötpunkt
soldering process	METAL	Lötverfahren
		= Lötprozess
soldering residues	MANUF	Lötmittelrückstand
soldering side (→solder side)	ELECTRON	(→Lötseite)
soldering sleeve	COMM.CABLE	Löthülse

soldering spot (→soldering point)	ELECTRON	(→Lötstelle)	
soldering station	ELECTRON	**Lötstation**	
soldering tab	COMPON	**Lötlasche**	
soldering tag 2 (→soldering lug)	COMPON	(→Lötfahne)	
soldering-tag terminal strip	COMPON	**Lötleiste**	
= tag block; tag-end terminal block			
soldering tail (→soldering lug)	COMPON	(→Lötfahne)	
soldering technique	METAL	**Löttechnik**	
soldering temperature	METAL	**Löttemperatur**	
soldering tin	METAL	**Lötzinn**	
= solder tin			
≈ tin solder		≈ Zinnlot	
soldering tip	ELECTRON	**Lötspitze**	
soldering-tip cleaner	ELECTRON	**Lötspitzenreiniger**	
soldering tool	METAL	**Löthilfe**	
= soldering aid			
soldering varnish	METAL	**Lötlack**	
= soldering fluid			
solder joint (→soldering point)	ELECTRON	(→Lötstelle)	
solder layer	ELECTRON	**Lötschicht**	
solderless	ELECTRON	**lötfrei**	
		= nichtverlötet	
solderless connection	EL.TECH	**lötfreie Verbindung**	
solderless lug	MECH	**Schraubklemme**	
= screw terminal			
solderless termination	EQUIP.ENG	**lötfreie Anschlußtechnik**	
solder lug (→soldering lug)	EL.INST	(→Lötschuh)	
solder mask	MANUF	**Lötmaske**	
		= Lötmittelmaske	
solder paste (→soldering paste)	MANUF	(→Lötpaste)	
solder pin	COMPON	**Lötstift**	
= soldering pin		= Lötanschlußstift	
↑ solder terminal		↑ Lötanschlußpunkt	
solder point (→soldering point)	ELECTRON	(→Lötstelle)	
solder post (→solder terminal)	COMPON	(→Lötanschlußpunkt)	
solder pot (→solder bath)	MANUF	(→Lötbad)	
solder projection	MANUF	**Lotvorsprung**	
solder reflow	MANUF	**Lötmittelrückfluß**	
solder resist	MANUF	**Lötstopplack**	
[PCB assembly]		[Leiterplattenfertigung]	
= solder stop-off		= Lötabdeckschicht; Lötabdecklack; Lötresist; Lackabdeckung	
solder resistance	METAL	**Lotbeständigkeit**	
solder-resist mask	MANUF	**Lötstoppmaske**	
solder side	ELECTRON	**Lötseite**	
= soldering side; opposite side; side two		[Leiterplatte]	
≠ component side		= Schwallseite	
		≈ Bauteileseite	
solder slinger	MANUF	**Lötmittelschleuder**	
		= Lotschleuder	
solder splash	ELECTRON	**Lotspritzer**	
↓ tin splash		= Lötspritzer	
		↓ Zinnspritzer	
solder stop-off (→solder resist)	MANUF	(→Lötstopplack)	
solder strap	ELECTRON	**Lötbrücke**	
= wiring strap			
solder-strap option	ELECTRON	**Lötbrückeneinstellung**	
= strapping option			
solder-strappable	ELECTRON	**mit Lötbrücke einstellbar**	
solder surface	METAL	**Lötfläche**	
solder tag 1	COMPON	**Löthülse**	
↑ solder terminal		= Lötanschlußhülse	
		↑ Lötanschlußpunkt	
solder terminal	COMPON	**Lötanschlußpunkt**	
= solder post; turred tag		= Lötanschluß; Lötstützpunkt	
↓ soldering lug; solder tag 1; soldering eyelet; solder pin		↓ Lötfahne; Löthülse; Lötöse; Lötstift	
solder tin (→soldering tin)	METAL	(→Lötzinn)	
solder-tin dispenser	ELECTRON	**Lötzinn-Abroller**	
= solder-tin holder			
solder-tin holder (→solder-tin dispenser)	ELECTRON	(→Lötzinn-Abroller)	
solder-up flow	MANUF	**Lotanstieg**	
		[Schwallbad]	
solder void	METAL	**Lotblase**	
solder wave	MANUF	**Lotwelle**	
solder wetting	METAL	**Lotbenetzung**	
sold-out	ECON	**vergriffen**	
		= ausverkauft	
solenoid	PHYS	**Solenoid**	
[a cylindric coil]		[zylindrische Spule]	
solenoid (→magnet coil)	COMPON	(→Magnetspule)	
solenoid (→cylindrical coil)	COMPON	(→Zylinderspule)	
solenoidal field (→source-free field)	PHYS	(→quellenfreies Feld)	
solenoidal field (→rotational field)	MATH	(→Wirbelfeld)	
solenoid valve	COMPON	**Magnetventil**	
sole selling rights (→exclusive sale franchise)	ECON	(→Alleinverkaufsrecht)	
solicitation (→invitation to tender)	ECON	(→Ausschreibung)	
solicitation (→advertising)	ECON	(→Werbung)	
solicitor (BRI) (→lawyer)	ECON	(→Rechtsanwalt)	
solid (adj.)	ENG.DRAW	**ausgezogen**	
[line, curve]		[Linie, Kurve]	
solid (adj.)	TECH	**massiv**	
solid	PHYS	**Festkörper**	
solid (→friction-locked)	TECH	(→kraftschlüßig)	
solid angle	MATH	**Raumwinkel**	
[SI unit: steradian; sr]		[SI-Einheit: Steradiant; sr]	
= Ω		= räumlicher Winkel; Ω	
solid-angle element	MATH	**Raumwinkelelement**	
solid complete angle	MATH	**räumlicher Vollwinkel**	
solid dielectric cable (→solid insulated cable)	HF	(→Vollkabel)	
solid fiber (→tight buffer tube)	OPT.COMM	(→Vollader)	
solid font printer (→type printer)	TERM&PER	(→Typendrucker)	
solidification	PHYS	**Erstarrung**	
		= Erstarren	
solidification point	PHYS	**Erstarrungspunkt**	
solidify	PHYS	**erstarren**	
		= erhärten	
solid insulated cable	HF	**Vollkabel**	
= solid dielectric cable		= vollisoliertes Kabel	
solid insulation	COMM.CABLE	**Vollisolierung**	
= full insulation			
solid laser (→solid-state laser)	OPTOEL	(→Festkörperlaser)	
solid leading rope	ANT	**Vollseil**	
≠ hollow leading rope		≠ Hohlseil	
solid line	ENG.DRAW	**ausgezogene Linie**	
solid model	DATA PROC	**Volumenmodell**	
[computer graphics; graphics without hidden lines]		[Computergraphik; Graphik ohne unsichtbaren Linien]	
= solid modelling		= Festkörpermodell; Körpermodell	
solid modelling (→solid model)	DATA PROC	(→Volumenmodell)	
solid-phase epitaxy	MICROEL	**Festphasen-Epitaxie**	
solid polyethylene	COMM.CABLE	**Vollpolyäthylen**	
solid rivet	MECH	**Vollniet**	
solid-state (adj.)	MICROEL	**Halbleiter-**	
= semiconductor		= halbleitend	
solid state amplifier	CIRC.ENG	**Halbleiterverstärker**	
= semiconductor amplifier			
solid-state cartridge	DATA PROC	(→Programmkarte)	
(→program card)			
solid-state cartridge (→ROM cartridge)	TERM&PER	(→Festspeicherkassette)	
solid-state component	MICROEL	(→Halbleiterbauelement)	
(→semiconductor device)			
solid state device	MICROEL	(→Halbleiterbauelement)	
(→semiconductor device)			
solid state electronics	ELECTRON	(→Halbleiterelektronik)	
(→semiconductor electronics)			

solid-state imager

solid-state imager (→solid-state image sensor)	TV	(→Festkörperbildsensor)	
solid-state image sensor = solid-state imager	TV	**Festkörperbildsensor** = Festkörpervidikon; Halbleiter-Bildsensor	
solid-state laser = solid laser	OPTOEL	**Festkörperlaser**	
solid state memory = semiconductor memory; semiconductor storage ↑ memory ↓ RAM; ROM; PROM; EPROM; dynamic solid state memory; volatile solid state memory; non-volatile solid state memory	MICROEL	**Halbleiterspeicher** = Festkörperspeicher; mikroelektronischer Speicher ↑ Speicher ↓ RAM; ROM; PROM; EPROM; dynamischer Halbleiterspeicher; flüchtiger Halbleiterspeicher; nichtflüchtiger Halbleiterspeicher	
solid state memory chip = semiconductor memory chip	MICROEL	**Halbleiterspeicherchip**	
solid state physics ≈ semiconductor physics	PHYS	**Festkörperphysik** ≈ Halbleiterphysik	
solid-state power amplifier = SSPA	MICROEL	**Halbleiterleistungsverstärker**	
solid-state relay (→transistor relay)	COMPON	(→Transistorrelais)	
solidus	METAL	**Soliduskurve**	
solidus (→slash)	LING	(→Schrägstrich)	
solid-walled	TECH	**vollwandig** ≠ durchlöchert	
solid wire	METAL	**Volldraht** = Massivdraht	
solubility	PHYS	**Löslichkeit**	
soluble = dissolvable ↓ water-soluble	CHEM	**löslich** = lösbar ↓ wasserlöslich	
solute (v.t.) (→solve)	MATH	(→lösen)	
solution	MATH	**Lösung**	
solution [substance]	CHEM	**Lösung 1** [Substanz]	
solution algorithm	MATH	**Lösungsalgorithmus**	
solution pressure	PHYS	**Lösungsdruck**	
solution principle	MATH	**Lösungsprinzip**	
solution trial	MATH	**Ansatz**	
solve (v.t.) = solute (v.t.)	MATH	**lösen**	
solve (→manage)	COLLOQ	(→bewältigen)	
solvent (n.) ≈ detergent	TECH	**Lösungsmittel** ≈ Reinigungsmittel	
SOM (→start of message)	DATA COMM	(→Nachrichtenbeginn)	
Sommerfeld's fine-structure constant	PHYS	**Sommerfeld'sche Feinstrukturkonstante**	
SONAR (→ultrasonic ranging)	INSTR	(→Ultraschallortung)	
son band [grandfather-father-son mode]	DATA PROC	**Sohn-Band** [Generationen-Prinzip]	
sonde	RADIO LOC	**Lot** = Lotung	
sonde (→probe)	MICROW	(→Sonde)	
son file	DATA PROC	**Sohn-Datei**	
sonic barrier (→sound barrier)	ACOUS	(→Schallgrenze)	
sonic depth finder (→echo sounding)	RADIO NAV	(→Echolot)	
soot (n.)	TECH	**Ruß**	
SOP (→standard operation procedure)	DATA PROC	(→Standardbetriebsverfahren)	
sophisticated = high-grade; fine	TECH	**hochwertig** = raffiniert	
sort (v.t.) = grade (quality)	TECH	**sortieren**	
sort (v.t.) [to sort data following an inherent criterion] ≈ collate	DATA PROC	**sortieren** [ordnen von Daten nach inhärentem Sortierbegriff] ≈ mischen	
sort (n.) (→sorting)	DATA PROC	(→Sortieren)	
sort check (→screening inspection)	QUAL	(→Sortierprüfung)	
sort effort	DATA PROC	**Sortierungsaufwand**	
sorter (→sorting device)	TERM&PER	(→Sortiergerät)	
sort generator	DATA PROC	**Sortierprogramm-Generator**	
sorting ↓ oscillating sorting; cascade sorting; merge sorting; polyphase sorting; bubble sorting; multipath sorting	DATA PROC	**Sortieren** = Sortierung ↓ oszillierendes Sortieren; Kaskadensortieren; Mischsortieren; Polyphasensortieren; Bubble-Sortieren; Mehrwege-Sortieren	
sorting	LING	**Sortierung**	
sorting (→traffic sorting)	TELEC	(→Dienstetrennung)	
sorting algorithm	DATA PROC	**Sortieralgorithmus**	
sorting device = sorter; sorting machine	TERM&PER	**Sortiergerät** = Sortiermaschine; Sortierer	
sorting key = sort key ≈ key; search key	DATA PROC	**Sortierschlüssel** = Sortierbegriff; Sortiermerkmal; Sortierargument; Sortierkriterium ≈ Kennbegriff; Suchkriterium	
sorting machine (→sorting device)	TERM&PER	(→Sortiergerät)	
sorting method = sort method	DATA PROC	**Sortierverfahren** = Sortiermethode	
sorting needle = sort needle; sorting rod; sort rod	TERM&PER	**Sortiernadel**	
sorting program = sort program ≈ sort-merge program	DATA PROC	**Sortierprogramm** ≈ Sortier-Misch-Programm	
sorting rod (→sorting needle)	TERM&PER	(→Sortiernadel)	
sorting sequence = sort sequence; collation sequence ≈ collating sequence	DATA PROC	**Sortierfolge** = Sortierreihenfolge	
sort key (→sorting key)	DATA PROC	(→Sortierschlüssel)	
sort-merge program ≈ merge program; sorting program	DATA PROC	**Sortier-Misch-Programm** ≈ Mischprogramm: Sortierprogramm	
sort method (→sorting method)	DATA PROC	(→Sortierverfahren)	
sort n. (→kind)	TECH	(→Sorte)	
sort needle (→sorting needle)	TERM&PER	(→Sortiernadel)	
sort program (→sorting program)	DATA PROC	(→Sortierprogramm)	
sort rod (→sorting needle)	TERM&PER	(→Sortiernadel)	
sort run	ECON	**Sortierdurchlauf** = Sortierlauf	
sort sequence (→sorting sequence)	DATA PROC	(→Sortierfolge)	
sort way [number of strings concatenated in a sorting run]	DATA PROC	**Wegezahl** [Zahl der in einem Mischlauf verketteten Strings]	
SOS technology [silicone on sapphire technology] ↑ semiconductor film technlogy	MICROEL	**SOS-Technologie** ↑ Halbleiterfilmtechnik	
sound (v.t.) [to measure the depth]	INSTR	**loten** [die Tiefe bestimmen]	
sound 3 [sensation by non-sinus acoustic wave] ≈ tone	ACOUS	**Klang** [Empfindung durch nichtsinusförmige Schwingung] ≈ Ton	
sound 1 [mechanical wave propagating in a medium] ↓ audible sound; infrasound; ultrasound	ACOUS	**Schall 1** [in einem Medium sich ausbreitende mechanische Schwingung] ↓ Hörschall; Infraschall; Ultraschall	
sound 2 [from 16 Hz to 24 kHz] = audible sound ↑ sound 1 ↓ sound 3; tone	ACOUS	**Hörschall** [von 16 Hz bis 24 kHz] = Schall 2; Laut ↑ Schall 1 ↓ Klang; Ton	
sound absorber (→deadener)	ACOUS	(→Schalldämpfer)	
sound-absorbing (→sound-insulating)	ACOUS	(→schalldämmend)	
sound absorption (→sound insulation)	ACOUS	(→Schalldämmung)	
sound absorption coefficient	ACOUS	**Schallabsorptionsgrad**	
sound-absorptive = noise-absorptive	ACOUS	**schallschluckend** = lärmschluckend	

sound analysis method	INSTR		(→Suchtonverfahren)
(→search-tone method)			
sound barrier	ACOUS	**Schallgrenze**	
= sonic barrier		= Schallmauer	
soundboard (→sounding board)	ACOUS	(→Resonanzboden)	
sound broadcast (→sound broadcast transmission)	BROADC	(→Hörfunksendung)	
sound broadcasting	BROADC	**Hörfunk**	
= radio broadcasting; radiophony		= Tonrundfunk; Hörrundfunk; Rundspruch (CH)	
↑ bradcasting		↑ Rundfunk	
sound broadcast satellite	SAT.COMM	**Hörfunksatellit**	
↑ broadcast satellite		↑ Rundfunksatellit	
sound broadcast transmission	BROADC	**Hörfunksendung**	
= sound broadcast		= Rundfunksendung 2	
≈ sound broadcast program		≈ Hörfunkprogramm	
↑ broadcast transmission		↑ Rundfunksendung 1	
sound broadcast transmitter	BROADC	**Hörfunksender**	
= sound transmitter		= Tonsender	
↑ broadcast transmitter		↑ Rundfunksender	
sound broadcast user	BROADC	**Hörfunkteilnehmer**	
= radio broadcast user		= Rundfunkhörer	
≈ sound broadcast listener		≈ Rundfunkteilnehmer	
↑ broadcast service user			
sound carrier	EL.ACOUS	**Tonträger**	
[storage medium]		[Speichermedium]	
= aural carrier			
sound carrier	TV	**Tonträger**	
sound carrier step	TV	**Tontreppe**	
sound carrier trap	TV	**Tonträgerfalle**	
= sound trap		= Tonfalle	
sound channel	BROADC	**Tonkanal**	
= audio channel; tone channel; program channel			
sound chip	MICROEL	**Tonbaustein**	
		= Tonchip	
sound circuit transformer	TRANS	**Tonleitungsübertrager**	
sound control	EL.ACOUS	**Klangregler**	
= sound corrector; tone control			
sound converter (→electroacoustic transducer)	EL.ACOUS	(→Schallwandler)	
sound corrector (→sound control)	EL.ACOUS	(→Klangregler)	
sound-deadening (→sound-insulating)	ACOUS	(→schalldämmend)	
sound distortion	ACOUS	**Klangverzerrung**	
sound effect	ACOUS	**Klangeffekt**	
sound engineer	ACOUS	**Toningenieur**	
= sound supervisor; audio control engineer; sound supervisor; monitor man		= Tontechniker; Tonmeister	
sound engineering	EL.ACOUS	**Tontechnik**	
= audio engineering		= Audiotechnik	
sound field	ACOUS	**Schallfeld**	
sound filter	EL.ACOUS	**Klangfilter**	
sound flux	ACOUS	**Schallfluß**	
sound generator	EL.ACOUS	**Schallgeber**	
= sound source; sound transmitter; acoustic source; acoustic radiator; sound projector		= Schallsender; Schallstrahler; Schallquelle	
↓ loudspeaker		↓ Lautsprecher	
sound head	EL.ACOUS	**Tonkopf**	
= tape head		= Magnetbandkopf	
≈ magnetic head [TERM&PER]		≈ Magnetkopf [TERM&PER]	
↓ recording sound head		↓ Aufzeichnungsmagnetkopf	
sounding board	ACOUS	**Resonanzboden**	
= soundboard		= Schallboden	
sounding lead	TECH	**Senklot**	
sound-insulating	ACOUS	**schalldämmend**	
= sound-absorbing; sound-deadening; noise-attenuating; quietized		= schalldämpfend	
sound-insulating cover (→noise-absorbing cover)	TERM&PER	(→Schallschluckhaube)	
sound insulation	ACOUS	**Schalldämmung**	
= sound absorption; sound proofing		= Schalldämpfung; Schallschutz; Lärmschutz	
sound-in-syncs transmission	TV	**SIS-Übertragung**	
= SIS transmission			
sound intensity	ACOUS	**Schallstärke**	
		= Schallintensität	
sound intensity level	ACOUS	**Schallintensitätspegel**	
sound level	ACOUS	**Schallpegel**	
sound-level calibrator	EL.ACOUS	**Eichschallquelle**	
sound level meter	INSTR	**Schallpegelmesser**	
= sound meter			
sound locating	ACOUS	**Schallortung**	
= sound location			
sound location (→sound locating)	ACOUS	(→Schallortung)	
sound locator	ACOUS	**Schallortungsgerät**	
= locator			
sound meter (→sound level meter)	INSTR	(→Schallpegelmesser)	
sound mixer 1 [person]	EL.ACOUS	**Tonmischmeister**
		= Tonmischer	
sound mixer 2 (→sound mixing panel)	EL.ACOUS	(→Tonmischpult)	
sound mixing	EL.ACOUS	**Tonmischung**	
[combination of several sound sources]		[Mischung mehrerer Tonquellen]	
= dubbing			
sound mixing panel	EL.ACOUS	**Tonmischpult**	
= sound mixer 2			
sound particle velocity	ACOUS	**Schallschnelle**	
= acoustic velocity; group velocity of sound; velocity of sound		= Schnelle	
sound pressure (→effective sound pressure)	ACOUS	(→Schalldruck)	
sound pressure level	ACOUS	**Schalldruckpegel**	
sound processing	EL.ACOUS	**Tonnachbearbeitung**	
sound-producing	ACOUS	**schallerzeugend**	
sound program circuit	TELEC	**Tonleitung**	
sound program line amplifier	TRANS	**Tonleitungsverstärker**	
sound programm	TELEC	**Tonprogramm**	
[type of signal]		[Signalart]	
= audio program			
sound program transmission	TRANS	**Tonprogrammübertragung**	
sound projector (→sound generator)	EL.ACOUS	(→Schallgeber)	
sound-proof	ACOUS	**schalldicht**	
= sound-proofed		= schallsicher; lärmsicher	
soundproof cover (→noise-absorbing cover)	TERM&PER	(→Schallschluckhaube)	
sound-proofed (→sound-proof)	ACOUS	(→schalldicht)	
sound proofing (→sound insulation)	ACOUS	(→Schalldämmung)	
sound propagation	ACOUS	**Schallausbreitung**	
= sound radiation			
sound radiation (→sound propagation)	ACOUS	(→Schallausbreitung)	
sound radiation impedance	ACOUS	**Schallwellenwiderstand**	
= characteristic sound impedance		= Schallkennimpedanz	
sound reason	COLLOQ	**triftiger Grund**	
sound receiver (→acoustic receiver)	EL.ACOUS	(→Schallempfänger)	
sound recorder	EL.ACOUS	**Tonaufnahmegerät**	
sound recording	EL.ACOUS	**Schallaufnahme**	
		= Schallaufzeichnung; Tonaufnahme; Tonaufzeichnung	
sound reflection	ACOUS	**Schallreflexion**	
= sound reflexion		↓ Echo; Nachhall	
↓ echo; reverberation			
sound reflexion (→sound reflection)	ACOUS	(→Schallreflexion)	
sound sensing detector system	SIGN.ENG	**Geräuschmeldesystem**	
= audio detection system			
sound sensor (→acoustic sensor)	COMPON	(→Schallsensor)	
sound signal	EL.ACOUS	**Schallsignal**	
= audio signal; audible signal		= Tonsignal; akustisches Signal; Schallzeichen	
↓ hooter		↓ Hupe	
sound source (→sound generator)	EL.ACOUS	(→Schallgeber)	
sound source (→acoustic source)	ACOUS	(→Schallquelle)	

sound supervisor (→sound engineer) EL.ACOUS (→Toningenieur)
sound supervisor (→sound engineer) EL.ACOUS (→Toningenieur)
sound tape EL.ACOUS **Tonband**
↑ magnetic tape [TERM&PER] ↑ Magnetband [TERM&PER]
sound test instrument INSTR **Schallmeßgerät**
sound track EL.ACOUS **Tonspur**
sound tracking EL.ACOUS **Vertonung**
↑ post production [BROADC] ↑ Nachbearbeitung [BROADC]
sound transmission factor EL.ACOUS **Schalltransmissionsgrad**
= acoustical transmission factor = Transmission
sound transmitter (→sound generator) EL.ACOUS (→Schallgeber)
sound transmitter (→sound broadcast transmitter) BROADC (→Hörfunksender)
sound trap (→sound carrier trap) TV (→Tonträgerfalle)
sound velocity ACOUS **Schallgeschwindigkeit**
sound volume ACOUS **Klangumfang**
sound wave ACOUS **Schallwelle**
sound wave diffraction ACOUS **Schallwellenbeugung**
sound wave length ACOUS **Schallwellenlänge**
source (v.t.) ECON **beziehen**
source (n.) PHYS **Quelle**
≠ dip (n.) ≠ Senke
source (n.) INF **Quelle**
≠ drain (n.) ≠ Senke
source (n.) MICROEL **Quelle**
[region or terminal of FET delivering the controlled current] [Zone oder Pol von FET, woraus der gesteuerte Strom fließt]
≠ drain = S-Pol; Source
≠ Senke
source (→generator) INSTR (→Generator)
sourcebook LING **Nachschlagwerk**
source code CODING **Primärcode**
= primary code = Quellencode
source code 2 (→source program) DATA PROC (→Quellprogramm)
source code 1 (→source language) DATA PROC (→Quellsprache)
source coding (→primary coding) CODING (→Quellencodierung)
source data DATA PROC **Ursprungsdaten**
= raw data = Originaldaten; Rohdaten
source-data acquisition DATA PROC (→Quelldatenerfassung)
(→real-time data acquisition)
source-data automation DATA PROC (→Quelldatenerfassung)
(→real-time data acquisition)
source-data entry (→real-time DATA PROC (→Quelldatenerfassung)
data acquisition)
source disk DATA PROC **Quelldiskette**
≠ target disk = Ausgangsdiskette
≠ Zieldiskette
source document (→original document) DATA PROC (→Originalbeleg)
source documents OFFICE **Schriftgut**
= documents
source-drain leakage current MICROEL **Source-drain-Leckstrom**
source encoding (→primary coding) CODING (→Quellencodierung)
source field PHYS **Quellenfeld**
source-free PHYS **quellenfrei**
[field] [Feld]
source-free field PHYS **quellenfreies Feld**
[with divergence equal zero] [mit Divergenz gleich Null]
= solenoidal field = solenoidales Feld
source impedance EL.TECH **Quellenimpedanz**
source impedance (→intrinsic resistance) NETW.TH (→Innenwiderstand)
source information INF **Quellinformation**
source language DATA PROC **Quellsprache**
[program language used for the formulation of a program] [für Formulierung verwendete Programmiersprache]
= source code 1 = Ursprungssprache;
≠ object language Quellcode 1; Ursprungscode
≠ Zielsprache

source language LING **Ursprungssprache**
[dictionary] [Wörterbuch]
≠ target language = Leitsprache
≠ Zielsprache
source listing DATA PROC **Programmliste**
[protocol generated by a translating program] [vom Übersetzungsprogramm erstelltes Protokoll]
= program listing; programme listing; source protocol = Programmprotokoll; Programmausdruck
source module DATA PROC **Quellmodul**
source program DATA PROC **Quellprogramm**
[written in a program language, not translated into machine language] [in Programmiersprache formuliert, noch nicht in Maschinensprache übersetzt]
= source code 2 = Ursprungsprogramm; Primärprogramm; Quellcode 2; Quellenprogramm
≈ source code 1 ≈ Quellsprache
source protocol (→source listing) DATA PROC (→Programmliste)
source register DATA PROC **Quellregister**
source regulation (→mains control) POWER ENG (→Netzregelung)
source voucher (→original document) DATA PROC (→Originalbeleg)
sourcing (→origin) ECON (→Ursprung)
south pole GEOPHYS **Südpol**
space (v.t.) TECH **räumlich aufteilen**
space (n.) PHYS **Raum**
space (n.) ASTROPHYS **Weltraum**
= Weltall
space (n.) (→blank) CODING (→Leerstelle)
space (→spacing pulse) TELEGR (→Pausenschritt)
space (→letterspace) TYPOGR (→spatiieren)
space (→blank space) TYPOGR (→Leerraum)
space administration AERON **Raumfahrtbehörde**
= space agency
space agency (→space administration) AERON (→Raumfahrtbehörde)
space bar TERM&PER **Leertaste**
[generates a blank space] [erzeugt eine Leerstelle]
= spacebar; spacing key = Zwischenraumtaste
spacebar (→space bar) TERM&PER (→Leertaste)
space between words (→interword spacing) TYPOGR (→Wortzwischenraum)
space character TELEGR **Zwischenraumzeichen**
= blank character; idle character = Leerzeichen
≈ filler ≈ Füllzeichen
space charge PHYS **Raumladung**
space-charge capacity PHYS **Raumladungskapazität**
space-charge cloud PHYS **Raumladungswolke**
space-charge control ELECTRON **Raumladungssteuerung**
space-charge-control tube ELECTRON **Raumladungsgitterröhre**
space-charge debunching ELECTRON **Raumladungsstreuung**
space-charge density PHYS **Raumladungsdichte**
space-charge depth ELECTRON **Raumladungsweite**
space-charge effect PHYS **Raumladungseffekt**
space-charge grid ELECTRON **Raumladungsgitter**
space-charge layer ELECTRON **Raumladungsschicht**
space-charge limited PHYS **raumladungsbegrenzt**
space-charge region PHYS **Raumladungsgebiet**
space-charge wave MICROW **Raumladungswelle**
space coordinate MATH **Raumkoordinate**
= räumliche Koordinate
space curve MATH **Raumkurve**
spaced-antenna diversity (→space diversity) RADIO (→Raumdiversity)
spaced characters TYPOGR **Sperrschrift**
= spaced type; spaced text = Sperrdruck
space diversity RADIO **Raumdiversity**
= spaced-antenna diversity; antenna diversity = Antennendiversity; Standortdiversity
space-division multiplex SWITCH **Raumvielfach**
= space multiplex = Raummultiplex
spaced text (→spaced characters) TYPOGR (→Sperrschrift)
spaced type (→spaced characters) TYPOGR (→Sperrschrift)
space economy (→space saving) TECH (→Raumersparnis)

space electronics	ELECTRON	**Raumfahrtelektronik**	
space factor [coil]	COMPON	**Füllfaktor** [Spule]	
space factor = array factor	ANT	**Gruppencharakteristik** = Gruppenfaktor	
space-flight center	AERON	**Raumfahrtzentrum**	
space-flight engineering ↑ aerospace engineering	AERON	**Raumfahrttechnik** ↑ Luft- und Raumfahrttechnik	
space interval (→interstice)	TECH	(→Zwischenraum)	
space lattice (→crystal lattice)	PHYS	(→Kristallgitter)	
space multiplex (→space-division multiplex)	SWITCH	(→Raumvielfach)	
space occupancy (→space requirement)	TECH	(→Raumbedarf)	
space operation service	SAT.COMM	**Weltraum-Fernwirkfunkdienst**	
space pulse	DATA COMM	**Leerschritt**	
spacer	MECH	**Abstandhalter** = Abstandstück	
spacer	ELECTRON	**Abstandsrolle** = Distanzrolle	
spacer (→lifter)	TERM&PER	(→Abstandhalter)	
space radiocommunication service = space radio service	RADIO	**Weltraumfunkdienst**	
space radio service (→space radiocommunication service)	RADIO	(→Weltraumfunkdienst)	
space radio station = space station ≈ communications satellite	SAT.COMM	**Weltraumfunkstelle** ≈ Nachrichtensatellit	
space radiowave (→sky wave)	RADIO PROP	(→Raumwelle)	
spacer bushing	MECH	**Abstandsbuchse**	
space requirement = space occupancy ≈ packaging density [EQUIP.ENG]; floor space	TECH	**Raumbedarf** = Platzbedarf ≈ Packungsdichte [EQUIP.ENG]; Stellfläche	
space research	SCIE	**Weltraumforschung**	
space research radio service	RADIO	**Weltraumforschungs-Funkdienst**	
space restriction (→lack of space)	TECH	(→Raummangel)	
spacer ring (→distance ring)	MECH	(→Abstandsring)	
spacer tube (→spacing tube)	MECH	(→Abstandsrohr)	
space saving (n.) = space economy	TECH	**Raumersparnis**	
space saving (adj.) ≈ compact	TECH	**raumsparend** ≈ kompakt	
space shuttle	AERON	**Raumgleiter** = Weltraumtransporter	
space stage (→space switch)	SWITCH	(→Raumlagenvielfach)	
space station (→space radio station)	SAT.COMM	(→Weltraumfunkstelle)	
space switch = space stage	SWITCH	**Raumlagenvielfach** = Raumstufe	
space switching matrix	SWITCH	**Raumkoppelfeld**	
space switching stage	SWITCH	**Raumkoppelstufe**	
space-tapered array	ANT	**gestufte Gruppenantenne**	
space-to-earth link (→down-link)	SAT.COMM	(→Abwärtsstrecke)	
space wave (→sky wave)	RADIO PROP	(→Raumwelle)	
spacial (→volumetric)	MATH	(→räumlich)	
spacing 1	TECH	**räumliche Aufteilung** = räumliche Einteilung	
spacing 2 (→separation 1)	TECH	(→Abstand)	
spacing 2 (→character spacing)	TYPOGR	(→Zeichenabstand)	
spacing 1 (→word spacing)	TYPOGR	(→Ausschluß 3)	
spacing bolt	MECH	**Abstandsbolzen**	
spacing current = spacing pulse ≠ open-circuit current	TELEGR	**Ruhestrom** = Trennstrom; Zeichenstrom ≈ Pausenschritt ≠ Arbeitsstrom	
spacing frequency	TELEGR	**Pausenfrequenz**	
spacing jig	MECH	**Abstandslehre**	
spacing key (→space bar)	TERM&PER	(→Leertaste)	
spacing pulse = spacing signal; space; pause; no-current condition ≈ condition Z; spacing current ≠ marking pulse ↑ unit interval; signal condition [DATA COMM]	TELEGR	**Pausenschritt** = Pause; Trennschritt; Trennlage; Kein-Strom-Schritt; Pausenpolarität; Pausenlage ≈ Startpolarität; Stoppolarität; Ruhestrom ≠ Stromschritt ↑ Zeichenelement; Zeichenlage [DATA COMM]	
spacing pulse (→spacing current)	TELEGR	(→Ruhestrom)	
spacing signal (→spacing pulse)	TELEGR	(→Pausenschritt)	
spacing tube = spacer tube	MECH	**Abstandsrohr**	
spacistor ↑ analog transistor	MICROEL	**Spacistor** ↑ Analog-Transistor	
SPADE [single channel per carrier PCM multiple-access demand assignment]	SAT.COMM	**SPADE**	
spaeker protector (→loudspeaker protector)	EL.ACOUS	(→Lautsprecher-Schutzschalter)	
spaghetti code	DATA PROC	**Spaghetti-Programm**	
span (v.t.) [fig.; e.g. distances]	TECH	**überbrücken** [fig.; z.B. Entfernungen]	
span (→span length)	OUTS.PLANT	(→Spannweite)	
span length [horizontally measured] = span ≈ suspension span	OUTS.PLANT	**Spannweite** [waagrecht gemessen] = Stützpunktabstand; Leitungsfeld; Mastspannweite ≈ Spannabschnitt	
spanned record	DATA PROC	**segmentierter Satz**	
spanner (→wrench)	MECH	(→Schraubenschlüssel)	
spanner wrench	MECH	**Vierkantschlüssel**	
spar (n.) = upright	MECH	**Holm**	
s-parameter (→scattering parameter)	NETW.TH	(→Streuparameter)	
spare (→stand-by)	TECH	(→Reserve)	
spare conductor	COMM.CABLE	**Reserveader** = Vorratsader	
spare equipment	TELEC	**Ersatzgerät 2**	
spare fiber	OPT.COMM	**Ersatzfaser**	
spare length	COMM.CABLE	**Reservelänge**	
spare module	ELECTRON	**Ersatzbaugruppe**	
spare module stock = spare module store	ELECTRON	**Ersatzbaugruppenvorrat** = Ersatzbaugruppenlager	
spare module store (→spare module stock)	ELECTRON	(→Ersatzbaugruppenvorrat)	
spare part = replacement part	TECH	**Ersatzteil**	
spare part list	DOC	**Ersatzteilliste**	
spare-parts service	ECON	**Ersatzteildienst**	
spark	PHYS	**Funke**	
spark discharge	PHYS	**Funkenentladung**	
spark extinction = spark quenching; quenching; spark quench; quench	PHYS	**Funkenlöschung**	
spark extinguisher	PHYS	**Funkenlöscher**	
spark gap	PHYS	**Funkenstrecke**	
sparking distance	PHYS	**Funkenschlagweite** = Schlagweite	
sparkle (→glitter) (v.i.)	TECH	(→glitzern)	
spark-over (n.) [breadown through gas, along the surface of a solid or liquid isolator] = flash-over; arc-over ↑ breakdown	PHYS	**Funkenüberschlag** [Entladung durch leitend gewordene Luft, längs der Oberfläche eines festen oder flüssigen Isolierstoffes] = Überschlag ↑ Durchschlag	
spark-over voltage	PHYS	**Überschlagspannung**	
spark quench (→spark extinction)	PHYS	(→Funkenlöschung)	
spark quenching (→spark extinction)	PHYS	(→Funkenlöschung)	
spark transmitter	RADIO	**Funkensender**	

English	Domain	German
spark welding (→arc welding)	METAL	(→Lichtbogenschweißung)
sparse array	DATA PROC	dünnbesiedeltes Feld
		= dünnbesiedelte Matrix
spatial (→volumetric)	MATH	(→räumlich)
spatial data management	DATA PROC	räumliche Datenverwaltung
spatial digitizer	DATA PROC	räumlicher Abtaster
spatial distribution	PHYS	geometrische Verteilung
= geometric distribution		= räumliche Verteilung
spatula	TECH	Spachtel
SPC (→stored-program control)	SWITCH	(→speicherprogrammierte Steuerung)
SPC switching system (→switching system with stored program control)	SWITCH	(→speicherprogrammiertes Vermittlungssystem)
SPDT (→single-pole double throw switch)	MICROEL	(→einpoliger Umschalter)
speaker (→loudspeaker)	EL.ACOUS	(→Lautsprecher)
speaker baffle (→loudspeaker baffle 1)	EL.ACOUS	(→Lautsprecher-Abdeckung)
speaker box (→loudspeaker box)	EL.ACOUS	(→Lautsprecherbox)
speaker-dependent	INF.TECH	sprecherabhängig
speaker-independent	INF.TECH	sprecherunabhängig
speaker set (→loudspeaker set)	EL.ACOUS	(→Lautsprecher-Set)
speaker system (→loudspeaker system)	EL.ACOUS	(→Lautsprechersystem)
speaker textile (→loudspeaker textile)	EL.ACOUS	(→Lautsprecher-Bespannungsgewebe)
special accessories	TECH	Sonderzubehör
		= Spezialzubehör
special amplifier tube	ELECTRON	Spezialverstärkerröhre
		= Spezialröhre
special applications software	DATA PROC	Anwendungssoftware
= application software		
special balancing network (→Hoyt balancing network)	TELEPH	(→Hoyt-Nachbildung)
special branch (→branch)	SCIE	(→Fachgebiet)
special cable	COMM.CABLE	Sonderkabel
special channel	RADIO	Sonderkanal
special character	DATA PROC	Sonderzeichen
[all non-alphanumeric characters]		[alle Zeichen die weder Buchstabe noch Ziffer sind]
= special sign; pi character		= Spezialzeichen
special data network	TELEC	Datensondernetz
special delivery length	COMM.CABLE	Sonderlänge
special direct costs	ECON	Sondereinzelkosten
special feature	TECH	Besonderheit
= pecularity		= Eigenartigkeit
≈ characteristic		≈ Merkmal
special feature (→added feature)	TECH	(→Komfortleistungsmerkmal)
special feature (→special function)	TERM&PER	(→Sonderfunktion)
special-feature telephone (→added-feature telephone)	TELEPH	(→Komfortfernsprecher)
special finish	TECH	Sonderausführung
≈ standard model		= Spezialausführung
		= Sonderanfertigung
special function	TERM&PER	Sonderfunktion
= special feature		= Spezialfunktion
special information tone	TELEPH	Hinweiston
["ask the operator or observe the written explanations"; a sequence of tones, e.g. 950, 1400 and 1800 Hz]		["fragen Sie die Auskunft oder achten Sie auf den Text"; 3 Frequenzen von 950, 1400 und 1800 Hz in unmittelbarer Folge]
specialist (→professional)	SCIE	(→Fachmann) (pl.-leute)
specialist publication	LING	Fachpublikation
specialists staff	ECON	Spezialistenstab
speciality (→branch)	SCIE	(→Fachgebiet)
specialization	SCIE	Spezialisierung
specialized computer (→special-purpose computer)	DATA PROC	(→Spezialrechner)
specialized fair	ECON	Fachmesse
specialized knowledge (→technical knowledge)	SCIE	(→Fachkenntnis)
specialized personnel	ECON	Fachpersonal
= skilled personnel		≈ Fachmann
specialized worker (→skilled worker)	ECON	(→Facharbeiter)
special key	TERM&PER	Sondertaste
= dedicated key		= Spezialtaste
special keyboard	TERM&PER	Sondertastatur
↓ added-feature keyboard		= Spezialtastatur; Sondertastenfeld; Spezialtastenfeld
		↓ Komforttastaur
special knowledge (→technical knowledge)	SCIE	(→Fachkenntnis)
special lot	ECON	Sonderposten
special make	TECH	Sonderanfertigung
= special version		= Spezialanfertigung; Sonderbauform
≈ special finish		= Sonderausführung
≠ standard model		≠ Serienmodell
special paper	TERM&PER	Spezialpapier
special price	ECON	Sonderpreis
special purpose	TECH	Spezialzweck
= special use		= spezielle Anwendung
special-purpose computer	DATA PROC	Spezialrechner
= dedicated computer; specialized computer		= Spezialcomputer; dedizierter Computer
≠ general-purpose computer		≠ Universalrechner
special radio service	RADIO	Sonderfunkdienst
special request	ECON	Sonderwunsch
special RF range	RADIO	Sonderbereich
special service	TELEC	Sonderdienst
= supplementary service		= ergänzender Dienst; Zusatzdienst
special-service circuit	SWITCH	Sonderdienstsatz
↑ junctor		↑ Verbindungssatz
special services radio network	TELEC	Funksondernetz
= dedicated radio network		
special sign (→special character)	DATA PROC	(→Sonderzeichen)
special telephone service	TELEPH	Fernsprechsonderdienst
= telephonic special service		= Telefonsonderdienst; Telephonsonderdienst
↓ directory assistance; fault complaint service		↓ Fernsprechauftragsdienst; Fernsprechauskunft; Störungsannahme; Telegrammaufnahme
special telephone set	TELEPH	Sonderfernsprecher
special thread	MECH	Spezialgewinde
special-use (adj.) (→application-specific)	TECH	(→anwendungsspezifisch)
special use (→special purpose)	TECH	(→Spezialzweck)
special version (→special make)	TECH	(→Sonderanfertigung)
specific address (→absolute address)	DATA PROC	(→absolute Adresse)
specific addressing (→absolute addressing)	DATA PROC	(→absolute Adressierung)
specification 2	TECH	Lastenheft
[commercial and technical terms]		[kommerzielle und technische Bedingungen]
≈ specification 1		= Ausschreibungsbedingungen
		≈ Spezifikation
specification 1	TECH	Spezifikation
[estipulated technical features]		[vorgegebene technische Bedingungen]
		= Pflichtenheft; technische Daten
		≈ Lastenheft
specification (→code number)	TECH	(→Sachnummer)
specification tree	TELEC	Spezifikationsbaum
specific code (→machine code 1)	DATA PROC	(→Maschinencode 1)
specific electric loading	POWER ENG	elektrischer Strombelag
[sum of currents crossing a hole to its circumference]		[Summe der Querströme zum Bohrumfang durch Bohrumfang]
specific electric resistance	PHYS	spezifischer elektrischer Widerstand
[resistance of a unitary volume 1 cm^2 x 1 cm]		[Widerstand einer Volumeneinheit 1 cm^2 x 1 cm]
= specific resistance; resistivity		= spezifischer Widerstand
≠ electric conductivity		≠ elektrische Leitfähigkeit

specific heat		PHYS	spezifische Wärme	speech coding (→speech encryption)	INF.TECH	(→Sprachverschlüsselung)
specific load = charge/mass ratio		PHYS	spezifische Ladung	speech communication (→voice communication)	TELEC	(→Sprachkommunikation)
specific magnetic resistance (→reluctivity)		PHYS	(→Reluktivität)	speech compression = voice compression	TELEC	Sprachkompression
specific resistance (→specific electric resistance)		PHYS	(→spezifischer elektrischer Widerstand)	speech connection (→telephone communication)	TELEC	(→Fernsprechverbindung)
specific thermal resistance		PHYS	spezifischer Wärmewiderstand	speech-controlled (→voice-controlled)	ELECTRON	(→sprachgesteuert)
specific weight		PHYS	spezifisches Gewicht	speech converter (→voice converter)	TELEC	(→Sprachwandler)
specified time (→delivery time)		ECON	(→Lieferfrist)	speech current	TELEPH	Sprechstrom
specify		TECH	spezifizieren	= telephone current; voice current		
specimen (→unit under measurement)		INSTR	(→Meßobjekt)	speech decoding = voice decoding	INF.TECH	Sprachentschlüsselung ≠ Sprachverschlüsselung
specimen (→test specimen)		QUAL	(→Prüfling)	≠ speech encryption		
specimen (→type model)		TECH	(→Ausführungsmuster)	speech digitization (→voice digitization)	TELEC	(→Sprachdigitalisierung)
speckle noise (→modal noise)		OPT.COMM	(→Modenrauschen)	speech digitizer (→voice digitizer)	TELEC	(→Sprachdigitalisierer)
specs (→spectacles)		TECH	(→Brille)	speech direction	TELEC	Sprechrichtung
spectacles (n.pl.t.) = glasses; specs		TECH	Brille	speech efficiency	TELEPH	Sprechwirkungsgrad
spectacle wearer		COLLOQ	Brillenträger	speech encryption	INF.TECH	Sprachverschlüsselung
spectral		PHYS	spektral	= voice coding; speech guard encryption; voice-guard encryption; speech coding; voice coding; cryptography; ciphony		= Sprachverschleierung ≠ Sprachentschlüsselung ↑ Kryptographie ↓ Sprachverwürfelung; Sprachbandinvertierung
spectral analysis (→spectrum analysis)		INSTR	(→Spektrumsanalyse)			
spectral characteristic		PHYS	Spektraleigenschaft			
spectral form (→spectral shape)		PHYS	(→Spektralform)			
spectral frequency		PHYS	Spektralfrequenz			
spectral line		PHYS	Spektrallinie = Linie	≠ speech decoding ↑ cryptology ↓ speech scrambling; speech inversion		
spectral photometer		INSTR	Spektralfotometer	speech filing	TELEC	Sprachspeicherung
spectral purity		PHYS	Spektralreinheit	= voice filing		
spectral range = spectral region		PHYS	Spektralbereich	speech frequency (→voice frequency)	TELEC	(→Niederfrequenz)
spectral region (→spectral range)		PHYS	(→Spektralbereich)	speech frequency band (→voice band)	TELEC	(→Sprachband)
spectral separation		PHYS	spektrale Zerlegung	speech-grade channel (→voice-grade channel)	DATA COMM	(→Sprechkanal)
spectral series		PHYS	Spektralserie			
spectral shape = spectral form		PHYS	Spektralform	speech guard (→speech protection)	TELEC	(→Sprachschutz)
spectral shifting		MODUL	Spektrumsverlagerung	speech-guard encryption (→speech encryption)	INF.TECH	(→Sprachverschlüsselung)
spectral window (→transmission window)		OPT.COMM	(→Übertragungsfenster)	speech highway = voice highway	SWITCH	Sprachmultiplexleitung = Sprach-Highway
spectrometer		INSTR	Spektrometer	speech input	SIGN.TH	akustische Spracheingabe
spectroscope		INSTR	Spektroskop	= voice input		
spectrum		PHYS	Spektrum	speech input (→voice input)	DATA PROC	(→Spracheingabe)
spectrum (→product range)		ECON	(→Produktspektrum)	speech interface (→voice interface)	TELEC	(→Sprachschnittstelle)
spectrum administration		RADIO	Spektrumsverwaltung			
spectrum analysis = spectral analysis		INSTR	Spektrumsanalyse = Spektralanalyse	speech interpolation = voice interpolation	TELEC	Sprachinterpolation
spectrum analyzer [selective, swept-tuned measurement and display of amplitude vs. frequency] = frequency analyzer ↑ signal analyzer		INSTR	Spektrumanalysator [selektive, gewobbelte Messung und Darstellung des Frequenzganges der Amplitude] = Frequenzanalysator; Frequenzspektrometer ↑ Signalanalysator	speech inversion = voice band inversion ↑ speech encryption	INF.TECH	Sprachbandinvertierung ↑ Sprachverschlüsselung
				speech level = voice level	TELEC	Sprachpegel
				speech memory = voice memory	TELEC	Sprachspeicher
spectrum efficiency		RADIO	Spektrumeffizienz = Spektrumausnutzung	speech mode (→voice mode)	TELEC	(→Sprachbetrieb)
				speech multiplexer = voice multiplexer	SWITCH	Sprachmultiplexer
spectrum limitation		EL.TECH	Spektrumsbegrenzung			
spectrum monitor		INSTR	Spektrum-Monitor	speech output (→voice response)	DATA PROC	(→Sprachausgabe)
spectrum occupancy recording		RADIO LOC	Frequenzbandbelegungs-Registrieranlage	speech output system (→voice output system)	TERM&PER	(→Sprachausgabesystem)
spectrum shaping		MODUL	Spektrumformung	speech path (→voice path)	TELEC	(→Fernsprechweg)
speech (n.) = address		COLLOQ	Ansprache [Rede]	speech-plus duplex = voice-plus duplex	TELEGR	Zwischenkanal-WT = KWT
speech (→voice)		INF.TECH	(→Sprache)	speech-plus signaling (→inband signaling)	TRANS	(→Inbandsignalisierung)
speech analysis (→voice analysis)		TELEC	(→Sprachanalyse)	speech-plus telegraphy (→inband telegraphy)	TELEGR	(→Inbandtelegraphie)
speech answer (→speech recording)		EL.ACOUS	(→Sprachantwort)	speech protection [against interferences by voice signals] = speech guard; voice protection; voice guard; guarding; guard action ≈ speech security	TELEC	Sprachschutz [gegen Störungen durch Sprachsignale] ≈ Sprachsicherheit
speech band (→voice band)		TELEC	(→Sprachband)			
speech-band limitation (→voice-band limitation)		TELEC	(→Sprachbegrenzung)			
speech-band limiting filter (→voice-band limiting filter)		TELEC	(→Sprachbegrenzungsfilter)			
speech channel (→telephone channel)		TELEC	(→Fernsprechkanal)			
speech circuit (→voice circuit)		TELEC	(→Sprechkreis)			

speech recognition INF.TECH **Spracherkennung**
= voice recognition = Worterkennung
↓ personal voice recognition ↓ Stimmerkennung
speech-recognizing INF.TECH (→sprachverstehend)
(→voice-recognizing)
speech recognizing dialog DATA PROC (→sprachverstehendes Dialogsystem)
system (→voice recognizing dialog system)
speech recorder (→call TERM&PER (→Sprachaufzeichnungsgerät)
recorder)
speech recording EL.ACOUS **Sprachantwort**
= speech answer; voice recording; voice = Sprachaufzeichnung
answer
speech response (→voice DATA PROC (→Sprachausgabe)
response)
speech scrambling TELEC **Sprachverwürfelung**
= voice band scrambling = Sprachbandverwürfelung
↑ speech encryption ↑ Sprachverschlüsselung
speech security TELEC **Sprachsicherheit**
= voice security ≈ Sprachschutz
≈ speech protection
speech signal (→voice TELEC (→Sprachsignal)
signal)
speech signal transmission TELEC (→Sprachübertragung)
(→voice transmission)
speech simulation INF.TECH **Sprachsimulation**
= voice simulation
speech sound (→voice) INF.TECH (→Sprache)
speech synthesis (→voice INF.TECH (→Sprachsynthese)
synthesis)
speech synthesizer (→voice TERM&PER (→Sprachsynthesizer)
synthesizer)
speech time (→conversation TELEPH (→Gesprächsdauer)
time)
speech-time limiter SWITCH **Gesprächszeitbegrenzer**
speech wire TELEC **Sprechader**
= voice wire ≠ Signalader
≠ signal wire
speed 1 (→rotational speed) PHYS (→Drehzahl)
speed 2 (→velocity) PHYS (→Geschwindigkeit)
speed (→fastness) COLLOQ (→Schnelligkeit)
speed adaptation DATA COMM **Geschwindigkeitsanpassung**
speed board DATA PROC **Beschleunigerkarte**
= tuning board = Beschleunigungskarte
speedboard (→magnetic OFFICE (→Magnettafel)
board)
speed calling (→abbreviated SWITCH (→Kurzwahl)
dialing)
speed control CONTROL **Drehzahlregelung**
speed controller CONTROL **Drehzahlregler**
speed conversion DATA COMM **Geschwindigkeitsumsetzung**
speed counter (→revolution INSTR (→Drehzahlmesser)
counter)
speed distortion (→telegraph TELEGR (→Schrittverzerrung)
signal distortion)
speedily (→fast) TECH (→schnell)
speed of propagation PHYS **Fortpflanzungsgeschwindigkeit**
= propagation speed; propagation velocity
speed-power product MICROEL **tdP-Produkt**
speed-up capacitor CIRC.ENG **Überhöhungskondensator**
[to increase the switching speed] [zur Erhöhung der Schaltgeschwindigkeit]
≈ bypass capacitor [NETW.TH] = Speed-up-Kondensator
≈ Überbrückungskondensator [NETW.TH]
spell (v.t.) INF.TECH **buchstabieren**
spell check (→spelling DATA PROC (→Rechtschreibprüfung)
check)
spellchecking (→spelling DATA PROC (→Rechtschreibprüfung)
check)
spelling LING **Schreibweise**
≈ orthography = Schreibung
≈ Rechtschreibung
spelling check DATA PROC **Rechtschreibprüfung**
= spell check; spellchecking = Orthographieprüfung
spelling checker (→spelling DATA PROC (→Rechtschreibprogramm)
check program)
spelling checker program DATA PROC (→Rechtschreibprogramm)
(→spelling check program)

spelling check program DATA PROC **Rechtschreibprogramm**
= spelling checker program; spelling = Orthographieprogramm
checker; dictionary program
spelling error LING **Rechtschreibfehler**
↑ grammatical error = Schreibfehler
↑ grammatischer Fehler
spending (→expense) ECON (→Ausgabe)
sphere MATH **Kugel**
[volume delimited by a spheric surface] [von einer Kugeloberfläche begrenzter Raum]
= globe; ball
sphere (→ball) TECH (→Kugel)
sphere cap (→spherical cup) MATH (→Kugelkalotte)
sphere cap diaphragm EL.ACOUS **Kalottenmembran**
= Kugelkappenmembran
sphere cap loudspeaker EL.ACOUS **Kalottenlautsprecher**
= Kugelkappenlautsprecher
spherical MATH **sphärisch**
= kugelförmig; kugelig; ballförmig
≈ rund
spherical antenna (→isotropic ANT (→Kugelstrahler)
radiator)
spherical armature contact COMPON **Kugelankerkontakt**
↑ Schutzrohrkontakt
spherical array ANT **Kugelgruppenantenne**
= sphärische Gruppenantenne; sphärische Gruppe
spherical capacitor PHYS **Kugelkondensator**
spherical cavity resonator MICROW **Kugelresonator**
spherical coordinate (→polar MATH (→Polarkoordinate)
coordinate)
spherical cup MATH **Kugelkalotte**
= sphere cap = Kugelkappe; Kalotte
spherical function MATH **Kugelfunktion**
= Legendresches Polynom
spherical geometry MATH **Kugelgeometrie**
spherical lens PHYS **sphärische Linse**
↓ converging lens; diverging lens = Kugellinse
↓ Sammellinse; Zerstreuungslinse
spherical reflector ANT **Kugelschalen-Reflektor**
spherical roller bearing MECH **Tonnenrollenlager**
spherical sector MATH **Kugelsektor**
= Kugelausschnitt
spherical segment MATH **Kugelsegment**
[delimited by a sphere cap and a [durch Kugelkappe und
plain] ebene Fläche begrenzt]
= Kugelabschnitt
spherical slice MATH **Kugelschicht**
[with two parallel surfaces] [mit zwei parallelen Schnittflächen]
spherical surface MATH **Kugeloberfläche**
spherical symmetry MATH **kugelsymmetrisch**
spherical wave PHYS **Kugelwelle**
= spheric wave = sphärische Welle
spherical waves horn EL.ACOUS **Kugelwellentrichter**
spheric wave (→spherical wave) PHYS (→Kugelwelle)
spheroid MATH **Sphäroid**
[ball-like form] [kugelähnliche geometrische Figur]
≈ Rotationsellipsoid
spheroidal (adj.) MATH **sphäroid**
= ball-like = kugelähnlich
spider-grid bonding MICROEL **Spider-grid-Kontaktierung**
spider spanner MECH **Kreuzschlüssel**
spider web antenna ANT **Spinnennetzantenne**
= Telerana-Antenne
spider-web coil COMPON **Korbbodenspule**
= Spinngewebespule; Spinnnetzspule
spigot (→guide pin) MECH (→Führungsstift)
spike ELECTRON **Zacken**
[a spurious short peak superimpost to [einem Impuls überlagerte
a pulse form] kurzzeitige und unerwünschte Spitze]
= impulse hit
≈ surge [EL.TECH] = Überschwingspitze
= Stoß [EL.TECH]
spike (→disturbance effect) ELECTRON (→Störeffekt)
spill over (v.t.) ANT **vorbeistrahlen**
= überstrahlen

spillover (→overflow 1)		SWITCH	(→Überlauf)	splicing plate	OPT.COMM	**Spleißplatte**
spillover (n.) (→overradiation)		ANT	(→Vorbeistrahlung)	splicing sleeve (→jointing sleeve)	OUTS.PLANT	(→Verbindungsmuffe)
spillover (n.) (→overflow)		ELECTRON	(→Überlauf)	splicing tape	OUTS.PLANT	**Klebeband**
spin		PHYS	**Spin**	splicing technique	OPT.COMM	**Spleißtechnik**
spin alignment		PHYS	**Spinausrichtung**	= jointing technique		
spin axis		PHYS	**Drehachse**	spline	DATA PROC	**Spline**
= rotation axis				[computer graphics; polynomial for pieceweise approximation]		[Computergraphik; Polynom für eine abschnittsweise Näherung]
spin direction		PHYS	**Spin-Richtung**			
spindle		MECH	**Spindel**	spline (→wedge)	MECH	(→Keil)
spindle coupling		MECH	**Doppelgelenk-Kupplung**	spline area [computer graphics]	DATA PROC	**Splinefläche** [Computergraphik]
spindle potentiometer		COMPON	**Spindeltrimmer**	spline courve [computer graphics]	DATA PROC	**Splinekurve** [Computergraphik]
spindle-shaped		TECH	**spindelförmig**			
spin doublet		PHYS	**Spin-Dublett**	spline hub	MECH	**Keilnabe**
spin effect		PHYS	**Spin-Effekt**	splinter	MECH	**Splitter**
spinel		CHEM	**Spinell**	split (v.t.)	TECH	**spalten**
spin electron		PHYS	**Spin-Elektron**	split (n.)	TECH	**Spalte**
spinel ferrite		CHEM	**Spinell-Ferrit**	[a narrow, deep and long separation of a solid object]		[schmale, tiefe und längliche Trennung eines festen Gegenstands]
spinel lattice		PHYS	**Spinell-Gitter**			
spin flip		PHYS	**Spin-Umklappung**	= cleft		
spin moment		PHYS	**Spin-Moment**	≈ crack; notch		≈ Sprung; Kerbe
spinning		METAL	**Metalldrücken**	split (v.t.) (→alternate)	TELEPH	(→makeln)
			= Drücken	split anode	ELECTRON	**Schlitzanode**
spinning tool		MECH	**Druckstahl** [Werkzeug]	split-anode magnetron	MICROEL	**Zweischlitz-Magnetron**
spin-off (n.)		TECH	**Nebeneffekt**	split-baud-rate mode [receives at one and transmits at another]	DATA COMM	**Zwei-Baudraten-Betrieb** [empfängt in einer und sendet in anderer]
spin orbit		PHYS	**Spinbahn**			
spin orientation		PHYS	**Spin-Orientierung**	split cable grip (→shackle)	OUTS.PLANT	(→Kabelziehschlauch)
spin quantum number ↑ quantum number		PHYS	**Spin-Quantenzahl** ↑ Quantenzahl	split lens	PHYS	**Halblinse**
				split pin	MECH	**Spaltkegelstift**
spin-spin interaction		PHYS	**Spin-Wechselwirkung**	split screen = screen splitting; split window	DATA PROC	**geteilter Bildschirm**
spinwriter (→thimble printer)		TERM&PER	(→Typenkorb-Drucker)			
				split-sheath balun (→slit-tube balun)	ANT	(→Schlitzbalun)
spiral (→helical)		TECH	(→schraubenförmig)			
spiral (→helix)		MATH	(→Schraubenlinie)	splitting	TECH	**Spaltung**
spiral antenna (→helix antenna)		ANT	(→Wendelantenne)	splitting (→alternation between lines)	TELEPH	(→Makeln)
spiral cable		EQUIP.ENG	**Spiralkabel**	splitting (→call splitting)	SWITCH	(→Verbindungsaufspaltung)
spiral coil		COMPON	**Spiralspule**			
spiral conductor (→helical conductor)		ELECTRON	(→Wendelleitung)	splitting jack = splitting socket	COMPON	**Trennbuchse**
				splitting plug = breaker plug	COMPON	**Trennstecker**
spiral cord (→retractile cord)		TERM&PER	(→Spiralschnur)			
spiral-eight		COMM.CABLE	**Doppelsternvierer**	splitting point	EQUIP.ENG	**Trennstelle**
spiral-eight cable		COMM.CABLE	**Doppelsternkabel**	splitting set	TELEPH	**Makleranlage**
spiral-four (→star quad)		COMM.CABLE	(→Sternvierer)	splitting socket (→splitting jack)	COMPON	(→Trennbuchse)
spiral-four cable 2		COMM.CABLE	**Feldfernkabel**			
= field cable			= Feldkabel	split tube balun	ANT	**Halbschalensymmetrierglied** = Schlitzübertrager ↑ Symmetrietransformator
spiral-four cable 1 (→star-quad cable)		COMM.CABLE	(→Sternviererkabel)			
				split tubular pin	MECH	**Spannstift**
spiral-four quad (→star quad)		COMM.CABLE	(→Sternvierer)	split window (→split screen)	DATA PROC	(→geteilter Bildschirm)
spiral of Archimedes = archimedian spiral		MATH	**archimedische Spirale**	split-wire type transformer	EL.TECH	**Anlegewandler**
spiral scanning		ELECTRON	**Spiralabtastung**	spoke (n.) (→wheel spoke)	MECH	(→Radspeiche)
spiral spring (→helical spring)		MECH	(→Spiralfeder)	spoken announcement (→spoken message)	TELEPH	(→Ansage)
spiral wrap		COMPON	**Spiralwicklung**			
spire (→turn)		EL.TECH	(→Windung)	spoken message = spoken announcement	TELEPH	**Ansage** = Durchsage
spiroband		COMM.CABLE	**Spiralschlauch**			
SPL (→string processing language)		DATA PROC	(→Stringverarbeitungssprache)	spokesman ↓ corporate spokesman; governmental spokesman; military spokesman	ECON	**Sprecher** ↓ Firmensprecher; Regierungssprecher; Militärsprecher
splashing water		TECH	**Spritzwasser**			
splash-proof (adj.) = protected against splashing water		TECH	**spritzwasserfest** = spritzwasserdicht	sponge rubber ≈ foam rubber	TECH	**Schwammgummi** ≈ Schaumgummi
splash ringing (→immediate ringing)		SWITCH	(→Sofortruf)	sponsored television = commecial TV	BROADC	**Werbefernsehen**
splice (v.t.) = joint (v.t.)		COMM.CABLE	**spleißen**	spontaneous (→voluntary)	COLLOQ	(→freiwillig)
				spontaneous emission	PHYS	**spontane Emission**
splice (→cable splice)		COMM.CABLE	(→Kabelspleiß)	spontaneous information = spontaneous message; report by exception ≠ inquiry information	TELECONTR	**Spontanmeldung** ≠ Abfragemeldung
splice attenuation		OPT.COMM	**Spleißdämpfung**			
spliceman		OUTS.PLANT	**Spleißmonteur**			
splicing (→cable splice)		COMM.CABLE	(→Kabelspleiß)			
splicing chamber (AM) (→cable vault)		OUTS.PLANT	(→Kabelkeller)	spontaneous magnetization	PHYS	**spontane Magnetisierung**
				spontaneous message (→spontaneous information)	TELECONTR	(→Spontanmeldung)
splicing chamber (BRI) (→cable chamber)		OUTS.PLANT	(→Kabelschacht)			
splicing device		OPT.COMM	**Spleißgerät** = Spleißvorrichtung	spontaneous transmission = spontaneous mode	TELECONTR	**Spontanbetrieb** = spontaner Betrieb

spoofing SIGN.ENG
  [of an alarm system by electonic means]
  ≈ circumvention
spool (v.t.) TECH
  ≈ wind; wrap-up
spool 1 (n.) DATA PROC
  [simultaneous peripheral operation via buffers]
  = spooling 1
spool 1 (→reel) TECH
spool 2 (n.) DATA PROC
  (→spooling 2)
spooled file DATA PROC
  = spooling file; spool file
spooler DATA PROC
  [software or hardware permitting print while doing something else]
spool file (→spooled file) DATA PROC
spool frame (→magnetic tape spool) TERM&PER
spooling 2 DATA PROC
  ["Simultaneous Peripheral Operation On Line"; temporary transfer to external memory, to free faster central unit]
  = spool 2 (n.)
spooling 1 (→spool 1) DATA PROC
spooling file (→spooled file) DATA PROC
spool-in mode DATA PROC
  ↑ spool mode
spool-out mode DATA PROC
  ↑ spool mode
spool-up (→wrap-up) TECH
sporadic (→intermittent) TECH
sporadic fault (→intermittent fault) QUAL
spot (n.) BROADC
  ↓ commercial spot
spot (n.) TECH
spot (n.) ELECTRON
spot- (adj.) TECH
spot (→illumination spot) SAT.COMM
spot analysis (→point-to-point measurement) INSTR
spot beam ELECTRON
spot beam SAT.COMM
spot-beam antenna SAT.COMM
spot check (→sample) MATH
spot measurement (→point-to-point measurement) INSTR
spot noise factor TELEC
spot protection (→object protection) SIGN.ENG
spot solution TECH
spot-weld (v.t.) METAL
spot weld (→spot welding) METAL
spot welding METAL
  = spot weld
spot wobble TV
  [a periodic motion of the scanning spot, transverse to the scanning direction]
spout (→squirt) TECH
SPP MICROEL
  [Signal Processing Peripheral]
spray 1 (v.t.) TECH
  ≈ squirt; sprinkle
spray 2 (v.t.) (→pulverize) TECH

Ausschaltung
  [einer Warnanlage durch elektrischen Eingriff]
  ≈ Umgehung
spulen
  ≈ wickeln; aufwickeln
Spool 1
  [gleichzeitiges Austeuern mehrerer Peripheriegeräte über Puffer]
(→Spule 1)
(→Spool-Betrieb)
Ausspuldatei
  = ausgespulte Datei; Spool-Datei
Spooler
  [Software oder Hardware die einen Druckvorgang ermöglicht während etwas anderes abgewickelt wird]
(→Ausspuldatei)
(→Magnetbandspule)
Spool-Betrieb
  [zeitweiliges Auslagern auf externe Speicher, um schnellere Zentraleinheit freizustellen]
  = Auslagern; Ausspulen
  ↓ Spool-in-Betrieb; Spool-out-Betrieb
(→Spool 1)
(→Ausspuldatei)
Spool-in-Betrieb
  [Dateneingabeverfahren]
  ↑ Spool-Betrieb
Spool-out-Betrieb
  [Datenausgabeverfahren]
  ↑ Spool-Betrieb
(→aufwickeln)
(→intermittierend)
(→intermittierender Fehler)

↓ Werbeeinblendung
Einblendung
Fleck
Leuchtfleck
punktuell
(→Ausleuchtzone)
(→punktweises Messen)
Punktstrahl
Punktbündel
Spot-Antenne
(→Stichprobe)
(→punktweises Messen)
Spektralrauschzahl
(→Objektschutz)
Insellösung
punktschweißen
(→Punktschweißen)
Punktschweißen
  = Punktschweißung
Zeilenwobbelung
  [eine überlagerte Wobbelung des Abtaststrahls, quer zur Ablenkrichtung]
  = Zeilenwobbeln
(→spritzen)
Zusatzrechenwerk
  = SPP
sprühen
  ≈ spritzen; sprengen 2
(→zerstäuben)

spray gun TECH **Spritzpistole**
spray-on TECH **aufspritzen**
spray-paint TECH **Spritzlackierung**
spread (→margin 1) ECON (→Spanne)
spread (→scatter propagation) RADIO PROP (→Streuausbreitung)
spread (→spread width) INSTR (→Streubreite)
spreader TECH **Rahe** (n.f.; pl.-en)
  [device holding two lines apart]   [Stange zwischen Seilen]
  = Rah (n.f.)
spreading loss (→free-space loss) SAT.COMM (→Freiraumdämpfung)
spreading resistance MICROEL **Ausbreitungswiderstand**
spreading-resistance temperature sensor MICROEL **Ausbreitungswiderstand-Temperatursensor**
  = Spreading-Widerstand-Temperatursensor
spreadsheet (→worksheet) OFFICE (→Arbeitsblatt)
spreadsheet analysis DATA PROC **Tabellenkalkulation**
  = spreadsheet processing   = Tabellenverarbeitung
spreadsheet processing DATA PROC (→Tabellenkalkulation)
  (→spreadsheet analysis)
spreadsheet program DATA PROC **Tabellenkalkulationsprogramm**
  = tabulation program; tab program; electronic spreadsheet
spread-spectrum communications RADIO **Streuspektrumfunk**
  = Streuspektrumübertragung
spread spectrum modulation MODUL **Streuspektrum-Modulation**
spread sprectum MODUL **Streuspektrum**
  = SS
spread width INSTR **Streubreite**
  = spread
spread-wire dipole (→triangular dipol) ANT (→Spreizdipol)
spring MECH **Feder**
spring assembly MECH **Federsatz**
  = springset
spring-back MECH **Rückfederung**
spring-back file OFFICE **Klemmappe**
spring climp MECH **Federklammer**
spring constant MECH **Federkonstante**
spring contact COMPON **Federkontakt**
spring contact strip COMPON **Federleiste**
  = female terminal strip   ≠ Steckerleiste
  ≠ connector strip   ↑ Kontaktleiste
  ↑ contact strip
spring force MECH **Federkraft**
spring-hard MECH **federhart**
spring load (→spring tension) MECH (→Federspannung)
spring-loaded TECH **federbelastet**
spring lock washer (→lock washer) MECH (→Sicherungsscheibe)
spring pressure gauge INSTR **Federdruckmesser**
  = Federmanometer
spring reverb EL.ACOUS **Nachhallspirale**
springset (→spring assembly) MECH (→Federsatz)
spring sheet steel METAL **Federstahlblech**
spring steel METAL **Federstahl**
spring-steel wire METAL **Federstahldraht**
spring tension MECH **Federspannung**
  = spring load   ≈ Federkraft
spring washer (→lock washer) MECH (→Sicherungsscheibe)
spring wire MECH **Federdraht**
sprinkle (v.t.) TECH **sprengen 3**
  [to distribute a liquid]   [eine Flüssigkeit verteilen]
sprite DATA PROC **Sprite**
  [small object that can be moved freely in a computer graphics]   [kleine, frei in Graphiken bewegbare Figur]
sprockered margin (→pin-feed edge) TERM&PER (→Führungsstreifen)
sprocket drum TERM&PER **Stachelwalze**
sprocket-feed device TERM&PER **Traktor-Einzug**
sprocket hole (→feed hole) TERM&PER (→Vorschubloch)
sprocket holes (→feed holes) TERM&PER (→Transportlochung)
sprocket wheel MECH **Kettenrad**
sprocket wheel (→pinfeed platen) TERM&PER (→Stachelrad)

spruce-up (v.t.) (→refurbish)		TECH	(→aufpolieren)	square-loop ferrite	METAL	Rechteck-Ferrit
spur (→entrance link)		TRANS	(→Zubringerstrecke)	= rectangular-hysteresis ferrite		
spur gear		MECH	Stirnrad	square matrix	TELEC	quadratische Matrix
↑ gear			↑ Zahnrad	square meter (NAM)	PHYS	Quadratmeter
spurious		INF.TECH	unerwünscht	= square metre (BRI); m^2		= m^2
= undesired; parasitic; incidental			= parasitär	square metre (BRI) (→square meter)	PHYS	(→Quadratmeter)
spurious emission		MODUL	Nebenaussendung			
			= Nebenwelle; parasitäre Aussendung	square nut	MECH	Vierkantmutter
				square of absolute value	MATH	Betragsquadrat
spurious emission (→unwanted emission)		RADIO	(→Störstrahlung)	square pulse	ELECTRON	Rechteckimpuls
				= rectangular pulse; square-wave pulse		≈ Rechteckwelle
spurious emission attenuation		MODUL	Nebenwellendämpfung	≈ square-wave reversal		
spurious error (→transit error)		ELECTRON	(→Zufallsfehler)	square root	MATH	Quadratwurzel
						= Wurzel 2
spurious irradiance (→spurious irradiation)		ELECTRON	(→Einstrahlung)	square signal (→square-wave reversals)	ELECTRON	(→Rechteckwelle)
spurious irradiation		ELECTRON	Einstrahlung	square thread	MECH	Flachgewinde
= spurious irradiance; irradiation; irradiance				square wave (→square-wave reversals)	ELECTRON	(→Rechteckwelle)
spurious modulation		MODUL	Störmodulation	squarewave characteristics	ELECTRON	Rechteck-Kenndaten
spurious oscillation		PHYS	Nebenschwingung	square-wave generator	CIRC.ENG	Rechteckgenerator
= parasitic oscillation; parasitic effect			= Parasitärschwingung; unerwünschte Schwingung; Parasitäreffekt	= rectangular-wave generator		= Rechteckwellengenerator
				↑ oscillator		↑ Oszillator
				square-wave measuring generator	INSTR	Rechteckwellen-Meßgenerator
spurious radiation (→unwanted emission)		RADIO	(→Störstrahlung)	= rectangular-wave measuring generator		
spurious resonance		COMPON	(→Störresonanz)	square-wave modulation	MODUL	Rechteckmodulation [HF]
(→unwanted response)						
spurious response		ELECTRON	Störsignal-Ansprechverhalten	square-wave pulse (→square pulse)	ELECTRON	(→Rechteckimpuls)
spurious signal (→interfering signal)		TELEC	(→Störsignal)	square-wave response	ELECTRON	Rechteckwellenantwort
spur link (→entrance link)		TRANS	(→Zubringerstrecke)	square-wave reversals	ELECTRON	Rechteckwelle
sputter (v.t.) (→pulverize)		TECH	(→zerstäuben)	= reversals; rectangular wave; square wave; rectangular repetition rate; square signal		= Rechteckfolge; Mäander; Rechteckpuls; Rechtecksignal
sputtering (→cathode sputtering)		ELECTRON	(→Kathodenzerstäubung)			
sputtering (→pulverization)		TECH	(→Zerstäubung)	↑ relaxation oscillation		↑ Kippschwingung
SQ decoder		CONS.EL	SQ-Decodierer	square-wave voltage	ELECTRON	Rechteckspannung
[stereophonic quadro sound]			[Stereo-Quadrofonie]	square well (→potential well)	PHYS	(→Potentialmulde)
square (v.t.)		MATH	quadrieren			
↑ power (v.t.)			↑ potenzieren	squaring circuit	CIRC.ENG	Rechteckformer
square (n.)		MATH	Quadrat	≈ schmitt trigger		≈ Schmitt-Trigger
[equilateral rectangle]			= gleichseitiges Rechteck	squaring element	COMPON	Quadrierglied
= quadrate (n.)			↑ Rechteck	squeezable waveguide	MICROW	Quetschhohlleiter
↑ rectangle						= Quetschmeßleitung
square (adj.)		MATH	quadratisch	squeezed (→crushed)	TECH	(→gequetscht)
= quadratic; square-law			↑ rechteckig	squeezer (→circuit engineer)	ELECTRON	(→Schaltungsentwickler)
↑ rectangular						
square antenna		ANT	Quadratantenne	squeg (v.i.)	ELECTRON	überoszillieren
= square-loop antenna			= Viereckschleife; Quadratrahmen; Square-loop-Antenne	[to oscillate in a highly irregular manner]		
				squegging	ELECTRON	Überoszillation
square bar		METAL	Vierkantstange	[irregular oscillation]		[unregelmäßige Schwingung]
square-bar aluminum		METAL	Vierkantaluminium			
square-bare iron		METAL	Vierkantstahl	squegging oscillator (→blocking oscillator)	CIRC.ENG	(→Sperrschwinger)
= square iron						
square bolt		MECH	Vierkantkopfschraube	squelch (n.)	HF	Stummabstimmung
square bracket		MATH	eckige Klammer	= squelching		= stumme Regelung
[[ ]]			[[ ]]	squelch	RADIO	Empfänger-Rauschabschaltung
= bracket 2 (n.)				[desactivates a radio receiver when noise exceeds a level]		
square-end horn (→pyramidal horn)		ANT	(→Pyramidenhorn)			= Squelch
				squelch (→squelch circuit)	RADIO	(→Rauschsperre)
square feather key		MECH	Vierkantpaßfeder	squelch circuit	RADIO	Rauschsperre
square file		TECH	Vierkantfeile	= squelch; interference suppressor		= Störsperre
square flange		MECH	Rechteckflansch	squelching (→squelch)	HF	(→Stummabstimmung)
			≠ Rundflansch	squint (n.)	ANT	Schielen
square formation (→star quad)		COMM.CABLE	(→Sternvierer)	squint angle	TECH	Fehlwinkel
				squint angle	ANT	Schielwinkel
square head		METAL	Vierkantkopf	squint antenna	ANT	Schielantenne
square interpolation		MATH	quadratische Interpolation	squirrel cage	PHYS	Käfiganker
square iron (→square-bare iron)		METAL	(→Vierkantstahl)	squirrel cage antenna (→cage antenna)	ANT	(→Käfigantenne)
square-law (→square)		MATH	(→quadratisch)	squirrel-cage magnetron	MICROW	Käfigmagnetron
square-law detection		HF	quadratische Gleichrichtung	squirrel-cage rotor	POWER SYS	Kurzschlußläufer
square-law load		MICROW	Quadratlast	squirrel cage winding	EL.TECH	Käfigwicklung
square-law modulator		MODUL	quadratischer Modulator	= cage winding		
square loop		PHYS	Rechteckschleife	squirt (v.t.)		TECH spritzen
[ferrite]			[Ferrit]	= syringe; spout; jet		≈ sprengen 3; sprühen
= right-angle curve			= Rechteck-Hystereseschleife	= sprinkle; spray		
square-loop antenna (→square antenna)		ANT	(→Quadratantenne)	sr (→steradian)	PHYS	(→Steradiant)
				Sr (→strontium)	CHEM	(→Strontium)

SRAM		DATA PROC
= static RAM		
≠ DRAM		
↑ memory chip; random access memory		

SRAM
= statischer Direktzugriffsspeicher; statischer RAM; statischer Schreib-Lese-Speicher
≠ dynamischer Direktzugriffsspeicher
↑ Speicherbaustein; Direktzugriffsspeicher

SRV (→surface recombination velocity) PHYS (→Oberflächenrekombinations-Geschwindigkeit)
SS (→spread sprectum) MODUL (→Streuspektrum)
SSB (→single sideband) MODUL (→Einseitenband)
SSB filter HF **SSB-Filter**
= Einseitenbandfilter
SSB modulator HF **SSB-Modulator**
SSB phase noise TELEC **Einseitenband-Phasenrauschen**
SSB system (→single-sideband system) RADIO (→Einseitenbandsystem)
ß cutoff MICROEL **Beta-Grenzfrequenz**
= beta cutoff ↓ Transitfrequenz
↓ gain-bandwidth product
SS disc (→SS disk) TERM&PER (→SS-Diskette)
SS disk TERM&PER **SS-Diskette**
= SS disc; SS diskette; single-sided disk; single-sided disc; single-sided diskette; single-sided floppy disk; single-sided floppy disc; one-sided disk; one-sided disc; one-sided diskette; 1-sided disk; one-sided disc; one-sided diskette; 1-sided diskette
= einseitig beschreibbare Diskette; einseitige Diskette
SS diskette (→SS disk) TERM&PER (→SS-Diskette)
SSI MICROEL **Kleinstintegration**
[1 to 10 components per IC] [1 bis 10 Komponenten pro IC]
= small scale integration
= niedriger Integrationsgrad
S signal (→synchronizing signal) TV (→Synchronsignal)
SSPA (→solid-state power amplifier) MICROEL (→Halbleiterleistungsverstärker)
SSR (→secondary surveillance radar) RADIO LOC (→Sekundärradar)
SSTV (→slow-scan TV) TV (→Slow-scan-TV)
St (→Stokes) PHYS (→Stokes)
stability PHYS **Stabilität**
≠ lability ≠ Labilität
stability criterion CONTROL **Stabilitätskriterium**
stability test TECH **Stabilitätsprüfung**
stabilization characteristics CIRC.ENG **Stabilisierungsfaktor**
stabilization circuit CIRC.ENG **Stabilisierungsschaltung**
stabilization of bias point ELECTRON **Arbeitspunktstabilisierung**
= stabilization of working point
stabilization of working point ELECTRON (→Arbeitspunktstabilisierung)
(→stabilization of bias point)
stabilization voltage CIRC.ENG **Stabilisatorspannung**
stabilize (v.t.) TECH **stabilisieren**
stabilized current regulator CIRC.ENG **Stromstabilisator**
= constant-current power supply
= Konstantstromquelle; Konstantstrom-Stromversorgung; Präzisionsstromquelle; Präzisionsstrom
stabilized power supply POWER SYS **stabilisierte Stromversorgung**
stabilizer CIRC.ENG **Konstanthalter**
= Stabilisator; Konstanter
↓ Spannungsstabilisator; Stromstabilisator
stable TECH **stabil**
≈ balanced
stable ECON **wertbeständig**
= stabil
stack (v.t.) TECH **häufen**
≈ pile ≈ stapeln
stack (n.) TERM&PER **Stapel**
[of paper]
= deck (of cards)
stack (n.) TECH **Haufen**
≈ pile (n.) ≈ Stapel

stack DATA PROC **Datenstapel**
[sequential list of data in main memory; data are retrieved from one of both ends]
[sequentielle Datenliste im Hauptspeicher; wird von einem der beiden Enden abgearbeitet]
≈ stack storage
= Stapel 2
≈ Stapelspeicher
stack (→data stack) DATA PROC (→Datenkeller)
stackable plug COMPON **Stapelstecker**
= Turmstecker
stack construction style EQUIP.ENG (→Stapelbauweise)
(→stack system)
stacked dipole (→broadside array) ANT (→Dipolebene)
stacked omnidirectional antenna ANT **aufgestockter Rundstrahler**
stacker (→output stacker) TERM&PER (→Ablagefach)
stacker (→pocket) TECH (→Fach)
stacking COLLOQ **Stapelung**
= piling; batching = Schubbildung
stacking ANT **Stockung**
stacking distance ANT **Stockungsabstand**
stack instruction DATA PROC **Kellerbefehl**
stack memory (→stack storage) DATA PROC (→Stapelspeicher)
stack mode DATA PROC **Kellerungsverfahren**
stack pointer DATA PROC **Stapelzeiger**
[register in a stack memory containing the address of the first or last entry]
[Register mit Adresse des ersten oder letzten Eintrages eines Kellerspeichers]
= Kellerzeiger; Kellerzähler
stack storage DATA PROC **Stapelspeicher**
[temporary storage with sequential input and output from one defined stack extremity]
[Zwischenspeicher mit sequentiellem Eintrag und Abruf vom vorgegebenen Stapelende]
= stack store; stack memory; program stack
= Stack-Speicher
≈ stack ≈ Datenstapel
↓ push-down storage; push-up storage ↓ Kellerspeicher; Silospeicher
stack store (→stack storage) DATA PROC (→Stapelspeicher)
stack style (→stack system) EQUIP.ENG (→Stapelbauweise)
stack system EQUIP.ENG **Stapelbauweise**
= stack construction style; stack style
staff 1 ECON **Stab**
[of an organization] [Organisation]
staff (→workforce) ECON (→Belegschaft)
staff 2 (→staff member) ECON (→Mitarbeiter)
staff (→personnel) ECON (→Personal)
staff (→bar 1) MECH (→Stab)
staffed TELEC **bemannt**
[station] [Station]
= attended; manned = beaufsichtigt; bedient; besetzt
≈ maintained
≠ unstaffed ≈ gewartet
≠ unbemannt
staff member ECON **Mitarbeiter**
= staff 2
≈ employee ≈ Angestellter
stage (→circuit stage) CIRC.ENG (→Stufe)
stage-by-stage switching system SWITCH (→schritthaltendes Vermittlungssystem)
(→step-by-step switching system)
stage gain CIRC.ENG **Stufenverstärkung**
stagger (n.) RADIO LOC **Impulsstaffelung**
stagger (→detune) ELECTRON (→verstimmen)
staggered (→graded) COLLOQ (→gestaffelt)
staggered (→time-shifted) TECH (→zeitlich versetzt)
staggered (→detuned) ELECTRON (→verstimmt)
staggering (→detuning) ELECTRON (→Verstimmung)
stagnation (→concentration) PHYS (→Stau)
stainless (→rustless) CHEM (→rostfrei)
stainless steel METAL **rostfreier Stahl**
staircase effect DATA PROC **Treppeneffekt**
[computer graphics] [Computergraphik]
↑ aliasing ↑ Bildunschönheit
staircase signal ELECTRON **Treppensignal**
= step signal = Stufensignal
staircase voltage ELECTRON **Treppenspannung**
stake (v.t.) MECH **vernieten 2**
[mit Nietmeißel]
= einnieten; meißelnieten
staked METAL **meißelgenietet**
staking METAL **Meißelnieten**

stall (n.) (→ booth)		ECON	(→ Messestand)	standard case	EQUIP.ENG Normgehäuse
stall (→ delay)		TECH	(→ verzögern)	standard cell	POWER SYS Normalelement
stamp (v.t.)		METAL	formstanzen		= Normalbatterie; Normalzelle
stamp (v.t.)		MECH	stanzen		
stamp (n.)		OFFICE	Stempel	standard cell	MICROEL Standardzelle
↓ postmark [POST]			↓ Poststempel [POST]	standard cell technology	MICROEL Standardzellenverfahren
stamp		ECON	Wertzeichen	standard color	PHYS Eichfarbe
≈ postage stamp 2 [POST]			≈ Postwertzeichen [POST]	standard configuration	TECH Standardkonfiguration
↓ adhesive stamp; imprinted stamp			↓ Wertmarke; Wertstempel	standard construction practice	EQUIP.ENG Standardbauweise
stamped		MECH	gestanzt		= Regelbauweise; Normbauweise
stamping 1		METAL	Formstanzen	= standard design	
			= Stanzen	standard converter	TELEC Normwandler
stamping 2		MECH	Stanzteil	= transcoder	= Transcoder
stamping die (→ punch die)		MECH	(→ Stanzwerkzeug)	standard costs	ECON Standardkosten
stamping machine		OFFICE	Stanzmaschine	standard design	TECH Regelbauart
= paper stamping machine			= Papierstanzmaschine	standard design (→ standard construction practice)	EQUIP.ENG (→ Standardbauweise)
stand (n.)		EQUIP.ENG	Fuß		
= base; foot			= Gerätefuß; Aufstellstütze	standard deviation	MATH Standardabweichung
= socket			≈ Sockel	[positive square root of variance]	[positive Quadratwurzel der Varianz]
↓ floor stand; swivel stand; tilt stand; tilt-swivel stand			↓ Standfuß; Drehfuß; Schwenkfuß; Dreh-Schwenk-Fuß	≈ variance	= mittlere quadratische Abweichung; quadratische Streuung; mittlerer quadratischer Fehler; Streuung 2
stand (n.)		TECH	Ständer		
[frame for support]			[Gestell zum Aufhängen]		≈ Varianz
≈ pedestal			≈ Sockel	standard dipole	ANT Normaldipol
stand (→ rack)			(→ Regal)	standard document	DOC Regelunterlage
stand (n.) (→ booth)		ECON	(→ Messestand)	standard drawings	ENG.DRAW Normzeichnung
stand (→ endure)		TECH	(→ überstehen)	standard equipment	TECH Grundausstattung
stand-alone		DATA PROC	allein operierend	= basic features	= Grundbestückung; Regelausstattung
= standalone; autonomous; on local			= selbstständig	≠ expansion equipment	≠ Ergänzungsausstattung
stand-alone		EQUIP.ENG	eigenständig	standard equipment	SWITCH Regelbelegung
[equipment]			[Gerät]	= standard assignment	
≠ integrated			≠ autonom	standard error	MATH mittlerer Fehler
			≠ integriert	standard etching	MANUF Normalätztechnik
standalone (→ stand-alone)		DATA PROC	(→ allein operierend)	standard failure rate	QUAL Standardausfallrate
standalone (→ independent)		TECH	(→ unabhängig)	standard finish	TECH Grundausführung
stand-alone concentrator		SWITCH	Wählsterneinrichtung	≈ standard version	[in Qualität und Ausstattung]
[pair-saving device with concentrating and expanding equipment, indepent from the central office switch]			[Vorfeldeinrichtung mit spiegelbildlicher Konzentration und Expansion, zur Einsparung von Anschlußleitung, ohne Zusammenwirken mit der Vermittlungsstelle]		= Standardausführung
					≈ Grundbauform
= line concentrator 2; trunk concentrator 2				standard form	OFFICE Formular
				= form; worksheet; blank	= Vordruck
≈ remote concentrator				standard form	MATH Normalform
↑ front-end equipment				standard form pageprinter	TERM&PER Formularblattschreiber
			= Leitungsdurchschalter		= Blattschreiber mit Formulardruck
			≈ Verkehrskonzentrator		
			≈ Vorfeldeinrichtung	standard frequency	INSTR Normalfrequenz
stand-alone-concentrator line		SWITCH	Wählsternanschluß	= reference frequency	= Standardfrequenz; Eichfrequenz; Normfrequenz; Vergleichsfrequenz; Bezugsfrequenz
standalone system		EQUIP.ENG	eigenständiges Gerät		
stand-alone system		DATA PROC	(→ Einplatzsystem)		
(→ single-user system)				standard frequency emitter	RADIO Normalfrequenzsender
standard (n.)		TECH	Norm		= Standardfrequenzsender
= norm			= Standard	standard frequency generator 2	INSTR Eichmarkengeber
standard (n.)		INSTR	Normal	standard frequency generator 1	INSTR Normalfrequenzgenerator
standard (adj.)		TECH	normgerecht		= Normalgenerator 1; Standardfrequenzgenerator
			= standardisiert		
standard		COMPON	Normwertreihe	standard frequency receiver	RADIO Normalfrequenzempfänger
			= Normreihe		= Standardfrequenzempfänger
standard		TECH	Standard-		
[adj.]				standard function	DATA PROC Standardfunktion
≈ regular				standard generator	INSTR Normalgenerator 2
standard		PHYS	Normalmaß	= one-milliwatt generator	↑ Pegelsender
			= Normal; Eichmaß; Etalon	standard hydrogen electrode	PHYS Normalwasserstoff-Elektrode
standard (→ medium)		TYPOGR	(→ normal)	standard IC	MICROEL Standardbaustein
standard antenna		ANT	Meßantenne	= standard integrated circuit	= Standardschaltkreis
= standard radiator			= Normalantenne; Normalstrahler	standard inch	PHYS Normalzoll
↓ reference dipole				[2.540 cm]	[2,540 cm]
			≈ Bezugsantenne	= English inch	= amerikanischer Zoll
			↓ Normdipol	↑ inch	≈ englischer Zoll
standard assignment (→ standard equipment)		SWITCH	(→ Regelbelegung)		↑ Zoll
				standard integrated circuit (→ standard IC)	MICROEL (→ Standardbaustein)
standard atmosphere		METEOR	Normalatmosphäre	standard interface	TELEC Standard-Schnittstelle
			= Standardatmosphäre	standard intermediate frequency	BROADC Norm-Zwischenfrequenz
standard atmosphere		QUAL	Normalklima		
standard attenuator		INSTR	Eichteiler	standard ion dose	PHYS Standardionendosis
			= Präzisionsteiler	= exposure	
standard cabling		SWITCH	Regelverkabelung		
standard capacitor		INSTR	Normalkondensator		
			= Eichkondensator; Kapazitätsnormal		

standardization

standardization	TECH	Standardisierung	
= normalization		= Normung; Normierung	
standardization body	TECH	Standardisierungsgremium	
= standards committee		= Normungsgremium; Normungsausschußes; Normengremium	
standardization works	TECH	Standardisierungsarbeiten	
standardize (→normalize)	TECH	(→standardisieren)	
standardized plug connection	COMPON	Normsteckverbindung	
standardized programming	DATA PROC	normierte Programmierung	
standardizing (→gauging)	INSTR	(→Eichung)	
standardizing generator	INSTR	Eichgenerator	
		= Prüfgenerator	
standard keyboard	TERM&PER	Volltastatur	
standard label	SWITCH	Standard-Adresse	
standard lamp	INSTR	Normalglühlampe	
		= Normallampe	
standard language	DATA PROC	Standardsprache	
[understandable by different processors or compilers]		[von verschiedenen Prozessoren oder Kompilierern interpretierbar]	
standard length	COMM.CABLE	Regellänge	
standard letter (→boilerplate letter)	OFFICE	(→Standardbrief)	
standard letter (→form letter)	DATA PROC	(→Standardbrief)	
standard light source	PHYS	Normallichtquelle	
standard mask	INSTR	Standardmaske	
= agency mask			
standard measure	INSTR	Eichnormal	
		= Meßnormal	
standard microphone		Meßmikrophon	
= measuring microphone		[EL.ACOUS]	
		= Eichmikrophon	
standard model (→basic model)	TECH	(→Grundbauform)	
standard molar volume	PHYS	molares Normvolumen	
standard operation procedure	DATA PROC	Standardbetriebsverfahren	
= SOP			
standard paper (→normal paper)	TERM&PER	(→Normalpapier)	
standard part	MANUF	Normteil	
standard performance	TECH	Normalleistung	
standard pitch (→standard tuning tone)	ACOUS	(→Stimmton)	
standard position (→normal position)	TECH	(→Normalstellung)	
standard price (→normal price)	ECON	(→Normalpreis)	
standard program	DATA PROC	Standardprogramm	
standard punched card	TERM&PER	Standardlochkarte	
standard quality	ECON	Standardqualität	
standard radiator (→standard antenna)	ANT	(→Meßantenne)	
standard resistance 2 (→precision resistor)	COMPON	(→Normalwiderstand)	
standard resistor 1	INSTR	Eichwiderstand	
[standard to calibrate precision resistors]		[Gebrauchsnormal zur Eichung von Normalwiderständen]	
= precision resistance 1		= Widerstandsnormal	
≈ precision resistor		≈ Normalwiderstand	
standard resistor 2 (→precision resistor)	COMPON	(→Normalwiderstand)	
standard route	TELECONTR	Regelweg	
[the used in normal conditions]		[der im Normalfall benutzte]	
standards committee	TECH	(→Standardisierungsgremium)	
(→standardization body)			
standards conversion (→TV standards conversion)	TV	(→Normwandlung)	
standards converter (→TV standards converter)	TV	(→Normwandler)	
standards enforcer	DATA PROC	Normenüberprüfer	
standard sheet-metal	METAL	Standardblech	
standard software	DATA PROC	Standard-Software	
standard software	DATA PROC	Standard-Software	
= canned software; precanned routines		= Massen-Software	
≠ custom software		≠ kundenspezifische Software	
standard steel	METAL	Normstahl	
		= Normalstahl	
standard technology	MICROEL	Standardtechnologie	
↓ standard buried collector technology			
standard telephone set	TELEPH	Normalfernsprecher	
		= Standardfernsprecher; Normaltelephon	
standard time	MANUF	Vorgabezeit	
Standard Time	INSTR	Standardzeit	
		= Normalzeit	
standard time (→local civil time)	PHYS	(→Ortszeit)	
standard-time board	DATA PROC	Funkuhrkarte	
		= Normalzeitkarte	
standard tool	MANUF	Normalwerkzeug	
standard traffic load	SWITCH	Regelverkehrslast	
standard transmission line	TELEPH	Eichleitung	
= reference circuit			
standard tuning tone	ACOUS	Stimmton	
[tone a1; 440Hz]		[Ton a1; 440 Hz]	
= pitch 2; tuning pitch; standard pitch		= Normstimmton	
standard value	INSTR	Eichwert	
standard value	DATA PROC	Standardwert	
standard version (→basic model)	TECH	(→Grundbauform)	
standard volumen	PHYS	Normvolumen	
stand-by (n.)	TELEC	Ersatz	
= standby; stand by; back-up; reserve		≠ Betrieb	
stand-by (n.)	TECH	Reserve	
= standby; stand by; spare		= Ersatz 2	
stand-by (n.)	CONS.EL	Schlummerschaltung	
= standby; stand by			
stand by (→stand-by)	TELEC	(→Ersatz)	
standby (→stand-by)	TELEC	(→Ersatz)	
standby (→stand-by)	TECH	(→Reserve)	
stand by (→stand-by)	TECH	(→Reserve)	
stand by (→stand-by)	CONS.EL	(→Schlummerschaltung)	
standby (→stand-by)	CONS.EL	(→Schlummerschaltung)	
stand-by button	CONS.EL	Bereitschaftstaste	
stand-by channel	TRANS	Ersatzkanal	
= protection channel			
stand-by credit	ECON	Stützungskredit	
= standby credit			
standby credit (→stand-by credit)	ECON	(→Stützungskredit)	
stand-by device	TECH	Ersatzvorrichtung	
stand-by indicator	CONS.EL	Bereitschaftsanzeige	
stand-by power plant	POWER SYS	(→Notstromversorgung)	
(→emergency power supply)			
stand-by power supply	POWER SYS	(→Notstromversorgung)	
(→emergency power supply)			
stand-by route	TELECONTR	Ersatzweg	
stand-by set (→emergency power supply)	POWER SYS	(→Notstromversorgung)	
stand-by system	DATA PROC	Bereitschaftssystem	
[reserve programs and data for emergency]		[für den Notfall bereitehende Ersatzprogramme und -daten]	
= fall-back system		= Ersatzsystem; Stand-by-System; Verbundanlage	
≈ fall-back		≈ Bereitschaftsmaßnahme	
stand-by unit	TELEC	Ersatzgerät 1	
standing (→reputation)	ECON	(→Ruf)	
standing wave	PHYS	Stehwelle	
= stationary wave		= stehende Welle	
standing-wave antenna	ANT	Stehwellenantenne	
standing wave factor (→voltage standing wave ratio)	LINE TH	(→Welligkeitsfaktor)	
standing wave ratio (→voltage standing wave ratio)	LINE TH	(→Welligkeitsfaktor)	
standing-wave-ratio meter (→SWR power meter)	INSTR	(→Stehwellenmeßgerät)	
stand-off ratio	MICROEL	Abstandsverhältnis	
standstill	TECH	Stillstand	
staple	TECH	Heftklammer 1	
[U-shaped metal loop to be applied by a stapling machine]		[mit einer Heftmaschine anzubringende U-förmige Drahtklammer]	
≈ paper clip [OFFICE]		≈ Büroklammer [OFFICE]	
staple (→pile)	TECH	(→Stapel)	
staple (v.t.) (→pile)	TECH	(→stapeln)	
stapler	TERM&PER	Heftgerät	
stapling machine	OFFICE	Heftmaschine	

star		TELEC	**Stern**	**starting** (→beginning)		TECH	(→beginnend) (adj.)

star [network]	TELEC	**Stern** [Netz]
star (→asterisc)	TYPOGR	(→Sternchen)
star (→star connection)	NETW.TH	(→Sternschaltung)
star bus	DATA COMM	**Sternbus**
star circuit (→star connection)	NETW.TH	(→Sternschaltung)
star connection	DATA COMM	**Sternanschluß**
star connection	NETW.TH	**Sternschaltung**
= wye connection; T connection; Y connection; star circuit; wye circuit; T circuit; Y circuit; star network; wye network; T network; Y network; star section; wye section; T section; Y section; star; wye; T; Y		= Sternglied; T-Schaltung; T-Glied; Y-Schaltung; Y-Glied
star-delta circuit (→star-delta connection)	POWER SYS	(→Stern-Dreieck-Schalter)
star-delta connection	POWER SYS	**Stern-Dreieck-Schalter**
= star-delta transformation; star-wye connection		
star-delta conversion	NETW.TH	**Stern-Dreieck-Umwandlung**
= star-delta transformation; wye-delta conversion; Tau-Pi transformation; Y-Delta transformation		
star-delta transformation (→star-delta conversion)	NETW.TH	(→Stern-Dreieck-Umwandlung)
star drill	TECH	**Steinbohrer**
star ground system	SYS.INST	**Sternerder**
star network	TELEC	**Sternnetz**
= radial network; star-type network; star-shaped network		= sternförmiges Netz
≈ star topology		≈ Sternstruktur
star network (→star connection)	NETW.TH	(→Sternschaltung)
star quad	COMM.CABLE	**Sternvierer**
= spiral-four quad; spiral-four; square formation		↑ Viererseil; Verseilelement
↑ quad; stranding element		
star-quad cable	COMM.CABLE	**Sternviererkabel**
= spiral-four cable 1; star-quadded cable		↑ Viererkabel
↑ quadded cable		
star-quadded cable (→star-quad cable)	COMM.CABLE	(→Sternviererkabel)
star resistance	EL.TECH	**Sternschenkelwiderstand**
star section (→star connection)	NETW.TH	(→Sternschaltung)
star-shaped (→stellate)	TECH	(→sternförmig)
star-shaped network (→star network)	TELEC	(→Sternnetz)
start (n.)	CONTROL	**Anlassen**
		= Anreizen
start (n.)	DATA PROC	**Anlauf**
= system start		= Start; Systemstart
≈ initialization; initiation		≈ Initialisierung; Einleitung
↓ recovery 2		↓ Wiederanlauf
start (n.)	TELEGR	**Start**
		= Anlauf
≠ stop		≠ Stop
start (n.)	TECH	**Start**
= beginning; commencement		= Beginn; Anlauf; Anfang; Anbeginn
start (v.t.) (→trigger)	DATA PROC	(→auslösen)
start (n.) (→begin)	TECH	(→Beginnen)
start (→begin)	TECH	(→beginnen)
start address	DATA PROC	**Startadresse**
[address of the first instruction to be executed of a program after loading]		[Adresse des ersten nach der Ladung eines Programms auszuführenden Befehls]
= starting address; boot address; initial address		= Anfangsadresse
start bit	DATA COMM	**Startbit**
start bit (→start pulse)	TELEGR	(→Anlaufschritt)
start block	DATA COMM	**Startblock**
start command (→start signal)	SWITCH	(→Anlaufsignal)
start element (→start pulse)	TELEGR	(→Anlaufschritt)
starter battery	POWER SYS	**Starterbatterie**
starter kit (→evaluation kit)	MICROEL	(→Einarbeitungskit)
starter resistance	PHYS	**Anlaßwiderstand**
= starting rheostat		
starting (→beginning)	TECH	(→beginnend) (adj.)
starting address (→start address)	DATA PROC	(→Startadresse)
starting condition (→initial condition)	MATH	(→Anfangsbedingung)
starting costs (→initial costs)	ECON	(→Anlaufkosten)
starting current	EL.TECH	**Anschwingstrom**
= pre-oscillation current		= Anlaufstrom
starting deviation (→initial deviation)	CONTROL	(→Anfangsabweichung)
starting key (→calling key)	TELEGR	(→Anruftaste)
starting point	COLLOQ	**Ausgangsposition**
		= Ausgangspunkt
starting posicion (→initial position)	MECH	(→Anfangsstellung)
starting rheostat (→starter resistance)	PHYS	(→Anlaßwiderstand)
starting thermistor	CIRC.ENG	**Anlaßheißleiter**
start level	DATA PROC	**Startebene**
start-of-block signal	DATA COMM	**Blockbeginnzeichen**
start of conversation (→beginning of conversation)	SWITCH	(→Gesprächsbeginn)
start of heading	DATA COMM	**Anfang des Kopfes**
= SOH		= SOH
start of message	DATA COMM	**Nachrichtenbeginn**
= SOM		
start-of-message signal	DATA COMM	**Meldungsbeginnzeichen**
start of test	DATA COMM	**Prüfbeginn**
≠ end of test		≠ Prüfende
start of text	DATA COMM	**Textanfang**
[code]		[Code]
= STX		= Textbeginn; STX; Anfang des Textes
star topology	TELEC	**Sternstruktur**
≈ star network		= Sterntopologie
		≈ Sternnetz
start polarity	TELEGR	**Startpolarität**
start polarity (→signal condition A)	DATA COMM	(→Startpolarität)
start program	DATA PROC	**Startprogramm**
start pulse	TELEGR	**Anlaufschritt**
= start signal; start element; start bit; open start pulse		= Startschritt; Startelement; Startbit; Startzeichen
≠ stop pulse; information pulse		≠ Sperrschritt; Informationsschritt
start response	CONTROL	**Anreizverhalten**
start signal	SWITCH	**Anlaufsignal**
= start command; recovery signal		= Anlaufbefehl
start signal (→start pulse)	TELEGR	(→Anlaufschritt)
start-stop	TELEGR	**Start-Stop**
		= Anlauf-Sperrung
start-stop distortion (→telegraph signal distortion)	TELEGR	(→Schrittverzerrung)
start-stop information	TELECONTR	**Start-Stop-Information**
start-stop mode	DATA PROC	**Start-/Stop-Betrieb**
[interruption of a data stream after each character or cluster of characters]		[Unterbrechung des Datenflusses nach jedem Zeichen oder Gruuppe von Zeichen]
≠ continuous mode		≠ Zügig-Betrieb
start-stop mode (→start-stop operation)	TELEGR	(→Start-Stop-Betrieb)
start-stop multivibrator (→one-shot multivibrator)	CIRC.ENG	(→stabile Kippschaltung)
start-stop operation	TELEGR	**Start-Stop-Betrieb**
[asynchronous mode for binary characters, with a start pulse and a stop pulse for each character]		[Asynchronverfahren für Binärzeichen, mit einem Anlaufschritt und einem Sperrschritt pro Zeichen]
= start-stop mode; start-stop transmission; start-stop principle		= Start-Stop-Verfahren; Start-Stop-Übertragung
↑ asynchronous operation		↑ Asynchronbetrieb
start-stop principle (→start-stop operation)	TELEGR	(→Start-Stop-Betrieb)
start/stop system	TELEGR	**Start-Stop-System**
≠ synchronous system		≠ Synchronsystem
start-stop teleprinter	TELEGR	**Springschreiber**
start-stop transmission (→start-stop operation)	TELEGR	(→Start-Stop-Betrieb)

start track	RAILW.SIGN	**Startgleis**	
start-up (v.t.)	DATA PROC	**hochfahren**	
[to power-up and establish operating readiness]		[einschalten und Betriebsbereitschaft herstellen]	
≈ power-up (v.t.)			
start-up (v.i.)	TECH	**anlaufen**	
≈ begin (v.i.)		[fig.]	
		≈ beginnen	
start-up (n.)	DATA PROC	**Hochfahren**	
= startup (n.); power-up (n.)			
start-up (v.t.)	EL.TECH	(→einschalten)	
(→connect 2)			
start-up (→commissioning)	TELEC	(→Einschaltung)	
startup (n.) (→start-up)	DATA PROC	(→Hochfahren)	
start-up disk	DATA PROC	**Startdiskette**	
start-up problem	TECH	**Anlaufschwierigkeit**	
start-up test	TELEC	**Inbetriebnahmetest**	
start-up time (→response time)	TECH	(→Reaktionszeit)	
star-type network (→star network)	TELEC	(→Sternnetz)	
star-wye connection	POWER SYS	(→Stern-Dreieck-Schalter)	
(→star-delta connection)			
state (n.)	ECON	**Staat**	
≈ government		≈ Regierung	
state (n.)	TECH	**Zustand**	
= order; fashion; condition; shape [COLLOQ]		= Verfassung [COLLOQ]; Status	
state- (adj.)	ECON	**staatlich**	
= state-owned; governmental		= Regierungs-	
state	PHYS	**Zustand**	
= status; condition			
state (→status)	DATA PROC	(→Zustand)	
state (→signal condition)	DATA COMM	(→Zeichenlage)	
state change	ELECTRON	**Zustandsübergang**	
= state transition			
state change	TELECONTR	**Zustandswechsel**	
stated below	OFFICE	**nachstehend**	
[in business letter]		[in Geschäftsbrief]	
≈ hereinafter			
state density (→energy state density)	PHYS	(→Zustandsdichte)	
state diagram	INF	**Graph**	
		= Zustandsfolgediagramm	
state diagram	DATA PROC	**Zustandsdiagramm**	
= phase diagram		= State-Diagramm	
stated value (→nominal value)	ECON	(→Nennwert)	
state equation	SYS.TH	**Zustandsgleichung**	
state estimation	CONTROL	(→Identifikation)	
(→identification)			
state function	PHYS	**Zustandsfunktion**	
state holding	ECON	**Staatsholding**	
state information (→status signal)	TELECONTR	(→Zustandsmeldung)	
state-information change	TELECONTR	**Anreiz**	
statement 1	DATA PROC	**Anweisung 1**	
[an operating instruction of a problem-oriented language; generally composed of several sub-statements or instructions; synonym to instruction in computer-oriented languages]		[Arbeitsvorschrift einer problemorientierten Sprache; i.a. aus mehreren Unteranweisungen oder Befehlen bestehend; bei maschinenorientierten Sprachen synonym zu Befehl]	
= action statement			
≈ instruction; command		= Arbeitsanweisung; Arbeitsvorschrift; Statement; Aussage	
↓ sub-statement; macro-instruction		≈ Befehl; Kommando	
		↓ Unteranweisung; Makroanweisung	
statement	ECON	**Stellungnahme**	
= opinion			
statement	ECON	**Kontoauszug**	
statement 2 (mach.orient. languages) (→instruction)	DATA PROC	(→Befehl)	
statement (→list)	DOC	(→Liste)	
statement label	DATA PROC	**Anweisungsetikett**	
statement list (→instruction list)	DATA PROC	(→Befehlsliste)	
statement normalization (→instruction normalization)	DATA PROC	(→Befehlsnormalisierung)	
statement register	DATA PROC	**Anweisungsregister**	
≈ instruction register		≈ Befehlsregister	
statement sequence	DATA PROC	**Anweisungsfolge**	
= command sequence		≈ Befehlsfolge; Befehlskette; Befehlsliste	
≈ instruction sequence; instruction chain; instruction list			
statements field	DATA PROC	**Anweisungsfeld**	
statements section	DATA PROC	**Anweisungsteil**	
= instructions section			
state of affairs	COLLOQ	**Sachlage**	
≈ circumstance		≈ Umstand	
state of aggregation	PHYS	**Aggregatzustand**	
state of the art	TECH	**Stand der Technik**	
≈ modern		≈ modern	
state-of-the-art (→modern)	TECH	(→modern)	
state-owned (→state-)	ECON	(→staatlich)	
state-owned company (→state-owned enterprise)	ECON	(→staatliches Unternehmen)	
state-owned enterprise	ECON	**staatliches Unternehmen**	
= state-run enterprise; state-owned company; state-run company		= Staatsunternehmen; staatlicher Betrieb; Staatsbetrieb; staatliche Firma; Staatsfirma	
↑ public entreprise		↑ öffentliches Unternehmen	
state-run carrier (→governmental telecommunications carrier)	TELEC	(→staatliche Fernmeldeverwaltung)	
state-run communication carrier (→governmental telecommunications carrier)	TELEC	(→staatliche Fernmeldeverwaltung)	
state-run company (→state-owned enterprise)	ECON	(→staatliches Unternehmen)	
state-run enterprise (→state-owned enterprise)	ECON	(→staatliches Unternehmen)	
state-run telecommunication carrier (→governmental telecommunications carrier)	TELEC	(→staatliche Fernmeldeverwaltung)	
state space	SYS.TH	**Zustandsraum**	
state table (→status table)	DATA PROC	(→Zustandstabelle)	
state tagging	INSTR	**Zustandskennung**	
		= Statuskennung	
state trajectory	SYS.TH	**Zustandskurve**	
		= Zustandstrajektorie; Trajektorie	
state transition (→state change)	ELECTRON	(→Zustandsübergang)	
state variable	SYS.TH	**Zustandsvariable**	
		= Zustandsgröße	
state-variable filter	NETW.TH	**Zuastandsvariablenfilter**	
		= State-variable-Filter	
static (adj.)	PHYS	**statisch**	
≠ dynamic (adj.)		≠ dynamisch	
static analysis	DATA PROC	**Trockenanalyse**	
static characteristic	ELECTRON	**statisches Verhalten**	
= static response		= statische Kennlinie	
≠ switching characteristics		≠ Schaltverhalten	
static charge	ELECTRON	**statische Auflading**	
static converter (→static power converter)	POWER SYS	(→Stromrichter)	
static data	DATA PROC	**feste Daten**	
[cannot be altered by the user]		[vom Anwender nicht veränderbar]	
= fixed data		= Festdaten	
static dump	DATA PROC	**statischer Speicherabzug**	
static electricity	PHYS	**Reibungselektrizität**	
= friction electricity			
static energy (→potential energy)	PHYS	(→potentielle Energie)	
static field	PHYS	**statisches Feld**	
static flip-flop	CIRC.ENG	**statische Kippschaltung**	
		= statisches Flipflop	
staticizer (→series-parallel converter)	CIRC.ENG	(→Serien-Parallel-Umsetzer)	
staticizing (→instruction staticizing)	DATA PROC	(→Befehlsübernahme)	
static memory	DATA PROC	**statischer Speicher**	
= static storage; static store		= Statikspeicher	
static noise immunity	ELECTRON	**statische Störsicherheit**	

static power converter	POWER SYS	**Stromrichter**
[converts the type of current, using electrical circuits]		[formt elektrische Energie um, unter Verwendung von elektrischen Schaltkreisen]
= static converter; converter; current changer		= statischer Umformer; Umformer 2
↓ rectifier; inverter; voltage system converter		↓ Gleichrichter; Wechselrichter; Umrichter
static power dissipation	MICROEL	**statische Verlustleistung**
= quiescent power dissipation		
static RAM (→SRAM)	DATA PROC	(→SRAM)
static response (→static characteristic)	ELECTRON	(→statisches Verhalten)
statics	RADIO	**atmosphärisches Geräusch**
= atmospheric noise		= atmosphärisches Rauschen
statics	MECH	**Statik**
≠ dynamics		≠ Dynamik
static storage (→static memory)	DATA PROC	(→statischer Speicher)
static store (→static memory)	DATA PROC	(→statischer Speicher)
station	DATA COMM	**Station**
= terminal		= Endstelle; Terminal 1
≈ subscriber; line		≈ Teilnehmer; Anschluß
↓ data station; workstation		↓ Datenstation; Arbeitsplatzsystem
station (→telephone set)	TELEPH	(→Fernsprechapparat)
station (→telephone station)	TELEC	(→Sprechstelle)
station alarm	EQUIP.ENG	**Stationsalarm**
station apparatus (→subscriber set)	TELEPH	(→Teilnehmerapparat)
stationary	TECH	**ortsfest**
= fixed		= stationär; ortsgebunden
≠ mobile		≠ mobil
stationary	PHYS	**stationär**
= steady		= unveränderlich
stationary battery	POWER SYS	**ortsfeste Batterie**
stationary current	PHYS	**stationärer Strom**
= steady-state current; steady current		
stationary reader	TERM&PER	**stationärer Belegleser**
≠ hand-held reader		≠ Handleser
↑ document reader		↑ Belegleser
stationary wave (→standing wave)	PHYS	(→Stehwelle)
stationary wave index (→voltage standing wave ratio)	LINE TH	(→Welligkeitsfaktor)
station battery (→central battery)	TELEPH	(→Zentralbatterie)
station equipment (→user terminal)	TELEC	(→Teilnehmergerät)
stationery 1	OFFICE	**Schreibwaren**
		= Büromaterial
stationery 2	DOC	**Briefbogen**
station finder (→station search)	CONS.EL	(→Sendersuchlauf)
station forced busy (→station guarding)	TELEPH	(→Anrufschutz)
station group (→cluster)	DATA COMM	(→Cluster)
station guarding	TELEPH	**Anrufschutz**
= station forced busy; do-not-disturb		= Ruhe vor dem Telefon
station identification	DATA COMM	**Stationskennung**
= identification character; station identifier; identifier; identifying code; answerback code		= Kennung; Stationskennzeichen; Kennzeichen; Stationsidentifizierung; Identifizierung; Stationsbezeichnung; Bezeichnung; Bezeichner
station identification request	DATA COMM	**Kennungsabfrage**
= answerback code request		= Kennungsanforderung
station identifier (→station identification)	DATA COMM	(→Stationskennung)
station inspection (→station survey)	SYS.INST	(→Stationsbesichtigung)
station installation (→office installation technique)	TELEC	(→Amtsbautechnik)
station installation	SYS.INST	**Stationsaufbau**
(→office installation 1)		
station installation technique	TELEC	(→Amtsbautechnik)
(→office installation technique)		
station of destination	TELEC	**Bestimmungsstelle**
station of origin	TELEC	**Ursprungsstelle**
station power (→station power supply)	TELEC	(→Amtsstromversorgung)
station power supply	TELEC	**Amtsstromversorgung**
= station power system; station power		= Ämterstromversorgung
≈ primary power		≈ Primärstromversorgung
station power system (→station power supply)	TELEC	(→Amtsstromversorgung)
station preset	CONS.EL	**Senderspeicher**
		= Stationsspeicher; Festsenderspeicher
station register	SWITCH	**Amtssatz**
station search	CONS.EL	**Sendersuchlauf**
= station finder; autoscanning; scan tuning; seek		= Suchlaufabstimmung
station set (→subscriber set)	TELEPH	(→Teilnehmerapparat)
station survey	SYS.INST	**Stationsbesichtigung**
= station inspection		= Stationsbegehung; Amtsbesichtigung; Amtsbegehung
↑ site survey		↑ Ortsbegehung
station test	SYS.INST	**Stationsmessung**
= in-station test; site test		
station-to-station calling	TELEPH	**Direktanruf**
station tuning	CONS.EL	**Senderabstimmung**
statistical	MATH	**statistisch**
= random		= zufällig
≈ stochastic		≈ stochastisch
statistical communication theory	SIGN.TH	(→Korrelationselektronik)
(→correlation electronics)		
statistical data	MATH	**statistische Daten**
= statistics 2		
statistical encoding	CODING	**statistische Codierung**
= entropy encoding		= Entropiecodierung
statistical error	MATH	**statistischer Fehler**
= random error; accidental error		= zufälliger Fehler
statistical fluctuation	MATH	**statistische Schwankung**
statistically distribuited	MATH	(→statistisch verteilt)
(→randomly distributed)		
statistical mechanics	PHYS	**statistische Mechanik**
statistical multiplexer	DATA COMM	**statistischer Multiplexer**
= statistical multiplexor		
statistical multiplexing	DATA COMM	**gepufferte Multiplexierung**
statistical multiplexor	DATA COMM	(→statistischer Multiplexer)
(→statistical multiplexer)		
statistical noise	TELEC	**statistisches Rauschen**
statistical physics	PHYS	**statistische Physik**
statistic overflow	DATA PROC	**Statistikanschlag**
		= Statistiküberlauf
statistics 1	MATH	**Statistik**
statistics 2 (→statistical data)	MATH	(→statistische Daten)
statistics counter	TERM&PER	**Statistikzähler**
statistics tape	DATA PROC	**Statistikband**
statistor	MICROEL	**Statistor**
↑ MOSFET		↑ MOSFET
stator	POWER SYS	**Ständer**
[the stationary portion of an electric engine]		[statischer Teil eines Elektromotors]
≠ rotor		= Stator
		≠ Läufer
stator	COMPON	**Stator**
stator lamination	POWER SYS	**Statorpaket**
status	DATA PROC	**Zustand**
= state; condition 2		= Status
status (→state)	PHYS	(→Zustand)
status bit (→flag)	DATA PROC	(→Merker)
status byte	DATA PROC	**Zustandsbyte**
= condition byte; functional byte		= Statusbyte; funktionales Byte
status data	DATA PROC	**Zustandsdaten**
status identifier	DATA PROC	**Zustandskennzeichen**
		= Statuskennzeichen
status identifier (→status signal)	SWITCH	(→Zustandskennzeichen)
status line	DATA PROC	**Meldungszeile**
status of expansion	TECH	**Ausbauzustand**
status register	DATA PROC	**Zustandsregister**
		= Statusregister
status signal	SWITCH	**Zustandskennzeichen**
= status identifier		= Statuszeichen
↓ continuous signal		↓ Dauerkennzeichen

status signal	TELECONTR	**Zustandsmeldung**		steel number	METAL	**Schlagzahl**
= state information		= Mehrfachmeldung		steel plate (→steel sheet)	METAL	(→Stahlblech)
status table	DATA PROC	**Zustandstabelle**		steel-sealed relay	COMPON	(→Stahlrelais)
= state table		= Statustabelle		(→metal-enclosed relay)		
status word	DATA PROC	**Zustandswort**		steel sheet	METAL	**Stahlblech**
		= Statuswort		= steel plate		
statute	ECON	**Satzung**		steel tube mast	ANT	**Stahlrohrmast**
		= Statuten		steel wire	METAL	**Stahldraht**
stay (n.)	OUTS.PLANT	**Anker**		steep	MATH	**steil**
= guy anchor; anchor				= steep-sloped		
stay (→guy rope)	OUTS.PLANT	(→Abspannseil)		steep (→precipitous)	TECH	(→abschüßig)
stay block	OUTS.PLANT	**Ankerpfahl**		steep filter	NETW.TH	**steiles Filter**
stay clamp	OUTS.PLANT	**Ankerdrahtklemme**		steepness	MATH	**Steilheit**
stay crutch	OUTS.PLANT	**Ankerstütze**		≈ slope		≈ Steigung
stayed pole (→terminal pole)	OUTS.PLANT	(→Abspannmast)		steep radiation	ANT	**Steilstrahlung**
				= high-angle radiation		
stayed terminal pole (→end pole)	OUTS.PLANT	(→Endgestänge)		steep radiator (→high-angle radiator)	ANT	(→Steilstrahler)
stay rope	ANT	**stehendes Gut**		steep rise	ELECTRON	**steiler Anstieg**
= staywire		[nicht bewegliches Seil]		steep skirt	ELECTRON	**steile Flanke**
stay rope (→guy rope)	OUTS.PLANT	(→Abspannseil)		steep-sloped (→steep)	MATH	(→steil)
stay thimble	OUTS.PLANT	**Ankerkausche**		steer (v.t.)	TECH	**lenken**
stay wire (→guy rope)	OUTS.PLANT	(→Abspannseil)		= guide		= steuern
staywire (→stay rope)	ANT	(→stehendes Gut)		≈ swivel		≈ schwenken
stay-wire isolator	ANT	**Pardunengehänge** [zur Isolation von Abspannseilen]		steerable	TECH	**lenkbar**
				≈ adjustable; hinged		≈ einstellbar; schwenkbar
				steerable aerial (→steerable antenna)	ANT	(→einstellbare Antenne)
STD (BRI) (→direct distance dialing)	SWITCH	(→Selbstwählferndienst)		steerable antenna [with steerable characteristic]	ANT	**einstellbare Antenne** [mit schwenkbarer Charakteristik]
STD (→synchronous time division multiplexing)	TELEC	(→synchrone Zeitmultiplextechnik)		= mobile antenna 2; steerable aerial; mobile aerial 2		= schwenkbare Antenne; Schwenkkeulenantenne
STD code [trunk prefix + trunk code]	SWITCH	**Inlandsvorwahlnummer** [Verkehrsausscheidungszahl + Ortskennzahl, z.B. 089 für München]		steering committee	ECON	**Lenkungsausschuß**
						= Steuerungsausschuß
				steering range	TECH	**Schwenkbereich**
				= swiveling range		
steady (→stationary)	PHYS	(→stationär)		Stefan-Boltzmann constant	PHYS	**Stefan-Boltzmann-Konstante**
steady (→continuous)	MATH	(→stetig)		stellate (adj.)	TECH	**sternförmig**
steady (→stepless)	TECH	(→stufenlos)		= stelliform; star-shaped		≈ strahlenförmig
steady current (→quiescent current)	ELECTRON	(→Ruhestrom)		≈ radial		
				stelliform (→stellate)	TECH	(→sternförmig)
steady current (→stationary current)	PHYS	(→stationärer Strom)		stem (n.)	TECH	**Stutzen**
				= nozzle; stub		= Stummel; Stumpf
steady discharge	PHYS	**Dauerentladung**		stencil (→template)	TECH	(→Schablone)
steady load (→continuous load)	TECH	(→Dauerbelastung)		stencil printer	ENG.DRAW	**Schablonendrucker**
steady plate current	ELECTRON	**Anodenruhestrom**		stenography	OFFICE	**Stenographie**
steady plate voltage	ELECTRON	**Anodenruhespannung**		= shorthand writing		= Stenografie; Kurzschrift
steady production	MANUF	**laufende Fertigung**				↓ Eilschrift
steady state	PHYS	**eingeschwungener Zustand** = Gleichgewichtszustand; Beharrungszustand; Dauerzustand		step (n.)	ELECTRON	**Sprung**
				= jump; discontinuity; skip		= Stufe
				step 1 (n.) [horizontal]	TECH	**Schritt**
steady-state condition	PHYS	**Gleichgewichtsbedingung**				
		= Beharrungsbedingung		step 2 (n.) [vertical discontinuity]	TECH	**Stufe**
steady-state current (→stationary current)	PHYS	(→stationärer Strom)				
				step (→job step)	DATA PROC	(→Auftragsschritt)
steady-state oscillation	PHYS	**stationäre Schwingung**		step (n.) (→knee)	MATH	(→Stufe)
↓ steady-state vibration				step action (→step response)	TECH	(→Schrittverhalten)
steady-state value	ELECTRON	**Beharrungswert**		step-and-repeat camera	MICROEL	**Step-repeat-Kamera**
		= eingeschwungener Wert		step-at-a-time code (→counting code)	CODING	(→Zählcode)
steady stop	DATA COMM	**Dauerstopp**				
steady tone	TELEPH	**Dauerton**		step attenuator	INSTR	**Stufendämpfungsglied**
steady voltage (→quiescent voltage)	ELECTRON	(→Ruhespannung)		= attenuation box		= Stufenabschwächung; Stufenabschwächer; Eichleitung
steam (→vapor)	PHYS	(→Dampf)				
steel (n.) [iron alloy with carbon content]	METAL	**Stahl** [Kohlenstoff enthaltende Eisenlegierung]		step-by-step	SWITCH	**schritthaltend**
				step-by-step (→stepwise)	TECH	(→schrittweise)
steel- (adj.)	METAL	**stählern**		step-by-step control	SWITCH	**Einzelschrittsteuerung**
= steel-made				= single-step control		= direkte Steuerung; Direktsteuerung
steel cabinet	TECH	**Stahlschrank**		step-by-step mode	TELECONTR	**Schrittschaltverfahren**
steel concrete (→reinforced concrete)	CIV.ENG	(→Stahlbeton)		step-by-step operation	SWITCH	**schritthaltender Verbindungsaufbau**
steel concrete tower (→reinforced concrete tower)	CIV.ENG	(→Stahlbetonturm)				= schritthaltende Verbindungsherstellung
steel-core cable	CONS.EL	**Stahlkabel**				
steel lattice mast	CIV.ENG	**Stahlgittermast**		step-by-step switch	ELECTRON	**Schrittschaltwerk**
steel letter	TECH	**Schlagbuchstabe**		= stepping switch; stepper switch		= Schrittschalter
steel lettice tower	CIV.ENG	**Stahlgitterturm**		step-by-step switching system	SWITCH	**schritthaltendes Vermittlungssystem**
steel-made (→steel-)	METAL	(→stählern)		= stage-by-stage switching system		
steel mill	METAL	**Stahlwerk**		↓ direct control system		= schritthaltend gesteuertes

			Vermittlungssystem; schritthaltendes Wählsystem; schritthaltend gesteuertes Wählsystem; Schrittschaltsystem
			↓ Direktwahlsystem
step capacitance		COMPON	Stufenkondensator
step-down (v.t.)		MECH	untersetzen
step-down (adj.)		TECH	schrittweise abwärts
step-down transformer		POWER ENG	Abspanntransformator
			= Abspanner; Abwärtstransformator
step function		MATH	Sprungfunktion
			= Stufenfunktion; Treppenfunktion
step-index fiber		OPT.COMM	Stufenindexfaser
= step-index optical waveguide			= Stufenfaser; Stufenprofil-Wellenleiter; Kern-Mantel-Wellenleiter
≈ multimode fiber			≈ Mehrmodenfaser
step-index optical waveguide		OPT.COMM	(→Stufenindexfaser)
(→step-index fiber)			
step-index profile		OPT.COMM	Stufenindexprofil
			= Stufenprofil
step junction (→abrupt junction)		MICROEL	(→abrupter Übergang)
step key		COMPON	Stufentaste
stepless		TECH	stufenlos
= smooth; continuous 3; uniform; steady			= stufenfrei; glatt; stetig; gleichförmig 2
≈ endless; continuous 1			≈ endlos; kontinuierlich
≠ stepped			≠ gestuft
step-like curve		MATH	Stufenkurve
stepped		TECH	gestuft
≈ stepwise			≈ schrittweise
≠ stepless			≠ stufenlos
stepped lens		ANT	Zonenlinse
			= Stufenlinse
			↑ Beschleunigungslinse
stepped sweep		INSTR	gestufte Wobbelung
			= Stufenwobbelung
stepper (→succesive sequential circuit)		CONTROL	(→Schrittschaltwerk)
stepper motor		POWER ENG	Schrittmotor
= stepping motor; pulse motor			= Schrittschaltmotor; Schrittantrieb; Steppermotor [DATA PROC]
stepper switch (→step-by-step switch)		ELECTRON	(→Schrittschaltwerk)
stepping drive		MECH	Schrittantrieb
stepping mode		TERM&PER	Schritttechnik
[magnetic disc memory]			[Magnetplattenspeicher]
			= Steptechnik; Stepper-Mechanik
≠ servo mode			≠ Servotechnik
stepping motor (→stepper motor)		POWER ENG	(→Schrittmotor)
stepping pawl		SWITCH	Schaltklinke
[drives an electromechanical switch]			[betätigt einen elektromechanischen Schalter]
≠ holding pawl			≠ Sperrklinke
↑ pawl			↑ Klinke
stepping relais		COMPON	Schrittschaltrelais
stepping relay		COMPON	Wählerrelais
stepping switch		COMPON	Stufenschalter
= tapping switch			
stepping switch		ELECTRON	(→Schrittschaltwerk)
(→step-by-step switch)			
step potentiometer		COMPON	Stufenpotentiometer
step rate (→increment)		DATA PROC	(→Schrittweite)
step-recovery diode		MICROEL	Step-recovery-Diode
↑ charge-storage diode			= Schritt-Wiederholungs-Diode
			↑ Speicherschaltdiode
step response		CONTROL	Sprungantwort
↑ transient response			= Sprungverhalten
			↑ Übergangsverhalten
step response		TECH	Schrittverhalten
= step action			

step signal (→staircase signal)		ELECTRON	(→Treppensignal)
step size (→increment)		DATA PROC	(→Schrittweite)
step table		SWITCH	Schrittabelle
step-up (adj.)		TECH	schrittweise aufwärts
step-up transformer		POWER ENG	Aufwärtstransformator
stepwise		TECH	schrittweise
= in steps; step-by-step; gradually; incrementally 2			= stufenweise 2
stepwise approximation		MATH	schrittweise Näherung
steradian		PHYS	Steradiant
[complementary SI unit for space angles]			[ergänzende SI-Einheit für Raumwinkel]
= sr			= sr
Sterba array (→Sterba curtain array)		ANT	(→Sterba-Antenne)
Sterba curtain (→Sterba curtain array)		ANT	(→Sterba-Antenne)
Sterba curtain array		ANT	Sterba-Antenne
= Sterba array; Sterba curtain			
stereo (→stereophonic)		EL.ACOUS	(→stereophonisch)
stereo (→stereoscopic)		PHYS	(→stereoskopisch)
stereo amplifier		EL.ACOUS	Stereoverstärker
stereo broadcast signal		EL.ACOUS	Stereo-Multiplexsignal
stereo cartridge (→stereo pickup)		EL.ACOUS	(→Stereotonabnehmer)
stereocasting (→stereophonic broadcast)		BROADC	(→Stereorundfunk)
stereo coder		BROADC	Stereocoder
stereo conductor		COMM.CABLE	Stereo-Leitung
stereo crosstalk		EL.ACOUS	Übersprechen
[between the two stereo channels]			[zwischen den zwei Stereokanälen]
stereo decoder		CONS.EL	Stereodecoder
stereographic		DOC	stereographisch
			[Kartographie]
stereo headphone		EL.ACOUS	Stereo-Kopfhörer
stereo microphone		EL.ACOUS	Stereo-Mikrophon
			= Stereo-Mikrofon
stereo mixer		BROADC	Stereo-Mischpult
= stereo mixing console			
stereo mixing console (→stereo mixer)		BROADC	(→Stereo-Mischpult)
stereophonic (adj.)		EL.ACOUS	stereophonisch
= stereo			= stereofonisch; stereophon; stereofon
≈ binaural [ACOUS]			≈ zweiohrig [ACOUS]
≠ monophonic			≠ monophonisch
stereophonic broadcast		BROADC	Stereorundfunk
= stereocasting			
stereophonic groove		EL.ACOUS	Stereo-Rille
stereophonic quadro sound		EL.ACOUS	Stereoquadrophonie
stereophonic sound system		EL.ACOUS	Stereophonie
			= Stereofonie
stereophonic television		TV	Stereofernsehen
= stereo TV			[mit stereophonem Ton]
stereophonic transmission		TELEC	Stereoübertragung
≠ monophonic transmission			= stereophone Übertragung
			≠ monophone Übertragung
stereo pickup		EL.ACOUS	Stereotonabnehmer
= stereo cartridge; stereo reproducer			
stereo preamplifier		EL.ACOUS	Stereovorverstärker
stereo record		CONS.EL	Stereoschallplatte
stereo reproducer (→stereo pickup)		EL.ACOUS	(→Stereotonabnehmer)
stereoscopic		PHYS	stereoskopisch
= stereo			
stereo separation		EL.ACOUS	Stereo-Übersprechdämpfung
= channel separation			= Kanaltrennung; Übersprechdämpfung
stereo set		CONS.EL	Stereoanlage
stereo signal		BROADC	Stereoton
stereo subcarrier		BROADC	Stereohilfsträger
stereo tape recorder		EL.ACOUS	Stereotonbandgerät
stereo TV (→stereophonic television)		TV	(→Stereofernsehen)
stethoset type phone		EL.ACOUS	Kennbügelhörer
Stibitz code (→excess-three code)		CODING	(→Drei-Exzeß-Code)

stick

stick (v.t.)	ELECTRON	**kleben** [Kontakt]
stick (v.t.) = get stuck	TECH	**klemmen 2** (v.i.)
stick (n.) [electromagnetic store] = rod	COMPON	**Stäbchen** [elektromagnetischer Speicher]
stick (n.) ↑ handle 1	MECH	**Stiel** ↑ Handgriff 1
stick (→bar 1)	MECH	(→Stab)
sticker = gummed label	TECH	**Aufkleber**
sticking glossy paper	OFFICE	**Glanzpapier**
stick-on foot	EQUIP.ENG	**selbstklebender Gerätefuß**
stick-on label ≈ adhesive label	OFFICE	**Klebeetikette** ≈ Haftetikette
sticky tape	TECH	**Klebestreifen**
stiff = rigid 1 ≈ immovable; rigid 2	TECH	**starr** = steif ≈ unbeweglich; formbeständig
stiffen ≈ reinforce	TECH	**versteifen** ≈ verstärken
stiffening ≈ reinforcement	TECH	**Versteifung** ≈ Verstärkung
stiffness = rigidity 1 ≈ permanence of form	TECH	**Starrheit** = Formbeständigkeit
stiffness (→rigidity)	MECH	(→Starrheit)
stilb [unit for brightness; = 10,000 cd/m^2] = sb	PHYS	**Stilb** [Maßeinheit für Leuchtdichte; = 10.000 cd/m^2] = sb
still image	TV	**Standbild**
stillstand costs (→downtime costs)	ECON	(→Stillstandskosten)
still store	TV	**Standbildspeicher**
stimulate (→excite)	PHYS	(→anregen)
stimulated emission	PHYS	**induzierte Emission** = stimulierte Emission
stimulation (→excitation)	PHYS	(→Anregung)
stimulus-response measurement	INSTR	**Stimulus-Antwort-Messung**
stipulate (→agree)	COLLOQ	(→vereinbaren)
stipulation (→agreement 1)	COLLOQ	(→Vereinbarung)
stir	TECH	**rühren**
stitch bonding = scissors bonding	MICROEL	**Stichkontaktierung** = Stichverfahren; Stich-Kontaktierung
STM (→synchronous transfer mode)	TELEC	(→synchrones Übertragungsverfahren)
stochastic [depending on chance] ≈ probabilistic ≈ random; statistic ≠ deterministic	MATH	**stochastisch** [vom Zufall abhängig] ≈ probabilistisch ≈ zufällig; statistisch ≠ determiniert
stochastic-ergodic converter	INSTR	**stochastisch-ergodischer Umsetzer**
stochastic error (→random error)	TECH	(→Zufallsfehler)
stochastic event (→random result)	MATH	(→Zufallsergebnis)
stochastic event generator (→random sequence generator)	DATA PROC	(→Zufallsgenerator)
stochastic number (→random number)	MATH	(→Zufallszahl)
stochastic process (→random process)	MATH	(→Zufallsprozeß)
stochastic result (→random result)	MATH	(→Zufallsergebnis)
stochastic signal (→random signal)	INF	(→Zufallssignal)
stochastic simulation	DATA PROC	**stochastische Simulation**
stochastic variable (→random variable)	MATH	(→Zufallsvariable)
stock (n.) = store (n.); supply (n.); inventory; reserve (n.); provision	ECON	**Vorrat** = Lagerbestand; Bestand
stock (AM) (→share)	ECON	(→Aktie)
stock (→semi-finished good)	MANUF	(→halbfertiges Erzeugnis)
stock corporation (AM) (→corporation 2)	ECON	(→Aktiengesellschaft)
stockholder (AM); shareholder; shareowner = shareholder (BR)	ECON	**Aktionär**
stockholding (→stockkeeping)	ECON	(→Lagerhaltung)
stockkeeping = stockholding; warehousing	ECON	**Lagerhaltung**
stockkeeping programm	DATA PROC	**Lagerverwaltungsprogramm**
stock list (→parts list)	MANUF	(→Stückliste)
stockpiling = provision; storage	ECON	**Bevorratung** = Vorratshaltung
stockturn	ECON	**Lagerumschlag**
stock valuation ≈ stocktaking ↑ inventory valuation	ECON	**Lagerbestandsbewertung** ≈ Lagerbestandsaufnahme
stoichiometric (adj.)	CHEM	**stöchiometrisch**
stoichiometry [theory of quantification of chemical compositions]	CHEM	**Stöchiometrie** [Lehre de Quantifizierung chemischer Zusammensetzungen]
Stokes = St	PHYS	**Stokes** = St
STOL (→short take-off and landing)	RADIO NAV	(→Kurzstart und -landung)
stop (n.) [limit of excursion] ≈ touch; detent	MECH	**Anschlag** [eines Ausschlags] = Stopp
stop (n.)	RAILW.SIGN	**Halt**
stop (n.) ≠ start	TELEGR	**Stopp** = Stop ≠ Start
stop (→arrest)	MECH	(→arretieren)
stop (→explosive)	LING	(→Explosivum)
stop band = stop-band; stopband; rejection band ≠ pass band	NETW.TH	**Sperrbereich** = Sperrband ≠ Durchlaßbereich
stopband (→stop band)	NETW.TH	(→Sperrbereich)
stop-band (→stop band)	NETW.TH	(→Sperrbereich)
stop-band attenuation = suppression loss; rejection 2	NETW.TH	**Sperrdämpfung**
stop bit (→stop pulse)	TELEGR	(→Sperrschritt)
stopcock	TECH	**Abschlußhahn**
stop distance [of a magnetic tape]	TERM&PER	**Stoppweg** [eines Magnetbands] = Stopweg
stop element (→stop pulse)	TELEGR	(→Sperrschritt)
stop filter (→blocking filter)	NETW.TH	(→Sperrfilter)
stop frequency = notch frequency	NETW.TH	**Sperrfrequenz**
stop instruction	DATA PROC	**Haltebefehl** = Haltebefehl; Stopbefehl
stop key	CONS.EL	**Stopptaste** = Stoptaste
stop knob	COMPON	**Arretierknopf**
stop list	DATA PROC	**Sperrliste**
stop-off lacquer	METAL	**Abdecklack** [Galvanotechnik]
stop pin	MECH	**Anschlagstift**
stopping (→breaking 1)	MECH	(→Bremsung)
stopping potential	PHYS	**Haltepotential**
stopping power (→breaking power)	MECH	(→Bremsvermögen)
stop polarity	TELEGR	**Stoppolarität**
stop polarity (→signal condition Z)	DATA COMM	(→Stoppolarität)
stop pulse = stop signal; stop bit; stop element; closed stop pulse ≠ start pulse; information pulse	TELEGR	**Sperrschritt** = Stoppschritt; Stopschritt; Stoppelement; Stopelement; Stoppbit; Stopbit; Stoppzeichen; Stopzeichen ≠ Anlaufschritt; Informationsschritt
stop screw	MECH	**Anschlagschraube**
stop signal (→stop pulse)	TELEGR	(→Sperrschritt)
stop spring	MECH	**Festhaltefeder**
stop time [of a magnetic tape]	TERM&PER	**Stoppzeit** [eines Magnetbands] = Stoppgeschwindigkeit; Stopzeit; Stopgeschwindigkeit

stop valve (→blocking valve)	TECH	(→Sperrventil)	storage device (→memory chip)	MICROEL	(→Speicherbaustein)
stop watch	INSTR	Stoppuhr	storage disk	DATA PROC	Speicherplatte
↑ chronometer		↑ Chronometer	= storage target		
storable	TECH	lagerfähig	storage display screen	TERM&PER	Speicherbildschirm
storage	ECON	Einlagerung	≈ image storage tube		= Bildspeicherschirm; Speicherschirm
storage	ELECTRON	Speicherung			≈ Bildspeicherröhre
= registration		↓ Speicherung [DATA PROC]	storage dump (→memory dump)	DATA PROC	(→Speicherabzug)
↓ storage [DATA PROC]					
storage	ECON	Lagerung	storage effect	MICROEL	Speichereffekt
≈ deposit 1		≈ Verwahrung	storage efficiency	DATA PROC	Speicherausnutzung
storage 1	DATA PROC	Speicherung	= store efficiency; memory efficiency		≈ Speichernutzung
[the process of memorizing]		= Abspeicherung; Einspeicherung	storage element (→memory element)	DATA PROC	(→Speicherelement)
= the process of memorization					
↓ short-term storage; intermediate storage; long-term storage; archival		↓ Kurzzeitspeicherung; Zwischenspeicherung; Langzeitspeicherung; Archivierung	storage element (→storage circuit)	CIRC.ENG	(→Speicherglied)
storage (→stockpiling)	ECON	(→Bevorratung)	storage expansion (→memory expansion)	DATA PROC	(→Speichererweiterung)
storage 2 (→memory)	DATA PROC	(→Speicher)			
storage (→stockpiling)	ECON	(→Vorratshaltung)	storage extension (→memory expansion)	DATA PROC	(→Speichererweiterung)
storage access (→memory access)	DATA PROC	(→Speicherzugriff)	storage factor (→factor of quality)	EL.TECH	(→Gütefaktor)
storage access time (→memory access time)	DATA PROC	(→Speicherzugriffszeit)	storage factor (→factor of merit)	PHYS	(→Gütewert)
storage address (→memory address)	DATA PROC	(→Speicheradresse)	storage hierarchy (→memory hierarchy)	DATA PROC	(→Speicherhierarchie)
storage address register (→memory address register)	DATA PROC	(→Speicheradressregister)	storage instruction (→memory instruction)	DATA PROC	(→Speicherbefehl)
storage allocation (→memory management)	DATA PROC	(→Speicherverwaltung)	storage keyboard	TERM&PER	Speichertastatur
storage altitude	TECH	Lagerungshöhe	storage life	TECH	Lagerzeit
		= Lagerhöhe	= shelf life		
storage amplifier	CIRC.ENG	Speicherverstärker	storage list (→memory list)	DATA PROC	(→Speicherliste)
storage area (→memory area)	DATA PROC	(→Speicherbereich)	storage location (→memory location)	DATA PROC	(→Speicherstelle)
storage area protection (→memory area protection)	DATA PROC	(→Speicherbereichsschutz)	storage management (→memory management)	DATA PROC	(→Speicherverwaltung)
storage bank (→memory bank)	DATA PROC	(→Speicherbank)	storage map (→memory map 1)	DATA PROC	(→Speicherabbild)
storage battery (→secondary battery)	POWER SYS	(→Sammelbatterie)	storage-mapped (→memory-mapped)	DATA PROC	(→speicherabbildgetreu)
storage box	ELECTRON	Sortimentbox	storage-mapped (→memory-mapped)	DATA PROC	(→speicherkonform)
storage byte (→memory byte)	DATA PROC	(→Speicherbyte)	storage matrix	COMPON	Speichermatrix
storage cabinet (→drawer storage cabinet)	TECH	(→Schubladenmagazin)	= store matrix; memory matrix		
storage camera tube	ELECTRON	Bildspeicherröhre	storage medium	DATA PROC	Speichermedium
storage capacity (→memory capacity)	DATA PROC	(→Speicherkapazität)	= store medium; memory medium; medium		≈ Datenträger
storage case	TECH	Aufbewahrungsbox	≈ data carrier		
storage cell (→accumulator)	POWER SYS	(→Akkumulator)	storage memory (→mass memory)	DATA PROC	(→Massenspeicher)
storage charge [diode]	MICROEL	Speicherladung [Diode]	storage modification (→memory modification)	DATA PROC	(→Speicheränderung)
storage chip (→memory chip)	MICROEL	(→Speicherbaustein)	storage module (→memory module)	EQUIP.ENG	(→Speicherbaugruppe)
storage circuit	CIRC.ENG	Speicherglied	storage normalizer [network analyzer]	INSTR	Speicher-Normalisierer [Netzwerkanalysator]
= store circuit; storage element; memory flipflop; Jg,Kg flipflop		= speicherndes Schaltglied; Speicherschaltung; Speicherelement; Speicher-Flipflop; Jg,Kg-Flipflop	storage occupancy (→memory occupancy)	DATA PROC	(→Speicherbelegung)
↑ flipflop		↑ Kippschaltung	storage organization (→memory organization)	DATA PROC	(→Speicherorganisation)
storage component (→memory chip)	MICROEL	(→Speicherbaustein)	storage oscilloscope	INSTR	Speicheroszilloskop
storage contents (→memory contents)	DATA PROC	(→Speicherinhalt)			= Speicheroszillograph
storage control (→memory control)	DATA PROC	(→Speichersteuerung)	storage partitioning (→partitioning)	DATA PROC	(→Speicherplatzaufteilung)
storage core	COMPON	Speicherkern	storage place requirement (→memory space requirement)	DATA PROC	(→Speicherplatzbedarf)
= magnetic storage core		= Magnetspeicherkern	storage pool	DATA PROC	Speicherpool
storage cycle (→memory cycle)	DATA PROC	(→Speicherzyklus)	storage process	CIRC.ENG	Speichervorgang
storage-cycle counter (→memory-cycle counter)	DATA PROC	(→Speicherzyklus-Zähler)			= Speicherungsvorgang
storage-cycle request (→memory cycle request)	DATA PROC	(→Speicherzyklus-Anforderung)	storage protection (→memory protection)	DATA PROC	(→Speicherschutz)
storage data (→memory data)	DATA PROC	(→Speicherdaten)	storage recovery time	DATA PROC	Speichererholzeit
storage data register (→memory data register)	DATA PROC	(→Speicherdatenregister)	= store recovery time		
			storage register (→memory register)	DATA PROC	(→Speicherregister)
			storage request (→memory request)	DATA PROC	(→Speicheranforderung)
			storage requirement (→memory space requirement)	DATA PROC	(→Speicherplatzbedarf)

storage-resident

storage-resident (→memory-resident)	DATA PROC	(→speicherresident)
storage room (→store-room)	TECH	(→Lagerraum)
storage scan (→memory scanning)	DATA PROC	(→Speicherabtastung)
storage scanning (→memory scanning)	DATA PROC	(→Speicherabtastung)
storage sector (→memory area)	DATA PROC	(→Speicherbereich)
storage size (→memory size)	DATA PROC	(→Speichergröße)
storage space 1 (→memory capacity)	DATA PROC	(→Speicherkapazität)
storage space 2 (→memory location)	DATA PROC	(→Speicherstelle)
storage structure (→memory structure)	DATA PROC	(→Speicherstruktur)
storage subsystem	DATA PROC	**Speicher-Subsystem**
storage target (→storage disk)	DATA PROC	(→Speicherplatte)
storage temperature	TECH	**Lagerungstemperatur** = Lagertemperatur
storage time (→retention time)	ELECTRON	(→Speicherzeit)
storage time (→carrier storage time)	MICROEL	(→Speicherzeit)
storage time constant	MICROEL	**Speicherzeitkonstante**
storage tube = storing tube; electrostatic storage tube; charge-storage tube; charge-storing tube	ELECTRON	**Speicherröhre** [mit kurzzeitiger Speicherung] = Ladungsspeicherröhre; Williamsröhre
storage unit (→memory unit)	DATA PROC	(→Speichereinheit)
storage utilization = store utilization; memory utilization	DATA PROC	**Speichernutzung** ≈ Speicherausnutzung
storage word (→memory word)	DATA PROC	(→Speicherwort 1)
storage zone (→memory area)	DATA PROC	(→Speicherbereich)
store (v.t.) ≈ save ↓ file (v.t.); archive (v.t.)	DATA PROC	**speichern** = abspeichern ≈ sichern ↓ ablegen; archivieren
store (→deposit 2)	ECON	(→Lager)
store (BRI) (→memory)	DATA PROC	(→Speicher)
store (n.) (→stock)	ECON	(→Vorrat)
store access (→memory access)	DATA PROC	(→Speicherzugriff)
store access time (→memory access time)	DATA PROC	(→Speicherzugriffszeit)
store address (→memory address)	DATA PROC	(→Speicheradresse)
store address register (→memory address register)	DATA PROC	(→Speicheradressregister)
store allocation (→memory management)	DATA PROC	(→Speicherverwaltung)
store-and-forward line	SWITCH	**Teilvermittlungsleitung**
store-and-forward mode [PABX]	SWITCH	**Speicherbetrieb** [PABX]
store-and-forward network ↓ packet switched network	DATA COMM	**speichervermitteltes Netz** = teilstreckenvermitteltes Netz; Teilstreckennetz ↓ Paketvermittlungsnetz
store-and-forward switching [retransmission from switching center to switching center, with intermediate storage] = message storage switching ≠ circuit switching ↓ packet switching; message switching	DATA COMM	**Speichervermittlung** [abschnittsweise Übermittlung von Vermittlungsstelle zu Vermittlungsstelle, mit Zwischenspeicherung] = Teilstreckenvermittlung; Speicher-Datenübermittlung ≠ Durchschaltevermittlung ↓ Paketvermittlung; Sendungsvermittlung
store area (→memory area)	DATA PROC	(→Speicherbereich)
store bank (→memory bank)	DATA PROC	(→Speicherbank)
store byte (→memory byte)	DATA PROC	(→Speicherbyte)
store capacity (→memory capacity)	DATA PROC	(→Speicherkapazität)
store chip (→memory chip)	MICROEL	(→Speicherbaustein)
store circuit (→storage circuit)	CIRC.ENG	(→Speicherglied)
store component (→memory chip)	MICROEL	(→Speicherbaustein)
store contents (→memory contents)	DATA PROC	(→Speicherinhalt)
store control (→memory control)	DATA PROC	(→Speichersteuerung)
store cycle (→memory cycle)	DATA PROC	(→Speicherzyklus)
store-cycle counter (→memory-cycle counter)	DATA PROC	(→Speicherzyklus-Zähler)
store-cycle request (→memory cycle request)	DATA PROC	(→Speicherzyklus-Anforderung)
stored	ELECTRON	**gespeichert** ≈ registriert
store data (→memory data)	DATA PROC	(→Speicherdaten)
store data register (→memory data register)	DATA PROC	(→Speicherdatenregister)
store device (→memory chip)	MICROEL	(→Speicherbaustein)
stored program ≠ fixed program	DATA PROC	**Speicherprogramm** ≠ Festprogramm
stored-program computer	DATA PROC	**speicherprogrammierter Computer** = speicherprogrammierter Rechner
stored-program control ≠ fixed-program control	DATA PROC	**Speicherprogrammierung** ≠ Festprogrammierung
stored-program control = SPC	SWITCH	**speicherprogrammierte Steuerung** = SPC
store dump (→memory dump)	DATA PROC	(→Speicherabzug)
store efficiency (→storage efficiency)	DATA PROC	(→Speicherausnutzung)
store element (→memory element)	DATA PROC	(→Speicherelement)
store expansion (→memory expansion)	DATA PROC	(→Speichererweiterung)
store extension (→memory expansion)	DATA PROC	(→Speichererweiterung)
store for transit goods	ECON	**Transitlager**
store hierarchy (→memory hierarchy)	DATA PROC	(→Speicherhierarchie)
store instruction (→memory instruction)	DATA PROC	(→Speicherbefehl)
store intermediately (→buffer)	DATA PROC	(→zwischenspeichern)
store list (→memory list)	DATA PROC	(→Speicherliste)
store location (→memory location)	DATA PROC	(→Speicherstelle)
store management (→memory management)	DATA PROC	(→Speicherverwaltung)
store map (→memory map 1)	DATA PROC	(→Speicherabbild)
store-mapped (→memory-mapped)	DATA PROC	(→speicherkonform)
store matrix (→storage matrix)	COMPON	(→Speichermatrix)
store medium (→storage medium)	DATA PROC	(→Speichermedium)
store mode	TERM&PER	**Storemodus**
store modification (→memory modification)	DATA PROC	(→Speicheränderung)
store module (→memory module)	EQUIP.ENG	(→Speicherbaugruppe)
store occupancy (→memory occupancy)	DATA PROC	(→Speicherbelegung)
store organization (→memory organization)	DATA PROC	(→Speicherorganisation)
store partitioning (→partitioning)	DATA PROC	(→Speicherplatzaufteilung)
store protection (→memory protection)	DATA PROC	(→Speicherschutz)

store recovery time	DATA PROC	(→Speichererholzeit)	
(→storage recovery time)			
store register (→memory register)	DATA PROC	(→Speicherregister)	
store request (→memory request)	DATA PROC	(→Speicheranforderung)	
store requirement	DATA PROC	(→Speicherplatzbedarf)	
(→memory space requirement)			
store-resident	DATA PROC	(→speicherresident)	
(→memory-resident)			
store-room	TECH	**Lagerraum**	
= storage room		= Abstellraum	
store scan (→memory scanning)	DATA PROC	(→Speicherabtastung)	
store scanning (→memory scanning)	DATA PROC	(→Speicherabtastung)	
store sector (→memory area)	DATA PROC	(→Speicherbereich)	
store space 1 (→memory capacity)	DATA PROC	(→Speicherkapazität)	
store space 2 (→memory location)	DATA PROC	(→Speicherstelle)	
store structure (→memory structure)	DATA PROC	(→Speicherstruktur)	
store temporarily (→buffer)	DATA PROC	(→zwischenspeichern)	
store unit (→memory unit)	DATA PROC	(→Speichereinheit)	
store utilization (→storage utilization)	DATA PROC	(→Speichernutzung)	
store zone (→memory area)	DATA PROC	(→Speicherbereich)	
storing tube (→storage tube)	ELECTRON	(→Speicherröhre)	
storm spy	RADIO LOC	**Sturmwarnungsradar**	
stove	TECH	**Ofen**	
= furnace		↓ Schmelzofen	
↓ smelter			
stove bolt	MECH	**Herdschraube**	
stow	ECON	**verstauen**	
stow position	SAT.COMM	**Verstauposition**	
straddle band	INSTR	**Spreizband**	
straight (→rectilinear)	MATH	(→geradlinig)	
straight detection (→homodyne reception)	HF	(→Homodynempfang)	
straightedge	ENG.DRAW	**Zeichendreieck**	
= set square; drawing triangle		≈ Winkelmaß	
≈ protractor			
straightening (→alignment)	TECH	(→Ausrichtung)	
straightforward	TECH	**problemlos**	
= problem-free			
straightforward amplification	CIRC.ENG	**Geradeausverstärkung**	
= straight-through amplification			
straight line (n.)	MATH	**Gerade**	
straight-line (adj.)	MATH	(→linear)	
(→linear)			
straight-line coding	DATA PROC	(→gestreckte Programmierung)	
(→straight-line programming)			
straight-lined	MATH	(→geradlinig)	
(→rectilinear)			
straight line motion	PHYS	**geradlinige Bewegung**	
straight-line programming	DATA PROC	**gestreckte Programmierung**	
[without program loops]		[ohne Programmschleifen]	
= straight-line coding; in-line programming; in-line coding		≠ zyklische Programmierung	
≠ cyclic programming			
straightness (→linearity 1)	MATH	(→Geradlinigkeit)	
straight pipe thread	MECH	**zylindrisches Rohrgewinde**	
straight receiver	RADIO	**Geradeausempfänger**	
straight reception (→homodyne reception)	HF	(→Homodynempfang)	
straight section	MICROW	**gerades Rechteckhohlleiterstück**	
straight sleeve (→jointing sleeve)	OUTS.PLANT	(→Verbindungsmuffe)	
straight-through amplification	CIRC.ENG	(→Geradeausverstärkung)	
(→straightforward amplification)			
strain (n.)	COLLOQ	**Überanstrengung**	
strain (→mechanical stress)	MECH	(→Spannung)	
strain (→deformation)	TECH	(→Verformung)	
strain (→traction)	MECH	(→Zug)	
strain (→deform)	TECH	(→verformen)	
strain-free	MECH	**spannungsfrei**	
= stress-free			
strain gage (→strain gauge)	INSTR	(→Dehnmeßstreifen)	
strain gauge	INSTR	**Dehnmeßstreifen**	
= strain gage		= Dehnungsmeßstreifen	
strain-gauge amplifier	INSTR	**Dehnmeßstreifen-Verstärker**	
strain hardening	METAL	**kaltverfestigen**	
= work hardening			
strain relief	COMM.CABLE	**Zugentlastung**	
strand (v.t.)	COMM.CABLE	**verseilen**	
strand (n.)	TECH	**Einzeldraht**	
		= Einzelleiter; Teilleiter	
		≈ Einzelader	
strand 1 (→wire pair)	COMM.CABLE	(→Aderpaar)	
strand 2 (→stranded wire)	COMM.CABLE	(→Litzendraht)	
stranded	EL.INST	**mehrdrähtig**	
stranded element	COMM.CABLE	(→Verseilelement)	
(→stranding element)			
stranded wire	COMM.CABLE	**Litzendraht**	
= strand 2; braided conductor; litz wire; flexible; flex		≈ Litze	
stranding cover	ECON	**Strandungsfalldeckung**	
stranding element	COMM.CABLE	**Verseilelement**	
= stranded element; stranding unit; component		= Aufbauelement	
↓ wire pair; quad; star quad; multiple twin quad; coaxial pair		↓ Paar; Viererseil; Sternvierer; DM-Vierer; Koaxialpaar	
stranding machine	COMM.CABLE	**Verseilmaschine**	
↓ pairing machine; twining machine		↓ Paarverseilmaschine; Viererverseilmaschine	
stranding radius	COMM.CABLE	**Verseilradius**	
stranding unit (→stranding element)	COMM.CABLE	(→Verseilelement)	
strangulate	TECH	**einschnüren**	
≈ lace		= abschnüren	
		≈ schnüren	
strangulation	TECH	**Einschnürung**	
= necking down		≈ Einbuchtung	
≈ recession			
strap (n.)	ELECTRON	**Brücke**	
= link; bridge		↓ Lötbrücke; Drahtbrücke; Kabelbrücke; Brückenstecker	
↓ solder strap; wire link; jumper; plug link			
strap (n.)	TECH	**Lasche**	
= lug (n.)			
strap (n.)	SYS.INST	**Rangierdraht**	
[for jumpering within the same terminal block]		[für Rangierungen innerhalb eines Verteilerblocks]	
= patching wire; jumping wire		= Brückendraht; Leiterdraht	
≈ cross-connect (n.)		≈ Rangierleitung	
strap (→strip 2)	TECH	(→Leiste)	
strap (v.t.)	EQUIP.ENG	(→rangieren)	
(→cross-connect)			
strap (→belt)	MECH	(→Riemen)	
strapping	TECH	**Laschung**	
= lashing			
strapping	EL.TECH	(→Zusammenschaltung)	
(→interconnection)			
strapping option	ELECTRON	(→Lötbrückeneinstellung)	
(→solder-strap option)			
strap v.	TECH	**anlaschen**	
strata chart	DOC	**Schichtlinienschaubild**	
strategic planning	ECON	**strategische Planung**	
stratification	TECH	**Schichtung**	
= striation; packing			
stratified	TECH	**geschichtet**	
= laminated; layered		= geblättert	
stratopause	GEOPHYS	**Stratopause**	
stratosphere	GEOPHYS	**Stratosphäre**	
↑ atmosphere		↑ Erdatmosphäre	
stray (v.t.)	TECH	**streuen**	
[something granular]		[etwas Körniges]	
stray (→scattering)	PHYS	(→Streuung)	
stray capacitance	PHYS	**Streukapazität**	
stray current	EL.TECH	**Streustrom**	
= vagabond current		= Irrstrom	
≈ leak current		≈ Leckstrom	

English	Domain	German
stray-current corrosion	EL.TECH	Streustromkorrosion
stray current drain	EL.TECH	Streustromabsaugung
stray current leakage	EL.TECH	Streustromableitung
stray field	PHYS	Streufeld
= dispersion field; leakage field		
stray flux [magnetism]	PHYS	Streufluß [Magnetismus]
stray-free	EL.TECH	streuungsfrei
stray-free transformer	COMPON	streuungsfreier Übertrager = streuungsfreier Transformator
stray inductance	PHYS	Streuinduktivität
= leakage inductance		
stray radiation (→scattered radiation)	PHYS	(→Streustrahlung)
streak	PHYS	Schliere
streak (→strip 1)	TECH	(→Streifen)
streaked (→striated)	TECH	(→streifig)
streaking	TV	Fahneneffekt
streamer	TERM&PER	Streamer-Magnetbandgerät
[records continuously without starts and stops]		[zügig ohne Start-Stopps durchlaufend]
= cartridge streamer; streaming tape drive; tape streamer; floppy tape ↑ magnetic tape memory		= Streaming-Magnetbandspeicher; Streamer; Cartridge Streamer ↑ Magnetbandspeicher
streaming mode (→continuous mode)	DATA PROC	(→Zügig-Betrieb)
streaming tape	TERM&PER	Streamer-Magnetband
streaming tape drive (→streamer)	TERM&PER	Streamer-Magnetbandgerät
streamline (→line of force)	PHYS	(→Kraftlinie)
streamline (v.t.) (→rationalize)	ECON	(→rationalisieren)
streamlined ↓ aerodynamic 2	TECH	stromlinienförmig formschön ↓ aerodynamisch 2
streamlined (→efficient)	ECON	(→rationell)
streamlining (→rationalization)	ECON	(→Rationalisieren)
streamy = sustained ≠ bursty	TELEC	kontinuierlich ≠ diskontinuierlich
strech (v.t.)	METAL	strecken
strech (v.t.) = extend (v.t.); elongate	PHYS	dehnen = strecken; spannen
streched	METAL	gestreckt
streching	METAL	Streckung
streching (→extension)	PHYS	(→Ausdehnung 1)
streetcar line (→tramway)	TECH	(→Straßenbahn)
street-side cabinet	OUTS.PLANT	Straßenschrank
strength ≈ compatibility; resistance	TECH	Festigkeit ≈ Kompatibilität; Widerstandsfähigkeit
strength ≈ Widerstandskraft	MECH	Stärke
strength (→resistance)	TECH	(→Widerstandsfähigkeit)
strengthen (→reinforce)	TECH	(→verstärken)
strengthening (→reinforcement)	MECH	(→Verstärkung)
stress = load; charge	QUAL	Beanspruchung = Belastung
stress (→mechanical stress)	MECH	(→Spannung)
stress cycle	QUAL	Beanspruchungszyklus
stress duration	QUAL	Beanspruchungsdauer
stress-free (→strain-free)	MECH	(→spannungsfrei)
stress level	QUAL	Beanspruchungsgrad
stress-relieving anneal (→normalizing)	METAL	(→Normalglühen)
stress resisting clamp	TECH	Zugentlastungsschelle
stress screening	QUAL	Verschleißaussonderung = Verschleißsortierung
stress tensor	MECH	Spannungstensor
stress test = stress testing	QUAL	Belastungsprobe = Belastungstest; Verschleißprüfung
stress testing (→stress test)	QUAL	(→Belastungsprobe)
stress vector	MECH	Spannungsvektor
stretch (n.) = extension; expansion; dilation; dilatation; lengthening ≈ strain (n.) ↓ extension	PHYS	Dehnung [eines Festkörpers in die Länge oder Breite] ↓ Ausdehnung 1
stretching [computer graphics; possibility to stretch a graphic with automatic correction of dimensions, hatching etc.] ≈ post-stretching	DATA PROC	Stretchen [Computergraphik; Möglichkeit des Auseinanderziehens einer Gaphik bei automatischer Maßkorrektur, Nachschraffur etc.] = graphische Editierfähigkeit ≈ Nachtransformation
striated = streaked	TECH	streifig = gestreift
striation (→stratification)	TECH	(→Schichtung)
strictly confidential	ECON	streng geheim = streng vertraulich
strike (n.) = walkout	ECON	Streik = Ausstand
strike (→hit)	TECH	(→schlagen)
string	MECH	Saite
string 1 [an ordered sequence of data] = sequence	DATA PROC	String 1 [geordnete Folge von Datensätzen] = Sequenz; sortierte Folge; vorsortierte Folge
string (→cord)	TECH	(→Schnur)
string (→character string)	DATA PROC	(→Zeichenfolge)
stringency (→shortage)	ECON	(→Knappheit)
stringend = severe	COLLOQ	streng ≈ scharf (fig.); rigurös
stringent specification	ECON	strenge Spezifikation
string formation	DATA PROC	Zeichenkettenbildung = Stringbildung
string galvanometer	INSTR	Saitengalvanometer
string handling = string manipulation	DATA PROC	Zeichenkettenhandhabung = Stringhandhabung
string length	DATA PROC	Zeichenkettenlänge
string manipulation (→string handling)	DATA PROC	(→Zeichenkettenhandhabung)
string-oriented data compression ↑ data compression	DATA PROC	stringorientierte Datenverdichtung = horizontale Datenverdichtung ↑ Datenverdichtung
string processing language = SPL	DATA PROC	Stringverarbeitungssprache
string variable	DATA PROC	Stringvariable
strip 2 (n.) = strap ↓ dummy strip; ornamental strip	TECH	Leiste ↓ Abdeckleiste; Zierleiste
strip 1 (n.) [long, short section of piece from something larger] = stripe; streak ≈ tape; strap	TECH	Streifen [langer, schmaler Abschnitt oder Stück von etwas größerem] ≈ Band; Leiste
strip	METAL	Flachschiene
strip (→strip the isolation)	EL.TECH	(→abisolieren)
strip cable (→ribbon cable)	COMM.CABLE	(→Bandkabel)
strip copper	METAL	Kupferband
stripe (→strip 1)	TECH	(→Streifen)
strip feed (→tape feed 2)	TERM&PER	(→Streifenvorschub)
strip line = stripline	MICROW	Streifenleitung = Bandleitung; Streifenleiter; Streifenwellenleiter; Stripline
stripline (→strip line)	MICROW	(→Streifenleitung)
strip line antenna	ANT	Streifenleiterantenne
strip line circulator	MICROW	Streifenleiterzirkulator = Stripline-Zirkulator
stripline dipole	ANT	Stripline-Dipol
strip line filter (→interdigital filter)	MICROW	(→Interdigitalfilter)
stripline laser	OPTOEL	Oxydstreifenlaser

stripped-down version (→crippled version)	TECH	(→abgemagerte Version)	
strip printer (→tape printer)	TERM&PER	(→Streifendrucker)	
strip reader	TERM&PER	**Streifenleser 2**	
↑ optical character reader		↑ optischer Leser	
strip the isolation	EL.TECH	**abisolieren**	
= strip			
strip without notching	EL.TECH	**kerbfrei abisolieren**	
strobe (v.t.)	DATA PROC	**freigeben**	
≈ deallocate		= bestätigen	
		≈ Zuweisung aufheben	
strobe (n.)	DATA PROC	**Hinweissignal**	
[signal confirming some content being transmitted on a bus]		[bestätigt einen Signalinhalt, der geräde auf einem Bus übertragen wird]	
↓ address strobe; data strobe		= Freigabesignal; Aktivierungssignal; Übernahmesignal; Auswahlsignal	
		↓ Adreßhinweissignal; Datenhinweissignal	
strobe (n.) (→enable signal)	ELECTRON	(→Freigabeimpuls)	
strobe (→strobe pulse)	ELECTRON	(→Ausblendimpuls)	
strobe 2 (n.) (→clock frequency)	DATA PROC	(→Taktfrequenz)	
strobe frequency (→clock frequency)	DATA PROC	(→Taktfrequenz)	
strobe input	ELECTRON	**Strobe-Eingang**	
strobe pulse (n.)	ELECTRON	**Ausblendimpuls**	
= strobe		= Austastimpuls; Abtastimpuls; Tastimpuls; Strobe-Impuls; Strobe	
strobe signal	ELECTRON	**Ausblendsignal**	
		= Strobesignal	
strobing	ELECTRON	**Impulsausblendung**	
[to sample a long signal by a pulse]		= Impulsaustastung; Strobing	
= pulse blanking			
stroboscopic relay tester	TELEGR	**Glimmlampen-Relaissender**	
stroke 2 (n.)	MECH	**Hub**	
= travel			
stroke 1 (n.)	MECH	**Schlag**	
= hit 1 (n.)		≈ Mittenversatz; Aufschlag	
≈ excentricity; impact			
stroke	TV	**Störstreifen**	
stroke keyboard	TERM&PER	**Hubtastatur**	
stroke pattern	TECH	**Strichmuster**	
stroke width	ENG.DRAW	**Strichstärke**	
		= Strichbreite	
stroke writer	TERM&PER	**Strichschreiber**	
strong (→resistant)	TECH	(→widerstandsfähig)	
strontium	CHEM	**Strontium**	
= Sr		= Sr	
Strowger switching system	SWITCH	**Strowger-Vermittlungssystem**	
structogram	DATA PROC	**Struktogramm**	
[representation of logic sequences]		[Darstellung logischer Abläufe]	
= structure chart; structure diagram		= Nassi-Schneidermann-Diagramm; Strukturdiagramm; Strukturbild	
structural balance return loss	NETW.TH	**komplexes Echodämpfungsmaß**	
= active balance return loss		= komplexes Rückflußdämpfungsmaß	
structural etching	MICROEL	**Strukturätzen**	
structurally dual	NETW.TH	**strukturdual**	
structurally dual quadripole (→structurally dual two-port)	NETW.TH	(→strukturdualer Vierpol)	
structurally dual two-port	NETW.TH	**strukturdualer Vierpol**	
= structurally dual quadripole			
structurally symmetrical	NETW.TH	**struktursymmetrisch**	
structurally symmetric quadripole (→structurally symmetric two-port)	NETW.TH	(→struktursymmetrischer Vierpol)	
structurally symmetric two-port	NETW.TH	**struktursymmetrischer Vierpol**	
= structurally symmetric quadripole			
structural parameter	SCIE	**Strukturparameter**	
structural part (→mechanical component)	EQUIP.ENG	(→Konstruktionsteil)	
structural programming language	DATA PROC	**strukturierte Programmiersprache**	
= top-down programming language		= strukturierte Sprache	
structural shape	METAL	**Profilstahl**
structural shape (→profile)	MECH	(→Profil)
structural steel	METAL	**Baustahl**
↑ acero de construcción		
structural-steel construction	CIV.ENG	**Stahlbau**
structure	SCIE	**Struktur**
≈ pattern; configuration; texture		= Gefüge
		≈ Muster; Konfiguration; Textur
structure (→configuration)	TECH	(→Aufbau 2)
structure-borne sound detector	SIGN.ENG	**Körperschallmelder**
structure chart (→structogram)	DATA PROC	(→Struktogramm)
structured coding (→structured programming)	DATA PROC	(→strukturierte Programmierung)
structured design	DATA PROC	**strukturierter Entwurf**
structure diagram (→structogram)	DATA PROC	(→Struktogramm)
structured program	DATA PROC	**strukturiertes Programm**
structured programming	DATA PROC	**strukturierte Programmierung**
= top-down programming; structured coding		= Top-down-Programmierung
≈ GO-TO-free programming		
structure memory	COMPON	**Strukturspeicher**
structure-oriented	DATA PROC	**strukturorientiert**
structure-oriented data compression	DATA PROC	**strukturorientierte Datenverdichtung**
↑ data compression		= vertikale Datenverdichtung
		↑ Datenverdichtung
strut	MECH	**Strebe**
↓ transversal strut		↓ Querstrebe
strutted terminal pole (→end pole)	OUTS.PLANT	(→Endgestänge)
strutting	CIV.ENG	**Gestänge**
STTL	MICROEL	**STTL**
[Schottky-clamped transistor-transistor logic]		
stub (n.)	LINE TH	**Blindleitung**
stub	COMM.CABLE	**Stichleitung**
stub	TECH	**Stumpf**
		[Verzweigung]
stub	ANT	**Stichleitung**
		= angezapfte Speiseleitung
stub (→stem)	TECH	(→Stutzen)
stub antenna	ANT	**Stichleitungsantenne**
		= Stub-Antenne
stub cable	OUTS.PLANT	**Stumpfkabel**
		= Abzweigkabel
stub card	TERM&PER	**Kurzlochkarte**
stub feedline (→feedline with matching stub)	ANT	(→angezapfte Speiseleitung)
stub guy	OUTS.PLANT	**Überweganker**
[horizontal guy to an auxiliary pole, used under special circumstances]		[horizontales Abspannseil zu einem Hilfsmast, für Sonderfälle]
stub tooth	MECH	**Stumpfzahn**
[gear]		[Zahnrad]
stud	MECH	**Stehbolzen**
stud (→nipple stud)	MECH	(→Pimpel)
stud bolt	MECH	**Gewindebolzen**
= threaded bolt		
stud bolt	MECH	**Stiftschraube**
Student's t distribution	MATH	**Student-Verteilung**
= t distribution		= t-Verteilung
↑ test distribution		↑ Testverteilung
studio	BROADC	**Studio**
↓ television studio; audio broadcast studio		↓ Aufnahmestudio
		↓ Fernsehstudio; Hörfunkstudio
studio conference	TELEC	**Studiokonferenz**
studio directional microphone	EL.ACOUS	**Studio-Richtmikrophon**
studio engineer	BROADC	**Studiotechniker**
studio mixer	TV	**Trickmischgerät**
studio quality level	BROADC	**Studioqualität**
= contribution level		
studio standard	BROADC	**Studionorm**
study (n.)	TECH	**Studie**
study commission	ECON	**Studienkommission**
study period	ECON	**Studienperiode**
stuff	TECH	**stopfen**

stuffing 572

stuffing (→ pulse stuffing) CODING (→Stopfen)
stuffing bit CODING **Stopfbit**
= justification bit; stuffing digit; justification digit; justifying digit; padding bit
stuffing box MECH **Stopfbuchse**
stuffing digit (→stuffing bit) CODING (→Stopfbit)
stuffing frame (→justification frame) CODING (→Stopfrahmen)
stuffing mode CODING **Stopfverfahren**
= justification mode
↑ bit rate matching ↑ Bitratenanpassung
STX (→start of text) DATA COMM (→Textanfang)
styling TECH **Gestaltung**
stylus TERM&PER **Schreibnadel**
[pencil-type screen scanner] [stiftähnliches Abtastgerät]
≈ light pen ≈ Stylus
≈ Lichtgriffel
stylus (→needle) EL.ACOUS (→Nadel)
stylus printer TERM&PER **Nadeldrucker**
= wire matrix printer; wire printer ≈ Drahtdrucker
↑ impact printer; matrix printer ↑ Anschlagdrucker; Rasterdrucker
s-type conductivity PHYS **s-Leitfähigkeit**
[from "soft" doping]
styroflex capacitor COMPON **Styroflexkondensator**
styroform CHEM **Styropor**
SUB (→substitute character) DATA COMM (→Substitution)
subaqueous cable COMM.CABLE (→Unterwasserkabel)
(→underwater cable)
subarea MATH **Teilbereich**
subassembly EQUIP.ENG **Untereinheit**
= subunit
subassembly (→module) EQUIP.ENG (→Baugruppe)
subassembly address (→module SWITCH (→Baugruppenadresse)
address)
subassembly frame (→module EQUIP.ENG (→Baugruppenrahmen)
frame)
subassembly guide bar EQUIP.ENG (→Baugruppenführung)
(→module guide bar)
subassembly labeling EQUIP.ENG (→Baugruppenbeschriftung)
(→module labeling)
subassembly panel (→module EQUIP.ENG (→Baugruppenabdeckung)
front panel)
subassembly position EQUIP.ENG (→Baugruppenposition)
(→board position)
subassembly replacement EQUIP.ENG (→Baugruppentausch)
(→module replacement)
subassembly shelf (→module EQUIP.ENG (→Baugruppenrahmen)
frame)
subassembly variant (→board EQUIP.ENG (→Baugruppenvariante)
variant)
sub-audio channel TELEC **Unterlagerungskanal**
sub-audio measurement INSTR **Subaudiomessung**
sub-audio telegraphy TELEGR **Unterlagerungstelegraphie**
subband TELEC **Teilband**
= Unterband
subband switch-over RADIO **Teilbereichsumschalter**
= area switch-over
sub-baseband RADIO REL **Subbasisband**
sub-board (→expansion EQUIP.ENG (→Erweiterungskarte)
board)
subcarrier MODUL **Hilfsträger**
≠ main carrier ≠ Nebenträger
≠ Hauptträger
sub-central (→hub unit) TELECONTR (→Unterzentrale)
subchannel TELEC **Unterkanal**
= tributary channel ≈ Subkanal
≈ auxiliary channel ≈ Hilfskanal
sub-channel (→auxiliary channel) TELEC (→Hilfskanal)
subcollector MICROEL **Subkollektor**
subcontract (n.) ECON **Unterauftrag**
= Untervertrag
subcontracting ECON **Untervergabe**
= Subkontraktierung
subcontracting agreement ECON **Zuliefervertrag**
subcontractor ECON **Unterauftragnehmer**
≈ supplier = Subunternehmer; Nebenunternehmer; Unterlieferant
≈ Lieferant

subcritical SCIE **unterkritisch**
= subkritisch
subdeterminant MATH **Subdeterminante**
subdimensioning (→subsizing) TECH (→Unterdimensionierung)
subdirectory DOC **Unterverzeichnis**
↑ directory ↑ Verzeichnis
subdirectory DATA PROC **Unterkatalog**
[cointained within a main directory] [in einem Hauptverzeichnis enthalten]
= Unterverzeichnis
subdivide TECH **unterteilen**
≈ partition (v.t.) = untergliedern
↑ divide ≈ aufteilen
↑ teilen
subdivision ECON **Unterabteilung**
= branch [Organisation]
subdivision TECH **Unterteilung**
= fractionation = Fraktionierung; Untergliederung
≈ partition (n.)
↑ division ≈ Aufteilung
↑ Teilung
subdued color (NAM) PHYS **gedämpfte Farbe**
= subdued colour (BRI) = Mattfarbe
subdued colour (BRI) (→subdued PHYS (→gedämpfte Farbe)
color)
subfield DATA PROC **Teilfeld**
↑ data field ↑ Datenfeld
subfield MATH **Teilkörper**
subfluvial cable COMM.CABLE **Flußkabel**
↑ underwater cable ↑ Unterwasserkabel
subframe TV **Teilbild**
[partial picture] [Unterteilung eines Vollbildes]
≠ frame
↓ field ↓ Halbbild
subframe CODING **Teilrahmen**
= Unterrahmen
subgoal COLLOQ **Zwischenziel**
subgroup MATH **Untergruppe**
subgroup (→grading group) SWITCH (→Staffelgruppe)
subharmonic (adj.) PHYS **subharmonisch**
subharmonic (n.) (→subharmonic PHYS (→subharmonische Schwingung)
oscillation)
subharmonic oscillation PHYS **subharmonische Schwingung**
= subharmonic (n.) = Subharmonische
subindex [MATH] TYPOGR (→Tiefstellung)
(→subscript)
subject LING **Subjekt**
= Satzgegenstand
subject (→subject field) LING (→Fachgebiet)
subject (→re) OFFICE (→Betreff)
subject area (→subject LING (→Fachgebiet)
field)
subject field LING **Fachgebiet**
[terminology] [Terminologie]
= field; subject area; subject = Sachgebiet
subject index LING **Sachwortverzeichnis**
= index = Sachregister; Sachverzeichnis; Sachwörterverzeichnis; Stichwortregister; Stichwortverzeichnis
subjective S/N ratio TV **visueller Störabstand**
subject to authorization ECON **genehmigungspflichtig**
subject to fee TELEC (→gebührenpflichtig)
(→chargeable)
subjunction LING **Subjunktion**
= Fügewort
subjunctive (→subjunctive LING (→Konjunktiv)
mood)
subjunctive mood (n.) LING **Konjunktiv**
[e.g.: if he go] [z.B. er gehe; er ginge]
≈ subjunctive = Möglichkeitsform; Heischeform
≈ conditional mood
↑ mood ≈ Konditional
≠ Indikativ
↑ Modus
↓ Konjunktiv I; Konjunktiv II
sublattice PHYS **Teilgitter**
sublayer DATA COMM **Teilschicht**
[OSI] [ISO-Schichtenmodell]
sublicence ECON **Unterlizenz**

English	Domain	German
sublimate (→sublime)	PHYS	(→sublimieren)
sublimation	PHYS	Sublimation
[direct transition from solid to gaseous state]		[direkter Übergang von festem in gasförmigen Zustand]
≈ evaporation		≈ Verdampfung
sublime (v.i.)	PHYS	sublimieren
= sublimate		
subliminal	COLLOQ	unterschwellig
submarine cable	COMM.CABLE	Seekabel
= seacable; undersea cable; sea cable		↑ Unterwasserkabel
↑ underwater cable		↓ Tiefseekabel
↓ deep-sea cable		
submarine cable connection	TELEC	Seekabelverbindung
submarine cable plow	OUTS.PLANT	Unterwasserkabelpflug
= underwater cable plow		
submarine fiber cable	COMM.CABLE	optisches Seekabel
= undersea fiber cable		
submenu	DATA PROC	Untermenü
submicron range	MICROEL	Submikronbereich
		= Submicronbereich
sub-miniature	TECH	Subminiatur
subminiature (→miniature)	TECH	(→Kleinst-)
subminiature tube (→acorn tube)	ELECTRON	(→Kleinströhre)
submission	OFFICE	Vorlage 2
[of a topic]		[eines Vorgangs]
= presentation		
submission and delivery entity	DATA COMM	Sende- und Empfangsinstanz
= SDE		
submodulation	MODUL	Untermodulation
submodule	EQUIP.ENG	Unterbaugruppe
		= Teileinschub
sub-mux (→subrate multiplexer)	DATA COMM	(→Subratenmultiplexer)
subnetwork	TELEC	Teilnetz
subnetwork connection	DATA COMM	Vermittlungsinstanzen-Verbindung
subnormal (n.)	MATH	Subnormale
[projection of a perpendicular on the abscissa axis]		[Projektion einer Senkrechten auf die Abszissenachse]
suboffer	ECON	Unterangebot 2
		[an einen Hauptanbieter]
suboptimum (adj.)	SCIE	suboptimal
subordinate (n.)	ECON	Untergebener
≠ superior		≠ Vorgesetzter
subordinate (adj.)	TECH	untergeordnet
= slave		
subordinate clause	LING	Nebensatz
= clause		= Gliedsatz; Konstituentensatz; Teilsatz
suboscillation (→partial oscillation)	PHYS	(→Partialschwingung)
subproblem	SCIE	Teilproblem
= partial problem		
subprogram (→subroutine)	DATA PROC	(→Unterprogramm)
subrack (→module frame)	EQUIP.ENG	(→Baugruppenrahmen)
subrack (→inset)	EQUIP.ENG	(→Geräteeinsatz)
subrack equipping (→module frame packaging)	EQUIP.ENG	(→Baugruppenträgerbestückung)
subrack packaging (→module frame packaging)	EQUIP.ENG	(→Baugruppenträgerbestückung)
subrange	TECH	Teilbereich
		= Unterbereich
subrate	TELEC	Subrate
[less than 56 kbit/s in the US hierarchy, less than 64 kbit/s in the CEPT hierarchy]		[in der US-Hierarchie kleiner 56 kbit/s, in der CEPT-Hierarchie kleiner 64 kbit/s]
subrate multiplexer	DATA COMM	Subratenmultiplexer
= sub-mux		= Submultiplexer; Submuxer; Geschwindigkeitsumsetzer
subrates bus	DATA COMM	Subratenbus
subreflector	ANT	Nebenreflektor
≠ main reflector		= Subreflektor; Hilfsreflektor; Fangreflektor
		≠ Hauptreflektor
subrepertoire	INF	Unter-Zeichenvorrat
subroutine	DATA PROC	Unterprogramm
[reusable fixed set of instructions for specific tasks]		[feste mehrfach abrufbare Befehlsfolge für spezielle Aufgaben]
= subprogram		= Subroutine; Unterroutine; Teilprogramm
≈ procedure; routine		≈ Prozedur; Routine
≠ main program		≠ Hauptprogramm
subroutine call	DATA PROC	Unterprogrammaufruf
subroutine library	DATA PROC	Unterprogrammbibliothek
subroutine register	DATA PROC	Unterprogrammregister
subsampling	CODING	Unterabtastung
subscheme	DATA PROC	Teilschema
subscriber	ECON	Zeichner
[of a security]		[eines Wertpapiers]
subscriber acceptance	TELEC	Teilnehmerakzeptanz
= costumer acceptance		
subscriber busy	SWITCH	teilnehmerbesetzt
subscriber-busy condition	SWITCH	Teilnehmerbesetztzustand
subscriber cable	OUTS.PLANT	Teilnehmerkabel
subscriber carrier system (→subscriber loop carrier)	TRANS	(→Teilnehmermultiplexsystem)
subscriber circuit (→subscriber-line circuit)	SWITCH	(→Teilnehmersatz)
subscriber data	SWITCH	Teilnehmerdaten
subscriber density	TELEC	Teilnehmerdichte
subscriber directory	TELEC	Teilnehmerverzeichnis
≈ communications directory		≈ Kommunikationsverzeichnis
↓ telephone directory; telex directory		↓ Fernsprechverzeichnis; Telexverzeichnis
subscriber file	SWITCH	Teilnehmerdatei
subscriber identification	DATA COMM	Teilnehmerkennung
= network user identification; NUI; user identification		= Teilnehmeridentifizierung; Teilnehmeridentifikation
subscriber line	TELEC	Teilnehmerleitung
= costumer line; access line; subscriber loop; costumer loop; local loop; line loop; loop		= Teilnehmeranschlußleitung; Teilnehmeranschluß; Anschlußleitung; Teilnehmerschleife; lokale Schleife
↓ local subscriber line; toll line; telephone line		↓ Ortsanschlußleitung; Fernanschlußleitung; Fernsprechleitung
subscriber line carrier (→subscriber loop carrier)	TRANS	(→Teilnehmermultiplexsystem)
subscriber-line circuit	SWITCH	Teilnehmersatz
[detects an off-hook condition]		[erkennt den Belegungsanreiz eines Teilnehmers]
= line circuit; subscriber circuit; subscriber line interface circuit; SLIC		= Teilnehmerleitungssatz; Teilnehmeranschlußschaltung; Teilnehmerschaltung; Anschlußschaltung
↑ circuit		↑ Satz
subscriber line interface circuit (→subscriber-line circuit)	SWITCH	(→Teilnehmersatz)
subscriber line measuring system	SWITCH	Teilnehmerleitungs-Meßeinrichtung
subscriber-line module	SWITCH	Teilnehmeranschluß-Baugruppe
		= Teilnehmerleitungsmodul
subscriber line supervision board	SWITCH	Teilnehmerüberwachungsplatz
subscriber loop (→subscriber line)	TELEC	(→Teilnehmerleitung)
subscriber loop carrier	TRANS	Teilnehmermultiplexsystem
= SLC; subscriber line carrier; subscriber loop system; loop carrier system; subscriber carrier system; loop carrier; pair gain system; added-main-line system		= Teilnehmermultiplex; Teilnehmersystem; Paarvervielfachungssystem; Pair-gain-System
≈ network access system		≈ Teilnehmerzugangssystem
subscriber loop system (→subscriber loop carrier)	TRANS	(→Teilnehmermultiplexsystem)
subscriber network	TELEC	Teilnehmernetz 1
[within the public communications network]		[Teil des öffentlichen Netzes]
= exchange access network; access network; distribution network 2; user network 2		

subscriber number TELEC
[code necessra to dial a subscriber within a local network]
= directory number; terminal number
↑ dial number
↓ telephone number [TELEPH]; telex number [TELEGR]; network address [DATA COMM]

Teilnehmerrufnummer
[Nummer zur Wahl eines Teilnehmers in einem Ortsnetz]
= Teilnehmeranschlußnummer; Teilnehmernummer; Anschlußnummer
↑ Rufnummer
↓ Telefonnummer [TELEPH]; Telex-Nummer [TELEGR]; Netzadresse [DATA COMM]

subscriber PCM system (→digital TRANS loop carrier system)
(→digitales Teilnehmermultiplexsystem)

subscriber-premises equipment TELEC (→user terminal)
(→Teilnehmergerät)

subscriber ringing signal TELEC (→ringing signal)
(→Rufsignal)

subscriber's account TELEC
Teilnehmerrechnung

subscribers' class (→user DATA COMM group)
(→Teilnehmerbetriebsklasse)

subscriber's control information SWITCH
Teilnehmerinformation

subscriber's drop (→drop OUTS.PLANT cable)
(→Hauseinführungskabel)

subscriber selection stage SWITCH
Teilnehmerwahlstufe
= subscriber stage
= Teilnehmerstufe

subscriber service SWITCH
Teilnehmerentstörung

subscriber set TELEPH
Teilnehmerapparat
[the telephone set at the subscriber]
= subscriber station; station set; subset; station apparatus; customer's apparatus
↑ subscriber premises equipment [TELEC]
[beim Teilnehmer stehender Fernsprechapparat]
≈ Fernsprechapparat
↑ Teilnehmergerät [TELEC]

subscriber's premises (→suscriber TELEC premises)
(→Teilnehmerbereich)

subscriber's service line OUTS.PLANT
Endstellenleitung

subscriber stage (→subscriber SWITCH selection stage)
(→Teilnehmerwahlstufe)

subscriber station TELEC
Teilnehmerstelle
= Teilnehmerstation

subscriber station (→subscriber TELEPH set)
(→Teilnehmerapparat)

subscriber switching unit SWITCH
Teinehmerkoppelfeld
↑ switching unit
↑ Koppelfeld

subscriber trunk dialling (BRI) SWITCH (→direct distance dialing)
(→Selbstwählferndienst)

subscript (n.) TYPOGR
Tiefstellung
= subscripted character; inferior figure; inferior; subindex [MATH]; subscripting
≠ superscript
= Index; tiefgesetztes Zeichen; tiefstehendes Zeichen; Tiefstellen
≠ Hochstellung

subscripted (→indexed) MATH
(→indexiert)

subscripted character TYPOGR (→subscript)
(→Tiefstellung)

subscripted variable DATA PROC (→indexed variable)
(→indexierte Variable)

subscript expression DATA PROC
Indexausdruck

subscription ECON
Zeichnung
[of securities]
[von Wertpapieren]

subscription (→application) TELEC
(→Anmeldung)

subscription call connection SWITCH
Ausnahmequerverbindung

subscription fee TELEC
Anschlußgebühr
= connecting charge; connection charge
= Anschlußgebühr

subscription television (→pay BROADC TV)
(→Abonnementsfernsehen)

subscription TV (→pay TV) BROADC
(→Abonnementsfernsehen)

subsequent TECH
nachträglich

subsequent delivery ECON
Nachlieferung

subset MATH
Teilmenge
= partial set

subset (→set) TERM&PER
(→Untermenge)
(→Gerät)

subset (→subscriber set) TELEPH
(→Teilnehmerapparat)

subshell PHYS
Unterschale

subsidiary ECON
Tochtergesellschaft
= affiliate company; affiliate
≠ parent company
≠ Stammhaus

subsidize ECON
subventionieren

subsizing TECH
Unterdimensionierung
= subdimensioning

subsound (→infrasound) ACOUS
(→Infraschall)

subsripting (→subscript) TYPOGR
(→Tiefstellung)

substance CHEM
Substanz
= Stoff

substandard (adj.) TECH
unterdurchschnittlich

substandard (→low-quality) ECON
(→minderwertig)

substandard keyboard TERM&PER
Schmaltastatur

substantive (n.) LING
Substantiv
= noun 1
= Hauptwort; Nomen 1; Nennwort; Dingwort
↑ Nomen 2

substation (→tributary TELECONTR station)
(→Unterstation)

substitutable TECH (→exchangeable)
(→auswechselbar)

substitute (→exchange) TECH
(→auswechseln)

substitute character DATA COMM
Substitution
[code]
[Code]
= SUB
= SUB

substituting address DATA PROC
Substitutionsadresse
[substituierende Adresse]

substitution MATH
Substitution

substitution DATA PROC
Substitution
↓ address substitution
= Austausch
↓ Adressensubstitution

substitution method INSTR
Substitutionsmethode

substitution-type mixed crystal PHYS
Substitutions-Mischkristall

substrate MICROEL
Substrat
= bulk silicon; supporting material; carrier; host material
= Trägermaterial; Träger; Grundmaterial; Wirtsmaterial

substrate bias MICROEL
Substratvorspannung
= back gate bias

substrate transistor MICROEL
Substrat-Transistor
= parasitic transistor
= parasitärer Transistor

substring DATA PROC
Teilkette

subsystem TECH
Untersystem
≈ ancillary subsystem
= Teilsystem; Subsystem
= Hilfssystem

subtangent MATH
Subtangente
[projection of a tangent on the abscissa axis]
[Projektion einer Tangente auf die Abszissenachse]

subterranean (→buried) TECH
(→unterirdisch)

subtotal MATH
Zwischensumme
= intermediate total
= Subtotal

subtract MATH
subtrahieren
= deduct
= abziehen

subtracter (→subtracter CIRC.ENG circuit)
(→Subtrahierschaltung)

subtracter circuit CIRC.ENG
Subtrahierschaltung
= subtracter
= Subtrahierer; Subtrahierwerk; Subtrahierglied; Subtraktor; Subtrakter

subtraction MATH
Subtraktion
≈ abatement; deduction
= Subtrahieren; Abzug; Abziehen

subtractive MATH
subtraktiv

subtractive color composition TV
subtraktive Farbmischung
= subtractive colour composition

subtractive colour composition TV (→subtractive color composition)
(→subtraktive Farbmischung)

subtrahend MATH
Subtrahend
[number to be deducted from the minuend]
≠ minuend
[vom Minuend abzuziehende Zahl]
≠ Minuend

subunit (→subassembly) EQUIP.ENG
(→Untereinheit)

suburb ECON
Vorort
= outskirt
= Vorstadt; Peripherie

subvoltage (→undervoltage) EL.TECH
(→Unterspannung)

subwave (→partial wave) PHYS
(→Teilwelle)

subwoofer EL.ACOUS
Tiefpaßlautsprecher
≈ woofer
= Subwoofer
≈ Tieftonlautsprecher

succeeding COLLOQ
nachfolgend
≈ consecutive
≈ aufeinanderfolgend

success MATH
Erfolg

successful call SWITCH
erfolgreiche Belegung

succession COLLOQ
Aufeinanderfolge
= Folge 2

succession (→sequence) SCIE
(→Sequenz)

English	Domain	German
successive (→stepwise)	TECH	(→schrittweise)
successive (→consecutive)	TECH	(→aufeinanderfolgend)
successive	COLLOQ	sukzessiv
successive (→consecutive)	TECH	(→aufeinanderfolgend)
successive sequential circuit	CONTROL	Schrittschaltwerk
= stepper		
successor program	DATA PROC	Folgeprogramm
success probability	MATH	Erfolgswahrscheinlichkeit
suck (v.t.)		absaugen
sucking action (→suction effect)	TECH	(→Saugwirkung)
sucking effect (→suction effect)	TECH	(→Saugwirkung)
suction (→wake)	PHYS	(→Sog)
suction anode	ELECTRON	Sauganode
suction effect	TECH	Saugwirkung
= sucking effect; sucking action		
suction opening	TECH	Absaugöffnung
suction power	TECH	Saugleistung
suction pump	TECH	Saugpumpe
sudden failure	QUAL	Sprungausfall
≠ degradation failure		≠ Driftausfall
suffix	LING	Suffix
↑ affix		= Nachsilbe
		↑ Affix
suffix (→file extension)	DATA PROC	(→Dateikennung)
suffix dialing	SWITCH	Nachwahl
= suffix dialling; post-selection		
suffix dialling (→suffix dialing)	SWITCH	(→Nachwahl)
suffix key	TERM&PER	Nachwahltaste
suitability	TECH	Eignung
= aptitude; qualification		≈ Brauchbarkeit
≈ serviceableness		
suite (→rack row)	SYS.INST	(→Gestellreihe)
suite (→program family)	DATA PROC	(→Programmfamilie)
suited (adj.)	TECH	geeignet
= appropriate; capable		= befähigt; zweckentsprechend
≈ serviceable		
		≈ brauchbar
suited to demand (→keeping with requirement)	TECH	(→bedarfsgerecht)
sulfatation (→sulphatation)	CHEM	(→Sulfatierung)
sulfate (→sulphate)	CHEM	(→Sulfat)
sulfating (→sulphatation)	CHEM	(→Sulfatierung)
sulfur (→sulphur)	CHEM	(→Schwefel)
sulfuric acid (→sulphuric acid)	CHEM	(→Schwefelsäure)
sulphatation	CHEM	Sulfatierung
= sulphating; sulfatation; sulfating		= Sulfatieren
sulphate	CHEM	Sulfat
= sulfate		
sulphating (→sulphatation)	CHEM	(→Sulfatierung)
sulphur	CHEM	Schwefel
= sulfur; S		= Sulfur; S
sulphuric acid	CHEM	Schwefelsäure
= sulfuric acid; oil of vitriol		
sum (n.)	MATH	Summe
[the result od addition]		[das Ergebnis einer Addition]
= addition 2; total (n.)		
		= Summa
sum (v.t.) (→add)	MATH	(→addieren)
sum-current transformer	POWER SYS	Summenstromwandler
sum level	TELEC	Summenpegel
summarize	LING	zusammenfassen
≈ abridge		≈ kürzen
summary	LING	Zusammenfassung
[of a text]		[eines Textes]
= resumé; resume		≈ Abriß
≈ compendium		
summary (→listing)	DATA PROC	(→Protokoll)
summary alarm (BRI) (→common alarm)	EQUIP.ENG	(→Sammelalarm)
summation	SCIE	Summation
summation (→addition 1)	MATH	(→Addition)
summation element	CONTROL	Summierglied
		= Vergleicher
summing amplifier (→adder)	CIRC.ENG	(→Addierer)
summing integrator	CIRC.ENG	Summenintegrator
[integrates the sum of various input signals]		[integriert die Summe mehrerer Eingangssignale]
≈ inverse integrator; summing amplifier; integrator		= Summationsintegrator
		≈ Umkehrintegrator; Summierverstärker; Integrator
summing junction	CIRC.ENG	Summenpunkt
[operating amplifier]		[Operationsverstärker]
summing point	CONTROL	Vergleichsstelle
[derives the actuating signal from reference input signal and feedback signal]		[bildet aus Führungsgröße und gemessene Regelgröße die Regeldifferenz]
sum pattern	ANT	Summendiagramm
sum sign	MATH	Summenzeichen
[Σ]		[Σ]
= sum symbol		
sum symbol (→sum sign)	MATH	(→Summenzeichen)
sum total (→grand total)	MATH	(→Gesamtsumme)
sum-up	MATH	aufsummieren
= cumulate; accumulate		
sum vector	MATH	Summenvektor
sun	ASTROPHYS	Sonne
sundial style (→gnomon)	PHYS	(→Gnomon)
sunk (adj.)	MECH	eingelassen
sunk key	MECH	Einlegepaßfeder
sunspot	ASTROPHYS	Sonnenfleck
superannuated	ECON	verjährt
= prescriptive		
super-audio channel	TELEC	Überlagerungskanal
super capacitor	NETW.TH	Superkondensator
super-cardioid directional characteristic	EL.ACOUS	Supernieren-Richtcharakteristik
super coil	NETW.TH	Superspule
supercomputer	DATA PROC	Größtrechner
= ultralarge computer; superlarge computer		= Größtcomputer
		= Großrechner
↑ mainframe computer		
superconductive	PHYS	supraleitend
superconductivity	PHYS	Supraleitung
		= Supraleitfähigkeit
superconductor	PHYS	Supraleiter
super directivity	ANT	Super-Richtfaktor
super-fast blowing	COMPON	(→superflink)
(→super-quick reacting)		
super-fast blowing fuse (→super-fast-blowing melting fuse)	COMPON	(→superflinke Feinsicherung)
super-fast-blowing melting fuse	COMPON	superflinke Feinsicherung
= super-fast blowing fuse; super-quick reacting fuse		= superflinke Schmelzsicherung
		↑ Stromfeinsicherung
superficial	TECH	oberflächlich
superficial corrosion	CHEM	Oberflächenkorrosion
= crevice		
superficial leakage current (→tracking current)	EL.TECH	(→Kriechstrom)
super-finishing	MECH	Feinstbearbeitung
superframe (→multiframe)	CODING	(→Mehrfachrahmen)
super gain antenna	ANT	Super-gain-Antenne
supergroup	TRANS	Sekundärgruppe
[FDM group of 60 channles]		[TF-Technik, Gruppe von 60 Kanälen]
		= SG; Übergruppe
supergroup modulator	TRANS	Sekundärgruppenumsetzer
		= SGU
superhet (→heterodyne receiver)	RADIO	(→Überlagerungsempfänger)
superheterodyne receiver (→heterodyne receiver)	RADIO	(→Überlagerungsempfänger)
super high frequency	EL.TECH	Höchstfrequenz
[range of frequencies at which e.m. energy can be reasonably conveyed in the inner space of waveguides; often considered as the upper subrange of high frequencies; from about 100 MHz to 100 GHz]		[Frequenzbereich in dem elektromagnetische Energie im Innenraum von Wellenleitern vorteilhaft übertragen werden kann; oft als oberer Teilbereich der Hochfrequenz betrachtet; von 100 MHz bis 100 GHz]
= SHF; microwave		= Mikrowelle
≈ high frequency		≈ Hochfrequenz
super-high frequency (→centimetric waves)	RADIO	(→Zentimeterwellen)
super high integration [more than 10,000 gates]	MICROEL	Superintegration [mehr als 10.000 Gatter]
superhighway (→freeway)	CIV.ENG	(→Autobahn)
supericonoscope	ELECTRON	Superikonoskop
		= Zwischenbild-Ikonoskop

superimpose	TECH	**überlagern**
superimposed ac	EL.TECH	**überlagerte Wechselspannung**
= ripple voltage		≈ Spannungswelligkeit
≈ ripple content		
superimposed dc	EL.TECH	**überlagerte Gleichspannung**
superimposed telegraphy	TELEGR	**Überlagerungstelegraphie**
		= ÜT; Überlagerungstelegrafie
superimposition	TECH	**Überlagerung**
superintendent (→foreman)	ECON	(→Vorarbeiter)
superior	ECON	**Vorgesetzter**
≠ subordinate		≠ Untergebener
superior (→superscript)	TYPOGR	(→Hochstellung)
superior figure	TYPOGR	(→Hochstellung)
(→superscript)		
superlarge computer	DATA PROC	(→Größtrechner)
(→supercomputer)		
superlative (n.)	LING	**Superlativ**
≈ elative form		= Höchststufe
		≈ Elativ
superlattice	PHYS	**Supergitter**
supermalloy	METAL	**Supermalloy**
supermaster group	TRANS	**Quartärgruppe**
		= QG
supermaster group modulator	TRANS	**Quartärgruppenumsetzer**
		= QGU
supermicro (→super microcomputer)	DATA PROC	(→Supermikrorechner)
super microcomputer	DATA PROC	**Supermikrorechner**
= supermicro		= Supermicro
supermini (→super minicomputer)	DATA PROC	(→Super-Minirechner)
super minicomputer	DATA PROC	**Super-Minirechner**
= superminicomputer; supermini		= Super-Minicomputer; Supermini
superminicomputer (→super minicomputer)	DATA PROC	(→Super-Minirechner)
superorthicon (→image orthicon)	ELECTRON	(→Superorthikon)
superphantom (→double-phantom circuit)	LINE TH	(→Achterleitung)
superposability (→congruence)	MATH	(→Kongruenz)
superposable (→congruent)	MATH	(→kongruent)
superposition	PHYS	**Überlagerung**
		= Superposition
superposition (→heterodyning)	HF	(→Überlagerung)
superposition principle	SYS.TH	**Superpositionsprinzip**
		= Überlagerungsprinzip
super-quick reacting	COMPON	**superflink**
= super-fast blowing		[Sicherung]
super-quick reacting fuse (→super-fast-blowing melting fuse)	COMPON	(→superflinke Feinsicherung)
superrefraction	RADIO PROP	**Superrefraktion**
supersaturation	TECH	**Übersättigung**
superscript (n.)	TYPOGR	**Hochstellung**
= superscripted character; superscripting; superior figure; superior		= hochgestelltes Zeichen; hochgesetztes Zeichen; hochstehendes Zeichen; Hochstellen
≠ subscript		≠ Tiefstellung
superscripted character (→superscript)	TYPOGR	(→Hochstellung)
superscripting (→superscript)	TYPOGR	(→Hochstellung)
supersede (→replace)	TECH	(→ersetzen)
supersensitive (→extremely sensitive)	TECH	(→höchstempfindlich)
supersonic sound (→ultrasound)	ACOUS	(→Ultraschall)
superstructure	TECH	**Überbau**
= extension		= Aufbau 3; Aufsatz
super-turnstile antenna	ANT	**Super-Turnstile-Antenne**
supervapotron	ELECTRON	**Supervapotron**
[vapour cooling method for power tubes]		[Dampfkühlungsverfahren für Leistungsröhren]
≈ vapotron; hypervapotron		≈ Vapotron; Hypervapotron
supervise	TECH	**überwachen**
= monitor (v.t.)		
supervising module	EQUIP.ENG	**Überwachungseinschub**
= supervision module; watchdog module		
supervising unit	EQUIP.ENG	**Überwachungseinheit**
= supervision unit; watchdog unit		
supervision	TECH	**Überwachung**
= monitoring		
supervision device (→supervisory device)	EQUIP.ENG	(→Überwachungsvorrichtung)
supervision module (→supervising module)	EQUIP.ENG	(→Überwachungseinschub)
supervision relay (→sensor relay)	COMPON	(→Überwachungsrelais)
supervision sensor	COMPON	**Überwachungssensor**
supervision threshold	ELECTRON	**Überwachungsschwelle**
= monitoring threshold; sensing threshold		= Überwachungsgrenze
supervision unit (→supervising unit)	EQUIP.ENG	(→Überwachungseinheit)
supervisor	DATA PROC	**Ablaufteil**
↑ executive routine		↑ Organisationsprogramm
supervisor call	DATA PROC	**Organisationsaufruf**
= control system call		
supervisor position (→operator's console)	SWITCH	(→Vermittlungsplatz)
supervisor routine (→control program)	DATA PROC	(→Organisationsprogramm)
supervisory channel	TELEC	**Überwachungskanal**
= monitoring channel		
supervisory circuit (→control circuit 2)	CIRC.ENG	(→Überwachungsschaltung)
supervisory device	EQUIP.ENG	**Überwachungsvorrichtung**
= monitoring device; supervision device		
supervisory instruction	DATA PROC	**Überwachungsbefehl**
supervized (→monitored)	TECH	(→überwacht)
supervoltage (→high voltage)	POWER ENG	(→Hochspannung)
superworkstation	DATA PROC	**Super-Arbeitsplatzrechner**
supple (→pliable)	TECH	(→geschmeidig)
supplement (v.t.)	TECH	**ergänzen**
≈ amend; update		≈ berichten; aktualisieren
supplement (n.)	LING	**Nachtrag**
= postscript; addendum		
supplement (n.)	TECH	**Zusatz**
= addition		= Ergänzung; Beifügung
supplementary-added carrrier	TELEGR	**Zusatzträger**
= locally added carrier		
supplementary air (→auxiliary air)	TECH	(→Hilfsluft)
supplementary character set	TELEGR	**Supplementärzeichensatz**
supplementary condition	TECH	**Zusatzbedingung**
= additional condition		= Ergänzungsbedingung
supplementary memory	DATA PROC	**Zusatzspeicher**
= supplementary storage		
supplementary panel	TERM&PER	**Vorsatzscheibe**
supplementary processing (→further processing)	MANUF	(→Weiterverarbeitung)
supplementary service	TELEC	**Zusatzdienst**
supplementary service (→special service)	TELEC	(→Sonderdienst)
supplementary SI unit	PHYS	**ergänzende SI-Einheit**
supplementary storage (→supplementary memory)	DATA PROC	(→Zusatzspeicher)
suppleness (→pliability)	TECH	(→Geschmeidigkeit)
supplied unmounted	ECON	**lose mitgeliefert**
supplier 2	ECON	**Auftragnehmer**
= contractor 3		≈ Lieferant
≈ supplier 1		≠ Auftraggeber
		↑ Vertragspartner
supplier 1	ECON	**Lieferant**
= vendor 2; purveyor		= Bezugsquelle; Zulieferant; Zulieferer; Lieferer
≈ dealer; subcontractor; manufacturer		≈ Händler; Unterauftragnehmer; Hersteller
supplies (→operating supplies)	MANUF	(→Betriebsmittel)
supply (v.t.)	ELECTRON	**versorgen**
= provide; support (v.t.)		= speisen
supply (n.)	ECON	**Versorgung**
= provision 1; support (n.)		≈ Nachschub
≈ ordnance 1		≠ Entsorgung
≠ disposal		
supply (n.)	ELECTRON	**Versorgung**
= feed (n.)		= Speisung
↓ power supply		↓ Stromversorgung

supply (n.) (→stock)	ECON	(→Vorrat)	
supply air	TECH	Versorgungsluft	
supply bin	TERM&PER	Vorratsbehälter	
= bin			
supply bridge (→feeding bridge)	TELEPH	(→Speisebrücke)	
supply company	DATA PROC	Zubehörfirma	
supply current	ELECTRON	Speisestrom	
= feed current		= Versorgungsstrom	
supply noise (→hum)	TELEC	(→Brumm)	
supply potential	MICROEL	Versorgungspotential	
supply reel	TERM&PER	Abwickelspule	
= feed reel			
supply transformer (→mains transformer)	POWER ENG	(→Netztransformator)	
supply voltage	TELEPH	Speisespannung	
= battery voltage			
supply voltage	ELECTRON	Speisespannung	
= feed voltage		= Versorgungsspannung	
support (v.t.)	INF.TECH	unterstützen	
[e.g. a feature]		[z.B. ein Leistungsmerkmal]	
support (n.)	TECH	Auflage	
≈ underlay (n.); receptacle		[Stütze]	
		≈ Unterlage; Aufnahme	
support (n.)	ECON	Betreuung	
= service			
support (n.)	TECH	Stütze	
≈ bracket		= Bügel; Tragstütze	
≈ mount		≈ Halterung	
support (n.)	CIV.ENG	Träger	
= beam; joist		↓ Profilträger	
↓ girder 2			
support (→after-sales service)	ECON	(→Kundendienst)	
support (→pedestal)	TECH	(→Sockel)	
support (→support point)	OUTS.PLANT	(→Stützpunkt)	
support (v.t.) (→supply)	ELECTRON	(→versorgen)	
support (n.) (→supply)	ECON	(→Versorgung)	
support capacitor	CIRC.ENG	Stützkondensator	
support chip	MICROEL	Unterstützungsbaustein	
support environment	DATA PROC	Unterstützungsumfeld	
supporting body	COMPON	Tragkörper	
supporting material (→substrate)	MICROEL	(→Substrat)	
supporting pipe	TECH	Stützrohr	
supporting rail	TECH	Tragschiene	
supporting structure	TECH	Traggestell	
supporting tower	CIV.ENG	Tragturm	
support lug	TECH	Stützlappen	
support mast	ANT	Tragemast	
support plate	MECH	Stützplatte	
		= Stützblech	
support point	OUTS.PLANT	Stützpunkt	
[open-wire line]		[Freileitung]	
= support			
supposition	SCIE	Annahme	
≈ hypothesis; assumption		≈ Hypothese	
≈ prerequisite		≈ Voraussetzung	
suppress	ELECTRON	unterdrücken	
[an interference, a signal]		[eine Störung, ein Signal]	
= reject		≈ sperren	
≈ inhibit			
suppressed carrier	MODUL	unterdrückter Träger	
= quiescent carrier			
suppressed range	ELECTRON	Unterdrückungsbereich	
suppression	ELECTRON	Unterdrückung	
≈ rejection		≈ Sperrung	
≈ inhibition			
suppression capacitor	COMPON	Funkentstörkondensator	
= feed-through capacitor		= Durchführungskondensator	
suppression loss (→stop-band attenuation)	NETW.TH	(→Sperrdämpfung)	
suppressor circuit	CIRC.ENG	TSE-Beschaltung	
[protective connection for semiconductor gate]		[Schutzbeschaltung für Halbleiterventil]	
suppressor grid	ELECTRON	Bremsgitter	
= decelerating grid			
supraregional	ECON	überregional	
surcharge (n.)	ECON	Zuschlag 1	
= extra charge		= Aufschlag	
surd (→irrational number)	MATH	(→Irrationalzahl)	
surety (→guaranty 2)	ECON	(→Bürgschaft)	
surface	PHYS	Oberfläche	
↑ area		↑ Fläche	
surface-acoustic wave	PHYS	akustische Oberflächenwelle	
= SAW			
surface-acoustic-wave filter	COMPON	Oberflächenwellenfilter	
= surface-wave filter; SAW filter		= OFW-Filter; SAW-Filter	
surface analyzer	MECH	Glattheitsprüfer	
surface area	MATH	Oberflächeninhalt	
surface barrier (→surface layer)	MICROEL	(→Randschicht)	
surface barrier transistor	MICROEL	Randschichttransistor	
surface charge	PHYS	Oberflächenladung	
surface-charge density	PHYS	Flächenladungsdichte	
surface-charge transistor	MICROEL	Oberflächenladungstransistor	
= SCT		= SCT	
surface conductivity	PHYS	Oberflächenleitfähigkeit	
surface contamination	PHYS	Oberflächenverunreinigung	
surface coupler	PHYS	Oberflächenkoppler	
surface current	PHYS	Oberflächenstrom	
surface damage	TECH	Oberflächenbeschädigung	
surface density	PHYS	Flächendichte	
= areal density			
surface density of electric charge	PHYS	elektrische Flächendichte	
surface duct	RADIO PROP	erdnaher Dukt	
= ground-based duct			
surface earth	SYS.INST	Oberflächenerder	
surface electric resistance (→surface impedance)	PHYS	(→Oberflächenwiderstand)	
surface element	MATH	Flächenelement	
surface error	TECH	Konturgenauigkeit	
surface film (→surface layer)	PHYS	(→Oberflächenschicht)	
surface finish (→surface termination)	TECH	(→Oberflächenbehandlung)	
surface gage	MECH	Parallelreißer	
surface grade	ENG.DRAW	Oberflächenausführung	
= surface quality		= Oberflächengüte	
surface grinding	MECH	Flächenschleifen	
surface grounding	SYS.INST	Flächenerdung	
surface hardness	TECH	Oberflächenhärte	
surface hologram	OPTOEL	Flächenhologramm	
surface impedance	PHYS	Oberflächenwiderstand	
= resistance per square; surface electric resistance		= Flächenwiderstand	
surface integral	MATH	Oberflächenintegral	
[integration over a piece of any surface]		[Integration über ein Stück einer beliebigen Fläche]	
≈ double integral		= Flächenintegral	
		≈ Doppelintegral	
surface ionization	PHYS	Oberflächenionisation	
surface layer	PHYS	Oberflächenschicht	
= surface film		= Oberflächenfilm	
surface layer	MICROEL	Randschicht	
[of a semiconductor in contact with a non-metallic]		[eines Halbleiters wenn an Nichtmetall angrenzend]	
= boundary layer; surface barrier		= Randzone; Grenzschicht; Unstetigkeitsschicht	
≈ depletion layer		≈ Sperrschicht	
surface leakage current	PHYS	Oberflächenleckstrom	
surface life time	PHYS	Oberflächen-Lebensdauer	
surface mail	POST	normale Post	
surface mark	ENG.DRAW	Oberflächenzeichen	
surface migration	PHYS	Oberflächenwanderung	
surface model	DATA PROC	Flächenmodell	
[computer graphics]		[Computergraphik]	
surface mount device test probe (→SMD test probe)	INSTR	(→SMD-Tastkopf)	
surface-mountable	ELECTRON	oberflächenmontierbar	
surface mounted	ELECTRON	oberflächenmontiert	
surface mounted device	COMPON	SMD-Bauelement	
= SMD		= oberflächenmontiertes Bauelement; SMD; SMT-Bauelement	
surface mounted technology (→SMD technology)	ELECTRON	(→SMD-Technik)	
surface mounting	ELECTRON	Oberflächenmontage	
≈ SMD-technology		≈ SMD-Technik	
surface noise	EL.ACOUS	Abspielgeräusch	
= scratch noise; scratch			

surface of revolution		MATH	**Rotationsfläche**	surplus production (→excess production)		ECON	(→Überschußproduktion)
surface oscillation		PHYS	**Oberflächenschwingung**	surreptitious listening device (→eavesdropping device)		SIGN.ENG	(→Abhörvorrichtung)
surface passivation		METAL	**Oberflächenpassivierung**				
= passivation; surface stabilization			= Oberflächenstabilisierung	surrounding temperature (→environmental temperature)		TECH	(→Umgebungstemperatur)
surface point		MATH	**Flächenpunkt**				
↓ peak point; saddle point			↓ Gipfelpunkt; Sattelpunkt	surveillance radar (→panoramic radar)		RADIO LOC	(→Panoramaradar)
surface potential		MICROEL	**Oberflächenpotential**				
surface preparation (→surface termination)		TECH	(→Oberflächenbehandlung)	surveillance receiver (→monitoring receiver)		RADIO	(→Überwachungsempfänger)
surface quality (→surface grade)		ENG.DRAW	(→Oberflächenausführung)	survey (v.t.)		TECH	**vermessen** [Land]
surface reaction		CHEM	**Wandreaktion**	survey (→site survey)		SYS.INST	(→Ortsbegehung)
surface recombination		PHYS	**Oberflächenrekombination**	survey (n.) (→overview)		LING	(→Übersicht)
surface recombination velocity = SRV		PHYS	**Oberflächenrekombinations-Geschwindigkeit**	survey (→directory)		DOC	(→Verzeichnis)
				survey (→census)		MATH	(→Erhebung)
surface refinement (→surface termination)		TECH	(→Oberflächenbehandlung)	survey article		DOC	**Übersichtsartikel**
				survey plan = ÜP		SWITCH	**Übersichtsplan**
surface resistivity		PHYS	spezifischer Oberflächenwiderstand				
				survey plan = layout plan		TECH	**Übersichtsplan**
surface roughness		TECH	**Oberflächenrauhigkeit**				
surface stabilization (→surface passivation)		METAL	(→Oberflächenpassivierung)	survivability		COLLOQ	**Überlebensfähigkeit**
				survivals		QUAL	**Bestand**
surface technology		METAL	**Oberflächentechnik**	susceptance		NETW.TH	**Blindleitwert**
surface temperature		PHYS	**Oberflächentemperatur**	[imaginary part of complex admittance; SI unit: Siemens]			[Imaginärteil des komplexen Scheinleitwert; SI-Einheit: Siemens]
surface tension		PHYS	**Oberflächenspannung**				
surface termination		TECH	**Oberflächenbehandlung**				= Suszeptanz
= surface finish; surface treatment; surface refinement; surface preparation ≈ finish			= Oberflächenbearbeitung; Oberflächenveredelung ≈ Fertigbearbeitung	susceptibility		PHYS	**Suszeptibilität**
				susceptibility = liability		QUAL	**Störanfälligkeit** = Fehleranfälligkeit
surface treatment (→surface termination)		TECH	(→Oberflächenbehandlung)	susceptibility (→interference sensibility)		TELEC	(→Störempfindlichkeit)
surface voltage gradient		EL.TECH	**Schrittspannung**				
surface wave		PHYS	**Oberflächenwelle**	susceptibility test set		INSTR	**Suszeptibilitätsmeßplatz**
↑ transversal wave			↑ Transversalwelle	susceptible ≈ susceptive; accident-sensitive ≈ sensitive		TECH	**störanfällig** ≈ empfindlich
surface wave (→direct wave)		RADIO PROP	(→Bodenwelle)				
surface wave acoustic amplifier		COMPON	**Oberflächenwellenverstärker**	susceptive (→susceptible)		TECH	(→störanfällig)
surface-wave antenna		ANT	**Oberflächenwellenantenne**	suscriber = costumer; party ≈ user ↓ private subscriber; bussines subscriber		TELEC	**Teilnehmer** ≈ Benutzer ↓ Privatteilnehmer; Geschäftsteilnehmer
surface-wave filter (→surface-acoustic-wave filter)		COMPON	(→Oberflächenwellenfilter)				
				suscriber address		SWITCH	**Teilnehmeradresse**
surface wave transmission line = G line; single-wire		LINE TH	**Oberflächenwellenleitung** = Drahtwellenleitung; Goubeau-Leitung; G-Leitung	suscriber premises = costumer premises; subscriber's premises; costumer's premises ≈ subscriber building		TELEC	**Teilnehmerbereich** = Teilnehmerstandort; Teilnehmergelände; Teilnehmergrundstück; Teilnehmerräumlichkeiten; Teilnehmerdependance; Kundenbereich; Kundenstandort; Kundengelände; Kundengrundstück; Kundenräumlichkeiten; Kundenlokalität (SWZ); Kundendependance ≈ Teilnehmergebäude
surface wrinkling		TECH	**Runzelbildung**				
surge-absorbing capacitor = commutating capacitance		POWER SYS	**Löschkondensator** = Kommutierungskondensator				
surge arrester (→overvoltage protector)		COMPON	(→Spannungssicherung)				
surge blockers		POWER ENG	**Wellensperre**				
surge current ≈ current surge		EL.TECH	**Stoßstrom** ≈ Stromstoß				
surge diverter (→overvoltage protector)		COMPON	(→Spannungssicherung)				
surge generator		EL.TECH	**Blitzgenerator**				
surge impedance (→characteristic impedance)		NETW.TH	(→Wellenwiderstand)	suspend (v.t.) = halt (v.t.)		ELECTRON	**anhalten**
surge oscilloscope = Impulsoszilloskop		INSTR	**Stoßspannungsoszilloskop**	suspend		MECH	**aufhängen**
				suspended drawing cabinet		ENG.DRAW	**Zeichnungshängeschrank**
surge protection (→overvoltage protection)		EL.TECH	(→Überspannungsschutz)	suspended pocket file		OFFICE	**Hängeregister**
				suspender (→suspension hook)		TECH	(→Traghaken)
surge protector (→line filter)		CIRC.ENG	(→Netzfilter)	suspending wire		OUTS.PLANT	**Entlastungsseil**
surge resistance = lightning resistance		QUAL	**Blitzfestigkeit**	suspension [of an activity, payment]		ECON	**vorübergehende Einstellung** [einer Tätigkeit, Zahlung]
surge voltage ≈ impulse voltage ≈ voltage surge ↑ surge		EL.TECH	**Stoßspannung** ≈ Spannungsstoß ↑ Stoß	suspension clamp		OUTS.PLANT	**Hängeklemme**
				suspension guy = messenger wire; messenger cable		CIV.ENG	**Tragseil**
surge-voltage protector (→overvoltage protector)		COMPON	(→Spannungssicherung)	suspension hook = suspender		TECH	**Traghaken**
surge voltmeter		INSTR	**Stoßspannungsvoltmeter**	suspension insulator		OUTS.PLANT	**Hängeisolator**
surge withstand test		QUAL	**Überspannungsprüfung**	suspension points [...]		TYPOGR	**Auslassungspunkte** [...]
surjection [set theory]		MATH	**Surjektion** [Mengenlehre]	suspension span ≈ span length		OUTS.PLANT	**Abspannabschnitt** [Freileitung] ≈ Spannweite
surpass (→transgres)		TECH	(→überschreiten)				
surpass (n.) (→transgression)		TECH	(→Überschreitung)	suspension strand (→messenger wire)		OUTS.PLANT	(→Tragseil)
surplus		ECON	**Überschuß**				
surplus price (→ addition price)		ECON	(→Aufpreis)	sustained (→undamped)		PHYS	(→ungedämpft)
				sustained (→streamy)		TELEC	(→kontinuierlich)

sustained oscillation (→undamped	PHYS	(→ungedämpfte Schwingung)	
oscillation)			
sustained short-circuit current	EL.TECH	**Dauerkurzschlußstrom**	
sustained wave (→undamped	PHYS	(→ungedämpfte Welle)	
wave)			
Sv (→Sievert)	PHYS	(→Sievert)	
Svoboda map	ENG.LOG	**Svoboda-Diagramm**	
		= Svoboda-Plan	
SVR (→voltage standing wave	LINE TH	(→Welligkeitsfaktor)	
ratio)			
SW (→service word)	TRANS	(→Meldewort)	
swaging (→cold forging)	METAL	(→Fließpressen)	
SW antenna (→short-wave	ANT	(→Kurzwellenantenne)	
antenna)			
swap file	DATA PROC	**Swap-Datei**	
swap gate	DATA PROC	**Austauschtor**	
[magnetic bubble memory]		[Magnetblasenspeicher]	
swapping	DATA PROC	**Swapping**	
[temporary transfer from main to ex-		[vorübergehendes Ausla-	
ternal memory]		gern vom Hauptspeicher]	
≈ overlay		≈ Überlagerung	
swapping (→paging)	DATA PROC	(→Seitenwechsel)	
swarm (n.)	DATA PROC	**Fehlerhäufung**	
[several bugs]			
swash (→swash letter)	TYPOGR	(→Schwungbuchstabe)	
swash letter	TYPOGR	**Schwungbuchstabe**	
[typeface with flourishes]		= Zierbuchstabe	
= swash			
swastika	ANT	**Swastika**	
sway (n.)	TECH	**Seitenschwankung**	
sway (→fluctuate)	TECH	(→schwanken)	
sway (→fluctuation)	TECH	(→Schwankung)	
SW connection (→short-wave	TELEC	(→Kurzwellenverbindung)	
link)			
sweat (v.t.)	METAL	**ofenlöten**	
		↑ löten	
sweating	TECH	**Ofenlötung**	
↑ soldering		↑ Lötung	
sweep (v.t.)	ELECTRON	**wobbeln**	
= wobble; vobulate; warble			
sweep (n.)	ELECTRON	**Wobbeldurchgang**	
= pass (n.)			
sweep (n.)	DATA PROC	**Lasso**	
[clustering of several graphical ele-		[Einbindung mehrerer Gra-	
ments for common processing]		phikelemente zwecks ein-	
		heitlicher Bearbeitung]	
sweep (→deflection)	ELECTRON	(→Ablenkung)	
sweep (→excursion)	MODUL	(→Hub)	
sweep circuit	CIRC.ENG	(→Kippschaltung)	
(→multivibrator 1)			
sweep diode (→trigger	MICROEL	(→Triggerdiode)	
diode)			
sweep flatness	ELECTRON	**Wobbelfrequenzgang**	
sweep frequency (→sweep	ELECTRON	(→Wobbelfrequenz)	
rate)			
sweep frequency generator	INSTR	(→Wobbelsender)	
(→sweep generator)			
sweep-frequency method	ELECTRON	**Wobbelverfahren**	
= sweeping procedure; vobulation			
method			
sweep generator	TV	**Ablenkgenerator**	
sweep generator	INSTR	**Wobbelsender**	
= sweep frequency generator; sweep os-		= Wobbelgenerator; Wobb-	
cillator; wobbler; swept source; vobu-		ler; Wobbelgerät; Wobbel-	
lator		einrichtung	
sweep generator	ELECTRON	**Zeitablenkgenerator**	
= time-base generator		[erzeugt Sägespannungen]	
		= Wobbelgenerator; Zeitbasi-	
		steil	
sweeping	ELECTRON	**Wobbelung**	
= wobbling; wobble; vobulating; warble		= Wobbeln; Wobbelbetrieb	
sweeping coil (→deflection	ELECTRON	(→Ablenkspule)	
coil)			
sweeping procedure	ELECTRON	(→Wobbelverfahren)	
(→sweep-frequency method)			
sweepings (→garbage)	COLLOQ	(→Müll)	
sweeping system (→deflection	ELECTRON	(→Ablenksystem)	
system)			
sweep-level measuring set	INSTR	(→Wobbelmeßplatz)	
(→sweep measuring set)			
sweep measuring set	INSTR	**Wobbelmeßplatz**	
= sweep-level measuring set; wobbling			
measuring set; wobble measuring set			
sweep oscillator (→sweep	INSTR	(→Wobbelsender)	
generator)			
sweep range	ELECTRON	**Wobbelbereich**	
= vobulating range; wobbling range;		= Wobbelhub	
scan width			
sweep rate	ELECTRON	**Wobbelfrequenz**	
= sweep frequency; wobbling rate; wob-		= Wobbelgeschwindigkeit	
bling frequency; vobulating rate; vo-			
bulating frequency			
sweep span	ELECTRON	**Wobbelbandbreite**	
= frequency span			
sweep-synchronized	ELECTRON	**im Wobbeltakt**	
sweep time	ELECTRON	**Wobbelzeit**	
= sweeptime			
sweeptime (→sweep time)	ELECTRON	(→Wobbelzeit)	
sweep triode	MICROEL	**Kipptriode**	
≈ thyristor		= Vierschichttriode	
		≈ Thyristor	
swept	INSTR	**gewobbelt**	
swept analysis (→swept	INSTR	(→Wobbelmessung)	
measurement)			
swept measurement	INSTR	**Wobbelmessung**	
= swept analysis		≠ punktweises Messen	
≠ point-to-point measurement			
swept scalar analysis	INSTR	**skalare Wobbelmessung**	
= swept scalar measurement			
swept scalar measurement (→swept	INSTR	(→skalare Wobbelmessung)	
scalar analysis)			
swept-sine mode	INSTR	**gewobbelter Sinusbetrieb**	
swept source (→sweep	INSTR	(→Wobbelsender)	
generator)			
swept-tuned frequency mode	INSTR	**gewobbelte Frequenzabstim-**	
		mung	
swim (n.)	TERM&PER	**Bidschirmschwankung**	
[undesired movement of picture on a			
display]			
swing (v.i.)	MECH	**schaukeln**	
= roll (v.t.); rock		≈ schwingen; pendeln;	
≈ oscillate; fluctuate		schwanken	
swing	PHYS	**pendeln**	
↑ oscillate		[um einen Befestigungs-	
		punkt schwingen]	
		≈ schaukeln	
		↑ schwingen	
swing (→excursion)	MODUL	(→Hub)	
swing (→vibration)	MECH	(→mechanische Schwingung)	
swing dash	TYPOGR	**Wiederholungszeichen**	
= tilde		= Tilde	
swinging arm (→pickup arm)	EL.ACOUS	(→Tonabnehmerarm)	
swinging choke	ELECTRON	**Siebdrossel**	
swing mechanism	TECH	**Drehvorrichtung**	
= turning mechanism; revolving mech-		= Drehmechanismus	
anism; swivel (n.)		≈ Schwenkmechanismus	
≈ pivoting mechanism			
swing-out transient	ELECTRON	**Ausschwingvorgang**	
swirl-free (→irrotational)	MATH	(→wirbelfrei)	
Swiss quad antenna	ANT	**Swiss-quad-Antenne**	
switch (v.t.)	EL.TECH	**schalten**	
≈ connect			
switch (v.t.)	TELEC	**vermitteln**	
switch (n.)	COMPON	**Schalter**	
≠ non-locking key		≠ Taster	
↓ contactor [POWER SYS]		↓ Schütz [POWER SYS]	
switch (→program	DATA PROC	(→Programmschalter)	
switch)			
switch (→exchange)	SWITCH	(→Vermittlungsstelle)	
switch (→selector)	SWITCH	(→Wähler)	
switchable	TECH	**schaltbar**	
switch amplifier	CIRC.ENG	**Schaltverstärker**	
= switching amplifier		[Treiberstufe einer Lei-	
		stungsschaltstufe]	
switchboard (AM) (→manual	SWITCH	(→Handvermittlung 1)	
exchange)			
switchboard (→switchboard	SWITCH	(→Handvermittlungsplatz)	
position)			
switchboard (→operator's	SWITCH	(→Vermittlungsplatz)	
console)			

switchboard cable

switchboard cable	COMM.CABLE	(→Schaltkabel)	
(→connecting cable)			
switchboard instrument (→panel instrument)	INSTR	**Schalttafelinstrument**	
switchboard jack (→jack)	TELEPH	(→Klinke)	
switchboard plug (→jack connector)	TELEPH	(→Klinkenstecker)	
switchboard position	SWITCH	**Handvermittlungsplatz**	
= switchboard			
switchboard unit	EL.ACOUS	**Umschaltpult**	
switch deck	COMPON	**Schalterebene**	
= deck			
switch-desk (→control desk)	TECH	(→Kontrollpult)	
switch driver	INSTR	**Schaltertreiber**	
switched	POWER SYS	**getaktet**	
= switched mode ...		= geschaltet	
switched-capacitor filter	NETW.TH	**Schalterfilter**	
= SC filter		= Schalter-C-Filter; CSF; SC-Filter; Schalter-Kondensator-Filter	
switched connection	TELEC	**Wählleitung**	
= switched line; dial-up line		= Wählverbindung	
≠ fixed line		≠ Standleitung	
↓ switched telephone line; telex line		↓ Fernsprechwählleitung; Fernschreibwählleitung	
switched data network	TELEC	**vermitteltes Datennetz**	
↑ switched network		= Datenwählnetz	
		↑ Wählnetz	
switched line (→switched connection)	TELEC	(→Wählleitung)	
switched mode	POWER SYS	**Schaltbetrieb**	
switched mode ... (→switched)	POWER SYS	(→getaktet)	
switched mode dc converter	POWER SYS	**getakteter Gleichspannungswandler** [Fernmeldestromversorgung]	
		= geschalteter DC-Wandler	
switched mode mains power supply	POWER SYS	**Schaltnetzteil**	
		= getaktetes Netzgerät	
↑ switched mode power supply		↑ Schaltstromversorgung	
switched mode power supply	POWER SYS	**Schaltstromversorgung**	
= switched power supply; switching power supply		= getaktete Stromversorgung	
		↓ Schaltnetzteil	
switched network	TELEC	**Wählnetz**	
= dialing network; automatic network		= Vermittlungsnetz; vermitteltes Netz	
≠ fixed network		≠ Festnetz	
↓ switched telephone network; telex network; switched data network		↓ Fernsprechwählnetz; Fernschreibwählnetz; vermitteltes Datennetz	
switched operation	TELEC	**Wählbetrieb**	
= switched service; automatic operation; automatic service		= Wähldienst	
switched power supply	POWER SYS	(→Schaltstromversorgung)	
(→switched mode power supply)			
switched service (→switched operation)	TELEC	(→Wählbetrieb)	
switched tailring counter code (→Libraw-Craig code)	CODING	(→Libraw-Craig-Code)	
switched telegraph connection (→telex line)	TELEC	(→Telexleitung)	
switched telegraph line (→telex line)	TELEC	(→Telexleitung)	
switched telephone connection (→switched telephone line)	TELEC	(→Fernsprechwählleitung)	
switched telephone line	TELEC	**Fernsprechwählleitung**	
= switched telephone connection		= Fernsprechwählverbindung	
		↑ Wählleitung	
switched telephone network	TELEC	**Fernsprechwählnetz**	
≈ automatic telephone network		↑ Wählnetz	
↑ switched network			
switch finder (→line finder)	SWITCH	(→Anrufsucher)	
switchgear	POWER ENG	**Schaltgerät**	
[switching device with peripherals, for electric energy]		= Schaltanlage	
switch group	SWITCH	**Koppelgruppe**	
switch group control	SWITCH	**Koppelgruppensteuerung**	
switchhook (→hookswitch)	TELEPH	(→Gabelumschalter)	
switching	TELEC	**Vermittlung**
[device or process of temporary interconnection, controlled by the subscriber]		[Einrichtung oder Vorgang zur Herstellung zeitweiliger, vom Teilnehmer gesteuerter Verbindungen]
= communications switching; telecommunications switching		= Nachrichtenvermittlung
↑ communication		≈ Vermittlungstechnik
↓ circuit switching [SWITCH]; store-and-forward switching [SWITCH]		↑ Übermittlung
		↓ Durchschaltevermittlung [SWITCH]; Speichervermittlung [SWITCH]
switching-...	SWITCH	**vermittlungstechnisch**
= call-processing ...; call-...		
switching (→switchover)	ELECTRON	(→Umschaltung)
switching algebra (→engineering logic)	INF	(→Schaltalgebra)
switching amplifier (→switch amplifier)	CIRC.ENG	(→Schaltverstärker)
switching capacity	COMPON	**Schaltkapazität**
[parasitic capacitance]		[parasitäre Kapazität]
switching center (AM) (→exchange)	SWITCH	(→Vermittlungsstelle)
switching centre (BRI) (→exchange)	SWITCH	(→Vermittlungsstelle)
switching characteristics	ELECTRON	**Schaltverhalten**
= dynamic characteristics		= dynamisches Verhalten; dynamische Kennlinie
≠ static characteristics		≠ statisches Verhalten
switching circuit (→combinatorial circuit)	CIRC.ENG	(→Schaltnetz)
switching command	TELECONTR	**Schaltbefehl**
= teleswitching command		[Befehl zum Fernschalten]
switching component	COMPON	**Schaltelement**
≈ switching element [SWITCH]		≈ Koppelelement [SWITCH]
switching controller (→switching regulator)	POWER SYS	(→Schaltregler)
switching cycle	ELECTRON	**Schaltzyklus**
switching device (→switching equipment)	SWITCH	(→Vermittlungseinrichtung)
switching diode	MICROEL	**Schaltdiode**
[optimized for switching functions]		[für Schaltfunktionen dimensionierte Diode]
= computer diode		= Computerdiode
switching element	SWITCH	**Koppelelement**
≈ switching component [COMPON]		[Elementarschaltung von Koppelanordnungen]
↓ crosspoint		≈ Schaltelement 2 [COMPON]
		↓ Koppelanordnung
switching element [in a digital circuit]	CIRC.ENG	**Schaltungselement** [einer Digitalschaltung]
switching element (→logic gate)	CIRC.ENG	(→Verknüpfungsglied)
switching engineering	TELEC	**Vermittlungstechnik**
= switching technique		= Wähltechnik
↓ telephone switching engineering [SWITCH]; data switching engineering [SWITCH]		≈ Vermittlung
		↓ Fernsprechvermittlungstechnik [SWITCH]; Datenvermittlungstechnik [SWITCH]
switching equipment	SWITCH	**Vermittlungseinrichtung**
= switching device		
switching equipment	SWITCH	**Vermittlungsanlage**
= dialling exchange equipment		≈ Vermittlungsstelle; Vermittlungssystem
switching ferrite	METAL	**Schalt-Ferrit**
↑ square-loop ferrite		↑ Rechteckferrit
switching frequency	ELECTRON	**Schaltfrequenz**
= toggle rate		
switching function	ENG.LOG	**Schaltfunktion**
switching galvanometer	INSTR	**Schaltgalvanometer**
switching information	SWITCH	**vermittlungstechnische Information**
= control information		= Steuerinformation
switching layer	TRANS	**Schaltebene**
= granularity		= Granularität
switching loss	MICROEL	**Umschaltverlust**
		= Schaltverlust
switching matrix	CIRC.ENG	**Koppelfeld**

switching matrix	SWITCH	**Koppelvielfach**	**switching signal**	ELECTRON	**Schaltsignal**
[a simple switching network, arranged by switching rows in matrix array, any input connectable to any output]		[Vielfachschaltung von Koppelreihen in Form einer Matrix, jeder Eingang mit jedem Ausgang verbindbar]	**switching signal exchange**	SWITCH	**Kennzeichenaustausch** = Zeichenaustausch
			switching speed	EL.TECH	**Schaltgeschwindigkeit**
			switching stage	SWITCH	**Koppelstufe**
= connecting matrix; matrix		= Schaltmatrix	[structurally related switch stages, serving a definite function]		[strukturell zusammenhängende Koppelvielfache]
↑ switching network		↑ Koppelanordnung			
switching mechanism	COMPON	**Schaltwerk** [elektromechanisch]	= connecting stage		↑ Koppelanordnung
switching module	SWITCH	**Koppelbaugruppe**	↑ switching network		↓ Wahlstufe
switching network 1	SWITCH	**Koppelanordnung**	↓ selection stage		
[space-multiplex arrangement of crosspoints to connect, on demand and simultaneously, many inputs with many outputs or viceversa]		[Raumvielfach aus Koppelelementen zum wahlweisen und gleichzeitigen Verbinden mehrerer Eingänge mit mehreren Ausgängen oder umgekehrt]	**switching state**	ELECTRON	**Schaltzustand**
			switching surge	ELECTRON	**Schaltüberspannung**
			switching system	SWITCH	**Vermittlungssystem** = Wählsystem
			switching system with stored program control	SWITCH	**speicherprogrammiertes Vermittlungssystem**
= connecting network 1		= Koppeleinrichtung	= SPC switching system		= SPC-Vermittlung; speicherprogrammiertes Wählsystem
↓ switching row; switching matrix; switching stage; switching unit; switching network 2		↓ Koppelreihe; Koppelvielfach; Koppelstufe; Koppelfeld; Koppelnetz	↑ indirect control switching system		
					↑ indirekt gesteuertes Vermittlungssystem
switching network 2	SWITCH	**Koppelnetz**	**switching task**	SWITCH	**vermittlungstechnische Aufgabe**
[the totality of switching facilities of a central office]		[Gesamtheit der Koppelanordnungen einer Vermittlung]	= call-processing function		
			switching technique (→switching engineering)	TELEC	(→Vermittlungstechnik)
= connecting network 2; switching network array		= Koppelnetzwerk	**switching theory** (→engineering logic)	INF	(→Schaltalgebra)
switching network	CIRC.ENG	(→Schaltnetz)	**switching threshold**	ELECTRON	**Schaltschwelle**
(→combinatorial circuit)			= switchover threshold; changeover threshold		= Umschaltschwelle
switching network array	SWITCH	(→Koppelnetz)			
(→switching network 2)			**switching threshold** (→protection switching threshold)	TRANS	(→Umschaltschwelle)
switching network clock	SWITCH	**Koppelnetztakt**			
switching-network extension	SWITCH	**Koppelnetz-Erweiterung**	**switching time**	ELECTRON	**Schaltzeit**
switching node	DATA COMM	**Vermittlungsknoten**	**switching transistor**	MICROEL	**Schalttransistor**
switching office (→exchange)	SWITCH	(→Vermittlungsstelle)	[designed for switching functions]		[für Schaltvorgänge dimensionierter Transistor]
switching operations	SWITCH	**Vermittlungsbetrieb**			
switching panel	SWITCH	**Vermittlungsfeld**	**switching tube**	ELECTRON	**Schaltröhre**
switching path	SWITCH	**Koppelweg**	= switching valve		↓ T/R-Röhre
switching plug	COMPON	**Schaltstecker**	↓ T/R tube		
switching point	BROADC	**Schaltstelle**	**switching unit**	SWITCH	**Koppelfeld**
switching position (→operator's console)	SWITCH	(→Vermittlungsplatz)	[functional unit of a switching network array, formed by several switching stages]		[strukturell zusammenhängende Teile des Koppelnetzes einer Vermittlung]
switching power supply (→switched mode power supply)	POWER SYS	(→Schaltstromversorgung)			
			↓ subscriber switching unit; group switching unit		↑ Koppelanordnung ↓ Teilnehmerkoppelfeld; Richtungskoppelfeld
switching process	ELECTRON	**Schaltvorgang**			
switching processor	SWITCH	**Vermittlungsprozessor**	**switching unit**	RADIO REL	**Schalteinsatz**
[executes central controlling functions for all subscribers]		[führt zentrale Steuerfunktionen für alle Teilnehmer durch]	**switching valve** (→switching tube)	ELECTRON	(→Schaltröhre)
= call processor; central control unit		= Vermittlungsrechner; Zentralsteuereinheit; Zentralsteuerwerk	**switching variable**	ENG.LOG	**Schaltvariable**
			switching voltage	ELECTRON	**Schaltspannung**
			switch lever (→control lever)	TECH	(→Schalthebel)
switching program	SWITCH	**Vermittlungsprogramm**	**switch manufacturer** (→switch vendor)	TELEC	(→Vermittlungslieferant)
= call processing program					
switching rack	RADIO REL	**Schaltgestell**	**switch-off behaviour** (→turn-off behaviour)	MICROEL	(→Ausschaltverhalten)
switching reactor coil	CIRC.ENG	**Einschaltdrossel** = Einschaltspule			
			switch-off loss (→turn-off loss)	MICROEL	(→Ausschaltverlust)
switching regulator	POWER SYS	**Schaltregler** = Spannungsregler			
= switching controller			**switch-off time** (→turn-off time)	MICROEL	(→Ausschaltzeit)
switching relay	SWITCH	**Koppelrelais**			
= matrix switching relay			**switch-on** (→circuit closer)	CIRC.ENG	(→EIN-Schalter)
switching row	SWITCH	**Koppelreihe**	**switch-on peak**	EL.TECH	**Einschaltstromspitze**
[a row of crosspoints]		[Reihung von Koppelpunkten]	= transient current; inrush current		= Einschaltstromstoß; Einschaltstrom; Übergangsstrom
= connecting row					
↑ switching network		↑ Koppelanordnung	**switch-out** (v.t.) (→disconnect)	EL.TECH	(→ausschalten)
switching section	SWITCH	**Koppelabschnitt**			
switching section	RADIO REL	**Schaltabschnitt**	**switchover** (v.t.) = changeover (v.t.)	ELECTRON	**umschalten**
switching sequence	ELECTRON	**Schaltfolge**			
switching signal	SWITCH	**Kennzeichen**	**switchover** (n.)	ELECTRON	**Umschaltung** = Schaltung
[control signal for switching operations]		[Steuersignal für Schaltvorgänge in Vermittlungen]	= changeover; switching		
			switch-over command	TRANS	**Umschaltbefehl**
= signal; signal mark		= Schaltkennzeichen; Vermittlungskennzeichen; vermittlungstechnisches Zeichen; Zeichen	**switchover contact** (→changeover contact)	COMPON	(→Umschaltekontakt)
≈ metering signal					
↓ line signal; register signal; forward signal; backward signal; acknowledgment signal; pulse signal; continuous signal		≈ Zählzeichen	**switchover criterion**	ELECTRON	**Umschaltkriterium**
		↓ Leitungszeichen; Registerzeichen; Vorwärtszeichen; Rückwärtszeichen; Quittierungszeichen; Impulskennzeichen; Dauerkennzeichen	= changeover criterion		≈ Umschaltlogik
			≈ switchover logic		
			switchover logic	ELECTRON	**Umschaltlogik**
			= changeover		≈ Umschaltkriterium
			≈ switchover criterion		

switchover threshold

switchover threshold (→switching threshold)	ELECTRON	(→Schaltschwelle)
switchover threshold (→protection switching threshold)	TRANS	(→Umschaltschwelle)
switchover time = changeover time; transit time	ELECTRON	**Umschaltzeit** = Umschaltezeit
switchover time (→transit time)	COMPON	(→Umschlagzeit)
switch panel (→control panel)	TECH	(→Schalttafel)
switch position = throw	COMPON	**Schalterstellung** = Schalterstufe; Stellung; Stufe
switchroom	SWITCH	**Wählerraum**
switch-selectable	ELECTRON	**umschaltbar** = durch Schalter wählbar; durch Schalter einstellbar
switch tower (AM) = signal cabin (BRI); signal box; interlocking installation; interlocking post; positioner	RAILW.SIGN	**Stellwerk**
switch transaction (→barter business)	ECON	(→Tauschgeschäft)
switch-up (v.t.) (→connect 2)	EL.TECH	(→einschalten)
switch vendor = switch manufacturer	TELEC	**Vermittlungslieferant** = Vermittlungshersteller
swivel (v.t.) = pivot ≈ steer	MECH	**schwenken** ≈ lenken
swivel [adjustable junction element]	TECH	**Wirbel** [einstellbares Verbindungselement]
swivel (n.) (→swing mechanism)	TECH	(→Drehvorrichtung)
swivel arm = swivelling arm	TECH	**Schwenkarm**
swivel arm for terminal = terminal swivel arm	TERM&PER	**Terminal-Schwenkarm**
swivel base	ANT	**Antennenkopf**
swivel chair	TECH	**Drehstuhl**
swiveling (→hinged)	MECH	(→schwenkbar)
swiveling mechanism (→pivoting mechanism)	MECH	(→Schwenkmechanismus)
swiveling range (→steering range)	TECH	(→Schwenkbereich)
swivelling (→hinged)	MECH	(→schwenkbar)
swivelling arm (→swivel arm)	TECH	(→Schwenkarm)
swivel motor	TECH	**Schwenkmotor**
swivel stand ≈ tilt stand	EQUIP.ENG	**Drehfuß** ≈ Schwenkfuß
SW link (→short-wave link)	TELEC	(→Kurzwellenverbindung)
sworn translator	ECON	**vereidigter Übersetzer**
SWR bridge [standing-wave-ratio bridge] = SWR meter	INSTR	**SWR-Meßbrücke** = Stehwellenverhältnis-Meßbrücke; SWR-Meßgerät
S wrench	TECH	**S-Schlüssel**
SWR meter (→SWR bridge)	INSTR	(→SWR-Meßbrücke)
SWR power meter = standing-wave-ratio meter; VSWR resistance bridge; VSWR meter	INSTR	**Stehwellenmeßgerät** = Stehwellen-Leitungsmesser; SWR-Wattmeter; Stehwellenmeßbrücke; Anpassungsmeßgerät
SW transmitter (→short-wave transmitter)	RADIO	(→Kurzwellensender)
syllabe intelligibility (→syllabic intelligibility)	TELEPH	(→Silbenverständlichkeit)
syllabic articulation (→syllabic intelligibility)	TELEPH	(→Silbenverständlichkeit)
syllabic companding	TELEPH	**Silbenkompandierung**
syllabic intelligibility = syllabic articulation; logotom articulation; syllabe intelligibility ≈ discrete words intelligibility	TELEPH	**Silbenverständlichkeit** = Wortverständlichkeit
syllable-oriented computer ↑ word-oriented computer	DATA PROC	**Silbenmaschine** ↑ Wortmaschine
sylvine (→potassium chloride)	CHEM	(→Kaliumchlorid)
symbol	MATH	**Formelzeichen** = Symbol
symbol ≈ pictograph [DATA PROC]	DATA PROC	**Symbol** = Sinnbild ≈ Piktogramm [DATA PROC]
symbol ' (→minute)	MATH	(→Minute)
symbol (→character)	INF	(→Zeichen)
symbol (→sign)	LING	(→Zeichen)
symbol alignment (→character alignment)	DATA COMM	(→Zeichensynchronisierung)
symbol-bound (→character-oriented)	INF.TECH	(→zeichenorientiert)
symbol-by-symbol (→character-by-character)	INF.TECH	(→zeichenweise)
symbol check (→character check)	DATA COMM	(→Zeichenprüfung)
symbol checking (→character check)	DATA COMM	(→Zeichenprüfung)
symbol-controlled	DATA PROC	**symbolgesteuert**
symbol counter (→character counter)	TELEGR	(→Zeichenzähler)
symbol error (→character error)	INF.TECH	(→Zeichenfehler)
symbol error frequency (→character error rate)	DATA COMM	(→Zeichenfehlerhäufigkeit)
symbol error probability (→character error probability)	DATA COMM	(→Zeichenfehlerwahrscheinlichkeit)
symbol error rate (→character error rate)	DATA COMM	(→Zeichenfehlerhäufigkeit)
symbol frequency (→character rate)	DATA COMM	(→Zeichengeschwindigkeit)
symbol generator (→character generator)	DATA PROC	(→Zeichengeber)
symbolic	COLLOQ	**symbolisch**
symbolic address [by names] = mnemotechnic address	DATA PROC	**symbolische Adresse** [mit Namen] = mnemotechnische Adresse; mnemonische Adresse
symbolic addressing	DATA PROC	**symbolische Adressierung**
symbolic algebra (→Boolean algebra)	MATH	(→Boolesche Algebra)
symbolic assembler	DATA PROC	**symbolischer Assemblierer**
symbolic code (→symbolic programming language)	DATA PROC	(→symbolische Programmiersprache)
symbolic coding ≠ absolute coding	DATA PROC	**symbolische Codierung** ≠ absolute Codierung
symbolic data	DATA PROC	**symbolische Daten**
symbolic device	DATA PROC	**symbolisches Gerät**
symbolic editor	DATA PROC	**symbolisches Editierprogramm**
symbolic instruction	DATA PROC	**symbolischer Befehl** = symbolische Instruktion
symbolic language (→symbolic programming language)	DATA PROC	(→symbolische Programmiersprache)
symbolic logic = formale Logik	MATH	**symbolische Logik**
symbolic program [written in a symbolic program language]	DATA PROC	**symbolisches Programm** [in symbolischer Programmiersprache geschrieben]
symbolic program language (→symbolic programming language)	DATA PROC	(→symbolische Programmiersprache)
symbolic programming language [uses symbolic instructions and addresses for ease of programming] = symbolic program language; symbolic language; symbolic code ≈ symbol-oriented program language ≠ machine language ↓ machine-oriented programming language; high-level programming language	DATA PROC	**symbolische Programmiersprache** [verwendet symbolische Befehle und Adressen zur Erleichterung des Programmierens] = Symbolsprache 1 ≈ symbolorientierte Programmiersprache ≠ Maschinensprache ↓ maschinenorientierte Programmiersprache; problemorientierte Programmiersprache
symbolize	SCIE	**symbolisieren**
symbol key	TERM&PER	**Symboltaste**
symbol multiplexer (→character multiplexer)	INF.TECH	(→Zeichenmultiplexer)
symbology	SCIE	**Symbologie**

synchronous data transmission

symbol-oriented	INF.TECH	(→zeichenorientiert)	
(→character-oriented)			
symbol-oriented programming language	DATA PROC	**symbolorientierte Programmiersprache**	
[descriptive language, based on symbols and relations; the languages of fifth generation, as LISP, PROLOG, SMALLTAK]		[deskriptive, mit Symbolen und Verknüpfungen operierende Sprache; die Sprachen der 5. Generation wie LISP, PROLOG, SMALLTAK]	
≠ procedure-oriented programming language		= Symbolsprache 2; nichtprozedurale Programmiersprache	
↑ symbolic programming language		≠ prozedurorientierte Programmiersprache	
		↑ symbolische Programmiersprache	
symbol position (→character position)	DATA COMM	(→Zeichenstelle)	
symbol processing	DATA PROC	**Symbolverarbeitung**	
symbol range (→character set)	INF.TECH	(→Zeichenvorrat)	
symbol rate (→character rate)	DATA COMM	(→Zeichengeschwindigkeit)	
symbol reader (→character reader)	TERM&PER	(→Klarschriftleser)	
symbol recognition (→character recognition)	TERM&PER	(→Klarschriftlesen)	
symbol repertoire (→character set)	INF.TECH	(→Zeichenvorrat)	
symbol scanner (→character reader)	TERM&PER	(→Klarschriftleser)	
symbol set (→character set)	INF.TECH	(→Zeichenvorrat)	
symbol stencil (→character template)	ENG.DRAW	(→Zeichenschablone)	
symbol string (→character string)	DATA PROC	(→Zeichenfolge)	
symbol supply (→character set)	INF.TECH	(→Zeichenvorrat)	
symbol table	DATA PROC	**Symboltabelle**	
symmetric	MATH	**symmetrisch**	
= symmetrical		≈ ebenmäßig; mittig	
symmetric (→balanced)	EL.TECH	(→symmetrisch)	
symmetrical (→symmetric)	MATH	(→symmetrisch)	
symmetrical (→balanced)	EL.TECH	(→symmetrisch)	
symmetrical broadband antenna	ANT	**symmetrische Breitbandantenne**	
; **symmetrical feed** (→center feed)	ANT	(→Mittelpunktspeisung)	
symmetrical feeding	ANT	**Gegentaktspeisung**	
symmetrical grading	SWITCH	**symmetrische Mischung**	
symmetrical heterostatic circuit	EL.TECH	**Nadelschaltung**	
symmetrical line (→balanced line)	LINE TH	(→erdsymmetrische Leitung)	
symmetrical pair	COMM.CABLE	(→symmetrisches Aderpaar)	
(→balanced pair)			
symmetric distribution	MATH	**symmetrische Verteilung**	
symmetric quadripole	NETW.TH	(→widerstandssymmetrischer Vierpol)	
(→electrically symmetric two-port)			
symmetric two-port	NETW.TH	(→widerstandssymmetrischer Vierpol)	
(→electrically symmetric two-port)			
symmetry	MATH	**Symmetrie**	
≠ asymmetry		= Ebenmaß; Gleichmaß; Mittigkeit	
		≠ Asymmetrie	
symmetry (→balance)	NETW.TH	(→Symmetrie)	
symmetry axis	MATH	**Symmetrieachse**	
symmetry measure	EL.TECH	**Symmetriemaß**	
symmetry plane	MATH	**Symmetrieebene**	
symptom	DATA PROC	**Indiz**	
= diagnostic fact		≈ Symptom	
SYN (→synchronous idle)	DATA COMM	(→Synchronisierung)	
sync (→synchronization)	ELECTRON	(→Synchronismus)	
sync cell (→synchronization cell)	TELEC	(→Synchronisationszelle)	
synchro (→rotary resolver)	INSTR	(→Drehmelder)	
synchrogenerator (→rotary resolver)	INSTR	(→Drehmelder)	
synchronisation error (→loss of synchronism)	TECH	(→Gleichlaufstörung)	
synchronisation trouble (→loss of synchronism)	TECH	(→Gleichlaufstörung)	
synchronism	TECH	**Synchronismus**	
= synchronization; synchronizing; alignment		= Synchronisation; Gleichlauf; Synchronlauf; Taktgleichheit; Synchronisierung	
		≈ zeitliche Abstimmung	
synchronism	ELECTRON	(→Synchronismus)	
(→synchronization)			
synchronization	ELECTRON	**Synchronismus**	
= synchronizing; synchronism; sync		= Taktung; Synchronisierung; Synchronisation; Gleichlauf	
synchronization	TECH	(→Synchronismus)	
(→synchronism)			
synchronization bit	TELEC	**Synchronisierbit**	
= alignment bit			
synchronization cell [ATM]	TELEC	**Synchronisationszelle** [ATM]	
= sync cell		= Sync-Zelle	
synchronization character	CODING	**Synchronisierzeichen**	
[coded character for synchronizing purposes]		[codiertes Zeichen zur Synchronisierung]	
= synchronizing character		= Synchronisationszeichen; Synchronzeichen	
≈ timing signal [TELEC]; clock pulse [ELECTRON]		≈ Taktsignal [TELEC]; Schrittpuls [ELECTRON]	
synchronization clock system	TV	(→Impulsgenerator)	
(→synchronization signal generator)			
synchronization failure	TELEC	**Synchronisationsfehler**	
= alignment error			
synchronization signal (→timing signal)	CODING	(→Taktsignal)	
synchronization signal generator	TV	**Impulsgenerator**	
= synchronization clock system		= Impulsgeber; Fernseh-Taktgeber; Taktgeber	
synchronize	TECH	**synchronisieren**	
		= aufsynchronisieren	
synchronized	TECH	**synchronisiert**	
= in step with		= im Gleichlauf; im Gleichtakt	
synchronizer	TV	**Synchronisator**	
synchronizer (→synchronizing unit)	EQUIP.ENG	(→Synchronisiereinrichtung)	
synchronizing (→synchronism)	TECH	(→Synchronismus)	
synchronizing	ELECTRON	(→Synchronismus)	
(→synchronization)			
synchronizing character	CODING	(→Synchronisierzeichen)	
(→synchronization character)			
synchronizing circuit	BROADC	**Synchronisierleitung**	
synchronizing information	TELECONTR	**Gleichlaufinformation**	
synchronizing pulse	TV	**Synchronisierpuls**	
synchronizing pulse (→clock pulse)	ELECTRON	(→Schrittpuls)	
synchronizing pulse delay circuit	TV	**Impulsverzögerer**	
synchronizing pulse distribution amplifier	TV	**Impulsverteiler**	
synchronizing signal	TV	**Synchronsignal**	
= sync signal; S signal		= S-Signal	
synchronizing signal (→timing signal)	CODING	(→Taktsignal)	
synchronizing signal level	TV	**Synchronwert**	
= sync level			
synchronizing unit	EQUIP.ENG	**Synchronisiereinrichtung**	
= synchronizer		= Synchronisiereinheit	
≈ timing generator [CIRC.ENG]		≈ Taktgeber [CIRC.ENG]	
synchronoscope	INSTR	**Synchronoskop**	
synchronous	TELEC	**synchron**	
[with exactly the same clock]		[absolut taktgleich]	
= correctly phased; in step with		= gleichlaufend; taktgleich	
≈ concurrent; homochronous; isochronous; mesochronous; plesiochronous		≈ zeitgleich; homochron; isochron; mesochron; plesiochron	
≠ asynchronous		≠ asynchron	
synchronous computer	DATA PROC	**Synchronrechner**	
synchronous counter	CIRC.ENG	**Synchronzähler**	
≠ asynchronous counter		= synchroner Zähler	
		≠ Asynchronzähler	
synchronous data transmission	DATA COMM	**synchrone Datenübertragung**	

synchronous demodulator

synchronous demodulator (→synchronous detector)	MODUL	(→Synchrondemodulator)			= synonymes Wort ≈ Quasisynonym ≠ Antonym
synchronous detector = synchronous demodulator	MODUL	**Synchrondemodulator**	**syntagma** [word combination with fixed meaning, e.g. "at least"]	LING	**Syntagma** [stehenden Begriff bildende Wortverbindung, z.B. „zu hause"]
synchronous digital hierarchy = SDH; synchronous hierarchy ≈ SONET	TELEC	**Synchronhierarchie** = SDH; synchrone Digitalhierarchie; Synchrondigitalhierarchie ≈ SONET	**syntagmatic** (adj.)	LING	**syntagmatisch**
			syntax [rules for the formation of sentences] ≈ Phraseology ↑ grammar	LING	**Syntax** [Regeln der Satzbildung] = Satzlehre ≈ Phraseologie ↑ Grammatik
synchronous hierarchy (→synchronous digital hierarchy)	TELEC	(→Synchronhierarchie)			
synchronous idle [code] = SYN	DATA COMM	**Synchronisierung** [Code] = SYN	**syntax** [rules for character sequences of a programming language]	DATA PROC	**Syntax** [Regeln für die Bildung von Zeichenfolgen einer Programmiersprache]
synchronous machine	POWER SYS	**Synchronmaschine**	**syntax check**	DATA PROC	**Syntaxprüfung** = Syntaxüberprüfung
synchronous mixer	RADIO	**Synchronmischstufe**	**syntax-controlled**	DATA PROC	**syntaxtgesteuert**
synchronous mode = synchronous processing	DATA PROC	**Synchronverarbeitung** = synchrone Verarbeitung; synchrone Arbeitsweise	**syntax error** = grammatical error; grammatical mistake ↑ software error	DATA PROC	**Syntaxfehler** = Formfehler; Syntaxverstoß; syntaktischer Fehler; grammatischer Fehler ↑ Softwarefehler
synchronous mode = synchronous operation; synchronous transmission	DATA COMM	**Synchronbetrieb** = Synchronübertragung			
synchronous motor	POWER SYS	**Synchronmotor**			
synchronous network	TELEC	**synchrones Netz**	**synthesized signal generator**	INSTR	**Synthesizer**
synchronous operation (→synchronous mode)	DATA COMM	(→Synchronbetrieb)	**synthesizer** (→frequency synthesizer)	CIRC.ENG	(→Frequenzsynthesizer)
synchronous operation (→synchronous transfer mode)	TELEC	(→synchrones Übertragungsverfahren)	**synthesizer sweeper**	INSTR	**Synthesizer-Wobbler**
			synthetic ≈ artificial	TECH	**synthetisch** ≈ künstlich
synchronous orbit = geostationary orbit	ASTROPHYS	**Synchronlaufbahn** = geostationäre Umlaufbahn	**synthetic** (→synthetic fiber)	CHEM	(→Kunstfaser)
synchronous processing (→synchronous mode)	DATA PROC	(→Synchronverarbeitung)	**synthetic** (→synthetic material)	CHEM	(→Kunststoff)
synchronous satellite = geostationary satellite; geosynchronous satellite; 24-hours satellite	SAT.COMM	**Synchronsatellit** = geostationärer Satellit	**synthetic address** = generated address	DATA PROC	**synthetische Adresse**
			synthetic fiber = synthetic	CHEM	**Kunstfaser**
synchronous scanning	ELECTRON	**Synchronabtastung**	**synthetic material** = synthetic; plastic material; plastic (n.)	CHEM	**Kunststoff** = Plastik; Plast (DDR)
synchronous sequential logic	CIRC.ENG	**synchrones Schaltwerk** = synchrone Logik ↑ Logikschaltung	**synthetic resin**	CHEM	**Kunstharz**
synchronous service (→synchronous transfer mode)	TELEC	(→synchrones Übertragungsverfahren)	**synthetic resin varnish**	TECH	**Kunstharzlack**
			synthetizable	INF.TECH	**synthetisierbar**
synchronous system ≠ start/stop system	TELEGR	**Synchronsystem** ≠ Start-Stop-System	**synthetized language** = synthetized voice; synthetized speech	TELEC	**synthetische Sprache** = synthetische Stimme
synchronous TDM (→synchronous time division multiplexing)	TELEC	(→synchrone Zeitmultiplextechnik)	**synthetized speech** (→synthetized language)	TELEC	(→synthetische Sprache)
synchronous time division multiplexing = STD; synchronous TDM	TELEC	**synchrone Zeitmultiplextechnik** = STD	**synthetized voice** (→synthetized language)	TELEC	(→synthetische Sprache)
			synthetizer	INSTR	**Steuergenerator** = Synthetizer
synchronous transfer mode = STM; synchronous operation; synchronous service ≈ synchronous transmission; synchronous operation [DATA COMM]	TELEC	**synchrones Übertragungsverfahren** = synchroner Übertragungsmodus; STM; Synchronbetrieb; Synchronverfahren ≈ Synchronübertragung, Synchronübermittlung; Synchronbetrieb [DATA COMM]	**syntony** (→tuning)	CIRC.ENG	(→Abgleich)
			syringe (→squirt)	TECH	(→spritzen)
			system	TECH	**System**
			system	SYS.TH	**System**
			system (→electric network)	NETW.TH	(→Netzwerk)
			system administration (→housekeeping)	DATA PROC	(→Systemverwaltung)
synchronous transmission (→synchronous mode)	DATA COMM	(→Synchronbetrieb)	**system administrator**	DATA COMM	**Systemverwalter** = Systemadministrator
synchronous vibrator	CIRC.ENG	**Synchronzerhacker**	**system adviser** (→system designer)	TECH	(→Systemplaner)
synchron separator (→amplitude separator)	TV	(→Amplitudensieb)	**system analysis** = systems analysis	DATA PROC	**Systemanalyse**
sync level (→synchronizing signal level)	TV	(→Synchronwert)	**system analyst** = systems analyst	DATA PROC	**Systemanalytiker**
sync mixer	TV	**Signalmischgerät**	**system analyst** (→system designer)	TECH	(→Systemplaner)
sync regeneration	TV	**Impulserneuerung**			
sync separator	TV	**Separator**	**system architecture** (→system structure)	TECH	(→Systemstruktur)
sync signal (→synchronizing signal)	TV	(→Synchronsignal)	**systematic** ≈ methodic	SCIE	**systematisch** ≈ methodisch
syndicate (→consortium)	ECON	(→Konsortium)	**systematic code**	CODING	**systematischer Code**
syndrom evaluation [ATM]	TELEC	**Syndrom-Auswertung** [ATM]	**systematic error** = bias (n.)	MATH	**systematischer Fehler**
synistor	MICROEL	**Synistor**	**systematic failure** ≠ random failure	QUAL	**systematischer Ausfall** ≠ Zufallsausfall
synonym (n.) [word of the same meaning; sometimes also applied for words with nearly the same meaning] ≠ antonym	LING	**Synonym** (pl.-e;-a) [bedeutungsgleiches Wort; vielfach werden sinnverwandte Wörter mit einbezogen]	**system board** (→main board)	DATA PROC	(→Hauptplatine)
			system bus	EQUIP.ENG	**Systembus**

system characteristic		TECH	Systemkennwert	system management	DATA PROC (→Systemverwaltung)
system characteristics		TECH	Systemdaten		(→housekeeping)
= system parameters				system management function	DATA PROC (→Existenzfunktion)
system check (→system test)		MANUF	(→Systemprüfung)		(→existence function)
system check (→system test)		DATA PROC	(→Systemtest)	system manual	DATA PROC Systemhandbuch
system command		DATA PROC	Systembedienungsbefehl	= systems manual; system handbook	
system component		TECH	Systemkomponente	system model	MATH Systemmodell
system configuration		TECH	Systemkonfiguration	system monitoring programm	DATA PROC (→Systemüberwachungpro-
system configuration menu		DATA PROC	(→Konfigurationsmenu)	(→system diagnostics)	gramm)
(→configuration menu)				system node	POWER SYS Netzknoten
system configuration program	DATA PROC		(→Konfigurationsprogramm)	system of equations	MATH Gleichungssystem
(→configuration program)				system of units (→measurement	PHYS (→Maßsystem)
system connection		SWITCH	Systemanschaltung	system)	
system constant		TECH	Systemkonstante	system organization	TECH Systemorganisation
system control		TECH	Systemsteuerung	system overhead	DATA PROC Systemverwaltungszeit
system conversion program		DATA PROC	Systemumstellungsprogramm	system panel	EQUIP.ENG Betriebsanzeige
system crash		DATA PROC	Systemabsturz	system parameter	TECH Systemparameter
↑ abnormal end			= Maschinestopp	system parameters (→system	TECH (→Systemdaten)
			↑ Absturz	characteristics)	
system description		DOC	Systembeschreibung	system peripherals	TECH Systemperipheriegeräte
system design (→system		TECH	(→Systemkonzept)	system periphery	TECH Systemperipherie
philosophy)				system philosophy	TECH Systemkonzept
system designer		TECH	Systemplaner	= system design	= Systementwurf
= system engineer; system analyst; sys-			= Systemingenieur; System-	≈ system structure	≈ Systemstruktur
tem adviser			analytiker; Systemberater	system programmer	DATA PROC Systemprogrammierer
≈ system developer			≈ Systementwickler	system programming	DATA PROC Systemprogrammmie-
system developer		TECH	Systementwickler	≠ application programming	≠ Applikationsprogrammie-
≈ system designer			≈ Systemplaner; Systement-		rung
			werfer	system programming language	DATA PROC (→Implementierungssprache)
system development		TECH	Systementwicklung	(→implementation language)	
system diagnostics		DATA PROC	Systemüberwachungpro-	system programs (→system	DATA PROC (→Systemsoftware)
= system monitoring programm			gramm	software)	
system diplexer		RADIO REL	Systemweiche	system promt	DATA PROC System-Promt
≈ polarization-frequency diplexer			≈ Polarisation-Frequenz-Wei-	system reactance	POWER ENG Netzreaktanz
			che	system reaction	POWER ENG Netzrückwirkung
system disk 1		DATA PROC	Systemplatte	system recovery	TECH Systemerholung
[stores seldom used parts of the oper-			[speichert seltener benötig-	system requirements	TECH Systemanforderung
ating system]			te Teile des Betriebssy-	system reserve	RADIO REL Systemreserve
↑ operating system residence			stems]	system reserves	TECH Systemreserven
			↑ Betriebssystem-Residenz	systems analysis (→system	DATA PROC (→Systemanalyse)
system disk 2		DATA PROC	Systemdiskette	analysis)	
[contains the operating system]			[enthält das Betriebssy-	systems analyst (→system	DATA PROC (→Systemanalytiker)
			stem]	analyst)	
system embedment		DATA COMM	Systemeinbettung	system security	DATA PROC Systemsicherheit
system emulator		DATA PROC	Systememulator	systems engineering	SYS.TH Systemtechnik
system engineer (→system		TECH	(→Systemplaner)		= angewandte Systemtheorie
designer)				system simulation	TECH Systemsimulation
system environment		DATA PROC	Systemumgebung	systems manual (→system	DATA PROC (→Systemhandbuch)
system error		TECH	Systemfehler	manual)	
system expansion		TECH	Systemerweiterung	system software	DATA PROC Systemsoftware
= system upgrade			= Systemausbau	[set of programs making possible to	[Satz von Programmen die
system failure		TECH	Systemausfall	run application programs on a hard-	Anwendersoftware auf ei-
system family		TECH	Systemfamilie	ware]	ner Anlage ablauffähig ma-
system file		DATA PROC	Systemdatei	= basic software; system programs	chen]
system flowchart		DATA PROC	Systemflußdiagramm	≠ application software	= Grundsoftware; Basissoft-
≈ program flowchart			= Systemablaufplan	↑ operating system	ware; Systemprogramme
			= Programmablaufplan		≠ Anwendersoftware 1
system gain		RADIO REL	Systemgewinn		↑ Betriebssystem
system generation		DATA PROC	Systemgeneration	system speed	DATA PROC Systemgeschwindigkeit
system handbook (→system		DATA PROC	(→Systemhandbuch)	[working speed of a computer, de-	[von Taktfrequenz und Pro-
manual)				pending on clock frequency and pro-	zessor abhängige Arbeits-
system-immanent		TECH	(→systemeigen)	cessor]	geschwindigkeit eines
(→system-inherent)					Computers]
system imperfection		TECH	Systemimperfektion	system start (→start)	DATA PROC (→Anlauf)
system implementation		DATA PROC	Systemimplementierung	system structure	TECH Systemstruktur
system-inherent		TECH	systemeigen	= system architecture	= Systemaufbau; Systemar-
= intra-system; system-immanent			= systemintern; systemimma-	≈ system philosophy	chitektur
			nent		≈ Systemkonzept
system installation		DATA PROC	Systeminstallation	system study	TECH Systemstudie
			= Systeminstallierung	system supplier	ECON Systemlieferant
system integration		DATA PROC	Systemintegration	system test	MANUF Systemprüfung
= integration			= Integration 1	= system check	= Systemtest
system loader		DATA PROC	Systemlader	system test	DATA PROC Systemtest
system loading		TRANS	Systembelastung	[an overall performance test of a data	[Prüfung des Gesamtver-
system log		DATA PROC	Systemprotokoll	processing system]	haltens einer Datenverar-
			= Systemlog; Systemnachweis	= system check; overall-performance	beitungsanlage]
system loss		RADIO REL	Systemdämpfung	test	≈ Integrationstest
= total transmission loss				≈ integration test	
system maintenance		DATA PROC	Systemwartung	system test	TECH Systemversuch

system theory   TECH   **Systemtheorie**
[theory of structure ans performance of systems]
[Theorie des Aufbaus und der Eigenschaften von Systemen]
≈ system engineering
≈ Systemtechnik

system track   DATA PROC   **Systemspur**

system unit   DATA PROC   **Systemeinheit**
[central unit of a PC, generally incorporating a hard disk unit, a floppy disk drive and power supply, as a desk top or tower equipment]
[Zentraleinheit eines PC, i.a. außerdem Festplatte, Diskettenlaufwerk und Stromversorgung enthaltendend, als Tisch- oder Standgerät ausgeführt]

system upgrade (→system expansion)   TECH   (→Systemerweiterung)

system value   RADIO REL   **Systemwert**
system variant   TECH   **Systemvariante**
sytematics   SCIE   **Systematik**
s zone (→soft-doped zone)   MICROEL   (→s-Zone)

T

T (→Tesla)   PHYS   (→Tesla)
T (→star connection)   NETW.TH   (→Sternschaltung)
t (→metric ton-force)   PHYS   (→Tonne-Kraft)
t (→metric ton)   PHYS   (→Tonne)
T1 (→basic rate)   TRANS   (→Basisrate)
TA (→terminal adaptor)   TELEC   (→Endgeräteanpassung)
Ta (→tantalum)   CHEM   (→Tantal)
tab (→rider)   OFFICE   (→Reiter)
tab (n.) (→tabulator)   TERM&PER   (→Tabulator)
tab (v.t.) (→tabulate)   DOC   (→tabellieren)
tab argument (→table argument)   DATA PROC   (→Tabellenargument)
tabbing (→tabulation)   DOC   (→Tabellierung)
tab character (→tabulator character)   DATA PROC   (→Tabulatorzeichen)
tab eraser (→tabulator eraser)   TERM&PER   (→Tabulator-Löscher)
tab feature (→tabulation feature)   DATA PROC   (→Tabelliervorrichtung)
tab key (→tabulator key)   TERM&PER   (→Tabulatortaste)
table   DOC   **Tabelle**
= schedule
= Tafel
table argument   DATA PROC   **Tabellenargument**
= tab argument
table base   TECH   **Tischfuß**
table instrument   INSTR   **Tischinstrument**
= desktop instrument; desk instrument; bench instrument
= Tischgerät
table look-up   DATA PROC   **Tabellensuche**
table model (→desktop model)   EQUIP.ENG   (→Tischgerät)
table mounting   TECH   **Tischaufstellung**
= desktop mounting
table of contents (→contents)   DOC   (→Inhaltsverzeichnis)
table set   TERM&PER   **Tischapparat**
= desktop set
↓ Fernsprech-Tischapparat
↓ desktop telephone
table socket   EL.INST   **Tischsteckdose**
table telephone (→desktop telephone)   TELEPH   (→Fernsprech-Tischapparat)
table tipoid (microphone)   EL.ACOUS   **Dreibein-Tischfuß** [Mikrofon]
tab mark (→tabulation stop)   TERM&PER   (→Tabulatorstopp)
tab mark (→tabulation stop)   TERM&PER   (→Tabulatormarke)
tab memory (→tabulator memory)   DATA PROC   (→Tabulatorspeicher)
tab program (→spreadsheet program)   DATA PROC   (→Tabellenkalkulationsprogramm)
tab set (→tabulator set)   TERM&PER   (→Tabulator-Setzer)
tab set-clear (→tabulator set-clear)   TERM&PER   (→Tabulator-Setz-Lösch-Vorrichtung)
tab stop (→tabulation stop)   TERM&PER   (→Tabulatorstopp)
tab storage (→tabulator memory)   DATA PROC   (→Tabulatorspeicher)
tabular   DOC   **tabellarisch**
= in tabular form
tabulate   DOC   **tabellieren**
= tab (v.t.)
= tabellarisieren

tabulation   DOC   **Tabellierung**
= tabbing
= Tabellarisierung; Tabulierung
tabulation feature   DATA PROC   **Tabelliervorrichtung**
= tab feature
tabulation machine   TERM&PER   **Tabelliermaschine**
[special printer of punched card technique]
[Spezialdrucker der Lochkartentechnik]
= punched card tabulator; tabulator
tabulation program   DATA PROC   (→Tabellenkalkulationsprogramm)
(→spreadsheet program)
tabulation stop   TERM&PER   **Tabulatorstopp**
= tabulator stop; tab stop; tabulator mark; tab mark
= Tabstopp; Tabzeichen; Tabulatormarke
tabulator   TERM&PER   **Tabulator**
= tab (n.)
tabulator (→tabulation machine)   TERM&PER   (→Tabelliermaschine)
tabulator character   DATA PROC   **Tabulatorzeichen**
= tab character
tabulator eraser   TERM&PER   **Tabulator-Löscher**
= tab eraser
= Tab-Löscher
tabulator key   TERM&PER   **Tabulatortaste**
[cursor runs to next tabulation stop]
[Schreibmarke springt bis zum nächsten Tabulatorstopp]
= tab key
= Tabtaste
tabulator mark (→tabulation stop)   TERM&PER   (→Tabulatorstopp)
tabulator memory   DATA PROC   **Tabulatorspeicher**
= tab memory; tabulator storage; tab storage
tabulator set   TERM&PER   **Tabulator-Setzer**
= tab set
tabulator set-clear   TERM&PER   **Tabulator-Setz-Lösch-Vorrichtung**
= tab set-clear
tabulator stop (→tabulation stop)   TERM&PER   (→Tabulatorstopp)
tabulator storage   DATA PROC   (→Tabulatorspeicher)
(→tabulator memory)
TACAN (→tactical air navigation)   RADIO NAV   (→TACAN)
tacho pulse generator   TV   **Tachosignalgeber**
tack   TECH   **Zwecke**
[short nail with a broad head]
[kurzer Nagel mit breitem Kopf]
↓ thumb tack
↓ Reißzwecke
tackle   TECH   **Flaschenzug**
= treble block; pulley 2
tack weld   METAL   **Heftschweißnaht**
tactical air navigation   RADIO NAV   **TACAN**
= TACAN
tactical antenna   ANT   **Feldantenne**
tactile sensor (→contact sensor)   COMPON   (→Berührungssensor)
T adapter (→T junction)   MICROW   (→T-Verzweigung)
T adapter (→T junction)   COMPON   (→T-Verzweigung)
T adaptor (→T junction)   MICROW   (→T-Verzweigung)
tag (→label identifier)   DATA PROC   (→Etikettenkennzeichen)
tag (→lug)   COMPON   (→Fahne)
tag (→separator)   DATA PROC   (→Trennzeichen)
tag block (→soldering-tag terminal strip)   COMPON   (→Lötleiste)
tag check (→label check)   DATA PROC   (→Etikettprüfung)
tag-end terminal block   COMPON   (→Lötleiste)
(→soldering-tag terminal strip)
tagging   INSTR   **Kennung**
= Kennzeichnung
tag group (→label group)   DATA PROC   (→Etikettfolge)
tag memory   DATA PROC   **Etikettenspeicher**
= tag storage
tag storage (→tag memory)   DATA PROC   (→Etikettenspeicher)
TAHA (→tapered aperture horn antenna)   ANT   (→TAHA)
tail (n.)   DATA PROC   **Listenende-Markierung**
[end-of-list mark]
tail (→pulse tail)   ELECTRON   (→Impulsschwanz)

tailored (→tailor-made)		TECH	(→maßgeschneidert)	tanh (→ hyperbolic tangent)	MATH	(→Tangens hyperbolicus)
tailor-made		TECH	maßgeschneidert	tank circuit	CIRC.ENG	offener Schwingungskreis
= tailored			≈ kundenspezifisch	= open oscillatory circuit		= Tankkreis
≈ custom-designed				tank coil	CIRC.ENG	Tankspule
take a course		COLLOQ	verlaufen	tantalum	CHEM	Tantal
[an issue]			[ein Vorgang]	= Ta		= Ta
take apart (→disassemble)		TECH	(→zerlegen)	tantalum capacitor	COMPON	Tantalkondensator
take effect (→come into force)		ECON	(→inkrafttreten)	tantalum electrolyte capacitor	COMPON	Tantal-Elektrolytkondensator
				T antenna	ANT	T-Antenne
take off (v.i.)		AERON	abheben	tap (v.t.)	EL.TECH	abgreifen
≈ start (v.i.)			≈ starten			= anzapfen
take off (n.)		AERON	Abheben	tap (v.t.)	MECH	gewindebohren
≈ Start			≈ Start	[to make a threaded hole]		
take-off		TELEPH	abnehmen	tap (v.t.)	TELEC	anzapfen
[the handset]			[den Handapparat]	[a communication line for eavesdropping purposes]		[einer Fernmeldeleitung zum Zwecke des Abhörens]
take-off angle (→elevation angle)		ANT	(→Elevationswinkel)	tap (n.)	EL.TECH	Abgriff
take-off climb surface		AERON	Abflugfläche			= Anzapfung
take on lease		ECON	pachten	tap (n.)	BROADC	Abzweiger
[use of real estate during an established time and for a fixed pay]			[Nutzung einer Immobilie für festgelegte Dauer und Entgelt]	[CATV]		[CATV]
				tap (→eavesdrop)	TELEPH	(→abhören)
= lease 2				tape	TECH	Band
≈ give on rent			≈ mieten	[something made in long, narrow format]		[etwas in langer, schmaler Form Hergestelltes]
≠ give on lease			≠ verpachten	= ribbon		≈ Streifen
take out of service		TECH	außerbetriebsetzen	≈ strip		
= remove from service; deinstall			= außer Betrieb setzen; außer Betrieb nehmen	tape-armored cable	COMM.CABLE	bandbewehrtes Kabel
				tape block	BROADC	Bandblock
takeover fee		TELEC	Übernahmegebühr	tape card (→marginal punched card)	TERM&PER	(→Randlochkarte)
take-up reel		TERM&PER	Aufwickelrolle			
= take-up spool			= aufnehmende Lochstreifenspule	tapecartridge (→magnetic tape cassette)	TERM&PER	(→Magnetbandkassette)
take-up spool (→take-up reel)		TERM&PER	(→Aufwickelrolle)	tape cassette (→magnetic tape cassette)	TERM&PER	(→Magnetbandkassette)
taking out of service (→diconnecting)		SWITCH	(→Aufheben)	tape change	TERM&PER	Bandwechsel
				= tape swapping		= Magnetbandwechsel
talk (→lecture)		SCIE	(→Vortrag)	tape cleaner	TERM&PER	Bandreiniger
talk-back (→duplex intercommunication system)		TELEPH	(→Gegensprechanlage)	tape-controlled (→tape-operated)	TERM&PER	(→bandgesteuert)
talk-back loudspeaker		TERM&PER	Mikrophonlautsprecher	tape core (→laminated core)	EL.TECH	(→Schnittbandkern)
talker		TELEPH	Sprecher	tape deck (→magnetic tape device)	TERM&PER	(→Magnetbandgerät)
≠ listener			≠ Hörer			
talker (→data source)		DATA COMM	(→Datenquelle)	tape device (→magnetic tape device)	TERM&PER	(→Magnetbandgerät)
talking key		TELEPH	Sprechschalter			
talk path (→voice path)		TELEC	(→Fernsprechweg)	tape drive (→magnetic tape drive)	TERM&PER	(→Magnetbandantrieb)
talk-through facility (→simplex intercommunication system)		TELEPH	(→Wechselsprechanlage)	tape drive servo (→magnetic tape drive)	TERM&PER	(→Magnetbandantrieb)
tally chart		MATH	Strichliste			
tamper (v.t.)		TECH	manipulieren	tape drum	TERM&PER	Bandführungstrommel
= manipulate			= fälschen; unsachgemäß behandeln	= head drum		[Magnetband]
				tape error (→magnetic tape error)	TERM&PER	(→Magnetbandfehler)
tamper device 1		SIGN.ENG	Gehäuseöffnungsmelder			
tamper switch		SIGN.ENG	Gehäusekontakt	tape exhaustion contact	TELEGR	Papierendekontakt
tan (→tangent 2)		MATH	(→Tangens)	tape feed 1	TERM&PER	Bandvorschub
tandem (→series connection)		EL.TECH	(→Reihenschaltung)	↑ feed (n.)		[Magnetband]
						↑ Vorschub
tandem acquisition terminal		TERM&PER	Tandemerfassungsplatz	tape feed 2	TERM&PER	Streifenvorschub
tandem center (→tandem exchange)		SWITCH	(→Durchgangsamt)	= strip feed		↑ Vorschub
				↑ feed		
tandem circuit (→series connection)		EL.TECH	(→Reihenschaltung)	tape file (→magnetic tape file)	DATA PROC	(→Magnetbanddatei)
tandem computer		DATA PROC	Tandemcomputer	tape gate	TERM&PER	Klappe
tandem connection (→series connection)		EL.TECH	(→Reihenschaltung)	[puncher]		[Locher]
				tape guide (→tape leader 1)	TERM&PER	(→Bandführung)
tandem exchange		SWITCH	Durchgangsamt	tape handler (→magnetic tape device)	TERM&PER	(→Magnetbandgerät)
= tandem center; tandem office						
tandem office (→tandem exchange)		SWITCH	(→Durchgangsamt)	tape head (→sound head)	EL.ACOUS	(→Tonkopf)
				tape head cleaner	EL.ACOUS	Tonkopfreiniger
tandem potentiometer		COMPON	Doppelpotentiometer	= head cleaner		
			= Tandempotentiometer	tape jig	TERM&PER	Lochstreifenlehre
tandem selection		SWITCH	Durchgangswahl	tape label	TERM&PER	Bandetikett
tandem transistor		COMPON	Tandemtransistor	= tape volume		= Bandkennsatz
			= Doppeltransistor	≈ tape section marker		≈ Bandabschnittsmarke
tangent 2		MATH	Tangens	↑ label		↑ Etikett
= tan; tg			= tan; tg	tape leader 1	TERM&PER	Bandführung
tangent 1		MATH	Tangente	= leader 1; tape guide		[Magnetband]
[a line touching a curve]				tape leader 2	TERM&PER	Vorspannband
tangential fan		TECH	Querstromlüfter	[piece of tape remaining in the recorder]		[im Magnetbandgerät verbleibendes Bandstück]
tangential plane		MATH	Tangentialebene			
tangential sensitivity		INSTR	tangentiale Empfindlichkeit	= leader 2		
tangential signal		MICROEL	Tangiersignal			

tape length

tape length [magnetic tape]	TERM&PER	**Bandlänge** [Magnetband]
tape librarian	DATA PROC	**Magnetbänderverwalter**
tape library = magnetic type library	DATA PROC	**Magnetbandarchiv** = Bandarchiv; Magnetbandbibliothek; Bandbibliothek
tape loading routine	DATA PROC	**Bandlader**
tape mark [reflecting strip on magnetic tape] = control mark; marker ≈ tape section mark [DATA PROC] ↓ beginning-of-tape mark; end-of-tape mark	TERM&PER	**Bandmarke** [reflektierender Metallstreifen auf Magnetband] = Reflektormarke; Marke ≈ Bandabschnittsmarke [DATA PROC] ↓ Bandanfangsmarke; Bandendmarke
tape memory (→magnetic tape memory)	TERM&PER	(→Magnetbandspeicher)
tape noise	EL.ACOUS	**Bandrauschen**
tape-operated = tape-controlled	TERM&PER	**bandgesteuert**
tape operating system = TOS	DATA PROC	**Magnetband-Betriebssystem**
tape perforator (→tape punch)	TERM&PER	(→Lochstreifenstanzer)
tape printer = strip printer	TERM&PER	**Streifendrucker**
tape printer [teleprinter with tape attachment]	TELEGR	**Streifenschreiber** [Fernschreiber mit Lochstreifenzusatz]
tape punch = paper tape punch; tape perforator	TERM&PER	**Lochstreifenstanzer** = Lochstreifenlocher; Streifenlocher; Perforator
taper (n.) = narrowing ≈ cone	MATH	**Verjüngung** ≈ Kegel
taper [reader and/or puncher]	TERM&PER	**Lochstreifengerät** [Leser und/oder Stanzer] = Taper
taper (→conical horn)	ANT	(→Konushorn)
taper coupler	OPTOEL	**Taper-Koppler**
tape reader (→punched tape reader)	TERM&PER	(→Lochstreifenleser)
tape recorder = magnetic tape recorder	TERM&PER	**Bandaufzeichnungsgerät** = Magnetband-Aufzeichnungsgerät
tape recorder ↑ sound recorder ↓ audio tape recorder; cassette recorder	CONS.EL	**Tonbandgerät** ↑ Tonaufnahmegerät ↓ Tonbandmaschine; Kassettenrecorder
tape recording (→magnetic tape recording)	ELECTRON	(→Magnetbandaufzeichnung)
tapered = beveled; folded ≈ chamfered	MECH	**abgeschrägt** = abgekantet ≈ abgefast
tapered [with gradual diminution of cross section or width] ≈ conical	MATH	**verjüngt** ≈ kegelförmig
tapered (→beveled)	MECH	(→abgeschrägt)
tapered aperture horn antenna = TAHA	ANT	**TAHA**
tapered balun	ANT	**Tapered-balun**
tapered roller bearing	MECH	**Kegelrollenlager**
tapered waveguide	MICROW	**sich erweiternder Hohlleiter**
tape reel = tape spool ↑ reel	TERM&PER	**Lochstreifenspule** ↑ Spule
taper key ≈ feathered key	MECH	**Wellenkeil** ≈ Paßfeder
taper pin	MECH	**Kegelstift**
taper pipe thread	MECH	**kegeliges Rohrgewinde**
taper thread	MECH	**kegeliges Gewinde**
tape section mark [an indicating data block] = section mark ↑ tape mark	TERM&PER	**Bandabschnittsmarke** [markierender Datenblock] = Abschnittsmarke; Bandschreibmarke ↑ Bandmarke
tape selector (→band selector)	CONS.EL	(→Bandsortenschalter)
tape skew [deviation from ideal line of the recording track on a tape] = skew	TERM&PER	**Bandschräglauf** [Abweichung der Aufzeichnungsspur auf einem Band von der Ideallinie] = Schräglauf
tape speed	TERM&PER	**Bandgeschwindigkeit**
tape splice	TERM&PER	**Bandklebstelle**
tape spool (→tape reel)	TERM&PER	(→Lochstreifenspule)
tape station (→magnetic tape device)	TERM&PER	(→Magnetbandgerät)
tape storage (→magnetic tape memory)	TERM&PER	(→Magnetbandspeicher)
tape streamer (→streamer)	TERM&PER	(→Streamer-Magnetbandgerät)
tape swapping (→tape change)	TERM&PER	(→Bandwechsel)
tape transmission	TELEGR	**Lochstreifensendung**
tape transmitter	TELEGR	**Lochstreifensender**
tape transport (→magnetic tape drive)	TERM&PER	(→Magnetbandantrieb)
tape unit (→magnetic tape device)	TERM&PER	(→Magnetbandgerät)
tape volume (→tape label)	TERM&PER	(→Bandetikett)
tape width	TERM&PER	**Bandbreite** [Breite des Magnetbandes]
tape-wound core (→laminated core)	EL.TECH	(→Schnittbandkern)
tape-wounded core	COMPON	**bandumwickelter Kern**
tape-wounded core (→laminated core)	EL.TECH	(→Schnittbandkern)
tap-line plug	BROADC	**Stichleitungs-Steckdose**
tapped capacitor	COMPON	**Mehrfachkondensator**
tapped hole = threaded hole	MECH	**Gewindeloch**
tapped resistor	COMPON	**Anzapfwiderstand**
tapped transformer	COMPON	**Abzweigtransformator**
tapping	MECH	**Gewindebohren**
tapping point = Abgreifpunkt	EL.TECH	**Anzapfpunkt**
tapping screw	MECH	**Schneidschraube**
tapping switch (→stepping switch)	COMPON	(→Stufenschalter)
tap-proof	TELEC	**abhörsicher**
tar	TECH	**Teer**
tardy (→defaulting)	ECON	(→säumig)
tare [weight of package]	ECON	**Tara** = Verpackungsgewicht
target (v.t.)	COLLOQ	**abzielen auf**
target (v.t.) [fig.]	SCIE	**ausrichten auf** [fig.]
target code (→object language)	DATA PROC	(→Zielsprache)
target disk = destination disk ≠ source disk	DATA PROC	**Zieldiskette** ≠ Quelldiskette
target drive	DATA PROC	**Ziellaufwerk**
target group	ECON	**Zielgruppe**
target information ≈ dialing code	SWITCH	**Zielinformation**
target language [traduction] ≠ source language	LING	**Zielsprache** [Übersetzung] ≠ Ursprungssprache
target language (→object language)	DATA PROC	(→Zielsprache)
target position	RADIO LOC	**Zielposition**
target price = guiding price	ECON	**Richtpreis**
target program (→object program)	DATA PROC	(→Objektprogramm)
target station	TELEC	**Zielstation**
tariff (a schedule of charges or rates) (→charge 2)	ECON	(→Gebühr)
tariff administration	SWITCH	**Gebührenbehandlung**
tariff anouncement	TELEPH	**Gebührenansage**
tariff changeover time (→tariff change time)	SWITCH	(→Tarifumschaltepunkt)
tariff change time = tariff changeover time	SWITCH	**Tarifumschaltepunkt**
tariff classification	ECON	**zolltarifliche Einstufung**

tariff discount (→allowance of charge) ECON (→Gebührenermäßigung)
tariff rate SWITCH **Gebührentakt**
= Taktzeit
tariff zone (→metering zone) SWITCH (→Gebührenzone)
tarification (→tax metering) SWITCH (→Gebührenzählung)
tarnished METAL **angelaufen**
= coated
tarred CIV.ENG **geteert**
tarred tape OUTS.PLANT **Teerband**
TASI TELEPH **TASI**
[time assignment speech interpolation]
task (n.) DATA PROC **Teilaufgabe**
[a work element of a job] [ein Arbeitsschritt eines Auftrags]
↑ job; programm segment = Teilauftrag; Task
↑ Auftrag; Programmteil
task TECH **Aufgabe**
= job ≈ Funktion
≈ function
task SWITCH **Aufgabe**
= function
task (→process) DATA PROC (→Prozeß)
task-dependent TECH (→anwendungsspezifisch)
(→application-specific)
task handling (→order processing) DATA PROC (→Auftragsabwicklung)
task interval (→float) DATA PROC (→Taskintervall)
task level (→job level) DATA PROC (→Auftragsebene)
task management DATA PROC **Task-Management**
task processing (→order processing) DATA PROC (→Auftragsabwicklung)
task sharing DATA PROC **Aufgabenteilung**
task-specific TECH (→anwendungsspezifisch)
(→application-specific)
Tau-Pi transformation NETW.TH (→Stern-Dreieck-Umwandlung)
(→star-delta conversion)
tautology SCIE **Tautologie**
tax ECON **Steuer**
[a public charge upon person, property income, service or goods] [öffentliche Abgabe für Personen, Eigentum, Einkommen, Dienstleistungen oder Lieferungen]
= duty ≈ Gebühr; Zollgebühr
taxcard (→phonecard) TERM&PER (→Telefonkarte)
tax computer SWITCH **Gebührenrechner**
= charge computer; call-charge computer = Gebührencomputer
taxi (v.t.) AERON **rollen**
[low-speed displacement on the ground] [langsame Flugzeugbewegung am Boden]
↓ anrollen; ausrollen
taxilane (→taxiway) AERON (→Rollbahn)
taxiway AERON **Rollbahn**
[for low-speed maneuvers] [für langsame Anfahrten]
= taxilane
tax law (→fiscal law) ECON (→Steuerrecht)
tax meter SWITCH **Gebührenzähler**
= rate meter; rate counter; charge meter; time pulse counter = Zeitimpulszähler; Zeittaktzähler
tax metering SWITCH **Gebührenzählung**
= call metering; charging; message accounting; metering; tarification = Gebührenerfassung; Zählung; Tarifierung; Vergebührung
tax-metering pulse SWITCH **Gebührenimpuls**
= call-charge pulse
taxpayer identification code ECON **Steuernummer**
tax printing SWITCH **Gebührenausdruck**
= rate printing; charge printing; fee printing
Taylor distribution ANT **Taylor-Verteilung**
Taylor series MATH **Taylorsche Reihe**
= Taylorreihe
Taylor tuning fork antenna ANT **Taylor-Stimmgabelantenne**
Tb (→terbium) CHEM (→Terbium)
tb (→terabyte) DATA PROC (→Terabyte)
TC (→trunk exchange) SWITCH (→Fernvermittlungsstelle)

TC (→trunk circuit) SWITCH (→Leitungssatz)
TC (→temperature coefficient) PHYS (→Temperaturkoeffizient)
TC (→transmission control) DATA COMM (→Datenübertragungssteuerung 2)
Tc (→technetium) CHEM (→Technetium)
T circuit (→star connection) NETW.TH (→Sternschaltung)
TCM (→trellis-coded modulation) MODUL (→trelliscodierte Modulation)
T connection (→star connection) NETW.TH (→Sternschaltung)
TDE (→time-domain equalizer) NETW.TH (→Zeitbereichsentzerrer)
TDF (→intermediate distribution frame) SYS.INST (→Zwischenverteiler)
t distribution (→Student's t distribution) MATH (→Student-Verteilung)
TDM 1 (→time division multiplex) TELEC (→Zeitmultiplex)
TDM 2 (→time division multiplexing) TELEC (→Zeitmultiplextechnik)
TDMA (→time division multiple access) TELEC (→Zeitvielfachzugriff)
Te (→tellurium) CHEM (→Tellur)
teach (→train) TECH (→schulen)
teaching (→training) TECH (→Schulung)
teaching field SCIE **Lehrgebiet**
teaching step SCIE **Lehrschritt**
≈ learning step ≈ Lernschritt
team ECON **Mannschaft**
team ECON **Trupp**
= Arbeitstrupp
= Team
team work COLLOQ **Gemeinschaftsarbeit**
= Teamarbeit
tear 1 (v.t.) TECH **zerreißen**
= tear-off ≈ abreißen
tear 2 (v.t.) TECH **abreißen**
= tear-off
tear down a call SWITCH **eine Verbindung abbauen**
= tear down a connection; tear down a link
tear down a connection (→tear down a call) SWITCH (→eine Verbindung abbauen)
tear down a link (→tear down a call) SWITCH (→eine Verbindung abbauen)
tearing TV **Zeilenreißen**
[erratic lateral displacement of individual or blocks of lines] [unregelmäßige seitliche Verschiebung von einzelnen oder Blöcken von Zeilen]
tearing resistance TECH **Reißfestigkeit**
= tearing strength = Zerreißfestigkeit
tearing strength (→tearing resistance) TECH (→Reißfestigkeit)
tear-off (→tear 2) TECH (→abreißen)
tear-off edge TERM&PER **Abrißkante**
tear-off facility TERM&PER **Abtrennvorrichtung**
tear-off pad DOC **Abreißblock**
TEC (→thermoelectric cooler) TECH (→thermoelektrischer Kühler)
technality TECH **technische Einzelheit**
technetium CHEM **Technetium**
= Tc = Tc
technic 2 (→engineering) TECH (→Technik 1)
technic 1 (→technology 2) TECH (→Technologie)
technical atmosphere PHYS **technische Atmosphäre**
[= 98,066.50 pascal] [= 98.066,50 Pascal]
= at = at
technical cybernetics CONTROL **technische Kybernetik**
technical department ECON **technische Abteilung**
technical description DOC **technische Beschreibung**
technical dictionary LING **Fachwörterbuch**
= Sachwörterbuch; technisches Lexikon; Reallexikon
technical director (→technical manager) ECON (→technischer Leiter)
technical documentation DOC **technisches Schrifttum**
technical documentation filing OFFICE **technische Registratur**
technical document microfilming TERM&PER **Zeichnungsverfilmung**

technical drawing

technical drawing (→engineering drawing 1) TECH (→Technisches Zeichnen)
technical expression (→technical term) LING (→Fachausdruck)
technical improvement TECH technische Verbesserung
technical journal (→professional journal) DOC (→Fachzeitschrift)
technical know-how TECH technische Kenntnisse
technical knowledge SCIE Fachkenntnis
= specialized knowledge; special knowledge; expert knowledge; know how; expertness; expertise
= Fachwissen; Sachkenntnis; Sachkunde; Fachkunde
technical language LING Fachsprache
≈ jargon; terminology
≠ colloquial
≈ Jargon; Terminology
≠ Umgangssprache
technical manager ECON technischer Leiter
= technical director
= technischer Direktor
technical physics PHYS technische Physik
technical progress (→technological progress) TECH (→technischer Fortschritt)
technical term LING Fachausdruck
= term; technical expression
= Fachwort; Fachbegriff; Terminus (pl.-i); Fachterminus; Sachwort
technical terms of delivery (→engineering terms of delivery) TECH (→technische Lieferbedingungen)
technical university (BRI) SCIE Technische Universität
= Technische Hochschule
technical writer DOC technischer Fachautor
= Fachautor
technician TECH Techniker
≈ craftsman
≈ Handwerker
technique 2 (→engineering) TECH (→Technik 1)
technique 1 (→technology 2) TECH (→Technologie)
technochemical CHEM chemotechnisch
technological advance TECH (→technischer Fortschritt)
(→technological progress)
technological gap TECH technologische Lücke
technological highschool SCIE Fachhochschule
≈ technological university
≈ technische Universität
technological innovation TECH technische Neuerung
technological lead TECH technologischer Vorsprung
technological progress TECH technologischer Fortschritt
= technological advance; technical progress; advance in the art
≈ technische Errungenschaft
technology 2 TECH Technologie
[special knowlwdges and procedures of a technical branch]
[spezielle Erkenntnisse und Verfahren eines technischen Zweiges]
= technique 1; technic 1; practice
≈ engineering
≈ Gewerbekunde; Technik 2
≈ Technik 1
technology 1 (→engineering) TECH (→Technik 1)
technology transfer ECON Technologie-Transfer
tecnetron MICROEL Tecnetron
↑ junction field effect transistor
↑ Sperrschicht-Feldeffekttransistor
TED MICROEL TED
[transferred electron device]
tee METAL T-Profil
Tee adapter (→T junction) MICROW (→T-Verzweigung)
Tee adaptor (→T junction) MICROW (→T-Verzweigung)
Tee junction (→T junction) MICROW (→T-Verzweigung)
tee square ENG.DRAW Zeichenschiene
teflon CHEM Teflon
= Polytetrafluoräthylen; PTFE
telco (AM) (→telecommunications carrier) TELEC (→Fernmeldegesellschaft)
telcomms (→telecommunication engineering) INF.TECH (→Telekommunikationstechnik)
telcos (→telecommunications) INF.TECH (→Telekommunikation)
tele... INF.TECH Fern...
= remote
≈ Tele...
teleadjusting TELECONTR Ferneinstellen
[telecommand by polyvalent magnitudes]
[Fernsteuern durch mehrwerige Größen]
↑ telecommand
↑ Fernsteuern
teleautograph (AM) TELEGR (→Handschriften-Übertragungsgerät)
(→telewriter)

teleautography (AM) TELEGR (→Bildfernschreiben)
(→telewriting)
telebanking TELEC Telebanking
= electronic banking; home banking
telebox (→electronic mailbox) TELEC (→elektronischer Briefkasten)
telecard (→phonecard) TERM&PER (→Telefonkarte)
telecast (→television broadcast program) BROADC (→Fernsehprogramm)
teleciopier (→telefax set) TERM&PER (→Telefax-Gerät)
telecommand TELECONTR Fernsteuern
[commanding remote objects]
[Beeinflussen ferner Objekte]
= remote command; telecommand
↑ Fernwirken
↑ telecontrol
↓ Fernschalten; Ferneinstellen
↓ teleswitching; teleadjusting
telecommand (→telecommand) TELECONTR (→Fernsteuern)
telecomms equipment (→telecommunications equipment) TELEC (→Telekommunikationsgerät)
telecommunication ... TELEC Fernmelde...
= Telekommunikations...
telecommunication act TELEC (→Fernmeldegesetz)
(→telecommunication regulations)
telecommunication administration TELEC Fernmeldeverwaltung
[public, mostly state-run]
[öffentlich, meist staatlich]
= public communications common carrier; public communication carrier
= öffentliche Fernmeldegesellschaft
↑ telecommunications common carrier
↑ Fernmeldegesellschaft
↓ telecommunications authority; telephone administration
↓ staatliche Fernmeldeverwaltung; Fernsprechverwaltung
telecommunication bill TELEC Fernmelderechnung
↓ telephone bill
= Gebührenrechnung
= Telefonrechnung
telecommunication capability TERM&PER Telekommunikationsfähigkeit
telecommunication center (AM) TELEC fernmeldetechnisches Zentrum
= telecommunication centre (BRI)
= Fernmeldezentrum
telecommunication centre (BRI) TELEC (→fernmeldetechnisches Zentrum)
(→telecommunication center)
telecommunication circuit TELEC Fernmeldeleitung
= communications line
≈ Fernmeldelinie
telecommunication engineering INF.TECH Telekommunikationstechnik
= telecommunication technique; telecommunication system engineering; communications engineering; communications electronics; telcomms
= Fernmeldetechnik; Kommunikationstechnik; Nachrichtenübermittlungstechnik; Informationsübermittlungstechnik; Kommunikationselektronik
↑ information technology
↑ Nachrichtentechnik
telecommunication facility TELEC Fernmeldeanlage
telecommunication line (→trunk 1) TELEC (→Fernmeldelinie)
telecommunication power system POWER SYS Fernmelde-Stromversorgungsanlage
= telecommunications energy
telecommunication regulations TELEC Fernmeldegesetz
= telecommunication act
telecommunications INF.TECH Telekommunikation
[theory, engineering and practice of message transfer over distances]
[Theorie, Technik und Praxis der Fernübermittlung von Nachrichten]
= telcos
= Fernmeldewesen
≈ telematics
≈ Telematik
↑ communications
↑ Kommunikation
↓ telecommunications engineering
↓ Telekommunikationstechnik; Sprachkommikation; Datenkommunikation; Textkommunikation; Bildkommunikation
telecommunication satellite SAT.COMM Fernmeldesatellit
↑ communications satellite
↑ Nachrichtensatellit
telecommunications authority TELEC Fernmeldebehörde
≈ Fernmeldeverwaltung; Fernmeldefirma
telecommunications cable (→communication cable) COMM.CABLE (→Nachrichtenkabel)

telecommunications carrier TELEC [public or private company furnishing telecommunication services] = commercial telecommunications carrier; communications common carrier; common carrier; telco (AM) ≈ operating company; telecommunications concessionaire ↑ telecommunications company; carrier ↓ telecommunications administration; telecommunications co-operative	Fernmeldegesellschaft [öffentliche oder private Gesellschaft die Fernmeldedienste bietet] ≈ Betriebsgesellschaft; Fernmeldekonzessionär ↑ Fernmeldefirma; Träger ↓ Fernmeldeverwaltung; Fernmeldegenossenschaft	teleconferencing TELEC = teleconference ↓ audio conferencing; video conferencing telecontrol TELECONTR [supervision and command of remote objects] ↓ telemonitoring; telecommand	Telekonferenz ↓ Audiokonferenz; Videokonferenz Fernwirken [Überwachen und Beeinflussen ferner Objekte] = Fernbedienen; Fernlenken ↓ Fernüberwachen; Fernsteuern
telecommunications charge TELEC (→telecommunication services tariff)	(→Fernmeldegebühr)	telecontrol command TELECONTR = remote control telecontrol engineering TELEC = remote action technique	Fernwirkbefehl Fernwirktechnik
telecommunications company ECON [manufacturing or operating telecommunication facilities] = telecoms company ↓ telecommunications carrier; telecommunications equipment supplier	Fernmeldefirma [Lieferant oder Betreiber von Fernmeldeanlagen] ↓ Fernmeldegesellschaft; Fernmeldegeräte-Lieferant	telecontrol functional unit TELECONTR telecontrol information TELECONTR telecontrol installation TELECONTR [totality of interworking stations with their interconnection links]	Fernwirkfunktionseinheit Fernwirkinformation Fernwirkanlage [Gesamtheit zusammenwirkender Stationen mit deren Verbindungswegen]
telecommunications concesionaire TELEC = telecommunications concessioner ≈ telecommunications carrier	Fernmeldekonzessionär ≈ Fernmeldegesellschaft	telecontrol link TELECONTR telecontrol location TELECONTR [location with telecontrol station] telecontrol malfunction TELECONTR telecontrol message TELECONTR (→telecontrol telegram)	Fernwirkverbindung Fernwirkstelle [Ort mit Fernwirkstation] Fernwirkstörung (→Fernwirktelegramm)
telecommunications concessioner TELEC (→telecommunications concesionaire)	(→Fernmeldekonzessionär)	telecontrol network TELECONTR ↓ telesupervision network	Fernwirknetz ↓ Fernüberwachungsnetz
telecommunications co-operative TELEC = communications co-operative ↑ telecommunications carrier	Fernmeldegenossenschaft = Telefonverein ↑ Fernmeldegesellschaft	telecontrol receiver TELECONTR telecontrol signal TELECONTR telecontrol station TELECONTR telecontrol system TELECONTR	Fernwirkempfänger Fernwirksignal Fernwirkstation Fernwirksystem
telecommunications coverage TELEC (→communications coverage)	(→Fernmeldeversorgung)	telecontrol telegram TELECONTR = telecontrol message telecontrol transmitter TELECONTR	Fernwirktelegramm Fernwirksender
telecommunications energy POWER SYS (→telecommunication power system)	(→Fernmelde-Stromversorgungsanlage)	telecopying TELEC ≈ facsimile	Fernkopieren = Telekopieren ≈ Faksimile
telecommunications engineering TELEC authority of the F.R.G	Fernmeldetechnisches Zentralamt [DBP] = FTZ	telecopy service TELEC [service of Deutsche Bundespost]	Telefax-Dienst [Dienst der DBP für Faksimile-Übertragung] = Telefax; Fax-Dienst; Fernkopieren
telecommunications equipment TELEC = communications equipment; telecomms equipment	Telekommunikationsgerät = Fernmeldegerät; fernmeldetechnisches Gerät	telefax set TERM&PER = telecopier	Telefax-Gerät = Fernkopierer
telecommunication service TELEC = commercial telecommunications; teleservice	Fernmeldedienst = Telekommunikationsdienst	telegram (n.) POST = cable (n.); wire (n.) telegram chain TELECONTR	Telegramm Telegrammkette
telecommunication services tariff TELEC = telecommunications charge	Fernmeldegebühr ↓ Fernsprechgebühr; Fernschreibgebühr	telegram charge (→telegram service POST tariff) telegram form POST	(→Telegrammgebühr) Telegrammformular
telecommunications market ECON	Fernmeldemarkt	telegram service TELEC	Telegrammdienst
telecommunications network TELEC = telecom network ↑ communications network ↓ telephone network	Fernmeldenetz = Telekommunikationsnetz ↑ Kommunikationsnetz ↓ Fernsprechnetz	telegram service tariff POST = telegram charge telegram text POST = telegram wording telegram wording (→telegram POST	Telegrammgebühr Telegrammtext (→Telegrammtext)
telecommunications regulations ECON	Telekommunikationsordnung = Fernmeldeordnung	text) telegraph (v.t.) TELEC = cable (v.t.); wire (v.t.)	telegrafieren = telegraphieren
telecommunications switching TELEC (→switching)	(→Vermittlung)	telegraph alphabet (→teleprinter CODING code)	(→Fernschreibalphabet)
telecommunications testing INSTR	Telekommunikations-Meßtechnik = Fernmeldemeßtechnik	telegraph channel TELEC = telex channel ≈ telegraph line	Fernschreibkanal = Telegraphiekanal; Telegrafiekanal; Telegraphenkanal; Telegrafenkanal ≈ Fernschreibleitung
telecommunication system INF.TECH engineering (→telecommunication engineering)	(→Telekommunikationstechnik)	telegraph code (→teleprinter CODING code)	(→Fernschreibalphabet)
telecommunication systems TELEC supplier	Amtsbaufirma	telegraph current (→line TELEGR current)	(→Fernschreibstrom)
telecommunication technique INF.TECH (→telecommunication engineering)	(→Telekommunikationstechnik)	telegraph distortion (→telegraph TELEGR signal distortion)	(→Schrittverzerrung)
telecommunication tower OUTS.PLANT (→communication tower)	(→Fernmeldeturm)	telegraph engineering TELEC (→telegraphy)	(→Fernschreibtechnik)
telecommunication traffic TELEC	Fernmeldeverkehr	telegraphic TELEC	telegrafisch
telecommunication vehicle TELEC	Fernmeldefahrzeug	telegraphic address POST	Telegrammadresse
telecommuting TELEC	Telearbeit [Heimarbeit]	= cable address	
telecom network TELEC (→telecommunications network)	(→Fernmeldenetz)		
telecom operator (→operating TELEC company)	(→Betriebsgesellschaft)		
telecoms company ECON (→telecommunications company)	(→Fernmeldefirma)		
teleconference TELEC (→teleconferencing)	(→Telekonferenz)		

telegraphic equation

telegraphic equation   LINE TH   (→Leitungsgleichung)
 (→transmission equation)
telegraphic keyboard   TERM&PER   (→Fernschreibtastatur)
 (→teletypewriter keyboard)
telegraphic line (→telegraph   TELEC   (→Fernschreibleitung)
 line)
telegraphic traffic (→telex traffic)   TELEC   (→Fernschreibverkehr)
telegraphic transmitter   RADIO   Telegrafiesender
telegraph line   TELEC   Fernschreibleitung
 = telegraphic line   = Telegrafieleitung; Telegra-
 ≈ telegraph channel   phieleitung; Telegraphenlei-
         tung; Telegraphenleitung
telegraph modulation   TELEGR   Telegrafiemodulation
        = Telegraphiemodulation
telegraph noise   TELEPH   Telegrafiegeräusch
 = thump   = Telegraphiegeräusch; Tele-
         grafiergeräusch; Telegra-
         phiergeräusch
telegraph office (→telex   TELEC   (→Fernschreibvermittlungs-
 office)   amt)
telegraph pole   OUTS.PLANT   Telegrafenmast
 = telegraph post   = Telegraphenmast; Telegra-
 ≈ telephone pole   fenstange; Telegraphen-
 ↑ fixture   stange
         ≈ Telefonmast
         ↑ Freileitungsmast
telegraph post (→telegraph   OUTS.PLANT   (→Telegrafenmast)
 pole)
telegraph signal (→teleprinter   CODING   (→Fernschreibzeichen)
 signal)
telegraph signal distortion   TELEGR   Schrittverzerrung
 = start-stop distortion; telegraph distor-   = Drehzahlverzerrung; Be-
 tion; speed distortion   zugsverzerrung; Start-
         Stopp-Verzerrungsgrad
telegraph speed   TELEC   Schrittgeschwindigkeit
 [reciprocal of shortest nominal unit   [Kehrwert der kürzesten
 interval; unit: Baud]   nominellen Schrittdauer;
 = signaling rate; modulation rate   Einheit: Baud]
 [TELEGR]; baud rate; pulse clock   = Telegrafiergeschwindigkeit
 ≈ transmission rate   ; Tastgeschwindigkeit; Mo-
         dulationsgeschwindigkeit
         [TELEGR]; Schritttakt;
         Baud-Rate
         ≈ Übertragungsgeschwindig-
         keit
telegraph speed generator   DATA COMM   (→Baudratengenerator)
 (→baud rate generator)
telegraph switchboard   TELEGR   Fernschreib-Vermittlungs-
 = teletypewriter switchboard   schrank
telegraph switching   SWITCH   Fernschreibvermittlung 1
 [function]   [Funktion]
         = Telexvermittlung 1
telegraph terminal equipment   TERM&PER   Fernschreibendgerät
 ↑ user terminal   ↑ Teilnehmergerät
 ↓ teletypewriter   ↓ Fernschreiber
telegraph transmission system   TELEGR   Fernschreib-Übertragungssy-
         stem
telegraphy   TELEC   Fernschreibtechnik
 = telegraph engineering   = Telegrafie; Telegraphie; Te-
         legrafentechnik; Telegra-
         phentechnik
telegraphy over multiplex   TELEGR   (→Wechselstromtelegrafie)
 (→voice frequency telegraphy)
teleindication   TELECONTR   Fernanzeigen
 [telesupervision by two-states magni-   [Fernüberwachen durch
 tudes]   Größen die nur zwei Zu-
 = remote indication   stände annehmen können]
 ↑ telemonitoring   = Fernmelden
         ≈ Fernüberwachen
telelens   PHYS   Teleobjektiv
 = telephoto lens
telematics   INF.TECH   Telematik
 [informatics with exploration of tele-   [Informatik unter Einbezie-
 communications means]   hung der Mittel der Tele-
 = compunication   kommunikation]
 ≈ information technology   ≈ Informationstechnik
 ↑ communications   ↑ Kommunikation
 ↓ telecommunications; informatics
telematic service   TELEC   Telematikdienst
         = telematischer Dienst

telematic system   TELEC   Telematikanlage
telematic terminal   TERM&PER   Telematik-Endgerät
telemeter (v.t.)   TELECONTR   fernmessen
telemetering   TELECONTR   Fernmessen
 = telemetry; remote measuring   = Telemetrie; Fernmessung;
 ↑ telemonitoring   Fernmesstechnik
         ↑ Fernüberwachen
telemetry   TELECONTR   (→Fernmessen)
 (→telemetering)
telemetry data   TELECONTR   Telemetriedaten
         = Fernmeßdaten
telemonitoring   TELECONTR   Fernüberwachen
 = remote supervision; remote supervi-   = Fernüberwachung
 sory; telesupervision   ↑ Fernwirken
 ↑ telecontrol   ↓ Fernanzeigen; Fernmessen
 ↓ teleindication; telemetering
telemonitoring technique   TELECONTR   Fernüberwachungstechnik
telepak   DATA COMM   Breitband-Standleitungs-
 [service of leased broadband lines]   dienst
telephon cord (→handset   TELEPH   (→Hörerschnur)
 cord)
telephone (v.i.)   TELEPH   telefonieren
 = phone (v.i.); call (v.t.); make a phone   = telephonieren; fernspre-
 call; ring (v.i.)   chen; anrufen
telephone (→telephone set)   TELEPH   (→Fernsprechapparat)
telephone adapter   TELEPH   Telefon-Saugadapter
         = Telephon-Saugadapter
telephone administration   TELEC   Fernsprechverwaltung
 = telephone operating company; tele-   = Telefonverwaltung; Telephone company; operating telephone   phonverwaltung
 company (NAM); OTC (NAM)   ↑ Fernmeldeverwaltung
 ↑ telecommunications administration
telephone answering machine   TERM&PER   (→Anrufbeantworter)
 (→automatic answering equipment)
telephone answering service   TELEPH   Abwesenheitsdienst
 = absent-subscriber service; answering   ↑ Fernsprechauftragsdienst
 service
 ↑ telephone message service
telephone bill   TELEC   Telefonrechnung
 ↑ telcommunication bill   = Telephonrechnung
         ↑ Fernmelderechnung
telephone book (→telephone   TELEC   (→Fernsprechverzeichnis)
 directory)
telephone booth   TELEPH   Fernsprechzelle
 [a free standing box]   [freistehendes Häuschen]
 = telephone kiosk; telephone call box;   = Telefonzelle; Telephonzelle
 telephone box; phone box; call box;   ≈ Fernsprechkabine
 ≈ telephone cabin
telephone box (→telephone   TELEPH   (→Fernsprechzelle)
 booth)
telephone cabin   TELEPH   Fernsprechkabine
 [a compartment within a building]   [kleiner abgeteilter Raum
 ≈ telephone booth   innerhalb eines Gebäudes]
         = Telefonkabine; Telephon-
         kabine
         ≈ Fernsprechzelle
telephone cable (→handset   TELEPH   (→Hörerschnur)
 cord)
telephone call   TELEPH   Telefongespräch
 = telephone conversation; telephone   = Telephongespräch
 message   ≈ Ferngespräch
telephone call box (→telephone   TELEPH   (→Fernsprechzelle)
 booth)
telephone central office   SWITCH   (→Fernsprechvermittlungs-
 (→central office)   stelle)
telephone central office (→local   SWITCH   (→Teilnehmervermittlungs-
 central office)   stelle)
telephone channel   TELEC   Fernsprechkanal
 = voice channel; speech channel   = Telefonkanal; Telephonka-
 ≈ voice-grade channel [DATA COMM]   nal; Sprachkanal; Sprechka-
         nal
telephone charge   TELEC   Fernsprechgebühr
 ↑ telecommunications service tariff   = Telefongebühr; Telephon-
         gebühr
         ↑ Fernmeldegebühr
telephone chip card (→chip   TERM&PER   (→Chip-Karte)
 card)
telephone circuit   TELEPH   Fernsprecherschaltung
         = Fernsprechschaltung;
         Sprechschaltung

telephone communication TELEC
= telephonic communication; telephone connection 2; speech connection

telephone company (→telephone administration) TELEC

telephone conference equipment TERM&PER

telephone connection 1 TELEC
= POT line
≈ telephone line

telephone connection 2 TELEC
(→telephone communication)

telephone console TELEPH
= telephone switchboard

telephone conversation TELEPH
(→telephone call)

telephone current (→speech current) TELEPH

telephone density TELEC

telephone directory TELEC
= directory; telephone book; phone book
↑ communications directory

telephoned telegram TELEPH

telephoned telegram service TELEPH
↑ special telephone service

telephone engineering TELEC
= telephone transmission engineering; telephone transmission technique
≈ telephony
↑ telecommunications

telephone equipment TELEPH
≈ telephone set

telephone exchange (BRI) SWITCH
(→central office)

telephone exchange (AM) SWITCH
(→local exchange area)

telephone exchange (BRI) SWITCH
(→local central office)

telephone hood TELEPH
= phone hood

telephone indicator COMPON

telephone information service TELEPH

telephone installer TELEPH

telephone instrument TELEPH
(→telephone set)

telephone interference factor TELEPH

telephone keyboard TELEPH
(→telephone keypad)

telephone keypad TELEPH
= telephone keyboard

telephone kiosk (→telephone booth) TELEPH

telephone line TELEC
≈ telephone connection

telephone message (→telephone call) TELEPH

telephone message service TELEPH
↑ special telephone service
↓ telephone answering service; alarm-clock calling; telephone notification service

telephone network TELEC
↑ telecommunications network

telephone notification service TELEPH

telephone number TELEPH
↑ subscriber number [TELEC]

telephone operating company TELEC
(→telephone administration)

telephone plug (→jack connector) TELEPH

telephone pole OUTS.PLANT
= telephone post
≈ telegraph pole
↑ fixture

telephone post (→telephone pole) OUTS.PLANT

telephone recorder TERM&PER
(→automatic answering equipment)

telephone repeater (→voice frequency amplifier) TRANS

telephone service TELEC
↓ POT service

telephone service fuse COMPON
↑ overcurrent protector
↓ heat-coil fuse; resolderable fuse; reversible fuse

telephone set TELEPH
= telephone; telephone instrument; station
≈ telephonic station; telephonic equipment
↑ user terminal; telephone terminal equipment

telephone shop TELEC

telephone sidetone (→sidetone 1) TELEPH

telephone signal TELEC
[voice signal with standard band limiting from 300 Hz to 3,400 Hz for telecommunications]
= telephonic signal
↑ voice signal

telephone socket COMPON

telephone stands OFFICE

telephone station TELEC
= station
≈ telephone set

telephone statistics TELEC

telephone subscriber TELEC

telephone switchboard TELEPH
(→telephone console)

telephone switching engineering SWITCH
↑ switching engineering

telephone switching exchange SWITCH
(→central office)

telephone swivel arm OFFICE

telephone terminal equipment TELEPH
↑ user terminal
↓ telephone set; executive telephone system

↑ Fernsprechsonderdienst
↓ Abwesenheitsdienst; Weckdienst; Benachrichtigungsdienst

Fernsprechnetz
= Telefonnetz; Telephonnetz
↑ Fernmeldenetz

Benachrichtigungsdienst
↑ Fernsprechauftragsdienst

Telefonnummer
= Telephonnummer; Fernsprechnummer
↑ Teilnehmerrufnummer [TELEC]

(→Fernsprechverwaltung)

(→Klinkenstecker)

Telefonmast
= Telephonmast; Telefonstange; Telephonstange
≈ Telegrafenmast
↑ Freileitungsmast

(→Telefonmast)

(→Anrufbeantworter)

(→NF-Verstärker)

Fernsprechdienst
↓ konventioneller Fernsprechdienst

Fernmeldesicherung
↑ Stromsicherung
↓ Hitzdrahtsicherung; Rücklötsicherung; Umkehrauslöser

Fernsprechapparat
= Telefon 1; Telephon 1; Fernsprecher
≈ Fernsprechstelle; Fernsprechanlage
↑ Teilnehmergerät; Fernsprechendgerät

Telefonladen
= Telephonladen

(→Rückhören)

Fernsprechsignal
[für Fernmeldezwecke auf 300 Hz bis 3.400 Hz bandbreitenbegrenztes Sprachsignal]
= Telefoniesignal; Telephoniesignal; Telefonsignal; Telephonsignal
↑ Sprachsignal

Telefonbuchse
= Telephonbuchse

Telefonständer

Sprechstelle
≈ Fernsprechapparat

Fernsprechstatistik
= Telefonstatistik; Telephonstatistik

Fernsprechteilnehmer
= Telephonabonnent (CH)

(→Fernsprech-Vermittlungsschrank)

Fernsprechvermittlungstechnik
↑ Vermittlungstechnik

(→Fernsprechvermittlungsstelle)

Telefonschwenkarm

Fernsprechendgerät
↑ Teilnehmergerät
↓ Fernsprechapparat; Chef-Fernsprechanlage

Fernsprechverbindung
= Sprechverbindung; Gesprächsverbindung; Telefonverbindung; Telephonverbindung

(→Fernsprechverwaltung)

Telefonkonferenzeinrichtung

Fernsprechanschluß
= Telefonanschluß
≈ Fernsprechleitung

(→Fernsprechverbindung)

Fernsprech-Vermittlungsschrank

(→Telefongespräch)

(→Sprechstrom)

Fernsprechdichte
= Telefondichte; Telephondichte

Fernsprechverzeichnis
= Telefonverzeichnis; Telephonverzeichnis; Fernsprechbuch; Telefonbuch; Telephonbuch
↑ Kommunikationsverzeichnis

zugesprochenes Telegramm
= telefonisch zugestelltes Telegramm

Telegrammannahme
= Telegrammaufnahme
↑ Fernsprechsonderdienst

Fernsprechtechnik
≈ Fernsprechwesen
↑ Telekommunikation

Fernsprechanlage
= Telefonanlage; Telephonanlage
≈ Fernsprechapparat

(→Fernsprechvermittlungsstelle)

(→Ortsanschlußbereich)

(→Teilnehmervermittlungsstelle)

Telefonhaube

Telefonlampe
= Telephonlampe

Fernsprechansagedienst
= Telefonansagedienst; Telephonansagedienst

Sprechstelleneinrichter

(→Fernsprechapparat)

Fernsprechformfaktor

(→Fernsprechertastatur)

Fernsprechertastatur
= Fernsprechtastatur; Telefontastatur; Telephontastatur

(→Fernsprechzelle)

Fernsprechleitung
= Telefonleitung; Telephonleitung
≈ Fernsprechanschluß

(→Telefongespräch)

Fernsprechauftragsdienst
= Telefonauftragsdienst; Telephonauftragsdienst

telephone traffic

telephone traffic TELEC	**Fernsprechverkehr**	
= telephonic traffic	= Telefonieverkehr; Telefonverkehr	
telephone transmission TELEC	**Fernsprechübertragung**	
≈ telephony	≈ Fernsprechwesen	
↑ voice transmission	↑ Sprachübertragung	
telephone transmission engineering (→ telephone engineering) TELEC	(→ Fernsprechtechnik)	
telephone transmission technique (→ telephone engineering) TELEC	(→ Fernsprechtechnik)	
telephone transmitter TELEPH (→ transmission capsule)	(→ Sprechkapsel)	
telephone volume amplifier TERM&PER	**Telefonlautverstärker**	
telephonic (adj.) TELEC	**fernmündlich** (adj.) = telefonisch; telephonisch	
telephonically (→ by phone) TELEC	(→ fernmündlich) (adv.)	
telephonic communication TELEC (→ telephone communication)	(→ Fernsprechverbindung)	
telephonic hybrid TELEPH	**Fernsprechgabel**	
telephonic path (→ voice path) TELEC	(→ Fernsprechweg)	
telephonic signal (→ telephone signal) TELEC	(→ Fernsprechsignal)	
telephonic special service TELEPH (→ special telephone service)	(→ Fernsprechsonderdienst)	
telephonic station TELEPH	**Fernsprechstelle**	
≈ telephone set	≈ Fernsprechapparat	
telephonic traffic (→ telephone traffic) TELEC	(→ Fernsprechverkehr)	
telephony TELEC	**Fernsprechwesen**	
≈ telephone engineering	= Fernsprechen; Telefonie; Telephonie	
↑ telecommunications	≈ Fernsprechtechnik	
	↑ Telekommunikation	
telephotography TELEGR	**Fotographie-Fernübertragung**	
telephoto lens (→ telelens) PHYS	(→ Teleobjektiv)	
telepoint TELEC	**Telepoint**	
teleport (→ telecommunication center) TELEC	(→ fernmeldetechnisches Zentrum)	
teleprinter (BRI) TERM&PER (→ teletypewriter)	(→ Fernschreiber)	
teleprinter code CODING	**Fernschreibalphabet**	
= telegraph code; teletypewriters code; telegraph alphabet	= Fernschreibcode; Telegrafiecode; Telegraphiecode; Telegrafenalphabet; Telegraphenalphabet; Fernschreibercode	
↓ international telegraph alphabet no.2; Baudot code	↓ internationales Telegraphenalphabet Nr.2; CCITT-Alphabet Nr.1	
teleprinter exchange (→ telex office) TELEC	(→ Fernschreibvermittlungsamt)	
teleprinter installation TELEGR	**Fernschreibanlage** = Telexanlage	
teleprinter keyboard TERM&PER (→ teletypewriter keyboard)	(→ Fernschreibtastatur)	
teleprinter network (→ telex network) TELEC	(→ Fernschreibwählnetz)	
teleprinter signal CODING = telegraph signal	**Fernschreibzeichen** = Telegrafiezeichen	
teleprinter station TELEGR = telex station	**Fernschreibstelle** = Telexstelle	
teleprinter traffic (→ telex traffic) TELEC	(→ Fernschreibverkehr)	
teleprinting TELEC	**Ferndrucken**	
teleprinting over radio TELEGR (→ radiotelegraphy)	(→ Funktelegrafie)	
teleprocessing (→ remote data processing) DATA PROC	(→ Datenfernverarbeitung)	
teleprocessing monitor DATA PROC [program to coordinate data teleprocessing] = TP monitor	**Fernverarbeitungsmonitor** [Programm zur Koordination von Datenfernverarbeitung] = TP-Monitor; Transaktionsmonitor	
telescope PHYS	**Fernrohr**	
telescopic antenna ANT = pop-up antenna	**Teleskopantenne**	
telescopic mast ANT = extension mast	**Teleskopmast** = Kurbelmast	
telescopic mast antenna ANT	**Teleskopmast-Antenne** = Kurbelantenne	
telescoping cell MICROEL	**Teleskopzelle** = Erweiterungszelle	
teleservice DATA PROC = remote meintenance	**Fernwartung** = Fernservice; Teleservice	
teleservice (→ telecommunication service) TELEC	(→ Fernmeldedienst)	
teleshopping TELEC	**Teleshopping**	
telesoftware DATA PROC [fed via data link]	**Telesoftware** [über Datenfernübertragung eingespeist]	
telespectator (→ televiewer) BROADC	(→ Fernsehzuschauer)	
telesupervision TELECONTR (→ telemonitoring)	(→ Fernüberwachen)	
telesupervision network TELECONTR ↑ telecontrol network	**Fernüberwachungsnetz** ↑ Fernwirknetz	
teleswitching TELECONTR [telecommand by a bivalent magnitude] ↑ telecommand (n.)	**Fernschalten** [Fernsteuern mittels zweiwertiger Größe] ↑ Fernsteuern	
teleswitching command TELECONTR (→ switching command)	(→ Schaltbefehl)	
teletax signal (→ metering signal) SWITCH	(→ Zählzeichen)	
teletex TELEC [switched teleprint service with higher transmission rate and extended character set] = TTX ↑ text communications	**Teletex** [vermittelter Fernschreibdienst höherer Übertragungsgeschwindigkeit mit erweitertem Zeichenvorrat] = Bürofernschreiben; TTX ↑ Textkommunikation	
teletex-compatible TELEC	**teletexfähig**	
teletex concentrator DATA COMM	**Teletex-Konzentrator**	
teletex private branch exchange DATA COMM	**Teletex-Nebenstellanlage**	
teletext TELEC [one-way transmission of texts and symbols during the horizontal blanking interval of the TV signal, displayed on the domestic receiver with a special decoder] = broadcast videotex; one-way videotex; Videotext (BRD); Ceefax (GBR) ≈ videotex	**Teletext** [Einwegübertragung von Texten und Symbolen in der Zeilenaustaktlücke des Fernsehsignals, durch spezielle Decoder am Heimempfänger darstellbar] = Bildschirmzeitung; Videotext (BRD); Ceefax (GBR) ≈ Bildschirmtext	
teletex-telex converter DATA COMM	**Teletex-Telex-Umsetzer** = TTU	
teletex terminal TERM&PER	**Teletexendgerät** = Bürofernschreiber	
teletype (→ teletypewriter) TERM&PER	(→ Fernschreiber)	
teletypewriter (AM) TERM&PER [a terminal converting from or into telegraphic code, for telex or computing] = teleprinter (BRI); teletype; TTY ↑ telegraph terminal equipment	**Fernschreiber** [Oberbegriff für in oder aus Fernschreibcode umsetzende Endgeräte, für Telex oder als Peripheriegerät einer DVA] = Fernschreibmaschine ↑ Fernschreibendgerät ↓ Telexfernschreiber	
teletypewriter conference TELEC ≈ computer conferencing [DATA COMM]	**Fernschreibkonferenz** ≈ Computer-Konferenz [DATA COMM]	
teletypewriter keyboard TERM&PER = teleprinter keyboard; telegraphic keyboard; telex keyboard	**Fernschreibtastatur**	
teletypewriters code CODING (→ teleprinter code)	(→ Fernschreibalphabet)	
teletypewriter switchboard TELEGR (→ telegraph switchboard)	(→ Fernschreib-Vermittlungsschrank)	
televiewer BROADC = viewer; telespectator = television broadcast service user ↑ radio broadcast user	**Fernsehzuschauer** = Fernsehzuseher (OES) = Fernsehteilnehmer ↑ Rundfunkteilnehmer	
television TELEC = TV	**Fernsehen** = TV	
television advertizing BROADC = television commercial; TV advertizing; TV commercial; TV ad	**Fernsehwerbung** = Fernsehpropaganda; Fernsehreklame	

television aerial 1 (→television ANT
antenna 1)
television antenna 1 ANT
= television aerial 1; TV antenna 1; TV
aerial 1
↓ TV emission antenna; TV reception
antenna
television antenna 2 (→television ANT
reception antenna)
television band (→television RADIO
broadcast band)
television broadcast band RADIO
= TV broadcast band; television band;
TV band
television broadcast emitter BROADC
(→television transmitter)
television broadcasting BROADC
= TV broadcasting; radiovision
↑ broadcasting
television broadcasting BROADC
company
= TV broadcasting company
television broadcasting SAT.COMM
satellite (→TV broadcasting satellite)
television broadcast program BROADC
(AM)
= television program; TV broadcast
program; TV program; telecast; television broadcast programme (BRI);
TV broadcast programme; television
programme ; TV programme
≈ television broadcast transmission
↑ broadcast program
television broadcast programme BROADC
(BRI) (→television broadcast program)
television broadcast satellite SAT.COMM
= TV broadcast satellite; TV satellite
↑ broadcast satellite
television broadcast service BROADC
user
= TV broadcast service user
≈ television spectator
↑ broadcast service user
television broadcast station BROADC
= TV broadcast station
television broadcast BROADC
transmission
= television transmission; TV broadcast
transmission; TV transmission
↑ broadcast transmission
television by satellite BROADC
= satellite TV
television camera TV
= TV camera
television car (→television van) TV
television channel RADIO
= TV channel
television channel signal BROADC
= TV channel signal
television channel spacing RADIO
television commercial BROADC
(→television advertizing)
television control system SIGN.ENG
= TV control system; video monitoring
system
television converter (→television TV
translater)
television coverage BROADC
= TV coverage
television distribution center BROADC
television emitter (→television BROADC
transmitter)
television engineering TV
= TV engineering
television headphone CONS.EL
= TV headphone

television interference RADIO PROP
= TVI
television licence BROADC
= TV licence
television line TRANS
= TV line
television link BROADC
= TV link
television measurement TV
= TV measurement
television measuring signal TV
= TV measuring signal
television network BROADC
= TV network
television picture (→TV TV
picture)
television picture display (→TV TV
picture display)
television picture tube ELECTRON
= TV picture tube; TV tube
↑ picture tube
↓ black-and-white TV tube; color TV
tube
television program (→television BROADC
broadcast program)
television program center (AM) BROADC
= TV program center (AM); television
programme centre (BRI); TV programme centre (BRI)
television programme BROADC
(→television broadcast program)
television programme centre BROADC
(BRI) (→television program center)
television receiver CONS.EL
(→television set)
television reception BROADC
= TV reception
television reception antenna ANT
= TV reception antenna; television antenna 2; TV antenna 2
↑ television antenna 1
television relay transmitter BROADC
(→television repater)
television repater BROADC
[relays actively oder passively a TV
channel signal, to illuminate]
= TV repeater; television relay transmitter; TV relay transmitter
≈ television transposer
≠ a shaded region
television screen (→video TV
screen)
television set CONS.EL
= TV set; televisor; television receiver;
TV receiver; TV home receiver;
video receiver (AM)
television signal TV
[video plus audio]
= TV signal
television signal BROADC
distribution
= TV signal distribution
television standard TV
= TV standard
television studio BROADC
= TV studio
television test signal TV
= TV test signal
television tower ANT
= TV tower
television translater TV
= TV translater; television converter;
TV converter
television transmission BROADC
(→television broadcast transmission)

Fernsehstörung
[Störung von TV-Empfang]
Fernsehgebühr
Fernsehleitung
= TV-Leitung
Fernsehleitungsverbindung
= TV-Leitungsverbindung
Fernsehmeßtechnik
= TV-Meßtechnik
Fernsehmeßsignal
= TV-Meßsignal
Fernsehleitungsnetz
= TV-Leitungsnetz
(→Fernsehbild)

(→Fernsehbildempfänger)

Fernsehbildröhre
= Fernsehröhre; TV-Röhre
↑ Bildwiedergaberöhre
↓ Schwarz-Weiß-Fernsehbildröhre; Farb-Fernsehbildröhre
(→Fernsehprogramm)

Fernsehschaltstelle
= TV-Schaltsstelle

(→Fernsehprogramm)

(→Fernsehschaltstelle)

(→Fernsehgerät)

Fernsehempfang
= TV-Empfang
Fernsehempfangsantenne
= Fernsehantenne 2; TV-Antenne
↑ Fernsehantenne 1
Fernsehumlenksender

Fernsehumlenksender
[nicht umsetzender oder
passiver Repeater zum
Ausleuchten eines abgeschatteten Gebiets]
= TV-Umlenksender
≈ Fernsehumsetzer
(→Bildschirm)

Fernsehgerät
= Fernsehempfänger; Fernseher; Fernsehheimempfänger; Fersehapparat;
TV-Gerät; TV-Empfänger;
TV-Apparat
Fernsehsignal
[Bild plus Ton]
= TV-Signal
Fernsehverteilung
= TV-Verteilung

Fernsehnorm
= Fernsehstandard; TV-
Norm; TV-Standard
Fernsehstudio
= TV-Studio
Fernsehprüfsignal
= TV-Prüfsignal
Fernsehturm
= TV-Turm
↑ Fernmeldturm
Bereichsumsetzer
[für Kabelempfanf]
= Bereichs-Converter
(→Fernsehsendung)

Fernsehantenne 1
= TV-Antenne 1
↓ Fernsehsendeantenne;
Fernsehempfangsantenne

(→Fernsehempfangsantenne)

(→Fernsehband)

Fernsehband
= Fernsehbereich; TV-Band;
TV-Bereich

(→Fernsehsender)

Fernsehrundfunk
↑ Rundfunk

Fernsehanstalt
= Fernsehgesellschaft

(→Fernsehrundfunk-Satellit)

Fernsehprogramm
= TV-Programm
≈ Fernsehsendung
↑ Rundfunkprogramm

(→Fernsehprogramm)

Fernsehsatellit
= TV-Satellit
↑ Rundfunksatellit
Fernsehteilnehmer
≈ Fernsehzuschauer
↑ Rundfunkteilnehmer

Fernsehsendeanlage
= TV-Sendeanlage
≈ Fernsehsender
Fernsehsendung
= Fernsehübertragung
≈ Fernsehprogramm
↑ Rundfunksendung 1

Satelliten-Fernsehen

Fernsehkamera
= Fernsehaufnahmekamera;
TV-Kamera
(→Fernsehwagen)
Fernsehkanal
= TV-Kanal
Fernsehkanalsignal
= TV-Signal-Kanal
Fernsehkanalabstand
(→Fernsehwerbung)

Fernsehüberwachungsanlage
= TV-Überwachungsanlage

(→Bereichsumsetzer)

Fernsehversorgung
= TV-Versorgung
Bildsternpunkt
(→Fernsehsender)

Fernsehtechnik
= TV-Technik
Fernsehkopfhörer
= Fernsehhörer; TV-Kopfhörer

television transmitter	BROADC	Fernsehsender
= television emitter; television broadcast emitter; TV transmitter; TV emitter; TV broadcast emitter		= Fernsehsendestation; TV-Sender; TV-Sendestation
↑ broadcast transmitter		≈ Fernsehsendeanlage
		↑ Rundfunksender
television transmitter feeding link	BROADC	**Fernsehmodulationsleitung**
[from studio to a transmitter]		[vom Studio zu einem Sender]
= TV transmitter feeding link		= TV-Modulationsleitung
↑ transmitter feeding link		↑ Modulationsleitung
television transmitting antenna	ANT	**Fernsehsendeantenne**
= TV transmitting antenna		= TV-Sendeantenne
↑ television antenna 1		↓ Fernsehantenne 1
television transposer	BROADC	**Fernsehumsetzer**
[frequency-changing relay to illuminate a shaded region]		[frequenzumsetzende Relaisstelle zum Ausleuchten abgeschatteter Gebiete]
= rebroadcast transmitter; TV transposer; transposer		= Umsetzer; TV-Umsetzer; TVU
≈ television repeater		≈ Fernsehumlenksender
television van	TV	**Fernsehwagen**
= TV van; television car; TV car		= TV-Wagen
↑ outside broadcast van		↑ Übertragungswagen
televisor (→television set)	CONS.EL	(→Fernsehgerät)
telewriter	TELEGR	**Handschriften-Übertragungsgerät**
= teleautograph (AM)		
telewriting	TELEGR	**Bildfernschreiben**
[transmission of handwriting]		[Übertragung von Handschriften]
= teleautography (AM)		= Handschriftenübertragung
≈ videotelegraphy		≈ Bildtelegrafie
telex 1	TELEC	**Telex 1**
["teleprinter exchange"; a switched teletypewriter service by international standards, with 50 Baud]		[international genormter, vermittelter Fernschreibdienst, mit 50 Baud]
= telex service		= Telexdienst; Fernschreibdienst
telex 2 (→telex message)	TELEC	(→Fernschreiben)
telex channel (→telegraph channel)	TELEC	(→Fernschreibkanal)
telex charge	TELEC	**Fernschreibgebühr**
↑ telecommunications tariff		= Telexgebühr
		↑ Fernmeldegebühr
telex communication (→telex traffic)	TELEC	(→Fernschreibverkehr)
telex connection (→telex line)	TELEC	(→Telexleitung)
telex connection unit	TELEGR	**Telex-Anschlußeinheit**
telex directory	TELEC	**Telexverzeichnis**
↑ communications directory		= Fernschreibteilnehmer-Verzeichnis
		↑ Kommunikationsverzeichnis
telex exchange (→telex office)	TELEC	(→Fernschreibvermittlungsamt)
telex keyboard	TERM&PER	(→Fernschreibtastatur)
(→teletypewriter keyboard)		
telex line	TELEC	**Telexleitung**
= telex connection; switched telegraph line; switched telegraph connection		= Telexverbindung; Fernschreibwählleitung; Fernschreibwählverbindung
↓ switched line		↑ Wählleitung
telex machine (→telex teleprinter)	TERM&PER	(→Telexfernschreiber)
telex message	TELEC	**Fernschreiben**
= telex 2		[ein telegrafisch übermitteltes Schreiben]
		= Fernschreibnachricht; Telex 2
telex network	TELEC	**Fernschreibwählnetz**
= teleprinter network		= Telexnetz; Fernschreibnetz
↑ switched network		↑ Wählnetz
telex number	TELEGR	**Telexnummer**
↑ subscriber number [TELEC]		= Fernschreibnummer
		↑ Teilnehmerrufnummer [TELEC]
telex office	TELEC	**Fernschreibvermittlungsamt**
= teleprinter exchange; telex exchange; telegraph office		= Fernschreibamt; Telexamt; Telexvermittlung 2; Fernschreibvermittlung 2; Telexvermittlungsamt; Telegrafenamt; Fernschreibwählvermittlung; Fernschreib-Wählvermittlungsamt
telex page printer	TERM&PER	**Telex-Blattschreiber**
telex service (→telex 1)	TELEC	(→Telex)
telex station (→teleprinter station)	TELEGR	(→Fernschreibstelle)
telex subscriber	TELEC	**Fernschreibteilnehmer**
		= Telexteilnehmer
telex subscriber line	TELEC	**Telexanschluß**
		= Fernschreibanschluß
telex teleprinter	TERM&PER	**Telexfernschreiber**
= telex machine		= Telexmaschine
↑ teletypewriter		↑ Fernschreiber
telex traffic	TELEC	**Fernschreibverkehr**
= teleprinter traffic; telex communication; telegraphic traffic		= Telexverkehr
telgraph battery	POWER SYS	**Telegrafenbatterie**
= line battery		= Telegraphenbatterie
teller (→dividend 2)	MATH	(→Dividend 2)
teller terminal	TERM&PER	**Bankenterminal**
tellurium	CHEM	**Tellur**
= Te		= Te
temex	TELEC	**Temex**
[service of German PTT for telecontrol]		[Dienst der DBP für Fernwirken]
TEMEX service	TELEC	**TEMEX-Dienst**
TEM mode (→transverse electromagnetic wave)	LINE TH	(→transversale elektromagnetische Welle)
TE mode (→transverse electric wave)	LINE TH	(→transversale elektrische Welle)
temper	METAL	**anlassen**
temperature	PHYS	**Temperatur**
↓ thermodynamic temperature; Celsius temperature; Fahrenheit temperature		↓ thermodynamische Temperatur; Celsius-Temperatur; Fahrenheit-Temperatur
temperature alarm	SIGN.ENG	**Temperaturalarm**
temperature alarm system	SIGN.ENG	**Temperaturmelder**
temperature ascent (→temperature rise)	PHYS	(→Temperaturanstieg)
temperature coefficient	PHYS	**Temperaturkoeffizient**
= TC		= TK; Temperaturbeiwert
temperature coefficient of resistance	PHYS	**Temperaturkoeffizient des Widerstandes**
= resistance coefficient		= Temperaturbeiwert des Widerstandes; Widerstandskoeffizient; Widerstandsbeiwert
↑ temperature coefficient		↑ Temperaturkoeffizient
temperature-compensated	ELECTRON	**temperaturkompensiert**
		= thermisch ausgeglichen
temperature compensating lead	INSTR	**Thermoausgleichsleitung**
= compensating lead		= Ausgleichleitung
temperature compensation	PHYS	**Temperaturausgleich**
temperature compensation	ELECTRON	**Temperaturkompensation**
= temperature correction		= Temperaturausgleich; Temperaturkorrektur
temperature constancy	PHYS	**Temperaturkonstanz**
temperature controller	CONTROL	**Temperaturregler**
		= Temperaturwächter
temperature converter	COMPON	**Temperaturwandler**
= temperature transducer		= Temperaturmeßumformer; Temperaturmeßwandler
temperature correction	ELECTRON	(→Temperaturkompensation)
(→temperature compensation)		
temperature cycle	QUAL	**Temperaturzyklus**
		= Temperaturschleife
temperature cycling	COMPON	**Temperaturwelligkeit**
[crystal thermostate]		[Quarzthermostat]
temperature dependence	PHYS	**Temperaturabhängigkeit**
= temperature drift; temperature sensitivity		= Temperaturempfindlichkeit 2
≈ thermal response [TECH]; thermal sensitivity; themperature drift		≈ Temperaturgang [TECH]; Wärmeempfindlichkeit; Temperaturwanderung
temperature-dependent	PHYS	**temperaturabhängig**
≈ temperature-sensitive		≈ temperaturempfindlich

temperature detector		COMPON	(→Temperaturfühler)	TEM wave (→transverse electromagnetic wave)		LINE TH	(→transversale elektromagnetische Welle)

temperature detector COMPON (→Temperaturfühler)
  (→temperature sensor)
temperature drift PHYS **Temperaturwanderung**
  ≈ temperature dependence   = Temperaturdrift
    ≈ Temperaturabhängigkeit
temperature drift (→temperature PHYS (→Temperaturabhängigkeit)
  dependence)
temperature drop PHYS **Temperaturabfall**
  = temperature fall
temperature fall (→temperature PHYS (→Temperaturabfall)
  drop)
temperature fluctuation TECH **Temperaturschwankung**
  ≈ temperature variation   ≈ Temperaturänderung
temperature inversion METEOR **Temperaturinversion**
  = inversion   = Inversion
temperature limit PHYS **Temperaturgrenze**
temperature measurement INSTR **Temperaturmessung**
temperature on-off controller CONTROL **Temperatur-Zweipunktregler**
temperature probe INSTR **Temperaturfühler**
temperature punch-through MICROEL **Temperaturdurchgriff**
  [Transistorkenngröße]
temperature radiation PHYS **Temperaturstrahlung**
temperature range PHYS **Temperaturbereich**
temperature response (→thermal TECH (→Temperaturgang)
  response)
temperature rise PHYS **Temperaturanstieg**
  = temperature ascent
temperature scale INSTR **Temperaturskala**
temperature-sensible PHYS (→temperaturempfindlich)
  (→temperature-sensitive)
temperature-sensitive PHYS **temperaturempfindlich**
  = thermally sensitive; temperature-sensible; thermally sensible; thermosensible   = thermosensibel; wärmeempfindlich
  ≈ temperature-variable   ≈ temperaturabhängig
temperature-sensitive paper TERM&PER **Thermopapier**
  = heat-sensitive paper   = wärmeempfindliches Papier
temperature sensitivity PHYS (→Temperaturabhängigkeit)
  (→temperature dependence)
temperature sensor COMPON **Temperaturfühler**
  = temperature detector   = Temperatursensor; Temperaturmeßfühler
temperature setting TECH **Temperatureinstellung**
temperature shock TECH **Temperaturschock**
temperature stability PHYS **Temperaturstabilität**
temperature stabilization PHYS **Temperaturstabilisierung**
temperature-stabilized ELECTRON **temperaturstabilisiert**
temperature transducer COMPON (→Temperaturwandler)
  (→temperature converter)
temperature variation TECH **Temperaturänderung**
  ≈ temperature fluctuation   ≈ Temperaturschwankung
tempering METAL **Aushärtung**
tempering tarnish METAL **Anlauffarbe**
tempest equipment DATA PROC **Tempest-Gerät**
  [conforming special security specifications to U.S. NSA, especially relating to EMC]   [besonderen US-Sicherheitsbestimmungen v.a. hinsichtlich Abstrahlung genügendes Gerät]
template TECH **Schablone**
  = former; stencil   = Lehre; Maske
temporal adverb LING **Temporaladverb**
temporal course TECH **zeitlicher Verlauf**
  = course 2   = Verlauf; Lauf
temporal function MATH **Zeitfunktion**
temporary TECH **vorübergehend**
  = transitory; nonpermanent; momentary   = zeitweilig; momentan 2; transitorisch; temporär
  ≈ instantaneous   ≈ sofortig
temporary file DATA PROC **temporäre Datei**
temporary importation ECON **vorübergehende Einfuhr**
  = Temporärimport; temporärer Import
temporary memory (→buffer DATA PROC (→Pufferspeicher)
  store)
temporary solution (→expedient) TECH (→Notbehelf)
temporary sound program BROADC **Tonzubringerleitung**
  circuit
temporary staff ECON **Aushilfspersonal**
temporary storage DATA PROC (→Zwischenspeicherung)
  (→intermediate storage 1)

TEM wave (→transverse LINE TH (→transversale elektromagnetische Welle)
  electromagnetic wave)
tenant ECON **Mieter**
  = lessee   ≠ Vermieter
  ≠ letter
tend COLLOQ **tendieren**
  = neigen
tendency COLLOQ **Tendenz**
  = trend   = Trend; Verlaufsrichtung
tendentious COLLOQ **tendenziös**
tender 1 (→offer) ECON (→Angebot)
tender 2 (→invitation to ECON (→Ausschreibung)
  tender)
tender 3 (→payment means) ECON (→Zahlungsmittel)
tender documentation ECON **Ausschreibungsunterlagen**
  ≈ specification   ≈ Spezifikation
tenderer (→bidder) ECON (→Anbieter)
tender offer (→offer) ECON (→Angebot)
tender opening (→opening of ECON (→Angebotseröffnung)
  tender)
ten-key pad (→numeric TERM&PER (→Zifferntastenblock)
  keypad)
ten-layer TECH **zehnlagig**
  = ten-part   = zehnschichtig
ten-part (→ten-layer) TECH (→zehnlagig)
ten-party line with selective TELEC **Zehneranschluß**
  ringing   ↑ Gemeinschaftsanschluß
  ↑ party line
ten quadrillions (AM) MATH **Trillion**
  [ten to the power of sixteen]   [zehn hoch sechzehn]
  = ten thousand billions (BRI)
ten's complement DATA PROC (→Zehnerkomplement)
  (→complement to ten)
tense (n.) LING **Tempusform**
  [time-expression inflection form of a verb]   = Tempus; Zeitform; Zeitstufe
  ↓ present tense; past tense; future tense   ↓ Gegenwart; Vergangenheit; Zukunft
tense (adj.) MECH **gespannt 1**
tensile load MECH **Zugbelastung**
tensile strength 1 MECH **Zugfestigkeit**
  = longitudinal strength
tensile strength 2 MECH **Zugspannung**
tensile test MECH **Zerreißprobe**
tension (→mechanical stress) MECH (→Spannung)
tension (→voltage) EL.TECH (→elektrische Spannung)
tension arm TECH **Spannbügel**
  = tie strap   = Spannhebel
tension roller (→idler MECH (→Spannrolle)
  pulley)
tension spring MECH **Spannfeder**
  = Zugfeder
tension test MECH **Zugprüfung**
  = pulling test
tensor MATH **Tensor**
tensor algebra MATH **Tensoralgebra**
tensor field PHYS **Tensorfeld**
tentative (→improvised) COLLOQ (→provisorisch)
ten thousand billions (BRI) (→ten MATH (→Trillion)
  quadrillions)
tenuis (pl.-es) LING **Tenuis (n.f.; pl.-es)**
  [unvoiced explosive]   [stimmloser Verschlußlaut, z.B. p,t,k]
  ↑ explosive   = Tenues
  ↑ Explosivum
TEOS process MICROEL **TEOS-Verfahren**
  = tetrahexilene-oxisilane process
terabit DATA PROC **Terabit**
  [10E12 bit]   [10E12 Bit]
terabyte DATA PROC **Terabyte**
  [2E40 = 1,006,511,627,776 bytes]   [2E40 = 1.009.511.627.776 Bytes]
  = tb
teracycles (→terahertz) PHYS (→Terahertz)
terahertz PHYS **Terahertz**
  [10E12 hertz]   [10E12 Hertz]
  = THZ; teracycles   = THz
terameter PHYS **Terameter**
  [10E12 meters]   [10E12 Meter]
teraohm EL.TECH **Teraohm**
  [10E12 ohm]   [10E12 Ohm]

teraohmmeter	INSTR	**Teraohmmeter**
= TO meter		↑ Ohmmeter
↑ ohmmeter		
terbium	CHEM	**Terbium**
= Tb		= Tb
terephtalate	CHEM	**Terephthalat**
term	ECON	**Laufzeit**
		≈ Termin
term	MATH	**Term**
[algebra]		[Algebra]
= expression		= Glied; Ausdruck
↓ numeric term		↓ Zahlenterm; Variablenterm
term	ENG.LOG	**Term**
[propositional calculus]		[Aussagenlogik]
↓ minterm; maxterm		↓ Vollkonjunktion; Volldisjunktion
term	COLLOQ	**Frist**
[established duration]		[festgesetzte Zeitdauer]
= period of time		≈ Termin
≈ appointed time		
term (→ technical term)	LING	(→Fachausdruck)
term (→ due time)	ECON	(→Fälligkeit)
termbank (→terminological data bank)	DATA PROC	(→Terminologie-Datenbank)
terminal (n.)	SWITCH	**Anschlußeinheit**
		= AE
terminal (n.)	OUTS.PLANT	**Endverschluß**
= pothead		
terminal	COMPON	**Anschlußklemme**
= binding post; grabber		= Klemme; Anschlußpunkt
terminal	CIRC.ENG	**Pol**
terminal (→data terminal)	TERM&PER	(→Datenendgerät)
terminal (→terminal equipment)	TELEC	(→Endgerät)
terminal (→terminal station)	TRANS	(→Endstelle)
terminal (→station)	DATA COMM	(→Station)
terminal adapter	DATA COMM	**Terminaladapter**
terminal adaptor	TELEC	**Endgeräteanpassung**
[ISDN]		[ISDN]
= TA		
terminal base (→base)	ANT	(→Fußpunkt)
terminal base capacity	ANT	**Fußpunktkapazität**
terminal block	OUTS.PLANT	(→Kabelverzweiger)
(→distributing box)		
terminal block (→terminal strip)	EL.INST	(→Klemmleiste)
terminal computer	DATA COMM	**Datenstationsrechner**
= front-end processor		= Stationsrechner
terminal connection	SWITCH	**Anschluß**
= connection (NAM); connexion (BRI)		
terminal device	OUTS.PLANT	**Endeinrichtung**
terminal distributor	OUTS.PLANT	**Endverteiler**
terminal echo suppressor	TELEPH	**Endechosperre**
terminal emulation	DATA PROC	**Terminalemulation**
terminal end (→terminal station)	TRANS	(→Endstelle)
terminal equipment	TELEC	**Endgerät**
= terminal		= Terminal
≈ user terminal		≈ Benutzerstation
terminal equipment (→line terminal equipment)	TRANS	(→Leitungsendgerät)
terminal error	DATA PROC	**Abbruchfehler**
terminal exchange	SWITCH	**Endamt**
= end exchange		
terminal extension card	TERM&PER	**Terminalerweiterungskarten**
terminal field	EQUIP.ENG	**Anschlußfeld**
= connector panel		
terminal for handicapped persons	TERM&PER	**Behindertenterminal**
terminal keyboard	TERM&PER	**Terminaltastatur**
terminal mobility	MOB.COMM	**Endgerätemobilität**
terminal number (→subscriber number)	TELEC	(→Teilnehmerrufnummer)
terminal pad	MICROEL	**Anschlußfleck**
= pad 1		
terminal pad (→soldering land)	ELECTRON	(→Lötauge)
terminal pin	COMPON	**Anschlußstift**
≈ terminal point; lead		= Pin; Sockelstift; Beinchen
↑ pin		≈ Stützpunkt; Anschlußdraht
		↑ Stift
terminal point	COMPON	**Stützpunkt**
= connecting point		≈ Anschlußstift
≈ terminal pin		
terminal pole	OUTS.PLANT	**Abspannmast**
= stayed pole; rigid stay		[Freileitung]
terminal regenerator	TRANS	**Endregenerator**
terminal repeater	TRANS	**Endverstärker**
≈ terminal regenerator		≈ Endregenerator
↑ line repeater		≠ Zwischenverstärker
		↑ Leitungsverstärker
terminal repeater box	TELEGR	**Teilnehmer-Anschlußkasten**
terminal repeater circuit	TELEGR	**Fernteilnehmeranschluß-Schaltung**
terminal resistance	EL.TECH	(→Abschlußwiderstand)
(→terminating resistor)		
terminal screen	TERM&PER	**Terminalbildschirm**
terminal-side echo path	TELEPH	**Endechoweg**
terminal statement	DATA PROC	**letzte Anweisung**
terminal station	DATA COMM	**Datenstation**
[terminal of a data communication system; consists of data terminal equipment and data communications equipment]		[Endstelle einer Datenübertragung, bestehend aus Datenendeinrichtung und Datenübertragungseinrichtung]
= data station; data terminal; communication terminal		= Datenendstelle; Endstation; Terminal 2
↑ station		↑ Station
↓ interactive terminal; batch terminal		↓ Dialogstation; Stapelstation
terminal station	TRANS	**Endstelle**
= terminal; terminal end		
terminal station demodulator	RADIO REL	**Endstellendemodulator**
terminal station modem	RADIO REL	**Endstellenmodem**
terminal station modulator	RADIO REL	**Endstellenmodulator**
terminal strip	EL.INST	**Klemmleiste**
= terminal block; connection block		= Lüsterklemme; Anschlußklemmleiste; Klemmenleiste; Klemmenstreifen
terminal strip	COMPON	**Stützpunktleiste**
terminal strip (→distribution strip)	COMPON	(→Verteilerleiste)
terminal swivel arm (→swivel arm for terminal)	TERM&PER	(→Terminal-Schwenkarm)
terminal voltage	EL.TECH	**Klemmenspannung**
terminate	NETW.TH	**abschließen**
[a circuit with a load]		[eine Schaltung mit einer Last]
terminate	TECH	**beenden**
= close		≈ verlassen
≈ quit		
terminate	DATA PROC	**beenden**
= quit		= beendigen; verlassen
terminated by matched load	EL.TECH	(→angepaßt)
(→matched)		
terminated folded antenna	ANT	**TFD-Antenne**
= TFD antenna		
terminated line	LINE TH	**abgeschlossene Leitung**
↓ matched line		↓ angepaßte Leitung
terminating exchange	SWITCH	**Zielvermittlung**
= destination exchange		= Zielvermittlungsstelle; Zielamt; Bestimmungsvermittlung; Bestimmungsvermittlungsstelle; Bestimmungsamt
≈ destination point		
≠ originating exchange		
		≈ Zielpunkt
		≠ Ursprungsvermittlung
terminating point (→destination point)	SWITCH	(→Zielpunkt)
terminating register	SWITCH	**Endverkehrsregister**
terminating resistor	EL.TECH	**Abschlußwiderstand**
= terminal resistance; termination		≈ Lastwiderstand
≈ load resistance		
terminating set (→termination hybrid)	TELEC	(→Gabelschaltung)
terminating station (→called suscriber)	SWITCH	(→gerufener Teilnehmer)
terminating traffic	SWITCH	**Endverkehr**
		≈ kommender Verkehr

termination [of a circuit with a load]	NETW.TH	Abschluß [eines Schaltkreises mit einer Last]
termination	INSTR	Widerstandsabschluß
termination (→terminating resistor)	EL.TECH	(→Abschlußwiderstand)
termination (→treatment)	TECH	(→Behandlung)
termination (→abort)	DATA PROC	(→Abbruch)
termination amplifier (→hybrid amplifier)	TRANS	(→Gabelverstärker)
termination hybrid = hybrid circuit; hybrid; terminating set; four-to-two-wire	TELEC	Gabelschaltung [Übergang von Zweidraht auf Vierdraht] = Gabel; Endschaltung ≈ Gabelübertrager
terminator	INSTR	Abschlußwiderstand
terminator (→final character)	DATA COMM	(→Schlußzeichen)
terminological data bank = termbank	DATA PROC	Terminologie-Datenbank
terminology [special terms of a branch of science, engineering or activity] = nomenclature ≈ technical language	LING	Terminologie [Fachwörter eines Fachgebiets] = Nomenklatur ≈ Fachsprache
termi-point connection	COMPON	Termi-point-Verdrahtung
terms = conditions	ECON	Konditionen
ternary	CODING	ternär
ternary code	CODING	Ternärcode = ternärer Code
ternary logic	ENG.LOG	Ternärlogik = dreiwertige Logik
ternary number system [number system with three digits] ↑ denominational number system	MATH	ternäres Zahlensystem [Zahlendarstellung mit drei Ziffern] ↑ Stellenwertsystem
ternary signal	CODING	ternäres Signal
terpentine	CHEM	Terpentin
terrain	TECH	Gelände
terrain conditions	TECH	Bodenverhältnisse
terrain factor	RADIO PROP	Geländefaktor
terrain irregularity = path irregularity	RADIO PROP	Geländeunregelmäßigkeit
terrain profile (→path profile)	RADIO PROP	(→Geländeschnitt)
terrain roughness	RADIO PROP	Geländerauhigkeit
terrain rugosity	RADIO PROP	Bodenwelligkeit
terrestial cable ≠ underwater cable	COMM.CABLE	Landkabel ≠ Unterwasserkabel
terrestrial	TELEC	erdgebunden = terrestrisch
terrestrial electric field	PHYS	elektrisches Erdfeld
terrestrial field	PHYS	Erdfeld
terrestrial magnetic field	PHYS	erdmagnetisches Feld = magnetisches Erdfeld
terrestrial radio relay link	RADIO	terrestrische Richtfunkverbindung
teriary group (→mastergroup 2)	TRANS	(→Tertiärgruppe)
tertiary memory (→tertiary storage)	DATA PROC	(→Tertiärspeicher 1)
tertiary school ↓ university; academy	SCIE	Hochschule ↑ Lehranstalt ↓ Universität; Fachhochschule; Akademie
tertiary storage [external memory without on-line access] = tertiary memory	DATA PROC	Tertiärspeicher [externer, online nicht zugreifbarer Speicher]
Tesla [unit for magnetic flux density; = 1 V s/m²] = T	PHYS	Tesla [SI-Einheit für magnetische Flußdichte; = 1 V s/m²] = T
Tesla transformer	EL.TECH	Tesla-Transformator
test (n.) = measurement	INSTR	Messung = Prüfung
test (v.t.) (→prooftest)	QUAL	(→erproben)
test [INSTR] (→mearument)	PHYS	(→Messung)
test (→program test)	DATA PROC	(→Programmtest)
test (→check)	TECH	(→prüfen)
test (→check)	TECH	(→Prüfung)
test (n.) (→inspection)	QUAL	(→Prüfung)
test (→trial)	TECH	(→Versuch)
testability	ELECTRON	Prüffreundlichkeit = Prüfbarkeit
testability rule	MICROEL	Prüfbarkeitsregel
test access = test connection	SWITCH	Prüfanschalter
test access point (→test point)	ELECTRON	(→Meßpunkt)
test adapter (→measuring adapter)	INSTR	(→Meßvorsatz)
test aid = testing aid ≈ measuring aid; test gage	INSTR	Prüfhilfe = Testhilfe; Prüfhilfsmittel; Prüfhilfseinrichtung ≈ Meßhilfe; Prüflehre
test aid = debugging aid; diagnostic aid ≈ diagnostic program	DATA PROC	Testhilfe = Prüfhilfe; Meßhilfe; Fehlersuchhilfe; Diagnosehilfe; Diagnosemittel ≈ Diagnoseprogramm
test arrangement = measurement setup; test setup; measurement device	INSTR	Meßaufbau = Meßanordnung
test atmosphere	QUAL	Prüfklima
test automat (→automatic tester)	INSTR	(→Prüfautomat)
test bay	MANUF	Referenzgestell = Prüfgestell
test bench = test desk	QUAL	Prüfplatz = Prüftisch
test bench (→test department)	MANUF	(→Prüffeld)
test bit (→check bit)	CODING	(→Prüfbit)
test bit pattern	DATA COMM	Simulationsbitmuster = Stimuli; Prüfbitmuster
test block	DATA COMM	Prüfblock
test board = test module; function test module	EQUIP.ENG	Testbaugruppe = Prüfbaugruppe
test body	PHYS	Prüfkörper
test bus = scan path	MICROEL	Prüfbus = Testbus
test by substitution	QUAL	Substitutionsprüfung
test byte	DATA COMM	Prüfbyte
test card [punched card] ≈ diagnostic card	TERM&PER	Prüfkarte [Lochkarte]
test case ≈ test instrument	INSTR	Meßkoffer ≈ Meßgerät
test cassette	EL.ACOUS	Test-Bandkassette
test cell [ATM]	TELEC	Testzelle [ATM]
test cell detection [ATM]	TELEC	Testzellenerkennung [ATM]
test certificate ≈ test protokoll	QUAL	Prüfurkunde = Prüfzertifikat ≈ Prüfprotokoll
test channel	TELEC	Testkanal = Prüfkanal
test chart (→test pattern)	TV	(→Testbild)
test chart generator (→test pattern generator)	TV	(→Bildmustergenerator)
test circuit ↑ circuit	SWITCH	Prüfsatz ↑ Satz
test circuit	CIRC.ENG	Prüfschaltung = Testschaltung
test clip ↓ crocodil clip	ELECTRON	Prüfklemme = Testclip ↓ Abgreifklemme
test code = test sequence	DATA COMM	Prüftext
test code (→test sequence)	DATA COMM	(→Prüffolge)
test condition	TECH	Prüfbedingung
test conductor	SWITCH	Prüfader
test connection = test link	TELEC	Prüfverbindung = Prüfschaltung
test connection (→test access)	SWITCH	(→Prüfanschalter)
test coverage	MICROEL	Fehlererkennungsgrad

test data	DATA PROC	**Testdaten**		**test instrument**	INSTR	**Prüfgerät**
		= Prüfdaten		= test set; tester		= Testgerät
test data generator	DATA PROC	**Testdatengenerator**		≈ test equipment; testing device; measuring instrument		≈ Prüfapparatur; Prüfvorrichtung; Meßgerät
		= Prüfdatengenerator				
test data set	DATA PROC	**Testdatensatz**		**test jack**	EQUIP.ENG	**Meßbuchse**
		= Prüfdatensatz		**test jack**	TELEPH	**Prüfklinke**
test deck	TERM&PER	**Prüfkartensatz**		**test key**	EQUIP.ENG	**Prüfschalter**
test department	MANUF	**Prüffeld**		**test lead** (→measuring line)	INSTR	(→Meßleitung)
= production test area; production test floor; test bench; bench				**test lead holder**	INSTR	**Meßleitungshalter**
				test lead wire (→measuring line)	INSTR	(→Meßleitung)
test desk	SWITCH	**Prüfschrank**		**test level**	TRANS	**Meßpegel**
≈ test position		≈ Prüfplatz		= through level		
test desk (→test bench)	QUAL	(→Prüfplatz)		**test line**	TV	**Prüfzeile**
test distribution	MATH	**Testverteilung**		= vertical interval		
[distribution for statistical tests]		[Verteilung für statistische Tests]		**test line** (→measuring line)	INSTR	(→Meßleitung)
↓ chi-square function; Student's-t distribution		= Prüfverteilung		**test line group**	DATA COMM	**Prüfbündel**
		↓ Chi-Quadrat-Verteilung; Student-Verteilung		**test line insertion equipment**	TV	**Prüfzeilen-Einblendgerät**
				test link (→test connection)	TELEC	(→Prüfverbindung)
test document	QUAL	**Prüfunterlage**		**test load**	QUAL	**Prüflast**
≈ test protocol		= Testunterlage; Prüfbeleg; Prüfdokument		**test log** (→test protocol)	QUAL	(→Prüfprotokoll)
		≈ Prüfprotokoll		**test loop**	TELEC	**Prüfschleife**
test driver	DATA PROC	**Prüftreiber**		= loopback; diagnostic loop		= Testschleife
test electrode (→measuring electrode)	INSTR	(→Meßelektrode)		≈ measuring loop [ELECTRON]		≈ Meßschleife [ELECTRON]
				testmobile	INSTR	**Meßwagen**
test environment	DATA PROC	**Testumgebung**				= Testwagen
		= Prüfumgebung; Testumfeld; Prüfumfeld		**test mode**	MICROEL	**Testmodus**
						= Prüfmodus
test equipment module	EQUIP.ENG	**Prüfeinrichtungsbaugruppe**		**test mode** (→test operation)	TECH	(→Probebetrieb)
tester	QUAL	**Prüfer**		**test model** (→experimental prototype)	TECH	(→Versuchsmuster)
tester (→test instrument)	INSTR	(→Prüfgerät)		**test module** (→test board)	EQUIP.ENG	(→Testbaugruppe)
test fixture (→measuring adapter)	INSTR	(→Meßvorsatz)		**test multiple**	SWITCH	**Prüfvielfach**
				test number	SWITCH	**Testnummer**
test form (→test plan)	DATA PROC	(→Testvordruck)				= Prüfnummer
test gage	INSTR	**Prüflehre**		**test object** (→test specimen)	QUAL	(→Prüfling)
≈ test aid		≈ Prüfhilfe		**test operation**	TECH	**Probebetrieb**
test generator (→signal generator 1)	INSTR	(→Universal-Meßsender)		= test mode; experimental operation; experimental mode		= Testbetrieb
test head (→test probe)	INSTR	(→Prüfspitze)		**test output** (→test port)	ELECTRON	(→Meßausgang)
test hole [puncher]	TERM&PER	**Prüflochung** [Stanzgerät]		**test panel**	EQUIP.ENG	**Testfeld**
				test path	SWITCH	**Prüfweg**
testimonial	ECON	**Dienstzeugnis**		**test pattern**	TV	**Testbild**
= certificate 2; testimonio letter		= Arbeitszeugnis; Zeugnis		= test chart; resolution pattern; resolution chart		= Bildmuster
testimonio letter (→testimonial)	ECON	(→Dienstzeugnis)		**test pattern** (→test sequence)	DATA COMM	(→Prüffolge)
testing (→program test)	DATA PROC	(→Programmtest)		**test pattern generator**	TV	**Bildmustergenerator**
testing aid (→test aid)	INSTR	(→Prüfhilfe)		= pattern generator; test chart generator; resolution pattern generator		= Streifengenerator; Testbildgenerator; Testbildgeber
testing appliance (→testing device)	MANUF	(→Prüfaufnahme)		**test pattern generator** (→pattern generator)	INSTR	(→Bitmustergenerator)
testing center	TECH	**Prüfanstalt**		**test pattern source** (→pattern generator)	INSTR	(→Bitmustergenerator)
		= Prüfinstitution; Prüfinstitut; Prüfzentrum		**test perforated tape**	TERM&PER	(→Prüflochstreifen)
testing crew	SYS.INST	**Meßtrupp**		(→diagnostic test tape)		
testing device	MANUF	**Prüfaufnahme**		**test period**	DATA PROC	**Testzeit**
= testing appliance				**test pin**	CIRC.ENG	**Meßstift**
testing device	INSTR	**Prüfvorrichtung**		**test plan**	DATA PROC	**Testvordruck**
= measuring device		≈ Prüfgerät; Prüfeinrichtung; Meßvorrichtung		= test form		= Testformular
≈ test equipment; test tools; measuring device				**test plan** (→test schedule)	QUAL	(→Prüfplan)
				test point	ELECTRON	**Meßpunkt**
testing method (→test program)	QUAL	(→Prüfprogramm)		= test access point; checkpoint; measuring point; measuring access point		= Prüfpunkt; Testpunkt; Meßstelle; Prüfstelle; Teststelle
testing purpose	TECH	**Prüfzweck**		≈ monitoring point; test port		
		= Testzweck; Meßzweck				≈ Überwachungspunkt; Meßausgang
testing stand	TECH	**Prüfstand**				
testing van	OUTS.PLANT	**Kabelmeßwagen**		**test point selector**	EQUIP.ENG	**Meßstellenwahlschalter**
test inhibit	SWITCH	**Prüfsperre**		= checkpoint selector		
test inhibit	DATA COMM	**Testsperre**		**test port**	ELECTRON	**Meßausgang**
		= Prüfsperre		= test output		= Meßanschluß; Testanschluß; Testausgang; Prüfanschluß; Prüfausgang
test inhibit acknowledgment	DATA COMM	**Testsperrquittung**		≈ test point		
= test inhibit confirmation						
test inhibit confirmation	DATA COMM	(→Testsperrquittung)				≈ Meßpunkt
(→test inhibit acknowledgment)				**test port isolation**	INSTR	**Testanschlußentkopplung**
test inhibit signal	DATA COMM	**Testsperrsignal**		**test port reciprocity**	INSTR	**Testanschluß-Reziprozität**
		= Prüfsperrsignal		**test position**	SWITCH	**Prüfplatz**
test inhibit signal comparision	DATA COMM	**Testsperrsignalvergleich**		≈ test desk		≈ Prüfschrank
		= Prüfsperrsignalvergleich		**test position** (→measuring set)	INSTR	(→Meßplatz)
test instruction	TECH	**Prüfanweisung**		**test print**	TERM&PER	**Probedruck**
		= Prüferläuterung		= proof copy		= Andruck
test instruction (→test statement)	DATA PROC	(→Testanweisung)				

test probe		INSTR	**Prüfspitze**	**test specimen**	QUAL	**Prüfling**
= measuring probe; scanner probe; probe tip; probe; test head; measuring head; scanner head; sensor; sensing head; mount; pod; sensing device; sensing element; pick-up			= Meßkopf; Meßfühler; Meßsonde; Meßsensor; Sensor; Meßwertgeber; Meßdetektor; Meßtaster; Tastkopf; Taster; Tastspitze	= test object; specimen; device under test; DUT		= Prüfstück; Prüfobjekt
				≈ unit under measurement [INSTR]		≈ Meßobjekt [INSTR]
				test statement	DATA PROC	**Testanweisung**
≈ attenuator probe; logic tester			↓ Testteiler; Logiktester	= test instruction		= Testbefehl
test procedure (→test program)		QUAL	(→Prüfprogramm)	**test structure**	MICROEL	(→Teststruktur)
test program		QUAL	**Prüfprogramm**	(→process-control monitor)		
= test procedure; test sequence; testing method			= Testprogramm; Prüfablauf; Testablauf; Prüfroutine; Testroutine; Prüfverfahren; Testverfahren	**test system**	DATA PROC	**Testsystem**
				= Testrahmen		
				test tape (→diagnostic test tape)	TERM&PER	(→Prüflochstreifen)
≈ measurement procedure [INSTR]			≈ Meßverfahren [INSTR]	**test terminal block**	EQUIP.ENG	**Prüfleiste**
				test tolerance	QUAL	**Meßtoleranz**
test program		DATA PROC	**Testprogramm**	≈ measuring inaccuracy		= Prüftoleranz
			= Prüfprogramm 1			≈ Meßunsicherheit
			↓ Wartungsprogramm			
test protocol		QUAL	**Prüfprotokoll**	**test tone**	TELEPH	**Meßton**
= test log			= Testprotokoll	**test tone**	EL.ACOUS	**Prüfton**
			≈ Prüfbeleg			= Testton
test range		ANT	**Antennenmeßfeld**	**test tools**	INSTR	**Prüfeinrichtung**
= pattern range			= Antennenmeßstrecke	≈ testing device; measuring set		= Prüfvorrichtung; Meßplatz
test receiver		INSTR	**Meßempfänger**	**test trunk group**	SWITCH	**Prüfbündel**
= measuring receiver				**test-value processing**	INSTR	(→Meßwertverarbeitung)
test record		EL.ACOUS	**Testplatte**	(→measured-value processing)		
			= Prüfplatte	**test value recording**	INSTR	**Meßwerterfassung**
test record (→measurements record)		INSTR	(→Meßprotokoll)	= test results recording		
				test-values memory	INSTR	**Meßwertspeicher**
test relay		CIRC.ENG	**Prüfrelais**	= test results memory		
test report		TECH	**Prüfbericht**	**test-value transmission**	TELECONTR	**Meßwertübertragung**
≈ test protocol			= Testbericht	= test result transmission		
			≈ Prüfprotokoll	**test vector**	MICROEL	**Testvektor**
test request		DATA COMM	**Prüfauftrag**	**test voltage**	ELECTRON	**Prüfspannung**
			= Testanforderung	≈ measuring voltage [INSTR]		≈ Meßspannung [INSTR]
test requirement (→measurement requirement)		TELEC	(→Meßauftrag)	**tetrachloride process**	MICROEL	**Tetrachloridprozeß**
				tetrad	CODING	**Tetrade**
test resources		TECH	**Prüfmittel**	[group of 4 bit positions, normally to represent decimal digits]		[Gruppe von 4 Binärstellen, i.a. zur Darstellung von Dezimalziffern]
test result		DATA PROC	**Testergebnis**			
test result		QUAL	**Prüfergebnis**	= half-byte; nibble; nybble		= Halbyte; Nibble
			= Testergebnis	↓ pseudotetrade		↓ Pseudotetrade
test result		INSTR	**Meßwert**			
= measurement result; measuring result; measured value			= Prüfwert; Meßergebnis; Prüfergebnis; gemessener Wert	**tetradic code**	CODING	**Tetradencode**
				tetragonal (→quadrilateral)	MATH	(→viereckig)
				tetrahedral	MATH	**vierflächig**
test result processing		INSTR	(→Meßwertverarbeitung)	**tetrahedron**	MATH	**Tetraeder**
(→measured-value processing)				[a pyramyd of three side]		[dreiseitige Pyramide]
test results memory (→test-values memory)		INSTR	(→Meßwertspeicher)	↑ regular polyhedron		= Vierflächner
						↑ reguläres Polyeder
test results recording (→test value recording)		INSTR	(→Meßwerterfassung)	**tetrahexilene-oxisilane process** (→TEOS process)	MICROEL	(→TEOS-Verfahren)
				tetrajunction transistor	MICROEL	**Tetrajunction-Transistor**
test result transmission (→test-value transmission)		TELECONTR	(→Meßwertübertragung)	**tetraode** (→double-grid tube)	ELECTRON	(→Doppelgitterröhre)
				tetrode	ELECTRON	**Tetrode**
test reversal		DATA COMM	**Prüfwechsel**	[tube with four electrodes]		= Vierpolröhre
test room		QUAL	**Prüfraum**	↑ sreen grid tube		↑ Schirmgitterröhre
≈ measuring chamber [INSTR]			= Testraum	↓ screen-grid tube		
			≈ Meßraum [INSTR]	**TE wave** (→transverse electric wave)	LINE TH	(→transversale elektrische Welle)
test run		DATA PROC	**Testlauf**			
			= Prüflauf	**text**	LING	**Text**
test schedule		QUAL	**Prüfplan**	≈ wording		= Wortlaut
= test plan			= Testplan	**text alignment**	DATA PROC	**Textausrichtung**
test scheme		DATA PROC	**Testschema**	**text analysis**	DATA PROC	**Textanalyse**
test sequence		DATA COMM	**Prüffolge**	**text analysis program**	DATA PROC	**Textanalyseprogramm**
= test pattern; retest signal; test code			= Prüfmuster; Testfolge; Testmuster; Prüftext	**text and data network**	DATA COMM	**Text- und Datennetz**
				text and fax server	TELEC	**Text- und Fax-Server**
test sequence (→test program)		QUAL	(→Prüfprogramm)	= TFS		
test sequence (→test code)		DATA COMM	(→Prüftext)	**textbook**	SCIE	**Lehrbuch**
test series		PHYS	**Versuchsserie**	≈ tutorial; schoolbook; monograph		≈ Schulbuch; Fachbuch
test series		MANUF	**Vorserie**	**text character**	DATA COMM	**Textzeichen**
= pilot series; preproduction series			= Pilotserie; Versuchsserie	**text column**	TYPOGR	**Textspalte**
test set (→test instrument)		INSTR	(→Prüfgerät)	**text communication**	TELEC	**Textkommunikation**
test setup (→test arrangement)		INSTR	(→Meßaufbau)	↑ telecommunications		= Textübermittlung
				↓ telegraphy; videotext		↑ Telekommunikation
test signal		ELECTRON	**Testsignal**			↓ Fernschreiben; Videotext
= measuring signal			= Prüfsignal; Meßsignal	**text cursor**	DATA PROC	**Textcursor**
test signal generator		INSTR	**Meßsender 1**	**text editing**	DATA PROC	**Textbearbeitung**
test signal generator		TELEC	**Prüfsender**	[to redact, input, modify or format a text]		[Redigieren, Erfassen, Korrigieren, Formatieren eines Textes]
test socket		MICROEL	**Testfassung**			
[for IC's]			[zur Prüfung von IC's]			
test specification		QUAL	**Prüfvorschrift**	**text editor**	DATA PROC	**Texteditor**
			= Prüfbedingung			

textfax	TELEC	**Textfax**	
[service of German PTT, combining teletex and telefax]		[Dienst der DBP, Verbindung von Teletex und Telefax]	
text file	DATA PROC	**Textdatei**	
= document			
text formatter	DATA PROC	**Textformatierer**	
text formatting	DATA COMM	**Textformatierung**	
textile ribbon	TERM&PER	**Gewebefarbband**	
text mail	TELEC	**Mitteilungssystem für Texte**	
text manager	DATA PROC	**Textverwalter**	
text network	DATA COMM	**Textnetz**	
text processing	DATA PROC	**Textverarbeitung**	
[manipulation of texts by computer]		[Rationalisierung von Texterstellung mit Textbausteinen]	
= word processing; text system			
text processing cluster	DATA COMM	**Textverarbeitungs-Cluster**	
text processing system	DATA PROC	(→Textsystem)	
(→word processing system)			
text recognition (→character recognition)	TERM&PER	(→Klarschriftlesen)	
text station (→text terminal equipment)	TERM&PER	(→Textendgerät)	
text system (→text processing)	DATA PROC	(→Textverarbeitung)	
text terminal equipment	TERM&PER	**Textendgerät**	
= text station		= Textstation	
texture	SCIE	**Textur**	
≈ structure		≈ Struktur	
texture-conversion effect	OPTOEL	**Texturumwandlungseffekt**	
text window	DATA PROC	**Textfenster**	
TFD antenna (→terminated folded antenna)	ANT	(→TFD-Antenne)	
TFFET (→thin-film field effect transistor)	MICROEL	(→Dünnschicht-Feldeffekttransistor)	
T flipflop (→toggle flipflop)	CIRC.ENG	(→T-Flipflop)	
TFS (→text and fax server)	TELEC	(→Text- und Fax-Server)	
tg (→tangent 2)	MATH	(→Tangens)	
T1 group (→DS1 group)	TELEC	(→DS1-Bündel)	
T2 group (→DS2 group)	TELEC	(→DS2-Bündel)	
T3 group (→DS3 group)	TELEC	(→DS3-Bündel)	
th (→hyperbolic tangent)	MATH	(→Tangens hyperbolicus)	
Th (→thorium)	CHEM	(→Thorium)	
thallium	CHEM	**Thallium**	
= Tl		= Tl	
THD (→harmonic content)	EL.ACOUS	(→Oberschwingungsgehalt)	
THD (→total harmonic distortion)	ELECTRON	(→harmonische Gesamtverzerrung)	
T-head bolt	MECH	**Hammerkopfschraube**	
		= Hammerschraube; Hakenbolzen	
theft protection	SIGN.ENG	**Diebstahlsicherung**	
thematic area	SCIE	**Themenkreis**	
themodynamics	PHYS	**Thermodynamik**	
		= Wärmelehre	
"the number is engaged at the moment"	TELEPH	„der Apparat ist zur Zeit belegt"	
		= „der Teilnehmer ist gerade besetzt"	
theorem	SCIE	**Theorem**	
		= Lehrsatz	
theorema egregium	MATH	**Theorema egregium**	
theorem of coding	INF	**Codierungstheorem**	
theoretical physics	PHYS	**theoretische Physik**	
theory	SCIE	**Theorie**	
		= Lehre	
theory of colors (AM)	PHYS	**Farbenlehre**	
= theory of colours (BRI)			
theory of colours (BRI) (→theory of colors)	PHYS	(→Farbenlehre)	
theory of fields (→field theory)	PHYS	(→Feldtheorie)	
theory of games	MATH	**Spieltheorie**	
= games theory			
theory of invariants	MATH	**Invariantentheorie**	
theory of numbers	MATH	**Zahlentheorie**	
theory of signs (→semiotics)	LING	(→Semiologie)	
the process of memorization (→storage 1)	DATA PROC	(→Speicherung)	
thermal	PHYS	**thermisch**	
= thermic			
thermal agitation (→thermal motion)	PHYS	(→Wärmebewegung)	
thermal balance (→heat balance)	PHYS	(→Wärmebilanz)	
thermal beam (→thermal ray)	PHYS	(→Wärmestrahl)	
thermal binding equipment	OFFICE	**Thermobindegerät**	
thermal breakdown	MICROEL	**thermischer Durchbruch**	
= thermal instability; thermal runaway		= thermische Instabilität; thermischer Selbstmord	
thermal capacity (→heat capacity)	PHYS	(→Wärmekapazität)	
thermal comb	TERM&PER	**Thermokamm**	
[thermal printer]		[Thermodrucker]	
thermal compound (→heat sink compound)	ELECTRON	(→Wärmeleitpaste)	
thermal conductance	PHYS	**Wärmeleitwert**	
		= thermischer Leitwert	
thermal conducting (→heat conducting)	PHYS	(→wärmeleitend)	
thermal conduction	PHYS	**Wärmeleitung**	
= heat conduction; heat transport		= thermische Leitung; Wärmetransport; Wärmeübertragung	
≈ heat transfer		≈ Wärmeableitung	
thermal conductivity (→heat conductivity)	PHYS	(→Wärmeleitfähigkeit)	
thermal conductor	PHYS	**Wärmeleiter**	
= heat conductor		= thermischer Leiter	
thermal contact	COMPON	**Thermokontakt**	
		= thermischer Kontakt; Wärmekontakt	
thermal content	PHYS	**Wärmeinhalt**	
= heat content			
thermal convection (→heat convection)	PHYS	(→Wärmekonvektion)	
thermal converter	INSTR	**Thermoumformer**	
= thermoconverter		≈ Thermokreuz [PHYS]	
≈ thermal cross [PHYS]			
thermal cross	PHYS	**Thermokreuz**	
≈ thermal converter [INSTR]		≈ Thermoumformer [INSTR]	
thermal cutout	COMPON	**Thermoauslöser**	
		= thermischer Auslöser	
thermal derating factor	QUAL	**Reduzierkoeffizient**	
thermal diffusion	PHYS	**Thermodiffusion**	
		= Wärmediffusion	
thermal drift	TECH	**thermisches Weglaufen**	
= thermal runaway			
thermal effect	PHYS	**thermischer Effekt**	
		= Wärmewirkung	
thermal energy (→thermic energy)	PHYS	(→thermische Energie)	
thermal equivalent (→equivalent of heath)	PHYS	(→Wärmeäquivalent)	
thermal equivalent circuit	MICROEL	**Wärmeersatzschaltung**	
thermal exchange constant	PHYS	**Wärmeübergangszahl**	
thermal expansion	PHYS	**Wärmeausdehnung**	
		= Wärmeexpansion	
thermal expansion coefficient	PHYS	**Wärmeausdehnungskoeffizient**	
thermal flux	PHYS	**Wärmefluß**	
= flux of heat; heat flow; heat flux		= thermischer Fluß; Wärmestrom	
thermal image camera	INSTR	**Wärmebildkamera**	
thermal instability (→thermal breakdown)	MICROEL	(→thermischer Durchbruch)	
thermal insulating	PHYS	**wärmeisolierend**	
= thermally insulated		= wärmeundurchlässig	
thermal insulation	PHYS	**Wärmeisolierung**	
= heat insulation			
thermal leakage coefficient	MICROEL	**Ableitungskoeffizient**	
[power to overtemperature of a NTC thermistor]		[Leistung zu Übertemperatur eines Heißleiters]	
≠ thermal leakage constant		≠ Ableitungskonstante	
thermal leakage constant	MICROEL	**Ableitungskonstante**	
≠ thermal leakage coefficient		≠ Ableitungskoeffizient	
thermally curable	METAL	**wärmehärtbar**	
thermally insulated (→thermal insulating)	PHYS	(→wärmeisolierend)	
thermally sensible (→temperature-sensitive)	PHYS	(→temperaturempfindlich)	

thermally sensitive (→temperature-sensitive)	PHYS	(→temperaturempfindlich)
thermally sensitive resistor (→thermistor)	COMPON	(→Thermistor)
thermal measuring instrument	INSTR	thermisches Meßinstrument
thermal motion = thermal agitation	PHYS	Wärmebewegung
thermal noise = thermoionic noise; electronic noise ≈ basic noise; white noise	TELEC	thermisches Rauschen = Wärmerauschen; Widerstandsrauschen; elektronisches Rauschen ≈ Grundgeräusch; weißes Rauschen
thermal plotter	TERM&PER	Thermoplotter
thermal power	PHYS	thermische Leistung = Wärmeleistung
thermal power meter	INSTR	thermischer Leistungsmesser
thermal power station	TECH	Wärmekraftwerk = kalorisches Kraftwerk (OES)
thermal printer ↑ non impact printer	TERM&PER	Thermodrucker = thermischer Drucker ↑ anschlagfreier Drucker
thermal printing	TERM&PER	Thermodruck = thermisches Druckverfahren
thermal radiation = heat radiation	PHYS	Wärmestrahlung
thermal rating	QUAL	thermische Belastbarkeit
thermal ray = thermal beam ↑ ray	PHYS	Wärmestrahl ↑ Strahl
thermal recording [facsimile]	TERM&PER	thermografische Aufzeichnung [Faksimile]
thermal relay = thermoelectric relay; electrothermal relay ↑ time-delay relay	COMPON	Thermorelais ↑ Verzögerungsrelais
thermal removal (→heat transfer)	PHYS	(→Wärmeableitung)
thermal resistance	PHYS	Wärmewiderstand = thermischer Widerstand
thermal response = temperature response ≈ temperature dependence [PHYS]	TECH	Temperaturgang = Temperaturverhalten ≈ Temperaturabhängigkeit [PHYS]
thermal runaway (→thermal breakdown)	MICROEL	(→thermischer Durchbruch)
thermal runaway (→thermal drift)	TECH	(→thermisches Weglaufen)
thermal sensitivity ≈ temperature dependence	PHYS	Wärmeempfindlichkeit = Temperaturempfindlichkeit 1 ≈ Temperaturabhängigkeit
thermal shock	QUAL	Thermoschock = Wärmestoß; thermischer Schock
thermal shock resistance	QUAL	Thermoschockfestigkeit
thermal source (→heat source)	PHYS	(→Wärmequelle)
thermal stability = heat resistance; heat resistivity	TECH	Wärmebeständigkeit = Hitzebeständigkeit; Wärmestabilität; thermische Stabilität
thermal tension (→thermoelectric force)	PHYS	(→Thermospannung)
thermal transfer printer (→thermo-transfer printer)	TERM&PER	(→Thermotransferdrucker)
thermal voltage [atomic physics, voltage equivalent of thermal energy]	PHYS	Temperaturspannung [Atomphysik, einer thermischen Energie äquivalente Spannung] = Boltzmann-Spannung
thermic (→thermal)	PHYS	(→thermisch)
thermic distribution		thermische Besetzung
thermic emission (→thermoionic emission)	PHYS	(→Glühemission)
thermic energy = thermal energy	PHYS	thermische Energie = Wärmeenergie
thermic pointer	INSTR	Heizzeiger
thermic simulation	TECH	Thermosimulation
thermistor = thermally sensitive resistor ↑ adjustable resistance ↓ NTC thermistor; PTC thermistor	COMPON	Thermistor = temperaturabhängiger Widerstand ↑ veränderbarer Widerstand ↓ Heißleiter; Kaltleiter
thermoammeter (→thermocouple instrument)	INSTR	(→Thermoumformer-Instrument)
thermocell (→thermocouple)	PHYS	(→Thermoelement)
thermocompression	METAL	Thermokompression
thermocompression bonding	MICROEL	Thermokompressionsverfahren = Thermokompressionsschweißen; Wärmedruckverfahren
thermocompression welding	METAL	Thermokompressionsschweißen
thermocontractible = thermocontractile; heat-shrinking	TECH	wärmeschrumpfend
thermocontractile (→thermocontractible)	TECH	(→wärmeschrumpfend)
thermoconverter (→thermal converter)	INSTR	(→Thermoumformer)
thermocouple = thermocell; thermoelectric couple; thermoelement	PHYS	Thermoelement = Thermopaar
thermocouple instrument = thermoammeter	INSTR	Thermoumformer-Instrument = Thermostrommesser
thermodynamic temperature [SI unit: kelvin] = absolute temperature	PHYS	thermodynamische Temperatur [SI-Einheit: Kelvin] = absolute Temperatur; Kelvin-Temperatur
thermoelastic	TECH	thermoelastisch
thermoelectric	PHYS	thermoelektrisch
thermoelectric cooler = TEC	TECH	thermoelektrischer Kühler
thermoelectric cooling	TECH	thermoelektrische Kühlung
thermoelectric couple (→thermocouple)	PHYS	(→Thermoelement)
thermoelectric current	PHYS	Thermostrom = thermoelektrischer Strom
thermoelectric effect ↓ Seebeck effect; Peltier effect; Benedicks effect	PHYS	thermoelektrischer Effekt = glühelektrischer Effekt; Richardson-Effekt ↓ Seebeck-Effekt; Peltier-Effekt; Benedicks-Effekt
thermoelectric force = thermal tension; thermoelectric potential	PHYS	Thermospannung = integrale Thermokraft; Thermokraft; thermoelektrische Spannung; Thermo-EMK; thermische Spannung ↓ differentiale Thermokraft
thermoelectricity	PHYS	Thermoelektrizität
thermoelectric pickup	INSTR	thermoelektrischer Meßfühler
thermoelectric pile = thermopile	PHYS	Thermosäule
thermoelectric potential (→thermoelectric force)	PHYS	(→Thermospannung)
thermoelectric relay (→thermal relay)	COMPON	(→Thermorelais)
thermoelectric series	PHYS	thermoelektrische Spannungsreihe
thermoelement (→thermocouple)	PHYS	(→Thermoelement)
thermofuse = thermo fuse	COMPON	Thermosicherung
thermo fuse (→thermofuse)	COMPON	(→Thermosicherung)
thermographic	TECH	thermographisch = thermografisch
thermoionic	PHYS	thermoionisch
thermoionic cathode = hot cathode; incandescent cathode	ELECTRON	Glühkathode = thermische Kathode
thermoionic electron emission	PHYS	thermische Elektronenemission
thermoionic emission = thermic emission	PHYS	Glühemission = thermische Emission
thermoionic noise (→thermal noise)	TELEC	(→thermisches Rauschen)

thermoionic tube (→electron tube)	ELECTRON	(→Elektronenröhre)	
thermoionic valve (BRI) (→electron tube)	ELECTRON	(→Elektronenröhre)	
thermoluminescence [induced by temperature rise] ↑ luminescence	PHYS	**Thermolumineszenz** [durch Temperaturerhöhung induziert] ↑ Lumineszenz	
thermomagnetic	PHYS	**thermomagnetisch**	
thermometer	INSTR	**Thermometer**	
thermooptical effect	PHYS	**thermooptischer Effekt**	
thermophone	EL.ACOUS	**Thermophon**	
thermopile (→thermoelectric pile)	PHYS	(→Thermosäule)	
thermoplast = thermoplastic ↑ plastic	CHEM	**Thermoplast** = wärmeempfindlicher Kunststoff ↑ Kunststoff	
thermoplastic (adj.)	CHEM	**thermoplastisch**	
thermoplastic (→thermoplast)	CHEM	(→Thermoplast)	
thermosensible (→temperature-sensitive)	PHYS	(→temperaturempfindlich)	
thermosetting	CHEM	**wärmeaushärtend**	
thermosphere	GEOPHYS	**Thermosphäre**	
thermostat ↑ temperature controller [CONTROL]	TECH	**Thermostat** ↑ Temperaturregler [CONTROL]	
thermostatic switch (→thermoswitch)	COMPON	(→Thermoschalter)	
thermoswitch = thermostatic switch	COMPON	**Thermoschalter**	
thermo-transfer printer [melts-off ink from a ribbon] = thermal transfer printer	TERM&PER	**Thermotransferdrucker** [schmilzt Druckfarbe von einem Farbband ab]	
thesaurus 1 [a lexicon with entries linkable by concepts]	DATA PROC	**Thesaurus 1** [über Deskriptoren verknüpfbares Lexikon]	
thesaurus 2 [file of synonyms]	DATA PROC	**Thesaurus 2** [Datei von Synonymen]	
thesaurus function	DATA PROC	**Thesaurusfunktion**	
thesaurus program	DATA PROC	**Thesaurusprogramm**	
thesis = dissertation	SCIE	**Doktorarbeit** = Dissertation	
Thévenins's theorem ↑ Helmholtz equivalent-source theorem	NETW.TH	**Theorem von Thévenin** = Théveninsches Theorem; Satz von den Ersatzspannungsquellen ↑ Helmholtzscher Satz	
THF (→tremendously high frequency)	EL.TECH	(→Submillimeterwelle)	
thick (→viscous)	TECH	(→dickflüssig)	
thick film ≠ thin film	MICROEL	**Dickfilm** = Dickschicht ≠ Dünnfilm	
thick film capacitor	MICROEL	**Dickschichtkondensator** = Dickfilmkondensator	
thick film circuit	MICROEL	**Dickfilmschaltung** = Dickschichtschaltung	
thick film hybrid circuit	MICROEL	**Dickschicht-Hybridschaltung** = Dickfilm-Hybridschaltung	
thick-film hybrid technology ↑ film technology	MICROEL	**Dickfilm-Hybridtechnik** = Dickschicht-Hybridtechnik ↑ Filmtechnik	
thick film resistor	MICROEL	**Dickfilmwiderstand** = Dickschichtwiderstand	
thick film technology ↑ film technology	MICROEL	**Dickfilmtechnik** = Dickschichttechnik ↑ Filmtechnik	
thickly liquid (→viscous)	TECH	(→dickflüssig)	
thickness	PHYS	**Dicke** = Dickte	
thickness shear	PHYS	**Dickenscherung**	
thickness shear crystal	COMPON	**Dickenscherungsschwinger** [Quarz] = Dickenschwinger	
thickness shear mode = thickness shear vibration	PHYS	**Dickenscherungsschwingung** = Dickenschermode	
thickness shear vibration (→thickness shear mode)	PHYS	(→Dickenscherungsschwingung)	
thick-wall	TECH	**dickwandig**	
thimble = cable socket; cable lug; cable grip	COMPON	**Kabelschuh** ↓ Quetschkabelschuh	
thimble [grooved ring to protect a rope loop] = eye	TECH	**Kausch** (n.f.) [rinnenförmiger Ring zur Verstärkung einer Seilschlaufe] = Kausche	
thimble = type basket	OFFICE	**Typenkorb**	
thimble printer = spinwriter	TERM&PER	**Typenkorb-Drucker** = Thimble-Drucker	
thin (v.t.) ≠ to make thin a solid body	TECH	**verdünnen 2** [einen Festkörper dünner machen]	
thin	TYPOGR	**fein**	
thin film	MICROEL	**Dünnfilm** = Dünnschicht	
thin-film capacitor	MICROEL	**Dünnfilmkondensator** = Dünnschichtkondensator	
thin-film electroluminescence	MICROEL	**Dünnfilm-Elektrolumineszenz** = Dünnschicht-Elektrolumineszenz	
thin-film field effect transistor = TFFET	MICROEL	**Dünnschicht-Feldeffekttransistor** = TFFET	
thin-film hybrid circuit	MICROEL	**Dünnfilm-Hybridschaltung**	
thin-film hybrid technology ↑ film technology	MICROEL	**Dünnfilm-Hybridtechnik** = Dünnschicht-Hybridtechnik ↑ Filmtechnik	
thin film memory	MICROEL	**Dünnfilmspeicher** = Dünnschichtspeicher; Magnetfilmspeicher; Filmspeicher	
thin film resistor	MICROEL	**Dünnfilmwiderstand** = Dünnschichtwiderstand	
thin film technique (→thin film technology)	MICROEL	(→Dünnfilmtechnik)	
thin film technology = thin film technique ↑ film technology	MICROEL	**Dünnfilmtechnik** = Dünnschichttechnik ↑ Filmtechnik	
thin-layer	PHYS	**Blättchen**	
thinly liquid	TECH	**dünnflüssig**	
thinned antenna array	ANT	**verdünnte Gruppenantenne**	
thinnest at the center (→diverging lens)	PHYS	(→Zerstreuungslinse)	
thinning [of a solid body]	TECH	**Verdünnung 2** [eines Festkörpers]	
thin-wall = thin-walled	TECH	**dünnwandig**	
thin-walled (→thin-wall)	TECH	(→dünnwandig)	
thin-window display	DATA PROC	**Schmalfensteranzeige**	
third (n.)	ACOUS	**Terz**	
third-angle projection	ENG.DRAW	**amerikanische Projektion**	
third-generation computer [IC's; approx. 1965 to 1972]	DATA PROC	**Rechner der dritten Generation** [ICs; ca. 1965 bis 1972] = Computer der dritten Generation	
third-order intercept = TOI	TELEC	**Intercept-Punkt dritter Ordnung**	
third party	ECON	**Dritte** (pl.t.)	
third-party charging	TELEPH	**Belastung an Drittperson**	
third party liability	ECON	**Haftpflicht**	
third-party product (→purchased part)	ECON	(→Fremdfabrikat)	
third root (→cube root)	MATH	(→Kubikwurzel)	
third shell = M shell	PHYS	**3.Schale** = M-Schale	
third wire (→neutral conductor)	POWER ENG	(→Nulleiter)	
thixotropic [liquefying by vibration]	PHYS	**thixotrop** [bei Vibration sich verflüssigend]	
Thomson bridge = Kelvin-Bridge	INSTR	**Thomson-Meßbrücke** = Thomson-Doppelbrücke; Thomson-Brücke; Kelvin-Brücke; Doppelbrücke	
Thomson's rule	NETW.TH	**Thomsonsche Schwingungsgleichung**	

thoriated		CHEM	thoriert	three input adder (→full adder)		CIRC.ENG	(→Volladdierer)
thorium		CHEM	Thorium	three-layer		TECH	dreilagig
= Th			= Th; Thor	≈ three-part			= dreischichtig
thousand mega (→G)		PHYS	(→G)				≈ dreiteilig
thousand trillions (BRI)		MATH	(→Trillarde)	three-level modulation		MODUL	Drei-Pegel-Modulation
(→sextillon)				three-part		TECH	dreiteilig
thrashing		DATA PROC	Zeitverschwendung	≈ three-layer			≈ dreilagig
= churning				↑ multisectional			↑ mehrteilig
thread (v.t.)		MECH	gewindeschneiden	three-party conversation		TELEPH	(→Dreierkonferenz)
thread 2 (n.)		MECH	Gang	(→tripartite conference)			
			[Gewinde]	three-path system		EL.ACOUS	Dreiwegebox
thread		TECH	Faden	= three-way box			= Dreiwegsystem
[a strand of twisted filaments]			[sehr dünn, aus mehreren	three-phase 1 (adj.)		POWER ENG	dreiphasig
= yarn			Fasern gebildet]	= triple-phase			
≈ cord; fiber			≈ Schnur; Zwirn; Faser	three-phase 2 (adj.)		POWER ENG	dreisträngig
↓ string			↓ Bindfaden	[winding]			[Windung]
thread 2		DATA PROC	Teilprozeß				≈ dreiphasig
[a routine running independently]			[eine unabhängig ablaufende Routine]	three-phase alternator (→three-phase generator)		POWER ENG	(→Drehstromgenerator)
≈ threading; multithreading			= Thread 2	three-phase bridge		POWER ENG	Drehstrombrückenschaltung
thread 3		DATA PROC	Thread 3				= Dreiphasenbrückenschaltung
[a task in OS/2]			[im System OS/2 ein Auftrag]	three-phase current		POWER ENG	Drehstrom
thread 1 (n.) (→screw thread)		MECH	(→Gewinde)	= rotary current			= Dreiphasenstrom
thread 1 (→threaded code)		DATA PROC	(→gereihter Code)	↑ alternating current; polyphase current			↑ Wechselstrom; Mehrphasenstrom
thread class		MECH	Gewindeklasse	three-phase current integrator		INSTR	Drehstromzähler
threaded bolt (→stud bolt)		MECH	(→Gewindebolzen)				= Dreiphasenzähler
threaded bushing		MECH	Gewindebuchse	three-phase diffusion		MICROEL	Dreifachdiffusion
threaded code		DATA PROC	gereihter Code	three-phase generator		POWER ENG	Drehstromgenerator
[program composed by independent subroutines]			[aus unabhängig voneinander ablaufenden Unterprogrammen bestehendes Programm]	= three-phase alternator			= Dreiphasengenerator
				three-phase line		POWER ENG	Drehstromleitung
= thread 1							= Dreiphasenleitung
threaded data base		DATA PROC	verkettete Datenbank	three-phase mains (→three-phase network)		POWER ENG	(→Drehstromnetz)
threaded file (→chained file)		DATA PROC	(→verkettete Datei)	three-phase motor		POWER ENG	Drehstrommotor
threaded hole (→tapped hole)		MECH	(→Gewindeloch)				= Dreiphasenmotor
threaded list (→chained list)		DATA PROC	(→verkettete Liste)	three-phase network		POWER ENG	Drehstromnetz
				= three-phase mains			= Dreiphasennetz
threaded plate		MECH	Gewindeplatte	three-phase rectifier		POWER ENG	Drehstromgleichrichter
threaded spindle		MECH	Gewindespindel				= Dreiphasengleichrichter
thread electrometer (→filament electrometer)		INSTR	(→Fadenelektrometer)	three-phase switch		EL.INST	Drehstromschalter
							= Dreiphasenschalter
threading		MECH	Gewindeschneiden	three-phase synchronous alternator		POWER ENG	Drehstromsynchrongenerator
threading (→thread 2)		DATA PROC	(→Teilprozeß)				= Dreiphasensynchrongenerator
threading die		MECH	Gewindeschneider	three-phase transformer		POWER ENG	Drehstromtransformator
thread representation		ENG.DRAW	Gewindedarstellung				= Dreiphasentransformator
thread ring		MECH	Gewindering	three-place		MATH	dreistellig
thread series		MECH	Gewindereihe	↑ of many places			↑ mehrstellig
three-address instruction		DATA PROC	Drei-Adreß-Befehl	three-plus-one instruction		DATA PROC	Drei-plus-Eins-Befehl
three-chip equipment		EQUIP.ENG	Drei-Chip-Gerät	↑ four-address instruction			↑ Vieradreßbefehl
threechromatic		PHYS	dreifarbig	three-point circuit		EL.TECH	(→Dreipunktschaltung)
= three-colored				(→three-point connection)			
three-circuit ...		NETW.TH	Dreikreis-	three-point connection		EL.TECH	Dreipunktschaltung
= three-section ...				= three-point circuit			
three-colored		PHYS	(→dreifarbig)	three-pole		EL.TECH	dreipolig
(→threechromatic)				= triple-pole			
three-color print		TYPOGR	Dreifarbendruck	three-pole network		NETW.TH	Dreipol
= three-colour print			↑ Mehrfarbendruck	three-port circulator		MICROW	Dreiarmzirkulator
↑ multicolor print				three-pulse mid-point circuit		POWER SYS	Dreipuls-Mittelpunkt-Schaltung
three-colour print (→three-color print)		TYPOGR	(→Dreifarbendruck)				
				three-range (→three-step)		TECH	(→dreistufig)
three-conductor plug		TELEPH	dreipoliger Klinkenstecker	three-section ... (→three-circuit ...)		NETW.TH	(→Dreikreis-)
			= dreipoliger Stöpsel				
				three-section filter		NETW.TH	Dreikreisbandfilter
three-core (→three-wire)		EL.TECH	(→dreiadrig)	three-sigma limit		MATH	Drei-Sigma-Grenze
three-dB coupler		MICROW	Drei-dB-Koppler	[statistics]			[Statistik]
= 3-dB coupler; hybrid coupler			= 3-dB-Koppler	three-square file		MECH	Dreikantfeile
three-dimensional		MATH	dreidimensional	= triangular file			
three-dimensional sound		ACOUS	Raumklang	three-step		TECH	dreistufig
three-dimensional storage		DATA PROC	Drei-D-Speicher	= triple; three-range			≈ dreifach
three-electrode tube (→triode vacuum tube)		ELECTRON	(→Triode)	↑ multi-step			↑ mehrstufig
				three-step controller		CONTROL	Dreipunktregler
three-fingers rule		EL.TECH	Dreifingerregel	three-step oscillator		CIRC.ENG	Dreipunktoszillator
= right-hand rule; right-hand screw rule			= Rechtehandregel; Korkenzieherregel	three-valued (→trivalent)		MATH	(→dreiwertig)
				three-way box (→three-path system)		EL.ACOUS	(→Dreiwegebox)
threefold		COLLOQ	dreifach				
= triple				three-way power splitter		MICROW	Drei-Wege-Leistungsteiler
three fourth lambda dipole		ANT	Dreiviertel-Lambda-Dipol				

three-way reflex (→3-way reflex)	EL.ACOUS	(→3-Wege-Baßreflex)
three-way switch	COMPON	Dreiwegschalter
three-wire (adj.)	EL.TECH	dreiadrig
= three-core		= dreidrähtig
three-wire machine	POWER ENG	Dreileitermaschine
three-wire plug (→Schuko plug)	EL.INST	(→Schukostecker)
three-wire socket (→Schuko socket)	EL.INST	(→Schukosteckdose)
three-wire system	POWER ENG	Dreileitersystem
thre-phase A.C. generating set	POWER ENG	Drehstromaggregat = Dreiphasenaggregat
threshold	ELECTRON	Schwelle
threshold	CODING	Schwelle
threshold (→threshold value)	TECH	(→Schwelle)
threshold current	ELECTRON	Schwellenstrom = Schwellwertstrom
threshold element	COMPON	Schwellwertelement
threshold extension demodulator	SAT.COMM	Schwellenerweiterungsdemodulator
threshold gate	CIRC.ENG	Schwellwertgatter
threshold laser current	OPTOEL	Laser-Schwellstrom
threshold logic	CIRC.ENG	Schwellwertlogik
threshold of hearing (→audibility threshold)	ACOUS	(→Hörbarkeitsschwelle)
threshold potential	PHYS	Schwellenpotential
threshold switch	COMPON	Schwellwertschalter
threshold value	ECON	Eckwert
threshold value	TECH	Schwelle
= threshold; barrier		= Schwellenwert; Schwellwert
threshold voltage [thyristor]	MICROEL	Schleusenspannung [Thyristor]
threshold voltage	ELECTRON	Schwellenspannung = Schwellwertspannung
threshold voltage detector	CIRC.ENG	Schwellwertspannungs-Detektor
throttle (v.t.)	TECH	drosseln
throttling	TECH	Drosselung
[reduction of preesure by obstruction]		[Druckminderung durch Hindernis]
through (n.)	TECH	Wanne
[long and shallow receptacle]		[länglicher und flacher Behälter]
through adapter	INSTR	Durchgangsadapter
through connect	TELEC	durchschalten
= connect		= eine Verbindung aufbauen
≈ set up a connection [SWITCH]		[SWITCH]
through connecting filter	TRANS	Durchschaltefilter
through-connecting signal	SWITCH	Durchschaltekennzeichen
through connection	TRANS	Durchschaltung
through connection (→through switching)	SWITCH	(→Durchschaltung)
through-connection aknowledge	SWITCH	Durchschaltungsquittung
through connect loss	SWITCH	Durchschalteverlust
through-connect point	SWITCH	Durchschaltepunkt
through-contacted [PCB]	ELECTRON	durchkontaktiert [Leiterplatte]
= plated-through; through-plated		= durchmetallisiert
through-drilling (n.)	MECH	Durchbohrung
through group filter [FDM]	TRANS	Primärgruppen-Durchschaltefilter [TF-Technik] = PGDFi
through-hole plated board (→through-hole plated PCB)	ELECTRON	(→durchkontaktierte Leiterplatte)
through-hole plated PCB	ELECTRON	durchkontaktierte Leiterplatte
= through-hole plated board; plated-through PCB; plated-through board		= durchmetallisierte Leiterplatte
through level (→test level)	TRANS	(→Meßpegel)
through mastergroup filter	TRANS	Tertiärgruppendurchschaltefilter = TGDFi
through-plated (→through-contacted)	ELECTRON	(→durchkontaktiert)
through-plated hole [PCB]	COMPON	durchkontaktierte Lochung [Leiterplatte]
through-plating (→feedthrough)	ELECTRON	(→Durchkontaktierung)
throughput	TECH	Durchsatz
= thruput		↑ Leistung
↑ performance		
throughput	DATA PROC	Durchsatz
[maximum number of characters per second which can be transferred into or from the main memory]		[maximale Anzahl von Zeichen pro Sekunde die in den Arbeitsspeicher geladen, oder aus ihm gelesen werden kann]
= data throughput; thruput; data thruput; transfer rate		= Arbeitsspeicherdurchsatzrate; Durchsatzrate; Datenrate; Datendurchsatz
≈ transmission rate [TELEC]		≈ Übertragungsgeschwindigkeit [TELEC]
throughput class	DATA COMM	Durchsatzklasse
throughput power meter (→directional power meter)	INSTR	(→Durchgangsleistungsmesser)
throughput time (→turnaround time)	DATA COMM	(→Durchlaufzeit)
through repeater	TRANS	Durchgangsregenerator
through supergroup filter	TRANS	Sekundärgruppen-Durchschaltefilter = SGDFi
through supermaster group filter	TRANS	Quartärgruppen-Durchschaltefilter = QGDFi
through switching	SWITCH	Durchschaltung
= call throughput; through connection		
through switching multiplexer	SWITCH	Durchschaltemultiplexer
through traffic (→transit traffic)	SWITCH	(→Durchgangsverkehr)
throw (n.) (→feed)	TERM&PER	(→Vorschub)
throw (→switch position)	COMPON	(→Schalterstellung)
throw-away item	ECON	Wegwurfposition
throw-out (→lead-out spiral)	CONS.EL	(→Auslaufrille)
thruput (→throughput)	TECH	(→Durchsatz)
thruput (→throughput)	DATA PROC	(→Durchsatz)
thrust	MECH	Axialdruck
thrust bearing	MECH	Axiallager = Drucklager
thrust collar	MECH	Druckring
thrust washer	MECH	Druckscheibe
thulium	CHEM	Thulium
= Tm		= Tm
thumb [mark on scroll bar]	DATA PROC	relative Positionsanzeige [in der Bildlauflinie]
thumb-nail [layout reduction for control purposes]	TYPOGR	Layout-Struktur [Verkleinerung für Kontrollzwecke]
thumbscrew (→pressing screw)	MECH	(→Druckschraube)
thumb tack	TECH	Reißzwecke
= drawing pin		= Heftzwecke; Reißnagel
↑ tack		↑ Zwecke
thumbwheel [device to position a cursor]	TERM&PER	Rändelscheibe [Vorrichtung zur Schreibmarkensteuerung] = Rändelrad
thumbwheel (→knurled disk)	MECH	(→Rändelscheibe)
thumbwheel coding switch (→coded rotary switch)	COMPON	(→Dreh-Codierschalter)
thumbwheel switch	COMPON	Fingerradschalter
≈ decade switch		≈ Dekadenschalter
thump (→telegraph noise)	TELEPH	(→Telegrafiegeräusch)
thunder	METEOR	Donner
≈ thunderclap		≈ Donnerschlag
thunderbolt (→atmospheric discharge)	METEOR	(→atmosphärische Entladung)
thunderclap	METEOR	Donnerschlag
≈ thunder		≈ Donner
thundercloud	METEOR	Gewitterwolke
thundergust (→thunderstorm)	METEOR	(→Gewitter)
thunderstone (→atmospheric discharge)	METEOR	(→atmosphärische Entladung)
thunderstorm	METEOR	Gewitter
≈ thundergust		≈ Gewittersturm
≈ storm; thunderstroke		≈ Sturm; Unwetter; atmosphärische Entladung
thunderstorm day	METEOR	Gewittertag

thunderstrike	METEOR	Blitzschlag	
≈ lightning		≈ Blitz	
thunderstroke (→atmospheric discharge)	METEOR	(→atmosphärische Entladung)	
thundreous	METEOR	gewittrig	
		= gewitterig	
thyrator	MICROEL	Thyrator	
thyratron	ELECTRON	Thyratron	
↓ ion tube; rectifier tube		↑ Ionenröhre; Gleichrichterröhre	
thyristor	MICROEL	Thyristor	
["thyratron-like transistor"]		= steuerbarer Siliziumgleichrichter	
= silicon controlled rectifier; SCR			
↓ triac		↓ Triac	
thyristor amplifier	POWER SYS	Thyristorverstärker	
thyristor control	POWER SYS	Thyristorsteuerung	
thyristor controlled	POWER SYS	thyristorgesteuert	
thyristor-controlled convertor	POWER SYS	(→thyristorgesteuerter Gleichrichter)	
(→thyristor-controlled rectifier)			
thyristor-controlled rectifier	POWER SYS	thyristorgesteuerter Gleichrichter	
= thyristor-controlled convertor; thyristor rectifier		= Thyristorgleichrichter	
thyristor latch-up	MICROEL	Thyristordurchschaltung	
= thyristor latch-up effect		= Thyristor Latch-up	
thyristor rectifier	POWER SYS	(→thyristorgesteuerter Gleichrichter)	
(→thyristor-controlled rectifier)			
thyristor regulator	POWER SYS	Thyristorregler	
thyristor switch	POWER SYS	Thyristorschalter	
thyristor tetrode	MICROEL	Thyristortetrode	
		= beiseitig steuerbarer Thyristor	
thyristor triode	MICROEL	Thyristortriode	
THZ (→terahertz)	PHYS	(→Terahertz)	
Ti (→titanium)	CHEM	(→Titan)	
tick (v.t.)	ECON	ankreuzen	
[to mark a field in a form]		[ein Kästchen in einem Formular]	
tickler (→reaction coil)	CIRC.ENG	(→Rückkopplungsspule)	
tick mark	INSTR	Skalenmarkierung	
tie breaker (→arbiter)	DATA PROC	(→Zuteiler)	
tie circuit (→fixed line)	TELEC	(→Standleitung)	
tie-cut ratio	TERM&PER	Steg-Schlitz-Verhältnis	
tied	ECON	zweckgebunden	
= dedicated; commited		= gebunden	
tie line (PABX) (→direct connection)	TELEC	(→Querverbindung)	
tie line (→fixed line)	TELEC	(→Standleitung)	
tie-line traffic	SWITCH	Verbindungsverkehr	
tie off	TECH	abschnüren	
tie-on label	OFFICE	Anhängeetikette	
tie point	COMPON	Sammelkontaktstelle	
tier (n.)	TECH	Rang	
[one of rows arranged above others]		[eine Reihe aus einer Staffelung]	
tie strap (→tension arm)	TECH	(→Spannbügel)	
tie trunk [DATA COMM]	TELEC	(→Querverbindung)	
(→direct connection)			
tie-up (→bind)	EQUIP.ENG	(→abbinden)	
tie wire	TECH	Bindedraht	
= binding wire			
tight (adj.)	TECH	dicht 2	
≈ impermeable; leak-proof		≈ undurchlässig; lecksicher	
≠ leak		≠ undicht	
tight	MECH	straff	
		= gespannt 2	
tight buffer tube	OPT.COMM	Vollader	
= solid fiber		≠ Hohlader	
≠ loose buffer tube			
tight coated structure (→tight packaged structure)	OPT.COMM	(→Volladeraufbau)	
tightest fit	ENG.DRAW	Kleinstsitz	
= minimum clerarance; allowance		= engster Sitz; Kleinstspiel; größte Durchdringung	
tightness	TECH	Dichtigkeit	
≈ leakproofness; impermeability		≈ Dichtheit; Dichte	
≠ leakage		≈ Lecksicherheit; Undurchlässigkeit	
↓ airtightness		≠ Undichtigkeit	
		↓ Luftdichtigkeit	
tightness (→shortage)	ECON	(→Knappheit)	
tight packaged structure	OPT.COMM	Volladeraufbau	
= tight coated structure			
tilde	LING	Tilde	
[diacritic to palatalize the n in Spanish, or to mark nasality of vowels in Portuguese, like in ñ]		[zeigt Palatisierung des n im Spanischen, oder die Nasalisierung von Vokalen im Portugiesischen an, wie in ñ]	
		↑ diakritisches Zeichen	
tilde	ENG.LOG	Negationsfunktor 1	
tilde (→swing dash)	TYPOGR	(→Wiederholungszeichen)	
tilt (→pulse tilt)	TV	(→Dachschräge)	
tilt (→inclined)	TECH	(→geneigt)	
tilt (→inclination)	MECH	(→Neigung)	
tilt (→incline)	MECH	(→schrägstellen)	
tilt angle	MECH	Kippwinkel	
tilt angle	ANT	Absenkungswinkel	
[of transmitting antenas]		[von Sendeantennen]	
tilted aerial (→tilted antenna)	ANT	(→geneigte Antenne)	
tilted antenna	ANT	geneigte Antenne	
= tilted aerial			
tilted terminated folded dipole	ANT	TTFD-Antenne	
= TTFD antenna			
tilt equalizer (→slope equalizer)	TRANS	(→Neigungsentzerrer)	
tilt handle	EQUIP.ENG	Aufstellbügel	
tilting (→hinged)	MECH	(→schwenkbar)	
tilting screen	TERM&PER	schwenkbarer Bildschirm	
tilt-pivoting stand (→tilt-swivel stand)	EQUIP.ENG	(→Dreh-Schwenk-Fuß)	
tilt regulator (→slope regulator)	TRANS	(→Neigungsregler)	
tilt stand	EQUIP.ENG	Schwenkfuß	
≈ swivel stand		≈ Drehfuß	
tilt-swivel stand	EQUIP.ENG	Dreh-Schwenk-Fuß	
= tilt-pivoting stand		= Schwenk-Neige-Fuß	
↑ stand		↑ Fuß	
TIM	EL.ACOUS	TIM-Verzerrung	
= transient intermodulation			
timber	TECH	Nutzholz	
timbre	ACOUS	Klangfarbe	
= pitch 1			
time	PHYS	Zeit	
↓ instant; time interval		↓ Zeitpunkt; Zeitintervall	
time-acceleration factor	QUAL	Raffungsfaktor	
time acquisition terminal	TERM&PER	Zeiterfassungsterminal	
time allowance	SWITCH	Zeitgutschrift	
time announcement	TELEPH	Zeitansage	
= time announcement service; time announcer; time-of-day			
time announcement service (→time announcement)	TELEPH	(→Zeitansage)	
time announcer (→time announcement)	TELEPH	(→Zeitansage)	
time average	TECH	Zeitmittel	
= time mean			
time axis	PHYS	Zeitachse	
time base	ELECTRON	Zeitbasis	
time-base generator (→sweep generator)	ELECTRON	(→Zeitablenkgenerator)	
time-basis enhacement	INSTR	Triggerbasiserweiterung	
time behaviour (→time response)	ELECTRON	(→Zeitverhalten)	
time cell	CIRC.ENG	Zeitfach	
time clause	LING	Temporalsatz	
↑ conjunctional sentence		= Zeitsatz	
		↑ Konjunktionalsatz	
time clock	SIGN.ENG	Stechuhr	
= punch clock		= Stempeluhr	
time-consistent busy hour	SWITCH	zeitlich festgelegte Hauptverkehrsstunde	
[main traffic hour evaluated during several days of observation]		[über einen mehrtägigen Beobachtungszeitraum ermittelte H.]	
time constant	PHYS	Zeitkonstante	
time-consuming	COLLOQ	zeitraubend	
time-continuous	INF	zeitkontinuierlich	
≠ time-discrete		≠ zeitdiskret	
time counter	INSTR	Stundenzähler	

time counting	SWITCH	**Zeitzählung**	
time-critical	TECH	**zeitkritisch**	
≈ real-time			
time delay (→ delay)	ELECTRON	(→ Verzögerungszeit)	
time-delay fuse (→ slow blowing fuse)	COMPON	(→ träge Schmelzsicherung)	
time-delay relay	COMPON	**Verzögerungsrelais**	
= time relay; timing relais; delay relay		= Zeitrelais	
↓ thermal relay		↓ Thermorelais	
time-delay stage	CIRC.ENG	**Verzögerungsstufe**	
		= Laufzeitstufe	
time-delay switch (→ time switch)	COMPON	(→ Zeitschalter)	
time dependence	TECH	**Zeitabhängigkeit**	
time-dependent (adj.)	TECH	**zeitabhängig**	
time derivative	MATH	**Zeitableitung**	
↑ derivative		↑ Ableitung	
time-detined (→ time-discrete)	INF	(→ zeitdiskret)	
time deviation	MODUL	**Zeithub**	
= time swing			
time-discrete	INF	**zeitdiskret**	
= time-detined		≠ zeitkontinuierlich	
≠ time-continuous			
time discriminator	CIRC.ENG	**Zeitdiskriminator**	
		= Zeitentscheider	
time-dislocated	TECH	**zeitlich versetzt**	
(→ time-shifted)			
time distribution	SIGN.ENG	**Zeitdienst**	
time distribution	PHYS	**Zeitverteilung**	
		= zeitliche Verteilung	
time distribution system	SIGN.ENG	**Zeitdienstanlage**	
time diversity	MIL.COMM	**Zeitdiversity**	
[repetition of emission]		[Wiederholung der Übertragung]	
time division	PHYS	**Zeiteinteilung**	
time-division mode	TELEC	**Zeitgetrenntlage-Verfahren**	
= time-division operation; grouped-time mode; grouped-time operation; burst mode; burst operation		≈ Zeitmultiplex	
≈ time division multiplex			
time division multiple access	TELEC	**Zeitvielfachzugriff**	
= TDMA		= Vielfachzugriff in der Zeitebene; TDMA	
time division multiplex	TELEC	**Zeitmultiplex**	
= TDM 1		[zeitliche Verschachtelung mehrerer Signale]	
≈ time division mode		= Zeitvielfach; Zeitteilung; TDM	
		≈ Zeitgetrenntlage-Verfahren	
time division multiplexing	TELEC	**Zeitmultiplextechnik**	
= TDM 2			
time-division operation	TELEC	(→ Zeitgetrenntlage-Verfahren)	
(→ time-division mode)			
time-division switch unit	SWITCH	**Zeitmultiplex-Koppeleinrichtung**	
time-division system	SWITCH	**Zeitvielfachsystem**	
time domain	PHYS	**Zeitbereich**	
≈ frequency domain		≈ Frequenzbereich	
time-domain equalizer	NETW.TH	**Zeitbereichsentzerrer**	
= TDE			
time-domain reflectometer	INSTR	(→ Impulsreflektometer)	
(→ impulse reflectometer)			
timed release (→ timeout)	ELECTRON	(→ Zeitabschaltung)	
time economy (→ time saving)	TECH	(→ Zeitersparnis)	
time equalization	TV	**Zeitfehlerausgleich**	
time equalization circuit	TV	**Zeitfehler-Ausgleichschaltung**	
time expenditure	ECON	**Zeitaufwand**	
= time spent			
time fill (→ filler)	DATA COMM	(→ Füllzeichen)	
time filter	CIRC.ENG	**Zeitfilter**	
time frame (→ graticule)	INSTR	(→ Zeitraster)	
time indication	TECH	**Zeitangabe**	
time integral	MATH	**Zeitintegral**	
time interval	PHYS	**Zeitintervall**	
= time segment; time span; time section; period 2		= Zeitabschnitt; Zeitspanne; Zeitraum; Zeitsegment	
≈ period 1; duration		≈ Periode; Dauer	
↑ time		↑ Zeit	
time interval analyzer	INSTR	**Zeitintervallanalysator**	
time interval calibrator	INSTR	**Zeitintervallkalibrator**	
time interval counter	INSTR	**Zeitintervallzähler**	
time interval error	TELEC	**Phasenzeitfehler**	
time-interval measurement	INSTR	**Zeitintervallmessung**	
time-interval method	INSTR	**Zeitverfahren**	
time-invariant	PHYS	**zeitinvariant**	
≈ invariable		≈ unveränderlich	
timekeeping	TECH	**Zeitkontrolle**	
time-lag	TECH	**Nacheilung**	
= lag (n.)		= Nachlauf	
≈ delay		≈ Verzug	
≠ time-lead		≠ Voreilung	
time lag (→ delay)	ELECTRON	(→ Verzögerungszeit)	
time lead	TECH	**Voreilung**	
= lead (n.); anticipation		= Vorlauf	
≠ time lag		≠ Nacheilung	
time limit	TECH	**Zeitbegrenzung**	
time limit	ELECTRON	**Zeitgrenze**	
time limit (→ appointed time)	COLLOQ	(→ Termin)	
time limiter	ELECTRON	**Zeitbegrenzer**	
time log	TECH	**Zeitprotokollierung**	
timely	TECH	**rechtzeitig**	
		≈ pünktlich	
timely (→ on time)	ECON	(→ fristgerecht)	
timely (→ modern)	TECH	(→ modern)	
time management	MANUF	**Zeitwirtschaft**	
= manufacturing scheduling			
time mean (→ time average)	TECH	(→ Zeitmittel)	
time measure	PHYS	**Zeitmaß**	
time measurement	INSTR	**Zeitmessung**	
= metering; time metering; chronometry		= Zeitstoppung	
time metering (→ time measurement)	INSTR	(→ Zeitmessung)	
time multiplex signal	TELEC	**Zeitmultiplexsignal**	
time normalization	NETW.TH	**Zeitnormierung**	
time-of-day (→ time announcement)	TELEPH	(→ Zeitansage)	
time-on (→ beginning of conversation)	SWITCH	(→ Gesprächsbeginn)	
time-optimized	TECH	**zeitoptimiert**	
= optimum-time		= zeitoptimal	
timeout 1	TECH	**Zeitablauf 2**	
[ending of a time]		[Ende der Zeit]	
timeout	ELECTRON	**Zeitabschaltung**	
= timed release			
timeout 2	TECH	**Zeitüberschreitung**	
[exceeding of a time]			
time-out (→ time-out circuit)	CIRC.ENG	(→ Zeitüberwachung)	
time-out circuit	CIRC.ENG	**Zeitüberwachung**	
= time-out; timing device; watch-dog; timer 3		≈ Zeitgeber	
≈ timing circuit			
time pattern	TELEC	**Zeitraster**	
time pay (→ time wage)	ECON	(→ Zeitlohn)	
time pulse	ELECTRON	**Zeitsteuertakt**	
= timing pulse; periodic pulse; time tick; metering pulse [TELEC]		= Zeitimpuls; Zeitpuls; Zeittakt	
≈ clock pulse		≈ Schrittpuls	
time pulse counter (→ tax meter)	SWITCH	(→ Gebührenzähler)	
time-pulse metering	SWITCH	**Zeitimpulszählung**	
= timing pulse metering; periodic-pulse metering		[Zeittaktzählung; Gebührenzählung]	
↑ multimetering		↑ Mehrfachzählung	
time-pulse metering	CIRC.ENG	**Zeitimpulszählung**	
		= Zeitintervallzählung; Zeitpulszählung	
time quantum (→ time slice)	DATA PROC	(→ Zeitscheibe)	
timer	EL.INST	**Schaltuhr**	
timer 1 (→ timing circuit 1)	CIRC.ENG	(→ Zeitgeber)	
timer 2 (→ timing circuit 2)	CIRC.ENG	(→ Zeitglied)	
timer 3 (→ time-out circuit)	CIRC.ENG	(→ Zeitüberwachung)	
timer computer	EL.INST	**Schaltcomputer**	
time recorder	INSTR	**Zeitschreiber**	
↑ recording instrument		↑ Registriergerät	
time reference point	INSTR	**Zeitreferenzpunkt**	
		= Zeitbezugspunkt	
time-related	TECH	**zeitbezogen**	
= time-relative			

time-relative (→time-related)	TECH		(→zeitbezogen)
time relay (→time-delay relay)	COMPON		(→Verzögerungsrelais)
time response	ELECTRON		**Zeitverhalten**
= time behaviour			
times	MATH		**mal**
[multiplication]			[Multiplikation]
Times	TYPOGR		**Times**
[a seriph type font; the type of this book]			[eine Seriphenschrift; die Schrift dieses Buches]
time saving (n.)	TECH		**Zeitersparnis**
= time economy			= Zeiteinsparung
time-saving (adj.)	TECH		**zeitsparend**
time scale	INSTR		**Zeitmaßstab**
time scaling	TECH		**Zeitmaßstabsänderung**
↓ acceleration; slow-down (n.)			↓ Zeitraffung; Zeitdehnung
time schedule	TECH		**Zeitplan**
= timetable; schedule			≈ Programm
≈ program			
time schedule (→timing diagram)	TECH		(→Ablaufdiagramm)
time scheduling	TECH		**Terminplanung**
			= Zeitplanung
time section (→time interval)	PHYS		(→Zeitintervall)
time segment (→time interval)	PHYS		(→Zeitintervall)
time sequence	TECH		**zeitlicher Ablauf**
time sharing (→time sharing operation)	DATA PROC		(→Teilnehmerbetrieb)
time sharing operating system	DATA PROC		**Teilnehmer-Betriebssystem**
time sharing operation	DATA PROC		**Teilnehmerbetrieb**
[apparently simultaneous working of several users on a computer, with indipendent programs]			[scheinbar gleichzeitiges Arbeiten mehrerer Benutzer an einer DVA, mit unabhängigen Programmen]
= time sharing system; time sharing; time slicing system; time slicing; TSS			= Teilnehmerrechnersystem; Teilnehmersystem; Zeitscheibenbetrieb; Zeitteilung; Time-sharing-Betrieb; Time-sharing-System; Time-sharing-Verfahren; TSS; Mehrbenutzersystem 1
≈ transaction processing; multi-user system; simultaneous operation			≈ Teilhabersystem; Mehrplatzsystem; Mehrbenutzersystem 2; Mehrprogrammbetrieb; Simultanbetrieb 1
↑ multi-user mode; multiprogramming			↑ Mehrbenutzerbetrieb; Mehrprogrammbetrieb
time sharing system (→time sharing operation)	DATA PROC		(→Teilnehmerbetrieb)
time-shifted	TECH		**zeitlich versetzt**
= time-dislocated; staggered			≈ verzögert
≈ delayed			
time-shifted	TECH		**zeitverschoben**
			= zeitversetzt
time signal	SIGN.ENG		**Zeitzeichen**
			= Zeitsignal
time slice	DATA PROC		**Zeitscheibe**
[time sharing systems]			[Teilnehmersysteme]
= time quantum			
time slice (→time slot)	TELEC		(→Zeitschlitz)
time slicing (→time sharing operation)	DATA PROC		(→Teilnehmerbetrieb)
time slicing system (→time sharing operation)	DATA PROC		(→Teilnehmerbetrieb)
time slot	TELEC		**Zeitschlitz**
= time slice; TS			= Zeitlage; Zeitkanal; Zeitabschnitt; Zeitscheibe
time slot assignment	TELEC		**Zeitschlitzzuordnung**
			= Zeitschlitzzuweisung; Zeitschlitzzuteilung
time-slot classified	TELEC		**zeitschlitzsortiert**
time-slot controlled	TELEC		**zeitschlitzgesteuert**
time-slot coupler	TELEC		**Zeitschlitzkoppler**
time slot interchange	TELEC		**Zeitlagenwechsel**
			= Zeitlagentausch; Zeitschlitzwechsel; Zeitschlitztausch
time-space stage (→time-space switch)	SWITCH		(→Kombinationsstufe)
time-space switch	SWITCH		**Kombinationsstufe**
= time-space stage			= Kombinationsvielfach
time span (→time interval)	PHYS		(→Zeitintervall)
time spent (→time expenditure)	ECON		(→Zeitaufwand)
time stage	SWITCH		**Zeitstufe**
= time switching stage			= Zeitkoppelstufe
≈ time switching network			≈ Zeitkoppelfeld
↑ switching stage			↑ Koppelstufe
time stage group	SWITCH		**Zeitstufengruppe**
time stage incoming	SWITCH		**Zeitstufe kommend**
time stage outgoing	SWITCH		**Zeitstufe gehend**
time stamping	DATA PROC		**Zuschreiben von Datum und Uhrzeit**
time standard	INSTR		**Zeitnormal**
time study man	ECON		**Zeitnehmer**
time swing (→time deviation)	MODUL		(→Zeithub)
time switch	COMPON		**Zeitschalter**
= time-delay switch; delay switch			
time-switch clock	SIGN.ENG		**Zeitschaltuhr**
time switching network	SWITCH		**Zeitkoppelfeld**
≈ time stage			= Zeitvielfach-Koppelfeld
			≈ Zeitstufe
time switching point	SWITCH		**Zeitkoppelpunkt**
time switching stage (→time stage)	SWITCH		(→Zeitstufe)
time synthesis	INSTR		**Zeitsynthese**
timetable (→time schedule)	TECH		(→Zeitplan)
time tagging	INSTR		**Zeitkennung**
time tick (→time pulse)	ELECTRON		(→Zeitsteuertakt)
time unit	PHYS		**Zeiteinheit**
time value	PHYS		**Zeitwert**
↓ instant value			↓ Augenblickswert
time-variable	TECH		**zeitvariabel**
= time-varying			= zeitveränderlich; zeitvariant
time variable	MATH		**Zeitvariable**
time-varying (→time-variable)	TECH		(→zeitvariabel)
time wage	ECON		**Zeitlohn**
= time pay			
time zone	GEOPHYS		**Zeitzone**
time zone	SWITCH		**Zeitzone**
time-zone meter	SWITCH		**Zeitzonenzähler**
time-zone metering	SWITCH		**Zeitzonenzählung**
timing 2	TECH		**zeitliche Abstimmung**
≈ synchronism			≈ Synchronisierung
↓ anticipation; lag			↓ Vorlauf; Nachlauf
timing 1	TECH		**Zeitablauf 1**
≈ sequence			≈ Sequenz
↑ time			↑ Zeit
timing alignment	CODING		**Taktanpassung**
= clock alignment			= Taktabgleich
timing chart (→timing diagram)	TECH		(→Ablaufdiagramm)
timing circuit 1	CIRC.ENG		**Zeitgeber**
[marks time intervals]			[markiert Zeitintervalle]
= timer 1			= Zeitgeberschaltung
≈ time-out circuit; timing generator			≈ Zeitüberwachung; Taktgeber
timing circuit 2	CIRC.ENG		**Zeitglied**
[modifies the temporal sequence]			[beinflußt einen zeitlichen Ablauf]
= timer 2			≈ Verzögerungsglied; Zeitgeber; Zeitüberwachung
≈ delay element; timing circuit 1; time-out circuit			
↑ flipflop			↑ Kippschaltung
timing device (→timing register)	SWITCH		(→Gesprächszeitmesser)
timing device (→time-out circuit)	CIRC.ENG		(→Zeitüberwachung)
timing diagram	TECH		**Ablaufdiagramm**
≈ timing chart; flow diagram; flowchart; flow sheet; activity plan; process chart; time schedule; schedule			≈ Fließplan; Flußdiagramm; Ablaufplan
			≈ Chronogramm
timing diagram	ELECTRON		**Impulsplan**
timing disk	TELEGR		**Taktscheibe**
timing extraction (→clock recovery)	CODING		(→Taktrückgewinnung)
timing extractor	CIRC.ENG		**Zeitextraktor**
timing frequency (→clock frequency)	ELECTRON		(→Taktfrequenz)
timing generator	CIRC.ENG		**Taktgeber**
= clock pulse generator; clock generator; master clock; clock; clock oscillator			= Taktgenerator; Taktimpulsgenerator; Taktimpulserzeuger;

timing mark

≈ pulse generator; clock pulse supply; timing circuit 1; time-out circuit

timing mark ELECTRON **Zeitmarke**
timing pattern ELECTRON **Taktraster**
timing pulse (→time pulse) ELECTRON (→Zeitsteuertakt)
timing pulse metering SWITCH (→Zeitimpulszählung)
(→time-pulse metering)
timing pulse rate CODING **Taktfolge**
= clock pulse rate
timing recovery (→clock CODING (→Taktrückgewinnung)
recovery)
timing register SWITCH **Gesprächszeitmesser**
= timing device; chargeable-time device
timing relais (→time-delay COMPON (→Verzögerungsrelais)
relay)
timing resolution ELECTRON **Zeitauflösung**
timing signal CODING **Taktsignal**
= clock pulse signal; clock signal; clock pulse; synchronizing signal; synchronization signal
≈ clock pulse [ELECTRON]; synchronization character [TELEC]
= Synchronisiersignal
= Schrittpuls [ELECTRON]; Synchronisierzeichen [TELEC]
timistor MICROEL **Timistor**
[thyristor cascade to repeat pulses, with controllable delay]
↑ thyristor
[Kettenthyristor für Impulswiederholungen, mit steuerbarer Verzögerungszeit]
↑ Thyristor
tin CHEM **Zinn**
= Sn = Sn
tin (v.t.) (→tin-plate) METAL (→verzinnen)
tin alloy METAL **Zinnlegierung**
tin bath METAL **Zinnbad**
tin bronze METAL **Zinnbronze**
tin-coat (v.t.) METAL **feuerverzinnen**
↑ verzinnen
tin coat (n.) METAL **Zinnüberzug**
= Verzinnung
tin-coated (→hot-tinned) METAL (→feuerverzinnt)
tin coating METAL **Feuerverzinnung**
tinfoil METAL **Zinnfolie**
= Stanniol; Stanniolfolie; Stanniolpapier
tinker (v.t.) TECH **basteln**
≈ bungle ≈ murksen
tinker (n.) TECH **Bastler**
≈ amateur = Hobbybastler
tinned METAL **verzinnt**
tinner TECH **Klempner**
tin-plate (v.t.) METAL **verzinnen**
= tin (v.t.)
↓ tin-coat (v.t.)
↓ feuerverzinnen
tin plate (n.) METAL **Weißblech**
= tin sheet; white plate
tinsel conductor COMM.CABLE **Lahnlitze**
[tinsel cord] [Spiralschnur]
≈ retractile cord [TERM&PER] [TERM&PER]]
tinsel cord (→retractile cord) TERM&PER (→Spiralschnur)
tinsel wire COMM.CABLE **Lahnlitzendraht**
[a ribbon wire used for flexible telephone cords] [Spiralschnur]
tin sheet (→tin plate) METAL (→Weißblech)
tin solder METAL **Zinnlot**
≈ soldering tin = Lötzinn
tin splash MANUF **Zinnspritzer**
↑ solder splash ↑ Lötspritzer
tint 1 PHYS **Helltönung**
[a light tone of a color] = hellgetönte Farbe; heller Farbton
= color tint (NAM); colour tint (BRI)
tint 2 PHYS **Farbnuance**
[a slight variation of a color] ≈ Farbton
= color tint (AM); colour tint (BRI)
≈ hue
tint (→hue) PHYS (→Farbton)
tiny business ECON **Zwergunternehmen**
tip (n.) TECH **Spitze**
≈ Kamm

tip 1 TELEPH **Spitze**
[one of the poles of a jack connector] [einer der Pole des Klinkensteckers]
≈ ring; sleeve ≈ Ring; Hals
tip 2 (→tip wire) TELEPH (→A-Ader)
tip 3 (→tip2 wire) TELEPH (→D-Ader)
tip wire 1 TELEPH **A-Ader**
= tip 2; T wire; A wire; A-lead
tip wire 2 TELEPH **C-Ader**
= sleeve wire; C wire; C lead; guard; guard wire
= Prüfader
tip wire 3 TELEPH **D-Ader**
= tip 3; D wire; D lead
Tirill voltage regulator POWER SYS (→Tirill-Spannungsregler)
(→vibrating-magnet regulator)
titanate CHEM **Titanat**
titanium CHEM **Titan**
= Ti = Ti; Titanium
title (→claim 1) ECON (→Anspruch)
title (→heading) LING (→Überschrift)
title access CONS.EL **Titelzugriff**
title block ENG.DRAW **Zeichnungskopf**
title field (→lettering TERM&PER (→Schriftfeld)
field)
title page TYPOGR **Titelseite**
= front page = Titelblatt
≈ half title ≈ Respektblatt
titling (→capital character font) TYPOGR (→Versalschrift)
T joint MECH **T-Verbinder**
T junction MICROW **T-Verzweigung**
= T adapter; T adaptor; Tee junction; Tee adapter; Tee adaptor
= T-Stück
T junction COMPON **T-Verzweigung**
= T adapter = T-Adapter
Tl (→thallium) CHEM (→Thallium)
T1 level (→DS1 level) TELEC (→DS1-Ebene)
T2 level (→DS2 level) TELEC (→DS2-Ebene)
T3 level (→DS3 level) TELEC (→DS3-Ebene)
Tm (→thulium) CHEM (→Thulium)
T mast ANT **T-Mast**
T match ANT **T-Anpassung**
TM-bend MICROW **E-Knick**
TM mode (→transversal EL.TECH (→transversale magnetische Welle)
magnetic wave)
TM-plane junction MICROW **E-Verzweigung**
TM-plane offset MICROW **E-Versatz**
TMUX (→transmultiplexer) TRANS (→Transmultiplexer)
TM wave (→transversal EL.TECH (→transversale magnetische Welle)
magnetic wave)
T network (→star NETW.TH (→Sternschaltung)
connection)
to be located COLLOQ **befinden, sich** (v.r.)
to be released TELEC **freiwerden**
toggle (v.t.) ELECTRON **hin- und herschalten**
toggle (→toggle flipflop) CIRC.ENG (→T-Flipflop)
toggle flipflop CIRC.ENG **T-Flipflop**
= T flipflop; toggle = Trigger-Flipflop
toggle generator CIRC.ENG **Kippgenerator**
↑ relaxation oscillator ↑ Relaxationsoszillator
toggle joint MECH **Kniehebel**
toggle press MECH **Kniehebelpresse**
toggle rate (→switching ELECTRON (→Schaltfrequenz)
frequency)
toggle spacer ELECTRON **Gelenkbolzen**
toggle switch COMPON **Kippschalter**
= snap switch; lever key ≈ Kipptaste
≈ toggle key
to give a bird's eye view COLLOQ (→einen Überblick geben)
(→overview)
to go to town (US slang) COLLOQ **kurzen Prozeß machen**
TOI (→third-order TELEC (→Intercept-Punkt dritter Ordnung)
intercept)
token (n.) TERM&PER **Einwurfmünze**
= slug ≈ Münze
≈ coin
token (n.) DATA COMM **Sendeberechtigung**
[symbol assigning authority to control the transmission medium] [Sendepriorität erteilende Bitkombination, token = engl. „Staffelholz"]
= control token = Token

token access (→token passing)	TELEC	(→Token-Verfahren)	**toll restriction**	SWITCH	**Fernwahlsperre**
token passing	TELEC	**Token-Verfahren**	**toll route** (AM) (→long-haul route)	TRANS	(→Ferntrasse)
[circuit access through a circulating bit combination caled token]		[Sendeberechting durch Empfang einer spezifischen Bitkombination]	**toll switchboard**	SWITCH	**Fernschrank**
= token access		= Sendeberechtigungsverfahren	= trunk switchboard; trunk position		= Fernplatz
token ring	DATA COMM	**Token-Ring**	**toll switching office** (→trunk exchange)	SWITCH	(→Fernvermittlungsstelle)
tolerance	MECH	**Maßtoleranz**	**toll switching position**	SWITCH	**Fernvermittlungsplatz**
≈ deviation		≈ Maßabweichung	**toll terminal** (→toll line)	TELEPH	(→Fernanschlußleitung)
tolerance	TECH	**Zulässigkeit**	**toll ticketing**	SWITCH	**Einzelgebührenerfassung**
		= Toleranz	= detailed registration; detailed message accounting; detailed accounting; itemized billing		= Einzelgebührenauflistung; Einzelgebührnachweis; Einzelgebührregistrierung; Einzelgesprächserfassung; Einzelgesprächsauflistung; Einzelgesprächsnachweis; Einzelgesprächsregistrierung; Einzelerfassung
tolerance analysis	NETW.TH	**Toleranzanalyse**	≠ bulk billing		
tolerance band (→tolerance range)	TECH	(→Toleranzbereich)			
tolerance characteristic	TECH	**Grenzkurve**			
		= Toleranzkurve			
		≈ Toleranzmaske			≠ Summengebührenerfassung
tolerance criterion	QUAL	**Toleranzkriterium**	**toll traffic**	TELEC	**Fernverkehr**
tolerance mask	TECH	**Toleranzmaske**	= long-haul traffic; long-range traffic; long-distance traffic; long-haul communications; long-range communications; long-distance communications		= Weitverkehr
= tolerance scheme		= Maske; Toleranzschema ≈ Grenzkurve			≠ Ortsverkehr; Nahverkehr
tolerance measuring bridge	INSTR	**Toleranzmeßbrücke**	≠ local traffic; short-distance traffic		
= limit measuring bridge			**toll transmission network** (→toll network)	TELEC	(→Fernnetz)
tolerance of form	ENG.DRAW	**Formtoleranz**	**toll transmission system** (AM) (→long-haul system)	TRANS	(→Weitverkehrssystem)
= form tolerance					
tolerance range	TECH	**Toleranzbereich**	**toll trunk** (AM)	TELEC	**Fernleitung**
= tolerance band; tolerance zone			= trunk circuit (BRI); intercity trunk (AM); long-distance line; trunk line 1		= Fernverbindungsleitung; Fernlinie
tolerance scheme (→tolerance mask)	TECH	(→Toleranzmaske)	**to load initial program** (→bootstrap)	DATA PROC	(→urladen)
tolerance zone	TECH	**Toleranzfeld**	**TOM** (→voice frequency telegraphy)	TELEGR	(→Wechselstromtelegrafie)
		= Toleranzbereich			
tolerance zone (→tolerance range)	TECH	(→Toleranzbereich)	**TO meter** (→teraohmmeter)	INSTR	(→Teraohmmeter)
tolerated (→permissible)	TECH	(→zulässig)	**ton** (→metric ton)	PHYS	(→Tonne)
tolerated stress (→maximum limited stress)	QUAL	(→Grenzbeanspruchung)	**tonality** (→audible signal)	TELEPH	(→Hörton)
toll (→long-distance call fee)	TELEC	(→Ferngebühr)	**tonal response**	ACOUS	**Klangbild**
			tone	ACOUS	**Ton**
toll- (→chargeable)	TELEC	(→gebührenpflichtig)	[acoustical sensation due to a sinus or harmonic wave]		[Schallempfindung durch sinusförmige oder harmonische Schwingung]
toll area (BRI) (→near zone)	TELEPH	(→Nahzone)	≈ sound		≈ Klang
toll cable	COMM.CABLE	**Fernkabel**	↑ sound 2		↑ Hörschall
= trunk cable (BRI)		= Fernverbindungskabel; Weitverkehrskabel	**tone** (→hue)	PHYS	(→Farbton)
↑ transmission cable		↑ Übertragungskabel	**tone and ringing machine**	SWITCH	**Ruf- und Signalmaschine**
toll call 1 (AM)	TELEPH	**Ferngespräch**			= RSM
= long-distance call; trunk call (BRI)		≈ Nahgespräch	**tone arm** (→pickup arm)	EL.ACOUS	(→Tonabnehmerarm)
≈ interzone call		↓ Telefongespräch	**tone channel** (→sound channel)	BROADC	(→Tonkanal)
toll call 2 (BRI) (→interzone call)	TELEPH	(→Nahgespräch)	**tone control** (→sound control)	EL.ACOUS	(→Klangregler)
toll center (→trunk exchange)	SWITCH	(→Fernvermittlungsstelle)	**tone control aperture** [facsimile]	TELEGR	**Tonwertblende** [Fakssimile]
toll center office (→trunk exchange)	SWITCH	(→Fernvermittlungsstelle)			= Lichtrelais
toll-free (→without charge)	TELEC	(→gebührenfrei)	**tone detector** (→signal receiver)	TRANS	(→Signalempfänger)
toll line	TELEPH	**Fernanschlußleitung**	**tone filter**	EL.ACOUS	**Tonfilter**
[direct connection to toll office or subscribers of other exchange area]		[Direktanschluß an Fernamt, oder mit Teilnehmern eines anderen Ortsnetzes]	**tone generator**	TELEPH	**Tongenerator**
= toll terminal		↑ Teilnehmerleitung	**tone generator** (→signaling transmitter)	ELECTRON	(→Signalgeber)
↑ subscriber line			**toneme** [specific intonation]	LING	**Tonem** [bedeutungsunterscheidende Intonation]
toll network (AM)	TELEC	**Fernnetz**			
= trunk network; trunking network; long-haul network; long-distance network; long-range network; intercapital network; toll plant; trunk plant; toll transmission network; backbone network		= Fernleitungsnetz; Weiverkehrsnetz	**tone-off idle**	TRANS	**Arbeitsstromverfahren**
			= open-circuit signaling		≠ Ruhestromverfahren
			≠ tone-on idle		
			tone-on idle	TRANS	**Ruhestromverfahren**
toll office (→trunk exchange)	SWITCH	(→Fernvermittlungsstelle)	= closed-circuit signaling		≠ Arbeitsstromverfahren
			≠ tone-off idle		
toll open-wire line	TRANS	**Weitverkehrsfreileitung**	**tone-only paging**	MOB.COMM	**Tonfunkruf**
		= Fernfreileitung	↑ radio paging		↑ Funkruf
toll operator	TELEPH	**Fernbeamtin**	**tone pulse**	TELEPH	**Tonimpuls**
↑ operator		= Fernbeamter	**toner**	TERM&PER	**Toner**
		↑ Beamtin	[copy machine, laser printer]		[Kopierer; Laserdrucker]
toll plant (→toll network)	TELEC	(→Fernnetz)	≈ printer's ink		= Tonerpulver
toll register	SWITCH	**Knotenregister**			≈ Druckerschwärze
toll-restricted	TELEPH	(→halbamtsberechtigt)	**toner bin**	TERM&PER	**Tonerkassette**
(→outward-restricted)			**tone receiver**	TELEPH	**Hörtonempfänger**

tone sequence mode	INSTR	Tonsequenzbetrieb
tone wedge	TELEGR	Tonwertskala
[facsimile]		[Faksimile]
ton-force (→metric ton-force)	PHYS	(→Tonne-Kraft)
tongs	TECH	Zange 1
↓ pincers; flat-nose tongs		[groß]
		↓ Beißzange; Flachzange 1
tongue (→reed)	TECH	(→Zunge)
Tonna bazooka	ANT	Tonna-Einspeisung
tool (v.t.)	MECH	bearbeiten
= machine (v.t.); work (v.t.)		≈ verarbeiten [TECH]
≈ process [TECH]		↓ vorbearbeiten; nachbearbeiten; fertigbearbeiten
↓ pre-tool; re-tool; finish		
tool (n.)	TECH	Werkzeug
= implement (n.); instrument; appliance 3; utensil		= Gerät 3
		↓ Handwerkzeug
↓ hand tool		
tool (n.)	DATA PROC	Programmierwerkzeug
[software facilitating development of programs]		[Software zur Erleichterung der Entwicklung von Programmen]
= toolkit software; toolkit; engineering tool; development tool; designer tool; designer kit; development software; programming aid; program design aid		= Programmentwicklungssystem; Tool; Programmentwicklungswerkzeug; Entwicklungswerkzeug; Software-Entwicklungswerkzeug; Entwicklungssoftware; Programmierhilfe
≈ utility program		≈ Dienstprogramm
tool (→aid)	TECH	(→Hilfsmittel)
tool (→machine tool)	MECH	(→Werkzeugmaschine)
tool and die making (→tool shop)	MANUF	(→Werkzeugbau)
tool bag	TECH	Werkzeugtasche
toolbox	DATA PROC	Werkzeugkiste
[fig.]		[fig.]
		= Toolbox
toolbox (→tool case)	TECH	(→Werkzeugkoffer)
tool case	TECH	Werkzeugkoffer
= toolbox		
tool changing	TECH	Werkzeugwechsel
tool engineering	TECH	Werkzeugbau
		[Fachgebiet]
		= Vorrichtungsbau
toolhead	TECH	Werkzeugaufnahme
[part of machine where tools or toll-holders are clamped]		≈ Werkzeughalter
≈ toolholder		
toolholder	TECH	Werkzeughalter
≈ toolhead		≈ Werkzeugaufnahme
tooling (→work)	MECH	(→Bearbeitung)
tooling accuracy	MECH	Bearbeitungsgenauigkeit
tool kit	TECH	Werkzeugsatz
= tool set		
tool kit (→tool)	DATA PROC	(→Programmierwerkzeug)
toolkit software (→tool)	DATA PROC	(→Programmierwerkzeug)
toolmaker	TECH	Werkzeugmacher
toolroom (→tool shop)	MANUF	(→Werkzeugbau)
tool set (→tool kit)	TECH	(→Werkzeugsatz)
tool setting	MANUF	Maschineneinrichten
tool shop	MANUF	Werkzeugbau
= tool and die making; toolroom		= Werkzeugmacherei
tooth	MECH	Zahn
[gear]		[Zahnrad]
tooth clutch	MECH	Zahnkupplung
toothed	MECH	gezahnt
toothed belt	MECH	Zahnriemen
toothed washer (→tooth lock washer)	MECH	(→Zahnscheibe)
toothed wheel (→gear)	MECH	(→Zahnrad)
tooth lock washer	MECH	Zahnscheibe
= toothed washer		
tooth pitch	MECH	Zahnteilung
tooth rack	MECH	Zahnstange
tooth space	MECH	Zahnlücke
[gear]		[Zahnrad]
tooth thickness	MECH	Zahndicke
[gear]		[Zahnrad]
top	TECH	Kopfteil
TO packing	MICROEL	TO-Gehäuse
["transistor outlines"]		
top antenna (→roof antenna)	ANT	(→Dachantenne 1)
to pay taxes on	ECON	versteuern
top coat	TECH	Decklackierung
		[Farbe]
top-down design	MICROEL	Top-down-Entwurf
top-down programming	DATA PROC	(→strukturierte Programmierung)
(→structured programming)		
top-down programming language (→structural programming language)	DATA PROC	(→strukturierte Programmiersprache)
top-fed aerial (→top-fed antenna)	ANT	(→obengespeiste Antenne)
top-fed antenna	ANT	obengespeiste Antenne
= top-fed aerial		
top feed	ANT	Obenspeisung
= shunt feed		
top-heavy	MECH	oberlastig
		= kopflastig
topic (→matter)	OFFICE	(→Vorgang)
topical (→actual)	COLLOQ	(→aktuell)
topline (→pulse top)	ELECTRON	(→Impulsdach)
top load	ANT	Endkapazität
= end capacitance; top-loading capacitance		= Dachkapazität
top loaded antenna (→umbrella antenna)	ANT	(→Schirmantenne)
top-loading capacitance (→top load)	ANT	(→Endkapazität)
topographic map	GEOPHYS	topographische Karte
= topo map		
topography	GEOPHYS	Topographie
[description of places]		[Beschreibung der Lage]
≈ orography		≈ Orographie
topography of terrain (→orography)	GEOPHYS	(→Orographie)
topological group (→topologic group)	MATH	(→topologische Gruppe)
topological network structure	NETW.TH	topologische Netzwerkstruktur
topological structure	MATH	topologische Struktur
topological surface	MATH	topologische Fläche
topologic group	MATH	topologische Gruppe
= topological group		↓ Liesche Gruppe; Halbgruppe
↓ Lie group; semi-group		
topology	MATH	Topologie
topology (→network topology)	NETW.TH	(→Netzwerktopologie)
topo map (→topographic map)	GEOPHYS	(→topographische Karte)
top panel	EQUIP.ENG	Kopfblende
top plate	MECH	Deckplatte
topple (v.i.)	MECH	umkippen (v.t.)
		= kippen 1; umfallen; umstürzen
top quality	ECON	Spitzenqualität
top side	TECH	Oberseite
= upside		= obere Seite
top surface	MECH	Kopffläche
		[Schraube]
top technology	TECH	Spitzentechnologie
= leading-edge technology; cutting-edge technology (BRI)		
top view	ENG.DRAW	Aufsicht
		= Draufsicht
TOR (→radiotelegraphy)	TELEGR	(→Funktelegrafie)
torch	TECH	Taschenlampe
torch (→welding torch)	METAL	(→Schweißbrenner)
torch battery	POWER SYS	Taschenlampenbatterie
		[enthält 2 bis 3 Trockenelemente]
torn (adj.)	TECH	gerissen
		= zerrissen
toroid	MATH	Toroid
[surface generated by rotation of a closed curve]		[durch Rotation einer geschlossenen Kurve gebildete Fläche]
↓ torus		↓ Torus
toroidal balun transformer	ANT	Ringkern-Balun
toroidal coil	COMPON	Ringspule
≈ toroidal core coil		= Toroid
		≈ Ringkernspule

toroidal condenser	COMPON	**Ringkondensator**
toroidal core	COMPON	**Ringkern**
= annular core		= ringförmiger Magnetkern; Toroidkern
↑ ferrite core		↑ Ferritkern
toroidal core coil	COMPON	**Ringkernspule**
≈ toroidal coil		≈ Ringspule
toroidal mains transformer	COMPON	**Ringkerntransformator**
		= Toroidtransformator
toroidal pattern	ANT	**Ringwulst**
toroidal transformer	COMPON	**Ringübertrager**
toroidal variable isolating transformer	COMPON	**Ringkern-Regeltrenntransformator**
torque	PHYS	**Drehmoment**
[SI unit: Newtonmeter, Joule]		[SI-Einheit: Newtonmeter, Joule]
= moment of force; M; turning moment		= Drehkraft; Moment; Torsionskraft; M; Drall 3
		≈ Kraftmoment
torque controller	TECH	**Drehmomentregler**
torque indication	TECH	**Drehmomentanzeige**
torque meter	INSTR	**Drehmomentmesser**
torque motor	POWER ENG	**Drehmoment-Motor**
= torquer		= Torque-Motor; Torquer
torquer (→torque motor)	POWER ENG	(→Drehmoment-Motor)
torque screw driver	MECH	**Drehmoment-Schraubenzieher**
torque screw wrench	MECH	**Drehmoment-Schraubenschlüssel**
torque transducer	TECH	**Drehmomentwandler**
torque wrench	MECH	**Drehmomentschlüssel**
Torr	PHYS	**Torr**
[= 1.333224 bar]		[= 1,333224 bar]
= Torr		= Torr
Torr (→Torr)	PHYS	(→Torr)
torsion (n.) (→twist 1)	PHYS	(→Verwindung)
torsional constant (→directional constant)	INSTR	(→Drehfederkonstante)
torsional oscillation (→torsional wave 1)	PHYS	(→Torsionsschwingung)
torsional rod (→torsion rod)	MECH	(→Torsionstab)
torsional spring (→torsion spring)	MECH	(→Drehfeder)
torsional stiffness (→torsional strength)	MECH	(→Torsionsfestigkeit)
torsional strength	MECH	**Torsionsfestigkeit**
= torsional stiffness		= Verdrehungsfestigkeit
torsional stress	MECH	**Verdrehungsspannung**
torsional vibration (→torsional wave 1)	PHYS	(→Torsionsschwingung)
torsional wave 1	PHYS	**Torsionsschwingung**
= torsional oscillation; torsional vibration		= Drillingsschwingung; Verdrehungsschwingung
torsional wave 2	PHYS	**Torsionswelle**
↑ transversal wave		↑ Transversalwelle
torsion balance	PHYS	**Drehwaage**
torsion load	MECH	**Verdrehungsbelastung**
torsion rod	MECH	**Torsionstab**
= torsional rod		= Drehstab
torsion spring	MECH	**Drehfeder**
= torsional spring		= Torsionsfeder
tortuous (→sinuous)	TECH	(→geschlängelt)
torus (pl.-i)	MATH	**Torus** (pl.-i)
[surface generated by rotation of a circle]		[Fläche die durch Rotation eines Kreises gebildet wird]
= ring 1 (n.)		= Kreiswulst; Kreisringfläche; Ringröhre; Ring
≈ annulus		≈ Kreisring
↑ toroid		↑ Toroid
torus reflector	ANT	**Torusreflektor**
TOS (→tape operating system)	DATA PROC	(→Magnetband-Betriebssystem)
to set parameter values	DATA PROC	**parametrieren**
		= Parameter einstellen
total (→overall)	TECH	(→Gesamt-)
total (→grand total)	MATH	(→Gesamtsumme)
total (v.t.) (→add)	MATH	(→addieren)
total (n.) (→sum)	MATH	(→Summe)
total amount (→grand total)	MATH	(→Gesamtsumme)
total depth	MECH	**Zahnhöhe**
[gear]		[Zahnrad]
total duration (→total time)	TECH	(→Gesamtzeit)
total error	TECH	**Gesamtfehler**
total failure (→catastrophic failure)	QUAL	(→Totalausfall)
total harmonic distortion	ELECTRON	**harmonische Gesamtverzerrung**
= THD		
total harmonic distortion (→harmonic content)	EL.ACOUS	(→Oberschwingungsgehalt)
total inpection	QUAL	**Vollprüfung**
		= Hundertprozentprüfung
totality (→population)	MATH	(→Gesamtheit)
totalize (→add)	MATH	(→addieren)
totalizer	INSTR	**Ereigniszähler**
total loss reference equivalent	TELEPH	**Gesamtbezugsdämpfung**
totally (→fully)	COLLOQ	(→völlig) (adv.)
total permeability	PHYS	**totale Permeabilität**
total reflection	PHYS	**Totalreflexion**
total removal	NETW.TH	**Vollabbau**
↑ removal of poles		↑ Polabspaltung
total time	TECH	**Gesamtzeit**
= total duration		
total transmission loss (→system loss)	RADIO REL	(→Systemdämpfung)
totem-pole amplifier	CIRC.ENG	**Totem-pole-Verstärker**
= totem-pole circuit		= Totem-pole-Schaltung
totem-pole circuit (→totem-pole amplifier)	CIRC.ENG	(→Totem-pole-Verstärker)
totem-pole output	MICROEL	**Totem-pole-Ausgang**
to the best knowledge and belief	COLLOQ	**nach bestem Wissen und Gewissen**
to the power of	MATH	**hoch**
touch (→stop)	MECH	(→Anschlag)
touch (→impact)	TERM&PER	(→Anschlag)
touch (→contact)	PHYS	(→Kontakt)
touch calling (→pushbutton dialing)	SWITCH	(→Tastwahl)
touch dialing (→pushbutton dialing)	SWITCH	(→Tastwahl)
touch dialling (→pushbutton dialing)	SWITCH	(→Tastwahl)
touch down (v.i.)	AERON	**aufsetzen**
≈ land (v.i.)		≈ landen
touch-down zone	AERON	**Aufsetzzone**
touch-guard	EQUIP.ENG	**Berührungsschutz**
touching key (→touch key)	COMPON	(→Berührungstaste)
touch key	COMPON	**Berührungstaste**
= touching key; sensor key; sensory key; soft pad 1		= Sensortaste; Tipptaste
touch panel 1	TERM&PER	**Fingerspitzen-Tablett**
touch panel 2 (→touch screen 2)	TERM&PER	(→Berührungsbildschirm)
touch potential	EL.TECH	**Fehlerspannung**
		≈ Berührungsspannung
touch screen 2	TERM&PER	**Berührungsbildschirm**
= touch-sensitive screen; touch-sensitive CRT; touch panel 2		= berührungssensitiver Bildschirm; drucksensitiver Bildschirm; Kontaktbildschirm; Tastbildschirm; Sensorbildschirm; Sensorfeld
≈ touch screen 2		
		≈ Berührungstablett
touch-screen input	TERM&PER	**Berührungseingabe**
touch selection (→pushbutton dialing)	SWITCH	(→Tastwahl)
touch-sensitive	TECH	**berührungsempfindlich**
≈ pressure-sensitive		≈ kontaktempfindlich
		≈ druckempfindlich
touch-sensitive CRT (→touch screen 2)	TERM&PER	(→Berührungsbildschirm)
touch-sensitive keyboard	TERM&PER	**Berührungstastatur**
≈ membrane keyboard		= Sensortastatur
		≈ Folientastatur
touch-sensitive panel (→touch-sensitive tablet)	TERM&PER	(→Berührungstablett)
touch-sensitive screen (→touch screen 2)	TERM&PER	(→Berührungsbildschirm)
touch-sensitive switch	COMPON	**Kontaktschalter**

touch-sensitive tablet	TERM&PER	Berührungstablett
= touch-sensitive panel		= Sensorfeld
≈ touch screen 2		≈ Berührungsbildschirm
touch sensor (→contact sensor)	COMPON	(→Berührungssensor)
touch-tone calling (→VF pushbutton dialing)	SWITCH	(→tonfrequente Tastwahl)
touch-tone dialing (→VF pushbutton dialing)	SWITCH	(→tonfrequente Tastwahl)
touch-tone dialling (→VF pushbutton dialing)	SWITCH	(→tonfrequente Tastwahl)
touch-tone selection (→VF pushbutton dialing)	SWITCH	(→tonfrequente Tastwahl)
touch-tone telephone	TELEPH	Tastwahltelefon
= pushbutton telephon; pushbutton station		= Tastwahltelephon; Tastwahlapparat; Tastefernsprecher; Tastentelefon; Tastentelephon
touch-typing	TERM&PER	Zehnfingerschreiben
		= Zehnfinger-Blindschreiben
tourmaline	CHEM	Turmalin
tour of duty	ECON	Dienstreise
≈ business trip		≈ Geschäftsreise
tower	CIV.ENG	Turm
[free standing building or structure, very high in relation to its cross section]		[im Verhältnis zum Querschnitt sehr hohes, freistehendes Gebäude oder Struktur]
≈ mast		≈ Mast
tower	DATA PROC	Turmgerät
= tower-style case		= Turm; Towergerät; Tower
↑ under-desk equipment [EQUIP.ENG]		↑ Unterbaugerät [EQUIP.ENG]
tower antenna	ANT	Mastantenne
= pylon antenna		= M-Antenne; Pylon-Antenne
≈ cylindrical antenna		≈ Zylinderantenne
tower loading	CIV.ENG	Turmlast
tower mount	ANT	Turmbefestigung
		[von Antennen]
tower-style case (→tower)	DATA PROC	(→Turmgerät)
Townsend discharge	PHYS	Townsend-Entladung
toxid agent (→polluant)	TECH	(→Schadstoff)
TPA (→transient program area)	DATA PROC	(→Übergangsbereich)
TPI	TERM&PER	TPI
[measure of storage density]		[Maßeinheit für Spurendichte]
= tpi; tracks per inch		= Spuren pro Zoll
tpi (→TPI)	TERM&PER	(→TPI)
TP monitor (→teleprocessing monitor)	DATA PROC	(→Fernverarbeitungsmonitor)
TP monitor (→transaction processing monitor)	DATA COMM	(→Transaktionsystem)
TR (→transmitting-receiving)	RADIO LOC	(→Empfänger-Sperröhre)
trace (v.t.)	ENG.DRAW	durchzeichnen
trace (v.t.)	DATA PROC	verfolgen
trace 1 (n.)	ELECTRON	Hinlauf
[beam on a display]		[Bildschirm]
≠ reverse action		≠ Rücklauf
trace (n.)	CHEM	Spur
≈ trace element		≈ Spurenelement
trace (n.)	INSTR	Kurve
[of measurement display]		[einer Meßanzeige]
= curve		
trace (n.)	COLLOQ	Spur
trace (n.) (→fault trace)	DATA PROC	(→Ablaufverfolgung)
trace 2 (n.) (→track conductor)	ELECTRON	(→Leiterbahn)
traceability	INSTR	Nachweisbarkeit
trace concentration	CHEM	Spurenkonzentration
trace element	CHEM	Spurenelement
trace math operation	INSTR	mathematische Kurvenoperation
trace operation	INSTR	Kurvenoperation
trace program (→tracer)	DATA PROC	(→Überwacher)
tracer	CHEM	Indikator
tracer	PHYS	Radioindikator
tracer	DATA PROC	Überwacher
[program to protocolize test runs]		[Programm zur Protokollierung von Testläufen]
= monitor (n.); trace program		= Überwachungsprogramm; Ablaufverfolger; Monitorprogramm; Monitor; Tracer
tracer (→drawer)	ENG.DRAW	(→technischer Zeichner)
tracing (→route tracing)	TELEC	(→Streckenführung)
tracing paper (→transparent paper)	ENG.DRAW	(→Transparentpapier)
tracing routine	DATA PROC	Überwachungsroutine
tracing switch	SWITCH	Fangschaltung
= annoyance call trap		
track 1 (n.)	RAILW.SIGN	Geleise
track (n.)	TERM&PER	Spur
[a path on a data carrier permitting sequential recording of bits]		[Bahn auf einem Datenträger zur sequentiellen Speicherung von Bits]
= recording track		= Speicherspur; Kanal
≈ row		≈ Sprosse
↓ information track; clock track; channel		↓ Informationsspur; Taktspur; Lochspur
track (→track conductor)	ELECTRON	(→Leiterbahn)
track 0 (→primary track)	DATA PROC	(→erste Spur)
track address	DATA PROC	Spuradresse
[magnetic disk memory]		[Magnetplattenspeicher]
track arrangement	TERM&PER	Spuranordnung
= track format		= Spureinteilung; Spuraufteilung
track ball	TERM&PER	Rollkugel
[ball turnable from top; "turned mouse"]		[Kugel von oben rollbar; „umgedrehte Maus"]
= trackball		= Trackball; Standmaus
≈ mouse		≈ Maus
trackball (→track ball)	TERM&PER	(→Rollkugel)
track conductor	ELECTRON	Leiterbahn
= PCB track; path conductor; printed wire; conduction path; conducting track; track; trace 2 (n.)		= Leiterstreifen
track density	TERM&PER	Spurdichte
track description block	TERM&PER	Spurkennblock
= track desription record; track identifier		[enthält Angabe der Spureinteilung]
≈ key field		= Spurkennsatz
		≈ Schlüsselfeld
track descriptor (→identifier)	TERM&PER	(→Spurenkennblock)
track desription record (→track description block)	TERM&PER	(→Spurkennblock)
40-track disk (→fourty-track disk)	TERM&PER	(→Vierzig-Spuren-Diskette)
40-track diskette (→fourty-track disk)	TERM&PER	(→Vierzig-Spuren-Diskette)
tracked vehicle	TECH	Raupenfahrzeug
track format (→track arrangement)	TERM&PER	(→Spuranordnung)
track identifier (→track description block)	TERM&PER	(→Spurkennblock)
track index	DATA PROC	Spurindex
[indexed sequential access method]		[indiziert sequentieller Zugriff]
tracking	TECH	Führung
[influencing and shaping of a course]		[Beeinflussung oder Auslegung eines Verlaufs]
= routing		
tracking	RADIO LOC	Zielverfolgung
tracking	DATA PROC	Nachführung
[of a cursor]		[einer Schreibmarke]
tracking	SAT.COMM	Bahnverfolgung
tracking	TERM&PER	Spureinstellung
tracking (→route tracing)	TELEC	(→Streckenführung)
tracking accuracy	RADIO LOC	Folgegenauigkeit
tracking antenna	ANT	Aufspürantenne
tracking beam	RADIO LOC	Verfolgungsstrahl
tracking control	TERM&PER	Spursteuerung
[magnetic tape]		[Magnetband]
tracking control (→follow-up control)	CONTROL	(→Folgeregelung)
tracking cross	TERM&PER	Spurkreuz
		= Markierungskreuz

tracking current		EL.TECH	**Kriechstrom**		
[a superficial leakage current]			[oberflächlicher Leckstrom]		
≈ creeping current; superficial leakage current			= Oberflächenverluste		
			↑ Streustrom		
tracking current resistance		EL.TECH	**Kriechstromfestigkeit**		
= tracking resistance					
tracking distance		EL.TECH	**Kriechstrecke**		
= creeping distance					
tracking error		EL.ACOUS	**Rillenfehler**		
tracking error		ELECTRON	**Nachlauffehler**		
			= Spurfehler		
tracking generator		INSTR	**Mitlaufgenerator**		
tracking method		RADIO LOC	**Suchverfahren**		
tracking pitch		TERM&PER	**Spurteilung**		
[separation of adjacent tracks]			[Abstand benachbarter Spuren]		
tracking radar		RADIO LOC	**Zielverfolgungsradar**		
			= Nachlaufradar; Folgeradar; Verfolgungsradar		
tracking rate		ELECTRON	**Nachlauffrequenz**		
tracking resistance (→tracking current resistance)		EL.TECH	(→Kriechstromfestigkeit)		
tracking speed		ELECTRON	**Nachlaufgeschwindigkeit**		
tracking symbol		DATA PROC	**Nachführungssymbol**		
≈ cursor			≈ Sxchreibmarke		
tracking synthesizer		INSTR	**Mitlauf-Syntheziser**		
trackink output		INSTR	**Mitlaufausgang**		
track plotter (→curve recorder)		INSTR	(→Kurvenschreiber)		
track recorder (→curve recorder)		INSTR	(→Kurvenschreiber)		
track safety		RAILW.SIGN	**Zugsicherung**		
track section		RAILW.SIGN	**Streckenabschnitt**		
tracks per inch (→TPI)		TERM&PER	(→TPI)		
track switching		TERM&PER	**Spurwechsel**		
			≈ Spurumschaltung		
track-to-train radio system		RADIO	**Schiene-Zug-Funksystem**		
track width		TERM&PER	**Spurbreite**		
traction		MECH	**Zug**		
= draft; draught (BRI); drawing; pull; strain			≈ Spannung		
≈ tension					
traction element		COMM.CABLE	**Zugelement**		
traction line (→overhead line)		POWER ENG	(→Fahrleitung)		
traction rope		TECH	**Zugseil**		
↓ hoisting rope; haulage rope			↓ Hebeseil; Schleppseil		
traction rope (→winch cable)		OUTS.PLANT	(→Windenseil)		
traction wire		COMM.CABLE	**Zugdraht**		
≈ traction rope			≈ Zugseil		
tractor		TECH	**Zugmaschine**		
			= Traktor		
tractor (→paper tractor)		TERM&PER	(→Vorschubraupe)		
tractor-fed printer		TERM&PER	**Raupenvorschubdrucker**		
tractor feed		TERM&PER	**Raupenvorschub**		
= tractor feed mechanism			= Traktor		
↓ pulling tractor; pushing tractor			↓ Zugtraktor; Schubtraktor		
tractor feed mechanism (→tractor feed)		TERM&PER	(→Raupenvorschub)		
tractrix		MATH	**Traktrix**		
			= Schleppkurve		
trade		ECON	**Handel**		
= commerce; business transaction			= Wirtschaftsverkehr		
↓ specialized trade; wholesale trade; retail trade			↓ Fachhandel; Großhandel; Einzelhandel		
trade directory		ECON	**Branchenverzeichnis**		
trade fair (→trade show)		ECON	(→Handelmesse)		
trademark		ECON	**Markenzeichen**		
= brand			= Warenzeichen; Schutzmarke; Zeichen		
≈ brand name			≈ Markenname		
↑ logotype			↑ Logotype		
trademark (→corporate logo)		ECON	(→Firmenzeichen)		
trademark owner		ECON	**Warenzeicheninhaber**		
trade name		ECON	**Handelsname**		
trade-off (v.t.)		COLLOQ	**abwägen**		
= weigh (fig.); ponder					
trade-off (n.)		COLLOQ	**Abwägung**		
= tradeoff; ponderation			= Pro und Kontra; Bilanz		
tradeoff (→trade-off)		COLLOQ	(→Abwägung)		
trader (→dealer)		ECON	(→Händler)		
trade show		ECON	**Handelsmesse**		
= trade fair			↑ Austellung		
↑ exposition					
trade union (→union)		ECON	(→Gewerkschaft)		
trading (→business)		ECON	(→Geschäft)		
trading company		ECON	**Handelsfirma**		
≈ dealer			= Handelsunternehmen		
↓ retail distribution company; wholesale distribution company			≈ Händler		
			↓ Einzelhandelsunternehmen; Großhandelsunternehmen		
traditional (→usual)		COLLOQ	(→üblich)		
traffic		SWITCH	**Verkehr**		
traffic amount (→traffic volume)		SWITCH	(→Verkehrsmenge)		
traffic announcement		BROADC	**Verkehrsfunk**		
traffic announcement decoder		CONS.EL	**Verkehrsfunkdecoder**		
traffic capacity (→traffic-handling capability)		SWITCH	(→Verkehrskapazität)		
traffic carried (→traffic load)		TELEC	(→Verkehrsbelastung)		
traffic category		SWITCH	**Verkehrskategorie**		
traffic compensation		SWITCH	**Verkehrsausgleich**		
			= Verkehrskompensation		
traffic computer (→traffic control computer)		SIGN.ENG	(→Verkehrsrechner)		
traffic control		SIGN.ENG	**Verkehrsregelung**		
= road traffic control					
traffic control computer		SIGN.ENG	**Verkehrsrechner**		
= road traffic control computer; traffic computer			= Verkehrsleitrechner		
traffic control direction finder		RADIO LOC	**Verkehrspeiler**		
traffic control engineering (→road traffic engineering)		TECH	(→Straßenverkehrstechnik)		
traffic control radar		RADIO LOC	**Verkehrsradar**		
traffic counter		SWITCH	**Verkehrsdatenzähler**		
traffic data		SWITCH	**Verkehrsdaten**		
traffic data administration		SWITCH	**Verkehrsdatenerfassung**		
= traffic recording; traffic data collection; traffic statistics			= Verkehrsaufzeichnung		
traffic data collection		SWITCH	(→Verkehrsdatenerfassung)		
(→traffic data administration)					
traffic density (→traffic intensity)		SWITCH	(→Verkehrswert)		
traffic destination		SWITCH	**Verkehrsziel**		
traffic flow (→traffic intensity)		SWITCH	(→Verkehrswert)		
traffic handling		SWITCH	**Verkehrsabwicklung**		
			= Verkehrsverwaltung		
traffic-handling capability		SWITCH	**Verkehrskapazität**		
= traffic capacity			= Verkehrsleistung		
traffic intensity		SWITCH	**Verkehrswert**		
= traffic flow; carried traffic; traffic density					
traffic light		SIGN.ENG	**Verkehrsampel**		
			= Ampel		
traffic load		TELEC	**Verkehrsbelastung**		
= traffic loading; traffic carried; traffic occupancy			= Verkehrsanfall; Belastung		
traffic loading (→traffic load)		TELEC	(→Verkehrsbelastung)		
traffic matrix		SWITCH	**Verkehrsmatrix**		
traffic measurement		SWITCH	**Verkehrsmessung**		
traffic measurement data		SWITCH	**Verkehrsmeßdaten**		
traffic measurement system		SWITCH	**Verkehrsmeßsystem**		
traffic mode (→traffic type)		TELEC	(→Verkehrsart)		
traffic model		SWITCH	**Verkehrsmodell**		
traffic observation		SWITCH	**Verkehrsbeobachtung**		
traffic occupancy (→traffic load)		TELEC	(→Verkehrsbelastung)		
traffic peak		SWITCH	**Verkehrsspitze**		
traffic recording (→traffic data administration)		SWITCH	(→Verkehrsdatenerfassung)		
traffic relation		TELEC	**Verkehrsbeziehung**		
traffic route		SWITCH	**Verkehrsweg**		
traffic route		CIV.ENG	**Verkehrsweg**		

traffic route searching

traffic route searching	SWITCH	**Verkehrswegsuche**
= route searching		
traffic routing (→alternate routing)	SWITCH	(→Leitweglenkung)
traffic signal system	SIGN.ENG	**Verkehrssignalanlage**
traffic simulation	SWITCH	**Verkehrssimulation**
traffic simulator	SWITCH	**Verkehrssimulator**
traffic sink	SWITCH	**Verkehrssenke**
traffic sorting	TELEC	**Dienstetrennung**
= grooming; sorting		= Dienstesortierung; Vorsortierung
traffic source	SWITCH	**Verkehrsquelle**
traffic statistics (→traffic data administration)	SWITCH	(→Verkehrsdatenerfassung)
traffic structure	SWITCH	**Verkehrsstruktur**
traffic supervision	SWITCH	**Verkehrsüberwachung**
traffic theory	TELEC	**Verkehrstheorie**
		= Nachrichtenverkehrstheorie; Bedienungstheorie
traffic type	TELEC	**Verkehrsart**
= traffic mode		
traffic unit	SWITCH	**Verkehrswert-Einheit**
traffic volume	SWITCH	**Verkehrsmenge**
= traffic amount		= Verkehrsvolumen; Verkehrsaufkommen; Gesprächsaufkommen
traffic volume incoming	SWITCH	**Verkehrsmenge kommend**
traffic volume outgoing	SWITCH	**Verkehrsmenge gehend**
trailer	BROADC	**Vorspann**
trailer (→trailer label)	DATA PROC	(→Dateiend-Etikett)
trailer card	DATA PROC	**Fortsetzungskarte**
		= Folgekarte
trailer label	DATA PROC	**Dateiend-Etikett**
[identifies end of file]		[kennzeichnet das Dateiende]
= trailer record; trailer; postamble		= Dateiend-Kennsatz; Schlußetikett; Schlußkennsatz; Nachsatz; Nachspann
≠ header label		≠ Dateianfangs-Etikett
↑ label		↑ Etikett
trailer record (→trailer label)	DATA PROC	(→Dateiend-Etikett)
trailer statement	DATA PROC	**Endanweisung**
trailing	TECH	**nacheilend**
= lagging		≠ voreilend
≠ leading		
trailing angle	PHYS	**Nacheilwinkel**
= lag angle; lagging angle		= nacheilender Winkel
trailing antenna (→drag antenna)	ANT	(→Schleppantenne)
trailing contact (→make-break contact)	COMPON	(→Folgekontakt)
trailing edge	TERM&PER	**Hinterkante**
[punched card]		[Lochkarte]
trailing edge	ELECTRON	**Abfallflanke**
= falling edge; negative edge; negative-going edge		[eines Signals, Impulses]
		= Hinterflanke; fallende Flanke; hintere Flanke; Rückflanke
≠ rising edge		≠ Anstiegflanke
↓ trailing pulse edge		↓ Impulsabfallflanke
trailing-edge overshoot (→baseline overshoot)	ELECTRON	(→Nachschwingung)
trailing end	TERM&PER	**Bandende**
[magnetic tape]		[Magnetband]
trailing pulse edge	ELECTRON	**Impulsabfallflanke**
= negative pulse edge; falling pulse edge; decaying pulse edge		= hintere Impulsflanke; fallende Impulsflanke; Impulsrückflanke; Pulsabfallflanke; Pulsrückflanke
≈ pulse tail		≈ Impulsschwanz
≠ leding pulse edge		≠ Impulsanstiegflanke
trailing transition time (→decay time)	ELECTRON	(→Abfallzeit)
train	TECH	**schulen**
= school; teach		= ausbilden
trainable (→learnable)	SCIE	(→erlernbar)
train destination indicator	RAILW.SIGN	**Zugzielanzeiger**
trainee 2	ECON	**Praktikant**
[on the job]		
trainee 3	ECON	**Schüler**
[participants in a course]		= Schulungsteilnehmer
		≈ Kursteilnehmer
trainee 1 (→apprentice)	ECON	(→Lehrling)
training	TECH	**Schulung**
= schooling; teaching; occupational education		= Ausbildung 1
		≈ Weiterbildung; Information
≈ education; information		
training center	ECON	**Schulungszentrum**
training center (AM) (→training plant)	TECH	(→Schulungsanlage)
training centre (BRI) (→training plant)	TECH	(→Schulungsanlage)
training documentation (→course material)	ECON	(→Schulungsunterlage)
training manual	TECH	**Lehrhandbuch**
= tutorial		
training material (→course material)	ECON	(→Schulungsunterlage)
training on the job	ECON	**Ausbildung am Arbeitsplatz**
= on-the-job training		
training plant	TECH	**Schulungsanlage**
= training center (AM); training centre (BRI)		= Schulungszentrum
training room (→classroom)	SCIE	(→Schulungsraum)
train performance processor	RAILW.SIGN	**Zugfahrtrechner**
trajectory	PHYS	**Flugbahn**
trajectory	MATH	**Trajektorie**
[a curve intersecting isogonally a family of curves]		[eine Kurvenschar gleichwinklig schneidende Kurve]
tram (BRI) (→tramway)	TECH	(→Straßenbahn)
tramline (BRI) (→tramway)	TECH	(→Straßenbahn)
tramway (BRI)	TECH	**Straßenbahn**
= tramline (BRI); tram (BRI); streetcar line; city rail		= Trambahn; Tram
≈ trolleybus		≈ Obus
tranche	ECON	**Tranche**
≈ part amount		≈ Teilbetrag
transact	ECON	**abwickeln 2**
[a business]		[ein Geschäft]
transaction	DATA PROC	**Transaktion**
[elementary input, processing and output step in an interactive mode]		[elementarer Eingabe-, Bearbeitungs- und Ausgabeschritt bei Dialogverarbeitung]
		= Vorgang; Interaktion
transaction (→business)	ECON	(→Geschäft)
transaction (→matter)	OFFICE	(→Vorgang)
transaction code	DATA PROC	**Transaktionscode**
transaction data	DATA PROC	**Bewegungsdaten**
= change data; movement data		= Änderungsdaten
transaction file	DATA PROC	**Bewegungsdatei**
[contains changes later used to update a master file]		[enthält Änderungen mit denen dann eine Stammdatei aktualisiert wird]
= activity file; change file; detail file; movement file		= Änderungsdatei
≠ master file		≠ Stammdatei
transaction handling (→handling)	OFFICE	(→Bearbeitung)
transaction mode (→interactive processing)	DATA PROC	(→Dialogbetrieb)
transaction processing	DATA PROC	**Teilhaberbetrieb**
[simultaneous working of several users on a program]		[gleichzeitiges Arbeiten mehrerer Benutzer mit einem Programm]
≈ time-sharing operation		= Teilhabersystem; Transaktionsbetrieb
		≈ Teilnehmerbetrieb
transaction processing (→interactive processing)	DATA PROC	(→Dialogbetrieb)
transaction processing monitor	DATA COMM	**Transaktionssystem**
[supervision and control programs for transaction processing]		[Überwachungs- und Steuerprogramme für Teihaberbetrieb]
= TP monitor		= TP-Monitor
transactions (n.plt.)	ECON	**Tagungsbericht**
[official written record]		= Kongreßbericht
= proceedings (n.plt.)		
transaction tape	DATA PROC	**Bewegungsband**

transadmittance NETW.TH
↑ transfer function
transatlantic cable TELEC
transatlantic fiber optic cable TELEC
transborder (→cross-border) ECON
transceiver TERM&PER
[facsimile]
= recording system
transceiver RADIO REL
= radio equipment
transceiver MICROEL
= transmitter/receiver circuit
transceiver TELEC
[equipment able to transmit and receive]
= transmitter-receiver
transceiver RADIO
[radio transmitter plus radio receiver, including all central parts]
= transmitter-receiver; Tx/Rx; transmit-receive set
↑ radio equipment
↓ HF transceiver
transceiver (→HF transceiver) HF
transceiver rack RADIO REL
transceiver switch TERM&PER
(→send/receive switch)
transceiver test equipment INSTR
(→radiocommunication tester)
transcendental equation MATH
transcendental function MATH
transcendental number MATH
≠ algebraic number
↑ irrational number
transcoder TRANS
[fits two 2 Mbit/s signals into one 2 Mbit/s stream and viceversa]
≈ ADPCM
transcoder (→standard TELEC
converter)
transcoder (→TV standards TV
converter)
transcoding (→code conversion) CODING
transconductance ELECTRON
[factor of merit for electronic tube, relation of plate-current change to grid-voltage change]
= grid-plate transconductance; mutual conductance
transconductance of ELECTRON
characteristic
transconductance operational CIRC.ENG
amplifier (→OTA)
transconductor CIRC.ENG
[for special mathematical operations]
↑ operational amplifier
transcribe LING
transcription LING
[reproduction by phonetically corresponding letters of another alphabet]
transducer INSTR
= primary detector; primary element
transducer COMPON
[changes the physical signal carrier]
= converter
transducer INSTR
↑ transformer [EL.TECH]
transducer (→mode MICROW
transformer)

Transadmittanz
= Gegenscheinleitwert
↑ Transferfunktion
Transatlantikkabel
LWL-Transatlantikkabel
(→grenzüberschreitend)
Aufzeichnungssystem
[Faksimile]
= Transceiver
Funkgerät
= RF-Gerät; Sender-Empfänger
Sender-/Empfänger-Schaltung
= Transceiver
Transceiver
[Gerät das sowohl senden als auch empfangen kann]
= Sender-Empfänger
Transceiver
[Funksender mit Funkempfänger, einschließlich aller Zentraleinheiten]
= Sender-Empfänger; S/E; Sende-Empfangs-Gerät
↑ Funkgerät
↓ HF-Transceiver
(→HF-Transceiver)
Funkgestell
(→Sende-Empfangs-Umschalter)
(→Funkmeßplatz)
transzendente Gleichung
transzendente Funktion
transzendente Zahl
≠ algebraische Zahl
↑ Irrationalzahl
Transcoder
[bildet aus zwei 2-Mbit/s-Signalen einen 2-Mbit/s-Strom und umgekehrt]
≈ ADPCM
(→Normwandler)
(→Normwandler)
(→Codeumsetzung)
Steilheit
[Kenngröße einer Röhre, Anodenstrom-Änderung zu Gitterspannungs-Änderung]
= Gegenleitwert; Transkonduktanz
Kennliniensteilheit
(→Transkonduktanzverstärker)
Transconductor
[für spezielle mathematische Operationen]
↑ Operationsverstärker
umschreiben
Transkription
[lautlich entsprechende Wiedergabe eines anderen Alphabets]
= Umschrift; Transkribieren
Meßaufnehmer
= Meßwertaufnehmer; Aufnehmer; Meßgeber
Wandler
[ändert den physikalischen Signalparameter]
= Umwandler; Umsetzer
≈ Sensor [COMPON]
Wandler
↑ Transformator [EL.TECH]
(→Modenwandler)

transducer gain NETW.TH
transducer loss ELECTRON
transducer sensitivity TELEPH
transductor EL.TECH
[device by means of which a current or voltage may be varied by varying the saturation of an inductor]
≈ transductor amplifier
transductor amplifier CIRC.ENG
[an amplifier whoes output power is controlled by transductors]
= magnetic amplifier
transductor regulator EL.TECH
[transductor to control an aelctrical quantity]
transfer (v.t.) ECON
[money]
= remit
transfer (v.t.) TELEPH
transfer 1 (n.) ECON
= cession; assignment
transfer (n.) ELECTRON
[PCB]
transfer 2 (n.) (→remittance) ECON
transfer (n.) DATA PROC
[of memory contents]
transfer (n.) COLLOQ
[figurative]
= assignment
transfer (n.) PHYS
= transport (n.)
transfer 3 ECON
[of personnel]
transfer (→carry) MATH
transfer (→remittance) ECON
transfer (→relay) TELEC
transfer (→delivery) TELEC
transfer (→transpose) SCIE
transfer (→transposition) SCIE
transferability COLLOQ
transferable COLLOQ
transfer address DATA PROC
= jump address
transfer admittance (→mutual NETW.TH
admittance)
transfer bit (→carry bit) DATA PROC
transfer capacity (→transmission INF
capacity)
transfer characteristic SYS.TH
transfer characteristic MICROEL
transfer element CONTROL
= element
↓ D-element; I-element
transfer-elements CONTROL
combination
transference number PHYS
= transport number
transfer flag (→carry DATA PROC
bit)
transfer function NETW.TH
[transforms input into output signal]
≈ transmission coefficient
↑ active function
↓ transimpedance; transadmittance

Übertragungs-Leistungsverstärkung
Wandlerverlust
Leistungsübertragungsfaktor
= Wandlerempfindlichkeit
Transduktor
[Vorrichtung mit der ein Strom oder eine Spannung durch Änderung der Vormagnetisierung von Drosseln verändert werden kann]
≈ Transduktorverstärker
Transduktorverstärker
[ein Verstärker dessen Ausgangsleistung durch Transduktoren gesteuert wird]
= magnetischer Verstärker; Magnetverstärker
Transduktorregler
[Transduktor zur Steuerung einer elektrischen Größe]
überweisen
[Geld]
= transferieren; übersenden 2
Übergeben
Abtretung
= Überführung
Transfer
[gedruckte Schaltung]
(→Überweisung)
Transfer
[Verlagerung von Speicherinhalten]
Übertragung
[figurativ]
Übertragung
= Transport
Versetzung
[von Personal]
(→Übertrag)
(→Überweisung)
(→weiterleiten)
(→Weitersendung)
(→transponieren)
(→Transponierung)
Übertragbarkeit
= Transferierbarkeit
übertragbar
= transferierbar
Verzweigungsadresse
= Sprungadresse
(→Kernleitwert)
(→Übertragbit)
(→Übertragungskapazität)
Transfercharakteristik
Transfercharakteristik
= Transferkennlinie
Übertragungsglied
= Glied
↓ D-Glied; I-Glied
Übertragungsglieder-Verknüpfung
= Verknüpfung 2
Überführungszahl
(→Übertragbit)
Transferfunktion
[wandelt Eingangssignal in Ausgangssignal]
= Übertragungsfunktion
≈ Übertragungsfaktor
↑ Wirkungsfunktion
↓ Transimpedanz; Transadmittanz

transfer function	CONTROL	Übertragungsfunktion	
transfer gate	MICROEL	Transfergatter	
= transmission gate			
transfer impedance (→mutual impedance)	NETW.TH	(→Kernwiderstand)	
transfer inefficiency [CCD]	MICROEL	Transferineffizienz [CCD]	
transfer instruction	DATA PROC	Transferbefehl	
transfer method (→transfer mode)	DATA COMM	(→Übermittlungsprinzip)	
transfer mode	DATA COMM	Übermittlungsprinzip	
= transfer method		= Übermittlungsverfahren	
transfer operation	DATA PROC	Transferoperation	
transfer point	TELEC	Transferpunkt	
transfer rate (→throughput)	DATA PROC	(→Durchsatz)	
transferred-charge call [called party takes over the bill after beeing asked] = collect call	TELEPH	R-Gespräch [nach Rückfrage übernimmt der Angerufene die Gebühr] ≈ Gebührenübernahme	
transferred information (→transinformation)	INF	(→Transinformation)	
transferred subscriber (→changed number)	TELEC	(→Rufnummer geändert)	
transfer register	DATA PROC	Transferregister	
transfer server	DATA COMM	Transferserver	
transfer system	DATA COMM	Übermittlungssystem	
transfer time = data transfer time	DATA COMM	Transferzeit = Datentransferzeit; Datentransferdauer; Datenübermittlungsdauer; Übermittlungsdauer	
transfer time	TELECONTR	Übermittlungszeit	
transfer to external storage	DATA PROC	auslagern	
transfer tree	SWITCH	Tannenbaumverzweigung	
transfer value [signal to interference ratio at output related to RF decoupling at input]	RADIO	Transferwert [Störabstand am Ausgang zu RF-Entkopplung am Eingang]	
transfinite	MATH	transfinit	
transfluxor [ferrite nucleus with several holes for read/write leads or control windings]	COMPON	Transfluxor [Ferritkern mit mehreren Löchern für Schreib-/Lesedrähte oder Steuerwicklungen]	
transfluxor ferrite ↑ square-loop ferrite	METAL	Transfluxor-Ferrit ↑ Rechteck-Ferrit	
transform (v.t.)	MATH	transformieren	
transform (n.)	MATH	Transformierte	
transformation	MATH	Transformation	
transformation = conversion ≈ change	TECH	Umwandlung = Transformation; Konversion ≈ Veränderung	
transformation [computer graphics; to displace a graphic]	DATA PROC	Transformation [Computergraphik; Verschieben einer Graphik]	
transformation (→conversion)	PHYS	(→Konversion)	
transformation loop	NETW.TH	Transformationsschleife	
transformation matrix	MATH	Transformationsmatrix	
transformation rate = conversion rate	TECH	Umwandlungsgeschwindigkeit	
transformation ratio (→turns ratio)	EL.TECH	(→Windungsverhältnis)	
transformator yoke [device conducting the manetic flux of a transformator] = yoke ↑ magnet yoke	EL.TECH	Transformatorjoch [Vorrichtung die den Magnetfluß eines Transformators leitet] ↑ Magnetjoch	
transform encoding	CODING	Transformationscodierung	
transformer 1 [component with several windings coupled by a closed magnetic circuit] ↓ transformer 2 [COMPON]; high-power transformer [POWER ENG]; transducer [INSTR]	EL.TECH	Transformator [Bauteil mit mehreren Induktionswicklungen. die über einen geschlossenen magnetischen Kreis gekoppelt sind; in der Starkstromtechnik zur Spannungs- oder Stromübersetzung eingesetzt; in der Schwachstromtechnik zur Impedanzanpassung oder galvanischen Trennung, dort Übertrager genannt] ↓ Übertrager [COMPON]; Umspanner [POWER ENG]; Wandler [INSTR]	
transformer 2 [deditaced to electronics or telecommunications applications, with function of impedance matching or isolating] ↑ transformer 1 [EL.TECH]	COMPON	Übertrager [Transformator für nachrichtentechnisches Zwecke, speziell für Impedanzanpassung oder Potentialtrennung] ↑ Transformator [EL.TECH]	
transformer amplifier (→transformer-coupled amplifier)	CIRC.ENG	(→Transformator-Verstärker)	
transformer bridge	CIRC.ENG	Übertragerbrücke = Transformatorbrücke	
transformer-coupled amplifier = transformer amplifier	CIRC.ENG	Transformator-Verstärker	
transformer coupling	EL.TECH	Transformatorkopplung	
transformer-free (→transformerless)	CIRC.ENG	(→übertragerlos)	
transformer-free telephone circuit	TELEPH	übertragerlose Fernsprechschaltung	
transformerless = transformer-free	CIRC.ENG	übertragerlos = transformatorlos	
transformer oil	EL.TECH	Transformatoröl	
transformer ratio (→turns ratio)	EL.TECH	(→Windungsverhältnis)	
transformer sheet	METAL	Transformatorblech = Elektroblech	
transformer winding ↑ winding [EL.TECH] ↓ primary winding; secondary winding	COMPON	Transformatorwicklung ↑ Wicklung [EL.TECH] ↓ Primärwicklung; Sekundärwicklung	
transgres = overstep; overrun (v.t.); surpass; exceed	TECH	überschreiten	
transgression = overstepping; overrun (n.); surpass (n.); excess (n.); overrange (n.)	TECH	Überschreitung = Überschuß	
tranship (v.t.) (→reload)	ECON	(→umladen)	
transhorizon communication (→transhorizon radiolink)	RADIO REL	(→Überhorizontverbindung)	
transhorizon link (→transhorizon radiolink)	RADIO REL	(→Überhorizontverbindung)	
transhorizon radiolink = transhorizon link; transhorizon communication ↓ troposcatter radiolink	RADIO REL	Überhorizontverbindung ↓ Streustrahl-Richtfunkverbindung	
transhybrid loss	TELEC	Gabelübergangsdämpfung = Gabelsperrdämpfung	
transient 1 (n.) [transition from one steady state to another] = settling process; build-up process ↓ preshoot; ringing; overshoot; baseline overshoot	ELECTRON	Übergangsvorgang [von einem stabilen Zustand in einen anderen] = Ausgleichvorgang; Transient ↓ Vorschwinger; Einschwinger; Überschwinger; Nachschwinger	
transient (adj.) [not permanently in the main memory] = loadable ≠ resident	DATA PROC	transient [nicht dauernd im Hauptspeicher] = ladbar; ladefähig ≠ resident	
transient ≈ ephemeral	SCIE	vorübergehend = transient ≈ ephemer	
transient (→building-up transient)	PHYS	(→Einschwingvorgang)	
transient 2 (→interfering pulse)	ELECTRON	(→Störimpuls)	
transient area (→transient program area)	DATA PROC	(→Übergangsbereich)	
transient behaviour (→transient response)	ELECTRON	(→Übergangsverhalten)	
transient characteristics (→transient response)	ELECTRON	(→Übergangsverhalten)	

transient command	DATA PROC	transientes Kommando
[stays in main memory only during execution]		[nur während der Ausführung im Hauptspeicher]
= external command		= externes Kommando
≠ resident command		≠ residentes Kommando
transient current (→switch-on peak)	EL.TECH	(→Einschaltstromspitze)
transient-current-ratio saturation	MICROEL	Umschaltstromverhältnis
transient data	DATA PROC	transiente Daten
transient decay current	EL.TECH	Nachwirkungsstrom
transient disturbance (→interfering pulse)	ELECTRON	(→Störimpuls)
transient effect	ELECTRON	Einschaltvorgang
= transient phenomenon		
transient information	TELECONTR	Wischermeldung
= transitory alarm		≈ Kurzzeitmeldung
≈ fleeting information		
transient intermodulation (→TIM)	EL.ACOUS	(→TIM-Verzerrung)
transient limiter	CIRC.ENG	Störspitzenbegrenzer
transient oscillation	PHYS	Einschwingung
transient period	PHYS	Einschwingzeit
= transient time		
transient period (→build-up time)	ELECTRON	(→Einschwingzeit)
transient phenomenon	PHYS	Ausgleichsvorgang
transient phenomenon (→transient effect)	ELECTRON	(→Einschaltvorgang)
transient program	DATA PROC	transientes Programm
[only when required temporarily in main memory]		[nur bei Bedarf vorübergehend im Hauptspeicher]
transient program area	DATA PROC	Übergangsbereich
[area of main memory reserved for seldom used parts of the executive routine]		[für selten eingesetzte Teile des Organisationsprogramms reservierter Bereich des Arbeitsspeichers]
= TPA; transient area		
≠ nucleus		≠ Nukleus
transient read (→read error)	TERM&PER	(→Lesefehler)
transient recorder	INSTR	Transienten-Recorder
transient response	ELECTRON	Übergangsverhalten
= transient behaviour; transient characteristics; settling characteristics; build-up response; build-up behaviour; build-up characteristics		≈ Einschwingverhalten
		≈ Sprungantwort [CONTROL]; Impulsantwort
≈ step response [CONTROL]; pulse response		
transient response	CONTROL	Übertragungsverhalten
↓ step response; rise response		↓ Sprungantwort; Anstiegantwort
transients fidelity	EL.ACOUS	Impulstreue
transient signal	ELECTRON	Einzelimpuls
transient state	TECH	Übergangszustand
transient suppressor (→line filter)	CIRC.ENG	(→Netzfilter)
transient time (→transient period)	PHYS	(→Einschwingzeit)
transient time (→build-up time)	ELECTRON	(→Einschwingzeit)
transient voltage	EL.TECH	Übergangsspannung
transimpedance	NETW.TH	Transimpedanz
↑ transfer function		↑ Transferfunktion
transinformation	INF	Transinformation
[effectivly transmitted information, after deduction of the information losses on the transmission path called "equivocation"]		[tatsächlich übertragene Information, unter Abzug der bei der Übertragung verlorengegangenen Information „Äquivokation"]
= transferred information; transmitted information; mutual information		= übertragene Information
transinformation content	INF	Transinformationsgehalt
transistor	MICROEL	Transistor
[semiconductor device with a minimum of three terminals, which permits to amplify the current, voltage or power of a signal]		[Halbleiterbauelement mit mindestens drei Anschlüssen, mit welchem der Strom, die Spannung oder die Leistung eines Signals verstärkt werden kann]
↓ point-contact transistor; junction transistor; pnp transistor; npn transistor; small-signal transistor; power transistor		↓ Spitzentransistor; Flächentransistor; PNP-Transistor; NPN-Transistor; Kleinsignaltransistor; Leistungstransistor
transistor base circuit	CIRC.ENG	(→Transistor-Grundschaltung)
(→transistor base connection)		
transistor base connection	CIRC.ENG	Transistor-Grundschaltung
= transistor base circuit		↓ Basisschaltung; Emitterschaltung; Kollektorschaltung
↓ common base circuit; common emitter connection; common collector emitter		
transistor chip	MICROEL	Transistorchip
↑ chip		= Transitorplättchen
		↑ Chip
transistor circuit	CIRC.ENG	Transistorschaltung
transistor function	MICROEL	Transistorfunktion
transistorized	ELECTRON	transistorisiert
= all-transistorized; fully transistorized		= volltransistorisiert; transistoriert; volltransistoriert
transistorized microphone	TELEPH	Transistorsprechkapsel
		= Transistormikrophon; Linearmikrophon
transistor microphone	EL.ACOUS	Transistormikrophon
transistor mounting insulator	COMPON	Transistorisolierstück
		= Antiwärmstück
transistor multimeter	INSTR	Transistorvielfachmesser
transistor noise	MICROEL	Transistorrauschen
transistor parameter	MICROEL	Transistorkenngröße
		= Transistorparameter; Transistor-Signalkenngrößen
transistor region	MICROEL	Transistorzone
↓ emitter; basis; collector		↓ Emitter; Basis; Kollektor
transistor relay	COMPON	Transistorrelais
= electronic relay; solid-state relay; semiconductor relay		= elektronisches Relais; Verstärkerrelais; Halbleiterrelais
transistor set	CONS.EL	Transistorradio
		= Transistorgerät; Transistor
transistor-transistor logic	MICROEL	Transistor-Transistor-Logik
= TTL		= TTL
transistor voltmeter	INSTR	Transistorvoltmeter
transit	ECON	Transit
transit call	DATA COMM	Durchgangsverbindung
		= Transitverbindung
transit case (→transport case)	TECH	(→Transportkoffer)
transit center (→transit exchange)	SWITCH	(→Durchgangsvermittlungstelle)
transit error	ELECTRON	Zufallsfehler
= spurious error		
transit exchange	SWITCH	Durchgangsvermittlungstelle
= transit switching center; transit center		= Durchgangsvermittlung; Transitvermittlung
↑ exchange		↑ Vermittlungsstelle
transit gate	CIRC.ENG	Durchgangsgatter
transitient limiter	ELECTRON	Störspannungsbegrenzer
= interference limiter		
transition	PHYS	Übergang
[between states]		= Zustandsänderung
transition	INF	Übergang
[from one value of the information parameter to another]		[zwischen Werten des Informationsparameters]
transition (→junction)	MICROEL	(→Übergang)
transitional surface	AERON	Übergangsfläche
transition band	NETW.TH	Übergangsband
		= Übergangsbereich
transition fit	ENG.DRAW	Übergangspassung
↑ class of fits		= Übergangssitz
		↑ Sitzklasse
transition frequency (energy level)	PHYS	Übergangsfrequenz
		[Energieniveau]
transition frequency	MICROEL	(→Transitfrequenz)
(→gain-bandwidth product)		
transition period	TECH	Übergangszeit
transition probability	PHYS	Übergangswahrscheinlichkeit
transition region 2	MICROEL	Übergangsgebiet 2
[variation range of the base voltage necessary to trigger a transistor]		[Änderungsbereich der Basisspannung um einen Transistor umzuschalten]
↑ transistor parameter		= Übergangsbereich 2
		↑ Transistorkenngröße

transition region 1 MICROEL (→Übergang)
(→junction)
transition region (→transition PHYS (→Übergangsbereich)
zone)
transition temperature PHYS **Sprungtemperatur**
[superconductivity] [Supraleitung]
= critical temperature
transition time (→rise time) ELECTRON (→Anstiegzeit)
transition zone PHYS **Übergangsbereich**
= transition region
transition zone ANT **Übergangsbereich**
[from near field to far field region] [zwischen Nahfeld und Fernfeld]
transition zone (→junction) MICROEL (→Übergang)
transitive (adj.) LING **transitiv**
= zielend
transit network DATA COMM **Transitnetz**
transit node DATA COMM **Durchgangsknoten**
= intermediate node = Transitknoten
transitory (→temporary) TECH (→vorübergehend)
transitory alarm (→transient TELECONTR (→Wischermeldung)
information)
transit primary exchange SWITCH **Durchgangs-Hauptvermitt-**
lungsstelle
transit register SWITCH **Durchgangsregister**
transit register DATA PROC **Transitregister**
transitron CIRC.ENG **Transitron**
↑ tube circuit ↑ Röhrenschaltung
transit storage CIRC.ENG **Durchlaufspeicher**
transit stove MANUF **Durchlaufofen**
transit switching center SWITCH (→Durchgangsvermittlungs-
(→transit exchange) telle)
transit system DATA COMM **Transitsystem**
= intermediate system ; relay system
transit time COMPON **Umschlagzeit**
[contact] [Kontakt]
= switchover time; changeover time = Umschaltezeit
transit time (→propagation time) PHYS (→Laufzeit)
transit time (→switchover ELECTRON (→Umschaltzeit)
time)
transit-time distortion TELEC **Laufzeitschräglage**
transit-time effect ELECTRON **Laufzeiteffekt**
transit-time tube MICROW (→Laufzeitröhre)
(→velocity-modulated tube)
transit traffic SWITCH **Durchgangsverkehr**
= through traffic = Transitverkehr; durchge-
hender Verkehr; durchlau-
fender Verkehr
translate LING **übersetzen**
[to repeaty literally in another lan- [wortgetreu in anderer
guage] Sprache wiedergeben]
↓ interprete ↓ dolmetschen
translate DATA PROC **übersetzen**
[program language] [Programmiersprachen]
↓ assemble; compile ↓ assemblieren; kompilieren
translate MODUL **umsetzen**
[translate the frequency of a signal] [ein Signal in seiner Fre-
quenzlage verändern]
translate (→transpose) SCIE (→transponieren)
translated program (→object DATA PROC (→Objektprogramm)
program)
translating program DATA PROC (→Übersetzer)
(→translator)
translation PHYS **Fortbewegung**
= Translation
translation LING **Übersetzung**
translation MODUL **Umsetzung**
= conversion
translation (→transposition) SCIE (→Transponierung)
translation instruction DATA PROC (→Übersetzungsanweisung)
(→directive)
translations agency ECON **Übersetzungsbüro**
translation stage TRANS **Umsetzerstufe**
[FDM] [TF-Technik]
≈ hierarchical level ≈ Hierarchiestufe
translation table (→conversion TECH (→Umrechnungstabelle)
table)
translation time DATA PROC **Übersetzungszeit**
↓ compilation time; assembly time ↓ Kompilierzeit; Assemblier-
zeit

translator LING **Übersetzer**
↓ interpreter ↓ Dolmetscher
translator 1 DATA PROC **Übersetzer**
[program to translate program lan- [Programm zur Überset-
guage or data format] zung von Programmierspra-
= language translator; translating pro- che oder Datenformat]
gram; language processor = Übersetzerprogramm;
↓ assembler; compiler; interpreter 1; Übersetzungsprogramm
processor 2 ↓ Assembler; Kompilierer;
Interpretierer 1; Prozessor
2
translator TRANS **Umsetzer**
translator 2 (→allocator) DATA PROC (→Zuordner)
translator listing DATA PROC **Übersetzungsprotokoll**
↑ listing ↑ Protokoll
↓ compiler listing; assembler listing ↓ Kompiliererprotokoll; Assemblerprotokoll
translator unit (→dial pulse TELEGR (→Wählnumsetzer)
storage)
transliteration LING **Transliteration**
[spelling of an alphabet by another [Umschreibung eines Alp-
one] habets in einem anderen]
translucent (→translucide) PHYS (→durchscheinend)
translucide PHYS **durchscheinend**
= translucent; diaphanous = lichtdurchlässig
≈ transparent ≈ durchsichtig
transmission TELEC **Übertragung**
= communication transmission = Nachrichtenübertragung
↑ communication ↑ Übermittlung
transmission (→drive) TECH (→Antrieb)
transmission (→motion MECH (→Bewegungsübertragung)
transmission)
transmission (→broadcast BROADC (→Rundfunksendung 1)
transmission)
transmission (→transmittance) PHYS (→Transmissionsgrad)
transmission band TELEC **Übertragungsband**
transmission budget TRANS **Leistungsbilanz**
transmission cable COMM.CABLE **Übertragungskabel**
↑ communication cable ↑ Nachrichtenkabel
↓ local junction cable; toll cable ↓ Ortskabel; Fernkabel
transmission capacity INF **Übertragungskapazität**
= communication capacity; communica-
tions capacity; transfer capacity
transmission capacity TELEC **Übertragungskapazität**
transmission capsule TELEPH **Sprechkapsel**
= microphone capsule; transmitter = Mikrophonkapsel
inset; microphone inset; telephone
transmitter; transmitter unit
transmission carrier TELEC (→Übertragungsmedium)
(→transmission medium)
transmission central CIRC.ENG **Sendezentrale**
transmission chain TELEC **Sendekette**
transmission channel INF (→Übertragungskanal)
(→communication channel)
transmission characteristic NETW.TH **Durchlaßcharakteristik**
transmission characteristic TELEC **Übertragungseigenschaft**
= transmission property = Übertragungscharakteri-
stik; Übertragungsverhalten
transmission code (→line code) CODING (→Leitungscode)
transmission coefficient NETW.TH **Übertragungsfaktor**
[output to input value; reciprocal of [Ausgangsgröße zu Ein-
the attenuation factor] gangsgröße; Kehrwert des
= transmission factor; gain factor Dämpfungsfaktors]
≈ transfer function = Übertragungsverhältnis;
≠ attenuation factor Übersetzungsverhältnis;
↓ image transmission coefficient; effec- Übersetzung
tive transmission coefficient; voltage ≈ Übertragungsfunktion
transmission coefficient; current ≠ Dämpfungsfaktor
transmission coefficient; power trans- ↓ Wellenübertragungsfaktor;
mission coefficient; transfer imped- Betriebsübetragungsfaktor;
ance; transfer admittance; gain Spannungsübertragungsfaktor; Stromübertragungsfak-
tor; Leistungsübertragungsfaktor; Kernwiderstand;
Kernleitwert; Verstärkung
transmission control DATA COMM **Datenübertragungssteue-**
[code] **rung 2**
= TC; communication control 2 [Code]
= Übertragungssteuerung 2;
TC

transmission control (→data DATA COMM (→Datenübertragungssteue-
transmission control)   rung 1)
transmission control   DATA COMM **Übertragungssteuerzeichen**
character
= communication control character
transmission data   DATA COMM **Sendedaten**
= transmittal data
transmission delay   TELEC **Durchlaufzeit**
transmission direction   TELEC **Übertragungsrichtung**
= transmission sense   = Übertragungssinn
↓ go direction; return direction   ↓ Hin-Richtung; Rückrich-
                                             tung
transmission efficiency   TELEC **Übertragungswirkungsgrad**
transmission electron   ELECTRON **Durchstrahlungs-Elektronen-**
microscope   **mikroskop**
transmission engineering   TELEC (→Nachrichtenübertragungs-
(→communications transmission en-   technik)
gineering)
transmission equation   LINE TH **Leitungsgleichung**
= telegraphic equation   = Telegrafengleichung
transmission equipment   TELEC **Übertragungseinrichtung**
= transmission facility   = Übertragungsgerät
≈ transmission system   ≈ Übertragungssystem
transmission error   TELEC **Übertragungsfehler**
transmission facility   TELEC (→Übertragungseinrichtung)
(→transmission equipment)
transmission factor   PHYS (→Transmissionsgrad)
(→transmittance)
transmission factor   NETW.TH (→Übertragungsfaktor)
(→transmission coefficient)
transmission gate (→transfer   MICROEL (→Transfergatter)
gate)
transmission hierarchy   TRANS **Übertragungshierarchie**
transmission hologram   PHYS **Transmissionshologramm**
transmission impairment test set INSTR **Übertragungsfehler-Meßplatz**
transmission input (→transmit   TRANS (→Sendeeingang)
input)
transmission interrupt   TELEC **Sendeunterbrechung**
transmission level   TRANS **Sendepegel**
= send level; transmit level
transmission line   TELEC **Übertragungsleitung**
transmission line   POWER ENG **Starkstrom-Freileitung**
= overhead power line   = Freileitung
↓ high voltage transmission line   ↓ Hochspannungsfreileitung
transmission line (→antenna   ANT (→Antennenspeiseleitung)
feeding line)
transmission line antenna   ANT (→DDRR-Antenne)
(→DDRR antenna)
transmission line chart   LINE TH **Leitungsdiagramm**
[for graphical determination of line   [zur graphischen Bestim-
impedances]   mung von Leitungsimped-
= impedance chart; line chart; circle   anzen]
chart   = Kreisdiagramm
↓ rectangular impedance chart; Smith   ↓ Schmidt-Buschbeck-Dia-
chart; Carter chart   gramm; Smith-Diagramm;
                                           Carter-Diagramm
transmission-line constant   LINE TH **Leitungskonstante**
↓ distributed leakage; distributed in-   = Leitungsbelag
ductance; distributed capacitance; dis-   ↓ Ableitungsbelag; Induktivi-
tributed resistance   tätsbelag; Kapazitätsbelag;
                                           Widerstandsbelag
transmission link (→link)   TRANS (→Strecke)
transmission loss   NETW.TH **Durchgangsdämpfung**
= via net loss
transmission loss   TELEC **Übertragungsdämpfung**
                                           = Übertragungsverlust
transmission loss   LINE TH (→Dämpfung)
(→attenuation)
transmission management   TELEC **Übertragungsverwaltung**
≈ path management   ≈ Wegewaltung
transmission mean   TELEC (→Übertragungsmedium)
(→transmission medium)
transmission measuring set   INSTR **TF-Pegelmeßplatz**
↓ level generator; level meter   ↓ Pegelsender; Pegelmesser
transmission medium   TELEC **Übertragungsmedium**
= transmitting medium; transmission   = Übertragungsmittel; Trans-
mean; transmission carrier; carrier   portmittel
transmission method   TELEC (→Übertragungsverfahren)
(→transmission mode)
transmission microscope   PHYS **Durchdringungsmikroskop**

transmission mode   TELEC **Übertragungsverfahren**
= transmission method   = Übertragungsmodus; Über-
                                           tragungsmethode
transmission mode   HF **Übertragungsart**
= transmission type
transmission on demand   TELECONTR (→Aufrufbetrieb)
(→polling mode)
transmission parameter   NETW.TH **Übertragungsparameter**
↓ attenuation factor; transmission coef-   = Übertragungsgröße
ficient; complex transfer constant;   ↓ Dämpfungsfaktor; Übertra-
complex attenuation constant; trans-   gungsfaktor; komplexes
fer constant; attenuation constant;   Übertragungsmaß; komple-
phase angle factor; phase angle;   xes Dämpfungsmaß; Über-
image parameter; effective trans-   tragungsmaß; Dämpfungs-
mission parameter   maß; Phasenmaß;
                                           Dämpfungswinkel; Wellen-
                                           parameter; Betriebsüber-
                                           tragungsparameter
transmission path   TELEC **Übertragungsweg**
≈ transmission link
transmission performance   TELEC (→Übertragungsgüte)
(→transmission quality)
transmission plan   TELEC **Dämpfungsplan**
= attenuation plan
transmission plant   RADIO **Sendeanlage**
= transmitting installation; transmitting ≈ Sendestation
equipment
≈ transmitting station
transmission power (→transmit   RADIO (→Sendeleistung)
power)
transmission procedure   DATA COMM **Übertragungsprozedur**
[rules for data communication]   [Regeln für den Ablauf
↓ HDLC procedure; SDLC procedure   von Datenübermittlungen]
                                           = Prozedur; Übertragungs-
                                           verfahren
                                           ↓ HDLC-Prozedur;
                                           SDLC-Prozedur
transmission program   DATA PROC (→Kommunikationspro-
(→communications program)   gramm)
transmission protocol   DATA COMM (→Kommunikationsproto-
(→communications protocol)   koll)
transmission quality   TELEC **Übertragungsgüte**
= transmission performance   = Übertragungsqualität
transmission range   TELEC **Übertragungsbereich**
≈ transmission band   ≈ Übertragungsband
transmission rate   TELEC **Übertragungsgeschwindigkeit**
[product of signaling rate with num-   [Schrittgeschwindigkeit
ber of bits per unit intervall, includ-   mal Anzahl Bits pro
ing pay load and auxiliary bits; unit: Schritt, einschließlich Nutz-
bit/s]   bits und Hilfsbits; Einheit:
= transmission speed; gross bit rate;   Bit/s]
digit rate; data signaling rate [DATA   = Datenrate [DATA
COMM]; data rate [DATA COMM];   COMM]; Datenübertra-
equivalent bit rate [DATA COMM];   gungsgeschwindigkeit [DA-
overall bit rate; composite bit rate; ag- TA COMM]; Bitrate; Bit-
gregate bit rate; bit rate   geschwindigkeit;
≈ throughput [DATA PROC]; signal-   Bitfrequenz; Bitfolgefre-
ing rate   quenz; Gesamtbitrate; Sum-
↓ data transfer rate [DATA COMM]   menbitrate
                                           ≈ Durchsatz [DATA
                                           PROC]; Schrittgeschwin-
                                           digkeit
                                           ↓ Transfergeschwindigkeit
                                           [DATA COMM]
transmission reference equiva- TELEPH (→Sendebezugsdämpfung)
lent (→sending reference equiva-
lent)
transmission reference system   TELEPH (→Ureichkreis)
(→master telephone system)
transmission reflection test set INSTR **Übertragungsreflexions-Meß-**
                                           platz
transmission route (→route)   TRANS (→Trasse)
transmission securing   TELECONTR **Betriebssicherung**
transmission sense   TELEC (→Übertragungsrichtung)
(→transmission direction)
transmission signal   TELEC **Übertragungssignal**
transmission software   DATA COMM (→Kommunikationssoftware)
(→communications software)
transmission speed   TELEC (→Übertragungsgeschwindig-
(→transmission rate)   keit)

transmission station

transmission station	RADIO	(→Sendestation)
(→transmitting station)		
transmission system	TRANS	**Übertragungssystem**
= interexchange carrier		≈ Übertragungseinrichtung
≈ transmission equipment		
transmission technique	TELEC	(→Nachrichtenübertragungstechnik)
(→communications transmission engineering)		
transmission test set	INSTR	**Übertragungsmeßplatz**
transmission time	TELEC	**Übertragungszeit**
		= Übertragungsdauer
transmission trunking engineering	TELEC	(→Weitverkehrstechnik)
(→long-haul communications engineering)		
transmission trunking equipment	TRANS	(→Weitverkehrsgerät)
(→long-haul transmission equipment)		
transmission trunking system	TRANS	(→Weitverkehrssystem)
(→long-haul system)		
transmission type (→transmission mode)	HF	(→Übertragungsart)
transmission vendor	TELEC	**Übertragungslieferant**
transmission window	OPT.COMM	**Übertragungsfenster**
= spectral window; window		= optisches Fenster; Fenster
transmission with information feedback (→echo principle)	TELECONTR	(→Echobetrieb)
transmissive (→transparent)	PHYS	(→durchsichtig)
transmisssion property	TELEC	(→Übertragungseigenschaft)
(→transmission characteristic)		
transmit (v.t.)	RADIO	**senden**
= emit; send		≈ übertragen
transmit 1	TELEC	**übertragen**
[to convey]		= durchgeben
≈ emit		≈ senden
transmit 2 (→send)	TELEC	(→senden)
transmit .. (→on the transmit side)	TELEC	(→sendeseitig)
transmit aerial (→transmitting antenna)	ANT	(→Sendeantenne)
transmit antenna (→transmitting antenna)	ANT	(→Sendeantenne)
transmit bit	CODING	**Sendebit**
transmit equipment	TELEC	**Sendegerät**
= sending equipment		
transmit frequency	RADIO	**Sendefrequenz**
= send frequency		
transmit input	TRANS	**Sendeeingang**
= transmission input		= F2an (DBP)
transmit level (→transmission level)	TRANS	(→Sendepegel)
transmit low-pass filter	TELEC	**Sendetiefpaß**
transmit mode	TELEC	**Sendebetrieb**
= send mode; transmit operation		
transmit operation (→transmit mode)	TELEC	(→Sendebetrieb)
transmit output	TRANS	**Sendeausgang**
= transmitting output		= F1ab (DBP)
transmit power	RADIO	**Sendeleistung**
= transmission power; emitting power		
transmit-receive set	RADIO	(→Transceiver)
(→transceiver)		
transmit/receive switch	TERM&PER	(→Sende-Empfangs-Umschalter)
(→send/receive switch)		
transmit side	TELEC	**Sendeseite**
= transmitting side		
transmit signal (→transmitted signal)	TELEC	(→Sendesignal)
transmittal data	DATA COMM	(→Sendedaten)
(→transmission data)		
transmittance	PHYS	**Transmissionsgrad**
[optics]		[Optik]
= transmission factor; transmission		= Durchlässigkeit 2
		≈ Transparenz
transmitted information	INF	(→Transinformation)
(→transinformation)		
transmitted light	PHYS	**durchgelassenes Licht**
transmitted sideband	MODUL	**Hauptseitenband**
transmitted signal	TELEC	**Sendesignal**
= transmit signal; send signal		≈ Ausgangssignal
≈ outgoing signal		
transmitted wave	PHYS	**durchgehende Welle**
transmitter	TELEC	**Sender**
transmitter	RADIO	**Sender**
= sender		
transmitter (→measurand transducer)	INSTR	(→Meßumformer)
transmitter (→microphone)	TELEPH	(→Mikrophon)
transmitter amplifier	TELEC	**Sendeverstärker**
transmitter combining filter	RADIO	(→Sendeantennenweiche)
(→antenna duplexer)		
transmitter current supply	TELEPH	**Mikrophonspeisung**
transmitter feeding link	BROADC	**Modulationsleitung**
[from distibution node to transmitter]		[von Schaltpunkt zu Fernsehsender]
↓ TV transmitter feeding link		↓ Fernsehmodulationsleitung
transmitter feeding network	BROADC	**Modulationsleitungsnetz**
		[DBP, Studio-Sender]
transmitter horizon	RADIO PROP	**Radiohorizont des Senders**
		= Funkhorizont des Senders
transmitter inset (→transmission capsule)	TELEPH	(→Sprechkapsel)
transmitter microphone	EL.ACOUS	**Sender-Mikrofon**
transmitter muting	RADIO REL	**Senderabschaltung**
transmitter noise	TELEPH	**Mikrophongeräusch**
= microphone noise; frying (BRI)		
transmitter power supply	RADIO	**Senderstromversorgung**
transmitter-receiver (→transceiver)	TELEC	(→Transceiver)
transmitter-receiver (→transceiver)	RADIO	(→Transceiver)
transmitter/receiver circuit (→transceiver)	MICROEL	(→Sender-/Empfänger-Schaltung)
transmitter stand-by	RADIO	**Senderersatz**
transmitter tube (→transmitting tube)	ELECTRON	(→Senderöhre)
transmitter unit (→transmission capsule)	TELEPH	(→Sprechkapsel)
transmitter valve (→transmitting tube)	ELECTRON	(→Senderöhre)
transmit time	TELECONTR	**Durchgabezeit**
transmitting aerial (→transmitting antenna)	ANT	(→Sendeantenne)
transmitting antenna	ANT	**Sendeantenne**
= transmit antenna; transmitting aerial; transmit aerial; emitting antenna; emitting aerial; sending antenna; sending aerial		
transmitting clock	TRANS	**Sendetakt**
transmitting equipment (→transmission plant)	RADIO	(→Sendeanlage)
transmitting installation (→transmission plant)	RADIO	(→Sendeanlage)
transmitting medium (→transmission medium)	TELEC	(→Übertragungsmedium)
transmitting output (→transmit output)	TRANS	(→Sendeausgang)
transmitting-receiving	RADIO LOC	**Empfänger-Sperröhre**
= TR		[Radar]
transmitting section	TELEC	**Sendeteil**
transmitting side (→transmit side)	TELEC	(→Sendeseite)
transmitting station	RADIO	**Sendestation**
= transmission station		≈ Sendeanlage
≈ transmission plant		
transmitting terminal	DATA COMM	**Sendestation**
= master station; control station		
transmitting tube	ELECTRON	**Senderöhre**
= transmitter tube; transmitting valve; transmitter valve		
transmitting valve (→transmitting tube)	ELECTRON	(→Senderöhre)
transmultiplexer	TRANS	**Transmultiplexer**
[converts a FDM bundle into PCM streams, e.g. a 60 channel supergroup into two 2 Mbit/s streams, and viceversa]		[wandelt TF-Grundgruppen in 2 Mbit/s Bündel, z.B. eine Sekundärgruppe in zwei Bündel 2 Mbit/s, und umgekehrt]
= transmux; TMUX		
transmux (→transmultiplexer)	TRANS	(→Transmultiplexer)
transnational (→international)	ECON	(→international)

transparency	PHYS	Transparenz	
[optics]		[Optik]	
≈ transmittance		= Durchsichtigkeit	
		≈ Transmissionsgrad	
transparency	OFFICE	Transparentfolie	
[for overhead proyector]		[für Tageslichtprojektor]	
= projection transparency; transparent foil; overhead transparency		= Klarsichtfolie; Folie	
transparency	SCIE	Transparenz	
[fig.]		[fig.]	
		= Durchschaubarkeit; Überschaubarkeit	
transparent (adj.)	SCIE	transparent	
[fig.]		[fig.]	
		= durchschaubar; überschaubar	
transparent	PHYS	durchsichtig	
= transmissive		= transparent	
≈ translucide		≈ lichtdurchlässig; durchscheinend	
≠ opaque		≠ undurchsichtig	
transparent	INF	transparent	
transparent cover	OFFICE	Sichthülle	
transparent foil	OFFICE	(→Transparentfolie)	
(→transparency)			
transparent mode	DATA COMM	transparenter Modus	
transparent paper	ENG.DRAW	Transparentpapier	
= tracing paper; detail paper			
transparent pocket	TECH	Klarsichttasche	
transparent transmission	TELEC	transparente Übertragung	
transphasor	OPTOEL	Transphasor	
[a sort of optical transistor]		[eine Art optischer Transistor]	
transponder	RADIO NAV	Antwortbake	
= responder beacon		= Transponder; Wiederholerbake; Antwortgerät	
transponder	SAT.COMM	Transponder	
[transmitter + responder; receives a signal in the satellite, converts the frequency, and sends the signal back to earth]		[empfängt im Satelliten ein Signal, setzt dessen Frequenz um, und sendet es zur Erde zurück]	
transport (n.)	ECON	Transport	
= transportation (AM); carriage		= Beförderung	
transport (→convey)	TECH	(→befördern)	
transport (n.) (→transfer)	PHYS	(→Übertragung)	
transportability	TECH	Transportierbarkeit	
≈ portability		≈ Tragbarkeit	
transportable	TECH	transportierbar	
= luggable		= transportfähig	
≈ mobile; portable		≈ beweglich; tragbar	
transportable cellular phone	MOB.COMM	transportierbares Mobiltelefon	
[can be moved from the car]		[kann dem Wagen entnommen werden]	
↑ cellular phone		↑ Zellulartelefon	
transport address	DATA COMM	Teilnehmeradresse	
= network user access; NUA			
transportation (AM)	ECON	(→Transport)	
(→transport)			
transportation costs	ECON	(→Frachtkosten)	
(→freight 1)			
transportation engineering	TECH	Verkehrstechnik	
transportation layer	DATA COMM	(→Transportschicht)	
(→transport layer)			
transportation mode (AM)	ECON	Transportart	
= kind of transport		= Frachtart	
transport case	TECH	Transportkoffer	
= transit case			
transport container	CODING	Transportcontainer	
[SDH]		[SDH]	
transport costs (→freight 1)	ECON	(→Frachtkosten)	
transport cover (→transport protection)	TECH	(→Transportschutz)	
transport direction	MECH	Transportrichtung	
= feed direction		= Vorschubrichtung	
transport frame	TRANS	Transportrahmen	
[SDH]		[SDH]	
transport hole (→feed hole)	TERM&PER	(→Vorschubloch)	
transport holes (→feed holes)	TERM&PER	(→Transportlochung)	
transport instruction	DATA PROC	Transportbefehl	
transport layer	DATA COMM	Transportschicht	
[4th layer of OSI; defines the parameter relevant for data transfer]		[4. Schicht des ISO-Schichtenmodells; legt die für die Datenübertragung relevanten Parameter fest]	
= transportation layer		= Transportebene; Transportsteuerung	
transport lock	TECH	Transportsicherung	
= transport securing device; shipping spacer			
transport means	TECH	Verkehrsmittel	
transport module	TRANS	Transportmodul	
[SDH]		[SDH]	
transport number (→transference number)	PHYS	(→Überführungszahl)	
transport protection	TECH	Transportschutz	
= transport cover			
transport securing device	TECH	(→Transportsicherung)	
(→transport lock)			
transpose (v.t.)	DATA PROC	umstellen	
transpose	TECH	versetzen	
≈ displace		≈ verlagern	
transpose	SCIE	transponieren	
= translate; transfer		= überführen; übertragen	
transposed band mode	TRANS	Getrenntlageverfahren	
≠ equal band mode		≠ Gleichlageverfahren	
transposed matrix	MATH	transponierte Matrix	
transposer (→television transposer)	BROADC	(→Fernsehumsetzer)	
transposition	OUTS.PLANT	Kreuzung	
[open-wire line]		[Freileitung]	
= crossing			
transposition	SAT.COMM	Transponierung	
= frequency transposition		= Frequenzumsetzung	
transposition	TECH	Versetzung	
≈ displacement		≈ Verlagerung	
transposition	SCIE	Transponierung	
= transfer; translation		= Übertragung; Überführung	
transposition error	TRANS	Transpositionsfehler	
transposition insulator	OUTS.PLANT	Kreuzungsisolator	
transposition interval	OUTS.PLANT	(→Kreuzungsfeld)	
(→transposition section)			
transposition point	OUTS.PLANT	Kreuzungspunkt	
[open-wire line]		[Freileitung]	
transposition pole	OUTS.PLANT	Kreuzungsgestänge	
transposition section	OUTS.PLANT	Kreuzungsfeld	
[open-wire line]		[Freileitung]	
= transposition interval			
transposition system	OUTS.PLANT	Kreuzungsschema	
= rolled transposition			
transputer	DATA PROC	Transputer	
[trans(more than) + computer; powerful computer with non-convential structure, e.g. with a multiple arithmetic-logic unit]		[leistungsstarker Computer mit unkonventioneller Struktur, z.B. mit mehrfachem Rechenwerk]	
transputer network	DATA COMM	Transputernetzwerk	
transshipment	ECON	Warenumschlag	
= handling		= Umschlag	
transshipment (→reloading)	ECON	(→Umladung)	
transversal (adj.)	TECH	transversal	
= transverse; cross; traverse		= querliegend; quer	
≈ diagonal; broadside		≈ diagonal; breitseitig	
transversal attenuation	TRANS	Querdämpfung	
transversal equalizer	TRANS	Transversalentzerrer	
transversal filter	NETW.TH	Transversalfilter	
		= nichtrekursives Digitalfilter; nichtrekursives Filter	
transversal fin	MECH	Querrippe	
= transversal rib			
transversality	SCIE	Transversalität	
transversal magnetic wave	EL.TECH	transversale magnetische Welle	
= TM wave; TM mode		= TM-Welle; E-Welle; magnetische Transversalwelle	
transversal recording	TV	Querspurverfahren	
= quadruplex recording			
transversal recording	TERM&PER	Transversalaufzeichnung	
≠ longitudinal recording		≠ Longitudinalaufzeichnung	

transversal rib (→transversal fin)		MECH	(→Querrippe)	**trapezoidal tooth antenna**	ANT	**Trapez-Zahn-Antenne**
transversal section		NETW.TH	**Transversalglied**	**trapping** (→program interrupt)	DATA PROC	(→Programmunterbrechung)
transversal slot		EQUIP.ENG	**Querschlitz**	**trash** (→scrap)	TECH	(→Abfall)
transversal strut		MECH	**Querstrebe**	**trash** (→trashy goods)	ECON	(→Schundware)
≈ crossarm			≈ Querträger	**trashy goods**	ECON	**Schundware**
↑ strut			↑ Strebe	= trash		= Schund
transversal vibration		PHYS	**Querschwingung**	**travel** (→stroke 2)	MECH	(→Hub)
≈ transverse wave			≈ Transversalwelle	**travel-agency terminal**	TERM&PER	**Reisebüro-Terminal**
transversal wave		PHYS	**Transversalwelle**	**traveling-wave electrode**	ELECTRON	**Lauffeldelektrode**
= transverse wave			= transversale Welle; Querwelle			= Wanderwellenelektrode
≈ transversal vibration			≈ Querschwingung	**travelling-field tube**	MICROEL	**Lauffeldröhre**
transverse		MECH	**querlaufend**	↑ velocity-modulated tube		↑ Laufzeitröhre
transverse (→transversal)		TECH	(→transversal)	↓ travelling-wave tube		↓ Wanderfeldröhre
transverse cable		OUTS.PLANT	**Quertragseil**	**travelling wave**	PHYS	**fortschreitende Welle**
transverse contraction		PHYS	**Querkontraktion**	= running wave		= wandernde Welle; Wanderwelle; laufende Welle
transverse contraction ratio		PHYS	**Querkontraktionskoeffizient**			
			= Poissonsche Konstante	**travelling-wave antenna**	ANT	**Wanderwellenantenne**
transverse electric wave		LINE TH	**transversale elektrische Welle**	↓ fishbone antenna; helix antenna; long-wire antenna; rhombic antenna; V antenna		↓ Fischgrätenantenne; Wendelantenne; Langdrahtantenne; Rhombusantenne; V-Antenne
= TE wave; TE mode			= TE-Welle; H-Welle; elektrische Transversalwelle			
transverse electromagnetic wave		LINE TH	**transversale elektromagnetische Welle**	**travelling-wave cathode-ray tube**	ELECTRON	**Wanderfeld-Kathodenstrahlröhre**
= guided wave; TEM wave; TEM mode			= Lecher-Welle; L-Welle; TEM-Welle; Leitungswelle; elektromagnetische Transversalwelle	= travelling-wave CRT		↑ Oszilloskopröhre
				↑ oscilloscope tube		
				travelling-wave CRT	ELECTRON	(→Wanderfeld-Kathodenstrahlröhre)
transverse field		PHYS	**Querfeld**	(→travelling-wave cathode-ray tube)		
			= Transversalfeld	**travelling-wave klystron**	MICROW	(→Wanderfeldklystron)
transverse force (→cross force)		PHYS	(→Querkraft)	(→travelling-wave multiple-beam klystron)		
transverse grind		MECH	**Querschliff**			
transverse magnetization		PHYS	**Quermagnetisierung**	**travelling-wave magnetron**	MICROW	**Wanderfeldmagnetron**
transverse motion		MECH	**Querbewegung**	= multicavity magnetron; multisegment magnetron		= Lauffeldmagnetron; Wanderfeld-Magnetfeldröhre; Vielschlitzmagnetron; Vielkammermagnetron
= transverse movement				↑ magnetron		
transverse movement (→transverse motion)		MECH	(→Querbewegung)			
						↑ Magnetron
transverse suspension		MECH	**Queraufhängung**	**travelling-wave multiple-beam klystron**	MICROW	**Wanderfeldklystron**
transverse-symmetrical		MATH	**quersymmetrisch**			↑ Klystron
transverse symmetric quadripole		NETW.TH	**quersymmetrischer Vierpol**	= travelling-wave klystron		
(→transverse symmetric two-port)				↑ klystron		
transverse symmetric two-port		NETW.TH	**quersymmetrischer Vierpol**	**travelling wave tube**	MICROW	**Wanderfeldröhre**
= transverse symmetric quadripole				= TWT		= TWT
transverse symmetry		MATH	**Quersymmetrie**			↑ Lauffeldröhre
transverse voltage (→cross voltage)		PHYS	(→Querspannung)	**travel time** (→propagation time)	PHYS	(→Laufzeit)
				traverse (v.t.)	TECH	**durchqueren**
transverse wave (→transversal wave)		PHYS	(→Transversalwelle)	≈ cross (v.t.)		≈ kreuzen
				traverse (→crossarm)	MECH	(→Querträger)
transzorb		MICROEL	**Transzorb**	**traverse** (→transversal)	TECH	(→transversal)
↑ Zener diode			↑ Zenerdiode	**tray** (→paper tray)	TERM&PER	(→Papiernachfüllmagazin)
trap (v.t.)		TECH	**einfangen**	**trayectory**	PHYS	**Bahn**
trap (n.)		TECH	**Falle**	= path line; path		= Bahnkurve
[device to catch]				↓ circular orbit; orbit		↓ Umlaufbahn; Kreisbahn
trap (n.)		MICROEL	**Haftstelle**	**T/R distance**	RADIO	**Sende-Empfangs-Abstand**
			= Falle; Einfangstelle; Fangstelle; Trap	**treasury**	ECON	**Fiskus**
						= Staatskasse
trap (n.)		TECH	**Abscheider**	**treasury secretary** (AM)	ECON	(→Finanzminister)
= separator				(→minister of finance)		
trap [HF] (→bandstop filter)		NETW.TH	(→Bandsperrfilter)	**treat**	TECH	**behandeln**
trap (→unprogrammed conditional program jump)		DATA PROC	(→nichtprogrammierter Programmsprung)	**treatise**	LING	**Abhandlung**
				treatment	TECH	**Behandlung**
trap (→wave trap)		ANT	(→Wellenfalle)	= termination; preparation		= Bearbeitung
trapatt diode		MICROEL	**Trapatt-Diode**	≈ improvement; processing		≈ Veredelung; Verarbeitung
↑ avalanche photodiode			↑ Lawinenphotodiode	**treatment** (→work)	MECH	(→Bearbeitung)
trapdoor		DATA PROC	**Falltür**	**treble**	EL.ACOUS	**Höhen**
[way to force acess to a system]			[Einbruchstelle in ein System]	**treble** (→triplicate)	MATH	(→verdreifachen)
				treble block (→tackle)	TECH	(→Flaschenzug)
trapezium (pl.-iums or -ia) (AM)		MATH	**Trapez**	**treble correction**	EL.ACOUS	**Höhenanhebung**
[quadrilateral with no two sides parallel]			[Viereck ohne Parallelität von Seiten]	= pre-emphasis		
				trebling (→triplication)	MATH	(→Verdreifachung)
= trapezoid (BRI)			≠ Parallelogramm	**tree**	NETW.TH	**Baum**
≠ parallelogram			↑ Viereck	**tree diagram**	TECH	**Baumdiagramm**
↑ quadrilateral				**tree-like**	TECH	**baumähnlich**
trapezium converter		POWER SYS	**Trapezumrichter**	≈ tree-shaped		= baumförmig
trapezium distortion		TV	**Trapezfehler**	≈ branched		≈ verzweigt
= keystone distortion			= Trapezverzeichnung	**tree-shaped** (→tree-like)	TECH	(→baumähnlich)
trapezoid (BRI) (→trapezium)		MATH	(→Trapez)	**tree-shaped network**	TELEC	**Baumnetz**
trapezoidal		MATH	**trapezförmig**	≈ tree structure		≈ Baumstruktur
trapezoidal field winding		POWER SYS	**Trapezwicklung**	**tree sort**	DATA PROC	**Baumsortierung**
			= Trapezfeldwicklung	= head sort		

tree structure (→tree topology)	TELEC	(→Baumstruktur)
tree topology	TELEC	**Baumstruktur**
= tree structure; hierarchical structure		= hierarchische Struktur
trellis-coded modulation	MODUL	**trelliscodierte Modulation**
= TCM		= TCM
trembler bell (→dc bell)	COMPON	(→Gleichstromwecker)
tremendously high frequency	EL.TECH	**Submillimeterwelle**
= THF		[300 GHz bis 3.000 GHz]
trench (n.)	CIV.ENG	**Graben**
= ditch		↓ Kabelgraben
trench (n.)	MICROEL	**Graben**
trencher	CIV.ENG	**Grabmaschine**
trenching	CIV.ENG	**Grabenöffnung**
trend (→tendency)	COLLOQ	(→Tendenz)
trend line	DATA PROC	**Trendkurve**
trend-setting	TECH	**zukunftsweisend**
≈ future-oriented		= richtungsweisend
		≈ zukunftsorientiert
trend setting costumer	ECON	**Leitkunde**
= leading costumer		
triac	MICROEL	**Triac**
["triode & alternating current"; thyristor pair with one gate]		[zwei parallel geschaltete Thyristoren mit einer Steuerelektrode]
= AC thyristor; bidirectional triode thyristor; triode AC semiconductor switch		= Wechselstromthyristor; Zweiweg-Thyristor; Zweirichtungs-Thyristordiode
↑ thyristor		↑ Thyristor
triad (n.)	DATA PROC	**Triade**
[three bits or characters]		[drei Bits oder Zeichen]
triadic bell	TERM&PER	**Dreiklangton**
trial (v.t.)	TECH	**ausloten**
		≈ erproben
trial (n.)	TECH	**Versuch**
= test; experiment; attempt		= Probe; Experiment
trial (v.t.) (→prooftest)	QUAL	(→erproben)
trial (→prooftesting)	QUAL	(→Erprobung)
trial and error	SCIE	**Trial-and-error**
		= Lernen am Erfolg
trial condition	TECH	**Versuchsbedingung**
trial equipment (→evaluation unit)	TECH	(→Vorführgerät)
trial product sample	ECON	**Produktmuster**
trial run	DATA PROC	**Versuchslauf**
triangle	MATH	**Dreieck**
↑ polygon		↑ Vieleck
triangle antenna	ANT	**Dreiecksantenne**
= multiwire-triatic antenna		= Dreieckantenne
triangle characteristics	ELECTRON	**Dreieck-Kenndaten**
triangle circuit (→delta section)	NETW.TH	(→Pi-Schaltung)
triangle connection (→delta section)	NETW.TH	(→Pi-Schaltung)
triangle generator	CIRC.ENG	**Dreieckgenerator**
triangle loop	ANT	**Dreieckschleife**
triangle pulse (→sawtooth wave)	ELECTRON	(→Sägezahnschwingung)
triangle-sine wave generator	CIRC.ENG	**Dreieck-Sinus-Generator**
triangle-square wave generator	CIRC.ENG	**Dreieck-Rechteck-Generator**
triangular	MATH	**dreieckig**
↑ polygonal		↑ vieleckig
triangular dipol	ANT	**Spreizdipol**
= spread-wire dipole		= Dreiecksdipol
triangular file (→three-square file)	MECH	(→Dreikantfeile)
triangular grid array	ANT	**Dreiecksgitter-Gruppe**
triangular lattice	PHYS	**Dreiecksgitter**
triangular wave (→sawtooth wave)	ELECTRON	(→Sägezahnschwingung)
triaxial	TECH	**dreiachsig**
triband	RADIO	**Dreiband**
triband antenna	ANT	**Dreibandantenne**
= tribander		
triband cubical quad antenna	ANT	**Dreiband-Cubical-Quad-Antenne**
triband delta loop antenna	ANT	**Dreiband-Delta-loop-Antenne**
triband element	ANT	**Dreiband-Strahlelement**
tribander (→triband antenna)	ANT	(→Dreibandantenne)
triband groundplane antenna	ANT	**Dreiband-Groundplane-Antenne**
triband trap antenna	ANT	**Dreiband-Trap-Antenne**
tribit	INF	**Tribit**
		= Drei-Bit; 3-Bit
triboluminescence	PHYS	**Triboluminszenz**
[by mechanic impact]		[durch mechanische Einwirkung induziert]
tributary	TRANS	**Unterbündel**
[multiplex technique]		[Multiplextechnik]
tributary (→tributary signal)	TRANS	(→Zubringersignal)
tributary channel	TELEC	(→Unterkanal)
(→subchannel)		
tributary signal	TRANS	**Zubringersignal**
[subsignal of a multiplex grouping]		[Untersignal einer Multiplexbündelung]
= tributary; constituent signal		= Zubringer 2
tributary station	TELECONTR	**Unterstation**
= secondary station; substation; slave station; outstation		= Trabantenstation; Fernwirk-Unterstation
≠ master station		≠ Zentrale
↓ passive station		↓ Wartestation
tributary unit	TELEC	**Zubringereinheit**
= TU		[SDH]
trichromatric	PHYS	**trichromatisch**
trickle-charge	POWER SYS	(→puffern)
(→float-charge)		
tricky (→ingenious)	TECH	(→ausgeklügelt)
triclinic	PHYS	**triklin**
[crystallography; three distinct symmetry axes at irregular angles]		[Kristallographie; drei ungleichwertige Achsen in beliebigen Winkeln]
		= triklinisch; asymmetrisch
tridimensional antenna array	ANT	(→räumliche Gruppe)
(→tridimensional array)		
tridimensional array	ANT	**räumliche Gruppe**
= tridimensional antenna array		= räumliche Gruppenantenne
trigger (v.t.)	DATA PROC	**auslösen**
= initiate; clear down; release (v.t.); enable (v.t.); start (v.t.); activate		= einleiten; starten
trigger (v.t.)	ELECTRON	**auslösen**
[to activate a process by a signal]		[einen Vorgang durch ein Signal auslösen]
= activate		= zünden; triggern
trigger 1 (n.)	ELECTRON	**Auslöser**
= ignitor; firing device		= Auslösevorrichtung; Zünder; Trigger
trigger 2 (n.) (→triggering)	ELECTRON	(→Auslösung)
trigger (→trigger circuit 1)	CIRC.ENG	(→Triggerschaltung)
trigger (n.) (→trigger pulse)	ELECTRON	(→Triggerimpuls)
trigger characteristics	ELECTRON	**Trigger-Kenndaten**
trigger circuit 1	CIRC.ENG	**Triggerschaltung**
[circuit triggering something]		[Schaltung die etwas auslöst]
= trigger		= Trigger; Auslöser; Auslöseschaltung
≈ multivibrator		
↑ control circuit 1		≈ Kippschaltung
		↑ Steuerschaltung
trigger circuit 2	CIRC.ENG	**Zündschaltung**
[to drive a thyristor]		[zum Anschalten von Thyristoren]
trigger current	ELECTRON	**Zündstrom**
= triggering current; firing current; activation current		= Auslösestrom; Triggerstrom
trigger delay (→trigger holdoff)	ELECTRON	(→Triggerverzögerung)
trigger diode	MICROEL	**Triggerdiode**
[corresponds to two antiparallel diodes, triggers in both directions]		[äquivalent zu zwei antiparallel geschalteten Dioden, zündet in beiden Richtung]
= diode AC switch; diac; biswitch; sweep diode		= Zweiweg-Schaltdiode; Kippdiode; Fünfschichtdiode; Diac; Biswitch; Auslösediode
trigger electrode	ELECTRON	**Triggerelektrode**
		= Auslöseelektrode
trigger enhacement	INSTR	**Triggererweiterung**
trigger generator (→trigger source)	ELECTRON	(→Triggerquelle)

trigger holdoff 626

trigger holdoff	ELECTRON	**Triggerverzögerung**	
= trigger delay		= Auslöseverzögerung	
triggering	ELECTRON	**Auslösung**	
= trigger 2 (n.); activation; firing; ignition; release (n.); clearing ≈ enabling		= Zündung; Zünden; Triggern; Triggerung; Aktivierung ≈ Einschaltung	
triggering (→drive 1)	ELECTRON	(→Ansteuerung)	
triggering current (→trigger current)	ELECTRON	(→Zündstrom)	
triggering level	ELECTRON	**Triggerschwelle**	
= triggering threshold; trigger threshold		= Triggerpegel; Auslöseschwelle; Auslösepegel	
triggering signal	ELECTRON	**Triggersignal**	
		= Auslösesignal	
triggering threshold (→triggering level)	ELECTRON	(→Triggerschwelle)	
trigger input	ELECTRON	**Triggereingang**	
		= Auslöseeingang	
trigger level range	INSTR	**Triggerpegelbereich**	
trigger logic (→control logic)	ELECTRON	(→Ansteuerlogik)	
trigger pulse	ELECTRON	**Triggerimpuls**	
[iniciates some action]		[löst einen Vorgang aus]	
= trigger (n.)		= Auslöseimpuls	
≈ drive pulse; triggering signal; control pulse		≈ Ansteuerimpuls; Triggersignal; Steuerimpuls	
trigger source	ELECTRON	**Triggerquelle**	
= trigger generator		= Auslösequelle	
trigger switch	ELECTRON	**Triggerschalter**	
		= Auslöseschalter	
trigger threshold (→triggering level)	ELECTRON	(→Triggerschwelle)	
trigger transformer	CIRC.ENG	**Ansteuerübertrager**	
= control transformer; drive transformer			
trigger voltage	ELECTRON	**Zündspannung**	
= firing voltage		= Triggerspannung; Auslösespannung	
trigger winding	ELECTRON	**Triggerwicklung**	
		= Auslösewicklung	
trigistor	MICROEL	**Trigistor**	
		[PNPN-Transistor mit Flipflop-Eigenschaften]	
trigonometric	MATH	**trigonometrisch**	
trigonometric function (→angular function)	MATH	(→Winkelfunktion)	
trigonometry	MATH	**Trigonometrie**	
trillion (AM)	MATH	**Billion**	
[a million of millions]		[eine Million Millionen]	
= billion (BRI)			
trim (→masking)	TECH	(→Verkleidung)	
trimester (→quarter)	ECON	(→Quartal)	
trimestral (→quarterly)	ECON	(→vierteljährlich)	
trimestrial (→quarterly)	ECON	(→vierteljährlich)	
trimmer 1 (→trimming capacitor)	COMPON	(→Trimmerkondensator)	
trimmer 2 (→trimming potentiometer)	COMPON	(→Trimmerwiderstand)	
trimmer capacitor (→trimming capacitor)	COMPON	(→Trimmerkondensator)	
trimmer resistor (→trimming potentiometer)	COMPON	(→Trimmerwiderstand)	
trimming	METAL	**Beschneiden**	
trimming capacitor	COMPON	**Trimmerkondensator**	
[variable for adjustment purposes]		[für Abgleich veränderbar]	
= trimming condenser; trimming capacitor; trimmer 1; balancing capacitor; ↑ variable capacitor		= Abgleichkondensator = Trimmer 1; Korrektionskondensator ↑ einstellbarer Kondensator	
trimming condenser (→trimming capacitor)	COMPON	(→Trimmerkondensator)	
trimming potentiometer	COMPON	**Trimmerwiderstand**	
= trimmer resistor; trim-pot; trimmer 2 ↑ potentiometer		= Trimmwiderstand; Trimmerpotentiometer; Trimmpotentiometer; Widerstandstrimmer; Trimmer 2 ↑ Regelwiderstand	
trimming resistor (→balancing resistor)	ELECTRON	(→Abgleichwiderstand)	
trimmtool (→alignment tool)	ELECTRON	(→Abgleichwerkzeug)	
trim-pot (→trimming potentiometer)	COMPON	(→Trimmerwiderstand)	
trinistor	MICROEL	**Trinistor**	
trinitron (→trinitron tube)	ELECTRON	(→Trinitronröhre)	
trinitron tube	ELECTRON	**Trinitronröhre**	
[the beam of one electron gun is splitted into three by an electronic prism]		[aus einer Elektronenkanone werden die Strahlen für die drei Farben über ein elektronisches Prisma erzeugt]	
= trinitron		= Trinitron	
triode (→triode vacuum tube)	ELECTRON	(→Triode)	
triode AC semiconductor switch	MICROEL	(→Triac)	
(→triac)			
triode vacuum tube	ELECTRON	**Triode**	
[with three electrodes]		[mit drei Elektroden]	
= triode; three-electrode tube		= Hochvakuumtriode; Dreielektrodenröhre	
↑ electron tube		↑ Elektronenröhre	
trip (v.t.)	POWER SYS	**auslösen**	
[release to initiate]			
= release			
trip (n.)	POWER SYS	**Auslösung**	
[release that initiates]			
= tripping; release			
trip (v.t.) (→unlock)	MECH	(→entriegeln)	
tripartite conference	TELEPH	**Dreierkonferenz**	
= three-party conversation; add-on conference		= Dreiergespräch	
triplate stripline	MICROW	**Triplate-Leitung**	
[screened stripline]		[abgeschirmter Streifenleiter]	
triple (n.)	EL.TECH	**Dreileiter**	
triple (n.)	MATH	**Tripel**	
[mathematical quantity of three elements]		[mathematische Größe aus drei Elementen]	
triple (→threefold)	COLLOQ	(→dreifach)	
triple (→three-step)	TECH	(→dreistufig)	
triple (→triplicate)	MATH	(→verdreifachen)	
triple conductor	COMM.CABLE	**Dreier**	
triple-C section	EQUIP.ENG	**Dreifach-C-Profil**	
triple eurocard size	ELECTRON	**Dreifacheuropaformat**	
		= Dreifach-Europakartenformat	
triple folded dipole	ANT	**Doppelschleifendipol**	
		= Dreifach-Faltdipol	
triple integral	MATH	**dreifaches Integral**	
↑ multiple integral		↑ mehrfaches Integral	
triple leg antenna	ANT	**Triple-leg-Antenne**	
triple mirror	PHYS	**Zentralspiegel**	
triple output	EL.TECH	**Dreifachausgang**	
triple-phase	POWER ENG	(→dreiphasig)	
(→three-phase 1)			
triple-pole (→three-pole)	EL.TECH	(→dreipolig)	
triple precision	DATA PROC	**Dreifachgenauigkeit**	
triple resonant circuit	ANT	**Dreiband-Resonanzkreis**	
triple-threaded	MECH	**dreigängig**	
		[Gewinde]	
trip level	POWER SYS	**Auslösepegel**	
triplexer	RADIO	**Triplexer**	
triple-zero key	TERM&PER	**Dreifachnulltaste**	
triplicate	MATH	**verdreifachen**	
= triple; treble		↑ vervielfachen	
↑ multiplicate			
triplication	MATH	**Verdreifachung**	
= trebling		↑ Vervielfachung	
↑ multiplication			
triplug	COMPON	**Dreifachstecker**	
tripod	TECH	**Dreifuß**	
tripole	ANT	**Tripol**	
↑ multipole		↑ Multipol	
tripole antenna	ANT	**Tripolantenne**	
tripping (→trip)	POWER SYS	(→Auslösung)	

tripping current (→blowing current)		COMPON	(→Auslösestrom)	troy weight (→troy system)	PHYS	(→Zwölf-Unzen-System)
				TR switch (→duplexer 2)	RADIO	(→Antennenumschalter)
trip point		POWER SYS	**Auslösepunkt**	T/R tube	RADIO	**T/R-Röhre**
trip signal		POWER SYS	**Auslösesignal**	[changes an antenna over between transmit and receive operation]		[schaltet Antenne von Sende- auf Empfangsbetrieb um]
= release signal						
trisistor		MICROEL	**Trisistor**	↑ duplexer; switching tube [ELECTRON]		↑ Antennenumschalter; Schaltröhre [ELECTRON]
[characteristic similar to thyristor]			[mit einem Thyristor ähnlichen Eigenschaften]			
↑ switching transistor			↑ Schalttransistor	truck	TECH	**Lastkraftwagen**
				= lorry (BRI)		= LKW
tristate		MICROEL	**Tristate**	true	MATH	**wahr**
tristate gate		MICROEL	**Tristate-Gatter**	true complement (→radix complement)	DATA PROC	(→Basiskomplement)
tri-state logic		MICROEL	**Dreizustands-Logik**			
tristate output		MICROEL	**Tristate-Ausgang**	true motion radar	RADIO	**Kursradar**
tristate technique		MICROEL	**Tristate-Technik**	true power (→active power)	EL.TECH	(→Wirkleistung)
tristimulus value (→chrominance)		TV	(→Farbwert)			
trisulfide		CHEM	**Trisulfid**	true power match (→active power matching)	EL.TECH	(→Wirkleistungsanpassung)
trivalent		MATH	**dreiwertig**			
= three-valued				true power matching (→active power matching)	EL.TECH	(→Wirkleistungsanpassung)
Troeger connection (→Morgan connection)		POWER SYS	(→Morgan-Schaltung)			
				true resistance	NETW.TH	**reeller Widerstand**
Trojan Horse		DATA PROC	**Trojanisches Pferd**	[current and voltage in phase]		[Gleichphasigkeit zwischen Strom und Spannung]
trolleybus		TECH	**Obus**	≈ effective resistance; ohmic resistance [EL.TECH]		
≈ tramway			= Oberleitungsomnibus			≈ Wirkwiderstand; ohmscher Widerstand [EL.TECH]
			≈ Straßenbahn			
trombone line		MICROW	**Posaune**	true to gage	TECH	**maßgerecht**
[of adjustable length]			≠ veränderbarer Länge	true to scale	ENG.DRAW	**maßstabgerecht**
tropicalized		QUAL	**tropenfest**	≈ scaled		= maßstabsgerecht; maßstabgetreu; maßstabgetreu
= tropi-proof			= tropentauglich			
↑ climate-resistant			↑ Klimabeständig			≈ maßstäblich
tropics		METEOR	**Tropen**	true to shape	MECH	**formgerecht**
tropi-proof (→tropicalized)		QUAL	(→tropenfest)	= shape-accurate		= formgenau
tropopause		GEOPHYS	**Tropopause**	truncate	MATH	**abstreichen**
troposcatter		RADIO PROP	**troposphärische Steuerausbreitung**	[of fractional digits]		[von Kommastellen]
= tropospheric scatter; tropospheric scatter propagation			= Überhorizontausbreitung; Troposcatter	= cut-off		≈ abrunden
				≈ round-off		
troposcatter link (→troposcatter radio link)		RADIO REL	(→Streustrahl-Richtfunkverbindung)	truncated	MECH	**abgeschnitten**
				truncation	MATH	**Abstrich**
troposcatter radio link		RADIO REL	**Streustrahl-Richtfunkverbindung**	≈ rounding-off		≈ Abrundung
= troposcatter link; scatter link			= Streustrahlverbindung; Troposcatter-Verbindung; Scatterverbindung	truncation (→abnormal end)	DATA PROC	(→Absturz)
↑ trans-horizon radio link						
				truncation error	MATH	**Abstreichfehler**
			↑ Überhorizontverbindung	trunk	SWITCH	**Amtsleitung**
troposphere		GEOPHYS	**Troposphäre**	trunk 1	TELEC	**Fernmeldelinie**
tropospheric scatter (→troposcatter)		RADIO PROP	(→troposphärische Steuerausbreitung)	[a single or multichannel connection, routed commonly on the same carrier system, by cable, open wire line or microwave line]		[zwischen entfernten Punkten eingerichteter Verbindungsweg, mit einem oder mehreren gemeinsam auf der gleichen Trasse geführten Kanälen, über Kabel, Freileitung oder Funk]
tropospheric scatter propagation (→troposcatter)		RADIO PROP	(→troposphärische Steuerausbreitung)			
trouble (→malfunction)		TECH	(→Funktionsstörung)	= telecommunication line; line		
trouble (→difficulty)		COLLOQ	(→Schwierigkeit)	≈ trunk group		
trouble diagnosis (→fault diagnosis)		QUAL	(→Fehlersuche)	↓ trunk line; local line; cable line; open-wire line; line-of-sight line		= Linie
						≈ Leitungsbündel
trouble-free		TECH	**störungsfrei**			↓ Fernlinie; Ortslinie; Kabellinie; Freileitungslinie; Richtfunklinie
= perturbation-free			≈ fehlerfrei			
≈ error-free						
trouble indication (→malfunction indication)		TELECONTR	(→Störungsanzeige)	trunk 2 (→interoffice trunk)	TELEC	(→Amtsverbindungsleitung)
				trunk (→trunk line)	BROADC	(→Strecke)
trouble localization (→fault diagnosis)		QUAL	(→Fehlersuche)	trunk buffer	SWITCH	**Leitungspuffer**
				trunk cable (BRI) (→toll cable)	COMM.CABLE	(→Fernkabel)
trouble report (→failure report)		QUAL	(→Fehlerbericht)			
troubleshoot (v.t.) (→debug)		DATA PROC	(→austesten)	trunk call (BRI) (→toll call 1)	TELEPH	(→Ferngespräch)
trouble-shooter (→fault-clearer)		TECH	(→Entstörer)	trunk carrier		**Fernnetzbetreiber**
troubleshooting (→debugging)		DATA PROC	(→Austesten)	= long-distance carrier; long-haul carrier; long-range carrier; interexchange carrier		= Fernverkehrsgesellschaft; Fernleitungsgesellschaft; Weitverkehrsgesellschaft
trouble shooting (→fault diagnosis)		QUAL	(→Fehlersuche)	↑ operating company		↑ Betriebsgesellschaft
				trunk charge (→long-distance call fee)	TELEC	(→Ferngebühr)
troubleshooting kit		INSTR	**Fehlersuchsatz**			
			= Fehlersuch-Kit	trunk circuit	SWITCH	**Leitungssatz**
trouble ticket		QUAL	**Fehlerformular**	[interface unit to trunk lines]		[Schnittstelleneinheit zu Verbindungsleitungen]
trouble tracking (→fault diagnosis)		QUAL	(→Fehlersuche)	= TC; line supplement		
trough		SYS.INST	**Wanne**	↑ circuit		↑ Satz
↓ cable trough			↓ Kabelwanne	trunk circuit (BRI) (→toll trunk)	TELEC	(→Fernleitung)
trough (→dip)		PHYS	(→Senke)			
trough-connecting path		SWITCH	**Durchschaltweg**	trunk circuit bothway	SWITCH	**wechselseitiger Leitungssatz**
troy system		PHYS	**Zwölf-Unzen-System**	trunk code (→area code)	SWITCH	(→Ortskennzahl)
[system of mass measures based on the pound of 12 ounces]			[auf ein Pound von 12 Unzen basierendes angelsächsisches Maßsystem für Masse]	trunk concentrator 1	SWITCH	**Verkehrskonzentrator**
				= line concentrator 1		= Leitungskonzentrator 1
= troy weight				≈ stand-alone concentrator		≈ Wählsterneinrichtung

trunk concentrator 2 SWITCH (→Wählsterneinrichtung)
  (→stand-alone concentrator)
trunk dialing (→long-distance SWITCH (→Fernwahl)
  dialing)
trunk distribution frame SYS.INST (→Zwischenverteiler)
  (→intermediate distribution frame)
trunk exchange SWITCH **Fernvermittlungsstelle**
  = toll office; long-distance exchange; = Fernamt; Fernwählamt
  toll center office; toll center; TC; toll ↑ Vermittlungsstelle
  switching office ↓ Endvermittlungsstelle;
  ↑ exchange Knotenvermittlungsstelle;
  ↓ local exchange; secondary exchange; Hauptvermittlungsstelle;
  primary exchange; regional ex- Zentralvermittlungsstelle;
  change; international gateway ex- Auslandskopfvermittlungs-
  change stelle; Auslandsvermitt-
   lungsstelle; internationale
   Durchgangsvermittlungs-
   stelle
trunk group SWITCH **Leitungsbündel**
  [group of trunks between equal end- [Leitungsbündel zwischen
  points] gleichen Endpunkten]
  = bunch (AM); trunking; group of = Bündel
  trunks; group
trunk group alarm SWITCH **Bündelalarm**
trunk group number SWITCH **Bündelnummer**
trunk hunting SWITCH **Ausgangssuche**
trunking MOB.COMM **Bündelfunk**
  [exploitation of a RF band by a [Verwendung eines Bandes
  closed user groups] durch geschlossene Benut-
  = multi-user-band radio zergruppen]
  ↑ mobile radiocommunications = Chekker (DBP)
   ↑ Mobilfunk
trunking (→grouping) SWITCH (→Bündelung)
trunking (→trunk group) SWITCH (→Leitungsbündel)
trunking network (→toll TELEC (→Fernnetz)
  network)
trunking route (→long-haul TRANS (→Ferntrasse)
  route)
trunking system (→long-haul TRANS (→Weitverkehrssystem)
  system)
trunk line BROADC **Strecke**
  [CATV] [CATV]
  = trunk
trunk line 2 (→interoffice TELEC (→Amtsverbindungsleitung)
  trunk)
trunk line 1 (→toll trunk) TELEC (→Fernleitung)
trunk module SWITCH **Leitungssatzbaugruppe**
trunk network (→toll network) TELEC (→Fernnetz)
trunk plant (→toll network) TELEC (→Fernnetz)
trunk position (→toll SWITCH (→Fernschrank)
  switchboard)
trunk prefix SWITCH **nationale Verkehrsausschei-**
  [digit to provide access to trunk ex- **dungszahl**
  change] [im Netz der DBP: 0]
trunk release (→clearing) SWITCH (→freischalten)
trunk switchboard (→toll SWITCH (→Fernschrank)
  switchboard)
trunk transmission engineering TELEC **Nachrichtenübertragungs-**
  (→communications transmission en- technik)
  gineering)
truss (n.) CIV.ENG **Hängewerk**
truss head MECH **Flachrundkopf**
  [screw] [Schraube]
truss-head rivet MECH **Flachrundkopfniet**
  = Rundkopfniet
truss-head screw MECH **Rundkopfschraube**
  = Flachrundkopfschraube
truth function MATH **Wahrheitsfunktion**
truth table MATH **Wahrheitstabelle**
  = Bewertungstabelle; Ver-
   knüpfungstafel
truth variable MATH **Wahrheitsvariable**
tryout (→prooftesting) QUAL (→Erprobung)
try square MECH **Anschlagwinkel**
  [zur Markierung und Prü-
   fung rechter Winkel]
TS (→time slot) TELEC (→Zeitschlitz)
T section (→star NETW.TH (→Sternschaltung)
  connection)
T section low-pass NETW.TH **T-Transformationsglied**

T-slot MECH **T-Nut**
T-square ENG.DRAW **Reißschiene**
TSR program DATA PROC **TSR-Programm**
  [Terminal Stay Resident]
TSS (→time sharing DATA PROC (→Teilnehmerbetrieb)
  operation)
T-switch DATA PROC **Umschaltbox**
t test MATH **t-Test**
TTFD antenna (→tilted terminated ANT (→TTFD-Antenne)
  folded dipole)
TTL (→transistor-transistor MICROEL (→Transistor-Transistor-Lo-
  logic) gik)
TTX (→teletex) TELEC (→Teletex)
TTY (→teletypewriter) TERM&PER (→Fernschreiber)
TU (→tributary unit) TELEC (→Zubringereinheit)
tube 1 TECH **Rohr**
  = pipe
tube (→electron tube) ELECTRON (→Elektronenröhre)
tube (→coaxial pair) COMM.CABLE (→Koaxialpaar)
tube 2 (→hose) TECH (→Schlauch)
tube (→well) MICROEL (→Wanne)
tube amplifier CIRC.ENG **Röhrenverstärker**
  = valve amplifier
tube base (→tube socket) ELECTRON (→Röhrenfassung)
tube characteristic ELECTRON **Röhrenkennlinie**
tube furnace PHYS **Rohrofen**
tube generator CIRC.ENG **Röhrengenerator**
  = valve generator
tube hum ELECTRON **Röhrenbrummen**
  ≈ tube noise ≈ Röhrenrauschen
tube measuring rectifier INSTR **Röhrenmeßgleichrichter**
tube noise ELECTRON **Röhrenrauschen**
  = valve noise ≈ Röhrenbrumm
  ≈ tube hum
tube of force PHYS **Kraftlinienröhre**
tube oscillator MICROW **Röhrenoszillator**
tube reference guide ELECTRON **Röhren-Vergleichstabelle**
tube socket ELECTRON **Röhrenfassung**
  = tube base; valve socket; valve base = Röhrensockel
tube stage CIRC.ENG **Röhrenstufe**
  = valve stage
tube system ELECTRON **Röhrensystem**
tube transmitter RADIO **Röhrensender**
  = valve transmitter
tube voltmeter INSTR **Röhrenvoltmeter**
tubing lattice pylon CIV.ENG **Rohrgittermast**
tubular TECH **rohrförmig**
  ≈ fistulous
tubular capacitor COMPON **Rohrkondensator**
  = Zylinderkondensator
tubular earthing electrode SYS.INST **Rohrerder**
tubular mast OUTS.PLANT **Rohrmast**
tubular pole OUTS.PLANT **Hohlmast**
tubular rivet MECH **Rohrniet**
tuition 2 SCIE **Lehrgeld**
  [charge fore instruction service] = Schulgeld
tuition 1 (→instruction) SCIE (→Lehre)
tumbling DATA PROC **Torkeln**
  [computer graphics; rotation of [Computergraphik; Rota-
  graphics elements] tion von Darstellungsele-
   menten]
tunability range (→tuning CIRC.ENG (→Ziehbereich)
  range)
tunable CIRC.ENG **ziehbar**
  = abstimmbar; durchstimmbar
tunable filter NETW.TH **durchstimmbares Filter**
tunable oscillator CIRC.ENG **ziehbarer Oszillator**
tune CIRC.ENG **ziehen**
  = abstimmen
tune (→sintonize) HF (→abstimmen)
tuned CIRC.ENG **abgestimmt**
  = sintonized ≠ unabgestimmt
  ≠ untuned
tuned aerial (→periodic ANT (→abgestimmte Antenne)
  antenna)
tuned amplifier CIRC.ENG **Resonanzverstärker**
  = resonance amplifier = abgestimmter Verstärker
tuned antenna (→periodic ANT (→abgestimmte Antenne)
  antenna)

tuned cavity (→coaxial cavity resonator)		MICROW	(→Topfkreis)	tunnel diode oscillator	CIRC.ENG	Tunneldiodenoszillator
tuned circuit		CIRC.ENG	abgestimmter Schaltkreis = Abstimmkreis	tunnel effect [electron surmounts a superior potential wall]	PHYS	Tunneleffekt [Elektron überwindet übermäßigen Potentialberg]
tuned feeder		ANT	abgestimmte Speiseleitung	tunnel transistor	MICROEL	Tunneltransistor
tuned-grid tuned-anode oscillator (→tuned-plate tuned-grid oscillator)		CIRC.ENG	(→Huth-Kühn-Oszillator)	tupel [consisting of two elements]	MATH	Tupel [aus zwei Elementen bestehend] = zweistellige Relation
tuned-plate tuned-grid oscillator = tuned-grid tuned-anode oscillator ↑ LC oscillator		CIRC.ENG	Huth-Kühn-Oszillator ↑ LC-Oszillator	turbid = dull 2 ≈ mat ≠ clear	TECH	trübe ≈ matt ≠ klar
tuned transformer		EL.TECH	Resonanzübertrager = Resonanztransformator	turbidity	TECH	Trübung
tuned trap [HF] (→bandstop filter)		NETW.TH	(→Bandsperrfilter)	turbulence (→turbulent flow)	PHYS	(→Wirbelströmung)
tuner		MICROW	Abstimmvorrichtung	turbulence inversion	RADIO PROP	Turbulenzinversion
tuner [HF preamp + oscillator + mixer]		HF	Tuner [HF-Vorstufe + Oszillator + Mischer]	turbulent	PHYS	turbulent
				turbulent flow = turbulence ≈ curl	PHYS	Wirbelströmung = turbulente Strömung; Turbulenzströmung; Turbulenz ≈ Wirbel
tuner [FM receiver without audio booster]		CONS.EL	Tuner 1 [UKW-Empfänger ohne NF-Leistungsverstärker]	Turing machine	INF	Turing-Maschine
				Turing test	DATA PROC	Turing-Test
tuner = matching unit; matchbox		ANT	Anpaßgerät = Anpassungseinrichtung	turn (v.t.) [to machine]	TECH	drehen [zerspanen]
tuner (→tuning coil)		HF	(→Abstimmspule)	turn 2 (v.i.) [to move around its axis] ≈ rotate; circulate	PHYS	drehen [sich um die eigene Achse bewegen] ≈ rotieren; kreisen
tuner 2 (→channel selector)		CONS.EL	(→Kanalwähler)			
tungsten = W; wolfram		CHEM	Wolfram = W			
tungsten filament lamp		EL.INST	Wolframlampe	turn (n.) = spire ≈ winding	EL.TECH	Windung ≈ Wicklung
tuning [to frequency] = syntony		CIRC.ENG	Abgleich [auf Frequenz] = Abstimmung; Trimmen ≈ Eichen	turn (→twist 3)	PHYS	(→Drehung)
				turn 1 (→rotate)	PHYS	(→rotieren)
tuning [increase of processing speed by an add-on board]		DATA PROC	Beschleunigung [mittels Zusatzkarte]	turnaround (v.t.)	DATA PROC	umkehren
				turnaround (n.) (→execution time)	DATA PROC	(→Ausführungszeit)
tuning board (→speed board)		DATA PROC	(→Beschleunigerkarte)	turnaround form	TERM&PER	Kreisverkehrsbeleg
tuning capacitor		HF	Abstimmkondensator	turnaround time = run time; throughput time	DATA COMM	Durchlaufzeit = Ablaufzeit; Rechenzeit
tuning capacitor		CIRC.ENG	Ziehkondensator	turned part	MECH	Drehteil
tuning coil = tuner		HF	Abstimmspule	turning mechanism (→swing mechanism)	TECH	(→Drehvorrichtung)
tuning control (→tuning indicator)		ELECTRON	(→Abstimmanzeige)	turning moment (→torque)	PHYS	(→Drehmoment)
tuning device ↓ tuning screw		MICROW	Frequenzabstimmelement = Abstimmelement ↓ Abgleichschraube	turning point [fig.]	COLLOQ	Wendepunkt [fig.]
				turning shop	MANUF	Dreherei
tuning device		CIRC.ENG	Selektionsmittel = Selektivitätsmittel	turning swivel (→pulley 1)	TECH	(→Umlenkrolle)
				turn-key	TECH	schlüsselfertig = Turn-Key-
tuning diode [with controllable capacitance]		MICROEL	Abstimmdiode [mit steuerbarer Kapazität]	turn key system	TECH	Gesamtanlage = Turn-key-System
tuning diode modulator = varicap modulator		MODUL	Kapazitätsdioden-Modulator	turn-off (v.t.) (→disconnect)	EL.TECH	(→ausschalten)
tuning figure		ELECTRON	Verstimmungsmaß	turn-off behaviour = switch-off behaviour	MICROEL	Ausschaltverhalten = Ausschalteverhalten; Abschaltverhalten
tuning figure		CIRC.ENG	Verstimmungsmaß			
tuning fork = diapason		ACOUS	Stimmgabel = Diapason	turn-off delay	ELECTRON	Abschaltverzögerung
				turn-off loss = breaking loss; switch-off loss	MICROEL	Ausschaltverlust = Ausschalteverlust; Abschaltverlust
tuning indicator = tuning control; tuning meter		ELECTRON	Abstimmanzeige			
tuning meter (→tuning indicator)		ELECTRON	(→Abstimmanzeige)	turn-off time [storage plus fall time] = switch-off time	MICROEL	Ausschaltzeit [Speicher- plus Abfallzeit] = Ausschaltezeit; Abschaltzeit
tuning pitch (→standard tuning tone)		ACOUS	(→Stimmton)			
tuning range = tunability range; pull-in range		CIRC.ENG	Ziehbereich = Abstimmbereich; Durchstimmbereich	turn-off time	ELECTRON	Ausschaltezeit = Freiwerdezeit; Löschzeit
				turn-on [transistor]	MICROEL	durchsteuern [Transistor]
tuning screw = slug ↑ tuning device		MICROW	Abgleichschraube ↑ Frequenzabstimmelement	turn-on (v.t.) (→connect 2)	EL.TECH	(→einschalten)
				turn-on characteristics	EL.TECH	Einschaltverhalten
tuning speed		CIRC.ENG	Abstimmgeschwindigkeit = Durchstimmgeschwindigkeit	turn-on delay	ELECTRON	Einschaltverzögerung
				turn-on level	TELEC	Einschaltpegel
tuning stub		MICROW	Abstimmblindleitung	turn-on loss	EL.TECH	Einschaltverlust
tunnel (→cable subway)		OUTS.PLANT	(→begehbarer Kabelkanal)	turn-on time (→on period)	ELECTRON	(→Einschaltzeit)
tunnel breakdown		MICROEL	Tunneldurchbruch	turn over (v.t.)	ECON	umsetzen = Umsatz legen
tunnel current		MICROEL	Tunnelstrom			
tunnel diode = Esaki diode ↑ junction diode		MICROEL	Tunneldiode = Esaki-Diode ↑ Flächendiode	turnover (n.) = sales (AM); sales volume (AM); filled orders; revenue	ECON	Umsatz = U
				turn-over (v.t.) (→invert)	TECH	(→invertieren)

turnover commission   ECON **Umsatzprovision**
turn-over voltage (→reverse   MICROEL (→Umkehrspannung)
  isolation)
turns factor   EL.TECH **Windungszahlfaktor**
turns ratio   EL.TECH **Windungsverhältnis**
  [transformer]   [Transformator, Übertra-
  = transformer ratio; transformation   ger]
    ratio   = Übersetzungsverhältnis;
                                               Übersetzung
turnstile antenna   ANT **Drehkreuzantenne**
  = crossed antenna   = Kreuzdipolantenne; Kreuz-
                                             dipol; Kreuzstrahler; Dreh-
                                             standantenne;
                                             Turnstile-Antenne; Quir-
                                             lantenne
turn switch (→rotary   COMPON (→Drehschalter)
  switch)
turntable   TECH **Drehtisch**
  = rotary table
turntable   CONS.EL **Plattenteller**
                                           = Schallplattenteller
turn-up (→commission)   TELEC (→einschalten)
turn-up (→commissioning)   TELEC (→Einschaltung)
turquoise (→turquoise blue)   PHYS (→türkisblau)
turquoise blue   PHYS **türkisblau**
  = turquoise   = türkis
turquoise green   PHYS **blaugrün**
turred tag (→solder   COMPON (→Lötanschlußpunkt)
  terminal)
turret lathe   MECH **Revolverdrehbank**
turtle   DATA PROC **Schildkröte**
  = screen turtle   = Igel
  ↑ icon   ↑ Bildschirmsymbol
turtle (→floor turtle)   TERM&PER (→Schildkröte)
turtle graphics   DATA PROC **Schildkrötengraphik**
  [computer-aided freehand drawing]   [rechnergestütztes Frei-
                                               handzeichen]
                                           = Igelgraphik;
                                             Turtle-Graphik
tutorial 1 (n.)   SCIE **Einzelunterricht**
  [individual instruction class]
tutorial (→tutorial   DATA PROC (→Lernprogramm)
  program)
tutorial (→training manual)   TECH (→Lehrhandbuch)
tutorial program   DATA PROC **Lernprogramm**
  [a training program]   = Tutorial
  = tutorial
TV (→television)   TELEC (→Fernsehen)
TV ad (→television   BROADC (→Fernsehwerbung)
  advertizing)
TV advertizing (→television   BROADC (→Fernsehwerbung)
  advertizing)
TV aerial 1 (→television   ANT (→Fernsehantenne 1)
  antenna 1)
TV antenna 1 (→television   ANT (→Fernsehantenne 1)
  antenna 1)
TV antenna 2 (→television reception ANT (→Fernsehempfangsantenne)
  antenna)
TV band (→television broadcast RADIO (→Fernsehband)
  band)
TV broadcast band (→television RADIO (→Fernsehband)
  broadcast band)
TV broadcast emitter   BROADC (→Fernsehsender)
  (→television transmitter)
TV broadcasting (→television   BROADC (→Fernsehrundfunk)
  broadcasting)
TV broadcasting company   BROADC (→Fernsehanstalt)
  (→television broadcasting company)
TV broadcasting satellite   SAT.COMM **Fernsehrundfunk-Satellit**
  = television broadcasting satellite
TV broadcast program   BROADC (→Fernsehprogramm)
  (→television broadcast program)
TV broadcast programme   BROADC (→Fernsehprogramm)
  (→television broadcast program)
TV broadcast satellite   SAT.COMM (→Fernsehsatellit)
  (→television broadcast satellite)
TV broadcast service user   BROADC (→Fernsehteilnehmer)
  (→television broadcast service user)
TV broadcast station   BROADC (→Fernsehsendeanlage)
  (→television broadcast station)

TV broadcast transmission   BROADC (→Fernsehsendung)
  (→television broadcast transmission)
TV camera (→television camera) TV (→Fernsehkamera)
TV car (→television van)   TV (→Fernsehwagen)
TV channel (→television   RADIO (→Fernsehkanal)
  channel)
TV channel signal (→television BROADC (→Fernsehkanalsignal)
  channel signal)
TV commercial (→television   BROADC (→Fernsehwerbung)
  advertizing)
TV control system (→television SIGN.ENG (→Fernsehüberwachungsan-
  control system)   lage)
TV converter (→television   TV (→Bereichsumsetzer)
  translater)
TV coverage (→television   BROADC (→Fernsehversorgung)
  coverage)
TV data line decoder   TV **TV-Datenzeilendecoder**
TV demodulator   INSTR **TV-Meßdemodulator**
                                           = Fernseh-Meßdemodulator
TV digital oscilloscope   INSTR **TV-Digitaloszilloskop**
TV distribution system   BROADC **Fernsehverteilanlage**
  ↓ common antenna system; CATV;   ↓ Gemeinschaftsantennenan-
    broadband cable system   lage; Groß-Gemeinschafts-
                                               antennenanlage;
                                               Breitband-Kabelanlage
TV emitter (→television   BROADC (→Fernsehsender)
  transmitter)
TV engineering (→television   TV (→Fernsehtechnik)
  engineering)
TV headphone (→television   CONS.EL (→Fernsehkopfhörer)
  headphone)
TV home receiver (→television CONS.EL (→Fernsehgerät)
  set)
TVI (→television   RADIO PROP (→Fernsehstörung)
  interference)
TV installer   TV **Fernsehinstallateur**
TV licence (→television   BROADC (→Fernsehgebühr)
  licence)
TV line (→television line)   TRANS (→Fernsehleitung)
TV link (→television link)   BROADC (→Fernsehleitungsverbin-
                                             dung)
TV measurement (→television   TV (→Fernsehmeßtechnik)
  measurement)
TV measuring signal (→television TV (→Fernsehmeßsignal)
  measuring signal)
TV monitoring receiver   RADIO LOC **TV-Überwachungsempfänger**
TV network (→television   BROADC (→Fernsehleitungsnetz)
  network)
TV oscilloscope   INSTR **TV-Oszilloskop**
TV picture   TV **Fernsehbild**
  = television picture; picture   = TV-Bild; Bild
TV picture display   TV **Fernsehbildempfänger**
  = television picture display   = TV-Empfänger
TV picture tube (→television   ELECTRON (→Fernsehbildröhre)
  picture tube)
TV program (→television   BROADC (→Fernsehprogramm)
  broadcast program)
TV program center (AM)   BROADC (→Fernsehschaltstelle)
  (→television program center)
TV program contribution line   BROADC **Fernsehzuführungsleitung**
  [in the network of German PTT a   [im Netz der DBP die Lei-
  line from a regional studio to the pro-   tung vom Regionalstudio
  gram exchange center]   zum Sternpunkt der ARD
                                               oder Sendezentrum des
                                               ZDF]
                                           = TV-Zuführungsleitung; Zu-
                                             führungsleitung
TV program distribution center BROADC **Sternpunkt**
  [of the German 1st TV program   [Programmaustauschzen-
  (ARD)]   trale der ARD, in Frank-
                                               furt a.M.]
                                           ↑ Sendezentrum
TV program distribution line   BROADC **Fernsehverteilleitung**
  [in the German TV network a line   [von der Programmzentra-
  from the national program center to   le (Sternpunkt ARD oder
  the node of a regional transmitter   Sendezentrum ZDF) zum
  feeding network]   Quellpunkt eines regiona-
  = program distribution line   len Modulationsleitungsnet-
                                               zes]
                                           = Verteilleitung

TV program exchange line	BROADC	**Fernsehaustauschleitung**	
[for international or national program exchange]		[für internationalen oder nationalen Programmaustausch]	
↑ program exchange line		= TV-Austauschleitung	
		↑ Austauschleitung	
TV programme (→ television broadcast program)	BROADC	(→ Fernsehprogramm)	
TV programme centre (BRI) (→ television program center)	BROADC	(→ Fernsehschaltstelle)	
TV receiver (→ television set)	CONS.EL	(→ Fernsehgerät)	
TV reception (→ television reception)	BROADC	(→ Fernsehempfang)	
TV reception antenna (→ television reception antenna)	ANT	(→ Fernsehempfangsantenne)	
TV relay transmitter (→ television repeater)	BROADC	(→ Fernsehumlenksender)	
TV repeater (→ television repeater)	BROADC	(→ Fernsehumlenksender)	
TV/RF test transmitter	BROADC	**TV-RF-Meßsender**	
TV satellite (→ television broadcast satellite)	SAT.COMM	(→ Fernsehsatellit)	
TV screen (→ video screen)	TV	(→ Bildschirm)	
TV set (→ television set)	CONS.EL	(→ Fernsehgerät)	
TV signal (→ television signal)	TV	(→ Fernsehsignal)	
TV signal distribution (→ television signal distribution)	BROADC	(→ Fernsehverteilung)	
TV standard (→ television standard)	TV	(→ Fernsehnorm)	
TV standards conversion	TV	**Normwandlung**	
= standards conversion; TV system conversion		= Fernsehnormwandlung	
TV standards converter	TV	**Normwandler**	
= TV system converter; standards converter; transcoder		= Fernsehnormwandler; Normkonverter; Fernsehnormkonverter; Transcoder	
TV studio (→ television studio)	BROADC	(→ Fernsehstudio)	
TV subscriber line	BROADC	**Fernsehanschlußleitung**	
TV system conversion (→ TV standards conversion)	TV	(→ Normwandlung)	
TV system converter (→ TV standards converter)	TV	(→ Normwandler)	
TV terminal	TERM&PER	**Fernseh-Bildschirmgerät**	
TV test signal (→ television test signal)	TV	(→ Fernsehprüfsignal)	
TV text	TELEC	**Fernsehtext**	
TV tower (→ television tower)	ANT	(→ Fernsehturm)	
TV translater (→ television translater)	TV	(→ Bereichsumsetzer)	
TV transmission (→ television broadcast transmission)	BROADC	(→ Fernsehsendung)	
TV transmitter (→ television transmitter)	BROADC	(→ Fernsehsender)	
TV transmitter feeding link (→ television transmitter feeding link)	BROADC	(→ Fernsehmodulationsleitung)	
TV transmitting antenna (→ television transmitting antenna)	ANT	(→ Fernsehsendeantenne)	
TV transposer (→ television transposer)	BROADC	(→ Fernsehumsetzer)	
TV tube (→ television picture tube)	ELECTRON	(→ Fernsehbildröhre)	
TV van (→ television van)	TV	(→ Fernsehwagen)	
tweak (v.t.)	DATA PROC	**trimmen**	
tweeter	EL.ACOUS	**Hochtonlautsprecher**	
		= Hochtöner	
tweezer (→ pressure clamp)	EL.INST	(→ Klemme)	
tweezers (→ pincers)	TECH	(→ Pinzette)	
twelve-punch	TERM&PER	**Zwölferlochung**	
twenty-four-hour ...	ECON	**Tag-/Nacht-**	
twice (adv.)	TECH	**zweimal** (adv.)	
		= zweimalig	
twilight switch	EL.INST	**Dämmerungsschalter 2**	
[switches automatically on at twilight]		[schaltet bei Dämmerung ein]	
twin (adj.)	TECH	**Zwillings-**	
twin antenna	ANT	**Doppelantenne 1**	
		= Zwillingsantenne	
twin-band antenna (→ two-band antenna)	ANT	(→ Zweibandantenne)	
twin computer system (→ dual computer system)	DATA PROC	(→ Doppelsystem)	
twin contact	COMPON	**Doppelkontakt**	
= dual contact			
twin-cross contact	COMPON	**Doppelkreuzkontakt**	
twin crystal	PHYS	**Zwillingskristall**	
twin delta loop (→ double delta loop)	ANT	(→ Doppel-Delta-Loop)	
twin dipole	ANT	**Doppeldipol**	
twining machine	COMM.CABLE	**Viererverseilmaschine**	
= twisting machine		↑ Verseilmaschine	
↑ stranding machine			
twinkle box	TERM&PER	**Winkelsensor**	
[an input device with angular sensing of light-emitting sources]			
twin lamp indicator	ANT	**Twin-Lamp**	
[of VSWR]			
twin lead (→ ribbon cable)	COMM.CABLE	(→ Bandkabel)	
twin lead (→ two-wire feeder)	ANT	(→ Zweidrahtleitung)	
twin-lead Marconi antenna	ANT	**Bandkabel-Marconi-Antenne**	
twin output (→ dual output)	EL.TECH	(→ Zweifachausgang)	
twin quad	ANT	**gestockte Cubical-quad-Antenne**	
= twin square			
twin square (→ twin quad)	ANT	(→ gestockte Cubical-quad-Antenne)	
twin-T-filter	NETW.TH	**Doppel-T-Filter**	
= notch filter		= Notch-Filter	
twin-T network (→ balanced T section)	NETW.TH	(→ Viereckschaltung)	
twin-well technology	MICROEL	**Twin-well-Technik**	
twin wire (→ wire pair)	COMM.CABLE	(→ Aderpaar)	
T wire (→ tip wire)	TELEPH	(→ A-Ader)	
twist (v.t.)	TECH	**verdrillen**	
[threads, wires]		[von Fäden, Drähten]	
≈ twist		= verdrallen; zusammendrehen	
		≈ verwinden	
twist 1 (v.t.)	PHYS	**verwinden**	
[to exert force in order to move one end along the longitudinal axis, while the other end is held fast or forced to the other direction]		[ein Ende um die Längsachse drehen, unter Ausübung einer Gegenkraft am anderen Ende]	
= wrench		= verdrehen; tordieren	
≈ twist [TECH]		≈ verdrillen [TECH]	
twist 2 (n.)	PHYS	**Drall 1**	
[circular movement on the own axis]		[Drehbewegung um eigene Achse]	
↑ circular movement		↑ Kreisbewegung	
twist 3 (n.)	PHYS	**Drehung**	
= turn; rotation		↓ Umdrehung	
twist (n.)	TECH	**Verdrillung**	
		= Verdrallung	
twist 1 (n.)	PHYS	**Verwindung**	
= wrench (n.); torsion (n.)		= Torsion; Drillung; Verdrillung	
twist drill bit	MECH	**Spiralbohrer**	
twisted	TECH	**verdrillt**	
		= verdrallt; gedrallt; verwunden; geschränkt	
twisted joint	OUTS.PLANT	**Würgerverbindung**	
= hand twist; Western Union joint			
twisted pair	COMM.CABLE	**verdrilltes Aderpaar**	
		= verdrilltes Paar	
twisted slot antenna	ANT	**Doppelschlitzstrahler**	
twisted thread	TECH	**Zwirn**	
= twisted yarn		[aus mehreren Fäden gedrehter dünner Faden]	
↑ thread		↑ Faden	
twisted yarn (→ twisted thread)	TECH	(→ Zwirn)	
twisting	COMM.CABLE	**Verseilung**	
= formation		= Adernverseilung; Verseilart	
↓ pair formation; quad formation		↓ Paarverseilung; Sternverseilung	
twisting by pairs (→ pairing)	COMM.CABLE	(→ Paarverseilung)	
twisting machine (→ twining machine)	COMM.CABLE	(→ Viererverseilmaschine)	

twistor

twistor COMPON
[storage matrix formed by twisted wires]

Twistor
[Speichermatrix aus tordierten Drähten]
= magnetischer Drahtspeicher

90° twist section MICROW **90°-Twist-Stück**

two-address instruction DATA PROC **Zwei-Adreß-Befehl**
↑ multi-address instruction
↑ Mehr-Adreß-Befehl

two-band antenna ANT **Zweibandantenne**
= dual-band antenna; twin-band antenna; duoband antenna; duo-bander
= Zweibereichsantenne; Doppelantenne 2

two-bit error CODING **Zwei-Bit-Fehler**

two-carrier mode RADIO REL **Zweiträgerverfahren**

two-cavity klystron MICROEL (→two-resonator klystron)
(→Zweikammerklystron)

two-channel synthesizer INSTR **Zweikanal-Synthesizer**

two-chip equipment EQUIP.ENG **Zwei-Chip-Gerät**

two-circuit NETW.TH **zweikreisig**
= double-tuned
↑ multi-circuit
= Zweikreis-
↑ mehrkreisig

two-color print TYPOGR **Zweifarbendruck**
= two-colour print
↑ multicolor print
↑ Mehrfarbendruck

two-colour print (→two-color print) TYPOGR (→Zweifarbendruck)

two-conductor plug TELEPH **zweipoliger Klinkenstecker**
= zweipoliger Stöpsel

two-core POWER SYS **zweiadrig**

two-dimensional MATH **zweidimensional**
= bivariate

two-dimensional field PHYS **ebenes Feld**

two-dimensional scanning RADIO LOC **Kreuzabtastung**

two-element cubical quad antenna ANT **Zwei-Element-Quad-Antenne**

two-element Yagi-Uda antenna ANT **Zwei-Element-Yagi-Uda-Antenne**

two-faced (→bilateral) TECH (→zweiseitig)

two-figure (→two-place) MATH (→zweistellig)

twofold TECH **zweifach**
= double; dual; duplex

twofold refraction PHYS (→Doppelbrechung)
(→birefringence)

two-frequency recording DATA PROC **Wechseltaktschrift**
= pulse-width recording; FM code
↑ recording mode
= Zwei-Frequenzen-Schrift; Zweifrequenzaufzeichnung; Zweifrequenzverfahren; FM-Code; Impulsbreiten-Aufzeichnung
↑ Aufzeichnungsverfahren

two-frequency signaling TELEPH **Zweifrequenztonwahl**
[zwei Frequenzen pro Ziffer]
= Zweifrequenzverfahren

two-gong bell TERM&PER **Zweischalenwecker**

two-handed TECH **zweihändig**

two-hole directional coupler MICROW **Zweilochrichtkoppler**

two-input adder (→half-adder) CIRC.ENG (→Halbaddierer)

two-layer (→two-layered) TECH (→zweischichtig)

two-layered TECH **zweischichtig**
= bipack; two-layer
≈ two-part
= doppelschichtig; Zweischicht-; zweilagig
≈ zweiteilig

two-layer PBC ELECTRON **Zweilagen-Leiterplatte**
= Zweiebenen-Leiterplatte

two-layer wiring MICROEL **Zweilagenverdrahtung**
[mit zwei Metallisierungsebenen]
= Zweiebenenverdrahtung

two-level action (→two-position control) CONTROL (→Zweipunktregelung)

two-level control (→two-position control) CONTROL (→Zweipunktregelung)

two-level phase shift keying (→2 PSK) MODUL (→2 PSK)

two-loci method CONTROL **Zwei-Ortskurven-Methode**
[stability analysis]
[Stabilitätskriterium]
= Zwei-Ortskurven-Verfahren

two-motion selector SWITCH **Hebdrehwähler**
= Koordinatenwähler

two-out-of-five code CODING **Zwei-aus-Fünf-Code**
↓ Walking code
↓ Walking-Code

two-package adhesive CHEM **Zweikomponentenkleber**

two-part TECH **zweiteilig**
= bipartite
≈ two-layered
≈ zweischichtig

two-party circuit SWITCH **Zweierverbindungssatz**

two-party line with selective ringing TELEC **Zweieranschluß**
[both parties can call simultaneously]
≈ two-party omnibus line
↑ party line
[gleichzeitig verfügbar, mit Selektivruf]
≈ Doppelanschluß
↑ Gemeinschaftsanschluß

two-party omnibus line TELEC **Doppelanschluß**
[nicht gleichzeitig verfügbar]
≈ Zweieranschluß
↑ Gemeinschaftsanschluß

two pass DATA PROC **Zweifachdurchlauf**

two-path fading RADIO PROP **Zweiwegeschwund**
= Zweiwegefading

two-phase (→diphase) TECH (→zweiphasig)

two-phase clock MICROEL **Zweiphasentakt**

two-phase current POWER SYS **Zweiphasenstrom**

two-phased (→diphase) TECH (→zweiphasig)

two-phase recording (→phase encoding) DATA PROC (→Richtungstaktschrift)

two-phase rotating field POWER SYS **Zweiphasen-Drehfeld**

two-place MATH **zweistellig**
= two-figure
↑ of many places
↑ vielstellig

two-point EL.TECH **Zweipunkt-**
= two-position; two-spot; double-spot; on-off

two-point action control CONTROL (→Zweipunktregelung)
(→two-position control)

two-point control CONTROL (→Zweipunktregelung)
(→two-position control)

two-pole (→bipolar) PHYS (→zweipolig)

two-pole armature EL.TECH **Doppel-T-Anker**
= shuttle armature

two-port (→two-port network) NETW.TH (→Vierpol)

two-port attenuation (→image attenuation) NETW.TH (→Wellendämpfung)

two-port attenuation coefficient NETW.TH (→Wellendämpfungsmaß)
(→image attenuation constant)

two-port attenuation constant NETW.TH (→Wellendämpfungsmaß)
(→image attenuation constant)

two-port attenuation factor NETW.TH (→Wellendämpfungsfaktor)
(→image attenuation factor)

two-port connection NETW.TH **Vierpolzusammenschaltung**
= quadripole connection; four-pole connection

two-port determinant NETW.TH (→Vierpoldeterminante)
(→network determinant)

two-port dissection NETW.TH **Vierpolzerlegung**
= quadripole dissection; four-pole dissection

two-port equation (→network equation) NETW.TH (→Vierpolgleichung)

two-port matrix (→network matrix) NETW.TH (→Vierpolmatrix)

two-port network NETW.TH **Vierpol**
[a network with two pairs of terminals (with two ports)]
= two-port; two-terminal pair network; four-terminal network; quadripole; four-pole; fourpole; four-pole network; fourpole network
↑ network
[Netzwerk mit zwei Anschlußklemmenpaaren (mit zwei Toren)]
= Zweitor; Vierpolnetzwerk
↑ Netzwerk

two-port network analysis NETW.TH **Vierpoltheorie**
= quadripole analysis; four-pole analysis

two-port network transformation NETW.TH **Vierpoltransformation**
= quadripole transformation; four-pole transformation

two-port parameter (→network parameter) NETW.TH (→Vierpolparameter)

two-port transmission coefficient (→image transfer constant) NETW.TH (→Wellenübertragungsmaß)

two-port transmission constant	NETW.TH	(→Wellenübertragungsmaß)
(→image transfer constant)		
two-port transmission factor	NETW.TH	(→Wellenübertragungsmaß)
(→image transfer constant)		
two-port wave	NETW.TH	**Vierpolwelle**
= four-pole wave		↓ Spannungswelle; Stromwelle
↓ voltage wave; current wave		
two-position (→two-point)	EL.TECH	(→Zweipunkt-)
two-position action	CONTROL	(→Zweipunktregelung)
(→two-position control)		
two-position control	CONTROL	**Zweipunktregelung**
= two-position regulation; two-position action; two-spot control; two-spot regulation; two-spot action; on-off control; on-off action; two-level control; two-level action; two-point action control; two-point control		
two-position regulation	CONTROL	(→Zweipunktregelung)
(→two-position control)		
two-pulse bridge	POWER ENG	**Zweipuls-Brückenschaltung**
two-pulse midpoint connection	POWER SYS	**Zweipuls-Mittelpunktschaltung**
= single-phase midpoint connection		
two-quadrant multiplier	COMPON	**Zweiquadranten-Multiplikator**
two-quadrant operation	POWER SYS	**Zweiquadrantenbetrieb**
two-resonator klystron	MICROEL	**Zweikammerklystron**
= two-cavity klystron		= Zweikreisklystron
two-row	EQUIP.ENG	(→zweizeilig)
(→double-height)		
two's complement	DATA PROC	(→Zweierkomplement)
(→complement to two)		
two-sided (→bilateral)	TECH	(→zweiseitig)
two-sided board (→double-face PCB)	ELECTRON	(→doppelt kaschierte Leiterplatte)
two-sided PCB (→double-face PCB)	ELECTRON	(→doppelt kaschierte Leiterplatte)
two-spot (→two-point)	EL.TECH	(→Zweipunkt-)
two-spot action (→two-position control)	CONTROL	(→Zweipunktregelung)
two-spot control (→two-position control)	CONTROL	(→Zweipunktregelung)
two-spot regulation (→two-position control)	CONTROL	(→Zweipunktregelung)
two-spot tuning (→double-spot tuning)	ELECTRON	(→Zweipunktabstimmung)
two-stage	CIRC.ENG	**zweistufig**
two-step	TECH	**zweistufig**
↑ multi-step		↑ mehrstufig
two-step relay	COMPON	**Zweistufenrelais**
= sequence action relay		= Stufenrelais
two-terminal [R,L,C]	NETW.TH	**Zweipol** [R,L,C]
= two-terminal network; one-port		= Eintor
↑ network		↑ Netzwerk
two-terminal network (→two-terminal)	NETW.TH	(→Zweipol)
two-terminal network synthesis	NETW.TH	**Zweipolsynthese**
two-terminal network theory	NETW.TH	**Zweipoltheorie**
two-terminal pair network (→two-port network)	NETW.TH	(→Vierpol)
two-thirds-of-line precision offset [TV]	BROADC	**Zweidrittelzeilen-Präzisionsoffset** [TV]
↑ offset		↑ Versatzbetrieb
two-to-four-wire transition point	TELEC	**Gabelpunkt**
two-tone measurement	INSTR	**Zweitonmessung**
two-tone signal	INSTR	**Zweitonsignal**
two-tone sweep	INSTR	**Zweitonwobbelung**
two-tone voice frequency telegraphy	TELEGR	**Doppelton-WT**
two-valued (→bivalent)	MATH	(→zweiwertig 2)
two-way (→both-way)	TELEC	(→doppeltgerichtet)
two-way alternate	TELEC	**wechselseitig**
= either-way		≈ Halbduplex-Betrieb; beidseitig
≈ half-duplex; two-way simultaneous		
two-way alternate communication	DATA COMM	**wechselseitige Datenübermittlung**
= half-duplex data communication		
two-way alternate mode (→half duplex)	TELEC	(→Halbduplexbetrieb)
two-way alternate operation (→half duplex)	TELEC	(→Halbduplexbetrieb)
two-way alternate transmission (→half duplex)	TELEC	(→Halbduplexbetrieb)
two-way converter (→two-way convertor)	POWER SYS	(→Umkehrstromrichter)
two-way convertor	POWER SYS	**Umkehrstromrichter**
= reversible convertor; double convertor; two-way converter		
two-way distributor	BROADC	**Zweifachverteiler**
two-way simultaneous	TELEC	**beidseitig**
≈ two-way alternate		≈ wechselseitig
two-way simultaneous communication	DATA COMM	**beidseitige Datenübermittlung**
= full duplex data communication		
two-way splitting (→alternation between lines)	TELEPH	(→Makeln)
two-way telephone system (→simplex intercommunication system)	TELEPH	(→Wechselsprechanlage)
two weeks (AM) (→fortnight)	COLLOQ	(→zwei Wochen)
two-wire	TELEC	**zweidrätig**
= 2-wire		= zweiadrig; Zweidraht-
two-wire amplifier	TRANS	**Zweidrahtverstärker**
↓ two-wire repeater		↓ Zweidraht-Zwischenverstärker
two-wire antenna	ANT	**Bifilarantenne**
two-wire carrier system	TRANS	**Zweidraht-Getrenntlage-System**
= N + N system		= Z-System
two-wire feeder	ANT	**Zweidrahtleitung**
= twin lead		
two-wire line	LINE TH	**Lecher-Leitung**
= Lecher line; Lecher wires; parallel-wire line		= Doppelleitung; Paralleldrahtleitung
two-wire line	TELEC	**Zweidrahtleitung**
= two-wire link		
two-wire link (→two-wire line)	TELEC	(→Zweidrahtleitung)
two-wire neutral current	TELEGR	**Zweidraht-Einfachstrom**
two-wire operation	TELEC	**Zweidrahtbetrieb**
two-wire repeater	TRANS	**Zweidraht-Zwischenverstärker**
↑ two-wire amplifier		↑ Zweidraht-Verstärker
two-wire switching	SWITCH	**Zweidraht-Durchschaltung**
= 2-wire switching		= 2-Draht-Durchschaltung
two-wire system	TRANS	**Zweidrahtsystem**
TWT (→travelling wave tube)	MICROW	(→Wanderfeldröhre)
Tx/Rx (→transceiver)	RADIO	(→Transceiver)
tying-up (→binding)	EQUIP.ENG	(→Abbindung)
type (v.t.)	OFFICE	**maschineschreiben**
		= maschinschreiben (OES)
type (n.)	TECH	**Ausführung 1**
= execution; make; model 3; version; configuration		= Ausführungsform; Ausführungsart; Bauform; Konfiguration; Version
≈ type model; grade; workmanship 2		≈ Ausführungsmuster; Gütegrad; Ausführungsqualität
type (n.)	TYPOGR	**Drucktype**
[a block with the relief of a character]		[Block mit reliefartigem Schriftzeichen, zu einem Satz zusammenfügbar]
= letter; metal character; type-cast letter		
≈ character font; type style		= Schrifttype; Type; Letter
		≈ Schriftzeichensatz; Schriftart
type (n.)	ECON	**Typ**
= model		= Modell
type (→typesetting)	TYPOGR	(→Satz)
type (v.t.) (→typewrite)	TERM&PER	(→tippen)
type (→type model)	TECH	(→Ausführungsmuster)
type acceptance (→type approval)	TECH	(→Typenabnahme)
type acceptance test	QUAL	(→Typprüfung)
(→homologation test)		
type-accepted	QUAL	**typgeprüft**
= type-proved		= typengeprüft
type approval	TECH	**Typenabnahme**
= type acceptance; homologation		= Homologation
≈ type acceptance test		≈ Typprüfung

type approval test QUAL (→Typprüfung)
  (→homologation test)
type area (→image area) TYPOGR (→Satzspiegel)
type attribute (→typeface TYPOGR (→Schriftschnitt)
  design)
type-ball printer TERM&PER **Kugelkopfdrucker**
  = golf-ball printer; ball printer
type bar OFFICE **Typenhebel**
  = typebar; print bar = Typenstab; Typenstange;
  ↑ type carrier   Druckstange
    ↑ Typenträger
typebar (→type bar) OFFICE (→Typenhebel)
typebar printer TERM&PER **Typenstabdrucker**
  = bar printer = Typenstangendrucker
    ↑ Zeilendrucker
type basket (→thimble) OFFICE (→Typenkorb)
type carrier TERM&PER **Typenträger**
  = type pallet ↑ Druckkopf
  ↑ print head ↓ Druckkette; Typenband;
  ↓ type chain; printer tape; print drum;   Typenwalze; Schreibkopf
  print hear; type pallet; print ball; ty-   1; Typenkopf; Kugelkopf;
  pewheel; type magazine; type bar   Typenrad; Typenmagazin;
    Typenhebel
type-cast letter (→type) TYPOGR (→Drucktype)
type chain (→print chain) TERM&PER (→Druckkette)
type design (→typeface design) TYPOGR (→Schriftschnitt)
typeface TYPOGR **Schriftbild**
  = face ≈ Schriftart; Schriftzeichen-
  ≈ type style; font   satz
typeface attribute (→typeface TYPOGR (→Schriftschnitt)
  design)
typeface design TYPOGR **Schriftschnitt**
  = typeface attribute; type design; type = Schriftattribut
  attribute; font design; font attribute; ↓ normal; mager; ultraleicht;
  lettering type; lettering attribute   halbfett; fett; extrafett; kur-
  ↓ standard; light; ultralight; bold; semi   siv; zart kursiv; halbfett
  bold; extra bold; italic; light italic;   kursiv
  semi bold italic
type family TYPOGR **Schriftfamilie**
  = font family; letter family
type font (→font) TYPOGR (→Schriftzeichensatz)
type head TERM&PER **Typenkopf**
  [cylindrical type carrier]   [zylindrischer Typenträger]
  ↑ print head; type carrier ↑ Druckkopf 1; Typenträger
typehead (→print head) TYPOGR (→Druckkopf)
type head (→print head) TERM&PER (→Druckkopf)
type height TYPOGR **Schrifthöhe**
  = font height; letter height
type label EQUIP.ENG **Typenschild**
  = designation label; nameplate
typematic TERM&PER **Anschlagwiederholfunktion**
type matter preparation DATA PROC **Satzaufbereitungsprogramm**
  program
type model TECH **Ausführungsmuster**
  = type; prototype; sample (n.); spe- = Typmuster; Muster 3; Proto-
  cimen   totyp
  ↑ model ↑ Modell
type of monitored binary TELECONTR **Meldungsart**
  information
type pallet (→type carrier) TERM&PER (→Typenträger)
type printer TERM&PER **Typendrucker**
  [with mechanically defined font]   [mit mechanisch vorgegebe-
  = solid font printer   nem Typenvorrat]
  ↑ printer = Ganzzeichendrucker
  ↓ type wheel printer; drum printer; ↑ Drucker
  chain printer ↓ Typenraddrucker; Walzen-
    drucker; Kettendrucker
type printing TYPOGR **Typendruck**
  = typographical printing ↑ Hochdruck
  ↑ letterprint
type-proved (→type-accepted) QUAL (→typgeprüft)
typescript OFFICE **getipptes Manuskript**
typesetting TYPOGR **Satz**
  = setting; type; composition; print = Schriftsatz; Setzen
  ≈ makeup ≈ Umbruch
  ↓ justified typesetting; ragged typeset- ↓ Blocksatz; Flattersatz
  ting
typesetting computer DATA PROC (→Satzrechner)
  (→composition computer)

typesetting machine TYPOGR **Setzmaschine**
type size TYPOGR **Schriftgröße**
  = font size; lettering size = Schriftgrad
type size 6 point ((AM)) TYPOGR **Schriftgröße 6 Punkt**
  = Nonpareil (BRI)
type size 5 1/2 point (AM) TYPOGR **Schriftgröße 5 1/2 Punkt**
  = Agate (BRI)
type size 7 point (AM) TYPOGR **Schriftgröße 7 Punkt**
  = Minion (BRI)
type size 8 point (AM) TYPOGR **Schriftgröße 8 Punkt**
  = Brevier (BRI)
type size 9 point (AM) TYPOGR **Schriftgröße 9 Punkt**
  = Burgeois (BRI)
type size 10 point (AM) TYPOGR **Schriftgröße 10 Punkt**
  = Long Primer (BRI)
type size 11 point (AM) TYPOGR **Schriftgröße 11 Punkt**
  = Small Pica (BRI)
type size 12 point (AM) TYPOGR **Schriftgröße 12 Punkt**
  = Pica (BRI)
type size 14 point (AM) TYPOGR **Schriftgröße 14 Punkt**
  = English (BRI)
type size 16 point (AM) TYPOGR **Schriftgröße 16 Punkt**
  = Columbian (BRI)
type size 18 point (AM) TYPOGR **Schriftgröße 18 Punkt**
  = Great Primer (BRI)
type size 4 1/2 point (AM) TYPOGR **Schriftgröße 4 1/2 Punkt**
  = Diamond (BRI)
type size 5 point (AM) TYPOGR **Schriftgröße 5 Punkt**
  = Pearl (BRI)
type style TYPOGR **Schriftart**
  = font style; letter type = Schrifttype; Schrift
  ≈ character font; typeface ≈ Schriftzeichensatz; Schrift-
    bild
typewheel TERM&PER **Typenrad**
  = daisy wheel; daisy; print wheel; char-   [daisy = engl.„Gänseblu-
  acter wheel   me"]
  ↑ type carrier = Typenscheibe; Scheibrad;
    Daisywheel
    ↑ Typenträger
typew-heel printer TERM&PER **Typenraddrucker**
  = daisy-wheel printer; character-wheel = Typenscheibendrucker;
  printer; print wheel-printer; wheel   Scheibenraddrucker
  printer ↑ Typendrucker
  ↑ type printer
typewrite TERM&PER **tippen**
  = type (v.t.)
typewriter OFFICE **Schreibmaschine**
typewriter composing machine OFFICE **Schreibsetzmaschine**
typewriter face document TERM&PER **Schreibmaschinenschrift-Le-**
  reader   ser
typewriter keyboard OFFICE **Schreibmaschinentastatur**
typewriter paper OFFICE **Schreibmaschinenpapier**
typewriter table OFFICE **Schreibmaschinentisch**
typewriter type OFFICE **Schreibmaschinentype**
typical equipment configuration TECH **Bestückungsbeispiel**
typing condition TELEGR **Schreibzustand**
  = Schreibbereitschaft
typing error (→clerical error) DOC (→Schreibfehler)
typing speed TELEGR **Schreibgeschwindigkeit**
typist OFFICE **Typistin**
  [female]
  ↓ shorthand typist 1; phonotypist 1; ↓ Stenotypistin; Phonotypi-
  data entry operator 1   stin; Datentypistin
typist OFFICE **Schreibkraft**
  = clerk-typist
typographical printing (→type TYPOGR (→Typendruck)
  printing)
typographic command DATA PROC (→Satzanweisung)
  (→typographic instruction)
typographic instruction DATA PROC **Satzanweisung**
  [word processing]   [Textverarbeitung]
  = typographic command = typographische Anwei-
    sung; Satzbefehl;
    typographischer Befehl
typography 1 DOC **Typographie 1**
  [art of letterpress printing] = Buchdruckerkunst
typography 2 DOC **Typographie 2**
  [arrangement of letterpress matter]   [Gestaltung von Druckseit-
    en]

U

U (→voltage)	EL.TECH	(→elektrische Spannung)	
U (→uranium)	CHEM	(→Uran)	
UA (→user agent)	DATA COMM	(→End-Systemteil)	
U Adcock antenna	ANT	U-Adcock-Antenne	
UAE (→user agent entity)	DATA COMM	(→Instanz des End-Systemteils)	
U antenna	ANT	U-Antenne	
ubication (→site)	ECON	(→Standort)	
ubiquitous	COLLOQ	allgegenwärtig	
= omnipresent			
UG antenna	ANT	UG-Antenne	
UHF (→decimetric waves)	RADIO	(→Dezimeterwellen)	
UHF engineering	RADIO	Dezimeterwellentechnik	
		= Dezitechnik	
UHF jack (→UHF socket)	COMPON	(→UHF-Buchse)	
UHF plug	COMPON	UHF-Stecker	
UHF socket	COMPON	UHF-Buchse	
= UHF jack		= UHF-Kupplung	
UIT (→ITU)	TELEC	(→UIT)	
UJT (→unijunction transistor)	MICROEL	(→Doppelbasisdiode)	
ULA (→gate array)	MICROEL	(→Gate Array)	
ULSI (→ultra-large-scale integration)	MICROEL	(→Ultrahöchstintegration)	
ultimate buyer (→end-consumer)	ECON	(→Endabnehmer)	
ultimate configuration	TECH	(→Endausbau)	
(→maximum capacity)			
ultimate consumer (→final consumer)	ECON	(→Endverbraucher)	
ultimate load (→breaking load)	MECH	(→Bruchlast)	
ultimate objective	COLLOQ	Endziel	
= ultimate goal			
ultimate strength	MECH	Bruchfestigkeit	
ultimate stress	MECH	Zerreißspannung	
		= Bruchspannung	
ultimate taker (→end-consumer)	ECON	(→Endabnehmer)	
ultimate value (→final value)	TECH	(→Endwert)	
ultor electrode	ELECTRON	Ultor-Hochspannungsanode	
ultrabold (→extra bold)	TYPOGR	(→extrafett)	
ultrafiche	TERM&PER	Ultrafiche	
[with reduction by hundred or more]		[mit Vergrößerungsfaktor Hundert und mehr]	
↑ microfilm		↑ Mikrofilm	
ultra-high frequencies	RADIO	(→Dezimeterwellen)	
(→decimetric waves)			
ultralarge computer	DATA PROC	(→Größtrechner)	
(→supercomputer)			
ultra-large-scale integration	MICROEL	Ultrahöchstintegration	
= ULSI; V²LSI; wafer scale integration; WSI		= ULSI; V²LSI; WSI; Waver-scale-Integration	
ultra light	TYPOGR	ultraleicht	
↑ typeface design		↑ Schriftschnitt	
ultralinear circuit	CIRC.ENG	Ultralinearschaltung	
ultramarine	PHYS	ultramarin	
		= ultramarinblau	
ultra-modern	TECH	hochmodern	
ultrarapid	TECH	superschnell	
ultrared radiation (→infrared radiation)	PHYS	(→Infrarotstrahlung)	
ultrasonic (→ultrasound)	ACOUS	(→Ultraschall)	
ultrasonic bonding	MICROEL	Ultraschallkontaktierung	
		= Ultraschallschweißen; Ultraschallbonden; Mikroschweißen; Ultraschallbondierung	
ultrasonic cleaning	MANUF	Ultraschallbad	
ultrasonic cleaning equipment	MANUF	Ultraschall-Reinigungsanlage	
ultrasonic communications	TELEC	Ultraschallkommunikation	
ultrasonic control	TERM&PER	Ultraschallfernbedienung	
ultrasonic delay line	COMPON	Ultraschall-Verzögerungsleitung	
ultrasonic generator	EL.ACOUS	Ultraschallgeber	
= ultrasonic source			
ultrasonic image converter		Ultraschallbildwandler	
ultrasonic ranging	INSTR	Ultraschallortung	
[Sound Navigation And Ranging]		= SONAR	
= SONAR; ultrasonic sounding			
ultrasonics	TECH	Ultraschalltechnik	
= ultrasound engineering			
ultrasonic sounding (→ultrasonic ranging)	INSTR	(→Ultraschallortung)	
ultrasonic source (→ultrasonic generator)	EL.ACOUS	(→Ultraschallgeber)	
ultrasonic stamping	METAL	Ultraschallprägen	
ultrasonic transducer	COMPON	Ultraschallwandler	
= ultrasound transducer		[Sensor]	
ultrasonic wave	PHYS	Ultraschallwelle	
ultrasonic welding	METAL	Ultraschallschweißen	
ultrasound (n.)	ACOUS	Ultraschall	
[oscillations above 24 Hz]		[Schwingungen über 24 kHz]	
= supersonic sound; ultrasonic			
≠ infrasound		≠ Überschall	
↑ sonic 1		≠ Infraschall	
		↑ Schall 1	
ultrasound engineering (→ultrasonics)	TECH	(→Ultraschalltechnik)	
ultrasound transducer (→ultrasonic transducer)	COMPON	(→Ultraschallwandler)	
ultraviolet	PHYS	ultraviolett	
		= UV	
ultraviolet light-erasable PROM (→EPROM)	MICROEL	(→EPROM)	
ultraviolet radiation	PHYS	Ultraviolettstrahlung	
[wavelength less than 0.4 μ]		[Wellenbereich kleiner 0,4 μ]	
		= ultraviolette Strahlung; Ultraviolett	
umber	PHYS	umbra	
umbilical point	PHYS	Nabelpunkt	
umbra	PHYS	Kernschatten	
umbrella	ANT	Schirm	
		= Umbrella	
umbrella antenna	ANT	Schirmantenne	
= top loaded antenna			
umbrella organization	ECON	Dachorganisation	
umlaut	LING	Umlautzeichen	
≈ diaresis		≈ Trema	
↑ diacritical mark		↑ diakritisches Zeichen	
umlaut 1	LING	Umlaut	
= mutated vowel			
unabridged	LING	ungekürzt	
[text]		[Text]	
unadulterated	TECH	unverfälscht	
= genuine			
unalloyed	METAL	unlegiert	
unaltered (→unchanged)	TECH	(→unverändert)	
unambiguity	MATH	Eindeutigkeit	
= uniqueness			
unambiguous	MATH	eindeutig	
= one-valued			
unambiguous function	MATH	eindeutige Funktion	
unamplified	ELECTRON	unverstärkt	
unanswered	COLLOQ	unbeantwortet	
unarmored cable	COMM. CABLE	unbewehrtes Kabel	
unary (adj.)	MATH	unär	
≈ one-place		= einwertig; monovalent	
		≈ einstellig	
unary (→monadic)	DATA PROC	(→monadisch)	
unassembled board	ELECTRON	(→unbestückte Leiterplatte)	
(→unassembled PCB)			
unassembled PCB	ELECTRON	unbestückte Leiterplatte	
= unpopulated PCB; bare PCB; unassembled board; unpopulated board; bare board			
unassignable (→not transferable)	ECON	(→unübertragbar)	
unattached (→unmounted)	TECH	(→lose)	
unattended (→unstaffed)	TELEC	(→unbemannt)	
unauthorized	COLLOQ	unbefugt	
≈ mischievous		= unerlaubt	
		≈ schädlich	
unauthorized (→unlawful)	ECON	(→unerlaubt)	
unaveraged measurement	INSTR	Messung ohne Mittelwertsbildung	
= unaveraged mode			
unaveraged mode (→unaveraged measurement)	INSTR	(→Messung ohne Mittelwertsbildung)	
unavoidable	COLLOQ	unabwendbar	
		= unvermeidbar	

unbalance

unbalance	NETW.TH	**Unsymmetrie**	
= dissymmertry; impedance unbalance; imbalance		≠ Symmetrie	
≠ balance			
unbalance attenuation	NETW.TH	**Unsymmetriedämpfung**	
= common-mode suppression			
unbalanced	NETW.TH	**unsymmetrisch**	
unbalanced aerial (→unbalanced antenna)	ANT	(→unsymmetrische Antenne)	
unbalanced antenna	ANT	**unsymmetrische Antenne**	
= unbalanced aerial			
unbalanced code (→unipolar code)	CODING	(→unipolarer Code)	
unbalance factor	NETW.TH	**Unsymmetriegrad**	
Unbehauen procedure	NETW.TH	**Unbehauen-Prozeß**	
unbiased	MATH	**erwartungstreu**	
[statistics]		[Statistik]	
unbiasedness	MATH	**Erwartungstreue**	
[statistics]		[Statistik]	
unbilled costs	ECON	**unverrechnete Lieferungen und Leistungen**	
[not yet accouted]		= UL	
unblock	DATA PROC	**entblocken**	
= deblock			
unblocking	SWITCH	**Entblockierung**	
= deblocking		= Entblocken	
unblocking	DATA PROC	**Entblocken**	
= deblocking		= Entblockierung	
unbranched	DATA PROC	(→**unverzweigt**)	
(→non-branched)			
unbreakable 2	TECH	**unzerbrechlich**	
[cannot be broken]		= bruchfest	
= break-proof		≈ bruchsicher	
≈ unbreakable 1		≠ zerbrechlich	
≠ fragile			
unbreakable 1	TECH	**bruchsicher**	
[protected against break]		[gegen Bruch gesichert]	
		≈ bruchfest	
unbuffered	CIRC.ENG	**ungepuffert**	
unbundled	DATA PROC	**nicht im Preis einbegriffen**	
= not included in the price		= getrennt verrechnet	
uncabled	EL.TECH	**unverkabelt**	
≈ unwired		≈ unverdrahtet	
≠ cabled		≠ verkabelt	
uncalibrated	INSTR	**ungeeicht**	
uncertainity	SCIE	**Unschärfe**	
= indeterminacy; unsharpness; fuzziness		= Unbestimmtheit	
uncertainity	INF	**Unsicherheit**	
uncertainity	SCIE	**Unschärfe**	
= indeterminacy; unsharpness; fuzziness		= Unbestimmtheit; Unbestimmbarkeit	
uncertainity relation	PHYS	**Unbestimmtheitsrelation**	
unchanged	TECH	**unverändert**	
= unaltered			
unchecked	TECH	**ungeprüft**	
≈ untested		= unkontrolliert	
≠ checked		≈ unerprobt	
		≠ geprüft	
uncial	TYPOGR	**Unzialschrift**	
[late roman style with rounded letters]		[spätrömische Schrift mit abgerundeten Buchstaben]	
uncoated (→bare)	COMM.CABLE	(→blank)	
uncoded	INF	**uncodiert**	
= non-coded		= unkodiert	
uncoded text	INF.TECH	(→**Klartext**)	
(→plaintext)			
uncoiling	MECH	**abspulen**	
= unwind		= abwickeln	
uncollectable	ECON	**uneinbringlich**	
≈ irretrievable			
uncombined	CHEM	**ungebunden**	
uncommited	ECON	**nicht zweckgebunden**	
		= ungebunden	
uncommited logic array (→gate array)	MICROEL	(→Gate Array)	
unconditional	COLLOQ	**unbedingt**	
= unconstrained		= nichtbedingt	
≈ unrestricted			
unconditional branch (→unconditional program branch)	DATA PROC	(→**unbedingte Programmverzweigung**)	
unconditional branch instruction (→unconditional jump instruction)	DATA PROC	(→unbedingter Sprungbefehl)	
unconditional jump (→unconditional program jump)	DATA PROC	(→unbedingter Programmsprung)	
unconditional jump instruction	DATA PROC	**unbedingter Sprungbefehl**	
= unconditional branch instruction		= unbedingter Verzweigungsbefehl	
unconditional program branch	DATA PROC	**unbedingte Programmverzweigung**	
= unconditional branch			
unconditional program jump	DATA PROC	**unbedingter Programmsprung**	
= unconditional jump		= unbedingter Sprung	
unconfirmed	COLLOQ	**unbestätigt**	
unconformity	TECH	**Nichtübereinstimmung**	
= non-conformity			
unconstrained (→unconditional)	COLLOQ	(→**unbedingt**)	
unconstrained (→unrestricted)	COLLOQ	(→**unbeschrankt**)	
uncontrollable	TECH	**unbeherrschbar**	
		= unkontrollierbar	
uncontrolled (→with uncontrolled ambient)	TECH	(→**unklimatisiert**)	
uncontrolled loop	DATA PROC	**unkontrollierte Schleife**	
uncorrected error (→residual error)	MATH	(→**Restfehler**)	
uncorrelated	MATH	**unkorreliert**	
uncostumed	ECON	**unverzollt**	
uncountable (n.)	LING	**unzählbares Nomen**	
uncouple (→decouple)	EL.TECH	(→**entkoppeln**)	
uncoupled two-port	NETW.TH	**kopplungsfreier Vierpol**	
		= koppelfreier Vierpol	
uncover (v.t.)	TECH	**abdecken 1**	
≠ cover 1 (v.t.)		[Deckung abnehmen]	
		≠ zudecken	
uncovered	ECON	**ungedeckt**	
undamaged	ECON	**unversehrt**	
= intact		= unbeschädigt; intakt	
undamped	PHYS	**ungedämpft**	
= sustained			
undamped oscillation	PHYS	**ungedämpfte Schwingung**	
= sustained oscillation			
undamped wave	PHYS	**ungedämpfte Welle**	
= sustained wave			
undated	COLLOQ	**undatiert**	
undecidability	MATH	**Unentscheidbarkeit**	
undecidable	MATH	**unentscheidbar**	
undefined	SCIE	**unbestimmt**	
= indeterminate		= undefiniert; indeterminiert	
undeformable	TECH	**formtreu**	
		= unverformbar	
undelayed	ELECTRON	**verzögerungsfrei**	
= quick-action; quick-response; quick-operating		= unverzögert	
underbid	ECON	**unterbieten**	
= undercut			
under-car antenna	ANT	**Chassisantenne**	
undercurrent	EL.TECH	**Unterstrom**	
≠ overcurrent		≠ Überstrom	
undercut (v.t.)	MICROEL	**unterätzen**	
undercut (→underbid)	ECON	(→unterbieten)	
underdamped	PHYS	**unterdämpft**	
underdamping	PHYS	**Unterdämpfung**	
		= unterkritische Dämpfung	
under-desk equipment	EQUIP.ENG	**Unterbaugerät**	
↓ tower [DATA PROC]		= Unter-Tisch-Gerät	
		↓ Turmgerät [DATA PROC]	
underdiffusion	MICROEL	**Unterdiffusion**	
underemploy (→underuse)	TECH	(→unterauslasten)	
underemployment	ECON	**Unterbeschäftigung**	
underemployment (→underuse)	TECH	(→Unterauslastung)	
underequip	TECH	**unterbestücken**	
≈ equip partially		≈ teilbestücken	
underequipping	TECH	**Unterbestückung**	
≈ partial equipping		≈ Teilbestückung	
underetching	MICROEL	**Unterätzung**	
= undermining			
underexposure	PHYS	**Unterbelichtung**	

underflow	DATA PROC	Unterlauf		undervoltage	EL.TECH	Unterspannung
≠ overflow		= Bereichsunterschreitung; Underflow		= subvoltage		≠ Überspannung
		≠ Überlauf		≠ overvoltage		
undergoing (→ongoing)	TECH	(→laufend 1)		underwater acoustics	ACOUS	Unterwasserakustik
underground (→buried)	TECH	(→unterirdisch)		underwater cable	COMM.CABLE	Unterwasserkabel
underground amplifier	TRANS	(→Unterflurverstärker)		= subaqueous cable		≠ Landkabel
(→underground repeater)				≠ terrestrial cable		↓ Seekabel; Tiefseekabel; Flußkabel
underground cable (BRI)	COMM.CABLE	(→Erdkabel)		↓ submarine cable; deep-see cable; subfluvial cable		
(→earth cable)				underwater cable plow	OUTS.PLANT	(→Unterwasserkabelpflug)
underground cable (BRI)	COMM.CABLE	(→unterirdisches Kabel)		(→submarine cable plow)		
(→below-ground cable)				underwater loudspeaker	EL.ACOUS	Unterwasserlautsprecher
underground cable (AM)	COMM.CABLE	(→Röhrenkabel)		underway (→ongoing)	TECH	(→laufend 1)
(→duct cable)				underweight	ECON	Untergewicht
underground construction	CIV.ENG	Tiefbau		= short weight		= Mindergewicht
↑ civil engineering		↑ Bauwesen		underwriter (→insurer)	ECON	(→Versicherer)
underground construction engineering	CIV.ENG	Tiefbautechnik		undesigned (→unintentional)	COLLOQ	(→unbeabsichtigt)
↑ civil engineering		↑ Bauwesen		undesired (→spurious)	INF.TECH	(→unerwünscht)
underground container	OUTS.PLANT	Unterflurbehälter		undetected error	DATA PROC	nichterfaßter Fehler
underground line	OUTS.PLANT	Erdleitung		undetected fault	QUAL	stiller Fehler
underground power cable	POWER ENG	Starkstrom-Erdkabel		undiluted	TECH	unverdünnt
underground regenerator	TRANS	Unterflurregenerator		undiminished	COLLOQ	unvermindert
≈ underground regenerator		≈ Unterflurverstärker		undisclosed (→unnamed)	COLLOQ	(→ungenannt)
underground repeater	TRANS	Unterflurverstärker		undisclosed reserves	ECON	stille Rücklage
= underground amplifier; buried repeater		≈ Unterflurregenerator				= stille Reserve
≈ underground regenerator				undistorted	TELEC	unverzerrt
underlay (v.t.)	ELECTRON	unterlegen		≠ distorted		≠ verzerrt
underlay (n.)	TECH	Unterlage		undisturbed	ELECTRON	ungestört
≈ support (n.)		≈ Auflage		undo (v.t.)	TECH	rückgängigmachen
underlay (n.)	ELECTRON	Unterlegung		undulate (→oscillate)	MECH	(→schwingen)
underlayed	ELECTRON	unterlegt		undulating (→undulatory)	PHYS	(→wellenförmig)
= underlaying				undulating (→wavy)	TECH	(→wellig)
underlaying	ELECTRON	(→unterlegt)		undulation	PHYS	Wellenbewegung
(→underlayed)				undulation (→vibration)	MECH	(→mechanische Schwingung)
underlaying coding	CODING	unterlegte Codierung		undulatory	PHYS	wellenförmig
= modulation-matched coding		= modulationsangepaßte Codierung		= undulating		= wellig
				undulatory mechanics (→wave mechanics)	PHYS	(→Wellenmechanik)
underline (v.t.)	LING	unterstreichen		undulatory optics (→wave optics)	PHYS	(→Wellenoptik)
= underscore				undulatory physics (→wave mechanics)	PHYS	(→Wellenmechanik)
underline (n.)	LING	Unterstreichung				
= underscore (n.)				undulatory theory (→wave theory)	PHYS	(→Wellentheorie)
underline character	TYPOGR	Unterstreichungszeichen		unearthed	EL.TECH	ungeerdet
[symbol –]		[Symbol –]		≠ earthed		≠ geerdet
= underscore character; break character		= Unterstrich		unearthed protective conductor	POWER ENG	ungeerdeter Schutzleiter
undermatching	NETW.TH	Stromanpassung		= protective conductor		
[intrinsic resistance very larger than load resistance]		[Innenwiderstand viel größer als Außenwiderstand]		uneconomic	ECON	unwirtschaftlich
≠ overmatching		= Unteranpassung				= unrationell
↑ matching 1		≠ Spannungsanpassung		unequal	MATH	ungleich
		↑ Anpassung 1		= inequal; not equal		≠ gleich
				≠ equal		
undermining (→underetching)	MICROEL	(→Unterätzung)		unequal (→different)	COLLOQ	(→verschieden)
underpass	OUTS.PLANT	Wegunterführung		unequaled (→unmatched)	COLLOQ	(→unerreicht)
underrun	TECH	Unterschreitung		unequality (→inequality)	MATH	(→Ungleichung)
undersaturation	ELECTRON	Untersättigung		unequalled (→unrivalled)	ECON	(→konkurrenzlos 1)
underscore (→underline)	LING	(→unterstreichen)		unequipped	TECH	unbestückt
underscore (n.) (→underline)	LING	(→Unterstreichung)				= nicht ausgebaut
underscore character	TYPOGR	(→Unterstreichungszeichen)		uneven	TECH	uneben
(→underline character)				≠ even		≠ eben
undersea cable	COMM.CABLE	(→Seekabel)		uneven	MATH	ungeradzahlig
(→submarine cable)				= uneven-numbered; odd		= ungerade
undersea fiber cable	COMM.CABLE	(→optisches Seekabel)		≠ even		≠ geradzahlig
(→submarine fiber cable)				unevenness (→inequality)	MATH	(→Ungleichung)
underserved	TELEC	unterversorgt		uneven number	MATH	ungerade Zahl
underside	TECH	Unterseite		= odd number		
		= untere Seite		uneven-numbered (→uneven)	MATH	(→ungeradzahlig)
undersize (v.t.)	TECH	unterdimensionieren		uneven parity (→odd parity)	CODING	(→ungerade Parität)
		= zu schwach bemessen		unfaded	RADIO PROP	schwundfrei
undertaker (BRI)	ECON	(→Unternehmer)		unfaulted (→fault-free)	TECH	(→fehlerfrei)
(→entrepreneur)				unfavorable	COLLOQ	ungünstig
undertaking (→company)	ECON	(→Gesellschaft)		= adverse		
underuse (v.t.)	TECH	unterauslasten		unfinished	MANUF	unfertig
= underemploy; underutilize		≈ unterbelasten [QUAL]		unfinished (→raw)	TECH	(→roh)
≈ underrate [QUAL]				unfinished product	MANUF	unfertiges Produkt
underuse (n.)	TECH	Unterauslastung		= work-in-process		≈ halbfertiges Erzeugnis
= underemployment; underutilization		≈ Unterbelastung [QUAL]		unfit	TECH	ungeeignet
≈ underrating [QUAL]				= unsuitable; inadequate		= untauglich; unangemessen; inadäquat; unzulänglich
underutilization (→underuse)	TECH	(→Unterauslastung)		≈ useless; insufficient		≈ unbrauchbar; unzureichend
underutilize (→underuse)	TECH	(→unterauslasten)				

unformatted

unformatted DATA PROC **formatfrei**
= nonformatted = unformatiert
≠ formatted ≠ formatiert
unformatted capacity DATA PROC **Brutto-Speicherkapazität**
= Bruttokapazität; unformatierte Speicherkapazität
unformatted input DATA PROC (→formatfreie Eingabe)
(→nonformatted input)
unforseeable COLLOQ **unabsehbar**
= unvorhersehbar; unvorausschaubar
unfriendly takeover (→hostile takeover) ECON (→feindliche Übernahme)
ungeordnete Datei DATA PROC **ungeordnete Datei**
unhatched (→unshaded) ENG.DRAW (→unschraffiert)
uniaxial TECH **einachsig**
unibus (→central bus) DATA PROC (→Zentralbus)
UNIC NETW.TH **spannungsumkehrender Negativ-Impedanzkonverter**
unidimensional MATH (→eindimensional)
(→one-dimensional)
unidirectional TECH **unidirektional**
≠ bidirectional = einseitig gerichtet
≠ bidirektional
unidirectional (→one-way) TELEC (→einfachgerichtet)
unidirectional antenna ANT **einseitige Richtantenne**
↓ Yagi-Uda antenna; parabolic antenna ↓ Yagi-Uda-Antenne; Parabolantenne
unidirectional current (→direct current) EL.TECH (→Gleichstrom)
unidirectional rhombic antenna ANT **unidirektionale Rhombusantenne**
unifet (→unipolar field-effect trransistor) MICROEL (→Unipolar-Feldeffekttransistor)
unification TECH **Vereinheitlichung**
= uniformity = Standardisierung; Vereinfachung
≈ standardization; simplification
unification COLLOQ **Vereinigung**
= union; merger = Zusammenschluß
unified (→uniform) TECH (→gleichförmig 1)
unifilar (→single-wired) EL.TECH (→einadrig)
unifilar electrometer INSTR **Einfadenelektrometer**
uniform (v.t.) TECH **vereinheitlichen**
≈ standardize; simplify ≈ standardisieren; vereinfachen
uniform (adj.) TECH **gleichförmig 1**
= unified; unique = gleichmäßig; einheitlich
uniform (→continuous) MATH (→stetig)
uniform (→stepless) TECH (→stufenlos)
uniform approximation MATH **gleichmäßige Näherung**
uniform distribution MATH **Gleichverteilung**
= equipartition = gleichförmige Verteilung; Rechteckverteilung
uniformity (→unification) TECH (→Vereinheitlichung)
uniform load MECH **Flächenlast**
uniform loading COMM.CABLE (→Krarupisierung)
(→continuous loading)
uniform numbering SWITCH **Feststellennumerierung**
= fixed numbering = einheitliche Numerierung
uniform quantization CODING **gleichförmige Quantisierung**
= gleichmäßige Quantisierung
unify (→unite) COLLOQ (→vereinigen)
unijunction transistor MICROEL **Doppelbasisdiode**
= UJT; double-base diode = Zweibasisdiode; Unijunction-Transistor; UJT; Zweizonen-Transistor; Doppelbasistransistor; Zweibasistransistor
unilateral (→single-sided) TECH (→einseitig)
unilateral quadripole NETW.TH (→übertragungsunsymmetrischer Vierpol)
(→unilateral two-port)
unilateral tolerance MECH **einseitiges Abmaß**
unilateral two-port NETW.TH **übertragungsunsymmetrischer Vierpol**
= unilateral quadripole
unimodal MATH **eingipflig**
uninsulated EL.TECH **unisoliert**
uninsured ECON **unversichert**
unintelligible TELEC **unverständlich**
≠ intelligible ≠ verständlich

unintelligible crosstalk TELEC **unverständliches Nebensprechen**
= inverted crosstalk
unintended COLLOQ (→unbeabsichtigt)
(→unintentional)
unintentional COLLOQ **unbeabsichtigt**
= unintended; undesigned = ungewollt
≠ intentional ≠ beabsichtigt
uninterruptable power supply POWER SYS **unterbrechungsfreie Stromversorgung**
= UPS; no-break power supply; interruption-free power supply
uninterrupted TECH (→kontinuierlich)
(→continuous 1)
uninverted crosstalk TELEC (→verständliches Nebensprechen)
(→intelligible crosstalk)
uninvertible (→irreversible) SCIE (→irreversibel)
union ECON **Gewerkschaft**
= labor union (AM); labour union (BRI); trade union
union (→sleeve) TECH (→Muffe)
union (→unification) COLLOQ (→Vereinigung)
union of sets MATH **Vereinigungsmenge**
unipolar PHYS **unipolar**
= monopole
unipolar (→single-pole) EL.TECH (→einpolig)
unipolar circuit CIRC.ENG **Unipolarschaltung**
unipolar code CODING **unipolarer Code**
= unbalanced code = unsymmetrischer Code
unipolar FET (→unipolar field-effect transistor) MICROEL (→Unipolar-Feldeffekttransistor)
unipolar field-effect transistor MICROEL **Unipolar-Feldeffekttransistor**
= unipolar FET; unifet = Unipolar-FET; Unifet
unipolar transistor MICROEL **Unipolartransistor**
≠ bipolar transistor = unipolarer Transistor
↓ field-effect transistor: MOS transistor ≠ Bipolartransistor
↓ Feldeffekttransistor:; MOS-Transistor
unipole (→monopole) ANT (→Monopol)
uniquad ANT **Uniquad**
unique (→uniform) TECH (→gleichförmig 1)
uniqueness (→unambiguity) MATH (→Eindeutigkeit)
uniselector SWITCH **Einwegwähler**
unit MATH **Einheit**
= unity
unit EQUIP.ENG **Einheit**
unit (→inset) EQUIP.ENG (→Geräteeinsatz)
unitary impulse ELECTRON **Einheitsimpuls**
= input impulse
unit assembly MANUF **Teilzusammenbau**
unit charge PHYS **Ladungseinheit**
unit charge (→basic rental) TELEC (→Grundgebühr)
unit cube PHYS **Elementarwürfel**
[crystal] [Kristall]
= fundamental cell = Elementarzelle; Gitterbaustein
unit-distance code (→cyclic code) CODING (→zyklischer Code)
unite COLLOQ **vereinigen**
= unify
unit fan-in (→unit load) CIRC.ENG (→Einheitslast)
unit-fee SWITCH **Gebühreneinheit**
= charging unit; charge unit
unit-fee metering (→single metering) SWITCH (→Einfachzählung)
unit interval TELEGR **Zeichenelement**
= signal element; unit signal element; signal interval; length element; significant interval; signal element length = Zeichenintervall; Zeichenschritt 1; Zeichenlänge; Zeichendauer; Signalelement; Signalintervall; Signaldauer; Signalschritt; Schrittelement; Schrittintervall; Schrittlänge; Schrittdauer; Kennabschnitt; Kennintervall; kennzeichnendes Intervall; Einheitsschritt; Übertragungsschritt; Schritt
↓ marking pulse; spacing pulse; start pulse; information pulse; stop pulse
↓ Stromschritt; Pausenschritt; Anlaufschritt; Informationsschritt; Sperrschritt

unit load	CIRC.ENG	Einheitslast
= unit fan-in		
unit matrix	MATH	Einheitsmatrix
unit position	DATA PROC	Einheitsfeld
unit production	MANUF	Einzelanfertigung
= manufacture to order; job work; single-part production		= Einzelfertigung
unit pulse	ELECTRON	Einheitspuls
unit rate	ECON	Satz
= rate		
unit separator [code]	DATA PROC	Teilgruppenzeichen [Code]
= US		= US
unit signal element (→unit interval)	TELEGR	(→Zeichenelement)
unit-step function	INSTR	Einheitssprung
unit under measurement	INSTR	Meßobjekt
= unit under test; device under test; specimen		= Meßgegenstand; Meßling; Probe
≈ test specimen [QUAL]		≈ Prüfling [QUAL]
unit under test (→unit under measurement)	INSTR	(→Meßobjekt)
uni tunnel diode (→backward diode)	MICROEL	(→Rückwärtsdiode)
unit vector	MATH	Einheitsektor
unit wiring (→intra-shelf wiring)	EQUIP.ENG	(→Einsatzverdrahtung)
unity (→unit)	MATH	(→Einheit)
univariate (→one-dimensional)	MATH	(→eindimensional)
universal	TECH	universell
= general-purpose		≈ Mehrzweck-; Universal-
≈ multi-purpose		
universal amplifier	TRANS	Allverstärker
universal closure (→universal sleeve)	OUTS.PLANT	(→Universalmuffe)
universal computer (→general-purpose computer)	DATA PROC	(→Universalrechner)
universal counter	INSTR	Universalzähler
universal document reader	TERM&PER	Universalschriftleser
universal identifier [for persons]	ECON	Personenkennzahl
universal joint	MECH	Kardangelenk
= Kreuzgelenk		
universally valid (→generally valid)	SCIE	(→allgemeingültig)
universal network	TELEC	Universalnetz
universal product code [a US bar code]	DATA PROC	UPC-Strichcode
= UPC		
universal register	DATA PROC	Universalregister
universal shunt (→Ayrton shunt)	INSTR	(→Ayrton-Nebenwiderstand)
universal sleeve	OUTS.PLANT	Universalmuffe
= universal closure		
universal validity (→general validity)	SCIE	(→Allgemeingültigkeit)
universal-wound coil	COMPON	Kreuzwickelspule
		≈ Honigwabenspule
univibrator (→monostable multivibrator)	CIRC.ENG	(→monostabile Kippstufe)
UNIX [UNIversal eXchange]	DATA PROC	UNIX
		= Unix
= Unix		↑ Betriebssystem
↑ operating system		
Unix (→UNIX)	DATA PROC	(→UNIX)
unjustified composition (→ragged typesetting)	TYPOGR	(→Flattersatz)
unjustified print (→ragged typesetting)	TYPOGR	(→Flattersatz)
unjustified setting (→ragged typesetting)	TYPOGR	(→Flattersatz)
unjustified type (→ragged typesetting)	TYPOGR	(→Flattersatz)
unjustified typesetting (→ragged typesetting)	TYPOGR	(→Flattersatz)
unjustified typesetting with hyphenation (→ragged typesetting with hyphenation)	TYPOGR	(→Rauhsatz)
unknown	COLLOQ	unbekannt
unknown quantity	MATH	Unbekannte
≈ variable (n.)		= unbekannte Größe; Unbestimmte; unbestimmte Größe
		≈ Variable
unlawful	ECON	unerlaubt
= unauthorized; illigal		= illegal; widerrechtlich
unleveled	CIRC.ENG	nicht pegelgeregelt
unlicensed listener	BROADC	Schwarzhörer
= blacklistener; illicit listener		
unlicensed transmitter	RADIO	Schwarzsender
= non-licensed transmitter; unlicensed transmitting station; pirate transmitting station; illicit transmitter		
unlicensed transmitting station (→unlicensed transmitter)	RADIO	(→Schwarzsender)
unlicensed TV viewer	BROADC	Schwarzfernseher
= illicit TV viewer		
unlike (→different)	COLLOQ	(→verschieden)
unlikeness (→difference)	COLLOQ	(→Verschiedenheit)
unlimited [in time]	ECON	unbefristet
unlimited (→unrestricted)	COLLOQ	(→unbeschränkt)
unlink (→remove)	SWITCH	(→aushängen)
unload	TECH	entladen
= discharge		
unloaded	TECH	unbelastet
		[ohne Gewicht]
unloaded	COMM.CABLE	unbespult
= nonloaded		= unpupinisiert
≠ coil-loaded		≠ bespult
unloaded cable	COMM.CABLE	unbespultes Kabel
= nonloaded cable		= unpupinisiertes Kabel
unloading [of goods]	ECON	Entladung [einer Ware]
= discharge		
unlock (v.t.)	MECH	entriegeln
= trip (v.t.)		= auslösen
unlocking	TECH	Ausrasten
unmachined (→raw)	TECH	(→roh)
unmachined part (→preform)	MANUF	(→Rohling)
unmanned (→unstaffed)	TELEC	(→unbemannt)
unmark (→erase)	ELECTRON	(→löschen)
unmatched	NETW.TH	unangepaßt
unmatched	COLLOQ	unerreicht
= unequaled		≈ unübertroffen
= unsurpassed		
unmistakable	COLLOQ	unverkennbar
		= unverwechselbar
unmonitored	TECH	unüberwacht
= unsupervized		≠ überwacht
≠ monitored		
unmounted	TECH	lose
= loose; unattached; slack		= locker; nicht fixiert
unnamed	COLLOQ	ungenannt
= undisclosed		= unbenannt
unnecesary seizure	SWITCH	unnötige Belegung
unnumbered	TECH	unnummeriert
unobjectionable	TECH	einwandfrei
≈ fault-free		≈ fehlerfrei
unobtainable (→disconnected)	SWITCH	(→unbeschaltet)
unofficial print	DOC	Informationspause
= courtesy copy		= Informationskopie
unordered	ECON	unbestellt
unordered	TECH	ungeordnet
unpack (v.t.)	DATA PROC	entpacken
≠ pack (v.t.)		≠ packen
unpack	TECH	auspacken
unpacked	ECON	unverpackt
≈ loose		≈ lose
unpacking instructions	TECH	Auspackanleitung
unpaid	ECON	unbezahlt
= outstanding		= offenstehend
unparity (→odd parity)	CODING	(→ungerade Parität)
unparity check (→odd parity check)	CODING	(→Imparitätskontrolle)
unparity control (→odd parity check)	CODING	(→Imparitätskontrolle)

unplanned	COLLOQ	**unplanmäßig**
= unscheduled; non-regular		= außerplanmäßig; unprogrammiert
≈ extraordinary		≈ außergewöhnlich
unpopulated board (→unassembled PCB)	ELECTRON	(→unbestückte Leiterplatte)
unpopulated PCB (→unassembled PCB)	ELECTRON	(→unbestückte Leiterplatte)
unprecedented (→novel)	COLLOQ	(→neuartig)
unpredictable	COLLOQ	**unvorhersagbar**
unproductive	ECON	**unproduktiv**
unprogrammed conditional program jump	DATA PROC	**nichtprogrammierter Programmsprung**
= trap		= nichtprogrammierter Sprung; Trap
unprotected	TELEC	**ungeschützt**
= single		≠ ersatzgeschützt
≠ protected		
unprotected contact	COMPON	**ungeschützter Kontakt**
		= offener Kontakt
unpublished	LING	**unveröffentlicht**
unpunched	TERM&PER	**ungelocht**
unpunched tape	TERM&PER	**ungelochter Lochstreifen**
= blank tape		
unquote (→unquote sign)	TYPOGR	(→Abführung)
unquote sign	TYPOGR	**Abführung**
= unquote		= schließendes Anführungszeichen
↑ quotation mark		↑ Anführungszeichen
unrecoverable (→irretrievable)	DATA PROC	(→unwiederbringlich)
unreliability	COLLOQ	**Unzuverlässigkeit**
unreliable	COLLOQ	**unzuverlässig**
unrestricted	COLLOQ	**unbeschränkt**
= unlimited; unconstrained; limitless		= uneingeschränkt; unbegrenzt; grenzenlos
≈ unconditional; free		≈ unbedingt; frei
unrestricted	OFFICE	**allgemein zugänglich**
≠ confidential; secret		≠ vertraulich; geheim
unrestricted data	DATA PROC	**freie Daten**
unrivalled	ECON	**konkurrenzlos 1**
= unequalled		[überlegen]
unsalable	ECON	**unverkäuflich**
unscheduled (→unplanned)	COLLOQ	(→unplanmäßig)
unscrew	MECH	**herausdrehen**
= screw-out		= ausschrauben; abschrauben; herausschrauben
unserviceable (→useless)	TECH	(→unbrauchbar)
unset (v.t.)	CODING	**auf Null setzen**
unshaded	ENG.DRAW	**unschraffiert**
= unhatched		≈ schraffiert
≠ shaded		
unsharpness (→uncertainty)	SCIE	(→Unschärfe)
unsigned	MATH	**vorzeichenlos**
unskilled	ECON	**ungelernt**
unsolder (→desolder)	ELECTRON	(→auslöten)
unsophisticated (→primitive)	TECH	(→primitiv)
unstable	TECH	**instabil**
= instable; labile		= labil; unbeständig
≈ variable		≈ veränderlich
unstableness (→instability)	TECH	(→Instabilität)
unstaffed	TELEC	**unbemannt**
[station]		[Station]
= unattended; unmanned		= unbedient; unbesetzt; unbeaufsichtigt
≠ staffed		≠ bemannt
unsteady (→non-steady)	TECH	(→nichtstationär)
unstress (v.t.)	MECH	**entspannen**
unstressed	TECH	**ungespannt**
unstressing	METAL	**Entspannung**
unsuccessful call	SWITCH	**erfolglose Belegung**
unsufficient	TECH	**ungenügend**
unsuitable (→unfit)	TECH	(→ungeeignet)
unsupervized (→unmonitored)	TECH	(→unüberwacht)
unsupported (→self-supporting)	TECH	(→selbsttragend)
unsurpassed	COLLOQ	**unübertroffen**
≈ unmatched		≈ unerreicht
untested	TECH	**unerprobt**
≈ unchecked		≈ ungeprüft
untie	TECH	**losbinden**
≈ undo		≈ aufbinden
		≈ lösen
untimely	COLLOQ	**unzeitig**
↓ premature; belated; delayed		↓ vorzeitig; verspätet; verzögert
untraceable	COLLOQ	**unauffindbar**
untune (→detune)	ELECTRON	(→verstimmen)
untuned (→detuned)	ELECTRON	(→verstimmt)
untuned antenna	ANT	**unabgestimmte Antenne**
untuned feedline	ANT	**unabgestimmte Speiseleitung**
untuning (→detuning)	ELECTRON	(→Verstimmung)
unusable (→useless)	TECH	(→unbrauchbar)
unused (→unutilized)	TECH	(→ungenutzt)
unutilized	TECH	**ungenutzt**
= unused; idle		= unausgenutzt; unausgelastet
unvalid	MATH	**ungültig**
≠ valid		≠ gültig
unveil (→present)	ECON	(→vorstellen)
unwanted emission	RADIO	**Störstrahlung**
= spurious emission; spurious radiation; radiated interference; radiated emission; electromagnetic interference; EMI		= unerwünschte Ausstrahlung
unwanted emission (→unwanted emission)	RADIO	(→Störstrahlung)
unwanted mode (→interfering mode)	MICROW	(→Störwelle)
unwanted response	COMPON	**Störresonanz**
[quartz]		[Quarz]
= spurious resonance		= Nebenresonanz
unweighted noise voltage (→interference voltage)	TELEC	(→Störspannung)
unwind (→uncoiling)	MECH	(→abspulen)
unwinding roller	OUTS.PLANT	**Abspulroller**
unwired	EL.TECH	**unverdrahtet**
≈ uncabled		≈ unverkabelt
≠ wired		≠ verdrahtet
unwrought (→raw)	TECH	(→roh)
up-and-running	DATA PROC	**auf Anhieb funktionierend**
UPC (→universal product code)	DATA PROC	(→UPC-Strichcode)
up-conversion	HF	**Aufwärtsmischung**
up-conversion mixer	HF	**Aufwärtsmischer**
up-converted	HF	**aufwärts gemischt**
up-converter	RADIO	**Sendeumsetzer**
up-counter	CIRC.ENG	**Vorwärtszähler**
= progressive counter		
update (v.t.)	DATA PROC	**aktualisieren**
		= fortschreiben
update (n.)	TECH	**Aktualisierung**
= actualization		= Fortschreibung
≈ enhancement		≈ Verbesserung
update (n.) (→program update)	DATA PROC	(→Programmaktualisierung)
update information	TECH	**Änderungsmitteilung**
= update notification; change information; change notification; document change; update message; change notice		= Änderungsinformation
update information service	TECH	**Änderungsdienst**
= update service		
update message (→update information)	TECH	(→Änderungsmitteilung)
update notification (→update information)	TECH	(→Änderungsmitteilung)
update record (→amendment record)	DATA PROC	(→Ergänzungseintrag)
update service (→update information service)	TECH	(→Änderungsdienst)
updating run	DATA PROC	**Aktualisierungslauf**
up-down counter	CIRC.ENG	**Vor-Rückwärts-Zähler**
= reversible counter; bidirectional counter		= Zweirichtungszähler; bidirektionaler Zähler
up-fading	RADIO PROP	**Aufwärtsschwund**
		= Aufwärtsfading
upgradability	DATA PROC	**Ausbaufähigkeit**
= Systemaktualisierbarkeit		= Aufrüstbarkeit; Systemaktualisierbarkeit

upgradability (→expansion capability)		TECH	(→Erweiterungsmöglichkeit)	**usage**	COLLOQ	**Gepflogenheit**
				= custom; practice		= Usance
upgradable		TECH	**ausbaufähig**	**usage**	LING	**Sprachgebrauch**
= expandable; open-ended			= erweiterbar; ausbaubar;	**usage** (→occupancy)	TELEC	(→Belegung)
≈ adaptable			aufrüstbar; offen (fig.)	**usage** (→use)	TECH	(→Benutzung)
			≈ umrüstbar	**usage of channels** (→channel occupancy)	TELEC	(→Kanalbelegung)
upgrade (→expand)		TECH	(→ausbauen 1)			
upgrade (n.) (→program upgrade)		DATA PROC	(→Programmerweiterung)	**USASCII code** (→ASCII code)	CODING	(→ASCII-Code)
				use (v.t.)	TECH	**benutzen**
upgrading (→expansion)		TECH	(→Erweiterung)	= employ		= verwenden
up-link		SAT.COMM	**Aufwärtsstrecke**	≈ utilize		≈ ausnutzen
= earth-to-space link			= Aufwärtsrichtung	**use** (n.)	TECH	**Benutzung**
upload		DATA PROC	**hinaufladen**	= usage		= Verwendung; Gebrauch; Nutzung
[transfer to a larger computer]			[auf einen größeren Computer transferieren]	≈ application; utilization		
≠ download			≠ herunterladen			≈ Anwendung; Ausnutzung
U plug (→plug link)		ELECTRON	(→Brückenstecker)	**use** (n.) (→operation 1)	TECH	(→Betrieb)
upper band		RADIO	**Oberband**	**use-dependent**	TECH	**anwendungsspezifisch**
≠ lower band			≠ Unterband	(→application-specific)		
upper band limit		RADIO	**Bandende 2**	**useful**	TECH	**nützlich**
			≠ Bandanfang	= helpful; utile		≈ brauchbar; nutzbar
			↑ Bandende 1	**useful bit**	CODING	**Nutzbit**
uppercase character (→capital character)	TYPOGR	(→Großbuchstabe)		= payload bit; data bit		
				useful byte	CODING	**Nutzbyte**
uppercase letter (→capital character)	TYPOGR	(→Großbuchstabe)		= data byte		
				useful field	EL.TECH	**Nutzfeld**
uppercase shift (→case shift)	TERM&PER	(→Groß-Klein-Umschaltung)		**useful field intensity**	RADIO	**Nutzfeldstärke**
				useful information	CODING	**Nutzinformation**
upper/lower case shift (→case shift)	TERM&PER	(→Groß-Klein-Umschaltung)		= wanted signal; payload; data		= Nutznachricht
				useful level	TELEC	**Nutzpegel**
upper mode		TERM&PER	**Umschaltebene**	**useful life**	QUAL	**Betriebsbrauchbarkeitsdauer**
[the upper symbols of a double assigned keuboard are valid]			[es gelten die oberen Zeichen einer doppelt belegten Tastatur]	= life utility		[bis zu Überschreitung festgelegter Grenzwerte]
				≈ utilization time; life time		≈ Brauchbarkeitsdauer; Gebrauchslebensdauer
= secondary allocation			= sekundäre Belegung			
≠ lower mode			≠ Grundebene			≈ Nutzungsdauer; Lebensdauer
upper part		TECH	**Oberteil**			
≠ lower part			≠ Unterteil	**usefulness**	TECH	**Nützlichkeit**
upper sideband		MODUL	**oberes Seitenband**	= utility		≈ Brauchbarkeit; Nutzbarkeit
= upright sideband				**useful signal** (→wanted signal)	TELEC	(→Nutzsignal)
upright		TECH	**aufrecht**	**use-independent**	TECH	**anwendungsneutral**
upright (→spar)		MECH	(→Holm)	(→application-independent)		
upright sideband (→upper sideband)	MODUL	(→oberes Seitenband)		**use instruction 1**	TECH	**Bedienungsanleitung**
				= instructions for use; operating instructions 2; directions for use; operating guide		= Gebrauchsanleitung; Bedienungsanweisung; Gebrauchsanweisung; Betriebsanleitung; Betriebsvorschrift; Bedienungshinweise
UPS (→uninterruptable power supply)	POWER SYS	(→unterbrechungsfreie Stromversorgung)				
up-scale (adj.)		INSTR	**skalenaufwärts**			
upset (v.t.)		METAL	**stauchen**	≈ operating instruction 1; notes for the operator		
upset (adj.)		METAL	**gestaucht**			
upsetting		METAL	**Stauchen**			
upside (→top side)		TECH	(→Oberseite)	**use instruction 2**	TECH	**Einsatzhinweis**
upstroke		TERM&PER	**Hub**	**useless**	TECH	**unbrauchbar**
[ink ribbon]			[Farbband]	= unserviceable; unusable		= unnütz; nutzlos
uptime		QUAL	**Betriebszeit**	≈ unfit		≈ ungeeignet
≠ downtime			≠ Ausfallzeit	**uselessness**	TECH	**Nutzlosigkeit**
up-time ratio		QUAL	(→Verfügbarkeit)			= Unbrauchbarkeit
(→availability)				**use-orientated**	DATA PROC	(→anwendungsorientiert)
up-to-date (→modern)		TECH	(→modern)	(→application-oriented)		
upward arrow		TYPOGR	**Pfeil nach oben**	**use-oriented**	DATA PROC	(→anwendungsorientiert)
[symbol ^]			[Symbol ^]	(→application-oriented)		
= caret				**user**	TELEC	**Benutzer**
upward compatible		DATA PROC	**aufwärtskompatibel**	≈ suscriber		≈ Teilnehmer
[with later versions]			[mit Nachfolgeausgabe kompatibel]	**user**	TECH	**Benutzer**
				≈ operator		= Nutzer; Anwender
uranium		CHEM	**Uran**			≈ Bedienperson
= U			= U	**user acceptance**	TECH	**Benutzerakzeptanz**
urban area		ECON	**Stadtbereich**	= user acceptation		= Nutzerakzeptanz; Anwenderakzeptanz; Bedienerakzeptanz
≈ metropolitan area			≈ Ballungsgebiet			
urban switch (→local switching center)	SWITCH	(→Ortsvermittlungsstelle)				
				user acceptation (→user acceptance)	TECH	(→Benutzerakzeptanz)
urgency		COLLOQ	**Dringlichkeit**			
≈ priority			≈ Priorität	**user action frame** (→response frame)	TELEC	(→Antwortseite)
urgent alarm		EQUIP.ENG	**dringender Alarm**			
= prompt alarm; mayor alarm				**user agent**	DATA COMM	**End-Systemteil**
us (→microsecond)		PHYS	(→Mikrosekunde)	[software to manage the storage of messages]		[Software zur Verwaltung der Nachrichtenspeicherung]
US (→unit separator)		DATA PROC	(→Teilgruppenzeichen)			
usability		TECH	**Nutzbarkeit**	= UA		
			≈ Brauchbarkeit; Nützlichkeit	**user agent entity**	DATA COMM	**Instanz des End-Systemteils**
usability (→serviceableness)	TECH	(→Brauchbarkeit)		[exchanges control information for message transfer]		[tauscht Steuerinformationen für die Nachrichtenübermittlung aus]
usable (→serviceable)		TECH	(→brauchbar)			
				= UAE		

user area

user area	DATA PROC	**Benutzerbereich**
		= Nutzerbereich; Anwenderbereich
user cluster	TELEC	**Teilnehmerkonzentration**
user code	DATA PROC	**Benutzercode**
		= Nutzercode; Anwendercode
user credential check	DATA PROC	**Berechtigungsprüfung**
user credential file	DATA PROC	**Berechtigungskatalog**
user data	DATA PROC	**Benutzerdaten**
		= Nutzerdaten; Anwenderdaten
user-definable	DATA PROC	**anwenderdefinierbar**
≈ user-programmable		= benutzerdefinierbar; anwenderfestlegbar; benutzerfestlegbar; freiprogrammierbar
		≈ anwenderprogrammierbar
user-definable	TECH	(→anwenderdefiniert)
(→user-defined)		
user-defined	TECH	**anwenderdefiniert**
= user-selectable; user-definable		= benutzerdefiniert
user-dependant	TECH	(→anwenderspezifisch)
(→user-specific)		
user-dependent	TECH	(→anwenderspezifisch)
(→user-specific)		
user documentation	TECH	**Benutzerdokumentation**
≈ user's manual		≈ Benutzerhandbuch
user-driven	TECH	**benutzergesteuert**
		= nutzergesteuert; anwendergesteuert
user environment	TELEC	**Teilnehmerumgebung**
user equipment (→user terminal)	TELEC	(→Teilnehmergerät)
user facility	TELEC	**Teilnehmerdienst**
user facility (→feature)	TECH	(→Leistungsmerkmal)
user friendliness	TECH	**Benutzerfreundlichkeit**
≈ applicability [COLLOQ]; operability; maintainability; serviceability		= Anwenderfreundlichkeit; Anwendungsfreundlichkeit; Bedienerfreundlichkeit; Nutzerfreundlichkeit; Betriebsfreundlichkeit
		≈ Anwendbarkeit [COLLOQ]; Betriebsfähigkeit; Wartbarkeit; Brauchbarkeit
user-friendly	TECH	**benutzerfreundlich**
= serviceable		= anwenderfreundlich; bedienerfreundlich
		≈ brauchbar
user function	ELECTRON	**Benutzerfunktion**
		= Anwenderfunktion
user group	DATA COMM	**Teilnehmerbetriebsklasse**
= class of service; COS; subscribers' class; class of line; service category		= Teilnehmerklasse; Benutzergruppe; Teilnehmergruppe; Anschlußklasse; Dienstklasse; Dienstart; Dienstkategorie
user group	DATA PROC	**Benutzergruppe**
		= Nutzergruppe; Anwendergruppe
user guidance (→user interface)	DATA PROC	(→Benutzeroberfläche)
user id (→keyword 1)	DATA PROC	(→Paßwort)
user identification (→keyword 1)	DATA PROC	(→Paßwort)
user identification (→subscriber identification)	DATA COMM	(→Teilnehmerkennung)
user-independent	TECH	**anwenderneutral**
≈ use-independent		≈ anwendungsneutral
≠ user-specific		≠ anwenderspezifisch
user-installable	TECH	**vom Benutzer installierbar**
= costumer-installable		
user interface	DATA PROC	**Benutzeroberfläche**
[complex of hardware and software facilities deployed for the user action]		[Gesamtheit der für die Benutzerbedienung angebotenen Hardware- und Softwareeinrichtungen]
= user surface; shell; user guidance; operator prompting; prompting 2; operator guidance; interactive operation		= Benutzerschnittstelle; Benutzerführung; Anwenderoberfläche; Anwenderschnittstelle; Anwenderführung; Bedieneroberfläche; Bedienerführung; Bedienerschnittstelle; Shell; Dialogführung; Nutzeroberfläche
↓ menu mode		↓ Menütechnik
user interrupt	DATA PROC	**Anwender-Interrupt**
user list	DATA PROC	**Verwenderliste**
		= Anwenderliste
user-near	TELEC	**teilnehmernahe**
= costumer-premises near		
user network 2 (→subscriber network)	TELEC	(→Teilnehmernetz 1)
user-oriented	TECH	**benutzerorientiert**
		= nutzerorientiert; anwenderorientiert
user part	DATA PROC	**Benutzerteil**
		= Bedienerteil; Anwenderteil
user profile	TECH	**Benutzerprofil**
		= Nutzerprofil; Anwenderprofil
user program	DATA PROC	**Anwenderprogramm 2**
= user's program; user-written program		[vom Anwender geschrieben]
user-programmable	DATA PROC	**anwenderprogrammierbar**
≈ user-definable		= benutzerprogrammierbar
		≈ anwenderdefinierbar
user requirement	TECH	**Benutzeranforderung**
		= Nutzeranforderung; Anwenderanforderung
user-selectable	TECH	(→anwenderdefiniert)
(→user-defined)		
user service category	TELEC	(→Anschlußberechtigung)
(→authorized class of service)		
user service class	SWITCH	**Berechtigung**
user's manual (→operating documentation)	TECH	(→Gebrauchsunterlagen)
user software	DATA PROC	**Anwendersoftware 2**
		[vom Anwender geschrieben]
user-specific	TECH	**anwenderspezifisch**
= user-dependent; user-dependant		= benutzerspezifisch
≈ application-specific; application-oriented		≈ anwendungsspezifisch; anwendungsorientiert
≠ user-independent		≠ anwenderneutral
user's program (→user program)	DATA PROC	(→Anwenderprogramm 2)
user surface (→user interface)	DATA PROC	(→Benutzeroberfläche)
user terminal	TERM&PER	**Benutzerstation**
≈ terminal equipment		= Bedienerstation; Bedienstation; Nutzerstation
		≈ Endgerät
user terminal	TELEC	**Teilnehmergerät**
= user equipment; subscriber-premises equipment; station equipment; costumer-premises equipment; end-user equipment; costumer equipment		[beim Teilnehmer stehendes Endgerät]
		= Teilnehmerendgerät; Teilnehmereinrichtung; Teilnehmerapparat
↑ terminal equipment		↑ Endgerät
↓ telephone terminal equipment; telegraph terminal equipment		↓ Fernsprechendgerät; Fernschreiber
user training	ECON	**Anwenderschulung**
user-written program (→user program)	DATA PROC	(→Anwenderprogramm 2)
use-specific (→application-specific)	TECH	(→anwendungsspezifisch)
U-shaped	TECH	**U-förmig**
= hairpin-; channeled		= haarnadelförmig
US-standard keyboard	TERM&PER	**amerikanische Tastatur**
= QWERTY keyboard		= QWERTY-Tastatur; amerikanische Schreibmaschinentastatur
usual	COLLOQ	**üblich**
= customary; established; conventional; regular; traditional		= herkömmlich; traditionell
		≈ regelmäßig
≈ regular		
usufruct	ECON	**Nutznießung**
		= Nießbrauch

utensil (→tool)	TECH	(→Werkzeug)		vacuum relay	COMPON	Vakuumrelais
utile (→useful)	TECH	(→nützlich)		vacuumtight	TECH	vakuumdicht
utilities (→utility provider)	ECON	Versorgungsunternehmen		vacuum tube	ELECTRON	Vakuumröhre
				↑ electron tube		↑ Elektronenröhre
utility (→utility program)	DATA PROC	(→Dienstprogramm)		vagabond current (→stray current)	EL.TECH	(→Streustrom)
utility (→usefulness)	TECH	(→Nützlichkeit)		vagueness	SCIE	Unklarheit
utility model (→registered design)	ECON	(→Gebrauchsmuster)		≈ ambiguity		≈ Mehrdeutigkeit
utility program	DATA PROC	Dienstprogramm		valence	PHYS	Valenz
[auxiliary program of an operating system, to support the user in routine activities]		[Hilfsprogramm eines Betriebssystems, um dem Anwender Routinetätigkeiten zu erleichtern]		= valency		= Wertigkeit
				valence band	PHYS	Valenzband
				valence binding (→valence bond)	CHEM	(→Valenzbindung)
= utility; service program; service routine		= Hilfsprogramm; Dienstleistungsprogramm; Softwarehilfe; Utility		valence bond	CHEM	Valenzbindung
				= valence binding; covalent binding		= Elektronenpaar-Bindung
↓ tools		≈ Programmierwerkzeug		valence electron (→bonding electron)	PHYS	(→Valenzelektron)
		↓ Tools				
				valency (→valence)	PHYS	(→Valenz)
				Valentine antenna	ANT	Valentine-Antenne
utility provider	ECON	Versorgungsunternehmen		valid (adj.)	DATA PROC	zulässig
= utilities		≈ Dienstleistungsunternehmen		= permissible; legal; correct (adj.)		= gültig; richtig
				valid (adj.)	MATH	gültig
utility routine	DATA PROC	Dienstroutine		≠ invalid		≠ ungültig
≈ utility program		≈ Dienstprogramm		validate 2	TECH	Gültigkeit feststellen
utility socket (→mains socket)	EL.INST	(→Netzsteckdose)		validate 1	TECH	gültig setzen
				validation	ECON	Legalisierung
utilizable (→serviceable)	TECH	(→brauchbar)		= legalization		
utilization	TECH	Ausnutzung		validation (→revision)	TECH	(→Nachprüfung)
= exploitation		= Verwertung; Ausbeutung		validation 1 (→data validation)	DATA PROC	(→Datenüberprüfung)
≈ use		≈ Benutzung		validation 2 (→validity check)	DATA PROC	(→Zulässigkeitsprüfung)
utilization density	TELEC	Nutzungsdichte				
utilization mode (→applicability)	TECH	(→Anwendbarkeit)		validity	MATH	Gültigkeit
utilization of capacity	MANUF	Kapazitätsauslastung		validity	DATA PROC	Zulässigkeit
		= Auslastung		= legitimacy		= Gültigkeit; Richtigkeit
utilization time	QUAL	Nutzungsdauer		≈ plausibility		≈ Plausibilität
≈ useful life; life time		[bis zur Außerdienststellung]		validity check	DATA PROC	Zulässigkeitsprüfung
				= validity checking; validation 2		= Gültigkeitsprüfung
		= Verwendungsdauer		≈ plausibility test		≈ Plausibilitätsprüfung
		≈ Gebrauchslebensdauer; Lebensdauer		validity checking (→validity check)	DATA PROC	(→Zulässigkeitsprüfung)
utilize	TECH	ausnutzen		validity range	MATH	Gültigkeitsbereich
≈ use (v.t.)		≈ verwerten		= domain		= Bereich
		≈ benutzen		valley	GEOPHYS	Tal
UTR (→availability)	QUAL	(→Verfügbarkeit)		= dale		
				valley point	MATH	Talpunkt
V				[curve]		[Kurve]
				valley point current	MICROEL	Talstrom
v (→velocity)	PHYS	(→Geschwindigkeit)		[tunnel diode]		[Tunneldiode]
V (→vanadium)	CHEM	(→Vanadium)		valley point voltage	MICROEL	Talspannung
V (→volt)	PHYS	(→Volt)		[tunnel diode]		[Tunneldiode]
V (→volume)	MATH	(→Volumen)		valuable (adj.)	ECON	geldwert
V²LSI (→ultra-large-scale integration)	MICROEL	(→Ultrahöchstintegration)		valuation	COLLOQ	Bewerung
				≈ evaluation		≈ Auswertung
VA (→volt-ampere)	EL.TECH	(→Volt-Ampere)		valuation (→estimate)	COLLOQ	(→Schätzung)
vacancy (→lattice vacancy)	PHYS	(→Gitterfehlstelle)		valuator	TERM&PER	Valuator
vacant pipe	OUTS.PLANT	Leerrohr		↑ input device		↑ Eingabegerät
vacation (AM)	ECON	Urlaub		value	ECON	Wert
= leave (AM); holiday (BRI)		= Ferien		[quantified commercial value]		[in Zahlen ausgedrückter Handelswert]
vacuous (→vacuum)	PHYS	(→luftleer)		≈ amount		≈ Betrag
vacuum (n.)	PHYS	Vakuum		value (→measured value)	PHYS	(→Meßwert)
≈ negative pressure		≈ Unterdruck		value (→numerical value)	MATH	(→Zahlenwert)
↓ high vacuum; extra-high vacuum		↓ Hochvakuum; Höchstvakuum		value added (→added value)	ECON	(→Wertschöpfung)
				value added network	TELEC	Mehrwertdienstnetz
vacuum (adj.)	PHYS	luftleer		[common carrier lines leased with enhancements]		[Standleitungen besonderer Leistungsmerkmale]
= vacuous		= gasleer; Vakuum-				
vacuum 2 (→negative pressure)	PHYS	(→Unterdruck)		value added network service (→value added service)	TELEC	(→Mehrwertdienst)
vacuum blower	TERM&PER	Unterdruckgebläse				
[ink jet printer]		[Tintendrucker]		value added service	TELEC	Mehrwertdienst
vacuum capacitor	POWER SYS	Vakuumkondensator		= VAS; value added network service; VANS		
vacuum cell	PHYS	Vakuumzelle				
vacuum column	TERM&PER	Unterdrucksäule		value analysis	ECON	Wertanalyse
vacuum container	TERM&PER	Unterdruckbehälter		= value engineering		
[ink jet printer]		[Tintendrucker]		value assignment	DATA PROC	Wertzuweisung
vacuum electronics	ELECTRON	Vakuumelektronik		value comparator	CIRC.ENG	(→Wertevergleicher)
vacuum guide	TV	Bandführungssegment		(→magnitude comparator)		
		= Kopfschuh		value-continuous	INF	wertkontinuierlich
vacuum packed	TECH	vakuumverpackt		≈ analog		= werstetig
vacuum pump	TECH	Vakuumpumpe		≠ value-discrete		≈ analog
↓ high vacuum pump		≈ Luftpumpe				≠ wertdiskret
		↓ Hochvakuumpumpe				

value-continuous signal	CODING	**wertkontinuierliches Signal**
≠ n-level signal		= wertstetiges Signal
		≠ wertstetiges Signal
value course	INF	**Werteverlauf**
value declaration	ECON	**Wertangabe**
value determination	ECON	**Wertfestsetzung**
value-discrete	INF	**wertdiskret**
= n-level; multilevel		= mehrstufig
≈ digital		≈ digital
≠ value-continuous		≠ wertkontinuierlich
value engineering (→value analysis)	ECON	(→Wertanalyse)
value of quality	QUAL	**Qualitätswert**
value parameter	DATA PROC	**Werteparameter**
valuer (→appraiser)	ECON	(→Schätzer)
valve	TECH	**Ventil**
= cock		
valve	POWER ENG	**Ventil**
valve (BRI) (→electron tube)	ELECTRON	(→Elektronenröhre)
valve amplifier (→tube amplifier)	CIRC.ENG	(→Röhrenverstärker)
valve base (→tube socket)	ELECTRON	(→Röhrenfassung)
valve detector (→rectifier tube)	ELECTRON	(→Gleichrichterröhre)
valve generator (→tube generator)	CIRC.ENG	(→Röhrengenerator)
valve noise (→tube noise)	ELECTRON	(→Röhrenrauschen)
valve receiver	RADIO	**Röhrenempfänger**
valve-side	POWER ENG	**ventilseitig**
valve socket (→tube socket)	ELECTRON	(→Röhrenfassung)
valve stage (→tube stage)	CIRC.ENG	(→Röhrenstufe)
valve transmitter (→tube transmitter)	RADIO	(→Röhrensender)
vanadium	CHEM	**Vanadium**
= V		= Vanadin; V
Van Allen belt	GEOPHYS	**Van-Allen-Gürtel**
Van Atta array	ANT	**Van-Atta-Antenne**
= Van Atta reflector array		= retroaktive Antenne
Van Atta reflector array (→Van Atta array)	ANT	(→Van-Atta-Antenne)
van de Graaff generator	PHYS	**elektrostatischer Bandgenerator**
		= Bandgenerator
van der Waals bonding	CHEM	**van-der-Waals-Bindung**
van Dycke brown	PHYS	**vandyckbraun**
vane switch	COMPON	**Magnetgabelschranke**
vanish (v.i.)	COLLOQ	**schwinden**
= fade (v.i.)		= verschwinden; entschwinden
VANS (→value added service)	TELEC	(→Mehrwertdienst)
V antenna	ANT	**V-Antenne**
= vee antenna; vee		↑ Wanderwellenantenne
↑ travelling-wave antenna		
vapor (AM)	PHYS	**Dampf**
= vapour (BRI); steam		= Dunst
≈ gas		≈ Gas
↓ water vapor		↓ Wasserdampf
vapor-deposition facility	MICROEL	**Aufdampfanlage**
vapor-deposition method	MICROEL	**Aufdampfverfahren**
= deposition method		
vapor flow soldering	MICROEL	**Dampfströmungsschweißung**
vaporize	PHYS	**verdampfen**
= evaporate		= abdampfen; verflüchtigen
≈ sublime; boil		≈ sublimieren; sieden
↓ volatilize		↓ verdunsten
vaporized	TECH	**aufgedampft**
vaporous	PHYS	**dampfförmig**
vapotron	ELECTRON	**Vapotron**
[vapour cooling method for power tubes]		[Dampfkühlungsverfahren für Leistungsröhren]
≈ supervapotron; hypervapotron		≈ Supervapotron; Hypervapotron
vapour (BRI) (→vapor)	PHYS	(→Dampf)
vapour pressure (AM)	PHYS	**Dampfdruck**
= vapour pressure (BRI)		
vapour pressure (BRI) (→vapour pression)	PHYS	(→Dampfdruck)
var	EL.TECH	**Volt-Ampere reaktiv**
[SI unit for reactive power]		[SI-Einheit für elektrische Blindleistung]
		= var
		≈ Watt [PHYS]
VAR (→visual/aural range)	RADIO NAV	(→optisch-akustischer Leitstrahlsender)
varactor	MICROEL	**Varaktor**
[variable capacitor; semiconductor device exploiting characteristics of reverse biased operation]		[Eigenschaften des Betriebs in Sperrichtung ausnutzendes Halbleiter-Bauelement]
↓ varactor diode; MIS varactor		= Varactor
		↓ Varaktordiode; MIS-Varaktor
varactor diode	MICROEL	**Varaktordiode**
[exploits the bias-voltage dependence of junction capacity]		[nutzt die Abhängigkeit der Sperrschichtkapazität von der Sperrspannung aus]
= variable-capacitance diode; capacitance diode; varicap		= Varactordiode; Kapazitätsdiode; Kapazitätsvariations-Diode; Reaktanzdiode; Varicap;
↓ junction varactor; charge-storage diode		↓ Sperrschichtvaraktor; Speichervaraktor
variability	TECH	**Veränderlichkeit**
= variableness; inconstancy		= Inkonstanz; Variabilität; Unbeständigkeit
≈ instability		≈ Instabilität
variable (n.)	MATH	**Variable**
= variable quantity		= variable Größe; Zahlenvariable; Veränderliche
≈ unknown quantity		≈ Konstante
≠ constant (n.)		
variable (n.)	DATA PROC	**Variable**
[operand of a computer instruction, containing the address of the procured information]		[Operand eines Rechnerbefehls für die Adresse der gesuchten Befehls enthält]
≈ parameter; argument		≈ Parameter; Argument
≠ literal constant		≠ Literalkonstante
variable (adj.)	TECH	**veränderlich**
= inconstant		[sich ändernd]
≈ unstable		≈ inkonstant
↑ alterable		≈ instabil
		↑ veränderbar
variable (adj.)	MATH	**variabel**
≠ constant		= veränderlich
		≠ konstant
variable (→characteristic quantity)	TECH	(→Kenngröße)
variable accessibility	SWITCH	**variable Erreichbarkeit**
= variable availability		= variable Verfügbarkeit
variable attenuator	CIRC.ENG	**Dämpfungsregler**
= variable pad		
variable availability (→variable accessibility)	SWITCH	(→variable Erreichbarkeit)
variable bit rate mode	TELEC	**variable-Bitrate-Betrieb**
variable-capacitance diode (→varactor diode)	MICROEL	(→Varaktordiode)
variable capacitor	COMPON	**einstellbarer Kondensator**
↓ rotatable capacitor; decade capacitance box; trimming capacitor		= veränderlicher Kondensator; regelbarer Kondensator; variabler Kondensator
		↓ Drehkondensator; Dekadenkondensator; Trimmerkondensator
variable data	DATA PROC	**variable Daten**
= current data; varying data		= Tagesdaten; Veränderungsdaten
≠ master data		≠ Stammdaten
variable declaration	DATA PROC	**Variablendeklaration**
		= Variablenvereinbarung
variable field (→alternating field)	PHYS	(→Wechselfeld)
variable film potentiometer	COMPON	**Schicht-Drehwiderstand**
		= Schichtpotentiometer
variable film resistor	COMPON	**Schicht-Schiebewiderstand**
≈ fader		
variable-frequency oscillator	CIRC.ENG	**durchstimmbarer Oszillator**
= VFO		= VFO

variable information length	DATA PROC	**variable Informationslänge**	
≈ variable word length		≈ variable Wortlänge	
variable isolating transformer	COMPON	**Regel-Trenntransformator**	
variable-length instruction	DATA PROC	(→Mehrbytebefehl)	
(→multiple-byte instruction)			
variable-length record	DATA PROC	**Datensatz variabler Länge**	
variable-length word	DATA PROC	**Wort variabler Länge**	
[can be varied instruction-wise]		[von Befehl zu Befehl veränderbar]	
variable memory	DATA PROC	**Variablenspeicher**	
= variable storage; variable store			
variable mu tube (→regulator tube)	ELECTRON	(→Regelröhre)	
variable name	DATA PROC	**Variablenname**	
variableness (→variability)	TECH	(→Veränderlichkeit)	
variable numbering	SWITCH	**variable Numerierung**	
variable pad (→variable attenuator)	CIRC.ENG	(→Dämpfungsregler)	
variable quantity (→variable)	MATH	(→Variable)	
variable-ratio transformer	POWER SYS	**Stelltransformator**	
= variable transformer		= Regeltransformator; regelbarer Transformator	
≈ variable transformer [COMPON]		≈ regelbarer Übertrager [COMPON]	
variable resistance (→adjustable resistor)	COMPON	(→veränderbarer Widerstand)	
variable resistor (→rheostat)	PHYS	(→Regelwiderstand)	
variable resistor (→adjustable resistor)	COMPON	(→veränderbarer Widerstand)	
variable-speed (adj.)	TECH	**drehzahlveränderbar**	
variable storage (→variable memory)	DATA PROC	(→Variablenspeicher)	
variable store (→variable memory)	DATA PROC	(→Variablenspeicher)	
variable-threshold logic (→VTL logic)	MICROEL	(→VTL-Logik)	
variable transformer	COMPON	**regelbarer Übertrager**	
≈ variable ratio transformer [POWER SYS]		≈ regelbarer Transformator ≈ Stelltransformator [POWER SYS]	
variable transformer (→variable-ratio transformer)	POWER SYS	(→Stelltransformator)	
variable wire-wound resistor (→wirewound potentiometer)	COMPON	(→Drahtpotentiometer)	
variable word length	DATA PROC	**variable Wortlänge**	
variance	MATH	**Varianz**	
[arithmetic mean of the squares of individual deviations from average]		[arithmetisches Mittel der Quadrate der Einzelabweichungen vom Mittelwert]	
≈ standard deviation		= Streuung 1; Dispersion ≈ Standardabweichung	
variance analysis	MATH	**Varianzanalyse**	
variance-ratio distribution (→F-distribution)	MATH	(→F-Verteilung)	
variant	EQUIP.ENG	**Variante**	
≈ option		= Gerätevariante ≈ Option	
variant of subrack	EQUIP.ENG	**Einsatzvariante**	
= inset variant		[Variante eines Geräteeinsatzes]	
variate (→random variable)	MATH	(→Zufallsvariable)	
variation	MATH	**Variation**	
↑ combination		= Kombination mit Berücksichtigung der Anordnung ↑ Kombination 1	
variation (→adjustment)	TECH	(→Einstellung)	
variation (→adjustment)	ELECTRON	(→Einstellung)	
variation (→deviation)	PHYS	(→Abweichung)	
variational calculus	MATH	**Variationsrechnung**	
= calculus of variations ↑ analysis		↑ Analysis	
variational resistance (→impedance)	NETW.TH	(→komplexer Scheinwiderstand)	
variation range	MATH	**Spannweite**	
[statistics; difference between maximum and minimum]		[Statistik; Differenz zwischen größtem und kleinstem Stichprobenwert]	
= range 2		= Streubreite; Variationsbreite	
variation range	PHYS	**Streubereich**	
varicap (→varactor diode)	MICROEL	(→Varaktordiode)	
varicap modulator (→tuning diode modulator)	MODUL	(→Kapazitätsdioden-Modulator)	
variety	SCIE	**Abart**	
[slightly different]		[geringfügig unterschiedlich] = Varietät	
variety	COLLOQ	**Vielfalt**	
variety (→kind)	TECH	(→Sorte)	
variety parts list (→parts list)	MANUF	(→Stückliste)	
variety plan (→parts list)	MANUF	(→Stückliste)	
variety production	MANUF	**Sortenfertigung**	
[serial production with product differentiation in the very last manufacturing steps]		[Serienfertigung mit Produktdifferenzierungen erst in den letzten Fertigungsschritten] = Sortenproduktion	
varilosser	ELECTRON	**pegelabhängiges Dämpfungsglied**	
variometer	COMPON	**Variometer**	
[continuously variable inductor, by]		[stetig veränderbare Induktivität]	
various (adj.)	COLLOQ	**vielfältig**	
[having many different qualities]		[mit vielen verschiedenen Eigenschaften]	
= diverse 2		= vielfach 2	
≈ multiple; manifold 1; different		≈ vielfach 1; mannigfaltig; verschieden	
variousness	COLLOQ	**Vielfältigkeit**	
variovent	EL.ACOUS	**Variovent**	
		[mit Dämpfungsmaterial gefüllte Öffnung]	
varistor	COMPON	**Varistor**	
= voltage-dependent resistor; VDR ↑ adjustable resistor		= spannungsabhängiger Widerstand; VDR-Widerstand ↑ veränderbarer Widerstand	
varmeter (→reactive power meter)	INSTR	(→Blindleistungsmesser)	
varnish	TECH	**Firnis**	
≈ lacquer		≈ Lack	
varnish 2 (n.) (→lacquer)	CHEM	(→Lack)	
varnish color	TECH	**Lackfarbe**	
= paint			
varnish-color coat	TECH	**Lacküberzg**	
varnished (→lacquered)	TECH	(→lackiert)	
varnished paper	TECH	**Lackpapier**	
varnished wire	COMM.CABLE	**Lackdraht**	
= enameled wire			
vary (→change)	TECH	(→ändern)	
varying data (→variable data)	DATA PROC	(→variable Daten)	
VAS (→value added service)	TELEC	(→Mehrwertdienst)	
vaseline	CHEM	**Vaseline**	
vat (→barrel)	TECH	(→Tonne)	
V-ATE process	MICROEL	**V-ATE-Verfahren**	
= vertical anisotropic etch process			
VAWS (→vector arbitrary waveform synthesizer)	INSTR	(→Vektor-arbitrary-waveform-Synthesizer)	
V belt	MECH	**Keilriemen**	
↑ driving belt		↑ Treibriemen	
V-belt drive	MECH	**Keilriemenantrieb**	
VC (→virtual container)	TELEC	(→virtueller Container)	
VCA (→voltage-controlled amplifier)	CIRC.ENG	(→spannungsgesteuerter Verstärker)	
VCF (→voltage-controlled filter)	NETW.TH	(→spannungsgesteuertes Filter)	
VCO (→voltage-controlled oscillator)	CIRC.ENG	(→spannungsgesteuerter Oszillator)	
VCO post-tuning drift	INSTR	**VCO-Ausgangsdrift**	
VCS (→voltage-controlled current source)	NETW.TH	(→spannungsgesteuerte Stromquelle)	
V dipole (→angular dipole)	ANT	(→Winkeldipol)	
VDR (→varistor)	COMPON	(→Varistor)	
VDT (→data display terminal)	TERM&PER	(→Datensichtgerät)	
VDU (→data display terminal)	TERM&PER	(→Datensichtgerät)	
vectographic	DATA PROC	**vektorgraphisch**	

vector | | MATH | **Vektor**
[magnitde characterized by a value and a direction] | | | [nicht nur durch Zahlen, sondern auch durch eine Richtung charakterisierte Größe]
= vectorial quantity | | | = vektorielle Größe
≠ scalar | | | ≠ Skalar
vector (→array 1) | | DATA PROC | (→Datenfeld 2)
vector algebra | | MATH | **Vektoralgebra**
vector alternating current | | EL.TECH | (→komplexer Wechselstrom)
(→complex alternating current) | | |
vector alternating voltage | | EL.TECH | (→komplexe Wechselspannung)
(→complex alternating voltage) | | |
vector analysis | | MATH | **Vektoranalysis**
vector analyzer | | INSTR | **Vektoranalysator**
≈ vector modulation analyzer | | | ≈ Vektormodulationsanalysator
↑ signal analyzer | | | ↑ Signalanalysator
vector arbitrary waveform synthesizer | | INSTR | **Vektor-arbitrary-waveform-Synthesizer**
= VAWS | | |
vector averaging | | INSTR | **Vektormittelwertbildung**
vector computer (→array processor) | | DATA PROC | (→Vektorrechner)
vector diagram | | INSTR | **Vektordiagramm**
 | | | = Zeigerdiagramm
vector diagram | | POWER ENG | **Zeigerbild**
vector display | | TERM&PER | **Vektorbildschirm**
[cathode ray trace controlled directly by a program, electrom beam moves randomly to trace] | | | [Kathodenstrahl wird direkt von einem Programm gesteuert, er bewegt sich dem Kurvenverlauf entsprechend]
= vector scan display; random-scan display | | | = Vektorsichtgerät
≠ raster display | | | ≠ Rasterbildschirm
↑ display terminal | | | ↑ Sichtgerät
vector display | | INSTR | **Vektordiagrammanzeige**
vectored (→directional) | | TECH | (→gerichtet)
vectored interrupt | | DATA PROC | **Vektorunterbrechung**
 | | | = vektorielle Unterbrechung; gerichtete Unterbrechung
vector field | | PHYS | **Vektorfeld**
vector function | | MATH | **Vektorfunktion**
vector graphics | | DATA PROC | **Vektorgraphik**
 | | | = Vektorgrafik
≠ raster graphics | | | ≠ Rastergraphik
vector group | | POWER ENG | **Schaltgruppe**
vectorial | | MATH | **vektoriell**
≠ scalar | | | ≠ skalar
vectorial calculus | | MATH | **Vektorrechnung**
vectorial potential | | PHYS | **vektorielles Potential**
vectorial quantity | | MATH | **vektorielle Größe**
vectorial quantity (→vector) | | MATH | (→Vektor)
vectorial representation | | MATH | **Vektordarstellung**
vector impedance (→impedance) | | NETW.TH | (→komplexer Scheinwiderstand)
vector impedance meter | | INSTR | **Vektor-Impedanzmesser**
vectoring | | DATA PROC | **Zwischenadressierung**
vector line | | MATH | **Vektorlinie**
≈ exhaust | | |
vector measurement | | INSTR | **Vektormessung**
[of amplitude and phase] | | | [von Amplitude und Phase]
≠ scalar measurement | | | ≠ skalare Messung
vector meter | | INSTR | **Vektormesser**
vector modulation analyzer | | INSTR | **Vektor-Modulationsanalysator**
vector network analyzer | | INSTR | **Vektor-Netzwerkanalysator**
= vector voltmeter | | | = Vektorvoltmeter
vector operator | | MATH | **Nablaoperator**
[vector analysis] | | | [Vektorrechnung]
= Hamilton operator | | | = Hamiltonoperator
vector-oriented | | DATA PROC | **vektororientiert**
vector pair | | DATA PROC | **Vektorpaar**
vector potential | | PHYS | **Vektorpotential**
vector power (→apparent power) | | EL.TECH | (→Scheinleistung)
vector processor (→array processor) | | DATA PROC | (→Vektorrechner)
vector product | | MATH | **Vektorprodukt**
vector scan display (→vector display) | | TERM&PER | (→Vektorbildschirm)
vectorscope | | ELECTRON | **Vektorskop**
[for polar representations] | | | [für Polarkoordinatendarstellung]
↑ oscilloscope | | | ↑ Oszilloskop
vector signal generator | | INSTR | **Vektorsignalgenerator**
vector space | | MATH | **Vektorraum**
vector sum | | MATH | **Vektorsumme**
vector-to-raster converter | | TERM&PER | **Vektor-Raster-Umsetzer**
= VRC | | |
vector voltmeter (→vector network analyzer) | | INSTR | (→Vektor-Netzwerkanalysator)
vee (→V antenna) | | ANT | (→V-Antenne)
vee antenna (→V antenna) | | ANT | (→V-Antenne)
vee-beam star antenna | | ANT | **V-Stern-Antenne**
vehicle | | TECH | **Fahrzeug**
= car | | | = Wagen
vehicle engineering | | TECH | **Fahrzeugbau**
vehicle-mounted aerial (→car antenna) | | ANT | (→Fahrzeugantenne)
vehicle-mounted antenna (→car antenna) | | ANT | (→Fahrzeugantenne)
velocity | | PHYS | **Geschwindigkeit**
[distance per unit of time; SI unit: meter per second] | | | [Länge pro Zeiteinheit; SI-Einheit: Meter durch Sekunde]
= v; speed 2 | | | = v
velocity (→fastness) | | COLLOQ | (→Schnelligkeit)
velocity microphone | | EL.ACOUS | **Schnellewandler**
 | | | = Druckgradientenmikrofon
velocity-modulated tube | | MICROW | **Laufzeitröhre**
= transit-time tube; v.m. tube | | | ↑ Kathodenstrahlröhre
↑ cathode ray tube | | | ↓ Triftröhre; Lauffeldröhre
↓ linear-beam tube; | | |
velocity modulation | | ELECTRON | **Geschwindigkeitsmodulation**
velocity of propagation | | PHYS | **Ausbreitungsgeschwindigkeit**
velocity of sound (→sound particle velocity) | | ACOUS | (→Schallschnelle)
vendable (→salable) | | ECON | (→verkäuflich)
vendee (→purchaser) | | ECON | (→Käufer)
vender (→seller) | | ECON | (→Verkäufer)
vendible (→salable) | | ECON | (→verkäuflich)
vending machine | | TECH | **Verkaufsautomat**
 | | | = Warenautomat
vendor 3 (→dealer) | | ECON | (→Händler)
vendor 2 (→supplier 1) | | ECON | (→Lieferant)
vendor 1 (→seller) | | ECON | (→Verkäufer)
veneer | | TECH | **Fournier**
[wood] | | | [Holz]
Venn diagram | | ENG.LOG | **Venn-Diagramm**
vent | | TECH | **Abzug**
 | | | [Lüftung]
vent (v.t.) (→ventilate) | | TECH | (→lüften)
vent hole (→ventilation slit) | | TECH | (→Lüftungsschlitz)
ventialting vane (→cooling vane) | | COMPON | (→Kühlfahne)
ventilate | | TECH | **lüften**
= vent (v.t.) | | | ≈ entlüften
≈ exhaust | | |
ventilation | | TECH | **Lüftung**
≈ exhaust | | | = Belüftung
 | | | ≈ Entlüftung
ventilation hole (→cooling hole) | | TECH | (→Lüftungsloch)
ventilation opening (→cooling opening) | | TECH | (→Lüftungsöffnung)
ventilation slit | | TECH | **Lüftungsschlitz**
= cooling slit; vent hole | | | = Ventilationsschlitz; Belüftungsschlitz
ventilator (→fan) | | TECH | (→Ventilator)
venture | | ECON | **Unternehmung 1**
[action] | | | [Aktion]
 | | | = Unternehmen 2
venue (→meeting place) | | ECON | (→Tagungsort)
verb 1 (n.) | | LING | **Verb**
 | | | = Verbum; Zeitwort; Tätigkeitswort; Tunwort
 | | | ≈ Prädikat
verb 2 (→predicate) | | LING | (→Prädikat)
verification (→revision) | | TECH | (→Nachprüfung)

verification kit	INSTR	**Prüfsatz**	
verifier (→punched card verifier)	TERM&PER	(→Kartenprüfer)	
vermilion = vermillion	PHYS	**zinnober**	
vermillion (→vermilion)	PHYS	(→zinnober)	
vernacular (→colloquial)	LING	(→Umgangssprache)	
vernier	MECH	**Nonius** = Vernier	
vernier	INSTR	**Feineinstellung** = Nonius	
vernier accuracy	INSTR	**Feineinstellungsgenauigkeit** = Noniusgenauigkeit	
vernier caliper (→slide gauge)	MECH	(→Schiebelehre)	
vernier dial	INSTR	**Noniusskala**	
vernier drive	MECH	**Feinantrieb**	
vernier rule (→slide gauge)	MECH	(→Schiebelehre)	
versatile	COLLOQ	**vielseitig**	
versatile (→rotatable)	MECH	(→drehbar)	
versatility	COLLOQ	**Vielseitigkeit**	
versed (→experienced)	SCIE	(→sachkundig)	
version (→type)	TECH	(→Ausführung 1)	
version (→release version)	DATA PROC	(→Software-Version)	
version number	TECH	**Versionsnummer**	
version number (→release number)	DATA PROC	(→Versionsnummer)	
verso (n.) = rear side	TYPOGR	**Rückseite**	
vertex (→apex)	MATH	(→Scheitelpunkt 1)	
vertex plate ↑ auxiliary reflector	ANT	**Scheitelplatte** ↑ Hilfsreflektor	
vertical (adj.) = perpendicular ≈ plumb (TECH); rectangular ≠ horizontal	MATH	**senkrecht** = vertikal ≈ lotrecht (TECH); rechtwinklig ≠ waagrecht	
vertical (n.) (→perpendicular)	MATH	(→Senkrechte)	
vertical advance (→line spacing)	TERM&PER	(→Zeilenabstand)	
vertical aerial (→vertical antenna)	ANT	(→Vertikalantenne)	
vertical amplifier = Y amplifier	ELECTRON	**Vertikalverstärker** = Y-Verstärker; Y-Ablenkverstärker	
vertical anisotropic etch metal-oxide semiconductor process (→ VMOS process)	MICROEL	**VMOS-Verfahren**	
vertical anisotropic etch process (→V-ATE process)	MICROEL	**V-ATE-Verfahren**	
vertical antenna = vertical aerial ≈ monopole ↓ rod antenna; discone antenna; ground plane antenna; Marconi antenna; cage monopole antenna	ANT	**Vertikalantenne** = Vertikalstrahler ≈ Monopol ↓ Stabantenne; Discone-Antenne; Ground-plane-Antenne; Marconi-Antenne; Reusen-Monopol	
vertical axis (→Z axis)	MATH	(→Z-Achse)	
vertical blanking (→frame suppression)	TV	(→Bildaustastung)	
vertical blanking interval = field-blanking interval ↑ vertical blanking interval	TV	**Bildaustastlücke** ↑ Austastlücke	
vertical cage antenna	ANT	**Vertikalreuse**	
vertical construction practice = vertical design ≠ horizontal construction practice	EQUIP.ENG	**Vertikalbauweise** ≠ Horizontalbauweise	
vertical deflection = vertical scan	ELECTRON	**Vertikalablenkung** = V-Ablenkung; Vertikalsteuerung	
vertical deflection final amplifier	TV	**Vertikalendstufe**	
vertical deflection oscillator = vertical scan oscillator; vertical oscillator	TV	**Vertikalablenkoszillator** = Vertikalablenkgenerator; Vertikalfrequenzgenerator; Vertikaloszillator; Bildkipposzillator; Bildablenkgenerator; Bildablenkoszillator; Vertikalsteueroszillator	
vertical deflection transformer = vertical scan transformer	TV	**Vertikalablenktransformator**	
vertical design (→vertical construction practice)	EQUIP.ENG	(→Vertikalbauweise)	
vertical diagram = vertical radiation pattern; elevation pattern	ANT	**Vertikaldiagramm**	
vertical distance (→line spacing)	TERM&PER	(→Zeilenabstand)	
vertical flyback = field flyback (AM); frame flyback	TV	**Vertikalrücklauf**	
vertical focusing	TV	**Vertikalfokussierung**	
vertical format buffer = format buffer	TERM&PER	**Formularformatspeicher**	
vertical frequency (→field frequency)	TV	(→Teilbildfrequenz)	
vertical gain = Y gain	ELECTRON	**Vertikalverstärkung** = Y-Verstärkung	
vertical half rhombic antenna	ANT	**halbe Rhombusantenne**	
vertical half-wave dipole	ANT	**Halbwellen-Vertikaldipol**	
vertical inset	EQUIP.ENG	**Schmaleinsatz**	
vertical interval (→frame suppression)	TV	(→Bildaustastung)	
vertical interval (→test line)	TV	(→Prüfzeile)	
vertical interval reference signal = VIRS	TV	**Prüfzeilen-Referenzsignal**	
vertical interval test signal = VITS	TV	**Prüfzeilen-Meßsignal**	
vertical motion	MECH	**Hebbewegung**	
vertical oscillator (→vertical deflection oscillator)	TV	(→Vertikalablenkoszillator)	
vertical parity bit track (→parity track)	TERM&PER	(→Paritätsbitspur)	
vertical parity check = vertical redundancy check	CODING	**Querparitätsprüfung** = Querprüfung	
vertical polarization = vertikale Polarisation	PHYS	**Vertikalpolarisation**	
vertical potter (→beltbed plotter)	TERM&PER	(→Vertikalplotter)	
vertical print density	TERM&PER	**Zeilendichte**	
vertical radiation pattern (→vertical diagram)	ANT	(→Vertikaldiagramm)	
vertical recording [magnetic particles vertically aligned]	TERM&PER	**Vertikalaufzeichnung** [Magnetpartikel senkrecht orientiert]	
vertical redundancy check (→vertical parity check)	CODING	(→Querparitätsprüfung)	
vertical scan (→vertical deflection)	ELECTRON	(→Vertikalablenkung)	
vertical scanning	TV	**Vertikalaustastung** = V-Austastung	
vertical scan oscillator (→vertical deflection oscillator)	TV	(→Vertikalablenkoszillator)	
vertical scan transformer (→vertical deflection transformer)	TV	(→Vertikalablenktransformator)	
vertical separation (→line spacing)	TERM&PER	(→Zeilenabstand)	
vertical size control	TV	**Teilbildhöhenregler**	
vertical spacing (→line spacing)	TERM&PER	(→Zeilenabstand)	
vertical step	SWITCH	**Hebschritt**	
vertical synchronization	TV	**Vertikalsynchronisierung** = Vertikalsynchronisation	
vertical synchronizing pulse = vertical sync pulse	TV	**Vertikalsynchronimpuls** = V-Synchronimpuls	
vertical sync pulse (→vertical synchronizing pulse)	TV	(→Vertikalsynchronimpuls)	
vertical tabulator = VT	DATA PROC	**Vertikaltabulator** = VT	
vertical take-of and landing = VTOL	AERON	**Senkrechtstart und -landung**	
vertical Zepp (→vertical Zeppelin antenna)	ANT	(→vertikale Zeppelinantenne)	
vertical Zeppelin antenna = vertical Zepp	ANT	**vertikale Zeppelinantenne** = Vertikal-Zepp	
very fast memory (→high-speed memory)	DATA PROC	(→Schnellspeicher 1)	
very fast storage (→high-speed memory)	DATA PROC	(→Schnellspeicher 1)	
very fast store (→high-speed memory)	DATA PROC	(→Schnellspeicher 1)	

very high frequency (→metric waves) RADIO (→Ultrakurzwellen)
very high performance computer Höchstleistungsrechner = Höchstleistungs-Computer
very high speed logic MICROEL Schnellstlogik = VHSL = VHSL
very large data base DATA PROC (→Größtdatenbank) (→VLDB)
very large scale integration MICROEL (→Größtintegration) (→VLSI)
very low frequencies RADIO (→Myriameter-Wellen) (→myriametric waves)
very low temperature physics PHYS (→Tiefsttemperaturphysik) (→cryophysics)
Very sincerely yours (AM) ECON Mit freundlichen Grüßen Ihr [personal form of complimentary close] [verbindlicher Briefschluß] = Beste Grüße von Ihrem; = Yours sincerely (BRI); Sincerely (AM); Cordially yours (AM) Mit freundschaftlichen Grüßen ≈ Sincerely yours ≈ Hochachtungsvoll
very small aperture terminal SAT.COMM (→VSAT) (→VSAT)
Very truly yours (AM) ECON (→Hochachtungsvoll) (→Sincerely yours)
vessel TECH Behälter = container; receptacle; box; bin; hopper = Box ≈ Tank; Gefäß ≈ tank; receptacle
vestigial sideband MODUL Restseitenband = VSB
vestigial-sideband filter TV Restseitenbandfilter
vestigial sideband modulation MODUL Restseitenband-Modulation
vestigial sideband transmission TV Restseitenband-Übertragung
VF (→voice frequency) TELEC (→Niederfrequenz)
VF amplifier (→voice frequency amplifier) TRANS (→NF-Verstärker)
VF amplifier (→booster amplifier) TRANS (→NF-Zwischenverstärker)
VF cable (→audio frequency cable) COMM.CABLE (→NF-Kabel)
VF calling (→VF pushbutton dialing) SWITCH (→tonfrequente Tastwahl)
VF channel TELEC NF-Kanal = voice frequency channel = Niederfrequenzkanal
VF circuit (→voice-frequency circuit) TELEC (→NF-Leitung)
VF dialling (→VF pushbutton dialing) SWITCH (→tonfrequente Tastwahl)
VF engineering (→voice-frequency engineering) TELEC (→NF-Technik)
VF level (→voice-frequency level) TELEC (→NF-Pegel)
VF link (→voice-frequency circuit) TELEC (→NF-Leitung)
VFO (→variable-frequency oscillator) CIRC.ENG (→durchstimmbarer Oszillator)
VF oscilloscope (→video-frequency oscilloscope) TV (→Fernseh-Kontrolloszilloskop)
VF pushbutton dialing SWITCH tonfrequente Tastwahl = touch-tone dialing; VF dialling; VF selection; VF calling; touch-tone dialling; touch-tone selection; touch-tone calling
VF selection (→VF pushbutton dialing) SWITCH (→tonfrequente Tastwahl)
VF signal (→voice frequency signal) TELEC (→NF-Signal)
VFT (→voice frequency telegraphy) TELEGR (→Wechselstromtelegrafie)
VFT channel TELEC WT-Kanal = voice frequency telegraphy channel = Wechselstromtelegrafie-Kanal
VFT circuit (→voice-frequency telegraph circuit) TELEGR (→WT-Verbindung)
VFTG (→voice frequency telegraphy) TELEGR (→Wechselstromtelegrafie)
VGA TERM&PER VGA [Video Graphics Adapter; graphics standard with 640x480 pixels, 16 to 256 out of 262,144 colors, 70 Hz; with special driver software also 1024x768 pixels] [Graphikstandard mit 640x480 Punkten, 16 bis 256 aus 262.144 Farben, 70Hz; durch Treibersoftware auch 1024x768 Punkte]

VGA board TERM&PER VGA-Karte ↑ color graphics board = VGA-Adapter ↑ Farbgraphikkarte
VGA monitor TERM&PER VGA-Monitor [640x480 pixels and 16 to 256 colors] [640x480 Bildpunkte mit 16 bis 256 Farben] ↑ color monitor ↑ Farbmonitor
VHDL MICROEL VHDL-Sprache [very high speed IC hardware description language]
VHF (→metric waves) RADIO (→Ultrakurzwellen)
VHF omnidirectional range RADIO NAV UKW-Drehfunkfeuer = VOR; VHF omnirange = VOR
VHF omnirange (→VHF omnidirectional range) RADIO NAV (→UKW-Drehfunkfeuer)
VHF radiotelephony RADIO UKW-Sprechfunk = UKW-Telefonie
VHF signal generator INSTR VHF-Signalgenerator
VHSIC MICROEL VHSIC [Very High Speed Integrated Circuit]
VHSL (→very high speed logic) MICROEL (→Schnellstlogik)
viability TECH Durchführbarkeit = feasibility; practicability; practicality; performability; realizability = Realisierbarkeit; Machbarkeit
viable (→feasible 1) TECH (→durchführbar)
viahole (→feedthrough) ELECTRON (→Durchkontaktierung)
via net loss (→transmission loss) NETW.TH (→Durchgangsdämpfung)
vibrate (→oscillate) MECH (→schwingen)
vibrating capacitor CIRC.ENG Schwingkreiskondensator = Schwingkondensator
vibrating-magnet regulator POWER SYS Tirill-Spannungsregler = Tirill voltage regulator = Tirill-Regler; Vibrationsregler
vibrating reed electrometer INSTR Vibrationselektrometer
vibrating-reed frequency meter INSTR (→Zungenfrequenzmesser) (→reed-type frequency meter)
vibrating-reed instrument INSTR (→Zungenfrequenzmesser) (→reed-type frequency meter)
vibrating reed magnetometer INSTR Vibrationsmagnetometer
vibrating-reed measuring system INSTR Vibrationsmeßwerk
vibrating reed rectifier EL.TECH Vibrationsgleichrichter
vibrating relay COMPON Vibrationsrelais
vibration MECH mechanische Schwingung = mechanical oscillation; oscillation; undulation; swing = Schwingung; Vibration ↓ Formschwingung; Körperschwingung ↓ contour vibration; bulk vibration
vibration antinode MECH Schwingungsbauch = oscillation antinode ≠ Schwingungsknoten ≠ vibration node
vibration detector SIGN.ENG Vibrationsmelder = vibration sensor = Schwingungsmelder
vibration endurance limit MECH Schwingungsfestigkeit
vibration energy PHYS Schwingungsenergie
vibration-free MECH schwingungsfrei = vibrationsfrei; erschütterungsfrei
vibration frequency MECH Schwingungsfrequenz = oscillating frequency ≈ Schwingfrequenz ≈ intrinsic frequency
vibration galvanometer INSTR Vibrationsgalvanometer ↑ moving-magnet galvanometer ↑ Drehmagnetgalvanometer
vibration measuring amplifier INSTR Schwingungsmeßverstärker
vibration node MECH Schwingungsknoten = oscillation node ≠ Schwingungsbauch ≠ vibration antinode
vibration pickup (→vibration transducer) INSTR (→Schwingungsaufnehmer)
vibration resistance TECH Schwingbeständigkeit = Vibrationsbeständigkeit
vibration resistant TECH vibrationsfest
vibration sensor (→vibration detector) SIGN.ENG (→Vibrationsmelder)
vibration test QUAL Schwingungsprüfung = Schwingprüfung; Vibrationsprüfung
vibration transducer INSTR Schwingungsaufnehmer = vibration pickup ≈ acceleration pickup

vibrator (AM)		POWER SYS	(→Wechselrichter)	video-frequency oscilloscope	TV	Fernseh-Kontrolloszilloskop
(→inverter)				= VF oscilloscope; video oscilloscope		= Videofrequenz-Oszilloskop; VF-Oszilloskop; VF-Oszillograf
vibrator (→chopper)		CIRC.ENG	(→Zerhacker)			
vibrometer		INSTR	Schwingungsmeßgerät			
vice (BRI) (→vise)		MECH	(→Schraubstock)	video game	DATA PROC	Videospiel
video		TV	Video	= computer game; computerized game		= Computerspiel
[TV picture signal]			[Fernsehsignal]	video generator	TERM&PER	Videogenerator
video amplifier		TV	Videoverstärker	video grabber (→frame grabber)	DATA PROC	(→Bildfangschaltung)
			= Bildverstärker			
video analyzer		INSTR	Video-Analysator	video grabbing (→frame grabbing)	DATA PROC	(→Bildeinfangung)
video animation		TELEC	Video-Animation			
video band (→video signal)		TV	(→Videosignal)	video head	TV	Videokopfrad
video bandwidth		TV	Videobandbreite			= Kopfrad
video board (→graphics board)		TERM&PER	(→Graphikkarte)	video information system	OFFICE	Video-Auskunftsanlagen
				video integration [RADAR]	RADIO LOC	Videointegration [RADAR]
video cable		COMM.CABLE	Videokabel			
video cable amplifier [CATV]		BROADC	VF-Entzerrerverstärker [CATV]	video memory (→graphics memory)	DATA PROC	(→Graphikspeicher)
video cable equalizer [CATV]		BROADC	VF-Kabelentzerrer [CATV]	video memory (→refresh memory)	DATA PROC	(→Bildwiederholspeicher)
video camera		CONS.EL	Videokamera	video mixer	TV	Bildmischgerät
[TV camera of consumer electronics]			[Farbfernsehkamera der Konsumelektronik]			= Bildmischpult
				video modulator	TV	Videogleichrichter
video card (→graphics board)		TERM&PER	(→Graphikkarte)			= Videomodulator
				video monitor (→video signal monitor)	TV	(→Videokontrollschirm)
video carrier		TV	Bildträger			
video carrier spacing		TV	Bildträgerabstand	video monitoring system (→television control system)	SIGN.ENG	(→Fernsehüberwachungsanlage)
video cassette recorder (→videotape recorder)		TV	(→Videorecorder)			
				video oscilloscope (→video-frequency oscilloscope)	TV	(→Fernseh-Kontrolloszilloskop)
video channel		TELEC	Videokanal			
video circuit section [CATV]		BROADC	VF-Leitungsabschnitt [CATV]	video output	TERM&PER	Videoausgang
				video port	TERM&PER	Bildschirmanschluß
video cleaning cassette ↑ cleaning cassette		TV	Video-Reinigungskassette ↑ Reinigungskassette	video presentation system	OFFICE	Video-Präsentationsanlage
				video RAM (→refresh memory)	DATA PROC	(→Bildwiederholspeicher)
video codec		TELEC	Videocodierer			
			= Videocodec	video receiver (AM) (→television set)	CONS.EL	(→Fernsehgerät)
video codification		TELEC	Videocodierung			
video communication service		TELEC	Videokommunikationsdienst	video recorder (→videotape recorder)	TV	(→Videorecorder)
video computer		DATA PROC	Bildschirmcomputer			
videoconference (→video conferencing)		TELEC	(→Bildkonferenz)	video recording	TV	Videoaufzeichnung
				video scanner (→image scanner)	TERM&PER	(→Bildabtaster)
video conferencing = videoconference ↑ teleconferencing		TELEC	Bildkonferenz = Bildschirmkonferenz; Videokonferenz ↑ Telekonferenz			
				videoscope	INSTR	Videoskop
				video screen = television screen; TV screen; screen ↑ luminescent screen	TV	Bildschirm = Fernsehbildschirm ↑ Leuchtschirm
videoconferencing system		TELEC	Videokonferenzsystem			
video connector		COMPON	Videosteckverbinder = Videostecker	video selector	TV	Videowahlschalter
				video signal	TV	Videosignal
video control and regenerator amplifier [CATV]		BROADC	VF-Regel-und Regenerierverstärker [CATV]	[the blanked picture signal delivered by the TV camera, plus the sync signal; the signal containing the picture]		[BA-Signal der Fernsehkamera mit Synchronsignal; das Signal des Bildinhalts]
video copying lead		CONS.EL	Videoüberspielkabel			
video crossbar system [CATV]		BROADC	VF-Koppelpunkt [CATV]	= composite picture signal; picture signal; video band ↓ composite color picture signal; blanked picture signal		= BAS-Signal; Fernsehsignalgemisch; Signalgemisch; Basisband; Bildsignal; B-Signal; Videoband 2
video demodulator		TV	Videodemodulator			
video detection		TV	Bildgleichrichtung			
video digitizer		TERM&PER	Video-Digitizer			↓ FBAS-Signal; BA-Signal
videodisc = videodisk		CONS.EL	Videospeicherplatte	video signal monitor = video monitor	TV	Videokontrollschirm = Videomonitor
video-disc player		CONS.EL	Bildplattenspieler	video switcher [CATV]	BROADC	VF-Schaltverteiler [CATV]
video disk (→optical disk)		TERM&PER	(→Bildplatte)			
				videotape ↑ magnetic tape	TV	Videoband 1 ↑ Magnetband
videodisk (→videodisc)		CONS.EL	(→Videospeicherplatte)			
video display (→display on screen)		TELECONTR	(→Bildschirmanzeige)	videotape recorder = VTR; video cassette recorder; video recorder	TV	Videorecorder = Bildaufnahmegerät; Videobandgerät
video display terminal (→data display terminal)		TERM&PER	(→Datensichtgerät)			
				video tape recording = VTR ↑ magnetic recording [ELECTRON]	TV	Magnetband-Fernsehaufzeichnung = magnetische Videosignallaufzeichnung ↑ Magnetaufzeichnung [ELECTRON]
video display unit (→data display terminal)		TERM&PER	(→Datensichtgerät)			
video distortion analyzer		TV	Prüfzeilen-Analysator			
video distribution amplifier		TV	Videotrennverstärker			
video engineer		TV	Bildingenieur	video techniques (→video engineering)	TV	(→Videofrequenztechnik)
video engineering = video techniques		TV	Videofrequenztechnik = VF-Technik			
				videotelegraphy [transmission of pictures]	TELEGR	Bildtelegrafie [Übertragung von Bildern]
video enhancer		CONS.EL	Videoüberspielverstärker			
video frequency [10 Hz to 5 MHz]		TV	Videofrequenz [10 Hz – 5 MHz]	= phototelegraphy; picture telegraphy ≈ telewriting ↑ facsimile telegraphy		= Fototelegrafie ≈ Bildfernschreiben ↑ Faksimiletelegrafie
video frequency (→frame frequency)		TV	(→Bildfolgefrequenz)			
				videotelephony	TELEC	Bildfernsprechen

video terminal (→data display terminal)		TERM&PER	(→Datensichtgerät)	**virtual connection** (→virtual call)	DATA COMM	(→virtuelle Verbindung)
video test signal generator		INSTR	**Videotestsignalgenerator**	**virtual container** [SDH/SONET] = VC	TELEC	**virtueller Container** [SDH/SONET] = VC
videotex (→interactive videotex)		TELEC	(→Bildschirmtext)	**virtual drive** (→RAM disk)	DATA PROC	(→virtuelles Laufwerk)
videotex agency		TELEC	**Bildschirmtext-Agentur** = Btx-Agentur; Videotex-Agentur	**virtual fixed connection** = virtual permanent circuit	DATA COMM	**virtuelle Standleitung** = virtuelle Standverbindung
videotex-compatible		TELEC	**bildschirmtextfähig**	**virtual hardware**	DATA PROC	**virtuelle Hardware**
videotex computer center		TELEC	**Bildschirmtext-Zentrale**	**virtual image**	PHYS	**virtuelles Bild**
videotex connection box		TELEC	**Btx-Anschlußbox**	**virtual link** (→virtual call)	DATA COMM	(→virtuelle Verbindung)
videotex control station		TELEC	**Btx-Leitzentrale**	**virtual memory** [addressable as unit beyond physical limits] = virtual storage; virtual store ≠ real memory	DATA PROC	**virtueller Speicher** [über physikalische Begrenzung hinaus als Einheit adressierbar] = Hintergrundspeicher 1; Seitenwechselspeicher ≠ realer Speicher
videotex information provider		TELEC	**Btx-Anbieter**			
videotex inquiry terminal		TERM&PER	**Bildschirmtext-Abfragegerät**			
videotex page		TELEC	**Bildschirmtext-Seite** = Btx-Seite			
Videotext (BRD) (→teletext)		TELEC	(→Teletext)			
video track		TV	**Videospur**	**virtual network**	DATA COMM	**virtuelles Netz**
video unit (→data display terminal)		TERM&PER	(→Datensichtgerät)	**virtual permanent circuit** (→virtual fixed connection)	DATA COMM	**virtuelle Standleitung**
videowall		CONS.EL	**Videowand**	**virtual storage**	DATA PROC	**virtuelle Speicherung**
vidicon ↑ camera tube		ELECTRON	**Vidikon** ↑ Bildaufnahmeröhre	**virtual storage** (→virtual memory)	DATA PROC	(→virtueller Speicher)
vidicon telecine		TV	**Vidikon-Filmabtaster**	**virtual store** (→virtual memory)	DATA PROC	(→virtueller Speicher)
view (n.)		DATA PROC	**Darstellungsart**			
view		ENG.DRAW	**Ansicht**	**virus** [hidden corrupting subroutine] = computer virus; software virus; virus program	DATA PROC	**Virus** = Computer-Virus; Software-Virus; Virus-Programm; Software-Fremdkörper
viewdata service (→interactive videotex)		TELEC	(→Bildschirmtext)			
viewer (→televiewer)		BROADC	(→Fernsehzuschauer)	**virus program** (→virus)	DATA PROC	(→Virus)
viewfinder		TECH	**Sucher** [optischer]	**visa**	ECON	**Visum** = Einreiseerlaubnis
viewing angle [of a sreen]		TERM&PER	**Beobachtungswinkel** [Bildschirm]	**viscosity** = sluggishness	PHYS	**Zähflüssigkeit** = Viskosität; Dickflüssigkeit; Schwerflüssigkeit
viewing distance		TERM&PER	**Betrachtungsabstand** [Bildschirm]			
viewing storage tube		ELECTRON	**Sichtspeicherröhre** ↑ Oszilloskopröhre	**viscous** = thick; thickly liquid; sluggish	TECH	**dickflüssig** = zähflüssig; schwerflüssig; viskos; viskös; leimartig
viewing transformation (→viewpoint transformation)		DATA PROC	(→Fensterabbildung)			
viewpoint transformation = viewing transformation		DATA PROC	**Fensterabbildung** = Fenstertransformation	**vise** (AM) = vice (BRI)	MECH	**Schraubstock**
viewport [computer graphics]		DATA PROC	**Arbeitsfläche** [Computergraphik] = Darstellungsfeld; Darstellungsbereich; Bildschirmbereich; Bildflächenausschnitt	**visibility limit**	PHYS	**Sichtbarkeitsgrenze**
				visibility range = visual range; optical range; line-of-sight distance	PHYS	**Sichtbereich** = Sichtweite; Sehweite
				visible ≈ optical; visual	PHYS	**sichtbar** ≈ optisch; visuell
view state (→display mode)		INSTR	(→Anzeigefunktion)	**visible outline**	ENG.DRAW	**sichtbare Kante**
vigilance buttom		RAILW.SIGN	**Wachsamkeitstaste**	**vision channel**	TELEC	**Bildkanal**
Villard connection ↑ converter connection		POWER SYS	**Villard-Schaltung** ↑ Stromrichterschaltung	**vision modulation** = image modulation	TV	**Bildmodulation**
vine red		PHYS	**weinrot**	**visit** (→site survey)	SYS.INST	(→Ortsbegehung)
violate		CODING	**verletzen**	**visiting card** = calling card (AM)	ECON	**Visitenkarte**
violation		CODING	**Verletzung**			
violation (→coding law violation)		CODING	(→Coderegelverletzung)	**visiting lecturer**	SCIE	**Gastprofessor**
				visitor location register ≠ home location register	MOB.COMM	**Besucherdatei** ≠ Heimatdatei
violation monitoring (→code violation monitoring)		CODING	(→Coderegelüberwachung)			
				visitors' cafeteria (AM) (→visitors' restaurant)	ECON	(→Gästekasino)
violet		PHYS	**violett**			
virgin cassette		TERM&PER	**Leerkassette** = unbeschriebene Kassette; Leercassette	**visitors' canteen** (→visitors' restaurant)	ECON	(→Gästekasino)
				visitors' casino (→visitors' restaurant)	ECON	(→Gästekasino)
virgin curve (→initial magnetization curve)		PHYS	(→Neukurve)			
				visitors' restaurant = visitors' canteen; visitors' casino; visitors' cafeteria (AM)	ECON	**Gästekasino** = Gästekantine
virgule (→slash)		LING	(→Schrägstrich)			
VIRS (→vertical interval reference signal)		TV	(→Prüfzeilen-Referenzsignal)			
virtual		SCIE	**virtuell**	**visual** ≈ visible; optical	TECH	**visuell** ≈ sichtbar; optisch
virtual ≠ real		DATA PROC	**virtuell** ≠ reell	**visual acuity**	PHYS	**Sehschärfe**
virtual address = logical address ≠ absolute address		DATA PROC	**virtuelle Adresse** = logische Adresse ≠ absolute Adresse	**visual alarm** = alarm indication	EQUIP.ENG	**optischer Alarm** = Alarmanzeige; Lichtzeichen
				visual angle (→angle of sight)	PHYS	(→Sehwinkel)
virtual call [data call in the packet mode] = virtual circuit; virtual link; virtual connection; datagram mode		DATA COMM	**virtuelle Verbindung** [paketvermittelter Datenaustausch] = virtuelle Wählvernbindung; Datagramm-Dienst	**visual/aural range** = VAR	RADIO NAV	**optisch-akustischer Leitstrahlsender** = VAR
				visual axis (→optical axis)	PHYS	(→Binormale)
				visual-card index	OFFICE	**Sichtkartei**
				visual check (→visual inspection)	QUAL	(→Sichtprüfung)
virtual carrier		MODUL	**virtueller Träger**	**visual display** (→display)	TERM&PER	(→Bildanzeige)
virtual circuit (→virtual call)		DATA COMM	(→virtuelle Verbindung)	**visual display** (→display 3)	TERM&PER	(→Bildanzeige)

visual display device TERM&PER (→Sichtgerät)
(→display terminal)
visual display mode (→screen TERM&PER (→Bildschirmdarstellung)
mode)
visual display unit (→display TERM&PER (→Sichtgerät)
terminal)
visual field PHYS **Gesichtsfeld**
= field
visual indicator EQUIP.ENG **Schauzeichen**
= visual signal
visual inspection QUAL **Sichtprüfung**
= visual check; macroscopic instruction = Sichtkontrolle; makroskopische Untersuchung
visualization TECH **Visualisierung**
visual monitoring SIGN.ENG **optische Überwachung**
visual range (→visibility range) PHYS (→Sichtbereich)
visual scanner (→optical TERM&PER (→optischer Leser)
character reader)
visual signal (→visual EQUIP.ENG (→Schauzeichen)
indicator)
Viterbi coding CODING **Viterbi-Codierung**
vitreous TECH **gläsern**
= glassy
VITS (→vertical interval test signal) TV (→Prüfzeilen-Meßsignal)
Vivaldi antenna ANT **Vivaldi-Antenne**
vivid color (AM) PHYS **lebhafte Farbe**
= vivid colour (BRI)
vivid colour (BRI) (→vivid color) PHYS (→lebhafte Farbe)
VLDB DATA PROC **Größtdatenbank**
= very large data base
VLF (→myriametric waves) RADIO (→Myriameter-Wellen)
VLF antenna ANT **Längstwellenantenne**
VLSI MICROEL **Größtintegration**
[10,000 to 100,000 components per IC] [10.000 bis 100.000 Komponenten pro IC]
= very large scale integration = Höchstintegration; sehr hoher Integrationsgrad; VLSI
VME bus MICROEL **VME-Bus**
[Versa Module Europe bus]
VMOS MICROEL **VMOS**
[vertical metal-oxide semiconductor]
VMOS process MICROEL **VMOS-Verfahren**
= vertical anisotropic etch metal-oxide semiconductor process
v.m. tube (→velocity-modulated MICROW (→Laufzeitröhre)
tube)
V notch laser OPTOEL **V-Nut-Laser**
vobulate (→sweep) ELECTRON (→wobbeln)
vobulating (→sweeping) ELECTRON (→Wobbelung)
vobulating frequency ELECTRON (→Wobbelfrequenz)
(→sweep rate)
vobulating range (→sweep ELECTRON (→Wobbelbereich)
range)
vobulating rate (→sweep rate) ELECTRON (→Wobbelfrequenz)
vobulation method ELECTRON (→Wobbelverfahren)
(→sweep-frequency method)
vobulator (→sweep generator) INSTR (→Wobbelsender)
vocabulary DATA PROC **Vokabular**
[set of codes and instruction for programming] [von Befehlen für eine Programmierung]
vocation (→profession) ECON (→Beruf)
vocational school SCIE **Berufsfachschule**
vocational training (→professional ECON (→Berufsausbildung)
education)
vocoder TELEPH **Vocoder**
[band saving tranmission of characteristic parameters, used on receive side for speech synthesis] [frequenzsparende Übertragung charakteristischer Sprachparameter, mit denen empfangsseitig künstliche Sprache erzeugt wird]
= voice-operated coder
≈ voice coder [TELEC] ≈ Sprachverschlüsselungsgerät [TELEC]
vocoder analyzer TELEPH **Vocoder-Analysator**
vocoder synthesizer TELEPH **Vocoder-Synthesizer**
voice INF.TECH **Sprache**
[information conveyed by voice] [über Stimme mitgeteilte Information]
= speech; voice sound; speech sound = Sprechen; Sprachlaut
≈ voice signal; conversation; voice [ACOUS] ≈ Sprachsignal; Gespräch; Stimme [ACOUS]

voice ACOUS **Stimme**
[sound produced by human vocal organism] [vom Sprachorgan abgegebener Laut]
≈ speech ≈ Sprache 1
voice analysis TELEC **Sprachanalyse**
= speech analysis
voice annotation TERM&PER **Sprachanmerkung**
[speech comments to a stored text] [gesprochener Kommentar zu gespeichertem Text]
= word annotation = Voice Annotation
≈ multi-media mail ≈ Mitteilungssystem für Text und Sprache
voice answer (→speech EL.ACOUS (→Sprachantwort)
recording)
voice band TELEC **Sprachband**
[in public telecommunications generally 0.3 to 3.4 kHz] [in öffentlichen Telekommunikation meist 0,3–3,4 kHz]
= voiceband; speech band; speech frequency band = Sprachfrequenzband; Sprechfrequenzbereich
voiceband (→voice band) TELEC (→Sprachband)
voiceband data transmission DATA COMM **Sprachband-Datenübertragung**
voice band inversion (→speech INF.TECH (→Sprachbandinvertierung)
inversion)
voice-band limitation TELEC **Sprachbegrenzung**
= speech-band limitation
voice-band limiting filter TELEC **Sprachbegrenzungsfilter**
= speech-band limiting filter
voice band scrambling (→speech TELEC (→Sprachverwürfelung)
scrambling)
voice calling TELEPH **Einzelansprechen**
voice channel (→telephone TELEC (→Fernsprechkanal)
channel)
voice circuit TELEC **Sprechkreis**
= speech circuit
voice coder TELEC **Sprachverschlüsselungsgerät**
[speech coding for secrecy purposes] [Sprachverschlüsselung zum Zwecke der Geheimhaltung]
≈ vocoder (TELEPH) ≈ Vocoder (TELEPH)
voice coding (→speech INF.TECH (→Sprachverschlüsselung)
encryption)
voice coil EL.ACOUS **Schwingspule**
= Lautsprecher-Schwingspule
voice communication TELEC **Sprachkommunikation**
= speech communication ≠ Nicht-Sprache-Kommunikation
≠ non-voice communication
↑ telecommunications ↑ Telekommunikation
↓ telephony; sound broadcasting ↓ Fernsprechen; Tonrundfunk
voice compression (→speech TELEC (→Sprachkompression)
compression)
voice-controlled ELECTRON **sprachgesteuert**
= speech-controlled
voice converter TELEC **Sprachwandler**
= speech converter ↓ Sprachdigitalisierer
↓ voice digitizer
voice current (→speech TELEPH (→Sprechstrom)
current)
voice decoding (→speech INF.TECH (→Sprachentschlüsselung)
decoding)
voice digitization TELEC **Sprachdigitalisierung**
= speech digitization
voice digitizer TELEC **Sprachdigitalisierer**
= speech digitizer ↑ Sprachwandler
↑ voice converter
voice encryption (→speech INF.TECH (→Sprachverschlüsselung)
encryption)
voice filing (→speech filing) TELEC (→Sprachspeicherung)
voice frequency TELEC **Niederfrequenz**
[frequency range for voice signal transmission in public telecommunications, usually 300 Hz to 3400 Hz] [Frequenzbereich für Spachsignalübertragung in der öffentlichen Telekommunikation, meist 300 Hz bis 3400 Hz]
= VF; audio frequency; AF; low frequency; LF; speech frequency = NF; Tonfrequenz; Sprachfrequenz

voice frequency amplifier

voice frequency amplifier	TRANS	**NF-Verstärker**	**voice-plus duplex** (→speech-plus duplex)	TELEGR (→Zwischenkanal-WT)
= VF amplifier; audio frequency amplifier; audio amplifier; telephone repeater		= Niederfrequenzverstärker	**voice protection** (→speech protection)	TELEC (→Sprachschutz)
voice frequency amplifier (→booster amplifier)	TRANS	(→NF-Zwischenverstärker)	**voice recognition** (→speech recognition)	INF.TECH (→Spracherkennung)
voice-frequency carrier telegraphy (→voice frequency telegraphy)	TELEGR	(→Wechselstromtelegrafie)	**voice-recognizing** = speech-recognizing	INF.TECH **sprachverstehend**
voice frequency channel (→VF channel)	TELEC	(→NF-Kanal)	**voice recognizing dialog system**	DATA PROC **sprachverstehendes Dialogsystem**
voice-frequency circuit = VF circuit; VF link	TELEC	**NF-Leitung** = Niederfrequenzleitung	= speech recognizing dialog system; interactive dialog system	
voice-frequency engineering = VF engineering; audio frequency engineering	TELEC	**NF-Technik** = Niederfrequenztechnik	**voice recorder** (→call recorder)	TERM&PER (→Sprachaufzeichnungsgerät)
voice-frequency level = VF level; audio frequency level	TELEC	**NF-Pegel** = Niederfrequenzpegel	**voice recording** (→speech recording)	EL.ACOUS (→Sprachantwort)
voice-frequency line section	TELEGR	**WT-Abschnitt**	**voice response** = voice output; speech response; speech output	DATA PROC **Sprachausgabe**
voice frequency signal = VF signal; audio frequency signal	TELEC	**NF-Signal** = Niederfrequenzsignal; Tonfrequenzsignal	**voice security** (→speech security)	TELEC (→Sprachsicherheit)
voice frequency signaling = low frequency signaling	TELEC	**Tonfrequenz-Signalisierung** = Niederfrequenz-Signalisierung	**voice service** (→POT service)	TELEC (→konventioneller Fernsprechdienst)
voice-frequency telegraph circuit = VFT circuit	TELEGR	**WT-Verbindung**	**voice signal** = speech signal ↓ telephone signal	TELEC **Sprachsignal** = Sprechsignal ↓ Fernsprechsignal
voice frequency telegraphy = VFT; VFTG; carrier telegraphy; telegraphy over multiplex; TOM; voice-frequency carrier telegraphy; ac telegraphy	TELEGR	**Wechselstromtelegrafie** = WT; Trägerfrequenztelegrafie	**voice signal transmission** (→voice transmission)	TELEC (→Sprachübertragung)
			voice simulation (→speech simulation)	INF.TECH (→Sprachsimulation)
			voice sound (→voice)	INF.TECH (→Sprache)
voice frequency telegraphy channel (→VFT channel)	TELEC	(→WT-Kanal)	**voice synthesis** = speech synthesis	INF.TECH **Sprachsynthese** = Sprachgenerierung
voice-grade channel = speech-grade channel	DATA COMM	**Sprechkanal** [mit Fernsprechqualität]	**voice synthesizer** = speech synthesizer	TERM&PER **Sprachsynthesizer** = Sprachgenerator; Synthesator
voice guard (→speech protection)	TELEC	(→Sprachschutz)	**voice transmission** = voice signal transmission; speech signal transmission ≈ telephone transmission	TELEC **Sprachübertragung** = Sprachsignalübertragung ↓ Fernsprechübertragung
voice-guard encryption (→speech encryption)	INF.TECH	(→Sprachverschlüsselung)		
voice highway (→speech highway)	SWITCH	(→Sprachmultiplexleitung)	**voice wire** (→speech wire)	TELEC (→Sprechader)
voice input = speech input	DATA PROC	**Spracheingabe**	**void** (n.)	MECH **Aussparung**
voice input (→speech input)	SIGN.TH	(→akustische Spracheingabe)	**void** (→lattice vacancy)	PHYS (→Gitterfehlstelle)
voice interface = speech interface	TELEC	**Sprachschnittstelle**	**void** (n.) (→blank)	CODING (→Leerstelle)
voice interpolation (→speech interpolation)	TELEC	(→Sprachinterpolation)	**void** (→invalid)	ECON (→ungültig)
			void (→ineffective)	TECH (→unwirksam)
voiceless	LING	**stimmlos**	**void** (→empty)	TERM&PER (→leer)
voice level (→speech level)	TELEC	(→Sprachpegel)	**void cell** [ATM]	TELEC **Leerzelle** [ATM]
voice mail [computer aided storage and transfer of telephone messages]	TELEC	**Sprachkommunikationssystem** [rechnergestützte Aufzeichnung und Übermittlung telefonischer Mitteilungen] = Sprachpost; Voice Mail; Mitteilungssystem für Sprache	**void list** (→empty list)	DATA PROC (→Leerliste)
			void tape = empty tape	DATA PROC **Leerband**
			volatile ≠ non-volatile	PHYS **flüchtig** ≠ nichtflüchtig
			volatile memory [information is lost if operating power is switched off] = volatile storage; volatile store ≠ non-volatile memory ↑ read/write memory ↓ volatile solid state memory	DATA PROC **flüchtiger Speicher** [Information geht bei Abschalten der Betriebsspannung verloren] ≠ nichtflüchtiger Speicher ↑ Lese-Schreib-Speicher ↓ flüchtiger Halbleiterspeicher; statischer Speicher; dynamischer Speicher
voice mail server [ISDN]	TELEC	**Sprachinformationsserver** [ISDN]		
voice memory (→speech memory)	TELEC	(→Sprachspeicher)		
voice mode = speech mode	TELEC	**Sprachbetrieb**		
voice multiplexer (→speech multiplexer)	SWITCH	(→Sprachmultiplexer)	**volatile semiconductor memory** (→volatile solid state memory)	MICROEL (→flüchtiger Halbleiterspeicher)
voice-operated coder (→vocoder)	TELEPH	(→Vocoder)	**volatile solid state memory** [looses its content when power is turned off] = volatile semiconductor memory ↑ volatile memory	MICROEL **flüchtiger Halbleiterspeicher** [Inhalt geht bei Abschalten der Stromversorgung verloren] ↑ flüchtiger Speicher
voice output (→voice response)	DATA PROC	(→Sprachausgabe)		
voice output system = speech output system	TERM&PER	**Sprachausgabesystem**	**volatile storage** (→volatile memory)	DATA PROC (→flüchtiger Speicher)
voice paging	TELEPH	**Sammelansprechen**	**volatile store** (→volatile memory)	DATA PROC (→flüchtiger Speicher)
voice paging ↑ radio paging	MOB.COMM	**Sprechfunkruf** ↑ Funkruf	**volatility** [change rate of a file]	DATA PROC **Dateiänderungshäufigkeit**
voice path = telephonic path; speech path; talk path	TELEC	**Fernsprechweg** = Sprechweg	**volatilization** = slow evaporation ↑ evaporation	PHYS **Verdunstung** [langsame Verdampfung] = Flüchtigkeit ↑ Verdampfung

volatilize [evaporate slowly] ↑ evaporate		PHYS	**verdunsten** [langsam verdampfen] = verflüchtigen ↑ verdampfen	**voltage driving magnetic amplifier** (→voltage driving transductor amplifier)	CIRC.ENG	(→spannungssteuernder Transduktorverstärker)
volt [derived SI unit for electric tension] = V		PHYS	**Volt** [abgeleitete SI-Einheit für elektrische Spannung] = V	**voltage driving transductor amplifier** = voltage driving magnetic amplifier ↑ transductor amplifier	CIRC.ENG	**spannungssteuernder Transduktorverstärker** = spannungssteuernder Magnetverstärker ↑ Transduktorverstärker
Volta effect		PHYS	**Volta-Effekt**	**voltage drop**	EL.TECH	**Spannungsabfall**
voltage [potential difference expressed in Volt; SI-unit: Volt] = tension; U ≈ potential difference [PHYS]		EL.TECH	**elektrische Spannung** [SI-Einheit: Volt] = Spannung; U ≈ Potentialdifferenz [PHYS]	= resistance drop; drop ≈ potential fall **voltage endurance** (→dielectric strength) **voltage-fed antenna** (→Zeppelin antenna)	EL.TECH ANT	= Spannungsverlust ≈ Potentialabfall (→Spannungsfestigkeit) (→Zeppelin-Antenne)
voltage adapter switch (→voltage selector)		EQUIP.ENG	(→Spannungswähler)	**voltage feed** = voltage source driving	CIRC.ENG	**Spannungssteuerung** = Spannungseinkopplung; Spannungseionspeisung
voltage amplification (→voltage gain)		NETW.TH	(→Spannungsverstärkung)	**voltage feed** ≈ power supply	EL.TECH	**Stromzuführung** = Spannungszuführung ≈ Stromversorgung
voltage amplifier		CIRC.ENG	**Spannungsverstärker**	**voltage feedback**	CIRC.ENG	**Spannungsrückkopplung**
voltage amplitude		EL.TECH	**Spannungsamplitude**	[part of output voltage is fed to input voltage] ↑ feedback ↓ positive voltage feedback; negative voltage feedback		[Teil der Ausgangsspannung wird auf den Eingang rückgekoppelt] ↑ Rückkopplung ↓ Spannungsmitkopplung; Spannungsgegenkopplung
voltage antinode (→voltage crest)		EL.TECH	(→Spannungsbauch)			
voltage attenuation		NETW.TH	**Spannungsdämpfung**			
voltage attenuation factor		NETW.TH	**Spannungsdämpfungsfaktor** [Spannung am Eingang zu Spannung am Ausgang]			
voltage brakdown (→black-out)		POWER SYS	(→Spannungsausfall)	**voltage fluctuation**	EL.TECH	**Spannungsschwankung**
				voltage follower [operational amplifier with extreme feedback and unitary amplification]	CIRC.ENG	**Spannungsfolger** [extrem stark gegengekoppelter Operationsverstärker mit Gesamtverstärkung Eins] ↑ Operationsverstärker
voltage-carrying ≠ voltage-free ↑ live		EL.TECH	**spannungsführend** ≠ spannungsfrei ↑ heiß			
voltage collapse		EL.TECH	**Spannungszusammenbruch**			
voltage comparator		CIRC.ENG	**Spannungskomparator**	**voltage-free** ≠ voltage-carrying ↑ dead	EL.TECH	**spannungsfrei** = spannungslos ≠ spannungsführend ↑ kalt
voltage comparison encoder (→incremental-step converter)		CIRC.ENG	(→Stufenkompensationsumsetzer)			
voltage compatibility		ELECTRON	**Spannungsverträglichkeit**	**voltage-frequency converter**	CIRC.ENG	**Spannungs-Frequenz-Umsetzer** = Spannungs-Frequenz-Wandler
voltage compensator		ELECTRON	**Spannungskompensator**			
voltage-controlled		ELECTRON	**spannungsgesteuert**	**voltage gain** = voltage amplification ↑ gain	NETW.TH	**Spannungsverstärkung** ↑ Verstärkung
voltage-controlled amplifier = VCA		CIRC.ENG	**spannungsgesteuerter Verstärker** = VCA			
voltage-controlled current source = VCS		NETW.TH	**spannungsgesteuerte Stromquelle** = VCS	**voltage generator** **voltage generator** (→voltage source)	NETW.TH PHYS	**Urspannungsquelle** (→Spannungsquelle)
voltage-controlled filter = VCF		NETW.TH	**spannungsgesteuertes Filter** = VCF	**voltage indication** = voltage coupling	ANT	**Spannungskopplung**
voltage-controlled oscillator = VCO		CIRC.ENG	**spannungsgesteuerter Oszillator**	**voltage indicator** = voltage tester	INSTR	**Spannungsprüfer** = Spannungsindikator
voltage-controlled voltage source = VVS		NETW.TH	**spannungsgesteuerte Spannungsquelle** = VVS	**voltage insufficiency** (→brown-out)	POWER SYS	(→Spannungsmangel)
voltage converter		CIRC.ENG	**Spannungsumsetzer** = Spannungswandler	**voltage jump**	EL.TECH	**Spannungssprung**
voltage converter = voltage transformer; potential transformer		INSTR	**Spannungswandler**	**voltage level** ≠ power level	TELEC	**Spannungspegel** ≠ Leistungspegel
voltage corrector (→voltage stabilizer)		POWER ENG	(→Spannungsstabilisator)	**voltage level difference**	CIRC.ENG	**Spannungshub**
voltage coupling (→voltage indication)		ANT	(→Spannungskopplung)	**voltage limit**	EL.TECH	**Spannungsgrenzwert**
				voltage limitation	EL.TECH	**Spannungsbegrenzung**
voltage crest = voltage antinode; voltage maximum ≠ voltage node		EL.TECH	**Spannungsbauch** ≠ Spannungsknoten	**voltage limiter**	CIRC.ENG	**Spannungsbegrenzer**
				voltage matrix	NETW.TH	**Spannungsmatrix**
				voltage maximum (→voltage crest)	EL.TECH	(→Spannungsbauch)
voltage-current characteristic		MICROEL	**Spannungs-Strom-Charakteristik** = Spannungs-Strom-Kennlinie	**voltage minimum** (→voltage node)	EL.TECH	(→Spannungsknoten)
				voltage multiplication	CIRC.ENG	**Spannungsverfielfachung**
voltage curve		PHYS	**Spannungsverlauf**	**voltage multiplicator** ↑ multiplicator	CIRC.ENG	**Spannungsvervielfacherschaltung** = Spannungsvervielfacher ↑ Vervielfacher
voltage-dependent resistor (→varistor)		COMPON	(→Varistor)			
voltage discharge gap (→discharger)		COMPON	(→Schutzfunkenstrecke)	**voltage multiplier**	CIRC.ENG	**Spannungsvervielfacher**
voltage distribution		EL.TECH	**Spannungsverteilung**	**voltage node** = voltage minimum ≠ voltage crest	EL.TECH	**Spannungsknoten** ≠ Spannungsbauch
voltage divider = potential divider		EL.TECH	**Spannungsteiler**			
voltage doubler ↑ doubler circuit		CIRC.ENG	**Spannungsverdoppler** = Spannungsdoppler ↑ Verdopplerschaltung	**voltage overshoot**	EL.TECH	**Spannungsüberhöhung**
				voltage path	INSTR	**Spannungspfad**
				voltage pulse (→voltage surge)	EL.TECH	(→Spannungsstoß)
voltage driving		EL.TECH	**spannungssteuernd**	**voltage range**	EL.TECH	**Spannungsbereich**

voltage rating (→nominal voltage) EL.TECH (→Nennspannung)

voltage ratio NETW.TH **Spannungs-Übersetzungsverhältnis**
= voltage transfer ratio
= Spannungsverhältnis

voltage reducer CIRC.ENG **Spannungsverkleinerer**

voltage reference diode MICROEL **Referenzdiode**
[diode with a flat voltage response over a broad current interval, thereby suited for the generation of reference voltages or for the stabilization of dc voltages; purely function-oriented definition]
[Diode mit fast konstanter Spannung über einen breiten Strombereich, u.a. für die Erzeugung von Bezugsspannungen oder zur Gleichspannungsstabilisierung geeignet; auf die Anwendungsmöglichkeit bezogene Bezeichnung]
= reference diode; voltage stabilisator diode
= Regulatordiode; Stabilisatordiode; Spannungsstabilisatordiode; Bezugsspannungsdiode
≈ Zener diode; avalanche diode
≈ Zenerdiode; Lawinendiode

voltage-regulated ELECTRON **spannungsgeregelt**

voltage regulation (→voltage stabilization) EL.TECH (→Spannungsstabilisierung)

voltage regulator CIRC.ENG **Spannungsregler**

voltage regulator (→voltage stabilizer) POWER ENG (→Spannungsstabilisator)

voltage regulator diode CIRC.ENG **Spannungsreglerdiode**

voltage regulator tube ELECTRON **Stabilisatorröhre**
= glow-discharge voltage regulator

voltage resonance (→series resonance) NETW.TH (→Serienresonanz)

voltage rise EL.TECH **Spannungsanstieg**
= voltage step-up
= Spannungserhöhung

voltage selection EQUIP.ENG **Spannungsumschaltung**

voltage selector EQUIP.ENG **Spannungswähler**
= voltage selector switch; voltage adapter switch
= Spannungsumschalter

voltage selector switch (→voltage selector) EQUIP.ENG (→Spannungswähler)

voltage source PHYS **Spannungsquelle**
= voltage generator

voltage source driving (→voltage feed) CIRC.ENG (→Spannungssteuerung)

voltage stabilisator diode (→voltage reference diode) MICROEL (→Referenzdiode)

voltage stabilization EL.TECH **Spannungsstabilisierung**
= voltage regulation
= Spannungsregelung

voltage stabilizer POWER ENG **Spannungsstabilisator**
= voltage regulator; voltage corrector; constant-voltage source; constant-voltage power supply
= Spannungskonstanthalter; Spannungskonstanter; Konstantspannungsquelle; Konstantspannungs-Stromversorgung

voltage standard INSTR **Spannungsnormal**

voltage standing wave ratio LINE TH **Welligkeitsfaktor**
[voltage maximum to voltage minimum; measure of impedance matching between a line and a load]
[Spannungsmaximum zu Spannungsminimum; Maß der Impedanzanpassung einer Last an einer Leitung]
= VSWR; standing wave ratio; SVR; standing wave factor; stationary wave index
= Stehwellenverhältnis; Stehwellen-Spannungsverhältnis; Stehwellenfaktor; Welligkeit; VSWR; SWR
≠ inverse standing wave ratio
≠ Anpassungsfaktor

voltage standing wave ratio (→VSWR) ANT (→Rückflußdämpfung)

voltage step-up (→voltage rise) EL.TECH (→Spannungsanstieg)

voltage supervision EL.TECH **Spannungsüberwachung**

voltage surge EL.TECH **Spannungsstoß**
= voltage pulse
= Spannungsimpuls
≈ surge voltage; line surge
≈ Stoßspannung; Netzstoß

voltage system converter POWER SYS **Umrichter**
[converts the voltage system, employing convertor valves]
[wandelt das Spannungssystem um, unter Verwendung von Gleichrichterventilen]
= voltage system convertor; converter 2 (IEC)
↑ static power converter
↑ Stromrichter

↓ DC converter; AC converter; frequency converter
↓ Gleichstrom-Umrichter; Wechselstrom-Umrichter; Frequenzumrichter

voltage system convertor POWER SYS (→Umrichter)
(→voltage system converter)

voltage tester (→voltage indicator) INSTR (→Spannungsprüfer)

voltage-to-current converter CIRC.ENG **Spannungs-Strom-Wandler**

voltage transducer POWER SYS **Spannungswandler**
[converts DC into AC]
[wandelt Gleichstrom in Wechselstrom]
= voltage transformer

voltage transfer ratio NETW.TH (→Spannungs-Übersetzungsverhältnis)
(→voltage ratio)

voltage transformer POWER SYS **Spannungswandler**
(→voltage transducer)

voltage transformer (→voltage converter) INSTR (→Spannungswandler)

voltage transmission coefficient NETW.TH **Spannungsübertragungsfaktor**
[voltage at output to voltage at input; reciprocal of voltage attenuation factor]
[Spannung am Ausgang zu Spannung am Eingang; Kehrwert des Spannungsdämpfungsfaktor]
≠ voltage attenuation factor
= Spannungsverstärkungsfaktor
↑ transmission coefficient
↓ image voltage transmission coefficient; effective voltage transmission coefficient
≠ Spannungsdämpfungsfaktor
↑ Übertragungsfaktor
↓ Wellenspannungsübertragungsfaktor; Betriebsspannungsübertragungsfaktor

voltaic arc PHYS **Lichtbogen**
= Flammenbogen (OES)

voltaic battery (→battery) POWER SYS (→Batterie)

voltaic cell (→galvanic cell) POWER SYS (→galvanisches Element)

voltaic couple (→galvanic cell) POWER SYS (→galvanisches Element)

voltaic electricity (→contact electricity) EL.TECH (→Berührungselektrizität)

voltaic potential PHYS **Volta-Spannung**

volt-ampere EL.TECH **Volt-Ampere**
[unit for apparent power]
[Einheit für elektrische Scheinleistung]
= VA
= VA
≈ Watt [PHYS]

voltmeter INSTR **Spannungsmesser**
= Voltmeter

Volt-second PHYS **Volt-Sekunde**
[unit for magnetic flux; = 1 Wb]
[Einheit für magnetischen Fluß; = 1 Wb]

volume TYPOGR **Band**
= fascicle

volume MATH **Volumen**
[SI unit: cubic meter]
[SI-Einheit: Kubikmeter]
= V
= Rauminhalt; V

volume (→data carrier) DATA PROC (→Datenträger)

volume (→volume of sound) ACOUS (→Lautstärkepegel)

volume compression TRANS **Dynamikkompression**
= dynamics compression

volume control (→level controller) EL.ACOUS (→Pegelregler)

volume discount (→quantity discount) ECON (→Mengenrabatt)

volume excitation ANT **Flächenanregung**
[to excite a body to radiate]
[einen Körper zum Strahlen anregen]

volume grating PHYS **Volumengitter**

volume hologram OPTOEL **Volumenhologramm**
= holographic image

volume integral MATH **Volumenintegral**

volume ionization PHYS **Volumenionisierung**

volume label DATA PROC **Datenträger-Etikett**
[identifies the data carrier]
[identifiziert den Datenträger]
= Volume-Etikett; Datenträgerkennung; Datenträgererkennung; Datenträger-Kennsatz

volume limiter		EL.ACOUS	**Lautstärkebegrenzer**	**W**		
≈ limiter						
volume meter		EL.ACOUS	**Lautstärkemesser**	W (→watt)	PHYS	(→Watt)
= volumen indicator				W (→tungsten)	CHEM	(→Wolfram)
volumen indicator (→volume meter)		EL.ACOUS	(→Lautstärkemesser)	wadding	TECH	**Watte**
				wafer (→wafer crystal)	MICROEL	(→Kristallscheibe)
volume of sound		ACOUS	**Lautstärkepegel**	wafer crystal	MICROEL	**Kristallscheibe**
= loudness level; volume				[disk of semiconducting material with circuits diffused on it, which are called chips when cutted into pieces]		[Halbleiterscheibe mit auf-diffundierten Schaltungen, die nach Zerteilen Chips genannt werden]
volume range		EL.ACOUS	**Dynamikbereich**			
= dynamic range						
volume recombination		PHYS	**Volumenrekombination**	= wafer; silicon wafer; silicon slice; slice		= Halbleiterscheibe; Träger-platte; Halbleiter-Wafer; Wafer; Siliziumscheibe; Scheibe
volume regulator (→level controller)		EL.ACOUS	(→Pegelregler)	≈ chip		
volumetric		MATH	**räumlich**			
= spatial; spacial						≈ Chip
volumetric expansion		PHYS	**Volumenausdehnung**	wafer probe	INSTR	**Wafer-Testsonde**
volumetric measure		PHYS	**Raummaß**	wafer scale integration (→ultra-large-scale integration)	MICROEL	(→Ultrahöchstintegration)
= measure of volume; measure of capacity			= Hohlmaß; Volumenmaß			
				wafer sort (→wafer test)	MICROEL	(→Scheibentest)
volumetric variation		PHYS	**Volumeninhalt**	wafer test	MICROEL	**Scheibentest**
volume-unit meter (→ VU meter)		EL.ACOUS	(→VU-Meter)	= wafer sort		≈ Wafertest
				wage	ECON	**Lohn**
voluminous (adj.)		TECH	**voluminös**	[pay for work done]		≈ Gehalt; Bezahlung
↑ large			↑ groß	≈ pay		↓ Stundenlohn
voluntary		COLLOQ	**freiwillig**	≈ salary; payment		
= self-imposed; spontaneous; free 2 (adj.)				↓ hourly wage		
				wage and salary earner (→employed)	ECON	(→Arbeitnehmer)
VOR (→ VHF omnidirectional range)		RADIO NAV	(→UKW-Drehfunkfeuer)			
				wage-earner	ECON	**Lohnempfänger**
vortex (→curl)		PHYS	(→Wirbel)	waggon (BRI) (→wagon)	ECON	(→Waggon)
vote		ECON	**Stimme**	**Wagner ground**	INSTR	**Wagnerscher Hilfszweig**
[election]			[Wahl]	wagon (NAM)	ECON	**Waggon**
voting		ECON	**stimmberechtigt**	= waggon (BRI); rail car		
voucher 1		ECON	**Beleg**	wait call	DATA PROC	**Warteaufruf**
[paper serving as evidence]			= Abrechnungsbeleg; Zahlungsbeleg	≈ wait instruction		≈ Wartebefehl
≈ document				wait condition	DATA PROC	(→Wartestellung)
			≈ Dokument	(→disconnected mode)		
voucher 2 (→receipt)		ECON	**Quittung**	wait condition (→waiting condition)	TECH	(→Wartezustand)
voucher (→document)		TERM&PER	(→Beleg)			
voucher copy		DOC	**Belegexemplar**	waiting (→queing)	SWITCH	(→Warten)
voucher feed (→document feed)		TERM&PER	(→Belegzufuhr)	waiting condition	TECH	**Wartezustand**
				= wait condition		
voucher feeder (→document feed)		TERM&PER	(→Belegzufuhr)	waiting delay (→call delay)	SWITCH	(→Wartedauer)
				waiting list	ECON	**Warteliste**
vowel		LING	**Vokal**			= Vormerkliste
↑ phoneme			= Selbstlaut	waiting period (→waiting time)	TECH	(→Wartezeit)
			↑ Phonem	waiting program	DATA PROC	**Warteprogramm**
VRAM (→refresh memory)		DATA PROC	(→Bildwiederholspeicher)			= wartendes Programm
VRC (→vector-to-raster converter)		TERM&PER	(→Vektor-Raster-Umsetzer)	waiting queue	SWITCH	**Warteschlange**
				= queue		= Schlange; Eintrittswarteschlange
V reflector (→corner reflector)		ANT	(→Winkelreflektor)			
VS1AA antenna		ANT	**Mehrband-Windom-Antenne**	waiting time	TECH	**Wartezeit**
VSAT		SAT.COMM	**VSAT**	= waiting period; wait time; wait period		
= very small aperture terminal						
VSB (→vestigial sideband)		MODUL	(→Restseitenband)	waiting traffic	SWITCH	**Wartebelastung**
VSWR		ANT	**Rückflußdämpfung**			= Warteverkehr
= voltage standing wave ratio			= VSWR	wait instruction	DATA PROC	**Wartebefehl**
VSWR (→voltage standing wave ratio)		LINE TH	(→Welligkeitsfaktor)	[to coordinate input/output operations with processing operations in the simultaneous operation]		[zur Koordinierung von Ein-/Ausgabe- mit Verarbeitungsvorgängen bei Simultanbetrieb]
VSWR meter (→SWR power meter)		INSTR	(→Stehwellenmeßgerät)			
				≈ wait call		≈ Warteaufruf
VSWR resistance bridge (→SWR power meter)		INSTR	(→Stehwellenmeßgerät)	wait loop	DATA PROC	**Warteschleife**
				≈ holding loop		≈ Halteschleife
VT (→vertical tabulator)		DATA PROC	(→Vertikaltabulator)	wait period (→waiting time)	TECH	(→Wartezeit)
				wait state (→disconnected mode)	DATA PROC	(→Wartestellung)
VTL logic		MICROEL	**VTL-Logik**			
= variable-threshold logic				wait time (→waiting time)	TECH	(→Wartezeit)
VTOL (→vertical take-of and landing)		AERON	(→Senkrechtstart und -landung)	waiver (→renunciation)	ECON	(→Verzicht)
				wake (n.)	PHYS	**Sog**
VTR (→video tape recording)		TV	(→Magnetband-Fernsehaufzeichnung)	≈ suction		= Luftsog
				walkie-talkie	RADIO	**Handfunksprechgerät**
VTR (→videotape recorder)		TV	(→Videorecorder)	= handheld radiotelephone; handheld transceiver; handi-talkie		= Handfunkgerät; Handsprechfunkgerät
vulnerability		TECH	**Anfälligkeit**			
vulnerable		TECH	**anfällig**	**Walking code**	CODING	**Walking Code**
VU meter		EL.ACOUS	**VU-Meter**	↑ two-out-of-five code		↑ Zwei-aus-Fünf-Code
= volume-unit meter			[Lautstärkemesser nach US-Norm]	walkman	CONS.EL	**Walkman**
				walkout (→strike)	ECON	(→Streik)
VVS (→voltage-controlled voltage source)		NETW.TH	(→spannungsgesteuerte Spannungsquelle)			

walk-through (v.t.)	TECH	**durchgehen**
		= analysieren
walkway	CIV	**Fußgängerübergang**
walkway	TECH	**Laufsteg**
wall	CIV.ENG	**Mauer**
wall	TECH	**Wand**
		= Wandung
wall bracket	TECH	**Wandarm**
wall break-through (→wall opening)	CIV.ENG	(→Wanddurchbruch)
wall cabinet (→wall-mounted cabinet)	EQUIP.ENG	(→Wandgehäuse)
wall current	MICROW	**Wandstrom**
wall fastening (→wall mounting)	EQUIP.ENG	(→Wandbefestigung)
wall feed-through (→wall opening)	CIV.ENG	(→Wanddurchbruch)
wall model	TERM&PER	**Wandgerät**
wall-mounted	EQUIP.ENG	**Wand-**
wall-mounted cabinet	EQUIP.ENG	**Wandgehäuse**
= wall cabinet; wall-mounted frame		= Wandrahmen; Wandschrank
wall-mounted frame (→wall-mounted cabinet)	EQUIP.ENG	(→Wandgehäuse)
wall mounting	SYS.INST	**Wandaufbau**
≠ floor mounting		≠ Reihenaufbau
wall mounting	EQUIP.ENG	**Wandbefestigung**
= wall fastening		
wall opening	CIV.ENG	**Wanddurchbruch**
= opening; wall break-through; break-through; wall feed-through		= Durchbruch; Wanddurchführung; Durchführung; Wandeinführung
wall telephone	TERM&PER	**Wandtelefon**
		= Wandfernsprecher
wall thickness	TECH	**Wandstärke**
Walsh function	MATH	**Walsh-Funktion**
↑ sequencial function		↑ Sequenzfunktion
WAN (→wide-area network)	DATA COMM	(→weiträumiges Netz)
wand (→code pen)	TERM&PER	(→Lesestift)
wander	ELECTRON	**Wander**
[very slow phase fluctuations]		[sehr langsame Phasenschwankungen]
≈ jitter		≈ Jitter
wanted band	TELEC	**Nutzband**
wanted emission	RADIO	**Nutzaussendung**
wanted signal	TELEC	**Nutzsignal**
= useful signal; desired signal		
wanted signal (→useful information)	CODING	(→Nutzinformation)
warble (→sweep)	ELECTRON	(→wobbeln)
warble (→sweeping)	ELECTRON	(→Wobbelung)
Ward-Leonard drive	POWER ENG	**Leonard-Satz**
[DC drive actuator]		[Stellglied für Gleichstromantrieb]
warehouse	ECON	**Lagerhaus**
warehousing (→stockkeeping)	ECON	(→Lagerhaltung)
warm (→heat)	PHYS	(→erwärmen)
warm boot (→warm start)	DATA PROC	(→Warmstart)
warmer (→heater)	TECH	(→Erhitzer)
warm start (n.)	TECH	**Warmstart**
warm start	DATA PROC	**Warmstart**
[continuation of a process interrupted by failure, from a restart point, pulling the test routines of a cold start]		[Fortsetzung eines durch Störung unterbrochenen Ablaufs, ab einem Wiederanlaufpunkt, ohne die bei Kaltstart üblichen Testroutinen]
= warm boot		
≠ cold start		≠ Kaltstart
warmth [TECH] (→heat)	PHYS	(→Wärme)
warm-up (v.t.)	TECH	**anwärmen**
= preheat (v.t.)		= vorwärmen; vorheizen
warm-up period (→heating time)	ELECTRON	(→Anheizzeit)
warm-up period (→warm-up time)	TECH	(→Aufwärmzeit)
warm-up time	TECH	**Aufwärmzeit**
= warm-up period		= Anwärmzeit; Warmlaufzeit; Einlaufzeit; Einlaufdauer
≈ preheating time		≈ Vorheizzeit

warning	COLLOQ	**Warnung**
= alert (n.)		≈ Alarm [TECH]
≈ alarm [TECH]		
warning beacon (→hazard bacon)	RADIO NAV	(→Warnungsbake)
warning message	DATA PROC	**Warnmeldung**
warning notice (→danger notice)	TECH	(→Warntafel)
warning paint	TECH	**Warnbemalung**
= alerting paint; obstruction paint; hazard paint		
warning signal	TECH	**Aufmerksamkeitszeichen**
warning system (→danger detection system)	SIGN.ENG	(→Gefahrenmeldeanlage)
warning tape	OUTS.PLANT	**Trassenband**
		= Warnband
warp (v.i.)	TECH	**verziehen**
≈ twist 1		= verwerfen 2; verwölben
		≈ verwinden4
warp (→shrink)	MECH	(→Verzug)
warped	TECH	**windschief**
= out of shape		= verzogen
≈ skew		≈ schief 1
warping	TECH	**Verwindung**
		= Verwerfung; Verwölbung
warrant (→guarantee)	ECON	(→garantieren)
warranted performance	TECH	**Garantiewert**
warranty (→guaranty 1)	ECON	(→Garantie)
washer (→plain washer)	MECH	(→Beilagscheibe)
washer (→disc)	MECH	(→Scheibe)
waste (v.t.)	ECON	**verschwenden**
		= vergeuden
waste (n.)	ECON	**Verschwendung**
		= Vergeudung
waste-basket	DATA PROC	**Papierkorb**
[pictograph]		[Piktogramm]
waste disposal (→disposal)	TECH	(→Entsorgung)
waste disposal facility	TECH	**Entsorgungsanlage**
waste gas	TECH	**Abgas**
↓ flue gas		↓ Rauchgas
waste-heat	TECH	**Abwärme**
≈ dissipated heat		≈ Verlustwärme
wastepaper	TYPOGR	**Makulatur**
[printed sheet not fit for use]		[beim Druck schlecht geratener Bogen]
waste-paper basket	OFFICE	**Papierkorb**
waste press	OFFICE	**Abfallpresse**
waster	MANUF	**Ausschußteil**
waste-water	TECH	**Abwasser**
= sewage		
wasting	TYPOGR	**Makulierung**
		= Einstampfung
watch-dog (→time-out circuit)	CIRC.ENG	(→Zeitüberwachung)
watchdog module (→supervising module)	EQUIP.ENG	(→Überwachungseinschub)
watchdog unit (→supervising unit)	EQUIP.ENG	(→Überwachungseinheit)
water (v.t.)	TECH	**verwässern**
↑ dilute		↑ verdünnen
water	CHEM	**Wasser**
water blocking	TECH	**Nässeschutz**
≈ humidity barrier		≈ Feuchteschutz
water consumption	TECH	**Wasserverbrauch**
water cooling	TECH	**Wasserkühlung**
water disposal	CIV.ENG	**Wasserentsorgung**
water ingress	TECH	**Wassereindringung**
water management	CIV.ENG	**Wasserwirtschaft**
waterproof	TECH	**wasserdicht**
= water-tight		= wasserfest
waterproof loudspeaker	EL.ACOUS	**wasserfester Lautsprecher**
water-proofness	TECH	**Wasserdichtigkeit**
= water-tightness		
water-soluble	CHEM	**wasserlöslich**
↑ soluble		↑ löslich
water supply	CIV.ENG	**Wasserversorgung**
water table depth	CIV.ENG	**Grundwasserspiegel**
water-tight (→waterproof)	TECH	(→wasserdicht)
water-tightness (→water-proofness)	TECH	(→Wasserdichtigkeit)

water vapor (AM) = water vapour (BRI)	PHYS	Wasserdampf
water vapour (BRI) (→water vapor)	PHYS	(→Wasserdampf)
WATS [an arrangement allowing unlimited number of long distance calls within a prescribed area, by a global monthly fee] = wide area telephone service	TELEC	Ferngesprächspauschale [in USA eingeführte Verrechnungsmodalität, die gegen eine Monatspauschale eine unbegrenzte Anzahl von Ferngesprächen innerhalb eines bestimmten Bereiches erlaubt]
Watson-Watt direction finder = dual-channel cathode ray direction finder	RADIO LOC	Watson-Watt-Sichtfunkpeiler = Wattson-Watt-Peiler; Zweikanalpeiler
watt [SI unit for power, energy flow, heat flow; = 1J/s] = W ≈ Joule-second; volt-ampere [EL.TECH]; Var [EL.TECH]	PHYS	Watt [SI-Einheit für Leistung, Energiestrom, Wärmestrom; = 1 J/s] = W ≈ Joule-Sekunde; Volt-Ampere [EL.TECH]; Volt-Ampere reaktiv [EL.TECH]
wattage (→electric power)	EL.TECH	(→elektrische Leistung)
wattage (→power input)	EL.TECH	(→Leistungsaufnahme)
wattage rating (→nominal power)	TECH	(→Nennleistung)
wattful loss (→ohmic loss)	EL.TECH	(→ohmscher Verlust)
watthour meter [integrating measurement of active, reactive or apparent power]	INSTR	Wechselstromzähler [intergrierende Messung von Wirk-, Blind- oder Scheinleistung] = Wechselstrom-Induktionszähler
wattmeter (→power meter)	INSTR	(→Leistungsmesser)
wave [a periodic process in the time and space domain] ≈ oscillation	PHYS	Welle [zeitlich und räumlich periodischer Vorgang] ≈ Schwingung
wave amplitude	PHYS	Wellenamplitude
wave analysis (→frequency analysis)	TELEC	(→Frequenzanalyse)
wave analyzer (→selective voltmeter)	INSTR	(→Selektivspannungsmesser)
wave angle (→angle of incidence)	PHYS	(→Einfallswinkel)
wave antenna 1 [horizontal receiving antenna, several wavelengths long] ↓ Beverage antenna	ANT	Wellenantenne [horizontale Empfangsantenne, mehrere Wellenlängen lang, Richtwirkung in Längsrichtung] ↓ Beverage-Antenne
wave antenna 2 (→Beverage antenna)	ANT	(→Beverage-Antenne)
waveband (→frequency band)	RADIO	(→Frequenzband)
wave band (→wavelength range)	PHYS	(→Wellenlängenbereich)
waveband selection (→wave selection)	RADIO	(→Wellenbereichswahl)
waveband switch (→wave-change switch)	RADIO	(→Wellenbereichsschalter)
wave-carrier modulation ≠ pulse modulation ↑ modulation	MODUL	Schwingungsmodulation = Sinusmodulation ≠ Pulsmodulation ↑ Modulation
wave change (→wave selection)	RADIO	(→Wellenbereichswahl)
wave-change switch = waveband switch; band-change switch	RADIO	Wellenbereichsschalter
wave crest = wave top ≠ wave trough	PHYS	Wellenberg = Wellenscheitel ≠ Wellental
wave equation	PHYS	Wellengleichung
wave filter	NETW.TH	Wellenfilter
waveform = wave form; waveshape; wave shape; oscillation mode; mode ≈ signal form [ELECTRON]		Schwingungsform = Schwingungsverlauf; Wellenform; Wellenverlauf; Verlauf ≈ Signalform [ELECTRON]
wave form (→waveform)	PHYS	(→Schwingungsform)
waveform (→signal form)	ELECTRON	(→Signalform)
waveform analyzer	INSTR	Signalformanalysator = Wellenformanalysator
waveform generator	INSTR	Signalformgenerator = Funktionsgenerator
waveform math (→waveform mathematics)	INSTR	(→Signalform-Mathematik)
waveform mathematics = waveform math	INSTR	Signalform-Mathematik
wave-form monitor	TV	Waveform-Monitor
waveform reconstruction = signal reconstruction	ELECTRON	Signalrekonstruktion
wave front [equal-phase surface] = wave surface	PHYS	Wellenfläche [Fläche gleicher Phase] = Wellenfront
wave function	PHYS	Wellenfunktion
wave group	PHYS	Wellengruppe
waveguide = W/G	MICROW	Hohlleiter = Wellenleiter [PHYS]; Leiter
waveguide antenna (→waveguide radiator)	ANT	(→Hohlleiterstrahler)
waveguide attenuator	MICROW	Hohlleiter-Dämpfungsglied = Hohlleiterabschwächer
waveguide band-rejection filter	MICROW	Hohlleitersperre
waveguide bend = waveguide elbow	MICROW	Hohlleiterkrümmung
waveguide bridge	ANT	Hohlleiterbrücke
waveguide circulator	MICROW	Hohlleiterzirkulator
waveguide component	MICROW	Hohlleiterelement
waveguide cutoff (→cutoff frequency)	MICROW	(→Grenzfrequenz)
waveguide directional coupler	MICROW	Hohlleiter-Richtkoppler
waveguide dispersion	OPT.COMM	Wellenlängendispersion
waveguide elbow (→waveguide bend)	MICROW	(→Hohlleiterkrümmung)
waveguide filter (→mode filter)	MICROW	(→Modenfilter)
waveguide flange	MICROW	Hohlleiterflansch
waveguide junction (→waveguide transition)	MICROW	(→Hohlleiterübergang)
waveguide lens	MICROW	Hohlleiterlinse
waveguide mode = propagation mode; mode	MICROW	Wellentyp = Schwingungsart; Feldtyp; Hohlleitermodus; Modus; Mode
waveguide phase shifter	MICROW	Hohlleiter-Phasenschieber
waveguide probe	MICROW	Hohlleitersonde
waveguide radiator = waveguide antenna ↓ horn radiator	ANT	Hohlleiterstrahler ↓ Hornstrahler
waveguide resonator	MICROW	Hohlleiterresonator = Hohlleiter-Resonanzkreis
waveguide run	RADIO REL	Hohlleiterzug
waveguide short	MICROW	Hohlleiterkurzschluß
waveguide shutter	MICROW	Hohlleiterblende
waveguide stub	MICROW	Hohlleiter-Stichleitung
waveguide switch	MICROW	Hohlleiterschalter = Wellenleiterschalter
waveguide system	ANT	Hohlleitersystem
waveguide transformer	MICROW	Hohlleitertransformator = Wellenleitertransformator
waveguide transition = waveguide junction	MICROW	Hohlleiterübergang = Übergang
waveguide tuner	MICROW	einstellbarer Hohlleitertransformator
waveguide window	MICROW	Hohlleiterfenster = Hohlleiterdurchführung
wavelength [spacial period of a wave] = wave length	PHYS	Wellenlänge [räumliche Periode einer Welle]
wave length (→wavelength)	PHYS	(→Wellenlänge)
wavelength band (→frequency band)	RADIO	(→Frequenzband)
wavelength dependency	PHYS	Wellenlängenabhängigkeit
wavelength division multiplex = WDM	OPT.COMM	Wellenlängenmultiplex = WDM
wavelength meter = wavemeter	INSTR	Wellenlängenmesser

wavelength range PHYS **Wellenlängenbereich**
= wave band = Wellenbereich; Wellenband
wavelenth tracker INSTR **Wellenlängennormal**
wave mechanics PHYS **Wellenmechanik**
= undulatory mechanics; wave physics; = Wellenphysik
undulatory physics ≈ Wellentheorie
≈ wave theory
wavemeter (→wavelength INSTR (→Wellenlängenmesser)
meter)
wave-mode conversion MICROW **Modenwandlung**
= wave-mode transformation = Wellenumwandlung
wave-mode transformation MICROW (→Modenwandlung)
(→wave-mode conversion)
wave number PHYS **Wellenzahl**
wave optics PHYS **Wellenoptik**
= undulatory optics
wave parameter (→alternating EL.TECH (→Wechselgröße)
value)
wave physics (→wave mechanics) PHYS (→Wellenmechanik)
wave radiation PHYS **Wellenstrahlung**
wave range (→frequency band) RADIO (→Frequenzband)
wave resistance NETW.TH (→Wellenwiderstand)
(→characteristic impedance)
waver neon lamp TECH **Flackerkerzen-Glühlampe**
wave rotation circulator MICROW (→Faraday-Zirkulator)
(→rotation circulator)
wave rotation isolator MICROW **Faraday-Richtungsleitung**
wave rotator (→phase rotator) MICROW (→Phasendreher)
wave selection RADIO **Wellenbereichswahl**
= waveband selection; band selection;
wave change; band change
wave shape (→waveform) PHYS (→Schwingungsform)
waveshape (→waveform) PHYS (→Schwingungsform)
waveshape (→signal form) ELECTRON (→Signalform)
wave soldering (→flow MANUF (→Schwallbadlötung)
soldering)
wave-soldering bath MANUF (→Schwallbad)
(→flow-soldering bath)
wave surface (→wave front) PHYS (→Wellenfläche)
wave synthesis (→frequency TELEC (→Frequenzsynthese)
synthesis)
wave theory PHYS **Wellentheorie**
= undulatory theory = Undulationstheorie
≈ wave mechanics ≈ Wellenmechanik
wave top (→wave crest) PHYS (→Wellenberg)
wave train PHYS **Wellenzug**
wave trap ANT **Wellenfalle**
= trap
wave trap (→parallel-resonant NETW.TH (→Parallelschwingkreis)
circuit)
wave trough PHYS **Wellental**
≠ wave crest ≠ Wellenberg
waviness MECH **Welligkeit**
[surface quality] [Oberflächengüte]
waviness height MECH **Welligkeitshöhe**
[surface quality] [Oberflächengüte]
waviness width MECH **Wellenabstand**
[surface quality] [Oberflächengüte]
wavy TECH **wellig**
= undulating
waw (n.) EL.ACOUS **Jaulen**
[fast pitch fluctuations] [schnelle Schwankungen
der Tonhöhe]
waxed paper TECH **Wachspapier**
waxed paper recording INSTR **Wachspapierschrift**
waybill (→shipping ECON (→Versandpapiere)
documents)
wayleave (→right of OUTS.PLANT (→Wegerecht)
way)
3-way reflex EL.ACOUS **3-Wege-Baßreflex**
= three-way reflex = Drei-Wege-Baßreflex
way-side derivation TRANS (→Unterwegsabzweig)
(→intermediate derivation)
way-side traffic TRANS **Unterwegsverkehr**
= add-drop traffic = Abzweigverkehr
Wb (→Weber) PHYS (→Weber)
WDM (→wavelength division OPT.COMM (→Wellenlängenmultiplex)
multiplex)
weak (adj.) TECH **schwach**
≈ fragile ≈ zerbrechlich

weak current (→low EL.TECH (→Schwachstrom)
current)
weak-current enginnering EL.TECH (→Schwachstromtechnik)
(→low-current engineering)
weaken TECH **schwächen**
weakness (→defect) QUAL (→Mangel)
weak point TECH **Schwachstelle**
wealth ECON **Vermögen**
≈ property ≈ Besitz
wealth of information COLLOQ **Informationsfülle**
wear (n.) TECH **Abnützung**
= wearout = Verschleiß
≈ fatigue ≈ Ermüdung
wear (→fatigue) QUAL (→Ermüdung)
wear and tear ECON **Wertminderung**
[depreciation by normal use] [durch normalen Gebrauch
= depreciation verursachte Minderung]
= Wertverzehr
wear down (→abrade) TECH (→abschleifen)
wearfree TECH **verschleißfrei**
wearout (→wear) TECH (→Abnützung)
wearout failure (→ageing QUAL (→Verschleißausfall)
failure)
wear-out period QUAL **Verschleißphase**
= Alterungsphase; Ermü-
dungsphase
wear resistant TECH **verschleißfest**
weather METEOR **Wetter**
[atmospheric conditions at a given [atmosphärische Bedin-
place and time] gung an einem bestimmten
≈ weather conditions; climate Ort und Zeitpunkt]
≈ Witterung; Klima
weatherbeaten (→weatherworn) TECH (→verwittert)
weather conditions METEOR **Witterung**
[weather conditions during a deter- [Wetterbedingungen wäh-
mined time span] rend eines bestimmten
≈ weather; climate Zeitraums]
= Wetterverhältnisse; Wetter-
bedingungen; Witterungs-
bedingungen;
Witterungsverhältnisse
≈ Wetter; Klima
weather-induced TECH **witterungsbedingt**
weatherproof TECH (→wetterfest)
(→weather-resistant)
weatherproof bell (→outdoor TERM&PER (→Außenwecker)
bell)
weatherproof enclosure EQUIP.ENG (→Wettergehäuse)
(→weatherproof housing)
weatherproof housing EQUIP.ENG **Wettergehäuse**
= weatherproof enclosure; outdoor = Wetterschutzgehäuse; Wet-
housing; outdoor enclosure; outdoor terschutzschrank; Freilift-
cabinet; outdoor case gehäuse; Freiluftschrank;
Freiraumgehäuse; Frei-
raumschrank
weather-protected (adj.) TECH **wettergeschützt**
= weathertight; outdoor- (adj.) = Freiluft-
≈ wheatherproof ≈ wetterfest
weather radar RADIO LOC (→Wetterradar)
(→meteorological radar)
weather-resistant TECH **wetterfest**
= weatherproof; wheatherproofed = wetterbeständig; witte-
(BRI) rungsbeständig
≈ wheather-protected ≈ wettergeschützt
weather station METEOR **Wetterstation**
weathertight TECH (→wettergeschützt)
(→weather-protected)
weatherworn TECH **verwittert**
= weatherbeaten
Weber PHYS **Weber**
[SI unit for magnetic flux; = 1 V s] [SI-Einheit für magneti-
= Wb schen Fluß; = 1 V s]
= Wb
wedge (v.t.) MECH **einkeilen**
wedge (n.) MECH **Keil**
= spline ↓ Wellenkeil; Querkeil
wedge bonding MICROEL **Keilbondierung**
= Keilschweißen; Schneiden-
kontaktierung
wedge shaped MECH **keilförmig**

weed (v.t.)	DATA PROC	ausmisten	
week	COLLOQ	Woche	
weekday (→working day)	COLLOQ	(→Werktag)	
weekly paper	COLLOQ	Wochenzeitung	
↑ newspaper		↑ Zeitung	
Wehnelt cylinder	ELECTRON	Wehneltzylinder	
= control cylinder			
Weibull distribution	MATH	Weibull-Verteilung	
weigh (fig.) (→trade-off)	COLLOQ	(→abwägen)	
weight	PHYS	Gewicht	
[force acting on masses by gravity; units: dyn, pond, kilogramm-force]		[von der Erdanziehung auf Massen wirkende Kraft; Maßeinheiten: Dyn, Pond, Kilogramm-Kraft]	
= force due to gravity; G		= Gewichtskraft; G	
≈ gravitational force; avoirdupois		≈ Gravitationskraft	
↑ force		↑ Kraft	
weight (→significance)	INF	(→Wertigkeit)	
weight (→weighting)	MATH	(→Wichtung)	
weighted (v.t.)	MATH	bewerten	
		= wichten; wiegen	
weighted	MATH	bewertet	
		= gewichtet; gewogen	
weighted average	MATH	gewogenes Mittel	
		= bewertetes Mittel; gewichtetes Mittel	
weighted code	CODING	gewichteter Code	
weighted noise	TELEC	Geräusch	
= psophometric noise		[psofometrisch bewertetes Rauschen]	
↑ noise		↑ Rauschen	
weighted noise voltage	TELEC	(→Geräuschspannung)	
(→psophometric noise voltage)			
weighted scoring	SCIE	Stufentest	
weighting	TELEC	Bewertung	
weighting	MATH	Wichtung	
= weight		= Gewicht; Wertigkeit; Bewertung	
weighting factor	TECH	Bewertungsziffer	
weighting filter	TELEC	Bewerungsfilter	
Weiss domain	PHYS	Weisscher Bezirk	
Weiss field	PHYS	Weissches Feld	
= molecular field			
welcome (n.)	ECON	Grußadresse	
		= Begrüßung	
weld (v.t.)	METAL	anschweißen	
weld (v.t.)	METAL	schweißen	
[union of metalic pieces by fusing contacting parts and possibly adding some material of similar composition]		[Verbindung durch Aufschmelzen der zu verbindenden Metallteile; eventuell zugefügtes Material ist von ähnlicher Zusammensetzung]	
≈ solder		≈ löten	
weld (n.)	METAL	Schweißnaht	
welded	METAL	geschweißt	
welded joint	METAL	Schweißverbindung	
welded part	MECH	Schweißteil	
welder	TECH	Schweißer	
welding	METAL	Schweißung	
≈ soldering		= Schweißen	
		= Löten	
welding appliance	METAL	Schweißvorrichtung	
= welding tool			
welding drawing	ENG.DRAW	Schweißzeichnung	
welding electrode	METAL	Schweißelektrode	
welding flux	METAL	Flußmittel 2	
= fluxing; flux 2		[Schweißen]	
welding practice (→welding process)	METAL	(→Schweißtechnik)	
welding process	METAL	Schweißtechnik	
= welding practice			
welding rod	METAL	Schweißdraht	
= welding wire			
welding symbol	ENG.DRAW	Schweißzeichen	
welding tool (→welding appliance)	METAL	(→Schweißvorrichtung)	
welding torch	METAL	Schweißbrenner	
= torch			
welding wire (→welding rod)	METAL	(→Schweißdraht)	
well (n.)	MICROEL	Wanne	
= tube			
well-known	COLLOQ	namhaft	
= reputable			
Western plug	TELEC	ISDN-Stecker	
= ISDN plug		= S0-Stecker	
Western Union joint (→twisted joint)	OUTS.PLANT	(→Würgerverbindung)	
Weston cell	POWER SYS	Weston-Element	
= Weston standard cell		= Weston-Normalelement	
Weston standard cell (→Weston cell)	POWER SYS	(→Weston-Element)	
wet (v.t.)	TECH	betzen	
		[mit Flüssigkeit]	
wet (v.t.)	COMPON	fritten	
wet (adj.)	PHYS	naß	
≈ humid		≈ feucht	
wet grinding	MECH	Naßschleifen	
wetted contact 1	COMPON	gefritteter Kontakt	
[with superimposed dc]		[mit Gleichstromüberlagerung]	
wetted contact 2 (→mercury contact)	COMPON	(→Quecksilberkontakt)	
wetting	TECH	Benetzung	
		[mit Flüssigkeit]	
wetting 1	COMPON	Frittung	
[superposition of direct current]		[Unterlagerung mit Gleichstrom]	
= fritting			
≈ dc underlay [ELECTRON]		≈ Gleichstromunterlegung [ELECTRON]	
wetting resistance	COMPON	Frittwiderstand	
= fritting resistance			
wetting temperature	METAL	Benetzungstemperatur	
[soldering]		[Lötung]	
wetting voltage	COMPON	Frittspannung	
= fritting voltage			
wetzel (n.)	TERM&PER	Retuschierungselement	
[picture element to improve sharpness of display]			
W/G (→waveguide)	MICROW	(→Hohlleiter)	
wheatherproofed (BRI) (→weather-resistant)	TECH	(→wetterfest)	
wheather protection	TECH	Witterungsschutz	
		= Wetterschutz	
Wheatstone bridge	INSTR	Wheatstonebrücke	
= Wheatstone measuring bridge; resistance bridge		= Wheatstone-Meßbrücke; Wheatstonesche Brücke	
Wheatstone measuring bridge (→Wheatstone bridge)	INSTR	(→Wheatstonebrücke)	
wheel	TECH	Rad	
wheel hub	MECH	Radnabe	
= nave		↑ Nabe	
↑ hub			
wheel printer (→typew-heel printer)	TERM&PER	(→Typenraddrucker)	
wheel rim	MECH	Radkranz	
wheel spoke	MECH	Radspeiche	
= spoke (n.)		= Speiche	
when using	COLLOQ	bei Einsatz von	
where-used list	DATA PROC	Verwendungsliste	
whichever is greater	MATH	größer der beiden	
whip antenna	ANT	Peitschenantenne	
↑ monopole antenna		↑ Monopol	
whir (v.i.)	COLLOQ	schwirren	
= whirr			
whirr (→whir)	COLLOQ	(→schwirren)	
white	PHYS	weiß	
white board (→magnetic board)	OFFICE	(→Magnettafel)	
white box	TECH	weißer Kasten	
[unit with known functioning]		[Einheit durchschaubarer Funktionsweise]	
≠ black box		≠ schwarzer Kasten	
white-hot	TECH	weißglühend	
white level (→reference white level)	TV	(→Weißpegel)	
white metal	METAL	Weißmetall	
white noise	TELEC	weißes Rauschen	
[equal energy per unit bandwidth]		[konstantes und frequenzunabhängiges Leistungsspektrum]	
≈ thermal noise			
		≈ thermisches Rauschen	

white peak	TV	**Weißspitze**
white plate (→tin plate)	METAL	(→Weißblech)
white printer	TERM&PER	**Weißdrucker**
= white writer		↑ Laserdrucker
↑ laser printer		
white signal (→reference white level)	TV	(→Weißpegel)
white wallboard	OFFICE	**Weißwandtafel**
white writer (→white printer)	TERM&PER	(→Weißdrucker)
white-write technique	TERM&PER	**White-write-Verfahren**
whiz (pl. whizzes)	ACOUS	**Säuseln**
= whizz		
whizz (→whiz)	ACOUS	(→Säuseln)
who are you ? (→identification please)	TELEGR	(→wer da ?)
WHO-ARE-YOU key	TELEGR	**WER-DA-Taste**
whole number (→integer)	MATH	(→Ganzzahl)
"who's that speaking please ?"	TELEPH	**„wer spricht bitte ?"**
wide (→extended)	TYPOGR	(→breit)
wide-angle-cone antenna	ANT	**Weitwinkelkonusantenne**
wide-angle lens	PHYS	**Weitwinkelobjektiv**
= wide lens		
wide-aperture direction finder	RADIO LOC	**Großbasispeiler**
		≈ **Wullenweverantenne**
≈ large base direction finder		
≈ wullenwever antenna		
wide-aperture Doppler direction finder	RADIO LOC	**Großbasis-Dopplerpeiler**
wide-aperture principle	RADIO LOC	**Großbasisverfahren**
= large-base principle		
wide-area network	DATA COMM	**weiträumiges Netz**
[data communication network based on limited long rage transmission capacity]		[auf die beschränkten Übertragungskapazitäten des Weitverkehrs abgestimmtes Datenübertragungsnetz]
= WAN		= WAN
wide area telephone service (→WATS)	TELEC	(→Ferngesprächspauschale)
wideband (n.) (→broadband)	TELEC	(→Breitband)
wideband adj. (→broadband)	TELEC	(→breitbandig)
wideband aerial (→broadband antenna)	ANT	(→Breitbandantenne)
wideband amplifier	CIRC.ENG	(→Breitbandverstärker)
(→broadband amplifier)		
wideband antenna (→broadband antenna)	ANT	(→Breitbandantenne)
wideband aperture antenna (→broadband aperture antenna)	ANT	(→Breitband-Flächenantenne)
wideband balun	ANT	**Breitband-Balun**
wideband cable network	TELEC	(→Breitband-Kabelnetz)
(→broadband cable network)		
wideband cage antenna (→broadband cage antenna)	ANT	(→Breitbandreuse)
wideband channel (→broadband channel)	TELEC	(→Breitbandkanal)
wideband communication (→broadband communication)	TELEC	(→Breitbandkommunikation)
wideband compensation (→broadband compensation)	ANT	(→Breitbandkompensation)
wideband conical monopole (→broadband conical monopole)	ANT	(→Breitband-Kegelantenne)
wideband corner reflector antenna	ANT	**Breitband-Winkelreflektorantenne**
wideband dipole (→broadband dipole)	ANT	(→Breitband-Dipol)
wideband feeding (→broadband feeding)	ANT	(→Breitbandspeisung)
wideband filter (→broadband filter)	NETW.TH	(→Breitbandfilter)
wideband ISDN (→broadband ISDN)	TELEC	(→Breitband-ISDN)
wideband link (→high capacity link)	TRANS	(→Breitbandstrecke)
wideband loudspeaker (→broadband loudspeaker)	EL.ACOUS	(→Breitbandlautsprecher)
wideband measuring (→broadband measuring)	INSTR	(→Breitbandmessung)
wideband modulation (→broadband modulation)	MODUL	(→Breitbandmodulation)
wideband network (→broadband network)	TELEC	(→Breitbandnetz)
wideband noise	TELEC	**Breitbandrauschen**
= broadband noise		
wideband polarization filter	RADIO REL	(→Breitband-Polarisationsweiche)
(→broadband polarization diplexer)		
wideband rhombic antenna	ANT	(→Breitband-Rhombusantenne)
(→broadband rhombic antenna)		
wideband route (→high capacity link)	TRANS	(→Breitbandstrecke)
wideband service (→broadband service)	TELEC	(→Breitbanddienst)
wideband sweep (→broadband sweep)	INSTR	(→Breitbandwobbeln)
wideband system (→high-capacity system)	TRANS	(→Breitbandsystem)
wide-bandwidth (→broadband)	TELEC	(→breitbandig)
wide carriage	OFFICE	**Breitwagen**
wide-coverage panoramic radar	RADIO LOC	**Großrundsichtradar**
wide-flange (adj.)	MECH	**breitflanschig**
wide font (→expanded font)	TYPOGR	(→Breitschrift)
wide lens (→wide-angle lens)	PHYS	(→Weitwinkelobjektiv)
wide lettering (→expanded font)	TYPOGR	(→Breitschrift)
widen	TECH	**verbreitern**
= broaden		
widening	TECH	**Verbreiterung**
= broadening		
wide punched tape	TERM&PER	**Breitlochstreifen**
wide-ranging	COLLOQ	(→weitreichend)
(→far-reaching)		
wide selection	COLLOQ	**reichhaltige Auswahl**
widow	TYPOGR	**Überhangzeile**
[last line of a chapter passing to next page]		[letzte Zeile eines Abschnitts, die auf die nächste Seite oder Spalte fällt]
≈ orphan		= Hurenkind; Witwe
		≈ Schusterjunge
width	PHYS	**Breite**
≈ lateral dimension		= Weite
↑ dimension		= Querabmessung
		↑ Dimension
width (→character width)	TYPOGR	(→Dickte)
Wien bridge	INSTR	**Wien-Brücke**
Wien-Robinson bridge	INSTR	**Wien-Robinson-Brücke**
Wien-Robinson oscillator	CIRC.ENG	**Wien-Robinson-Oszillator**
↑ RC oscillator		= Wien-Robinson-Brückenoszillator
		↑ RC-Oszillator
Wien's displacement law	PHYS	**Wiensches Verschiebungsgesetz**
[black body]		[schwarzer Strahler]
wild card	DATA PROC	**Stellvertreterzeichen**
[a symbol, like in MS-DOS, substituting any character in a file name]		[Symbol, z.B. in MS-DOS, welches beliebige Zeichen in einem Dateinamen ersetzt]
= global character		= Ersatzzeichen; Joker-Zeichen; Joker
Wilson cloud chamber	PHYS	**Wilsonsche Nebelkammer**
winch	TECH	**Winde**
≈ windlass		= Zugwinde; Seilwinde; Haspel
≈ reel		≈ Haspel
winch cable	OUTS.PLANT	**Windenseil**
= draw cable; traction rope; pulling-in line		= Zugseil
Winchester disk	DATA PROC	**Winchesterplatte**
= lubricated magnetic disk		[Magnetplatte mit Gleitschicht]
↑ fixed disk		↑ Festplatte
Winchester disk drive 1	TERM&PER	**Winchester-Laufwerk**
= Winchester drive		
Winchester disk drive 2	TERM&PER	(→Winchesterplattenspeicher)
(→Winchester disk memory)		
Winchester disk memory	TERM&PER	**Winchesterplattenspeicher**
= Winchester file; Winchester disk drive 2		= Winchesterspeicher
↑ fixed disk memory		↑ Magnetplattenspeicher

Winchester drive	TERM&PER	(→Winchester-Laufwerk)	**wiper** (→wiping contact)	COMPON	(→Wischkontakt)
(→Winchester disk drive 1)			**wiper switch**	COMPON	**Schiebeschalter**
Winchester file (→Winchester	TERM&PER	(→Winchesterplattenspeicher)	= slide switch		
disk memory)			**wiping contact**	COMPON	**Wischkontakt**
wind	METEOR	**Wind**	= wiper; slip contact; passing contact; momentary contact		= Schleifkontakt
wind (→wrap)	TECH	(→wickeln)			
wind energy	POWER SYS	**Windenergie**	**wiping relay**	COMPON	**Wischrelais**
winder	TECH	**Aufwickelvorrichtung**	[with short contact times only]		[mit nur kurzen Kontaktschließungen]
= winding device					
windflaw	METEOR	**Windstoß**	**wiping wire** (→slide wire)	EL.TECH	(→Schleifdraht)
= gust; wind rush; blast; flaw		≈ Windsturm	**wire 1** (v.t.)	EQUIP.ENG	**verdrahten**
≈ wind storm			≈ cable 1 (v.t.); connect		≈ verkabeln; anschließen
winding	TECH	**Wicklung**	**wire**	COMM.CABLE	**Ader**
= wrapping		= Umwicklung; Bewicklung	= lead; conductor		= Einzelader; Kabelader
winding	EL.TECH	**Wicklung**	**wire 2**	EQUIP.ENG	**beschalten**
= coil		↓ Spulenwicklung [COMPON]; Relaiswicklung [COMPON]	= cable 2		
≈ turn			**wire**	METAL	**Draht**
↓ coil winding [COMPON]; relay winding [COMPON]			**wire**	EL.TECH	**Draht**
			= conductor		= Leiter
winding (→sinuous)	TECH	(→geschlängelt)	**wire** (v.t.) (→telegraph)	TELEC	(→telegrafieren)
winding (→wrapping 1)	TECH	(→Wickel)	**wire** (n.) (→telegram)	POST	(→Telegramm)
winding capacitance	COMPON	**Wicklungskapazität**	**2-wire** (→two-wire)	TELEC	(→zweidrähtig)
[capacitance between turns]		[Kapazität zwischen Windungen]	**wire antenna**	ANT	**Drahtantenne**
= interwinding capacity			**wire-armored cable**	COMM.CABLE	**drahtbewehrtes Kabel**
winding device (→winder)	TECH	(→Aufwickelvorrichtung)	**wire assignment**	EQUIP.ENG	**Adernbelegung**
winding diagram	MANUF	**Wickelplan**	**wire bonding**	MICROEL	**Drahtkontaktierung**
winding factor	EL.TECH	**Wicklungsfaktor**			= Drahtanschluß
winding layer	COMPON	**Wickellage**	**wire-bound**	TELEC	**drahtgebunden**
winding machine	MANUF	**Wickelmaschine**	= wire-conducted		≠ drahtlos
= coil winder		≈ Wickelautomat	≠ wireless		↑ leitergebunden
≈ automatic winding machine			↑ conducted		
winding mandrel	ELECTRON	**Wickeldorn**	**wire-bound communication**	TELEC	**drahtgebundene Kommunikation**
winding shop	MANUF	**Wickelei**	= wire-conducted communication		
winding surface	PHYS	**Windungsfläche**	**wire breakage**	COMM.CABLE	**Drahtbruch**
winding wire (→wrap wire)	ELECTRON	(→Wickeldraht)	= broken wire		= Aderbruch
windlass (→winch)	TECH	(→Winde)	**wire broadcasting** (→wired broadcasting)	BROADC	(→Kabelrundfunk)
wind load	CIV.ENG	**Windlast**			
Windom antenna	ANT	**Windom-Antenne**	**wire brush**	TECH	**Drahtbürste**
= off-center fed antenna			**wire bundle**	COMM.CABLE	**Aderbündel**
window	DATA PROC	**Fenster**	= lead bundle; conductor bundle		= Adernbündel
= screen window; display window		= Bildschirmausschnitt; Ausschnitt; Sichtfenster; Anzeigefenster; Ausgabefenster; Bildfenster	**wire cable end** (→wire-end sleeve)	COMPON	(→Aderendhülse)
			wire-conducted (→wire-bound)	TELEC	(→drahtgebunden)
			wire-conducted communication (→wire-bound communication)	TELEC	(→drahtgebundene Kommunikation)
window (→radio window)	RADIO PROP	(→Radiofenster)			
window (→transmission window)	OPT.COMM	(→Übertragungsfenster)	**wire-conducted wave**	LINE TH	**Drahtwelle**
			wire-crimp connection (→crimp connection)	COMPON	(→Quetschverbindung)
window antenna	ANT	**Fensterantenne**	**wire cutter** (→side cutting pliers)	TECH	(→Seitenschneider)
= window-sill antenna			**wired**	EQUIP.ENG	**verdrahtet**
window comparator	CIRC.ENG	**Fensterdetektor**	≈ cabled		≈ verkabelt
		[Impulsabgabe wenn Eingangssignal im Toleranzbereich]	≠ unwired		≠ unverdrahtet
			wired (→hardwired)	DATA PROC	(→festverdrahtet)
		= Fensterdiskriminator	**wired broadcasting**	BROADC	**Kabelrundfunk**
window envelope	OFFICE	**Fensterumschlag**	= wire broadcasting; line broadcasting; wired radio		= Drahtfunk
windowing	DATA PROC	**Fenstertechnik**			↓ Kabelfernsehen; Kabelhörfunk
window monitoring system	SIGN.ENG	**Fensterüberwachungseinrichtung**	↓ cable TV; wired sound broadcasting		
window overlapping	DATA PROC	**Fensterüberlappung**	**wired-in** (→hardwired)	DATA PROC	(→festverdrahtet)
= overlapping			**wire dispenser** (→cable dispenser)	OUTS.PLANT	(→Abrollvorrichtung)
window-sill antenna (→window antenna)	ANT	(→Fensterantenne)	**wired logic**	DATA PROC	**verdrahtete Logik**
			wired OR	MICROEL	**verdrahtetes ODER**
window technique	TERM&PER	**Ausschnittseinblendung**			= Phantom-ODER
[on the screen]		[am Bildschirm]	**wired program** (→hardwired program)	DATA PROC	(→festverdrahtetes Programm)
wind pressure	TECH	**Winddruck**			
wind rush (→windflaw)	METEOR	(→Windstoß)	**wired program computer** (→hardwired-program computer)	DATA PROC	(→festprogrammierter Computer)
windscreen	EL.ACOUS	**Windschutzkorb**			
windshield mounting aerial	ANT	**Frontscheibenantenne**	**wired radio** (→wired broadcasting)	BROADC	(→Kabelrundfunk)
wind storm	METEOR	**Windsturm**			
≈ windflaw		≈ Windsturm	**wired sound broadcasting**	BROADC	**Kabelhörfunk**
wind tunnel	TECH	**Windkanal**	↑ wired broadcasting		↑ Kabelrundfunk
wind-up (v.t.)	MECH	**aufziehen**	**wire-end sleeve**	COMPON	**Aderendhülse**
[a spring]		[Feder]	= wire cable end		
wind-up (n.)	TERM&PER	**Aufzug**	**wire-end stripper**	EL.TECH	**Abisolierzange**
[of the dialing disk]		[Nummernscheibe]	= wire stripper; skinning tool		= Abisolierwerkzeug
wind-up (v.t.) (→wrap-up)	TECH	(→aufwickeln)	**wire frame**	DATA PROC	**Drahtmodell**
wing screw	MECH	**Flügelschraube**	[graphic with hidden lines]		[Graphik mit verdeckten Linien]
wink (→acknowlegment signal)	TELEC	(→Quittungszeichen)			
wink pulse (→acknowlegment signal)	TELEC	(→Quittungszeichen)			= Drahtgittermodell; Linienmodell; Kantenmodell

wire gauge	METAL	**Drahtmaß**
↓ AWG		= Drahtlehre; Aderdurchmesser; Aderndurchmesser; Aderdicke; Aderndicke
		↓ amerikanische Drahtlehre
wire-grid lens antenna	ANT	**Maschendraht-Linsenantenne**
wire guide	EQUIP.ENG	**Drahtführung**
wire junction	ANT	**Drahtverbindung**
wire lead (→lead)	COMPON	(→Anschlußdraht)
wireless	TELEC	**drahtlos**
= cordless		≠ drahtgebunden
≠ wire-bound		
wireless (BRI) (→radio)	TELEC	(→Funk)
wireless communications	TELEC	**drahtlose Kommunikation**
wireless loop	TELEC	**drahtlose Anschlußleitung**
wireless message 1 (→radio message 1)	TELEC	(→Funkspruch)
wireless message 2 (→radiogram)	TELEC	(→Funktelegramm)
wireless microphone	EL.ACOUS	**drahtloses Mikrophon**
wireless network	TELEC	(→Funknetz)
(→radio communications network)		
wireless phone call announcer	TERM&PER	**Anruf-Meldesystem**
wireless picture transmission (→videotelegraphy)	TELEC	(→Bildtelegrafie)
wireless plant (→radio installation)	RADIO	(→Funkanlage)
wireless telephone (→cordless telephone)	TELEPH	(→schnurloser Fernsprechapparat)
wire link	ELECTRON	**Drahtbrücke**
= jumper wire; wire strap		= Drahtbügel
≈ plug link; jumper		≈ Steckbrücke; Jumper
wire list	EQUIP.ENG	**Drahtliste**
= wiring list		= Drahtfarbliste; Drahtführungsliste; Verdrahtungsliste
wire list	COMM.CABLE	**Farbliste**
wire matrix printer (→stylus printer)	TERM&PER	(→Nadeldrucker)
wire matrix printing mechanism	TERM&PER	**Nadeldruckwerk**
		= Nadeldruckkopf
wire netting	TECH	**Drahtgeflecht**
		= Gewebedraht
wire pair	COMM.CABLE	**Aderpaar**
= pair; twin wire; strand 1; lead pair; conductor pair		= Adernpaar; Doppelader; Paar
↑ stranding element		↑ Verseilelement
wire printer (→stylus printer)	TERM&PER	(→Nadeldrucker)
wire resistor	COMPON	**Drahtwiderstand**
= wirewound resistor		
wire rod	METAL	**Walzdraht**
= rod		≈ Rundstahl
≈ round-bar iron		
wire rope	METAL	**Drahtseil**
wire scraper tool	ELECTRON	**Blankmacher**
		= Lackabkratzer
wire strap (→wire link)	ELECTRON	(→Drahtbrücke)
wire stripper (→wire-end stripper)	EL.TECH	(→Abisolierzange)
2-wire switching (→two-wire switching)	SWITCH	(→Zweidraht-Durchschaltung)
wire turn	COMPON	**Drahtwindung**
wirewound potentiometer	COMPON	**Drahtpotentiometer**
= variable wire-wound resistor		= Drahtdrehwiderstand
wirewound resistor (→wire resistor)	COMPON	(→Drahtwiderstand)
wirewound strain gauge	INSTR	**Drahtdehnmeßstreifen**
wire-wrap connection (→wrapped connection)	ELECTRON	(→Wickelverbindung)
wire-wrap technique	ELECTRON	**Drahtwickeltechnik**
		= Wire-wrap-Technik; Wrap-Technik
wiring	EQUIP.ENG	**Verdrahtung**
≈ cabling		≈ Verkabelung
wiring	MICROEL	**Verdrahtung**
= interconnect (n.); interconnection		
wiring board	EQUIP.ENG	**Verdrahtungsfeld**
= wiring matrix		= Verdrahtungsplatte
wiring channel	MICROEL	**Verdrahtungskanal**
= interconnect channel		
wiring diagram	EL.TECH	**Bauschaltplan**
[with physically correct positions]		[lagerichtige Darstellung einer Schaltung]
≈ assembly diagram; connection diagram; wiring scheme		= Montageschaltplan; Montageschaltbild; Montagestromlauf; MS; Verdrahtungsplan; Verdrahtungsunterlage; Verkabelungsplan
≈ circuit diagramm		≈ Stromlaufplan
wiring error	EQUIP.ENG	**Verdrahtungsfehler**
wiring frame	MANUF	**Verdrahtungsrahmen**
wiring layer	MICROEL	**Verdrahtungslage**
= interconnect layer		
wiring level	MICROEL	**Verdrahtungsebene**
= interconnect level		
wiring list (→wire list)	EQUIP.ENG	(→Drahtliste)
wiring matrix (→wiring board)	EQUIP.ENG	(→Verdrahtungsfeld)
wiring pattern [PCB]	ELECTRON	**Verdrahtungsmuster** [Leiterplatte]
= conductive pattern		
wiring plan	MICROEL	**Beschaltungsplan**
wiring scheme (→wiring diagram)	EL.TECH	(→Bauschaltplan)
wiring scheme (→circuit diagram)	ELECTRON	(→Stromlaufplan)
wiring strap (→solder strap)	ELECTRON	(→Lötbrücke)
wiring test	MANUF	**Verdrahtungsprüfung**
with all trunks busy (→congested)	SWITCH	(→gassenbesetzt)
with a narrow opening	TECH	**engmündig**
with congestion (→congested)	SWITCH	(→gassenbesetzt)
with controlled ambient (→airconditioned)	TECH	(→klimatisiert)
with costs	ECON	**kostenpflichtig**
withdraw	TECH	**zurückziehen**
= retract		
withdraw (→cancel)	ECON	(→rückgängigmachen)
withdrawal	COLLOQ	**Entnahme**
withdrawal [of an order]	ECON	**Widerruf 2** [einer Bestellung]
≈ countermand (n.); revocation; redemption; cancellation		= Annullierung; Rücknahme
with equality of access (→equal-access)	TELEC	(→gleichberechtigt)
with general authorization	ECON	**generalbevollmächtigt**
with global coverage (→nationwide)	TELEC	(→landesweit)
with high definition (→high-definition)	TV	(→hochauflösend)
within costumer reach	ECON	**kundennah**
within-series adapter (→in-series adapter)	COMPON	(→Kupplung 1)
with internal clock (→self-clocked)	ELECTRON	(→selbsttaktend)
with low runtime	EL.TECH	**laufzeitarm**
with matched load (→matched)	EL.TECH	(→angepaßt)
with moderate costs	ECON	**kostengünstig** [mit geringen Kosten verbunden]
with multiple recording	TERM&PER	**mehrfachbeschreibbar**
with narrow pores	TECH	**engporig**
with nine faces	MATH	**neunflächig**
without charge	TELEC	**gebührenfrei**
= no-chatge; free-code; free of charge; toll-free		
without claiming completeness	COLLOQ	**ohne Anspruch auf Vollständigkeit**
without competition	ECON	**konkurrenzlos 2** [ohne Wettbewerb]
without costs (→free of charge)	ECON	(→kostenlos)
without engagement (→non-binding)	ECON	(→unverbindlich)
without notice	ECON	**fristlos**
without obligation (→non-binding)	ECON	(→unverbindlich)

with priority (→prior-ranking)	COLLOQ	(→vorrangig)
withstand voltage (→dielectric strength)	EL.TECH	(→Spannungsfestigkeit)
with two reference magnitudes and two controlled magnitudes (→double control)	CONTROL	(→Zweifachregelung)
with uncontrolled ambient = uncontrolled	TECH	unklimatisiert
wizard [an experienced hacker 1] ≈ hacker 1	DATA PROC	Wizard [ein erfahrener Hacker 1] ≈ Hacker 1
wobble (→sweep)	ELECTRON	(→wobbeln)
wobble (→sweeping)	ELECTRON	(→Wobbelung)
wobble measuring set (→sweep measuring set)	INSTR	(→Wobbelmeßplatz)
wobbler (→sweep generator)	INSTR	(→Wobbelsender)
wobbling (→sweeping)	ELECTRON	(→Wobbelung)
wobbling frequency (→sweep rate)	ELECTRON	(→Wobbelfrequenz)
wobbling measuring set (→sweep measuring set)	INSTR	(→Wobbelmeßplatz)
wobbling range (→sweep range)	ELECTRON	(→Wobbelbereich)
wobbling rate (→sweep rate)	ELECTRON	(→Wobbelfrequenz)
wolfram (→tungsten)	CHEM	(→Wolfram)
wood = lumber	TECH	Holz
wooden crate	TECH	Holzkiste
wooden mast (→wooden pool)	OUTS.PLANT	(→Holzmast)
wooden pool = wooden mast	OUTS.PLANT	Holzmast
Woodruff key	MECH	Scheibenfeder
wood screw	MECH	Holzschraube
Wood's metal ↑ solder	METAL	Woodsches Metall = Wood-Metall ↑ Lot
wood wool = excelsior (AM)	TECH	Holzwolle
woofer ≈ subwoofer	EL.ACOUS	Tieftonlautsprecher = Tieftöner; Tiefenkonus ≈ Tiefpaßlautsprecher
word [smallest unit which can be addressed and processed by a computer; can be a sequence of signals, bits or bytes] ≈ byte ↑ unit of information	DATA PROC	Wort [kleinste vom Computer adressierbare und verarbeitbare Einheit; kann eine Folge von Zeichen, Bits oder Bytes sein] = Rechnerwort ≈ Byte ↑ Informationseinheit
word [sequence of characters considered as unit]	CODING	Wort [als Einheit betrachtete Zeichenfolge]
word 1 [smallest significance bearing unit of language]	LING	Wort 1 (pl. Wörter) [kleinste Bedeutung tragende Spracheinheit]
word 2 [word with special significance]	LING	Wort 2 (pl. Worte) [bedeutungsvolles Wort]
word (→machine word)	DATA PROC	(→Maschinenwort)
word address	DATA PROC	Wortadresse
word annotation (→voice annotation)	TERM&PER	(→Sprachanmerkung)
word class = part of speech ↓ noun; verb; adjective; adverb; pronoun; preposition; article; numeral	LING	Wortart ↓ Nomen; Verb; Adjektive; Adverb; Pronomen; Präposition; Artikel; Numerale
word clipping	TELEPH	Wortverstümmelung
word clock	DATA PROC	Worttakt
word computer (→word-oriented computer)	DATA PROC	(→Wortmaschine)
word counting contact	TELEGR	Wortzählkontakt
word generator [test generator for sequences of logic states]	INSTR	Wortgenerator [Meßgenerator für Folgen von Logikzuständen] = Wortgeber; Datenmustergenerator; Datensignalgenerator
wording	LING	Wortlaut = Textformulierung
word length [number of bits, characters or bytes, which are processed by the computer always together as smallest data unit] = word size; data word size ≈ data capacity	DATA PROC	Wortlänge [Anzahl der Bits, Zeichen oder Bytes, die von einem Computer als kleinste Dateneinheit immer zusammenhängend verarbeitet werden] = Wortbreite ≈ Datenbreite
word line = wordline	MICROEL	Wortleitung
wordline (→word line)	MICROEL	(→Wortleitung)
wordline driver	MICROEL	Worttreiber
word machine (→word-oriented computer)	DATA PROC	(→Wortmaschine)
word mark [for variable-length words]	DATA PROC	Wortmarke [für variable Wortlängen]
word marker	DATA PROC	Wortmarkierer
word order	LING	Wortstellung
word-organized = word-oriented	DATA PROC	wortorganisiert = wortorientiert
word-oriented (→word-organized)	DATA PROC	(→wortorganisiert)
word-oriented computer [computer with words as smallest adressable units] = word-oriented machine; word computer; word machine ≠ character-oriented computer ↓ syllable-oriented computer	DATA PROC	Wortmaschine [Computer dessen kleinste adressierbare Einheit ein Wort ist] ≠ Stellenmaschine ↓ Silbenmaschine
word-oriented machine (→word-oriented computer)	DATA PROC	(→Wortmaschine)
word plane	DATA PROC	Wortebene
word processing (→text processing)	DATA PROC	(→Textverarbeitung)
word-processing bureau	ECON	Textverarbeitungsbüro
word processing center	DATA PROC	Textverarbeitungszentrum
word processing equipment (→word processing system)	DATA PROC	(→Textsystem)
word processing program (→word processing system)	DATA PROC	(→Textsystem)
word processing system = text processing system; word processor 2; word processing equipment; word processing program ≈ desktop publishing ↑ workstation	DATA PROC	Textsystem = Textautomat; Wortprozessor; Textverarbeitungssystem; Textverarbeitungsprogramm; Textprogramm ≈ Desktop Publishing ↑ Arbeitsplatzsystem
word processor 2 (→word processing system)	DATA PROC	(→Textsystem)
words articulation (→discrete words intelligibility)	TELEPH	(→Wortverständlichkeit)
words intelligibility (→discrete words intelligibility)	TELEPH	(→Wortverständlichkeit)
word size (→word length)	DATA PROC	(→Wortlänge)
word spacing = spacing 1	TYPOGR	Ausschluß 3 [Leerraum zwischen Wörtern] = Wortabstand
word stress	LING	Wortbetonung
word wrap (→line break)	TYPOGR	(→Zeilenumbruch)
word wrap (→automatic line break)	DATA PROC	(→automatischer Zeilenumbruch)
word wrapping (→line break)	TYPOGR	(→Zeilenumbruch)
work (v.t.; past participle: worked or wrought) ≈ deform	TECH	bearbeiten ≈ verformen
work (n.) [force by path; SI unit: Joule] ≈ energy	PHYS	Arbeit [Kraft mal Verschiebung; SI-Einheit: Joule] ≈ Energie
work (n.) = treatment; processing; tooling ≈ processing [TECH]; deformation	MECH	Bearbeitung ≈ Verarbeitung [TECH]; Verformung
work (v.t.) (→tool)	MECH	(→bearbeiten)
work (v.i.) (→function)	TECH	(→funktionieren)

workability

workability	TECH	**Bearbeitbarkeit**
≈ deformability; plasticity		≈ Formbarkeit; Verformbarkeit
workable	TECH	**bearbeitbar**
≈ deformable		≈ verformbar
work accident	ECON	**Arbeitsunfall**
workbench	DATA PROC	**Mehrbenutzeranlage**
work conditions	ECON	**Arbeitsbedingungen**
work disk	DATA PROC	**Arbeitsdiskette**
≠ back-up disk		≠ Sicherungsdiskette
work effort (→expenditure of work)	ECON	(→Arbeitsaufwand)
worker	ECON	**Arbeiter**
= workman		
work file	DATA PROC	**Arbeitsdatei**
workflow (→job sequence)	TECH	(→Arbeitsablauf)
workforce	ECON	**Belegschaft**
= personnel; staff; employees; labor force		= Mitarbeiterzahl; Personalbestand
≈ headcount		≈ Kopfzahl
workforce (→labour force)	ECON	(→Arbeitskraft)
work function	PHYS	**Ablösearbeit**
= metallic work function		= Abtrennarbeit; Metallaustrittsarbeit; Austrittsarbeit; Ablöseenergie
≈ ionization energy; activation energy		≈ Ionisierungsenergie; Aktivierungsenergie
work function voltage	MICROEL	**Austrittsspannung**
work hardening (→strain hardening)	METAL	(→kaltverfestigen)
working (adj.)	TECH	**funktionstüchtig**
= operating		= funktionsfähig; funktionierend
≈ functional 2; operational		≈ funktionell; betriebsfähig
working	ECON	**berufstätig**
≈ occupied		≈ beschäftigt
working (→non-cutting shaping)	MECH	(→spanlose Bearbeitung)
working (→operation 1)	TECH	(→Betrieb)
working (→on-line)	TELEC	(→in Betrieb)
working (→operating)	TECH	(→Betriebs-)
working (→active)	DATA PROC	(→aktiv)
working (→active)	TECH	(→aktiv)
working area	DATA PROC	(→Arbeitsbereich)
(→workspace)		
working attenuation	NETW.TH	(→Betriebsdämpfungsmaß)
(→effective attenuation constant)		
working copy	DATA PROC	**Arbeitskopie**
[of an original program]		[eines Originalprogramms]
working costs (→operating costs)	ECON	(→Betriebskosten)
working current	ELECTRON	**Arbeitsstrom**
≠ quiescent current		≠ Ruhestrom
working current (→operating current)	EL.TECH	(→Betriebsstrom)
working day	COLLOQ	**Werktag**
[all days without Sunday and holidays, sometimes also without Saturday]		[Tag, an dem allgemein gearbeitet wird]
= weekday		= Wochentag; Arbeitstag
≠ sunday; holiday		≠ Sonntag; Feiertag
working depth	MECH	**Eingriffstiefe**
		[Zahnrad]
working drawing	ENG.DRAW	**Werkstattzeichnung**
working engineer	TECH	**berufstätiger Ingenieur**
working example	MATH	**Anwendungsbeispiel**
		[einer Formel]
working frequency	ELECTRON	**Arbeitsfrequenz**
= actual frequency		[Oszillator]
working frequency	RADIO	**Betriebsfrequenz**
working frequency	HF	**Verkehrsfrequenz**
working group	ECON	**Arbeitsgruppe**
= working party		= Arbeitskreis
working meeting	ECON	**Arbeitssitzung**
working memory	DATA PROC	**Arbeitsspeicher 3**
[part of main memory 1, for data in current use]		[Sektor des Hauptspeichers, für aktuelle Daten]
= working storage; workspace		↑ Hauptspeicher 1
↑ main memory 1		
working mode (→operating mode)	TECH	(→Betriebsart)
working party (→working group)	ECON	(→Arbeitsgruppe)
working plate	MICROEL	**Arbeitsmaske**
working platform	OUTS.PLANT	**Arbeitsplattform**
working point (→operating point)	TECH	(→Arbeitspunkt)
working point (→operating point)	ELECTRON	(→Arbeitspunkt)
working polarization	RADIO	**Nutzpolarisation**
≈ co-polarization		= Arbeitspolarisation
		≈ Kopolarisation
working principle (→operating mode)	TECH	(→Betriebsart)
working range (→operating range)	TECH	(→Betriebsbereich)
working region (→operating range)	TECH	(→Betriebsbereich)
working sequence	TECH	**Betriebsablauf**
working standard	TELEC	**Arbeitseichkreis**
working standard	INSTR	**Arbeitseichgröße**
[a standard for regular use]		= Maßverkörperung
working storage (→working memory)	DATA PROC	(→Arbeitsspeicher 3)
working stroke	MECH	**Arbeitshub**
working temperature (→operating temperature)	TECH	(→Betriebstemperatur)
working time	ECON	**Arbeitszeit**
working time recording system	SIGN.ENG	**Anwesenheitszeiterfassung**
		= AZE; Personalzeiterfassung
working tool	MECH	**Bearbeitungswerkzeug**
working voltage (→operating voltage)	EL.TECH	(→Betriebsspannung)
work-in-process (→unfinished product)	MANUF	(→unfertiges Produkt)
workload	COLLOQ	**Arbeitsbelastung**
workman (→worker)	ECON	(→Arbeiter)
workmanship 2	TECH	**Ausführungsqualität**
[imparted quality]		= Ausführung 2; Verarbeitungsqualität; Verarbeitung 2
= finish 2 (n.); fabric		
workmanship 1	TECH	**Handwerkskunst**
= craftmanship		
workpiece	TECH	**Werkstück**
workplace	ECON	**Arbeitsplatz**
work preparation	MANUF	**Arbeitsvorbereitung**
≈ production control		= Avo
		≈ Fertigungslenkung
workroom (→operating room)	TECH	(→Betriebsraum)
works	TECH	**Baustelle**
= project site; building site [CIV.ENG]		
works (→plant)	TECH	(→Anlage)
works (→factory)	MANUF	(→Fabrik)
works employee	ECON	**Werkangehöriger**
		= Werksangehöriger (ÖS)
worksheet	OFFICE	**Arbeitsblatt**
= spreadsheet; plansheet		= Kalkulationstabelle
worksheet (→standard form)	OFFICE	(→Formular)
works holidays	ECON	**Werkferien**
= holiday shutdown		= Werksferien (ÖS); Betriebsferien; Kollektivurlaub; Kollektivferien; Fabrikferien
workshop	MANUF	**Werkhalle**
		= Fertigungshalle; Werkshalle (ÖS); Betrieb; Fertigungsbetrieb
workshop	TECH	**Werkstatt**
= shop (AM)		= Werkstätte
works manager	TECH	**Bauleiter**
= site engineer		≈ Projektleiter
≈ project manager		
works manager (→manufacturing manager)	MANUF	(→Fertigungsleiter)
workspace	DATA PROC	**Arbeitsbereich**
[in a store]		[eines Speichers]
= working area		
workspace (→working memory)	DATA PROC	(→Arbeitsspeicher 3)
works superintendent (→manufacturing manager)	MANUF	(→Fertigungsleiter)

workstation	DATA PROC	**Arbeitsplatzrechner**	**wrapper** (→wrapping 1)	TECH	(→**Wickel**)
[autonomous high power computer in the Mips range and RAM greater 2 Mbyte]		[autonomer Computer hoher Leistung im Mips-Bereich und RAM größer 2 MByte]	**wrap pin**	COMPON	**Wrap-Stift** = Wickelstift
= workstation computer			**wrapping 2**	TECH	**Hülle 2**
≈ workstation [DATA COMM]; personal computer		= APR; Arbeitsplatzcomputer; APC; Workstation; Arbeitsstation; Arbeitsplatzsystem	[something covering] ↓ protective wrapping		[aus faltbarem Material] = Umhüllung 1 ↓ Schutzhülle 2
≠ server			**wrapping 1**	TECH	**Wickel**
↑ computer		≈ Workstation [DATA COMM]; Personal-Computer	[something wrapped] = wrapper; winding		[etwas Gewickeltes] = Bewicklung; Wicklung; Umwicklung
↓ word processing system; graphical workstation		↑ Computer ↓ Textsystem; graphisches Arbeitsplatzsystem			≈ Knäuel
			wrapping 3	TECH	**Umhüllung 2**
			≈ cladding		≈ Kaschierung
			wrapping (→serving)	COMM.CABLE	(→**Umwicklung**)
workstation	DATA COMM	**Workstation**	**wrapping** (→line break)	TYPOGR	(→**Zeilenumbruch**)
[terminal for a user of a computer network, mostly with local processing capability]		[Arbeitsplatz für einen Benutzer eines Mehrrechnersystems, meist mit lokaler Verarbeitungskapazität]	**wrapping paper**	TECH	**Packpapier** = Einschlagpapier
			wrapping sleeve	ELECTRON	**Wickelhülse**
≈ workstation [DATA PROC]			**wrap tool**	ELECTRON	**Wickelwerkzeug**
≠ server		= Arbeitsplatzstation; Arbeitsplatzrechner			= Wrap-Werkzeug
↑ station		≈ Arbeitsplatzsystem [DATA PROC]	**wrap-up** (v.t.) = wind-up (v.t.); spool-up; coil (v.t.)	TECH	**aufwickeln** = aufspulen ≈ spulen
		≠ Server ↑ Station	≈ spool		
			wrap wire	ELECTRON	**Wickeldraht**
workstation computer (→workstation)	DATA PROC	(→**Arbeitsplatzrechner**)	= winding wire **wrench** (n.)	MECH	**Schraubenschlüssel**
work-year (→man-year)	ECON	(→**Mann-Jahr**)	= spanner		↑ Werkzeug
world air route	RADIO NAV	**Weltflugverbindung**	↑ tool		
world band receiver (→world receiver)	CONS.EL	(→**Weltempfänger**)	**wrench** (→driver) **wrench** (→twist 1)	TECH PHYS	(→**Schlüssel 2**) (→**verwinden**)
World Bank	ECON	**Weltbank**	**wrench** (n.) (→twist 1)	PHYS	(→**Verwindung**)
world debut	ECON	**Weltpremiere**	**wring** (v.t.)	TECH	**wringen**
world economy = international economy	ECON	**Weltwirtschaft** = internationale Wirtschaft	[to squeeze by twisting]		[durch Gegendrehung pressen]
world market	ECON	**Weltmarkt**	**wring nut**	MECH	**Flügelmutter**
= internationaL market		= internationaler Markt	**wrist-watch paging**	MOB.COMM	**Armbanduhr-Funkruf**
world receiver	CONS.EL	**Weltempfänger**	↑ radio paging		
= world band receiver			**write** (v.t.)	LING	**schreiben**
world's vanguard	COLLOQ	**Weltspitze**	≈ redact; record		≈ verfassen; aufzeichnen
world trade = international trade	ECON	**Welthandel** = internationaler Handel	**write** (v.t.) [data into a memory]	DATA PROC	**schreiben** [Daten in einen Speicher]
worm [gear]	MECH	**Schnecke** [Zahnrad]	≠ read ↓ overwrite		= einspeicher ≠ lesen ↓ überschreiben
WORM	TERM&PER	**WORM**	**write** (→program)	MICROEL	(→**schießen**)
["write once read multiple"] = DRAW		[einmal beschreibbarer, dann nur noch lesbarer optischer Speicher]	**write** (→write line) **write address**	ELECTRON DATA PROC	(→**Schreibleitung**) **Schreibadresse**
↑ optical disk		= DRAW	**write buffer**	DATA PROC	**Schreibpuffer**
		↑ Bildplatte	**write element** (→print head)	TERM&PER	(→**Druckkopf**)
WORM disk	TERM&PER	**WORM-Platte**	**write-enable ring**	TERM&PER	**Schreibsicherungsring**
[optical disk] = write-once-read-mostly disk		[optische Speicherplatte]	[magnetic tape] = write-permit ring; file protection ring		[Magnetband] = Schreibring
worm drive [gear]	MECH	**Schneckentrieb** [Zahnrad]	**write error** = print error; recording error	TERM&PER	**Druckfehler** ≈ Schreibfehler
worm shaft	TELEPH	**Spindel**	**write head**	TERM&PER	**Schreibkopf 2**
[rotary dial switch]		[Nummernschalter]	= record head; recording head		≠ Lesekopf
worm wheel	MECH	**Schneckenrad**	≠ read head		↑ Magnetkopf
↑ gear		↑ Zahnrad	↑ magnetic head		
worn	TECH	**abgenutzt** ≈ verschlissen	**write head** (→print head) **writehead** (→print head)	TERM&PER TERM&PER	(→**Druckkopf**) (→**Druckkopf**)
worshop	ECON	**Arbeitstagung**	**write-in** (v.t.)	DATA PROC	**einschreiben**
worst-case condition	QUAL	**Worst-Case-Fall**	**write-in** (n.)	DATA PROC	**Einschreiben**
"would you like to leave a message?"	TELEPH	„möchten Sie eine Nachricht hinterlassen?"	≠ read-out (v.t.) **write-inhibit ring** (→write-protect ring)	TERM&PER	≠ Auslesen (→**Schreibschutzring**)
woven tape (→fabric tape)	POWER SYS	(→**Gewebeband**)	**write instruction**	DATA PROC	**Schreibbefehl**
wow (n.)	EL.ACOUS	**Tonhöhenschwankung**	**write keyboard**	TERM&PER	**Schreibtastatur**
[unwanted pitch variations] ≈ flutter [TERM&PER]			**write line** = write	ELECTRON	**Schreibleitung**
wrap (v.t.)	TECH	**wickeln**	**write lock** (→write protect)	TERM&PER	(→**Schreibschutz**)
= wind		= umwickeln; bewickeln	**write lockout** (→write protect)	TERM&PER	(→**Schreibschutz**)
wraparound (→automatic line break)	DATA PROC	(→**automatischer Zeilenumbruch**)	**write-once-read-mostly disk** (→WORM disk)	TERM&PER	(→**WORM-Platte**)
wrapped connection = wire-wrap connection	ELECTRON	**Wickelverbindung** = Wrap-Verbindung; Drahtwickelverbindung	**write-permit ring** (→write-enable ring)	TERM&PER	(→**Schreibsicherungsring**)
wrapper (→book jacket)	TYPOGR	(→**Buchumschlag**)			

write protect [floppy disk] = write protection; write lockout; write lock; file protection ↓ write-protect notch; write-protect ring	TERM&PER	**Schreibschutz** [Diskette] = Schreibsperre; Löschschutz ↓ Schreibschutzkerbe; Schreibschutzring	
write-protected	TERM&PER	**schreibgeschützt**	
write-protected disk (→locked disk)	TERM&PER	(→Diskette mit Schreibschutz)	
write-protect hole [floppy disk]	TERM&PER	**Schreibschutzloch** [Diskette]	
write protection (→write protect)		(→Schreibschutz)	
write-protect notch [floppy disk]	TERM&PER	**Schreibschutzkerbe** [Diskette]	
write-protect ring = write-inhibit ring ↑ write protect	TERM&PER	**Schreibschutzring** ↑ Schreibschutz	
write protect sensor	TERM&PER	**Schreibschutzschranke**	
write pulse	ELECTRON	**Schreibimpuls**	
write-read ... (→read-write ...)	TERM&PER	(→Schreib-Lese-)	
write velocity = writing speed; recording speed	TERM&PER	**Schreibgeschwindigkeit**	
write wire [magnetic core memory] = set wire ≈ read-write wire ≠ read wire ↓ Y write wire; X write wire	TERM&PER	**Schreibdraht** [Magnetkernspeicher] = Setzdraht ≈ Schreib-Lese-Draht ≠ Lesedraht ↓ Spaltendraht; Zeilendraht	
writing 1 [the process of]	LING	**Schreiben 1** [Vorgang des Schreibens]	
writing 2 [system for the readable reproduction of language]	LING	**Schrift 1** [System zur lesbaren Wiedergabe einer Sprache]	
writing amplifier	ELECTRON	**Schreibverstärker**	
writing beam (→recording beam)	ELECTRON	(→Schreibstrahl)	
writing head (→print head)	TERM&PER	(→Druckkopf)	
writing magnet	TERM&PER	**Schreibmagnet**	
writing speed (→write velocity)	TERM&PER	(→Schreibgeschwindigkeit)	
writing table (→desk)	OFFICE	(→Schreibtisch)	
written (adj.) = in writing ≠ spoken	LING	**schriftlich** (adj.) ≠ mündlich	
written form (→in writing)	ECON	(→Schriftform)	
wrong connection = fallacy	TELEC	**Falschverbindung** = Fehlverbindung	
wrong inference ≈ fallacy	SCIE	**Fehlschluß** ≈ Trugschluß	
"wrong number"	TELEPH	**„falsch verbunden"**	
wrong position (→malposition)	TECH	(→Fehlstellung)	
wrought (→forged)	METAL	(→geschmiedet)	
wrought alloy	METAL	**Knetlegierung**	
wrought iron	METAL	**Schweißstahl**	
WSI (→ultra-large-scale integration)	MICROEL	(→Ultrahöchstintegration)	
Wullenweber antenna ≈ circular array antenna	ANT	**Wullenweber-Antenne** = Wullenweberantenne ≈ Kreisgruppenantenne	
wye (→star connection)	NETW.TH	(→Sternschaltung)	
wye circuit (→star connection)	NETW.TH	(→Sternschaltung)	
wye connection (→star connection)	NETW.TH	(→Sternschaltung)	
wye-delta conversion (→star-delta conversion)	NETW.TH	(→Stern-Dreieck-Umwandlung)	
wye network (→star connection)	NETW.TH	(→Sternschaltung)	
wye section (→star connection)	NETW.TH	(→Sternschaltung)	
WYSIWYG (→WYSIWYG display)	DATA PROC	(→WYSIWYG-Darstellung)	
WYSIWYG display ["what you see is what you get"; word processing] = WYSIWYG	DATA PROC	**WYSIWYG-Darstellung** [Textverarbeitung]; = druckgleiche Bildschirmdarstellung; druckkonforme Bilddarstellung	

X

X (→reactance)	NETW.TH	(→Blindwiderstand)	
X amplifier = horizontal amplifier	ELECTRON	**Horizontalverstärker** = X-Verstärker; X-Ablenkverstärker	
X-axis (→abscissa)	MATH	(→Abszisse)	
X-coordinate (→abscissa)	MATH	(→Abszisse)	
Xe (→xenon)	CHEM	(→Xenon)	
xenon = Xe	CHEM	**Xenon** = Xe	
xerographic printer [with optical principle] ↓ laser printer	TERM&PER	**xerographischer Drucker** [mit optischem Prinzip] ↓ Laserdrucker	
xerography (→electrostatic copying process)	OFFICE	(→elektrostatisches Kopierverfahren)	
X gain (→horizontal gain)	ELECTRON	(→Horizontalverstärkung)	
x height [distance between top and bottom of a character, without ascender or descender]	TYPOGR	**Mittellänge** [Höhe eines Kleinbuchstabens ohne Unter- oder Oberlängen]	
XLR audio connector	COMPON	**XLR-Steckverbinder**	
Xmastree antenna (→fishbone antenna)	ANT	(→Fischgrätenantenne)	
x modulation (→cross modulation)	MODUL	(→Kreuzmodulation)	
XOFF ≈ log-off [DATA PROC] ≠ XON	DATA COMM	**XOFF** ≈ Abmeldung [DATA PROC] ≠ XON	
XON ≈ log-in [DATA PROC] ≠ XOFF	DATA COMM	**XON** ≈ Anmeldung [DATA PROC] ≠ XOFF	
XOR (→EXCLUSIVE-OR operation)	ENG.LOG	(→EXKLUSIV-ODER-Verknüpfung)	
XOR circuit (→EXCLUSIVE OR gate)	CIRC.ENG	(→EXKLUSIV-ODER-Glied)	
XOR element (→EXCLUSIVE OR gate)	CIRC.ENG	(→EXKLUSIV-ODER-Glied)	
XOR function (→EXCLUSIVE-OR operation)	ENG.LOG	(→EXKLUSIV-ODER-Verknüpfung)	
XOR gate (→EXCLUSIVE OR gate)	CIRC.ENG	(→EXKLUSIV-ODER-Glied)	
XOR operation (→EXCLUSIVE-OR operation)	ENG.LOG	(→EXKLUSIV-ODER-Verknüpfung)	
XPD (→cross-polar discrimination)	RADIO	(→Kreuzpolarisationsentkopplung)	
XPIC (→cross-polar-interference canceler)	RADIO REL	(→Depolarisationskompensator)	
X radiation = Röntgenbremsstrahlung	PHYS	**Röntgenstrahlung** = Röntgenbremsstrahlung	
X ray ↑ ray	PHYS	**Röntgenstrahl** ↑ Strahl	
X-ray absorption edge	PHYS	**Röntgenabsorptionskante**	
X-ray analysis	INSTR	**Röntgenanalytik**	
X-ray image converter	ELECTRON	**Röntgenbildwandler**	
X-ray lithography	MICROEL	**Röntgenstrahl-Lithographie**	
X-ray microscope	ELECTRON	**Röntgen-Mikroskop**	
X-ray photography	PHYS	**Röntgenfotografie**	
X ray specxtrometer	PHYS	**Röntgen-Spektrometer**	
X ray tube	ELECTRON	**Röntgenröhre**	
X read-write wire ≠ Y read-write wire	DATA PROC	**Zeilendraht** [Kernspeicher] ≠ Spaltendraht	
X-switch	DATA PROC	**Kreuzschalter**	
XT computer [eXtended Technology; a PC generation based on 16-bit processors 8086 or 8088 and equipped with hard disk drive] = PCXT; PC/XT	DATA PROC	**XT-Computer** [PC-Generation basierend auf 16-Bit-Prozessoren 8086 und 8088 und mit Festplatte ausgerüstet] = XT-Rechner; PCXT; PC/XT	
XY display	INSTR	**XY-Anzeige**	
X-Y plotter = graph plotter	TERM&PER	**Koordinatenschreiber** = X-Y-Schreiber; X-Y-Recorder ↑ Plotter	
XYT recorder	INSTR	**XYT-Schreiber**	

Y

Y (→admittance)	NETW.TH	(→komplexer Scheinleitwert)
Y (→star connection)	NETW.TH	(→Sternschaltung)
Y (→yttrium)	CHEM	(→Yttrium)
yagi (→Yagi antenna)	ANT	(→Yagi-Antenne)
Yagi aerial (→Yagi antenna)	ANT	(→Yagi-Antenne)
Yagi antenna	ANT	Yagi-Antenne
= Yagi aerial; yagi		= Yagi-Uda-Antenne
Y amplifier (→vertical amplifier)	ELECTRON	(→Vertikalverstärker)
yard	PHYS	Yard
[3 ft = 36 in = 0.914 40 m]		[angelsächsisches Längenmaß; = 3 Fuß = 36 Zoll = 0,914 40 m]
= yd		= yd
yard good	ECON	Meterware
		= Schnittware
yardstick	INSTR	Zollstock
		= Gliedermaßstab (DDR)
yarn (→thread)	TECH	(→Faden)
yaw (v.i.)	TECH	gieren
[angular oscillation along vertical axis]		[um Vertikalachse schwingen]
yaw axis	TECH	Gierachse
Y axis (→ordinate)	MATH	(→Ordinate)
Yb (→ytterbium)	CHEM	(→Ytterbium)
Y circuit (→star connection)	NETW.TH	(→Sternschaltung)
Y connection (→star connection)	NETW.TH	(→Sternschaltung)
yd (→yard)	PHYS	(→Yard)
Y-Delta transformation (→star-delta conversion)	NETW.TH	(→Stern-Dreieck-Umwandlung)
year	PHYS	Jahr
[8,765.8 h]		[8.765,8 h]
= a		= a
year of manufacturing	MANUF	Fertigungsjahr
= year of production		= Herstellungsjahr
year of production (→year of manufacturing)	MANUF	(→Fertigungsjahr)
year under review	ECON	Berichtsjahr
yellow	PHYS	gelb
yellow cable	DATA COMM	gelbes Kabel
[a special coaxial cable for LAN]		[spezielles Koaxialkabel für LAN]
yellowish	PHYS	gelblich
yellow ocher	PHYS	goldocker
= yellow ochre		
yellow ochre (→yellow ocher)	PHYS	(→goldocker)
yellow orange	PHYS	gelborange
Yellow Pages	TELEPH	Branchen-Telefonbuch
		= Branchen-Fernsprechbuch; Gelbe Seiten
yellow-passivize	METAL	gelbchromatisieren
Y gain (→vertical gain)	ELECTRON	(→Vertikalverstärkung)
yield (v.t.)	MECH	fließen
[of a solid]		[Festkörper]
yield (v.i.)	MECH	nachgeben
yield (v.t.)	ECON	einbringen
= profit (v.t.)		= abwerfen
yield (n.) (→profit)	ECON	(→Gewinn)
yield	QUAL	Ausbeute
yield (n.) (→return)	ECON	(→Rendite)
yield point	MECH	Fließgrenze
		= Streckgrenze
yield strength	MECH	Fließfestigkeit
YIG filter	MICROW	YIG-Filter
[resonance of Yttrium-iron-garnet]		[Resonanz von Yttrium-Eisengranat]
		= YIG-Nachlauffilter
YIG oscillator	MICROW	YIG-Oszillator
YIG-tuned	MICROW	YIG-abgestimmt
Y matrix (→admittance matrix)	NETW.TH	(→Leitwertmatrix)
Y network (→star connection)	NETW.TH	(→Sternschaltung)
yoke	EL.TECH	Joch
[closes the magnetic loop of a transformer]		[schließt den magnetischen Kreis eines Transformators]
yoke	TERM&PER	Zugriffskamm
[fixed disk]		[Festplatte]
yoke (→magnet yoke)	EL.TECH	(→Magnetjoch)
yoke (→transformator yoke)	EL.TECH	(→Transformatorjoch)
Yours faithfully (BRI) (→Sincerely yours)	ECON	(→Hochachtungsvoll)
Yours sincerely (BRI) (→Very sincerely yours)	ECON	(→Mit freundlichen Grüßen Ihr)
Yours truly (BRI) (→Sincerely yours)	ECON	(→Hochachtungsvoll)
Yours very truly (BRI) (→Sincerely yours)	ECON	(→Hochachtungsvoll)
y parameter (→conductance parameter)	NETW.TH	(→Leitwertparameter)
Y read-write wire	DATA PROC	Spaltendraht
[magnetic core memory]		[Magnetkernspeicher]
≠ X read-write wire		≠ Zeilendraht
↑ set wire		↑ Setzdraht
Y section (→star connection)	NETW.TH	(→Sternschaltung)
Y-shaped branching circulator	MICROW	Y-Verzweigungszirkulator
ytterbium	CHEM	Ytterbium
= Yb		= Yb
yttrium	CHEM	Yttrium
= Y		= Y
yttrium garnet	CHEM	Yttriumgranat

Z

Z (→impedance)	NETW.TH	(→komplexer Scheinwiderstand)
zap (v.t.)	DATA PROC	versehentlich löschen
= delete accidentally		
zap (n.)	DATA PROC	Tabelleneintrag-Löschanweisung
zapon-varnish (v.t.)	TECH	zaponieren
		↑ lackieren
zapon varnish (n.)	TECH	Zaponlack
Z axis	MATH	Z-Achse
= vertical axis		= Vertikalachse
Z clipping	DATA PROC	Z-Abschneiden
[computer graphics; suppression of too near or far situated image elements]		[Computergraphik; Unterdrückung zu nah oder fern liegender Bildteile]
Z diode (→Zener diode)	MICROEL	(→Zenerdiode)
ZDR (→zoned density recording)	TERM&PER	(→angepaßte Schreibdichte)
Zener breakdown	MICROEL	Zenerdurchbruch
Zener diode	MICROEL	Zenerdiode
[reverse biased diode exploiting the Zener effect or the tunnel effect]		[in Sperrichtung betriebene Diode unter Ausnutzung des Zener-Effekts oder des Lawineneffekts]
= Z diode		= Z-Diode
≈ voltage reference diode		≈ Referenzdiode
Zener effect	PHYS	Zenereffekt
Zener resistance	MICROEL	Zenerwiderstand
Zener voltage	MICROEL	Zenerspannung
= Z voltage		↑ Durchbruchspannung
↑ breakdown voltage		
Zenneck wave	RADIO PROP	Zenneck-Welle
zepp antenna (→Zeppelin antenna)	ANT	(→Zeppelin-Antenne)
Zeppelin antenna	ANT	Zeppelin-Antenne
= zepp antenna; end-fed antenna; voltage-fed antenna		= Zepp-Antenne; Begerow-Antenne; Fuchsantenne; endgespeiste Antenne
zero (n.)	MATH	Null
= nil; null; nought		
zero-access memory (→high-speed memory)	DATA PROC	(→Schnellspeicher 1)
zero-access storage (→high-speed memory)	DATA PROC	(→Schnellspeicher 1)
zero-access store (→high-speed memory)	DATA PROC	(→Schnellspeicher 1)
zero adjuster	INSTR	Nullpunkteinsteller
		= Nullsteller
zero adjustment (→zero balancing)	ELECTRON	(→Nullabgleich)
zero balance	TELEGR	Nullvergleich

zero balancing

zero balancing	ELECTRON	**Nullabgleich**
= zero adjustment; nulling		= Nullpunkteinstellung; Nulleinstellung; Nullung
zero balancing motor	POWER ENG	**Nullmotor**
zero branch	EL.TECH	**Nullzweig**
zero carryover	MATH	**Nullübertrag**
zero compression	DATA PROC	**Nullunterdrückung**
[suppression of leading zeros]		[Entfernen führender Nullen]
= ripple blanking		
zero condition ($\rightarrow$zero state)	ELECTRON	($\rightarrow$Nullzustand)
zero conductor ($\rightarrow$neutral conductor)	POWER ENG	($\rightarrow$Nulleiter)
zero correction ($\rightarrow$zero-point correction)	INSTR	($\rightarrow$Nullpunktkorrektur)
zero crossing	MATH	**Nulldurchgang**
$\approx$ zero point		$\approx$ Nullstelle
zero crossing	EL.TECH	**Nulldurchgang**
		= Zeichenwechsel
zero-current amplifier	CIRC.ENG	**Nullstromverstärker**
		= Nullverstärker
zero-defects principle	QUAL	**Null-Fehler-Prinzip**
zero deviation ($\rightarrow$balance error)	INSTR	($\rightarrow$Nullpunktfehler)
zero-dissipation ($\rightarrow$loss-free)	EL.TECH	($\rightarrow$verlustfrei)
zero-distortion	TELEC	($\rightarrow$verzerrungsfrei)
zero-distortion line ($\rightarrow$distortionless line)	LINE TH	($\rightarrow$verzerrungsfreie Leitung)
zero divisor	MATH	**Nullteiler**
zero drift	ELECTRON	**Nullpunktdrift**
		= Nullpunktwanderung
zero-energy level ($\rightarrow$ionization level)	MICROEL	($\rightarrow$Außenraum-Niveau)
zero error ($\rightarrow$balance error)	INSTR	($\rightarrow$Nullpunktfehler)
zerofill (v.t.)	DATA PROC	($\rightarrow$nullen)
($\rightarrow$zeroize)		
zero flag	DATA PROC	**Null-Hinweiszeichen**
zero-flux leakage ($\rightarrow$absence of flux leakage)	EL.TECH	($\rightarrow$Streuungsfreiheit)
zero-fraction sign [,-]	ECON	**Nullstrich** [,-]
zero frequency	TRANS	**Nullfrequenz**
zero-insertion loss ($\rightarrow$overall loss)	TELEC	($\rightarrow$Restdämpfung)
zero instrument	INSTR	**Nullinstrument**
= null indicator; null detector		= Nullindikator
zeroize (v.t.)	DATA PROC	**nullen**
= zerofill (v.t.)		= auf Null stellen
zero key	TERM&PER	**Nulltaste**
zero level	TELEC	**Nullpegel**
zero line	MATH	**Nullinie**
zero-loss ($\rightarrow$loss-free)	EL.TECH	($\rightarrow$verlustfrei)
zero-loss aerial ($\rightarrow$zero-loss antenna)	ANT	($\rightarrow$verlustlose Antenne)
zero-loss antenna	ANT	**verlustlose Antenne**
= zero-loss aerial		
zero-loss line	LINE TH	**verlustlose Leitung**
= lossless line		= verlustfreie Leitung
zero-loss transformer	COMPON	**verlustloser Übertrager**
		= verlustfreier Transformator
zero mark	ELECTRON	**Nullpunkt**
zero mark resistance	MICROEL	**Nullpunktwiderstand**
zero method	INSTR	**Nullmethode**
= null method		
zero order hold circuit	CIRC.ENG	**Halteglied nullter Ordnung**
zero phase angle	EL.TECH	**Nullphasenwinkel**
		= Nullphase; Anfangsphasenwinkel
zero point	MATH	**Nullstelle**
= null		= Nullpunkt
$\approx$ zero crossing		$\approx$ Nulldurchgang
zero point	NETW.TH	**Nullstelle**
zero point ($\rightarrow$directional null)	ANT	($\rightarrow$Nullstelle)
zero-point angle	ANT	**Nullwertswinkel**
[between maximum and null]		[zwischen Maximum und Nullstelle]
zero-point beamwidth	ANT	**Nullwertsbreite**
zero-point correction	INSTR	**Nullpunktkorrektur**
= zero correction		
zero position	INSTR	**Nullstellung**
zero potential ($\rightarrow$earth potential)	EL.TECH	($\rightarrow$Erdpotential)
zero potential ($\rightarrow$zero voltage)	POWER ENG	($\rightarrow$Nullspannung)
zero resistivity	INSTR	**Nullwiderstand**
zero shift	INSTR	**Nullpunktsverschiebung**
zero span	ELECTRON	**Nullwobbelung**
zero stability	INSTR	**Nullpunktstabilität**
zero stage	CIRC.ENG	**Differenzstufe**
zero state	ELECTRON	**Nullzustand**
= zero condition		
zero stuffing	CODING	**Nullstopfen**
zero substitution	MICROEL	**Null-Substitution**
zero suppression	INSTR	**Nullpunktunterdrückung**
zero variation ($\rightarrow$balance error)	INSTR	($\rightarrow$Nullpunktfehler)
zero voltage	POWER ENG	**Nullspannung**
= zero potential		
Z fold ($\rightarrow$fan folding)	TERM&PER	($\rightarrow$Leporellofalzung)
Z-folded ($\rightarrow$fanfold)	TERM&PER	($\rightarrow$zickzackgefaltet)
Z folding ($\rightarrow$fan folding)	TERM&PER	($\rightarrow$Leporellofalzung)
zigzag	TECH	**Zickzack**
zigzag antenna	ANT	**Zickzackantenne**
= Chireix-Mesny antenna		= Chireix-Mesny-Antenne; Sägezahnantenne
$\uparrow$ helix antenna		$\uparrow$ Wendelantenne
zigzag connection	POWER SYS	**Zickzackverbindung**
$\uparrow$ three-phase transformer		[Drehstromtransformator]
zigzag dipole	ANT	**Zickzack-Dipol**
zigzag filter	NETW.TH	**Zickzackfilter**
$\uparrow$ bandpass filter		$\uparrow$ Bandpaßfilter
zigzag resonator	OPTOEL	**Zickzackresonator**
zigzag routing	RADIO REL	**Zickzackführung**
		= Zickzackkurs
zinc	CHEM	**Zink**
= Zn		= Zn
zinc alloy	METAL	**Zinklegierung**
zinc coat	METAL	**Zinküberzug**
		= Verzinkung
zinc die-casting	METAL	**Zinkspritzgießen**
zinc sulfide	CHEM	**Zinksulfid**
		= Zinkblende
zink-coat	METAL	**verzinken**
$\downarrow$ galvanize; hot-galvanize		$\downarrow$ galvanisch verzinken; feuerverzinken
zip code (AM) ($\rightarrow$postcode)	POST	($\rightarrow$Postleitzahl)
zirconium	CHEM	**Zirkon**
= Zr		= Zr
Z matrix ($\rightarrow$impedance matrix)	NETW.TH	($\rightarrow$Widerstandsmatrix)
Zn ($\rightarrow$zinc)	CHEM	($\rightarrow$Zink)
zone (v.t.)	SWITCH	**verzonen**
zone (n.)	MICROEL	**Zone**
= region		= Gebiet
$\downarrow$ junction		$\downarrow$ Übergang
zone ($\rightarrow$region)	COLLOQ	($\rightarrow$Gebiet)
zone bit	DATA PROC	**Zone-Bit**
zone bit recording	TERM&PER	**Zone-Bit-Recording**
[variable number of sectors/track]		[veränderliche Zahl von Sektoren/Spur]
zoned antenna	ANT	**gezonte Antenne**
zoned density recording	TERM&PER	**angepaßte Schreibdichte**
[magnetic store]		[Magnetspeicher]
= ZDR		
zone melting	MICROEL	**Zonenschmelzen**
		= Zonenziehen
zoner	SWITCH	**Verzoner**
[evaluates the dialled area code and adjusts the metering pulse generator correspondingly]		[wertet die Ortskennzahl aus und stellt den Zählimpulsgeber auf entsprechenden Takt ein]
zone refining	MICROEL	**Zonenreinigen**
zone restriction	SWITCH	**Zonenbeschränkung**
zone sequence	MICROEL	**Zonenfolge**
= region sequence		
zoning	SWITCH	**Verzonung**
zoning list ($\rightarrow$charging zone list)	SWITCH	($\rightarrow$Verzonungsliste)
zoning table	SWITCH	**Verzonungstabelle**
zoom (n.)	DATA PROC	**Zoom**
[graphics program]		[Graphikprogramm]

zoom	INSTR	**Zoom**
zooming (→scaling)	DATA PROC	(→Skalierung)
zoom lens	TV	**Gummilinse**
		= Varioptic
zoom measurement	INSTR	**Zoom-Messung**
z parameter	NETW.TH	**Z-Vierpolparameter**
		= Z-Parameter

Zr (→zirconium)	CHEM	(→Zirkon)
Z/3 star	COMPON	**Z/3-Stern**
z transformation	MATH	**Z-Transformation**
Zulu time (→mean Greewich time)	PHYS	(→Weltzeit)
Z voltage (→Zener voltage)	MICROEL	(→Zenerspannung)